Prepare for Class "Read the Book"

Feature	Description	Benefit	Page
Every chapter begins with....			
Chapter Opening Article & Project	Each chapter begins with a current article and ends with a related project.	The Article describes a real situation. The Project lets you apply what you learned to solve a related problem.	193, 273
Every section begins with....			
Learning Objectives ✓1	Each section begins with a list of objectives. Objectives also appear in the text where the objective is covered.	These focus your studying by emphasizing what's most important and where to find it.	296
Most sections contain...			
PREPARING FOR THIS SECTION	Most sections begin with a list of key concepts to review with page numbers.	Ever forget what you've learned? This feature highlights previously learned material to be used in this section. Review it, and you'll always be prepared to move forward.	296
Now Work the 'Are You Prepared?' Problems	Problems that assess whether you have the prerequisite knowledge for the upcoming section.	Not sure you need the Preparing for This Section review? Work the 'Are You Prepared?' problems. If you get one wrong, you'll know exactly what you need to review and where to review it!	296, 307
"Now Work" PROBLEMS	These follow most examples and direct you to a related exercise.	We learn best by doing. You'll solidify your understanding of examples if you try a similar problem right away, to be sure you understand what you've just read.	305
WARNING	Warnings are provided in the text.	These point out common mistakes and help you to avoid them.	329
Explorations and **Seeing the Concept**	These represent graphing utility activities to foreshadow a concept or solidify a concept just presented.	You will obtain a deeper and more intuitive understanding of theorems and definitions.	203, 301
In Words	These provide alternative descriptions of select definitions and theorems.	Does math ever look foreign to you? This feature translates math into plain English.	298
Calculus Icon	These appear next to information essential for the study of calculus.	Pay attention – if you spend extra time now, you'll do better later!	70
NEW! Showcase **EXAMPLES**	These examples provide "how-to" instruction by offering a guided, step-by-step approach to solving a problem.	With each step presented on the left and the mathematics displayed on the right, students can immediately see how each step is employed.	206–207
NEW! **Model It!** Examples and Problems	Marked with ⬤. These are examples and problems that require you to build a mathematical model from either a verbal description or data. The homework Model It! problems are marked by purple numbers.	It is rare for a problem to come in the form, "Solve the following equation". Rather, the equation must be developed based on an explanation of the problem. These problems require you to develop models that will allow you to describe the problem mathematically and suggest a solution to the problem.	341, 348

Practice "Work the Problems"

Feature	Description	Benefit	Page
"Assess Your Understanding" contains a variety of problems at the end of each section.			
'Are You Prepared?' Problems	These assess your retention of the prerequisite material you'll need. Answers are given at the end of the section exercises. This feature is related to the Preparing for This Section feature.	Do you always remember what you've learned? Working these problems is the best way to find out. If you get one wrong, you'll know exactly what you need to review and where to review it!	296, 307
Concepts and Vocabulary	These short-answer questions, mainly Fill-in-the-Blank and True/False items, assess your understanding of key definitions and concepts in the current section.	It is difficult to learn math without knowing the language of mathematics. These problems test your understanding of the formulas and vocabulary.	307
Skill Building	Correlated to section examples, these problems provide straightforward practice.	It's important to dig in and develop your skills. These problems provide you with ample practice to do so.	307–309
NEW! **Mixed Practice**	These problems offer comprehensive assessment of the skills learned in the section by asking problems that relate to more than one concept or objective. These problems may also require you to utilize skills learned in previous sections.	Learning mathematics is a building process. Many concepts are interrelated. These problems help you see how mathematics builds on itself and also see how the concepts tie together.	309–310
Applications and Extensions	These problems allow you to apply your skills to real-world problems. These problems also allow you to extend concepts learned in the section.	You will see that the material learned within the section has many uses in everyday life.	310–312
Explaining Concepts: Discussion and Writing	"Discussion and Writing" problems are colored red. These support class discussion, verbalization of mathematical ideas, and writing and research projects.	To verbalize an idea, or to describe it clearly in writing, shows real understanding. These problems nurture that understanding. Many are challenging but you'll get out what you put in.	312
"Now Work" PROBLEMS	Many examples refer you to a related homework problem. These related problems are marked by a pencil and yellow numbers.	If you get stuck while working problems, look for the closest Now Work problem and refer back to the related example to see if it helps.	303
Graphing Calculator	These optional problems require the use of a graphing utility, and are marked by a special icon and green numbers.	Your instructor will usually provide guidance on whether or not to do these problems. If so, these problems help to verify and visualize your analytical results.	333

Review "Study for Quizzes and Tests"

Feature	Description	Benefit	Page
Chapter Reviews at the end of each chapter contain...			
"Things to Know"	A detailed list of important theorems, formulas, and definitions from the chapter.	Review these and you'll know the most important material in the chapter!	468–470
"You Should Be Able to..."	Contains a complete list of objectives by section, examples that illustrate the object, and practice exercises that test your understanding of the objective.	Do the recommended exercises and you'll have mastery over the key material. If you get something wrong, review the suggested page numbers and try again.	470–471
Review Exercises	These provide comprehensive review and practice of key skills, matched to the Learning Objectives for each section.	Practice makes perfect. These problems combine exercises from all sections, giving you a comprehensive review in one place.	471–474
CHAPTER TEST	About 15–20 problems that can be taken as a Chapter Test. Be sure to take the Chapter Test under test conditions—no notes!	Be prepared. Take the sample practice test under test conditions. This will get you ready for your instructor's test. If you get a problem wrong, watch the Chapter Test Prep video.	474–475
CUMULATIVE REVIEW	These problem sets appear at the end of each chapter, beginning with Chapter 2. They combine problems from previous chapters, providing an ongoing cumulative review.	These are really important. They will ensure that you are not forgetting anything as you go. These will go a long way toward keeping you constantly primed for the final exam.	475
CHAPTER PROJECTS	The Chapter Project applies what you've learned in the chapter. Additional projects are available on the Instructor's Resource Center (IRC).	The Project gives you an opportunity to apply what you've learned in the chapter to solve a problem related to the opening article. If your instructor allows, these make excellent opportunities to work in a group, which is often the best way of learning math.	476

STUDY SMARTER

CHAPTER
Test Prep
V I D E O S

Step-by-step solutions on video for all chapter test exercises from the text

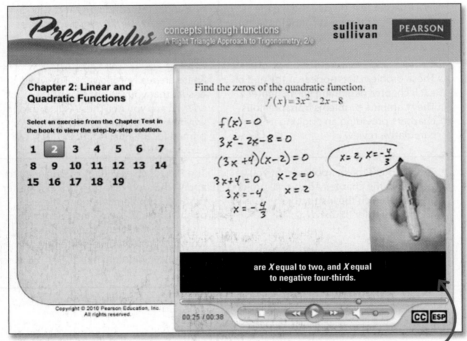

Precalculus concepts through functions
A Right Triangle Approach to Trigonometry, 2/e

sullivan sullivan   PEARSON

Chapter 2: Linear and Quadratic Functions

Select an exercise from the Chapter Test in the book to view the step-by-step solution.

1 2 3 4 5 6 7
8 9 10 11 12 13 14
15 16 17 18 19

Find the zeros of the quadratic function.
$$f(x) = 3x^2 - 2x - 8$$

$f(x) = 0$

$3x^2 - 2x - 8 = 0$

$(3x + 4)(x - 2) = 0$ $x = 2, x = -\frac{4}{3}$

$3x + 4 = 0$ $x - 2 = 0$

$3x = -4$ $x = 2$

$x = -\frac{4}{3}$

are *X* equal to two, and *X* equal
to negative four-thirds.

00:25 / 00:38   CC ESP

English and Spanish Subtitles Available

CHAPTER TEST PREP VIDEOS ARE ACCESSIBLE THROUGH THE FOLLOWING:

Video Resources on DVD

MyMathLab

You Tube™

Precalculus

Concepts Through Functions

A RIGHT TRIANGLE APPROACH TO

TRIGONOMETRY

Second Edition

Michael Sullivan

Chicago State University

Michael Sullivan, III

Joliet Junior College

Prentice Hall

Boston   Columbus   Indianapolis   New York   San Francisco   Upper Saddle River
Amsterdam   Cape Town   Dubai   London   Madrid   Milan   Munich   Paris   Montréal   Toronto
Delhi   Mexico City   São Paulo   Sydney   Hong Kong   Seoul   Singapore   Taipei   Tokyo

Editorial Director, Mathematics: *Christine Hoag*
Executive Editor: *Adam Jaworski*
Sponsoring Editor: *Dawn Murrin*
Assistant Editor: *Joseph Colella*
Senior Managing Editor: *Karen Wernholm*
Associate Managing Editor: *Tamela Ambush*
Art Editor: *Thomas Benfatti*
Senior Design Supervisor: *Andrea Nix*
Art Director: *Heather Scott*
Interior Designer: *Tamara Newnam*
Cover Designer: *Tamara Newnam*
Manager, Cover Visual Research and Permissions: *Karen Sanatar*
Permissions Project Supervisor: *Michael Joyce*
Photo Researcher: *Regalle Jaramillo*
Digital Assets Manager: *Marianne Groth*
Media Producer: *Audra Walsh*
Associate Content Manager: *Tanya Farber*
Senior Content Developer: *Mary Durnwald*
Senior Author Support/Technology Specialist: *Joe Vetere*
Senior Marketing Manager: *Katherine Greig*
Marketing Assistant: *Katherine Minton*
Senior Manufacturing Buyer: *Carol Melville*
Production Coordination: *Bob Walters, Prepress Management, Inc.*
Composition: *MPS Limited, A Macmillan Company*
Illustrations: *Precision Graphics and Laserwords Private Ltd.*

Cover Image: *Corbis*

Many of the designations used by manufacturers and sellers to distinguish their products are claimed as trademarks. Where those designations appear in this book, and Pearson Education was aware of a trademark claim, the designations have been printed in initial caps or all caps.

Library of Congress Cataloging-in-Publication Data

Sullivan, Michael, 1942-
 Precalculus :concepts through functions, a right triangle approach to trigonometry / Michael Sullivan, Michael Sullivan, III. —2nd ed.
  p. cm.
 Includes bibliographical references and index.
 ISBN 0-321-64508-1 (alk. paper)
 1. Functions—Textbooks. 2. Trigonometry—Textbooks. I. Sullivan, Michael, 1967 July 2– II. Title.
 QA331.3.S9238 2011
 515—dc22

                                    2009040217

    2 3 4 5 6 7 8 9 10—CRK—14 13 12 11 10

Prentice Hall
is an imprint of

www.pearsonhighered.com

ISBN-10: 0-321-64508-1
ISBN-13: 978-0-321-64508-1

For Michael S., Kevin, and Marissa (Sullivan)

Shannon, Patrick, and Ryan (Murphy)

Maeve, Sean, and Nolan (Sullivan)

Kaleigh, Billy, and Timmy (O'Hara)

The Next Generation

Contents

Three Distinct Series

Students have different goals, learning styles, and levels of preparation. Instructors have different teaching philosophies, styles, and techniques. Rather than write one series to fit all, the Sullivans have written three distinct series. All share the same goal—to develop a high level of mathematical understanding and an appreciation for the way mathematics can describe the world around us. The manner of reaching that goal, however, differs from series to series.

Concepts through Functions Series, Second Edition

This series differs from the others, utilizing a functions approach that serves as the organizing principle tying concepts together. Functions are introduced early in various formats. This approach supports the Rule of Four, which states that functions are represented symbolically, numerically, graphically, and verbally. Each chapter introduces a new type of function and then develops all concepts pertaining to that particular function. The solutions of equations and inequalities, instead of being developed as stand-alone topics, are developed in the context of the underlying functions. Graphing utility coverage is optional and can be included or excluded at the discretion of the instructor: *College Algebra; Precalculus, with a Unit Circle Approach to Trigonometry; Precalculus, with a Right Triangle Approach to Trigonometry.*

Enhanced with Graphing Utilities Series, Fifth Edition

This series provides a more thorough integration of graphing utilities into topics, allowing students to explore mathematical concepts and foreshadow ideas usually studied in later courses. Using technology, the approach to solving certain problems differs from the Contemporary series, while the emphasis on understanding concepts and building strong skills does not: *College Algebra, Algebra & Trigonometry, Trigonometry, Precalculus.*

Contemporary Series, Eighth Edition

The Contemporary Series is the most traditional in approach, yet modern in its treatment of precalculus mathematics. Graphing utility coverage is optional and can be included or excluded at the discretion of the instructor: *College Algebra, College Algebra Essentials, Algebra & Trigonometry, Trigonometry, Precalculus.*

Preface to the Instructor

As professors at both an urban university and a community college, Michael Sullivan and Michael Sullivan, III, are aware of the varied needs of Precalculus students, ranging from those who have little mathematical background and a fear of mathematics courses, to those having a strong mathematical education and a high level of motivation. For some of your students, this will be their last course in mathematics, whereas others will further their mathematical education. This text is written for both groups.

As a teacher, and as an author of precalculus, engineering calculus, finite mathematics, and business calculus texts, Michael Sullivan understands what students must know if they are to be focused and successful in upper-level math courses. However, as a father of four, he also understands the realities of college life. As an author of a developmental mathematics series, Michael's co-author and son, Michael Sullivan, III, understands the trepidations and skills students bring to the Precalculus course. Michael, III also believes in the value of technology as a tool for learning that enhances understanding without sacrificing math skills. Together, both authors have taken great pains to ensure that the text contains solid, student-friendly examples and problems, as well as a clear and seamless writing style.

A tremendous benefit of authoring a successful series is the broad-based feedback we receive from teachers and students. We are sincerely grateful for their support. Virtually every change in this edition is the result of their thoughtful comments and suggestions. We are sincerely grateful for this support and hope that we have been able to take these ideas and, building upon a successful first edition, make this series an even better tool for learning and teaching. We continue to encourage you to share with us your experiences teaching from this text.

About This Book

This book utilizes a functions approach to Precalculus. Functions are introduced early (Chapter 1) in various formats: maps, tables, sets of ordered pairs, equations, and graphs. Our approach to functions illustrates the symbolic, numeric, graphic, and verbal representations of functions. This allows students to make connections between the visual representation of a function and its algebraic representation.

It is our belief that students need to "hit the ground running" so that they do not become complacent in their studies. After all, it is highly likely that students have been exposed to solving equations and inequalities prior to entering this class. By spending precious time reviewing these concepts, students are likely to think of the course as a rehash of material learned in other courses and say to themselves, "I know this material, so I don't have to

study." This may result in the students developing poor study habits for this course. By introducing functions early in the course, students are less likely to develop bad habits.

Another advantage of the early introduction of functions is that the discussion of equations and inequalities can focus around the concept of a function. For example, rather than asking students to solve an equation such as $2x^2 + 5x + 2 = 0$, we ask students to find the zeros of $f(x) = 2x^2 + 5x + 2$ or solve $f(x) = 0$ when $f(x) = 2x^2 + 5x + 2$. While the technique used to solve this type of problem is the same, the fact that the problem looks different to the student means the student is less apt to say, "Oh, I already have seen this problem before, and I know how to solve it." In addition, in Calculus students are going to be asked to solve equations such as $f'(x) = 0$, so solving $f(x) = 0$ is a logical prerequisite skill to practice in Precalculus. Another advantage to solving equations through the eyes of a function is that the properties of functions can be included in the solution. For example, the linear function $f(x) = 2x - 3$ has one real zero because the function f is increasing on its domain.

Features in the Second Edition

Rather than provide a list of new features here, that information can be found on the endpapers in the front of this book. This places the new features in their proper context, as building blocks of an overall learning system that has been carefully crafted over the years to help students get the most out of the time they put into studying. Please take the time to review the features listed in the front of this book, and to discuss them with your students at the beginning of your course. Our experience has been that when students utilize these features, they are more successful in the course.

New to the Second Edition

- **Showcase Examples** are used to present examples in a guided, step-by-step format. Students can immediately see how each of the steps in a problem are employed. The "How To" examples have a two-column format in which the left column describes the step in solving the problem and the right column displays the algebra complete with annotations.
- **Model It** examples and exercises are clearly ma with a 🌐 icon. These examples and exerc meant to develop the student's ability to bu from both verbal descriptions and data. problems involving data require the stud termine the appropriate model (line so on) to fit to the data and justify th

- **Exercise Sets** at the end of each section have been classified according to purpose. The **Concepts and Vocabulary** exercises have been updated. These fill-in-the-blank and True/False problems have been written to serve as reading quizzes. For example, see Section 4.5, page 332. We also added **Mixed Practice** exercises where appropriate. These problems offer a comprehensive assessment of the skills learned in the section by asking problems that relate to more than one objective. In addition, we may present problems from previous sections so students must utilize skills learned throughout the course. For example, see Section 1.3, page 78; Section 3.4, page 246; and Section 4.6, page 339. **Applications and Extension** problems have been updated and many new problems involving sourced information and data have been added to bring relevance and timeliness to the exercises. The **Explaining Concepts: Discussion and Writing** exercises have been updated and reworded to stimulate discussion of concepts in online discussion forums. These can also be used to spark classroom discussion.

Changes in the Second Edition

- **CONTENT**
  - The discussion of finding zeros of a function in Chapter 1, Section 2, *The Graph of a Function*, has been expanded.
  - A new Objective, *Use a Graph to Locate the Absolute Maximum and the Absolute Minimum*, appears in Chapter 1, Section 3, *Properties of Functions*. The Extreme Value Theorem is included here as well.
  - The section *Solving Equations* in Appendix A has been expanded to include a discussion of solving radical equations.

- **ORGANIZATION**
  - The discussion of Variation now appears in Chapter 1, Section 7, *Building Mathematical Models Using Variation*.
  - *Problem Solving: Interest, Mixture, Uniform Motion, Constant Rate Jobs* now appears in Appendix A, Section 9, before *Interval Notation; Solving Inequalities* so that students are exposed to problem-solving skills before utilizing these skills in building models ~~~near inequalities.~~~

- More applications of decomposing vectors have been added to Chapter 8, Section 5 on the Dot Product.

Using this Book Effectively and Efficiently with Your Syllabus

To meet the varied needs of diverse syllabi, this book contains more content than is likely to be covered in a typical Precalculus course. As the chart illustrates, this book has been organized with flexibility of use in mind. Even within a given chapter, certain sections are optional and can be omitted without loss of continuity. (See the detail following the flow chart.)

Foundations   A Prelude to Functions

Quick coverage of this chapter, which is mainly review material, will enable you to get to Chapter 1, *Functions and Their Graphs*, earlier.

Chapter 1   Functions and Their Graphs

Perhaps the most important chapter. Section 1.6 is optional.

Chapter 2   Linear and Quadratic Functions

Topic selection depends on your syllabus. Section 2.2, 2.6, and 2.7 may be omitted without a loss of continuity.

Chapter 3   Polynomial and Rational Functions

Topic selection depends on your syllabus. Section 3.6 is optional.

Chapter 4   Exponential and Logarithmic Functions

Sections 4.1–4.6 follow in sequence. Sections 4.7–4.9 are optional.

Chapter 5   Trigonometric Functions

The sections follow in sequence. Section 5.8 is optional.

Chapter 6   Analytic Trigonometry

Sections 6.2 and 6.7 may be omitted in a brief course.

Chapter 7   Applications of Trigonometric Functions

Sections 7.4 and 7.5 may be omitted in a brief course.

Chapter 8   Polar Coordinates; Vectors

Sections 8.1–8.3 and Sections 8.4–8.7 are independent and may be covered separately.

Chapter 9   Analytic Geometry

Sections 9.2–9.4 follow in sequence. Sections 9.5, 9.6, and 9.7, are independent of each other, but each requires Sections 9.1–9.4.

Chapter 10 Systems of Equations and Inequalities

Sections 10.2–10.7 may be covered in any order. Section 10.8 requires Section 10.7.

Chapter 11 Sequences; Induction; the Binomial Theorem

There are three independent parts: Sections 11.1–11.3, Section 11.4, and Section 11.5.

Chapter 12 Counting and Probability

The sections follow in sequence.

Appendix A Review

This review material may be covered at the start of a course or used as a just-in-time review. Specific references to this material occur throughout the text to assist in the review process.

Appendix B Graphing Utilities

Reference is made to these sections at the appropriate place in the text.

Acknowledgments

Textbooks are written by authors but evolve from an idea to final form through the efforts of many people. It was Don Dellen who first suggested this book and series. Don is remembered for his extensive contributions to publishing and mathematics.

Thanks are due the following people for their assistance and encouragement in the preparation of this edition:

- From Pearson Education: Adam Jaworski for his substantial contributions, ideas, and enthusiasm; Katherine Greig, who remains a huge fan and supporter; Dawn Murrin, for her unmatched talent at getting the details right; Bob Walters for his superb organizational skills in directing production; Chris Hoag for her continued support and genuine interest; Greg Tobin for his leadership and commitment to excellence; and the Pearson Math and Science Sales team, for their continued confidence and personal support of our books.

- Accuracy checkers: C. Brad Davis, who read the entire manuscript and accuracy checked answers. His attention to detail is amazing; Timothy Britt, for creating the Solutions Manuals and accuracy checking answers.

- Reviewers: Richard Andrews—Florida A&M University; Rebecca Berthiaume—Edison State College; Michael Brook—University of Delaware; Tim Chappell—Penn Valley Community College; Alicia Collins—Mesa Community College; Jerry DeGroot—Purdue North Central; Joanna DelMonaco—Middlesex Community College; Stephanie Deacon—Liberty University; Jerrett Dumouchel—Florida Community College at Jacksonville; Nina Girard—University of Pittsburgh at Johnstown; Mary Beth Grayson—Liberty University; Celeste Hernandez—Richland College; Gloria P. Hernandez—Louisiana State University at Eunice; Glenn Johnson—Middlesex Community College; Susitha Karunaratne—Purdue University North Central; Debra Kopcso—Louisiana State University; Yelena Kravchuk—University of Alabama at Birmingham; Lynn Marecek—Santa Ana College; James McLaughlin—West Chester University; Kathleen Miranda—SUNY at Old Westbury; Chris Mirbaha—The Community College of Baltimore County; Karla Neal—Louisiana State University; Denise Nunley—Maricopa Community Colleges; Leticia Oropesa—University of Miami; Laura Pyzdrowski—West Virginia University; Mike Rosenthal—Florida International University; Phoebe Rouse—Louisiana State University; Brenda Santistevan—Salt Lake Community College; Ingrid Scott—Montgomery College; Charlotte Smedberg—University of Tampa; Leslie Soltis—Mercyhurst College; Katrina Staley—North Carolina Agricultural and Technical State University; Sonya Stephens—Florida A&M University; John Sumner—University of Tampa; Marilyn Toscano—University of Wisconsin, Superior; Timothy L. Warkentin—Cloud County Community College; Hayat Weiss—Middlesex Community College; Larissa Williamson—University of Florida

Finally, we offer our grateful thanks to the dedicated users and reviewers of our books, whose collective insights form the backbone of each textbook revision.

Our list of indebtedness just grows and grows. And, if we've forgotten anyone, please accept our apology. Thank you all.

STUDENT RESOURCES

Available to students are the following supplements:

- **_Student Solutions Manual (ISBN 10: 0321645154; ISBN 13: 9780321645159)_**
  Fully worked solutions to odd-numbered exercises.

- **_Algebra Review (ISBN 10: 0131480065; ISBN 13: 9780131480063)_**
  Four chapters of Intermediate Algebra Review. Perfect for a slower-paced course or for individual review.

- **_Videos on DVD (ISBN 10: 0321645197; ISBN 13: 9780321645197)_**
  The Videos on DVD contain short video clips of the author of the book working key book examples. Chapter Test Prep Videos (also included) provide fully worked solutions to the Chapter Test exercises. The Chapter Test Prep Videos are also available within MyMathLab® or on YouTube™ (go to **www.youtube.com/SullivanConceptsRT2e**). Videos have optional subtitles.

MathXL® Online Course (access code required)

MathXL® is a powerful online homework, tutorial, and assessment system that accompanies Pearson Education's textbooks in mathematics or statistics. With MathXL, instructors can:

- Create, edit, and assign online homework and tests using algorithmically generated exercises correlated at the objective level to the textbook.
- Create and assign their own online exercises and import TestGen tests for added flexibility.
- Maintain records of all student work tracked in MathXL's online gradebook.

With MathXL, students can:

- Take chapter tests in MathXL and receive personalized study plans based on their test results.
- Use the study plan to link directly to tutorial exercises for the objectives they need to study and retest.
- Access supplemental animations and video clips directly from selected exercises.

MathXL is available to qualified adopters. For more information, visit our website at **www.mathxl.com** <http://www.mathxl.com>, or contact your Pearson sales representative.

MyMathLab® Online Course (access code required)

MyMathLab is a series of text-specific, easily customizable online courses for Pearson Education's textbooks in mathematics and statistics. Powered by CourseCompass™ (our online teaching and learning environment) and MathXL® (our online homework, tutorial, and assessment system). MyMathLab gives you the tools you need to deliver all or a portion of your course online, whether your students are in a lab setting or working from home. MyMathLab provides a rich and flexible set of course materials, featuring free-response exercises that are algorithmically generated for unlimited practice and mastery. Students can also use online tools, such as video lectures, animations, and a multimedia textbook, to independently improve their understanding and performance. Instructors can use MyMathLab's homework and test managers to select and assign online exercises correlated directly to the textbook, and they can also create and assign their own online exercises and import TestGen tests for added flexibility. MyMathLab's online gradebook—designed specifically for mathematics and statistics—automatically tracks students' homework and test results and gives the instructor control over how to calculate final grades. Instructors can also add offline (paper-and-pencil) grades to the gradebook. MyMathLab also includes access to the **Pearson Tutor Center** (**www.pearsontutorservices.com** <http://www.pearsontutorservices.com>). The Tutor Center is staffed by qualified mathematics instructors who provide textbook-specific tutoring for students via toll-free phone, fax, email, and interactive Web sessions. MyMathLab is available to qualified adopters. For more information, visit our website at **www.mymathlab.com** or contact your sales representative.

MathXL® Tutorials on CD

This interactive tutorial CD-ROM provides algorithmically generated practice exercises that are correlated at the objective lev
the exercises in the textbook. Every practice exercise is accompanied by an example and a guided solution designed to invo
dents in the solution process. Selected exercises may also include a video clip to help students visualize concepts. The soft
vides helpful feedback for incorrect answers and can generate printed summaries of students' progress.

el to
ve stu-
ware pro-

Applications Index

Photo Credits

To the Student

As you begin, you may feel anxious about the number of theorems, definitions, procedures, and equations. You may wonder if you can learn it all in time. Don't worry, your concerns are normal. This textbook was written with you in mind. If you attend class, work hard, and read and study this book, you will build the knowledge and skills you need to be successful. Here's how you can use the book to your benefit.

Read Carefully

When you get busy, it's easy to skip reading and go right to the problems. Don't. . . the book has a large number of examples and clear explanations to help you break down the mathematics into easy-to-understand steps. Reading will provide you with a clearer understanding, beyond simple memorization. Read sections to be covered before class (not after) so you can ask questions about anything you didn't understand. You'll be amazed at how much more you'll get out of class if you do this.

Getting the Most from This Book

We've provided a brief guide to getting the most from this book (refer to the "Prepare for Class," "Practice," and "Review" pages on the inside front cover of this book). Spend 15 minutes reviewing the guide and familiarizing yourself with the features by flipping to the page numbers provided. Then, as you read, use them. This is the best way to make the most of your textbook.

Do not hesitate to contact us, through Prentice Hall, with any questions, suggestions or comments that would improve this text. We look forward to hearing from you, and best of luck with all your studies.

Best Wishes!

Michael Sullivan
Michael Sullivan, III

Foundations:
A Prelude to Functions

Outline

◁ A Look Back   Appendix A reviews skills from Intermediate Algebra.

A Look Ahead ▷   Here we make the connection between algebra and geometry through the rectangular coordinate system. The idea of using a system of rectangular coordinates dates back to ancient times, when such a system was used for surveying and city planning. Apollonius of Perga, in 200 BC, used a form of rectangular coordinates in his work on conics, although this use does not stand out as clearly as it does in modern treatments. Sporadic use of rectangular coordinates continued until the 1600s. By that time, algebra had developed sufficiently so that René Descartes (1596–1650) and Pierre de Fermat (1601–1665) could take the crucial step, which was the use of rectangular coordinates to translate geometry problems into algebra problems, and vice versa. This step was important for two reasons. First, it allowed both geometers and algebraists to gain new insights into their subjects, which previously had been regarded as separate, but now were seen to be connected in many important ways. Second, these insights made the development of calculus possible, which greatly enlarged the number of areas in which mathematics could be applied and made possible a much deeper understanding of these areas.

1 The Distance and Midpoint Formulas

PREPARING FOR THIS SECTION *Before getting started, review the following:*

- Algebra Essentials (Appendix A, Section A.1, pp. A1–A10)
- Geometry Essentials (Appendix A, Section A.2, pp. A14–A19)

Now Work the **'Are You Prepared?'** problems on page 6.

OBJECTIVES 1 Use the Distance Formula (p. 3)
2 Use the Midpoint Formula (p. 5)

Figure 1

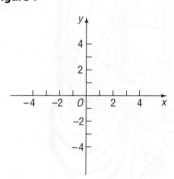

Rectangular Coordinates

We locate a point on the real number line by assigning it a single real number, called the *coordinate of the point*. For work in a two-dimensional plane, we locate points by using two numbers.

We begin with two real number lines located in the same plane: one horizontal and the other vertical. We call the horizontal line the **x-axis,** the vertical line the **y-axis,** and the point of intersection the **origin O.** See Figure 1. We assign coordinates to every point on these number lines using a convenient scale. Recall that the scale of a number line is the distance between 0 and 1. In mathematics, we usually use the same scale on each axis; in applications, a different scale is often used on each axis.

The origin *O* has a value of 0 on both the *x*-axis and *y*-axis. Points on the *x*-axis to the right of *O* are associated with positive real numbers, and those to the left of *O* are associated with negative real numbers. Points on the *y*-axis above *O* are associated with positive real numbers, and those below *O* are associated with negative real numbers. In Figure 1, the *x*-axis and *y*-axis are labeled as *x* and *y*, respectively, and we have used an arrow at the end of each axis to denote the positive direction.

The coordinate system described here is called a **rectangular** or **Cartesian*** **coordinate system.** The plane formed by the *x*-axis and *y*-axis is sometimes called the **xy-plane,** and the *x*-axis and *y*-axis are referred to as the **coordinate axes.**

Figure 2

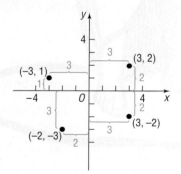

Any point *P* in the *xy*-plane can be located by using an **ordered pair** (x, y) of real numbers. Let *x* denote the signed distance of *P* from the *y*-axis (*signed* means that, if *P* is to the right of the *y*-axis, then $x > 0$, and if *P* is to the left of the *y*-axis, then $x < 0$); and let *y* denote the signed distance of *P* from the *x*-axis. The ordered pair (x, y), also called the **coordinates** of *P*, then gives us enough information to locate the point *P* in the plane.

For example, to locate the point whose coordinates are $(-3, 1)$, go 3 units along the *x*-axis to the left of *O* and then go straight up 1 unit. We **plot** this point by placing a dot at this location. See Figure 2, in which the points with coordinates $(-3, 1)$, $(-2, -3)$, $(3, -2)$, and $(3, 2)$ are plotted.

The origin has coordinates $(0, 0)$. Any point on the *x*-axis has coordinates of the form $(x, 0)$, and any point on the *y*-axis has coordinates of the form $(0, y)$.

If (x, y) are the coordinates of a point *P*, then *x* is called the **x-coordinate,** or **abscissa,** of *P* and *y* is the **y-coordinate,** or **ordinate,** of *P*. We identify the point *P* by its coordinates (x, y) by writing $P = (x, y)$. Usually, we will simply say "the point (x, y)" rather than "the point whose coordinates are (x, y)."

The coordinate axes divide the *xy*-plane into four sections called **quadrants,** as shown in Figure 3. In quadrant I, both the *x*-coordinate and the *y*-coordinate of all points are positive; in quadrant II, *x* is negative and *y* is positive; in quadrant III, both *x* and *y* are negative; and in quadrant IV, *x* is positive and *y* is negative. Points on the coordinate axes belong to no quadrant.

Figure 3

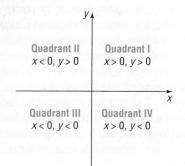

Now Work PROBLEM 11

*Named after René Descartes (1596–1650), a French mathematician, philosopher, and theologian.

Figure 4

COMMENT On a graphing calculator, you can set the scale on each axis. Once this has been done, you obtain the **viewing rectangle.** See Figure 4 for a typical viewing rectangle. You should now read Section B.1, *The Viewing Rectangle*, in Appendix B. ∎

1 Use the Distance Formula

If the same units of measurement, such as inches, centimeters, and so on, are used for both the x-axis and y-axis, then all distances in the xy-plane can be measured using this unit of measurement.

EXAMPLE 1	

Finding the Distance between Two Points

Find the distance d between the points $(1, 3)$ and $(5, 6)$.

Solution First we plot the points $(1, 3)$ and $(5, 6)$ and connect them with a straight line. See Figure 5(a). We are looking for the length d. We begin by drawing a horizontal line from $(1, 3)$ to $(5, 3)$ and a vertical line from $(5, 3)$ to $(5, 6)$, forming a right triangle, as shown in Figure 5(b). One leg of the triangle is of length 4 (since $|5 - 1| = 4$), and the other is of length 3 (since $|6 - 3| = 3$). By the Pythagorean Theorem, the square of the distance d that we seek is

$$d^2 = 4^2 + 3^2 = 16 + 9 = 25$$
$$d = \sqrt{25} = 5$$

Figure 5

(a)   (b)

The **distance formula** provides a straightforward method for computing the distance between two points.

THEOREM

Distance Formula

The distance between two points $P_1 = (x_1, y_1)$ and $P_2 = (x_2, y_2)$, denoted by $d(P_1, P_2)$, is

$$d(P_1, P_2) = \sqrt{(x_2 - x_1)^2 + (y_2 - y_1)^2} \qquad (1)$$

In Words
To compute the distance between two points, find the difference of the x-coordinates, square it, and add this to the square of the difference of the y-coordinates. The square root of this sum is the distance.

Proof of the Distance Formula Let (x_1, y_1) denote the coordinates of point P_1 and let (x_2, y_2) denote the coordinates of point P_2. Assume that the line joining P_1 and P_2 is neither horizontal nor vertical. Refer to Figure 6(a). The coordinates of P_3 are (x_2, y_1). The horizontal distance from P_1 to P_3 is the absolute value of the difference of the x-coordinates, $|x_2 - x_1|$. The vertical distance from P_3 to P_2 is the

Figure 6

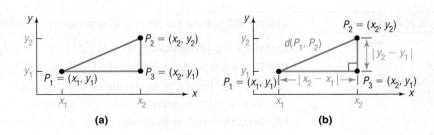

(a)   (b)

absolute value of the difference of the y-coordinates, $|y_2 - y_1|$. See Figure 6(b). The distance $d(P_1, P_2)$ that we seek is the length of the hypotenuse of the right triangle, so, by the Pythagorean Theorem, it follows that

$$[d(P_1, P_2)]^2 = |x_2 - x_1|^2 + |y_2 - y_1|^2$$
$$= (x_2 - x_1)^2 + (y_2 - y_1)^2$$
$$d(P_1, P_2) = \sqrt{(x_2 - x_1)^2 + (y_2 - y_1)^2}$$

Now, if the line joining P_1 and P_2 is horizontal, then the y-coordinate of P_1 equals the y-coordinate of P_2; that is, $y_1 = y_2$. Refer to Figure 7(a). In this case, the distance formula (1) still works, because, for $y_1 = y_2$, it reduces to

$$d(P_1, P_2) = \sqrt{(x_2 - x_1)^2 + 0^2} = \sqrt{(x_2 - x_1)^2} = |x_2 - x_1|$$

Figure 7

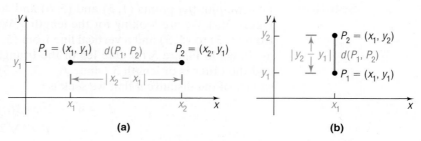

(a)   (b)

A similar argument holds if the line joining P_1 and P_2 is vertical. See Figure 7(b). The distance formula is valid in all cases.   ■

EXAMPLE 2

Using the Distance Formula

Find the distance d between the points $(-3, 5)$ and $(3, 2)$.

Solution

Use the distance formula, equation (1), with $P_1 = (x_1, y_1) = (-3, 5)$ and $P_2 = (x_2, y_2) = (3, 2)$. Then

$$d = \sqrt{[3 - (-3)]^2 + (2 - 5)^2} = \sqrt{6^2 + (-3)^2}$$
$$= \sqrt{36 + 9}$$
$$= \sqrt{45}$$
$$= 3\sqrt{5} \approx 6.71$$

●

Now Work PROBLEMS **15** AND **19**

The distance between two points $P_1 = (x_1, y_1)$ and $P_2 = (x_2, y_2)$ is never a negative number. Furthermore, the distance between two points is 0 only when the points are identical, that is, when $x_1 = x_2$ and $y_1 = y_2$. Also, because $(x_2 - x_1)^2 = (x_1 - x_2)^2$ and $(y_2 - y_1)^2 = (y_1 - y_2)^2$, it makes no difference whether the distance is computed from P_1 to P_2 or from P_2 to P_1; that is, $d(P_1, P_2) = d(P_2, P_1)$.

The introduction to this chapter mentioned that rectangular coordinates enable us to translate geometry problems into algebra problems, and vice versa. The next example shows how algebra (the distance formula) can be used to solve geometry problems.

EXAMPLE 3

Using Algebra to Solve Geometry Problems

Consider the three points $A = (-2, 1)$, $B = (2, 3)$, and $C = (3, 1)$.

(a) Plot each point and form the triangle ABC.
(b) Find the length of each side of the triangle.
(c) Verify that the triangle is a right triangle.
(d) Find the area of the triangle.

Solution

(a) Points A, B, C and triangle ABC are plotted in Figure 8.

(b) We use the distance formula, equation (1).

Figure 8

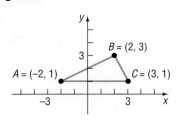

$$d(A, B) = \sqrt{[2 - (-2)]^2 + (3 - 1)^2} = \sqrt{16 + 4} = \sqrt{20} = 2\sqrt{5}$$
$$d(B, C) = \sqrt{(3 - 2)^2 + (1 - 3)^2} = \sqrt{1 + 4} = \sqrt{5}$$
$$d(A, C) = \sqrt{[3 - (-2)]^2 + (1 - 1)^2} = \sqrt{25 + 0} = 5$$

(c) To show that the triangle is a right triangle, we need to show that the sum of the squares of the lengths of two of the sides equals the square of the length of the third side. (Why is this sufficient?) Looking at Figure 8, it seems reasonable to conjecture that the right angle is at vertex B. We shall check to see whether

$$[d(A, B)]^2 + [d(B, C)]^2 = [d(A, C)]^2$$

We find that

$$[d(A, B)]^2 + [d(B, C)]^2 = \left(2\sqrt{5}\right)^2 + \left(\sqrt{5}\right)^2$$
$$= 20 + 5 = 25 = [d(A, C)]^2$$

so it follows from the converse of the Pythagorean Theorem that triangle ABC is a right triangle.

(d) Because the right angle is at vertex B, the sides AB and BC form the base and height of the triangle. Its area is

$$\text{Area} = \frac{1}{2}(\text{Base})(\text{Height}) = \frac{1}{2}\left(2\sqrt{5}\right)\left(\sqrt{5}\right) = 5 \text{ square units}$$

●

Now Work PROBLEM 29

2 Use the Midpoint Formula

Figure 9

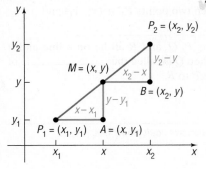

We now derive a formula for the coordinates of the **midpoint of a line segment.** Let $P_1 = (x_1, y_1)$ and $P_2 = (x_2, y_2)$ be the endpoints of a line segment, and let $M = (x, y)$ be the point on the line segment that is the same distance from P_1 as it is from P_2. See Figure 9. The triangles P_1AM and MBP_2 are congruent. [Do you see why? Angle $AP_1M = $ angle BMP_2,* angle $P_1MA = $ angle MP_2B, and $d(P_1, M) = d(M, P_2)$ is given. So, we have angle–side–angle.] Hence, corresponding sides are equal in length. That is,

$$x - x_1 = x_2 - x \qquad \text{and} \qquad y - y_1 = y_2 - y$$
$$2x = x_1 + x_2 \qquad\qquad\qquad 2y = y_1 + y_2$$
$$x = \frac{x_1 + x_2}{2} \qquad\qquad\qquad y = \frac{y_1 + y_2}{2}$$

THEOREM

In Words

To find the midpoint of a line segment, average the x-coordinates, and average the y-coordinates of the endpoints.

Midpoint Formula

The midpoint $M = (x, y)$ of the line segment from $P_1 = (x_1, y_1)$ to $P_2 = (x_2, y_2)$ is

$$M = (x, y) = \left(\frac{x_1 + x_2}{2}, \frac{y_1 + y_2}{2} \right) \qquad (2)$$

*A postulate from geometry states that the transversal $\overline{P_1P_2}$ forms congruent corresponding angles with the parallel line segments $\overline{P_1A}$ and $\overline{MB}$.

EXAMPLE 4

Finding the Midpoint of a Line Segment

Find the midpoint of the line segment from $P_1 = (-5, 5)$ to $P_2 = (3, 1)$. Plot the points P_1 and P_2 and the midpoint.

Solution

We apply the midpoint formula (2) using $x_1 = -5$, $y_1 = 5$, $x_2 = 3$, and $y_2 = 1$. Then the coordinates (x, y) of the midpoint M are

$$x = \frac{x_1 + x_2}{2} = \frac{-5 + 3}{2} = -1 \quad \text{and} \quad y = \frac{y_1 + y_2}{2} = \frac{5 + 1}{2} = 3$$

That is, $M = (-1, 3)$. See Figure 10.

Figure 10

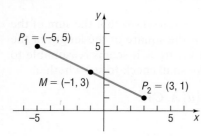

━━━ **Now Work** PROBLEM 35

1 Assess Your Understanding

'Are You Prepared?' *Answers are given at the end of these exercises. If you get a wrong answer, read the pages listed in red.*

1. On the real number line the origin is assigned the number _____. (p. A4)

2. If -3 and 5 are the coordinates of two points on the real number line, the distance between these points is _____. (p. A6)

3. If 3 and 4 are the legs of a right triangle, the hypotenuse is _____. (pp. A14–A15)

4. Use the converse of the Pythagorean Theorem to show that a triangle whose sides are of lengths 11, 60, and 61 is a right triangle. (p. A14)

5. State the formula for the area A of a triangle whose base is b and whose altitude is h. (p. A15)

6. State the three cases for which two triangles are congruent. (p. A17)

Concepts and Vocabulary

7. If (x, y) are the coordinates of a point P in the xy-plane, then x is called the _____ of P and y is the _____ of P.

8. The coordinate axes divide the xy-plane into four sections called _____.

9. The distance d between two points $P_1 = (x_1, y_1)$ and $P_2 = (x_2, y_2)$ is $d =$ _____.

10. If three distinct points P, Q, and R all lie on a line and if $d(P, Q) = d(Q, R)$, then Q is called the _____ of the line segment from P to R.

Skill Building

In Problems 11 and 12, plot each point in the xy-plane. Tell in which quadrant or on what coordinate axis each point lies.

11. (a) $A = (-3, 2)$
    (b) $B = (6, 0)$
    (c) $C = (-2, -2)$
    (d) $D = (6, 5)$
    (e) $E = (0, -3)$
    (f) $F = (6, -3)$

12. (a) $A = (1, 4)$
    (b) $B = (-3, -4)$
    (c) $C = (-3, 4)$
    (d) $D = (4, 1)$
    (e) $E = (0, 1)$
    (f) $F = (-3, 0)$

13. Plot the points $(2, 0), (2, -3), (2, 4), (2, 1),$ and $(2, -1)$. Describe the set of all points of the form $(2, y)$, where y is a real number.

14. Plot the points $(0, 3), (1, 3), (-2, 3), (5, 3),$ and $(-4, 3)$. Describe the set of all points of the form $(x, 3)$, where x is a real number.

In Problems 15–28, find the distance $d(P_1, P_2)$ between the points P_1 and P_2.

15.

16.

17.

18.

19. $P_1 = (3, -4)$; $P_2 = (5, 4)$

20. $P_1 = (-1, 0)$; $P_2 = (2, 4)$

21. $P_1 = (-3, 2)$; $P_2 = (6, 0)$

22. $P_1 = (2, -3)$; $P_2 = (4, 2)$

23. $P_1 = (4, -2)$; $P_2 = (-2, -5)$

24. $P_1 = (-4, -3)$; $P_2 = (6, 2)$

25. $P_1 = (-0.2, 0.3)$; $P_2 = (2.3, 1.1)$

26. $P_1 = (1.2, 2.3)$; $P_2 = (-0.3, 1.1)$

27. $P_1 = (a, b)$; $P_2 = (0, 0)$

28. $P_1 = (a, a)$; $P_2 = (0, 0)$

In Problems 29–34, plot each point and form the triangle ABC. Verify that the triangle is a right triangle. Find its area.

29. $A = (-2, 5)$; $B = (1, 3)$; $C = (-1, 0)$

30. $A = (-2, 5)$; $B = (12, 3)$; $C = (10, -11)$

31. $A = (-5, 3)$; $B = (6, 0)$; $C = (5, 5)$

32. $A = (-6, 3)$; $B = (3, -5)$; $C = (-1, 5)$

33. $A = (4, -3)$; $B = (0, -3)$; $C = (4, 2)$

34. $A = (4, -3)$; $B = (4, 1)$; $C = (2, 1)$

In Problems 35–44, find the midpoint of the line segment joining the points P_1 and P_2.

35. $P_1 = (3, -4)$; $P_2 = (5, 4)$

36. $P_1 = (-2, 0)$; $P_2 = (2, 4)$

37. $P_1 = (-3, 2)$; $P_2 = (6, 0)$

38. $P_1 = (2, -3)$; $P_2 = (4, 2)$

39. $P_1 = (4, -2)$; $P_2 = (-2, -5)$

40. $P_1 = (-4, -3)$; $P_2 = (2, 2)$

41. $P_1 = (-0.2, 0.3)$; $P_2 = (2.3, 1.1)$

42. $P_1 = (1.2, 2.3)$; $P_2 = (-0.3, 1.1)$

43. $P_1 = (a, b)$; $P_2 = (0, 0)$

44. $P_1 = (a, a)$; $P_2 = (0, 0)$

Applications and Extensions

45. Find all points having an x-coordinate of 2 whose distance from the point $(-2, -1)$ is 5.

46. Find all points having a y-coordinate of -3 whose distance from the point $(1, 2)$ is 13.

47. Find all points on the x-axis that are 5 units from the point $(4, -3)$.

48. Find all points on the y-axis that are 5 units from the point $(4, 4)$.

49. Geometry The **medians** of a triangle are the line segments from each vertex to the midpoint of the opposite side (see the figure). Find the lengths of the medians of the triangle with vertices at $A = (0, 0)$, $B = (6, 0)$, and $C = (4, 4)$.

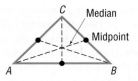

50. Geometry An **equilateral triangle** is one in which all three sides are of equal length. If two vertices of an equilateral triangle are $(0, 4)$ and $(0, 0)$, find the third vertex. How many of these triangles are possible?

51. Geometry Find the midpoint of each diagonal of a square with side of length s. Draw the conclusion that the diagonals of a square intersect at their midpoints.

[**Hint:** Use $(0, 0)$, $(0, s)$, $(s, 0)$, and (s, s) as the vertices of the square.]

52. Geometry Verify that the points $(0, 0)$, $(a, 0)$, and $\left(\dfrac{a}{2}, \dfrac{\sqrt{3}a}{2}\right)$ are the vertices of an equilateral triangle. Then show that the midpoints of the three sides are the vertices of a second equilateral triangle (refer to Problem 50).

*In Problems 53–56, find the length of each side of the triangle determined by the three points P_1, P_2, and P_3. State whether the triangle is an isosceles triangle, a right triangle, neither of these, or both. (An **isosceles triangle** is one in which at least two of the sides are of equal length.)*

53. $P_1 = (2, 1)$; $P_2 = (-4, 1)$; $P_3 = (-4, -3)$

54. $P_1 = (-1, 4)$; $P_2 = (6, 2)$; $P_3 = (4, -5)$

55. $P_1 = (-2, -1)$; $P_2 = (0, 7)$; $P_3 = (3, 2)$

56. $P_1 = (7, 2)$; $P_2 = (-4, 0)$; $P_3 = (4, 6)$

57. Baseball A major league baseball "diamond" is actually a square, 90 feet on a side (see the figure). What is the distance directly from home plate to second base (the diagonal of the square)?

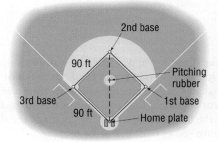

58. Little League Baseball The layout of a Little League playing field is a square, 60 feet on a side.* How far is it directly from home plate to second base (the diagonal of the square)?

Source: Little League Baseball, Official Regulations and Playing Rules, 2005.

59. Baseball Refer to Problem 57. Overlay a rectangular coordinate system on a major league baseball diamond so that the origin is at home plate, the positive x-axis lies in the direction from home plate to first base, and the positive y-axis lies in the direction from home plate to third base.
  (a) What are the coordinates of first base, second base, and third base? Use feet as the unit of measurement.
  (b) If the right fielder is located at $(310, 15)$, how far is it from there to second base?
  (c) If the center fielder is located at $(300, 300)$, how far is it from there to third base?

60. Little League Baseball Refer to Problem 58. Overlay a rectangular coordinate system on a Little League baseball diamond so that the origin is at home plate, the positive x-axis lies in the direction from home plate to first base, and the positive y-axis lies in the direction from home plate to third base.
  (a) What are the coordinates of first base, second base, and third base? Use feet as the unit of measurement.
  (b) If the right fielder is located at $(180, 20)$, how far is it from there to second base?
  (c) If the center fielder is located at $(220, 220)$, how far is it from there to third base?

61. Distance between Moving Objects A Dodge Neon and a Mack truck leave an intersection at the same time. The Neon heads east at an average speed of 30 miles per hour, while the truck heads south at an average speed of 40 miles per hour. Find an expression for their distance apart d (in miles) at the end of t hours.

62. Distance of a Moving Object from a Fixed Point A hot-air balloon, headed due east at an average speed of 15 miles per hour and at a constant altitude of 100 feet, passes over an intersection (see the figure). Find an expression for the distance d (measured in feet) from the balloon to the intersection t seconds later.

100 ft

15 mph

63. Drafting Error When a draftsman draws three lines that are to intersect at one point, the lines may not intersect as intended and subsequently will form an **error triangle.** If this error triangle is long and thin, one estimate for the location of the desired point is the midpoint of the shortest side. The figure shows one such error triangle.

(a) Find an estimate for the desired intersection point.
(b) Find the length of the median for the midpoint found in part (a). See Problem 49.

Source: www.uwgb.edu/DutchS/structge/s100.htm

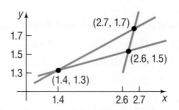

64. Net Sales The figure illustrates how net sales of Wal-Mart Stores, Inc., have grown from 2002 through 2008. Use the midpoint formula to estimate the net sales of Wal-Mart Stores, Inc., in 2005. How does your result compare to the reported value of $282 billion?

Source: Wal-Mart Stores, Inc., 2008 Annual Report

Wal-Mart Stores, Inc.
Net sales (in $ billions)

65. Poverty Threshold Poverty thresholds are determined by the U.S. Census Bureau. A poverty threshold represents the minimum annual household income for a family not to be considered poor. In 1998, the poverty threshold for a family of four with two children under the age of 18 years was $16,530. In 2008, the poverty threshold for a family of four with two children under the age of 18 years was $21,834. Assuming poverty thresholds increase in a straight-line fashion, use the midpoint formula to estimate the poverty threshold of a family of four with two children under the age of 18 in 2003. How does your result compare to the actual poverty threshold in 2003 of $18,660?

Source: U.S. Census Bureau

66. Horizontal and Vertical Shifts Suppose that $A = (2, 5)$ are the coordinates of a point in the xy-plane.
  (a) Find the coordinates of the point if A is shifted 3 units to the right and 2 units down.
  (b) Find the coordinates of the point if A is shifted 2 units to the left and 8 units up.

67. Completing a Line Segment Plot the points $A = (-1, 8)$ and $M = (2, 3)$ in the xy-plane. If M is the midpoint of a line segment AB, find the coordinates of B.

Explaining Concepts: Discussion and Writing

68. Write a paragraph that describes a Cartesian plane. Then write a second paragraph that describes how to plot points in the Cartesian plane. Your paragraphs should include the terms "coordinate axes," "ordered pair," "coordinates," "plot," "x-coordinate," and "y-coordinate."

'Are You Prepared?' Answers

1. 0   **2.** 8   **3.** 5   **4.** $11^2 + 60^2 = 61^2$ **5.** $A = \frac{1}{2}bh$ **6.** Angle–Side–Angle; Side–Side–Side; Side–Angle–Side

---

2 Graphs of Equations in Two Variables; Intercepts; Symmetry

PREPARING FOR THIS SECTION   *Before getting started, review the following:*

- Solving Equations (Appendix A, Section A.8, pp. A65–A71)

Now Work the *'Are You Prepared?'* problems on page 16.

OBJECTIVES   **1** Graph Equations by Plotting Points (p. 9)
  2 Find Intercepts from a Graph (p. 11)
  3 Find Intercepts from an Equation (p. 12)
  4 Test an Equation for Symmetry (p. 12)
  5 Know How to Graph Key Equations (p. 14)

1 Graph Equations by Plotting Points

An **equation in two variables,** say x and y, is a statement in which two expressions involving x and y are equal. The expressions are called the **sides** of the equation. Since an equation is a statement, it may be true or false, depending on the value of the variables. Any values of x and y that result in a true statement are said to **satisfy** the equation.

For example, the following are all equations in two variables x and y:

$$x^2 + y^2 = 5 \qquad 2x - y = 6 \qquad y = 2x + 5 \qquad x^2 = y$$

The first of these, $x^2 + y^2 = 5$, is satisfied for $x = 1, y = 2$, since $1^2 + 2^2 = 1 + 4 = 5$. Other choices of x and y, such as $x = -1, y = -2$, also satisfy this equation. It is not satisfied for $x = 2$ and $y = 3$, since $2^2 + 3^2 = 4 + 9 = 13 \neq 5$.

The **graph of an equation in two variables** x and y consists of the set of points in the xy-plane whose coordinates (x, y) satisfy the equation.

EXAMPLE 1   **Determining Whether a Point Is on the Graph of an Equation**

Determine if the following points are on the graph of the equation $2x - y = 6$.

(a) $(2, 3)$ (b) $(2, -2)$

Solution   (a) For the point $(2, 3)$, we check to see if $x = 2, y = 3$ satisfies the equation $2x - y = 6$.

$$2x - y = 2(2) - 3 = 4 - 3 = 1 \neq 6$$

The equation is not satisfied, so the point $(2, 3)$ is not on the graph.

(b) For the point $(2, -2)$, we have

$$2x - y = 2(2) - (-2) = 4 + 2 = 6$$

The equation is satisfied, so the point $(2, -2)$ is on the graph.

━━━━▶ **Now Work** PROBLEM 9

EXAMPLE 2 | **How to Graph an Equation by Plotting Points**

Graph the equation: $y = -2x + 3$

Step-by-Step Solution

Step 1: Find points (x, y) that satisfy the equation. To determine these points, choose values of x and use the equation to find the corresponding values for y. See Table 1.

Table 1

x	$y = -2x + 3$	(x, y)
-2	$-2(-2) + 3 = 7$	$(-2, 7)$
-1	$-2(-1) + 3 = 5$	$(-1, 5)$
0	$-2(0) + 3 = 3$	$(0, 3)$
1	$-2(1) + 3 = 1$	$(1, 1)$
2	$-2(2) + 3 = -1$	$(2, -1)$

Step 2: Plot the points found in the table as shown in Figure 11(a). Now connect the points to obtain the graph of the equation (a *line*), as shown in Figure 11(b).

Figure 11

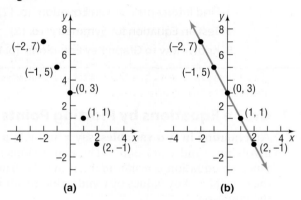

(a)   (b)

EXAMPLE 3 | **Graphing an Equation by Plotting Points**

Graph the equation: $y = x^2$

Solution   Table 2 provides several points on the graph. In Figure 12 we plot these points and connect them with a smooth curve to obtain the graph (a *parabola*).

Table 2

x	$y = x^2$	(x, y)
-4	16	$(-4, 16)$
-3	9	$(-3, 9)$
-2	4	$(-2, 4)$
-1	1	$(-1, 1)$
0	0	$(0, 0)$
1	1	$(1, 1)$
2	4	$(2, 4)$
3	9	$(3, 9)$
4	16	$(4, 16)$

Figure 12

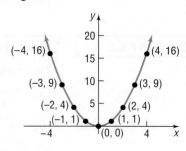

The graphs of the equations shown in Figures 11 and 12 do not show all the points that are on the graph. For example, in Figure 11, the point $(20, -37)$ is a part of the graph of $y = -2x + 3$, but it is not shown. Since the graph of $y = -2x + 3$ could be extended out as far as we please, we use arrows to indicate that the pattern shown continues. It is important when illustrating a graph to present enough of the graph so that any viewer of the illustration will "see" the rest of it as an obvious continuation of what is actually there. This is referred to as a **complete graph.**

One way to obtain a complete graph of an equation is to plot a sufficient number of points on the graph until a pattern becomes evident. Then these points are connected with a smooth curve following the suggested pattern. But how many points are sufficient? Sometimes knowledge about the equation tells us. For example, we will learn in the next section that, if an equation is of the form $y = mx + b$, then its graph is a line. In this case, only two points are needed to obtain the graph.

One purpose of this book is to investigate the properties of equations in order to decide whether a graph is complete. Sometimes we shall graph equations by plotting points. Shortly, we shall investigate various techniques that will enable us to graph an equation without plotting so many points.

🖳 COMMENT   Another way to obtain the graph of an equation is to use a graphing utility. Read Section B.2, *Using a Graphing Utility to Graph Equations*, in Appendix B.   ∎

Two techniques that sometimes reduce the number of points required to graph an equation involve finding *intercepts* and checking for *symmetry*.

2 Find Intercepts from a Graph

The points, if any, at which a graph crosses or touches the coordinate axes are called the **intercepts.** See Figure 13. The x-coordinate of a point at which the graph crosses or touches the x-axis is an **x-intercept,** and the y-coordinate of a point at which the graph crosses or touches the y-axis is a **y-intercept.** For a graph to be complete, all its intercepts must be displayed.

Figure 13

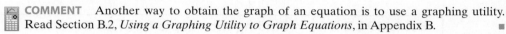

EXAMPLE 4   **Finding Intercepts from a Graph**

Find the intercepts of the graph in Figure 14. What are its x-intercepts? What are its y-intercepts?

Solution   The intercepts of the graph are the points

$$(-3, 0), \quad (0, 3), \quad \left(\frac{3}{2}, 0\right), \quad \left(0, -\frac{4}{3}\right), \quad (0, -3.5), \quad (4.5, 0)$$

The x-intercepts are -3, $\dfrac{3}{2}$, and 4.5; the y-intercepts are -3.5, $-\dfrac{4}{3}$, and 3.   ●

Figure 14

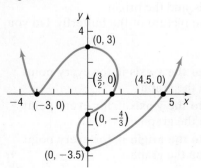

In Example 4, you should notice the following usage: If we do not specify the type of intercept (x- versus y-), then we report the intercept as an ordered pair. However, if we specify the type of intercept, then we only report the coordinate of the intercept. For x-intercepts, we report the x-coordinate of the intercept; for y-intercepts, we report the y-coordinate of the intercept.

⟜━━━━ **Now Work** PROBLEM 37(a)

3 Find Intercepts from an Equation

The intercepts of a graph can be found from its equation by using the fact that points on the x-axis have y-coordinates equal to 0 and points on the y-axis have x-coordinates equal to 0.

Procedure for Finding Intercepts

1. To find the x-intercept(s), if any, of the graph of an equation, let $y = 0$ in the equation and solve for x.

2. To find the y-intercept(s), if any, of the graph of an equation, let $x = 0$ in the equation and solve for y.

Because the x-intercepts of the graph of an equation are those x-values for which $y = 0$, they are also called the **zeros** (or **roots**) of the equation.

EXAMPLE 5 | **Finding Intercepts from an Equation**

Find the x-intercept(s) and the y-intercept(s) of the graph of $y = x^2 - 4$.

Solution To find the x-intercept(s), we let $y = 0$ and obtain the equation

$$x^2 - 4 = 0$$
$$(x + 2)(x - 2) = 0 \quad \text{Factor.}$$
$$x + 2 = 0 \quad \text{or} \quad x - 2 = 0 \quad \text{Zero-Product Property}$$
$$x = -2 \quad \text{or} \quad x = 2 \quad \text{Solve.}$$

The equation has two solutions, -2 and 2. The x-intercepts (the zeros) are -2 and 2. To find the y-intercept(s), we let $x = 0$ in the equation.

$$y = x^2 - 4$$
$$= 0^2 - 4$$
$$= -4$$

The y-intercept is -4.

 Now Work PROBLEM 19

COMMENT For many equations, finding intercepts may not be so easy. In such cases, a graphing utility can be used. Read the first part of Section B.3, *Using a Graphing Utility to Locate Intercepts and Check for Symmetry*, in Appendix B to find out how a graphing utility locates intercepts. ∎

4 Test an Equation for Symmetry

Another helpful tool for graphing equations by hand involves *symmetry*, particularly symmetry with respect to the x-axis, the y-axis, and the origin.

Symmetry often occurs in nature. Consider the picture of the butterfly. Do you see the symmetry?

DEFINITION A graph is said to be **symmetric with respect to the x-axis** if, for every point (x, y) on the graph, the point $(x, -y)$ is also on the graph.

A graph is said to be **symmetric with respect to the y-axis** if, for every point (x, y) on the graph, the point $(-x, y)$ is also on the graph.

A graph is said to be **symmetric with respect to the origin** if, for every point (x, y) on the graph, the point $(-x, -y)$ is also on the graph.

Figure 15 illustrates the definition. When a graph is symmetric with respect to the *x*-axis, notice that the part of the graph above the *x*-axis is a reflection or mirror image of the part below it, and vice versa. When a graph is symmetric with respect to the *y*-axis, the part of the graph to the right of the *y*-axis is a reflection of the part to the left of it, and vice versa. Symmetry with respect to the origin may be viewed in two ways:

1. As a reflection about the *y*-axis, followed by a reflection about the *x*-axis
2. As a projection along a line through the origin so that the distances from the origin are equal

Figure 15

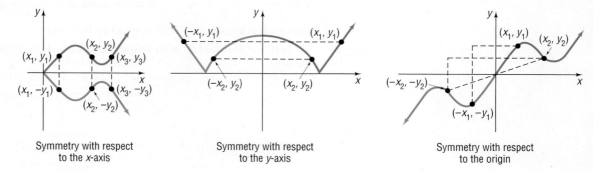

Symmetry with respect to the *x*-axis   Symmetry with respect to the *y*-axis   Symmetry with respect to the origin

EXAMPLE 6   **Symmetric Points**

(a) If a graph is symmetric with respect to the *x*-axis and the point $(4, 2)$ is on the graph, then the point $(4, -2)$ is also on the graph.
(b) If a graph is symmetric with respect to the *y*-axis and the point $(4, 2)$ is on the graph, then the point $(-4, 2)$ is also on the graph.
(c) If a graph is symmetric with respect to the origin and the point $(4, 2)$ is on the graph, then the point $(-4, -2)$ is also on the graph.

Now Work PROBLEM 27

When the graph of an equation is symmetric with respect to the *x*-axis, the *y*-axis, or the origin, the number of points that you need to plot in order to see the pattern is reduced. For example, if the graph of an equation is symmetric with respect to the *y*-axis, then, once points to the right of the *y*-axis are plotted, an equal number of points on the graph can be obtained by reflecting them about the *y*-axis. Because of this, before we graph an equation, we first want to determine whether it has any symmetry. The following tests are used for this purpose.

Tests for Symmetry

To test the graph of an equation for symmetry with respect to the

x-Axis   Replace *y* by $-y$ in the equation. If an equivalent equation results, the graph of the equation is symmetric with respect to the *x*-axis.

y-Axis   Replace *x* by $-x$ in the equation. If an equivalent equation results, the graph of the equation is symmetric with respect to the *y*-axis.

Origin   Replace *x* by $-x$ and *y* by $-y$ in the equation. If an equivalent equation results, the graph of the equation is symmetric with respect to the origin.

| EXAMPLE 7 | **Testing an Equation for Symmetry** |

Test $y = \dfrac{4x^2}{x^2 + 1}$ for symmetry.

Solution   *x-Axis:*   To test for symmetry with respect to the x-axis, replace y by −y. Since $-y = \dfrac{4x^2}{x^2 + 1}$ is not equivalent to $y = \dfrac{4x^2}{x^2 + 1}$, the graph of the equation is not symmetric with respect to the x-axis.

y-Axis: To test for symmetry with respect to the y-axis, replace x by −x. Since $y = \dfrac{4(-x)^2}{(-x)^2 + 1} = \dfrac{4x^2}{x^2 + 1}$ is equivalent to $y = \dfrac{4x^2}{x^2 + 1}$, the graph of the equation is symmetric with respect to the y-axis.

Origin: To test for symmetry with respect to the origin, replace x by −x and y by −y.

$$-y = \frac{4(-x)^2}{(-x)^2 + 1} \qquad \text{Replace x by } -x \text{ and y by } -y.$$

$$-y = \frac{4x^2}{x^2 + 1} \qquad \text{Simplify.}$$

$$y = -\frac{4x^2}{x^2 + 1} \qquad \text{Multiply both sides by } -1.$$

Since the result is not equivalent to the original equation, the graph of the equation $y = \dfrac{4x^2}{x^2 + 1}$ is not symmetric with respect to the origin.

Figure 16

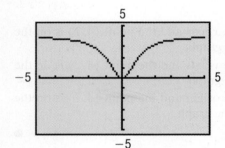

🖩 Seeing the Concept

Figure 16 shows the graph of $y = \dfrac{4x^2}{x^2 + 1}$ using a graphing utility. Do you see the symmetry with respect to the y-axis?

✎━━━ **Now Work** PROBLEM 57

5 Know How to Graph Key Equations

The next three examples use intercepts, symmetry, and point plotting to obtain the graphs of key equations. It is important to know the graphs of these key equations because we use them later. The first of these is $y = x^3$.

| EXAMPLE 8 | **Graphing the Equation $y = x^3$ by Finding Intercepts and Checking for Symmetry** |

Graph the equation $y = x^3$ by plotting points. Find any intercepts and check for symmetry first.

Solution   First, we seek the intercepts. When $x = 0$, then $y = 0$; and when $y = 0$, then $x = 0$. The origin $(0, 0)$ is the only intercept. Now we test for symmetry.

x-Axis: Replace y by −y. Since $-y = x^3$ is not equivalent to $y = x^3$, the graph is not symmetric with respect to the x-axis.

y-Axis: Replace x by −x. Since $y = (-x)^3 = -x^3$ is not equivalent to $y = x^3$, the graph is not symmetric with respect to the y-axis.

Origin: Replace x by −x and y by −y. Since $-y = (-x)^3 = -x^3$ is equivalent to $y = x^3$ (multiply both sides by −1), the graph is symmetric with respect to the origin.

To graph $y = x^3$, we use the equation to obtain several points on the graph. Because of the symmetry, we only need to locate points on the graph for which $x \geq 0$. See Table 3. Since $(1, 1)$ is on the graph, and the graph is symmetric with respect to the origin, the point $(-1, -1)$ is also on the graph. Figure 17 shows the graph.

Table 3

x	$y = x^3$	(x, y)
0	0	$(0, 0)$
1	1	$(1, 1)$
2	8	$(2, 8)$
3	27	$(3, 27)$

Figure 17

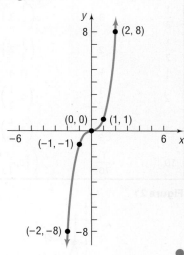

EXAMPLE 9

Graphing the Equation $x = y^2$

(a) Graph the equation $x = y^2$. Find any intercepts and check for symmetry first.

(b) Graph $x = y^2$, $y \geq 0$.

Solution

(a) The lone intercept is $(0, 0)$. The graph is symmetric with respect to the x-axis. (Do you see why? Replace y by $-y$.) Figure 18 shows the graph.

(b) If we restrict y so that $y \geq 0$, the equation $x = y^2$, $y \geq 0$, may be written equivalently as $y = \sqrt{x}$. The portion of the graph of $x = y^2$ in quadrant I is therefore the graph of $y = \sqrt{x}$. See Figure 19.

Figure 20

Figure 18

Figure 19

COMMENT   To see the graph of the equation $x = y^2$ on a graphing calculator, you will need to graph two equations: $Y_1 = \sqrt{x}$ and $Y_2 = -\sqrt{x}$. We discuss why in Chapter 1. See Figure 20. ∎

EXAMPLE 10

Graphing the Equation $y = \dfrac{1}{x}$

Graph the equation $y = \dfrac{1}{x}$. Find any intercepts and check for symmetry first.

Solution

We check for intercepts first. If we let $x = 0$, we obtain 0 in the denominator, which makes y undefined. We conclude that there is no y-intercept. If we let $y = 0$, we get the equation $\dfrac{1}{x} = 0$, which has no solution. We conclude that there is no x-intercept. The graph of $y = \dfrac{1}{x}$ does not cross or touch the coordinate axes.

Table 4

x	$y = \dfrac{1}{x}$	(x, y)
$\dfrac{1}{10}$	10	$\left(\dfrac{1}{10}, 10\right)$
$\dfrac{1}{3}$	3	$\left(\dfrac{1}{3}, 3\right)$
$\dfrac{1}{2}$	2	$\left(\dfrac{1}{2}, 2\right)$
1	1	$(1, 1)$
2	$\dfrac{1}{2}$	$\left(2, \dfrac{1}{2}\right)$
3	$\dfrac{1}{3}$	$\left(3, \dfrac{1}{3}\right)$
10	$\dfrac{1}{10}$	$\left(10, \dfrac{1}{10}\right)$

Figure 21

Next we check for symmetry:

x-Axis: Replacing y by $-y$ yields $-y = \dfrac{1}{x}$, which is not equivalent to $y = \dfrac{1}{x}$.

y-Axis: Replacing x by $-x$ yields $y = \dfrac{1}{-x} = -\dfrac{1}{x}$, which is not equivalent to $y = \dfrac{1}{x}$.

Origin: Replacing x by $-x$ and y by $-y$ yields $-y = -\dfrac{1}{x}$, which is equivalent to $y = \dfrac{1}{x}$. The graph is symmetric with respect to the origin.

Now we set up Table 4, listing several points on the graph. Because of the symmetry with respect to the origin, we use only positive values of x. From Table 4 we infer that if x is a large and positive number then $y = \dfrac{1}{x}$ is a positive number close to 0. We also infer that if x is a positive number close to 0 then $y = \dfrac{1}{x}$ is a large and positive number. Armed with this information, we can graph the equation.

Figure 21 illustrates some of these points and the graph of $y = \dfrac{1}{x}$. Observe how the absence of intercepts and the existence of symmetry with respect to the origin are utilized.

COMMENT Refer to Example 3 in Appendix B, Section B.3, for the graph of $y = \dfrac{1}{x}$ using a graphing utility. ∎

2 Assess Your Understanding

'Are You Prepared?' *Answers are given at the end of these exercises. If you get a wrong answer, read the pages listed in red.*

1. Solve the equation $2(x + 3) - 1 = -7$. (pp. A66–A68)

2. Solve the equation $x^2 - 4x - 12 = 0$. (p. A69)

Concepts and Vocabulary

3. The points, if any, at which a graph crosses or touches the coordinate axes are called _____.

4. Because the x-intercepts of the graph of an equation are those x-values for which $y = 0$, they are also called _____ or _____.

5. If for every point (x, y) on the graph of an equation the point $(-x, y)$ is also on the graph, then the graph is symmetric with respect to the _____.

6. *True or False* To find the y-intercepts of the graph of an equation, let $x = 0$ and solve for y.

7. *True or False* The graph of an equation in two variables x and y consists of the set of points in the xy-plane whose coordinates (x, y) satisfy the equation.

8. *True or False* If a graph is symmetric with respect to the x-axis, then it cannot be symmetric with respect to the y-axis.

Skill Building

In Problems 9–14, tell whether the given points are on the graph of the equation.

9. Equation: $y = x^4 - \sqrt{x}$
Points: $(0, 0)$; $(1, 1)$; $(-1, 0)$

10. Equation: $y = x^3 - 2\sqrt{x}$
Points: $(0, 0)$; $(1, 1)$; $(1, -1)$

11. Equation: $y^2 = x^2 + 9$
Points: $(0, 3)$; $(3, 0)$; $(-3, 0)$

12. Equation: $y^3 = x + 1$
Points: $(1, 2)$; $(0, 1)$; $(-1, 0)$

13. Equation: $x^2 + y^2 = 4$
Points: $(0, 2)$; $(-2, 2)$; $\left(\sqrt{2}, \sqrt{2}\right)$

14. Equation: $x^2 + 4y^2 = 4$
Points: $(0, 1)$; $(2, 0)$; $\left(2, \dfrac{1}{2}\right)$

In Problems 15–26, find the intercepts and graph each equation by plotting points. Be sure to label the intercepts.

15. $y = x + 2$

16. $y = x - 6$

17. $y = 2x + 8$

18. $y = 3x - 9$

19. $y = x^2 - 1$

20. $y = x^2 - 9$

21. $y = -x^2 + 4$

22. $y = -x^2 + 1$

23. $2x + 3y = 6$

24. $5x + 2y = 10$

25. $9x^2 + 4y = 36$

26. $4x^2 + y = 4$

In Problems 27–36, plot each point. Then plot the point that is symmetric to it with respect to (a) the x-axis; (b) the y-axis; (c) the origin.

27. $(3, 4)$ **28.** $(5, 3)$ **29.** $(-2, 1)$ **30.** $(4, -2)$ **31.** $(5, -2)$

32. $(-1, -1)$ **33.** $(-3, -4)$ **34.** $(4, 0)$ **35.** $(0, -3)$ **36.** $(-3, 0)$

In Problems 37–48, the graph of an equation is given. (a) Find the intercepts. (b) Indicate whether the graph is symmetric with respect to the x-axis, the y-axis, or the origin.

37.

38.

39.

40.

41.

42.

43.

44.

45.

46.

47.

48.

In Problems 49–52, draw a complete graph so that it has the type of symmetry indicated.

49. *y*-axis

50. *x*-axis

51. Origin

52. *y*-axis

In Problems 53–68, list the intercepts and test for symmetry.

53. $y^2 = x + 4$

54. $y^2 = x + 9$

55. $y = \sqrt[3]{x}$

56. $y = \sqrt[5]{x}$

57. $y = x^4 - 8x^2 - 9$

58. $y = x^4 - 2x^2 - 8$

59. $9x^2 + 4y^2 = 36$

60. $4x^2 + y^2 = 4$

61. $y = x^3 + x^2 - 9x - 9$

62. $y = x^3 + 2x^2 - 4x - 8$

63. $y = |x| - 4$

64. $y = |x| - 2$

65. $y = \dfrac{3x}{x^2 + 9}$

66. $y = \dfrac{x^2 - 4}{2x}$

67. $y = \dfrac{-x^3}{x^2 - 9}$

68. $y = \dfrac{x^4 + 1}{2x^5}$

In Problems 69–72, draw a quick sketch of each equation.

69. $y = x^3$

70. $x = y^2$

71. $y = \sqrt{x}$

72. $y = \dfrac{1}{x}$

73. If $(3, b)$ is a point on the graph of $y = 4x + 1$, what is b?

74. If $(-2, b)$ is a point on the graph of $2x + 3y = 2$, what is b?

75. If $(a, 4)$ is a point on the graph of $y = x^2 + 3x$, what is a?

76. If $(a, -5)$ is a point on the graph of $y = x^2 + 6x$, what is a?

Mixed Practice

In Problems 77–84, (a) find the intercepts of the graph of each equation, (b) test each equation for symmetry with respect to the x-axis, the y-axis, and the origin, and (c) graph each equation by plotting points. Be sure to label the intercepts on the graph and use any symmetry to assist in drawing the graph.

77. $y = x^2 - 5$ **78.** $y = x^2 - 8$ **79.** $x - y^2 = -9$ **80.** $x + y^2 = 4$

81. $x^2 + y^2 = 9$ **82.** $x^2 + y^2 = 16$ **83.** $y = x^3 - 4x$ **84.** $y = x^3 - x$

Applications and Extensions

85. Given that the point $(1, 2)$ is on the graph of an equation that is symmetric with respect to the origin, what other point is on the graph?

86. If the graph of an equation is symmetric with respect to the y-axis and 6 is an x-intercept of this graph, name another x-intercept.

87. If the graph of an equation is symmetric with respect to the origin and -4 is an x-intercept of this graph, name another x-intercept.

88. If the graph of an equation is symmetric with respect to the x-axis and 2 is a y-intercept, name another y-intercept.

89. Microphones In studios and on stages, cardioid microphones are often preferred for the richness they add to voices and for their ability to reduce the level of sound from the sides and rear of the microphone. Suppose one such cardioid pattern is given by the equation $(x^2 + y^2 - x)^2 = x^2 + y^2$.

(a) Find the intercepts of the graph of the equation.

(b) Test for symmetry with respect to the x-axis, y-axis, and origin.

Source: www.notaviva.com

90. Solar Energy The solar electric generating systems at Kramer Junction, California, use parabolic troughs to heat a heat-transfer fluid to a high temperature. This fluid is used to generate steam that drives a power conversion system to produce electricity. For troughs 7.5 feet wide, an equation for the cross section is $16y^2 = 120x - 225$.

(a) Find the intercepts of the graph of the equation.

(b) Test for symmetry with respect to the x-axis, y-axis, and origin.

Source: U.S. Department of Energy

Explaining Concepts: Discussion and Writing

In Problem 91, use a graphing utility.

91. (a) Graph $y = \sqrt{x^2}$, $y = x$, $y = |x|$, and $y = (\sqrt{x})^2$, noting which graphs are the same.

(b) Explain why the graphs of $y = \sqrt{x^2}$ and $y = |x|$ are the same.

(c) Explain why the graphs of $y = x$ and $y = (\sqrt{x})^2$ are not the same.

(d) Explain why the graphs of $y = \sqrt{x^2}$ and $y = x$ are not the same.

92. Explain what is meant by a complete graph.

93. Draw a graph of an equation that contains two x-intercepts; at one the graph crosses the x-axis, and at the other the graph touches the x-axis.

94. Make up an equation with the intercepts $(2, 0)$, $(4, 0)$, and $(0, 1)$. Compare your equation with a friend's equation. Comment on any similarities.

95. Draw a graph that contains the points $(-2, -1)$, $(0, 1)$, $(1, 3)$, and $(3, 5)$. Compare your graph with those of other students. Are most of the graphs almost straight lines? How many are "curved"? Discuss the various ways that these points might be connected.

96. An equation is being tested for symmetry with respect to the x-axis, the y-axis, and the origin. Explain why, if two of these symmetries are present, the remaining one must also be present.

97. Draw a graph that contains the points $(-2, 5)$, $(-1, 3)$, and $(0, 2)$ that is symmetric with respect to the y-axis. Compare your graph with those of other students; comment on any similarities. Can a graph contain these points and be symmetric with respect to the x-axis? the origin? Why or why not?

'Are You Prepared?' Answers

1. $\{-6\}$ **2.** $\{-2, 6\}$

3 Lines

In this section we study a certain type of equation that contains two variables, called a *linear equation*, and its graph, a *line*.

1 Calculate and Interpret the Slope of a Line

Figure 22

Consider the staircase illustrated in Figure 22. Each step contains exactly the same horizontal **run** and the same vertical **rise.** The ratio of the rise to the run, called the *slope*, is a numerical measure of the steepness of the staircase. For example, if the run is increased and the rise remains the same, the staircase becomes less steep. If the run is kept the same, but the rise is increased, the staircase becomes more steep. This important characteristic of a line is best defined using rectangular coordinates.

DEFINITION

Let $P = (x_1, y_1)$ and $Q = (x_2, y_2)$ be two distinct points. If $x_1 \neq x_2$, the **slope m** of the nonvertical line L containing P and Q is defined by the formula

$$m = \frac{y_2 - y_1}{x_2 - x_1} \qquad x_1 \neq x_2 \qquad \textbf{(1)}$$

If $x_1 = x_2$, L is a **vertical line** and the slope m of L is **undefined** (since this results in division by 0).

Figure 23(a) provides an illustration of the slope of a nonvertical line; Figure 23(b) illustrates a vertical line.

Figure 23

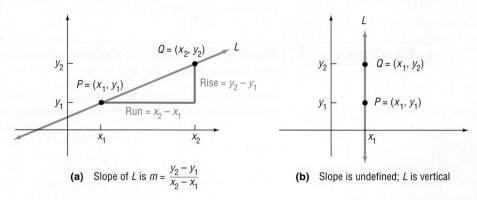

(a)   Slope of L is $m = \dfrac{y_2 - y_1}{x_2 - x_1}$ (b)   Slope is undefined; L is vertical

As Figure 23(a) illustrates, the slope m of a nonvertical line may be viewed as

$$m = \frac{y_2 - y_1}{x_2 - x_1} = \frac{\text{Rise}}{\text{Run}} \quad \text{or} \quad m = \frac{y_2 - y_1}{x_2 - x_1} = \frac{\text{Change in } y}{\text{Change in } x} = \frac{\Delta y}{\Delta x}$$

That is, the slope m of a nonvertical line L measures the amount that y changes as x changes from x_1 to x_2. This is called the **average rate of change** of y with respect to x.

Two comments about computing the slope of a nonvertical line may prove helpful:

1. Any two distinct points on the line can be used to compute the slope of the line. (See Figure 24 for justification.)

Figure 24 Triangles ABC and PQR are similar (equal angles). Hence, ratios of corresponding sides are proportional so that

Slope using P and $Q = \dfrac{y_2 - y_1}{x_2 - x_1}$

$= \dfrac{d(B, C)}{d(A, C)} = $ Slope using A and B

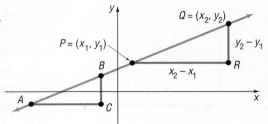

2. The slope of a line may be computed from $P = (x_1, y_1)$ to $Q = (x_2, y_2)$ or from Q to P because

$$\frac{y_2 - y_1}{x_2 - x_1} = \frac{y_1 - y_2}{x_1 - x_2}$$

EXAMPLE 1

Finding and Interpreting the Slope of a Line Containing Two Points

The slope m of the line containing the points $(1, 2)$ and $(5, -3)$ may be computed as

$$m = \frac{-3 - 2}{5 - 1} = \frac{-5}{4} = -\frac{5}{4} \quad \text{or as} \quad m = \frac{2 - (-3)}{1 - 5} = \frac{5}{-4} = -\frac{5}{4}$$

For every 4-unit change in x, y will change by -5 units. That is, if x increases by 4 units, then y will decrease by 5 units. The average rate of change of y with respect to x is $-\dfrac{5}{4}$.

●

Now Work PROBLEMS 11 AND 17

To get a better idea of the meaning of the slope m of a line L, consider the following example.

EXAMPLE 2

Finding the Slopes of Various Lines Containing the Same Point (2, 3)

Compute the slopes of the lines L_1, L_2, L_3, and L_4 containing the following pairs of points. Graph all four lines on the same set of coordinate axes.

$$L_1: \quad P = (2, 3) \qquad Q_1 = (-1, -2)$$
$$L_2: \quad P = (2, 3) \qquad Q_2 = (3, -1)$$
$$L_3: \quad P = (2, 3) \qquad Q_3 = (5, 3)$$
$$L_4: \quad P = (2, 3) \qquad Q_4 = (2, 5)$$

Solution Let m_1, m_2, m_3, and m_4 denote the slopes of the lines L_1, L_2, L_3, and L_4, respectively. Then

Figure 25

$$m_1 = \frac{-2 - 3}{-1 - 2} = \frac{-5}{-3} = \frac{5}{3} \qquad \text{A rise of 5 divided by a run of 3}$$

$$m_2 = \frac{-1 - 3}{3 - 2} = \frac{-4}{1} = -4$$

$$m_3 = \frac{3 - 3}{5 - 2} = \frac{0}{3} = 0$$

m_4 is undefined because $x_1 = x_2 = 2$

The graphs of these lines are given in Figure 25.

Figure 25 illustrates the following facts:

1. When the slope of a line is positive, the line slants upward from left to right (L_1).
2. When the slope of a line is negative, the line slants downward from left to right (L_2).
3. When the slope is 0, the line is horizontal (L_3).
4. When the slope is undefined, the line is vertical (L_4).

Seeing the Concept

On the same screen, graph the following equations:

$Y_1 = 0$	Slope of line is 0.
$Y_2 = \frac{1}{4}x$	Slope of line is $\frac{1}{4}$.
$Y_3 = \frac{1}{2}x$	Slope of line is $\frac{1}{2}$.
$Y_4 = x$	Slope of line is 1.
$Y_5 = 2x$	Slope of line is 2.
$Y_6 = 6x$	Slope of line is 6.

See Figure 26.

Figure 26

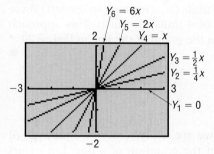

Seeing the Concept

On the same screen, graph the following equations:

$Y_1 = 0$	Slope of line is 0.
$Y_2 = -\frac{1}{4}x$	Slope of line is $-\frac{1}{4}$.
$Y_3 = -\frac{1}{2}x$	Slope of line is $-\frac{1}{2}$.
$Y_4 = -x$	Slope of line is −1.
$Y_5 = -2x$	Slope of line is −2.
$Y_6 = -6x$	Slope of line is −6.

See Figure 27.

Figure 27

Figures 26 and 27 illustrate that the closer the line is to the vertical position, the greater the absolute value of the slope.

2 Graph Lines Given a Point and the Slope

The next example illustrates how the slope of a line can be used to graph the line.

EXAMPLE 3

Graphing a Line Given a Point and the Slope

Draw a graph of the line that contains the point $(3, 2)$ and has a slope of:

(a) $\dfrac{3}{4}$
(b) $-\dfrac{4}{5}$

Solution

(a) Slope $= \dfrac{\text{Rise}}{\text{Run}}$. The fact that the slope is $\dfrac{3}{4}$ means that for every horizontal movement (run) of 4 units to the right there will be a vertical movement (rise) of 3 units. If we start at the given point $(3, 2)$ and move 4 units to the right and 3 units up, we reach the point $(7, 5)$. By drawing the line through this point and the point $(3, 2)$, we have the graph. See Figure 28.

(b) The fact that the slope is

$$-\frac{4}{5} = \frac{-4}{5} = \frac{\text{Rise}}{\text{Run}}$$

Figure 28

means that for every horizontal movement of 5 units to the right there will be a corresponding vertical movement of -4 units (a downward movement). If we start at the given point $(3, 2)$ and move 5 units to the right and then 4 units down, we arrive at the point $(8, -2)$. By drawing the line through these points, we have the graph. See Figure 29.

Alternatively, we can set

$$-\frac{4}{5} = \frac{4}{-5} = \frac{\text{Rise}}{\text{Run}}$$

Figure 29

so that for every horizontal movement of -5 units (a movement to the left) there will be a corresponding vertical movement of 4 units (upward). This approach brings us to the point $(-2, 6)$, which is also on the graph of the line in Figure 29.

Now Work PROBLEM 23

3 Find the Equation of a Vertical Line

Now that we have discussed the slope of a line, we are ready to derive equations of lines. As we shall see, there are several forms of the equation of a line. Let's start with an example.

EXAMPLE 4

Graphing a Line

Graph the equation: $x = 3$

Solution

To graph $x = 3$, recall that we are looking for all points (x, y) in the plane for which $x = 3$. No matter what y-coordinate is used, the corresponding x-coordinate always equals 3. Consequently, the graph of the equation $x = 3$ is a vertical line with x-intercept 3 and undefined slope. See Figure 30.

Figure 30

As suggested by Example 4, we have the following result:

THEOREM

Equation of a Vertical Line

A vertical line is given by an equation of the form

$$x = a$$

where a is the x-intercept.

COMMENT To graph an equation using a graphing utility, we need to express the equation in the form $y = \{expression\ in\ x\}$. But $x = 3$ cannot be put in this form. To overcome this, most graphing utilities have special commands for drawing vertical lines. LINE, PLOT, and VERT are among the more common ones. Consult your manual to determine the correct methodology for your graphing utility. ∎

Figure 31

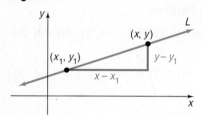

4 Use the Point–Slope Form of a Line; Identify Horizontal Lines

Now let L be a nonvertical line with slope m and containing the point (x_1, y_1). See Figure 31. For any other point (x, y) on L, we have

$$m = \frac{y - y_1}{x - x_1} \quad \text{or} \quad y - y_1 = m(x - x_1)$$

THEOREM

Point–Slope Form of an Equation of a Line

An equation of a nonvertical line with slope m that contains the point (x_1, y_1) is

$$y - y_1 = m(x - x_1) \qquad (2)$$

EXAMPLE 5 Using the Point–Slope Form of a Line

Figure 32

An equation of the line with slope 4 and containing the point $(1, 2)$ can be found by using the point–slope form with $m = 4$, $x_1 = 1$, and $y_1 = 2$.

$$y - y_1 = m(x - x_1)$$
$$y - 2 = 4(x - 1) \qquad m = 4, x_1 = 1, y_1 = 2$$
$$y = 4x - 2 \qquad \text{Solve for } y.$$

See Figure 32 for the graph.

EXAMPLE 6 Finding the Equation of a Horizontal Line

Find an equation of the horizontal line containing the point $(3, 2)$.

Solution Because all the y-values are equal on a horizontal line, the slope of a horizontal line is 0. To get an equation, we use the point–slope form with $m = 0$, $x_1 = 3$, and $y_1 = 2$.

$$y - y_1 = m(x - x_1)$$
$$y - 2 = 0 \cdot (x - 3) \qquad m = 0, x_1 = 3, \text{ and } y_1 = 2$$
$$y - 2 = 0$$
$$y = 2$$

Figure 33

See Figure 33 for the graph.

As suggested by Example 6, we have the following result:

THEOREM

Equation of a Horizontal Line

A horizontal line is given by an equation of the form

$$y = b$$

where b is the y-intercept.

5 Find the Equation of a Line Given Two Points

Given two points, we use the slope formula and the point–slope form of a line to find the equation.

> **EXAMPLE 7**

Finding an Equation of a Line Given Two Points

Find an equation of the line containing the points $(2, 3)$ and $(-4, 5)$. Graph the line.

Solution

We first compute the slope of the line with $(x_1, y_1) = (2, 3)$ and $(x_2, y_2) = (-4, 5)$.

$$m = \frac{5 - 3}{-4 - 2} = \frac{2}{-6} = -\frac{1}{3} \qquad m = \frac{y_2 - y_1}{x_2 - x_1}$$

Figure 34

We use the point $(x_1, y_1) = (2, 3)$ and the slope $m = -\frac{1}{3}$ to get the point–slope form of the equation of the line.

$$y - 3 = -\frac{1}{3}(x - 2) \qquad y - y_1 = m(x - x_1)$$

See Figure 34 for the graph.

In the solution to Example 7, we could have used the other point, $(-4, 5)$, instead of the point $(2, 3)$. The equation that results, although it looks different, is equivalent to the equation that we obtained in the example. (Try it for yourself.)

━━━━▶ **Now Work** PROBLEM 37

6 Write the Equation of a Line in Slope–Intercept Form

Another useful equation of a line is obtained when the slope m and y-intercept b are known. In this event, we know both the slope m of the line and a point $(0, b)$ on the line; then we may use the point–slope form, equation (2), to obtain the following equation:

$$y - b = m(x - 0) \quad \text{or} \quad y = mx + b$$

THEOREM

Slope–Intercept Form of an Equation of a Line

An equation of a line with slope m and y-intercept b is

$$y = mx + b \tag{3}$$

━━━━▶ **Now Work** PROBLEM 45 (EXPRESS ANSWER
IN SLOPE–INTERCEPT FORM)

Figure 35 $y = mx + 2$

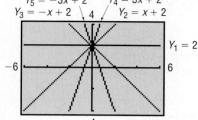

$Y_5 = -3x + 2$ $Y_4 = 3x + 2$
$Y_3 = -x + 2$ $Y_2 = x + 2$
$Y_1 = 2$

Seeing the Concept

To see the role that the slope m plays, graph the following lines on the same square screen.

$Y_1 = 2$	$m = 0$
$Y_2 = x + 2$	$m = 1$
$Y_3 = -x + 2$	$m = -1$
$Y_4 = 3x + 2$	$m = 3$
$Y_5 = -3x + 2$	$m = -3$

See Figure 35. What do you conclude about the lines $y = mx + 2$?

Figure 36 $y = 2x + b$

$Y_4 = 2x + 4$ $Y_2 = 2x + 1$
$Y_1 = 2x$
$Y_3 = 2x - 1$
$Y_5 = 2x - 4$

Seeing the Concept

To see the role of the y-intercept b, graph the following lines on the same square screen.

$Y_1 = 2x$	$b = 0$
$Y_2 = 2x + 1$	$b = 1$
$Y_3 = 2x - 1$	$b = -1$
$Y_4 = 2x + 4$	$b = 4$
$Y_5 = 2x - 4$	$b = -4$

See Figure 36. What do you conclude about the lines $y = 2x + b$?

7 Identify the Slope and y-Intercept of a Line from Its Equation

When the equation of a line is written in slope–intercept form, it is easy to find the slope m and y-intercept b of the line. For example, suppose that the equation of a line is

$$y = -2x + 3$$

Compare it to $y = mx + b$.

$$y = -2x + 3$$
$$y = mx + b$$

The slope of this line is -2 and its y-intercept is 3.

Now Work PROBLEM 71

EXAMPLE 8   **Finding the Slope and y-Intercept**

Find the slope m and y-intercept b of the equation $2x + 4y = 8$. Graph the equation.

Solution   To obtain the slope and y-intercept, we write the equation in slope–intercept form by solving for y.

$$2x + 4y = 8$$
$$4y = -2x + 8$$
$$y = -\frac{1}{2}x + 2 \qquad y = mx + b$$

Figure 37
$2x + 4y = 8$

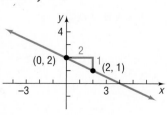

The coefficient of x, $-\frac{1}{2}$, is the slope, and the y-intercept is 2. We can graph the line using the fact that the y-intercept is 2 and the slope is $-\frac{1}{2}$. Then, starting at the point $(0, 2)$, go to the right 2 units and then down 1 unit to the point $(2, 1)$. See Figure 37.

Now Work PROBLEM 77

8 Graph Lines Written in General Form Using Intercepts

The form of the equation of the line in Example 8, $2x + 4y = 8$, is called the *general form*.

DEFINITION

The equation of a line L is in **general form*** when it is written as

$$Ax + By = C \qquad \textbf{(4)}$$

where A, B, and C are real numbers and A and B are not both 0.

When we want to graph an equation that is written in general form, we can solve the equation for y (if $B \neq 0$) and write the equation in slope–intercept form as we did in Example 8. Another approach to graphing the equation would be to find its intercepts. Remember, the intercepts of the graph of an equation are the points where the graph crosses or touches a coordinate axis.

EXAMPLE 9

Graphing an Equation in General Form Using Its Intercepts

Graph the equation $2x + 4y = 8$ by finding its intercepts.

Solution

To obtain the x-intercept, let $y = 0$ in the equation and solve for x.

$$2x + 4y = 8$$
$$2x + 4(0) = 8 \qquad \text{Let } y = 0.$$
$$2x = 8$$
$$x = 4 \qquad \text{Divide both sides by 2.}$$

The x-intercept is 4 and the point $(4, 0)$ is on the graph of the equation.
To obtain the y-intercept, let $x = 0$ in the equation and solve for y.

$$2x + 4y = 8$$
$$2(0) + 4y = 8 \qquad \text{Let } x = 0.$$
$$4y = 8$$
$$y = 2 \qquad \text{Divide both sides by 4.}$$

Figure 38

The y-intercept is 2 and the point $(0, 2)$ is on the graph of the equation.
We plot the points $(4, 0)$ and $(0, 2)$ in a Cartesian plane and draw a line through the points. See Figure 38.

Now Work PROBLEM 91

Every line has an equation that is equivalent to an equation written in general form. For example, a vertical line whose equation is

$$x = a$$

can be written in the general form

$$1 \cdot x + 0 \cdot y = a \qquad A = 1, B = 0, C = a$$

A horizontal line whose equation is

$$y = b$$

can be written in the general form

$$0 \cdot x + 1 \cdot y = b \qquad A = 0, B = 1, C = b$$

*Some books use the term **standard form.**

Lines that are neither vertical nor horizontal have general equations of the form

$$Ax + By = C \qquad A \neq 0 \text{ and } B \neq 0$$

Because the equation of every line can be written in general form, any equation equivalent to equation (4) is called a **linear equation.**

9 Find Equations of Parallel Lines

When two lines (in the plane) do not intersect (that is, they have no points in common), they are said to be **parallel.** Look at Figure 39. There we have drawn two lines and have constructed two right triangles by drawing sides parallel to the coordinate axes. These lines are parallel if and only if the right triangles are similar. (Do you see why? Two angles are equal.) And the triangles are similar if and only if the ratios of corresponding sides are equal.

Figure 39
The distinct lines are parallel if and only if their slopes are equal.

This suggests the following result:

THEOREM

Criterion for Parallel Lines

Two nonvertical lines are parallel if and only if their slopes are equal and they have different y-intercepts.

The use of the words "if and only if" in the preceding theorem means that actually two statements are being made, one the converse of the other.

If two nonvertical lines are parallel, then their slopes are equal and they have different y-intercepts.

If two nonvertical lines have equal slopes and they have different y-intercepts, then they are parallel.

EXAMPLE 10

Showing That Two Lines Are Parallel

Show that the lines given by the following equations are parallel:

$$L_1: \quad 2x + 3y = 6, \qquad L_2: \quad 4x + 6y = 0$$

Solution

To determine whether these lines have equal slopes and different y-intercepts, we write each equation in slope–intercept form:

Figure 40 Parallel lines

$$
\begin{array}{ll}
L_1: \quad 2x + 3y = 6 & L_2: \quad 4x + 6y = 0 \\
3y = -2x + 6 & 6y = -4x \\
y = -\dfrac{2}{3}x + 2 & y = -\dfrac{2}{3}x \\
\text{Slope} = -\dfrac{2}{3};\ y\text{-intercept} = 2 & \text{Slope} = -\dfrac{2}{3};\ y\text{-intercept} = 0
\end{array}
$$

Because these lines have the same slope, $-\dfrac{2}{3}$, but different y-intercepts, the lines are parallel. See Figure 40.

EXAMPLE 11 | **Finding a Line That Is Parallel to a Given Line**

Find an equation for the line that contains the point $(2, -3)$ and is parallel to the line $2x + y = 6$.

Solution

Since the two lines are to be parallel, the slope of the line that we seek equals the slope of the line $2x + y = 6$. We begin by writing the equation of the line $2x + y = 6$ in slope–intercept form.

$$2x + y = 6$$
$$y = -2x + 6$$

The slope is -2. Since the line that we seek contains the point $(2, -3)$, we use the point–slope form to obtain

$$y - y_1 = m(x - x_1) \qquad \text{Point–slope form}$$
$$y - (-3) = -2(x - 2) \qquad m = -2, x_1 = 2, y_1 = -3$$
$$y + 3 = -2x + 4 \qquad \text{Simplify.}$$
$$y = -2x + 1 \qquad \text{Slope–intercept form}$$
$$2x + y = 1 \qquad \text{General form}$$

This line is parallel to the line $2x + y = 6$ and contains the point $(2, -3)$. See Figure 41.

Figure 41

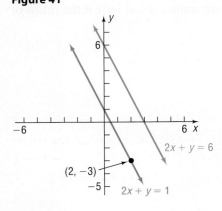

$2x + y = 6$
$(2, -3)$
$2x + y = 1$

─────Now Work PROBLEM 59

10 Find Equations of Perpendicular Lines

When two lines intersect at a right angle ($90°$), they are said to be **perpendicular.** See Figure 42.

Figure 42
Perpendicular lines

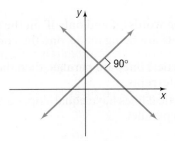

$90°$

The following result gives a condition, in terms of their slopes, for two lines to be perpendicular.

THEOREM

Criterion for Perpendicular Lines

Two nonvertical lines are perpendicular if and only if the product of their slopes is -1.

Here we shall prove the "only if" part of the statement:

If two nonvertical lines are perpendicular, then the product of their slopes is -1.

In Problem 128, you are asked to prove the "if" part of the theorem; that is:

If two nonvertical lines have slopes whose product is -1, then the lines are perpendicular.

Proof Let m_1 and m_2 denote the slopes of the two lines. There is no loss in generality (that is, neither the angle nor the slopes are affected) if we situate the lines so that they meet at the origin. See Figure 43. The point $A = (1, m_2)$ is on the line having slope m_2, and the point $B = (1, m_1)$ is on the line having slope m_1. (Do you see why this must be true?)

Suppose that the lines are perpendicular. Then triangle OAB is a right triangle. As a result of the Pythagorean Theorem, it follows that

$$[d(O, A)]^2 + [d(O, B)]^2 = [d(A, B)]^2 \qquad (5)$$

By the distance formula, we can write the squares of these distances as

$$[d(O, A)]^2 = (1 - 0)^2 + (m_2 - 0)^2 = 1 + m_2^2$$
$$[d(O, B)]^2 = (1 - 0)^2 + (m_1 - 0)^2 = 1 + m_1^2$$
$$[d(A, B)]^2 = (1 - 1)^2 + (m_2 - m_1)^2 = m_2^2 - 2m_1m_2 + m_1^2$$

Using these facts in equation (5), we get

$$(1 + m_2^2) + (1 + m_1^2) = m_2^2 - 2m_1m_2 + m_1^2$$

which, upon simplification, can be written as

$$m_1m_2 = -1$$

If the lines are perpendicular, the product of their slopes is -1. ∎

You may find it easier to remember the condition for two nonvertical lines to be perpendicular by observing that the equality $m_1m_2 = -1$ means that m_1 and m_2 are negative reciprocals of each other; that is, either $m_1 = -\dfrac{1}{m_2}$ or $m_2 = -\dfrac{1}{m_1}$.

Figure 43

EXAMPLE 12	**Finding the Slope of a Line Perpendicular to Another Line**

If a line has slope $\dfrac{3}{2}$, any line having slope $-\dfrac{2}{3}$ is perpendicular to it.

EXAMPLE 13	**Finding the Equation of a Line Perpendicular to a Given Line**

Find an equation of the line that contains the point $(1, -2)$ and is perpendicular to the line $x + 3y = 6$. Graph the two lines.

Solution We first write the equation of the given line in slope–intercept form to find its slope.

$$x + 3y = 6$$
$$3y = -x + 6 \qquad \text{\textit{Proceed to solve for y.}}$$
$$y = -\frac{1}{3}x + 2 \qquad \text{\textit{Place in the form } } y = mx + b.$$

The given line has slope $-\dfrac{1}{3}$. Any line perpendicular to this line will have slope 3.

Because we require the point $(1, -2)$ to be on this line with slope 3, we use the point–slope form of the equation of a line.

$$y - y_1 = m(x - x_1) \qquad \text{\textit{Point–slope form}}$$
$$y - (-2) = 3(x - 1) \qquad \text{\textit{m = 3, } x_1 = 1, y_1 = -2}$$

To obtain other forms of the equation, we proceed as follows:

$$y + 2 = 3(x - 1)$$
$$y + 2 = 3x - 3 \qquad \text{\textit{Simplify.}}$$
$$y = 3x - 5 \qquad \text{\textit{Slope–intercept form}}$$
$$3x - y = 5 \qquad \text{\textit{General form}}$$

Figure 44 shows the graphs.

Figure 44

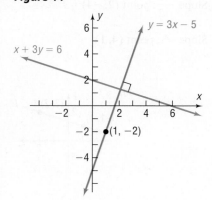

Now Work PROBLEM 65

3 Assess Your Understanding

Concepts and Vocabulary

1. The slope of a vertical line is _____; the slope of a horizontal line is _____.

2. Let $P = (x_1, y_1)$ and $Q = (x_2, y_2)$ be two distinct points. If $x_1 \neq x_2$, the slope m of the nonvertical line containing P and Q is defined by the formula _____.

3. A horizontal line is given by an equation of the form _____, where b is the _____.

4. **True or False** Vertical lines have an undefined slope.

5. **True or False** The slope of the line $2y = 3x + 5$ is 3.

6. **True or False** The point $(1, 2)$ is on the line $2x + y = 4$.

7. Two nonvertical lines have slopes m_1 and m_2, respectively. The lines are parallel if _____ and the _____ are unequal; the lines are perpendicular if _____ or _____.

8. The lines $y = 2x + 3$ and $y = ax + 5$ are parallel if $a =$ _____.

9. The lines $y = 2x - 1$ and $y = ax + 2$ are perpendicular if $a =$ _____.

10. A slope of $\dfrac{3}{5}$ means that for every horizontal movement of 5 units to the right there will be a vertical movement of _____ units up.

Skill Building

In Problems 11–14, (a) find the slope of the line and (b) interpret the slope.

11.

12.

13.

14.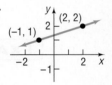

In Problems 15–22, plot each pair of points and determine the slope of the line containing them. Graph the line.

15. $(2, 3); (4, 0)$

16. $(4, 2); (3, 4)$

17. $(-2, 3); (2, 1)$

18. $(-1, 1); (2, 3)$

19. $(-3, -1); (2, -1)$

20. $(4, 2); (-5, 2)$

21. $(-1, 2); (-1, -2)$

22. $(2, 0); (2, 2)$

In Problems 23–30, graph the line containing the point P and having slope m.

23. $P = (1, 2); m = 3$

24. $P = (2, 1); m = 4$

25. $P = (2, 4); m = -\dfrac{3}{4}$

26. $P = (1, 3); m = -\dfrac{2}{5}$

27. $P = (-1, 3); m = 0$

28. $P = (2, -4); m = 0$

29. $P = (0, 3)$; slope undefined

30. $P = (-2, 0)$; slope undefined

In Problems 31–36, the slope and a point on a line are given. Use this information to locate three additional points on the line. Answers may vary.

[**Hint:** *It is not necessary to find the equation of the line. See Example 2.*]

31. Slope 4; point $(1, 2)$

32. Slope 2; point $(-2, 3)$

33. Slope $-\dfrac{3}{2}$; point $(2, -4)$

34. Slope $\dfrac{4}{3}$; point $(-3, 2)$

35. Slope -2; point $(-2, -3)$

36. Slope -1; point $(4, 1)$

In Problems 37–44, find an equation of the line L.

37.

38.

39.

40.

41.

L is parallel to y = 2x

42.

L is parallel to y = −x

43.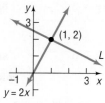

L is perpendicular
to y = 2x

44.

L is perpendicular
to y = −x

In Problems 45–70, find an equation for the line with the given properties. Express your answer using either the general form or the slope–intercept form of the equation of a line, whichever you prefer.

45. Slope $= 3$; containing the point $(-2, 3)$

46. Slope $= 2$; containing the point $(4, -3)$

47. Slope $= -\dfrac{2}{3}$; containing the point $(1, -1)$

48. Slope $= \dfrac{1}{2}$; containing the point $(3, 1)$

49. Containing the points $(1, 3)$ and $(-1, 2)$

50. Containing the points $(-3, 4)$ and $(2, 5)$

51. Slope $= -3$; y-intercept $= 3$

52. Slope $= -2$; y-intercept $= -2$

53. x-intercept $= 2$; y-intercept $= -1$

54. x-intercept $= -4$; y-intercept $= 4$

55. Slope undefined; containing the point $(2, 4)$

56. Slope undefined; containing the point $(3, 8)$

57. Horizontal; containing the point $(-3, 2)$

58. Vertical; containing the point $(4, -5)$

59. Parallel to the line $y = 2x$; containing the point $(-1, 2)$

60. Parallel to the line $y = -3x$; containing the point $(-1, 2)$

61. Parallel to the line $2x - y = -2$; containing the point $(0, 0)$

62. Parallel to the line $x - 2y = -5$; containing the point $(0, 0)$

63. Parallel to the line $x = 5$; containing the point $(4, 2)$

64. Parallel to the line $y = 5$; containing the point $(4, 2)$

65. Perpendicular to the line $y = \dfrac{1}{2}x + 4$; containing the point $(1, -2)$

66. Perpendicular to the line $y = 2x - 3$; containing the point $(1, -2)$

67. Perpendicular to the line $2x + y = 2$; containing the point $(-3, 0)$

68. Perpendicular to the line $x - 2y = -5$; containing the point $(0, 4)$

69. Perpendicular to the line $x = 8$; containing the point $(3, 4)$

70. Perpendicular to the line $y = 8$; containing the point $(3, 4)$

In Problems 71–90, find the slope and y-intercept of each line. Graph the line.

71. $y = 2x + 3$

72. $y = -3x + 4$

73. $\dfrac{1}{2}y = x - 1$

74. $\dfrac{1}{3}x + y = 2$

75. $y = \dfrac{1}{2}x + 2$

76. $y = 2x + \dfrac{1}{2}$

77. $x + 2y = 4$

78. $-x + 3y = 6$

79. $2x - 3y = 6$

80. $3x + 2y = 6$

81. $x + y = 1$ **82.** $x - y = 2$ **83.** $x = -4$ **84.** $y = -1$ **85.** $y = 5$

86. $x = 2$ **87.** $y - x = 0$ **88.** $x + y = 0$ **89.** $2y - 3x = 0$ **90.** $3x + 2y = 0$

In Problems 91–100, (a) find the intercepts of the graph of each equation and (b) graph the equation.

91. $2x + 3y = 6$ **92.** $3x - 2y = 6$ **93.** $-4x + 5y = 40$ **94.** $6x - 4y = 24$

95. $7x + 2y = 21$ **96.** $5x + 3y = 18$ **97.** $\frac{1}{2}x + \frac{1}{3}y = 1$ **98.** $x - \frac{2}{3}y = 4$

99. $0.2x - 0.5y = 1$ **100.** $-0.3x + 0.4y = 1.2$

101. Find an equation of the *x*-axis.   **102.** Find an equation of the *y*-axis.

In Problems 103–106, the equations of two lines are given. Determine if the lines are parallel, perpendicular, or neither.

103. $y = 2x - 3$
$y = 2x + 4$

104. $y = \frac{1}{2}x - 3$
$y = -2x + 4$

105. $y = 4x + 5$
$y = -4x + 2$

106. $y = -2x + 3$
$y = -\frac{1}{2}x + 2$

In Problems 107–110, match each graph with the correct equation:

(a) $y = x$ (b) $y = 2x$ (c) $y = \frac{x}{2}$ (d) $y = 4x$

107.

108.

109.

110.

Applications and Extensions

111. Geometry Use slopes to show that the triangle whose vertices are $(-2, 5)$, $(1, 3)$, and $(-1, 0)$ is a right triangle.

112. Geometry Use slopes to show that the quadrilateral whose vertices are $(1, -1)$, $(4, 1)$, $(2, 2)$, and $(5, 4)$ is a parallelogram.

113. Geometry Use slopes to show that the quadrilateral whose vertices are $(-1, 0)$, $(2, 3)$, $(1, -2)$, and $(4, 1)$ is a rectangle.

114. Geometry Use slopes and the distance formula to show that the quadrilateral whose vertices are $(0, 0)$, $(1, 3)$, $(4, 2)$, and $(3, -1)$ is a square.

115. Truck Rentals A truck rental company rents a moving truck for one day by charging $29 plus $0.20 per mile. Write a linear equation that relates the cost C, in dollars, of renting the truck to the number x of miles driven. What is the cost of renting the truck if the truck is driven 110 miles? 230 miles?

116. Cost Equation The **fixed costs** of operating a business are the costs incurred regardless of the level of production. Fixed costs include rent, fixed salaries, and costs of leasing machinery. The **variable costs** of operating a business are the costs that change with the level of output. Variable costs include raw materials, hourly wages, and electricity. Suppose that a manufacturer of jeans has fixed daily costs of $500 and variable costs of $8 for each pair of jeans manufactured. Write a linear equation that relates the daily cost C, in dollars, of manufacturing the jeans to the number x of jeans manufactured. What is the cost of manufacturing 400 pairs of jeans? 740 pairs?

117. Cost of Driving a Car The annual fixed costs for owning a small sedan are $1289, assuming the car is completely paid for. The cost to drive the car is approximately $0.15 per mile. Write a linear equation that relates the cost C and the number x of miles driven annually.
Source: www.pacebus.com

118. **Wages of a Car Salesperson** Dan receives $375 per week for selling new and used cars at a car dealership in Oak Lawn, Illinois. In addition, he receives 5% of the profit on any sales that he generates. Write a linear equation that relates Dan's weekly salary S when he has sales that generate a profit of x dollars.

119. **Electricity Rates in Illinois** Commonwealth Edison Company supplies electricity to residential customers for a monthly customer charge of $10.47 plus 10.701 cents per kilowatt-hour for up to 400 kilowatt-hours.

(a) Write a linear equation that relates the monthly charge C, in dollars, to the number x of kilowatt-hours used in a month, $0 \leq x \leq 400$.
(b) Graph this equation.
(c) What is the monthly charge for using 100 kilowatt-hours?
(d) What is the monthly charge for using 300 kilowatt-hours?
(e) Interpret the slope of the line.

Source: Commonwealth Edison Company, March, 2008.

120. **Electricity Rates in Florida** Florida Power & Light Company supplies electricity to residential customers for a monthly customer charge of $5.33 plus 10 cents per kilowatt-hour for up to 1000 kilowatt-hours.
(a) Write a linear equation that relates the monthly charge C, in dollars, to the number x of kilowatt-hours used in a month, $0 \leq x \leq 1000$.
(b) Graph this equation.
(c) What is the monthly charge for using 200 kilowatt-hours?
(d) What is the monthly charge for using 500 kilowatt-hours?
(e) Interpret the slope of the line.

Source: Florida Power & Light Company, April 2009.

121. **Measuring Temperature** The relationship between Celsius (°C) and Fahrenheit (°F) degrees for measuring temperature is linear. Find a linear equation relating °C and °F if 0°C corresponds to 32°F and 100°C corresponds to 212°F. Use the equation to find the Celsius measure of 70°F.

122. **Measuring Temperature** The Kelvin (K) scale for measuring temperature is obtained by adding 273 to the Celsius temperature.
(a) Write a linear equation relating K and °C.
(b) Write a linear equation relating K and °F (see Problem 121).

123. **Access Ramp** A wooden access ramp is being built to reach a platform that sits 30 inches above the floor. The proposed ramp drops 2 inches for every 25-inch run.

(a) Write a linear equation that relates the height y above the floor to the horizontal distance x from the platform.
(b) Find and interpret the x-intercept of the graph of your equation.
(c) Design requirements stipulate that the maximum run be 30 feet and that the maximum slope be a drop of 1 inch for each 12 inches of run. Will this ramp meet the requirements? Explain.
(d) What slopes could be used to obtain the 30-inch rise and still meet design requirements?

Source: www.adaptiveaccess.com/wood_ramps.php

124. **Cigarette Use** A report in the Child Trends DataBase indicated that, in 1996, 22.2% of twelfth grade students reported daily use of cigarettes. In 2006, 12.2% of twelfth grade students reported daily use of cigarettes.
(a) Write a linear equation that relates the percent y of twelfth grade students who smoke cigarettes daily to the number x of years after 1996.
(b) Find the intercepts of the graph of your equation.
(c) Do the intercepts have any meaningful interpretation?
(d) Use your equation to predict the percent for the year 2016. Is this result reasonable?

Source: www.childtrendsdatabank.org

125. **Product Promotion** A cereal company finds that the number of people who will buy one of its products in the first month that it is introduced is linearly related to the amount of money it spends on advertising. If it spends $40,000 on advertising, then 100,000 boxes of cereal will be sold, and if it spends $60,000, then 200,000 boxes will be sold.
(a) Write a linear equation that relates the amount A spent on advertising to the number x of boxes the company aims to sell.
(b) How much advertising is needed to sell 300,000 boxes of cereal?
(c) Interpret the slope.

126. Show that the line containing the points (a, b) and (b, a), $a \neq b$, is perpendicular to the line $y = x$. Also show that the midpoint of (a, b) and (b, a) lies on the line $y = x$.

127. The equation $2x - y = C$ defines a **family of lines,** one line for each value of C. On one set of coordinate axes, graph the members of the family when $C = -4$, $C = 0$, and $C = 2$. Can you draw a conclusion from the graph about each member of the family?

128. Prove that if two nonvertical lines have slopes whose product is -1 then the lines are perpendicular. [**Hint:** Refer to Figure 43 and use the converse of the Pythagorean Theorem.]

Explaining Concepts: Discussion and Writing

129. Which of the following equations might have the graph shown? (More than one answer is possible.)
  (a) $2x + 3y = 6$
  (b) $-2x + 3y = 6$
  (c) $3x - 4y = -12$
  (d) $x - y = 1$
  (e) $x - y = -1$
  (f) $y = 3x - 5$
  (g) $y = 2x + 3$
  (h) $y = -3x + 3$

130. Which of the following equations might have the graph shown? (More than one answer is possible.)
  (a) $2x + 3y = 6$
  (b) $2x - 3y = 6$
  (c) $3x + 4y = 12$
  (d) $x - y = 1$
  (e) $x - y = -1$
  (f) $y = -2x - 1$
  (g) $y = -\dfrac{1}{2}x + 10$
  (h) $y = x + 4$

131. The figure shows the graph of two parallel lines. Which of the following pairs of equations might have such a graph?
  (a) $x - 2y = 3$
    $x + 2y = 7$
  (b) $x + y = 2$
    $x + y = -1$
  (c) $x - y = -2$
    $x - y = 1$
  (d) $x - y = -2$
    $2x - 2y = -4$
  (e) $x + 2y = 2$
    $x + 2y = -1$

132. The figure shows the graph of two perpendicular lines. Which of the following pairs of equations might have such a graph?
  (a) $y - 2x = 2$
    $y + 2x = -1$
  (b) $y - 2x = 0$
    $2y + x = 0$
  (c) $2y - x = 2$
    $2y + x = -2$
  (d) $y - 2x = 2$
    $x + 2y = -2$
  (e) $2x + y = -2$
    $2y + x = -2$

133. **m Is for Slope** The accepted symbol used to denote the slope of a line is the letter m. Investigate the origin of this symbolism. Begin by consulting a French dictionary and looking up the French word *monter*. Write a brief essay on your findings.

134. **Grade of a Road** The term *grade* is used to describe the inclination of a road. How does this term relate to the notion of slope of a line? Is a 4% grade very steep? Investigate the grades of some mountainous roads and determine their slopes. Write a brief essay on your findings.

135. **Carpentry** Carpenters use the term *pitch* to describe the steepness of staircases and roofs. How does pitch relate to slope? Investigate typical pitches used for stairs and for roofs. Write a brief essay on your findings.

136. Can the equation of every line be written in slope–intercept form? Why?

137. Does every line have exactly one x-intercept and one y-intercept? Are there any lines that have no intercepts? Explain.

138. What can you say about two lines that have equal slopes and equal y-intercepts?

139. What can you say about two lines with the same x-intercept and the same y-intercept? Assume that the x-intercept is not 0.

140. If two distinct lines have the same slope, but different x-intercepts, can they have the same y-intercept?

141. If two distinct lines have the same y-intercept, but different slopes, can they have the same x-intercept?

142. **What Went Wrong?** A student is asked to find the slope of the line joining $(-3, 2)$ and $(1, -4)$. He states that the slope is $\dfrac{3}{2}$. Is he correct? If not, what went wrong?

4 Circles

PREPARING FOR THIS SECTION *Before getting started, review the following:*

- Completing the Square (Appendix A, Section A.4, p. A40)
- Solving Equations (Appendix A, Section A.8, pp. A65–A71)

Now Work the 'Are You Prepared?' problems on page 39.

OBJECTIVES **1** Write the Standard Form of the Equation of a Circle (p. 35)
2 Graph a Circle (p. 36)
3 Work with the General Form of the Equation of a Circle (p. 37)

1 Write the Standard Form of the Equation of a Circle

One advantage of a coordinate system is that it enables us to translate a geometric statement into an algebraic statement, and vice versa. Consider, for example, the following geometric statement that defines a circle.

DEFINITION

A **circle** is a set of points in the xy-plane that are a fixed distance r from a fixed point (h, k). The fixed distance r is called the **radius,** and the fixed point (h, k) is called the **center** of the circle.

Figure 45

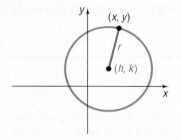

Figure 45 shows the graph of a circle. To find the equation, we let (x, y) represent the coordinates of any point on a circle with radius r and center (h, k). Then the distance between the points (x, y) and (h, k) must always equal r. That is, by the distance formula

$$\sqrt{(x - h)^2 + (y - k)^2} = r$$

After squaring both sides, we obtain

$$(x - h)^2 + (y - k)^2 = r^2$$

DEFINITION

The **standard form of an equation of a circle** with radius r and center (h, k) is

$$(x - h)^2 + (y - k)^2 = r^2 \qquad \textbf{(1)}$$

THEOREM

The standard form of an equation of a circle of radius r with center at the origin $(0, 0)$ is

$$x^2 + y^2 = r^2$$

DEFINITION

If the radius $r = 1$, the circle whose center is at the origin is called the **unit circle** and has the equation

$$x^2 + y^2 = 1$$

See Figure 46.

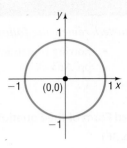

Figure 46
Unit circle $x^2 + y^2 = 1$

EXAMPLE 1   **Writing the Standard Form of the Equation of a Circle**

Write the standard form of the equation of the circle with radius 5 and center $(-3, 6)$.

Solution   Using the form of equation (1) and substituting the values $r = 5$, $h = -3$, and $k = 6$, we have

$$(x - h)^2 + (y - k)^2 = r^2$$
$$(x - (-3))^2 + (y - 6)^2 = 5^2$$
$$(x + 3)^2 + (y - 6)^2 = 25$$

—**Now Work** PROBLEM 7

2 Graph a Circle

The graph of any equation of the form $(x - h)^2 + (y - k)^2 = r^2$ is that of a circle with radius r and center (h, k).

EXAMPLE 2   **Graphing a Circle**

Graph the equation: $(x + 3)^2 + (y - 2)^2 = 16$

Solution   Since the equation is in the form of equation (1), its graph is a circle. To graph the equation, we first compare the given equation to the standard form of the equation of a circle. The comparison yields information about the circle.

Figure 47

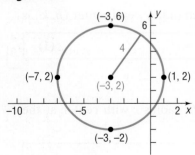

$$(x + 3)^2 + (y - 2)^2 = 16$$
$$(x - (-3))^2 + (y - 2)^2 = 4^2$$
$$\uparrow \qquad\qquad \uparrow \qquad \uparrow$$
$$(x - h)^2 + (y - k)^2 = r^2$$

We see that $h = -3$, $k = 2$, and $r = 4$. The circle has center $(-3, 2)$ and a radius of 4 units. To graph this circle, we first plot the center $(-3, 2)$. Since the radius is 4, we can locate four points on the circle by plotting points 4 units to the left, to the right, and up and down from the center. These four points can then be used as guides to obtain the graph. See Figure 47.

—**Now Work** PROBLEMS 23(a) AND (b)

EXAMPLE 3   **Finding the Intercepts of a Circle**

For the circle $(x + 3)^2 + (y - 2)^2 = 16$, find the intercepts, if any, of its graph.

Solution   This is the equation discussed and graphed in Example 2. To find the x-intercepts, let $y = 0$ and solve for x. Then

$$(x + 3)^2 + (y - 2)^2 = 16$$
$$(x + 3)^2 + (0 - 2)^2 = 16 \qquad \text{\textit{y = 0}}$$
$$(x + 3)^2 + 4 = 16 \qquad \text{\textit{Simplify.}}$$
$$(x + 3)^2 = 12 \qquad \text{\textit{Subtract 4 from both sides.}}$$
$$x + 3 = \pm\sqrt{12} \qquad \text{\textit{Solve for x + 3.}}$$
$$x = -3 \pm 2\sqrt{3} \qquad \text{\textit{Solve for x.}}$$

The *x*-intercepts are $-3 - 2\sqrt{3} \approx -6.46$ and $-3 + 2\sqrt{3} \approx 0.46$.
To find the *y*-intercepts, we let $x = 0$ and solve for *y*. Then

$$(x + 3)^2 + (y - 2)^2 = 16$$
$$(0 + 3)^2 + (y - 2)^2 = 16 \qquad \text{\textit{x = 0}}$$
$$9 + (y - 2)^2 = 16$$
$$(y - 2)^2 = 7$$
$$y - 2 = \pm\sqrt{7} \qquad \text{\textit{Solve for y - 2.}}$$
$$y = 2 \pm \sqrt{7} \qquad \text{\textit{Solve for y.}}$$

The *y*-intercepts are $2 - \sqrt{7} \approx -0.65$ and $2 + \sqrt{7} \approx 4.65$.
Look back at Figure 47 to verify the approximate locations of the intercepts.

Now Work PROBLEM 23(C)

3 Work with the General Form of the Equation of a Circle

If we eliminate the parentheses from the standard form of the equation of the circle given in Example 3, we get

$$(x + 3)^2 + (y - 2)^2 = 16$$
$$x^2 + 6x + 9 + y^2 - 4y + 4 = 16$$

which we find, upon simplifying, is equivalent to

$$x^2 + y^2 + 6x - 4y - 3 = 0$$

It can be shown that any equation of the form

$$x^2 + y^2 + ax + by + c = 0$$

has a graph that is a circle, or is a point, or has no graph at all. For example, the graph of the equation $x^2 + y^2 = 0$ is the single point $(0, 0)$. The equation $x^2 + y^2 + 5 = 0$, or $x^2 + y^2 = -5$, has no graph, because sums of squares of real numbers are never negative.

DEFINITION

When its graph is a circle, the equation

$$x^2 + y^2 + ax + by + c = 0$$

is referred to as the **general form of the equation of a circle.**

If an equation of a circle is in the general form, we use the method of completing the square to put the equation in standard form so that we can identify its center and radius.

EXAMPLE 4 Graphing a Circle Whose Equation Is in General Form

Graph the equation: $x^2 + y^2 + 4x - 6y + 12 = 0$

Solution

We complete the square in both x and y to put the equation in standard form. Group the expression involving x, group the expression involving y, and put the constant on the right side of the equation. The result is

$$(x^2 + 4x) + (y^2 - 6y) = -12$$

Next, complete the square of each expression in parentheses. Remember that any number added on the left side of the equation must be added on the right.

Figure 48

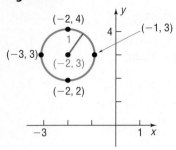

$$(x^2 + 4x + 4) + (y^2 - 6y + 9) = -12 + 4 + 9$$

$$\left(\tfrac{4}{2}\right)^2 = 4 \qquad \left(\tfrac{-6}{2}\right)^2 = 9$$

$$(x + 2)^2 + (y - 3)^2 = 1 \qquad \text{Factor}$$

We recognize this equation as the standard form of the equation of a circle with radius 1 and center $(-2, 3)$. To graph the equation, use the center $(-2, 3)$ and the radius 1. See Figure 48.

Now Work PROBLEM 27

EXAMPLE 5   **Finding the General Equation of a Circle**

Find the general equation of the circle whose center is $(1, -2)$ and whose graph contains the point $(4, -2)$.

Solution

To find the equation of a circle, we need to know its center and its radius. Here, we know that the center is $(1, -2)$. Since the point $(4, -2)$ is on the graph, the radius r will equal the distance from $(4, -2)$ to the center $(1, -2)$. See Figure 49. Thus,

Figure 49

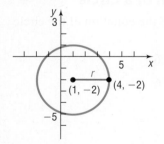

$$r = \sqrt{(4 - 1)^2 + [-2 - (-2)]^2}$$

$$= \sqrt{9} = 3$$

The standard form of the equation of the circle is

$$(x - 1)^2 + (y + 2)^2 = 9$$

Eliminating the parentheses and rearranging terms, we get the general equation

$$x^2 + y^2 - 2x + 4y - 4 = 0$$

Now Work PROBLEM 13

 COMMENT Example 6 requires information from Section B.5, *Square Screens*, in Appendix B. ∎

EXAMPLE 6   **Using a Graphing Utility to Graph a Circle**

Graph the equation: $x^2 + y^2 = 4$

Solution

This is the equation of a circle with center at the origin and radius 2. To graph this equation, we must first solve for y.

Figure 50

$$x^2 + y^2 = 4$$

$$y^2 = 4 - x^2 \qquad \text{Subtract } x^2 \text{ from each side.}$$

$$y = \pm\sqrt{4 - x^2} \qquad \text{Solve for } y.$$

There are two equations to graph: first, we graph $Y_1 = \sqrt{4 - x^2}$ and then $Y_2 = -\sqrt{4 - x^2}$ on the same square screen. (Your circle will appear oval if you do not use a square screen.) See Figure 50.

Overview

The discussion in Sections 3 and 4 about lines and circles dealt with two main types of problems that can be generalized as follows:

1. Given an equation, classify it and graph it.
2. Given a graph, or information about a graph, find its equation.

This text deals with both types of problems. We shall study various equations, classify them, and graph them. The second type of problem is usually more difficult to solve than the first.

4 Assess Your Understanding

Are You Prepared? *Answers are given at the end of these exercises. If you get a wrong answer, read the pages listed in red.*

1. To complete the square of $x^2 + 10x$, you would add the number _____. (p. A40)

2. Solve the equation $(x - 2)^2 = 9$. (pp. A65–A71)

Concepts and Vocabulary

3. *True or False* Every equation of the form
$$x^2 + y^2 + ax + by + c = 0$$
has a circle as its graph.

4. For a circle, the _____ is the distance from the center to any point on the circle.

5. The standard form of an equation of a circle with radius r and center (h, k) is _____.

6. The circle whose equation is given by $x^2 + y^2 = 1$ is called the _____ circle.

Skill Building

In Problems 7–10, find the center and radius of each circle. Write the standard form of the equation.

7.

8.

9.

10.

In Problems 11–20, write the standard form of the equation and the general form of the equation of each circle of radius r and center (h, k). Graph each circle.

11. $r = 2$; $(h, k) = (0, 0)$

12. $r = 3$; $(h, k) = (0, 0)$

13. $r = 2$; $(h, k) = (0, 2)$

14. $r = 3$; $(h, k) = (1, 0)$

15. $r = 5$; $(h, k) = (4, -3)$

16. $r = 4$; $(h, k) = (2, -3)$

17. $r = 4$; $(h, k) = (-2, 1)$

18. $r = 7$; $(h, k) = (-5, -2)$

19. $r = \dfrac{1}{2}$; $(h, k) = \left(\dfrac{1}{2}, 0\right)$

20. $r = \dfrac{1}{2}$; $(h, k) = \left(0, -\dfrac{1}{2}\right)$

In Problems 21–34, (a) find the center (h, k) and radius r of each circle; (b) graph each circle; (c) find the intercepts, if any.

21. $x^2 + y^2 = 4$

22. $x^2 + (y - 1)^2 = 1$

23. $2(x - 3)^2 + 2y^2 = 8$

24. $3(x + 1)^2 + 3(y - 1)^2 = 6$

25. $x^2 + y^2 - 2x - 4y - 4 = 0$

26. $x^2 + y^2 + 4x + 2y - 20 = 0$

27. $x^2 + y^2 + 4x - 4y - 1 = 0$

28. $x^2 + y^2 - 6x + 2y + 9 = 0$

29. $x^2 + y^2 - x + 2y + 1 = 0$

30. $x^2 + y^2 + x + y - \dfrac{1}{2} = 0$

31. $2x^2 + 2y^2 - 12x + 8y - 24 = 0$

32. $2x^2 + 2y^2 + 8x + 7 = 0$

33. $2x^2 + 8x + 2y^2 = 0$

34. $3x^2 + 3y^2 - 12y = 0$

In Problems 35–42, find the standard form of the equation of each circle.

35. Center at the origin and containing the point $(-2, 3)$

36. Center $(1, 0)$ and containing the point $(-3, 2)$

37. Center $(2, 3)$ and tangent to the *x*-axis

38. Center $(-3, 1)$ and tangent to the *y*-axis

39. With endpoints of a diameter at $(1, 4)$ and $(-3, 2)$

40. With endpoints of a diameter at $(4, 3)$ and $(0, 1)$

41. Center $(-1, 3)$ and tangent to the line $y = 2$

42. Center $(4, -2)$ and tangent to the line $x = 1$

In Problems 43–46, match each graph with the correct equation.

(a) $(x - 3)^2 + (y + 3)^2 = 9$ (b) $(x + 1)^2 + (y - 2)^2 = 4$ (c) $(x - 1)^2 + (y + 2)^2 = 4$ (d) $(x + 3)^2 + (y - 3)^2 = 9$

43.

44.

45.

46.

Applications and Extensions

47. Find the area of the square in the figure.

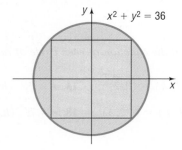

$x^2 + y^2 = 9$

48. Find the area of the blue shaded region in the figure, assuming the quadrilateral inside the circle is a square.

$x^2 + y^2 = 36$

49. Ferris Wheel The original Ferris wheel was built in 1893 by Pittsburgh, Pennsylvania bridge-builder George W. Ferris. The Ferris wheel was originally built for the 1893 World's Fair in Chicago, but was also later reconstructed for the 1904 World's Fair in St. Louis. It had a maximum height of 264 feet and a wheel diameter of 250 feet. Find an equation for the wheel if the center of the wheel is on the *y*-axis.

Source: inventors.about.com

50. Ferris Wheel In 2008, The Flyer opened as the world's largest Ferris wheel. It has a maximum height of 165 meters and a diameter of 150 meters, with one full rotation taking approximately 30 minutes. Find an equation for the wheel if the center of the wheel is on the *y*-axis.

Source: Wikipedia

51. Weather Satellites Earth is represented on a map of a portion of the solar system so that its surface is the circle with equation $x^2 + y^2 + 2x + 4y - 4091 = 0$. A weather satellite circles 0.6 unit above Earth with the center of its circular orbit at the center of Earth. Find the equation for the orbit of the satellite on this map.

52. The **tangent line** to a circle may be defined as the line that intersects the circle in a single point, called the **point of tangency.** See the figure.

If the equation of the circle is $x^2 + y^2 = r^2$ and the equation of the tangent line is $y = mx + b$, show that:

(a) $r^2(1 + m^2) = b^2$

[**Hint:** The quadratic equation $x^2 + (mx + b)^2 = r^2$ has exactly one solution.]

(b) The point of tangency is $\left(\dfrac{-r^2 m}{b}, \dfrac{r^2}{b} \right)$.

(c) The tangent line is perpendicular to the line containing the center of the circle and the point of tangency.

53. The Greek Method The Greek method for finding the equation of the tangent line to a circle uses the fact that at any point on a circle the lines containing the center and the tangent line are perpendicular (see Problem 52). Use this method to find an equation of the tangent line to the circle $x^2 + y^2 = 9$ at the point $(1, 2\sqrt{2})$.

54. Use the Greek method described in Problem 53 to find an equation of the tangent line to the circle $x^2 + y^2 - 4x + 6y + 4 = 0$ at the point $(3, 2\sqrt{2} - 3)$.

55. Refer to Problem 52. The line $x - 2y + 4 = 0$ is tangent to a circle at $(0, 2)$. The line $y = 2x - 7$ is tangent to the same circle at $(3, -1)$. Find the center of the circle.

56. Find an equation of the line containing the centers of the two circles

$$x^2 + y^2 - 4x + 6y + 4 = 0$$

and

$$x^2 + y^2 + 6x + 4y + 9 = 0$$

57. If a circle of radius 2 is made to roll along the x-axis, what is an equation for the path of the center of the circle?

58. If the circumference of a circle is 6π, what is its radius?

Explaining Concepts: Discussion and Writing

59. Which of the following equations might have the graph shown? (More than one answer is possible.)

(a) $(x - 2)^2 + (y + 3)^2 = 13$

(b) $(x - 2)^2 + (y - 2)^2 = 8$

(c) $(x - 2)^2 + (y - 3)^2 = 13$

(d) $(x + 2)^2 + (y - 2)^2 = 8$

(e) $x^2 + y^2 - 4x - 9y = 0$

(f) $x^2 + y^2 + 4x - 2y = 0$

(g) $x^2 + y^2 - 9x - 4y = 0$

(h) $x^2 + y^2 - 4x - 4y = 4$

60. Which of the following equations might have the graph shown? (More than one answer is possible.)

(a) $(x - 2)^2 + y^2 = 3$

(b) $(x + 2)^2 + y^2 = 3$

(c) $x^2 + (y - 2)^2 = 3$

(d) $(x + 2)^2 + y^2 = 4$

(e) $x^2 + y^2 + 10x + 16 = 0$

(f) $x^2 + y^2 + 10x - 2y = 1$

(g) $x^2 + y^2 + 9x + 10 = 0$

(h) $x^2 + y^2 - 9x - 10 = 0$

61. Explain how the center and radius of a circle can be used to graph a circle.

62. What Went Wrong? A student stated that the center and radius of the graph whose equation is $(x + 3)^2 + (y - 2)^2 = 16$ are $(3, -2)$ and 4, respectively. Why is this incorrect?

'Are You Prepared?' Answers

1. 25   **2.** $\{-1, 5\}$

1 Functions and Their Graphs

One Call Started It All: Industry Marks 25th Anniversary of First Commercial Cellular Call

One of the most significant events in the history of Chicago's storied Soldier Field had nothing to do with football legends Dick Butkus and Walter Payton and certainly didn't draw a crowd near the size of your typical Bears home game.

On Oct. 13, 1983, a small group gathered outside the stadium as Bob Barnett, President of Ameritech Mobile Communications Inc., placed a call from inside a car to the grandson of famed telephone inventor Alexander Graham Bell in Berlin. It was the inaugural call for the nation's first citywide, commercial cellular system and marked the arrival of cellular telephone service 36 years after the concept was first introduced.

Twenty-five years after that call, industry executives find it hard to believe that cellular technology has redefined the communications landscape in so short a period of time.

Even as cell phones shrank in size and became more affordable, they faced an uphill battle for acceptance. In 1995, a poll by Gallup revealed that 53 percent of Americans thought that people who had cell phones didn't really need them and were just showing off or acting self-important.

Attitudes certainly have changed in the 13 years since that poll, with 84 percent of Americans now having wireless phone service, including the 15.8 percent of U.S. homes that are wireless-only households, according to figures released earlier this year by the National Center for Health Statistics. The industry's growth has indeed been exponential. There were only 7.5 million U.S. subscribers in 1992, a figure that swelled to more than 158 million in 2004 and has almost doubled in the four years since.

Source: The CTIA Semi-Annual Wireless Industry Indices, December 2004 Edition. Used with the permission of CTIA—The Wireless Association®

—See Chapter Project I—

◁ **A Look Back** Up to now, our discussion has focused on graphing equations containing two variables.

A Look Ahead ▷ In this chapter, we look at a special type of equation involving two variables called a *function*. This chapter deals with what a function is, how to graph functions, properties of functions, and how functions are used in applications. The word *function* apparently was introduced by René Descartes in 1637. For him, a function simply meant any positive integral power of a variable *x*. Gottfried Wilhelm Leibniz (1646–1716), who always emphasized the geometric side of mathematics, used the word *function* to denote any quantity associated with a curve, such as the coordinates of a point on the curve. Leonhard Euler (1707–1783) employed the word to mean any equation or formula involving variables and constants. His idea of a function is similar to the one most often seen in courses that precede calculus. Later, the use of functions in investigating heat flow equations led to a very broad definition, due to Lejeune Dirichlet (1805–1859), which describes a function as a rule or correspondence between two sets. It is his definition that we use here.

1.1 Functions

PREPARING FOR THIS SECTION   *Before getting started, review the following:*

- Intervals (Appendix A, Section A.10, pp. A83–A84)
- Evaluating Algebraic Expressions, Domain of a Variable (Appendix A, Section A.1, pp. A6–A7)
- Solving Inequalities (Appendix A, Section A.10, pp. A86–A88)
- Polynomials (Appendix A, Section A.3, pp. A23–A31)
- Rational Expressions (Appendix A, Section A.6, pp. A46–A54)

Now Work the **'Are You Prepared?'** problems on page 53.

OBJECTIVES   **1** Determine Whether a Relation Represents a Function (p. 43)
   2 Find the Value of a Function (p. 46)
   3 Find the Domain of a Function Defined by an Equation (p. 49)
   4 Form the Sum, Difference, Product, and Quotient of Two Functions (p. 51)

1 Determine Whether a Relation Represents a Function

We often see situations where one variable is somehow linked to another variable. For example, an individual's level of education is linked to annual income. Engine size is linked to gas mileage. When the value of one variable is related to the value of a second variable, we have a *relation*. A **relation** is a correspondence between two sets. If x and y are two elements in these sets and if a relation exists between x and y, then we say that x **corresponds** to y or that y **depends on** x, and we write $x \rightarrow y$.

We have a number of ways to express relations between two sets. For example, the equation $y = 3x - 1$ shows a relation between x and y. It says that if we take some number x, multiply it by 3, and then subtract 1, we obtain the corresponding value of y. In this sense, x serves as the **input** to the relation and y is the **output** of the relation. We can also express this relation as a graph, as shown in Figure 1.

Not only can a relation be expressed through an equation or graph, but we can also express relations using through a technique called *mapping*. A **map** illustrates a relation by using a set of inputs and drawing arrows to the corresponding element in the set of outputs. Ordered pairs can also be used to represent $x \rightarrow y$ as (x, y). We illustrate these two concepts in Example 1.

Figure 1

EXAMPLE 1   **Maps and Ordered Pairs as Relations**

Figure 2 shows a relation between states and the number of representatives each state has in the House of Representatives. The relation might be named "number of representatives."

Figure 2

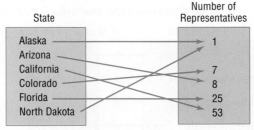

In this relation, Alaska corresponds to 1, Arizona corresponds to 8, and so on. Using ordered pairs, this relation would be expressed as

{(Alaska, 1), (Arizona, 8), (California, 53), (Colorado, 7), (Florida, 25), (North Dakota, 1)}

We now present one of the most important concepts in algebra—the *function*. A function is a special type of relation. To understand the idea behind a function,

Figure 3

Weight (grams)   IQ

let's revisit the relation presented in Example 1. If we were to ask, "How many representatives does Alaska have?," you would respond 1. In fact, each input *state* corresponds to a single output *number of representatives.*

Let's consider a second relation where we have a correspondence between the weight of a brain (in grams) and IQ. See Figure 3. Notice a brain that weighs 1342 grams corresponds to two different IQs: 104 and 111. So, if someone asked, "What is the IQ of an individual whose brain weighs 1342 grams?," you cannot provide a single answer.

Let's look at one more relation. Figure 4 is a relation that shows a correspondence between *animals* and *life expectancy*. If asked to determine the life expectancy of a dog, we would all respond "11 years." If asked to determine the life expectancy of a rabbit, we would all respond "7 years."

Figure 4

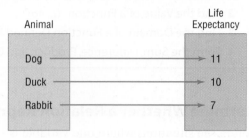

Notice that the relations presented in Figures 2 and 4 have something in common. What is it? The common link between these two relations is that each input corresponds to exactly one output. This leads to the definition of a *function.*

DEFINITION

Figure 5

Let X and Y be two nonempty sets.* A **function** from X into Y is a relation that associates with each element of X exactly one element of Y.

The set X is called the **domain** of the function. For each element x in X, the corresponding element y in Y is called the **value** of the function at x, or the **image** of x. The set of all images of the elements in the domain is called the **range** of the function. See Figure 5.

Since there may be some elements in Y that are not the image of some x in X, it follows that the range of a function may be a subset of Y, as shown in Figure 5.

Not all relations between two sets are functions. The next example shows how to determine whether a relation is a function.

EXAMPLE 2   **Determining Whether a Relation Represents a Function**

Determine which of the following relations represent a function. If the relation is a function, then state its domain and range.

(a) See Figure 6. For this relation, the domain represents the level of education and the range represents the unemployment rate.

Figure 6
Source: Statistical Abstract of the United States, 2008

Level of Education   Unemployment Rate

No High School Diploma   →   7.1.%
High School Diploma   →   4.4%
Some College   →   3.6%
College Graduate   →   2.0%

(b) See Figure 7. For this relation, the domain represents the number of calories in a sandwich from a fast-food restaurant and the range represents the fat content (in grams).

*The sets X and Y will usually be sets of real numbers, in which case a (real) function results. The two sets can also be sets of complex numbers, and then we have defined a complex function. In the broad definition (due to Lejeune Dirichlet), X and Y can be any two sets.

Figure 7
Source: Each company's Web site

Figure 8
Source: Used with permission of Diamonds.com

(c) See Figure 8. For this relation, the domain represents the weight (in carats) of pear-cut diamonds and the range represents the price (in dollars).

Solution

(a) The relation in Figure 6 is a function because each element in the domain corresponds to exactly one element in the range. The domain of the relation is {No High School Diploma, High School Diploma, Some College, College Graduate}, and the range of the relation is {7.1%, 4.4%, 3.6%, 2.0%}.

(b) The relation in Figure 7 is a function because each element in the domain corresponds to exactly one element in the range. The domain of the relation is {410, 580, 540, 750, 600, 430}. The range of the relation is {19, 29, 24, 33, 23}. Notice that it is okay for more than one element in the domain to correspond to the same element in the range (McDonald's and Burger King's chicken sandwich both have 23 grams of fat).

(c) The relation in Figure 8 is not a function because each element in the domain does not correspond to exactly one element in the range. If a 0.71-carat diamond is chosen from the domain, a single price cannot be assigned to it.

Now Work PROBLEM 15

The idea behind a function is its predictability. If the input is known, we can use the function to determine the output. With "nonfunctions," we don't have this predictability. Look back at Figure 7. The inputs are {410, 580, 540, 750, 600, 430}. The correspondence is *number of fat grams*, and the outputs are {19, 29, 24, 33, 23}. If asked, "How many grams of fat are in a 410-calorie sandwich?," we can use the correspondence to answer "19." Now consider Figure 8. If asked, "What is the price of a 0.71-carat diamond?," we could not give a single response because two outputs result from the single input "0.71." For this reason, the relation in Figure 8 is not a function.

We may also think of a function as a set of ordered pairs (x, y) in which no ordered pairs have the same first element and different second elements. The set of all first elements x is the domain of the function, and the set of all second elements y is its range. Each element x in the domain corresponds to exactly one element y in the range.

> **In Words**
> For a function, no input has more than one output. The domain of a function is the set of all inputs; the range is the set of all outputs.

EXAMPLE 3

Determining Whether a Relation Represents a Function

Determine whether each relation represents a function. If it is a function, state the domain and range.
(a) $\{(1, 4), (2, 5), (3, 6), (4, 7)\}$
(b) $\{(1, 4), (2, 4), (3, 5), (6, 10)\}$
(c) $\{(-3, 9), (-2, 4), (0, 0), (1, 1), (-3, 8)\}$

Solution

(a) This relation is a function because there are no ordered pairs with the same first element and different second elements. The domain of this function is $\{1, 2, 3, 4\}$, and its range is $\{4, 5, 6, 7\}$.

(b) This relation is a function because there are no ordered pairs with the same first element and different second elements. The domain of this function is $\{1, 2, 3, 6\}$, and its range is $\{4, 5, 10\}$.

(c) This relation is not a function because there are two ordered pairs, $(-3, 9)$ and $(-3, 8)$, that have the same first element and different second elements.

In Example 3(b), notice that 1 and 2 in the domain each have the same image in the range. This does not violate the definition of a function; two different first elements can have the same second element. A violation of the definition occurs when two ordered pairs have the same first element and different second elements, as in Example 3(c).

Now Work PROBLEM 19

Up to now we have shown how to identify when a relation is a function for relations defined by mappings (Example 2) and ordered pairs (Example 3). We know that relations can also be expressed as equations. We discuss next the circumstances under which equations are functions.

To determine whether an equation, where y depends on x, is a function, it is often easiest to solve the equation for y. If any value of x in the domain corresponds to more than one y, the equation does not define a function; otherwise, it does define a function.

EXAMPLE 4 **Determining Whether an Equation Is a Function**

Determine if the equation $y = 2x - 5$ defines y as a function of x.

Solution The equation tells us to take an input x, multiply it by 2, and then subtract 5. For any input x, these operations yield only one output y. For example, if $x = 1$, then $y = 2(1) - 5 = -3$. If $x = 3$, then $y = 2(3) - 5 = 1$. For this reason, the equation is a function.

EXAMPLE 5 **Determining Whether an Equation Is a Function**

Determine if the equation $x^2 + y^2 = 1$ defines y as a function of x.

Solution To determine whether the equation $x^2 + y^2 = 1$, which defines the unit circle, is a function, we need to solve the equation for y.

$$x^2 + y^2 = 1$$
$$y^2 = 1 - x^2$$
$$y = \pm\sqrt{1 - x^2}$$

For values of x between -1 and 1, two values of y result. For example, if $x = 0$, then $y = \pm 1$, so two different outputs result from the same input. This means that the equation $x^2 + y^2 = 1$ does not define a function.

Now Work PROBLEM 33

2 Find the Value of a Function

Functions are often denoted by letters such as f, F, g, G, and others. If f is a function, then for each number x in its domain the corresponding image in the range is designated by the symbol $f(x)$, read as "f of x" or as "f at x." We refer to $f(x)$ as the **value of f at the number x**; $f(x)$ is the number that results when x is given and the function f is applied; $f(x)$ is the output corresponding to x or the image of x; $f(x)$ does *not* mean "f times x." For example, the function given in Example 4 may be written as $y = f(x) = 2x - 5$. Then $f\left(\dfrac{3}{2}\right) = -2$.

Figure 9 illustrates some other functions. Notice that, in every function, for each x in the domain there is one value in the range.

Figure 9

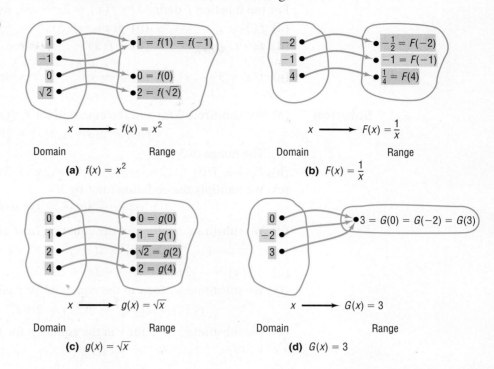

(a) $f(x) = x^2$

(b) $F(x) = \dfrac{1}{x}$

(c) $g(x) = \sqrt{x}$

(d) $G(x) = 3$

Figure 10

Input x

Output
$y = f(x)$

Sometimes it is helpful to think of a function f as a machine that receives as input a number from the domain, manipulates it, and outputs the value. See Figure 10.

The restrictions on this input/output machine are as follows:

1. It only accepts numbers from the domain of the function.
2. For each input, there is exactly one output (which may be repeated for different inputs).

For a function $y = f(x)$, the variable x is called the **independent variable,** because it can be assigned any of the permissible numbers from the domain. The variable y is called the **dependent variable,** because its value depends on x.

Any symbol can be used to represent the independent and dependent variables. For example, if f is the *cube function,* then f can be given by $f(x) = x^3$ or $f(t) = t^3$ or $f(z) = z^3$. All three functions are the same. Each tells us to cube the independent variable to get the output. In practice, the symbols used for the independent and dependent variables are based on common usage, such as using C for cost in business.

The independent variable is also called the **argument** of the function. Thinking of the independent variable as an argument can sometimes make it easier to find the value of a function. For example, if f is the function defined by $f(x) = x^3$, then f tells us to cube the argument. Thus, $f(2)$ means to cube 2, $f(a)$ means to cube the number a, and $f(x + h)$ means to cube the quantity $x + h$.

EXAMPLE 6

Illustrating Language Used with Functions

Consider the function $f(p) = p^2$. The name of the function is f. The *independent variable* or *argument* of f is p. If $q = f(p)$, then q is the *dependent variable.* Since $f(3) = 9$, we say that 9 is the *value* of f at 3 or 9 is the *image* of 3. Because we can square any real number, the *domain* of the function is the set of all real numbers, or $(-\infty, \infty)$ using interval notation. Because the square of any real number is greater than or equal to zero, the *range* of f is $[0, \infty)$. In words, the function f tells us to square the argument.

EXAMPLE 7 | **Finding Values of a Function**

For the function f defined by $f(x) = 2x^2 - 3x$, evaluate

(a) $f(3)$ (b) $f(x) + f(3)$ (c) $3f(x)$

(d) $f(-x)$ (e) $-f(x)$ (f) $f(3x)$

(g) $f(x + 3)$ (h) $\dfrac{f(x + h) - f(x)}{h}$ $h \neq 0$

Solution

(a) We substitute 3 for x in the equation for f, $f(x) = 2x^2 - 3x$, to get
$$f(3) = 2(3)^2 - 3(3) = 18 - 9 = 9$$
The image of 3 is 9.

(b) $f(x) + f(3) = (2x^2 - 3x) + (9) = 2x^2 - 3x + 9$

(c) We multiply the equation for f by 3.
$$3f(x) = 3(2x^2 - 3x) = 6x^2 - 9x$$

(d) We substitute $-x$ for x in the equation for f and simplify.
$$f(-x) = 2(-x)^2 - 3(-x) = 2x^2 + 3x$$

(e) $-f(x) = -(2x^2 - 3x) = -2x^2 + 3x$

(f) We substitute $3x$ for x in the equation for f and simplify.
$$f(3x) = 2(3x)^2 - 3(3x) = 2(9x^2) - 9x = 18x^2 - 9x$$

(g) We substitute $x + 3$ for x in the equation for f and simplify.
$$f(x + 3) = 2(x + 3)^2 - 3(x + 3) \quad \text{Notice the use of parentheses here.}$$
$$= 2(x^2 + 6x + 9) - 3x - 9$$
$$= 2x^2 + 12x + 18 - 3x - 9$$
$$= 2x^2 + 9x + 9$$

(h) $\dfrac{f(x + h) - f(x)}{h} = \dfrac{[2(x + h)^2 - 3(x + h)] - [2x^2 - 3x]}{h}$

$\quad\quad\quad\quad\quad\quad\quad\quad\uparrow$
$\quad\quad\quad f(x + h) = 2(x + h)^2 - 3(x + h)$

$$= \dfrac{2(x^2 + 2xh + h^2) - 3x - 3h - 2x^2 + 3x}{h} \quad \text{Simplify.}$$

$$= \dfrac{2x^2 + 4xh + 2h^2 - 3h - 2x^2}{h} \quad \text{Distribute and combine like terms.}$$

$$= \dfrac{4xh + 2h^2 - 3h}{h} \quad \text{Combine like terms.}$$

$$= \dfrac{h(4x + 2h - 3)}{h} \quad \text{Factor out } h.$$

$$= 4x + 2h - 3 \quad \text{Divide out the } h\text{'s.}$$

●

Notice in Example 7 that $f(x + 3) \neq f(x) + f(3)$, $f(-x) \neq -f(x)$, and $3f(x) \neq f(3x)$.

The expression in part (h) is called the **difference quotient** of f, an important expression in calculus.

Now Work   PROBLEMS 39 AND 75

Most calculators have special keys that allow you to find the value of certain commonly used functions. For example, you should be able to find the square function $f(x) = x^2$, the square root function $f(x) = \sqrt{x}$, the reciprocal function $f(x) = \dfrac{1}{x} = x^{-1}$, and many others that will be discussed later in this book (such as $\ln x$ and $\log x$). Verify the results of Example 8, which follows, on your calculator.

EXAMPLE 8	**Finding Values of a Function on a Calculator**

(a) $f(x) = x^2$ $f(1.234) = 1.234^2 = 1.522756$

(b) $F(x) = \dfrac{1}{x}$ $F(1.234) = \dfrac{1}{1.234} \approx 0.8103727715$

(c) $g(x) = \sqrt{x}$ $g(1.234) = \sqrt{1.234} \approx 1.110855526$

COMMENT Graphing calculators can be used to evaluate any function that you wish. Figure 11 shows the result obtained in Example 7(a) on a TI-84 Plus graphing calculator with the function to be evaluated, $f(x) = 2x^2 - 3x$, in Y_1.

Figure 11

Implicit Form of a Function

In general, when a function f is defined by an equation in x and y, we say that the function f is given **implicitly**. If it is possible to solve the equation for y in terms of x, then we write $y = f(x)$ and say that the function is given **explicitly.** For example,

Implicit Form	**Explicit Form**
$3x + y = 5$	$y = f(x) = -3x + 5$
$x^2 - y = 6$	$y = f(x) = x^2 - 6$
$xy = 4$	$y = f(x) = \dfrac{4}{x}$

We list next a summary of some important facts to remember about a function f.

SUMMARY

Important Facts about Functions

(a) For each x in a domain of a function f, there is exactly one image $f(x)$ in the range; however, an element in the range can result from more than one x in the domain.
(b) f is the symbol that we use to denote the function. It is symbolic of the equation (rule) that we use to get from an x in the domain to $f(x)$ in the range.
(c) If $y = f(x)$, then x is called the independent variable or argument of f, and y is called the dependent variable or the value of f at x.

3 Find the Domain of a Function Defined by an Equation

Often the domain of a function f is not specified; instead, only the equation defining the function is given. In such cases, we agree that the **domain of f** is the largest set of real numbers for which the value $f(x)$ is a real number. The domain of a function f is the same as the domain of the variable x in the expression $f(x)$.

EXAMPLE 9	**Finding the Domain of a Function**

Find the domain of each of the following functions:

(a) $f(x) = x^2 + 5x$ (b) $g(x) = \dfrac{3x}{x^2 - 4}$

(c) $h(t) = \sqrt{4 - 3t}$ (d) $F(x) = \dfrac{\sqrt{3x + 12}}{x - 5}$

Solution

(a) The function tells us to square a number and then add five times the number. Since these operations can be performed on any real number, we conclude that the domain of f is the set of all real numbers.

(b) The function g tells us to divide $3x$ by $x^2 - 4$. Since division by 0 is not defined, the denominator $x^2 - 4$ can never be 0, so x can never equal -2 or 2. The domain of the function g is $\{x \mid x \neq -2, x \neq 2\}$.

(c) The function h tells us to take the square root of $4 - 3t$. But only nonnegative numbers have real square roots, so the expression under the square root (the radicand) must be nonnegative (greater than or equal to zero). This requires that

$$4 - 3t \geq 0$$
$$-3t \geq -4$$
$$t \leq \frac{4}{3}$$

The domain of h is $\left\{ t \mid t \leq \dfrac{4}{3} \right\}$ or the interval $\left(-\infty, \dfrac{4}{3} \right]$.

(d) The function F tells us to take the square root of $3x + 12$ and divide this result by $x - 5$. This requires that $3x + 12 \geq 0$, so $x \geq -4$, and also that $x - 5 \neq 0$, so $x \neq 5$. Combining these two restrictions, the domain of F is $\{x \mid x \geq -4, x \neq 5\}$.

●

> **In Words**
> The domain of g found in Example 9(b) is $\{x \mid x \neq -2, x \neq 2\}$. This notation is read, "The domain of the function g is the set of all real numbers x such that x does not equal -2 and x does not equal 2."

For the functions that we will encounter in this book, the following steps may prove helpful for finding the domain of a function that is defined by an equation whose domain is a subset of the real numbers.

Finding the Domain of a Function Defined by an Equation

1. Start with the domain as the set of real numbers.
2. If the equation has a denominator, exclude any numbers that give a zero denominator.
3. If the equation has a radical of even index, exclude any numbers that cause the expression inside the radical to be negative.

Now Work PROBLEM 51

If x is in the domain of a function f, we shall say that *f* **is defined at x,** or *f(x)* **exists.** If x is not in the domain of f, we say that *f* **is not defined at x,** or *f(x)* **does not exist.** For example, if $f(x) = \dfrac{x}{x^2 - 1}$, then $f(0)$ exists, but $f(1)$ and $f(-1)$ do not exist. (Do you see why?)

We have not said much about finding the range of a function. We will say more about finding the range when we look at the graph of a function in the next section. When a function is defined by an equation, it can be difficult to find the range.* Therefore, we shall usually be content to find just the domain of a function when the function is defined by an equation. We shall express the domain of a function using inequalities, interval notation, set notation, or words, whichever is most convenient.

When we use functions in applications, the domain may be restricted by physical or geometric considerations. For example, the domain of the function f defined by $f(x) = x^2$ is the set of all real numbers. However, if f is used to obtain the area of a square when the length x of a side is known, then we must restrict the domain of f to the positive real numbers, since the length of a side can never be 0 or negative.

---

*In Section 4.2, we discuss a way to find the range for a special class of functions.

EXAMPLE 10   **Finding the Domain in an Application**

Express the area of a circle as a function of its radius. Find the domain.

Solution   See Figure 12. We know that the formula for the area A of a circle of radius r is $A = \pi r^2$. If we use r to represent the independent variable and A to represent the dependent variable, the function expressing this relationship is

Figure 12

$$A(r) = \pi r^2$$

In this setting, the domain is $\{r \mid r > 0\}$. (Do you see why?)

Observe in the solution to Example 10 that we used the symbol A in two ways: It is used to name the function, and it is used to symbolize the dependent variable. This double use is common in applications and should not cause any difficulty.

⟶**Now Work** PROBLEM **89**

4 Form the Sum, Difference, Product, and Quotient of Two Functions

Next we introduce some operations on functions. We shall see that functions, like numbers, can be added, subtracted, multiplied, and divided. For example, if $f(x) = x^2 + 9$ and $g(x) = 3x + 5$, then

$$f(x) + g(x) = (x^2 + 9) + (3x + 5) = x^2 + 3x + 14$$

The new function $y = x^2 + 3x + 14$ is called the *sum function* $f + g$. Similarly,

$$f(x) \cdot g(x) = (x^2 + 9)(3x + 5) = 3x^3 + 5x^2 + 27x + 45$$

The new function $y = 3x^3 + 5x^2 + 27x + 45$ is called the *product function* $f \cdot g$.
The general definitions are given next.

DEFINITION   If f and g are functions:
The **sum** $f + g$ is the function defined by

$$(f + g)(x) = f(x) + g(x)$$

In Words
Remember, the symbol ∩ stands for intersection. It means you should find the elements that are common to two sets.

The domain of $f + g$ consists of the numbers x that are in the domains of both f and g. That is, domain of $f + g =$ domain of f ∩ domain of g.

DEFINITION   The **difference** $f - g$ is the function defined by

$$(f - g)(x) = f(x) - g(x)$$

The domain of $f - g$ consists of the numbers x that are in the domains of both f and g. That is, domain of $f - g =$ domain of f ∩ domain of g.

DEFINITION   The **product** $f \cdot g$ is the function defined by

$$(f \cdot g)(x) = f(x) \cdot g(x)$$

The domain of $f \cdot g$ consists of the numbers x that are in the domains of both f and g. That is, domain of $f \cdot g =$ domain of f ∩ domain of g.

DEFINITION

The **quotient** $\dfrac{f}{g}$ is the function defined by

$$\left(\frac{f}{g}\right)(x) = \frac{f(x)}{g(x)} \qquad g(x) \neq 0$$

The domain of $\dfrac{f}{g}$ consists of the numbers x for which $g(x) \neq 0$ and that are in the domains of both f and g. That is,

$$\text{domain of } \frac{f}{g} = \{x \,|\, g(x) \neq 0\} \cap \text{ domain of } f \cap \text{ domain of } g$$

EXAMPLE 11   **Operations on Functions**

Let f and g be two functions defined as

$$f(x) = \frac{1}{x+2} \quad \text{and} \quad g(x) = \frac{x}{x-1}$$

Find the following, and determine the domain in each case.

(a) $(f + g)(x)$ (b) $(f - g)(x)$ (c) $(f \cdot g)(x)$ (d) $\left(\dfrac{f}{g}\right)(x)$

Solution   The domain of f is $\{x \,|\, x \neq -2\}$ and the domain of g is $\{x \,|\, x \neq 1\}$.

(a) $(f + g)(x) = f(x) + g(x) = \dfrac{1}{x+2} + \dfrac{x}{x-1}$

$$= \frac{x-1}{(x+2)(x-1)} + \frac{x(x+2)}{(x+2)(x-1)} = \frac{x^2 + 3x - 1}{(x+2)(x-1)}$$

The domain of $f + g$ consists of those numbers x that are in the domains of both f and g. Therefore, the domain of $f + g$ is $\{x \,|\, x \neq -2, x \neq 1\}$.

(b) $(f - g)(x) = f(x) - g(x) = \dfrac{1}{x+2} - \dfrac{x}{x-1}$

$$= \frac{x-1}{(x+2)(x-1)} - \frac{x(x+2)}{(x+2)(x-1)} = \frac{-(x^2 + x + 1)}{(x+2)(x-1)}$$

The domain of $f - g$ consists of those numbers x that are in the domains of both f and g. Therefore, the domain of $f - g$ is $\{x \,|\, x \neq -2, x \neq 1\}$.

(c) $(f \cdot g)(x) = f(x) \cdot g(x) = \dfrac{1}{x+2} \cdot \dfrac{x}{x-1} = \dfrac{x}{(x+2)(x-1)}$

The domain of $f \cdot g$ consists of those numbers x that are in the domains of both f and g. Therefore, the domain of $f \cdot g$ is $\{x \,|\, x \neq -2, x \neq 1\}$.

(d) $\left(\dfrac{f}{g}\right)(x) = \dfrac{f(x)}{g(x)} = \dfrac{\dfrac{1}{x+2}}{\dfrac{x}{x-1}} = \dfrac{1}{x+2} \cdot \dfrac{x-1}{x} = \dfrac{x-1}{x(x+2)}$

The domain of $\dfrac{f}{g}$ consists of the numbers x for which $g(x) \neq 0$ and that are in the domains of both f and g. Since $g(x) = 0$ when $x = 0$, we exclude 0 as well as -2 and 1 from the domain. The domain of $\dfrac{f}{g}$ is $\{x \,|\, x \neq -2, x \neq 0, x \neq 1\}$.

━━━━━ **Now Work** PROBLEM 63

In calculus, it is sometimes helpful to view a complicated function as the sum, difference, product, or quotient of simpler functions. For example,

$$F(x) = x^2 + \sqrt{x} \text{ is the sum of } f(x) = x^2 \text{ and } g(x) = \sqrt{x}.$$

$$H(x) = \frac{x^2 - 1}{x^2 + 1} \text{ is the quotient of } f(x) = x^2 - 1 \text{ and } g(x) = x^2 + 1.$$

SUMMARY

We list here some of the important vocabulary introduced in this section, with a brief description of each term.

Function	A relation between two sets of real numbers so that each number x in the first set, the domain, has corresponding to it exactly one number y in the second set
	A set of ordered pairs (x, y) or $(x, f(x))$ in which no first element is paired with two different second elements
	The range is the set of y values of the function that are the images of the x values in the domain.
	A function f may be defined implicitly by an equation involving x and y or explicitly by writing $y = f(x)$.
Unspecified domain	If a function f is defined by an equation and no domain is specified, then the domain will be taken to be the largest set of real numbers for which the equation defines a real number.
Function notation	$y = f(x)$
	f is a symbol for the function.
	x is the independent variable or argument.
	y is the dependent variable.
	$f(x)$ is the value of the function at x, or the image of x.

1.1 Assess Your Understanding

'Are You Prepared?' *Answers are given at the end of these exercises. If you get a wrong answer, read the pages listed in* red.

1. If $x = -2$, the value of the expression $3x^2 - 5x + \dfrac{1}{x}$ is

_____. (pp. A6–A7)

2. The domain of the variable in the expression $\dfrac{x - 3}{x + 4}$ is

_____. (p. A7)

3. Solve the inequality: $3 - 2x > 5$. Write the solution in interval notation and graph the solution set. (pp. A83–A87)

4. Perform the indicated operation.

(a) $(x^2 + x - 1) - (x^2 + 4x - 3)$ (pp. A25–A26)

(b) $\dfrac{x + 1}{x - 2} + \dfrac{x}{x - 3}$ (pp. A50–A52)

(c) $\dfrac{2x + 3}{x - 4} \cdot \dfrac{x + 1}{2x + 3}$ (pp. A47–A48)

Concepts and Vocabulary

5. If f is a function defined by the equation $y = f(x)$, then x is called the _____ variable and y is the _____ variable.

6. The set of all images of the elements in the domain of a function is called the _____.

7. If the domain of f is all real numbers in the interval $[0, 7]$ and the domain of g is all real numbers in the interval $[-2, 5]$, the domain of $f + g$ is all real numbers in the interval _____.

8. If $f(5) = 17$, then 17 is the _____ of 5.

9. If $f(x) = x + 1$ and $g(x) = x^3$, then _____ $=$ $x^3 - (x + 1)$.

10. *True or False* Every relation is a function.

11. *True or False* The domain of $(f \cdot g)(x)$ consists of the numbers x that are in the domains of both f and g.

12. *True or False* The independent variable is sometimes referred to as the argument of the function.

13. *True or False* If no domain is specified for a function f, then the domain of f is taken to be the set of real numbers.

14. *True or False* The domain of the function $f(x) = \dfrac{x^2 - 4}{x}$ is $\{x \mid x \neq \pm 2\}$.

Skill Building

In Problems 15–26, determine whether each relation represents a function. For each function, state the domain and range.

15.

16.

17. Hours Worked   Salary

20 Hours	$200
	$300
30 Hours	$350
40 Hours	$425

18.

Level of Education	Average Income
Less than 9th grade	$18,120
9th-12th grade	$23,251
High School Graduate	$36,055
Some College	$45,810
College Graduate	$67,165

19. $\{(2, 8), (-3, 6), (4, 9), (2, 10)\}$

20. $\{(-2, 5), (-1, 3), (3, 7), (4, 12)\}$

21. $\{(1, 3), (2, 3), (3, 3), (4, 3)\}$

22. $\{(0, -2), (1, 3), (2, 3), (3, 7)\}$

23. $\{(-2, 4), (-2, 6), (0, 3), (3, 7)\}$

24. $\{(-4, 4), (-3, 3), (-2, 2), (-1, 1), (-4, 0)\}$

25. $\{(-2, 4), (-1, 1), (0, 0), (1, 1)\}$

26. $\{(-2, 16), (-1, 4), (0, 3), (1, 4)\}$

In Problems 27–38, determine whether the equation defines y as a function of x.

27. $y = x^2$

28. $y = x^3$

29. $y = \dfrac{1}{x}$

30. $y = |x|$

31. $y = \pm\sqrt{4 - x^2}$

32. $y = \pm\sqrt{1 - 2x}$

33. $x = y^2$

34. $x + y^2 = 1$

35. $y = 2x^2 - 3x + 4$

36. $y = \dfrac{3x - 1}{x + 2}$

37. $2x^2 + 3y^2 = 1$

38. $x^2 - 4y^2 = 1$

In Problems 39–46, find the following for each function:

(a) $f(0)$ (b) $f(1)$ (c) $f(-1)$ (d) $f(-x)$ (e) $-f(x)$ (f) $f(x + 1)$ (g) $f(2x)$ (h) $f(x + h)$

39. $f(x) = 3x^2 + 2x - 4$

40. $f(x) = -2x^2 + x - 1$

41. $f(x) = \dfrac{x}{x^2 + 1}$

42. $f(x) = \dfrac{x^2 - 1}{x + 4}$

43. $f(x) = |x| + 4$

44. $f(x) = \sqrt{x^2 + x}$

45. $f(x) = \dfrac{2x + 1}{3x - 5}$

46. $f(x) = 1 - \dfrac{1}{(x + 2)^2}$

In Problems 47–62, find the domain of each function.

47. $f(x) = -5x + 4$

48. $f(x) = x^2 + 2$

49. $f(x) = \dfrac{x}{x^2 + 1}$

50. $f(x) = \dfrac{x^2}{x^2 + 9}$

51. $g(x) = \dfrac{x}{x^2 - 16}$

52. $h(x) = \dfrac{2x}{x^2 - 4}$

53. $F(x) = \dfrac{x - 2}{x^3 - 2x^2 - 9x + 18}$

54. $G(x) = \dfrac{x + 4}{x^3 + 4x^2 - 4x - 16}$

55. $h(x) = \sqrt{3x - 12}$

56. $G(x) = \sqrt{1 - x}$

57. $f(x) = \dfrac{4}{\sqrt{x - 9}}$

58. $f(x) = \dfrac{x}{\sqrt{x - 4}}$

59. $P(x) = \sqrt{-5x + 30}$

60. $q(x) = \sqrt{-x - 2}$

61. $P(t) = \dfrac{\sqrt{t - 4}}{3t - 21}$

62. $h(z) = \dfrac{\sqrt{z + 3}}{z - 2}$

In Problems 63–72, for the given functions f and g, find the following. For parts (a)–(d), also find the domain.

(a) $(f + g)(x)$ (b) $(f - g)(x)$ (c) $(f \cdot g)(x)$ (d) $\left(\dfrac{f}{g}\right)(x)$ (e) $(f + g)(3)$ (f) $(f - g)(4)$ (g) $(f \cdot g)(2)$ (h) $\left(\dfrac{f}{g}\right)(1)$

63. $f(x) = 3x + 4;\quad g(x) = 2x - 3$

64. $f(x) = 2x + 1;\quad g(x) = 3x - 2$

65. $f(x) = x - 1;\quad g(x) = 2x^2 + 1$

66. $f(x) = 2x^2 + 3;\quad g(x) = 4x^3 + 1$

67. $f(x) = \sqrt{x + 1};\ g(x) = \sqrt{x - 1}$

68. $f(x) = \sqrt{x - 1};\quad g(x) = \sqrt{4 - x}$

69. $f(x) = \dfrac{2x + 3}{3x - 2};\quad g(x) = \dfrac{4x}{3x - 2}$

70. $f(x) = \dfrac{x - 3}{2x + 1};\ g(x) = \dfrac{2x}{2x + 1}$

71. $f(x) = \dfrac{x - 1}{x + 3};\ g(x) = \dfrac{3x - 1}{x - 1}$

72. $f(x) = \dfrac{2x}{x - 2};\ g(x) = \dfrac{x - 2}{x + 4}$

73. Given $f(x) = 3x + 1$ and $(f + g)(x) = 6 - \dfrac{1}{2}x$, find the function g.

74. Given $f(x) = \dfrac{1}{x}$ and $\left(\dfrac{f}{g}\right)(x) = \dfrac{x + 1}{x^2 - x}$, find the function g.

✍ *In Problems 75–82, find the difference quotient of f; that is, find* $\dfrac{f(x + h) - f(x)}{h}$, $h \neq 0$, *for each function. Be sure to simplify.*

75. $f(x) = 4x + 3$

76. $f(x) = -3x + 1$

77. $f(x) = x^2 - x + 4$

78. $f(x) = x^2 + 5x - 1$

79. $f(x) = 3x^2 - 2x + 6$

80. $f(x) = 4x^2 + 5x - 7$

81. $f(x) = x^3 - 2$

82. $f(x) = \dfrac{1}{x + 3}$

Applications and Extensions

83. If $f(x) = 2x^3 + Ax^2 + 4x - 5$ and $f(2) = 5$, what is the value of A?

84. If $f(x) = 3x^2 - Bx + 4$ and $f(-1) = 12$, what is the value of B?

85. If $f(x) = \dfrac{3x + 8}{2x - A}$ and $f(0) = 2$, what is the value of A?

86. If $f(x) = \dfrac{2x - B}{3x + 4}$ and $f(2) = \dfrac{1}{2}$, what is the value of B?

87. If $f(x) = \dfrac{2x - A}{x - 3}$ and $f(4) = 0$, what is the value of A? Where is f not defined?

88. If $f(x) = \dfrac{x - B}{x - A}$, $f(2) = 0$ and $f(1)$ is undefined, what are the values of A and B?

89. **Geometry** Express the area A of a rectangle as a function of the length x if the length of the rectangle is twice its width.

90. **Geometry** Express the area A of an isosceles right triangle as a function of the length x of one of the two equal sides.

91. **Constructing Functions** Express the gross salary G of a person who earns \$10 per hour as a function of the number x of hours worked.

92. **Constructing Functions** Tiffany, a commissioned salesperson, earns \$100 base pay plus \$10 per item sold. Express her gross salary G as a function of the number x of items sold.

93. **Population as a Function of Age** The function
$$P(a) = 0.015a^2 - 4.962a + 290.580$$
represents the population P (in millions) of Americans that were a years of age or older in 2005.
Source: U.S. Census Bureau
(a) Identify the dependent and independent variable.
(b) Evaluate $P(20)$. Provide a verbal explanation of the meaning of $P(20)$.
(c) Evaluate $P(0)$. Provide a verbal explanation of the meaning of $P(0)$.

94. **Number of Rooms** The function
$$N(r) = -1.44r^2 + 14.52r - 14.96$$
represents the number N of housing units (in millions) in 2005 that had r rooms, where r is an integer and $2 \leq r \leq 9$.
Source: U.S. Census Bureau
(a) Identify the dependent and independent variable.
(b) Evaluate $N(3)$. Provide a verbal explanation of the meaning of $N(3)$.

95. **Effect of Gravity on Earth** If a rock falls from a height of 20 meters on Earth, the height H (in meters) after x seconds is approximately
$$H(x) = 20 - 4.9x^2$$
(a) What is the height of the rock when $x = 1$ second? $x = 1.1$ seconds? $x = 1.2$ seconds? $x = 1.3$ seconds?
(b) When is the height of the rock 15 meters? When is it 10 meters? When is it 5 meters?
(c) When does the rock strike the ground?

96. **Effect of Gravity on Jupiter** If a rock falls from a height of 20 meters on the planet Jupiter, its height H (in meters) after x seconds is approximately
$$H(x) = 20 - 13x^2$$

(a) What is the height of the rock when $x = 1$ second? $x = 1.1$ seconds? $x = 1.2$ seconds?

(b) When is the height of the rock 15 meters? When is it 10 meters? When is it 5 meters?

(c) When does the rock strike the ground?

97. Cost of Trans-Atlantic Travel   A Boeing 747 crosses the Atlantic Ocean (3000 miles) with an airspeed of 500 miles per hour. The cost C (in dollars) per passenger is given by

$$C(x) = 100 + \frac{x}{10} + \frac{36,000}{x}$$

where x is the ground speed (airspeed ± wind).

(a) What is the cost per passenger for quiescent (no wind) conditions?

(b) What is the cost per passenger with a head wind of 50 miles per hour?

(c) What is the cost per passenger with a tail wind of 100 miles per hour?

(d) What is the cost per passenger with a head wind of 100 miles per hour?

98. Cross-sectional Area   The cross-sectional area of a beam cut from a log with radius 1 foot is given by the function $A(x) = 4x\sqrt{1 - x^2}$, where x represents the length, in feet, of half the base of the beam. See the figure. Determine the cross-sectional area of the beam if the length of half the base of the beam is as follows:

(a) One-third of a foot

(b) One-half of a foot

(c) Two-thirds of a foot

$$A(x) = 4x\sqrt{1 - x^2}$$

99. Economics   The **participation rate** is the number of people in the labor force divided by the civilian population (excludes military). Let $L(x)$ represent the size of the labor force in year x and $P(x)$ represent the civilian population in year x. Determine a function that represents the participation rate R as a function of x.

100. Crimes   Suppose that $V(x)$ represents the number of violent crimes committed in year x and $P(x)$ represents the number of property crimes committed in year x. Determine a function T that represents the combined total of violent crimes and property crimes in year x.

101. Health Care   Suppose that $P(x)$ represents the percentage of income spent on health care in year x and $I(x)$ represents income in year x. Determine a function H that represents total health care expenditures in year x.

102. Income Tax   Suppose that $I(x)$ represents the income of an individual in year x before taxes and $T(x)$ represents the individual's tax bill in year x. Determine a function N that represents the individual's net income (income after taxes) in year x.

103. Profit Function   Suppose that the revenue R, in dollars, from selling x cell phones, in hundreds, is $R(x) = -1.2x^2 + 220x$. The cost C, in dollars, of selling x cell phones, in hundreds, is $C(x) = 0.05x^3 - 2x^2 + 65x + 500$.

(a) Find the profit function, $P(x) = R(x) - C(x)$.

(b) Find the profit if $x = 15$ hundred cell phones are sold.

(c) Interpret $P(15)$.

104. Profit Function   Suppose that the revenue R, in dollars, from selling x clocks is $R(x) = 30x$. The cost C, in dollars, of selling x clocks is $C(x) = 0.1x^2 + 7x + 400$.

(a) Find the profit function, $P(x) = R(x) - C(x)$.

(b) Find the profit if $x = 30$ clocks are sold.

(c) Interpret $P(30)$.

105. Some functions f have the property that $f(a + b) = f(a) + f(b)$ for all real numbers a and b. Which of the following functions have this property?

(a) $h(x) = 2x$ (b) $g(x) = x^2$

(c) $F(x) = 5x - 2$ (d) $G(x) = \dfrac{1}{x}$

Explaining Concepts: Discussion and Writing

106. Are the functions $f(x) = x - 1$ and $g(x) = \dfrac{x^2 - 1}{x + 1}$ the same? Explain.

107. Investigate when, historically, the use of the function notation $y = f(x)$ first appeared.

108. Write the equation of the following function f. The square root of three times a number x increased by 5.

109. Write the following function in words: $g(t) = t^2 - 2$.

110. Will the equation of a circle define a function? Explain.

'Are You Prepared?' Answers

1. 21.5   **2.** $\{x | x \neq -4\}$ **3.** $(-\infty, -1)$ $\xleftarrow{\qquad}$ $\underset{-1 \quad 0 \quad 1}{\longmapsto}$ **4.** (a) $-3x + 2$ (b) $\dfrac{2x^2 - 4x - 3}{(x - 2)(x - 3)}$ (c) $\dfrac{x + 1}{x - 4}$

1.2 The Graph of a Function

PREPARING FOR THIS SECTION *Before getting started, review the following:*

- Graphs of Equations (Foundations, Section 2, pp. 9–11)
- Intercepts (Foundations, Section 2, pp. 11–12)

Now Work the 'Are You Prepared?' problems on page 62.

OBJECTIVES **1** Identify the Graph of a Function (p. 58)
           2 Obtain Information from or about the Graph of a Function (p. 58)

In applications, a graph often demonstrates more clearly the relationship between two variables than, say, an equation or table would. For example, Table 1 shows the average price of gasoline at a particular gas station in Texas (for the years 1979–2008 adjusted for inflation, based on 2008 dollars). If we plot these data and then connect the points, we obtain Figure 13.

Table 1

Year	Price	Year	Price	Year	Price	Year	Price
1979	3.04	1987	1.90	1995	1.68	2003	1.79
1980	3.41	1988	1.77	1996	1.87	2004	2.13
1981	3.26	1989	1.83	1997	1.65	2005	2.60
1982	3.15	1990	2.25	1998	1.50	2006	2.62
1983	2.51	1991	1.90	1999	1.73	2007	3.29
1984	2.51	1992	1.82	2000	1.85	2008	1.60
1985	2.46	1993	1.70	2001	1.40		
1986	1.63	1994	1.85	2002	1.86		

Source: http://www.randomuseless.info/gasprice/gasprice.html

Figure 13
Average retail price of gasoline (2008 dollars)

Source: http://www.randomuseless.info/gasprice/gasprice.html

We can see from the graph that the price of gasoline (adjusted for inflation) fell from 1980 to 1986 and rose rapidly from 2003 to 2007. The graph also shows that the lowest price occurred in 2001. To learn information such as this from an equation requires that some calculations be made.

Look again at Figure 13. The graph shows that for each date on the horizontal axis there is only one price on the vertical axis. The graph represents a function, although the exact rule for getting from date to price is not given.

When a function is defined by an equation in *x* and *y*, the **graph of the function** is the graph of the equation, that is, the set of points (x, y) in the *xy*-plane that satisfy the equation.

1 Identify the Graph of a Function

Not every collection of points in the xy-plane represents the graph of a function. Remember, for a function, each number x in the domain has exactly one image y in the range. This means that the graph of a function cannot contain two points with the same x-coordinate and different y-coordinates. Therefore, the graph of a function must satisfy the following **vertical-line test.**

THEOREM   **Vertical-line Test**

A set of points in the xy-plane is the graph of a function if and only if every vertical line intersects the graph in at most one point.

In other words, if any vertical line intersects a graph at more than one point, the graph is not the graph of a function.

EXAMPLE 1   **Identifying the Graph of a Function**

Which of the graphs in Figure 14 are graphs of functions?

Figure 14

(a) $y = x^2$ (b) $y = x^3$ (c) $x = y^2$ (d) $x^2 + y^2 = 1$

Solution   The graphs in Figures 14(a) and 14(b) are graphs of functions, because every vertical line intersects each graph in at most one point. The graphs in Figures 14(c) and 14(d) are not graphs of functions, because there is a vertical line that intersects each graph in more than one point. Notice in Figure 14(c) that the input 1 corresponds to two outputs, -1 and 1. This is why the graph does not represent a function.

━━━━ **Now Work** PROBLEM 15

2 Obtain Information from or about the Graph of a Function

If (x, y) is a point on the graph of a function f, then y is the value of f at x; that is, $y = f(x)$. Also if $y = f(x)$, then (x, y) is a point on the graph of f. For example, if $(-2, 7)$ is on the graph of f, then $f(-2) = 7$, and if $f(5) = 8$, then the point $(5, 8)$ is on the graph of $y = f(x)$. The next example illustrates how to obtain information about a function if its graph is given.

EXAMPLE 2   **Obtaining Information from the Graph of a Function**

Let f be the function whose graph is given in Figure 15. (The graph of f might represent the distance that the bob of a pendulum is from its *at-rest* position. Negative values of y mean that the pendulum is to the left of the at-rest position, and positive values of y mean that the pendulum is to the right of the at-rest position.)

Figure 15

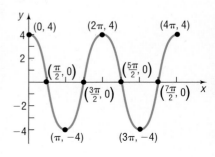

(a) What are $f(0)$, $f\left(\dfrac{3\pi}{2}\right)$, and $f(3\pi)$?

(b) What is the domain of f?

(c) What is the range of f?

(d) List the intercepts. (Recall that these are the points, if any, where the graph crosses or touches the coordinate axes.)

(e) How often does the line $y = 2$ intersect the graph?

(f) For what values of x does $f(x) = -4$?

(g) For what values of x is $f(x) > 0$?

Solution

(a) Since $(0, 4)$ is on the graph of f, the y-coordinate 4 is the value of f at the x-coordinate 0; that is, $f(0) = 4$. In a similar way, we find that when $x = \dfrac{3\pi}{2}$, then $y = 0$, so $f\left(\dfrac{3\pi}{2}\right) = 0$. When $x = 3\pi$, then $y = -4$, so $f(3\pi) = -4$.

(b) To determine the domain of f, we notice that the points on the graph of f have x-coordinates between 0 and 4π, inclusive; and for each number x between 0 and 4π, there is a point $(x, f(x))$ on the graph. The domain of f is $\{x \mid 0 \le x \le 4\pi\}$ or the interval $[0, 4\pi]$.

(c) The points on the graph all have y-coordinates between -4 and 4, inclusive; and for each such number y, there is at least one number x in the domain. The range of f is $\{y \mid -4 \le y \le 4\}$ or the interval $[-4, 4]$.

(d) The intercepts are

$$(0, 4), \left(\dfrac{\pi}{2}, 0\right), \left(\dfrac{3\pi}{2}, 0\right), \left(\dfrac{5\pi}{2}, 0\right), \text{ and } \left(\dfrac{7\pi}{2}, 0\right)$$

(e) If we draw the horizontal line $y = 2$ on the graph in Figure 15, we find that it intersects the graph four times.

(f) Since $(\pi, -4)$ and $(3\pi, -4)$ are the only points on the graph for which $y = f(x) = -4$, we have $f(x) = -4$ when $x = \pi$ and $x = 3\pi$.

(g) To determine where $f(x) > 0$, we look at Figure 15 and determine the x-values from 0 to 4π for which the y-coordinate is positive. This occurs on $\left[0, \dfrac{\pi}{2}\right) \cup \left(\dfrac{3\pi}{2}, \dfrac{5\pi}{2}\right) \cup \left(\dfrac{7\pi}{2}, 4\pi\right]$. Using inequality notation, $f(x) > 0$ for

$$0 \le x < \dfrac{\pi}{2} \text{ or } \dfrac{3\pi}{2} < x < \dfrac{5\pi}{2} \text{ or } \dfrac{7\pi}{2} < x \le 4\pi.$$

●

When the graph of a function is given, its domain may be viewed as the shadow created by the graph on the x-axis by vertical beams of light. Its range can be viewed as the shadow created by the graph on the y-axis by horizontal beams of light. Try this technique with the graph given in Figure 15.

The Zeros of a Function

If $f(r) = 0$ for a real number r, then r is called a real **zero** of f. For example, the zeros of the function $f(x) = x^2 - 4$ are -2 and 2 since $f(-2) = 0$ and $f(2) = 0$. We can identify the zeros of a function from its graph. To see how, we must remember that $y = f(x)$ means that the point (x, y) is on the graph of f. So, if r is a zero of f, then $f(r) = 0$, which means the point $(r, 0)$ is on the graph of f. That is, r is an x-intercept. We conclude that the x-intercepts of the graph of a function are also the zeros of the function.

EXAMPLE 3   **Finding the Zeros of a Function from Its Graph**

Find the zeros of the function f whose graph is shown in Figure 15.

Solution The zeros of a function are the x-intercepts of the graph of the function. Because the x-intercepts of the function shown in Figure 15 are $\dfrac{\pi}{2}$, $\dfrac{3\pi}{2}$, $\dfrac{5\pi}{2}$, and $\dfrac{7\pi}{2}$, the zeros of f are $\dfrac{\pi}{2}$, $\dfrac{3\pi}{2}$, $\dfrac{5\pi}{2}$, and $\dfrac{7\pi}{2}$.

●

Now Work PROBLEMS 9 AND 13

EXAMPLE 4   **Obtaining Information about the Graph of a Function**

Consider the function: $f(x) = \dfrac{x^2 - 1}{x + 2}$

(a) Is the point $(2, 1)$ on the graph of f?
(b) If $x = 4$, what is $f(x)$? What point is on the graph of f?
(c) If $f(x) = -8$, what is x? What point(s) are on the graph of f?
(d) What are the zeros of f (if any)? What point(s) are on the graph of f?

Solution   (a) When $x = 2$, then

$$f(x) = \frac{x^2 - 1}{x + 2}$$

$$f(2) = \frac{2^2 - 1}{2 + 2} = \frac{3}{4}$$

The point $\left(2, \dfrac{3}{4}\right)$ is on the graph of f; the point $(2, 1)$ is not.

(b) If $x = 4$, then

$$f(x) = \frac{x^2 - 1}{x + 2}$$

$$f(4) = \frac{4^2 - 1}{4 + 2} = \frac{15}{6} = \frac{5}{2}$$

The point $\left(4, \dfrac{5}{2}\right)$ is on the graph of f.

(c) If $f(x) = -8$, then

$$\frac{x^2 - 1}{x + 2} = -8$$

$$x^2 - 1 = -8(x + 2) \quad \text{Multiply both sides by } x + 2.$$

$$x^2 - 1 = -8x - 16 \quad \text{Remove parentheses.}$$

$$x^2 + 8x + 15 = 0 \qquad \text{Set equal to 0.}$$

$$(x + 5)(x + 3) = 0 \qquad \text{Factor.}$$

$$x + 5 = 0 \quad \text{or} \quad x + 3 = 0 \qquad \text{Zero-product property}$$

$$x = -5 \quad \text{or} \qquad x = -3 \qquad \text{Solve for } x.$$

If $f(x) = -8$, then $x = -5$ or $x = -3$. The points $(-5, -8)$ and $(-3, -8)$ are on the graph of f.

(d) The zeros of f are the real solutions of the equation $f(x) = 0$. The real solutions of the equation $f(x) = \dfrac{x^2 - 1}{x + 2} = 0$ are $x = -1$ or $x = 1$, so -1 and 1 are the zeros of f. Since $f(-1) = 0$ and $f(1) = 0$, the points $(-1, 0)$ and $(1, 0)$ are on the graph of f.

●

Now Work PROBLEM 25

EXAMPLE 5

Average Cost Function

The average cost $\overline{C}$ of manufacturing x computers per day is given by the function

$$\overline{C}(x) = 0.56x^2 - 34.39x + 1212.57 + \dfrac{20{,}000}{x}$$

Determine the average cost of manufacturing:

(a) 30 computers in a day
(b) 40 computers in a day
(c) 50 computers in a day
(d) Graph the function $\overline{C} = \overline{C}(x), 0 < x \le 80$.
(e) Create a TABLE with TblStart = 1 and ΔTbl = 1. Which value of x minimizes the average cost?

Solution

(a) The average cost of manufacturing $x = 30$ computers is

$$\overline{C}(30) = 0.56(30)^2 - 34.39(30) + 1212.57 + \dfrac{20{,}000}{30} = \$1351.54$$

(b) The average cost of manufacturing $x = 40$ computers is

$$\overline{C}(40) = 0.56(40)^2 - 34.39(40) + 1212.57 + \dfrac{20{,}000}{40} = \$1232.97$$

(c) The average cost of manufacturing $x = 50$ computers is

$$\overline{C}(50) = 0.56(50)^2 - 34.39(50) + 1212.57 + \dfrac{20{,}000}{50} = \$1293.07$$

(d) See Figure 16 for the graph of $\overline{C} = \overline{C}(x)$.

(e) With the function $\overline{C} = \overline{C}(x)$ in Y_1, we create Table 2. We scroll down until we find a value of x for which Y_1 is smallest. Table 3 shows that manufacturing $x = 41$ computers minimizes the average cost at $\$1231.74$ per computer.

Figure 16

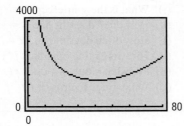

Table 2

X	Y1
1	21179
2	11146
3	7781.1
4	6084
5	5054.6
6	4359.7
7	3856.4

$Y_1 \blacksquare .56X^2 - 34.39X\ldots$

Table 3

X	Y1
38	1240.7
39	1235.9
40	1233
41	**1231.7**
42	1232.2
43	1234.4
44	1238.1

$Y_1 = 1231.74487805$

●

Now Work PROBLEM 31

SUMMARY

Graph of a function	The collection of points (x, y) that satisfy the equation $y = f(x)$
Vertical Line Test	A collection of points is the graph of a function provided that every vertical line intersects the graph in at most one point.

1.2 Assess Your Understanding

'Are You Prepared?' *Answers are given at the end of these exercises. If you get a wrong answer, read the pages listed in red.*

1. The intercepts of the equation $x^2 + 4y^2 = 16$ are _____.
(p. 12)

2. *True or False* The point $(-2, -6)$ is on the graph of the equation $x = 2y - 2$. (pp. 9–10)

Concepts and Vocabulary

3. A set of points in the xy-plane is the graph of a function if and only if every _____ line intersects the graph in at most one point.

4. If the point $(5, -3)$ is a point on the graph of f, then $f(\_) = \_$.

5. Find a so that the point $(-1, 2)$ is on the graph of $f(x) = ax^2 + 4$.

6. *True or False* A function can have more than one y-intercept.

7. *True or False* The graph of a function $y = f(x)$ always crosses the y-axis.

8. *True or False* The y-intercept of the graph of the function $y = f(x)$, whose domain is all real numbers, is $f(0)$.

Skill Building

9. Use the given graph of the function f to answer parts (a)–(o).

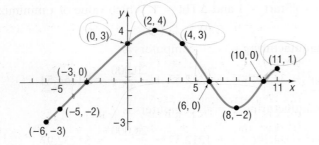

(a) Find $f(0)$ and $f(-6)$.
(b) Find $f(6)$ and $f(11)$.
(c) Is $f(3)$ positive or negative?
(d) Is $f(-4)$ positive or negative?
(e) For what values of x is $f(x) = 0$?
(f) For what values of x is $f(x) > 0$?
(g) What is the domain of f?
(h) What is the range of f?
(i) What are the x-intercepts?
(j) What is the y-intercept?
(k) How often does the line $y = \dfrac{1}{2}$ intersect the graph?
(l) How often does the line $x = 5$ intersect the graph?
(m) For what values of x does $f(x) = 3$?
(n) For what values of x does $f(x) = -2$?
(o) What are the zeros of f?

10. Use the given graph of the function f to answer parts (a)–(o).

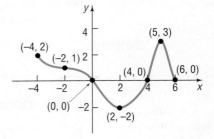

(a) Find $f(0)$ and $f(6)$.
(b) Find $f(2)$ and $f(-2)$.
(c) Is $f(3)$ positive or negative?
(d) Is $f(-1)$ positive or negative?
(e) For what values of x is $f(x) = 0$?
(f) For what values of x is $f(x) < 0$?
(g) What is the domain of f?
(h) What is the range of f?
(i) What are the x-intercepts?
(j) What is the y-intercept?
(k) How often does the line $y = -1$ intersect the graph?
(l) How often does the line $x = 1$ intersect the graph?
(m) For what value of x does $f(x) = 3$?
(n) For what value of x does $f(x) = -2$?
(o) What are the zeros of f?

In Problems 11–22, determine whether the graph is that of a function by using the vertical-line test. If it is, use the graph to find:
(a) The domain and range   (b) The intercepts, if any   (c) Any symmetry with respect to the x-axis, the y-axis, or the origin

11.

12.

13.

14.

15.

16.

17.

18.

19.

20.

21.

22.

In Problems 23–28, answer the questions about the given function.

23. $f(x) = 2x^2 - x - 1$
   (a) Is the point $(-1, 2)$ on the graph of f?
   (b) If $x = -2$, what is $f(x)$? What point is on the graph of f?
   (c) If $f(x) = -1$, what is x? What point(s) are on the graph of f?
   (d) What is the domain of f?
   (e) List the x-intercepts, if any, of the graph of f.
   (f) List the y-intercept, if there is one, of the graph of f.
   (g) What are the zeros of f?

24. $f(x) = -3x^2 + 5x$
   (a) Is the point $(-1, 2)$ on the graph of f?
   (b) If $x = -2$, what is $f(x)$? What point is on the graph of f?
   (c) If $f(x) = -2$, what is x? What point(s) are on the graph of f?
   (d) What is the domain of f?
   (e) List the x-intercepts, if any, of the graph of f.
   (f) List the y-intercept, if there is one, of the graph of f.
   (g) What are the zeros of f?

25. $f(x) = \dfrac{x + 2}{x - 6}$
   (a) Is the point $(3, 14)$ on the graph of f?
   (b) If $x = 4$, what is $f(x)$? What point is on the graph of f?
   (c) If $f(x) = 2$, what is x? What point(s) are on the graph of f?
   (d) What is the domain of f?
   (e) List the x-intercepts, if any, of the graph of f.
   (f) List the y-intercept, if there is one, of the graph of f.
   (g) What are the zeros of f?

26. $f(x) = \dfrac{x^2 + 2}{x + 4}$
   (a) Is the point $\left(1, \dfrac{3}{5}\right)$ on the graph of f?

   (b) If $x = 0$, what is $f(x)$? What point is on the graph of f?
   (c) If $f(x) = \dfrac{1}{2}$, what is x? What point(s) are on the graph of f?
   (d) What is the domain of f?
   (e) List the x-intercepts, if any, of the graph of f.
   (f) List the y-intercept, if there is one, of the graph of f.
   (g) What are the zeros of f?

27. $f(x) = \dfrac{2x^2 + 3x}{x^2 + 1}$
   (a) Is the point $\left(-1, -\dfrac{1}{2}\right)$ on the graph of f?
   (b) If $x = -2$, what is $f(x)$? What point is on the graph of f?
   (c) If $f(x) = \dfrac{5}{2}$, what is x? What point(s) are on the graph of f?
   (d) What is the domain of f?
   (e) List the x-intercepts, if any, of the graph of f.
   (f) List the y-intercept, if there is one, of the graph of f.
   (g) What are the zeros of f?

28. $f(x) = \dfrac{x^2 + 4x}{x - 2}$
   (a) Is the point $(1, 5)$ on the graph of f?
   (b) If $x = 3$, what is $f(x)$? What point is on the graph of f?
   (c) If $f(x) = 1$, what is x? What point(s) are on the graph of f?
   (d) What is the domain of f?
   (e) List the x-intercepts, if any, of the graph of f.
   (f) List the y-intercept, if there is one, of the graph of f.
   (g) What are the zeros of f?

Applications and Extensions

29. Free-throw Shots   According to physicist Peter Brancazio, the key to a successful foul shot in basketball lies in the arc of the shot. Brancazio determined the optimal angle of the arc from the free-throw line to be 45 degrees. The arc also depends on the velocity with which the ball is shot. If a player shoots a foul shot, releasing the ball at a 45-degree angle from a position 6 feet above the floor, then the path of the ball can be modeled by the function

$$h(x) = -\frac{44x^2}{v^2} + x + 6$$

where h is the height of the ball above the floor, x is the forward distance of the ball in front of the foul line, and v is the initial velocity with which the ball is shot in feet per second. Suppose a player shoots a ball with an initial velocity of 28 feet per second.

(a) Determine the height of the ball after it has traveled 8 feet in front of the foul line.

(b) Determine the height of the ball after it has traveled 12 feet in front of the foul line.

(c) Find additional points and graph the path of the basketball.

(d) The center of the hoop is 10 feet above the floor and 15 feet in front of the foul line. Will the ball go through the hoop? Why or why not? If not, with what initial velocity must the ball be shot in order for the ball to go through the hoop?

Source: The Physics of Foul Shots, Discover, *Vol. 21, No. 10, October 2000*

30. Granny Shots   The last player in the NBA to use an underhand foul shot (a "granny" shot) was Hall of Fame forward Rick Barry who retired in 1980. Barry believes that current NBA players could increase their free-throw percentage if they were to use an underhand shot. Since underhand shots are released from a lower position, the angle of the shot must be increased. If a player shoots an underhand foul shot, releasing the ball at a 70-degree angle from a position 3.5 feet above the floor, then the path of the ball can be modeled by the function $h(x) = -\frac{136x^2}{v^2} + 2.7x + 3.5$, where h is the height of the ball above the floor, x is the forward distance of the ball in front of the foul line, and v is the initial velocity with which the ball is shot in feet per second.

(a) The center of the hoop is 10 feet above the floor and 15 feet in front of the foul line. Determine the initial velocity with which the ball must be shot in order for the ball to go through the hoop.

(b) Write the function for the path of the ball using the velocity found in part (a).

(c) Determine the height of the ball after it has traveled 9 feet in front of the foul line.

(d) Find additional points and graph the path of the basketball.

Source: The Physics of Foul Shots, Discover, *Vol. 21, No. 10, October 2000*

31. Cost of Trans-Atlantic Travel   A Boeing 747 crosses the Atlantic Ocean (3000 miles) with an airspeed of 500 miles per hour. The cost C (in dollars) per passenger is given by

$$C(x) = 100 + \frac{x}{10} + \frac{36{,}000}{x}$$

where x is the ground speed (airspeed ± wind).

(a) Determine the cost if the ground speed is 400 miles per hour; 450 miles per hour.

(b) Find the domain of C.

(c) Use a graphing utility to graph the function $C = C(x)$.

(d) Create a TABLE with TblStart = 0 and ΔTbl = 50.

(e) To the nearest 50 miles per hour, what ground speed minimizes the cost per passenger?

32. Cross-sectional Area   The cross-sectional area of a beam cut from a log with radius 1 foot is given by the function $A(x) = 4x\sqrt{1 - x^2}$, where x represents the length, in feet, of half the base of the beam. See the figure.

$$A(x) = 4x\sqrt{1 - x^2}$$

(a) Find the domain of A.

(b) Use a graphing utility to graph the function $A = A(x)$.

(c) Create a TABLE with TblStart = 0 and ΔTbl = 0.1 for $0 \le x \le 1$. Which value of x maximizes the cross-sectional area? What should be the length of the base of the beam to maximize the cross-sectional area?

33. The graphs of two functions, f and g, are illustrated. Use the graphs to answer parts (a)–(f).

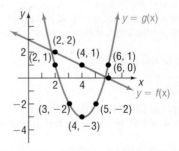

(a) $(f + g)(2)$

(b) $(f + g)(4)$

(c) $(f - g)(6)$

(d) $(g - f)(6)$

(e) $(f \cdot g)(2)$

(f) $\left(\dfrac{f}{g}\right)(4)$

34. Reading and Interpreting Graphs   Let C be the function whose graph is given on the following page. This graph represents the cost C of manufacturing q computers in a day.

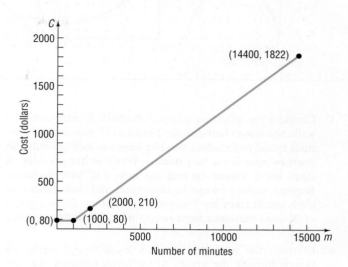

35. Reading and Interpreting Graphs   Let C be the function whose graph is given below. This graph represents the cost C of using m anytime cell phone minutes in a month for a five-person family plan.

(a) Determine $C(0)$. Interpret this value.
(b) Determine $C(10)$. Interpret this value.
(c) Determine $C(50)$. Interpret this value.
(d) What is the domain of C? What does this domain imply in terms of daily production?
(e) Describe the shape of the graph.
(f) The point $(30, 32000)$ is called an *inflection point*. Describe the behavior of the graph around the inflection point.

(a) Determine $C(0)$. Interpret this value.
(b) Determine $C(1000)$. Interpret this value.
(c) Determine $C(2000)$. Interpret this value.
(d) What is the domain of C? What does this domain imply in terms of the number of anytime minutes?
(e) Describe the shape of the graph.

Explaining Concepts: Discussion and Writing

36. Describe how you would proceed to find the domain and range of a function if you were given its graph. How would your strategy change if you were given the equation defining the function instead of its graph?

37. How many x-intercepts can the graph of a function have? How many y-intercepts can the graph of a function have?

38. Is a graph that consists of a single point the graph of a function? Can you write the equation of such a function?

39. Match each of the following functions with the graph that best describes the situation.
(a) The cost of building a house as a function of its square footage
(b) The height of an egg dropped from a 300-foot building as a function of time
(c) The height of a human as a function of time
(d) The demand for Big Macs as a function of price
(e) The height of a child on a swing as a function of time

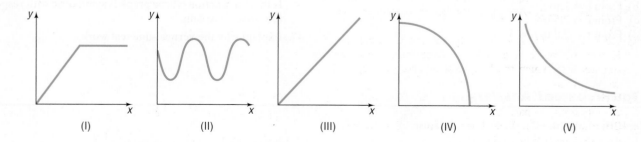

40. Match each of the following functions with the graph shown on the following page that best describes the situation.
(a) The temperature of a bowl of soup as a function of time
(b) The number of hours of daylight per day over a 2-year period
(c) The population of Florida as a function of time
(d) The distance traveled by a car going at a constant velocity as a function of time
(e) The height of a golf ball hit with a 7-iron as a function of time

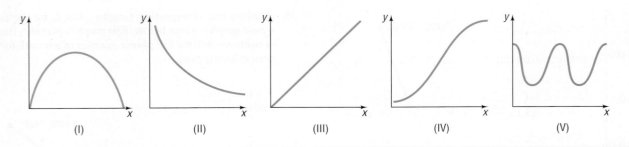

(I)   (II)   (III)   (IV)   (V)

41. Consider the following scenario: Barbara decides to take a walk. She leaves home, walks 2 blocks in 5 minutes at a constant speed, and realizes that she forgot to lock the door. So Barbara runs home in 1 minute. While at her doorstep, it takes her 1 minute to find her keys and lock the door. Barbara walks 5 blocks in 15 minutes and then decides to jog home. It takes her 7 minutes to get home. Draw a graph of Barbara's distance from home (in blocks) as a function of time.

42. Consider the following scenario: Jayne enjoys riding her bicycle through the woods. At the forest preserve, she gets on her bicycle and rides up a 2000-foot incline in 10 minutes. She then travels down the incline in 3 minutes. The next 5000 feet are level terrain and she covers the distance in 20 minutes. She rests for 15 minutes. Jayne then travels 10,000 feet in 30 minutes. Draw a graph of Jayne's distance traveled (in feet) as a function of time.

43. The sketch below represents the distance d (in miles) that Kevin is from home as a function of time t (in hours). Answer the questions based on the graph. In parts (a)–(g), how many hours elapsed and how far was Kevin from home during this time?

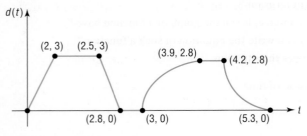

(a) From $t = 0$ to $t = 2$
(b) From $t = 2$ to $t = 2.5$
(c) From $t = 2.5$ to $t = 2.8$
(d) From $t = 2.8$ to $t = 3$

(e) From $t = 3$ to $t = 3.9$
(f) From $t = 3.9$ to $t = 4.2$
(g) From $t = 4.2$ to $t = 5.3$
(h) What is the farthest distance that Kevin is from home?
(i) How many times did Kevin return home?

44. The following sketch represents the speed v (in miles per hour) of Michael's car as a function of time t (in minutes).

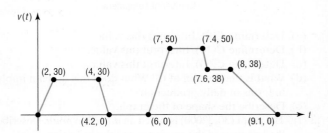

(a) Over what interval of time is Michael traveling fastest?
(b) Over what interval(s) of time is Michael's speed zero?
(c) What is Michael's speed between 0 and 2 minutes?
(d) What is Michael's speed between 4.2 and 6 minutes?
(e) What is Michael's speed between 7 and 7.4 minutes?
(f) When is Michael's speed constant?

45. Draw the graph of a function whose domain is $\{x \mid -3 \le x \le 8, \ x \ne 5\}$ and whose range is $\{y \mid -1 \le y \le 2, \ y \ne 0\}$. What point(s) in the rectangle $-3 \le x \le 8, -1 \le y \le 2$ cannot be on the graph? Compare your graph with those of other students. What differences do you see?

46. Is there a function whose graph is symmetric with respect to the x-axis? Explain.

47. Explain why the vertical-line test works.

'Are You Prepared?' Answers

1. $(-4, 0), (4, 0), (0, -2), (0, 2)$ **2.** False

1.3 Properties of Functions

PREPARING FOR THIS SECTION *Before getting started, review the following:*

- Intervals (Appendix A, Section A.10, pp. A83–A84)
- Intercepts (Foundations, Section 2, pp. 11–12)
- Slope of a Line (Foundations, Section 3, pp. 19–21)
- Point–Slope Form of a Line (Foundations, Section 3, pp. 23–24)
- Symmetry (Foundations, Section 2, pp. 12–14)

Now Work the *'Are You Prepared?'* problems on page 75.

OBJECTIVES
1 Determine Even and Odd Functions from a Graph (p. 67)
2 Identify Even and Odd Functions from the Equation (p. 68)
3 Use a Graph to Determine Where a Function Is Increasing, Decreasing, or Constant (p. 69)
4 Use a Graph to Locate Local Maxima and Local Minima (p. 70)
5 Use a Graph to Locate the Absolute Maximum and the Absolute Minimum (p. 71)
6 Use a Graphing Utility to Approximate Local Maxima and Local Minima and to Determine Where a Function Is Increasing or Decreasing (p. 73)
7 Find the Average Rate of Change of a Function (p. 73)

To obtain the graph of a function $y = f(x)$, it is often helpful to know certain properties that the function has and the impact of these properties on the way that the graph will look.

1 Determine Even and Odd Functions from a Graph

The words *even* and *odd*, when applied to a function f, describe the symmetry that exists for the graph of the function.

A function f is even, if and only if, whenever the point (x, y) is on the graph of f then the point $(-x, y)$ is also on the graph. Using function notation, we define an even function as follows:

DEFINITION
A function f is **even** if, for every number x in its domain, the number $-x$ is also in the domain and
$$f(-x) = f(x)$$

A function f is odd, if and only if, whenever the point (x, y) is on the graph of f then the point $(-x, -y)$ is also on the graph. Using function notation, we define an odd function as follows:

DEFINITION
A function f is **odd** if, for every number x in its domain, the number $-x$ is also in the domain and
$$f(-x) = -f(x)$$

Refer to page 13, where the tests for symmetry are listed. The following results are then evident.

THEOREM
A function is even if and only if its graph is symmetric with respect to the y-axis. A function is odd if and only if its graph is symmetric with respect to the origin.

EXAMPLE 1   **Determining Even and Odd Functions from the Graph**

Determine whether each graph given in Figure 17 is the graph of an even function, an odd function, or a function that is neither even nor odd.

Figure 17

(a)   (b)   (c)

Solution   (a) The graph in Figure 17(a) is that of an even function, because the graph is symmetric with respect to the y-axis.

(b) The function whose graph is given in Figure 17(b) is neither even nor odd, because the graph is neither symmetric with respect to the y-axis nor symmetric with respect to the origin.

(c) The function whose graph is given in Figure 17(c) is odd, because its graph is symmetric with respect to the origin.

➤ **Now Work** PROBLEMS 21(a), (b), AND (d)

2 Identify Even and Odd Functions from the Equation

In the next example, we use algebraic techniques to verify whether a given function is even, odd, or neither.

EXAMPLE 2   **Identifying Even and Odd Functions Algebraically**

Determine whether each of the following functions is even, odd, or neither. Then determine whether the graph is symmetric with respect to the y-axis or with respect to the origin.

(a) $f(x) = x^2 - 5$ (b) $g(x) = x^3 - 1$
(c) $h(x) = 5x^3 - x$ (d) $F(x) = |x|$

Solution   (a) To determine whether f is even, odd, or neither, we replace x by $-x$ in $f(x) = x^2 - 5$. Then

$$f(-x) = (-x)^2 - 5 = x^2 - 5 = f(x)$$

Since $f(-x) = f(x)$, we conclude that f is an even function, and the graph of f is symmetric with respect to the y-axis.

(b) We replace x by $-x$ in $g(x) = x^3 - 1$. Then

$$g(-x) = (-x)^3 - 1 = -x^3 - 1$$

Since $g(-x) \neq g(x)$ and $g(-x) \neq -g(x) = -(x^3 - 1) = -x^3 + 1$, we conclude that g is neither even nor odd. The graph of g is not symmetric with respect to the y-axis nor is it symmetric with respect to the origin.

(c) We replace x by $-x$ in $h(x) = 5x^3 - x$. Then

$$h(-x) = 5(-x)^3 - (-x) = -5x^3 + x = -(5x^3 - x) = -h(x)$$

Since $h(-x) = -h(x)$, h is an odd function, and the graph of h is symmetric with respect to the origin.

(d) We replace x by $-x$ in $F(x) = |x|$. Then

$$F(-x) = |-x| = |-1| \cdot |x| = |x| = F(x)$$

Since $F(-x) = F(x)$, F is an even function, and the graph of F is symmetric with respect to the y-axis.

──────── **Now Work** PROBLEM 41

3 Use a Graph to Determine Where a Function Is Increasing, Decreasing, or Constant

Consider the graph given in Figure 18. If you look from left to right along the graph of the function, you will notice that parts of the graph are going up, parts are going down, and parts are horizontal. In such cases, the function is described as *increasing*, *decreasing*, or *constant*, respectively.

Figure 18

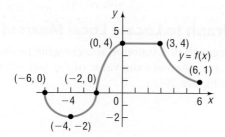

EXAMPLE 3

Determining Where a Function Is Increasing, Decreasing, or Constant from Its Graph

Where is the function in Figure 18 increasing? Where is it decreasing? Where is it constant?

Solution

To answer the question of where a function is increasing, where it is decreasing, and where it is constant, we use strict inequalities involving the independent variable x, or we use open intervals* of x-coordinates. The function whose graph is given in Figure 18 is increasing on the open interval $(-4, 0)$ or for $-4 < x < 0$. The function is decreasing on the open intervals $(-6, -4)$ and $(3, 6)$ or for $-6 < x < -4$ and $3 < x < 6$. The function is constant on the open interval $(0, 3)$ or for $0 < x < 3$.

WARNING We describe the behavior of a graph in terms of its x-values. Do not say the graph in Figure 18 is increasing from the point $(-4, -2)$ to the point $(0, 4)$. Rather, say it is increasing on the interval $(-4, 0)$. ■

More precise definitions follow:

DEFINITIONS

A function f is **increasing** on an open interval I if, for any choice of x_1 and x_2 in I, with $x_1 < x_2$, we have $f(x_1) < f(x_2)$.

A function f is **decreasing** on an open interval I if, for any choice of x_1 and x_2 in I, with $x_1 < x_2$, we have $f(x_1) > f(x_2)$.

A function f is **constant** on an open interval I if, for all choices of x in I, the values $f(x)$ are equal.

In Words
If a function is increasing, then, as the values of x get bigger, the values of the function also get bigger. If a function is decreasing, then, as the values of x get bigger, the values of the function get smaller. If a function is constant, then, as the values of x get bigger, the values of the function remain unchanged.

Figure 19 on the following page illustrates the definitions. The graph of an increasing function goes up from left to right, the graph of a decreasing function goes down from left to right, and the graph of a constant function remains at a fixed height.

*The open interval (a, b) consists of all real numbers x for which $a < x < b$.

Figure 19

(a) For $x_1 < x_2$ in I, $f(x_1) < f(x_2)$; f is increasing on I

(b) For $x_1 < x_2$ in I, $f(x_1) > f(x_2)$; f is decreasing on I

(c) For all x in I, the values of f are equal; f is constant on I

—**Now Work** PROBLEMS **11, 13, 15**, AND **21(c)**

4 Use a Graph to Locate Local Maxima and Local Minima

When the graph of a function is increasing to the left of $x = c$ and decreasing to the right of $x = c$, then at c the value of f is largest. This value is called a *local maximum* of f.* See Figure 20(a).

Figure 20

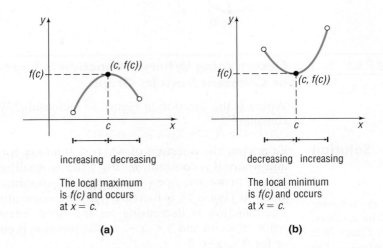

increasing decreasing

The local maximum is $f(c)$ and occurs at $x = c$.

(a)

decreasing increasing

The local minimum is $f(c)$ and occurs at $x = c$.

(b)

When the graph of a function is decreasing to the left of $x = c$ and is increasing to the right of $x = c$, then at c the value of f is the smallest. This value is called a *local minimum* of f. See Figure 20(b).

DEFINITIONS

> A function f has a **local maximum** at c if there is an open interval I containing c so that, for all x not equal to c in I, $f(x) < f(c)$. We call $f(c)$ a **local maximum of f.**
>
> A function f has a **local minimum** at c if there is an open interval I containing c so that, for all x not equal to c in I, $f(x) > f(c)$. We call $f(c)$ a **local minimum of f.**

If f has a local maximum at c, then the value of f at c is greater than the values of f near c. If f has a local minimum at c, then the value of f at c is less than the values of f near c. The word *local* is used to suggest that it is only near c that the value $f(c)$ is largest or smallest.

*Some texts will use the term *relative* instead of *local*.

Finding Local Maxima and Local Minima from the Graph of a Function and Determining Where the Function Is Increasing, Decreasing, or Constant

Figure 21

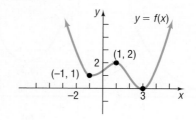

Figure 21 shows the graph of a function f.
(a) At what value(s) of x, if any, does f have a local maximum? What are the local maxima?
(b) At what value(s) of x, if any, does f have a local minimum? What are the local minima?
(c) List the intervals on which f is increasing. List the intervals on which f is decreasing.

Solution

WARNING The y-value is the local maximum or local minimum and it occurs at some x-value. For Figure 21, we say the local maximum is 2 and that the local maximum occurs at $x = 1$. ∎

The domain of f is the set of real numbers.
(a) f has a local maximum at 1, since for all x close to 1, $x \neq 1$, we have $f(x) < f(1)$. The local maximum is $f(1) = 2$.
(b) f has a local minimum at -1 and at 3. The local minima are $f(-1) = 1$ and $f(3) = 0$.
(c) The function whose graph is given in Figure 21 is increasing for all values of x between -1 and 1 and for all values of x greater than 3. That is, the function is increasing on the intervals $(-1, 1)$ and $(3, \infty)$ or for $-1 < x < 1$ and $x > 3$. The function is decreasing for all values of x less than -1 and for all values of x between 1 and 3. That is, the function is decreasing on the intervals $(-\infty, -1)$ and $(1, 3)$ or for $x < -1$ and $1 < x < 3$.

Now Work PROBLEMS 17 AND 19

Figure 22

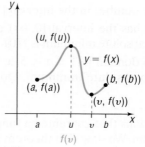

domain: $[a, b]$
for all x in $[a, b]$, $f(x) \leq f(u)$
for all x in $[a, b]$, $f(x) \geq f(v)$
absolute maximum: $f(u)$
absolute minimum: $f(v)$

5 Use a Graph to Locate the Absolute Maximum and the Absolute Minimum

 Look at the graph of the function f given in Figure 22. We see that the domain of f is the closed interval $[a, b]$. We also see that the largest value of f is $f(u)$ and the smallest value of f is $f(v)$. These are called, respectively, the *absolute maximum* and the *absolute minimum* of f on $[a, b]$. A precise definition is given next.

DEFINITION

Let f denote a function defined on some interval I. If there is a number u in I for which $f(x) \leq f(u)$ for all x in I, then $f(u)$ is the **absolute maximum of f** on I and we say **the absolute maximum of f occurs at u.**

If there is a number v in I for which $f(x) \geq f(v)$ for all x in I, then $f(v)$ is the **absolute minimum of f** on I and we say **the absolute minimum of f occurs at v.**

The absolute maximum and absolute minimum of a function f are sometimes called the **extreme values** of f on I.

The absolute maximum or absolute minimum of a function f may not exist. Let's look at some examples.

EXAMPLE 5

Finding the Absolute Maximum and the Absolute Minimum from the Graph of a Function

For each graph of a function $y = f(x)$ in Figure 23 on the following page, find the absolute maximum and the absolute minimum, if they exist.

Figure 23

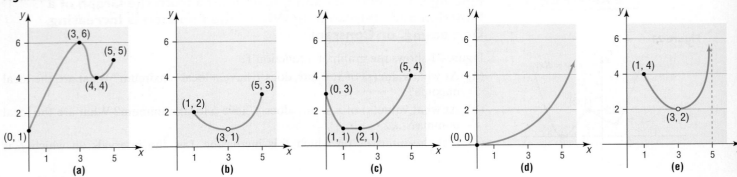

Solution

(a) The function f whose graph is given in Figure 23(a) has the closed interval $[0, 5]$ as its domain. The largest value of f is $f(3) = 6$, the absolute maximum. The smallest value of f is $f(0) = 1$, the absolute minimum.

(b) The function f whose graph is given in Figure 23(b) has the domain $\{x | 1 \leq x \leq 5, x \neq 3\}$. Note that we exclude 3 from the domain because of the "hole" at $(3, 1)$. The largest value of f on its domain is $f(5) = 3$, the absolute maximum. There is no absolute minimum. Do you see why? As you trace the graph, getting closer to the point $(3, 1)$, there is no single smallest value. [As soon as you claim a smallest value, we can trace closer to $(3, 1)$ and get a smaller value!]

(c) The function f whose graph is given in Figure 23(c) has the interval $[0, 5]$ as its domain. The absolute maximum of f is $f(5) = 4$. The absolute minimum is 1. Notice that the absolute minimum 1 occurs at any number in the interval $[1, 2]$.

(d) The graph of the function f given in Figure 23(d) has the interval $[0, \infty)$ as its domain. The function has no absolute maximum; the absolute minimum is $f(0) = 0$.

(e) The graph of the function f in Figure 23(e) has the domain $\{x | 1 \leq x < 5, x \neq 3\}$. The function f has no absolute maximum and no absolute minimum. Do you see why?

In calculus, there is a theorem that provides conditions that guarantee a function will have an absolute maximum and an absolute minimum. We state that theorem next.

THEOREM

Extreme Value Theorem

If f is a continuous function* whose domain is a closed interval $[a, b]$, then f has an absolute maximum and an absolute minimum on $[a, b]$.

The absolute maximum (minimum) can be found by selecting the largest (smallest) value of f from the following list:

1. The value of f at any local maxima and local minima of f in $[a, b]$.
2. The value of f at each endpoint of $[a, b]$, that is, $f(a)$ and $f(b)$.
3. The value of f on any interval in $[a, b]$ on which f is constant.

For example, the graph of the function f given in Figure 23(a) is continuous on the closed interval $[0, 5]$. The Extreme Value Theorem guarantees f has extreme values on $[0, 5]$. To find them, we list:

1. The value of f at the local extrema: $f(3) = 6$, $f(4) = 4$
2. The value of f at the endpoints: $f(0) = 1$, $f(5) = 5$

The largest of these, 6, is the absolute maximum; the smallest of these, 1, is the absolute minimum.

Now Work PROBLEM 33

*Although it requires calculus for a precise definition, we'll agree for now that a continuous function is one whose graph has no gaps or holes and can be traced without lifting the pencil from the paper.

6 Use a Graphing Utility to Approximate Local Maxima and Local Minima and to Determine Where a Function Is Increasing or Decreasing

To locate the exact value at which a function f has a local maximum or a local minimum usually requires calculus. However, a graphing utility may be used to approximate these values by using the MAXIMUM and MINIMUM features.

EXAMPLE 6   **Using a Graphing Utility to Approximate Local Maxima and Minima and to Determine Where a Function Is Increasing or Decreasing**

(a) Use a graphing utility to graph $f(x) = 6x^3 - 12x + 5$ for $-2 < x < 2$. Approximate where f has a local maximum and where f has a local minimum.

(b) Determine where f is increasing and where it is decreasing.

Solution   (a) Graphing utilities have a feature that finds the maximum or minimum point of a graph within a given interval. Graph the function f for $-2 < x < 2$. The MAXIMUM and MINIMUM commands require us to first determine the open interval I. The graphing utility will then approximate the maximum or minimum value in the interval. Using MAXIMUM we find that the local maximum is 11.53 and it occurs at $x = -0.82$, rounded to two decimal places. See Figure 24(a). Using MINIMUM, we find that the local minimum is -1.53 and it occurs at $x = 0.82$, rounded to two decimal places. See Figure 24(b).

Figure 24

(a)   (b)

(b) Looking at Figures 24(a) and (b), we see that the graph of f is increasing from $x = -2$ to $x = -0.82$ and from $x = 0.82$ to $x = 2$, so f is increasing on the intervals $(-2, -0.82)$ and $(0.82, 2)$ or for $-2 < x < -0.82$ and $0.82 < x < 2$. The graph is decreasing from $x = -0.82$ to $x = 0.82$, so f is decreasing on the interval $(-0.82, 0.82)$ or for $-0.82 < x < 0.82$.

Now Work   PROBLEM 53

7 Find the Average Rate of Change of a Function

In Foundations, Section 3, we said that the slope of a line could be interpreted as the average rate of change. To find the average rate of change of a function between any two points on its graph, we calculate the slope of the line containing the two points.

DEFINITION   If a and b, $a \neq b$, are in the domain of a function $y = f(x)$, the **average rate of change of f** from a to b is defined as

$$\text{Average rate of change} = \frac{\Delta y}{\Delta x} = \frac{f(b) - f(a)}{b - a} \qquad a \neq b \qquad \textbf{(1)}$$

The symbol Δy in (1) is the "change in y," and Δx is the "change in x." The average rate of change of f is the change in y divided by the change in x.

| EXAMPLE 7 | Finding the Average Rate of Change |

Find the average rate of change of $f(x) = 3x^2$:

(a) From 1 to 3   (b) From 1 to 5   (c) From 1 to 7

Solution   (a) The average rate of change of $f(x) = 3x^2$ from 1 to 3 is

$$\frac{\Delta y}{\Delta x} = \frac{f(3) - f(1)}{3 - 1} = \frac{27 - 3}{3 - 1} = \frac{24}{2} = 12$$

(b) The average rate of change of $f(x) = 3x^2$ from 1 to 5 is

$$\frac{\Delta y}{\Delta x} = \frac{f(5) - f(1)}{5 - 1} = \frac{75 - 3}{5 - 1} = \frac{72}{4} = 18$$

(c) The average rate of change of $f(x) = 3x^2$ from 1 to 7 is

$$\frac{\Delta y}{\Delta x} = \frac{f(7) - f(1)}{7 - 1} = \frac{147 - 3}{7 - 1} = \frac{144}{6} = 24$$

Figure 25

See Figure 25 for a graph of $f(x) = 3x^2$. The function f is increasing for $x > 0$. The fact that the average rate of change is positive for any $x_1, x_2, x_1 \neq x_2$ in the interval $(1, 7)$ indicates that the graph is increasing on $1 < x < 7$. In addition, since the average rate of change is consistently getting larger for $1 < x < 7$, indicates that the graph is increasing at an increasing rate.

➤ **Now Work** PROBLEM 61

The Secant Line

The average rate of change of a function has an important geometric interpretation. Look at the graph of $y = f(x)$ in Figure 26. We have labeled two points on the graph: $(a, f(a))$ and $(b, f(b))$. The line containing these two points is called the **secant line;** its slope is

$$m_{\text{sec}} = \frac{f(b) - f(a)}{b - a}$$

Figure 26

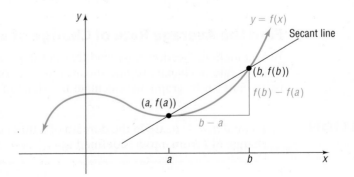

| THEOREM | **Slope of the Secant Line** |

The average rate of change of a function from a to b equals the slope of the secant line containing the two points $(a, f(a))$ and $(b, f(b))$ on its graph.

EXAMPLE 8   **Finding the Equation of a Secant Line**

Suppose that $g(x) = 3x^2 - 2x + 3$.

(a) Find the average rate of change of g from -2 to 1.

(b) Find an equation of the secant line containing $(-2, g(-2))$ and $(1, g(1))$.

(c) Using a graphing utility, draw the graph of g and the secant line obtained in part (b) on the same screen.

Solution   (a) The average rate of change of $g(x) = 3x^2 - 2x + 3$ from -2 to 1 is

$$\text{Average rate of change} = \frac{g(1) - g(-2)}{1 - (-2)}$$

$$= \frac{4 - 19}{3} \qquad \begin{array}{l} g(1) = 3(1)^2 - 2(1) + 3 = 4 \\ g(-2) = 3(-2)^2 - 2(-2) + 3 = 19 \end{array}$$

$$= -\frac{15}{3} = -5$$

(b) The slope of the secant line containing $(-2, g(-2)) = (-2, 19)$ and $(1, g(1)) = (1, 4)$ is $m_{\text{sec}} = -5$. We use the point–slope form to find an equation of the secant line.

$$y - y_1 = m_{\text{sec}}(x - x_1) \qquad \text{Point–slope form of the secant line}$$
$$y - 19 = -5(x - (-2)) \qquad x_1 = -2, y_1 = g(-2) = 19, m_{\text{sec}} = -5$$
$$y - 19 = -5x - 10 \qquad \text{Simplify.}$$
$$y = -5x + 9 \qquad \text{Slope–intercept form of the secant line}$$

(c) Figure 27 shows the graph of g along with the secant line $y = -5x + 9$.

Figure 27

➤ **Now Work** PROBLEM 67

1.3 Assess Your Understanding

'Are You Prepared?' *Answers are given at the end of these exercises. If you get a wrong answer, read the pages listed in red.*

1. The interval $(2, 5)$ can be written as the inequality _____. (pp. A83–A84)

2. The slope of the line containing the points $(-2, 3)$ and $(3, 8)$ is _____. (pp. 19–21)

3. Test the equation $y = 5x^2 - 1$ for symmetry with respect to the x-axis, the y-axis, and the origin. (pp. 12–14)

4. Write the point–slope form of the line with slope 5 containing the point $(3, -2)$. (p. 23)

5. The intercepts of the equation $y = x^2 - 9$ are _____. (p. 12)

Concepts and Vocabulary

6. A function f is _____ on an open interval I if, for any choice of x_1 and x_2 in I, with $x_1 < x_2$, we have $f(x_1) < f(x_2)$.

7. An _____ function f is one for which $f(-x) = f(x)$ for every x in the domain of f; an _____ function f is one for which $f(-x) = -f(x)$ for every x in the domain of f.

8. Let f denote a function defined on some interval I. If there is a number u in I for which $f(x) \le f(u)$ for all x in I, then $f(u)$ is the _____ _____ of f on I.

9. A function f has a _____ _____ at c if there is an open interval I containing c so that, for all x not equal to c in I, $f(x) < f(c)$.

10. Even functions have graphs that are symmetric with respect to the _____.

Skill Building

In Problems 11–20, use the given graph of the function f.

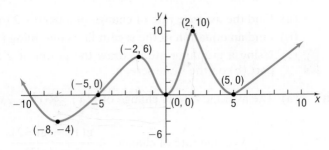

11. Is f increasing on the interval $(-8, -2)$?

12. Is f decreasing on the interval $(-8, -4)$?

13. Is f increasing on the interval $(2, 10)$?

14. Is f decreasing on the interval $(2, 5)$?

15. List the interval(s) on which f is increasing.

16. List the interval(s) on which f is decreasing.

17. Is there a local maximum at 2? If yes, what is it?

18. Is there a local maximum at 5? If yes, what is it?

19. List the numbers at which f has a local maximum. What are these local maxima?

20. List the numbers at which f has a local minimum. What are these local minima?

In Problems 21–28, the graph of a function is given. Use the graph to find:
  (a) *The intercepts, if any*   (b) *The domain and range*   (c) *The intervals on which the function is increasing, decreasing, or constant*
  (d) *Whether the function is even, odd, or neither*

21.

22.

23.

24.

25.

26.

27.

28.

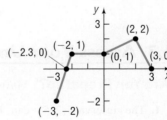

In Problems 29–32, the graph of a function f is given. Use the graph to find:
  (a) *The values, if any, at which f has a local maximum. What are these local maxima?*
  (b) *The values, if any, at which f has a local minimum. What are these local minima?*

29.

30.

31.

32.

In Problems 33–40, for each graph of a function $y = f(x)$, *find the absolute maximum and the absolute minimum, if it exists.*

33.

34.

35.

36.

37.

38.

39.

40.

In Problems 41–52, determine algebraically whether each function is even, odd, or neither.

41. $F(x) = 4x^3 + x$

42. $f(x) = 2x^4 - x^2$

43. $g(x) = -3x^2 - 5$

44. $h(x) = 3x^5 + 5x$

45. $F(x) = \sqrt[3]{x}$

46. $G(x) = \sqrt{x}$

47. $f(x) = x + |x|$

48. $f(x) = \sqrt[3]{2x^2 + 1}$

49. $g(x) = \dfrac{1}{x^2}$

50. $h(x) = \dfrac{x}{x^2 - 1}$

51. $h(x) = \dfrac{-x^3}{3x^2 - 9}$

52. $F(x) = \dfrac{2x}{|x|}$

In Problems 53–60, use a graphing utility to graph each function over the indicated interval and approximate any local maxima and local minima. Determine where the function is increasing and where it is decreasing. Round answers to two decimal places.

53. $f(x) = x^3 - 3x + 2$ $(-2, 2)$

54. $f(x) = x^3 - 3x^2 + 5$ $(-1, 3)$

55. $f(x) = x^5 - x^3$ $(-2, 2)$

56. $f(x) = x^4 - x^2$ $(-2, 2)$

57. $f(x) = -0.2x^3 - 0.6x^2 + 4x - 6$ $(-6, 4)$

58. $f(x) = -0.4x^3 + 0.6x^2 + 3x - 2$ $(-4, 5)$

59. $f(x) = 0.25x^4 + 0.3x^3 - 0.9x^2 + 3$ $(-3, 2)$

60. $f(x) = -0.4x^4 - 0.5x^3 + 0.8x^2 - 2$ $(-3, 2)$

61. Find the average rate of change of $f(x) = -2x^2 + 4$
   (a) From 0 to 2
   (b) From 1 to 3
   (c) From 1 to 4

62. Find the average rate of change of $f(x) = -x^3 + 1$
   (a) From 0 to 2
   (b) From 1 to 3
   (c) From −1 to 1

63. Find the average rate of change of $g(x) = x^3 - 2x + 1$
   (a) From −3 to −2
   (b) From −1 to 1
   (c) From 1 to 3

64. Find the average rate of change of $h(x) = x^2 - 2x + 3$
   (a) From −1 to 1
   (b) From 0 to 2
   (c) From 2 to 5

65. $f(x) = 5x - 2$
   (a) Find the average rate of change from 1 to 3.
   (b) Find an equation of the secant line containing $(1, f(1))$ and $(3, f(3))$.

66. $f(x) = -4x + 1$
   (a) Find the average rate of change from 2 to 5.
   (b) Find an equation of the secant line containing $(2, f(2))$ and $(5, f(5))$.

67. $g(x) = x^2 - 2$
   (a) Find the average rate of change from −2 to 1.
   (b) Find an equation of the secant line containing $(-2, g(-2))$ and $(1, g(1))$.

68. $g(x) = x^2 + 1$
   (a) Find the average rate of change from −1 to 2.
   (b) Find an equation of the secant line containing $(-1, g(-1))$ and $(2, g(2))$.

69. $h(x) = x^2 - 2x$
   (a) Find the average rate of change from 2 to 4.
   (b) Find an equation of the secant line containing $(2, h(2))$ and $(4, h(4))$.

70. $h(x) = -2x^2 + x$
   (a) Find the average rate of change from 0 to 3.
   (b) Find an equation of the secant line containing $(0, h(0))$ and $(3, h(3))$.

Mixed Practice

71. $g(x) = x^3 - 27x$
   (a) Determine whether g is even, odd, or neither.
   (b) There is a local minimum of -54 at 3. Determine the local maximum.

72. $f(x) = -x^3 + 12x$
   (a) Determine whether f is even, odd, or neither.
   (b) There is a local maximum of 16 at 2. Determine the local minimum.

73. $F(x) = -x^4 + 8x^2 + 8$
   (a) Determine whether F is even, odd, or neither.
   (b) There is a local maximum of 24 at $x = 2$. Determine a second local maximum.

⚠(c) Suppose the area under the graph of F between $x = 0$ and $x = 3$ that is bounded below by the x-axis is 47.4 square units. Using the results from part (a), determine the area under the graph of F between $x = -3$ and $x = 0$ bounded below by the x-axis.

74. $G(x) = -x^4 + 32x^2 + 144$
   (a) Determine whether G is even, odd, or neither.
   (b) There is a local maximum of 400 at $x = 4$. Determine a second local maximum.

⚠(c) Suppose the area under the graph of G between $x = 0$ and $x = 6$ that is bounded below by the x-axis is 1612.8 square units. Using the results from part (a), determine the area under the graph of F between $x = -6$ and $x = 0$ bounded below by the x-axis.

Applications and Extensions

75. Minimum Average Cost The average cost per hour in dollars, $\overline{C}$, of producing x riding lawn mowers can be modeled by the function

$$\overline{C}(x) = 0.3x^2 + 21x - 251 + \frac{2500}{x}$$

   (a) Use a graphing utility to graph $\overline{C} = \overline{C}(x)$.
   (b) Determine the number of riding lawn mowers to produce in order to minimize average cost.
   (c) What is the minimum average cost?

76. Medicine Concentration The concentration C of a medication in the bloodstream t hours after being administered is modeled by the function

$$C(t) = -0.002x^4 + 0.039t^3 - 0.285t^2 + 0.766t + 0.085$$

   (a) After how many hours will the concentration be highest?
   (b) A woman nursing a child must wait until the concentration is below 0.5 before she can feed him. After taking the medication, how long must she wait before feeding her child?

77. E-coli Growth A strain of E-coli Beu 397-recA441 is placed into a nutrient broth at 30° Celsius and allowed to grow. The data shown in the table are collected. The population is measured in grams and the time in hours. Since population P depends on time t and each input corresponds to exactly one output, we can say that population is a function of time; so $P(t)$ represents the population at time t.

   (a) Find the average rate of change of the population from 0 to 2.5 hours.
   (b) Find the average rate of change of the population from 4.5 to 6 hours.

   (c) What is happening to the average rate of change as time passes?

Time (hours), t	Population (grams), P
0	0.09
2.5	0.18
3.5	0.26
4.5	0.35
6	0.50

78. e-Filing Tax Returns The Internal Revenue Service Restructuring and Reform Act (RRA) was signed into law by President Bill Clinton in 1998. A major objective of the RRA was to promote electronic filing of tax returns. The data in the table on the following page show the percentage of individual income tax returns filed electronically for filing years 2000–2008. Since the percentage P of returns filed electronically depends on the filing year y and each input corresponds to exactly one output, the percentage of returns filed electronically is a function of the filing year; so $P(y)$ represents the percentage of returns filed electronically for filing year y.
   (a) Find the average rate of change of the percentage of e-filed returns from 2000 to 2002.
   (b) Find the average rate of change of the percentage of e-filed returns from 2004 to 2006.
   (c) Find the average rate of change of the percentage of e-filed returns from 2006 to 2008.
   (d) What is happening to the average rate of change as time passes?

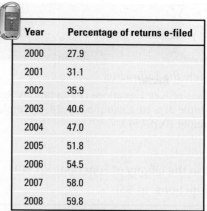

Year	Percentage of returns e-filed
2000	27.9
2001	31.1
2002	35.9
2003	40.6
2004	47.0
2005	51.8
2006	54.5
2007	58.0
2008	59.8

SOURCE: Internal Revenue Service

79. For the function $f(x) = x^2$, compute each average rate of change:
    (a) From 0 to 1
    (b) From 0 to 0.5
    (c) From 0 to 0.1
    (d) From 0 to 0.01
    (e) From 0 to 0.001
    (f) Use a graphing utility to graph each of the secant lines along with f.
    (g) What do you think is happening to the secant lines?
    (h) What is happening to the slopes of the secant lines? Is there some number that they are getting closer to? What is that number?

80. For the function $f(x) = x^2$, compute each average rate of change:
    (a) From 1 to 2
    (b) From 1 to 1.5
    (c) From 1 to 1.1
    (d) From 1 to 1.01
    (e) From 1 to 1.001
    (f) Use a graphing utility to graph each of the secant lines along with f.
    (g) What do you think is happening to the secant lines?
    (h) What is happening to the slopes of the secant lines? Is there some number that they are getting closer to? What is that number?

Problems 81–88 require the following discussion of a secant line. The slope of the secant line containing the two points $(x, f(x))$ and $(x + h, f(x + h))$ on the graph of a function $y = f(x)$ may be given as

$$m_{sec} = \frac{f(x + h) - f(x)}{(x + h) - x} = \frac{f(x + h) - f(x)}{h} \qquad h \neq 0$$

*In calculus, this expression is called the **difference quotient of f.***
(a) *Express the slope of the secant line of each function in terms of x and h. Be sure to simplify your answer.*
(b) *Find m_{sec} for $h = 0.5, 0.1$, and 0.01 at $x = 1$. What value does m_{sec} approach as h approaches 0?*
(c) *Find the equation for the secant line at $x = 1$ with $h = 0.01$.*
(d) *Use a graphing utility to graph f and the secant line found in part (c) on the same viewing window.*

81. $f(x) = 2x + 5$

82. $f(x) = -3x + 2$

83. $f(x) = x^2 + 2x$

84. $f(x) = 2x^2 + x$

85. $f(x) = 2x^2 - 3x + 1$

86. $f(x) = -x^2 + 3x - 2$

87. $f(x) = \dfrac{1}{x}$

88. $f(x) = \dfrac{1}{x^2}$

Explaining Concepts: Discussion and Writing

89. Draw the graph of a function that has the following properties: domain: all real numbers; range: all real numbers; intercepts: $(0, -3)$ and $(3, 0)$; a local maximum of -2 is at -1; a local minimum of -6 is at 2. Compare your graph with those of others. Comment on any differences.

90. Redo Problem 89 with the following additional information: increasing on $(-\infty, -1), (2, \infty)$; decreasing on $(-1, 2)$. Again compare your graph with others and comment on any differences.

91. How many x-intercepts can a function defined on an interval have if it is increasing on that interval? Explain.

92. Suppose that a friend of yours does not understand the idea of increasing and decreasing functions. Provide an explanation, complete with graphs, that clarifies the idea.

93. Can a function be both even and odd? Explain.

94. Using a graphing utility, graph $y = 5$ on the interval $(-3, 3)$. Use MAXIMUM to find the local maxima on $(-3, 3)$. Comment on the result provided by the calculator.

95. A function f has a positive average rate of change on the interval $[2, 5]$. Is f increasing on $[2, 5]$? Explain.

96. Show that a constant function $f(x) = b$ has an average rate of change of 0. Compute the average rate of change of $y = \sqrt{4 - x^2}$ on the interval $[-2, 2]$. Explain how this can happen.

'Are You Prepared?' Answers

1. $2 < x < 5$ 2. 1 3. symmetric with respect to the y-axis   4. $y + 2 = 5(x - 3)$ 5. $(-3, 0), (3, 0), (0, -9)$

1.4 Library of Functions; Piecewise-defined Functions

PREPARING FOR THIS SECTION *Before getting started, review the following:*

- Intercepts (Foundations, Section 2, pp. 11–12)
- Graphs of Key Equations (Foundations, Section 2: Example 3, p. 10, Example 8, p. 14; Example 9, p. 15; Example 10, p. 15)

Now Work the 'Are You Prepared?' problems on page 86.

OBJECTIVES **1** Graph the Functions Listed in the Library of Functions (p. 80)
2 Graph Piecewise-defined Functions (p. 85)

1 Graph the Functions Listed in the Library of Functions

We now introduce a few more functions to add to our list of important functions. We begin with the *square root function*.

On page 15, we graphed the equation $y = \sqrt{x}$. Figure 28 shows a graph of the function $f(x) = \sqrt{x}$. Based on the graph, we have the following properties:

Figure 28

Properties of $f(x) = \sqrt{x}$

1. The domain and the range are the set of nonnegative real numbers.
2. The x-intercept of the graph of $f(x) = \sqrt{x}$ is 0. The y-intercept of the graph of $f(x) = \sqrt{x}$ is also 0.
3. The function is neither even nor odd.
4. It is increasing on the interval $(0, \infty)$.
5. It has an absolute minimum of 0 at $x = 0$.

EXAMPLE 1 | **Graphing the Cube Root Function**

(a) Determine whether $f(x) = \sqrt[3]{x}$ is even, odd, or neither. State whether the graph of f is symmetric with respect to the y-axis or symmetric with respect to the origin.
(b) Determine the intercepts, if any, of the graph of $f(x) = \sqrt[3]{x}$.
(c) Graph $f(x) = \sqrt[3]{x}$.

Solution | (a) Because

$$f(-x) = \sqrt[3]{-x} = -\sqrt[3]{x} = -f(x)$$

the function is odd. The graph of f is symmetric with respect to the origin.

(b) The y-intercept is $f(0) = \sqrt[3]{0} = 0$. The x-intercept is found by solving the equation $f(x) = 0$.

$$f(x) = 0$$
$$\sqrt[3]{x} = 0 \quad \text{f(x)} = \sqrt[3]{x}$$
$$x = 0 \quad \text{Cube both sides of the equation.}$$

The x-intercept is also 0.

(c) We use the function to form Table 4 and obtain some points on the graph. Because of the symmetry with respect to the origin, we need to find only points (x, y) for which $x \geq 0$. Figure 29 shows the graph of $f(x) = \sqrt[3]{x}$.

Table 4

x	$y = f(x) = \sqrt[3]{x}$	(x, y)
0	0	(0, 0)
$\frac{1}{8}$	$\frac{1}{2}$	$\left(\frac{1}{8}, \frac{1}{2}\right)$
1	1	(1, 1)
2	$\sqrt[3]{2} \approx 1.26$	$\left(2, \sqrt[3]{2}\right)$
8	2	(8, 2)

Figure 29

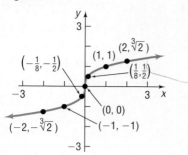

From the results of Example 1 and Figure 29, we have the following properties of the cube root function.

Properties of $f(x) = \sqrt[3]{x}$

1. The domain and the range are the set of all real numbers.
2. The x-intercept of the graph of $f(x) = \sqrt[3]{x}$ is 0. The y-intercept of the graph of $f(x) = \sqrt[3]{x}$ is also 0.
3. The graph is symmetric with respect to the origin. The function is odd.
4. The function is increasing on the interval $(-\infty, \infty)$.
5. The function does not have any local minima or any local maxima.

EXAMPLE 2 **Graphing the Absolute Value Function**

(a) Determine whether $f(x) = |x|$ is even, odd, or neither. State whether the graph of f is symmetric with respect to the y-axis or symmetric with respect to the origin.
(b) Determine the intercepts, if any, of the graph of $f(x) = |x|$.
(c) Graph $f(x) = |x|$.

Solution (a) Because

$$f(-x) = |-x|$$
$$= |x| = f(x)$$

the function is even. The graph of f is symmetric with respect to the y-axis.
(b) The y-intercept is $f(0) = |0| = 0$. The x-intercept is found by solving the equation $f(x) = 0$ or $|x| = 0$. So the x-intercept is 0.
(c) We use the function to form Table 5 and obtain some points on the graph. Because of the symmetry with respect to the y-axis, we need to find only points (x, y) for which $x \geq 0$. Figure 30 shows the graph of $f(x) = |x|$.

Table 5

x	$y = f(x) = |x|$	(x, y)
0	0	(0, 0)
1	1	(1, 1)
2	2	(2, 2)
3	3	(3, 3)

Figure 30

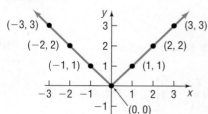

From the results of Example 2 and Figure 30, we have the following properties of the absolute value function.

Properties of $f(x) = |x|$

1. The domain is the set of all real numbers. The range of f is $\{y | y \geq 0\}$.
2. The x-intercept of the graph of $f(x) = |x|$ is 0. The y-intercept of the graph of $f(x) = |x|$ is also 0.
3. The graph is symmetric with respect to the y-axis. The function is even.
4. The function is decreasing on the interval $(-\infty, 0)$. It is increasing on the interval $(0, \infty)$.
5. The function has a local minimum of 0 at $x = 0$.

Seeing the Concept

Graph $y = |x|$ on a square screen and compare what you see with Figure 30. Note that some graphing calculators use abs(x) for absolute value.

We now provide a summary of the key functions that we have encountered. In going through this list, pay special attention to the properties of each function, particularly to the shape of each graph. Knowing these graphs along with key points on the graph will lay the foundation for further graphing techniques.

Constant Function

$$f(x) = b \qquad b \text{ is a real number}$$

Figure 31 Constant Function

See Figure 31.
The domain of a **constant function** is the set of all real numbers; its range is the set consisting of a single number b. Its graph is a horizontal line whose y-intercept is b. The constant function is an even function whose graph is a horizontal line over its domain.

Identity Function

$$f(x) = x$$

Figure 32 Identity Function

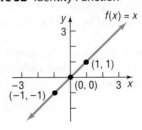

See Figure 32.
The domain and the range of the **identity function** are the set of all real numbers. Its graph is a line whose slope is $m = 1$ and whose y-intercept is 0. The line consists of all points for which the x-coordinate equals the y-coordinate. The identity function is an odd function that is increasing over its domain. Note that the graph bisects quadrants I and III.

Square Function

$$f(x) = x^2$$

Figure 33 Square Function

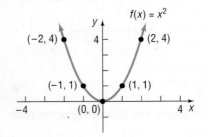

See Figure 33.
The domain of the **square function** f is the set of all real numbers; its range is the set of nonnegative real numbers. The graph of this function is a parabola whose intercept is at $(0, 0)$. The square function is an even function that is decreasing on the interval $(-\infty, 0)$ and increasing on the interval $(0, \infty)$.

Figure 34 Cube Function

Cube Function

$$f(x) = x^3$$

See Figure 34.

The domain and the range of the **cube function** are the set of all real numbers. The intercept of the graph is at $(0, 0)$. The cube function is odd and is increasing on the interval $(-\infty, \infty)$.

Square Root Function

$$f(x) = \sqrt{x}$$

Figure 35 Square Root Function

See Figure 35.

The domain and the range of the **square root function** are the set of nonnegative real numbers. The intercept of the graph is at $(0, 0)$. The square root function is neither even nor odd and is increasing on the interval $(0, \infty)$.

Cube Root Function

$$f(x) = \sqrt[3]{x}$$

Figure 36 Cube Root Function

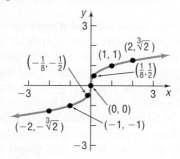

See Figure 36.

The domain and the range of the **cube root function** are the set of all real numbers. The intercept of the graph is at $(0, 0)$. The cube root function is an odd function that is increasing on the interval $(-\infty, \infty)$.

Reciprocal Function

$$f(x) = \frac{1}{x}$$

Figure 37 Reciprocal Function

Refer to Example 10, page 15, for a discussion of the equation $y = \dfrac{1}{x}$. See Figure 37.

The domain and the range of the **reciprocal function** are the set of all nonzero real numbers. The graph has no intercepts. The reciprocal function is decreasing on the intervals $(-\infty, 0)$ and $(0, \infty)$ and is an odd function.

Absolute Value Function

$$f(x) = |x|$$

Figure 38 Absolute Value Function

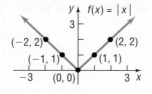

See Figure 38.

The domain of the **absolute value function** is the set of all real numbers; its range is the set of nonnegative real numbers. The intercept of the graph is at $(0, 0)$. If $x \geq 0$, then $f(x) = x$, and the graph of f is part of the line $y = x$; if $x < 0$, then $f(x) = -x$, and the graph of f is part of the line $y = -x$. The absolute value function is an even function; it is decreasing on the interval $(-\infty, 0)$ and increasing on the interval $(0, \infty)$.

The notation int(x) stands for the largest integer less than or equal to x. For example,

$$\text{int}(1) = 1, \quad \text{int}(2.5) = 2, \quad \text{int}\left(\frac{1}{2}\right) = 0, \quad \text{int}\left(-\frac{3}{4}\right) = -1, \quad \text{int}(\pi) = 3$$

This type of correspondence occurs frequently enough in mathematics that we give it a name.

DEFINITION

Greatest Integer Function

$$f(x) = \text{int}(x)^* = \text{greatest integer less than or equal to } x$$

Table 6

x	$y = f(x)$ $= \text{int}(x)$	(x, y)
-1	-1	$(-1, -1)$
$-\frac{1}{2}$	-1	$\left(-\frac{1}{2}, -1\right)$
$-\frac{1}{4}$	-1	$\left(-\frac{1}{4}, -1\right)$
0	0	$(0, 0)$
$\frac{1}{4}$	0	$\left(\frac{1}{4}, 0\right)$
$\frac{1}{2}$	0	$\left(\frac{1}{2}, 0\right)$
$\frac{3}{4}$	0	$\left(\frac{3}{4}, 0\right)$

We obtain the graph of $f(x) = \text{int}(x)$ by plotting several points. See Table 6. For values of x, $-1 \le x < 0$, the value of $f(x) = \text{int}(x)$ is -1; for values of x, $0 \le x < 1$, the value of f is 0. See Figure 39 for the graph.

Figure 39 Greatest Integer Function

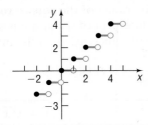

The domain of the **greatest integer function** is the set of all real numbers; its range is the set of integers. The y-intercept of the graph is 0. The x-intercepts lie in the interval $[0, 1)$. The greatest integer function is neither even nor odd. It is constant on every interval of the form $[k, k + 1)$, for k an integer. In Figure 39, we use a solid dot to indicate, for example, that at $x = 1$ the value of f is $f(1) = 1$; we use an open circle to illustrate that the function does not assume the value of 0 at $x = 1$.

Although a precise definition requires the idea of a limit, discussed in calculus, in a rough sense, a function is said to be **continuous** if its graph has no gaps or holes and can be drawn without lifting a pencil from the paper on which the graph is drawn. We contrast this with a *discontinuous* function. A function is **discontinuous** if its graph has gaps or holes so that its graph cannot be drawn without lifting a pencil from the paper.

From the graph of the greatest integer function, we can see why it is also called a **step function.** At $x = 0$, $x = \pm1$, $x = \pm2$, and so on, this function is discontinuous because, at integer values, the graph suddenly "steps" from one value to another without taking on any of the intermediate values. For example, to the immediate left of $x = 3$, the y-coordinates of the points on the graph are 2, and at $x = 3$ and to the immediate right of $x = 3$, the y-coordinates of the points on the graph are 3. So, the graph has gaps in it.

Figure 40 $f(x) = \text{int}(x)$

(a) Connected mode

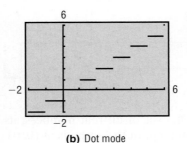

(b) Dot mode

COMMENT When graphing a function using a graphing utility, you can choose either the **connected mode,** in which points plotted on the screen are connected, making the graph appear without any breaks, or the **dot mode,** in which only the points plotted appear. When graphing the greatest integer function with a graphing utility, it may be necessary to be in the dot mode. This is to prevent the utility from "connecting the dots" when $f(x)$ changes from one integer value to the next. See Figure 40. ∎

The functions that we have discussed so far are basic. Whenever you encounter one of them, you should see a mental picture of its graph. For example, if you encounter the function $f(x) = x^2$, you should see in your mind's eye a picture like Figure 33.

-**Now Work** PROBLEMS **9** THROUGH **16**

*Some books use the notation $f(x) = [x]$ instead of int(x).

2 Graph Piecewise-defined Functions

Sometimes a function is defined differently on different parts of its domain. For example, the absolute value function $f(x) = |x|$ is actually defined by two equations: $f(x) = x$ if $x \geq 0$ and $f(x) = -x$ if $x < 0$. For convenience, we generally combine these equations into one expression as

$$f(x) = |x| = \begin{cases} x & \text{if } x \geq 0 \\ -x & \text{if } x < 0 \end{cases}$$

When a function is defined by different equations on different parts of its domain, it is called a **piecewise-defined** function.

Let's look at another example of a piecewise-defined function.

EXAMPLE 3   **Analyzing a Piecewise-defined Function**

The function f is defined as

$$f(x) = \begin{cases} -2x + 1 & \text{if } -3 \leq x < 1 \\ 2 & \text{if } x = 1 \\ x^2 & \text{if } x > 1 \end{cases}$$

(a) Find $f(-2)$, $f(1)$, and $f(2)$.   (b) Determine the domain of f.
(c) Locate any intercepts.   (d) Graph f.
(e) Use the graph to find the range of f.   (f) Is f continuous on its domain?

Solution   (a) To find $f(-2)$, we observe that when $x = -2$ the equation for f is given by $f(x) = -2x + 1$. So we have

$$f(-2) = -2(-2) + 1 = 5$$

When $x = 1$, the equation for f is $f(x) = 2$. Thus,

$$f(1) = 2$$

When $x = 2$, the equation for f is $f(x) = x^2$. So

$$f(2) = 2^2 = 4$$

(b) To find the domain of f, we look at its definition. Since f is defined for all x greater than or equal to -3, the domain of f is $\{x \mid x \geq -3\}$, or the interval $[-3, \infty)$.

(c) The y-intercept of the graph of the function is $f(0)$. Because the equation for f when $x = 0$ is $f(x) = -2x + 1$, the y-intercept is $f(0) = -2(0) + 1 = 1$. The x-intercepts of the graph of a function f are the real solutions to the equation $f(x) = 0$. To find the x-intercepts of f, we solve $f(x) = 0$ for each "piece" of the function and then determine if the value of x satisfies the condition that defines the piece.

$f(x) = 0$	$f(x) = 0$	$f(x) = 0$
$-2x + 1 = 0 \quad -3 \leq x < 1$	$2 = 0 \quad x = 1$	$x^2 = 0 \quad x > 1$
$-2x = -1$	No solution	$x = 0$
$x = \dfrac{1}{2}$		

The first potential x-intercept, $x = \dfrac{1}{2}$, satisfies the condition $-3 \leq x < 1$, so $x = \dfrac{1}{2}$ is an x-intercept. The second potential x-intercept, $x = 0$, does not satisfy the condition $x > 1$, so $x = 0$ is not an x-intercept. The only x-intercept is $\dfrac{1}{2}$. The intercepts are $(0, 1)$ and $\left(\dfrac{1}{2}, 0\right)$.

(d) To graph f, we graph "each piece." First we graph the line $y = -2x + 1$ and keep only the part for which $-3 \leq x < 1$. Then we plot the point $(1, 2)$ because, when $x = 1$, $f(x) = 2$. Finally, we graph the parabola $y = x^2$ and keep only the part for which $x > 1$. See Figure 41.

Figure 41

(e) From the graph, we conclude that the range of f is $\{y | y > -1\}$, or the interval $(-1, \infty)$.

(f) The function f is not continuous because there is a "jump" in the graph at $x = 1$.

Now Work PROBLEM 29

EXAMPLE 4   **Cost of Electricity**

In the summer of 2009, Duke Energy supplied electricity to residences of Ohio for a monthly customer charge of $4.50 plus 4.2345¢ per kilowatt-hour (kWhr) for the first 1000 kWhr supplied in the month and 5.3622¢ per kWhr for all usage over 1000 kWhr in the month.

(a) What is the charge for using 300 kWhr in a month?

(b) What is the charge for using 1500 kWhr in a month?

(c) If C is the monthly charge for x kWhr, develop a model relating the monthly charge and kilowatt-hours used. That is, express C as a function of x.

Source: Duke Energy, 2009.

Solution   (a) For 300 kWhr, the charge is $4.50 plus 4.2345¢ = $0.042345 per kWhr. That is,

$$\text{Charge} = \$4.50 + \$0.042345(300) = \$17.20$$

(b) For 1500 kWhr, the charge is $4.50 plus 4.2345¢ per kWhr for the first 1000 kWhr plus 5.3622¢ per kWhr for the 500 kWhr in excess of 1000. That is,

$$\text{Charge} = \$4.50 + \$0.042345(1000) + \$0.053622(500) = \$73.66$$

(c) Let x represent the number of kilowatt-hours used. If $0 \leq x \leq 1000$, the monthly charge C (in dollars) can be found by multiplying x times $0.042345 and adding the monthly customer charge of $4.50. So, if $0 \leq x \leq 1000$, then $C(x) = 0.042345x + 4.50$.

For $x > 1000$, the charge is $0.042345(1000) + 4.50 + 0.053622(x - 1000)$, since $x - 1000$ equals the usage in excess of 1000 kWhr, which costs $0.053622 per kWhr. That is, if $x > 1000$, then

$$C(x) = 0.042345(1000) + 4.50 + 0.053622(x - 1000)$$

$$= 46.845 + 0.053622(x - 1000)$$

$$= 0.053622x - 6.777$$

The rule for computing C follows two equations:

$$C(x) = \begin{cases} 0.042345x + 4.50 & \text{if } 0 \leq x \leq 1000 \\ 0.053622x - 6.777 & \text{if } x > 1000 \end{cases} \qquad \textit{The Model}$$

See Figure 42 for the graph.

Figure 42

1.4 Assess Your Understanding

'Are You Prepared?'   *Answers are given at the end of these exercises. If you get a wrong answer, read the pages listed in red.*

1. Sketch the graph of $y = \sqrt{x}$. (p. 15)

2. Sketch the graph of $y = \dfrac{1}{x}$. (pp. 15–16)

3. List the intercepts of the equation $y = x^3 - 8$. (p. 12)

Concepts and Vocabulary

4. The function $f(x) = x^2$ is decreasing on the interval _____.

5. When functions are defined by more than one equation, they are called _____ functions.

6. *True or False* The cube function is odd and is increasing on the interval $(-\infty, \infty)$.

7. *True or False* The cube root function is odd and is decreasing on the interval $(-\infty, \infty)$.

8. *True or False* The domain and the range of the reciprocal function are the set of all real numbers.

Skill Building

In Problems 9–16, match each graph to its function.
   A. *Constant function*   B. *Identity function*   C. *Square function*
   D. *Cube function*   E. *Square root function*   F. *Reciprocal function*
   G. *Absolute value function* H. *Cube root function*

9.   **10.**   **11.**   **12.**

13.   **14.**   **15.**   **16.**

In Problems 17–24, sketch the graph of each function. Be sure to label three points on the graph.

17. $f(x) = x$ **18.** $f(x) = x^2$ **19.** $f(x) = x^3$ **20.** $f(x) = \sqrt{x}$

21. $f(x) = \dfrac{1}{x}$ **22.** $f(x) = |x|$ **23.** $f(x) = \sqrt[3]{x}$ **24.** $f(x) = 3$

25. If $f(x) = \begin{cases} x^2 & \text{if } x < 0 \\ 2 & \text{if } x = 0 \\ 2x + 1 & \text{if } x > 0 \end{cases}$

   find: (a) $f(-2)$ (b) $f(0)$ (c) $f(2)$

26. If $f(x) = \begin{cases} -3x & \text{if } x < -1 \\ 0 & \text{if } x = -1 \\ 2x^2 + 1 & \text{if } x > -1 \end{cases}$

   find: (a) $f(-2)$ (b) $f(-1)$ (c) $f(0)$

27. If $f(x) = \begin{cases} 2x - 4 & \text{if } -1 \le x \le 2 \\ x^3 - 2 & \text{if } 2 < x \le 3 \end{cases}$

   find: (a) $f(0)$ (b) $f(1)$ (c) $f(2)$ (d) $f(3)$

28. If $f(x) = \begin{cases} x^3 & \text{if } -2 \le x < 1 \\ 3x + 2 & \text{if } 1 \le x \le 4 \end{cases}$

   find: (a) $f(-1)$ (b) $f(0)$ (c) $f(1)$ (d) $f(3)$

In Problems 29–40:
   (a) *Find the domain of each function.*   (b) *Locate any intercepts.*   (c) *Graph each function.*
   (d) *Based on the graph, find the range.*   (e) *Is f continuous on its domain?*

29. $f(x) = \begin{cases} 2x & \text{if } x \ne 0 \\ 1 & \text{if } x = 0 \end{cases}$

30. $f(x) = \begin{cases} 3x & \text{if } x \ne 0 \\ 4 & \text{if } x = 0 \end{cases}$

31. $f(x) = \begin{cases} -2x + 3 & \text{if } x < 1 \\ 3x - 2 & \text{if } x \ge 1 \end{cases}$

32. $f(x) = \begin{cases} x + 3 & \text{if } x < -2 \\ -2x - 3 & \text{if } x \ge -2 \end{cases}$

33. $f(x) = \begin{cases} x + 3 & \text{if } -2 \le x < 1 \\ 5 & \text{if } x = 1 \\ -x + 2 & \text{if } x > 1 \end{cases}$

34. $f(x) = \begin{cases} 2x + 5 & \text{if } -3 \le x < 0 \\ -3 & \text{if } x = 0 \\ -5x & \text{if } x > 0 \end{cases}$

35. $f(x) = \begin{cases} 1 + x & \text{if } x < 0 \\ x^2 & \text{if } x \ge 0 \end{cases}$

36. $f(x) = \begin{cases} \dfrac{1}{x} & \text{if } x < 0 \\ \sqrt[3]{x} & \text{if } x \ge 0 \end{cases}$

37. $f(x) = \begin{cases} |x| & \text{if } -2 \le x < 0 \\ x^3 & \text{if } x > 0 \end{cases}$

38. $f(x) = \begin{cases} 2 - x & \text{if } -3 \le x < 1 \\ \sqrt{x} & \text{if } x > 1 \end{cases}$

39. $f(x) = 2\,\text{int}(x)$

40. $f(x) = \text{int}(2x)$

In Problems 41–44, the graph of a piecewise-defined function is given. Write a definition for each function.

41.

42.

43.

44.
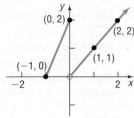

45. If $f(x) = \text{int}(2x)$, find
  (a) $f(1.2)$ (b) $f(1.6)$ (c) $f(-1.8)$

46. If $f(x) = \text{int}\left(\dfrac{x}{2}\right)$, find
  (a) $f(1.2)$ (b) $f(1.6)$ (c) $f(-1.8)$

Applications and Extensions

47. Cell Phone Service Sprint PCS offers a monthly cellular phone plan for $39.99. It includes 450 anytime minutes and charges $0.45 per minute for additional minutes. The following function is used to compute the monthly cost for a subscriber:

$$C(x) = \begin{cases} 39.99 & \text{if } 0 \le x \le 450 \\ 0.45x - 162.51 & \text{if } x > 450 \end{cases}$$

where x is the number of anytime minutes used. Compute the monthly cost of the cellular phone for use of the following anytime minutes:
  (a) 200   (b) 465   (c) 451
Source: Sprint PCS

48. Parking at O'Hare International Airport The short-term (no more than 24 hours) parking fee F (in dollars) for parking x hours at O'Hare International Airport's main parking garage can be modeled by the function

$$F(x) = \begin{cases} 3 & \text{if } 0 < x \le 3 \\ 5\,\text{int}(x+1) + 1 & \text{if } 3 < x < 9 \\ 50 & \text{if } 9 \le x \le 24 \end{cases}$$

Determine the fee for parking in the short-term parking garage for
  (a) 2 hours   (b) 7 hours   (c) 15 hours
  (d) 8 hours and 24 minutes
Source: O'Hare International Airport

49. Cost of Natural Gas In April 2009, Peoples Energy had the following rate schedule for natural gas usage in single-family residences:

Monthly service charge	$15.95
Per therm service charge	
1st 50 therms	$0.33606/therm
Over 50 therms	$0.10580/therm
Gas charge	$0.3940/therm

  (a) What is the charge for using 50 therms in a month?
  (b) What is the charge for using 500 therms in a month?
  (c) Develop a model that relates the monthly charge C for x therms of gas.
  (d) Graph the function found in part (c).
Source: Peoples Energy, Chicago, Illinois, 2009

50. Cost of Natural Gas In April 2009, Nicor Gas had the following rate schedule for natural gas usage in single-family residences:

Monthly customer charge	$8.40
Distribution charge	
1st 20 therms	$0.1473/therm
Next 30 therms	$0.0579/therm
Over 50 therms	$0.0519/therm
Gas supply charge	$0.43/therm

  (a) What is the charge for using 40 therms in a month?
  (b) What is the charge for using 150 therms in a month?
  (c) Develop a model that gives the monthly charge C for x therms of gas.
  (d) Graph the function found in part (c).
Source: Nicor Gas, Aurora, Illinois, 2009

51. Federal Income Tax Two 2009 Tax Rate Schedules are given in the accompanying table. If x equals taxable income and y equals the tax due, construct a function $y = f(x)$ for Schedule X.

REVISED 2009 TAX RATE SCHEDULES										
Schedule X—Single						**Schedule Y-1—Married Filing jointly or qualifying Widow(er)**				
If Taxable Income Is Over	But Not Over	The Tax Is This Amount	Plus This %	Of the Excess Over		If Taxable Income Is Over	But Not Over	The Tax Is This Amount	Plus This %	Of The Excess Over
$0	$8,350	–	+ 10%	$0		$0	$16,700	–	+ 10%	$0
8,350	33,950	$835.00	+ 15%	8,350		16,700	67,900	$1,670.00	+ 15%	16,700
33,950	82,250	4,675.00	+ 25%	33,950		67,900	137,050	9,350.00	+ 25%	67,900
82,250	171,550	16,750.00	+ 28%	82,250		137,050	208,850	26,637.50	+ 28%	137,050
171,550	372,950	41,754.00	+ 33%	171,550		208,850	372,950	46,741.50	+ 33%	208,850
372,950	–	108,216.00	+ 35%	372,950		372,950	–	100,894.50	+ 35%	372,950

Source: Internal Revenue Service

52. Federal Income Tax Refer to the revised 2009 tax rate schedules. If x equals taxable income and y equals the tax due, construct a function $y = f(x)$ for Schedule Y-1.

53. Cost of Transporting Goods A trucking company transports goods between Chicago and New York, a distance of 960 miles. The company's policy is to charge, for each pound, $0.50 per mile for the first 100 miles, $0.40 per mile for the next 300 miles, $0.25 per mile for the next 400 miles, and no charge for the remaining 160 miles.
(a) Graph the relationship between the cost of transportation in dollars and mileage over the entire 960-mile route.
(b) Find the cost as a function of mileage for hauls between 100 and 400 miles from Chicago.
(c) Find the cost as a function of mileage for hauls between 400 and 800 miles from Chicago.

54. Car Rental Costs An economy car rented in Florida from National Car Rental® on a weekly basis costs $95 per week. Extra days cost $24 per day until the day rate exceeds the weekly rate, in which case the weekly rate applies. Also, any part of a day used counts as a full day. Find the cost C of renting an economy car as a function of the number x of days used, where $7 \leq x \leq 14$. Graph this function.

55. Minimum Payments for Credit Cards Holders of credit cards issued by banks, department stores, oil companies, and so on, receive bills each month that state minimum amounts that must be paid by a certain due date. The minimum due depends on the total amount owed. One such credit card company uses the following rules: For a bill of less than $10, the entire amount is due. For a bill of at least $10 but less than $500, the minimum due is $10. A minimum of $30 is due on a bill of at least $500 but less than $1000, a minimum of $50 is due on a bill of at least $1000 but less than $1500, and a minimum of $70 is due on bills of $1500 or more. Find the function f that describes the minimum payment due on a bill of x dollars. Graph f.

56. Interest Payments for Credit Cards Refer to Problem 55. The card holder may pay any amount between the minimum due and the total owed. The organization issuing the card charges the card holder interest of 1.5% per month for the first $1000 owed and 1% per month on any unpaid balance over $1000. Find the function g that gives the amount of interest charged per month on a balance of x dollars. Graph g.

57. Wind Chill The wind chill factor represents the equivalent air temperature at a standard wind speed that would produce the same heat loss as the given temperature and wind speed. One formula for computing the equivalent temperature is

$$W = \begin{cases} t & 0 \leq v < 1.79 \\ 33 - \dfrac{(10.45 + 10\sqrt{v} - v)(33 - t)}{22.04} & 1.79 \leq v \leq 20 \\ 33 - 1.5958(33 - t) & v > 20 \end{cases}$$

where v represents the wind speed (in meters per second) and t represents the air temperature (°C). Compute the wind chill for the following:
(a) An air temperature of 10°C and a wind speed of 1 meter per second (m/sec)
(b) An air temperature of 10°C and a wind speed of 5 m/sec
(c) An air temperature of 10°C and a wind speed of 15 m/sec
(d) An air temperature of 10°C and a wind speed of 25 m/sec
(e) Explain the physical meaning of the equation corresponding to $0 \leq v < 1.79$.
(f) Explain the physical meaning of the equation corresponding to $v > 20$.

58. Wind Chill Redo Problem 57(a)–(d) for an air temperature of -10°C.

59. First-class Mail In 2009 the U.S. Postal Service charged $1.17 postage for first-class mail retail flats (such as an 8.5" by 11" envelope) weighing up to 1 ounce, plus $0.17 for each additional ounce up to 13 ounces. First-class rates do not apply to flats weighing more than 13 ounces. Develop a model that relates C, the first-class postage charged, for a flat weighing x ounces. Graph the function.
Source: United States Postal Service

Explaining Concepts: Discussion and Writing

In Problems 60–67, use a graphing utility.

60. Exploration Graph $y = x^2$. Then on the same screen graph $y = x^2 + 2$, followed by $y = x^2 + 4$, followed by $y = x^2 - 2$. What pattern do you observe? Can you predict the graph of $y = x^2 - 4$? Of $y = x^2 + 5$?

61. Exploration Graph $y = x^2$. Then on the same screen graph $y = (x - 2)^2$, followed by $y = (x - 4)^2$, followed by $y = (x + 2)^2$. What pattern do you observe? Can you predict the graph of $y = (x + 4)^2$? Of $y = (x - 5)^2$?

62. Exploration Graph $y = |x|$. Then on the same screen graph $y = 2|x|$, followed by $y = 4|x|$, followed by $y = \frac{1}{2}|x|$. What pattern do you observe? Can you predict the graph of $y = \frac{1}{4}|x|$? Of $y = 5|x|$?

63. Exploration Graph $y = x^2$. Then on the same screen graph $y = -x^2$. What pattern do you observe? Now try $y = |x|$ and $y = -|x|$. What do you conclude?

64. Exploration Graph $y = \sqrt{x}$. Then on the same screen graph $y = \sqrt{-x}$. What pattern do you observe? Now try $y = 2x + 1$ and $y = 2(-x) + 1$. What do you conclude?

65. Exploration Graph $y = x^3$. Then on the same screen graph $y = (x - 1)^3 + 2$. Could you have predicted the result?

66. Exploration Graph $y = x^2$, $y = x^4$, and $y = x^6$ on the same screen. What do you notice is the same about each graph? What do you notice that is different?

67. Exploration Graph $y = x^3$, $y = x^5$, and $y = x^7$ on the same screen. What do you notice is the same about each graph? What do you notice that is different?

68. Consider the equation

$$y = \begin{cases} 1 & \text{if } x \text{ is rational} \\ 0 & \text{if } x \text{ is irrational} \end{cases}$$

Is this a function? What is its domain? What is its range? What is its y-intercept, if any? What are its x-intercepts, if any? Is it even, odd, or neither? How would you describe its graph?

69. Define some functions that pass through $(0, 0)$ and $(1, 1)$ and are increasing for $x \geq 0$. Begin your list with $y = \sqrt{x}$, $y = x$, and $y = x^2$. Can you propose a general result about such functions?

'Are You Prepared?' Answers

1.

2.

3. $(0, -8)$, $(2, 0)$

1.5 Graphing Techniques: Transformations

OBJECTIVES 1 Graph Functions Using Vertical and Horizontal Shifts (p. 90)
2 Graph Functions Using Compressions and Stretches (p. 93)
3 Graph Functions Using Reflections about the x-Axis and the y-Axis (p. 96)

At this stage, if you were asked to graph any of the functions defined by $y = x$, $y = x^2$, $y = x^3$, $y = \sqrt{x}$, $y = \sqrt[3]{x}$, $y = \dfrac{1}{x}$, or $y = |x|$, your response should be, "Yes, I recognize these functions and know the general shapes of their graphs." (If this is not your answer, review the previous section, Figures 32 through 38.)

Sometimes we are asked to graph a function that is "almost" like one that we already know how to graph. In this section, we look at some of these functions and develop techniques for graphing them. Collectively, these techniques are referred to as **transformations.**

1 Graph Functions Using Vertical and Horizontal Shifts

EXAMPLE 1

Vertical Shift Up

Use the graph of $f(x) = x^2$ to obtain the graph of $g(x) = x^2 + 3$.

Solution

We begin by obtaining some points on the graphs of f and g. For example, when $x = 0$, then $y = f(0) = 0$ and $y = g(0) = 3$. When $x = 1$, then $y = f(1) = 1$ and $y = g(1) = 4$. Table 7 lists these and a few other points on each graph. Notice that each y-coordinate of a point on the graph of g is 3 units larger than the y-coordinate of the corresponding point on the graph of f. We conclude that the graph of g is identical to that of f, except that it is shifted vertically up 3 units. See Figure 43.

Table 7

x	$y = f(x)$ $= x^2$	$y = g(x)$ $= x^2 + 3$
−2	4	7
−1	1	4
0	0	3
1	1	4
2	4	7

Figure 43

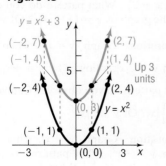

Let's look at another example.

EXAMPLE 2   **Vertical Shift Down**

Use the graph of $f(x) = x^2$ to obtain the graph of $g(x) = x^2 - 4$.

Solution   Table 8 lists some points on the graphs of f and g. Notice that each y-coordinate of g is 4 units less than the corresponding y-coordinate of f.

To obtain the graph of g from the graph of f, subtract 4 from each y-coordinate on the graph of f. So the graph of g is identical to that of f, except that it is shifted down 4 units. See Figure 44.

Table 8

x	$y = f(x)$ $= x^2$	$y = g(x)$ $= x^2 - 4$
-2	4	0
-1	1	-3
0	0	-4
1	1	-3
2	4	0

Figure 44

Figure 45

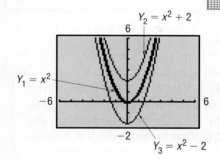

Exploration

On the same screen, graph each of the following functions:
$$Y_1 = x^2$$
$$Y_2 = x^2 + 2$$
$$Y_3 = x^2 - 2$$

Figure 45 illustrates the graphs. You should have observed a general pattern. With $Y_1 = x^2$ on the screen, the graph of $Y_2 = x^2 + 2$ is identical to that of $Y_1 = x^2$, except that it is shifted vertically up 2 units. The graph of $Y_3 = x^2 - 2$ is identical to that of $Y_1 = x^2$, except that it is shifted vertically down 2 units.

We are led to the following conclusions:

> If a positive real number k is added to the output of a function $y = f(x)$, the graph of the new function $y = f(x) + k$ is the graph of f **shifted vertically up** k units.

> If a positive real number k is subtracted from the output of a function $y = f(x)$, the graph of the new function $y = f(x) - k$ is the graph of f **shifted vertically down** k units.

Now Work   PROBLEM 39

EXAMPLE 3   **Horizontal Shift to the Right**

Use the graph of $f(x) = x^2$ to obtain the graph of $g(x) = (x - 2)^2$.

Solution   The function $g(x) = (x - 2)^2$ is basically a square function. Table 9 lists some points on the graphs of f and g. Note that when $f(x) = 0$ then $x = 0$, and when $g(x) = 0$, then $x = 2$. Also, when $f(x) = 4$, then $x = -2$ or 2, and when $g(x) = 4$, then $x = 0$ or 4. Notice that the x-coordinates on the graph of g are two units larger than the corresponding x-coordinates on the graph of f for any given y-coordinate. We conclude that the graph of g is identical to that of f, except that it is shifted horizontally 2 units to the right. See Figure 46.

Table 9

x	$y = f(x)$ $= x^2$	$y = g(x)$ $= (x - 2)^2$
-2	4	16
0	0	4
2	4	0
4	16	4

Figure 46

Figure 47

Exploration

On the same screen, graph each of the following functions:

$$Y_1 = x^2$$
$$Y_2 = (x - 3)^2$$
$$Y_3 = (x + 2)^2$$

Result Figure 47 illustrates the graphs.

You should have observed the following pattern. With the graph of $Y_1 = x^2$ on the screen, the graph of $Y_2 = (x - 3)^2$ is identical to that of $Y_1 = x^2$, except that it is shifted horizontally to the right 3 units. The graph of $Y_3 = (x + 2)^2$ is identical to that of $Y_1 = x^2$, except that it is shifted horizontally to the left 2 units.

In Other Words

If a positive number h is subtracted from x in $y = f(x)$, the graph of the new function $y = f(x - h)$ is the graph of $y = f(x)$ shifted horizontally right h units. If h is added to x, shift horizontally left h units.

We are led to the following conclusion.

If the argument x of a function f is replaced by $x - h, h > 0$, the graph of the new function $y = f(x - h)$ is the graph of f **shifted horizontally right** h units.
If the argument x of a function f is replaced by $x + h, h > 0$, the graph of the new function $y = f(x + h)$ is the graph of f **shifted horizontally left** h units.

EXAMPLE 4

Horizontal Shift to the Left

Use the graph of $f(x) = x^2$ to obtain the graph of $g(x) = (x + 4)^2$.

Solution

Again, the function $g(x) = (x + 4)^2$ is basically a square function. Its graph is the same as that of f, except that it is shifted horizontally 4 units to the left. See Figure 48.

Figure 48

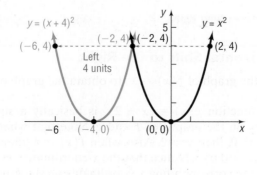

Now Work PROBLEM 43

Notice the distinction between vertical and horizontal shifts. The graph of $f(x) = \sqrt{x} + 3$ is obtained by shifting the graph of $y = \sqrt{x}$ *up* 3 units, because we evaluate the square root function first and then add 3. The graph of $g(x) = \sqrt{x + 3}$ is obtained by shifting the graph of $y = \sqrt{x}$ *left* 3 units, because we add 3 to x before we evaluate the square root function.

Vertical and horizontal shifts are sometimes combined.

EXAMPLE 5   **Combining Vertical and Horizontal Shifts**

Graph the function: $f(x) = (x + 3)^2 - 5$

Solution   We graph f in steps. First, we note that the rule for f is basically a square function, so we begin with the graph of $y = x^2$ as shown in Figure 49(a). Next, to get the graph of $y = (x + 3)^2$, we shift the graph of $y = x^2$ horizontally 3 units to the left. See Figure 49(b). Finally, to get the graph of $y = (x + 3)^2 - 5$, we shift the graph of $y = (x + 3)^2$ vertically down 5 units. See Figure 49(c). Note the points plotted on each graph. Using key points can be helpful in keeping track of the transformation that has taken place.

Figure 49

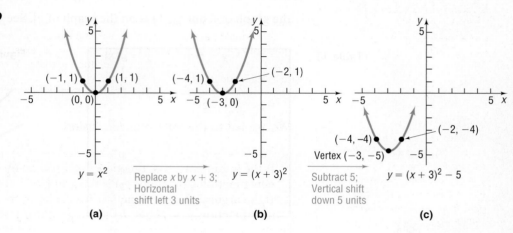

(a)   (b)   (c)

✓ Check: Graph $Y_1 = f(x) = (x + 3)^2 - 5$ and compare the graph to Figure 49(c).

In Example 5, if the vertical shift had been done first, followed by the horizontal shift, the final graph would have been the same. Try it for yourself.

Now Work PROBLEM 45

---

2 Graph Functions Using Compressions and Stretches

EXAMPLE 6   **Vertical Stretch**

Use the graph of $f(x) = |x|$ to obtain the graph of $g(x) = 2|x|$.

Solution   To see the relationship between the graphs of f and g, we form Table 10, listing points on each graph. For each x, the y-coordinate of a point on the graph of g is 2 times as large as the corresponding y-coordinate on the graph of f. The graph of $f(x) = |x|$ is vertically stretched by a factor of 2 to obtain the graph of $g(x) = 2|x|$ [for example, $(1, 1)$ is on the graph of f, but $(1, 2)$ is on the graph of g]. See Figure 50.

Table 10

x	$y = f(x)$ $= \|x\|$	$y = g(x)$ $= 2\|x\|$
−2	2	4
−1	1	2
0	0	0
1	1	2
2	2	4

Figure 50

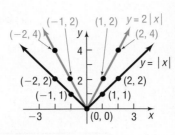

EXAMPLE 7 **Vertical Compression**

Use the graph of $f(x) = |x|$ to obtain the graph of $g(x) = \frac{1}{2}|x|$.

Solution For each x, the y-coordinate of a point on the graph of g is $\frac{1}{2}$ as large as the corresponding y-coordinate on the graph of f. The graph of $f(x) = |x|$ is vertically compressed by a factor of $\frac{1}{2}$ to obtain the graph of $g(x) = \frac{1}{2}|x|$ [for example, $(2, 2)$ is on the graph of f, but $(2, 1)$ is on the graph of g]. See Table 11 and Figure 51.

Table 11

x	$y = f(x)$ $= \|x\|$	$y = g(x)$ $= \frac{1}{2}\|x\|$
−2	2	1
−1	1	$\frac{1}{2}$
0	0	0
1	1	$\frac{1}{2}$
2	2	1

Figure 51

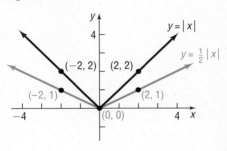

> When the right side of a function $y = f(x)$ is multiplied by a positive number a, the graph of the new function $y = af(x)$ is obtained by multiplying each y-coordinate on the graph of $y = f(x)$ by a. The new graph is a **vertically compressed** (if $0 < a < 1$) or a **vertically stretched** (if $a > 1$) version of the graph of $y = f(x)$.

 Now Work PROBLEM 47

What happens if the argument x of a function $y = f(x)$ is multiplied by a positive number a, creating a new function $y = f(ax)$? To find the answer, we look at the following Exploration.

Exploration

On the same screen, graph each of the following functions:

$$Y_1 = f(x) = \sqrt{x} \qquad Y_2 = f(2x) = \sqrt{2x} \qquad Y_3 = f\left(\frac{1}{2}x\right) = \sqrt{\frac{1}{2}x} = \sqrt{\frac{x}{2}}$$

Create a table of values to explore the relation between the x- and y-coordinates of each function.

Result You should have obtained the graphs in Figure 52. Look at Table 12(a). Notice that $(1, 1)$, $(4, 2)$, and $(9, 3)$ are points on the graph of $Y_1 = \sqrt{x}$. Also, $(0.5, 1)$, $(2, 2)$, and $(4.5, 3)$ are points on the graph of $Y_2 = \sqrt{2x}$. For a given y-coordinate, the x-coordinate on the graph of Y_2 is $\frac{1}{2}$ of the x-coordinate on Y_1.

Figure 52

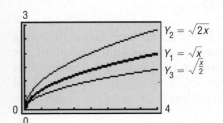

Table12

X	Y₁	Y₂
0	0	0
.5	.70711	1
1	1	1.4142
2	1.4142	2
4	2	2.8284
4.5	2.1213	3
9	3	4.2426

Y₂ ◼√(2X)

(a)

X	Y₁	Y₃
0	0	0
1	1	.70711
2	1.4142	1
4	2	1.4142
8	2.8284	2
9	3	2.1213
18	4.2426	3

Y₃ ◼√(X/2)

(b)

We conclude that the graph of $Y_2 = \sqrt{2x}$ is obtained by multiplying the *x*-coordinate of each point on the graph of $Y_1 = \sqrt{x}$ by $\frac{1}{2}$. The graph of $Y_2 = \sqrt{2x}$ is the graph of $Y_1 = \sqrt{x}$ *compressed* horizontally.

Look at Table 12(b). Notice that (1, 1), (4, 2), and (9, 3) are points on the graph of $Y_1 = \sqrt{x}$. Also notice that (2, 1), (8, 2), and (18, 3) are points on the graph of $Y_3 = \sqrt{\dfrac{x}{2}}$. For a given *y*-coordinate, the *x*-coordinate on the graph of Y_3 is 2 times the *x*-coordinate on Y_1. We conclude that the graph of $Y_3 = \sqrt{\dfrac{x}{2}}$ is obtained by multiplying the *x*-coordinate of each point on the graph of $Y_1 = \sqrt{x}$ by 2. The graph of $Y_3 = \sqrt{\dfrac{x}{2}}$ is the graph of $Y_1 = \sqrt{x}$ *stretched* horizontally.

Based on the results of the Exploration, we have the following result:

> If the argument *x* of a function $y = f(x)$ is multiplied by a positive number *a*, the graph of the new function $y = f(ax)$ is obtained by multiplying each *x*-coordinate of $y = f(x)$ by $\frac{1}{a}$. A **horizontal compression** results if $a > 1$, and a **horizontal stretch** occurs if $0 < a < 1$.

Let's look at an example.

EXAMPLE 8   **Graphing Using Stretches and Compressions**

The graph of $y = f(x)$ is given in Figure 53. Use this graph to find the graphs of

(a) $y = 2f(x)$ (b) $y = f(3x)$

Solution   (a) The graph of $y = 2f(x)$ is obtained by multiplying each *y*-coordinate of $y = f(x)$ by 2. See Figure 54.

(b) The graph of $y = f(3x)$ is obtained from the graph of $y = f(x)$ by multiplying each *x*-coordinate of $y = f(x)$ by $\frac{1}{3}$. See Figure 55.

Figure 53

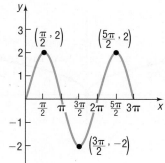

Figure 54

$y = 2f(x)$

Figure 55

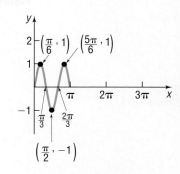

$y = f(3x)$

Now Work PROBLEMS 69(e) AND (g)

3 Graph Functions Using Reflections about the *x*-Axis and the *y*-Axis

EXAMPLE 9

Reflection about the *x*-Axis

Graph the function: $f(x) = -x^2$

Solution

We begin with the graph of $y = x^2$, as shown in black in Figure 56. For each point (x, y) on the graph of $y = x^2$, the point $(x, -y)$ is on the graph of $y = -x^2$, as indicated in Table 13. We can draw the graph of $y = -x^2$ by reflecting the graph of $y = x^2$ about the *x*-axis. See Figure 56.

Figure 56

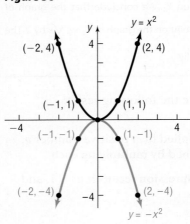

Table 13

x	$y = x^2$	$y = -x^2$
−2	4	−4
−1	1	−1
0	0	0
1	1	−1
2	4	−4

When the right side of the function $y = f(x)$ is multiplied by −1, the graph of the new function $y = -f(x)$ is the **reflection about the *x*-axis** of the graph of the function $y = f(x)$.

━━━ **Now Work** PROBLEM 51

EXAMPLE 10

Reflection about the *y*-Axis

Graph the function: $f(x) = \sqrt{-x}$

Solution

First, notice that the domain of f consists of all real numbers x for which $-x \geq 0$ or, equivalently, $x \leq 0$. To get the graph of $f(x) = \sqrt{-x}$, we begin with the graph of $y = \sqrt{x}$, as shown in Figure 57. For each point (x, y) on the graph of $y = \sqrt{x}$, the point $(-x, y)$ is on the graph of $y = \sqrt{-x}$. We obtain the graph of $y = \sqrt{-x}$ by reflecting the graph of $y = \sqrt{x}$ about the *y*-axis. See Figure 57.

Figure 57

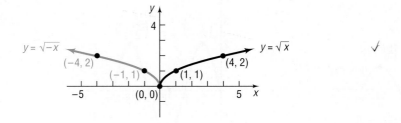

When the graph of the function $y = f(x)$ is known, the graph of the new function $y = f(-x)$ is the **reflection about the *y*-axis** of the graph of the function $y = f(x)$.

SUMMARY OF GRAPHING TECHNIQUES

To Graph:	Draw the Graph of f and:	Functional Change to $f(x)$
Vertical shifts		
$y = f(x) + k, \quad k > 0$	Raise the graph of f by k units.	Add k to $f(x)$.
$y = f(x) - k, \quad k > 0$	Lower the graph of f by k units.	Subtract k from $f(x)$.
Horizontal shifts		
$y = f(x + h), \quad h > 0$	Shift the graph of f to the left h units.	Replace x by $x + h$.
$y = f(x - h), \quad h > 0$	Shift the graph of f to the right h units.	Replace x by $x - h$.
Compressing or stretching		
$y = af(x), \quad a > 0$	Multiply each y-coordinate of $y = f(x)$ by a. Stretch the graph of f vertically if $a > 1$. Compress the graph of f vertically if $0 < a < 1$.	Multiply $f(x)$ by a.
$y = f(ax), \quad a > 0$	Multiply each x-coordinate of $y = f(x)$ by $\dfrac{1}{a}$. Stretch the graph of f horizontally if $0 < a < 1$. Compress the graph of f horizontally if $a > 1$.	Replace x by ax.
Reflection about the x-axis		
$y = -f(x)$	Reflect the graph of f about the x-axis.	Multiply $f(x)$ by -1.
Reflection about the y-axis		
$y = f(-x)$	Reflect the graph of f about the y-axis.	Replace x by $-x$.

EXAMPLE 11 **Determining the Function Obtained from a Series of Transformations**

Find the function that is finally graphed after the following three transformations are applied to the graph of $y = |x|$.

1. Shift left 2 units
2. Shift up 3 units
3. Reflect about the y-axis

Solution

1. Shift left 2 units: Replace x by $x + 2$. $y = |x + 2|$
2. Shift up 3 units: Add 3. $y = |x + 2| + 3$
3. Reflect about the y-axis: Replace x by $-x$. $y = |-x + 2| + 3$

Now Work PROBLEM 27

The examples that follow combine some of the procedures outlined in this section to get the required graph.

EXAMPLE 12 **Combining Graphing Procedures**

Graph the function $f(x) = \dfrac{3}{x - 2} + 1$. Find the domain and the range of f.

Solution It is helpful to write f as $f(x) = 3\left(\dfrac{1}{x-2}\right) + 1$. Now we use the following steps to obtain the graph of f:

STEP 1: $y = \dfrac{1}{x}$ Reciprocal function

STEP 2: $y = 3 \cdot \left(\dfrac{1}{x}\right) = \dfrac{3}{x}$ Multiply by 3; vertical stretch of the graph of $y = \dfrac{1}{x}$ by a factor of 3.

STEP 3: $y = \dfrac{3}{x-2}$ Replace x by $x-2$; horizontal shift to the right 2 units.

STEP 4: $y = \dfrac{3}{x-2} + 1$ Add 1; vertical shift up 1 unit.

See Figure 58.

Figure 58

(a) $y = \dfrac{1}{x}$ Multiply by 3; Vertical stretch   (b) $y = \dfrac{3}{x}$ Replace x by $x-2$; Horizontal shift right 2 units   (c) $y = \dfrac{3}{x-2}$ Add 1; Vertical shift up 1 unit   (d) $y = \dfrac{3}{x-2} + 1$

The domain of $y = \dfrac{1}{x}$ is $\{x \mid x \neq 0\}$ and its range is $\{y \mid y \neq 0\}$. Because we shifted right 2 units and up 1 unit to obtain f, the domain of f is $\{x \mid x \neq 2\}$ and its range is $\{y \mid y \neq 1\}$.

Other orderings of the steps shown in Example 12 would also result in the graph of f. For example, try this one:

STEP 1: $y = \dfrac{1}{x}$ Reciprocal function

STEP 2: $y = \dfrac{1}{x-2}$ Replace x by $x-2$; horizontal shift to the right 2 units.

STEP 3: $y = \dfrac{3}{x-2}$ Multiply by 3; vertical stretch of the graph of $y = \dfrac{1}{x-2}$ by a factor of 3.

STEP 4: $y = \dfrac{3}{x-2} + 1$ Add 1; vertical shift up 1 unit.

Hint: Although the order in which transformations are performed can be altered, you may consider using the following order for consistency:

1. Reflections

2. Compressions and stretches

3. Shifts

EXAMPLE 13   **Combining Graphing Procedures**

Graph the function $f(x) = \sqrt{1-x} + 2$. Find the domain and the range of f.

Solution   Because horizontal shifts require the form $x - h$, we begin by rewriting $f(x)$ as $f(x) = \sqrt{1 - x} + 2 = \sqrt{-(x - 1)} + 2$. Now use the following steps:

STEP 1: $y = \sqrt{x}$ Square root function
STEP 2: $y = \sqrt{-x}$ Replace x by $-x$; reflect about the y-axis.
STEP 3: $y = \sqrt{-(x - 1)} = \sqrt{1 - x}$ Replace x by x − 1; horizontal shift to the right 1 unit.
STEP 4: $y = \sqrt{1 - x} + 2$ Add 2; vertical shift up 2 units.

See Figure 59.

Figure 59

(a) $y = \sqrt{x}$ Replace x by $-x$; Reflect about y-axis
(b) $y = \sqrt{-x}$ Replace x by x − 1; Horizontal shift right 1 unit
(c) $y = \sqrt{-(x - 1)}$ $= \sqrt{-x + 1}$ $= \sqrt{1 - x}$ Add 2; Vertical shift up 2 units
(d) $y = \sqrt{1 - x} + 2$

The domain of f is $(-\infty, 1]$ and the range is $[2, \infty)$.

Now Work PROBLEM 61

1.5 Assess Your Understanding

Concepts and Vocabulary

1. Suppose that the graph of a function f is known. Then the graph of $y = f(x - 2)$ may be obtained by a(n) _____ shift of the graph of f to the _____ a distance of 2 units.

2. Suppose that the graph of a function f is known. Then the graph of $y = f(-x)$ may be obtained by a reflection about the _____-axis of the graph of the function $y = f(x)$.

3. Suppose that the graph of a function g is known. The graph of $y = g(x) + 2$ may be obtained by a _____ shift of the graph of g _____ a distance of 2 units.

4. *True or False* The graph of $y = -f(x)$ is the reflection about the x-axis of the graph of $y = f(x)$.

5. *True or False* To obtain the graph of $f(x) = \sqrt{x + 2}$, shift the graph of $y = \sqrt{x}$ horizontally to the right 2 units.

6. *True or False* To obtain the graph of $f(x) = x^3 + 5$, shift the graph of $y = x^3$ vertically up 5 units.

Skill Building

In Problems 7–18, match each graph to one of the following functions:

A. $y = x^2 + 2$

B. $y = -x^2 + 2$

C. $y = |x| + 2$

D. $y = -|x| + 2$

E. $y = (x - 2)^2$

F. $y = -(x + 2)^2$

G. $y = |x - 2|$

H. $y = -|x + 2|$

I. $y = 2x^2$

J. $y = -2x^2$

K. $y = 2|x|$

L. $y = -2|x|$

7.

8.

9.

10.

11.

12.

13.

14.

15.

16.

17.

18.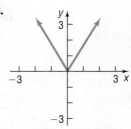

In Problems 19–26, write the function whose graph is the graph of $y = x^3$, but is:

19. Shifted to the right 4 units

20. Shifted to the left 4 units

21. Shifted up 4 units

22. Shifted down 4 units

23. Reflected about the y-axis

24. Reflected about the x-axis

25. Vertically stretched by a factor of 4

26. Horizontally stretched by a factor of 4

In Problems 27–30, find the function that is finally graphed after each of the following transformations is applied to the graph of $y = \sqrt{x}$ in the order stated.

27. (1) Shift up 2 units
   (2) Reflect about the x-axis
   (3) Reflect about the y-axis

28. (1) Reflect about the x-axis
   (2) Shift right 3 units
   (3) Shift down 2 units

29. (1) Reflect about the x-axis
   (2) Shift up 2 units
   (3) Shift left 3 units

30. (1) Shift up 2 units
   (2) Reflect about the y-axis
   (3) Shift left 3 units

31. If $(3, 6)$ is a point on the graph of $y = f(x)$, which of the following points must be on the graph of $y = -f(x)$?
   (a) $(6, 3)$ (b) $(6, -3)$
   (c) $(3, -6)$ (d) $(-3, 6)$

32. If $(3, 6)$ is a point on the graph of $y = f(x)$, which of the following points must be on the graph of $y = f(-x)$?
   (a) $(6, 3)$ (b) $(6, -3)$
   (c) $(3, 6)$ (d) $(-3, 6)$

33. If $(1, 3)$ is a point on the graph of $y = f(x)$, which of the following points must be on the graph of $y = 2f(x)$?
   (a) $(1, 3)$ (b) $(2, 3)$
   (c) $(1, 6)$ (d) $\left(\dfrac{1}{2}, 3\right)$

34. If $(4, 2)$ is a point on the graph of $y = f(x)$, which of the following points must be on the graph of $y = f(2x)$?
   (a) $(4, 1)$ (b) $(8, 2)$
   (c) $(2, 2)$ (d) $(4, 4)$

35. Suppose that the x-intercepts of the graph of $y = f(x)$ are -5 and 3.
   (a) What are the x-intercepts of the graph of $y = f(x + 2)$?
   (b) What are the x-intercepts of the graph of $y = f(x - 2)$?
   (c) What are the x-intercepts of the graph of $y = 4f(x)$?
   (d) What are the x-intercepts of the graph of $y = f(-x)$?

36. Suppose that the x-intercepts of the graph of $y = f(x)$ are -8 and 1.
   (a) What are the x-intercepts of the graph of $y = f(x + 4)$?
   (b) What are the x-intercepts of the graph of $y = f(x - 3)$?
   (c) What are the x-intercepts of the graph of $y = 2f(x)$?
   (d) What are the x-intercepts of the graph of $y = f(-x)$?

37. Suppose that the function $y = f(x)$ is increasing on the interval $(-1, 5)$.
   (a) Over what interval is the graph of $y = f(x + 2)$ increasing?
   (b) Over what interval is the graph of $y = f(x - 5)$ increasing?
   (c) What can be said about the graph of $y = -f(x)$?
   (d) What can be said about the graph of $y = f(-x)$?

38. Suppose that the function $y = f(x)$ is decreasing on the interval $(-2, 7)$.
   (a) Over what interval is the graph of $y = f(x + 2)$ decreasing?
   (b) Over what interval is the graph of $y = f(x - 5)$ decreasing?
   (c) What can be said about the graph of $y = -f(x)$?
   (d) What can be said about the graph of $y = f(-x)$?

In Problems 39–68, graph each function using the techniques of shifting, compressing, stretching, and/or reflecting. Start with the graph of the basic function (for example, $y = x^2$) and show all stages. Be sure to show at least three key points. Find the domain and the range of each function.

39. $f(x) = x^2 - 1$

40. $f(x) = x^2 + 4$

41. $g(x) = x^3 + 1$

42. $g(x) = x^3 - 1$

43. $h(x) = \sqrt{x - 2}$

44. $h(x) = \sqrt{x + 1}$

45. $f(x) = (x - 1)^3 + 2$

46. $f(x) = (x + 2)^3 - 3$

47. $g(x) = 4\sqrt{x}$

48. $g(x) = \frac{1}{2}\sqrt{x}$

49. $h(x) = \frac{1}{2x}$

50. $h(x) = \sqrt[3]{2x}$

51. $f(x) = -\sqrt[3]{x}$

52. $f(x) = -\sqrt{x}$

53. $g(x) = \sqrt[3]{-x}$

54. $g(x) = \frac{1}{-x}$

55. $h(x) = -x^3 + 2$

56. $h(x) = \frac{1}{-x} + 2$

57. $f(x) = 2(x + 1)^2 - 3$

58. $f(x) = 3(x - 2)^2 + 1$

59. $g(x) = 2\sqrt{x - 2} + 1$

60. $g(x) = 3|x + 1| - 3$

61. $h(x) = \sqrt{-x} - 2$

62. $h(x) = \frac{4}{x} + 2$

63. $f(x) = -(x + 1)^3 - 1$

64. $f(x) = -4\sqrt{x - 1}$

65. $g(x) = 2|1 - x|$

66. $g(x) = 4\sqrt{2 - x}$

67. $h(x) = 2 \operatorname{int}(x - 1)$

68. $h(x) = \operatorname{int}(-x)$

In Problems 69–72, the graph of a function f is illustrated. Use the graph of f as the first step toward graphing each of the following functions:

(a) $F(x) = f(x) + 3$

(b) $G(x) = f(x + 2)$

(c) $P(x) = -f(x)$

(d) $H(x) = f(x + 1) - 2$

(e) $Q(x) = \frac{1}{2}f(x)$

(f) $g(x) = f(-x)$

(g) $h(x) = f(2x)$

69.

70.

71.

72.

Mixed Practice

In Problems 73–82, complete the square of each quadratic expression. Then graph each function using the technique of shifting. (If necessary, refer to Appendix A, Section A.4 to review completing the square.)

73. $f(x) = x^2 + 2x$

74. $f(x) = x^2 - 6x$

75. $f(x) = x^2 - 8x + 1$

76. $f(x) = x^2 + 4x + 2$

77. $f(x) = x^2 + x + 1$

78. $f(x) = x^2 - x + 1$

79. $f(x) = 2x^2 - 12x + 19$

80. $f(x) = 3x^2 + 6x + 1$

81. $f(x) = -3x^2 - 12x - 17$

82. $f(x) = -2x^2 - 12x - 13$

Applications and Extensions

83. The equation $y = (x - c)^2$ defines a *family of parabolas*, one parabola for each value of c. On one set of coordinate axes, graph the members of the family for $c = 0, c = 3$, and $c = -2$.

84. Repeat Problem 83 for the family of parabolas $y = x^2 + c$.

85. Thermostat Control Energy conservation experts estimate that homeowners can save 5% to 10% on winter heating bills by programming their thermostats 5 to 10 degrees lower while sleeping. In the given graph, the temperature T (in degrees Fahrenheit) of a home is given as a function of time t (in hours after midnight) over a 24-hour period.

(a) At what temperature is the thermostat set during daytime hours? At what temperature is the thermostat set overnight?

(b) The homeowner reprograms the thermostat to $y = T(t) - 2$. Explain how this affects the temperature in the house. Graph this new function.

(c) The homeowner reprograms the thermostat to $y = T(t + 1)$. Explain how this affects the temperature in the house. Graph this new function.

Source: Roger Albright, 547 Ways to Be Fuel Smart, 2000

86. **Digital Music Revenues** The total projected worldwide digital music revenues R, in millions of dollars, for the years 2005 through 2010 can be estimated by the function

$$R(x) = 170.7x^2 + 1373x + 1080$$

where x is the number of years after 2005.

(a) Find $R(0)$, $R(3)$, and $R(5)$ and explain what each value represents.

(b) Find $r = R(x - 5)$.

(c) Find $r(5)$, $r(8)$, and $r(10)$ and explain what each value represents.

(d) In the model r, what does x represent?

(e) Would there be an advantage in using the model r when estimating the projected revenues for a given year instead of the model R?

Source: eMarketer.com, May 2006

87. **Temperature Measurements** The relationship between the Celsius (°C) and Fahrenheit (°F) scales for measuring temperature is given by the equation

$$F = \frac{9}{5}C + 32$$

The relationship between the Celsius (°C) and Kelvin (K) scales is $K = C + 273$. Graph the equation $F = \frac{9}{5}C + 32$ using degrees Fahrenheit on the y-axis and degrees Celsius on the x-axis. Use the techniques introduced in this section to obtain the graph showing the relationship between Kelvin and Fahrenheit temperatures.

88. **Period of a Pendulum** The period T (in seconds) of a simple pendulum is a function of its length l (in feet) defined by the equation

$$T = 2\pi\sqrt{\frac{l}{g}}$$

where $g \approx 32.2$ feet per second per second is the acceleration of gravity.

(a) Use a graphing utility to graph the function $T = T(l)$.

(b) Now graph the functions $T = T(l + 1)$, $T = T(l + 2)$, and $T = T(l + 3)$.

(c) Discuss how adding to the length l changes the period T.

(d) Now graph the functions $T = T(2l)$, $T = T(3l)$, and $T = T(4l)$.

(e) Discuss how multiplying the length l by factors of 2, 3, and 4 changes the period T.

89. **Cigar Company Profits** The daily profits of a cigar company from selling x cigars are given by

$$p(x) = -0.05x^2 + 100x - 2000$$

The government wishes to impose a tax on cigars (sometimes called a *sin tax*) that gives the company the option of either paying a flat tax of $10,000 per day or a tax of 10% on profits. As chief financial officer (CFO) of the company, you need to decide which tax is the better option for the company.

(a) On the same screen, graph $Y_1 = p(x) - 10,000$ and $Y_2 = (1 - 0.10)p(x)$.

(b) Based on the graph, which option would you select? Why?

(c) Using the terminology learned in this section, describe each graph in terms of the graph of $p(x)$.

(d) Suppose that the government offered the options of a flat tax of $4800 or a tax of 10% on profits. Which would you select? Why?

90. The graph of a function f is illustrated in the figure.

(a) Draw the graph of $y = |f(x)|$.

(b) Draw the graph of $y = f(|x|)$.

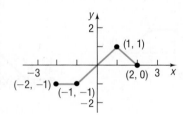

91. The graph of a function f is illustrated in the figure.

(a) Draw the graph of $y = |f(x)|$.

(b) Draw the graph of $y = f(|x|)$.

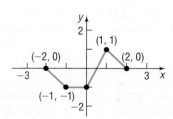

92. Suppose $(1, 3)$ is a point on the graph of $y = f(x)$.

(a) What point is on the graph of $y = f(x + 3) - 5$?

(b) What point is on the graph of $y = -2f(x - 2) + 1$?

(c) What point is on the graph of $y = f(2x + 3)$?

93. Suppose $(-3, 5)$ is a point on the graph of $y = g(x)$.

(a) What point is on the graph of $y = g(x + 1) - 3$?

(b) What point is on the graph of $y = -3g(x - 4) + 3$?

(c) What point is on the graph of $y = g(3x + 9)$?

Explaining Concepts: Discussion and Writing

94. Suppose that the graph of a function f is known. Explain how the graph of $y = 4f(x)$ differs from the graph of $y = f(4x)$.

95. Suppose that the graph of a function f is known. Explain how the graph of $y = f(x) - 2$ differs from the graph of $y = f(x - 2)$.

96. The area under the curve $y = \sqrt{x}$ bounded below by the x-axis and on the right by $x = 4$ is $\dfrac{16}{3}$ square units. Using the ideas presented in this section, what do you think is the area under the curve of $y = \sqrt{-x}$ bounded below by the x-axis and on the left by $x = -4$? Justify your answer.

1.6 Mathematical Models: Building Functions

OBJECTIVE **1** Build and Analyze Functions (p. 103)

1 Build and Analyze Functions

Real-world problems often result in mathematical models that involve functions. These functions need to be constructed or built based on the information given. In building functions, we must be able to translate the verbal description into the language of mathematics. We do this by assigning symbols to represent the independent and dependent variables and then by finding the function or rule that relates these variables.

EXAMPLE 1

Finding the Distance from the Origin to a Point on a Graph

Let $P = (x, y)$ be a point on the graph of $y = x^2 - 1$.

(a) Express the distance d from P to the origin O as a function of x.

(b) What is d if $x = 0$?

(c) What is d if $x = 1$?

(d) What is d if $x = \dfrac{\sqrt{2}}{2}$?

 (e) Use a graphing utility to graph the function $d = d(x)$, $x \geq 0$. Rounded to two decimal places, find the value(s) of x at which d has a local minimum. [This gives the point(s) on the graph of $y = x^2 - 1$ closest to the origin.]

Solution

(a) Figure 60 illustrates the graph of $y = x^2 - 1$. The distance d from P to O is

$$d = \sqrt{(x - 0)^2 + (y - 0)^2} = \sqrt{x^2 + y^2}$$

Figure 60

Since P is a point on the graph of $y = x^2 - 1$, we substitute $x^2 - 1$ for y. Then

$$d(x) = \sqrt{x^2 + (x^2 - 1)^2} = \sqrt{x^4 - x^2 + 1}$$

We have expressed the distance d as a function of x.

(b) If $x = 0$, the distance d is

$$d(0) = \sqrt{0^4 - 0^2 + 1} = \sqrt{1} = 1$$

(c) If $x = 1$, the distance d is

$$d(1) = \sqrt{1^4 - 1^2 + 1} = 1$$

(d) If $x = \dfrac{\sqrt{2}}{2}$, the distance d is

$$d\left(\frac{\sqrt{2}}{2}\right) = \sqrt{\left(\frac{\sqrt{2}}{2}\right)^4 - \left(\frac{\sqrt{2}}{2}\right)^2 + 1} = \sqrt{\frac{1}{4} - \frac{1}{2} + 1} = \frac{\sqrt{3}}{2}$$

Figure 61

(e) Figure 61 shows the graph of $Y_1 = \sqrt{x^4 - x^2 + 1}$. Using the MINIMUM feature on a graphing utility, we find that when $x \approx 0.71$ the value of d is smallest. The local minimum is $d \approx 0.87$ rounded to two decimal places. Since $d(x)$ is even, by symmetry, it follows that when $x \approx -0.71$ the value of d is also a local minimum. Since $(\pm 0.71)^2 - 1 = -0.50$, the points $(-0.71, -0.50)$ and $(0.71, -0.50)$ on the graph of $y = x^2 - 1$ are closest to the origin.

━━━━━ **Now Work** PROBLEM 1

EXAMPLE 2 **Area of a Rectangle**

Figure 62

A rectangle has one corner in quadrant I on the graph of $y = 25 - x^2$, another at the origin, a third on the positive y-axis, and the fourth on the positive x-axis. See Figure 62.

(a) Express the area A of the rectangle as a function of x.

(b) What is the domain of A?

(c) Graph $A = A(x)$.

(d) For what value of x is the area largest?

Solution

(a) The area A of the rectangle is $A = xy$, where $y = 25 - x^2$. Substituting this expression for y, we obtain $A(x) = x(25 - x^2) = 25x - x^3$.

(b) Since (x, y) is in quadrant I, we have $x > 0$. Also, $y = 25 - x^2 > 0$, which implies that $x^2 < 25$, so $-5 < x < 5$. Combining these restrictions, we have the domain of A as $\{x | 0 < x < 5\}$, or $(0, 5)$ using interval notation.

(c) See Figure 63 for the graph of $A = A(x)$.

(d) Using MAXIMUM, we find that the maximum area is 48.11 square units at $x = 2.89$ inches, each rounded to two decimal places. See Figure 64.

Figure 63

Figure 64

━━━━━ **Now Work** PROBLEM 7

EXAMPLE 3 **Making a Playpen***

A manufacturer of children's playpens makes a square model that can be opened at one corner and attached at right angles to a wall or, perhaps, the side of a house. If each side is 3 feet in length, the open configuration doubles the available area in which the child can play from 9 square feet to 18 square feet. See Figure 65.

Now suppose that we place hinges at the outer corners to allow for a configuration like the one shown in Figure 66.

*Adapted from *Proceedings, Summer Conference for College Teachers on Applied Mathematics* (University of Missouri, Rolla), 1971.

Figure 65

Figure 66

(a) Build a model that expresses the area A of the configuration shown in Figure 66 as a function of the distance x between the two parallel sides.

(b) Find the domain of A.

(c) Find A if $x = 5$.

(d) Graph $A = A(x)$. For what value of x is the area largest? What is the maximum area?

Solution   (a) Refer to Figure 66. The area A that we seek consists of the area of a rectangle (with width 3 and length x) and the area of an isosceles triangle (with base x and two equal sides of length 3). The height h of the triangle may be found using the Pythagorean Theorem.

$$h^2 + \left(\frac{x}{2}\right)^2 = 3^2$$

$$h^2 = 3^2 - \left(\frac{x}{2}\right)^2 = 9 - \frac{x^2}{4} = \frac{36 - x^2}{4}$$

$$h = \frac{1}{2}\sqrt{36 - x^2}$$

The area A enclosed by the playpen is

$$A = \text{area of rectangle} + \text{area of triangle} = 3x + \frac{1}{2}x\left(\frac{1}{2}\sqrt{36 - x^2}\right)$$

The area A expressed as a function of x is

$$A(x) = 3x + \frac{x\sqrt{36 - x^2}}{4} \qquad \text{The Model}$$

(b) To find the domain of A, we note first that $x > 0$, since x is a length. Also, the expression under the square root must be positive, so

$$36 - x^2 > 0$$
$$x^2 < 36$$
$$-6 < x < 6$$

Combining these restrictions, we find that the domain of A is $0 < x < 6$, or $(0, 6)$ using interval notation.

Figure 67

(c) If $x = 5$, the area is

$$A(5) = 3(5) + \frac{5}{4}\sqrt{36 - (5)^2} \approx 19.15 \text{ square feet}$$

If the length of the playpen is 5 feet, its area is 19.15 square feet.

(d) See Figure 67. The maximum area is about 19.82 square feet, obtained when x is about 5.58 feet.

1.6 Assess Your Understanding

Applications and Extensions

1. Let $P = (x, y)$ be a point on the graph of $y = x^2 - 8$.
   (a) Express the distance d from P to the origin as a function of x.
   (b) What is d if $x = 0$?
   (c) What is d if $x = 1$?
   (d) Use a graphing utility to graph $d = d(x)$.
   (e) For what values of x is d smallest?

2. Let $P = (x, y)$ be a point on the graph of $y = x^2 - 8$.
   (a) Express the distance d from P to the point $(0, -1)$ as a function of x.
   (b) What is d if $x = 0$?
   (c) What is d if $x = -1$?
   (d) Use a graphing utility to graph $d = d(x)$.
   (e) For what values of x is d smallest?

3. Let $P = (x, y)$ be a point on the graph of $y = \sqrt{x}$.
   (a) Express the distance d from P to the point $(1, 0)$ as a function of x.
   (b) Use a graphing utility to graph $d = d(x)$.
   (c) For what values of x is d smallest?

4. Let $P = (x, y)$ be a point on the graph of $y = \dfrac{1}{x}$.
   (a) Express the distance d from P to the origin as a function of x.
   (b) Use a graphing utility to graph $d = d(x)$.
   (c) For what values of x is d smallest?

5. A right triangle has one vertex on the graph of $y = x^3, x > 0$, at (x, y), another at the origin, and the third on the positive y-axis at $(0, y)$, as shown in the figure. Express the area A of the triangle as a function of x.

6. A right triangle has one vertex on the graph of $y = 9 - x^2, x > 0$, at (x, y), another at the origin, and the third on the positive x-axis at $(x, 0)$. Express the area A of the triangle as a function of x.

7. A rectangle has one corner in quadrant I on the graph of $y = 16 - x^2$, another at the origin, a third on the positive y-axis, and the fourth on the positive x-axis. See the figure.

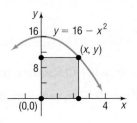

(a) Express the area A of the rectangle as a function of x.
(b) What is the domain of A?
(c) Graph $A = A(x)$. For what value of x is A largest?

8. A rectangle is inscribed in a semicircle of radius 2. See the figure. Let $P = (x, y)$ be the point in quadrant I that is a vertex of the rectangle and is on the circle.

(a) Express the area A of the rectangle as a function of x.
(b) Express the perimeter p of the rectangle as a function of x.
(c) Graph $A = A(x)$. For what value of x is A largest?
(d) Graph $p = p(x)$. For what value of x is p largest?

9. A rectangle is inscribed in a circle of radius 2. See the figure. Let $P = (x, y)$ be the point in quadrant I that is a vertex of the rectangle and is on the circle.

(a) Express the area A of the rectangle as a function of x.
(b) Express the perimeter p of the rectangle as a function of x.
(c) Graph $A = A(x)$. For what value of x is A largest?
(d) Graph $p = p(x)$. For what value of x is p largest?

10. A circle of radius r is inscribed in a square. See the figure.

(a) Express the area A of the square as a function of the radius r of the circle.
(b) Express the perimeter p of the square as a function of r.

11. **Geometry** A wire 10 meters long is to be cut into two pieces. One piece will be shaped as a square, and the other piece will be shaped as a circle. See the figure.

(a) Express the total area A enclosed by the pieces of wire as a function of the length x of a side of the square.

(b) What is the domain of A?

(c) Graph $A = A(x)$. For what value of x is A smallest?

12. Geometry   A wire 10 meters long is to be cut into two pieces. One piece will be shaped as an equilateral triangle, and the other piece will be shaped as a circle.

(a) Express the total area A enclosed by the pieces of wire as a function of the length x of a side of the equilateral triangle.

(b) What is the domain of A?

(c) Graph $A = A(x)$. For what value of x is A smallest?

13. A wire of length x is bent into the shape of a circle.

(a) Express the circumference C of the circle as a function of x.

(b) Express the area A of the circle as a function of x.

14. A wire of length x is bent into the shape of a square.

(a) Express the perimeter p of the square as a function of x.

(b) Express the area A of the square as a function of x.

15. Geometry   A semicircle of radius r is inscribed in a rectangle so that the diameter of the semicircle is the length of the rectangle. See the figure.

(a) Express the area A of the rectangle as a function of the radius r of the semicircle.

(b) Express the perimeter p of the rectangle as a function of r.

16. Geometry   An equilateral triangle is inscribed in a circle of radius r. See the figure. Express the circumference C of the circle as a function of the length x of a side of the triangle.

[**Hint:** First show that $r^2 = \dfrac{x^2}{3}$.]

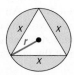

17. Geometry   An equilateral triangle is inscribed in a circle of radius r. See the figure in Problem 16. Express the area A within the circle, but outside the triangle, as a function of the length x of a side of the triangle.

18. Uniform Motion   Two cars leave an intersection at the same time. One is headed south at a constant speed of 30 miles per hour, and the other is headed west at a constant speed of 40 miles per hour (see the figure). Build a model that expresses the distance d between the cars as a function of the time t.

[**Hint:** At $t = 0$, the cars leave the intersection.]

19. Uniform Motion   Two cars are approaching an intersection. One is 2 miles south of the intersection and is moving at a constant speed of 30 miles per hour. At the same time, the other car is 3 miles east of the intersection and is moving at a constant speed of 40 miles per hour.

(a) Build a model that expresses the distance d between the cars as a function of time t.

[**Hint:** At $t = 0$, the cars are 2 miles south and 3 miles east of the intersection, respectively.]

(b) Use a graphing utility to graph $d = d(t)$. For what value of t is d smallest?

20. Inscribing a Cylinder in a Sphere   Inscribe a right circular cylinder of height h and radius r in a sphere of fixed radius R. See the illustration. Express the volume V of the cylinder as a function of h.

[**Hint:** $V = \pi r^2 h$. Note also the right triangle.]

Sphere

21. Inscribing a Cylinder in a Cone   Inscribe a right circular cylinder of height h and radius r in a cone of fixed radius R and fixed height H. See the illustration. Express the volume V of the cylinder as a function of r.

[**Hint:** $V = \pi r^2 h$. Note also the similar triangles.]

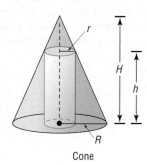

Cone

22. Installing Cable TV MetroMedia Cable is asked to provide service to a customer whose house is located 2 miles from the road along which the cable is buried. The nearest connection box for the cable is located 5 miles down the road. See the figure.

(a) If the installation cost is $500 per mile along the road and $700 per mile off the road, build a model that expresses the total cost C of installation as a function of the distance x (in miles) from the connection box to the point where the cable installation turns off the road. Give the domain.
(b) Compute the cost if $x = 1$ mile.
(c) Compute the cost if $x = 3$ miles.
(d) Graph the function $C = C(x)$. Use TRACE to see how the cost C varies as x changes from 0 to 5.
(e) What value of x results in the least cost?

23. Time Required to Go from an Island to a Town An island is 2 miles from the nearest point P on a straight shoreline. A town is 12 miles down the shore from P. See the illustration.

(a) If a person can row a boat at an average speed of 3 miles per hour and the same person can walk 5 miles per hour, build a model that expresses the time T that it takes to go from the island to town as a function of the distance x from P to where the person lands the boat.
(b) What is the domain of T?

(c) How long will it take to travel from the island to town if the person lands the boat 4 miles from P?
(d) How long will it take if the person lands the boat 8 miles from P?

24. Filling a Conical Tank Water is poured into a container in the shape of a right circular cone with radius 4 feet and height 16 feet. See the figure. Express the volume V of the water in the cone as a function of the height h of the water.

[**Hint:** The volume V of a cone of radius r and height h is $V = \dfrac{1}{3}\pi r^2 h$.]

25. Constructing an Open Box An open box with a square base is to be made from a square piece of cardboard 24 inches on a side by cutting out a square from each corner and turning up the sides. See the figure.

(a) Express the volume V of the box as a function of the length x of the side of the square cut from each corner.
(b) What is the volume if a 3-inch square is cut out?
(c) What is the volume if a 10-inch square is cut out?
(d) Graph $V = V(x)$. For what value of x is V largest?

26. Constructing an Open Box An open box with a square base is required to have a volume of 10 cubic feet.
(a) Express the amount A of material used to make such a box as a function of the length x of a side of the square base.
(b) How much material is required for a base 1 foot by 1 foot?
(c) How much material is required for a base 2 feet by 2 feet?
(d) Use a graphing utility to graph $A = A(x)$. For what value of x is A smallest?

1.7 Building Mathematical Models Using Variation

OBJECTIVES 1 Construct a Model Using Direct Variation (p. 109)
2 Construct a Model Using Inverse Variation (p. 110)
3 Construct a Model Using Joint or Combined Variation (p. 110)

When a mathematical model is developed for a real-world problem, it often involves relationships between quantities that are expressed in terms of proportionality:

Force is proportional to acceleration.

When an ideal gas is held at a constant temperature, pressure and volume are inversely proportional.

The force of attraction between two objects is inversely proportional to the square of the distance between them.

Revenue is directly proportional to sales.

Each of these statements illustrates the idea of **variation,** or how one quantity varies in relation to another quantity. Quantities may vary *directly*, *inversely*, or *jointly*.

1 Construct a Model Using Direct Variation

DEFINITION

Let x and y denote two quantities. Then y **varies directly** with x, or y is **directly proportional to** x, if there is a nonzero number k such that

$$y = kx$$

The number k is called the **constant of proportionality.**

Figure 68
$y = kx; k > 0, x \geq 0$

The graph in Figure 68 illustrates the relationship between y and x if y varies directly with x and $k > 0$, $x \geq 0$. Note that the constant of proportionality is, in fact, the slope of the line.

If we know that two quantities vary directly, then knowing the value of each quantity in one instance enables us to write a model that is true in all cases.

EXAMPLE 1 | **Mortgage Payments**

The monthly payment p on a mortgage varies directly with the amount B borrowed. If the monthly payment on a 30-year mortgage is $6.65 for every $1000 borrowed, find a model that relates the monthly payment p to the amount B borrowed for a mortgage with these terms. Then find the monthly payment p when the amount B borrowed is $120,000.

Solution Because p varies directly with B, we know that

$$p = kB$$

for some constant k. Because $p = 6.65$ when $B = 1000$, it follows that

$$6.65 = k(1000)$$
$$k = 0.00665 \quad \text{Solve for } k.$$

Since $p = kB$, we have

$$p = 0.00665B \quad \text{The model}$$

In particular, when $B = \$120,000$, we find that

$$p = 0.00665(\$120,000) = \$798$$

Figure 69 illustrates the relationship between the monthly payment p and the amount B borrowed.

Figure 69

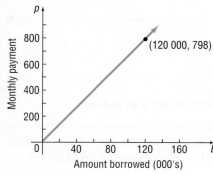

(120 000, 798)

Amount borrowed (000's)

Now Work PROBLEMS 3 AND 21

2 Construct a Model Using Inverse Variation

DEFINITION

Figure 70

$y = \dfrac{k}{x}; k > 0, x > 0$

Let x and y denote two quantities. Then y **varies inversely** with x, or y is **inversely proportional to** x, if there is a nonzero constant k such that

$$y = \frac{k}{x}$$

The graph in Figure 70 illustrates the relationship between y and x if y varies inversely with x and $k > 0$, $x > 0$.

EXAMPLE 2

Maximum Weight That Can Be Supported by a Piece of Pine

Figure 71

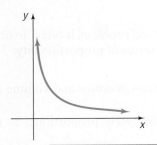

See Figure 71. The maximum weight W that can be safely supported by a 2-inch by 4-inch piece of pine varies inversely with its length l. Experiments indicate that the maximum weight that a 10-foot-long 2-by-4 piece of pine can support is 500 pounds. Write a model relating the maximum weight W (in pounds) to length l (in feet). Find the maximum weight W that can be safely supported by a length of 25 feet.

Solution Because W varies inversely with l, we know that

$$W = \frac{k}{l}$$

for some constant k. Because $W = 500$ when $l = 10$, we have

$$500 = \frac{k}{10}$$
$$k = 5000$$

Figure 72

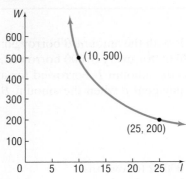

Since $W = \dfrac{k}{l}$, we have

$$W = \frac{5000}{l} \quad \text{The Model}$$

In particular, the maximum weight W that can be safely supported by a piece of pine 25 feet in length is

$$W = \frac{5000}{25} = 200 \text{ pounds}$$

Figure 72 illustrates the relationship between the weight W and the length l.

◆

Now Work PROBLEM 31

3 Construct a Model Using Joint or Combined Variation

When a variable quantity Q is proportional to the product of two or more other variables, we say that Q **varies jointly** with these quantities. Combinations of direct and/or inverse variation may also occur. This is usually referred to as **combined variation.**

Let's look at an example.

EXAMPLE 3

Loss of Heat Through a Wall

The loss of heat through a wall varies jointly with the area of the wall and the difference between the inside and outside temperatures and varies inversely with the thickness of the wall. Write an equation that relates these quantities.

Solution

We begin by assigning symbols to represent the quantities:

$$L = \text{Heat loss} \qquad T = \text{Temperature difference}$$
$$A = \text{Area of wall} \qquad d = \text{Thickness of wall}$$

Then

$$L = k\frac{AT}{d}$$

where k is the constant of proportionality.

In direct or inverse variation, the quantities that vary may be raised to powers. For example, in the early seventeenth century, Johannes Kepler (1571–1630) discovered that the square of the period of revolution T of a planet around the Sun varies directly with the cube of its mean distance a from the Sun. That is, $T^2 = ka^3$, where k is the constant of proportionality.

EXAMPLE 4

Force of the Wind on a Window

The force F of the wind on a flat surface positioned at a right angle to the direction of the wind varies jointly with the area A of the surface and the square of the speed v of the wind. A wind of 30 miles per hour blowing on a window measuring 4 feet by 5 feet has a force of 150 pounds. See Figure 73. What is the force on a window measuring 3 feet by 4 feet caused by a wind of 50 miles per hour?

Solution

Since F varies jointly with A and v^2, we have

$$F = kAv^2$$

Figure 73

Wind

where k is the constant of proportionality. We are told that $F = 150$ pounds when $A = 4 \cdot 5 = 20$ feet^2 and $v = 30$ miles per hour. Then we have

$$150 = k(20)(900) \qquad F = kAv^2, F = 150, A = 20, v = 30$$
$$k = \frac{1}{120}$$

Since $F = kAv^2$, we have

$$F = \frac{1}{120}Av^2$$

For a wind of 50 miles per hour blowing on a window whose area is $A = 3 \cdot 4 = 12$ square feet, the force F is

$$F = \frac{1}{120}(12)(2500) = 250 \text{ pounds}$$

Now Work PROBLEM 39

1.7 Assess Your Understanding

Concepts and Vocabulary

1. If x and y are two quantities, then y is directly proportional to x if there is a nonzero number k such that _____ .

2. *True or False* If y varies directly with x, then $y = \dfrac{k}{x}$, where k is a constant.

Skill Building

In Problems 3–14, write a general formula to describe each variation.

3. y varies directly with x; $\quad y = 2$ when $x = 10$

4. v varies directly with t; $\quad v = 16$ when $t = 2$

5. A varies directly with x^2; $\quad A = 4\pi$ when $x = 2$

6. V varies directly with x^3; $\quad V = 36\pi$ when $x = 3$

7. F varies inversely with d^2; $\quad F = 10$ when $d = 5$

8. y varies inversely with $\sqrt{x}$; $\quad y = 4$ when $x = 9$

9. z varies directly with the sum of the squares of x and y; $\quad z = 5$ when $x = 3$ and $y = 4$

10. T varies jointly with the cube root of x and the square of d; $\quad T = 18$ when $x = 8$ and $d = 3$

11. M varies directly with the square of d and inversely with the square root of x; $\quad M = 24$ when $x = 9$ and $d = 4$

12. z varies directly with the sum of the cube of x and the square of y; $\quad z = 1$ when $x = 2$ and $y = 3$

13. The square of T varies directly with the cube of a and inversely with the square of d; $\quad T = 2$ when $a = 2$ and $d = 4$

14. The cube of z varies directly with the sum of the squares of x and y; $\quad z = 2$ when $x = 9$ and $y = 4$

Applications and Extensions

In Problems 15–20, write an equation that relates the quantities.

15. Geometry The volume V of a sphere varies directly with the cube of its radius r. The constant of proportionality is $\dfrac{4\pi}{3}$.

16. Geometry The square of the length of the hypotenuse c of a right triangle varies jointly with the sum of the squares of the lengths of its legs a and b. The constant of proportionality is 1.

17. Geometry The area A of a triangle varies jointly with the lengths of the base b and the height h. The constant of proportionality is $\dfrac{1}{2}$.

18. Geometry The perimeter p of a rectangle varies jointly with the sum of the lengths of its sides l and w. The constant of proportionality is 2.

19. Physics: Newton's Law The force F (in newtons) of attraction between two objects varies jointly with their masses m and M (in kilograms) and inversely with the square of the distance d (in meters) between them. The constant of proportionality is $G = 6.67 \times 10^{-11}$.

20. Physics: Simple Pendulum The **period** of a pendulum is the time required for one oscillation; the pendulum is usually referred to as **simple** when the angle made to the vertical is less than $5°$. The period T of a simple pendulum (in seconds) varies directly with the square root of its length l (in feet). The constant of proportionality is $\dfrac{2\pi}{\sqrt{32}}$.

21. Mortgage Payments The monthly payment p on a mortgage varies directly with the amount B borrowed. If the monthly payment on a 30-year mortgage is $6.49 for every $1000 borrowed, find a model that relates the monthly payment p to the amount B borrowed for a mortgage with the same terms. Then find the monthly payment p when the amount B borrowed is $145,000.

22. Mortgage Payments The monthly payment p on a mortgage varies directly with the amount B borrowed. If the monthly payment on a 15-year mortgage is $8.99 for every $1000 borrowed, find a model that relates the monthly payment p to the amount B borrowed for a mortgage with the same terms. Then find the monthly payment p when the amount B borrowed is $175,000.

23. Physics: Falling Objects The distance s that an object falls is directly proportional to the square of the time t of the fall. If an object falls 16 feet in 1 second, how far will it fall in 3 seconds? How long will it take an object to fall 64 feet?

24. Physics: Falling Objects The velocity v of a falling object is directly proportional to the time t of the fall. If, after 2 seconds, the velocity of the object is 64 feet per second, what will its velocity be after 3 seconds?

25. **Physics: Stretching a Spring** The elongation E of a spring balance varies directly with the applied weight W (see the figure). If $E = 3$ when $W = 20$, find E when $W = 15$.

26. **Physics: Vibrating String** The rate of vibration of a string under constant tension varies inversely with the length of the string. If a string is 48 inches long and vibrates 256 times per second, what is the length of a string that vibrates 576 times per second?

27. **Revenue Equation** At the corner Shell station, the revenue R varies directly with the number g of gallons of gasoline sold. If the revenue is $47.40 when the number of gallons sold is 12, find a model that relates revenue R to the number g of gallons of gasoline. Then find the revenue R when the number of gallons of gasoline sold is 10.5.

28. **Cost Equation** The cost C of chocolate-covered almonds varies directly with the number A of pounds of almonds purchased. If the cost is $23.75 when the number of pounds of chocolate-covered almonds purchased is 5, find a model that relates the cost C to the number A of pounds of almonds purchased. Then find the cost C when the number of pounds of almonds purchased is 3.5.

29. **Demand** Suppose that the demand D for candy at the movie theater is inversely related to the price p.
    (a) When the price of candy is $2.75 per bag, the theater sells 156 bags of candy. Express the demand for candy in terms of its price.
    (b) Determine the number of bags of candy that will be sold if the price is raised to $3 a bag.

30. **Driving to School** The time t that it takes to get to school varies inversely with your average speed s.
    (a) Suppose that it takes you 40 minutes to get to school when your average speed is 30 miles per hour. Express the driving time to school in terms of average speed.
    (b) Suppose that your average speed to school is 40 miles per hour. How long will it take you to get to school?

31. **Pressure** The volume V of a gas held at a constant temperature in a closed container varies inversely with its pressure P. If the volume of a gas is 600 cubic centimeters (cm^3) when the pressure is 150 millimeters of mercury (mm Hg), find the volume when the pressure is 200 mm Hg.

32. **Resistance** The current i in a circuit is inversely proportional to its resistance Z measured in ohms. Suppose that when the current in a circuit is 30 amperes the resistance is 8 ohms. Find the current in the same circuit when the resistance is 10 ohms.

33. **Weight** The weight of an object above the surface of Earth varies inversely with the square of the distance from the center of Earth. If Maria weighs 125 pounds when she is on the surface of Earth (3960 miles from the center), determine

Maria's weight if she is at the top of Mount McKinley (3.8 miles from the surface of Earth).

34. **Intensity of Light** The intensity I of light (measured in foot-candles) varies inversely with the square of the distance from the bulb. Suppose that the intensity of a 100-watt light bulb at a distance of 2 meters is 0.075 foot-candle. Determine the intensity of the bulb at a distance of 5 meters.

35. **Geometry** The volume V of a right circular cylinder varies jointly with the square of its radius r and its height h. The constant of proportionality is π. See the figure. Write an equation for V.

36. **Geometry** The volume V of a right circular cone varies jointly with the square of its radius r and its height h. The constant of proportionality is $\dfrac{\pi}{3}$. See the figure. Write an equation for V.

37. **Weight of a Body** The weight of a body above the surface of Earth varies inversely with the square of the distance from the center of Earth. If a certain body weighs 55 pounds when it is 3960 miles from the center of Earth, how much will it weigh when it is 3965 miles from the center?

38. **Force of the Wind on a Window** The force exerted by the wind on a plane surface varies jointly with the area of the surface and the square of the velocity of the wind. If the force on an area of 20 square feet is 11 pounds when the wind velocity is 22 miles per hour, find the force on a surface area of 47.125 square feet when the wind velocity is 36.5 miles per hour.

39. **Horsepower** The horsepower (hp) that a shaft can safely transmit varies jointly with its speed (in revolutions per minute, rpm) and the cube of its diameter. If a shaft of a certain material 2 inches in diameter can transmit 36 hp at 75 rpm, what diameter must the shaft have in order to transmit 45 hp at 125 rpm?

40. **Chemistry: Gas Laws** The volume V of an ideal gas varies directly with the temperature T and inversely with the pressure P. Write a model relating V, T, and P using k as the constant of proportionality. If a cylinder contains oxygen at a temperature of 300 K and a pressure of 15 atmospheres in a volume of 100 liters, what is the constant of proportionality k? If a piston is lowered into the cylinder, decreasing the volume occupied by the gas to 80 liters and raising the temperature to 310 K, what is the gas pressure?

41. **Physics: Kinetic Energy**   The kinetic energy K of a moving object varies jointly with its mass m and the square of its velocity v. If an object weighing 25 kilograms and moving with a velocity of 10 meters per second has a kinetic energy of 1250 joules, find its kinetic energy when the velocity is 15 meters per second.

42. **Electrical Resistance of a Wire**   The electrical resistance of a wire varies directly with the length of the wire and inversely with the square of the diameter of the wire. If a wire 432 feet long and 4 millimeters in diameter has a resistance of 1.24 ohms, find the length of a wire of the same material whose resistance is 1.44 ohms and whose diameter is 3 millimeters.

43. **Measuring the Stress of Materials**   The stress in the material of a pipe subject to internal pressure varies jointly with the internal pressure and the internal diameter of the pipe and inversely with the thickness of the pipe. The stress is 100 pounds per square inch when the diameter is 5 inches, the thickness is 0.75 inch, and the internal pressure is 25 pounds per square inch. Find the stress when the internal pressure is 40 pounds per square inch if the diameter is 8 inches and the thickness is 0.50 inch.

44. **Safe Load for a Beam**   The maximum safe load for a horizontal rectangular beam varies jointly with the width of the beam and the square of the thickness of the beam and inversely with its length. If an 8-foot beam will support up to 750 pounds when the beam is 4 inches wide and 2 inches thick, what is the maximum safe load in a similar beam 10 feet long, 6 inches wide, and 2 inches thick?

Explaining Concepts: Discussion and Writing

45. In the early 17th century, Johannes Kepler discovered that the square of the period T for a planet to orbit the Sun varies directly with the cube of its mean distance a from the Sun. Go to the library and research this law and Kepler's other two laws. Write a brief paper about these laws and Kepler's place in history.

46. Using a situation that has not been discussed in the text, write a real-world problem that you think involves two variables that vary directly. Exchange your problem with another student to solve and critique.

47. Using a situation that has not been discussed in the text, write a real-world problem that you think involves two variables that vary inversely. Exchange your problem with another student to solve and critique.

48. Using a situation that has not been discussed in the text, write a real-world problem that you think involves three variables that vary jointly. Exchange your problem with another student to solve and critique.

CHAPTER REVIEW

Library of Functions

Constant function (p. 82)

$$f(x) = b$$

The graph is a horizontal line with y-intercept b.

Identity function (p. 82)

$$f(x) = x$$

The graph is a line with slope 1 and y-intercept 0.

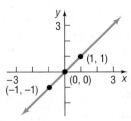

Square function (p. 82)

$$f(x) = x^2$$

The graph is a parabola with intercept at $(0, 0)$.

Cube function (p. 83)

$$f(x) = x^3$$

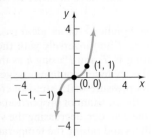

Square root function (p. 83)

$$f(x) = \sqrt{x}$$

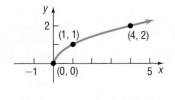

Cube root function (p. 83)

$$f(x) = \sqrt[3]{x}$$

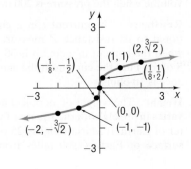

Reciprocal function (p. 83)

$$f(x) = \frac{1}{x}$$

Absolute value function (p. 83)

$$f(x) = |x|$$

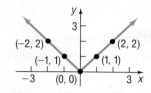

Greatest integer function (p. 84)

$$f(x) = \text{int}(x)$$

Things to Know

Function (pp. 43–46)

A relation between two sets so that each element x in the first set, the domain, has corresponding to it exactly one element y in the second set. The range is the set of y values of the function for the x values in the domain.

A function can also be characterized as a set of ordered pairs (x, y) in which no first element is paired with two different second elements.

Function notation (pp. 46–49)

$y = f(x)$

f is a symbol for the function.

x is the argument, or independent variable.

y is the dependent variable.

$f(x)$ is the value of the function at x, or the image of x.

A function f may be defined implicitly by an equation involving x and y or explicitly by writing $y = f(x)$.

Difference quotient of f (pp. 48 and 79)

$$\frac{f(x + h) - f(x)}{h} \quad h \neq 0$$

Domain (pp. 44 and 49)

If unspecified, the domain of a function f is the largest set of real numbers for which $f(x)$ is a real number.

Vertical-line test (p. 58)

A set of points in the plane is the graph of a function if and only if every vertical line intersects the graph in at most one point.

Even function f (p. 67)

$f(-x) = f(x)$ for every x in the domain ($-x$ must also be in the domain).

Odd function f (p. 67)

$f(-x) = -f(x)$ for every x in the domain ($-x$ must also be in the domain).

Increasing function (p. 69)

A function f is increasing on an open interval I if, for any choice of x_1 and x_2 in I, with $x_1 < x_2$, we have $f(x_1) < f(x_2)$.

Decreasing function (p. 69)

A function f is decreasing on an open interval I if, for any choice of x_1 and x_2 in I, with $x_1 < x_2$, we have $f(x_1) > f(x_2)$.

Constant function (p. 69)

A function f is constant on an open interval I if, for all choices of x in I, the values of $f(x)$ are equal.

Local maximum (p. 70)

A function f has a local maximum at c if there is an open interval I containing c so that, for all x not equal to c in I, $f(x) < f(c)$.

Local minimum (p. 70)

A function f has a local minimum at c if there is an open interval I containing c so that, for all x not equal to c in I, $f(x) > f(c)$.

Absolute maximum and Absolute minimum (p. 71)

Let f denote a function defined on some interval I.

If there is a number u in I for which $f(x) \leq f(u)$ for all x in I, then $f(u)$ is the absolute maximum of f on I and we say the absolute maximum of f occurs at u.

If there is a number v in I for which $f(x) \geq f(v)$, for all x in I, then $f(v)$ is the absolute minimum of f on I and we say the absolute minimum of f occurs at v.

Average rate of change of a function (p. 73)

The average rate of change of f from a to b is

$$\frac{\Delta y}{\Delta x} = \frac{f(b) - f(a)}{b - a} \quad a \neq b$$

Direct variation (p. 109)

$y = kx, \ k \neq 0$

Inverse variation (p. 110)

$y = \dfrac{k}{x}, \ k \neq 0$

Objectives

Review Exercises

In Problems 1 and 2, determine whether each relation represents a function. For each function, state the domain and the range.

1. $\{(-1, 0), (2, 3), (4, 0)\}$

2. $\{(4, -1), (2, 1), (4, 2)\}$

In Problems 3–8, find the following for each function:

  (a) $f(2)$ *(b)* $f(-2)$ *(c)* $f(-x)$ *(d)* $-f(x)$ *(e)* $f(x - 2)$ *(f)* $f(2x)$

3. $f(x) = \dfrac{3x}{x^2 - 1}$

4. $f(x) = \dfrac{x^2}{x + 1}$

5. $f(x) = \sqrt{x^2 - 4}$

6. $f(x) = |x^2 - 4|$

7. $f(x) = \dfrac{x^2 - 4}{x^2}$

8. $f(x) = \dfrac{x^3}{x^2 - 9}$

In Problems 9–16, find the domain of each function.

9. $f(x) = \dfrac{x}{x^2 - 9}$

10. $f(x) = \dfrac{3x^2}{x - 2}$

11. $f(x) = \sqrt{2 - x}$

12. $f(x) = \sqrt{x + 2}$

13. $h(x) = \dfrac{\sqrt{x}}{|x|}$

14. $g(x) = \dfrac{|x|}{x}$

15. $f(x) = \dfrac{x}{x^2 + 2x - 3}$

16. $F(x) = \dfrac{1}{x^2 - 3x - 4}$

In Problems 17–22, find $f + g, f - g, f \cdot g,$ and $\dfrac{f}{g}$ for each pair of functions. State the domain of each.

17. $f(x) = 2 - x; \quad g(x) = 3x + 1$

18. $f(x) = 2x - 1; \quad g(x) = 2x + 1$

19. $f(x) = 3x^2 + x + 1; \quad g(x) = 3x$

20. $f(x) = 3x; \quad g(x) = 1 + x + x^2$

21. $f(x) = \dfrac{x + 1}{x - 1}; \quad g(x) = \dfrac{1}{x}$

22. $f(x) = \dfrac{1}{x - 3}; \quad g(x) = \dfrac{3}{x}$

In Problems 23 and 24, find the difference quotient of each function f; that is, find

$$\frac{f(x + h) - f(x)}{h} \qquad h \neq 0$$

23. $f(x) = -2x^2 + x + 1$

24. $f(x) = 3x^2 - 2x + 4$

25. Using the graph of the function f shown:

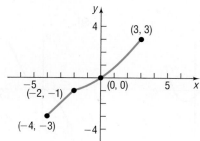

(a) Find the domain and the range of f.
(b) List the intercepts.
(c) Find $f(-2)$.
(d) For what value of x does $f(x) = -3$?
(e) Solve $f(x) > 0$.
(f) Graph $y = f(x - 3)$.
(g) Graph $y = f\left(\dfrac{1}{2}x\right)$.
(h) Graph $y = -f(x)$.

26. Using the graph of the function g shown:

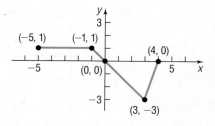

(a) Find the domain and the range of g.
(b) Find $g(-1)$.
(c) List the intercepts.
(d) For what value of x does $g(x) = -3$?
(e) Solve $g(x) > 0$.
(f) Graph $y = g(x - 2)$.
(g) Graph $y = g(x) + 1$.
(h) Graph $y = 2g(x)$.

In Problems 27 and 28, use the graph of the function f to find:
(a) The domain and the range of f.
(b) The intervals on which f is increasing, decreasing, or constant.
(c) The local minima and local maxima.
(d) The absolute maximum and absolute minimum.
(e) Whether the graph is symmetric with respect to the x-axis, the y-axis, or the origin.
(f) Whether the function is even, odd, or neither.
(g) The intercepts, if any.

27.

28.

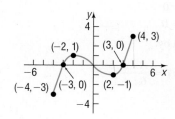

In Problems 29–36, determine (algebraically) whether the given function is even, odd, or neither.

29. $f(x) = x^3 - 4x$

30. $g(x) = \dfrac{4 + x^2}{1 + x^4}$

31. $h(x) = \dfrac{1}{x^4} + \dfrac{1}{x^2} + 1$

32. $F(x) = \sqrt{1 - x^3}$

33. $G(x) = 1 - x + x^3$

34. $H(x) = 1 + x + x^2$

35. $f(x) = \dfrac{x}{1 + x^2}$

36. $g(x) = \dfrac{1 + x^2}{x^3}$

In Problems 37–40, use a graphing utility to graph each function over the indicated interval. Approximate any local maxima and local minima. Determine where the function is increasing and where it is decreasing.

37. $f(x) = 2x^3 - 5x + 1$ $(-3, 3)$

38. $f(x) = -x^3 + 3x - 5$ $(-3, 3)$

39. $f(x) = 2x^4 - 5x^3 + 2x + 1$ $(-2, 3)$

40. $f(x) = -x^4 + 3x^3 - 4x + 3$ $(-2, 3)$

In Problems 41 and 42, find the average rate of change of f:
(a) From 1 to 2   *(b) From 0 to 1*   *(c) From 2 to 4*

41. $f(x) = 8x^2 - x$

42. $f(x) = 2x^3 + x$

In Problems 43–46, find the average rate of change from 2 to 3 for each function f. Be sure to simplify.

43. $f(x) = 2 - 5x$ **44.** $f(x) = 2x^2 + 7$ **45.** $f(x) = 3x - 4x^2$ **46.** $f(x) = x^2 - 3x + 2$

In Problems 47–50, is the graph shown the graph of a function?

47.

48.

49.

50.

In Problems 51–54, sketch the graph of each function. Be sure to label at least three points.

51. $f(x) = |x|$ **52.** $f(x) = \sqrt[3]{x}$ **53.** $f(x) = \sqrt{x}$ **54.** $f(x) = \dfrac{1}{x}$

In Problems 55–66, graph each function using the techniques of shifting, compressing or stretching, and reflections. Identify any intercepts on the graph. State the domain and, based on the graph, find the range.

55. $F(x) = |x| - 4$ **56.** $f(x) = |x| + 4$ **57.** $g(x) = -2|x|$ **58.** $g(x) = \dfrac{1}{2}|x|$

59. $h(x) = \sqrt{x - 1}$ **60.** $h(x) = \sqrt{x} - 1$ **61.** $f(x) = \sqrt{1 - x}$ **62.** $f(x) = -\sqrt{x + 3}$

63. $h(x) = (x - 1)^2 + 2$ **64.** $h(x) = (x + 2)^2 - 3$ **65.** $g(x) = 3(x - 1)^3 + 1$ **66.** $g(x) = -2(x + 2)^3 - 8$

In Problems 67–70,
(a) Find the domain of each function.   *(c) Graph each function.*   *(e) Is f continuous on its domain?*
(b) Locate any intercepts.   *(d) Based on the graph, find the range.*

67. $f(x) = \begin{cases} 3x & \text{if } -2 < x \le 1 \\ x + 1 & \text{if } x > 1 \end{cases}$

68. $f(x) = \begin{cases} x - 1 & \text{if } -3 < x < 0 \\ 3x - 1 & \text{if } x \ge 0 \end{cases}$

69. $f(x) = \begin{cases} x & \text{if } -4 \le x < 0 \\ 1 & \text{if } x = 0 \\ 3x & \text{if } x > 0 \end{cases}$

70. $f(x) = \begin{cases} x^2 & \text{if } -2 \le x \le 2 \\ 2x - 1 & \text{if } x > 2 \end{cases}$

71. A function f is defined by

$$f(x) = \frac{Ax + 5}{6x - 2}$$

If $f(1) = 4$, find A.

72. A function g is defined by

$$g(x) = \frac{A}{x} + \frac{8}{x^2}$$

If $g(-1) = 0$, find A.

73. Page Design   A page with dimensions of $8\dfrac{1}{2}$ inches by 11 inches has a border of uniform width x surrounding the printed matter of the page, as shown in the figure.
(a) Develop a model that expresses the area A of the printed part of the page as a function of the width x of the border.
(b) Give the domain and the range of A.
(c) Find the area of the printed page for borders of widths 1 inch, 1.2 inches, and 1.5 inches.
(d) Graph the function $A = A(x)$.

74. Constructing a Closed Box   A closed box with a square base is required to have a volume of 10 cubic feet.
   (a) Build a model that expresses the amount A of material used to make such a box as a function of the length x of a side of the square base.
   (b) How much material is required for a base 1 foot by 1 foot?
   (c) How much material is required for a base 2 feet by 2 feet?
   (d) Graph $A = A(x)$. For what value of x is A smallest?

75. A rectangle has one vertex in quadrant I on the graph of $y = 10 - x^2$, another at the origin, one on the positive x-axis, and one on the positive y-axis.
   (a) Express the area A of the rectangle as a function of x.
   (b) Find the largest area A that can be enclosed by the rectangle.

76. Mortgage Payments   The monthly payment p on a mortgage varies directly with the amount B borrowed. If the monthly payment on a 30-year mortgage is $854.00 when $130,000 is borrowed, find a model that relates the monthly payment p to the amount B borrowed for a mortgage with the same terms. Then find the monthly payment p when the amount borrowed is $165,000

77. Revenue Function   At the corner Esso station, the revenue R varies directly with the number g of gallons of gasoline sold. If the revenue is $46.67 when the number of gallons sold is 13, find a model that relates revenue R to the number g of gallons of gasoline. Then find the revenue R when the number of gallons of gasoline sold is 11.2.

78. Weight of a Body   The weight of a body varies inversely with the square of its distance from the center of Earth. Assuming that the radius of Earth is 3960 miles, how much would a man weigh at an altitude of 1 mile above Earth's surface if he weighs 200 pounds on Earth's surface?

79. Kepler's Third Law of Planetary Motion   Kepler's Third Law of Planetary Motion states that the square of the period of revolution T of a planet varies directly with the cube of its mean distance a from the Sun. If the mean distance of Earth from the Sun is 93 million miles, what is the mean distance of the planet Mercury from the Sun, given that Mercury has a "year" of 88 days?

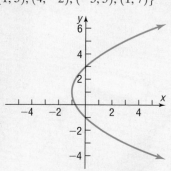

CHAPTER TEST

CHAPTER **Test Prep** VIDEOS

Step-by-step test solutions are found on the Chapter Test Prep Videos available via the Video Resources on DVD, in *MyMathLab* or on YouTube (search "Sullivan College Algebra, Concepts Through Functions 2e).

1. Determine whether each relation represents a function. For each function, state the domain and the range.
   (a) $\{(2, 5), (4, 6), (6, 7), (8, 8)\}$
   (b) $\{(1, 3), (4, -2), (-3, 5), (1, 7)\}$
   (c)

(d)

In Problems 2–4, find the domain of each function and evaluate each function at $x = -1$.

2. $f(x) = \sqrt{4 - 5x}$

3. $g(x) = \dfrac{x + 2}{|x + 2|}$

4. $h(x) = \dfrac{x - 4}{x^2 + 5x - 36}$

5. Using the graph of the function f:

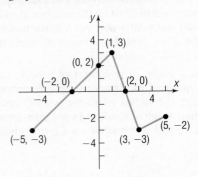

(a) Find the domain and the range of f.
(b) List the intercepts.
(c) Find $f(1)$.
(d) For what value(s) of x does $f(x) = -3$?
(e) Solve $f(x) < 0$.

6. Use a graphing utility to graph the function $f(x) = -x^4 + 2x^3 + 4x^2 - 2$ on the interval $(-5, 5)$. Approximate any local maxima and local minima rounded to two decimal places. Determine where the function is increasing and where it is decreasing.

7. Consider the function $g(x) = \begin{cases} 2x + 1 & \text{if } x < -1 \\ x - 4 & \text{if } x \geq -1 \end{cases}$

(a) Graph the function.
(b) List the intercepts.
(c) Find $g(-5)$.
(d) Find $g(2)$.

8. For the function $f(x) = 3x^2 - 2x + 4$, find the average rate of change of f from 3 to 4.

9. For the functions $f(x) = 2x^2 + 1$ and $g(x) = 3x - 2$, find the following and simplify:

(a) $f - g$
(b) $f \cdot g$
(c) $f(x + h) - f(x)$

10. Graph each function using the techniques of shifting, compressing or stretching, and reflections. Start with the graph of the basic function and show all stages.

(a) $h(x) = -2(x + 1)^3 + 3$
(b) $g(x) = |x + 4| + 2$

11. The variable interest rate on a student loan changes each July 1 based on the bank prime loan rate. For the years 1992–2007, this rate can be approximated by the model $r(x) = -0.115x^2 + 1.183x + 5.623$, where x is the number of years since 1992 and r is the interest rate as a percent.

(a) Use a graphing utility to estimate the highest rate during this time period. During which year was the interest rate the highest?
(b) Use the model to estimate the rate in 2010. Does this value seem reasonable?

Source: U.S. Federal Reserve

12. A community skating rink is in the shape of a rectangle with semicircles attached at the ends. The length of the rectangle is 20 feet less than twice the width. The thickness of the ice is 0.75 inch.

(a) Build a model that expresses the ice volume, V, as a function of the width, x.
(b) How much ice is in the rink if the width is 90 feet?

13. The resistance (in ohms) of a circular conductor varies directly with the length of the conductor and inversely with the square of the radius of the conductor. If 50 feet of wire with a radius of 6×10^{-3} inch has a resistance of 10 ohms, what would be the resistance of 100 feet of the same wire if the radius is increased to 7×10^{-3} inch?

CHAPTER PROJECTS

I. Cell Phone Service This month, as you were paying your bills, you noticed that your cell phone service contract had expired. Since cell phone numbers are now portable, you decided to investigate different companies. Since you regularly travel outside your local calling area, you decide to look only at plans with no roaming charges. Here are your findings:

	Anytime Minutes Included	Charge for Each Extra Minute	Mobile-to-Mobile Minutes	National Long Distance	Nights and Weekends
Company A:					
Plan A1: $39.99	450 with rollover	$0.45	Unlimited	Included	5000 (9 PM–6 AM)
Plan A2: $59.99	900 with rollover	$0.40	Unlimited	Included	Unlimited (9 PM–6 AM)
Company B:					
Plan B1: $39.99	600	$0.40	None	Included	Unlimited (9 PM–7 AM)
Plan B2: $49.99	1000	$0.40	None	Included	Unlimited (9 PM–7 AM)
Company C:					
Plan C1: $59.99	550	551–1050, $5 for each 50 minutes; above 1050, $0.10 each minute	Unlimited	Included	Unlimited (7 PM–7 AM)
Plan C2: $69.99	800	801–1300, $5 for each 50 minutes; above 1300, $0.10 each minute	Unlimited	Included	Unlimited (7 PM–7 AM)

Source: Based on rates from the Web sites of the companies Cingular, T Mobile, and Sprint PCS for the zip code 76201 on May 20, 2006. (*www.att.com, www.t-mobile.com, www.sprintpcs.com*)

Each plan requires a 2-year contract.

1. Determine the total cost of each plan for the life of the contract, assuming that you stay within the allotted anytime minutes provided by each contract.

2. If you expect to use 400 anytime, 200 mobile-to-mobile, and 4500 night and weekend minutes per month, which plan provides the best deal? If you expect to use 400 anytime, 200 mobile-to-mobile, and 5500 night and weekend minutes per month, which plan provides the best deal? If you expect to use 500 anytime, and 1000 mobile-to-mobile, and 2000 night and weekend minutes per month, which plan provides the best deal?

3. Ignoring any mobile-to-mobile and night and weekend usage, if you expect to use 850 anytime minutes each month, which option provides the best deal? What if you use 1050 anytime minutes per month?

4. Each monthly charge includes a specific number of anytime minutes in the monthly fee. Write a function for each option, where C is the monthly cost and x is the number of anytime minutes used.

5. Graph each function from part 4.

6. For each of the companies A, B, and C, determine the average price per minute for each plan, based on no extra minutes used. For each company, which plan is better?

7. Looking at the three plans that you found to be the best for companies A, B, and C in part 6, which of these three seems to be the best deal?

8. Based on your own cell phone usage, which plan would be the best for you?

The following projects are available on the Instructor's Resource Center (IRC):

II. Project at Motorola: *Wireless Internet Service* Use functions and their graphs to analyze the total cost of various wireless Internet service plans.

III. Cost of Cable When government regulations and customer preference influence the path of a new cable line, the Pythagorean Theorem can be used to assess the cost of installation.

IV. Oil Spill Functions are used to analyze the size and spread of an oil spill from a leaking tanker.

2 Linear and Quadratic Functions

The Use of Statistics in Making Investment Decisions: Using Beta

One of the most common statistics used by investors is beta, which measures a security or portfolio's movement relative to the S&P 500. When determining beta graphically, returns are plotted on a graph, with the horizontal axis representing the S&P returns and the vertical axis representing the security or portfolio returns; the points that are plotted represent the return of the security versus the return of the S&P 500 at the same time period. A regression line is then drawn, which is the straight line that most closely approximates the points in the graph. Beta represents the slope of that line; a steep slope (45 degrees or greater) indicates that when the stock market moved up or down the security or portfolio moved up or down on average to the same degree or greater, and a flatter slope indicates that when the stock market moved up or down the security or portfolio moved to a lesser degree. Since beta measures variability, it is used as a measure of risk; the greater the beta is, the greater the risk.

Source: Paul E. Hoffman, Journal of the American Association of Individual Investors, September, 1991; www.aaii.com

—See Chapter Project I—

Outline

◁ A Look Back Up to now, our discussion has focused on graphs of equations and the idea of a function. We have learned how to graph equations using the point-plotting method, intercepts, and the tests for symmetry. In addition, we have learned what a function is and how to identify whether a relation represents a function. We also introduced properties of functions, such as domain/range, increasing/decreasing, even/odd, and zeros.

A Look Ahead ▷ Going forward, we will look at specific types of functions. In this chapter, we focus on linear and quadratic functions, their properties, and applications.

2.1 Properties of Linear Functions and Linear Models

PREPARING FOR THIS SECTION *Before getting started, review the following:*

- Lines (Foundations, Section 3, pp. 19–27)
- Solve Linear Equations (Appendix A, Section A.8, pp. A66–68)
- Interval Notation; Solving Inequalities (Appendix A, Section A.10, pp. A83–A87)
- Functions (Section 1.1, pp. 46–51)
- The Graph of a Function (Section 1.2, pp. 57–61)
- Properties of Functions (Section 1.3, pp. 67–75)

Now Work the 'Are You Prepared?' problems on page 130.

OBJECTIVES 1 Graph Linear Functions (p. 123)
2 Use Average Rate of Change to Identify Linear Functions (p. 123)
3 Determine Whether a Linear Function Is Increasing, Decreasing, or Constant (p. 126)
4 Find the Zero of a Linear Function (p. 127)
5 Build Linear Models from Verbal Descriptions (p. 127)

1 Graph Linear Functions

In Foundations, Section 3, we discussed lines. In particular, we developed the slope–intercept form of the equation of a line $y = mx + b$. When we write the slope–intercept form of a line using function notation, we have a *linear function*.

DEFINITION

A **linear function** is a function of the form

$$f(x) = mx + b$$

The graph of a linear function is a line with slope m and y-intercept b. Its domain is the set of all real numbers. If $m \neq 0$, the range of a linear function is the set of all real numbers. If $m = 0$, the range is $\{y \,|\, y = b\}$.

EXAMPLE 1

Graphing a Linear Function

Graph the linear function $f(x) = -3x + 7$. What is the domain and the range of f?

Solution

This is a linear function with slope $m = -3$ and y-intercept $b = 7$. To graph this function, we start by plotting the point $(0, 7)$ (the y-intercept is 7), and use the slope to find an additional point by moving right 1 unit and down 3 units. See Figure 1. The domain and the range of f is the set of all real numbers.

Alternatively, we could have found an additional point by evaluating the function at some $x \neq 0$. For $x = 1$, we find $f(1) = -3(1) + 7 = 4$ and obtain the point $(1, 4)$ on the graph.

Figure 1

Now Work PROBLEMS 13(a), (b), AND (c)

2 Use Average Rate of Change to Identify Linear Functions

Look at Table 1, which shows certain values of the independent variable x and corresponding values of the dependent variable y for the function $y = f(x) = -3x + 7$. Notice that, as the value of the independent variable x increases by 1, the value of the dependent variable y decreases by 3. That is, the average rate of change of y with respect to x is a constant, -3.

Table 1

x	$y = f(x) = -3x + 7$	Average Rate of Change $= \dfrac{\Delta y}{\Delta x}$
-2	13	
		$\dfrac{10 - 13}{-1 - (-2)} = \dfrac{-3}{1} = -3$
-1	10	
		$\dfrac{7 - 10}{0 - (-1)} = \dfrac{-3}{1} = -3$
0	7	
		-3
1	4	
		-3
2	1	
		-3
3	-2	

It is not a coincidence that the average rate of change of the linear function $f(x) = -3x + 7$ is the slope of the linear function. That is, $\dfrac{\Delta y}{\Delta x} = m = -3$. The following theorem states this fact.

THEOREM

Average Rate of Change of a Linear Function

Linear functions have a constant average rate of change. That is, the average rate of change of a linear function $f(x) = mx + b$ is

$$\frac{\Delta y}{\Delta x} = m$$

Proof The average rate of change of $f(x) = mx + b$ from x_1 to x_2, $x_1 \neq x_2$, is

$$\frac{\Delta y}{\Delta x} = \frac{f(x_2) - f(x_1)}{x_2 - x_1} = \frac{(mx_2 + b) - (mx_1 + b)}{x_2 - x_1}$$

$$= \frac{mx_2 - mx_1}{x_2 - x_1} = \frac{m(x_2 - x_1)}{x_2 - x_1} = m$$

∎

Based on the theorem just proved, the average rate of change of the function $g(x) = -\dfrac{2}{5}x + 5$ is $-\dfrac{2}{5}$.

Now Work PROBLEM 13(d)

As it turns out, only linear functions have a constant average rate of change. Because of this, we can use the average rate of change to determine whether a particular function is linear or not. Functions that are not linear are said to be **nonlinear.**

EXAMPLE 2

Using the Average Rate of Change to Identify Linear Functions

(a) A strain of E-coli Beu 397-recA441 is placed into a Petri dish at 30° Celsius and allowed to grow. The data shown in Table 2 are collected. The population is measured in grams and the time in hours. Plot the ordered pairs (x, y) in the Cartesian plane, and use the average rate of change to determine whether the function is linear.

(b) The data in Table 3 represent the maximum number of heartbeats that a healthy individual should have during a 15-second interval of time while exercising for different ages. Plot the ordered pairs (x, y) in the Cartesian plane and use the average rate of change to determine whether the function is linear.

Table 2

Time (hours), x	Population (grams), y	(x, y)
0	0.09	(0, 0.09)
1	0.12	(1, 0.12)
2	0.16	(2, 0.16)
3	0.22	(3, 0.22)
4	0.29	(4, 0.29)
5	0.39	(5, 0.39)

Table 3

Age, x	Maximum Number of Heartbeats, y	(x, y)
20	50	(20, 50)
30	47.5	(30, 47.5)
40	45	(40, 45)
50	42.5	(50, 42.5)
60	40	(60, 40)
70	37.5	(70, 37.5)

Source: American Heart Association

Figure 2

Time (hours), *x*

Solution

We compute the average rate of change of each function. If the average rate of change is constant, the function is linear. If the average rate of change is not constant, the function is nonlinear.

(a) Figure 2 shows the points listed in Table 2 plotted in the Cartesian plane. Notice that it is impossible to draw a straight line that contains all the points. Table 4 displays the average rate of change of population.

Because the average rate of change is not constant, we know that the function is not linear. In fact, because the average rate of change is increasing as the value of the independent variable increases, we say that the function is increasing at an increasing rate. So not only is the population increasing over time, but it is also growing more rapidly as time passes.

Table 4

Time (hours), x	Population (grams), y	Average Rate of Change $= \dfrac{\Delta y}{\Delta x}$
0	0.09	$\dfrac{0.12 - 0.09}{1 - 0} = 0.03$
1	0.12	0.04
2	0.16	0.06
3	0.22	0.07
4	0.29	0.10
5	0.39	

(b) Figure 3 on the following page shows the points listed in Table 3 plotted in the Cartesian plane. We can see that the data in Figure 3 lie on a straight line. Table 5 shows the average rate of change of the maximum number of heartbeats. The average rate of change of the heartbeat data is constant, -0.25 beat for every 15-second interval per year, so the function is linear. We can find the linear function using the point–slope formula with $x_1 = 20$, $y_1 = 50$, and $m = -0.25$.

$$y - y_1 = m(x - x_1)$$
$$y - 50 = -0.25(x - 20)$$
$$y - 50 = -0.25x + 5$$
$$y = -0.25x + 55$$

Figure 3

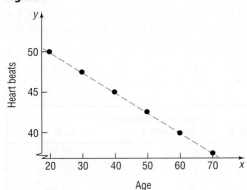

Age

Table 5

Age, x	Maximum Number of Heartbeats, y	Average Rate of Change $= \dfrac{\Delta y}{\Delta x}$
20	50	
		$\dfrac{47.5 - 50}{30 - 20} = -0.25$
30	47.5	
		-0.25
40	45	
		-0.25
50	42.5	
		-0.25
60	40	
		-0.25
70	37.5	

Now Work PROBLEM 27

3 Determine Whether a Linear Function Is Increasing, Decreasing, or Constant

Look back at the Seeing the Concept on page 21. We know that when the slope m of a linear function is positive ($m > 0$), the line slants upward from left to right. When the slope m of a linear function is negative ($m < 0$), the line slants downward from left to right. When the slope m of a linear function is zero ($m = 0$), the line is horizontal.

Based on these results, we have the following theorem.

THEOREM

Increasing, Decreasing, and Constant Linear Functions

A linear function $f(x) = mx + b$ is increasing over its domain if its slope, m, is positive. It is decreasing over its domain if its slope, m, is negative. It is constant over its domain if its slope, m, is zero.

EXAMPLE 3

Determining Whether a Linear Function Is Increasing, Decreasing, or Constant

Determine whether the following linear functions are increasing, decreasing, or constant.

(a) $f(x) = 5x - 2$ (b) $g(x) = -2x + 8$

(c) $s(t) = \dfrac{3}{4}t - 4$ (d) $h(z) = 7$

Solution

(a) For the linear function $f(x) = 5x - 2$, the slope is 5, which is positive. The function f is increasing on the interval $(-\infty, \infty)$.

(b) For the linear function $g(x) = -2x + 8$, the slope is -2, which is negative. The function g is decreasing on the interval $(-\infty, \infty)$.

(c) For the linear function $s(t) = \dfrac{3}{4}t - 4$, the slope is $\dfrac{3}{4}$, which is positive. The function s is increasing on the interval $(-\infty, \infty)$.

(d) We can write the linear function h as $h(z) = 0z + 7$. Because the slope is 0, the function h is constant on the interval $(-\infty, \infty)$.

Now Work PROBLEM 13(e)

4 Find the Zero of a Linear Function

Recall from Section 1.2 that a real number r is a zero of a function f if $f(r) = 0$. So, to find the zeros of a function $y = f(x)$, we solve the equation $f(x) = 0$. The solution(s) represent the zero(s) of f. Any linear function $f(x) = mx + b$ has exactly one zero provided that $m \neq 0$, since, if $m \neq 0$, the equation $mx + b = 0$ has exactly one solution. In other words, a linear function has exactly one zero if it is increasing or decreasing.

EXAMPLE 4   **Finding the Zero of a Linear Function**

(a) Does $f(x) = -\dfrac{1}{2}x + 2$ have a zero?

(b) If it does, find the zero.

(c) Use the zero along with the y-intercept to graph f.

(d) Solve $f(x) > 0$.

Solution

(a) The linear function f is decreasing (do you see why?), so it has one zero.

(b) We solve the equation $f(x) = 0$ to find the zero.

$$f(x) = 0$$

$$-\frac{1}{2}x + 2 = 0 \qquad f(x) = -\frac{1}{2}x + 2$$

$$-\frac{1}{2}x = -2 \qquad \text{Subtract 2 from each side of the equation.}$$

$$x = 4 \qquad \text{Multiply both sides of the equation by } -2.$$

✓ **Check:** $f(4) = -\dfrac{1}{2}(4) + 2 = 0$.

The zero of f is 4.

(c) The y-intercept is 2, so the point $(0, 2)$ is on the graph of the function. The zero of f is 4, so the x-intercept is 4 and the point $(4, 0)$ is on the graph. See Figure 4 for the graph of f.

(d) From Figure 4, we see $f(x) > 0$ (above the x-axis) when $x < 4$. Alternatively, we can solve the inequality.

$$-\frac{1}{2}x + 2 > 0 \qquad f(x) > 0 \text{ where } f(x) = -\frac{1}{2}x + 2$$

$$-\frac{1}{2}x > -2 \qquad \text{Add } -2 \text{ to both sides.}$$

$$-2\left(-\frac{1}{2}x\right) < -2 \cdot (-2) \qquad \begin{array}{l} \text{Multiply both sides by } -2. \text{ Change the} \\ \text{direction of the inequality.} \end{array}$$

$$x < 4 \qquad \text{Simplify.}$$

Figure 4

 Now Work PROBLEM 21

---

5 Build Linear Models from Verbal Descriptions

When the average rate of change of a function is constant, we can use a linear function to model the relation between the two variables. For example, if your phone company charges you $0.07 per minute to talk regardless of the number of minutes used, we can model the relation between the cost C and minutes used x as the linear function $C(x) = 0.07x$, with slope $m = \dfrac{0.07 \, \text{dollar}}{1 \, \text{minute}}$.

Modeling with a Linear Function

If the average rate of change of a function is a constant m, a linear function f can be used to model the relation between the two variables as follows:

$$f(x) = mx + b$$

where b is the value of f at 0, that is, $b = f(0)$.

We illustrate how to develop a linear model in the next example.

EXAMPLE 5 **Modeling Straight-line Depreciation**

Book value is the value of an asset that a company uses to create its balance sheet. Some companies depreciate their assets using straight-line depreciation so that the value of the asset declines by a fixed amount each year. The amount of the decline depends on the useful life that the company places on the asset. Suppose that a company just purchased a fleet of new cars for its sales force at a cost of $28,000 per car. The company chooses to depreciate each vehicle using the straight-line method over 7 years. This means that each car will depreciate by $\dfrac{\$28{,}000}{7} = \$4000$ per year.

(a) Build a linear model that expresses the book value V of each car as a function of its age, x.
(b) What is the implied domain of the function found in part (a)?
(c) Graph the linear function.
(d) What is the book value of each car after 3 years?
(e) When will the book value of each car be $8000?
   [**Hint:** Solve the equation $V(x) = 8000$.]

Solution

(a) If we let $V(x)$ represent the value of each car after x years, then $V(0)$ represents the original value of each car, so $V(0) = \$28{,}000$. The y-intercept of the linear function is $28,000. Because each car depreciates by $4000 per year, the slope of the linear function is -4000. The linear function that represents the book value V of each car after x years is

$$V(x) = -4000x + 28{,}000 \quad \text{The Model}$$

(b) Because x represents the age of the car, we have $x \geq 0$. Also, the value of the car cannot be negative. Since $V(7) = 0$, we have $x \leq 7$. Therefore, the domain of the function V is $\{x \,|\, 0 \leq x \leq 7\}$ or $[0, 7]$ using interval notation.

(c) Figure 5 shows the graph of V.

(d) The book value of each car after 3 years is

$$V(3) = -4000(3) + 28{,}000$$
$$= \$16{,}000$$

(e) To find when the book value is $8000, we solve the equation

$$V(x) = 8000$$
$$-4000x + 28{,}000 = 8000$$
$$-4000x = -20{,}000 \quad \text{Subtract 28,000 from each side.}$$
$$x = \frac{-20{,}000}{-4000} = 5 \quad \text{Divide by } -4000.$$

The car will have a book value of $8000 when it is 5 years old.

Figure 5

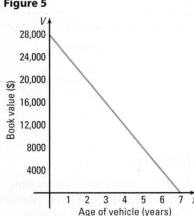

Age of vehicle (years)

Now Work PROBLEM 51

Next we look at an application from economics.

EXAMPLE 6 **Supply and Demand**

The **quantity supplied** of a good is the amount of a product that a company is willing to make available for sale at a given price. The **quantity demanded** of a good is the amount of a product that consumers are willing to purchase at a given price. Suppose that the quantity supplied, S, and quantity demanded, D, of cellular telephones each month are given by the following functions:

$$S(p) = 60p - 900$$
$$D(p) = -15p + 2850$$

where p is the price (in dollars) of the telephone.

(a) The **equilibrium price** of a product is defined as the price at which quantity supplied equals quantity demanded. That is, the equilibrium price is the price at which $S(p) = D(p)$. Find the equilibrium price of cellular telephones. What is the **equilibrium quantity,** the amount demanded (or supplied) at the equilibrium price?

(b) Determine the prices for which quantity supplied is greater than quantity demanded. That is, solve the inequality $S(p) > D(p)$.

(c) Graph $S = S(p)$, $D = D(p)$ and label the equilibrium point.

Solution

(a) To find the equilibrium price, we solve the equation $S(p) = D(p)$.

$$60p - 900 = -15p + 2850 \quad \text{\small $S(p) = 60p - 900$;}$$
$$\text{\small $D(p) = -15p + 2850$}$$
$$60p = -15p + 3750 \quad \text{\small Add 900 to each side.}$$
$$75p = 3750 \quad \text{\small Add 15p to each side.}$$
$$p = 50 \quad \text{\small Divide each side by 75.}$$

The equilibrium price is \$50 per cellular phone. To find the equilibrium quantity, we evaluate either $S(p)$ or $D(p)$ at $p = 50$.

$$S(50) = 60(50) - 900 = 2100$$

The equilibrium quantity is 2100 cellular phones. At a price of \$50 per phone, the company will produce and sell 2100 phones each month and have no shortages or excess inventory.

(b) We solve the inequality $S(p) > D(p)$.

$$60p - 900 > -15p + 2850 \quad \text{\small $S(p) > D(p)$}$$
$$60p > -15p + 3750 \quad \text{\small Add 900 to each side.}$$
$$75p > 3750 \quad \text{\small Add 15p to each side.}$$
$$p > 50 \quad \text{\small Divide each side by 75.}$$

If the company charges more than \$50 per phone, quantity supplied will exceed quantity demanded. In this case the company will have excess phones in inventory.

(c) Figure 6 shows the graphs of $S = S(p)$ and $D = D(p)$ with the equilibrium point labeled. Notice that we only graph each function in quadrant I. Do you know why?

Figure 6

2.1 Assess Your Understanding

'Are You Prepared?' *Answers are given at the end of these exercises. If you get a wrong answer, read the pages listed in red.*

1. Graph $y = 2x - 3$. (p. 25)
2. Find the slope of the line joining the points $(2, 5)$ and $(-1, 3)$. (pp. 19–21)
3. Find the average rate of change of $f(x) = 3x^2 - 2$ from 2 to 4. (pp. 73–74)

4. (a) Solve: $60x - 900 = -15x + 2850$ (pp. A66–A68)
   (b) Solve: $5x - 4 \le 3x + 6$ (pp. A86–A87)
5. If $f(x) = -5x - 14$, find $f(0)$. (pp. 46–49)
6. **True or False** The graph of the function $f(x) = x^2$ is increasing on the interval $(0, \infty)$. (pp. 69–70)

Concepts and Vocabulary

7. For the graph of the linear function $f(x) = mx + b$, m is the _____ and b is the ——————.
8. For the graph of the linear function $H(z) = -4z + 3$, the slope is _____ and the y-intercept is _____.
9. If the slope m of the graph of a linear function is ——————, the function is increasing over its domain.

10. **True or False** The slope of a line is the average rate of change of the linear function.
11. **True or False** If the average rate of change of a linear function is $\dfrac{2}{3}$ then, if y increases by 3, x will increase by 2.
12. **True or False** The average rate of change of $f(x) = 2x + 8$ is 8.

Skill Building

In Problems 13–20, a linear function is given.
    (a) Determine the slope and y-intercept of each function. (b) Use the slope and y-intercept to graph the linear function. (c) Determine the domain and the range of each function. (d) Determine the average rate of change of each function. (e) Determine whether the linear function is increasing, decreasing, or constant.

13. $f(x) = 2x + 3$ 14. $g(x) = 5x - 4$ 15. $h(x) = -3x + 4$ 16. $p(x) = -x + 6$

17. $f(x) = \dfrac{1}{4}x - 3$ 18. $h(x) = -\dfrac{2}{3}x + 4$ 19. $F(x) = 4$ 20. $G(x) = -2$

In Problems 21–26, (a) find the zero of each linear function and (b) graph each function using the zero and y-intercept.

21. $g(x) = 2x - 8$ 22. $g(x) = 3x + 12$ 23. $f(x) = -5x + 10$

24. $f(x) = -6x + 12$ 25. $H(x) = -\dfrac{1}{2}x + 4$ 26. $G(x) = \dfrac{1}{3}x - 4$

In Problems 27–34, determine whether the given function is linear or nonlinear. If it is linear, determine the equation of the line.

27.
x	y
-2	4
-1	1
0	-2
1	-5
2	-8

28.
x	y
-2	$\dfrac{1}{4}$
-1	$\dfrac{1}{2}$
0	1
1	2
2	4

29.
x	y
-20	-8
-10	-3
0	0
10	1
20	0

30.
x	y
-20	-40
-10	0
0	40
10	80
20	120

31.
x	y
-2	-26
-1	-4
0	2
1	-2
2	-10

32.
x	y
-2	-4
-1	-3.5
0	-3
1	-2.5
2	-2

33.
x	y
-8	41
-5	29
0	9
2	1
6	-15

34.
x	y
-10	0
-5	2
0	4
3	9
8	16

Applications and Extensions

35. Suppose that $f(x) = 4x - 1$ and $g(x) = -2x + 5$.
(a) Solve $f(x) = 0$.
(b) Solve $f(x) > 0$.
(c) Solve $f(x) = g(x)$.
(d) Solve $f(x) \le g(x)$.
(e) Graph $y = f(x)$ and $y = g(x)$ and label the point that represents the solution to the equation $f(x) = g(x)$.

36. Suppose that $f(x) = 3x + 5$ and $g(x) = -2x + 15$.
(a) Solve $f(x) = 0$.
(b) Solve $f(x) < 0$.
(c) Solve $f(x) = g(x)$.
(d) Solve $f(x) \ge g(x)$.
(e) Graph $y = f(x)$ and $y = g(x)$ and label the point that represents the solution to the equation $f(x) = g(x)$.

37. In parts (a)–(f), use the following figure.

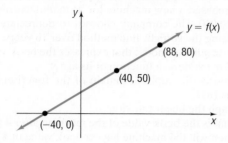

(a) Solve $f(x) = 50$.
(b) Solve $f(x) = 80$.
(c) Solve $f(x) = 0$.
(d) Solve $f(x) > 50$.
(e) Solve $f(x) \le 80$.
(f) Solve $0 < f(x) < 80$.

38. In parts (a)–(f), use the following figure.

(a) Solve $g(x) = 20$.
(b) Solve $g(x) = 60$.
(c) Solve $g(x) = 0$.
(d) Solve $g(x) > 20$.
(e) Solve $g(x) \le 60$.
(f) Solve $0 < g(x) < 60$.

39. In parts (a) and (b) use the following figure.

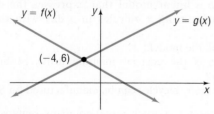

(a) Solve the equation: $f(x) = g(x)$.
(b) Solve the inequality: $f(x) > g(x)$.

40. In parts (a) and (b), use the following figure.

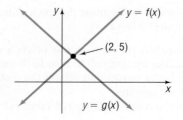

(a) Solve the equation: $f(x) = g(x)$.
(b) Solve the inequality: $f(x) \le g(x)$.

41. In parts (a) and (b), use the following figure.

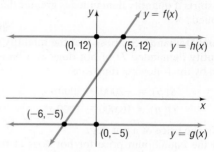

(a) Solve the equation: $f(x) = g(x)$.
(b) Solve the inequality: $g(x) \le f(x) < h(x)$.

42. In parts (a) and (b), use the following figure.

(a) Solve the equation: $f(x) = g(x)$.
(b) Solve the inequality: $g(x) < f(x) \le h(x)$.

43. **Car Rentals** The cost C, in dollars, of renting a moving truck for a day is modeled by the function $C(x) = 0.25x + 35$, where x is the number of miles driven.
(a) What is the cost if you drive $x = 40$ miles?
(b) If the cost of renting the moving truck is $80, how many miles did you drive?
(c) Suppose that you want the cost to be no more than $100. What is the maximum number of miles that you can drive?
(d) What is the implied domain of C?

44. **Phone Charges** The monthly cost C, in dollars, for international calls on a certain cellular phone plan is modeled by the function $C(x) = 0.38x + 5$, where x is the number of minutes used.
(a) What is the cost if you talk on the phone for $x = 50$ minutes?
(b) Suppose that your monthly bill is $29.32. How many minutes did you use the phone?
(c) Suppose that you budget yourself $60 per month for the phone. What is the maximum number of minutes that you can talk?
(d) What is the implied domain of C if there are 30 days in the month?

45. **Supply and Demand** Suppose that the quantity supplied S and quantity demanded D of T-shirts at a concert are given by the following functions:

$$S(p) = -200 + 50p$$
$$D(p) = 1000 - 25p$$

where p is the price of a T-shirt.

(a) Find the equilibrium price for T-shirts at this concert. What is the equilibrium quantity?

(b) Determine the prices for which quantity demanded is greater than quantity supplied.

(c) What do you think will eventually happen to the price of T-shirts if quantity demanded is greater than quantity supplied?

46. **Supply and Demand** Suppose that the quantity supplied S and quantity demanded D of hot dogs at a baseball game are given by the following functions:

$$S(p) = -2000 + 3000p$$
$$D(p) = 10,000 - 1000p$$

where p is the price of a hot dog.

(a) Find the equilibrium price for hot dogs at the baseball game. What is the equilibrium quantity?

(b) Determine the prices for which quantity demanded is less than quantity supplied.

(c) What do you think will eventually happen to the price of hot dogs if quantity demanded is less than quantity supplied?

47. **Taxes** The function $T(x) = 0.15(x - 8350) + 835$ represents the tax bill T of a single person whose adjusted gross income is x dollars for income between $8350 and $33,950, inclusive, in 2009.

Source: Internal Revenue Service

(a) What is the domain of this linear function?

(b) What is a single filer's tax bill if adjusted gross income is $20,000?

(c) Which variable is independent and which is dependent?

(d) Graph the linear function over the domain specified in part (a).

(e) What is a single filer's adjusted gross income if the tax bill is $3707.50?

48. **Luxury Tax** In 2002, major league baseball signed a labor agreement with the players. In this agreement, any team whose payroll exceeded $136.5 million in 2006 had to pay a luxury tax of 40% (for second offenses). The linear function $T(p) = 0.40(p - 136.5)$ describes the luxury tax T of a team whose payroll was p (in millions of dollars).

Source: Major League Baseball

(a) What is the implied domain of this linear function?

(b) What was the luxury tax for the New York Yankees whose 2006 payroll was $171.1 million?

(c) Graph the linear function.

(d) What is the payroll of a team that pays a luxury tax of $11.7 million?

*The point at which a company's profits equal zero is called the company's **break-even point**. For Problems 49 and 50, let R represent a company's revenue, let C represent the company's costs, and let x represent the number of units produced and sold each day.*

(a) Find the firm's break-even point; that is, find x so that R = C.

(b) Find the values of x such that R(x) > C(x). This represents the number of units that the company must sell to earn a profit.

49. $R(x) = 8x$
    $C(x) = 4.5x + 17,500$

50. $R(x) = 12x$
    $C(x) = 10x + 15,000$

51. **Straight-line Depreciation** Suppose that a company has just purchased a new computer for $3000. The company chooses to depreciate the computer using the straight-line method over 3 years.

(a) Write a linear model that expresses the book value V of the computer as a function of its age x.

(b) What is the implied domain of the function found in part (a)?

(c) Graph the linear function.

(d) What is the book value of the computer after 2 years?

(e) When will the computer have a book value of $2000?

52. **Straight-line Depreciation** Suppose that a company has just purchased a new machine for its manufacturing facility for $120,000. The company chooses to depreciate the machine using the straight-line method over 10 years.

(a) Write a linear model that expresses the book value V of the machine as a function of its age x.

(b) What is the implied domain of the function found in part (a)?

(c) Graph the linear function.

(d) What is the book value of the machine after 4 years?

(e) When will the machine have a book value of $72,000?

53. **Cost Function** The simplest cost function is the linear cost function, $C(x) = mx + b$, where the y-intercept b represents the fixed costs of operating a business and the slope m represents the cost of each item produced. Suppose that a small bicycle manufacturer has daily fixed costs of $1800 and each bicycle costs $90 to manufacture.

(a) Write a linear model that expresses the cost C of manufacturing x bicycles in a day.

(b) Graph the model.

(c) What is the cost of manufacturing 14 bicycles in a day?

(d) How many bicycles could be manufactured for $3780?

54. **Cost Function** Refer to Problem 53. Suppose that the landlord of the building increases the bicycle manufacturer's rent by $100 per month.

(a) Assuming that the manufacturer is open for business 20 days per month, what are the new daily fixed costs?

(b) Write a linear model that expresses the cost C of manufacturing x bicycles in a day with the higher rent.

(c) Graph the model.

(d) What is the cost of manufacturing 14 bicycles in a day?

(e) How many bicycles can be manufactured for $3780?

55. **Truck Rentals** A truck rental company rents a truck for one day by charging $29 plus $0.07 per mile.

(a) Write a linear model that relates the cost C, in dollars, of renting the truck to the number x of miles driven.

(b) What is the cost of renting the truck if the truck is driven 110 miles? 230 miles?

56. **Long Distance** A phone company offers a domestic long distance package by charging $5 plus $0.05 per minute.

(a) Write a linear model that relates the cost C, in dollars, of talking x minutes.

(b) What is the cost of talking 105 minutes? 180 minutes?

Mixed Practice

57. Developing a Linear Model from Data The following data represent the price p and quantity demanded per day q of an MP3 player.

Price, p (in dollars)	Quantity Demanded, q
150	100
200	80
250	60
300	40

(a) Plot the ordered pairs (p, q) in a Cartesian plane.
(b) Show that quantity demanded q is a linear function of the price p.
(c) Determine the linear function that describes the relation between p and q.
(d) What is the implied domain of the linear function?
(e) Graph the linear function in the Cartesian plane drawn in part (a).
(f) Interpret the slope.
(g) Interpret the value of the intercepts.

58. Developing a Linear Model from Data The following data represent the various combinations of soda and hot dogs that Yolanda can buy at a baseball game with $60.

Soda, s	Hot Dogs, h
20	0
15	3
10	6
5	9

(a) Plot the ordered pairs (s, h) in a Cartesian plane.
(b) Show that the number of hot dogs purchased h is a linear function of the number of sodas purchased s.
(c) Determine the linear function that describes the relation between s and h.
(d) What is the implied domain of the linear function?
(e) Graph the linear function in the Cartesian plane drawn in part (a).
(f) Interpret the slope.
(g) Interpret the value of the intercepts.

Explaining Concepts: Discussion and Writing

59. Which of the following functions might have the graph shown? (More than one answer is possible.)

(a) $f(x) = 2x - 7$
(b) $g(x) = -3x + 4$
(c) $H(x) = 5$
(d) $F(x) = 3x + 4$
(e) $G(x) = \dfrac{1}{2}x + 2$

60. Which of the following functions might have the graph shown? (More than one answer is possible.)

(a) $f(x) = 3x + 1$
(b) $g(x) = -2x + 3$
(c) $H(x) = 3$
(d) $F(x) = -4x - 1$
(e) $G(x) = -\dfrac{2}{3}x + 3$

61. Under what circumstances is a linear function $f(x) = mx + b$ odd? Can a linear function ever be even?

62. Explain how the graph of $f(x) = mx + b$ can be used to solve $mx + b > 0$.

'Are You Prepared?' Answers

1.

2. $\dfrac{2}{3}$ **3.** 18   **4.** (a) $\{50\}$ (b) $\{x \,|\, x \le 5\}$ or $(-\infty, 5]$ **5.** -14 **6.** True

2.2 Building Linear Models from Data

PREPARING FOR THIS SECTION *Before getting started, review the following:*

- Rectangular Coordinates (Foundations, Section 1, pp. 2–3)
- Functions (Section 1.1, pp. 43–46)
- Lines (Foundations, Section 3, pp. 19–27)

Now Work the 'Are You Prepared?' problems on page 137.

OBJECTIVES 1 Draw and Interpret Scatter Diagrams (p. 134)
   2 Distinguish between Linear and Nonlinear Relations (p. 135)
   3 Use a Graphing Utility to Find the Line of Best Fit (p. 136)

1 Draw and Interpret Scatter Diagrams

In Section 2.1, we built linear models from verbal descriptions. Linear models can also be constructed by fitting a linear function to data. The first step in finding this relation is to plot the ordered pairs using rectangular coordinates. The resulting graph is called a **scatter diagram.**

EXAMPLE 1   **Drawing and Interpreting a Scatter Diagram**

In baseball, the on-base percentage for a team represents the percentage of time that the players safely reach base. The data given in Table 6 represent the number of runs scored y and the on-base percentage x for teams in the National League during the 2008 baseball season.

Table 6

Team	On-Base Percentage, x	Runs Scored, y	(x, y)
Atlanta	34.5	753	(34.5, 753)
St. Louis	35.0	779	(35.0, 779)
Colorado	33.6	747	(33.6, 747)
Houston	32.3	712	(32.3, 712)
Philadelphia	33.2	799	(33.2, 799)
San Francisco	32.1	640	(32.1, 640)
Pittsburgh	32.0	735	(32.0, 735)
Florida	32.6	770	(32.6, 770)
Chicago Cubs	35.4	855	(35.4, 855)
Arizona	32.7	720	(32.7, 720)
Milwaukee	32.5	750	(32.5, 750)
Washington	32.3	641	(32.3, 641)
Cincinnati	32.1	704	(32.1, 704)
San Diego	31.7	637	(31.7, 637)
NY Mets	34.0	799	(34.0, 799)
Los Angeles	33.3	700	(33.3, 700)

Source: Based on data from http://www.baseball-reference.com.
A Sports Reference, LLC, web site.

(a) Draw a scatter diagram of the data, treating on-base percentage as the independent variable.

 (b) Use a graphing utility to draw a scatter diagram.

(c) Describe what happens to runs scored as the on-base percentage increases.

Solution

(a) To draw a scatter diagram, we plot the ordered pairs listed in Table 6, with the on-base percentage as the *x*-coordinate and the runs scored as the *y*-coordinate. See Figure 7(a). Notice that the points in the scatter diagram are not connected.

(b) Figure 7(b) shows a scatter diagram using a TI-84 Plus graphing calculator.

(c) We see from the scatter diagrams that, as the on-base percentage increases, the trend is that the number of runs scored also increases.

Figure 7

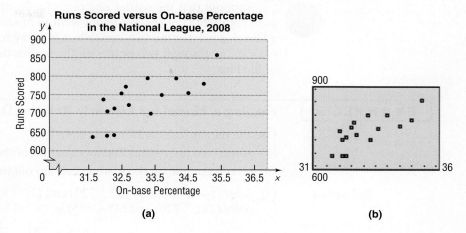

(a)

(b)

Now Work PROBLEM 11(a)

2 Distinguish between Linear and Nonlinear Relations

You should notice that the points in Figure 7 do not follow a perfect linear relation (as they do in Figure 3 in Section 2.1). However, the data do exhibit a linear pattern. There are numerous explanations as to why the data are not perfectly linear, but one easy explanation is the fact that other variables besides on-base percentage play a role in determining runs scored, such as number of home runs hit.

Scatter diagrams are used to help us to see the type of relation that exists between two variables. In this text, we will discuss a variety of different relations that may exist between two variables. For now, we concentrate on distinguishing between linear and nonlinear relations. See Figure 8.

Figure 8

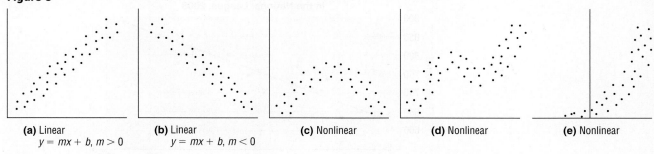

(a) Linear
$y = mx + b, m > 0$

(b) Linear
$y = mx + b, m < 0$

(c) Nonlinear

(d) Nonlinear

(e) Nonlinear

EXAMPLE 2   **Distinguishing between Linear and Nonlinear Relations**

Determine whether the relation between the two variables in Figure 9 is linear or nonlinear.

Figure 9

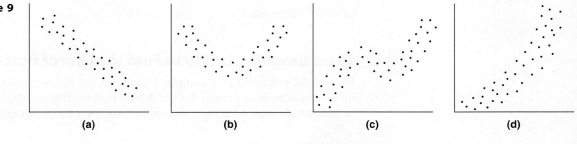

(a)

(b)

(c)

(d)

Solution (a) Linear (b) Nonlinear (c) Nonlinear (d) Nonlinear

━ **Now Work** PROBLEM 5

In this section we will study data whose scatter diagrams imply that a linear relation exists between the two variables.

Suppose that the scatter diagram of a set of data appears to be linearly related as in Figure 8(a) or (b). We might want to model the data by finding an equation of a line that relates the two variables. One way to obtain a model for such data is to draw a line through two points on the scatter diagram and determine the equation of the line.

EXAMPLE 3   **Finding a Model for Linearly Related Data**

Use the data in Table 6 from Example 1 to:

(a) Select two points and find an equation of the line containing the points.
(b) Graph the line on the scatter diagram obtained in Example 1(b).

Solution (a) Select two points, say $(32.7, 720)$ and $(35.4, 855)$. The slope of the line joining the points $(32.7, 720)$ and $(35.4, 855)$ is

$$m = \frac{855 - 720}{35.4 - 32.7} = \frac{135}{2.7} = 50$$

The equation of the line with slope 50 and passing through $(32.7, 720)$ is found using the point–slope form with $m = 50$, $x_1 = 32.7$, and $y_1 = 720$.

$$y - y_1 = m(x - x_1) \qquad \text{Point–slope form of a line}$$
$$y - 720 = 50(x - 32.7) \qquad x_1 = 32.7, \ y_1 = 720, \ m = 50$$
$$y - 720 = 50x - 1635$$
$$y = 50x - 915 \qquad \text{The Model}$$

(b) Figure 10 shows the scatter diagram with the graph of the line found in part (a).

Figure 10

Select two other points and complete the solution. Graph the line on the scatter diagram obtained in Figure 7.

━ **Now Work** PROBLEMS 11(b) AND (c)

3 Use a Graphing Utility to Find the Line of Best Fit

The model obtained in Example 3 depends on the selection of points, which will vary from person to person. So the model that we found might be different from the model you found. Although the model we found in Example 3 appears to fit the data well, there may be a model that "fits it better." Do you think your model fits the data better? Is there a *line of best fit*? As it turns out, there is a method for finding a model that best fits linearly related data (called the **line of best fit**).*

EXAMPLE 4   **Finding a Model for Linearly Related Data**

Use the data in Table 6 from Example 1.

(a) Use a graphing utility to find the line of best fit that models the relation between on-base percentage and runs scored.

(b) Graph the line of best fit on the scatter diagram obtained in Example 1(b).

(c) Interpret the slope.

(d) Use the line of best fit to predict the number of runs a team will score if their on-base percentage is 34.1.

Solution

(a) Graphing utilities contain built-in programs that find the line of best fit for a collection of points in a scatter diagram. Upon executing the LINear REGression program, we obtain the results shown in Figure 11. The output that the utility provides shows us the equation $y = ax + b$, where a is the slope of the line and b is the y-intercept. The line of best fit that relates on-base percentage to runs scored may be expressed as the line

$$y = 40.85x - 617.66 \quad \text{The Model}$$

Figure 11

(b) Figure 12 shows the graph of the line of best fit, along with the scatter diagram.

(c) The slope of the line of best fit is 40.85, which means that, for every 1 percent increase in the on-base percentage, runs scored increase 40.85, on average.

(d) Letting $x = 34.1$ in the equation of the line of best fit, we obtain $y = 40.85(34.1) - 617.66 \approx 775$ runs.

Figure 12

Now Work PROBLEMS 11(d) AND (e)

Does the line of best fit appear to be a good fit? In other words, does the line appear to accurately describe the relation between on-base percentage and runs scored?

And just how "good" is this line of best fit? Look again at Figure 11. The last line of output is $r = 0.751$. This number, called the **correlation coefficient,** r, $-1 \leq r \leq 1$, is a measure of the strength of the linear relation that exists between two variables. The closer that $|r|$ is to 1, the more perfect the linear relationship is. If r is close to 0, there is little or no linear relationship between the variables. A negative value of r, $r < 0$, indicates that as x increases y decreases; a positive value of r, $r > 0$, indicates that as x increases y does also. The data given in Table 6, having a correlation coefficient of 0.751, are indicative of a linear relationship with positive slope.

2.2 Assess Your Understanding

'Are You Prepared?' *Answers are given at the end of these exercises. If you get a wrong answer, read the pages listed in red.*

1. Plot the points $(1, 5)$, $(2, 6)$, $(3, 9)$, $(1, 12)$ in the Cartesian plane. Is the relation $\{(1, 5), (2, 6), (3, 9), (1, 12)\}$ a function? Why? (pp. 2 and 43–46)

2. Find an equation of the line containing the points $(1, 4)$ and $(3, 8)$. (p. 24)

Concepts and Vocabulary

3. A _____ _____ is used to help us to see the type of relation, if any, that may exist between two variables.

4. *True or False* The correlation coefficient is a measure of the strength of a linear relation between two variables and must lie between -1 and 1, inclusive.

---

*We shall not discuss the underlying mathematics of lines of best fit in this book.

Skill Building

In Problems 5–10, examine the scatter diagram and determine whether the type of relation is linear or nonlinear.

5.

6.

7.

8.

9.

10.

In Problems 11–16,
> *(a) Draw a scatter diagram.*
> *(b) Select two points from the scatter diagram and find the equation of the line containing the points selected.*
> *(c) Graph the line found in part (b) on the scatter diagram.*
> *(d) Use a graphing utility to find the line of best fit.*
> *(e) Use a graphing utility to draw the scatter diagram and graph the line of best fit on it.*

11.

x	3	4	5	6	7	8	9
y	4	6	7	10	12	14	16

12.

x	3	5	7	9	11	13
y	0	2	3	6	9	11

13.

x	−2	−1	0	1	2
y	−4	0	1	4	5

14.

x	−2	−1	0	1	2
y	7	6	3	2	0

15.

x	−20	−17	−15	−14	−10
y	100	120	118	130	140

16.

x	−30	−27	−25	−20	−14
y	10	12	13	13	18

Applications and Extensions

17. Candy The following data represent the weight (in grams) of various candy bars and the corresponding number of calories.

Candy Bar	Weight, x	Calories, y
Hershey's Milk Chocolate®	44.28	230
Nestle's Crunch®	44.84	230
Butterfinger®	61.30	270
Baby Ruth®	66.45	280
Almond Joy®	47.33	220
Twix® (with Caramel)	58.00	280
Snickers®	61.12	280
Heath®	39.52	210

Source: Megan Pocius, Student at Joliet Junior College

(a) Draw a scatter diagram of the data treating weight as the independent variable.
(b) What type of relation appears to exist between the weight of a candy bar and the number of calories?
(c) Select two points and find a linear model that contains the points.

(d) Graph the line on the scatter diagram drawn in part (a).
(e) Use the linear model to predict the number of calories in a candy bar that weighs 62.3 grams.
(f) Interpret the slope of the line found in part (c).

18. Raisins The following data represent the weight (in grams) of a box of raisins and the number of raisins in the box.

Weight (in grams), w	Number of Raisins, N
42.3	87
42.7	91
42.8	93
42.4	87
42.6	89
42.4	90
42.3	82
42.5	86
42.7	86
42.5	86

Source: Jennifer Maxwell, Student at Joliet Junior College

(a) Draw a scatter diagram of the data treating weight as the independent variable.
(b) What type of relation appears to exist between the weight of a box of raisins and the number of raisins?
(c) Select two points and find a linear model that contains the points.
(d) Graph the line on the scatter diagram drawn in part (b).
(e) Use the linear model to predict the number of raisins in a box that weighs 42.5 grams.
(f) Interpret the slope of the line found in part (c).

19. Video Games and Grade-Point Average Professor Grant Alexander wanted to find a linear model that relates the number of hours a student plays video games each week, h, to the cumulative grade-point average, G, of the student. He obtained a random sample of 10 full-time students at his college and asked each student to disclose the number of hours spent playing video games and the student's cumulative grade-point average.

Hours of Video Games per Week, h	Grade-point Average, G
0	3.49
0	3.05
2	3.24
3	2.82
3	3.19
5	2.78
8	2.31
8	2.54
10	2.03
12	2.51

(a) Explain why the number of hours spent playing video games is the independent variable and cumulative grade-point average is the dependent variable.

(b) Use a graphing utility to draw a scatter diagram.
(c) Use a graphing utility to find the line of best fit that models the relation between number of hours of video game playing each week and grade-point average. Express the model using function notation.
(d) Interpret the slope
(e) Predict the grade-point average of a student who plays video games for 8 hours each week.
(f) How many hours of video game playing do you think a student plays whose grade-point average is 2.40?

20. Height versus Head Circumference A pediatrician wanted to find a linear model that relates a child's height, H, to head circumference, C. She randomly selects nine children from her practice, measures their height and head circumference, and obtains the data shown. Let H represent the independent variable and C the dependent variable.

(a) Use a graphing utility to draw a scatter diagram.
(b) Use a graphing utility to find the line of best fit that models the relation between height and head circumference. Express the model using function notation.
(c) Interpret the slope.
(d) Predict the head circumference of a child that is 26 inches tall.
(e) What is the height of a child whose head circumference is 17.4 inches?

Height, H (inches)	Head Circumference, C (inches)
25.25	16.4
25.75	16.9
25	16.9
27.75	17.6
26.5	17.3
27	17.5
26.75	17.3
26.75	17.5
27.5	17.5

Source: Denise Slucki, Student at Joliet Junior College

Mixed Practice

21. Demand for Jeans The marketing manager at Levi-Strauss wishes to find a function that relates the demand D for men's jeans and p, the price of the jeans. The following data were obtained based on a price history of the jeans.

Price ($/Pair), p	Demand (Pairs of Jeans Sold per Day), D
20	60
22	57
23	56
23	53
27	52
29	49
30	44

(a) Does the relation defined by the set of ordered pairs (p, D) represent a function?
(b) Draw a scatter diagram of the data.
(c) Using a graphing utility, find the line of best fit that models the relation between price and quantity demanded.
(d) Interpret the slope.
(e) Express the relationship found in part (c) using function notation.
(f) What is the domain of the function?
(g) How many jeans will be demanded if the price is $28 a pair?

22. Advertising and Sales Revenue A marketing firm wishes to find a function that relates the sales S of a product and A, the amount spent on advertising the product. The data are obtained from past experience. Advertising and sales are measured in thousands of dollars.

Advertising Expenditures, A	Sales, S
20	335
22	339
22.5	338
24	343
24	341
27	350
28.3	351

(a) Does the relation defined by the set of ordered pairs (A, S) represent a function?
(b) Draw a scatter diagram of the data.
(c) Using a graphing utility, find the line of best fit that models the relation between advertising expenditures and sales.
(d) Interpret the slope.
(e) Express the relationship found in part (c) using function notation.
(f) What is the domain of the function?
(g) Predict sales if advertising expenditures are $25,000.

Explaining Concepts: Discussion and Writing

23. **Maternal Age versus Down Syndrome**   A biologist would like to know how the age of the mother affects the incidence rate of Down syndrome. The data to the right represent the age of the mother and the incidence rate of Down syndrome per 1000 pregnancies.

    Draw a scatter diagram treating age of the mother as the independent variable. Would it make sense to find the line of best fit for these data? Why or why not?

24. Find the line of best fit for the ordered pairs $(1, 5)$ and $(3, 8)$. What is the correlation coefficient for these data? Why is this result reasonable?

25. What does a correlation coefficient of 0 imply?

26. Refer to Problem 17. Explain why it does not make sense to interpret the y-intercept.

27. Refer to Problem 19. Solve $G(h) = 0$. Provide an interpretation of this result. Find $G(0)$. Provide an interpretation of this result.

Age of Mother, x	Incidence of Down Syndrome, y
33	2.4
34	3.1
35	4
36	5
37	6.7
38	8.3
39	10
40	13.3
41	16.7
42	22.2
43	28.6
44	33.3
45	50

Source: E. B. Hook, P. K. Cross, and D. M. Schreinemachers, "Chromosomal abnormality rates at amniocentesis and in live-born infants," *Journal of the American Medical Association,* 249(15), 1983, pp. 2034–2038.

'Are You Prepared?' Answers

1.

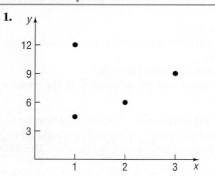

2. $y = 2x + 2$

No, because the input, 1, corresponds to two different outputs.

2.3 Quadratic Functions and Their Zeros

PREPARING FOR THIS SECTION *Before getting started, review the following:*

- Zero-Product Property (Appendix A, Section A.1, p. A4)
- Factoring Polynomials (Appendix A, Section A.4, pp. A33–A39)
- Square Roots (Appendix A, Section A.7, pp. A57–A58)

- Completing the Square (Appendix A, Section A.4, p. A40)
- The Graph of a Function (Section 1.2, pp. 57–59)
- Zeros of a Function (Section 1.2, pp. 59–60)

Now work the 'Are You Prepared?' problems on page 149.

OBJECTIVES **1** Find the Zeros of a Quadratic Function by Factoring (p. 142)
2 Find the Zeros of a Quadratic Function Using the Square Root Method (p. 142)
3 Find the Zeros of a Quadratic Function by Completing the Square (p. 144)
4 Find the Zeros of a Quadratic Function Using the Quadratic Formula (p. 145)
5 Find the Point of Intersection of Two Functions (p. 147)
6 Solve Equations That Are Quadratic in Form (p. 148)

Here are some examples of quadratic functions.

$$F(x) = 3x^2 - 5x + 1 \qquad g(x) = -6x^2 + 1 \qquad H(x) = \frac{1}{2}x^2 + \frac{2}{3}x$$

A **quadratic function** is a function of the form

$$f(x) = ax^2 + bx + c$$

where a, b, and c are real numbers and $a \neq 0$. The domain of a quadratic function consists of all real numbers.

We may also say that a quadratic function is a function defined by a second-degree polynomial in one variable.

Remember, solving the equation $f(x) = 0$ is equivalent to finding the zeros of f and is also equivalent to finding the x-intercepts of the graph of $y = f(x)$. So, to find the zeros, or x-intercepts of the graph, of a quadratic function $f(x) = ax^2 + bx + c$, we must be able to solve the equation $ax^2 + bx + c = 0$. These types of equations are called **quadratic equations.**

A **quadratic equation** is an equation equivalent to one of the form

$$ax^2 + bx + c = 0 \qquad\qquad (1)$$

where a, b, and c are real numbers and $a \neq 0$.

Examples of quadratic equations are

$$3x^2 - 5x + 1 = 0 \qquad -6x^2 + 1 = 0 \qquad \frac{1}{2}x^2 + \frac{2}{3}x = 0$$

A quadratic equation written in the form $ax^2 + bx + c = 0$ is said to be in **standard form.**

Sometimes a quadratic equation is called a **second-degree equation,** because the left side is a polynomial of degree 2. We shall discuss four ways of solving quadratic equations: by factoring, by the square root method, by completing the square, and by using the quadratic formula.

1 Find the Zeros of a Quadratic Function by Factoring

When a quadratic equation is written in standard form, $ax^2 + bx + c = 0$, it may be possible to factor the expression on the left side as the product of two first-degree polynomials. Then, we use the Zero-Product Property and set each factor equal to 0 and solve the resulting linear equations to obtain the exact solutions of the quadratic equation.

EXAMPLE 1   **Finding the Zeros of a Quadratic Function by Factoring**

Find the zeros of $f(x) = x^2 + x - 12$. List any x-intercepts of the graph of f.

Solution   To find the zeros of a function f, we solve the equation $f(x) = 0$.

$$f(x) = 0$$
$$x^2 + x - 12 = 0 \quad \text{f(x)} = x^2 + x - 12$$
$$(x + 4)(x - 3) = 0 \quad \text{Factor.}$$
$$x + 4 = 0 \quad \text{or} \quad x - 3 = 0 \quad \text{Zero-Product Property}$$
$$x = -4 \quad \text{or} \quad x = 3$$

✓ **Check:** $f(x) = x^2 + x - 12$

$x = -4: f(-4) = (-4)^2 + (-4) - 12 \qquad x = 3: f(3) = (3)^2 + (3) - 12$
$\qquad\qquad\quad = 16 - 4 - 12 \qquad\qquad\qquad\qquad = 9 + 3 - 12$
$\qquad\qquad\quad = 0 \qquad\qquad\qquad\qquad\qquad\qquad = 0$

The zeros of $f(x) = x^2 + x - 12$ are -4 and 3. The x-intercepts of the graph of $f(x) = x^2 + x - 12$ are -4 and 3.

●

Now Work PROBLEM 11

Sometimes, the quadratic expression factors into two linear equations with the same solution. When this happens, the quadratic equation is said to have a **repeated solution.** We also call this solution a **root of multiplicity 2,** or a **double root.**

EXAMPLE 2   **Finding the Zeros of a Quadratic Function by Factoring**

Find the zeros of $f(x) = x^2 - 6x + 9$. List any x-intercepts of the graph of f.

Solution   To find the zeros of a function f, we solve the equation $f(x) = 0$.

$$f(x) = 0$$
$$x^2 - 6x + 9 = 0 \quad \text{f(x)} = x^2 - 6x + 9$$
$$(x - 3)(x - 3) = 0 \quad \text{Factor.}$$
$$x - 3 = 0 \quad \text{or} \quad x - 3 = 0 \quad \text{Zero-Product Property}$$
$$x = 3 \quad \text{or} \quad x = 3$$

We leave the check to you. The only zero of $f(x) = x^2 - 6x + 9$ is 3. The only x-intercept of the graph of $f(x) = x^2 - 6x + 9$ is 3.

●

Now Work PROBLEM 23

---

2 Find the Zeros of a Quadratic Function Using the Square Root Method

Suppose that we wish to solve the quadratic equation

$$x^2 = p \tag{2}$$

where p is a nonnegative number ($p \geq 0$). We proceed as in the earlier examples.

$$x^2 - p = 0 \qquad \text{Put in standard form.}$$
$$\left(x - \sqrt{p}\right)\left(x + \sqrt{p}\right) = 0 \qquad \text{Factor (over the real numbers).}$$
$$x = \sqrt{p} \quad \text{or} \quad x = -\sqrt{p} \qquad \text{Solve.}$$

We have the following result:

If $x^2 = p$ and $p \geq 0$, then $x = \sqrt{p}$ or $x = -\sqrt{p}$. **(3)**

When statement (3) is used, it is called the **Square Root Method.** In statement (3), note that if $p > 0$ the equation $x^2 = p$ has two solutions, $x = \sqrt{p}$ and $x = -\sqrt{p}$. We usually abbreviate these solutions as $x = \pm\sqrt{p}$, read as "x equals plus or minus the square root of p." For example, the two solutions of the equation

$$x^2 = 4$$

are

$$x = \pm\sqrt{4}$$

and, since $\sqrt{4} = 2$, we have

$$x = \pm 2$$

EXAMPLE 3

Finding the Zeros of a Quadratic Function Using the Square Root Method

Find the zeros of each function.

(a) $f(x) = x^2 - 12$ 　　　　　　(b) $g(x) = (x - 2)^2 - 16$

List any x-intercepts of the graph.

Solution　(a) To find the zeros of a function f, we solve the equation $f(x) = 0$.

$$f(x) = 0$$
$$x^2 - 12 = 0 \qquad f(x) = x^2 - 12$$
$$x^2 = 12 \qquad \text{Add 12 to both sides.}$$
$$x = \pm\sqrt{12} \qquad \text{Use the Square Root Method.}$$
$$x = \pm 2\sqrt{3}$$
$$x = 2\sqrt{3} \quad \text{or} \quad x = -2\sqrt{3}$$

The zeros of $f(x) = x^2 - 12$ are $-2\sqrt{3}$ and $2\sqrt{3}$. The x-intercepts of the graph of $f(x) = x^2 - 12$ are $-2\sqrt{3}$ and $2\sqrt{3}$.

(b) To find the zeros of a function g, we solve the equation $g(x) = 0$.

$$g(x) = 0$$
$$(x - 2)^2 - 16 = 0 \qquad g(x) = (x - 2)^2 - 16$$
$$(x - 2)^2 = 16 \qquad \text{Add 16 to both sides.}$$
$$x - 2 = \pm\sqrt{16} \qquad \text{Use the Square Root Method.}$$
$$x - 2 = \sqrt{16} \quad \text{or} \quad x - 2 = -\sqrt{16}$$
$$x - 2 = 4 \quad \text{or} \quad x - 2 = -4$$
$$x = 6 \quad \text{or} \quad x = -2$$

The zeros of $g(x) = (x - 2)^2 - 16$ are -2 and 6. The x-intercepts of the graph of $g(x) = (x - 2)^2 - 16$ are -2 and 6.

Now Work PROBLEM 27

3 Find the Zeros of a Quadratic Function by Completing the Square

The next example illustrates how the procedure of completing the square can be used to find the zeros of a quadratic function.

EXAMPLE 4

Finding the Zeros of a Quadratic Function by Completing the Square

Find the zeros of $f(x) = x^2 + 5x + 4$ by completing the square. List any x-intercepts of the graph of f.

Solution

To find the zeros of a function f, we solve the equation $f(x) = 0$.

$$f(x) = 0$$
$$x^2 + 5x + 4 = 0 \quad f(x) = x^2 + 5x + 4$$

We always begin this procedure by rearranging the equation so that the constant is on the right side.

$$x^2 + 5x + 4 = 0$$
$$x^2 + 5x = -4$$

Since the coefficient of x^2 is 1, we can complete the square on the left side by adding $\left(\frac{1}{2} \cdot 5\right)^2 = \frac{25}{4}$. Of course, in an equation, whatever we add to the left side must also be added to the right side. We add $\frac{25}{4}$ to both sides.

$$x^2 + 5x + \frac{25}{4} = -4 + \frac{25}{4} \quad \text{Add } \frac{25}{4} \text{ to both sides.}$$

$$\left(x + \frac{5}{2}\right)^2 = \frac{9}{4} \quad \text{Factor; simplify.}$$

$$x + \frac{5}{2} = \pm\sqrt{\frac{9}{4}} \quad \text{Use the Square Root Method.}$$

$$x + \frac{5}{2} = \pm\frac{3}{2}$$

$$x = -\frac{5}{2} \pm \frac{3}{2}$$

$$x = -\frac{5}{2} + \frac{3}{2} = -1 \quad \text{or} \quad x = -\frac{5}{2} - \frac{3}{2} = -4$$

We leave the check to you. The zeros of $f(x) = x^2 + 5x + 4$ are -4 and -1. The x-intercepts of the graph of $f(x) = x^2 + 5x + 4$ are -4 and -1.

●

The zeros of the function in Example 4 can also be obtained by factoring. Rework Example 4 using factoring.

NOTE If the coefficient of the square term is not 1, divide through by the coefficient of the square term before attempting to complete the square. For example, to find the zeros of $f(x) = 2x^2 - 8x - 5$ by completing the square, we solve $2x^2 - 8x - 5 = 0$. Start by dividing both sides of the equation by 2 and obtain

$$x^2 - 4x - \frac{5}{2} = 0$$

■

Now Work PROBLEM 31

4 Find the Zeros of a Quadratic Function Using the Quadratic Formula

We can use the method of completing the square to obtain a general formula for solving the quadratic equation

$$ax^2 + bx + c = 0, \qquad a \neq 0$$

As in Example 4, we begin by rearranging the terms as

$$ax^2 + bx = -c, \qquad a > 0$$

Since $a > 0$, we can divide both sides by a to get

$$x^2 + \frac{b}{a}x = -\frac{c}{a}$$

Now the coefficient of x^2 is 1. To complete the square on the left side, add the square of $\frac{1}{2}$ the coefficient of x: that is, add

$$\left(\frac{1}{2} \cdot \frac{b}{a}\right)^2 = \frac{b^2}{4a^2}$$

to each side. Then

$$x^2 + \frac{b}{a}x + \frac{b^2}{4a^2} = \frac{b^2}{4a^2} - \frac{c}{a}$$

$$\left(x + \frac{b}{2a}\right)^2 = \frac{b^2 - 4ac}{4a^2} \qquad \frac{b^2}{4a^2} - \frac{c}{a} = \frac{b^2}{4a^2} - \frac{4ac}{4a^2} = \frac{b^2 - 4ac}{4a^2} \qquad \textbf{(4)}$$

Provided that $b^2 - 4ac \geq 0$, we now can use the Square Root Method to get

$$x + \frac{b}{2a} = \pm\sqrt{\frac{b^2 - 4ac}{4a^2}}$$

$$x + \frac{b}{2a} = \frac{\pm\sqrt{b^2 - 4ac}}{2a} \qquad \text{The square root of a quotient equals the quotient of the square roots. Also, } \sqrt{4a^2} = 2a \text{ since } a > 0.$$

$$x = -\frac{b}{2a} \pm \frac{\sqrt{b^2 - 4ac}}{2a} \qquad \text{Add } -\frac{b}{2a} \text{ to both sides.}$$

$$x = \frac{-b \pm \sqrt{b^2 - 4ac}}{2a} \qquad \text{Combine the quotients on the right.}$$

What if $b^2 - 4ac$ is negative? Then equation (4) states that the left expression (a real number squared) equals the right expression (a negative number). Since this occurrence is impossible for real numbers, we conclude that if $b^2 - 4ac < 0$ the quadratic equation has no real solution.*

We now state the quadratic formula.

THEOREM

Quadratic Formula

Consider the quadratic equation

$$ax^2 + bx + c = 0, \qquad a \neq 0$$

If $b^2 - 4ac < 0$, this equation has no real solution.

If $b^2 - 4ac \geq 0$, the real solution(s) of this equation is (are) given by the **quadratic formula:**

$$x = \frac{-b \pm \sqrt{b^2 - 4ac}}{2a}$$

*If $b^2 - 4ac < 0$, the quadratic function will have complex zeros. We discuss the complex zeros of a quadratic function in Section 2.7.

The quantity $b^2 - 4ac$ is called the **discriminant** of the quadratic equation because its value tells us whether the equation has real solutions and how many solutions to expect. The discriminant also tells us how many real zeros, or x-intercepts, the quadratic function $f(x) = ax^2 + bx + c$ will have.

> **Discriminant of a Quadratic Equation**
>
> For a quadratic equation $ax^2 + bx + c = 0$:
> 1. If $b^2 - 4ac > 0$, there are two unequal real solutions. Therefore, the graph of $f(x) = ax^2 + bx + c$ will have two distinct x-intercepts.
> 2. If $b^2 - 4ac = 0$, there is a repeated real solution, a root of multiplicity 2. Therefore, the graph of $f(x) = ax^2 + bx + c$ will have one x-intercept, at which the graph touches the x-axis.
> 3. If $b^2 - 4ac < 0$, there is no real solution. Therefore, the graph of $f(x) = ax^2 + bx + c$ will have no x-intercept.

If you want to know the number and nature of the zeros of a quadratic function, evaluate the discriminant first.

EXAMPLE 5 | **Finding the Zeros of a Quadratic Function Using the Quadratic Formula**

Find the real zeros, if any, of the function $f(x) = 3x^2 - 5x + 1$. List any x-intercepts of the graph of f.

Solution To find the real zeros, we solve the equation $f(x) = 0$.

$$f(x) = 0$$
$$3x^2 - 5x + 1 = 0 \quad f(x) = 3x^2 - 5x + 1$$

We compare $3x^2 - 5x + 1 = 0$ to $ax^2 + bx + c = 0$ to find a, b, and c.

$$3x^2 - 5x + 1 = 0$$
$$ax^2 + bx + c = 0, \quad a = 3, b = -5, c = 1$$

With $a = 3$, $b = -5$, and $c = 1$, we evaluate the discriminant $b^2 - 4ac$.

$$b^2 - 4ac = (-5)^2 - 4(3)(1) = 25 - 12 = 13$$

Since $b^2 - 4ac > 0$, there are two unequal real solutions.
We use the quadratic formula with $a = 3$, $b = -5$, $c = 1$, and $b^2 - 4ac = 13$.

$$x = \frac{-b \pm \sqrt{b^2 - 4ac}}{2a} = \frac{-(-5) \pm \sqrt{13}}{2(3)} = \frac{5 \pm \sqrt{13}}{6}$$

The real zeros of $f(x) = 3x^2 - 5x + 1$ are $\dfrac{5 - \sqrt{13}}{6}$ and $\dfrac{5 + \sqrt{13}}{6}$. The x-intercepts of the graph of $f(x) = 3x^2 - 5x + 1$ are $\dfrac{5 - \sqrt{13}}{6}$ and $\dfrac{5 + \sqrt{13}}{6}$.

●

EXAMPLE 6 | **Finding the Zeros of a Quadratic Function Using the Quadratic Formula**

Find the real zeros, if any, of the function $f(x) = 3x^2 - 4x + 2$. List any x-intercepts of the graph.

Solution To find the real zeros, we solve the equation $f(x) = 0$.

$$f(x) = 0$$
$$3x^2 - 4x + 2 = 0 \quad \text{f(x) = 3x}^2 \text{ − 4x + 2}$$
$$ax^2 + bx + c = 0 \quad \text{Compare to standard form.}$$

With $a = 3, b = -4$, and $c = 2$, we find that

$$b^2 - 4ac = (-4)^2 - 4(3)(2) = 16 - 24 = -8$$

Since $b^2 - 4ac < 0$, the equation has no real solution. Therefore, the function $f(x) = 3x^2 - 4x + 2$ has no real zeros. The graph of $f(x) = 3x^2 - 4x + 2$ has no x-intercept.

●

━━━━▶ **Now Work** PROBLEMS 41 AND 43

SUMMARY

Finding the Zeros, or x-intercepts of the Graph, of a Quadratic Function

STEP 1: Given that $f(x) = ax^2 + bx + c$, write the equation $ax^2 + bx + c = 0$.
STEP 2: Identify a, b, and c.
STEP 3: Evaluate the discriminant, $b^2 - 4ac$.
STEP 4: (a) If the discriminant is negative, the equation has no real solution, so the function has no real zeros. The graph of the function has no x-intercepts.

(b) If the discriminant is nonnegative (greater than or equal to zero), determine whether the left side can be factored. If you can easily spot factors, use the factoring method to solve the equation. Otherwise, use the quadratic formula or the method of completing the square. The solution(s) to the equation is(are) the real zero(s) of the function. The solution(s) also represent(s) the x-intercept(s) of the graph of the function.

---

5 Find the Point of Intersection of Two Functions

We are often interested in knowing where two functions are equal to each other. For example, if $R = R(x)$ represents the revenue for selling x units of a good and $C = C(x)$ represents the cost of manufacturing x units of a good, we might want to know the number of goods for which revenue equals cost. This requires us to solve $R(x) = C(x)$.

If f and g are two functions, the solution to the equation $f(x) = g(x)$ has a geometric interpretation—it represents the x-coordinate at which the graphs of $y = f(x)$ and $y = g(x)$ intersect.

EXAMPLE 7 **Finding the Point of Intersection of the Graphs of Two Functions**

If $f(x) = x^2 + 5x - 3$ and $g(x) = 2x + 1$, solve $f(x) = g(x)$. At what point(s) do the graphs of the two functions intersect?

Solution
$$f(x) = g(x)$$
$$x^2 + 5x - 3 = 2x + 1 \quad \text{f(x) = x}^2 \text{ + 5x − 3; g(x) = 2x + 1}$$
$$x^2 + 3x - 4 = 0 \quad \text{Put in standard form.}$$
$$(x + 4)(x - 1) = 0 \quad \text{Factor.}$$
$$x + 4 = 0 \quad \text{or} \quad x - 1 = 0 \quad \text{Zero-Product Property}$$
$$x = -4 \quad \text{or} \quad x = 1$$

The x-coordinates of the points of intersection are -4 and 1. To find the y-coordinates of the points of intersection, we evaluate either f or g at $x = -4$ and $x = 1$. It is easier to evaluate g.

$$g(x) = 2x + 1 \qquad\qquad g(x) = 2x + 1$$
$$x = -4: \quad g(-4) = 2(-4) + 1 \qquad x = 1: \quad g(1) = 2(1) + 1$$
$$= -7 \qquad\qquad\qquad = 3$$

The graphs of the two functions intersect at the points $(-4, -7)$ and $(1, 3)$.

●

── **Now Work** PROBLEM 53

6 Solve Equations That Are Quadratic in Form

So far, this section has been dedicated to finding zeros of a quadratic function $f(x) = ax^2 + bx + c$ by solving the equation $ax^2 + bx + c = 0$. Often, rather than being asked to find the zeros of a quadratic function, we might be asked to find the zeros of a function that can be transformed into a quadratic function with an appropriate substitution.

For example, the function $f(x) = x^4 + x^2 - 12$ is not a quadratic function, but if we let $u = x^2$, we get $f(u) = u^2 + u - 12$, which is a quadratic function. We can find the zeros of this function by solving the equation $u^2 + u - 12 = 0$. Then we can find x by solving $u = x^2$.

In general, if an appropriate substitution u transforms an equation into one of the form

$$au^2 + bu + c = 0, \qquad a \neq 0$$

the original equation is called an **equation of the quadratic type** or an **equation quadratic in form.**

The difficulty of solving such an equation lies in the determination that the equation is, in fact, quadratic in form. After you are told an equation is quadratic in form, it is easy enough to see it, but some practice is needed to enable you to recognize them on your own.

EXAMPLE 8

Finding the Real Zeros of a Function

Find the real zeros of the function $f(x) = (x + 2)^2 + 11(x + 2) - 12$. Find any x-intercepts of the graph of f.

Solution To find the real zeros, we solve the equation $f(x) = 0$.

$$f(x) = 0$$
$$(x + 2)^2 + 11(x + 2) - 12 = 0 \quad {\scriptstyle f(x) = (x + 2)^2 + 11(x + 2) - 12}$$

For this equation, let $u = x + 2$. Then $u^2 = (x + 2)^2$, and the original equation

$$(x + 2)^2 + 11(x + 2) - 12 = 0$$

becomes

$$u^2 + 11u - 12 = 0 \quad {\scriptstyle \text{Let } u = x + 2.}$$
$$(u + 12)(u - 1) = 0 \quad {\scriptstyle \text{Factor.}}$$
$$u = -12 \quad \text{or} \quad u = 1 \quad {\scriptstyle \text{Solve.}}$$

But we want to solve for x. Because $u = x + 2$, we have

$$x + 2 = -12 \quad \text{or} \quad x + 2 = 1$$
$$x = -14 \qquad\qquad x = -1$$

The real zeros of f are -14 and -1. The x-intercepts of the graph of f are -14 and -1.

●

EXAMPLE 9   **Finding the Real Zeros of a Function**

Find the real zeros of $g(x) = x + 2\sqrt{x} - 3$. Find any x-intercepts of the graph of g.

Solution   To find the real zeros, we solve the equation $g(x) = 0$.

$$g(x) = 0$$

$$x + 2\sqrt{x} - 3 = 0 \quad g(x) = x + 2\sqrt{x} - 3$$

For the equation $x + 2\sqrt{x} - 3 = 0$, let $u = \sqrt{x}$. Then $u^2 = x$, and the original equation

$$x + 2\sqrt{x} - 3 = 0$$

becomes

$$u^2 + 2u - 3 = 0 \quad \text{Let } u = \sqrt{x}.$$

$$(u + 3)(u - 1) = 0 \quad \text{Factor.}$$

$$u = -3 \quad \text{or} \quad u = 1 \quad \text{Solve.}$$

Since $u = \sqrt{x}$, we have $\sqrt{x} = -3$ or $\sqrt{x} = 1$. The first of these, $\sqrt{x} = -3$, has no real solution, since the square root of a real number is never negative. The second one, $\sqrt{x} = 1$, has the solution $x = 1$.

Check: $g(x) = x + 2\sqrt{x} - 3$

$g(1) = 1 + 2\sqrt{1} - 3 = 1 + 2 - 3 = 0$

The only real zero of the function is 1. The only x-intercept of the graph of the function is 1.

Now Work   PROBLEM 69

2.3 Assess Your Understanding

'Are You Prepared?'   *Answers are given at the end of these exercises. If you get a wrong answer, read the pages listed in red.*

1. (a) Factor: $x^2 - 5x - 6$ (pp. A36–A37)
   (b) Factor: $2x^2 - x - 3$ (pp. A38–A39)
2. Simplify: $\sqrt{8^2 - 4 \cdot 2 \cdot 3}$ (pp. A57–A58)
3. Solve: $(x - 3)(3x + 5) = 0$ (p. A4)

4. To complete the square of: $x^2 + 6x$ _____ (add/subtract) _____ . (p. A40)
5. If $f(4) = 10$, what point is on the graph of f? (pp. 58–59)
6. Is -3 a zero of $f(x) = x^2 + 4x + 3$? (pp. 59–60)

Concepts and Vocabulary

7. When a quadratic equation has a repeated solution, it is called a(n) _____ root or a root of _____ .

8. The quantity $b^2 - 4ac$ is called the _____ of a quadratic equation. If it is _____ , the equation has no real solution.

9. How many real zeros can a quadratic function have?
10. State the quadratic formula.

Skill Building

In Problems 11–24, find the zeros of each quadratic function by factoring. What are the x-intercepts of the graph of the function?

11. $f(x) = x^2 - 9x$ **12.** $f(x) = x^2 + 4x$ **13.** $g(x) = x^2 - 25$ **14.** $G(x) = x^2 - 9$

15. $F(x) = x^2 + x - 6$ **16.** $H(x) = x^2 + 7x + 6$ **17.** $g(x) = 2x^2 - 5x - 3$ **18.** $f(x) = 3x^2 + 5x + 2$

19. $P(x) = 3x^2 - 48$ **20.** $H(x) = 2x^2 - 50$ **21.** $g(x) = x(x + 8) + 12$ **22.** $f(x) = x(x - 4) - 12$

23. $G(x) = 4x^2 + 9 - 12x$ **24.** $F(x) = 25x^2 + 16 - 40x$

In Problems 25–30, find the zeros of each quadratic function using the Square Root Method. What are the x-intercepts of the graph of the function?

25. $f(x) = x^2 - 8$ **26.** $g(x) = x^2 - 18$ **27.** $g(x) = (x - 1)^2 - 4$
28. $G(x) = (x + 2)^2 - 1$ **29.** $F(x) = (2x + 3)^2 - 32$ **30.** $g(x) = (3x - 2)^2 - 75$

In Problems 31–36, find the zeros of each quadratic function by completing the square. What are the x-intercepts of the graph of the function?

31 $f(x) = x^2 + 4x - 8$ **32.** $f(x) = x^2 - 6x - 9$ **33.** $g(x) = x^2 - \dfrac{1}{2}x - \dfrac{3}{16}$
34. $g(x) = x^2 + \dfrac{2}{3}x - \dfrac{1}{3}$ **35.** $F(x) = 3x^2 + x - \dfrac{1}{2}$ **36.** $G(x) = 2x^2 - 3x - 1$

In Problems 37–50, find the real zeros, if any, of each quadratic function using the quadratic formula. What are the x-intercepts, if any, of the graph of the function?

37. $f(x) = x^2 - 4x + 2$ **38.** $f(x) = x^2 + 4x + 2$ **39.** $g(x) = x^2 - 4x - 1$ **40.** $g(x) = x^2 + 6x + 1$
41. $F(x) = 2x^2 - 5x + 3$ **42.** $g(x) = 2x^2 + 5x + 3$ **43.** $P(x) = 4x^2 - x + 2$ **44.** $H(x) = 4x^2 + x + 1$
45. $f(x) = 4x^2 - 1 + 2x$ **46.** $f(x) = 2x^2 - 1 + 2x$ **47.** $G(x) = 2x(x + 2) - 3$ **48.** $F(x) = 3x(x + 2) - 1$
49. $P(x) = 9x^2 - 6x + 1$ **50.** $g(x) = 4x^2 + 20x + 25$

In Problems 51–56, solve $f(x) = g(x)$. What are the points of intersection of the graphs of the two functions?

51. $f(x) = x^2 + 6x + 3$ **52.** $f(x) = x^2 - 4x + 3$ **53.** $f(x) = -2x^2 + 1$
   $g(x) = 3$ $g(x) = 3$ $g(x) = 3x + 2$

54. $f(x) = 3x^2 - 7$ **55.** $f(x) = x^2 - x + 1$ **56.** $f(x) = x^2 + 5x - 3$
   $g(x) = 10x + 1$ $g(x) = 2x^2 - 3x - 14$ $g(x) = 2x^2 + 7x - 27$

In Problems 57–74, find the real zeros of each function. What are the x-intercepts of the graph of the function?

57. $P(x) = x^4 - 6x^2 - 16$ **58.** $H(x) = x^4 - 3x^2 - 4$ **59.** $f(x) = x^4 - 5x^2 + 4$
60. $f(x) = x^4 - 10x^2 + 24$ **61.** $G(x) = 3x^4 - 2x^2 - 1$ **62.** $F(x) = 2x^4 - 5x^2 - 12$
63. $g(x) = x^6 + 7x^3 - 8$ **64.** $g(x) = x^6 - 7x^3 - 8$ **65.** $G(x) = (x + 2)^2 + 7(x + 2) + 12$
66. $f(x) = (2x + 5)^2 - (2x + 5) - 6$ **67.** $f(x) = (3x + 4)^2 - 6(3x + 4) + 9$ **68.** $H(x) = (2 - x)^2 + (2 - x) - 20$
69. $P(x) = 2(x + 1)^2 - 5(x + 1) - 3$ **70.** $H(x) = 3(1 - x)^2 + 5(1 - x) + 2$ **71.** $G(x) = x - 4\sqrt{x}$
72. $f(x) = x + 8\sqrt{x}$ **73.** $g(x) = x + \sqrt{x} - 20$ **74.** $f(x) = x + \sqrt{x} - 2$

Mixed Practice

In Problems 75–86, find the real zeros of each quadratic function using any method you wish. What are the x-intercepts, if any, of the graph of the function?

75. $f(x) = x^2 - 50$ **76.** $f(x) = x^2 - 20$ **77.** $g(x) = 16x^2 - 8x + 1$

78. $F(x) = 4x^2 - 12x + 9$ **79.** $G(x) = 10x^2 - 19x - 15$ **80.** $f(x) = 6x^2 + 7x - 20$

81. $P(x) = 6x^2 - x - 2$ **82.** $H(x) = 6x^2 + x - 2$ **83.** $G(x) = x^2 + \sqrt{2}x - \dfrac{1}{2}$

84. $F(x) = \dfrac{1}{2}x^2 - \sqrt{2}x - 1$ **85.** $f(x) = x^2 + x - 4$ **86.** $g(x) = x^2 + x - 1$

In Problems 87–92, (a) graph each function using transformations, (b) find the real zeros of each function, and (c) label the x-intercepts on the graph of the function.

87. $g(x) = (x - 1)^2 - 4$

88. $F(x) = (x + 3)^2 - 9$

89. $f(x) = 2(x + 4)^2 - 8$

90. $h(x) = 3(x - 2)^2 - 12$

91. $H(x) = -3(x - 3)^2 + 6$

92. $f(x) = -2(x + 1)^2 + 12$

In Problems 93–98, solve $f(x) = g(x)$. What are the points of intersection of the graphs of the two functions?

93. $f(x) = 5x(x - 1)$
$g(x) = -7x^2 + 2$

94. $f(x) = 10x(x + 2)$
$g(x) = -3x + 5$

95. $f(x) = 3(x^2 - 4)$
$g(x) = 3x^2 + 2x + 4$

96. $f(x) = 4(x^2 + 1)$
$g(x) = 4x^2 - 3x - 8$

97. $f(x) = \dfrac{3x}{x + 2} - \dfrac{5}{x + 1}$
$g(x) = \dfrac{-5}{x^2 + 3x + 2}$

98. $f(x) = \dfrac{2x}{x - 3} - \dfrac{3}{x + 1}$
$g(x) = \dfrac{2x + 18}{x^2 - 2x - 3}$

99. Suppose that $f(x) = x^2 + 5x - 14$ and $g(x) = x^2 + 3x - 4$.
(a) Find the zeros of $(f + g)(x)$.
(b) Find the zeros of $(f - g)(x)$.
(c) Find the zeros of $(f \cdot g)(x)$.

100. Suppose that $f(x) = x^2 - 3x - 18$ and $g(x) = x^2 + 2x - 3$.
(a) Find the zeros of $(f + g)(x)$.
(b) Find the zeros of $(f - g)(x)$
(c) Find the zeros of $(f \cdot g)(x)$.

Applications and Extensions

101. Dimensions of a Window   The function $A(x) = x(x + 2)$ describes the area A of the opening of a rectangular window in which the length is 2 feet more than the width, x. Find the dimensions of the window if the area of the opening is to be 143 square feet by solving $A(x) = 143$.

102. Dimensions of a Window   The function $A(x) = x(x + 1)$ describes the area A of the opening of a rectangular window in which the length is 1 centimeter more than the width, x. Find the dimensions of the window if the area of the opening is to be 306 square centimeters by solving $A(x) = 306$.

103. Constructing a Box   An open box is to be constructed from a square sheet of sheet metal with dimensions x feet by x feet by removing a square of side 1 foot from each corner and turning up the edges. The volume V of the box is $V(x) = (x - 2)^2$. Find the dimensions of the sheet metal needed to make a box that will hold 4 cubic feet by solving the equation $V(x) = 4$.

104. Constructing a Box   Rework Problem 103 if the box needs to hold 16 cubic feet.

105. Physics   A ball is thrown vertically upward from the top of a building 96 feet tall with an initial velocity of 80 feet per second. The distance s (in feet) of the ball from the ground after t seconds is $s(t) = 96 + 80t - 16t^2$.
(a) After how many seconds does the ball strike the ground?
(b) After how many seconds will the ball pass the top of the building on its way down?

106. Physics   An object is propelled vertically upward with an initial velocity of 20 meters per second. The distance s (in meters) of the object from the ground after t seconds is $s(t) = -4.9t^2 + 20t$.

(a) When will the object be 15 meters above the ground?
(b) When will it strike the ground?
(c) Will the object reach a height of 100 meters?

107. Consecutive Integers   The sum S of the consecutive integers $1, 2, 3, \ldots, n$ is given by the function $S(n) = \dfrac{1}{2}n(n + 1)$. Determine the number of consecutive integers, starting with 1, that must be added to get a sum of 210.

108. Geometry   The number of diagonals D of a polygon with n sides is given by the function $D(n) = \dfrac{1}{2}n(n - 3)$. Determine the number of sides a polygon with 65 diagonals will have. Is there a polygon with 80 diagonals?

109. Show that the sum of the roots of a quadratic equation is $-\dfrac{b}{a}$.

110. Show that the product of the roots of a quadratic equation is $\dfrac{c}{a}$.

111. Find k such that the function $f(x) = kx^2 + x + k = 0$ has a repeated real zero.

112. Find k such that the function $f(x) = x^2 - kx + 4 = 0$ has a repeated real zero.

113. Show that the real zeros of the function $f(x) = ax^2 + bx + c$ are the negatives of the real zeros of the function $g(x) = ax^2 - bx + c$. Assume that $b^2 - 4ac \geq 0$.

114. Show that the real zeros of the function $f(x) = ax^2 + bx + c$ are the reciprocals of the real zeros of the function $g(x) = cx^2 + bx + a$. Assume that $b^2 - 4ac \geq 0$.

Explaining Concepts: Discussion and Writing

115. Which of the following pairs of equations are equivalent? Explain.
  (a) $x^2 = 9$; $x = 3$
  (b) $x = \sqrt{9}$; $x = 3$
  (c) $(x - 1)(x - 2) = (x - 1)^2$; $x - 2 = x - 1$

116. Describe four methods you might use to find the zeros of a quadratic function. State your preferred method. Explain why you chose it.

117. Explain the benefits of evaluating the discriminant of a quadratic equation before attempting to solve it.

118. Create three quadratic functions, one having two distinct zeros, one having no real zeros, and one having exactly one real zero.

119. The word quadratic seems to imply four (*quad*), yet a quadratic equation is an equation that involves a polynomial of degree 2. Investigate the origin of the term quadratic as it is used in the expression quadratic equation. Write a brief essay on your findings.

120. How many times can the graphs of two different quadratic functions intersect?

'Are You Prepared?' Answers

1. (a) $(x - 6)(x + 1)$ (b) $(2x - 3)(x + 1)$ **2.** $2\sqrt{10}$ **3.** $\left\{-\dfrac{5}{3}, 3\right\}$ **4.** add; 9 **5.** (4, 10) **6.** Yes

2.4 Properties of Quadratic Functions

PREPARING FOR THIS SECTION *Before getting started, review the following:*

- Intercepts (Foundations, Section 2, pp. 11–12)
- Graphing Techniques: Transformations (Section 1.5, pp. 90–99)
- Completing the Square (Appendix A, Section A.4, p. A40)
- Solving Equations (Appendix A, Section A.8, pp. A69–A70)

Now Work the 'Are You Prepared?' problems on page 161.

OBJECTIVES **1** Graph a Quadratic Function Using Transformations (p. 153)
  2 Identify the Vertex and Axis of Symmetry of a Quadratic Function (p. 155)
  3 Graph a Quadratic Function Using Its Vertex, Axis, and Intercepts (p. 156)
  4 Find the Maximum or Minimum Value of a Quadratic Function (p. 160)

Many applications require a knowledge of quadratic functions. For example, suppose that Texas Instruments collects the data shown in Table 7, which relate the number of calculators sold at the price p (in dollars) per calculator. Since the price of a product determines the quantity that will be purchased, we treat price as the independent variable. The relationship between the number x of calculators sold and the price p per calculator may be approximated by the linear equation

$$x = 21,000 - 150p$$

Table 7

Price per Calculator, p (Dollars)	Number of Calculators, x
60	11,100
65	10,115
70	9,652
75	8,731
80	8,087
85	7,205
90	6,439

Then the revenue R derived from selling x calculators at the price p per calculator is equal to the unit selling price p times the number x of units actually sold. That is,

$$R = xp$$
$$R(p) = (21{,}000 - 150p)p$$
$$= -150p^2 + 21{,}000p$$

So the revenue R is a quadratic function of the price p. Figure 13 illustrates the graph of this revenue function, whose domain is $0 \leq p \leq 140$, since both x and p must be nonnegative.

Figure 13

A second situation in which a quadratic function appears involves the motion of a projectile. Based on Newton's second law of motion (force equals mass times acceleration, $F = ma$), it can be shown that, ignoring air resistance, the path of a projectile propelled upward at an inclination to the horizontal is the graph of a quadratic function. See Figure 14 for an illustration.

Figure 14 Path of a cannonball

1 Graph a Quadratic Function Using Transformations

We know how to graph the square function $f(x) = x^2$. Figure 15 shows the graph of three functions of the form $f(x) = ax^2$, $a > 0$, for $a = 1$, $a = \dfrac{1}{2}$, and $a = 3$. Notice that the larger the value of a, the "narrower" the graph is, and the smaller the value of a, the "wider" the graph is.

Figure 16 shows graphs of $f(x) = ax^2$ for $a < 0$. Notice that these graphs are reflections about the x-axis of the graphs in Figure 15. Based on the results of

Figure 15

Figure 16

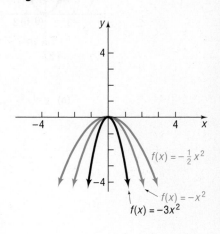

Figure 17
Graphs of a quadratic function,
$f(x) = ax^2 + bx + c, a \neq 0$

Axis of
symmetry

Vertex is
highest point

Vertex is
lowest point

Axis of
symmetry

(a) Opens up

$a > 0$

(b) Opens down

$a < 0$

these two figures, we can draw some general conclusions about the graph of $f(x) = ax^2$. First, as $|a|$ increases, the graph becomes "taller" (a vertical stretch), and as $|a|$ gets closer to zero, the graph gets "shorter" (a vertical compression). Second, if a is positive, the graph opens "up," and if a is negative, the graph opens "down."

The graphs in Figures 15 and 16 are typical of the graphs of all quadratic functions, which we call **parabolas.*** Refer to Figure 17, where two parabolas are pictured. The one on the left **opens up** and has a lowest point; the one on the right **opens down** and has a highest point. The lowest or highest point of a parabola is called the **vertex.** The vertical line passing through the vertex in each parabola in Figure 17 is called the **axis of symmetry** (usually abbreviated to **axis**) of the parabola. Because the parabola is symmetric about its axis, the axis of symmetry of a parabola can be used to find additional points on the parabola.

The parabolas shown in Figure 17 are the graphs of a quadratic function $f(x) = ax^2 + bx + c, a \neq 0$. Notice that the coordinate axes are not included in the figure. Depending on the values of a, b, and c, the axes could be placed anywhere. The important fact is that the shape of the graph of a quadratic function will look like one of the parabolas in Figure 17.

In the following example, we use techniques from Section 1.5 to graph a quadratic function $f(x) = ax^2 + bx + c, a \neq 0$. In so doing, we shall complete the square and write the function f in the form $f(x) = a(x - h)^2 + k$.

EXAMPLE 1   **Graphing a Quadratic Function Using Transformations**

Graph the function $f(x) = 2x^2 + 8x + 5$. Find the vertex and axis of symmetry.

Solution   We begin by completing the square on the right side.

$$f(x) = 2x^2 + 8x + 5$$
$$= 2(x^2 + 4x) + 5 \qquad \text{Factor out the 2 from } 2x^2 + 8x.$$
$$= 2(x^2 + 4x + 4) + 5 - 8 \quad \text{Complete the square of } x^2 + 4x \text{ by adding 4.}$$
$$\qquad\qquad\qquad\qquad\quad \text{Notice that the factor of 2 requires that}$$
$$= 2(x + 2)^2 - 3 \qquad\quad 8 \text{ be added and subtracted.}$$

The graph of f can be obtained in three stages, as shown in Figure 18. Now compare this graph to the graph in Figure 17(a). The graph of $f(x) = 2x^2 + 8x + 5$ is a parabola that opens up and has its vertex (lowest point) at $(-2, -3)$. Its axis of symmetry is the line $x = -2$.

Figure 18

(a) $y = x^2$ Multiply by 2; Vertical stretch

(b) $y = 2x^2$ Replace x by $x + 2$; Shift left 2 units

(c) $y = 2(x + 2)^2$ Subtract 3; Shift down 3 units

(d) $y = 2(x + 2)^2 - 3$ Vertex $(-2, -3)$

━━━ **Now Work** PROBLEM 21

*We shall study parabolas using a geometric definition later in this book.

The method used in Example 1 can be used to graph any quadratic function $f(x) = ax^2 + bx + c, a \neq 0$, as follows:

$$f(x) = ax^2 + bx + c$$

$$= a\left(x^2 + \frac{b}{a}x\right) + c \qquad \text{Factor out } a \text{ from } ax^2 + bx.$$

$$= a\left(x^2 + \frac{b}{a}x + \frac{b^2}{4a^2}\right) + c - a\left(\frac{b^2}{4a^2}\right) \qquad \begin{array}{l}\text{Complete the square by adding } \frac{b^2}{4a^2}.\\ \text{Look closely at this step!}\end{array}$$

$$= a\left(x + \frac{b}{2a}\right)^2 + c - \frac{b^2}{4a} \qquad \text{Factor; simplify.}$$

$$= a\left(x + \frac{b}{2a}\right)^2 + \frac{4ac - b^2}{4a} \qquad c - \frac{b^2}{4a} = c \cdot \frac{4a}{4a} - \frac{b^2}{4a} = \frac{4ac - b^2}{4a}$$

$$= a\left(x - \left(-\frac{b}{2a}\right)\right)^2 + \frac{4ac - b^2}{4a}$$

Based on these results, we conclude the following:

If $h = -\dfrac{b}{2a}$ and $k = \dfrac{4ac - b^2}{4a}$, then

$$f(x) = ax^2 + bx + c = a(x - h)^2 + k \qquad \qquad \textbf{(1)}$$

The graph of $f(x) = a(x - h)^2 + k$ is the parabola $y = ax^2$ shifted horizontally h units (replace x by $x - h$) and vertically k units (add k). As a result, the vertex is at (h, k), and the graph opens up if $a > 0$ and down if $a < 0$. The axis of symmetry is the vertical line $x = h$.

For example, compare equation (1) with the solution given in Example 1.

$$f(x) = 2(x + 2)^2 - 3$$
$$= 2(x - (-2))^2 + (-3)$$
$$\qquad\quad \uparrow \qquad\qquad \uparrow \qquad\quad \uparrow$$
$$= a(x - \quad h)^2 + \quad k$$

We conclude that $a = 2$, so the graph opens up. Also, we find that $h = -2$ and $k = -3$, so its vertex is at $(-2, -3)$.

2 Identify the Vertex and Axis of Symmetry of a Quadratic Function

We do not need to complete the square to obtain the vertex. In almost every case, it is easier to obtain the vertex of a quadratic function f by remembering that its x-coordinate is $h = -\dfrac{b}{2a}$. The y-coordinate k can then be found by evaluating f at $-\dfrac{b}{2a}$. That is, $k = f\left(-\dfrac{b}{2a}\right)$.

We summarize these properties of quadratic functions below.

Properties of the Graph of a Quadratic Function

$$f(x) = ax^2 + bx + c \qquad a \neq 0$$

$$\text{Vertex} = \left(-\frac{b}{2a}, f\left(-\frac{b}{2a}\right)\right) \qquad \text{Axis of symmetry: the line } x = -\frac{b}{2a} \qquad \textbf{(2)}$$

Parabola opens up if $a > 0$; the vertex is a minimum point.
Parabola opens down if $a < 0$; the vertex is a maximum point.

EXAMPLE 2   **Locating the Vertex without Graphing**

Without graphing, locate the vertex and axis of symmetry of the parabola defined by $f(x) = -3x^2 + 6x + 1$. Does it open up or down?

Solution   For this quadratic function, $a = -3, b = 6$, and $c = 1$. The x-coordinate of the vertex is

$$h = -\frac{b}{2a} = -\frac{6}{-6} = 1$$

The y-coordinate of the vertex is

$$k = f\left(-\frac{b}{2a}\right) = f(1) = -3(1)^2 + 6(1) + 1 = 4$$

The vertex is located at the point $(1, 4)$. The axis of symmetry is the line $x = 1$. Because $a = -3 < 0$, the parabola opens down.

●

3 Graph a Quadratic Function Using Its Vertex, Axis, and Intercepts

The location of the vertex and the location of the intercepts, along with knowledge as to whether the graph of the quadratic function opens up or down, usually provide enough information to graph $f(x) = ax^2 + bx + c, a \neq 0$.

The y-intercept is the value of f at $x = 0$; that is, the y-intercept is $f(0) = c$.

The x-intercepts, if there are any, are found by solving the quadratic equation

$$ax^2 + bx + c = 0$$

This equation has two, one, or no real solutions, depending on whether the discriminant $b^2 - 4ac$ is positive, 0, or negative. Depending on the value of the discriminant, the graph of f has x-intercepts, as follows:

> **The x-Intercepts of a Quadratic Function**
>
> **1.** If the discriminant $b^2 - 4ac > 0$, the graph of $f(x) = ax^2 + bx + c$ has two distinct x-intercepts, so it crosses the x-axis in two places.
> **2.** If the discriminant $b^2 - 4ac = 0$, the graph of $f(x) = ax^2 + bx + c$ has one x-intercept, so it touches the x-axis at its vertex.
> **3.** If the discriminant $b^2 - 4ac < 0$, the graph of $f(x) = ax^2 + bx + c$ has no x-intercept, so it does not cross or touch the x-axis.

Figure 19 illustrates these possibilities for parabolas that open up.

Figure 19
$f(x) = ax^2 + bx + c, a > 0$

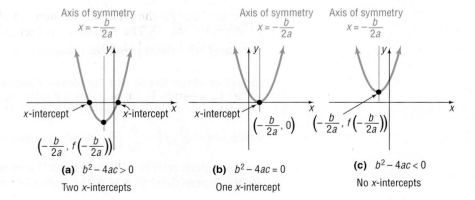

EXAMPLE 3

How to Graph a Quadratic Function Using Its Properties

Graph $f(x) = -3x^2 + 6x + 1$ using its properties. Determine the domain and the range of f. Determine where f is increasing and where it is decreasing.

Step-by-Step Solution

Step 1: Determine whether the graph of f opens up or down.

In Example 2, we determined that the graph of $f(x) = -3x^2 + 6x + 1$ opens down because $a = -3 < 0$.

Step 2: Determine the vertex and axis of symmetry of the graph of f.

In Example 2, we found the vertex to be at the point whose coordinates are $(1, 4)$. The axis of symmetry is the line $x = 1$.

Step 3: Determine the intercepts of the graph of f.

The y-intercept is found by letting $x = 0$. The y-intercept is $f(0) = 1$. The x-intercepts are found by solving the equation $f(x) = 0$.

$$f(x) = 0$$
$$-3x^2 + 6x + 1 = 0 \quad a = -3, b = 6, c = 1$$

The discriminant $b^2 - 4ac = 6^2 - 4(-3)(1) = 36 + 12 = 48 > 0$, so the equation has two real solutions and the graph has two x-intercepts. Using the quadratic formula, we find that

$$x = \frac{-b + \sqrt{b^2 - 4ac}}{2a} = \frac{-6 + \sqrt{48}}{2(-3)} = \frac{-6 + 4\sqrt{3}}{-6} \approx -0.15$$

and

$$x = \frac{-b - \sqrt{b^2 - 4ac}}{2a} = \frac{-6 - \sqrt{48}}{2(-3)} = \frac{-6 - 4\sqrt{3}}{-6} \approx 2.15$$

The x-intercepts are approximately -0.15 and 2.15.

Step 4: Use the information in Steps 1 through 3 to graph f.

The graph is illustrated in Figure 20. Notice how we use the y-intercept and the axis of symmetry, $x = 1$, to obtain the additional point $(2, 1)$ on the graph.

Figure 20

The domain of f is the set of all real numbers. Based on the graph, the range of f is the interval $(-\infty, 4]$. The function f is increasing on the interval $(-\infty, 1)$ and decreasing on the interval $(1, \infty)$.

●

Graph the function in Example 3 by completing the square and using transformations. Which method do you prefer?

Now Work PROBLEM 29

If the graph of a quadratic function has only one x-intercept or no x-intercepts, it is usually necessary to plot an additional point to obtain the graph.

EXAMPLE 4

Graphing a Quadratic Function Using Its Vertex, Axis, and Intercepts

(a) Graph $f(x) = x^2 - 6x + 9$ by determining whether the graph opens up or down and by finding its vertex, axis of symmetry, y-intercept, and x-intercepts, if any.

(b) Determine the domain and the range of f.

(c) Determine where f is increasing and where it is decreasing.

Solution

(a) **STEP 1:** For $f(x) = x^2 - 6x + 9$, we have $a = 1, b = -6$, and $c = 9$. Since $a = 1 > 0$, the parabola opens up.

STEP 2: The x-coordinate of the vertex is

$$h = -\frac{b}{2a} = -\frac{-6}{2(1)} = 3$$

The y-coordinate of the vertex is

$$k = f(3) = (3)^2 - 6(3) + 9 = 0$$

Figure 21

So the vertex is at $(3, 0)$. The axis of symmetry is the line $x = 3$.

STEP 3: The y-intercept is $f(0) = 9$. Since the vertex $(3, 0)$ lies on the x-axis, the graph touches the x-axis at the x-intercept.

STEP 4: By using the axis of symmetry and the y-intercept at $(0, 9)$, we can locate the additional point $(6, 9)$ on the graph. See Figure 21.

(b) The domain of f is the set of all real numbers. Based on the graph, the range of f is the interval $[0, \infty)$.

(c) The function f is decreasing on the interval $(-\infty, 3)$ and increasing on the interval $(3, \infty)$.

●

━━━━ **Now Work** PROBLEM 35

EXAMPLE 5

Graphing a Quadratic Function Using Its Vertex, Axis, and Intercepts

(a) Graph $f(x) = 2x^2 + x + 1$ by determining whether the graph opens up or down and by finding its vertex, axis of symmetry, y-intercept, and x-intercepts, if any.

(b) Determine the domain and the range of f.

(c) Determine where f is increasing and where it is decreasing.

Solution

(a) **STEP 1:** For $f(x) = 2x^2 + x + 1$, we have $a = 2, b = 1$, and $c = 1$. Since $a = 2 > 0$, the parabola opens up.

STEP 2: The x-coordinate of the vertex is

$$h = -\frac{b}{2a} = -\frac{1}{4}$$

The y-coordinate of the vertex is

$$k = f\left(-\frac{1}{4}\right) = 2\left(\frac{1}{16}\right) + \left(-\frac{1}{4}\right) + 1 = \frac{7}{8}$$

So the vertex is at $\left(-\frac{1}{4}, \frac{7}{8}\right)$. The axis of symmetry is the line $x = -\frac{1}{4}$.

NOTE In Example 5, since the vertex is above the x-axis and the parabola opens up, we can conclude that the graph of the quadratic function will have no x-intercepts. ∎

Figure 22

STEP 3: The y-intercept is $f(0) = 1$. The x-intercept(s), if any, obey the equation $2x^2 + x + 1 = 0$. Since the discriminant $b^2 - 4ac = (1)^2 - 4(2)(1) = -7 < 0$, this equation has no real solutions, and therefore the graph has no x-intercepts.

STEP 4: We use the point $(0, 1)$ and the axis of symmetry $x = -\dfrac{1}{4}$ to locate the additional point $\left(-\dfrac{1}{2}, 1\right)$ on the graph. See Figure 22.

(b) The domain of f is the set of all real numbers. Based on the graph, the range of f is the interval $\left[\dfrac{7}{8}, \infty\right)$.

(c) The function f is decreasing on the interval $\left(-\infty, -\dfrac{1}{4}\right)$ and is increasing on the interval $\left(-\dfrac{1}{4}, \infty\right)$.

Now Work PROBLEM 39

SUMMARY

Steps for Graphing a Quadratic Function $f(x) = ax^2 + bx + c, a \neq 0$

Option 1

STEP 1: Complete the square in x to write the quadratic function in the form $f(x) = a(x - h)^2 + k$.

STEP 2: Graph the function in stages using transformations.

Option 2

STEP 1: Determine whether the graph of f opens up or down.

STEP 2: Determine the vertex $\left(-\dfrac{b}{2a}, f\left(-\dfrac{b}{2a}\right)\right)$ and the axis of symmetry, $x = -\dfrac{b}{2a}$.

STEP 3: Determine the y-intercept, $f(0)$. Determine the x-intercept(s), if any.

    (a) If $b^2 - 4ac > 0$, then the graph of the quadratic function has two x-intercepts, which are found by solving the equation $ax^2 + bx + c = 0$.

    (b) If $b^2 - 4ac = 0$, the vertex is the x-intercept.

    (c) If $b^2 - 4ac < 0$, there are no x-intercepts.

STEP 4: Determine an additional point by using the y-intercept and the axis of symmetry. Plot the points and draw the graph.

Finding a Quadratic Function from Its Graph

Given the vertex (h, k) and one additional point on the graph of a quadratic function $f(x) = ax^2 + bx + c, a \neq 0$, we can use

$$f(x) = a(x - h)^2 + k \qquad (3)$$

to obtain the quadratic function.

EXAMPLE 6   **Finding the Quadratic Function Given Its Vertex and One Other Point**

Determine the quadratic function whose vertex is $(1, -5)$ and whose y-intercept is -3. The graph of the parabola is shown in Figure 23.

Solution

The vertex is $(1, -5)$, so $h = 1$ and $k = -5$. Substitute these values into equation (3).

$$f(x) = a(x - h)^2 + k \quad \text{Equation (3)}$$
$$f(x) = a(x - 1)^2 - 5 \quad h = 1, k = -5$$

To determine the value of a, we use the fact that $f(0) = -3$ (the y-intercept).

$$f(x) = a(x - 1)^2 - 5$$
$$-3 = a(0 - 1)^2 - 5 \quad x = 0, y = f(0) = -3$$
$$-3 = a - 5$$
$$a = 2$$

The quadratic function whose graph is shown in Figure 23 is

$$f(x) = a(x - h)^2 + k = 2(x - 1)^2 - 5$$
$$= 2x^2 - 4x - 3$$

Figure 23

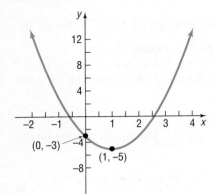

(0, -3)

(1, -5)

━━━━ **Now Work** PROBLEM 43

4 Find the Maximum or Minimum Value of a Quadratic Function

The graph of a quadratic function

$$f(x) = ax^2 + bx + c, \quad a \neq 0$$

is a parabola with vertex at $\left(-\dfrac{b}{2a}, f\left(-\dfrac{b}{2a}\right)\right)$. This vertex is the highest point on the graph if $a < 0$ and the lowest point on the graph if $a > 0$. If the vertex is the highest point $(a < 0)$, then $f\left(-\dfrac{b}{2a}\right)$ is the **maximum value** of f. If the vertex is the lowest point $(a > 0)$, then $f\left(-\dfrac{b}{2a}\right)$ is the **minimum value** of f.

| EXAMPLE 7 | **Finding the Maximum or Minimum Value of a Quadratic Function** |

Determine whether the quadratic function

$$f(x) = x^2 - 4x - 5$$

has a maximum or minimum value. Then find the maximum or minimum value.

Solution

We compare $f(x) = x^2 - 4x - 5$ to $f(x) = ax^2 + bx + c$. We conclude that $a = 1, b = -4$, and $c = -5$. Since $a > 0$, the graph of f opens up, so the vertex is a minimum point. The minimum value occurs at

$$x = -\frac{b}{2a} = -\frac{-4}{2(1)} = \frac{4}{2} = 2$$
$$a = 1, b = -4$$

The minimum value is

$$f\left(-\frac{b}{2a}\right) = f(2) = 2^2 - 4(2) - 5 = 4 - 8 - 5 = -9$$

━━━━ **Now Work** PROBLEM 51

2.4 Assess Your Understanding

'Are You Prepared?' *Answers are given at the end of these exercises. If you get a wrong answer, read the pages listed in* red.

1. List the intercepts of the graph of $y = x^2 - 9$. (p. 12)
2. Find the real solutions of the equation $2x^2 + 7x - 4 = 0$. (p. A69–A70)

3. To complete the square of $x^2 - 5x$, you add the number _____ . (p. A40)
4. To graph $y = (x - 4)^2$, you shift the graph of $y = x^2$ to the _____ a distance of _____ units. (pp. 91–93)

Concepts and Vocabulary

5. The graph of a quadratic function is called a(n) _____ .

6. The vertical line passing through the vertex of a parabola is called the _____ .

7. The x-coordinate of the vertex of $f(x) = ax^2 + bx + c$, $a \neq 0$, is _____ .

8. **True or False** The graph of $f(x) = 2x^2 + 3x - 4$ opens up.

9. **True or False** The y-coordinate of the vertex of $f(x) = -x^2 + 4x + 5$ is $f(2)$.

10. **True or False** If the discriminant $b^2 - 4ac = 0$, the graph of $f(x) = ax^2 + bx + c, a \neq 0$, will touch the x-axis at its vertex.

Skill Building

In Problems 11–18, match graphs A–H to one of the following functions.

11. $f(x) = x^2 - 1$

12. $f(x) = -x^2 - 1$

13. $f(x) = x^2 - 2x + 1$

14. $f(x) = x^2 + 2x + 1$

15. $f(x) = x^2 - 2x + 2$

16. $f(x) = x^2 + 2x$

17. $f(x) = (x - 1)^2 - 1$

18. $f(x) = (x + 1)^2 + 1$

A.

B.

C.

D.

E.

F.

G.

H.
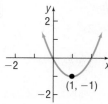

In Problems 19–28, graph the function f by starting with the graph of $y = x^2$ and using transformations (shifting, compressing, stretching, and/or reflection).

[Hint: If necessary, write f in the form $f(x) = a(x - h)^2 + k$.]

19. $f(x) = \dfrac{1}{4}x^2 - 2$

20. $f(x) = 2x^2 + 4$

21. $f(x) = (x + 2)^2 - 2$

22. $f(x) = (x - 3)^2 - 10$

23. $f(x) = 2(x - 1)^2 - 1$

24. $f(x) = 3(x + 1)^2 - 3$

25. $f(x) = -x^2 - 2x$

26. $f(x) = -2x^2 + 6x + 2$

27. $f(x) = 2x^2 + 4x - 2$

28. $f(x) = 3x^2 + 12x + 5$

In Problems 29–42, (a) graph each quadratic function by determining whether its graph opens up or down and by finding its vertex, axis of symmetry, y-intercept, and x-intercepts, if any. (b) Determine the domain and the range of the function. (c) Determine where the function is increasing and where it is decreasing.

29. $f(x) = x^2 + 2x$ 30. $f(x) = x^2 - 4x$ 31. $f(x) = -x^2 - 6x$ 32. $f(x) = -x^2 + 4x$

33. $f(x) = x^2 + 2x - 8$ 34. $f(x) = x^2 - 2x - 3$ 35. $f(x) = x^2 + 2x + 1$ 36. $f(x) = x^2 + 6x + 9$

37. $f(x) = 2x^2 - x + 2$ 38. $f(x) = 4x^2 - 2x + 1$ 39. $f(x) = -2x^2 + 2x - 3$ 40. $f(x) = -3x^2 + 3x - 2$

41. $f(x) = -4x^2 - 6x + 2$ 42. $f(x) = 3x^2 - 8x + 2$

In Problems 43–48, determine the quadratic function whose graph is given.

43.

44.

45.

Vertex: (−3, 5)

46.

Vertex: (2, 3)

47.
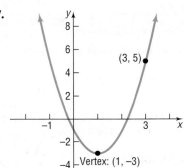
(3, 5)
Vertex: (1, −3)

48.

Vertex: (−2, 6)

In Problems 49–56, determine, without graphing, whether the given quadratic function has a maximum value or a minimum value and then find the value.

49. $f(x) = 2x^2 + 12x$ 50. $f(x) = -2x^2 + 12x$ 51. $f(x) = 2x^2 + 12x - 3$ 52. $f(x) = 4x^2 - 8x + 3$

53. $f(x) = -x^2 + 10x - 4$ 54. $f(x) = -2x^2 + 8x + 3$ 55. $f(x) = -3x^2 + 12x + 1$ 56. $f(x) = 4x^2 - 4x$

Mixed Practice

In Problems 57–68, (a) graph each function. (b) Determine the domain and the range of the function. (c) Determine where the function is increasing and where it is decreasing.

57. $f(x) = x^2 - 2x - 15$ 58. $g(x) = x^2 - 2x - 8$ 59. $F(x) = 2x - 5$ 60. $f(x) = \frac{3}{2}x - 2$

61. $g(x) = -2(x - 3)^2 + 2$ 62. $h(x) = -3(x + 1)^2 + 4$ 63. $f(x) = 2x^2 + x + 1$ 64. $G(x) = 3x^2 + 2x + 5$

65. $h(x) = -\frac{2}{5}x + 4$ 66. $f(x) = -3x + 2$ 67. $H(x) = -4x^2 - 4x - 1$ 68. $F(x) = -4x^2 + 20x - 25$

Applications and Extensions

69. The graph of the function $f(x) = ax^2 + bx + c$ has vertex at $(0, 2)$ and passes through the point $(1, 8)$. Find a, b, and c.

70. The graph of the function $f(x) = ax^2 + bx + c$ has vertex at $(1, 4)$ and passes through the point $(-1, -8)$. Find a, b, and c.

In Problems 71–76, for the given functions f and g,
(a) Graph f and g on the same Cartesian plane.   (b) Solve $f(x) = g(x)$.   (c) Use the result of part (b) to label the points of intersection of the graphs of f and g.   (d) Shade the region between the points of intersection.

71. $f(x) = 2x - 1$; $g(x) = x^2 - 4$

72. $f(x) = -2x - 1$; $g(x) = x^2 - 9$

73. $f(x) = -x^2 + 4$; $g(x) = -2x + 1$

74. $f(x) = -x^2 + 9$; $g(x) = 2x + 1$

75. $f(x) = -x^2 + 5x$; $g(x) = x^2 + 3x - 4$

76. $f(x) = -x^2 + 7x - 6$; $g(x) = x^2 + x - 6$

Answer Problems 77 and 78 using the following: A quadratic function of the form $f(x) = ax^2 + bx + c$ with $b^2 - 4ac > 0$ may also be written in the form $f(x) = a(x - r_1)(x - r_2)$, where r_1 and r_2 are the x-intercepts of the graph of the quadratic function.

77. (a) Find a quadratic function whose x-intercepts are -3 and 1 with $a = 1$; $a = 2$; $a = -2$; $a = 5$.
(b) How does the value of a affect the intercepts?
(c) How does the value of a affect the axis of symmetry?
(d) How does the value of a affect the vertex?
(e) Compare the x-coordinate of the vertex with the midpoint of the x-intercepts. What might you conclude?

78. (a) Find a quadratic function whose x-intercepts are -5 and 3 with $a = 1$; $a = 2$; $a = -2$; $a = 5$.
(b) How does the value of a affect the intercepts?
(c) How does the value of a affect the axis of symmetry?
(d) How does the value of a affect the vertex?
(e) Compare the x-coordinate of the vertex with the midpoint of the x-intercepts. What might you conclude?

---

79. Suppose that $f(x) = x^2 + 4x - 21$.
(a) What is the vertex of f?
(b) What are the x-intercepts of the graph of f?
(c) Solve $f(x) = -21$ for x. What points are on the graph of f?
(d) Use the information obtained in parts (a)–(c) to graph $f(x) = x^2 + 4x - 21$.

80. Suppose that $f(x) = x^2 + 2x - 8$.
(a) What is the vertex of f?
(b) What are the x-intercepts of the graph of f?
(c) Solve $f(x) = -8$ for x. What points are on the graph of f?
(d) Use the information obtained in parts (a)–(c) to graph $f(x) = x^2 + 2x - 8$.

81. Maximizing Revenue Suppose that the manufacturer of a gas clothes dryer has found that, when the unit price is p dollars, the revenue R (in dollars) is

$$R(p) = -4p^2 + 4000p$$

What unit price should be established for the dryer to maximize revenue? What is the maximum revenue?

82. Maximizing Revenue The John Deere company has found that the revenue, in dollars, from sales of heavy-duty tractors is a function of the unit price p, in dollars, that it charges. If the revenue R is

$$R(p) = -\frac{1}{2}p^2 + 1900p$$

what unit price p should be charged to maximize revenue? What is the maximum revenue?

83. Minimizing Marginal Cost The **marginal cost** of a product can be thought of as the cost of producing one additional unit of output. For example, if the marginal cost of producing the 50th product is $6.20, it cost $6.20 to increase production from 49 to 50 units of output. Suppose the marginal cost C (in dollars) to produce x thousand MP3 players is given by the function

$$C(x) = x^2 - 140x + 7400$$

(a) How many players should be produced to minimize the marginal cost?
(b) What is the minimum marginal cost?

84. Minimizing Marginal Cost (See Problem 83.) The marginal cost C (in dollars) of manufacturing x cell phones (in thousands) is given by

$$C(x) = 5x^2 - 200x + 4000$$

(a) How many cell phones should be manufactured to minimize the marginal cost?
(b) What is the minimum marginal cost?

85. Stopping Distance An accepted relationship between stopping distance, d (in feet), and the speed of a car, v (in mph), is $d = 1.1v + 0.06v^2$ on dry, level concrete.
(a) How many feet will it take a car traveling 45 mph to stop on dry, level concrete?
(b) If an accident occurs 200 feet ahead of you, what is the maximum speed you can be traveling to avoid being involved?
(c) What might the term $1.1v$ represent?
Source: www2.nsta.org/Energy/fn_braking.html

86. Birthrate of Unmarried Women In the United States, the birthrate B of unmarried women (births per 1000 unmarried women) for women whose age is a is modeled by the function $B(a) = -0.27a^2 + 14.23a - 120.16$.
(a) What is the age of unmarried women with the highest birthrate?
(b) What is the highest birthrate of unmarried women?
(c) Evaluate and interpret $B(40)$.
Source: United States Statistical Abstract, 2009

87. Business The monthly revenue R achieved by selling x wristwatches is modeled by $R(x) = 75x - 0.2x^2$. The monthly cost C of selling x wristwatches is $C(x) = 32x + 1750$.
(a) How many wristwatches must the firm sell to maximize revenue? What is the maximum revenue?
(b) Profit is given as $P(x) = R(x) - C(x)$. What is the profit function?

(c) How many wristwatches must the firm sell to maximize profit? What is the maximum profit?

(d) Provide a reasonable explanation as to why the answers found in parts (a) and (c) differ. Explain why a quadratic function is a reasonable model for revenue.

88. Business The daily revenue R achieved by selling x boxes of candy is modeled by $R(x) = 9.5x - 0.04x^2$. The daily cost C of selling x boxes of candy is $C(x) = 1.25x + 250$.

(a) How many boxes of candy must the firm sell to maximize revenue? What is the maximum revenue?

(b) Profit is given as $P(x) = R(x) - C(x)$. What is the profit function?

(c) How many boxes of candy must the firm sell to maximize profit? What is the maximum profit?

(d) Provide a reasonable explanation as to why the answers found in parts (a) and (c) differ. Explain why a quadratic function is a reasonable model for revenue.

89. See Problems 77 and 78. Find a quadratic function whose x-intercepts are -4 and 2 and whose range is $[-18, \infty)$.

90. See Problems 77 and 78. Find a quadratic function whose x-intercepts are -1 and 5 and whose range is $(-\infty, 9]$.

91. Let $f(x) = ax^2 + bx + c$, where $a, b,$ and c are odd integers. If x is an integer, show that $f(x)$ must be an odd integer.

[**Hint:** x is either an even integer or an odd integer.]

Explaining Concepts: Discussion and Writing

92. Make up a quadratic function that opens down and has only one x-intercept. Compare yours with others in the class. What are the similarities? What are the differences?

93. On one set of coordinate axes, graph the family of parabolas $f(x) = x^2 + 2x + c$ for $c = -3, c = 0,$ and $c = 1$. Describe the characteristics of a member of this family.

94. On one set of coordinate axes, graph the family of parabolas $f(x) = x^2 + bx + 1$ for $b = -4, b = 0,$ and $b = 4$. Describe the general characteristics of this family.

95. State the circumstances that cause the graph of a quadratic function $f(x) = ax^2 + bx + c$ to have no x-intercepts.

96. Why does the graph of a quadratic function open up if $a > 0$ and down if $a < 0$?

97. Can a quadratic function have a range of $(-\infty, \infty)$? Justify your answer.

98. What are the possibilities for the number of times the graphs of two different quadratic functions intersect? Justify your answer.

'Are You Prepared?' Answers

1. $(0, -9), (-3, 0), (3, 0)$ **2.** $\left\{-4, \dfrac{1}{2}\right\}$ **3.** $\dfrac{25}{4}$ **4.** right; 4

2.5 Inequalities Involving Quadratic Functions

PREPARING FOR THIS SECTION *Before getting started, review the following:*

- Solve Inequalities (Appendix A, Section A.10, pp. A86–A87)
- Use Interval Notation (Appendix A, Section A.10, pp. A83–A84)

Now Work the 'Are You Prepared?' problems on page 166.

OBJECTIVES 1 Solve Inequalities Involving a Quadratic Function (p. 164)

1 Solve Inequalities Involving a Quadratic Function

In this section we solve inequalities that involve quadratic functions. We will accomplish this by using their graphs. For example, to solve the inequality

$$ax^2 + bx + c > 0 \qquad a \neq 0$$

we graph the function $f(x) = ax^2 + bx + c$ and, from the graph, determine where it is above the x-axis, that is, where $f(x) > 0$. To solve the inequality $ax^2 + bx + c < 0, a \neq 0$, we graph the function $f(x) = ax^2 + bx + c$ and determine where the graph is below the x-axis. If the inequality is not strict, we include the x-intercepts in the solution.

Let's look at an example.

EXAMPLE 1   **Solving an Inequality**

Solve the inequality $x^2 - 4x - 12 \leq 0$ and graph the solution set.

Solution

We graph the function $f(x) = x^2 - 4x - 12$. The intercepts are

Figure 24

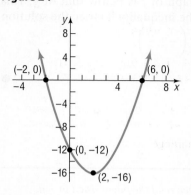

y-intercept: $\qquad\qquad\qquad f(0) = -12$ Evaluate f at 0.

x-intercepts (if any): $\qquad x^2 - 4x - 12 = 0$ Solve $f(x) = 0$.

$$(x - 6)(x + 2) = 0 \qquad \text{Factor.}$$

$$x - 6 = 0 \quad \text{or} \quad x + 2 = 0 \qquad \text{Apply the Zero-Product Property.}$$

$$x = 6 \quad \text{or} \qquad x = -2$$

The y-intercept is -12; the x-intercepts are -2 and 6.

The vertex is at $x = -\dfrac{b}{2a} = -\dfrac{-4}{2(1)} = 2$. Since $f(2) = -16$, the vertex is $(2, -16)$. See Figure 24 for the graph.

The graph is below the x-axis for $-2 < x < 6$. Since the original inequality is not strict, we include the x-intercepts. The solution set is $\{x | -2 \le x \le 6\}$ or, using interval notation, $[-2, 6]$.

Figure 25

See Figure 25 for the graph of the solution set.

$\blacksquare$

Now Work PROBLEM 9

EXAMPLE 2	**Solving an Inequality**

Solve the inequality $2x^2 < x + 10$ and graph the solution set.

Solution

Option 1 We rearrange the inequality so that 0 is on the right side.

Figure 26

$$2x^2 < x + 10$$

$$2x^2 - x - 10 < 0 \qquad \text{Subtract } x + 10 \text{ from both sides.}$$

This inequality is equivalent to the one that we wish to solve.

Next we graph the function $f(x) = 2x^2 - x - 10$. We want to determine where the graph is below the x-axis. The intercepts are

y-intercept: $\qquad\qquad\qquad\qquad f(0) = -10$ Evaluate f at 0.

x-intercepts (if any): $\qquad 2x^2 - x - 10 = 0$ Solve $f(x) = 0$.

$$(2x - 5)(x + 2) = 0 \qquad \text{Factor.}$$

$$2x - 5 = 0 \quad \text{or} \quad x + 2 = 0 \qquad \begin{array}{l}\text{Apply the Zero-Product}\\\text{Property.}\end{array}$$

$$x = \frac{5}{2} \quad \text{or} \qquad x = -2$$

The y-intercept is -10; the x-intercepts are -2 and $\dfrac{5}{2}$.

Figure 27

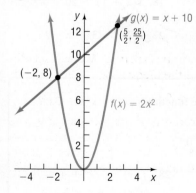

The vertex is at $x = -\dfrac{b}{2a} = -\dfrac{-1}{2(2)} = \dfrac{1}{4}$. Since $f\left(\dfrac{1}{4}\right) = -10.125$, the vertex is $\left(\dfrac{1}{4}, -10.125\right)$. See Figure 26 for the graph.

The graph is below the x-axis between $x = -2$ and $x = \dfrac{5}{2}$. Since the inequality is strict, the solution set is $\left\{x \Big| -2 < x < \dfrac{5}{2}\right\}$ or, using interval notation, $\left(-2, \dfrac{5}{2}\right)$.

Option 2 If $f(x) = 2x^2$ and $g(x) = x + 10$, the inequality that we wish to solve is $f(x) < g(x)$. We graph the functions $f(x) = 2x^2$ and $g(x) = x + 10$. See Figure 27. The graphs intersect where $f(x) = g(x)$. Then

$$2x^2 = x + 10 \qquad\qquad f(x) = g(x)$$

$$2x^2 - x - 10 = 0$$

$$(2x - 5)(x + 2) = 0 \qquad\qquad \text{Factor.}$$

$$2x - 5 = 0 \quad \text{or} \quad x + 2 = 0 \qquad \text{Apply the Zero-Product Property.}$$

$$x = \frac{5}{2} \quad \text{or} \qquad x = -2$$

The graphs intersect at the points $(-2, 8)$ and $\left(\dfrac{5}{2}, \dfrac{25}{2}\right)$. The inequality that we wish to solve is $f(x) < g(x)$, so we seek where the graph of f is below that of g. This happens between the points of intersection. Since the inequality is strict, the solution set is $\left\{x \mid -2 < x < \dfrac{5}{2}\right\}$ or, using interval notation, $\left(-2, \dfrac{5}{2}\right)$.

See Figure 28 for the graph of the solution set.

Figure 28

A number line from -4 to 4 showing the solution set.

──**Now Work** PROBLEMS 5 AND 11

2.5 Assess Your Understanding

'Are You Prepared?' *Answers are given at the end of these exercises. If you get a wrong answer, read the pages listed in red.*

1. Solve the inequality $-3x - 2 < 7$. (pp. A86–A87)

2. Write $(-2, 7]$ using inequality notation and graph the inequality. (pp. A83–A84)

Skill Building

In Problems 3–6, use the figure to solve each inequality.

3.

(a) $f(x) > 0$
(b) $f(x) \leq 0$

4.

(a) $g(x) < 0$
(b) $g(x) \geq 0$

5.

(a) $g(x) \geq f(x)$
(b) $f(x) > g(x)$

6.

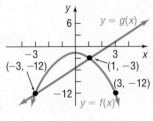

(a) $f(x) < g(x)$
(b) $f(x) \geq g(x)$

In Problems 7–22, solve each inequality.

7. $x^2 - 3x - 10 \leq 0$

8. $x^2 + 3x - 10 > 0$

9. $x^2 - 4x > 0$

10. $x^2 + 8x \leq 0$

11. $x^2 + x > 12$

12. $x^2 + 7x < -12$

13. $x^2 + 6x + 9 \leq 0$

14. $x^2 - 4x + 4 \leq 0$

15. $2x^2 \leq 5x + 3$

16. $6x^2 \leq 6 + 5x$

17. $x(x - 7) > 8$

18. $x(x + 1) > 20$

19. $4x^2 + 9 < 6x$

20. $25x^2 + 16 < 40x$

21. $6(x^2 + 1) \geq 5x$

22. $2(2x^2 - 3x) \geq -9$

Mixed Practice

23. What is the domain of the function $f(x) = \sqrt{x^2 - 16}$?

24. What is the domain of the function $f(x) = \sqrt{x - 3x^2}$?

In Problems 25–32, use the given functions f and g.

(a) Solve $f(x) = 0$. (b) Solve $g(x) = 0$. (c) Solve $f(x) = g(x)$. (d) Solve $f(x) > 0$.
(e) Solve $g(x) \leq 0$. (f) Solve $f(x) > g(x)$. (g) Solve $f(x) \geq 1$.

25. $f(x) = x^2 - 1$
    $g(x) = 3x + 3$

26. $f(x) = -x^2 + 3$
    $g(x) = -3x + 3$

27. $f(x) = -x^2 + 1$
    $g(x) = 4x + 1$

28. $f(x) = -x^2 + 4$
    $g(x) = -x - 2$

29. $f(x) = x^2 - 4$
    $g(x) = -x^2 + 4$

30. $f(x) = x^2 - 2x + 1$
    $g(x) = -x^2 + 1$

31. $f(x) = x^2 - x - 2$
    $g(x) = x^2 + x - 2$

32. $f(x) = -x^2 - x + 1$
    $g(x) = -x^2 + x + 6$

Applications and Extensions

33. Physics   A ball is thrown vertically upward with an initial velocity of 80 feet per second. The distance s (in feet) of the ball from the ground after t seconds is $s(t) = 80t - 16t^2$.

$s = 80t - 16t^2$

96 ft

(a) At what time t will the ball strike the ground?
(b) For what time t is the ball more than 96 feet above the ground?

34. Physics   A ball is thrown vertically upward with an initial velocity of 96 feet per second. The distance s (in feet) of the ball from the ground after t seconds is $s(t) = 96t - 16t^2$.

(a) At what time t will the ball strike the ground?
(b) For what time t is the ball more than 128 feet above the ground?

35. Revenue   Suppose that the manufacturer of a gas clothes dryer has found that, when the unit price is p dollars, the revenue R (in dollars) is

$$R(p) = -4p^2 + 4000p$$

(a) At what prices p is revenue zero?
(b) For what range of prices will revenue exceed $800,000?

36. Revenue   The John Deere company has found that the revenue from sales of heavy-duty tractors is a function of the unit price p, in dollars, that it charges. The revenue R, in dollars, is

$$R(p) = -\frac{1}{2}p^2 + 1900p$$

(a) At what prices p is revenue zero?
(b) For what range of prices will revenue exceed $1,200,000?

Explaining Concepts: Discussion and Writing

37. Explain why the inequality $(x - 4)^2 \leq 0$ has exactly one solution.

38. Explain why the inequality $(x - 2)^2 > 0$ has one real number that is not a solution.

39. Explain why the inequality $x^2 + x + 1 > 0$ has all real numbers as the solution set.

40. Explain why the inequality $x^2 - x + 1 < 0$ has the empty set as solution set.

41. Explain the circumstances under which the x-intercepts of the graph of a quadratic function are included in the solution set of a quadratic inequality.

'Are You Prepared?' Answers

1. $\{x | x > -3\}$ or $(-3, \infty)$ **2.** $-2 < x \leq 7$

-3   0   3   6   8 x

2.6 Building Quadratic Models from Verbal Descriptions and from Data

PREPARING FOR THIS SECTION   *Before getting started, review the following:*

• Problem Solving (Appendix A, Section A.9, pp. A74–A79)

• Building Linear Models from Data (Section 2.2, pp. 134–137)

Now Work the 'Are You Prepared?' problems on page 173.

OBJECTIVES   **1** Build Quadratic Models from Verbal Descriptions (p. 168)
              2 Build Quadratic Models from Data (p. 172)

In this section we will first discuss models in the form of a quadratic function when a verbal description of the problem is given. We end the section by fitting a quadratic function to data, which is another form of modeling.

When a mathematical model is in the form of a quadratic function, the properties of the graph of the quadratic function can provide important information about the model. In particular, we can use the quadratic function to determine the maximum or minimum value of the function. The fact that the graph of a quadratic function has a maximum or minimum value enables us to answer questions involving **optimization,** that is, finding the maximum or minimum values in models.

1 Build Quadratic Models from Verbal Descriptions

In economics, revenue R, in dollars, is defined as the amount of money received from the sale of an item and is equal to the unit selling price p, in dollars, of the item times the number x of units actually sold. That is,

$$R = xp$$

In economics, the Law of Demand states that p and x are related: As one increases, the other decreases. The equation that relates p and x is called the **demand equation.** When the demand equation is linear, the revenue model is a quadratic function.

EXAMPLE 1

Maximizing Revenue

The marketing department at Texas Instruments has found that, when certain calculators are sold at a price of p dollars per unit, the number x of calculators sold is given by the demand equation

$$x = 21{,}000 - 150p$$

(a) Find a model that expresses the revenue R as a function of the price p.
(b) What is the domain of R?
(c) What unit price should be used to maximize revenue?
(d) If this price is charged, what is the maximum revenue?
(e) How many units are sold at this price?
(f) Graph R.
(g) What price should Texas Instruments charge to collect at least $675,000 in revenue?

Solution

(a) The revenue R is $R = xp$, where $x = 21{,}000 - 150p$.

$$R = xp = (21{,}000 - 150p)p = -150p^2 + 21{,}000p \quad \text{The Model}$$

(b) Because x represents the number of calculators sold, we have $x \geq 0$, so $21{,}000 - 150p \geq 0$. Solving this linear inequality, we find that $p \leq 140$. In addition, Texas Instruments will only charge a positive price for the calculator, so $p > 0$. Combining these inequalities, we have that the domain of R is $\{p \mid 0 < p \leq 140\}$.

(c) The function R is a quadratic function with $a = -150, b = 21{,}000$, and $c = 0$. Because $a < 0$, the vertex is the highest point on the parabola. The revenue R is therefore a maximum when the price p is

$$p = -\frac{b}{2a} = -\frac{21{,}000}{2(-150)} = -\frac{21{,}000}{-300} = \$70.00$$

$$\uparrow$$
$$a = -150, b = 21{,}000$$

(d) The maximum revenue R is

$$R(70) = -150(70)^2 + 21{,}000(70) = \$735{,}000$$

(e) The number of calculators sold is given by the demand equation $x = 21{,}000 - 150p$. At a price of $p = \$70$,

$$x = 21{,}000 - 150(70) = 10{,}500$$

calculators are sold.

(f) We plot the intercepts, $(0, 0)$ and $(140, 0)$, and the vertex $(70, 735000)$. See Figure 29 for the graph of R.

Figure 29

(g) We graph $R = 675,000$ and $R(p) = -150p^2 + 21,000p$ on the same Cartesian plane. See Figure 30. Next we find where the graphs intersect by solving

$$675,000 = -150p^2 + 21,000p$$

$150p^2 - 21,000p + 675,000 = 0$	Add $150p^2 - 21,000p$ to both sides.
$p^2 - 140p + 4500 = 0$	Divide both sides by 150.
$(p - 50)(p - 90) = 0$	Factor.
$p = 50$ or $p = 90$	Use the Zero-Product Property.

Figure 30

Price p per calculator (dollars)

The graphs intersect at $(50, 675,000)$ and $(90, 675,000)$. Based on the graph in Figure 30, Texas Instruments should charge between $50 and $90 to earn at least $675,000 in revenue.

Now Work PROBLEM 3

EXAMPLE 2

Maximizing the Area Enclosed by a Fence

A farmer has 2000 yards of fence to enclose a rectangular field. What are the dimensions of the rectangle that encloses the most area?

Figure 31

Solution Figure 31 illustrates the situation. The available fence represents the perimeter of the rectangle. If x is the length and w is the width, then

$$2x + 2w = 2000 \qquad \textbf{(1)}$$

The area A of the rectangle is

$$A = xw$$

To express A in terms of a single variable, we solve equation (1) for w and substitute the result in $A = xw$. Then A involves only the variable x. [You could also solve equation (1) for x and express A in terms of w alone. Try it!]

$$2x + 2w = 2000$$
$$2w = 2000 - 2x$$
$$w = \frac{2000 - 2x}{2} = 1000 - x$$

Then the area A is

$$A = xw = x(1000 - x) = -x^2 + 1000x$$

Now, A is a quadratic function of x.

$$A(x) = -x^2 + 1000x \quad a = -1, b = 1000, c = 0$$

Figure 32

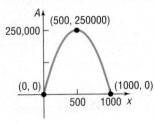

Figure 32 shows a graph of $A(x) = -x^2 + 1000x$. Since $a < 0$, the vertex is a maximum point on the graph of A. The maximum value occurs at

$$x = -\frac{b}{2a} = -\frac{1000}{2(-1)} = 500$$

The maximum value of A is

$$A\left(-\frac{b}{2a}\right) = A(500) = -500^2 + 1000(500) = -250{,}000 + 500{,}000 = 250{,}000$$

The largest rectangle that can be enclosed by 2000 yards of fence has an area of 250,000 square yards. Its dimensions are 500 yards by 500 yards.

━━━━━━━ **Now Work** PROBLEM 7

EXAMPLE 3

Analyzing the Motion of a Projectile

A projectile is fired from a cliff 500 feet above the water at an inclination of 45° to the horizontal, with a muzzle velocity of 400 feet per second. In physics, it is established that the height h of the projectile above the water can be modeled by

$$h(x) = \frac{-32x^2}{(400)^2} + x + 500$$

where x is the horizontal distance of the projectile from the base of the cliff. See Figure 33.

Figure 33

(a) Find the maximum height of the projectile.
(b) How far from the base of the cliff will the projectile strike the water?

Solution

(a) The height of the projectile is given by a quadratic function.

$$h(x) = \frac{-32x^2}{(400)^2} + x + 500 = \frac{-1}{5000}x^2 + x + 500$$

We are looking for the maximum value of h. Since $a < 0$, the maximum value is obtained at the vertex. We compute

$$x = -\frac{b}{2a} = -\frac{1}{2\left(-\dfrac{1}{5000}\right)} = \frac{5000}{2} = 2500$$

The maximum height of the projectile is

$$h(2500) = \frac{-1}{5000}(2500)^2 + 2500 + 500 = -1250 + 2500 + 500 = 1750 \text{ ft}$$

(b) The projectile will strike the water when the height is zero. To find the distance x traveled, we need to solve the equation

$$h(x) = \frac{-1}{5000}x^2 + x + 500 = 0$$

We find the discriminant first.

$$b^2 - 4ac = 1^2 - 4\left(\frac{-1}{5000}\right)(500) = 1.4$$

Then

$$x = \frac{-b \pm \sqrt{b^2 - 4ac}}{2a} = \frac{-1 \pm \sqrt{1.4}}{2\left(-\dfrac{1}{5000}\right)} \approx \begin{cases} -458 \\ 5458 \end{cases}$$

We discard the negative solution and find that the projectile will strike the water at a distance of about 5458 feet from the base of the cliff.

●

── **Now Work** PROBLEM 11

Seeing the Concept

Graph

$$h(x) = \frac{-1}{5000}x^2 + x + 500$$

$$0 \le x \le 5500$$

Use MAXIMUM to find the maximum height of the projectile, and use ROOT or ZERO to find the distance from the base of the cliff to where it strikes the water. Compare your results with those obtained in Example 3.

EXAMPLE 4

The Golden Gate Bridge

The Golden Gate Bridge, a suspension bridge, spans the entrance to San Francisco Bay. Its 746-foot-tall towers are 4200 feet apart. The bridge is suspended from two huge cables more than 3 feet in diameter; the 90-foot-wide roadway is 220 feet above the water. The cables are parabolic in shape* and touch the road surface at the center of the bridge. Find the height of the cable above the road at a distance of 1000 feet from the center.

Solution

See Figure 34. We begin by choosing the placement of the coordinate axes so that the x-axis coincides with the road surface and the origin coincides with the center of the bridge. As a result, the twin towers will be vertical (height $746 - 220 = 526$ feet above the road) and located 2100 feet from the center. Also, the cable, which has the shape of a parabola, will extend from the towers, open up, and have its vertex at $(0, 0)$. The choice of placement of the axes enables us to identify the equation of the parabola as $y = ax^2$, $a > 0$. We can also see that the points $(-2100, 526)$ and $(2100, 526)$ are on the graph.

Figure 34

*A cable suspended from two towers is in the shape of a **catenary,** but when a horizontal roadway is suspended from the cable, the cable takes the shape of a parabola.

Based on these facts, we can find the value of a in $y = ax^2$.

$$y = ax^2$$
$$526 = a(2100)^2 \quad \text{\small{$x = 2100, y = 526$}}$$
$$a = \frac{526}{(2100)^2}$$

The equation of the parabola is therefore

$$y = \frac{526}{(2100)^2}x^2$$

The height of the cable when $x = 1000$ is

$$y = \frac{526}{(2100)^2}(1000)^2 \approx 119.3 \text{ feet}$$

The cable is 119.3 feet above the road at a distance of 1000 feet from the center of the bridge.

●

━━━━**Now Work** PROBLEM 13

 2 Build Quadratic Models from Data

In Section 2.2, we found the line of best fit for data that appeared to be linearly related. It was noted that data may also follow a nonlinear relation. Figures 35(a) and (b) show scatter diagrams of data that follow a quadratic relation.

Figure 35

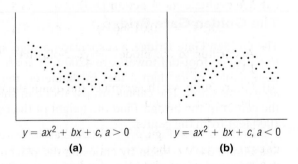

$y = ax^2 + bx + c, a > 0$ $y = ax^2 + bx + c, a < 0$

(a)   **(b)**

EXAMPLE 5   **Fitting a Quadratic Function to Data**

The data in Table 8 represent the percentage D of the population that is divorced for various ages a in 2007.

(a) Draw a scatter diagram of the data treating age as the independent variable. Comment on the type of relation that may exist between age and percentage of the population divorced.

(b) Use a graphing utility to find the quadratic function of best fit that models the relation between age and percentage of the population divorced.

(c) Use the model found in part (b) to approximate the age at which the percentage of the population divorced is greatest.

(d) Use the model found in part (b) to approximate the highest percentage of the population that is divorced.

(e) Use a graphing utility to draw the quadratic function of best fit on the scatter diagram.

Table 8

Age, a	Percentage Divorced, D
22	0.8
27	2.8
32	6.4
37	8.7
42	12.3
50	14.5
60	13.8
70	9.6
80	4.9

Source: United States Statistical Abstract, 2009

Solution

(a) Figure 36 shows the scatter diagram, from which it appears that the data follow a quadratic relation, with $a < 0$.

Figure 36

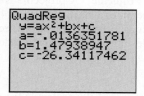

(b) Upon executing the QUADratic REGression program, we obtain the results shown in Figure 37. The output of the utility shows us the equation $y = ax^2 + bx + c$. The quadratic function of best fit that models the relation between age and percentage divorced is

$$D(a) = -0.0136a^2 + 1.4794a - 26.3412 \quad \text{The Model}$$

where a represents age and D represents the percentage divorced.

(c) Based on the quadratic function of best fit, the age with the greatest percentage divorced is

$$a = -\frac{b}{2a} = -\frac{1.4794}{2(-0.0136)} \approx 54 \text{ years}$$

Figure 37

```
QuadReg
y=ax²+bx+c
a=-.0136351781
b=1.47938947
c=-26.34117462
```

(d) We evaluate the function $D(a)$ at $a = 54.4$.

$$D(54) = -0.0136(54)^2 + 1.4794(54) - 26.3412 \approx 13.9 \text{ percent}$$

According to the model, 54-year-olds have the highest percentage divorced at 13.9 percent.

(e) Figure 38 shows the graph of the quadratic function found in part (b) drawn on the scatter diagram.

Figure 38

Look again at Figure 37. Notice that the output given by the graphing calculator does not include r, the correlation coefficient. Recall that the correlation coefficient is a measure of the strength of a linear relation that exists between two variables. The graphing calculator does not provide an indication of how well the function fits the data in terms of r since a quadratic function cannot be expressed as a linear function.

Now Work PROBLEM 25

2.6 Assess Your Understanding

'Are You Prepared?' *Answers are given at the end of these exercises. If you get a wrong answer, read the pages listed in red.*

1. Translate the following sentence to a mathematical equation: The total revenue R from selling x hot dogs is $3 times the number of hot dogs sold. (p. A74)

2. Use a graphing utility to find the line of best fit for the following data: (pp. 134–137)

x	3	5	5	6	7	8
y	10	13	12	15	16	19

Applications and Extensions

3. **Maximizing Revenue** The price p (in dollars) and the quantity x sold of a certain product obey the demand equation

$$p = -\frac{1}{6}x + 100$$

(a) Find a model that expresses the revenue R as a function of x. (Remember, $R = xp$.)
(b) What is the domain of R?
(c) What is the revenue if 200 units are sold?
(d) What quantity x maximizes revenue? What is the maximum revenue?
(e) What price should the company charge to maximize revenue?

4. **Maximizing Revenue** The price p (in dollars) and the quantity x sold of a certain product obey the demand equation

$$p = -\frac{1}{3}x + 100$$

(a) Find a model that expresses the revenue R as a function of x.
(b) What is the domain of R?
(c) What is the revenue if 100 units are sold?
(d) What quantity x maximizes revenue? What is the maximum revenue?
(e) What price should the company charge to maximize revenue?

5. **Maximizing Revenue** The price p (in dollars) and the quantity x sold of a certain product obey the demand equation

$$x = -5p + 100 \qquad 0 < p \le 20$$

(a) Express the revenue R as a function of x.
(b) What is the revenue if 15 units are sold?
(c) What quantity x maximizes revenue? What is the maximum revenue?
(d) What price should the company charge to maximize revenue?
(e) What price should the company charge to earn at least $480 in revenue?

6. **Maximizing Revenue** The price p (in dollars) and the quantity x sold of a certain product obey the demand equation

$$x = -20p + 500 \qquad 0 < p \le 25$$

(a) Express the revenue R as a function of x.
(b) What is the revenue if 20 units are sold?
(c) What quantity x maximizes revenue? What is the maximum revenue?
(d) What price should the company charge to maximize revenue?
(e) What price should the company charge to earn at least $3000 in revenue?

7. **Enclosing a Rectangular Field** David has 400 yards of fencing and wishes to enclose a rectangular area.
(a) Express the area A of the rectangle as a function of the width w of the rectangle.
(b) For what value of w is the area largest?
(c) What is the maximum area?

8. **Enclosing a Rectangular Field** Beth has 3000 feet of fencing available to enclose a rectangular field.
(a) Express the area A of the rectangle as a function of x, where x is the length of the rectangle.
(b) For what value of x is the area largest?
(c) What is the maximum area?

9. **Enclosing the Most Area with a Fence** A farmer with 4000 meters of fencing wants to enclose a rectangular plot that borders on a river. If the farmer does not fence the side along the river, what is the largest area that can be enclosed? (See the figure.)

$$4000 - 2x$$

10. **Enclosing the Most Area with a Fence** A farmer with 2000 meters of fencing wants to enclose a rectangular plot that borders on a straight highway. If the farmer does not fence the side along the highway, what is the largest area that can be enclosed?

11. **Analyzing the Motion of a Projectile** A projectile is fired from a cliff 200 feet above the water at an inclination of 45° to the horizontal, with a muzzle velocity of 50 feet per second. The height h of the projectile above the water is modeled by

$$h(x) = \frac{-32x^2}{(50)^2} + x + 200$$

where x is the horizontal distance of the projectile from the face of the cliff.
(a) At what horizontal distance from the face of the cliff is the height of the projectile a maximum?
(b) Find the maximum height of the projectile.
(c) At what horizontal distance from the face of the cliff will the projectile strike the water?
(d) Using a graphing utility, graph the function h, $0 \le x \le 200$.
(e) Use a graphing utility to verify the solutions found in parts (b) and (c).
(f) When the height of the projectile is 100 feet above the water, how far is it from the cliff?

12. **Analyzing the Motion of a Projectile** A projectile is fired at an inclination of 45° to the horizontal, with a muzzle velocity of 100 feet per second. The height h of the projectile is modeled by

$$h(x) = \frac{-32x^2}{(100)^2} + x$$

where x is the horizontal distance of the projectile from the firing point.
(a) At what horizontal distance from the firing point is the height of the projectile a maximum?

(b) Find the maximum height of the projectile.

(c) At what horizontal distance from the firing point will the projectile strike the ground?

(d) Using a graphing utility, graph the function h, $0 \leq x \leq 350$.

(e) Use a graphing utility to verify the results obtained in parts (b) and (c).

(f) When the height of the projectile is 50 feet above the ground, how far has it traveled horizontally?

13. Suspension Bridge A suspension bridge with weight uniformly distributed along its length has twin towers that extend 75 meters above the road surface and are 400 meters apart. The cables are parabolic in shape and are suspended from the tops of the towers. The cables touch the road surface at the center of the bridge. Find the height of the cables at a point 100 meters from the center. (Assume that the road is level.)

14. Architecture A parabolic arch has a span of 120 feet and a maximum height of 25 feet. Choose suitable rectangular coordinate axes and find the equation of the parabola. Then calculate the height of the arch at points 10 feet, 20 feet, and 40 feet from the center.

15. Constructing Rain Gutters A rain gutter is to be made of aluminum sheets that are 12 inches wide by turning up the edges 90°. See the illustration.

(a) What depth will provide maximum cross-sectional area and hence allow the most water to flow?

(b) What depths will allow at least 16 square inches of water to flow?

16. Norman Windows A **Norman window** has the shape of a rectangle surmounted by a semicircle of diameter equal to the width of the rectangle. See the figure. If the perimeter of the window is 20 feet, what dimensions will admit the most light (maximize the area)?

[**Hint:** Circumference of a circle $= 2\pi r$; area of a circle $= \pi r^2$, where r is the radius of the circle.]

17. Constructing a Stadium A track and field playing area is in the shape of a rectangle with semicircles at each end. See the figure. The inside perimeter of the track is to be 1500 meters. What should the dimensions of the rectangle be so that the area of the rectangle is a maximum?

18. Architecture A special window has the shape of a rectangle surmounted by an equilateral triangle. See the figure. If the perimeter of the window is 16 feet, what dimensions will admit the most light?

[**Hint:** Area of an equilateral triangle $= \left(\dfrac{\sqrt{3}}{4}\right)x^2$, where x is the length of a side of the triangle.]

19. Chemical Reactions A self-catalytic chemical reaction results in the formation of a compound that causes the formation ratio to increase. If the reaction rate V is modeled by

$$V(x) = kx(a - x), \qquad 0 \leq x \leq a$$

where k is a positive constant, a is the initial amount of the compound, and x is the variable amount of the compound, for what value of x is the reaction rate a maximum?

20. Calculus: Simpson's Rule The figure shows the graph of $y = ax^2 + bx + c$. Suppose that the points $(-h, y_0)$, $(0, y_1)$, and (h, y_2) are on the graph. It can be shown that the area enclosed by the parabola, the x-axis, and the lines $x = -h$ and $x = h$ is

$$\text{Area} = \frac{h}{3}(2ah^2 + 6c)$$

Show that this area may also be given by

$$\text{Area} = \frac{h}{3}(y_0 + 4y_1 + y_2)$$

21. Use the result obtained in Problem 20 to find the area enclosed by $f(x) = -5x^2 + 8$, the x-axis, and the lines $x = -1$ and $x = 1$.

22. Use the result obtained in Problem 20 to find the area enclosed by $f(x) = 2x^2 + 8$, the x-axis, and the lines $x = -2$ and $x = 2$.

23. Use the result obtained in Problem 20 to find the area enclosed by $f(x) = x^2 + 3x + 5$, the x-axis, and the lines $x = -4$ and $x = 4$.

24. Use the result obtained in Problem 20 to find the area enclosed by $f(x) = -x^2 + x + 4$, the x-axis, and the lines $x = -1$ and $x = 1$.

25. **Life Cycle Hypothesis** An individual's income varies with his or her age. The following table shows the median income I of males of different age groups within the United States for 2006. For each age group, let the class midpoint represent the independent variable, x. For the class "65 years and older," we will assume that the class midpoint is 69.5.

Age	Class Midpoint, x	Median Income, I
15–24 years	19.5	$10,964
25–34 years	29.5	$32,131
35–44 years	39.5	$42,637
45–54 years	49.5	$45,693
55–64 years	59.5	$41,477
65 years and older	69.5	$23,500

Source: U.S. Census Bureau

(a) Use a graphing utility to draw a scatter diagram of the data. Comment on the type of relation that may exist between the two variables.
(b) Use a graphing utility to find the quadratic function of best fit that models the relation between age and median income.
(c) Use the function found in part (b) to determine the age at which an individual can expect to earn the most income.

(d) Use the function found in part (b) to predict the peak income earned.
(e) With a graphing utility, graph the quadratic function of best fit on the scatter diagram.

26. **Height of a Ball** A shot-putter throws a ball at an inclination of 45° to the horizontal. The following data represent the height of the ball h at the instant that it has traveled x feet horizontally.

Distance, x	Height, h
20	25
40	40
60	55
80	65
100	71
120	77
140	77
160	75
180	71
200	64

(a) Use a graphing utility to draw a scatter diagram of the data. Comment on the type of relation that may exist between the two variables.
(b) Use a graphing utility to find the quadratic function of best fit that models the relation between distance and height.
(c) Use the function found in part (b) to determine how far the ball will travel before it reaches its maximum height.
(d) Use the function found in part (b) to find the maximum height of the ball.
(e) With a graphing utility, graph the quadratic function of best fit on the scatter diagram.

Mixed Practice

27. **Which Model?** The following data represent the square footage and rents (dollars per month) for apartments in the Del Mar area of San Diego, California.

Square Footage, x	Rent per Month, R
686	1600
770	1665
817	1750
800	1685
809	1700
901	1770
803	1725

Source: apartments.com

(a) Using a graphing utility, draw a scatter diagram of the data treating square footage as the independent variable. What type of relation appears to exist between square footage and rent?
(b) Based on your response to part (a), find either a linear or quadratic model that describes the relation between square footage and rent.
(c) Use your model to predict the rent of an apartment in San Diego that is 850 square feet.

28. **Which Model?** An engineer collects the following data showing the speed s of a Toyota Camry and its average miles per gallon, M.
(a) Using a graphing utility, draw a scatter diagram of the data treating speed as the independent variable. What type of relation appears to exist between speed and miles per gallon?

Speed, s	Miles per Gallon, M
30	18
35	20
40	23
40	25
45	25
50	28
55	30
60	29
65	26
65	25
70	25

(b) Based on your response to part (a), find either a linear or quadratic model that describes the relation between speed and miles per gallon.
(c) Use your model to predict the miles per gallon for a Camry that is traveling 63 miles per hour.

29. **Which Model?** The following data represent the percentage of the U.S. population aged 25 years or older whose age is x who do not have a high school diploma as of March 2005.

Age, a	Percentage without a High School Diploma, P
30	13.3
40	11.6
50	10.9
60	13.7
70	22.3
80	30.2

Source: U.S. Census Bureau

(a) Using a graphing utility, draw a scatter diagram of the data treating age as the independent variable. What type of relation appears to exist between age and percentage of the population without a high school diploma?
(b) Based on your response to part (a), find either a linear or quadratic model that describes the relation between age and percentage of the population that do not have a high school diploma.
(c) Use your model to predict the percentage of 35-year-olds that do not have a high school diploma.

30. **Which Model?** A cricket makes a chirping noise by sliding its wings together rapidly. Perhaps you have noticed that the number of chirps seems to increase with the temperature. The following data list the temperature (in degrees Fahrenheit) and the number of chirps per second for the striped ground cricket.

Temperature, x	Chirps per Second, C
88.6	20.0
93.3	19.8
80.6	17.1
69.7	14.7
69.4	15.4
79.6	15.0
80.6	16.0
76.3	14.4
75.2	15.5

Source: Pierce, George W. *The Songs of Insects.* Cambridge, MA Harvard University Press, 1949, pp. 12 – 21

(a) Using a graphing utility, draw a scatter diagram of the data treating temperature as the independent variable. What type of relation appears to exist between temperature and chirps per second?
(b) Based on your response to part (a), find either a linear or quadratic model that best describes the relation between temperature and chirps per second.
(c) Use your model to predict the chirps per second if the temperature is 80°F.

Explaining Concepts: Discussion and Writing

31. Refer to Example 1 on page 168. Notice that if the price charged for the calculators is $0 or $140 the revenue is $0. It is easy to explain why revenue would be $0 if the price charged is $0, but how can revenue be $0 if the price charged is $140?

'Are You Prepared?' Answers

1. $R = 3x$ 2. $y = 1.7826x + 4.0652$

2.7 Complex Zeros of a Quadratic Function*

PREPARING FOR THIS SECTION *Before getting started, review the following:*

- Real numbers (Appendix A, Section A.1, pp. A3–A4)
- Complex numbers (Appendix A, Section A.11, pp. A92–A97)

Now Work the *'Are You Prepared?'* problems on page 180.

OBJECTIVE **1** Find the Complex Zeros of a Quadratic Function (p. 178)

1 Find the Complex Zeros of a Quadratic Function

For a function f, the solutions of the equation $f(x) = 0$ are called the zeros of f. If we are working in the real number system, these zeros are called *real* zeros of f. If we are working in the complex number system, these zeros are called **complex zeros of f.**

EXAMPLE 1 **Finding Complex Zeros of a Quadratic Function**

Find the complex zeros of each of the following quadratic functions. Graph each function and label the intercepts.

(a) $f(x) = x^2 - 4$ (b) $g(x) = x^2 + 9$

Solution (a) To find the complex zeros of $f(x) = x^2 - 4$, we solve the equation $f(x) = 0$.

Figure 39

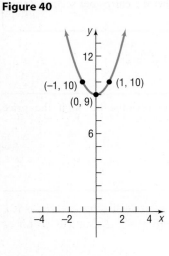

$$f(x) = 0$$
$$x^2 - 4 = 0 \qquad \text{\small $f(x) = x^2 - 4$}$$
$$x^2 = 4 \qquad \text{\small Add 4 to both sides.}$$
$$x = \pm\sqrt{4} \qquad \text{\small Use the Square Root Method.}$$
$$x = \pm 2$$

The complex zeros of f are -2 and 2. Because the zeros are real numbers, we can also say that the x-intercepts of the graph of f are -2 and 2. Figure 39 shows the graph of $f(x) = x^2 - 4$ with the intercepts labeled.

(b) To find the complex zeros of $g(x) = x^2 + 9$, we solve the equation $g(x) = 0$.

Figure 40

$$g(x) = 0$$
$$x^2 + 9 = 0 \qquad \text{\small $g(x) = x^2 + 9$}$$
$$x^2 = -9 \qquad \text{\small Subtract 9 from both sides.}$$
$$x = \pm\sqrt{-9} \qquad \text{\small Use the Square Root Method.}$$
$$x = \pm 3i$$

The complex zeros of g are $-3i$ and $3i$. Because neither of the zeros is a real number, there are no x-intercepts for the graph of g. Figure 40 shows the graph of $g(x) = x^2 + 9$.

●

Now Work PROBLEM 9

In Appendix A, Section A.11, we defined the square root of a negative number. We can now restate the quadratic formula without restriction.

---

*This section is optional and may be omitted without loss of continuity.

THEOREM

In the complex number system, the solutions of the quadratic equation $ax^2 + bx + c = 0$, where a, b, and c are real numbers and $a \neq 0$, are given by the formula

$$x = \frac{-b \pm \sqrt{b^2 - 4ac}}{2a} \qquad \text{(1)}$$

So, in the complex number system, a quadratic equation has either one solution or two solutions.

EXAMPLE 2

Finding Complex Zeros of a Quadratic Function

Find the complex zeros of each of the following quadratic functions. Graph each function and label the intercepts.

(a) $f(x) = x^2 + 6x + 2$ (b) $g(x) = x^2 - 4x + 8$

Solution

(a) To find the complex zeros of $f(x) = x^2 + 6x + 2$, we solve the equation $f(x) = 0$.

$$f(x) = 0$$

$$x^2 + 6x + 2 = 0 \quad \text{\small{$f(x) = x^2 + 6x + 2$}}$$

Here $a = 1$, $b = 6$, $c = 2$, and $b^2 - 4ac = 6^2 - 4(1)(2) = 28$. Using equation (1), we find that

$$x = \frac{-6 \pm \sqrt{28}}{2} = \frac{-6 \pm 2\sqrt{7}}{2} = -3 \pm \sqrt{7}$$

The complex zeros of f are $-3 - \sqrt{7} \approx -5.65$ and $-3 + \sqrt{7} \approx -0.35$. Because the zeros are real numbers, we can also say that the x-intercepts of the graph of f are $-3 - \sqrt{7} \approx -5.65$ and $-3 + \sqrt{7} \approx -0.35$. Figure 41 shows the graph of $f(x) = x^2 + 6x + 2$ with the x-intercepts labeled.

(b) To find the complex zeros of $g(x) = x^2 - 4x + 8$, we solve the equation $g(x) = 0$.

$$g(x) = 0$$

$$x^2 - 4x + 8 = 0 \quad \text{\small{$g(x) = x^2 - 4x + 8$}}$$

Here $a = 1$, $b = -4$, $c = 8$, and $b^2 - 4ac = (-4)^2 - 4(1)(8) = -16$. Using equation (1), we find that

$$x = \frac{-(-4) \pm \sqrt{-16}}{2(1)} = \frac{4 \pm 4i}{2} = 2 \pm 2i$$

✓**Check:**

$$2 + 2i: g(2 + 2i) = (2 + 2i)^2 - 4(2 + 2i) + 8 = 4 + 8i + 4i^2 - 8 - 8i + 8$$
$$= 4 - 4 = 0$$

$$2 - 2i: g(2 - 2i) = (2 - 2i)^2 - 4(2 - 2i) + 8 = 4 - 8i + 4i^2 - 8 + 8i + 8$$
$$= 4 - 4 = 0$$

The complex zeros of f are $2 - 2i$ and $2 + 2i$. Because neither of the zeros is a real number, the graph of g has no x-intercepts. Figure 42 shows the graph of $g(x) = x^2 - 4x + 8$.

Figure 41

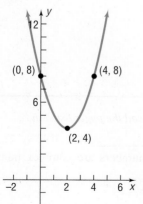

Figure 42

Now Work PROBLEM **15**

The discriminant $b^2 - 4ac$ still serves as a way to determine the character of the solutions of a quadratic equation.

> **Character of the Solutions of a Quadratic Equation**
>
> In the complex number system, consider a quadratic equation $ax^2 + bx + c = 0$ with real coefficients.
>
> 1. If $b^2 - 4ac > 0$, the equation has two unequal real solutions.
> 2. If $b^2 - 4ac = 0$, the equation has a repeated real solution, a double root.
> 3. If $b^2 - 4ac < 0$, the equation has two complex solutions that are not real. The complex solutions are conjugates of each other.

The third conclusion in the display is a consequence of the fact that if $b^2 - 4ac = -N < 0$, then, by the quadratic formula, the solutions are

$$x = \frac{-b + \sqrt{b^2 - 4ac}}{2a} = \frac{-b + \sqrt{-N}}{2a} = \frac{-b + \sqrt{N}\,i}{2a} = \frac{-b}{2a} + \frac{\sqrt{N}}{2a}\,i$$

and

$$x = \frac{-b - \sqrt{b^2 - 4ac}}{2a} = \frac{-b - \sqrt{-N}}{2a} = \frac{-b - \sqrt{N}\,i}{2a} = \frac{-b}{2a} - \frac{\sqrt{N}}{2a}\,i$$

which are conjugates of each other.

Therefore, when $b^2 - 4ac < 0$, the zeros of the quadratic function $f(x) = ax^2 + bx + c$ will be conjugates of each other.

EXAMPLE 3

Determining the Character of the Solutions of a Quadratic Equation

Without solving, determine the character of the solutions of each equation.

(a) $3x^2 + 4x + 5 = 0$ (b) $2x^2 + 4x + 1 = 0$ (c) $9x^2 - 6x + 1 = 0$

Solution

(a) Here $a = 3$, $b = 4$, and $c = 5$, so $b^2 - 4ac = 16 - 4(3)(5) = -44$. The solutions are two complex numbers that are not real and are conjugates of each other.

(b) Here $a = 2$, $b = 4$, and $c = 1$, so $b^2 - 4ac = 16 - 8 = 8$. The solutions are two unequal real numbers.

(c) Here $a = 9$, $b = -6$, and $c = 1$, so $b^2 - 4ac = 36 - 4(9)(1) = 0$. The solution is a repeated real number, that is, a double root.

●

➤**Now Work** PROBLEM 25

2.7 Assess Your Understanding

'Are You Prepared?' *Answers are given at the end of these exercises. If you get a wrong answer, read the pages listed in* red.

1. Name the integers and the rational numbers in the set $\left\{-3, 0, \sqrt{2}, \frac{6}{5}, \pi\right\}$. (pp. A3–A4)

2. **True or False** Rational numbers and irrational numbers are in the set of real numbers. (pp. A3–A4)

3. $(2 + i)(3 - 4i) =$ _____ (pp. A92–A97)

4. The conjugate of $2 + 5i$ is _____. (pp. A92–A97)

5. **True or False** All real numbers are complex numbers. (pp. A92–A97)

6. $\sqrt{-81} =$ _____ . (pp. A92–A97)

Concepts and Vocabulary

7. If $2 - 3i$ is a zero of a quadratic function with real coefficients, then _____ is also a zero.

8. *True or False* Consider the quadratic function $f(x) = ax^2 + bx + c$. If $b^2 - 4ac > 0$, the graph of f will have two unequal x-intercepts.

Skill Building

In Problems 9–24, find the complex zeros of each quadratic function. Graph each function and label the intercepts.

9. $f(x) = x^2 + 4$

10. $f(x) = x^2 - 9$

11. $f(x) = x^2 - 16$

12. $f(x) = x^2 + 25$

13. $f(x) = x^2 - 6x + 13$

14. $f(x) = x^2 + 4x + 8$

15. $f(x) = x^2 - 6x + 10$

16. $f(x) = x^2 - 2x + 5$

17. $f(x) = x^2 - 4x + 1$

18. $f(x) = x^2 + 6x + 1$

19. $f(x) = 2x^2 + 2x + 1$

20. $f(x) = 3x^2 + 6x + 4$

21. $f(x) = x^2 + x + 1$

22. $f(x) = x^2 - x + 1$

23. $f(x) = -2x^2 + 8x + 1$

24. $f(x) = -3x^2 + 6x + 1$

In Problems 25–30, without solving, determine the character of the solutions of each equation in the complex number system.

25. $3x^2 - 3x + 4 = 0$

26. $2x^2 - 4x + 1 = 0$

27. $2x^2 + 3x - 4 = 0$

28. $x^2 + 2x + 6 = 0$

29. $9x^2 - 12x + 4 = 0$

30. $4x^2 + 12x + 9 = 0$

Mixed Practice

In Problems 31–34, find all complex zeros of each function.

31. $f(t) = t^4 - 16$

32. $G(y) = y^4 - 81$

33. $F(x) = x^6 - 9x^3 + 8$

34. $P(z) = z^6 + 28z^3 + 27$

'Are You Prepared?' Answers

1. Integers: $\{-3, 0\}$; rational numbers: $\left\{-3, 0, \dfrac{6}{5}\right\}$ **2.** True   **3.** $10 - 5i$ **4.** $2 - 5i$ **5.** True   **6.** $9i$

2.8 Equations and Inequalities Involving the Absolute Value Function

PREPARING FOR THIS SECTION *Before getting started, review the following:*

- Algebra Essentials (Appendix A, Section A.1, pp. A4–A5)
- Solving Equations (Appendix A, Section A.8, pp. A65–A71)
- Interval Notation; Solving Inequalities (Appendix A, Section A.10, pp. A83–A88)
- Absolute Value Function (Section 1.4, pp. 81–82)
- Graphing Techniques: Transformations (Section 1.5, pp. 90–99)

Now Work the 'Are You Prepared?' problems on page 184.

OBJECTIVES **1** Solve Absolute Value Equations (p. 181)

2 Solve Absolute Value Inequalities (p. 182)

1 Solve Absolute Value Equations

Recall that, on the real number line, the absolute value of a equals the distance from the origin to the point whose coordinate is a. For example, there are two points whose distance from the origin is 5 units, -5 and 5. See Figure 43. Thus the equation $|x| = 5$ will have the solution set $\{-5, 5\}$.

Figure 43

The equation $|u| = a$ leads to two equations depending on whether u is nonnegative (greater than or equal to zero) or negative.

If $u < 0$	**If $u \geq 0$**
$|u| = a$	$|u| = a$
$-u = a$ $|u| = -u$ when $u < 0$	$u = a$ $|u| = u$ when $u \geq 0$
$u = -a$ Multiply both sides by -1.	

So we have the following result.

> **Equations Involving Absolute Value**
>
> If a is a positive real number and if u is any algebraic expression, then
>
> $$|u| = a \quad \text{is equivalent to} \quad u = a \quad \text{or} \quad u = -a \qquad \textbf{(1)}$$

EXAMPLE 1

Solving an Equation Involving Absolute Value

Solve the equation: $|x + 2| = 5$

Solution

Figure 44

The equation follows the form of equation (1), where $u = x + 2$. There are two possibilities.

$$x + 2 = 5 \quad \text{or} \quad x + 2 = -5$$
$$x = 3 \qquad\qquad x = -7$$

The solution set is $\{-7, 3\}$.

We can visualize the solution by graphing $f(x) = |x + 2|$ and $g(x) = 5$. Figure 44 shows the graphs of $f(x) = |x + 2|$ and $g(x) = 5$. The x-coordinates of the points of intersection are -7 and 3, the solutions of the equation $f(x) = g(x)$.

━━━ **Now Work** PROBLEMS **13(a)** AND **19**

Figure 45

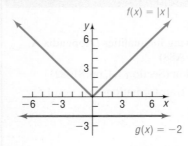

In equation (1), we required that a be a positive number. If $a = 0$, equation (1) becomes $|u| = 0$, which is equivalent to $u = 0$. If a is less than zero, the equation has no real solution. To see why, consider the equation $|x| = -2$. Figure 45 shows the graphs of $f(x) = |x|$ and $g(x) = -2$. Notice that the graphs do not intersect, which implies that the equation $|x| = -2$ has no solution. The solution set is the empty set, $\varnothing$ or $\{\ \}$.

2 Solve Absolute Value Inequalities

Let's look at an inequality involving absolute value. Once again we will use the graph of the absolute value function to help us visualize the results of the algebra.

EXAMPLE 2

Solving an Inequality Involving Absolute Value

Solve the inequality: $|x| < 4$

Solution

Figure 46

Less than 4 units from origin

-5 -4 -3 -2 -1 0 1 2 3 4

We are looking for all points whose coordinate x is a distance less than 4 units from the origin. See Figure 46 for an illustration.

Because any x between -4 and 4 satisfies the condition $|x| < 4$, the solution set consists of all numbers x for which $-4 < x < 4$, that is, all x in the interval $(-4, 4)$.

Figure 47

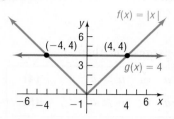

To visualize the results of Example 2, we graph $f(x) = |x|$ and $g(x) = 4$. See Figure 47. Because we are solving $f(x) < g(x)$, we look for all x-coordinates such that the graph of $f(x)$ is below the graph of $g(x)$. From the graph, we can see that the graph of $f(x) = |x|$ is below the graph of $g(x) = 4$ for all x between -4 and 4. So the solution set consists of all x for which $-4 < x < 4$, or all x in the interval $(-4, 4)$.

The results of Example 2 lead us to the following:

Inequalities Involving Absolute Value

If a is any positive number and if u is any algebraic expression, then

$$|u| < a \quad \text{is equivalent to} \quad -a < u < a \qquad \textbf{(2)}$$

$$|u| \leq a \quad \text{is equivalent to} \quad -a \leq u \leq a \qquad \textbf{(3)}$$

In other words, $|u| < a$ is equivalent to $-a < u$ and $u < a$.

EXAMPLE 3

Solving an Inequality Involving Absolute Value

Solve the inequality $|2x + 4| \leq 3$, and graph the solution set.

Solution

$$|2x + 4| \leq 3 \qquad \text{This follows the form of statement (3);}$$
$$\text{the expression } u = 2x + 4 \text{ is inside the}$$
$$\text{absolute value bars.}$$

$$-3 \leq \quad 2x + 4 \quad \leq 3 \qquad \text{Apply statement (3).}$$

$$-3 - 4 \leq \quad 2x + 4 - 4 \leq 3 - 4 \qquad \text{Subtract 4 from each part.}$$

$$-7 \leq \quad 2x \quad \leq -1 \qquad \text{Simplify.}$$

$$\frac{-7}{2} \leq \quad \frac{2x}{2} \quad \leq \frac{-1}{2} \qquad \text{Divide each part by 2.}$$

$$-\frac{7}{2} \leq \quad x \quad \leq -\frac{1}{2} \qquad \text{Simplify.}$$

Figure 48

The solution set is $\left\{ x \mid -\dfrac{7}{2} \leq x \leq -\dfrac{1}{2} \right\}$, that is, all x in the interval $\left[-\dfrac{7}{2}, -\dfrac{1}{2} \right]$. See Figure 48 for a graph of the solution set.

Now Work PROBLEMS 13(b) AND 47

EXAMPLE 4

Solving an Inequality Involving Absolute Value

Solve the inequality: $\ |x| > 3$

Solution

Figure 49

We are looking for all points whose coordinate x is a distance greater than 3 units from the origin. Figure 49 illustrates the situation.

We conclude that any x less than -3 or greater than 3 satisfies the condition $|x| > 3$. Consequently, the solution set consists of all numbers x for which $x < -3$ or $x > 3$, that is, all x in the intervals $(-\infty, -3)$ or $(3, \infty)$.

Figure 50

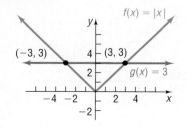

To visualize these results, we graph $f(x) = |x|$ and $g(x) = 3$. See Figure 50. Because we are solving $f(x) > g(x)$, we look for all x-coordinates such that the graph of $f(x)$ is above the graph of $g(x)$. From the graph, we can see that the graph of $f(x) = |x|$ is above the graph of $g(x) = 3$ for all x less than -3 or all x greater than 3. So the solution set consists of all x for which $x < -3$ or $x > 3$, that is, all x in the intervals $(-\infty, -3)$ or $(3, \infty)$.

WARNING A common error to be avoided is to attempt to write the solution $x < 1$ or $x > 4$ as $1 > x > 4$, which is incorrect, since there are no numbers x for which $1 > x$ and $x > 4$. Another common error is to "mix" the symbols and write $1 < x > 4$, which makes no sense. ∎

The results of Example 4 lead us to the following:

Inequalities Involving Absolute Value

If a is any positive number and u is any algebraic expression, then

$$|u| > a \quad \text{is equivalent to} \quad u < -a \quad \text{or} \quad u > a \qquad \textbf{(4)}$$

$$|u| \geq a \quad \text{is equivalent to} \quad u \leq -a \quad \text{or} \quad u \geq a \qquad \textbf{(5)}$$

EXAMPLE 5   **Solving an Inequality Involving Absolute Value**

Solve the inequality $|2x - 5| > 3$, and graph the solution set.

Solution

$|2x - 5| > 3$ This follows the form of statement (4); the expression $u = 2x - 5$ is inside the absolute value bars.

$$
\begin{array}{ccll}
2x - 5 < -3 & \text{or} & 2x - 5 > 3 & \text{Apply statement (4).} \\
2x - 5 + 5 < -3 + 5 & \text{or} & 2x - 5 + 5 > 3 + 5 & \text{Add 5 to each part.} \\
2x < 2 & \text{or} & 2x > 8 & \text{Simplify.} \\
\dfrac{2x}{2} < \dfrac{2}{2} & \text{or} & \dfrac{2x}{2} > \dfrac{8}{2} & \text{Divide each part by 2.} \\
x < 1 & \text{or} & x > 4 & \text{Simplify.}
\end{array}
$$

Figure 51

The solution set is $\{x \mid x < 1 \text{ or } x > 4\}$, that is, all x in the intervals $(-\infty, 1)$ or $(4, \infty)$.

See Figure 51 for a graph of the solution set.

━━━ **Now Work**   PROBLEMS **13(C)** AND **51**

2.8 Assess Your Understanding

'Are You Prepared?'   *Answers are given at the end of these exercises. If you get a wrong answer, read the pages listed in red.*

1. Graph the inequality $x \geq -2$. (pp. A4–A5)

2. Use a real number line to define $|a|$. (p. A5)

3. Solve $4x - 3 = 9$. (pp. A66–A67)

4. Solve $3x - 2 > 7$. (pp. A86–A87)

5. Solve $-1 < 2x + 5 < 13$. (pp. A87–A88)

6. Use transformations to graph the function $f(x) = |x - 3|$. (pp. 91–93)

Concepts and Vocabulary

7. $|u| = a$ is equivalent to $u =$ _____ or $u =$ _____.

8. $|u| < a$ is equivalent to _____.

9. $|u| < a$ will have no solution if a _____ 0.

10. *True or False* $|x| = -4$ has no real solution.

11. *True or False* $|x| > -2$ has no real solution.

12. *True or False* $|u| > a$ is equivalent to $-a < u < a$.

Skill Building

In Problems 13–16, use the graphs of the functions given to solve each problem.

13. $f(x) = |x + 3|$; $g(x) = 6$
(a) $f(x) = g(x)$
(b) $f(x) \leq g(x)$
(c) $f(x) > g(x)$

14. $f(x) = |x - 2|$; $g(x) = 2$
(a) $f(x) = g(x)$
(b) $f(x) \leq g(x)$
(c) $f(x) > g(x)$

15. $f(x) = |2x - 1|$; $g(x) = 5$
  (a) $f(x) = g(x)$
  (b) $f(x) \geq g(x)$
  (c) $f(x) < g(x)$

16. $f(x) = |2x + 1|$; $g(x) = 7$
  (a) $f(x) = g(x)$
  (b) $f(x) \geq g(x)$
  (c) $f(x) < g(x)$

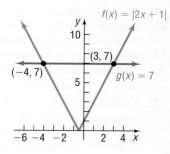

In Problems 17–38, solve each equation.

17. $|x| = 6$

18. $|x| = 12$

19. $|2x + 3| = 5$

20. $|3x - 1| = 2$

21. $|1 - 4t| + 8 = 13$

22. $|1 - 2z| + 6 = 9$

23. $|-2x| = 8$

24. $|-x| = 1$

25. $4 - |2x| = 3$

26. $5 - \left|\frac{1}{2}x\right| = 3$

27. $\frac{2}{3}|x| = 9$

28. $\frac{3}{4}|x| = 9$

29. $\left|\frac{x}{3} + \frac{2}{5}\right| = 2$

30. $\left|\frac{x}{2} - \frac{1}{3}\right| = 1$

31. $|u - 2| = -\frac{1}{2}$

32. $|2 - v| = -1$

33. $|x^2 - 9| = 0$

34. $|x^2 - 16| = 0$

35. $|x^2 - 2x| = 3$

36. $|x^2 + x| = 12$

37. $|x^2 + x - 1| = 1$

38. $|x^2 + 3x - 2| = 2$

In Problems 39–62, solve each absolute value inequality. Express your answer using set-builder or interval notation. Graph the solution set.

39. $|x| < 6$

40. $|x| < 9$

41. $|x| > 4$

42. $|x| > 1$

43. $|2x| < 8$

44. $|3x| < 15$

45. $|3x| > 12$

46. $|2x| > 6$

47. $|x - 2| + 2 < 3$

48. $|x + 4| + 3 < 5$

49. $|3t - 2| \leq 4$

50. $|2u + 5| \leq 7$

51. $|x - 3| \geq 2$

52. $|x + 4| \geq 2$

53. $|1 - 4x| - 7 < -2$

54. $|1 - 2x| - 4 < -1$

55. $|1 - 2x| > |-3|$

56. $|2 - 3x| > |-1|$

57. $|2x + 1| < -1$

58. $|3x - 4| \geq 0$

59. $|(3x - 2) - 7| < \frac{1}{2}$

60. $|(4x - 1) - 11| < \frac{1}{4}$

61. $5 - |x - 1| > 2$

62. $6 - |x + 3| \geq 2$

Mixed Practice

63. If $f(x) = -3|5x - 2|$ and $g(x) = -9$, solve:
  (a) $f(x) = g(x)$
  (b) $f(x) > g(x)$
  (c) $f(x) \leq g(x)$

64. If $f(x) = -2|2x - 3|$ and $g(x) = -12$, solve:
  (a) $f(x) = g(x)$
  (b) $f(x) \geq g(x)$
  (c) $f(x) < g(x)$

65. If $f(x) = |-3x + 2|$ and $g(x) = x + 10$, solve:
  (a) $f(x) = g(x)$
  (b) $f(x) \geq g(x)$
  (c) $f(x) < g(x)$

66. If $f(x) = |4x - 3|$ and $g(x) = x + 2$, solve:
  (a) $f(x) = g(x)$
  (b) $f(x) > g(x)$
  (c) $f(x) \leq g(x)$

Applications and Extensions

67. Express the fact that x differs from 10 by less than 2 as an inequality involving absolute value. Solve for x.

68. Express the fact that x differs from -6 by less than 3 as an inequality involving absolute value. Solve for x.

69. Express the fact that twice x differs from -1 by more than 5 as an inequality involving absolute value. Solve for x.

70. Express the fact that twice x differs from 3 by more than 1 as an inequality involving absolute value. Solve for x.

71. Tolerance A certain rod in an internal combustion engine is supposed to be 5.7 inches long. The tolerance on the rod is 0.0005 inch. If x represents the length of the rod, the acceptable lengths can be expressed as $|x - 5.7| \leq 0.0005$. Determine the acceptable lengths of the rod.

Source: Wiseco Piston

72. Tolerance A certain rod in an internal combustion engine is supposed to be 6.125 inches long. The tolerance on the rod is 0.0005 inch. If x represents the length of the rod, the acceptable lengths can be expressed as $|x - 6.125| \leq 0.0005$. Determine the acceptable lengths of the rod.

73. IQ Scores According to the Stanford–Binet IQ test, a normal IQ score is 100. It can be shown that anyone with an IQ x that satisfies the inequality $\left|\dfrac{x - 100}{15}\right| > 1.96$ has an unusual IQ score. Determine the IQ scores that would be considered unusual.

74. Gestation Period The length of human pregnancy is about 266 days. It can be shown that a mother whose gestation period x satisfies the inequality $\left|\dfrac{x - 266}{16}\right| > 1.96$ has an unusual length of pregnancy. Determine the lengths of pregnancy that would be considered unusual.

Explaining Concepts: Discussion and Writing

75. The equation $|5x + 1| + 7 = 5$ has no real solution. Explain why.

76. The inequality $|2x + 5| + 3 > 1$ has all real numbers as the solution set. Explain why.

77. The inequality $|2x - 1| \leq 0$ has $\left\{\dfrac{1}{2}\right\}$ as the solution set. Explain why.

'Are You Prepared?' Answers

1.

2. The distance from the origin to a is $|a|$ for any real number a.

3. $\{3\}$　　　　**4.** $x > 3$　　　　**5.** $-3 < x < 4$　　　　**6.**
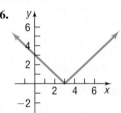

CHAPTER REVIEW

Things to Know

Linear function (p. 123)

$f(x) = mx + b$ 　　　　Average rate of change $= m$

Graph is a line with slope m and y-intercept b.

Quadratic function (pp. 141, 155–156)

$f(x) = ax^2 + bx + c, a \neq 0$ 　　　　Graph is a parabola that opens up if $a > 0$ and opens down if $a < 0$.

Vertex: $\left(-\dfrac{b}{2a}, f\left(-\dfrac{b}{2a}\right)\right)$

Axis of symmetry: $x = -\dfrac{b}{2a}$

y-intercept: $f(0)$

x-intercept(s): if any, found by finding the real solutions of the equation $ax^2 + bx + c = 0$

Quadratic equation and quadratic formula

The real solutions of the equation $ax^2 + bx + c = 0, a \neq 0$, are given by $x = \dfrac{-b \pm \sqrt{b^2 - 4ac}}{2a}$, provided $b^2 - 4ac \geq 0$. If $b^2 - 4ac < 0$, there are no real solutions. (p. 145)

In the complex number system, the solutions of the equation $ax^2 + bx + c = 0, a \neq 0$, are given by $x = \dfrac{-b \pm \sqrt{b^2 - 4ac}}{2a}$. (p. 179)

Discriminant (pp. 146 and 180)

If $b^2 - 4ac > 0$, there are two distinct real solutions.

If $b^2 - 4ac = 0$, there is one repeated real solution.

If $b^2 - 4ac < 0$, there are no real solutions, but there are two distinct complex solutions that are not real; the complex solutions are conjugates of each other.

Absolute value

If $|u| = a, a > 0$, then $u = -a$ or $u = a$. (p. 182)

If $|u| \le a, a > 0$, then $-a \le u \le a$. (p. 183)

If $|u| \ge a, a > 0$, then $u \le -a$ or $u \ge a$. (p. 184)

Objectives

Section		You should be able to ...	Example(s)	Review Exercises
2.1	1	Graph linear functions (p. 123)	1	1(a)–6(a), 1(b)–6(b), 7, 8
	2	Use average rate of change to identify linear functions (p. 123)	2	9, 10
	3	Determine whether a linear function is increasing, decreasing, or constant (p. 126)	3	1(e)–6(e)
	4	Find the zero of a linear function (p. 127)	4	7, 8
	5	Build linear models from verbal descriptions (p. 127)	5, 6	79, 80
2.2	1	Draw and interpret scatter diagrams (p. 134)	1	91(a), 92(a)
	2	Distinguish between linear and nonlinear relations (p. 135)	2	91(b) 92(a)
	3	Use a graphing utility to find the line of best fit (p. 136)	4	91(c)
2.3	1	Find the zeros of a quadratic function by factoring (p. 142)	1, 2	11–14, 17, 18
	2	Find the zeros of a quadratic function using the Square Root Method (p. 142)	3	15, 16
	3	Find the zeros of a quadratic function by completing the square (p. 144)	4	11–22
	4	Find the zeros of a quadratic function using the quadratic formula (p. 145)	5, 6	11–22
	5	Find the point of intersection of two functions (p. 147)	7	23–26
	6	Solve equations that are quadratic in form (p. 148)	8, 9	27–32
2.4	1	Graph a quadratic function using transformations (p. 153)	1	33–38
	2	Identify the vertex and axis of symmetry of a quadratic function (p. 155)	2	39–48
	3	Graph a quadratic function using its vertex, axis, and intercepts (p. 156)	3–6	39–48
	4	Find the maximum or minimum value of a quadratic function (p. 160)	7	49–54, 86, 87, 89
2.5	1	Solve inequalities involving a quadratic function (p. 164)	1, 2	55–58, 81(f)
2.6	1	Build quadratic models from verbal descriptions (p. 168)	1–4	79–90
	2	Build quadratic models from data (p. 172)	5	92
2.7	1	Find the complex zeros of a quadratic function (p. 178)	1–3	59–66
2.8	1	Solve absolute value equations (p. 181)	1	67–70
	2	Solve absolute value inequalities (p. 182)	2–5	71–78

Review Exercises

In Problems 1–6:
   (a) Determine the slope and y-intercept of each linear function.
   (b) Graph each function. Label the intercepts.
   (c) Determine the domain and the range of each function.
   (d) Determine the average rate of change of each function.
   (e) Determine whether the function is increasing, decreasing, or constant.

1. $f(x) = 2x - 5$

2. $g(x) = -4x + 7$

3. $h(x) = \dfrac{4}{5}x - 6$

4. $F(x) = -\dfrac{1}{3}x + 1$

5. $G(x) = 4$

6. $H(x) = -3$

7. Find the zero and y-intercept of $f(x) = 2x + 14$. Use the zero and y-intercept to graph f.

8. Find the zero and y-intercept of $f(x) = -\dfrac{1}{4}x + 2$. Use the zero and y-intercept to graph f.

In Problems 9 and 10, determine whether the function is linear or nonlinear. If the function is linear, determine the equation that defines $y = f(x)$.

9.

x	$y = f(x)$
-2	-7
0	3
1	8
3	18
6	33

10.

x	$y = g(x)$
-1	-3
0	4
1	7
2	6
3	1

In Problems 11–22, find the zeros of each quadratic function. What are the x-intercepts of the graph of the function?

11. $f(x) = x^2 + x - 72$

12. $G(x) = x^2 - 3x - 40$

13. $P(x) = 6x^2 - 13x - 5$

14. $H(x) = 3x^2 - 17x + 20$

15. $g(x) = (x - 3)^2 - 4$

16. $F(x) = (x + 5)^2 - 9$

17. $h(x) = 9x^2 + 6x + 1$

18. $f(x) = 4x^2 - 4x + 1$

19. $G(x) = 2x^2 - 4x - 1$

20. $P(x) = 3x^2 + 5x + 1$

21. $f(x) = -2x^2 + x + 1$

22. $H(x) = -x^2 - 2x + 5$

In Problems 23–26, solve $f(x) = g(x)$. Graph each function and label the intersection points.

23. $f(x) = (x - 3)^2$; $g(x) = 16$

24. $f(x) = x^2 + 6x$; $g(x) = 16$

25. $f(x) = x^2 + 4x - 5$; $g(x) = 4x - 1$

26. $f(x) = -2x^2 + 3x + 20$; $g(x) = 3x + 12$.

In Problems 27–32, find the real zeros of each function. What are the x-intercepts of the graph of the function?

27. $f(x) = x^4 - 5x^2 + 4$

28. $g(x) = 3x^4 + 4x^2 + 1$

29. $F(x) = (x - 3)^2 - 2(x - 3) - 48$

30. $G(x) = 2(x + 4)^2 + 3(x + 4)^2 - 14$

31. $h(x) = 3x - 13\sqrt{x} - 10$

32. $f(x) = \left(\dfrac{1}{x}\right)^2 - 4\left(\dfrac{1}{x}\right) - 12$

In Problems 33–38, graph each function using transformations (shifting, compressing, stretching, and reflection).

33. $f(x) = (x - 2)^2 + 2$

34. $f(x) = (x + 1)^2 - 4$

35. $f(x) = -(x - 4)^2$

36. $f(x) = (x - 1)^2 - 3$

37. $f(x) = 2(x + 1)^2 + 4$

38. $f(x) = -3(x + 2)^2 + 1$

In Problems 39–48, (a) graph each quadratic function by determining whether its graph opens up or down and by finding its vertex, axis of symmetry, y-intercept, and x-intercepts, if any. (b) Determine the domain and the range of the function. (c) Determine where the function is increasing and where it is decreasing.

39. $f(x) = (x - 2)^2 + 2$

40. $f(x) = (x + 1)^2 - 4$

41. $f(x) = \dfrac{1}{4}x^2 - 16$

42. $f(x) = -\dfrac{1}{2}x^2 + 2$

43. $f(x) = -4x^2 + 4x$

44. $f(x) = 9x^2 - 6x + 3$

45. $f(x) = \dfrac{9}{2}x^2 + 3x + 1$

46. $f(x) = -x^2 + x + \dfrac{1}{2}$

47. $f(x) = 3x^2 + 4x - 1$

48. $f(x) = -2x^2 - x + 4$

In Problems 49–54, determine whether the given quadratic function has a maximum value or a minimum value, and then find the value.

49. $f(x) = 3x^2 - 6x + 4$

50. $f(x) = 2x^2 + 8x + 5$

51. $f(x) = -x^2 + 8x - 4$

52. $f(x) = -x^2 - 10x - 3$

53. $f(x) = -3x^2 + 12x + 4$

54. $f(x) = -2x^2 + 4$

In Problems 55–58, solve each quadratic inequality.

55. $x^2 + 6x - 16 < 0$

56. $3x^2 - 2x - 1 \geq 0$

57. $3x^2 \geq 14x + 5$

58. $4x^2 < 13x - 3$

In Problems 59–66, find the complex zeros of each quadratic function. Graph each function and label the intercepts.

59. $f(x) = x^2 + 8$

60. $f(x) = x^2 - 8$

61. $g(x) = x^2 + 2x - 4$

62. $h(x) = x^2 + 4x + 5$

63. $p(x) = -2x^2 + 4x - 3$

64. $F(x) = -3x^2 + 6x - 5$

65. $f(x) = 4x^2 + 4x + 3$

66. $G(x) = 5x^2 + 5x + 2$

In Problems 67–70, solve each absolute value equation.

67. $|2x + 3| = 7$

68. $|3x - 1| = 5$

69. $|2 - 3x| + 2 = 9$

70. $|1 - 2x| + 1 = 4$

In Problems 71–78, solve each absolute value inequality. Express your answer using set-builder notation or interval notation. Graph the solution set.

71. $|3x + 4| < \dfrac{1}{2}$

72. $|1 - 2x| < \dfrac{1}{3}$

73. $|2x - 5| \geq 9$

74. $|3x + 1| \geq 10$

75. $2 + |2 - 3x| \leq 4$

76. $\dfrac{1}{2} + \left|\dfrac{2x - 1}{3}\right| \leq 1$

77. $1 - |2 - 3x| < -4$

78. $1 - \left|\dfrac{2x - 1}{3}\right| < -2$

---

79. Comparing Phone Companies Marissa must decide between one of two companies as her long-distance phone provider. Company A charges a monthly fee of $7.00 plus $0.06 per minute, while Company B does not have a monthly fee, but charges $0.08 per minute.
(a) Find a linear model that relates cost, C, to total minutes on the phone, x, for each company.
(b) Determine the number of minutes x for which the bill from Company A will equal the bill from Company B.
(c) For how many minutes x will the bill from Company B be less than the bill from Company A?

80. Sales Commissions Bill's sales position salary is $15,000 per year plus 1% of his total annual sales.
(a) Find a linear model that relates Bill's annual salary, S, to his total annual sales, x.
(b) In 2005, Bill had total annual sales of $1,000,000. What was Bill's salary?
(c) What would Bill have to sell to earn $100,000?
(d) Determine the sales required of Bill for his salary to exceed $150,000.

81. Demand Equation The price p (in dollars) and the quantity x sold of a certain product obey the demand equation

$$x = 1500 - 10p$$

(a) Find a model that expresses the revenue R as a function of the price p.
(b) What is the domain of R?
(c) What unit price should be used to maximize revenue?
(d) If this price is charged, what is the maximum revenue?
(e) How many units are sold at this price?
(f) What price should be charged to collect at least $56,000 in revenue?

82. Business: Determining the Cost of a Charter A group of 20 senior citizens can charter a bus for a one-day excursion trip for $15 per person. The charter company agrees to reduce the price of each ticket by 10¢ for each additional passenger in excess of 20 who goes on the trip, up to a maximum of 44 passengers (the capacity of the bus). If the final bill from the charter company was $482.40, how many seniors went on the trip, and how much did each pay?

83. Pleasing Proportions One formula stating the relationship between the length l and width w of a rectangle of "pleasing proportion" is $l^2 = w(l + w)$. How should a 4-foot by 8-foot sheet of plasterboard be cut so that the result is a rectangle of "pleasing proportion" with a width of 4 feet?

84. Geometry The hypotenuse of a right triangle measures 13 centimeters. Find the lengths of the legs if their sum is 17 centimeters.

85. Geometry The diagonal of a rectangle measures 10 inches. If the length is 2 inches more than the width, find the dimensions of the rectangle.

86. Minimizing Marginal Cost Callaway Golf Company has determined that the marginal cost C of manufacturing x Big Bertha golf clubs may be expressed by the quadratic function

$$C(x) = 4.9x^2 - 617.4x + 19,600$$

(a) How many clubs should be manufactured to minimize the marginal cost?
(b) At this level of production, what is the marginal cost?

87. Landscaping A landscape engineer has 200 feet of border to enclose a rectangular pond. What dimensions will result in the largest pond?

88. Geometry Find the length and width of a rectangle whose perimeter is 20 feet and whose area is 16 square feet.

89. A rectangle has one vertex on the line $y = 10 - x$, $x > 0$, another at the origin, one on the positive x-axis, and one on the positive y-axis. Find the largest area A that can be enclosed by the rectangle.

90. Parabolic Arch Bridge A horizontal bridge is in the shape of a parabolic arch. Given the information shown in the figure, what is the height h of the arch 2 feet from shore?

91. Bone Length Research performed at NASA, led by Dr. Emily R. Morey-Holton, measured the lengths of the right humerus and right tibia in 11 rats that were sent to space on Spacelab Life Sciences 2. The data on the following page were collected.
(a) Draw a scatter diagram of the data treating length of the right humerus as the independent variable.

(b) Based on the scatter diagram, do you think that there is a linear relation between the length of the right humerus and the length of the right tibia?

Right Humerus (mm), x	Right Tibia (mm), y
24.80	36.05
24.59	35.57
24.59	35.57
24.29	34.58
23.81	34.20
24.87	34.73
25.90	37.38
26.11	37.96
26.63	37.46
26.31	37.75
26.84	38.50

SOURCE: NASA Life Sciences Data Archive

(c) If the two variables appear to be linearly related, use a graphing utility to find the line of best fit that models the relation between length of the right humerus and length of the right tibia.

(d) Predict the length of the right tibia on a rat whose right humerus is 26.5 mm.

92. Advertising A small manufacturing firm collected the following data on advertising expenditures A (in thousands of dollars) and total revenue R (in thousands of dollars).

Advertising	Total Revenue
20	$6101
22	$6222
25	$6350
25	$6378
27	$6453
28	$6423
29	$6360
31	$6231

(a) Draw a scatter diagram of the data. Comment on the type of relation that may exist between the two variables.

(b) Use a graphing utility to find the quadratic function of best fit that models the relation between advertising expenditures and total revenue. Use this model to determine the optimal level of advertising.

(c) Use the model to predict the total revenue when the optimal level of advertising is spent.

(d) Use a graphing utility to draw a scatter diagram of the data and then graph the quadratic function of best fit on the scatter diagram.

CHAPTER TEST   CHAPTER Test Prep VIDEOS

Step-by-step test solutions are found on the Chapter Test Prep Videos available via the Video Resources on DVD, in *MyMathLab* or on *You Tube* (search "Sullivan College Algebra, Concepts Through Functions 2e).

1. For the linear function $f(x) = -4x + 3$,
   (a) Find the slope and y-intercept.
   (b) Determine whether f is increasing, decreasing, or constant.
   (c) Graph f.

2. Determine whether the given function is linear or nonlinear. If it is linear, determine the equation that defines $y = f(x)$.

x	y
−2	12
−1	7
0	2
1	−3
2	−8

In Problems 3 and 4, find the zeros of each quadratic function.

3. $f(x) = 3x^2 - 2x - 8$

4. $G(x) = -2x^2 + 4x + 1$

5. Given that $f(x) = x^2 + 3x$ and $g(x) = 5x + 3$, solve $f(x) = g(x)$. Graph each function and label the points of intersection.

6. Find the real zeros of $f(x) = (x - 1)^2 + 5(x - 1) + 4$.

7. Graph $f(x) = (x - 3)^2 - 2$ using transformations.

In Problems 8 and 9,
   (a) Determine whether the graph opens up or down.
   (b) Determine the vertex of the graph of the quadratic function.
   (c) Determine the axis of symmetry of the graph of the quadratic function.
   (d) Determine the intercepts of the graph of the quadratic function.
   (e) Use the information in parts (a)–(d) to graph the quadratic function.
   (f) Based on the graph, determine the domain and the range of the quadratic function.
   (g) Based on the graph, determine where the function is increasing and where it is decreasing.

8. $f(x) = 3x^2 - 12x + 4$

9. $g(x) = -2x^2 + 4x - 5$

10. Determine the quadratic function for the given graph.

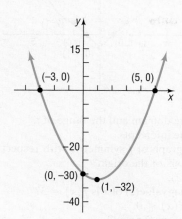

11. Determine whether $f(x) = -2x^2 + 12x + 3$ has a maximum or minimum value. Then find the maximum or minimum value.

12. Solve: $x^2 - 10x + 24 \geq 0$

13. Find the complex zeros of $f(x) = 2x^2 + 4x + 5$.

14. Solve: $|3x + 1| = 8$

In Problems 15 and 16, solve each absolute value inequality. Express your answer using set-builder notation or interval notation. Graph the solution set.

15. $\left| \dfrac{x + 3}{4} \right| < 2$

16. $|2x + 3| - 4 \geq 3$

17. **RV Rental** The weekly rental cost of a 20-foot recreational vehicle is $129.50 plus $0.15 per mile.
    Source: westernrv.com
    (a) Find a linear model that expresses the cost C as a function of miles driven m.

(b) What is the rental cost if 860 miles are driven?

(c) How many miles were driven if the rental cost is $213.80?

18. The price p (in dollars) and the quantity x sold of a certain product obey the demand equation $p = -\dfrac{1}{10}x + 1000$.
    (a) Find a model that expresses the revenue R as a function of x.
    (b) What is the revenue if 400 units are sold?
    (c) What quantity x maximizes revenue? What is the maximum revenue?
    (d) What price should the company charge to maximize revenue?

19. Consider these two data sets:

Set A

x	-2	-1	0	0	1	2	2
y	5	2	1	-3	-8	-12	-10

Set B

x	-2	-1	0	0	1	2	2
y	10	4	2	3	5	10	12

One data set follows a linear pattern and one data set follows a quadratic relation.
(a) Draw a scatter diagram of each data set. Determine which is linear and which is quadratic. For the linear data, indicate whether the relation shows a positive or negative slope. For the quadratic relation, indicate whether the quadratic function of best fit will open up or down.
(b) For the linear data set, find the line of best fit.
(c) For the quadratic data set, find the quadratic function of best fit.

CUMULATIVE REVIEW

1. Find the distance between the points $P = (-1, 3)$ and $Q = (4, -2)$. Find the midpoint of the line segment P to Q.

2. Which of the following points are on the graph of $y = x^3 - 3x + 1$?
   (a) $(-2, -1)$ (b) $(2, 3)$ (c) $(3, 1)$

3. Solve the inequality $5x + 3 \geq 0$ and graph the solution set.

4. Find the equation of the line containing the points $(-1, 4)$ and $(2, -2)$. Express your answer in slope–intercept form and graph the line.

5. Find the equation of the line perpendicular to the line $y = 2x + 1$ and containing the point $(3, 5)$. Express your answer in slope–intercept form and graph the line.

6. Graph the equation $x^2 + y^2 - 4x + 8y - 5 = 0$.

7. Determine whether the following relation represents a function: $\{(-3, 8), (1, 3), (2, 5), (3, 8)\}$.

8. For the function f defined by $f(x) = x^2 - 4x + 1$, evaluate:
   (a) $f(2)$ (b) $f(x) + f(2)$
   (c) $f(-x)$ (d) $-f(x)$
   (e) $f(x + 2)$ (f) $\dfrac{f(x + h) - f(x)}{h}, h \neq 0$

9. Find the domain of $h(z) = \dfrac{3z - 1}{6z - 7}$

10. Determine whether the following graph is the graph of a function.

11. Consider the function $f(x) = \dfrac{x}{x+4}$.

(a) Is the point $\left(1, \dfrac{1}{4}\right)$ on the graph of f?

(b) If $x = -2$, what is $f(x)$? What point is on the graph of f?

(c) If $f(x) = 2$, what is x? What point is on the graph of f?

12. Is the function $f(x) = \dfrac{x^2}{2x+1}$ even, odd, or neither?

13. Approximate the local maxima and local minima of $f(x) = x^3 - 5x + 1$ on $(-4, 4)$. Determine where the function is increasing and where it is decreasing.

14. If $f(x) = 3x + 5$ and $g(x) = 2x + 1$,

(a) Solve $f(x) = g(x)$.

(b) Solve $f(x) > g(x)$.

15. For the graph of the function f,

(a) Find the domain and the range of f.

(b) Find the intercepts.

(c) Is the graph of f symmetric with respect to the x-axis, the y-axis, or the origin?

(d) Find $f(2)$.

(e) For what value(s) of x is $f(x) = 3$?

(f) Solve $f(x) < 0$.

(g) Graph $y = f(x) + 2$.

(h) Graph $y = f(-x)$.

(i) Graph $y = 2f(x)$.

(j) Is f even, odd, or neither?

(k) Find the interval(s) on which f is increasing.

CHAPTER PROJECTS

I. **Analyzing a Stock** The beta, β, of a stock represents the relative risk of a stock compared with a market basket of stocks, such as Standard and Poor's 500 Index of stocks. Beta is computed by finding the slope of the line of best fit between the rate of return of the stock and the rate of return of the S&P 500. The rates of return are computed on a weekly basis.

(a) Find the weekly closing price of your favorite stock and the S&P 500 for 20 weeks. One good source is on the Internet at /finance.yahoo.com.

(b) Compute the rate of return by computing the weekly percentage change in the closing price of your stock and the weekly percentage change in the S&P 500 using the following formula:

$$\text{Weekly \% change} = \frac{P_2 - P_1}{P_1}$$

where P_1 is last week's price and P_2 is this week's price.

(c) Using a graphing utility, find the line of best fit, treating the weekly rate of return of the S&P 500 as the independent variable and the weekly rate of return of your stock as the dependent variable.

(d) What is the beta of your stock?

(e) Compare your result with that of the Value Line Investment Survey found in your library. What might account for any differences?

The following projects are available on the Instructor's Resource Center (IRC):

II. **First and Second Differences** Finite differences provide a numerical method that is used to estimate the graph of an unknown function.

III. Cannons A battery commander uses the weight of a missile, its initial velocity, and the position of its gun to determine where the missile will travel.

IV. CBL Experiment Computer simulation is used to study the physical properties of a bouncing ball.

Polynomial and Rational Functions

Outline

Day Length

Day length refers to the time each day from the moment the upper limb of the sun's disk appears above the horizon during sunrise to the moment when the upper limb disappears below the horizon during sunset. Due to the diffusion and refraction of sunlight by the atmosphere, there is actually daylight even when the sun is slightly below the horizon. The period when it is still light even though the sun is below the horizon is called twilight.

The length of a day varies throughout the year and depends on latitude. This variation is caused by the tilt of Earth's axis of rotation with respect to the ecliptic plane of Earth around the sun. At the solstice occurring about June 20–22, the north pole is tilted toward the sun, and therefore the northern hemisphere has days ranging in length from just over 12 hours in the southern portion of the Tropic of Cancer to 24 hours in the Arctic Circle, while the southern hemisphere has days ranging in length from just under 12 hours in the northern portion of the Tropic of Capricorn to zero in the Antarctic Circle. At the equinox occurring about September 22–23, the poles are neither tilted toward nor away from the sun, and the length of a day is generally about 12 hours all over Earth. At the solstice occurring about December 20–22, the south pole is tilted toward the sun, and therefore the southern hemisphere has days ranging in length from just over 12 hours in the northern portion of the Tropic of Capricorn to 24 hours in the Antarctic Circle, whereas the northern hemisphere has days ranging in length from just under 12 hours in the southern portion of the Tropic of Cancer to zero in the Arctic Circle. At the equinox occurring about March 19–21, the poles are again aligned so that the length of a day is generally about 12 hours all over Earth.

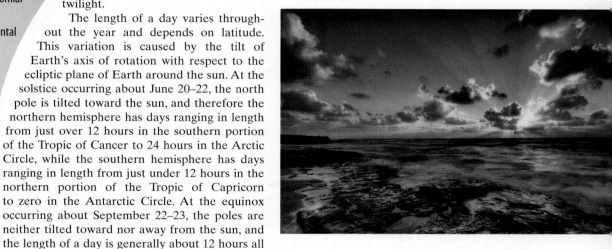

Source: Day length. (2009, May 19). In *Wikipedia, The Free Encyclopedia*. Retrieved May 27, 2009, from http://en.wikipedia.org/w/index.php?title=Day_length&oldid=290991965

—See Chapter Project I—

◁ **A Look Back** In Chapter 1, we began our discussion of functions. We defined domain and range and independent and dependent variables; we found the value of a function and graphed functions. We continued our study of functions by listing the properties that a function might have, such as being even or odd, and we created a library of functions, naming key functions and listing their properties, including their graphs.

In Chapter 2, we discussed linear functions and quadratic functions, which belong to the class of *polynomial functions*.

A Look Ahead ▷ In this chapter, we look at two general classes of functions, polynomial functions and rational functions, and examine their properties. Polynomial functions are arguably the simplest expressions in algebra. For this reason, they are often used to approximate other, more complicated functions. Rational functions are ratios of polynomial functions.

3.1 Polynomial Functions and Models

PREPARING FOR THIS SECTION   *Before getting started, review the following:*

- Polynomials (Appendix A, Section A.3, pp. A23–A31)
- Using a Graphing Utility to Approximate Local Maxima and Local Minima (Section 1.3, p. 73)
- Graphing Techniques: Transformations (Section 1.5, pp. 90–99)
- Intercepts (Foundations, Section 2, pp. 11–12)
- Real Zeros of a Function (Section 1.2, pp. 59–60)

Now Work the 'Are You Prepared?' problems on page 210.

> **OBJECTIVES**   1 Identify Polynomial Functions and Their Degree (p. 194)
> 2 Graph Polynomial Functions Using Transformations (p. 198)
> 3 Identify the Real Zeros of a Polynomial Function and Their Multiplicity (p. 199)
> 4 Analyze the Graph of a Polynomial Function (p. 206)
> 5 Build Cubic Models from Data (p. 209)

1 Identify Polynomial Functions and Their Degree

In Chapter 2 we studied the linear function $f(x) = mx + b$, which we can write as

$$f(x) = a_1x + a_0$$

and we studied the quadratic function $f(x) = ax^2 + bx + c, a \neq 0$, which we can write as

$$f(x) = a_2x^2 + a_1x + a_0 \qquad a_2 \neq 0$$

Each of these functions is an example of a *polynomial function*.

DEFINITION

> A **polynomial function** is a function of the form
>
> $$f(x) = a_nx^n + a_{n-1}x^{n-1} + \cdots + a_1x + a_0 \qquad (1)$$
>
> where $a_n, a_{n-1}, \ldots, a_1, a_0$ are real numbers and n is a nonnegative integer. The domain of a polynomial function is the set of all real numbers.

In Words
A polynomial function is the sum of monomials.

A polynomial function is a function whose rule is given by a polynomial in one variable. The **degree** of a polynomial function is the largest power of x that appears. The zero polynomial function $f(x) = 0 + 0x + 0x^2 + \cdots + 0x^n$ is not assigned a degree.

Polynomial functions are among the simplest expressions in algebra. They are easy to evaluate: only addition and repeated multiplication are required. Because of this, they are often used to approximate other, more complicated functions. In this section, we investigate properties of this important class of functions.

EXAMPLE 1   **Identifying Polynomial Functions**

Determine which of the following are polynomial functions. For those that are, state the degree; for those that are not, tell why not.

(a) $f(x) = 2 - 3x^4$ (b) $g(x) = \sqrt{x}$ (c) $h(x) = \dfrac{x^2 - 2}{x^3 - 1}$

(d) $F(x) = 0$ (e) $G(x) = 8$ (f) $H(x) = -2x^3(x - 1)^2$

Solution (a) f is a polynomial function of degree 4.

(b) g is not a polynomial function because $g(x) = \sqrt{x} = x^{\frac{1}{2}}$, so the variable x is raised to the $\dfrac{1}{2}$ power, which is not a nonnegative integer.

(c) h is not a polynomial function. It is the ratio of two distinct polynomials, and the polynomial in the denominator is of positive degree.

(d) F is the zero polynomial function; it is not assigned a degree.

(e) G is a nonzero constant function. It is a polynomial function of degree 0 since $G(x) = 8 = 8x^0$.

(f) $H(x) = -2x^3(x-1)^2 = -2x^3(x^2 - 2x + 1) = -2x^5 + 4x^4 - 2x^3$. So H is a polynomial function of degree 5. Do you see a way to find the degree of H without multiplying out?

●

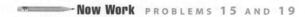 **Now Work** PROBLEMS **15** AND **19**

We have already discussed in detail polynomial functions of degrees 0, 1, and 2. See Table 1 for a summary of the properties of the graphs of these polynomial functions.

Table 1

Degree	Form	Name	Graph
No degree	$f(x) = 0$	Zero function	The x-axis
0	$f(x) = a_0, \quad a_0 \neq 0$	Constant function	Horizontal line with y-intercept a_0
1	$f(x) = a_1 x + a_0, \quad a_1 \neq 0$	Linear function	Nonvertical, nonhorizontal line with slope a_1 and y-intercept a_0
2	$f(x) = a_2 x^2 + a_1 x + a_0, \quad a_2 \neq 0$	Quadratic function	Parabola: graph opens up if $a_2 > 0$; graph opens down if $a_2 < 0$

One objective of this section is to analyze the graph of a polynomial function. If you take a course in calculus, you will learn that the graph of every polynomial function is both smooth and continuous. By **smooth,** we mean that the graph contains no sharp corners or cusps; by **continuous,** we mean that the graph has no gaps or holes and can be drawn without lifting pencil from paper. See Figures 1(a) and (b).

Figure 1

(a) Graph of a polynomial function: smooth, continuous

(b) Cannot be the graph of a polynomial function

Power Functions

We begin the analysis of the graph of a polynomial function by discussing *power functions*, a special kind of polynomial function.

DEFINITION

A **power function of degree n** is a monomial function of the form

$$f(x) = ax^n \tag{2}$$

where a is a real number, $a \neq 0$, and $n > 0$ is an integer.

Examples of power functions are

$$f(x) = 3x \qquad f(x) = -5x^2 \qquad f(x) = 8x^3 \qquad f(x) = -5x^4$$
$$\text{degree 1} \qquad\quad \text{degree 2} \qquad\quad \text{degree 3} \qquad\quad \text{degree 4}$$

The graph of a power function of degree 1, $f(x) = ax$, is a straight line, with slope a, that passes through the origin. The graph of a power function of degree 2, $f(x) = ax^2$, is a parabola, with vertex at the origin, that opens up if $a > 0$ and down if $a < 0$.

If we know how to graph a power function of the form $f(x) = x^n$, a compression or stretch and, perhaps, a reflection about the x-axis will enable us to obtain the graph of $g(x) = ax^n$. Consequently, we shall concentrate on graphing power functions of the form $f(x) = x^n$.

We begin with power functions of even degree of the form $f(x) = x^n$, $n \geq 2$ and n even. The domain of f is the set of all real numbers, and the range is the set of nonnegative real numbers. Such a power function is an even function (do you see why?), so its graph is symmetric with respect to the y-axis. Its graph always contains the origin and the points $(-1, 1)$ and $(1, 1)$.

If $n = 2$, the graph is the familiar parabola $y = x^2$ that opens up, with vertex at the origin. If $n \geq 4$, the graph of $f(x) = x^n$, n even, will be closer to the x-axis than the parabola $y = x^2$ if $-1 < x < 1$, $x \neq 0$, and farther from the x-axis than the parabola $y = x^2$ if $x < -1$ or if $x > 1$. Figure 2(a) illustrates this conclusion. Figure 2(b) shows the graphs of $y = x^4$ and $y = x^8$ for comparison.

Figure 2

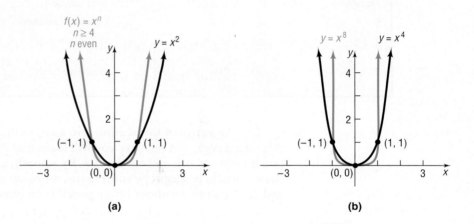

(a)   (b)

From Figure 2, we can see that as n increases the graph of $f(x) = x^n$, $n \geq 2$ and n even, tends to flatten out near the origin and to increase very rapidly when x is far from 0. For large n, it may appear that the graph coincides with the x-axis near the origin, but it does not; the graph actually touches the x-axis only at the origin (see Table 2). Also, for large n, it may appear that for $x < -1$ or for $x > 1$ the graph is vertical, but it is not; it is only increasing very rapidly in these intervals. If the graphs were enlarged many times, these distinctions would be clear.

Table 2

	$x = 0.1$	$x = 0.3$	$x = 0.5$
$f(x) = x^8$	10^{-8}	0.0000656	0.0039063
$f(x) = x^{20}$	10^{-20}	$3.487 \cdot 10^{-11}$	0.000001
$f(x) = x^{40}$	10^{-40}	$1.216 \cdot 10^{-21}$	$9.095 \cdot 10^{-13}$

Seeing the Concept

Graph $Y_1 = x^4$, $Y_2 = x^8$, and $Y_3 = x^{12}$ using the viewing rectangle $-2 \leq x \leq 2$, $-4 \leq y \leq 16$. Then graph each again using the viewing rectangle $-1 \leq x \leq 1$, $0 \leq y \leq 1$. See Figure 3. TRACE along one of the graphs to confirm that for x close to 0 the graph is above the x-axis and that for $x > 0$ the graph is increasing.

Figure 3

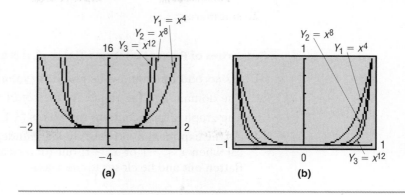

(a)   (b)

Properties of Power Functions, $f(x) = x^n$, n Is a Positive Even Integer

1. f is an even function, so its graph is symmetric with respect to the y-axis.
2. The domain is the set of all real numbers. The range is the set of non-negative real numbers.
3. The graph always contains the points $(-1, 1)$, $(0, 0)$, and $(1, 1)$.
4. As the exponent n increases in magnitude, the graph becomes more vertical when $x < -1$ or $x > 1$; but for x near the origin, the graph tends to flatten out and lie closer to the x-axis.

Figure 4

Now we consider power functions of odd degree of the form $f(x) = x^n$, $n \geq 3$ and n odd. The domain and the range of f are the set of real numbers. Such a power function is an odd function (do you see why?), so its graph is symmetric with respect to the origin. Its graph always contains the origin and the points $(-1, -1)$ and $(1, 1)$.

The graph of $f(x) = x^n$ when $n = 3$ has been shown several times and is repeated in Figure 4. If $n \geq 5$, the graph of $f(x) = x^n$, n odd, will be closer to the x-axis than that of $y = x^3$ if $-1 < x < 1$ and farther from the x-axis than that of $y = x^3$ if $x < -1$ or if $x > 1$. Figure 4 also illustrates this conclusion.

Figure 5 shows the graph of $y = x^5$ and the graph of $y = x^9$ for further comparison.

Figure 5

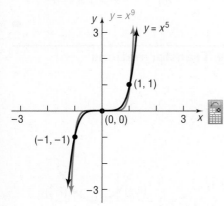

It appears that each graph coincides with the x-axis near the origin, but it does not; each graph actually crosses the x-axis at the origin. Also, it appears that as x increases the graph becomes vertical, but it does not; each graph is increasing very rapidly.

Seeing the Concept

Graph $Y_1 = x^3$, $Y_2 = x^7$, and $Y_3 = x^{11}$ using the viewing rectangle $-2 \leq x \leq 2$, $-16 \leq y \leq 16$. Then graph each again using the viewing rectangle $-1 \leq x \leq 1$, $-1 \leq y \leq 1$. See Figure 6. TRACE along one of the graphs to confirm that the graph is increasing and crosses the x-axis at the origin.

Figure 6

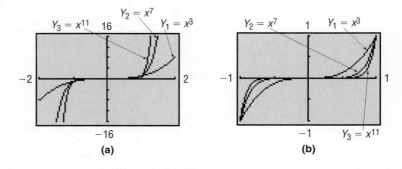

(a)   (b)

To summarize:

> **Properties of Power Functions, $f(x) = x^n$, n Is a Positive Odd Integer**
>
> 1. f is an odd function, so its graph is symmetric with respect to the origin.
> 2. The domain and the range are the set of all real numbers.
> 3. The graph always contains the points $(-1, -1)$, $(0, 0)$, and $(1, 1)$.
> 4. As the exponent n increases in magnitude, the graph becomes more vertical when $x < -1$ or $x > 1$; but for x near the origin, the graph tends to flatten out and lie closer to the x-axis.

2 Graph Polynomial Functions Using Transformations

The methods of shifting, compression, stretching, and reflection studied in Section 1.5, when used with the facts just presented, will enable us to graph polynomial functions that are transformations of power functions.

EXAMPLE 2 **Graphing a Polynomial Function Using Transformations**

Graph: $f(x) = 1 - x^5$

Solution It is helpful to rewrite f as $f(x) = -x^5 + 1$. Figure 7 shows the required stages.

Figure 7

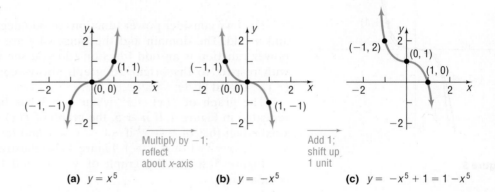

(a) $y = x^5$ **(b)** $y = -x^5$ **(c)** $y = -x^5 + 1 = 1 - x^5$

EXAMPLE 3 **Graphing a Polynomial Function Using Transformations**

Graph: $f(x) = \dfrac{1}{2}(x - 1)^4$

Solution Figure 8 shows the required stages.

Figure 8

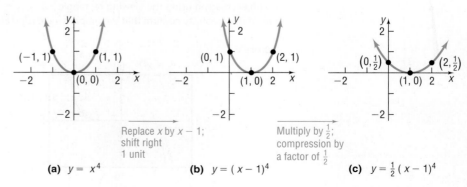

(a) $y = x^4$ **(b)** $y = (x - 1)^4$ **(c)** $y = \frac{1}{2}(x - 1)^4$

══**Now Work** PROBLEMS 27 AND 33

3 Identify the Real Zeros of a Polynomial Function and Their Multiplicity

Figure 9 shows the graph of a polynomial function with four x-intercepts. Notice that at the x-intercepts the graph must either cross the x-axis or touch the x-axis. Consequently, between consecutive x-intercepts the graph is either above the x-axis or below the x-axis. We will make use of this property of the graph of a polynomial function shortly.

Figure 9

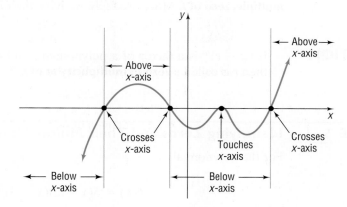

If a polynomial function f is factored completely, it is easy to locate the x-intercepts of the graph by solving the equation $f(x) = 0$ using the Zero-Product Property. For example, if $f(x) = (x - 1)^2(x + 3)$, then the solutions of the equation

$$f(x) = (x - 1)^2(x + 3) = 0$$

are identified as 1 and -3. Because $f(1) = f(-3) = 0$, 1 and -3 are real zeros of f. Based on this analysis, the following statements are equivalent.

1. r is a real zero of a polynomial function f.
2. r is an x-intercept of the graph of f.
3. $x - r$ is a factor of f.
4. r is a solution to the equation $f(x) = 0$.

EXAMPLE 4   **Finding a Polynomial Function from Its Zeros**

(a) Find a polynomial function of degree 3 whose zeros are -3, 2, and 5.

 (b) Use a graphing utility to graph the polynomial found in part (a) to verify your result.

Solution (a) If r is a real zero of a polynomial function f, then $x - r$ is a factor of f. This means that $x - (-3) = x + 3$, $x - 2$, and $x - 5$ are factors of f. As a result, any polynomial function of the form

$$f(x) = a(x + 3)(x - 2)(x - 5)$$

where a is a nonzero real number, qualifies. The value of a causes a stretch, compression, or reflection, but does not affect the x-intercepts of the graph. Do you know why?

(b) We choose to graph f with $a = 1$. Then

$$f(x) = (x + 3)(x - 2)(x - 5) = x^3 - 4x^2 - 11x + 30$$

Figure 10 shows the graph of f. Notice that the x-intercepts are -3, 2, and 5.

Figure 10

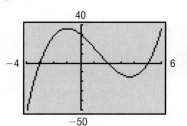

Seeing the Concept

Graph the function found in Example 4 for $a = 2$ and $a = -1$. Does the value of a affect the zeros of f? How does the value of a affect the graph of f?

Now Work PROBLEM 41

If the same factor $x - r$ occurs more than once, r is called a **repeated,** or **multiple, zero of f.** More precisely, we have the following definition.

DEFINITION

If $(x - r)^m$ is a factor of a polynomial f and $(x - r)^{m+1}$ is not a factor of f, then r is called a **zero of multiplicity m of f.** *

EXAMPLE 5

Identifying Zeros and Their Multiplicities

For the polynomial

$$f(x) = 5(x - 2)(x + 3)^2\left(x - \frac{1}{2}\right)^4$$

2 is a zero of multiplicity 1 because the exponent on the factor $x - 2$ is 1.

-3 is a zero of multiplicity 2 because the exponent on the factor $x + 3$ is 2.

$\dfrac{1}{2}$ is a zero of multiplicity 4 because the exponent on the factor $x - \dfrac{1}{2}$ is 4.

●

Now Work PROBLEM 49(a)

Suppose that it is possible to factor completely a polynomial function and, as a result, locate all the x-intercepts of its graph (the real zeros of the function). These x-intercepts then divide the x-axis into open intervals and, on each such interval, the graph of the polynomial will be either above or below the x-axis. Let's look at an example.

EXAMPLE 6

Graphing a Polynomial Using Its x-Intercepts

For the polynomial: $f(x) = x^2(x - 2)$

(a) Find the x- and y-intercepts of the graph of f.

(b) Use the x-intercepts to find the intervals on which the graph of f is above the x-axis and the intervals on which the graph of f is below the x-axis.

(c) Locate other points on the graph and connect all the points plotted with a smooth, continuous curve.

Solution

(a) The y-intercept is $f(0) = 0^2(0 - 2) = 0$. The x-intercepts satisfy the equation

$$f(x) = x^2(x - 2) = 0$$

from which we find

$$x^2 = 0 \quad \text{or} \quad x - 2 = 0$$
$$x = 0 \quad \text{or} \quad x = 2$$

The x-intercepts are 0 and 2.

---

*Some books use the terms **multiple root** and **root of multiplicity m**.

(b) The two x-intercepts divide the x-axis into three intervals:

$$(-\infty, 0) \qquad (0, 2) \qquad (2, \infty)$$

Since the graph of f crosses or touches the x-axis only at $x = 0$ and $x = 2$, it follows that the graph of f is either above the x-axis $[f(x) > 0]$ or below the x-axis $[f(x) < 0]$ on each of these three intervals. To see where the graph lies, we only need to pick a number in each interval, evaluate f there, and see whether the value is positive (above the x-axis) or negative (below the x-axis). See Table 3.

(c) In constructing Table 3, we obtained three additional points on the graph: $(-1, -3), (1, -1),$ and $(3, 9)$. Figure 11 illustrates these points, the intercepts, and a smooth, continuous curve (the graph of f) connecting them.

Table 3

Interval	$(-\infty, 0)$	$(0, 2)$	$(2, \infty)$
Number chosen	-1	1	3
Value of f	$f(-1) = -3$	$f(1) = -1$	$f(3) = 9$
Location of graph	Below x-axis	Below x-axis	Above x-axis
Point on graph	$(-1, -3)$	$(1, -1)$	$(3, 9)$

Figure 11

Look again at Table 3. Since the graph of $f(x) = x^2(x - 2)$ is below the x-axis on both sides of 0, the graph of f *touches* the x-axis at $x = 0$, a *zero of multiplicity 2*. Since the graph of f is below the x-axis for $x < 2$ and above the x-axis for $x > 2$, the graph of f *crosses* the x-axis at $x = 2$, a *zero of multiplicity 1*.

This suggests the following results:

If r Is a Zero of Even Multiplicity

The sign of $f(x)$ does not change from one side to the other side of r.   The graph of f **touches** the x-axis at r.

If r Is a Zero of Odd Multiplicity

The sign of $f(x)$ changes from one side to the other side of r.   The graph of f **crosses** the x-axis at r.

Now Work PROBLEM 49(b)

Behavior Near a Zero

We have just learned how the multiplicity of a zero can be used to determine whether the graph of a function touches or crosses the x-axis at the zero. However, we can learn more about the behavior of the graph near its zeros than just whether the graph crosses or touches the x-axis. Consider the function $f(x) = x^2(x - 2)$ whose graph is drawn in Figure 11. The zeros of f are 0 and 2. Table 4 shows the values of $f(x) = x^2(x - 2)$ and $y = -2x^2$ for x near 0. Figure 12 shows the points $(-0.1, -0.021), (-0.05, -0.0051),$

Table 4

x	$f(x) = x^2(x - 2)$	$y = -2x^2$
−0.1	−0.021	−0.02
−0.05	−0.005125	−0.005
−0.03	−0.001827	−0.0018
−0.01	−0.000201	−0.0002
0	0	0
0.01	−0.000199	−0.0002
0.03	−0.001773	−0.0018
0.05	−0.004875	−0.005
0.1	−0.019	−0.02

Figure 12

and so on, that are on the graph of $f(x) = x^2(x - 2)$ along with the graph of $y = -2x^2$ on the same Cartesian plane. From the table and graph, we can see that the points on the graph of $f(x) = x^2(x - 2)$ and the points on the graph of $y = -2x^2$ are indistinguishable near $x = 0$. So $y = -2x^2$ describes the behavior of the graph of $f(x) = x^2(x - 2)$ near $x = 0$.

But how did we know that the function $f(x) = x^2(x - 2)$ behaves like $y = -2x^2$ when x is close to 0? In other words, where did $y = -2x^2$ come from? Because the zero, 0, comes from the factor x^2, we evaluate all factors in the function f at 0 with the exception of x^2.

$$f(x) = x^2(x - 2)$$
$$\approx x^2(0 - 2)$$
$$= -2x^2$$

The factor x^2 gives rise to the zero, so we keep the factor x^2 and let $x = 0$ in the remaining factors to find the behavior near 0.

This tells us that the graph of $f(x) = x^2(x - 2)$ will behave like the graph of $y = -2x^2$ near $x = 0$.

Now let's discuss the behavior of $f(x) = x^2(x - 2)$ near $x = 2$, the other zero. Because the zero, 2, comes from the factor $x - 2$, we evaluate all factors of the function f at 2 with the exception of $x - 2$.

$$f(x) = x^2(x - 2)$$
$$\approx 2^2(x - 2)$$
$$= 4(x - 2)$$

The factor $x - 2$ gives rise to the zero, so we keep the factor $x - 2$ and let $x = 2$ in the remaining factors to find the behavior near 2.

So the graph of $f(x) = x^2(x - 2)$ will behave like the graph of $y = 4(x - 2)$ near $x = 2$. Table 5 verifies that $f(x) = x^2(x - 2)$ and $y = 4(x - 2)$ have similar values for x near 2. Figure 13 shows the points $(1.9, -0.361)$, $(1.99, -0.0396)$, and so on, that are on the graph of $f(x) = x^2(x - 2)$ along with the graph of $y = 4(x - 2)$

Table 5

x	$f(x) = x^2(x - 2)$	$y = 4(x - 2)$
1.9	−0.361	−0.4
1.99	−0.0396	−0.04
1.999	−0.003996	−0.004
2	0	0
2.001	0.004004	0.004
2.01	0.0404	0.04
2.1	0.441	0.4

Figure 13

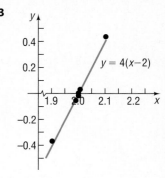

on the same Cartesian plane. We can see that the points on the graph of $f(x) = x^2(x - 2)$ and the points on the graph of $y = 4(x - 2)$ are indistinguishable near $x = 2$. So $y = 4(x - 2)$, a line with slope 4, describes the behavior of the graph of $f(x) = x^2(x - 2)$ near $x = 2$.

Figure 14 illustrates how we would use this information to begin to graph $f(x) = x^2(x - 2)$.

Figure 14

- By determining the multiplicity of a real zero, we determine whether the graph crosses or touches the x-axis at the zero.
- By determining the behavior of the graph near the real zero, we determine how the graph touches or crosses the x-axis.

 Now Work PROBLEM 49(c)

Turning Points

Look again at Figure 11 on page 201. We cannot be sure just how low the graph actually goes between $x = 0$ and $x = 2$. But we do know that somewhere in the interval $(0, 2)$ the graph of f must change direction (from decreasing to increasing). The points at which a graph changes direction are called **turning points.** In calculus, such points are called **local maxima** or **local minima,** and techniques for locating them are given. So we shall not ask for the location of turning points in our graphs. Instead, we will use the following result from calculus, which tells us the maximum number of turning points that the graph of a polynomial function can have.

THEOREM

> **Turning Points**
>
> If f is a polynomial function of degree n, then the graph of f has at most $n - 1$ turning points.
>
> If the graph of a polynomial function f has $n - 1$ turning points, the degree of f is at least n.

For example, the graph of $f(x) = x^2(x - 2)$ shown in Figure 11 is the graph of a polynomial function of degree 3 and has $3 - 1 = 2$ turning points: one at $(0, 0)$ and the other somewhere between $x = 0$ and $x = 2$.

If the graph of a polynomial function has three turning points, then the degree of the function must be at least 4.

 Exploration

A graphing utility can be used to locate the turning points of a graph. Graph $Y_1 = x^2(x - 2)$. Use MINIMUM to find the location of the turning point for $0 < x < 2$. See Figure 15.

Figure 15

 Now Work PROBLEM 49(d)

| EXAMPLE 7 | Identifying the Graph of a Polynomial Function |

Which of the graphs in Figure 16 could be the graph of a polynomial function? For those that could, list the real zeros and state the least degree the polynomial can have. For those that could not, say why not.

Figure 16

(a)   (b)   (c)   (d)

Solution

(a) The graph in Figure 16(a) cannot be the graph of a polynomial function because of the gap that occurs at $x = -1$. Remember, the graph of a polynomial function is continuous—no gaps or holes.

(b) The graph in Figure 16(b) could be the graph of a polynomial function because the graph is smooth and continuous. It has three real zeros, at -2, at 1, and at 2. Since the graph has two turning points, the degree of the polynomial function must be at least 3.

(c) The graph in Figure 16(c) cannot be the graph of a polynomial function because of the cusp at $x = 1$. Remember, the graph of a polynomial function is smooth.

(d) The graph in Figure 16(d) could be the graph of a polynomial function. It has two real zeros, at -2 and at 1. Since the graph has three turning points, the degree of the polynomial function is at least 4.

 Now Work PROBLEM 61

End Behavior

 One last remark about Figure 11. For very large values of x, either positive or negative, the graph of $f(x) = x^2(x - 2)$ looks like the graph of $y = x^3$. To see why, we write f in the form

$$f(x) = x^2(x - 2) = x^3 - 2x^2 = x^3\left(1 - \frac{2}{x}\right)$$

Now, for large values of x, either positive or negative, the term $\frac{2}{x}$ is close to 0, so for large values of x

$$f(x) = x^3 - 2x^2 = x^3\left(1 - \frac{2}{x}\right) \approx x^3$$

The behavior of the graph of a function for large values of x, either positive or negative, is referred to as its **end behavior.**

THEOREM

End Behavior

For large values of x, either positive or negative, the graph of the polynomial function

$$f(x) = a_n x^n + a_{n-1} x^{n-1} + \cdots + a_1 x + a_0$$

resembles the graph of the power function

$$y = a_n x^n$$

For example, if $f(x) = -2x^3 + 5x^2 + x - 4$, then the graph of f will behave like the graph of $y = -2x^3$ for very large values of x, either positive or negative. We can see that the graphs of f and $y = -2x^3$ "behave" the same by considering Table 6 and Figure 17.

Table 6

x	$f(x)$	$y = -2x^3$
10	$-1{,}494$	$-2{,}000$
100	$-1{,}949{,}904$	$-2{,}000{,}000$
500	$-248{,}749{,}504$	$-250{,}000{,}000$
1,000	$-1{,}994{,}999{,}004$	$-2{,}000{,}000{,}000$

Figure 17

Notice that, as x becomes a larger and larger positive number, the values of f become larger and larger negative numbers. When this happens, we say that f is **unbounded in the negative direction.** Rather than using words to describe the behavior of the graph of the function, we explain its behavior using notation. We can symbolize "the value of f becomes a larger and larger negative number as x becomes a larger and larger positive number" by writing $f(x) \rightarrow -\infty$ as $x \rightarrow \infty$ (read "the values of f approach negative infinity as x approaches infinity"). In calculus, **limits** are used to convey these ideas. There we use the symbolism $\lim_{x \to \infty} f(x) = -\infty$, read "the limit of $f(x)$ as x approaches infinity equals negative infinity," to mean that $f(x) \rightarrow -\infty$ as $x \rightarrow \infty$. When the value of a limit equals infinity, we mean that the values of the function are unbounded in the positive or negative direction and call the limit an **infinite limit.** Also based on Figure 17, we see that $\lim_{x \to -\infty} f(x) = \infty$. What does this mean in words?

When we discuss limits as x becomes unbounded in the negative direction or unbounded in the positive direction, we are discussing **limits at infinity.**

Look back at Figures 2 and 4. Based on the preceding theorem and the previous discussion on power functions, the end behavior of a polynomial function can only be of four types. See Figure 18.

Figure 18

End behavior of $f(x) = a_n x^n + a_{n-1} x^{n-1} + \cdots + a_1 x + a_0$

$f(x) \rightarrow \infty$ as $x \rightarrow -\infty$ $f(x) \rightarrow \infty$ as $x \rightarrow \infty$

$f(x) \rightarrow \infty$ as $x \rightarrow \infty$ $f(x) \rightarrow \infty$ as $x \rightarrow -\infty$

$f(x) \rightarrow -\infty$ as $x \rightarrow -\infty$ $f(x) \rightarrow -\infty$ as $x \rightarrow \infty$ $f(x) \rightarrow -\infty$ as $x \rightarrow -\infty$ $f(x) \rightarrow -\infty$ as $x \rightarrow \infty$

(a)
$n \geq 2$ even; $a_n > 0$

(b)
$n \geq 2$ even; $a_n < 0$

(c)
$n \geq 3$ odd; $a_n > 0$

(d)
$n \geq 3$ odd; $a_n < 0$

For example, if $f(x) = -2x^4 + x^3 + 4x^2 - 7x + 1$, the graph of f will resemble the graph of the power function $y = -2x^4$ for large $|x|$. The graph of f will behave like Figure 18(b) for large $|x|$.

➤ **Now Work** PROBLEM 49(e)

EXAMPLE 8

Identifying the Graph of a Polynomial Function

Which of the graphs in Figure 19 could be the graph of

$$f(x) = x^4 + 5x^3 + 5x^2 - 5x - 6?$$

Figure 19

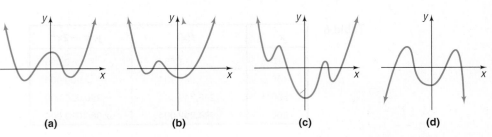

(a)   (b)   (c)   (d)

Solution

The y-intercept of f is $f(0) = -6$. We can eliminate the graph in Figure 19(a), whose y-intercept is positive.

We don't have any methods for finding the x-intercepts of f, so we move on to investigate the turning points of each graph. Since f is of degree 4, the graph of f has at most 3 turning points. We eliminate the graph in Figure 19(c) since that graph has 5 turning points.

Now we look at end behavior. For large values of x, the graph of f will behave like the graph of $y = x^4$. This eliminates the graph in Figure 19(d), whose end behavior is like the graph of $y = -x^4$.

Only the graph in Figure 19(b) could be (and, in fact, is) the graph of $f(x) = x^4 + 5x^3 + 5x^2 - 5x - 6$.

●

─── **Now Work** PROBLEM 65

SUMMARY

Graph of a Polynomial Function $f(x) = a_n x^n + a_{n-1} x^{n-1} + \cdots + a_1 x + a_0$ $\quad a_n \neq 0$

Degree of the polynomial function f: n

Graph is smooth and continuous

Maximum number of turning points: $n - 1$

At a zero of even multiplicity: The graph of f touches the x-axis.

At a zero of odd multiplicity: The graph of f crosses the x-axis.

Between zeros, the graph of f is either above or below the x-axis.

End behavior: For large $|x|$, the graph of f behaves like the graph of $y = a_n x^n$.

4 Analyze the Graph of a Polynomial Function

EXAMPLE 9

How to Analyze the Graph of a Polynomial Function

Analyze the graph of the polynomial function $f(x) = (2x + 1)(x - 3)^2$.

Step-by-Step Solution

Step 1: Determine the end behavior of the graph of the function.

First, we expand the polynomial to write it in the form

$$f(x) = a_n x^n + a_{n-1} x^{n-1} + \cdots + a_1 x + a_0$$

$$f(x) = (2x + 1)(x - 3)^2$$
$$= (2x + 1)(x^2 - 6x + 9)$$
$$= 2x^3 - 12x^2 + 18x + x^2 - 6x + 9 \quad \text{Multiply.}$$
$$= 2x^3 - 11x^2 + 12x + 9 \quad \text{Combine like terms.}$$

The polynomial function f is of degree 3. The graph of f behaves like $y = 2x^3$ for large values of $|x|$.

Step 2: Find the x- and y-intercepts of the graph of the function.

The y-intercept is $f(0) = 9$. To find the x-intercepts, we solve $f(x) = 0$.

$$f(x) = 0$$
$$(2x + 1)(x - 3)^2 = 0$$
$$2x + 1 = 0 \quad \text{or} \quad (x - 3)^2 = 0$$
$$x = -\frac{1}{2} \quad \text{or} \quad x - 3 = 0$$
$$x = 3$$

The x-intercepts are $-\frac{1}{2}$ and 3.

Step 3: Determine the zeros of the function and their multiplicity. Use this information to determine whether the graph crosses or touches the x-axis at each x-intercept.

The zeros of f are $-\frac{1}{2}$ and 3. The zero $-\frac{1}{2}$ is a zero of multiplicity 1, so the graph of f crosses the x-axis at $x = -\frac{1}{2}$. The zero 3 is a zero of multiplicity 2, so the graph of f touches the x-axis at $x = 3$.

Step 4: Determine the maximum number of turning points on the graph of the function.

Because the polynomial function is of degree 3 (Step 1), the graph of the function will have at most $3 - 1 = 2$ turning points.

Step 5: Determine the behavior of the graph of f near each x-intercept.

The two x-intercepts are $-\frac{1}{2}$ and 3.

$$\text{Near} -\frac{1}{2}: \quad f(x) = (2x + 1)(x - 3)^2$$
$$\approx (2x + 1)\left(-\frac{1}{2} + 3\right)^2$$
$$= (2x + 1)\left(\frac{25}{4}\right)$$
$$= \frac{25}{2}x + \frac{25}{4} \qquad \text{A line with slope } \frac{25}{2}$$

$$\text{Near } 3: \quad f(x) = (2x + 1)(x - 3)^2$$
$$\approx (2 \cdot 3 + 1)(x + 3)^2$$
$$= 7(x + 3)^2 \qquad \text{A parabola that opens up}$$

Step 6: Put all the information from Steps 1 through 5 together to obtain the graph of f.

Figure 20(a) illustrates the information obtained from Steps 1 through 5. We evaluate f at -1, 1, and 4 to help establish the scale on the y-axis. The graph of f is given in Figure 20(b).

Figure 20

(a)

(b)

SUMMARY

> **Analyzing the Graph of a Polynomial Function**
>
> **STEP 1:** Determine the end behavior of the graph of the function.
>
> **STEP 2:** Find the x- and y-intercepts of the graph of the function.
>
> **STEP 3:** Determine the zeros of the function and their multiplicity. Use this information to determine whether the graph crosses or touches the x-axis at each x-intercept.
>
> **STEP 4:** Determine the maximum number of turning points on the graph of the function.
>
> **STEP 5:** Determine the behavior of the graph near each x-intercept.
>
> **STEP 6:** Use the information in Steps 1 through 5 to draw a complete graph of the function.

━━━━━━━━ **Now Work**   PROBLEM **69**

For polynomial functions that have noninteger coefficients and for polynomials that are not easily factored, we utilize the graphing utility early in the analysis. This is because the amount of information that can be obtained from algebraic analysis is limited.

EXAMPLE 10

How to Use a Graphing Utility to Analyze the Graph of a Polynomial Function

Analyze the graph of the polynomial function

$$f(x) = x^3 + 2.48x^2 - 4.3155x + 2.484406$$

Step-by-Step Solution

Step 1: Determine the end behavior of the graph of the function.

The polynomial function f is of degree 3. The graph of f behaves like $y = x^3$ for large values of $|x|$.

Step 2: Graph the function using a graphing utility.

See Figure 21 for the graph of f.

Figure 21

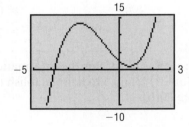

Step 3: Use a graphing utility to approximate the x- and y-intercepts of the graph.

The y-intercept is $f(0) = 2.484406$. In Example 9, the polynomial function was factored, so it was easy to find the x-intercepts algebraically. However, it is not readily apparent how to factor f in this example. Therefore, we use a graphing utility's ZERO (or ROOT or SOLVE) feature and find the lone x-intercept to be -3.79, rounded to two decimal places.

Step 4: Use a graphing utility to create a TABLE to find points on the graph around each x-intercept.

Table 7 shows values of x on each side of the x-intercept. The points $(-4, -4.57)$ and $(-2, 13.04)$ are on the graph.

Table 7

X	Y1
-4	-4.574
-2	13.035

Y1■X^3+2.48X²-4...

Step 5: Approximate the turning points of the graph.

From the graph of f shown in Figure 21, we can see that f has two turning points. Using MAXIMUM, one turning point is at $(-2.28, 13.36)$, rounded to two decimal places. Using MINIMUM, the other turning point is at $(0.63, 1)$, rounded to two decimal places.

Step 6: Use the information in Steps 1 through 5 to draw a complete graph of the function by hand.

Figure 22 shows a graph of f using the information in Steps 1 through 5.

Figure 22

Step 7: Find the domain and the range of the function.

The domain and the range of f are the set of all real numbers.

Step 8: Use the graph to determine where the function is increasing and where it is decreasing.

Based on the graph, f is increasing on the intervals $(-\infty, -2.28)$ and $(0.63, \infty)$. Also, f is decreasing on the interval $(-2.28, 0.63)$.

SUMMARY

Using a Graphing Utility to Analyze the Graph of a Polynomial Function

STEP 1: Determine the end behavior of the graph of the function.

STEP 2: Graph the function using a graphing utility.

STEP 3: Use a graphing utility to approximate the x- and y-intercepts of the graph.

STEP 4: Use a graphing utility to create a TABLE to find points on the graph around each x-intercept.

STEP 5: Approximate the turning points of the graph.

STEP 6: Use the information in Steps 1 through 5 to draw a complete graph of the function by hand.

STEP 7: Find the domain and the range of the function.

STEP 8: Use the graph to determine where the function is increasing and where it is decreasing.

━━━ **Now Work** PROBLEM 87

5 Build Cubic Models from Data

In Section 2.2 we found the line of best fit from data, and in Section 2.6 we found the quadratic function of best fit. It is also possible to find polynomial functions of best fit. However, most statisticians do not recommend finding polynomials of best fit of degree higher than 3.

Data that follow a cubic relation should look like Figure 23(a) or (b).

Figure 23

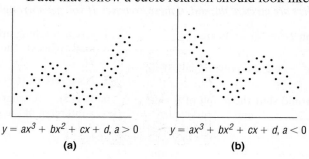

$y = ax^3 + bx^2 + cx + d, a > 0$ $\qquad$ $y = ax^3 + bx^2 + cx + d, a < 0$

(a) $\qquad\qquad\qquad\qquad$ (b)

EXAMPLE 11   **A Cubic Function of Best Fit**

The data in Table 8 represent the weekly cost C (in thousands of dollars) of printing x thousand textbooks.

(a) Draw a scatter diagram of the data using x as the independent variable and C as the dependent variable. Comment on the type of relation that may exist between the two variables x and C.

(b) Using a graphing utility, find the cubic function of best fit $C = C(x)$ that models the relation between number of texts and cost.

(c) Graph the cubic function of best fit on your scatter diagram.

(d) Use the function found in part (b) to predict the cost of printing 22 thousand texts per week.

Table 8

Number of Text Books, x	Cost, C
0	100
5	128.1
10	144
13	153.5
17	161.2
18	162.6
20	166.3
23	178.9
25	190.2
27	221.8

Solution

(a) Figure 24 shows the scatter diagram. A cubic relation may exist between the two variables.

(b) Upon executing the CUBIC REGression program, we obtain the results shown in Figure 25. The output that the utility provides shows us the equation $y = ax^3 + bx^2 + cx + d$. The cubic function of best fit to the data is $C(x) = 0.0155x^3 - 0.5951x^2 + 9.1502x + 98.4327$.

(c) Figure 26 shows the graph of the cubic function of best fit on the scatter diagram. The function fits the data reasonably well.

Figure 24

Figure 25

Figure 26

(d) We evaluate the function $C(x)$ at $x = 22$.

$$C(22) = 0.0155(22)^3 - 0.5951(22)^2 + 9.1502(22) + 98.4327 \approx 176.8$$

The model predicts that the cost of printing 22 thousand textbooks in a week will be 176.8 thousand dollars, that is $176,800.

3.1 Assess Your Understanding

'Are You Prepared?' *Answers are given at the end of these exercises. If you get a wrong answer, read the pages listed in red.*

1. The intercepts of the equation $9x^2 + 4y = 36$ are _____. (pp. 11–12)

2. Is the expression $4x^3 - 3.6x^2 - \sqrt{2}$ is a polynomial? If so, what is its degree? (pp. A23–A25)

3. To graph $y = x^2 - 4$, you would shift the graph of $y = x^2$ _____ a distance of _____ units. (pp. 90–91)

4. Use a graphing utility to approximate (rounded to two decimal places) the local maxima and local minima of $f(x) = x^3 - 2x^2 - 4x + 5$, for $-3 < x < 3$. (p. 73)

5. Is 4 a zero of $f(x) = x^3 - 4x^2 + 2x - 8$? Why? (pp. 59–60)

6. If $g(5) = 0$, what point is on the graph of g? What is the x-intercept of the graph of g? (pp. 59–60)

Concepts and Vocabulary

7. The graph of every polynomial function is both _____ and _____.

8. If r is a real zero of even multiplicity of a function f, then the graph of f _____ (crosses/touches) the x-axis at r.

9. The graphs of power functions of the form $f(x) = x^n$, where n is an even integer, always contain the points _____, _____, and _____.

10. If r is a solution to the equation $f(x) = 0$, name three additional statements that can be made about f and r assuming f is a polynomial function.

11. The points at which a graph changes direction (from increasing to decreasing or decreasing to increasing) are called _____.

12. The graph of the function $f(x) = 3x^4 - x^3 + 5x^2 - 2x - 7$ will behave like the graph of _____ for large values of $|x|$.

13. If $f(x) = -2x^5 + x^3 - 5x^2 + 7$, then $\lim\limits_{x \to -\infty} f(x) =$ _____ and $\lim\limits_{x \to \infty} f(x) =$ _____.

14. Explain what the notation $\lim\limits_{x \to \infty} f(x) = -\infty$ means.

Skill Building

In Problems 15–26, determine which functions are polynomial functions. For those that are, state the degree. For those that are not, tell why not.

15. $f(x) = 4x + x^3$

16. $f(x) = 5x^2 + 4x^4$

17. $g(x) = \dfrac{1 - x^2}{2}$

18. $h(x) = 3 - \dfrac{1}{2}x$

19. $f(x) = 1 - \dfrac{1}{x}$

20. $f(x) = x(x - 1)$

21. $g(x) = x^{3/2} - x^2 + 2$

22. $h(x) = \sqrt{x}(\sqrt{x} - 1)$

23. $F(x) = 5x^4 - \pi x^3 + \dfrac{1}{2}$

24. $F(x) = \dfrac{x^2 - 5}{x^3}$

25. $G(x) = 2(x - 1)^2(x^2 + 1)$

26. $G(x) = -3x^2(x + 2)^3$

In Problems 27–40, use transformations of the graph of $y = x^4$ or $y = x^5$ to graph each function.

27. $f(x) = (x + 1)^4$

28. $f(x) = (x - 2)^5$

29. $f(x) = x^5 - 3$

30. $f(x) = x^4 + 2$

31. $f(x) = \dfrac{1}{2}x^4$

32. $f(x) = 3x^5$

33. $f(x) = -x^5$

34. $f(x) = -x^4$

35. $f(x) = (x - 1)^5 + 2$

36. $f(x) = (x + 2)^4 - 3$

37. $f(x) = 2(x + 1)^4 + 1$

38. $f(x) = \dfrac{1}{2}(x - 1)^5 - 2$

39. $f(x) = 4 - (x - 2)^5$

40. $f(x) = 3 - (x + 2)^4$

In Problems 41–48, form a polynomial function whose real zeros and degree are given. Answers will vary depending on the choice of a leading coefficient.

41. Zeros: $-1, 1, 3$; degree 3

42. Zeros: $-2, 2, 3$; degree 3

43. Zeros: $-3, 0, 4$; degree 3

44. Zeros: $-4, 0, 2$; degree 3

45. Zeros: $-4, -1, 2, 3$; degree 4

46. Zeros: $-3, -1, 2, 5$; degree 4

47. Zeros: -1, multiplicity 1; 3, multiplicity 2; degree 3

48. Zeros: -2, multiplicity 2; 4, multiplicity 1; degree 3

In Problems 49–60, for each polynomial function:
(a) List each real zero and its multiplicity.
(b) Determine whether the graph crosses or touches the x-axis at each x-intercept.
(c) Determine the behavior of the graph near each x-intercept (zero).
(d) Determine the maximum number of turning points on the graph.
(e) Determine the end behavior; that is, find the power function that the graph of f resembles for large values of $|x|$.

49. $f(x) = 3(x - 7)(x + 3)^2$

50. $f(x) = 4(x + 4)(x + 3)^3$

51. $f(x) = 4(x^2 + 1)(x - 2)^3$

52. $f(x) = 2(x - 3)(x^2 + 4)^3$

53. $f(x) = -2\left(x + \dfrac{1}{2}\right)^2(x + 4)^3$

54. $f(x) = \left(x - \dfrac{1}{3}\right)^2(x - 1)^3$

55. $f(x) = (x - 5)^3(x + 4)^2$

56. $f(x) = (x + \sqrt{3})^2(x - 2)^4$

57. $f(x) = 3(x^2 + 8)(x^2 + 9)^2$

58. $f(x) = -2(x^2 + 3)^3$

59. $f(x) = -2x^2(x^2 - 2)$

60. $f(x) = 4x(x^2 - 3)$

In Problems 61–64, identify which of the graphs could be the graph of a polynomial function. For those that could, list the real zeros and state the least degree the polynomial can have. For those that could not, say why not.

61.

62.

63.

64.

In Problems 65–68, construct a polynomial function that might have the given graph. (More than one answer may be possible.)

65.

66.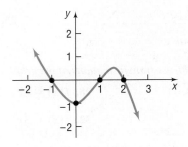

67.

68.

In Problems 69–86, analyze each polynomial function by following Steps 1 through 6 on page 208.

69. $f(x) = x^2(x - 3)$

70. $f(x) = x(x + 2)^2$

71. $f(x) = (x + 4)(x - 2)^2$

72. $f(x) = (x - 1)(x + 3)^2$

73. $f(x) = -2(x + 2)(x - 2)^3$

74. $f(x) = -\frac{1}{2}(x + 4)(x - 1)^3$

75. $f(x) = (x + 1)(x - 2)(x + 4)$

76. $f(x) = (x - 1)(x + 4)(x - 3)$

77. $f(x) = x^2(x - 2)(x + 2)$

78. $f(x) = x^2(x - 3)(x + 4)$

79. $f(x) = (x + 1)^2(x - 2)^2$

80. $f(x) = (x + 1)^3(x - 3)$

81. $f(x) = x^2(x - 3)(x + 1)$

82. $f(x) = x^2(x - 3)(x - 1)$

83. $f(x) = (x + 2)^2(x - 4)^2$

84. $f(x) = (x - 2)^2(x + 2)(x + 4)$

85. $f(x) = x^2(x - 2)(x^2 + 3)$

86. $f(x) = x^2(x^2 + 1)(x + 4)$

In Problems 87–94, analyze each polynomial function f by following Steps 1 through 8 on page 209.

87. $f(x) = x^3 + 0.2x^2 - 1.5876x - 0.31752$

88. $f(x) = x^3 - 0.8x^2 - 4.6656x + 3.73248$

89. $f(x) = x^3 + 2.56x^2 - 3.31x + 0.89$

90. $f(x) = x^3 - 2.91x^2 - 7.668x - 3.8151$

91. $f(x) = x^4 - 2.5x^2 + 0.5625$

92. $f(x) = x^4 - 18.5x^2 + 50.2619$

93. $f(x) = 2x^4 - \pi x^3 + \sqrt{5}x - 4$

94. $f(x) = -1.2x^4 + 0.5x^2 - \sqrt{3}x + 2$

Mixed Practice

In Problems 95–102, analyze each polynomial function by following Steps 1 through 6 on page 208.
[**Hint:** You will need to first factor the polynomial].

95. $f(x) = 4x - x^3$

96. $f(x) = x - x^3$

97. $f(x) = x^3 + x^2 - 12x$

98. $f(x) = x^3 + 2x^2 - 8x$

99. $f(x) = 2x^4 + 12x^3 - 8x^2 - 48x$

100. $f(x) = 4x^3 + 10x^2 - 4x - 10$

101. $f(x) = -x^5 - x^4 + x^3 + x^2$

102. $f(x) = -x^5 + 5x^4 + 4x^3 - 20x^2$

In Problems 103–106, construct a polynomial function f with the given characteristics.

103. Zeros: $-3, 1, 4$; degree 3; y-intercept: 36

104. Zeros: $-4, -1, 2$; degree 3; y-intercept: 16

105. Zeros: -5 (multiplicity 2); 2(multiplicity 1); 4 (multiplicity 1); degree 4; contains the point $(3, 128)$

106. Zeros: -4 (multiplicity 1); 0 (multiplicity 3); 2 (multiplicity 1); degree 5; contains the point $(-2, 64)$

107. $G(x) = (x + 3)^2(x - 2)$

   (a) Identify the x-intercepts of the graph of G. -3
   (b) What are the x-intercepts of the graph of $y = G(x + 3)$?

108. $h(x) = (x + 2)(x - 4)^3$

   (a) Identify the x-intercepts of the graph of h.
   (b) What are the x-intercepts of the graph of $y = h(x - 2)$?

Applications and Extensions

109. Future Value of Money Suppose that you make deposits of $500 at the beginning of every year into an Individual Retirement Account (IRA) earning interest r. At the beginning of the first year, the value of the account will be $500; at the beginning of the second year, the value of the account, will be

$$\underbrace{\$500 + \$500r}_{\text{Value of 1st deposit}} + \underbrace{500}_{\text{Value of 2nd deposit}} = \$500(1 + r) + \$500 = 500r + 1000$$

   (a) Verify that the value of the account at the beginning of the third year is $T(r) = 500r^2 + 1500r + 1500$.

   (b) The account value at the beginning of the fourth year is $F(r) = 500r^3 + 2000r^2 + 3000r + 2000$. If the annual rate of interest is 5% $= 0.05$, what will be the value of the account at the beginning of the fourth year?

110. A Geometric Series In calculus, you will learn that certain functions can be approximated by polynomial functions. We will explore one such function now.

   (a) Using a graphing utility, create a table of values with
$$Y_1 = f(x) = \frac{1}{1-x} \text{ and } Y_2 = g_2(x) = 1 + x + x^2 + x^3$$
for $-1 < x < 1$ with $\Delta\text{Tbl} = 0.1$.

   (b) Using a graphing utility, create a table of values with
$$Y_1 = f(x) = \frac{1}{1-x} \text{ and } Y_3 = g_3(x) = 1 + x + x^2 +$$
$x^3 + x^4$ for $-1 < x < 1$ with $\Delta\text{Tbl} = 0.1$.

   (c) Using a graphing utility, create a table of values with
$$Y_1 = f(x) = \frac{1}{1-x} \text{ and } Y_4 = g_4(x) = 1 + x + x^2 +$$
$x^3 + x^4 + x^5$ for $-1 < x < 1$ with $\Delta\text{Tbl} = 0.1$.

   (d) What do you notice about the values of the function as more terms are added to the polynomial? Are there some values of x for which the approximations are better?

111. Hurricanes In 2005, hurricane Katrina struck the Gulf Coast of the United States, killing 1289 people and causing an estimated $200 billion in damage. The following data represent the number of major hurricane strikes in the United States (category 3, 4, or 5) each decade from 1921 to 2000.

Decade, x	Major Hurricanes Striking United States, H
1921–1930, 1	5
1931–1940, 2	8
1941–1950, 3	10
1951–1960, 4	8
1961–1970, 5	6
1971–1980, 6	4
1981–1990, 7	5
1991–2000, 8	5

Source: National Oceanic & Atmospheric Administration

   (a) Draw a scatter diagram of the data. Comment on the type of relation that may exist between the two variables.

   (b) Use a graphing utility to find the cubic function of best fit that models the relation between decade and number of major hurricanes.

   (c) Use the model found in part (b) to predict the number of major hurricanes that struck the United States between 1961 and 1970.

(d) With a graphing utility, draw a scatter diagram of the data and then graph the cubic function of best fit on the scatter diagram.

(e) Concern has risen about the increase in the number and intensity of hurricanes, but some scientists believe this is just a natural fluctuation that could last another decade or two. Use your model to predict the number of major hurricanes that will strike the United States between 2001 and 2010. Does your result appear to agree with what these scientists believe?

(f) From 2001 through 2008, 10 major hurricanes struck the United States. Does this support or contradict your prediction in part (e)?

112. Cost of Manufacturing The following data represent the cost C (in thousands of dollars) of manufacturing Chevy Cobalts and the number x of Cobalts produced.

(a) Draw a scatter diagram of the data using x as the independent variable and C as the dependent variable. Comment on the type of relation that may exist between the two variables C and x.

(b) Use a graphing utility to find the cubic function of best fit $C = C(x)$.

(c) Graph the cubic function of best fit on the scatter diagram.

Number of Cobalts Produced, x	Cost, C
0	10
1	23
2	31
3	38
4	43
5	50
6	59
7	70
8	85
9	105
10	135

(d) Use the function found in part (b) to predict the cost of manufacturing 11 Cobalts.

(e) Interpret the y-intercept.

Mixed Practice

113. Temperature The following data represent the temperature T (°Fahrenheit) in Kansas City, Missouri, x hours after midnight on May 15, 2005.

Hours after Midnight, x	Temperature (°F), T
3	45.0
6	44.1
9	51.1
12	57.9
15	63.0
18	63.0
21	59.0
24	54.0

Source: The Weather Underground

(a) Draw a scatter diagram of the data. Comment on the type of relation that may exist between the two variables.

(b) Find the average rate of change in temperature from 9 AM to 12 noon.

(c) What is the average rate of change in temperature from 3 PM to 6 PM?

(d) Decide on a function of best fit to these data (linear, quadratic, or cubic) and use this function to predict the temperature at 5 PM.

(e) With a graphing utility, draw a scatter diagram of the data and then graph the function of best fit on the scatter diagram.

(f) Interpret the y-intercept.

Explaining Concepts: Discussion and Writing

114. Can the graph of a polynomial function have no y-intercept? Can it have no x-intercepts? Explain.

115. Write a few paragraphs that provide a general strategy for graphing a polynomial function. Be sure to mention the following: degree, intercepts, end behavior, and turning points.

116. Make up a polynomial that has the following characteristics: crosses the x-axis at -1 and 4, touches the x-axis at 0 and 2, and is above the x-axis between 0 and 2. Give your polynomial to a fellow classmate and ask for a written critique.

117. Make up two polynomials, not of the same degree, with the following characteristics: crosses the x-axis at -2, touches the x-axis at 1, and is above the x-axis between -2 and 1. Give your polynomials to a fellow classmate and ask for a written critique.

118. The graph of a polynomial function is always smooth and continuous. Name a function studied earlier that is smooth and not continuous. Name one that is continuous, but not smooth.

119. Which of the following statements are true regarding the graph of the cubic polynomial $f(x) = x^3 + bx^2 + cx + d$? (Give reasons for your conclusions.)

(a) It intersects the y-axis in one and only one point.

(b) It intersects the x-axis in at most three points.

(c) It intersects the x-axis at least once.

(d) For $|x|$ very large, it behaves like the graph of $y = x^3$.

(e) It is symmetric with respect to the origin.

(f) It passes through the origin.

120. The illustration shows the graph of a polynomial function.

(a) Is the degree of the polynomial even or odd?
(b) Is the leading coefficient positive or negative?
(c) Is the function even, odd, or neither?
(d) Why is x^2 necessarily a factor of the polynomial?

(e) What is the minimum degree of the polynomial?
(f) Formulate five different polynomials whose graphs could look like the one shown. Compare yours to those of other students. What similarities do you see? What differences?

121. Design a polynomial function with the following characteristics: degree 6; four distinct real zeros, one of multiplicity 3; y-intercept 3; behaves like $y = -5x^6$ for large values of $|x|$. Is this polynomial unique? Compare your polynomial with those of other students. What terms will be the same as everyone else's? Add some more characteristics, such as symmetry or naming the real zeros. How does this modify the polynomial?

'Are You Prepared?' Answers

1. $(-2,0), (2,0), (0,9)$ **2.** Yes; 3   **3.** Down; 4   **4.** Local maximum 6.48 at $x = -0.67$; local minimum -3 at $x = 2$
5. Yes, because $f(4) = 0$ **6.** $(5,0); 5$

3.2 Properties of Rational Functions

PREPARING FOR THIS SECTION *Before getting started, review the following:*

- Rational Expressions (Appendix A, Section A.6, pp. A46–A47)
- Polynomial Division (Appendix A, Section A.3, pp. A28–A31)
- Graph of $f(x) = \dfrac{1}{x}$ (Foundations, Section 2, Example 10, p. 15)
- Graphing Techniques: Transformations (Section 1.5, pp. 90–99)

Now Work the 'Are You Prepared?' problems on page 223.

OBJECTIVES **1** Find the Domain of a Rational Function (p. 215)
2 Find the Vertical Asymptotes of a Rational Function (p. 219)
3 Find the Horizontal or Oblique Asymptote of a Rational Function (p. 220)

Ratios of integers are called *rational numbers*. Similarly, ratios of polynomial functions are called *rational functions*. Examples of rational functions are

$$R(x) = \frac{x^2 - 4}{x^2 + x + 1} \qquad F(x) = \frac{x^3}{x^2 - 4} \qquad G(x) = \frac{3x^2}{x^4 - 1}$$

DEFINITION

A **rational function** is a function of the form

$$R(x) = \frac{p(x)}{q(x)}$$

where p and q are polynomial functions and q is not the zero polynomial. The domain of a rational function is the set of all real numbers except those for which the denominator q is 0.

1 Find the Domain of a Rational Function

EXAMPLE 1   Finding the Domain of a Rational Function

(a) The domain of $R(x) = \dfrac{2x^2 - 4}{x + 5}$ is the set of all real numbers x except -5; that is, the domain is $\{x \mid x \neq -5\}$.

(b) The domain of $R(x) = \dfrac{1}{x^2 - 4}$ is the set of all real numbers x except -2 and 2; that is, the domain is $\{x \mid x \neq -2, x \neq 2\}$.

(c) The domain of $R(x) = \dfrac{x^3}{x^2 + 1}$ is the set of all real numbers.

(d) The domain of $R(x) = \dfrac{x^2 - 1}{x - 1}$ is the set of all real numbers x except 1; that is, the domain is $\{x \mid x \neq 1\}$.

It is important to observe that the functions

$$R(x) = \frac{x^2 - 1}{x - 1} \quad \text{and} \quad f(x) = x + 1$$

are not equal, since the domain of R is $\{x \mid x \neq 1\}$ and the domain of f is the set of all real numbers.

━━━━━**Now Work** PROBLEM 15

If $R(x) = \dfrac{p(x)}{q(x)}$ is a rational function and if p and q have no common factors, then the rational function R is said to be in **lowest terms.** For a rational function $R(x) = \dfrac{p(x)}{q(x)}$ in lowest terms, the real zeros, if any, of the numerator are the x-intercepts of the graph of R and so will play a major role in the graph of R. The real zeros of the denominator of R [that is, the numbers x, if any, for which $q(x) = 0$], although not in the domain of R, also play a major role in the graph of R. We will discuss this role shortly.

We have already discussed the properties of the rational function $y = \dfrac{1}{x}$. (Refer to Example 10, page 15). The next rational function that we take up is $H(x) = \dfrac{1}{x^2}$.

EXAMPLE 2

Graphing $y = \dfrac{1}{x^2}$

Analyze the graph of $H(x) = \dfrac{1}{x^2}$.

Table 9

x	$H(x) = \dfrac{1}{x^2}$
$\dfrac{1}{2}$	4
$\dfrac{1}{100}$	$10{,}000$
$\dfrac{1}{10{,}000}$	$100{,}000{,}000$
1	1
2	$\dfrac{1}{4}$
100	$\dfrac{1}{10{,}000}$
$10{,}000$	$\dfrac{1}{100{,}000{,}000}$

Solution The domain of $H(x) = \dfrac{1}{x^2}$ is the set of all real numbers x except 0. The graph has no y-intercept, because x can never equal 0. The graph has no x-intercept because the equation $H(x) = 0$ has no solution. Therefore, the graph of H will not cross or touch either of the coordinate axes. Because

$$H(-x) = \frac{1}{(-x)^2} = \frac{1}{x^2} = H(x)$$

H is an even function, so its graph is symmetric with respect to the y-axis.

Table 9 shows the behavior of $H(x) = \dfrac{1}{x^2}$ for selected positive numbers x. (We will use symmetry to obtain the graph of H when $x < 0$.) From the first three rows of Table 9, we see that, as the values of x approach (get closer to) 0, the values of $H(x)$ become larger and larger positive numbers, so H is unbounded in the positive direction. In calculus, limits are used to convey these ideas. We use the symbolism $\lim\limits_{x \to 0} H(x) = \infty$, read "the limit of $H(x)$ as x approaches zero equals infinity," to mean that $H(x) \to \infty$ as $x \to 0$.

Look at the last four rows of Table 9. As $x \to \infty$, the values of $H(x)$ approach 0 (the end behavior of the graph). In calculus, this is symbolized by writing $\lim_{x \to \infty} H(x) = 0$. Figure 27 shows the graph. Notice the use of red dashed lines to convey the ideas discussed above.

Figure 27

$H(x) = \dfrac{1}{x^2}$

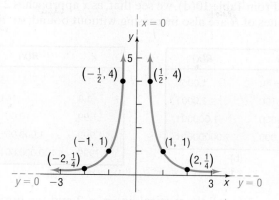

Sometimes transformations (shifting, compressing, stretching, and reflection) can be used to graph a rational function.

EXAMPLE 3   **Using Transformations to Graph a Rational Function**

Graph the rational function: $R(x) = \dfrac{1}{(x-2)^2} + 1$

Solution   First, we take note of the fact that the domain of R is the set of all real numbers except $x = 2$. To graph R, we start with the graph of $y = \dfrac{1}{x^2}$. See Figure 28 for the steps.

Figure 28

(a) $y = \dfrac{1}{x^2}$ (b) $y = \dfrac{1}{(x-2)^2}$ (c) $y = \dfrac{1}{(x-2)^2} + 1$

Replace x by $x - 2$; shift right 2 units

Add 1; shift up 1 unit

Now Work PROBLEM 33

Asymptotes

Let's investigate the roles of the vertical line $x = 2$ and the horizontal line $y = 1$ in Figure 28(c).

First, let's consider the end behavior of $R(x) = \dfrac{1}{(x-2)^2} + 1$. Table 10(a) shows the values of R at $x = 10, 100, 1000, 10,000$. Notice that, as x becomes unbounded in the positive direction, the values of R approach 1, so $\lim_{x \to \infty} R(x) = 1$. From Table 10(b) we see that, as x becomes unbounded in the negative direction, the values of R also approach 1, so $\lim_{x \to -\infty} R(x) = 1$.

Notice that $x = 2$ is not in the domain of R. However, we want to know the behavior of the graph of R near $x = 2$. Table 10(c) shows the values of R at $x = 1.5, 1.9, 1.99, 1.999,$ and 1.9999. We see that, as x approaches 2 for $x < 2$, denoted $x \rightarrow 2^-$, the values of R are increasing without bound, so $\lim\limits_{x \rightarrow 2^-} R(x) = \infty$. From Table 10(d), we see that, as x approaches 2 for $x > 2$, denoted $x \rightarrow 2^+$, the values of R are also increasing without bound, so $\lim\limits_{x \rightarrow 2^+} R(x) = \infty$.

Table 10

x	R(x)
10	1.0156
100	1.0001
1,000	1.000001
10,000	1.00000001

(a)

x	R(x)
−10	1.0069
−100	1.0001
−1000	1.000001
−10,000	1.00000001

(b)

x	R(x)
1.5	5
1.9	101
1.99	10,001
1.999	1,000,001
1.9999	100,000,001

(c)

x	R(x)
2.5	5
2.1	101
2.01	10,001
2.001	1,000,001
2.0001	100,000,001

(d)

We call the vertical line $x = 2$ and the horizontal line $y = 1$ *asymptotes* of the graph of R, which we define as follows:

DEFINITION

Let R denote a function.

If, as $x \rightarrow -\infty$ or as $x \rightarrow \infty$, the values of $R(x)$ approach some fixed number L, then the line $y = L$ is a **horizontal asymptote** of the graph of R. [Refer to Figures 29(a) and (b).]

If, as x approaches some number c, the values $|R(x)| \rightarrow \infty [R(x) \rightarrow -\infty$ or $R(x) \rightarrow \infty]$, then the line $x = c$ is a **vertical asymptote** of the graph of R. The graph of R never intersects a vertical asymptote. [Refer to Figures 29(c) and (d).]

Figure 29

(a) End behavior:
As $x \rightarrow \infty$, the values of $R(x)$ approach L [$\lim\limits_{x \rightarrow \infty} R(x) = L$]. That is, the points on the graph of R are getting closer to the line $y = L$; $y = L$ is a horizontal asymptote.

(b) End behavior:
As $x \rightarrow -\infty$, the values of $R(x)$ approach L [$\lim\limits_{x \rightarrow -\infty} R(x) = L$]. That is, the points on the graph of R are getting closer to the line $y = L$; $y = L$ is a horizontal asymptote.

(c) As x approaches c, the values of $|R(x)| \rightarrow \infty$ [$\lim\limits_{x \rightarrow c^-} R(x) = \infty$; $\lim\limits_{x \rightarrow c^+} R(x) = \infty$]. That is, the points on the graph of R are getting closer to the line $x = c$; $x = c$ is a vertical asymptote.

(d) As x approaches c, the values of $|R(x)| \rightarrow \infty$ [$\lim\limits_{x \rightarrow c^-} R(x) = -\infty$; $\lim\limits_{x \rightarrow c^+} R(x) = \infty$]. That is, the points on the graph of R are getting closer to the line $x = c$; $x = c$ is a vertical asymptote.

Figure 30

A horizontal asymptote, when it occurs, describes the **end behavior** of the graph as $x \rightarrow \infty$ or as $x \rightarrow -\infty$. **The graph of a function may intersect a horizontal asymptote.**

A vertical asymptote, when it occurs, describes the behavior of the graph when x is close to some number c. **The graph of a function will never intersect a vertical asymptote.**

There is a third possibility. If, as $x \rightarrow -\infty$ or as $x \rightarrow \infty$, the value of a rational function $R(x)$ approaches a linear expression $ax + b$, $a \neq 0$, then the line $y = ax + b$, $a \neq 0$, is an **oblique asymptote** of R. Figure 30 shows an oblique asymptote. An oblique asymptote, when it occurs, describes the end behavior of the graph. **The graph of a function may intersect an oblique asymptote.**

Now Work PROBLEM 25

2 Find the Vertical Asymptotes of a Rational Function

The vertical asymptotes of a rational function $R(x) = \dfrac{p(x)}{q(x)}$, in lowest terms, are located at the real zeros of the denominator of $q(x)$. Suppose that r is a real zero of q, so $x - r$ is a factor of q. As x approaches r, symbolized as $x \rightarrow r$, the values of $x - r$ approach 0, causing the ratio to become unbounded, that is, $|R(x)| \rightarrow \infty$. Based on the definition, we conclude that the line $x = r$ is a vertical asymptote.

THEOREM

WARNING If a rational function is not in lowest terms, an application of this theorem may result in an incorrect listing of vertical asymptotes. ∎

> **Locating Vertical Asymptotes**
>
> A rational function $R(x) = \dfrac{p(x)}{q(x)}$, *in lowest terms*, will have a vertical asymptote $x = r$ if r is a real zero of the *denominator* q. That is, if $x - r$ is a factor of the denominator q of a rational function $R(x) = \dfrac{p(x)}{q(x)}$, in lowest terms, R will have the vertical asymptote $x = r$.

EXAMPLE 4 **Finding Vertical Asymptotes**

Find the vertical asymptotes, if any, of the graph of each rational function.

(a) $F(x) = \dfrac{x + 3}{x - 1}$ (b) $R(x) = \dfrac{x}{x^2 - 4}$

(c) $H(x) = \dfrac{x^2}{x^2 + 1}$ (d) $G(x) = \dfrac{x^2 - 9}{x^2 + 4x - 21}$

Solution

WARNING In Example 4(a), the vertical asymptote is $x = 1$. Do not say that the vertical asymptote is 1. ∎

(a) F is in lowest terms and the only zero of the denominator is 1. The line $x = 1$ is the vertical asymptote of the graph of F.

(b) R is in lowest terms and the zeros of the denominator $x^2 - 4$ are -2 and 2. The lines $x = -2$ and $x = 2$ are the vertical asymptotes of the graph of R.

(c) H is in lowest terms and the denominator has no real zeros, because the equation $x^2 + 1 = 0$ has no real solutions. The graph of H has no vertical asymptotes.

(d) Factor $G(x)$ to determine if it is in lowest terms.

$$G(x) = \frac{x^2 - 9}{x^2 + 4x - 21} = \frac{(x + 3)(x - 3)}{(x + 7)(x - 3)} = \frac{x + 3}{x + 7} \qquad x \neq 3$$

The only zero of the denominator of $G(x)$ in lowest terms is -7. The line $x = -7$ is the only vertical asymptote of the graph of G.

●

As Example 4 points out, rational functions can have no vertical asymptotes, one vertical asymptote, or more than one vertical asymptote.

 Exploration

Graph each of the following rational functions:

$$R(x) = \frac{1}{x - 1} \qquad R(x) = \frac{1}{(x - 1)^2} \qquad R(x) = \frac{1}{(x - 1)^3} \qquad R(x) = \frac{1}{(x - 1)^4}$$

Each has the vertical asymptote $x = 1$. What happens to the value of $R(x)$ as x approaches 1 from the right side of the vertical asymptote; that is, what is $\lim\limits_{x \to 1^+} R(x)$? What happens to the value of $R(x)$ as x approaches 1 from the left side of the vertical asymptote; that is, what is $\lim\limits_{x \to 1^-} R(x)$? How does the multiplicity of the zero in the denominator affect the graph of R?

Now Work PROBLEM 47 (FIND THE VERTICAL ASYMPTOTES, IF ANY.)

3 Find the Horizontal or Oblique Asymptote of a Rational Function

The procedure for finding horizontal and oblique asymptotes is somewhat more involved. To find such asymptotes, we need to know how the values of a function behave as $x \to -\infty$ or as $x \to \infty$. That is, we need to find the end behavior of the rational function.

If a rational function $R(x)$ is **proper,** that is, if the degree of the numerator is less than the degree of the denominator, then as $x \to -\infty$ or as $x \to \infty$ the value of $R(x)$ approaches 0. Consequently, the line $y = 0$ (the x-axis) is a horizontal asymptote of the graph.

THEOREM   If a rational function is proper, the line $y = 0$ is a horizontal asymptote of its graph.

EXAMPLE 5   Finding a Horizontal Asymptote

Find the horizontal asymptote, if one exists, of the graph of

$$R(x) = \frac{x - 12}{4x^2 + x + 1}$$

Solution   Since the degree of the numerator, 1, is less than the degree of the denominator, 2, the rational function R is proper. We conclude that the line $y = 0$ is a horizontal asymptote of the graph of R.

                                            ●

To see why $y = 0$ is a horizontal asymptote of the function R in Example 5, we need to investigate the behavior of R as $x \to -\infty$ and $x \to \infty$. When $|x|$ is very large, the numerator of R, which is $x - 12$, can be approximated by the power function $y = x$, while the denominator of R, which is $4x^2 + x + 1$, can be approximated by the power function $y = 4x^2$. Applying these ideas to $R(x)$, we find

$$R(x) = \frac{x - 12}{4x^2 + x + 1} \underset{\substack{\uparrow \\ \text{For } |x| \text{ very large}}}{\approx} \frac{x}{4x^2} = \frac{1}{4x} \underset{\substack{\uparrow \\ \text{As } x \to -\infty \text{ or } x \to \infty}}{\to 0}$$

This shows that the line $y = 0$ is a horizontal asymptote of the graph of R.

If a rational function $R(x) = \dfrac{p(x)}{q(x)}$ is **improper,** that is, if the degree of the numerator is greater than or equal to the degree of the denominator, we must use long division to write the rational function as the sum of a polynomial $f(x)$ (the quotient) plus a proper rational function $\dfrac{r(x)}{q(x)}$ ($r(x)$ is the remainder). That is, we write

$$R(x) = \frac{p(x)}{q(x)} = f(x) + \frac{r(x)}{q(x)}$$

where $f(x)$ is a polynomial and $\dfrac{r(x)}{q(x)}$ is a proper rational function. Since $\dfrac{r(x)}{q(x)}$ is proper, $\dfrac{r(x)}{q(x)} \to 0$ as $x \to -\infty$ or as $x \to \infty$. As a result,

$$R(x) = \frac{p(x)}{q(x)} \to f(x) \quad \text{as } x \to -\infty \text{ or as } x \to \infty$$

The possibilities are listed next.

1. If $f(x) = b$, a constant, the line $y = b$ is a horizontal asymptote of the graph of R.

2. If $f(x) = ax + b$, $a \neq 0$, the line $y = ax + b$ is an oblique asymptote of the graph of R.

3. In all other cases, the graph of R approaches the graph of f, and there are no horizontal or oblique asymptotes.

We illustrate each of the possibilities in Examples 6 through 8.

EXAMPLE 6   **Finding a Horizontal or Oblique Asymptote**

Find the horizontal or oblique asymptote, if one exists, of the graph of

$$H(x) = \frac{3x^4 - x^2}{x^3 - x^2 + 1}$$

Solution   Since the degree of the numerator, 4, is greater than the degree of the denominator, 3, the rational function H is improper. To find a horizontal or oblique asymptote, we use long division.

$$
\begin{array}{r}
3x + 3 \\
x^3 - x^2 + 1 \overline{) 3x^4 - x^2 } \\
\underline{3x^4 - 3x^3 + 3x} \\
3x^3 - x^2 - 3x \\
\underline{3x^3 - 3x^2 + 3} \\
2x^2 - 3x - 3
\end{array}
$$

As a result,

$$H(x) = \frac{3x^4 - x^2}{x^3 - x^2 + 1} = 3x + 3 + \frac{2x^2 - 3x - 3}{x^3 - x^2 + 1}$$

As $x \to -\infty$ or as $x \to \infty$,

$$\frac{2x^2 - 3x - 3}{x^3 - x^2 + 1} \approx \frac{2x^2}{x^3} = \frac{2}{x} \to 0$$

As $x \to -\infty$ or as $x \to \infty$, we have $H(x) \to 3x + 3$. We conclude that the graph of the rational function H has an oblique asymptote $y = 3x + 3$.

●

EXAMPLE 7   **Finding a Horizontal or Oblique Asymptote**

Find the horizontal or oblique asymptote, if one exists, of the graph of

$$R(x) = \frac{8x^2 - x + 2}{4x^2 - 1}$$

Solution   Since the degree of the numerator, 2, equals the degree of the denominator, 2, the rational function R is improper. To find a horizontal or oblique asymptote, we use long division.

$$
\begin{array}{r}
2 \\
4x^2 - 1 \overline{) 8x^2 - x + 2} \\
\underline{8x^2 - 2} \\
- x + 4
\end{array}
$$

As a result,

$$R(x) = \frac{8x^2 - x + 2}{4x^2 - 1} = 2 + \frac{-x + 4}{4x^2 - 1}$$

Then, as $x \to -\infty$ or as $x \to \infty$,

$$\frac{-x + 4}{4x^2 - 1} \approx \frac{-x}{4x^2} = \frac{-1}{4x} \to 0$$

As $x \to -\infty$ or as $x \to \infty$, we have $R(x) \to 2$. We conclude that $y = 2$ is a horizontal asymptote of the graph.

●

In Example 7, we note that the quotient 2 obtained by long division is the quotient of the leading coefficients of the numerator polynomial and the denominator polynomial $\left(\dfrac{8}{4}\right)$. This means that we can avoid the long division process for rational functions where the numerator and denominator *are of the same degree* and conclude that the quotient of the leading coefficients will give us the horizontal asymptote.

Now Work PROBLEMS 43 AND 45

EXAMPLE 8 **Finding a Horizontal or Oblique Asymptote**

Find the horizontal or oblique asymptote, if one exists, of the graph of

$$G(x) = \frac{2x^5 - x^3 + 2}{x^3 - 1}$$

Solution Since the degree of the numerator, 5, is greater than the degree of the denominator, 3, the rational function G is improper. To find a horizontal or oblique asymptote, we use long division.

$$
\begin{array}{r}
2x^2 - 1 \\
x^3 - 1 \overline{) 2x^5 - x^3 \qquad + 2} \\
\underline{2x^5 \qquad - 2x^2} \\
-x^3 + 2x^2 + 2 \\
\underline{-x^3 \qquad + 1} \\
2x^2 + 1
\end{array}
$$

As a result,

$$G(x) = \frac{2x^5 - x^3 + 2}{x^3 - 1} = 2x^2 - 1 + \frac{2x^2 + 1}{x^3 - 1}$$

Then, as $x \to -\infty$ or as $x \to \infty$,

$$\frac{2x^2 + 1}{x^3 - 1} \approx \frac{2x^2}{x^3} = \frac{2}{x} \to 0$$

As $x \to -\infty$ or as $x \to \infty$, we have $G(x) \to 2x^2 - 1$. We conclude that, for large values of $|x|$, the graph of G approaches the graph of $y = 2x^2 - 1$. That is, the graph of G will look like the graph of $y = 2x^2 - 1$ as $x \to -\infty$ or $x \to \infty$. Since $y = 2x^2 - 1$ is not a linear function, G has no horizontal or oblique asymptotes.

●

We now summarize the procedure for finding horizontal and oblique asymptotes.

SUMMARY

Finding a Horizontal or Oblique Asymptote of a Rational Function

Consider the rational function

$$R(x) = \frac{p(x)}{q(x)} = \frac{a_n x^n + a_{n-1} x^{n-1} + \cdots + a_1 x + a_0}{b_m x^m + b_{m-1} x^{m-1} + \cdots + b_1 x + b_0}$$

in which the degree of the numerator is n and the degree of the denominator is m.

1. If $n < m$ (the degree of the numerator is less than the degree of the denominator), then R is a proper rational function, and the graph of R will have the horizontal asymptote $y = 0$ (the x-axis).

2. If $n \geq m$ (the degree of the numerator is greater than or equal to the degree of the denominator), then R is improper. Here long division is used.

   (a) If $n = m$ (the degree of the numerator equals the degree of the denominator), the quotient obtained will be the number $\dfrac{a_n}{b_m}$, and the line $y = \dfrac{a_n}{b_m}$ is a horizontal asymptote.

(b) If $n = m + 1$ (the degree of the numerator is one more than the degree of the denominator), the quotient obtained is of the form $ax + b$ (a polynomial of degree 1), and the line $y = ax + b$ is an oblique asymptote.

(c) If $n \geq m + 2$ (the degree of the numerator is two or more greater than the degree of the denominator), the quotient obtained is a polynomial of degree 2 or higher, and R has neither a horizontal nor an oblique asymptote. In this case, for very large values of $|x|$, the graph of R will behave like the graph of the quotient.

Note: The graph of a rational function either has one horizontal or one oblique asymptote or else has no horizontal and no oblique asymptote. ∎

3.2 Assess Your Understanding

'Are You Prepared?' *Answers are given at the end of these exercises. If you get a wrong answer, read the pages listed in red.*

1. *True or False* The quotient of two polynomial expressions is a rational expression. (p. A46)

2. What is the quotient and remainder when $3x^4 - x^2$ is divided by $x^3 - x^2 + 1$. (pp. A29–A31)

3. Graph $y = \dfrac{1}{x}$. (p. 15)

4. Graph $y = 2(x + 1)^2 - 3$ using transformations. (pp. 90–95)

Concepts and Vocabulary

5. *True or False* The domain of every rational function is the set of all real numbers.

6. If, as $x \to -\infty$ or as $x \to \infty$, the values of $R(x)$ approach some fixed number L, then the line $y = L$ is a _____ of the graph of R.

7. If, as x approaches some number c, the values of $|R(x)| \to \infty$, then the line $x = c$ is a _____ _____ of the graph of R.

8. For a rational function R, if the degree of the numerator is less than the degree of the denominator, then R is _____.

9. *True or False* The graph of a rational function may intersect a horizontal asymptote.

10. *True or False* The graph of a rational function may intersect a vertical asymptote.

11. If a rational function is proper, then _____ is a horizontal asymptote.

12. *True or False* If the degree of the numerator of a rational function equals the degree of the denominator, then the ratio of the leading coefficients gives rise to the horizontal asymptote.

Skill Building

In Problems 13–24, find the domain of each rational function.

13. $R(x) = \dfrac{4x}{x - 3}$

14. $R(x) = \dfrac{5x^2}{3 + x}$

15. $H(x) = \dfrac{-4x^2}{(x - 2)(x + 4)}$

16. $G(x) = \dfrac{6}{(x + 3)(4 - x)}$

17. $F(x) = \dfrac{3x(x - 1)}{2x^2 - 5x - 3}$

18. $Q(x) = \dfrac{-x(1 - x)}{3x^2 + 5x - 2}$

19. $R(x) = \dfrac{x}{x^3 - 8}$

20. $R(x) = \dfrac{x}{x^4 - 1}$

21. $H(x) = \dfrac{3x^2 + x}{x^2 + 4}$

22. $G(x) = \dfrac{x - 3}{x^4 + 1}$

23. $R(x) = \dfrac{3(x^2 - x - 6)}{4(x^2 - 9)}$

24. $F(x) = \dfrac{-2(x^2 - 4)}{3(x^2 + 4x + 4)}$

In Problems 25–30, use the graph shown to find:
- (a) *The domain and range of each function*
- (b) *The intercepts, if any*
- (c) *Horizontal asymptotes, if any*
- (d) *Vertical asymptotes, if any*
- (e) *Oblique asymptotes, if any*

25.

26.

27.

28.

29.

30.

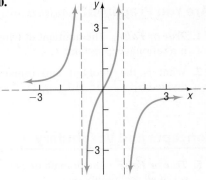

In Problems 31–42, graph each rational function using transformations.

31. $F(x) = 2 + \dfrac{1}{x}$

32. $Q(x) = 3 + \dfrac{1}{x^2}$

33. $R(x) = \dfrac{1}{(x-1)^2}$

34. $R(x) = \dfrac{3}{x}$

35. $H(x) = \dfrac{-2}{x+1}$

36. $G(x) = \dfrac{2}{(x+2)^2}$

37. $R(x) = \dfrac{-1}{x^2 + 4x + 4}$

38. $R(x) = \dfrac{1}{x-1} + 1$

39. $G(x) = 1 + \dfrac{2}{(x-3)^2}$

40. $F(x) = 2 - \dfrac{1}{x+1}$

41. $R(x) = \dfrac{x^2 - 4}{x^2}$

42. $R(x) = \dfrac{x-4}{x}$

In Problems 43–54, find the vertical, horizontal, and oblique asymptotes, if any, of each rational function.

43. $R(x) = \dfrac{3x}{x+4}$

44. $R(x) = \dfrac{3x+5}{x-6}$

45. $H(x) = \dfrac{x^3 - 8}{x^2 - 5x + 6}$

46. $G(x) = \dfrac{x^3 + 1}{x^2 - 5x - 14}$

47. $T(x) = \dfrac{x^3}{x^4 - 1}$

48. $P(x) = \dfrac{4x^2}{x^3 - 1}$

49. $Q(x) = \dfrac{2x^2 - 5x - 12}{3x^2 - 11x - 4}$

50. $F(x) = \dfrac{x^2 + 6x + 5}{2x^2 + 7x + 5}$

51. $R(x) = \dfrac{6x^2 + 7x - 5}{3x + 5}$

52. $R(x) = \dfrac{8x^2 + 26x - 7}{4x - 1}$

53. $G(x) = \dfrac{x^4 - 1}{x^2 - x}$

54. $F(x) = \dfrac{x^4 - 16}{x^2 - 2x}$

Applications and Extensions

55. Gravity In physics, it is established that the acceleration due to gravity, g (in meters/sec^2), at a height h meters above sea level is given by

$$g(h) = \dfrac{3.99 \times 10^{14}}{(6.374 \times 10^6 + h)^2}$$

where 6.374×10^6 is the radius of Earth in meters.
- (a) What is the acceleration due to gravity at sea level?
- (b) The Willis Tower in Chicago, Illinois, is 443 meters tall. What is the acceleration due to gravity at the top of the Willis Tower?

(c) The peak of Mount Everest is 8848 meters above sea level. What is the acceleration due to gravity on the peak of Mount Everest?

(d) Find the horizontal asymptote of $g(h)$.

(e) Solve $g(h) = 0$. How do you interpret your answer?

56. Population Model   A rare species of insect was discovered in the Amazon Rain Forest. To protect the species, environmentalists declare the insect endangered and transplant the insect into a protected area. The population P of the insect t months after being transplanted is

$$P(t) = \frac{50(1 + 0.5t)}{(2 + 0.01t)}$$

(a) How many insects were discovered? In other words, what was the population when $t = 0$?

(b) What will the population be after 5 years?

(c) Determine the horizontal asymptote of $P(t)$. What is the largest population that the protected area can sustain?

57. Resistance in Parallel Circuits   From Ohm's law for circuits, it follows that the total resistance R_{tot} of two components hooked in parallel is given by the equation

$$R_{tot} = \frac{R_1 R_2}{R_1 + R_2}$$

where R_1 and R_2 are the individual resistances.

(a) Let $R_1 = 10$ ohms, and graph R_{tot} as a function of R_2.

(b) Find and interpret any asymptotes of the graph obtained in part (a).

(c) If $R_2 = 2\sqrt{R_1}$, what value of R_1 will yield an R_{tot} of 17 ohms?

Source: en.wikipedia.org/wiki/Series_and_parallel_circuits

58. Newton's Method   In calculus you will learn that, if

$$p(x) = a_n x^n + a_{n-1} x^{n-1} + \cdots + a_1 x + a_0$$

is a polynomial function, then the *derivative* of $p(x)$ is

$$p'(x) = n a_n x^{n-1} + (n - 1) a_{n-1} x^{n-2} + \cdots + 2 a_2 x + a_1$$

Newton's Method is an efficient method for finding the x-intercepts (or real zeros) of a function, such as $p(x)$. The following steps outline Newton's Method.

STEP 1: Select an initial value x_0 that is somewhat close to the x-intercept being sought.

STEP 2: Find values for x using the relation

$$x_{n+1} = x_n - \frac{p(x_n)}{p'(x_n)} \quad n = 1, 2, \ldots$$

until you get two consecutive values x_n and x_{n+1} that agree to whatever decimal place accuracy you desire.

STEP 3: The approximate zero will be x_{n+1}.

Consider the polynomial $p(x) = x^3 - 7x - 40$.

(a) Evaluate $p(5)$ and $p(-3)$.

(b) What might we conclude about a zero of p? Explain.

(c) Use Newton's Method to approximate an x-intercept, r, $-3 < r < 5$, of $p(x)$ to four decimal places.

(d) Use a graphing utility to graph $p(x)$ and verify your answer in part (c).

(e) Using a graphing utility, evaluate $p(r)$ to verify your result.

Explaining Concepts: Discussion and Writing

59. If the graph of a rational function R has the vertical asymptote $x = 4$, the factor $x - 4$ must be present in the denominator of R. Explain why.

60. If the graph of a rational function R has the horizontal asymptote $y = 2$, the degree of the numerator of R equals the degree of the denominator of R. Explain why.

61. Can the graph of a rational function have both a horizontal and an oblique asymptote? Explain.

62. Make up a rational function that has $y = 2x + 1$ as an oblique asymptote. Explain the methodology that you used.

'Are You Prepared?' Answers

1. True   **2.** Quotient: $3x + 3$; remainder: $2x^2 - 3x - 3$ **3.**

4.

3.3 The Graph of a Rational Function

PREPARING FOR THIS SECTION *Before getting started, review the following:*

- Intercepts (Foundations, Section 2, pp. 11–12)
- Even/Odd Functions (Section 1.3, pp. 67–69)
- Solving Rational Equations (Appendix A, Section A.8, pp. A68–A69)

Now Work the 'Are You Prepared?' problems on page 237.

OBJECTIVES 1 Analyze the Graph of a Rational Function (p. 226)
            2 Solve Applied Problems Involving Rational Functions (p. 237)

1 Analyze the Graph of a Rational Function

We commented earlier that calculus provides the tools required to graph a polynomial function accurately. The same holds true for rational functions. However, we can gather together quite a bit of information about their graphs to get an idea of the general shape and position of the graph.

EXAMPLE 1   **How to Analyze the Graph of a Rational Function**

Analyze the graph of the rational function: $R(x) = \dfrac{x-1}{x^2-4}$

Step-by-Step Solution

Step 1: Factor the numerator and denominator of R. Find the domain of the rational function.

$$R(x) = \frac{x-1}{x^2-4} = \frac{x-1}{(x+2)(x-2)}$$

The domain of R is $\{x \mid x \neq -2, x \neq 2\}$.

Step 2: Write R in lowest terms.

Because there are no common factors between the numerator and denominator, R is in lowest terms.

Step 3: Locate the intercepts of the graph. Determine the behavior of the graph of R near each x-intercept using the same procedure as for polynomial functions. Plot each x-intercept and indicate the behavior of the graph near it.

Since 0 is in the domain of R, the y-intercept is $R(0) = \dfrac{1}{4}$. The x-intercepts are found by determining the real zeros of the numerator of R written in lowest terms. By solving $x - 1 = 0$, we find that the only real zero of the numerator is 1, so the only x-intercept of the graph of R is 1. We now analyze the behavior of the graph of R near $x = 1$:

Near 1: $R(x) = \dfrac{x-1}{(x+2)(x-2)} \approx \dfrac{x-1}{(1+2)(1-2)} = -\dfrac{1}{3}(x-1)$

Plot the point $(1, 0)$ and draw a line through $(1, 0)$ with a negative slope.

Step 4: Locate the vertical asymptotes. Graph each vertical asymptote using a dashed line.

The vertical asymptotes are the zeros of the denominator with the rational function in lowest terms. With R written in lowest terms, we find that the graph of R has two vertical asymptotes: the lines $x = -2$ and $x = 2$. See Figure 31(a).

Step 5: Locate the horizontal or oblique asymptote, if one exists. Determine points, if any, at which the graph of R intersects this asymptote. Graph the asymptotes using a dashed line. Plot any points at which the graph of R intersects the asymptote.

Because the degree of the numerator is less than the degree of the denominator, R is proper and the line $y = 0$ (the x-axis) is a horizontal asymptote of the graph. To determine if the graph of R intersects the horizontal asymptote, we solve the equation $R(x) = 0$:

$$\frac{x-1}{x^2-4} = 0$$
$$x - 1 = 0$$
$$x = 1$$

The only solution is $x = 1$, so the graph of R intersects the horizontal asymptote at $(1, 0)$.

Step 6: Use the zeros of the numerator and denominator of R to divide the x-axis into intervals. Determine where the graph of R is above or below the x-axis by choosing a number in each interval and evaluating R there. Plot the points found.

The zero of the numerator, 1, and the zeros of the denominator, -2 and 2, divide the x-axis into four intervals:

$$(-\infty, -2) \qquad (-2, 1) \qquad (1, 2) \qquad (2, \infty)$$

Now construct Table 11.

Table 11

Interval	$(-\infty, -2)$	$(-2, 1)$	$(1, 2)$	$(2, \infty)$
Number chosen	-3	0	$\dfrac{3}{2}$	3
Value of R	$R(-3) = -0.8$	$R(0) = \dfrac{1}{4}$	$R\left(\dfrac{3}{2}\right) = -\dfrac{2}{7}$	$R(3) = 0.4$
Location of graph	Below x-axis	Above x-axis	Below x-axis	Above x-axis
Point on graph	$(-3, -0.8)$	$\left(0, \dfrac{1}{4}\right)$	$\left(\dfrac{3}{2}, -\dfrac{2}{7}\right)$	$(3, 0.4)$

Figure 31(a) shows the asymptotes, the points from Table 11, the y-intercept, the x-intercept, and the behavior of the graph near the x-intercept, 1.

Step 7: Analyze the behavior of the graph of R near each asymptote and indicate this behavior on the graph.

- Since $y = 0$ (the x-axis) is a horizontal asymptote and the graph lies below the x-axis for $x < -2$, we can sketch a portion of the graph by placing a small arrow to the far left and under the x-axis.
- Since the line $x = -2$ is a vertical asymptote and the graph lies below the x-axis for $x < -2$, we place an arrow well below the x-axis and approaching the line $x = -2$ from the left $\left(\lim\limits_{x \to -2^-} R(x) = -\infty\right)$.
- Since the graph is above the x-axis for $-2 < x < 1$ and $x = -2$ is a vertical asymptote, the graph will continue on the right of $x = -2$ at the top $\left(\lim\limits_{x \to -2^+} R(x) = +\infty\right)$.

  Similar explanations account for the other arrows shown in Figure 31(b).

Step 8: Use the results obtained in Steps 1 through 7 to graph R.

Figure 31(c) shows the graph of R.

Figure 31

(a)   (b)   (c)

Exploration

Graph $R(x) = \dfrac{x-1}{x^2-4}$

Result The analysis just completed in Example 1 helps us to set the viewing rectangle to obtain a complete graph. Figure 32(a) shows the graph of $R(x) = \dfrac{x-1}{x^2-4}$ in connected mode, and Figure 32(b) shows it in dot mode. Notice in Figure 32(a) that the graph has vertical lines at $x = -2$ and $x = 2$. This is due to the fact that, when the graphing utility is in connected mode, it will connect the dots between consecutive pixels and vertical lines may occur. We know that the graph of R does not cross the lines $x = -2$ and $x = 2$, since R is not defined at $x = -2$ or $x = 2$. So, when graphing rational functions, dot mode should be used to avoid extraneous vertical lines that are not part of the graph. See Figure 32(b).

Figure 32

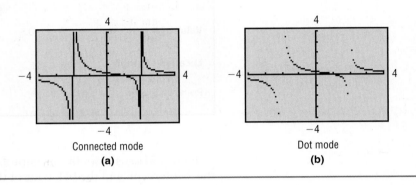

Connected mode
(a)

Dot mode
(b)

Now Work   PROBLEM 7

Next we summarize the steps to be used when analyzing a rational function.

SUMMARY

Analyzing the Graph of a Rational Function R

STEP 1: Factor the numerator and denominator of R. Find the domain of the rational function.

STEP 2: Write R in lowest terms.

STEP 3: Locate the intercepts of the graph. Determine the behavior of the graph of R near each x-intercept.

STEP 4: Determine the vertical asymptotes. Graph each vertical asymptote using a dashed line.

STEP 5: Determine the horizontal or oblique asymptote, if one exists. Determine points, if any, at which the graph of R intersects this asymptote. Graph the asymptote using a dashed line. Plot any points at which the graph of R intersects the asymptote.

STEP 6: Use the zeros of the numerator and denominator of R to divide the x-axis into intervals. Determine where the graph of R is above or below the x-axis by choosing a number in each interval and evaluating R there. Plot the points found.

STEP 7: Analyze the behavior of the graph of R near each asymptote and indicate this behavior on the graph.

STEP 8: Use the results obtained in Steps 1 through 7 to graph R.

EXAMPLE 2   **Analyzing the Graph of a Rational Function**

Analyze the graph of the rational function: $R(x) = \dfrac{x^2-1}{x}$

Solution   **STEP 1:** $R(x) = \dfrac{(x+1)(x-1)}{x}$. The domain of R is $\{x \mid x \neq 0\}$.

STEP 2: R is in lowest terms.

STEP 3: Because x cannot equal 0, there is no y-intercept. The graph has two x-intercepts: -1 and 1.

$$\text{Near } -1: \quad R(x) = \frac{(x+1)(x-1)}{x} \approx \frac{(x+1)(-1-1)}{-1} = 2(x+1)$$

$$\text{Near } 1: \quad R(x) = \frac{(x+1)(x-1)}{x} \approx \frac{(1+1)(x-1)}{1} = 2(x-1)$$

Plot the point $(-1, 0)$ and indicate a line with positive slope there. Plot the point $(1, 0)$ and indicate a line with positive slope there.

STEP 4: The real zero of the denominator with R in lowest terms is 0, so the graph of R has the line $x = 0$ (the y-axis) as a vertical asymptote. Graph $x = 0$ using a dashed line.

STEP 5: Since the degree of the numerator, 2, is one greater than the degree of the denominator, 1, the rational function will have an oblique asymptote. To find the oblique asymptote, we use long division.

NOTE Because the denominator of the rational function is a monomial, we can also find the oblique asymptote as follows:

$$\frac{x^2 - 1}{x} = \frac{x^2}{x} - \frac{1}{x} = x - \frac{1}{x}$$

Since $\frac{1}{x} \to 0$ as $x \to \infty$, $y = x$ is the oblique asymptote. ∎

$$
\begin{array}{r}
x \phantom{{}-1} \\
x{\overline{\smash{\big)}\,x^2 - 1}} \\
\underline{x^2} \\
-1
\end{array}
$$

The quotient is x, so the line $y = x$ is an oblique asymptote of the graph. Graph $y = x$ using a dashed line.

To determine whether the graph of R intersects the asymptote $y = x$, we solve the equation $R(x) = x$.

$$R(x) = \frac{x^2 - 1}{x} = x$$

$$x^2 - 1 = x^2$$

$$-1 = 0 \qquad \text{Impossible}$$

We conclude that the equation $\frac{x^2 - 1}{x} = x$ has no solution, so the graph of R does not intersect the line $y = x$.

STEP 6: The zeros of the numerator are -1 and 1; the zero of the denominator is 0. We use these values to divide the x-axis into four intervals:

$$(-\infty, -1) \qquad (-1, 0) \qquad (0, 1) \qquad (1, \infty)$$

Now we construct Table 12. Plot the points from Table 12. You should now have Figure 33(a).

Table 12

		-1		0		1	$\longrightarrow x$
Interval	$(-\infty, -1)$		$(-1, 0)$		$(0, 1)$		$(1, \infty)$
Number chosen	-2		$-\dfrac{1}{2}$		$\dfrac{1}{2}$		2
Value of R	$R(-2) = -\dfrac{3}{2}$		$R\left(-\dfrac{1}{2}\right) = \dfrac{3}{2}$		$R\left(\dfrac{1}{2}\right) = -\dfrac{3}{2}$		$R(2) = \dfrac{3}{2}$
Location of graph	Below x-axis		Above x-axis		Below x-axis		Above x-axis
Point on graph	$\left(-2, -\dfrac{3}{2}\right)$		$\left(-\dfrac{1}{2}, \dfrac{3}{2}\right)$		$\left(\dfrac{1}{2}, -\dfrac{3}{2}\right)$		$\left(2, \dfrac{3}{2}\right)$

STEP 7: Since the graph of R is below the x-axis for $x < -1$ and is above the x-axis for $x > 1$, and since the graph of R does not intersect the oblique asymptote $y = x$, the graph of R will approach the line $y = x$ as shown in Figure 33(b).

Since the graph of R is above the x-axis for $-1 < x < 0$, the graph of R will approach the vertical asymptote $x = 0$ at the top to the left of $x = 0$ [$\lim_{x \to 0^-} R(x) = \infty$]; since the graph of R is below the x-axis for $0 < x < 1$, the graph of R will approach the vertical asymptote $x = 0$ at the bottom to the right of $x = 0$ [$\lim_{x \to 0^+} R(x) = -\infty$]. See Figure 33(b).

STEP 8: The complete graph is given in Figure 33(c).

Figure 33

(a)

(b)   (c)

Seeing the Concept

Graph $R(x) = \dfrac{x^2 - 1}{x}$ and compare what you see with Figure 33(c). Could you have predicted from the graph that $y = x$ is an oblique asymptote? Graph $y = x$ and ZOOM-OUT. What do you observe?

Now Work PROBLEM 15

EXAMPLE 3 | **Analyzing the Graph of a Rational Function**

Analyze the graph of the rational function: $R(x) = \dfrac{x^4 + 1}{x^2}$

Solution
STEP 1: R is completely factored. The domain of R is $\{x \mid x \neq 0\}$.
STEP 2: R is in lowest terms.
STEP 3: There is no y-intercept. Since $x^4 + 1 = 0$ has no real solutions, there are no x-intercepts.
STEP 4: R is in lowest terms, so $x = 0$ (the y-axis) is a vertical asymptote of R. Graph the line $x = 0$ using dashes.
STEP 5: Since the degree of the numerator, 4, is two more than the degree of the denominator, 2, the rational function will not have a horizontal or oblique asymptote. We use long division to find the end behavior of R.

$$\begin{array}{r} x^2 \\ x^2 \overline{)\,x^4 + 1} \\ \underline{x^4 } \\ 1 \end{array}$$

The quotient is x^2, so the graph of R will approach the graph of $y = x^2$ as $x \to -\infty$ and as $x \to \infty$. The graph of R does not intersect $y = x^2$. Do you know why? Graph $y = x^2$ using dashes.
STEP 6: The numerator has no real zeros, and the denominator has one real zero at 0. We divide the x-axis into the two intervals

$$(-\infty, 0) \qquad (0, \infty)$$

and construct Table 13.

Table 13

	$(-\infty, 0)$	$(0, \infty)$
Interval	$(-\infty, 0)$	$(0, \infty)$
Number chosen	-1	1
Value of R	$R(-1) = 2$	$R(1) = 2$
Location of graph	Above x-axis	Above x-axis
Point on graph	$(-1, 2)$	$(1, 2)$

Plot the points $(-1, 2)$ and $(1, 2)$.

STEP 7: Since the graph of R is above the x-axis and does not intersect $y = x^2$, we place arrows above $y = x^2$ as shown in Figure 34(a). Also, since the graph of R is above the x-axis, it will approach the vertical asymptote $x = 0$ at the top to the left of $x = 0$ and at the top to the right of $x = 0$. See Figure 34(a).

STEP 8: Figure 34(b) shows the complete graph.

Figure 34

NOTE Notice that R in Example 3 is an even function. Do you see the symmetry about the y-axis in the graph of R? ∎

(a)   (b)

Seeing the Concept

Graph $R(x) = \dfrac{x^4 + 1}{x^2}$ and compare what you see with Figure 34(b). Use MINIMUM to find the two turning points. Enter $Y_2 = x^2$ and ZOOM-OUT. What do you see?

Now Work PROBLEM 13

EXAMPLE 4   **Analyzing the Graph of a Rational Function**

Analyze the graph of the rational function: $R(x) = \dfrac{3x^2 - 3x}{x^2 + x - 12}$

Solution   **STEP 1:** We factor R to get

$$R(x) = \frac{3x(x - 1)}{(x + 4)(x - 3)}$$

The domain of R is $\{x \mid x \neq -4, x \neq 3\}$.

STEP 2: R is in lowest terms.

STEP 3: The y-intercept is $R(0) = 0$. Plot the point $(0, 0)$. Since the real solutions of the equation $3x(x - 1) = 0$ are $x = 0$ and $x = 1$, the graph has two x-intercepts, 0 and 1. We determine the behavior of the graph of R near each x-intercept.

$$\text{Near } 0: \quad R(x) = \frac{3x(x - 1)}{(x + 4)(x - 3)} \approx \frac{3x(0 - 1)}{(0 + 4)(0 - 3)} = \frac{1}{4}x$$

$$\text{Near } 1: \quad R(x) = \frac{3x(x - 1)}{(x + 4)(x - 3)} \approx \frac{3(1)(x - 1)}{(1 + 4)(1 - 3)} = -\frac{3}{10}(x - 1)$$

Plot the point $(0, 0)$ and show a line with positive slope there. Plot the point $(1, 0)$ and show a line with negative slope there.

STEP 4: R is in lowest terms. The real solutions of the equation $(x + 4)(x - 3) = 0$ are $x = -4$ and $x = 3$, so the graph of R has two vertical asymptotes, the lines $x = -4$ and $x = 3$. Graph these lines using dashes.

STEP 5: Since the degree of the numerator equals the degree of the denominator, the graph has a horizontal asymptote. To find it, we form the quotient of the leading coefficient of the numerator, 3, and the leading coefficient of the denominator, 1. The graph of R has the horizontal asymptote $y = 3$.

To find out whether the graph of R intersects the asymptote, we solve the equation $R(x) = 3$.

$$R(x) = \frac{3x^2 - 3x}{x^2 + x - 12} = 3$$
$$3x^2 - 3x = 3x^2 + 3x - 36$$
$$-6x = -36$$
$$x = 6$$

The graph intersects the line $y = 3$ at $x = 6$, and $(6, 3)$ is a point on the graph of R. Plot the point $(6, 3)$ and graph the line $y = 3$ using dashes.

STEP 6: The real zeros of the numerator, 0 and 1, and the real zeros of the denominator, -4 and 3, divide the x-axis into five intervals:

$$(-\infty, -4) \quad (-4, 0) \quad (0, 1) \quad (1, 3) \quad (3, \infty)$$

Construct Table 14. Plot the points from Table 14. Figure 35(a) shows the graph we have so far.

Table 14

	-4	0	1	3	$\rightarrow x$
Interval	$(-\infty, -4)$	$(-4, 0)$	$(0, 1)$	$(1, 3)$	$(3, \infty)$
Number chosen	-5	-2	$\frac{1}{2}$	2	4
Value of R	$R(-5) = 11.25$	$R(-2) = -1.8$	$R\left(\frac{1}{2}\right) = \frac{1}{15}$	$R(2) = -1$	$R(4) = 4.5$
Location of graph	Above x-axis	Below x-axis	Above x-axis	Below x-axis	Above x-axis
Point on graph	$(-5, 11.25)$	$(-2, -1.8)$	$\left(\frac{1}{2}, \frac{1}{15}\right)$	$(2, -1)$	$(4, 4.5)$

STEP 7: Since the graph of R is above the x-axis for $x < -4$ and only crosses the line $y = 3$ at $(6, 3)$, as x approaches $-\infty$ the graph of R will approach the horizontal asymptote $y = 3$ from above ($\lim\limits_{x \to -\infty} R(x) = 3$). The graph of R will approach the vertical asymptote $x = -4$ at the top to the left of $x = -4$ ($\lim\limits_{x \to -4^-} R(x) = +\infty$) and at the bottom to the right of $x = -4$ ($\lim\limits_{x \to -4^+} R(x) = -\infty$). The graph of R will approach the vertical asymptote

$x = 3$ at the bottom to the left of $x = 3$ ($\lim\limits_{x \to 3^-} R(x) = -\infty$) and at the top to the right of $x = 3$ ($\lim\limits_{x \to 3^+} R(x) = +\infty$).

We do not know whether the graph of R crosses or touches the line $y = 3$ at $(6, 3)$. To see whether the graph, in fact, crosses or touches the line $y = 3$, we plot an additional point to the right of $(6, 3)$. We use $x = 7$ to find $R(7) = \dfrac{63}{22} < 3$. The graph crosses $y = 3$ at $x = 6$. Because $(6, 3)$ is the only point where the graph of R intersects the asymptote $y = 3$, the graph must approach the line $y = 3$ from below as $x \to \infty$ ($\lim\limits_{x \to \infty} R(x) = 3$). See Figure 35(b).

STEP 8: The complete graph is shown in Figure 35(c).

Figure 35

(a)

(b)

(c)

Exploration

Graph $R(x) = \dfrac{3x^2 - 3x}{x^2 + x - 12}$

Result Figure 36 shows the graph in connected mode, and Figure 37(a) shows it in dot mode. Neither graph displays clearly the behavior of the function between the two x-intercepts, 0 and 1. Nor do they clearly display the fact that the graph crosses the horizontal asymptote at $(6, 3)$. To see these parts better, we graph R for $-1 \leq x \leq 2$ [Figure 37(b)] and for $4 \leq x \leq 60$ [Figure 38(b)].

Figure 36

Figure 37

Connected mode

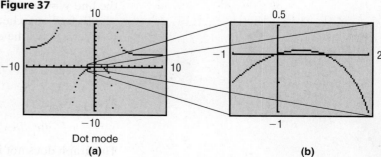

Dot mode

(a)

(b)

Figure 38

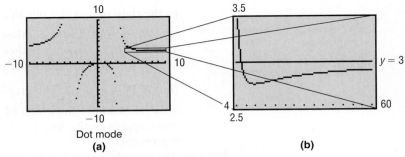

Dot mode
(a)   (b)

The new graphs reflect the behavior produced by the analysis. Furthermore, we observe two turning points, one between 0 and 1 and the other to the right of 6. Rounded to two decimal places, these turning points are (0.52, 0.07) and (11.48, 2.75).

EXAMPLE 5   **Analyzing the Graph of a Rational Function with a Hole**

Analyze the graph of the rational function: $R(x) = \dfrac{2x^2 - 5x + 2}{x^2 - 4}$

Solution   **STEP 1:** We factor R and obtain

$$R(x) = \frac{(2x - 1)(x - 2)}{(x + 2)(x - 2)}$$

The domain of R is $\{x \mid x \ne -2, x \ne 2\}$.

STEP 2: In lowest terms,

$$R(x) = \frac{2x - 1}{x + 2} \qquad x \ne -2, x \ne 2$$

STEP 3: The y-intercept is $R(0) = -\dfrac{1}{2}$. Plot the point $\left(0, -\dfrac{1}{2}\right)$. The graph has one

x-intercept: $\dfrac{1}{2}$.

Near $\dfrac{1}{2}$: $R(x) = \dfrac{2x - 1}{x + 2} \approx \dfrac{2x - 1}{\dfrac{1}{2} + 2} = \dfrac{2}{5}(2x - 1)$

Plot the point $\left(\dfrac{1}{2}, 0\right)$ showing a line with positive slope.

STEP 4: Since $x + 2$ is the only factor of the denominator of $R(x)$ *in lowest terms*, the graph has one vertical asymptote, $x = -2$. However, the rational function is undefined at both $x = 2$ and $x = -2$. Graph the line $x = -2$ using dashes.

STEP 5: Since the degree of the numerator equals the degree of the denominator, the graph has a horizontal asymptote. To find it, we form the quotient of the leading coefficient of the numerator, 2, and the leading coefficient of the denominator, 1. The graph of R has the horizontal asymptote $y = 2$. Graph the line $y = 2$ using dashes.

To find out whether the graph of R intersects the horizontal asymptote $y = 2$, we solve the equation $R(x) = 2$.

$$R(x) = \frac{2x - 1}{x + 2} = 2$$
$$2x - 1 = 2(x + 2)$$
$$2x - 1 = 2x + 4$$
$$-1 = 4 \qquad \textit{Impossible}$$

The graph does not intersect the line $y = 2$.

STEP 6: Look at the factored expression for R in Step 1. The real zeros of the numerator and denominator, -2, $\frac{1}{2}$, and 2, divide the x-axis into four intervals:

$$(-\infty, -2) \qquad \left(-2, \frac{1}{2}\right) \qquad \left(\frac{1}{2}, 2\right) \qquad (2, \infty)$$

Construct Table 15. Plot the points in Table 15.

Table 15

	-2	$1/2$	2	$\rightarrow x$
Interval	$(-\infty, -2)$	$\left(-2, \frac{1}{2}\right)$	$\left(\frac{1}{2}, 2\right)$	$(2, \infty)$
Number chosen	-3	-1	1	3
Value of R	$R(-3) = 7$	$R(-1) = -3$	$R(1) = \frac{1}{3}$	$R(3) = 1$
Location of graph	Above x-axis	Below x-axis	Above x-axis	Above x-axis
Point on graph	$(-3, 7)$	$(-1, -3)$	$\left(1, \frac{1}{3}\right)$	$(3, 1)$

STEP 7: From Table 15 we know that the graph of R is above the x-axis for $x < -2$. From Step 5 we know that the graph of R does not intersect the asymptote $y = 2$. Therefore, the graph of R will approach $y = 2$ from above as $x \to -\infty$ and will approach the vertical asymptote $x = -2$ at the top from the left.

Since the graph of R is below the x-axis for $-2 < x < \frac{1}{2}$, the graph of R will approach $x = -2$ at the bottom from the right. Finally, since the graph of R is above the x-axis for $x > \frac{1}{2}$ and does not intersect the horizontal asymptote $y = 2$, the graph of R will approach $y = 2$ from below as $x \to \infty$. See Figure 39(a).

STEP 8: See Figure 39(b) for the complete graph. Since R is not defined at 2, there is a hole at the point $\left(2, \frac{3}{4}\right)$.

NOTE The coordinates of the hole were obtained by evaluating R in lowest terms at 2. R in lowest terms is $\frac{2x - 1}{x + 2}$, which, at $x = 2$, is $\frac{2(2) - 1}{2 + 2} = \frac{3}{4}$. ∎

Figure 39

(a)   (b)

Exploration

Graph $R(x) = \dfrac{2x^2 - 5x + 2}{x^2 - 4}$. Do you see the hole at $\left(2, \dfrac{3}{4}\right)$? TRACE along the graph. Did you obtain an ERROR at $x = 2$? Are you convinced that an algebraic analysis of a rational function is required in order to accurately interpret the graph obtained with a graphing utility?

As Example 5 shows, **the zeros of the denominator of a rational function give rise to either vertical asymptotes or holes on the graph.**

Now Work PROBLEM 33

We now discuss the problem of finding a rational function from its graph.

EXAMPLE 6 **Constructing a Rational Function from Its Graph**

Find a rational function that might have the graph shown in Figure 40.

Figure 40

Solution The numerator of a rational function $R(x) = \dfrac{p(x)}{q(x)}$ in lowest terms determines the x-intercepts of its graph. The graph shown in Figure 40 has x-intercepts -2 (even multiplicity; graph touches the x-axis) and 5 (odd multiplicity; graph crosses the x-axis). So one possibility for the numerator is $p(x) = (x + 2)^2(x - 5)$.

The denominator of a rational function in lowest terms determines the vertical asymptotes of its graph. The vertical asymptotes of the graph are $x = -5$ and $x = 2$. Since $R(x)$ approaches ∞ to the left of $x = -5$ and $R(x)$ approaches $-\infty$ to the right of $x = -5$, we know that $(x + 5)$ is a factor of odd multiplicity in $q(x)$. Also, $R(x)$ approaches $-\infty$ on both sides of $x = 2$, so $(x - 2)$ is a factor of even multiplicity in $q(x)$. A possibility for the denominator is $q(x) = (x + 5)(x - 2)^2$. So far we have
$$R(x) = \frac{(x + 2)^2(x - 5)}{(x + 5)(x - 2)^2}.$$

However, the horizontal asymptote of the graph given in Figure 40 is $y = 2$, so we know that the degree of the numerator must equal the degree of the denominator and the quotient of leading coefficients must be $\dfrac{2}{1}$. This leads to

$$R(x) = \frac{2(x + 2)^2(x - 5)}{(x + 5)(x - 2)^2}$$

●

Figure 41

 Check: Figure 41 shows the graph of R on a graphing utility. Since Figure 41 looks similar to Figure 40, we have found a rational function R for the graph in Figure 40.

Now Work PROBLEM 51

2 Solve Applied Problems Involving Rational Functions

EXAMPLE 7 Finding the Least Cost of a Can

Reynolds Metal Company manufactures aluminum cans in the shape of a cylinder with a capacity of 500 cubic centimeters $\left(\frac{1}{2} \text{ liter}\right)$. The top and bottom of the can are made of a special aluminum alloy that costs 0.05¢ per square centimeter. The sides of the can are made of material that costs 0.02¢ per square centimeter.

(a) Express the cost of material for the can as a function of the radius r of the can.
(b) Use a graphing utility to graph the function $C = C(r)$.
(c) What value of r will result in the least cost?
(d) What is this least cost?

Solution

(a) Figure 42 illustrates the components of a can in the shape of a right circular cylinder. Notice that the material required to produce a cylindrical can of height h and radius r consists of a rectangle of area $2\pi rh$ and two circles, each of area πr^2. The total cost C (in cents) of manufacturing the can is therefore

Figure 42

Top
Area $= \pi r^2$

Lateral Surface
Area $= 2\pi rh$

h h

Area $= \pi r^2$

Bottom

$$C = \text{Cost of the top and bottom} + \text{Cost of the side}$$
$$= \underbrace{2(\pi r^2)}_{\substack{\text{Total area} \\ \text{of top and} \\ \text{bottom}}} \ \underbrace{(0.05)}_{\substack{\text{Cost/unit} \\ \text{area}}} \ + \ \underbrace{(2\pi rh)}_{\substack{\text{Total} \\ \text{area of} \\ \text{side}}} \ \underbrace{(0.02)}_{\substack{\text{Cost/unit} \\ \text{area}}}$$

$$= 0.10\pi r^2 + 0.04\pi rh$$

But we have the additional restriction that the height h and radius r must be chosen so that the volume V of the can is 500 cubic centimeters. Since $V = \pi r^2 h$, we have

$$500 = \pi r^2 h \quad \text{so} \quad h = \frac{500}{\pi r^2}$$

Substituting this expression for h, the cost C, in cents, as a function of the radius r is

$$C(r) = 0.10\pi r^2 + 0.04\pi r \cdot \frac{500}{\pi r^2} = 0.10\pi r^2 + \frac{20}{r} = \frac{0.10\pi r^3 + 20}{r}$$

Figure 43

(b) See Figure 43 for the graph of $C = C(r)$.
(c) Using the MINIMUM command, the cost is least for a radius of about 3.17 centimeters.
(d) The least cost is $C(3.17) \approx 9.47¢$.

➤ **Now Work** PROBLEM 61

3.3 Assess Your Understanding

'Are You Prepared?' *The answers are given at the end of these exercises. If you get a wrong answer, read the pages listed in red.*

1. (a) Find the intercepts of the graph of the equation $y = \frac{x^2 - 1}{x^2 - 4}$. (pp. 11–12)

   (b) Is the function $y = f(x)$ in part (a) even, odd, or neither? (pp. 67–69)

2. Solve $\frac{x - 3}{x^2 + 1} = -2$. (pp. A68–A69)

Concepts and Vocabulary

3. If the numerator and the denominator of a rational function have no common factors, the rational function is _____.

4. The graph of a rational function never intersects a _____ asymptote.

5. *True or False* The graph of a rational function sometimes intersects an oblique asymptote.

6. *True or False* The graph of a rational function sometimes has a hole.

Skill Building

In Problems 7–44, follow Steps 1 through 8 on page 228 to analyze the graph of each function.

7. $R(x) = \dfrac{x+1}{x(x+4)}$

8. $R(x) = \dfrac{x}{(x-1)(x+2)}$

9. $R(x) = \dfrac{3x+3}{2x+4}$

10. $R(x) = \dfrac{2x+4}{x-1}$

11. $R(x) = \dfrac{3}{x^2-4}$

12. $R(x) = \dfrac{6}{x^2-x-6}$

13. $P(x) = \dfrac{x^4+x^2+1}{x^2-1}$

14. $Q(x) = \dfrac{x^4-1}{x^2-4}$

15. $H(x) = \dfrac{x^3-1}{x^2-9}$

16. $G(x) = \dfrac{x^3+1}{x^2+2x}$

17. $R(x) = \dfrac{x^2}{x^2+x-6}$

18. $R(x) = \dfrac{x^2+x-12}{x^2-4}$

19. $G(x) = \dfrac{x}{x^2-4}$

20. $G(x) = \dfrac{3x}{x^2-1}$

21. $R(x) = \dfrac{3}{(x-1)(x^2-4)}$

22. $R(x) = \dfrac{-4}{(x+1)(x^2-9)}$

23. $H(x) = \dfrac{x^2-1}{x^4-16}$

24. $H(x) = \dfrac{x^2+4}{x^4-1}$

25. $F(x) = \dfrac{x^2-3x-4}{x+2}$

26. $F(x) = \dfrac{x^2+3x+2}{x-1}$

27. $R(x) = \dfrac{x^2+x-12}{x-4}$

28. $R(x) = \dfrac{x^2-x-12}{x+5}$

29. $F(x) = \dfrac{x^2+x-12}{x+2}$

30. $G(x) = \dfrac{x^2-x-12}{x+1}$

31. $R(x) = \dfrac{x(x-1)^2}{(x+3)^3}$

32. $R(x) = \dfrac{(x-1)(x+2)(x-3)}{x(x-4)^2}$

33. $R(x) = \dfrac{x^2+x-12}{x^2-x-6}$

34. $R(x) = \dfrac{x^2+3x-10}{x^2+8x+15}$

35. $R(x) = \dfrac{6x^2-7x-3}{2x^2-7x+6}$

36. $R(x) = \dfrac{8x^2+26x+15}{2x^2-x-15}$

37. $R(x) = \dfrac{x^2+5x+6}{x+3}$

38. $R(x) = \dfrac{x^2+x-30}{x+6}$

39. $f(x) = x + \dfrac{1}{x}$

40. $f(x) = 2x + \dfrac{9}{x}$

41. $f(x) = x^2 + \dfrac{1}{x}$

42. $f(x) = 2x^2 + \dfrac{16}{x}$

43. $f(x) = x + \dfrac{1}{x^3}$

44. $f(x) = 2x + \dfrac{9}{x^3}$

In Problems 45–50, graph each function using a graphing utility; then use MINIMUM to obtain the minimum value, rounded to two decimal places.

45. $f(x) = x + \dfrac{1}{x}, \quad x > 0$

46. $f(x) = 2x + \dfrac{9}{x}, \quad x > 0$

47. $f(x) = x^2 + \dfrac{1}{x}, \quad x > 0$

48. $f(x) = 2x^2 + \dfrac{9}{x}, \quad x > 0$

49. $f(x) = x + \dfrac{1}{x^3}, \quad x > 0$

50. $f(x) = 2x + \dfrac{9}{x^3}, \quad x > 0$

In Problems 51–54, find a rational function that might have the given graph. (More than one answer might be possible.)

51.

52.

53.

54.

Applications and Extensions

55. Drug Concentration The concentration C of a certain drug in a patient's bloodstream t hours after injection is given by

$$C(t) = \frac{t}{2t^2 + 1}$$

(a) Find the horizontal asymptote of $C(t)$. What happens to the concentration of the drug as t increases?
(b) Using your graphing utility, graph $C = C(t)$.
(c) Determine the time at which the concentration is highest.

56. Drug Concentration The concentration C of a certain drug in a patient's bloodstream t minutes after injection is given by

$$C(t) = \frac{50t}{t^2 + 25}$$

(a) Find the horizontal asymptote of $C(t)$. What happens to the concentration of the drug as t increases?
(b) Using your graphing utility, graph $C = C(t)$.
(c) Determine the time at which the concentration is highest.

57. Minimum Cost A rectangular area adjacent to a river is to be fenced in; no fence is needed on the river side. The enclosed area is to be 1000 square feet. Fencing for the side parallel to the river is $5 per linear foot, and fencing for the other two sides is $8 per linear foot; the four corner posts

are $25 apiece. Let x be the length of one of the sides perpendicular to the river.
(a) Write a function $C(x)$ that describes the cost of the project.
(b) What is the domain of C?
(c) Use a graphing utility to graph $C = C(x)$.
(d) Find the dimensions of the cheapest enclosure.

Source: www.uncwil.edu/courses/math111hb/PandR/rational/rational.html

58. Doppler Effect The Doppler effect (named after Christian Doppler) is the change in the pitch (frequency) of the sound from a source (s) as heard by an observer (o) when one or both are in motion. If we assume both the source and the observer are moving in the same direction, the relationship is

$$f' = f_a\left(\frac{v - v_o}{v - v_s}\right)$$

where f' = perceived pitch by the observer

f_a = actual pitch of the source

v = speed of sound in air (assume 772.4 mph)

v_o = speed of the observer

v_s = speed of the source

Suppose that you are traveling down the road at 45 mph and you hear an ambulance (with siren) coming toward you from the rear. The actual pitch of the siren is 600 hertz (Hz).

(a) Write a function $f'(v_s)$ that describes this scenario.
(b) If $f' = 620$ Hz, find the speed of the ambulance.
(c) Use a graphing utility to graph the function.
(d) Verify your answer from part (b).
Source: www.kettering.edu/~drussell/

59. **Minimizing Surface Area** United Parcel Service has contracted you to design a closed box with a square base that has a volume of 10,000 cubic inches. See the illustration.

(a) Express the surface area S of the box as a function of x.
(b) Using a graphing utility, graph the function found in part (a).
(c) What is the minimum amount of cardboard that can be used to construct the box?
(d) What are the dimensions of the box that minimize the surface area?
(e) Why might UPS be interested in designing a box that minimizes the surface area?

60. **Minimizing Surface Area** United Parcel Service has contracted you to design an open box with a square base that has a volume of 5000 cubic inches. See the illustration.

(a) Express the surface area S of the box as a function of x.
(b) Using a graphing utility, graph the function found in part (a).
(c) What is the minimum amount of cardboard that can be used to construct the box?
(d) What are the dimensions of the box that minimize the surface area?
(e) Why might UPS be interested in designing a box that minimizes the surface area?

61. **Cost of a Can** A can in the shape of a right circular cylinder is required to have a volume of 500 cubic centimeters. The top and bottom are made of material that costs 6¢ per square centimeter, while the sides are made of material that costs 4¢ per square centimeter.

(a) Express the total cost C of the material as a function of the radius r of the cylinder. (Refer to Figure 42.)
(b) Graph $C = C(r)$. For what value of r is the cost C a minimum?

62. **Material Needed to Make a Drum** A steel drum in the shape of a right circular cylinder is required to have a volume of 100 cubic feet.

(a) Express the amount A of material required to make the drum as a function of the radius r of the cylinder.
(b) How much material is required if the drum's radius is 3 feet?
(c) How much material is required if the drum's radius is 4 feet?
(d) How much material is required if the drum's radius is 5 feet?
(e) Graph $A = A(r)$. For what value of r is A smallest?

Explaining Concepts: Discussion and Writing

63. Graph each of the following functions:
$$y = \frac{x^2 - 1}{x - 1} \qquad y = \frac{x^3 - 1}{x - 1}$$
$$y = \frac{x^4 - 1}{x - 1} \qquad y = \frac{x^5 - 1}{x - 1}$$
Is $x = 1$ a vertical asymptote? Why not? What is happening for $x = 1$? What do you conjecture about $y = \frac{x^n - 1}{x - 1}$, $n \geq 1$ an integer, for $x = 1$?

64. Graph each of the following functions:
$$y = \frac{x^2}{x - 1} \quad y = \frac{x^4}{x - 1} \quad y = \frac{x^6}{x - 1} \quad y = \frac{x^8}{x - 1}$$
What similarities do you see? What differences?

65. Write a few paragraphs that provide a general strategy for graphing a rational function. Be sure to mention the following: proper, improper, intercepts, and asymptotes.

66. Create a rational function that has the following characteristics: crosses the x-axis at 2; touches the x-axis at -1; one vertical asymptote at $x = -5$ and another at $x = 6$; and one horizontal asymptote, $y = 3$. Compare your function to a fellow classmate's. How do they differ? What are their similarities?

67. Create a rational function that has the following characteristics: crosses the x-axis at 3; touches the x-axis at -2; one vertical asymptote, $x = 1$; and one horizontal asymptote, $y = 2$. Give your rational function to a fellow classmate and ask for a written critique of your rational function.

68. Create a rational function with the following characteristics: three real zeros, one of multiplicity 2; y-intercept 1; vertical asymptotes, $x = -2$ and $x = 3$; oblique asymptote, $y = 2x + 1$. Is this rational function unique? Compare your function with those of other students. What will be the same as everyone else's? Add some more characteristics, such as symmetry or naming the real zeros. How does this modify the rational function?

69. Explain the circumstances under which the graph of a rational function will have a hole.

3.4 Polynomial and Rational Inequalities

PREPARING FOR THIS SECTION *Before getting started, review the following:*

- Solving Linear Inequalities (Appendix A, Section A.10, pp. A86–A88)
- Solving Quadratic Inequalities (Section 2.5, pp. 164–166)

Now Work the **'Are You Prepared?'** problems on page 245.

OBJECTIVES **1** Solve Polynomial Inequalities (p. 241)
2 Solve Rational Inequalities (p. 243)

1 Solve Polynomial Inequalities

In this section we solve inequalities that involve polynomials of degree 3 and higher, along with inequalities that involve rational functions. To help understand the algebraic procedure for solving such inequalities, we use the information obtained in the previous three sections about the graphs of polynomial and rational functions. The approach follows the same logic that we used to solve inequalities involving quadratic functions.

EXAMPLE 1

Solving a Polynomial Inequality Using Its Graph

Solve $(x + 3)(x - 1)^2 > 0$ by graphing $f(x) = (x + 3)(x - 1)^2$.

Solution

We graph $f(x) = (x + 3)(x - 1)^2$ and determine the intervals of x for which the graph is above the x-axis. Do you see why? These values of x result in $f(x)$ being positive. Using Steps 1 through 8 on page 208, we obtain the graph shown in Figure 44.

Figure 44

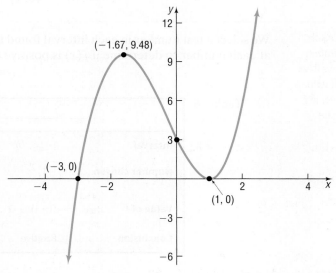

From the graph, we can see that $f(x) > 0$ for $-3 < x < 1$ or $x > 1$. The solution set is $\{x \mid -3 < x < 1 \text{ or } x > 1\}$ or, using interval notation, $(-3, 1) \cup (1, \infty)$.

Now Work PROBLEM **9**

The results of Example 1 lead to the following approach to solving polynomial inequalities algebraically. Suppose that the polynomial or rational inequality is in one of the forms

$$f(x) < 0 \qquad f(x) > 0 \qquad f(x) \leq 0 \qquad f(x) \geq 0$$

Locate the zeros (x-intercepts of the graph) of the polynomial function f. We know that the sign of f can change on either side of an x-intercept, so we use these zeros to divide the real number line into intervals. Then we know that on each interval the graph of f is either above $[f(x) > 0]$ or below $[f(x) < 0]$ the x-axis.

EXAMPLE 2

How to Solve a Polynomial Inequality Algebraically

Solve the inequality $x^4 > x$ algebraically, and graph the solution set.

Step-by-Step Solution

Step 1: Write the inequality so that a polynomial expression f is on the left side and zero is on the right side.

Rearrange the inequality so that 0 is on the right side.

$$x^4 > x$$
$$x^4 - x > 0 \quad \text{Subtract } x \text{ from both sides of the inequality.}$$

This inequality is equivalent to the one we wish to solve.

Step 2: Determine the real zeros (x-intercepts of the graph) of f.

Find the real zeros of $f(x) = x^4 - x$ by solving $x^4 - x = 0$.

$$x^4 - x = 0$$
$$x(x^3 - 1) = 0 \quad \text{Factor out } x.$$
$$x(x - 1)(x^2 + x + 1) = 0 \quad \text{Factor the difference of two cubes.}$$
$$x = 0 \quad \text{or} \quad x - 1 = 0 \quad \text{or} \quad x^2 + x + 1 = 0 \quad \text{Set each factor to zero and solve.}$$
$$x = 0 \quad \text{or} \qquad x = 1$$

The equation $x^2 + x + 1 = 0$ has no real solutions. Do you see why?

Step 3: Use the zeros found in Step 2 to divide the real number line into intervals.

We use the real zeros to separate the real number line into three intervals:

$$(-\infty, 0) \qquad (0, 1) \qquad (1, \infty)$$

Step 4: Select a number in each interval, evaluate f at the number, and determine whether f is positive or negative. If f is positive, all values of x in the interval are positive. If f is negative, all values of x in the interval are negative.

We select a test number in each interval found in Step 3 and evaluate $f(x) = x^4 - x$ at each number to determine if $f(x)$ is positive or negative. See Table 16.

Table 16

Interval	$(-\infty, 0)$	$(0, 1)$	$(1, \infty)$
Number chosen	-1	$\dfrac{1}{2}$	2
Value of f	$f(-1) = 2$	$f\left(\dfrac{1}{2}\right) = -\dfrac{7}{16}$	$f(2) = 14$
Conclusion	Positive	Negative	Positive

NOTE If the inequality is not strict ($\leq$ or $\geq$), include the solutions of $f(x) = 0$ in the solution set. ∎

Since we want to know where $f(x)$ is positive, we conclude that $f(x) > 0$ for all numbers x for which $x < 0$ or $x > 1$. Because the original inequality is strict,

Figure 45

numbers x that satisfy the equation $x^4 = x$ are not solutions. The solution set of the inequality $x^4 > x$ is $\{x \mid x < 0 \text{ or } x > 1\}$ or, using interval notation, $(-\infty, 0) \cup (1, \infty)$.

Figure 45 shows the graph of the solution set.

Now Work PROBLEM 21

2 Solve Rational Inequalities

Just as we presented a graphical approach to help us understand the algebraic procedure for solving inequalities involving polynomials, we present a graphical approach to help us understand the algebraic procedure for solving inequalities involving rational expressions.

EXAMPLE 3 **Solving a Rational Inequality Using Its Graph**

Solve $\dfrac{x-1}{x^2-4} \geq 0$ by graphing $R(x) = \dfrac{x-1}{x^2-4}$.

Solution We graph $R(x) = \dfrac{x-1}{x^2-4}$ and determine the intervals of x such that the graph is above or on the x-axis. Do you see why? These values of x result in $R(x)$ being positive or zero. We graphed $R(x) = \dfrac{x-1}{x^2-4}$ in Example 1 from Section 3.3 (p. 226). We reproduce the graph in Figure 46.

Figure 46

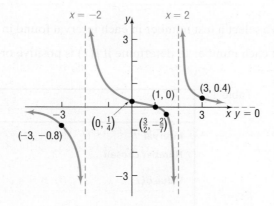

From the graph, we can see that $R(x) \geq 0$ for $-2 < x \leq 1$ or $x > 2$. The solution set is $\{x \mid -2 < x \leq 1 \text{ or } x > 2\}$ or, using interval notation, $(-2, 1] \cup (2, \infty)$.

Now Work PROBLEM 33

To solve a rational inequality algebraically, we follow the same approach that we used to solve a polynomial inequality algebraically. However, we must also identify the zeros of the denominator of the rational function, because the sign of a rational function may change on either side of a vertical asymptote. Convince yourself of this by looking at Figure 46. Notice that the function values are negative for $x < -2$ and are positive for $x > -2$ (but less than 1).

EXAMPLE 4   How to Solve a Rational Inequality Algebraically

Solve the inequality $\dfrac{4x+5}{x+2} \geq 3$ algebraically, and graph the solution set.

Step-by-Step Solution

Step 1: Write the inequality so that a rational expression f is on the left side and zero is on the right side.

Rearrange the inequality so that 0 is on the right side.

$$\frac{4x+5}{x+2} \geq 3$$

$$\frac{4x+5}{x+2} - 3 \geq 0 \quad \text{Subtract 3 from both sides of the inequality.}$$

$$\frac{4x+5}{x+2} - 3 \cdot \frac{x+2}{x+2} \geq 0 \quad \text{Multiply 3 by}$$

$$\frac{4x+5-3x-6}{x+2} \geq 0 \quad \text{Write as a single quotient.}$$

$$\frac{x-1}{x+2} \geq 0 \quad \text{Combine like terms.}$$

Step 2: Determine the real zeros (x-intercepts of the graph) of f and the real numbers for which f is undefined.

The zero of $f(x) = \dfrac{x-1}{x+2}$ is 1. Also, f is undefined for $x = -2$.

Step 3: Use the zeros and undefined values found in Step 2 to divide the real number line into intervals.

We use the zeros and undefined values to separate the real number line into three intervals:

$$(-\infty, -2) \qquad (-2, 1) \qquad (1, \infty)$$

Step 4: Select a number in each interval, evaluate f at the number, and determine whether f is positive or negative. If f is positive, all values of f in the interval are positive. If f is negative, all values of f in the interval are negative.

We select a test number in each interval found in Step 3 and evaluate $f(x) = \dfrac{x-1}{x+2}$ at each number to determine if $f(x)$ is positive or negative. See Table 17.

Table 17

Interval	$(-\infty, -2)$	$(-2, 1)$	$(1, \infty)$
Number chosen	-3	0	2
Value of f	$f(-3) = 4$	$f(0) = -\dfrac{1}{2}$	$f(2) = \dfrac{1}{4}$
Conclusion	Positive	Negative	Positive

NOTE If the inequality is not strict ($\leq$ or $\geq$), include the solutions of $f(x) = 0$ in the solution set. ∎

Since we want to know where $f(x)$ is positive or zero, we conclude that $f(x) \geq 0$ for all numbers x for which $x < -2$ or $x \geq 1$. Notice we do not include -2 in the solution because -2 is not in the domain of f. The solution set of the inequality $\dfrac{4x+5}{x+2} \geq 3$ is $\{x \,|\, x < -2 \text{ or } x \geq 1\}$ or, using interval notation, $(-\infty, -2) \cup [1, \infty)$. Figure 47 shows the graph of the solution set.

Figure 47

Now Work PROBLEM 39

Next we summarize the steps for solving a polynomial or rational inequality algebraically.

SUMMARY

Steps for Solving Polynomial and Rational Inequalities Algebraically

STEP 1: Write the inequality so that a polynomial or rational expression f is on the left side and zero is on the right side in one of the following forms:

$$f(x) > 0 \qquad f(x) \geq 0 \qquad f(x) < 0 \qquad f(x) \leq 0$$

For rational expressions, be sure that the left side is written as a single quotient and find the domain of f.

STEP 2: Determine the real numbers at which the expression f equals zero and, if the expression is rational, the real numbers at which the expression f is undefined.

STEP 3: Use the numbers found in Step 2 to separate the real number line into intervals.

STEP 4: Select a number in each interval and evaluate f at the number.

(a) If the value of f is positive, then $f(x) > 0$ for all numbers x in the interval.

(b) If the value of f is negative, then $f(x) < 0$ for all numbers x in the interval.

If the inequality is not strict ($\geq$ or $\leq$), include the solutions of $f(x) = 0$ in the solution set. Be careful to exclude values of x where f is undefined.

3.4 Assess Your Understanding

'Are You Prepared?' *Answers are given at the end of these exercises. If you get a wrong answer, read the pages listed in red.*

1. Solve the inequality $3 - 4x > 5$. Graph the solution set. (pp. A86–A87)

2. Solve the inequality $x^2 - 5x \leq 24$. Graph the solution set. (pp. 164–166)

Concepts and Vocabulary

3. *True or False* A test number for the interval $-2 < x < 5$ could be 4.

4. *True or False* The graph of $f(x) = \dfrac{x}{x - 3}$ is above the x-axis for $x < 0$ or $x > 3$, so the solution set of the inequality $\dfrac{x}{x - 3} \geq 0$ is $\{x \mid x \leq 0 \text{ or } x \geq 3\}$.

Skill Building

In Problems 5–8, use the graph of the function f to solve the inequality.

5. (a) $f(x) > 0$
    (b) $f(x) \leq 0$

6. (a) $f(x) < 0$
    (b) $f(x) \geq 0$

7. (a) $f(x) < 0$
    (b) $f(x) \geq 0$

8. (a) $f(x) > 0$
    (b) $f(x) \leq 0$

In Problems 9–14, solve the inequality by using the graph of the function.
[**Hint:** The graphs were drawn in Problems 69–74 of Section 3.1.]

9. Solve $f(x) < 0$, where $f(x) = x^2(x - 3)$.

10. Solve $f(x) \leq 0$, where $f(x) = x(x + 2)^2$.

11. Solve $f(x) \geq 0$, where $f(x) = (x + 4)(x - 2)^2$.

12. Solve $f(x) > 0$, where $f(x) = (x - 1)(x + 3)^2$.

13. Solve $f(x) \leq 0$, where $f(x) = -2(x + 2)(x - 2)^3$.

14. Solve $f(x) < 0$, where $f(x) = -\frac{1}{2}(x + 4)(x - 1)^3$.

In Problems 15–18, solve the inequality by using the graph of the function.
[**Hint:** The graphs were drawn in Problems 7–10 of Section 3.3.]

15. Solve $R(x) > 0$, where $R(x) = \dfrac{x + 1}{x(x + 4)}$.

16. Solve $R(x) < 0$, where $R(x) = \dfrac{x}{(x - 1)(x + 2)}$.

17. Solve $R(x) \leq 0$, where $R(x) = \dfrac{3x + 3}{2x + 4}$.

18. Solve $R(x) \geq 0$, where $R(x) = \dfrac{2x + 4}{x - 1}$.

In Problems 19–48, solve each inequality algebraically.

19. $(x - 5)^2(x + 2) < 0$

20. $(x - 5)(x + 2)^2 > 0$

21. $x^3 - 4x^2 > 0$

22. $x^3 + 8x^2 < 0$

23. $2x^3 > -8x^2$

24. $3x^3 < -15x^2$

25. $(x - 1)(x - 2)(x - 3) \leq 0$

26. $(x + 1)(x + 2)(x + 3) \leq 0$

27. $x^3 - 2x^2 - 3x > 0$

28. $x^3 + 2x^2 - 3x > 0$

29. $x^4 > x^2$

30. $x^4 < 9x^2$

31. $x^4 > 1$

32. $x^3 > 1$

33. $\dfrac{x + 1}{x - 1} > 0$

34. $\dfrac{x - 3}{x + 1} > 0$

35. $\dfrac{(x - 1)(x + 1)}{x} \leq 0$

36. $\dfrac{(x - 3)(x + 2)}{x - 1} \leq 0$

37. $\dfrac{(x - 2)^2}{x^2 - 1} \geq 0$

38. $\dfrac{(x + 5)^2}{x^2 - 4} \geq 0$

39. $\dfrac{x + 4}{x - 2} \leq 1$

40. $\dfrac{x + 2}{x - 4} \geq 1$

41. $\dfrac{3x - 5}{x + 2} \leq 2$

42. $\dfrac{x - 4}{2x + 4} \geq 1$

43. $\dfrac{1}{x - 2} < \dfrac{2}{3x - 9}$

44. $\dfrac{5}{x - 3} > \dfrac{3}{x + 1}$

45. $\dfrac{x^2(3 + x)(x + 4)}{(x + 5)(x - 1)} \geq 0$

46. $\dfrac{x(x^2 + 1)(x - 2)}{(x - 1)(x + 1)} \geq 0$

47. $\dfrac{(3 - x)^3(2x + 1)}{x^3 - 1} < 0$

48. $\dfrac{(2 - x)^3(3x - 2)}{x^3 + 1} < 0$

Mixed Practice

In Problems 49–60, solve each inequality algebraically.

49. $(x + 1)(x - 3)(x - 5) > 0$

50. $(2x - 1)(x + 2)(x + 5) < 0$

51. $7x - 4 \geq -2x^2$

52. $x^2 + 3x \geq 10$

53. $\dfrac{x + 1}{x - 3} \leq 2$

54. $\dfrac{x - 1}{x + 2} \geq -2$

55. $3(x^2 - 2) < 2(x - 1)^2 + x^2$

56. $(x - 3)(x + 2) < x^2 + 3x + 5$

57. $6x - 5 < \dfrac{6}{x}$

58. $x + \dfrac{12}{x} < 7$

59. $x^3 - 9x \leq 0$

60. $x^3 - x \geq 0$

Application and Extensions

61. For what positive numbers will the cube of a number exceed four times its square?

62. For what positive numbers will the cube of a number be less than the number?

63. What is the domain of the function $f(x) = \sqrt{x^4 - 16}$?

64. What is the domain of the function $f(x) = \sqrt{x^3 - 3x^2}$?

65. What is the domain of the function $f(x) = \sqrt{\dfrac{x - 2}{x + 4}}$?

66. What is the domain of the function $f(x) = \sqrt{\dfrac{x - 1}{x + 4}}$?

In Problems 67–70, determine where the graph of f is below the graph of g by solving the inequality $f(x) \le g(x)$. Graph f and g together.

67. $f(x) = x^4 - 1$
$g(x) = -2x^2 + 2$

68. $f(x) = x^4 - 1$
$g(x) = x - 1$

69. $f(x) = x^4 - 4$
$g(x) = 3x^2$

70. $f(x) = x^4$
$g(x) = 2 - x^2$

71. Average Cost Suppose that the daily cost C of manufacturing bicycles is given by $C(x) = 80x + 5000$. Then the average daily cost $\overline{C}$ is given by $\overline{C}(x) = \dfrac{80x + 5000}{x}$. How many bicycles must be produced each day for the average cost to be no more than $100?

72. Average Cost See Problem 71. Suppose that the government imposes a $1000 per day tax on the bicycle manufacturer so that the daily cost C of manufacturing x bicycles is now given by $C(x) = 80x + 6000$. Now the average daily cost $\overline{C}$ is given by $\overline{C}(x) = \dfrac{80x + 6000}{x}$. How many bicycles must be produced each day for the average cost to be no more than $100?

73. Bungee Jumping Originating on Pentecost Island in the Pacific, the practice of a person jumping from a high place harnessed to a flexible attachment was introduced to western culture in 1979 by the Oxford University Dangerous Sport Club. One important parameter to know before attempting a bungee jump is the amount the cord will stretch

at the bottom of the fall. The stiffness of the cord is related to the amount of stretch by the equation

$$K = \frac{2W(S + L)}{S^2}$$

where W = weight of the jumper (pounds)
 K = cord's stiffness (pounds per foot)
 L = free length of the cord (feet)
 S = stretch (feet)

(a) A 150-pound person plans to jump off a ledge attached to a cord of length 42 feet. If the stiffness of the cord is no less than 16 pounds per foot, how much will the cord stretch?

(b) If safety requirements will not permit the jumper to get any closer than 3 feet to the ground, what is the minimum height required for the ledge in part (a)?
Source: American Institute of Physics, Physics News Update, No. 150, November 5, 1993

74. Gravitational Force According to Newton's Law of universal gravitation, the attractive force F between two bodies is given by

$$F = G\frac{m_1 m_2}{r^2}$$

where m_1, m_2 = the masses of the two bodies
 r = distance between the two bodies
 G = gravitational constant 6.6742×10^{-11} newtons meter^2 kilogram^{-2}

Suppose an object is traveling directly from Earth to the moon. The mass of Earth is 5.9742×10^{24} kilograms, the mass of the moon is 7.349×10^{22} kilograms, and the mean distance from Earth to the moon is 384,400 kilometers. For an object between Earth and the moon, how far from Earth is the force on the object due to the moon greater than the force on the object due to Earth?
Source: www.solarviews.com;en.wikipedia.org

75. Field Trip Mrs. West has decided to take her fifth grade class to a play. The manager of the theater agreed to discount the regular $40 price of the ticket by $0.20 for each ticket sold. The cost of the bus, $500, will be split equally among each of the students. How many students must attend to keep the cost per student at or below $40?

Explaining Concepts: Discussion and Writing

76. Make up an inequality that has no solution. Make up one that has exactly one solution.

77. The inequality $x^4 + 1 < -5$ has no solution. Explain why.

78. A student attempted to solve the inequality $\dfrac{x + 4}{x - 3} \le 0$ by multiplying both sides of the inequality by $x - 3$ to get

$x + 4 \le 0$. This led to a solution of $\{x \mid x \le -4\}$. Is the student correct? Explain.

79. Write a rational inequality whose solution set is $\{x \mid -3 < x \le 5\}$.

'Are You Prepared?'

1. $\left\{x \mid x < -\dfrac{1}{2}\right\}$ or $\left(-\infty, -\dfrac{1}{2}\right)$

2. $\{x \mid -3 \le x \le 8\}$ or $[-3, 8]$

3.5 The Real Zeros of a Polynomial Function

PREPARING FOR THIS SECTION *Before getting started, review the following:*

- Evaluating Functions (Section 1.1, pp. 46–49)
- Factoring Polynomials (Appendix A, Section A.4, pp. A33–A39)
- Synthetic Division (Appendix A, Section A.5, pp. A42–A45)

- Polynomial Division (Appendix A, Section A.3, pp. A28–A31)
- Zeros of a Quadratic Function (Section 2.3, pp. 141–149)

Now Work the *'Are You Prepared?'* problems on page 259.

OBJECTIVES 1 Use the Remainder and Factor Theorems (p. 248)
 2 Use Descartes' Rule of Signs to Determine the Number of Positive and the Number of Negative Real Zeros of a Polynomial Function (p. 251)
 3 Use the Rational Zeros Theorem to List the Potential Rational Zeros of a Polynomial Function (p. 252)
 4 Find the Real Zeros of a Polynomial Function (p. 253)
 5 Solve Polynomial Equations (p. 255)
 6 Use the Theorem for Bounds on Zeros (p. 256)
 7 Use the Intermediate Value Theorem (p. 257)

In Section 3.1, we were able to identify the real zeros of a polynomial function because either the polynomial function was in factored form or it could be easily factored. But how do we find the real zeros of a polynomial function if it is not factored or cannot be easily factored?

Recall that if r is a real zero of a polynomial function f then $f(r) = 0$, r is an x-intercept of the graph of f, $x - r$ is a factor of f, and r is a solution of the equation $f(x) = 0$. For example, if $x - 4$ is a factor of f, then 4 is a real zero of f and 4 is a solution to the equation $f(x) = 0$. For polynomial and rational functions, we have seen the importance of the real zeros for graphing. In most cases, however, the real zeros of a polynomial function are difficult to find using algebraic methods. No nice formulas like the quadratic formula are available to help us find zeros for polynomials of degree 3 or higher. Formulas do exist for solving any third- or fourth-degree polynomial equation, but they are somewhat complicated. No general formulas exist for polynomial equations of degree 5 or higher. Refer to the Historical Feature at the end of this section for more information.

1 Use the Remainder and Factor Theorems

When we divide one polynomial (the dividend) by another (the divisor), we obtain a quotient polynomial and a remainder, the remainder being either the zero polynomial or a polynomial whose degree is less than the degree of the divisor. To check our work, we verify that

$$(\text{Quotient})(\text{Divisor}) + \text{Remainder} = \text{Dividend}$$

This checking routine is the basis for a famous theorem called the **division algorithm* for polynomials,** which we now state without proof.

---

*A systematic process in which certain steps are repeated a finite number of times is called an **algorithm.** For example, long division is an algorithm.

THEOREM | **Division Algorithm for Polynomials**

If $f(x)$ and $g(x)$ denote polynomial functions and if $g(x)$ is a polynomial whose degree is greater than zero, then there are unique polynomial functions $q(x)$ and $r(x)$ such that

$$\frac{f(x)}{g(x)} = q(x) + \frac{r(x)}{g(x)} \quad \text{or} \quad f(x) = q(x)g(x) + r(x) \qquad \text{(1)}$$

dividend   quotient   divisor   remainder

where $r(x)$ is either the zero polynomial or a polynomial of degree less than that of $g(x)$.

In equation (1), $f(x)$ is the **dividend,** $g(x)$ is the **divisor,** $q(x)$ is the **quotient,** and $r(x)$ is the **remainder.**

If the divisor $g(x)$ is a first-degree polynomial of the form

$$g(x) = x - c \qquad c \text{ a real number}$$

then the remainder $r(x)$ is either the zero polynomial or a polynomial of degree 0. As a result, for such divisors, the remainder is some number, say R, and we may write

$$f(x) = (x - c)q(x) + R \qquad \text{(2)}$$

This equation is an identity in x and is true for all real numbers x. Suppose that $x = c$. Then equation (2) becomes

$$f(c) = (c - c)q(c) + R$$
$$f(c) = R$$

Substitute $f(c)$ for R in equation (2) to obtain

$$f(x) = (x - c)q(x) + f(c) \qquad \text{(3)}$$

We have now proved the **Remainder Theorem.**

REMAINDER THEOREM | Let f be a polynomial function. If $f(x)$ is divided by $x - c$, then the remainder is $f(c)$.

EXAMPLE 1   **Using the Remainder Theorem**

Find the remainder if $f(x) = x^3 - 4x^2 - 5$ is divided by

(a) $x - 3$ (b) $x + 2$

Solution (a) We could use long division or synthetic division, but it is easier to use the Remainder Theorem, which says that the remainder is $f(3)$.

$$f(3) = (3)^3 - 4(3)^2 - 5 = 27 - 36 - 5 = -14$$

The remainder is -14.

(b) To find the remainder when $f(x)$ is divided by $x + 2 = x - (-2)$, we evaluate $f(-2)$.

$$f(-2) = (-2)^3 - 4(-2)^2 - 5 = -8 - 16 - 5 = -29$$

The remainder is -29.

●

Compare the method used in Example 1(a) with the method used in Example 1 of Appendix A, Section A.5. Which method do you prefer? Give reasons.

COMMENT A graphing utility provides another way to find the value of a function using the eVALUEate feature. Consult your manual for details. Then check the results of Example 1. ∎

An important and useful consequence of the Remainder Theorem is the **Factor Theorem.**

FACTOR THEOREM

Let f be a polynomial function. Then $x - c$ is a factor of $f(x)$ if and only if $f(c) = 0$.

The Factor Theorem actually consists of two separate statements:

1. If $f(c) = 0$, then $x - c$ is a factor of $f(x)$.
2. If $x - c$ is a factor of $f(x)$, then $f(c) = 0$.

The proof requires two parts.

Proof

1. Suppose that $f(c) = 0$. Then, by equation (3), we have
$$f(x) = (x - c)q(x)$$
for some polynomial $q(x)$. That is, $x - c$ is a factor of $f(x)$.

2. Suppose that $x - c$ is a factor of $f(x)$. Then there is a polynomial function q such that
$$f(x) = (x - c)q(x)$$
Replacing x by c, we find that
$$f(c) = (c - c)q(c) = 0 \cdot q(c) = 0$$

This completes the proof. ∎

One use of the Factor Theorem is to determine whether a polynomial has a particular factor.

EXAMPLE 2   **Using the Factor Theorem**

Use the Factor Theorem to determine whether the function
$$f(x) = 2x^3 - x^2 + 2x - 3$$
has the factor

(a) $x - 1$ (b) $x + 3$

Solution   The Factor Theorem states that if $f(c) = 0$ then $x - c$ is a factor.

(a) Because $x - 1$ is of the form $x - c$ with $c = 1$, we find the value of $f(1)$. We choose to use substitution.
$$f(1) = 2(1)^3 - (1)^2 + 2(1) - 3 = 2 - 1 + 2 - 3 = 0$$
By the Factor Theorem, $x - 1$ is a factor of $f(x)$.

(b) To test the factor $x + 3$, we first need to write it in the form $x - c$. Since $x + 3 = x - (-3)$, we find the value of $f(-3)$. We choose to use synthetic division.

$$
\begin{array}{r|rrrr}
-3 & 2 & -1 & 2 & -3 \\
   &   & -6 & 21 & -69 \\
\hline
   & 2 & -7 & 23 & -72
\end{array}
$$

Because $f(-3) = -72 \neq 0$, we conclude from the Factor Theorem that $x - (-3) = x + 3$ is not a factor of $f(x)$.

━━━━ **Now Work** PROBLEM 11

In Example 2(a), we found that $x - 1$ was a factor of f. To write f in factored form, we use long division or synthetic division. Using synthetic division, we find that

$$
\begin{array}{r}
1\overline{)2 \quad -1 \quad 2 \quad -3} \\
2 \quad\;\; 1 \quad\;\; 3 \\
\hline
2 \quad\;\; 1 \quad\;\; 3 \quad\;\; 0
\end{array}
$$

The quotient is $q(x) = 2x^2 + x + 3$ with a remainder of 0, as expected. We can write f in factored form as

$$f(x) = 2x^3 - x^2 + 2x - 3 = (x - 1)(2x^2 + x + 3)$$

The next theorem concerns the number of real zeros that a polynomial function may have. In counting the zeros of a polynomial, we count each zero as many times as its multiplicity.

THEOREM   **Number of Real Zeros**

A polynomial function cannot have more real zeros than its degree.

Proof   The proof is based on the Factor Theorem. If r is a real zero of a polynomial function f, then $f(r) = 0$ and, hence, $x - r$ is a factor of $f(x)$. Each real zero corresponds to a factor of degree 1. Because f cannot have more first-degree factors than its degree, the result follows. ∎

2 Use Descartes' Rule of Signs to Determine the Number of Positive and the Number of Negative Real Zeros of a Polynomial Function

Descartes' Rule of Signs provides information about the number and location of the real zeros of a polynomial function written in standard form (descending powers of x). It requires that we count the number of variations in the sign of the coefficients of $f(x)$ and $f(-x)$.

For example, the following polynomial function has two variations in the signs of the coefficients.

$$
\begin{aligned}
f(x) &= -3x^7 + 4x^4 + 3x^2 - 2x - 1 \\
&= -3x^7 + 0x^6 + 0x^5 + 4x^4 + 0x^3 + 3x^2 - 2x - 1
\end{aligned}
$$

$$- \text{ to } + \qquad\qquad\qquad + \text{ to } -$$

Notice that we ignored the zero coefficients in $0x^6, 0x^5$, and $0x^3$ in counting the number of variations in the sign of $f(x)$. Replacing x by $-x$, we get

$$
\begin{aligned}
f(-x) &= -3(-x)^7 + 4(-x)^4 + 3(-x)^2 - 2(-x) - 1 \\
&= 3x^7 + 4x^4 + 3x^2 + 2x - 1
\end{aligned}
$$

$$+ \text{ to } -$$

which has one variation in sign.

THEOREM   **Descartes' Rule of Signs**

Let f denote a polynomial function written in standard form.

The number of positive real zeros of f either equals the number of variations in the sign of the nonzero coefficients of $f(x)$ or else equals that number less an even integer.

The number of negative real zeros of f either equals the number of variations in the sign of the nonzero coefficients of $f(-x)$ or else equals that number less an even integer.

We shall not prove Descartes' Rule of Signs. Let's see how it is used.

EXAMPLE 3	**Using the Number of Real Zeros Theorem and Descartes' Rule of Signs**

Discuss the real zeros of $f(x) = 3x^6 - 4x^4 + 3x^3 + 2x^2 - x - 3$.

Solution   Because the polynomial is of degree 6, by the Number of Real Zeros Theorem there are at most six real zeros. Since there are three variations in the sign of the nonzero coefficients of $f(x)$, by Descartes' Rule of Signs we expect either three positive real zeros or one positive real zero. To continue, we look at $f(-x)$.

$$f(-x) = 3x^6 - 4x^4 - 3x^3 + 2x^2 + x - 3$$

There are three variations in sign, so we expect either three negative real zeros or one negative real zero. Equivalently, we now know that the graph of f has either three positive x-intercepts or one positive x-intercept and three negative x-intercepts or one negative x-intercept.

───────── **Now Work** PROBLEM 21

3 Use the Rational Zeros Theorem to List the Potential Rational Zeros of a Polynomial Function

The next result, called the **Rational Zeros Theorem,** provides information about the rational zeros of a polynomial *with integer coefficients*.

THEOREM

Rational Zeros Theorem

Let f be a polynomial function of degree 1 or higher of the form

$$f(x) = a_n x^n + a_{n-1} x^{n-1} + \cdots + a_1 x + a_0 \qquad a_n \neq 0, \quad a_0 \neq 0$$

where each coefficient is an integer. If $\dfrac{p}{q}$, in lowest terms, is a rational zero of f, then p must be a factor of a_0, and q must be a factor of a_n.

EXAMPLE 4	**Listing Potential Rational Zeros**

List the potential rational zeros of

$$f(x) = 2x^3 + 11x^2 - 7x - 6$$

Solution   Because f has integer coefficients, we may use the Rational Zeros Theorem. First, we list all the integers p that are factors of the constant term $a_0 = -6$ and all the integers q that are factors of the leading coefficient $a_3 = 2$.

$$p: \quad \pm1, \pm2, \pm3, \pm6 \quad \text{Factors of } -6$$
$$q: \quad \pm1, \pm2 \qquad\qquad \text{Factors of } 2$$

Now we form all possible ratios $\dfrac{p}{q}$.

$$\frac{p}{q}: \quad \pm1, \pm2, \pm3, \pm6, \pm\frac{1}{2}, \pm\frac{3}{2}$$

If f has a rational zero, it will be found in this list, which contains 12 possibilities.

In Words
For the polynomial function
$f(x) = 2x^3 + 11x^2 - 7x - 6$, we
know 5 is not a zero, because 5 is
not in the list of potential
rational zeros. However, −1 may
or may not be a zero.

───────── **Now Work** PROBLEM 33

Be sure that you understand what the Rational Zeros Theorem says: For a polynomial with integer coefficients, *if* there is a rational zero, it is one of those listed. It may be the case that the function does not have any rational zeros.

Long division, synthetic division, or substitution can be used to test each potential rational zero to determine whether it is indeed a zero. To make the work easier, integers are usually tested first.

4 Find the Real Zeros of a Polynomial Function

| EXAMPLE 5 | **How to Find the Real Zeros of a Polynomial Function** |

Find the real zeros of the polynomial function $f(x) = 2x^3 + 11x^2 - 7x - 6$. Write f in factored form.

Step-by-Step Solution

Step 1: Use the degree of the polynomial to determine the maximum number of zeros.

Since f is a polynomial of degree 3, there are at most three real zeros.

Step 2: Use Descartes' Rule of Signs to determine the possible number of positive zeros and negative zeros.

By Descartes' Rule of Signs, there is one positive real zero. Also, because

$$f(-x) = -2x^3 + 11x^2 + 7x - 6$$

there are two negative zeros or no negative zeros.

Step 3: If the polynomial has integer coefficients, use the Rational Zeros Theorem to identify those rational numbers that potentially can be zeros. Use the Factor Theorem to determine if each potential rational zero is a zero. If it is, use synthetic division or long division to factor the polynomial function. Repeat Step 3 until all the zeros of the polynomial function have been identified and the polynomial function is completely factored.

We list the potential rational zeros obtained in Example 4:

$$\pm 1, \pm 2, \pm 3, \pm 6, \pm \frac{1}{2}, \pm \frac{3}{2}$$

From our list of potential rational zeros, we will test 6 to determine if it is a zero of f. Because $f(6) = 780 \neq 0$, we know that 6 is not a zero of f. Now, let's test if -6 is a zero. Because $f(-6) = 0$, we know that -6 is a zero and $x - (-6) = x + 6$ is a factor of f. We can use long division or synthetic division to factor f. (We will not show the division here, but you are encouraged to verify the results shown.) After dividing f by $x + 6$, we find that the quotient is $2x^2 - x - 1$, so

$$f(x) = 2x^3 + 11x^2 - 7x - 6$$
$$= (x + 6)(2x^2 - x - 1)$$

Now any solution of the equation $2x^2 - x - 1 = 0$ will be a zero of f. We call the equation $2x^2 - x - 1 = 0$ a **depressed equation** of f. Because any solution to the equation $2x^2 - x - 1 = 0$ is a zero of f, we work with the depressed equation to find the remaining zeros of f.

The depressed equation $2x^2 - x - 1 = 0$ is a quadratic equation with discriminant $b^2 - 4ac = (-1)^2 - 4(2)(-1) = 9 > 0$. The equation has two real solutions, which can be found by factoring.

$$2x^2 - x - 1 = (2x + 1)(x - 1) = 0$$
$$2x + 1 = 0 \quad \text{or} \quad x - 1 = 0$$
$$x = -\frac{1}{2} \quad \text{or} \quad x = 1$$

The zeros of f are -6, $-\frac{1}{2}$, and 1.

We completely factor f as follows:

$$f(x) = 2x^3 + 11x^2 - 7x - 6 = (x + 6)(2x^2 - x - 1)$$
$$= (x + 6)(2x + 1)(x - 1)$$

Notice that the three zeros of f are in the list of potential rational zeros.

We summarize the steps for finding the real zeros of a polynomial function.

SUMMARY

Steps for Finding the Real Zeros of a Polynomial Function

STEP 1: Use the degree of the polynomial to determine the maximum number of real zeros.

STEP 2: Use Descartes' Rule of Signs to determine the possible number of positive zeros and negative zeros.

STEP 3: (a) If the polynomial has integer coefficients, use the Rational Zeros Theorem to identify those rational numbers that potentially could be zeros.
(b) Use substitution, synthetic division, or long division to test each potential rational zero. Each time that a zero (and thus a factor) is found, repeat Step 3 on the depressed equation.

In attempting to find the zeros, remember to use (if possible) the factoring techniques that you already know (special products, factoring by grouping, and so on).

EXAMPLE 6   **Finding the Real Zeros of a Polynomial Function**

Find the real zeros of $f(x) = x^5 - 5x^4 + 12x^3 - 24x^2 + 32x - 16$. Write f in factored form.

Solution   We gather all the information that we can about the zeros.

STEP 1: There are at most five real zeros.

STEP 2: By Descartes' Rule of Signs, there are five or three positive zeros or one positive zero. There are no negative zeros because

$$f(-x) = -x^5 - 5x^4 - 12x^3 - 24x^2 - 32x - 16$$

has no variation in sign.

STEP 3: Because the leading coefficient $a_5 = 1$ and there are no negative zeros, the potential rational zeros are the integers 1, 2, 4, 8, and 16, the positive factors of the constant term, 16. We test the potential rational zero 1 first, using synthetic division.

$$\begin{array}{r|rrrrr} 1 & 1 & -5 & 12 & -24 & 32 & -16 \\ & & 1 & -4 & 8 & -16 & 16 \\ \hline & 1 & -4 & 8 & -16 & 16 & 0 \end{array}$$

The remainder is $f(1) = 0$, so 1 is a zero and $x - 1$ is a factor of f. Using the entries in the bottom row of the synthetic division, we can begin to factor f.

$$f(x) = x^5 - 5x^4 + 12x^3 - 24x^2 + 32x - 16$$
$$= (x - 1)(x^4 - 4x^3 + 8x^2 - 16x + 16)$$

We now work with the first depressed equation:

$$q_1(x) = x^4 - 4x^3 + 8x^2 - 16x + 16 = 0$$

REPEAT STEP 3: The potential rational zeros of q_1 are still 1, 2, 4, 8, and 16. We test 1 first, since it may be a repeated zero of f.

$$\begin{array}{r|rrrrr} 1 & 1 & -4 & 8 & -16 & 16 \\ & & 1 & -3 & 5 & -11 \\ \hline & 1 & -3 & 5 & -11 & 5 \end{array}$$

Since the remainder is 5, 1 is not a repeated zero. We try 2 next.

$$\begin{array}{r} 2\overline{\smash{)}\,1 \quad -4 \quad\ \ 8 \quad -16 \quad\ \ 16} \\ 2 \quad -4 \quad\ \ 8 \quad -16 \\ \hline 1 \quad -2 \quad\ \ 4 \quad\ \ -8 \qquad 0 \end{array}$$

The remainder is $f(2) = 0$, so 2 is a zero and $x - 2$ is a factor of f. Again using the bottom row, we find

$$f(x) = x^5 - 5x^4 + 12x^3 - 24x^2 + 32x - 16$$
$$= (x - 1)(x - 2)(x^3 - 2x^2 + 4x - 8)$$

The remaining zeros satisfy the new depressed equation

$$q_2(x) = x^3 - 2x^2 + 4x - 8 = 0$$

Notice that $q_2(x)$ can be factored using grouping. (Alternatively, you could repeat Step 3 and check the potential rational zero 2.) Then

$$x^3 - 2x^2 + 4x - 8 = 0$$
$$x^2(x - 2) + 4(x - 2) = 0$$
$$(x^2 + 4)(x - 2) = 0$$
$$x^2 + 4 = 0 \quad \text{or} \quad x - 2 = 0$$
$$x = 2$$

Since $x^2 + 4 = 0$ has no real solutions, the real zeros of f are 1 and 2, with 2 being a zero of multiplicity 2. The factored form of f is

$$f(x) = x^5 - 5x^4 + 12x^3 - 24x^2 + 32x - 16$$
$$= (x - 1)(x - 2)^2(x^2 + 4)$$

━━━**Now Work** PROBLEM 45

5 Solve Polynomial Equations

EXAMPLE 7

Solving a Polynomial Equation

Find the real solutions of the equation: $x^5 - 5x^4 + 12x^3 - 24x^2 + 32x - 16 = 0$

Solution

The real solutions of this equation are the real zeros of the polynomial function

$$f(x) = x^5 - 5x^4 + 12x^3 - 24x^2 + 32x - 16$$

Using the result of Example 6, the real zeros of f are 1 and 2. These are the real solutions of the equation

$$x^5 - 5x^4 + 12x^3 - 24x^2 + 32x - 16 = 0$$

━━━**Now Work** PROBLEM 57

In Example 6, the quadratic factor $x^2 + 4$ that appears in the factored form of f is called *irreducible*, because the polynomial $x^2 + 4$ cannot be factored over the real numbers. In general, we say that a quadratic factor $ax^2 + bx + c$ is **irreducible** if it cannot be factored over the real numbers, that is, if it is prime over the real numbers.

Refer to Examples 5 and 6. The polynomial function of Example 5 has three real zeros, and its factored form contains three linear factors. The polynomial function of Example 6 has two distinct real zeros, and its factored form contains two distinct linear factors and one irreducible quadratic factor.

THEOREM

THEOREM
Every polynomial function with real coefficients can be uniquely factored into a product of linear factors and/or irreducible quadratic factors.

We shall prove this result in Section 3.6, and, in fact, we shall draw several additional conclusions about the zeros of a polynomial function. One conclusion is worth noting now. If a polynomial with real coefficients is of odd degree, it must contain at least one linear factor. (Do you see why? Consider the end behavior of polynomial functions of odd degree.) This means that it must have at least one real zero.

THEOREM
A polynomial function of odd degree that has real coefficients has at least one real zero.

6 Use the Theorem for Bounds on Zeros

The search for the real zeros of a polynomial function can be reduced somewhat if *bounds* on the zeros are found. A number M is a **bound** on the zeros of a polynomial if every zero lies between $-M$ and M, inclusive. That is, M is a bound on the zeros of a polynomial f if

$$-M \leq \text{any real zero of } f \leq M$$

THEOREM

Bounds on Zeros

Let f denote a polynomial function whose leading coefficient is 1.

$$f(x) = x^n + a_{n-1}x^{n-1} + \cdots + a_1x + a_0$$

A bound M on the real zeros of f is the smaller of the two numbers

$$\text{Max}\{1, |a_0| + |a_1| + \cdots + |a_{n-1}|\}, \quad 1 + \text{Max}\{|a_0|, |a_1|, \ldots, |a_{n-1}|\} \quad \textbf{(4)}$$

where $\text{Max}\{\ \}$ means "choose the largest entry in $\{\ \}$."

An example will help to make the theorem clear.

EXAMPLE 8

Using the Theorem for Finding Bounds on Zeros

Find a bound on the real zeros of each polynomial function.

(a) $f(x) = x^5 + 3x^3 - 9x^2 + 5$
(b) $g(x) = 4x^5 - 2x^3 + 2x^2 + 1$

Solution
(a) The leading coefficient of f is 1.

$$f(x) = x^5 + 3x^3 - 9x^2 + 5 \quad a_4 = 0, a_3 = 3, a_2 = -9, a_1 = 0, a_0 = 5$$

We evaluate the two expressions in (4).

$$\text{Max}\{1, |a_0| + |a_1| + \cdots + |a_{n-1}|\} = \text{Max}\{1, |5| + |0| + |-9| + |3| + |0|\}$$
$$= \text{Max}\{1, 17\} = 17$$
$$1 + \text{Max}\{|a_0|, |a_1|, \ldots, |a_{n-1}|\} = 1 + \text{Max}\{|5|, |0|, |-9|, |3|, |0|\}$$
$$= 1 + 9 = 10$$

The smaller of the two numbers, 10, is the bound. Every real zero of f lies between -10 and 10.

(b) First we write g so that it is the product of a constant times a polynomial whose leading coefficient is 1 by factoring out the leading coefficient of g, 4.

$$g(x) = 4x^5 - 2x^3 + 2x^2 + 1 = 4\left(x^5 - \frac{1}{2}x^3 + \frac{1}{2}x^2 + \frac{1}{4}\right)$$

Next we evaluate the two expressions in (4) with $a_4 = 0$, $a_3 = -\frac{1}{2}$, $a_2 = \frac{1}{2}$, $a_1 = 0$, and $a_0 = \frac{1}{4}$.

$$\text{Max}\{1, |a_0| + |a_1| + \cdots + |a_{n-1}|\} = \text{Max}\left\{1, \left|\frac{1}{4}\right| + |0| + \left|\frac{1}{2}\right| + \left|-\frac{1}{2}\right| + |0|\right\}$$

$$= \text{Max}\left\{1, \frac{5}{4}\right\} = \frac{5}{4}$$

$$1 + \text{Max}\{|a_0|, |a_1|, \ldots, |a_{n-1}|\} = 1 + \text{Max}\left\{\left|\frac{1}{4}\right|, |0|, \left|\frac{1}{2}\right|, \left|-\frac{1}{2}\right|, |0|\right\}$$

$$= 1 + \frac{1}{2} = \frac{3}{2}$$

COMMENT The bounds on the zeros of a polynomial provide good choices for setting Xmin and Xmax of the viewing rectangle. With these choices, all the x-intercepts of the graph can be seen. ■

The smaller of the two numbers, $\frac{5}{4}$, is the bound. Every real zero of g lies between $-\frac{5}{4}$ and $\frac{5}{4}$.

-Now Work PROBLEM 69

7 Use the Intermediate Value Theorem

The next result, called the **Intermediate Value Theorem**, is based on the fact that the graph of a polynomial function is continuous; that is, it contains no "holes" or "gaps."

THEOREM

Intermediate Value Theorem

Let f denote a polynomial function. If $a < b$ and if $f(a)$ and $f(b)$ are of opposite sign, there is at least one real zero of f between a and b.

Although the proof of this result requires advanced methods in calculus, it is easy to "see" why the result is true. Look at Figure 48.

Figure 48
If $f(a) < 0$ and $f(b) > 0$, there is a zero between a and b.

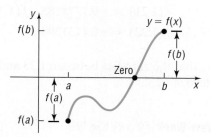

EXAMPLE 9 | **Using the Intermediate Value Theorem to Locate a Real Zero**

Show that $f(x) = x^5 - x^3 - 1$ has a zero between 1 and 2.

Solution We evaluate f at 1 and at 2.

$$f(1) = -1 \quad \text{and} \quad f(2) = 23$$

Because $f(1) < 0$ and $f(2) > 0$, it follows from the Intermediate Value Theorem that the polynomial function f has at least one zero between 1 and 2.

●

───**Now Work** PROBLEM 77

Let's look at the polynomial f of Example 9 more closely. Based on Descartes' Rule of Signs, f has exactly one positive real zero. Based on the Rational Zeros Theorem, 1 is the only potential positive rational zero. Since $f(1) \neq 0$, we conclude that the zero between 1 and 2 is irrational. We can use the Intermediate Value Theorem to approximate it.

Approximating the Real Zeros of a Polynomial Function

STEP 1: Find two consecutive integers a and $a + 1$ such that f has a zero between them.

STEP 2: Divide the interval $[a, a + 1]$ into 10 equal subintervals.

STEP 3: Evaluate f at each endpoint of the subintervals until the Intermediate Value Theorem applies; this interval then contains a zero.

STEP 4: Repeat the process starting at Step 2 until the desired accuracy is achieved.

EXAMPLE 10

Approximating a Real Zero of a Polynomial Function

Find the positive zero of $f(x) = x^5 - x^3 - 1$ correct to two decimal places.

Solution

From Example 9 we know that the positive zero is between 1 and 2. We divide the interval $[1, 2]$ into 10 equal subintervals: $[1, 1.1], [1.1, 1.2], [1.2, 1.3], [1.3, 1.4], [1.4, 1.5], [1.5, 1.6], [1.6, 1.7], [1.7, 1.8], [1.8, 1.9], [1.9, 2]$. Now we find the value of f at each endpoint until the Intermediate Value Theorem applies.

$$f(x) = x^5 - x^3 - 1$$
$$f(1.0) = -1 \qquad\qquad f(1.2) = -0.23968$$
$$f(1.1) = -0.72049 \qquad f(1.3) = 0.51593$$

We can stop here and conclude that the zero is between 1.2 and 1.3. Now we divide the interval $[1.2, 1.3]$ into 10 equal subintervals and proceed to evaluate f at each endpoint.

$$f(1.20) = -0.23968 \qquad f(1.23) \approx -0.0455613$$
$$f(1.21) \approx -0.1778185 \qquad f(1.24) \approx 0.025001$$
$$f(1.22) \approx -0.1131398$$

We conclude that the zero lies between 1.23 and 1.24, and so, correct to two decimal places, the zero is 1.23.

───**Now Work** PROBLEM 89

Exploration

We examine the polynomial f given in Example 10. The Theorem on Bounds of Zeros tells us that every zero is between -2 and 2. If we graph f using $-2 \leq x \leq 2$ (see Figure 49), we see that f has exactly one x-intercept. Using ZERO or ROOT, we find this zero to be 1.24 rounded to two decimal places. Correct to two decimal places, the zero is 1.23.

Figure 49

COMMENT The TABLE feature of a graphing calculator makes the computations in the solution to Example 10 a lot easier. ∎

There are many other numerical techniques for approximating the zeros of a polynomial. The one outlined in Example 10 (a variation of the *bisection method*) has the advantages that it will always work, it can be programmed rather easily on a computer, and each time it is used another decimal place of accuracy is achieved. See Problem 115 for the bisection method, which places the zero in a succession of intervals, with each new interval being half the length of the preceding one.

Historical Feature

Formulas for the solution of third- and fourth-degree polynomial equations exist, and, while not very practical, they do have an interesting history.

In the 1500s in Italy, mathematical contests were a popular pastime, and persons possessing methods for solving problems kept them secret. (Solutions that were published were already common knowledge.) Niccolo of Brescia (1499–1557), commonly referred to as Tartaglia ("the stammerer"), had the secret for solving cubic (third-degree) equations, which gave him a decided advantage in the contests. Girolamo Cardano (1501–1576) found out that Tartaglia had the secret, and, being interested in cubics, he requested it from Tartaglia. The reluctant Tartaglia hesitated for some time, but finally, swearing Cardano to secrecy with midnight oaths by candlelight, told him the secret. Cardano then published the solution in his book *Ars Magna* (1545), giving Tartaglia the credit but rather compromising the secrecy. Tartaglia exploded into bitter recriminations, and each wrote pamphlets that reflected on the other's mathematics, moral character, and ancestry.

The quartic (fourth-degree) equation was solved by Cardano's student Lodovico Ferrari, and this solution also was included, with credit and this time with permission, in the *Ars Magna*.

Attempts were made to solve the fifth-degree equation in similar ways, all of which failed. In the early 1800s, P. Ruffini, Niels Abel, and Evariste Galois all found ways to show that it is not possible to solve fifth-degree equations by formula, but the proofs required the introduction of new methods. Galois's methods eventually developed into a large part of modern algebra.

Historical Problems

Problems 1–8 develop the Tartaglia–Cardano solution of the cubic equation and show why it is not altogether practical.

1. Show that the general cubic equation $y^3 + by^2 + cy + d = 0$ can be transformed into an equation of the form $x^3 + px + q = 0$ by using the substitution $y = x - \dfrac{b}{3}$.

2. In the equation $x^3 + px + q = 0$, replace x by $H + K$. Let $3HK = -p$, and show that $H^3 + K^3 = -q$.

3. Based on Problem 2, we have the two equations

$$3HK = -p \quad \text{and} \quad H^3 + K^3 = -q$$

   Solve for K in $3HK = -p$ and substitute into $H^3 + K^3 = -q$. Then show that

$$H = \sqrt[3]{\dfrac{-q}{2} + \sqrt{\dfrac{q^2}{4} + \dfrac{p^3}{27}}}$$

   [**Hint:** Look for an equation that is quadratic in form.]

4. Use the solution for H from Problem 3 and the equation $H^3 + K^3 = -q$ to show that

$$K = \sqrt[3]{\dfrac{-q}{2} - \sqrt{\dfrac{q^2}{4} + \dfrac{p^3}{27}}}$$

5. Use the results from Problems 2 to 4 to show that the solution of $x^3 + px + q = 0$ is

$$x = \sqrt[3]{\dfrac{-q}{2} + \sqrt{\dfrac{q^2}{4} + \dfrac{p^3}{27}}} + \sqrt[3]{\dfrac{-q}{2} - \sqrt{\dfrac{q^2}{4} + \dfrac{p^3}{27}}}$$

6. Use the result of Problem 5 to solve the equation $x^3 - 6x - 9 = 0$.

7. Use a calculator and the result of Problem 5 to solve the equation $x^3 + 3x - 14 = 0$.

8. Use the methods of this section to solve the equation $x^3 + 3x - 14 = 0$.

3.5 Assess Your Understanding

'Are You Prepared?' *Answers are given at the end of these exercises. If you get a wrong answer, read the pages listed in* red.

1. Find $f(-1)$ if $f(x) = 2x^2 - x$. (pp. 46–48)

2. Factor the expression $6x^2 + x - 2$. (pp. A38–A39)

3. Find the quotient and remainder if $3x^4 - 5x^3 + 7x - 4$ is divided by $x - 3$. (pp. A29–A30 or A42–A44)

4. Find the zeros of $f(x) = x^2 + x - 3$. (pp. 145–147)

Concepts and Vocabulary

5. In the process of polynomial division, (Divisor)(Quotient) + _____ = _____.

6. When a polynomial function f is divided by $x - c$, the remainder is _____.

7. If a function f, whose domain is all real numbers, is even and if 4 is a zero of f, then _____ is also a zero.

8. **True or False** Every polynomial function of degree 3 with real coefficients has exactly three real zeros.

9. If f is a polynomial function and $x - 4$ is a factor of f, then $f(4) =$ _____.

10. **True or False** If f is a polynomial function of degree 4 and if $f(2) = 5$, then

$$\dfrac{f(x)}{x - 2} = p(x) + \dfrac{5}{x - 2}$$

   where $p(x)$ is a polynomial of degree 3.

Skill Building

In Problems 11–20, use the Remainder Theorem to find the remainder when $f(x)$ is divided by $x - c$. Then use the Factor Theorem to determine whether $x - c$ is a factor of $f(x)$.

11. $f(x) = 4x^3 - 3x^2 - 8x + 4; \ x - 2$

12. $f(x) = -4x^3 + 5x^2 + 8; \ x + 3$

13. $f(x) = 3x^4 - 6x^3 - 5x + 10; \ x - 2$

14. $f(x) = 4x^4 - 15x^2 - 4; \ x - 2$

15. $f(x) = 3x^6 + 82x^3 + 27; \ x + 3$

16. $f(x) = 2x^6 - 18x^4 + x^2 - 9; \ x + 3$

17. $f(x) = 4x^6 - 64x^4 + x^2 - 15; \ x + 4$

18. $f(x) = x^6 - 16x^4 + x^2 - 16; \ x + 4$

19. $f(x) = 2x^4 - x^3 + 2x - 1; \ x - \dfrac{1}{2}$

20. $f(x) = 3x^4 + x^3 - 3x + 1; \ x + \dfrac{1}{3}$

In Problems 21–32, tell the maximum number of real zeros that each polynomial function may have. Then use Descartes' Rule of Signs to determine how many positive and how many negative zeros each polynomial function may have. Do not attempt to find the zeros.

21. $f(x) = -4x^7 + x^3 - x^2 + 2$

22. $f(x) = 5x^4 + 2x^2 - 6x - 5$

23. $f(x) = 2x^6 - 3x^2 - x + 1$

24. $f(x) = -3x^5 + 4x^4 + 2$

25. $f(x) = 3x^3 - 2x^2 + x + 2$

26. $f(x) = -x^3 - x^2 + x + 1$

27. $f(x) = -x^4 + x^2 - 1$

28. $f(x) = x^4 + 5x^3 - 2$

29. $f(x) = x^5 + x^4 + x^2 + x + 1$

30. $f(x) = x^5 - x^4 + x^3 - x^2 + x - 1$

31. $f(x) = x^6 - 1$

32. $f(x) = x^6 + 1$

In Problems 33–44, list the potential rational zeros of each polynomial function. Do not attempt to find the zeros.

33. $f(x) = 3x^4 - 3x^3 + x^2 - x + 1$

34. $f(x) = x^5 - x^4 + 2x^2 + 3$

35. $f(x) = x^5 - 6x^2 + 9x - 3$

36. $f(x) = 2x^5 - x^4 - x^2 + 1$

37. $f(x) = -4x^3 - x^2 + x + 2$

38. $f(x) = 6x^4 - x^2 + 2$

39. $f(x) = 6x^4 - x^2 + 9$

40. $f(x) = -4x^3 + x^2 + x + 6$

41. $f(x) = 2x^5 - x^3 + 2x^2 + 12$

42. $f(x) = 3x^5 - x^2 + 2x + 18$

43. $f(x) = 6x^4 + 2x^3 - x^2 + 20$

44. $f(x) = -6x^3 - x^2 + x + 10$

In Problems 45–56, use the Rational Zeros Theorem to find all the real zeros of each polynomial function. Use the zeros to factor f over the real numbers.

45. $f(x) = x^3 + 2x^2 - 5x - 6$

46. $f(x) = x^3 + 8x^2 + 11x - 20$

47. $f(x) = 2x^3 - x^2 + 2x - 1$

48. $f(x) = 2x^3 + x^2 + 2x + 1$

49. $f(x) = 2x^3 - 4x^2 - 10x + 20$

50. $f(x) = 3x^3 + 6x^2 - 15x - 30$

51. $f(x) = 2x^4 + x^3 - 7x^2 - 3x + 3$

52. $f(x) = 2x^4 - x^3 - 5x^2 + 2x + 2$

53. $f(x) = x^4 + x^3 - 3x^2 - x + 2$

54. $f(x) = x^4 - x^3 - 6x^2 + 4x + 8$

55. $f(x) = 4x^4 + 5x^3 + 9x^2 + 10x + 2$

56. $f(x) = 3x^4 + 4x^3 + 7x^2 + 8x + 2$

In Problems 57–68, solve each equation in the real number system.

57. $x^4 - x^3 + 2x^2 - 4x - 8 = 0$

58. $2x^3 + 3x^2 + 2x + 3 = 0$

59. $3x^3 + 4x^2 - 7x + 2 = 0$

60. $2x^3 - 3x^2 - 3x - 5 = 0$

61. $3x^3 - x^2 - 15x + 5 = 0$

62. $2x^3 - 11x^2 + 10x + 8 = 0$

63. $x^4 + 4x^3 + 2x^2 - x + 6 = 0$

64. $x^4 - 2x^3 + 10x^2 - 18x + 9 = 0$

65. $x^3 - \dfrac{2}{3}x^2 + \dfrac{8}{3}x + 1 = 0$

66. $x^3 + \dfrac{3}{2}x^2 + 3x - 2 = 0$

67. $2x^4 - 19x^3 + 57x^2 - 64x + 20 = 0$

68. $2x^4 + x^3 - 24x^2 + 20x + 16 = 0$

In Problems 69–76, find bounds on the real zeros of each polynomial function.

69. $f(x) = x^4 - 3x^2 - 4$

70. $f(x) = x^4 - 5x^2 - 36$

71. $f(x) = x^4 + x^3 - x - 1$

72. $f(x) = x^4 - x^3 + x - 1$

73. $f(x) = 3x^4 + 3x^3 - x^2 - 12x - 12$

74. $f(x) = 3x^4 - 3x^3 - 5x^2 + 27x - 36$

75. $f(x) = 4x^5 - x^4 + 2x^3 - 2x^2 + x - 1$

76. $f(x) = 4x^5 + x^4 + x^3 + x^2 - 2x - 2$

In Problems 77–82, use the Intermediate Value Theorem to show that each polynomial function has a zero in the given interval.

77. $f(x) = 8x^4 - 2x^2 + 5x - 1; [0, 1]$

78. $f(x) = x^4 + 8x^3 - x^2 + 2; [-1, 0]$

79. $f(x) = 2x^3 + 6x^2 - 8x + 2; [-5, -4]$

80. $f(x) = 3x^3 - 10x + 9; [-3, -2]$

81. $f(x) = x^5 - x^4 + 7x^3 - 7x^2 - 18x + 18; [1.4, 1.5]$

82. $f(x) = x^5 - 3x^4 - 2x^3 + 6x^2 + x + 2; [1.7, 1.8]$

In Problems 83–86, each equation has a solution r in the interval indicated. Use the method of Example 10 to approximate this solution correct to two decimal places.

83. $8x^4 - 2x^2 + 5x - 1 = 0; 0 \le r \le 1$

84. $x^4 + 8x^3 - x^2 + 2 = 0; -1 \le r \le 0$

85. $2x^3 + 6x^2 - 8x + 2 = 0; -5 \le r \le -4$

86. $3x^3 - 10x + 9 = 0; -3 \le r \le -2$

In Problems 87–90, each polynomial function has exactly one positive zero. Use the method of Example 10 to approximate the zero correct to two decimal places.

87. $f(x) = x^3 + x^2 + x - 4$

88. $f(x) = 2x^4 + x^2 - 1$

89. $f(x) = 2x^4 - 3x^3 - 4x^2 - 8$

90. $f(x) = 3x^3 - 2x^2 - 20$

Mixed Practice

In Problems 91–102, graph each polynomial function.

91. $f(x) = x^3 + 2x^2 - 5x - 6$

92. $f(x) = x^3 + 8x^2 + 11x - 20$

93. $f(x) = 2x^3 - x^2 + 2x - 1$

94. $f(x) = 2x^3 + x^2 + 2x + 1$

95. $f(x) = x^4 + x^2 - 2$

96. $f(x) = x^4 - 3x^2 - 4$

97. $f(x) = 4x^4 + 7x^2 - 2$

98. $f(x) = 4x^4 + 15x^2 - 4$

99. $f(x) = x^4 + x^3 - 3x^2 - x + 2$

100. $f(x) = x^4 - x^3 - 6x^2 + 4x + 8$

101. $f(x) = 4x^5 - 8x^4 - x + 2$

102. $f(x) = 4x^5 + 12x^4 - x - 3$

Applications and Extensions

103. Find k such that $f(x) = x^3 - kx^2 + kx + 2$ has the factor $x - 2$.

104. Find k such that $f(x) = x^4 - kx^3 + kx^2 + 1$ has the factor $x + 2$.

105. What is the remainder when $f(x) = 2x^{20} - 8x^{10} + x - 2$ is divided by $x - 1$?

106. What is the remainder when $f(x) = -3x^{17} + x^9 - x^5 + 2x$ is divided by $x + 1$?

107. Use the Factor Theorem to prove that $x - c$ is a factor of $x^n - c^n$ for any positive integer n.

108. Use the Factor Theorem to prove that $x + c$ is a factor of $x^n + c^n$ if $n \ge 1$ is an odd integer.

109. One solution of the equation $x^3 - 8x^2 + 16x - 3 = 0$ is 3. Find the sum of the remaining solutions.

110. One solution of the equation $x^3 + 5x^2 + 5x - 2 = 0$ is -2. Find the sum of the remaining solutions.

111. Geometry What is the length of the edge of a cube if, after a slice 1 inch thick is cut from one side, the volume remaining is 294 cubic inches?

112. Geometry What is the length of the edge of a cube if its volume could be doubled by an increase of 6 centimeters in one edge, an increase of 12 centimeters in a second edge, and a decrease of 4 centimeters in the third edge?

113. Let $f(x)$ be a polynomial function whose coefficients are integers. Suppose that r is a real zero of f and that the leading coefficient of f is 1. Use the Rational Zeros Theorem to show that r is either an integer or an irrational number.

114. Prove the Rational Zeros Theorem.

[**Hint:** Let $\dfrac{p}{q}$, where p and q have no common factors except 1 and -1, be a zero of the polynomial function

$$f(x) = a_n x^n + a_{n-1} x^{n-1} + \cdots + a_1 x + a_0$$

whose coefficients are all integers. Show that

$$a_n p^n + a_{n-1} p^{n-1} q + \cdots + a_1 pq^{n-1} + a_0 q^n = 0$$

Now, because p is a factor of the first n terms of this equation, p must also be a factor of the term $a_0 q^n$. Since p is not a factor of q (why?), p must be a factor of a_0. Similarly, q must be a factor of a_n.]

ection Method for Approximating Zeros of a Function f
We begin with two consecutive integers, a and $a + 1$, such that $f(a)$ and $f(a + 1)$ are of opposite sign. Evaluate f at the midpoint m_1 of a and $a + 1$. If $f(m_1) = 0$, then m_1 is the zero of f, and we are finished. Otherwise, $f(m_1)$ is of opposite sign to either $f(a)$ or $f(a + 1)$. Suppose that it is $f(a)$ and $f(m_1)$ that are of opposite sign. Now evaluate f at the midpoint m_2 of a and m_1. Repeat this process until the desired degree of accuracy is obtained. Note that each iteration places the zero in an interval whose length is half that of the previous interval. Use the bisection method to approximate the zero of $f(x) = 8x^4 - 2x^2 + 5x - 1$ in the interval $[0, 1]$ correct to three decimal places.

[**Hint:** The process ends when both endpoints agree to the desired number of decimal places.]

Explaining Concepts: Discussion and Writing

116. Is $\frac{1}{3}$ a zero of $f(x) = 2x^3 + 3x^2 - 6x + 7$? Explain.

117. Is $\frac{1}{3}$ a zero of $f(x) = 4x^3 - 5x^2 - 3x + 1$? Explain.

118. Is $\frac{3}{5}$ a zero of $f(x) = 2x^6 - 5x^4 + x^3 - x + 1$? Explain.

119. Is $\frac{2}{3}$ a zero of $f(x) = x^7 + 6x^5 - x^4 + x + 2$? Explain.

'Are You Prepared?' Answers

1. 3　　　**2.** $(3x + 2)(2x - 1)$　　　**3.** Quotient: $3x^3 + 4x^2 + 12x + 43$; Remainder: 125　　　**4.** $\dfrac{-1 - \sqrt{13}}{2}, \dfrac{-1 + \sqrt{13}}{2}$

3.6 Complex Zeros; Fundamental Theorem of Algebra

PREPARING FOR THIS SECTION *Before getting started, review the following:*

- Complex Numbers (Appendix A, Section A.11, pp. A92–A97)
- Complex Zeros of a Quadratic Function (Section 2.7, pp. 178–180)

Now Work the 'Are You Prepared?' problems on page 266.

OBJECTIVES **1** Use the Conjugate Pairs Theorem (p. 263)
　　　　　　 2 Find a Polynomial Function with Specified Zeros (p. 264)
　　　　　　 3 Find the Complex Zeros of a Polynomial Function (p. 265)

In Section 2.3, we found the real zeros of a quadratic function. That is, we found the real zeros of a polynomial function of degree 2. Then, in Section 2.7 we found the complex zeros of a quadratic function. That is, we found the complex zeros of a polynomial function of degree 2.

In Section 3.5, we found the real zeros of polynomial functions of degree 3 or higher. In this section we will find the *complex zeros* of polynomial functions of degree 3 or higher.

DEFINITION
A variable in the complex number system is referred to as a **complex variable.** A **complex polynomial function** f of degree n is a function of the form

$$f(x) = a_n x^n + a_{n-1} x^{n-1} + \cdots + a_1 x + a_0 \qquad (1)$$

where $a_n, a_{n-1}, \ldots, a_1, a_0$ are complex numbers, $a_n \neq 0$, n is a nonnegative integer, and x is a complex variable. As before, a_n is called the **leading coefficient** of f. A complex number r is called a **complex zero** of f if $f(r) = 0$.

In most of our work the coefficients in (1) will be real numbers.

We have learned that some quadratic equations have no real solutions, but that in the complex number system every quadratic equation has a solution, either real or complex. The next result, proved by Karl Friedrich Gauss (1777–1855) when he was 22 years old,* gives an extension to complex polynomials. In fact, this result is so important and useful that it has become known as the **Fundamental Theorem of Algebra.**

*In all, Gauss gave four different proofs of this theorem, the first one in 1799 being the subject of his doctoral dissertation.

FUNDAMENTAL THEOREM OF ALGEBRA

Every complex polynomial function $f(x)$ of degree $n \geq 1$ has at least one complex zero.

We shall not prove this result, as the proof is beyond the scope of this book. However, using the Fundamental Theorem of Algebra and the Factor Theorem, we can prove the following result:

THEOREM

Every complex polynomial function $f(x)$ of degree $n \geq 1$ can be factored into n linear factors (not necessarily distinct) of the form

$$f(x) = a_n(x - r_1)(x - r_2) \cdot \cdots \cdot (x - r_n) \qquad (2)$$

where $a_n, r_1, r_2, \ldots, r_n$ are complex numbers. That is, every complex polynomial function of degree $n \geq 1$ has exactly n complex zeros, some of which may repeat.

Proof Let

$$f(x) = a_n x^n + a_{n-1}x^{n-1} + \cdots + a_1 x + a_0$$

By the Fundamental Theorem of Algebra, f has at least one zero, say r_1. Then, by the Factor Theorem, $x - r_1$ is a factor, and

$$f(x) = (x - r_1)q_1(x)$$

where $q_1(x)$ is a complex polynomial of degree $n - 1$ whose leading coefficient is a_n. Repeating this argument n times, we arrive at

$$f(x) = (x - r_1)(x - r_2) \cdot \cdots \cdot (x - r_n)q_n(x)$$

where $q_n(x)$ is a complex polynomial of degree $n - n = 0$ whose leading coefficient is a_n. That is, $q_n(x) = a_n x^0 = a_n$, and so

$$f(x) = a_n(x - r_1)(x - r_2) \cdot \cdots \cdot (x - r_n)$$

We conclude that every complex polynomial function $f(x)$ of degree $n \geq 1$ has exactly n (not necessarily distinct) zeros. ∎

1 Use the Conjugate Pairs Theorem

We can use the Fundamental Theorem of Algebra to obtain valuable information about the complex zeros of polynomial functions whose coefficients are real numbers.

CONJUGATE PAIRS THEOREM

Let $f(x)$ be a polynomial function whose coefficients are real numbers. If $r = a + bi$ is a zero of f, the complex conjugate $\bar{r} = a - bi$ is also a zero of f.

In other words, for polynomial functions whose coefficients are real numbers, the complex zeros occur in conjugate pairs. This result should not be all that surprising since the complex zeros of a quadratic function occurred in conjugate pairs.

Proof Let

$$f(x) = a_n x^n + a_{n-1}x^{n-1} + \cdots + a_1 x + a_0$$

where $a_n, a_{n-1}, \ldots, a_1, a_0$ are real numbers and $a_n \neq 0$. If $r = a + bi$ is a zero of f, then $f(r) = f(a + bi) = 0$, so

$$a_n r^n + a_{n-1}r^{n-1} + \cdots + a_1 r + a_0 = 0$$

We take the conjugate of both sides to get

$$\overline{a_n r^n + a_{n-1} r^{n-1} + \cdots + a_1 r + a_0} = \overline{0}$$

$$\overline{a_n r^n} + \overline{a_{n-1} r^{n-1}} + \cdots + \overline{a_1 r} + \overline{a_0} = \overline{0}$$ *The conjugate of a sum equals the sum of the conjugates (see Section A.11).*

$$\overline{a_n} (\overline{r})^n + \overline{a_{n-1}} (\overline{r})^{n-1} + \cdots + \overline{a_1} \overline{r} + \overline{a_0} = \overline{0}$$ *The conjugate of a product equals the product of the conjugates.*

$$a_n (\overline{r})^n + a_{n-1} (\overline{r})^{n-1} + \cdots + a_1 \overline{r} + a_0 = 0$$ *The conjugate of a real number equals the real number.*

This last equation states that $f(\overline{r}) = 0$; that is, $\overline{r} = a - bi$ is a zero of f.   ■

The importance of this result should be clear. Once we know that, say, $3 + 4i$ is a zero of a polynomial function with real coefficients, then we know that $3 - 4i$ is also a zero. This result has an important corollary.

COROLLARY   A polynomial function f of odd degree with real coefficients has at least one real zero.

Proof   Because complex zeros occur as conjugate pairs in a polynomial function with real coefficients, there will always be an even number of zeros that are not real numbers. Consequently, since f is of odd degree, one of its zeros has to be a real number.   ■

For example, the polynomial function $f(x) = x^5 - 3x^4 + 4x^3 - 5$ has at least one zero that is a real number, since f is of degree 5 (odd) and has real coefficients.

EXAMPLE 1   **Using the Conjugate Pairs Theorem**

A polynomial function f of degree 5 whose coefficients are real numbers has the zeros $1, 5i$, and $1 + i$. Find the remaining two zeros.

Solution   Since f has coefficients that are real numbers, complex zeros appear as conjugate pairs. It follows that $-5i$, the conjugate of $5i$, and $1 - i$, the conjugate of $1 + i$, are the two remaining zeros.   ●

Now Work PROBLEM 7

2 Find a Polynomial Function with Specified Zeros

EXAMPLE 2   **Finding a Polynomial Function Whose Zeros Are Given**

Find a polynomial function f of degree 4 whose coefficients are real numbers that has the zeros $1, 1$, and $-4 + i$.

Solution   Since $-4 + i$ is a zero, by the Conjugate Pairs Theorem, $-4 - i$ must also be a zero of f. Because of the Factor Theorem, if $f(c) = 0$, then $x - c$ is a factor of $f(x)$. So we can now write f as

$$f(x) = a(x - 1)(x - 1)[x - (-4 + i)][x - (-4 - i)]$$

where a is any real number. Then

$$
\begin{aligned}
f(x) &= a(x - 1)(x - 1)[x - (-4 + i)][x - (-4 - i)] \\
&= a(x^2 - 2x + 1)[x^2 - (-4 + i)x - (-4 - i)x + (-4 + i)(-4 - i)] \\
&= a(x^2 - 2x + 1)(x^2 + 4x - ix + 4x + ix + 16 + 4i - 4i - i^2) \\
&= a(x^2 - 2x + 1)(x^2 + 8x + 17) \\
&= a(x^4 + 8x^3 + 17x^2 - 2x^3 - 16x^2 - 34x + x^2 + 8x + 17) \\
&= a(x^4 + 6x^3 + 2x^2 - 26x + 17)
\end{aligned}
$$

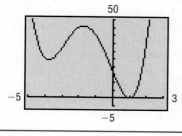

Figure 50

Exploration

Graph the function f found in Example 2 for $a = 1$. Does the value of a affect the zeros of f? How does the value of a affect the graph of f? What information about f is sufficient to uniquely determine a?

Result A quick analysis of the polynomial function f tells us what to expect:

At most three turning points.

For large $|x|$, the graph will behave like $y = x^4$.

A repeated real zero at 1, so the graph will touch the x-axis at 1.

The only x-intercept is at 1; the y-intercept is 17.

Figure 50 shows the complete graph. (Do you see why? The graph has exactly three turning points.) The value of a causes a stretch or compression; a reflection also occurs if $a < 0$. The zeros are not affected.

If any point other than an x-intercept on the graph of f is known, then a can be determined. For example, if $(2, 3)$ is on the graph, then $f(2) = 3 = a(37)$, so $a = 3/37$. Why won't an x-intercept work?

Now we can prove the theorem we conjectured in Section 3.5.

THEOREM

> Every polynomial function with real coefficients can be uniquely factored over the real numbers into a product of linear factors and/or irreducible quadratic factors.

Proof Every complex polynomial function f of degree n has exactly n zeros and can be factored into a product of n linear factors. If its coefficients are real, those zeros that are complex numbers will always occur as conjugate pairs. As a result, if $r = a + bi$ is a complex zero, then so is $\bar{r} = a - bi$. Consequently, when the linear factors $x - r$ and $x - \bar{r}$ of f are multiplied, we have

$$(x - r)(x - \bar{r}) = x^2 - (r + \bar{r})x + r\bar{r} = x^2 - 2ax + a^2 + b^2$$

This second-degree polynomial has real coefficients and is irreducible (over the real numbers). Thus, the factors of f are either linear or irreducible quadratic factors. ■

Now Work   PROBLEM 17

3 Find the Complex Zeros of a Polynomial Function

The steps for finding the complex zeros of a polynomial function are the same as those for finding the real zeros.

EXAMPLE 3   **Finding the Complex Zeros of a Polynomial Function**

Find the complex zeros of the polynomial function

$$f(x) = 3x^4 + 5x^3 + 25x^2 + 45x - 18$$

Write f in factored form.

Solution   **STEP 1:** The degree of f is 4. So f will have four complex zeros.

STEP 2: Descartes' Rule of Signs provides information about the real zeros. For this polynomial function, there is one positive real zero. There are three negative real zeros or one negative real zero because

$$f(-x) = 3x^4 - 5x^3 + 25x^2 - 45x - 18$$

has three variations in sign.

STEP 3: The Rational Zeros Theorem provides information about the potential rational zeros of polynomial functions with integer coefficients. For this polynomial function (which has integer coefficients), the potential rational zeros are

$$\pm\frac{1}{3}, \pm\frac{2}{3}, \pm 1, \pm 2, \pm 3, \pm 6, \pm 9, \pm 18$$

We test 1 first:

$$1)\overline{\begin{array}{rrrrr} 3 & 5 & 25 & 45 & -18 \\ & 3 & 8 & 33 & 78 \\ \hline 3 & 8 & 33 & 78 & 60 \end{array}}$$ 1 is not a zero.

We test −1:

$$-1)\overline{\begin{array}{rrrrr} 3 & 5 & 25 & 45 & -18 \\ & -3 & -2 & -23 & -22 \\ \hline 3 & 2 & 23 & 22 & -40 \end{array}}$$ −1 is not a zero.

We test 2:

$$2)\overline{\begin{array}{rrrrr} 3 & 5 & 25 & 45 & -18 \\ & 6 & 22 & 94 & 278 \\ \hline 3 & 11 & 47 & 139 & 260 \end{array}}$$ 2 is not a zero.

We test −2:

$$-2)\overline{\begin{array}{rrrrr} 3 & 5 & 25 & 45 & -18 \\ & -6 & 2 & -54 & 18 \\ \hline 3 & -1 & 27 & -9 & 0 \end{array}}$$

Since $f(-2) = 0$, then -2 is a zero and $x + 2$ is a factor of f. The depressed equation is

$$3x^3 - x^2 + 27x - 9 = 0$$

REPEAT STEP 3: We factor the depressed equation by grouping.

$$3x^3 - x^2 + 27x - 9 = 0$$
$$x^2(3x - 1) + 9(3x - 1) = 0 \quad \text{Factor } x^2 \text{ from } 3x^3 - x^2 \text{ and 9 from } 27x - 9.$$
$$(x^2 + 9)(3x - 1) = 0 \quad \text{Factor out the common factor } 3x - 1.$$
$$x^2 + 9 = 0 \quad \text{or} \quad 3x - 1 = 0 \quad \text{Apply the Zero-Product Property.}$$
$$x^2 = -9 \quad \text{or} \quad x = \frac{1}{3}$$
$$x = -3i, \quad x = 3i \quad \text{or} \quad x = \frac{1}{3}$$

The four complex zeros of f are $\left\{-3i, 3i, -2, \dfrac{1}{3}\right\}$.

The factored form of f is

$$f(x) = 3x^4 + 5x^3 + 25x^2 + 45x - 18$$
$$= 3(x + 3i)(x - 3i)(x + 2)\left(x - \frac{1}{3}\right)$$

━━━━ **Now Work** PROBLEM 33

3.6 Assess Your Understanding

'Are You Prepared?' *Answers are given at the end of these exercises. If you get the wrong answer, read the pages listed in red.*

1. Find the sum and the product of the complex numbers $3 - 2i$ and $-3 + 5i$. (pp. A92–A94)

2. In the complex number system, find the complex zeros of $f(x) = x^2 + 2x + 2$. (pp. 178–180)

Concepts and Vocabulary

3. Every polynomial function of odd degree with real coefficients will have at least _____ real zero(s).

4. If $3 + 4i$ is a zero of a polynomial function of degree 5 with real coefficients, then so is _____.

5. *True or False* A polynomial function of degree n with real coefficients has exactly n complex zeros. At most n of them are real zeros.

6. *True or False* A polynomial function of degree 4 with real coefficients could have $-3, 2 + i, 2 - i$, and $-3 + 5i$ as its zeros.

Skill Building

In Problems 7–16, information is given about a polynomial function f(x) whose coefficients are real numbers. Find the remaining zeros of f.

7. Degree 3; zeros: $3, 4 - i$

8. Degree 3; zeros: $4, 3 + i$

9. Degree 4; zeros: $i, 1 + i$

10. Degree 4; zeros: $1, 2, 2 + i$

11. Degree 5; zeros: $1, i, 2i$

12. Degree 5; zeros: $0, 1, 2, i$

13. Degree 4; zeros: $i, 2, -2$

14. Degree 4; zeros: $2 - i, -i$

15. Degree 6; zeros: $2, 2 + i, -3 - i, 0$

16. Degree 6; zeros: $i, 3 - 2i, -2 + i$

In Problems 17–22, form a polynomial function f(x) with real coefficients having the given degree and zeros. Answers will vary depending on the choice of leading coefficient.

17. Degree 4; zeros: $3 + 2i$; 4, multiplicity 2

18. Degree 4; zeros: $i, 1 + 2i$

19. Degree 5; zeros: $2; -i; 1 + i$

20. Degree 6; zeros: $i, 4 - i; 2 + i$

21. Degree 4; zeros: 3, multiplicity 2; $-i$

22. Degree 5; zeros: 1, multiplicity 3; $1 + i$

In Problems 23–30, use the given zero to find the remaining zeros of each function.

23. $f(x) = x^3 - 4x^2 + 4x - 16$; zero: $2i$

24. $g(x) = x^3 + 3x^2 + 25x + 75$; zero: $-5i$

25. $f(x) = 2x^4 + 5x^3 + 5x^2 + 20x - 12$; zero: $-2i$

26. $h(x) = 3x^4 + 5x^3 + 25x^2 + 45x - 18$; zero: $3i$

27. $h(x) = x^4 - 9x^3 + 21x^2 + 21x - 130$; zero: $3 - 2i$

28. $f(x) = x^4 - 7x^3 + 14x^2 - 38x - 60$; zero: $1 + 3i$

29. $h(x) = 3x^5 + 2x^4 + 15x^3 + 10x^2 - 528x - 352$; zero: $-4i$

30. $g(x) = 2x^5 - 3x^4 - 5x^3 - 15x^2 - 207x + 108$; zero: $3i$

In Problems 31–40, find the complex zeros of each polynomial function. Write f in factored form.

31. $f(x) = x^3 - 1$

32. $f(x) = x^4 - 1$

33. $f(x) = x^3 - 8x^2 + 25x - 26$

34. $f(x) = x^3 + 13x^2 + 57x + 85$

35. $f(x) = x^4 + 5x^2 + 4$

36. $f(x) = x^4 + 13x^2 + 36$

37. $f(x) = x^4 + 2x^3 + 22x^2 + 50x - 75$

38. $f(x) = x^4 + 3x^3 - 19x^2 + 27x - 252$

39. $f(x) = 3x^4 - x^3 - 9x^2 + 159x - 52$

40. $f(x) = 2x^4 + x^3 - 35x^2 - 113x + 65$

Explaining Concepts: Discussion and Writing

In Problems 41 and 42, explain why the facts given are contradictory.

41. $f(x)$ is a polynomial function of degree 3 whose coefficients are real numbers; its zeros are $4 + i, 4 - i$, and $2 + i$.

42. $f(x)$ is a polynomial function of degree 3 whose coefficients are real numbers; its zeros are $2, i$, and $3 + i$.

43. $f(x)$ is a polynomial function of degree 4 whose coefficients are real numbers; three of its zeros are $2, 1 + 2i$, and $1 - 2i$. Explain why the remaining zero must be a real number.

44. $f(x)$ is a polynomial function of degree 4 whose coefficients are real numbers; two of its zeros are -3 and $4 - i$. Explain why one of the remaining zeros must be a real number. Write down one of the missing zeros.

'Are You Prepared?' Answers

1. Sum: $3i$; product: $1 + 21i$ **2.** $-1 - i, \ -1 + i$

CHAPTER REVIEW

Things to Know

Power function (pp. 195–198)

$f(x) = x^n, \quad n \geq 2$ even

Domain: all real numbers   Range: nonnegative real numbers

Passes through $(-1, 1), (0, 0), (1, 1)$

Even function

Decreasing on $(-\infty, 0)$, increasing on $(0, \infty)$

$f(x) = x^n, \quad n \geq 3$ odd

Domain: all real numbers   Range: all real numbers

Passes through $(-1, -1), (0, 0), (1, 1)$

Odd function

Increasing on $(-\infty, \infty)$

Polynomial function (pp. 194, 199–206)

$f(x) = a_n x^n + a_{n-1} x^{n-1}$
$\quad + \cdots + a_1 x + a_0, \quad a_n \neq 0$

Domain: all real numbers

At most $n - 1$ turning points

End behavior: Behaves like $y = a_n x^n$ for large $|x|$

Real zeros of a polynomial function f (p. 199)

Real numbers for which $f(x) = 0$; the real zeros of f are the x-intercepts of the graph of f.

Rational function (pp. 215–223)

$R(x) = \dfrac{p(x)}{q(x)}$

p, q are polynomial functions and q is not the zero polynomial.

Domain: $\{x \,|\, q(x) \neq 0\}$

Vertical asymptotes: With $R(x)$ in lowest terms, if $q(r) = 0$ for some real number, then $x = r$ is a vertical asymptote.

Horizontal or oblique asymptote: See the summary on pages 222–223.

Remainder Theorem (p. 249)

If a polynomial function $f(x)$ is divided by $x - c$, then the remainder is $f(c)$.

Factor Theorem (p. 250)

$x - c$ is a factor of a polynomial function $f(x)$ if and only if $f(c) = 0$.

Descartes' Rule of Signs (p. 251)

Let f denote a polynomial function written in standard form. The number of positive zeros of f either equals the number of variations in sign of the nonzero coefficients of $f(x)$ or else equals that number less some even integer. The number of negative zeros of f either equals the number of variations in sign of the nonzero coefficients of $f(-x)$ or else equals that number less some even integer.

Rational Zeros Theorem (p. 252)

Let f be a polynomial function of degree 1 or higher of the form

$$f(x) = a_n x^n + a_{n-1} x^{n-1} + \cdots + a_1 x + a_0 \qquad a_n \neq 0, a_0 \neq 0$$

where each coefficient is an integer. If $\dfrac{p}{q}$, in lowest terms, is a rational zero of f, then p must be a factor of a_0, and q must be a factor of a_n.

Intermediate Value Theorem (p. 257)

Let f be a polynomial function. If $a < b$ and $f(a)$ and $f(b)$ are of opposite sign, then there is at least one real zero of f between a and b.

Fundamental Theorem of Algebra (p. 263)

Every complex polynomial function $f(x)$ of degree $n \geq 1$ has at least one complex zero.

Conjugate Pairs Theorem (p. 263)

Let $f(x)$ be a polynomial whose coefficients are real numbers. If $r = a + bi$ is a zero of f, then its complex conjugate $\bar{r} = a - bi$ is also a zero of f.

Objectives

Section		You should be able to . . .	Example(s)	Review Exercises
3.1	1	Identify polynomial functions and their degree (p. 194)	1	1–4
	2	Graph polynomial functions using transformations (p. 198)	2, 3	5–10
	3	Identify the real zeros of a polynomial function and their multiplicity (p. 199)	4–6	11–18 (a) and (b)
	4	Analyze the graph of a polynomial function (p. 206)	9, 10	11–18
	5	Build cubic models from data (p. 209)	11	90
3.2	1	Find the domain of a rational function (p. 215)	1	19–22
	2	Find the vertical asymptotes of a rational function (p. 219)	4	19–22
	3	Find the horizontal or oblique asymptote of a rational function (p. 220)	5–8	19–22
3.3	1	Analyze the graph of a rational function (p. 226)	1–5	23–34
	2	Solve applied problems involving rational functions (p. 237)	7	89
3.4	1	Solve polynomial inequalities (p. 241)	1, 2	35, 36
	2	Solve rational inequalities (p. 243)	3, 4	37–44
3.5	1	Use the Remainder and Factor Theorems (p. 248)	1, 2	45–50
	2	Use Descartes' Rule of Signs to determine the number of positive and the number of negative real zeros of a polynomial function (p. 251)	3	51, 52
	3	Use the Rational Zeros Theorem to list the potential rational zeros of a polynomial function (p. 252)	4	53–60
	4	Find the real zeros of a polynomial function (p. 253)	5, 6	55–60
	5	Solve polynomial equations (p. 255)	7	61–64
	6	Use the Theorem for Bounds on Zeros (p. 256)	8	65–68
	7	Use the Intermediate Value Theorem (p. 257)	9	69–76
3.6	1	Use the Conjugate Pairs Theorem (p. 263)	1	77–80
	2	Find a polynomial function with specified zeros (p. 264)	2	77–80
	3	Find the complex zeros of a polynomial (p. 265)	3	81–88

Review Exercises

In Problems 1–4, determine whether the function is a polynomial function, rational function, or neither. For those that are polynomial functions, state the degree. For those that are not polynomial functions, tell why not.

1. $f(x) = 4x^5 - 3x^2 + 5x - 2$ **2.** $f(x) = \dfrac{3x^5}{2x+1}$ **3.** $f(x) = 3x^2 + 5x^{1/2} - 1$ **4.** $f(x) = 3$

In Problems 5–10, graph each function using transformations (shifting, compressing, stretching, and reflection). Show all the stages.

5. $f(x) = (x+2)^3$ **6.** $f(x) = -x^3 + 3$ **7.** $f(x) = -(x-1)^4$

8. $f(x) = (x-1)^4 - 2$ **9.** $f(x) = (x-1)^4 + 2$ **10.** $f(x) = (1-x)^3$

In Problems 11–18, analyze each polynomial function by following Steps 1 through 6 on page 208.

11. $f(x) = x(x+2)(x+4)$ **12.** $f(x) = x(x-2)(x-4)$ **13.** $f(x) = (x-2)^2(x+4)$

14. $f(x) = (x-2)(x+4)^2$ **15.** $f(x) = -2x^3 + 4x^2$ **16.** $f(x) = -4x^3 + 4x$

17. $f(x) = (x-1)^2(x+3)(x+1)$ **18.** $f(x) = (x-4)(x+2)^2(x-2)$

In Problems 19–22, find the domain of each rational function. Find any horizontal, vertical, or oblique asymptotes.

19. $R(x) = \dfrac{x+2}{x^2-9}$ **20.** $R(x) = \dfrac{x^2+4}{x-2}$ **21.** $R(x) = \dfrac{x^2+3x+2}{(x+2)^2}$ **22.** $R(x) = \dfrac{x^3}{x^3-1}$

In Problems 23–34, discuss each rational function following the eight steps given on page 228.

23. $R(x) = \dfrac{2x - 6}{x}$

24. $R(x) = \dfrac{4 - x}{x}$

25. $H(x) = \dfrac{x + 2}{x(x - 2)}$

26. $H(x) = \dfrac{x}{x^2 - 1}$

27. $R(x) = \dfrac{x^2 + x - 6}{x^2 - x - 6}$

28. $R(x) = \dfrac{x^2 - 6x + 9}{x^2}$

29. $F(x) = \dfrac{x^3}{x^2 - 4}$

30. $F(x) = \dfrac{3x^3}{(x - 1)^2}$

31. $R(x) = \dfrac{2x^4}{(x - 1)^2}$

32. $R(x) = \dfrac{x^4}{x^2 - 9}$

33. $G(x) = \dfrac{x^2 - 4}{x^2 - x - 2}$

34. $F(x) = \dfrac{(x - 1)^2}{x^2 - 1}$

In Problems 35–44, solve each inequality. Graph the solution set.

35. $x^3 + x^2 < 4x + 4$

36. $x^3 + 4x^2 \geq x + 4$

37. $\dfrac{6}{x + 3} \geq 1$

38. $\dfrac{-2}{1 - 3x} < 1$

39. $\dfrac{2x - 6}{1 - x} < 2$

40. $\dfrac{3 - 2x}{2x + 5} \geq 2$

41. $\dfrac{(x - 2)(x - 1)}{x - 3} \geq 0$

42. $\dfrac{x + 1}{x(x - 5)} \leq 0$

43. $\dfrac{x^2 - 8x + 12}{x^2 - 16} > 0$

44. $\dfrac{x(x^2 + x - 2)}{x^2 + 9x + 20} \leq 0$

In Problems 45–48, find the remainder R when f(x) is divided by g(x). Is g a factor of f?

45. $f(x) = 8x^3 - 3x^2 + x + 4$; $g(x) = x - 1$

46. $f(x) = 2x^3 + 8x^2 - 5x + 5$; $g(x) = x - 2$

47. $f(x) = x^4 - 2x^3 + 15x - 2$; $g(x) = x + 2$

48. $f(x) = x^4 - x^2 + 2x + 2$; $g(x) = x + 1$

49. Find the value of $f(x) = 12x^6 - 8x^4 + 1$ at $x = 4$.

50. Find the value of $f(x) = -16x^3 + 18x^2 - x + 2$ at $x = -2$.

In Problems 51 and 52, use Descartes' Rule of Signs to determine how many positive and negative zeros each polynomial function may have. Do not attempt to find the zeros.

51. $f(x) = 12x^8 - x^7 + 8x^4 - 2x^3 + x + 3$

52. $f(x) = -6x^5 + x^4 + 5x^3 + x + 1$

53. List all the potential rational zeros of $f(x) = 12x^8 - x^7 + 6x^4 - x^3 + x - 3$.

54. List all the potential rational zeros of $f(x) = -6x^5 + x^4 + 2x^3 - x + 1$.

In Problems 55–60, use the Rational Zeros Theorem to find all the real zeros of each polynomial function. Use the zeros to factor f over the real numbers.

55. $f(x) = x^3 - 3x^2 - 6x + 8$

56. $f(x) = x^3 - x^2 - 10x - 8$

57. $f(x) = 4x^3 + 4x^2 - 7x + 2$

58. $f(x) = 4x^3 - 4x^2 - 7x - 2$

59. $f(x) = x^4 - 4x^3 + 9x^2 - 20x + 20$

60. $f(x) = x^4 + 6x^3 + 11x^2 + 12x + 18$

In Problems 61–64, solve each equation in the real number system.

61. $2x^4 + 2x^3 - 11x^2 + x - 6 = 0$

62. $3x^4 + 3x^3 - 17x^2 + x - 6 = 0$

63. $2x^4 + 7x^3 + x^2 - 7x - 3 = 0$

64. $2x^4 + 7x^3 - 5x^2 - 28x - 12 = 0$

In Problems 65–68, find bounds to the real zeros of each polynomial function.

65. $f(x) = x^3 - x^2 - 4x + 2$

66. $f(x) = x^3 + x^2 - 10x - 5$

67. $f(x) = 2x^3 - 7x^2 - 10x + 35$

68. $f(x) = 3x^3 - 7x^2 - 6x + 14$

In Problems 69–72, use the Intermediate Value Theorem to show that each polynomial function has a zero in the given interval.

69. $f(x) = 3x^3 - x - 1;$ $[0, 1]$

70. $f(x) = 2x^3 - x^2 - 3;$ $[1, 2]$

71. $f(x) = 8x^4 - 4x^3 - 2x - 1;$ $[0, 1]$

72. $f(x) = 3x^4 + 4x^3 - 8x - 2;$ $[1, 2]$

In Problems 73–76, each polynomial function has exactly one positive zero. Approximate the zero correct to two decimal places.

73. $f(x) = x^3 - x - 2$

74. $f(x) = 2x^3 - x^2 - 3$

75. $f(x) = 8x^4 - 4x^3 - 2x - 1$

76. $f(x) = 3x^4 + 4x^3 - 8x - 2$

In Problems 77–80, information is given about a complex polynomial $f(x)$ whose coefficients are real numbers. Find the remaining zeros of f. Then find a polynomial function with real coefficients that has the zeros.

77. Degree 3; zeros: $4 + i, 6$

78. Degree 3; zeros: $3 + 4i, 5$

79. Degree 4; zeros: $i, 1 + i$

80. Degree 4; zeros: $1, 2, 1 + i$

In Problems 81–88, find the complex zeros of each polynomial function $f(x)$. Write f in factored form.

81. $f(x) = x^3 - 3x^2 - 6x + 8$

82. $f(x) = x^3 - x^2 - 10x - 8$

83. $f(x) = 4x^3 + 4x^2 - 7x + 2$

84. $f(x) = 4x^3 - 4x^2 - 7x - 2$

85. $f(x) = x^4 - 4x^3 + 9x^2 - 20x + 20$

86. $f(x) = x^4 + 6x^3 + 11x^2 + 12x + 18$

87. $f(x) = 2x^4 + 2x^3 - 11x^2 + x - 6$

88. $f(x) = 3x^4 + 3x^3 - 17x^2 + x - 6$

---

89. Making a Can A can in the shape of a right circular cylinder is required to have a volume of 250 cubic centimeters.

(a) Express the amount A of material to make the can as a function of the radius r of the cylinder.

(b) How much material is required if the can is of radius 3 centimeters?

(c) How much material is required if the can is of radius 5 centimeters?

(d) Graph $A = A(r)$. For what value of r is A smallest?

90. Model It: Poverty Rates The following data represent the percentage of families in the United States whose income is below the poverty level.

Year, t	Percent below Poverty Level, p
1990, 1	10.9
1991, 2	11.5
1992, 3	11.9
1993, 4	12.3
1994, 5	11.6
1995, 6	10.8
1996, 7	11.0
1997, 8	10.3
1998, 9	10.0
1999, 10	9.3
2000, 11	8.7
2001, 12	9.2
2002, 13	9.6
2003, 14	10.0
2004, 15	10.2

Source: U.S. Census Bureau

(a) With a graphing utility, draw a scatter diagram of the data. Comment on the type of relation that appears to exist between the two variables.

(b) Decide on a function of best fit to these data (linear, quadratic, or cubic), and use this function to predict the percentage of U.S. families that were below the poverty level in 2005 ($t = 16$).

(c) Draw the function of best fit on the scatter diagram drawn in part (a).

91. Design a polynomial function with the following characteristics: degree 6; four real zeros, one of multiplicity 3; y-intercept 3; behaves like $y = -5x^6$ for large values of $|x|$. Is this polynomial unique? Compare your polynomial with those of other students. What terms will be the same as everyone else's? Add some more characteristics, such as symmetry or naming the real zeros. How does this modify the polynomial?

92. Design a rational function with the following characteristics: three real zeros, one of multiplicity 2; y-intercept 1; vertical asymptotes $x = -2$ and $x = 3$; oblique asymptote $y = 2x + 1$. Is this rational function unique? Compare yours with those of other students. What will be the same as everyone else's? Add some more characteristics, such as symmetry or naming the real zeros. How does this modify the rational function?

93. The illustration shows the graph of a polynomial function.

(a) Is the degree of the polynomial even or odd?

(b) Is the leading coefficient positive or negative?

(c) Is the function even, odd, or neither?

(d) Why is x^2 necessarily a factor of the polynomial?

(e) What is the minimum degree of the polynomial?

(f) Formulate five different polynomials whose graphs could look like the one shown. Compare yours to those of other students. What similarities do you see? What differences?

CHAPTER TEST

Step-by-step test solutions are found on the Chapter Test Prep Videos available via the Video Resources on DVD, in *MyMathLab* or on You Tube (search "Sullivan College Algebra, Concepts Through Functions 2e).

1. Graph $f(x) = (x - 3)^4 - 2$ using transformations.

2. For the polynomial function $g(x) = 2x^3 + 5x^2 - 28x - 15$,
   (a) Determine the maximum number of real zeros that the function may have.
   (b) Find bounds to the zeros of the function.
   (c) List the potential rational zeros.
   (d) Determine the real zeros of g. Factor g over the reals.
   (e) Find the x- and y-intercepts of the graph of g.
   (f) Determine whether the graph crosses or touches the x-axis at each x-intercept.
   (g) Find the power function that the graph of g resembles for large values of $|x|$.
   (h) Determine the behavior of the graph of g near each x-intercept.
   (i) Put all the information together to obtain the graph of g.

3. Find the complex zeros of $f(x) = x^3 - 4x^2 + 25x - 100$.

4. Solve $3x^3 + 2x - 1 = 8x^2 - 4$ in the complex number system.

In Problems 5 and 6, find the domain of each function. Find any horizontal, vertical, or oblique asymptotes.

5. $g(x) = \dfrac{2x^2 - 14x + 24}{x^2 + 6x - 40}$

6. $r(x) = \dfrac{x^2 + 2x - 3}{x + 1}$

7. Sketch the graph of the function in Problem 6. Label all intercepts, vertical asymptotes, horizontal asymptotes, and oblique asymptotes.

In Problems 8 and 9, write a function that meets the given conditions.

8. Fourth-degree polynomial with real coefficients; zeros: $-2, 0, 3 + i$

9. Rational function; asymptotes: $y = 2, x = 4$; domain: $\{x \mid x \neq 4, x \neq 9\}$

10. Use the Intermediate Value Theorem to show that the function $f(x) = -2x^2 - 3x + 8$ has at least one real zero on the interval $[0, 4]$.

11. Solve: $\dfrac{x + 2}{x - 3} < 2$

CUMULATIVE REVIEW

1. Find the distance between the points $P = (1, 3)$ and $Q = (-4, 2)$.

2. Solve the inequality $x^2 \geq x$ and graph the solution set.

3. Solve the inequality $x^2 - 3x < 4$ and graph the solution set.

4. Find a linear function with slope -3 that contains the point $(-1, 4)$. Graph the function.

5. Find the equation of the line parallel to the line $y = 2x + 1$ and containing the point $(3, 5)$. Express your answer in slope–intercept form and graph the line.

6. Graph the equation $y = x^3$.

7. Does the relation $\{(3, 6), (1, 3), (2, 5), (3, 8)\}$ represent a function? Why or why not?

8. Solve the equation $x^3 - 6x^2 + 8x = 0$.

9. Solve the inequality $3x + 2 \leq 5x - 1$ and graph the solution set.

10. Find the center and radius of the circle $x^2 + 4x + y^2 - 2y - 4 = 0$. Graph the circle.

11. For the equation $y = x^3 - 9x$, determine the intercepts and test for symmetry.

12. Find an equation of the line perpendicular to $3x - 2y = 7$ that contains the point $(1, 5)$.

13. Is the following the graph of a function? Why or why not?

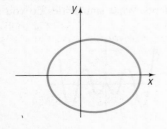

14. For the function $f(x) = x^2 + 5x - 2$, find
    (a) $f(3)$
    (b) $f(-x)$
    (c) $-f(x)$
    (d) $f(3x)$
    (e) $\dfrac{f(x + h) - f(x)}{h} \quad h \neq 0$

15. Answer the following questions regarding the function
    $$f(x) = \frac{x + 5}{x - 1}$$
    (a) What is the domain of f?
    (b) Is the point $(2, 6)$ on the graph of f?
    (c) If $x = 3$, what is $f(x)$? What point is on the graph of f?
    (d) If $f(x) = 9$, what is x? What point is on the graph of f?
    (e) If f a polynomial or rational function?

16. Graph the function $f(x) = -3x + 7$.

17. Graph $f(x) = 2x^2 - 4x + 1$ by determining whether its graph opens up or down and by finding its vertex, axis of symmetry, y-intercept, and x-intercepts, if any.

18. Find the average rate of change of $f(x) = x^2 + 3x + 1$ from 1 to 2. Use this result to find the equation of the secant line containing $(1, f(1))$ and $(2, f(2))$.

19. In parts (a) to (f), use the following graph.

(a) Determine the intercepts.

(b) Based on the graph, tell whether the graph is symmetric with respect to the x-axis, the y-axis, and/or the origin.

(c) Based on the graph, tell whether the function is even, odd, or neither.

(d) List the intervals on which f is increasing. List the intervals on which f is decreasing.

(e) List the numbers, if any, at which f has a local maximum. What are these local maxima?

(f) List the numbers, if any, at which f has a local minimum. What are these local minima?

20. Determine algebraically whether the function

$$f(x) = \frac{5x}{x^2 - 9}$$

is even, odd, or neither.

21. For the function $f(x) = \begin{cases} 2x + 1 & \text{if } -3 < x < 2 \\ -3x + 4 & \text{if } x \geq 2 \end{cases}$

(a) Find the domain of f.

(b) Locate any intercepts.

(c) Graph the function.

(d) Based on the graph, find the range.

22. Graph the function $f(x) = -3(x + 1)^2 + 5$ using transformations.

23. Suppose that $f(x) = x^2 - 5x + 1$ and $g(x) = -4x - 7$.

(a) Find $f + g$ and state its domain.

(b) Find $\dfrac{f}{g}$ and state its domain.

24. **Demand Equation** The price p (in dollars) and the quantity x sold of a certain product obey the demand equation

$$p = -\frac{1}{10}x + 150,$$

(a) Express the revenue R as a function of x.

(b) What is the revenue if 100 units are sold?

(c) What quantity x maximizes revenue? What is the maximum revenue?

(d) What price should the company charge to maximize revenue?

CHAPTER PROJECTS

Latitude (degrees)	Length of Day (hours)
0	12.117
10	12.2
20	12.283
30	12.383
40	12.517
50	12.7
60	13.0
70	13.567
80	15.267

Source: http://encarta.msn.com/media_701500905_761563211_-1_1/hours_of_daylight_by_latitude.html

I. **Length of Day** The table shows the length of a day for various latitudes in the northern hemisphere on September 15.

1. Let the independent variable, x, represent the latitude. Let the dependent variable, y, represent the length of day. Draw a scatter diagram of the data using your graphing utility. Comment on the shape of the data.

2. Use a graphing utility to determine the linear function of best fit. Graph the linear function of best fit on the scatter diagram. Do you think the function accurately describes the relation between latitude and length of the day?

3. Use a graphing utility to determine the quadratic function of best fit. Graph the quadratic function of best fit on the scatter diagram. Do you think the function accurately describes the relation between latitude and length of the day?

4. Use a graphing utility to determine the cubic function of best fit. Graph the cubic function of best fit on the scatter diagram. Do you think the function accurately describes the relation between latitude and length of the day?

5. Which of the three models seems to fit the data best? Explain your reasoning.

6. Use your model to predict the hours of daylight on September 15 for Chicago, (41.85 degrees north latitude)

7. Investigate various Web sites that claim to provide hours of daylight for various latitudes. Compare the predictions of these Web sites to your prediction. Provide some reasons why the different Web sites have different results for hours of daylight. Do any of the Web sites have the same prediction as yours?

The following project is available at the Instructor's Resource Center (IRC):

II. **Theory of Equations** The coefficients of a polynomial function can be found if its zeros are known, an advantage of using polynomials in modeling.

4

Exponential and Logarithmic Functions

The McDonald's Scalding Coffee Case

April 3, 1996 – There is a lot of hype about the McDonald's scalding coffee case. No one is in favor of frivolous cases or outlandish results; however, it is important to understand some points that were not reported in most of the stories about the case. McDonald's coffee

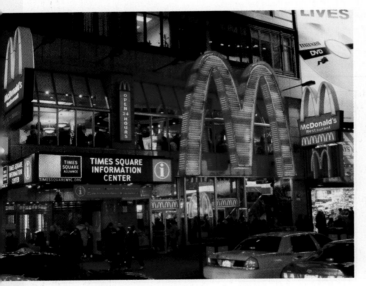

was not only hot, it was scalding, capable of almost instantaneous destruction of skin, flesh, and muscle.

Plaintiff's expert, a scholar in thermodynamics applied to human skin burns, testified that liquids at 180 degrees will cause a full thickness burn to human skin in two to seven seconds. Other testimony showed that, as the temperature decreases toward 155 degrees, the extent of the burn relative to that temperature decreases exponentially. Thus, if (the) spill had involved coffee at 155 degrees, the liquid would have cooled and given her time to avoid a serious burn.

Source: "McDonald's Scalding Coffee Case," ATLA Press Room, http://www.atla.org/pressroom/ FACTS/frivolous/McdonaldsCoffeecase.aspx, accessed September 30, 2005. Used with permission.

—See Chapter Project I—

◁ A Look Back Until now, our study of functions has concentrated on polynomial and rational functions. These functions belong to the class of **algebraic functions,** that is, functions that can be expressed in terms of sums, differences, products, quotients, powers, or roots of polynomials. Functions that are not algebraic are termed **transcendental** (they transcend, or go beyond, algebraic functions).

A Look Ahead ▷ In this chapter, we study two transcendental functions: the exponential function and the logarithmic function. These functions occur frequently in a wide variety of applications, such as biology, chemistry, economics, and psychology.

The chapter begins with a discussion of composite, one-to-one, and inverse functions, concepts needed to see the relationship between exponential and logarithmic functions.

4.1 Composite Functions

PREPARING FOR THIS SECTION *Before getting started, review the following:*

• Find the Value of a Function (Section 1.1, pp. 46–49)   • Domain of a Function (Section 1.1, pp. 49–51)

Now Work the *'Are You Prepared?'* problems on page 280.

OBJECTIVES   **1** Form a Composite Function (p. 275)
                       2 Find the Domain of a Composite Function (p. 276)

1 Form a Composite Function

Figure 1

Suppose that an oil tanker is leaking oil and you want to determine the area of the circular oil patch around the ship. See Figure 1. It is determined that the oil is leaking from the tanker in such a way that the radius of the circular patch of oil around the ship is increasing at a rate of 3 feet per minute. Therefore, the radius r of the oil patch at any time t, in minutes, is given by $r(t) = 3t$. So after 20 minutes the radius of the oil patch is $r(20) = 3(20) = 60$ feet.

The area A of a circle as a function of the radius r is given by $A(r) = \pi r^2$. The area of the circular patch of oil after 20 minutes is $A(60) = \pi(60)^2 = 3600\pi$ square feet. Notice that $60 = r(20)$, so $A(60) = A(r(20))$. The argument of the function A is itself a function!

In general, we can find the area of the oil patch as a function of time t by evaluating $A(r(t))$ and obtaining $A(r(t)) = A(3t) = \pi(3t)^2 = 9\pi t^2$. The function $A(r(t))$ is a special type of function called a *composite function*.

As another example, consider the function $y = (2x + 3)^2$. If we write $y = f(u) = u^2$ and $u = g(x) = 2x + 3$, then, by a substitution process, we can obtain the original function: $y = f(u) = f(g(x)) = (2x + 3)^2$.

In general, suppose that f and g are two functions and that x is a number in the domain of g. By evaluating g at x, we get $g(x)$. If $g(x)$ is in the domain of f, then we may evaluate f at $g(x)$ and obtain the expression $f(g(x))$. The correspondence from x to $f(g(x))$ is called a *composite function* $f \circ g$.

DEFINITION

Given two functions f and g, the **composite function,** denoted by $f \circ g$ (read as "f composed with g"), is defined by

$$(f \circ g)(x) = f(g(x))$$

The domain of $f \circ g$ is the set of all numbers x in the domain of g such that $g(x)$ is in the domain of f.

Look carefully at Figure 2. Only those x's in the domain of g for which $g(x)$ is in the domain of f can be in the domain of $f \circ g$. The reason is that if $g(x)$ is not in the domain of f then $f(g(x))$ is not defined. Because of this, the domain of $f \circ g$ is a subset of the domain of g; the range of $f \circ g$ is a subset of the range of f.

Figure 2

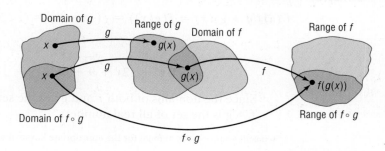

Figure 3 provides a second illustration of the definition. Here x is the input to the function g, yielding $g(x)$. Then $g(x)$ is the input to the function f, yielding $f(g(x))$. Notice that the "inside" function g in $f(g(x))$ is done first.

Figure 3

Let's look at some examples.

EXAMPLE 1   **Evaluating a Composite Function**

Suppose that $f(x) = 2x^2 - 3$ and $g(x) = 4x$. Find:

(a) $(f \circ g)(1)$ (b) $(g \circ f)(1)$ (c) $(f \circ f)(-2)$ (d) $(g \circ g)(-1)$

Solution   (a) $(f \circ g)(1) = f(g(1)) = f(4) = 2 \cdot 4^2 - 3 = 29$

$$g(x) = 4x \quad f(x) = 2x^2 - 3$$
$$g(1) = 4$$

(b) $(g \circ f)(1) = g(f(1)) = g(-1) = 4 \cdot (-1) = -4$

$$f(x) = 2x^2 - 3 \quad g(x) = 4x$$
$$f(1) = -1$$

(c) $(f \circ f)(-2) = f(f(-2)) = f(5) = 2 \cdot 5^2 - 3 = 47$

$$f(-2) = 2(-2)^2 - 3 = 5$$

(d) $(g \circ g)(-1) = g(g(-1)) = g(-4) = 4 \cdot (-4) = -16$

$$g(-1) = -4$$

Figure 4

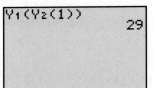

COMMENT   Graphing calculators can be used to evaluate composite functions.* Let $Y_1 = f(x) = 2x^2 - 3$ and $Y_2 = g(x) = 4x$. Then, using a TI-84 Plus graphing calculator, $(f \circ g)(1)$ is found as shown in Figure 4. Notice that this is the result obtained in Example 1(a).   ∎

——**Now Work** PROBLEM 11

2 Find the Domain of a Composite Function

EXAMPLE 2   **Finding a Composite Function and Its Domain**

Suppose that $f(x) = x^2 + 3x - 1$ and $g(x) = 2x + 3$.

Find: (a) $f \circ g$ (b) $g \circ f$

Then find the domain of each composite function.

Solution   The domain of f and the domain of g are the set of all real numbers.

(a) $(f \circ g)(x) = f(g(x)) = f(2x + 3) = (2x + 3)^2 + 3(2x + 3) - 1$

$$f(x) = x^2 + 3x - 1$$

$$= 4x^2 + 12x + 9 + 6x + 9 - 1 = 4x^2 + 18x + 17$$

Since the domains of both f and g are the set of all real numbers, the domain of $f \circ g$ is the set of all real numbers.

*Consult your owner's manual for the appropriate keystrokes.

(b) $(g \circ f)(x) = g(f(x)) = g(x^2 + 3x - 1) = 2(x^2 + 3x - 1) + 3$

$$\uparrow$$
$$g(x) = 2x + 3$$

$$= 2x^2 + 6x - 2 + 3 = 2x^2 + 6x + 1$$

Since the domains of both f and g are the set of all real numbers, the domain of $g \circ f$ is the set of all real numbers.

Look back at Figure 2. In determining the domain of the composite function $(f \circ g)(x) = f(g(x))$, keep the following two thoughts in mind about the input x.

1. Any x not in the domain of g must be excluded.
2. Any x for which $g(x)$ is not in the domain of f must be excluded.

EXAMPLE 3   **Finding the Domain of $f \circ g$**

Find the domain of $f \circ g$ if $f(x) = \dfrac{1}{x + 2}$ and $g(x) = \dfrac{4}{x - 1}$.

Solution   For $(f \circ g)(x) = f(g(x))$, we first note that the domain of g is $\{x \mid x \neq 1\}$, so we exclude 1 from the domain of $f \circ g$. Next we note that the domain of f is $\{x \mid x \neq -2\}$, which means that $g(x)$ cannot equal -2. We solve the equation $g(x) = -2$ to determine what additional value(s) of x to exclude.

$$\frac{4}{x - 1} = -2 \qquad g(x) = -2$$
$$4 = -2(x - 1)$$
$$4 = -2x + 2$$
$$2x = -2$$
$$x = -1$$

We also exclude -1 from the domain of $f \circ g$. The domain of $f \circ g$ is $\{x \mid x \neq -1, x \neq 1\}$.

✓**Check:** For $x = 1$, $g(x) = \dfrac{4}{x - 1}$ is not defined, so $(f \circ g)(x) = f(g(x))$ is not defined.

For $x = -1$, $g(-1) = \dfrac{4}{-2} = -2$, and $(f \circ g)(-1) = f(g(-1)) = f(-2)$ is not defined.

Now Work PROBLEM 21

EXAMPLE 4   **Finding a Composite Function and Its Domain**

Suppose that $f(x) = \dfrac{1}{x + 2}$ and $g(x) = \dfrac{4}{x - 1}$.

Find: (a) $f \circ g$ (b) $f \circ f$

Then find the domain of each composite function.

Solution   The domain of f is $\{x \mid x \neq -2\}$ and the domain of g is $\{x \mid x \neq 1\}$.

(a) $(f \circ g)(x) = f(g(x)) = f\left(\dfrac{4}{x - 1}\right) = \dfrac{1}{\dfrac{4}{x - 1} + 2} = \dfrac{x - 1}{4 + 2(x - 1)} = \dfrac{x - 1}{2x + 2} = \dfrac{x - 1}{2(x + 1)}$

$$\uparrow \qquad\qquad\qquad\qquad \uparrow$$
$$f(x) = \frac{1}{x + 2} \qquad \text{Multiply by } \frac{x - 1}{x - 1}.$$

In Example 3, we found the domain of $f \circ g$ to be $\{x \mid x \neq -1, x \neq 1\}$.

We could also find the domain of $f \circ g$ by first looking at the domain of g: $\{x \mid x \neq 1\}$. We exclude 1 from the domain of $f \circ g$ as a result. Then we look at $f \circ g$ and notice that x cannot equal -1, since $x = -1$ results in division by 0. So we also exclude -1 from the domain of $f \circ g$. Therefore, the domain of $f \circ g$ is $\{x \mid x \neq -1, x \neq 1\}$.

(b) $(f \circ f)(x) = f(f(x)) = f\left(\dfrac{1}{x+2}\right) = \dfrac{1}{\underset{\uparrow}{\dfrac{1}{x+2}} + 2} = \dfrac{x+2}{1+2(x+2)} = \dfrac{x+2}{2x+5}$

$$f(x) = \dfrac{1}{x+2} \qquad\qquad \text{Multiply by } \dfrac{x+2}{x+2}.$$

The domain of $f \circ f$ consists of those x in the domain of f, $\{x \mid x \neq -2\}$, for which

$$f(x) = \dfrac{1}{x+2} \neq -2 \qquad \begin{aligned}\dfrac{1}{x+2} &= -2 \\ 1 &= -2(x+2) \\ 1 &= -2x - 4 \\ 2x &= -5 \\ x &= -\dfrac{5}{2}\end{aligned}$$

or, equivalently,

$$x \neq -\dfrac{5}{2}$$

The domain of $f \circ f$ is $\left\{x \mid x \neq -\dfrac{5}{2}, x \neq -2\right\}$.

We could also find the domain of $f \circ f$ by recognizing that -2 is not in the domain of f and so should be excluded from the domain of $f \circ f$. Then, looking at $f \circ f$, we see that x cannot equal $-\dfrac{5}{2}$. Do you see why? Therefore, the domain of $f \circ f$ is $\left\{x \mid x \neq -\dfrac{5}{2}, x \neq -2\right\}$. ●

Now Work PROBLEMS 33 AND 35

Look back at Example 2, which illustrates that, in general, $f \circ g \neq g \circ f$. Sometimes $f \circ g$ does equal $g \circ f$, as shown in the next example.

EXAMPLE 5   **Showing That Two Composite Functions Are Equal**

If $f(x) = 3x - 4$ and $g(x) = \dfrac{1}{3}(x + 4)$, show that

$$(f \circ g)(x) = (g \circ f)(x) = x$$

for every x in the domain of $f \circ g$ and $g \circ f$.

Solution $(f \circ g)(x) = f(g(x))$

$$= f\left(\dfrac{x+4}{3}\right) \qquad g(x) = \dfrac{1}{3}(x+4) = \dfrac{x+4}{3}$$

$$= 3\left(\dfrac{x+4}{3}\right) - 4 \qquad \text{Substitute } g(x) \text{ into the rule for } f, f(x) = 3x - 4.$$

$$= x + 4 - 4 = x$$

Seeing the Concept

Using a graphing calculator, let

$$Y_1 = f(x) = 3x - 4$$

$$Y_2 = g(x) = \frac{1}{3}(x + 4)$$

$$Y_3 = f \circ g, Y_4 = g \circ f$$

Using the viewing window $-3 \leq x \leq 3$, $-2 \leq y \leq 2$, graph only Y_3 and Y_4. What do you see? TRACE to verify that $Y_3 = Y_4$.

$$(g \circ f)(x) = g(f(x))$$

$$= g(3x - 4) \qquad f(x) = 3x - 4$$

$$= \frac{1}{3}[(3x - 4) + 4] \qquad \text{Substitute } f(x) \text{ into the rule for } g, g(x) = \frac{1}{3}(x + 4).$$

$$= \frac{1}{3}(3x) = x$$

We conclude that $(f \circ g)(x) = (g \circ f)(x) = x$.

In Section 4.2, we shall see that there is an important relationship between functions f and g for which $(f \circ g)(x) = (g \circ f)(x) = x$.

━━━━━ **Now Work** PROBLEM 45

Calculus Application

 Some techniques in calculus require that we be able to determine the components of a composite function. For example, the function $H(x) = \sqrt{x + 1}$ is the composition of the functions f and g, where $f(x) = \sqrt{x}$ and $g(x) = x + 1$, because $H(x) = (f \circ g)(x) = f(g(x)) = f(x + 1) = \sqrt{x + 1}$.

| EXAMPLE 6 | **Finding the Components of a Composite Function** |

Find functions f and g such that $f \circ g = H$ if $H(x) = (x^2 + 1)^{50}$.

Solution The function H takes $x^2 + 1$ and raises it to the power 50. A natural way to decompose H is to raise the function $g(x) = x^2 + 1$ to the power 50. If we let $f(x) = x^{50}$ and $g(x) = x^2 + 1$, then

Figure 5

$$(f \circ g)(x) = f(g(x))$$

$$= f(x^2 + 1)$$

$$= (x^2 + 1)^{50} = H(x)$$

See Figure 5.

Other functions f and g may be found for which $f \circ g = H$ in Example 6. For example, if $f(x) = x^2$ and $g(x) = (x^2 + 1)^{25}$, then

$$(f \circ g)(x) = f(g(x)) = f((x^2 + 1)^{25}) = [(x^2 + 1)^{25}]^2 = (x^2 + 1)^{50}$$

Although the functions f and g found as a solution to Example 6 are not unique, there is usually a "natural" selection for f and g that comes to mind first.

| EXAMPLE 7 | **Finding the Components of a Composite Function** |

Find functions f and g such that $f \circ g = H$ if $H(x) = \dfrac{1}{x + 1}$.

Solution Here H is the reciprocal of $g(x) = x + 1$. If we let $f(x) = \dfrac{1}{x}$ and $g(x) = x + 1$, we find that

$$(f \circ g)(x) = f(g(x)) = f(x + 1) = \frac{1}{x + 1} = H(x)$$

━━━━━ **Now Work** PROBLEM 53

4.1 Assess Your Understanding

'Are You Prepared?' *Answers are given at the end of these exercises. If you get a wrong answer, read the pages listed in red.*

1. Find $f(3)$ if $f(x) = -4x^2 + 5x$. (pp. 46–49)
2. Find $f(3x)$ if $f(x) = 4 - 2x^2$. What is the argument of f? (pp. 46–49)
3. Find the domain of the function $f(x) = \dfrac{x^2 - 1}{x^2 - 25}$. (pp. 49–50)

Concepts and Vocabulary

4. Given two functions f and g, the _____ _____, denoted $f \circ g$, is defined by $f \circ g(x) = $ _____.
5. **True or False** $f(g(x)) = f(x) \cdot g(x)$.
6. **True or False** The domain of the composite function $(f \circ g)(x)$ is the same as the domain of $g(x)$.

Skill Building

In Problems 7 and 8, evaluate each expression using the values given in the table.

7.

x	−3	−2	−1	0	1	2	3
f(x)	−7	−5	−3	−1	3	5	7
g(x)	8	3	0	−1	0	3	8

(a) $(f \circ g)(1)$ (b) $(f \circ g)(-1)$
(c) $(g \circ f)(-1)$ (d) $(g \circ f)(0)$
(e) $(g \circ g)(-2)$ (f) $(f \circ f)(-1)$

8.

x	−3	−2	−1	0	1	2	3
f(x)	11	9	7	5	3	1	−1
g(x)	−8	−3	0	1	0	−3	−8

(a) $(f \circ g)(1)$ (b) $(f \circ g)(2)$
(c) $(g \circ f)(2)$ (d) $(g \circ f)(3)$
(e) $(g \circ g)(1)$ (f) $(f \circ f)(3)$

In Problems 9 and 10, evaluate each expression using the graphs of $y = f(x)$ and $y = g(x)$ shown in the figure.

9. (a) $(g \circ f)(-1)$ (b) $(g \circ f)(0)$
   (c) $(f \circ g)(-1)$ (d) $(f \circ g)(4)$

10. (a) $(g \circ f)(1)$ (b) $(g \circ f)(5)$
    (c) $(f \circ g)(0)$ (d) $(f \circ g)(2)$

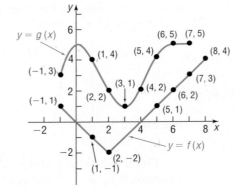

In Problems 11–20, for the given functions f and g, find:

(a) $(f \circ g)(4)$ (b) $(g \circ f)(2)$ (c) $(f \circ f)(1)$ (d) $(g \circ g)(0)$

11. $f(x) = 2x$; $g(x) = 3x^2 + 1$
12. $f(x) = 3x + 2$; $g(x) = 2x^2 - 1$

13. $f(x) = 4x^2 - 3$; $g(x) = 3 - \dfrac{1}{2}x^2$
14. $f(x) = 2x^2$; $g(x) = 1 - 3x^2$

15. $f(x) = \sqrt{x}$; $g(x) = 2x$
16. $f(x) = \sqrt{x + 1}$; $g(x) = 3x$

17. $f(x) = |x|$; $g(x) = \dfrac{1}{x^2 + 1}$
18. $f(x) = |x - 2|$; $g(x) = \dfrac{3}{x^2 + 2}$

19. $f(x) = \dfrac{3}{x + 1}$; $g(x) = \sqrt[3]{x}$
20. $f(x) = x^{3/2}$; $g(x) = \dfrac{2}{x + 1}$

In Problems 21–28, find the domain of the composite function $f \circ g$.

21. $f(x) = \dfrac{3}{x - 1}$; $g(x) = \dfrac{2}{x}$
22. $f(x) = \dfrac{1}{x + 3}$; $g(x) = -\dfrac{2}{x}$

23. $f(x) = \dfrac{x}{x-1}$; $g(x) = -\dfrac{4}{x}$

24. $f(x) = \dfrac{x}{x+3}$; $g(x) = \dfrac{2}{x}$

25. $f(x) = \sqrt{x}$; $g(x) = 2x + 3$

26. $f(x) = x - 2$; $g(x) = \sqrt{1-x}$

27. $f(x) = x^2 + 1$; $g(x) = \sqrt{x-1}$

28. $f(x) = x^2 + 4$; $g(x) = \sqrt{x-2}$

In Problems 29–44, for the given functions f and g, find:
  (a) f ∘ g   *(b) g ∘ f*   *(c) f ∘ f*   *(d) g ∘ g*
State the domain of each composite function.

29. $f(x) = 2x + 3$; $g(x) = 3x$

30. $f(x) = -x$; $g(x) = 2x - 4$

31. $f(x) = 3x + 1$; $g(x) = x^2$

32. $f(x) = x + 1$; $g(x) = x^2 + 4$

33. $f(x) = x^2$; $g(x) = x^2 + 4$

34. $f(x) = x^2 + 1$; $g(x) = 2x^2 + 3$

35. $f(x) = \dfrac{3}{x-1}$; $g(x) = \dfrac{2}{x}$

36. $f(x) = \dfrac{1}{x+3}$; $g(x) = -\dfrac{2}{x}$

37. $f(x) = \dfrac{x}{x-1}$; $g(x) = -\dfrac{4}{x}$

38. $f(x) = \dfrac{x}{x+3}$; $g(x) = \dfrac{2}{x}$

39. $f(x) = \sqrt{x}$; $g(x) = 2x + 3$

40. $f(x) = \sqrt{x-2}$; $g(x) = 1 - 2x$

41. $f(x) = x^2 + 1$; $g(x) = \sqrt{x-1}$

42. $f(x) = x^2 + 4$; $g(x) = \sqrt{x-2}$

43. $f(x) = \dfrac{x-5}{x+1}$; $g(x) = \dfrac{x+2}{x-3}$

44. $f(x) = \dfrac{2x-1}{x-2}$; $g(x) = \dfrac{x+4}{2x-5}$

In Problems 45–52, show that $(f \circ g)(x) = (g \circ f)(x) = x$.

45. $f(x) = 2x$; $g(x) = \dfrac{1}{2}x$

46. $f(x) = 4x$; $g(x) = \dfrac{1}{4}x$

47. $f(x) = x^3$; $g(x) = \sqrt[3]{x}$

48. $f(x) = x + 5$; $g(x) = x - 5$

49. $f(x) = 2x - 6$; $g(x) = \dfrac{1}{2}(x + 6)$

50. $f(x) = 4 - 3x$; $g(x) = \dfrac{1}{3}(4 - x)$

51. $f(x) = ax + b$; $g(x) = \dfrac{1}{a}(x - b)$ $a \neq 0$

52. $f(x) = \dfrac{1}{x}$; $g(x) = \dfrac{1}{x}$

In Problems 53–58, find functions f and g so that $f \circ g = H$.

53. $H(x) = (2x + 3)^4$

54. $H(x) = (1 + x^2)^3$

55. $H(x) = \sqrt{x^2 + 1}$

56. $H(x) = \sqrt{1 - x^2}$

57. $H(x) = |2x + 1|$

58. $H(x) = |2x^2 + 3|$

Applications and Extensions

59. If $f(x) = 2x^3 - 3x^2 + 4x - 1$ and $g(x) = 2$, find $(f \circ g)(x)$ and $(g \circ f)(x)$.

60. If $f(x) = \dfrac{x+1}{x-1}$, find $(f \circ f)(x)$.

61. If $f(x) = 2x^2 + 5$ and $g(x) = 3x + a$, find a so that the graph of $f \circ g$ crosses the y-axis at 23.

62. If $f(x) = 3x^2 - 7$ and $g(x) = 2x + a$, find a so that the graph of $f \circ g$ crosses the y-axis at 68.

In Problems 63 and 64, use the functions f and g to find:
  (a) f ∘ g   *(b) g ∘ f*
  (c) the domain of f ∘ g and of g ∘ f
  (d) the conditions for which f ∘ g = g ∘ f

63. $f(x) = ax + b$; $g(x) = cx + d$

64. $f(x) = \dfrac{ax + b}{cx + d}$; $g(x) = mx$

65. Surface Area of a Balloon The surface area S (in square meters) of a hot-air balloon is given by

$$S(r) = 4\pi r^2$$

where r is the radius of the balloon (in meters). If the radius r is increasing with time t (in seconds) according to the formula $r(t) = \dfrac{2}{3}t^3$, $t \geq 0$, find the surface area S of the balloon as a function of the time t.

66. Volume of a Balloon The volume V (in cubic meters) of the hot-air balloon described in Problem 65 is given by $V(r) = \dfrac{4}{3}\pi r^3$. If the radius r is the same function of t as in Problem 65, find the volume V as a function of the time t.

67. Automobile Production The number N of cars produced at a certain factory in one day after t hours of operation is given by $N(t) = 100t - 5t^2, 0 \leq t \leq 10$. If the cost C

(in dollars) of producing N cars is $C(N) = 15{,}000 + 8000N$, find the cost C as a function of the time t of operation of the factory.

68. Environmental Concerns The spread of oil leaking from a tanker is in the shape of a circle. If the radius r (in feet) of the spread after t hours is $r(t) = 200\sqrt{t}$, find the area A of the oil slick as a function of the time t.

69. Production Cost The price p, in dollars of a certain product and the quantity x sold obey the demand equation

$$p = -\frac{1}{4}x + 100 \quad 0 \le x \le 400$$

Suppose that the cost C, in dollars, of producing x units is

$$C = \frac{\sqrt{x}}{25} + 600$$

Assuming that all items produced are sold, find the cost C as a function of the price p.

[Hint: Solve for x in the demand equation and then form the composite.]

70. Cost of a Commodity The price p, in dollars, of a certain commodity and the quantity x sold obey the demand equation

$$p = -\frac{1}{5}x + 200 \quad 0 \le x \le 1000$$

Suppose that the cost C, in dollars, of producing x units is

$$C = \frac{\sqrt{x}}{10} + 400$$

Assuming that all items produced are sold, find the cost C as a function of the price p.

71. Volume of a Cylinder The volume V of a right circular cylinder of height h and radius r is $V = \pi r^2 h$. If the height is twice the radius, express the volume V as a function of r.

72. Volume of a Cone The volume V of a right circular cone is $V = \frac{1}{3}\pi r^2 h$. If the height is twice the radius, express the volume V as a function of r.

73. Foreign Exchange Traders often buy foreign currency in hope of making money when the currency's value changes. For example, on June 5, 2009, one U.S. dollar could purchase 0.7143 Euros, and one Euro could purchase 137.402 yen. Let $f(x)$ represent the number of Euros you can buy with x dollars, and let $g(x)$ represent the number of yen you can buy with x Euros.
 (a) Find a function that relates dollars to Euros.
 (b) Find a function that relates Euros to yen.
 (c) Use the results of parts (a) and (b) to find a function that relates dollars to yen. That is, find $(g \circ f)(x) = g(f(x))$.
 (d) What is $g(f(1000))$?

74. Temperature Conversion The function $C(F) = \frac{5}{9}(F - 32)$ converts a temperature in degrees Fahrenheit, F, to a temperature in degrees Celsius, C. The function $K(C) = C + 273$, converts a temperature in degrees Celsius to a temperature in kelvins, K.
 (a) Find a function that converts a temperature in degrees Fahrenheit to a temperature in kelvins.
 (b) Determine 80 degrees Fahrenheit in kelvins.

75. Discounts The manufacturer of a computer is offering two discounts on last year's model computer. The first discount is a \$200 rebate and the second discount is 20% off the regular price, p.
 (a) Write a function f that represents the sale price if only the rebate applies.
 (b) Write a function g that represents the sale price if only the 20% discount applies.
 (c) Find $f \circ g$ and $g \circ f$. What does each of these functions represent? Which combination of discounts represents a better deal for the consumer? Why?

76. If f and g are odd functions, show that the composite function $f \circ g$ is also odd.

77. If f is an odd function and g is an even function, show that the composite functions $f \circ g$ and $g \circ f$ are both even.

'Are You Prepared?' Answers

1. -21　　**2.** $4 - 18x^2; 3x$　　**3.** $\{x \mid x \ne -5, x \ne 5\}$

4.2 One-to-One Functions; Inverse Functions

PREPARING FOR THIS SECTION *Before getting started, review the following:*

- Functions (Section 1.1, pp. 43–51)
- Increasing/Decreasing Functions (Section 1.3, pp. 69–71)
- Rational Expressions (Appendix A, Section A.6, pp. A52–A53)

Now Work the 'Are You Prepared?' problems on page 292.

OBJECTIVES　**1** Determine Whether a Function Is One-to-One　(p. 283)
　　　　　　　　2 Determine the Inverse of a Function Defined by a Map or a Set of Ordered Pairs　(p. 285)
　　　　　　　　3 Obtain the Graph of the Inverse Function from the Graph of the Function　(p. 287)
　　　　　　　　4 Find the Inverse of a Function Defined by an Equation　(p. 288)

1 Determine Whether a Function Is One-to-One

In Section 1.1, we presented four different ways to represent a function: as (1) a map, (2) a set of ordered pairs, (3) a graph, and (4) an equation. For example, Figures 6 and 7 illustrate two different functions represented as mappings. The function in Figure 6 shows the correspondence between states and their population (in millions). The function in Figure 7 shows a correspondence between animals and life expectancy (in years).

Figure 6

Figure 7

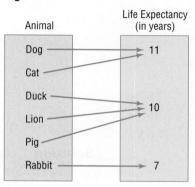

Suppose that we asked a group of people to name the state that has a population of 0.8 million based on the function in Figure 6. Everyone in the group would respond South Dakota. Now, if we asked the same group of people to name the animal whose life expectancy is 11 years based on the function in Figure 7, some would respond dog, while others would respond cat. What is the difference between the functions in Figures 6 and 7? In Figure 6, we can see that no two elements in the domain correspond to the same element in the range. In Figure 7, this is not the case: two different elements in the domain correspond to the same element in the range. We give functions such as the one in Figure 6 a special name.

DEFINITION

A function is **one-to-one** if any two different inputs in the domain correspond to two different outputs in the range. That is, if x_1 and x_2 are two different inputs of a function f, then f is one-to-one if $f(x_1) \neq f(x_2)$.

In Words

A function is not one-to-one if two different inputs correspond to the same output.

Put another way, a function f is one-to-one if no y in the range is the image of more than one x in the domain. A function is not one-to-one if two different elements in the domain correspond to the same element in the range. So the function in Figure 7 is not one-to-one because two different elements in the domain, *dog* and *cat*, both correspond to 11. Figure 8 illustrates the distinction of one-to-one functions, functions that are not one-to-one, and relations that are not functions.

Figure 8

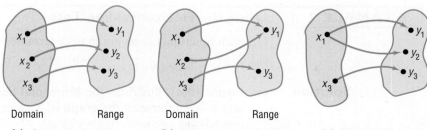

(a) One-to-one function: Each *x* in the domain has one and only one image in the range.

(b) Not a one-to-one function: y_1 is the image of both x_1 and x_2.

(c) Not a function: x_1 has two images, y_1 and y_2.

EXAMPLE 1   Determining Whether a Function Is One-to-One

Determine whether the following functions are one-to-one.

(a) For the following function, the domain represents the age of five males and the range represents their HDL (good) cholesterol (mg/dL).

(b) $\{(-2, 6), (-1, 3), (0, 2), (1, 5), (2, 8)\}$

Solution   (a) The function is not one-to-one because there are two different inputs, 55 and 61, that correspond to the same output, 38.

(b) The function is one-to-one because there are no two distinct inputs that correspond to the same output.

●

━━━ **Now Work** PROBLEMS 11 AND 15

For functions defined by an equation $y = f(x)$ and for which the graph of f is known, there is a simple test, called the **horizontal-line test,** to determine whether f is one-to-one.

THEOREM   **Horizontal-line Test**

Figure 9
$f(x_1) = f(x_2) = h$ and $x_1 \neq x_2$; f is not a one-to-one function.

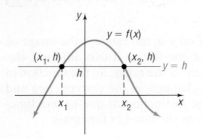

If every horizontal line intersects the graph of a function f in at most one point, then f is one-to-one.

The reason that this test works can be seen in Figure 9, where the horizontal line $y = h$ intersects the graph at two distinct points, (x_1, h) and (x_2, h). Since h is the image of both x_1 and x_2 and $x_1 \neq x_2$, f is not one-to-one. Based on Figure 9, we can state the horizontal-line test in another way: If the graph of any horizontal line intersects the graph of a function f at more than one point, then f is not one-to-one.

EXAMPLE 2   Using the Horizontal-line Test

For each function, use its graph to determine whether the function is one-to-one.
(a) $f(x) = x^2$ (b) $g(x) = x^3$

Solution   (a) Figure 10(a) illustrates the horizontal-line test for $f(x) = x^2$. The horizontal line $y = 1$ intersects the graph of f twice, at $(1, 1)$ and at $(-1, 1)$, so f is not one-to-one.

(b) Figure 10(b) illustrates the horizontal-line test for $g(x) = x^3$. Because every horizontal line intersects the graph of g exactly once, it follows that g is one-to-one.

Figure 10

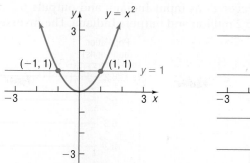

(a) A horizontal line intersects the graph twice; *f* is not one-to-one

(b) Every horizontal line intersects the graph exactly once; *g* is one-to-one

Now Work PROBLEM 19

Let's look more closely at the one-to-one function $g(x) = x^3$. This function is an increasing function. Because an increasing (or decreasing) function will always have different *y*-values for unequal *x*-values, it follows that a function that is increasing (or decreasing) over its domain is also a one-to-one function.

THEOREM

A function that is increasing on an interval *I* is a one-to-one function on *I*.
A function that is decreasing on an interval *I* is a one-to-one function on *I*.

2 Determine the Inverse of a Function Defined by a Map or a Set of Ordered Pairs

We begin with the definition.

DEFINITION

Suppose that *f* is a one-to-one function. Then, to each *x* in the domain of *f*, there is exactly one *y* in the range (because *f* is a function); and to each *y* in the range of *f*, there is exactly one *x* in the domain (because *f* is one-to-one). The correspondence from the range of *f* back to the domain of *f* is called the **inverse function of f.** We use the symbol f^{-1} to denote the inverse of *f*.

In Words

Suppose that we have a one-to-one function *f* where the input 5 corresponds to the output 10. In the inverse function f^{-1}, the input 10 would correspond to the output 5.

Recall that we have a variety of ways of representing functions. We will discuss how to find inverses for all four representations of functions: (1) maps, (2) sets of ordered pairs, (3) graphs, and (4) equations. We begin with finding inverses of functions represented by maps or sets of ordered pairs.

EXAMPLE 3 Finding the Inverse of a Function Defined by a Map

Find the inverse of the following function. Let the domain of the function represent certain states, and let the range represent the state's population (in millions). State the domain and the range of the inverse function.

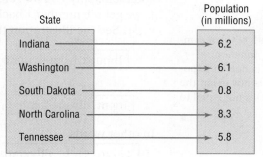

Solution The function is one-to-one. To find the inverse function, we interchange the elements in the domain with the elements in the range. For example, the function

receives as input Indiana and outputs 6.2 million. So the inverse receives as input 6.2 million and outputs Indiana. The inverse function is shown next.

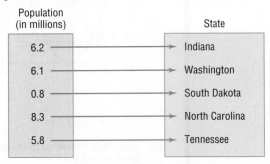

The domain of the inverse function is {6.2, 6.1, 0.8, 8.3, 5.8}. The range of the inverse function is {Indiana, Washington, South Dakota, North Carolina, Tennessee}.

If the function f is a set of ordered pairs (x, y), then the inverse of f, denoted f^{-1}, is the set of ordered pairs (y, x).

EXAMPLE 4

Finding the Inverse of a Function Defined by a Set of Ordered Pairs

Find the inverse of the following one-to-one function:

$$\{(-3, -27), (-2, -8), (-1, -1), (0, 0), (1, 1), (2, 8), (3, 27)\}$$

State the domain and the range of the function and its inverse.

Solution

The inverse of the given function is found by interchanging the entries in each ordered pair and so is given by

$$\{(-27, -3), (-8, -2), (-1, -1), (0, 0), (1, 1), (8, 2), (27, 3)\}$$

The domain of the function is $\{-3, -2, -1, 0, 1, 2, 3\}$. The range of the function is $\{-27, -8, -1, 0, 1, 8, 27\}$. The domain of the inverse function is $\{-27, -8, -1, 0, 1, 8, 27\}$. The range of the inverse function is $\{-3, -2, -1, 0, 1, 2, 3\}$.

$\blacktriangleright$ **Now Work** PROBLEMS 25 AND 29

Remember, if f is a one-to-one function, it has an inverse function, f^{-1}. See Figure 11.

Based on the results of Example 4 and Figure 11, two facts are now apparent about a one-to-one function f and its inverse f^{-1}.

Figure 11

Domain of f Range of f

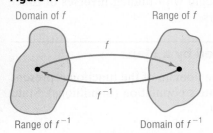

Range of f^{-1} Domain of f^{-1}

WARNING Be careful! f^{-1} is a symbol for the inverse function of f. The -1 used in f^{-1} is not an exponent. That is, f^{-1} does not mean the reciprocal of f; $f^{-1}(x)$ is not equal to $\dfrac{1}{f(x)}$. ∎

Domain of f = Range of f^{-1}	Range of f = Domain of f^{-1}

Look again at Figure 11 to visualize the relationship. If we start with x, apply f, and then apply f^{-1}, we get x back again. If we start with x, apply f^{-1}, and then apply f, we get the number x back again. To put it simply, what f does, f^{-1} undoes, and vice versa. See the illustration that follows.

| Input x from domain of f | $\xrightarrow{\text{Apply } f}$ | $f(x)$ | $\xrightarrow{\text{Apply } f^{-1}}$ | $f^{-1}(f(x)) = x$ |

| Input x from domain of f^{-1} | $\xrightarrow{\text{Apply } f^{-1}}$ | $f^{-1}(x)$ | $\xrightarrow{\text{Apply } f}$ | $f(f^{-1}(x)) = x$ |

In other words,

$$f^{-1}(f(x)) = x \quad \text{where } x \text{ is in the domain of } f$$
$$f(f^{-1}(x)) = x \quad \text{where } x \text{ is in the domain of } f^{-1}$$

Consider the function $f(x) = 2x$, which multiplies the argument x by 2. The inverse function f^{-1} undoes whatever f does. So the inverse function of f is is $f^{-1}(x) = \dfrac{1}{2}x$, which divides the argument by 2. For example, $f(3) = 2(3) = 6$ and $f^{-1}(6) = \dfrac{1}{2}(6) = 3$, so f^{-1} undoes what f did. We can verify this by showing that

$$f^{-1}(f(x)) = f^{-1}(2x) = \frac{1}{2}(2x) = x \quad \text{and} \quad f(f^{-1}(x)) = f\left(\frac{1}{2}x\right) = 2\left(\frac{1}{2}x\right) = x$$

See Figure 12.

Figure 12

$f^{-1}(2x) = \frac{1}{2}(2x) = x$

EXAMPLE 5 **Verifying Inverse Functions**

(a) We verify that the inverse of $g(x) = x^3$ is $g^{-1}(x) = \sqrt[3]{x}$ by showing that

$$g^{-1}(g(x)) = g^{-1}(x^3) = \sqrt[3]{x^3} = x \quad \text{for all } x \text{ in the domain of } g$$
$$g(g^{-1}(x)) = g(\sqrt[3]{x}) = (\sqrt[3]{x})^3 = x \quad \text{for all } x \text{ in the domain of } g^{-1}$$

(b) We verify that the inverse of $f(x) = 2x + 3$ is $f^{-1}(x) = \dfrac{1}{2}(x - 3)$ by showing that

$$f^{-1}(f(x)) = f^{-1}(2x + 3) = \frac{1}{2}[(2x + 3) - 3] = \frac{1}{2}(2x) = x \qquad \text{for all } x \text{ in the domain of } f$$

$$f(f^{-1}(x)) = f\left(\frac{1}{2}(x - 3)\right) = 2\left[\frac{1}{2}(x - 3)\right] + 3 = (x - 3) + 3 = x \quad \text{for all } x \text{ in the domain of } f^{-1}$$

EXAMPLE 6 **Verifying Inverse Functions**

Verify that the inverse of $f(x) = \dfrac{1}{x - 1}$ is $f^{-1}(x) = \dfrac{1}{x} + 1$. For what values of x is $f^{-1}(f(x)) = x$? For what values of x is $f(f^{-1}(x)) = x$?

Solution The domain of f is $\{x \mid x \neq 1\}$ and the domain of f^{-1} is $\{x \mid x \neq 0\}$. Now

$$f^{-1}(f(x)) = f^{-1}\left(\frac{1}{x - 1}\right) = \frac{1}{\dfrac{1}{x - 1}} + 1 = x - 1 + 1 = x \qquad \text{provided } x \neq 1$$

$$f(f^{-1}(x)) = f\left(\frac{1}{x} + 1\right) = \frac{1}{\dfrac{1}{x} + 1 - 1} = \frac{1}{\dfrac{1}{x}} = x \qquad \text{provided } x \neq 0$$

Now Work PROBLEM 33

3 Obtain the Graph of the Inverse Function from the Graph of the Function

Figure 13

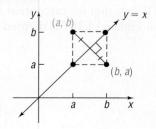

Suppose that (a, b) is a point on the graph of a one-to-one function f defined by $y = f(x)$. Then $b = f(a)$. This means that $a = f^{-1}(b)$, so (b, a) is a point on the graph of the inverse function f^{-1}. The relationship between the point (a, b) on f and the point (b, a) on f^{-1} is shown in Figure 13. The line segment with endpoints (a, b) and (b, a) is perpendicular to the line $y = x$ and is bisected by the line $y = x$. (Do you see why?) It follows that the point (b, a) on f^{-1} is the reflection about the line $y = x$ of the point (a, b) on f.

THEOREM

> The graph of a one-to-one function f and the graph of its inverse f^{-1} are symmetric with respect to the line $y = x$.

Figure 14 illustrates this result. Notice that, once the graph of f is known, the graph of f^{-1} may be obtained by reflecting the graph of f about the line $y = x$.

Figure 14

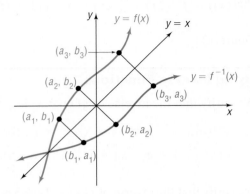

EXAMPLE 7   **Graphing the Inverse Function**

The graph in Figure 15(a) is that of a one-to-one function $y = f(x)$. Draw the graph of its inverse.

Solution   We begin by adding the graph of $y = x$ to Figure 15(a). Since the points $(-2, -1), (-1, 0)$, and $(2, 1)$ are on the graph of f, we know that the points $(-1, -2), (0, -1)$, and $(1, 2)$ must be on the graph of f^{-1}. Keeping in mind that the graph of f^{-1} is the reflection about the line $y = x$ of the graph of f, we can draw f^{-1}. See Figure 15(b).

Figure 15

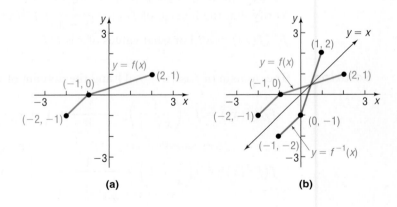

(a)   (b)

Now Work   PROBLEM 43

4 Find the Inverse of a Function Defined by an Equation

The fact that the graphs of a one-to-one function f and its inverse function f^{-1} are symmetric with respect to the line $y = x$ tells us more. It says that we can obtain f^{-1} by interchanging the roles of x and y in f. Look again at Figure 14. If f is defined by the equation

$$y = f(x)$$

then f^{-1} is defined by the equation

$$x = f(y)$$

The equation $x = f(y)$ defines f^{-1} *implicitly*. If we can solve this equation for y, we will have the *explicit* form of f^{-1}, that is,

$$y = f^{-1}(x)$$

Let's use this procedure to find the inverse of $f(x) = 2x + 3$. (Since f is a linear function and is increasing, we know that f is one-to-one and so has an inverse function.)

EXAMPLE 8

How to Find the Inverse Function

Find the inverse of $f(x) = 2x + 3$. Graph f and f^{-1} on the same coordinate axes.

Step-by-Step Solution

Step 1: Replace $f(x)$ with y. In $y = f(x)$, interchange the variables x and y to obtain $x = f(y)$. This equation defines the inverse function f^{-1} implicitly.

We replace $f(x)$ with y in $f(x) = 2x + 3$ and obtain $y = 2x + 3$. Now we interchange the variables x and y to obtain

$$x = 2y + 3$$

This equation defines the inverse f^{-1} implicitly.

Step 2: If possible, solve the implicit equation for y in terms of x to obtain the explicit form of f^{-1}, $y = f^{-1}(x)$.

To find the explicit form of the inverse, we solve $x = 2y + 3$ for y.

$$x = 2y + 3$$
$$2y + 3 = x \qquad \text{Reflexive Property; If } a = b, \text{ then } b = a.$$
$$2y = x - 3 \qquad \text{Subtract 3 from both sides.}$$
$$y = \frac{1}{2}(x - 3) \qquad \text{Divide both sides by 2.}$$

The explicit form of the inverse f^{-1} is therefore

$$f^{-1}(x) = \frac{1}{2}(x - 3)$$

Step 3: Check the result by showing that $f^{-1}(f(x)) = x$ and $f(f^{-1}(x)) = x$.

We verified that f and f^{-1} are inverses in Example 5(b).

The graphs of $f(x) = 2x + 3$ and its inverse $f^{-1}(x) = \frac{1}{2}(x - 3)$ are shown in Figure 16. Note the symmetry of the graphs with respect to the line $y = x$.

Figure 16

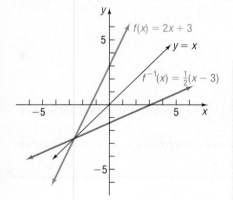

We now list the steps to follow for finding the inverse of a one-to-one function.

Procedure for Finding the Inverse of a One-to-One Function

STEP 1: In $y = f(x)$, interchange the variables x and y to obtain

$$x = f(y)$$

This equation defines the inverse function f^{-1} implicitly.

STEP 2: If possible, solve the implicit equation for y in terms of x to obtain the explicit form of f^{-1}:

$$y = f^{-1}(x)$$

STEP 3: Check the result by showing that

$$f^{-1}(f(x)) = x \quad \text{and} \quad f(f^{-1}(x)) = x$$

EXAMPLE 9   **Finding the Inverse Function**

The function

$$f(x) = \frac{2x + 1}{x - 1} \quad x \neq 1$$

is one-to-one. Find its inverse and check the result.

Solution   **STEP 1:** Replace $f(x)$ with y and interchange the variables x and y in

$$y = \frac{2x + 1}{x - 1}$$

to obtain

$$x = \frac{2y + 1}{y - 1}$$

STEP 2: Solve for y.

$$x = \frac{2y + 1}{y - 1}$$

$x(y - 1) = 2y + 1$ Multiply both sides by $y - 1$.

$xy - x = 2y + 1$ Apply the Distributive Property.

$xy - 2y = x + 1$ Subtract $2y$ from both sides; add x to both sides.

$(x - 2)y = x + 1$ Factor.

$$y = \frac{x + 1}{x - 2}$$ Divide by $x - 2$.

The inverse is

$$f^{-1}(x) = \frac{x + 1}{x - 2} \quad x \neq 2 \quad \text{Replace } y \text{ by } f^{-1}(x).$$

STEP 3: ✓Check:

$$f^{-1}(f(x)) = f^{-1}\left(\frac{2x + 1}{x - 1}\right) = \frac{\dfrac{2x + 1}{x - 1} + 1}{\dfrac{2x + 1}{x - 1} - 2} = \frac{2x + 1 + x - 1}{2x + 1 - 2(x - 1)} = \frac{3x}{3} = x \quad x \neq 1$$

$$f(f^{-1}(x)) = f\left(\frac{x + 1}{x - 2}\right) = \frac{2\left(\dfrac{x + 1}{x - 2}\right) + 1}{\dfrac{x + 1}{x - 2} - 1} = \frac{2(x + 1) + x - 2}{x + 1 - (x - 2)} = \frac{3x}{3} = x \quad x \neq 2$$

●

Exploration

In Example 9, we found that, if $f(x) = \dfrac{2x + 1}{x - 1}$, then $f^{-1}(x) = \dfrac{x + 1}{x - 2}$. Compare the vertical and horizontal asymptotes of f and f^{-1}.

Result The vertical asymptote of f is $x = 1$, and the horizontal asymptote is $y = 2$. The vertical asymptote of f^{-1} is $x = 2$, and the horizontal asymptote is $y = 1$.

━━━ **Now Work** PROBLEM 51

We said in Chapter 1 that finding the range of a function f is not easy. However, if f is one-to-one, we can find its range by finding the domain of the inverse function f^{-1}.

EXAMPLE 10   **Finding the Range of a One-to-One Function**

Find the domain and the range of

$$f(x) = \frac{2x + 1}{x - 1}$$

Solution   The domain of f is $\{x | x \neq 1\}$. To find the range of f, we use the fact that the domain of f^{-1} equals the range of f. Based on Example 9, we have

$$f^{-1}(x) = \frac{x + 1}{x - 2}$$

The domain of f^{-1} is $\{x | x \neq 2\}$, so the range of f is $\{y | y \neq 2\}$. Also, because the domain of f is $\{x | x \neq 1\}$, the range of f^{-1} is $\{y | y \neq 1\}$.

 Now Work   PROBLEM 65

If a function is not one-to-one, it has no inverse function. Sometimes, though, an appropriate restriction on the domain of such a function will yield a new function that is one-to-one. Then the function defined on the restricted domain has an inverse function. Let's look at an example of this common practice.

EXAMPLE 11   **Finding the Inverse of a Domain-restricted Function**

Find the inverse of $y = f(x) = x^2$ if $x \geq 0$.

Solution   The function $y = x^2$ is not one-to-one. [Refer to Example 2(a).] However, if we restrict the domain of this function to $x \geq 0$, as indicated, we have a new function that is increasing and therefore is one-to-one. As a result, the function defined by $y = f(x) = x^2$, $x \geq 0$, has an inverse function, f^{-1}.
We follow the steps given previously to find f^{-1}.

STEP 1:   In the equation $y = x^2$, $x \geq 0$, interchange the variables x and y. The result is

$$x = y^2 \qquad y \geq 0$$

This equation defines (implicitly) the inverse function.

Figure 17

STEP 2:   We solve for y to get the explicit form of the inverse. Since $y \geq 0$, only one solution for y is obtained: $y = \sqrt{x}$. So $f^{-1}(x) = \sqrt{x}$.

STEP 3:   ✓ **Check:** $f^{-1}(f(x)) = f^{-1}(x^2) = \sqrt{x^2} = |x| = x$ since $x \geq 0$
$$f(f^{-1}(x)) = f(\sqrt{x}) = (\sqrt{x})^2 = x$$

Figure 17 illustrates the graphs of $f(x) = x^2$, $x \geq 0$, and $f^{-1}(x) = \sqrt{x}$.

SUMMARY

1. If a function f is one-to-one, then it has an inverse function f^{-1}.
2. Domain of f = Range of f^{-1}; Range of f = Domain of f^{-1}.
3. To verify that f^{-1} is the inverse of f, show that $f^{-1}(f(x)) = x$ for every x in the domain of f and $f(f^{-1}(x)) = x$ for every x in the domain of f^{-1}.
4. The graphs of f and f^{-1} are symmetric with respect to the line $y = x$.
5. To find the range of a one-to-one function f, find the domain of its inverse function f^{-1}.

4.2 Assess Your Understanding

1. Is the set of ordered pairs $\{(1, 3), (2, 3), (-1, 2)\}$ a function? Why or why not? (pp. 43–46)

2. Where is the function $f(x) = x^2$ increasing? Where is it decreasing? (pp. 69–70)

3. What is the domain of $f(x) = \dfrac{x + 5}{x^2 + 3x - 18}$? (pp. 49–50)

4. Simplify: $\dfrac{\dfrac{1}{x} + 1}{\dfrac{1}{x^2} - 1}$ (pp. A52–A53)

Concepts and Vocabulary

5. If x_1 and x_2 are two different inputs of a function f, then f is one-to-one if _____.

6. If every horizontal line intersects the graph of a function f at no more than one point, f is a(n) _____ function.

7. If f is a one-to-one function and $f(3) = 8$, then $f^{-1}(8) =$ _____.

8. If f^{-1} denotes the inverse of a function f, then the graphs of f and f^{-1} are symmetric with respect to the line _____.

9. If the domain of a one-to-one function f is $[4, \infty)$, the range of its inverse, f^{-1}, is _____.

10. *True or False* If f and g are inverse functions, the domain of f is the same as the range of g.

Skill Building

In Problems 11–18, determine whether the function is one-to-one.

11.

12.

13.

14.

15. $\{(2, 6), (-3, 6), (4, 9), (1, 10)\}$

16. $\{(-2, 5), (-1, 3), (3, 7), (4, 12)\}$

17. $\{(0, 0), (1, 1), (2, 16), (3, 81)\}$

18. $\{(1, 2), (2, 8), (3, 18), (4, 32)\}$

In Problems 19–24, the graph of a function f is given. Use the horizontal-line test to determine whether f is one-to-one.

19.

20.

21.

22.

23.

24.

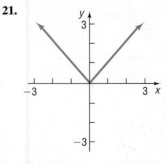

In Problems 25–32, find the inverse of each one-to-one function. State the domain and the range of each inverse function.

25.

Location	Annual Rainfall (inches)
Mt Waialeale, Hawaii	460.00
Monrovia, Liberia	202.01
Pago Pago, American Samoa	196.46
Moulmein, Burma	191.02
Lae, Papua New Guinea	182.87

Source: Information Please Almanac

26.

Title	Domestic Gross (in millions)
Star Wars	$461
Star Wars: Episode One – The Phantom Menace	$431
E.T. the Extra Terrestrial	$400
Jurassic Park	$357
Forrest Gump	$330

Source: Information Please Almanac

27.

Age	Monthly Cost of Life Insurance
30	$7.09
40	$8.40
45	$11.29

Source: eterm.com

28.

State	Unemployment Rate
Virginia	11%
Nevada	5.5%
Tennessee	5.1%
Texas	6.3%

Source: United States Statistical Abstract

29. $\{(-3, 5), (-2, 9), (-1, 2), (0, 11), (1, -5)\}$

30. $\{(-2, 2), (-1, 6), (0, 8), (1, -3), (2, 9)\}$

31. $\{(-2, 1), (-3, 2), (-10, 0), (1, 9), (2, 4)\}$

32. $\{(-2, -8), (-1, -1), (0, 0), (1, 1), (2, 8)\}$

In Problems 33–42, verify that the functions f and g are inverses of each other by showing that $f(g(x)) = x$ and $g(f(x)) = x$. Give any values of x that need to be excluded from the domain of f and the domain of g.

33. $f(x) = 3x + 4; \quad g(x) = \frac{1}{3}(x - 4)$

34. $f(x) = 3 - 2x; \quad g(x) = -\frac{1}{2}(x - 3)$

35. $f(x) = 4x - 8; \quad g(x) = \frac{x}{4} + 2$

36. $f(x) = 2x + 6; \quad g(x) = \frac{1}{2}x - 3$

37. $f(x) = x^3 - 8; \quad g(x) = \sqrt[3]{x + 8}$

38. $f(x) = (x - 2)^2, x \geq 2; \quad g(x) = \sqrt{x} + 2$

39. $f(x) = \frac{1}{x}; \quad g(x) = \frac{1}{x}$

40. $f(x) = x; \quad g(x) = x$

41. $f(x) = \frac{2x + 3}{x + 4}; \quad g(x) = \frac{4x - 3}{2 - x}$

42. $f(x) = \frac{x - 5}{2x + 3}; \quad g(x) = \frac{3x + 5}{1 - 2x}$

In Problems 43–48, the graph of a one-to-one function f is given. Draw the graph of the inverse function f^{-1}. For convenience (and as a hint), the graph of $y = x$ is also given.

43.

44.

45.

46.

47.

48.

In Problems 49–60, the function f is one-to-one. Find its inverse and check your answer. Graph f, f^{-1}, and $y = x$ on the same coordinate axes.

49. $f(x) = 3x$

50. $f(x) = -4x$

51. $f(x) = 4x + 2$

52. $f(x) = 1 - 3x$

53. $f(x) = x^3 - 1$

54. $f(x) = x^3 + 1$

55. $f(x) = x^2 + 4 \quad x \geq 0$

56. $f(x) = x^2 + 9 \quad x \geq 0$

57. $f(x) = \dfrac{4}{x}$

58. $f(x) = -\dfrac{3}{x}$

59. $f(x) = \dfrac{1}{x - 2}$

60. $f(x) = \dfrac{4}{x + 2}$

In Problems 61–72, the function f is one-to-one. Find its inverse and check your answer. State the domain and the range of f and f^{-1}.

61. $f(x) = \dfrac{2}{3 + x}$

62. $f(x) = \dfrac{4}{2 - x}$

63. $f(x) = \dfrac{3x}{x + 2}$

64. $f(x) = -\dfrac{2x}{x - 1}$

65. $f(x) = \dfrac{2x}{3x - 1}$

66. $f(x) = -\dfrac{3x + 1}{x}$

67. $f(x) = \dfrac{3x + 4}{2x - 3}$

68. $f(x) = \dfrac{2x - 3}{x + 4}$

69. $f(x) = \dfrac{2x + 3}{x + 2}$

70. $f(x) = \dfrac{-3x - 4}{x - 2}$

71. $f(x) = \dfrac{x^2 - 4}{2x^2} \quad x > 0$

72. $f(x) = \dfrac{x^2 + 3}{3x^2} \quad x > 0$

Applications and Extensions

73. Use the graph of $y = f(x)$ given in Problem 43 to evaluate the following:

    (a) $f(-1)$ (b) $f(1)$ (c) $f^{-1}(1)$ (d) $f^{-1}(2)$

74. Use the graph of $y = f(x)$ given in Problem 44 to evaluate the following:

    (a) $f(2)$ (b) $f(1)$ (c) $f^{-1}(0)$ (d) $f^{-1}(-1)$

75. If $f(7) = 13$ and f is one-to-one, what is $f^{-1}(13)$?

76. If $g(-5) = 3$ and g is one-to-one, what is $g^{-1}(3)$?

77. The domain of a one-to-one function f is $[5, \infty)$, and its range is $[-2, \infty)$. State the domain and the range of f^{-1}.

78. The domain of a one-to-one function f is $[0, \infty)$, and its range is $[5, \infty)$. State the domain and the range of f^{-1}.

79. The domain of a one-to-one function g is $(-\infty, 0]$, and its range is $[0, \infty)$. State the domain and the range of g^{-1}.

80. The domain of a one-to-one function g is $[0, 15]$, and its range is $(0, 8)$. State the domain and the range of g^{-1}.

81. A function $y = f(x)$ is increasing on the interval $(0, 5)$. What conclusions can you draw about the graph of $y = f^{-1}(x)$?

82. A function $y = f(x)$ is decreasing on the interval $(0, 5)$. What conclusions can you draw about the graph of $y = f^{-1}(x)$?

83. Find the inverse of the linear function

$$f(x) = mx + b \quad m \neq 0$$

84. Find the inverse of the function

$$f(x) = \sqrt{r^2 - x^2} \quad 0 \leq x \leq r$$

85. A function f has an inverse function. If the graph of f lies in quadrant I, in which quadrant does the graph of f^{-1} lie?

86. A function f has an inverse function. If the graph of f lies in quadrant II, in which quadrant does the graph of f^{-1} lie?

87. The function $f(x) = |x|$ is not one-to-one. Find a suitable restriction on the domain of f so that the new function that results is one-to-one. Then find the inverse of f.

88. The function $f(x) = x^4$ is not one-to-one. Find a suitable restriction on the domain of f so that the new function that results is one-to-one. Then find the inverse of f.

In applications, the symbols used for the independent and dependent variables are often based on common usage. So, rather than using y = f(x) to represent a function, an applied problem might use C = C(q) to represent the cost C of manufacturing q units of a good since, in economics, q is used for output. Because of this, the inverse notation f⁻¹ used in a pure mathematics problem is not used when finding inverses of applied problems. Rather, the inverse of a function such as C = C(q) will be q = q(C). So C = C(q) is a function that represents the cost C as a function of the output q, while q = q(C) is a function that represents the output q as a function of the cost C. Problems 89–92 illustrate this idea.

89. Vehicle Stopping Distance Taking into account reaction time, the distance d (in feet) that a car requires to come to a complete stop while traveling r miles per hour is given by the function

$$d(r) = 6.97r - 90.39$$

(a) Express the speed r at which the car is traveling as a function of the distance d required to come to a complete stop.
(b) Verify that $r = r(d)$ is the inverse of $d = d(r)$ by showing that $r(d(r)) = r$ and $d(r(d)) = d$.
(c) Predict the speed that a car was traveling if the distance required to stop was 300 feet.

90. Height and Head Circumference The head circumference C of a child is related to the height H of the child (both in inches) through the function

$$H(C) = 2.15C - 10.53$$

(a) Express the head circumference C as a function of height H.
(b) Verify that $C = C(H)$ is the inverse of $H = H(C)$ by showing that $H(C(H)) = H$ and $C(H(C)) = C$.
(c) Predict the head circumference of a child who is 26 inches tall.

91. Ideal Body Weight One model for the ideal body weight W for men (in kilograms) as a function of height h (in inches) is given by the function

$$W(h) = 50 + 2.3(h - 60)$$

(a) What is the ideal weight of a 6-foot male?
(b) Express the height h as a function of weight W.
(c) Verify that $h = h(W)$ is the inverse of $W = W(h)$ by showing that $h(W(h)) = h$ and $W(h(W)) = W$.
(d) What is the height of a male who is at his ideal weight of 80 kilograms?

[**Note:** The ideal body weight W for women (in kilograms) as a function of height h (in inches) is given by $W(h) = 45.5 + 2.3(h - 60)$.]

92. Temperature Conversion The function $F(C) = \dfrac{9}{5}C + 32$ converts a temperature from C degrees Celsius to F degrees Fahrenheit.

(a) Express the temperature in degrees Celsius C as a function of the temperature in degrees Fahrenheit F.
(b) Verify that $C = C(F)$ is the inverse of $F = F(C)$ by showing that $C(F(C)) = C$ and $F(C(F)) = F$.
(c) What is the temperature in degrees Celsius if it is 70 degrees Fahrenheit?

93. Income Taxes The function

$$T(g) = 4675 + 0.25(g - 33,950)$$

represents the 2009 federal income tax T (in dollars) due for a "single" filer whose modified adjusted gross income is g dollars, where $33,950 \le g \le 82,250$.

(a) What is the domain of the function T?
(b) Given that the tax due T is an increasing linear function of modified adjusted gross income g, find the range of the function T.
(c) Find adjusted gross income g as a function of federal income tax T. What are the domain and the range of this function?

94. Income Taxes The function

$$T(g) = 1670 + 0.15(g - 16,700)$$

represents the 2009 federal income tax T (in dollars) due for a "married filing jointly" filer whose modified adjusted gross income is g dollars, where $16,700 \le g \le 67,900$.

(a) What is the domain of the function T?
(b) Given that the tax due T is an increasing linear function of modified adjusted gross income g, find the range of the function T.
(c) Find adjusted gross income g as a function of federal income tax T. What are the domain and the range of this function?

95. Gravity on Earth If a rock falls from a height of 100 meters on Earth, the height H (in meters) after t seconds is approximately

$$H(t) = 100 - 4.9t^2$$

(a) In general, quadratic functions are not one-to-one. However, the function H is one-to-one. Why?
(b) Find the inverse of H and verify your result.
(c) How long will it take a rock to fall 80 meters?

96. Period of a Pendulum The period T (in seconds) of a simple pendulum as a function of its length l (in feet) is given by

$$T(l) = 2\pi\sqrt{\dfrac{l}{32.2}}$$

(a) Express the length l as a function of the period T.
(b) How long is a pendulum whose period is 3 seconds?

97. Given

$$f(x) = \dfrac{ax + b}{cx + d}$$

find $f^{-1}(x)$. If $c \ne 0$, under what conditions on a, b, c, and d is $f = f^{-1}$?

Explaining Concepts: Discussion and Writing

98. Can a one-to-one function and its inverse be equal? What must be true about the graph of f for this to happen? Give some examples to support your conclusion.

99. Draw the graph of a one-to-one function that contains the points $(-2, -3)$, $(0, 0)$, and $(1, 5)$. Now draw the graph of its inverse. Compare your graph to those of other students. Discuss any similarities. What differences do you see?

100. Give an example of a function whose domain is the set of real numbers and that is neither increasing nor decreasing on its domain, but is one-to-one.

[**Hint:** Use a piecewise-defined function.]

101. Is every odd function one-to-one? Explain.

102. Suppose that $C(g)$ represents the cost C, in dollars, of manufacturing g cars. Explain what $C^{-1}(800,000)$ represents.

103. Explain why the horizontal-line test can be used to identify one-to-one functions from a graph.

'Are You Prepared?' Answers

1. Yes; for each input x there is one output y.

2. Increasing on $(0, \infty)$; decreasing on $(-\infty, 0)$

3. $\{x \mid x \neq -6, x \neq 3\}$

4. $\dfrac{x}{1 - x}, x \neq 0, x \neq -1$

4.3 Exponential Functions

PREPARING FOR THIS SECTION *Before getting started, review the following:*

- Exponents (Appendix A, Section A.1, pp. A7–A9, and Section A.7, pp. A60–A61)
- Graphing Techniques: Transformations (Section 1.5, pp. 90–99)
- Linear Equations (Appendix A, Section A.8, pp. A66–A68)

- Average Rate of Change (Section 1.3, pp. 73–75)
- Quadratic Functions (Section 2.3, pp. 141–149)
- Linear Functions (Section 2.1, pp. 123–127)
- Horizontal Asymptotes (Section 3.2, pp. 217–223)

Now Work the 'Are You Prepared?' problems on page 307.

OBJECTIVES **1** Evaluate Exponential Functions (p. 296)

2 Graph Exponential Functions (p. 300)

3 Define the Number e (p. 303)

4 Solve Exponential Equations (p. 305)

1 Evaluate Exponential Functions

In Appendix A, Section A.7, we give a definition for raising a real number a to a rational power. Based on that discussion, we gave meaning to expressions of the form

$$a^r$$

where the base a is a positive real number and the exponent r is a rational number.

 But what is the meaning of a^x, where the base a is a positive real number and the exponent x is an irrational number? Although a rigorous definition requires methods discussed in calculus, the basis for the definition is easy to follow: Select a rational number r that is formed by truncating (removing) all but a finite number of digits from the irrational number x. Then it is reasonable to expect that

$$a^x \approx a^r$$

For example, take the irrational number $\pi = 3.14159\ldots$. Then an approximation to a^π is

$$a^\pi \approx a^{3.14}$$

where the digits after the hundredths position have been removed from the value for π. A better approximation would be

$$a^\pi \approx a^{3.14159}$$

where the digits after the hundred-thousandths position have been removed. Continuing in this way, we can obtain approximations to a^π to any desired degree of accuracy.

Most calculators have an $\boxed{x^y}$ key or a caret key $\boxed{\wedge}$ for working with exponents. To evaluate expressions of the form a^x, enter the base a, then press the $\boxed{x^y}$ key (or the $\boxed{\wedge}$ key), enter the exponent x, and press $\boxed{=}$ (or $\boxed{\text{ENTER}}$).

EXAMPLE 1	Using a Calculator to Evaluate Powers of 2

Using a calculator, evaluate:

(a) $2^{1.4}$ (b) $2^{1.41}$ (c) $2^{1.414}$ (d) $2^{1.4142}$ (e) $2^{\sqrt{2}}$

Solution   (a) $2^{1.4} \approx 2.639015822$ (b) $2^{1.41} \approx 2.657371628$

(c) $2^{1.414} \approx 2.66474965$ (d) $2^{1.4142} \approx 2.665119089$

(e) $2^{\sqrt{2}} \approx 2.665144143$

●

Now Work PROBLEM 15

It can be shown that the familiar laws for rational exponents hold for real exponents.

THEOREM	**Laws of Exponents**

If s, t, a, and b are real numbers with $a > 0$ and $b > 0$, then

$$a^s \cdot a^t = a^{s+t} \qquad (a^s)^t = a^{st} \qquad (ab)^s = a^s \cdot b^s$$

$$1^s = 1 \qquad a^{-s} = \frac{1}{a^s} = \left(\frac{1}{a}\right)^s \qquad a^0 = 1 \qquad \qquad \textbf{(1)}$$

Introduction to Exponential Growth

Let's examine a function f that has the following two properties:

1. The value of f doubles with every 1-unit increase in the independent variable x.
2. The value of f at $x = 0$ is 5, so $f(0) = 5$.

Table 1

x	$f(x)$
0	5
1	10
2	20
3	40
4	80

Table 1 shows values of the function f for $x = 0, 1, 2, 3$, and 4.

We seek an equation $y = f(x)$ that describes this function f. The key is the fact that the value of f doubles for every 1-unit increase in x.

$$f(0) = 5$$
$$f(1) = 2f(0) = 2 \cdot 5 = 5 \cdot 2^1 \qquad \textit{We double the value of f at 0 to get the value at 1.}$$
$$f(2) = 2f(1) = 2(5 \cdot 2) = 5 \cdot 2^2 \qquad \textit{We double the value of f at 1 to get the value at 2.}$$
$$f(3) = 2f(2) = 2(5 \cdot 2^2) = 5 \cdot 2^3$$
$$f(4) = 2f(3) = 2(5 \cdot 2^3) = 5 \cdot 2^4$$

The pattern leads us to

$$f(x) = 2f(x - 1) = 2(5 \cdot 2^{x-1}) = 5 \cdot 2^x$$

We are now ready for the following definition.

DEFINITION	An **exponential function** is a function of the form

$$f(x) = Ca^x$$

where a is a positive real number $(a > 0)$ and $a \neq 1$, and $C \neq 0$ is a real number. The domain of f is the set of all real numbers. The base a is the **growth factor**, and because $f(0) = Ca^0 = C$, we call C the **initial value**.

WARNING It is important to distinguish a power function, $g(x) = ax^n$, $n \geq 2$, an integer, from an exponential function, $f(x) = C \cdot a^x$, $a \neq 1$, $a > 0$. In a power function, the base is a variable and the exponent is a constant. In an exponential function, the base is a constant and the exponent is a variable. ∎

In the definition of an exponential function, we exclude the base $a = 1$ because this function is simply the constant function $f(x) = C \cdot 1^x = C$. We also need to exclude bases that are negative; otherwise, we would have to exclude many values of x from the domain, such as $x = \dfrac{1}{2}$ and $x = \dfrac{3}{4}$. [Recall that $(-2)^{1/2} = \sqrt{-2}$, $(-3)^{3/4} = \sqrt[4]{(-3)^3} = \sqrt[4]{-27}$, and so on, are not defined in the set of real numbers.] Finally, transformations (vertical shifts, horizontal shifts, reflections, and so on) of a function of the form $f(x) = Ca^x$ also represent exponential functions.

Some examples of exponential functions are

$$f(x) = 2^x \qquad F(x) = \left(\frac{1}{3}\right)^x + 5 \qquad G(x) = 2 \cdot 3^{x-3}$$

Notice for each function that the base is a constant and the exponent contains a variable.

In the function $f(x) = 5 \cdot 2^x$, notice that the ratio of consecutive outputs is constant for 1-unit increases in the input. This ratio equals the constant 2, the base of the exponential function. In other words,

$$\frac{f(1)}{f(0)} = \frac{5 \cdot 2^1}{5} = 2 \qquad \frac{f(2)}{f(1)} = \frac{5 \cdot 2^2}{5 \cdot 2^1} = 2 \qquad \frac{f(3)}{f(2)} = \frac{5 \cdot 2^3}{5 \cdot 2^2} = 2 \quad \text{and so on}$$

This leads to the following result.

THEOREM

For an exponential function $f(x) = Ca^x$, $a > 0$, $a \neq 1$, if x is any real number, then

$$\frac{f(x + 1)}{f(x)} = a \quad \text{or} \quad f(x + 1) = af(x)$$

In Words For 1-unit changes in the input x of an exponential function $f(x) = C \cdot a^x$, the ratio of consecutive outputs is the constant a.

Proof

$$\frac{f(x + 1)}{f(x)} = \frac{Ca^{x+1}}{Ca^x} = a^{x+1-x} = a^1 = a \qquad ∎$$

EXAMPLE 2 | **Identifying Linear or Exponential Functions**

Determine whether the given function is linear, exponential, or neither. For those that are linear, find a linear function that models the data. For those that are exponential, find an exponential function that models the data.

(a)

x	y
−1	5
0	2
1	−1
2	−4
3	−7

(b)

x	y
−1	32
0	16
1	8
2	4
3	2

(c)

x	y
−1	2
0	4
1	7
2	11
3	16

Solution

For each function we compute the average rate of change of y with respect to x. If the average rate of change is constant, then the function is linear. Also, we compute the ratio of consecutive outputs. If the ratio is constant, then the function is exponential.

(a) See Table 2(a). The average rate of change for every 1-unit increase in x is -3. Therefore, the function is a linear function. In a linear function the average rate of change is the slope m, so $m = -3$. The y-intercept b is the value of the function at $x = 0$, so $b = 2$. The linear function that models the data is $f(x) = mx + b = -3x + 2$.

Table 2

x	y	Average Rate of Change	Ratio of Consecutive Outputs
−1	5	$\dfrac{\Delta y}{\Delta x}=\dfrac{2-5}{0-(-1)}=-3$	$\dfrac{2}{5}$
0	2	−3	$-\dfrac{1}{2}$
1	−1	−3	4
2	−4	−3	$\dfrac{7}{4}$
3	−7		

(a)

x	y	Average Rate of Change	Ratio of Consecutive Outputs
−1	32	$\dfrac{\Delta y}{\Delta x}=\dfrac{16-32}{0-(-1)}=-16$	$\dfrac{16}{32}=\dfrac{1}{2}$
0	16	−8	$\dfrac{8}{16}=\dfrac{1}{2}$
1	8	−4	$\dfrac{4}{8}=\dfrac{1}{2}$
2	4	−2	$\dfrac{2}{4}=\dfrac{1}{2}$
3	2		

(b)

x	y	Average Rate of Change	Ratio of Consecutive Outputs
−1	2	$\dfrac{\Delta y}{\Delta x}=\dfrac{4-2}{0-(-1)}=2$	2
0	4	3	$\dfrac{7}{4}$
1	7	4	$\dfrac{11}{7}$
2	11	5	$\dfrac{16}{11}$
3	16		

(c)

(b) See Table 2(b). In this function the average rate of change from −1 to 0 is −16, and the average rate of change from 0 to 1 is −8. Because the average rate of change is not constant, the function is not a linear function. The ratio of consecutive outputs for a 1-unit increase in the inputs is a constant, $\dfrac{1}{2}$. Because the ratio of consecutive outputs is constant, the function is an exponential function with growth factor $a = \dfrac{1}{2}$. The initial value of the exponential function is $C = 16$. Therefore, the exponential function that models the data is
$$g(x) = Ca^x = 16 \cdot \left(\dfrac{1}{2}\right)^x.$$

(c) See Table 2(c). In this function the average rate of change from −1 to 0 is 2, and the average rate of change from 0 to 1 is 3. Because the average rate of change is not constant, the function is not a linear function. In this function, the ratio of consecutive outputs from −1 to 0 is 2, and the ratio of consecutive outputs from 0 to 1 is $\dfrac{7}{4}$. Because the ratio of consecutive outputs is not a constant, the function is not an exponential function.

●

⟶ **Now Work** PROBLEM 25

2 Graph Exponential Functions

If we know how to graph an exponential function of the form $f(x) = a^x$, then we could use transformations (shifting, stretching, and so on) to obtain the graph of any exponential function.

First, we graph the exponential function $f(x) = 2^x$.

EXAMPLE 3 **Graphing an Exponential Function**

Graph the exponential function: $f(x) = 2^x$

Solution The domain of $f(x) = 2^x$ is the set of all real numbers. We begin by locating some points on the graph of $f(x) = 2^x$, as listed in Table 3.

Since $2^x > 0$ for all x, the range of f is $(0, \infty)$. From this, we conclude that the graph has no x-intercepts, and, in fact, the graph will lie above the x-axis for all x. As Table 3 indicates, the y-intercept is 1. Table 3 also indicates that as $x \to -\infty$ the values of $f(x) = 2^x$ get closer and closer to 0. We conclude that the x-axis ($y = 0$) is a horizontal asymptote to the graph as $x \to -\infty$. This gives us the end behavior for x large and negative.

To determine the end behavior for x large and positive, look again at Table 3. As $x \to \infty$, $f(x) = 2^x$ grows very quickly, causing the graph of $f(x) = 2^x$ to rise very rapidly. It is apparent that f is an increasing function and hence is one-to-one.

Using all this information, we plot some of the points from Table 3 and connect them with a smooth, continuous curve, as shown in Figure 18.

Table 3

x	$f(x) = 2^x$
-10	$2^{-10} \approx 0.00098$
-3	$2^{-3} = \dfrac{1}{8}$
-2	$2^{-2} = \dfrac{1}{4}$
-1	$2^{-1} = \dfrac{1}{2}$
0	$2^0 = 1$
1	$2^1 = 2$
2	$2^2 = 4$
3	$2^3 = 8$
10	$2^{10} = 1024$

Figure 18

●

As we shall see, graphs that look like the one in Figure 18 occur very frequently in a variety of situations. For example, look at the graph in Figure 19, which illustrates the number of cellular telephone subscribers at the end of each year from 1985 to 2008. We might conclude from this graph that the number of cellular telephone subscribers is growing *exponentially*.

Figure 19

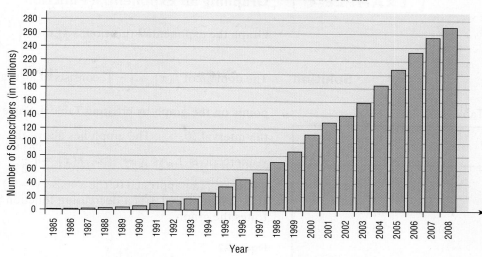

Number of Cellular Phone Subscribers at Year End

Source: ©2009CTIA–The Wireless Association®. All Rights Reserved.

We shall have more to say about situations that lead to exponential growth later in this chapter. For now, we continue to seek properties of exponential functions.

The graph of $f(x) = 2^x$ in Figure 18 is typical of all exponential functions of the form $f(x) = a^x$ with $a > 1$. Such functions are increasing functions and hence are one-to-one. Their graphs lie above the x-axis, pass through the point $(0, 1)$, and thereafter rise rapidly as $x \to \infty$. As $x \to -\infty$, the x-axis $(y = 0)$ is a horizontal asymptote. There are no vertical asymptotes. Finally, the graphs are smooth and continuous with no corners or gaps.

Figure 20 illustrates the graphs of two more exponential functions whose bases are larger than 1. Notice that the larger the base, the steeper the graph is when $x > 0$, and when $x < 0$, the larger the base, the closer the graph of the equation is to the x-axis.

Figure 20

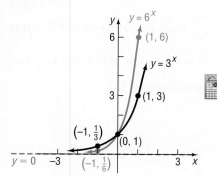

Seeing the Concept

Graph $Y_1 = 2^x$ and compare what you see to Figure 18. Clear the screen and graph $Y_1 = 3^x$ and $Y_2 = 6^x$ and compare what you see to Figure 20. Clear the screen and graph $Y_1 = 10^x$ and $Y_2 = 100^x$.

The following display summarizes the information that we have about $f(x) = a^x$, $a > 1$.

Properties of the Exponential Function $f(x) = a^x$, $a > 1$

1. The domain is the set of all real numbers or $(-\infty, \infty)$ using interval notation; the range is the set of positive real numbers or $(0, \infty)$ using interval notation.

2. There are no x-intercepts; the y-intercept is 1.

3. The x-axis $(y = 0)$ is a horizontal asymptote as $x \to -\infty$. $\left[\lim\limits_{x \to -\infty} a^x = 0\right]$.

4. $f(x) = a^x$, $a > 1$, is an increasing function and is one-to-one.

5. The graph of f contains the points $(0, 1)$, $(1, a)$, and $\left(-1, \dfrac{1}{a}\right)$.

6. The graph of f is smooth and continuous, with no corners or gaps. See Figure 21.

Figure 21

Now we consider $f(x) = a^x$ when $0 < a < 1$.

<table>
</table>

EXAMPLE 4	**Graphing an Exponential Function**

Graph the exponential function: $f(x) = \left(\dfrac{1}{2}\right)^x$

Solution

The domain of $f(x) = \left(\dfrac{1}{2}\right)^x$ consists of all real numbers. As before, we locate some points on the graph by creating Table 4. Since $\left(\dfrac{1}{2}\right)^x > 0$ for all x, the range of f is the interval $(0, \infty)$. The graph lies above the x-axis and so has no x-intercepts. The y-intercept is 1. As $x \to -\infty$, $f(x) = \left(\dfrac{1}{2}\right)^x$ grows very quickly. As $x \to \infty$, the values of $f(x)$ approach 0. The x-axis $(y = 0)$ is a horizontal asymptote as $x \to \infty$. It is apparent that f is a decreasing function and so is one-to-one. Figure 22 illustrates the graph.

Table 4

x	$f(x) = \left(\dfrac{1}{3}\right)^x$
-10	$\left(\dfrac{1}{2}\right)^{-10} = 1024$
-3	$\left(\dfrac{1}{2}\right)^{-3} = 8$
-2	$\left(\dfrac{1}{2}\right)^{-2} = 4$
-1	$\left(\dfrac{1}{2}\right)^{-1} = 2$
0	$\left(\dfrac{1}{2}\right)^{0} = 1$
1	$\left(\dfrac{1}{2}\right)^{1} = \dfrac{1}{2}$
2	$\left(\dfrac{1}{2}\right)^{2} = \dfrac{1}{4}$
3	$\left(\dfrac{1}{2}\right)^{3} = \dfrac{1}{8}$
10	$\left(\dfrac{1}{2}\right)^{10} \approx 0.00098$

Figure 22

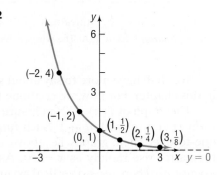

We could have obtained the graph of $y = \left(\dfrac{1}{2}\right)^x$ from the graph of $y = 2^x$ using transformations. The graph of $y = \left(\dfrac{1}{2}\right)^x = 2^{-x}$ is a reflection about the y-axis of the graph of $y = 2^x$ (replace x by $-x$). See Figures 23(a) and (b).

Figure 23

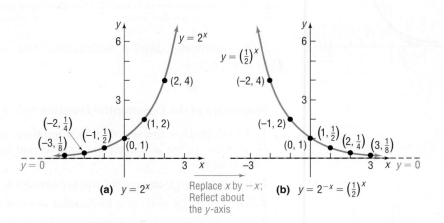

(a) $y = 2^x$

Replace x by $-x$;
Reflect about
the y-axis

(b) $y = 2^{-x} = \left(\dfrac{1}{2}\right)^x$

 Seeing the Concept

Using a graphing utility, simultaneously graph:

(a) $Y_1 = 3^x, Y_2 = \left(\dfrac{1}{3}\right)^x$

(b) $Y_1 = 6^x, Y_2 = \left(\dfrac{1}{6}\right)^x$

Conclude that the graph of $Y_2 = \left(\dfrac{1}{a}\right)^x$, for $a > 0$, is the reflection about the y-axis of the graph of $Y_1 = a^x$.

The graph of $f(x) = \left(\dfrac{1}{2}\right)^x$ in Figure 22 is typical of all exponential functions of the form $f(x) = a^x$ with $0 < a < 1$. Such functions are decreasing and one-to-one. Their graphs lie above the x-axis and pass through the point $(0, 1)$. The graphs rise rapidly as $x \to -\infty$. As $x \to \infty$, the x-axis $(y = 0)$ is a horizontal asymptote. There are no vertical asymptotes. Finally, the graphs are smooth and continuous, with no corners or gaps.

Figure 24

Figure 24 illustrates the graphs of two more exponential functions whose bases are between 0 and 1. Notice that the smaller base results in a graph that is steeper when $x < 0$. When $x > 0$, the graph of the equation with the smaller base is closer to the x-axis.

The following display summarizes the information that we have about the function $f(x) = a^x, 0 < a < 1$.

Properties of the Exponential Function $f(x) = a^x$, $0 < a < 1$

1. The domain is the set of all real numbers or $(-\infty, \infty)$ using interval notation; the range is the set of positive real numbers or $(0, \infty)$ using interval notation.

2. There are no x-intercepts; the y-intercept is 1.

3. The x-axis ($y = 0$) is a horizontal asymptote as $x \to \infty$ $\left[\lim\limits_{x \to \infty} a^x = 0 \right]$.

4. $f(x) = a^x, 0 < a < 1$, is a decreasing function and is one-to-one.

5. The graph of f contains the points $\left(-1, \dfrac{1}{a} \right)$, $(0, 1)$, and $(1, a)$.

6. The graph of f is smooth and continuous, with no corners or gaps. See Figure 25.

Figure 25

EXAMPLE 5

Graphing Exponential Functions Using Transformations

Graph $f(x) = 2^{-x} - 3$ and determine the domain, range, and horizontal asymptote of f.

Solution

We begin with the graph of $y = 2^x$. Figure 26 shows the stages.

Figure 26

(a) $y = 2^x$ Replace x by $-x$; reflect about y-axis   (b) $y = 2^{-x}$ Subtract 3; shift down 3 units   (c) $y = 2^{-x} - 3$

As Figure 26(c) illustrates, the domain of $f(x) = 2^{-x} - 3$ is the interval $(-\infty, \infty)$ and the range is the interval $(-3, \infty)$. The horizontal asymptote of f is the line $y = -3$.

━━━**Now Work** PROBLEM 41

3 Define the Number e

As we shall see shortly, many problems that occur in nature require the use of an exponential function whose base is a certain irrational number, symbolized by the letter e.

Let's look at one way of arriving at this important number e.

DEFINITION

The **number e** is defined as the number that the expression

$$\left(1 + \frac{1}{n}\right)^n \tag{2}$$

approaches as $n : \infty$. In calculus, this is expressed using limit notation as

$$e = \lim_{n : \infty} \left(1 + \frac{1}{n}\right)^n$$

Table 5 illustrates what happens to the defining expression (2) as n takes on increasingly large values. The last number in the right column in the table is correct to nine decimal places and is the same as the entry given for e on your calculator (if expressed correctly to nine decimal places).

The exponential function $f(x) = e^x$, whose base is the number e, occurs with such frequency in applications that it is usually referred to as *the* exponential function. Indeed, most calculators have the key $\boxed{e^x}$ or $\boxed{\exp(x)}$, which may be used to evaluate the exponential function for a given value of x. *

Table 5

n	$\dfrac{1}{n}$	$1 + \dfrac{1}{n}$	$\left(1 + \dfrac{1}{n}\right)^n$
1	1	2	2
2	0.5	1.5	2.25
5	0.2	1.2	2.48832
10	0.1	1.1	2.59374246
100	0.01	1.01	2.704813829
1,000	0.001	1.001	2.716923932
10,000	0.0001	1.0001	2.718145927
100,000	0.00001	1.00001	2.718268237
1,000,000	0.000001	1.000001	2.718280469
1,000,000,000	10^{-9}	$1 + 10^{-9}$	2.718281827

Table 6

x	e^x
-2	$e^{-2} \approx 0.14$
-1	$e^{-1} \approx 0.37$
0	$e^0 \approx 1$
1	$e^1 \approx 2.72$
2	$e^2 \approx 7.39$

Figure 27
$y = e^x$

Now use your calculator to approximate e^x for $x = -2$, $x = -1$, $x = 0$, $x = 1$, and $x = 2$, as we have done to create Table 6. The graph of the exponential function $f(x) = e^x$ is given in Figure 27. Since $2 < e < 3$, the graph of $y = e^x$ lies between the graphs of $y = 2^x$ and $y = 3^x$. Do you see why? (Refer to Figures 18 and 20.)

 Seeing the Concept

Graph $Y_1 = e^x$ and compare what you see to Figure 27. Use eVALUEate or TABLE to verify the points on the graph shown in Figure 27. Now graph $Y_2 = 2^x$ and $Y_3 = 3^x$ on the same screen as $Y_1 = e^x$. Notice that the graph of $Y_1 = e^x$ lies between these two graphs.

EXAMPLE 6 **Graphing Exponential Functions Using Transformations**

Graph $f(x) = -e^{x-3}$ and determine the domain, range, and horizontal asymptote of f.

Solution We begin with the graph of $y = e^x$. Figure 28 shows the stages.

*If your calculator does not have one of these keys, refer to your Owner's Manual.

Figure 28

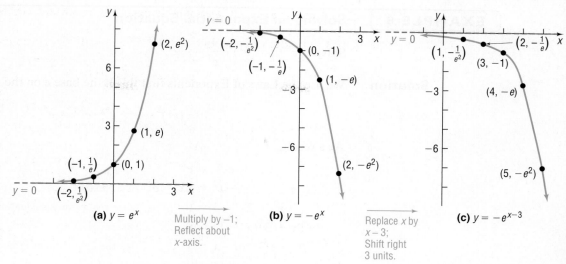

(a) $y = e^x$

Multiply by −1;
Reflect about
x-axis.

(b) $y = -e^x$

Replace x by
x − 3;
Shift right
3 units.

(c) $y = -e^{x-3}$

As Figure 28(c) illustrates, the domain of $f(x) = -e^{x-3}$ is the interval $(-\infty, \infty)$, and the range is the interval $(-\infty, 0)$. The horizontal asymptote is the line $y = 0$.

━━━━ **Now Work** PROBLEM 53

4 Solve Exponential Equations

Equations that involve terms of the form $a^x, a > 0, a \neq 1$, are referred to as **exponential equations.** Such equations can sometimes be solved by appropriately applying the Laws of Exponents and property (3):

> **In Words**
> When two exponential expressions with the same base are equal, then their exponents are equal.

$$\text{If } \quad a^u = a^v, \quad \text{then} \quad u = v. \qquad (3)$$

Property (3) is a consequence of the fact that exponential functions are one-to-one. To use property (3), each side of the equality must be written with the same base.

EXAMPLE 7   **Solving Exponential Equations**

Solve each exponential equation.

(a) $3^{x+1} = 81$ (b) $4^{2x-1} = 8^{x+3}$

Solution   (a) Since $81 = 3^4$, we can write the equation as

$$3^{x+1} = 81 = 3^4$$

Now we have the same base, 3, on each side, so we can set the exponents equal to each other to obtain

$$x + 1 = 4$$
$$x = 3$$

The solution set is {3}.

(b)
$$4^{2x-1} = 8^{x+3}$$
$$(2^2)^{(2x-1)} = (2^3)^{(x+3)} \qquad 4 = 2^2; 8 = 2^3$$
$$2^{2(2x-1)} = 2^{3(x+3)} \qquad (a^r)^s = a^{rs}$$
$$2(2x - 1) = 3(x + 3) \qquad \text{If } a^u = a^v, \text{ then } u = v.$$
$$4x - 2 = 3x + 9$$
$$x = 11$$

The solution set is {11}.

━━━━ **Now Work** PROBLEM 63

EXAMPLE 8

Solving an Exponential Equation

Solve: $e^{-x^2} = (e^x)^2 \cdot \dfrac{1}{e^3}$

Solution   We use the Laws of Exponents first to get the base e on the right side.

$$(e^x)^2 \cdot \frac{1}{e^3} = e^{2x} \cdot e^{-3} = e^{2x-3}$$

As a result,

$$e^{-x^2} = e^{2x-3}$$
$$-x^2 = 2x - 3 \qquad \text{Apply property (3).}$$
$$x^2 + 2x - 3 = 0 \qquad \text{Place the quadratic equation in standard form.}$$
$$(x + 3)(x - 1) = 0 \qquad \text{Factor.}$$
$$x = -3 \quad \text{or} \quad x = 1 \qquad \text{Use the Zero-Product Property.}$$

The solution set is $\{-3, 1\}$.

Many applications involve the exponential function. Let's look at one.

EXAMPLE 9

Exponential Probability

Between 9:00 PM and 10:00 PM cars arrive at Burger King's drive-thru at the rate of 12 cars per hour (0.2 car per minute). The following formula from statistics can be used to determine the probability that a car will arrive within t minutes of 9:00 PM.

$$F(t) = 1 - e^{-0.2t}$$

(a) Determine the probability that a car will arrive within 5 minutes of 9 PM (that is, before 9:05 PM).

(b) Determine the probability that a car will arrive within 30 minutes of 9 PM (before 9:30 PM).

 (c) Graph F using your graphing utility.

(d) What value does F approach as t increases without bound in the positive direction?

Solution   (a) The probability that a car will arrive within 5 minutes is found by evaluating $F(t)$ at $t = 5$.

$$F(5) = 1 - e^{-0.2(5)} \approx 0.63212$$
$$\uparrow$$
$$\text{Use a calculator.}$$

We conclude that there is a 63% probability that a car will arrive within 5 minutes.

(b) The probability that a car will arrive within 30 minutes is found by evaluating $F(t)$ at $t = 30$.

$$F(30) = 1 - e^{-0.2(30)} \approx 0.9975$$
$$\uparrow$$
$$\text{Use a calculator.}$$

There is a 99.75% probability that a car will arrive within 30 minutes.

(c) See Figure 29 for the graph of F.

(d) As time passes, the probability that a car will arrive increases. The value that F approaches can be found by letting $t \to \infty$. Since $e^{-0.2t} = \dfrac{1}{e^{0.2t}}$, it follows that $e^{-0.2t} \to 0$ as $t \to \infty$. We conclude that F approaches 1 as t gets large. The algebraic analysis is confirmed by Figure 29.

Figure 29

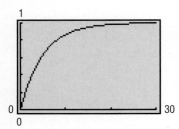

Now Work PROBLEM 107

SUMMARY

> **Properties of the Exponential Function**
>
> $f(x) = a^x, \quad a > 1$ Domain: the interval $(-\infty, \infty)$; range: the interval $(0, \infty)$
> x-intercepts: none; y-intercept: 1
> Horizontal asymptote: x-axis $(y = 0)$ as $x : -\infty$
> Increasing; one-to-one; smooth; continuous
> See Figure 21 for a typical graph.
>
> $f(x) = a^x, \quad 0 < a < 1$ Domain: the interval $(-\infty, \infty)$; range: the interval $(0, \infty)$
> x-intercepts: none; y-intercept: 1
> Horizontal asymptote: x-axis $(y = 0)$ as $x \to \infty$
> Decreasing; one-to-one; smooth; continuous
> See Figure 25 for a typical graph.
>
> If $a^u = a^v$, then $u = v$.

4.3 Assess Your Understanding

'Are You Prepared?' *Answers are given at the end of these exercises. If you get a wrong answer, read the pages listed in red.*

1. $4^3 = $ _____; $8^{2/3} = $ _____; $3^{-2} = $ _____ . (pp. A7–A9 and pp. A60–A61)

2. Solve: $x^2 + 3x = 4$ (p. 142)

3. *True or False* To graph $y = (x - 2)^3$, shift the graph of $y = x^3$ to the left 2 units. (pp. 90–93)

4. Find the average rate of change of $f(x) = 3x - 5$ from $x = 0$ to $x = 4$. (pp. 73–74; 123–124)

5. *True or False* The function $f(x) = \dfrac{2x}{x - 3}$ has $y = 2$ as a horizontal asymptote. (pp. 220–223)

Concepts and Vocabulary

6. A(n) _____ is a function of the form $f(x) = Ca^x, a > 0, a \neq 1$ and $C \neq 0$ are real numbers. The base a is the _____ and C is the _____ _____.

7. For an exponential function $f(x) = Ca^x, a > 0, a \neq 1$, and $C \neq 0$ are real numbers, $\dfrac{f(x + 1)}{f(x)} = $ _____ or $f(x + 1) = $ _____ $\cdot f(x)$.

8. *True or False* The domain of the exponential function $f(x) = a^x, a > 0, a \neq 1$, is the set of all real numbers.

9. *True or False* The range of the exponential function $f(x) = a^x, a > 0, a \neq 1$, is the set of all real numbers.

10. *True or False* The graph of the exponential function $f(x) = a^x, a > 0, a \neq 1$, has no x-intercept.

11. The graph of every exponential function $f(x) = a^x, a > 0, a \neq 1$, passes through three points: _____, _____, and _____.

12. If the graph of the exponential function $f(x) = a^x, a > 0, a \neq 1$, is decreasing, then a must be less than _____.

13. If $3^x = 3^4$, then $x = $ _____.

14. *True or False* The graphs of $y = 3^x$ and $y = \left(\dfrac{1}{3}\right)^x$ are identical.

Skill Building

In Problems 15–24, approximate each number using a calculator. Express your answer rounded to three decimal places.

15. (a) $3^{2.2}$ (b) $3^{2.23}$ (c) $3^{2.236}$ (d) $3^{\sqrt{5}}$

16. (a) $5^{1.7}$ (b) $5^{1.73}$ (c) $5^{1.732}$ (d) $5^{\sqrt{3}}$

17. (a) $2^{3.14}$ (b) $2^{3.141}$ (c) $2^{3.1415}$ (d) 2^{π}

18. (a) $2^{2.7}$ (b) $2^{2.71}$ (c) $2^{2.718}$ (d) 2^{e}

19. (a) $3.1^{2.7}$ (b) $3.14^{2.71}$ (c) $3.141^{2.718}$ (d) π^{e}

20. (a) $2.7^{3.1}$ (b) $2.71^{3.14}$ (c) $2.718^{3.141}$ (d) e^{π}

21. $e^{1.2}$ **22.** $e^{-1.3}$ **23.** $e^{-0.85}$ **24.** $e^{2.1}$

In Problems 25–32, determine whether the given function is linear, exponential, or neither. For those that are linear functions, find a linear function that models the data; for those that are exponential, find an exponential function that models the data.

25.

x	f(x)
−1	3
0	6
1	12
2	18
3	30

26.

x	g(x)
−1	2
0	5
1	8
2	11
3	14

27.

x	H(x)
−1	$\frac{1}{4}$
0	1
1	4
2	16
3	64

28.

x	F(x)
−1	$\frac{2}{3}$
0	1
1	$\frac{3}{2}$
2	$\frac{9}{4}$
3	$\frac{27}{8}$

29.

x	f(x)
−1	$\frac{3}{2}$
0	3
1	6
2	12
3	24

30.

x	g(x)
−1	6
0	1
1	0
2	3
3	10

31.

x	H(x)
−1	2
0	4
1	6
2	8
3	10

32.

x	F(x)
−1	$\frac{1}{2}$
0	$\frac{1}{4}$
1	$\frac{1}{8}$
2	$\frac{1}{16}$
3	$\frac{1}{32}$

In Problems 33–40, the graph of an exponential function is given. Match each graph to one of the following functions.

(a) $y = 3^x$ (b) $y = 3^{-x}$ (c) $y = -3^x$ (d) $y = -3^{-x}$

(e) $y = 3^x - 1$ (f) $y = 3^{x-1}$ (g) $y = 3^{1-x}$ (h) $y = 1 - 3^x$

33.

34.

35.

36.

37.

38.

39.

40.

In Problems 41–52, use transformations to graph each function. Determine the domain, range, and horizontal asymptote of each function.

41. $f(x) = 2^x + 1$ **42.** $f(x) = 3^x - 2$ **43.** $f(x) = 3^{x-1}$ **44.** $f(x) = 2^{x+2}$

45. $f(x) = 3 \cdot \left(\frac{1}{2}\right)^x$ **46.** $f(x) = 4 \cdot \left(\frac{1}{3}\right)^x$ **47.** $f(x) = 3^{-x} - 2$ **48.** $f(x) = -3^x + 1$

49. $f(x) = 2 + 4^{x-1}$ **50.** $f(x) = 1 - 2^{x+3}$ **51.** $f(x) = 2 + 3^{x/2}$ **52.** $f(x) = 1 - 2^{-x/3}$

In Problems 53–60, begin with the graph of $y = e^x$ [Figure 27] and use transformations to graph each function. Determine the domain, range, and horizontal asymptote of each function.

53. $f(x) = e^{-x}$ **54.** $f(x) = -e^x$ **55.** $f(x) = e^{x+2}$ **56.** $f(x) = e^x - 1$

57. $f(x) = 5 - e^{-x}$ **58.** $f(x) = 9 - 3e^{-x}$ **59.** $f(x) = 2 - e^{-x/2}$ **60.** $f(x) = 7 - 3e^{2x}$

In Problems 61–80, solve each equation.

61. $7^x = 7^3$

62. $5^x = 5^{-6}$

63. $2^{-x} = 16$

64. $3^{-x} = 81$

65. $\left(\dfrac{1}{5}\right)^x = \dfrac{1}{25}$

66. $\left(\dfrac{1}{4}\right)^x = \dfrac{1}{64}$

67. $2^{2x-1} = 4$

68. $5^{x+3} = \dfrac{1}{5}$

69. $3^{x^3} = 9^x$

70. $4^{x^2} = 2^x$

71. $8^{-x+14} = 16^x$

72. $9^{-x+15} = 27^x$

73. $3^{x^2-7} = 27^{2x}$

74. $5^{x^2+8} = 125^{2x}$

75. $4^x \cdot 2^{x^2} = 16^2$

76. $9^{2x} \cdot 27^{x^2} = 3^{-1}$

77. $e^x = e^{3x+8}$

78. $e^{3x} = e^{2-x}$

79. $e^{x^2} = e^{3x} \cdot \dfrac{1}{e^2}$

80. $(e^4)^x \cdot e^{x^2} = e^{12}$

81. If $4^x = 7$, what does 4^{-2x} equal?

82. If $2^x = 3$, what does 4^{-x} equal?

83. If $3^{-x} = 2$, what does 3^{2x} equal?

84. If $5^{-x} = 3$, what does 5^{3x} equal?

In Problems 85–88, determine the exponential function whose graph is given.

85.

86.

87.

88.

89. Find an exponential function with horizontal asymptote $y = 2$ whose graph contains the points $(0, 3)$ and $(1, 5)$.

90. Find an exponential function with horizontal asymptote $y = -3$ whose graph contains the points $(0, 2)$ and $(-2, 1)$.

Mixed Practice

91. Suppose that $f(x) = 2^x$.
  (a) What is $f(4)$? What point is on the graph of f?
  (b) If $f(x) = \dfrac{1}{16}$, what is x? What point is on the graph of f?

92. Suppose that $f(x) = 3^x$.
  (a) What is $f(4)$? What point is on the graph of f?
  (b) If $f(x) = \dfrac{1}{9}$, what is x? What point is on the graph of f?

93. Suppose that $g(x) = 4^x + 2$.
  (a) What is $g(-1)$? What point is on the graph of g?
  (b) If $g(x) = 66$, what is x? What point is on the graph of g?

94. Suppose that $g(x) = 5^x - 3$.
  (a) What is $g(-1)$? What point is on the graph of g?
  (b) If $g(x) = 122$, what is x? What point is on the graph of g?

95. Suppose that $H(x) = \left(\dfrac{1}{2}\right)^x - 4$.
  (a) What is $H(-6)$? What point is on the graph of H?
  (b) If $H(x) = 12$, what is x? What point is on the graph of H?
  (c) Find the zero of H.

96. Suppose that $F(x) = \left(\dfrac{1}{3}\right)^x - 3$.
  (a) What is $F(-5)$? What point is on the graph of F?
  (b) If $F(x) = 24$, what is x? What point is on the graph of F?
  (c) Find the zero of F.

In Problems 97–100, graph each function. Based on the graph, state the domain and the range and find any intercepts.

97. $f(x) = \begin{cases} e^{-x} & \text{if } x < 0 \\ e^x & \text{if } x \geq 0 \end{cases}$

98. $f(x) = \begin{cases} e^x & \text{if } x < 0 \\ e^{-x} & \text{if } x \geq 0 \end{cases}$

99. $f(x) = \begin{cases} -e^x & \text{if } x < 0 \\ -e^{-x} & \text{if } x \geq 0 \end{cases}$

100. $f(x) = \begin{cases} -e^{-x} & \text{if } x < 0 \\ -e^x & \text{if } x \geq 0 \end{cases}$

Applications and Extensions

101. Optics If a single pane of glass obliterates 3% of the light passing through it, the percent p of light that passes through n successive panes is given approximately by the function

$$p(n) = 100(0.97)^n$$

(a) What percent of light will pass through 10 panes?
(b) What percent of light will pass through 25 panes?

102. Atmospheric Pressure The atmospheric pressure p on a balloon or plane decreases with increasing height. This pressure, measured in millimeters of mercury, is related to the height h (in kilometers) above sea level by the function

$$p(h) = 760e^{-0.145h}$$

(a) Find the atmospheric pressure at a height of 2 kilometers (over a mile).
(b) What is it at a height of 10 kilometers (over 30,000 feet)?

103. Depreciation The price p, in dollars, of a Honda Civic DX Sedan that is x years old is modeled by

$$p(x) = 16{,}630(0.90)^x$$

(a) How much should a 3-year-old Civic DX Sedan cost?
(b) How much should a 9-year-old Civic DX Sedan cost?

104. Healing of Wounds The normal healing of wounds can be modeled by an exponential function. If A_0 represents the original area of the wound and if A equals the area of the wound, then the function

$$A(n) = A_0 e^{-0.35n}$$

describes the area of a wound after n days following an injury when no infection is present to retard the healing. Suppose that a wound initially had an area of 100 square millimeters.
(a) If healing is taking place, how large will the area of the wound be after 3 days?
(b) How large will it be after 10 days?

105. Drug Medication The function

$$D(h) = 5e^{-0.4h}$$

can be used to find the number of milligrams D of a certain drug that is in a patient's bloodstream h hours after the drug has been administered. How many milligrams will be present after 1 hour? After 6 hours?

106. Spreading of Rumors A model for the number N of people in a college community who have heard a certain rumor is

$$N = P(1 - e^{-0.15d})$$

where P is the total population of the community and d is the number of days that have elapsed since the rumor began. In a community of 1000 students, how many students will have heard the rumor after 3 days?

107. Exponential Probability Between 12:00 PM and 1:00 PM, cars arrive at Citibank's drive-thru at the rate of 6 cars per hour (0.1 car per minute). The following formula from probability can be used to determine the probability that a car will arrive within t minutes of 12:00 PM:

$$F(t) = 1 - e^{-0.1t}$$

(a) Determine the probability that a car will arrive within 10 minutes of 12:00 PM (that is, before 12:10 PM).
(b) Determine the probability that a car will arrive within 40 minutes of 12:00 PM (before 12:40 PM).
(c) What value does F approach as t becomes unbounded in the positive direction?
(d) Graph F using a graphing utility.
(e) Using INTERSECT, determine how many minutes are needed for the probability to reach 50%.

108. Exponential Probability Between 5:00 PM and 6:00 PM, cars arrive at Jiffy Lube at the rate of 9 cars per hour (0.15 car per minute). The following formula from probability can be used to determine the probability that a car will arrive within t minutes of 5:00 PM:

$$F(t) = 1 - e^{-0.15t}$$

(a) Determine the probability that a car will arrive within 15 minutes of 5:00 PM (that is, before 5:15 PM).
(b) Determine the probability that a car will arrive within 30 minutes of 5:00 PM (before 5:30 PM).
(c) What value does F approach as t becomes unbounded in the positive direction?
(d) Graph F using a graphing utility.
(e) Using INTERSECT, determine how many minutes are needed for the probability to reach 60%.

109. Poisson Probability Between 5:00 PM and 6:00 PM, cars arrive at McDonald's drive-thru at the rate of 20 cars per hour. The following formula from probability can be used to determine the probability that x cars will arrive between 5:00 PM and 6:00 PM.

$$P(x) = \frac{20^x e^{-20}}{x!}$$

where

$$x! = x \cdot (x - 1) \cdot (x - 2) \cdot \cdots \cdot 3 \cdot 2 \cdot 1$$

(a) Determine the probability that $x = 15$ cars will arrive between 5:00 PM and 6:00 PM.
(b) Determine the probability that $x = 20$ cars will arrive between 5:00 PM and 6:00 PM.

110. Poisson Probability People enter a line for the *Demon Roller Coaster* at the rate of 4 per minute. The following formula from probability can be used to determine the probability that x people will arrive within the next minute.

$$P(x) = \frac{4^x e^{-4}}{x!}$$

where

$$x! = x \cdot (x-1) \cdot (x-2) \cdots 3 \cdot 2 \cdot 1$$

(a) Determine the probability that $x = 5$ people will arrive within the next minute.
(b) Determine the probability that $x = 8$ people will arrive within the next minute.

111. Relative Humidity The relative humidity is the ratio (expressed as a percent) of the amount of water vapor in the air to the maximum amount that it can hold at a specific temperature. The relative humidity, R, is found using the following formula:

$$R = 10^{\left(\frac{4221}{T+459.4} - \frac{4221}{D+459.4} + 2\right)}$$

where T is the air temperature (in °F) and D is the dew point temperature (in °F).
(a) Determine the relative humidity if the air temperature is 50° Fahrenheit and the dew point temperature is 41° Fahrenheit.
(b) Determine the relative humidity if the air temperature is 68° Fahrenheit and the dew point temperature is 59° Fahrenheit.
(c) What is the relative humidity if the air temperature and the dew point temperature are the same?

112. Learning Curve Suppose that a student has 500 vocabulary words to learn. If the student learns 15 words after 5 minutes, the function

$$L(t) = 500(1 - e^{-0.0061t})$$

approximates the number of words L that the student will learn after t minutes.
(a) How many words will the student learn after 30 minutes?
(b) How many words will the student learn after 60 minutes?

113. Current in a RL Circuit The equation governing the amount of current I (in amperes) after time t (in seconds) in a single RL circuit consisting of a resistance R (in ohms), an inductance L (in henrys), and an electromotive force E (in volts) is

$$I = \frac{E}{R}[1 - e^{-(R/L)t}]$$

(a) If $E = 120$ volts, $R = 10$ ohms, and $L = 5$ henrys, how much current I_1 is flowing after 0.3 second? After 0.5 second? After 1 second?
(b) What is the maximum current?
(c) Graph this function $I = I_1(t)$, measuring I along the y-axis and t along the x-axis.
(d) If $E = 120$ volts, $R = 5$ ohms, and $L = 10$ henrys, how much current I_2 is flowing after 0.3 second? After 0.5 second? After 1 second?
(e) What is the maximum current?
(f) Graph the function $I = I_2(t)$ on the same coordinate axes as $I_1(t)$.

114. Current in a RC Circuit The equation governing the amount of current I (in amperes) after time t (in microseconds) in a single RC circuit consisting of a resistance R (in ohms), a capacitance C (in microfarads), and an electromotive force E (in volts) is

$$I = \frac{E}{R}e^{-t/(RC)}$$

(a) If $E = 120$ volts, $R = 2000$ ohms, and $C = 1.0$ microfarad, how much current I_1 is flowing initially $(t = 0)$? After 1000 microseconds? After 3000 microseconds?
(b) What is the maximum current?
(c) Graph the function $I = I_1(t)$, measuring I along the y-axis and t along the x-axis.
(d) If $E = 120$ volts, $R = 1000$ ohms, and $C = 2.0$ microfarads, how much current I_2 is flowing initially? After 1000 microseconds? After 3000 microseconds?
(e) What is the maximum current?
(f) Graph the function $I = I_2(t)$ on the same coordinate axes as $I_1(t)$.

115. If f is an exponential function of the form $f(x) = C \cdot a^x$ with growth factor 3 and $f(6) = 12$, what is $f(7)$?

116. Another Formula for e Use a calculator to compute the values of

$$2 + \frac{1}{2!} + \frac{1}{3!} + \cdots + \frac{1}{n!}$$

for $n = 4, 6, 8,$ and 10. Compare each result with e.
[**Hint:** $1! = 1, 2! = 2 \cdot 1, 3! = 3 \cdot 2 \cdot 1,$
$n! = n(n-1) \cdots (3)(2)(1).$]

117. Another Formula for e Use a calculator to compute the various values of the expression. Compare the values to e.

$$2 + \cfrac{1}{1 + \cfrac{1}{2 + \cfrac{2}{3 + \cfrac{3}{4 + \cfrac{4}{\text{etc.}}}}}}$$

118. Difference Quotient If $f(x) = a^x$, show that

$$\frac{f(x+h) - f(x)}{h} = a^x \cdot \frac{a^h - 1}{h} \quad h \neq 0$$

119. If $f(x) = a^x$, show that $f(A + B) = f(A) \cdot f(B)$.

120. If $f(x) = a^x$, show that $f(-x) = \dfrac{1}{f(x)}$.

121. If $f(x) = a^x$, show that $f(\alpha x) = [f(x)]^{\alpha}$.

Problems 122 and 123 provide definitions for two other transcendental functions.

122. The **hyperbolic sine function,** designated by sinh x, is defined as

$$\sinh x = \frac{1}{2}(e^x - e^{-x})$$

(a) Show that $f(x) = \sinh x$ is an odd function.

(b) Graph $f(x) = \sinh x$ using a graphing utility.

123. The **hyperbolic cosine function,** designated by cosh x, is defined as

$$\cosh x = \frac{1}{2}(e^x + e^{-x})$$

(a) Show that $f(x) = \cosh x$ is an even function.

(b) Graph $f(x) = \cosh x$ using a graphing utility.

(c) Refer to Problem 122. Show that, for every x,

$$(\cosh x)^2 - (\sinh x)^2 = 1$$

124. Historical Problem Pierre de Fermat (1601–1665) conjectured that the function

$$f(x) = 2^{(2^x)} + 1$$

for $x = 1, 2, 3, \ldots$, would always have a value equal to a prime number. But Leonhard Euler (1707–1783) showed that this formula fails for $x = 5$. Use a calculator to determine the prime numbers produced by f for $x = 1, 2, 3, 4$. Then show that $f(5) = 641 \times 6,700,417$, which is not prime.

Explaining Concepts: Discussion and Writing

125. The bacteria in a 4-liter container double every minute. After 60 minutes the container is full. How long did it take to fill half the container?

126. Explain in your own words what the number e is. Provide at least two applications that use this number.

127. Do you think that there is a power function that increases more rapidly than an exponential function whose base is greater than 1? Explain.

128. As the base a of an exponential function $f(x) = a^x, a > 1$, increases, what happens to the behavior of its graph for $x > 0$? What happens to the behavior of its graph for $x < 0$?

129. The graphs of $y = a^{-x}$ and $y = \left(\dfrac{1}{a}\right)^x$ are identical. Why?

'Are You Prepared?' Answers

1. 64; 4; $\dfrac{1}{9}$ **2.** $\{-4, 1\}$ **3.** False   **4.** 3   **5.** True

4.4 Logarithmic Functions

PREPARING FOR THIS SECTION *Before getting started, review the following:*

- Solving Inequalities (Appendix A, Section A.10, pp. A86–A87)
- Quadratic Inequalities (Section 2.5, pp. 164–166)
- Polynomial and Rational Inequalities (Section 3.4, pp. 241–245)
- Solve Linear Equations (Appendix A, Section A.8, pp. A65–A68)

Now Work the *'Are You Prepared?'* problems on page 321.

OBJECTIVES **1** Change Exponential Statements to Logarithmic Statements and Logarithmic Statements to Exponential Statements (p. 313)
   2 Evaluate Logarithmic Expressions (p. 313)
   3 Determine the Domain of a Logarithmic Function (p. 314)
   4 Graph Logarithmic Functions (p. 315)
   5 Solve Logarithmic Equations (p. 319)

Recall that a one-to-one function $y = f(x)$ has an inverse function that is defined (implicitly) by the equation $x = f(y)$. In particular, the exponential function $y = f(x) = a^x, a > 0, a \neq 1$, is one-to-one and hence has an inverse function that is defined implicitly by the equation

$$x = a^y, \qquad a > 0, \qquad a \neq 1$$

This inverse function is so important that it is given a name, the *logarithmic function.*

DEFINITION

The **logarithmic function to the base a,** where $a > 0$ and $a \neq 1$, is denoted by $y = \log_a x$ (read as "y is the logarithm to the base a of x") and is defined by

$$y = \log_a x \quad \text{if and only if} \quad x = a^y$$

The domain of the logarithmic function $y = \log_a x$ is $x > 0$.

In Words
When you read $\log_a x$, think to yourself "a raised to what power gives me x."

As this definition illustrates, **a logarithm is a name for a certain exponent.** So, $\log_a x$ represents the exponent to which a must be raised to obtain x.

EXAMPLE 1   **Relating Logarithms to Exponents**

(a) If $y = \log_3 x$, then $x = 3^y$. For example, the logarithmic statement $4 = \log_3 81$ is equivalent to the exponential statement $81 = 3^4$.

(b) If $y = \log_5 x$, then $x = 5^y$. For example, $-1 = \log_5\left(\dfrac{1}{5}\right)$ is equivalent to $\dfrac{1}{5} = 5^{-1}$.

---

1 Change Exponential Statements to Logarithmic Statements and Logarithmic Statements to Exponential Statements

We can use the definition of a logarithm to convert from exponential form to logarithmic form, and vice versa, as the following two examples illustrate.

EXAMPLE 2   **Changing Exponential Statements to Logarithmic Statements**

Change each exponential statement to an equivalent statement involving a logarithm.

(a) $1.2^3 = m$ (b) $e^b = 9$ (c) $a^4 = 24$

Solution   We use the fact that $y = \log_a x$ and $x = a^y$, $a > 0$, $a \neq 1$, are equivalent.

(a) If $1.2^3 = m$, then $3 = \log_{1.2} m$.
(b) If $e^b = 9$, then $b = \log_e 9$.
(c) If $a^4 = 24$, then $4 = \log_a 24$.

━━━ **Now Work** PROBLEM 9

EXAMPLE 3   **Changing Logarithmic Statements to Exponential Statements**

Change each logarithmic statement to an equivalent statement involving an exponent.

(a) $\log_a 4 = 5$ (b) $\log_e b = -3$ (c) $\log_3 5 = c$

Solution   (a) If $\log_a 4 = 5$, then $a^5 = 4$.
(b) If $\log_e b = -3$, then $e^{-3} = b$.
(c) If $\log_3 5 = c$, then $3^c = 5$.

━━━ **Now Work** PROBLEM 17

---

2 Evaluate Logarithmic Expressions

To find the exact value of a logarithm, we write the logarithm in exponential notation using the fact that $y = \log_a x$ is equivalent to $a^y = x$ and use the fact that if $a^u = a^v$ then $u = v$.

EXAMPLE 4	**Finding the Exact Value of a Logarithmic Expression**

Find the exact value of:

(a) $\log_2 16$

(b) $\log_3 \dfrac{1}{27}$

Solution

(a) To evaluate $\log_2 16$, think "2 raised to what power yields 16." So,

$$y = \log_2 16$$
$$2^y = 16 \qquad \textit{Change to exponential form.}$$
$$2^y = 2^4 \qquad 16 = 2^4$$
$$y = 4 \qquad \textit{Equate exponents.}$$

Therefore, $\log_2 16 = 4$.

(b) To evaluate $\log_3 \dfrac{1}{27}$, think "3 raised to what power yields $\dfrac{1}{27}$." So,

$$y = \log_3 \dfrac{1}{27}$$
$$3^y = \dfrac{1}{27} \qquad \textit{Change to exponential form.}$$
$$3^y = 3^{-3} \qquad \dfrac{1}{27} = \dfrac{1}{3^3} = 3^{-3}$$
$$y = -3 \qquad \textit{Equate exponents.}$$

Therefore, $\log_3 \dfrac{1}{27} = -3$.

●

━━ **Now Work** PROBLEM 25

3 Determine the Domain of a Logarithmic Function

The logarithmic function $y = \log_a x$ has been defined as the inverse of the exponential function $y = a^x$. That is, if $f(x) = a^x$, then $f^{-1}(x) = \log_a x$. Based on the discussion given in Section 4.2 on inverse functions, for a function f and its inverse f^{-1}, we have

$$\text{Domain of } f^{-1} = \text{Range of } f \quad \text{and} \quad \text{Range of } f^{-1} = \text{Domain of } f$$

Consequently, it follows that

> Domain of the logarithmic function = Range of the exponential function = $(0, \infty)$
>
> Range of the logarithmic function = Domain of the exponential function = $(-\infty, \infty)$

In the next box, we summarize some properties of the logarithmic function:

> $y = \log_a x$ (defining equation: $x = a^y$)
>
> Domain: $0 < x < \infty$ Range: $-\infty < y < \infty$

The domain of a logarithmic function consists of the *positive* real numbers, so the argument of a logarithmic function must be greater than zero.

EXAMPLE 5	**Finding the Domain of a Logarithmic Function**

Find the domain of each logarithmic function.

(a) $F(x) = \log_2(x + 3)$ (b) $g(x) = \log_5\left(\dfrac{1 + x}{1 - x}\right)$ (c) $h(x) = \log_{1/2}|x|$

Solution

(a) The domain of F consists of all x for which $x + 3 > 0$, that is, $x > -3$. Using interval notation, the domain of f is $(-3, \infty)$.

(b) The domain of g is restricted to

$$\dfrac{1 + x}{1 - x} > 0$$

Solving this inequality, we find that the domain of g consists of all x between -1 and 1, that is, $-1 < x < 1$ or, using interval notation, $(-1, 1)$.

(c) Since $|x| > 0$, provided that $x \neq 0$, the domain of h consists of all real numbers except zero or, using interval notation, $(-\infty, 0) \cup (0, \infty)$.

──── **Now Work** PROBLEMS 39 AND 45

4 Graph Logarithmic Functions

Since exponential functions and logarithmic functions are inverses of each other, the graph of the logarithmic function $y = \log_a x$ is the reflection about the line $y = x$ of the graph of the exponential function $y = a^x$, as shown in Figure 30.

Figure 30

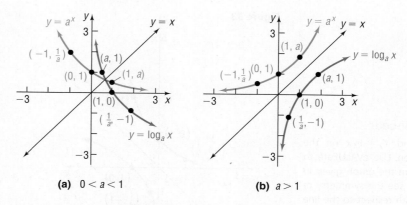

(a) $0 < a < 1$ **(b)** $a > 1$

For example, to graph $y = \log_2 x$, graph $y = 2^x$ and reflect it about the line $y = x$. See Figure 31. To graph $y = \log_{1/3} x$, graph $y = \left(\dfrac{1}{3}\right)^x$ and reflect it about the line $y = x$. See Figure 32.

Figure 31

Figure 32

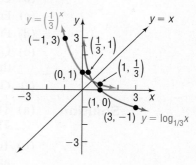

──── **Now Work** PROBLEM 59

The graphs of $y = \log_a x$ in Figures 30(a) and (b) lead to the following properties.

> **Properties of the Logarithmic Function $f(x) = \log_a x$**
>
> 1. The domain is the set of positive real numbers or $(0, \infty)$ using interval notation; the range is the set of all real numbers or $(-\infty, \infty)$ using interval notation.
> 2. The x-intercept of the graph is 1. There is no y-intercept.
> 3. The y-axis ($x = 0$) is a vertical asymptote of the graph.
> 4. A logarithmic function is decreasing if $0 < a < 1$ and increasing if $a > 1$.
> 5. The graph of f contains the points $(1, 0)$, $(a, 1)$, and $\left(\dfrac{1}{a}, -1\right)$.
> 6. The graph is smooth and continuous, with no corners or gaps.

If the base of a logarithmic function is the number e, then we have the **natural logarithm function.** This function occurs so frequently in applications that it is given a special symbol, **ln** (from the Latin, *logarithmus naturalis*). That is,

$$y = \ln x \quad \text{if and only if} \quad x = e^y \qquad \qquad \textbf{(1)}$$

Since $y = \ln x$ and the exponential function $y = e^x$ are inverse functions, we can obtain the graph of $y = \ln x$ by reflecting the graph of $y = e^x$ about the line $y = x$. See Figure 33.

Using a calculator with an $\boxed{\text{ln}}$ key, we can obtain other points on the graph of $f(x) = \ln x$. See Table 7.

Figure 33

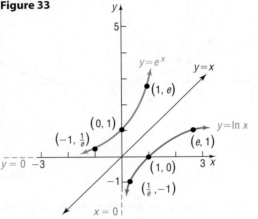

Table 7	x	$\ln x$
	$\dfrac{1}{2}$	-0.69
	2	0.69
	3	1.10

Seeing the Concept

Graph $Y_1 = e^x$ and $Y_2 = \ln x$ on the same square screen. Use eVALUEate to verify the points on the graph given in Figure 33. Do you see the symmetry of the two graphs with respect to the line $y = x$?

EXAMPLE 6

Graphing a Logarithmic Function and Its Inverse

(a) Find the domain of the logarithmic function $f(x) = -\ln(x - 2)$.
(b) Graph f.
(c) From the graph, determine the range and vertical asymptote of f.
(d) Find f^{-1}, the inverse of f.
(e) Use f^{-1} to confirm the range of f found in part (c). From the domain of f, find the range of f^{-1}.
(f) Graph f^{-1}.

Solution

(a) The domain of f consists of all x for which $x - 2 > 0$ or, equivalently, $x > 2$. The domain of f is $\{x \mid x > 2\}$ or $(2, \infty)$ in interval notation.

(b) To obtain the graph of $y = -\ln(x - 2)$, we begin with the graph of $y = \ln x$ and use transformations. See Figure 34.

Figure 34

(a) $y = \ln x$

Multiply by -1; reflect about x-axis

(b) $y = -\ln x$

Replace x by $x - 2$; shift right 2 units.

(c) $y = -\ln(x - 2)$

(c) The range of $f(x) = -\ln(x - 2)$ is the set of all real numbers. The vertical asymptote is $x = 2$. [Do you see why? The original asymptote $(x = 0)$ is shifted to the right 2 units.]

(d) To find f^{-1}, we begin with $y = -\ln(x - 2)$. The inverse function is defined (implicitly) by the equation

$$x = -\ln(y - 2)$$

We proceed to solve for y.

$$-x = \ln(y - 2) \quad \text{Isolate the logarithm.}$$
$$e^{-x} = y - 2 \quad \text{Change to an exponential equation.}$$
$$y = e^{-x} + 2 \quad \text{Solve for y.}$$

The inverse of f is $f^{-1}(x) = e^{-x} + 2$.

(e) The range of f is the domain of f^{-1}, which is the set of all real numbers, confirming what we found from the graph of f. The range of f^{-1} is the domain of f, which is $(2, \infty)$ in interval notation.

(f) To graph f^{-1}, we use the graph of f in Figure 34(c) and reflect it about the line $y = x$. See Figure 35. We could also graph $f^{-1}(x) = e^{-x} + 2$ using transformations.

Figure 35

Figure 36

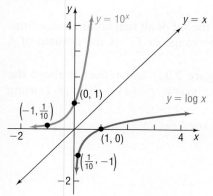

Now Work PROBLEM **71**

If the base of a logarithmic function is the number 10, then we have the **common logarithm function.** If the base a of the logarithmic function is not indicated, it is understood to be 10. That is,

$$y = \log x \quad \text{if and only if} \quad x = 10^y$$

Since $y = \log x$ and the exponential function $y = 10^x$ are inverse functions, we can obtain the graph of $y = \log x$ by reflecting the graph of $y = 10^x$ about the line $y = x$. See Figure 36.

EXAMPLE 7   **Graphing a Logarithmic Function and Its Inverse**

(a) Find the domain of the logarithmic function $f(x) = 3 \log (x - 1)$.
(b) Graph f.
(c) From the graph, determine the range and vertical asymptote of f.
(d) Find f^{-1}, the inverse of f.
(e) Use f^{-1} to confirm the range of f found in part (c). From the domain of f, find the range of f^{-1}.
(f) Graph f^{-1}.

Solution   (a) The domain of f consists of all x for which $x - 1 > 0$ or, equivalently, $x > 1$. The domain of f is $\{x | x > 1\}$ or $(1, \infty)$ in interval notation.

(b) To obtain the graph of $y = 3 \log(x - 1)$, we begin with the graph of $y = \log x$ and use transformations. See Figure 37.

Figure 37

(a) $y = \log x$ **(b)** $y = \log(x - 1)$ **(c)** $y = 3\log(x - 1)$

(c) The range of $f(x) = 3\log(x - 1)$ is the set of all real numbers. The vertical asymptote is $x = 1$.

(d) We begin with $y = 3\log(x - 1)$. The inverse function is defined (implicitly) by the equation

$$x = 3\log(y - 1)$$

We proceed to solve for y.

$$\frac{x}{3} = \log(y - 1) \quad \text{Isolate the logarithm.}$$

$$10^{x/3} = y - 1 \quad \text{Change to an exponential equation.}$$

$$y = 10^{x/3} + 1 \quad \text{Solve for } y.$$

The inverse of f is $f^{-1}(x) = 10^{x/3} + 1$.

(e) The range of f is the domain of f^{-1}, which is the set of all real numbers, confirming what we found from the graph of f. The range of f^{-1} is the domain of f, which is $(1, \infty)$ in interval notation.

(f) To graph f^{-1}, we use the graph of f in Figure 37(c) and reflect it about the line $y = x$. See Figure 38. We could also graph $f^{-1}(x) = 10^{x/3} + 1$ using transformations.

Figure 38

5 Solve Logarithmic Equations

Equations that contain logarithms are called **logarithmic equations.** Care must be taken when solving logarithmic equations algebraically. In the expression $\log_a M$, remember that a and M are positive and $a \neq 1$. Be sure to check each apparent solution in the original equation and discard any that are extraneous.

Some logarithmic equations can be solved by changing the logarithmic equation to exponential form using the fact that $y = \log_a x$ means $a^y = x$.

EXAMPLE 8   **Solving Logarithmic Equations**

Solve:

(a) $\log_3(4x - 7) = 2$ (b) $\log_x 64 = 2$

Solution   (a) We can obtain an exact solution by changing the logarithmic equation to exponential form.

$$\log_3(4x - 7) = 2$$
$$4x - 7 = 3^2 \qquad \text{Change to exponential form using } y = \log_a x$$
$$\text{means } a^y = x.$$
$$4x - 7 = 9$$
$$4x = 16$$
$$x = 4$$

✓ Check: $\log_3(4x - 7) = \log_3(4 \cdot 4 - 7) = \log_3 9 = 2 \quad 3^2 = 9$

The solution set is $\{4\}$.

(b) We can obtain an exact solution by changing the logarithmic equation to exponential form.

$$\log_x 64 = 2$$
$$x^2 = 64 \qquad \text{Change to exponential form.}$$
$$x = \pm\sqrt{64} = \pm 8 \quad \text{Square Root Method}$$

The base of a logarithm is always positive. As a result, we discard -8. We check the solution 8.

✓ Check: $\log_8 64 = 2 \quad 8^2 = 64$

The solution set is $\{8\}$.

●

EXAMPLE 9   **Using Logarithms to Solve an Exponential Equation**

Solve: $e^{2x} = 5$

Solution   We can obtain an exact solution by changing the exponential equation to logarithmic form.

$$e^{2x} = 5$$
$$\ln 5 = 2x \qquad \text{Change to logarithmic form using the}$$
$$\text{fact that if } e^y = x \text{ then } y = \ln x.$$
$$x = \frac{\ln 5}{2} \qquad \text{Exact solution}$$
$$\approx 0.805 \qquad \text{Approximate solution}$$

The solution set is $\left\{ \dfrac{\ln 5}{2} \right\}$.

●

━━━**Now Work** PROBLEMS 87 AND 99

EXAMPLE 10   **Alcohol and Driving**

Blood alcohol concentration (BAC) is a measure of the amount of alcohol in a person's bloodstream. A BAC of 0.04% means that a person has 4 parts alcohol per 10,000 parts blood in the body. Relative risk is defined as the likelihood of one event occurring divided by the likelihood of a second event occurring. For example, if an individual with a BAC of 0.02% is 1.4 times as likely to have a car accident as an individual that has not been drinking, the relative risk of an accident with a BAC of 0.02% is 1.4. Recent medical research suggests that the relative risk R of having an accident while driving a car can be modeled by an equation of the form

$$R = e^{kx}$$

where x is the percent of concentration of alcohol in the bloodstream and k is a constant.

(a) Research indicates that the relative risk of a person having an accident with a BAC of 0.02% is 1.4. Find the constant k in the equation.

(b) Using this value of k, what is the relative risk if the concentration is 0.17%?

(c) Using this same value of k, what BAC corresponds to a relative risk of 100?

(d) If the law asserts that anyone with a relative risk of 5 or more should not have driving privileges, at what concentration of alcohol in the bloodstream should a driver be arrested and charged with a DUI (driving under the influence)?

Solution

(a) For a concentration of alcohol in the blood of 0.02% and a relative risk of 1.4, we let $x = 0.02$ and $R = 1.4$ in the equation and solve for k.

$$R = e^{kx}$$
$$1.4 = e^{k(0.02)} \qquad \text{\color{gray}{$R = 1.4$; $x = 0.02$}}$$
$$0.02k = \ln 1.4 \qquad \text{\color{gray}{Change to a logarithmic equation.}}$$
$$k = \frac{\ln 1.4}{0.02} \approx 16.82 \qquad \text{\color{gray}{Solve for k.}}$$

(b) For a concentration of 0.17%, we have $x = 0.17$. Using $k = 16.82$ in the equation, we find the relative risk R to be

$$R = e^{kx} = e^{(16.82)(0.17)} \approx 17.5$$

For a concentration of alcohol in the blood of 0.17%, the relative risk of an accident is about 17.5. That is, a person with a BAC of 0.17% is 17.5 times as likely to have a car accident as a person with no alcohol in the bloodstream.

(c) For a relative risk of 100, we have $R = 100$. Using $k = 16.82$ in the equation $R = e^{kx}$, we find the concentration x of alcohol in the blood obeys

$$100 = e^{16.82x} \qquad \text{\color{gray}{$R = e^{kx}$, $R = 100$; $k = 16.82$}}$$
$$16.82x = \ln 100 \qquad \text{\color{gray}{Change to a logarithmic equation.}}$$
$$x = \frac{\ln 100}{16.82} \approx 0.27 \qquad \text{\color{gray}{Solve for x.}}$$

NOTE A BAC of 0.30% results in a loss of consciousness in most people. ∎

For a concentration of alcohol in the blood of 0.27%, the relative risk of an accident is 100.

(d) For a relative risk of 5, we have $R = 5$. Using $k = 16.82$ in the equation $R = e^{kx}$, we find the concentration x of alcohol in the bloodstream obeys

$$5 = e^{16.82x}$$
$$16.82x = \ln 5$$
$$x = \frac{\ln 5}{16.82} \approx 0.096$$

NOTE Most states use 0.08% or 0.10% as the blood alcohol content at which a DUI citation is given. ∎

A driver with a BAC of 0.096% or more should be arrested and charged with DUI.

●

SUMMARY

Properties of the Logarithmic Function

$f(x) = \log_a x, \quad a > 1$
($y = \log_a x$ means $x = a^y$)

Domain: the interval $(0, \infty)$; Range: the interval $(-\infty, \infty)$

x-intercept: 1; y-intercept: none; vertical asymptote: $x = 0$ (y-axis); increasing; one-to-one

See Figure 39(a) for a typical graph.

$f(x) = \log_a x, \quad 0 < a < 1$
($y = \log_a x$ means $x = a^y$)

Domain: the interval $(0, \infty)$; Range: the interval $(-\infty, \infty)$

x-intercept: 1; y-intercept: none; vertical asymptote: $x = 0$ (y-axis); decreasing; one-to-one

See Figure 39(b) for a typical graph.

Figure 39

(a) $a > 1$ **(b)** $0 < a < 1$

4.4 Assess Your Understanding

'Are You Prepared?' *Answers are given at the end of these exercises. If you get a wrong answer, read the pages listed in red.*

1. Solve each inequality:
   (a) $3x - 7 \leq 8 - 2x$ (pp. A86–A87)
   (b) $x^2 - x - 6 > 0$ (pp. 164–166)

2. Solve the inequality: $\dfrac{x - 1}{x + 4} > 0$ (pp. 243–245)

3. Solve: $2x + 3 = 9$ (pp. A65–A68)

Concepts and Vocabulary

4. The domain of the logarithmic function $f(x) = \log_a x$ is _____.

5. The graph of every logarithmic function $f(x) = \log_a x, a > 0$, $a \neq 1$, passes through three points: _____, _____, and _____.

6. If the graph of a logarithmic function $f(x) = \log_a x, a > 0$, $a \neq 1$, is increasing, then its base must be larger than _____.

7. *True or False* If $y = \log_a x$, then $y = a^x$.

8. *True or False* The graph of $f(x) = \log_a x, a > 0, a \neq 1$, has an x-intercept equal to 1 and no y-intercept.

Skill Building

In Problems 9–16, change each exponential statement to an equivalent statement involving a logarithm.

9. $9 = 3^2$

10. $16 = 4^2$

11. $a^2 = 1.6$

12. $a^3 = 2.1$

13. $2^x = 7.2$

14. $3^x = 4.6$

15. $e^x = 8$

16. $e^{2.2} = M$

In Problems 17–24, change each logarithmic statement to an equivalent statement involving an exponent.

17. $\log_2 8 = 3$

18. $\log_3\left(\dfrac{1}{9}\right) = -2$

19. $\log_a 3 = 6$

20. $\log_b 4 = 2$

21. $\log_3 2 = x$

22. $\log_2 6 = x$

23. $\ln 4 = x$

24. $\ln x = 4$

In Problems 25–36, find the exact value of each logarithm without using a calculator.

25. $\log_2 1$

26. $\log_8 8$

27. $\log_5 25$

28. $\log_3\left(\dfrac{1}{9}\right)$

29. $\log_{1/2} 16$

30. $\log_{1/3} 9$

31. $\log_{10}\sqrt{10}$

32. $\log_5\sqrt[3]{25}$

33. $\log_{\sqrt{2}} 4$

34. $\log_{\sqrt{3}} 9$

35. $\ln\sqrt{e}$

36. $\ln e^3$

In Problems 37–48, find the domain of each function.

37. $f(x) = \ln(x - 3)$

38. $g(x) = \ln(x - 1)$

39. $F(x) = \log_2 x^2$

40. $H(x) = \log_5 x^3$

41. $f(x) = 3 - 2\log_4\left[\dfrac{x}{2} - 5\right]$

42. $g(x) = 8 + 5\ln(2x + 3)$

43. $f(x) = \ln\left(\dfrac{1}{x + 1}\right)$

44. $g(x) = \ln\left(\dfrac{1}{x - 5}\right)$

45. $g(x) = \log_5\left(\dfrac{x + 1}{x}\right)$

46. $h(x) = \log_3\left(\dfrac{x}{x - 1}\right)$

47. $f(x) = \sqrt{\ln x}$

48. $g(x) = \dfrac{1}{\ln x}$

In Problems 49–56, use a calculator to evaluate each expression. Round your answer to three decimal places.

49. $\ln\dfrac{5}{3}$

50. $\dfrac{\ln 5}{3}$

51. $\dfrac{\ln\dfrac{10}{3}}{0.04}$

52. $\dfrac{\ln\dfrac{2}{3}}{-0.1}$

53. $\dfrac{\ln 4 + \ln 2}{\log 4 + \log 2}$

54. $\dfrac{\log 15 + \log 20}{\ln 15 + \ln 20}$

55. $\dfrac{2\ln 5 + \log 50}{\log 4 - \ln 2}$

56. $\dfrac{3\log 80 - \ln 5}{\log 5 + \ln 20}$

57. Find a so that the graph of $f(x) = \log_a x$ contains the point $(2, 2)$.

58. Find a so that the graph of $f(x) = \log_a x$ contains the point $\left(\dfrac{1}{2}, -4\right)$.

In Problems 59–62, graph each function and its inverse on the same Cartesian plane.

59. $f(x) = 3^x;\ f^{-1}(x) = \log_3 x$

60. $f(x) = 4^x;\ f^{-1}(x) = \log_4 x$

61. $f(x) = \left(\dfrac{1}{2}\right)^x;\ f^{-1}(x) = \log_{\frac{1}{2}} x$

62. $f(x) = \left(\dfrac{1}{3}\right)^x;\ f^{-1}(x) = \log_{\frac{1}{3}} x$

In Problems 63–70, the graph of a logarithmic function is given. Match each graph to one of the following functions:

(a) $y = \log_3 x$

(b) $y = \log_3(-x)$

(c) $y = -\log_3 x$

(d) $y = -\log_3(-x)$

(e) $y = \log_3 x - 1$

(f) $y = \log_3(x - 1)$

(g) $y = \log_3(1 - x)$

(h) $y = 1 - \log_3 x$

63.

64.

65.

66.

67.

68.

69.

70.

In Problems 71–86, use the given function f to:

(a) *Find the domain of f.* (b) *Graph f.* (c) *From the graph, determine the range and any asymptotes of f.* (d) *Find f^{-1}, the inverse of f.* (e) *Use f^{-1} to confirm the range of f found in part (c). From the domain of f, find the range of f^{-1}.* (f) *Graph f^{-1}.*

71. $f(x) = \ln(x + 4)$

72. $f(x) = \ln(x - 3)$

73. $f(x) = 2 + \ln x$

74. $f(x) = -\ln(-x)$

75. $f(x) = \ln(2x) - 3$ **76.** $f(x) = -2\ln(x + 1)$ **77.** $f(x) = \log(x - 4) + 2$ **78.** $f(x) = \frac{1}{2}\log x - 5$

79. $f(x) = \frac{1}{2}\log(2x)$ **80.** $f(x) = \log(-2x)$ **81.** $f(x) = 3 + \log_3(x + 2)$ **82.** $f(x) = 2 - \log_3(x + 1)$

83. $f(x) = e^{x+2} - 3$ **84.** $f(x) = 3e^x + 2$ **85.** $f(x) = 2^{x/3} + 4$ **86.** $f(x) = -3^{x+1}$

In Problems 87–110, solve each equation.

87. $\log_3 x = 2$ **88.** $\log_5 x = 3$ **89.** $\log_2(2x + 1) = 3$ **90.** $\log_3(3x - 2) = 2$

91. $\log_x 4 = 2$ **92.** $\log_x\left(\frac{1}{8}\right) = 3$ **93.** $\ln e^x = 5$ **94.** $\ln e^{-2x} = 8$

95. $\log_4 64 = x$ **96.** $\log_5 625 = x$ **97.** $\log_3 243 = 2x + 1$ **98.** $\log_6 36 = 5x + 3$

99. $e^{3x} = 10$ **100.** $e^{-2x} = \frac{1}{3}$ **101.** $e^{2x+5} = 8$ **102.** $e^{-2x+1} = 13$

103. $\log_3(x^2 + 1) = 2$ **104.** $\log_5(x^2 + x + 4) = 2$ **105.** $\log_2 8^x = -3$ **106.** $\log_3 3^x = -1$

107. $5e^{0.2x} = 7$ **108.** $8 \cdot 10^{2x-7} = 3$ **109.** $2 \cdot 10^{2-x} = 5$ **110.** $4e^{x+1} = 5$

Mixed Practice

111. Suppose that $G(x) = \log_3(2x + 1) - 2$.
   (a) What is the domain of G?
   (b) What is $G(40)$? What point is on the graph of G?
   (c) If $G(x) = 3$, what is x? What point is on the graph of G?
   (d) What is the zero of G?

112. Suppose that $F(x) = \log_2(x + 1) - 3$.
   (a) What is the domain of F?
   (b) What is $F(7)$? What point is on the graph of F?
   (c) If $F(x) = -1$, what is x? What point is on the graph of F?
   (d) What is the zero of F?

In Problems 113–116, graph each function. Based on the graph, state the domain and the range and find any intercepts.

113. $f(x) = \begin{cases} \ln(-x) & \text{if } x < 0 \\ \ln x & \text{if } x > 0 \end{cases}$

114. $f(x) = \begin{cases} \ln(-x) & \text{if } x \le -1 \\ -\ln(-x) & \text{if } -1 < x < 0 \end{cases}$

115. $f(x) = \begin{cases} -\ln x & \text{if } 0 < x < 1 \\ \ln x & \text{if } x \ge 1 \end{cases}$

116. $f(x) = \begin{cases} \ln x & \text{if } 0 < x < 1 \\ -\ln x & \text{if } x \ge 1 \end{cases}$

Applications and Extensions

117. Chemistry   The pH of a chemical solution is given by the formula

$$\text{pH} = -\log_{10}[\text{H}^+]$$

where $[\text{H}^+]$ is the concentration of hydrogen ions in moles per liter. Values of pH range from 0 (acidic) to 14 (alkaline).
   (a) What is the pH of a solution for which $[\text{H}^+]$ is 0.1?
   (b) What is the pH of a solution for which $[\text{H}^+]$ is 0.01?
   (c) What is the pH of a solution for which $[\text{H}^+]$ is 0.001?
   (d) What happens to pH as the hydrogen ion concentration decreases?
   (e) Determine the hydrogen ion concentration of an orange (pH = 3.5).
   (f) Determine the hydrogen ion concentration of human blood (pH = 7.4).

118. Diversity Index   **Shannon's diversity index** is a measure of the diversity of a population. The diversity index is given by the formula

$$H = -(p_1 \log p_1 + p_2 \log p_2 + \cdots + p_n \log p_n)$$

where p_1 is the proportion of the population that is species 1, p_2 is the proportion of the population that is species 2, and so on.
   (a) According to the U.S. Census Bureau, the distribution of race in the United States in 2007 was as follows:

Race	Proportion
American Indian or Native Alaskan	0.015
Asian	0.042
Black or African American	0.129
Hispanic	0.125
Native Hawaiian or Pacific Islander	0.003
White	0.686

Source: U.S. Census Bureau

Compute the diversity index of the United States in 2007.

(b) The largest value of the diversity index is given by $H_{max} = \log(S)$, where S is the number of categories of race. Compute H_{max}.

(c) The **evenness ratio** is given by $E_H = \dfrac{H}{H_{max}}$, where $0 \le E_H \le 1$. If $E_H = 1$, there is complete evenness. Compute the evenness ratio for the United States.

(d) Obtain the distribution of race for the United States in 2000 from the Census Bureau. Compute Shannon's diversity index. Is the United States becoming more diverse? Why?

119. **Atmospheric Pressure** The atmospheric pressure p on an object decreases with increasing height. This pressure, measured in millimeters of mercury, is related to the height h (in kilometers) above sea level by the function

$$p(h) = 760e^{-0.145h}$$

(a) Find the height of an aircraft if the atmospheric pressure is 320 millimeters of mercury.

(b) Find the height of a mountain if the atmospheric pressure is 667 millimeters of mercury.

120. **Healing of Wounds** The normal healing of wounds can be modeled by an exponential function. If A_0 represents the original area of the wound and if A equals the area of the wound, then the function

$$A(n) = A_0 e^{-0.35n}$$

describes the area of a wound after n days following an injury when no infection is present to retard the healing. Suppose that a wound initially had an area of 100 square millimeters.

(a) If healing is taking place, after how many days will the wound be one-half its original size?

(b) How long before the wound is 10% of its original size?

121. **Exponential Probability** Between 12:00 PM and 1:00 PM, cars arrive at Citibank's drive-thru at the rate of 6 cars per hour (0.1 car per minute). The following formula from statistics can be used to determine the probability that a car will arrive within t minutes of 12:00 PM.

$$F(t) = 1 - e^{-0.1t}$$

(a) Determine how many minutes are needed for the probability to reach 50%.

(b) Determine how many minutes are needed for the probability to reach 80%.

(c) Is it possible for the probability to equal 100%? Explain.

122. **Exponential Probability** Between 5:00 PM and 6:00 PM, cars arrive at Jiffy Lube at the rate of 9 cars per hour (0.15 car per minute). The following formula from statistics can be used to determine the probability that a car will arrive within t minutes of 5:00 PM.

$$F(t) = 1 - e^{-0.15t}$$

(a) Determine how many minutes are needed for the probability to reach 50%.

(b) Determine how many minutes are needed for the probability to reach 80%.

123. **Drug Medication** The formula

$$D = 5e^{-0.4h}$$

can be used to find the number of milligrams D of a certain drug that is in a patient's bloodstream h hours after the drug was administered. When the number of milligrams reaches 2, the drug is to be administered again. What is the time between injections?

124. **Spreading of Rumors** A model for the number N of people in a college community who have heard a certain rumor is

$$N = P(1 - e^{-0.15d})$$

where P is the total population of the community and d is the number of days that have elapsed since the rumor began. In a community of 1000 students, how many days will elapse before 450 students have heard the rumor?

125. **Current in a RL Circuit** The equation governing the amount of current I (in amperes) after time t (in seconds) in a simple RL circuit consisting of a resistance R (in ohms), an inductance L (in henrys), and an electromotive force E (in volts) is

$$I = \frac{E}{R}\left[1 - e^{-(R/L)t}\right]$$

If $E = 12$ volts, $R = 10$ ohms, and $L = 5$ henrys, how long does it take to obtain a current of 0.5 ampere? Of 1.0 ampere? Graph the equation.

126. **Learning Curve** Psychologists sometimes use the function

$$L(t) = A(1 - e^{-kt})$$

to measure the amount L learned at time t. The number A represents the amount to be learned, and the number k measures the rate of learning. Suppose that a student has an amount A of 200 vocabulary words to learn. A psychologist determines that the student learned 20 vocabulary words after 5 minutes.

(a) Determine the rate of learning k.

(b) Approximately how many words will the student have learned after 10 minutes?

(c) After 15 minutes?

(d) How long does it take for the student to learn 180 words?

Loudness of Sound *Problems 127–130 use the following discussion: The* **loudness** $L(x)$*, measured in decibels (dB), of a sound of intensity x, measured in watts per square meter, is defined as $L(x) = 10 \log \dfrac{x}{I_0}$, where $I_0 = 10^{-12}$ watt per square meter is the least intense sound that a human ear can detect. Determine the loudness, in decibels, of each of the following sounds.*

127. Normal conversation: intensity of $x = 10^{-7}$ watt per square meter.

128. Amplified rock music: intensity of 10^{-1} watt per square meter.

129. Heavy city traffic: intensity of $x = 10^{-3}$ watt per square meter.

130. Diesel truck traveling 40 miles per hour 50 feet away: intensity 10 times that of a passenger car traveling 50 miles per hour 50 feet away whose loudness is 70 decibels.

The Richter Scale *Problems 131 and 132 use the following discussion: The **Richter scale** is one way of converting seismographic readings into numbers that provide an easy reference for measuring the magnitude M of an earthquake. All earthquakes are compared to a **zero-level earthquake** whose seismographic reading measures 0.001 millimeter at a distance of 100 kilometers from the epicenter. An earthquake whose seismographic reading measures x millimeters has **magnitude M(x)**, given by*

$$M(x) = \log\left(\frac{x}{x_0}\right)$$

where $x_0 = 10^{-3}$ is the reading of a zero-level earthquake the same distance from its epicenter. In Problems 131 and 132, determine the magnitude of each earthquake.

131. Magnitude of an Earthquake Mexico City in 1985: seismographic reading of 125,892 millimeters 100 kilometers from the center

132. Magnitude of an Earthquake San Francisco in 1906: seismographic reading of 50,119 millimeters 100 kilometers from the center

133. Alcohol and Driving The concentration of alcohol in a person's bloodstream is measurable. Suppose that the relative risk R of having an accident while driving a car can be modeled by an equation of the form

$$R = e^{kx}$$

where x is the percent of concentration of alcohol in the bloodstream and k is a constant.

(a) Suppose that a concentration of alcohol in the bloodstream of 0.03 percent results in a relative risk of an accident of 1.4. Find the constant k in the equation.

(b) Using this value of k, what is the relative risk if the concentration is 0.17 percent?

(c) Using the same value of k, what concentration of alcohol corresponds to a relative risk of 100?

(d) If the law asserts that anyone with a relative risk of having an accident of 5 or more should not have driving privileges, at what concentration of alcohol in the bloodstream should a driver be arrested and charged with a DUI?

(e) Compare this situation with that of Example 10. If you were a lawmaker, which situation would you support? Give your reasons.

Explaining Concepts: Discussion and Writing

134. Is there any function of the form $y = x^\alpha, 0 < \alpha < 1$, that increases more slowly than a logarithmic function whose base is greater than 1? Explain.

135. In the definition of the logarithmic function, the base a is not allowed to equal 1. Why?

136. Critical Thinking In buying a new car, one consideration might be how well the price of the car holds up over time. Different makes of cars have different depreciation rates. One way to compute a depreciation rate for a car is given here. Suppose that the current prices of a certain automobile are as shown in the table.

	Age in Years				
New	**1**	**2**	**3**	**4**	**5**
$38,000	$36,600	$32,400	$28,750	$25,400	$21,200

Use the formula New = Old(e^{Rt}) to find R, the annual depreciation rate, for a specific time t. When might be the best time to trade in the car? Consult the NADA ("blue") book and compare two like models that you are interested in. Which has the better depreciation rate?

'Are You Prepared?' Answers

1. (a) $x \le 3$ (b) $x < -2$ or $x > 3$ **2.** $x < -4$ or $x > 1$ **3.** {3}

4.5 Properties of Logarithms

OBJECTIVES **1** Work with the Properties of Logarithms (p. 325)

2 Write a Logarithmic Expression as a Sum or Difference of Logarithms (p. 327)

3 Write a Logarithmic Expression as a Single Logarithm (p. 328)

4 Evaluate Logarithms Whose Base Is Neither 10 Nor e (p. 329)

1 Work with the Properties of Logarithms

Logarithms have some very useful properties that can be derived directly from the definition and the laws of exponents.

| EXAMPLE 1 | **Establishing Properties of Logarithms** |

(a) Show that $\log_a 1 = 0$.   (b) Show that $\log_a a = 1$.

Solution   (a) This fact was established when we graphed $y = \log_a x$ (see Figure 30 on page 315). To show the result algebraically, let $y = \log_a 1$. Then

$$y = \log_a 1$$
$$a^y = 1 \qquad \textit{Change to an exponential equation.}$$
$$a^y = a^0 \qquad \textit{a}^0 = 1 \textit{ since } a > 0, a \neq 1$$
$$y = 0 \qquad \textit{Solve for y.}$$
$$\log_a 1 = 0 \qquad y = \log_a 1$$

(b) Let $y = \log_a a$. Then

$$y = \log_a a$$
$$a^y = a \qquad \textit{Change to an exponential equation.}$$
$$a^y = a^1 \qquad \textit{a} = a^1$$
$$y = 1 \qquad \textit{Solve for y.}$$
$$\log_a a = 1 \qquad y = \log_a a$$

To summarize:

$$\log_a 1 = 0 \qquad \log_a a = 1$$

THEOREM   **Properties of Logarithms**

In the properties given next, M and a are positive real numbers, $a \neq 1$, and r is any real number.

The number $\log_a M$ is the exponent to which a must be raised to obtain M. That is,

$$a^{\log_a M} = M \tag{1}$$

The logarithm to the base a of a raised to a power equals that power. That is,

$$\log_a a^r = r \tag{2}$$

The proof uses the fact that $y = a^x$ and $y = \log_a x$ are inverses.

Proof of Property (1)   For inverse functions,

$$f(f^{-1}(x)) = x \quad \text{for all } x \text{ in the domain of } f^{-1}$$

Using $f(x) = a^x$ and $f^{-1}(x) = \log_a x$, we find

$$f(f^{-1}(x)) = a^{\log_a x} = x \quad \text{for } x > 0$$

Now let $x = M$ to obtain $a^{\log_a M} = M$, where $M > 0$. ∎

Proof of Property (2)   For inverse functions,

$$f^{-1}(f(x)) = x \quad \text{for all } x \text{ in the domain of } f$$

Using $f(x) = a^x$ and $f^{-1}(x) = \log_a x$, we find

$$f^{-1}(f(x)) = \log_a a^x = x \quad \text{for all real numbers } x$$

Now let $x = r$ to obtain $\log_a a^r = r$, where r is any real number. ∎

EXAMPLE 2   **Using Properties (1) and (2)**

(a) $2^{\log_2 \pi} = \pi$ (b) $\log_{0.2} 0.2^{-\sqrt{2}} = -\sqrt{2}$ (c) $\ln e^{kt} = kt$

●

Now Work PROBLEM 15

Other useful properties of logarithms are given next.

THEOREM | **Properties of Logarithms**

In the following properties, M, N, and a are positive real numbers, $a \neq 1$, and r is any real number.

The Log of a Product Equals the Sum of the Logs

$$\log_a(MN) = \log_a M + \log_a N \tag{3}$$

The Log of a Quotient Equals the Difference of the Logs

$$\log_a\left(\frac{M}{N}\right) = \log_a M - \log_a N \tag{4}$$

The Log of a Power Equals the Product of the Power and the Log

$$\log_a M^r = r \log_a M \tag{5}$$

We shall derive properties (3) and (5) and leave the derivation of property (4) as an exercise (see Problem 109).

Proof of Property (3)   Let $A = \log_a M$ and let $B = \log_a N$. These expressions are equivalent to the exponential expressions

$$a^A = M \quad \text{and} \quad a^B = N$$

Now

$$\begin{aligned}
\log_a(MN) = \log_a(a^A a^B) &= \log_a a^{A+B} &&\text{Law of Exponents}\\
&= A + B &&\text{Property (2) of logarithms}\\
&= \log_a M + \log_a N
\end{aligned}$$

■

Proof of Property (5)   Let $A = \log_a M$. This expression is equivalent to

$$a^A = M$$

Now

$$\begin{aligned}
\log_a M^r = \log_a(a^A)^r &= \log_a a^{rA} &&\text{Law of Exponents}\\
&= rA &&\text{Property (2) of logarithms}\\
&= r \log_a M
\end{aligned}$$

■

Now Work PROBLEM 19

2 Write a Logarithmic Expression as a Sum or Difference of Logarithms

Logarithms can be used to transform products into sums, quotients into differences, and powers into factors. Such transformations prove useful in certain types of calculus problems.

EXAMPLE 3 **Writing a Logarithmic Expression as a Sum of Logarithms**

Write $\log_a\left(x\sqrt{x^2 + 1}\right)$, $x > 0$, as a sum of logarithms. Express all powers as factors.

Solution
$$\log_a\left(x\sqrt{x^2 + 1}\right) = \log_a x + \log_a \sqrt{x^2 + 1} \qquad \log_a(M \cdot N) = \log_a M + \log_a N$$
$$= \log_a x + \log_a(x^2 + 1)^{1/2}$$
$$= \log_a x + \frac{1}{2}\log_a(x^2 + 1) \qquad \log_a M^r = r\log_a M$$

EXAMPLE 4 **Writing a Logarithmic Expression as a Difference of Logarithms**

Write

$$\ln\frac{x^2}{(x - 1)^3} \qquad x > 1$$

as a difference of logarithms. Express all powers as factors.

Solution
$$\ln\frac{x^2}{(x - 1)^3} = \ln x^2 - \ln(x - 1)^3 = 2\ln x - 3\ln(x - 1)$$
$$\uparrow \qquad\qquad\qquad \uparrow$$
$$\log_a\left(\frac{M}{N}\right) = \log_a M - \log_a N \quad \log_a M^r = r\log_a M$$

EXAMPLE 5 **Writing a Logarithmic Expression as a Sum and Difference of Logarithms**

Write

$$\log_a\frac{\sqrt{x^2 + 1}}{x^3(x + 1)^4} \qquad x > 0$$

as a sum and difference of logarithms. Express all powers as factors.

Solution
$$\log_a\frac{\sqrt{x^2 + 1}}{x^3(x + 1)^4} = \log_a\sqrt{x^2 + 1} - \log_a[x^3(x + 1)^4] \qquad \text{Property (4)}$$
$$= \log_a\sqrt{x^2 + 1} - [\log_a x^3 + \log_a(x + 1)^4] \qquad \text{Property (3)}$$
$$= \log_a(x^2 + 1)^{1/2} - \log_a x^3 - \log_a(x + 1)^4$$
$$= \frac{1}{2}\log_a(x^2 + 1) - 3\log_a x - 4\log_a(x + 1) \qquad \text{Property (5)}$$

WARNING In using properties (3) through (5), be careful about the values that the variable may assume. For example, the domain of the variable for $\log_a x$ is $x > 0$ and for $\log_a(x - 1)$ it is $x > 1$. If we add these functions, the domain is $x > 1$. That is, the equality

$$\log_a x + \log_a(x - 1) = \log_a[x(x - 1)]$$

is true only for $x > 1$. ∎

━━━ **Now Work** PROBLEM 51

3 Write a Logarithmic Expression as a Single Logarithm

Another use of properties (3) through (5) is to write sums and/or differences of logarithms with the same base as a single logarithm. This skill will be needed to solve certain logarithmic equations discussed in the next section.

EXAMPLE 6 **Writing Expressions as a Single Logarithm**

Write each of the following as a single logarithm.

(a) $\log_a 7 + 4\log_a 3$ (b) $\frac{2}{3}\ln 8 - \ln(5^2 - 1)$

(c) $\log_a x + \log_a 9 + \log_a(x^2 + 1) - \log_a 5$

Solution   (a) $\log_a 7 + 4 \log_a 3 = \log_a 7 + \log_a 3^4$ $r \log_a M = \log_a M^r$

$$= \log_a 7 + \log_a 81$$

$$= \log_a(7 \cdot 81)$$ $\log_a M + \log_a N = \log_a(M \cdot N)$

$$= \log_a 567$$

(b) $\dfrac{2}{3}\ln 8 - \ln(5^2 - 1) = \ln 8^{2/3} - \ln(25 - 1)$ $r \log_a M = \log_a M^r$

$$= \ln 4 - \ln 24$$ $8^{2/3} = (\sqrt[3]{8})^2 = 2^2 = 4$

$$= \ln\left(\frac{4}{24}\right)$$ $\log_a M - \log_a N = \log_a\left(\dfrac{M}{N}\right)$

$$= \ln\left(\frac{1}{6}\right)$$

$$= \ln 1 - \ln 6$$

$$= -\ln 6$$ $\ln 1 = 0$

(c) $\log_a x + \log_a 9 + \log_a(x^2 + 1) - \log_a 5 = \log_a(9x) + \log_a(x^2 + 1) - \log_a 5$

$$= \log_a[9x(x^2 + 1)] - \log_a 5$$

$$= \log_a\left[\frac{9x(x^2 + 1)}{5}\right]$$

WARNING   A common error made by some students is to express the logarithm of a sum as the sum of logarithms.

$$\log_a(M + N) \quad \textit{is not equal to} \quad \log_a M + \log_a N$$

Correct statement $\log_a(MN) = \log_a M + \log_a N$ Property (3)

Another common error is to express the difference of logarithms as the quotient of logarithms.

$$\log_a M - \log_a N \quad \textit{is not equal to} \quad \frac{\log_a M}{\log_a N}$$

Correct statement $\log_a M - \log_a N = \log_a\left(\dfrac{M}{N}\right)$ Property (4)

A third common error is to express a logarithm raised to a power as the product of the power times the logarithm.

$$(\log_a M)^r \quad \textit{is not equal to} \quad r \log_a M$$

Correct statement $\log_a M^r = r \log_a M$ Property (5) ∎

🖉 **Now Work**   PROBLEM **57**

Two other properties of logarithms that we need to know are consequences of the fact that the logarithmic function $y = \log_a x$ is a one-to-one function.

THEOREM

Properties of Logarithms

In the following properties, M, N, and a are positive real numbers, $a \neq 1$.

If $M = N$, then $\log_a M = \log_a N$.   **(6)**
If $\log_a M = \log_a N$, then $M = N$.   **(7)**

When property (6) is used, we start with the equation $M = N$ and say "take the logarithm of both sides" to obtain $\log_a M = \log_a N$.

Properties (6) and (7) are useful for solving *exponential and logarithmic equations*, a topic discussed in the next section.

4 Evaluate Logarithms Whose Base Is Neither 10 Nor *e*

Logarithms to the base 10, common logarithms, were used to facilitate arithmetic computations before the widespread use of calculators. (See the Historical Feature at the end of this section.) Natural logarithms, that is, logarithms whose base is the number *e*, remain very important because they arise frequently in the study of natural phenomena.

Common logarithms are usually abbreviated by writing **log,** with the base understood to be 10, just as natural logarithms are abbreviated by **ln,** with the base understood to be *e*.

Most calculators have both $\boxed{\log}$ and $\boxed{\ln}$ keys to calculate the common logarithm and natural logarithm of a number. Let's look at an example to see how to approximate logarithms having a base other than 10 or *e*.

EXAMPLE 7   **Approximating a Logarithm Whose Base Is Neither 10 Nor *e***

Approximate $\log_2 7$. Round the answer to four decimal places.

Solution   Remember, $\log_2 7$ means "2 raised to what exponent equals 7." If we let $y = \log_2 7$, then $2^y = 7$. Because $2^2 = 4$ and $2^3 = 8$, we expect $\log_2 7$ to be between 2 and 3.

$$2^y = 7$$
$$\ln 2^y = \ln 7 \quad \text{Property (6)}$$
$$y \ln 2 = \ln 7 \quad \text{Property (5)}$$
$$y = \frac{\ln 7}{\ln 2} \quad \text{Exact value}$$
$$y \approx 2.8074 \quad \text{Approximate value rounded to four decimal places}$$

Example 7 shows how to approximate a logarithm whose base is 2 by changing to logarithms involving the base *e*. In general, we use the **Change-of-Base Formula.**

THEOREM

Change-of-Base Formula

If $a \neq 1$, $b \neq 1$, and M are positive real numbers, then

$$\log_a M = \frac{\log_b M}{\log_b a} \qquad (8)$$

Proof   We derive this formula as follows: Let $y = \log_a M$. Then

$$a^y = M$$
$$\log_b a^y = \log_b M \quad \text{Property (6)}$$
$$y \log_b a = \log_b M \quad \text{Property (5)}$$
$$y = \frac{\log_b M}{\log_b a} \quad \text{Solve for y.}$$
$$\log_a M = \frac{\log_b M}{\log_b a} \quad y = \log_a M$$

Since calculators have keys only for $\boxed{\log}$ and $\boxed{\ln}$, in practice, the Change-of-Base Formula uses either $b = 10$ or $b = e$. That is,

$$\log_a M = \frac{\log M}{\log a} \quad \text{and} \quad \log_a M = \frac{\ln M}{\ln a} \qquad (9)$$

EXAMPLE 8   **Using the Change-of-Base Formula**

Approximate:

(a) $\log_5 89$ (b) $\log_{\sqrt{2}} \sqrt{5}$

Round answers to four decimal places.

Solution   (a) $\log_5 89 = \dfrac{\log 89}{\log 5} \approx \dfrac{1.949390007}{0.6989700043}$
≈ 2.7889

or

$\log_5 89 = \dfrac{\ln 89}{\ln 5} \approx \dfrac{4.48863637}{1.609437912}$
≈ 2.7889

(b) $\log_{\sqrt{2}} \sqrt{5} = \dfrac{\log \sqrt{5}}{\log \sqrt{2}} = \dfrac{\frac{1}{2}\log 5}{\frac{1}{2}\log 2}$

$= \dfrac{\log 5}{\log 2} \approx 2.3219$

or

$\log_{\sqrt{2}} \sqrt{5} = \dfrac{\ln \sqrt{5}}{\ln \sqrt{2}} = \dfrac{\frac{1}{2}\ln 5}{\frac{1}{2}\ln 2}$

$= \dfrac{\ln 5}{\ln 2} \approx 2.3219$

Now Work PROBLEMS 23 AND 71

 COMMENT To graph logarithmic functions when the base is different from e or 10 requires the Change-of-Base Formula. For example, to graph $y = \log_2 x$, we would instead graph $y = \dfrac{\ln x}{\ln 2}$. Try it.

Now Work PROBLEM 79

SUMMARY

Properties of Logarithms

In the list that follows, a, b, M, N, and r are real numbers. Also, $a > 0$, $a \neq 1$, $b > 0$, $b \neq 1$, $M > 0$, and $N > 0$.

Definition $y = \log_a x$ means $x = a^y$

Properties of logarithms $\log_a 1 = 0$; $\log_a a = 1$ $\log_a M^r = r \log_a M$

$a^{\log_a M} = M$; $\log_a a^r = r$ If $M = N$, then $\log_a M = \log_a N$.

$\log_a(MN) = \log_a M + \log_a N$ If $\log_a M = \log_a N$, then $M = N$.

$\log_a\left(\dfrac{M}{N}\right) = \log_a M - \log_a N$

Change-of-Base Formula $\log_a M = \dfrac{\log_b M}{\log_b a}$

Historical Feature

*John Napier
(1550–1617)*

Logarithms were invented about 1590 by John Napier (1550–1617) and Joost Bürgi (1552–1632), working independently. Napier, whose work had the greater influence, was a Scottish lord, a secretive man whose neighbors were inclined to believe him to be in league with the devil. His approach to logarithms was very different from ours; it was based on the relationship between arithmetic and geometric sequences, discussed in a later chapter, and not on the inverse function relationship of logarithms to exponential functions (described in Section 4.4).

Napier's tables, published in 1614, listed what would now be called *natural logarithms* of sines and were rather difficult to use. A London professor, Henry Briggs, became interested in the tables and visited Napier. In their conversations, they developed the idea of common logarithms, which were published in 1617. Their importance for calculation was immediately recognized, and by 1650 they were being printed as far away as China. They remained an important calculation tool until the advent of the inexpensive handheld calculator about 1972, which has decreased their calculational, but not their theoretical, importance.

A side effect of the invention of logarithms was the popularization of the decimal system of notation for real numbers.

4.5 Assess Your Understanding

Concepts and Vocabulary

1. $\log_a 1 = $ _____

2. $\log_a a = $ _____

3. $a^{\log_a M} = $ _____

4. $\log_a a^r = $ _____

5. $\log_a(MN) = $ _____ + _____

6. $\log_a\left(\dfrac{M}{N}\right) = $ _____ − _____

7. $\log_a M^r = $ _____

8. If $\log_a x = \log_a 6$, then $x = $ _____.

9. If $\log_8 M = \dfrac{\log_5 7}{\log_5 8}$, then $M = $ _____.

10. *True or False* $\ln(x + 3) - \ln(2x) = \dfrac{\ln(x + 3)}{\ln(2x)}$

11. *True or False* $\log_2(3x^4) = 4\log_2(3x)$

12. *True or False* $\dfrac{\ln 8}{\ln 4} = 2$

Skill Building

In Problems 13–28, use properties of logarithms to find the exact value of each expression. Do not use a calculator.

13. $\log_3 3^{71}$

14. $\log_2 2^{-13}$

15. $\ln e^{-4}$

16. $\ln e^{\sqrt{2}}$

17. $2^{\log_2 7}$

18. $e^{\ln 8}$

19. $\log_8 2 + \log_8 4$

20. $\log_6 9 + \log_6 4$

21. $\log_6 18 - \log_6 3$

22. $\log_8 16 - \log_8 2$

23. $\log_2 6 \cdot \log_6 8$

24. $\log_3 8 \cdot \log_8 9$

25. $3^{\log_3 5 - \log_3 4}$

26. $5^{\log_5 6 + \log_5 7}$

27. $e^{\log_{e^2} 16}$

28. $e^{\log_{e^2} 9}$

In Problems 29–36, suppose that $\ln 2 = a$ and $\ln 3 = b$. Use properties of logarithms to write each logarithm in terms of a and b.

29. $\ln 6$

30. $\ln \dfrac{2}{3}$

31. $\ln 1.5$

32. $\ln 0.5$

33. $\ln 8$

34. $\ln 27$

35. $\ln \sqrt[5]{6}$

36. $\ln \sqrt[4]{\dfrac{2}{3}}$

In Problems 37–56, write each expression as a sum and/or difference of logarithms. Express powers as factors.

37. $\log_5(25x)$

38. $\log_3 \dfrac{x}{9}$

39. $\log_2 z^3$

40. $\log_7 x^5$

41. $\ln(ex)$

42. $\ln \dfrac{e}{x}$

43. $\ln \dfrac{x}{e^x}$

44. $\ln(xe^x)$

45. $\log_a(u^2 v^3)$ $u > 0, v > 0$

46. $\log_2\left(\dfrac{a}{b^2}\right)$ $a > 0, b > 0$

47. $\ln\left(x^2\sqrt{1 - x}\right)$ $0 < x < 1$

48. $\ln\left(x\sqrt{1 + x^2}\right)$ $x > 0$

49. $\log_2\left(\dfrac{x^3}{x - 3}\right)$ $x > 3$

50. $\log_5\left(\dfrac{\sqrt[3]{x^2 + 1}}{x^2 - 1}\right)$ $x > 1$

51. $\log\left[\dfrac{x(x + 2)}{(x + 3)^2}\right]$ $x > 0$

52. $\log\left[\dfrac{x^3\sqrt{x + 1}}{(x - 2)^2}\right]$ $x > 2$

53. $\ln\left[\dfrac{x^2 - x - 2}{(x + 4)^2}\right]^{1/3}$ $x > 2$

54. $\ln\left[\dfrac{(x - 4)^2}{x^2 - 1}\right]^{2/3}$ $x > 4$

55. $\ln\dfrac{5x\sqrt{1 + 3x}}{(x - 4)^3}$ $x > 4$

56. $\ln\left[\dfrac{5x^2\sqrt[3]{1 - x}}{4(x + 1)^2}\right]$ $0 < x < 1$

In Problems 57–70, write each expression as a single logarithm.

57. $3\log_5 u + 4\log_5 v$

58. $2\log_3 u - \log_3 v$

59. $\log_3 \sqrt{x} - \log_3 x^3$

60. $\log_2\left(\dfrac{1}{x}\right) + \log_2\left(\dfrac{1}{x^2}\right)$

61. $\log_4(x^2 - 1) - 5\log_4(x + 1)$

62. $\log(x^2 + 3x + 2) - 2\log(x + 1)$

63. $\ln\left(\dfrac{x}{x - 1}\right) + \ln\left(\dfrac{x + 1}{x}\right) - \ln(x^2 - 1)$

64. $\log\left(\dfrac{x^2 + 2x - 3}{x^2 - 4}\right) - \log\left(\dfrac{x^2 + 7x + 6}{x + 2}\right)$

65. $8\log_2\sqrt{3x - 2} - \log_2\left(\dfrac{4}{x}\right) + \log_2 4$

66. $21 \log_3 \sqrt[3]{x} + \log_3(9x^2) - \log_3 9$

67. $2 \log_a(5x^3) - \dfrac{1}{2} \log_a(2x + 3)$

68. $\dfrac{1}{3} \log(x^3 + 1) + \dfrac{1}{2} \log(x^2 + 1)$

69. $2 \log_2(x + 1) - \log_2(x + 3) - \log_2(x - 1)$

70. $3 \log_5(3x + 1) - 2 \log_5(2x - 1) - \log_5 x$

In Problems 71–78, use the Change-of-Base Formula and a calculator to evaluate each logarithm. Round your answer to three decimal places.

71. $\log_3 21$

72. $\log_5 18$

73. $\log_{1/3} 71$

74. $\log_{1/2} 15$

75. $\log_{\sqrt{2}} 7$

76. $\log_{\sqrt{5}} 8$

77. $\log_{\pi} e$

78. $\log_{\pi} \sqrt{2}$

In Problems 79–84, graph each function using a graphing utility and the Change-of-Base Formula.

79. $y = \log_4 x$

80. $y = \log_5 x$

81. $y = \log_2(x + 2)$

82. $y = \log_4(x - 3)$

83. $y = \log_{x-1}(x + 1)$

84. $y = \log_{x+2}(x - 2)$

Mixed Practice

85. If $f(x) = \ln x$, $g(x) = e^x$, and $h(x) = x^2$, find:
(a) $(f \circ g)(x)$. What is the domain of $f \circ g$?
(b) $(g \circ f)(x)$. What is the domain of $g \circ f$?
(c) $(f \circ g)(5)$
(d) $(f \circ h)(x)$. What is the domain of $f \circ h$?
(e) $(f \circ h)(e)$

86. If $f(x) = \log_2 x$, $g(x) = 2^x$, and $h(x) = 4x$, find:
(a) $(f \circ g)(x)$. What is the domain of $f \circ g$?
(b) $(g \circ f)(x)$. What is the domain of $g \circ f$?
(c) $(f \circ g)(3)$
(d) $(f \circ h)(x)$. What is the domain of $f \circ h$?
(e) $(f \circ h)(8)$

Applications and Extensions

In Problems 87–96, express y as a function of x. The constant C is a positive number.

87. $\ln y = \ln x + \ln C$

88. $\ln y = \ln(x + C)$

89. $\ln y = \ln x + \ln(x + 1) + \ln C$

90. $\ln y = 2 \ln x - \ln(x + 1) + \ln C$

91. $\ln y = 3x + \ln C$

92. $\ln y = -2x + \ln C$

93. $\ln(y - 3) = -4x + \ln C$

94. $\ln(y + 4) = 5x + \ln C$

95. $3 \ln y = \dfrac{1}{2} \ln(2x + 1) - \dfrac{1}{3} \ln(x + 4) + \ln C$

96. $2 \ln y = -\dfrac{1}{2} \ln x + \dfrac{1}{3} \ln(x^2 + 1) + \ln C$

97. Find the value of $\log_2 3 \cdot \log_3 4 \cdot \log_4 5 \cdot \log_5 6 \cdot \log_6 7 \cdot \log_7 8$.

98. Find the value of $\log_2 4 \cdot \log_4 6 \cdot \log_6 8$.

99. Find the value of $\log_2 3 \cdot \log_3 4 \cdot \cdots \cdot \log_n(n + 1) \cdot \log_{n+1} 2$.

100. Find the value of $\log_2 2 \cdot \log_2 4 \cdot \cdots \cdot \log_2 2^n$.

101. Show that $\log_a\left(x + \sqrt{x^2 - 1}\right) + \log_a\left(x - \sqrt{x^2 - 1}\right) = 0$.

102. Show that $\log_a\left(\sqrt{x} + \sqrt{x - 1}\right) + \log_a\left(\sqrt{x} - \sqrt{x - 1}\right) = 0$.

103. Show that $\ln(1 + e^{2x}) = 2x + \ln(1 + e^{-2x})$.

104. Difference Quotient If $f(x) = \log_a x$, show that $\dfrac{f(x + h) - f(x)}{h} = \log_a\left(1 + \dfrac{h}{x}\right)^{1/h}$, $h \neq 0$.

105. If $f(x) = \log_a x$, show that $-f(x) = \log_{1/a} x$.

106. If $f(x) = \log_a x$, show that $f(AB) = f(A) + f(B)$.

107. If $f(x) = \log_a x$, show that $f\left(\dfrac{1}{x}\right) = -f(x)$.

108. If $f(x) = \log_a x$, show that $f(x^\alpha) = \alpha f(x)$.

109. Show that $\log_a\left(\dfrac{M}{N}\right) = \log_a M - \log_a N$, where a, M, and N are positive real numbers and $a \neq 1$.

110. Show that $\log_a\left(\dfrac{1}{N}\right) = -\log_a N$, where a and N are positive real numbers and $a \neq 1$.

Explaining Concepts: Discussion and Writing

111. Graph $Y_1 = \log(x^2)$ and $Y_2 = 2 \log(x)$ using a graphing utility. Are they equivalent? What might account for any differences in the two functions?

112. Write an example that illustrates why $(\log_a x)^r \neq r \log_a x$.

113. Write an example that illustrates why $\log_2(x + y) \neq \log_2 x + \log_2 y$.

114. Does $3^{\log_3(-5)} = -5$? Why or why not?

4.6 Logarithmic and Exponential Equations

PREPARING FOR THIS SECTION *Before getting started, review the following:*

- Solving Equations Using a Graphing Utility (Appendix B, Section B.4, pp. B6–B7)
- Solving Quadratic Equations (Section 2.3, pp. 141–147)
- Solving Equations Quadratic in Form (Section 2.3, pp. 148–149)

Now Work the *'Are You Prepared?'* problems on page 338.

OBJECTIVES **1** Solve Logarithmic Equations (p. 334)
 2 Solve Exponential Equations (p. 336)
 3 Solve Logarithmic and Exponential Equations Using a Graphing Utility (p. 337)

1 Solve Logarithmic Equations

In Section 4.4 we solved logarithmic equations by changing a logarithmic expression to an exponential expression. That is, we used the definition of a logarithm:

$$y = \log_a x \quad \text{is equivalent to} \quad x = a^y \qquad a > 0, a \neq 1$$

For example, to solve the equation $\log_2(1 - 2x) = 3$, we write the logarithmic equation as an equivalent exponential equation $1 - 2x = 2^3$ and solve for x.

$$\log_2(1 - 2x) = 3$$
$$1 - 2x = 2^3 \qquad \text{\textit{Change to an exponential equation.}}$$
$$-2x = 7 \qquad \text{\textit{Simplify.}}$$
$$x = -\frac{7}{2} \qquad \text{\textit{Solve.}}$$

You should check this solution for yourself.

For most logarithmic equations, some manipulation of the equation (usually using properties of logarithms) is required to obtain a solution. Also, to avoid extraneous solutions with logarithmic equations, we determine the domain of the variable first.

We begin with an example of a logarithmic equation that requires using the fact that a logarithmic function is a one-to-one function.

$$\text{If } \log_a M = \log_a N, \text{ then } M = N \qquad M, N, \text{ and } a \text{ are positive and } a \neq 1.$$

EXAMPLE 1 **Solving a Logarithmic Equation**

Solve: $2 \log_5 x = \log_5 9$

Solution The domain of the variable in this equation is $x > 0$. Because each logarithm is to the same base, 5, we can obtain an exact solution as follows:

$$2 \log_5 x = \log_5 9$$
$$\log_5 x^2 = \log_5 9 \qquad \text{\textit{log}}_a M^r = r \log_a M$$
$$x^2 = 9 \qquad \text{\textit{If log}}_a M = \log_a N, \text{ then } M = N.$$
$$x = 3 \quad \text{or} \quad x = -3$$

Recall that the domain of the variable is $x > 0$. Therefore, -3 is extraneous and we discard it.

✓ **Check:** $2 \log_5 3 \overset{?}{=} \log_5 9$

$\log_5 3^2 \overset{?}{=} \log_5 9$ $r \log_a M = \log_a M^r$

$\log_5 9 = \log_5 9$

The solution set is $\{3\}$.

━━▶ **Now Work** PROBLEM 13

Often we need to use one or more properties of logarithms to rewrite the equation as a single logarithm. In the next example we employ the log of a product property to solve a logarithmic equation.

EXAMPLE 2   **Solving a Logarithmic Equation**

Solve: $\log_5(x + 6) + \log_5(x + 2) = 1$

Solution   The domain of the variable requires that $x + 6 > 0$ and $x + 2 > 0$, so $x > -6$ and $x > -2$. This means any solution must satisfy $x > -2$. To obtain an exact solution, we need to express the left side as a single logarithm. Then we will change the equation to an equivalent exponential equation.

$\log_5(x + 6) + \log_5(x + 2) = 1$

$\log_5[(x + 6)(x + 2)] = 1$ $\log_a M + \log_a N = \log_a(MN)$

$(x + 6)(x + 2) = 5^1 = 5$ Change to an exponential equation.

$x^2 + 8x + 12 = 5$ Simplify.

$x^2 + 8x + 7 = 0$ Place the quadratic equation in standard form.

$(x + 7)(x + 1) = 0$ Factor.

$x = -7$ or $x = -1$ Zero-Product Property

WARNING A negative solution is not automatically extraneous. You must determine whether the potential solution causes the argument of any logarithmic expression in the equation to be negative. ∎

Only $x = -1$ satisfies the restriction that $x > -2$, so $x = -7$ is extraneous. The solution set is $\{-1\}$, which you should check.

━━▶ **Now Work** PROBLEM 21

EXAMPLE 3   **Solving a Logarithmic Equation**

Solve: $\ln x = \ln(x + 6) - \ln(x - 4)$

Solution   The domain of the variable requires that $x > 0$, $x + 6 > 0$, and $x - 4 > 0$. As a result, the domain of the variable here is $x > 4$. We begin the solution using the log of a difference property.

$\ln x = \ln(x + 6) - \ln(x - 4)$

$\ln x = \ln\left(\dfrac{x + 6}{x - 4}\right)$ $\ln M - \ln N = \ln\left(\dfrac{M}{N}\right)$

$x = \dfrac{x + 6}{x - 4}$ If $\ln M = \ln N$, then $M = N$.

$x(x - 4) = x + 6$ Multiply both sides by $x - 4$

$x^2 - 4x = x + 6$ Simplify.

$x^2 - 5x - 6 = 0$ Place the quadratic equation in standard form.

$(x - 6)(x + 1) = 0$ Factor.

$x = 6$ or $x = -1$ Zero-Product Property

Since the domain of the variable is $x > 4$, we discard -1 as extraneous. The solution set is $\{6\}$, which you should check.

━━▶ **Now Work** PROBLEM 31

2 Solve Exponential Equations

In Sections 4.3 and 4.4, we solved exponential equations algebraically by expressing each side of the equation using the same base. That is, we used the one-to-one property of the exponential function:

$$\text{If } a^u = a^v, \quad \text{then } u = v \qquad a > 0, a \neq 1$$

For example, to solve the exponential equation $4^{2x+1} = 16$, we notice that $16 = 4^2$ and apply the property above to obtain $2x + 1 = 2$, from which we find $x = \dfrac{1}{2}$.

For most exponential equations, we cannot express each side of the equation using the same base. In such cases, algebraic techniques can sometimes be used to obtain exact solutions.

In the next example we solve two exponential equations by changing the exponential expression to a logarithmic expression.

EXAMPLE 4

Solving Exponential Equations

Solve: (a) $2^x = 5$ (b) $8 \cdot 3^x = 5$

Solution

(a) Since 5 cannot be written as an integer power of 2 ($2^2 = 4$ and $2^3 = 8$), we write the exponential equation as the equivalent logarithmic equation.

$$2^x = 5$$
$$x = \log_2 5 = \frac{\ln 5}{\ln 2}$$

↑
Change-of-Base Formula (9), Section 4.5

Alternatively, we can solve the equation $2^x = 5$ by taking the natural logarithm (or common logarithm) of each side. Taking the natural logarithm,

$$2^x = 5$$
$$\ln 2^x = \ln 5 \qquad \text{If } M = N, \text{ then } \ln M = \ln N.$$
$$x \ln 2 = \ln 5 \qquad \ln M^r = r \ln M$$
$$x = \frac{\ln 5}{\ln 2} \qquad \text{Exact solution}$$
$$\approx 2.322 \qquad \text{Approximate solution}$$

The solution set is $\left\{ \dfrac{\ln 5}{\ln 2} \right\}$.

(b) $8 \cdot 3^x = 5$
$$3^x = \frac{5}{8} \qquad \text{Solve for } 3^x.$$

$$x = \log_3\left(\frac{5}{8}\right) = \frac{\ln\left(\frac{5}{8}\right)}{\ln 3} \qquad \text{Exact solution}$$
$$\approx -0.428 \qquad \text{Approximate solution}$$

The solution set is $\left\{ \dfrac{\ln\left(\frac{5}{8}\right)}{\ln 3} \right\}$.

Now Work PROBLEM 35

EXAMPLE 5	**Solving an Exponential Equation**

Solve: $5^{x-2} = 3^{3x+2}$

Solution   Because the bases are different, we first apply property (6), Section 4.5 (take the natural logarithm of each side), and then use a property of logarithms. The result is an equation in x that we can solve.

$$5^{x-2} = 3^{3x+2}$$

$$\ln 5^{x-2} = \ln 3^{3x+2} \qquad \text{If } M = N, \ln M = \ln N.$$

$$(x-2)\ln 5 = (3x+2)\ln 3 \qquad \ln M^r = r\ln M$$

$$(\ln 5)x - 2\ln 5 = (3\ln 3)x + 2\ln 3 \qquad \text{Distribute.}$$

$$(\ln 5)x - (3\ln 3)x = 2\ln 3 + 2\ln 5 \qquad \text{Place terms involving x on the left.}$$

$$(\ln 5 - 3\ln 3)x = 2(\ln 3 + \ln 5) \qquad \text{Factor.}$$

$$x = \frac{2(\ln 3 + \ln 5)}{\ln 5 - 3\ln 3} \qquad \text{Exact solution}$$

$$\approx -3.212 \qquad \text{Approximate solution}$$

The solution set is $\left\{\dfrac{2(\ln 3 + \ln 5)}{\ln 5 - 3\ln 3}\right\}$.

═══► **Now Work**   PROBLEM 45

The next example deals with an exponential equation that is quadratic in form.

EXAMPLE 6	**Solving an Exponential Equation That Is Quadratic in Form**

Solve: $4^x - 2^x - 12 = 0$

Solution   We note that $4^x = (2^2)^x = 2^{(2x)} = (2^x)^2$, so the equation is quadratic in form, and we can rewrite it as

$$(2^x)^2 - 2^x - 12 = 0 \qquad \text{Let } u = 2^x; \text{ then } u^2 - u - 12 = 0.$$

Now we can factor as usual.

$$(2^x - 4)(2^x + 3) = 0 \qquad (u - 4)(u + 3) = 0$$

$$2^x - 4 = 0 \quad \text{or} \quad 2^x + 3 = 0 \qquad u - 4 = 0 \quad \text{or} \quad u + 3 = 0$$

$$2^x = 4 \qquad\qquad 2^x = -3 \qquad u = 2^x = 4 \qquad u = 2^x = -3$$

The equation on the left has the solution $x = 2$, since $2^x = 4 = 2^2$; the equation on the right has no solution, since $2^x > 0$ for all x. The only solution is 2. The solution set is $\{2\}$.

═══► **Now Work**   PROBLEM 53

 3 Solve Logarithmic and Exponential Equations Using a Graphing Utility

 The algebraic techniques introduced in this section to obtain exact solutions apply only to certain types of logarithmic and exponential equations. Solutions for other types are usually studied in calculus, using numerical methods. For such types, we can use a graphing utility to approximate the solution.

| EXAMPLE 7 | **Solving Equations Using a Graphing Utility** |

Solve: $x + e^x = 2$

Express the solution(s) rounded to two decimal places.

Figure 40

Solution   The solution is found by graphing $Y_1 = x + e^x$ and $Y_2 = 2$. Since Y_1 is an increasing function (do you know why?), there is only one point of intersection for Y_1 and Y_2. Figure 40 shows the graphs of Y_1 and Y_2. Using the INTERSECT command, the solution is 0.44 rounded to two decimal places.

━━━ **Now Work** PROBLEM 63

4.6 Assess Your Understanding

'Are You Prepared?' *Answers are given at the end of these exercises. If you get a wrong answer, read the pages listed in red.*

1. Solve $x^2 - 7x - 30 = 0$. (pp. 141–147)

2. Solve $(x + 3)^2 - 4(x + 3) + 3 = 0$. (pp. 148–149)

3. Approximate the solution(s) to $x^3 = x^2 - 5$ using a graphing utility. (pp. B6–B7)

4. Approximate the solution(s) to $x^3 - 2x + 2 = 0$ using a graphing utility. (pp. B6–B7)

Skill Building

In Problems 5–32, solve each logarithmic equation. Express irrational solutions in exact form and as a decimal rounded to 3 decimal places.

5. $\log_4 x = 2$

6. $\log(x + 6) = 1$

7. $\log_2(5x) = 4$

8. $\log_3(3x - 1) = 2$

9. $\log_4(x + 2) = \log_4 8$

10. $\log_5(2x + 3) = \log_5 3$

11. $\dfrac{1}{2}\log_3 x = 2\log_3 2$

12. $-2\log_4 x = \log_4 9$

13. $3\log_2 x = -\log_2 27$

14. $2\log_5 x = 3\log_5 4$

15. $3\log_2(x - 1) + \log_2 4 = 5$

16. $2\log_3(x + 4) - \log_3 9 = 2$

17. $\log x + \log(x + 15) = 2$

18. $\log x + \log(x - 21) = 2$

19. $\log(2x + 1) = 1 + \log(x - 2)$

20. $\log(2x) - \log(x - 3) = 1$

21. $\log_2(x + 7) + \log_2(x + 8) = 1$

22. $\log_6(x + 4) + \log_6(x + 3) = 1$

23. $\log_8(x + 6) = 1 - \log_8(x + 4)$

24. $\log_5(x + 3) = 1 - \log_5(x - 1)$

25. $\ln x + \ln(x + 2) = 4$

26. $\ln(x + 1) - \ln x = 2$

27. $\log_3(x + 1) + \log_3(x + 4) = 2$

28. $\log_2(x + 1) + \log_2(x + 7) = 3$

29. $\log_{1/3}(x^2 + x) - \log_{1/3}(x^2 - x) = -1$

30. $\log_4(x^2 - 9) - \log_4(x + 3) = 3$

31. $\log_a(x - 1) - \log_a(x + 6) = \log_a(x - 2) - \log_a(x + 3)$

32. $\log_a x + \log_a(x - 2) = \log_a(x + 4)$

In Problems 33–60, solve each exponential equation. Express irrational solutions in exact form and as a decimal rounded to 3 decimal places.

33. $2^{x-5} = 8$

34. $5^{-x} = 25$

35. $2^x = 10$

36. $3^x = 14$

37. $8^{-x} = 1.2$

38. $2^{-x} = 1.5$

39. $5(2^{3x}) = 8$

40. $0.3(4^{0.2x}) = 0.2$

41. $3^{1-2x} = 4^x$

42. $2^{x+1} = 5^{1-2x}$

43. $\left(\dfrac{3}{5}\right)^x = 7^{1-x}$

44. $\left(\dfrac{4}{3}\right)^{1-x} = 5^x$

45. $1.2^x = (0.5)^{-x}$

46. $0.3^{1+x} = 1.7^{2x-1}$

47. $\pi^{1-x} = e^x$

48. $e^{x+3} = \pi^x$

49. $2^{2x} + 2^x - 12 = 0$ **50.** $3^{2x} + 3^x - 2 = 0$ **51.** $3^{2x} + 3^{x+1} - 4 = 0$ **52.** $2^{2x} + 2^{x+2} - 12 = 0$

53. $16^x + 4^{x+1} - 3 = 0$ **54.** $9^x - 3^{x+1} + 1 = 0$ **55.** $25^x - 8 \cdot 5^x = -16$ **56.** $36^x - 6 \cdot 6^x = -9$

57. $3 \cdot 4^x + 4 \cdot 2^x + 8 = 0$ **58.** $2 \cdot 49^x + 11 \cdot 7^x + 5 = 0$ **59.** $4^x - 10 \cdot 4^{-x} = 3$ **60.** $3^x - 14 \cdot 3^{-x} = 5$

In Problems 61–74, use a graphing utility to solve each equation. Express your answer rounded to two decimal places.

61. $\log_5(x + 1) - \log_4(x - 2) = 1$ **62.** $\log_2(x - 1) - \log_6(x + 2) = 2$

63. $e^x = -x$ **64.** $e^{2x} = x + 2$ **65.** $e^x = x^2$ **66.** $e^x = x^3$

67. $\ln x = -x$ **68.** $\ln(2x) = -x + 2$ **69.** $\ln x = x^3 - 1$ **70.** $\ln x = -x^2$

71. $e^x + \ln x = 4$ **72.** $e^x - \ln x = 4$ **73.** $e^{-x} = \ln x$ **74.** $e^{-x} = -\ln x$

In Problems 75–86, solve each equation. Express irrational solutions in exact form and as a decimal rounded to 3 decimal places.

75. $\log_2(x + 1) - \log_4 x = 1$
[**Hint:** Change $\log_4 x$ to base 2.]

76. $\log_2(3x + 2) - \log_4 x = 3$ **77.** $\log_{16} x + \log_4 x + \log_2 x = 7$

78. $\log_9 x + 3 \log_3 x = 14$ **79.** $\left(\sqrt[3]{2}\right)^{2-x} = 2^{x^2}$ **80.** $\log_2 x^{\log_2 x} = 4$

81. $\dfrac{e^x + e^{-x}}{2} = 1$
[**Hint:** Multiply each side by e^x.]

82. $\dfrac{e^x + e^{-x}}{2} = 3$ **83.** $\dfrac{e^x - e^{-x}}{2} = 2$

84. $\dfrac{e^x - e^{-x}}{2} = -2$ **85.** $\log_5 x + \log_3 x = 1$ **86.** $\log_2 x + \log_6 x = 3$
[**Hint:** Use the Change-of-Base Formula.]

Mixed Practice

87. $f(x) = \log_2(x + 3)$ and $g(x) = \log_2(3x + 1)$.
  (a) Solve $f(x) = 3$. What point is on the graph of f?
  (b) Solve $g(x) = 4$. What point is on the graph of g?
  (c) Solve $f(x) = g(x)$. Do the graphs of f and g intersect? If so, where?
  (d) Solve $(f + g)(x) = 7$.
  (e) Solve $(f - g)(x) = 2$.

88. $f(x) = \log_3(x + 5)$ and $g(x) = \log_3(x - 1)$.
  (a) Solve $f(x) = 2$. What point is on the graph of f?
  (b) Solve $g(x) = 3$. What point is on the graph of g?
  (c) Solve $f(x) = g(x)$. Do the graphs of f and g intersect? If so, where?
  (d) Solve $(f + g)(x) = 3$.
  (e) Solve $(f - g)(x) = 2$.

89. (a) If $f(x) = 3^{x+1}$ and $g(x) = 2^{x+2}$, graph f and g on the same Cartesian plane.
  (b) Find the point(s) of intersection of the graphs of f and g by solving $f(x) = g(x)$. Round answers to three decimal places. Label this point on the graph drawn in part (a).
  (c) Based on the graph, solve $f(x) > g(x)$.

90. (a) If $f(x) = 5^{x-1}$ and $g(x) = 2^{x+1}$, graph f and g on the same Cartesian plane.
  (b) Find the point(s) of intersection of the graphs of f and g by solving $f(x) = g(x)$. Label this point on the graph drawn in part (a).
  (c) Based on the graph, solve $f(x) > g(x)$.

91. (a) Graph $f(x) = 3^x$ and $g(x) = 10$ on the same Cartesian plane.

  (b) Shade the region bounded by the y-axis, $f(x) = 3^x$, and $g(x) = 10$ on the graph drawn in part (a).
  (c) Solve $f(x) = g(x)$ and label the point of intersection on the graph drawn in part (a).

92. (a) Graph $f(x) = 2^x$ and $g(x) = 12$ on the same Cartesian plane.
  (b) Shade the region bounded by the y-axis, $f(x) = 2^x$, and $g(x) = 12$ on the graph drawn in part (a).
  (c) Solve $f(x) = g(x)$ and label the point of intersection on the graph drawn in part (a).

93. (a) Graph $f(x) = 2^{x+1}$ and $g(x) = 2^{-x+2}$ on the same Cartesian plane.
  (b) Shade the region bounded by the y-axis, $f(x) = 2^{x+1}$, and $g(x) = 2^{-x+2}$ on the graph draw in part (a).
  (c) Solve $f(x) = g(x)$ and label the point of intersection on the graph drawn in part (a).

94. (a) Graph $f(x) = 3^{-x+1}$ and $g(x) = 3^{x-2}$ on the same Cartesian plane.
  (b) Shade the region bounded by the y-axis, $f(x) = 3^{-x+1}$, and $g(x) = 3^{x-2}$ on the graph draw in part (a).
  (c) Solve $f(x) = g(x)$ and label the point of intersection on the graph drawn in part (a).

95. (a) Graph $f(x) = 2^x - 4$.
  (b) Find the zero of f.
  (c) Based on the graph, solve $f(x) < 0$.

96. (a) Graph $g(x) = 3^x - 9$.
  (b) Find the zero of g.
  (c) Based on the graph, solve $g(x) > 0$.

Applications and Extensions

97. A Population Model The resident population of the United States in 2008 was 304 million people and was growing at a rate of 0.9% per year. Assuming that this growth rate continues, the model $P(t) = 304(1.009)^{t-2008}$ represents the population P (in millions of people) in year t.
  (a) According to this model, when will the population of the United States be 354 million people?
  (b) According to this model, when will the population of the United States be 416 million people?
  Source: Statistical Abstract of the United States, 125th ed., 2009.

98. A Population Model The population of the world in 2009 was 6.78 billion people and was growing at a rate of 1.14% per year. Assuming that this growth rate continues, the model $P(t) = 6.78(1.0114)^{t-2009}$ represents the population P (in billions of people) in year t.
  (a) According to this model, when will the population of the world be 8.7 billion people?

  (b) According to this model, when will the population of the world be 14 billion people?
  Source: U.S. Census Bureau.

99. Depreciation The value V of a Chevy Cobalt that is t years old can be modeled by $V(t) = 16,500(0.82)^t$.
  (a) According to the model, when will the car be worth $9000?
  (b) According to the model, when will the car be worth $4000?
  (c) According to the model, when will the car be worth $2000?
  Source: Kelley Blue Book

100. Depreciation The value V of a Honda Civic DX that is t years old can be modeled by $V(t) = 16,775(0.905)^t$.
  (a) According to the model, when will the car be worth $15,000?
  (b) According to the model, when will the car be worth $8000?
  (c) According to the model, when will the car be worth $4000?
  Source: Kelley Blue Book

Explaining Concepts: Discussion and Writing

101. Fill in reasons for each step in the following two solutions.
Solve: $\log_3(x-1)^2 = 2$

Solution A

$\log_3(x-1)^2 = 2$
$(x-1)^2 = 3^2 = 9$ _____
$(x-1) = \pm 3$ _____
$x - 1 = -3$ or $x - 1 = 3$ _____
$x = -2$ or $x = 4$ _____

Solution B

$\log_3(x-1)^2 = 2$
$2\log_3(x-1) = 2$ _____
$\log_3(x-1) = 1$ _____
$x - 1 = 3^1 = 3$ _____
$x = 4$ _____

Both solutions given in Solution A check. Explain what caused the solution $x = -2$ to be lost in Solution B.

'Are You Prepared?' Answers

1. $\{-3, 10\}$ **2.** $\{-2, 0\}$ **3.** $\{-1.43\}$ **4.** $\{-1.77\}$

4.7 Financial Models

PREPARING FOR THIS SECTION *Before getting started, review the following:*

- Simple Interest (Appendix A, Section A.9, p. A75)

Now Work the 'Are You Prepared?' problems on page 347.

OBJECTIVES   **1** Determine the Future Value of a Lump Sum of Money (p. 341)
                   2 Calculate Effective Rates of Return (p. 344)
                   3 Determine the Present Value of a Lump Sum of Money (p. 345)
                   4 Determine the Rate of Interest or Time Required to Double a Lump Sum of Money (p. 346)

1 Determine the Future Value of a Lump Sum of Money

Interest is money paid for the use of money. The total amount borrowed (whether by an individual from a bank in the form of a loan or by a bank from an individual in the form of a savings account) is called the **principal.** The **rate of interest,** expressed as a percent, is the amount charged for the use of the principal for a given period of time, usually on a yearly (that is, per annum) basis.

THEOREM

> **Simple Interest Formula**
>
> If a principal of P dollars is borrowed for a period of t years at a per annum interest rate r, expressed as a decimal, the interest I charged is
>
> $$I = Prt \qquad\qquad (1)$$
>
> Interest charged according to formula (1) is called **simple interest.**

In working with problems involving interest, we define the term **payment period** as follows:

Annually:	Once per year	**Monthly:**	12 times per year
Semiannually:	Twice per year	**Daily:**	365 times per year*
Quarterly:	Four times per year		

When the interest due at the end of a payment period is added to the principal so that the interest computed at the end of the next payment period is based on this new principal amount (old principal + interest), the interest is said to have been **compounded. Compound interest** is interest paid on the principal and previously earned interest.

EXAMPLE 1   **Computing Compound Interest**

A credit union pays interest of 8% per annum compounded quarterly on a certain savings plan. If $1000 is deposited in such a plan and the interest is left to accumulate, how much is in the account after 1 year?

Solution   We use the simple interest formula, $I = Prt$. The principal P is $1000 and the rate of interest is 8% = 0.08. After the first quarter of a year, the time t is $\frac{1}{4}$ year, so the interest earned is

$$I = Prt = (\$1000)(0.08)\left(\frac{1}{4}\right) = \$20$$

*Most banks use a 360-day "year." Why do you think they do?

The new principal is $P + I = \$1000 + \$20 = \$1020$. At the end of the second quarter, the interest on this principal is

$$I = (\$1020)(0.08)\left(\frac{1}{4}\right) = \$20.40$$

At the end of the third quarter, the interest on the new principal of $\$1020 + \$20.40 = \$1040.40$ is

$$I = (\$1040.40)(0.08)\left(\frac{1}{4}\right) = \$20.81$$

Finally, after the fourth quarter, the interest is

$$I = (\$1061.21)(0.08)\left(\frac{1}{4}\right) = \$21.22$$

After 1 year the account contains $\$1061.21 + \$21.22 = \$1082.43$.

The pattern of the calculations performed in Example 1 leads to a general formula for compound interest. To fix our ideas, let P represent the principal to be invested at a per annum interest rate r that is compounded n times per year, so the time of each compounding period is $\frac{1}{n}$ years. (For computing purposes, r is expressed as a decimal.) The interest earned after each compounding period is given by formula (1).

$$\text{Interest} = \text{principal} \times \text{rate} \times \text{time} = P \cdot r \cdot \frac{1}{n} = P \cdot \left(\frac{r}{n}\right)$$

The amount A after one compounding period is

$$A = P + P \cdot \left(\frac{r}{n}\right) = P \cdot \left(1 + \frac{r}{n}\right)$$

After two compounding periods, the amount A, based on the new principal $P \cdot \left(1 + \frac{r}{n}\right)$, is

$$A = \underbrace{P \cdot \left(1 + \frac{r}{n}\right)}_{\substack{\text{New} \\ \text{principal}}} + \underbrace{P \cdot \left(1 + \frac{r}{n}\right)\left(\frac{r}{n}\right)}_{\substack{\text{Interest on} \\ \text{new principal}}} = \underset{\substack{\uparrow \\ \text{Factor out } P \cdot (1 + \frac{r}{n})}}{P \cdot \left(1 + \frac{r}{n}\right)\left(1 + \frac{r}{n}\right)} = P \cdot \left(1 + \frac{r}{n}\right)^2$$

After three compounding periods, the amount A is

$$A = P \cdot \left(1 + \frac{r}{n}\right)^2 + P \cdot \left(1 + \frac{r}{n}\right)^2\left(\frac{r}{n}\right) = P \cdot \left(1 + \frac{r}{n}\right)^2 \cdot \left(1 + \frac{r}{n}\right) = P \cdot \left(1 + \frac{r}{n}\right)^3$$

Continuing this way, after n compounding periods (1 year), the amount A is

$$A = P \cdot \left(1 + \frac{r}{n}\right)^n$$

Because t years will contain $n \cdot t$ compounding periods, after t years we have

$$A = P \cdot \left(1 + \frac{r}{n}\right)^{nt}$$

THEOREM

Compound Interest Formula

The amount A after t years due to a principal P invested at an annual interest rate r compounded n times per year is

$$A = P \cdot \left(1 + \frac{r}{n}\right)^{nt} \tag{2}$$

For example, to rework Example 1, we would use $P = \$1000$, $r = 0.08$, $n = 4$ (quarterly compounding), and $t = 1$ year to obtain

$$A = P \cdot \left(1 + \frac{r}{n}\right)^{nt} = 1000\left(1 + \frac{0.08}{4}\right)^{4 \cdot 1} = \$1082.43$$

In equation (2), the amount A is typically referred to as the **future value** of the account, while P is called the **present value.**

Now Work PROBLEM 7

EXAMPLE 2

Comparing Investments Using Different Compounding Periods

Investing \$1000 at an annual rate of 10% compounded annually, semiannually, quarterly, monthly, and daily will yield the following amounts after 1 year:

Annual compounding ($n = 1$): $\quad A = P \cdot (1 + r)$
$$= (\$1000)(1 + 0.10) = \$1100.00$$

Semiannual compounding ($n = 2$): $\quad A = P \cdot \left(1 + \dfrac{r}{2}\right)^2$
$$= (\$1000)(1 + 0.05)^2 = \$1102.50$$

Quarterly compounding ($n = 4$): $\quad A = P \cdot \left(1 + \dfrac{r}{4}\right)^4$
$$= (\$1000)(1 + 0.025)^4 = \$1103.81$$

Monthly compounding ($n = 12$): $\quad A = P \cdot \left(1 + \dfrac{r}{12}\right)^{12}$
$$= (\$1000)\left(1 + \frac{0.10}{12}\right)^{12} = \$1104.71$$

Daily compounding ($n = 365$): $\quad A = P \cdot \left(1 + \dfrac{r}{365}\right)^{365}$
$$= (\$1000)\left(1 + \frac{0.10}{365}\right)^{365} = \$1105.16$$

From Example 2, we can see that the effect of compounding more frequently is that the amount after 1 year is higher: \$1000 compounded 4 times a year at 10% results in \$1103.81, \$1000 compounded 12 times a year at 10% results in \$1104.71, and \$1000 compounded 365 times a year at 10% results in \$1105.16. This leads to the following question: What would happen to the amount after 1 year if the number of times that the interest is compounded were increased without bound?

Let's find the answer. Suppose that P is the principal, r is the per annum interest rate, and n is the number of times that the interest is compounded each year. The amount after 1 year is

$$A = P \cdot \left(1 + \frac{r}{n}\right)^n$$

Rewrite this expression as follows:

$$A = P \cdot \left(1 + \frac{r}{n}\right)^n = P \cdot \left(1 + \frac{1}{\frac{n}{r}}\right)^n = P \cdot \left[\left(1 + \frac{1}{\frac{n}{r}}\right)^{n/r}\right]^r = P \cdot \left[\left(1 + \frac{1}{h}\right)^h\right]^r \quad \text{(3)}$$
$$h = \frac{n}{r}$$

Now suppose that the number n of times that the interest is compounded per year gets larger and larger; that is, suppose that $n \to \infty$. Then $h = \dfrac{n}{r} \to \infty$, and the expression in brackets in Equation (3) equals e. That is, $A \to Pe^r$.

Table 8 compares $\left(1 + \dfrac{r}{n}\right)^n$, for large values of n, to e^r for $r = 0.05$, $r = 0.10$, $r = 0.15$, and $r = 1$. The larger that n gets, the closer $\left(1 + \dfrac{r}{n}\right)^n$ gets to e^r. No matter how frequent the compounding, the amount after 1 year has the definite ceiling Pe^r.

Table 8

	$\left(1 + \frac{r}{n}\right)^n$			
	$n = 100$	$n = 1000$	$n = 10{,}000$	e^r
$r = 0.05$	1.0512580	1.0512698	1.051271	1.0512711
$r = 0.10$	1.1051157	1.1051654	1.1051704	1.1051709
$r = 0.15$	1.1617037	1.1618212	1.1618329	1.1618342
$r = 1$	2.7048138	2.7169239	2.7181459	2.7182818

When interest is compounded so that the amount after 1 year is Pe^r, we say that the interest is **compounded continuously.**

THEOREM

Continuous Compounding

The amount A after t years due to a principal P invested at an annual interest rate r compounded continuously is

$$A = Pe^{rt} \qquad\qquad (4)$$

EXAMPLE 3   **Using Continuous Compounding**

The amount A that results from investing a principal P of $1000 at an annual rate r of 10% compounded continuously for a time t of 1 year is

$$A = \$1000e^{0.10} = (\$1000)(1.10517) = \$1105.17$$

●

── **Now Work** PROBLEM 13

2 Calculate Effective Rates of Return

Suppose that you have $1000 and a bank offers to pay you 3% annual interest on a savings account compounded monthly. What annual interest rate do you need to earn to have the same amount at the end of the year if the interest is compounded annually (once per year)? To answer this question, we first determine the value of the $1000 in the account that earns 3% compounded monthly.

$$A = \$1000\left(1 + \frac{0.03}{12}\right)^{12} \quad \text{Use } A = P\left(1 + \frac{r}{n}\right)^n \text{ with } P = \$1000, r = 0.03, n = 12.$$

$$= \$1030.42$$

So the interest earned is $30.42. Using $I = Prt$ with $t = 1$, $I = \$30.42$, and $P = \$1000$, we find the annual simple interest rate is $0.03042 = 3.042\%$. This interest rate is known as the *effective rate of interest*.

The **effective rate of interest** is the equivalent annual simple interest rate that would yield the same amount as compounding n times per year, or continuously, after 1 year. We can determine the effective rate of interest using the following.

THEOREM

> **Effective Rate of Interest**
>
> The effective rate of interest r_e of an investment earning an annual interest rate r is given by
>
> Compounding n times per year: $r_e = \left(1 + \dfrac{r}{n}\right)^n - 1$
>
> Continuous compounding: $r_e = e^r - 1$

EXAMPLE 4 **Computing the Effective Rate of Interest—Which Is the Best Deal?**

Suppose that you want to open a money market account. You visit three banks to determine their money market rates. Bank A offers you 6% annual interest compounded daily and Bank B offers you 6.02% compounded quarterly. Bank C offers 5.98% compounded continuously. Determine which bank is offering the best deal.

Solution The bank that offers the best deal is the one with the highest effective interest rate.

Bank A	**Bank B**	**Bank C**
$r_e = \left(1 + \dfrac{0.06}{365}\right)^{365} - 1$	$r_e = \left(1 + \dfrac{0.0602}{4}\right)^{4} - 1$	$r_e = e^{0.0598} - 1$
$\approx 1.06183 - 1$	$\approx 1.06157 - 1$	$\approx 1.06162 - 1$
$= 0.06183$	$= 0.06157$	$= 0.06162$
$= 6.183\%$	$= 6.157\%$	$= 6.162\%$

Since the effective rate of interest is highest for Bank A, Bank A is offering the best deal.

──➤**Now Work** PROBLEM 23

3 Determine the Present Value of a Lump Sum of Money

When people in finance speak of the "time value of money," they are usually referring to the *present value* of money. The **present value** of A dollars to be received at a future date is the principal that you would need to invest now so that it will grow to A dollars in the specified time period. The present value of money to be received at a future date is always less than the amount to be received, since the amount to be received will equal the present value (money invested now) *plus* the interest accrued over the time period.

   We use the compound interest formula (2) to get a formula for present value. If P is the present value of A dollars to be received after t years at a per annum interest rate r compounded n times per year, then, by formula (2),

$$A = P \cdot \left(1 + \frac{r}{n}\right)^{nt}$$

To solve for P, we divide both sides by $\left(1 + \dfrac{r}{n}\right)^{nt}$. The result is

$$\frac{A}{\left(1 + \dfrac{r}{n}\right)^{nt}} = P \quad \text{or} \quad P = A \cdot \left(1 + \frac{r}{n}\right)^{-nt}$$

THEOREM **Present Value Formulas**

The present value P of A dollars to be received after t years, assuming a per annum interest rate r compounded n times per year, is

$$P = A \cdot \left(1 + \frac{r}{n}\right)^{-nt} \tag{5}$$

If the interest is compounded continuously,

$$P = Ae^{-rt} \tag{6}$$

To prove (6), solve formula (4) for P.

EXAMPLE 5 **Computing the Value of a Zero-coupon Bond**

A zero-coupon (noninterest-bearing) bond can be redeemed in 10 years for $1000. How much should you be willing to pay for it now if you want a return of
(a) 8% compounded monthly?   (b) 7% compounded continuously?

Solution (a) We are seeking the present value of $1000. We use formula (5) with $A = \$1000$, $n = 12$, $r = 0.08$, and $t = 10$.

$$P = A \cdot \left(1 + \frac{r}{n}\right)^{-nt} = \$1000\left(1 + \frac{0.08}{12}\right)^{-12(10)} = \$450.52$$

For a return of 8% compounded monthly, you should pay $450.52 for the bond.

(b) Here we use formula (6) with $A = \$1000$, $r = 0.07$, and $t = 10$.

$$P = Ae^{-rt} = \$1000e^{-(0.07)(10)} = \$496.59$$

For a return of 7% compounded continuously, you should pay $496.59 for the bond.

━━━━ **Now Work** PROBLEM 15

4 Determine the Rate of Interest or Time Required to Double a Lump Sum of Money

EXAMPLE 6 **Rate of Interest Required to Double an Investment**

What annual rate of interest compounded annually should you seek if you want to double your investment in 5 years?

Solution If P is the principal and we want P to double, the amount A will be $2P$. We use the compound interest formula with $n = 1$ and $t = 5$ to find r.

$$A = P \cdot \left(1 + \frac{r}{n}\right)^{nt}$$

$$2P = P \cdot (1 + r)^5 \qquad A = 2P, n = 1, t = 5$$

$$2 = (1 + r)^5 \qquad \text{Divide both sides by } P.$$

$$1 + r = \sqrt[5]{2} \qquad \text{Take the fifth root of each side.}$$

$$r = \sqrt[5]{2} - 1 \approx 1.148698 - 1 = 0.148698$$

The annual rate of interest needed to double the principal in 5 years is 14.87%.

━━━━ **Now Work** PROBLEM 31

| EXAMPLE 7 | Time Required to Double or Triple an Investment |

(a) How long will it take for an investment to double in value if it earns 5% compounded continuously?

(b) How long will it take to triple at this rate?

Solution (a) If P is the initial investment and we want P to double, the amount A will be $2P$. We use formula (4) for continuously compounded interest with $r = 0.05$. Then

$$A = Pe^{rt}$$
$$2P = Pe^{0.05t} \qquad A = 2P, r = 0.05$$
$$2 = e^{0.05t} \qquad \text{Cancel the P's.}$$
$$0.05t = \ln 2 \qquad \text{Rewrite as a logarithm.}$$
$$t = \frac{\ln 2}{0.05} \approx 13.86 \qquad \text{Solve for t.}$$

It will take about 14 years to double the investment.

(b) To triple the investment, we set $A = 3P$ in formula (4).

$$A = Pe^{rt}$$
$$3P = Pe^{0.05t} \qquad A = 3P, r = 0.05$$
$$3 = e^{0.05t}. \qquad \text{Cancel the P's.}$$
$$0.05t = \ln 3 \qquad \text{Rewrite as a logarithm.}$$
$$t = \frac{\ln 3}{0.05} \approx 21.97 \qquad \text{Solve for t.}$$

It will take about 22 years to triple the investment.

Now Work PROBLEM 35

4.7 Assess Your Understanding

'Are You Prepared?' *Answers are given at the end of these exercises. If you get a wrong answer, read the pages listed in* red.

1. What is the interest due if $500 is borrowed for 6 months at a simple interest rate of 6% per annum? (p. A75)

2. If you borrow $5000 and, after 9 months, pay off the loan in the amount of $5500, what per annum rate of interest was charged? (p. A75)

Concepts and Vocabulary

3. The total amount borrowed (whether by an individual from a bank in the form of a loan or by a bank from an individual in the form of a savings account) is called the _____.

4. If a principal of P dollars is borrowed for a period of t years at a per annum interest rate r, expressed as a decimal, the interest I charged is _____ = _____. Interest charged according to this formula is called _____.

5. In working problems involving interest, if the payment period of the interest is quarterly, then interest is paid _____ times per year.

6. The _____ _____ ___ _____ is the equivalent annual simple interest rate that would yield the same amount as compounding n times per year, or continuously, after 1 year.

Skill Building

In Problems 7–14, find the amount that results from each investment.

7. $100 invested at 4% compounded quarterly after a period of 2 years

8. $50 invested at 6% compounded monthly after a period of 3 years

9. $500 invested at 8% compounded quarterly after a period of $2\frac{1}{2}$ years

10. $300 invested at 12% compounded monthly after a period of $1\frac{1}{2}$ years

11. $600 invested at 5% compounded daily after a period of 3 years

12. $700 invested at 6% compounded daily after a period of 2 years

13. $1000 invested at 11% compounded continuously after a period of 2 years

14. $400 invested at 7% compounded continuously after a period of 3 years

In Problems 15–22, find the principal needed now to get each amount; that is, find the present value.

15. To get $100 after 2 years at 6% compounded monthly

16. To get $75 after 3 years at 8% compounded quarterly

17. To get $1000 after $2\frac{1}{2}$ years at 6% compounded daily

18. To get $800 after $3\frac{1}{2}$ years at 7% compounded monthly

19. To get $600 after 2 years at 4% compounded quarterly

20. To get $300 after 4 years at 3% compounded daily

21. To get $80 after $3\frac{1}{4}$ years at 9% compounded continuously

22. To get $800 after $2\frac{1}{2}$ years at 8% compounded continuously

In Problems 23–26, find the effective rate of interest.

23. For 5% compounded quarterly

24. For 6% compounded monthly

25. For 5% compounded continuously

26. For 6% compounded continuously

In Problems 27–30, determine the rate that represents the better deal.

27. 6% compounded quarterly or $6\frac{1}{4}$% compounded annually

28. 9% compounded quarterly or $9\frac{1}{4}$% compounded annually

29. 9% compounded monthly or 8.8% compounded daily

30. 8% compounded semiannually or 7.9% compounded daily

31. What rate of interest compounded annually is required to double an investment in 3 years?

32. What rate of interest compounded annually is required to double an investment in 6 years?

33. What rate of interest compounded annually is required to triple an investment in 5 years?

34. What rate of interest compounded annually is required to triple an investment in 10 years?

35. (a) How long does it take for an investment to double in value if it is invested at 8% compounded monthly?
(b) How long does it take if the interest is compounded continuously?

36. (a) How long does it take for an investment to triple in value if it is invested at 6% compounded monthly?
(b) How long does it take if the interest is compounded continuously?

37. What rate of interest compounded quarterly will yield an effective interest rate of 7%?

38. What rate of interest compounded continuously will yield an effective interest rate of 6%?

Applications and Extensions

39. Time Required to Reach a Goal If Tanisha has $100 to invest at 8% per annum compounded monthly, how long will it be before she has $150? If the compounding is continuous, how long will it be?

40. Time Required to Reach a Goal If Angela has $100 to invest at 10% per annum compounded monthly, how long will it be before she has $175? If the compounding is continuous, how long will it be?

41. Time Required to Reach a Goal How many years will it take for an initial investment of $10,000 to grow to $25,000? Assume a rate of interest of 6% compounded continuously.

42. Time Required to Reach a Goal How many years will it take for an initial investment of $25,000 to grow to $80,000? Assume a rate of interest of 7% compounded continuously.

43. Price Appreciation of Homes What will a $90,000 condominium cost 5 years from now if the price appreciation for condos over that period averages 3% compounded annually?

44. Credit Card Interest A department store charges 1.25% per month on the unpaid balance for customers with charge accounts (interest is compounded monthly). A customer charges $200 and does not pay her bill for 6 months. What is the bill at that time?

45. Saving for a Car Jerome will be buying a used car for $15,000 in 3 years. How much money should he ask his parents for now so that, if he invests it at 5% compounded continuously, he will have enough to buy the car?

46. Paying off a Loan John requires $3000 in 6 months to pay off a loan that has no prepayment privileges. If he has the $3000 now, how much of it should he save in an account paying 3% compounded monthly so that in 6 months he will have exactly $3000?

47. Return on a Stock George contemplates the purchase of 100 shares of a stock selling for $15 per share. The stock pays no dividends. The history of the stock indicates that it should grow at an annual rate of 15% per year.

How much should the 100 shares of stock be worth in 5 years?

48. **Return on an Investment** A business purchased for $650,000 in 2005 is sold in 2008 for $850,000. What is the annual rate of return for this investment?

49. **Comparing Savings Plans** Jim places $1000 in a bank account that pays 5.6% compounded continuously. After 1 year, will he have enough money to buy a computer system that costs $1060? If another bank will pay Jim 5.9% compounded monthly, is this a better deal?

50. **Savings Plans** On January 1, Kim places $1000 in a certificate of deposit that pays 6.8% compounded continuously and matures in 3 months. Then Kim places the $1000 and the interest in a passbook account that pays 5.25% compounded monthly. How much does Kim have in the passbook account on May 1?

51. **Comparing IRA Investments** Will invests $2000 in his IRA in a bond trust that pays 9% interest compounded semiannually. His friend Henry invests $2000 in his IRA in a certificate of deposit that pays $8\frac{1}{2}$% compounded continuously. Who has more money after 20 years, Will or Henry?

52. **Comparing Two Alternatives** Suppose that April has access to an investment that will pay 10% interest compounded continuously. Which is better: to be given $1000 now so that she can take advantage of this investment opportunity or to be given $1325 after 3 years?

53. **College Costs** The average annual cost of college at 4-year private colleges was $25,143 in the 2008–2009 academic year. This was a 5.9% increase from the previous year.
   Source: The College Board
   (a) If the cost of college increases by 5.9% each year, what will be the average cost of college at a 4-year private college for the 2028–2029 academic year?

(b) College savings plans, such as a 529 plan, allow individuals to put money aside now to help pay for college later. If one such plan offers a rate of 4% compounded continuously, how much should be put in a college savings plan in 2010 to pay for 1 year of the cost of college for an incoming freshman in 2028?

54. **Analyzing Interest Rates on a Mortgage** Colleen and Bill have just purchased a house for $650,000, with the seller holding a second mortgage of $100,000. They promise to pay the seller $100,000 plus all accrued interest 5 years from now. The seller offers them three interest options on the second mortgage:
   (a) Simple interest at 12% per annum
   (b) $11\frac{1}{2}$% interest compounded monthly
   (c) $11\frac{1}{4}$% interest compounded continuously
   Which option is best; that is, which results in the least interest on the loan?

55. **2009 Federal Stimulus Package** In February 2009, President Obama signed into law a $787 billion federal stimulus package. At that time, 20-year Series EE bonds had a fixed rate of 1.3% compounded semiannually. If the federal government financed the stimulus through EE bonds, how much would it have to pay back in 2029? How much interest was paid to finance the stimulus?
   Source: U.S. Treasury Department

56. **Per Capita Federal Debt** In 2008, the federal debt was about $10 trillion. In 2008, the U.S., population was about 304 million. Assuming that the federal debt is increasing about 7.8% per year and the U.S. population is increasing about 0.9% per year, determine the per capita debt (total debt divided by population) in 2020.

Inflation *Problems 57–62 require the following discussion.* **Inflation** *is a term used to describe the erosion of the purchasing power of money. For example, if the annual inflation rate is 3%, then $1000 worth of purchasing power now will have only $970 worth of purchasing power in 1 year because 3% of the original $1000 (0.03 × 1000 = 30) has been eroded due to inflation. In general, if the rate of inflation averages r% over n years, the amount A that $P will purchase after n years is*

$$A = P \cdot (1 - r)^n$$

where r is expressed as a decimal.

57. **Inflation** If the inflation rate averages 3%, how much will $1000 purchase in 2 years?

58. **Inflation** If the inflation rate averages 2%, how much will $1000 purchase in 3 years?

59. **Inflation** If the amount that $1000 will purchase is only $950 after 2 years, what was the average inflation rate?

60. **Inflation** If the amount that $1000 will purchase is only $930 after 2 years, what was the average inflation rate?

61. **Inflation** If the average inflation rate is 2%, how long is it until purchasing power is cut in half?

62. **Inflation** If the average inflation rate is 4%, how long is it until purchasing power is cut in half?

---

*Problems 63–66 involve zero-coupon bonds. A **zero-coupon bond** is a bond that is sold now at a discount and will pay its face value at the time when it matures; no interest payments are made.*

63. **Zero-Coupon Bonds** A zero-coupon bond can be redeemed in 20 years for $10,000. How much should you be willing to pay for it now if you want a return of:
   (a) 10% compounded monthly?
   (b) 10% compounded continuously?

64. **Zero-Coupon Bonds** A child's grandparents are considering buying a $40,000 face-value, zero-coupon bond at birth so that she will have enough money for her college education 17 years later. If they want a rate of return of 8% compounded annually, what should they pay for the bond?

65. Zero-Coupon Bonds How much should a $10,000 face-value, zero-coupon bond, maturing in 10 years, be sold for now if its rate of return is to be 8% compounded annually?

66. Zero-Coupon Bonds If Pat pays $12,485.52 for a $25,000 face-value, zero-coupon bond that matures in 8 years, what is his annual rate of return?

67. Time to Double or Triple an Investment The formula

$$t = \frac{\ln m}{n \ln\left(1 + \dfrac{r}{n}\right)}$$

can be used to find the number of years t required to multiply an investment m times when r is the per annum interest rate compounded n times a year.
(a) How many years will it take to double the value of an IRA that compounds annually at the rate of 12%?
(b) How many years will it take to triple the value of a savings account that compounds quarterly at an annual rate of 6%?
(c) Give a derivation of this formula.

68. Time to Reach an Investment Goal The formula

$$t = \frac{\ln A - \ln P}{r}$$

can be used to find the number of years t required for an investment P to grow to a value A when compounded continuously at an annual rate r.
(a) How long will it take to increase an initial investment of $1000 to $8000 at an annual rate of 10%?
(b) What annual rate is required to increase the value of a $2000 IRA to $30,000 in 35 years?
(c) Give a derivation of this formula.

*Problems 69–72 require the following discussion. The **Consumer Price Index (CPI)** indicates the relative change in price over time for a fixed basket of goods and services. It is a cost of living index that helps measure the effect of inflation on the cost of goods and services. The CPI uses the base period 1982–1984 for comparison (the CPI for this period is 100). The CPI for January 2006 was 198.3. This means that $100 in the period 1982–1984 had the same purchasing power as $198.30 in January 2006. In general, if the rate of inflation averages r% over n years, then the CPI index after n years is*

$$CPI = CPI_0\left(1 + \frac{r}{100}\right)^n$$

where CPI_0 is the CPI index at the beginning of the n-year period.
Source: U.S. Bureau of Labor Statistics

69. Consumer Price Index
(a) The CPI was 163.0 for 1998 and 215.3 for 2008. Assuming that annual inflation remained constant for this time period, determine the average annual inflation rate.
(b) Using the inflation rate from part (a), in what year will the CPI reach 300?

70. Consumer Price Index If the current CPI is 234.2 and the average annual inflation rate is 2.8%, what will be the CPI in 5 years?

71. Consumer Price Index If the average annual inflation rate is 3.1%, how long will it take for the CPI index to double? (A doubling of the CPI index means purchasing power is cut in half).

72. Consumer Price Index The base period for the CPI changed in 1998. Under the previous weight and item structure, the CPI for 1995 was 456.5. If the average annual inflation rate was 5.57%, what year was used as the base period for the CPI?

Explaining Concepts: Discussion and Writing

73. Explain in your own words what the term *compound interest* means. What does *continuous compounding* mean?

74. Explain in your own words the meaning of *present value*.

75. Critical Thinking You have just contracted to buy a house and will seek financing in the amount of $100,000. You go to several banks. Bank 1 will lend you $100,000 at the rate of 8.75% amortized over 30 years with a loan origination fee of 1.75%. Bank 2 will lend you $100,000 at the rate of 8.375% amortized over 15 years with a loan origination fee of 1.5%. Bank 3 will lend you $100,000 at the rate of 9.125% amortized over 30 years with no loan origination fee. Bank 4 will lend you $100,000 at the rate of 8.625% amortized over 15 years with no loan origination fee. Which loan would you take? Why? Be sure to have sound reasons for your choice.

Use the information in the table to assist you. If the amount of the monthly payment does not matter to you, which loan would you take? Again, have sound reasons for your choice. Compare your final decision with others in the class. Discuss.

	Monthly Payment	Loan Origination Fee
Bank 1	$786.70	$1,750.00
Bank 2	$977.42	$1,500.00
Bank 3	$813.63	$0.00
Bank 4	$992.08	$0.00

'Are You Prepared?' Answers

1. $15
2. $13\frac{1}{3}\%$

4.8 Exponential Growth and Decay Models; Newton's Law; Logistic Growth and Decay Models

OBJECTIVES
1 Find Equations of Populations That Obey the Law of Uninhibited Growth (p. 351)
2 Find Equations of Populations That Obey the Law of Decay (p. 353)
3 Use Newton's Law of Cooling (p. 354)
4 Use Logistic Models (p. 356)

1 Find Equations of Populations That Obey the Law of Uninhibited Growth

Many natural phenomena have been found to follow the law that an amount A varies with time t according to the function

Figure 41

(a) $A(t) = A_0 e^{kt}, k > 0$
Exponential growth

(b) $A(t) = A_0 e^{kt}, k < 0$
Exponential decay

$$A(t) = A_0 e^{kt} \qquad (1)$$

Here A_0 is the original amount ($t = 0$) and $k \neq 0$ is a constant.

If $k > 0$, then equation (1) states that the amount A is increasing over time; if $k < 0$, the amount A is decreasing over time. In either case, when an amount A varies over time according to equation (1), it is said to follow the **exponential law** or the **law of uninhibited growth** ($k > 0$) **or decay** ($k < 0$). See Figure 41.

For example, we saw in Section 4.7 that continuously compounded interest follows the law of uninhibited growth. In this section we shall look at some additional phenomena that follow the exponential law.

Cell division is the growth process of many living organisms, such as amoebas, plants, and human skin cells. Based on an ideal situation in which no cells die and no by-products are produced, the number of cells present at a given time follows the law of uninhibited growth. Actually, however, after enough time has passed, growth at an exponential rate will cease due to the influence of factors such as lack of living space and dwindling food supply. The law of uninhibited growth accurately models only the early stages of the cell division process.

The cell division process begins with a culture containing N_0 cells. Each cell in the culture grows for a certain period of time and then divides into two identical cells. We assume that the time needed for each cell to divide in two is constant and does not change as the number of cells increases. These new cells then grow, and eventually each divides in two, and so on.

Uninhibited Growth of Cells

A model that gives the number N of cells in a culture after a time t has passed (in the early stages of growth) is

$$N(t) = N_0 e^{kt} \qquad k > 0 \qquad (2)$$

where N_0 is the initial number of cells and k is a positive constant that represents the growth rate of the cells.

In using formula (2) to model the growth of cells, we are using a function that yields positive real numbers, even though we are counting the number of cells, which must be an integer. This is a common practice in many applications.

EXAMPLE 1 **Bacterial Growth**

A colony of bacteria that grows according to the law of uninhibited growth is modeled by the function $N(t) = 100e^{0.045t}$, where N is measured in grams and t is measured in days.

(a) Determine the initial amount of bacteria.

(b) What is the growth rate of the bacteria?

(c) What is the population after 5 days?

(d) How long will it take for the population to reach 140 grams?

(e) What is the doubling time for the population?

Solution (a) The initial amount of bacteria, N_0, is obtained when $t = 0$, so

$$N_0 = N(0) = 100e^{0.045(0)} = 100 \text{ grams}$$

(b) Compare $N(t) = 100e^{0.045t}$ to $N(t) = 100e^{kt}$. The value of k, 0.045, indicates a growth rate of 4.5%.

(c) The population after 5 days is $N(5) = 100e^{0.045(5)} \approx 125.2$ grams.

(d) To find how long it takes for the population to reach 140 grams, we solve the equation $N(t) = 140$.

$$100e^{0.045t} = 140$$

$e^{0.045t} = 1.4$ *Divide both sides of the equation by 100.*

$0.045t = \ln 1.4$ *Rewrite as a logarithm.*

$t = \dfrac{\ln 1.4}{0.045}$ *Divide both sides of the equation by 0.045.*

≈ 7.5 days

(e) The population doubles when $N(t) = 200$ grams, so we find the doubling time by solving the equation $200 = 100e^{0.045t}$ for t.

$$200 = 100e^{0.045t}$$

$2 = e^{0.045t}$ *Divide both sides of the equation by 100.*

$\ln 2 = 0.045t$ *Rewrite as a logarithm.*

$t = \dfrac{\ln 2}{0.045}$ *Divide both sides of the equation by 0.045.*

≈ 15.4 days

The population doubles approximately every 15.4 days.

●

━━ **Now Work** PROBLEM 1

EXAMPLE 2 **Bacterial Growth**

A colony of bacteria increases according to the law of uninhibited growth.

(a) If N is the number of cells and t is the time in hours, express N as a function of t.

(b) If the number of bacteria doubles in 3 hours, find the function that gives the number of cells in the culture.

(c) How long will it take for the size of the colony to triple?

(d) How long will it take for the population to double a second time (that is, increase four times)?

Solution (a) Using formula (2), the number N of cells at a time t is

$$N(t) = N_0 e^{kt}$$

where N_0 is the initial number of bacteria present and k is a positive number.

(b) We seek the number k. The number of cells doubles in 3 hours, so we have

$$N(3) = 2N_0$$

But $N(3) = N_0e^{k(3)}$, so

$$N_0e^{k(3)} = 2N_0$$
$$e^{3k} = 2 \qquad \text{Divide both sides by } N_0.$$
$$3k = \ln 2 \qquad \text{Write the exponential equation as a logarithm.}$$
$$k = \frac{1}{3}\ln 2 \approx 0.23105$$

The function that models this growth process is therefore

$$N(t) = N_0e^{0.23105t}$$

(c) The time t needed for the size of the colony to triple requires that $N = 3N_0$. We substitute $3N_0$ for N to get

$$3N_0 = N_0e^{0.23105t}$$
$$3 = e^{0.23105t}$$
$$0.23105t = \ln 3$$
$$t = \frac{\ln 3}{0.23105} \approx 4.755 \text{ hours}$$

It will take about 4.755 hours or 4 hours, 45 minutes for the size of the colony to triple.

(d) If a population doubles in 3 hours, it will double a second time in 3 more hours, for a total time of 6 hours.

●

2 Find Equations of Populations That Obey the Law of Decay

Radioactive materials follow the law of uninhibited decay.

Uninhibited Radioactive Decay

The amount A of a radioactive material present at time t is given by

$$A(t) = A_0e^{kt} \qquad k < 0 \tag{3}$$

where A_0 is the original amount of radioactive material and k is a negative number that represents the rate of decay.

All radioactive substances have a specific **half-life,** which is the time required for half of the radioactive substance to decay. In **carbon dating,** we use the fact that all living organisms contain two kinds of carbon, carbon 12 (a stable carbon) and carbon 14 (a radioactive carbon with a half-life of 5600 years). While an organism is living, the ratio of carbon 12 to carbon 14 is constant. But when an organism dies, the original amount of carbon 12 present remains unchanged, whereas the amount of carbon 14 begins to decrease. This change in the amount of carbon 14 present relative to the amount of carbon 12 present makes it possible to calculate when the organism died.

EXAMPLE 3   **Estimating the Age of Ancient Tools**

Traces of burned wood along with ancient stone tools in an archeological dig in Chile were found to contain approximately 1.67% of the original amount of carbon 14. If the half-life of carbon 14 is 5600 years, approximately when was the tree cut and burned?

Solution   Using formula (3), the amount A of carbon 14 present at time t is

$$A(t) = A_0 e^{kt}$$

where A_0 is the original amount of carbon 14 present and k is a negative number. We first seek the number k. To find it, we use the fact that after 5600 years half of the original amount of carbon 14 remains, so $A(5600) = \frac{1}{2} A_0$. Then

$$\frac{1}{2} A_0 = A_0 e^{k(5600)}$$

$$\frac{1}{2} = e^{5600k} \qquad \textit{Divide both sides of the equation by } A_0.$$

$$5600k = \ln \frac{1}{2} \qquad \textit{Rewrite as a logarithm.}$$

$$k = \frac{1}{5600} \ln \frac{1}{2} \approx -0.000124$$

Formula (3) therefore becomes

$$A(t) = A_0 e^{-0.000124t}$$

If the amount A of carbon 14 now present is 1.67% of the original amount, it follows that

$$0.0167 A_0 = A_0 e^{-0.000124t}$$

$$0.0167 = e^{-0.000124t} \qquad \textit{Divide both sides of the equation by } A_0.$$

$$-0.000124t = \ln 0.0167 \qquad \textit{Rewrite as a logarithm.}$$

$$t = \frac{\ln 0.0167}{-0.000124} \approx 33{,}003 \text{ years}$$

The tree was cut and burned about 33,003 years ago. Some archeologists use this conclusion to argue that humans lived in the Americas 33,000 years ago, much earlier than is generally accepted.

●

─ **Now Work** PROBLEM 3

3 Use Newton's Law of Cooling

Newton's Law of Cooling* states that the temperature of a heated object decreases exponentially over time toward the temperature of the surrounding medium.

> **Newton's Law of Cooling**
>
> The temperature u of a heated object at a given time t can be modeled by the following function:
>
> $$u(t) = T + (u_0 - T)e^{kt} \qquad k < 0 \qquad \textbf{(4)}$$
>
> where T is the constant temperature of the surrounding medium, u_0 is the initial temperature of the heated object, and k is a negative constant.

EXAMPLE 4   **Using Newton's Law of Cooling**

An object is heated to 100°C (degrees Celsius) and is then allowed to cool in a room whose air temperature is 30°C.

(a) If the temperature of the object is 80°C after 5 minutes, when will its temperature be 50°C?

(b) Determine the elapsed time before the temperature of the object is 35°C.

(c) What do you notice about the temperature as time passes?

*Named after Sir Isaac Newton (1643–1727), one of the cofounders of calculus.

Solution (a) Using formula (4) with $T = 30$ and $u_0 = 100$, the temperature $u(t)$ (in degrees Celsius) of the object at time t (in minutes) is

$$u(t) = 30 + (100 - 30)e^{kt} = 30 + 70e^{kt}$$

where k is a negative constant. To find k, we use the fact that $u = 80$ when $t = 5$. Then

$$u(t) = 30 + 70e^{kt}$$
$$80 = 30 + 70e^{k(5)} \qquad u(5) = 80$$
$$50 = 70e^{5k} \qquad \text{Simplify.}$$
$$e^{5k} = \frac{50}{70} \qquad \text{Solve for } e^{5k}.$$
$$5k = \ln\frac{5}{7} \qquad \text{Take ln of both sides.}$$
$$k = \frac{1}{5}\ln\frac{5}{7} \approx -0.0673 \qquad \text{Solve for } k.$$

Formula (4) therefore becomes

$$u(t) = 30 + 70e^{-0.0673t} \qquad\qquad \textbf{(5)}$$

We want to find t when $u = 50°C$, so

$$50 = 30 + 70e^{-0.0673t}$$
$$20 = 70e^{-0.0673t} \qquad \text{Simplify.}$$
$$e^{-0.0673t} = \frac{20}{70}$$
$$-0.0673t = \ln\frac{2}{7} \qquad \text{Take ln of both sides.}$$
$$t = \frac{\ln\dfrac{2}{7}}{-0.0673} \approx 18.6 \text{ minutes} \qquad \text{Solve for } t.$$

The temperature of the object will be 50°C after about 18.6 minutes or 18 minutes, 36 seconds.

(b) If $u = 35°C$, then, based on equation (5), we have

$$35 = 30 + 70e^{-0.0673t}$$
$$5 = 70e^{-0.0673t} \qquad \text{Simplify.}$$
$$e^{-0.0673t} = \frac{5}{70}$$
$$-0.0673t = \ln\frac{5}{70} \qquad \text{Take ln of both sides.}$$
$$t = \frac{\ln\dfrac{5}{70}}{-0.0673} \approx 39.2 \text{ minutes} \qquad \text{Solve for } t.$$

The object will reach a temperature of 35°C after about 39.2 minutes.

(c) Look at equation (5). As t increases, the exponent $-0.0673t$ becomes unbounded in the negative direction. As a result, the value of $e^{-0.0673t}$ approaches zeros so the value of u, the temperature of the object, approaches 30°C, the air temperature of the room.

Now Work PROBLEM 13

4 Use Logistic Models

The exponential growth model $A(t) = A_0 e^{kt}$, $k > 0$, assumes uninhibited growth, meaning that the value of the function grows without limit. Recall that we stated that cell division could be modeled using this function, assuming that no cells die and no by-products are produced. However, cell division eventually is limited by factors such as living space and food supply. The **logistic model,** given next, can describe situations where the growth or decay of the dependent variable is limited.

Logistic Model

In a logistic growth model, the population P after time t is given by the function

$$P(t) = \frac{c}{1 + ae^{-bt}} \qquad \text{(6)}$$

where a, b, and c are constants with $c > 0$. The model is a growth model if $b > 0$; the model is a decay model if $b < 0$.

The number c is called the **carrying capacity** (for growth models) because the value $P(t)$ approaches c as t approaches infinity; that is, $\lim_{t \to \infty} P(t) = c$. The number $|b|$ is the growth rate for $b > 0$ and the decay rate for $b < 0$. Figure 42(a) shows the graph of a typical logistic growth function, and Figure 42(b) shows the graph of a typical logistic decay function.

Figure 42

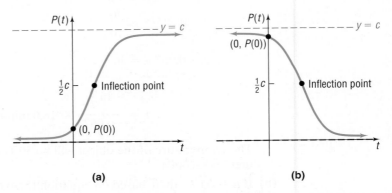

(a)   (b)

Based on the figures, we have the following properties of logistic growth functions.

Properties of the Logistic Growth Model, Equation (6)

1. The domain is the set of all real numbers. The range is the interval $(0, c)$, where c is the carrying capacity.
2. There are no x-intercepts; the y-intercept is $P(0)$.
3. There are two horizontal asymptotes: $y = 0$ and $y = c$.
4. $P(t)$ is an increasing function if $b > 0$ and a decreasing function if $b < 0$.
5. There is an **inflection point** where $P(t)$ equals $\frac{1}{2}$ of the carrying capacity. The inflection point is the point on the graph where the graph changes from being curved upward to curved downward for growth functions and the point where the graph changes from being curved downward to curved upward for decay functions.
6. The graph is smooth and continuous, with no corners or gaps.

EXAMPLE 5   **Fruit Fly Population**

Fruit flies are placed in a half-pint milk bottle with a banana (for food) and yeast plants (for food and to provide a stimulus to lay eggs). Suppose that the fruit fly population after t days is given by

$$P(t) = \frac{230}{1 + 56.5e^{-0.37t}}$$

(a) State the carrying capacity and the growth rate.

(b) Determine the initial population.

(c) What is the population after 5 days?

(d) How long does it take for the population to reach 180?

 (e) Use a graphing utility to determine how long it takes for the population to reach one-half of the carrying capacity by graphing $Y_1 = P(t)$ and $Y_2 = 115$ and using INTERSECT.

Solution   (a) As $t \to \infty$, $e^{-0.37t} \to 0$ and $P(t) \to \dfrac{230}{1}$. The carrying capacity of the half-pint bottle is 230 fruit flies. The growth rate is $|b| = |0.37| = 37\%$ per day.

(b) To find the initial number of fruit flies in the half-pint bottle, we evaluate $P(0)$.

$$
\begin{aligned}
P(0) &= \frac{230}{1 + 56.5e^{-0.37(0)}} \\
&= \frac{230}{1 + 56.5} \\
&= 4
\end{aligned}
$$

So, initially, there were 4 fruit flies in the half-pint bottle.

(c) To find the number of fruit flies in the half-pint bottle after 5 days, we evaluate $P(5)$.

$$P(5) = \frac{230}{1 + 56.5e^{-0.37(5)}} \approx 23 \text{ fruit flies}$$

After 5 days, there are approximately 23 fruit flies in the bottle.

(d) To determine when the population of fruit flies will be 180, we solve the equation $P(t) = 180$.

$$
\begin{aligned}
\frac{230}{1 + 56.5e^{-0.37t}} &= 180 & \\
230 &= 180(1 + 56.5e^{-0.37t}) & \\
1.2778 &= 1 + 56.5e^{-0.37t} & \text{Divide both sides by 180.} \\
0.2778 &= 56.5e^{-0.37t} & \text{Subtract 1 from both sides.} \\
0.0049 &= e^{-0.37t} & \text{Divide both sides by 56.5.} \\
\ln(0.0049) &= -0.37t & \text{Rewrite as a logarithmic expression.} \\
t &\approx 14.4 \text{ days} & \text{Divide both sides by } -0.37.
\end{aligned}
$$

It will take approximately 14.4 days (14 days, 10 hours) for the population to reach 180 fruit flies.

(e) One-half of the carrying capacity is 115 fruit flies. We solve $P(t) = 115$ by graphing $Y_1 = \dfrac{230}{1 + 56.5e^{-0.37t}}$ and $Y_2 = 115$ and using INTERSECT. See Figure 43.

The population will reach one-half of the carrying capacity in about 10.9 days (10 days, 22 hours).

Figure 43

Look back at Figure 43. Notice the point where the graph reaches 115 fruit flies (one-half of the carrying capacity): the graph changes from being curved upward to being curved downward. Using the language of calculus, we say the graph changes from increasing at an increasing rate to increasing at a decreasing rate. For any logistic growth function, when the population reaches one-half the carrying capacity, the population growth starts to slow down.

> **Now Work** PROBLEM 23

Exploration

On the same viewing rectangle, graph

$$Y_1 = \frac{500}{1 + 24e^{-0.03t}} \text{ and } Y_2 = \frac{500}{1 + 24e^{-0.08t}}$$

What effect does the growth rate $|b|$ have on the logistic growth function?

EXAMPLE 6   **Wood Products**

The EFISCEN wood product model classifies wood products according to their life-span. There are four classifications: short (1 year), medium short (4 years), medium long (16 years), and long (50 years). Based on data obtained from the European Forest Institute, the percentage of remaining wood products after t years for wood products with long life-spans (such as those used in the building industry) is given by

$$P(t) = \frac{100.3952}{1 + 0.0316e^{0.0581t}}$$

(a) What is the decay rate?
(b) What is the percentage of remaining wood products after 10 years?
(c) How long does it take for the percentage of remaining wood products to reach 50%?
(d) Explain why the numerator given in the model is reasonable.

Solution (a) The decay rate is $|b| = |-0.0581| = 5.81\%$.

(b) Evaluate $P(10)$.

$$P(10) = \frac{100.3952}{1 + 0.0316e^{0.0581(10)}} \approx 95.0$$

So 95% of long-life-span wood products remain after 10 years.

(c) Solve the equation $P(t) = 50$.

$$\frac{100.3952}{1 + 0.0316e^{0.0581t}} = 50$$

$$100.3952 = 50(1 + 0.0316e^{0.0581t})$$

$$2.0079 = 1 + 0.0316e^{0.0581t} \qquad \text{Divide both sides by 50.}$$

$$1.0079 = 0.0316e^{0.0581t} \qquad \text{Subtract 1 from both sides.}$$

$$31.8957 = e^{0.0581t} \qquad \text{Divide both sides by 0.0316.}$$

$$\ln(31.8957) = 0.0581t \qquad \text{Rewrite as a logarithmic expression.}$$

$$t \approx 59.6 \text{ years} \qquad \text{Divide both sides by 0.0581.}$$

It will take approximately 59.6 years for the percentage of long-life-span wood products remaining to reach 50%.

(d) The numerator of 100.3952 is reasonable because the maximum percentage of wood products remaining that is possible is 100%.

4.8 Assess Your Understanding

Applications and Extensions

1. **Growth of an Insect Population** The size P of a certain insect population at time t (in days) obeys the function $P(t) = 500e^{0.02t}$.
   (a) Determine the number of insects at $t = 0$ days.
   (b) What is the growth rate of the insect population?
   (c) What is the population after 10 days?
   (d) When will the insect population reach 800?
   (e) When will the insect population double?

2. **Growth of Bacteria** The number N of bacteria present in a culture at time t (in hours) obeys the law of uninhibited growth $N(t) = 1000e^{0.01t}$.
   (a) Determine the number of bacteria at $t = 0$ hours.
   (b) What is the growth rate of the bacteria?
   (c) What is the population after 4 hours?
   (d) When will the number of bacteria reach 1700?
   (e) When will the number of bacteria double?

3. **Radioactive Decay** Strontium 90 is a radioactive material that decays according to the function $A(t) = A_0 e^{-0.0244t}$, where A_0 is the initial amount present and A is the amount present at time t (in years). Assume that a scientist has a sample of 500 grams of strontium 90.
   (a) What is the decay rate of strontium 90?
   (b) How much strontium 90 is left after 10 years?
   (c) When will 400 grams of strontium 90 be left?
   (d) What is the half-life of strontium 90?

4. **Radioactive Decay** Iodine 131 is a radioactive material that decays according to the function $A(t) = A_0 e^{-0.087t}$, where A_0 is the initial amount present and A is the amount present at time t (in days). Assume that a scientist has a sample of 100 grams of iodine 131.
   (a) What is the decay rate of iodine 131?
   (b) How much iodine 131 is left after 9 days?
   (c) When will 70 grams of iodine 131 be left?
   (d) What is the half-life of iodine 131?

5. **Growth of a Colony of Mosquitoes** The population of a colony of mosquitoes obeys the law of uninhibited growth.
   (a) If N is the population of the colony and t is the time in days, express N as a function of t.
   (b) If there are 1000 mosquitoes initially and there are 1800 after 1 day, what is the size of the colony after 3 days?
   (c) How long is it until there are 10,000 mosquitoes?

6. **Bacterial Growth** A culture of bacteria obeys the law of uninhibited growth.
   (a) If N is the number of bacteria in the culture and t is the time in hours, express N as a function of t.
   (b) If 500 bacteria are present initially and there are 800 after 1 hour, how many will be present in the culture after 5 hours?
   (c) How long is it until there are 20,000 bacteria?

7. **Population Growth** The population of a southern city follows the exponential law.
   (a) If N is the population of the city and t is the time in years, express N as a function of t.

   (b) If the population doubled in size over an 18-month period and the current population is 10,000, what will the population be 2 years from now?

8. **Population Decline** The population of a midwestern city follows the exponential law.
   (a) If N is the population of the city and t is the time in years, express N as a function of t.
   (b) If the population decreased from 900,000 to 800,000 from 2003 to 2005, what was the population in 2007?

9. **Radioactive Decay** The half-life of radium is 1690 years. If 10 grams is present now, how much will be present in 50 years?

10. **Radioactive Decay** The half-life of radioactive potassium is 1.3 billion years. If 10 grams is present now, how much will be present in 100 years? In 1000 years?

11. **Estimating the Age of a Tree** A piece of charcoal is found to contain 30% of the carbon 14 that it originally had. When did the tree die from which the charcoal came? Use 5600 years as the half-life of carbon 14.

12. **Estimating the Age of a Fossil** A fossilized leaf contains 70% of its normal amount of carbon 14. How old is the fossil?

13. **Cooling Time of a Pizza Pan** A pizza pan is removed at 5:00 PM from an oven whose temperature is fixed at 450°F into a room that is a constant 70°F. After 5 minutes, the pan is 300°F.
    (a) At what time is the temperature of the pan 135°F?
    (b) Determine the time that needs to elapse before the pan is 160°F.
    (c) What do you notice about the temperature as time passes?

14. **Newton's Law of Cooling** A thermometer reading 72°F is placed in a refrigerator where the temperature is a constant 38°F.
    (a) If the thermometer reads 60°F after 2 minutes, what will it read after 7 minutes?
    (b) How long will it take before the thermometer reads 39°F?
    (c) Determine the time needed to elapse before the thermometer reads 45°F.
    (d) What do you notice about the temperature as time passes?

15. **Newton's Law of Heating** A thermometer reading 8°C is brought into a room with a constant temperature of 35°C. If the thermometer reads 15°C after 3 minutes, what will it read after being in the room for 5 minutes? For 10 minutes?

    [**Hint:** You need to construct a formula similar to equation (4).]

16. **Warming Time of a Beer Stein** A beer stein has a temperature of 28°F. It is placed in a room with a constant temperature of 70°F. After 10 minutes, the temperature of the stein has risen to 35°F. What will the temperature of the stein be after 30 minutes? How long will it take the stein to reach a temperature of 45°F? (See the hint given for Problem 15.)

17. **Decomposition of Chlorine in a Pool** Under certain water conditions, the free chlorine (hypochlorous acid, HOCl) in a swimming pool decomposes according to the law of uninhibited decay. After shocking his pool, Ben tested the water and found the amount of free chlorine to be 2.5 parts per million (ppm). Twenty-four hours later, Ben tested the water again and found the amount of free chlorine to be 2.2 ppm. What will be the reading after 3 days (that is, 72 hours)? When the chlorine level reaches 1.0 ppm, Ben must shock the pool again. How long can Ben go before he must shock the pool again?

18. **Decomposition of Dinitrogen Pentoxide** At 45°C, dinitrogen pentoxide (N_2O_5) decomposes into nitrous dioxide (NO_2) and oxygen (O_2) according to the law of uninhibited decay. An initial amount of 0.25 M of dinitrogen pentoxide decomposes to 0.15 M in 17 minutes. How much dinitrogen pentoxide will remain after 30 minutes? How long will it take until 0.01 M of dinitrogen pentoxide remains?

19. **Decomposition of Sucrose** Reacting with water in an acidic solution at 35°C, sucrose ($C_{12}H_{22}O_{11}$) decomposes into glucose ($C_6H_{12}O_6$) and fructose ($C_6H_{12}O_6$)* according to the law of uninhibited decay. An initial amount of 0.40 M of sucrose decomposes to 0.36 M in 30 minutes. How much sucrose will remain after 2 hours? How long will it take until 0.10 M of sucrose remains?

20. **Decomposition of Salt in Water** Salt (NaCl) decomposes in water into sodium (Na^+) and chloride (Cl^-) ions according to the law of uninhibited decay. If the initial amount of salt is 25 kilograms and, after 10 hours, 15 kilograms of salt is left, how much salt is left after 1 day? How long does it take until $\frac{1}{2}$ kilogram of salt is left?

21. **Radioactivity from Chernobyl** After the release of radioactive material into the atmosphere from a nuclear power plant at Chernobyl (Ukraine) in 1986, the hay in Austria was contaminated by iodine 131 (half-life 8 days). If it is safe to feed the hay to cows when 10% of the iodine 131 remains, how long did the farmers need to wait to use this hay?

22. **Pig Roasts** The hotel Bora-Bora is having a pig roast. At noon, the chef put the pig in a large earthen oven. The pig's original temperature was 75°F. At 2:00 PM the chef checked the pig's temperature and was upset because it had reached only 100°F. If the oven's temperature remains a constant 325°F, at what time may the hotel serve its guests, assuming that pork is done when it reaches 175°F?

23. **Population of a Bacteria Culture** The logistic growth model

    $$P(t) = \frac{1000}{1 + 32.33e^{-0.439t}}$$

    represents the population (in grams) of a bacterium after t hours.
    (a) Determine the carrying capacity of the environment.
    (b) What is the growth rate of the bacteria?
    (c) Determine the initial population size.
    (d) What is the population after 9 hours?
    (e) When will the population be 700 grams?
    (f) How long does it take for the population to reach one-half the carrying capacity?

24. **Population of an Endangered Species** Often environmentalists capture an endangered species and transport the species to a controlled environment where the species can produce offspring and regenerate its population. Suppose that six American bald eagles are captured, transported to Montana, and set free. Based on experience, the environmentalists expect the population to grow according to the model

    $$P(t) = \frac{500}{1 + 83.33e^{-0.162t}}$$

    where t is measured in years.

(a) Determine the carrying capacity of the environment.
(b) What is the growth rate of the bald eagle?
(c) What is the population after 3 years?
(d) When will the population be 300 eagles?
(e) How long does it take for the population to reach one-half of the carrying capacity?

25. **The *Challenger* Disaster** After the *Challenger* disaster in 1986, a study was made of the 23 launches that preceded the fatal flight. A mathematical model was developed involving the relationship between the Fahrenheit temperature x around the O-rings and the number y of eroded or leaky primary O-rings. The model stated that

    $$y = \frac{6}{1 + e^{-(5.085 - 0.1156x)}}$$

    where the number 6 indicates the 6 primary O-rings on the spacecraft.

*Author's Note: Surprisingly, the chemical formulas for glucose and fructose are the same: This is not a typo.

(a) What is the predicted number of eroded or leaky primary O-rings at a temperature of 100°F?

(b) What is the predicted number of eroded or leaky primary O-rings at a temperature of 60°F?

(c) What is the predicted number of eroded or leaky primary O-rings at a temperature of 30°F?

(d) Graph the equation using a graphing utility. At what temperature is the predicted number of eroded or leaky O-rings 1? 3? 5?

Source: Linda Tappin, "Analyzing Data Relating to the Challenger Disaster," Mathematics Teacher, *Vol. 87, No. 6, September 1994, pp. 423–426.*

4.9 Building Exponential, Logarithmic, and Logistic Models from Data

PREPARING FOR THIS SECTION *Before getting started, review the following:*

- Building Linear Models from Data (Section 2.2, pp. 134–137)
- Building Cubic Models from Data (Section 3.1, pp. 209–210)
- Building Quadratic Models from Data (Section 2.6, pp. 172–173)

OBJECTIVES **1** Build an Exponential Model from Data (p. 362)
2 Build a Logarithmic Model from Data (p. 363)
3 Build a Logistic Model from Data (p. 364)

In Section 2.2, we discussed how to find the linear function of best fit ($y = ax + b$), in Section 2.6, we discussed how to find the quadratic function of best fit ($y = ax^2 + bx + c$). and in Section 3.1, we discussed how to find the cubic function of best fit ($y = ax^3 + bx^2 + cx + d$).

In this section we will discuss how to use a graphing utility to find equations of best fit that describe the relation between two variables when the relation is thought to be exponential ($y = ab^x$), logarithmic ($y = a + b \ln x$), or logistic $\left(y = \dfrac{c}{1 + ae^{-bx}} \right)$. As before, we draw a scatter diagram of the data to help to determine the appropriate model to use.

Figure 44 shows scatter diagrams that will typically be observed for the three models. Below each scatter diagram are any restrictions on the values of the parameters.

Figure 44

$y = ab^x$, $a > 0$, $b > 1$	$y = ab^x$, $0 < b < 1$, $a > 0$	$y = a + b \ln x$, $a > 0$, $b < 0$	$y = a + b \ln x$, $a > 0$, $b > 0$	$y = \dfrac{c}{1 + ae^{-bx}}$, $a > 0$, $b > 0$, $c > 0$
Exponential	Exponential	Logarithmic	Logarithmic	Logistic

Most graphing utilities have REGression options that fit data to a specific type of curve. Once the data have been entered and a scatter diagram obtained, the type of curve that you want to fit to the data is selected. Then that REGression option is used to obtain the curve of *best fit* of the type selected.

The correlation coefficient r will appear only if the model can be written as a linear expression. As it turns out, r will appear for the linear, power, exponential, and logarithmic models, since these models can be written as a linear expression. Remember, the closer $|r|$ is to 1, the better the fit.

Let's look at some examples.

 1 Build an Exponential Model from Data

We saw in Section 4.7 that the future value of money behaves exponentially, and we saw in Section 4.8 that growth and decay models also behave exponentially. The next example shows how data can lead to an exponential model.

EXAMPLE 1	**Fitting an Exponential Function to Data**

Kathleen is interested in finding a function that explains the growth of cell phone usage in the United States. She gathers data on the number (in millions) of U.S. cell phone subscribers from 1985 through 2008. The data are shown in Table 9.

(a) Using a graphing utility, draw a scatter diagram with year as the independent variable.

(b) Using a graphing utility, build an exponential model from the data.

(c) Express the function found in part (b) in the form $A = A_0 e^{kt}$.

(d) Graph the exponential function found in part (b) or (c) on the scatter diagram.

(e) Using the solution to part (b) or (c), predict the number of U.S. cell phone subscribers in 2009.

(f) Interpret the value of k found in part (c).

Solution

(a) Enter the data into the graphing utility, letting 1 represent 1985, 2 represent 1986, and so on. We obtain the scatter diagram shown in Figure 45.

(b) A graphing utility fits the data in Figure 45 to an exponential function of the form $y = ab^x$ using the EXPonential REGression option. From Figure 46 we find that $y = ab^x = 0.86498(1.31855)^x$. Notice that $|r|$ is close to 1, indicating a good fit.

Table 9

Year, x	Number of Subscribers (in millions), y
1985 ($x = 1$)	0.34
1986 ($x = 2$)	0.68
1987 ($x = 3$)	1.23
1988 ($x = 4$)	2.07
1989 ($x = 5$)	3.51
1990 ($x = 6$)	5.28
1991 ($x = 7$)	7.56
1992 ($x = 8$)	11.03
1993 ($x = 9$)	16.01
1994 ($x = 10$)	24.13
1995 ($x = 11$)	33.76
1996 ($x = 12$)	44.04
1997 ($x = 13$)	55.31
1998 ($x = 14$)	69.21
1999 ($x = 15$)	86.05
2000 ($x = 16$)	109.48
2001 ($x = 17$)	128.37
2002 ($x = 18$)	140.77
2003 ($x = 19$)	158.72
2004 ($x = 20$)	182.14
2005 ($x = 21$)	207.90
2006 ($x = 22$)	233.00
2007 ($x = 23$)	255.40
2008 ($x = 24$)	270.33

Source: ©2009 CTIA–The Wireless Association®. All Rights Reserved.

Figure 45

Figure 46

```
ExpReg
y=a*b^x
a=.8649775385
b=1.318554023
r²=.9310544659
r=.9649116363
```

(c) To express $y = ab^x$ in the form $A = A_0 e^{kt}$, where $x = t$ and $y = A$, we proceed as follows:

$$ab^x = A_0 e^{kt} \quad x = t$$

When $x = t = 0$, we find that $a = A_0$. This leads to

$$a = A_0 \qquad b^x = e^{kt}$$
$$b^x = (e^k)^t$$
$$b = e^k \qquad {\scriptstyle x \,=\, t}$$

Since $y = ab^x = 0.86498(1.31855)^x$, we find that $a = 0.86498$ and $b = 1.31855$.

$$a = A_0 = 0.86498 \quad \text{and} \quad b = e^k = 1.31855$$

We want to find k, so we rewrite $e^k = 1.31855$ as a logarithm and obtain

$$k = \ln(1.31855) \approx 0.2765$$

As a result, $A = A_0 e^{kt} = 0.86498 e^{0.2765t}$.

Figure 47

275

0

0 ————————————— 25

(d) See Figure 47 for the graph of the exponential function of best fit.

(e) Let $t = 25$ (end of 2009) in the function found in part (c). The predicted number (in millions) of cell phone subscribers in the United States in 2009 is

$$A_0 e^{kt} = 0.86498 e^{0.2765(25)} \approx 869$$

This prediction (869 millon) far exceeds what the U.S. population was in 2009 (currently the U.S. population is about 304 million). See the answer in part (f).

(f) The value of $k = 0.2765$ represents the growth rate of the number of cell phone subscribers in the United States. Over the period 1985 through 2008, the number of cell phone subscribers has grown at an annual rate of 27.65% compounded continuously. This growth rate is not sustainable as we learned in part (e). In Problem 10 you are asked to build a better model from these data.

●

━━━━━**Now Work** PROBLEM 1

 2 Build a Logarithmic Model from Data

Many relations between variables do not follow an exponential model; instead, the independent variable is related to the dependent variable using a logarithmic model.

| EXAMPLE 2 | **Fitting a Logarithmic Function to Data** |

Table 10

Atmospheric Pressure, p	Height, h
760	0
740	0.184
725	0.328
700	0.565
650	1.079
630	1.291
600	1.634
580	1.862
550	2.235

Jodi, a meteorologist, is interested in finding a function that explains the relation between the height of a weather balloon (in kilometers) and the atmospheric pressure (measured in millimeters of mercury) on the balloon. She collects the data shown in Table 10.

(a) Using a graphing utility, draw a scatter diagram of the data with atmospheric pressure as the independent variable.

(b) It is known that the relation between atmospheric pressure and height follows a logarithmic model. Using a graphing utility, build a logarithmic model from the data.

(c) Draw the logarithmic function found in part (b) on the scatter diagram.

(d) Use the function found in part (b) to predict the height of the weather balloon if the atmospheric pressure is 560 millimeters of mercury.

Solution

(a) After entering the data into the graphing utility, we obtain the scatter diagram shown in Figure 48.

(b) A graphing utility fits the data in Figure 48 to a logarithmic function of the form $y = a + b \ln x$ by using the LOGarithm REGression option. See Figure 49. The logarithmic model from the data is

$$h(p) = 45.7863 - 6.9025 \ln p$$

where h is the height of the weather balloon and p is the atmospheric pressure. Notice that $|r|$ is close to 1, indicating a good fit.

(c) Figure 50 shows the graph of $h(p) = 45.7863 - 6.9025 \ln p$ on the scatter diagram.

Figure 48

Figure 49

```
LnReg
y=a+blnx
a=45.78632064
b=-6.902524299
r=-.9999946336
```

Figure 50

(d) Using the function found in part (b), Jodi predicts the height of the weather balloon when the atmospheric pressure is 560 to be

$$h(560) = 45.7863 - 6.9025 \ln 560$$

$$\approx 2.108 \text{ kilometers}$$

Now Work PROBLEM 5

3 Build a Logistic Model from Data

Logistic growth models can be used to model situations for which the value of the dependent variable is limited. Many real-world situations conform to this scenario. For example, the population of the human race is limited by the availability of natural resources such as food and shelter. When the value of the dependent variable is limited, a logistic growth model is often appropriate.

EXAMPLE 3

Fitting a Logistic Function to Data

The data in Table 11 represent the amount of yeast biomass in a culture after t hours.

Table 11

Time (in hours)	Yeast Biomass	Time (in hours)	Yeast Biomass
0	9.6	10	513.3
1	18.3	11	559.7
2	29.0	12	594.8
3	47.2	13	629.4
4	71.1	14	640.8
5	119.1	15	651.1
6	174.6	16	655.9
7	257.3	17	659.6
8	350.7	18	661.8
9	441.0		

Source: Tor Carlson (Über Geschwindigkeit und Grösse der Hefevermehrung in Würze, Biochemische Zeitschrift, Bd. 57, pp. 313–334, 1913)

(a) Using a graphing utility, draw a scatter diagram of the data with time as the independent variable.

(b) Using a graphing utility, build a logistic model from the data.

(c) Using a graphing utility, graph the function found in part (b) on the scatter diagram.

(d) What is the predicted carrying capacity of the culture?

(e) Use the function found in part (b) to predict the population of the culture at $t = 19$ hours.

Solution

(a) See Figure Figure 51 for a scatter diagram of the data.

(b) A graphing utility fits a logistic growth model of the form $y = \dfrac{c}{1 + ae^{-bx}}$ by using the LOGISTIC regression option. See Figure 52. The logistic model from the data is

$$y = \frac{663.0}{1 + 71.6e^{-0.5470x}}$$

where y is the amount of yeast biomass in the culture and x is the time.

(c) See Figure 53 for the graph of the logistic model.

Figure 51

Figure 52

Figure 53

(d) Based on the logistic growth model found in part (b), the carrying capacity of the culture is 663.

(e) Using the logistic growth model found in part (b), the predicted amount of yeast biomass at $t = 19$ hours is

$$y = \frac{663.0}{1 + 71.6e^{-0.5470(19)}} \approx 661.5$$

Now Work PROBLEM 7

4.9 Assess Your Understanding

Applications and Extensions

1. Biology   A strain of E-coli Beu 397-recA441 is placed into a nutrient broth at 30° Celsius and allowed to grow. The following data are collected. Theory states that the number of bacteria in the petri dish will initially grow according to the law of uninhibited growth. The population is measured using an optical device in which the amount of light that passes through the petri dish is measured.

Time (hours), x	Population, y
0	0.09
2.5	0.18
3.5	0.26
4.5	0.35
6	0.50

Source: Dr. Polly Lavery, Joliet Junior College

(a) Draw a scatter diagram treating time as the independent variable.

(b) Using a graphing utility, build an exponential model from the data.

(c) Express the function found in part (b) in the form $N(t) = N_0 e^{kt}$.

(d) Graph the exponential function found in part (b) or (c) on the scatter diagram.

(e) Use the exponential function from part (b) or (c) to predict the population at $x = 7$ hours.

(f) Use the exponential function from part (b) or (c) to predict when the population will reach 0.75.

2. Biology   A strain of E-coli SC18del-recA718 is placed into a nutrient broth at 30° Celsius and allowed to grow. The data on the following page are collected. Theory states that the number of bacteria in the petri dish will initially grow according to the law of uninhibited growth. The population is measured using an optical device in which the amount of light that passes through the petri dish is measured.

Time (hours), x	Population, y
2.5	0.175
3.5	0.38
4.5	0.63
4.75	0.76
5.25	1.20

Source: Dr. Polly Lavery, Joliet Junior College

(a) Draw a scatter diagram treating time as the independent variable.
(b) Using a graphing utility, build an exponential model from the data.
(c) Express the function found in part (b) in the form $N(t) = N_0 e^{kt}$.
(d) Graph the exponential function found in part (b) or (c) on the scatter diagram.
(e) Use the exponential function from part (b) or (c) to predict the population at $x = 6$ hours.
(f) Use the exponential function from part (b) or (c) to predict when the population will reach 2.1.

3. **Chemistry** A chemist has a 100-gram sample of a radioactive material. He records the amount of radioactive material every week for 6 weeks and obtains the following data:

Week	Weight (in Grams)
0	100.0
1	88.3
2	75.9
3	69.4
4	59.1
5	51.8
6	45.5

(a) Using a graphing utility, draw a scatter diagram with week as the independent variable.
(b) Using a graphing utility, build an exponential model from the data.
(c) Express the function found in part (b) in the form $A(t) = A_0 e^{kt}$.
(d) Graph the exponential function found in part (b) or (c) on the scatter diagram.
(e) From the result found in part (b), determine the half-life of the radioactive material.
(f) How much radioactive material will be left after 50 weeks?
(g) When will there be 20 grams of radioactive material?

4. **Cigarette Exports** The following data represent the number of cigarettes (in billions) exported from the United States by year.

Year	Cigarette Exports (in billions of pieces)
1995	231.1
1998	201.3
1999	151.4
2000	147.9
2001	133.9
2002	127.4
2003	121.5
2004	118.7

Source: Statistical Abstract of the United States, 2006

(a) Let $t =$ the number of years since 1995. Using a graphing utility, draw a scatter diagram of the data using t as the independent variable and number of cigarettes as the dependent variable.
(b) Using a graphing utility, build an exponential model from the data.
(c) Express the function found in part (b) in the form $A(t) = A_0 e^{kt}$.
(d) Graph the exponential function found in part (b) or (c) on the scatter diagram.
(e) Use the exponential function from part (b) or (c) to predict the number of cigarettes that will be exported from the United States in 2010.
(f) Use the exponential function from part (b) or (c) to predict when the number of cigarettes exported from the United States will decrease to 50 billion.

5. **Economics and Marketing** The following data represent the price and quantity demanded in 2009 for Dell personal computers.

Price ($/Computer)	Quantity Demanded
2300	152
2000	159
1700	164
1500	171
1300	176
1200	180
1000	189

(a) Using a graphing utility, draw a scatter diagram of the data with price as the dependent variable.
(b) Using a graphing utility, build a logarithmic model from the data.
(c) Using a graphing utility, draw the logarithmic function found in part (b) on the scatter diagram.
(d) Use the function found in part (b) to predict the number of Dell personal computers that will be demanded if the price is $1650.

6. **Economics and Marketing** The following data represent the price and quantity supplied in 2009 for Dell personal computers.

Price ($/Computer)	Quantity Supplied
2300	180
2000	173
1700	160
1500	150
1300	137
1200	130
1000	113

(a) Using a graphing utility, draw a scatter diagram of the data with price as the dependent variable.
(b) Using a graphing utility, build a logarithmic model from the data.
(c) Using a graphing utility, draw the logarithmic function found in part (b) on the scatter diagram.
(d) Use the function found in part (b) to predict the number of Dell personal computers that will be supplied if the price is $1650.

7. **Population Model** The following data represent the population of the United States. An ecologist is interested in building a model that describes the population of the United States.

Year	Population
1900	76,212,168
1910	92,228,496
1920	106,021,537
1930	123,202,624
1940	132,164,569
1950	151,325,798
1960	179,323,175
1970	203,302,031
1980	226,542,203
1990	248,709,873
2000	281,421,906

Source: U.S. Census Bureau

(a) Using a graphing utility, draw a scatter diagram of the data using years since 1900 as the independent variable and population as the dependent variable.
(b) Using a graphing utility, build a logistic model from the data.
(c) Using a graphing utility, draw the function found in part (b) on the scatter diagram.
(d) Based on the function found in part (b), what is the carrying capacity of the United States?
(e) Use the function found in part (b) to predict the population of the United States in 2004.
(f) When will the United States population be 300,000,000?
(g) Compare actual U.S. Census figures to the predictions found in parts (e) and (f). Discuss any differences.

8. **Population Model** The following data represent the world population. An ecologist is interested in building a model that describes the world population.

Year	Population (in Billions)
2001	6.17
2002	6.25
2003	6.32
2004	6.40
2005	6.48
2006	6.55
2007	6.63
2008	6.71
2009	6.79

Source: U.S. Census Bureau

(a) Using a graphing utility, draw a scatter diagram of the data using years since 2000 as the independent variable and population as the dependent variable.
(b) Using a graphing utility, build a logistic model from the data.
(c) Using a graphing utility, draw the function found in part (b) on the scatter diagram.
(d) Based on the function found in part (b), what is the carrying capacity of the world?
(e) Use the function found in part (b) to predict the population of the world in 2015.
(f) When will world population be 10 billion?

9. **Cable Subscribers** The following data represent the number of basic cable TV subscribers in the United States. A market researcher believes that external factors, such as satellite TV, have affected the growth of cable subscribers. She is interested in building a model that can be used to describe the number of cable TV subscribers in the United States.

Year	Subscribers (1,000)
1975 ($t = 5$)	9,800
1980 ($t = 10$)	17,500
1985 ($t = 15$)	35,440
1990 ($t = 20$)	50,520
1992 ($t = 22$)	54,300
1994 ($t = 24$)	58,373
1996 ($t = 26$)	62,300
1998 ($t = 28$)	64,650
2000 ($t = 30$)	66,250
2002 ($t = 32$)	66,472
2004 ($t = 34$)	65,727
2006 ($t = 36$)	65,319

Source: Statistical Abstract of the United States, 2009

(a) Using a graphing utility, draw a scatter diagram of the data using the number of years after 1970, t, as the independent variable and number of subscribers as the dependent variable.
(b) Using a graphing utility, build a logistic model from the data.

(c) Using a graphing utility, draw the function found in part (b) on the scatter diagram.

(d) Based on the model found in part (b), what is the maximum number of cable TV subscribers in the United States?

(e) Use the model found in part (b) to predict the number of cable TV subscribers in the United States in 2015.

10. **Cell Phone Users** Refer to the data in Table 9 (p. 362).

(a) Using a graphing utility, build a logistic model from the data.

(b) Graph the logistic function found in part (b) on a scatter diagram of the data.

(c) What is the predicted carrying capacity of U.S. cell phone subscribers?

(d) Use the model found in part (b) to predict the number of U.S. cell phone subscribers at the end of 2009.

(e) Compare the answer to part (d) above with the answer to Example 1, part (e). How do you explain the different predictions?

Mixed Practice

11. **Age versus Total Cholesterol** The following data represent the age and average total cholesterol for adult males at various ages.

Age	Total Cholesterol
27	189
40	205
50	215
60	210
70	210
80	194

(a) Using a graphing utility, draw a scatter diagram of the data using age, x, as the independent variable and total cholesterol, y, as the dependent variable.

(b) Based on the scatter diagram drawn in part (a), decide on a model (linear, quadratic, cubic, exponential, logarithmic, or logistic) that you think best describes the relation between age and total cholesterol. Be sure to justify your choice of model.

(c) Using a graphing utility, find the model of best fit.

(d) Using a graphing utility, draw the model of best fit on the scatter diagram drawn in part (a).

(e) Use your model to predict the total cholesterol of a 35-year-old male.

12. **Income versus Crime Rate** The following data represent crime rate against individuals (crimes per 1000 households) and their income in the United States in 2006.

Income	Crime Rate
$5000	217.3
11,250	195.7
20,000	183.1
30,000	179.4
42,500	166.2
62,500	166.8
85,000	162.0

Source: Statistical Abstract of the United States, 2009

(a) Using a graphing utility, draw a scatter diagram of the data using income, x, as the independent variable and crime rate, y, as the dependent variable.

(b) Based on the scatter diagram drawn in part (a), decide on a model (linear, quadratic, cubic, exponential, logarithmic, or logistic) that you think best describes the relation between income and crime rate. Be sure to justify your choice of model.

(c) Using a graphing utility, find the model of best fit.

(d) Using a graphing utility, draw the model of best fit on the scatter diagram drawn in part (a).

(e) Use your model to predict the crime rate of a household whose income is $55,000.

13. **Depreciation of a Chevrolet Impala** The following data represent the asking price and age of a Chevrolet Impala SS.

Age	Asking Price
1	$27,417
1	26,750
2	22,995
2	23,195
3	17,999
4	16,995
4	16,490

Source: cars.com

(a) Using a graphing utility, draw a scatter diagram of the data using age, x, as the independent variable and asking price, y, as the dependent variable.

(b) Based on the scatter diagram drawn in part (a), decide on a model (linear, quadratic, cubic, exponential, logarithmic, or logistic) that you think best describes the relation between age and asking price. Be sure to justify your choice of model.

(c) Using a graphing utility, find the model of best fit.

(d) Using a graphing utility, draw the model of best fit on the scatter diagram drawn in part (a).

(e) Use your model to predict the asking price of a Chevrolet Impala SS that is 5 years old.

CHAPTER REVIEW

Things to Know

Composite function (p. 275)

$(f \circ g)(x) = f(g(x))$ The domain of $f \circ g$ is the set of all numbers x in the domain of g for which $g(x)$ is in the domain of f.

One-to-one function f (p. 283)

A function for which any two different inputs in the domain correspond to two different outputs in the range
For any choice of elements x_1, x_2 in the domain of f, if $x_1 \neq x_2$, then $f(x_1) \neq f(x_2)$.

Horizontal-line test (p. 284)

If every horizontal line intersects the graph of a function f in at most one point, f is one-to-one.

Inverse function f^{-1} of f (pp. 285–288)

Domain of f = range of f^{-1}; range of f = domain of f^{-1}
$f^{-1}(f(x)) = x$ for all x in the domain of f
$f(f^{-1}(x)) = x$ for all x in the domain of f^{-1}
The graphs of f and f^{-1} are symmetric with respect to the line $y = x$.

Properties of the exponential function (pp. 300–303)

$f(x) = Ca^x, \quad a > 1$

Domain: the interval $(-\infty, \infty)$

Range: the interval $(0, \infty)$

x-intercepts: none; y-intercept: C

Horizontal asymptote: x-axis ($y = 0$) as $x : -\infty$

Increasing; one-to-one; smooth; continuous

See Figure 21 for a typical graph.

$f(x) = Ca^x, \quad 0 < a < 1$

Domain: the interval $(-\infty, \infty)$

Range: the interval $(0, \infty)$

x-intercepts: none; y-intercept: C

Horizontal asymptote: x-axis ($y = 0$) as $x : \infty$

Decreasing; one-to-one; smooth; continuous

See Figure 25 for a typical graph.

Number e (p. 304)

Value approached by the expression $\left(1 + \dfrac{1}{n}\right)^n$ as $n : \infty$; that is, $\lim\limits_{n:\infty}\left(1 + \dfrac{1}{n}\right)^n = e$.

Property of exponents (p. 305)

If $a^u = a^v$, then $u = v$.

Properties of the logarithmic function (pp. 313–315)

$f(x) = \log_a x, \quad a > 1$
($y = \log_a x$ means $x = a^y$)

Domain: the interval $(0, \infty)$

Range: the interval $(-\infty, \infty)$

x-intercept: 1; y-intercept: none

Vertical asymptote: $x = 0$ (y-axis)

Increasing; one-to-one; smooth; continuous

See Figure 39(a) for a typical graph.

$f(x) = \log_a x, \quad 0 < a < 1$
($y = \log_a x$ means $x = a^y$)

Domain: the interval $(0, \infty)$

Range: the interval $(-\infty, \infty)$

x-intercept: 1; y-intercept: none

Vertical asymptote: $x = 0$ (y-axis)

Decreasing; one-to-one; smooth; continuous

See Figure 39(b) for a typical graph.

Natural logarithm (p. 316)

$y = \ln x$ means $x = e^y$.

Properties of logarithms (pp. 326–327, 329)

$\log_a 1 = 0 \qquad \log_a a = 1 \qquad a^{\log_a M} = M \qquad \log_a a^r = r$

$\log_a(MN) = \log_a M + \log_a N \qquad \log_a\left(\dfrac{M}{N}\right) = \log_a M - \log_a N$
$\log_a M^r = r \log_a M$
If $M = N$, then $\log_a M = \log_a N$.
If $\log_a M = \log_a N$, then $M = N$.

Formulas

Change-of-Base Formula (p. 330)	$\log_a M = \dfrac{\log_b M}{\log_b a}$
Compound Interest Formula (p. 342)	$A = P \cdot \left(1 + \dfrac{r}{n}\right)^{nt}$
Continuous compounding (p. 344)	$A = Pe^{rt}$
Effective rate of interest (p. 345)	Compounding n times per year: $r_e = \left(1 + \dfrac{r}{n}\right)^n - 1$
	Continuous compounding: $r_e = e^r - 1$
Present Value Formulas (p. 346)	$P = A \cdot \left(1 + \dfrac{r}{n}\right)^{-nt}$ or $P = Ae^{-rt}$
Growth and decay (p. 351)	$A(t) = A_0 e^{kt}$
Newton's Law of Cooling (p. 354)	$u(t) = T + (u_0 - T)e^{kt} \quad k < 0$
Logistic model (p. 356)	$P(t) = \dfrac{c}{1 + ae^{-bt}}$

Objectives

Section		You should be able to ...	Example(s)	Review Exercises
4.1	1	Form a composite function (p. 275)	1, 4, 5	1–12
	2	Find the domain of a composite function (p. 276)	2–4	7–12
4.2	1	Determine whether a function is one-to-one (p. 283)	1, 2	13(a), 14(a), 15, 16
	2	Determine the inverse of a function defined by a map or a set of ordered pairs (p. 285)	3, 4	13(b), 14(b)
	3	Obtain the graph of the inverse function from the graph of the function (p. 287)	7	15, 16
	4	Find the inverse of a function defined by an equation (p. 288)	8, 9, 11	17–22
4.3	1	Evaluate exponential functions (p. 296)	1	23(a), (c), 24(a), (c), 87(a)
	2	Graph exponential functions (p. 300)	3–6	55–60
	3	Define the number e (p. 303)	pg. 304	59, 60
	4	Solve exponential equations (p. 305)	7, 8	63–66, 71, 72, 74–76, 87(b)
4.4	1	Change exponential statements to logarithmic statements and logarithmic statements to exponential statements (p. 313)	2, 3	25–28
	2	Evaluate logarithmic expressions (p. 313)	4	23(b), (d), 24(b), (d), 33, 34, 83(b), 84(b), 85, 86, 88(a), 89
	3	Determine the domain of a logarithmic function (p. 314)	5	29–32, 61(a), 62(a)
	4	Graph logarithmic functions (p. 315)	6, 7	61, 62, 83(a), 84(a)
	5	Solve logarithmic equations (p. 319)	8, 9	67, 68, 73, 83(c), 84(c), 88(b)
4.5	1	Work with the properties of logarithms (p. 325)	1, 2	35–38
	2	Write a logarithmic expression as a sum or difference of logarithms (p. 327)	3–5	39–44
	3	Write a logarithmic expression as a single logarithm (p. 328)	6	45–50
	4	Evaluate logarithms whose base is neither 10 nor e (p. 329)	7, 8	51, 52
4.6	1	Solve logarithmic equations (p. 334)	1–3	67, 77, 78
	2	Solve exponential equations (p. 336)	4–6	63–66, 69, 70, 74–76, 79–82
	3	Solve logarithmic and exponential equations using a graphing utility (p. 337)	7	69–82
4.7	1	Determine the future value of a lump sum of money (p. 341)	1–3	90, 92, 97
	2	Calculate effective rates of return (p. 344)	4	90
	3	Determine the present value of a lump sum of money (p. 345)	5	91
	4	Determine the rate of interest or time required to double a lump sum of money (p. 346)	6, 7	90

4.8	**1** Find equations of populations that obey the law of uninhibited growth (p. 351)	1, 2	95
	2 Find equations of populations that obey the law of decay (p. 353)	3	93, 96
	3 Use Newton's Law of Cooling (p. 354)	4	94
	4 Use logistic models (p. 356)	5, 6	98
4.9	**1** Build an exponential model from data (p. 362)	1	99
	2 Build a logarithmic model from data (p. 363)	2	100
	3 Build a logistic model from data (p. 364)	3	101

Review Exercises

In Problems 1–6, for the given functions f and g find:

(a) $(f \circ g)(2)$ *(b)* $(g \circ f)(-2)$ *(c)* $(f \circ f)(4)$ *(d)* $(g \circ g)(-1)$

1. $f(x) = 3x - 5$; $g(x) = 1 - 2x^2$

2. $f(x) = 4 - x$; $g(x) = 1 + x^2$

3. $f(x) = \sqrt{x + 2}$; $g(x) = 2x^2 + 1$

4. $f(x) = 1 - 3x^2$; $g(x) = \sqrt{4 - x}$

5. $f(x) = e^x$; $g(x) = 3x - 2$

6. $f(x) = \dfrac{2}{1 + 2x^2}$; $g(x) = 3x$

In Problems 7–12, find $f \circ g$, $g \circ f$, $f \circ f$, and $g \circ g$ for each pair of functions. State the domain of each composite function.

7. $f(x) = 2 - x$; $g(x) = 3x + 1$

8. $f(x) = 2x - 1$; $g(x) = 2x + 1$

9. $f(x) = 3x^2 + x + 1$; $g(x) = |3x|$

10. $f(x) = \sqrt{3x}$; $g(x) = 1 + x + x^2$

11. $f(x) = \dfrac{x + 1}{x - 1}$; $g(x) = \dfrac{1}{x}$

12. $f(x) = \sqrt{x - 3}$; $g(x) = \dfrac{3}{x}$

In Problems 13 and 14, (a) verify that the function is one-to-one, and (b) find the inverse of the given function.

13. $\{(1, 2), (3, 5), (5, 8), (6, 10)\}$

14. $\{(-1, 4), (0, 2), (1, 5), (3, 7)\}$

In Problems 15 and 16, state why the graph of the function is one-to-one. Then draw the graph of the inverse function f^{-1}. For convenience (and as a hint), the graph of $y = x$ is also given.

15.

16.

In Problems 17–22, the function f is one-to-one. Find the inverse of each function and check your answer. Find the domain and the range of f and f^{-1}.

17. $f(x) = \dfrac{2x + 3}{5x - 2}$

18. $f(x) = \dfrac{2 - x}{3 + x}$

19. $f(x) = \dfrac{1}{x - 1}$

20. $f(x) = \sqrt{x - 2}$

21. $f(x) = \dfrac{3}{x^{1/3}}$

22. $f(x) = x^{1/3} + 1$

In Problems 23 and 24, $f(x) = 3^x$ and $g(x) = \log_3 x$.

23. Evaluate: (a) $f(4)$ (b) $g(9)$ (c) $f(-2)$ (d) $g\left(\dfrac{1}{27}\right)$

24. Evaluate: (a) $f(1)$ (b) $g(81)$ (c) $f(-4)$ (d) $g\left(\dfrac{1}{243}\right)$

In Problems 25 and 26, convert each exponential statement to an equivalent statement involving a logarithm. In Problems 27 and 28, convert each logarithmic statement to an equivalent statement involving an exponent.

25. $5^2 = z$ **26.** $a^5 = m$ **27.** $\log_5 u = 13$ **28.** $\log_a 4 = 3$

In Problems 29–32, find the domain of each logarithmic function.

29. $f(x) = \log(3x - 2)$ **30.** $F(x) = \log_5(2x + 1)$ **31.** $H(x) = \log_2(x^2 - 3x + 2)$ **32.** $F(x) = \ln(x^2 - 9)$

In Problems 33–38, evaluate each expression. Do not use a calculator.

33. $\log_2\left(\dfrac{1}{8}\right)$ **34.** $\log_3 81$ **35.** $\ln e^{\sqrt{2}}$ **36.** $e^{\ln 0.1}$ **37.** $2^{\log_2 0.4}$ **38.** $\log_2 2^{\sqrt{3}}$

In Problems 39–44, write each expression as the sum and/or difference of logarithms. Express powers as factors.

39. $\log_3\left(\dfrac{uv^2}{w}\right)$, $u > 0, v > 0, w > 0$ **40.** $\log_2\left(a^2\sqrt{b}\right)^4$, $a > 0, b > 0$ **41.** $\log\left(x^2\sqrt{x^3 + 1}\right)$, $x > 0$

42. $\log_5\left(\dfrac{x^2 + 2x + 1}{x^2}\right)$, $x > 0$ **43.** $\ln\left(\dfrac{x\sqrt[3]{x^2 + 1}}{x - 3}\right)$, $x > 3$ **44.** $\ln\left(\dfrac{2x + 3}{x^2 - 3x + 2}\right)^2$, $x > 2$

In Problems 45–50, write each expression as a single logarithm.

45. $3\log_4 x^2 + \dfrac{1}{2}\log_4\sqrt{x}$ **46.** $-2\log_3\left(\dfrac{1}{x}\right) + \dfrac{1}{3}\log_3\sqrt{x}$

47. $\ln\left(\dfrac{x - 1}{x}\right) + \ln\left(\dfrac{x}{x + 1}\right) - \ln(x^2 - 1)$ **48.** $\log(x^2 - 9) - \log(x^2 + 7x + 12)$

49. $2\log 2 + 3\log x - \dfrac{1}{2}[\log(x + 3) + \log(x - 2)]$ **50.** $\dfrac{1}{2}\ln(x^2 + 1) - 4\ln\dfrac{1}{2} - \dfrac{1}{2}[\ln(x - 4) + \ln x]$

In Problems 51 and 52, use the Change-of-Base Formula and a calculator to evaluate each logarithm. Round your answer to three decimal places.

51. $\log_4 19$ **52.** $\log_2 21$

In Problems 53 and 54, graph each function using a graphing utility and the Change-of-Base Formula.

53. $y = \log_3 x$ **54.** $y = \log_7 x$

In Problems 55–62, use the given function f to:
   (a) Find the domain of f.   *(b) Graph f.*   *(c) From the graph, determine the range and any asymptotes of f.*
   (d) Find f^{-1}, the inverse of f.   *(e) Use f^{-1} to find the range of f.*   *(f) Graph f^{-1}.*

55. $f(x) = 2^{x-3}$ **56.** $f(x) = -2^x + 3$ **57.** $f(x) = \dfrac{1}{2}(3^{-x})$ **58.** $f(x) = 1 + 3^{-x}$

59. $f(x) = 1 - e^{-x}$ **60.** $f(x) = 3e^{x-2}$ **61.** $f(x) = \dfrac{1}{2}\ln(x + 3)$ **62.** $f(x) = 3 + \ln(2x)$

In Problems 63–82, solve each equation. Express irrational solutions in exact form and as a decimal rounded to 3 decimal places.

63. $4^{1-2x} = 2$ **64.** $8^{6+3x} = 4$ **65.** $3^{x^2+x} = \sqrt{3}$ **66.** $4^{x-x^2} = \dfrac{1}{2}$

67. $\log_x 64 = -3$ **68.** $\log_{\sqrt{2}} x = -6$ **69.** $5^x = 3^{x+2}$ **70.** $5^{x+2} = 7^{x-2}$

71. $9^{2x} = 27^{3x-4}$ **72.** $25^{2x} = 5^{x^2-12}$ **73.** $\log_3\sqrt{x - 2} = 2$ **74.** $2^{x+1}\cdot 8^{-x} = 4$

75. $8 = 4^{x^2} \cdot 2^{5x}$

76. $2^x \cdot 5 = 10^x$

77. $\log_6(x + 3) + \log_6(x + 4) = 1$

78. $\log(7x - 12) = 2 \log x$

79. $e^{1-x} = 5$

80. $e^{1-2x} = 4$

81. $9^x + 4 \cdot 3^x - 3 = 0$

82. $4^x - 14 \cdot 4^{-x} = 5$

83. Suppose that $f(x) = \log_2(x - 2) + 1$.
(a) Graph f.
(b) What is $f(6)$? What point is on the graph of f?
(c) Solve $f(x) = 4$. What point is on the graph of f?
(d) Based on the graph drawn in part (a), solve $f(x) > 0$.
(e) Find $f^{-1}(x)$. Graph f^{-1} on the same Cartesian plane as f.

84. Suppose that $f(x) = \log_3(x + 1) - 4$.
(a) Graph f.
(b) What is $f(8)$? What point is on the graph of f?
(c) Solve $f(x) = -3$. What point is on the graph of f?
(d) Based on the graph drawn in part (a), solve $f(x) < 0$.
(e) Find $f^{-1}(x)$. Graph f^{-1} on the same Cartesian plane as f.

In Problems 85 and 86, use the following result: If x is the atmospheric pressure (measured in millimeters of mercury), then the formula for the altitude h(x) (measured in meters above sea level) is

$$h(x) = (30T + 8000) \log\left(\frac{P_0}{x}\right)$$

where T is the temperature (in degrees Celsius) and P_0 is the atmospheric pressure at sea level, which is approximately 760 millimeters of mercury.

85. Finding the Altitude of an Airplane At what height is a Piper Cub whose instruments record an outside temperature of 0°C and a barometric pressure of 300 millimeters of mercury?

86. Finding the Height of a Mountain How high is a mountain if instruments placed on its peak record a temperature of 5°C and a barometric pressure of 500 millimeters of mercury?

87. Amplifying Sound An amplifier's power output P (in watts) is related to its decibel voltage gain d by the formula
$$P = 25e^{0.1d}$$

(a) Find the power output for a decibel voltage gain of 4 decibels.
(b) For a power output of 50 watts, what is the decibel voltage gain?

88. Limiting Magnitude of a Telescope A telescope is limited in its usefulness by the brightness of the star that it is aimed at and by the diameter of its lens. One measure of a star's brightness is its *magnitude;* the dimmer the star, the larger its magnitude. A formula for the limiting magnitude L of a telescope, that is, the magnitude of the dimmest star that it can be used to view, is given by
$$L = 9 + 5.1 \log d$$
where d is the diameter (in inches) of the lens.
(a) What is the limiting magnitude of a 3.5-inch telescope?
(b) What diameter is required to view a star of magnitude 14?

89. Salvage Value The number of years n for a piece of machinery to depreciate to a known salvage value can be found using the formula
$$n = \frac{\log s - \log i}{\log(1 - d)}$$
where s is the salvage value of the machinery, i is its initial value, and d is the annual rate of depreciation.
(a) How many years will it take for a piece of machinery to decline in value from \$90,000 to \$10,000 if the annual rate of depreciation is 0.20 (20%)?

(b) How many years will it take for a piece of machinery to lose half of its value if the annual rate of depreciation is 15%?

90. Funding a College Education A child's grandparents purchase a \$10,000 bond fund that matures in 18 years to be used for her college education. The bond fund pays 4% interest compounded semiannually. How much will the bond fund be worth at maturity? What is the effective rate of interest? How long will it take the bond to double in value under these terms?

91. Funding a College Education A child's grandparents wish to purchase a bond that matures in 18 years to be used for her college education. The bond pays 4% interest compounded semiannually. How much should they pay so that the bond will be worth \$85,000 at maturity?

92. Funding an IRA First Colonial Bankshares Corporation advertised the following IRA investment plans.

Target IRA Plans

For each \$5000 Maturity Value Desired	
Deposit:	At a Term of:
\$620.17	20 Years
\$1045.02	15 Years
\$1760.92	10 Years
\$2967.26	5 Years

(a) Assuming continuous compounding, what annual rate of interest did they offer?
(b) First Colonial Bankshares claims that \$4000 invested today will have a value of over \$32,000 in 20 years. Use the answer found in part (a) to find the actual value of \$4000 in 20 years. Assume continuous compounding.

93. Estimating the Date That a Prehistoric Man Died The bones of a prehistoric man found in the desert of New Mexico contain approximately 5% of the original amount of carbon 14. If the half-life of carbon 14 is 5600 years, approximately how long ago did the man die?

94. **Temperature of a Skillet** A skillet is removed from an oven whose temperature is 450°F and placed in a room whose temperature is 70°F. After 5 minutes, the temperature of the skillet is 400°F. How long will it be until its temperature is 150°F?

95. **World Population** The annual growth rate of the world's population in 2005 was $k = 1.15\% = 0.0115$. The population of the world in 2005 was 6,451,058,790. Letting $t = 0$ represent 2005, use the uninhibited growth model to predict the world's population in the year 2015.

Source: U.S. Census Bureau

96. **Radioactive Decay** The half-life of radioactive cobalt is 5.27 years. If 100 grams of radioactive cobalt is present now, how much will be present in 20 years? In 40 years?

97. **Federal Deficit** In fiscal year 2005, the federal deficit was $319 billion. At that time, 10-year treasury notes were paying 4.25% interest per annum. If the federal government financed this deficit through 10-year notes, how much would it have to pay back in 2015?

Source: U.S. Treasury Department

98. **Logistic Growth** The logistic growth model

$$P(t) = \frac{0.8}{1 + 1.67e^{-0.16t}}$$

represents the proportion of new cars with a global positioning system (GPS). Let $t = 0$ represent 2006, $t = 1$ represent 2007, and so on.

(a) What proportion of new cars in 2006 had a GPS?
(b) Determine the maximum proportion of new cars that have a GPS.
(c) Using a graphing utility, graph $P = P(t)$.
(d) When will 75% of new cars have a GPS?

99. **CBL Experiment** The following data were collected by placing a temperature probe in a portable heater, removing the probe, and then recording temperature over time.

Time (sec.)	Temperature (°F)
0	165.07
1	164.77
2	163.99
3	163.22
4	162.82
5	161.96
6	161.20
7	160.45
8	159.35
9	158.61
10	157.89
11	156.83
12	156.11
13	155.08
14	154.40
15	153.72

According to Newton's Law of Cooling, these data should follow an exponential model.

(a) Using a graphing utility, draw a scatter diagram for the data.
(b) Using a graphing utility, build an exponential model from the data.
(c) Graph the exponential function found in part (b) on the scatter diagram.
(d) Predict how long it will take for the probe to reach a temperature of 110°F.

100. **Wind Chill Factor** The following data represent the wind speed (mph) and wind chill factor at an air temperature of 15°F.

Wind Speed (mph)	Wind Chill Factor (°F)
5	7
10	3
15	0
20	−2
25	−4
30	−5
35	−7

Source: U.S. National Weather Service

(a) Using a graphing utility, draw a scatter diagram with wind speed as the independent variable.
(b) Using a graphing utility, build a logarithmic model from the data.
(c) Using a graphing utility, draw the logarithmic function found in part (b) on the scatter diagram.
(d) Use the function found in part (b) to predict the wind chill factor if the air temperature is 15°F and the wind speed is 23 mph.

101. **Spreading of a Disease** Jack and Diane live in a small town of 50 people. Unfortunately, both Jack and Diane have a cold. Those who come in contact with someone who has this cold will themselves catch the cold. The following data represent the number of people in the small town who have caught the cold after t days.

Days, t	Number of People with Cold, C
0	2
1	4
2	8
3	14
4	22
5	30
6	37
7	42
8	44

(a) Using a graphing utility, draw a scatter diagram of the data. Comment on the type of relation that appears to exist between the days and number of people with a cold.
(b) Using a graphing utility, build a logistic model from the data.

(c) Graph the function found in part (b) on the scatter diagram.

(d) According to the function found in part (b), what is the maximum number of people who will catch the cold? In reality, what is the maximum number of people who could catch the cold?

(e) Sometime between the second and third day, 10 people in the town had a cold. According to the model found in part (b), when did 10 people have a cold?

(f) How long will it take for 46 people to catch the cold?

CHAPTER TEST

Step-by-step test solutions are found on the Chapter Test Prep Videos available via the Video Resources on DVD, in *MyMathLab* or on You Tube (search "Sullivan College Algebra, Concepts Through Functions 2e).

1. Given $f(x) = \dfrac{x+2}{x-2}$ and $g(x) = 2x + 5$, find:
   (a) $f \circ g$ and state its domain
   (b) $(g \circ f)(-2)$
   (c) $(f \circ g)(-2)$

2. Determine whether the function is one-to-one.
   (a) $y = 4x^2 + 3$
   (b) $y = \sqrt{x+3} - 5$

3. Find the inverse of $f(x) = \dfrac{2}{3x-5}$ and check your answer. State the domain and the range of f and f^{-1}.

4. If the point $(3, -5)$ is on the graph of a one-to-one function f, what point must be on the graph of f^{-1}?

In Problems 5–7, solve each equation.

5. $3^x = 243$ 6. $\log_b 16 = 2$

7. $\log_5 x = 4$

In Problems 8–11, use a calculator to evaluate each expression. Round your answer to three decimal places.

8. $e^3 + 2$ 9. $\log 20$

10. $\log_3 21$ 11. $\ln 133$

In Problems 12 and 13, use the given function f to:
   (a) *Find the domain of f.*
   (b) *Graph f.*
   (c) *From the graph, determine the range and any asymptotes of f.*
   (d) *Find f^{-1}, the inverse of f.*
   (e) *Use f^{-1} to find the range of f.*
   (f) *Graph f^{-1}.*

12. $f(x) = 4^{x+1} - 2$

13. $f(x) = 1 - \log_5(x - 2)$

In Problems 14–19, solve each equation.

14. $5^{x+2} = 125$ 15. $\log(x + 9) = 2$

16. $8 - 2e^{-x} = 4$ 17. $\log(x^2 + 3) = \log(x + 6)$

18. $7^{x+3} = e^x$ 19. $\log_2(x - 4) + \log_2(x + 4) = 3$

20. Write $\log_2\left(\dfrac{4x^3}{x^2 - 3x - 18}\right)$ as the sum and/or difference of logarithms. Express powers as factors.

21. A 50-mg sample of a radioactive substance decays to 34 mg after 30 days. How long will it take for there to be 2 mg remaining?

22. (a) If $1000 is invested at 5% compounded monthly, how much is there after 8 months?
   (b) If you want to have $1000 in 9 months, how much do you need to place in a savings account now that pays 5% compounded quarterly?
   (c) How long does it take to double your money if you can invest it at 6% compounded annually?

23. The decibel level, D, of sound is given by the equation $D = 10 \log\left(\dfrac{I}{I_0}\right)$, where I is the intensity of the sound and $I_0 = 10^{-12}$ watt per square meter.
   (a) If the shout of a single person measures 80 decibels, how loud will the sound be if two people shout at the same time? That is, how loud would the sound be if the intensity doubled?
   (b) The pain threshold for sound is 125 decibels. If the Athens Olympic Stadium 2004 (Olympiako Stadio Athinas 'Spyros Louis') can seat 74,400 people, how many people in the crowd need to shout at the same time for the resulting sound level to meet or exceed the pain threshold? (Ignore any possible sound dampening.)

CUMULATIVE REVIEW

1. Is the following graph the graph of a function? If it is, is the function one-to-one?

2. For the function $f(x) = 2x^2 - 3x + 1$, find the following:
   (a) $f(3)$ (b) $f(-x)$ (c) $f(x + h)$

3. Determine which of the following points are on the graph of $x^2 + y^2 = 1$.
   (a) $\left(\dfrac{1}{2}, \dfrac{1}{2}\right)$ (b) $\left(\dfrac{1}{2}, \dfrac{\sqrt{3}}{2}\right)$

4. Solve the equation $3(x - 2) = 4(x + 5)$.

5. Graph the line $2x - 4y = 16$.

6. (a) Graph the quadratic function $f(x) = -x^2 + 2x - 3$ by determining whether its graph opens up or down and by

finding its vertex, axis of symmetry, y-intercept, and x-intercept(s), if any.

(b) Solve $f(x) \leq 0$.

7. Determine the quadratic function whose graph is given in the figure.

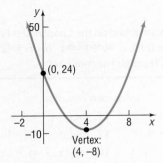

(0, 24)

Vertex:
(4, −8)

8. Graph $f(x) = 3(x + 1)^3 - 2$ using transformations.

9. Given that $f(x) = x^2 + 2$ and $g(x) = \dfrac{2}{x - 3}$, find $f(g(x))$ and state its domain. What is $f(g(5))$?

10. For the polynomial function $f(x) = 4x^3 + 9x^2 - 30x - 8$:

(a) Find the real zeros of f.

(b) Determine the intercepts of the graph of f.

(c) Use a graphing utility to approximate the local maxima and local minima.

(d) Draw a complete graph of f. Be sure to label the intercepts and turning points.

11. For the function $g(x) = 3^x + 2$:

(a) Graph g using transformations. State the domain, range, and horizontal asymptote of g.

(b) Determine the inverse of g. State the domain, range, and vertical asymptote of g^{-1}.

(c) On the same graph as g, graph g^{-1}.

12. Solve the equation $4^{x-3} = 8^{2x}$.

13. Solve the equation: $\log_3(x + 1) + \log_3(2x - 3) = \log_9 9$.

14. Suppose that $f(x) = \log_3(x + 2)$. Solve:

(a) $f(x) = 0$ (b) $f(x) > 0$ (c) $f(x) = 3$

15. Data Analysis The following data represent the percent of all drivers by age that have been stopped by the police for any reason within the past year. The median age represents the midpoint of the upper and lower limit for the age range.

Age Range	Median Age, x	Percentage Stopped, y
16–19	17.5	18.2
20–29	24.5	16.8
30–39	34.5	11.3
40–49	44.5	9.4
50–59	54.5	7.7
≥60	69.5	3.8

(a) Using your graphing utility, draw a scatter diagram of the data treating median age, x, as the independent variable.

(b) Determine a model that you feel best describes the relation between median age and percentage stopped. You may choose from among linear, quadratic, cubic, exponential, logarithmic, or logistic models.

(c) Provide a justification for the model that you selected in part (b).

CHAPTER PROJECTS

I. **Hot Coffee** A fast-food restaurant wants a special container to hold coffee. The restaurant wishes the container to quickly cool the coffee from 200° to 130°F and keep the liquid between 110° and 130°F as long as possible. The restaurant has three containers to select from.

   1. The CentiKeeper Company has a container that reduces the temperature of a liquid from 200° to 100°F in 30 minutes by maintaining a constant temperature of 70°F.

   2. The TempControl Company has a container that reduces the temperature of a liquid from 200° to 110°F in 25 minutes by maintaining a constant temperature of 60°F.

   3. The Hot'n'Cold Company has a container that reduces the temperature of a liquid from 200° to 120°F in 20 minutes by maintaining a constant temperature of 65°F.

You need to recommend which container the restaurant should purchase.

(a) Use Newton's Law of Cooling to find a model relating the temperature of the liquid over time for each container.

(b) How long does it take each container to lower the coffee temperature from 200° to 130°F?

(c) How long will the coffee temperature remain between 110° and 130°F? This temperature is considered the optimal drinking temperature.

(d) Graph each function using a graphing utility.

(e) Which company would you recommend to the restaurant? Why?

(f) How might the cost of the container affect your decision?

The following projects are available on the Instructor's Resource Center (IRC):

II. **Project at Motorola** *Thermal Fatigue of Solder Connections* Product reliability is a major concern of a manufacturer. Here a logarithmic transformation is used to simplify the analysis of a cell phone's ability to withstand temperature change.

III. **Depreciation of a New Car** Resale value is a factor to consider when purchasing a car, and exponential functions provide a way to compare the depreciation rates of different makes and models.

Trigonometric Functions

Outline

Surf's Up: Professor Using Models To Predict Huge Waves

GALVESTON, Feb. 15, 2005 — If you are a ship captain and there might be 50-foot waves headed your way, you would appreciate some information about them, right?

That's the idea behind a wave model system a Texas A&M University at Galveston professor has developed. His detailed wave prediction system is currently in use in the Gulf of Mexico and the Gulf of Maine.

Vijay Panchang, head of the Department of Maritime Systems Engineering, doesn't make waves; he predicts what they will do, when they will do it, and how high they will get.

Using data provided daily from NOAA [National Oceanic and Atmospheric Administration] and his

own complex mathematical models, Panchang and research engineer Doncheng Li provide daily wave model predictions for much of the Texas coast, the Gulf of Mexico, and the Gulf of Maine. Their simulations, updated every 12 hours, provide a forecast for two days ahead.

Because the models use wind data, tsunamis that are created by undersea earthquakes cannot be predicted. But that is not to say his modeling system does not come up with some big waves.

His wave model predicted big waves in November 2003 in the Gulf of Maine, and it was accurate; waves as high as 30 feet were recorded during one storm even in coastal regions.

Last summer during Hurricane Ivan, a buoy located 60 miles south of the Alabama coast recorded a whopping 60-foot wave. "There may have been higher waves because right after recording the 60-foot wave, the buoy snapped and stopped functioning," he says.

Source: Texas A&M University News & Information, February 8, 2005. Accessed at http://communications.tamu.edu/ newsarchives/05/020805-18.html, February 2007. Adapted with permission.

—See Chapter Project I—

◁ **A Look Back** In Chapter 1, we began our discussion of functions. We defined domain and range and independent and dependent variables; we found the value of a function and graphed functions. We continued our study of functions by listing properties that a function might have, like being even or odd, and we created a library of functions, naming key functions and listing their properties, including the graph.

A Look Ahead ▷ In this chapter we define the trigonometric functions, six functions that have wide application. We shall talk about their domain and range, see how to find values, graph them, and develop a list of their properties.

There are two widely accepted approaches to the development of the trigonometric functions: one uses right triangles; the other uses circles, especially the unit circle. In this book, we develop the trigonometric functions using right triangles. In Section 5.5, we introduce trigonometric functions using the unit circle and show that this approach leads to the definition using right triangles.

5.1 Angles and Their Measure

PREPARING FOR THIS SECTION *Before getting started, review the following:*

- Circumference and Area of a Circle (Appendix A, Section A.2, p. A16)
- Uniform Motion (Appendix A, Section A.9, pp. A77–A78)

Now Work the 'Are You Prepared?' problems on page 387.

OBJECTIVES 1 Convert between Decimals and Degrees, Minutes, Seconds Measures for Angles (p. 380)
       2 Find the Length of an Arc of a Circle (p. 382)
       3 Convert from Degrees to Radians and from Radians to Degrees (p. 382)
       4 Find the Area of a Sector of a Circle (p. 385)
       5 Find the Linear Speed of an Object Traveling in Circular Motion (p. 386)

Figure 1

A **ray,** or **half-line,** is that portion of a line that starts at a point V on the line and extends indefinitely in one direction. The starting point V of a ray is called its **vertex.** See Figure 1.

If two rays are drawn with a common vertex, they form an **angle.** We call one ray of an angle the **initial side** and the other the **terminal side.** The angle formed is identified by showing the direction and amount of rotation from the initial side to the terminal side. If the rotation is in the counterclockwise direction, the angle is **positive;** if the rotation is clockwise, the angle is **negative.** See Figure 2.

Figure 2

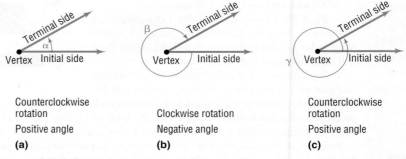

Lowercase Greek letters, such as α (alpha), β (beta), γ (gamma), and θ (theta), will often be used to denote angles. Notice in Figure 2(a) that the angle α is positive because the direction of the rotation from the initial side to the terminal side is counterclockwise. The angle β in Figure 2(b) is negative because the rotation is clockwise. The angle γ in Figure 2(c) is positive. Notice that the angle α in Figure 2(a) and the angle γ in Figure 2(c) have the same initial side and the same terminal side. However, α and γ are unequal, because the amount of rotation required to go from the initial side to the terminal side is greater for angle γ than for angle α.

An angle θ is said to be in **standard position** if its vertex is at the origin of a rectangular coordinate system and its initial side coincides with the positive x-axis. See Figure 3.

Figure 3

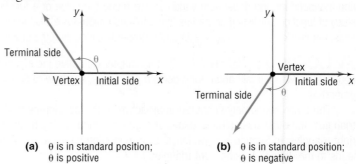

When an angle θ is in standard position, the terminal side will lie either in a quadrant, in which case we say that θ **lies in that quadrant,** or the terminal side will lie on the x-axis or the y-axis, in which case we say that θ is a **quadrantal angle.** For example, the angle θ in Figure 4(a) lies in quadrant II, the angle θ in Figure 4(b) lies in quadrant IV, and the angle θ in Figure 4(c) is a quadrantal angle.

Figure 4

(a) θ lies in quadrant II   (b) θ lies in quadrant IV   (c) θ is a quadrantal angle

We measure angles by determining the amount of rotation needed for the initial side to become coincident with the terminal side. The two commonly used measures for angles are *degrees* and *radians.*

Degrees

HISTORICAL NOTE One counterclock-wise rotation is 360° due to the Babylonian year, which had 360 days.   ■

The angle formed by rotating the initial side exactly once in the counterclockwise direction until it coincides with itself (1 revolution) is said to measure 360 degrees, abbreviated 360°. **One degree, 1°,** is $\frac{1}{360}$ revolution. A **right angle** is an angle that measures 90°, or $\frac{1}{4}$ revolution; a **straight angle** is an angle that measures 180°, or $\frac{1}{2}$ revolution. See Figure 5. As Figure 5(b) shows, it is customary to indicate a right angle by using the symbol ⌐.

Figure 5

(a)   1 revolution counterclockwise, 360°   (b)   right angle, $\frac{1}{4}$ revolution counterclockwise, 90°   (c)   straight angle, $\frac{1}{2}$ revolution counterclockwise, 180°

It is also customary to refer to an angle that measures θ degrees as an angle *of* θ degrees.

EXAMPLE 1

Drawing an Angle

Draw each angle.

(a) 45°   (b) −90°   (c) 225°   (d) 405°

Solution   (a) An angle of 45° is $\frac{1}{2}$ of a right angle.   (b) An angle of −90° is $\frac{1}{4}$ revolution in See Figure 6.   the clockwise direction. See Figure 7.

Figure 6   **Figure 7**

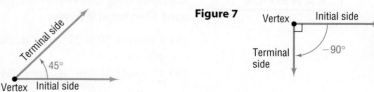

(c) An angle of 225° consists of a rotation through 180° followed by a rotation through 45°. See Figure 8.

(d) An angle of 405° consists of 1 revolution (360°) followed by a rotation through 45°. See Figure 9.

Figure 8

Figure 9

Now Work PROBLEM 11

1 Convert between Decimals and Degrees, Minutes, Seconds Measures for Angles

Although subdivisions of a degree may be obtained by using decimals, we also may use the notion of *minutes* and *seconds*. **One minute,** denoted by **1′**, is defined as $\frac{1}{60}$ degree. **One second,** denoted by **1″**, is defined as $\frac{1}{60}$ minute, or equivalently, $\frac{1}{3600}$ degree. An angle of, say, 30 degrees, 40 minutes, 10 seconds is written compactly as 30°40′10″. To summarize:

$$1 \text{ counterclockwise revolution} = 360°$$
$$1° = 60' \qquad 1' = 60'' \qquad \textbf{(1)}$$

It is sometimes necessary to convert from the degree, minute, second notation (D°M′S″) to a decimal form, and vice versa. Check your calculator; it should be capable of doing the conversion for you.

Before using your calculator, though, you must set the mode to degrees because there are two common ways to measure angles: degree mode and radian mode. (We will define radians shortly.) Usually, a menu is used to change from one mode to another. Check your owner's manual to find out how your particular calculator works.

Now let's see how to convert from the degree, minute, second notation (D°M′S″) to a decimal form, and vice versa, by looking at some examples:

$$15°30' = 15.5° \quad \text{because} \quad 30' = 30 \cdot 1' = 30 \cdot \left(\frac{1}{60}\right)° = 0.5°$$
$$\underset{\uparrow}{}$$
$$1' = \left(\frac{1}{60}\right)°$$

$$32.25° = 32°15' \quad \text{because} \quad 0.25° = \left(\frac{1}{4}\right)° = \frac{1}{4} \cdot 1° = \frac{1}{4}(60') = 15'$$
$$\underset{\uparrow}{}$$
$$1° = 60'$$

EXAMPLE 2   **Converting between Degrees, Minutes, Seconds, and Decimal Forms**

(a) Convert 50°6′21″ to a decimal in degrees. Round the answer to four decimal places.

(b) Convert 21.256° to the D°M′S″ form. Round the answer to the nearest second.

Solution (a) Because $1' = \left(\dfrac{1}{60}\right)^\circ$ and $1'' = \left(\dfrac{1}{60}\right)' = \left(\dfrac{1}{60} \cdot \dfrac{1}{60}\right)^\circ$, we convert as follows:

$$50°6'21'' = 50° + 6' + 21''$$
$$= 50° + 6 \cdot \left(\frac{1}{60}\right)^\circ + 21 \cdot \left(\frac{1}{60} \cdot \frac{1}{60}\right)^\circ$$
$$\approx 50° + 0.1° + 0.0058°$$
$$= 50.1058°$$

(b) We proceed as follows:

$$21.256° = 21° + 0.256°$$
$$= 21° + (0.256)(60')$$ Convert fraction of degree to minutes; $1° = 60'$.
$$= 21° + 15.36'$$
$$= 21° + 15' + 0.36'$$
$$= 21° + 15' + (0.36)(60'')$$ Convert fraction of minute to seconds; $1' = 60''$.
$$= 21° + 15' + 21.6''$$
$$\approx 21°15'22''$$ Round to the nearest second.

Now Work PROBLEMS 23 AND 29

 In many applications, such as describing the exact location of a star or the precise position of a ship at sea, angles measured in degrees, minutes, and even seconds are used. For calculation purposes, these are transformed to decimal form. In other applications, especially those in calculus, angles are measured using *radians*.

Radians

A **central angle** is a positive angle whose vertex is at the center of a circle. The rays of a central angle subtend (intersect) an arc on the circle. If the radius of the circle is r and the length of the arc subtended by the central angle is also r, then the measure of the angle is **1 radian.** See Figure 10(a).

For a circle of radius 1, the rays of a central angle with measure 1 radian subtend an arc of length 1. For a circle of radius 3, the rays of a central angle with measure 1 radian subtend an arc of length 3. See Figure 10(b).

Figure 10

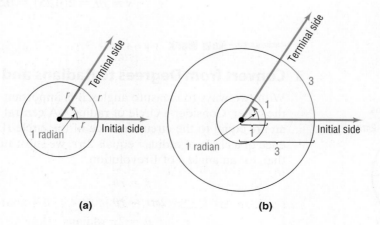

(a)　　　(b)

Figure 11

$$\frac{\theta}{\theta_1} = \frac{s}{s_1}$$

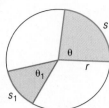

2 Find the Length of an Arc of a Circle

Now consider a circle of radius r and two central angles, θ and θ_1, measured in radians. Suppose that these central angles subtend arcs of lengths s and s_1, respectively, as shown in Figure 11. From geometry, we know that the ratio of the measures of the angles equals the ratio of the corresponding lengths of the arcs subtended by these angles; that is,

$$\frac{\theta}{\theta_1} = \frac{s}{s_1} \qquad \text{(2)}$$

Suppose that $\theta_1 = 1$ radian. Refer again to Figure 10(a). The length s_1 of the arc subtended by the central angle $\theta_1 = 1$ radian equals the radius r of the circle. Then $s_1 = r$, so equation (2) reduces to

$$\frac{\theta}{1} = \frac{s}{r} \quad \text{or} \quad s = r\theta \qquad \text{(3)}$$

THEOREM

> **Arc Length**
>
> For a circle of radius r, a central angle of θ radians subtends an arc whose length s is
>
> $$s = r\theta \qquad \text{(4)}$$

NOTE Formulas must be consistent with regard to the units used. In equation (4), we write

$$s = r\theta$$

To see the units, however, we must go back to equation (3) and write

$$\frac{\theta \text{ radians}}{1 \text{ radian}} = \frac{s \text{ length units}}{r \text{ length units}}$$

$$s \text{ length units} = r \text{ length units} \frac{\theta \text{ radians}}{1 \text{ radian}}$$

Since the radians divide out, we are left with

$$s \text{ length units} = (r \text{ length units})\theta \qquad s = r\theta$$

where θ appears to be "dimensionless" but, in fact, is measured in radians. So, in using the formula $s = r\theta$, the dimension for θ is radians, and any convenient unit of length (such as inches or meters) may be used for s and r. ∎

EXAMPLE 3

Finding the Length of an Arc of a Circle

Find the length of the arc of a circle of radius 2 meters subtended by a central angle of 0.25 radian.

Solution

We use equation (4) with $r = 2$ meters and $\theta = 0.25$. The length s of the arc is

$$s = r\theta = 2(0.25) = 0.5 \text{ meter}$$

●

—**Now Work** PROBLEM 71

3 Convert from Degrees to Radians and from Radians to Degrees

Figure 12

1 revolution = 2π radians

With two ways to measure angles, it is important to be able to convert from one to the other. Consider a circle of radius r. A central angle of 1 revolution will subtend an arc equal to the circumference of the circle (Figure 12). Because the circumference of a circle of radius r equals $2\pi r$, we substitute $2\pi r$ for s in equation (4) to find that, for an angle θ of 1 revolution,

$$s = r\theta$$

$$2\pi r = r\theta \qquad \theta = 1 \text{ revolution; } s = 2\pi r$$

$$\theta = 2\pi \text{ radians} \qquad \text{Solve for } \theta.$$

From this, we have

$$1 \text{ revolution} = 2\pi \text{ radians} \qquad \textbf{(5)}$$

Since 1 revolution = 360°, we have

$$360° = 2\pi \text{ radians}$$

Dividing both sides by 2 yields

$$180° = \pi \text{ radians} \qquad \textbf{(6)}$$

Divide both sides of equation (6) by 180. Then

$$1 \text{ degree} = \frac{\pi}{180} \text{ radian}$$

Divide both sides of (6) by π. Then

$$\frac{180}{\pi} \text{ degrees} = 1 \text{ radian}$$

We have the following two conversion formulas:*

$$1 \text{ degree} = \frac{\pi}{180} \text{ radian} \qquad 1 \text{ radian} = \frac{180}{\pi} \text{ degrees} \qquad \textbf{(7)}$$

EXAMPLE 4

Converting from Degrees to Radians

Convert each angle in degrees to radians.

(a) 60° (b) 150° (c) −45° (d) 90° (e) 107°

Solution

(a) $60° = 60 \cdot 1 \text{ degree} = 60 \cdot \dfrac{\pi}{180} \text{ radian} = \dfrac{\pi}{3} \text{ radians}$

(b) $150° = 150 \cdot 1° = 150 \cdot \dfrac{\pi}{180} \text{ radian} = \dfrac{5\pi}{6} \text{ radians}$

(c) $-45° = -45 \cdot \dfrac{\pi}{180} \text{ radian} = -\dfrac{\pi}{4} \text{ radian}$

(d) $90° = 90 \cdot \dfrac{\pi}{180} \text{ radian} = \dfrac{\pi}{2} \text{ radians}$

(e) $107° = 107 \cdot \dfrac{\pi}{180} \text{ radian} \approx 1.868 \text{ radians}$

Example 4, parts (a)–(d), illustrates that angles that are "nice" fractions of a revolution are expressed in radian measure as fractional multiples of π, rather than as decimals. For example, a right angle, as in Example 4(d), is left in the form $\dfrac{\pi}{2}$ radians, which is exact, rather than using the approximation $\dfrac{\pi}{2} \approx \dfrac{3.1416}{2} = 1.5708$ radians. When the fractions are not "nice," we use the decimal approximation of the angle, as in Example 4(e).

━━━ **Now Work** PROBLEMS 35 AND 61

---

*Some students prefer instead to use the proportion $\dfrac{\text{Degree}}{180°} = \dfrac{\text{Radian}}{\pi}$. Then substitute for what is given and solve for the measurement sought.

> **EXAMPLE 5**

Converting Radians to Degrees

Convert each angle in radians to degrees.

(a) $\dfrac{\pi}{6}$ radian   (b) $\dfrac{3\pi}{2}$ radians   (c) $-\dfrac{3\pi}{4}$ radians

(d) $\dfrac{7\pi}{3}$ radians   (e) 3 radians

Solution (a) $\dfrac{\pi}{6}$ radian $= \dfrac{\pi}{6} \cdot 1$ radian $= \dfrac{\pi}{6} \cdot \dfrac{180}{\pi}$ degrees $= 30°$

(b) $\dfrac{3\pi}{2}$ radians $= \dfrac{3\pi}{2} \cdot \dfrac{180}{\pi}$ degrees $= 270°$

(c) $-\dfrac{3\pi}{4}$ radians $= -\dfrac{3\pi}{4} \cdot \dfrac{180}{\pi}$ degrees $= -135°$

(d) $\dfrac{7\pi}{3}$ radians $= \dfrac{7\pi}{3} \cdot \dfrac{180}{\pi}$ degrees $= 420°$

(e) 3 radians $= 3 \cdot \dfrac{180}{\pi}$ degrees $\approx 171.89°$

─── **Now Work** PROBLEM 47

Table 1 lists the degree and radian measures of some commonly encountered angles. You should learn to feel equally comfortable using degree or radian measure for these angles.

Table 1

Degrees	0°	30°	45°	60°	90°	120°	135°	150°	180°
Radians	0	$\dfrac{\pi}{6}$	$\dfrac{\pi}{4}$	$\dfrac{\pi}{3}$	$\dfrac{\pi}{2}$	$\dfrac{2\pi}{3}$	$\dfrac{3\pi}{4}$	$\dfrac{5\pi}{6}$	π
Degrees		210°	225°	240°	270°	300°	315°	330°	360°
Radians		$\dfrac{7\pi}{6}$	$\dfrac{5\pi}{4}$	$\dfrac{4\pi}{3}$	$\dfrac{3\pi}{2}$	$\dfrac{5\pi}{3}$	$\dfrac{7\pi}{4}$	$\dfrac{11\pi}{6}$	2π

> **EXAMPLE 6**

Finding the Distance between Two Cities

The latitude of a location L is the measure of the angle formed by a ray drawn from the center of Earth to the Equator and a ray drawn from the center of Earth to L. See Figure 13(a). Glasgow, Montana, is due north of Albuquerque, New Mexico. Find the distance between Glasgow (48°9′ north latitude) and Albuquerque (35°5′ north latitude). See Figure 13(b). Assume that the radius of Earth is 3960 miles.

Figure 13

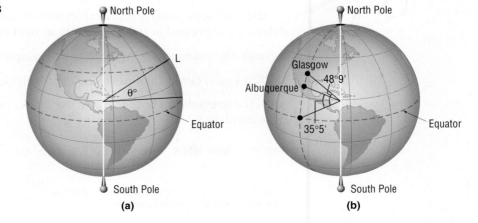

(a)   (b)

Solution

The measure of the central angle between the two cities is $48°9' - 35°5' = 13°4'$. We use equation (4), $s = r\theta$, but first we must convert the angle of $13°4'$ to radians.

$$\theta = 13°4' \approx 13.0667° = 13.0667 \cdot \frac{\pi}{180} \text{ radian} \approx 0.228 \text{ radian}$$

$$\uparrow$$
$$4' = 4\left(\frac{1}{60}\right)°$$

We use $\theta = 0.228$ radian and $r = 3960$ miles in equation (4). The distance between the two cities is

$$s = r\theta = 3960 \cdot 0.228 \approx 903 \text{ miles}$$

●

NOTE If the measure of an angle is given as 5, it is understood to mean 5 radians; if the measure of an angle is given as 5°, it means 5 degrees. ∎

When an angle is measured in degrees, the degree symbol will always be shown. However, when an angle is measured in radians, we will follow the usual practice and omit the word *radians*. So, if the measure of an angle is given as $\frac{\pi}{6}$, it is understood to mean $\frac{\pi}{6}$ radian.

──**Now Work** PROBLEM 101

Figure 14

Figure 15

$$\frac{\theta}{\theta_1} = \frac{A}{A_1}$$

4 Find the Area of a Sector of a Circle

Consider a circle of radius r. Suppose that θ, measured in radians, is a central angle of this circle. See Figure 14. We seek a formula for the area A of the sector (shown in blue) formed by the angle θ.

Now consider a circle of radius r and two central angles θ and θ_1, both measured in radians. See Figure 15. From geometry, we know that the ratio of the measures of the angles equals the ratio of the corresponding areas of the sectors formed by these angles. That is,

$$\frac{\theta}{\theta_1} = \frac{A}{A_1}$$

Suppose that $\theta_1 = 2\pi$ radians. Then $A_1 =$ area of the circle $= \pi r^2$. Solving for A, we find

$$A = A_1 \frac{\theta}{\theta_1} = \pi r^2 \frac{\theta}{2\pi} = \frac{1}{2}r^2\theta$$

$$\uparrow$$
$$A_1 = \pi r^2$$
$$\theta_1 = 2\pi$$

THEOREM

Area of a Sector

The area A of the sector of a circle of radius r formed by a central angle of θ radians is

$$A = \frac{1}{2}r^2\theta \qquad (8)$$

EXAMPLE 7

Finding the Area of a Sector of a Circle

Find the area of the sector of a circle of radius 2 feet formed by an angle of 30°. Round the answer to two decimal places.

Solution

We use equation (8) with $r = 2$ feet and $\theta = 30° = \frac{\pi}{6}$ radian. [Remember, in equation (8), θ must be in radians.]

$$A = \frac{1}{2}r^2\theta = \frac{1}{2}(2)^2\frac{\pi}{6} = \frac{\pi}{3} \approx 1.05$$

The area A of the sector is 1.05 square feet, rounded to two decimal places.

●

──**Now Work** PROBLEM 79

5 Find the Linear Speed of an Object Traveling in Circular Motion

In Appendix A, Section A.9, we defined the average speed of an object as the distance traveled divided by the elapsed time.

Figure 16
$$v = \frac{s}{t}$$

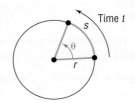

DEFINITION

Suppose that an object moves around a circle of radius r at a constant speed. If s is the distance traveled in time t around this circle, then the **linear speed** v of the object is defined as

$$v = \frac{s}{t} \tag{9}$$

As this object travels around the circle, suppose that θ (measured in radians) is the central angle swept out in time t. See Figure 16.

DEFINITION

The **angular speed** ω (the Greek letter omega) of this object is the angle θ (measured in radians) swept out, divided by the elapsed time t; that is,

$$\omega = \frac{\theta}{t} \tag{10}$$

Angular speed is the way the turning rate of an engine is described. For example, an engine idling at 900 rpm (revolutions per minute) is one that rotates at an angular speed of

$$900 \frac{\text{revolutions}}{\text{minute}} = 900 \frac{\text{revolutions}}{\text{minute}} \cdot 2\pi \frac{\text{radians}}{\text{revolution}} = 1800\pi \frac{\text{radians}}{\text{minute}}$$

There is an important relationship between linear speed and angular speed:

$$\text{linear speed} = v = \underset{\underset{(9)}{\uparrow}}{\frac{s}{t}} = \underset{\underset{s\,=\,r\theta}{\uparrow}}{\frac{r\theta}{t}} = r\left(\frac{\theta}{t}\right) = \underset{\underset{(10)}{\uparrow}}{r \cdot \omega}$$

So

$$v = r\omega \tag{11}$$

where ω is measured in radians per unit time.

When using equation (11), remember that $v = \dfrac{s}{t}$ (the linear speed) has the dimensions of length per unit of time (such as feet per second or miles per hour), r (the radius of the circular motion) has the same length dimension as s, and ω (the angular speed) has the dimensions of radians per unit of time. If the angular speed is given in terms of *revolutions* per unit of time (as is often the case), be sure to convert it to *radians* per unit of time using the fact that 1 revolution $= 2\pi$ radians before attempting to use equation (11).

EXAMPLE 8   **Finding Linear Speed**

A child is spinning a rock at the end of a 2-foot rope at the rate of 180 revolutions per minute (rpm). Find the linear speed of the rock when it is released.

Solution   Look at Figure 17. The rock is moving around a circle of radius $r = 2$ feet. The angular speed ω of the rock is

$$\omega = 180 \frac{\text{revolutions}}{\text{minute}} = 180 \frac{\text{revolutions}}{\text{minute}} \cdot 2\pi \frac{\text{radians}}{\text{revolution}} = 360\pi \frac{\text{radians}}{\text{minute}}$$

Figure 17

From equation (11), the linear speed v of the rock is

$$v = r\omega = 2 \text{ feet} \cdot 360\pi \frac{\text{radians}}{\text{minute}} = 720\pi \frac{\text{feet}}{\text{minute}} \approx 2262 \frac{\text{feet}}{\text{minute}}$$

The linear speed of the rock when it is released is 2262 ft/min $\approx$ 25.7 mi/hr.

━━━━ **Now Work** PROBLEM 97

Historical Feature

Trigonometry was developed by Greek astronomers, who regarded the sky as the inside of a sphere, so it was natural that triangles on a sphere were investigated early (by Menelaus of Alexandria about AD 100) and that triangles in the plane were studied much later. The first book containing a systematic treatment of plane and spherical trigonometry was written by the Persian astronomer Nasir Eddin (about AD 1250).

Regiomontanus (1436–1476) is the person most responsible for moving trigonometry from astronomy into mathematics. His work was improved by Copernicus (1473–1543) and Copernicus's student

Rhaeticus (1514–1576). Rhaeticus's book was the first to define the six trigonometric functions as ratios of sides of triangles, although he did not give the functions their present names. Credit for this is due to Thomas Finck (1583), but Finck's notation was by no means universally accepted at the time. The notation was finally stabilized by the textbooks of Leonhard Euler (1707–1783).

Trigonometry has since evolved from its use by surveyors, navigators, and engineers to present applications involving ocean tides, the rise and fall of food supplies in certain ecologies, brain wave patterns, and many other phenomena.

5.1 Assess Your Understanding

'Are You Prepared?' *Answers are given at the end of these exercises. If you get a wrong answer, read the pages listed in* red.

1. What is the formula for the circumference C of a circle of radius r? What is the formula for the area A of a circle of radius r? (p. A16)

2. If a particle has a speed of r feet per second and travels a distance d (in feet) in time t (in seconds), then $d =$ _____. (p. A77)

Concepts and Vocabulary

3. An angle θ is in _____ _____ if its vertex is at the origin of a rectangular coordinate system and its initial side coincides with the positive x-axis.

4. A _____ _____ is a positive angle whose vertex is at the center of a circle.

5. If the radius of a circle is r and the length of the arc subtended by a central angle is also r, then the measure of the angle is 1 _____.

6. On a circle of radius r, a central angle of θ radians subtends an arc of length $s =$ _____; the area of the sector formed by this angle θ is $A =$ _____.

7. $180° =$ _____ radians

8. An object travels around a circle of radius r with constant speed. If s is the distance traveled in time t around the circle and θ is the central angle (in radians) swept out in time t, then the linear speed of the object is $v =$ _____ and the angular speed of the object is $\omega =$ _____.

9. **True or False** The angular speed ω of an object traveling around a circle of radius r is the angle θ (measured in radians) swept out, divided by the elapsed time t.

10. **True or False** For circular motion on a circle of radius r, linear speed equals angular speed divided by r.

Skill Building

In Problems 11–22, draw each angle.

11. $30°$
12. $60°$
13. $135°$
14. $-120°$
15. $450°$
16. $540°$

17. $\dfrac{3\pi}{4}$
18. $\dfrac{4\pi}{3}$
19. $-\dfrac{\pi}{6}$
20. $-\dfrac{2\pi}{3}$
21. $\dfrac{16\pi}{3}$
22. $\dfrac{21\pi}{4}$

In Problems 23–28, convert each angle to a decimal in degrees. Round your answer to two decimal places.

23. $40°10'25''$
24. $61°42'21''$
25. $1°2'3''$
26. $73°40'40''$
27. $9°9'9''$
28. $98°22'45''$

In Problems 29–34, convert each angle to D°M'S" form. Round your answer to the nearest second.

29. 40.32°   **30.** 61.24°   **31.** 18.255°   **32.** 29.411°   **33.** 19.99°   **34.** 44.01°

In Problems 35–46, convert each angle in degrees to radians. Express your answer as a multiple of π.

35. 30°   **36.** 120°   **37.** 240°   **38.** 330°   **39.** −60°   **40.** −30°

41. 180°   **42.** 270°   **43.** −135°   **44.** −225°   **45.** −90°   **46.** −180°

In Problems 47–58, convert each angle in radians to degrees.

47. $\dfrac{\pi}{3}$ **48.** $\dfrac{5\pi}{6}$ **49.** $-\dfrac{5\pi}{4}$ **50.** $-\dfrac{2\pi}{3}$ **51.** $\dfrac{\pi}{2}$ **52.** 4π

53. $\dfrac{\pi}{12}$ **54.** $\dfrac{5\pi}{12}$ **55.** $-\dfrac{\pi}{2}$ **56.** $-\pi$ **57.** $-\dfrac{\pi}{6}$ **58.** $-\dfrac{3\pi}{4}$

In Problems 59–64, convert each angle in degrees to radians. Express your answer in decimal form, rounded to two decimal places.

59. 17°   **60.** 73°   **61.** −40°   **62.** −51°   **63.** 125°   **64.** 350°

In Problems 65–70, convert each angle in radians to degrees. Express your answer in decimal form, rounded to two decimal places.

65. 3.14   **66.** 0.75   **67.** 2   **68.** 3   **69.** 6.32   **70.** $\sqrt{2}$

In Problems 71–78, s denotes the length of the arc of a circle of radius r subtended by the central angle θ. Find the missing quantity. Round answers to three decimal places.

71. $r = 10$ meters, $\theta = \dfrac{1}{2}$ radian, $s = ?$

72. $r = 6$ feet, $\theta = 2$ radians, $s = ?$

73. $\theta = \dfrac{1}{3}$ radian, $s = 2$ feet, $r = ?$

74. $\theta = \dfrac{1}{4}$ radian, $s = 6$ centimeters, $r = ?$

75. $r = 5$ miles, $s = 3$ miles, $\theta = ?$

76. $r = 6$ meters, $s = 8$ meters, $\theta = ?$

77. $r = 2$ inches, $\theta = 30°$, $s = ?$

78. $r = 3$ meters, $\theta = 120°$, $s = ?$

In Problems 79–86, A denotes the area of the sector of a circle of radius r formed by the central angle θ. Find the missing quantity. Round answers to three decimal places.

79. $r = 10$ meters, $\theta = \dfrac{1}{2}$ radian, $A = ?$

80. $r = 6$ feet, $\theta = 2$ radians, $A = ?$

81. $\theta = \dfrac{1}{3}$ radian, $A = 2$ square feet, $r = ?$

82. $\theta = \dfrac{1}{4}$ radian, $A = 6$ square centimeters, $r = ?$

83. $r = 5$ miles, $A = 3$ square miles, $\theta = ?$

84. $r = 6$ meters, $A = 8$ square meters, $\theta = ?$

85. $r = 2$ inches, $\theta = 30°$, $A = ?$

86. $r = 3$ meters, $\theta = 120°$, $A = ?$

In Problems 87–90, find the length s and area A. Round answers to three decimal places.

87.

88.

89.

90.

Applications and Extensions

91. Movement of a Minute Hand The minute hand of a clock is 6 inches long. How far does the tip of the minute hand move in 15 minutes? How far does it move in 25 minutes? Round answers to two decimal places.

92. Movement of a Pendulum A pendulum swings through an angle of 20° each second. If the pendulum is 40 inches long, how far does its tip move each second? Round answers to two decimal places.

93. Area of a Sector Find the area of the sector of a circle of radius 4 meters formed by an angle of 45°. Round the answer to two decimal places.

94. Area of a Sector Find the area of the sector of a circle of radius 3 centimeters formed by an angle of 60°. Round the answer to two decimal places.

95. Watering a Lawn A water sprinkler sprays water over a distance of 30 feet while rotating through an angle of 135°. What area of lawn receives water?

96. Designing a Water Sprinkler An engineer is asked to design a water sprinkler that will cover a field of 100 square yards that is in the shape of a sector of a circle of radius 15 yards. Through what angle should the sprinkler rotate?

97. Motion on a Circle An object is traveling around a circle with a radius of 5 centimeters. If in 20 seconds a central angle of $\frac{1}{3}$ radian is swept out, what is the angular speed of the object? What is its linear speed?

98. Motion on a Circle An object is traveling around a circle with a radius of 2 meters. If in 20 seconds the object travels 5 meters, what is its angular speed? What is its linear speed?

99. Bicycle Wheels The diameter of each wheel of a bicycle is 26 inches. If you are traveling at a speed of 35 miles per hour on this bicycle, through how many revolutions per minute are the wheels turning?

100. Car Wheels The radius of each wheel of a car is 15 inches. If the wheels are turning at the rate of 3 revolutions per second, how fast is the car moving? Express your answer in inches per second and in miles per hour.

In Problems 101–104, the latitude of a location L is the angle formed by a ray drawn from the center of Earth to the Equator and a ray drawn from the center of Earth to L. See the figure.

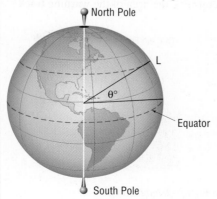

101. Distance between Cities Memphis, Tennessee, is due north of New Orleans, Louisiana. Find the distance between Memphis (35°9′ north latitude) and New Orleans (29°57′ north latitude). Assume that the radius of Earth is 3960 miles.

102. Distance between Cities Charleston, West Virginia, is due north of Jacksonville, Florida. Find the distance between Charleston (38°21′ north latitude) and Jacksonville (30°20′ north latitude). Assume that the radius of Earth is 3960 miles.

103. Linear Speed on Earth Earth rotates on an axis through its poles. The distance from the axis to a location on Earth 30° north latitude is about 3429.5 miles. Therefore, a location on Earth at 30° north latitude is spinning on a circle of radius 3429.5 miles. Compute the linear speed on the surface of Earth at 30° north latitude.

104. Linear Speed on Earth Earth rotates on an axis through its poles. The distance from the axis to a location on Earth 40° north latitude is about 3033.5 miles. Therefore, a location on Earth at 40° north latitude is spinning on a circle of radius 3033.5 miles. Compute the linear speed on the surface of Earth at 40° north latitude.

105. Speed of the Moon The mean distance of the moon from Earth is 2.39×10^5 miles. Assuming that the orbit of the moon around Earth is circular and that 1 revolution takes 27.3 days, find the linear speed of the moon. Express your answer in miles per hour.

106. Speed of Earth The mean distance of Earth from the Sun is 9.29×10^7 miles. Assuming that the orbit of Earth around the Sun is circular and that 1 revolution takes 365 days, find the linear speed of Earth. Express your answer in miles per hour.

107. Pulleys Two pulleys, one with radius 2 inches and the other with radius 8 inches, are connected by a belt. (See the figure.) If the 2-inch pulley is caused to rotate at 3 revolutions per minute, determine the revolutions per minute of the 8-inch pulley.

[Hint: The linear speeds of the pulleys are the same; both equal the speed of the belt.]

108. Ferris Wheels A neighborhood carnival has a Ferris wheel whose radius is 30 feet. You measure the time it takes for one revolution to be 70 seconds. What is the linear speed (in feet per second) of this Ferris wheel? What is the angular speed in radians per second?

109. Computing the Speed of a River Current To approximate the speed of the current of a river, a circular paddle wheel with radius 4 feet is lowered into the water. If the current causes the wheel to rotate at a speed of 10 revolutions per minute, what is the speed of the current? Express your answer in miles per hour.

110. Spin Balancing Tires   A spin balancer rotates the wheel of a car at 480 revolutions per minute. If the diameter of the wheel is 26 inches, what road speed is being tested? Express your answer in miles per hour. At how many revolutions per minute should the balancer be set to test a road speed of 80 miles per hour?

111. The Cable Cars of San Francisco   At the Cable Car Museum you can see the four cable lines that are used to pull cable cars up and down the hills of San Francisco. Each cable travels at a speed of 9.55 miles per hour, caused by a rotating wheel whose diameter is 8.5 feet. How fast is the wheel rotating? Express your answer in revolutions per minute.

112. Difference in Time of Sunrise   Naples, Florida, is approximately 90 miles due west of Ft. Lauderdale. How much sooner would a person in Ft. Lauderdale first see the rising Sun than a person in Naples? See the hint.

[**Hint:** Consult the figure. When a person at Q sees the first rays of the Sun, a person at P is still in the dark. The person at P sees the first rays after Earth has rotated so that P is at the location Q. Now use the fact that at the latitude of Ft. Lauderdale in 24 hours an arc of length $2\pi(3559)$ miles is subtended.]

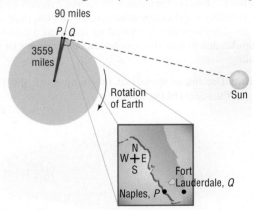

113. Keeping Up with the Sun   How fast would you have to travel on the surface of Earth at the equator to keep up with the Sun (that is, so that the Sun would appear to remain in the same position in the sky)?

114. Nautical Miles   A **nautical mile** equals the length of arc subtended by a central angle of 1 minute on a great circle* on the surface of Earth. See the figure. If the radius of Earth is taken as 3960 miles, express 1 nautical mile in terms of ordinary, or **statute,** miles.

*Any circle drawn on the surface of Earth that divides Earth into two equal hemispheres.

115. Approximating the Circumference of Earth   Eratosthenes of Cyrene (276–195 BC) was a Greek scholar who lived and worked in Cyrene and Alexandria. One day while visiting in Syene he noticed that the Sun's rays shone directly down a well. On this date 1 year later, in Alexandria, which is 500 miles due north of Syene he measured the angle of the Sun to be about 7.2 degrees. See the figure. Use this information to approximate the radius and circumference of Earth.

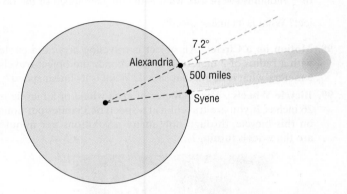

116. Designing a Little League Field   For a 60-foot Little League Baseball field, the distance from home base to the nearest fence (or other obstruction) in fair territory should be a minimum of 200 feet. The commissioner of parks and recreation is making plans for a new 60-foot field. Because of limited ground availability, he will use the minimum required distance to the outfield fence. To increase safety, however, he plans to include a 10-foot wide warning track on the inside of the fence. To further increase safety, the fence and warning track will extend both directions into foul territory. In total, the arc formed by the outfield fence (including the extensions into the foul territories) will be subtended by a central angle at home plate measuring 96°, as illustrated.
(a) Determine the length of the outfield fence.
(b) Determine the area of the warning track.

Source: www.littleleague.org

[**Note:** There is a 90° angle between the two foul lines. Then there are two 3° angles between the foul lines and the dotted lines shown. The angle between the two dotted lines outside the 200-foot foul lines is 96°.]

117. Pulleys Two pulleys, one with radius r_1 and the other with radius r_2, are connected by a belt. The pulley with radius r_1 rotates at ω_1 revolutions per minute, whereas the pulley with radius r_2 rotates at ω_2 revolutions per minute. Show that
$$\frac{r_1}{r_2} = \frac{\omega_2}{\omega_1}.$$

Explaining Concepts: Discussion and Writing

118. Do you prefer to measure angles using degrees or radians? Provide justification and a rationale for your choice.

119. What is 1 radian? What is 1°?

120. Which angle has the larger measure: 1 degree or 1 radian? Or are they equal?

121. Explain the difference between linear speed and angular speed.

122. For a circle of radius r, a central angle of θ degrees subtends an arc whose length s is $s = \dfrac{\pi}{180} r\theta$. Discuss whether this is a true or false statement. Give reasons to defend your position.

123. Discuss why ships and airplanes use nautical miles to measure distance. Explain the difference between a nautical mile and a statute mile.

124. Investigate the way that speed bicycles work. In particular, explain the differences and similarities between 5-speed and 9-speed derailleurs. Be sure to include a discussion of linear speed and angular speed.

125. In Example 6, we found that the distance between Albuquerque, New Mexico, and Glasgow, Montana, is approximately 903 miles. According to *mapquest.com*, the distance is approximately 1300 miles. What might account for the difference?

'Are You Prepared?' Answers

1. $C = 2\pi r;\ A = \pi r^2$ **2.** $r \cdot t$

5.2 Right Triangle Trigonometry

PREPARING FOR THIS SECTION *Before getting started, review the following:*

- Geometry Essentials (Appendix A, Section A.2, pp. A14–A19)
- Functions (Section 1.1, pp. 43–51)

Now Work the 'Are You Prepared?' problems on page 399.

OBJECTIVES   **1** Find the Values of Trigonometric Functions of Acute Angles (p. 391)
            2 Use the Fundamental Identities (p. 393)
            3 Find the Values of the Remaining Trigonometric Functions, Given the Value of One of Them (p. 395)
            4 Use the Complementary Angle Theorem (p. 397)

1 Find the Values of Trigonometric Functions of Acute Angles

A triangle in which one angle is a right angle (90°) is called a **right triangle.** Recall that the side opposite the right angle is called the **hypotenuse,** and the remaining two sides are called the **legs** of the triangle. In Figure 18 we have labeled the hypotenuse as c to indicate that its length is c units, and, in a like manner, we have labeled the legs as a and b. Because the triangle is a right triangle, the Pythagorean Theorem tells us that

$$c^2 = a^2 + b^2$$

Now, suppose that θ is an **acute angle;** that is, $0° < \theta < 90°$ (if θ is measured in degrees) and $0 < \theta < \dfrac{\pi}{2}$ (if θ is measured in radians). See Figure 19(a) on page 392. Using this acute angle θ, we can form a right triangle, like the one illustrated in

Figure 18

Figure 19(b), with hypotenuse of length c and legs of lengths a and b. Using the three sides of this triangle, we can form exactly six ratios:

$$\frac{b}{c}, \quad \frac{a}{c}, \quad \frac{b}{a}, \quad \frac{c}{b}, \quad \frac{c}{a}, \quad \frac{a}{b}$$

Figure 19

(a) Acute angle   (b) Right triangle   (c) Similar triangles

In fact, these ratios depend only on the size of the angle θ and not on the triangle formed. To see why, look at Figure 19(c). Any two right triangles formed using the angle θ will be similar and, hence, corresponding ratios will be equal. As a result,

$$\frac{b}{c} = \frac{b'}{c'} \quad \frac{a}{c} = \frac{a'}{c'} \quad \frac{b}{a} = \frac{b'}{a'} \quad \frac{c}{b} = \frac{c'}{b'} \quad \frac{c}{a} = \frac{c'}{a'} \quad \frac{a}{b} = \frac{a'}{b'}$$

Because the ratios depend only on the angle θ and not on the triangle itself, we give each ratio a name that involves θ: sine of θ, cosine of θ, tangent of θ, cosecant of θ, secant of θ, and cotangent of θ.

DEFINITION

The six ratios of the lengths of the sides of a right triangle are called **trigonometric functions of acute angles** and are defined as follows:

Function Name	Abbreviation	Value	Function Name	Abbreviation	Value
sine of θ	$\sin \theta$	$\dfrac{b}{c}$	cosecant of θ	$\csc \theta$	$\dfrac{c}{b}$
cosine of θ	$\cos \theta$	$\dfrac{a}{c}$	secant of θ	$\sec \theta$	$\dfrac{c}{a}$
tangent of θ	$\tan \theta$	$\dfrac{b}{a}$	cotangent of θ	$\cot \theta$	$\dfrac{a}{b}$

Figure 20

As an aid to remembering these definitions, it may be helpful to refer to the lengths of the sides of the triangle by the names *hypotenuse (c), opposite (b),* and *adjacent (a)*. See Figure 20. In terms of these names, we have the following ratios:

$$\sin \theta = \frac{\text{opposite}}{\text{hypotenuse}} = \frac{b}{c} \qquad \cos \theta = \frac{\text{adjacent}}{\text{hypotenuse}} = \frac{a}{c} \qquad \tan \theta = \frac{\text{opposite}}{\text{adjacent}} = \frac{b}{a}$$

$$\csc \theta = \frac{\text{hypotenuse}}{\text{opposite}} = \frac{c}{b} \qquad \sec \theta = \frac{\text{hypotenuse}}{\text{adjacent}} = \frac{c}{a} \qquad \cot \theta = \frac{\text{adjacent}}{\text{opposite}} = \frac{a}{b}$$

(1)

Since a, b, and c are positive, each of the trigonometric functions of an acute angle θ is positive.

EXAMPLE 1   **Finding the Value of Trigonometric Functions**

Find the value of each of the six trigonometric functions of the angle θ in Figure 21.

Solution   We see in Figure 21 that the two given sides of the triangle are

$$c = \text{hypotenuse} = 5 \qquad a = \text{adjacent} = 3$$

Figure 21

To find the length of the opposite side, we use the Pythagorean Theorem.

$$(\text{adjacent})^2 + (\text{opposite})^2 = (\text{hypotenuse})^2$$
$$3^2 + (\text{opposite})^2 = 5^2$$
$$(\text{opposite})^2 = 25 - 9 = 16$$
$$\text{opposite} = 4$$

WARNING When writing the values of the trigonometric functions, do not forget the argument of the function

$$\sin \theta = \frac{4}{5} \quad \textit{correct}$$

$$\sin = \frac{4}{5} \quad \textit{incorrect} \quad \blacksquare$$

Now that we know the lengths of the three sides, we use the ratios in (1) to find the value of each of the six trigonometric functions:

$$\sin \theta = \frac{\text{opposite}}{\text{hypotenuse}} = \frac{4}{5} \qquad \cos \theta = \frac{\text{adjacent}}{\text{hypotenuse}} = \frac{3}{5} \qquad \tan \theta = \frac{\text{opposite}}{\text{adjacent}} = \frac{4}{3}$$

$$\csc \theta = \frac{\text{hypotenuse}}{\text{opposite}} = \frac{5}{4} \qquad \sec \theta = \frac{\text{hypotenuse}}{\text{adjacent}} = \frac{5}{3} \qquad \cot \theta = \frac{\text{adjacent}}{\text{opposite}} = \frac{3}{4}$$

●

Now Work PROBLEM 11

2 Use the Fundamental Identities

You may have observed some relationships that exist among the six trigonometric functions of acute angles. For example, $\csc \theta = \dfrac{c}{b} = \dfrac{1}{\dfrac{b}{c}} = \dfrac{1}{\sin \theta}$. The **reciprocal identities** are

Reciprocal Identities

$$\csc \theta = \frac{1}{\sin \theta} \qquad \sec \theta = \frac{1}{\cos \theta} \qquad \cot \theta = \frac{1}{\tan \theta} \tag{2}$$

Two other fundamental identities that are easy to see are the **quotient identities.**

Quotient Identities

$$\tan \theta = \frac{\sin \theta}{\cos \theta} \qquad \cot \theta = \frac{\cos \theta}{\sin \theta} \tag{3}$$

If $\sin \theta$ and $\cos \theta$ are known, the reciprocal and quotient identities in (2) and (3) make it easy to find the values of the remaining trigonometric functions.

EXAMPLE 2

Finding the Values of the Remaining Trigonometric Functions, Given $\sin \theta$ and $\cos \theta$

Given $\sin \theta = \dfrac{\sqrt{5}}{5}$ and $\cos \theta = \dfrac{2\sqrt{5}}{5}$, find the value of each of the four remaining trigonometric functions of θ.

Solution

Based on the quotient identity in formula (3), we have

$$\tan \theta = \frac{\sin \theta}{\cos \theta} = \frac{\dfrac{\sqrt{5}}{5}}{\dfrac{2\sqrt{5}}{5}} = \frac{1}{2}$$

Then we use the reciprocal identities from formula (2) to get

$$\csc\theta = \frac{1}{\sin\theta} = \frac{1}{\dfrac{\sqrt{5}}{5}} = \frac{5}{\sqrt{5}} = \sqrt{5} \qquad \sec\theta = \frac{1}{\cos\theta} = \frac{1}{\dfrac{2\sqrt{5}}{5}} = \frac{5}{2\sqrt{5}} = \frac{\sqrt{5}}{2} \qquad \cot\theta = \frac{1}{\tan\theta} = \frac{1}{\dfrac{1}{2}} = 2$$

●

━━━━ **Now Work** PROBLEM 21

Figure 22

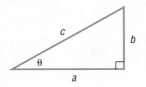

Refer now to the right triangle in Figure 22. The Pythagorean Theorem states that $a^2 + b^2 = c^2$, which we can write as

$$b^2 + a^2 = c^2$$

Dividing each side by c^2, we get

$$\frac{b^2}{c^2} + \frac{a^2}{c^2} = 1 \quad \text{or} \quad \left(\frac{b}{c}\right)^2 + \left(\frac{a}{c}\right)^2 = 1$$

In terms of trigonometric functions of the angle θ, this equation states that

$$(\sin\theta)^2 + (\cos\theta)^2 = 1 \tag{4}$$

Equation (4) is an identity, since the equation is true for any acute angle θ.

It is customary to write $\sin^2\theta$ instead of $(\sin\theta)^2$, $\cos^2\theta$ instead of $(\cos\theta)^2$, and so on. With this notation, we can rewrite equation (4) as

$$\boxed{\sin^2\theta + \cos^2\theta = 1} \tag{5}$$

Another identity can be obtained from equation (5) by dividing each side by $\cos^2\theta$.

$$\frac{\sin^2\theta}{\cos^2\theta} + 1 = \frac{1}{\cos^2\theta}$$

Now use formulas (2) and (3) to get

$$\boxed{\tan^2\theta + 1 = \sec^2\theta} \tag{6}$$

Similarly, by dividing each side of equation (5) by $\sin^2\theta$, we get $1 + \cot^2\theta = \csc^2\theta$, which we write as

$$\boxed{\cot^2\theta + 1 = \csc^2\theta} \tag{7}$$

Collectively, the identities in equations (5), (6), and (7) are referred to as the **Pythagorean Identities.**

Let's pause here to summarize the fundamental identities.

Fundamental Identities

$$\tan\theta = \frac{\sin\theta}{\cos\theta} \qquad \cot\theta = \frac{\cos\theta}{\sin\theta}$$

$$\csc\theta = \frac{1}{\sin\theta} \qquad \sec\theta = \frac{1}{\cos\theta} \qquad \cot\theta = \frac{1}{\tan\theta}$$

$$\sin^2\theta + \cos^2\theta = 1 \qquad \tan^2\theta + 1 = \sec^2\theta \qquad \cot^2\theta + 1 = \csc^2\theta$$

EXAMPLE 3

Finding the Exact Value of a Trigonometric Expression Using Identities

Find the exact value of each expression. Do not use a calculator.

(a) $\tan 20° - \dfrac{\sin 20°}{\cos 20°}$

(b) $\sin^2 \dfrac{\pi}{12} + \dfrac{1}{\sec^2 \dfrac{\pi}{12}}$

Solution

(a) $\tan 20° - \underset{\underset{\dfrac{\sin \theta}{\cos \theta} = \tan \theta}{\uparrow}}{\dfrac{\sin 20°}{\cos 20°}} = \tan 20° - \tan 20° = 0$

(b) $\sin^2 \dfrac{\pi}{12} + \underset{\underset{\cos \theta = \dfrac{1}{\sec \theta}}{\uparrow}}{\dfrac{1}{\sec^2 \dfrac{\pi}{12}}} = \underset{\underset{\sin^2 \theta + \cos^2 \theta = 1}{\uparrow}}{\sin^2 \dfrac{\pi}{12} + \cos^2 \dfrac{\pi}{12}} = 1$

Now Work **PROBLEM 39**

3 Find the Values of the Remaining Trigonometric Functions, Given the Value of One of Them

Once the value of one trigonometric function is known, it is possible to find the value of each of the remaining five trigonometric functions.

EXAMPLE 4

Finding the Values of the Remaining Trigonometric Functions, Given sin θ, θ Acute

Given that $\sin \theta = \dfrac{1}{3}$ and θ is an acute angle, find the exact value of each of the remaining five trigonometric functions of θ.

Solution

We solve this problem in two ways: The first way uses the definition of the trigonometric functions; the second method uses the fundamental identities.

Solution 1 Using the Definition

Because $\sin \theta = \dfrac{1}{3} = \dfrac{b}{c}$ we draw a right triangle with acute angle θ, opposite side of length $b = 1$, and hypotenuse of length $c = 3$. See Figure 23. The adjacent side a can be found by using the Pythagorean Theorem.

Figure 23

$$a^2 + 1^2 = 3^2 \qquad a^2 + b^2 = c^2; b = 1, c = 3$$
$$a^2 + 1 = 9$$
$$a^2 = 8$$
$$a = 2\sqrt{2}$$

Now the definitions given in equation (1) can be used to find the value of each of the remaining five trigonometric functions. (Refer back to the method used in Example 1.) Using $a = 2\sqrt{2}, b = 1$, and $c = 3$, we have

$$\cos \theta = \frac{a}{c} = \frac{2\sqrt{2}}{3} \qquad \tan \theta = \frac{b}{a} = \frac{1}{2\sqrt{2}} = \frac{\sqrt{2}}{4}$$

$$\csc \theta = \frac{c}{b} = \frac{3}{1} = 3 \quad \sec \theta = \frac{c}{a} = \frac{3}{2\sqrt{2}} = \frac{3\sqrt{2}}{4} \quad \cot \theta = \frac{a}{b} = \frac{2\sqrt{2}}{1} = 2\sqrt{2}$$

Solution 2
Using Identities

We begin by seeking $\cos \theta$, which can be found by using the Pythagorean Identity from equation (5).

$$\sin^2 \theta + \cos^2 \theta = 1 \qquad \text{Formula (5)}$$

$$\frac{1}{9} + \cos^2 \theta = 1 \qquad \sin \theta = \frac{1}{3}$$

$$\cos^2 \theta = 1 - \frac{1}{9} = \frac{8}{9}$$

Recall that the trigonometric functions of an acute angle are positive. In particular, $\cos \theta > 0$ for an acute angle θ, so we have

$$\cos \theta = \sqrt{\frac{8}{9}} = \frac{2\sqrt{2}}{3}$$

Now we know that $\sin \theta = \frac{1}{3}$ and $\cos \theta = \frac{2\sqrt{2}}{3}$, so we can proceed as we did in Example 2.

$$\tan \theta = \frac{\sin \theta}{\cos \theta} = \frac{\dfrac{1}{3}}{\dfrac{2\sqrt{2}}{3}} = \frac{1}{2\sqrt{2}} = \frac{\sqrt{2}}{4} \qquad \cot \theta = \frac{1}{\tan \theta} = \frac{1}{\dfrac{\sqrt{2}}{4}} = \frac{4}{\sqrt{2}} = 2\sqrt{2}$$

$$\sec \theta = \frac{1}{\cos \theta} = \frac{1}{\dfrac{2\sqrt{2}}{3}} = \frac{3}{2\sqrt{2}} = \frac{3\sqrt{2}}{4} \qquad \csc \theta = \frac{1}{\sin \theta} = \frac{1}{\dfrac{1}{3}} = 3$$

●

Finding the Values of the Trigonometric Functions When One Is Known

Given the value of one trigonometric function of an acute angle θ, the exact value of each of the remaining five trigonometric functions of θ can be found in either of two ways.

Method 1 Using the Definition
STEP 1: Draw a right triangle showing the acute angle θ.
STEP 2: Two of the sides can then be assigned values based on the value of the given trigonometric function.
STEP 3: Find the length of the third side by using the Pythagorean Theorem.
STEP 4: Use the definitions in equation (1) to find the value of each of the remaining trigonometric functions.

Method 2 Using Identities
Use appropriately selected identities to find the value of each of the remaining trigonometric functions.

EXAMPLE 5

Given One Value of a Trigonometric Function, Find the Values of the Remaining Ones

Given $\tan \theta = \dfrac{1}{2}$, θ an acute angle, find the exact value of each of the remaining five trigonometric functions of θ.

Solution 1
Using the Definition

Figure 24 shows a right triangle with acute angle θ, where

$$\tan \theta = \frac{1}{2} = \frac{\text{opposite}}{\text{adjacent}} = \frac{b}{a}$$

Figure 24

$\tan \theta = \dfrac{1}{2}$

With $b = 1$ and $a = 2$, the hypotenuse c can be found by using the Pythagorean Theorem.

$$c^2 = a^2 + b^2 = 2^2 + 1^2 = 5$$

$$c = \sqrt{5}$$

Now apply the definitions using $a = 2$, $b = 1$, and $c = \sqrt{5}$.

$$\sin \theta = \frac{b}{c} = \frac{1}{\sqrt{5}} = \frac{\sqrt{5}}{5} \qquad \cos \theta = \frac{a}{c} = \frac{2}{\sqrt{5}} = \frac{2\sqrt{5}}{5}$$

$$\csc \theta = \frac{c}{b} = \frac{\sqrt{5}}{1} = \sqrt{5} \qquad \sec \theta = \frac{c}{a} = \frac{\sqrt{5}}{2} \qquad \cot \theta = \frac{a}{b} = \frac{2}{1} = 2$$

●

Solution 2
Using Identities

Because we know the value of $\tan \theta$, we use the Pythagorean Identity that involves $\tan \theta$:

$$\tan^2 \theta + 1 = \sec^2 \theta \qquad \textit{Formula (6)}$$

$$\left(\frac{1}{2}\right)^2 + 1 = \sec^2 \theta \qquad \textit{tan } \theta = \frac{1}{2}$$

$$\sec^2 \theta = \frac{1}{4} + 1 = \frac{5}{4} \qquad \textit{Proceed to solve for sec } \theta.$$

$$\sec \theta = \frac{\sqrt{5}}{2} \qquad \textit{sec } \theta > 0 \quad \textit{since } \theta \textit{ is acute.}$$

Now we know $\tan \theta = \dfrac{1}{2}$ and $\sec \theta = \dfrac{\sqrt{5}}{2}$. Using reciprocal identities, we find

$$\cos \theta = \frac{1}{\sec \theta} = \frac{1}{\dfrac{\sqrt{5}}{2}} = \frac{2}{\sqrt{5}} = \frac{2\sqrt{5}}{5}$$

$$\cot \theta = \frac{1}{\tan \theta} = \frac{1}{\dfrac{1}{2}} = 2$$

To find $\sin \theta$, we use the following reasoning:

$$\tan \theta = \frac{\sin \theta}{\cos \theta}, \quad \text{so} \quad \sin \theta = (\tan \theta)(\cos \theta) = \frac{1}{2} \cdot \frac{2\sqrt{5}}{5} = \frac{\sqrt{5}}{5}$$

$$\csc \theta = \frac{1}{\sin \theta} = \frac{1}{\dfrac{\sqrt{5}}{5}} = \sqrt{5}$$

●

 Now Work   PROBLEM **25**

4 Use the Complementary Angle Theorem

Two acute angles are called **complementary** if their sum is a right angle, or 90°. Because the sum of the angles of any triangle is 180°, it follows that, for a right triangle, the two acute angles are complementary, so the sum of the acute angles in a right triangle is 90°.

Refer now to Figure 25; we have labeled the angle opposite side b as B and the angle opposite side a as A. Notice that side a is adjacent to angle B and is opposite angle A. Similarly, side b is opposite angle B and is adjacent to angle A. As a result,

Figure 25

Adjacent to A
opposite B

Adjacent to B
opposite A

$$\sin B = \frac{b}{c} = \cos A \qquad \cos B = \frac{a}{c} = \sin A \qquad \tan B = \frac{b}{a} = \cot A$$

$$\csc B = \frac{c}{b} = \sec A \qquad \sec B = \frac{c}{a} = \csc A \qquad \cot B = \frac{a}{b} = \tan A$$

(8)

Because of these relationships, the functions sine and cosine, tangent and cotangent, and secant and cosecant are called **cofunctions** of each other. The identities (8) may be expressed in words as follows:

THEOREM

Complementary Angle Theorem

Cofunctions of complementary angles are equal.

Here are examples of this theorem.

Complementary angles
$$\sin 30° = \cos 60°$$
Cofunctions

Complementary angles
$$\tan 40° = \cot 50°$$
Cofunctions

Complementary angles
$$\sec 80° = \csc 10°$$
Cofunctions

If an angle θ is measured in degrees, we will use the degree symbol when writing a trigonometric function of θ, as, for example, in $\sin 30°$ and $\tan 45°$. If an angle θ is measured in radians, then no symbol is used when writing a trigonometric function of θ, as, for example, in $\cos \pi$ and $\sec \dfrac{\pi}{3}$.

If θ is an acute angle measured in degrees, the angle $90° - \theta$ (or $\dfrac{\pi}{2} - \theta$, if θ is in radians) is the angle complementary to θ. Table 2 restates the preceding theorem on cofunctions.

Table 2

θ (Degrees)	θ (Radians)
$\sin \theta = \cos(90° - \theta)$	$\sin \theta = \cos\left(\dfrac{\pi}{2} - \theta\right)$
$\cos \theta = \sin(90° - \theta)$	$\cos \theta = \sin\left(\dfrac{\pi}{2} - \theta\right)$
$\tan \theta = \cot(90° - \theta)$	$\tan \theta = \cot\left(\dfrac{\pi}{2} - \theta\right)$
$\csc \theta = \sec(90° - \theta)$	$\csc \theta = \sec\left(\dfrac{\pi}{2} - \theta\right)$
$\sec \theta = \csc(90° - \theta)$	$\sec \theta = \csc\left(\dfrac{\pi}{2} - \theta\right)$
$\cot \theta = \tan(90° - \theta)$	$\cot \theta = \tan\left(\dfrac{\pi}{2} - \theta\right)$

The angle θ in Table 2 is acute. We will see later (Section 6.5) that these results are valid for any angle θ.

EXAMPLE 6

Using the Complementary Angle Theorem

(a) $\sin 62° = \cos(90° - 62°) = \cos 28°$

(b) $\tan \dfrac{\pi}{12} = \cot\left(\dfrac{\pi}{2} - \dfrac{\pi}{12}\right) = \cot \dfrac{5\pi}{12}$

(c) $\cos \dfrac{\pi}{4} = \sin\left(\dfrac{\pi}{2} - \dfrac{\pi}{4}\right) = \sin \dfrac{\pi}{4}$

(d) $\csc \dfrac{\pi}{6} = \sec\left(\dfrac{\pi}{2} - \dfrac{\pi}{6}\right) = \sec \dfrac{\pi}{3}$

EXAMPLE 7   **Using the Complementary Angle Theorem**

Find the exact value of each expression. Do not use a calculator.

(a) $\sec 28° - \csc 62°$

(b) $\dfrac{\sin 35°}{\cos 55°}$

Solution   (a) $\sec 28° - \csc 62° = \csc(90° - 28°) - \csc 62°$

$= \csc 62° - \csc 62° = 0$

(b) $\dfrac{\sin 35°}{\cos 55°} = \dfrac{\cos(90° - 35°)}{\cos 55°} = \dfrac{\cos 55°}{\cos 55°} = 1$

━━━━ **Now Work** PROBLEM 43

Historical Feature

The name *sine* for the sine function is due to a medieval confusion. The name comes from the Sanskrit word *jīva*, (meaning *chord*), first used in India by Araybhata the Elder (AD 510). He really meant half-chord, but abbreviated it. This was brought into Arabic as *jība*, which was meaningless. Because the proper Arabic word *jaib* would be written the same way (short vowels are not written out in Arabic), *jība*, was pronounced as *jaib*, which meant bosom or hollow, and *jaib* remains as the Arabic word for sine to this day. Scholars translating the Arabic works into Latin found that the word *sinus* also meant bosom or hollow, and from *sinus* we get the word *sine*.

The name *tangent*, due to Thomas Finck (1583), can be understood by looking at Figure 26. The line segment $\overline{DC}$ is tangent to the circle at C. If $d(O, B) = d(O, C) = 1$, then the length of the line segment $\overline{DC}$ is

$$d(D, C) = \frac{d(D, C)}{1} = \frac{d(D, C)}{d(O, C)} = \tan \alpha$$

The old name for the tangent is *umbra versa* (meaning turned shadow), referring to the use of the tangent in solving height problems with shadows.

Figure 26

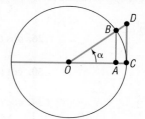

The names of the cofunctions came about as follows. If *A* and *B* are complementary angles, then cos *A* = sin *B*. Because *B* is the complement of *A*, it was natural to write the cosine of *A* as *sin co A*. Probably for reasons involving ease of pronunciation, the *co* migrated to the front, and then cosine received a three-letter abbreviation to match sin, sec, and tan. The two other cofunctions were similarly treated, except that the long forms *cotan* and *cosec* survive to this day in some countries.

5.2 Assess Your Understanding

'Are You Prepared?'   *Answers are given at the end of these exercises. If you get a wrong answer, read the pages listed in red.*

1. In a right triangle with legs $a = 6$ and $b = 10$, the Pythagorean Theorem tells us that the hypotenuse $c =$ _____.
(pp. A14–A15)

2. The value of the function $f(x) = 3x - 7$ at 5 is _____.
(p. 48)

Concepts and Vocabulary

3. Two acute angles whose sum is a right angle are called _____.

4. The sine and _____ functions are cofunctions.

5. $\tan 28° = \cot$ _____.

6. For any angle θ, $\sin^2 \theta + \cos^2 \theta =$ _____.

7. *True or False* $\tan \theta = \dfrac{\sin \theta}{\cos \theta}$.

8. *True or False* $1 + \tan^2 \theta = \csc^2 \theta$.

9. *True or False*   If θ is an acute angle and $\sec \theta = 3$, then $\cos \theta = \dfrac{1}{3}$.

10. *True or False* $\tan \dfrac{\pi}{5} = \cot \dfrac{4\pi}{5}$.

Skill Building

In Problems 11–20, find the value of the six trigonometric functions of the angle θ in each figure.

11.

12.

13.

14.

15.

16.

17.

18.

19.

20.

In Problems 21–24, use identities to find the exact value of each of the four remaining trigonometric functions of the acute angle θ.

21. $\sin\theta = \dfrac{1}{2}$, $\cos\theta = \dfrac{\sqrt{3}}{2}$ **22.** $\sin\theta = \dfrac{\sqrt{3}}{2}$, $\cos\theta = \dfrac{1}{2}$ **23.** $\sin\theta = \dfrac{2}{3}$, $\cos\theta = \dfrac{\sqrt{5}}{3}$ **24.** $\sin\theta = \dfrac{1}{3}$, $\cos\theta = \dfrac{2\sqrt{2}}{3}$

In Problems 25–36, use the definition or identities to find the exact value of each of the remaining five trigonometric functions of the acute angle θ.

25. $\sin\theta = \dfrac{\sqrt{2}}{2}$ **26.** $\cos\theta = \dfrac{\sqrt{2}}{2}$ **27.** $\cos\theta = \dfrac{1}{3}$ **28.** $\sin\theta = \dfrac{\sqrt{3}}{4}$

29. $\tan\theta = \dfrac{1}{2}$ **30.** $\cot\theta = \dfrac{1}{2}$ **31.** $\sec\theta = 3$ **32.** $\csc\theta = 5$

33. $\tan\theta = \sqrt{2}$ **34.** $\sec\theta = \dfrac{5}{3}$ **35.** $\csc\theta = 2$ **36.** $\cot\theta = 2$

In Problems 37–54, use Fundamental Identities and/or the Complementary Angle Theorem to find the exact value of each expression. Do not use a calculator.

37. $\sin^2 20° + \cos^2 20°$ **38.** $\sec^2 28° - \tan^2 28°$ **39.** $\sin 80° \csc 80°$ **40.** $\tan 10° \cot 10°$

41. $\tan 50° - \dfrac{\sin 50°}{\cos 50°}$ **42.** $\cot 25° - \dfrac{\cos 25°}{\sin 25°}$ **43.** $\sin 38° - \cos 52°$ **44.** $\tan 12° - \cot 78°$

45. $\dfrac{\cos 10°}{\sin 80°}$ **46.** $\dfrac{\cos 40°}{\sin 50°}$ **47.** $1 - \cos^2 20° - \cos^2 70°$ **48.** $1 + \tan^2 5° - \csc^2 85°$

49. $\tan 20° - \dfrac{\cos 70°}{\cos 20°}$ **50.** $\cot 40° - \dfrac{\sin 50°}{\sin 40°}$ **51.** $\tan 35° \cdot \sec 55° \cdot \cos 35°$ **52.** $\cot 25° \cdot \csc 65° \cdot \sin 25°$

53. $\cos 35° \sin 55° + \cos 55° \sin 35°$ **54.** $\sec 35° \csc 55° - \tan 35° \cot 55°$

---

55. Given $\sin 30° = \dfrac{1}{2}$, use trigonometric identities to find the exact value of
(a) $\cos 60°$
(b) $\cos^2 30°$
(c) $\csc\dfrac{\pi}{6}$
(d) $\sec\dfrac{\pi}{3}$

56. Given $\sin 60° = \dfrac{\sqrt{3}}{2}$, use trigonometric identities to find the exact value of
(a) $\cos 30°$
(b) $\cos^2 60°$
(c) $\sec\dfrac{\pi}{6}$
(d) $\csc\dfrac{\pi}{3}$

57. Given $\tan\theta = 4$, use trigonometric identities to find the exact value of
(a) $\sec^2\theta$
(b) $\cot\theta$
(c) $\cot\left(\dfrac{\pi}{2} - \theta\right)$
(d) $\csc^2\theta$

58. Given $\sec\theta = 3$, use trigonometric identities to find the exact value of
(a) $\cos\theta$ (b) $\tan^2\theta$ (c) $\csc(90° - \theta)$ (d) $\sin^2\theta$

59. Given $\csc\theta = 4$, use trigonometric identities to find the exact value of
(a) $\sin\theta$ (b) $\cot^2\theta$ (c) $\sec(90° - \theta)$ (d) $\sec^2\theta$

60. Given $\cot\theta = 2$, use trigonometric identities to find the exact value of
(a) $\tan\theta$
(b) $\csc^2\theta$
(c) $\tan\left(\dfrac{\pi}{2} - \theta\right)$
(d) $\sec^2\theta$

61. Given the approximation $\sin 38° \approx 0.62$, use trigonometric identities to find the approximate value of
(a) $\cos 38°$
(b) $\tan 38°$
(c) $\cot 38°$
(d) $\sec 38°$
(e) $\csc 38°$
(f) $\sin 52°$
(g) $\cos 52°$
(h) $\tan 52°$

62. Given the approximation $\cos 21° \approx 0.93$, use trigonometric identities to find the approximate value of
(a) $\sin 21°$ (b) $\tan 21°$
(c) $\cot 21°$ (d) $\sec 21°$
(e) $\csc 21°$ (f) $\sin 69°$
(g) $\cos 69°$ (h) $\tan 69°$

63. If $\sin \theta = 0.3$, find the exact value of $\sin \theta + \cos\left(\dfrac{\pi}{2} - \theta\right)$.

64. If $\tan \theta = 4$, find the exact value of $\tan \theta + \tan\left(\dfrac{\pi}{2} - \theta\right)$.

65. Find an acute angle θ that satisfies the equation
$$\sin \theta = \cos(2\theta + 30°)$$

66. Find an acute angle θ that satisfies the equation
$$\tan \theta = \cot(\theta + 45°)$$

Applications and Extensions

67. Calculating the Time of a Trip From a parking lot you want to walk to a house on the ocean. The house is located 1500 feet down a paved path that parallels the beach, which is 500 feet wide. Along the path you can walk 300 feet per minute, but in the sand on the beach you can only walk 100 feet per minute. See the illustration.

(a) Calculate the time T if you walk 1500 feet along the paved path and then 500 feet in the sand to the house.
(b) Calculate the time T if you walk in the sand directly toward the ocean for 500 feet and then turn left and walk along the beach for 1500 feet to the house.
(c) Express the time T to get from the parking lot to the beachhouse as a function of the angle θ shown in the illustration.
(d) Calculate the time T if you walk directly from the parking lot to the house.

$$\left[\textbf{Hint: } \tan \theta = \frac{500}{1500}\right]$$

(e) Calculate the time T if you walk 1000 feet along the paved path and then walk directly to the house.
(f) Graph $T = T(\theta)$. For what angle θ is T least? What is x for this angle? What is the minimum time?
(g) Explain why $\tan \theta = \dfrac{1}{3}$ gives the smallest angle θ that is possible.

68. Carrying a Ladder around a Corner Two hallways, one of width 3 feet, the other of width 4 feet, meet at a right angle. See the illustration.
(a) Express the length L of the line segment shown as a function of the angle θ.
(b) Discuss why the length of the longest ladder that can be carried around the corner is equal to the smallest value of L.

69. Electrical Engineering A resistor and an inductor connected in a series network impede the flow of an alternating current. This impedance Z is determined by the reactance X of the inductor and the resistance R of the resistor. The three quantities, all measured in ohms, can be represented by the sides of a right triangle as illustrated, so $Z^2 = X^2 + R^2$. The angle ϕ is called the **phase angle.** Suppose a series network has an inductive reactance of $X = 400$ ohms and a resistance of $R = 600$ ohms.
(a) Find the impedance Z.
(b) Find the values of the six trigonometric functions of the phase angle ϕ.

70. Electrical Engineering Refer to Problem 69. A series network has a resistance of $R = 588$ ohms. The phase angle ϕ is such that $\tan \phi = \dfrac{5}{12}$.
(a) Determine the inductive reactance X and the impedance Z.
(b) Determine the values of the remaining five trigonometric functions of the phase angle ϕ.

71. Geometry Suppose that the angle θ is a central angle of a circle of radius 1 (see the figure). Show that
(a) Angle $OAC = \dfrac{\theta}{2}$
(b) $|CD| = \sin \theta$ and $|OD| = \cos \theta$
(c) $\tan \dfrac{\theta}{2} = \dfrac{\sin \theta}{1 + \cos \theta}$

72. Geometry Show that the area A of an isosceles triangle is $A = a^2 \sin \theta \cos \theta$, where a is the length of one of the two equal sides and θ is the measure of one of the two equal angles (see the figure).

73. Geometry Let $n \geq 1$ be any real number and let θ be any angle for which $0 < n\theta < \dfrac{\pi}{2}$. Then we can draw a triangle with the angles θ and $n\theta$ and included side of length 1 (do you see why?) and place it on the unit circle as illustrated.

Now, drop the perpendicular from C to $D = (x, 0)$ and show that

$$x = \frac{\tan(n\theta)}{\tan\theta + \tan(n\theta)}$$

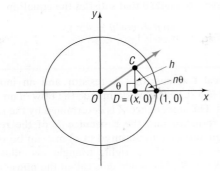

74. Geometry Refer to the figure. The smaller circle, whose radius is a, is tangent to the larger circle, whose radius is b. The ray $\overrightarrow{OA}$ contains a diameter of each circle, and the ray $\overrightarrow{OB}$ is tangent to each circle. Show that

$$\cos\theta = \frac{\sqrt{ab}}{\dfrac{a+b}{2}}$$

(This shows that $\cos\theta$ equals the ratio of the geometric mean of a and b to the arithmetic mean of a and b.)

[Hint: First show that $\sin\theta = (b - a)/(b + a)$.**]**

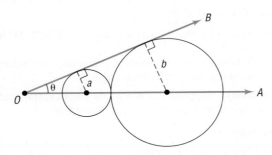

75. Geometry Refer to the figure. If $|OA| = 1$, show that

(a) Area $\triangle OAC = \dfrac{1}{2}\sin\alpha\cos\alpha$

(b) Area $\triangle OCB = \dfrac{1}{2}|OB|^2\sin\beta\cos\beta$

(c) Area $\triangle OAB = \dfrac{1}{2}|OB|\sin(\alpha + \beta)$

(d) $|OB| = \dfrac{\cos\alpha}{\cos\beta}$

(e) $\sin(\alpha + \beta) = \sin\alpha\cos\beta + \cos\alpha\sin\beta$

[Hint: Area $\triangle OAB = $ Area $\triangle OAC + $ Area $\triangle OCB$**]**

76. Geometry Refer to the figure, where a unit circle is drawn. The line segment $\overline{DB}$ is tangent to the circle.

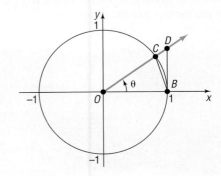

(a) Express the area of $\triangle OBC$ in terms of $\sin\theta$ and $\cos\theta$.

   [Hint: Use the altitude from C to the base $\overline{OB} = 1$.**]**

(b) Express the area of $\triangle OBD$ in terms of $\sin\theta$ and $\cos\theta$.

(c) The area of the sector OBC of the circle is $\dfrac{1}{2}\theta$, where θ is measured in radians. Use the results of parts (a) and (b) and the fact that

$$\text{Area } \triangle OBC < \text{Area of sector } OBC < \text{Area } \triangle OBD$$

to show that

$$1 < \frac{\theta}{\sin\theta} < \frac{1}{\cos\theta}$$

77. If $\cos\alpha = \tan\beta$ and $\cos\beta = \tan\alpha$, where α and β are acute angles, show that

$$\sin\alpha = \sin\beta = \sqrt{\frac{3 - \sqrt{5}}{2}}$$

78. If θ is an acute angle and $\tan\theta = x$, $x \neq 0$, express the remaining five trigonometric functions in terms of x.

Explaining Concepts: Discussion and Writing

79. If θ is an acute angle, explain why $\sec\theta > 1$.

80. If θ is an acute angle, explain why $0 < \sin\theta < 1$.

81. How would you explain the meaning of the sine function to a fellow student who has just completed College Algebra?

82. Look back at Example 5. Which of the two solutions do you prefer? Explain your reasoning.

'Are You Prepared?' Answers

1. $2\sqrt{34}$ **2.** $f(5) = 8$

5.3 Computing the Values of Trigonometric Functions of Acute Angles

In the previous section, we developed ways to find the value of each trigonometric function of an acute angle when one of the functions is known. In this section, we discuss the problem of finding the value of each trigonometric function of an acute angle when the angle is given.

For three special acute angles, we can use some results from plane geometry to find the exact value of each of the six trigonometric functions.

1 Find the Exact Values of the Trigonometric Functions of $\frac{\pi}{4} = 45°$

EXAMPLE 1   **Finding the Exact Values of the Trigonometric Functions of $\frac{\pi}{4} = 45°$**

Find the exact values of the six trigonometric functions of $\frac{\pi}{4} = 45°$.

Solution   Using the right triangle in Figure 27(a), in which one of the angles is $\frac{\pi}{4} = 45°$, it follows that the other acute angle is also $\frac{\pi}{4} = 45°$, and hence the triangle is isosceles. As a result, side a and side b are equal in length. Since the values of the trigonometric functions of an angle depend only on the angle and not on the size of the triangle, we may assign any values to a and b for which $a = b > 0$. We decide to use the triangle for which

$$a = b = 1$$

Then, by the Pythagorean Theorem,

$$c^2 = a^2 + b^2 = 1 + 1 = 2$$
$$c = \sqrt{2}$$

As a result, we have the triangle in Figure 27(b), from which we find

$$\sin\frac{\pi}{4} = \sin 45° = \frac{b}{c} = \frac{1}{\sqrt{2}} = \frac{\sqrt{2}}{2} \qquad \cos\frac{\pi}{4} = \cos 45° = \frac{a}{c} = \frac{1}{\sqrt{2}} = \frac{\sqrt{2}}{2}$$

Figure 27

(a)

(b)

Using Quotient and Reciprocal Identities, we find

$$\tan \frac{\pi}{4} = \tan 45° = \frac{\sin 45°}{\cos 45°} = \frac{\frac{\sqrt{2}}{2}}{\frac{\sqrt{2}}{2}} = 1 \qquad \cot \frac{\pi}{4} = \cot 45° = \frac{1}{\tan 45°} = \frac{1}{1} = 1$$

$$\sec \frac{\pi}{4} = \sec 45° = \frac{1}{\cos 45°} = \frac{1}{\frac{1}{\sqrt{2}}} = \sqrt{2} \qquad \csc \frac{\pi}{4} = \csc 45° = \frac{1}{\sin 45°} = \frac{1}{\frac{1}{\sqrt{2}}} = \sqrt{2}$$

●

EXAMPLE 2   **Finding the Exact Value of a Trigonometric Expression**

Find the exact value of each expression.

(a) $(\sin 45°)(\tan 45°)$

(b) $\left(\sec \frac{\pi}{4} \right)\left(\cot \frac{\pi}{4} \right)$

Solution   We use the results obtained in Example 1.

(a) $(\sin 45°)(\tan 45°) = \frac{\sqrt{2}}{2} \cdot 1 = \frac{\sqrt{2}}{2}$

(b) $\left(\sec \frac{\pi}{4} \right)\left(\cot \frac{\pi}{4} \right) = \sqrt{2} \cdot 1 = \sqrt{2}$

●

━━━ **Now Work**   PROBLEMS 5 AND 17

---

2 Find the Exact Values of the Trigonometric Functions
of $\frac{\pi}{6} = 30°$ and $\frac{\pi}{3} = 60°$

EXAMPLE 3   **Finding the Exact Values of the Trigonometric Functions**
of $\frac{\pi}{6} = 30°$ and $\frac{\pi}{3} = 60°$

Find the exact values of the six trigonometric functions of $\frac{\pi}{6} = 30°$ and $\frac{\pi}{3} = 60°$.

Solution   Form a right triangle in which one of the angles is $\frac{\pi}{6} = 30°$. It then follows that the third angle is $\frac{\pi}{3} = 60°$. Figure 28(a) illustrates such a triangle with hypotenuse of length 2. Our problem is to determine a and b.

We begin by placing next to the triangle in Figure 28(a) another triangle congruent to the first, as shown in Figure 28(b). Notice that we now have a triangle whose angles are each 60°. This triangle is therefore equilateral, so each side is of length 2. In particular, the base is $2a = 2$, so $a = 1$. By the Pythagorean Theorem, b satisfies the equation $a^2 + b^2 = c^2$, so we have

$$a^2 + b^2 = c^2$$
$$1^2 + b^2 = 2^2 \qquad a = 1, c = 2$$
$$b^2 = 4 - 1 = 3$$
$$b = \sqrt{3}$$

Figure 28

(a)

(b)

(c)

Using the triangle in Figure 28(c) and the fact that $\dfrac{\pi}{6} = 30°$ and $\dfrac{\pi}{3} = 60°$ are complementary angles, we find

$$\sin\frac{\pi}{6} = \sin 30° = \frac{\text{opposite}}{\text{hypotenuse}} = \frac{1}{2} \qquad\qquad \cos\frac{\pi}{3} = \cos 60° = \frac{1}{2}$$

$$\cos\frac{\pi}{6} = \cos 30° = \frac{\text{adjacent}}{\text{hypotenuse}} = \frac{\sqrt{3}}{2} \qquad\qquad \sin\frac{\pi}{3} = \sin 60° = \frac{\sqrt{3}}{2}$$

$$\tan\frac{\pi}{6} = \tan 30° = \frac{\sin 30°}{\cos 30°} = \frac{\dfrac{1}{2}}{\dfrac{\sqrt{3}}{2}} = \frac{1}{\sqrt{3}} = \frac{\sqrt{3}}{3} \qquad\qquad \cot\frac{\pi}{3} = \cot 60° = \frac{\sqrt{3}}{3}$$

$$\csc\frac{\pi}{6} = \csc 30° = \frac{1}{\sin 30°} = \frac{1}{\dfrac{1}{2}} = 2 \qquad\qquad \sec\frac{\pi}{3} = \sec 60° = 2$$

$$\sec\frac{\pi}{6} = \sec 30° = \frac{1}{\cos 30°} = \frac{1}{\dfrac{\sqrt{3}}{2}} = \frac{2}{\sqrt{3}} = \frac{2\sqrt{3}}{3} \qquad\qquad \csc\frac{\pi}{3} = \csc 60° = \frac{2\sqrt{3}}{3}$$

$$\cot\frac{\pi}{6} = \cot 30° = \frac{1}{\tan 30°} = \frac{1}{\dfrac{\sqrt{3}}{3}} = \frac{3}{\sqrt{3}} = \sqrt{3} \qquad\qquad \tan\frac{\pi}{3} = \tan 60° = \sqrt{3}$$

Table 3 summarizes the information just derived for the angles $\dfrac{\pi}{6} = 30°$, $\dfrac{\pi}{4} = 45°$, and $\dfrac{\pi}{3} = 60°$. Rather than memorize the entries in Table 3, you can draw the appropriate triangle to determine the values given in the table.

Table 3

θ (Radians)	θ (Degrees)	$\sin\theta$	$\cos\theta$	$\tan\theta$	$\csc\theta$	$\sec\theta$	$\cot\theta$
$\dfrac{\pi}{6}$	30°	$\dfrac{1}{2}$	$\dfrac{\sqrt{3}}{2}$	$\dfrac{\sqrt{3}}{3}$	2	$\dfrac{2\sqrt{3}}{3}$	$\sqrt{3}$
$\dfrac{\pi}{4}$	45°	$\dfrac{\sqrt{2}}{2}$	$\dfrac{\sqrt{2}}{2}$	1	$\sqrt{2}$	$\sqrt{2}$	1
$\dfrac{\pi}{3}$	60°	$\dfrac{\sqrt{3}}{2}$	$\dfrac{1}{2}$	$\sqrt{3}$	$\dfrac{2\sqrt{3}}{3}$	2	$\dfrac{\sqrt{3}}{3}$

EXAMPLE 4   **Finding the Exact Value of a Trigonometric Expression**

Find the exact value of each expression.

(a) $\sin 45° \cos 30°$ (b) $\tan\dfrac{\pi}{4} - \sin\dfrac{\pi}{3}$ (c) $\tan^2\dfrac{\pi}{6} + \sin^2\dfrac{\pi}{4}$

Solution   (a) $\sin 45° \cos 30° = \dfrac{\sqrt{2}}{2} \cdot \dfrac{\sqrt{3}}{2} = \dfrac{\sqrt{6}}{4}$

(b) $\tan\dfrac{\pi}{4} - \sin\dfrac{\pi}{3} = 1 - \dfrac{\sqrt{3}}{2} = \dfrac{2 - \sqrt{3}}{2}$

(c) $\tan^2\dfrac{\pi}{6} + \sin^2\dfrac{\pi}{4} = \left(\dfrac{\sqrt{3}}{3}\right)^2 + \left(\dfrac{\sqrt{2}}{2}\right)^2 = \dfrac{1}{3} + \dfrac{1}{2} = \dfrac{5}{6}$

━━━━ **Now Work** PROBLEMS 9 AND 19

3 Use a Calculator to Approximate the Values of Trigonometric Functions of Acute Angles

Before getting started, you must first decide whether to enter the angle in the calculator using radians or degrees and then set the calculator to the correct MODE. (Check your instruction manual to find out how your calculator handles degrees and radians.) Your calculator has the keys marked $\boxed{\text{sin}}$, $\boxed{\text{cos}}$, and $\boxed{\text{tan}}$. To find the values of the remaining three trigonometric functions (secant, cosecant, and cotangent), we use the reciprocal identities.

WARNING On your calculator the second functions $\sin^{-1}$, $\cos^{-1}$, and $\tan^{-1}$ do not represent the reciprocal of sin, cos, and tan. ∎

$$\sec \theta = \frac{1}{\cos \theta} \qquad \csc \theta = \frac{1}{\sin \theta} \qquad \cot \theta = \frac{1}{\tan \theta}$$

EXAMPLE 5

Using a Calculator to Approximate the Value of Trigonometric Functions

Use a calculator to find the approximate value of:

(a) $\cos 48°$ (b) $\csc 21°$ (c) $\tan \dfrac{\pi}{12}$

Express your answer rounded to two decimal places.

Solution

(a) First, we set the MODE to receive degrees. Rounded to two decimal places,

$$\cos 48° = 0.67$$

(b) Most calculators do not have a csc key. The manufacturers assume the user knows some trigonometry. To find the value of $\csc 21°$, we use the fact that $\csc 21° = \dfrac{1}{\sin 21°}$. Rounded to two decimal places,

$$\csc 21° = 2.79$$

Figure 29

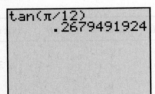

(c) Set the MODE to receive radians. Figure 29 shows the solution using a TI-84 Plus graphing calculator. Rounded to two decimal places,

$$\tan \frac{\pi}{12} = 0.27$$

➤ **Now Work** PROBLEM 29

---

4 Model and Solve Applied Problems Involving Right Triangles

Right triangles can be used to model many types of situations, such as the optimal design of a rain gutter.*

EXAMPLE 6

Constructing a Rain Gutter

A rain gutter is to be constructed of aluminum sheets 12 inches wide. After marking off a length of 4 inches from each edge, the sides are bent up at an angle θ. See Figure 30.

Figure 30

(a) Express the area A of the opening as a function of θ.

[**Hint:** Let b denote the vertical height of the bend.]

 (b) Find the area A of the opening for $\theta = 30°$, $\theta = 45°$, $\theta = 60°$, and $\theta = 75°$.

(c) Graph $A = A(\theta)$. Find the angle θ that makes A largest. (This bend will allow the most water to flow through the gutter.)

*In applied problems, it is important that answers be reported with both justifiable accuracy and appropriate significant figures. Throughout the text, we shall assume that angles are measured to the nearest tenth and sides are measured to the nearest hundredth, resulting in sides being rounded to two decimal places and angles being rounded to one decimal place.

Solution   (a) Look again at Figure 30. The area A of the opening is the sum of the areas of two congruent right triangles and one rectangle. Look at Figure 31, which shows the triangle on the right in Figure 30 redrawn. We see that

Figure 31

$$\cos \theta = \frac{a}{4}, \quad \text{so} \quad a = 4 \cos \theta \qquad \sin \theta = \frac{b}{4}, \quad \text{so} \quad b = 4 \sin \theta$$

The area of the triangle is

$$\text{area of triangle} = \frac{1}{2}(\text{base})(\text{height}) = \frac{1}{2}ab = \frac{1}{2}(4 \cos \theta)(4 \sin \theta) = 8 \sin \theta \cos \theta$$

So the area of the two congruent triangles is $16 \sin \theta \cos \theta$.
The rectangle has length 4 and height b, so its area is

$$\text{area of rectangle} = 4b = 4(4 \sin \theta) = 16 \sin \theta$$

The area A of the opening is

$$A = \text{area of the two triangles} + \text{area of the rectangle}$$
$$A(\theta) = 16 \sin \theta \cos \theta + 16 \sin \theta = 16 \sin \theta(\cos \theta + 1)$$

(b) For $\theta = 30°$: $A(30°) = 16 \sin 30°(\cos 30° + 1)$

$$= 16\left(\frac{1}{2}\right)\left(\frac{\sqrt{3}}{2} + 1\right) = 4\sqrt{3} + 8 \approx 14.93$$

The area of the opening for $\theta = 30°$ is about 14.93 square inches.

For $\theta = 45°$: $A(45°) = 16 \sin 45°(\cos 45° + 1)$

$$= 16\left(\frac{\sqrt{2}}{2}\right)\left(\frac{\sqrt{2}}{2} + 1\right) = 8 + 8\sqrt{2} \approx 19.31$$

The area of the opening for $\theta = 45°$ is about 19.31 square inches.

For $\theta = 60°$: $A(60°) = 16 \sin 60°(\cos 60° + 1)$

$$= 16\left(\frac{\sqrt{3}}{2}\right)\left(\frac{1}{2} + 1\right) = 12\sqrt{3} \approx 20.78$$

The area of the opening for $\theta = 60°$ is about 20.78 square inches.

Figure 32

```
 24
  │        ⁕
  │    ⁀⁀⁀⁀⁀⁀⁀
  │  ⁀            
0°│⁀                90°
  Maximum
  X=60.000002  Y=20.78461
 -8
```

For $\theta = 75°$: $A(75°) = 16 \sin 75°(\cos 75° + 1) \approx 19.45$

The area of the opening for $\theta = 75°$ is about 19.45 square inches.

(c) Figure 32 shows the graph of $A = A(\theta)$. Using MAXIMUM, the angle θ that makes A largest is $60°$.

───── **Now Work** PROBLEM 61

In addition to developing models using right triangles, we can use right triangle trigonometry to measure heights and distances that are either awkward or impossible to measure by ordinary means. When using right triangles to solve these problems, pay attention to the known measures. This will indicate what trigonometric function to use. For example, if we know the measure of an angle and the length of the side adjacent to the angle, and we wish to find the length of the opposite side, we would use the tangent function. Do you know why?

EXAMPLE 7   **Finding the Width of a River**

A surveyor can measure the width of a river by setting up a transit at a point C on one side of the river and taking a sighting of a point A on the other side. Refer to Figure 33. After turning through an angle of 90° at C, the surveyor walks a distance of 200 meters to point B. Using the transit at B, the angle θ is measured and found to be 20°. What is the width of the river rounded to the nearest meter?

Figure 33

Solution   We seek the length of side b. We know a and θ. So we use the fact that b is opposite θ and a is adjacent to θ and write

$$\tan \theta = \frac{b}{a}$$

which leads to

$$\tan 20° = \frac{b}{200}$$

$$b = 200 \tan 20° \approx 72.79 \text{ meters}$$

The width of the river is 73 meters, rounded to the nearest meter.

➤ **Now Work** PROBLEM 69

Vertical heights can sometimes be measured using either the *angle of elevation* or the *angle of depression*. If a person is looking up at an object, the acute angle measured from the horizontal to a line of sight to the object is called the **angle of elevation.** See Figure 34(a).

Figure 34

(a)   (b)

If a person is standing on a cliff looking down at an object, the acute angle made by the line of sight to the object and the horizontal is called the **angle of depression.** See Figure 34(b).

EXAMPLE 8   Finding the Height of a Cloud

Meteorologists find the height of a cloud using an instrument called a **ceilometer.** A ceilometer consists of a **light projector** that directs a vertical light beam up to the cloud base and a **light detector** that scans the cloud to detect the light beam. See Figure 35(a). On December 8, 2010, at Midway Airport in Chicago, a ceilometer was employed to find the height of the cloud cover. It was set up with its light detector 300 feet from its light projector. If the angle of elevation from the light detector to the base of the cloud was 75°, what was the height of the cloud cover?

Figure 35

(a)   (b)

Solution   Figure 35(b) illustrates the situation. To find the height h, we use the fact that $\tan 75° = \dfrac{h}{300}$, so

$$h = 300 \tan 75° \approx 1120 \text{ feet}$$

The ceiling (height to the base of the cloud cover) was approximately 1120 feet.

Now Work   PROBLEM 71

The idea behind Example 8 can also be used to find the height of an object with a base that is not accessible to the horizontal.

EXAMPLE 9   Finding the Height of a Statue on a Building

Adorning the top of the Board of Trade building in Chicago is a statue of Ceres, the Roman goddess of wheat. From street level, two observations are taken 400 feet from the center of the building. The angle of elevation to the base of the statue is found to be 55.1°, and the angle of elevation to the top of the statue is 56.5°. See Figure 36(a). What is the height of the statue?

Figure 36

(a)   (b)

Solution   Figure 36(b) shows two triangles that replicate Figure 36(a). The height of the statue of Ceres will be $b' - b$. To find b and b', we refer to Figure 36(b).

$$\tan 55.1° = \frac{b}{400} \qquad\qquad \tan 56.5° = \frac{b'}{400}$$

$$b = 400 \tan 55.1° \approx 573.39 \qquad b' = 400 \tan 56.5° \approx 604.33$$

The height of the statue is approximately $604.33 - 573.39 = 30.94$ feet ≈ 31 feet.

●

---- **Now Work** PROBLEM **77**

5.3 Assess Your Understanding

Concepts and Vocabulary

1. $\tan \dfrac{\pi}{4} + \sin 30° =$ _____.

2. Using a calculator, $\sin 2 =$ _____, rounded to two decimal places.

3. *True or False* Exact values can be found for the trigono-metric functions of 60°.

4. *True or False* Exact values can be found for the sine of any angle.

Skill Building

5. Write down the exact value of each of the six trigonometric functions of 45°.

6. Write down the exact value of each of the six trigonometric functions of 30° and of 60°.

In Problems 7–16, $f(\theta) = \sin \theta$ and $g(\theta) = \cos \theta$. Find the exact value of each expression if $\theta = 60°$. Do not use a calculator.

7. $f(\theta)$

8. $g(\theta)$

9. $f\left(\dfrac{\theta}{2}\right)$

10. $g\left(\dfrac{\theta}{2}\right)$

11. $[f(\theta)]^2$

12. $[g(\theta)]^2$

13. $2f(\theta)$

14. $2g(\theta)$

15. $\dfrac{f(\theta)}{2}$

16. $\dfrac{g(\theta)}{2}$

In Problems 17–28, find the exact value of each expression. Do not use a calculator.

17. $4 \cos 45° - 2 \sin 45°$

18. $2 \sin 45° + 4 \cos 30°$

19. $6 \tan 45° - 8 \cos 60°$

20. $\sin 30° \cdot \tan 60°$

21. $\sec \dfrac{\pi}{4} + 2 \csc \dfrac{\pi}{3}$

22. $\tan \dfrac{\pi}{4} + \cot \dfrac{\pi}{4}$

23. $\sec^2 \dfrac{\pi}{6} - 4$

24. $4 + \tan^2 \dfrac{\pi}{3}$

25. $\sin^2 30° + \cos^2 60°$

26. $\sec^2 60° - \tan^2 45°$

27. $1 - \cos^2 30° - \cos^2 60°$

28. $1 + \tan^2 30° - \csc^2 45°$

In Problems 29–46, use a calculator to find the approximate value of each expression. Round the answer to two decimal places.

29. $\sin 28°$

30. $\cos 14°$

31. $\tan 21°$

32. $\cot 70°$

33. $\sec 41°$

34. $\csc 55°$

35. $\sin \dfrac{\pi}{10}$

36. $\cos \dfrac{\pi}{8}$

37. $\tan \dfrac{5\pi}{12}$

38. $\cot \dfrac{\pi}{18}$

39. $\sec \dfrac{\pi}{12}$

40. $\csc \dfrac{5\pi}{13}$

41. $\sin 1$

42. $\tan 1$

43. $\sin 1°$

44. $\tan 1°$

45. $\tan 0.3$

46. $\tan 0.1$

Mixed Practice

In Problems 47–56, $f(x) = \sin x$, $g(x) = \cos x$, $h(x) = 2x$, and $p(x) = \dfrac{x}{2}$. Find the value of each of the following:

47. $(f + g)(30°)$

48. $(f - g)(60°)$

49. $(f \cdot g)\left(\dfrac{\pi}{4}\right)$

50. $(f \cdot g)\left(\dfrac{\pi}{3}\right)$

51. $(f \circ h)\left(\dfrac{\pi}{6}\right)$

52. $(g \circ p)(60°)$

53. $(p \circ g)(45°)$

54. $(h \circ f)\left(\dfrac{\pi}{6}\right)$

55. (a) Find $f\left(\dfrac{\pi}{4}\right)$. What point is on the graph of f?

(b) Assuming f is one-to-one, use the result of part (a), to find a point on the graph of f^{-1}.

(c) What point is on the graph of $y = f\left(x + \dfrac{\pi}{4}\right) - 3$ if $x = \dfrac{\pi}{4}$?

56. (a) Find $g\left(\dfrac{\pi}{6}\right)$. What point is on the graph of g?

(b) Assuming g is one-to-one, use the result of part (a), to find a point on the graph of g^{-1}.

(c) What point is on the graph of $y = 2g\left(x - \dfrac{\pi}{6}\right)$ if $x = \dfrac{\pi}{6}$?

Applications and Extensions

Problems 57–60 require the following discussion.

Projectile Motion   The path of a projectile fired at an inclination θ to the horizontal with initial speed v_0 is a parabola. See the figure. The **range R** of the projectile, that is, the horizontal distance that the projectile travels, is found by using the function

$$R(\theta) = \frac{2v_0^2 \sin\theta \cos\theta}{g}$$

where $g \approx 32.2$ feet per second per second ≈ 9.8 meters per second per second is the acceleration due to gravity. The maximum height H of the projectile is given by the function

$$H(\theta) = \frac{v_0^2 \sin^2\theta}{2g}$$

In Problems 57– 60, find the range R and maximum height H of the projectile. Round answers to two decimal places.

57. The projectile is fired at an angle of 45° to the horizontal with an initial speed of 100 feet per second.

58. The projectile is fired at an angle of 30° to the horizontal with an initial speed of 150 meters per second.

59. The projectile is fired at an angle of 25° to the horizontal with an initial speed of 500 meters per second.

60. The projectile is fired at an angle of 50° to the horizontal with an initial speed of 200 feet per second.

61. Inclined Plane   See the illustration. If friction is ignored, the time t (in seconds) required for a block to slide down an inclined plane is modeled by the function

$$t(\theta) = \sqrt{\frac{2a}{g \sin\theta \cos\theta}}$$

where a is the length (in feet) of the base and $g \approx 32$ feet per second per second is the acceleration due to gravity. How long does it take a block to slide down an inclined plane with base $a = 10$ feet when

(a) $\theta = 30°$?   **(b)** $\theta = 45°$?

(c) $\theta = 60°$?

62. Piston Engines   See the illustration. In a certain piston engine, the distance x (in inches) from the center of the drive shaft to the head of the piston is modeled by the function

$$x(\theta) = \cos\theta + \sqrt{16 + 0.5(2\cos^2\theta - 1)}$$

where θ is the angle between the crank and the path of the piston head. Find x when $\theta = 30°$ and when $\theta = 45°$.

63. Calculating the Time of a Trip   Two oceanfront homes are located 8 miles apart on a straight stretch of beach, each a distance of 1 mile from a paved path that parallels the ocean. Sally can jog 8 miles per hour along the paved path, but only 3 miles per hour in the sand on the beach. Because a river flows between the two houses, it is necessary to jog on the sand to the path, continue on the path, and then jog on the sand to get from one house to the other. See the illustration.

(a) Express the time T to get from one house to the other as a function of the angle θ shown in the illustration.

(b) Calculate the time T for $\theta = 30°$. How long is Sally on the paved path?

(c) Calculate the time T for $\theta = 45°$. How long is Sally on the paved path?

(d) Calculate the time T for $\theta = 60°$. How long is Sally on the paved path?

(e) Calculate the time T for $\theta = 90°$. Describe the route taken.

(f) Calculate the time T for $\tan \theta = \dfrac{1}{4}$. Describe the route taken. Explain why θ must be larger than 14°.

(g) Graph $T = T(\theta)$. What angle θ results in the least time? What is the least time? How long is Sally on the paved path?

64. **Designing Fine Decorative Pieces**   A designer of decorative art plans to market solid gold spheres encased in clear crystal cones. Each sphere is of fixed radius R and will be enclosed in a cone of height h and radius r. See the illustration. Many cones can be used to enclose the sphere, each having a different slant angle θ.

(a) Express the volume V of the cone as a function of the slant angle θ of the cone.

   [**Hint:** The volume V of a cone of height h and radius r is $V = \dfrac{1}{3}\pi r^2 h$.]

(b) What volume V is required to enclose a sphere of radius 2 centimeters in a cone whose slant angle θ is 30°? 45°? 60°?

(c) What slant angle θ should be used for the volume V of the cone to be a minimum? (This choice minimizes the amount of crystal required and gives maximum emphasis to the gold sphere.)

65. **Geometry**   A right triangle has a hypotenuse of length 8 inches. If one angle is 35°, find the length of each leg.

66. **Geometry**   A right triangle has a hypotenuse of length 10 centimeters. If one angle is 40°, find the length of each leg.

67. **Geometry**   A right triangle contains a 25° angle.
(a) If one leg is of length 5 inches, what is the length of the hypotenuse?
(b) There are two answers. How is this possible?

68. **Geometry**   A right triangle contains an angle of $\dfrac{\pi}{8}$ radian.
(a) If one leg is of length 3 meters, what is the length of the hypotenuse?
(b) There are two answers. How is this possible?

69. **Finding the Width of a Gorge**   Find the distance from A to C across the gorge illustrated in the figure.

70. **Finding the Distance across a Pond**   Find the distance from A to C across the pond illustrated in the figure.

71. **The Eiffel Tower**   The tallest tower built before the era of television masts, the Eiffel Tower was completed on March 31, 1889. Find the height of the Eiffel Tower (before a television mast was added to the top) using the information given in the illustration.

72. **Finding the Distance of a Ship from Shore**   A person in a small boat, offshore from a vertical cliff known to be 100 feet in height, takes a sighting of the top of the cliff. If the angle of elevation is found to be 30°, how far offshore is the boat?

73. **Finding the Distance to a Plateau**   Suppose that you are headed toward a plateau 50 meters high. If the angle of elevation to the top of the plateau is 60°, how far are you from the base of the plateau?

74. **Finding the Reach of a Ladder**   A 22-foot extension ladder leaning against a building makes a 70° angle with the ground. How far up the building does the ladder touch?

75. **Finding the Distance between Two Objects**   A blimp flying at an altitude of 500 feet, lies directly over a line from Soldier Field to the Adler Planetarium on Lake Michigan (see the figure). If the angle of depression from the blimp to the stadium is 32° and from the blimp to the planetarium is 23°, find the distance between Soldier Field and the Adler Planetarium.

76. Hot-air Balloon While taking a ride in a hot-air balloon in Napa Valley, Francisco wonders how high he is. To find out, he chooses a landmark that is to the east of the balloon and measures the angle of depression to be 54°. A few minutes later, after traveling 100 feet east, the angle of depression to the same landmark is determined to be 61°. Use this information to determine the height of the balloon.

77. Mt. Rushmore To measure the height of Lincoln's caricature on Mt. Rushmore, two sightings 800 feet from the base of the mountain are taken. If the angle of elevation to the bottom of Lincoln's face is 32° and the angle of elevation to the top is 35°, what is the height of Lincoln's face?

78. The CN Tower The CN Tower, located in Toronto, Canada, is the tallest structure in the Americas. While visiting Toronto, a tourist wondered what the height of the tower above the top of the Sky Pod is. While standing 4000 feet from the tower, she measured the angle to the top of the Sky Pod to be 20.1°. At this same distance, the angle of elevation to the top of the tower was found to be 24.4°. Use this information to determine the height of the tower above the Sky Pod.

79. Finding the Length of a Guy Wire A radio transmission tower is 200 feet high. How long should a guy wire be if it is to be attached to the tower 10 feet from the top and is to make an angle of 45° with the ground?

80. Finding the Height of a Tower A guy wire 80 feet long is attached to the top of a radio transmission tower, making an angle of 45° with the ground. How high is the tower?

81. Washington Monument The angle of elevation of the Sun is 35.1° at the instant the shadow cast by the Washington Monument is 789 feet long. Use this information to calculate the height of the monument.

82. Finding the Length of a Mountain Trail A straight trail with an inclination of 17° leads from a hotel at an elevation of 9000 feet to a mountain lake at an elevation of 11,200 feet. What is the length of the trail?

83. Constructing a Highway A highway whose primary directions are north–south is being constructed along the west coast

of Florida. Near Naples, a bay obstructs the straight path of the road. Since the cost of a bridge is prohibitive, engineers decide to go around the bay. The illustration shows the path that they decide on and the measurements taken. What is the length of highway needed to go around the bay?

84. Photography A camera is mounted on a tripod 4 feet high at a distance of 10 feet from George, who is 6 feet tall. See the illustration. If the camera lens has angles of depression and elevation of 20°, will George's feet and head be seen by the lens? If not, how far back will the camera need to be moved to include George's feet and head?

85. Calculating Pool Shots A pool player located at X wants to shoot the white ball off the top cushion and hit the red ball dead center. He knows from physics that the white ball will come off a cushion at the same angle as it hits a cushion. Where on the top cushion should he hit the white ball?

86. The Freedom Tower The Freedom Tower is to be the centerpiece of the rebuilding of the World Trade Center in New York City. The tower will be 1776 feet tall (not including a broadcast antenna). The angle of elevation from the base of an office building to the top of the tower is 34°. The angle of elevation from the helipad on the roof of the office building to the top of the tower is 20°.

(a) How far away is the office building from the Freedom Tower? Assume the side of the tower is vertical. Round to the nearest foot.

(b) How tall is the office building? Round to the nearest foot.

△ **87.** Use a calculator set in radian mode to complete the following table. What can you conclude about the value of $f(\theta) = \dfrac{\sin\theta}{\theta}$ as θ approaches 0?

θ	0.5	0.4	0.2	0.1	0.01	0.001	0.0001	0.00001
$f(\theta) = \dfrac{\sin\theta}{\theta}$								

△ **88.** Use a calculator set in radian mode to complete the following table. What can you conclude about the value of $g(\theta) = \dfrac{\cos\theta - 1}{\theta}$ as θ approaches 0?

θ	0.5	0.4	0.2	0.1	0.01	0.001	0.0001	0.00001
$g(\theta) = \dfrac{\cos\theta - 1}{\theta}$								

89. Find the exact value of $\tan 1° \cdot \tan 2° \cdot \tan 3° \cdot \ldots \cdot \tan 89°$.

90. Find the exact value of $\cot 1° \cdot \cot 2° \cdot \cot 3° \cdot \ldots \cdot \cot 89°$.

91. Find the exact value of $\cos 1° \cdot \cos 2° \cdot \ldots \cdot \cos 45° \cdot \csc 46° \cdot \ldots \cdot \csc 89°$.

92. Find the exact value of $\sin 1° \cdot \sin 2° \cdot \ldots \cdot \sin 45° \cdot \sec 46° \cdot \ldots \cdot \sec 89°$.

Explaining Concepts: Discussion and Writing

93. Write a brief paragraph that explains how to quickly compute the trigonometric functions of 30°, 45°, and 60°.

94. Explain how you would measure the width of the Grand Canyon from a point on its ridge.

95. Explain how you would measure the height of a TV tower that is on the roof of a tall building.

5.4 Trigonometric Functions of any Angle

OBJECTIVES   **1** Find the Exact Values of the Trigonometric Functions for any Angle (p. 414)
   2 Use Coterminal Angles to Find the Exact Value of a Trigonometric Function (p. 417)
   3 Determine the Signs of the Trigonometric Functions of an Angle in a Given Quadrant (p. 418)
   4 Find the Reference Angle of an Angle (p. 419)
   5 Use a Reference Angle to Find the Exact Value of a Trigonometric Function (p. 420)
   6 Find the Exact Values of Trigonometric Functions of an Angle, Given Information about the Functions (p. 421)

Figure 37

1 Find the Exact Values of the Trigonometric Functions for any Angle

To extend the definition of the trigonometric functions to include angles that are not acute, we employ a rectangular coordinate system and place the angle in the standard position so that its vertex is at the origin and its initial side is along the positive x-axis. See Figure 37.

DEFINITION

Let θ be any angle in standard position, and let (a, b) denote the coordinates of any point, except the origin $(0, 0)$, on the terminal side of θ. If $r = \sqrt{a^2 + b^2}$ denotes the distance from $(0, 0)$ to (a, b), then the **six trigonometric functions of θ** are defined as the ratios

$$\sin \theta = \frac{b}{r} \qquad \cos \theta = \frac{a}{r} \qquad \tan \theta = \frac{b}{a}$$

$$\csc \theta = \frac{r}{b} \qquad \sec \theta = \frac{r}{a} \qquad \cot \theta = \frac{a}{b}$$

provided no denominator equals 0. If a denominator equals 0, that trigonometric function of the angle θ is not defined.

Figure 38

Notice in the preceding definitions that if $a = 0$, that is, if the point (a, b) is on the y-axis, then the tangent function and the secant function are undefined. Also, if $b = 0$, that is, if the point (a, b) is on the x-axis, then the cosecant function and the cotangent function are undefined.

By constructing similar triangles, you should be convinced that the values of the six trigonometric functions of an angle θ do not depend on the selection of the point (a, b) on the terminal side of θ, but rather depend only on the angle θ itself. See Figure 38 for an illustration of this when θ lies in quadrant II. Since the triangles are similar, the ratio $\dfrac{b}{r}$ equals the ratio $\dfrac{b'}{r'}$, which equals $\sin \theta$. Also, the ratio $\dfrac{|a|}{r}$ equals

Figure 39

the ratio $\dfrac{|a'|}{r'}$, so $\dfrac{a}{r} = \dfrac{a'}{r'}$, which equals $\cos \theta$. And so on.

Also, observe that if θ is acute these definitions reduce to the right triangle definitions given in Section 5.2, as illustrated in Figure 39.

Finally, from the definition of the six trigonometric functions of any angle, we see that the Quotient and Reciprocal Identities hold. Also, using $r^2 = a^2 + b^2$ and dividing each side by r^2, we can derive the Pythagorean Identities for any angle.

EXAMPLE 1

Finding the Exact Values of the Six Trigonometric Functions of θ, Given a Point on the Terminal Side

Find the exact value of each of the six trigonometric functions of a positive angle θ if $(4, -3)$ is a point on its terminal side.

Solution

Figure 40 illustrates the situation. For the point $(a, b) = (4, -3)$, we have $a = 4$ and $b = -3$. Then $r = \sqrt{a^2 + b^2} = \sqrt{16 + 9} = 5$, so

Figure 40

$$\sin \theta = \frac{b}{r} = -\frac{3}{5} \qquad \cos \theta = \frac{a}{r} = \frac{4}{5} \qquad \tan \theta = \frac{b}{a} = -\frac{3}{4}$$

$$\csc \theta = \frac{r}{b} = -\frac{5}{3} \qquad \sec \theta = \frac{r}{a} = \frac{5}{4} \qquad \cot \theta = \frac{a}{b} = -\frac{4}{3}$$

● **Now Work** PROBLEM 11

In the next example, we find the exact value of each of the six trigonometric functions at the quadrantal angles 0, $\dfrac{\pi}{2}$, π, and $\dfrac{3\pi}{2}$.

Figure 41
$\theta = 0 = 0°$

Figure 42
$\theta = \dfrac{\pi}{2} = 90°$

Figure 43
$\theta = \pi = 180°$

Figure 44
$\theta = \dfrac{3\pi}{2} = 270°$

EXAMPLE 2 — **Finding the Exact Values of the Six Trigonometric Functions of Quadrantal Angles**

Find the exact value of each of the six trigonometric functions of

(a) $\theta = 0 = 0°$ (b) $\theta = \dfrac{\pi}{2} = 90°$ (c) $\theta = \pi = 180°$ (d) $\theta = \dfrac{3\pi}{2} = 270°$

Solution

(a) We can choose any point on the terminal side of $\theta = 0 = 0°$. For convenience, we choose the point $P = (1, 0) = (a, b)$, which is a distance of $r = 1$ unit from the origin. See Figure 41. Then

$$\sin 0 = \sin 0° = \frac{b}{r} = \frac{0}{1} = 0 \qquad \cos 0 = \cos 0° = \frac{a}{r} = \frac{1}{1} = 1$$

$$\tan 0 = \tan 0° = \frac{b}{a} = \frac{0}{1} = 0 \qquad \sec 0 = \sec 0° = \frac{r}{a} = \frac{1}{1} = 1$$

Since the y-coordinate of P is 0, $\csc 0$ and $\cot 0$ are not defined.

(b) The point $P = (0, 1) = (a, b)$ is on the terminal side of $\theta = \dfrac{\pi}{2} = 90°$ and is a distance of $r = 1$ unit from the origin. See Figure 42. Then

$$\sin \frac{\pi}{2} = \sin 90° = \frac{b}{r} = \frac{1}{1} = 1 \qquad \cos \frac{\pi}{2} = \cos 90° = \frac{a}{r} = \frac{0}{1} = 0$$

$$\csc \frac{\pi}{2} = \csc 90° = \frac{r}{b} = \frac{1}{1} = 1 \qquad \cot \frac{\pi}{2} = \cot 90° = \frac{a}{b} = \frac{0}{1} = 0$$

Since the x-coordinate of P is 0, $\tan \dfrac{\pi}{2}$ and $\sec \dfrac{\pi}{2}$ are not defined.

(c) The point $P = (-1, 0)$ is on the terminal side of $\theta = \pi = 180°$ and is a distance of $r = 1$ unit from the origin. See Figure 43. Then

$$\sin \pi = \sin 180° = \frac{0}{1} = 0 \qquad \cos \pi = \cos 180° = \frac{-1}{1} = -1$$

$$\tan \pi = \tan 180° = \frac{0}{-1} = 0 \qquad \sec \pi = \sec 180° = \frac{1}{-1} = -1$$

Since the y-coordinate of P is 0, $\csc \pi$ and $\cot \pi$ are not defined.

(d) The point $P = (0, -1)$ is on the terminal side of $\theta = \dfrac{3\pi}{2} = 270°$ and is a distance of $r = 1$ unit from the origin. See Figure 44. Then

$$\sin \frac{3\pi}{2} = \sin 270° = \frac{-1}{1} = -1 \qquad \cos \frac{3\pi}{2} = \cos 270° = \frac{0}{1} = 0$$

$$\csc \frac{3\pi}{2} = \csc 270° = \frac{1}{-1} = -1 \qquad \cot \frac{3\pi}{2} = \cot 270° = \frac{0}{-1} = 0$$

Since the x-coordinate of P is 0, $\tan \dfrac{3\pi}{2}$ and $\sec \dfrac{3\pi}{2}$ are not defined.

Table 4 summarizes the values of the trigonometric functions found in Example 2.

Table 4

θ (Radians)	θ (Degrees)	$\sin \theta$	$\cos \theta$	$\tan \theta$	$\csc \theta$	$\sec \theta$	$\cot \theta$
0	$0°$	0	1	0	Not defined	1	Not defined
$\dfrac{\pi}{2}$	$90°$	1	0	Not defined	1	Not defined	0
π	$180°$	0	-1	0	Not defined	-1	Not defined
$\dfrac{3\pi}{2}$	$270°$	-1	0	Not defined	-1	Not defined	0

There is no need to memorize Table 4. To find the value of a trigonometric function of a quadrantal angle, draw the angle and apply the definition, as we did in Example 2.

2 Use Coterminal Angles to Find the Exact Value of a Trigonometric Function

DEFINITION

Two angles in standard position are said to be **coterminal** if they have the same terminal side.

See Figure 45.

Figure 45

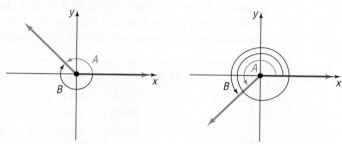

(a) *A* and *B* are coterminal   **(b)** *A* and *B* are coterminal

For example, the angles 60° and 420° are coterminal, as are the angles −40° and 320°.

In general, if θ is an angle measured in degrees, then $\theta + 360°k$, where k is any integer, is an angle coterminal with θ. If θ is measured in radians, then $\theta + 2\pi k$, where k is any integer, is an angle coterminal with θ.

Because coterminal angles have the same terminal side, it follows that the values of the trigonometric functions of coterminal angles are equal. We use this fact in the next example.

EXAMPLE 3

Using a Coterminal Angle to Find the Exact Value of a Trigonometric Function

Find the exact value of each of the following:

(a) $\sin 390°$ (b) $\cos 420°$ (c) $\tan \dfrac{9\pi}{4}$ (d) $\sec\left(-\dfrac{7\pi}{4}\right)$ (e) $\csc(-270°)$

Figure 46

Solution

(a) It is best to sketch the angle first. See Figure 46. The angle 390° is coterminal with 30°.

$$\sin 390° = \sin(30° + 360°)$$
$$= \sin 30° = \frac{1}{2}$$

Figure 47

(b) See Figure 47. The angle 420° is coterminal with 60°.

$$\cos 420° = \cos(60° + 360°)$$
$$= \cos 60° = \frac{1}{2}$$

Figure 48

(c) See Figure 48. The angle $\dfrac{9\pi}{4}$ is coterminal with $\dfrac{\pi}{4}$.

$$\tan \frac{9\pi}{4} = \tan\left(\frac{\pi}{4} + 2\pi\right)$$
$$= \tan \frac{\pi}{4} = 1$$

Figure 49

Figure 50

(d) See Figure 49. The angle $-\dfrac{7\pi}{4}$ is coterminal with $\dfrac{\pi}{4}$.

$$\sec\left(-\frac{7\pi}{4}\right) = \sec\left(\frac{\pi}{4} + 2\pi(-1)\right) = \sec\frac{\pi}{4} = \sqrt{2}$$

(e) See Figure 50. The angle $-270°$ is coterminal with $90°$.

$$\csc(-270°) = \csc(90° + 360°(-1))$$
$$= \csc 90° = 1$$

As Example 3 illustrates, the value of a trigonometric function of any angle is equal to the value of the same trigonometric function of an angle θ coterminal to it, where $0° \le \theta < 360°$ (or $0 \le \theta < 2\pi$). Because the angles θ and $\theta + 360°k$ (or $\theta + 2\pi k$), where k is any integer, are coterminal, and because the values of the trigonometric functions are equal for coterminal angles, it follows that

θ degrees	θ radians
$\sin(\theta + 360°k) = \sin\theta$	$\sin(\theta + 2\pi k) = \sin\theta$
$\cos(\theta + 360°k) = \cos\theta$	$\cos(\theta + 2\pi k) = \cos\theta$
$\tan(\theta + 360°k) = \tan\theta$	$\tan(\theta + 2\pi k) = \tan\theta$
$\csc(\theta + 360°k) = \csc\theta$	$\csc(\theta + 2\pi k) = \csc\theta$
$\sec(\theta + 360°k) = \sec\theta$	$\sec(\theta + 2\pi k) = \sec\theta$
$\cot(\theta + 360°k) = \cot\theta$	$\cot(\theta + 2\pi k) = \cot\theta$ **(1)**

where k is any integer.

These formulas show that the values of the trigonometric functions repeat themselves every $360°$ (or 2π radians).

Now Work PROBLEM 21

3 Determine the Signs of the Trigonometric Functions of an Angle in a Given Quadrant

If θ is not a quadrantal angle, then it will lie in a particular quadrant. In such a case, the signs of the x-coordinate and y-coordinate of a point (a, b) on the terminal side of θ are known. Because $r = \sqrt{a^2 + b^2} > 0$, it follows that the signs of the trigonometric functions of an angle θ can be found if we know in which quadrant θ lies.

For example, if θ lies in quadrant II, as shown in Figure 51, then a point (a, b) on the terminal side of θ has a negative x-coordinate and a positive y-coordinate. Then

Figure 51
θ in quadrant II, $a < 0, b > 0, r > 0$

$$\sin\theta = \frac{b}{r} > 0 \qquad \cos\theta = \frac{a}{r} < 0 \qquad \tan\theta = \frac{b}{a} < 0$$

$$\csc\theta = \frac{r}{b} > 0 \qquad \sec\theta = \frac{r}{a} < 0 \qquad \cot\theta = \frac{a}{b} < 0$$

Table 5 lists the signs of the six trigonometric functions for each quadrant. Figure 52 provides two illustrations.

Table 5

Quadrant of θ	$\sin\theta$, $\csc\theta$	$\cos\theta$, $\sec\theta$	$\tan\theta$, $\cot\theta$
I	Positive	Positive	Positive
II	Positive	Negative	Negative
III	Negative	Negative	Positive
IV	Negative	Positive	Negative

Figure 52

(a)

(b)

EXAMPLE 4 Finding the Quadrant in Which an Angle Lies

If $\sin \theta < 0$ and $\cos \theta < 0$, name the quadrant in which the angle θ lies.

Solution If $\sin \theta < 0$, then θ lies in quadrant III or IV. If $\cos \theta < 0$, then θ lies in quadrant II or III. Therefore, θ lies in quadrant III.

━━━ **Now Work** PROBLEM 33

4 Find the Reference Angle of an Angle

Once we know in which quadrant an angle lies, we know the sign of each trigonometric function of this angle. This information, along with the *reference angle*, will allow us to evaluate the trigonometric functions of such an angle.

DEFINITION Let θ denote an angle that lies in a quadrant. The acute angle formed by the terminal side of θ and the x-axis is called the **reference angle** for θ.

Figure 53 illustrates the reference angle for some general angles θ. Note that a reference angle is always an acute angle. That is, a reference angle has a measure between $0°$ and $90°$.

Figure 53

(a)   (b)   (c)   (d)

Although formulas can be given for calculating reference angles, usually it is easier to find the reference angle for a given angle by making a quick sketch of the angle.

EXAMPLE 5 Finding Reference Angles

Find the reference angle for each of the following angles:

(a) $150°$ (b) $-45°$ (c) $\dfrac{9\pi}{4}$ (d) $-\dfrac{5\pi}{6}$

Solution

(a) Refer to Figure 54. The reference angle for 150° is 30°.

Figure 54

(b) Refer to Figure 55. The reference angle for −45° is 45°.

Figure 55

(c) Refer to Figure 56. The reference angle for $\dfrac{9\pi}{4}$ is $\dfrac{\pi}{4}$.

Figure 56

(d) Refer to Figure 57. The reference angle for $-\dfrac{5\pi}{6}$ is $\dfrac{\pi}{6}$.

Figure 57

━━ **Now Work** PROBLEM 41

5 Use a Reference Angle to Find the Exact Value of a Trigonometric Function

The advantage of using reference angles is that, except for the correct sign, the values of the trigonometric functions of any angle θ equal the values of the trigonometric functions of its reference angle.

THEOREM

Reference Angles

If θ is an angle that lies in a quadrant and if α is its reference angle, then

$$\sin \theta = \pm\sin \alpha \qquad \cos \theta = \pm\cos \alpha \qquad \tan \theta = \pm\tan \alpha$$
$$\csc \theta = \pm\csc \alpha \qquad \sec \theta = \pm\sec \alpha \qquad \cot \theta = \pm\cot \alpha \qquad \textbf{(2)}$$

where the $+$ or $-$ sign depends on the quadrant in which θ lies.

Figure 58

For example, suppose that θ lies in quadrant II and α is its reference angle. See Figure 58. If (a, b) is a point on the terminal side of θ and if $r = \sqrt{a^2 + b^2}$, we have

$$\sin \theta = \frac{b}{r} = \sin \alpha \qquad \cos \theta = \frac{a}{r} = \frac{-|a|}{r} = -\cos \alpha$$

and so on.

The next example illustrates how the theorem on reference angles is used.

━━━━━━━━━━━━━━━━━━━━━━━━━━━━

| EXAMPLE 6 | **Using the Reference Angle to Find the Exact Value of a Trigonometric Function** |

Find the exact value of each of the following trigonometric functions using reference angles.

(a) $\sin 135°$ (b) $\cos 600°$ (c) $\cos \dfrac{17\pi}{6}$ (d) $\tan\left(-\dfrac{\pi}{3}\right)$

Solution

(a) Refer to Figure 59. The reference angle for 135° is 45° and $\sin 45° = \dfrac{\sqrt{2}}{2}$. The angle 135° is in quadrant II, where the sine function is positive, so

$$\sin 135° = \sin 45° = \dfrac{\sqrt{2}}{2}$$

Figure 59

(b) Refer to Figure 60. The reference angle for 600° is 60° and $\cos 60° = \dfrac{1}{2}$. The angle 600° is in quadrant III, where the cosine function is negative, so

$$\cos 600° = -\cos 60° = -\dfrac{1}{2}$$

Figure 60

(c) Refer to Figure 61. The reference angle for $\dfrac{17\pi}{6}$ is $\dfrac{\pi}{6}$ and $\cos \dfrac{\pi}{6} = \dfrac{\sqrt{3}}{2}$. The angle $\dfrac{17\pi}{6}$ is in quadrant II, where the cosine function is negative, so

$$\cos \dfrac{17\pi}{6} = -\cos \dfrac{\pi}{6} = -\dfrac{\sqrt{3}}{2}$$

Figure 61

(d) Refer to Figure 62. The reference angle for $-\dfrac{\pi}{3}$ is $\dfrac{\pi}{3}$ and $\tan \dfrac{\pi}{3} = \sqrt{3}$. The angle $-\dfrac{\pi}{3}$ is in quadrant IV, where the tangent function is negative, so

$$\tan\left(-\dfrac{\pi}{3}\right) = -\tan \dfrac{\pi}{3} = -\sqrt{3}$$

Figure 62

Finding the Values of the Trigonometric Functions of any Angle

- If the angle θ is a quadrantal angle, draw the angle, pick a point on its terminal side, and apply the definition of the trigonometric functions.
- If the angle θ lies in a quadrant:
  1. Find the reference angle α of θ.
  2. Find the value of the trigonometric function at α.
  3. Adjust the sign ($+$ or $-$) of the value of the trigonometric function based on the quadrant in which θ lies.

Now Work PROBLEMS **59** AND **61**

---

6 Find the Exact Values of Trigonometric Functions of an Angle, Given Information about the Functions

EXAMPLE 7 **Finding the Exact Values of Trigonometric Functions**

Given that $\cos \theta = -\dfrac{2}{3}, \dfrac{\pi}{2} < \theta < \pi$, find the exact value of each of the remaining trigonometric functions.

Solution The angle θ lies in quadrant II, so we know that $\sin \theta$ and $\csc \theta$ are positive and the other four trigonometric functions are negative. If α is the reference angle for θ, then $\cos \alpha = \dfrac{2}{3} = \dfrac{\text{adjacent}}{\text{hypotenuse}}$. The values of the remaining trigonometric functions

Figure 63

$\cos \alpha = \dfrac{2}{3}$

of the reference angle α can be found by drawing the appropriate triangle and using the Pythagorean Theorem to determine that the side opposite α is $\sqrt{5}$. We use Figure 63 to obtain

$$\sin \alpha = \frac{\sqrt{5}}{3} \qquad \cos \alpha = \frac{2}{3} \qquad \tan \alpha = \frac{\sqrt{5}}{2}$$

$$\csc \alpha = \frac{3}{\sqrt{5}} = \frac{3\sqrt{5}}{5} \qquad \sec \alpha = \frac{3}{2} \qquad \cot \alpha = \frac{2}{\sqrt{5}} = \frac{2\sqrt{5}}{5}$$

Now we assign the appropriate signs to each of these values to find the values of the trigonometric functions of θ.

$$\sin \theta = \frac{\sqrt{5}}{3} \qquad \cos \theta = -\frac{2}{3} \qquad \tan \theta = -\frac{\sqrt{5}}{2}$$

$$\csc \theta = \frac{3\sqrt{5}}{5} \qquad \sec \theta = -\frac{3}{2} \qquad \cot \theta = -\frac{2\sqrt{5}}{5}$$

●

Now Work PROBLEM 83

EXAMPLE 8   **Finding the Exact Values of Trigonometric Functions**

If $\tan \theta = -4$ and $\sin \theta < 0$, find the exact value of each of the remaining trigonometric functions of θ.

Solution   Since $\tan \theta = -4 < 0$ and $\sin \theta < 0$, it follows that θ lies in quadrant IV. If α is the reference angle for θ, then $\tan \alpha = 4 = \dfrac{4}{1} = \dfrac{b}{a}$. With $a = 1$ and $b = 4$, we find $r = \sqrt{1^2 + 4^2} = \sqrt{17}$. See Figure 64. Then

$$\sin \alpha = \frac{4}{\sqrt{17}} = \frac{4\sqrt{17}}{17} \qquad \cos \alpha = \frac{1}{\sqrt{17}} = \frac{\sqrt{17}}{17} \qquad \tan \alpha = \frac{4}{1} = 4$$

$$\csc \alpha = \frac{\sqrt{17}}{4} \qquad \sec \alpha = \frac{\sqrt{17}}{1} = \sqrt{17} \qquad \cot \alpha = \frac{1}{4}$$

Figure 64

$\tan \alpha = 4$

We assign the appropriate sign to each of these to obtain the values of the trigonometric functions of θ.

$$\sin \theta = -\frac{4\sqrt{17}}{17} \qquad \cos \theta = \frac{\sqrt{17}}{17} \qquad \tan \theta = -4$$

$$\csc \theta = -\frac{\sqrt{17}}{4} \qquad \sec \theta = \sqrt{17} \qquad \cot \theta = -\frac{1}{4}$$

●

Now Work PROBLEM 93

5.4 Assess Your Understanding

Concepts and Vocabulary

1. For an angle θ that lies in quadrant III, the trigonometric functions _____ and _____ are positive.

2. Two angles in standard position that have the same terminal side are _____.

3. The reference angle of 240° is _____.

4. *True or False* $\sin 182° = \cos 2°$.

5. *True or False* $\tan \dfrac{\pi}{2}$ is not defined.

6. *True or False*   The reference angle is always an acute angle.

7. What is the reference angle of 600°?

8. In which quadrants is the cosine function positive?

9. If $0 \le \theta < 2\pi$, for what angles θ, if any, is $\tan \theta$ undefined?

10. What is the reference angle of $\dfrac{13\pi}{3}$?

Skill Building

In Problems 11–20, a point on the terminal side of an angle θ in standard position is given. Find the exact value of each of the six trigonometric functions of θ.

11. $(-3, 4)$ **12.** $(5, -12)$ **13.** $(2, -3)$ **14.** $(-1, -2)$ **15.** $(-3, -3)$

16. $(2, -2)$ **17.** $\left(\dfrac{\sqrt{3}}{2}, \dfrac{1}{2} \right)$ **18.** $\left(-\dfrac{1}{2}, \dfrac{\sqrt{3}}{2} \right)$ **19.** $\left(\dfrac{\sqrt{2}}{2}, -\dfrac{\sqrt{2}}{2} \right)$ **20.** $\left(-\dfrac{\sqrt{2}}{2}, -\dfrac{\sqrt{2}}{2} \right)$

In Problems 21–32, use a coterminal angle to find the exact value of each expression. Do not use a calculator.

21. $\sin 405°$ **22.** $\cos 420°$ **23.** $\tan 405°$ **24.** $\sin 390°$ **25.** $\csc 450°$ **26.** $\sec\ 540°$

27. $\cot 390°$ **28.** $\sec 420°$ **29.** $\cos \dfrac{33\pi}{4}$ **30.** $\sin \dfrac{9\pi}{4}$ **31.** $\tan(21\pi)$ **32.** $\csc \dfrac{9\pi}{2}$

In Problems 33–40, name the quadrant in which the angle θ lies.

33. $\sin \theta > 0, \quad \cos \theta < 0$ **34.** $\sin \theta < 0, \quad \cos \theta > 0$ **35.** $\sin \theta < 0, \quad \tan \theta < 0$

36. $\cos \theta > 0, \quad \tan \theta > 0$ **37.** $\cos \theta > 0, \quad \cot \theta < 0$ **38.** $\sin \theta < 0, \quad \cot \theta > 0$

39. $\sec \theta < 0, \quad \tan \theta > 0$ **40.** $\csc \theta > 0, \quad \cot \theta < 0$

In Problems 41–58, find the reference angle of each angle.

41. $-30°$ **42.** $-60°$ **43.** $120°$ **44.** $210°$ **45.** $300°$ **46.** $330°$

47. $\dfrac{5\pi}{4}$ **48.** $\dfrac{5\pi}{6}$ **49.** $\dfrac{8\pi}{3}$ **50.** $\dfrac{7\pi}{4}$ **51.** $-135°$ **52.** $-240°$

53. $-\dfrac{2\pi}{3}$ **54.** $-\dfrac{7\pi}{6}$ **55.** $440°$ **56.** $490°$ **57.** $\dfrac{15\pi}{4}$ **58.** $\dfrac{19\pi}{6}$

In Problems 59–82, use the reference angle to find the exact value of each expression. Do not use a calculator.

59. $\sin 150°$ **60.** $\cos 210°$ **61.** $\sin 510°$ **62.** $\cos 600°$ **63.** $\cos(-45°)$ **64.** $\sin(-240°)$

65. $\sec 240°$ **66.** $\csc 300°$ **67.** $\cot 330°$ **68.** $\tan 225°$ **69.** $\sin \dfrac{3\pi}{4}$ **70.** $\cos \dfrac{2\pi}{3}$

71. $\cos \dfrac{13\pi}{4}$ **72.** $\tan \dfrac{8\pi}{3}$ **73.** $\sin\left(-\dfrac{2\pi}{3} \right)$ **74.** $\cot\left(-\dfrac{\pi}{6} \right)$ **75.** $\tan \dfrac{14\pi}{3}$ **76.** $\sec \dfrac{11\pi}{4}$

77. $\sin(8\pi)$ **78.** $\cos(-2\pi)$ **79.** $\tan(7\pi)$ **80.** $\cot(5\pi)$ **81.** $\sec(-3\pi)$ **82.** $\csc\left(-\dfrac{5\pi}{2} \right)$

In Problems 83–100, find the exact value of each of the remaining trigonometric functions of θ.

83. $\sin \theta = \dfrac{12}{13}, \quad \theta$ in quadrant II   **84.** $\cos \theta = \dfrac{3}{5}, \quad \theta$ in quadrant IV   **85.** $\cos \theta = -\dfrac{4}{5}, \quad \theta$ in quadrant III

86. $\sin \theta = -\dfrac{5}{13}, \quad \theta$ in quadrant III   **87.** $\sin \theta = \dfrac{5}{13}, \quad 90° < \theta < 180°$ **88.** $\cos \theta = \dfrac{4}{5}, \quad 270° < \theta < 360°$

89. $\cos \theta = -\dfrac{1}{3}, \quad 180° < \theta < 270°$ **90.** $\sin \theta = -\dfrac{2}{3}, \quad 180° < \theta < 270°$ **91.** $\sin \theta = \dfrac{2}{3}, \quad \tan \theta < 0$

92. $\cos \theta = -\dfrac{1}{4}, \quad \tan \theta > 0$ **93.** $\sec \theta = 2, \quad \sin \theta < 0$ **94.** $\csc \theta = 3, \quad \cot \theta < 0$

95. $\tan \theta = \dfrac{3}{4}, \quad \sin \theta < 0$ **96.** $\cot \theta = \dfrac{4}{3}, \quad \cos \theta < 0$ **97.** $\tan \theta = -\dfrac{1}{3}, \quad \sin \theta > 0$

98. $\sec \theta = -2, \quad \tan \theta > 0$ **99.** $\csc \theta = -2, \quad \tan \theta > 0$ **100.** $\cot \theta = -2, \quad \sec \theta > 0$

101. Find the exact value of $\sin 40° + \sin 130° + \sin 220° + \sin 310°$.

102. Find the exact value of $\tan 40° + \tan 140°$.

Mixed Practice

In Problems 103–106, $f(x) = \sin x$, $g(x) = \cos x$, $h(x) = \tan x$, $F(x) = \csc x$, $G(x) = \sec x$, and $H(x) = \cot x$.

103. (a) Find $f(315°)$. What point is on the graph of f?

    (b) Find $G(315°)$. What point is on the graph of G?

    (c) Find $h(315°)$. What point is on the graph of h?

104. (a) Find $g(120°)$. What point is on the graph of g?

    (b) Find $F(120°)$. What point is on the graph of F?

    (c) Find $H(120°)$. What point is on the graph of H?

105. (a) Find $g\left(\dfrac{7\pi}{6}\right)$. What point is on the graph of g?

    (b) Find $F\left(\dfrac{7\pi}{6}\right)$. What point is on the graph of F?

    (c) Find $H(-315°)$. What point is on the graph of H?

106. (a) Find $f\left(\dfrac{7\pi}{4}\right)$. What point is on the graph of f?

    (b) Find $G\left(\dfrac{7\pi}{4}\right)$. What point is on the graph of G?

    (c) Find $F(-225°)$. What point is on the graph of F?

Applications and Extensions

107. If $f(\theta) = \sin \theta = 0.2$, find $f(\theta + \pi)$.

108. If $g(\theta) = \cos \theta = 0.4$, find $g(\theta + \pi)$.

109. If $F(\theta) = \tan \theta = 3$, find $F(\theta + \pi)$.

110. If $G(\theta) = \cot \theta = -2$, find $G(\theta + \pi)$.

111. If $\sin \theta = \dfrac{1}{5}$, find $\csc(\theta + \pi)$.

112. If $\cos \theta = \dfrac{2}{3}$, find $\sec(\theta + \pi)$.

113. Find the exact value of
$\sin 1° + \sin 2° + \sin 3° + \cdots + \sin 358° + \sin 359°$

114. Find the exact value of
$\cos 1° + \cos 2° + \cos 3° + \cdots + \cos 358° + \cos 359°$

115. **Projectile Motion** An object is propelled upward at an angle θ, $45° < \theta < 90°$, to the horizontal with an initial velocity of v_0 feet per second from the base of a plane that makes an angle of $45°$ with the horizontal. See the illustration. If air resistance is ignored, the distance R that it travels up the inclined plane is given by the function

$$R(\theta) = \frac{v_0^2 \sqrt{2}}{32}[\sin(2\theta) - \cos(2\theta) - 1]$$

(a) Find the distance R that the object travels along the inclined plane if the initial velocity is 32 feet per second and $\theta = 60°$.

(b) Graph $R = R(\theta)$ if the initial velocity is 32 feet per second.

(c) What value of θ makes R largest?

Explaining Concepts: Discussion and Writing

116. Give three examples that demonstrate how to use the theorem on reference angles.

117. Write a brief paragraph that explains how to quickly compute the value of the trigonometric functions of $0°, 90°, 180°$, and $270°$.

118. Explain what a reference angle is. What role does it play in finding the value of a trigonometric function?

5.5 Unit Circle Approach; Properties of the Trigonometric Functions

PREPARING FOR THIS SECTION *Before getting started, review the following:*

- Unit Circle (Foundations, Section 4, p. 35–36)
- Functions (Section 1.1, pp. 43–51)
- Even and Odd Functions (Section 1.3, pp. 67–69)

Now Work the 'Are You Prepared?' problems on page 433.

OBJECTIVES **1** Find the Exact Values of the Trigonometric Functions Using the Unit Circle (p. 425)

2 Know the Domain and Range of the Trigonometric Functions (p. 429)

3 Use the Periodic Properties to Find the Exact Values of the Trigonometric Functions (p. 430)

4 Use Even–Odd Properties to Find the Exact Values of the Trigonometric Functions (p. 432)

In this section, we develop important properties of the trigonometric functions. We begin by introducing the trigonometric functions using the unit circle. This approach will lead to the definition given earlier of the trigonometric functions of any angle (page 415).

1 Find the Exact Values of the Trigonometric Functions Using the Unit Circle

Recall that the unit circle is a circle whose radius is 1 and whose center is at the origin of a rectangular coordinate system. Also recall that any circle of radius r has circumference of length $2\pi r$. Therefore, the unit circle (radius $= 1$) has a circumference of length 2π. So, for 1 revolution around the unit circle the length of the arc is 2π units.

The following discussion sets the stage for defining the trigonometric functions using the unit circle.

Draw a vertical real number line with the origin of the number line at $(1, 0)$ in the Cartesian Plane. Let $t \geq 0$ be any real number and let s be the distance from the origin to t on the real number line. See the red portion of Figure 65(a). Now look at the unit circle in Figure 65(a). Beginning at the point $(1, 0)$ on the unit circle, travel $s = t$ units in the counterclockwise direction along the circle to arrive at the point $P = (a, b)$. In this sense, the length $s = t$ units is being **wrapped** around the unit circle.

If $t < 0$, we begin at the point $(1, 0)$ on the unit circle and travel $s = |t|$ units in the clockwise direction to arrive at the point $P = (a, b)$. See Figure 65(b).

Figure 65

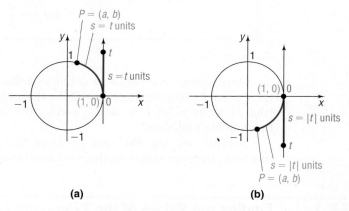

(a)   **(b)**

If $t > 2\pi$ or if $t < -2\pi$, it will be necessary to travel around the unit circle more than once before arriving at point P. Do you see why?

Let's describe this process another way. Picture a string of length $s = |t|$ units being wrapped around a circle of radius 1 unit. We start wrapping the string around the circle at the point $(1, 0)$. If $t \geq 0$, we wrap the string in the counterclockwise direction; if $t < 0$, we wrap the string in the clockwise direction. The point $P = (a, b)$ is the point where the string ends.

This discussion tells us that, for any real number t, we can locate a unique point $P = (a, b)$ on the unit circle. We call this point **the point P on the unit circle that corresponds to t.** This is the important idea here. No matter what real number t is chosen, there is a unique point P on the unit circle corresponding to it. We use the coordinates of the point $P = (a, b)$ on the unit circle corresponding to the real number t to define the **six trigonometric functions of t.** Be sure to consult Figures 65(a) and (b) as you read these definitions.

DEFINITION

Let t be a real number and let $P = (a, b)$ be the point on the unit circle that corresponds to t.

The **sine function** associates with t the y-coordinate of P and is denoted by

$$\sin t = b$$

In Words
The sine function takes as input a real number *t* that corresponds to a point *P* = (*a*, *b*) on the unit circle and outputs the *y*-coordinate, *b*. The cosine function takes as input a real number *t* that corresponds to a point *P* = (*a*, *b*) on the unit circle and outputs the *x*-coordinate, *a*.

The **cosine function** associates with *t* the *x*-coordinate of *P* and is denoted by

$$\cos t = a$$

If $a \neq 0$, the **tangent function** associates with *t* the ratio of the *y*-coordinate to the *x*-coordinate of *P* and is denoted by

$$\tan t = \frac{b}{a}$$

If $b \neq 0$, the **cosecant function** is defined as

$$\csc t = \frac{1}{b}$$

If $a \neq 0$, the **secant function** is defined as

$$\sec t = \frac{1}{a}$$

If $b \neq 0$, the **cotangent function** is defined as

$$\cot t = \frac{a}{b}$$

Once again, notice in these definitions that if $a = 0$ (that is, if the point *P* is on the *y*-axis) the tangent function and the secant function are undefined. Also, if $b = 0$ (that is, if the point *P* is on the *x*-axis), the cosecant function and the cotangent function are undefined.

Because we use the unit circle in these definitions of the trigonometric functions, they are also sometimes referred to as **circular functions.**

EXAMPLE 1

Finding the Values of the Trigonometric Functions Using a Point on the Unit Circle

Find the values of $\sin t, \cos t, \tan t, \csc t, \sec t$, and $\cot t$ if $P = \left(-\frac{1}{2}, \frac{\sqrt{3}}{2}\right)$ is the point on the unit circle that corresponds to the real number *t*.

Solution

See Figure 66. We follow the definition of the six trigonometric functions using $P = \left(-\frac{1}{2}, \frac{\sqrt{3}}{2}\right) = (a, b)$. Then, with $a = -\frac{1}{2}$ and $b = \frac{\sqrt{3}}{2}$, we have

Figure 66

$$\sin t = b = \frac{\sqrt{3}}{2} \qquad \cos t = a = -\frac{1}{2} \qquad \tan t = \frac{b}{a} = \frac{\frac{\sqrt{3}}{2}}{-\frac{1}{2}} = -\sqrt{3}$$

$$\csc t = \frac{1}{b} = \frac{1}{\frac{\sqrt{3}}{2}} = \frac{2}{\sqrt{3}} = \frac{2\sqrt{3}}{3} \qquad \sec t = \frac{1}{a} = \frac{1}{-\frac{1}{2}} = -2 \qquad \cot t = \frac{a}{b} = \frac{-\frac{1}{2}}{\frac{\sqrt{3}}{2}} = -\frac{1}{\sqrt{3}} = -\frac{\sqrt{3}}{3}$$

Now Work PROBLEM 9

Trigonometric Functions of Angles

Let $P = (a, b)$ be the point on the unit circle corresponding to the real number t. See Figure 67(a). Let θ be the angle in standard position, measured in radians, whose terminal side is the ray from the origin through P and whose arc length is $|t|$. See Figure 67(b). Since the unit circle has radius 1 unit, if $s = |t|$ units, then from the arc length formula $s = r\theta$, we have $\theta = t$ radians. See Figures 67(c) and (d).

Figure 67

(a)

(b)

(c)

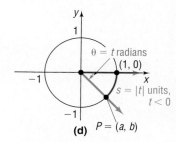
(d)

The point $P = (a, b)$ on the unit circle that corresponds to the real number t is also the point P on the terminal side of the angle $\theta = t$ radians. As a result, we can say that

$$\underset{\underset{\text{Real number}}{\uparrow}}{\sin t} = \underset{\underset{\theta = t \text{ radians}}{\uparrow}}{\sin \theta}$$

and so on. We can now define the trigonometric functions of the angle θ.

DEFINITION

If $\theta = t$ radians, the six **trigonometric functions of the angle θ** are defined as

$\sin \theta = \sin t$	$\cos \theta = \cos t$	$\tan \theta = \tan t$
$\csc \theta = \csc t$	$\sec \theta = \sec t$	$\cot \theta = \cot t$

Even though the trigonometric functions can be viewed both as functions of real numbers and as functions of angles, it is customary to refer to trigonometric functions of real numbers and trigonometric functions of angles collectively as *the trigonometric functions*. We will follow this practice from now on.

Since the values of the trigonometric functions of an angle θ are determined by the coordinates of the point $P = (a, b)$ on the unit circle corresponding to θ, the units used to measure the angle θ are irrelevant. For example, it does not matter whether we write $\theta = \dfrac{\pi}{2}$ radians or $\theta = 90°$. In either case, the point on the unit circle corresponding to this angle is $P = (0, 1)$. As a result,

$$\sin \frac{\pi}{2} = \sin 90° = 1 \quad \text{and} \quad \cos \frac{\pi}{2} = \cos 90° = 0$$

The discussion based on Figure 67 implies the following: To find the exact value of a trigonometric function of an angle θ requires that we locate the corresponding point $P^* = (a^*, b^*)$ on the unit circle. In fact, though, any circle whose center is at the origin can be used.

Let θ be any nonquadrantal angle placed in standard position. Let $P = (a, b)$ be the point on the circle $x^2 + y^2 = r^2$ that corresponds to θ, and let $P^* = (a^*, b^*)$ be the point on the unit circle that corresponds to θ. See Figure 68.

Figure 68

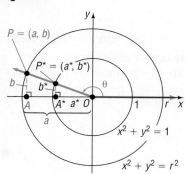

Notice that the triangles OA^*P^* and OAP are similar, so ratios of corresponding sides are equal.

$\dfrac{b^*}{1} = \dfrac{b}{r}$	$\dfrac{a^*}{1} = \dfrac{a}{r}$	$\dfrac{b^*}{a^*} = \dfrac{b}{a}$
$\dfrac{1}{b^*} = \dfrac{r}{b}$	$\dfrac{1}{a^*} = \dfrac{r}{a}$	$\dfrac{a^*}{b^*} = \dfrac{a}{b}$

These results lead us to formulate the following theorem:

THEOREM

For an angle θ in standard position, let $P = (a, b)$ be any point on the terminal side of θ that is also on the circle $x^2 + y^2 = r^2$. Then

$$\sin \theta = \frac{b}{r} \qquad \cos \theta = \frac{a}{r} \qquad \tan \theta = \frac{b}{a} \quad a \neq 0$$

$$\csc \theta = \frac{r}{b} \quad b \neq 0 \qquad \sec \theta = \frac{r}{a} \quad a \neq 0 \qquad \cot \theta = \frac{a}{b} \quad b \neq 0$$

This result coincides with the definition given in Section 5.4 for the six trigonometric functions of any angle θ (page 415).

Consider Figure 69, which shows a circle of radius $\sqrt{2}$. Inside the circle we have drawn a right triangle whose acute angles both measure 45°. The lengths of the legs are then 1 unit each. This is the right triangle in Figure 27(b) on page 403. Therefore, the point on the circle that corresponds to an angle of 45° is $(1, 1) = (a, b)$. From this,

$$\sin 45° = \frac{b}{c} = \frac{b}{r} = \frac{1}{\sqrt{2}} = \frac{\sqrt{2}}{2}, \quad \cos 45° = \frac{a}{c} = \frac{a}{r} = \frac{1}{\sqrt{2}} = \frac{\sqrt{2}}{2}, \text{and so on.}$$

Now consider Figure 70(a), which shows a circle of radius 2. Inside the circle we have drawn a right triangle whose acute angles measure 30° and 60°. Refer to Figure 28(c) on page 405. If we place the 30° angle at the origin, we find that the point on the circle that corresponds to an angle of 30° is $(\sqrt{3}, 1)$. From this,

$$\sin 30° = \frac{b}{c} = \frac{b}{r} = \frac{1}{2}, \quad \cos 30° = \frac{a}{c} = \frac{a}{r} = \frac{\sqrt{3}}{2}, \text{and so on. Figure 70(b) also shows}$$

a circle of radius 2, but the angle at the origin is 60°. Now we can see that

$$\sin 60° = \frac{b}{c} = \frac{b}{r} = \frac{\sqrt{3}}{2}, \text{and so on.}$$

Figure 69

Figure 70

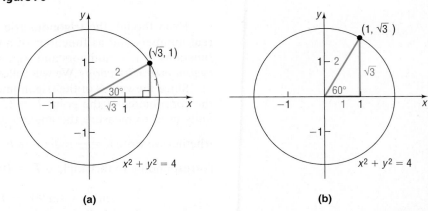

(a)   (b)

EXAMPLE 2 **Finding the Exact Values of the Six Trigonometric Functions**

Find the exact values of each of the six trigonometric functions of an angle θ if $(4, -3)$ is a point on its terminal side.

Figure 71

Solution See Figure 71. The point $(4, -3)$ is on a circle of radius $r = \sqrt{4^2 + (-3)^2} = \sqrt{16 + 9} = \sqrt{25} = 5$ with the center at the origin.

For the point $(a, b) = (4, -3)$, we have $a = 4$ and $b = -3$. Since $r = 5$, we find

$$\sin \theta = \frac{b}{r} = -\frac{3}{5} \qquad \cos \theta = \frac{a}{r} = \frac{4}{5} \qquad \tan \theta = \frac{b}{a} = -\frac{3}{4}$$

$$\csc \theta = \frac{r}{b} = -\frac{5}{3} \qquad \sec \theta = \frac{r}{a} = \frac{5}{4} \qquad \cot \theta = \frac{a}{b} = -\frac{4}{3}$$

Now Work PROBLEM 15

2 Know the Domain and Range of the Trigonometric Functions

Let θ be an angle in standard position, and let $P = (a, b)$ be the point on the unit circle that corresponds to θ. See Figure 72. Then, by the definition given earlier:

Figure 72

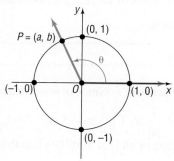

$$\sin \theta = b \qquad \cos \theta = a \qquad \tan \theta = \frac{b}{a} \quad a \neq 0$$

$$\csc \theta = \frac{1}{b} \quad b \neq 0 \qquad \sec \theta = \frac{1}{a} \quad a \neq 0 \qquad \cot \theta = \frac{a}{b} \quad b \neq 0$$

For $\sin \theta$ and $\cos \theta$, there is no concern about dividing by 0, so θ can be any angle. It follows that the domain of the sine function and cosine function is the set of all real numbers.

> The domain of the sine function is the set of all real numbers.
>
> The domain of the cosine function is the set of all real numbers.

For the tangent function and secant function, the x-coordinate of $P = (a, b)$ cannot be 0 since this results in division by 0. See Figure 72. On the unit circle, there are two such points, $(0, 1)$ and $(0, -1)$. These two points correspond to the angles $\frac{\pi}{2}(90°)$ and $\frac{3\pi}{2}(270°)$ or, more generally, to any angle that is an odd integer multiple of $\frac{\pi}{2}(90°)$, such as $\pm\frac{\pi}{2}(\pm90°)$, $\pm\frac{3\pi}{2}(\pm270°)$, and $\pm\frac{5\pi}{2}(\pm450°)$. Such angles must therefore be excluded from the domain of the tangent function and secant function.

> The domain of the tangent function is the set of all real numbers, except odd integer multiples of $\frac{\pi}{2}(90°)$.
>
> The domain of the secant function is the set of all real numbers, except odd integer multiples of $\frac{\pi}{2}(90°)$.

For the cotangent function and cosecant function, the y-coordinate of $P = (a, b)$ cannot be 0 since this results in division by 0. See Figure 72. On the unit circle, there are two such points, $(1, 0)$ and $(-1, 0)$. These two points correspond to the angles $0 \ (0°)$ and $\pi(180°)$ or, more generally, to any angle that is an integer multiple of $\pi(180°)$, such as $0(0°)$, $\pm\pi(\pm180°)$, $\pm2\pi(\pm360°)$, and $\pm3\pi(\pm540°)$. Such angles must be excluded from the domain of the cotangent function and cosecant function.

> The domain of the cotangent function is the set of all real numbers, except integer multiples of $\pi(180°)$.
>
> The domain of the cosecant function is the set of all real numbers, except integer multiples of $\pi(180°)$.

Next, we determine the range of each of the six trigonometric functions. Refer again to Figure 72. Let $P = (a, b)$ be the point on the unit circle that corresponds to the angle θ. It follows that $-1 \leq a \leq 1$ and $-1 \leq b \leq 1$. Since $\sin \theta = b$ and $\cos \theta = a$, we have

$$-1 \leq \sin \theta \leq 1 \quad \text{and} \quad -1 \leq \cos \theta \leq 1$$

The range of both the sine function and the cosine function consists of all real numbers between -1 and 1, inclusive. Using absolute value notation, we have $|\sin \theta| \leq 1$ and $|\cos \theta| \leq 1$.

If θ is not an integer multiple of $\pi(180°)$, then $\csc\theta = \dfrac{1}{b}$. Since $b = \sin\theta$ and $|b| = |\sin\theta| \le 1$, it follows that $|\csc\theta| = \dfrac{1}{|\sin\theta|} = \dfrac{1}{|b|} \ge 1 \left(\dfrac{1}{b} \le -1 \text{ or } \dfrac{1}{b} \ge 1 \right)$. Since $\csc\theta = \dfrac{1}{b}$, the range of the cosecant function consists of all real numbers less than or equal to -1 or greater than or equal to 1. That is,

$$\csc\theta \le -1 \quad \text{or} \quad \csc\theta \ge 1$$

If θ is not an odd integer multiple of $\dfrac{\pi}{2}(90°)$, then $\sec\theta = \dfrac{1}{a}$. Since $a = \cos\theta$ and $|a| = |\cos\theta| \le 1$, it follows that $|\sec\theta| = \dfrac{1}{|\cos\theta|} = \dfrac{1}{|a|} \ge 1 \left(\dfrac{1}{a} \le -1 \text{ or } \dfrac{1}{a} \ge 1 \right)$. Since $\sec\theta = \dfrac{1}{a}$, the range of the secant function consists of all real numbers less than or equal to -1 or greater than or equal to 1. That is,

$$\sec\theta \le -1 \quad \text{or} \quad \sec\theta \ge 1$$

The range of both the tangent function and the cotangent function consists of all real numbers. That is,

$$-\infty < \tan\theta < \infty \quad \text{and} \quad -\infty < \cot\theta < \infty$$

You are asked to prove this in Problems 91 and 92.
Table 6 summarizes these results.

Table 6

Function	Symbol	Domain	Range
sine	$f(\theta) = \sin\theta$	All real numbers	All real numbers from -1 to 1, inclusive
cosine	$f(\theta) = \cos\theta$	All real numbers	All real numbers from -1 to 1, inclusive
tangent	$f(\theta) = \tan\theta$	All real numbers, except odd integer multiples of $\dfrac{\pi}{2}(90°)$	All real numbers
cosecant	$f(\theta) = \csc\theta$	All real numbers, except integer multiples of $\pi(180°)$	All real numbers greater than or equal to 1 or less than or equal to -1
secant	$f(\theta) = \sec\theta$	All real numbers, except odd integer multiples of $\dfrac{\pi}{2}(90°)$	All real numbers greater than or equal to 1 or less than or equal to -1
cotangent	$f(\theta) = \cot\theta$	All real numbers, except integer multiples of $\pi(180°)$	All real numbers

Now Work PROBLEMS 61 AND 65

3 Use the Periodic Properties to Find the Exact Values of the Trigonometric Functions

Look at Figure 73. This figure shows that for an angle of $\dfrac{\pi}{3}$ radians the corresponding point P on the unit circle is $\left(\dfrac{1}{2}, \dfrac{\sqrt{3}}{2} \right)$. Notice that for an angle of $\dfrac{\pi}{3} + 2\pi$ radians the corresponding point P on the unit circle is also $\left(\dfrac{1}{2}, \dfrac{\sqrt{3}}{2} \right)$. As a result,

$$\sin\frac{\pi}{3} = \frac{\sqrt{3}}{2} \quad \text{and} \quad \sin\left(\frac{\pi}{3} + 2\pi\right) = \frac{\sqrt{3}}{2}$$

$$\cos\frac{\pi}{3} = \frac{1}{2} \quad \text{and} \quad \cos\left(\frac{\pi}{3} + 2\pi\right) = \frac{1}{2}$$

Figure 73

Figure 74

This example illustrates a more general situation. For a given angle θ, measured in radians, suppose that we know the corresponding point $P = (a, b)$ on the unit circle. Now add 2π to θ. The point on the unit circle corresponding to $\theta + 2\pi$ is identical to the point P on the unit circle corresponding to θ. See Figure 74. The values of the trigonometric functions of $\theta + 2\pi$ are equal to the values of the corresponding trigonometric functions of θ.

If we add (or subtract) integer multiples of 2π to θ, the values of the sine and cosine function remain unchanged. That is, for all θ

$$\sin(\theta + 2\pi k) = \sin\theta \qquad \cos(\theta + 2\pi k) = \cos\theta$$
$$\text{where } k \text{ is any integer} \qquad \qquad \textbf{(1)}$$

Functions that exhibit this kind of behavior are called *periodic functions*.

DEFINITION

A function f is called **periodic** if there is a positive number p such that, whenever θ is in the domain of f, so is $\theta + p$, and

$$f(\theta + p) = f(\theta)$$

If there is a smallest such number p, this smallest value is called the **(fundamental) period** of f.

Figure 75

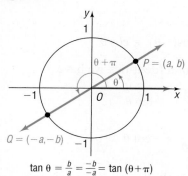

$$\tan\theta = \frac{b}{a} = \frac{-b}{-a} = \tan(\theta + \pi)$$

Based on equation (1), the sine and cosine functions are periodic. In fact, the sine, cosine, secant, and cosecant functions have period 2π. You are asked to prove this in Problems 93 through 96.

The tangent and cotangent functions are periodic with period π. See Figure 75 for a partial justification. You are asked to prove this statement in Problems 97 and 98.

These facts are summarized as follows:

Periodic Properties

$$\sin(\theta + 2\pi) = \sin\theta \qquad \cos(\theta + 2\pi) = \cos\theta \qquad \tan(\theta + \pi) = \tan\theta$$
$$\csc(\theta + 2\pi) = \csc\theta \qquad \sec(\theta + 2\pi) = \sec\theta \qquad \cot(\theta + \pi) = \cot\theta$$

In Words
Tangent and cotangent have period π; the others have period 2π.

Because the sine, cosine, secant, and cosecant functions have period 2π, once we know their values for $0 \le \theta < 2\pi$, we know all their values; similarly, since the tangent and cotangent functions have period π, once we know their values for $0 \le \theta < \pi$, we know all their values.

EXAMPLE 3

Using Periodic Properties to Find Exact Values

Find the exact value of:

(a) $\sin 420°$ 　　　　(b) $\tan \dfrac{5\pi}{4}$ 　　　　(c) $\cos \dfrac{11\pi}{4}$

Solution

(a) $\sin 420° = \sin(60° + 360°) = \sin 60° = \dfrac{\sqrt{3}}{2}$

(b) $\tan \dfrac{5\pi}{4} = \tan\left(\dfrac{\pi}{4} + \pi\right) = \tan \dfrac{\pi}{4} = 1$

(c) $\cos \dfrac{11\pi}{4} = \cos\left(\dfrac{3\pi}{4} + \dfrac{8\pi}{4}\right) = \cos\left(\dfrac{3\pi}{4} + 2\pi\right) = \cos \dfrac{3\pi}{4} = -\dfrac{\sqrt{2}}{2}$

The periodic properties of the trigonometric functions will be very helpful to us when we study their graphs in the next section.

Now Work PROBLEMS 21 AND 79

4 Use Even–Odd Properties to Find the Exact Values of the Trigonometric Functions

Recall that a function f is even if $f(-\theta) = f(\theta)$ for all θ in the domain of f; a function f is odd if $f(-\theta) = -f(\theta)$ for all θ in the domain of f. We will now show that the trigonometric functions sine, tangent, cotangent, and cosecant are odd functions and the functions cosine and secant are even functions.

THEOREM

Even–Odd Properties

$$\sin(-\theta) = -\sin\theta \qquad \cos(-\theta) = \cos\theta \qquad \tan(-\theta) = -\tan\theta$$
$$\csc(-\theta) = -\csc\theta \qquad \sec(-\theta) = \sec\theta \qquad \cot(-\theta) = -\cot\theta$$

Proof Let $P = (a, b)$ be the point on the unit circle that corresponds to the angle θ. See Figure 76. The point Q on the unit circle that corresponds to the angle $-\theta$ will have coordinates $(a, -b)$. Using the definition for the trigonometric functions, we have

$$\sin\theta = b \qquad \sin(-\theta) = -b \qquad \cos\theta = a \qquad \cos(-\theta) = a$$

Figure 76

so

$$\sin(-\theta) = -\sin\theta \qquad\qquad \cos(-\theta) = \cos\theta$$

Now, using these results and some of the Fundamental Identities, we have

$$\tan(-\theta) = \frac{\sin(-\theta)}{\cos(-\theta)} = \frac{-\sin\theta}{\cos\theta} = -\tan\theta$$

$$\cot(-\theta) = \frac{1}{\tan(-\theta)} = \frac{1}{-\tan\theta} = -\cot\theta$$

$$\sec(-\theta) = \frac{1}{\cos(-\theta)} = \frac{1}{\cos\theta} = \sec\theta$$

$$\csc(-\theta) = \frac{1}{\sin(-\theta)} = \frac{1}{-\sin\theta} = -\csc\theta \qquad\blacksquare$$

In Words
Cosine and secant are even functions; the others are odd functions.

EXAMPLE 4 **Finding Exact Values Using Even–Odd Properties**

Find the exact value of:

(a) $\sin(-45°)$ (b) $\cos(-\pi)$ (c) $\cot\left(-\dfrac{3\pi}{2}\right)$ (d) $\tan\left(-\dfrac{37\pi}{4}\right)$

Solution (a) $\sin(-45°) \underset{\substack{\uparrow \\ \text{Odd function}}}{=} -\sin 45° = -\dfrac{\sqrt{2}}{2}$

(b) $\cos(-\pi) \underset{\substack{\uparrow \\ \text{Even function}}}{=} \cos\pi = -1$

(c) $\cot\left(-\dfrac{3\pi}{2}\right) \underset{\substack{\uparrow \\ \text{Odd function}}}{=} -\cot\dfrac{3\pi}{2} = 0$

(d) $\tan\left(-\dfrac{37\pi}{4}\right) \underset{\substack{\uparrow \\ \text{Odd function}}}{=} -\tan\dfrac{37\pi}{4} = -\tan\left(\dfrac{\pi}{4} + 9\pi\right) \underset{\substack{\uparrow \\ \text{Period is }\pi.}}{=} -\tan\dfrac{\pi}{4} = -1$

Now Work PROBLEMS 37 AND 73

5.5 Assess Your Understanding

'Are You Prepared?' *Answers are given at the end of these exercises. If you get a wrong answer, read the pages listed in* red.

1. What is the equation of the unit circle? (p. 35)

2. The domain of the function $f(x) = \dfrac{3x - 6}{x - 4}$ is _____. (pp. 49–50)

3. A function for which $f(x) = f(-x)$ for all x in the domain of f is called a(n) _____ function. (pp. 68–69)

Concepts and Vocabulary

4. The sine, cosine, cosecant, and secant functions have period _____; the tangent and cotangent functions have period _____.

5. Let t be a real number and let $p = (a, b)$ be the point on the unit circle that corresponds to t. Then $\sin t =$ ____ and $\cos t =$ ____.

6. For any angle θ in standard position, let $p = (a, b)$ be any point on the terminal side of θ that is also on the circle $x^2 + y^2 = r^2$. Then $\sin\theta =$ _____ and $\cos\theta =$ _____.

7. If $\sin\theta = 0.2$, then $\sin(-\theta) =$ _____ and $\sin(\theta + 2\pi) =$ _____.

8. **True or False** The only even trigonometric functions are the cosine and secant functions.

Skill Building

In Problems 9–14, the point P on the unit circle that corresponds to a real number t is given. Find $\sin t, \cos t, \tan t, \csc t, \sec t,$ *and* $\cot t$.

9. $\left(\dfrac{\sqrt{3}}{2}, -\dfrac{1}{2}\right)$ 10. $\left(-\dfrac{\sqrt{3}}{2}, -\dfrac{1}{2}\right)$ 11. $\left(-\dfrac{\sqrt{2}}{2}, -\dfrac{\sqrt{2}}{2}\right)$ 12. $\left(\dfrac{\sqrt{2}}{2}, -\dfrac{\sqrt{2}}{2}\right)$ 13. $\left(\dfrac{\sqrt{5}}{3}, \dfrac{2}{3}\right)$ 14. $\left(-\dfrac{\sqrt{5}}{5}, \dfrac{2\sqrt{5}}{5}\right)$

In Problems 15–20, the point P on the circle $x^2 + y^2 = r^2$ that is also on the terminal side of an angle θ in standard position is given. Find $\sin\theta, \cos\theta, \tan\theta, \csc\theta, \sec\theta,$ *and* $\cot\theta$.

15. $(3, -4)$ 16. $(5, -12)$ 17. $(-2, 3)$ 18. $(2, -4)$ 19. $(-1, -1)$ 20. $(-3, 1)$

In Problems 21–36, use the fact that the trigonometric functions are periodic to find the exact value of each expression. Do not use a calculator.

21. $\sin 405°$
22. $\cos 420°$
23. $\tan 405°$
24. $\sin 390°$
25. $\csc 450°$
26. $\sec 540°$
27. $\cot 390°$
28. $\sec 420°$
29. $\cos\dfrac{33\pi}{4}$
30. $\sin\dfrac{9\pi}{4}$
31. $\tan(21\pi)$
32. $\csc\dfrac{9\pi}{2}$
33. $\sec\dfrac{17\pi}{4}$
34. $\cot\dfrac{17\pi}{4}$
35. $\tan\dfrac{19\pi}{6}$
36. $\sec\dfrac{25\pi}{6}$

In Problems 37–54, use the even–odd properties to find the exact value of each expression. Do not use a calculator.

37. $\sin(-60°)$
38. $\cos(-30°)$
39. $\tan(-30°)$
40. $\sin(-135°)$
41. $\sec(-60°)$
42. $\csc(-30°)$
43. $\sin(-90°)$
44. $\cos(-270°)$
45. $\tan\left(-\dfrac{\pi}{4}\right)$
46. $\sin(-\pi)$
47. $\cos\left(-\dfrac{\pi}{4}\right)$
48. $\sin\left(-\dfrac{\pi}{3}\right)$
49. $\tan(-\pi)$
50. $\sin\left(-\dfrac{3\pi}{2}\right)$
51. $\csc\left(-\dfrac{\pi}{4}\right)$
52. $\sec(-\pi)$
53. $\sec\left(-\dfrac{\pi}{6}\right)$
54. $\csc\left(-\dfrac{\pi}{3}\right)$

In Problems 55–60, find the exact value of each expression. Do not use a calculator.

55. $\sin(-\pi) + \cos(5\pi)$
56. $\tan\left(-\dfrac{5\pi}{6}\right) - \cot\dfrac{7\pi}{2}$
57. $\sec(-\pi) + \csc\left(-\dfrac{\pi}{2}\right)$
58. $\tan(-6\pi) + \cos\dfrac{9\pi}{4}$
59. $\sin\left(-\dfrac{9\pi}{4}\right) - \tan\left(-\dfrac{9\pi}{4}\right)$
60. $\cos\left(-\dfrac{17\pi}{4}\right) - \sin\left(-\dfrac{3\pi}{2}\right)$

61. What is the domain of the sine function?
62. What is the domain of the cosine function?
63. For what numbers θ is $f(\theta) = \tan\theta$ not defined?
64. For what numbers θ is $f(\theta) = \cot\theta$ not defined?
65. For what numbers θ is $f(\theta) = \sec\theta$ not defined?
66. For what numbers θ is $f(\theta) = \csc\theta$ not defined?
67. What is the range of the sine function?
68. What is the range of the cosine function?
69. What is the range of the tangent function?
70. What is the range of the cotangent function?
71. What is the range of the secant function?
72. What is the range of the cosecant function?
73. Is the sine function even, odd, or neither? Is its graph symmetric? With respect to what?
74. Is the cosine function even, odd, or neither? Is its graph symmetric? With respect to what?

75. Is the tangent function even, odd, or neither? Is its graph symmetric? With respect to what?

76. Is the cotangent function even, odd, or neither? Is its graph symmetric? With respect to what?

77. Is the secant function even, odd, or neither? Is its graph symmetric? With respect to what?

78. Is the cosecant function even, odd, or neither? Is its graph symmetric? With respect to what?

79. If $\sin \theta = 0.3$, find the value of:
$$\sin \theta + \sin(\theta + 2\pi) + \sin(\theta + 4\pi)$$

80. If $\cos \theta = 0.2$, find the value of:
$$\cos \theta + \cos(\theta + 2\pi) + \cos(\theta + 4\pi)$$

81. If $\tan \theta = 3$, find the value of:
$$\tan \theta + \tan(\theta + \pi) + \tan(\theta + 2\pi)$$

82. If $\cot \theta = -2$, find the value of:
$$\cot \theta + \cot(\theta - \pi) + \cot(\theta - 2\pi)$$

In Problems 83–88, use the periodic and even–odd properties.

83. If $f(x) = \sin x$ and $f(a) = \dfrac{1}{3}$, find the exact value of:
(a) $f(-a)$ (b) $f(a) + f(a + 2\pi) + f(a + 4\pi)$

84. If $f(x) = \cos x$ and $f(a) = \dfrac{1}{4}$, find the exact value of:
(a) $f(-a)$ (b) $f(a) + f(a + 2\pi) + f(a - 2\pi)$

85. If $f(x) = \tan x$ and $f(a) = 2$, find the exact value of:
(a) $f(-a)$ (b) $f(a) + f(a + \pi) + f(a + 2\pi)$

86. If $f(x) = \cot x$ and $f(a) = -3$, find the exact value of:
(a) $f(-a)$ (b) $f(a) + f(a + \pi) + f(a + 4\pi)$

87. If $f(x) = \sec x$ and $f(a) = -4$, find the exact value of:
(a) $f(-a)$ (b) $f(a) + f(a + 2\pi) + f(a + 4\pi)$

88. If $f(x) = \csc x$ and $f(a) = 2$, find the exact value of:
(a) $f(-a)$ (b) $f(a) + f(a + 2\pi) + f(a + 4\pi)$

Applications and Extensions

In Problems 89–90, use the figure to approximate the value of the six trigonometric functions at t to the nearest tenth. Then use a calculator to approximate each of the six trigonometric functions at t.

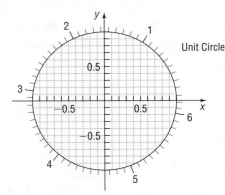

Unit Circle

89. (a) $t = 1$ (b) $t = 5.1$

90. (a) $t = 2$ (b) $t = 4$

91. Show that the range of the tangent function is the set of all real numbers.

92. Show that the range of the cotangent function is the set of all real numbers.

93. Show that the period of $f(\theta) = \sin \theta$ is 2π.

[**Hint:** Assume that $0 < p < 2\pi$ exists so that $\sin(\theta + p) = \sin \theta$ for all θ. Let $\theta = 0$ to find p. Then let $\theta = \dfrac{\pi}{2}$ to obtain a contradiction.]

94. Show that the period of $f(\theta) = \cos \theta$ is 2π.

95. Show that the period of $f(\theta) = \sec \theta$ is 2π.

96. Show that the period of $f(\theta) = \csc \theta$ is 2π.

97. Show that the period of $f(\theta) = \tan \theta$ is π.

98. Show that the period of $f(\theta) = \cot \theta$ is π.

99. If $\theta, 0 < \theta < \pi$, is the angle between a horizontal ray directed to the right (say, the positive x-axis) and a nonhorizontal, nonvertical line L, show that the slope m of L equals $\tan \theta$. The angle θ is called the **inclination** of L.

[**Hint:** See the illustration, where we have drawn the line M parallel to L and passing through the origin. Use the fact that M intersects the unit circle at the point $(\cos \theta, \sin \theta)$.]

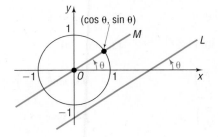

Explaining Concepts: Discussion and Writing

100. Explain how you would find the value of $\sin 390°$ using periodic properties.

101. Explain how you would find the value of $\cos(-45°)$ using even–odd properties.

102. Write down five properties of the tangent function. Explain the meaning of each.

103. Describe your understanding of the meaning of a periodic function.

'Are You Prepared?' Answers

1. $x^2 + y^2 = 1$ **2.** $\{x \mid x \neq 4\}$ **3.** even

5.6 Graphs of the Sine and Cosine Functions*

PREPARING FOR THIS SECTION *Before getting started, review the following:*

- Graphing Techniques: Transformations (Section 1.5, pp. 90–99)

Now Work the 'Are You Prepared?' problems on page 445.

OBJECTIVES **1** Graph Functions of the Form $y = A \sin(\omega x)$ Using Transformations (p. 436)
2 Graph Functions of the Form $y = A \cos(\omega x)$ Using Transformations (p. 438)
3 Determine the Amplitude and Period of Sinusoidal Functions (p. 439)
4 Graph Sinusoidal Functions Using Key Points (p. 440)
5 Find an Equation for a Sinusoidal Graph (p. 444)

Since we want to graph the trigonometric functions in the xy-plane, we shall use the traditional symbols x for the independent variable (or argument) and y for the dependent variable (or value at x) for each function. So we write the six trigonometric functions as

$$y = f(x) = \sin x \qquad y = f(x) = \cos x \qquad y = f(x) = \tan x$$
$$y = f(x) = \csc x \qquad y = f(x) = \sec x \qquad y = f(x) = \cot x$$

 Here the independent variable x represents an angle, measured in radians. In calculus, x will usually be treated as a real number. As we said earlier, these are equivalent ways of viewing x.

Table 7

x	$y = \sin x$	(x, y)
0	0	$(0, 0)$
$\dfrac{\pi}{6}$	$\dfrac{1}{2}$	$\left(\dfrac{\pi}{6}, \dfrac{1}{2}\right)$
$\dfrac{\pi}{2}$	1	$\left(\dfrac{\pi}{2}, 1\right)$
$\dfrac{5\pi}{6}$	$\dfrac{1}{2}$	$\left(\dfrac{5\pi}{6}, \dfrac{1}{2}\right)$
π	0	$(\pi, 0)$
$\dfrac{7\pi}{6}$	$-\dfrac{1}{2}$	$\left(\dfrac{7\pi}{6}, -\dfrac{1}{2}\right)$
$\dfrac{3\pi}{2}$	-1	$\left(\dfrac{3\pi}{2}, -1\right)$
$\dfrac{11\pi}{6}$	$-\dfrac{1}{2}$	$\left(\dfrac{11\pi}{6}, -\dfrac{1}{2}\right)$
2π	0	$(2\pi, 0)$

The Graph of the Sine Function $y = \sin x$

Since the sine function has period 2π, we only need to graph $y = \sin x$ on the interval $[0, 2\pi]$. The remainder of the graph will consist of repetitions of this portion of the graph.

We begin by constructing Table 7, which lists some points on the graph of $y = \sin x$, $0 \le x \le 2\pi$. As the table shows, the graph of $y = \sin x$, $0 \le x \le 2\pi$, begins at the origin. As x increases from 0 to $\dfrac{\pi}{2}$, the value of $y = \sin x$ increases from 0 to 1; as x increases from $\dfrac{\pi}{2}$ to π to $\dfrac{3\pi}{2}$, the value of y decreases from 1 to 0 to -1; as x increases from $\dfrac{3\pi}{2}$ to 2π, the value of y increases from -1 to 0. If we plot the points listed in Table 7 and connect them with a smooth curve, we obtain the graph shown in Figure 77.

Figure 77
$y = \sin x, 0 \le x \le 2\pi$

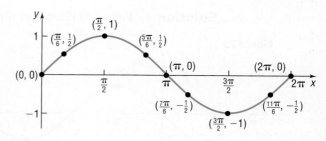

The graph in Figure 77 is one period, or **cycle,** of the graph of $y = \sin x$. To obtain a more complete graph of $y = \sin x$, we continue the graph in each direction, as shown in Figure 78.

*For those who wish to include phase shifts here, Section 5.8 can be covered immediately after Section 5.6 without loss of continuity.

Figure 78
$y = \sin x, -\infty < x < \infty$

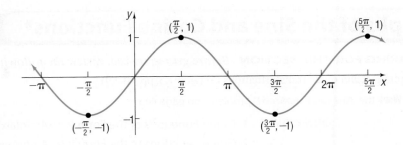

The graph of $y = \sin x$ illustrates some of the facts that we already know about the sine function.

Properties of the Sine Function $y = \sin x$

1. The domain is the set of all real numbers.
2. The range consists of all real numbers from -1 to 1, inclusive.
3. The sine function is an odd function, as the symmetry of the graph with respect to the origin indicates.
4. The sine function is periodic, with period 2π.
5. The x-intercepts are $\ldots, -2\pi, -\pi, 0, \pi, 2\pi, 3\pi, \ldots$; the y-intercept is 0.
6. The maximum value is 1 and occurs at $x = \ldots, -\dfrac{3\pi}{2}, \dfrac{\pi}{2}, \dfrac{5\pi}{2}, \dfrac{9\pi}{2}, \ldots$; the minimum value is -1 and occurs at $x = \ldots, -\dfrac{\pi}{2}, \dfrac{3\pi}{2}, \dfrac{7\pi}{2}, \dfrac{11\pi}{2}, \ldots$.

━━━▶ **Now Work** PROBLEM 9

1 Graph Functions of the Form $y = A \sin(\omega x)$ Using Transformations

EXAMPLE 1

Graphing Functions of the Form $y = A \sin(\omega x)$ Using Transformations

Graph $y = 3 \sin x$ using transformations.

Solution   Figure 79 illustrates the steps.

Figure 79

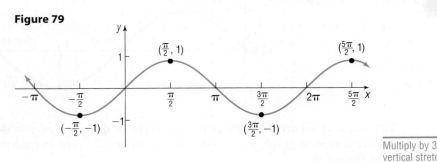

(a) $y = \sin x$

Multiply by 3
vertical stretch
by a factor of 3

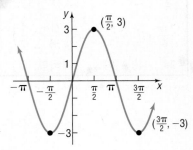

(b) $y = 3 \sin x$

EXAMPLE 2

Graphing Functions of the Form $y = A \sin(\omega x)$ Using Transformations

Graph $y = -\sin(2x)$ using transformations.

Solution Figure 80 illustrates the steps.

Figure 80

Multiply by -1;
Reflect about the
x-axis

Replace x by $2x$;
Horizontal compression
by a factor of $\frac{1}{2}$

(a) $y = \sin x$

(b) $y = -\sin x$

(c) $y = -\sin(2x)$

Notice in Figure 80(c) that the period of the function $y = -\sin(2x)$ is π due to the horizontal compression of the original period 2π by a factor of $\frac{1}{2}$.

 Now Work PROBLEM 37 USING TRANSFORMATIONS

The Graph of the Cosine Function

The cosine function also has period 2π. We proceed as we did with the sine function by constructing Table 8, which lists some points on the graph of $y = \cos x$, $0 \leq x \leq 2\pi$. As the table shows, the graph of $y = \cos x$, $0 \leq x \leq 2\pi$, begins at the point $(0, 1)$. As x increases from 0 to $\frac{\pi}{2}$ to π, the value of y decreases from 1 to 0 to -1; as x increases from π to $\frac{3\pi}{2}$ to 2π, the value of y increases from -1 to 0 to 1. As before, we plot the points in Table 8 to get one period or cycle of the graph. See Figure 81.

Table 8

x	$y = \cos x$	(x, y)
0	1	$(0, 1)$
$\dfrac{\pi}{3}$	$\dfrac{1}{2}$	$\left(\dfrac{\pi}{3}, \dfrac{1}{2}\right)$
$\dfrac{\pi}{2}$	0	$\left(\dfrac{\pi}{2}, 0\right)$
$\dfrac{2\pi}{3}$	$-\dfrac{1}{2}$	$\left(\dfrac{2\pi}{3}, -\dfrac{1}{2}\right)$
π	-1	$(\pi, -1)$
$\dfrac{4\pi}{3}$	$-\dfrac{1}{2}$	$\left(\dfrac{4\pi}{3}, -\dfrac{1}{2}\right)$
$\dfrac{3\pi}{2}$	0	$\left(\dfrac{3\pi}{2}, 0\right)$
$\dfrac{5\pi}{3}$	$\dfrac{1}{2}$	$\left(\dfrac{5\pi}{3}, \dfrac{1}{2}\right)$
2π	1	$(2\pi, 1)$

Figure 81
$y = \cos x$, $0 \leq x \leq 2\pi$

A more complete graph of $y = \cos x$ is obtained by continuing the graph in each direction, as shown in Figure 82.

Figure 82
$y = \cos x$, $-\infty < x < \infty$

The graph of $y = \cos x$ illustrates some of the facts that we already know about the cosine function.

Properties of the Cosine Function

1. The domain is the set of all real numbers.
2. The range consists of all real numbers from -1 to 1, inclusive.
3. The cosine function is an even function, as the symmetry of the graph with respect to the y-axis indicates.
4. The cosine function is periodic, with period 2π.
5. The x-intercepts are $\ldots, -\dfrac{3\pi}{2}, -\dfrac{\pi}{2}, \dfrac{\pi}{2}, \dfrac{3\pi}{2}, \dfrac{5\pi}{2}, \ldots$; the y-intercept is 1.
6. The maximum value is 1 and occurs at $x = \ldots, -2\pi, 0, 2\pi, 4\pi, 6\pi, \ldots$; the minimum value is -1 and occurs at $x = \ldots, -\pi, \pi, 3\pi, 5\pi, \ldots$.

2 Graph Functions of the Form $y = A\cos(\omega x)$ Using Transformations

EXAMPLE 3 **Graphing Functions of the Form $y = A\cos(\omega x)$ Using Transformations**

Graph $y = 2\cos(3x)$ using transformations.

Solution Figure 83 shows the steps.

Figure 83

(a) $y = \cos x$

Multiply by 2; Vertical stretch by a factor of 2

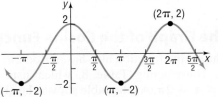

(b) $y = 2\cos x$

Replace x by 3x; Horizontal compression by a factor of $\frac{1}{3}$

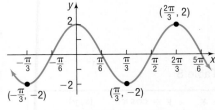

(c) $y = 2\cos(3x)$

Notice in Figure 83(c) that the period of the function $y = 2\cos(3x)$ is $\dfrac{2\pi}{3}$ due to the compression of the original period 2π by a factor of $\dfrac{1}{3}$.

Now Work PROBLEM **45** USING TRANSFORMATIONS

Sinusoidal Graphs

Shift the graph of $y = \cos x$ to the right $\dfrac{\pi}{2}$ units to obtain the graph of $y = \cos\left(x - \dfrac{\pi}{2}\right)$. See Figure 84(a). Now look at the graph of $y = \sin x$ in Figure 84(b). We see that the graph of $y = \sin x$ is the same as the graph of $y = \cos\left(x - \dfrac{\pi}{2}\right)$.

Figure 84

(a) $y = \cos x$ $y = \cos\left(x - \frac{\pi}{2}\right)$

(b) $y = \sin x$

Based on Figure 84, we conjecture that

$$\sin x = \cos\left(x - \frac{\pi}{2}\right)$$

Seeing the Concept

Graph $Y_1 = \sin x$ and $Y_2 = \cos\left(x - \frac{\pi}{2}\right)$.

How many graphs do you see?

(We shall prove this fact in Chapter 6.) Because of this relationship, the graphs of functions of the form $y = A \sin(\omega x)$ or $y = A \cos(\omega x)$ are referred to as **sinusoidal graphs.**

Let's look at some general properties of sinusoidal graphs.

3 Determine the Amplitude and Period of Sinusoidal Functions

In Figure 85(b) we show the graph of $y = 2 \cos x$. Notice that the values of $y = 2 \cos x$ lie between -2 and 2, inclusive.

Figure 85

(a) $y = \cos x$

Multiply by 2; Vertical stretch by a factor of 2

(b) $y = 2 \cos x$

In general, the values of the functions $y = A \sin x$ and $y = A \cos x$, where $A \neq 0$, will always satisfy the inequalities

$$-|A| \leq A \sin x \leq |A| \quad \text{and} \quad -|A| \leq A \cos x \leq |A|$$

respectively. The number $|A|$ is called the **amplitude** of $y = A \sin x$ or $y = A \cos x$. See Figure 86.

Figure 86

$y = A \sin x, A > 0$
Period $= 2\pi$

In Figure 87(b), we show the graph of $y = \cos(3x)$. Notice that the period of this function is $\dfrac{2\pi}{3}$, due to the horizontal compression of the original period 2π by a factor of $\dfrac{1}{3}$.

Figure 87

(a) $y = \cos x$

Replace x by $3x$; Horizontal compression by a factor of $\frac{1}{3}$

(b) $y = \cos(3x)$

In general, if $\omega > 0$, the functions $y = \sin(\omega x)$ and $y = \cos(\omega x)$ will have period $T = \dfrac{2\pi}{\omega}$. To see why, recall that the graph of $y = \sin(\omega x)$ is obtained from the graph of $y = \sin x$ by performing a horizontal compression or stretch by a factor $\dfrac{1}{\omega}$. This horizontal compression replaces the interval $[0, 2\pi]$, which contains one period of the graph of $y = \sin x$, by the interval $\left[0, \dfrac{2\pi}{\omega}\right]$, which contains one period of the graph of $y = \sin(\omega x)$. So, the function $y = \cos(3x)$, graphed in Figure 87(b), $\omega = 3$, has period $\dfrac{2\pi}{\omega} = \dfrac{2\pi}{3}$.

One period of the graph of $y = \sin(\omega x)$ or $y = \cos(\omega x)$ is called a **cycle**. Figure 88 illustrates the general situation. The blue portion of the graph is one cycle.

Figure 88

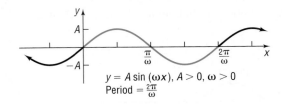

$y = A \sin(\omega x),\ A > 0,\ \omega > 0$
Period $= \frac{2\pi}{\omega}$

NOTE Recall that a function f is even if $f(-x) = f(x)$; a function f is odd if $f(-x) = -f(x)$. Since the sine function is odd, $\sin(-x) = -\sin x$; since the cosine function is even, $\cos(-x) = \cos x$. ∎

When graphing $y = \sin(\omega x)$ or $y = \cos(\omega x)$, we want ω to be positive. To graph $y = \sin(-\omega x), \omega > 0$ or $y = \cos(-\omega x), \omega > 0$, we use the Even–Odd Properties of the sine and cosine functions as follows:

$$\sin(-\omega x) = -\sin(\omega x) \quad \text{and} \quad \cos(-\omega x) = \cos(\omega x)$$

This gives us an equivalent form in which the coefficient of x in the argument is positive. For example,

$$\sin(-2x) = -\sin(2x) \quad \text{and} \quad \cos(-\pi x) = \cos(\pi x)$$

Because of this, we can assume that $\omega > 0$.

THEOREM

If $\omega > 0$, the amplitude and period of $y = A \sin(\omega x)$ and $y = A \cos(\omega x)$ are given by

$$\text{Amplitude} = |A| \qquad \text{Period} = T = \frac{2\pi}{\omega} \qquad \textbf{(1)}$$

EXAMPLE 4

Finding the Amplitude and Period of a Sinusoidal Function

Determine the amplitude and period of $y = 3 \sin(4x)$.

Solution

Comparing $y = 3 \sin(4x)$ to $y = A \sin(\omega x)$, we find that $A = 3$ and $\omega = 4$. From equation (1),

$$\text{Amplitude} = |A| = 3 \qquad \text{Period} = T = \frac{2\pi}{\omega} = \frac{2\pi}{4} = \frac{\pi}{2}$$

●

━━ **Now Work** PROBLEM 15

4 Graph Sinusoidal Functions Using Key Points

So far, we have graphed functions of the form $y = A \sin(\omega x)$ or $y = A \cos(\omega x)$ using transformations. We now introduce another method that can be used to graph these functions.

Figure 89 shows one cycle of the graphs of $y = \sin x$ and $y = \cos x$ on the interval $[0, 2\pi]$. Notice that each graph consists of four parts corresponding to the four subintervals:

$$\left[0, \frac{\pi}{2}\right], \quad \left[\frac{\pi}{2}, \pi\right], \quad \left[\pi, \frac{3\pi}{2}\right], \quad \left[\frac{3\pi}{2}, 2\pi\right]$$

Each subinterval is of length $\frac{\pi}{2}$ (the period 2π divided by 4, the number of parts), and the endpoints of these intervals $x = 0$, $x = \frac{\pi}{2}$, $x = \pi$, $x = \frac{3\pi}{2}$, $x = 2\pi$ give rise to five key points on each graph:

For $y = \sin x$: $(0, 0), \left(\frac{\pi}{2}, 1\right), (\pi, 0), \left(\frac{3\pi}{2}, -1\right), (2\pi, 0)$

For $y = \cos x$: $(0, 1), \left(\frac{\pi}{2}, 0\right), (\pi, -1), \left(\frac{3\pi}{2}, 0\right), (2\pi, 1)$

Look again at Figure 89.

Figure 89

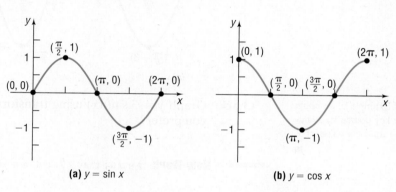

(a) $y = \sin x$ **(b)** $y = \cos x$

Let's look at an example that illustrates how to graph a sinusoidal function using key points.

EXAMPLE 5

How to Graph a Sinusoidal Function Using Key Points

Graph $y = 3 \sin(4x)$ using key points.

Step-by-Step Solution

Step 1: Determine the amplitude and period of the sinusoidal function.

Comparing $y = 3 \sin(4x)$ to $y = A \sin(\omega x)$, we see that $A = 3$ and $\omega = 4$, so the amplitude is $|A| = 3$ and the period is $\frac{2\pi}{\omega} = \frac{2\pi}{4} = \frac{\pi}{2}$. Because the amplitude is 3, the graph of $y = 3 \sin(4x)$ will lie between -3 and 3 on the y-axis. Because the period is $\frac{\pi}{2}$, one cycle will begin at $x = 0$ and end at $x = \frac{\pi}{2}$.

Step 2: Divide the interval $\left[0, \frac{2\pi}{\omega}\right]$ into four subintervals of the same length.

We divide the interval $\left[0, \frac{\pi}{2}\right]$ into four subintervals, each of length $\frac{\pi}{2} \div 4 = \frac{\pi}{8}$, as follows:

$$\left[0, \frac{\pi}{8}\right] \quad \left[\frac{\pi}{8}, \frac{\pi}{8} + \frac{\pi}{8}\right] = \left[\frac{\pi}{8}, \frac{\pi}{4}\right] \quad \left[\frac{\pi}{4}, \frac{\pi}{4} + \frac{\pi}{8}\right] = \left[\frac{\pi}{4}, \frac{3\pi}{8}\right] \quad \left[\frac{3\pi}{8}, \frac{3\pi}{8} + \frac{\pi}{8}\right] = \left[\frac{3\pi}{8}, \frac{\pi}{2}\right]$$

The endpoints of the subintervals are $0, \frac{\pi}{8}, \frac{\pi}{4}, \frac{3\pi}{8}, \frac{\pi}{2}$. These values represent the x-coordinates of the five key points on the graph.

Step 3: Use the endpoints of these subintervals to obtain five key points on the graph.

To obtain the y-coordinates of the five key points of $y = 3 \sin(4x)$, we multiply the y-coordinates of the five key points for $y = \sin x$ in Figure 89(a) by $A = 3$. The five key points are

$$(0, 0) \quad \left(\frac{\pi}{8}, 3\right) \quad \left(\frac{\pi}{4}, 0\right) \quad \left(\frac{3\pi}{8}, -3\right) \quad \left(\frac{\pi}{2}, 0\right)$$

Step 4: Plot the five key points and draw a sinusoidal graph to obtain the graph of one cycle. Extend the graph in each direction to make it complete.

We plot the five key points obtained in Step 3 and fill in the graph of the sine curve as shown in Figure 90(a). We extend the graph in each direction to obtain the complete graph shown in Figure 90(b). Notice that additional key points appear every $\frac{\pi}{8}$ radian.

Figure 90

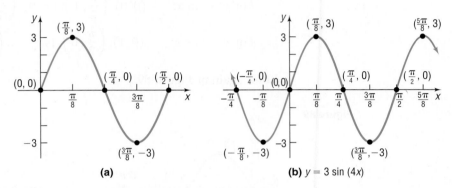

(a)

(b) $y = 3 \sin (4x)$

NOTE In Step 3 of Example 5, we could also obtain the five key points by evaluating $y = 3 \sin(4x)$ at each value of x. ■

✓**Check:** Graph $y = 3 \sin(4x)$ using transformations. Which graphing method do you prefer?

━━━ **Now Work** PROBLEM 37 USING KEY POINTS

SUMMARY

Steps for Graphing a Sinusoidal Function of the Form $y = A \sin(\omega x)$ or $y = A \cos(\omega x)$ Using Key Points

STEP 1: Determine the amplitude and period of the sinusoidal function.

STEP 2: Divide the interval $\left[0, \frac{2\pi}{\omega}\right]$ into four subintervals of the same length.

STEP 3: Use the endpoints of these subintervals to obtain five key points on the graph.

STEP 4: Plot the five key points and draw a sinusoidal graph to obtain the graph of one cycle. Extend the graph in each direction to make it complete.

EXAMPLE 6 **Graphing a Sinusoidal Function Using Key Points**

Graph $y = 2 \sin\left(-\frac{\pi}{2}x\right)$ using key points.

Solution Since the sine function is odd, we can use the equivalent form:

$$y = -2 \sin\left(\frac{\pi}{2}x\right)$$

STEP 1: Comparing $y = -2 \sin\left(\frac{\pi}{2}x\right)$ to $y = A \sin(\omega x)$, we find that $A = -2$ and $\omega = \frac{\pi}{2}$. The amplitude is $|A| = |-2| = 2$, and the period is $T = \frac{2\pi}{\omega} = \frac{2\pi}{\frac{\pi}{2}} = 4$.

The graph of $y = -2\sin\left(\dfrac{\pi}{2}x\right)$ will lie between -2 and 2 on the y-axis. One cycle will begin at $x = 0$ and end at $x = 4$.

STEP 2: We divide the interval $[0, 4]$ into four subintervals, each of length $4 \div 4 = 1$. The x-coordinates of the five key points are

0	$0 + 1 = 1$	$1 + 1 = 2$	$2 + 1 = 3$	$3 + 1 = 4$
1st x-coordinate	2nd x-coordinate	3rd x-coordinate	4th x-coordinate	5th x-coordinate

> **COMMENT** To graph a sinusoidal function of the form $y = A\sin(\omega x)$ or $y = A\cos(\omega x)$ using a graphing utility, we use the amplitude to set $Y_{\min}$ and $Y_{\max}$ and use the period to set $X_{\min}$ and $X_{\max}$. ∎

STEP 3: Since $y = -2\sin\left(\dfrac{\pi}{2}x\right)$, we multiply the y-coordinates of the five key points in Figure 89(a) by $A = -2$. The five key points on the graph are

$$(0, 0) \quad (1, -2) \quad (2, 0) \quad (3, 2) \quad (4, 0)$$

STEP 4: We plot these five points and fill in the graph of the sine function as shown in Figure 91(a). Extending the graph in each direction, we obtain Figure 91(b).

Figure 91

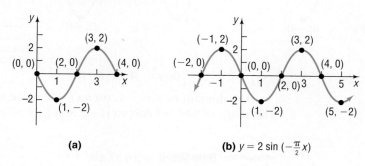

(a)

(b) $y = 2\sin\left(-\dfrac{\pi}{2}x\right)$

✓**Check:** Graph $y = 2\sin\left(-\dfrac{\pi}{2}x\right)$ using transformations. Which graphing method do you prefer?

●

━━ **Now Work** PROBLEM 41 USING KEY POINTS

If the function to be graphed is of the form $y = A\sin(\omega x) + B$ [or $y = A\cos(\omega x) + B$], first graph $y = A\sin(\omega x)$ [or $y = A\cos(\omega x)$] and then use a vertical shift.

EXAMPLE 7

Graphing a Sinusoidal Function Using Key Points

Graph $y = -4\cos(\pi x) - 2$ using key points. Use the graph to determine the domain and the range of $y = -4\cos(\pi x) - 2$.

Solution

We begin by graphing the function $y = -4\cos(\pi x)$. Comparing $y = -4\cos(\pi x)$ with $y = A\cos(\omega x)$, we find that $A = -4$ and $\omega = \pi$. The amplitude is $|A| = |-4| = 4$, and the period is $T = \dfrac{2\pi}{\omega} = \dfrac{2\pi}{\pi} = 2$.

The graph of $y = -4\cos(\pi x)$ will lie between -4 and 4 on the y-axis. One cycle will begin at $x = 0$ and end at $x = 2$.

We divide the interval $[0, 2]$ into four subintervals, each of length $2 \div 4 = \dfrac{1}{2}$. The x-coordinates of the five key points are

0	$0 + \dfrac{1}{2} = \dfrac{1}{2}$	$\dfrac{1}{2} + \dfrac{1}{2} = 1$	$1 + \dfrac{1}{2} = \dfrac{3}{2}$	$\dfrac{3}{2} + \dfrac{1}{2} = 2$
1st x-coordinate	2nd x-coordinate	3rd x-coordinate	4th x-coordinate	5th x-coordinate

Since $y = -4\cos(\pi x)$, we multiply the y-coordinates of the five key points of $y = \cos x$ shown in Figure 89(b) by $A = -4$ to obtain the five key points on the graph of $y = -4\cos(\pi x)$:

$$(0, -4) \quad \left(\frac{1}{2}, 0\right) \quad (1, 4) \quad \left(\frac{3}{2}, 0\right) \quad (2, -4)$$

We plot these five points and fill in the graph of the cosine function as shown in Figure 92(a). Extending the graph in each direction, we obtain Figure 92(b), the graph of $y = -4\cos(\pi x)$.

A vertical shift down 2 units gives the graph of $y = -4\cos(\pi x) - 2$, as shown in Figure 92(c).

Figure 92

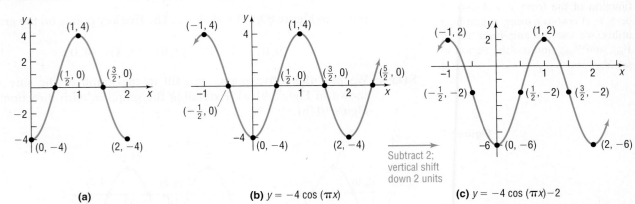

(a)   **(b)** $y = -4\cos(\pi x)$ **(c)** $y = -4\cos(\pi x) - 2$

The domain of $y = -4\cos(\pi x) - 2$ is the set of all real numbers or $(-\infty, \infty)$. The range of $y = -4\cos(\pi x) - 2$ is $\{y \mid -6 \le y \le 2\}$ or $[-6, 2]$.

Now Work   PROBLEM 51

5 Find an Equation for a Sinusoidal Graph

We can also use the ideas of amplitude and period to identify a sinusoidal function when its graph is given.

EXAMPLE 8

Finding an Equation for a Sinusoidal Graph

Find an equation for the graph shown in Figure 93.

Figure 93

Solution   The graph has the characteristics of a cosine function. Do you see why? The maximum value, 3, occurs at $x = 0$. So we view the equation as a cosine function $y = A\cos(\omega x)$ with $A = 3$ and period $T = 1$. Then $\dfrac{2\pi}{\omega} = 1$, so $\omega = 2\pi$. The cosine function whose graph is given in Figure 93 is

$$y = A\cos(\omega x) = 3\cos(2\pi x)$$

✓ **Check:**   Graph $Y_1 = 3\cos(2\pi x)$ and compare the result with Figure 93.

EXAMPLE 9   **Finding an Equation for a Sinusoidal Graph**

Find an equation for the graph shown in Figure 94.

Figure 94

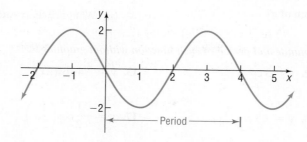

Solution   The graph is sinusoidal, with amplitude $|A| = 2$. The period is 4, so $\dfrac{2\pi}{\omega} = 4$ or $\omega = \dfrac{\pi}{2}$. Since the graph passes through the origin, it is easiest to view the equation as a sine function,[†] but notice that the graph is actually the reflection of a sine function about the x-axis (since the graph is decreasing near the origin). This requires that $A = -2$. The sine function whose graph is given in Figure 94 is

$$y = A \sin(\omega x) = -2 \sin\left(\frac{\pi}{2} x\right)$$

 ✓**Check:** Graph $Y_1 = -2 \sin\left(\dfrac{\pi}{2} x\right)$ and compare the result with Figure 94.

●

Now Work PROBLEMS 59 AND 63

5.6 Assess Your Understanding

'Are You Prepared?' *Answers are given at the end of these exercises. If you get a wrong answer, read the pages listed in red.*

1. Use transformations to graph $y = 3x^2$. (pp. 93–94)

2. Use transformations to graph $y = \sqrt{2x}$. (p. 96)

Concepts and Vocabulary

3. The maximum value of $y = \sin x$, $0 \le x \le 2\pi$, is _____ and occurs at $x =$ _____.

4. The function $y = A \sin(\omega x)$, $A > 0$, has amplitude 3 and period 2; then $A =$ _____ and $\omega =$ _____.

5. The function $y = 3 \cos(6x)$ has amplitude _____ and period _____.

6. *True or False* The graphs of $y = \sin x$ and $y = \cos x$ are identical except for a horizontal shift.

7. *True or False* For $y = 2 \sin(\pi x)$, the amplitude is 2 and the period is $\dfrac{\pi}{2}$.

8. *True or False* The graph of the sine function has infinitely many x-intercepts.

Skill Building

9. $f(x) = \sin x$

(a) What is the y-intercept of the graph of f?

(b) For what numbers x, $-\pi \le x \le \pi$, is the graph of f increasing?

(c) What is the absolute maximum of f?

(d) For what numbers x, $0 \le x \le 2\pi$, does $f(x) = 0$?

(e) For what numbers x, $-2\pi \le x \le 2\pi$, does $f(x) = 1$? Where does $f(x) = -1$?

(f) For what numbers x, $-2\pi \le x \le 2\pi$, does $f(x) = -\dfrac{1}{2}$?

(g) What are the x-intercepts of f?

[†]The equation could also be viewed as a cosine function with a horizontal shift, but viewing it as a sine function is easier.

10. $g(x) = \cos x$

(a) What is the y-intercept of the graph of g?

(b) For what numbers x, $-\pi \le x \le \pi$, is the graph of g decreasing?

(c) What is the absolute minimum of g?

(d) For what numbers x, $0 \le x \le 2\pi$, does $g(x) = 0$?

(e) For what numbers x, $-2\pi \le x \le 2\pi$, does $g(x) = 1$? Where does $g(x) = -1$?

(f) For what numbers x, $-2\pi \le x \le 2\pi$, does $g(x) = \dfrac{\sqrt{3}}{2}$?

(g) What are the x-intercepts of g?

In Problems 11–20, determine the amplitude and period of each function without graphing.

11. $y = 2 \sin x$

12. $y = 3 \cos x$

13. $y = -4 \cos(2x)$

14. $y = -\sin\left(\dfrac{1}{2}x\right)$

15. $y = 6 \sin(\pi x)$

16. $y = -3 \cos(3x)$

17. $y = -\dfrac{1}{2}\cos\left(\dfrac{3}{2}x\right)$

18. $y = \dfrac{4}{3}\sin\left(\dfrac{2}{3}x\right)$

19. $y = \dfrac{5}{3}\sin\left(-\dfrac{2\pi}{3}x\right)$

20. $y = \dfrac{9}{5}\cos\left(-\dfrac{3\pi}{2}x\right)$

In Problems 21–30, match the given function to one of the graphs (A)–(J).

(A)

(B)

(C)

(D)

(E)

(F)

(G)

(H)

(I)

(J)

21. $y = 2 \sin\left(\dfrac{\pi}{2}x\right)$

22. $y = 2 \cos\left(\dfrac{\pi}{2}x\right)$

23. $y = 2 \cos\left(\dfrac{1}{2}x\right)$

24. $y = 3 \cos(2x)$

25. $y = -3 \sin(2x)$

26. $y = 2 \sin\left(\dfrac{1}{2}x\right)$

27. $y = -2 \cos\left(\dfrac{1}{2}x\right)$

28. $y = -2 \cos\left(\dfrac{\pi}{2}x\right)$

29. $y = 3 \sin(2x)$

30. $y = -2 \sin\left(\dfrac{1}{2}x\right)$

In Problems 31–34, match the given function to one of the graphs (A)–(D).

(A)

(B)

(C)

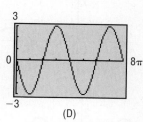

(D)

31. $y = 3 \sin\left(\dfrac{1}{2}x\right)$

32. $y = -3 \sin(2x)$

33. $y = 3 \sin(2x)$

34. $y = -3 \sin\left(\dfrac{1}{2}x\right)$

In Problems 35–58, graph each function. Be sure to label key points and show at least two cycles. Use the graph to determine the domain and the range of each function.

35. $y = 4 \cos x$

36. $y = 3 \sin x$

37. $y = -4 \sin x$

38. $y = -3 \cos x$

39. $y = \cos(4x)$

40. $y = \sin(3x)$

41. $y = \sin(-2x)$

42. $y = \cos(-2x)$

43. $y = 2 \sin\left(\dfrac{1}{2}x\right)$

44. $y = 2 \cos\left(\dfrac{1}{4}x\right)$

45. $y = -\dfrac{1}{2}\cos(2x)$

46. $y = -4 \sin\left(\dfrac{1}{8}x\right)$

47. $y = 2 \sin x + 3$

48. $y = 3 \cos x + 2$

49. $y = 5 \cos(\pi x) - 3$

50. $y = 4 \sin\left(\dfrac{\pi}{2}x\right) - 2$

51. $y = -6 \sin\left(\dfrac{\pi}{3}x\right) + 4$

52. $y = -3 \cos\left(\dfrac{\pi}{4}x\right) + 2$

53. $y = 5 - 3 \sin(2x)$

54. $y = 2 - 4 \cos(3x)$

55. $y = \dfrac{5}{3}\sin\left(-\dfrac{2\pi}{3}x\right)$

56. $y = \dfrac{9}{5}\cos\left(-\dfrac{3\pi}{2}x\right)$

57. $y = -\dfrac{3}{2}\cos\left(\dfrac{\pi}{4}x\right) + \dfrac{1}{2}$

58. $y = -\dfrac{1}{2}\sin\left(\dfrac{\pi}{8}x\right) + \dfrac{3}{2}$

In Problems 59–62, write the equation of a sine function that has the given characteristics.

59. Amplitude: 3
Period: π

60. Amplitude: 2
Period: 4π

61. Amplitude: 3
Period: 2

62. Amplitude: 4
Period: 1

In Problems 63–76, find an equation for each graph.

63.

64.

65.

66.

67.

68.

69.

70.

71.

72.

73.

74.

75.

76.

Mixed Practice

In Problems 77–80, find the average rate of change of f from 0 to $\dfrac{\pi}{2}$.

77. $f(x) = \sin x$ **78.** $f(x) = \cos x$ **79.** $f(x) = \sin\left(\dfrac{x}{2}\right)$ **80.** $f(x) = \cos(2x)$

In Problems 81–84, find $(f \circ g)(x)$ and $(g \circ f)(x)$ and graph each of these functions.

81. $f(x) = \sin x$
    $g(x) = 4x$

82. $f(x) = \cos x$
    $g(x) = \dfrac{1}{2}x$

83. $f(x) = -2x$
    $g(x) = \cos x$

84. $f(x) = -3x$
    $g(x) = \sin x$

Applications and Extensions

85. Alternating Current (ac) Circuits The current I, in amperes, flowing through an ac (alternating current) circuit at time t in seconds, is

$$I(t) = 220 \sin(60\pi t) \qquad t \geq 0$$

What is the period? What is the amplitude? Graph this function over two periods.

86. Alternating Current (ac) Circuits The current I, in amperes, flowing through an ac (alternating current) circuit at time t in seconds, is

$$I(t) = 120 \sin(30\pi t) \qquad t \geq 0$$

What is the period? What is the amplitude? Graph this function over two periods.

87. Alternating Current (ac) Generators The voltage V, in volts, produced by an ac generator at time t, in seconds, is

$$V(t) = 220 \sin(120\pi t)$$

(a) What is the amplitude? What is the period?
(b) Graph V over two periods, beginning at $t = 0$.
(c) If a resistance of $R = 10$ ohms is present, what is the current I?
     [**Hint:** Use Ohm's Law, $V = IR$.]
(d) What is the amplitude and period of the current I?
(e) Graph I over two periods, beginning at $t = 0$.

88. Alternating Current (ac) Generators The voltage V, in volts, produced by an ac generator at time t, in seconds, is

$$V(t) = 120 \sin(120\pi t)$$

(a) What is the amplitude? What is the period?
(b) Graph V over two periods, beginning at $t = 0$.
(c) If a resistance of $R = 20$ ohms is present, what is the current I?
     [**Hint:** Use Ohm's Law, $V = IR$.]
(d) What is the amplitude and period of the current I?
(e) Graph I over two periods, beginning at $t = 0$.

89. Alternating Current (ac) Generators The voltage V produced by an ac generator is sinusoidal. As a function of time, the voltage V is

$$V(t) = V_0 \sin(2\pi f t)$$

where f is the **frequency,** the number of complete oscillations (cycles) per second. [In the United States and Canada, f is 60 hertz (Hz).] The **power** P delivered to a resistance R at any time t is defined as

$$P(t) = \frac{[V(t)]^2}{R}$$

(a) Show that $P(t) = \dfrac{V_0^2}{R}\sin^2(2\pi f t)$.

(b) The graph of P is shown in the figure. Express P as a sinusoidal function.

Power in an ac generator

(c) Deduce that
$$\sin^2(2\pi ft) = \frac{1}{2}[1 - \cos(4\pi ft)]$$

90. Bridge Clearance A one-lane highway runs through a tunnel in the shape of one-half a sine curve cycle. The opening is 28 feet wide at road level and is 15 feet tall at its highest point.

(a) Find an equation for the sine curve that fits the opening. Place the origin at the left end of the sine curve.
(b) If the road is 14 feet wide with 7-foot shoulders on each side, what is the height of the tunnel at the edge of the road?

Source: en.wikipedia.org/wiki/Interstate_Highway_standards and *Ohio Revised Code*

91. Biorhythms In the theory of biorhythms, a sine function of the form
$$P(t) = 50\sin(\omega t) + 50$$
is used to measure the percent P of a person's potential at time t, where t is measured in days and $t = 0$ is the person's birthday. Three characteristics are commonly measured:

Physical potential: period of 23 days
Emotional potential: period of 28 days
Intellectual potential: period of 33 days

(a) Find ω for each characteristic.
(b) Using a graphing utility, graph all three functions on the same screen.
(c) Is there a time t when all three characteristics have 100% potential? When is it?
(d) Suppose that you are 20 years old today ($t = 7305$ days). Describe your physical, emotional, and intellectual potential for the next 30 days.

92. Graph $y = |\cos x|$, $-2\pi \le x \le 2\pi$.

93. Graph $y = |\sin x|$, $-2\pi \le x \le 2\pi$.

In Problems 94–97, the graphs of the given pairs of functions intersect infinitely many times. Find four of these points of intersection.

94. $y = \sin x$
$y = \frac{1}{2}$

95. $y = \cos x$
$y = \frac{1}{2}$

96. $y = 2\sin x$
$y = -2$

97. $y = \tan x$
$y = 1$

Explaining Concepts: Discussion and Writing

98. Explain how you would scale the x-axis and y-axis before graphing $y = 3\cos(\pi x)$.

99. Explain the term *amplitude* as it relates to the graph of a sinusoidal function.

100. Explain how the amplitude and period of a sinusoidal graph are used to establish the scale on each coordinate axis.

101. Find an application in your major field that leads to a sinusoidal graph. Write a paper about your findings.

'Are You Prepared?' Answers

1. Vertical stretch by a factor of 3

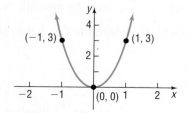

2. Horizontal compression by a factor of $\frac{1}{2}$

5.7 Graphs of the Tangent, Cotangent, Cosecant, and Secant Functions

PREPARING FOR THIS SECTION *Before getting started, review the following:*

• Vertical Asymptotes (Section 3.2, pp. 217–219)

Now Work the 'Are You Prepared?' problems on page 456.

> **OBJECTIVES** 1 Graph Functions of the Form $y = A \tan(\omega x) + B$
> and $y = A \cot(\omega x) + B$ (p. 452)
> 2 Graph Functions of the Form $y = A \csc(\omega x) + B$
> and $y = A \sec(\omega x) + B$ (p. 454)

The Graph of the Tangent Function

Because the tangent function has period π, we only need to determine the graph over some interval of length π. The rest of the graph will consist of repetitions of that graph. Because the tangent function is not defined at $\ldots$, $-\dfrac{3\pi}{2}, -\dfrac{\pi}{2}, \dfrac{\pi}{2}, \dfrac{3\pi}{2}, \ldots$, we will concentrate on the interval $\left(-\dfrac{\pi}{2}, \dfrac{\pi}{2}\right)$, of length π, and construct Table 9, which lists some points on the graph of $y = \tan x$, $-\dfrac{\pi}{2} < x < \dfrac{\pi}{2}$. We plot the points in the table and connect them with a smooth curve. See Figure 95 for a partial graph of $y = \tan x$, where $-\dfrac{\pi}{3} \le x \le \dfrac{\pi}{3}$.

Table 9

x	$y = \tan x$	(x, y)
$-\dfrac{\pi}{3}$	$-\sqrt{3} \approx -1.73$	$\left(-\dfrac{\pi}{3}, -\sqrt{3}\right)$
$-\dfrac{\pi}{4}$	-1	$\left(-\dfrac{\pi}{4}, -1\right)$
$-\dfrac{\pi}{6}$	$-\dfrac{\sqrt{3}}{3} \approx -0.58$	$\left(-\dfrac{\pi}{6}, -\dfrac{\sqrt{3}}{3}\right)$
0	0	$(0, 0)$
$\dfrac{\pi}{6}$	$\dfrac{\sqrt{3}}{3} \approx 0.58$	$\left(\dfrac{\pi}{6}, \dfrac{\sqrt{3}}{3}\right)$
$\dfrac{\pi}{4}$	1	$\left(\dfrac{\pi}{4}, 1\right)$
$\dfrac{\pi}{3}$	$\sqrt{3} \approx 1.73$	$\left(\dfrac{\pi}{3}, \sqrt{3}\right)$

Figure 95
$y = \tan x, -\dfrac{\pi}{3} \le x \le \dfrac{\pi}{3}$

To complete one period of the graph of $y = \tan x$, we need to investigate the behavior of the function as x approaches $-\dfrac{\pi}{2}$ and $\dfrac{\pi}{2}$. We must be careful, though, because $y = \tan x$ is not defined at these numbers. To determine this behavior, we use the identity

$$\tan x = \frac{\sin x}{\cos x}$$

See Table 10. If x is close to $\dfrac{\pi}{2} \approx 1.5708$, but remains less than $\dfrac{\pi}{2}$, then $\sin x$ will be close to 1 and $\cos x$ will be positive and close to 0. (To see this, refer back to the graphs of the sine function and the cosine function.) So the ratio $\dfrac{\sin x}{\cos x}$ will be positive and large. In fact, the closer x gets to $\dfrac{\pi}{2}$, the closer $\sin x$ gets to 1 and $\cos x$ gets to 0, so $\tan x$ approaches ∞ $\left(\displaystyle\lim_{x \to \frac{\pi}{2}^-} \tan x = \infty\right)$. In other words, the vertical line $x = \dfrac{\pi}{2}$ is a vertical asymptote to the graph of $y = \tan x$.

Table 10

x	$\sin x$	$\cos x$	$y = \tan x$
$\dfrac{\pi}{3} \approx 1.05$	$\dfrac{\sqrt{3}}{2}$	$\dfrac{1}{2}$	$\sqrt{3} \approx 1.73$
1.5	0.9975	0.0707	14.1
1.57	0.9999	7.96×10^{-4}	1255.8
1.5707	0.9999	9.6×10^{-5}	10,381
$\dfrac{\pi}{2} \approx 1.5708$	1	0	Undefined

If x is close to $-\dfrac{\pi}{2}$, but remains greater than $-\dfrac{\pi}{2}$, then $\sin x$ will be close to -1 and $\cos x$ will be positive and close to 0. The ratio $\dfrac{\sin x}{\cos x}$ approaches $-\infty$ $\left(\displaystyle\lim_{x \to -\frac{\pi}{2}^+} \tan x = -\infty\right)$. In other words, the vertical line $x = -\dfrac{\pi}{2}$ is also a vertical asymptote to the graph.

With these observations, we can complete one period of the graph. We obtain the complete graph of $y = \tan x$ by repeating this period, as shown in Figure 96.

Figure 96
$y = \tan x$, $-\infty < x < \infty$, x not equal to odd multiples of $\dfrac{\pi}{2}$, $-\infty < y < \infty$

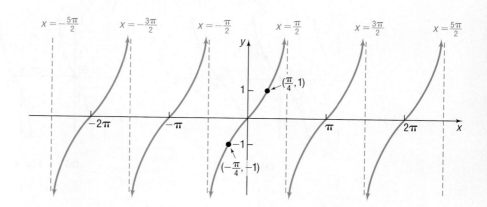

Check: Graph $Y_1 = \tan x$ and compare the result with Figure 96. Use TRACE to see what happens as x gets close to $\dfrac{\pi}{2}$, but is less than $\dfrac{\pi}{2}$.

The graph of $y = \tan x$ in Figure 96 illustrates the following properties.

Properties of the Tangent Function

1. The domain is the set of all real numbers, except odd multiples of $\dfrac{\pi}{2}$.
2. The range is the set of all real numbers.
3. The tangent function is an odd function, as the symmetry of the graph with respect to the origin indicates.

4. The tangent function is periodic, with period π.

5. The x-intercepts are $\ldots, -2\pi, -\pi, 0, \pi, 2\pi, 3\pi, \ldots$; the y-intercept is 0.

6. Vertical asymptotes occur at $x = \ldots, -\dfrac{3\pi}{2}, -\dfrac{\pi}{2}, \dfrac{\pi}{2}, \dfrac{3\pi}{2}, \ldots$.

─── **Now Work** PROBLEMS 7 AND 15

1 Graph Functions of the Form $y = A\tan(\omega x) + B$

For tangent functions, there is no concept of amplitude since the range of the tangent function is $(-\infty, \infty)$. The role of A in $y = A\tan(\omega x) + B$ is to provide the magnitude of the vertical stretch. The period of $y = \tan x$ is π, so the period of $y = A\tan(\omega x) + B$ is $\dfrac{\pi}{\omega}$, caused by the horizontal compression of the graph by a factor of $\dfrac{1}{\omega}$. Finally, the presence of B indicates that a vertical shift is required.

EXAMPLE 1 | **Graphing Functions of the Form $y = A\tan(\omega x) + B$**

Graph: $y = 2\tan x - 1$. Use the graph to determine the domain and the range of $y = 2\tan x - 1$.

Solution Figure 97 shows the steps using transformations.

Figure 97

The domain of $y = 2\tan x - 1$ is $\left\{x \middle| x \neq \dfrac{k\pi}{2}, k \text{ is an odd integer}\right\}$, and the range is the set of all real numbers, or $(-\infty, \infty)$.

✓ **Check:** Graph $Y_1 = 2\tan x - 1$ to verify the graph shown in Figure 97(c).

EXAMPLE 2 | **Graphing Functions of the Form $y = A\tan(\omega x) + B$**

Graph $y = 3\tan(2x)$. Use the graph to determine the domain and the range of $y = 3\tan(2x)$.

Solution Figure 98 shows the steps using transformations.

Figure 98

(a) $y = \tan x$

Multiply by 3;
Vertical stretch by
a factor of 3

(b) $y = 3 \tan x$

Replace x by $2x$;
Horizontal compression
by a factor of $\frac{1}{2}$

(c) $y = 3 \tan (2x)$

The domain of $y = 3 \tan (2x)$ is $\{x \mid x \neq \dfrac{k\pi}{4}, k \text{ is an odd integer}\}$, and the range is the set of all real numbers or $(-\infty, \infty)$.

Check: Graph $Y_1 = 3 \tan(2x)$ to verify the graph in Figure 98(c).

Notice in Figure 98(c) that the period of $y = 3 \tan(2x)$ is $\dfrac{\pi}{2}$ due to the compression of the original period π by a factor of $\dfrac{1}{2}$. Notice that the asymptotes are $x = -\dfrac{\pi}{4}, x = \dfrac{\pi}{4}, x = \dfrac{3\pi}{4}$, and so on, also due to the compression.

Now Work PROBLEM 21

The Graph of the Cotangent Function

We obtain the graph of $y = \cot x$ as we did the graph of $y = \tan x$. The period of $y = \cot x$ is π. Because the cotangent function is not defined for integer multiples of π, we will concentrate on the interval $(0, \pi)$. Table 11 lists some points on the graph of $y = \cot x, 0 < x < \pi$. As x approaches 0, but remains greater than 0, the value of $\cos x$ will be close to 1 and the value of $\sin x$ will be positive and close to 0. Hence, the ratio $\dfrac{\cos x}{\sin x} = \cot x$ will be positive and large; so as x approaches 0, with $x > 0$, $\cot x$ approaches $\infty (\lim\limits_{x \to 0^+} \cot x = \infty)$. Similarly, as x approaches π, but remains less than π, the value of $\cos x$ will be close to -1, and the value of $\sin x$ will be positive and close to 0. So the ratio $\dfrac{\cos x}{\sin x} = \cot x$ will be negative and will approach $-\infty$ as x approaches $\pi (\lim\limits_{x \to \pi^-} \cot x = -\infty)$. Figure 99 shows the graph.

Table 11

x	$y = \cot x$	(x, y)
$\dfrac{\pi}{6}$	$\sqrt{3}$	$\left(\dfrac{\pi}{6}, \sqrt{3}\right)$
$\dfrac{\pi}{4}$	1	$\left(\dfrac{\pi}{4}, 1\right)$
$\dfrac{\pi}{3}$	$\dfrac{\sqrt{3}}{3}$	$\left(\dfrac{\pi}{3}, \dfrac{\sqrt{3}}{3}\right)$
$\dfrac{\pi}{2}$	0	$\left(\dfrac{\pi}{2}, 0\right)$
$\dfrac{2\pi}{3}$	$-\dfrac{\sqrt{3}}{3}$	$\left(\dfrac{2\pi}{3}, -\dfrac{\sqrt{3}}{3}\right)$
$\dfrac{3\pi}{4}$	-1	$\left(\dfrac{3\pi}{4}, -1\right)$
$\dfrac{5\pi}{6}$	$-\sqrt{3}$	$\left(\dfrac{5\pi}{6}, -\sqrt{3}\right)$

Figure 99
$y = \cot x, -\infty < x < \infty, x$ not equal to integer multiples of π,
$-\infty < y < \infty$

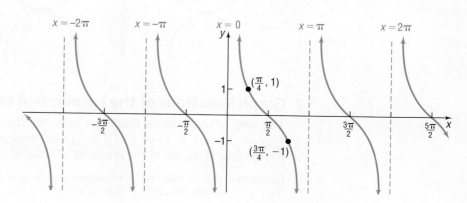

The graph of $y = A \cot(\omega x) + B$ has similar characteristics to those of the tangent function. The cotangent function $y = A \cot(\omega x) + B$ has period $\dfrac{\pi}{\omega}$. The cotangent function has no amplitude. The role of A is to provide the magnitude of the vertical stretch; the presence of B indicates a vertical shift is required.

━━━━━━ **Now Work** PROBLEM 23

The Graphs of the Cosecant Function and the Secant Function

The cosecant and secant functions, sometimes referred to as **reciprocal functions,** are graphed by making use of the reciprocal identities

$$\csc x = \frac{1}{\sin x} \quad \text{and} \quad \sec x = \frac{1}{\cos x}$$

For example, the value of the cosecant function $y = \csc x$ at a given number x equals the reciprocal of the corresponding value of the sine function, provided that the value of the sine function is not 0. If the value of $\sin x$ is 0, then x is an integer multiple of π. At such numbers, the cosecant function is not defined. In fact, the graph of the cosecant function has vertical asymptotes at integer multiples of π. Figure 100 shows the graph.

Figure 100
$y = \csc x$, $-\infty < x < \infty$, x not equal to integer multiples of π, $|y| \geq 1$

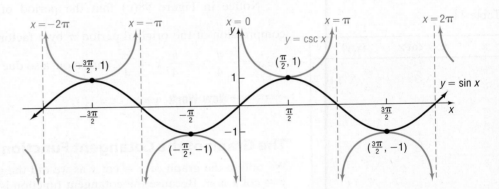

Using the idea of reciprocals, we can similarly obtain the graph of $y = \sec x$. See Figure 101.

Figure 101
$y = \sec x$, $-\infty < x < \infty$, x not equal to odd multiples of $\dfrac{\pi}{2}$, $|y| \geq 1$

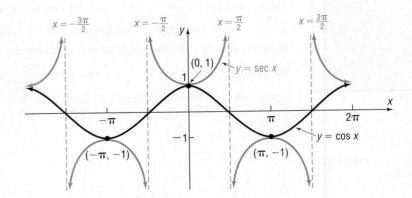

2 Graph Functions of the Form $y = A \csc(\omega x) + B$

The role of A in these functions is to set the range. The range of $y = \csc x$ is $\{y | y \leq -1 \text{ or } y \geq 1\}$ or $\{y | |y| \geq 1\}$; the range of $y = A \csc x$ is $\{y | |y| \geq |A|\}$, due to the vertical stretch of the graph by a factor of $|A|$. Just as with the sine and cosine functions, the period of $y = \csc(\omega x)$ and $y = \sec(\omega x)$ becomes $\dfrac{2\pi}{\omega}$, due to the horizontal compression of the graph by a factor of $\frac{1}{\omega}$. The presence of B indicates that a vertical shift is required.

We shall graph these functions in two ways: using transformations and using the reciprocal function.

EXAMPLE 3   **Graphing Functions of the Form $y = A \csc(\omega x) + B$**

Graph $y = 2 \csc x - 1$. Use the graph to determine the domain and the range of $y = 2 \csc x - 1$.

**Solution
Using Transformations**

Figure 102 shows the required steps.

Figure 102

(a) $y = \csc x$ Multiply by 2; Vertical stretch by a factor of 2   (b) $y = 2 \csc x$ Subtract 1; Vertical shift down 1 unit   (c) $y = 2 \csc x - 1$

**Solution
Using the Reciprocal
Function**

We graph $y = 2 \csc x - 1$ by first graphing the function $y = 2 \sin x - 1$ and then filling in the graph of $y = 2 \csc x - 1$, using the idea of reciprocals. See Figure 103.

Figure 103

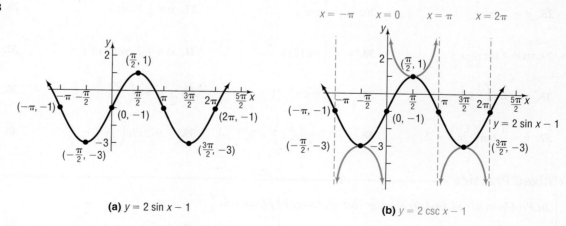

(a) $y = 2 \sin x - 1$ (b) $y = 2 \csc x - 1$

The domain of $y = 2 \csc x - 1$ is $\{x \mid x \neq k\pi, k \text{ is an integer}\}$ and the range is $\{y \mid y \leq -3 \text{ or } y \geq 1\}$ or, using interval notation, $(-\infty, -3] \cup [1, \infty)$.

✓ **Check:** Graph $Y_1 = 2 \csc x - 1$ to verify the graph shown in Figure 102 or 103.

━━━**Now Work** PROBLEM **29**

5.7 Assess Your Understanding

'Are You Prepared?' *Answers are given at the end of these exercises. If you get a wrong answer, read the pages listed in red.*

1. The graph of $y = \dfrac{3x - 6}{x - 4}$ has a vertical asymptote. What is it? (p. 219)

2. **True or False**   If $x = 3$ is a vertical asymptote of a function R, then $\lim\limits_{x \to 3} |R(x)| = \infty$. (pp. 217–219)

Concepts and Vocabulary

3. The graph of $y = \tan x$ is symmetric with respect to the _____ and has vertical asymptotes at _____.

4. The graph of $y = \sec x$ is symmetric with respect to the _____ and has vertical asymptotes at _____ .

5. It is easiest to graph $y = \sec x$ by first sketching the graph of _____.

6. **True or False**   The graphs of $y = \tan x, y = \cot x$, $y = \sec x$, and $y = \csc x$ each have infinitely many vertical asymptotes.

Skill Building

In Problems 7–16, if necessary, refer to the graphs to answer each question.

7. What is the y-intercept of $y = \tan x$?

8. What is the y-intercept of $y = \cot x$?

9. What is the y-intercept of $y = \sec x$?

10. What is the y-intercept of $y = \csc x$?

11. For what numbers x, $-2\pi \le x \le 2\pi$, does $\sec x = 1$? For what numbers x does $\sec x = -1$?

12. For what numbers x, $-2\pi \le x \le 2\pi$, does $\csc x = 1$? For what numbers x does $\csc x = -1$?

13. For what numbers x, $-2\pi \le x \le 2\pi$, does the graph of $y = \sec x$ have vertical asymptotes?

14. For what numbers x, $-2\pi \le x \le 2\pi$, does the graph of $y = \csc x$ have vertical asymptotes?

15. For what numbers x, $-2\pi \le x \le 2\pi$, does the graph of $y = \tan x$ have vertical asymptotes?

16. For what numbers x, $-2\pi \le x \le 2\pi$, does the graph of $y = \cot x$ have vertical asymptotes?

In Problems 17–40, graph each function. Be sure to label key points and show at least two cycles. Use the graph to determine the domain and the range of each function.

17. $y = 3 \tan x$

18. $y = -2 \tan x$

19. $y = 4 \cot x$

20. $y = -3 \cot x$

21. $y = \tan\left(\dfrac{\pi}{2}x\right)$

22. $y = \tan\left(\dfrac{1}{2}x\right)$

23. $y = \cot\left(\dfrac{1}{4}x\right)$

24. $y = \cot\left(\dfrac{\pi}{4}x\right)$

25. $y = 2 \sec x$

26. $y = \dfrac{1}{2} \csc x$

27. $y = -3 \csc x$

28. $y = -4 \sec x$

29. $y = 4 \sec\left(\dfrac{1}{2}x\right)$

30. $y = \dfrac{1}{2} \csc(2x)$

31. $y = -2 \csc(\pi x)$

32. $y = -3 \sec\left(\dfrac{\pi}{2}x\right)$

33. $y = \tan\left(\dfrac{1}{4}x\right) + 1$

34. $y = 2 \cot x - 1$

35. $y = \sec\left(\dfrac{2\pi}{3}x\right) + 2$

36. $y = \csc\left(\dfrac{3\pi}{2}x\right)$

37. $y = \dfrac{1}{2} \tan\left(\dfrac{1}{4}x\right) - 2$

38. $y = 3 \cot\left(\dfrac{1}{2}x\right) - 2$

39. $y = 2 \csc\left(\dfrac{1}{3}x\right) - 1$

40. $y = 3 \sec\left(\dfrac{1}{4}x\right) + 1$

Mixed Practice

In Problems 41–44, find the average rate of change of f from 0 to $\dfrac{\pi}{6}$.

41. $f(x) = \tan x$

42. $f(x) = \sec x$

43. $f(x) = \tan(2x)$

44. $f(x) = \sec(2x)$

In Problems 45–48, find $(f \circ g)(x)$ and $(g \circ f)(x)$ and graph each of these functions.

45. $f(x) = \tan x$
    $g(x) = 4x$

46. $f(x) = 2 \sec x$
    $g(x) = \dfrac{1}{2}x$

47. $f(x) = -2x$
    $g(x) = \cot x$

48. $f(x) = \dfrac{1}{2}x$
    $g(x) = 2 \csc x$

Applications and Extensions

49. Carrying a Ladder around a Corner Two hallways, one of width 3 feet, the other of width 4 feet, meet at a right angle. See the illustration.

(a) Show that the length L of the line segment shown as a function of the angle θ is
$$L(\theta) = 3 \sec \theta + 4 \csc \theta$$

(b) Graph $L = L(\theta), 0 < \theta < \dfrac{\pi}{2}$.

(c) For what value of θ is L the least?

(d) What is the length of the longest ladder that can be carried around the corner? Why is this also the least value of L?

50. A Rotating Beacon Suppose that a fire truck is parked in front of a building as shown in the figure.

The beacon light on top of the fire truck is located 10 feet from the wall and has a light on each side. If the beacon light rotates 1 revolution every 2 seconds, then a model for determining the distance d, in feet, that the beacon of light is from point A on the wall after t seconds is given by
$$d(t) = |10 \tan(\pi t)|$$

(a) Graph $d(t) = |10 \tan(\pi t)|$ for $0 \le t \le 2$.

(b) For what values of t is the function undefined? Explain what this means in terms of the beam of light on the wall.

(c) Fill in the following table.

t	0	0.1	0.2	0.3	0.4
$d(t) = 10\tan(\pi t)$					

(d) Compute $\dfrac{d(0.1) - d(0)}{0.1 - 0}, \dfrac{d(0.2) - d(0.1)}{0.2 - 0.1}$, and so on, for each consecutive value of t. These are called **first differences.**

(e) Interpret the first differences found in part (d). What is happening to the speed of the beam of light as d increases?

51. Exploration Graph
$$y = \tan x \quad \text{and} \quad y = -\cot\left(x + \frac{\pi}{2}\right)$$

Do you think that $\tan x = -\cot\left(x + \dfrac{\pi}{2}\right)$?

'Are You Prepared?' Answers

1. $x = 4$ **2.** True

5.8 Phase Shift; Sinusoidal Curve Fitting

> **OBJECTIVES** 1 Graph Sinusoidal Functions of the Form $y = A\sin(\omega x - \phi) + B$ (p. 457)
> 2 Build Sinusoidal Models from Data (p. 461)

Figure 104
One cycle of
$y = A\sin(\omega x), A > 0, \omega > 0$

1 Graph Sinusoidal Functions of the Form $y = A\sin(\omega x - \phi) + B$

We have seen that the graph of $y = A\sin(\omega x), \omega > 0$, has amplitude $|A|$ and period $T = \dfrac{2\pi}{\omega}$. One cycle can be drawn as x varies from 0 to $\dfrac{2\pi}{\omega}$ or, equivalently, as ωx varies from 0 to 2π. See Figure 104.

We now want to discuss the graph of

$$y = A \sin(\omega x - \phi)$$

which may also be written as

$$y = A \sin\left[\omega\left(x - \frac{\phi}{\omega}\right)\right]$$

where $\omega > 0$ and ϕ (the Greek letter phi) are real numbers. The graph will be a sine curve with amplitude $|A|$. As $\omega x - \phi$ varies from 0 to 2π, one period will be traced out. This period will begin when

NOTE We can also find the beginning and end of the period by solving the inequality:

$$0 \leq \omega x - \phi \leq 2\pi$$
$$\phi \leq \omega x \leq 2\pi + \phi$$
$$\frac{\phi}{\omega} \leq x \leq \frac{2\pi}{\omega} + \frac{\phi}{\omega}$$ ∎

$$\omega x - \phi = 0 \quad \text{or} \quad x = \frac{\phi}{\omega}$$

and will end when

$$\omega x - \phi = 2\pi \quad \text{or} \quad x = \frac{\phi}{\omega} + \frac{2\pi}{\omega}$$

See Figure 105.

We see that the graph of $y = A \sin(\omega x - \phi) = A \sin\left[\omega\left(x - \frac{\phi}{\omega}\right)\right]$ is the same as the graph of $y = A \sin(\omega x)$, except that it has been shifted $\frac{\phi}{\omega}$ units (to the right if $\phi > 0$ and to the left if $\phi < 0$). This number $\frac{\phi}{\omega}$ is called the **phase shift** of the graph of $y = A \sin(\omega x - \phi)$.

Figure 105
One cycle of $y = A \sin(\omega x - \phi)$, $A > 0$, $\omega > 0$, $\phi > 0$

For the graphs of $y = A \sin(\omega x - \phi)$ or $y = A \cos(\omega x - \phi)$, $\omega > 0$,

$$\text{Amplitude} = |A| \qquad \text{Period} = T = \frac{2\pi}{\omega} \qquad \text{Phase shift} = \frac{\phi}{\omega}$$

The phase shift is to the left if $\phi < 0$ and to the right if $\phi > 0$.

EXAMPLE 1 **Finding the Amplitude, Period, and Phase Shift of a Sinusoidal Function and Graphing It**

Find the amplitude, period, and phase shift of $y = 3 \sin(2x - \pi)$ and graph the function.

Solution We will use the same four steps used to graph sinusoidal functions of the form $y = A \sin(\omega x)$ or $y = A \cos(\omega x)$ given on page 442.

STEP 1: Comparing

$$y = 3 \sin(2x - \pi) = 3 \sin\left[2\left(x - \frac{\pi}{2}\right)\right]$$

to

$$y = A \sin(\omega x - \phi) = A \sin\left[\omega\left(x - \frac{\phi}{\omega}\right)\right]$$

we find that $A = 3$, $\omega = 2$, and $\phi = \pi$. The graph is a sine curve with amplitude $|A| = 3$, period $T = \frac{2\pi}{\omega} = \frac{2\pi}{2} = \pi$, and phase shift $= \frac{\phi}{\omega} = \frac{\pi}{2}$.

NOTE We also can find the interval defining one cycle by solving the inequality

$$0 \le 2x - \pi \le 2\pi$$

Then

$$\pi \le 2x \le 3\pi$$

$$\frac{\pi}{2} \le x \le \frac{3\pi}{2}$$ ∎

STEP 2: The graph of $y = 3\sin(2x - \pi)$ will lie between -3 and 3 on the y-axis. One cycle will begin at $x = \dfrac{\phi}{\omega} = \dfrac{\pi}{2}$ and end at $x = \dfrac{\phi}{\omega} + \dfrac{2\pi}{\omega} = \dfrac{\pi}{2} + \pi = \dfrac{3\pi}{2}$.
To find the five key points, we divide the interval $\left[\dfrac{\pi}{2}, \dfrac{3\pi}{2}\right]$ into four subintervals, each of length $\pi \div 4 = \dfrac{\pi}{4}$, by finding the following values of x:

$$\underset{\text{1st }x\text{-coordinate}}{\frac{\pi}{2}} \qquad \underset{\text{2nd }x\text{-coordinate}}{\frac{\pi}{2} + \frac{\pi}{4} = \frac{3\pi}{4}} \qquad \underset{\text{3rd }x\text{-coordinate}}{\frac{3\pi}{4} + \frac{\pi}{4} = \pi} \qquad \underset{\text{4th }x\text{-coordinate}}{\pi + \frac{\pi}{4} = \frac{5\pi}{4}} \qquad \underset{\text{5th }x\text{-coordinate}}{\frac{5\pi}{4} + \frac{\pi}{4} = \frac{3\pi}{2}}$$

STEP 3: Use these values of x to determine the five key points on the graph:

$$\left(\frac{\pi}{2}, 0\right) \quad \left(\frac{3\pi}{4}, 3\right) \quad (\pi, 0) \quad \left(\frac{5\pi}{4}, -3\right) \quad \left(\frac{3\pi}{2}, 0\right)$$

STEP 4: We plot these five points and fill in the graph of the sine function as shown in Figure 106(a). Extending the graph in each direction, we obtain Figure 106(b).

Figure 106

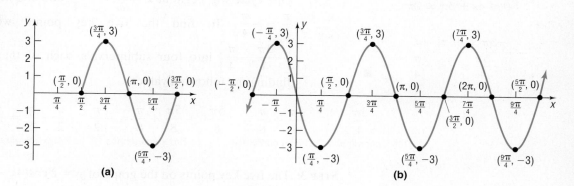

(a)

(b)

The graph of $y = 3\sin(2x - \pi) = 3\sin\left[2\left(x - \dfrac{\pi}{2}\right)\right]$ may also be obtained using transformations. See Figure 107.

Figure 107

(a) $y = \sin x$

Multiply by 3;
Vertical stretch
by a factor of 3

(b) $y = 3\sin x$

Replace x by $2x$;
Horizontal compression
by a factor of $\frac{1}{2}$

(c) $y = 3\sin(2x)$

Replace x by $x - \frac{\pi}{2}$;
Shift right
$\frac{\pi}{2}$ units

(d) $y = 3\sin\left[2\left(x - \frac{\pi}{2}\right)\right]$
$= 3\sin(2x - \pi)$

To graph a sinusoidal function of the form $y = A\sin(\omega x - \phi) + B$, we first graph the function $y = A\sin(\omega x - \phi)$ and then apply a vertical shift.

EXAMPLE 2   **Finding the Amplitude, Period, and Phase Shift of a Sinusoidal Function and Graphing It**

Find the amplitude, period, and phase shift of $y = 2\cos(4x + 3\pi) + 1$ and graph the function.

Solution   **STEP 1:** We begin by graphing $y = 2\cos(4x + 3\pi)$. Comparing

$$y = 2\cos(4x + 3\pi) = 2\cos\left[4\left(x + \frac{3\pi}{4}\right)\right]$$

to

$$y = A\cos(\omega x - \phi) = A\cos\left[\omega\left(x - \frac{\phi}{\omega}\right)\right]$$

we see that $A = 2$, $\omega = 4$, and $\phi = -3\pi$. The graph is a cosine curve with amplitude $|A| = 2$, period $T = \dfrac{2\pi}{\omega} = \dfrac{2\pi}{4} = \dfrac{\pi}{2}$, and phase shift $= \dfrac{\phi}{\omega} = -\dfrac{3\pi}{4}$.

NOTE We can also find the interval defining one cycle by solving the inequality

$$0 \le 4x + 3\pi \le 2\pi$$

Then

$$-3\pi \le 4x \le -\pi$$

$$-\frac{3\pi}{4} \le x \le -\frac{\pi}{4}$$ ∎

STEP 2: The graph of $y = 2\cos(4x + 3\pi)$ will lie between -2 and 2 on the y-axis. One cycle will begin at $x = \dfrac{\phi}{\omega} = -\dfrac{3\pi}{4}$ and end at $x = \dfrac{\phi}{\omega} + \dfrac{2\pi}{\omega} = -\dfrac{3\pi}{4} + \dfrac{\pi}{2} = -\dfrac{\pi}{4}$. To find the five key points, we divide the interval $\left[-\dfrac{3\pi}{4}, -\dfrac{\pi}{4}\right]$ into four subintervals, each of the length $\dfrac{\pi}{2} \div 4 = \dfrac{\pi}{8}$, by finding the following values.

$$-\frac{3\pi}{4} \qquad -\frac{3\pi}{4} + \frac{\pi}{8} = -\frac{5\pi}{8} \qquad -\frac{5\pi}{8} + \frac{\pi}{8} = -\frac{\pi}{2} \qquad -\frac{\pi}{2} + \frac{\pi}{8} = -\frac{3\pi}{8} \qquad -\frac{3\pi}{8} + \frac{\pi}{8} = -\frac{\pi}{4}$$

1st x-coordinate   2nd x-coordinate   3rd x-coordinate   4th x-coordinate   5th x-coordinate

STEP 3: The five key points on the graph of $y = 2\cos(4x + 3\pi)$ are

$$\left(-\frac{3\pi}{4}, 2\right) \quad \left(-\frac{5\pi}{8}, 0\right) \quad \left(-\frac{\pi}{2}, -2\right) \quad \left(-\frac{3\pi}{8}, 0\right) \quad \left(-\frac{\pi}{4}, 2\right)$$

STEP 4: We plot these five points and fill in the graph of the cosine function as shown in Figure 108(a). Extending the graph in each direction, we obtain Figure 108(b), the graph of $y = 2\cos(4x + 3\pi)$.

STEP 5: A vertical shift up 1 unit gives the final graph. See Figure 108(c).

Figure 108

(a)

(b) $y = 2\cos(4x + 3\pi)$

Add 1;
Vertical shift
up 1 unit

(c) $y = 2\cos(4x + 3\pi) + 1$

The graph of $y = 2\cos(4x + 3\pi) + 1 = 2\cos\left[4\left(x + \dfrac{3\pi}{4}\right)\right] + 1$ may also be obtained using transformations. See Figure 109.

Figure 109

(a) $y = 2 \cos x$

Replace x by $4x$;
Horizontal compression
by a factor of $\frac{1}{4}$

(b) $y = 2 \cos (4x)$

Replace x by $x + \frac{3\pi}{4}$;
Shift left $\frac{3\pi}{4}$ units

(c) $y = 2 \cos \left[4 \left(x + \frac{3\pi}{4} \right) \right]$
$= 2 \cos (4x + 3\pi)$

Add 1;
Vertical shift
up 1 unit

(d) $y = 2 \cos (4x + 3\pi) + 1$

──── **Now Work** PROBLEM 3

SUMMARY

> **Steps for Graphing Sinusoidal Functions $y = A \sin(\omega x - \phi) + B$ or $y = A \cos(\omega x - \phi) + B$**
>
> **STEP 1:** Determine the amplitude $|A|$ and period $T = \dfrac{2\pi}{\omega}$.
>
> **STEP 2:** Determine the starting point of one cycle of the graph, $\dfrac{\phi}{\omega}$. Determine the ending point of one cycle of the graph, $\dfrac{\phi}{\omega} + \dfrac{2\pi}{\omega}$. Divide the interval $\left[\dfrac{\phi}{\omega}, \dfrac{\phi}{\omega} + \dfrac{2\pi}{\omega} \right]$ into four subintervals, each of length $\dfrac{2\pi}{\omega} \div 4$.
>
> **STEP 3:** Use the endpoints of the subintervals to find the five key points on the graph.
>
> **STEP 4:** Plot the five key points with a sinusoidal graph to obtain one cycle of the graph. Extend the graph in each direction to make it complete.
>
> **STEP 5:** If $B \neq 0$, apply a vertical shift.

2 Build Sinusoidal Models from Data

Scatter diagrams of data sometimes take the form of a sinusoidal function. Let's look at an example.

The data given in Table 12 represent the average monthly temperatures in Denver, Colorado. Since the data represent *average* monthly temperatures collected over many years, the data will not vary much from year to year and so will essentially repeat each year. In other words, the data are periodic. Figure 110 shows the scatter diagram of these data repeated over 2 years, where $x = 1$ represents January, $x = 2$ represents February, and so on.

Notice that the scatter diagram looks like the graph of a sinusoidal function. We choose to fit the data to a sine function of the form

$$y = A \sin(\omega x - \phi) + B$$

where A, B, ω, and ϕ are constants.

Table 12

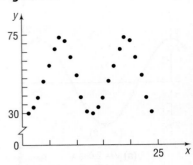

Figure 110

Month, x	Average Monthly Temperature, °F
January, 1	29.7
February, 2	33.4
March, 3	39.0
April, 4	48.2
May, 5	57.2
June, 6	66.9
July, 7	73.5
August, 8	71.4
September, 9	62.3
October, 10	51.4
November, 11	39.0
December, 12	31.0

Source: U.S. National Oceanic and Atmospheric Administration

EXAMPLE 3

Finding a Sinusoidal Function from Temperature Data

Fit a sine function to the data in Table 12.

Figure 111

Solution

We begin with a scatter diagram of the data for one year. See Figure 111. The data will be fitted to a sine function of the form

$$y = A \sin(\omega x - \phi) + B$$

STEP 1: To find the amplitude A, we compute

$$\text{Amplitude} = \frac{\text{largest data value} - \text{smallest data value}}{2}$$

$$= \frac{73.5 - 29.7}{2} = 21.9$$

To see the remaining steps in this process, we superimpose the graph of the function $y = 21.9 \sin x$, where x represents months, on the scatter diagram.

Figure 112 shows the two graphs. To fit the data, the graph needs to be shifted vertically, shifted horizontally, and stretched horizontally.

STEP 2: We determine the vertical shift by finding the average of the highest and lowest data values.

$$\text{Vertical shift} = \frac{73.5 + 29.7}{2} = 51.6$$

Now we superimpose the graph of $y = 21.9 \sin x + 51.6$ on the scatter diagram. See Figure 113.

Figure 112

Figure 113

Figure 114

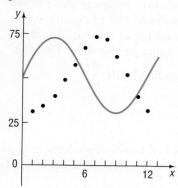

We see that the graph needs to be shifted horizontally and stretched horizontally.

STEP 3: It is easier to find the horizontal stretch factor first. Since the temperatures repeat every 12 months, the period of the function is $T = 12$. Since $T = \dfrac{2\pi}{\omega} = 12$, we find

$$\omega = \frac{2\pi}{12} = \frac{\pi}{6}$$

Now we superimpose the graph of $y = 21.9 \sin\left(\dfrac{\pi}{6}x\right) + 51.6$ on the scatter diagram. See Figure 114. We see that the graph still needs to be shifted horizontally.

STEP 4: To determine the horizontal shift, we use the period $T = 12$ and divide the interval $[0, 12]$ into four subintervals of length $12 \div 4 = 3$:

$$[0, 3], \quad [3, 6], \quad [6, 9], \quad [9, 12]$$

The sine curve is increasing on the interval $(0, 3)$ and is decreasing on the interval $(3, 9)$, so a local maximum occurs at $x = 3$. The data indicate that a maximum occurs at $x = 7$ (corresponding to July's temperature), so we must shift the graph of the function 4 units to the right by replacing x by $x - 4$. Doing this, we obtain

$$y = 21.9 \sin\left(\frac{\pi}{6}(x - 4)\right) + 51.6$$

Multiplying out, we find that a sine function of the form $y = A \sin(\omega x - \phi) + B$ that fits the data is

$$y = 21.9 \sin\left(\frac{\pi}{6}x - \frac{2\pi}{3}\right) + 51.6$$

Figure 115

The graph of $y = 21.9 \sin\left(\dfrac{\pi}{6}x - \dfrac{2\pi}{3}\right) + 51.6$ and the scatter diagram of the data are shown in Figure 115.

The steps to fit a sine function

$$y = A \sin(\omega x - \phi) + B$$

to sinusoidal data follow:

Steps for Fitting Data to a Sine Function $y = A \sin(\omega x - \phi) + B$

STEP 1: Determine A, the amplitude of the function.

$$\text{Amplitude} = \frac{\text{largest data value} - \text{smallest data value}}{2}$$

STEP 2: Determine B, the vertical shift of the function.

$$\text{Vertical shift} = \frac{\text{largest data value} + \text{smallest data value}}{2}$$

STEP 3: Determine ω. Since the period T, the time it takes for the data to repeat, is $T = \dfrac{2\pi}{\omega}$, we have

$$\omega = \frac{2\pi}{T}$$

STEP 4: Determine the horizontal shift of the function by using the period of the data. Divide the period into four subintervals of equal length. Determine the x-coordinate for the maximum of the sine function and the x-coordinate for the maximum value of the data. Use this information to determine the value of the phase shift, $\dfrac{\phi}{\omega}$.

────────**Now Work**   PROBLEMS 29(a)–(c)

Let's look at another example. Since the number of hours of sunlight in a day cycles annually, the number of hours of sunlight in a day for a given location can be modeled by a sinusoidal function.

The longest day of the year (in terms of hours of sunlight) occurs on the day of the summer solstice. For locations in the northern hemisphere, the summer solstice is the time when the sun is farthest north. In 2010, the summer solstice occurred on June 21 (the 172nd day of the year) at 6:28 AM EDT. The shortest day of the year occurs on the day of the winter solstice. The winter solstice is the time when the Sun is farthest south (again, for locations in the northern hemisphere). In 2010, the winter solstice occurred on December 21 (the 355th day of the year) at 6:38 PM (EST).

EXAMPLE 4   **Finding a Sinusoidal Function for Hours of Daylight**

According to the *Old Farmer's Almanac*, the number of hours of sunlight in Boston on the summer solstice is 15.30 and the number of hours of sunlight on the winter solstice is 9.08.

(a) Find a sinusoidal function of the form $y = A \sin(\omega x - \phi) + B$ that fits the data.

(b) Use the function found in part (a) to predict the number of hours of sunlight on April 1, the 91st day of the year.

(c) Draw a graph of the function found in part (a).

(d) Look up the number of hours of sunlight for April 1 in the *Old Farmer's Almanac* and compare it to the results found in part (b).

Source: The Old Farmer's Almanac, www.almanac.com/rise

Solution   (a) **STEP 1:** Amplitude $= \dfrac{\text{largest data value} - \text{smallest data value}}{2}$

$$= \dfrac{15.30 - 9.08}{2} = 3.11$$

STEP 2: Vertical shift $= \dfrac{\text{largest data value} + \text{smallest data value}}{2}$

$$= \dfrac{15.30 + 9.08}{2} = 12.19$$

STEP 3: The data repeat every 365 days. Since $T = \dfrac{2\pi}{\omega} = 365$, we find

$$\omega = \dfrac{2\pi}{365}$$

So far, we have $y = 3.11 \sin\left(\dfrac{2\pi}{365}x - \phi\right) + 12.19$.

STEP 4: To determine the horizontal shift, we use the period $T = 365$ and divide the interval $[0, 365]$ into four subintervals of length $365 \div 4 = 91.25$:

$$[0, 91.25], \quad [91.25, 182.5], \quad [182.5, 273.75], \quad [273.75, 365]$$

The sine curve is increasing on the interval $(0, 91.25)$ and is decreasing on the interval $(91.25, 273.75)$, so a local maximum occurs at

$x = 91.25$. Since the maximum occurs on the summer solstice at $x = 172$, we must shift the graph of the function $172 - 91.25 = 80.75$ units to the right by replacing x by $x - 80.75$. Doing this, we obtain

$$y = 3.11 \sin\left(\frac{2\pi}{365}(x - 80.75)\right) + 12.19$$

Multiplying out, we find that a sine function of the form $y = A\sin(\omega x - \phi) + B$ that fits the data is

$$y = 3.11 \sin\left(\frac{2\pi}{365}x - \frac{323\pi}{730}\right) + 12.19$$

(b) To predict the number of hours of daylight on April 1, we let $x = 91$ in the function found in part (a) and obtain

$$y = 3.11 \sin\left(\frac{2\pi}{365}\cdot 91 - \frac{323}{730}\pi\right) + 12.19$$

$$\approx 12.74$$

So we predict that there will be about 12.74 hours = 12 hours, 44 minutes of sunlight on April 1 in Boston.

 (c) The graph of the function found in part (a) is given in Figure 116.

(d) According to the *Old Farmer's Almanac*, there will be 12 hours 45 minutes of sunlight on April 1 in Boston.

Figure 116

────Now Work PROBLEM 35

Certain graphing utilities (such as a TI-83, TI-84 Plus, and TI-86) have the capability of finding the sine function of best fit for sinusoidal data. At least four data points are required for this process.

EXAMPLE 5 **Finding the Sine Function of Best Fit**

Use a graphing utility to find the sine function of best fit for the data in Table 12. Graph this function with the scatter diagram of the data.

Solution Enter the data from Table 12 and execute the SINe REGression program. The result is shown in Figure 117.

The output that the utility provides shows the equation

$$y = a\sin(bx + c) + d$$

The sinusoidal function of best fit is

$$y = 21.15 \sin(0.55x - 2.35) + 51.19$$

where x represents the month and y represents the average temperature.

Figure 118 shows the graph of the sinusoidal function of best fit on the scatter diagram.

Figure 117

```
SinReg
y=a*sin(bx+c)+d
a=21.14682796
b=.5494591199
c=-2.35007307
d=51.19288889
```

Figure 118

────Now Work PROBLEMS 29(d) AND (e)

5.8 Assess Your Understanding

Concepts and Vocabulary

1. For the graph of $y = A \sin(\omega x - \phi)$, the number $\dfrac{\phi}{\omega}$ is called the _____ _____.

2. **True or False** Only two data points are required by a graphing utility to find the sine function of best fit.

Skill Building

In Problems 3–14, find the amplitude, period, and phase shift of each function. Graph each function. Be sure to label key points. Show at least two periods.

3. $y = 4\sin(2x - \pi)$

4. $y = 3\sin(3x - \pi)$

5. $y = 2\cos\left(3x + \dfrac{\pi}{2}\right)$

6. $y = 3\cos(2x + \pi)$

7. $y = -3\sin\left(2x + \dfrac{\pi}{2}\right)$

8. $y = -2\cos\left(2x - \dfrac{\pi}{2}\right)$

9. $y = 4\sin(\pi x + 2) - 5$

10. $y = 2\cos(2\pi x + 4) + 4$

11. $y = 3\cos(\pi x - 2) + 5$

12. $y = 2\cos(2\pi x - 4) - 1$

13. $y = -3\sin\left(-2x + \dfrac{\pi}{2}\right)$

14. $y = -3\cos\left(-2x + \dfrac{\pi}{2}\right)$

In Problems 15–18, write the equation of a sine function that has the given characteristics.

15. Amplitude: 2
    Period: π
    Phase shift: $\dfrac{1}{2}$

16. Amplitude: 3
    Period: $\dfrac{\pi}{2}$
    Phase shift: 2

17. Amplitude: 3
    Period: 3π
    Phase shift: $-\dfrac{1}{3}$

18. Amplitude: 2
    Period: π
    Phase shift: -2

Mixed Practice

In Problems 19–26, apply the methods of this and the previous section to graph each function. Be sure to label key points and show at least two periods.

19. $y = 2\tan(4x - \pi)$

20. $y = \dfrac{1}{2}\cot(2x - \pi)$

21. $y = 3\csc\left(2x - \dfrac{\pi}{4}\right)$

22. $y = \dfrac{1}{2}\sec(3x - \pi)$

23. $y = -\cot\left(2x + \dfrac{\pi}{2}\right)$

24. $y = -\tan\left(3x + \dfrac{\pi}{2}\right)$

25. $y = -\sec(2\pi x + \pi)$

26. $y = -\csc\left(-\dfrac{1}{2}\pi x + \dfrac{\pi}{4}\right)$

Applications and Extensions

27. **Alternating Current (ac) Circuits** The current I, in amperes, flowing through an ac (alternating current) circuit at time t, in seconds, is

    $$I(t) = 120\sin\left(30\pi t - \dfrac{\pi}{3}\right) \qquad t \geq 0$$

    What is the period? What is the amplitude? What is the phase shift? Graph this function over two periods.

28. **Alternating Current (ac) Circuits** The current I, in amperes, flowing through an ac (alternating current) circuit at time t, in seconds, is

    $$I(t) = 220\sin\left(60\pi t - \dfrac{\pi}{6}\right) \qquad t \geq 0$$

    What is the period? What is the amplitude? What is the phase shift? Graph this function over two periods.

29. **Monthly Temperature** The following data represent the average monthly temperatures for Juneau, Alaska.

Month, x	Average Monthly Temperature, °F
January, 1	24.2
February, 2	28.4
March, 3	32.7
April, 4	39.7
May, 5	47.0
June, 6	53.0
July, 7	56.0
August, 8	55.0
September, 9	49.4
October, 10	42.2
November, 11	32.0
December, 12	27.1

Source: U.S. National Oceanic and Atmospheric Administration

(a) Draw a scatter diagram of the data for one period.
(b) Find a sinusoidal function of the form
$y = A \sin(\omega x - \phi) + B$ that models the data.
(c) Draw the sinusoidal function found in part (b) on the scatter diagram.
(d) Use a graphing utility to find the sinusoidal function of best fit.
(e) Draw the sinusoidal function of best fit on a scatter diagram of the data.

30. **Monthly Temperature** The following data represent the average monthly temperatures for Washington, D.C.
(a) Draw a scatter diagram of the data for one period.
(b) Find a sinusoidal function of the form
$y = A \sin(\omega x - \phi) + B$ that models the data.
(c) Draw the sinusoidal function found in part (b) on the scatter diagram.
(d) Use a graphing utility to find the sinusoidal function of best fit.
(e) Graph the sinusoidal function of best fit on a scatter diagram of the data.

Month, x	Average Monthly Temperature, °F
January, 1	34.6
February, 2	37.5
March, 3	47.2
April, 4	56.5
May, 5	66.4
June, 6	75.6
July, 7	80.0
August, 8	78.5
September, 9	71.3
October, 10	59.7
November, 11	49.8
December, 12	39.4

Source: U.S. National Oceanic and Atmospheric Administration

31. **Monthly Temperature** The following data represent the average monthly temperatures for Indianapolis, Indiana.

Month, x	Average Monthly Temperature, °F
January, 1	25.5
February, 2	29.6
March, 3	41.4
April, 4	52.4
May, 5	62.8
June, 6	71.9
July, 7	75.4
August, 8	73.2
September, 9	66.6
October, 10	54.7
November, 11	43.0
December, 12	30.9

Source: U.S. National Oceanic and Atmospheric Administration

(a) Draw a scatter diagram of the data for one period.
(b) Find a sinusoidal function of the form
$y = A \sin(\omega x - \phi) + B$ that models the data.
(c) Draw the sinusoidal function found in part (b) on the scatter diagram.
(d) Use a graphing utility to find the sinusoidal function of best fit.
(e) Graph the sinusoidal function of best fit on a scatter diagram of the data.

32. **Monthly Temperature** The following data represent the average monthly temperatures for Baltimore, Maryland.
(a) Draw a scatter diagram of the data for one period.
(b) Find a sinusoidal function of the form
$y = A \sin(\omega x - \phi) + B$ that models the data.
(c) Draw the sinusoidal function found in part (b) on the scatter diagram.
(d) Use a graphing utility to find the sinusoidal function of best fit.
(e) Graph the sinusoidal function of best fit on a scatter diagram of the data.

Month, x	Average Monthly Temperature, °F
January, 1	31.8
February, 2	34.8
March, 3	44.1
April, 4	53.4
May, 5	63.4
June, 6	72.5
July, 7	77.0
August, 8	75.6
September, 9	68.5
October, 10	56.6
November, 11	46.8
December, 12	36.7

Source: U.S. National Oceanic and Atmospheric Administration

33. **Tides** The length of time between consecutive high tides is 12 hours and 25 minutes. According to the National Oceanic and Atmospheric Administration, on Saturday, July 25, 2009, in Charleston, South Carolina, high tide occurred at 11:30 AM (11.5 hours) and low tide occurred at 5:31 PM (17.5167 hours). Water heights are measured as the amounts above or below the mean lower low water. The height of the water at high tide was 5.84 feet, and the height of the water at low tide was −0.37 foot.
(a) Approximately when will the next high tide occur?
(b) Find a sinusoidal function of the form
$y = A \sin(\omega x - \phi) + B$ that models the data.
(c) Use the function found in part (b) to predict the height of the water at 3 PM on July 25, 2009.

34. **Tides** The length of time between consecutive high tides is 12 hours and 25 minutes. According to the National Oceanic and Atmospheric Administration, on Saturday, July 25, 2009, in Sitka Sound, Alaska, high tide occurred at 2:37 AM (2.6167 hours) and low tide occurred at 9:12 PM (9.2 hours). Water heights are measured as the amounts above or below the mean lower low water. The height of the water at high tide was 11.09 feet, and the height of the water at low tide was −2.49 feet.

(a) Approximately when will the next high tide occur?
(b) Find a sinusoidal function of the form
   $y = A \sin(\omega x - \phi) + B$ that models the data.
(c) Use the function found in part (b) to predict the height of the water at 6 PM.

35. **Hours of Daylight** According to the *Old Farmer's Almanac*, in Miami, Florida, the number of hours of sunlight on the summer solstice of 2010 was 13.75, and the number of hours of sunlight on the winter solstice was 10.55.
   (a) Find a sinusoidal function of the form
      $y = A \sin(\omega x - \phi) + B$ that models the data.
   (b) Use the function found in part (a) to predict the number of hours of sunlight on April 1, the 91st day of the year.
   (c) Draw a graph of the function found in part (a).
   (d) Look up the number of hours of sunlight for April 1 in the *Old Farmer's Almanac*, and compare the actual hours of daylight to the results found in part (c).

36. **Hours of Daylight** According to the *Old Farmer's Almanac*, in Detroit, Michigan, the number of hours of sunlight on the summer solstice of 2010 was 15.30, and the number of hours of sunlight on the winter solstice was 9.10.
   (a) Find a sinusoidal function of the form
      $y = A \sin(\omega x - \phi) + B$ that models the data.
   (b) Use the function found in part (a) to predict the number of hours of sunlight on April 1, the 91st day of the year.
   (c) Draw a graph of the function found in part (a).

(d) Look up the number of hours of sunlight for April 1 in the *Old Farmer's Almanac*, and compare the actual hours of daylight to the results found in part (c).

37. **Hours of Daylight** According to the *Old Farmer's Almanac*, in Anchorage, Alaska, the number of hours of sunlight on the summer solstice of 2010 was 19.42 and the number of hours of sunlight on the winter solstice was 5.48.
   (a) Find a sinusoidal function of the form
      $y = A \sin(\omega x - \phi) + B$ that models the data.
   (b) Use the function found in part (a) to predict the number of hours of sunlight on April 1, the 91st day of the year.
   (c) Draw a graph of the function found in part (a).
   (d) Look up the number of hours of sunlight for April 1 in the *Old Farmer's Almanac*, and compare the actual hours of daylight to the results found in part (c).

38. **Hours of Daylight** According to the *Old Farmer's Almanac*, in Honolulu, Hawaii, the number of hours of sunlight on the summer solstice of 2010 was 13.43 and the number of hours of sunlight on the winter solstice was 10.85.
   (a) Find a sinusoidal function of the form
      $y = A \sin(\omega x - \phi) + B$ that models the data.
   (b) Use the function found in part (a) to predict the number of hours of sunlight on April 1, the 91st day of the year.
   (c) Draw a graph of the function found in part (a).
   (d) Look up the number of hours of sunlight for April 1 in the *Old Farmer's Almanac*, and compare the actual hours of daylight to the results found in part (c).

Explaining Concepts: Discussion and Writing

39. Explain how the amplitude and period of a sinusoidal graph are used to establish the scale on each coordinate axis.

40. Find an application in your major field that leads to a sinusoidal graph. Write a paper about your findings.

CHAPTER REVIEW

Things to Know

Definitions

Angle in standard position (p. 378)	Vertex is at the origin; initial side is along the positive x-axis.
1 Degree (1°) (p. 379)	$1° = \dfrac{1}{360}$ revolution
1 Radian (p. 381)	The measure of a central angle of a circle whose rays subtend an arc whose length is the radius of the circle
Acute angle (p. 391)	An angle θ whose measure is $0° < \theta < 90°$ or $0 < \theta < \dfrac{\pi}{2}$, θ in radians
Complementary angles (p. 397)	Two acute angles whose sum is $90° \left(\dfrac{\pi}{2} \text{ radians} \right)$
Cofunction (p. 398)	The following pairs of functions are cofunctions of each other: sine and cosine; tangent and cotangent; secant and cosecant.
Trigonometric functions of any angle (p. 415)	$P = (a, b)$ is the point on the terminal side of θ a distance r from the origin:

$$\sin \theta = \frac{b}{r} \qquad \cos \theta = \frac{a}{r} \qquad \tan \theta = \frac{b}{a}, \quad a \neq 0$$

$$\csc \theta = \frac{r}{b}, \quad b \neq 0 \qquad \sec \theta = \frac{r}{a}, \quad a \neq 0 \qquad \cot \theta = \frac{a}{b}, \quad b \neq 0$$

| Reference angle of θ (p. 419) | The acute angle formed by the terminal side of θ and either the positive or negative x-axis |
| Periodic function (p. 431) | $f(\theta + p) = f(\theta)$, for all θ, $p > 0$, where the smallest such p is the fundamental period |

Formulas

1 counterclockwise
revolution $= 360°$ (p. 380)
$\qquad = 2\pi$ radians (p. 383)

$s = r\theta$ (p. 382) — θ is measured in radians; s is the length of the arc subtended by the central angle θ of the circle of radius r.

$A = \dfrac{1}{2}r^2\theta$ (p. 385) — A is the area of the sector of a circle of radius r formed by a central angle of θ radians.

$v = r\omega$ (p. 386) — v is the linear speed along the circle of radius r; ω is the angular speed (measured in radians per unit time).

Table of Values

θ (Radians)	θ (Degrees)	$\sin\theta$	$\cos\theta$	$\tan\theta$	$\csc\theta$	$\sec\theta$	$\cot\theta$
0	0°	0	1	0	Not defined	1	Not defined
$\dfrac{\pi}{6}$	30°	$\dfrac{1}{2}$	$\dfrac{\sqrt{3}}{2}$	$\dfrac{\sqrt{3}}{3}$	2	$\dfrac{2\sqrt{3}}{3}$	$\sqrt{3}$
$\dfrac{\pi}{4}$	45°	$\dfrac{\sqrt{2}}{2}$	$\dfrac{\sqrt{2}}{2}$	1	$\sqrt{2}$	$\sqrt{2}$	1
$\dfrac{\pi}{3}$	60°	$\dfrac{\sqrt{3}}{2}$	$\dfrac{1}{2}$	$\sqrt{3}$	$\dfrac{2\sqrt{3}}{3}$	2	$\dfrac{\sqrt{3}}{3}$
$\dfrac{\pi}{2}$	90°	1	0	Not defined	1	Not defined	0
π	180°	0	−1	0	Not defined	−1	Not defined
$\dfrac{3\pi}{2}$	270°	−1	0	Not defined	−1	Not defined	0

Fundamental identities (p. 394)

$$\tan\theta = \frac{\sin\theta}{\cos\theta} \qquad \cot\theta = \frac{\cos\theta}{\sin\theta}$$

$$\csc\theta = \frac{1}{\sin\theta} \qquad \sec\theta = \frac{1}{\cos\theta} \qquad \cot\theta = \frac{1}{\tan\theta}$$

$$\sin^2\theta + \cos^2\theta = 1 \qquad \tan^2\theta + 1 = \sec^2\theta \qquad \cot^2\theta + 1 = \csc^2\theta$$

Properties of the trigonometric functions

$y = \sin x$ (p. 436)

Domain: $-\infty < x < \infty$
Range: $-1 \le y \le 1$
Periodic: period $= 2\pi$ (360°)
Odd function

$y = \cos x$ (p. 438)

Domain: $-\infty < x < \infty$
Range: $-1 \le y \le 1$
Periodic: period $= 2\pi$ (360°)
Even function

$y = \tan x$ (pp. 451–452)   Domain: $-\infty < x < \infty$, except odd integer multiples of $\dfrac{\pi}{2}$ (90°)

Range: $-\infty < y < \infty$

Periodic: period $= \pi$ (180°)

Odd function

Vertical asymptotes at odd integer multiples of $\dfrac{\pi}{2}$

$y = \cot x$ (pp. 453–454)   Domain: $-\infty < x < \infty$, except integer multiples of π (180°)

Range: $-\infty < y < \infty$

Periodic: period $= \pi$ (180°)

Odd function

Vertical asymptotes at integer multiples of π

$y = \csc x$ (p. 454)   Domain: $-\infty < x < \infty$, except integer multiples of π (180°)

Range: $|y| \geq 1$

Periodic: period $= 2\pi$ (360°)

Odd function

Vertical asymptotes at integer multiples of π

$y = \sec x$ (p. 454)   Domain: $-\infty < x < \infty$, except odd integer multiples of $\dfrac{\pi}{2}$ (90°)

Range: $|y| \geq 1$

Periodic: period $= 2\pi$ (360°)

Even function

Vertical asymptotes at odd integer multiples of $\dfrac{\pi}{2}$

Sinusoidal graphs

$y = A \sin(\omega x) + B, \quad \omega > 0$

$y = A \cos(\omega x) + B, \quad \omega > 0$

$y = A \sin(\omega x - \phi) + B = A \sin\left[\omega\left(x - \dfrac{\phi}{\omega}\right)\right] + B$

$y = A \cos(\omega x - \phi) + B = A \cos\left[\omega\left(x - \dfrac{\phi}{\omega}\right)\right] + B$

Period $= \dfrac{2\pi}{\omega}$ (pp. 440, 458)

Amplitude $= |A|$ (pp. 440, 458)

Phase shift $= \dfrac{\phi}{\omega}$ (p. 458)

Objectives

Section		You should be able to:	Example(s)	Review Exercises
5.1	1	Convert between decimals and degrees, minutes, seconds forms for angles (p. 380)	2	86
	2	Find the length of an arc of a circle (p. 382)	3	87, 88
	3	Convert from degrees to radians and from radians to degrees (p. 382)	4–6	1–8
	4	Find the area of a sector of a circle (p. 385)	7	87
	5	Find the linear speed of an object traveling in circular motion (p. 386)	8	89–92
5.2	1	Find the values of trigonometric functions of acute angles (p. 391)	1	79
	2	Use the Fundamental Identities (p. 393)	2, 3	21–24
	3	Find the values of the remaining trigonometric functions, given the value of one of them (p. 395)	4, 5	31–32
	4	Use the Complementary Angle Theorem (p. 397)	6, 7	25–26

5.3	1	Find the exact values of the trigonometric functions of $\frac{\pi}{4} = 45°$ (p. 403)	1, 2	9, 11
	2	Find the exact values of the trigonometric functions of $\frac{\pi}{6} = 30°$ and $\frac{\pi}{3} = 60°$ (p. 404)	3, 4	9–12
	3	Use a calculator to approximate the values of the trigonometric functions of acute angles (p. 406)	5	80
	4	Model and solve applied problems involving right triangles (p. 406)	6–9	93–97
5.4	1	Find the exact values of the trigonometric functions for any angle (p. 414)	1, 2	84
	2	Use coterminal angles to find the exact value of a trigonometric function (p. 417)	3	19, 20
	3	Determine the signs of the trigonometric functions of an angle in a given quadrant (p. 418)	4	81
	4	Find the reference angle of an angle (p. 419)	5	82
	5	Use a reference angle to find the exact value of a trigonometric function (p. 420)	6	13–16, 19
	6	Find the exact values of the trigonometric functions of an angle, given information about the functions (p. 421)	7, 8	31–46
5.5	1	Find the exact values of the trigonometric functions using the unit circle (p. 425)	1	83
	2	Know the domain and range of the trigonometric functions (p. 429)	pp. 429–430	85
	3	Use the periodic properties to find the exact values of the trigonometric functions (p. 430)	3	19, 20
	4	Use even–odd properties to find the exact values of the trigonometric functions (p. 432)	4	13, 15, 16, 18–20, 27–30
5.6	1	Graph functions of the form $y = A \sin(\omega x)$ using transformations (p. 436)	1, 2	47
	2	Graph functions of the form $y = A \cos(\omega x)$ using transformations (p. 438)	3	48
	3	Determine the amplitude and period of sinusoidal functions (p. 439)	4	63–68
	4	Graph sinusoidal functions using key points (p. 440)	5–7	47, 48, 67, 68
	5	Find an equation for a sinusoidal graph (p. 444)	8, 9	75–78
5.7	1	Graph functions of the form $y = A \tan(\omega x) + B$ and $y = A \cot(\omega x) + B$ (p. 452)	1, 2	53, 54, 56
	2	Graph functions of the form $y = A \csc(\omega x) + B$ and $y = A \sec(\omega x) + B$ (p. 454)	3	57
5.8	1	Graph sinusoidal functions of the form $y = A \sin(\omega x - \phi) + B$ (p. 457)	1, 2	49, 50, 59, 60, 98
	2	Build sinusoidal models from data (p. 461)	3–5	99, 100

Review Exercises

In Problems 1–4, convert each angle in degrees to radians. Express your answer as a multiple of π.

1. 135°   **2.** 210°   **3.** 18°   **4.** 15°

In Problems 5–8, convert each angle in radians to degrees.

5. $\dfrac{3\pi}{4}$ **6.** $\dfrac{2\pi}{3}$ **7.** $-\dfrac{5\pi}{2}$ **8.** $-\dfrac{3\pi}{2}$

In Problems 9–30, find the exact value of each expression. Do not use a calculator.

9. $\tan \dfrac{\pi}{4} - \sin \dfrac{\pi}{6}$

10. $\cos \dfrac{\pi}{3} + \sin \dfrac{\pi}{2}$

11. $3 \sin 45° - 4 \tan \dfrac{\pi}{6}$

12. $4 \cos 60° + 3 \tan \dfrac{\pi}{3}$

13. $6 \cos \dfrac{3\pi}{4} + 2 \tan\left(-\dfrac{\pi}{3}\right)$

14. $3 \sin \dfrac{2\pi}{3} - 4 \cos \dfrac{5\pi}{2}$

15. $\sec\left(-\dfrac{\pi}{3}\right) - \cot\left(-\dfrac{5\pi}{4}\right)$

16. $4\csc\dfrac{3\pi}{4} - \cot\left(-\dfrac{\pi}{4}\right)$

17. $\tan\pi + \sin\pi$

18. $\cos\dfrac{\pi}{2} - \csc\left(-\dfrac{\pi}{2}\right)$

19. $\cos 540° - \tan(-405°)$

20. $\sin 270° + \cos(-180°)$

21. $\sin^2 20° + \dfrac{1}{\sec^2 20°}$

22. $\dfrac{1}{\cos^2 40°} - \dfrac{1}{\cot^2 40°}$

23. $\sec 50° \cos 50°$

24. $\tan 10° \cot 10°$

25. $\dfrac{\sin 50°}{\cos 40°}$

26. $\dfrac{\tan 20°}{\cot 70°}$

27. $\dfrac{\sin(-40°)}{\cos 50°}$

28. $\tan(-20°)\cot 20°$

29. $\sin 400° \sec(-50°)$

30. $\cot 200° \cot(-70°)$

In Problems 31–46, find the exact value of each of the remaining trigonometric functions.

31. $\sin\theta = \dfrac{4}{5}, \quad \theta$ is acute

32. $\tan\theta = \dfrac{1}{4}, \quad \theta$ is acute

33. $\tan\theta = \dfrac{12}{5}, \quad \sin\theta < 0$

34. $\cot\theta = \dfrac{12}{5}, \quad \cos\theta < 0$

35. $\sec\theta = -\dfrac{5}{4}, \quad \tan\theta < 0$

36. $\csc\theta = -\dfrac{5}{3}, \quad \cot\theta < 0$

37. $\sin\theta = \dfrac{12}{13}, \quad \theta$ in quadrant II

38. $\cos\theta = -\dfrac{3}{5}, \quad \theta$ in quadrant III

39. $\sin\theta = -\dfrac{5}{13}, \quad \dfrac{3\pi}{2} < \theta < 2\pi$

40. $\cos\theta = \dfrac{12}{13}, \quad \dfrac{3\pi}{2} < \theta < 2\pi$

41. $\tan\theta = \dfrac{1}{3}, \quad 180° < \theta < 270°$

42. $\tan\theta = -\dfrac{2}{3}, \quad 90° < \theta < 180°$

43. $\sec\theta = 3, \quad \dfrac{3\pi}{2} < \theta < 2\pi$

44. $\csc\theta = -4, \quad \pi < \theta < \dfrac{3\pi}{2}$

45. $\cot\theta = -2, \quad \dfrac{\pi}{2} < \theta < \pi$

46. $\tan\theta = -2, \quad \dfrac{3\pi}{2} < \theta < 2\pi$

In Problems 47–62, graph each function. Each graph should contain at least two periods. Use the graph to determine the domain and the range of each function.

47. $y = 2\sin(4x)$

48. $y = -3\cos(2x)$

49. $y = -2\cos\left(x + \dfrac{\pi}{2}\right)$

50. $y = 3\sin(x - \pi)$

51. $y = \tan(x + \pi)$

52. $y = -\tan\left(x - \dfrac{\pi}{2}\right)$

53. $y = -2\tan(3x)$

54. $y = 4\tan(2x)$

55. $y = \cot\left(x + \dfrac{\pi}{4}\right)$

56. $y = -4\cot(2x)$

57. $y = 4\sec(2x)$

58. $y = \csc\left(x + \dfrac{\pi}{4}\right)$

59. $y = 4\sin(2x + 4) - 2$

60. $y = 3\cos(4x + 2) + 1$

61. $y = 4\tan\left(\dfrac{x}{2} + \dfrac{\pi}{4}\right)$

62. $y = 5\cot\left(\dfrac{x}{3} - \dfrac{\pi}{4}\right)$

In Problems 63–66, determine the amplitude and period of each function without graphing.

63. $y = 4\cos x$

64. $y = \sin(2x)$

65. $y = -8\sin\left(\dfrac{\pi}{2}x\right)$

66. $y = -2\cos(3\pi x)$

In Problems 67–74, find the amplitude, period, and phase shift of each function. Graph each function. Show at least two periods.

67. $y = 4\sin(3x)$

68. $y = 2\cos\left(\dfrac{1}{3}x\right)$

69. $y = 2\sin(2x - \pi)$

70. $y = -\cos\left(\dfrac{1}{2}x + \dfrac{\pi}{2}\right)$

71. $y = \dfrac{1}{2}\sin\left(\dfrac{3}{2}x - \pi\right)$

72. $y = \dfrac{3}{2}\cos(6x + 3\pi)$

73. $y = -\dfrac{2}{3}\cos(\pi x - 6)$

74. $y = -7\sin\left(\dfrac{\pi}{3}x - \dfrac{4}{3}\right)$

In Problems 75–78, find a function whose graph is given.

75.

76.

77.

78.

79. Find the value of each of the six trigonometric functions of the acute angle θ in a right triangle where the length of the side opposite θ is 3 and the length of the hypotenuse is 7.

80. Use a calculator to approximate $\sin\dfrac{\pi}{8}$. Use a calculator to approximate $\sec 10°$. Round answers to two decimal places.

81. Determine the signs of the six trigonometric functions of an angle θ whose terminal side is in quadrant III.

82. Find the reference angle of $750°$.

83. Find the exact values of the six trigonometric functions of t if $P = \left(-\dfrac{1}{3}, \dfrac{2\sqrt{2}}{3}\right)$ is the point on the unit circle that corresponds to t.

84. Find the exact value of $\sin t$, $\cos t$, and $\tan t$ if $P = (-2, 5)$ is the point on the circle that corresponds to t.

85. What is the domain and the range of the secant function? What is the period?

86. (a) Convert the angle $32°20'35''$ to a decimal in degrees. Round the answer to two decimal places.
(b) Convert the angle $63.18°$ to D°M′S″ form. Express the answer to the nearest second.

87. Find the length of the arc subtended by a central angle of $30°$ on a circle of radius 2 feet. What is the area of the sector?

88. The minute hand of a clock is 8 inches long. How far does the tip of the minute hand move in 30 minutes? How far does it move in 20 minutes?

89. Angular Speed of a Race Car A race car is driven around a circular track at a constant speed of 180 miles per hour. If the diameter of the track is $\dfrac{1}{2}$ mile, what is the angular speed of the car? Express your answer in revolutions per hour (which is equivalent to laps per hour).

90. Merry-Go-Rounds A neighborhood carnival has a merry-go-round whose radius is 25 feet. If the time for one revolution is 30 seconds, how fast is the merry-go-round going?

91. Lighthouse Beacons The Montauk Point Lighthouse on Long Island has dual beams (two light sources opposite each other). Ships at sea observe a blinking light every 5 seconds. What rotation speed is required to do this?

92. Spin Balancing Tires The radius of each wheel of a car is 16 inches. At how many revolutions per minute should a spin balancer be set to balance the tires at a speed of 90 miles per hour? Is the setting different for a wheel of radius 14 inches? If so, what is this setting?

93. Measuring the Length of a Lake From a stationary hot-air balloon 500 feet above the ground, two sightings of a lake are made (see the figure). How long is the lake?

94. Finding the Speed of a Glider From a glider 200 feet above the ground, two sightings of a stationary object directly in front are taken 1 second apart (see the figure). What is the speed of the glider?

95. Finding the Width of a River Find the distance from A to C across the river illustrated in the figure.

96. Finding the Height of a Building Find the height of the building shown in the figure.

97. Finding the Distance to Shore The Willis Tower in Chicago is 1454 feet tall and is situated about 1 mile inland from the shore of Lake Michigan, as indicated in the figure. An observer in a pleasure boat on the lake directly in front of the Willis Tower looks at the top of the tower and measures the angle of elevation as 5°. How far offshore is the boat?

98. Alternating Current The current I, in amperes, flowing through an ac (alternating current) circuit at time t is

$$I = 220 \sin\left(30\pi t + \frac{\pi}{6}\right) \qquad t \geq 0$$

(a) What is the period?
(b) What is the amplitude?
(c) What is the phase shift?
(d) Graph this function over two periods.

99. Monthly Temperature The following data represent the average monthly temperatures for Phoenix, Arizona.
(a) Draw a scatter diagram of the data for one period.

(b) Find a sinusoidal function of the form $y = A \sin(\omega x - \phi) + B$ that models the data.
(c) Draw the sinusoidal function found in part (b) on the scatter diagram.
(d) Use a graphing utility to find the sinusoidal function of best fit.
(e) Graph the sinusoidal function of best fit on the scatter diagram.

Month, m	Average Monthly Temperature, T
January, 1	51
February, 2	55
March, 3	63
April, 4	67
May, 5	77
June, 6	86
July, 7	90
August, 8	90
September, 9	84
October, 10	71
November, 11	59
December, 12	52

Source: U.S. National Oceanic and Atmospheric Administration

100. Hours of Daylight According to the *Old Farmer's Almanac*, in Las Vegas, Nevada, the number of hours of sunlight on the summer solstice is 14.63 and the number of hours of sunlight on the winter solstice is 9.72.
(a) Find a sinusoidal function of the form $y = A \sin(\omega x - \phi) + B$ that models the data.
(b) Use the function found in part (a) to predict the number of hours of sunlight on April 1, the 91st day of the year.
(c) Draw a graph of the function found in part (a).
(d) Look up the number of hours of sunlight for April 1 in the *Old Farmer's Almanac* and compare the actual hours of daylight to the results found in part (c).

CHAPTER TEST

CHAPTER **Test Prep** VIDEOS

Step-by-step test solutions are found on the Chapter Test Prep Videos available via the Video Resources on DVD, in **MyMathLab** or on You Tube (search "Sullivan Precalculus: A Right Triangle Approach, Concepts Through Functions 2e).

In Problems 1–3, convert each angle in degrees to radians. Express your answer as a multiple of π.

1. 260°   **2.** −400°   **3.** 13°

In Problems 4–6 convert each angle in radians to degrees.

4. $-\dfrac{\pi}{8}$ **5.** $\dfrac{9\pi}{2}$ **6.** $\dfrac{3\pi}{4}$

In Problems 7–12, find the exact value of each expression.

7. $\sin \dfrac{\pi}{6}$

8. $\cos\left(-\dfrac{5\pi}{4}\right) - \cos \dfrac{3\pi}{4}$

9. $\cos(-120°)$

10. $\tan 330°$

11. $\sin \dfrac{\pi}{2} - \tan \dfrac{19\pi}{4}$

12. $2\sin^2 60° - 3\cos 45°$

In Problems 13–16, use a calculator to evaluate each expression. Round your answer to three decimal places.

13. $\sin 17°$ **14.** $\cos \dfrac{2\pi}{5}$ **15.** $\sec 229°$ **16.** $\cot \dfrac{28\pi}{9}$

17. Fill in each table entry with the sign of each function.

	$\sin\theta$	$\cos\theta$	$\tan\theta$	$\sec\theta$	$\csc\theta$	$\cot\theta$
θ in QI						
θ in QII						
θ in QIII						
θ in QIV						

18. If $f(x) = \sin x$ and $f(a) = \dfrac{3}{5}$, find $f(-a)$.

In Problems 19–21 find the value of the remaining five trigonometric functions of θ.

19. $\sin \theta = \dfrac{5}{7}$, θ in quadrant II   **20.** $\cos \theta = \dfrac{2}{3}$, $\dfrac{3\pi}{2} < \theta < 2\pi$

21. $\tan \theta = -\dfrac{12}{5}$, $\dfrac{\pi}{2} < \theta < \pi$

In Problems 22–24, the point (x, y) is on the terminal side of angle θ in standard position. Find the exact value of the given trigonometric function.

22. $(2, 7)$, $\sin \theta$ **23.** $(-5, 11)$, $\cos \theta$

24. $(6, -3)$, $\tan \theta$

In Problems 25 and 26, graph the function.

25. $y = 2 \sin\left(\dfrac{x}{3} - \dfrac{\pi}{6}\right)$ **26.** $y = \tan\left(-x + \dfrac{\pi}{4}\right) + 2$

27. Write an equation for a sinusoidal graph with the following properties:

$$A = -3 \qquad \text{period} = \dfrac{2\pi}{3} \qquad \text{phase shift} = -\dfrac{\pi}{4}$$

28. Logan has a garden in the shape of a sector of a circle; the outer rim of the garden is 25 feet long and the central angle of the sector is 50°. She wants to add a 3-foot wide walk to the outer rim; how many square feet of paving blocks will she need to build the walk?

29. Hungarian Adrian Annus won the gold medal for the hammer throw at the 2004 Olympics in Athens with a winning distance of 83.19 meters.* The event consists of swinging a 16-pound weight attached to a wire 190 centimeters long in a circle and then releasing it. Assuming his release is at a 45° angle to the ground, the hammer will travel a distance of $\dfrac{v_0^2}{g}$ meters, where $g = 9.8$ meters/second² and v_0 is the linear speed of the hammer when released. At what rate (rpm) was he swinging the hammer upon release?

30. A ship is just offshore of New York City. A sighting is taken of the Statue of Liberty, which is about 305 feet tall. If the angle of elevation to the top of the statue is 20°, how far is the ship from the base of the statue?

31. To measure the height of a building, two sightings are taken a distance of 50 feet apart. If the first angle of elevation is 40° and the second is 32°, what is the height of the building?

*Annus was stripped of his medal after refusing to cooperate with postmedal drug testing.

CUMULATIVE REVIEW

1. Find the real solutions, if any, of the equation $2x^2 + x - 1 = 0$.

2. Find an equation for the line with slope -3 containing the point $(-2, 5)$.

3. Find an equation for a circle of radius 4 and center at the point $(0, -2)$.

4. Discuss the equation $2x - 3y = 12$. Graph it.

5. Discuss the equation $x^2 + y^2 - 2x + 4y - 4 = 0$. Graph it.

6. Use transformations to graph the function $y = (x - 3)^2 + 2$.

7. Sketch a graph of each of the following functions. Label at least three points on each graph.
   (a) $y = x^2$ (b) $y = x^3$ (c) $y = e^x$
   (d) $y = \ln x$ (e) $y = \sin x$ (f) $y = \tan x$

8. Find the inverse function of $f(x) = 3x - 2$.

9. Find the exact value of $(\sin 14°)^2 + (\cos 14°)^2 - 3$.

10. Graph $y = 3 \sin(2x)$.

11. Find the exact value of $\tan \dfrac{\pi}{4} - 3 \cos \dfrac{\pi}{6} + \csc \dfrac{\pi}{6}$.

12. Find an exponential function for the following graph. Express your answer in the form $y = Ab^x$.

13. Find a sinusoidal function for the following graph.

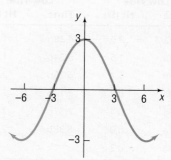

14. (a) Find a linear function that contains the points $(-2, 3)$ and $(1, -6)$. What is the slope? What are the intercepts of the function? Graph the function. Be sure to label the intercepts.
   (b) Find a quadratic function that contains the point $(-2, 3)$ with vertex $(1, -6)$. What are the intercepts of the function? Graph the function.
   (c) Show that there is no exponential function of the form $f(x) = ae^x$ that contains the points $(-2, 3)$ and $(1, -6)$.

15. (a) Find a polynomial function of degree 3 whose y-intercept is 5 and whose x-intercepts are $-2, 3$, and 5. Graph the function.
   (b) Find a rational function whose y-intercept is 5 and whose x-intercepts are $-2, 3$, and 5 that has the line $x = 2$ as a vertical asymptote. Graph the function.

CHAPTER PROJECTS

I. Tides The given table is a partial tide table for November 2006 for the Sabine Bank Lighthouse, a shoal located offshore from Texas where the Sabine River empties into the Gulf of Mexico.

1. On November 15, when was the tide high? This is called *high tide*. On November 19, when was the tide low? This is called *low tide*. Most days will have two high tides and two low tides.

2. Why do you think there is a negative height for the low tide on November 20? What is the height measured against?

3. On your graphing utility, draw a scatter diagram for the data in the table. Let t (time, in hours) be the independent variable, with $t = 0$ being 12:00 AM on November 14, $t = 24$ being 12:00 AM on November 15, and so on. Let h be the height in feet. Remember that there are 60 minutes in an hour. Also, make sure your graphing utility is in radian mode.

4. What shape does the data take? What is the period of the data? What is the amplitude? Is the amplitude constant? Explain.

5. Find a sine curve that models the data. Is there a vertical shift? Is there a phase shift?

6. Using your graphing utility, find the sinusoidal function of best fit. How does this model compare to the model from part 5?

7. Using the model found in part 5 and the sinusoidal equation of best fit found in part 6, predict the high tides and the low tides on November 21.

8. Looking at the times of day that the low tides occur, what do you think causes the low tides to vary so much each day? Explain. Does this seem to have the same type of effect on the high tides? Explain.

Nov	Low Tide Time	Ht (ft)	Low Tide Time	Ht (ft)	High Tide Time	Ht (ft)	High Tide Time	Ht (ft)	Sun/Moon Phase Sunrise/set	Moonrise/set
14	6:26a	2.0	4:38p	1.4	9:29a	2.2	11:14p	2.8	6:40a/5:20p	1:05a/2:02p
15	6:22a	1.6	5:34p	1.8	11:18a	2.4	11:15p	2.6	6:41a/5:20p	1:58a/2:27p
16	6:28a	1.2	6:25p	2.0	12:37p	2.6	11:16p	2.6	6:41a/5:19p	2:50a/2:52p
17	6:40a	0.8	7:12p	2.4	1:38p	2.8	11:16p	2.6	6:42a/5:19p	3:43a/3:19p
18	6:56a	0.4	7:57p	2.6	2:27p	3.0	11:14p	2.8	6:43a/5:19p	4:38a/3:47p
19	7:17a	0.0	8:38p	2.6	3:10p	3.2	11:05p	2.8	6:44a/5:18p	5:35a/4:20p
20	7:43a	−0.2			3:52p	3.4			6:45a/5:18p	6:34a/4:57p

[Note: a, AM; p, PM.]

Sources: National Oceanic and Atmospheric Administration (http://tidesandcurrents.noaa.gov) and U.S. Naval Observatory (http://aa.usno.navy.mil)

The following projects are available on the Instructor's Resource Center (IRC):

II. Project at Motorola ***Digital Transmission over the Air*** Learn how Motorola Corporation transmits digital sequences by modulating the phase of the carrier waves.

III. Identifying Mountain Peaks in Hawaii The visibility of a mountain is affected by its altitude, distance from the viewer, and the curvature of Earth's surface. Trigonometry can be used to determine whether a distant object can be seen.

IV. CBL Experiment Technology is used to model and study the effects of damping on sound waves.

Analytic Trigonometry

6

Outline

In an Instant, the Land Merged with the Sea

John Lancaster and Peter S. Goodman—Washington Post Foreign Service—Sunday, January 2, 2005—Although the tsunami would have been barely perceptible to passengers on a ship at sea, it rolled outward from the fault zone at about 420 miles per hour, roughly the speed of a commercial jetliner. Even at that speed, however, the surge took about two hours to travel to the Indian mainland.

The tsunami hit coastal areas in different ways. In some places, the first event was the recession of water; in others, it was a flood. The difference was determined at the origin of the tsunami: Along the fault line, the sea floor rose in some places and fell in others. Some coastal areas felt the effect of the sinking first and saw the water recede before it returned, often with a vengeance; elsewhere, the opposite happened. Places farther from the origin of the tsunami, such as Sri Lanka, experienced more waves than did closer areas, such as Sumatra.

As the tsunami approached shallower water near land, its behavior began to change. Its velocity slowed and the distance between its crests, known as wavelength, shortened. The stored energy in the wave was partly transformed into an increase in wave height. What happened next varied widely according to local bottom contours. In some areas, the tsunami broke like a massive beach wave; in others, it caused a sudden, rapid increase in sea level, like water pouring over the edge of a tub.

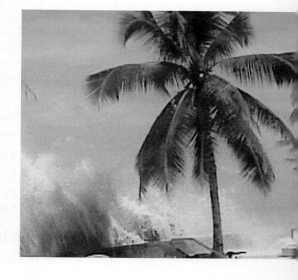

Source: Excerpt from *"In an Instant, the Land Merged with the Sea,"* by John Lancaster and Peter S. Goodman. Washington Post, *January 2, 2005, p. A1. © 2005, The Washington Post. Reprinted with Permission.*

—See Chapter Project I—

◁ A Look Back In Chapter 4, we defined inverse functions and developed their properties, particularly the relationship between the domain and range of a function and its inverse. We learned that the graphs of a function and its inverse are symmetric with respect to the line $y = x$.

We continued in Chapter 4 by defining the exponential function and the inverse of the exponential function, the logarithmic function. In Chapter 5, we defined the six trigonometric functions and looked at their properties.

A Look Ahead ▷ In the first two sections of this chapter, we define the six inverse trigonometric functions and investigate their properties. In Section 6.3, we discuss equations that contain trigonometric functions. In Sections 6.4 through 6.7, we continue the derivation of identities. These identities play an important role in calculus, the physical and life sciences, and economics, where they are used to simplify complicated expressions.

6.1 The Inverse Sine, Cosine, and Tangent Functions

PREPARING FOR THIS SECTION *Before getting started, review the following:*

- Inverse Functions (Section 4.2, pp. 282–291)
- Values of the Trigonometric Functions (Section 5.3, pp. 403–405; Section 5.4, pp. 416, 420–421)
- Properties of the Sine, Cosine, and Tangent Functions (Section 5.5, pp. 429–432)
- Graphs of the Sine, Cosine, and Tangent Functions (Sections 5.6, pp. 435–438 and 5.7, pp. 450–452)

Now Work the **'Are You Prepared?'** problems on page 487.

OBJECTIVES **1** Find the Exact Value of an Inverse Sine, Cosine, or Tangent Function (p. 479)
2 Find an Approximate Value of an Inverse Sine Function (p. 480)
3 Use Properties of Inverse Functions to Find Exact Values of Certain Composite Functions (p. 481)
4 Find the Inverse Function of a Trigonometric Function (p. 486)
5 Solve Equations Involving Inverse Trigonometric Functions (p. 487)

In Section 4.2 we discussed inverse functions, and we concluded that if a function is one-to-one it will have an inverse function. We also observed that if a function is not one-to-one it may be possible to restrict its domain in some suitable manner so that the restricted function is one-to-one For example, the function $y = x^2$ is not one-to-one; however, if we restrict the domain to $x \geq 0$, the function is one-to-one.

Other properties of a one-to-one function f and its inverse function f^{-1} that we discussed in Section 4.2 are summarized next.

1. $f^{-1}(f(x)) = x$ for every x in the domain of f and $f(f^{-1}(x)) = x$ for every x in the domain of f^{-1}.
2. Domain of f = range of f^{-1} and range of f = domain of f^{-1}.
3. The graph of f and the graph of f^{-1} are symmetric with respect to the line $y = x$.
4. If a function $y = f(x)$ has an inverse function, the implicit equation of the inverse function is $x = f(y)$. If we solve this equation for y, we obtain the explicit equation $y = f^{-1}(x)$.

The Inverse Sine Function

In Figure 1, we show the graph of $y = \sin x$. Because every horizontal line $y = b$, where b is between -1 and 1, inclusive, intersects the graph of $y = \sin x$ infinitely many times, it follows from the horizontal-line test that the function $y = \sin x$ is not one-to-one.

Figure 1
$y = \sin x, -\infty < x < \infty, -1 \leq y \leq 1$

Figure 2
$y = \sin x, -\dfrac{\pi}{2} \leq x \leq \dfrac{\pi}{2}, -1 \leq y \leq 1$

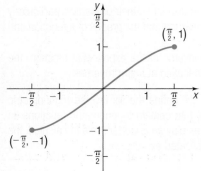

However, if we restrict the domain of $y = \sin x$ to the interval $\left[-\dfrac{\pi}{2}, \dfrac{\pi}{2}\right]$, the restricted function

$$y = \sin x \qquad -\frac{\pi}{2} \leq x \leq \frac{\pi}{2}$$

is one-to-one and so will have an inverse function.* See Figure 2.

*Although there are many other ways to restrict the domain and obtain a one-to-one function, mathematicians have agreed to use the interval $\left[-\dfrac{\pi}{2}, \dfrac{\pi}{2}\right]$ to define the inverse of $y = \sin x$.

NOTE Remember, the domain of a function f equals the range of its inverse, f^{-1}, and the range of a function f equals the domain of its inverse, f^{-1}. Because the restricted domain of the sine function is $\left[-\dfrac{\pi}{2}, \dfrac{\pi}{2}\right]$, the range of the inverse sine function is $\left[-\dfrac{\pi}{2}, \dfrac{\pi}{2}\right]$ and because the range of the sine function is $[-1, 1]$, the domain of the inverse sine function is $[-1, 1]$. ∎

An equation for the inverse of $y = f(x) = \sin x$ is obtained by interchanging x and y. The implicit form of the inverse function is $x = \sin y, -\dfrac{\pi}{2} \le y \le \dfrac{\pi}{2}$. The explicit form is called the **inverse sine** of x and is symbolized by $y = f^{-1}(x) = \sin^{-1} x$.

$$y = \sin^{-1} x \quad \text{means} \quad x = \sin y$$
$$\text{where} \quad -1 \le x \le 1 \quad \text{and} \quad -\dfrac{\pi}{2} \le y \le \dfrac{\pi}{2} \tag{1}$$

Because $y = \sin^{-1} x$ means $x = \sin y$, we read $y = \sin^{-1} x$ as "y is the angle or real number whose sine equals x." Alternatively, we can say that "y is the inverse sine of x." Be careful about the notation used. The superscript -1 that appears in $y = \sin^{-1} x$ is not an exponent, but is the symbolism used to denote the inverse function f^{-1} of f. (To avoid this notation, some books use the notation $y = \text{Arcsin } x$ instead of $y = \sin^{-1} x$.)

The inverse of a function f receives as input an element from the range of f and returns as output an element in the domain of f. The restricted sine function, $y = f(x) = \sin x$, receives as input an angle or real number x in the interval $\left[-\dfrac{\pi}{2}, \dfrac{\pi}{2}\right]$ and outputs a real number in the interval $[-1, 1]$. Therefore, the inverse sine function $y = \sin^{-1} x$ receives as input a real number in the interval $[-1, 1]$ or $-1 \le x \le 1$, its domain, and outputs an angle or real number in the interval $\left[-\dfrac{\pi}{2}, \dfrac{\pi}{2}\right]$ or $-\dfrac{\pi}{2} \le y \le \dfrac{\pi}{2}$, its range.

The graph of the inverse sine function can be obtained by reflecting the restricted portion of the graph of $y = f(x) = \sin x$ about the line $y = x$, as shown in Figure 3.

✔**Check:** Graph $Y_1 = \sin x$ and $Y_2 = \sin^{-1} x$. Compare the result with Figure 3.

Figure 3

$y = \sin^{-1} x, -1 \le x \le 1, -\dfrac{\pi}{2} \le y \le \dfrac{\pi}{2}$

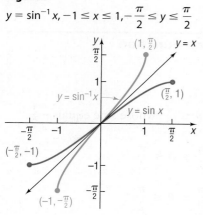

1 Find the Exact Value of an Inverse Sine Function

For some numbers x, it is possible to find the exact value of $y = \sin^{-1} x$.

EXAMPLE 1 **Finding the Exact Value of an Inverse Sine Function**

Find the exact value of: $\sin^{-1} 1$

Solution Let $\theta = \sin^{-1} 1$. We seek the angle $\theta, -\dfrac{\pi}{2} \le \theta \le \dfrac{\pi}{2}$, whose sine equals 1.

$$\theta = \sin^{-1} 1 \qquad -\dfrac{\pi}{2} \le \theta \le \dfrac{\pi}{2}$$

$$\sin \theta = 1 \qquad -\dfrac{\pi}{2} \le \theta \le \dfrac{\pi}{2} \quad \textit{By definition of } y = \sin^{-1} x$$

Now look at Table 1 and Figure 4.

Table 1

θ	$-\dfrac{\pi}{2}$	$-\dfrac{\pi}{3}$	$-\dfrac{\pi}{4}$	$-\dfrac{\pi}{6}$	0	$\dfrac{\pi}{6}$	$\dfrac{\pi}{4}$	$\dfrac{\pi}{3}$	$\dfrac{\pi}{2}$
$\sin \theta$	-1	$-\dfrac{\sqrt{3}}{2}$	$-\dfrac{\sqrt{2}}{2}$	$-\dfrac{1}{2}$	0	$\dfrac{1}{2}$	$\dfrac{\sqrt{2}}{2}$	$\dfrac{\sqrt{3}}{2}$	1

We see that the only angle θ within the interval $\left[-\dfrac{\pi}{2}, \dfrac{\pi}{2}\right]$ whose sine is 1 is $\dfrac{\pi}{2}$.

(Note that $\sin \dfrac{5\pi}{2}$ also equals 1, but $\dfrac{5\pi}{2}$ lies outside the interval $\left[-\dfrac{\pi}{2}, \dfrac{\pi}{2}\right]$ and hence

Figure 4

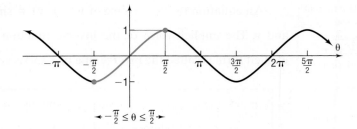

$$-\frac{\pi}{2} \le \theta \le \frac{\pi}{2}$$

is not admissible.) So,

$$\sin^{-1} 1 = \frac{\pi}{2}$$

━━━ **Now Work**   PROBLEM 13

EXAMPLE 2   **Finding the Exact Value of an Inverse Sine Function**

Find the exact value of: $\sin^{-1}\left(-\frac{1}{2}\right)$

Solution   Let $\theta = \sin^{-1}\left(-\frac{1}{2}\right)$. We seek the angle θ, $-\frac{\pi}{2} \le \theta \le \frac{\pi}{2}$, whose sine equals $-\frac{1}{2}$.

$$\theta = \sin^{-1}\left(-\frac{1}{2}\right) \qquad -\frac{\pi}{2} \le \theta \le \frac{\pi}{2}$$

$$\sin \theta = -\frac{1}{2} \qquad -\frac{\pi}{2} \le \theta \le \frac{\pi}{2}$$

(Refer to Table 1 and Figure 4, if necessary.) The only angle within the interval $\left[-\frac{\pi}{2}, \frac{\pi}{2}\right]$ whose sine is $-\frac{1}{2}$ is $-\frac{\pi}{6}$. So,

$$\sin^{-1}\left(-\frac{1}{2}\right) = -\frac{\pi}{6}$$

━━━ **Now Work**   PROBLEM 19

---

2 Find an Approximate Value of an Inverse Sine Function

For most numbers x, the value $y = \sin^{-1} x$ must be approximated.

EXAMPLE 3   **Finding an Approximate Value of an Inverse Sine Function**

Find an approximate value of:

(a) $\sin^{-1}\dfrac{1}{3}$ (b) $\sin^{-1}\left(-\dfrac{1}{4}\right)$

Express the answer in radians rounded to two decimal places.

Solution   (a) Because we want the angle measured in radians, we first set the mode of our calculator to radians.* Rounded to two decimal places, we have

$$\sin^{-1}\frac{1}{3} = 0.34$$

---

*On most calculators, the inverse sine is obtained by pressing SHIFT or 2nd, followed by sin. On some calculators, $\boxed{\sin^{-1}}$ is pressed first, then 1/3 is entered; on others, this sequence is reversed. Consult your owner's manual for the correct sequence.

Figure 5

```
sin⁻¹(-1/4)
    -.2526802551
```

(b) Figure 5 shows the solution using a TI-84 Plus graphing calculator in radian mode. We have $\sin^{-1}\left(-\dfrac{1}{4}\right) = -0.25$, rounded to two decimal places.

Now Work PROBLEM 25

3 Use Properties of Inverse Functions to Find Exact Values of Certain Composite Functions

When we discussed functions and their inverses in Section 4.2, we found that $f^{-1}(f(x)) = x$ for all x in the domain of f and $f(f^{-1}(x)) = x$ for all x in the domain of f^{-1}. In terms of the sine function and its inverse, these properties are of the form

$$f^{-1}(f(x)) = \sin^{-1}(\sin x) = x \qquad \text{where } -\frac{\pi}{2} \le x \le \frac{\pi}{2} \qquad \textbf{(2a)}$$

$$f(f^{-1}(x)) = \sin(\sin^{-1} x) = x \qquad \text{where } -1 \le x \le 1 \qquad \textbf{(2b)}$$

EXAMPLE 4 **Finding the Exact Value of Certain Composite Functions**

Find the exact value of each of the following composite functions:

(a) $\sin^{-1}\left(\sin \dfrac{\pi}{8}\right)$

(b) $\sin^{-1}\left(\sin \dfrac{5\pi}{8}\right)$

Solution (a) The composite function $\sin^{-1}\left(\sin \dfrac{\pi}{8}\right)$ follows the form of equation (2a). Because $\dfrac{\pi}{8}$ is in the interval $\left[-\dfrac{\pi}{2}, \dfrac{\pi}{2}\right]$, we can use (2a). Then

$$\sin^{-1}\left(\sin \frac{\pi}{8}\right) = \frac{\pi}{8}$$

(b) The composite function $\sin^{-1}\left(\sin \dfrac{5\pi}{8}\right)$ follows the form of equation (2a), but $\dfrac{5\pi}{8}$ is not in the interval $\left[-\dfrac{\pi}{2}, \dfrac{\pi}{2}\right]$. To use (2a), we need to find an angle θ in the interval $\left[-\dfrac{\pi}{2}, \dfrac{\pi}{2}\right]$ for which $\sin \theta = \sin \dfrac{5\pi}{8}$. Then, using (2a), $\sin^{-1}\left(\sin \dfrac{5\pi}{8}\right) = \sin^{-1}(\sin \theta) = \theta$, and we are finished.

Look at Figure 6. The angle $\dfrac{5\pi}{8}$ is in quadrant II. The reference angle of $\dfrac{5\pi}{8}$ is $\dfrac{3\pi}{8}$ and $\sin \dfrac{5\pi}{8} = \dfrac{b}{r} = \sin \dfrac{3\pi}{8}$. Since $\dfrac{3\pi}{8}$ is in the interval $\left[-\dfrac{\pi}{2}, \dfrac{\pi}{2}\right]$, we have

$$\sin^{-1}\left(\sin \frac{5\pi}{8}\right) = \sin^{-1}\left(\sin \frac{3\pi}{8}\right) = \frac{3\pi}{8}$$

Apply (2a).

Figure 6

Now Work PROBLEM 41

EXAMPLE 5 **Finding the Exact Value of Certain Composite Functions**

Find the exact value, if any, of each composite function.

(a) $\sin(\sin^{-1} 0.5)$

(b) $\sin(\sin^{-1} 1.8)$

Solution   (a) The composite function $\sin(\sin^{-1} 0.5)$ follows the form of equation (2b) and 0.5 is in the interval $[-1, 1]$. So we use (2b):

$$\sin(\sin^{-1} 0.5) = 0.5$$

(b) The composite function $\sin(\sin^{-1} 1.8)$ follows the form of equation (2b), but 1.8 is not in the domain of the inverse sine function. This composite function is not defined.

●

Now Work   PROBLEM 45

The Inverse Cosine Function

In Figure 7 we show the graph of $y = \cos x$. Because every horizontal line $y = b$, where b is between -1 and 1, inclusive, intersects the graph of $y = \cos x$ infinitely many times, it follows that the cosine function is not one-to-one.

Figure 7
$y = \cos x, -\infty < x < \infty,$
$-1 \leq y \leq 1$

Figure 8
$y = \cos x, 0 \leq x \leq \pi, -1 \leq y \leq 1$

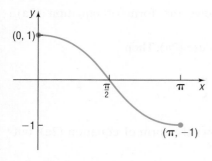

However, if we restrict the domain of $y = \cos x$ to the interval $[0, \pi]$, the restricted function

$$y = \cos x \qquad 0 \leq x \leq \pi$$

is one-to-one and hence will have an inverse function.* See Figure 8.

An equation for the inverse of $y = f(x) = \cos x$ is obtained by interchanging x and y. The implicit form of the inverse function is $x = \cos y, 0 \leq y \leq \pi$. The explicit form is called the **inverse cosine** of x and is symbolized by $y = f^{-1}(x) = \cos^{-1} x$ (or by $y = \text{Arccos } x$).

DEFINITION

$$y = \cos^{-1} x \quad \text{means} \quad x = \cos y$$
$$\text{where} \quad -1 \leq x \leq 1 \quad \text{and} \quad 0 \leq y \leq \pi \tag{3}$$

Figure 9
$y = \cos^{-1} x, -1 \leq x \leq 1, 0 \leq y \leq \pi$

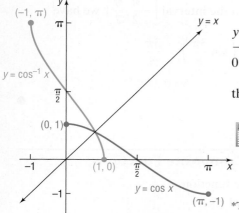

Here y is the angle whose cosine is x. Because the range of the cosine function, $y = \cos x$, is $-1 \leq y \leq 1$, the domain of the inverse function $y = \cos^{-1} x$ is $-1 \leq x \leq 1$. Because the restricted domain of the cosine function, $y = \cos x$, is $0 \leq x \leq \pi$, the range of the inverse function $y = \cos^{-1} x$ is $0 \leq y \leq \pi$.

The graph of $y = \cos^{-1} x$ can be obtained by reflecting the restricted portion of the graph of $y = \cos x$ about the line $y = x$, as shown in Figure 9.

✓ **Check:** Graph $Y_1 = \cos x$ and $Y_2 = \cos^{-1} x$. Compare the result with Figure 9.

*This is the generally accepted restriction to define the inverse cosine function.

EXAMPLE 6   **Finding the Exact Value of an Inverse Cosine Function**

Find the exact value of: $\cos^{-1} 0$

Solution   Let $\theta = \cos^{-1} 0$. We seek the angle $\theta, 0 \le \theta \le \pi$, whose cosine equals 0.

$$\theta = \cos^{-1} 0 \qquad 0 \le \theta \le \pi$$
$$\cos \theta = 0 \qquad 0 \le \theta \le \pi$$

Look at Table 2 and Figure 10.

Table 2

θ	$\cos \theta$
0	1
$\dfrac{\pi}{6}$	$\dfrac{\sqrt{3}}{2}$
$\dfrac{\pi}{4}$	$\dfrac{\sqrt{2}}{2}$
$\dfrac{\pi}{3}$	$\dfrac{1}{2}$
$\dfrac{\pi}{2}$	0
$\dfrac{2\pi}{3}$	$-\dfrac{1}{2}$
$\dfrac{3\pi}{4}$	$-\dfrac{\sqrt{2}}{2}$
$\dfrac{5\pi}{6}$	$-\dfrac{\sqrt{3}}{2}$
π	-1

Figure 10

We see that the only angle θ within the interval $[0, \pi]$ whose cosine is 0 is $\dfrac{\pi}{2}$. [Note that $\cos \dfrac{3\pi}{2}$ and $\cos\left(-\dfrac{\pi}{2}\right)$ also equal 0, but they lie outside the interval $[0, \pi]$ and hence are not admissible.] We conclude that

$$\cos^{-1} 0 = \frac{\pi}{2}$$

EXAMPLE 7   **Finding the Exact Value of an Inverse Cosine Function**

Find the exact value of: $\cos^{-1}\left(-\dfrac{\sqrt{2}}{2}\right)$

Solution   Let $\theta = \cos^{-1}\left(-\dfrac{\sqrt{2}}{2}\right)$. We seek the angle $\theta, \ 0 \le \theta \le \pi$, whose cosine equals $-\dfrac{\sqrt{2}}{2}$.

$$\theta = \cos^{-1}\left(-\frac{\sqrt{2}}{2}\right) \qquad 0 \le \theta \le \pi$$
$$\cos \theta = -\frac{\sqrt{2}}{2} \qquad 0 \le \theta \le \pi$$

Look at Table 2 and Figure 11.

Figure 11

We see that the only angle θ within the interval $[0, \pi]$ whose cosine is $-\dfrac{\sqrt{2}}{2}$ is $\dfrac{3\pi}{4}$. So,

$$\cos^{-1}\left(-\frac{\sqrt{2}}{2}\right) = \frac{3\pi}{4}$$

━━━━━ **Now Work** PROBLEM 23

For the cosine function and its inverse, the following properties hold:

$f^{-1}(f(x)) = \cos^{-1}(\cos x) = x$ where $0 \le x \le \pi$	**(4a)**
$f(f^{-1}(x)) = \cos(\cos^{-1} x) = x$ where $-1 \le x \le 1$	**(4b)**

EXAMPLE 8 **Using Properties of Inverse Functions to Find the Exact Value of Certain Composite Functions**

Find the exact value of:

(a) $\cos^{-1}\left(\cos \dfrac{\pi}{12}\right)$ (b) $\cos[\cos^{-1}(-0.4)]$ (c) $\cos^{-1}\left[\cos\left(-\dfrac{2\pi}{3}\right)\right]$ (d) $\cos(\cos^{-1}\pi)$

Solution (a) $\cos^{-1}\left(\cos \dfrac{\pi}{12}\right) = \dfrac{\pi}{12}$ $\dfrac{\pi}{12}$ is in the interval $[0, \pi]$; use Property (4a).

(b) $\cos[\cos^{-1}(-0.4)] = -0.4$ -0.4 is in the interval $[-1, 1]$; use Property (4b).

(c) The angle $-\dfrac{2\pi}{3}$ is not in the interval $[0, \pi]$ so we cannot use (4a). However, because the cosine function is even, $\cos\left(-\dfrac{2\pi}{3}\right) = \cos \dfrac{2\pi}{3}$. Since $\dfrac{2\pi}{3}$ is in the interval $[0, \pi]$, we have

$$\cos^{-1}\left[\cos\left(-\frac{2\pi}{3}\right)\right] = \cos^{-1}\left(\cos \frac{2\pi}{3}\right) = \frac{2\pi}{3} \qquad \frac{2\pi}{3} \text{ is in the interval } [0, \pi]; \text{ apply (4a)}.$$

(d) Because π is not in the interval $[-1, 1]$, the domain of the inverse cosine function, $\cos^{-1} \pi$ is not defined. This means the composite function $\cos(\cos^{-1}\pi)$ is also not defined.

━━━━━ **Now Work** PROBLEMS 37 AND 49

The Inverse Tangent Function

In Figure 12 we show the graph of $y = \tan x$. Because every horizontal line intersects the graph infinitely many times, it follows that the tangent function is not one-to-one.

However, if we restrict the domain of $y = \tan x$ to the interval $\left(-\dfrac{\pi}{2}, \dfrac{\pi}{2}\right)$, the restricted function

$$y = \tan x \qquad -\frac{\pi}{2} < x < \frac{\pi}{2}$$

is one-to-one and hence has an inverse function.* See Figure 13.

*This is the generally accepted restriction.

Figure 12
$y = \tan x, -\infty < x < \infty, x$ not equal to odd multiples of $\dfrac{\pi}{2}, -\infty < y < \infty$

Figure 13
$y = \tan x, -\dfrac{\pi}{2} < x < \dfrac{\pi}{2}, -\infty < y < \infty$

An equation for the inverse of $y = f(x) = \tan x$ is obtained by interchanging x and y. The implicit form of the inverse function is $x = \tan y, -\dfrac{\pi}{2} < y < \dfrac{\pi}{2}$. The explicit form is called the **inverse tangent** of x and is symbolized by $y = f^{-1}(x) = \tan^{-1} x$ (or by $y = \text{Arctan } x$).

DEFINITION

$$y = \tan^{-1} x \quad \text{means} \quad x = \tan y$$
$$\text{where} \quad -\infty < x < \infty \quad \text{and} \quad -\frac{\pi}{2} < y < \frac{\pi}{2} \qquad (5)$$

Here y is the angle whose tangent is x. The domain of the function $y = \tan^{-1} x$ is $-\infty < x < \infty$, and its range is $-\dfrac{\pi}{2} < y < \dfrac{\pi}{2}$. The graph of $y = \tan^{-1} x$ can be obtained by reflecting the restricted portion of the graph of $y = \tan x$ about the line $y = x$, as shown in Figure 14.

Figure 14
$y = \tan^{-1} x, -\infty < x < \infty, -\dfrac{\pi}{2} < y < \dfrac{\pi}{2}$

 ✓ **Check:** Graph $Y_1 = \tan x$ and $Y_2 = \tan^{-1} x$. Compare the result with Figure 14.

EXAMPLE 9 **Finding the Exact Value of an Inverse Tangent Function**

Find the exact value of:

(a) $\tan^{-1} 1$ (b) $\tan^{-1}\left(-\sqrt{3}\right)$

Solution (a) Let $\theta = \tan^{-1} 1$. We seek the angle $\theta, -\dfrac{\pi}{2} < \theta < \dfrac{\pi}{2}$, whose tangent equals 1.

$$\theta = \tan^{-1} 1 \qquad -\frac{\pi}{2} < \theta < \frac{\pi}{2}$$
$$\tan \theta = 1 \qquad -\frac{\pi}{2} < \theta < \frac{\pi}{2}$$

Table 3

θ	$\tan \theta$
$-\dfrac{\pi}{2}$	Undefined
$-\dfrac{\pi}{3}$	$-\sqrt{3}$
$-\dfrac{\pi}{4}$	-1
$-\dfrac{\pi}{6}$	$-\dfrac{\sqrt{3}}{3}$
0	0
$\dfrac{\pi}{6}$	$\dfrac{\sqrt{3}}{3}$
$\dfrac{\pi}{4}$	1
$\dfrac{\pi}{3}$	$\sqrt{3}$
$\dfrac{\pi}{2}$	Undefined

Look at Table 3. The only angle θ within the interval $\left(-\dfrac{\pi}{2}, \dfrac{\pi}{2}\right)$ whose tangent is 1 is $\dfrac{\pi}{4}$. So,

$$\tan^{-1} 1 = \dfrac{\pi}{4}$$

(b) Let $\theta = \tan^{-1}\left(-\sqrt{3}\right)$. We seek the angle $\theta, -\dfrac{\pi}{2} < \theta < \dfrac{\pi}{2}$, whose tangent equals $-\sqrt{3}$.

$$\theta = \tan^{-1}\left(-\sqrt{3}\right) \qquad -\dfrac{\pi}{2} < \theta < \dfrac{\pi}{2}$$
$$\tan \theta = -\sqrt{3} \qquad -\dfrac{\pi}{2} < \theta < \dfrac{\pi}{2}$$

Look at Table 3. The only angle θ within the interval $\left(-\dfrac{\pi}{2}, \dfrac{\pi}{2}\right)$ whose tangent is $-\sqrt{3}$ is $-\dfrac{\pi}{3}$. So,

$$\tan^{-1}\left(-\sqrt{3}\right) = -\dfrac{\pi}{3}$$

Now Work PROBLEM 17

For the tangent function and its inverse, the following properties hold:

$$f^{-1}(f(x)) = \tan^{-1}(\tan x) = x \qquad \text{where } -\dfrac{\pi}{2} < x < \dfrac{\pi}{2}$$
$$f(f^{-1}(x)) = \tan(\tan^{-1} x) = x \qquad \text{where } -\infty < x < \infty$$

Now Work PROBLEM 43

4 Find the Inverse Function of a Trigonometric Function

EXAMPLE 10 **Finding the Inverse Function of a Trigonometric Function**

Find the inverse function f^{-1} of $f(x) = 2 \sin x - 1, -\dfrac{\pi}{2} \le x \le \dfrac{\pi}{2}$. Find the range of f and the domain and range of f^{-1}.

Solution The function f is one-to-one and so has an inverse function. We follow the steps on page 289 for finding the inverse function.

$$y = 2 \sin x - 1$$
$$x = 2 \sin y - 1 \qquad \text{Interchange } x \text{ and } y.$$
$$x + 1 = 2 \sin y \qquad \text{Proceed to solve for } y.$$
$$\sin y = \dfrac{x + 1}{2}$$
$$y = \sin^{-1} \dfrac{x + 1}{2} \qquad \text{Apply the definition (1).}$$

The inverse function is $f^{-1}(x) = \sin^{-1} \dfrac{x + 1}{2}$.

To find the range of f, we solve $y = 2 \sin x - 1$ for $\sin x$ and use the fact that $-1 \le \sin x \le 1$.

$$y = 2 \sin x - 1$$
$$\sin x = \dfrac{y + 1}{2}$$

$$-1 \le \frac{y+1}{2} \le 1$$
$$-2 \le y + 1 \le 2$$
$$-3 \le y \le 1$$

The range of f is $\{y \mid -3 \le y \le 1\}$ or $[-3, 1]$ using internal notation.
The domain of f^{-1} equals the range of f, $[-3, 1]$.

The range of f^{-1} equals the domain of f, $\left[-\dfrac{\pi}{2}, \dfrac{\pi}{2}\right]$.

━━━━ **Now Work** PROBLEM 55

5 Solve Equations Involving Inverse Trigonometric Functions

Equations that contain inverse trigonometric functions are called **inverse trigono-metric equations**.

EXAMPLE 11 **Solving an Equation Involving an Inverse Trigonometric Function**

Solve the equation: $3 \sin^{-1} x = \pi$

Solution To solve an equation involving a single inverse trigonometric function, first isolate the inverse trigonometric function.

$$3 \sin^{-1} x = \pi$$
$$\sin^{-1} x = \frac{\pi}{3} \qquad \text{Divide both sides by 3.}$$
$$x = \sin \frac{\pi}{3} \qquad y = \sin^{-1} x \text{ means } x = \sin y.$$
$$x = \frac{\sqrt{3}}{2}$$

The solution set is $\left\{\dfrac{\sqrt{3}}{2}\right\}$.

━━━━ **Now Work** PROBLEM 61

6.1 Assess Your Understanding

'Are You Prepared?' *Answers are given at the end of these exercises. If you get a wrong answer, read the pages listed in red.*

1. What is the domain and the range of $y = \sin x$? (p. 429)

2. A suitable restriction on the domain of the function $f(x) = (x - 1)^2$ to make it one-to-one would be _____. (p. 291)

3. If the domain of a one-to-one function is $[3, \infty)$, the range of its inverse is _____. (p. 286)

4. **True or False** The graph of $y = \cos x$ is decreasing on the interval $[0, \pi]$. (p. 437)

5. $\tan \dfrac{\pi}{4} =$ _____ ; $\sin \dfrac{\pi}{3} =$ _____ (pp. 403–405)

6. $\sin\left(-\dfrac{\pi}{6}\right) =$ _____ ; $\cos \pi =$ _____. (pp. 416 and 420–421)

Concepts and Vocabulary

7. $y = \sin^{-1} x$ means _____, where $-1 \le x \le 1$ and $-\dfrac{\pi}{2} \le y \le \dfrac{\pi}{2}$.

8. $\cos^{-1}(\cos x) = x$ where _____.

9. $\tan(\tan^{-1} x) = x$ where _____.

10. **True or False** The domain of $y = \sin^{-1} x$ is $-\dfrac{\pi}{2} \le x \le \dfrac{\pi}{2}$.

11. **True or False** $\sin(\sin^{-1} 0) = 0$ and $\cos(\cos^{-1} 0) = 0$.

12. **True or False** $y = \tan^{-1} x$ means $x = \tan y$, where $-\infty < x < \infty$ and $-\dfrac{\pi}{2} < y < \dfrac{\pi}{2}$.

Skill Building

In Problems 13–24, find the exact value of each expression.

13. $\sin^{-1} 0$
14. $\cos^{-1} 1$
15. $\sin^{-1}(-1)$
16. $\cos^{-1}(-1)$

17. $\tan^{-1} 0$
18. $\tan^{-1}(-1)$
19. $\sin^{-1} \dfrac{\sqrt{2}}{2}$
20. $\tan^{-1} \dfrac{\sqrt{3}}{3}$

21. $\tan^{-1} \sqrt{3}$
22. $\sin^{-1}\left(-\dfrac{\sqrt{3}}{2}\right)$
23. $\cos^{-1}\left(-\dfrac{\sqrt{3}}{2}\right)$
24. $\sin^{-1}\left(-\dfrac{\sqrt{2}}{2}\right)$

In Problems 25–36, use a calculator to find the value of each expression rounded to two decimal places.

25. $\sin^{-1} 0.1$
26. $\cos^{-1} 0.6$
27. $\tan^{-1} 5$
28. $\tan^{-1} 0.2$

29. $\cos^{-1} \dfrac{7}{8}$
30. $\sin^{-1} \dfrac{1}{8}$
31. $\tan^{-1}(-0.4)$
32. $\tan^{-1}(-3)$

33. $\sin^{-1}(-0.12)$
34. $\cos^{-1}(-0.44)$
35. $\cos^{-1} \dfrac{\sqrt{2}}{3}$
36. $\sin^{-1} \dfrac{\sqrt{3}}{5}$

In Problems 37–44, find the exact value of each expression. Do not use a calculator.

37. $\cos^{-1}\left(\cos \dfrac{4\pi}{5}\right)$
38. $\sin^{-1}\left[\sin\left(-\dfrac{\pi}{10}\right)\right]$
39. $\tan^{-1}\left[\tan\left(-\dfrac{3\pi}{8}\right)\right]$
40. $\sin^{-1}\left[\sin\left(-\dfrac{3\pi}{7}\right)\right]$

41. $\sin^{-1}\left(\sin \dfrac{9\pi}{8}\right)$
42. $\cos^{-1}\left[\cos\left(-\dfrac{5\pi}{3}\right)\right]$
43. $\tan^{-1}\left(\tan \dfrac{4\pi}{5}\right)$
44. $\tan^{-1}\left[\tan\left(-\dfrac{2\pi}{3}\right)\right]$

In Problems 45–52, find the exact value, if any, of each composite function. If there is no value, say it is "not defined." Do not use a calculator.

45. $\sin\left(\sin^{-1} \dfrac{1}{4}\right)$
46. $\cos\left[\cos^{-1}\left(-\dfrac{2}{3}\right)\right]$
47. $\tan\left(\tan^{-1} 4\right)$
48. $\tan\left[\tan^{-1}(-2)\right]$

49. $\cos\left(\cos^{-1} 1.2\right)$
50. $\sin\left[\sin^{-1}(-2)\right]$
51. $\tan\left(\tan^{-1} \pi\right)$
52. $\sin\left[\sin^{-1}(-1.5)\right]$

In Problems 53–60, find the inverse function f^{-1} of each function f. Find the range of f and the domain and range of f^{-1}.

53. $f(x) = 5 \sin x + 2; -\dfrac{\pi}{2} \le x \le \dfrac{\pi}{2}$
54. $f(x) = 2 \tan x - 3; -\dfrac{\pi}{2} < x < \dfrac{\pi}{2}$

55. $f(x) = -2 \cos(3x); 0 \le x \le \dfrac{\pi}{3}$
56. $f(x) = 3 \sin(2x); -\dfrac{\pi}{4} \le x \le \dfrac{\pi}{4}$

57. $f(x) = -\tan(x + 1) - 3; -1 - \dfrac{\pi}{2} < x < \dfrac{\pi}{2} - 1$
58. $f(x) = \cos(x + 2) + 1; -2 \le x \le \pi - 2$

59. $f(x) = 3 \sin(2x + 1); -\dfrac{1}{2} - \dfrac{\pi}{4} \le x \le -\dfrac{1}{2} + \dfrac{\pi}{4}$
60. $f(x) = 2 \cos(3x + 2); -\dfrac{2}{3} \le x \le -\dfrac{2}{3} + \dfrac{\pi}{3}$

In Problems 61–68, find the exact solution of each equation.

61. $4 \sin^{-1} x = \pi$
62. $2 \cos^{-1} x = \pi$
63. $3 \cos^{-1}(2x) = 2\pi$

64. $-6 \sin^{-1}(3x) = \pi$
65. $3 \tan^{-1} x = \pi$
66. $-4 \tan^{-1} x = \pi$

67. $4 \cos^{-1} x - 2\pi = 2 \cos^{-1} x$
68. $5 \sin^{-1} x - 2\pi = 2 \sin^{-1} x - 3\pi$

Applications and Extensions

In Problems 69–74, use the following discussion. The formula

$$D = 24\left[1 - \dfrac{\cos^{-1}(\tan i \tan \theta)}{\pi}\right]$$

can be used to approximate the number of hours of daylight D when the declination of the Sun is $i°$ at a location $\theta°$ north latitude for any date between the vernal equinox and autumnal equinox. The declination of the Sun is defined as the angle i between the equatorial plane and any ray of light from the Sun. The latitude of a location

is the angle θ between the Equator and the location on the surface of Earth, with the vertex of the angle located at the center of Earth. See the figure. To use the formula, $\cos^{-1}(\tan i \tan \theta)$ *must be expressed in radians.*

69. Approximate the number of hours of daylight in Houston, Texas (29°45′ north latitude), for the following dates:
 (a) Summer solstice ($i = 23.5°$)
 (b) Vernal equinox ($i = 0°$)
 (c) July 4 ($i = 22°48′$)

70. Approximate the number of hours of daylight in New York, New York (40°45′ north latitude), for the following dates:
 (a) Summer solstice ($i = 23.5°$)
 (b) Vernal equinox ($i = 0°$)
 (c) July 4 ($i = 22°48′$)

71. Approximate the number of hours of daylight in Honolulu, Hawaii (21°18′ north latitude), for the following dates:
 (a) Summer solstice ($i = 23.5°$)
 (b) Vernal equinox ($i = 0°$)
 (c) July 4 ($i = 22°48′$)

72. Approximate the number of hours of daylight in Anchorage, Alaska (61°10′ north latitude), for the following dates:
 (a) Summer solstice ($i = 23.5°$)
 (b) Vernal equinox ($i = 0°$)
 (c) July 4 ($i = 22°48′$)

73. Approximate the number of hours of daylight at the Equator (0° north latitude) for the following dates:
 (a) Summer solstice ($i = 23.5°$)
 (b) Vernal equinox ($i = 0°$)
 (c) July 4 ($i = 22°48′$)
 (d) What do you conclude about the number of hours of daylight throughout the year for a location at the Equator?

74. Approximate the number of hours of daylight for any location that is 66°30′ north latitude for the following dates:
 (a) Summer solstice ($i = 23.5°$)
 (b) Vernal equinox ($i = 0°$)
 (c) July 4 ($i = 22°48′$)
 (d) The number of hours of daylight on the winter solstice may be found by computing the number of hours of daylight on the summer solstice and subtracting this result from 24 hours, due to the symmetry of the orbital path of Earth around the Sun. Compute the number of hours of daylight for this location on the winter solstice. What do you conclude about daylight for a location at 66°30′ north latitude?

75. Being the First to See the Rising Sun   Cadillac Mountain, elevation 1530 feet, is located in Acadia National Park, Maine, and is the highest peak on the east coast of the United States. It is said that a person standing on the summit will be the first person in the United States to see the rays of the rising Sun. How much sooner would a person atop Cadillac Mountain see the first rays than a person standing below, at sea level?

[**Hint:** Consult the figure. When the person at D sees the first rays of the Sun, the person at P does not. The person at P sees the first rays of the Sun only after Earth has rotated so that P is at location Q. Compute the length of the arc subtended by the central angle θ. Then use the fact that, at the latitude of Cadillac Mountain, in 24 hours a length of $2\pi(2710) \approx 17027.4$ miles is subtended, and find the time that it takes to subtend this length.]

76. Movie Theater Screens   Suppose that a movie theater has a screen that is 28 feet tall. When you sit down, the bottom of the screen is 6 feet above your eye level. The angle formed by drawing a line from your eye to the bottom of the screen and your eye and the top of the screen is called the **viewing angle.** In the figure, θ is the viewing angle. Suppose that you sit x feet from the screen. The viewing angle θ is given by the function

$$\theta(x) = \tan^{-1}\left(\frac{34}{x}\right) - \tan^{-1}\left(\frac{6}{x}\right)$$

(a) What is your viewing angle if you sit 10 feet from the screen? 15 feet? 20 feet?
(b) If there is 5 feet between the screen and the first row of seats and there is 3 feet between each row, which row results in the largest viewing angle?
(c) Using a graphing utility, graph

$$\theta(x) = \tan^{-1}\left(\frac{34}{x}\right) - \tan^{-1}\left(\frac{6}{x}\right)$$

What value of x results in the largest viewing angle?

77. Area under a Curve   The area under the graph of $y = \dfrac{1}{1 + x^2}$ and above the x-axis between $x = a$ and $x = b$ is given by

$$\tan^{-1} b - \tan^{-1} a$$

See the figure.

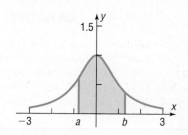

(a) Find the exact area under the graph of $y = \dfrac{1}{1 + x^2}$ and above the x-axis between $x = 0$ and $x = \sqrt{3}$.

(b) Find the exact area under the graph of $y = \dfrac{1}{1 + x^2}$ and above the x-axis between $x = -\dfrac{\sqrt{3}}{3}$ and $x = 1$.

△ **78. Area under a Curve** The area under the graph of $y = \dfrac{1}{\sqrt{1 - x^2}}$ and above the x-axis between $x = a$ and $x = b$ is given by

$$\sin^{-1} b - \sin^{-1} a$$

See the figure.

(a) Find the exact area under the graph of $y = \dfrac{1}{\sqrt{1 - x^2}}$ and above the x-axis between $x = 0$ and $x = \dfrac{\sqrt{3}}{2}$.

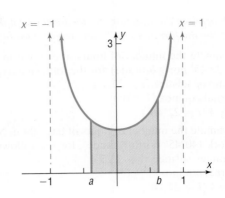

(b) Find the exact area under the graph of $y = \dfrac{1}{\sqrt{1 - x^2}}$ and above the x-axis between $x = -\dfrac{1}{2}$ and $x = \dfrac{1}{2}$.

Problems 79 and 80 require the following discussion:
The shortest distance between two points on Earth's surface can be determined from the latitude and longitude of the two locations. For example, if location 1 has (lat, lon) = (α_1, β_1) and location 2 has (lat, lon) = (α_2, β_2), the shortest distance between the two locations is approximately

$$d = r\cos^{-1}[(\cos \alpha_1 \cos \beta_1 \cos \alpha_2 \cos \beta_2) + (\cos \alpha_1 \sin \beta_1 \cos \alpha_2 \sin \beta_2) + (\sin \alpha_1 \sin \alpha_2)]$$

where r = radius of Earth ≈ 3960 miles and the inverse cosine function is expressed in degrees. Also N latitude and E longitude are positive angles while S latitude and W longitude are negative angles.

Source: www.infoplease.com

City	Latitude	Longitude
Chicago, IL	41°50′N	87°37′W
Honolulu, HI	21°18′N	157°50′W
Melbourne, Australia	37°47′S	144°58′E

79. Shortest Distance from Chicago to Honolulu Find the shortest distance from Chicago, latitude 41°50′N, longitude 87°37′W to Honolulu, latitude 21°18′N, longitude 157°50′W. Round your answer to the nearest mile.

80. Shortest Distance from Honolulu to Melbourne, Australia Find the shortest distance from Honolulu to Melbourne, Australia, latitude 37°47′S, longitude 144°58′E. Round your answer to the nearest mile.

'Are You Prepared?' Answers

1. domain: the set of all real numbers; range: $-1 \le y \le 1$

2. Two answers are possible: $x \le 1$ or $x \ge 1$

3. $[3, \infty)$ **4.** True   **5.** $1; \dfrac{\sqrt{3}}{2}$ **6.** $-\dfrac{1}{2}; -1$

6.2 The Inverse Trigonometric Functions (Continued)

PREPARING FOR THIS SECTION *Before getting started, review the following concepts:*

- Finding Exact Values Given the Value of a Trigonometric Function and the Quadrant of the Angle (Section 5.4, pp. 421–422)
- Graphs of the Secant, Cosecant, and Cotangent Functions (Section 5.7, pp. 453–455)
- Domain and Range of the Secant, Cosecant, and Cotangent Functions (Section 5.5, pp. 429–430)

Now Work the 'Are You Prepared?' problems on page 494.

OBJECTIVES   **1** Find the Exact Value of Expressions Involving the Inverse Sine, Cosine, and Tangent Functions (p. 491)

2 Define the Inverse Secant, Cosecant, and Cotangent Functions (p. 492)

3 Use a Calculator to Evaluate $\sec^{-1} x$, $\csc^{-1} x$, and $\cot^{-1} x$ (p. 492)

4 Write a Trigonometric Expression as an Algebraic Expression (p. 493)

1 Find the Exact Value of Expressions Involving the Inverse Sine, Cosine, and Tangent Functions

EXAMPLE 1

Finding the Exact Value of an Expression Involving Inverse Trigonometric Functions

Find the exact value of: $\sin\left(\tan^{-1}\dfrac{1}{2}\right)$

Solution

Let $\theta = \tan^{-1}\dfrac{1}{2}$. Then $\tan\theta = \dfrac{1}{2}$, where $-\dfrac{\pi}{2} < \theta < \dfrac{\pi}{2}$. We seek $\sin\theta$. Because $\tan\theta > 0$, it follows that $0 < \theta < \dfrac{\pi}{2}$, so θ lies in quadrant I. Now in Figure 15 we draw a triangle in quadrant I. Because $\tan\theta = \dfrac{b}{a} = \dfrac{1}{2}$, the side opposite θ is $b = 1$ and the side adjacent to θ is $a = 2$. The hypotenuse of this triangle is found using $a^2 + b^2 = c^2$ or $2^2 + 1^2 = c^2$, so $c = \sqrt{5}$. Then $\sin\theta = \dfrac{b}{c} = \dfrac{1}{\sqrt{5}}$, and

$$\sin\left(\tan^{-1}\dfrac{1}{2}\right) = \sin\theta = \dfrac{1}{\sqrt{5}} = \dfrac{\sqrt{5}}{5}$$

Figure 15

$\tan\theta = \dfrac{1}{2}$

EXAMPLE 2

Finding the Exact Value of an Expression Involving Inverse Trigonometric Functions

Find the exact value of: $\cos\left[\sin^{-1}\left(-\dfrac{1}{3}\right)\right]$

Solution

Let $\theta = \sin^{-1}\left(-\dfrac{1}{3}\right)$. Then $\sin\theta = -\dfrac{1}{3}$ and $-\dfrac{\pi}{2} \le \theta \le \dfrac{\pi}{2}$. We seek $\cos\theta$. Because $\sin\theta < 0$, it follows that $-\dfrac{\pi}{2} \le \theta < 0$, so θ lies in quadrant IV. Figure 16 illustrates $\sin\theta = -\dfrac{1}{3}$ for θ in quadrant IV. Then

$$\cos\left[\sin^{-1}\left(-\dfrac{1}{3}\right)\right] = \cos\theta = \dfrac{2\sqrt{2}}{3}$$

$\uparrow$
$\cos\theta = \dfrac{a}{r}$

Figure 16

$\sin\theta = -\dfrac{1}{3}$

EXAMPLE 3

Finding the Exact Value of an Expression Involving Inverse Trigonometric Functions

Find the exact value of: $\tan\left[\cos^{-1}\left(-\dfrac{1}{3}\right)\right]$

Solution

Let $\theta = \cos^{-1}\left(-\dfrac{1}{3}\right)$. Then $\cos\theta = -\dfrac{1}{3}$ and $0 \le \theta \le \pi$. We seek $\tan\theta$. Because $\cos\theta < 0$, it follows that $\dfrac{\pi}{2} < \theta \le \pi$, so θ lies in quadrant II. Figure 17 illustrates $\cos\theta = -\dfrac{1}{3}$ for in quadrant II. Then

$$\tan\left[\cos^{-1}\left(-\dfrac{1}{3}\right)\right] = \tan\theta = \dfrac{2\sqrt{2}}{-1} = -2\sqrt{2}$$

Figure 17

$\cos\theta = -\dfrac{1}{3}$

Now Work PROBLEMS 9 AND 27

2 Define the Inverse Secant, Cosecant, and Cotangent Functions

The inverse secant, inverse cosecant, and inverse cotangent functions are defined as follows:

DEFINITION

$$y = \sec^{-1} x \quad \text{means} \quad x = \sec y \qquad (1)$$
$$\text{where} \quad |x| \geq 1 \quad \text{and} \quad 0 \leq y \leq \pi, \quad y \neq \frac{\pi}{2}^*$$

$$y = \csc^{-1} x \quad \text{means} \quad x = \csc y \qquad (2)$$
$$\text{where} \quad |x| \geq 1 \quad \text{and} \quad -\frac{\pi}{2} \leq y \leq \frac{\pi}{2}, \quad y \neq 0^\dagger$$

$$y = \cot^{-1} x \quad \text{means} \quad x = \cot y \qquad (3)$$
$$\text{where} \quad -\infty < x < \infty \quad \text{and} \quad 0 < y < \pi$$

You are encouraged to review the graphs of the cotangent, cosecant, and secant functions in Figures 99, 100, and 101 in Section 5.7 to help you to see the basis for these definitions.

EXAMPLE 4

Finding the Exact Value of an Inverse Cosecant Function

Find the exact value of: $\csc^{-1} 2$

Solution   Let $\theta = \csc^{-1} 2$. We seek the angle $\theta, -\dfrac{\pi}{2} \leq \theta \leq \dfrac{\pi}{2}, \theta \neq 0$, whose cosecant equals 2 $\left(\text{or, equivalently, whose sine equals } \dfrac{1}{2}\right)$.

$$\theta = \csc^{-1} 2 \qquad -\frac{\pi}{2} \leq \theta \leq \frac{\pi}{2}, \quad \theta \neq 0$$
$$\csc \theta = 2 \qquad -\frac{\pi}{2} \leq \theta \leq \frac{\pi}{2}, \quad \theta \neq 0 \quad \sin\theta = \frac{1}{2}$$

The only angle θ in the interval $-\dfrac{\pi}{2} \leq \theta \leq \dfrac{\pi}{2}, \theta \neq 0$, whose cosecant is 2 $\left[\sin\theta = \dfrac{1}{2}\right]$ is $\dfrac{\pi}{6}$, so $\csc^{-1} 2 = \dfrac{\pi}{6}$.

●

━━━**Now Work** PROBLEM 39

---

3 Use a Calculator to Evaluate $\sec^{-1} x$, $\csc^{-1} x$, and $\cot^{-1} x$

Most calculators do not have keys for evaluating the inverse cotangent, cosecant, and secant functions. The easiest way to evaluate them is to convert to an inverse trigonometric function whose range is the same as the one to be evaluated. In this regard, notice that $y = \cot^{-1} x$ and $y = \sec^{-1} x$, except where undefined, each have the same range as $y = \cos^{-1} x$; $y = \csc^{-1} x$, except where undefined, has the same range as $y = \sin^{-1} x$.

NOTE Remember, the range of $y = \sin^{-1} x$ is $\left[-\dfrac{\pi}{2}, \dfrac{\pi}{2}\right]$; the range of $y = \cos^{-1} x$ is $[0, \pi]$ ■

---

*Most books use this definition. A few use the restriction $0 \leq y < \dfrac{\pi}{2}, \pi \leq y < \dfrac{3\pi}{2}$.

†Most books use this definition. A few use the restriction $-\pi < y \leq -\dfrac{\pi}{2}, 0 < y \leq \dfrac{\pi}{2}$.

EXAMPLE 5 | **Approximating the Value of Inverse Trigonometric Functions**

Use a calculator to approximate each expression in radians rounded to two decimal places.

(a) $\sec^{-1} 3$ (b) $\csc^{-1}(-4)$ (c) $\cot^{-1} \dfrac{1}{2}$ (d) $\cot^{-1}(-2)$

Solution First, set your calculator to radian mode.

(a) Let $\theta = \sec^{-1} 3$. Then $\sec \theta = 3$ and $0 \le \theta \le \pi, \theta \ne \dfrac{\pi}{2}$. We seek $\cos \theta$ because $y = \cos^{-1} x$ has the same range as $y = \sec^{-1} x$, except where undefined. Since $\sec \theta = \dfrac{1}{\cos \theta} = 3$, we have $\cos \theta = \dfrac{1}{3}$. Then $\theta = \cos^{-1} \dfrac{1}{3}$ and

$$\sec^{-1} 3 = \theta = \cos^{-1} \frac{1}{3} \approx 1.23$$
$$\uparrow$$
$$\textit{Use a calculator.}$$

(b) Let $\theta = \csc^{-1}(-4)$. Then $\csc \theta = -4, -\dfrac{\pi}{2} \le \theta \le \dfrac{\pi}{2}, \theta \ne 0$. We seek $\sin \theta$ because $y = \sin^{-1} x$ has the same range as $y = \csc^{-1} x$, except where undefined. Since $\csc \theta = \dfrac{1}{\sin \theta} = -4$, we have $\sin \theta = -\dfrac{1}{4}$. Then $\theta = \sin^{-1}\left(-\dfrac{1}{4}\right)$, and

$$\csc^{-1}(-4) = \theta = \sin^{-1}\left(-\frac{1}{4}\right) \approx -0.25$$

Figure 18

$\cot \theta = \dfrac{1}{2}, 0 < \theta < \pi$

(c) We proceed as before. Let $\theta = \cot^{-1} \dfrac{1}{2}$. Then $\cot \theta = \dfrac{1}{2}, 0 < \theta < \pi$. From these facts we know that θ lies in quadrant I. We seek $\cos \theta$ because $y = \cos^{-1} x$ has the same range as $y = \cot^{-1} x$, except where undefined. To find $\cos \theta$ we use Figure 18. We find that $\cos \theta = \dfrac{1}{\sqrt{5}}, 0 < \theta < \dfrac{\pi}{2}$, so $\theta = \cos^{-1}\left(\dfrac{1}{\sqrt{5}}\right)$. Then

$$\cot^{-1} \frac{1}{2} = \theta = \cos^{-1}\left(\frac{1}{\sqrt{5}}\right) \approx 1.11$$

Figure 19

$\cot \theta = -2, 0 < \theta < \pi$

(d) Let $\theta = \cot^{-1}(-2)$. Then $\cot \theta = -2, 0 < \theta < \pi$. From these facts we know that θ lies in quadrant II. We seek $\cos \theta$. To find it, we use Figure 19. We find that $\cos \theta = -\dfrac{2}{\sqrt{5}}, \dfrac{\pi}{2} < \theta < \pi$, so $\theta = \cos^{-1}\left(-\dfrac{2}{\sqrt{5}}\right)$. Then

$$\cot^{-1}(-2) = \theta = \cos^{-1}\left(-\frac{2}{\sqrt{5}}\right) \approx 2.68$$

Now Work PROBLEM 45

4 Write a Trigonometric Expression as an Algebraic Expression

EXAMPLE 6 | **Writing a Trigonometric Expression as an Algebraic Expression**

Write $\sin(\tan^{-1} u)$ as an algebraic expression containing u.

Solution Let $\theta = \tan^{-1} u$ so that $\tan \theta = u, -\dfrac{\pi}{2} < \theta < \dfrac{\pi}{2}, -\infty < u < \infty$. As a result, we know that $\sec \theta > 0$. Then

$$\sin(\tan^{-1} u) = \sin \theta = \sin \theta \cdot \frac{\cos \theta}{\cos \theta} = \tan \theta \cos \theta = \frac{\tan \theta}{\sec \theta} = \frac{\tan \theta}{\sqrt{1 + \tan^2 \theta}} = \frac{u}{\sqrt{1 + u^2}}$$

Multiply by 1: $\dfrac{\cos \theta}{\cos \theta}$. $\dfrac{\sin \theta}{\cos \theta} = \tan \theta$ $\sec^2 \theta = 1 + \tan^2 \theta$

$\sec \theta > 0$

Now Work PROBLEM 57

6.2 Assess Your Understanding

'Are You Prepared?' *Answers are given at the end of these exercises. If you get a wrong answer, read the pages listed in* red.

1. What is the domain and the range of $y = \sec x$? (pp. 429–430)

2. *True or False* The graph of $y = \sec x$ is increasing on the interval $\left[0, \dfrac{\pi}{2}\right)$ and on the interval $\left(\dfrac{\pi}{2}, \pi\right]$. (p. 454)

3. If $\tan \theta = \dfrac{1}{2}, -\dfrac{\pi}{2} < \theta < \dfrac{\pi}{2}$, then $\sin \theta =$ _____. (pp. 421–422)

Concepts and Vocabulary

4. $y = \sec^{-1} x$ means _____, where $|x|$ _____ and _____ $\leq y \leq$ _____, $y \neq \dfrac{\pi}{2}$.

5. To find the inverse secant of a real number x such that $|x| \geq 1$, convert the inverse secant to an inverse _____.

6. *True or False* It is impossible to obtain exact values for the inverse secant function.

7. *True or False* $\csc^{-1} 0.5$ is not defined.

8. *True or False* The domain of the inverse cotangent function is the set of real numbers.

Skill Building

In Problems 9–36, find the exact value of each expression.

9. $\cos\left(\sin^{-1} \dfrac{\sqrt{2}}{2}\right)$

10. $\sin\left(\cos^{-1} \dfrac{1}{2}\right)$

11. $\tan\left[\cos^{-1}\left(-\dfrac{\sqrt{3}}{2}\right)\right]$

12. $\tan\left[\sin^{-1}\left(-\dfrac{1}{2}\right)\right]$

13. $\sec\left(\cos^{-1} \dfrac{1}{2}\right)$

14. $\cot\left[\sin^{-1}\left(-\dfrac{1}{2}\right)\right]$

15. $\csc(\tan^{-1} 1)$

16. $\sec\left(\tan^{-1} \sqrt{3}\right)$

17. $\sin[\tan^{-1}(-1)]$

18. $\cos\left[\sin^{-1}\left(-\dfrac{\sqrt{3}}{2}\right)\right]$

19. $\sec\left[\sin^{-1}\left(-\dfrac{1}{2}\right)\right]$

20. $\csc\left[\cos^{-1}\left(-\dfrac{\sqrt{3}}{2}\right)\right]$

21. $\cos^{-1}\left(\cos \dfrac{5\pi}{4}\right)$

22. $\tan^{-1}\left(\tan \dfrac{2\pi}{3}\right)$

23. $\sin^{-1}\left[\sin\left(-\dfrac{7\pi}{6}\right)\right]$

24. $\cos^{-1}\left[\cos\left(-\dfrac{\pi}{3}\right)\right]$

25. $\tan\left(\sin^{-1} \dfrac{1}{3}\right)$

26. $\tan\left(\cos^{-1} \dfrac{1}{3}\right)$

27. $\sec\left(\tan^{-1} \dfrac{1}{2}\right)$

28. $\cos\left(\sin^{-1} \dfrac{\sqrt{2}}{3}\right)$

29. $\cot\left[\sin^{-1}\left(-\dfrac{\sqrt{2}}{3}\right)\right]$

30. $\csc[\tan^{-1}(-2)]$

31. $\sin[\tan^{-1}(-3)]$

32. $\cot\left[\cos^{-1}\left(-\dfrac{\sqrt{3}}{3}\right)\right]$

33. $\sec\left(\sin^{-1} \dfrac{2\sqrt{5}}{5}\right)$

34. $\csc\left(\tan^{-1} \dfrac{1}{2}\right)$

35. $\sin^{-1}\left(\cos \dfrac{3\pi}{4}\right)$

36. $\cos^{-1}\left(\sin \dfrac{7\pi}{6}\right)$

In Problems 37–44, find the exact value of each expression.

37. $\cot^{-1} \sqrt{3}$

38. $\cot^{-1} 1$

39. $\csc^{-1}(-1)$

40. $\csc^{-1} \sqrt{2}$

41. $\sec^{-1} \dfrac{2\sqrt{3}}{3}$

42. $\sec^{-1}(-2)$

43. $\cot^{-1}\left(-\dfrac{\sqrt{3}}{3}\right)$

44. $\csc^{-1}\left(-\dfrac{2\sqrt{3}}{3}\right)$

In Problems 45–56, use a calculator to find the value of each expression rounded to two decimal places.

45. $\sec^{-1} 4$ **46.** $\csc^{-1} 5$ **47.** $\cot^{-1} 2$ **48.** $\sec^{-1}(-3)$

49. $\csc^{-1}(-3)$ **50.** $\cot^{-1}\left(-\dfrac{1}{2}\right)$ **51.** $\cot^{-1}(-\sqrt{5})$ **52.** $\cot^{-1}(-8.1)$

53. $\csc^{-1}\left(-\dfrac{3}{2}\right)$ **54.** $\sec^{-1}\left(-\dfrac{4}{3}\right)$ **55.** $\cot^{-1}\left(-\dfrac{3}{2}\right)$ **56.** $\cot^{-1}(-\sqrt{10})$

In Problems 57–66, write each trigonometric expression as an algebraic expression in u.

57. $\cos(\tan^{-1} u)$ **58.** $\sin(\cos^{-1} u)$ **59.** $\tan(\sin^{-1} u)$ **60.** $\tan(\cos^{-1} u)$

61. $\sin(\sec^{-1} u)$ **62.** $\sin(\cot^{-1} u)$ **63.** $\cos(\csc^{-1} u)$ **64.** $\cos(\sec^{-1} u)$

65. $\tan(\cot^{-1} u)$ **66.** $\tan(\sec^{-1} u)$

Mixed Practice

In Problems 67–78, $f(x) = \sin x, -\dfrac{\pi}{2} \le x \le \dfrac{\pi}{2}, g(x) = \cos x, 0 \le x \le \pi$, and $h(x) = \tan x, -\dfrac{\pi}{2} < x < \dfrac{\pi}{2}$. Find the exact value of each composite function.

67. $g\left(f^{-1}\left(\dfrac{12}{13}\right)\right)$ **68.** $f\left(g^{-1}\left(\dfrac{5}{13}\right)\right)$ **69.** $g^{-1}\left(f\left(\dfrac{7\pi}{4}\right)\right)$ **70.** $f^{-1}\left(g\left(\dfrac{5\pi}{6}\right)\right)$

71. $h\left(f^{-1}\left(-\dfrac{3}{5}\right)\right)$ **72.** $h\left(g^{-1}\left(-\dfrac{4}{5}\right)\right)$ **73.** $g\left(h^{-1}\left(\dfrac{12}{5}\right)\right)$ **74.** $f\left(h^{-1}\left(\dfrac{5}{12}\right)\right)$

75. $g^{-1}\left(f\left(-\dfrac{4\pi}{3}\right)\right)$ **76.** $g^{-1}\left(f\left(-\dfrac{5\pi}{6}\right)\right)$ **77.** $h\left(g^{-1}\left(-\dfrac{1}{4}\right)\right)$ **78.** $h\left(f^{-1}\left(-\dfrac{2}{5}\right)\right)$

Applications and Extensions

*Problems 79 and 80 require the following discussion: When granular materials are allowed to fall freely, they form conical (cone-shaped) piles. The naturally occurring angle of slope, measured from the horizontal, at which the loose material comes to rest is called the **angle of repose** and varies for different materials. The angle of repose θ is related to the height h and base radius r of the conical pile by the equation $\theta = \cot^{-1}\dfrac{r}{h}$. See the illustration.*

79. Angle of Repose: Deicing Salt Due to potential transportation issues (for example, frozen waterways) deicing salt used by highway departments in the Midwest must be ordered early and stored for future use. When deicing salt is stored in a pile 14 feet high, the diameter of the base of the pile is 45 feet.
(a) Find the angle of repose for deicing salt.
(b) What is the base diameter of a pile that is 17 feet high?
(c) What is the height of a pile that has a base diameter of approximately 122 feet?
Source: Salt Institute, *The Salt Storage Handbook,* 2006

80. Angle of Repose: Bunker Sand The steepness of sand bunkers on a golf course is affected by the angle of repose of the sand (a larger angle of repose allows for steeper bunkers). A freestanding pile of loose sand from a United States Golf Association (USGA) bunker had a height of 4 feet and a base diameter of approximately 6.68 feet.
(a) Find the angle of repose for USGA bunker sand.
(b) What is the height of such a pile if the diameter of the base is 8 feet?
(c) A 6-foot high pile of loose Tour Grade 50/50 sand has a base diameter of approximately 8.44 feet. Which type of sand (USGA or Tour Grade 50/50) would be better suited for steep bunkers?
Source: 2004 Annual Report, Purdue University Turfgrass Science Program

81. Artillery A projectile fired into the first quadrant from the origin of a coordinate system will pass through the point (x, y) at time t according to the relationship $\cot \theta = \dfrac{2x}{2y + gt^2}$, where θ = the angle of elevation of the launcher and g = the acceleration due to gravity = 32.2 feet/second^2. An artilleryman is firing at an enemy bunker located 2450 feet up the side of a hill that is 6175 feet away. He fires a round, and exactly 2.27 seconds later he scores a direct hit.

(a) What angle of elevation did he use?
(b) If the angle of elevation is also given by $\sec \theta = \dfrac{v_0 t}{x}$, where v_0 is the muzzle velocity of the weapon, find the muzzle velocity of the artillery piece he used.
Source: www.egwald.com/geometry/projectile3d.php

82. Using a graphing utility, graph $y = \cot^{-1} x$.

83. Using a graphing utility, graph $y = \sec^{-1} x$.

84. Using a graphing utility, graph $y = \csc^{-1} x$.

Explaining Concepts: Discussion and Writing

85. Explain in your own words how you would use your calculator to find the value of $\cot^{-1} 10$.

86. Consult three books on calculus and write down the definition in each of $y = \sec^{-1} x$ and $y = \csc^{-1} x$. Compare these with the definitions given in this book.

'Are You Prepared?' Answers

1. Domain: $\left\{ x \mid x \neq \text{odd integer multiples of } \dfrac{\pi}{2} \right\}$; range: $\{ y \leq -1 \text{ or } y \geq 1 \}$ 2. True   3. $\dfrac{\sqrt{5}}{5}$

6.3 Trigonometric Equations

PREPARING FOR THIS SECTION *Before getting started, review the following:*

- Solving Equations (Appendix A, Section A.8 pp. A65–A72)
- Values of the Trigonometric Functions (Section 5.3, pp. 403–405; Section 5.4, pp. 416, 420–421)

- Solving Quadratic Equations (Section 2.3, pp. 141–149)
- Using a Graphing Utility to Solve Equations (Appendix B, Section B.4, pp. B6–B7)

Now Work the 'Are You Prepared?' Problems on page 502.

OBJECTIVES 1 Solve Equations Involving a Single Trigonometric Function (p. 496)
2 Solve Trigonometric Equations Using a Calculator (p. 499)
3 Solve Trigonometric Equations Quadratic in Form (p. 500)
4 Solve Trigonometric Equations Using Fundamental Identities (p. 500)
5 Solve Trigonometric Equations Using a Graphing Utility (p. 501)

1 Solve Equations Involving a Single Trigonometric Function

In this section, we discuss **trigonometric equations,** that is, equations involving trigonometric functions that are satisfied only by some values of the variable (or, possibly, are not satisfied by any values of the variable). The values that satisfy the equation are called **solutions** of the equation.

EXAMPLE 1

Checking Whether a Given Number Is a Solution of a Trigonometric Equation

Determine whether $\theta = \dfrac{\pi}{4}$ is a solution of the equation $2 \sin \theta - 1 = 0$. Is $\theta = \dfrac{\pi}{6}$ a solution?

Solution   Replace θ by $\dfrac{\pi}{4}$ in the given equation. The result is

$$2 \sin \frac{\pi}{4} - 1 = 2 \cdot \frac{\sqrt{2}}{2} - 1 = \sqrt{2} - 1 \neq 0$$

We conclude that $\dfrac{\pi}{4}$ is not a solution.

Next replace θ by $\dfrac{\pi}{6}$ in the equation. The result is

$$2 \sin \frac{\pi}{6} - 1 = 2 \cdot \frac{1}{2} - 1 = 0$$

We conclude that $\dfrac{\pi}{6}$ is a solution of the given equation.

The equation given in Example 1 has other solutions besides $\theta = \dfrac{\pi}{6}$. For example, $\theta = \dfrac{5\pi}{6}$ is also a solution, as is $\theta = \dfrac{13\pi}{6}$. (You should check this for yourself.) In fact, the equation has an infinite number of solutions due to the periodicity of the sine function, as can be seen in Figure 20 where we graph $y = 2 \sin x - 1$. Each x-intercept of the graph represents a solution to the equation $2 \sin x - 1 = 0$.

Figure 20

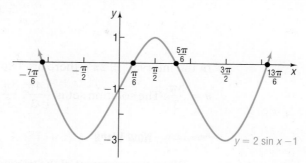

Unless the domain of the variable is restricted, we need to find *all* the solutions of a trigonometric equation. As the next example illustrates, finding all the solutions can be accomplished by first finding solutions over an interval whose length equals the period of the function and then adding multiples of that period to the solutions found.

EXAMPLE 2

Finding All the Solutions of a Trigonometric Equation

Solve the equation: $\cos \theta = \dfrac{1}{2}$

Give a general formula for all the solutions. List eight of the solutions.

Solution The period of the cosine function is 2π. In the interval $[0, 2\pi)$, there are two angles θ for which $\cos \theta = \dfrac{1}{2}$: $\theta = \dfrac{\pi}{3}$ and $\theta = \dfrac{5\pi}{3}$. See Figure 21. Because the cosine function has period 2π, all the solutions of $\cos \theta = \dfrac{1}{2}$ may be given by the general formula

$$\theta = \frac{\pi}{3} + 2k\pi \quad \text{or} \quad \theta = \frac{5\pi}{3} + 2k\pi \quad k \text{ any integer}$$

Eight of the solutions are

$$\underbrace{-\frac{5\pi}{3},}_{k\,=\,-1} \quad \underbrace{-\frac{\pi}{3},}_{} \quad \underbrace{\frac{\pi}{3},}_{k\,=\,0} \quad \underbrace{\frac{5\pi}{3},}_{} \quad \underbrace{\frac{7\pi}{3},}_{k\,=\,1} \quad \underbrace{\frac{11\pi}{3},}_{} \quad \underbrace{\frac{13\pi}{3},}_{k\,=\,2} \quad \underbrace{\frac{17\pi}{3}}_{}$$

✓Check: We can verify the solutions by graphing $Y_1 = \cos x$ and $Y_2 = \dfrac{1}{2}$ to determine where the graphs intersect. (Be sure to graph in radian mode.) See Figure 22. The graph of Y_1 intersects the graph of Y_2 at $x = 1.05 \left(\approx \dfrac{\pi}{3}\right)$, $5.24 \left(\approx \dfrac{5\pi}{3}\right)$, $7.33 \left(\approx \dfrac{7\pi}{3}\right)$, and $11.52 \left(\approx \dfrac{11\pi}{3}\right)$, rounded to two decimal places.

Now Work PROBLEM 35

In most of our work, we shall be interested only in finding solutions of trigonometric equations for $0 \le \theta < 2\pi$.

In Words
The equation $\cos \theta = \dfrac{1}{2}$ means find all the angles θ whose cosine is $\dfrac{1}{2}$.

Figure 21

Figure 22

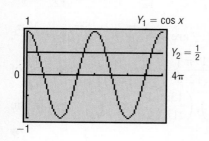

EXAMPLE 3	**Solving a Linear Trigonometric Equation**

Solve the equation: $2 \sin \theta + \sqrt{3} = 0$, $0 \leq \theta < 2\pi$

Solution We solve the equation for $\sin \theta$.

$$2 \sin \theta + \sqrt{3} = 0$$

$$2 \sin \theta = -\sqrt{3} \quad \text{Subtract } \sqrt{3} \text{ from both sides.}$$

$$\sin \theta = -\frac{\sqrt{3}}{2} \quad \text{Divide both sides by 2.}$$

In the interval $[0, 2\pi)$, there are two angles θ for which $\sin \theta = -\frac{\sqrt{3}}{2}$: $\theta = \frac{4\pi}{3}$ and $\theta = \frac{5\pi}{3}$. The solution set is $\left\{ \frac{4\pi}{3}, \frac{5\pi}{3} \right\}$.

━━▶ **Now Work** PROBLEM 11

When the argument of the trigonometric function in an equation is a multiple of θ, the general formula must be used to solve the equation.

EXAMPLE 4	**Solving a Trigonometric Equation**

Solve the equation: $\sin(2\theta) = \frac{1}{2}$, $0 \leq \theta < 2\pi$

Solution In the interval $[0, 2\pi)$, the sine function equals $\frac{1}{2}$ at $\frac{\pi}{6}$ and $\frac{5\pi}{6}$. See Figure 23(a). So, we know that 2θ must equal $\frac{\pi}{6}$ and $\frac{5\pi}{6}$. Here's the problem, however. The period of $y = \sin(2\theta)$ is $\frac{2\pi}{2} = \pi$. So, in the interval $[0, 2\pi)$, the graph of $y = \sin(2\theta)$ will complete two cycles, and the graph of $y = \sin(2\theta)$ will intersect the graph of $y = \frac{1}{2}$ four times. See Figure 23(b). So there are four solutions to the equation $\sin(2\theta) = \frac{1}{2}$ in $[0, 2\pi)$. To find these solution, we write the general formula that gives all the solutions.

Figure 23

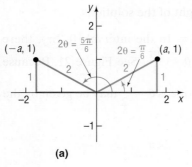

(a)

(b)

$$2\theta = \frac{\pi}{6} + 2k\pi \quad \text{or} \quad 2\theta = \frac{5\pi}{6} + 2k\pi \quad \text{k any integer}$$

$$\theta = \frac{\pi}{12} + k\pi \quad \text{or} \quad \theta = \frac{5\pi}{12} + k\pi \quad \text{Divide by 2.}$$

Then

$\theta = \frac{\pi}{12} + (-1)\pi = \frac{-11\pi}{12}$	$k = -1$	$\theta = \frac{5\pi}{12} + (-1)\pi = \frac{-7\pi}{12}$	
$\theta = \frac{\pi}{12} + (0)\pi = \frac{\pi}{12}$	$k = 0$	$\theta = \frac{5\pi}{12} + (0)\pi = \frac{5\pi}{12}$	
$\theta = \frac{\pi}{12} + (1)\pi = \frac{13\pi}{12}$	$k = 1$	$\theta = \frac{5\pi}{12} + (1)\pi = \frac{17\pi}{12}$	
$\theta = \frac{\pi}{12} + (2)\pi = \frac{25\pi}{12}$	$k = 2$	$\theta = \frac{5\pi}{12} + (2)\pi = \frac{29\pi}{12}$	

In the interval $[0, 2\pi)$, the solutions of $\sin(2\theta) = \frac{1}{2}$ are $\theta = \frac{\pi}{12}$, $\theta = \frac{5\pi}{12}$, $\theta = \frac{13\pi}{12}$, and $\theta = \frac{17\pi}{12}$. The solution set is $\left\{ \frac{\pi}{12}, \frac{5\pi}{12}, \frac{13\pi}{12}, \frac{17\pi}{12} \right\}$. We now know the graph of $y = \sin(2\theta)$ intersects $y = \frac{1}{2}$ at $\left(\frac{\pi}{12}, \frac{1}{2} \right)$, $\left(\frac{5\pi}{12}, \frac{1}{2} \right)$, $\left(\frac{13\pi}{12}, \frac{1}{2} \right)$, and $\left(\frac{17\pi}{12}, \frac{1}{2} \right)$ in the interval $[0, 2\pi)$.

✓**Check:** Verify these solutions by graphing $Y_1 = \sin(2x)$ and $Y_2 = \dfrac{1}{2}$ for $0 \le x \le 2\pi$.

WARNING In solving a trigonometric equation for $\theta, 0 \le \theta < 2\pi$, in which the argument is not θ (as in Example 4), you must write down all the solutions first and then list those that are in the interval $[0, 2\pi)$. Otherwise, solutions may be lost. For example, in solving $\sin(2\theta) = \dfrac{1}{2}$, if you merely write the solutions $2\theta = \dfrac{\pi}{6}$ and $2\theta = \dfrac{5\pi}{6}$, you will find only $\theta = \dfrac{\pi}{12}$ and $\theta = \dfrac{5\pi}{12}$ and miss the other solutions. ∎

EXAMPLE 5	**Solving a Trigonometric Equation**

Solve the equation: $\quad \tan\left(\theta - \dfrac{\pi}{2}\right) = 1, \quad 0 \le \theta < 2\pi$

Solution

The period of the tangent function is π. In the interval $[0, \pi)$, the tangent function has the value 1 when the argument is $\dfrac{\pi}{4}$. Because the argument is $\theta - \dfrac{\pi}{2}$ in the given equation, we write the general formula that gives all the solutions.

$$\theta - \frac{\pi}{2} = \frac{\pi}{4} + k\pi \qquad k \text{ any integer}$$

$$\theta = \frac{3\pi}{4} + k\pi$$

In the interval $[0, 2\pi), \theta = \dfrac{3\pi}{4}$ and $\theta = \dfrac{3\pi}{4} + \pi = \dfrac{7\pi}{4}$ are the only solutions. The solution set is $\left\{ \dfrac{3\pi}{4}, \dfrac{7\pi}{4} \right\}$.

●

─ **Now Work** PROBLEM 17

2 Solve Trigonometric Equations Using a Calculator

The next example illustrates how to solve trigonometric equations using a calculator. Remember that the function keys on a calculator will only give values consistent with the definition of the function.

EXAMPLE 6	**Solving a Trigonometric Equation with a Calculator**

Use a calculator to solve the equation $\sin \theta = 0.3, 0 \le \theta < 2\pi$. Express any solutions in radians, rounded to two decimal places.

Solution

Figure 24

To solve $\sin \theta = 0.3$ on a calculator, first set the mode to radians. Then use the $\boxed{\sin^{-1}}$ key to obtain

$$\theta = \sin^{-1}(0.3) \approx 0.3046927$$

Rounded to two decimal places, $\theta = \sin^{-1}(0.3) = 0.30$ radian. Because of the definition of $y = \sin^{-1} x$, the angle θ that we obtain is the angle $-\dfrac{\pi}{2} \le \theta \le \dfrac{\pi}{2}$ for which $\sin \theta = 0.3$. Another angle for which $\sin \theta = 0.3$ is $\pi - 0.30$. See Figure 24. The angle $\pi - 0.30$ is the angle in quadrant II, where $\sin \theta = 0.3$. The solutions for $\sin \theta = 0.3, 0 \le \theta < 2\pi$, are

$$\theta = 0.30 \text{ radian} \quad \text{and} \quad \theta = \pi - 0.30 \approx 2.84 \text{ radians}$$

The solution set is $\{0.30, 2.84\}$.

●

WARNING Example 6 illustrates that caution must be exercised when solving trigonometric equations on a calculator. Remember that the calculator supplies an angle only within the restrictions of the definition of the inverse trigonometric function. To find the remaining solutions, you must identify other quadrants, if any, in which a solution may be located. ∎

──────**Now Work** PROBLEM 45

3 Solve Trigonometric Equations Quadratic in Form

Many trigonometric equations can be solved by applying techniques that we already know, such as applying the quadratic formula (if the equation is a second-degree polynomial) or factoring.

| EXAMPLE 7 | **Solving a Trigonometric Equation Quadratic in Form** |

Solve the equation: $2 \sin^2 \theta - 3 \sin \theta + 1 = 0, \quad 0 \le \theta < 2\pi$

Solution   This equation is a quadratic equation (in $\sin \theta$) that can be factored.

$$2 \sin^2 \theta - 3 \sin \theta + 1 = 0 \qquad 2x^2 - 3x + 1 = 0, \quad x = \sin \theta$$
$$(2 \sin \theta - 1)(\sin \theta - 1) = 0 \qquad (2x - 1)(x - 1) = 0$$
$$2 \sin \theta - 1 = 0 \quad \text{or} \quad \sin \theta - 1 = 0 \quad \text{Use the Zero-Product Property.}$$
$$\sin \theta = \frac{1}{2} \quad \text{or} \quad \sin \theta = 1$$

Solving each equation in the interval $[0, 2\pi)$, we obtain

$$\theta = \frac{\pi}{6}, \qquad \theta = \frac{5\pi}{6}, \qquad \theta = \frac{\pi}{2}$$

The solution set is $\left\{ \dfrac{\pi}{6}, \dfrac{5\pi}{6}, \dfrac{\pi}{2} \right\}$.

──────**Now Work** PROBLEM 59

4 Solve Trigonometric Equations Using Fundamental Identities

When a trigonometric equation contains more than one trigonometric function, identities sometimes can be used to obtain an equivalent equation that contains only one trigonometric function.

| EXAMPLE 8 | **Solving a Trigonometric Equation Using Identities** |

Solve the equation: $3 \cos \theta + 3 = 2 \sin^2 \theta, \quad 0 \le \theta < 2\pi$

Solution   The equation in its present form contains a sine and a cosine. However, a form of the Pythagorean Identity, $\sin^2 \theta + \cos^2 \theta = 1$, can be used to transform the equation into an equivalent expression containing only cosines.

$$3 \cos \theta + 3 = 2 \sin^2 \theta$$
$$3 \cos \theta + 3 = 2(1 - \cos^2 \theta) \qquad \sin^2 \theta = 1 - \cos^2 \theta$$
$$3 \cos \theta + 3 = 2 - 2 \cos^2 \theta$$
$$2 \cos^2 \theta + 3 \cos \theta + 1 = 0 \qquad \text{Quadratic in } \cos \theta$$
$$(2 \cos \theta + 1)(\cos \theta + 1) = 0 \qquad \text{Factor.}$$
$$2 \cos \theta + 1 = 0 \quad \text{or} \quad \cos \theta + 1 = 0 \qquad \text{Use the Zero-Product Property.}$$
$$\cos \theta = -\frac{1}{2} \quad \text{or} \quad \cos \theta = -1$$

Solving each equation in the interval $[0, 2\pi)$, we obtain

$$\theta = \frac{2\pi}{3}, \qquad \theta = \frac{4\pi}{3}, \qquad \theta = \pi$$

The solution set is $\left\{\dfrac{2\pi}{3}, \pi, \dfrac{4\pi}{3}\right\}$.

 ✓**Check:** Graph $Y_1 = 3 \cos x + 3$ and $Y_2 = 2 \sin^2 x, 0 \le x \le 2\pi$, and find the points of intersection. How close are your approximate solutions to the exact ones found in Example 8?

EXAMPLE 9 **Solving a Trigonometric Equation Using Identities**

Solve the equation: $\cos^2 \theta + \sin \theta = 2, \quad 0 \le \theta < 2\pi$

Solution This equation involves two trigonometric functions, sine and cosine. We use a form of the Pythagorean Identity, $\sin^2 \theta + \cos^2 \theta = 1$, to rewrite the equation in terms of $\sin \theta$.

$$\cos^2 \theta + \sin \theta = 2$$
$$(1 - \sin^2 \theta) + \sin \theta = 2 \quad {\scriptstyle \cos^2 \theta = 1 - \sin^2 \theta}$$
$$\sin^2 \theta - \sin \theta + 1 = 0$$

This is a quadratic equation in $\sin \theta$. The discriminant is $b^2 - 4ac = 1 - 4 = -3 < 0$. Therefore, the equation has no real solution. The solution set is the empty set, $\varnothing$.

 ✓**Check:** Graph $Y_1 = \cos^2 x + \sin x$ and $Y_2 = 2$ to see that the two graphs never intersect, so the equation $Y_1 = Y_2$ has no real solution.

---

 5 Solve Trigonometric Equations Using a Graphing Utility

⚠ The techniques introduced in this section apply only to certain types of trigonometric equations. Solutions for other types are usually studied in calculus, using numerical methods.

EXAMPLE 10 **Solving Trigonometric Equations Using a Graphing Utility**

Solve: $5 \sin x + x = 3$

Express the solution(s) rounded to two decimal places.

Solution This type of trigonometric equation cannot be solved by previous methods. A graphing utility, though, can be used here. Each solution of this equation is the x-coordinate of a point of intersection of the graphs of $Y_1 = 5 \sin x + x$ and $Y_2 = 3$. See Figure 25.

Figure 25

There are three points of intersection; the x-coordinates are the solutions that we seek. Using INTERSECT, we find

$$x = 0.52, \qquad x = 3.18, \qquad x = 5.71$$

The solution set is $\{0.52, 3.18, 5.71\}$.

—**Now Work** PROBLEM 81

6.3 Assess Your Understanding

'Are You Prepared?' *Answers are given at the end of these exercises. If you get a wrong answer, read the pages listed in* red.

1. Solve: $3x - 5 = -x + 1$ (pp. A65–A68)

2. $\sin\left(\dfrac{\pi}{4}\right) = $ _____; $\cos\left(\dfrac{8\pi}{3}\right) = $ _____
(pp. 403–404 and 420–421)

3. Find the real solutions of $4x^2 - x - 5 = 0$. (pp. 141–142)

4. Find the real solutions of $x^2 - x - 1 = 0$. (pp. 145–147)

5. Find the real solutions of $(2x - 1)^2 - 3(2x - 1) - 4 = 0$.
(pp. 148–149)

6. Use a graphing utility to solve $5x^3 - 2 = x - x^2$. Round answers to two decimal places. (pp. B6–B7)

Concepts and Vocabulary

7. Two solutions of the equation $\sin\theta = \dfrac{1}{2}$ are _____ and

_____.

8. All the solutions of the equation $\sin\theta = \dfrac{1}{2}$ are _____.

9. *True or False* Most trigonometric equations have unique solutions.

10. *True or False* The equation $\sin\theta = 2$ has a real solution that can be found using a calculator.

Skill Building

In Problems 11–34, solve each equation on the interval $0 \le \theta < 2\pi$.

11. $2\sin\theta + 3 = 2$

12. $1 - \cos\theta = \dfrac{1}{2}$

13. $4\cos^2\theta = 1$

14. $\tan^2\theta = \dfrac{1}{3}$

15. $2\sin^2\theta - 1 = 0$

16. $4\cos^2\theta - 3 = 0$

17. $\sin(3\theta) = -1$

18. $\tan\dfrac{\theta}{2} = \sqrt{3}$

19. $\cos(2\theta) = -\dfrac{1}{2}$

20. $\tan(2\theta) = -1$

21. $\sec\dfrac{3\theta}{2} = -2$

22. $\cot\dfrac{2\theta}{3} = -\sqrt{3}$

23. $2\sin\theta + 1 = 0$

24. $\cos\theta + 1 = 0$

25. $\tan\theta + 1 = 0$

26. $\sqrt{3}\cot\theta + 1 = 0$

27. $4\sec\theta + 6 = -2$

28. $5\csc\theta - 3 = 2$

29. $3\sqrt{2}\cos\theta + 2 = -1$

30. $4\sin\theta + 3\sqrt{3} = \sqrt{3}$

31. $\cos\left(2\theta - \dfrac{\pi}{2}\right) = -1$

32. $\sin\left(3\theta + \dfrac{\pi}{18}\right) = 1$

33. $\tan\left(\dfrac{\theta}{2} + \dfrac{\pi}{3}\right) = 1$

34. $\cos\left(\dfrac{\theta}{3} - \dfrac{\pi}{4}\right) = \dfrac{1}{2}$

In Problems 35–44, solve each equation. Give a general formula for all the solutions. List six solutions.

35. $\sin\theta = \dfrac{1}{2}$

36. $\tan\theta = 1$

37. $\tan\theta = -\dfrac{\sqrt{3}}{3}$

38. $\cos\theta = -\dfrac{\sqrt{3}}{2}$

39. $\cos\theta = 0$

40. $\sin\theta = \dfrac{\sqrt{2}}{2}$

41. $\cos(2\theta) = -\dfrac{1}{2}$

42. $\sin(2\theta) = -1$

43. $\sin\dfrac{\theta}{2} = -\dfrac{\sqrt{3}}{2}$

44. $\tan\dfrac{\theta}{2} = -1$

In Problems 45–56, use a calculator to solve each equation on the interval $0 \le \theta < 2\pi$. Round answers to two decimal places.

45. $\sin\theta = 0.4$

46. $\cos\theta = 0.6$

47. $\tan\theta = 5$

48. $\cot\theta = 2$

49. $\cos\theta = -0.9$

50. $\sin\theta = -0.2$

51. $\sec\theta = -4$

52. $\csc\theta = -3$

53. $5\tan\theta + 9 = 0$

54. $4\cot\theta = -5$

55. $3\sin\theta - 2 = 0$

56. $4\cos\theta + 3 = 0$

In Problems 57–80, solve each equation on the interval $0 \le \theta < 2\pi$.

57. $2 \cos^2 \theta + \cos \theta = 0$

58. $\sin^2 \theta - 1 = 0$

59. $2 \sin^2 \theta - \sin \theta - 1 = 0$

60. $2 \cos^2 \theta + \cos \theta - 1 = 0$

61. $(\tan \theta - 1)(\sec \theta - 1) = 0$

62. $(\cot \theta + 1)\left(\csc \theta - \dfrac{1}{2}\right) = 0$

63. $\sin^2 \theta - \cos^2 \theta = 1 + \cos \theta$

64. $\cos^2 \theta - \sin^2 \theta + \sin \theta = 0$

65. $\sin^2 \theta = 6(\cos(-\theta) + 1)$

66. $2 \sin^2 \theta = 3(1 - \cos(-\theta))$

67. $\cos \theta = -\sin(-\theta)$

68. $\cos \theta - \sin(-\theta) = 0$

69. $\tan \theta = 2 \sin \theta$

70. $\tan \theta = \cot \theta$

71. $1 + \sin \theta = 2 \cos^2 \theta$

72. $\sin^2 \theta = 2 \cos \theta + 2$

73. $2 \sin^2 \theta - 5 \sin \theta + 3 = 0$

74. $2 \cos^2 \theta - 7 \cos \theta - 4 = 0$

75. $3(1 - \cos \theta) = \sin^2 \theta$

76. $4(1 + \sin \theta) = \cos^2 \theta$

77. $\tan^2 \theta = \dfrac{3}{2} \sec \theta$

78. $\csc^2 \theta = \cot \theta + 1$

79. $\sec^2 \theta + \tan \theta = 0$

80. $\sec \theta = \tan \theta + \cot \theta$

In Problems 81–92, use a graphing utility to solve each equation. Express the solution(s) rounded to two decimal places.

81. $x + 5 \cos x = 0$

82. $x - 4 \sin x = 0$

83. $22x - 17 \sin x = 3$

84. $19x + 8 \cos x = 2$

85. $\sin x + \cos x = x$

86. $\sin x - \cos x = x$

87. $x^2 - 2 \cos x = 0$

88. $x^2 + 3 \sin x = 0$

89. $x^2 - 2 \sin(2x) = 3x$

90. $x^2 = x + 3 \cos(2x)$

91. $6 \sin x - e^x = 2, \quad x > 0$

92. $4 \cos(3x) - e^x = 1, \quad x > 0$

Mixed Practice

93. What are the zeros of $f(x) = 4 \sin^2 x - 3$ on the interval $[0, 2\pi]$?

94. What are the zeros of $f(x) = 2 \cos(3x) + 1$ on the interval $[0, \pi]$?

95. $f(x) = 3 \sin x$
  (a) Find the zeros of f on the interval $[-2\pi, 4\pi]$.
  (b) Graph $f(x) = 3 \sin x$ on the interval $[-2\pi, 4\pi]$.
  (c) Solve $f(x) = \dfrac{3}{2}$ on the interval $[-2\pi, 4\pi]$. What points are on the graph of f? Label these points on the graph drawn in part (b).
  (d) Use the graph drawn in part (b) along with the results of part (c) to determine the values of x such that $f(x) > \dfrac{3}{2}$ on the interval $[-2\pi, 4\pi]$.

96. $f(x) = 2 \cos x$
  (a) Find the zeros of f on the interval $[-2\pi, 4\pi]$.
  (b) Graph $f(x) = 2 \cos x$ on the interval $[-2\pi, 4\pi]$.

  (c) Solve $f(x) = -\sqrt{3}$ on the interval $[-2\pi, 4\pi]$. What points are on the graph of f? Label these points on the graph drawn in part (b).
  (d) Use the graph drawn in part (b) along with the results of part (c) to determine the values of x such that $f(x) < -\sqrt{3}$ on the interval $[-2\pi, 4\pi]$.

97. $f(x) = 4 \tan x$
  (a) Solve $f(x) = -4$.
  (b) For what values of x is $f(x) < -4$ on the interval $\left(-\dfrac{\pi}{2}, \dfrac{\pi}{2}\right)$?

98. $f(x) = \cot x$
  (a) Solve $f(x) = -\sqrt{3}$.
  (b) For what values of x is $f(x) > -\sqrt{3}$ on the interval $(0, \pi)$?

99. (a) Graph $f(x) = 3 \sin(2x) + 2$ and $g(x) = \dfrac{7}{2}$ on the same Cartesian plane for the interval $[0, \pi]$.

(b) Solve $f(x) = g(x)$ on the interval $[0, \pi]$ and label the points of intersection on the graph drawn in part (b).

(c) Solve $f(x) > g(x)$ on the interval $[0, \pi]$.

(d) Shade the region bounded by $f(x) = 3 \sin(2x) + 2$ and $g(x) = \dfrac{7}{2}$ between the two points found in part (b) on the graph drawn in part (a).

100. (a) Graph $f(x) = 2 \cos \dfrac{x}{2} + 3$ and $g(x) = 4$ on the same Cartesian plane for the interval $[0, 4\pi]$.

(b) Solve $f(x) = g(x)$ on the interval $[0, 4\pi]$ and label the points of intersection on the graph drawn in part (b).

(c) Solve $f(x) < g(x)$ on the interval $[0, 4\pi]$.

(d) Shade the region bounded by $f(x) = 2 \cos \dfrac{x}{2} + 3$ and $g(x) = 4$ between the two points found in part (b) on the graph drawn in part (a).

101. (a) Graph $f(x) = -4 \cos x$ and $g(x) = 2 \cos x + 3$ on the same Cartesian plane for the interval $[0, 2\pi]$.

(b) Solve $f(x) = g(x)$ on the interval $[0, 2\pi]$ and label the points of intersection on the graph drawn in part (b).

(c) Solve $f(x) > g(x)$ on the interval $[0, 2\pi]$.

(d) Shade the region bounded by $f(x) = -4 \cos x$ and $g(x) = 2 \cos x + 3$ between the two points found in part (b) on the graph drawn in part (a).

102. (a) Graph $f(x) = 2 \sin x$ and $g(x) = -2 \sin x + 2$ on the same Cartesian plane for the interval $[0, 2\pi]$.

(b) Solve $f(x) = g(x)$ on the interval $[0, 2\pi]$ and label the points of intersection on the graph drawn in part (b).

(c) Solve $f(x) > g(x)$ on the interval $[0, 2\pi]$.

(d) Shade the region bounded by $f(x) = 2 \sin x$ and $g(x) = -2 \sin x + 2$ between the two points found in part (b) on the graph drawn in part (a).

Applications and Extensions

103. Blood Pressure Blood pressure is a way of measuring the amount of force exerted on the walls of blood vessels. It is measured using two numbers: systolic (as the heart beats) blood pressure and diastolic (as the heart rests) blood pressure. Blood pressures vary substantially from person to person, but a typical blood pressure is 120/80, which means the systolic blood pressure is 120 mmHg and the diastolic blood pressure is 80 mmHg. Assuming that a person's heart beats 70 times per minute, the blood pressure of an individual P after t seconds can be modeled by the function

$$P(t) = 100 + 20 \sin\left(\frac{7\pi}{3} t\right)$$

(a) In the interval $[0, 1]$, determine the times at which the blood pressure is 100 mmHg.

(b) In the interval $[0, 1]$, determine the times at which the blood pressure is 120 mmHg.

(c) In the interval $[0, 1]$, determine the times at which the blood pressures is between 100 and 105 mmHg.

104. The Ferris Wheel In 1893, George Ferris engineered the Ferris Wheel. It was 250 feet in diameter. If the wheel makes 1 revolution every 40 seconds, then the function

$$h(t) = 125 \sin\left(0.157t - \frac{\pi}{2}\right) + 125$$

represents the height h, in feet, of a seat on the wheel as a function of time t, where t is measured in seconds. The ride begins when $t = 0$.

(a) During the first 40 seconds of the ride, at what time t is an individual on the Ferris Wheel exactly 125 feet above the ground?

(b) During the first 80 seconds of the ride, at what time t is an individual on the Ferris Wheel exactly 250 feet above the ground?

(c) During the first 40 seconds of the ride, over what interval of time t is an individual on the Ferris Wheel more than 125 feet above the ground?

105. Holding Pattern An airplane is asked to stay within a holding pattern near Chicago's O'Hare International Airport. The function $d(x) = 70 \sin(0.65x) + 150$ represents the distance d, in miles, that the airplane is from the airport at time x, in minutes.

(a) When the plane enters the holding pattern, $x = 0$, how far is it from O'Hare?

(b) During the first 20 minutes after the plane enters the holding pattern, at what time x is the plane exactly 100 miles from the airport?

(c) During the first 20 minutes after the plane enters the holding pattern, at what time x is the plane more than 100 miles from the airport?

(d) While the plane is in the holding pattern, will it ever be within 70 miles of the airport? Why?

106. Projectile Motion A golfer hits a golf ball with an initial velocity of 100 miles per hour. The range R of the ball as a function of the angle θ to the horizontal is given by $R(\theta) = 672 \sin(2\theta)$, where R is measured in feet.

(a) At what angle θ should the ball be hit if the golfer wants the ball to travel 450 feet (150 yards)?

(b) At what angle θ should the ball be hit if the golfer wants the ball to travel 540 feet (180 yards)?

(c) At what angle θ should the ball be hit if the golfer wants the ball to travel at least 480 feet (160 yards)?

(d) Can the golfer hit the ball 720 feet (240 yards)?

107. Heat Transfer In the study of heat transfer, the equation $x + \tan x = 0$ occurs. Graph $Y_1 = -x$ and $Y_2 = \tan x$ for $x \geq 0$. Conclude that there are an infinite number of points of intersection of these two graphs. Now find the first two positive solutions of $x + \tan x = 0$ rounded to two decimal places.

108. Carrying a Ladder Around a Corner Two hallways, one of width 3 feet, the other of width 4 feet, meet at a right angle. See the illustration. It can be shown that the length L of the ladder as a function of θ is $L(\theta) = 4 \csc \theta + 3 \sec \theta$.

(a) In calculus, you will be asked to find the length of the longest ladder that can turn the corner by solving the equation

$$3 \sec \theta \tan \theta - 4 \csc \theta \cot \theta = 0, \quad 0° < \theta < 90°$$

Solve this equation for θ.

(b) What is the length of the longest ladder that can be carried around the corner?

🖥 (c) Graph $L = L(\theta), 0° \le \theta \le 90°$, and find the angle θ that minimizes the length L.

(d) Compare the result with the one found in part (b). Explain why the two answers are the same.

109. Projectile Motion The horizontal distance that a projectile will travel in the air (ignoring air resistance) is given by the equation

$$R(\theta) = \frac{v_0^2 \sin(2\theta)}{g}$$

where v_0 is the initial velocity of the projectile, θ is the angle of elevation, and g is acceleration due to gravity (9.8 meters per second squared).

(a) If you can throw a baseball with an initial speed of 34.8 meters per second, at what angle of elevation θ should you direct the throw so that the ball travels a distance of 107 meters before striking the ground?

(b) Determine the maximum distance that you can throw the ball.

🖥 (c) Graph $R = R(\theta)$, with $v_0 = 34.8$ meters per second.

(d) Verify the results obtained in parts (a) and (b) using a graphing utility.

110. Projectile Motion Refer to Problem 109.

(a) If you can throw a baseball with an initial speed of 40 meters per second, at what angle of elevation θ should you direct the throw so that the ball travels a distance of 110 meters before striking the ground?

(b) Determine the maximum distance that you can throw the ball.

🖥 (c) Graph $R = R(\theta)$, with $v_0 = 40$ meters per second.

(d) Verify the results obtained in parts (a) and (b) using a graphing utility.

⚠ *The following discussion of Snell's Law of Refraction* (named after Willebrord Snell, 1580–1626) is needed for Problems 111–118. Light, sound, and other waves travel at different speeds, depending on the media (air, water, wood, and so on) through which they pass. Suppose that light travels from a point A in one medium, where its speed is v_1, to a point B in another medium, where its speed is v_2. Refer to the figure, where the angle θ_1 is called the* **angle of incidence** *and the angle θ_2 is the* **angle of refraction.** *Snell's Law, which can be proved using calculus, states that*

$$\frac{\sin \theta_1}{\sin \theta_2} = \frac{v_1}{v_2}$$

The ratio $\dfrac{v_1}{v_2}$ is called the **index of refraction.** *Some values are given in the table shown to the right.*

Some Indexes of Refraction	
Medium	**Index of Refraction[†]**
Water	1.33
Ethyl alcohol (20°C)	1.36
Carbon disulfide	1.63
Air (1 atm and 0°C)	1.00029
Diamond	2.42
Fused quartz	1.46
Glass, crown	1.52
Glass, dense flint	1.66
Sodium chloride	1.54

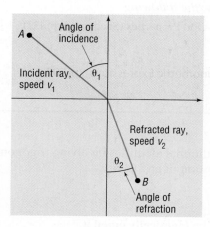

111. The index of refraction of light in passing from a vacuum into water is 1.33. If the angle of incidence is 40°, determine the angle of refraction.

112. The index of refraction of light in passing from a vacuum into dense flint glass is 1.66. If the angle of incidence is 50°, determine the angle of refraction.

113. Ptolemy, who lived in the city of Alexandria in Egypt during the second century AD, gave the measured values in the following table for the angle of incidence θ_1 and the angle of refraction θ_2 for a light beam passing from air into water. Do these values agree with Snell's Law? If so, what index of refraction results? (These data are of interest as the oldest recorded physical measurements.)[†]

θ_1	θ_2	θ_1	θ_2
10°	8°	50°	35°0′
20°	15°30′	60°	40°30′
30°	22°30′	70°	45°30′
40°	29°0′	80°	50°0′

---

*Because this law was also deduced by René Descartes in France, it is also known as Descartes's Law.

[†]For light of wavelength 589 nanometers, measured with respect to a vacuum. The index with respect to air is negligibly different in most cases.

114. Bending Light The speed of yellow sodium light (wavelength, 589 nanometers) in a certain liquid is measured to be 1.92×10^8 meters per second. What is the index of refraction of this liquid, with respect to air, for sodium light?*

[**Hint:** The speed of light in air is approximately 2.998×10^8 meters per second.]

115. Bending Light A beam of light with a wavelength of 589 nanometers traveling in air makes an angle of incidence of 40° on a slab of transparent material, and the refracted beam makes an angle of refraction of 26°. Find the index of refraction of the material.*

116. Bending Light A light ray with a wavelength of 589 nanometers (produced by a sodium lamp) traveling through air makes an angle of incidence of 30° on a smooth, flat slab of crown glass. Find the angle of refraction.*

117. A light beam passes through a thick slab of material whose index of refraction is n_2. Show that the emerging beam is parallel to the incident beam.[†]

118. Brewster's Law If the angle of incidence and the angle of refraction are complementary angles, the angle of incidence is referred to as the Brewster angle θ_B. The Brewster angle is related to the index of refractions of the two media, n_1 and n_2, by the equation $n_1 \sin \theta_B = n_2 \cos \theta_B$, where n_1 is the index of refraction of the incident medium and n_2 is the index of refraction of the refractive medium. Determine the Brewster angle for a light beam traveling through water (at 20°C) that makes an angle of incidence with a smooth, flat slab of crown glass.

[†]Adapted from Halliday and Resnick, *Fundamentals of Physics*, 7th ed., 2005, John Wiley & Sons.

Explaining Concepts: Discussion and Writing

119. Explain in your own words how you would use your calculator to solve the equation $\cos x = -0.6, 0 \le x < 2\pi$. How would you modify your approach to solve the equation $\cot x = 5, 0 < x < 2\pi$?

120. Provide a justification as to why no further points of intersection (and therefore solutions) exist in Figure 25 for $x < -\pi$ or $x > 4\pi$.

'Are You Prepared?' Answers

1. $\left\{\dfrac{3}{2}\right\}$ **2.** $\dfrac{\sqrt{2}}{2}; -\dfrac{1}{2}$ **3.** $\left\{-1, \dfrac{5}{4}\right\}$ **4.** $\left\{\dfrac{1 - \sqrt{5}}{2}, \dfrac{1 + \sqrt{5}}{2}\right\}$ **5.** $\left\{0, \dfrac{5}{2}\right\}$ **6.** $\{0.76\}$

6.4 Trigonometric Identities

PREPARING FOR THIS SECTION *Before getting started, review the following:*

• Fundamental Identities (Section 5.2, p. 394)   • Even–Odd Properties (Section 5.5, p. 432)

Now Work the 'Are You Prepared?' problems on page 511.

OBJECTIVES **1** Use Algebra to Simplify Trigonometric Expressions (p. 507)
            2 Establish Identities (p. 508)

In this chapter we establish some additional identities involving trigonometric functions. But first, let's review the definition of an *identity*.

DEFINITION

Two functions f and g are said to be **identically equal** if

$$f(x) = g(x)$$

for every value of x for which both functions are defined. Such an equation is referred to as an **identity**. An equation that is not an identity is called a **conditional equation**.

For example, the following are identities:

$$(x + 1)^2 = x^2 + 2x + 1 \qquad \sin^2 x + \cos^2 x = 1 \qquad \csc x = \frac{1}{\sin x}$$

The following are conditional equations:

$$2x + 5 = 0 \qquad \text{True only if } x = -\frac{5}{2}$$

$$\sin x = 0 \qquad \text{True only if } x = k\pi, \text{ } k \text{ an integer}$$

$$\sin x = \cos x \qquad \text{True only if } x = \frac{\pi}{4} + 2k\pi \text{ or } x = \frac{5\pi}{4} + 2k\pi, \text{ } k \text{ an integer}$$

The following summarizes the trigonometric identities that we have established thus far.

Quotient Identities

$$\tan \theta = \frac{\sin \theta}{\cos \theta} \qquad \cot \theta = \frac{\cos \theta}{\sin \theta}$$

Reciprocal Identities

$$\csc \theta = \frac{1}{\sin \theta} \qquad \sec \theta = \frac{1}{\cos \theta} \qquad \cot \theta = \frac{1}{\tan \theta}$$

Pythagorean Identities

$$\sin^2 \theta + \cos^2 \theta = 1 \qquad \tan^2 \theta + 1 = \sec^2 \theta$$

$$\cot^2 \theta + 1 = \csc^2 \theta$$

Even–Odd Identities

$$\sin(-\theta) = -\sin \theta \qquad \cos(-\theta) = \cos \theta \qquad \tan(-\theta) = -\tan \theta$$

$$\csc(-\theta) = -\csc \theta \qquad \sec(-\theta) = \sec \theta \qquad \cot(-\theta) = -\cot \theta$$

This list of identities comprises what we shall refer to as the **basic trigonometric identities.** These identities should not merely be memorized, but should be *known* (just as you know your name rather than have it memorized). In fact, minor variations of a basic identity are often used. For example, we might want to use

$$\sin^2 \theta = 1 - \cos^2 \theta \quad \text{or} \quad \cos^2 \theta = 1 - \sin^2 \theta$$

instead of $\sin^2 \theta + \cos^2 \theta = 1$. For this reason, among others, you need to know these relationships and be comfortable with variations of them.

1 Use Algebra to Simplify Trigonometric Expressions

The ability to use algebra to manipulate trigonometric expressions is a key skill that one must have to establish identities. Some of the techniques that are used in establishing identities are multiplying by a "well-chosen 1," writing a trigonometric expression over a common denominator, rewriting a trigonometric expression in terms of sine and cosine only, and factoring.

| EXAMPLE 1 | **Using Algebraic Techniques to Simplify Trigonometric Expressions** |

(a) Simplify $\dfrac{\cot \theta}{\csc \theta}$ by rewriting each trigonometric function in terms of sine and cosine functions.

(b) Show that $\dfrac{\cos \theta}{1 + \sin \theta} = \dfrac{1 - \sin \theta}{\cos \theta}$ by multiplying the numerator and denominator by $1 - \sin \theta$.

(c) Simplify $\dfrac{1 + \sin u}{\sin u} + \dfrac{\cot u - \cos u}{\cos u}$ by rewriting the expression over a common denominator.

(d) Simplify $\dfrac{\sin^2 v - 1}{\tan v \sin v - \tan v}$ by factoring.

Solution

(a) $\dfrac{\cot \theta}{\csc \theta} = \dfrac{\dfrac{\cos \theta}{\sin \theta}}{\dfrac{1}{\sin \theta}} = \dfrac{\cos \theta}{\sin \theta} \cdot \dfrac{\sin \theta}{1} = \cos \theta$

(b) $\dfrac{\cos \theta}{1 + \sin \theta} = \dfrac{\cos \theta}{1 + \sin \theta} \cdot \dfrac{1 - \sin \theta}{1 - \sin \theta} = \dfrac{\cos \theta(1 - \sin \theta)}{1 - \sin^2 \theta}$

$\uparrow$ Multiply by a well-chosen 1: $\dfrac{1 - \sin \theta}{1 - \sin \theta}$.

$= \dfrac{\cos \theta(1 - \sin \theta)}{\cos^2 \theta} = \dfrac{1 - \sin \theta}{\cos \theta}$

(c) $\dfrac{1 + \sin u}{\sin u} + \dfrac{\cot u - \cos u}{\cos u} = \dfrac{1 + \sin u}{\sin u} \cdot \dfrac{\cos u}{\cos u} + \dfrac{\cot u - \cos u}{\cos u} \cdot \dfrac{\sin u}{\sin u}$

$= \dfrac{\cos u + \sin u \cos u + \cot u \sin u - \cos u \sin u}{\sin u \cos u} = \dfrac{\cos u + \dfrac{\cos u}{\sin u} \cdot \sin u}{\sin u \cos u}$

$\uparrow$ $\cot u = \dfrac{\cos u}{\sin u}$

$= \dfrac{\cos u + \cos u}{\sin u \cos u} = \dfrac{2 \cos u}{\sin u \cos u} = \dfrac{2}{\sin u}$

(d) $\dfrac{\sin^2 v - 1}{\tan v \sin v - \tan v} = \dfrac{(\sin v + 1)(\sin v - 1)}{\tan v(\sin v - 1)} = \dfrac{\sin v + 1}{\tan v}$

●

Now Work PROBLEMS **9, 11,** AND **13**

2 Establish Identities

In the examples that follow, the directions will read "Establish the identity. . . . " As you will see, this is accomplished by starting with one side of the given equation (usually the one containing the more complicated expression) and, using appropriate basic identities and algebraic manipulations, arriving at the other side. The selection of appropriate basic identities to obtain the desired result is learned only through experience and lots of practice.

| EXAMPLE 2 | **Establishing an Identity** |

Establish the identity: $\csc \theta \cdot \tan \theta = \sec \theta$

Solution

NOTE A graphing utility can be used to provide evidence of an identity. For example, if we graph $Y_1 = \csc\theta \cdot \tan\theta$ and $Y_2 = \sec\theta$, the graphs appear to be the same. This provides evidence that $Y_1 = Y_2$. However, it does not prove their equality. A graphing utility cannot be used to establish an identity—identities must be established algebraically. ∎

We start with the left side, because it contains the more complicated expression, and apply a reciprocal identity and a quotient identity.

$$\csc\theta \cdot \tan\theta = \frac{1}{\sin\theta} \cdot \frac{\sin\theta}{\cos\theta} = \frac{1}{\cos\theta} = \sec\theta$$

Having arrived at the right side, the identity is established.

Now Work PROBLEM 19

EXAMPLE 3 | **Establishing an Identity**

Establish the identity: $\sin^2(-\theta) + \cos^2(-\theta) = 1$

Solution

We begin with the left side and, because the arguments are $-\theta$, apply Even–Odd Identities.

$$\sin^2(-\theta) + \cos^2(-\theta) = [\sin(-\theta)]^2 + [\cos(-\theta)]^2$$
$$= (-\sin\theta)^2 + (\cos\theta)^2 \qquad \text{Even–Odd Identities}$$
$$= (\sin\theta)^2 + (\cos\theta)^2$$
$$= 1 \qquad \text{Pythagorean Identity}$$

EXAMPLE 4 | **Establishing an Identity**

Establish the identity: $\dfrac{\sin^2(-\theta) - \cos^2(-\theta)}{\sin(-\theta) - \cos(-\theta)} = \cos\theta - \sin\theta$

Solution

We begin with two observations: The left side contains the more complicated expression. Also, the left side contains expressions with the argument $-\theta$, whereas the right side contains expressions with the argument θ. We decide, therefore, to start with the left side and apply Even–Odd Identities.

$$\frac{\sin^2(-\theta) - \cos^2(-\theta)}{\sin(-\theta) - \cos(-\theta)} = \frac{[\sin(-\theta)]^2 - [\cos(-\theta)]^2}{\sin(-\theta) - \cos(-\theta)}$$
$$= \frac{(-\sin\theta)^2 - (\cos\theta)^2}{-\sin\theta - \cos\theta} \qquad \text{Even–Odd Identities}$$
$$= \frac{(\sin\theta)^2 - (\cos\theta)^2}{-\sin\theta - \cos\theta} \qquad \text{Simplify.}$$
$$= \frac{(\sin\theta - \cos\theta)(\sin\theta + \cos\theta)}{-(\sin\theta + \cos\theta)} \qquad \text{Factor.}$$
$$= \cos\theta - \sin\theta \qquad \text{Cancel and simplify.}$$

EXAMPLE 5 | **Establishing an Identity**

Establish the identity: $\dfrac{1 + \tan u}{1 + \cot u} = \tan u$

Solution

$$\frac{1 + \tan u}{1 + \cot u} = \frac{1 + \tan u}{1 + \dfrac{1}{\tan u}} = \frac{1 + \tan u}{\dfrac{\tan u + 1}{\tan u}}$$
$$= \frac{\tan u(1 + \tan u)}{\tan u + 1} = \tan u$$

Now Work PROBLEMS 23 AND 27

When sums or differences of quotients appear, it is usually best to rewrite them as a single quotient, especially if the other side of the identity consists of only one term.

EXAMPLE 6   **Establishing an Identity**

Establish the identity: $\dfrac{\sin \theta}{1 + \cos \theta} + \dfrac{1 + \cos \theta}{\sin \theta} = 2 \csc \theta$

Solution   The left side is more complicated, so we start with it and proceed to add.

$$\dfrac{\sin \theta}{1 + \cos \theta} + \dfrac{1 + \cos \theta}{\sin \theta} = \dfrac{\sin^2 \theta + (1 + \cos \theta)^2}{(1 + \cos \theta)(\sin \theta)} \qquad \text{Add the quotients.}$$

$$= \dfrac{\sin^2 \theta + 1 + 2 \cos \theta + \cos^2 \theta}{(1 + \cos \theta)(\sin \theta)} \qquad \begin{array}{l}\text{Remove parentheses in}\\\text{the numerator.}\end{array}$$

$$= \dfrac{(\sin^2 \theta + \cos^2 \theta) + 1 + 2 \cos \theta}{(1 + \cos \theta)(\sin \theta)} \qquad \text{Regroup.}$$

$$= \dfrac{2 + 2 \cos \theta}{(1 + \cos \theta)(\sin \theta)} \qquad \text{Pythagorean Identity}$$

$$= \dfrac{2(1 + \cos \theta)}{(1 + \cos \theta)(\sin \theta)} \qquad \text{Factor and cancel.}$$

$$= \dfrac{2}{\sin \theta}$$

$$= 2 \csc \theta \qquad \text{Reciprocal Identity}$$

Now Work PROBLEM 49

Sometimes it helps to write one side in terms of sine and cosine functions only.

EXAMPLE 7   **Establishing an Identity**

Establish the identity: $\dfrac{\tan v + \cot v}{\sec v \csc v} = 1$

Solution $\dfrac{\tan v + \cot v}{\sec v \csc v} = \dfrac{\dfrac{\sin v}{\cos v} + \dfrac{\cos v}{\sin v}}{\dfrac{1}{\cos v} \cdot \dfrac{1}{\sin v}} = \dfrac{\dfrac{\sin^2 v + \cos^2 v}{\cos v \sin v}}{\dfrac{1}{\cos v \sin v}}$

$\qquad\qquad\qquad\quad \uparrow \qquad\qquad\qquad \uparrow$

$\qquad\qquad\quad$ Change to sines $\qquad$ Add the quotients
$\qquad\qquad\quad$ and cosines. $\qquad\quad$ in the numerator.

$$= \dfrac{1}{\cos v \sin v} \cdot \dfrac{\cos v \sin v}{1} = 1$$

$\qquad\qquad\quad \uparrow$

$\qquad\qquad$ Divide the quotients;
$\qquad\qquad\quad \sin^2 v + \cos^2 v = 1.$

Now Work PROBLEM 69

Sometimes, multiplying the numerator and denominator by an appropriate factor will result in a simplification.

EXAMPLE 8   Establishing an Identity

Establish the identity: $\dfrac{1 - \sin \theta}{\cos \theta} = \dfrac{\cos \theta}{1 + \sin \theta}$

Solution   We start with the left side and multiply the numerator and the denominator by $1 + \sin \theta$. (Alternatively, we could multiply the numerator and denominator of the right side by $1 - \sin \theta$.)

$$\dfrac{1 - \sin \theta}{\cos \theta} = \dfrac{1 - \sin \theta}{\cos \theta} \cdot \dfrac{1 + \sin \theta}{1 + \sin \theta} \qquad \text{Multiply the numerator and denominator by } 1 + \sin \theta.$$

$$= \dfrac{1 - \sin^2 \theta}{\cos \theta (1 + \sin \theta)}$$

$$= \dfrac{\cos^2 \theta}{\cos \theta (1 + \sin \theta)} \qquad 1 - \sin^2 \theta = \cos^2 \theta$$

$$= \dfrac{\cos \theta}{1 + \sin \theta} \qquad \text{Cancel.}$$

━━ **Now Work** PROBLEM 53

Although a lot of practice is the only real way to learn how to establish identities, the following guidelines should prove helpful.

WARNING Be careful not to handle identities to be established as if they were conditional equations. You *cannot* establish an identity by such methods as adding the same expression to each side and obtaining a true statement. This practice is not allowed, because the original statement is precisely the one that you are trying to establish. You do not know until it has been established that it is, in fact, true. ■

Guidelines for Establishing Identities

1. It is almost always preferable to start with the side containing the more complicated expression.
2. Rewrite sums or differences of quotients as a single quotient.
3. Sometimes rewriting one side in terms of sine and cosine functions only will help.
4. Always keep your goal in mind. As you manipulate one side of the expression, you must keep in mind the form of the expression on the other side.

6.4 Assess Your Understanding

'Are You Prepared?' *Answers are given at the end of these exercises. If you get a wrong answer, read the pages listed in red.*

1. *True or False* $\sin^2 \theta = 1 - \cos^2 \theta$. (p. 394)

2. *True or False* $\sin(-\theta) + \cos(-\theta) = \cos \theta - \sin \theta$. (p. 432)

Concepts and Vocabulary

3. Suppose that f and g are two functions with the same domain. If $f(x) = g(x)$ for every x in the domain, the equation is called a(n) _____. Otherwise, it is called a(n) _____ equation.

4. $\tan^2 \theta - \sec^2 \theta =$ _____.

5. $\cos(-\theta) - \cos \theta =$ _____.

6. *True or False* $\sin(-\theta) + \sin \theta = 0$ for any value of θ.

7. *True or False*   In establishing an identity, it is often easiest to just multiply both sides by a well-chosen nonzero expression involving the variable.

8. *True or False* $\tan \theta \cdot \cos \theta = \sin \theta$ for any $\theta \neq (2k + 1)\dfrac{\pi}{2}$.

Skill Building

In Problems 9–18, simplify each trigonometric expression by following the indicated direction.

9. Rewrite in terms of sine and cosine functions:
$$\tan \theta \cdot \csc \theta.$$

10. Rewrite in terms of sine and cosine functions:
$$\cot \theta \cdot \sec \theta.$$

11. Multiply $\dfrac{\cos \theta}{1 - \sin \theta}$ by $\dfrac{1 + \sin \theta}{1 + \sin \theta}$.

12. Multiply $\dfrac{\sin \theta}{1 + \cos \theta}$ by $\dfrac{1 - \cos \theta}{1 - \cos \theta}$.

13. Rewrite over a common denominator:

$$\frac{\sin \theta + \cos \theta}{\cos \theta} + \frac{\cos \theta - \sin \theta}{\sin \theta}$$

14. Rewrite over a common denominator:

$$\frac{1}{1 - \cos v} + \frac{1}{1 + \cos v}$$

15. Multiply and simplify: $\dfrac{(\sin \theta + \cos \theta)(\sin \theta + \cos \theta) - 1}{\sin \theta \cos \theta}$

16. Multiply and simplify: $\dfrac{(\tan \theta + 1)(\tan \theta + 1) - \sec^2 \theta}{\tan \theta}$

17. Factor and simplify: $\dfrac{3 \sin^2 \theta + 4 \sin \theta + 1}{\sin^2 \theta + 2 \sin \theta + 1}$

18. Factor and simplify: $\dfrac{\cos^2 \theta - 1}{\cos^2 \theta - \cos \theta}$

In Problems 19–98, establish each identity.

19. $\csc \theta \cdot \cos \theta = \cot \theta$

20. $\sec \theta \cdot \sin \theta = \tan \theta$

21. $1 + \tan^2(-\theta) = \sec^2 \theta$

22. $1 + \cot^2(-\theta) = \csc^2 \theta$

23. $\cos \theta(\tan \theta + \cot \theta) = \csc \theta$

24. $\sin \theta(\cot \theta + \tan \theta) = \sec \theta$

25. $\tan u \cot u - \cos^2 u = \sin^2 u$

26. $\sin u \csc u - \cos^2 u = \sin^2 u$

27. $(\sec \theta - 1)(\sec \theta + 1) = \tan^2 \theta$

28. $(\csc \theta - 1)(\csc \theta + 1) = \cot^2 \theta$

29. $(\sec \theta + \tan \theta)(\sec \theta - \tan \theta) = 1$

30. $(\csc \theta + \cot \theta)(\csc \theta - \cot \theta) = 1$

31. $\cos^2 \theta(1 + \tan^2 \theta) = 1$

32. $(1 - \cos^2 \theta)(1 + \cot^2 \theta) = 1$

33. $(\sin \theta + \cos \theta)^2 + (\sin \theta - \cos \theta)^2 = 2$

34. $\tan^2 \theta \cos^2 \theta + \cot^2 \theta \sin^2 \theta = 1$

35. $\sec^4 \theta - \sec^2 \theta = \tan^4 \theta + \tan^2 \theta$

36. $\csc^4 \theta - \csc^2 \theta = \cot^4 \theta + \cot^2 \theta$

37. $\sec u - \tan u = \dfrac{\cos u}{1 + \sin u}$

38. $\csc u - \cot u = \dfrac{\sin u}{1 + \cos u}$

39. $3 \sin^2 \theta + 4 \cos^2 \theta = 3 + \cos^2 \theta$

40. $9 \sec^2 \theta - 5 \tan^2 \theta = 5 + 4 \sec^2 \theta$

41. $1 - \dfrac{\cos^2 \theta}{1 + \sin \theta} = \sin \theta$

42. $1 - \dfrac{\sin^2 \theta}{1 - \cos \theta} = -\cos \theta$

43. $\dfrac{1 + \tan v}{1 - \tan v} = \dfrac{\cot v + 1}{\cot v - 1}$

44. $\dfrac{\csc v - 1}{\csc v + 1} = \dfrac{1 - \sin v}{1 + \sin v}$

45. $\dfrac{\sec \theta}{\csc \theta} + \dfrac{\sin \theta}{\cos \theta} = 2 \tan \theta$

46. $\dfrac{\csc \theta - 1}{\cot \theta} = \dfrac{\cot \theta}{\csc \theta + 1}$

47. $\dfrac{1 + \sin \theta}{1 - \sin \theta} = \dfrac{\csc \theta + 1}{\csc \theta - 1}$

48. $\dfrac{\cos \theta + 1}{\cos \theta - 1} = \dfrac{1 + \sec \theta}{1 - \sec \theta}$

49. $\dfrac{1 - \sin v}{\cos v} + \dfrac{\cos v}{1 - \sin v} = 2 \sec v$

50. $\dfrac{\cos v}{1 + \sin v} + \dfrac{1 + \sin v}{\cos v} = 2 \sec v$

51. $\dfrac{\sin \theta}{\sin \theta - \cos \theta} = \dfrac{1}{1 - \cot \theta}$

52. $1 - \dfrac{\sin^2 \theta}{1 + \cos \theta} = \cos \theta$

53. $\dfrac{1 - \sin \theta}{1 + \sin \theta} = (\sec \theta - \tan \theta)^2$

54. $\dfrac{1 - \cos \theta}{1 + \cos \theta} = (\csc \theta - \cot \theta)^2$

55. $\dfrac{\cos \theta}{1 - \tan \theta} + \dfrac{\sin \theta}{1 - \cot \theta} = \sin \theta + \cos \theta$

56. $\dfrac{\cot \theta}{1 - \tan \theta} + \dfrac{\tan \theta}{1 - \cot \theta} = 1 + \tan \theta + \cot \theta$

57. $\tan \theta + \dfrac{\cos \theta}{1 + \sin \theta} = \sec \theta$

58. $\dfrac{\sin \theta \cos \theta}{\cos^2 \theta - \sin^2 \theta} = \dfrac{\tan \theta}{1 - \tan^2 \theta}$

59. $\dfrac{\tan \theta + \sec \theta - 1}{\tan \theta - \sec \theta + 1} = \tan \theta + \sec \theta$

60. $\dfrac{\sin \theta - \cos \theta + 1}{\sin \theta + \cos \theta - 1} = \dfrac{\sin \theta + 1}{\cos \theta}$

61. $\dfrac{\tan \theta - \cot \theta}{\tan \theta + \cot \theta} = \sin^2 \theta - \cos^2 \theta$

62. $\dfrac{\sec \theta - \cos \theta}{\sec \theta + \cos \theta} = \dfrac{\sin^2 \theta}{1 + \cos^2 \theta}$

63. $\dfrac{\tan u - \cot u}{\tan u + \cot u} + 1 = 2 \sin^2 u$

64. $\dfrac{\tan u - \cot u}{\tan u + \cot u} + 2 \cos^2 u = 1$

65. $\dfrac{\sec \theta + \tan \theta}{\cot \theta + \cos \theta} = \tan \theta \sec \theta$

66. $\dfrac{\sec \theta}{1 + \sec \theta} = \dfrac{1 - \cos \theta}{\sin^2 \theta}$

67. $\dfrac{1 - \tan^2 \theta}{1 + \tan^2 \theta} + 1 = 2 \cos^2 \theta$

68. $\dfrac{1 - \cot^2 \theta}{1 + \cot^2 \theta} + 2 \cos^2 \theta = 1$

69. $\dfrac{\sec \theta - \csc \theta}{\sec \theta \csc \theta} = \sin \theta - \cos \theta$

70. $\dfrac{\sin^2 \theta - \tan \theta}{\cos^2 \theta - \cot \theta} = \tan^2 \theta$

71. $\sec \theta - \cos \theta = \sin \theta \tan \theta$

72. $\tan \theta + \cot \theta = \sec \theta \csc \theta$

73. $\dfrac{1}{1 - \sin \theta} + \dfrac{1}{1 + \sin \theta} = 2 \sec^2 \theta$

74. $\dfrac{1 + \sin \theta}{1 - \sin \theta} - \dfrac{1 - \sin \theta}{1 + \sin \theta} = 4 \tan \theta \sec \theta$

75. $\dfrac{\sec \theta}{1 - \sin \theta} = \dfrac{1 + \sin \theta}{\cos^3 \theta}$

76. $\dfrac{1 + \sin \theta}{1 - \sin \theta} = (\sec \theta + \tan \theta)^2$

77. $\dfrac{(\sec v - \tan v)^2 + 1}{\csc v(\sec v - \tan v)} = 2 \tan v$

78. $\dfrac{\sec^2 v - \tan^2 v + \tan v}{\sec v} = \sin v + \cos v$

79. $\dfrac{\sin \theta + \cos \theta}{\cos \theta} - \dfrac{\sin \theta - \cos \theta}{\sin \theta} = \sec \theta \csc \theta$

80. $\dfrac{\sin \theta + \cos \theta}{\sin \theta} - \dfrac{\cos \theta - \sin \theta}{\cos \theta} = \sec \theta \csc \theta$

81. $\dfrac{\sin^3 \theta + \cos^3 \theta}{\sin \theta + \cos \theta} = 1 - \sin \theta \cos \theta$

82. $\dfrac{\sin^3 \theta + \cos^3 \theta}{1 - 2\cos^2 \theta} = \dfrac{\sec \theta - \sin \theta}{\tan \theta - 1}$

83. $\dfrac{\cos^2 \theta - \sin^2 \theta}{1 - \tan^2 \theta} = \cos^2 \theta$

84. $\dfrac{\cos \theta + \sin \theta - \sin^3 \theta}{\sin \theta} = \cot \theta + \cos^2 \theta$

85. $\dfrac{(2\cos^2 \theta - 1)^2}{\cos^4 \theta - \sin^4 \theta} = 1 - 2\sin^2 \theta$

86. $\dfrac{1 - 2\cos^2 \theta}{\sin \theta \cos \theta} = \tan \theta - \cot \theta$

87. $\dfrac{1 + \sin \theta + \cos \theta}{1 + \sin \theta - \cos \theta} = \dfrac{1 + \cos \theta}{\sin \theta}$

88. $\dfrac{1 + \cos \theta + \sin \theta}{1 + \cos \theta - \sin \theta} = \sec \theta + \tan \theta$

89. $(a \sin \theta + b \cos \theta)^2 + (a \cos \theta - b \sin \theta)^2 = a^2 + b^2$

90. $(2a \sin \theta \cos \theta)^2 + a^2(\cos^2 \theta - \sin^2 \theta)^2 = a^2$

91. $\dfrac{\tan \alpha + \tan \beta}{\cot \alpha + \cot \beta} = \tan \alpha \tan \beta$

92. $(\tan \alpha + \tan \beta)(1 - \cot \alpha \cot \beta) + (\cot \alpha + \cot \beta)(1 - \tan \alpha \tan \beta) = 0$

93. $(\sin \alpha + \cos \beta)^2 + (\cos \beta + \sin \alpha)(\cos \beta - \sin \alpha) = 2\cos \beta(\sin \alpha + \cos \beta)$

94. $(\sin \alpha - \cos \beta)^2 + (\cos \beta + \sin \alpha)(\cos \beta - \sin \alpha) = -2\cos \beta(\sin \alpha - \cos \beta)$

95. $\ln |\sec \theta| = -\ln |\cos \theta|$

96. $\ln |\tan \theta| = \ln |\sin \theta| - \ln |\cos \theta|$

97. $\ln |1 + \cos \theta| + \ln |1 - \cos \theta| = 2 \ln |\sin \theta|$

98. $\ln |\sec \theta + \tan \theta| + \ln |\sec \theta - \tan \theta| = 0$

In Problems 99–102, show that the functions f and g are identically equal.

99. $f(x) = \sin x \cdot \tan x \qquad g(x) = \sec x - \cos x$

100. $f(x) = \cos x \cdot \cot x \qquad g(x) = \csc x - \sin x$

101. $f(\theta) = \dfrac{1 - \sin \theta}{\cos \theta} - \dfrac{\cos \theta}{1 + \sin \theta} \qquad g(\theta) = 0$

102. $f(\theta) = \tan \theta + \sec \theta \qquad g(\theta) = \dfrac{\cos \theta}{1 - \sin \theta}$

Applications and Extensions

103. Searchlights A searchlight at the grand opening of a new car dealership casts a spot of light on a wall located 75 meters from the searchlight. The acceleration $\ddot{r}$ of the spot of light is found to be $\ddot{r} = 1200 \sec \theta (2 \sec^2 \theta - 1)$. Show that this is equivalent to $\ddot{r} = 1200 \left(\dfrac{1 + \sin^2 \theta}{\cos^3 \theta} \right)$.

Source: Adapted from Hibbeler, *Engineering Mechanics: Dynamics*, 10th ed. © 2004.

104. Optical Measurement Optical methods of measurement often rely on the interference of two light waves. If two light waves, identical except for a phase lag, are mixed together, the resulting intensity, or irradiance, is given by $I_t = 4A^2 \dfrac{(\csc \theta - 1)(\sec \theta + \tan \theta)}{\csc \theta \sec \theta}$. Show that this is equivalent to $I_t = (2A \cos \theta)^2$.

Source: Experimental Techniques, July/August 2002

Explaining Concepts: Discussion and Writing

105. Write a few paragraphs outlining your strategy for establishing identities.

106. Write down the three Pythagorean Identities.

107. Why do you think it is usually preferable to start with the side containing the more complicated expression when establishing an identity?

108. Make up an identity that is not a Fundamental Identity.

'Are You Prepared?' Answers

1. True **2.** True

6.5 Sum and Difference Formulas

PREPARING FOR THIS SECTION *Before getting started, review the following:*

- Distance Formula (Foundations, Section 1, p. 3)
- Values of the Trigonometric Functions (Section 5.3, pp. 403–405; Section 5.4, pp. 416, 420–421)
- Finding Exact Values Given the Value of a Trigonometric Function and the Quadrant of the Angle (Section 5.4, pp. 421–422)

Now Work the 'Are You Prepared?' problems on page 523.

OBJECTIVES 1 Use Sum and Difference Formulas to Find Exact Values (p. 515)
2 Use Sum and Difference Formulas to Establish Identities (p. 516)
3 Use Sum and Difference Formulas Involving Inverse Trigonometric Functions (p. 520)
4 Solve Trigonometric Equations Linear in Sine and Cosine (p. 521)

In this section, we continue our derivation of trigonometric identities by obtaining formulas that involve the sum or difference of two angles, such as $\cos(\alpha + \beta)$, $\cos(\alpha - \beta)$, or $\sin(\alpha + \beta)$. These formulas are referred to as the **sum and difference formulas.** We begin with the formulas for $\cos(\alpha + \beta)$ and $\cos(\alpha - \beta)$.

THEOREM

Sum and Difference Formulas for the Cosine Function

$$\cos(\alpha + \beta) = \cos \alpha \cos \beta - \sin \alpha \sin \beta \qquad (1)$$

$$\cos(\alpha - \beta) = \cos \alpha \cos \beta + \sin \alpha \sin \beta \qquad (2)$$

In Words
Formula (1) states that the cosine of the sum of two angles equals the cosine of the first angle times the cosine of the second angle minus the sine of the first angle times the sine of the second angle.

Proof We will prove formula (2) first. Although this formula is true for all numbers α and β, we shall assume in our proof that $0 < \beta < \alpha < 2\pi$. We begin with the unit circle and place the angles α and β in standard position, as shown in Figure 26(a). The point P_1 lies on the terminal side of β, so its coordinates are $(\cos \beta, \sin \beta)$; and the point P_2 lies on the terminal side of α, so its coordinates are $(\cos \alpha, \sin \alpha)$.

Figure 26

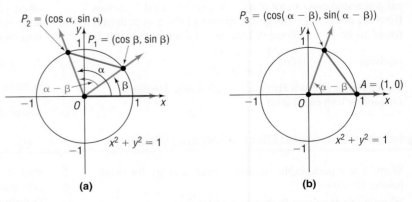

(a)   (b)

Now place the angle $\alpha - \beta$ in standard position, as shown in Figure 26(b). The point A has coordinates $(1, 0)$, and the point P_3 is on the terminal side of the angle $\alpha - \beta$, so its coordinates are $(\cos(\alpha - \beta), \sin(\alpha - \beta))$.

Looking at triangle OP_1P_2 in Figure 26(a) and triangle OAP_3 in Figure 26(b), we see that these triangles are congruent. (Do you see why? We have SAS: two sides and the included angle, $\alpha - \beta$, are equal.) As a result, the unknown side of each triangle must be equal; that is,

$$d(A, P_3) = d(P_1, P_2)$$

Using the distance formula, we find that

$$\sqrt{[\cos(\alpha - \beta) - 1]^2 + [\sin(\alpha - \beta) - 0]^2} = \sqrt{(\cos\alpha - \cos\beta)^2 + (\sin\alpha - \sin\beta)^2} \quad \textit{d(A, P}_3) = d(P_1, P_2)$$

$$[\cos(\alpha - \beta) - 1]^2 + \sin^2(\alpha - \beta) = (\cos\alpha - \cos\beta)^2 + (\sin\alpha - \sin\beta)^2 \quad \textit{Square both sides.}$$

$$\cos^2(\alpha - \beta) - 2\cos(\alpha - \beta) + 1 + \sin^2(\alpha - \beta) = \cos^2\alpha - 2\cos\alpha\cos\beta + \cos^2\beta \quad \textit{Multiply out the squared terms.}$$
$$+ \sin^2\alpha - 2\sin\alpha\sin\beta + \sin^2\beta$$

$$2 - 2\cos(\alpha - \beta) = 2 - 2\cos\alpha\cos\beta - 2\sin\alpha\sin\beta \quad \textit{Apply a Pythagorean Identity (3 times).}$$

$$-2\cos(\alpha - \beta) = -2\cos\alpha\cos\beta - 2\sin\alpha\sin\beta \quad \textit{Subtract 2 from each side.}$$

$$\cos(\alpha - \beta) = \cos\alpha\cos\beta + \sin\alpha\sin\beta \quad \textit{Divide each side by }-2.$$

This is formula (2). ∎

The proof of formula (1) follows from formula (2) and the Even–Odd Identities. We use the fact that $\alpha + \beta = \alpha - (-\beta)$. Then

$$\cos(\alpha + \beta) = \cos[\alpha - (-\beta)]$$
$$= \cos\alpha\cos(-\beta) + \sin\alpha\sin(-\beta) \quad \textit{Use formula (2).}$$
$$= \cos\alpha\cos\beta - \sin\alpha\sin\beta \quad \textit{Even–Odd Identities}$$

1 Use Sum and Difference Formulas to Find Exact Values

One use of formulas (1) and (2) is to obtain the exact value of the cosine of an angle that can be expressed as the sum or difference of angles whose sine and cosine are known exactly.

EXAMPLE 1 **Using the Sum Formula to Find an Exact Value**

Find the exact value of $\cos 75°$.

Solution Since $75° = 45° + 30°$, we use formula (1) to obtain

$$\cos 75° = \cos(45° + 30°) = \cos 45° \cos 30° - \sin 45° \sin 30°$$
$$\uparrow$$
$$\textit{Formula (1)}$$

$$= \frac{\sqrt{2}}{2} \cdot \frac{\sqrt{3}}{2} - \frac{\sqrt{2}}{2} \cdot \frac{1}{2} = \frac{1}{4}(\sqrt{6} - \sqrt{2})$$

EXAMPLE 2 **Using the Difference Formula to Find an Exact Value**

Find the exact value of $\cos\dfrac{\pi}{12}$.

Solution
$$\cos\frac{\pi}{12} = \cos\left(\frac{3\pi}{12} - \frac{2\pi}{12}\right) = \cos\left(\frac{\pi}{4} - \frac{\pi}{6}\right)$$

$$= \cos\frac{\pi}{4}\cos\frac{\pi}{6} + \sin\frac{\pi}{4}\sin\frac{\pi}{6} \quad \textit{Use formula (2).}$$

$$= \frac{\sqrt{2}}{2} \cdot \frac{\sqrt{3}}{2} + \frac{\sqrt{2}}{2} \cdot \frac{1}{2} = \frac{1}{4}(\sqrt{6} + \sqrt{2})$$

Now Work PROBLEM 13

2 Use Sum and Difference Formulas to Establish Identities

Another use of formulas (1) and (2) is to establish other identities. Two important identities we conjectured earlier in Section 5.2 are given next.

$$\cos\left(\frac{\pi}{2} - \theta\right) = \sin\theta \qquad \textbf{(3a)}$$

$$\sin\left(\frac{\pi}{2} - \theta\right) = \cos\theta \qquad \textbf{(3b)}$$

Proof To prove formula (3a), we use the formula for $\cos(\alpha - \beta)$ with $\alpha = \dfrac{\pi}{2}$ and $\beta = \theta$.

$$\cos\left(\frac{\pi}{2} - \theta\right) = \cos\frac{\pi}{2}\cos\theta + \sin\frac{\pi}{2}\sin\theta$$
$$= 0\cdot\cos\theta + 1\cdot\sin\theta$$
$$= \sin\theta$$

To prove formula (3b), we make use of the identity (3a) just established.

$$\sin\left(\frac{\pi}{2} - \theta\right) = \cos\left[\frac{\pi}{2} - \left(\frac{\pi}{2} - \theta\right)\right] = \cos\theta$$
$$\uparrow$$
$$\text{Use (3a).}$$

Also, since

$$\cos\left(\frac{\pi}{2} - \theta\right) = \cos\left[-\left(\theta - \frac{\pi}{2}\right)\right] = \cos\left(\theta - \frac{\pi}{2}\right)$$
$$\uparrow$$
$$\text{Even Property of Cosine}$$

and since

$$\cos\left(\frac{\pi}{2} - \theta\right) = \sin\theta$$
$$\uparrow$$
$$\text{3(a)}$$

it follows that $\cos\left(\theta - \dfrac{\pi}{2}\right) = \sin\theta$. The graphs of $y = \cos\left(\theta - \dfrac{\pi}{2}\right)$ and $y = \sin\theta$ are identical.

Having established the identities in formulas (3a) and (3b), we now can derive the sum and difference formulas for $\sin(\alpha + \beta)$ and $\sin(\alpha - \beta)$.

Proof

$$\sin(\alpha + \beta) = \cos\left[\frac{\pi}{2} - (\alpha + \beta)\right] \qquad \text{Formula (3a)}$$

$$= \cos\left[\left(\frac{\pi}{2} - \alpha\right) - \beta\right]$$

$$= \cos\left(\frac{\pi}{2} - \alpha\right)\cos\beta + \sin\left(\frac{\pi}{2} - \alpha\right)\sin\beta \qquad \text{Formula (2)}$$

$$= \sin\alpha\cos\beta + \cos\alpha\sin\beta \qquad \text{Formulas (3a) and (3b)}$$

$$\sin(\alpha - \beta) = \sin[\alpha + (-\beta)]$$

$$= \sin\alpha\cos(-\beta) + \cos\alpha\sin(-\beta) \qquad \text{Use the sum formula for sine just obtained.}$$

$$= \sin\alpha\cos\beta + \cos\alpha(-\sin\beta) \qquad \text{Even–Odd Identities}$$

$$= \sin\alpha\cos\beta - \cos\alpha\sin\beta$$

THEOREM

Sum and Difference Formulas for the Sine Function

$$\sin(\alpha + \beta) = \sin\alpha\cos\beta + \cos\alpha\sin\beta \qquad \textbf{(4)}$$

$$\sin(\alpha - \beta) = \sin\alpha\cos\beta - \cos\alpha\sin\beta \qquad \textbf{(5)}$$

EXAMPLE 3 **Using the Sum Formula to Find an Exact Value**

Find the exact value of $\sin\dfrac{7\pi}{12}$.

Solution

$$\sin\frac{7\pi}{12} = \sin\left(\frac{3\pi}{12} + \frac{4\pi}{12}\right) = \sin\left(\frac{\pi}{4} + \frac{\pi}{3}\right)$$

$$= \sin\frac{\pi}{4}\cos\frac{\pi}{3} + \cos\frac{\pi}{4}\sin\frac{\pi}{3} \qquad \text{Formula (4)}$$

$$= \frac{\sqrt{2}}{2}\cdot\frac{1}{2} + \frac{\sqrt{2}}{2}\cdot\frac{\sqrt{3}}{2} = \frac{1}{4}\left(\sqrt{2} + \sqrt{6}\right)$$

Now Work PROBLEM 19

EXAMPLE 4 **Using the Difference Formula to Find an Exact Value**

Find the exact value of $\sin 80°\cos 20° - \cos 80°\sin 20°$.

Solution The form of the expression $\sin 80°\cos 20° - \cos 80°\sin 20°$ is that of the right side of the formula (5) for $\sin(\alpha - \beta)$ with $\alpha = 80°$ and $\beta = 20°$. That is,

$$\sin 80°\cos 20° - \cos 80°\sin 20° = \sin(80° - 20°) = \sin 60° = \frac{\sqrt{3}}{2}$$

Now Work PROBLEMS 25 AND 29

EXAMPLE 5 **Finding Exact Values**

If it is known that $\sin\alpha = \dfrac{4}{5}$, $\dfrac{\pi}{2} < \alpha < \pi$, and that $\sin\beta = -\dfrac{2}{\sqrt{5}} = -\dfrac{2\sqrt{5}}{5}$, $\pi < \beta < \dfrac{3\pi}{2}$, find the exact value of

(a) $\cos\alpha$ (b) $\cos\beta$ (c) $\cos(\alpha + \beta)$ (d) $\sin(\alpha + \beta)$

Solution

(a) Since $\sin\alpha = \dfrac{4}{5} = \dfrac{b}{r}$ and $\dfrac{\pi}{2} < \alpha < \pi$, we let $b = 4$ and $r = 5$ and place α in quadrant II. See Figure 27. Since the point $P = (a, b) = (a, 4)$ is in quadrant II, we have $a < 0$. The distance from $(a, 4)$ to $(0, 0)$ is 5, so

$$a^2 + b^2 = 25$$
$$a^2 + 16 = 25 \qquad b = 4$$
$$a^2 = 25 - 16 = 9$$
$$a = -3 \qquad a < 0$$

Then

$$\cos\alpha = \frac{a}{r} = -\frac{3}{5}$$

Figure 27

$\sin\alpha = \dfrac{4}{5}, \dfrac{\pi}{2} < \alpha < \pi$

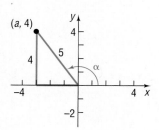

Alternatively, we can find $\cos \alpha$ using identities, as follows:

$$\cos \alpha = -\sqrt{1 - \sin^2 \alpha} = -\sqrt{1 - \frac{16}{25}} = -\sqrt{\frac{9}{25}} = -\frac{3}{5}$$

$\uparrow$ α in quadrant II, $\cos \alpha < 0$

Figure 28

$\sin \beta = \dfrac{-2}{\sqrt{5}}, \pi < \beta < \dfrac{3\pi}{2}$

$(a, -2)$

(b) Since $\sin \beta = \dfrac{-2}{\sqrt{5}} = \dfrac{b}{r}$ and $\pi < \beta < \dfrac{3\pi}{2}$, we let $b = -2$ and $r = \sqrt{5}$ and place β in quadrant III. See Figure 28. Since the point $P = (a, b) = (a, -2)$ is in quadrant III, we have that $a < 0$. The distance from $(a, -2)$ to $(0, 0)$ is $\sqrt{5}$, so

$$a^2 + b^2 = 5$$
$$a^2 + 4 = 5 \qquad b = -2$$
$$a^2 = 1$$
$$a = -1 \qquad a < 0$$

Then

$$\cos \beta = \frac{a}{r} = \frac{-1}{\sqrt{5}} = -\frac{\sqrt{5}}{5}$$

Alternatively, we can find $\cos \beta$ using identities, as follows:

$$\cos \beta = -\sqrt{1 - \sin^2 \beta} = -\sqrt{1 - \frac{4}{5}} = -\sqrt{\frac{1}{5}} = -\frac{\sqrt{5}}{5}$$

(c) Using the results found in parts (a) and (b) and formula (1), we have

$$\cos(\alpha + \beta) = \cos \alpha \cos \beta - \sin \alpha \sin \beta$$
$$= -\frac{3}{5}\left(-\frac{\sqrt{5}}{5}\right) - \frac{4}{5}\left(-\frac{2\sqrt{5}}{5}\right) = \frac{11\sqrt{5}}{25}$$

(d) $\sin(\alpha + \beta) = \sin \alpha \cos \beta + \cos \alpha \sin \beta$
$$= \frac{4}{5}\left(-\frac{\sqrt{5}}{5}\right) + \left(-\frac{3}{5}\right)\left(-\frac{2\sqrt{5}}{5}\right) = \frac{2\sqrt{5}}{25}$$

═══ **Now Work** PROBLEMS 33(a), (b), AND (c)

EXAMPLE 6 | **Establishing an Identity**

Establish the identity: $\dfrac{\cos(\alpha - \beta)}{\sin \alpha \sin \beta} = \cot \alpha \cot \beta + 1$

Solution

$$\frac{\cos(\alpha - \beta)}{\sin \alpha \sin \beta} = \frac{\cos \alpha \cos \beta + \sin \alpha \sin \beta}{\sin \alpha \sin \beta}$$

$$= \frac{\cos \alpha \cos \beta}{\sin \alpha \sin \beta} + \frac{\sin \alpha \sin \beta}{\sin \alpha \sin \beta}$$

$$= \frac{\cos \alpha}{\sin \alpha} \cdot \frac{\cos \beta}{\sin \beta} + 1$$

$$= \cot \alpha \cot \beta + 1$$

═══ **Now Work** PROBLEMS 47 AND 59

We use the identity $\tan \theta = \dfrac{\sin \theta}{\cos \theta}$ and the sum formulas for $\sin(\alpha + \beta)$ and $\cos(\alpha + \beta)$ to derive a formula for $\tan(\alpha + \beta)$.

Proof $\tan(\alpha + \beta) = \dfrac{\sin(\alpha + \beta)}{\cos(\alpha + \beta)} = \dfrac{\sin \alpha \cos \beta + \cos \alpha \sin \beta}{\cos \alpha \cos \beta - \sin \alpha \sin \beta}$

Now we divide the numerator and denominator by $\cos \alpha \cos \beta$.

$$\tan(\alpha + \beta) = \dfrac{\dfrac{\sin \alpha \cos \beta + \cos \alpha \sin \beta}{\cos \alpha \cos \beta}}{\dfrac{\cos \alpha \cos \beta - \sin \alpha \sin \beta}{\cos \alpha \cos \beta}} = \dfrac{\dfrac{\sin \alpha \cancel{\cos \beta}}{\cos \alpha \cancel{\cos \beta}} + \dfrac{\cancel{\cos \alpha} \sin \beta}{\cancel{\cos \alpha} \cos \beta}}{\dfrac{\cancel{\cos \alpha} \cancel{\cos \beta}}{\cancel{\cos \alpha} \cancel{\cos \beta}} - \dfrac{\sin \alpha \sin \beta}{\cos \alpha \cos \beta}}$$

$$= \dfrac{\dfrac{\sin \alpha}{\cos \alpha} + \dfrac{\sin \beta}{\cos \beta}}{1 - \dfrac{\sin \alpha}{\cos \alpha} \cdot \dfrac{\sin \beta}{\cos \beta}} = \dfrac{\tan \alpha + \tan \beta}{1 - \tan \alpha \tan \beta}$$ ∎

Proof We use the sum formula for $\tan(\alpha + \beta)$ and Even–Odd Properties to get the difference formula.

$$\tan(\alpha - \beta) = \tan[\alpha + (-\beta)] = \dfrac{\tan \alpha + \tan(-\beta)}{1 - \tan \alpha \tan(-\beta)} = \dfrac{\tan \alpha - \tan \beta}{1 + \tan \alpha \tan \beta}$$ ∎

We have proved the following results:

THEOREM

> **Sum and Difference Formulas for the Tangent Function**
>
> $$\tan(\alpha + \beta) = \dfrac{\tan \alpha + \tan \beta}{1 - \tan \alpha \tan \beta} \qquad (6)$$
>
> $$\tan(\alpha - \beta) = \dfrac{\tan \alpha - \tan \beta}{1 + \tan \alpha \tan \beta} \qquad (7)$$

In Words
Formula (6) states that the tangent of the sum of two angles equals the tangent of the first angle plus the tangent of the second angle, all divided by 1 minus their product.

―**Now Work** PROBLEM 33(d)

EXAMPLE 7 **Establishing an Identity**

Prove the identity: $\tan(\theta + \pi) = \tan \theta$

Solution $\tan(\theta + \pi) = \dfrac{\tan \theta + \tan \pi}{1 - \tan \theta \tan \pi} = \dfrac{\tan \theta + 0}{1 - \tan \theta \cdot 0} = \tan \theta$ ●

The result obtained in Example 7 verifies that the tangent function is periodic with period π, a fact that we discussed earlier.

EXAMPLE 8 **Establishing an Identity**

Prove the identity: $\tan\left(\theta + \dfrac{\pi}{2}\right) = -\cot \theta$

Solution We cannot use formula (6), since $\tan \dfrac{\pi}{2}$ is not defined. Instead, we proceed as follows:

WARNING Be careful when using formulas (6) and (7). These formulas can be used only for angles α and β for which $\tan \alpha$ and $\tan \beta$ are defined, that is, all angles except odd integer multiples of $\dfrac{\pi}{2}$. ∎

$$\tan\left(\theta + \dfrac{\pi}{2}\right) = \dfrac{\sin\left(\theta + \dfrac{\pi}{2}\right)}{\cos\left(\theta + \dfrac{\pi}{2}\right)} = \dfrac{\sin \theta \cos \dfrac{\pi}{2} + \cos \theta \sin \dfrac{\pi}{2}}{\cos \theta \cos \dfrac{\pi}{2} - \sin \theta \sin \dfrac{\pi}{2}}$$

$$= \dfrac{(\sin \theta)(0) + (\cos \theta)(1)}{(\cos \theta)(0) - (\sin \theta)(1)} = \dfrac{\cos \theta}{-\sin \theta} = -\cot \theta$$ ●

3 Use Sum and Difference Formulas Involving Inverse Trigonometric Functions

EXAMPLE 9

Finding the Exact Value of an Expression Involving Inverse Trigonometric Functions

Find the exact value of: $\sin\left(\cos^{-1}\dfrac{1}{2} + \sin^{-1}\dfrac{3}{5}\right)$

Solution

We seek the sine of the sum of two angles, $\alpha = \cos^{-1}\dfrac{1}{2}$ and $\beta = \sin^{-1}\dfrac{3}{5}$. Then

$$\cos\alpha = \frac{1}{2} \quad 0 \le \alpha \le \pi \quad \text{and} \quad \sin\beta = \frac{3}{5} \quad -\frac{\pi}{2} \le \beta \le \frac{\pi}{2}$$

We use Pythagorean Identities to obtain $\sin\alpha$ and $\cos\beta$. Since $\sin\alpha \ge 0$ and $\cos\beta \ge 0$ (do you know why?), we find

$$\sin\alpha = \sqrt{1 - \cos^2\alpha} = \sqrt{1 - \frac{1}{4}} = \sqrt{\frac{3}{4}} = \frac{\sqrt{3}}{2}$$

$$\cos\beta = \sqrt{1 - \sin^2\beta} = \sqrt{1 - \frac{9}{25}} = \sqrt{\frac{16}{25}} = \frac{4}{5}$$

NOTE In Example 9, we could also find $\sin\alpha$ by using $\cos\alpha = \dfrac{1}{2} = \dfrac{a}{r}$, so $a = 1$ and $r = 2$. Then $y = \sqrt{3}$ and $\sin\alpha = \dfrac{b}{r} = \dfrac{\sqrt{3}}{2}$. We could find $\cos\beta$ in a similar fashion. ∎

As a result,

$$\sin\left(\cos^{-1}\frac{1}{2} + \sin^{-1}\frac{3}{5}\right) = \sin(\alpha + \beta) = \sin\alpha\cos\beta + \cos\alpha\sin\beta$$

$$= \frac{\sqrt{3}}{2}\cdot\frac{4}{5} + \frac{1}{2}\cdot\frac{3}{5} = \frac{4\sqrt{3} + 3}{10}$$

➤ **Now Work** PROBLEM 75

EXAMPLE 10

Writing a Trigonometric Expression as an Algebraic Expression

Write $\sin(\sin^{-1} u + \cos^{-1} v)$ as an algebraic expression containing u and v (that is, without any trigonometric functions). Give the restrictions on u and v.

Solution

First, for $\sin^{-1} u$, we have $-1 \le u \le 1$, and for $\cos^{-1} v$, we have $-1 \le v \le 1$. Now let $\alpha = \sin^{-1} u$ and $\beta = \cos^{-1} v$. Then

$$\sin\alpha = u \quad -\frac{\pi}{2} \le \alpha \le \frac{\pi}{2} \quad -1 \le u \le 1$$
$$\cos\beta = v \quad 0 \le \beta \le \pi \quad -1 \le v \le 1$$

Since $-\dfrac{\pi}{2} \le \alpha \le \dfrac{\pi}{2}$, we know that $\cos\alpha \ge 0$. As a result,

$$\cos\alpha = \sqrt{1 - \sin^2\alpha} = \sqrt{1 - u^2}$$

Similarly, since $0 \le \beta \le \pi$, we know that $\sin\beta \ge 0$. Then

$$\sin\beta = \sqrt{1 - \cos^2\beta} = \sqrt{1 - v^2}$$

As a result,

$$\sin(\sin^{-1} u + \cos^{-1} v) = \sin(\alpha + \beta) = \sin\alpha\cos\beta + \cos\alpha\sin\beta$$

$$= uv + \sqrt{1 - u^2}\sqrt{1 - v^2}$$

➤ **Now Work** PROBLEM 85

4 Solve Trigonometric Equations Linear in Sine and Cosine

Sometimes it is necessary to square both sides of an equation to obtain expressions that allow the use of identities. Remember, squaring both sides of an equation may introduce extraneous solutions. As a result, apparent solutions must be checked.

| EXAMPLE 11 | Solving a Trigonometric Equation Linear in Sine and Cosine |

Solve the equation: $\sin \theta + \cos \theta = 1$, $0 \le \theta < 2\pi$

Solution A

Attempts to use available identities do not lead to equations that are easy to solve. (Try it yourself.) Given the form of this equation, we decide to square each side.

$$\sin \theta + \cos \theta = 1$$

$$(\sin \theta + \cos \theta)^2 = 1 \quad \text{Square each side.}$$

$$\sin^2 \theta + 2 \sin \theta \cos \theta + \cos^2 \theta = 1 \quad \text{Remove parentheses.}$$

$$2 \sin \theta \cos \theta = 0 \quad \sin^2 \theta + \cos^2 \theta = 1$$

$$\sin \theta \cos \theta = 0$$

Setting each factor equal to zero, we obtain

$$\sin \theta = 0 \quad \text{or} \quad \cos \theta = 0$$

The apparent solutions are

$$\theta = 0, \quad \theta = \pi, \quad \theta = \frac{\pi}{2}, \quad \theta = \frac{3\pi}{2}$$

Because we squared both sides of the original equation, we must check these apparent solutions to see if any are extraneous.

$$\theta = 0: \quad \sin 0 + \cos 0 = 0 + 1 = 1 \qquad \text{A solution}$$

$$\theta = \pi: \quad \sin \pi + \cos \pi = 0 + (-1) = -1 \qquad \text{Not a solution}$$

$$\theta = \frac{\pi}{2}: \quad \sin \frac{\pi}{2} + \cos \frac{\pi}{2} = 1 + 0 = 1 \qquad \text{A solution}$$

$$\theta = \frac{3\pi}{2}: \quad \sin \frac{3\pi}{2} + \cos \frac{3\pi}{2} = -1 + 0 = -1 \qquad \text{Not a solution}$$

The values $\theta = \pi$ and $\theta = \frac{3\pi}{2}$ are extraneous. The solution set is $\left\{ 0, \frac{\pi}{2} \right\}$.

●

Solution B

We start with the equation

$$\sin \theta + \cos \theta = 1$$

and divide each side by $\sqrt{2}$. (The reason for this choice will become apparent shortly.) Then

$$\frac{1}{\sqrt{2}} \sin \theta + \frac{1}{\sqrt{2}} \cos \theta = \frac{1}{\sqrt{2}}$$

The left side now resembles the formula for the sine of the sum of two angles, one of which is θ. The other angle is unknown (call it ϕ.) Then

$$\sin(\theta + \phi) = \sin \theta \cos \phi + \cos \theta \sin \phi = \frac{1}{\sqrt{2}} = \frac{\sqrt{2}}{2} \qquad \textbf{(8)}$$

where

$$\cos \phi = \frac{1}{\sqrt{2}} = \frac{\sqrt{2}}{2}, \qquad \sin \phi = \frac{1}{\sqrt{2}} = \frac{\sqrt{2}}{2}, \qquad 0 \le \phi < 2\pi$$

Figure 29

The angle ϕ is therefore $\dfrac{\pi}{4}$. As a result, equation (8) becomes

$$\sin\left(\theta + \frac{\pi}{4}\right) = \frac{\sqrt{2}}{2}$$

In the interval $[0, 2\pi)$, there are two angles whose sine is $\dfrac{\sqrt{2}}{2}$: $\dfrac{\pi}{4}$ and $\dfrac{3\pi}{4}$. See Figure 29. As a result,

$$\theta + \frac{\pi}{4} = \frac{\pi}{4} \quad \text{or} \quad \theta + \frac{\pi}{4} = \frac{3\pi}{4}$$

$$\theta = 0 \quad \text{or} \quad \theta = \frac{\pi}{2}$$

The solution set is $\left\{0, \dfrac{\pi}{2}\right\}$.

This second method of solution can be used to solve any linear equation in the variables $\sin\theta$ and $\cos\theta$. Let's look at an example.

EXAMPLE 12 **Solving a Trigonometric Equation Linear in $\sin\theta$ and $\cos\theta$**

Solve:

$$a\sin\theta + b\cos\theta = c \tag{9}$$

where a, b, and c are constants and either $a \neq 0$ or $b \neq 0$.

Solution We divide each side of equation (9) by $\sqrt{a^2 + b^2}$. Then

$$\frac{a}{\sqrt{a^2 + b^2}}\sin\theta + \frac{b}{\sqrt{a^2 + b^2}}\cos\theta = \frac{c}{\sqrt{a^2 + b^2}} \tag{10}$$

There is a unique angle ϕ, $0 \le \phi < 2\pi$, for which

$$\cos\phi = \frac{a}{\sqrt{a^2 + b^2}} \quad \text{and} \quad \sin\phi = \frac{b}{\sqrt{a^2 + b^2}} \tag{11}$$

Figure 30

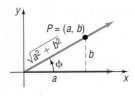

Figure 30 shows the situation for $a > 0$ and $b > 0$. Equation (10) may be written as

$$\sin\theta\cos\phi + \cos\theta\sin\phi = \frac{c}{\sqrt{a^2 + b^2}}$$

or, equivalently,

$$\sin(\theta + \phi) = \frac{c}{\sqrt{a^2 + b^2}} \tag{12}$$

where ϕ satisfies equation (11).

If $|c| > \sqrt{a^2 + b^2}$, then $\sin(\theta + \phi) > 1$ or $\sin(\theta + \phi) < -1$, and equation (12) has no solution.

If $|c| \le \sqrt{a^2 + b^2}$, then the solutions of equation (12) are

$$\theta + \phi = \sin^{-1}\frac{c}{\sqrt{a^2 + b^2}} \quad \text{or} \quad \theta + \phi = \pi - \sin^{-1}\frac{c}{\sqrt{a^2 + b^2}}$$

Because the angle ϕ is determined by equations (11), these give the solutions to equation (9).

Now Work PROBLEM 93

SUMMARY

Sum and Difference Formulas

$$\cos(\alpha + \beta) = \cos \alpha \cos \beta - \sin \alpha \sin \beta \qquad \cos(\alpha - \beta) = \cos \alpha \cos \beta + \sin \alpha \sin \beta$$

$$\sin(\alpha + \beta) = \sin \alpha \cos \beta + \cos \alpha \sin \beta \qquad \sin(\alpha - \beta) = \sin \alpha \cos \beta - \cos \alpha \sin \beta$$

$$\tan(\alpha + \beta) = \frac{\tan \alpha + \tan \beta}{1 - \tan \alpha \tan \beta} \qquad \tan(\alpha - \beta) = \frac{\tan \alpha - \tan \beta}{1 + \tan \alpha \tan \beta}$$

6.5 Assess Your Understanding

'Are You Prepared?' *Answers are given at the end of these exercises. If you get a wrong answer, read the pages listed in red.*

1. The distance d from the point $(2, -3)$ to the point $(5, 1)$ is. _____ . (pp. 3–4)

2. If $\sin \theta = \dfrac{4}{5}$ and θ is in quadrant II, then $\cos \theta =$ _____ . (pp. 421–422)

3. (a) $\sin \dfrac{\pi}{4} \cdot \cos \dfrac{\pi}{3} =$ _____ . (pp. 403–405)

   (b) $\tan \dfrac{\pi}{4} - \sin \dfrac{\pi}{6} =$ _____ . (pp. 403–405)

4. If $\sin \alpha = -\dfrac{4}{5}$, $\pi < \alpha < \dfrac{3\pi}{2}$, then $\cos \alpha =$ _____ . (pp. 421–422)

Concepts and Vocabulary

5. $\cos(\alpha + \beta) = \cos \alpha \cos \beta$ _____ $\sin \alpha \sin \beta$

6. $\sin(\alpha - \beta) = \sin \alpha \cos \beta$ _____ $\cos \alpha \sin \beta$

7. **True or False** $\sin(\alpha + \beta) = \sin \alpha + \sin \beta + 2 \sin \alpha \sin \beta$

8. **True or False** $\tan 75° = \tan 30° + \tan 45°$

9. **True or False** $\cos\left(\dfrac{\pi}{2} - \theta\right) = \cos \theta$

10. **True or False** If $f(x) = \sin x$ and $g(x) = \cos x$, then $g(\alpha + \beta) = g(\alpha)g(\beta) - f(\alpha)f(\beta)$

Skill Building

In Problems 11–22, find the exact value of each expression.

11. $\sin \dfrac{5\pi}{12}$

12. $\sin \dfrac{\pi}{12}$

13. $\cos \dfrac{7\pi}{12}$

14. $\tan \dfrac{7\pi}{12}$

15. $\cos 165°$

16. $\sin 105°$

17. $\tan 15°$

18. $\tan 195°$

19. $\sin \dfrac{17\pi}{12}$

20. $\tan \dfrac{19\pi}{12}$

21. $\sec\left(-\dfrac{\pi}{12}\right)$

22. $\cot\left(-\dfrac{5\pi}{12}\right)$

In Problems 23–32, find the exact value of each expression.

23. $\sin 20° \cos 10° + \cos 20° \sin 10°$

24. $\sin 20° \cos 80° - \cos 20° \sin 80°$

25. $\cos 70° \cos 20° - \sin 70° \sin 20°$

26. $\cos 40° \cos 10° + \sin 40° \sin 10°$

27. $\dfrac{\tan 20° + \tan 25°}{1 - \tan 20° \tan 25°}$

28. $\dfrac{\tan 40° - \tan 10°}{1 + \tan 40° \tan 10°}$

29. $\sin \dfrac{\pi}{12} \cos \dfrac{7\pi}{12} - \cos \dfrac{\pi}{12} \sin \dfrac{7\pi}{12}$

30. $\cos \dfrac{5\pi}{12} \cos \dfrac{7\pi}{12} - \sin \dfrac{5\pi}{12} \sin \dfrac{7\pi}{12}$

31. $\cos \dfrac{\pi}{12} \cos \dfrac{5\pi}{12} + \sin \dfrac{5\pi}{12} \sin \dfrac{\pi}{12}$

32. $\sin \dfrac{\pi}{18} \cos \dfrac{5\pi}{18} + \cos \dfrac{\pi}{18} \sin \dfrac{5\pi}{18}$

In Problems 33–38, find the exact value of each of the following under the given conditions:

(a) $\sin(\alpha + \beta)$ (b) $\cos(\alpha + \beta)$ (c) $\sin(\alpha - \beta)$ (d) $\tan(\alpha - \beta)$

33. $\sin \alpha = \dfrac{3}{5}$, $0 < \alpha < \dfrac{\pi}{2}$; $\cos \beta = \dfrac{2\sqrt{5}}{5}$, $-\dfrac{\pi}{2} < \beta < 0$

34. $\cos \alpha = \dfrac{\sqrt{5}}{5}$, $0 < \alpha < \dfrac{\pi}{2}$; $\sin \beta = -\dfrac{4}{5}$, $-\dfrac{\pi}{2} < \beta < 0$

35. $\tan \alpha = -\dfrac{4}{3}$, $\dfrac{\pi}{2} < \alpha < \pi$; $\cos \beta = \dfrac{1}{2}$, $0 < \beta < \dfrac{\pi}{2}$

36. $\tan \alpha = \dfrac{5}{12}$, $\pi < \alpha < \dfrac{3\pi}{2}$; $\sin \beta = -\dfrac{1}{2}$, $\pi < \beta < \dfrac{3\pi}{2}$

37. $\sin \alpha = \dfrac{5}{13}$, $-\dfrac{3\pi}{2} < \alpha < -\pi$; $\tan \beta = -\sqrt{3}$, $\dfrac{\pi}{2} < \beta < \pi$

38. $\cos \alpha = \dfrac{1}{2}$, $-\dfrac{\pi}{2} < \alpha < 0$; $\sin \beta = \dfrac{1}{3}$, $0 < \beta < \dfrac{\pi}{2}$

39. If $\sin \theta = \dfrac{1}{3}$, θ in quadrant II, find the exact value of:

(a) $\cos \theta$

(b) $\sin\left(\theta + \dfrac{\pi}{6}\right)$

(c) $\cos\left(\theta - \dfrac{\pi}{3}\right)$

(d) $\tan\left(\theta + \dfrac{\pi}{4}\right)$

40. If $\cos \theta = \dfrac{1}{4}$, θ in quadrant IV, find the exact value of:

(a) $\sin \theta$

(b) $\sin\left(\theta - \dfrac{\pi}{6}\right)$

(c) $\cos\left(\theta + \dfrac{\pi}{3}\right)$

(d) $\tan\left(\theta - \dfrac{\pi}{4}\right)$

In Problems 41–46, use the figures to evaluate each function if $f(x) = \sin x$, $g(x) = \cos x$, *and* $h(x) = \tan x$.

41. $f(\alpha + \beta)$

42. $g(\alpha + \beta)$

43. $g(\alpha - \beta)$

44. $f(\alpha - \beta)$

45. $h(\alpha + \beta)$

46. $h(\alpha - \beta)$

In Problems 47–72, establish each identity.

47. $\sin\left(\dfrac{\pi}{2} + \theta\right) = \cos \theta$

48. $\cos\left(\dfrac{\pi}{2} + \theta\right) = -\sin \theta$

49. $\sin(\pi - \theta) = \sin \theta$

50. $\cos(\pi - \theta) = -\cos \theta$

51. $\sin(\pi + \theta) = -\sin \theta$

52. $\cos(\pi + \theta) = -\cos \theta$

53. $\tan(\pi - \theta) = -\tan \theta$

54. $\tan(2\pi - \theta) = -\tan \theta$

55. $\sin\left(\dfrac{3\pi}{2} + \theta\right) = -\cos \theta$

56. $\cos\left(\dfrac{3\pi}{2} + \theta\right) = \sin \theta$

57. $\sin(\alpha + \beta) + \sin(\alpha - \beta) = 2\sin \alpha \cos \beta$

58. $\cos(\alpha + \beta) + \cos(\alpha - \beta) = 2\cos \alpha \cos \beta$

59. $\dfrac{\sin(\alpha + \beta)}{\sin \alpha \cos \beta} = 1 + \cot \alpha \tan \beta$

60. $\dfrac{\sin(\alpha + \beta)}{\cos \alpha \cos \beta} = \tan \alpha + \tan \beta$

61. $\dfrac{\cos(\alpha + \beta)}{\cos \alpha \cos \beta} = 1 - \tan \alpha \tan \beta$

62. $\dfrac{\cos(\alpha - \beta)}{\sin \alpha \cos \beta} = \cot \alpha + \tan \beta$

63. $\dfrac{\sin(\alpha + \beta)}{\sin(\alpha - \beta)} = \dfrac{\tan \alpha + \tan \beta}{\tan \alpha - \tan \beta}$

64. $\dfrac{\cos(\alpha + \beta)}{\cos(\alpha - \beta)} = \dfrac{1 - \tan \alpha \tan \beta}{1 + \tan \alpha \tan \beta}$

65. $\cot(\alpha + \beta) = \dfrac{\cot \alpha \cot \beta - 1}{\cot \beta + \cot \alpha}$

66. $\cot(\alpha - \beta) = \dfrac{\cot \alpha \cot \beta + 1}{\cot \beta - \cot \alpha}$

67. $\sec(\alpha + \beta) = \dfrac{\csc \alpha \csc \beta}{\cot \alpha \cot \beta - 1}$

68. $\sec(\alpha - \beta) = \dfrac{\sec \alpha \sec \beta}{1 + \tan \alpha \tan \beta}$

69. $\sin(\alpha - \beta)\sin(\alpha + \beta) = \sin^2 \alpha - \sin^2 \beta$

70. $\cos(\alpha - \beta)\cos(\alpha + \beta) = \cos^2 \alpha - \sin^2 \beta$

71. $\sin(\theta + k\pi) = (-1)^k \sin \theta$, k any integer

72. $\cos(\theta + k\pi) = (-1)^k \cos \theta$, k any integer

In Problems 73–84, find the exact value of each expression.

73. $\sin\left(\sin^{-1}\dfrac{1}{2} + \cos^{-1} 0\right)$

74. $\sin\left(\sin^{-1}\dfrac{\sqrt{3}}{2} + \cos^{-1} 1\right)$

75. $\sin\left[\sin^{-1}\dfrac{3}{5} - \cos^{-1}\left(-\dfrac{4}{5}\right)\right]$

76. $\sin\left[\sin^{-1}\left(-\dfrac{4}{5}\right) - \tan^{-1}\dfrac{3}{4}\right]$

77. $\cos\left(\tan^{-1}\dfrac{4}{3} + \cos^{-1}\dfrac{5}{13}\right)$

78. $\cos\left[\tan^{-1}\dfrac{5}{12} - \sin^{-1}\left(-\dfrac{3}{5}\right)\right]$

79. $\cos\left(\sin^{-1}\dfrac{5}{13} - \tan^{-1}\dfrac{3}{4}\right)$

80. $\cos\left(\tan^{-1}\dfrac{4}{3} + \cos^{-1}\dfrac{12}{13}\right)$

81. $\tan\left(\sin^{-1}\dfrac{3}{5} + \dfrac{\pi}{6}\right)$

82. $\tan\left(\dfrac{\pi}{4} - \cos^{-1}\dfrac{3}{5}\right)$

83. $\tan\left(\sin^{-1}\dfrac{4}{5} + \cos^{-1} 1\right)$

84. $\tan\left(\cos^{-1}\dfrac{4}{5} + \sin^{-1} 1\right)$

In Problems 85–90, write each trigonometric expression as an algebraic expression containing u and v. Give the restrictions required on u and v.

85. $\cos(\cos^{-1} u + \sin^{-1} v)$

86. $\sin(\sin^{-1} u - \cos^{-1} v)$

87. $\sin(\tan^{-1} u - \sin^{-1} v)$

88. $\cos(\tan^{-1} u + \tan^{-1} v)$

89. $\tan(\sin^{-1} u - \cos^{-1} v)$

90. $\sec(\tan^{-1} u + \cos^{-1} v)$

In Problems 91–96, solve each equation on the interval $0 \le \theta < 2\pi$.

91. $\sin \theta - \sqrt{3} \cos \theta = 1$

92. $\sqrt{3} \sin \theta + \cos \theta = 1$

93. $\sin \theta + \cos \theta = \sqrt{2}$

94. $\sin \theta - \cos \theta = -\sqrt{2}$

95. $\tan \theta + \sqrt{3} = \sec \theta$

96. $\cot \theta + \csc \theta = -\sqrt{3}$

Applications and Extensions

97. Show that $\sin^{-1} v + \cos^{-1} v = \dfrac{\pi}{2}$.

98. Show that $\tan^{-1} v + \cot^{-1} v = \dfrac{\pi}{2}$.

99. Show that $\tan^{-1}\left(\dfrac{1}{v}\right) = \dfrac{\pi}{2} - \tan^{-1} v$, if $v > 0$.

100. Show that $\cot^{-1} e^v = \tan^{-1} e^{-v}$.

101. Show that $\sin(\sin^{-1} v + \cos^{-1} v) = 1$.

102. Show that $\cos(\sin^{-1} v + \cos^{-1} v) = 0$.

103. Calculus Show that the difference quotient for $f(x) = \sin x$ is given by

$$\frac{f(x + h) - f(x)}{h} = \frac{\sin(x + h) - \sin x}{h}$$

$$= \cos x \cdot \frac{\sin h}{h} - \sin x \cdot \frac{1 - \cos h}{h}$$

104. Calculus Show that the difference quotient for $f(x) = \cos x$ is given by

$$\frac{f(x + h) - f(x)}{h} = \frac{\cos(x + h) - \cos x}{h}$$

$$= -\sin x \cdot \frac{\sin h}{h} - \cos x \cdot \frac{1 - \cos h}{h}$$

105. One, Two, Three
(a) Show that $\tan(\tan^{-1} 1 + \tan^{-1} 2 + \tan^{-1} 3) = 0$.
(b) Conclude from part (a) that

$$\tan^{-1} 1 + \tan^{-1} 2 + \tan^{-1} 3 = \pi$$

Source: College Mathematics Journal, Vol. 37, No. 3, May 2006

106. Electric Power In an alternating current (ac) circuit, the instantaneous power p at time t is given by

$$p(t) = V_m I_m \cos \phi \sin^2(\omega t) - V_m I_m \sin \phi \sin(\omega t) \cos(\omega t)$$

Show that this is equivalent to

$$p(t) = V_m I_m \sin(\omega t) \sin(\omega t - \phi)$$

Source: HyperPhysics, hosted by Georgia State University

107. Geometry: Angle between Two Lines Let L_1 and L_2 denote two nonvertical intersecting lines, and let θ denote the acute angle between L_1 and L_2 (see the figure). Show that

$$\tan \theta = \frac{m_2 - m_1}{1 + m_1 m_2}$$

where m_1 and m_2 are the slopes of L_1 and L_2, respectively. [**Hint:** Use the facts that $\tan \theta_1 = m_1$ and $\tan \theta_2 = m_2$.]

108. If $\alpha + \beta + \gamma = 180°$ and

$$\cot \theta = \cot \alpha + \cot \beta + \cot \gamma, \quad 0 < \theta < 90°$$

show that

$$\sin^3 \theta = \sin(\alpha - \theta) \sin(\beta - \theta) \sin(\gamma - \theta)$$

109. If $\tan \alpha = x + 1$ and $\tan \beta = x - 1$, show that

$$2 \cot(\alpha - \beta) = x^2$$

Explaining Concepts: Discussion and Writing

110. Discuss the following derivation:

$$\tan\left(\theta + \frac{\pi}{2}\right) = \frac{\tan \theta + \tan \dfrac{\pi}{2}}{1 - \tan \theta \tan \dfrac{\pi}{2}} = \frac{\dfrac{\tan \theta}{\tan \dfrac{\pi}{2}} + 1}{\dfrac{1}{\tan \dfrac{\pi}{2}} - \tan \theta} = \frac{0 + 1}{0 - \tan \theta} = \frac{1}{-\tan \theta} = -\cot \theta$$

Can you justify each step?

111. Explain why formula (7) cannot be used to show that

$$\tan\left(\frac{\pi}{2} - \theta\right) = \cot\theta$$

Establish this identity by using formulas (3a) and (3b).

'Are You Prepared?' Answers

1. 5   **2.** $-\dfrac{3}{5}$ **3.** (a) $\dfrac{\sqrt{2}}{4}$ (b) $\dfrac{1}{2}$ **4.** $-\dfrac{3}{5}$

6.6 Double-angle and Half-angle Formulas

OBJECTIVES **1** Use Double-angle Formulas to Find Exact Values (p. 526)
   2 Use Double-angle Formulas to Establish Identities (p. 527)
   3 Use Half-angle Formulas to Find Exact Values (p. 530)

In this section we derive formulas for $\sin(2\theta)$, $\cos(2\theta)$, $\sin\left(\dfrac{1}{2}\theta\right)$, and $\cos\left(\dfrac{1}{2}\theta\right)$ in terms of $\sin\theta$ and $\cos\theta$. They are derived using the sum formulas.

In the sum formulas for $\sin(\alpha + \beta)$ and $\cos(\alpha + \beta)$, let $\alpha = \beta = \theta$. Then

$$\sin(\alpha + \beta) = \sin\alpha\cos\beta + \cos\alpha\sin\beta$$
$$\sin(\theta + \theta) = \sin\theta\cos\theta + \cos\theta\sin\theta$$
$$\sin(2\theta) = 2\sin\theta\cos\theta$$

and

$$\cos(\alpha + \beta) = \cos\alpha\cos\beta - \sin\alpha\sin\beta$$
$$\cos(\theta + \theta) = \cos\theta\cos\theta - \sin\theta\sin\theta$$
$$\cos(2\theta) = \cos^2\theta - \sin^2\theta$$

An application of the Pythagorean Identity $\sin^2\theta + \cos^2\theta = 1$ results in two other ways to express $\cos(2\theta)$.

$$\cos(2\theta) = \cos^2\theta - \sin^2\theta = (1 - \sin^2\theta) - \sin^2\theta = 1 - 2\sin^2\theta$$

and

$$\cos(2\theta) = \cos^2\theta - \sin^2\theta = \cos^2\theta - (1 - \cos^2\theta) = 2\cos^2\theta - 1$$

We have established the following **Double-angle Formulas:**

THEOREM   **Double-angle Formulas**

$\sin(2\theta) = 2\sin\theta\cos\theta$	**(1)**
$\cos(2\theta) = \cos^2\theta - \sin^2\theta$	**(2)**
$\cos(2\theta) = 1 - 2\sin^2\theta$	**(3)**
$\cos(2\theta) = 2\cos^2\theta - 1$	**(4)**

1 Use Double-angle Formulas to Find Exact Values

EXAMPLE 1   **Finding Exact Values Using the Double-angle Formula**

If $\sin\theta = \dfrac{3}{5}$, $\dfrac{\pi}{2} < \theta < \pi$, find the exact value of:

(a) $\sin(2\theta)$ (b) $\cos(2\theta)$

Solution   (a) Because $\sin(2\theta) = 2\sin\theta\cos\theta$ and we already know that $\sin\theta = \dfrac{3}{5}$, we only need to find $\cos\theta$. Since $\sin\theta = \dfrac{3}{5} = \dfrac{b}{r}$, $\dfrac{\pi}{2} < \theta < \pi$, we let $b = 3$ and $r = 5$ and place θ in quadrant II. See Figure 31. Since the point $P = (a, b) = (a, 3)$ is in quadrant II, we know that $a < 0$. The distance from $(a, 3)$ to $(0, 0)$ is 5, so

Figure 31

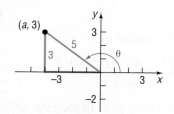

$$a^2 + 3^2 = 5^2 \qquad a^2 + b^2 = r^2$$

$$a^2 = 25 - 9 = 16$$

$$a = -4 \qquad a < 0$$

We find that $\cos\theta = \dfrac{a}{r} = \dfrac{-4}{5}$. Now we use formula (1) to obtain

$$\sin(2\theta) = 2\sin\theta\cos\theta = 2\left(\frac{3}{5}\right)\left(-\frac{4}{5}\right) = -\frac{24}{25}$$

(b) Because we are given $\sin\theta = \dfrac{3}{5}$, it is easiest to use formula (3) to get $\cos(2\theta)$.

$$\cos(2\theta) = 1 - 2\sin^2\theta = 1 - 2\left(\frac{9}{25}\right) = 1 - \frac{18}{25} = \frac{7}{25}$$

WARNING In finding $\cos(2\theta)$ in Example 1(b), we chose to use a version of the Double-angle Formula, formula (3). Note that we are unable to use the Pythagorean Identity $\cos(2\theta) = \pm\sqrt{1 - \sin^2(2\theta)}$, with $\sin(2\theta) = -\dfrac{24}{25}$, because we have no way of knowing which sign to choose. ∎

Now Work   PROBLEMS 7(a) AND (b)

2 Use Double-angle Formulas to Establish Identities

EXAMPLE 2   **Establishing Identities**

(a) Develop a formula for $\tan(2\theta)$ in terms of $\tan\theta$.
(b) Develop a formula for $\sin(3\theta)$ in terms of $\sin\theta$ and $\cos\theta$.

Solution   (a) In the sum formula for $\tan(\alpha + \beta)$, let $\alpha = \beta = \theta$. Then

$$\tan(\alpha + \beta) = \frac{\tan\alpha + \tan\beta}{1 - \tan\alpha\tan\beta}$$

$$\tan(\theta + \theta) = \frac{\tan\theta + \tan\theta}{1 - \tan\theta\tan\theta}$$

$$\tan(2\theta) = \frac{2\tan\theta}{1 - \tan^2\theta} \qquad \textbf{(5)}$$

(b) To get a formula for $\sin(3\theta)$, we use the sum formula and write 3θ as $2\theta + \theta$.

$$\sin(3\theta) = \sin(2\theta + \theta) = \sin(2\theta)\cos\theta + \cos(2\theta)\sin\theta$$

Now use the Double-angle Formulas to get

$$\sin(3\theta) = (2 \sin \theta \cos \theta)(\cos \theta) + (\cos^2 \theta - \sin^2 \theta)(\sin \theta)$$
$$= 2 \sin \theta \cos^2 \theta + \sin \theta \cos^2 \theta - \sin^3 \theta$$
$$= 3 \sin \theta \cos^2 \theta - \sin^3 \theta$$

The formula obtained in Example 2(b) can also be written as

$$\sin(3\theta) = 3 \sin \theta \cos^2 \theta - \sin^3 \theta = 3 \sin \theta(1 - \sin^2 \theta) - \sin^3 \theta$$
$$= 3 \sin \theta - 4 \sin^3 \theta$$

That is, $\sin(3\theta)$ is a third-degree polynomial in the variable $\sin \theta$. In fact, $\sin(n\theta)$, n a positive odd integer, can always be written as a polynomial of degree n in the variable $\sin \theta$.*

━━━▶ **Now Work** PROBLEM 65

By rearranging the Double-angle Formulas (3) and (4), we obtain other formulas that we will use later in this section.

We begin with formula (3) and proceed to solve for $\sin^2 \theta$.

$$\cos(2\theta) = 1 - 2 \sin^2 \theta$$
$$2 \sin^2 \theta = 1 - \cos(2\theta)$$

$$\sin^2 \theta = \frac{1 - \cos(2\theta)}{2} \tag{6}$$

Similarly, using formula (4), we proceed to solve for $\cos^2 \theta$.

$$\cos(2\theta) = 2 \cos^2 \theta - 1$$
$$2 \cos^2 \theta = 1 + \cos(2\theta)$$

$$\cos^2 \theta = \frac{1 + \cos(2\theta)}{2} \tag{7}$$

Formulas (6) and (7) can be used to develop a formula for $\tan^2 \theta$.

$$\tan^2 \theta = \frac{\sin^2 \theta}{\cos^2 \theta} = \frac{\dfrac{1 - \cos(2\theta)}{2}}{\dfrac{1 + \cos(2\theta)}{2}}$$

$$\tan^2 \theta = \frac{1 - \cos(2\theta)}{1 + \cos(2\theta)} \tag{8}$$

Formulas (6) through (8) do not have to be memorized since their derivations are so straightforward.

Formulas (6) and (7) are important in calculus. The next example illustrates a problem that arises in calculus requiring the use of formula (7).

EXAMPLE 3   **Establishing an Identity**

Write an equivalent expression for $\cos^4 \theta$ that does not involve any powers of sine or cosine greater than 1.

---

*Because of the work done by P. L. Chebyshëv, these polynomials are sometimes called *Chebyshëv polynomials*.

Solution   The idea here is to apply formula (7) twice.

$$\cos^4 \theta = (\cos^2 \theta)^2 = \left(\frac{1 + \cos(2\theta)}{2} \right)^2 \qquad \text{Formula (7)}$$

$$= \frac{1}{4} [1 + 2\cos(2\theta) + \cos^2(2\theta)]$$

$$= \frac{1}{4} + \frac{1}{2}\cos(2\theta) + \frac{1}{4}\cos^2(2\theta)$$

$$= \frac{1}{4} + \frac{1}{2}\cos(2\theta) + \frac{1}{4}\left\{ \frac{1 + \cos[2(2\theta)]}{2} \right\} \qquad \text{Formula (7)}$$

$$= \frac{1}{4} + \frac{1}{2}\cos(2\theta) + \frac{1}{8}[1 + \cos(4\theta)]$$

$$= \frac{3}{8} + \frac{1}{2}\cos(2\theta) + \frac{1}{8}\cos(4\theta)$$

Now Work PROBLEM 41

EXAMPLE 4   **Solving a Trigonometric Equation Using Identities**

Solve the equation: $\sin \theta \cos \theta = -\dfrac{1}{2}, \quad 0 \leq \theta < 2\pi$

Solution   The left side of the given equation is in the form of the Double-angle Formula $2 \sin \theta \cos \theta = \sin(2\theta)$, except for a factor of 2. We multiply each side by 2.

$$\sin \theta \cos \theta = -\frac{1}{2}$$

$$2 \sin \theta \cos \theta = -1 \qquad \text{Multiply each side by 2.}$$

$$\sin(2\theta) = -1 \qquad \text{Double-angle Formula}$$

The argument here is 2θ. So we need to write all the solutions of this equation and then list those that are in the interval $[0, 2\pi)$. Because $\sin\left(\dfrac{3\pi}{2} + 2\pi k \right) = -1$, for any integer k we have

$$2\theta = \frac{3\pi}{2} + 2k\pi \qquad k \text{ any integer}$$

$$\theta = \frac{3\pi}{4} + k\pi$$

$$\underset{\substack{\uparrow \\ k = -1}}{\theta = \frac{3\pi}{4} + (-1)\pi = -\frac{\pi}{4}}, \quad \underset{\substack{\uparrow \\ k = 0}}{\theta = \frac{3\pi}{4} + (0)\pi = \frac{3\pi}{4}}, \quad \underset{\substack{\uparrow \\ k = 1}}{\theta = \frac{3\pi}{4} + (1)\pi = \frac{7\pi}{4}}, \quad \underset{\substack{\uparrow \\ k = 2}}{\theta = \frac{3\pi}{4} + (2)\pi = \frac{11\pi}{4}}$$

The solutions in the interval $[0, 2\pi)$ are

$$\theta = \frac{3\pi}{4}, \qquad \theta = \frac{7\pi}{4}$$

The solution set is $\left\{ \dfrac{3\pi}{4}, \dfrac{7\pi}{4} \right\}$.

Now Work PROBLEM 69

EXAMPLE 5   **Projectile Motion**

Figure 32

An object is propelled upward at an angle θ to the horizontal with an initial velocity of v_0 feet per second. See Figure 32. If air resistance is ignored, the **range** R, the horizontal distance that the object travels, is given by the function

$$R(\theta) = \frac{1}{16} v_0^2 \sin \theta \cos \theta$$

(a) Show that $R(\theta) = \dfrac{1}{32} v_0^2 \sin(2\theta)$.

(b) Find the angle θ for which R is a maximum.

Solution (a) We rewrite the given expression for the range using the Double-angle Formula $\sin(2\theta) = 2\sin\theta\cos\theta$. Then

$$R(\theta) = \frac{1}{16}v_0^2\sin\theta\cos\theta = \frac{1}{16}v_0^2\frac{2\sin\theta\cos\theta}{2} = \frac{1}{32}v_0^2\sin(2\theta)$$

(b) In this form, the largest value for the range R can be found. For a fixed initial speed v_0, the angle θ of inclination to the horizontal determines the value of R. Since the largest value of a sine function is 1, occurring when the argument 2θ is $90°$, it follows that for maximum R we must have

$$2\theta = 90°$$
$$\theta = 45°$$

An inclination to the horizontal of $45°$ results in the maximum range.

3 Use Half-angle Formulas to Find Exact Values

Another important use of formulas (6) through (8) is to prove the *Half-angle Formulas*. In formulas (6) through (8), let $\theta = \dfrac{\alpha}{2}$. Then

$$\sin^2\frac{\alpha}{2} = \frac{1 - \cos\alpha}{2} \qquad \cos^2\frac{\alpha}{2} = \frac{1 + \cos\alpha}{2} \qquad \tan^2\frac{\alpha}{2} = \frac{1 - \cos\alpha}{1 + \cos\alpha} \qquad \textbf{(9)}$$

 The identities in box (9) will prove useful in integral calculus.

If we solve for the trigonometric functions on the left sides of equations (9), we obtain the Half-angle Formulas.

THEOREM

Half-angle Formulas

$$\sin\frac{\alpha}{2} = \pm\sqrt{\frac{1 - \cos\alpha}{2}} \qquad\qquad \textbf{(10a)}$$

$$\cos\frac{\alpha}{2} = \pm\sqrt{\frac{1 + \cos\alpha}{2}} \qquad\qquad \textbf{(10b)}$$

$$\tan\frac{\alpha}{2} = \pm\sqrt{\frac{1 - \cos\alpha}{1 + \cos\alpha}} \qquad\qquad \textbf{(10c)}$$

where the $+$ or $-$ sign is determined by the quadrant of the angle $\dfrac{\alpha}{2}$.

EXAMPLE 6 **Finding Exact Values Using Half-angle Formulas**

Use a Half-angle Formula to find the exact value of:
(a) $\cos 15°$ (b) $\sin(-15°)$

Solution (a) Because $15° = \dfrac{30°}{2}$, we can use the Half-angle Formula for $\cos\dfrac{\alpha}{2}$ with $\alpha = 30°$. Also, because $15°$ is in quadrant I, $\cos 15° > 0$, we choose the $+$ sign in using formula (10b):

$$\cos 15° = \cos\frac{30°}{2} = \sqrt{\frac{1 + \cos 30°}{2}}$$

$$= \sqrt{\frac{1 + \sqrt{3}/2}{2}} = \sqrt{\frac{2 + \sqrt{3}}{4}} = \frac{\sqrt{2 + \sqrt{3}}}{2}$$

(b) We use the fact that $\sin(-15°) = -\sin 15°$ and then apply formula (10a).

$$\sin(-15°) = -\sin\frac{30°}{2} = -\sqrt{\frac{1 - \cos 30°}{2}}$$

$$= -\sqrt{\frac{1 - \sqrt{3}/2}{2}} = -\sqrt{\frac{2 - \sqrt{3}}{4}} = -\frac{\sqrt{2 - \sqrt{3}}}{2}$$

It is interesting to compare the answer found in Example 6(a) with the answer to Example 2 of Section 6.5. There we calculated

$$\cos\frac{\pi}{12} = \cos 15° = \frac{1}{4}\left(\sqrt{6} + \sqrt{2}\right)$$

Based on this and the result of Example 6(a), we conclude that

$$\frac{1}{4}\left(\sqrt{6} + \sqrt{2}\right) \quad \text{and} \quad \frac{\sqrt{2 + \sqrt{3}}}{2}$$

are equal. (Since each expression is positive, you can verify this equality by squaring each expression.) Two very different looking, yet correct, answers can be obtained, depending on the approach taken to solve a problem.

Now Work PROBLEM 19

EXAMPLE 7

Finding Exact Values Using Half-angle Formulas

If $\cos\alpha = -\dfrac{3}{5}$, $\pi < \alpha < \dfrac{3\pi}{2}$, find the exact value of:

(a) $\sin\dfrac{\alpha}{2}$ (b) $\cos\dfrac{\alpha}{2}$ (c) $\tan\dfrac{\alpha}{2}$

Solution

First, we observe that if $\pi < \alpha < \dfrac{3\pi}{2}$ then $\dfrac{\pi}{2} < \dfrac{\alpha}{2} < \dfrac{3\pi}{4}$. As a result, $\dfrac{\alpha}{2}$ lies in quadrant II.

(a) Because $\dfrac{\alpha}{2}$ lies in quadrant II, $\sin\dfrac{\alpha}{2} > 0$, so we use the $+$ sign in formula (10a) to get

$$\sin\frac{\alpha}{2} = \sqrt{\frac{1 - \cos\alpha}{2}} = \sqrt{\frac{1 - \left(-\frac{3}{5}\right)}{2}}$$

$$= \sqrt{\frac{\frac{8}{5}}{2}} = \sqrt{\frac{4}{5}} = \frac{2}{\sqrt{5}} = \frac{2\sqrt{5}}{5}$$

(b) Because $\dfrac{\alpha}{2}$ lies in quadrant II, $\cos\dfrac{\alpha}{2} < 0$, so we use the $-$ sign in formula (10b) to get

$$\cos\frac{\alpha}{2} = -\sqrt{\frac{1 + \cos\alpha}{2}} = -\sqrt{\frac{1 + \left(-\frac{3}{5}\right)}{2}}$$

$$= -\sqrt{\frac{\frac{2}{5}}{2}} = -\frac{1}{\sqrt{5}} = -\frac{\sqrt{5}}{5}$$

(c) Because $\dfrac{\alpha}{2}$ lies in quadrant II, $\tan\dfrac{\alpha}{2} < 0$, so we use the $-$ sign in formula (10c) to get

$$\tan\frac{\alpha}{2} = -\sqrt{\frac{1-\cos\alpha}{1+\cos\alpha}} = -\sqrt{\frac{1-\left(-\dfrac{3}{5}\right)}{1+\left(-\dfrac{3}{5}\right)}} = -\sqrt{\frac{\dfrac{8}{5}}{\dfrac{2}{5}}} = -2$$

Another way to solve Example 7(c) is to use the results of parts (a) and (b).

$$\tan\frac{\alpha}{2} = \frac{\sin\dfrac{\alpha}{2}}{\cos\dfrac{\alpha}{2}} = \frac{\dfrac{2\sqrt{5}}{5}}{-\dfrac{\sqrt{5}}{5}} = -2$$

─ **Now Work** PROBLEMS 7(c) AND (d)

There is a formula for $\tan\dfrac{\alpha}{2}$ that does not contain $+$ and $-$ signs, making it more useful than Formula 10(c). To derive it, we use the formulas

$$1 - \cos\alpha = 2\sin^2\frac{\alpha}{2} \qquad \text{Formula (9)}$$

and

$$\sin\alpha = \sin\left[2\left(\frac{\alpha}{2}\right)\right] = 2\sin\frac{\alpha}{2}\cos\frac{\alpha}{2} \qquad \text{Double-angle Formula}$$

Then

$$\frac{1-\cos\alpha}{\sin\alpha} = \frac{2\sin^2\dfrac{\alpha}{2}}{2\sin\dfrac{\alpha}{2}\cos\dfrac{\alpha}{2}} = \frac{\sin\dfrac{\alpha}{2}}{\cos\dfrac{\alpha}{2}} = \tan\frac{\alpha}{2}$$

Since it also can be shown that

$$\frac{1-\cos\alpha}{\sin\alpha} = \frac{\sin\alpha}{1+\cos\alpha}$$

we have the following two Half-angle Formulas:

Half-angle Formulas for $\tan\dfrac{\alpha}{2}$

$$\tan\frac{\alpha}{2} = \frac{1-\cos\alpha}{\sin\alpha} = \frac{\sin\alpha}{1+\cos\alpha} \qquad \textbf{(11)}$$

With this formula, the solution to Example 7(c) can be obtained as follows:

$$\cos\alpha = -\frac{3}{5} \quad \pi < \alpha < \frac{3\pi}{2}$$

$$\sin\alpha = -\sqrt{1-\cos^2\alpha} = -\sqrt{1-\frac{9}{25}} = -\sqrt{\frac{16}{25}} = -\frac{4}{5}$$

Then, by equation (11),

$$\tan\frac{\alpha}{2} = \frac{1-\cos\alpha}{\sin\alpha} = \frac{1-\left(-\dfrac{3}{5}\right)}{-\dfrac{4}{5}} = \frac{\dfrac{8}{5}}{-\dfrac{4}{5}} = -2$$

6.6 Assess Your Understanding

Concepts and Vocabulary

1. $\cos(2\theta) = \cos^2 \theta -$ _____ $=$ _____ $- 1 = 1 -$ _____.

2. $\sin^2 \dfrac{\theta}{2} = \dfrac{\rule{2cm}{0.4pt}}{2}$.

3. $\tan \dfrac{\theta}{2} = \dfrac{1 - \cos \theta}{\rule{1cm}{0.4pt}}$.

4. *True or False* $\tan(2\theta) = \dfrac{2 \tan \theta}{1 - \tan^2 \theta}$

5. *True or False* $\sin(2\theta)$ has two equivalent forms:
$$2 \sin \theta \cos \theta \quad \text{and} \quad \sin^2 \theta - \cos^2 \theta$$

6. *True or False* $\tan(2\theta) + \tan(2\theta) = \tan(4\theta)$

Skill Building

In Problems 7–18, use the information given about the angle θ, $0 \leq \theta < 2\pi$, to find the exact value of

    (a) $\sin(2\theta)$ (b) $\cos(2\theta)$ (c) $\sin \dfrac{\theta}{2}$ (d) $\cos \dfrac{\theta}{2}$

7. $\sin \theta = \dfrac{3}{5}, \quad 0 < \theta < \dfrac{\pi}{2}$

8. $\cos \theta = \dfrac{3}{5}, \quad 0 < \theta < \dfrac{\pi}{2}$

9. $\tan \theta = \dfrac{4}{3}, \quad \pi < \theta < \dfrac{3\pi}{2}$

10. $\tan \theta = \dfrac{1}{2}, \quad \pi < \theta < \dfrac{3\pi}{2}$

11. $\cos \theta = -\dfrac{\sqrt{6}}{3}, \quad \dfrac{\pi}{2} < \theta < \pi$

12. $\sin \theta = -\dfrac{\sqrt{3}}{3}, \quad \dfrac{3\pi}{2} < \theta < 2\pi$

13. $\sec \theta = 3, \quad \sin \theta > 0$

14. $\csc \theta = -\sqrt{5}, \quad \cos \theta < 0$

15. $\cot \theta = -2, \quad \sec \theta < 0$

16. $\sec \theta = 2, \quad \csc \theta < 0$

17. $\tan \theta = -3, \quad \sin \theta < 0$

18. $\cot \theta = 3, \quad \cos \theta < 0$

In Problems 19–28, use the Half-angle Formulas to find the exact value of each expression.

19. $\sin 22.5°$

20. $\cos 22.5°$

21. $\tan \dfrac{7\pi}{8}$

22. $\tan \dfrac{9\pi}{8}$

23. $\cos 165°$

24. $\sin 195°$

25. $\sec \dfrac{15\pi}{8}$

26. $\csc \dfrac{7\pi}{8}$

27. $\sin\left(-\dfrac{\pi}{8}\right)$

28. $\cos\left(-\dfrac{3\pi}{8}\right)$

In Problems 29–40, use the figures to evaluate each function given that $f(x) = \sin x$, $g(x) = \cos x$, and $h(x) = \tan x$.

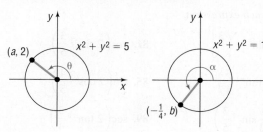

29. $f(2\theta)$

30. $g(2\theta)$

31. $g\left(\dfrac{\theta}{2}\right)$

32. $f\left(\dfrac{\theta}{2}\right)$

33. $h(2\theta)$

34. $h\left(\dfrac{\theta}{2}\right)$

35. $g(2\alpha)$

36. $f(2\alpha)$

37. $f\left(\dfrac{\alpha}{2}\right)$

38. $g\left(\dfrac{\alpha}{2}\right)$

39. $h\left(\dfrac{\alpha}{2}\right)$

40. $h(2\alpha)$

41. Show that $\sin^4 \theta = \dfrac{3}{8} - \dfrac{1}{2}\cos(2\theta) + \dfrac{1}{8}\cos(4\theta)$.

42. Show that $\sin(4\theta) = (\cos \theta)(4 \sin \theta - 8 \sin^3 \theta)$.

43. Develop a formula for $\cos(3\theta)$ as a third-degree polynomial in the variable $\cos \theta$.

44. Develop a formula for $\cos(4\theta)$ as a fourth-degree polynomial in the variable $\cos \theta$.

45. Find an expression for $\sin(5\theta)$ as a fifth-degree polynomial in the variable $\sin \theta$.

46. Find an expression for $\cos(5\theta)$ as a fifth-degree polynomial in the variable $\cos \theta$.

In Problems 47–68, establish each identity.

47. $\cos^4 \theta - \sin^4 \theta = \cos(2\theta)$

48. $\dfrac{\cot \theta - \tan \theta}{\cot \theta + \tan \theta} = \cos(2\theta)$

49. $\cot(2\theta) = \dfrac{\cot^2 \theta - 1}{2 \cot \theta}$

50. $\cot(2\theta) = \dfrac{1}{2}(\cot \theta - \tan \theta)$

51. $\sec(2\theta) = \dfrac{\sec^2 \theta}{2 - \sec^2 \theta}$

52. $\csc(2\theta) = \dfrac{1}{2} \sec \theta \csc \theta$

53. $\cos^2(2u) - \sin^2(2u) = \cos(4u)$

54. $(4 \sin u \cos u)(1 - 2 \sin^2 u) = \sin(4u)$

55. $\dfrac{\cos(2\theta)}{1 + \sin(2\theta)} = \dfrac{\cot \theta - 1}{\cot \theta + 1}$

56. $\sin^2 \theta \cos^2 \theta = \dfrac{1}{8}[1 - \cos(4\theta)]$

57. $\sec^2 \dfrac{\theta}{2} = \dfrac{2}{1 + \cos \theta}$

58. $\csc^2 \dfrac{\theta}{2} = \dfrac{2}{1 - \cos \theta}$

59. $\cot^2 \dfrac{v}{2} = \dfrac{\sec v + 1}{\sec v - 1}$

60. $\tan \dfrac{v}{2} = \csc v - \cot v$

61. $\cos \theta = \dfrac{1 - \tan^2 \dfrac{\theta}{2}}{1 + \tan^2 \dfrac{\theta}{2}}$

62. $1 - \dfrac{1}{2} \sin(2\theta) = \dfrac{\sin^3 \theta + \cos^3 \theta}{\sin \theta + \cos \theta}$

63. $\dfrac{\sin(3\theta)}{\sin \theta} - \dfrac{\cos(3\theta)}{\cos \theta} = 2$

64. $\dfrac{\cos \theta + \sin \theta}{\cos \theta - \sin \theta} - \dfrac{\cos \theta - \sin \theta}{\cos \theta + \sin \theta} = 2 \tan(2\theta)$

65. $\tan(3\theta) = \dfrac{3 \tan \theta - \tan^3 \theta}{1 - 3 \tan^2 \theta}$

66. $\tan \theta + \tan(\theta + 120°) + \tan(\theta + 240°) = 3 \tan(3\theta)$

67. $\ln |\sin \theta| = \dfrac{1}{2}(\ln |1 - \cos(2\theta)| - \ln 2)$

68. $\ln |\cos \theta| = \dfrac{1}{2}(\ln |1 + \cos(2\theta)| - \ln 2)$

In Problems 69–78, solve each equation on the interval $0 \le \theta < 2\pi$.

69. $\cos(2\theta) + 6 \sin^2 \theta = 4$

70. $\cos(2\theta) = 2 - 2 \sin^2 \theta$

71. $\cos(2\theta) = \cos \theta$

72. $\sin(2\theta) = \cos \theta$

73. $\sin(2\theta) + \sin(4\theta) = 0$

74. $\cos(2\theta) + \cos(4\theta) = 0$

75. $3 - \sin \theta = \cos(2\theta)$

76. $\cos(2\theta) + 5 \cos \theta + 3 = 0$

77. $\tan(2\theta) + 2 \sin \theta = 0$

78. $\tan(2\theta) + 2 \cos \theta = 0$

Mixed Practice

In Problems 79–90, find the exact value of each expression.

79. $\sin\left(2 \sin^{-1} \dfrac{1}{2}\right)$

80. $\sin\left[2 \sin^{-1} \dfrac{\sqrt{3}}{2}\right]$

81. $\cos\left(2 \sin^{-1} \dfrac{3}{5}\right)$

82. $\cos\left(2 \cos^{-1} \dfrac{4}{5}\right)$

83. $\tan\left[2 \cos^{-1}\left(-\dfrac{3}{5}\right)\right]$

84. $\tan\left(2 \tan^{-1} \dfrac{3}{4}\right)$

85. $\sin\left(2 \cos^{-1} \dfrac{4}{5}\right)$

86. $\cos\left[2 \tan^{-1}\left(-\dfrac{4}{3}\right)\right]$

87. $\sin^2\left(\dfrac{1}{2} \cos^{-1} \dfrac{3}{5}\right)$

88. $\cos^2\left(\dfrac{1}{2} \sin^{-1} \dfrac{3}{5}\right)$

89. $\sec\left(2 \tan^{-1} \dfrac{3}{4}\right)$

90. $\csc\left[2 \sin^{-1}\left(-\dfrac{3}{5}\right)\right]$

In Problems 91–93, find the real zeros of each trigonometric function on the interval $0 \le \theta < 2\pi$.

91. $f(x) = \sin(2x) - \sin x$

92. $f(x) = \cos(2x) + \cos x$

93. $f(x) = \cos(2x) + \sin^2 x$

Applications and Extensions

94. Constructing a Rain Gutter A rain gutter is to be constructed of aluminum sheets 12 inches wide. After marking off a length of 4 inches from each edge, this length is bent up at an angle θ. See the illustration. The area A of the opening as a function of θ is given by

$$A(\theta) = 16 \sin \theta (\cos \theta + 1) \quad 0° < \theta < 90°$$

$\measuredangle$ (a) In calculus, you will be asked to find the angle θ that maximizes A by solving the equation

$$\cos(2\theta) + \cos\theta = 0, \quad 0° < \theta < 90°$$

Solve this equation for θ.

(b) What is the maximum area A of the opening?

(c) Graph $A = A(\theta), 0° \le \theta \le 90°$, and find the angle θ that maximizes the area A. Also find the maximum area. Compare the results to the answer found earlier.

95. Laser Projection   In a laser projection system, the **optical** or **scanning angle** θ is related to the throw distance D from the scanner to the screen and the projected image width W by the equation

$$D = \frac{\frac{1}{2}W}{\csc\theta - \cot\theta}$$

(a) Show that the projected image width is given by

$$W = 2D\tan\frac{\theta}{2}$$

(b) Find the optical angle if the throw distance is 15 feet and the projected image width is 6.5 feet.

Source: Pangolin Laser Systems, Inc.

96. Product of Inertia   The **product of inertia** for an area about inclined axes is given by the formula

$$I_{uv} = I_x \sin\theta\cos\theta - I_y \sin\theta\cos\theta + I_{xy}(\cos^2\theta - \sin^2\theta)$$

Show that this is equivalent to

$$I_{uv} = \frac{I_x - I_y}{2}\sin(2\theta) + I_{xy}\cos(2\theta)$$

Source: Adapted from Hibbeler, Engineering Mechanics: Statics, 10th ed., Prentice Hall © 2004.

97. Projectile Motion   An object is propelled upward at an angle $\theta, 45° < \theta < 90°$, to the horizontal with an initial velocity of v_0 feet per second from the base of a plane that makes an angle of $45°$ with the horizontal. See the illustration. If air resistance is ignored, the distance R that it travels up the inclined plane is given by the function

$$R(\theta) = \frac{v_0^2\sqrt{2}}{16}\cos\theta(\sin\theta - \cos\theta)$$

(a) Show that

$$R(\theta) = \frac{v_0^2\sqrt{2}}{32}[\sin(2\theta) - \cos(2\theta) - 1]$$

$\measuredangle$ (b) In calculus, you will be asked to find the angle θ that maximizes R by solving the equation

$$\sin(2\theta) + \cos(2\theta) = 0$$

Solve this equation for θ.

(c) What is the maximum distance R if $v_0 = 32$ feet per second?

(d) Graph $R = R(\theta), 45° \le \theta \le 90°$, and find the angle θ that maximizes the distance R. Also find the maximum distance. Use $v_0 = 32$ feet per second. Compare the results with the answers found earlier.

98. Sawtooth Curve   An oscilloscope often displays a sawtooth curve. This curve can be approximated by sinusoidal curves of varying periods and amplitudes. A first approximation to the sawtooth curve is given by

$$y = \frac{1}{2}\sin(2\pi x) + \frac{1}{4}\sin(4\pi x)$$

Show that $y = \sin(2\pi x)\cos^2(\pi x)$.

99. Area of an Isosceles Triangle   Show that the area A of an isosceles triangle whose equal sides are of length s and θ is the angle between them is

$$A = \frac{1}{2}s^2\sin\theta$$

[**Hint:** See the illustration. The height h bisects the angle θ and is the perpendicular bisector of the base.]

100. Geometry   A rectangle is inscribed in a semicircle of radius 1. See the illustration.

(a) Express the area A of the rectangle as a function of the angle θ shown in the illustration.

(b) Show that $A(\theta) = \sin(2\theta)$.

(c) Find the angle θ that results in the largest area A.

(d) Find the dimensions of this largest rectangle.

101. If $x = 2\tan\theta$, express $\sin(2\theta)$ as a function of x.

102. If $x = 2\tan\theta$, express $\cos(2\theta)$ as a function of x.

103. Find the value of the number C:

$$\frac{1}{2}\sin^2 x + C = -\frac{1}{4}\cos(2x)$$

104. Find the value of the number C:

$$\frac{1}{2}\cos^2 x + C = \frac{1}{4}\cos(2x)$$

105. If $z = \tan\dfrac{\alpha}{2}$, show that $\sin\alpha = \dfrac{2z}{1 + z^2}$.

106. If $z = \tan\dfrac{\alpha}{2}$, show that $\cos\alpha = \dfrac{1 - z^2}{1 + z^2}$.

107. Graph $f(x) = \sin^2 x = \dfrac{1 - \cos(2x)}{2}$ for $0 \le x \le 2\pi$ by using transformations.

108. Repeat Problem 107 for $g(x) = \cos^2 x$.

109. Use the fact that

$$\cos\frac{\pi}{12} = \frac{1}{4}\left(\sqrt{6} + \sqrt{2}\right)$$

to find $\sin\dfrac{\pi}{24}$ and $\cos\dfrac{\pi}{24}$.

110. Show that

$$\cos\frac{\pi}{8} = \frac{\sqrt{2 + \sqrt{2}}}{2}$$

and use it to find $\sin\dfrac{\pi}{16}$ and $\cos\dfrac{\pi}{16}$.

111. Show that

$$\sin^3\theta + \sin^3(\theta + 120°) + \sin^3(\theta + 240°) = -\frac{3}{4}\sin(3\theta)$$

112. If $\tan\theta = a\tan\dfrac{\theta}{3}$, express $\tan\dfrac{\theta}{3}$ in terms of a.

Explaining Concepts: Discussion and Writing

113. Go to the library and research Chebyshёv polynomials. Write a report on your findings.

6.7 Product-to-Sum and Sum-to-Product Formulas

> **OBJECTIVES** **1** Express Products as Sums (p. 536)
> **2** Express Sums as Products (p. 537)

1 Express Products as Sums

Sum and difference formulas can be used to derive formulas for writing the products of sines and/or cosines as sums or differences. These identities are usually called the **Product-to-Sum Formulas.**

THEOREM

Product-to-Sum Formulas

$$\sin\alpha\sin\beta = \frac{1}{2}[\cos(\alpha - \beta) - \cos(\alpha + \beta)] \qquad (1)$$

$$\cos\alpha\cos\beta = \frac{1}{2}[\cos(\alpha - \beta) + \cos(\alpha + \beta)] \qquad (2)$$

$$\sin\alpha\cos\beta = \frac{1}{2}[\sin(\alpha + \beta) + \sin(\alpha - \beta)] \qquad (3)$$

These formulas do not have to be memorized. Instead, you should remember how they are derived. Then, when you want to use them, either look them up or derive them, as needed.

To derive formulas (1) and (2), write down the sum and difference formulas for the cosine:

$$\cos(\alpha - \beta) = \cos\alpha\cos\beta + \sin\alpha\sin\beta \qquad (4)$$

$$\cos(\alpha + \beta) = \cos\alpha\cos\beta - \sin\alpha\sin\beta \qquad (5)$$

Subtract equation (5) from equation (4) to get

$$\cos(\alpha - \beta) - \cos(\alpha + \beta) = 2\sin\alpha\sin\beta$$

from which

$$\sin\alpha\sin\beta = \frac{1}{2}[\cos(\alpha - \beta) - \cos(\alpha + \beta)]$$

Now add equations (4) and (5) to get

$$\cos(\alpha - \beta) + \cos(\alpha + \beta) = 2\cos\alpha\cos\beta$$

from which

$$\cos\alpha\cos\beta = \frac{1}{2}[\cos(\alpha - \beta) + \cos(\alpha + \beta)]$$

To derive Product-to-Sum Formula (3), use the sum and difference formulas for sine in a similar way. (You are asked to do this in Problem 47.)

EXAMPLE 1   **Expressing Products as Sums**

Express each of the following products as a sum containing only sines or only cosines.

(a) $\sin(6\theta)\sin(4\theta)$ (b) $\cos(3\theta)\cos\theta$ (c) $\sin(3\theta)\cos(5\theta)$

Solution   (a) We use formula (1) to get

$$\sin(6\theta)\sin(4\theta) = \frac{1}{2}[\cos(6\theta - 4\theta) - \cos(6\theta + 4\theta)]$$

$$= \frac{1}{2}[\cos(2\theta) - \cos(10\theta)]$$

(b) We use formula (2) to get

$$\cos(3\theta)\cos\theta = \frac{1}{2}[\cos(3\theta - \theta) + \cos(3\theta + \theta)]$$

$$= \frac{1}{2}[\cos(2\theta) + \cos(4\theta)]$$

(c) We use formula (3) to get

$$\sin(3\theta)\cos(5\theta) = \frac{1}{2}[\sin(3\theta + 5\theta) + \sin(3\theta - 5\theta)]$$

$$= \frac{1}{2}[\sin(8\theta) + \sin(-2\theta)] = \frac{1}{2}[\sin(8\theta) - \sin(2\theta)]$$

Now Work PROBLEM 1

2 Express Sums as Products

The **Sum-to-Product Formulas** are given next.

THEOREM   **Sum-to-Product Formulas**

$$\sin\alpha + \sin\beta = 2\sin\frac{\alpha + \beta}{2}\cos\frac{\alpha - \beta}{2} \qquad (6)$$

$$\sin\alpha - \sin\beta = 2\sin\frac{\alpha - \beta}{2}\cos\frac{\alpha + \beta}{2} \qquad (7)$$

$$\cos\alpha + \cos\beta = 2\cos\frac{\alpha + \beta}{2}\cos\frac{\alpha - \beta}{2} \qquad (8)$$

$$\cos\alpha - \cos\beta = -2\sin\frac{\alpha + \beta}{2}\sin\frac{\alpha - \beta}{2} \qquad (9)$$

We will derive formula (6) and leave the derivations of formulas (7) through (9) as exercises (see Problems 48 through 50).

Proof

$$2 \sin \frac{\alpha + \beta}{2} \cos \frac{\alpha - \beta}{2} = 2 \cdot \frac{1}{2} \left[\sin\left(\frac{\alpha + \beta}{2} + \frac{\alpha - \beta}{2} \right) + \sin\left(\frac{\alpha + \beta}{2} - \frac{\alpha - \beta}{2} \right) \right]$$

$\uparrow$
Product-to-Sum Formula (3)

$$= \sin \frac{2\alpha}{2} + \sin \frac{2\beta}{2} = \sin \alpha + \sin \beta \qquad \blacksquare$$

EXAMPLE 2 **Expressing Sums (or Differences) as a Product**

Express each sum or difference as a product of sines and/or cosines.

(a) $\sin(5\theta) - \sin(3\theta)$ 　　　　 (b) $\cos(3\theta) + \cos(2\theta)$

Solution (a) We use formula (7) to get

$$\sin(5\theta) - \sin(3\theta) = 2 \sin \frac{5\theta - 3\theta}{2} \cos \frac{5\theta + 3\theta}{2}$$

$$= 2 \sin \theta \cos(4\theta)$$

(b) $\cos(3\theta) + \cos(2\theta) = 2 \cos \dfrac{3\theta + 2\theta}{2} \cos \dfrac{3\theta - 2\theta}{2}$ 　 Formula (8)

$$= 2 \cos \frac{5\theta}{2} \cos \frac{\theta}{2}$$

Now Work PROBLEM 11

6.7 Assess Your Understanding

Skill Building

In Problems 1–10, express each product as a sum containing only sines or only cosines.

1. $\sin(4\theta) \sin(2\theta)$ 　　 **2.** $\cos(4\theta) \cos(2\theta)$ 　　 **3.** $\sin(4\theta) \cos(2\theta)$ 　　 **4.** $\sin(3\theta) \sin(5\theta)$ 　　 **5.** $\cos(3\theta) \cos(5\theta)$

6. $\sin(4\theta) \cos(6\theta)$ 　　 **7.** $\sin \theta \sin(2\theta)$ 　　 **8.** $\cos(3\theta) \cos(4\theta)$ 　　 **9.** $\sin \dfrac{3\theta}{2} \cos \dfrac{\theta}{2}$ 　　 **10.** $\sin \dfrac{\theta}{2} \cos \dfrac{5\theta}{2}$

In Problems 11–18, express each sum or difference as a product of sines and/or cosines.

11. $\sin(4\theta) - \sin(2\theta)$ 　　　 **12.** $\sin(4\theta) + \sin(2\theta)$ 　　　 **13.** $\cos(2\theta) + \cos(4\theta)$ 　　　 **14.** $\cos(5\theta) - \cos(3\theta)$

15. $\sin \theta + \sin(3\theta)$ 　　　 **16.** $\cos \theta + \cos(3\theta)$ 　　　 **17.** $\cos \dfrac{\theta}{2} - \cos \dfrac{3\theta}{2}$ 　　　 **18.** $\sin \dfrac{\theta}{2} - \sin \dfrac{3\theta}{2}$

In Problems 19–36, establish each identity.

19. $\dfrac{\sin \theta + \sin(3\theta)}{2 \sin(2\theta)} = \cos \theta$ 　　 **20.** $\dfrac{\cos \theta + \cos(3\theta)}{2 \cos(2\theta)} = \cos \theta$ 　　 **21.** $\dfrac{\sin(4\theta) + \sin(2\theta)}{\cos(4\theta) + \cos(2\theta)} = \tan(3\theta)$

22. $\dfrac{\cos \theta - \cos(3\theta)}{\sin(3\theta) - \sin \theta} = \tan(2\theta)$ 　　 **23.** $\dfrac{\cos \theta - \cos(3\theta)}{\sin \theta + \sin(3\theta)} = \tan \theta$ 　　 **24.** $\dfrac{\cos \theta - \cos(5\theta)}{\sin \theta + \sin(5\theta)} = \tan(2\theta)$

25. $\sin \theta [\sin \theta + \sin(3\theta)] = \cos \theta [\cos \theta - \cos(3\theta)]$ 　　　　 **26.** $\sin \theta [\sin(3\theta) + \sin(5\theta)] = \cos \theta [\cos(3\theta) - \cos(5\theta)]$

27. $\dfrac{\sin(4\theta) + \sin(8\theta)}{\cos(4\theta) + \cos(8\theta)} = \tan(6\theta)$ 　　　 **28.** $\dfrac{\sin(4\theta) - \sin(8\theta)}{\cos(4\theta) - \cos(8\theta)} = -\cot(6\theta)$

29. $\dfrac{\sin(4\theta) + \sin(8\theta)}{\sin(4\theta) - \sin(8\theta)} = -\dfrac{\tan(6\theta)}{\tan(2\theta)}$ 　　　 **30.** $\dfrac{\cos(4\theta) - \cos(8\theta)}{\cos(4\theta) + \cos(8\theta)} = \tan(2\theta) \tan(6\theta)$

31. $\dfrac{\sin \alpha + \sin \beta}{\sin \alpha - \sin \beta} = \tan \dfrac{\alpha + \beta}{2} \cot \dfrac{\alpha - \beta}{2}$

32. $\dfrac{\cos \alpha + \cos \beta}{\cos \alpha - \cos \beta} = -\cot \dfrac{\alpha + \beta}{2} \cot \dfrac{\alpha - \beta}{2}$

33. $\dfrac{\sin \alpha + \sin \beta}{\cos \alpha + \cos \beta} = \tan \dfrac{\alpha + \beta}{2}$

34. $\dfrac{\sin \alpha - \sin \beta}{\cos \alpha - \cos \beta} = -\cot \dfrac{\alpha + \beta}{2}$

35. $1 + \cos(2\theta) + \cos(4\theta) + \cos(6\theta) = 4 \cos \theta \cos(2\theta) \cos(3\theta)$

36. $1 - \cos(2\theta) + \cos(4\theta) - \cos(6\theta) = 4 \sin \theta \cos(2\theta) \sin(3\theta)$

In Problems 37–40, solve each equation on the interval $0 \le \theta < 2\pi$.

37. $\sin(2\theta) + \sin(4\theta) = 0$

38. $\cos(2\theta) + \cos(4\theta) = 0$

39. $\cos(4\theta) - \cos(6\theta) = 0$

40. $\sin(4\theta) - \sin(6\theta) = 0$

Applications and Extensions

41. Touch-Tone Phones On a Touch-Tone phone, each button produces a unique sound. The sound produced is the sum of two tones, given by

$$y = \sin(2\pi l t) \quad \text{and} \quad y = \sin(2\pi h t)$$

where l and h are the low and high frequencies (cycles per second) shown on the illustration. For example, if you touch 7, the low frequency is $l = 852$ cycles per second and the high frequency is $h = 1209$ cycles per second. The sound emitted by touching 7 is

$$y = \sin[2\pi(852)t] + \sin[2\pi(1209)t]$$

Touch-Tone phone

(a) Write this sound as a product of sines and/or cosines.
(b) Determine the maximum value of y.
(c) Graph the sound emitted by touching 7.

42. Touch-Tone Phones
(a) Write the sound emitted by touching the # key as a product of sines and/or cosines.
(b) Determine the maximum value of y.
(c) Graph the sound emitted by touching the # key.

43. Moment of Inertia The moment of inertia I of an object is a measure of how easy it is to rotate the object about some fixed point. In engineering mechanics, it is sometimes necessary to compute moments of inertia with respect to

a set of rotated axes. These moments are given by the equations

$$I_u = I_x \cos^2 \theta + I_y \sin^2 \theta - 2I_{xy} \sin \theta \cos \theta$$
$$I_v = I_x \sin^2 \theta + I_y \cos^2 \theta + 2I_{xy} \sin \theta \cos \theta$$

Use Product-to-Sum Formulas to show that

$$I_u = \dfrac{I_x + I_y}{2} + \dfrac{I_x - I_y}{2} \cos(2\theta) - I_{xy} \sin(2\theta)$$

and

$$I_v = \dfrac{I_x + I_y}{2} - \dfrac{I_x - I_y}{2} \cos(2\theta) + I_{xy} \sin(2\theta)$$

Source: Adapted from Hibbeler, *Engineering Mechanics: Statics,* 10th ed., Prentice Hall © 2004.

44. Projectile Motion The range R of a projectile propelled downward from the top of an inclined plane at an angle θ to the inclined plane is given by

$$R(\theta) = \dfrac{2v_0^2 \sin \theta \cos(\theta - \phi)}{g \cos^2 \phi}$$

where v_0 is the initial velocity of the projectile, ϕ is the angle the plane makes with respect to the horizontal, and g is acceleration due to gravity.
(a) Show that for fixed v_0 and ϕ the maximum range down the incline is given by $R_{\max} = \dfrac{v_0^2}{g(1 - \sin \phi)}$.
(b) Determine the maximum range if the projectile has an initial velocity of 50 meters/second, the angle of the plane is $\phi = 35°$, and $g = 9.8$ meters/second^2.

45. If $\alpha + \beta + \gamma = \pi$, show that

$$\sin(2\alpha) + \sin(2\beta) + \sin(2\gamma) = 4 \sin \alpha \sin \beta \sin \gamma$$

46. If $\alpha + \beta + \gamma = \pi$, show that

$$\tan \alpha + \tan \beta + \tan \gamma = \tan \alpha \tan \beta \tan \gamma$$

47. Derive formula (3).

48. Derive formula (7).

49. Derive formula (8).

50. Derive formula (9).

CHAPTER REVIEW

Things to Know

Definitions of the six inverse trigonometric functions

$y = \sin^{-1} x$ means $x = \sin y$ where $-1 \le x \le 1,$ $-\dfrac{\pi}{2} \le y \le \dfrac{\pi}{2}$ (p. 479)

$y = \cos^{-1} x$ means $x = \cos y$ where $-1 \le x \le 1,$ $0 \le y \le \pi$ (p. 482)

$y = \tan^{-1} x$ means $x = \tan y$ where $-\infty < x < \infty,$ $-\dfrac{\pi}{2} < y < \dfrac{\pi}{2}$ (p. 485)

$y = \sec^{-1} x$ means $x = \sec y$ where $|x| \ge 1,$ $0 \le y \le \pi,$ $y \ne \dfrac{\pi}{2}$ (p. 492)

$y = \csc^{-1} x$ means $x = \csc y$ where $|x| \ge 1,$ $-\dfrac{\pi}{2} \le y \le \dfrac{\pi}{2},$ $y \ne 0$ (p. 492)

$y = \cot^{-1} x$ means $x = \cot y$ where $-\infty < x < \infty,$ $0 < y < \pi$ (p. 492)

Sum and Difference Formulas (pp. 514, 517, and 519)

$$\cos(\alpha + \beta) = \cos\alpha \cos\beta - \sin\alpha \sin\beta \qquad \cos(\alpha - \beta) = \cos\alpha \cos\beta + \sin\alpha \sin\beta$$

$$\sin(\alpha + \beta) = \sin\alpha \cos\beta + \cos\alpha \sin\beta \qquad \sin(\alpha - \beta) = \sin\alpha \cos\beta - \cos\alpha \sin\beta$$

$$\tan(\alpha + \beta) = \frac{\tan\alpha + \tan\beta}{1 - \tan\alpha \tan\beta} \qquad\qquad \tan(\alpha - \beta) = \frac{\tan\alpha - \tan\beta}{1 + \tan\alpha \tan\beta}$$

Double-angle Formulas (pp. 526 and 527)

$$\sin(2\theta) = 2\sin\theta \cos\theta \qquad\qquad \cos(2\theta) = \cos^2\theta - \sin^2\theta \qquad\qquad \tan(2\theta) = \frac{2\tan\theta}{1 - \tan^2\theta}$$

$$\cos(2\theta) = 2\cos^2\theta - 1 \qquad\qquad \cos(2\theta) = 1 - 2\sin^2\theta$$

Half-angle Formulas (pp. 530 and 532)

$$\sin^2\frac{\alpha}{2} = \frac{1 - \cos\alpha}{2} \qquad\qquad \cos^2\frac{\alpha}{2} = \frac{1 + \cos\alpha}{2} \qquad\qquad \tan^2\frac{\alpha}{2} = \frac{1 - \cos\alpha}{1 + \cos\alpha}$$

$$\sin\frac{\alpha}{2} = \pm\sqrt{\frac{1 - \cos\alpha}{2}} \qquad\qquad \cos\frac{\alpha}{2} = \pm\sqrt{\frac{1 + \cos\alpha}{2}} \qquad\qquad \tan\frac{\alpha}{2} = \pm\sqrt{\frac{1 - \cos\alpha}{1 + \cos\alpha}} = \frac{1 - \cos\alpha}{\sin\alpha} = \frac{\sin\alpha}{1 + \cos\alpha}$$

where the $+$ or $-$ is determined by the quadrant of $\dfrac{\alpha}{2}$.

Product-to-Sum Formulas (p. 536)

$$\sin\alpha \sin\beta = \frac{1}{2}[\cos(\alpha - \beta) - \cos(\alpha + \beta)]$$

$$\cos\alpha \cos\beta = \frac{1}{2}[\cos(\alpha - \beta) + \cos(\alpha + \beta)]$$

$$\sin\alpha \cos\beta = \frac{1}{2}[\sin(\alpha + \beta) + \sin(\alpha - \beta)]$$

Sum-to-Product Formulas (p. 537)

$$\sin\alpha + \sin\beta = 2\sin\frac{\alpha + \beta}{2}\cos\frac{\alpha - \beta}{2} \qquad\qquad \sin\alpha - \sin\beta = 2\sin\frac{\alpha - \beta}{2}\cos\frac{\alpha + \beta}{2}$$

$$\cos\alpha + \cos\beta = 2\cos\frac{\alpha + \beta}{2}\cos\frac{\alpha - \beta}{2} \qquad\qquad \cos\alpha - \cos\beta = -2\sin\frac{\alpha + \beta}{2}\sin\frac{\alpha - \beta}{2}$$

Objectives

Section		You should be able to ...	Example(s)	Review Exercises
6.1	1	Find the exact value of an inverse sine, cosine, or tangent function (p. 479)	1, 2, 6, 7, 9	1–6
	2	Find an approximate value of an inverse sine function (p. 480)	3	121–124
	3	Use properties of inverse functions to find exact values of certain composite functions (p. 481)	4, 5, 8	9–20
	4	Find the inverse function of a trigonometric function (p. 486)	10	33–36
	5	Solve equations involving inverse trigonometric functions (p. 487)	11	133, 134
6.2	1	Find the exact value of expressions involving the inverse sine, cosine, and tangent functions (p. 491)	1–3	21–32
	2	Define the inverse secant, cosecant, and cotangent functions (p. 492)	4	7, 8, 29, 30
	3	Use a calculator to evaluate $\sec^{-1} x$, $\csc^{-1} x$, and $\cot^{-1} x$ (p. 492)	5	125, 126
	4	Write a trigonometric expression as an algebraic expression (p. 493)	6	37–40
6.3	1	Solve equations involving a single trigonometric function (p. 496)	1–5	97–106
	2	Solve trigonometric equations using a calculator (p. 499)	6	107, 108
	3	Solve trigonometric equations quadratic in form (p. 500)	7	113, 114
	4	Solve trigonometric equations using fundamental identities (p. 500)	8, 9	109–112, 115–118
	5	Solve trigonometric equations using a graphing utility (p. 501)	10	127–132
6.4	1	Use algebra to simplify trigonometric expressions (p. 507)	1	41–72
	2	Establish identities (p. 508)	2–8	41–58
6.5	1	Use sum and difference formulas to find exact values (p. 515)	1, 2	73–80, 81–90(a)–(d), 135
	2	Use sum and difference formulas to establish identities (p. 516)	3–8	59–62
	3	Use sum and difference formulas involving inverse trigonometric functions (p. 520)	9, 10	91–94
	4	Solve trigonometric equations linear in sine and cosine (p. 521)	11, 12	119, 120
6.6	1	Use double-angle formulas to find exact values (p. 526)	1	81–90(e), (f), 95, 96
	2	Use double-angle formulas to establish identities (p. 527)	2–5	65–67
	3	Use half-angle formulas to find exact values (p. 530)	6, 7	81–90(g), (h), 135
6.7	1	Express products as sums (p. 536)	1	68
	2	Express sums as products (p. 537)	2	69–72

Review Exercises

In Problems 1–8, find the exact value of each expression. Do not use a calculator.

1. $\sin^{-1} 1$

2. $\cos^{-1} 0$

3. $\tan^{-1} 1$

4. $\sin^{-1}\left(-\dfrac{1}{2}\right)$

5. $\cos^{-1}\left(-\dfrac{\sqrt{3}}{2}\right)$

6. $\tan^{-1}\left(-\sqrt{3}\right)$

7. $\sec^{-1}\sqrt{2}$

8. $\cot^{-1}(-1)$

In Problems 9–32, find the exact value, if any, of each composite function. If there is no value, say it is "not defined." Do not use a calculator.

9. $\sin^{-1}\left(\sin \dfrac{3\pi}{8}\right)$

10. $\cos^{-1}\left(\cos \dfrac{3\pi}{4}\right)$

11. $\tan^{-1}\left(\tan \dfrac{2\pi}{3}\right)$

12. $\sin^{-1}\left[\sin\left(-\dfrac{\pi}{8}\right)\right]$

13. $\cos^{-1}\left(\cos \dfrac{15\pi}{7}\right)$

14. $\sin^{-1}\left[\sin\left(-\dfrac{8\pi}{9}\right)\right]$

15. $\sin(\sin^{-1} 0.9)$

16. $\cos(\cos^{-1} 0.6)$

17. $\cos[\cos^{-1}(-0.3)]$

18. $\tan[\tan^{-1} 5]$

19. $\cos[\cos^{-1}(-1.6)]$

20. $\sin(\sin^{-1} 1.6)$

21. $\sin^{-1}\left(\cos \dfrac{2\pi}{3}\right)$

22. $\cos^{-1}\left(\tan \dfrac{3\pi}{4}\right)$

23. $\tan^{-1}\left(\tan \dfrac{7\pi}{4}\right)$

24. $\cos^{-1}\left(\cos \dfrac{7\pi}{6}\right)$

25. $\tan\left[\sin^{-1}\left(-\dfrac{\sqrt{3}}{2}\right)\right]$ **26.** $\tan\left[\cos^{-1}\left(-\dfrac{1}{2}\right)\right]$ **27.** $\sec\left(\tan^{-1}\dfrac{\sqrt{3}}{3}\right)$ **28.** $\csc\left(\sin^{-1}\dfrac{\sqrt{3}}{2}\right)$

29. $\sin\left(\cot^{-1}\dfrac{3}{4}\right)$ **30.** $\cos\left(\csc^{-1}\dfrac{5}{3}\right)$ **31.** $\tan\left[\sin^{-1}\left(-\dfrac{4}{5}\right)\right]$ **32.** $\tan\left[\cos^{-1}\left(-\dfrac{3}{5}\right)\right]$

In Problems 33–36, find the inverse function f^{-1} of each function f. Find the range of f and the domain and range of f^{-1}.

33. $f(x) = 2\,\sin(3x)$
$-\dfrac{\pi}{6} \le x \le \dfrac{\pi}{6}$

34. $f(x) = \tan(2x + 3) - 1$
$-\dfrac{3}{2} - \dfrac{\pi}{4} < x < -\dfrac{3}{2} + \dfrac{\pi}{4}$

35. $f(x) = -\cos x + 3$
$0 \le x \le \pi$

36. $f(x) = 2\,\sin(-x + 1)$
$1 - \dfrac{\pi}{2} \le x \le 1 + \dfrac{\pi}{2}$

In Problems 37–40, write each trigonometric expression as an algebraic expression in u.

37. $\cos(\sin^{-1} u)$ **38.** $\cos(\csc^{-1} u)$ **39.** $\sin(\csc^{-1} u)$ **40.** $\tan(\csc^{-1} u)$

In Problems 41–72, establish each identity.

41. $\tan\theta\,\cot\theta - \sin^2\theta = \cos^2\theta$ **42.** $\sin\theta\,\csc\theta - \sin^2\theta = \cos^2\theta$ **43.** $\sin^2\theta(1 + \cot^2\theta) = 1$

44. $(1 - \sin^2\theta)(1 + \tan^2\theta) = 1$ **45.** $5\cos^2\theta + 3\sin^2\theta = 3 + 2\cos^2\theta$ **46.** $4\sin^2\theta + 2\cos^2\theta = 4 - 2\cos^2\theta$

47. $\dfrac{1 - \cos\theta}{\sin\theta} + \dfrac{\sin\theta}{1 - \cos\theta} = 2\csc\theta$ **48.** $\dfrac{\sin\theta}{1 + \cos\theta} + \dfrac{1 + \cos\theta}{\sin\theta} = 2\csc\theta$ **49.** $\dfrac{\cos\theta}{\cos\theta - \sin\theta} = \dfrac{1}{1 - \tan\theta}$

50. $1 - \dfrac{\sin^2\theta}{1 + \cos\theta} = \cos\theta$ **51.** $\dfrac{\csc\theta}{1 + \csc\theta} = \dfrac{1 - \sin\theta}{\cos^2\theta}$ **52.** $\dfrac{1 + \sec\theta}{\sec\theta} = \dfrac{\sin^2\theta}{1 - \cos\theta}$

53. $\csc\theta - \sin\theta = \cos\theta\,\cot\theta$ **54.** $\dfrac{\csc\theta}{1 - \cos\theta} = \dfrac{1 + \cos\theta}{\sin^3\theta}$ **55.** $\dfrac{1 - \sin\theta}{\sec\theta} = \dfrac{\cos^3\theta}{1 + \sin\theta}$

56. $\dfrac{1 - \cos\theta}{1 + \cos\theta} = (\csc\theta - \cot\theta)^2$ **57.** $\dfrac{1 - 2\sin^2\theta}{\sin\theta\,\cos\theta} = \cot\theta - \tan\theta$ **58.** $\dfrac{(2\sin^2\theta - 1)^2}{\sin^4\theta - \cos^4\theta} = 1 - 2\cos^2\theta$

59. $\dfrac{\cos(\alpha + \beta)}{\cos\alpha\,\sin\beta} = \cot\beta - \tan\alpha$ **60.** $\dfrac{\sin(\alpha - \beta)}{\sin\alpha\,\cos\beta} = 1 - \cot\alpha\,\tan\beta$ **61.** $\dfrac{\cos(\alpha - \beta)}{\cos\alpha\,\cos\beta} = 1 + \tan\alpha\,\tan\beta$

62. $\dfrac{\cos(\alpha + \beta)}{\sin\alpha\,\cos\beta} = \cot\alpha - \tan\beta$ **63.** $(1 + \cos\theta)\tan\dfrac{\theta}{2} = \sin\theta$ **64.** $\sin\theta\,\tan\dfrac{\theta}{2} = 1 - \cos\theta$

65. $2\cot\theta\,\cot(2\theta) = \cot^2\theta - 1$ **66.** $2\sin(2\theta)(1 - 2\sin^2\theta) = \sin(4\theta)$ **67.** $1 - 8\sin^2\theta\,\cos^2\theta = \cos(4\theta)$

68. $\dfrac{\sin(3\theta)\cos\theta - \sin\theta\,\cos(3\theta)}{\sin(2\theta)} = 1$ **69.** $\dfrac{\sin(2\theta) + \sin(4\theta)}{\cos(2\theta) + \cos(4\theta)} = \tan(3\theta)$ **70.** $\dfrac{\sin(2\theta) + \sin(4\theta)}{\sin(2\theta) - \sin(4\theta)} + \dfrac{\tan(3\theta)}{\tan\theta} = 0$

71. $\dfrac{\cos(2\theta) - \cos(4\theta)}{\cos(2\theta) + \cos(4\theta)} - \tan\theta\,\tan(3\theta) = 0$ **72.** $\cos(2\theta) - \cos(10\theta) = \tan(4\theta)[\sin(2\theta) + \sin(10\theta)]$

In Problems 73–80, find the exact value of each expression.

73. $\sin 165°$ **74.** $\tan 105°$ **75.** $\cos\dfrac{5\pi}{12}$ **76.** $\sin\left(-\dfrac{\pi}{12}\right)$

77. $\cos 80°\cos 20° + \sin 80°\sin 20°$ **78.** $\sin 70°\cos 40° - \cos 70°\sin 40°$

79. $\tan\dfrac{\pi}{8}$ **80.** $\sin\dfrac{5\pi}{8}$

In Problems 81–90, use the information given about the angles α and β to find the exact value of:

(a) $\sin(\alpha + \beta)$ (b) $\cos(\alpha + \beta)$ (c) $\sin(\alpha - \beta)$ (d) $\tan(\alpha + \beta)$

(e) $\sin(2\alpha)$ (f) $\cos(2\beta)$ (g) $\sin\dfrac{\beta}{2}$ (h) $\cos\dfrac{\alpha}{2}$

81. $\sin\alpha = \dfrac{4}{5}, 0 < \alpha < \dfrac{\pi}{2}; \sin\beta = \dfrac{5}{13}, \dfrac{\pi}{2} < \beta < \pi$ **82.** $\cos\alpha = \dfrac{4}{5}, 0 < \alpha < \dfrac{\pi}{2}; \cos\beta = \dfrac{5}{13}, -\dfrac{\pi}{2} < \beta < 0$

83. $\sin\alpha = -\dfrac{3}{5}, \pi < \alpha < \dfrac{3\pi}{2}; \cos\beta = \dfrac{12}{13}, \dfrac{3\pi}{2} < \beta < 2\pi$ **84.** $\sin\alpha = -\dfrac{4}{5}, -\dfrac{\pi}{2} < \alpha < 0; \cos\beta = -\dfrac{5}{13}, \dfrac{\pi}{2} < \beta < \pi$

85. $\tan \alpha = \frac{3}{4}, \pi < \alpha < \frac{3\pi}{2}; \tan \beta = \frac{12}{5}, 0 < \beta < \frac{\pi}{2}$

86. $\tan \alpha = -\frac{4}{3}, \frac{\pi}{2} < \alpha < \pi; \cot \beta = \frac{12}{5}, \pi < \beta < \frac{3\pi}{2}$

87. $\sec \alpha = 2, -\frac{\pi}{2} < \alpha < 0; \sec \beta = 3, \frac{3\pi}{2} < \beta < 2\pi$

88. $\csc \alpha = 2, \frac{\pi}{2} < \alpha < \pi; \sec \beta = -3, \frac{\pi}{2} < \beta < \pi$

89. $\sin \alpha = -\frac{2}{3}, \pi < \alpha < \frac{3\pi}{2}; \cos \beta = -\frac{2}{3}, \pi < \beta < \frac{3\pi}{2}$

90. $\tan \alpha = -2, \frac{\pi}{2} < \alpha < \pi; \cot \beta = -2, \frac{\pi}{2} < \beta < \pi$

In Problems 91–96, find the exact value of each expression.

91. $\cos\left(\sin^{-1} \frac{3}{5} - \cos^{-1} \frac{1}{2} \right)$

92. $\sin\left(\cos^{-1} \frac{5}{13} - \cos^{-1} \frac{4}{5} \right)$

93. $\tan\left[\sin^{-1}\left(-\frac{1}{2}\right) - \tan^{-1} \frac{3}{4} \right]$

94. $\cos\left[\tan^{-1}(-1) + \cos^{-1}\left(-\frac{4}{5}\right) \right]$

95. $\sin\left[2 \cos^{-1}\left(-\frac{3}{5}\right) \right]$

96. $\cos\left(2 \tan^{-1} \frac{4}{3} \right)$

In Problems 97–120, solve each equation on the interval $0 \le \theta < 2\pi$.

97. $\cos \theta = \frac{1}{2}$

98. $\sin \theta = -\frac{\sqrt{3}}{2}$

99. $2 \cos \theta + \sqrt{2} = 0$

100. $\tan \theta + \sqrt{3} = 0$

101. $\sin(2\theta) + 1 = 0$

102. $\cos(2\theta) = 0$

103. $\tan(2\theta) = 0$

104. $\sin(3\theta) = 1$

105. $\sec^2 \theta = 4$

106. $\csc^2 \theta = 1$

107. $0.2 \sin \theta = 0.05$

108. $0.9 \cos(2\theta) = 0.7$

109. $\sin \theta + \sin(2\theta) = 0$

110. $\cos(2\theta) = \sin \theta$

111. $\sin(2\theta) - \cos \theta - 2 \sin \theta + 1 = 0$

112. $\sin(2\theta) - \sin \theta - 2 \cos \theta + 1 = 0$

113. $2 \sin^2 \theta - 3 \sin \theta + 1 = 0$

114. $2 \cos^2 \theta + \cos \theta - 1 = 0$

115. $4 \sin^2 \theta = 1 + 4 \cos \theta$

116. $8 - 12 \sin^2 \theta = 4 \cos^2 \theta$

117. $\sin(2\theta) = \sqrt{2} \cos \theta$

118. $1 + \sqrt{3} \cos \theta + \cos(2\theta) = 0$

119. $\sin \theta - \cos \theta = 1$

120. $\sin \theta - \sqrt{3} \cos \theta = 2$

In Problems 121–126, use a calculator to find an approximate value for each expression, rounded to two decimal places.

121. $\sin^{-1} 0.7$

122. $\cos^{-1} \frac{4}{5}$

123. $\tan^{-1}(-2)$

124. $\cos^{-1}(-0.2)$

125. $\sec^{-1} 3$

126. $\cot^{-1}(-4)$

In Problems 127–132, use a graphing utility to solve each equation on the interval $0 \le x \le 2\pi$. Approximate any solutions rounded to two decimal places.

127. $2x = 5 \cos x$

128. $2x = 5 \sin x$

129. $2 \sin x + 3 \cos x = 4x$

130. $3 \cos x + x = \sin x$

131. $\sin x = \ln x$

132. $\sin x = e^{-x}$

In Problems 133 and 134, find the exact solution of each equation.

133. $-3 \sin^{-1} x = \pi$

134. $2 \cos^{-1} x + \pi = 4 \cos^{-1} x$

135. Use a half-angle formula to find the exact value of $\sin 15°$. Then use a difference formula to find the exact value of $\sin 15°$. Show that the answers found are the same.

136. If you are given the value of $\cos \theta$ and want the exact value of $\cos(2\theta)$, what form of the double-angle formula for $\cos(2\theta)$ is most efficient to use?

CHAPTER TEST

Step-by-step test solutions are found on the Chapter Test Prep Videos available via the Video Resources on DVD, in **MyMathLab** or on You Tube (search "Sullivan Precalculus: A Right Triangle Approach, Concepts Through Functions 2e).

In Problems 1–6, find the exact value of each expression. Express angles in radians.

1. $\sec^{-1}\left(\dfrac{2}{\sqrt{3}}\right)$

2. $\sin^{-1}\left(-\dfrac{\sqrt{2}}{2}\right)$

3. $\sin^{-1}\left(\sin\dfrac{11\pi}{5}\right)$

4. $\tan\left(\tan^{-1}\dfrac{7}{3}\right)$

5. $\cot\left(\csc^{-1}\sqrt{10}\right)$

6. $\sec\left(\cos^{-1}\left(-\dfrac{3}{4}\right)\right)$

In Problems 7–10, use a calculator to evaluate each expression. Express angles in radians rounded to two decimal places.

7. $\sin^{-1}0.382$

8. $\sec^{-1}1.4$

9. $\tan^{-1}3$

10. $\cot^{-1}5$

In Problems 11–16 establish each identity.

11. $\dfrac{\csc\theta+\cot\theta}{\sec\theta+\tan\theta}=\dfrac{\sec\theta-\tan\theta}{\csc\theta-\cot\theta}$

12. $\sin\theta\tan\theta+\cos\theta=\sec\theta$

13. $\tan\theta+\cot\theta=2\csc(2\theta)$

14. $\dfrac{\sin(\alpha+\beta)}{\tan\alpha+\tan\beta}=\cos\alpha\cos\beta$

15. $\sin(3\theta)=3\sin\theta-4\sin^3\theta$

16. $\dfrac{\tan\theta-\cot\theta}{\tan\theta+\cot\theta}=1-2\cos^2\theta$

In Problems 17–24 use sum, difference, product or half-angle formulas to find the exact value of each expression.

17. $\cos 15°$

18. $\tan 75°$

19. $\sin\left(\dfrac{1}{2}\cos^{-1}\dfrac{3}{5}\right)$

20. $\tan\left(2\sin^{-1}\dfrac{6}{11}\right)$

21. $\cos\left(\sin^{-1}\dfrac{2}{3}+\tan^{-1}\dfrac{3}{2}\right)$

22. $\sin 75°\cos 15°$

23. $\sin 75°+\sin 15°$

24. $\cos 65°\cos 20°+\sin 65°\sin 20°$

In Problems 25–29, solve each equation on $0\le\theta<2\pi$.

25. $4\sin^2\theta-3=0$

26. $-3\cos\left(\dfrac{\pi}{2}-\theta\right)=\tan\theta$

27. $\cos^2\theta+2\sin\theta\cos\theta-\sin^2\theta=0$

28. $\sin(\theta+1)=\cos\theta$

29. $4\sin^2\theta+7\sin\theta=2$

30. Stage 16 of the 2004 Tour de France was a time trial from Bourg d'Oisans to L'Alpe d'Huez. The average grade (slope as a percent) for most of the 15-kilometer mountainous trek was 7.9%. What was the change in elevation from the beginning to the end of the route?

CUMULATIVE REVIEW

1. Find the real solutions, if any, of the equation $3x^2+x-1=0$.

2. Find an equation for the line containing the points $(-2,5)$ and $(4,-1)$. What is the distance between these points? What is their midpoint?

3. Test the equation $3x+y^2=9$ for symmetry with respect to the x-axis, y-axis, and origin. List the intercepts.

4. Use transformations to graph the equation $y=|x-3|+2$.

5. Use transformations to graph the equation $y=3e^x-2$.

6. Use transformations to graph the equation $y=\cos\left(x-\dfrac{\pi}{2}\right)-1$.

7. Sketch a graph of each of the following functions. Label at least three points on each graph. Name the inverse function of each and show its graph.
(a) $y=x^3$
(b) $y=e^x$
(c) $y=\sin x,\quad -\dfrac{\pi}{2}\le x\le\dfrac{\pi}{2}$
(d) $y=\cos x,\quad 0\le x\le\pi$

8. If $\sin\theta=-\dfrac{1}{3}$ and $\pi<\theta<\dfrac{3\pi}{2}$, find the exact value of:
(a) $\cos\theta$ (b) $\tan\theta$ (c) $\sin(2\theta)$
(d) $\cos(2\theta)$ (e) $\sin\left(\dfrac{1}{2}\theta\right)$ (f) $\cos\left(\dfrac{1}{2}\theta\right)$

9. Find the exact value of $\cos(\tan^{-1}2)$.

10. If $\sin \alpha = \dfrac{1}{3}, \dfrac{\pi}{2} < \alpha < \pi$, and $\cos \beta = -\dfrac{1}{3}, \pi < \beta < \dfrac{3\pi}{2}$, find the exact value of:

(a) $\cos \alpha$ (b) $\sin \beta$ (c) $\cos(2\alpha)$

(d) $\cos(\alpha + \beta)$ (e) $\sin \dfrac{\beta}{2}$

11. For the function
$$f(x) = 2x^5 - x^4 - 4x^3 + 2x^2 + 2x - 1:$$

(a) Find the real zeros and their multiplicity.
(b) Find the intercepts.
(c) Find the power function that the graph of f resembles for large $|x|$.

(d) Graph f using a graphing utility.
(e) Approximate the turning points, if any exist.
(f) Use the information obtained in parts (a)–(e) to sketch a graph of f by hand.
(g) Identify the intervals on which f is increasing, decreasing, or constant.

12. If $f(x) = 2x^2 + 3x + 1$ and $g(x) = x^2 + 3x + 2$, solve:

(a) $f(x) = 0$ (b) $f(x) = g(x)$

(c) $f(x) > 0$ (d) $f(x) \geq g(x)$

CHAPTER PROJECTS

I. **Waves** A stretched string that is attached at both ends, pulled in a direction perpendicular to the string, and released has motion that is described as wave motion. If we assume no friction and a length such that there are no "echoes" (that is, the wave doesn't bounce back), the transverse motion (motion perpendicular to the string) can be described by the equation

$$y = y_m \sin(kx - \omega t)$$

where y_m is the amplitude measured in meters and k and ω are constants. The height of the sound wave depends on the distance x from one endpoint of the string and on the time t, so a typical wave has horizontal and vertical motion over time.

(a) What is the amplitude of the wave $y = 0.00421 \sin(68.3x - 2.68t)$?

(b) The value of ω is the angular frequency measured in radians per second. What is the angular frequency of the wave given in part (a)?

(c) The frequency f is the number of vibrations per second (hertz) made by the wave as it passes a certain point. Its value is found using the formula $f = \dfrac{\omega}{2\pi}$. What is the frequency of the wave given in part (a)?

(d) The wavelength, λ, of a wave is the shortest distance at which the wave pattern repeats itself for a constant t. Thus, $\lambda = \dfrac{2\pi}{k}$. What is the wavelength of the wave given in part (a)?

(e) Graph the height of the string a distance $x = 1$ meter from an endpoint.

(f) If two waves travel simultaneously along the same stretched string, the vertical displacement of the string when both waves act is $y = y_1 + y_2$, where y_1 is the vertical displacement of the first wave and y_2 is the vertical

displacement of the second wave. This result is called the Principle of Superposition and was analyzed by the French mathematician Jean Baptiste Fourier (1768–1830). When two waves travel along the same string, one wave will differ from the other wave by a phase constant ϕ. That is,

$$y_1 = y_m \sin(kx - \omega t)$$
$$y_2 = y_m \sin(kx - \omega t + \phi)$$

assuming that each wave has the same amplitude. Write $y_1 + y_2$ as a product using the Sum-to-Product Formulas.

(g) Suppose that two waves are moving in the same direction along a stretched string. The amplitude of each wave is 0.0045 meter, and the phase difference between them is 2.5 radians. The wavelength, λ, of each wave is 0.09 meter and the frequency, f, is 2.3 hertz. Find y_1, y_2, and $y_1 + y_2$.

(h) Using a graphing utility, graph y_1, y_2, and $y_1 + y_2$ on the same viewing window.

(i) Redo parts (g) and (h) with the phase difference between the waves being 0.4 radian.

(j) What effect does the phase difference have on the amplitude of $y_1 + y_2$?

The following projects are available on the Instructor's Resource Center (IRC):

II. **Project at Motorola** *Sending Pictures Wirelessly* The electronic transmission of pictures is made practical by image compression, mathematical methods that greatly reduce the number of bits of data used to compose the picture.

III. Calculus of Differences Finding consecutive difference quotients is called finding finite differences and is used to analyze the graph of an unknown function.

Applications of Trigonometric Functions

From Lewis and Clark to Landsat

For $140, you can buy a handheld Global Positioning System receiver that will gauge your latitude and longitude to within a couple of meters. But in 1804, when Meriwether Lewis and William Clark ventured across the Louisiana Territory, a state of the art positioning system consisted of an octant, a pocket chronometer, and a surveyor's compass.

But somehow, Clark—the cartographer in the group—made do. When San Francisco map collector David Rumsey took his copy of Lewis and Clark's published map of their journey, scanned

it into a computer, and matched landmarks such as river junctions against corresponding features on today's maps, he found that it took only a slight amount of digital stretching and twisting to make Clark's map conform to modern coordinates. In fact, Rumsey was able to combine Clark's depiction of his party's route to the Pacific with pages from government atlases from the 1870s and 1970s and photos from NASA Landsat satellites, creating a digital composite that documents not only a historic adventure, but also the history of mapmaking itself.

Source: Wade Roush, "From Lewis and Clark to Landsat: David Rumsey's Digital Maps Marry Past and Present," *Technology Review*, 108, no. 7 (July 2005): 26–27.

—*See Chapter Project II—*

◁ **A Look Back** In Chapter 5, we defined the six trigonometric functions using right triangles and then extended this definition to include any angle. In particular, we learned to evaluate the trigonometric functions. We also learned how to graph sinusoidal functions. In Chapter 6, we defined the inverse trigonometric functions and solved equations involving the trigonometric functions.

A Look Ahead ▷ In this chapter, we use the trigonometric functions to solve applied problems. The first four sections deal with applications involving right triangles and *oblique triangles*, triangles that do not have a right angle. To solve problems involving oblique triangles, we will develop the Law of Sines and the Law of Cosines. We will also develop formulas for finding the area of a triangle.

The final section deals with applications of sinusoidal functions involving simple harmonic motion and damped motion.

7.1 Applications Involving Right Triangles

PREPARING FOR THIS SECTION *Before getting started, review the following:*

- Pythagorean Theorem (Appendix A, Section A.2, pp. A14–A15)
- Trigonometric Equations (I) (Section 6.3, pp. 496–501)
- Complementary Angles (Section 5.2, pp. 397–399)

Now Work the 'Are You Prepared?' problems on page 550.

OBJECTIVES 1 Solve Right Triangles (p. 547)
2 Solve Applied Problems (p. 548)

1 Solve Right Triangles

Figure 1

In the discussion that follows, we will always label a right triangle so that side a is opposite angle A, side b is opposite angle B, and side c is the hypotenuse, as shown in Figure 1. **To solve a right triangle** means to find the missing lengths of its sides and the measurements of its angles. We shall follow the practice of expressing the lengths of the sides rounded to two decimal places and expressing angles in degrees rounded to one decimal place. (Be sure that your calculator is in degree mode.)

To solve a right triangle, we need to know one of the acute angles A or B and a side, or else two sides. Then we make use of the Pythagorean Theorem and the fact that the sum of the angles of a triangle is $180°$. The sum of the angles A and B in a right triangle is therefore $90°$.

THEOREM

For the right triangle shown in Figure 1, we have

$$c^2 = a^2 + b^2 \qquad A + B = 90°$$

EXAMPLE 1 **Solving a Right Triangle**

Use Figure 2. If $b = 2$ and $A = 40°$, find a, c, and B.

Solution Since $A = 40°$ and $A + B = 90°$, we find that $B = 50°$. To find the sides a and c, we use the facts that

Figure 2

$$\tan 40° = \frac{a}{2} \quad \text{and} \quad \cos 40° = \frac{2}{c}$$

Now solve for a and c.

$$a = 2 \tan 40° \approx 1.68 \quad \text{and} \quad c = \frac{2}{\cos 40°} \approx 2.61$$

Now Work PROBLEM 9

EXAMPLE 2 **Solving a Right Triangle**

Use Figure 3. If $a = 3$ and $b = 2$, find c, A, and B.

Figure 3

Solution Since $a = 3$ and $b = 2$, then, by the Pythagorean Theorem, we have

$$c^2 = a^2 + b^2 = 3^2 + 2^2 = 9 + 4 = 13$$
$$c = \sqrt{13} \approx 3.61$$

NOTE To avoid round-off errors when using a calculator, we will store un-rounded values in memory for use in subsequent calculations. ∎

To find angle A, we use the fact that

$$\tan A = \frac{3}{2} \quad \text{so} \quad A = \tan^{-1}\frac{3}{2}$$

Set the mode on your calculator to degrees. Then, rounded to one decimal place, we find that $A = 56.3°$. Since $A + B = 90°$, we find that $B = 33.7°$.

━━━━**Now Work** PROBLEM 19

2 Solve Applied Problems

In Section 5.3, we used right triangle trigonometry to find the lengths of unknown sides of a right triangle, given the measure of an angle and the length of a side. Now that we understand the concept of inverse trigonometric functions and know how to solve trigonometric equations, we can solve applied problems that require finding the measure of an angle given the lengths of two sides in a right triangle.

EXAMPLE 3 | **Finding the Inclination of a Mountain Trail**

A straight trail leads from the Alpine Hotel, elevation 8000 feet, to a scenic over-look, elevation 11,100 feet. The length of the trail is 14,100 feet. What is the inclination (grade) of the trail? That is, what is the angle B in Figure 4?

Solution   As we can see in Figure 4, we know the length of the side opposite angle B is $11,100 - 8000 = 3100$ feet and the length of the hypotenuse is 14,100 feet. The angle B obeys the equation

$$\sin B = \frac{3100}{14,100}$$

Using a calculator,

$$B = \sin^{-1}\frac{3100}{14,100} \approx 12.7°$$

The inclination (grade) of the trail is approximately $12.7°$.

Figure 4

━━━━**Now Work** PROBLEM 25

EXAMPLE 4 | **The Gibb's Hill Lighthouse, Southampton, Bermuda**

In operation since 1846, the Gibb's Hill Lighthouse stands 117 feet high on a hill 245 feet high, so its beam of light is 362 feet above sea level. A brochure states that the light can be seen on the horizon about 26 miles distant. Verify the accuracy of this statement.

Solution   Figure 5 illustrates the situation. The central angle θ, positioned at the center of Earth, radius 3960 miles, obeys the equation

$$\cos \theta = \frac{3960}{3960 + \dfrac{362}{5280}} \approx 0.999982687 \quad \text{1 mile = 5280 feet}$$

Solving for θ, we find

$$\theta = \cos^{-1} 0.999982687 \approx 0.33715° \approx 20.23'$$

The brochure does not indicate whether the distance is measured in nautical miles or statute miles. Let's calculate both distances.

The distance s in nautical miles (refer to Problem 114, p. 390) is the measure of the angle θ in minutes, so $s \approx 20.23$ nautical miles.

Figure 5

The distance s in statute miles is given by the formula $s = r\theta$, where θ is measured in radians. Then, since

$$\theta \approx 20.23' \approx 0.33715° \approx 0.00588 \text{ radian}$$

$$1' = \frac{1°}{60} \qquad 1° = \frac{\pi}{180} \text{ radian}$$

we find that

$$s = r\theta \approx (3960)(0.00588) \approx 23.3 \text{ miles}$$

In either case, it would seem that the brochure overstated the distance somewhat.

Figure 6

In navigation and surveying, the **direction** or **bearing** from a point O to a point P equals the acute angle θ between the ray OP and the vertical line through O, the north–south line.

Figure 6 illustrates some bearings. Notice that the bearing from O to P_1 is denoted by the symbolism N30°E, indicating that the bearing is 30° east of north. In writing the bearing from O to P, the direction north or south always appears first, followed by an acute angle, followed by east or west. In Figure 6, the bearing from O to P_2 is S50°W, and from O to P_3 it is N70°W.

EXAMPLE 5   **Finding the Bearing of an Object**

In Figure 6, what is the bearing from O to an object at P_4?

Solution   The acute angle between the ray OP_4 and the north–south line through O is 20°. The bearing from O to P_4 is S20°E.

EXAMPLE 6   **Finding the Bearing of an Airplane**

A Boeing 777 aircraft takes off from Nashville International Airport on runway 2 LEFT, which has a bearing of N20°E.* After flying for 1 mile, the pilot of the aircraft requests permission to turn 90° and head toward the northwest. The request is granted. After the plane goes 2 miles in this direction, what bearing should the control tower use to locate the aircraft?

Solution   Figure 7 illustrates the situation. After flying 1 mile from the airport O (the control tower), the aircraft is at P. After turning 90° toward the northwest and flying 2 miles, the aircraft is at the point Q. In triangle OPQ, the angle θ obeys the equation

Figure 7

$$\tan \theta = \frac{2}{1} = 2 \quad \text{so} \quad \theta = \tan^{-1} 2 \approx 63.4°$$

The acute angle between north and the ray OQ is 63.4° − 20° = 43.4°. The bearing of the aircraft from O to Q is N43.4°W.

━━━━━**Now Work**   PROBLEM 31

---

*In air navigation, the term **azimuth** denotes the positive angle measured clockwise from the north (N) to a ray OP. In Figure 6, the azimuth from O to P_1 is 30°; the azimuth from O to P_2 is 230°; the azimuth from O to P_3 is 290°. In naming runways, the units digit is left off the azimuth. Runway 2 LEFT means the left runway with a direction of azimuth 20° (bearing N20°E). Runway 23 is the runway with azimuth 230° and bearing S50°W.

7.1 Assess Your Understanding

'Are You Prepared?' *Answers are given at the end of these exercises. If you get a wrong answer, read the pages listed in* red.

1. In a right triangle, if the length of the hypotenuse is 5 and the length of one of the other sides is 3, what is the length of the third side? (pp. A14–A15)

2. **True or False** $\sin 52° = \cos 48°$. (pp. 397–398)

3. If θ is an acute angle, solve the equation $\tan \theta = \dfrac{1}{2}$. Express your answer in degrees, rounded to one decimal place. (pp. 496–498)

4. If θ is an acute angle, solve the equation $\sin \theta = \dfrac{1}{2}$. (pp. 496–498)

Concepts and Vocabulary

5. In a right triangle, the sum of the two acute angles is _____ degrees.

6. In navigation or surveying, the _____ or _____ from a point O to a point P equals the acute angle θ between ray OP and the vertical line through O, the north–south line.

7. **True or False** In a right triangle, if two sides are known, we can solve the triangle.

8. **True or False** In a right triangle, if we know the two acute angles, we can solve the triangle.

Skill Building

In Problems 9–22, use the right triangle shown below. Then, using the given information, solve the triangle.

9. $b = 5$, $B = 20°$;   find a, c, and A

10. $b = 4$, $B = 10°$;   find a, c, and A

11. $a = 6$, $B = 40°$;   find b, c, and A

12. $a = 7$, $B = 50°$;   find b, c, and A

13. $b = 4$, $A = 10°$;   find a, c, and B

14. $b = 6$, $A = 20°$;   find a, c, and B

15. $a = 5$, $A = 25°$;   find b, c, and B

16. $a = 6$, $A = 40°$;   find b, c, and B

17. $c = 9$, $B = 20°$;   find b, a, and A

18. $c = 10$, $A = 40°$;   find b, a, and B

19. $a = 5$, $b = 3$;   find c, A, and B

20. $a = 2$, $b = 8$;   find c, A, and B

21. $a = 2$, $c = 5$;   find b, A, and B

22. $b = 4$, $c = 6$;   find a, A, and B

Applications and Extensions

23. **Geometry** The hypotenuse of a right triangle is 5 inches. If one leg is 2 inches, find the degree measure of each angle.

24. **Geometry** The hypotenuse of a right triangle is 3 feet. If one leg is 1 foot, find the degree measure of each angle.

25. **Finding the Angle of Elevation of the Sun** At 10 AM on April 26, 2009, a building 300 feet high casts a shadow 50 feet long. What is the angle of elevation of the Sun?

26. **Directing a Laser Beam** A laser beam is to be directed through a small hole in the center of a circle of radius 10 feet. The origin of the beam is 35 feet from the circle (see the figure). At what angle of elevation should the beam be aimed to ensure that it goes through the hole?

27. **Finding the Speed of a Truck** A state trooper is hidden 30 feet from a highway. One second after a truck passes, the angle θ between the highway and the line of observation from the patrol car to the truck is measured. See the illustration.

(a) If the angle measures 15°, how fast is the truck traveling? Express the answer in feet per second and in miles per hour.

(b) If the angle measures 20°, how fast is the truck traveling? Express the answer in feet per second and in miles per hour.

(c) If the speed limit is 55 miles per hour and a speeding ticket is issued for speeds of 5 miles per hour or more over the limit, for what angles should the trooper issue a ticket?

28. **Security** A security camera in a neighborhood bank is mounted on a wall 9 feet above the floor. What angle of depression should be used if the camera is to be directed to a spot 6 feet above the floor and 12 feet from the wall?

29. Parallax One method of measuring the distance from Earth to a star is the parallax method. The idea behind computing this distance is to measure the angle formed between the Earth and the star at two different points in time. Typically, the measurements are taken so that the side opposite the angle is as large as possible. Therefore, the optimal approach is to measure the angle when Earth is on opposite sides of the Sun, as shown in the figure.

(a) Proxima Centauri is 4.22 light-years from Earth. If 1 light-year is about 5.9 trillion miles, how many miles is Proxima Centauri from Earth?

(b) The mean distance from Earth to the Sun is 93,000,000 miles. What is the parallax of Proxima Centauri?

30. Parallax See Problem 29. 61 Cygni, sometimes called Bessel's Star (after Friedrich Bessel, who measured the distance from Earth to the star in 1838), is a star in the constellation Cygnus.

(a) 61 Cygni is 11.14 light-years from Earth. If 1 light-year is about 5.9 trillion miles, how many miles is 61 Cygni from Earth?

(b) The mean distance from Earth to the Sun is 93,000,000 miles. What is the parallax of 61 Cygni?

31. Finding the Bearing of an Aircraft A DC-9 aircraft leaves Midway Airport from runway 4 RIGHT, whose bearing is N40°E. After flying for $\frac{1}{2}$ mile, the pilot requests permission to turn 90° and head toward the southeast. The permission is granted. After the airplane goes 1 mile in this direction, what bearing should the control tower use to locate the aircraft?

32. Finding the Bearing of a Ship A ship leaves the port of Miami with a bearing of S80°E and a speed of 15 knots. After 1 hour, the ship turns 90° toward the south. After 2 hours, maintaining the same speed, what is the bearing to the ship from the port?

33. Niagara Falls Incline Railway Situated between Portage Road and the Niagara Parkway directly across from the Canadian Horseshoe Falls, the Falls Incline Railway is a funicular that carries passengers up an embankment to Table Rock Observation Point. If the length of the track is 51.8 meters and the angle of inclination is 36°2′, determine the height of the embankment.

Source: www.niagaraparks.com

34. Willis Tower The Willis Tower in Chicago is the third tallest building in the world and is topped by a high antenna. A surveyor on the ground makes the following measurement:

1. The angle of elevation from her position to the top of the building is 34°.
2. The distance from her position to the top of the building is 2593 feet.
3. The distance from her position to the top of the antenna is 2743 feet.

(a) How far away from the base of the building is the surveyor located?

(b) How tall is the building?

(c) What is the angle of elevation from the surveyor to the top of the antenna?

(d) How tall is the antenna?

Source: www.infoplease.com/ce6/us/A0844218.html

35. Chicago Skyscrapers The angle of inclination from the base of the John Hancock Center to the top of the main structure of the Willis Tower is approximately 10.3°. If the main structure of the Willis Tower is 1451 feet tall, how far apart are the two skyscrapers? Assume the bases of the two buildings are at the same elevation.

Source: www.emporis.com

36. Estimating the Width of the Mississippi River A tourist at the top of the Gateway Arch (height, 630 feet) in St. Louis, Missouri, observes a boat moored on the Illinois side of the Mississippi River 2070 feet directly across from the Arch. She also observes a boat moored on the Missouri side directly across from the first boat (see diagram). Given that

$$B = \cot^{-1}\frac{67}{55},$$ estimate the width of the Mississippi River at the St. Louis riverfront.

Source: U.S. Army Corps of Engineers

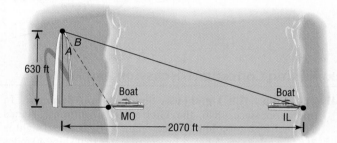

37. Finding the Pitch of a Roof A carpenter is preparing to put a roof on a garage that is 20 feet by 40 feet by 20 feet. A steel support beam 46 feet in length is positioned in the center of the garage. To support the roof, another beam will be attached to the top of the center beam (see the figure). At what angle of elevation is the new beam? In other words, what is the pitch of the roof?

38. Shooting Free Throws in Basketball The eyes of a basketball player are 6 feet above the floor. The player is at the free-throw line, which is 15 feet from the center of the basket rim (see the figure). What is the angle of elevation from the player's eyes to the center of the rim?

[**Hint:** The rim is 10 feet above the floor.]

39. Geometry Find the value of the angle θ (see the figure) in degrees rounded to the nearest tenth of a degree.

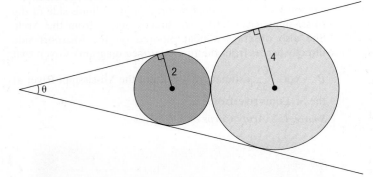

40. Surveillance Satellites A surveillance satellite circles Earth at a height of h miles above the surface. Suppose that d is the distance, in miles, on the surface of Earth that can be observed from the satellite. See the illustration.
 (a) Find an equation that relates the central angle θ to the height h.
 (b) Find an equation that relates the observable distance d and θ.
 (c) Find an equation that relates d and h.
 (d) If d is to be 2500 miles, how high must the satellite orbit above Earth?
 (e) If the satellite orbits at a height of 300 miles, what distance d on the surface can be observed?

Explaining Concepts: Discussion and Writing

41. The Gibb's Hill Lighthouse, Southampton, Bermuda In operation since 1846, the Gibb's Hill Lighthouse stands 117 feet high on a hill 245 feet high, so its beam of light is 362 feet above sea level. A brochure states that ships 40 miles away can see the light and planes flying at 10,000 feet can see it 120 miles away. Verify the accuracy of these statements. What assumption did the brochure make about the height of the ship?

'Are You Prepared?' Answers

 1. 4   **2.** False   **3.** 26.6°   **4.** 30°

7.2 The Law of Sines

PREPARING FOR THIS SECTION *Before getting started, review the following:*

- Trigonometric Equations (Section 6.3, pp. 496–501)
- Difference Formula for the Sine Function (Section 6.5, p. 517)
- Geometry Essentials (Appendix A, Section A.2, pp. A16–A19)

Now Work the 'Are You Prepared?' problems on page 559.

OBJECTIVES **1** Solve SAA or ASA Triangles (p. 553)
 2 Solve SSA Triangles (p. 554)
 3 Solve Applied Problems (p. 557)

If none of the angles of a triangle is a right angle, the triangle is called **oblique.** An oblique triangle will have either three acute angles or two acute angles and one obtuse angle (an angle between 90° and 180°). See Figure 8.

Figure 8

(a) All angles are acute   **(b)** Two acute angles and one obtuse angle

Figure 9

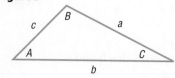

In the discussion that follows, we will always label an oblique triangle so that side a is opposite angle A, side b is opposite angle B, and side c is opposite angle C, as shown in Figure 9.

To **solve an oblique triangle** means to find the lengths of its sides and the measurements of its angles. To do this, we shall need to know the length of one side* along with (i) two angles; (ii) one angle and one other side; or (iii) the other two sides. There are four possibilities to consider:

Case 1:   One side and two angles are known (ASA or SAA).
Case 2:   Two sides and the angle opposite one of them are known (SSA).
Case 3:   Two sides and the included angle are known (SAS).
Case 4:   Three sides are known (SSS).

Figure 10 illustrates the four cases.

Figure 10

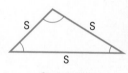

Case 1: ASA   Case 1: SAA   Case 2: SSA   Case 3: SAS   Case 4: SSS

WARNING Oblique triangles cannot be solved using the methods of Section 7.1. Do you know why? ∎

The **Law of Sines** is used to solve triangles for which Case 1 or 2 holds. Cases 3 and 4 are considered when we study the Law of Cosines in the next section.

THEOREM

Law of Sines

For a triangle with sides a, b, c and opposite angles A, B, C, respectively,

$$\frac{\sin A}{a} = \frac{\sin B}{b} = \frac{\sin C}{c} \tag{1}$$

A proof of the Law of Sines is given at the end of this section. The Law of Sines actually consists of three equalities:

$$\frac{\sin A}{a} = \frac{\sin B}{b} \qquad \frac{\sin A}{a} = \frac{\sin C}{c} \qquad \frac{\sin B}{b} = \frac{\sin C}{c}$$

Formula (1) is a compact way to write these three equations.

In applying the Law of Sines to solve triangles, we use the fact that the sum of the angles of any triangle equals 180°; that is,

$$A + B + C = 180° \tag{2}$$

1 Solve SAA or ASA Triangles

Our first two examples show how to solve a triangle when one side and two angles are known (Case 1: SAA or ASA).

*The reason we need to know the length of one side is that, if we only know the angles, this will result in a family of *similar triangles*.

EXAMPLE 1

Using the Law of Sines to Solve an SAA Triangle

Solve the triangle: $A = 40°, B = 60°, a = 4$

Solution

Figure 11 shows the triangle that we want to solve. The third angle C is found using equation (2).

$$A + B + C = 180°$$
$$40° + 60° + C = 180°$$
$$C = 80°$$

Figure 11

NOTE Although not a check, we can verify the reasonableness of our answer by determining if the longest side is opposite the largest angle and the shortest side is opposite the smallest angle. ■

Now we use the Law of Sines (twice) to find the unknown sides b and c.

$$\frac{\sin A}{a} = \frac{\sin B}{b} \qquad \frac{\sin A}{a} = \frac{\sin C}{c}$$

Because $a = 4$, $A = 40°$, $B = 60°$, and $C = 80°$, we have

$$\frac{\sin 40°}{4} = \frac{\sin 60°}{b} \qquad \frac{\sin 40°}{4} = \frac{\sin 80°}{c}$$

Solving for b and c, we find that

$$b = \frac{4 \sin 60°}{\sin 40°} \approx 5.39 \qquad c = \frac{4 \sin 80°}{\sin 40°} \approx 6.13$$

Notice in Example 1 that we found b and c by working with the given side a. This is better than finding b first and working with a rounded value of b to find c.

━━━━━➤ **Now Work** PROBLEM 9

EXAMPLE 2

Using the Law of Sines to Solve an ASA Triangle

Solve the triangle: $A = 35°, B = 15°, c = 5$

Solution

Figure 12 illustrates the triangle that we want to solve. Because we know two angles ($A = 35°$ and $B = 15°$), we find the third angle using equation (2).

$$A + B + C = 180°$$
$$35° + 15° + C = 180°$$
$$C = 130°$$

Figure 12

Now we know the three angles and one side ($c = 5$) of the triangle. To find the remaining two sides a and b, we use the Law of Sines (twice).

$$\frac{\sin A}{a} = \frac{\sin C}{c} \qquad\qquad \frac{\sin B}{b} = \frac{\sin C}{c}$$

$$\frac{\sin 35°}{a} = \frac{\sin 130°}{5} \qquad\qquad \frac{\sin 15°}{b} = \frac{\sin 130°}{5}$$

$$a = \frac{5 \sin 35°}{\sin 130°} \approx 3.74 \qquad\qquad b = \frac{5 \sin 15°}{\sin 130°} \approx 1.69$$

━━━━━➤ **Now Work** PROBLEM 23

Figure 13

2 Solve SSA Triangles

Case 2 (SSA), which applies to triangles for which two sides and the angle opposite one of them are known, is referred to as the **ambiguous case,** because the known information may result in one triangle, two triangles, or no triangle at all. Suppose that we are given sides a and b and angle A, as illustrated in Figure 13. The key to

$\sin A = \frac{h}{b}$

determining the possible triangles, if any, that may be formed from the given information lies primarily with the relative size of side a, the height h, and the fact that $h = b \sin A$.

No Triangle If $a < h = b \sin A$, then side a is not sufficiently long to form a triangle. See Figure 14.

One Right Triangle If $a = h = b \sin A$, then side a is just long enough to form a right triangle. See Figure 15.

Figure 14
$a < h = b \sin A$

Figure 15
$a = h = b \sin A$

Two Triangles If $h = b \sin A < a$, and $a < b$, two distinct triangles can be formed from the given information. See Figure 16.

One Triangle If $a \geq b$, only one triangle can be formed. See Figure 17.

Figure 16
$b \sin A < a$ and $a < b$

Figure 17
$a \geq b$

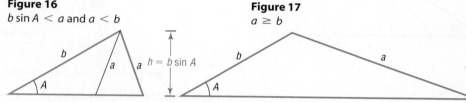

Fortunately, we do not have to rely on an illustration or complicated relationships to draw the correct conclusion in the ambiguous case. The Law of Sines will lead us to the correct determination. Let's see how.

EXAMPLE 3   **Using the Law of Sines to Solve an SSA Triangle (One Solution)**

Solve the triangle: $a = 3, b = 2, A = 40°$

Solution

See Figure 18(a). Because we know an angle ($A = 40°$), the side opposite the known angle ($a = 3$), and the side opposite angle B ($b = 2$), we use the Law of Sines to find the angle B.

Figure 18(a)

Then

$$\frac{\sin A}{a} = \frac{\sin B}{b}$$

$$\frac{\sin 40°}{3} = \frac{\sin B}{2}$$

$$\sin B = \frac{2 \sin 40°}{3} \approx 0.43$$

There are two angles B, $0° < B < 180°$, for which $\sin B \approx 0.43$.

$$B_1 \approx 25.4° \quad \text{and} \quad B_2 \approx 180° - 25.4° = 154.6°$$

NOTE Here we computed B_1 by determining the value of $\sin^{-1}\left(\frac{2 \sin 40°}{3}\right)$. If you use the rounded value and evaluate $\sin^{-1}(0.43)$, you will obtain a slightly different result.   ∎

The second possibility, $B_2 \approx 154.6°$, is ruled out, because $A = 40°$ makes $A + B_2 \approx 194.6° > 180°$. Now, using $B_1 \approx 25.4°$, we find that

$$C = 180° - A - B_1 \approx 180° - 40° - 25.4° = 114.6°$$

The third side c may now be determined using the Law of Sines.

$$\frac{\sin A}{a} = \frac{\sin C}{c}$$

$$\frac{\sin 40°}{3} = \frac{\sin 114.6°}{c}$$

Figure 18(b)

$$c = \frac{3 \sin 114.6°}{\sin 40°} \approx 4.24$$

Figure 18(b) illustrates the solved triangle.

| **EXAMPLE 4** | **Using the Law of Sines to Solve an SSA Triangle (Two Solutions)** |

Solve the triangle: $a = 6, b = 8, A = 35°$

Solution   See Figure 19(a). Because $a = 6$, $b = 8$, and $A = 35°$ are known, we use the Law of Sines to find the angle B.

Figure 19(a)

$$\frac{\sin A}{a} = \frac{\sin B}{b}$$

Then

$$\frac{\sin 35°}{6} = \frac{\sin B}{8}$$

$$\sin B = \frac{8 \sin 35°}{6} \approx 0.76$$

$$B_1 \approx 49.9° \quad \text{or} \quad B_2 \approx 180° - 49.9° = 130.1°$$

For both choices of B, we have $A + B < 180°$. There are two triangles, one containing the angle $B_1 \approx 49.9°$ and the other containing the angle $B_2 \approx 130.1°$. The third angle C is either

$$C_1 = 180° - A - B_1 \approx 95.1° \quad \text{or} \quad C_2 = 180° - A - B_2 \approx 14.9°$$

$$\uparrow \qquad\qquad\qquad\qquad\qquad\qquad \uparrow$$

$$A = 35° \qquad\qquad\qquad\qquad\qquad A = 35°$$
$$B_1 = 49.9° \qquad\qquad\qquad\qquad B_2 = 130.1°$$

The third side c obeys the Law of Sines, so we have

Figure 19(b)

$$\frac{\sin A}{a} = \frac{\sin C_1}{c_1} \qquad\qquad \frac{\sin A}{a} = \frac{\sin C_2}{c_2}$$

$$\frac{\sin 35°}{6} = \frac{\sin 95.1°}{c_1} \qquad\qquad \frac{\sin 35°}{6} = \frac{\sin 14.9°}{c_2}$$

$$c_1 = \frac{6 \sin 95.1°}{\sin 35°} \approx 10.42 \qquad\qquad c_2 = \frac{6 \sin 14.9°}{\sin 35°} \approx 2.69$$

The two solved triangles are illustrated in Figure 19(b).

| **EXAMPLE 5** | **Using the Law of Sines to Solve an SSA Triangle (No Solution)** |

Solve the triangle: $a = 2, c = 1, C = 50°$

Solution   Because $a = 2$, $c = 1$, and $C = 50°$ are known, we use the Law of Sines to find the angle A.

$$\frac{\sin A}{a} = \frac{\sin C}{c}$$

$$\frac{\sin A}{2} = \frac{\sin 50°}{1}$$

$$\sin A = 2 \sin 50° \approx 1.53$$

Figure 20

Since there is no angle A for which $\sin A > 1$, there can be no triangle with the given measurements. Figure 20 illustrates the measurements given. Notice that, no matter how we attempt to position side c, it will never touch side b to form a triangle.

━━━━ **Now Work** PROBLEMS 25 AND 31

 3 Solve Applied Problems

The Law of Sines is particularly useful for solving certain applied problems.

EXAMPLE 6   **Finding the Height of a Mountain**

To measure the height of a mountain, a surveyor takes two sightings of the peak at a distance 900 meters apart on a direct line to the mountain.* See Figure 21(a). The first observation results in an angle of elevation of 47°, and the second results in an angle of elevation of 35°. If the transit is 2 meters high, what is the height h of the mountain?

Figure 21

(a)   (b)

Solution   Figure 21(b) shows the triangles that replicate the illustration in Figure 21(a). Since $C + 47° = 180°$, we find that $C = 133°$. Also, since $A + C + 35° = 180°$, we find that $A = 180° - 35° - C = 145° - 133° = 12°$. We use the Law of Sines to find c.

$$\frac{\sin A}{a} = \frac{\sin C}{c} \qquad A = 12°, C = 133°, a = 900$$

$$c = \frac{900 \sin 133°}{\sin 12°} \approx 3165.86$$

Using the larger right triangle, we have

$$\sin 35° = \frac{b}{c} \qquad c = 3165.86$$

$$b = 3165.86 \sin 35° \approx 1815.86 \approx 1816 \text{ meters}$$

The height of the peak from ground level is approximately $1816 + 2 = 1818$ meters.

━━━━ **Now Work** PROBLEM 39

EXAMPLE 7   **Rescue at Sea**

Coast Guard Station Zulu is located 120 miles due west of Station X-ray. A ship at sea sends an SOS call that is received by each station. The call to Station Zulu indicates that the bearing of the ship from Zulu is N40°E (40° east of north). The call to Station X-ray indicates that the bearing of the ship from X-ray is N30°W (30° west of north).

(a) How far is each station from the ship?

(b) If a helicopter capable of flying 200 miles per hour is dispatched from the nearest station to the ship, how long will it take to reach the ship?

*For simplicity, we assume that these sightings are at the same level.

Figure 22

Solution

(a) Figure 22 illustrates the situation. The angle C is found to be

$$C = 180° - 50° - 60° = 70°$$

The Law of Sines can now be used to find the two distances a and b that we seek.

$$\frac{\sin 50°}{a} = \frac{\sin 70°}{120}$$

$$a = \frac{120 \sin 50°}{\sin 70°} \approx 97.82 \text{ miles}$$

$$\frac{\sin 60°}{b} = \frac{\sin 70°}{120}$$

$$b = \frac{120 \sin 60°}{\sin 70°} \approx 110.59 \text{ miles}$$

Station Zulu is about 111 miles from the ship, and Station X-ray is about 98 miles from the ship.

(b) The time t needed for the helicopter to reach the ship from Station X-ray is found by using the formula

$$(\text{Rate}, r)(\text{Time}, t) = \text{Distance}, a$$

Then

$$t = \frac{a}{r} = \frac{97.82}{200} \approx 0.49 \text{ hour} \approx 29 \text{ minutes}$$

It will take about 29 minutes for the helicopter to reach the ship.

●

──**Now Work** PROBLEM **37**

Figure 23

(a)

(b)

Proof of the Law of Sines To prove the Law of Sines, we construct an altitude of length h from one of the vertices of a triangle. Figure 23(a) shows h for a triangle with three acute angles, and Figure 23(b) shows h for a triangle with an obtuse angle. In each case, the altitude is drawn from the vertex at B. Using either illustration, we have

$$\sin C = \frac{h}{a}$$

from which

$$h = a \sin C \tag{3}$$

From Figure 23(a), it also follows that

$$\sin A = \frac{h}{c}$$

from which

$$h = c \sin A \tag{4}$$

From Figure 23(b), it follows that

$$\sin(180° - A) = \sin A = \frac{h}{c}$$

$$\underset{\uparrow}{}$$

$$\sin (180° - A) = \sin 180° \cos A - \cos 180° \sin A = \sin A$$

which again gives

$$h = c \sin A$$

So, whether the triangle has three acute angles or has two acute angles and one obtuse angle, equations (3) and (4) hold. As a result, we may equate the expressions for h in equations (3) and (4) to get

$$a \sin C = c \sin A$$

Figure 24

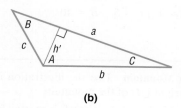

(a)

(b)

from which

$$\frac{\sin A}{a} = \frac{\sin C}{c} \qquad (5)$$

In a similar manner, by constructing the altitude h' from the vertex of angle A as shown in Figure 24, we can show that

$$\sin B = \frac{h'}{c} \quad \text{and} \quad \sin C = \frac{h'}{b}$$

Equating the expressions for h', we find that

$$h' = c \sin B = b \sin C$$

from which

$$\frac{\sin B}{b} = \frac{\sin C}{c} \qquad (6)$$

When equations (5) and (6) are combined, we have equation (1), the Law of Sines. ■

7.2 Assess Your Understanding

'Are You Prepared?' *Answers are given at the end of these exercises. If you get a wrong answer, read the pages listed in red.*

1. The difference formula for the sine function is $\sin(A - B) =$ ____. (p. 517)

2. If θ is an acute angle, solve the equation $\cos \theta = \dfrac{\sqrt{3}}{2}$. (p. 497)

3. The two triangles shown are similar. Find the missing length. (p. A18)

Concepts and Vocabulary

4. If none of the angles of a triangle is a right angle, the triangle is called _____.

5. For a triangle with sides a, b, c and opposite angles A, B, C, the Law of Sines states that _____.

6. *True or False* An oblique triangle in which two sides and an angle are given always results in at least one triangle.

7. *True or False* The Law of Sines can be used to solve triangles where three sides are known.

8. Triangles for which two sides and the angle opposite one of them are known (SSA) are referred to as the _____ _____.

Skill Building

In Problems 9–16, solve each triangle.

9.

10.

11.

12.

13.

14.

15.

16.

In Problems 17–24, solve each triangle.

17. $A = 40°$, $B = 20°$, $a = 2$

18. $A = 50°$, $C = 20°$, $a = 3$

19. $B = 70°$, $C = 10°$, $b = 5$

20. $A = 70°$, $B = 60°$, $c = 4$

21. $A = 110°$, $C = 30°$, $c = 3$

22. $B = 10°$, $C = 100°$, $b = 2$

23. $A = 40°$, $B = 40°$, $c = 2$

24. $B = 20°$, $C = 70°$, $a = 1$

In Problems 25–36, two sides and an angle are given. Determine whether the given information results in one triangle, two triangles, or no triangle at all. Solve any triangle(s) that results.

25. $a = 3$, $b = 2$, $A = 50°$

26. $b = 4$, $c = 3$, $B = 40°$

27. $b = 5$, $c = 3$, $B = 100°$

28. $a = 2$, $c = 1$, $A = 120°$

29. $a = 4$, $b = 5$, $A = 60°$

30. $b = 2$, $c = 3$, $B = 40°$

31. $b = 4$, $c = 6$, $B = 20°$

32. $a = 3$, $b = 7$, $A = 70°$

33. $a = 2$, $c = 1$, $C = 100°$

34. $b = 4$, $c = 5$, $B = 95°$

35. $a = 2$, $c = 1$, $C = 25°$

36. $b = 4$, $c = 5$, $B = 40°$

Applications and Extensions

37. **Rescue at Sea** Coast Guard Station Able is located 150 miles due south of Station Baker. A ship at sea sends an SOS call that is received by each station. The call to Station Able indicates that the ship is located N55°E; the call to Station Baker indicates that the ship is located S60°E.
    (a) How far is each station from the ship?
    (b) If a helicopter capable of flying 200 miles per hour is dispatched from the station nearest the ship, how long will it take to reach the ship?

38. **Distance to the Moon** At exactly the same time, Tom and Alice measured the angle of elevation to the moon while standing exactly 300 km apart. The angle of elevation to the moon for Tom was 49.8974° and the angle of elevation to the moon for Alice was 49.9312°. See the figure. To the nearest 1000 km, how far was the moon from Earth when the measurement was obtained?

39. **Finding the Length of a Ski Lift** Consult the figure. To find the length of the span of a proposed ski lift from P to Q, a surveyor measures $\angle DPQ$ to be 25° and then walks off a distance of 1000 feet to R and measures $\angle PRQ$ to be 15°. What is the distance from P to Q?

40. **Finding the Height of a Mountain** Use the illustration in Problem 39 to find the height QD of the mountain.

41. **Finding the Height of an Airplane** An aircraft is spotted by two observers who are 1000 feet apart. As the airplane passes over the line joining them, each observer takes a sighting of the angle of elevation to the plane, as indicated in the figure. How high is the airplane?

42. **Finding the Height of the Bridge over the Royal Gorge** The highest bridge in the world is the bridge over the Royal Gorge of the Arkansas River in Colorado. Sightings to the same point at water level directly under the bridge are taken from each side of the 880-foot-long bridge, as indicated in the figure. How high is the bridge?

Source: Guinness Book of World Records

43. **Landscaping** Pat needs to determine the height of a tree before cutting it down to be sure that it will not fall on a nearby fence. The angle of elevation of the tree from one position on a flat path from the tree is 30°, and from a second position 40 feet farther along this path it is 20°. What is the height of the tree?

44. **Construction** A loading ramp 10 feet long that makes an angle of 18° with the horizontal is to be replaced by one that makes an angle of 12° with the horizontal. How long is the new ramp?

45. **Commercial Navigation** Adam must fly home to St. Louis from a business meeting in Oklahoma City. One flight option

flies directly to St. Louis, a distance of about 461.1 miles. A second flight option flies first to Kansas City and then connects to St. Louis. The bearing from Oklahoma City to Kansas City is N29.6°E, and the bearing from Oklahoma City to St. Louis is N57.7°E. The bearing from St. Louis to Oklahoma City is S62.1°W, and the bearing from St. Louis to Kansas City is N79.4°W. How many more frequent flyer miles will Adam receive if he takes the connecting flight rather than the direct flight?

Source: www.landings.com

46. **Time Lost due to a Navigation Error** In attempting to fly from city P to city Q, an aircraft followed a course that was 10° in error, as indicated in the figure. After flying a distance of 50 miles, the pilot corrected the course by turning at point R and flying 70 miles farther. If the constant speed of the aircraft was 250 miles per hour, how much time was lost due to the error?

47. **Finding the Lean of the Leaning Tower of Pisa** The famous Leaning Tower of Pisa was originally 184.5 feet high.* At a distance of 123 feet from the base of the tower, the angle of elevation to the top of the tower is found to be 60°. Find $\angle RPQ$ indicated in the figure. Also, find the perpendicular distance from R to PQ.

48. **Crankshafts on Cars** On a certain automobile, the crankshaft is 3 inches long and the connecting rod is 9 inches long (see the figure). At the time when $\angle OPQ$ is 15°, how far is the piston (P) from the center (O) of the crankshaft?

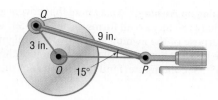

49. **Constructing a Highway** U.S. 41, a highway whose primary directions are north–south, is being constructed along the west coast of Florida. Near Naples, a bay obstructs the straight path of the road. Since the cost of a bridge is prohibitive, engineers decide to go around the bay. The illustration shows the path that they decide on and the measurements taken. What is the length of highway needed to go around the bay?

50. **Calculating Distances at Sea** The navigator of a ship at sea spots two lighthouses that she knows to be 3 miles apart along a straight seashore. She determines that the angles formed between two line-of-sight observations of the lighthouses and the line from the ship directly to shore are 15° and 35°. See the illustration.
(a) How far is the ship from lighthouse P?
(b) How far is the ship from lighthouse Q?
(c) How far is the ship from shore?

*On February 27, 1964, the government of Italy requested aid in preventing the tower from toppling. A multinational task force of engineers, mathematicians, and historians was assigned and met on the Azores islands to discuss stabilization methods. After over two decades of work on the subject, the tower was closed to the public in January 1990. During the time that the tower was closed, the bells were removed to relieve some weight, and cables were cinched around the third level and anchored several hundred meters away. Apartments and houses in the path of the tower were vacated for safety concerns. After a decade of corrective reconstruction and stabilization efforts, the tower was reopened to the public on December 15, 2001. Many methods were proposed to stabilize the tower, including the addition of 800 metric tons of lead counterweights to the raised end of the base. The final solution to correcting the lean was to remove 38 cubic meters of soil from underneath the raised end. The tower has been declared stable for at least another 300 years.

Source: http://en.wikipedia.org/wiki/Leaning_Tower_of_Pisa, page last modified June 28, 2006

51. Designing an Awning An awning that covers a sliding glass door that is 88 inches tall forms an angle of 50° with the wall. The purpose of the awning is to prevent sunlight from entering the house when the angle of elevation of the Sun is more than 65°. See the figure. Find the length L of the awning.

52. Finding Distances A forest ranger is walking on a path inclined at 5° to the horizontal directly toward a 100-foot-tall fire observation tower. The angle of elevation from the path to the top of the tower is 40°. How far is the ranger from the tower at this time?

53. Great Pyramid of Cheops One of the original Seven Wonders of the World, the Great Pyramid of Cheops was built about 2580 BC. Its original height was 480 feet 11 inches, but owing to the loss of its topmost stones, it is now shorter. Find the current height of the Great Pyramid using the information given in the illustration.

Source: Guinness Book of World Records

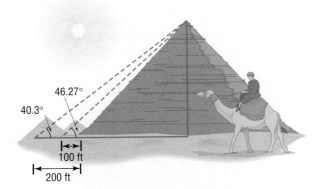

54. Determining the Height of an Aircraft Two sensors are spaced 700 feet apart along the approach to a small airport. When an aircraft is nearing the airport, the angle of elevation from the first sensor to the aircraft is 20°, and from the second sensor to the aircraft it is 15°. Determine how high the aircraft is at this time.

55. Mercury The distance from the Sun to Earth is approximately 149,600,000 kilometers (km). The distance from the Sun to Mercury is approximately 57,910,000 km. The **elongation angle** α is the angle formed between the line of sight from Earth to the Sun and the line of sight from Earth to Mercury. See the figure. Suppose that the elongation angle for Mercury is 15°. Use this information to find the possible distances between Earth and Mercury.

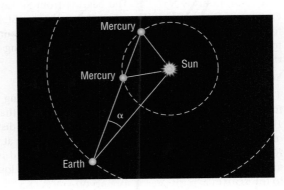

56. Venus The distance from the Sun to Earth is approximately 149,600,000 km. The distance from the Sun to Venus is approximately 108,200,000 km. The elongation angle α is the angle formed between the line of sight from Earth to the Sun and the line of sight from Earth to Venus. Suppose that the elongation angle for Venus is 10°. Use this information to find the possible distances between Earth and Venus.

57. The Original Ferris Wheel George Washington Gale Ferris, Jr., designed the original Ferris wheel for the 1893 World's Columbian Exposition in Chicago, Illinois. The wheel had 36 equally spaced cars each the size of a school bus. The distance between adjacent cars was approximately 22 feet. Determine the diameter of the wheel to the nearest foot.

Source: Carnegie Library of Pittsburgh, www.clpgh.org

58. Mollweide's Formula For any triangle, Mollweide's Formula (named after Karl Mollweide, 1774–1825) states that

$$\frac{a+b}{c} = \frac{\cos\left[\frac{1}{2}(A-B)\right]}{\sin\left(\frac{1}{2}C\right)}$$

Derive it.

[**Hint:** Use the Law of Sines and then a Sum-to-Product Formula. Notice that this formula involves all six parts of a triangle. As a result, it is sometimes used to check the solution of a triangle.]

59. Mollweide's Formula Another form of Mollweide's Formula is

$$\frac{a-b}{c} = \frac{\sin\left[\frac{1}{2}(A-B)\right]}{\cos\left(\frac{1}{2}C\right)}$$

Derive it.

60. For any triangle, derive the formula

$$a = b\cos C + c\cos B$$

[**Hint:** Use the fact that $\sin A = \sin(180° - B - C)$.]

61. Law of Tangents For any triangle, derive the Law of Tangents.

$$\frac{a - b}{a + b} = \frac{\tan\left[\frac{1}{2}(A - B)\right]}{\tan\left[\frac{1}{2}(A + B)\right]}$$

[**Hint:** Use Mollweide's Formula.]

62. Circumscribing a Triangle Show that

$$\frac{\sin A}{a} = \frac{\sin B}{b} = \frac{\sin C}{c} = \frac{1}{2r}$$

where r is the radius of the circle circumscribing the triangle PQR whose sides are a, b, and c, as shown in the figure.

[**Hint:** Draw the diameter PP'. Then $B = \angle PQR = \angle PP'R$, and angle $\angle PRP' = 90°$.]

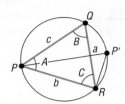

Explaining Concepts: Discussion and Writing

63. Make up three problems involving oblique triangles. One should result in one triangle, the second in two triangles, and the third in no triangle.

64. What do you do first if you are asked to solve a triangle and are given one side and two angles?

65. What do you do first if you are asked to solve a triangle and are given two sides and the angle opposite one of them?

'Are You Prepared?' Answers

1. $\sin A \cos B - \cos A \sin B$ **2.** $30°$ **3.** $\dfrac{15}{2}$

7.3 The Law of Cosines

PREPARING FOR THIS SECTION *Before getting started, review the following:*

- Trigonometric Equations (Section 6.3, pp. 496–501)
- Distance Formula (Foundations, Section 1, p. 3)

Now Work the 'Are You Prepared?' problems on page 566.

OBJECTIVES **1** Solve SAS Triangles (p. 564)
  2 Solve SSS Triangles (p. 565)
  3 Solve Applied Problems (p. 565)

In the previous section, we used the Law of Sines to solve Case 1 (SAA or ASA) and Case 2 (SSA) of an oblique triangle. In this section, we derive the Law of Cosines and use it to solve the remaining cases, 3 and 4.

Case 3: Two sides and the included angle are known (SAS).
Case 4: Three sides are known (SSS).

THEOREM

Law of Cosines

For a triangle with sides a, b, c and opposite angles A, B, C, respectively,

$$c^2 = a^2 + b^2 - 2ab \cos C \qquad \text{(1)}$$
$$b^2 = a^2 + c^2 - 2ac \cos B \qquad \text{(2)}$$
$$a^2 = b^2 + c^2 - 2bc \cos A \qquad \text{(3)}$$

Figure 25

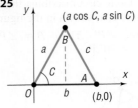

(a) Angle *C* is acute

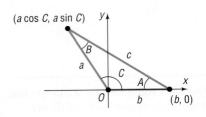

(b) Angle *C* is obtuse

Proof We will prove only formula (1) here. Formulas (2) and (3) may be proved using the same argument.

We begin by strategically placing a triangle on a rectangular coordinate system so that the vertex of angle *C* is at the origin and side *b* lies along the positive *x*-axis. Regardless of whether *C* is acute, as in Figure 25(a), or obtuse, as in Figure 25(b), the vertex of angle *B* has coordinates $(a \cos C, a \sin C)$. The vertex of angle *A* has coordinates $(b, 0)$.

We can now use the distance formula to compute c^2.

$$c^2 = (b - a \cos C)^2 + (0 - a \sin C)^2$$
$$= b^2 - 2ab \cos C + a^2 \cos^2 C + a^2 \sin^2 C$$
$$= b^2 - 2ab \cos C + a^2(\cos^2 C + \sin^2 C)$$
$$= a^2 + b^2 - 2ab \cos C \qquad ■$$

Each of formulas (1), (2), and (3) may be stated in words as follows:

THEOREM

Law of Cosines

The square of one side of a triangle equals the sum of the squares of the other two sides minus twice their product times the cosine of their included angle.

Observe that if the triangle is a right triangle (so that, say, $C = 90°$), formula (1) becomes the familiar Pythagorean Theorem: $c^2 = a^2 + b^2$. The Pythagorean Theorem is a special case of the Law of Cosines!

1 Solve SAS Triangles

Let's see how to use the Law of Cosines to solve Case 3 (SAS), which applies to triangles for which two sides and the included angle are known.

EXAMPLE 1 **Using the Law of Cosines to Solve an SAS Triangle**

Solve the triangle: $a = 2$, $b = 3$, $C = 60°$

Solution See Figure 26. Because we know two sides, *a* and *b*, and the included angle, $C = 60°$, the Law of Cosines makes it easy to find the third side, *c*.

Figure 26

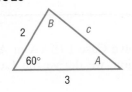

$$c^2 = a^2 + b^2 - 2ab \cos C$$
$$= 2^2 + 3^2 - 2 \cdot 2 \cdot 3 \cdot \cos 60° \quad {\scriptstyle a = 2, b = 3, C = 60°}$$
$$= 13 - \left(12 \cdot \frac{1}{2}\right) = 7$$
$$c = \sqrt{7}$$

Side *c* is of length $\sqrt{7}$. To find the angles *A* and *B*, we may use either the Law of Sines or the Law of Cosines. It is preferable to use the Law of Cosines, since it will lead to an equation with one solution. Using the Law of Sines would lead to an equation with two solutions that would need to be checked to determine which solution fits the given data. We choose to use formulas (2) and (3) of the Law of Cosines to find *A* and *B*.

For *A*:

$$a^2 = b^2 + c^2 - 2bc \cos A$$
$$2bc \cos A = b^2 + c^2 - a^2$$
$$\cos A = \frac{b^2 + c^2 - a^2}{2bc} = \frac{9 + 7 - 4}{2 \cdot 3\sqrt{7}} = \frac{12}{6\sqrt{7}} = \frac{2\sqrt{7}}{7}$$
$$A = \cos^{-1} \frac{2\sqrt{7}}{7} \approx 40.9°$$

For *B*:

$$b^2 = a^2 + c^2 - 2ac \cos B$$

$$\cos B = \frac{a^2 + c^2 - b^2}{2ac} = \frac{4 + 7 - 9}{4\sqrt{7}} = \frac{2}{4\sqrt{7}} = \frac{\sqrt{7}}{14}$$

$$B = \cos^{-1} \frac{\sqrt{7}}{14} \approx 79.1°$$

NOTE We could also have found *B* using the fact that the sum $A + B + C =$ 180°, so $B = 180° - 40.9° - 60° =$ 79.1°. However, using the Law of Cosines twice allows for a check. ∎

Notice that $A + B + C = 40.9° + 79.1° + 60° = 180°$, as required.

━━━━ **Now Work** PROBLEM 9

2 Solve SSS Triangles

The next example illustrates how the Law of Cosines is used when three sides of a triangle are known, Case 4 (SSS).

EXAMPLE 2

Using the Law of Cosines to Solve an SSS Triangle

Solve the triangle: $a = 4, b = 3, c = 6$

Solution

See Figure 27. To find the angles *A*, *B*, and *C*, we proceed as we did to find the angles in the solution to Example 1.

Figure 27

For *A*:

$$\cos A = \frac{b^2 + c^2 - a^2}{2bc} = \frac{9 + 36 - 16}{2 \cdot 3 \cdot 6} = \frac{29}{36}$$

$$A = \cos^{-1} \frac{29}{36} \approx 36.3°$$

For *B*:

$$\cos B = \frac{a^2 + c^2 - b^2}{2ac} = \frac{16 + 36 - 9}{2 \cdot 4 \cdot 6} = \frac{43}{48}$$

$$B = \cos^{-1} \frac{43}{48} \approx 26.4°$$

Since we know *A* and *B*,

$$C = 180° - A - B \approx 180° - 36.3° - 26.4° = 117.3°$$

━━━━ **Now Work** PROBLEM 15

3 Solve Applied Problems

EXAMPLE 3

Correcting a Navigational Error

A motorized sailboat leaves Naples, Florida, bound for Key West, 150 miles away. Maintaining a constant speed of 15 miles per hour, but encountering heavy cross-winds and strong currents, the crew finds, after 4 hours, that the sailboat is off course by 20°.

(a) How far is the sailboat from Key West at this time?

(b) Through what angle should the sailboat turn to correct its course?

(c) How much time has been added to the trip because of this? (Assume that the speed remains at 15 miles per hour.)

Solution   See Figure 28. With a speed of 15 miles per hour, the sailboat has gone 60 miles after 4 hours. We seek the distance x of the sailboat from Key West. We also seek the angle θ that the sailboat should turn through to correct its course.

(a) To find x, we use the Law of Cosines, since we know two sides and the included angle.

Figure 28

$$x^2 = 150^2 + 60^2 - 2(150)(60) \cos 20° \approx 9185.53$$
$$x \approx 95.8$$

The sailboat is about 96 miles from Key West.

(b) We now know three sides of the triangle, so we can use the Law of Cosines again to find the angle A opposite the side of length 150 miles.

$$150^2 = 96^2 + 60^2 - 2(96)(60) \cos A$$
$$9684 = -11{,}520 \cos A$$
$$\cos A \approx -0.8406$$
$$A \approx 147.2°$$

The sailboat should turn through an angle of

$$\theta = 180° - A \approx 180° - 147.2° = 32.8°$$

The sailboat should turn through an angle of about 33° to correct its course.

(c) The total length of the trip is now $60 + 96 = 156$ miles. The extra 6 miles will only require about 0.4 hour or 24 minutes more if the speed of 15 miles per hour is maintained.

────── **Now Work** PROBLEM 45

Historical Feature

The Law of Sines was known vaguely long before it was explicitly stated by Nasir Eddin (about AD 1250). Ptolemy (about AD 150) was aware of it in a form using a chord function instead of the sine function. But it was first clearly stated in Europe by Regiomontanus, writing in 1464.

The Law of Cosines appears first in Euclid's *Elements* (Book II), but in a well-disguised form in which squares built on the sides of triangles are added and a rectangle representing the cosine term is subtracted. It was thus known to all mathematicians because of their familiarity with Euclid's work. An early modern form of the Law of Cosines, that for finding the angle when the sides are known, was stated by François Viète (in 1593).

The Law of Tangents (see Problem 61 of Exercise 7.2) has become obsolete. In the past it was used in place of the Law of Cosines, because the Law of Cosines was very inconvenient for calculation with logarithms or slide rules. Mixing of addition and multiplication is now very easy on a calculator, however, and the Law of Tangents has been shelved along with the slide rule.

7.3 Assess Your Understanding

'Are You Prepared?'   *Answers are given at the end of these exercises. If you get a wrong answer, read the pages listed in* red.

1. Write the formula for the distance d from $P_1 = (x_1, y_1)$ to $P_2 = (x_2, y_2)$. (p. 3)

2. If θ is an acute angle, solve the equation $\cos \theta = \dfrac{\sqrt{2}}{2}$. (p. 497)

Concepts and Vocabulary

3. If three sides of a triangle are given, the Law of _____ is used to solve the triangle.

4. If one side and two angles of a triangle are given, the Law of _____ is used to solve the triangle.

5. If two sides and the included angle of a triangle are given, the Law of _____ is used to solve the triangle.

6. **True or False**   Given only the three sides of a triangle, there is insufficient information to solve the triangle.

7. **True or False**   Given two sides and the included angle, the first thing to do to solve the triangle is to use the Law of Sines.

8. **True or False**   A special case of the Law of Cosines is the Pythagorean Theorem.

Skill Building

In Problems 9–16, solve each triangle.

9.

10.

11.

12.

13.

14.

15.

16.

In Problems 17–32, solve each triangle.

17. $a = 3$, $b = 4$, $C = 40°$

18. $a = 2$, $c = 1$, $B = 10°$

19. $b = 1$, $c = 3$, $A = 80°$

20. $a = 6$, $b = 4$, $C = 60°$

21. $a = 3$, $c = 2$, $B = 110°$

22. $b = 4$, $c = 1$, $A = 120°$

23. $a = 2$, $b = 2$, $C = 50°$

24. $a = 3$, $c = 2$, $B = 90°$

25. $a = 12$, $b = 13$, $c = 5$

26. $a = 4$, $b = 5$, $c = 3$

27. $a = 2$, $b = 2$, $c = 2$

28. $a = 3$, $b = 3$, $c = 2$

29. $a = 5$, $b = 8$, $c = 9$

30. $a = 4$, $b = 3$, $c = 6$

31. $a = 10$, $b = 8$, $c = 5$

32. $a = 9$, $b = 7$, $c = 10$

Mixed Practice

In Problems 33–42, solve each triangle using either the Law of Sines or the Law of Cosines.

33. $B = 20°, C = 75°, b = 5$

34. $A = 50°, B = 55°, c = 9$

35. $a = 6, b = 8, c = 9$

36. $a = 14, b = 7, A = 85°$

37. $B = 35°, C = 65°, a = 15$

38. $a = 4, c = 5, B = 55°$

39. $A = 10°, a = 3, b = 10$

40. $A = 65°, B = 72°, b = 7$

41. $b = 5, c = 12, A = 60°$

42. $a = 10, b = 10, c = 15$

Applications and Extensions

43. Distance to the Green   A golfer hits an errant tee shot that lands in the rough. A marker in the center of the fairway is 150 yards from the center of the green. While standing on the marker and facing the green, the golfer turns 110° toward his ball. He then paces off 35 yards to his ball. See the figure. How far is the ball from the center of the green?

44. Navigation   An airplane flies due north from Ft. Myers to Sarasota, a distance of 150 miles, and then turns through an angle of 50° and flies to Orlando, a distance of 100 miles. See the figure.

(a) How far is it directly from Ft. Myers to Orlando?
(b) What bearing should the pilot use to fly directly from Ft. Myers to Orlando?

45. **Avoiding a Tropical Storm** A cruise ship maintains an average speed of 15 knots in going from San Juan, Puerto Rico, to Barbados, West Indies, a distance of 600 nautical miles. To avoid a tropical storm, the captain heads out of San Juan in a direction of 20° off a direct heading to Barbados. The captain maintains the 15-knot speed for 10 hours, after which time the path to Barbados becomes clear of storms.
    (a) Through what angle should the captain turn to head directly to Barbados?
    (b) Once the turn is made, how long will it be before the ship reaches Barbados if the same 15-knot speed is maintained?

46. **Revising a Flight Plan** In attempting to fly from Chicago to Louisville, a distance of 330 miles, a pilot inadvertently took a course that was 10° in error, as indicated in the figure.
    (a) If the aircraft maintains an average speed of 220 miles per hour and if the error in direction is discovered after 15 minutes, through what angle should the pilot turn to head toward Louisville?
    (b) What new average speed should the pilot maintain so that the total time of the trip is 90 minutes?

47. **Major League Baseball Field** A Major League baseball diamond is actually a square 90 feet on a side. The pitching rubber is located 60.5 feet from home plate on a line joining home plate and second base.
    (a) How far is it from the pitching rubber to first base?
    (b) How far is it from the pitching rubber to second base?
    (c) If a pitcher faces home plate, through what angle does he need to turn to face first base?

48. **Little League Baseball Field** According to Little League baseball official regulations, the diamond is a square 60 feet on a side. The pitching rubber is located 46 feet from home plate on a line joining home plate and second base.
    (a) How far is it from the pitching rubber to first base?
    (b) How far is it from the pitching rubber to second base?
    (c) If a pitcher faces home plate, through what angle does he need to turn to face first base?

49. **Finding the Length of a Guy Wire** The height of a radio tower is 500 feet, and the ground on one side of the tower slopes upward at an angle of 10° (see the figure).
    (a) How long should a guy wire be if it is to connect to the top of the tower and be secured at a point on the sloped side 100 feet from the base of the tower?
    (b) How long should a second guy wire be if it is to connect to the middle of the tower and be secured at a point 100 feet from the base on the flat side?

50. **Finding the Length of a Guy Wire** A radio tower 500 feet high is located on the side of a hill with an inclination to the horizontal of 5°. See the figure. How long should two guy wires be if they are to connect to the top of the tower and be secured at two points 100 feet directly above and directly below the base of the tower?

51. **Wrigley Field, Home of the Chicago Cubs** The distance from home plate to the fence in dead center in Wrigley Field is 400 feet (see the figure). How far is it from the fence in dead center to third base?

52. Little League Baseball The distance from home plate to the fence in dead center at the Oak Lawn Little League field is 280 feet. How far is it from the fence in dead center to third base?

[**Hint:** The distance between the bases in Little League is 60 feet.]

53. Building a Swing Set Clint is building a wooden swing set for his children. Each supporting end of the swing set is to be an A-frame constructed with two 10-foot-long 4 by 4s joined at a 45° angle. To prevent the swing set from tipping over, Clint wants to secure the base of each A-frame to concrete footings. How far apart should the footings for each A-frame be?

54. Rods and Pistons Rod *OA* rotates about the fixed point *O* so that point *A* travels on a circle of radius *r*. Connected to point *A* is another rod *AB* of length *L > 2r*, and point *B* is connected to a piston. See the figure. Show that the distance *x* between point *O* and point *B* is given by

$$x = r \cos \theta + \sqrt{r^2 \cos^2 \theta + L^2 - r^2}$$

where θ is the angle of rotation of rod *OA*.

55. Geometry Show that the length *d* of a chord of a circle of radius *r* is given by the formula

$$d = 2r \sin \frac{\theta}{2}$$

where θ is the central angle formed by the radii to the ends of the chord. See the figure. Use this result to derive the fact that $\sin \theta < \theta$, where $\theta > 0$ is measured in radians.

56. For any triangle, show that

$$\cos \frac{C}{2} = \sqrt{\frac{s(s - c)}{ab}}$$

where $s = \frac{1}{2}(a + b + c)$.

[**Hint:** Use a Half-angle Formula and the Law of Cosines.]

57. For any triangle show that

$$\sin \frac{C}{2} = \sqrt{\frac{(s - a)(s - b)}{ab}}$$

where $s = \frac{1}{2}(a + b + c)$.

58. Use the Law of Cosines to prove the identity

$$\frac{\cos A}{a} + \frac{\cos B}{b} + \frac{\cos C}{c} = \frac{a^2 + b^2 + c^2}{2abc}$$

Explaining Concepts: Discussion and Writing

59. What do you do first if you are asked to solve a triangle and are given two sides and the included angle?

60. What do you do first if you are asked to solve a triangle and are given three sides?

61. Make up an applied problem that requires using the Law of Cosines.

62. Write down your strategy for solving an oblique triangle.

63. State the Law of Cosines in words.

'Are You Prepared?' Answers

1. $d = \sqrt{(x_2 - x_1)^2 + (y_2 - y_1)^2}$ **2.** $\theta = 45°$

7.4 Area of a Triangle

PREPARING FOR THIS SECTION *Before getting started, review the following:*

- Geometry Essentials (Appendix A, Section A.2, p. A15)

Now Work the 'Are You Prepared?' problem on page 572.

OBJECTIVES **1** Find the Area of SAS Triangles (p. 570)

2 Find the Area of SSS Triangles (p. 571)

In this section, we will derive several formulas for calculating the area of a triangle. The most familiar of these is the following:

THEOREM

NOTE Typically A is used for area, but we use A as the measure of an angle, so we use K for area to avoid confusion. ∎

The area K of a triangle is

$$K = \frac{1}{2}bh \tag{1}$$

where b is the base and h is an altitude drawn to that base.

Proof The derivation of this formula is rather easy once a rectangle of base b and height h is constructed around the triangle. See Figures 29 and 30.

Triangles 1 and 2 in Figure 30 are equal in area, as are triangles 3 and 4. Consequently, the area of the triangle with base b and altitude h is exactly half the area of the rectangle, which is bh.

Figure 29

Figure 30

∎

1 Find the Area of SAS Triangles

Figure 31

If the base b and altitude h to that base are known, then we can find the area of such a triangle using formula (1). Usually, though, the information required to use formula (1) is not given. Suppose, for example, that we know two sides a and b and the included angle C. See Figure 31. Then the altitude h can be found by noting that

$$\frac{h}{a} = \sin C$$

so that

$$h = a \sin C$$

Using this fact in formula (1) produces

$$K = \frac{1}{2}bh = \frac{1}{2}b(a \sin C) = \frac{1}{2}ab \sin C$$

We now have the formula

$$K = \frac{1}{2}ab \sin C \tag{2}$$

By dropping altitudes from the other two vertices of the triangle, we obtain the following corresponding formulas:

$$K = \frac{1}{2}bc \sin A \tag{3}$$

$$K = \frac{1}{2}ac \sin B \tag{4}$$

It is easiest to remember these formulas using the following wording:

THEOREM

The area K of a triangle equals one-half the product of two of its sides times the sine of their included angle.

EXAMPLE 1 | Finding the Area of an SAS Triangle

Find the area K of the triangle for which $a = 8$, $b = 6$, and $C = 30°$.

Solution See Figure 32. We use formula (2) to get

Figure 32

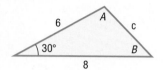

$$K = \frac{1}{2}ab \sin C = \frac{1}{2} \cdot 8 \cdot 6 \cdot \sin 30° = 12 \text{ square units}$$

━━━ **Now Work** PROBLEM 5

2 Find the Area of SSS Triangles

If the three sides of a triangle are known, another formula, called **Heron's Formula** (named after Heron of Alexandria), can be used to find the area of a triangle.

THEOREM

Heron's Formula

The area K of a triangle with sides a, b, and c is

$$K = \sqrt{s(s-a)(s-b)(s-c)} \qquad \text{(5)}$$

where $s = \frac{1}{2}(a + b + c)$.

EXAMPLE 2 | Finding the Area of an SSS Triangle

Find the area of a triangle whose sides are 4, 5, and 7.

Solution We let $a = 4$, $b = 5$, and $c = 7$. Then

$$s = \frac{1}{2}(a + b + c) = \frac{1}{2}(4 + 5 + 7) = 8$$

Heron's Formula then gives the area K as

$$K = \sqrt{s(s-a)(s-b)(s-c)} = \sqrt{8 \cdot 4 \cdot 3 \cdot 1} = \sqrt{96} = 4\sqrt{6} \text{ square units}$$

━━━ **Now Work** PROBLEM 11

Proof of Heron's Formula The proof that we shall give uses the Law of Cosines and is quite different from the proof given by Heron.

From the Law of Cosines,

$$c^2 = a^2 + b^2 - 2ab \cos C$$

and the Half-angle Formula

$$\cos^2 \frac{C}{2} = \frac{1 + \cos C}{2}$$

we find that

$$\cos^2 \frac{C}{2} = \frac{1 + \cos C}{2} = \frac{1 + \dfrac{a^2 + b^2 - c^2}{2ab}}{2}$$

$$= \frac{a^2 + 2ab + b^2 - c^2}{4ab} = \frac{(a + b)^2 - c^2}{4ab}$$

$$= \underbrace{\frac{(a + b - c)(a + b + c)}{4ab}}_{\text{Factor}} = \underbrace{\frac{2(s - c) \cdot 2s}{4ab}}_{\substack{a + b - c = a + b + c - 2c \\ = 2s - 2c = 2(s - c)}} = \frac{s(s - c)}{ab} \quad \textbf{(6)}$$

Similarly, using $\sin^2 \dfrac{C}{2} = \dfrac{1 - \cos C}{2}$, we find that

$$\sin^2 \frac{C}{2} = \frac{(s - a)(s - b)}{ab} \quad \textbf{(7)}$$

Now we use formula (2) for the area.

$$K = \frac{1}{2} ab \sin C$$

$$= \frac{1}{2} ab \cdot 2 \sin \frac{C}{2} \cos \frac{C}{2} \qquad \sin C = \sin\left[2\left(\frac{C}{2}\right)\right] = 2 \sin \frac{C}{2} \cos \frac{C}{2}$$

$$= ab \sqrt{\frac{(s - a)(s - b)}{ab}} \sqrt{\frac{s(s - c)}{ab}} \qquad \text{Use equations (6) and (7).}$$

$$= \sqrt{s(s - a)(s - b)(s - c)}$$

Historical Feature

eron's Formula (also known as *Hero's Formula*) is due to Heron of Alexandria (first century AD), who had, besides his mathematical talents, a good deal of engineering skills. In various temples his mechanical devices produced effects that seemed supernatural, and visitors presumably were thus influenced to generosity. Heron's book *Metrica*, on making such devices, has survived and was discovered in 1896 in the city of Constantinople.

Heron's Formulas for the area of a triangle caused some mild discomfort in Greek mathematics, because a product with two factors was an area, while one with three factors was a volume, but four factors seemed contradictory in Heron's time.

7.4 Assess Your Understanding

'Are You Prepared?'   *Answer is given at the end of these exercises. If you get the wrong answer, read the page listed in red.*

1. The area K of a triangle, whose base is b and whose height is h is _____ . (p. A15)

Concepts and Vocabulary

2. If two sides a and b and the included angle C are known in a triangle, then the area K is found using the formula

$K =$ _____ .

3. The area K of a triangle with sides a, b, and c is

$K =$ _____ where $s =$ _____ .

4. *True or False*   Heron's formula is used to find the area of SSS triangles.

Skill Building

In Problems 5–12, find the area of each triangle. Round answers to two decimal places.

5.

6.

7.

8.

9.

10.

11.

12.
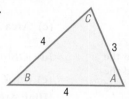

In Problems 13–24, find the area of each triangle. Round answers to two decimal places.

13. $a = 3$, $b = 4$, $C = 40°$

14. $a = 2$, $c = 1$, $B = 10°$

15. $b = 1$, $c = 3$, $A = 80°$

16. $a = 6$, $b = 4$, $C = 60°$

17. $a = 3$, $c = 2$, $B = 110°$

18. $b = 4$, $c = 1$, $A = 120°$

19. $a = 12$, $b = 13$, $c = 5$

20. $a = 4$, $b = 5$, $c = 3$

21. $a = 2$, $b = 2$, $c = 2$

22. $a = 3$, $b = 3$, $c = 2$

23. $a = 5$, $b = 8$, $c = 9$

24. $a = 4$, $b = 3$, $c = 6$

Applications and Extensions

25. Area of an ASA Triangle If two angles and the included side are given, the third angle is easy to find. Use the Law of Sines to show that the area K of a triangle with side a and angles A, B, and C is

$$K = \frac{a^2 \sin B \sin C}{2 \sin A}$$

26. Area of a Triangle Prove the two other forms of the formula given in Problem 25.

$$K = \frac{b^2 \sin A \sin C}{2 \sin B} \quad \text{and} \quad K = \frac{c^2 \sin A \sin B}{2 \sin C}$$

In Problems 27–32, use the results of Problem 25 or 26 to find the area of each triangle. Round answers to two decimal places.

27. $A = 40°$, $B = 20°$, $a = 2$

28. $A = 50°$, $C = 20°$, $a = 3$

29. $B = 70°$, $C = 10°$, $b = 5$

30. $A = 70°$, $B = 60°$, $c = 4$

31. $A = 110°$, $C = 30°$, $c = 3$

32. $B = 10°$, $C = 100°$, $b = 2$

33. Area of a Segment Find the area of the segment (shaded in blue in the figure) of a circle whose radius is 8 feet, formed by a central angle of 70°.

[**Hint:** Subtract the area of the triangle from the area of the sector to obtain the area of the segment.]

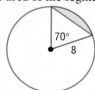

34. Area of a Segment Find the area of the segment of a circle whose radius is 5 inches, formed by a central angle of 40°.

35. Cost of a Triangular Lot The dimensions of a triangular lot are 100 feet by 50 feet by 75 feet. If the price of such land is $3 per square foot, how much does the lot cost?

36. Amount of Material to Make a Tent A cone-shaped tent is made from a circular piece of canvas 24 feet in diameter by

removing a sector with central angle 100° and connecting the ends. What is the surface area of the tent?

37. Dimensions of Home Plate The dimensions of home plate at any major league baseball stadium are shown. Find the area of home plate.

38. Computing Areas Find the area of the shaded region enclosed in a semicircle of diameter 10 inches. The length of the chord PQ is 8 inches.

[**Hint:** Triangle PQR is a right triangle.]

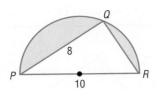

39. Geometry Consult the figure, which shows a circle of radius r with center at O. Find the area K of the shaded region as a function of the central angle θ.

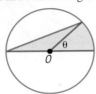

40. Approximating the Area of a Lake To approximate the area of a lake, a surveyor walks around the perimeter of the lake, taking the measurements shown in the illustration. Using this technique, what is the approximate area of the lake?

[**Hint:** Use the Law of Cosines on the three triangles shown and then find the sum of their areas.]

41. The Flatiron Building Completed in 1902 in New York City, the Flatiron Building is triangular shaped and bounded by 22nd Street, Broadway, and 5th Avenue. The building measures approximately 87 feet on the 22nd Street side, 190 feet on the Broadway side, and 173 feet on the 5th Avenue side. Approximate the ground area covered by the building.

Source: Sarah Bradford Landau and Carl W. Condit, *Rise of the New York Skyscraper: 1865–1913*. New Haven, CT: Yale University Press, 1996

42. Bermuda Triangle The Bermuda Triangle is roughly defined by Hamilton, Bermuda; San Juan, Puerto Rico; and Fort Lauderdale, Florida. The distances from Hamilton to Fort Lauderdale, Fort Lauderdale to San Juan, and San Juan to Hamilton are approximately 1028, 1046, and 965 miles respectively. Ignoring the curvature of Earth, approximate the area of the Bermuda Triangle.

Source: www.worldatlas.com

43. Geometry Refer to the figure. If $|OA| = 1$, show that:

(a) Area $\triangle OAC = \dfrac{1}{2}\sin\alpha\cos\alpha$

(b) Area $\triangle OCB = \dfrac{1}{2}|OB|^2\sin\beta\cos\beta$

(c) Area $\triangle OAB = \dfrac{1}{2}|OB|\sin(\alpha+\beta)$

(d) $|OB| = \dfrac{\cos\alpha}{\cos\beta}$

(e) $\sin(\alpha+\beta) = \sin\alpha\cos\beta + \cos\alpha\sin\beta$

[**Hint:** Area $\triangle OAB =$ Area $\triangle OAC +$ Area $\triangle OCB$.]

44. Geometry Refer to the figure, in which a unit circle is drawn. The line segment DB is tangent to the circle and θ is acute.

(a) Express the area of $\triangle OBC$ in terms of $\sin\theta$ and $\cos\theta$.

(b) Express the area of $\triangle OBD$ in terms of $\sin\theta$ and $\cos\theta$.

(c) The area of the sector $\overset{\frown}{OBC}$ of the circle is $\dfrac{1}{2}\theta$, where θ is measured in radians. Use the results of parts (a) and (b) and the fact that

$$\text{Area } \triangle OBC < \text{Area } \overset{\frown}{OBC} < \text{Area } \triangle OBD$$

to show that

$$1 < \dfrac{\theta}{\sin\theta} < \dfrac{1}{\cos\theta}$$

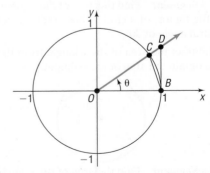

45. The Cow Problem* A cow is tethered to one corner of a square barn, 10 feet by 10 feet, with a rope 100 feet long. What is the maximum grazing area for the cow?

[**Hint:** See the illustration on the following page.]

*Suggested by Professor Teddy Koukounas of Suffolk Community College, who learned of it from an old farmer in Virginia. Solution provided by Professor Kathleen Miranda of SUNY at Old Westbury.

perimeter. Show that the triangles with the given side lengths are perfect.
(a) 9, 10, 17   (b) 6, 25, 29

Source: M.V. Bonsangue, G. E. Gannon, E. Buchman, and N. Gross, "In Search of Perfect Triangles," *Mathematics Teacher,* Vol. 92, No. 1, 1999: 56–61

48. If h_1, h_2, and h_3 are the altitudes dropped from P, Q, and R, respectively, in a triangle (see the figure), show that

$$\frac{1}{h_1} + \frac{1}{h_2} + \frac{1}{h_3} = \frac{s}{K}$$

where K is the area of the triangle and $s = \frac{1}{2}(a + b + c)$.
[**Hint:** $h_1 = \dfrac{2K}{a}$.]

46. Another Cow Problem If the barn in Problem 45 is rectangular, 10 feet by 20 feet, what is the maximum grazing area for the cow?

47. Perfect Triangles A *perfect triangle* is one having natural number sides for which the area is numerically equal to the

49. Show that a formula for the altitude h from a vertex to the opposite side a of a triangle is

$$h = \frac{a \sin B \sin C}{\sin A}$$

Inscribed Circle *For Problems 50–53, the lines that bisect each angle of a triangle meet in a single point O, and the perpendicular distance r from O to each side of the triangle is the same. The circle with center at O and radius r is called the **inscribed circle** of the triangle (see the figure).*

50. Apply the formula from Problem 49 to triangle OPQ to show that

$$r = \frac{c \sin \dfrac{A}{2} \sin \dfrac{B}{2}}{\cos \dfrac{C}{2}}$$

51. Use the result of Problem 50 and the result of Problems 56 and 57 in Section 7.3 to show that

$$\cot \frac{C}{2} = \frac{s - c}{r}$$

where $s = \frac{1}{2}(a + b + c)$.

52. Show that

$$\cot \frac{A}{2} + \cot \frac{B}{2} + \cot \frac{C}{2} = \frac{s}{r}$$

53. Show that the area K of triangle PQR is $K = rs$, where $s = \frac{1}{2}(a + b + c)$. Then show that

$$r = \sqrt{\frac{(s - a)(s - b)(s - c)}{s}}$$

Explaining Concepts: Discussion and Writing

54. What do you do first if you are asked to find the area of a triangle and are given two sides and the included angle?

55. What do you do first if you are asked to find the area of a triangle and are given three sides?

56. State the area of a SAS triangle in words.

'Are You Prepared?' Answers

1. $K = \dfrac{1}{2}bh$

7.5 Simple Harmonic Motion; Damped Motion; Combining Waves

PREPARING FOR THIS SECTION *Before getting started, review the following:*

- Sinusoidal Graphs (Section 5.6, pp. 438–445)

Now Work the 'Are You Prepared?' problem on page 582.

OBJECTIVES 1 Build a Model for an Object in Simple Harmonic Motion (p. 576)
2 Analyze Simple Harmonic Motion (p. 578)
3 Analyze an Object in Damped Motion (p. 579)
4 Graph the Sum of Two Functions (p. 580)

1 Build a Model for an Object in Simple Harmonic Motion

Many physical phenomena can be described as simple harmonic motion. Radio and television waves, light waves, sound waves, and water waves exhibit motion that is simple harmonic.

The swinging of a pendulum, the vibrations of a tuning fork, and the bobbing of a weight attached to a coiled spring are examples of vibrational motion. In this type of motion, an object swings back and forth over the same path. In Figure 33, the point B is the **equilibrium (rest) position** of the vibrating object. The **amplitude** is the distance from the object's rest position to its point of greatest displacement (either point A or point C in Figure 33). The **period** is the time required to complete one vibration, that is, the time it takes to go from, say, point A through B to C and back to A.

> **Simple harmonic motion** is a special kind of vibrational motion in which the acceleration a of the object is directly proportional to the negative of its displacement d from its rest position. That is, $a = -kd$, $k > 0$.

For example, when the mass hanging from the spring in Figure 33 is pulled down from its rest position B to the point C, the force of the spring tries to restore the mass to its rest position. Assuming that there is no frictional force* to retard the motion, the amplitude will remain constant. The force increases in direct proportion to the distance that the mass is pulled from its rest position. Since the force increases directly, the acceleration of the mass of the object must do likewise, because (by Newton's Second Law of Motion) force is directly proportional to acceleration. As a result, the acceleration of the object varies directly with its displacement, and the motion is an example of simple harmonic motion.

Simple harmonic motion is related to circular motion. To see this relationship, consider a circle of radius a, with center at $(0, 0)$. See Figure 34. Suppose that an

Vibrating tuning fork

Figure 33

Coiled spring

Figure 34

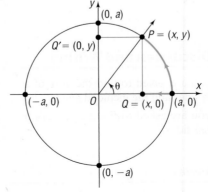

*If friction is present, the amplitude will decrease with time to 0. This type of motion is an example of **damped motion**, which is discussed later in this section.

object initially placed at $(a, 0)$ moves counterclockwise around the circle at a constant angular speed ω. Suppose further that after time t has elapsed the object is at the point $P = (x, y)$ on the circle. The angle θ, in radians, swept out by the ray $\overrightarrow{OP}$ in this time t is

$$\theta = \omega t$$

The coordinates of the point P at time t are

$$x = a \cos \theta = a \cos(\omega t)$$
$$y = a \sin \theta = a \sin(\omega t)$$

Corresponding to each position $P = (x, y)$ of the object moving about the circle, there is the point $Q = (x, 0)$, called the **projection of P on the x-axis**. As P moves around the circle at a constant rate, the point Q moves back and forth between the points $(a, 0)$ and $(-a, 0)$ along the x-axis with a motion that is simple harmonic. Similarly, for each point P there is a point $Q' = (0, y)$, called the **projection of P on the y-axis.** As P moves around the circle, the point Q' moves back and forth between the points $(0, a)$ and $(0, -a)$ on the y-axis with a motion that is simple harmonic. Simple harmonic motion can be described as the projection of constant circular motion on a coordinate axis.

To put it another way, again consider a mass hanging from a spring where the mass is pulled down from its rest position to the point C and then released. See Figure 35(a). The graph shown in Figure 35(b) describes the displacement d of the object from its rest position as a function of time t, assuming that no frictional force is present.

Figure 35

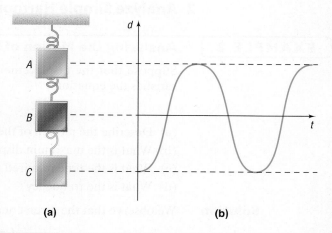

(a)   (b)

THEOREM

Simple Harmonic Motion

An object that moves on a coordinate axis so that the distance d from its rest position at time t is given by either

$$d = a \cos(\omega t) \quad \text{or} \quad d = a \sin(\omega t)$$

where a and $\omega > 0$ are constants, moves with simple harmonic motion. The motion has amplitude $|a|$ and period $\dfrac{2\pi}{\omega}$.

The **frequency** f of an object in simple harmonic motion is the number of oscillations per unit time. Since the period is the time required for one oscillation, it follows that the frequency is the reciprocal of the period; that is,

$$f = \frac{\omega}{2\pi} \qquad \omega > 0$$

EXAMPLE 1 **Build a Model for an Object in Harmonic Motion**

Figure 36

Suppose that an object attached to a coiled spring is pulled down a distance of 5 inches from its rest position and then released. If the time for one oscillation is 3 seconds, develop a model that relates the displacement d of the object from its rest position after time t (in seconds). Assume no friction.

Solution The motion of the object is simple harmonic. See Figure 36. When the object is released ($t = 0$), the displacement of the object from the rest position is -5 units (since the object was pulled down). Because $d = -5$ when $t = 0$, it is easier to use the cosine function*

$$d = a \cos(\omega t)$$

to describe the motion. Now the amplitude is $|-5| = 5$ and the period is 3, so

$$a = -5 \quad \text{and} \quad \frac{2\pi}{\omega} = \text{period} = 3, \quad \text{so} \quad \omega = \frac{2\pi}{3}$$

An equation that models the motion of the object is

$$d = -5 \cos\left[\frac{2\pi}{3}t\right]$$

NOTE In the solution to Example 1, we let $a = -5$, since the initial motion is down. If the initial direction were up, we would let $a = 5$. ∎

━━ **Now Work** PROBLEM 5

2 Analyze Simple Harmonic Motion

EXAMPLE 2 **Analyzing the Motion of an Object**

Suppose that the displacement d (in meters) of an object at time t (in seconds) satisfies the equation

$$d = 10 \sin(5t)$$

(a) Describe the motion of the object.
(b) What is the maximum displacement from its resting position?
(c) What is the time required for one oscillation?
(d) What is the frequency?

Solution We observe that the given equation is of the form

$$d = a \sin(\omega t) \quad d = 10 \sin(5t)$$

where $a = 10$ and $\omega = 5$.

(a) The motion is simple harmonic.

(b) The maximum displacement of the object from its resting position is the amplitude: $|a| = 10$ meters.

(c) The time required for one oscillation is the period:

$$\text{Period} = \frac{2\pi}{\omega} = \frac{2\pi}{5} \text{ seconds}$$

(d) The frequency is the reciprocal of the period. Thus,

$$\text{Frequency} = f = \frac{5}{2\pi} \text{ oscillation per second}$$

━━ **Now Work** PROBLEM 13

*No phase shift is required if a cosine function is used.

3 Analyze an Object in Damped Motion

Most physical phenomena are affected by friction or other resistive forces. These forces remove energy from a moving system and thereby damp its motion. For example, when a mass hanging from a spring is pulled down a distance a and released, the friction in the spring causes the distance that the mass moves from its at-rest position to decrease over time. As a result, the amplitude of any real oscillating spring or swinging pendulum decreases with time due to air resistance, friction, and so forth. See Figure 37.

Figure 37

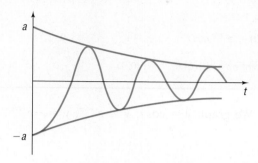

A model that describes this phenomenon maintains a sinusoidal component, but the amplitude of this component will decrease with time to account for the damping effect. In addition, the period of the oscillating component will be affected by the damping. The next result, from physics, describes damped motion.

THEOREM

Damped Motion

The displacement d of an oscillating object from its at-rest position at time t is given by

$$d(t) = ae^{-bt/(2m)} \cos\left(\sqrt{\omega^2 - \frac{b^2}{4m^2}}\, t\right)$$

where b is the **damping factor** or **damping coefficient** and m is the mass of the oscillating object. Here $|a|$ is the displacement at $t = 0$, and $\dfrac{2\pi}{\omega}$ is the period under simple harmonic motion (no damping).

Notice for $b = 0$ (zero damping) that we have the formula for simple harmonic motion with amplitude $|a|$ and period $\dfrac{2\pi}{\omega}$.

EXAMPLE 3   **Analyzing a Damped Vibration Curve**

Analyze the damped vibration curve

$$d(t) = e^{-t/\pi} \cos t, \quad t \ge 0$$

Solution   The displacement d is the product of $y = e^{-t/\pi}$ and $y = \cos t$. Using properties of absolute value and the fact that $|\cos t| \le 1$, we find that

$$|d(t)| = |e^{-t/\pi} \cos t| = |e^{-t/\pi}||\cos t| \le |e^{-t/\pi}| \underset{\uparrow}{=} e^{-t/\pi}$$
$$e^{-t/\pi} > 0$$

As a result,

$$-e^{-t/\pi} \le d(t) \le e^{-t/\pi}$$

This means that the graph of d will lie between the graphs of $y = e^{-t/\pi}$ and $y = -e^{-t/\pi}$, the **bounding curves** of d.

Also, the graph of d will touch these graphs when $|\cos t| = 1$, that is, when $t = 0, \pi, 2\pi$, and so on. The x-intercepts of the graph of d occur when $\cos t = 0$, that is, at $\dfrac{\pi}{2}, \dfrac{3\pi}{2}, \dfrac{5\pi}{2}$, and so on. See Table 1.

Table 1

t	0	$\dfrac{\pi}{2}$	π	$\dfrac{3\pi}{2}$	2π
$e^{-t/\pi}$	1	$e^{-1/2}$	e^{-1}	$e^{-3/2}$	e^{-2}
$\cos t$	1	0	-1	0	1
$d(t) = e^{-t/\pi} \cos t$	1	0	$-e^{-1}$	0	e^{-2}
Point on graph of d	$(0, 1)$	$\left(\dfrac{\pi}{2}, 0\right)$	$(\pi, -e^{-1})$	$\left(\dfrac{3\pi}{2}, 0\right)$	$(2\pi, e^{-2})$

We graph $y = \cos t$, $y = e^{-t/\pi}$, $y = -e^{-t/\pi}$, and $d(t) = e^{-t/\pi} \cos t$ in Figure 38.

Figure 38

Figure 39

Exploration

Graph $Y_1 = e^{-x/\pi} \cos x$, along with $Y_2 = e^{-x/\pi}$, and $Y_3 = -e^{-x/\pi}$, for $0 \le x \le 2\pi$. Determine where Y_1 has its first turning point (local minimum). Compare this to where Y_1 intersects Y_3.

Result Figure 39 shows the graphs of $Y_1 = e^{-x/\pi} \cos x$, $Y_2 = e^{-x/\pi}$, and $Y_3 = -e^{-x/\pi}$. Using MINIMUM, the first turning point occurs at $x \approx 2.83$; Y_1 INTERSECTS Y_3 at $x = \pi \approx 3.14$.

Now Work   PROBLEM 21

4 Graph the Sum of Two Functions

Many physical and biological applications require the graph of the sum of two functions, such as

$$f(x) = x + \sin x \quad \text{or} \quad g(x) = \sin x + \cos(2x)$$

For example, if two tones are emitted, the sound produced is the sum of the waves produced by the two tones. See Problem 57 for an explanation of Touch-Tone phones.

To graph the sum of two (or more) functions, we can use the method of adding y-coordinates described next.

EXAMPLE 4

Graphing the Sum of Two Functions

Use the method of adding y-coordinates to graph $f(x) = x + \sin x$.

Solution   First, we graph the component functions,

$$y = f_1(x) = x \qquad y = f_2(x) = \sin x$$

in the same coordinate system. See Figure 40(a). Now, select several values of x, say, $x = 0$, $x = \dfrac{\pi}{2}$, $x = \pi$, $x = \dfrac{3\pi}{2}$, and $x = 2\pi$, at which we compute $f(x) = f_1(x) + f_2(x)$. Table 2 shows the computation. We plot these points and connect them to get the graph, as shown in Figure 40(b).

Table 2

x	0	$\dfrac{\pi}{2}$	π	$\dfrac{3\pi}{2}$	2π
$y = f_1(x) = x$	0	$\dfrac{\pi}{2}$	π	$\dfrac{3\pi}{2}$	2π
$y = f_2(x) = \sin x$	0	1	0	-1	0
$f(x) = x + \sin x$	0	$\dfrac{\pi}{2} + 1 \approx 2.57$	π	$\dfrac{3\pi}{2} - 1 \approx 3.71$	2π
Point on graph of f	$(0, 0)$	$\left(\dfrac{\pi}{2}, 2.57\right)$	(π, π)	$\left(\dfrac{3\pi}{2}, 3.71\right)$	$(2\pi, 2\pi)$

Figure 40

(a)   (b)

In Figure 40(b), notice that the graph of $f(x) = x + \sin x$ intersects the line $y = x$ whenever $\sin x = 0$. Also, notice that the graph of f is not periodic. ●

✓**Check:** Graph $Y_1 = x$, $Y_2 = \sin x$, and $Y_3 = x + \sin x$ and compare the result with Figure 40(b). Use INTERSECT to verify that the graphs of Y_1 and Y_3 intersect when $\sin x = 0$.

The next example shows a periodic graph.

EXAMPLE 5   **Graphing the Sum of Two Sinusoidal Functions**

Use the method of adding y-coordinates to graph

$$f(x) = \sin x + \cos(2x)$$

Solution   Table 3 shows the steps for computing several points on the graph of f. Figure 41 illustrates the graphs of the component functions, $y = f_1(x) = \sin x$ and $y = f_2(x) = \cos(2x)$, and the graph of $f(x) = \sin x + \cos(2x)$, which is shown in red.

Table 3

x	$-\dfrac{\pi}{2}$	0	$\dfrac{\pi}{2}$	π	$\dfrac{3\pi}{2}$	2π
$y = f_1(x) = \sin x$	-1	0	1	0	-1	0
$y = f_2(x) = \cos(2x)$	-1	1	-1	1	-1	1
$f(x) = \sin x + \cos(2x)$	-2	1	0	1	-2	1
Point on graph of f	$\left(-\dfrac{\pi}{2}, -2\right)$	$(0, 1)$	$\left(\dfrac{\pi}{2}, 0\right)$	$(\pi, 1)$	$\left(\dfrac{3\pi}{2}, -2\right)$	$(2\pi, 1)$

Figure 41

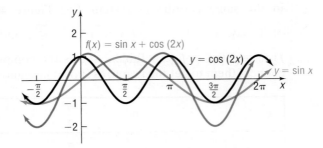

Notice that f is periodic, with period 2π.

 ✓ **Check:** Graph $Y_1 = \sin x$, $Y_2 = \cos(2x)$, and $Y_3 = \sin x + \cos(2x)$ and compare the result with Figure 41.

→ **Now Work** PROBLEM 25

7.5 Assess Your Understanding

'Are You Prepared?' *Answer given at the end of these exercises. If you get a wrong answer, read the pages listed in* red.

1. The amplitude A and period T of $f(x) = 5 \sin(4x)$ are _____ and _____. (pp. 439–440)

Concepts and Vocabulary

2. The motion of an object obeys the equation $d = 4 \cos(6t)$. Such motion is described as _____ _____. The number 4 is called the _____.

3. When a mass hanging from a spring is pulled down and then released, the motion is called _____ _____ if there is no frictional force to retard the motion, and the motion is called _____ if there is friction.

4. **True or False** If the distance d of an object from its rest position at time t is given by a sinusoidal graph, the motion of the object is simple harmonic motion.

Skill Building

In Problems 5–8, an object attached to a coiled spring is pulled down a distance a from its rest position and then released. Assuming that the motion is simple harmonic with period T, write an equation that relates the displacement d of the object from its rest position after t seconds. Also assume that the positive direction of the motion is up.

5. $a = 5$; $T = 2$ seconds

6. $a = 10$; $T = 3$ seconds

7. $a = 6$; $T = \pi$ seconds

8. $a = 4$; $T = \dfrac{\pi}{2}$ seconds

9. Rework Problem 5 under the same conditions except that, at time $t = 0$, the object is at its resting position and moving down.

10. Rework Problem 6 under the same conditions except that, at time $t = 0$, the object is at its resting position and moving down.

11. Rework Problem 7 under the same conditions except that, at time $t = 0$, the object is at its resting position and moving down.

12. Rework Problem 8 under the same conditions except that, at time $t = 0$, the object is at its resting position and moving down.

In Problems 13–20, the displacement d (in meters) of an object at time t (in seconds) is given.
  (a) *Describe the motion of the object.*
  (b) *What is the maximum displacement from its resting position?*
  (c) *What is the time required for one oscillation?*
  (d) *What is the frequency?*

13. $d = 5 \sin(3t)$

14. $d = 4 \sin(2t)$

15. $d = 6 \cos(\pi t)$

16. $d = 5 \cos\left(\dfrac{\pi}{2} t\right)$

17. $d = -3 \sin\left(\dfrac{1}{2} t\right)$

18. $d = -2 \cos(2t)$

19. $d = 6 + 2 \cos(2\pi t)$

20. $d = 4 + 3 \sin(\pi t)$

In Problems 21–24, graph each damped vibration curve for $0 \le t \le 2\pi$.

21. $d(t) = e^{-t/\pi} \cos(2t)$ **22.** $d(t) = e^{-t/2\pi} \cos(2t)$ **23.** $d(t) = e^{-t/2\pi} \cos t$ **24.** $d(t) = e^{-t/4\pi} \cos t$

In Problems 25–32, use the method of adding y-coordinates to graph each function.

25. $f(x) = x + \cos x$ **26.** $f(x) = x + \cos(2x)$ **27.** $f(x) = x - \sin x$

28. $f(x) = x - \cos x$ **29.** $f(x) = \sin x + \cos x$ **30.** $f(x) = \sin(2x) + \cos x$

31. $g(x) = \sin x + \sin(2x)$ **32.** $g(x) = \cos(2x) + \cos x$

Mixed Practice

In Problems 33–38, (a) use the Product-to-Sum Formulas to express each product as a sum, and (b) use the method of adding y-coordinates to graph each function on the interval $[0, 2\pi]$.

33. $f(x) = \sin(2x) \sin x$ **34.** $F(x) = \sin(3x) \sin x$ **35.** $G(x) = \cos(4x) \cos(2x)$

36. $h(x) = \cos(2x) \cos(x)$ **37.** $H(x) = 2 \sin(3x) \cos(x)$ **38.** $g(x) = 2 \sin x \cos(3x)$

Applications and Extensions

In Problems 39–44, an object of mass m (in grams) attached to a coiled spring with damping factor b (in grams per second) is pulled down a distance a (in centimeters) from its rest position and then released. Assume that the positive direction of the motion is up and the period is T (in seconds) under simple harmonic motion.

    (a) Write an equation that relates the distance d of the object from its rest position after t seconds.

    (b) Graph the equation found in part (a) for 5 oscillations using a graphing utility.

39. $m = 25$, $a = 10$, $b = 0.7$, $T = 5$ **40.** $m = 20$, $a = 15$, $b = 0.75$, $T = 6$

41. $m = 30$, $a = 18$, $b = 0.6$, $T = 4$ **42.** $m = 15$, $a = 16$, $b = 0.65$, $T = 5$

43. $m = 10$, $a = 5$, $b = 0.8$, $T = 3$ **44.** $m = 10$, $a = 5$, $b = 0.7$, $T = 3$

In Problems 45–50, the distance d (in meters) of the bob of a pendulum of mass m (in kilograms) from its rest position at time t (in seconds) is given. The bob is released from the left of its rest position and represents a negative direction.

    (a) Describe the motion of the object. Be sure to give the mass and damping factor.

    (b) What is the initial displacement of the bob? That is, what is the displacement at $t = 0$?

    (c) Graph the motion using a graphing utility.

    (d) What is the displacement of the bob at the start of the second oscillation?

    (e) What happens to the displacement of the bob as time increases without bound?

45. $d = -20e^{-0.7t/40} \cos\left(\sqrt{\left(\dfrac{2\pi}{5}\right)^2 - \dfrac{0.49}{1600}}\, t \right)$ **46.** $d = -20e^{-0.8t/40} \cos\left(\sqrt{\left(\dfrac{2\pi}{5}\right)^2 - \dfrac{0.64}{1600}}\, t \right)$

47. $d = -30e^{-0.6t/80} \cos\left(\sqrt{\left(\dfrac{2\pi}{7}\right)^2 - \dfrac{0.36}{6400}}\, t \right)$ **48.** $d = -30e^{-0.5t/70} \cos\left(\sqrt{\left(\dfrac{\pi}{2}\right)^2 - \dfrac{0.25}{4900}}\, t \right)$

49. $d = -15e^{-0.9t/30} \cos\left(\sqrt{\left(\dfrac{\pi}{3}\right)^2 - \dfrac{0.81}{900}}\, t \right)$ **50.** $d = -10e^{-0.8t/50} \cos\left(\sqrt{\left(\dfrac{2\pi}{3}\right)^2 - \dfrac{0.64}{2500}}\, t \right)$

51. Loudspeaker A loudspeaker diaphragm is oscillating in simple harmonic motion described by the equation $d = a \cos(\omega t)$ with a frequency of 520 hertz (cycles per second) and a maximum displacement of 0.80 millimeter. Find ω and then determine the equation that describes the movement of the diaphragm.

52. Colossus Added to Six Flags St. Louis in 1986, the Colossus is a giant Ferris wheel. Its diameter is 165 feet, it rotates at a rate of about 1.6 revolutions per minute, and the bottom of the wheel is 15 feet above the ground. Determine an equation that

relates a rider's height above the ground at time t. Assume the passenger begins the ride at the bottom of the wheel.

Source: Six Flags Theme Parks, Inc.

53. Tuning Fork The end of a tuning fork moves in simple harmonic motion described by the equation $d = a \sin(\omega t)$. If a tuning fork for the note A above middle C on an even-tempered scale (A_4, the tone by which an orchestra tunes itself) has a frequency of 440 hertz (cycles per second), find ω. If the maximum displacement of the end of the tuning fork is

0.01 millimeter, determine the equation that describes the movement of the tuning fork.

Source: David Lapp. *Physics of Music and Musical Instruments.* Medford, MA: Tufts University, 2003

54. **Tuning Fork** The end of a tuning fork moves in simple harmonic motion described by the equation $d = a \sin(\omega t)$. If a tuning fork for the note E above middle C on an even-tempered scale (E_4) has a frequency of approximately 329.63 hertz (cycles per second), find ω. If the maximum displacement of the end of the tuning fork is 0.025 millimeter, determine the equation that describes the movement of the tuning fork.

Source: David Lapp. *Physics of Music and Musical Instruments.* Medford, MA: Tufts University, 2003

55. **Charging a Capacitor** See the illustration. If a charged capacitor is connected to a coil by closing a switch, energy is transferred to the coil and then back to the capacitor in an oscillatory motion. The voltage V (in volts) across the capacitor will gradually diminish to 0 with time t (in seconds).

    (a) Graph the function relating V and t:

    $$V(t) = e^{-t/3} \cos(\pi t), \qquad 0 \le t \le 3$$

    (b) At what times t will the graph of V touch the graph of $y = e^{-t/3}$? When does the graph of V touch the graph of $y = -e^{-t/3}$?

    (c) When will the voltage V be between -0.4 and 0.4 volt?

Switch

Capacitor   Coil

56. **The Sawtooth Curve** An oscilloscope often displays a *sawtooth curve*. This curve can be approximated by sinusoidal curves of varying periods and amplitudes.

    (a) Use a graphing utility to graph the following function, which can be used to approximate the sawtooth curve.

    $$f(x) = \frac{1}{2}\sin(2\pi x) + \frac{1}{4}\sin(4\pi x), \qquad 0 \le x \le 4$$

    (b) A better approximation to the sawtooth curve is given by

    $$f(x) = \frac{1}{2}\sin(2\pi x) + \frac{1}{4}\sin(4\pi x) + \frac{1}{8}\sin(8\pi x)$$

    Use a graphing utility to graph this function for $0 \le x \le 4$ and compare the result to the graph obtained in part (a).

    (c) A third and even better approximation to the sawtooth curve is given by

    $$f(x) = \frac{1}{2}\sin(2\pi x) + \frac{1}{4}\sin(4\pi x) + \frac{1}{8}\sin(8\pi x) + \frac{1}{16}\sin(16\pi x)$$

Use a graphing utility to graph this function for $0 \le x \le 4$ and compare the result to the graphs obtained in parts (a) and (b).

    (d) What do you think the next approximation to the sawtooth curve is?

57. **Touch-Tone Phones** On a Touch-Tone phone, each button produces a unique sound. The sound produced is the sum of two tones, given by

$$y = \sin(2\pi l t) \quad \text{and} \quad y = \sin(2\pi h t)$$

where l and h are the low and high frequencies (cycles per second) shown in the illustration. For example, if you touch 7, the low frequency is $l = 852$ cycles per second and the high frequency is $h = 1209$ cycles per second. The sound emitted by touching 7 is

$$y = \sin[2\pi(852)t] + \sin[2\pi(1209)t]$$

Use a graphing utility to graph the sound emitted by touching 7.

Touch-Tone phone

58. Use a graphing utility to graph the sound emitted by the * key on a Touch-Tone phone. See Problem 57.

59. **CBL Experiment** Pendulum motion is analyzed to estimate simple harmonic motion. A plot is generated with the position of the pendulum over time. The graph is used to find a sinusoidal curve of the form $y = A \cos[B(x - C)] + D$. Determine the amplitude, period, and frequency. (Activity 16, Real-World Math with the CBL System.)

60. **CBL Experiment** The sound from a tuning fork is collected over time. Determine the amplitude, frequency, and period of the graph. A model of the form $y = A \cos[B(x - C)]$ is fitted to the data. (Activity 23, Real-World Math with the CBL System.)

Explaining Concepts: Discussion and Writing

61. Use a graphing utility to graph the function $f(x) = \dfrac{\sin x}{x}$, $x > 0$. Based on the graph, what do you conjecture about the value of $\dfrac{\sin x}{x}$ for x close to 0?

62. Use a graphing utility to graph $y = x \sin x$, $y = x^2 \sin x$, and $y = x^3 \sin x$ for $x > 0$. What patterns do you observe?

63. Use a graphing utility to graph $y = \dfrac{1}{x}\sin x$, $y = \dfrac{1}{x^2}\sin x$, and $y = \dfrac{1}{x^3}\sin x$ for $x > 0$. What patterns do you observe?

64. How would you explain to a friend what simple harmonic motion is? How would you explain damped motion?

'Are You Prepared?' Answers

1. $A = 5$; $T = \dfrac{\pi}{2}$

CHAPTER REVIEW

Things to Know

Formulas

Law of Sines (p. 553) $\dfrac{\sin A}{a} = \dfrac{\sin B}{b} = \dfrac{\sin C}{c}$

Law of Cosines (p. 563)
$$c^2 = a^2 + b^2 - 2ab \cos C$$
$$b^2 = a^2 + c^2 - 2ac \cos B$$
$$a^2 = b^2 + c^2 - 2bc \cos A$$

Area of a triangle (pp. 570–571)
$$K = \tfrac{1}{2}bh \quad K = \tfrac{1}{2}ab \sin C \quad K = \tfrac{1}{2}bc \sin A \quad K = \tfrac{1}{2}ac \sin B$$
$$K = \sqrt{s(s-a)(s-b)(s-c)} \quad \text{where} \quad s = \tfrac{1}{2}(a+b+c)$$

Objectives

Section		You should be able to …	Example(s)	Review Exercises
7.1	1	Solve right triangles (p. 547)	1, 2	1–4
	2	Solve applied problems (p. 548)	3–6	35, 36, 45–47
7.2	1	Solve SAA or ASA triangles (p. 553)	1, 2	5, 6, 22
	2	Solve SSA triangles (p. 554)	3–5	7–10, 12, 17, 18, 21
	3	Solve applied problems (p. 557)	6, 7	37–39
7.3	1	Solve SAS triangles (p. 564)	1	11, 15, 16, 23, 24
	2	Solve SSS triangles (p. 565)	2	13, 14, 19, 20
	3	Solve applied problems (p. 565)	3	40, 41
7.4	1	Find the area of SAS triangles (p. 570)	1	25–28, 43, 44
	2	Find the area of SSS triangles (p. 571)	2	29–32, 42
7.5	1	Build a model for an object in simple harmonic motion (p. 576)	1	48, 49
	2	Analyze simple harmonic motion (p. 578)	2	50–53
	3	Analyze an object in damped motion (p. 579)	3	54–57
	4	Graph the sum of two functions (p. 580)	4, 5	58, 59

Review Exercises

In Problems 1–4, solve each triangle.

1.

2.

3.

4.

In Problems 5–24, find the remaining angle(s) and side(s) of each triangle, if it (they) exists. If no triangle exists, say "No triangle."

5. $A = 50°$, $B = 30°$, $a = 1$

6. $A = 10°$, $C = 40°$, $c = 2$

7. $A = 100°$, $a = 5$, $c = 2$

8. $a = 2$, $c = 5$, $A = 60°$

9. $a = 3$, $c = 1$, $C = 110°$

10. $a = 3$, $c = 1$, $C = 20°$

11. $a = 3$, $c = 1$, $B = 100°$

12. $a = 3$, $b = 5$, $B = 80°$

13. $a = 2$, $b = 3$, $c = 1$

14. $a = 10$, $b = 7$, $c = 8$

15. $a = 1$, $b = 3$, $C = 40°$

16. $a = 4$, $b = 1$, $C = 100°$

17. $a = 5$, $b = 3$, $A = 80°$

18. $a = 2$, $b = 3$, $A = 20°$

19. $a = 1$, $b = \dfrac{1}{2}$, $c = \dfrac{4}{3}$

20. $a = 3$, $b = 2$, $c = 2$

21. $a = 3$, $A = 10°$, $b = 4$

22. $a = 4$, $A = 20°$, $B = 100°$

23. $c = 5$, $b = 4$, $A = 70°$

24. $a = 1$, $b = 2$, $C = 60°$

In Problems 25–34, find the area of each triangle.

25. $a = 2$, $b = 3$, $C = 40°$

26. $b = 5$, $c = 5$, $A = 20°$

27. $b = 4$, $c = 10$, $A = 70°$

28. $a = 2$, $b = 1$, $C = 100°$

29. $a = 4$, $b = 3$, $c = 5$

30. $a = 10$, $b = 7$, $c = 8$

31. $a = 4$, $b = 2$, $c = 5$

32. $a = 3$, $b = 2$, $c = 2$

33. $A = 50°$, $B = 30°$, $a = 1$

34. $A = 10°$, $C = 40°$, $c = 3$

---

35. Finding the Grade of a Mountain Trail   A straight trail with a uniform inclination leads from a hotel, elevation 5000 feet, to a lake in a valley, elevation 4100 feet. The length of the trail is 4100 feet. What is the inclination (grade) of the trail?

36. Geometry   The hypotenuse of a right triangle is 12 feet. If one leg is 8 feet, find the degree measure of each angle.

37. Finding the Height of a Helicopter   Two observers simultaneously measure the angle of elevation of a helicopter. One angle is measured as 25°, the other as 40° (see the figure). If the observers are 100 feet apart and the helicopter lies over the line joining them, how high is the helicopter?

38. Determining Distances at Sea   Rebecca, the navigator of a ship at sea, spots two lighthouses that she knows to be 2 miles apart along a straight shoreline. She determines that the angles formed between two line-of-sight observations of the lighthouses and the line from the ship directly to shore are 12° and 30°. See the illustration.
(a) How far is the ship from lighthouse L_1?
(b) How far is the ship from lighthouse L_2?
(c) How far is the ship from shore?

39. Constructing a Highway   A highway whose primary directions are north–south is being constructed along the west coast of Florida. Near Naples, a bay obstructs the straight path of the road. Since the cost of a bridge is prohibitive, engineers decide to go around the bay. The illustration shows the path that they decide on and the measurements taken. What is the length of highway needed to go around the bay?

40. Correcting a Navigational Error A sailboat leaves St. Thomas bound for an island in the British West Indies, 200 miles away. Maintaining a constant speed of 18 miles per hour, but encountering heavy crosswinds and strong currents, the crew finds after 4 hours that the sailboat is off course by 15°.
   (a) How far is the sailboat from the island at this time?
   (b) Through what angle should the sailboat turn to correct its course?
   (c) How much time has been added to the trip because of this? (Assume that the speed remains at 18 miles per hour.)

41. Surveying Two homes are located on opposite sides of a small hill. See the illustration. To measure the distance between them, a surveyor walks a distance of 50 feet from house P to point R, uses a transit to measure $\angle PRQ$, which is found to be 80°, and then walks to house Q, a distance of 60 feet. How far apart are the houses?

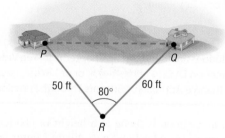

42. Approximating the Area of a Lake To approximate the area of a lake, Cindy walks around the perimeter of the lake, taking the measurements shown in the illustration. Using this technique, what is the approximate area of the lake?

[**Hint:** Use the Law of Cosines on the three triangles shown and then find the sum of their areas.]

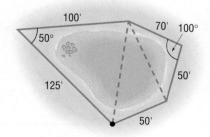

43. Calculating the Cost of Land The irregular parcel of land shown in the figure is being sold for $100 per square foot. What is the cost of this parcel?

44. Area of a Segment Find the area of the segment of a circle whose radius is 6 inches formed by a central angle of 50°.

45. Finding the Bearing of a Ship The *Majesty* leaves the Port at Boston for Bermuda with a bearing of S80°E at an average speed of 10 knots. After 1 hour, the ship turns 90° toward the southwest. After 2 hours at an average speed of 20 knots, what is the bearing of the ship from Boston?

46. Drive Wheels of an Engine The drive wheel of an engine is 13 inches in diameter, and the pulley on the rotary pump is 5 inches in diameter. If the shafts of the drive wheel and the pulley are 2 feet apart, what length of belt is required to join them as shown in the figure?

47. Rework Problem 46 if the belt is crossed, as shown in the figure.

In Problems 48 and 49, an object attached to a coiled spring is pulled down a distance a from its rest position and then released. Assuming that the motion is simple harmonic with period T, develop a model that relates the displacement d of the object from its rest position after t seconds. Also assume that the positive direction of the motion is up.

48. $a = 3$; $T = 4$ seconds

49. $a = 5$; $T = 6$ seconds

In Problems 50–53, the distance d (in feet) that an object travels in time t (in seconds) is given.
(a) *Describe the motion of the object.*
(b) *What is the maximum displacement from its rest position?*
(c) *What is the time required for one oscillation?*
(d) *What is the frequency?*

50. $d = 6 \sin(2t)$

51. $d = 2 \cos(4t)$

52. $d = -2 \cos(\pi t)$

53. $d = -3 \sin\left[\dfrac{\pi}{2}t\right]$

In Problems 54 and 55, an object of mass m attached to a coiled spring with damping factor b is pulled down a distance a from its rest position and then released. Assume that the positive direction of the motion is up and the period is T under simple harmonic motion.
(a) *Develop a model that relates the distance d of the object from its rest position after t seconds.*
(b) *Graph the equation found in part (a) for 5 oscillations.*

54. $m = 40$ grams; $a = 15$ centimeters; $b = 0.75$ gram/second; $T = 5$ seconds

55. $m = 25$ grams; $a = 13$ centimeters; $b = 0.65$ gram/second; $T = 4$ seconds*

In Problems 56 and 57, the distance d (in meters) of the bob of a pendulum of mass m (in kilograms) from its rest position at time t (in seconds) is given.
(a) *Describe the motion of the object.*
(b) *What is the initial displacement of the bob? That is, what is the displacement at t = 0?*
(c) *Graph the motion using a graphing utility.*
(d) *What is the displacement of the bob at the start of the second oscillation?*
(e) *What happens to the displacement of the bob as time increases without bound?*

56. $d = -15e^{-0.6t/40}\cos\left(\sqrt{\left(\dfrac{2\pi}{5}\right)^2 - \dfrac{0.36}{1600}}\, t\right)$

57. $d = -20e^{-0.5t/60}\cos\left(\sqrt{\left(\dfrac{2\pi}{3}\right)^2 - \dfrac{0.25}{3600}}\, t\right)$

In Problems 58 and 59, use the method of adding y-coordinates to graph each function.

58. $y = 2 \sin x + \cos(2x)$

59. $y = 2 \cos(2x) + \sin\dfrac{x}{2}$

CHAPTER TEST

CHAPTER **Test Prep** VIDEOS

Step-by-step test solutions are found on the Chapter Test Prep Videos available via the Video Resources on DVD, in *MyMathLab* or on You Tube (search "Sullivan Precalculus: A Right Triangle Approach, Concepts Through Functions 2e).

1. A 12-foot ladder leans against a building. The top of the ladder leans against the wall 10.5 feet from the ground. What is the angle formed by the ground and the ladder?

2. A hot-air balloon is flying at a height of 600 feet and is directly above the Marshall Space Flight Center in Huntsville, Alabama. The pilot of the balloon looks down at the airport that is known to be 5 miles from the Marshall Space Flight Center. What is the angle of depression from the balloon to the airport?

In Problems 3–5, use the given information to determine the three remaining parts of each triangle.

3.

4.

5.

In Problems 6–8, solve each triangle.

6. $A = 55°$, $C = 20°$, $a = 4$

7. $a = 3$, $b = 7$, $A = 40°$

8. $a = 8$, $b = 4$, $C = 70°$

---

9. Find the area of the triangle described in Problem 8.

10. Find the area of the triangle described in Problem 5.

11. Find the area of the shaded region enclosed in a semicircle of diameter 8 centimeters. The length of the chord AB is 6 centimeters.

[**Hint:** Triangle ABC is a right triangle.]

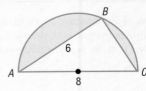

12. Find the area of the quadrilateral shown.

13. Madison wants to swim across Lake William from the fishing lodge (*A*) to the boat ramp (*B*), but she wants to know the distance first. Highway 20 goes right past the boat ramp, and County Road 3 goes to the lodge. The two roads intersect at point (*C*), 4.2 miles from the ramp and 3.5 miles from the lodge. Madison uses a transit to measure the angle of intersection of the two roads to be 32°. How far will she need to swim?

14. Given that △*OAB* is an isosceles triangle and the shaded sector is a semicircle, find the area of the entire region. Express your answer as a decimal rounded to two places.

15. The area of the triangle shown below is $54\sqrt{6}$ square units; find the lengths of the sides.

16. Logan is playing on her swing. One full swing (front to back to front) takes 6 seconds and at the peak of her swing she is at an angle of 42° with the vertical. If her swing is 5 feet long, and we ignore all resistive forces, build a model that relates her horizontal displacement (from the rest position) after time *t*.

CUMULATIVE REVIEW

1. Find the real solutions, if any, of the equation $3x^2 + 1 = 4x$.

2. Find an equation for the circle with center at the point $(-5, 1)$ and radius 3. Graph this circle.

3. Determine the domain of the function
$$f(x) = \sqrt{x^2 - 3x - 4}$$

4. Graph the function $y = 3\sin(\pi x)$.

5. Graph the function $y = -2\cos(2x - \pi)$.

6. If $\tan\theta = -2$ and $\dfrac{3\pi}{2} < \theta < 2\pi$, find the exact value of:

(a) $\sin\theta$ (b) $\cos\theta$ (c) $\sin(2\theta)$

(d) $\cos(2\theta)$ (e) $\sin\left(\dfrac{1}{2}\theta\right)$ (f) $\cos\left(\dfrac{1}{2}\theta\right)$

7. Graph each of the following functions on the interval $[0, 4]$:

(a) $y = e^x$ (b) $y = \sin x$

(c) $y = e^x\sin x$ (d) $y = 2x + \sin x$

8. Sketch the graph of each of the following functions:

(a) $y = x$ (b) $y = x^2$ (c) $y = \sqrt{x}$

(d) $y = x^3$ (e) $y = e^x$ (f) $y = \ln x$

(g) $y = \sin x$ (h) $y = \cos x$ (i) $y = \tan x$

9. Solve the triangle:

10. In the complex number system, solve the equation
$$3x^5 - 10x^4 + 21x^3 - 42x^2 + 36x - 8 = 0$$

11. Analyze the graph of the rational function
$$R(x) = \frac{2x^2 - 7x - 4}{x^2 + 2x - 15}$$

12. Solve $3^x = 12$. Round your answer to two decimal places.

13. Solve $\log_3(x + 8) + \log_3 x = 2$.

14. Suppose that $f(x) = 4x + 5$ and $g(x) = x^2 + 5x - 24$.

(a) Solve $f(x) = 0$.   (b) Solve $f(x) = 13$.

(c) Solve $f(x) = g(x)$.   (d) Solve $f(x) > 0$.

(e) Solve $g(x) \le 0$.   (f) Graph $y = f(x)$.

(g) Graph $y = g(x)$.

CHAPTER PROJECTS

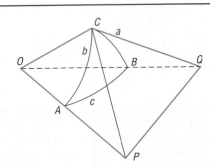

I. Spherical Trigonometry When the distance between two locations on the surface of Earth is small, we can compute the distance in statutory miles. Using this assumption, we can use the Law of Sines and the Law of Cosines to approximate distances and angles. However, if you look at a globe, you notice that Earth is a sphere, so, as the distance between two points on its surface increases, the linear distance is less accurate because of curvature. Under this circumstance, we need to take into account the curvature of Earth when using the Law of Sines and the Law of Cosines.

1. Draw a spherical triangle and label each vertex by A, B, and C. Then connect each vertex by a radius to the center O of the sphere. Now, draw tangent lines to the sides a and b of the triangle that go through C. Extend the lines OA and OB to intersect the tangent lines at P and Q, respectively. See the figure. List the plane right triangles. Determine the measures of the central angles.

2. Apply the Law of Cosines to triangles OPQ and CPQ to find two expressions for the length of PQ.

3. Subtract the expressions in part (2) from each other. Solve for the term containing $\cos c$.

4. Use the Pythagorean Theorem to find another value for $OQ^2 - CQ^2$ and $OP^2 - CP^2$. Now solve for $\cos c$.

5. Replacing the ratios in part (4) by the cosines of the sides of the spherical triangle, you should now have the Law of Cosines for spherical triangles:

$$\cos c = \cos a \cos b + \sin a \sin b \cos C$$

Source: For the spherical Law of Cosines; see *Mathematics from the Birth of Numbers* by Jan Gullberg. W. W. Norton & Co., Publishers, 1996, pp. 491–494.

II. The Lewis and Clark Expedition Lewis and Clark followed several rivers in their trek from what is now Great Falls, Montana, to the Pacific coast. First, they went down the Missouri and Jefferson rivers from Great Falls to Lemhi, Idaho. Because the two cities are on different longitudes and different latitudes, we must account for the curvature of Earth when computing the distance that they traveled. Assume that the radius of Earth is 3960 miles.

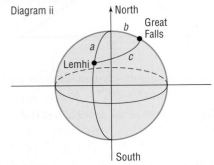

1. Great Falls is at approximately 47.5°N and 111.3°W. Lemhi is at approximately 45.5°N and 113.5°W. (We will assume

Diagram ii

that the rivers flow straight from Great Falls to Lemhi on the surface of Earth.) This line is called a geodesic line. Apply the Law of Cosines for a spherical triangle to find the angle between Great Falls and Lemhi. (The central angles are found by using the differences in the latitudes and longitudes of the towns. See the diagram.) Then find the length of the arc joining the two towns. (Recall $s = r\theta$.)

2. From Lemhi, they went up the Bitteroot River and the Snake River to what is now Lewiston and Clarkston on the border of Idaho and Washington. Although this is not really a side to a triangle, we will make a side that goes from Lemhi to Lewiston and Clarkston. If Lewiston and Clarkston are at about 46.5°N 117.0°W, find the distance from Lemhi using the Law of Cosines for a spherical triangle and the arc length.

3. How far did the explorers travel just to get that far?

4. Draw a plane triangle connecting the three towns. If the distance from Lewiston to Great Falls is 282 miles and the angle at Great Falls is 42° and the angle at Lewiston is 48.5°, find the distance from Great Falls to Lemhi and from Lemhi to Lewiston. How do these distances compare with the ones computed in parts (1) and (2)?

Source: For Lewis and Clark Expedition: *American Journey: The Quest for Liberty to 1877, Texas Edition.* Prentice Hall, 1992, p. 345.

Source: For map coordinates: *National Geographic Atlas of the World,* published by National Geographic Society, 1981, pp. 74–75.

The following projects are available at the Instructor's Resource Center (IRC):

III. Project at Motorola: ***How Can You Build or Analyze a Vibration Profile?*** Fourier functions are not only important to analyze vibrations, but they are also what a mathematician would call interesting. Complete the project to see why.

IV. Leaning Tower of Pisa Trigonometry is used to analyze the apparent height and tilt of the Leaning Tower of Pisa.

V. Locating Lost Treasure Clever treasure seekers who know the Law of Sines are able to efficiently find a buried treasure.

VI. Jacob's Field Angles of elevation and the Law of Sines are used to determine the height of the stadium wall and the distance from home plate to the top of the wall.

Polar Coordinates; Vectors

8

Outline

How Do Airplanes Fly?

Have you ever watched a big jetliner lumber into position on the runway for takeoff and wonder, "How does that thing ever get off the ground?" You know it's because of the wing that it stays up in the air, but how does it really work?

When air flows around a wing, it creates lift. The way it creates lift is based on the wing's movement through the air and the air pressure created around the wing. An airplane's wing, in varying degrees depending on the type and design of the airplane, is curved over the top of the wing and straighter underneath the wing. As air hits the wing, it is "split in two," with air moving both over and under the wing. Since the top of the wing has more curve than the bottom of the wing, the air moving over the top of the wing has farther to travel, and thus must move faster than the air moving underneath the wing. The air moving over the top of the wing now exerts less air pressure on the wing than the slower-moving air under the wing. Lift is created.

The difference in air pressure is the primary force creating lift on a wing, but one other force exerted on the wing also helps to produce lift. This is the force of deflection. Air moving along the underside of the wing is deflected downward. Remember the Newtonian principle: For every action, there is an equal and opposite reaction. The air that is deflected downward (action) helps to push the wing upward (reaction), producing more lift.

These two natural forces on the wing, pressure and deflection, produce lift. The faster the wing moves through the air, the greater the forces become, and the greater the lift.

Source: Thomas Schueneman. How do airplanes fly? http://meme.essortment.com/howdoairplane_rlmi.htm, accessed August 2006. © 2002 by Pagewise, Inc. Used with permission.

—See Chapter Project I—

◁ A Look Back, A Look Ahead ▷

This chapter is in two parts: Polar Coordinates, Sections 8.1–8.3, and Vectors, Sections 8.4–8.7. They are independent of each other and may be covered in any order.

Sections 8.1–8.3: In Foundations we introduced rectangular coordinates x and y and discussed the graph of an equation in two variables involving x and y. In Sections 8.1 and 8.2, we introduce polar coordinates, an alternative to rectangular coordinates, and discuss graphing equations that involve polar coordinates. In Section 4.3, we discussed raising a real number to a real power. In Section 8.3 we extend this idea by raising a complex number to a real power. As it turns out, polar coordinates are useful for the discussion.

Sections 8.4–8.7: We have seen in many chapters that we are often required to solve an equation to obtain a solution to applied problems. In the last four sections of this chapter, we develop the notion of a vector and show how it can be used to model applied problems in physics and engineering.

591

8.1 Polar Coordinates

PREPARING FOR THIS SECTION *Before getting started, review the following:*

- Rectangular Coordinates (Foundations, Section 1, pp. 2–3)
- Definitions of the Sine and Cosine Functions (Section 5.4, pp. 414–415)
- Inverse Tangent Function (Section 6.1, pp. 484–486)
- Completing the Square (Appendix A, Section A.4, p. A40)

Now Work the 'Are You Prepared?' problems on page 599.

OBJECTIVES **1** Plot Points Using Polar Coordinates (p. 592)

           2 Convert from Polar Coordinates to Rectangular Coordinates (p. 594)

           3 Convert from Rectangular Coordinates to Polar Coordinates (p. 596)

           4 Transform Equations between Polar and Rectangular Forms (p. 598)

So far, we have always used a system of rectangular coordinates to plot points in the plane. Now we are ready to describe another system, called *polar coordinates*. As we shall soon see, in many instances polar coordinates offer certain advantages over rectangular coordinates.

Figure 1

In a rectangular coordinate system, you will recall, a point in the plane is represented by an ordered pair of numbers (x, y), where x and y equal the signed distance of the point from the y-axis and x-axis, respectively. In a polar coordinate system, we select a point, called the **pole,** and then a ray with vertex at the pole, called the **polar axis.** See Figure 1. Comparing the rectangular and polar coordinate systems, we see that the origin in rectangular coordinates coincides with the pole in polar coordinates, and the positive x-axis in rectangular coordinates coincides with the polar axis in polar coordinates.

1 Plot Points Using Polar Coordinates

A point P in a polar coordinate system is represented by an ordered pair of numbers (r, θ). If $r > 0$, then r is the distance of the point from the pole; θ is an angle (in degrees or radians) formed by the polar axis and a ray from the pole through the point. We call the ordered pair (r, θ) the **polar coordinates** of the point. See Figure 2.

As an example, suppose that a point P has polar coordinates $\left(2, \dfrac{\pi}{4}\right)$. We locate P by first drawing an angle of $\dfrac{\pi}{4}$ radian, placing its vertex at the pole and its initial side along the polar axis. Then we go out a distance of 2 units along the terminal side of the angle to reach the point P. See Figure 3.

Figure 2

$P = (r, \theta)$

r

θ

O Pole   Polar axis

Figure 3

$P = \left(2, \frac{\pi}{4}\right)$

2

$\frac{\pi}{4}$

O Pole   Polar axis

In using polar coordinates (r, θ), it is possible for r to be negative. When this happens, instead of the point being on the terminal side of θ, it is on the ray from the pole extending in the direction *opposite* the terminal side of θ at a distance $|r|$ units from the pole. See Figure 4 for an illustration.

For example, to plot the point $\left(-3, \dfrac{2\pi}{3}\right)$, we use the ray in the opposite direction of $\dfrac{2\pi}{3}$ and go out $|-3| = 3$ units along that ray. See Figure 5.

Figure 4

Figure 5

EXAMPLE 1

Plotting Points Using Polar Coordinates

Plot the points with the following polar coordinates:

(a) $\left(3, \dfrac{5\pi}{3}\right)$ 　　(b) $\left(2, -\dfrac{\pi}{4}\right)$ 　　(c) $(3, 0)$ 　　(d) $\left(-2, \dfrac{\pi}{4}\right)$

Solution   Figure 6 shows the points.

Figure 6

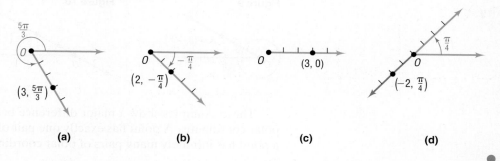

(a)　　　　　(b)　　　　　(c)　　　　　(d)

Now Work PROBLEMS **9, 17,** AND **27**

Recall that an angle measured counterclockwise is positive and an angle measured clockwise is negative. This convention has some interesting consequences relating to polar coordinates. Let's see what these consequences are.

EXAMPLE 2

Finding Several Polar Coordinates of a Single Point

Consider again the point P with polar coordinates $\left(2, \dfrac{\pi}{4}\right)$, as shown in Figure 7(a). Because $\dfrac{\pi}{4}$, $\dfrac{9\pi}{4}$, and $-\dfrac{7\pi}{4}$ all have the same terminal side, we also could have located this point P by using the polar coordinates $\left(2, \dfrac{9\pi}{4}\right)$ or $\left(2, -\dfrac{7\pi}{4}\right)$, as shown in Figures 7(b) and (c). The point $\left(2, \dfrac{\pi}{4}\right)$ can also be represented by the polar coordinates $\left(-2, \dfrac{5\pi}{4}\right)$. See Figure 7(d).

Figure 7

(a)　　　　　(b)　　　　　(c)　　　　　(d)

EXAMPLE 3   **Finding Other Polar Coordinates of a Given Point**

Plot the point P with polar coordinates $\left(3, \dfrac{\pi}{6}\right)$, and find other polar coordinates (r, θ) of this same point for which:

(a) $r > 0, \quad 2\pi \le \theta < 4\pi$ (b) $r < 0, \quad 0 \le \theta < 2\pi$

(c) $r > 0, \quad -2\pi \le \theta < 0$

Solution   The point $\left(3, \dfrac{\pi}{6}\right)$ is plotted in Figure 8.

(a) We add 1 revolution (2π radians) to the angle $\dfrac{\pi}{6}$ to get $P = \left(3, \dfrac{\pi}{6} + 2\pi\right) = \left(3, \dfrac{13\pi}{6}\right)$. See Figure 9.

Figure 8

$P = \left(3, \frac{\pi}{6}\right)$

(b) We add $\dfrac{1}{2}$ revolution (π radians) to the angle $\dfrac{\pi}{6}$ and replace 3 by -3 to get $P = \left(-3, \dfrac{\pi}{6} + \pi\right) = \left(-3, \dfrac{7\pi}{6}\right)$. See Figure 10.

(c) We subtract 2π from the angle $\dfrac{\pi}{6}$ to get $P = \left(3, \dfrac{\pi}{6} - 2\pi\right) = \left(3, -\dfrac{11\pi}{6}\right)$. See Figure 11.

Figure 9	**Figure 10**	**Figure 11**
$P = \left(3, \frac{13\pi}{6}\right)$	$P = \left(-3, \frac{7\pi}{6}\right)$	$P = \left(3, -\frac{11\pi}{6}\right)$

These examples show a major difference between rectangular coordinates and polar coordinates. A point has exactly one pair of rectangular coordinates; however, a point has infinitely many pairs of polar coordinates.

SUMMARY

A point with polar coordinates (r, θ) can also be represented by either of the following:

$$(r, \theta + 2\pi k) \quad \text{or} \quad (-r, \theta + \pi + 2\pi k) \qquad k \text{ any integer}$$

The polar coordinates of the pole are $(0, \theta)$, where θ can be any angle.

Now Work   PROBLEM 31

---

2 Convert from Polar Coordinates to Rectangular Coordinates

Sometimes we need to convert coordinates or equations in rectangular form to polar form, and vice versa. To do this, we recall that the origin in rectangular coordinates is the pole in polar coordinates and that the positive x-axis in rectangular coordinates is the polar axis in polar coordinates.

THEOREM   **Conversion from Polar Coordinates to Rectangular Coordinates**

If P is a point with polar coordinates (r, θ), the rectangular coordinates (x, y) of P are given by

$$x = r \cos \theta \qquad y = r \sin \theta \tag{1}$$

Figure 12

Proof Suppose that P has the polar coordinates (r, θ). We seek the rectangular coordinates (x, y) of P. Refer to Figure 12.

If $r = 0$, then, regardless of θ, the point P is the pole, for which the rectangular coordinates are $(0, 0)$. Formula (1) is valid for $r = 0$.

If $r > 0$, the point P is on the terminal side of θ, and $r = d(O, P) = \sqrt{x^2 + y^2}$. Since

$$\cos \theta = \frac{x}{r} \qquad \sin \theta = \frac{y}{r}$$

we have

$$x = r \cos \theta \qquad y = r \sin \theta$$

If $r < 0$, the point $P = (r, \theta)$ can be represented as $(-r, \pi + \theta)$, where $-r > 0$. Since

$$\cos(\pi + \theta) = -\cos \theta = \frac{x}{-r} \qquad \sin(\pi + \theta) = -\sin \theta = \frac{y}{-r}$$

we have

$$x = r \cos \theta \qquad y = r \sin \theta \qquad \blacksquare$$

EXAMPLE 4 **Converting from Polar Coordinates to Rectangular Coordinates**

Find the rectangular coordinates of the points with the following polar coordinates:

(a) $\left(6, \dfrac{\pi}{6}\right)$ (b) $\left(-4, -\dfrac{\pi}{4}\right)$

Solution We use formula (1): $x = r \cos \theta$ and $y = r \sin \theta$.

Figure 13

(a)

(b)

(a) Figure 13(a) shows $\left(6, \dfrac{\pi}{6}\right)$ plotted. Notice that $\left(6, \dfrac{\pi}{6}\right)$ lies in quadrant I of the rectangular coordinate system. So we expect both the x-coordinate and the y-coordinate to be positive. With $r = 6$ and $\theta = \dfrac{\pi}{6}$, we have

$$x = r \cos \theta = 6 \cos \frac{\pi}{6} = 6 \cdot \frac{\sqrt{3}}{2} = 3\sqrt{3}$$

$$y = r \sin \theta = 6 \sin \frac{\pi}{6} = 6 \cdot \frac{1}{2} = 3$$

The rectangular coordinates of the point $\left(6, \dfrac{\pi}{6}\right)$ are $\left(3\sqrt{3}, 3\right)$, which lies in quadrant I, as expected.

(b) Figure 13(b) shows $\left(-4, -\dfrac{\pi}{4}\right)$ plotted. Notice that $\left(-4, -\dfrac{\pi}{4}\right)$ lies in quadrant II of the rectangular coordinate system. With $r = -4$ and $\theta = -\dfrac{\pi}{4}$, we have

$$x = r \cos \theta = -4 \cos\left(-\frac{\pi}{4}\right) = -4 \cdot \frac{\sqrt{2}}{2} = -2\sqrt{2}$$

$$y = r \sin \theta = -4 \sin\left(-\frac{\pi}{4}\right) = -4\left(-\frac{\sqrt{2}}{2}\right) = 2\sqrt{2}$$

The rectangular coordinates of the point $\left(-4, -\dfrac{\pi}{4}\right)$ are $\left(-2\sqrt{2}, 2\sqrt{2}\right)$, which lies in quadrant II, as expected.

COMMENT Many calculators have the capability of converting from polar coordinates to rectangular coordinates. Consult your owner's manual for the proper key strokes. Since in most cases this procedure is tedious, you will find that using formula (1) is faster.

━━━ **Now Work** PROBLEMS 39 AND 51

3 Convert from Rectangular Coordinates to Polar Coordinates

Converting from rectangular coordinates (x, y) to polar coordinates (r, θ) is a little more complicated. Notice that we begin each example by plotting the given rectangular coordinates.

EXAMPLE 5

How to Convert from Rectangular Coordinates to Polar Coordinates with the Point on a Coordinate Axis

Find polar coordinates of a point whose rectangular coordinates are $(0, 3)$.

Step-by-Step Solution

Step 1: Plot the point (x, y) and note the quadrant the point lies in or the coordinate axis the point lies on.

Figure 14

We plot the point $(0, 3)$ in a rectangular coordinate system. See Figure 14. The point lies on the positive y-axis.

Step 2: Determine the distance r from the origin to the point.

The point $(0, 3)$ lies on the y-axis a distance of 3 units from the origin (pole), so $r = 3$.

Step 3: Determine θ.

A ray with vertex at the pole through $(0, 3)$ forms an angle $\theta = \dfrac{\pi}{2}$ with the polar axis.

Polar coordinates for this point can be given by $\left(3, \dfrac{\pi}{2}\right)$. Other possible representations include $\left(-3, -\dfrac{\pi}{2}\right)$ and $\left(3, \dfrac{5\pi}{2}\right)$.

 COMMENT Most graphing calculators have the capability of converting from rectangular coordinates to polar coordinates. Consult your owner's manual for the proper keystrokes. ∎

Figure 15 shows polar coordinates of points that lie on either the x-axis or the y-axis. In each illustration, $a > 0$.

Figure 15

(a) $(x, y) = (a, 0)$, $a > 0$ **(b)** $(x, y) = (0, a)$, $a > 0$ **(c)** $(x, y) = (-a, 0)$, $a > 0$ **(d)** $(x, y) = (0, -a)$, $a > 0$

Now Work PROBLEM 55

EXAMPLE 6

How to Convert from Rectangular Coordinates to Polar Coordinates with the Point in a Quadrant

Find the polar coordinates of a point whose rectangular coordinates are $(2, -2)$.

Step-by-Step Solution

Step 1: Plot the point (x, y) and note the quadrant the point lies in or the coordinate axis the point lies on.

Figure 16

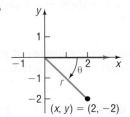

We plot the point $(2, -2)$ in a rectangular coordinate system. See Figure 16. The point lies in quadrant IV.

Step 2: Determine the distance r from the origin to the point using $r = \sqrt{x^2 + y^2}$.

$$r = \sqrt{x^2 + y^2} = \sqrt{(2)^2 + (-2)^2} = \sqrt{8} = 2\sqrt{2}$$

Step 3: Determine θ.

We find θ by recalling that $\tan\theta = \dfrac{y}{x}$, so $\theta = \tan^{-1}\dfrac{y}{x}, -\dfrac{\pi}{2} < \theta < \dfrac{\pi}{2}$. Since $(2, -2)$ lies in quadrant IV, we know that $-\dfrac{\pi}{2} < \theta < 0$. As a result,

$$\theta = \tan^{-1}\frac{y}{x} = \tan^{-1}\left(\frac{-2}{2}\right) = \tan^{-1}(-1) = -\frac{\pi}{4}$$

A set of polar coordinates for the point $(2,-2)$ is $\left(2\sqrt{2}, -\dfrac{\pi}{4}\right)$. Other possible representations include $\left(2\sqrt{2}, \dfrac{7\pi}{4}\right)$ and $\left(-2\sqrt{2}, \dfrac{3\pi}{4}\right)$.

●

EXAMPLE 7 **Converting from Rectangular Coordinates to Polar Coordinates**

Find polar coordinates of a point whose rectangular coordinates are $\left(-1, -\sqrt{3}\right)$.

Figure 17

$(x, y) = (-1, -\sqrt{3})$

Solution **STEP 1:** See Figure 17. The point lies in quadrant III.

STEP 2: The distance r from the origin to the point $\left(-1, -\sqrt{3}\right)$ is

$$r = \sqrt{(-1)^2 + \left(-\sqrt{3}\right)^2} = \sqrt{4} = 2$$

STEP 3: To find θ, we use $\alpha = \tan^{-1}\dfrac{y}{x} = \tan^{-1}\dfrac{-\sqrt{3}}{-1} = \tan^{-1}\sqrt{3}, -\dfrac{\pi}{2} < \alpha < \dfrac{\pi}{2}$. Since the point $\left(-1, -\sqrt{3}\right)$ lies in quadrant III and the inverse tangent function gives an angle in quadrant I, we add π to the result to obtain an angle in quadrant III.

$$\theta = \pi + \tan^{-1}\left(\frac{-\sqrt{3}}{-1}\right) = \pi + \tan^{-1}\sqrt{3} = \pi + \frac{\pi}{3} = \frac{4\pi}{3}$$

A set of polar coordinates for this point is $\left(2, \dfrac{4\pi}{3}\right)$. Other possible representations include $\left(-2, \dfrac{\pi}{3}\right)$ and $\left(2, -\dfrac{2\pi}{3}\right)$.

●

Figure 18 shows how to find polar coordinates of a point that lies in a quadrant when its rectangular coordinates (x, y) are given.

Figure 18

(a) $r = \sqrt{x^2 + y^2}$
$\theta = \tan^{-1}\dfrac{y}{x}$

(b) $r = \sqrt{x^2 + y^2}$
$\theta = \pi + \tan^{-1}\dfrac{y}{x}$

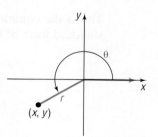

(c) $r = \sqrt{x^2 + y^2}$
$\theta = \pi + \tan^{-1}\dfrac{y}{x}$

(d) $r = \sqrt{x^2 + y^2}$
$\theta = \tan^{-1}\dfrac{y}{x}$

Based on the preceding discussion, we have the formulas

$$r^2 = x^2 + y^2 \qquad \tan\theta = \frac{y}{x} \qquad \text{if } x \neq 0 \qquad\qquad (2)$$

To use formula (2) effectively, follow these steps:

> **Steps for Converting from Rectangular to Polar Coordinates**
>
> **STEP 1:** Always plot the point (x, y) first, as we did in Examples 5, 6, and 7. Note the quadrant the point lies in or the coordinate axis the point lies on.
>
> **STEP 2:** If $x = 0$ or $y = 0$, use your illustration to find r. See Figure 15. If $x \neq 0$ and $y \neq 0$, then $r = \sqrt{x^2 + y^2}$.
>
> **STEP 3:** Find θ. If $x = 0$ or $y = 0$, use your illustration to find θ. See Figure 15. If $x \neq 0$ and $y \neq 0$, note the quadrant in which the point lies.
>
> $$\text{Quadrant I or IV:} \quad \theta = \tan^{-1} \frac{y}{x}$$
>
> $$\text{Quadrant II or III:} \quad \theta = \pi + \tan^{-1} \frac{y}{x}$$
>
> See Figure 18.

⟍════ **Now Work** PROBLEM 59

4 Transform Equations between Polar and Rectangular Forms

Formulas (1) and (2) may also be used to transform equations from polar form to rectangular form, and vice versa. Two common techniques for transforming an equation from polar form to rectangular form are the following:

1. Multiplying both sides of the equation by r
2. Squaring both sides of the equation

EXAMPLE 8	**Transforming an Equation from Polar to Rectangular Form**

Transform the equation $r = 6 \cos \theta$ from polar coordinates to rectangular coordinates, and identify the graph.

Solution   If we multiply each side by r, it will be easier to apply formulas (1) and (2).

$$r = 6 \cos \theta$$
$$r^2 = 6r \cos \theta \quad \text{\small Multiply each side by } r.$$
$$x^2 + y^2 = 6x \quad \text{\small } r^2 = x^2 + y^2; x = r \cos \theta$$

This is the equation of a circle, so we proceed to complete the square to obtain the standard form of the equation.

$$x^2 + y^2 = 6x$$
$$(x^2 - 6x) + y^2 = 0 \quad \text{\small General form}$$
$$(x^2 - 6x + 9) + y^2 = 9 \quad \text{\small Complete the square in x.}$$
$$(x - 3)^2 + y^2 = 9 \quad \text{\small Factor.}$$

This is the standard form of the equation of a circle with center $(3, 0)$ and radius 3.

    ●

⟍════ **Now Work** PROBLEM 75

EXAMPLE 9	**Transforming an Equation from Rectangular to Polar Form**

Transform the equation $4xy = 9$ from rectangular coordinates to polar coordinates.

Solution   We use formula (1): $x = r \cos \theta$ and $y = r \sin \theta$.

$$4xy = 9$$
$$4(r \cos \theta)(r \sin \theta) = 9 \quad x = r\cos\theta, y = r\sin\theta$$
$$4r^2 \cos \theta \sin \theta = 9$$

This is the polar form of the equation. It can be simplified as shown next:

$$2r^2(2 \sin \theta \cos \theta) = 9 \quad \text{Factor out } 2r^2.$$
$$2r^2 \sin(2\theta) = 9 \quad \text{Double-angle Formula}$$

●

Now Work PROBLEM 69

8.1 Assess Your Understanding

'Are You Prepared?' *Answers are given at the end of these exercises. If you get a wrong answer, read the pages listed in red.*

1. Plot the point whose rectangular coordinates are $(3, -1)$. What quadrant does the point lie in? (pp. 2–3)

2. To complete the square of $x^2 + 6x$, add _____. (p. A40)

3. If $P = (a, b)$ is a point on the terminal side of the angle θ at a distance r from the origin, then $\tan \theta =$ _____. (pp. 414–415)

4. $\tan^{-1}(-1) =$ _____ . (pp. 485–486)

Concepts and Vocabulary

5. The origin in rectangular coordinates coincides with the _____ in polar coordinates; the positive x-axis in rectangular coordinates coincides with the _____ _____ in polar coordinates.

6. True or False In the polar coordinates (r, θ), r can be negative.

7. True or False The polar coordinates of a point are unique.

8. If P is a point with polar coordinates (r, θ), the rectangular coordinates (x, y) of P are given by $x =$ _____ and $y =$ _____.

Skill Building

In Problems 9–16, match each point in polar coordinates with either A, B, C, or D on the graph.

9. $\left(2, -\dfrac{11\pi}{6}\right)$

10. $\left(-2, -\dfrac{\pi}{6}\right)$

11. $\left(-2, \dfrac{\pi}{6}\right)$

12. $\left(2, \dfrac{7\pi}{6}\right)$

13. $\left(2, \dfrac{5\pi}{6}\right)$

14. $\left(-2, \dfrac{5\pi}{6}\right)$

15. $\left(-2, \dfrac{7\pi}{6}\right)$

16. $\left(2, \dfrac{11\pi}{6}\right)$

In Problems 17–30, plot each point given in polar coordinates.

17. $(3, 90°)$

18. $(4, 270°)$

19. $(-2, 0)$

20. $(-3, \pi)$

21. $\left(6, \dfrac{\pi}{6}\right)$

22. $\left(5, \dfrac{5\pi}{3}\right)$

23. $(-2, 135°)$

24. $(-3, 120°)$

25. $\left(4, -\dfrac{2\pi}{3}\right)$

26. $\left(2, -\dfrac{5\pi}{4}\right)$

27. $\left(-1, -\dfrac{\pi}{3}\right)$

28. $\left(-3, -\dfrac{3\pi}{4}\right)$

29. $(-2, -\pi)$

30. $\left(-3, -\dfrac{\pi}{2}\right)$

In Problems 31–38, plot each point given in polar coordinates, and find other polar coordinates (r, θ) of the point for which:

(a) $r > 0$, $-2\pi \le \theta < 0$ 　　(b) $r < 0$, $0 \le \theta < 2\pi$ 　　(c) $r > 0$, $2\pi \le \theta < 4\pi$

31. $\left(5, \dfrac{2\pi}{3}\right)$

32. $\left(4, \dfrac{3\pi}{4}\right)$

33. $(-2, 3\pi)$

34. $(-3, 4\pi)$

35. $\left(1, \dfrac{\pi}{2}\right)$

36. $(2, \pi)$

37. $\left(-3, -\dfrac{\pi}{4}\right)$

38. $\left(-2, -\dfrac{2\pi}{3}\right)$

In Problems 39–54, the polar coordinates of a point are given. Find the rectangular coordinates of each point.

39. $\left(3, \dfrac{\pi}{2}\right)$ **40.** $\left(4, \dfrac{3\pi}{2}\right)$ **41.** $(-2, 0)$ **42.** $(-3, \pi)$

43. $(6, 150°)$ **44.** $(5, 300°)$ **45.** $\left(-2, \dfrac{3\pi}{4}\right)$ **46.** $\left(-2, \dfrac{2\pi}{3}\right)$

47. $\left(-1, -\dfrac{\pi}{3}\right)$ **48.** $\left(-3, -\dfrac{3\pi}{4}\right)$ **49.** $(-2, -180°)$ **50.** $(-3, -90°)$

51. $(7.5, 110°)$ **52.** $(-3.1, 182°)$ **53.** $(6.3, 3.8)$ **54.** $(8.1, 5.2)$

In Problems 55–66, the rectangular coordinates of a point are given. Find polar coordinates for each point.

55. $(3, 0)$ **56.** $(0, 2)$ **57.** $(-1, 0)$ **58.** $(0, -2)$

59. $(1, -1)$ **60.** $(-3, 3)$ **61.** $\left(\sqrt{3}, 1\right)$ **62.** $\left(-2, -2\sqrt{3}\right)$

63. $(1.3, -2.1)$ **64.** $(-0.8, -2.1)$ **65.** $(8.3, 4.2)$ **66.** $(-2.3, 0.2)$

In Problems 67–74, the letters x and y represent rectangular coordinates. Write each equation using polar coordinates (r, θ).

67. $2x^2 + 2y^2 = 3$ **68.** $x^2 + y^2 = x$ **69.** $x^2 = 4y$ **70.** $y^2 = 2x$

71. $2xy = 1$ **72.** $4x^2y = 1$ **73.** $x = 4$ **74.** $y = -3$

In Problems 75–82, the letters r and θ represent polar coordinates. Write each equation using rectangular coordinates (x, y).

75. $r = \cos \theta$ **76.** $r = \sin \theta + 1$ **77.** $r^2 = \cos \theta$ **78.** $r = \sin \theta - \cos \theta$

79. $r = 2$ **80.** $r = 4$ **81.** $r = \dfrac{4}{1 - \cos \theta}$ **82.** $r = \dfrac{3}{3 - \cos \theta}$

Applications and Extensions

83. Chicago In Chicago, the road system is set up like a Cartesian plane, where streets are indicated by the number of blocks they are from Madison Street and State Street. For example, Wrigley Field in Chicago is located at 1060 West Addison, which is 10 blocks west of State Street and 36 blocks north of Madison Street. Treat the intersection of Madison Street and State Street as the origin of a coordinate system, with east being the positive x-axis.
  (a) Write the location of Wrigley Field using rectangular coordinates.
  (b) Write the location of Wrigley Field using polar coordinates. Use the east direction for the polar axis. Express θ in degrees.
  (c) U.S. Cellular Field, home of the White Sox, is located at 35th and Princeton, which is 3 blocks west of State Street and 35 blocks south of Madison. Write the location of U.S. Cellular Field using rectangular coordinates.
  (d) Write the location of U.S. Cellular Field using polar coordinates. Use the east direction for the polar axis. Express θ in degrees.

84. Show that the formula for the distance d between two points $P_1 = (r_1, \theta_1)$ and $P_2 = (r_2, \theta_2)$ is

$$d = \sqrt{r_1^2 + r_2^2 - 2r_1 r_2 \cos(\theta_2 - \theta_1)}$$

City of Chicago, Illinois

Addison Street

1 mile
1 km

Madison Street

35th Street

State Street

Addison Street

Wrigley Field
1060 West Addison

U.S. Cellular Field
35th and Princeton

35th Street

Explaining Concepts: Discussion and Writing

85. In converting from polar coordinates to rectangular coordinates, what formulas will you use?

86. Explain how you proceed to convert from rectangular coordinates to polar coordinates.

87. Is the street system in your town based on a rectangular coordinate system, a polar coordinate system, or some other system? Explain.

'Are You Prepared?' Answers

1. ; quadrant IV   **2.** 9   **3.** $\dfrac{b}{a}$ **4.** $-\dfrac{\pi}{4}$

(graph showing point $(3, -1)$ in quadrant IV)

8.2 Polar Equations and Graphs

PREPARING FOR THIS SECTION *Before getting started, review the following:*

- Symmetry (Foundations, Section 2, pp. 12–14)
- Circles (Foundations, Section 4, pp. 35–39)
- Even–Odd Properties of Trigonometric Functions (Section 5.5, p. 432)

- Difference Formulas for Sine and Cosine (Section 6.5, pp. 514 and 617
- Value of the Sine and Cosine Functions at Certain Angles (Section 5.3, pp. 403–405, Section 5.4, pp. 416, 420–421)

Now Work the 'Are You Prepared?' problems on page 613.

OBJECTIVES   **1** Graph and Identify Polar Equations by Converting to Rectangular Equations (p. 602)
     2 Test Polar Equations for Symmetry (p. 605)
     3 Graph Polar Equations by Plotting Points (p. 606)

Just as a rectangular grid may be used to plot points given by rectangular coordinates, as in Figure 19(a), we can use a grid consisting of concentric circles (with centers at the pole) and rays (with vertices at the pole) to plot points given by polar coordinates, as shown in Figure 19(b). We shall use such **polar grids** to graph *polar equations*.

Figure 19

(a) Rectangular grid

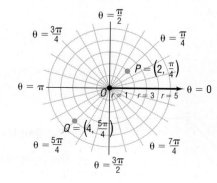

(b) Polar grid

DEFINITION

> An equation whose variables are polar coordinates is called a **polar equation.** The **graph of a polar equation** consists of all points whose polar coordinates satisfy the equation.

1 Graph and Identify Polar Equations by Converting to Rectangular Equations

One method that we can use to graph a polar equation is to convert the equation to rectangular coordinates. In the discussion that follows, (x, y) represent the rectangular coordinates of a point P, and (r, θ) represent polar coordinates of the point P.

EXAMPLE 1 **Identifying and Graphing a Polar Equation (Circle)**

Identify and graph the equation: $r = 3$

Solution We convert the polar equation to a rectangular equation.

$$r = 3$$
$$r^2 = 9 \quad \text{Square both sides.}$$
$$x^2 + y^2 = 9 \quad r^2 = x^2 + y^2$$

The graph of $r = 3$ is a circle, with center at the pole and radius 3. See Figure 20.

Figure 20
$r = 3$ or $x^2 + y^2 = 9$

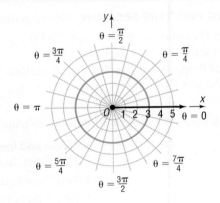

Now Work PROBLEM 13

EXAMPLE 2 **Identifying and Graphing a Polar Equation (Line)**

Identify and graph the equation: $\theta = \dfrac{\pi}{4}$

Solution We convert the polar equation to a rectangular equation.

Figure 21
$\theta = \dfrac{\pi}{4}$ or $y = x$

$$\theta = \frac{\pi}{4}$$
$$\tan \theta = \tan \frac{\pi}{4} \quad \text{Take the tangent of both sides.}$$
$$\frac{y}{x} = 1 \quad \tan \theta = \frac{y}{x}; \tan \frac{\pi}{4} = 1$$
$$y = x$$

The graph of $\theta = \dfrac{\pi}{4}$ is a line passing through the pole making an angle of $\dfrac{\pi}{4}$ with the polar axis. See Figure 21.

Now Work PROBLEM 15

EXAMPLE 3   **Identifying and Graphing a Polar Equation (Horizontal Line)**

Identify and graph the equation: $r \sin \theta = 2$

Solution   Since $y = r \sin \theta$, we can write the equation as

$$y = 2$$

We conclude that the graph of $r \sin \theta = 2$ is a horizontal line 2 units above the pole. See Figure 22.

Figure 22
$r \sin \theta = 2$ or $y = 2$

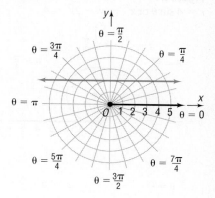

COMMENT   A graphing utility can be used to graph polar equations. Read Using a Graphing Utility to Graph a Polar Equation, Appendix B, Section B.8.   ∎

EXAMPLE 4   **Identifying and Graphing a Polar Equation (Vertical Line)**

Identify and graph the equation: $r \cos \theta = -3$

Solution   Since $x = r \cos \theta$, we can write the equation as

$$x = -3$$

We conclude that the graph of $r \cos \theta = -3$ is a vertical line 3 units to the left of the pole. See Figure 23.

Figure 23
$r \cos \theta = -3$ or $x = -3$

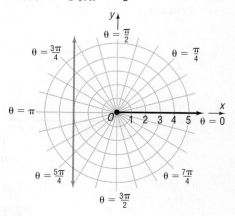

Based on Examples 3 and 4, we are led to the following results. (The proofs are left as exercises. See Problems 81 and 82.)

THEOREM

Let a be a nonzero real number. Then the graph of the equation

$$r \sin \theta = a$$

is a horizontal line a units above the pole if $a > 0$ and $|a|$ units below the pole if $a < 0$.
   The graph of the equation

$$r \cos \theta = a$$

is a vertical line a units to the right of the pole if $a > 0$ and $|a|$ units to the left of the pole if $a < 0$.

─**Now Work** PROBLEM **19**

EXAMPLE 5   **Identifying and Graphing a Polar Equation (Circle)**

Identify and graph the equation: $r = 4 \sin \theta$

Solution   To transform the equation to rectangular coordinates, we multiply each side by r.

$$r^2 = 4r \sin \theta$$

Now we use the facts that $r^2 = x^2 + y^2$ and $y = r \sin \theta$. Then

$$x^2 + y^2 = 4y$$

$$x^2 + (y^2 - 4y) = 0$$

$$x^2 + (y^2 - 4y + 4) = 4 \qquad \text{Complete the square in } y.$$

$$x^2 + (y - 2)^2 = 4 \qquad \text{Factor.}$$

This is the standard equation of a circle with center at $(0, 2)$ in rectangular coordinates and radius 2. See Figure 24.

Figure 24
$r = 4 \sin \theta$ or $x^2 + (y - 2)^2 = 4$

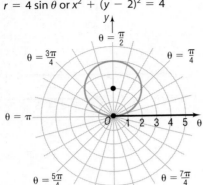

EXAMPLE 6   **Identifying and Graphing a Polar Equation (Circle)**

Identify and graph the equation: $r = -2 \cos \theta$

Solution   We proceed as in Example 5.

$$r^2 = -2r \cos \theta \qquad \text{Multiply both sides by } r.$$

$$x^2 + y^2 = -2x \qquad r^2 = x^2 + y^2; \ x = r \cos \theta$$

$$x^2 + 2x + y^2 = 0$$

$$(x^2 + 2x + 1) + y^2 = 1 \qquad \text{Complete the square in } x.$$

$$(x + 1)^2 + y^2 = 1 \qquad \text{Factor.}$$

This is the standard equation of a circle with center at $(-1, 0)$ in rectangular coordinates and radius 1. See Figure 25.

Figure 25
$r = -2 \cos \theta$ or $(x + 1)^2 + y^2 = 1$

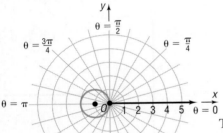

Exploration

Using a square screen, graph $r_1 = \sin \theta$, $r_2 = 2 \sin \theta$, and $r_3 = 3 \sin \theta$. Do you see the pattern? Clear the screen and graph $r_1 = -\sin \theta$, $r_2 = -2 \sin \theta$, and $r_3 = -3 \sin \theta$. Do you see the pattern? Clear the screen and graph $r_1 = \cos \theta$, $r_2 = 2 \cos \theta$, and $r_3 = 3 \cos \theta$. Do you see the pattern? Clear the screen and graph $r_1 = -\cos \theta$, $r_2 = -2 \cos \theta$, and $r_3 = -3 \cos \theta$. Do you see the pattern?

Based on Examples 5 and 6 and the preceding Exploration, we are led to the following results. (The proofs are left as exercises. See Problems 83–86.)

THEOREM   Let a be a positive real number. Then

Equation	Description
(a) $r = 2a \sin \theta$	Circle: radius a; center at $(0, a)$ in rectangular coordinates
(b) $r = -2a \sin \theta$	Circle: radius a; center at $(0, -a)$ in rectangular coordinates
(c) $r = 2a \cos \theta$	Circle: radius a; center at $(a, 0)$ in rectangular coordinates
(d) $r = -2a \cos \theta$	Circle: radius a; center at $(-a, 0)$ in rectangular coordinates

Each circle passes through the pole.

━━━━━━**Now Work** PROBLEM 21

The method of converting a polar equation to an identifiable rectangular equation to obtain the graph is not always helpful, nor is it always necessary. Usually, we set up a table that lists several points on the graph. By checking for symmetry, it may be possible to reduce the number of points needed to draw the graph.

2 Test Polar Equations for Symmetry

In polar coordinates, the points (r, θ) and $(r, -\theta)$ are symmetric with respect to the polar axis (and to the x-axis). See Figure 26(a). The points (r, θ) and $(r, \pi - \theta)$ are symmetric with respect to the line $\theta = \dfrac{\pi}{2}$ (the y-axis). See Figure 26(b). The points (r, θ) and $(-r, \theta)$ are symmetric with respect to the pole (the origin). See Figure 26(c).

Figure 26

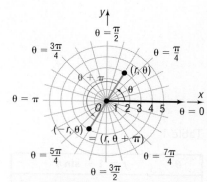

(a) Points symmetric with respect to the polar axis

(b) Points symmetric with respect to the line $\theta = \dfrac{\pi}{2}$

(c) Points symmetric with respect to the pole

The following tests are a consequence of these observations.

THEOREM

Tests for Symmetry

Symmetry with Respect to the Polar Axis (x-Axis)

In a polar equation, replace θ by $-\theta$. If an equivalent equation results, the graph is symmetric with respect to the polar axis.

Symmetry with Respect to the Line $\theta = \dfrac{\pi}{2}$ (y-Axis)

In a polar equation, replace θ by $\pi - \theta$. If an equivalent equation results, the graph is symmetric with respect to the line $\theta = \dfrac{\pi}{2}$.

Symmetry with Respect to the Pole (Origin)

In a polar equation, replace r by $-r$ or θ by $\theta + \pi$. If an equivalent equation results, the graph is symmetric with respect to the pole.

The three tests for symmetry given here are *sufficient* conditions for symmetry, but they are not *necessary* conditions. That is, an equation may fail these tests and still have a graph that is symmetric with respect to the polar axis, the line $\theta = \dfrac{\pi}{2}$, or the pole. For example, the graph of $r = \sin(2\theta)$ turns out to be symmetric with respect to the polar axis, the line $\theta = \dfrac{\pi}{2}$, and the pole, but only the test for symmetry with respect to the pole (replace θ by $\theta + \pi$) works. See also Problems 87–89.

3 Graph Polar Equations by Plotting Points

EXAMPLE 7 — **Graphing a Polar Equation (Cardioid)**

Graph the equation: $r = 1 - \sin\theta$

Solution

We check for symmetry first.

Polar Axis: Replace θ by $-\theta$. The result is

$$r = 1 - \sin(-\theta) = 1 + \sin\theta \qquad \sin(-\theta) = -\sin\theta$$

The test fails, so the graph may or may not be symmetric with respect to the polar axis.

The Line $\theta = \dfrac{\pi}{2}$: Replace θ by $\pi - \theta$. The result is

$$r = 1 - \sin(\pi - \theta) = 1 - (\sin\pi\cos\theta - \cos\pi\sin\theta)$$
$$= 1 - [0 \cdot \cos\theta - (-1)\sin\theta] = 1 - \sin\theta$$

The test is satisfied, so the graph is symmetric with respect to the line $\theta = \dfrac{\pi}{2}$.

The Pole: Replace r by $-r$. Then the result is $-r = 1 - \sin\theta$, so $r = -1 + \sin\theta$. The test fails. Replace θ by $\theta + \pi$. The result is

$$r = 1 - \sin(\theta + \pi)$$
$$= 1 - [\sin\theta\cos\pi + \cos\theta\sin\pi]$$
$$= 1 - [\sin\theta\cdot(-1) + \cos\theta\cdot 0]$$
$$= 1 + \sin\theta$$

This test also fails. So the graph may or may not be symmetric with respect to the pole.

Next, we identify points on the graph by assigning values to the angle θ and calculating the corresponding values of r. Due to the symmetry with respect to the line $\theta = \dfrac{\pi}{2}$, we only need to assign values to θ from $-\dfrac{\pi}{2}$ to $\dfrac{\pi}{2}$, as given in Table 1.

Now we plot the points (r, θ) from Table 1 and trace out the graph, beginning at the point $\left(2, -\dfrac{\pi}{2}\right)$ and ending at the point $\left(0, \dfrac{\pi}{2}\right)$. Then we reflect this portion of the graph about the line $\theta = \dfrac{\pi}{2}$ (the y-axis) to obtain the complete graph. See Figure 27.

Table 1

θ	$r = 1 - \sin\theta$
$-\dfrac{\pi}{2}$	$1 - (-1) = 2$
$-\dfrac{\pi}{3}$	$1 - \left(-\dfrac{\sqrt{3}}{2}\right) \approx 1.87$
$-\dfrac{\pi}{6}$	$1 - \left(-\dfrac{1}{2}\right) = \dfrac{3}{2}$
0	$1 - 0 = 1$
$\dfrac{\pi}{6}$	$1 - \dfrac{1}{2} = \dfrac{1}{2}$
$\dfrac{\pi}{3}$	$1 - \dfrac{\sqrt{3}}{2} \approx 0.13$
$\dfrac{\pi}{2}$	$1 - 1 = 0$

Figure 27
$r = 1 - \sin\theta$

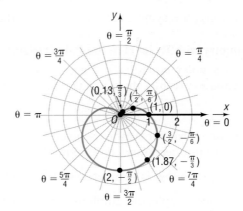

Exploration

Graph $r_1 = 1 + \sin\theta$. Clear the screen and graph $r_1 = 1 - \cos\theta$. Clear the screen and graph $r_1 = 1 + \cos\theta$. Do you see a pattern?

The curve in Figure 27 is an example of a *cardioid* (a heart-shaped curve).

DEFINITION

Cardioids are characterized by equations of the form

$$r = a(1 + \cos \theta) \qquad r = a(1 + \sin \theta)$$
$$r = a(1 - \cos \theta) \qquad r = a(1 - \sin \theta)$$

where $a > 0$. The graph of a cardioid passes through the pole.

━━━ Now Work PROBLEM 37

EXAMPLE 8

Graphing a Polar Equation (Limaçon without an Inner Loop)

Graph the equation: $r = 3 + 2 \cos \theta$

Solution

We check for symmetry first.

Polar Axis: Replace θ by $-\theta$. The result is

$$r = 3 + 2 \cos(-\theta) = 3 + 2 \cos \theta \qquad \cos(-\theta) = \cos \theta$$

The test is satisfied, so the graph is symmetric with respect to the polar axis.

The Line $\theta = \dfrac{\pi}{2}$: Replace θ by $\pi - \theta$. The result is

$$r = 3 + 2 \cos(\pi - \theta) = 3 + 2(\cos \pi \cos \theta + \sin \pi \sin \theta)$$
$$= 3 - 2 \cos \theta$$

The test fails, so the graph may or may not be symmetric with respect to the line $\theta = \dfrac{\pi}{2}$.

The Pole: Replace r by $-r$. The test fails, so the graph may or may not be symmetric with respect to the pole. Replace θ by $\theta + \pi$. The test fails, so the graph may or may not be symmetric with respect to the pole.

Next, we identify points on the graph by assigning values to the angle θ and calculating the corresponding values of r. Due to the symmetry with respect to the polar axis, we only need to assign values to θ from 0 to π, as given in Table 2.

Now we plot the points (r, θ) from Table 2 and trace out the graph, beginning at the point $(5, 0)$ and ending at the point $(1, \pi)$. Then we reflect this portion of the graph about the polar axis (the x-axis) to obtain the complete graph. See Figure 28.

Table 2

θ	$r = 3 + 2 \cos \theta$
0	$3 + 2(1) = 5$
$\dfrac{\pi}{6}$	$3 + 2\left(\dfrac{\sqrt{3}}{2}\right) \approx 4.73$
$\dfrac{\pi}{3}$	$3 + 2\left(\dfrac{1}{2}\right) = 4$
$\dfrac{\pi}{2}$	$3 + 2(0) = 3$
$\dfrac{2\pi}{3}$	$3 + 2\left(-\dfrac{1}{2}\right) = 2$
$\dfrac{5\pi}{6}$	$3 + 2\left(-\dfrac{\sqrt{3}}{2}\right) \approx 1.27$
π	$3 + 2(-1) = 1$

Figure 28
$r = 3 + 2 \cos \theta$

Exploration

Graph $r_1 = 3 - 2 \cos \theta$. Clear the screen and graph $r_1 = 3 + 2 \sin \theta$. Clear the screen and graph $r_1 = 3 - 2 \sin \theta$. Do you see a pattern?

The curve in Figure 28 is an example of a *limaçon* (a French word for *snail*) *without an inner loop*.

DEFINITION

Limaçons without an inner loop are characterized by equations of the form

$$r = a + b \cos \theta \qquad r = a + b \sin \theta$$
$$r = a - b \cos \theta \qquad r = a - b \sin \theta$$

where $a > 0, b > 0$, and $a > b$. The graph of a limaçon without an inner loop does not pass through the pole.

─────────**Now Work** PROBLEM 43

EXAMPLE 9 **Graphing a Polar Equation (Limaçon with an Inner Loop)**

Graph the equation: $r = 1 + 2 \cos \theta$

Solution First, we check for symmetry.

Polar Axis: Replace θ by $-\theta$. The result is

$$r = 1 + 2 \cos(-\theta) = 1 + 2 \cos \theta$$

The test is satisfied, so the graph is symmetric with respect to the polar axis.

The Line $\theta = \dfrac{\pi}{2}$: Replace θ by $\pi - \theta$. The result is

$$r = 1 + 2 \cos(\pi - \theta) = 1 + 2(\cos \pi \cos \theta + \sin \pi \sin \theta)$$
$$= 1 - 2 \cos \theta$$

The test fails, so the graph may or may not be symmetric with respect to the line $\theta = \dfrac{\pi}{2}$.

The Pole: Replace r by $-r$. The test fails, so the graph may or may not be symmetric with respect to the pole. Replace θ by $\theta + \pi$. The test fails, so the graph may or may not be symmetric with respect to the pole.

Next, we identify points on the graph of $r = 1 + 2 \cos \theta$ by assigning values to the angle θ and calculating the corresponding values of r. Due to the symmetry with respect to the polar axis, we only need to assign values to θ from 0 to π, as given in Table 3.

Now we plot the points (r, θ) from Table 3, beginning at $(3, 0)$ and ending at $(-1, \pi)$. See Figure 29(a). Finally, we reflect this portion of the graph about the polar axis (the x-axis) to obtain the complete graph. See Figure 29(b).

Table 3

θ	$r = 1 + 2 \cos \theta$
0	$1 + 2(1) = 3$
$\dfrac{\pi}{6}$	$1 + 2\left(\dfrac{\sqrt{3}}{2}\right) \approx 2.73$
$\dfrac{\pi}{3}$	$1 + 2\left(\dfrac{1}{2}\right) = 2$
$\dfrac{\pi}{2}$	$1 + 2(0) = 1$
$\dfrac{2\pi}{3}$	$1 + 2\left(-\dfrac{1}{2}\right) = 0$
$\dfrac{5\pi}{6}$	$1 + 2\left(-\dfrac{\sqrt{3}}{2}\right) \approx -0.73$
π	$1 + 2(-1) = -1$

Figure 29

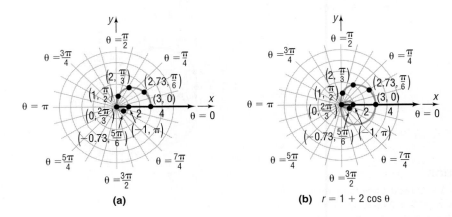

(a)

(b) $r = 1 + 2 \cos \theta$

Exploration

Graph $r_1 = 1 - 2 \cos \theta$. Clear the screen and graph $r_1 = 1 + 2 \sin \theta$. Clear the screen and graph $r_1 = 1 - 2 \sin \theta$. Do you see a pattern?

The curve in Figure 29(b) is an example of a *limaçon with an inner loop*.

DEFINITION

Limaçons with an inner loop are characterized by equations of the form

$$r = a + b \cos \theta \qquad r = a + b \sin \theta$$
$$r = a - b \cos \theta \qquad r = a - b \sin \theta$$

where $a > 0, b > 0$, and $a < b$. The graph of a limaçon with an inner loop will pass through the pole twice.

→ **Now Work** PROBLEM 45

EXAMPLE 10   **Graphing a Polar Equation (Rose)**

Graph the equation: $r = 2 \cos(2\theta)$

Solution   We check for symmetry.

Polar Axis: If we replace θ by $-\theta$, the result is

$$r = 2 \cos[2(-\theta)] = 2 \cos(2\theta)$$

The test is satisfied, so the graph is symmetric with respect to the polar axis.

The Line $\theta = \dfrac{\pi}{2}$:   If we replace θ by $\pi - \theta$, we obtain

$$r = 2 \cos[2(\pi - \theta)] = 2 \cos(2\pi - 2\theta) = 2 \cos(2\theta)$$

The test is satisfied, so the graph is symmetric with respect to the line $\theta = \dfrac{\pi}{2}$.

The Pole: Since the graph is symmetric with respect to both the polar axis and the line $\theta = \dfrac{\pi}{2}$, it must be symmetric with respect to the pole.

Next, we construct Table 4. Due to the symmetry with respect to the polar axis, the line $\theta = \dfrac{\pi}{2}$, and the pole, we consider only values of θ from 0 to $\dfrac{\pi}{2}$.

We plot and connect these points in Figure 30(a). Finally, because of symmetry, we reflect this portion of the graph first about the polar axis (the x-axis) and then about the line $\theta = \dfrac{\pi}{2}$ (the y-axis) to obtain the complete graph. See Figure 30(b).

Table 4

θ	$r = 2 \cos(2\theta)$
0	$2(1) = 2$
$\dfrac{\pi}{6}$	$2\left(\dfrac{1}{2}\right) = 1$
$\dfrac{\pi}{4}$	$2(0) = 0$
$\dfrac{\pi}{3}$	$2\left(-\dfrac{1}{2}\right) = -1$
$\dfrac{\pi}{2}$	$2(-1) = -2$

Figure 30

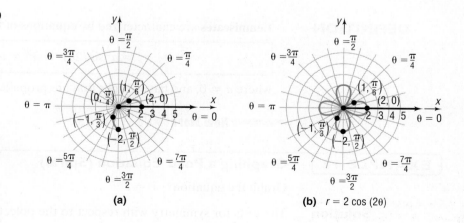

(a)

(b) $r = 2 \cos(2\theta)$

Exploration

Graph $r_1 = 2 \cos(4\theta)$; clear the screen and graph $r_1 = 2 \cos(6\theta)$. How many petals did each of these graphs have?

Clear the screen and graph, in order, each on a clear screen, $r_1 = 2 \cos(3\theta)$, $r_1 = 2 \cos(5\theta)$, and $r_1 = 2 \cos(7\theta)$. What do you notice about the number of petals?

The curve in Figure 30(b) is called a *rose* with four petals.

DEFINITION

Rose curves are characterized by equations of the form

$$r = a\cos(n\theta), \qquad r = a\sin(n\theta), \qquad a \neq 0$$

and have graphs that are rose shaped. If $n \neq 0$ is even, the rose has $2n$ petals; if $n \neq \pm 1$ is odd, the rose has n petals.

Now Work PROBLEM 49

EXAMPLE 11 **Graphing a Polar Equation (Lemniscate)**

Graph the equation: $r^2 = 4\sin(2\theta)$

Solution We leave it to you to verify that the graph is symmetric with respect to the pole. Because of the symmetry with respect to the pole, we only need to consider values of θ between $\theta = 0$ and $\theta = \pi$. Note that there are no points on the graph for $\frac{\pi}{2} < \theta < \pi$ (quadrant II), since $\sin(2\theta) < 0$ for such values. Table 5 lists points on the graph for values of $\theta = 0$ through $\theta = \frac{\pi}{2}$. The points from Table 5 where $r \geq 0$ are plotted in Figure 31(a). The remaining points on the graph may be obtained by using symmetry. Figure 31(b) shows the final graph drawn.

Table 5

θ	$r^2 = 4\sin(2\theta)$	r
0	$4(0) = 0$	0
$\frac{\pi}{6}$	$4\left(\frac{\sqrt{3}}{2}\right) = 2\sqrt{3}$	± 1.9
$\frac{\pi}{4}$	$4(1) = 4$	± 2
$\frac{\pi}{3}$	$4\left(\frac{\sqrt{3}}{2}\right) = 2\sqrt{3}$	± 1.9
$\frac{\pi}{2}$	$4(0) = 0$	0

Figure 31

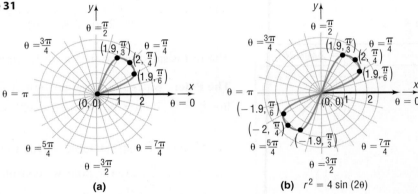

(a)

(b) $r^2 = 4\sin(2\theta)$

The curve in Figure 31(b) is an example of a *lemniscate* (from the Greek word *ribbon*).

DEFINITION

Lemniscates are characterized by equations of the form

$$r^2 = a^2\sin(2\theta) \qquad r^2 = a^2\cos(2\theta)$$

where $a \neq 0$, and have graphs that are propeller shaped.

Now Work PROBLEM 53

EXAMPLE 12 **Graphing a Polar Equation (Spiral)**

Graph the equation: $r = e^{\theta/5}$

Solution The tests for symmetry with respect to the pole, the polar axis, and the line $\theta = \frac{\pi}{2}$ fail. Furthermore, there is no number θ for which $r = 0$, so the graph does not pass through the pole. We observe that r is positive for all θ, r increases as θ increases,

Table 6

θ	$r = e^{\theta/5}$
$-\dfrac{3\pi}{2}$	0.39
$-\pi$	0.53
$-\dfrac{\pi}{2}$	0.73
$-\dfrac{\pi}{4}$	0.85
0	1
$\dfrac{\pi}{4}$	1.17
$\dfrac{\pi}{2}$	1.37
π	1.87
$\dfrac{3\pi}{2}$	2.57
2π	3.51

$r \to 0$ as $\theta \to -\infty$, and $r \to \infty$ as $\theta \to \infty$. With the help of a calculator, we obtain the values in Table 6. See Figure 32.

Figure 32
$r = e^{\theta/5}$

The curve in Figure 32 is called a **logarithmic spiral**, since its equation may be written as $\theta = 5 \ln r$ and it spirals infinitely both toward the pole and away from it.

Classification of Polar Equations

The equations of some lines and circles in polar coordinates and their corresponding equations in rectangular coordinates are given in Table 7. Also included are the names and graphs of a few of the more frequently encountered polar equations.

Table 7

Lines			
Description	Line passing through the pole making an angle α with the polar axis	Vertical line	Horizontal line
Rectangular equation	$y = (\tan \alpha)x$	$x = a$	$y = b$
Polar equation	$\theta = \alpha$	$r \cos \theta = a$	$r \sin \theta = b$
Typical graph			

Circles			
Description	Center at the pole, radius a	Passing through the pole, tangent to the line $\theta = \dfrac{\pi}{2}$, center on the polar axis, radius a	Passing through the pole, tangent to the polar axis, center on the line $\theta = \dfrac{\pi}{2}$, radius a
Rectangular equation	$x^2 + y^2 = a^2, \quad a > 0$	$x^2 + y^2 = \pm 2ax, \quad a > 0$	$x^2 + y^2 = \pm 2ay, \quad a > 0$
Polar equation	$r = a, \quad a > 0$	$r = \pm 2a \cos \theta, \quad a > 0$	$r = \pm 2a \sin \theta, \quad a > 0$
Typical graph			

Table 7 (Continued)

Other Equations		
Name	Cardioid	Limaçon without inner loop	Limaçon with inner loop
Polar equations	$r = a \pm a \cos \theta, \quad a > 0$ $r = a \pm a \sin \theta, \quad a > 0$	$r = a \pm b \cos \theta, \quad 0 < b < a$ $r = a \pm b \sin \theta, \quad 0 < b < a$	$r = a \pm b \cos \theta, \quad 0 < a < b$ $r = a \pm b \sin \theta, \quad 0 < a < b$
Typical graph			
Name	Lemniscate	Rose with three petals	Rose with four petals
Polar equations	$r^2 = a^2 \cos(2\theta), \quad a > 0$ $r^2 = a^2 \sin(2\theta), \quad a > 0$	$r = a \sin(3\theta), \quad a > 0$ $r = a \cos(3\theta), \quad a > 0$	$r = a \sin(2\theta), \quad a > 0$ $r = a \cos(2\theta), \quad a > 0$
Typical graph			

Sketching Quickly

If a polar equation involves only a sine (or cosine) function, you can quickly obtain a sketch of its graph by making use of Table 7, periodicity, and a short table.

EXAMPLE 13 **Sketching the Graph of a Polar Equation Quickly**

Graph the equation: $r = 2 + 2 \sin \theta$

Solution   We recognize the polar equation: Its graph is a cardioid. The period of $\sin \theta$ is 2π, so we form a table using $0 \le \theta \le 2\pi$, compute r, plot the points (r, θ), and sketch the graph of a cardioid as θ varies from 0 to 2π. See Table 8 and Figure 33.

Table 8

θ	$r = 2 + 2 \sin \theta$
0	$2 + 2(0) = 2$
$\dfrac{\pi}{2}$	$2 + 2(1) = 4$
π	$2 + 2(0) = 2$
$\dfrac{3\pi}{2}$	$2 + 2(-1) = 0$
2π	$2 + 2(0) = 2$

Figure 33
$r = 2 + 2 \sin \theta$

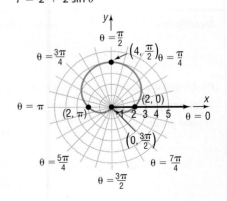

△ **Calculus Comment** For those of you who are planning to study calculus, a comment about one important role of polar equations is in order.

In rectangular coordinates, the equation $x^2 + y^2 = 1$, whose graph is the unit circle, is not the graph of a function. In fact, it requires two functions to obtain the graph of the unit circle:

$$y_1 = \sqrt{1 - x^2} \quad \text{Upper semicircle} \qquad y_2 = -\sqrt{1 - x^2} \quad \text{Lower semicircle}$$

In polar coordinates, the equation $r = 1$, whose graph is also the unit circle, does define a function. For each choice of θ, there is only one corresponding value of r, that is, $r = 1$. Since many problems in calculus require the use of functions, the opportunity to express nonfunctions in rectangular coordinates as functions in polar coordinates becomes extremely useful.

Note also that the vertical-line test for functions is valid only for equations in rectangular coordinates.

Historical Feature

Jakob Bernoulli (1654–1705)

Polar coordinates seem to have been invented by Jakob Bernoulli (1654–1705) in about 1691, although, as with most such ideas, earlier traces of the notion exist. Early users of calculus remained committed to rectangular coordinates, and polar coordinates did not become widely used until the early 1800s. Even then, it was mostly geometers who used them for describing odd curves. Finally, about the mid-1800s, applied mathematicians realized the tremendous simplification that polar coordinates make possible in the description of objects with circular or cylindrical symmetry. From then on their use became widespread.

8.2 Assess Your Understanding

'Are You Prepared?' *Answers are given at the end of these exercises. If you get a wrong answer, read the pages listed in* red.

1. If the rectangular coordinates of a point are $(4, -6)$, the point symmetric to it with respect to the origin is _____. (pp. 12–13)

2. The difference formula for cosine is $\cos(A - B) =$ _____. (p. 514)

3. The standard equation of a circle with center at $(-2, 5)$ and radius 3 is _____. (pp. 35–36)

4. Is the sine function even, odd, or neither? (p. 432)

5. $\sin \dfrac{5\pi}{4} =$ _____ . (pp. 420–421)

6. $\cos \dfrac{2\pi}{3} =$ _____. (pp. 420–421)

Concepts and Vocabulary

7. An equation whose variables are polar coordinates is called a(n) _____ _____.

8. *True or False* The tests for symmetry in polar coordinates are necessary, but not sufficient.

9. To test if a polar equation may be symmetric with respect to the polar axis, replace θ by _____.

10. To test if a polar equation may be symmetric with respect to the line $\theta = \dfrac{\pi}{2}$, replace θ by _____.

11. *True or False* The graph of a cardiod passes through the pole.

12. Rose curves are characterized by equations of the form $r = a \cos(n\theta)$ or $r = a \sin(n\theta), a \neq 0$. If $n \neq 0$ is even, the rose has _____ petals; if $n \neq \pm 1$ is odd, the rose has _____ petals.

Skill Building

In Problems 13–28, transform each polar equation to an equation in rectangular coordinates. Then identify and graph the equation.

13. $r = 4$

14. $r = 2$

15. $\theta = \dfrac{\pi}{3}$

16. $\theta = -\dfrac{\pi}{4}$

17. $r \sin \theta = 4$

18. $r \cos \theta = 4$

19. $r \cos \theta = -2$

20. $r \sin \theta = -2$

21. $r = 2 \cos \theta$ **22.** $r = 2 \sin \theta$ **23.** $r = -4 \sin \theta$ **24.** $r = -4 \cos \theta$

25. $r \sec \theta = 4$ **26.** $r \csc \theta = 8$ **27.** $r \csc \theta = -2$ **28.** $r \sec \theta = -4$

In Problems 29–36, match each of the graphs (A) through (H) to one of the following polar equations.

29. $r = 2$ **30.** $\theta = \dfrac{\pi}{4}$ **31.** $r = 2 \cos \theta$ **32.** $r \cos \theta = 2$

33. $r = 1 + \cos \theta$ **34.** $r = 2 \sin \theta$ **35.** $\theta = \dfrac{3\pi}{4}$ **36.** $r \sin \theta = 2$

(A)   (B)   (C)   (D)

(E)   (F)   (G)   (H)

In Problems 37–60, identify and graph each polar equation.

37. $r = 2 + 2 \cos \theta$ **38.** $r = 1 + \sin \theta$ **39.** $r = 3 - 3 \sin \theta$ **40.** $r = 2 - 2 \cos \theta$

41. $r = 2 + \sin \theta$ **42.** $r = 2 - \cos \theta$ **43.** $r = 4 - 2 \cos \theta$ **44.** $r = 4 + 2 \sin \theta$

45. $r = 1 + 2 \sin \theta$ **46.** $r = 1 - 2 \sin \theta$ **47.** $r = 2 - 3 \cos \theta$ **48.** $r = 2 + 4 \cos \theta$

49. $r = 3 \cos(2\theta)$ **50.** $r = 2 \sin(3\theta)$ **51.** $r = 4 \sin(5\theta)$ **52.** $r = 3 \cos(4\theta)$

53. $r^2 = 9 \cos(2\theta)$ **54.** $r^2 = \sin(2\theta)$ **55.** $r = 2^\theta$ **56.** $r = 3^\theta$

57. $r = 1 - \cos \theta$ **58.** $r = 3 + \cos \theta$ **59.** $r = 1 - 3 \cos \theta$ **60.** $r = 4 \cos(3\theta)$

Mixed Practice

In Problems 61–66, graph each pair of polar equations on the same polar grid. Find the polar coordinates of the point(s) of intersection and label the point(s) on the graph.

61. $r = 8 \cos \theta;\ r = 2 \sec \theta$ **62.** $r = 8 \sin \theta;\ r = 4 \csc \theta$ **63.** $r = \sin \theta;\ r = 1 + \cos \theta$

64. $r = 3;\ r = 2 + 2 \cos \theta$ **65.** $r = 1 + \sin \theta;\ r = 1 + \cos \theta$ **66.** $r = 1 + \cos \theta;\ r = 3 \cos \theta$

Applications and Extensions

In Problems 67–70, the polar equation for each graph is either $r = a + b \cos \theta$ or $r = a + b \sin \theta$, $a > 0$. Select the correct equation and find the values of a and b.

67.

68.

69.

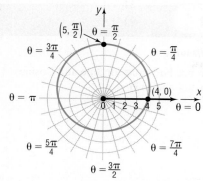

70.

In Problems 71–80, graph each polar equation by hand.

71. $r = \dfrac{2}{1 - \cos \theta}$ *(parabola)*

72. $r = \dfrac{2}{1 - 2 \cos \theta}$ *(hyperbola)*

73. $r = \dfrac{1}{3 - 2 \cos \theta}$ *(ellipse)*

74. $r = \dfrac{1}{1 - \cos \theta}$ *(parabola)*

75. $r = \theta, \quad \theta \geq 0$ *(spiral of Archimedes)*

76. $r = \dfrac{3}{\theta}$ *(reciprocal spiral)*

77. $r = \csc \theta - 2, \quad 0 < \theta < \pi$ *(conchoid)*

78. $r = \sin \theta \tan \theta$ *(cissoid)*

79. $r = \tan \theta, \quad -\dfrac{\pi}{2} < \theta < \dfrac{\pi}{2}$ *(kappa curve)*

80. $r = \cos \dfrac{\theta}{2}$

81. Show that the graph of the equation $r \sin \theta = a$ is a horizontal line a units above the pole if $a > 0$ and $|a|$ units below the pole if $a < 0$.

82. Show that the graph of the equation $r \cos \theta = a$ is a vertical line a units to the right of the pole if $a > 0$ and $|a|$ units to the left of the pole if $a < 0$.

83. Show that the graph of the equation $r = 2a \sin \theta, a > 0$, is a circle of radius a with center at $(0, a)$ in rectangular coordinates.

84. Show that the graph of the equation $r = -2a \sin \theta, a > 0$, is a circle of radius a with center at $(0, -a)$ in rectangular coordinates.

85. Show that the graph of the equation $r = 2a \cos \theta, a > 0$, is a circle of radius a with center at $(a, 0)$ in rectangular coordinates.

86. Show that the graph of the equation $r = -2a \cos \theta, a > 0$, is a circle of radius a with center at $(-a, 0)$ in rectangular coordinates.

Explaining Concepts: Discussion and Writing

87. Explain why the following test for symmetry is valid: Replace r by $-r$ and θ by $-\theta$ in a polar equation. If an equivalent equation results, the graph is symmetric with respect to the line $\theta = \dfrac{\pi}{2}$ (y-axis).

 (a) Show that the test on page 605 fails for $r^2 = \cos \theta$, yet this new test works.

 (b) Show that the test on page 605 works for $r^2 = \sin \theta$, yet this new test fails.

88. Write down two different tests for symmetry with respect to the polar axis. Find examples in which one test works and the other fails. Which test do you prefer to use? Justify your answer.

89. The tests for symmetry given on page 605 are sufficient, but not necessary. Explain what this means.

90. Explain why the vertical-line test used to identify functions in rectangular coordinates does not work for equations expressed in polar coordinates.

'Are You Prepared?' Answers

1. $(-4, 6)$ **2.** $\cos A \cos B + \sin A \sin B$ **3.** $(x + 2)^2 + (y - 5)^2 = 9$ **4.** Odd   **5.** $-\dfrac{\sqrt{2}}{2}$ **6.** $-\dfrac{1}{2}$

8.3 The Complex Plane; De Moivre's Theorem

PREPARING FOR THIS SECTION *Before getting started, review the following:*

- Complex Numbers (Appendix A, Section A.11, pp. A92–A97)
- Value of the Sine and Cosine Functions at Certain Angles (Section 5.3, pp. 403–405; Section 5.4, pp. 416, 420–421)

- Sum and Difference Formulas for Sine and Cosine (Section 6.5, pp. 514 and 517)

Now Work the 'Are You Prepared?' problems on page 622.

OBJECTIVES **1** Plot Points in the Complex Plane (p. 616)

2 Convert a Complex Number between Rectangular Form and Polar Form (p. 617)

3 Find Products and Quotients of Complex Numbers in Polar Form (p. 618)

4 Use De Moivre's Theorem (p. 619)

5 Find Complex Roots (p. 620)

Figure 34
Complex plane

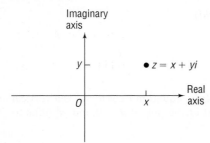

1 Plot Points in the Complex Plane

Complex numbers are discussed in Appendix A, Section A.11. In that discussion, we were not prepared to give a geometric interpretation of a complex number. Now we are ready.

A complex number $z = x + yi$ can be interpreted geometrically as the point (x, y) in the xy-plane. Each point in the plane corresponds to a complex number and, conversely, each complex number corresponds to a point in the plane. We shall refer to the collection of such points as the **complex plane.** The x-axis will be referred to as the **real axis,** because any point that lies on the real axis is of the form $z = x + 0i = x$, a real number. The y-axis is called the **imaginary axis,** because any point that lies on it is of the form $z = 0 + yi = yi$, a pure imaginary number. See Figure 34.

EXAMPLE 1   **Plotting a Point in the Complex Plane**

Plot the point corresponding to $z = \sqrt{3} - i$ in the complex plane.

Solution   The point corresponding to $z = \sqrt{3} - i$ has the rectangular coordinates $(\sqrt{3}, -1)$. The point, located in quadrant IV, is plotted in Figure 35.

Figure 35

DEFINITION

Let $z = x + yi$ be a complex number. The **magnitude** or **modulus** of z, denoted by $|z|$, is defined as the distance from the origin to the point (x, y). That is,

$$|z| = \sqrt{x^2 + y^2} \qquad (1)$$

Figure 36

See Figure 36 for an illustration.

This definition for $|z|$ is consistent with the definition for the absolute value of a real number: If $z = x + yi$ is real, then $z = x + 0i$ and

$$|z| = \sqrt{x^2 + 0^2} = \sqrt{x^2} = |x|$$

For this reason, the magnitude of z is sometimes called the **absolute value of z.**

Recall that if $z = x + yi$ then its **conjugate,** denoted by $\bar{z}$, is $\bar{z} = x - yi$. Because $z\bar{z} = x^2 + y^2$, which is a nonnegative real number, it follows from equation (1) that the magnitude of z can be written as

$$|z| = \sqrt{z\bar{z}} \qquad (2)$$

2 Convert a Complex Number between Rectangular Form and Polar Form

When a complex number is written in the standard form $z = x + yi$, we say that it is in **rectangular,** or **Cartesian, form,** because (x, y) are the rectangular coordinates of the corresponding point in the complex plane. Suppose that (r, θ) are the polar coordinates of this point. Then

$$x = r\cos\theta, \qquad y = r\sin\theta \qquad (3)$$

DEFINITION

If $r \geq 0$ and $0 \leq \theta < 2\pi$, the complex number $z = x + yi$ may be written in **polar form** as

$$z = x + yi = (r\cos\theta) + (r\sin\theta)i = r(\cos\theta + i\sin\theta) \qquad (4)$$

Figure 37

$z = x + yi = r(\cos\theta + i\sin\theta)$,
$r \geq 0, 0 \leq \theta < 2\pi$

See Figure 37.

If $z = r(\cos\theta + i\sin\theta)$ is the polar form of a complex number, the angle θ, $0 \leq \theta < 2\pi$, is called the **argument of z.**

Also, because $r \geq 0$, we have $r = \sqrt{x^2 + y^2}$. From equation (1), it follows that the magnitude of $z = r(\cos\theta + i\sin\theta)$ is

$$|z| = r$$

EXAMPLE 2

Writing a Complex Number in Polar Form

Write an expression for $z = \sqrt{3} - i$ in polar form.

Solution

The point, located in quadrant IV, is plotted in Figure 35. Because $x = \sqrt{3}$ and $y = -1$, it follows that

$$r = \sqrt{x^2 + y^2} = \sqrt{\left(\sqrt{3}\right)^2 + (-1)^2} = \sqrt{4} = 2$$

So

$$\sin\theta = \frac{y}{r} = \frac{-1}{2}, \qquad \cos\theta = \frac{x}{r} = \frac{\sqrt{3}}{2}, \qquad 0 \leq \theta < 2\pi$$

The angle $\theta, 0 \leq \theta < 2\pi$, that satisfies both equations is $\theta = \dfrac{11\pi}{6}$. With $\theta = \dfrac{11\pi}{6}$ and $r = 2$, the polar form of $z = \sqrt{3} - i$ is

$$z = r(\cos\theta + i\sin\theta) = 2\left(\cos\frac{11\pi}{6} + i\sin\frac{11\pi}{6}\right)$$

═══ **Now Work** PROBLEM 11

EXAMPLE 3

Plotting a Point in the Complex Plane and Converting from Polar to Rectangular Form

Plot the point corresponding to $z = 2(\cos 30° + i\sin 30°)$ in the complex plane, and write an expression for z in rectangular form.

Figure 38

Solution To plot the complex number $z = 2(\cos 30° + i \sin 30°)$, we plot the point whose polar coordinates are $(r, \theta) = (2, 30°)$, as shown in Figure 38. In rectangular form,

$$z = 2(\cos 30° + i \sin 30°) = 2\left(\frac{\sqrt{3}}{2} + \frac{1}{2}i\right) = \sqrt{3} + i$$

Now Work PROBLEM 23

3 Find Products and Quotients of Complex Numbers in Polar Form

The polar form of a complex number provides an alternative method for finding products and quotients of complex numbers.

THEOREM Let $z_1 = r_1(\cos \theta_1 + i \sin \theta_1)$ and $z_2 = r_2(\cos \theta_2 + i \sin \theta_2)$ be two complex numbers. Then

$$z_1 z_2 = r_1 r_2[\cos(\theta_1 + \theta_2) + i \sin(\theta_1 + \theta_2)] \tag{5}$$

If $z_2 \neq 0$, then

$$\frac{z_1}{z_2} = \frac{r_1}{r_2}[\cos(\theta_1 - \theta_2) + i \sin(\theta_1 - \theta_2)] \tag{6}$$

In Words
The magnitude of a complex number z is r and its argument is θ, so when
$$z = r(\cos \theta + i \sin \theta)$$
the magnitude of the product (quotient) of two complex numbers equals the product (quotient) of their magnitudes; the argument of the product (quotient) of two complex numbers is determined by the sum (difference) of their arguments.

Proof We will prove formula (5). The proof of formula (6) is left as an exercise (see Problem 66).

$$z_1 z_2 = [r_1(\cos \theta_1 + i \sin \theta_1)][r_2(\cos \theta_2 + i \sin \theta_2)]$$
$$= r_1 r_2[(\cos \theta_1 + i \sin \theta_1)(\cos \theta_2 + i \sin \theta_2)]$$
$$= r_1 r_2[(\cos \theta_1 \cos \theta_2 - \sin \theta_1 \sin \theta_2) + i(\sin \theta_1 \cos \theta_2 + \cos \theta_1 \sin \theta_2)]$$
$$= r_1 r_2[\cos(\theta_1 + \theta_2) + i \sin(\theta_1 + \theta_2)] \qquad ∎$$

Let's look at an example of how this theorem can be used.

EXAMPLE 4 **Finding Products and Quotients of Complex Numbers in Polar Form**

If $z = 3(\cos 20° + i \sin 20°)$ and $w = 5(\cos 100° + i \sin 100°)$, find the following (leave your answers in polar form):

(a) zw 　　　　(b) $\dfrac{z}{w}$

Solution (a) $zw = [3(\cos 20° + i \sin 20°)][5(\cos 100° + i \sin 100°)]$
$$= (3 \cdot 5)[\cos(20° + 100°) + i \sin(20° + 100°)]$$
$$= 15(\cos 120° + i \sin 120°) \qquad \text{Apply equation (5).}$$

(b) $\dfrac{z}{w} = \dfrac{3(\cos 20° + i \sin 20°)}{5(\cos 100° + i \sin 100°)}$

$$= \frac{3}{5}[\cos(20° - 100°) + i \sin(20° - 100°)] \quad \text{Apply equation (6).}$$

$$= \frac{3}{5}[\cos(-80°) + i \sin(-80°)]$$

$$= \frac{3}{5}(\cos 280° + i \sin 280°) \qquad \text{The argument must lie between } 0° \text{ and } 360°.$$

Now Work PROBLEM 33

4 Use De Moivre's Theorem

De Moivre's Theorem, stated by Abraham De Moivre (1667–1754) in 1730, but already known to many people by 1710, is important for the following reason: The fundamental processes of algebra are the four operations of addition, subtraction, multiplication, and division, together with powers and the extraction of roots. De Moivre's Theorem allows these latter fundamental algebraic operations to be applied to complex numbers.

De Moivre's Theorem, in its most basic form, is a formula for raising a complex number z to the power n, where $n \geq 1$ is a positive integer. Let's see if we can conjecture the form of the result.

Let $z = r(\cos \theta + i \sin \theta)$ be a complex number. Then, based on equation (5), we have

$$n = 2: \quad z^2 = r^2[\cos(2\theta) + i \sin(2\theta)] \qquad \text{Equation (5)}$$
$$n = 3: \quad z^3 = z^2 \cdot z$$
$$= \{r^2[\cos(2\theta) + i \sin(2\theta)]\}[r(\cos \theta + i \sin \theta)]$$
$$= r^3[\cos(3\theta) + i \sin(3\theta)] \qquad \text{Equation (5)}$$
$$n = 4: \quad z^4 = z^3 \cdot z$$
$$= \{r^3[\cos(3\theta) + i \sin(3\theta)]\}[r(\cos \theta + i \sin \theta)]$$
$$= r^4[\cos(4\theta) + i \sin(4\theta)] \qquad \text{Equation (5)}$$

Do you see the pattern?

THEOREM

> **De Moivre's Theorem**
>
> If $z = r(\cos \theta + i \sin \theta)$ is a complex number, then
>
> $$z^n = r^n[\cos(n\theta) + i \sin(n\theta)] \qquad \textbf{(7)}$$
>
> where $n \geq 1$ is a positive integer.

The proof of De Moivre's Theorem requires mathematical induction (which is not discussed until Section 11.4), so it is omitted here.

Let's look at some examples.

EXAMPLE 5

Using De Moivre's Theorem

Write $[2(\cos 20° + i \sin 20°)]^3$ in the standard form $a + bi$.

Solution

$[2(\cos 20° + i \sin 20°)]^3 = 2^3[\cos(3 \cdot 20°) + i \sin(3 \cdot 20°)]$ Apply De Moivre's Theorem.

$$= 8(\cos 60° + i \sin 60°)$$

$$= 8\left(\frac{1}{2} + \frac{\sqrt{3}}{2}i\right) = 4 + 4\sqrt{3}i$$

━━━━▶ **Now Work** PROBLEM 41

EXAMPLE 6

Using De Moivre's Theorem

Write $(1 + i)^5$ in the standard form $a + bi$.

Solution

To apply De Moivre's Theorem, we must first write the complex number in polar form. Since the magnitude of $1 + i$ is $\sqrt{1^2 + 1^2} = \sqrt{2}$, we begin by writing

NOTE We could also write $1 + i$ in polar form by following the approach used in Example 2. ∎

$$1 + i = \sqrt{2}\left(\frac{1}{\sqrt{2}} + \frac{1}{\sqrt{2}}i\right) = \sqrt{2}\left(\cos \frac{\pi}{4} + i \sin \frac{\pi}{4}\right)$$

Now

$$(1 + i)^5 = \left[\sqrt{2}\left(\cos\frac{\pi}{4} + i\sin\frac{\pi}{4}\right)\right]^5$$

$$= \left(\sqrt{2}\right)^5\left[\cos\left(5\cdot\frac{\pi}{4}\right) + i\sin\left(5\cdot\frac{\pi}{4}\right)\right]$$

$$= 4\sqrt{2}\left(\cos\frac{5\pi}{4} + i\sin\frac{5\pi}{4}\right)$$

$$= 4\sqrt{2}\left[-\frac{1}{\sqrt{2}} + \left(-\frac{1}{\sqrt{2}}\right)i\right] = -4 - 4i$$

5 Find Complex Roots

Let w be a given complex number, and let $n \geq 2$ denote a positive integer. Any complex number z that satisfies the equation

$$z^n = w$$

is called a **complex nth root** of w. In keeping with previous usage, if $n = 2$, the solutions of the equation $z^2 = w$ are called **complex square roots** of w, and if $n = 3$, the solutions of the equation $z^3 = w$ are called **complex cube roots** of w.

THEOREM

Finding Complex Roots

Let $w = r(\cos\theta_0 + i\sin\theta_0)$ be a complex number, and let $n \geq 2$ be an integer. If $w \neq 0$, there are n distinct complex roots of w, given by the formula

$$z_k = \sqrt[n]{r}\left[\cos\left(\frac{\theta_0}{n} + \frac{2k\pi}{n}\right) + i\sin\left(\frac{\theta_0}{n} + \frac{2k\pi}{n}\right)\right] \qquad (8)$$

where $k = 0, 1, 2, \ldots, n - 1$.

Proof (Outline)   We will not prove this result in its entirety. Instead, we shall show only that each z_k in equation (8) satisfies the equation $z_k^n = w$, proving that each z_k is a complex nth root of w.

$$z_k^n = \left\{\sqrt[n]{r}\left[\cos\left(\frac{\theta_0}{n} + \frac{2k\pi}{n}\right) + i\sin\left(\frac{\theta_0}{n} + \frac{2k\pi}{n}\right)\right]\right\}^n$$

$$= \left(\sqrt[n]{r}\right)^n\left\{\cos\left[n\left(\frac{\theta_0}{n} + \frac{2k\pi}{n}\right)\right] + i\sin\left[n\left(\frac{\theta_0}{n} + \frac{2k\pi}{n}\right)\right]\right\} \quad \text{Apply De Moivre's Theorem.}$$

$$= r[\cos(\theta_0 + 2k\pi) + i\sin(\theta_0 + 2k\pi)] \qquad \text{Simplify.}$$

$$= r(\cos\theta_0 + i\sin\theta_0) = w \qquad \text{The Periodic Property}$$

So each z_k, $k = 0, 1, \ldots, n - 1$, is a complex nth root of w. To complete the proof, we would need to show that each z_k, $k = 0, 1, \ldots, n - 1$, is, in fact, distinct and that there are no complex nth roots of w other than those given by equation (8). ∎

EXAMPLE 7

Finding Complex Cube Roots

Find the complex cube roots of $-1 + \sqrt{3}i$. Leave your answers in polar form, with the argument in degrees.

Solution   First, we express $-1 + \sqrt{3}i$ in polar form using degrees.

$$-1 + \sqrt{3}i = 2\left(-\frac{1}{2} + \frac{\sqrt{3}}{2}i\right) = 2(\cos 120° + i\sin 120°)$$

The three complex cube roots of $-1 + \sqrt{3}i = 2(\cos 120° + i \sin 120°)$ are

$$z_k = \sqrt[3]{2}\left[\cos\left(\frac{120°}{3} + \frac{360°k}{3}\right) + i \sin\left(\frac{120°}{3} + \frac{360°k}{3}\right)\right]$$

$$= \sqrt[3]{2}[\cos(40° + 120°k) + i \sin(40° + 120°k)] \qquad k = 0, 1, 2$$

So

WARNING Most graphing utilities will only provide the answer z_0 to the calculation $(-1 + \sqrt{3}i) \wedge \left(\dfrac{1}{3}\right)$. The paragraph following Example 7 explains how to obtain z_1 and z_2 from z_0. ∎

$$z_0 = \sqrt[3]{2}[\cos(40° + 120°\cdot 0) + i \sin(40° + 120°\cdot 0)] = \sqrt[3]{2}(\cos 40° + i \sin 40°)$$

$$z_1 = \sqrt[3]{2}[\cos(40° + 120°\cdot 1) + i \sin(40° + 120°\cdot 1)] = \sqrt[3]{2}(\cos 160° + i \sin 160°)$$

$$z_2 = \sqrt[3]{2}[\cos(40° + 120°\cdot 2) + i \sin(40° + 120°\cdot 2)] = \sqrt[3]{2}(\cos 280° + i \sin 280°)$$

●

Notice that each of the three complex roots of $-1 + \sqrt{3}i$ has the same magnitude, $\sqrt[3]{2}$. This means that the points corresponding to each cube root lie the same distance from the origin; that is, the three points lie on a circle with center at the origin and radius $\sqrt[3]{2}$. Furthermore, the arguments of these cube roots are 40°, 160°, and 280°, the difference of consecutive pairs being $120° = \dfrac{360°}{3}$. This means that the three points are equally spaced on the circle, as shown in Figure 39. These results are not coincidental. In fact, you are asked to show that these results hold for complex nth roots in Problems 63 through 65.

Figure 39

$z_1 = \sqrt[3]{2}(\cos 160° + i \sin 160°)$

$z_0 = \sqrt[3]{2}(\cos 40° + i \sin 40°)$

$x^2 + y^2 = (\sqrt[3]{2})^2$

$z_2 = \sqrt[3]{2}(\cos 280° + i \sin 280°)$

━━━▶ **Now Work** PROBLEM 53

Historical Feature

John Wallis

The Babylonians, Greeks, and Arabs considered square roots of negative quantities to be impossible and equations with complex solutions to be unsolvable. The first hint that there was some connection between real solutions of equations and complex numbers came when Girolamo Cardano (1501–1576) and Tartaglia (1499–1557) found *real* roots of cubic equations by taking cube roots of *complex* quantities. For centuries thereafter, mathematicians

worked with complex numbers without much belief in their actual existence. In 1673, John Wallis appears to have been the first to suggest the graphical representation of complex numbers, a truly significant idea that was not pursued further until about 1800. Several people, including Karl Friedrich Gauss (1777–1855), then rediscovered the idea, and graphical representation helped to establish complex numbers as equal members of the number family. In practical applications, complex numbers have found their greatest uses in the study of alternating current, where they are a commonplace tool, and in the field of subatomic physics.

Historical Problems

1. The quadratic formula will work perfectly well if the coefficients are complex numbers. Solve the following. [**Hint:** The answers are "nice."]

(a) $z^2 - (2 + 5i)z - 3 + 5i = 0$

(b) $z^2 - (1 + i)z - 2 - i = 0$

8.3 Assess Your Understanding

'Are You Prepared?' *Answers are given at the end of these exercises. If you get a wrong answer, read the pages listed in red.*

1. The conjugate of $-4 - 3i$ is _____. (p. A94)

2. The sum formula for the sine function is $\sin(A + B) =$ _____. (p. 517)

3. The sum formula for the cosine function is $\cos(A + B) =$ _____. (p. 514)

4. $\sin 120° =$ _____; $\cos 240° =$ _____. (pp. 394–401)

Concepts and Vocabulary

5. In the complex plane, the x-axis is referred to as the _____ axis and the y-axis is called the _____ axis.

6. When a complex number z is written in the polar form $z = r(\cos \theta + i \sin \theta)$, the nonnegative number r is the _____ or _____ of z, and the angle θ, $0 \le \theta < 2\pi$, is the _____ of z.

7. Let $z_1 = r_1(\cos \theta_1 + i \sin \theta_1)$ and $z_2 = r_2(\cos \theta_2 + i \sin \theta_2)$ be two complex numbers. Then
$z_1 z_2 =$ _____ $[\cos ($ _____ $) + i \sin ($ _____ $)]$.

8. If $z = r(\cos \theta + i \sin \theta)$ is a complex number, then $z^n =$ ____ $[\cos($ ____ $) + i \sin($ ____ $)]$

9. Every nonzero complex number will have exactly _____ distinct cube roots.

10. *True or False* The polar form of a complex number is unique.

Skill Building

In Problems 11–22, plot each complex number in the complex plane and write it in polar form. Express the argument in degrees.

11. $1 + i$

12. $-1 + i$

13. $\sqrt{3} - i$

14. $1 - \sqrt{3}i$

15. $-3i$

16. -2

17. $4 - 4i$

18. $9\sqrt{3} + 9i$

19. $3 - 4i$

20. $2 + \sqrt{3}i$

21. $-2 + 3i$

22. $\sqrt{5} - i$

In Problems 23–32, write each complex number in rectangular form.

23. $2(\cos 120° + i \sin 120°)$

24. $3(\cos 210° + i \sin 210°)$

25. $4\left(\cos \dfrac{7\pi}{4} + i \sin \dfrac{7\pi}{4}\right)$

26. $2\left(\cos \dfrac{5\pi}{6} + i \sin \dfrac{5\pi}{6}\right)$

27. $3\left(\cos \dfrac{3\pi}{2} + i \sin \dfrac{3\pi}{2}\right)$

28. $4\left(\cos \dfrac{\pi}{2} + i \sin \dfrac{\pi}{2}\right)$

29. $0.2(\cos 100° + i \sin 100°)$

30. $0.4(\cos 200° + i \sin 200°)$

31. $2\left(\cos \dfrac{\pi}{18} + i \sin \dfrac{\pi}{18}\right)$

32. $3\left(\cos \dfrac{\pi}{10} + i \sin \dfrac{\pi}{10}\right)$

In Problems 33–40, find zw and $\dfrac{z}{w}$. Leave your answers in polar form.

33. $z = 2(\cos 40° + i \sin 40°)$
$w = 4(\cos 20° + i \sin 20°)$

34. $z = \cos 120° + i \sin 120°$
$w = \cos 100° + i \sin 100°$

35. $z = 3(\cos 130° + i \sin 130°)$
$w = 4(\cos 270° + i \sin 270°)$

36. $z = 2(\cos 80° + i \sin 80°)$
$w = 6(\cos 200° + i \sin 200°)$

37. $z = 2\left(\cos \dfrac{\pi}{8} + i \sin \dfrac{\pi}{8}\right)$
$w = 2\left(\cos \dfrac{\pi}{10} + i \sin \dfrac{\pi}{10}\right)$

38. $z = 4\left(\cos \dfrac{3\pi}{8} + i \sin \dfrac{3\pi}{8}\right)$
$w = 2\left(\cos \dfrac{9\pi}{16} + i \sin \dfrac{9\pi}{16}\right)$

39. $z = 2 + 2i$
$w = \sqrt{3} - i$

40. $z = 1 - i$
$w = 1 - \sqrt{3}i$

In Problems 41–52, write each expression in the standard form $a + bi$.

41. $[4(\cos 40° + i \sin 40°)]^3$

42. $[3(\cos 80° + i \sin 80°)]^3$

43. $\left[2\left(\cos \dfrac{\pi}{10} + i \sin \dfrac{\pi}{10}\right)\right]^5$

44. $\left[\sqrt{2}\left(\cos\dfrac{5\pi}{16}+i\sin\dfrac{5\pi}{16}\right)\right]^4$ **45.** $\left[\sqrt{3}\,(\cos 10° + i\sin 10°)\right]^6$ **46.** $\left[\dfrac{1}{2}(\cos 72° + i\sin 72°)\right]^5$

47. $\left[\sqrt{5}\left(\cos\dfrac{3\pi}{16}+i\sin\dfrac{3\pi}{16}\right)\right]^4$ **48.** $\left[\sqrt{3}\left(\cos\dfrac{5\pi}{18}+i\sin\dfrac{5\pi}{18}\right)\right]^6$ **49.** $(1-i)^5$

50. $\left(\sqrt{3}-i\right)^6$ **51.** $\left(\sqrt{2}-i\right)^6$ **52.** $\left(1-\sqrt{5}i\right)^8$

In Problems 53–60, find all the complex roots. Leave your answers in polar form with the argument in degrees.

53. The complex cube roots of $1+i$

54. The complex fourth roots of $\sqrt{3}-i$

55. The complex fourth roots of $4-4\sqrt{3}i$

56. The complex cube roots of $-8-8i$

57. The complex fourth roots of $-16i$

58. The complex cube roots of -8

59. The complex fifth roots of i

60. The complex fifth roots of $-i$

Applications and Extensions

61. Find the four complex fourth roots of unity (1) and plot them.

62. Find the six complex sixth roots of unity (1) and plot them.

63. Show that each complex nth root of a nonzero complex number w has the same magnitude.

64. Use the result of Problem 63 to draw the conclusion that each complex nth root lies on a circle with center at the origin. What is the radius of this circle?

65. Refer to Problem 64. Show that the complex nth roots of a nonzero complex number w are equally spaced on the circle.

66. Prove formula (6).

67. Mandelbrot Sets

(a) Consider the expression $a_n=(a_{n-1})^2+z$, where z is some complex number (called the **seed**) and $a_0=z$. Compute a_1 $(=a_0^2+z)$, a_2 $(=a_1^2+z)$, a_3 $(=a_2^2+z)$, a_4, a_5, and a_6 for the following seeds: $z_1=0.1-0.4i$, $z_2=0.5+0.8i$, $z_3=-0.9+0.7i$, $z_4=-1.1+0.1i$, $z_5=0-1.3i$, and $z_6=1+1i$.

(b) The dark portion of the graph represents the set of all values $z=x+yi$ that are in the Mandelbrot set. Determine

which complex numbers in part (a) are in this set by plotting them on the graph. Do the complex numbers that are not in the Mandelbrot set have any common characteristics regarding the values of a_6 found in part (a)?

(c) Compute $|z|=\sqrt{x^2+y^2}$ for each of the complex numbers in part (a). Now compute $|a_6|$ for each of the complex numbers in part (a). For which complex numbers is $|a_6|\le|z|$ and $|z|\le2$? Conclude that the criterion for a complex number to be in the Mandelbrot set is that $|a_n|\le|z|$ and $|z|\le2$.

'Are You Prepared?' Answers

1. $-4+3i$ **2.** $\sin A\cos B + \cos A\sin B$ **3.** $\cos A\cos B - \sin A\sin B$ **4.** $\dfrac{\sqrt{3}}{2};-\dfrac{1}{2}$

8.4 Vectors

In simple terms, a **vector** (derived from the Latin *vehere*, meaning "to carry") is a quantity that has both magnitude and direction. It is customary to represent a vector by using an arrow. The length of the arrow represents the **magnitude** of the vector, and the arrowhead indicates the **direction** of the vector.

Many quantities in physics can be represented by vectors. For example, the velocity of an aircraft can be represented by an arrow that points in the direction of movement; the length of the arrow represents speed. If the aircraft speeds up, we lengthen the arrow; if the aircraft changes direction, we introduce an arrow in the new direction. See Figure 40. Based on this representation, it is not surprising that vectors and *directed line segments* are somehow related.

Figure 40

Geometric Vectors

If P and Q are two distinct points in the xy-plane, there is exactly one line containing both P and Q [Figure 41(a)]. The points on that part of the line that joins P to Q, including P and Q, form what is called the **line segment** $\overline{PQ}$ [Figure 41(b)]. If we order the points so that they proceed from P to Q, we have a **directed line segment** from P to Q, or a **geometric vector,** which we denote by $\overrightarrow{PQ}$. In a directed line segment $\overrightarrow{PQ}$, we call P the **initial point** and Q the **terminal point,** as indicated in Figure 41(c).

Figure 41

(a) Line containing P and Q **(b)** Line segment $\overline{PQ}$ **(c)** Directed line segment $\overrightarrow{PQ}$

The magnitude of the directed line segment $\overrightarrow{PQ}$ is the distance from the point P to the point Q; that is, it is the length of the line segment. The direction of $\overrightarrow{PQ}$ is from P to Q. If a vector $\mathbf{v}^*$ has the same magnitude and the same direction as the directed line segment $\overrightarrow{PQ}$, we write

$$\mathbf{v} = \overrightarrow{PQ}$$

The vector $\mathbf{v}$ whose magnitude is 0 is called the **zero vector, 0.** The zero vector is assigned no direction.

Two vectors $\mathbf{v}$ and $\mathbf{w}$ are **equal,** written

$$\mathbf{v} = \mathbf{w}$$

if they have the same magnitude and the same direction.

For example, the three vectors shown in Figure 42 have the same magnitude and the same direction, so they are equal, even though they have different initial points and different terminal points. As a result, we find it useful to think of a vector simply as an arrow, keeping in mind that two arrows (vectors) are equal if they have the same direction and the same magnitude (length).

Figure 42

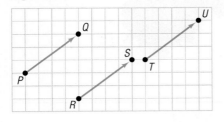

Adding Vectors Geometrically

The **sum** $\mathbf{v} + \mathbf{w}$ of two vectors is defined as follows: We position the vectors $\mathbf{v}$ and $\mathbf{w}$ so that the terminal point of $\mathbf{v}$ coincides with the initial point of $\mathbf{w}$, as shown in Figure 43. The vector $\mathbf{v} + \mathbf{w}$ is then the unique vector whose initial point coincides with the initial point of $\mathbf{v}$ and whose terminal point coincides with the terminal point of $\mathbf{w}$.

Figure 43

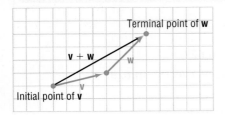

Vector addition is **commutative.** That is, if $\mathbf{v}$ and $\mathbf{w}$ are any two vectors, then

$$\mathbf{v} + \mathbf{w} = \mathbf{w} + \mathbf{v}$$

*Boldface letters will be used to denote vectors, to distinguish them from numbers. For handwritten work, an arrow is placed over the letter to signify a vector. For example, we write a vector by hand as $\vec{v}$.

Figure 44

Figure 44 illustrates this fact. (Observe that the commutative property is another way of saying that opposite sides of a parallelogram are equal and parallel.)

Vector addition is also **associative.** That is, if **u**, **v**, and **w** are vectors, then

$$\mathbf{u} + (\mathbf{v} + \mathbf{w}) = (\mathbf{u} + \mathbf{v}) + \mathbf{w}$$

Figure 45
$(\mathbf{u} + \mathbf{v}) + \mathbf{w} = \mathbf{u} + (\mathbf{v} + \mathbf{w})$

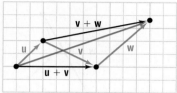

Figure 45 illustrates the associative property for vectors.

The zero vector **0** has the property that

$$\mathbf{v} + \mathbf{0} = \mathbf{0} + \mathbf{v} = \mathbf{v}$$

for any vector **v**.

Figure 46

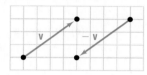

If **v** is a vector, then $-\mathbf{v}$ is the vector having the same magnitude as **v**, but whose direction is opposite to **v**, as shown in Figure 46.
Furthermore,

$$\mathbf{v} + (-\mathbf{v}) = \mathbf{0}$$

Figure 47

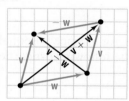

If **v** and **w** are two vectors, we define the **difference** $\mathbf{v} - \mathbf{w}$ as

$$\mathbf{v} - \mathbf{w} = \mathbf{v} + (-\mathbf{w})$$

Figure 47 illustrates the relationships among **v**, **w**, $\mathbf{v} + \mathbf{w}$, and $\mathbf{v} - \mathbf{w}$.

Multiplying Vectors by Numbers Geometrically

When dealing with vectors, we refer to real numbers as **scalars.** Scalars are quantities that have only magnitude. Examples from physics of scalar quantities are temperature, speed, and time. We now define how to multiply a vector by a scalar.

DEFINITION

If α is a scalar and **v** is a vector, the **scalar multiple** $\alpha\mathbf{v}$ is defined as follows:

1. If $\alpha > 0$, $\alpha\mathbf{v}$ is the vector whose magnitude is α times the magnitude of **v** and whose direction is the same as **v**.
2. If $\alpha < 0$, $\alpha\mathbf{v}$ is the vector whose magnitude is $|\alpha|$ times the magnitude of **v** and whose direction is opposite that of **v**.
3. If $\alpha = 0$ or if $\mathbf{v} = \mathbf{0}$, then $\alpha\mathbf{v} = \mathbf{0}$.

Figure 48

See Figure 48 for some illustrations.
For example, if **a** is the acceleration of an object of mass m due to a force **F** being exerted on it, then, by Newton's second law of motion, $\mathbf{F} = m\mathbf{a}$. Here, $m\mathbf{a}$ is the product of the scalar m and the vector **a**.
Scalar multiples have the following properties:

$$0\mathbf{v} = \mathbf{0} \qquad 1\mathbf{v} = \mathbf{v} \qquad -1\mathbf{v} = -\mathbf{v}$$
$$(\alpha + \beta)\mathbf{v} = \alpha\mathbf{v} + \beta\mathbf{v} \qquad \alpha(\mathbf{v} + \mathbf{w}) = \alpha\mathbf{v} + \alpha\mathbf{w}$$
$$\alpha(\beta\mathbf{v}) = (\alpha\beta)\mathbf{v}$$

1 Graph Vectors

EXAMPLE 1   **Graphing Vectors**

Use the vectors illustrated in Figure 49 to graph each of the following vectors:

(a) $\mathbf{v} - \mathbf{w}$ (b) $2\mathbf{v} + 3\mathbf{w}$ (c) $2\mathbf{v} - \mathbf{w} + \mathbf{u}$

Solution   Figure 50 illustrates each graph.

Figure 49

Figure 50

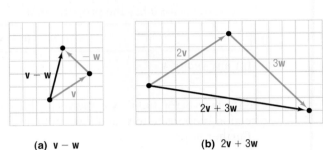

(a) $\mathbf{v} - \mathbf{w}$ (b) $2\mathbf{v} + 3\mathbf{w}$ (c) $2\mathbf{v} - \mathbf{w} + \mathbf{u}$

━━━ **Now Work** PROBLEMS 9 AND 11

Magnitude of Vectors

We use the symbol $\|\mathbf{v}\|$ to represent the **magnitude** of a vector $\mathbf{v}$. Since $\|\mathbf{v}\|$ equals the length of a directed line segment, it follows that $\|\mathbf{v}\|$ has the following properties:

THEOREM   **Properties of $\|\mathbf{v}\|$**

If $\mathbf{v}$ is a vector and if α is a scalar, then

(a) $\|\mathbf{v}\| \geq 0$ (b) $\|\mathbf{v}\| = 0$ if and only if $\mathbf{v} = \mathbf{0}$

(c) $\|-\mathbf{v}\| = \|\mathbf{v}\|$ (d) $\|\alpha\mathbf{v}\| = |\alpha|\|\mathbf{v}\|$

Property (a) is a consequence of the fact that distance is a nonnegative number. Property (b) follows, because the length of the directed line segment $\overrightarrow{PQ}$ is positive unless P and Q are the same point, in which case the length is 0. Property (c) follows because the length of the line segment $\overline{PQ}$ equals the length of the line segment $\overline{QP}$. Property (d) is a direct consequence of the definition of a scalar multiple.

DEFINITION   A vector $\mathbf{u}$ for which $\|\mathbf{u}\| = 1$ is called a **unit vector.**

2 Find a Position Vector

To compute the magnitude and direction of a vector, we need an algebraic way of representing vectors.

DEFINITION   An **algebraic vector** $\mathbf{v}$ is represented as

$$\mathbf{v} = \langle a, b \rangle$$

where a and b are real numbers (scalars) called the **components** of the vector $\mathbf{v}$.

Figure 51

We use a rectangular coordinate system to represent algebraic vectors in the plane. If $\mathbf{v} = \langle a, b \rangle$ is an algebraic vector whose initial point is at the origin, then $\mathbf{v}$ is called a **position vector**. See Figure 51. Notice that the terminal point of the position vector $\mathbf{v} = \langle a, b \rangle$ is $P = (a, b)$.

The next result states that any vector whose initial point is not at the origin is equal to a unique position vector.

THEOREM

Suppose that $\mathbf{v}$ is a vector with initial point $P_1 = (x_1, y_1)$, not necessarily the origin, and terminal point $P_2 = (x_2, y_2)$. If $\mathbf{v} = \overrightarrow{P_1P_2}$, then $\mathbf{v}$ is equal to the position vector

$$\mathbf{v} = \langle x_2 - x_1, y_2 - y_1 \rangle \qquad \textbf{(1)}$$

In Words

An algebraic vector represents "driving directions" to get from the initial point to the terminal point of a vector. So, if $\mathbf{v} = \langle 5, 4 \rangle$, travel 5 units right and 4 units up from the initial point to arrive at the terminal point.

To see why this is true, look at Figure 52.

Figure 52
$\mathbf{v} = \langle a, b \rangle = \langle x_2 - x_1, y_2 - y_1 \rangle$

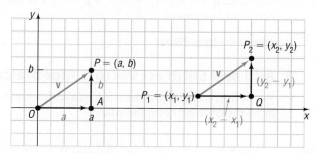

Triangle OPA and triangle P_1P_2Q are congruent. [Do you see why? The line segments have the same magnitude, so $d(O, P) = d(P_1, P_2)$; and they have the same direction, so $\angle POA = \angle P_2P_1Q$. Since the triangles are right triangles, we have angle–side–angle.] It follows that corresponding sides are equal. As a result, $x_2 - x_1 = a$ and $y_2 - y_1 = b$, so $\mathbf{v}$ may be written as

$$\mathbf{v} = \langle a, b \rangle = \langle x_2 - x_1, y_2 - y_1 \rangle$$

Because of this result, we can replace any algebraic vector by a unique position vector, and vice versa. This flexibility is one of the main reasons for the wide use of vectors.

EXAMPLE 2

Finding a Position Vector

Find the position vector of the vector $\mathbf{v} = \overrightarrow{P_1P_2}$ if $P_1 = (-1, 2)$ and $P_2 = (4, 6)$.

Solution

By equation (1), the position vector equal to $\mathbf{v}$ is

$$\mathbf{v} = \langle 4 - (-1), 6 - 2 \rangle = \langle 5, 4 \rangle$$

See Figure 53.

Figure 53

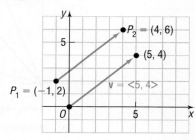

Two position vectors **v** and **w** are equal if and only if the terminal point of **v** is the same as the terminal point of **w**. This leads to the following result:

THEOREM

Equality of Vectors

Two vectors **v** and **w** are equal if and only if their corresponding components are equal. That is,

> If $\mathbf{v} = \langle a_1, b_1 \rangle$ and $\mathbf{w} = \langle a_2, b_2 \rangle$
>
> then $\mathbf{v} = \mathbf{w}$ if and only if $a_1 = a_2$ and $b_1 = b_2$.

Figure 54

We now present an alternative representation of a vector in the plane that is common in the physical sciences. Let **i** denote the unit vector whose direction is along the positive x-axis; let **j** denote the unit vector whose direction is along the positive y-axis. Then $\mathbf{i} = \langle 1, 0 \rangle$ and $\mathbf{j} = \langle 0, 1 \rangle$, as shown in Figure 54. Any vector $\mathbf{v} = \langle a, b \rangle$ can be written using the unit vectors **i** and **j** as follows:

$$\mathbf{v} = \langle a, b \rangle = a\langle 1, 0 \rangle + b\langle 0, 1 \rangle = a\mathbf{i} + b\mathbf{j}$$

We call a and b the **horizontal** and **vertical components** of **v**, respectively. For example, if $\mathbf{v} = \langle 5, 4 \rangle = 5\mathbf{i} + 4\mathbf{j}$, then 5 is the horizontal component and 4 is the vertical component.

━━━ **Now Work** PROBLEM 29

3 Add and Subtract Vectors Algebraically

We define addition, subtraction, scalar multiple, and magnitude of algebraic vectors in terms of their components.

DEFINITION

Let $\mathbf{v} = a_1\mathbf{i} + b_1\mathbf{j} = \langle a_1, b_1 \rangle$ and $\mathbf{w} = a_2\mathbf{i} + b_2\mathbf{j} = \langle a_2, b_2 \rangle$ be two vectors, and let α be a scalar. Then

> $$\mathbf{v} + \mathbf{w} = (a_1 + a_2)\mathbf{i} + (b_1 + b_2)\mathbf{j} = \langle a_1 + a_2, b_1 + b_2 \rangle \qquad \textbf{(2)}$$
> $$\mathbf{v} - \mathbf{w} = (a_1 - a_2)\mathbf{i} + (b_1 - b_2)\mathbf{j} = \langle a_1 - a_2, b_1 - b_2 \rangle \qquad \textbf{(3)}$$
> $$\alpha\mathbf{v} = (\alpha a_1)\mathbf{i} + (\alpha b_1)\mathbf{j} = \langle \alpha a_1, \alpha b_1 \rangle \qquad \textbf{(4)}$$
> $$\|\mathbf{v}\| = \sqrt{a_1^2 + b_1^2} \qquad \textbf{(5)}$$

In Words

To add two vectors, add corresponding components. To subtract two vectors, subtract corresponding components.

These definitions are compatible with the geometric definitions given earlier in this section. See Figure 55.

Figure 55

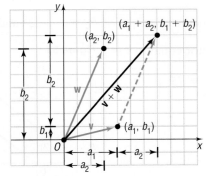

(a) Illustration of property (2)

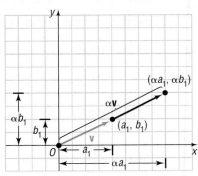

(b) Illustration of property (4), $\alpha > 0$

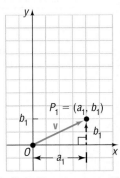

(c) Illustration of property (5):
$\|\mathbf{v}\| =$ Distance from O to P_1
$\|\mathbf{v}\| = \sqrt{a_1^2 + b_1^2}$

EXAMPLE 3   **Adding and Subtracting Vectors**

If $\mathbf{v} = 2\mathbf{i} + 3\mathbf{j} = \langle 2, 3 \rangle$ and $\mathbf{w} = 3\mathbf{i} - 4\mathbf{j} = \langle 3, -4 \rangle$, find:

(a) $\mathbf{v} + \mathbf{w}$ (b) $\mathbf{v} - \mathbf{w}$

Solution   (a) $\mathbf{v} + \mathbf{w} = (2\mathbf{i} + 3\mathbf{j}) + (3\mathbf{i} - 4\mathbf{j}) = (2 + 3)\mathbf{i} + (3 - 4)\mathbf{j} = 5\mathbf{i} - \mathbf{j}$

or

$\mathbf{v} + \mathbf{w} = \langle 2, 3 \rangle + \langle 3, -4 \rangle = \langle 2 + 3, 3 + (-4) \rangle = \langle 5, -1 \rangle$

(b) $\mathbf{v} - \mathbf{w} = (2\mathbf{i} + 3\mathbf{j}) - (3\mathbf{i} - 4\mathbf{j}) = (2 - 3)\mathbf{i} + [3 - (-4)]\mathbf{j} = -\mathbf{i} + 7\mathbf{j}$

or

$\mathbf{v} - \mathbf{w} = \langle 2, 3 \rangle - \langle 3, -4 \rangle = \langle 2 - 3, 3 - (-4) \rangle = \langle -1, 7 \rangle$

4 Find a Scalar Multiple and the Magnitude of a Vector

EXAMPLE 4   **Finding Scalar Multiples and Magnitudes of Vectors**

If $\mathbf{v} = 2\mathbf{i} + 3\mathbf{j} = \langle 2, 3 \rangle$ and $\mathbf{w} = 3\mathbf{i} - 4\mathbf{j} = \langle 3, -4 \rangle$, find:

(a) $3\mathbf{v}$ (b) $2\mathbf{v} - 3\mathbf{w}$ (c) $\|\mathbf{v}\|$

Solution   (a) $3\mathbf{v} = 3(2\mathbf{i} + 3\mathbf{j}) = 6\mathbf{i} + 9\mathbf{j}$

or

$3\mathbf{v} = 3\langle 2, 3 \rangle = \langle 6, 9 \rangle$

(b) $2\mathbf{v} - 3\mathbf{w} = 2(2\mathbf{i} + 3\mathbf{j}) - 3(3\mathbf{i} - 4\mathbf{j}) = 4\mathbf{i} + 6\mathbf{j} - 9\mathbf{i} + 12\mathbf{j}$

$= -5\mathbf{i} + 18\mathbf{j}$

or

$2\mathbf{v} - 3\mathbf{w} = 2\langle 2, 3 \rangle - 3\langle 3, -4 \rangle = \langle 4, 6 \rangle - \langle 9, -12 \rangle$

$= \langle 4 - 9, 6 - (-12) \rangle = \langle -5, 18 \rangle$

(c) $\|\mathbf{v}\| = \|2\mathbf{i} + 3\mathbf{j}\| = \sqrt{2^2 + 3^2} = \sqrt{13}$

⟶ **Now Work** PROBLEMS 35 AND 41

For the remainder of the section, we will express a vector $\mathbf{v}$ in the form $a\mathbf{i} + b\mathbf{j}$.

5 Find a Unit Vector

Recall that a unit vector $\mathbf{u}$ is a vector for which $\|\mathbf{u}\| = 1$. In many applications, it is useful to be able to find a unit vector $\mathbf{u}$ that has the same direction as a given vector $\mathbf{v}$.

THEOREM

Unit Vector in the Direction of v

For any nonzero vector $\mathbf{v}$, the vector

$$\mathbf{u} = \frac{\mathbf{v}}{\|\mathbf{v}\|}$$

is a unit vector that has the same direction as $\mathbf{v}$.

Proof   Let $\mathbf{v} = a\mathbf{i} + b\mathbf{j}$. Then $\|\mathbf{v}\| = \sqrt{a^2 + b^2}$ and

$$\mathbf{u} = \frac{\mathbf{v}}{\|\mathbf{v}\|} = \frac{a\mathbf{i} + b\mathbf{j}}{\sqrt{a^2 + b^2}} = \frac{a}{\sqrt{a^2 + b^2}}\mathbf{i} + \frac{b}{\sqrt{a^2 + b^2}}\mathbf{j}$$

The vector **u** is in the same direction as **v**, since $\|\mathbf{v}\| > 0$. Furthermore,

$$\|\mathbf{u}\| = \sqrt{\frac{a^2}{a^2 + b^2} + \frac{b^2}{a^2 + b^2}} = \sqrt{\frac{a^2 + b^2}{a^2 + b^2}} = 1$$

That is, **u** is a unit vector in the direction of **v**.   ∎

As a consequence of this theorem, if **u** is a unit vector in the same direction as a vector **v**, then **v** may be expressed as

$$\mathbf{v} = \|\mathbf{v}\|\mathbf{u} \tag{6}$$

This way of expressing a vector is useful in many applications.

EXAMPLE 5   **Finding a Unit Vector**

Find a unit vector in the same direction as $\mathbf{v} = 4\mathbf{i} - 3\mathbf{j}$.

Solution   We find $\|\mathbf{v}\|$ first.

$$\|\mathbf{v}\| = \|4\mathbf{i} - 3\mathbf{j}\| = \sqrt{16 + 9} = 5$$

Now we multiply **v** by the scalar $\dfrac{1}{\|\mathbf{v}\|} = \dfrac{1}{5}$. A unit vector in the same direction as **v** is

$$\frac{\mathbf{v}}{\|\mathbf{v}\|} = \frac{4\mathbf{i} - 3\mathbf{j}}{5} = \frac{4}{5}\mathbf{i} - \frac{3}{5}\mathbf{j}$$

✓ **Check:** This vector is, in fact, a unit vector because

$$\left(\frac{4}{5}\right)^2 + \left(-\frac{3}{5}\right)^2 = \frac{16}{25} + \frac{9}{25} = \frac{25}{25} = 1$$

───── **Now Work**   PROBLEM 51

---

6 Find a Vector from Its Direction and Magnitude

If a vector represents the speed and direction of an object, it is called a **velocity vector.** If a vector represents the direction and amount of a force acting on an object, it is called a **force vector.** In many applications, a vector is described in terms of its magnitude and direction, rather than in terms of its components. For example, a ball thrown with an initial speed of 25 miles per hour at an angle 30° to the horizontal is a velocity vector.

Suppose that we are given the magnitude $\|\mathbf{v}\|$ of a nonzero vector **v** and the **direction angle** α, $0° \le \alpha < 360°$, between **v** and **i**. To express **v** in terms of $\|\mathbf{v}\|$ and α, we first find the unit vector **u** having the same direction as **v**.

$$\mathbf{u} = \frac{\mathbf{v}}{\|\mathbf{v}\|} \quad \text{or} \quad \mathbf{v} = \|\mathbf{v}\|\mathbf{u} \tag{7}$$

Figure 56

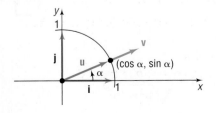

Look at Figure 56. The coordinates of the terminal point of **u** are $(\cos\alpha, \sin\alpha)$. Then $\mathbf{u} = \cos\alpha\,\mathbf{i} + \sin\alpha\,\mathbf{j}$ and, from (7),

$$\mathbf{v} = \|\mathbf{v}\|(\cos\alpha\,\mathbf{i} + \sin\alpha\,\mathbf{j}) \tag{8}$$

where α is the direction angle between **v** and **i**.

EXAMPLE 6 **Writing a Vector When Its Magnitude and Direction Are Given**

A ball is thrown with an initial speed of 25 miles per hour in a direction that makes an angle of 30° with the positive x-axis. Express the velocity vector **v** in terms of **i** and **j**. What is the initial speed in the horizontal direction? What is the initial speed in the vertical direction?

Solution The magnitude of **v** is $\|\mathbf{v}\| = 25$ miles per hour, and the angle between the direction of **v** and **i**, the positive x-axis, is $\alpha = 30°$. By equation (8),

Figure 57

$$\mathbf{v} = \|\mathbf{v}\|(\cos \alpha \mathbf{i} + \sin \alpha \mathbf{j}) = 25(\cos 30°\mathbf{i} + \sin 30°\mathbf{j}) = 25\left(\frac{\sqrt{3}}{2}\mathbf{i} + \frac{1}{2}\mathbf{j}\right) = \frac{25\sqrt{3}}{2}\mathbf{i} + \frac{25}{2}\mathbf{j}$$

The initial speed of the ball in the horizontal direction is the horizontal component of **v**, $\frac{25\sqrt{3}}{2} \approx 21.65$ miles per hour. The initial speed in the vertical direction is the vertical component of **v**, $\frac{25}{2} = 12.5$ miles per hour. See Figure 57.

━━━━━ **Now Work** PROBLEM 57

EXAMPLE 7 **Finding the Direction of a Vector**

Find the direction angle α between **i** and $\mathbf{v} = 4\mathbf{i} - 4\mathbf{j}$.

Figure 58

Solution See Figure 58. We see the direction angle α between **i** and $\mathbf{v} = 4\mathbf{i} - 4\mathbf{j}$ can be found by solving

$$\tan \alpha = \frac{-4}{4} = -1$$

so that $\alpha = -45°$. Because $0° \le \alpha < 360°$, the direction angle is $-45° + 360° = 315°$.

━━━━━ **Now Work** PROBLEM 63

Figure 59

7 Model with Vectors

Because forces can be represented by vectors, two forces "combine" the way that vectors "add." If $\mathbf{F}_1$ and $\mathbf{F}_2$ are two forces simultaneously acting on an object, the vector sum $\mathbf{F}_1 + \mathbf{F}_2$ is the **resultant force**. The resultant force produces the same effect on the object as that obtained when the two forces $\mathbf{F}_1$ and $\mathbf{F}_2$ act on the object. See Figure 59.

EXAMPLE 8 **Finding the Actual Speed and Direction of an Aircraft**

A Boeing 737 aircraft maintains a constant airspeed of 500 miles per hour headed due south. The jet stream is 80 miles per hour in the northeasterly direction.

(a) Express the velocity $\mathbf{v}_a$ of the 737 relative to the air and the velocity $\mathbf{v}_w$ of the jet stream in terms of **i** and **j**.

(b) Find the velocity of the 737 relative to the ground.

(c) Find the actual speed and direction of the 737 relative to the ground.

Solution

Figure 60

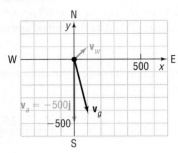

(a) We set up a coordinate system in which north (N) is along the positive y-axis. See Figure 60. The velocity of the 737 relative to the air is $\mathbf{v}_a = -500\mathbf{j}$. The velocity of the jet stream $\mathbf{v}_w$ has magnitude 80 and direction NE (northeast), so the angle between $\mathbf{v}_w$ and $\mathbf{i}$ is 45°. We express $\mathbf{v}_w$ in terms of $\mathbf{i}$ and $\mathbf{j}$ as

$$\mathbf{v}_w = 80(\cos 45°\mathbf{i} + \sin 45°\mathbf{j}) = 80\left(\frac{\sqrt{2}}{2}\mathbf{i} + \frac{\sqrt{2}}{2}\mathbf{j}\right) = 40\sqrt{2}\,(\mathbf{i} + \mathbf{j})$$

(b) The velocity of the 737 relative to the ground $\mathbf{v}_g$ is

$$\mathbf{v}_g = \mathbf{v}_a + \mathbf{v}_w = -500\mathbf{j} + 40\sqrt{2}\,(\mathbf{i} + \mathbf{j}) = 40\sqrt{2}\mathbf{i} + \left(40\sqrt{2} - 500\right)\mathbf{j}$$

(c) The actual speed of the 737 is

$$\|\mathbf{v}_g\| = \sqrt{\left(40\sqrt{2}\right)^2 + \left(40\sqrt{2} - 500\right)^2} \approx 447 \text{ miles per hour}$$

To find the actual direction of the 737 relative to the ground, we need to determine the angle between $\mathbf{v}_g$ and $\mathbf{i}$. The direction angle is found by solving

$$\tan \alpha = \frac{40\sqrt{2} - 500}{40\sqrt{2}}$$

so that $\alpha \approx -82.7°$. The 737 is traveling S 7.3° E.

Now Work PROBLEM 77

EXAMPLE 9

Finding the Weight of a Piano

Two movers require a magnitude of force of 300 pounds to push a piano up a ramp inclined at an angle 20° from the horizontal. How much does the piano weigh?

Solution

Figure 61

Let $\mathbf{F}_1$ represent the force of gravity, $\mathbf{F}_2$ represent the force required to move the piano up the ramp, and $\mathbf{F}_3$ represent the force of the piano against the ramp. See Figure 61. The angle between the ground and the ramp is the same as the angle between $\mathbf{F}_1$ and $\mathbf{F}_3$ because triangles ABC and BDE are similar, so $\angle BAC = \angle DBE = 20°$. We wish to find the magnitude of $\mathbf{F}_1$. So,

$$\sin 20° = \frac{\|\mathbf{F}_2\|}{\|\mathbf{F}_1\|} = \frac{300}{\|\mathbf{F}_1\|}$$

$$\|\mathbf{F}_1\| = \frac{300 \text{ lb}}{\sin 20°} \approx 877 \text{ lb}$$

The piano weighs approximately 877 pounds.

An object is said to be in **static equilibrium** if (1) the object is at rest and (2) the sum of all forces acting on the object is zero, that is, if the resultant force is 0.

EXAMPLE 10

An Object in Static Equilibrium

A box of supplies that weighs 1200 pounds is suspended by two cables attached to the ceiling, as shown in Figure 62. What are the tensions in the two cables?

Solution

We draw a force diagram using the vectors shown in Figure 63. The tensions in the cables are the magnitudes $\|\mathbf{F}_1\|$ and $\|\mathbf{F}_2\|$ of the force vectors $\mathbf{F}_1$ and $\mathbf{F}_2$. The magnitude of the force vector $\mathbf{F}_3$ equals 1200 pounds, the weight of the box. Now write each force vector in terms of the unit vectors $\mathbf{i}$ and $\mathbf{j}$. For $\mathbf{F}_1$ and $\mathbf{F}_2$, we use equation (8). Remember that α is the angle between the vector and the positive x-axis.

Figure 62

Figure 63

$$F_1 = \|F_1\|(\cos 150°i + \sin 150°j) = \|F_1\|\left(-\frac{\sqrt{3}}{2}i + \frac{1}{2}j\right) = -\frac{\sqrt{3}}{2}\|F_1\|i + \frac{1}{2}\|F_1\|j$$

$$F_2 = \|F_2\|(\cos 45°i + \sin 45°j) = \|F_2\|\left(\frac{\sqrt{2}}{2}i + \frac{\sqrt{2}}{2}j\right) = \frac{\sqrt{2}}{2}\|F_2\|i + \frac{\sqrt{2}}{2}\|F_2\|j$$

$$F_3 = -1200j$$

For static equilibrium, the sum of the force vectors must equal zero.

$$F_1 + F_2 + F_3 = -\frac{\sqrt{3}}{2}\|F_1\|i + \frac{1}{2}\|F_1\|j + \frac{\sqrt{2}}{2}\|F_2\|i + \frac{\sqrt{2}}{2}\|F_2\|j - 1200j = 0$$

The **i** component and **j** component will each equal zero. This results in the two equations

$$-\frac{\sqrt{3}}{2}\|F_1\| + \frac{\sqrt{2}}{2}\|F_2\| = 0 \tag{9}$$

$$\frac{1}{2}\|F_1\| + \frac{\sqrt{2}}{2}\|F_2\| - 1200 = 0 \tag{10}$$

We solve equation (9) for $\|F_2\|$ and obtain

$$\|F_2\| = \frac{\sqrt{3}}{\sqrt{2}}\|F_1\| \tag{11}$$

Substituting into equation (10) and solving for $\|F_1\|$, we obtain

$$\frac{1}{2}\|F_1\| + \frac{\sqrt{2}}{2}\left(\frac{\sqrt{3}}{\sqrt{2}}\|F_1\|\right) - 1200 = 0$$

$$\frac{1}{2}\|F_1\| + \frac{\sqrt{3}}{2}\|F_1\| - 1200 = 0$$

$$\frac{1 + \sqrt{3}}{2}\|F_1\| = 1200$$

$$\|F_1\| = \frac{2400}{1 + \sqrt{3}} \approx 878.5 \text{ pounds}$$

Substituting this value into equation (11) yields $\|F_2\|$.

$$\|F_2\| = \frac{\sqrt{3}}{\sqrt{2}}\|F_1\| = \frac{\sqrt{3}}{\sqrt{2}} \cdot \frac{2400}{1 + \sqrt{3}} \approx 1075.9 \text{ pounds}$$

The left cable has tension of approximately 878.5 pounds and the right cable has tension of approximately 1075.9 pounds.

━━━━▶ **Now Work** PROBLEM **85**

Historical Feature

Josiah Gibbs (1839–1903)

The history of vectors is surprisingly complicated for such a natural concept. In the xy-plane, complex numbers do a good job of imitating vectors. About 1840, mathematicians became interested in finding a system that would do for three dimensions what the complex numbers do for two dimensions. Hermann Grassmann (1809–1877), in Germany, and William Rowan Hamilton (1805–1865), in Ireland, both attempted to find solutions.

Hamilton's system was the *quaternions*, which are best thought of as a real number plus a vector, and do for four dimensions what complex numbers do for two dimensions. In this system the order of multiplication matters; that is, **ab** ≠ **ba**. Also, two products of

vectors emerged, the scalar (or dot) product and the vector (or cross) product.

Grassmann's abstract style, although easily read today, was almost impenetrable during the previous century, and only a few of his ideas were appreciated. Among those few were the same scalar and vector products that Hamilton had found.

About 1880, the American physicist Josiah Willard Gibbs (1839–1903) worked out an algebra involving only the simplest concepts: the vectors and the two products. He then added some calculus, and the resulting system was simple, flexible, and well adapted to expressing a large number of physical laws. This system remains in use essentially unchanged. Hamilton's and Grassmann's more extensive systems each gave birth to much interesting mathematics, but little of this mathematics is seen at elementary levels.

8.4 Assess Your Understanding

Concepts and Vocabulary

1. A _____ is a quantity that has both magnitude and direction.

2. If **v** is a vector, then **v** + (−**v**) = _____.

3. A vector **u** for which ‖**u**‖ = 1 is called a(n) _____ vector.

4. If **v** = <*a*, *b*> is an algebraic vector whose initial point is the origin, then **v** is called a(n) _____ vector.

5. If **v** = *a***i** + *b***j**, then *a* is called the _____ component of **v** and *b* is called the _____ component of **v**.

6. If **F**₁ and **F**₂ are two forces simultaneously acting on an object, the vector sum **F**₁ + **F**₂ is called the _____ force.

7. *True or False* Force is an example of a vector.

8. *True or False* Mass is an example of a vector.

Skill Building

In Problems 9–16, use the vectors in the figure at the right to graph each of the following vectors.

9. **v** + **w**

10. **u** + **v**

11. 3**v**

12. 4**w**

13. **v** − **w**

14. **u** − **v**

15. 3**v** + **u** − 2**w**

16. 2**u** − 3**v** + **w**

In Problems 17–24, use the figure at the right. Determine whether the given statement is true or false.

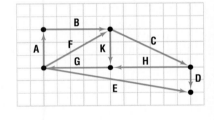

17. **A** + **B** = **F**

18. **K** + **G** = **F**

19. **C** = **D** − **E** + **F**

20. **G** + **H** + **E** = **D**

21. **E** + **D** = **G** + **H**

22. **H** − **C** = **G** − **F**

23. **A** + **B** + **K** + **G** = **0**

24. **A** + **B** + **C** + **H** + **G** = **0**

25. If ‖**v**‖ = 4, what is ‖3**v**‖?

26. If ‖**v**‖ = 2, what is ‖−4**v**‖?

*In Problems 27–34, the vector **v** has initial point P and terminal point Q. Write **v** in the form a**i** + b**j**; that is, find its position vector.*

27. $P = (0, 0)$; $Q = (3, 4)$

28. $P = (0, 0)$; $Q = (-3, -5)$

29. $P = (3, 2)$; $Q = (5, 6)$

30. $P = (-3, 2)$; $Q = (6, 5)$

31. $P = (-2, -1)$; $Q = (6, -2)$

32. $P = (-1, 4)$; $Q = (6, 2)$

33. $P = (1, 0)$; $Q = (0, 1)$

34. $P = (1, 1)$; $Q = (2, 2)$

*In Problems 35–40, find ‖**v**‖.*

35. **v** = 3**i** − 4**j**

36. **v** = −5**i** + 12**j**

37. **v** = **i** − **j**

38. **v** = −**i** − **j**

39. **v** = −2**i** + 3**j**

40. **v** = 6**i** + 2**j**

*In Problems 41–46, find each quantity if **v** = 3**i** − 5**j** and **w** = −2**i** + 3**j**.*

41. 2**v** + 3**w**

42. 3**v** − 2**w**

43. ‖**v** − **w**‖

44. ‖**v** + **w**‖

45. ‖**v**‖ − ‖**w**‖

46. ‖**v**‖ + ‖**w**‖

*In Problems 47–52, find the unit vector in the same direction as **v**.*

47. **v** = 5**i**

48. **v** = −3**j**

49. **v** = 3**i** − 4**j**

50. **v** = −5**i** + 12**j**

51. **v** = **i** − **j**

52. **v** = 2**i** − **j**

53. Find a vector **v** whose magnitude is 4 and whose component in the **i** direction is twice the component in the **j** direction.

54. Find a vector **v** whose magnitude is 3 and whose component in the **i** direction is equal to the component in the **j** direction.

55. If $\mathbf{v} = 2\mathbf{i} - \mathbf{j}$ and $\mathbf{w} = x\mathbf{i} + 3\mathbf{j}$, find all numbers x for which $\|\mathbf{v} + \mathbf{w}\| = 5$.

56. If $P = (-3, 1)$ and $Q = (x, 4)$, find all numbers x such that the vector represented by $\overrightarrow{PQ}$ has length 5.

*In Problems 57–62, write the vector **v** in the form $a\mathbf{i} + b\mathbf{j}$, given its magnitude $\|\mathbf{v}\|$ and the angle α it makes with the positive x-axis.*

57. $\|\mathbf{v}\| = 5, \quad \alpha = 60°$

58. $\|\mathbf{v}\| = 8, \quad \alpha = 45°$

59. $\|\mathbf{v}\| = 14, \quad \alpha = 120°$

60. $\|\mathbf{v}\| = 3, \quad \alpha = 240°$

61. $\|\mathbf{v}\| = 25, \quad \alpha = 330°$

62. $\|\mathbf{v}\| = 15, \quad \alpha = 315°$

*In Problems 63–70, find the direction angle between **i** and **v** for each vector.*

63. $\mathbf{v} = 3\mathbf{i} + 3\mathbf{j}$

64. $\mathbf{v} = \mathbf{i} + \sqrt{3}\mathbf{j}$

65. $\mathbf{v} = -3\sqrt{3}\mathbf{i} + 3\mathbf{j}$

66. $\mathbf{v} = -5\mathbf{i} - 5\mathbf{j}$

67. $\mathbf{v} = 4\mathbf{i} - 2\mathbf{j}$

68. $\mathbf{v} = 6\mathbf{i} - 4\mathbf{j}$

69. $\mathbf{v} = -\mathbf{i} - 5\mathbf{j}$

70. $\mathbf{v} = -\mathbf{i} + 3\mathbf{j}$

Applications and Extensions

71. Computer Graphics   The field of computer graphics utilizes vectors to compute translations of points. For example, if the point $(-3, 2)$ is to be translated by $\mathbf{v} = \langle 5, 2 \rangle$, then the new location will be $\mathbf{u'} = \mathbf{u} + \mathbf{v} = \langle -3, 2 \rangle + \langle 5, 2 \rangle = \langle 2, 4 \rangle$. As illustrated in the figure, the point $(-3, 2)$ is translated to $(2, 4)$ by **v**.

Source: Phil Dadd. Vectors and Matrices: A Primer. www.gamedev.net/reference/articles/article1832.asp

(a) Determine the new coordinates of $(3, -1)$ if it is translated by $\mathbf{v} = \langle -4, 5 \rangle$.

(b) Illustrate this translation graphically.

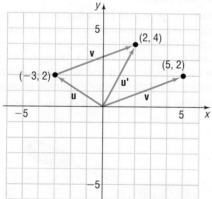

72. Computer Graphics   Refer to Problem 61. The points $(-3, 0), (-1, -2), (3, 1),$ and $(1, 3)$ are the vertices of a parallelogram $ABCD$.

(a) Find the new vertices of a parallelogram $A'B'C'D'$ if it is translated by $\mathbf{v} = \langle 3, -2 \rangle$.

(b) Find the new vertices of a parallelogram $A'B'C'D'$ if it is translated by $-\dfrac{1}{2}\mathbf{v}$.

73. Force Vectors   A child pulls a wagon with a force of 40 pounds. The handle of the wagon makes an angle of 30° with the ground. Express the force vector **F** in terms of **i** and **j**.

74. Force Vectors   A man pushes a wheelbarrow up an incline of 20° with a force of 100 pounds. Express the force vector **F** in terms of **i** and **j**.

75. Resultant Force   Two forces of magnitude 40 newtons (N) and 60 N act on an object at angles of 30° and −45° with the positive x-axis, as shown in the figure. Find the direction and magnitude of the resultant force; that is, find $\mathbf{F}_1 + \mathbf{F}_2$.

76. Resultant Force   Two forces of magnitude 30 newtons (N) and 70 N act on an object at angles of 45° and 120° with the positive x-axis, as shown in the figure. Find the direction and magnitude of the resultant force; that is, find $\mathbf{F}_1 + \mathbf{F}_2$.

77. Finding the Actual Speed and Direction of an Aircraft   A Boeing 747 jumbo jet maintains a constant airspeed of 550 miles per hour (mi/hr) headed due north. The jet stream is 100 mi/hr in the northeasterly direction.

(a) Express the velocity $\mathbf{v}_a$ of the 747 relative to the air and the velocity $\mathbf{v}_w$ of the jet stream in terms of **i** and **j**

(b) Find the velocity of the 747 relative to the ground.

(c) Find the acutal speed and direction of the 747 relative to the ground.

78. Finding the Actual Speed and Direction of an Aircraft   An Airbus A320 jet maintains a constant airspeed of 500 mi/hr headed due west. The jet stream is 100 mi/hr in the southeasterly direction.

(a) Express the velocity $\mathbf{v}_a$ of the A320 relative to the air and the velocity $\mathbf{v}_w$ of the jet stream in terms of **i** and **j**.

(b) Find the velocity of the A320 relative to the ground.
(c) Find the actual speed and direction of the A320 relative to the ground.

79. **Ground Speed and Direction of an Airplane** An airplane has an airspeed of 500 kilometers per hour (km/hr) bearing N45°E. The wind velocity is 60 km/hr in the direction N30°W. Find the resultant vector representing the path of the plane relative to the ground. What is the ground speed of the plane? What is its direction?

80. **Ground Speed and Direction of an Airplane** An airplane has an airspeed of 600 km/hr bearing S30°E. The wind velocity is 40 km/hr in the direction S45°E. Find the resultant vector representing the path of the plane relative to the ground. What is the ground speed of the plane? What is its direction?

81. **Weight of a Boat** A magnitude of 700 pounds of force is required to hold a boat and its trailer in place on a ramp whose incline is 10° to the horizontal. What is the combined weight of the boat and its trailer?

82. **Weight of a Car** A magnitude of 1200 pounds of force is required to prevent a car from rolling down a hill whose incline is 15° to the horizontal. What is the weight of the car?

83. **Correct Direction for Crossing a River** A river has a constant current of 3 km/hr. At what angle to a boat dock should a motorboat capable of maintaining a constant speed of 20 km/hr be headed in order to reach a point directly opposite the dock? If the river is $\frac{1}{2}$ kilometer wide, how long will it take to cross?

84. **Finding the Correct Compass Heading** The pilot of an aircraft wishes to head directly east, but is faced with a wind speed of 40 mi/hr from the northwest. If the pilot maintains an airspeed of 250 mi/hr, what compass heading should be maintained to head directly east? What is the actual speed of the aircarft

85. **Static Equilibrium** A weight of 1000 pounds is suspended from two cables as shown in the figure. What are the tensions in the two cables?

86. **Static Equilibrium** A weight of 800 pounds is suspended from two cables, as shown in the figure. What are the tensions in the two cables?

87. **Static Equilibrium** A tightrope walker located at a certain point deflects the rope as indicated in the figure. If the weight of the tightrope walker is 150 pounds, how much tension is in each part of the rope?

88. **Static Equilibrium** Repeat Problem 87 if the left angle is 3.8°, the right angle is 2.6°, and the weight of the tightrope walker is 135 pounds.

89. **Truck Pull** At a county fair truck pull, two pickup trucks are attached to the back end of a monster truck as illustrated in the figure. One of the pickups pulls with a force of 2000 pounds and the other pulls with a force of 3000 pounds with an angle of 45° between them. With how much force must the monster truck pull in order to remain unmoved?

[**Hint:** Find the resultant force of the two trucks.]

90. **Removing a Stump** A farmer wishes to remove a stump from a field by pulling it out with his tractor. Having removed many stumps before, he estimates that he will need 6 tons (12,000 pounds) of force to remove the stump. However, his tractor is only capable of pulling with a force of 7000 pounds, so he asks his neighbor to help. His neighbor's tractor can pull with a force of 5500 pounds. They attach the two tractors to the stump with a 40° angle between the forces as shown in the figure.
(a) Assuming the farmer's estimate of a needed 6-ton force is correct, will the farmer be successful in removing the stump? Explain.

(b) Had the farmer arranged the tractors with a 25° angle between the forces, would he have been successful in removing the stump? Explain.

5500 lb
40°
7000 lb

91. Static Equilibrium Show on the following graph the force needed for the object at P to be in static equilibrium.

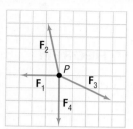

Explaining Concepts: Discussion and Writing

92. Explain in your own words what a vector is. Give an example of a vector.

93. Write a brief paragraph comparing the algebra of complex numbers and the algebra of vectors.

94. Explain the difference between an algebraic vector and a position vector.

8.5 The Dot Product

PREPARING FOR THIS SECTION *Before getting started, review the following:*

• Law of Cosines (Section 7.3, p. 563)

Now Work the '**Are You Prepared?**' problem on page 643.

OBJECTIVES 1 Find the Dot Product of Two Vectors (p. 637)
2 Find the Angle between Two Vectors (p. 638)
3 Determine Whether Two Vectors Are Parallel (p. 639)
4 Determine Whether Two Vectors Are Orthogonal (p. 639)
5 Decompose a Vector into Two Orthogonal Vectors (p. 640)
6 Compute Work (p. 642)

1 Find the Dot Product of Two Vectors

The definition for a product of two vectors is somewhat unexpected. However, such a product has meaning in many geometric and physical applications.

DEFINITION

If $\mathbf{v} = a_1\mathbf{i} + b_1\mathbf{j}$ and $\mathbf{w} = a_2\mathbf{i} + b_2\mathbf{j}$ are two vectors, the **dot product** $\mathbf{v} \cdot \mathbf{w}$ is defined as

$$\mathbf{v} \cdot \mathbf{w} = a_1a_2 + b_1b_2 \qquad (1)$$

EXAMPLE 1 **Finding Dot Products**

If $\mathbf{v} = 2\mathbf{i} - 3\mathbf{j}$ and $\mathbf{w} = 5\mathbf{i} + 3\mathbf{j}$, find:

(a) $\mathbf{v} \cdot \mathbf{w}$ (b) $\mathbf{w} \cdot \mathbf{v}$ (c) $\mathbf{v} \cdot \mathbf{v}$

(d) $\mathbf{w} \cdot \mathbf{w}$ (e) $\|\mathbf{v}\|$ (f) $\|\mathbf{w}\|$

Solution (a) $\mathbf{v} \cdot \mathbf{w} = 2(5) + (-3)3 = 1$ (b) $\mathbf{w} \cdot \mathbf{v} = 5(2) + 3(-3) = 1$

(c) $\mathbf{v} \cdot \mathbf{v} = 2(2) + (-3)(-3) = 13$ (d) $\mathbf{w} \cdot \mathbf{w} = 5(5) + 3(3) = 34$

(e) $\|\mathbf{v}\| = \sqrt{2^2 + (-3)^2} = \sqrt{13}$ (f) $\|\mathbf{w}\| = \sqrt{5^2 + 3^2} = \sqrt{34}$

COMMENT A scalar multiple $\alpha\mathbf{v}$ is a vector. A dot product $\mathbf{u}\cdot\mathbf{v}$ is a scalar (real number). ∎

Since the dot product $\mathbf{v}\cdot\mathbf{w}$ of two vectors $\mathbf{v}$ and $\mathbf{w}$ is a real number (scalar), we sometimes refer to it as the **scalar product.**

The results obtained in Example 1 suggest some general properties.

THEOREM

> **Properties of the Dot Product**
>
> If $\mathbf{u}$, $\mathbf{v}$, and $\mathbf{w}$ are vectors, then
>
> **Commutative Property**
>
> $$\mathbf{u}\cdot\mathbf{v} = \mathbf{v}\cdot\mathbf{u} \qquad (2)$$
>
> **Distributive Property**
>
> $$\mathbf{u}\cdot(\mathbf{v}+\mathbf{w}) = \mathbf{u}\cdot\mathbf{v} + \mathbf{u}\cdot\mathbf{w} \qquad (3)$$
>
> $$\mathbf{v}\cdot\mathbf{v} = \|\mathbf{v}\|^2 \qquad (4)$$
>
> $$\mathbf{0}\cdot\mathbf{v} = 0 \qquad (5)$$

Proof We will prove properties (2) and (4) here and leave properties (3) and (5) as exercises (see Problems 34 and 35).

To prove property (2), we let $\mathbf{u} = a_1\mathbf{i} + b_1\mathbf{j}$ and $\mathbf{v} = a_2\mathbf{i} + b_2\mathbf{j}$. Then

$$\mathbf{u}\cdot\mathbf{v} = a_1 a_2 + b_1 b_2 = a_2 a_1 + b_2 b_1 = \mathbf{v}\cdot\mathbf{u}$$

To prove property (4), we let $\mathbf{v} = a\mathbf{i} + b\mathbf{j}$. Then

$$\mathbf{v}\cdot\mathbf{v} = a^2 + b^2 = \|\mathbf{v}\|^2 \qquad ∎$$

2 Find the Angle between Two Vectors

One use of the dot product is to calculate the angle between two vectors. We proceed as follows.

Let $\mathbf{u}$ and $\mathbf{v}$ be two vectors with the same initial point A. Then the vectors $\mathbf{u}$, $\mathbf{v}$, and $\mathbf{u} - \mathbf{v}$ form a triangle. The angle θ at vertex A of the triangle is the angle between the vectors $\mathbf{u}$ and $\mathbf{v}$. See Figure 64. We wish to find a formula for calculating the angle θ.

The sides of the triangle have lengths $\|\mathbf{v}\|$, $\|\mathbf{u}\|$, and $\|\mathbf{u} - \mathbf{v}\|$, and θ is the included angle between the sides of length $\|\mathbf{v}\|$ and $\|\mathbf{u}\|$. The Law of Cosines (Section 7.3) can be used to find the cosine of the included angle.

$$\|\mathbf{u} - \mathbf{v}\|^2 = \|\mathbf{u}\|^2 + \|\mathbf{v}\|^2 - 2\|\mathbf{u}\|\|\mathbf{v}\|\cos\theta$$

Now we use property (4) to rewrite this equation in terms of dot products.

$$(\mathbf{u} - \mathbf{v})\cdot(\mathbf{u} - \mathbf{v}) = \mathbf{u}\cdot\mathbf{u} + \mathbf{v}\cdot\mathbf{v} - 2\|\mathbf{u}\|\|\mathbf{v}\|\cos\theta \qquad (6)$$

Then we apply the distributive property (3) twice on the left side of (6) to obtain

$$(\mathbf{u} - \mathbf{v})\cdot(\mathbf{u} - \mathbf{v}) = \mathbf{u}\cdot(\mathbf{u} - \mathbf{v}) - \mathbf{v}\cdot(\mathbf{u} - \mathbf{v})$$
$$= \mathbf{u}\cdot\mathbf{u} - \mathbf{u}\cdot\mathbf{v} - \mathbf{v}\cdot\mathbf{u} + \mathbf{v}\cdot\mathbf{v}$$
$$= \mathbf{u}\cdot\mathbf{u} + \mathbf{v}\cdot\mathbf{v} - 2\,\mathbf{u}\cdot\mathbf{v} \qquad (7)$$
$$\uparrow$$
$$\text{Property (2)}$$

Combining equations (6) and (7), we have

$$\mathbf{u}\cdot\mathbf{u} + \mathbf{v}\cdot\mathbf{v} - 2\,\mathbf{u}\cdot\mathbf{v} = \mathbf{u}\cdot\mathbf{u} + \mathbf{v}\cdot\mathbf{v} - 2\,\|\mathbf{u}\|\|\mathbf{v}\|\cos\theta$$
$$\mathbf{u}\cdot\mathbf{v} = \|\mathbf{u}\|\|\mathbf{v}\|\cos\theta$$

Figure 64

We have proved the following result:

THEOREM

Angle between Vectors

If $\mathbf{u}$ and $\mathbf{v}$ are two nonzero vectors, the angle θ, $0 \leq \theta \leq \pi$, between $\mathbf{u}$ and $\mathbf{v}$ is determined by the formula

$$\cos \theta = \frac{\mathbf{u} \cdot \mathbf{v}}{\|\mathbf{u}\|\|\mathbf{v}\|} \qquad (8)$$

EXAMPLE 2 Finding the Angle θ between Two Vectors

Find the angle θ between $\mathbf{u} = 4\mathbf{i} - 3\mathbf{j}$ and $\mathbf{v} = 2\mathbf{i} + 5\mathbf{j}$.

Solution We compute the quantities $\mathbf{u} \cdot \mathbf{v}$, $\|\mathbf{u}\|$, and $\|\mathbf{v}\|$.

Figure 65

$$\mathbf{u} \cdot \mathbf{v} = 4(2) + (-3)(5) = -7$$

$$\|\mathbf{u}\| = \sqrt{4^2 + (-3)^2} = 5$$

$$\|\mathbf{v}\| = \sqrt{2^2 + 5^2} = \sqrt{29}$$

By formula (8), if θ is the angle between $\mathbf{u}$ and $\mathbf{v}$, then

$$\cos \theta = \frac{\mathbf{u} \cdot \mathbf{v}}{\|\mathbf{u}\|\|\mathbf{v}\|} = \frac{-7}{5\sqrt{29}} \approx -0.26$$

We find that $\theta \approx 105°$. See Figure 65.

Now Work PROBLEMS 7(a) AND (b)

3 Determine Whether Two Vectors Are Parallel

Two vectors $\mathbf{v}$ and $\mathbf{w}$ are said to be **parallel** if there is a nonzero scalar α so that $\mathbf{v} = \alpha\mathbf{w}$. In this case, the angle θ between $\mathbf{v}$ and $\mathbf{w}$ is 0 or π.

EXAMPLE 3 Determining Whether Vectors Are Parallel

The vectors $\mathbf{v} = 3\mathbf{i} - \mathbf{j}$ and $\mathbf{w} = 6\mathbf{i} - 2\mathbf{j}$ are parallel, since $\mathbf{v} = \frac{1}{2}\mathbf{w}$. Furthermore, since

$$\cos \theta = \frac{\mathbf{v} \cdot \mathbf{w}}{\|\mathbf{v}\|\|\mathbf{w}\|} = \frac{18 + 2}{\sqrt{10}\sqrt{40}} = \frac{20}{\sqrt{400}} = 1$$

the angle θ between $\mathbf{v}$ and $\mathbf{w}$ is 0.

4 Determine Whether Two Vectors Are Orthogonal

Figure 66

v is orthogonal to **w**.

If the angle θ between two nonzero vectors $\mathbf{v}$ and $\mathbf{w}$ is $\frac{\pi}{2}$, the vectors $\mathbf{v}$ and $\mathbf{w}$ are called **orthogonal**.* See Figure 66.

Since $\cos \frac{\pi}{2} = 0$, it follows from formula (8) that if $\mathbf{v}$ and $\mathbf{w}$ are orthogonal then $\mathbf{v} \cdot \mathbf{w} = 0$.

*Orthogonal, perpendicular, and normal are all terms that mean "meet at a right angle." It is customary to refer to two vectors as being *orthogonal*, two lines as being *perpendicular*, and a line and a plane or a vector and a plane as being *normal*.

On the other hand, if $\mathbf{v} \cdot \mathbf{w} = 0$, then either $\mathbf{v} = 0$ or $\mathbf{w} = 0$ or $\cos \theta = 0$. In the latter case, $\theta = \dfrac{\pi}{2}$, and $\mathbf{v}$ and $\mathbf{w}$ are orthogonal. If $\mathbf{v}$ or $\mathbf{w}$ is the zero vector, then, since the zero vector has no specific direction, we adopt the convention that the zero vector is orthogonal to every vector.

THEOREM

Two vectors $\mathbf{v}$ and $\mathbf{w}$ are orthogonal if and only if

$$\mathbf{v} \cdot \mathbf{w} = 0$$

EXAMPLE 4   **Determining Whether Two Vectors Are Orthogonal**

The vectors

$$\mathbf{v} = 2\mathbf{i} - \mathbf{j} \quad \text{and} \quad \mathbf{w} = 3\mathbf{i} + 6\mathbf{j}$$

are orthogonal, since

$$\mathbf{v} \cdot \mathbf{w} = 6 - 6 = 0$$

See Figure 67.

Figure 67

━━━**Now Work** PROBLEM 7(C)

5 Decompose a Vector into Two Orthogonal Vectors

In the last section, we learned how to add two vectors to find the resultant vector. Now, we discuss the reverse problem, decomposing a vector into the sum of two components.

In many physical applications, it is necessary to find "how much" of a vector is applied in a given direction. Look at Figure 68. The force $\mathbf{F}$ due to gravity is pulling straight down (toward the center of Earth) on the block. To study the effect of gravity on the block, it is necessary to determine how much of $\mathbf{F}$ is actually pushing the block down the incline ($\mathbf{F}_1$) and how much is pressing the block against the incline ($\mathbf{F}_2$), at a right angle to the incline. Knowing the **decomposition** of $\mathbf{F}$ often will allow us to determine when friction (the force holding the block in place on the incline) is overcome and the block will slide down the incline.

Suppose that $\mathbf{v}$ and $\mathbf{w}$ are two nonzero vectors with the same initial point P. We seek to decompose $\mathbf{v}$ into two vectors: $\mathbf{v}_1$, which is parallel to $\mathbf{w}$, and $\mathbf{v}_2$, which is orthogonal to $\mathbf{w}$. See Figure 69(a) and (b). The vector $\mathbf{v}_1$ is called the **vector projection of v onto w.**

The vector $\mathbf{v}_1$ is obtained as follows: From the terminal point of $\mathbf{v}$, drop a perpendicular to the line containing $\mathbf{w}$. The vector $\mathbf{v}_1$ is the vector from P to the foot of this perpendicular. The vector $\mathbf{v}_2$ is given by $\mathbf{v}_2 = \mathbf{v} - \mathbf{v}_1$. Note that $\mathbf{v} = \mathbf{v}_1 + \mathbf{v}_2$, the vector $\mathbf{v}_1$ is parallel to $\mathbf{w}$, and the vector $\mathbf{v}_2$ is orthogonal to $\mathbf{w}$. This is the decomposition of $\mathbf{v}$ that we wanted.

Now we seek a formula for $\mathbf{v}_1$ that is based on a knowledge of the vectors $\mathbf{v}$ and $\mathbf{w}$. Since $\mathbf{v} = \mathbf{v}_1 + \mathbf{v}_2$, we have

$$\mathbf{v} \cdot \mathbf{w} = (\mathbf{v}_1 + \mathbf{v}_2) \cdot \mathbf{w} = \mathbf{v}_1 \cdot \mathbf{w} + \mathbf{v}_2 \cdot \mathbf{w} \qquad (9)$$

Since $\mathbf{v}_2$ is orthogonal to $\mathbf{w}$, we have $\mathbf{v}_2 \cdot \mathbf{w} = 0$. Since $\mathbf{v}_1$ is parallel to $\mathbf{w}$, we have $\mathbf{v}_1 = \alpha \mathbf{w}$ for some scalar α. Equation (9) can be written as

$$\mathbf{v} \cdot \mathbf{w} = \alpha \mathbf{w} \cdot \mathbf{w} = \alpha \|\mathbf{w}\|^2 \quad \text{\footnotesize $\mathbf{v}_1 = \alpha \mathbf{w}; \mathbf{v}_2 \cdot \mathbf{w} = 0$}$$

$$\alpha = \frac{\mathbf{v} \cdot \mathbf{w}}{\|\mathbf{w}\|^2}$$

Then

$$\mathbf{v}_1 = \alpha \mathbf{w} = \frac{\mathbf{v} \cdot \mathbf{w}}{\|\mathbf{w}\|^2} \mathbf{w}$$

Figure 68

Figure 69

(a)

(b)

THEOREM

If **v** and **w** are two nonzero vectors, the vector projection of **v** onto **w** is

$$\mathbf{v}_1 = \frac{\mathbf{v} \cdot \mathbf{w}}{\|\mathbf{w}\|^2}\mathbf{w} \tag{10}$$

The decomposition of **v** into $\mathbf{v}_1$ and $\mathbf{v}_2$, where $\mathbf{v}_1$ is parallel to **w** and $\mathbf{v}_2$ is orthogonal to **w**, is

$$\mathbf{v}_1 = \frac{\mathbf{v} \cdot \mathbf{w}}{\|\mathbf{w}\|^2}\mathbf{w} \qquad \mathbf{v}_2 = \mathbf{v} - \mathbf{v}_1 \tag{11}$$

EXAMPLE 5   Decomposing a Vector into Two Orthogonal Vectors

Find the vector projection of $\mathbf{v} = \mathbf{i} + 3\mathbf{j}$ onto $\mathbf{w} = \mathbf{i} + \mathbf{j}$. Decompose **v** into two vectors, $\mathbf{v}_1$ and $\mathbf{v}_2$, where $\mathbf{v}_1$ is parallel to **w** and $\mathbf{v}_2$ is orthogonal to **w**.

Solution   We use formulas (10) and (11).

Figure 70

$$\mathbf{v}_1 = \frac{\mathbf{v} \cdot \mathbf{w}}{\|\mathbf{w}\|^2}\mathbf{w} = \frac{1 + 3}{\left(\sqrt{2}\right)^2}\mathbf{w} = 2\mathbf{w} = 2(\mathbf{i} + \mathbf{j})$$

$$\mathbf{v}_2 = \mathbf{v} - \mathbf{v}_1 = (\mathbf{i} + 3\mathbf{j}) - 2(\mathbf{i} + \mathbf{j}) = -\mathbf{i} + \mathbf{j}$$

See Figure 70.

 Now Work PROBLEM 19

EXAMPLE 6   Finding the Force Required to Hold a Wagon on a Hill

A wagon with two small children as occupants that weighs 100 pounds is on a hill with a grade of 20°. What is the magnitude of the force that is required to keep the wagon from rolling down the hill?

Solution   See Figure 71. We wish to find the magnitude of the force **v** that is causing the wagon to roll down the hill. A force with the same magnitude in the opposite direction of **v** will keep the wagon from rolling down the hill. The force of gravity is orthogonal to the level ground, so we can represent the force of the wagon due to gravity by the vector

Figure 71

$$\mathbf{F}_w = -100\mathbf{j}$$

We need to determine the vector projection of $\mathbf{F}_w$ onto **w**, which is the force parallel to the hill. The vector **w** is given by

$$\mathbf{w} = \cos 20°\mathbf{i} + \sin 20°\mathbf{j}$$

The vector projection of $\mathbf{F}_w$ onto **w** is

$$\mathbf{v} = \frac{\mathbf{F}_w \cdot \mathbf{w}}{\|\mathbf{w}\|^2}\mathbf{w}$$

$$= \frac{-100(\sin 20°)}{\left(\sqrt{\cos^2 20° + \sin^2 20°}\right)^2}(\cos 20°\mathbf{i} + \sin 20°\mathbf{j})$$

$$= -34.2(\cos 20°\mathbf{i} + \sin 20°\mathbf{j})$$

The magnitude of **v** is 34.2 pounds, so the magnitude of the force required to keep the wagon from rolling down the hill is 34.2 pounds.

6 Compute Work

In elementary physics, the **work** W done by a constant force $\mathbf{F}$ in moving an object from a point A to a point B is defined as

$$W = (\text{magnitude of force})(\text{distance}) = \|\mathbf{F}\|\|\overrightarrow{AB}\|$$

Work is commonly measured in foot-pounds or in newton-meters (joules).

In this definition, it is assumed that the force $\mathbf{F}$ is applied along the line of motion. If the constant force $\mathbf{F}$ is not along the line of motion, but, instead, is at an angle θ to the direction of the motion, as illustrated in Figure 72, the **work** W **done by $\mathbf{F}$** in moving an object from A to B is defined as

$$W = \mathbf{F} \cdot \overrightarrow{AB} \qquad (12)$$

This definition is compatible with the force times distance definition, since

$$W = (\text{amount of force in the direction of } \overrightarrow{AB})(\text{distance})$$

$$= \|\text{projection of } \mathbf{F} \text{ on } AB\|\|\overrightarrow{AB}\| = \frac{\mathbf{F} \cdot \overrightarrow{AB}}{\|\overrightarrow{AB}\|^2}\|\overrightarrow{AB}\|\|\overrightarrow{AB}\| = \mathbf{F} \cdot \overrightarrow{AB}$$

$\uparrow$
Use formula (10).

Figure 72

| EXAMPLE 7 | **Computing Work** |

Figure 73(a) shows a girl pulling a wagon with a force of 50 pounds. How much work is done in moving the wagon 100 feet if the handle makes an angle of $30°$ with the ground?

Figure 73

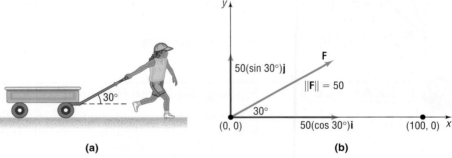

(a)　　　　　　(b)

Solution　We position the vectors in a coordinate system in such a way that the wagon is moved from $(0, 0)$ to $(100, 0)$. The motion is from $A = (0, 0)$ to $B = (100, 0)$, so $\overrightarrow{AB} = 100\mathbf{i}$. The force vector $\mathbf{F}$, as shown in Figure 73(b), is

$$\mathbf{F} = 50(\cos 30°\mathbf{i} + \sin 30°\mathbf{j}) = 50\left(\frac{\sqrt{3}}{2}\mathbf{i} + \frac{1}{2}\mathbf{j}\right) = 25\left(\sqrt{3}\mathbf{i} + \mathbf{j}\right)$$

By formula (12), the work done is

$$W = \mathbf{F} \cdot \overrightarrow{AB} = 25\left(\sqrt{3}\mathbf{i} + \mathbf{j}\right) \cdot 100\mathbf{i} = 2500\sqrt{3} \text{ foot-pounds}$$

━━━**Now Work** PROBLEM 25

Historical Feature

We stated in an earlier Historical Feature that complex numbers were used as vectors in the plane before the general notion of a vector was clarified. Suppose that we make the correspondence

Vector $\leftrightarrow$ Complex number
$a\mathbf{i} + b\mathbf{j} \leftrightarrow a + bi$
$c\mathbf{i} + d\mathbf{j} \leftrightarrow c + di$

Show that

$$(a\mathbf{i} + b\mathbf{j}) \cdot (c\mathbf{i} + d\mathbf{j}) = \text{real part } [\overline{(a + bi)}(c + di)]$$

This is how the dot product was found originally. The imaginary part is also interesting. It is a determinant (see Section 10.3) and represents the area of the parallelogram whose edges are the vectors. This is close to some of Hermann Grassmann's ideas and is also connected with the scalar triple product of three-dimensional vectors.

8.5 Assess Your Understanding

'Are You Prepared?' *The answer is given at the end of these exercises. If you get the wrong answer, read the page listed in red.*

1. In a triangle with sides a, b, c and angles A, B, C, the Law of Cosines states that _____. (p. 563)

Concepts and Vocabulary

2. If $\mathbf{v} = a_1\mathbf{i} + b_1\mathbf{j}$ and $\mathbf{w} = a_2\mathbf{i} + b_2\mathbf{j}$ are two vectors, the _____ _____ is defined as $\mathbf{v} \cdot \mathbf{w} = a_1a_2 + b_1b_2$.

3. If $\mathbf{v} \cdot \mathbf{w} = 0$, then the two vectors $\mathbf{v}$ and $\mathbf{w}$ are _____.

4. If $\mathbf{v} = 3\mathbf{w}$, then the two vectors $\mathbf{v}$ and $\mathbf{w}$ are _____.

5. **True or False** Given two nonzero vectors $\mathbf{v}$ and $\mathbf{w}$, it is always possible to decompose $\mathbf{v}$ into two vectors, one parallel to $\mathbf{w}$ and the other perpendicular to $\mathbf{w}$.

6. **True or False** Work is a physical example of a vector.

Skill Building

In Problems 7–16, (a) find the dot product $\mathbf{v} \cdot \mathbf{w}$; (b) find the angle between $\mathbf{v}$ and $\mathbf{w}$;(c) state whether the vectors are parallel, orthogonal, or neither.

7. $\mathbf{v} = \mathbf{i} - \mathbf{j}, \quad \mathbf{w} = \mathbf{i} + \mathbf{j}$

8. $\mathbf{v} = \mathbf{i} + \mathbf{j}, \quad \mathbf{w} = -\mathbf{i} + \mathbf{j}$

9. $\mathbf{v} = 2\mathbf{i} + \mathbf{j}, \quad \mathbf{w} = \mathbf{i} - 2\mathbf{j}$

10. $\mathbf{v} = 2\mathbf{i} + 2\mathbf{j}, \quad \mathbf{w} = \mathbf{i} + 2\mathbf{j}$

11. $\mathbf{v} = \sqrt{3}\mathbf{i} - \mathbf{j}, \quad \mathbf{w} = \mathbf{i} + \mathbf{j}$

12. $\mathbf{v} = \mathbf{i} + \sqrt{3}\mathbf{j}, \quad \mathbf{w} = \mathbf{i} - \mathbf{j}$

13. $\mathbf{v} = 3\mathbf{i} + 4\mathbf{j}, \quad \mathbf{w} = -6\mathbf{i} - 8\mathbf{j}$

14. $\mathbf{v} = 3\mathbf{i} - 4\mathbf{j}, \quad \mathbf{w} = 9\mathbf{i} - 12\mathbf{j}$

15. $\mathbf{v} = 4\mathbf{i}, \quad \mathbf{w} = \mathbf{j}$

16. $\mathbf{v} = \mathbf{i}, \quad \mathbf{w} = -3\mathbf{j}$

17. Find a so that the vectors $\mathbf{v} = \mathbf{i} - a\mathbf{j}$ and $\mathbf{w} = 2\mathbf{i} + 3\mathbf{j}$ are orthogonal.

18. Find b so that the vectors $\mathbf{v} = \mathbf{i} + \mathbf{j}$ and $\mathbf{w} = \mathbf{i} + b\mathbf{j}$ are orthogonal.

In Problems 19–24, decompose $\mathbf{v}$ into two vectors $\mathbf{v}_1$ and $\mathbf{v}_2$, where $\mathbf{v}_1$ is parallel to $\mathbf{w}$ and $\mathbf{v}_2$ is orthogonal to $\mathbf{w}$.

19. $\mathbf{v} = 2\mathbf{i} - 3\mathbf{j}, \quad \mathbf{w} = \mathbf{i} - \mathbf{j}$

20. $\mathbf{v} = -3\mathbf{i} + 2\mathbf{j}, \quad \mathbf{w} = 2\mathbf{i} + \mathbf{j}$

21. $\mathbf{v} = \mathbf{i} - \mathbf{j}, \quad \mathbf{w} = -\mathbf{i} - 2\mathbf{j}$

22. $\mathbf{v} = 2\mathbf{i} - \mathbf{j}, \quad \mathbf{w} = \mathbf{i} - 2\mathbf{j}$

23. $\mathbf{v} = 3\mathbf{i} + \mathbf{j}, \quad \mathbf{w} = -2\mathbf{i} - \mathbf{j}$

24. $\mathbf{v} = \mathbf{i} - 3\mathbf{j}, \quad \mathbf{w} = 4\mathbf{i} - \mathbf{j}$

Applications and Extensions

25. **Computing Work** Find the work done by a force of 3 pounds acting in the direction $60°$ to the horizontal in moving an object 6 feet from $(0, 0)$ to $(6, 0)$.

26. **Computing Work** A wagon is pulled horizontally by exerting a force of 20 pounds on the handle at an angle of $30°$ with the horizontal. How much work is done in moving the wagon 100 feet?

27. **Solar Energy** The amount of energy collected by a solar panel depends on the intensity of the sun's rays and the area of the panel. Let the vector $\mathbf{I}$ represent the intensity, in watts per square centimeter, having the direction of the sun's rays. Let the vector $\mathbf{A}$ represent the area, in square centimeters, whose direction is the orientation of a solar panel. See the figure. The total number of watts collected by the panel is given by $W = |\mathbf{I} \cdot \mathbf{A}|$.

   Suppose that $\mathbf{I} = \langle -0.02, -0.01 \rangle$ and $\mathbf{A} = \langle 300, 400 \rangle$.

   (a) Find $\|\mathbf{I}\|$ and $\|\mathbf{A}\|$ and interpret the meaning of each.

   (b) Compute W and interpret its meaning.

   (c) If the solar panel is to collect the maximum number of watts, what must be true about $\mathbf{I}$ and $\mathbf{A}$?

28. **Rainfall Measurement** Let the vector $\mathbf{R}$ represent the amount of rainfall, in inches, whose direction is the inclination of the rain to a rain gauge. Let the vector $\mathbf{A}$ represent the area, in square inches, whose direction is the orientation of the opening of the rain gauge. See the figure. The volume of rain collected in the gauge, in cubic inches, is given by $V = |\mathbf{R} \cdot \mathbf{A}|$, even when the rain falls in a slanted direction or the gauge is not perfectly vertical.

   Suppose that $\mathbf{R} = \langle 0.75, -1.75 \rangle$ and $\mathbf{A} = \langle 0.3, 1 \rangle$.

   (a) Find $\|\mathbf{R}\|$ and $\|\mathbf{A}\|$ and interpret the meaning of each.

   (b) Compute V and interpret its meaning.

   (c) If the gauge is to collect the maximum volume of rain, what must be true about $\mathbf{R}$ and $\mathbf{A}$?

29. **Braking Load** A Toyota Sienna with a gross weight of 5300 pounds is parked on a street with an $8°$ grade. See the figure.

Find the magnitude of the force required to keep the Sienna from rolling down the hill. What is the magnitude of the force perpendicular to the hill?

Weight = 5300 pounds

30. Braking Load A Pontiac Bonneville with a gross weight of 4500 pounds is parked on a street with a 10° grade. Find the magnitude of the force required to keep the Bonneville from rolling down the hill. What is the magnitude of the force perpendicular to the hill?

31. Ramp Angle Billy and Timmy are using a ramp to load furniture into a truck. While rolling a 250-pound piano up the ramp, they discover that the truck is too full of other furniture for the piano to fit. Timmy holds the piano in place on the ramp while Billy repositions other items to make room for it in the truck. If the angle of inclination of the ramp is 20°, how many pounds of force must Timmy exert to hold the piano in position?

20°

250 lb

32. Incline Angle A bulldozer exerts 1000 pounds of force to prevent a 5000-pound boulder from rolling down a hill. Determine the angle of inclination of the hill.

33. Find the acute angle that a constant unit force vector makes with the positive x-axis if the work done by the force in moving a particle from $(0, 0)$ to $(4, 0)$ equals 2.

34. Prove the distributive property:

$$\mathbf{u} \cdot (\mathbf{v} + \mathbf{w}) = \mathbf{u} \cdot \mathbf{v} + \mathbf{u} \cdot \mathbf{w}$$

35. Prove property (5), $\mathbf{0} \cdot \mathbf{v} = 0$.

36. If $\mathbf{v}$ is a unit vector and the angle between $\mathbf{v}$ and $\mathbf{i}$ is α, show that $\mathbf{v} = \cos \alpha \mathbf{i} + \sin \alpha \mathbf{j}$.

37. Suppose that $\mathbf{v}$ and $\mathbf{w}$ are unit vectors. If the angle between $\mathbf{v}$ and $\mathbf{i}$ is α and that between $\mathbf{w}$ and $\mathbf{i}$ is β, use the idea of the dot product $\mathbf{v} \cdot \mathbf{w}$ to prove that

$$\cos(\alpha - \beta) = \cos \alpha \cos \beta + \sin \alpha \sin \beta$$

38. Show that the projection of $\mathbf{v}$ onto $\mathbf{i}$ is $(\mathbf{v} \cdot \mathbf{i})\mathbf{i}$. Then show that we can always write a vector $\mathbf{v}$ as

$$\mathbf{v} = (\mathbf{v} \cdot \mathbf{i})\mathbf{i} + (\mathbf{v} \cdot \mathbf{j})\mathbf{j}$$

39. (a) If $\mathbf{u}$ and $\mathbf{v}$ have the same magnitude, show that $\mathbf{u} + \mathbf{v}$ and $\mathbf{u} - \mathbf{v}$ are orthogonal.
  (b) Use this to prove that an angle inscribed in a semicircle is a right angle (see the figure).

40. Let $\mathbf{v}$ and $\mathbf{w}$ denote two nonzero vectors. Show that the vector $\mathbf{v} - \alpha\mathbf{w}$ is orthogonal to $\mathbf{w}$ if $\alpha = \dfrac{\mathbf{v} \cdot \mathbf{w}}{\|\mathbf{w}\|^2}$.

41. Let $\mathbf{v}$ and $\mathbf{w}$ denote two nonzero vectors. Show that the vectors $\|\mathbf{w}\|\mathbf{v} + \|\mathbf{v}\|\mathbf{w}$ and $\|\mathbf{w}\|\mathbf{v} - \|\mathbf{v}\|\mathbf{w}$ are orthogonal.

42. In the definition of work given in this section, what is the work done if $\mathbf{F}$ is orthogonal to $\overrightarrow{AB}$?

43. Prove the **polarization identity**,

$$\|\mathbf{u} + \mathbf{v}\|^2 - \|\mathbf{u} - \mathbf{v}\|^2 = 4(\mathbf{u} \cdot \mathbf{v})$$

Explaining Concepts: Discussion and Writing

44. Create an application different from any found in the text that requires a dot product.

'Are You Prepared?' Answers

1. $c^2 = a^2 + b^2 - 2ab \cos C$

8.6 Vectors in Space

PREPARING FOR THIS SECTION *Before getting started, review the following:*

- Distance Formula (Foundations, Section 1, p. 3)

Now Work the 'Are You Prepared?' problem on page 652.

OBJECTIVES **1** Find the Distance between Two Points in Space (p. 646)
  2 Find Position Vectors in Space (p. 646)
  3 Perform Operations on Vectors (p. 647)
  4 Find the Dot Product (p. 648)
  5 Find the Angle between Two Vectors (p. 649)
  6 Find the Direction Angles of a Vector (p. 649)

Figure 74

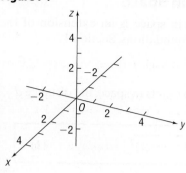

Rectangular Coordinates in Space

In the plane, each point is associated with an ordered pair of real numbers. In space, each point is associated with an ordered triple of real numbers. Through a fixed point, called the **origin** O, draw three mutually perpendicular lines, the x-axis, the y-axis, and the z-axis. On each of these axes, select an appropriate scale and the positive direction. See Figure 74.

The direction chosen for the positive z-axis in Figure 74 makes the system *right-handed*. This conforms to the *right-hand rule*, which states that, if the index finger of the right hand points in the direction of the positive x-axis and the middle finger points in the direction of the positive y-axis, then the thumb will point in the direction of the positive z-axis. See Figure 75.

We associate with each point P an ordered triple (x, y, z) of real numbers, the **coordinates of P.** For example, the point $(2, 3, 4)$ is located by starting at the origin and moving 2 units along the positive x-axis, 3 units in the direction of the positive y-axis, and 4 units in the direction of the positive z-axis. See Figure 76.

Figure 75

Figure 76

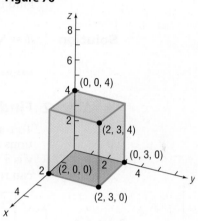

Figure 76 also shows the location of the points $(2, 0, 0)$, $(0, 3, 0)$, $(0, 0, 4)$, and $(2, 3, 0)$. Points of the form $(x, 0, 0)$ lie on the x-axis, while points of the form $(0, y, 0)$ and $(0, 0, z)$ lie on the y-axis and z-axis, respectively. Points of the form $(x, y, 0)$ lie in a plane, called the **xy-plane.** Its equation is $z = 0$. Similarly, points of the form $(x, 0, z)$ lie in the **xz-plane** (equation $y = 0$) and points of the form $(0, y, z)$ lie in the **yz-plane** (equation $x = 0$). See Figure 77(a). By extension of these ideas, all points obeying the equation $z = 3$ will lie in a plane parallel to and 3 units above the xy-plane. The equation $y = 4$ represents a plane parallel to the xz-plane and 4 units to the right of the plane $y = 0$. See Figure 77(b).

Figure 77

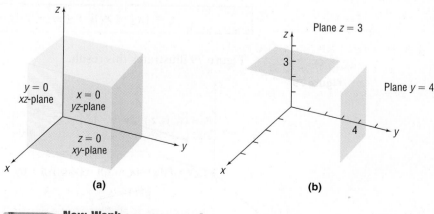

(a)   (b)

═══ **Now Work** PROBLEM 9

1 Find the Distance between Two Points in Space

The formula for the distance between two points in space is an extension of the Distance Formula for points in the plane given in Foundations, Section 1.

THEOREM

Distance Formula in Space

If $P_1 = (x_1, y_1, z_1)$ and $P_2 = (x_2, y_2, z_2)$ are two points in space, the distance d from P_1 to P_2 is

$$d = \sqrt{(x_2 - x_1)^2 + (y_2 - y_1)^2 + (z_2 - z_1)^2} \tag{1}$$

The proof, which we omit, utilizes a double application of the Pythagorean Theorem.

EXAMPLE 1

Using the Distance Formula

Find the distance from $P_1 = (-1, 3, 2)$ to $P_2 = (4, -2, 5)$.

Solution $d = \sqrt{[4 - (-1)]^2 + [-2 - 3]^2 + [5 - 2]^2} = \sqrt{25 + 25 + 9} = \sqrt{59}$

━━━━ **Now Work** PROBLEM 15

2 Find Position Vectors in Space

To represent vectors in space, we introduce the unit vectors $\mathbf{i}$, $\mathbf{j}$, and $\mathbf{k}$ whose directions are along the positive x-axis, positive y-axis, and positive z-axis, respectively. If $\mathbf{v}$ is a vector with initial point at the origin O and terminal point at $P = (a, b, c)$, we can represent $\mathbf{v}$ in terms of the vectors $\mathbf{i}$, $\mathbf{j}$, and $\mathbf{k}$ as

Figure 78

$$\mathbf{v} = a\mathbf{i} + b\mathbf{j} + c\mathbf{k}$$

See Figure 78.

The scalars a, b, and c are called the **components** of the vector $\mathbf{v} = a\mathbf{i} + b\mathbf{j} + c\mathbf{k}$, with a being the component in the direction $\mathbf{i}$, b the component in the direction $\mathbf{j}$, and c the component in the direction $\mathbf{k}$.

A vector whose initial point is at the origin is called a **position vector**. The next result states that any vector whose initial point is not at the origin is equal to a unique position vector.

THEOREM

Suppose that $\mathbf{v}$ is a vector with initial point $P_1 = (x_1, y_1, z_1)$, not necessarily the origin, and terminal point $P_2 = (x_2, y_2, z_2)$. If $\mathbf{v} = \overrightarrow{P_1 P_2}$, then $\mathbf{v}$ is equal to the position vector

$$\mathbf{v} = (x_2 - x_1)\mathbf{i} + (y_2 - y_1)\mathbf{j} + (z_2 - z_1)\mathbf{k} \tag{2}$$

Figure 79 illustrates this result.

Figure 79

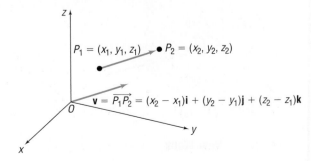

EXAMPLE 2 **Finding a Position Vector**

Find the position vector of the vector $\mathbf{v} = \overrightarrow{P_1 P_2}$ if $P_1 = (-1, 2, 3)$ and $P_2 = (4, 6, 2)$.

Solution By equation (2), the position vector equal to $\mathbf{v}$ is

$$\mathbf{v} = [4 - (-1)]\mathbf{i} + (6 - 2)\mathbf{j} + (2 - 3)\mathbf{k} = 5\mathbf{i} + 4\mathbf{j} - \mathbf{k}$$

Now Work PROBLEM 29

3 Perform Operations on Vectors

We can define equality, addition, subtraction, scalar product, and magnitude in terms of the components of a vector.

DEFINITION Let $\mathbf{v} = a_1\mathbf{i} + b_1\mathbf{j} + c_1\mathbf{k}$ and $\mathbf{w} = a_2\mathbf{i} + b_2\mathbf{j} + c_2\mathbf{k}$ be two vectors, and let α be a scalar. Then

$$\mathbf{v} = \mathbf{w} \quad \text{if and only if} \quad a_1 = a_2, b_1 = b_2, \text{ and } c_1 = c_2$$
$$\mathbf{v} + \mathbf{w} = (a_1 + a_2)\mathbf{i} + (b_1 + b_2)\mathbf{j} + (c_1 + c_2)\mathbf{k}$$
$$\mathbf{v} - \mathbf{w} = (a_1 - a_2)\mathbf{i} + (b_1 - b_2)\mathbf{j} + (c_1 - c_2)\mathbf{k}$$
$$\alpha\mathbf{v} = (\alpha a_1)\mathbf{i} + (\alpha b_1)\mathbf{j} + (\alpha c_1)\mathbf{k}$$
$$\|\mathbf{v}\| = \sqrt{a_1^2 + b_1^2 + c_1^2}$$

These definitions are compatible with the geometric definitions given for vectors in the plane earlier in Section 8.4.

EXAMPLE 3 **Adding and Subtracting Vectors**

If $\mathbf{v} = 2\mathbf{i} + 3\mathbf{j} - 2\mathbf{k}$ and $\mathbf{w} = 3\mathbf{i} - 4\mathbf{j} + 5\mathbf{k}$, find:

(a) $\mathbf{v} + \mathbf{w}$ (b) $\mathbf{v} - \mathbf{w}$

Solution (a) $\mathbf{v} + \mathbf{w} = (2\mathbf{i} + 3\mathbf{j} - 2\mathbf{k}) + (3\mathbf{i} - 4\mathbf{j} + 5\mathbf{k})$
$$= (2 + 3)\mathbf{i} + (3 - 4)\mathbf{j} + (-2 + 5)\mathbf{k}$$
$$= 5\mathbf{i} - \mathbf{j} + 3\mathbf{k}$$

(b) $\mathbf{v} - \mathbf{w} = (2\mathbf{i} + 3\mathbf{j} - 2\mathbf{k}) - (3\mathbf{i} - 4\mathbf{j} + 5\mathbf{k})$
$$= (2 - 3)\mathbf{i} + [3 - (-4)]\mathbf{j} + [-2 - 5]\mathbf{k}$$
$$= -\mathbf{i} + 7\mathbf{j} - 7\mathbf{k}$$

EXAMPLE 4 **Finding Scalar Products and Magnitudes**

If $\mathbf{v} = 2\mathbf{i} + 3\mathbf{j} - 2\mathbf{k}$ and $\mathbf{w} = 3\mathbf{i} - 4\mathbf{j} + 5\mathbf{k}$, find:

(a) $3\mathbf{v}$ (b) $2\mathbf{v} - 3\mathbf{w}$ (c) $\|\mathbf{v}\|$

Solution (a) $3\mathbf{v} = 3(2\mathbf{i} + 3\mathbf{j} - 2\mathbf{k}) = 6\mathbf{i} + 9\mathbf{j} - 6\mathbf{k}$

(b) $2\mathbf{v} - 3\mathbf{w} = 2(2\mathbf{i} + 3\mathbf{j} - 2\mathbf{k}) - 3(3\mathbf{i} - 4\mathbf{j} + 5\mathbf{k})$
$$= 4\mathbf{i} + 6\mathbf{j} - 4\mathbf{k} - 9\mathbf{i} + 12\mathbf{j} - 15\mathbf{k} = -5\mathbf{i} + 18\mathbf{j} - 19\mathbf{k}$$

(c) $\|\mathbf{v}\| = \|2\mathbf{i} + 3\mathbf{j} - 2\mathbf{k}\| = \sqrt{2^2 + 3^2 + (-2)^2} = \sqrt{17}$

Now Work PROBLEMS 33 AND 39

Recall that a unit vector $\mathbf{u}$ is one for which $\|\mathbf{u}\| = 1$. In many applications, it is useful to be able to find a unit vector $\mathbf{u}$ that has the same direction as a given vector $\mathbf{v}$.

THEOREM

> **Unit Vector in the Direction of v**
>
> For any nonzero vector $\mathbf{v}$, the vector
>
> $$\mathbf{u} = \frac{\mathbf{v}}{\|\mathbf{v}\|}$$
>
> is a unit vector that has the same direction as $\mathbf{v}$.

As a consequence of this theorem, if $\mathbf{u}$ is a unit vector in the same direction as a vector $\mathbf{v}$, then $\mathbf{v}$ may be expressed as

$$\mathbf{v} = \|\mathbf{v}\|\mathbf{u}$$

EXAMPLE 5

Finding a Unit Vector

Find the unit vector in the same direction as $\mathbf{v} = 2\mathbf{i} - 3\mathbf{j} - 6\mathbf{k}$.

Solution We find $\|\mathbf{v}\|$ first.

$$\|\mathbf{v}\| = \|2\mathbf{i} - 3\mathbf{j} - 6\mathbf{k}\| = \sqrt{4 + 9 + 36} = \sqrt{49} = 7$$

Now we multiply $\mathbf{v}$ by the scalar $\dfrac{1}{\|\mathbf{v}\|} = \dfrac{1}{7}$. The result is the unit vector

$$\mathbf{u} = \frac{\mathbf{v}}{\|\mathbf{v}\|} = \frac{2\mathbf{i} - 3\mathbf{j} - 6\mathbf{k}}{7} = \frac{2}{7}\mathbf{i} - \frac{3}{7}\mathbf{j} - \frac{6}{7}\mathbf{k}$$

⬤

Now Work PROBLEM 47

4 Find the Dot Product

The definition of *dot product* is an extension of the definition given for vectors in the plane.

DEFINITION

> If $\mathbf{v} = a_1\mathbf{i} + b_1\mathbf{j} + c_1\mathbf{k}$ and $\mathbf{w} = a_2\mathbf{i} + b_2\mathbf{j} + c_2\mathbf{k}$ are two vectors, the **dot product** $\mathbf{v} \cdot \mathbf{w}$ is defined as
>
> $$\mathbf{v} \cdot \mathbf{w} = a_1 a_2 + b_1 b_2 + c_1 c_2 \qquad \textbf{(3)}$$

EXAMPLE 6

Finding Dot Products

If $\mathbf{v} = 2\mathbf{i} - 3\mathbf{j} + 6\mathbf{k}$ and $\mathbf{w} = 5\mathbf{i} + 3\mathbf{j} - \mathbf{k}$, find:

(a) $\mathbf{v} \cdot \mathbf{w}$ (b) $\mathbf{w} \cdot \mathbf{v}$ (c) $\mathbf{v} \cdot \mathbf{v}$

(d) $\mathbf{w} \cdot \mathbf{w}$ (e) $\|\mathbf{v}\|$ (f) $\|\mathbf{w}\|$

Solution

(a) $\mathbf{v} \cdot \mathbf{w} = 2(5) + (-3)3 + 6(-1) = -5$

(b) $\mathbf{w} \cdot \mathbf{v} = 5(2) + 3(-3) + (-1)(6) = -5$

(c) $\mathbf{v} \cdot \mathbf{v} = 2(2) + (-3)(-3) + 6(6) = 49$

(d) $\mathbf{w} \cdot \mathbf{w} = 5(5) + 3(3) + (-1)(-1) = 35$

(e) $\|\mathbf{v}\| = \sqrt{2^2 + (-3)^2 + 6^2} = \sqrt{49} = 7$

(f) $\|\mathbf{w}\| = \sqrt{5^2 + 3^2 + (-1)^2} = \sqrt{35}$

⬤

The dot product in space has the same properties as the dot product in the plane.

THEOREM

Properties of the Dot Product

If **u**, **v**, and **w** are vectors, then

Commutative Property

$$\mathbf{u} \cdot \mathbf{v} = \mathbf{v} \cdot \mathbf{u}$$

Distributive Property

$$\mathbf{u} \cdot (\mathbf{v} + \mathbf{w}) = \mathbf{u} \cdot \mathbf{v} + \mathbf{u} \cdot \mathbf{w}$$

$$\mathbf{v} \cdot \mathbf{v} = \|\mathbf{v}\|^2$$
$$\mathbf{0} \cdot \mathbf{v} = 0$$

5 Find the Angle between Two Vectors

The angle θ between two vectors in space follows the same formula as for two vectors in the plane.

THEOREM

Angle between Vectors

If **u** and **v** are two nonzero vectors, the angle θ, $0 \le \theta \le \pi$, between **u** and **v** is determined by the formula

$$\cos \theta = \frac{\mathbf{u} \cdot \mathbf{v}}{\|\mathbf{u}\| \, \|\mathbf{v}\|} \tag{4}$$

EXAMPLE 7

Finding the Angle θ between Two Vectors

Find the angle θ between $\mathbf{u} = 2\mathbf{i} - 3\mathbf{j} + 6\mathbf{k}$ and $\mathbf{v} = 2\mathbf{i} + 5\mathbf{j} - \mathbf{k}$.

Solution

We compute the quantities $\mathbf{u} \cdot \mathbf{v}$, $\|\mathbf{u}\|$, and $\|\mathbf{v}\|$.

$$\mathbf{u} \cdot \mathbf{v} = 2(2) + (-3)(5) + 6(-1) = -17$$
$$\|\mathbf{u}\| = \sqrt{2^2 + (-3)^2 + 6^2} = \sqrt{49} = 7$$
$$\|\mathbf{v}\| = \sqrt{2^2 + 5^2 + (-1)^2} = \sqrt{30}$$

By formula (4), if θ is the angle between **u** and **v**, then

$$\cos \theta = \frac{\mathbf{u} \cdot \mathbf{v}}{\|\mathbf{u}\| \, \|\mathbf{v}\|} = \frac{-17}{7\sqrt{30}} \approx -0.443$$

We find that $\theta \approx 116.3°$.

━━━━ **Now Work** PROBLEM 51

6 Find the Direction Angles of a Vector

A nonzero vector **v** in space can be described by specifying its magnitude and its three **direction angles** α, β, and γ. These direction angles are defined as

α = the angle between **v** and **i**, the positive x-axis, $0 \le \alpha \le \pi$

β = the angle between **v** and **j**, the positive y-axis, $0 \le \beta \le \pi$

γ = the angle between **v** and **k**, the positive z-axis, $0 \le \gamma \le \pi$

See Figure 80.

Figure 80

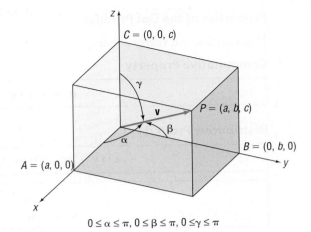

$$0 \leq \alpha \leq \pi, 0 \leq \beta \leq \pi, 0 \leq \gamma \leq \pi$$

Our first goal is to find expressions for α, β, and γ in terms of the components of a vector. Let $\mathbf{v} = a\mathbf{i} + b\mathbf{j} + c\mathbf{k}$ denote a nonzero vector. The angle α between $\mathbf{v}$ and $\mathbf{i}$, the positive x-axis, obeys

$$\cos \alpha = \frac{\mathbf{v} \cdot \mathbf{i}}{\|\mathbf{v}\| \, \|\mathbf{i}\|} = \frac{a}{\|\mathbf{v}\|}$$

Similarly,

$$\cos \beta = \frac{b}{\|\mathbf{v}\|} \qquad \cos \gamma = \frac{c}{\|\mathbf{v}\|}$$

Since $\|\mathbf{v}\| = \sqrt{a^2 + b^2 + c^2}$, we have the following result:

THEOREM

Direction Angles

If $\mathbf{v} = a\mathbf{i} + b\mathbf{j} + c\mathbf{k}$ is a nonzero vector in space, the direction angles α, β, and γ obey

$$\cos \alpha = \frac{a}{\sqrt{a^2 + b^2 + c^2}} = \frac{a}{\|\mathbf{v}\|} \qquad \cos \beta = \frac{b}{\sqrt{a^2 + b^2 + c^2}} = \frac{b}{\|\mathbf{v}\|}$$

$$\cos \gamma = \frac{c}{\sqrt{a^2 + b^2 + c^2}} = \frac{c}{\|\mathbf{v}\|} \qquad \qquad \textbf{(5)}$$

The numbers $\cos \alpha$, $\cos \beta$, and $\cos \gamma$ are called the **direction cosines** of the vector $\mathbf{v}$.

EXAMPLE 8

Finding the Direction Angles of a Vector

Find the direction angles of $\mathbf{v} = -3\mathbf{i} + 2\mathbf{j} - 6\mathbf{k}$.

Solution

$$\|\mathbf{v}\| = \sqrt{(-3)^2 + 2^2 + (-6)^2} = \sqrt{49} = 7$$

Using the formulas in equation (5), we have

$$\cos \alpha = \frac{-3}{7} \qquad \cos \beta = \frac{2}{7} \qquad \cos \gamma = \frac{-6}{7}$$

$$\alpha \approx 115.4° \qquad \beta \approx 73.4° \qquad \gamma \approx 149.0°$$

THEOREM

> **Property of the Direction Cosines**
>
> If α, β, and γ are the direction angles of a nonzero vector $\mathbf{v}$ in space, then
>
> $$\cos^2 \alpha + \cos^2 \beta + \cos^2 \gamma = 1 \qquad \text{(6)}$$

The proof is a direct consequence of the equations in (5).

Based on equation (6), when two direction cosines are known, the third is determined up to its sign. Knowing two direction cosines is not sufficient to uniquely determine the direction of a vector in space.

EXAMPLE 9 **Finding the Direction Angle of a Vector**

The vector $\mathbf{v}$ makes an angle of $\alpha = \dfrac{\pi}{3}$ with the positive x-axis, an angle of $\beta = \dfrac{\pi}{3}$ with the positive y-axis, and an acute angle γ with the positive z-axis. Find γ.

Solution By equation (6), we have

$$\cos^2\left(\frac{\pi}{3}\right) + \cos^2\left(\frac{\pi}{3}\right) + \cos^2 \gamma = 1 \qquad 0 < \gamma < \frac{\pi}{2}$$

$$\left(\frac{1}{2}\right)^2 + \left(\frac{1}{2}\right)^2 + \cos^2 \gamma = 1$$

$$\cos^2 \gamma = \frac{1}{2}$$

$$\cos \gamma = \frac{\sqrt{2}}{2} \quad \text{or} \quad \cos \gamma = -\frac{\sqrt{2}}{2}$$

$$\gamma = \frac{\pi}{4} \quad \text{or} \quad \gamma = \frac{3\pi}{4}$$

Since we are requiring that γ be acute, $\gamma = \dfrac{\pi}{4}$.

The direction cosines of a vector give information about only the direction of the vector; they provide no information about its magnitude. For example, *any* vector parallel to the xy-plane and making an angle of $\dfrac{\pi}{4}$ radian with the positive x-axis and y-axis has direction cosines

$$\cos \alpha = \frac{\sqrt{2}}{2} \qquad \cos \beta = \frac{\sqrt{2}}{2} \qquad \cos \gamma = 0$$

However, if the direction angles *and* the magnitude of a vector are known, the vector is uniquely determined.

EXAMPLE 10 **Writing a Vector in Terms of Its Magnitude and Direction Cosines**

Show that any nonzero vector $\mathbf{v}$ in space can be written in terms of its magnitude and direction cosines as

$$\mathbf{v} = \|\mathbf{v}\|[(\cos \alpha)\mathbf{i} + (\cos \beta)\mathbf{j} + (\cos \gamma)\mathbf{k}] \qquad \text{(7)}$$

Solution Let $\mathbf{v} = a\mathbf{i} + b\mathbf{j} + c\mathbf{k}$. From the equations in (5), we see that

$$a = \|\mathbf{v}\| \cos \alpha \qquad b = \|\mathbf{v}\| \cos \beta \qquad c = \|\mathbf{v}\| \cos \gamma$$

Substituting, we find that

$$\mathbf{v} = a\mathbf{i} + b\mathbf{j} + c\mathbf{k} = \|\mathbf{v}\|(\cos\alpha)\mathbf{i} + \|\mathbf{v}\|(\cos\beta)\mathbf{j} + \|\mathbf{v}\|(\cos\gamma)\mathbf{k}$$
$$= \|\mathbf{v}\|[(\cos\alpha)\mathbf{i} + (\cos\beta)\mathbf{j} + (\cos\gamma)\mathbf{k}]$$

——**Now Work** PROBLEM 59

Example 10 shows that the direction cosines of a vector **v** are also the components of the unit vector in the direction of **v**.

8.6 Assess Your Understanding

'Are You Prepared?' *Answer is given at the end of these exercises. If you get the wrong answer, read the page listed in red.*

1. The distance d from $P_1 = (x_1, y_1)$ to $P_2 = (x_2, y_2)$ is $d = $ _____ (p. 3)

Concepts and Vocabulary

2. In space, points of the form $(x, y, 0)$ lie in a plane called the _____.

3. If $\mathbf{v} = a\mathbf{i} + b\mathbf{j} + c\mathbf{k}$ is a vector in space, the scalars a, b, c are called the _____ of **v**.

4. The sum of the squares of the direction cosines of a vector in space add up to _____.

5. **True or False** In space, the dot product of two vectors is a positive number.

6. **True or False** A vector in space may be described by specifying its magnitude and its direction angles.

Skill Building

In Problems 7– 14, describe the set of points (x, y, z) defined by the equation(s).

7. $y = 0$

8. $x = 0$

9. $z = 2$

10. $y = 3$

11. $x = -4$

12. $z = -3$

13. $x = 1$ and $y = 2$

14. $x = 3$ and $z = 1$

In Problems 15– 20, find the distance from P_1 to P_2.

15. $P_1 = (0, 0, 0)$ and $P_2 = (4, 1, 2)$

16. $P_1 = (0, 0, 0)$ and $P_2 = (1, -2, 3)$

17. $P_1 = (-1, 2, -3)$ and $P_2 = (0, -2, 1)$

18. $P_1 = (-2, 2, 3)$ and $P_2 = (4, 0, -3)$

19. $P_1 = (4, -2, -2)$ and $P_2 = (3, 2, 1)$

20. $P_1 = (2, -3, -3)$ and $P_2 = (4, 1, -1)$

In Problems 21– 26, opposite vertices of a rectangular box whose edges are parallel to the coordinate axes are given. List the coordinates of the other six vertices of the box.

21. $(0, 0, 0);\ \ (2, 1, 3)$

22. $(0, 0, 0);\ \ (4, 2, 2)$

23. $(1, 2, 3);\ \ (3, 4, 5)$

24. $(5, 6, 1);\ \ (3, 8, 2)$

25. $(-1, 0, 2);\ \ (4, 2, 5)$

26. $(-2, -3, 0);\ \ (-6, 7, 1)$

*In Problems 27– 32, the vector **v** has initial point P and terminal point Q. Write **v** in the form $a\mathbf{i} + b\mathbf{j} + c\mathbf{k}$; that is, find its position vector.*

27. $P = (0, 0, 0);\ \ Q = (3, 4, -1)$

28. $P = (0, 0, 0);\ \ Q = (-3, -5, 4)$

29. $P = (3, 2, -1);\ \ Q = (5, 6, 0)$

30. $P = (-3, 2, 0);\ \ Q = (6, 5, -1)$

31. $P = (-2, -1, 4);\ \ Q = (6, -2, 4)$

32. $P = (-1, 4, -2);\ \ Q = (6, 2, 2)$

In Problems 33– 38, find $\|\mathbf{v}\|$.

33. $\mathbf{v} = 3\mathbf{i} - 6\mathbf{j} - 2\mathbf{k}$

34. $\mathbf{v} = -6\mathbf{i} + 12\mathbf{j} + 4\mathbf{k}$

35. $\mathbf{v} = \mathbf{i} - \mathbf{j} + \mathbf{k}$

36. $\mathbf{v} = -\mathbf{i} - \mathbf{j} + \mathbf{k}$

37. $\mathbf{v} = -2\mathbf{i} + 3\mathbf{j} - 3\mathbf{k}$

38. $\mathbf{v} = 6\mathbf{i} + 2\mathbf{j} - 2\mathbf{k}$

In Problems 39– 44, find each quantity if $\mathbf{v} = 3\mathbf{i} - 5\mathbf{j} + 2\mathbf{k}$ and $\mathbf{w} = -2\mathbf{i} + 3\mathbf{j} - 2\mathbf{k}$.

39. $2\mathbf{v} + 3\mathbf{w}$

40. $3\mathbf{v} - 2\mathbf{w}$

41. $\|\mathbf{v} - \mathbf{w}\|$

42. $\|\mathbf{v} + \mathbf{w}\|$

43. $\|\mathbf{v}\| - \|\mathbf{w}\|$

44. $\|\mathbf{v}\| + \|\mathbf{w}\|$

In Problems 45–50, find the unit vector in the same direction as **v**.

45. $\mathbf{v} = 5\mathbf{i}$

46. $\mathbf{v} = -3\mathbf{j}$

47. $\mathbf{v} = 3\mathbf{i} - 6\mathbf{j} - 2\mathbf{k}$

48. $\mathbf{v} = -6\mathbf{i} + 12\mathbf{j} + 4\mathbf{k}$

49. $\mathbf{v} = \mathbf{i} + \mathbf{j} + \mathbf{k}$

50. $\mathbf{v} = 2\mathbf{i} - \mathbf{j} + \mathbf{k}$

In Problems 51–58, find the dot product **v · w** *and the angle between* **v** *and* **w**.

51. $\mathbf{v} = \mathbf{i} - \mathbf{j}, \quad \mathbf{w} = \mathbf{i} + \mathbf{j} + \mathbf{k}$

52. $\mathbf{v} = \mathbf{i} + \mathbf{j}, \quad \mathbf{w} = -\mathbf{i} + \mathbf{j} - \mathbf{k}$

53. $\mathbf{v} = 2\mathbf{i} + \mathbf{j} - 3\mathbf{k}, \quad \mathbf{w} = \mathbf{i} + 2\mathbf{j} + 2\mathbf{k}$

54. $\mathbf{v} = 2\mathbf{i} + 2\mathbf{j} - \mathbf{k}, \quad \mathbf{w} = \mathbf{i} + 2\mathbf{j} + 3\mathbf{k}$

55. $\mathbf{v} = 3\mathbf{i} - \mathbf{j} + 2\mathbf{k}, \quad \mathbf{w} = \mathbf{i} + \mathbf{j} - \mathbf{k}$

56. $\mathbf{v} = \mathbf{i} + 3\mathbf{j} + 2\mathbf{k}, \quad \mathbf{w} = \mathbf{i} - \mathbf{j} + \mathbf{k}$

57. $\mathbf{v} = 3\mathbf{i} + 4\mathbf{j} + \mathbf{k}, \quad \mathbf{w} = 6\mathbf{i} + 8\mathbf{j} + 2\mathbf{k}$

58. $\mathbf{v} = 3\mathbf{i} - 4\mathbf{j} + \mathbf{k}, \quad \mathbf{w} = 6\mathbf{i} - 8\mathbf{j} + 2\mathbf{k}$

In Problems 59–66, find the direction angles of each vector. Write each vector in the form of equation (7).

59. $\mathbf{v} = 3\mathbf{i} - 6\mathbf{j} - 2\mathbf{k}$

60. $\mathbf{v} = -6\mathbf{i} + 12\mathbf{j} + 4\mathbf{k}$

61. $\mathbf{v} = \mathbf{i} + \mathbf{j} + \mathbf{k}$

62. $\mathbf{v} = \mathbf{i} - \mathbf{j} - \mathbf{k}$

63. $\mathbf{v} = \mathbf{i} + \mathbf{j}$

64. $\mathbf{v} = \mathbf{j} + \mathbf{k}$

65. $\mathbf{v} = 3\mathbf{i} - 5\mathbf{j} + 2\mathbf{k}$

66. $\mathbf{v} = 2\mathbf{i} + 3\mathbf{j} - 4\mathbf{k}$

Applications and Extensions

67. Robotic Arm Consider the double-jointed robotic arm shown in the figure. Let the lower arm be modeled by $\mathbf{a} = \langle 2, 3, 4 \rangle$, the middle arm be modeled by $\mathbf{b} = \langle 1, -1, 3 \rangle$, and the upper arm by $\mathbf{c} = \langle 4, -1, -2 \rangle$, where units are in feet.

  (a) Find a vector **d** that represents the position of the hand.
  (b) Determine the distance of the hand from the origin.

68. The Sphere In space, the collection of all points that are the same distance from some fixed point is called a **sphere**. See the illustration. The constant distance is called the **radius,** and the fixed point is the **center** of the sphere. Show that the equation of a sphere with center at (x_0, y_0, z_0) and radius r is

$$(x - x_0)^2 + (y - y_0)^2 + (z - z_0)^2 = r^2$$

[Hint: Use the Distance Formula (1).]

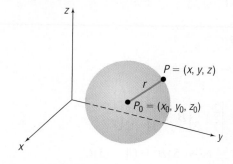

In Problems 69 and 70, find the equation of a sphere with radius r and center P_0.

69. $r = 1; \quad P_0 = (3, 1, 1)$

70. $r = 2; \quad P_0 = (1, 2, 2)$

In Problems 71–76, find the radius and center of each sphere.
[Hint: *Complete the square in each variable.]*

  71. $x^2 + y^2 + z^2 + 2x - 2y = 2$

  72. $x^2 + y^2 + z^2 + 2x - 2z = -1$

  73. $x^2 + y^2 + z^2 - 4x + 4y + 2z = 0$

  74. $x^2 + y^2 + z^2 - 4x = 0$

  75. $2x^2 + 2y^2 + 2z^2 - 8x + 4z = -1$

  76. $3x^2 + 3y^2 + 3z^2 + 6x - 6y = 3$

*The **work** W done by a constant force* **F** *in moving an object from a point A in space to a point B in space is defined as $W = \mathbf{F} \cdot \overrightarrow{AB}$. Use this definition in Problems 77–79.*

77. Work Find the work done by a force of 3 newtons acting in the direction $2\mathbf{i} + \mathbf{j} + 2\mathbf{k}$ in moving an object 2 meters from $(0, 0, 0)$ to $(0, 2, 0)$.

78. Work Find the work done by a force of 1 newton acting in the direction $2\mathbf{i} + 2\mathbf{j} + \mathbf{k}$ in moving an object 3 meters from $(0, 0, 0)$ to $(1, 2, 2)$.

79. Work Find the work done in moving an object along a vector $\mathbf{u} = 3\mathbf{i} + 2\mathbf{j} - 5\mathbf{k}$ if the applied force is $\mathbf{F} = 2\mathbf{i} - \mathbf{j} - \mathbf{k}$. Use meters for distance and newtons for force.

'Are You Prepared?' Answer

1. $\sqrt{(x_2 - x_1)^2 + (y_2 - y_1)^2}$

8.7 The Cross Product

1 Find the Cross Product of Two Vectors

For vectors in space, and only for vectors in space, a second product of two vectors is defined, called the *cross product*. The cross product of two vectors in space is, in fact, also a vector that has applications in both geometry and physics.

DEFINITION

If $\mathbf{v} = a_1\mathbf{i} + b_1\mathbf{j} + c_1\mathbf{k}$ and $\mathbf{w} = a_2\mathbf{i} + b_2\mathbf{j} + c_2\mathbf{k}$ are two vectors in space, the **cross product $\mathbf{v} \times \mathbf{w}$** is defined as the vector

$$\mathbf{v} \times \mathbf{w} = (b_1c_2 - b_2c_1)\mathbf{i} - (a_1c_2 - a_2c_1)\mathbf{j} + (a_1b_2 - a_2b_1)\mathbf{k} \quad \textbf{(1)}$$

Notice that the cross product $\mathbf{v} \times \mathbf{w}$ of two vectors is a vector. Because of this, it is sometimes referred to as the **vector product.**

EXAMPLE 1

Finding Cross Products Using Equation (1)

If $\mathbf{v} = 2\mathbf{i} + 3\mathbf{j} + 5\mathbf{k}$ and $\mathbf{w} = \mathbf{i} + 2\mathbf{j} + 3\mathbf{k}$, an application of equation (1) gives

$$\begin{aligned}
\mathbf{v} \times \mathbf{w} &= (3 \cdot 3 - 2 \cdot 5)\mathbf{i} - (2 \cdot 3 - 1 \cdot 5)\mathbf{j} + (2 \cdot 2 - 1 \cdot 3)\mathbf{k} \\
&= (9 - 10)\mathbf{i} - (6 - 5)\mathbf{j} + (4 - 3)\mathbf{k} \\
&= -\mathbf{i} - \mathbf{j} + \mathbf{k}
\end{aligned}$$

●

Determinants* may be used as an aid in computing cross products. A **2 by 2 determinant,** symbolized by

$$\begin{vmatrix} a_1 & b_1 \\ a_2 & b_2 \end{vmatrix}$$

has the value $a_1b_2 - a_2b_1$; that is,

$$\begin{vmatrix} a_1 & b_1 \\ a_2 & b_2 \end{vmatrix} = a_1b_2 - a_2b_1$$

A **3 by 3 determinant** has the value

$$\begin{vmatrix} A & B & C \\ a_1 & b_1 & c_1 \\ a_2 & b_2 & c_2 \end{vmatrix} = \begin{vmatrix} b_1 & c_1 \\ b_2 & c_2 \end{vmatrix} A - \begin{vmatrix} a_1 & c_1 \\ a_2 & c_2 \end{vmatrix} B + \begin{vmatrix} a_1 & b_1 \\ a_2 & b_2 \end{vmatrix} C$$

EXAMPLE 2

Evaluating Determinants

(a) $\begin{vmatrix} 2 & 3 \\ 1 & 2 \end{vmatrix} = 2 \cdot 2 - 1 \cdot 3 = 4 - 3 = 1$

(b) $\begin{vmatrix} A & B & C \\ 2 & 3 & 5 \\ 1 & 2 & 3 \end{vmatrix} = \begin{vmatrix} 3 & 5 \\ 2 & 3 \end{vmatrix} A - \begin{vmatrix} 2 & 5 \\ 1 & 3 \end{vmatrix} B + \begin{vmatrix} 2 & 3 \\ 1 & 2 \end{vmatrix} C$

$= (9 - 10)A - (6 - 5)B + (4 - 3)C$

$= -A - B + C$

●

Now Work PROBLEM **7**

*Determinants are discussed in detail in Section 10.3.

The cross product of the vectors $\mathbf{v} = a_1\mathbf{i} + b_1\mathbf{j} + c_1\mathbf{k}$ and $\mathbf{w} = a_2\mathbf{i} + b_2\mathbf{j} + c_2\mathbf{k}$, that is,

$$\mathbf{v} \times \mathbf{w} = (b_1 c_2 - b_2 c_1)\mathbf{i} - (a_1 c_2 - a_2 c_1)\mathbf{j} + (a_1 b_2 - a_2 b_1)\mathbf{k}$$

may be written symbolically using determinants as

$$\mathbf{v} \times \mathbf{w} = \begin{vmatrix} \mathbf{i} & \mathbf{j} & \mathbf{k} \\ a_1 & b_1 & c_1 \\ a_2 & b_2 & c_2 \end{vmatrix} = \begin{vmatrix} b_1 & c_1 \\ b_2 & c_2 \end{vmatrix}\mathbf{i} - \begin{vmatrix} a_1 & c_1 \\ a_2 & c_2 \end{vmatrix}\mathbf{j} + \begin{vmatrix} a_1 & b_1 \\ a_2 & b_2 \end{vmatrix}\mathbf{k}$$

EXAMPLE 3   **Using Determinants to Find Cross Products**

If $\mathbf{v} = 2\mathbf{i} + 3\mathbf{j} + 5\mathbf{k}$ and $\mathbf{w} = \mathbf{i} + 2\mathbf{j} + 3\mathbf{k}$, find:

(a) $\mathbf{v} \times \mathbf{w}$ (b) $\mathbf{w} \times \mathbf{v}$ (c) $\mathbf{v} \times \mathbf{v}$ (d) $\mathbf{w} \times \mathbf{w}$

Solution   (a) $\mathbf{v} \times \mathbf{w} = \begin{vmatrix} \mathbf{i} & \mathbf{j} & \mathbf{k} \\ 2 & 3 & 5 \\ 1 & 2 & 3 \end{vmatrix} = \begin{vmatrix} 3 & 5 \\ 2 & 3 \end{vmatrix}\mathbf{i} - \begin{vmatrix} 2 & 5 \\ 1 & 3 \end{vmatrix}\mathbf{j} + \begin{vmatrix} 2 & 3 \\ 1 & 2 \end{vmatrix}\mathbf{k} = -\mathbf{i} - \mathbf{j} + \mathbf{k}$

(b) $\mathbf{w} \times \mathbf{v} = \begin{vmatrix} \mathbf{i} & \mathbf{j} & \mathbf{k} \\ 1 & 2 & 3 \\ 2 & 3 & 5 \end{vmatrix} = \begin{vmatrix} 2 & 3 \\ 3 & 5 \end{vmatrix}\mathbf{i} - \begin{vmatrix} 1 & 3 \\ 2 & 5 \end{vmatrix}\mathbf{j} + \begin{vmatrix} 1 & 2 \\ 2 & 3 \end{vmatrix}\mathbf{k} = \mathbf{i} + \mathbf{j} - \mathbf{k}$

(c) $\mathbf{v} \times \mathbf{v} = \begin{vmatrix} \mathbf{i} & \mathbf{j} & \mathbf{k} \\ 2 & 3 & 5 \\ 2 & 3 & 5 \end{vmatrix}$

$= \begin{vmatrix} 3 & 5 \\ 3 & 5 \end{vmatrix}\mathbf{i} - \begin{vmatrix} 2 & 5 \\ 2 & 5 \end{vmatrix}\mathbf{j} + \begin{vmatrix} 2 & 3 \\ 2 & 3 \end{vmatrix}\mathbf{k} = 0\mathbf{i} - 0\mathbf{j} + 0\mathbf{k} = \mathbf{0}$

(d) $\mathbf{w} \times \mathbf{w} = \begin{vmatrix} \mathbf{i} & \mathbf{j} & \mathbf{k} \\ 1 & 2 & 3 \\ 1 & 2 & 3 \end{vmatrix}$

$= \begin{vmatrix} 2 & 3 \\ 2 & 3 \end{vmatrix}\mathbf{i} - \begin{vmatrix} 1 & 3 \\ 1 & 3 \end{vmatrix}\mathbf{j} + \begin{vmatrix} 1 & 2 \\ 1 & 2 \end{vmatrix}\mathbf{k} = 0\mathbf{i} - 0\mathbf{j} + 0\mathbf{k} = \mathbf{0}$

—**Now Work**   PROBLEM 15

2 Know Algebraic Properties of the Cross Product

Notice in Examples 3(a) and (b) that $\mathbf{v} \times \mathbf{w}$ and $\mathbf{w} \times \mathbf{v}$ are negatives of one another. From Examples 3(c) and (d), we might conjecture that the cross product of a vector with itself is the zero vector. These and other algebraic properties of the cross product are given next.

THEOREM   **Algebraic Properties of the Cross Product**

If $\mathbf{u}$, $\mathbf{v}$, and $\mathbf{w}$ are vectors in space and if α is a scalar, then

$\mathbf{u} \times \mathbf{u} = \mathbf{0}$	(2)
$\mathbf{u} \times \mathbf{v} = -(\mathbf{v} \times \mathbf{u})$	(3)
$\alpha(\mathbf{u} \times \mathbf{v}) = (\alpha\mathbf{u}) \times \mathbf{v} = \mathbf{u} \times (\alpha\mathbf{v})$	(4)
$\mathbf{u} \times (\mathbf{v} + \mathbf{w}) = (\mathbf{u} \times \mathbf{v}) + (\mathbf{u} \times \mathbf{w})$	(5)

Proof   We will prove properties (2) and (4) here and leave properties (3) and (5) as exercises (see Problems 60 and 61).

To prove property (2), we let $\mathbf{u} = a_1\mathbf{i} + b_1\mathbf{j} + c_1\mathbf{k}$. Then

$$\mathbf{u} \times \mathbf{u} = \begin{vmatrix} \mathbf{i} & \mathbf{j} & \mathbf{k} \\ a_1 & b_1 & c_1 \\ a_1 & b_1 & c_1 \end{vmatrix} = \begin{vmatrix} b_1 & c_1 \\ b_1 & c_1 \end{vmatrix}\mathbf{i} - \begin{vmatrix} a_1 & c_1 \\ a_1 & c_1 \end{vmatrix}\mathbf{j} + \begin{vmatrix} a_1 & b_1 \\ a_1 & b_1 \end{vmatrix}\mathbf{k}$$

$$= 0\mathbf{i} - 0\mathbf{j} + 0\mathbf{k} = \mathbf{0}$$

To prove property (4), we let $\mathbf{u} = a_1\mathbf{i} + b_1\mathbf{j} + c_1\mathbf{k}$ and $\mathbf{v} = a_2\mathbf{i} + b_2\mathbf{j} + c_2\mathbf{k}$. Then

$$\alpha(\mathbf{u} \times \mathbf{v}) = \alpha[(b_1c_2 - b_2c_1)\mathbf{i} - (a_1c_2 - a_2c_1)\mathbf{j} + (a_1b_2 - a_2b_1)\mathbf{k}]$$

$\uparrow$
Apply (1).

$$= \alpha(b_1c_2 - b_2c_1)\mathbf{i} - \alpha(a_1c_2 - a_2c_1)\mathbf{j} + \alpha(a_1b_2 - a_2b_1)\mathbf{k} \quad (6)$$

Since $\alpha\mathbf{u} = \alpha a_1\mathbf{i} + \alpha b_1\mathbf{j} + \alpha c_1\mathbf{k}$, we have

$$(\alpha\mathbf{u}) \times \mathbf{v} = (\alpha b_1c_2 - b_2\alpha c_1)\mathbf{i} - (\alpha a_1c_2 - a_2\alpha c_1)\mathbf{j} + (\alpha a_1b_2 - a_2\alpha b_1)\mathbf{k}$$
$$= \alpha(b_1c_2 - b_2c_1)\mathbf{i} - \alpha(a_1c_2 - a_2c_1)\mathbf{j} + \alpha(a_1b_2 - a_2b_1)\mathbf{k} \quad (7)$$

Based on equations (6) and (7), the first part of property (4) follows. The second part can be proved in like fashion. ∎

═══ **Now Work** PROBLEM **17**

3 Know Geometric Properties of the Cross Product

The cross product has several interesting geometric properties.

THEOREM

Geometric Properties of the Cross Product

Let $\mathbf{u}$ and $\mathbf{v}$ be vectors in space.

$\mathbf{u} \times \mathbf{v}$ is orthogonal to both $\mathbf{u}$ and $\mathbf{v}$.	(8)
$\|\mathbf{u} \times \mathbf{v}\| = \|\mathbf{u}\| \, \|\mathbf{v}\| \sin\theta$, where θ is the angle between $\mathbf{u}$ and $\mathbf{v}$.	(9)
$\|\mathbf{u} \times \mathbf{v}\|$ is the area of the parallelogram having $\mathbf{u} \neq \mathbf{0}$ and $\mathbf{v} \neq \mathbf{0}$ as adjacent sides.	(10)
$\mathbf{u} \times \mathbf{v} = \mathbf{0}$ if and only if $\mathbf{u}$ and $\mathbf{v}$ are parallel.	(11)

Figure 81

Figure 82

Proof of Property (8) Let $\mathbf{u} = a_1\mathbf{i} + b_1\mathbf{j} + c_1\mathbf{k}$ and $\mathbf{v} = a_2\mathbf{i} + b_2\mathbf{j} + c_2\mathbf{k}$. Then

$$\mathbf{u} \times \mathbf{v} = (b_1c_2 - b_2c_1)\mathbf{i} - (a_1c_2 - a_2c_1)\mathbf{j} + (a_1b_2 - a_2b_1)\mathbf{k}$$

Now we compute the dot product $\mathbf{u} \cdot (\mathbf{u} \times \mathbf{v})$.

$$\mathbf{u} \cdot (\mathbf{u} \times \mathbf{v}) = (a_1\mathbf{i} + b_1\mathbf{j} + c_1\mathbf{k}) \cdot [(b_1c_2 - b_2c_1)\mathbf{i} - (a_1c_2 - a_2c_1)\mathbf{j} + (a_1b_2 - a_2b_1)\mathbf{k}]$$
$$= a_1(b_1c_2 - b_2c_1) - b_1(a_1c_2 - a_2c_1) + c_1(a_1b_2 - a_2b_1) = 0$$

Since two vectors are orthogonal if their dot product is zero, it follows that $\mathbf{u}$ and $\mathbf{u} \times \mathbf{v}$ are orthogonal. Similarly, $\mathbf{v} \cdot (\mathbf{u} \times \mathbf{v}) = 0$, so $\mathbf{v}$ and $\mathbf{u} \times \mathbf{v}$ are orthogonal. ∎

4 Find a Vector Orthogonal to Two Given Vectors

As long as the vectors $\mathbf{u}$ and $\mathbf{v}$ are not parallel, they will form a plane in space. See Figure 81. Based on property (8), the vector $\mathbf{u} \times \mathbf{v}$ is normal to this plane. As Figure 81 illustrates, there are essentially (without regard to magnitude) two vectors normal to the plane containing $\mathbf{u}$ and $\mathbf{v}$. It can be shown that the vector $\mathbf{u} \times \mathbf{v}$ is the one determined by the thumb of the right hand when the other fingers of the right hand are cupped so that they point in a direction from **u** to **v**. See Figure 82.*

EXAMPLE 4

Finding a Vector Orthogonal to Two Given Vectors

Find a vector that is orthogonal to $\mathbf{u} = 3\mathbf{i} - 2\mathbf{j} + \mathbf{k}$ and $\mathbf{v} = -\mathbf{i} + 3\mathbf{j} - \mathbf{k}$.

Solution   Based on property (8), such a vector is $\mathbf{u} \times \mathbf{v}$.

$$\mathbf{u} \times \mathbf{v} = \begin{vmatrix} \mathbf{i} & \mathbf{j} & \mathbf{k} \\ 3 & -2 & 1 \\ -1 & 3 & -1 \end{vmatrix} = (2 - 3)\mathbf{i} - [-3 - (-1)]\mathbf{j} + (9 - 2)\mathbf{k} = -\mathbf{i} + 2\mathbf{j} + 7\mathbf{k}$$

The vector $-\mathbf{i} + 2\mathbf{j} + 7\mathbf{k}$ is orthogonal to both **u** and **v**.

✓**Check:** Two vectors are orthogonal if their dot product is zero.

$$\mathbf{u} \cdot (-\mathbf{i} + 2\mathbf{j} + 7\mathbf{k}) = (3\mathbf{i} - 2\mathbf{j} + \mathbf{k}) \cdot (-\mathbf{i} + 2\mathbf{j} + 7\mathbf{k}) = -3 - 4 + 7 = 0$$
$$\mathbf{v} \cdot (-\mathbf{i} + 2\mathbf{j} + 7\mathbf{k}) = (-\mathbf{i} + 3\mathbf{j} - \mathbf{k}) \cdot (-\mathbf{i} + 2\mathbf{j} + 7\mathbf{k}) = 1 + 6 - 7 = 0$$

Now Work PROBLEM 41

The proof of property (9) is left as an exercise. See Problem 62.

5 Find the Area of a Parallelogram

Figure 83

Proof of Property (10)   Suppose that **u** and **v** are adjacent sides of a parallelogram. See Figure 83. Then the lengths of these sides are $\|\mathbf{u}\|$ and $\|\mathbf{v}\|$. If θ is the angle between **u** and **v**, then the height of the parallelogram is $\|\mathbf{v}\| \sin \theta$ and its area is

$$\text{Area of parallelogram} = \text{Base} \times \text{Height} = \|\mathbf{u}\|[\|\mathbf{v}\| \sin \theta] = \|\mathbf{u} \times \mathbf{v}\|$$
$$\uparrow$$
$$\text{Property (9)}$$

■

EXAMPLE 5

Finding the Area of a Parallelogram

Find the area of the parallelogram whose vertices are $P_1 = (0, 0, 0)$, $P_2 = (3, -2, 1)$, $P_3 = (-1, 3, -1)$, and $P_4 = (2, 1, 0)$.

Solution   Two adjacent sides of this parallelogram are

$$\mathbf{u} = \overrightarrow{P_1P_2} = 3\mathbf{i} - 2\mathbf{j} + \mathbf{k} \quad \text{and} \quad \mathbf{v} = \overrightarrow{P_1P_3} = -\mathbf{i} + 3\mathbf{j} - \mathbf{k}$$

WARNING Not all pairs of vertices give rise to a side. For example, $\overrightarrow{P_1P_4}$ is a diagonal of the parallelogram since $\overrightarrow{P_1P_3} + \overrightarrow{P_3P_4} = \overrightarrow{P_1P_4}$. Also, $\overrightarrow{P_1P_3}$ and $\overrightarrow{P_2P_4}$ are not adjacent sides; they are parallel sides. ■

Since $\mathbf{u} \times \mathbf{v} = -\mathbf{i} + 2\mathbf{j} + 7\mathbf{k}$ (Example 4), the area of the parallelogram is

$$\text{Area of parallelogram} = \|\mathbf{u} \times \mathbf{v}\| = \sqrt{1 + 4 + 49} = \sqrt{54} = 3\sqrt{6} \text{ square units}$$

Now Work PROBLEM 49

Proof of Property (11)   The proof requires two parts. If **u** and **v** are parallel, then there is a scalar α such that $\mathbf{u} = \alpha\mathbf{v}$. Then

$$\mathbf{u} \times \mathbf{v} = (\alpha\mathbf{v}) \times \mathbf{v} = \alpha(\mathbf{v} \times \mathbf{v}) = 0$$
$$\uparrow \qquad\qquad \uparrow$$
$$\text{Property (4)} \quad \text{Property (2)}$$

If $\mathbf{u} \times \mathbf{v} = \mathbf{0}$, then, by property (9), we have

$$\|\mathbf{u} \times \mathbf{v}\| = \|\mathbf{u}\| \|\mathbf{v}\| \sin \theta = 0$$

Since $\mathbf{u} \neq \mathbf{0}$ and $\mathbf{v} \neq \mathbf{0}$, we must have $\sin \theta = 0$, so $\theta = 0$ or $\theta = \pi$. In either case, since θ is the angle between **u** and **v**, then **u** and **v** are parallel. ■

*This is a consequence of using a right-handed coordinate system.

8.7 Assess Your Understanding

Concepts and Vocabulary

1. **True or False** If **u** and **v** are parallel vectors, then $\mathbf{u} \times \mathbf{v} = \mathbf{0}$.

2. **True or False** For any vector **v**, $\mathbf{v} \times \mathbf{v} = \mathbf{0}$.

3. **True or False** If **u** and **v** are vectors, then $\mathbf{u} \times \mathbf{v} + \mathbf{v} \times \mathbf{u} = \mathbf{0}$.

4. **True or False** $\mathbf{u} \times \mathbf{v}$ is a vector that is parallel to both **u** and **v**.

5. **True or False** $\|\mathbf{u} \times \mathbf{v}\| = \|\mathbf{u}\| \|\mathbf{v}\| \cos \theta$, where θ is the angle between **u** and **v**.

6. **True or False** The area of the parallelogram having **u** and **v** as adjacent sides is the magnitude of the cross product of **u** and **v**.

Skill Building

In Problems 7–14, find the value of each determinant.

7. $\begin{vmatrix} 3 & 4 \\ 1 & 2 \end{vmatrix}$

8. $\begin{vmatrix} -2 & 5 \\ 2 & -3 \end{vmatrix}$

9. $\begin{vmatrix} 6 & 5 \\ -2 & -1 \end{vmatrix}$

10. $\begin{vmatrix} -4 & 0 \\ 5 & 3 \end{vmatrix}$

11. $\begin{vmatrix} A & B & C \\ 2 & 1 & 4 \\ 1 & 3 & 1 \end{vmatrix}$

12. $\begin{vmatrix} A & B & C \\ 0 & 2 & 4 \\ 3 & 1 & 3 \end{vmatrix}$

13. $\begin{vmatrix} A & B & C \\ -1 & 3 & 5 \\ 5 & 0 & -2 \end{vmatrix}$

14. $\begin{vmatrix} A & B & C \\ 1 & -2 & -3 \\ 0 & 2 & -2 \end{vmatrix}$

In Problems 15–22, find (a) $\mathbf{v} \times \mathbf{w}$, (b) $\mathbf{w} \times \mathbf{v}$, (c) $\mathbf{w} \times \mathbf{w}$, and (d) $\mathbf{v} \times \mathbf{v}$.

15. $\mathbf{v} = 2\mathbf{i} - 3\mathbf{j} + \mathbf{k}$
    $\mathbf{w} = 3\mathbf{i} - 2\mathbf{j} - \mathbf{k}$

16. $\mathbf{v} = -\mathbf{i} + 3\mathbf{j} + 2\mathbf{k}$
    $\mathbf{w} = 3\mathbf{i} - 2\mathbf{j} - \mathbf{k}$

17. $\mathbf{v} = \mathbf{i} + \mathbf{j}$
    $\mathbf{w} = 2\mathbf{i} + \mathbf{j} + \mathbf{k}$

18. $\mathbf{v} = \mathbf{i} - 4\mathbf{j} + 2\mathbf{k}$
    $\mathbf{w} = 3\mathbf{i} + 2\mathbf{j} + \mathbf{k}$

19. $\mathbf{v} = 2\mathbf{i} - \mathbf{j} + 2\mathbf{k}$
    $\mathbf{w} = \mathbf{j} - \mathbf{k}$

20. $\mathbf{v} = 3\mathbf{i} + \mathbf{j} + 3\mathbf{k}$
    $\mathbf{w} = \mathbf{i} - \mathbf{k}$

21. $\mathbf{v} = \mathbf{i} - \mathbf{j} - \mathbf{k}$
    $\mathbf{w} = 4\mathbf{i} - 3\mathbf{k}$

22. $\mathbf{v} = 2\mathbf{i} - 3\mathbf{j}$
    $\mathbf{w} = 3\mathbf{j} - 2\mathbf{k}$

*In Problems 23–44, use the given vectors **u**, **v**, and **w** to find each expression.*

$$\mathbf{u} = 2\mathbf{i} - 3\mathbf{j} + \mathbf{k} \qquad \mathbf{v} = -3\mathbf{i} + 3\mathbf{j} + 2\mathbf{k} \qquad \mathbf{w} = \mathbf{i} + \mathbf{j} + 3\mathbf{k}$$

23. $\mathbf{u} \times \mathbf{v}$

24. $\mathbf{v} \times \mathbf{w}$

25. $\mathbf{v} \times \mathbf{u}$

26. $\mathbf{w} \times \mathbf{v}$

27. $\mathbf{v} \times \mathbf{v}$

28. $\mathbf{w} \times \mathbf{w}$

29. $(3\mathbf{u}) \times \mathbf{v}$

30. $\mathbf{v} \times (4\mathbf{w})$

31. $\mathbf{u} \times (2\mathbf{v})$

32. $(-3\mathbf{v}) \times \mathbf{w}$

33. $\mathbf{u} \cdot (\mathbf{u} \times \mathbf{v})$

34. $\mathbf{v} \cdot (\mathbf{v} \times \mathbf{w})$

35. $\mathbf{u} \cdot (\mathbf{v} \times \mathbf{w})$

36. $(\mathbf{u} \times \mathbf{v}) \cdot \mathbf{w}$

37. $\mathbf{v} \cdot (\mathbf{u} \times \mathbf{w})$

38. $(\mathbf{v} \times \mathbf{u}) \cdot \mathbf{w}$

39. $\mathbf{u} \times (\mathbf{v} \times \mathbf{v})$

40. $(\mathbf{w} \times \mathbf{w}) \times \mathbf{v}$

41. Find a vector orthogonal to both **u** and **v**.

42. Find a vector orthogonal to both **u** and **w**.

43. Find a vector orthogonal to both **u** and $\mathbf{i} + \mathbf{j}$.

44. Find a vector orthogonal to both **u** and $\mathbf{j} + \mathbf{k}$.

In Problems 45–48, find the area of the parallelogram with one corner at P_1 and adjacent sides $\overrightarrow{P_1P_2}$ and $\overrightarrow{P_1P_3}$.

45. $P_1 = (0, 0, 0)$, $P_2 = (1, 2, 3)$, $P_3 = (-2, 3, 0)$

46. $P_1 = (0, 0, 0)$, $P_2 = (2, 3, 1)$, $P_3 = (-2, 4, 1)$

47. $P_1 = (1, 2, 0)$, $P_2 = (-2, 3, 4)$, $P_3 = (0, -2, 3)$

48. $P_1 = (-2, 0, 2)$, $P_2 = (2, 1, -1)$, $P_3 = (2, -1, 2)$

In Problems 49–52, find the area of the parallelogram with vertices P_1, P_2, P_3, and P_4.

49. $P_1 = (1, 1, 2)$, $P_2 = (1, 2, 3)$, $P_3 = (-2, 3, 0)$,
    $P_4 = (-2, 4, 1)$

50. $P_1 = (2, 1, 1)$, $P_2 = (2, 3, 1)$, $P_3 = (-2, 4, 1)$,
    $P_4 = (-2, 6, 1)$

51. $P_1 = (1, 2, -1)$, $P_2 = (4, 2, -3)$, $P_3 = (6, -5, 2)$,
    $P_4 = (9, -5, 0)$

52. $P_1 = (-1, 1, 1)$, $P_2 = (-1, 2, 2)$, $P_3 = (-3, 4, -5)$,
    $P_4 = (-3, 5, -4)$

Applications and Extensions

53. Find a unit vector normal to the plane containing $\mathbf{v} = \mathbf{i} + 3\mathbf{j} - 2\mathbf{k}$ and $\mathbf{w} = -2\mathbf{i} + \mathbf{j} + 3\mathbf{k}$.

54. Find a unit vector normal to the plane containing $\mathbf{v} = 2\mathbf{i} + 3\mathbf{j} - \mathbf{k}$ and $\mathbf{w} = -2\mathbf{i} - 4\mathbf{j} - 3\mathbf{k}$.

55. Volume of a Parallelepiped A **parallelepiped** is a prism whose faces are all parallelograms. Let $\mathbf{A}$, $\mathbf{B}$, and $\mathbf{C}$ be the vectors that define the parallelepiped shown in the figure. The volume V of the parallelepiped is given by the formula $V = |(\mathbf{A} \times \mathbf{B}) \cdot \mathbf{C}|$.

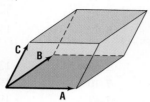

Find the volume of a parallelepiped if the defining vectors are $\mathbf{A} = 3\mathbf{i} - 2\mathbf{j} + 4\mathbf{k}$, $\mathbf{B} = 2\mathbf{i} + \mathbf{j} - 2\mathbf{k}$, and $\mathbf{C} = 3\mathbf{i} - 6\mathbf{j} - 2\mathbf{k}$.

56. Volume of a Parallelepiped Refer to Problem 55. Find the volume of a parallelepiped whose defining vectors are $\mathbf{A} = \langle 1, 0, 6 \rangle$, $\mathbf{B} = \langle 2, 3, -8 \rangle$, and $\mathbf{C} = \langle 8, -5, 6 \rangle$.

57. Prove for vectors $\mathbf{u}$ and $\mathbf{v}$ that

$$\|\mathbf{u} \times \mathbf{v}\|^2 = \|\mathbf{u}\|^2 \|\mathbf{v}\|^2 - (\mathbf{u} \cdot \mathbf{v})^2$$

[**Hint:** Proceed as in the proof of property (4), computing first the left side and then the right side.]

58. Show that if $\mathbf{u}$ and $\mathbf{v}$ are orthogonal then

$$\|\mathbf{u} \times \mathbf{v}\| = \|\mathbf{u}\| \|\mathbf{v}\|$$

59. Show that if $\mathbf{u}$ and $\mathbf{v}$ are orthogonal unit vectors then $\mathbf{u} \times \mathbf{v}$ is also a unit vector.

60. Prove property (3).

61. Prove property (5).

62. Prove property (9).

[**Hint:** Use the result of Problem 57 and the fact that if θ is the angle between $\mathbf{u}$ and $\mathbf{v}$ then $\mathbf{u} \cdot \mathbf{v} = \|\mathbf{u}\| \|\mathbf{v}\| \cos \theta$.]

Explaining Concepts: Discussion and Writing

63. If $\mathbf{u} \cdot \mathbf{v} = 0$ and $\mathbf{u} \times \mathbf{v} = \mathbf{0}$, what, if anything, can you conclude about $\mathbf{u}$ and $\mathbf{v}$?

CHAPTER REVIEW

Things to Know

Polar Coordinates (pp. 594–621)

Relationship between polar coordinates (r, θ) and rectangular coordinates (x, y) (pp. 594 and 597)	$x = r \cos \theta$, $y = r \sin \theta$ $r^2 = x^2 + y^2$, $\tan \theta = \dfrac{y}{x}$, $x \neq 0$
Polar form of a complex number (p. 617)	If $z = x + yi$, then $z = r(\cos \theta + i \sin \theta)$, where $r = |z| = \sqrt{x^2 + y^2}$, $\sin \theta = \dfrac{y}{r}$, $\cos \theta = \dfrac{x}{r}$, $0 \leq \theta < 2\pi$.
De Moivre's Theorem (p. 619)	If $z = r(\cos \theta + i \sin \theta)$, then $z^n = r^n[\cos(n\theta) + i \sin(n\theta)]$, where $n \geq 1$ is a positive integer.
nth root of a complex number $w = r(\cos \theta_0 + i \sin \theta_0)$ (p. 620)	$z_k = \sqrt[n]{r}\left[\cos\left(\dfrac{\theta_0}{n} + \dfrac{2k\pi}{n}\right) + i \sin\left(\dfrac{\theta_0}{n} + \dfrac{2k\pi}{n}\right)\right]$, $k = 0, \ldots, n - 1$, where $n \geq 2$ is an integer

Vectors (pp. 623–657)

	Quantity having magnitude and direction; equivalent to a directed line segment $\overrightarrow{PQ}$
Position vector (pp. 627 and 646)	Vector whose initial point is at the origin
Unit vector (pp. 626 and 648)	Vector whose magnitude is 1
Dot product (pp. 637 and 648)	If $\mathbf{v} = a_1\mathbf{i} + b_1\mathbf{j}$ and $\mathbf{w} = a_2\mathbf{i} + b_2\mathbf{j}$, then $\mathbf{v} \cdot \mathbf{w} = a_1 a_2 + b_1 b_2$. If $\mathbf{v} = a_1\mathbf{i} + b_1\mathbf{j} + c_1\mathbf{k}$ and $\mathbf{w} = a_2\mathbf{i} + b_2\mathbf{j} + c_2\mathbf{k}_2$, then $\mathbf{v} \cdot \mathbf{w} = a_1 a_2 + b_1 b_2 + c_1 c_2$.
Angle θ between two nonzero vectors $\mathbf{u}$ and $\mathbf{v}$ (pp. 639 and 649)	$\cos \theta = \dfrac{\mathbf{u} \cdot \mathbf{v}}{\|\mathbf{u}\| \|\mathbf{v}\|}$
Direction angles of a vector in space (p. 650)	If $\mathbf{v} = a\mathbf{i} + b\mathbf{j} + c\mathbf{k}$, then $\mathbf{v} = \|\mathbf{v}\|[(\cos \alpha)\mathbf{i} + (\cos \beta)\mathbf{j} + (\cos \gamma)\mathbf{k}]$, where $\cos \alpha = \dfrac{a}{\|\mathbf{v}\|}$, $\cos \beta = \dfrac{b}{\|\mathbf{v}\|}$, $\cos \gamma = \dfrac{c}{\|\mathbf{v}\|}$.
Cross product (p. 654)	If $\mathbf{v} = a_1\mathbf{i} + b_1\mathbf{j} + c_1\mathbf{k}$ and $\mathbf{w} = a_2\mathbf{i} + b_2\mathbf{j} + c_2\mathbf{k}$, then $\mathbf{v} \times \mathbf{w} = [b_1 c_2 - b_2 c_1]\mathbf{i} - [a_1 c_2 - a_2 c_1]\mathbf{j} + [a_1 b_2 - a_2 b_1]\mathbf{k}$.
Area of a parallelogram (p. 656)	$\|\mathbf{u} \times \mathbf{v}\| = \|\mathbf{u}\| \|\mathbf{v}\| \sin \theta$, where θ is the angle between $\mathbf{u}$ and $\mathbf{v}$.

Objectives

Section		You should be able to . . .	Example(s)	Review Exercises
8.1	1	Plot points using polar coordinates (p. 592)	1–3	1–6
	2	Convert from polar coordinates to rectangular coordinates (p. 594)	4	1–6
	3	Convert from rectangular coordinates to polar coordinates (p. 596)	5–7	7–12
	4	Transform equations between polar and rectangular forms (p. 598)	8, 9	13(a)–18(a)
8.2	1	Graph and identify polar equations by converting to rectangular equations (p. 602)	1–6	13(b)–18(b)
	2	Test polar equations for symmetry (p. 605)	7–10	19–24
	3	Graph polar equations by plotting points (p. 606)	7–12	19–24
8.3	1	Plot points in the complex plane (p. 616)	1	29–34
	2	Convert a complex number between rectangular form and polar form (p. 617)	2, 3	25–34
	3	Find products and quotients of complex numbers in polar form (p. 618)	4	35–40
	4	Use De Moivre's Theorem (p. 619)	5, 6	41–48
	5	Find complex roots (p. 620)	7	49–50
8.4	1	Graph vectors (p. 626)	1	51–54
	2	Find a position vector (p. 626)	2	55–58
	3	Add and subtract vectors algebraically (p. 628)	3	59, 60
	4	Find a scalar multiple and the magnitude of a vector (p. 629)	4	55–58, 61–66
	5	Find a unit vector (p. 629)	5	67, 68
	6	Find a vector from its direction and magnitude (p. 630)	6	69, 70
	7	Model with vectors (p. 631)	8–10	111–113
8.5	1	Find the dot product of two vectors (p. 637)	1	87–90
	2	Find the angle between two vectors (p. 638)	2	87–90
	3	Determine whether two vectors are parallel (p. 639)	3	95–100
	4	Determine whether two vectors are orthogonal (p. 639)	4	95–100
	5	Decompose a vector into two orthogonal vectors (p. 640)	5, 6	101–104, 115
	6	Compute work (p. 642)	7	114
8.6	1	Find the distance between two points in space (p. 646)	1	73, 74
	2	Find position vectors in space (p. 646)	2	75, 76
	3	Perform operations on vectors (p. 647)	3–5	77–82, 85
	4	Find the dot product (p. 648)	6	91–94
	5	Find the angle between two vectors (p. 649)	7	91–94
	6	Find the direction angles of a vector (p. 649)	8–10	105, 106
8.7	1	Find the cross product of two vectors (p. 654)	1–3	83, 84
	2	Know algebraic properties of the cross product (p. 655)	p. 649	109, 110
	3	Know geometric properties of the cross product (p. 656)	p. 650	107, 108
	4	Find a vector orthogonal to two given vectors (p. 656)	4	86
	5	Find the area of a parallelogram (p. 657)	5	107, 108

Review Exercises

In Problems 1–6, plot each point given in polar coordinates, and find its rectangular coordinates.

1. $\left(3, \frac{\pi}{6}\right)$ **2.** $\left(4, \frac{2\pi}{3}\right)$ **3.** $\left(-2, \frac{4\pi}{3}\right)$ **4.** $\left(-1, \frac{5\pi}{4}\right)$ **5.** $\left(-3, -\frac{\pi}{2}\right)$ **6.** $\left(-4, -\frac{\pi}{4}\right)$

In Problems 7–12, the rectangular coordinates of a point are given. Find two pairs of polar coordinates (r, θ) for each point, one with $r > 0$ and the other with $r < 0$. Express θ in radians.

7. $(-3, 3)$ **8.** $(1, -1)$ **9.** $(0, -2)$ **10.** $(2, 0)$ **11.** $(3, 4)$ **12.** $(-5, 12)$

In Problems 13– 18, the variables r and θ represent polar coordinates. (a) Write each polar equation as an equation in rectangular coordinates (x, y). (b) Identify the equation and graph it.

13. $r = 2 \sin \theta$

14. $3r = \sin \theta$

15. $r = 5$

16. $\theta = \dfrac{\pi}{4}$

17. $r \cos \theta + 3r \sin \theta = 6$

18. $r^2 + 4r \sin \theta - 8r \cos \theta = 5$

In Problems 19– 24, sketch the graph of each polar equation. Be sure to test for symmetry.

19. $r = 4 \cos \theta$

20. $r = 3 \sin \theta$

21. $r = 3 - 3 \sin \theta$

22. $r = 2 + \cos \theta$

23. $r = 4 - \cos \theta$

24. $r = 1 - 2 \sin \theta$

In Problems 25– 28, write each complex number in polar form. Express each argument in degrees.

25. $-1 - i$

26. $-\sqrt{3} + i$

27. $4 - 3i$

28. $3 - 2i$

In Problems 29– 34, write each complex number in the standard form a + bi and plot each in the complex plane.

29. $2(\cos 150° + i \sin 150°)$

30. $3(\cos 60° + i \sin 60°)$

31. $3\left(\cos \dfrac{2\pi}{3} + i \sin \dfrac{2\pi}{3} \right)$

32. $4\left(\cos \dfrac{3\pi}{4} + i \sin \dfrac{3\pi}{4} \right)$

33. $0.1(\cos 350° + i \sin 350°)$

34. $0.5(\cos 160° + i \sin 160°)$

In Problems 35– 40, find zw and $\dfrac{z}{w}$. Leave your answers in polar form.

35. $z = \cos 80° + i \sin 80°$
$w = \cos 50° + i \sin 50°$

36. $z = \cos 205° + i \sin 205°$
$w = \cos 85° + i \sin 85°$

37. $z = 3\left(\cos \dfrac{9\pi}{5} + i \sin \dfrac{9\pi}{5} \right)$
$w = 2\left(\cos \dfrac{\pi}{5} + i \sin \dfrac{\pi}{5} \right)$

38. $z = 2\left(\cos \dfrac{5\pi}{3} + i \sin \dfrac{5\pi}{3} \right)$
$w = 3\left(\cos \dfrac{\pi}{3} + i \sin \dfrac{\pi}{3} \right)$

39. $z = 5(\cos 10° + i \sin 10°)$
$w = \cos 355° + i \sin 355°$

40. $z = 4(\cos 50° + i \sin 50°)$
$w = \cos 340° + i \sin 340°$

In Problems 41– 48, write each expression in the standard form a + bi.

41. $[3(\cos 20° + i \sin 20°)]^3$

42. $[2(\cos 50° + i \sin 50°)]^3$

43. $\left[\sqrt{2} \left(\cos \dfrac{5\pi}{8} + i \sin \dfrac{5\pi}{8} \right) \right]^4$

44. $\left[2\left(\cos \dfrac{5\pi}{16} + i \sin \dfrac{5\pi}{16} \right) \right]^4$

45. $\left(1 - \sqrt{3}i \right)^6$

46. $(2 - 2i)^8$

47. $(3 + 4i)^4$

48. $(1 - 2i)^4$

49. Find all the complex cube roots of 27.

50. Find all the complex fourth roots of -16.

In Problems 51–54, use the figure to graph each of the following:

51. $\mathbf{u} + \mathbf{v}$

52. $\mathbf{v} + \mathbf{w}$

53. $2\mathbf{u} + 3\mathbf{v}$

54. $5\mathbf{v} - 2\mathbf{w}$

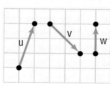

*In Problems 55–58, the vector **v** is represented by the directed line segment $\overrightarrow{PQ}$. Write **v** in the form a**i** + b**j** and find $\|\mathbf{v}\|$.*

55. $P = (1, -2); \quad Q = (3, -6)$

56. $P = (-3, 1); \quad Q = (4, -2)$

57. $P = (0, -2); \quad Q = (-1, 1)$

58. $P = (3, -4); \quad Q = (-2, 0)$

In Problems 59– 68, use the vectors $\mathbf{v} = -2\mathbf{i} + \mathbf{j}$ and $\mathbf{w} = 4\mathbf{i} - 3\mathbf{j}$ to find:

59. $\mathbf{v} + \mathbf{w}$

60. $\mathbf{v} - \mathbf{w}$

61. $4\mathbf{v} - 3\mathbf{w}$

62. $-\mathbf{v} + 2\mathbf{w}$

63. $\|\mathbf{v}\|$

64. $\|\mathbf{v} + \mathbf{w}\|$

65. $\|\mathbf{v}\| + \|\mathbf{w}\|$

66. $\|2\mathbf{v}\| - 3\|\mathbf{w}\|$

67. Find a unit vector in the same direction as **v**.

68. Find a unit vector in the opposite direction of **w**.

69. Find the vector $\mathbf{v}$ in the xy-plane with magnitude 3 if the angle between $\mathbf{v}$ and $\mathbf{i}$ is $60°$.

70. Find the vector $\mathbf{v}$ in the xy-plane with magnitude 5 if the angle between $\mathbf{v}$ and $\mathbf{i}$ is $150°$.

71. Find the direction angle α between $\mathbf{i}$ and $\mathbf{v} = -\mathbf{i} + \sqrt{3}\,\mathbf{j}$.

72. Find the direction angle between $\mathbf{i}$ and $\mathbf{v} = 2\mathbf{i} - 6\mathbf{j}$.

73. Find the distance from $P_1 = (1, 3, -2)$ to $P_2 = (4, -2, 1)$.

74. Find the distance from $P_1 = (0, -4, 3)$ to $P_2 = (6, -5, -1)$.

75. A vector $\mathbf{v}$ has initial point $P = (1, 3, -2)$ and terminal point $Q = (4, -2, 1)$. Write $\mathbf{v}$ in the form $\mathbf{v} = a\mathbf{i} + b\mathbf{j} + c\mathbf{k}$.

76. A vector $\mathbf{v}$ has initial point $P = (0, -4, 3)$ and terminal point $Q = (6, -5, -1)$. Write $\mathbf{v}$ in the form $\mathbf{v} = a\mathbf{i} + b\mathbf{j} + c\mathbf{k}$.

In Problems 77–86, use the vectors $\mathbf{v} = 3\mathbf{i} + \mathbf{j} - 2\mathbf{k}$ and $\mathbf{w} = -3\mathbf{i} + 2\mathbf{j} - \mathbf{k}$ to find each expression.

77. $4\mathbf{v} - 3\mathbf{w}$

78. $-\mathbf{v} + 2\mathbf{w}$

79. $\|\mathbf{v} - \mathbf{w}\|$

80. $\|\mathbf{v} + \mathbf{w}\|$

81. $\|\mathbf{v}\| - \|\mathbf{w}\|$

82. $\|\mathbf{v}\| + \|\mathbf{w}\|$

83. $\mathbf{v} \times \mathbf{w}$

84. $\mathbf{v} \cdot (\mathbf{v} \times \mathbf{w})$

85. Find a unit vector in the same direction as $\mathbf{v}$ and then in the opposite direction of $\mathbf{v}$.

86. Find a unit vector orthogonal to both $\mathbf{v}$ and $\mathbf{w}$.

In Problems 87–94, find the dot product $\mathbf{v} \cdot \mathbf{w}$ and the angle between $\mathbf{v}$ and $\mathbf{w}$.

87. $\mathbf{v} = -2\mathbf{i} + \mathbf{j}, \quad \mathbf{w} = 4\mathbf{i} - 3\mathbf{j}$

88. $\mathbf{v} = 3\mathbf{i} - \mathbf{j}, \quad \mathbf{w} = \mathbf{i} + \mathbf{j}$

89. $\mathbf{v} = \mathbf{i} - 3\mathbf{j}, \quad \mathbf{w} = -\mathbf{i} + \mathbf{j}$

90. $\mathbf{v} = \mathbf{i} + 4\mathbf{j}, \quad \mathbf{w} = 3\mathbf{i} - 2\mathbf{j}$

91. $\mathbf{v} = \mathbf{i} + \mathbf{j} + \mathbf{k}, \quad \mathbf{w} = \mathbf{i} - \mathbf{j} + \mathbf{k}$

92. $\mathbf{v} = \mathbf{i} - \mathbf{j} + \mathbf{k}, \quad \mathbf{w} = 2\mathbf{i} + \mathbf{j} + \mathbf{k}$

93. $\mathbf{v} = 4\mathbf{i} - \mathbf{j} + 2\mathbf{k}, \quad \mathbf{w} = \mathbf{i} - 2\mathbf{j} - 3\mathbf{k}$

94. $\mathbf{v} = -\mathbf{i} - 2\mathbf{j} + 3\mathbf{k}, \quad \mathbf{w} = 5\mathbf{i} + \mathbf{j} + \mathbf{k}$

In Problems 95–100, determine whether $\mathbf{v}$ and $\mathbf{w}$ are parallel, orthogonal, or neither.

95. $\mathbf{v} = 2\mathbf{i} + 3\mathbf{j}; \quad \mathbf{w} = -4\mathbf{i} - 6\mathbf{j}$

96. $\mathbf{v} = -2\mathbf{i} - \mathbf{j}; \quad \mathbf{w} = 2\mathbf{i} + \mathbf{j}$

97. $\mathbf{v} = 3\mathbf{i} - 4\mathbf{j}; \quad \mathbf{w} = -3\mathbf{i} + 4\mathbf{j}$

98. $\mathbf{v} = -2\mathbf{i} + 2\mathbf{j}; \quad \mathbf{w} = -3\mathbf{i} + 2\mathbf{j}$

99. $\mathbf{v} = 3\mathbf{i} - 2\mathbf{j}; \quad \mathbf{w} = 4\mathbf{i} + 6\mathbf{j}$

100. $\mathbf{v} = -4\mathbf{i} + 2\mathbf{j}; \quad \mathbf{w} = 2\mathbf{i} + 4\mathbf{j}$

In Problems 101–104, decompose $\mathbf{v}$ into two vectors, one parallel to $\mathbf{w}$ and the other orthogonal to $\mathbf{w}$.

101. $\mathbf{v} = 2\mathbf{i} + \mathbf{j}; \quad \mathbf{w} = -4\mathbf{i} + 3\mathbf{j}$

102. $\mathbf{v} = -3\mathbf{i} + 2\mathbf{j}; \quad \mathbf{w} = -2\mathbf{i} + \mathbf{j}$

103. $\mathbf{v} = 2\mathbf{i} + 3\mathbf{j}; \quad \mathbf{w} = 3\mathbf{i} + \mathbf{j}$

104. $\mathbf{v} = -\mathbf{i} + 2\mathbf{j}; \quad \mathbf{w} = 3\mathbf{i} - \mathbf{j}$

---

105. Find the direction angles of the vector $\mathbf{v} = 3\mathbf{i} - 4\mathbf{j} + 2\mathbf{k}$.

106. Find the direction angles of the vector $\mathbf{v} = \mathbf{i} - \mathbf{j} + 2\mathbf{k}$.

107. Find the area of the parallelogram with vertices $P_1 = (1, 1, 1)$, $P_2 = (2, 3, 4)$, $P_3 = (6, 5, 2)$, and $P_4 = (7, 7, 5)$.

108. Find the area of the parallelogram with vertices $P_1 = (2, -1, 1)$, $P_2 = (5, 1, 4)$, $P_3 = (0, 1, 1)$, and $P_4 = (3, 3, 4)$.

109. If $\mathbf{u} \times \mathbf{v} = 2\mathbf{i} - 3\mathbf{j} + \mathbf{k}$, what is $\mathbf{v} \times \mathbf{u}$?

110. Suppose that $\mathbf{u} = 3\mathbf{v}$. What is $\mathbf{u} \times \mathbf{v}$?

111. Actual Speed and Direction of a Swimmer   A swimmer can maintain a constant speed of 5 miles per hour. If the swimmer heads directly across a river that has a current moving at the rate of 2 miles per hour, what is the actual speed of the swimmer? (See the figure.) If the river is 1 mile wide, how far down-

stream will the swimmer end up from the point directly across the river from the starting point?

112. Actual Speed and Direction of an Airplane   An airplane has an airspeed of 500 kilometers per hour (k/hr) in a northerly direction. The wind velocity is 60 (k/hr) in a southeasterly direction. Find the actual speed and direction of the plane relative to the ground.

113. Static Equilibrium   A weight of 2000 pounds is suspended from two cables, as shown in the figure. What are the tensions in each cable?

114. Computing Work   Find the work done by a force of 5 pounds acting in the direction $60°$ to the horizontal in moving an object 20 feet from $(0, 0)$ to $(20, 0)$.

115. Braking Load   A moving van with a gross weight of 8000 pounds is parked on a street with a $5°$ grade. Find the magnitude of the force required to keep the van from rolling down the hill. What is the magnitude of the force perpendicular to the hill?

CHAPTER TEST

CHAPTER
Test Prep
VIDEOS

Step-by-step test solutions are found on the Chapter Test Prep Videos available via the Video Resources on DVD, in *MyMathLab* or on YouTube (search "Sullivan Precalculus: A Right Triangle Approach, Concepts Through Functions 2e).

In Problems 1–3, plot each point given in polar coordinates.

1. $\left(2, \dfrac{3\pi}{4}\right)$

2. $\left(3, -\dfrac{\pi}{6}\right)$

3. $\left(-4, \dfrac{\pi}{3}\right)$

4. Convert $\left(2, 2\sqrt{3}\right)$ from rectangular coordinates to polar coordinates (r, θ), where $r > 0$ and $0 \le \theta < 2\pi$.

In Problems 5–7, convert the polar equation to a rectangular equation. Graph the equation.

5. $r = 7$

6. $\tan \theta = 3$

7. $r \sin^2 \theta + 8 \sin \theta = r$

In Problems 8 and 9, test the polar equation for symmetry with respect to the pole, the polar axis, and the line $\theta = \dfrac{\pi}{2}$.

8. $r^2 \cos \theta = 5$

9. $r = 5 \sin \theta \cos^2 \theta$

In Problems 10–12, perform the given operation, where $z = 2(\cos 85° + i \sin 85°)$ and $w = 3(\cos 22° + i \sin 22°)$. Write your answer in polar form.

10. $z \cdot w$

11. $\dfrac{w}{z}$

12. w^5

13. Find all the complex cube roots of $-8 + 8\sqrt{3}i$. Then plot them in rectangular coordinates.

In Problems 14–18, $P_1 = \left(3\sqrt{2}, 7\sqrt{2}\right)$ and $P_2 = \left(8\sqrt{2}, 2\sqrt{2}\right)$.

14. Find the position vector $\mathbf{v}$ equal to $\overrightarrow{P_1 P_2}$.

15. Find $\|\mathbf{v}\|$.

16. Find the unit vector in the direction of $\mathbf{v}$.

17. Find the angle between $\mathbf{v}$ and $\mathbf{i}$.

18. Decompose $\mathbf{v}$ into its vertical and horizontal components.

In Problems 19–22, $\mathbf{v}_1 = \langle 4, 6 \rangle$, $\mathbf{v}_2 = \langle -3, -6 \rangle$, $\mathbf{v}_3 = \langle -8, 4 \rangle$, and $\mathbf{v}_4 = \langle 10, 15 \rangle$.

19. Find the vector $\mathbf{v}_1 + 2\mathbf{v}_2 - \mathbf{v}_3$.

20. Which two vectors are parallel?

21. Which two vectors are orthogonal?

22. Find the angle between vectors $\mathbf{v}_1$ and $\mathbf{v}_2$.

In Problems 23–25, use the vectors $\mathbf{u} = 2\mathbf{i} - 3\mathbf{j} + \mathbf{k}$ and $\mathbf{v} = -\mathbf{i} + 3\mathbf{j} + 2\mathbf{k}$.

23. Find $\mathbf{u} \times \mathbf{v}$.

24. Find the direction angles for $\mathbf{u}$.

25. Find the area of the parallelogram that has $\mathbf{u}$ and $\mathbf{v}$ as adjacent sides.

26. A 1200-pound chandelier is to be suspended over a large ballroom; the chandelier will be hung on two cables of equal length whose ends will be attached to the ceiling, 16 feet apart. The chandelier will be free hanging so that the ends of the cable will make equal angles with the ceiling. If the top of the chandelier is to be 16 feet from the ceiling, what is the minimum tension each cable must be able to endure?

CUMULATIVE REVIEW

1. Find the real solutions, if any, of the equation $e^{x^2-9} = 1$.

2. Find an equation for the line containing the origin that makes an angle of $30°$ with the positive x-axis.

3. Find an equation for the circle with center at the point $(0, 1)$ and radius 3. Graph this circle.

4. What is the domain of the function $f(x) = \ln(1 - 2x)$?

5. Test the equation $x^2 + y^3 = 2x^4$ for symmetry with respect to the x-axis, the y-axis, and the origin.

6. Graph the function $y = |\ln x|$.

7. Graph the function $y = |\sin x|$.

8. Graph the function $y = \sin|x|$.

9. Find the exact value of $\sin^{-1}\left(-\dfrac{1}{2}\right)$.

10. Graph the equations $x = 3$ and $y = 4$ on the same set of rectangular coordinates.

11. Graph the equations $r = 2$ and $\theta = \dfrac{\pi}{3}$ on the same set of polar coordinates.

12. What is the amplitude and period of $y = -4\cos(\pi x)$?

CHAPTER PROJECTS

Source: *www.aeromuseum.org/eduHowtoFly.html*

I. **Modeling Aircraft Motion** Four aerodynamic forces act on an airplane in flight lift, weight, thrust, and drag. While an aircraft is in flight, these four forces continuously battle each other. Weight opposes lift and drag opposes thrust. See the figure. In balanced flight at constant speed, both the lift and weight are equal and the thrust and drag are equal.

1. What will happen to the aircraft if the lift is held constant while the weight is decreased (say from burning off fuel)?

2. What will happen to the aircraft if the lift is decreased while the weight is held constant?

3. What will happen to the aircraft if the thrust is increased while the drag is held constant?

4. What will happen to the aircraft if the drag is increased while the thrust is held constant?

In 1903 the Wright brothers made the first controlled powered flight. The weight of their plane was approximately 700 pounds (lb). Newton's Second Law of motion states that force = mass × acceleration ($F = ma$). If the mass is measured in kilograms (kg) and acceleration in meters per second squared (m/sec^2), then the force will be measured in newtons (N). [**Note:** $1N = 1\text{kg} \cdot \text{m/sec}^2$.]

5. If 1 kg = 2.205 lb, convert the weight of the Wright brother's plane to kilograms.

6. If acceleration due to gravity is $a = 9.80$ m/sec^2, determine the force due to weight on the Wright brothers' plane.

7. What must be true about the lift force of the Wright brothers, plane for it to get off the ground?

8. The weight of a fully loaded Cessna 170B is 2200 lb. What lift force is required to get this plane off the ground?

9. The maximum gross weight of a Boeing 747 is 255,000 lb. What lift force is required to get this jet off the ground?

The following projects are available at the Instructors' Resource Center (IRC):

II. **Project at Motorola** *Signal Fades due to Interference* Complex trigonometric functions are used to assure that a cellphone has optimal reception as the user travels up and down an elevator.

III. **Compound Interest** The effect of continuously compounded interest is analyzed using polar coordinates.

IV. **Complex Equations** Analysis of complex equations illustrates the connections between complex and real equations. At times, using complex equations is more efficient for proving mathematical theorems.

Analytic Geometry

9

Outline

Pluto's Unusual Orbit

Pluto is about 39 times as far from the Sun as Earth is. Its average distance from the Sun is about 3,647,240,000 miles (5,869,660,000 kilometers). Pluto travels around the Sun in an elliptical (oval-shaped) orbit. At some point in its orbit, it comes closer to the Sun than Neptune, the outermost planet. It stays inside Neptune's orbit for about 20 Earth-years. This event occurs every 248 Earth-years, which is about the same number of Earth-years it takes Pluto to travel once around the Sun. Pluto entered Neptune's orbit on January 23, 1979, and remained there until February 11, 1999.

Source: Adapted from: Gulbis, Amanda A. S. "Pluto." World Book Online Reference Center. © 2007. http://www.worldbookonline.com. By permission of the publisher. All rights reserved.

—See Chapter Project I—

◁ **A Look Back** In the Foundations chapter, we introduced rectangular coordinates and showed how geometry problems can be solved algebraically. We defined a circle geometrically and then used the distance formula and rectangular coordinates to obtain an equation for a circle.

A Look Ahead ▷ In this chapter, we give geometric definitions for the conics and use the distance formula and rectangular coordinates to obtain their equations.

Historically, Apollonius (200 BC) was among the first to study *conics* and discover some of their interesting properties. Today, conics are still studied because of their many uses. *Paraboloids of revolution* (parabolas rotated about their axes of symmetry) are used as signal collectors (the satellite dishes used with radar and dish TV, for example), as solar energy collectors, and as reflectors (telescopes, light projection, and so on). The planets circle the Sun in approximately *elliptical* orbits. Elliptical surfaces can be used to reflect signals such as light and sound from one place to another. A third conic, the *hyperbola,* can be used to determine the location of lightning strikes.

The Greeks used the methods of Euclidean geometry to study conics. However, we shall use the more powerful methods of analytic geometry, which uses both algebra and geometry, for our study of conics.

9.1 Conics

1 Know the Names of the Conics

The word *conic* derives from the word *cone*, which is a geometric figure that can be constructed in the following way: Let *a* and *g* be two distinct lines that intersect at a point *V*. Keep the line *a* fixed. Now rotate the line *g* about *a* while maintaining the same angle between *a* and *g*. The collection of points swept out (generated) by the line *g* is called a **(right circular) cone.** See Figure 1. The fixed line *a* is called the **axis** of the cone; the point *V* is its **vertex;** the lines that pass through *V* and make the same angle with *a* as *g* are **generators** of the cone. Each generator is a line that lies entirely on the cone. The cone consists of two parts, called **nappes,** that intersect at the vertex.

Figure 1

Axis, *a*

Generators

Vertex, *V*

g

Conics, an abbreviation for **conic sections,** are curves that result from the intersection of a right circular cone and a plane. The conics we shall study arise when the plane does not contain the vertex, as shown in Figure 2. These conics are **circles** when the plane is perpendicular to the axis of the cone and intersects each generator; **ellipses** when the plane is tilted slightly so that it intersects each generator, but intersects only one nappe of the cone; **parabolas** when the plane is tilted farther so that it is parallel to one (and only one) generator and intersects only one nappe of the cone; and **hyperbolas** when the plane intersects both nappes.

If the plane does contain the vertex, the intersection of the plane and the cone is a point, a line, or a pair of intersecting lines. These are usually called **degenerate conics.**

Figure 2

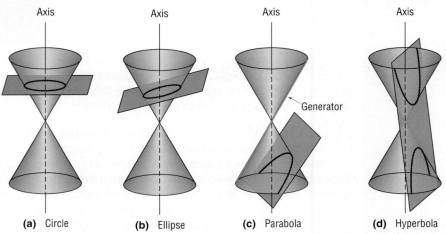

Axis   Axis   Axis   Axis

Generator

(a) Circle   **(b)** Ellipse   **(c)** Parabola   **(d)** Hyperbola

Conic sections are used in modeling many different applications. For example parabolas are used in describing satellite dishes and telescopes (see Figures 14 and 15 on page 672). Ellipses are used to model the orbits of planets and whispering chambers (see pages 682–683). And hyperbolas are used to model lightning strikes and nuclear cooling towers (see Problems 76 and 77 in Section 9.4).

9.2 The Parabola

PREPARING FOR THIS SECTION *Before getting started, review the following:*

- Distance Formula (Foundations, Section 1, p. 3)
- Symmetry (Foundations, Section 2, pp. 12–14)
- Square Root Method (Section 2.3, pp. 142–143)
- Completing the Square (Appendix A, Section A.4, p. A40)
- Graphing Techniques: Transformations (Section 1.5, pp. 90–99)

Now Work the 'Are You Prepared?' problems on page 673.

OBJECTIVES 1 Analyze Parabolas with Vertex at the Origin (p. 667)
2 Analyze Parabolas with Vertex at (h, k) (p. 670)
3 Solve Applied Problems Involving Parabolas (p. 672)

We stated earlier (Section 2.4) that the graph of a quadratic function is a parabola. In this section, we begin with a geometric definition of a parabola and use it to obtain an equation.

DEFINITION

A **parabola** is the collection of all points P in the plane that are the same distance d from a fixed point F as they are from a fixed line D. The point F is called the **focus** of the parabola, and the line D is its **directrix.** As a result, a parabola is the set of points P for which

$$d(F, P) = d(P, D) \qquad \text{(1)}$$

Figure 3

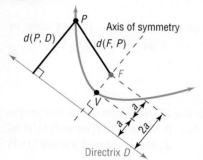

Figure 3 shows a parabola (in blue). The line through the focus F and perpendicular to the directrix D is called the **axis of symmetry** of the parabola. The point of intersection of the parabola with its axis of symmetry is called the **vertex** V.

Because the vertex V lies on the parabola, it must satisfy equation (1): $d(F, V) = d(V, D)$. The vertex is midway between the focus and the directrix. We shall let a equal the distance $d(F, V)$ from F to V. Now we are ready to derive an equation for a parabola. To do this, we use a rectangular system of coordinates positioned so that the vertex V, focus F, and directrix D of the parabola are conveniently located.

1 Analyze Parabolas with Vertex at the Origin

If we choose to locate the vertex V at the origin $(0, 0)$, we can conveniently position the focus F on either the x-axis or the y-axis. First, we consider the case where the focus F is on the positive x-axis, as shown in Figure 4. Because the distance from F to V is a, the coordinates of F will be $(a, 0)$ with $a > 0$. Similarly, because the distance from V to the directrix D is also a and, because D must be perpendicular to the x-axis (since the x-axis is the axis of symmetry), the equation of the directrix D must be $x = -a$.

Now, if $P = (x, y)$ is any point on the parabola, P must obey equation (1):

$$d(F, P) = d(P, D)$$

Figure 4

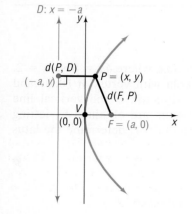

So we have

$$\sqrt{(x - a)^2 + (y - 0)^2} = |x + a| \qquad \text{Use the Distance Formula.}$$
$$(x - a)^2 + y^2 = (x + a)^2 \qquad \text{Square both sides.}$$
$$x^2 - 2ax + a^2 + y^2 = x^2 + 2ax + a^2 \qquad \text{Remove parentheses.}$$
$$y^2 = 4ax \qquad \text{Simplify.}$$

THEOREM

Equation of a Parabola: Vertex at (0, 0), Focus at (a, 0), a > 0

The equation of a parabola with vertex at $(0, 0)$, focus at $(a, 0)$, and directrix $x = -a, a > 0$, is

$$y^2 = 4ax \qquad\qquad (2)$$

Recall that a is the distance from the vertex to the focus of a parabola. When graphing the parabola $y^2 = 4ax$ it is helpful to determine the "opening" by finding the points that lie directly above or below the focus $(a, 0)$. This is done by letting $x = a$ in $y^2 = 4ax$, so $y^2 = 4a(a) = 4a^2$, or $y = \pm 2a$. The line segment joining these two points, $(a, 2a)$ and $(a, -2a)$, is called the **latus rectum**; its length is $4a$.

EXAMPLE 1 | **Finding the Equation of a Parabola and Graphing It**

Find an equation of the parabola with vertex at $(0, 0)$ and focus at $(3, 0)$. Graph the equation.

Solution

Figure 5

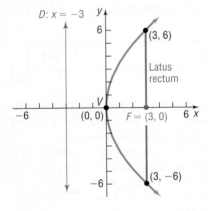

The distance from the vertex $(0, 0)$ to the focus $(3, 0)$ is $a = 3$. Based on equation (2), the equation of this parabola is

$$y^2 = 4ax$$
$$y^2 = 12x \quad a = 3$$

To graph this parabola, we find the two points that determine the latus rectum by letting $x = 3$. Then

$$y^2 = 12x = 12(3) = 36$$
$$y = \pm 6 \qquad \text{Solve for y.}$$

The points $(3, 6)$ and $(3, -6)$ determine the latus rectum. These points help in graphing the parabola because they determine the "opening." See Figure 5.

Now Work PROBLEM 19

COMMENT To graph the parabola $y^2 = 12x$ discussed in Example 1, we need to graph the two functions $Y_1 = \sqrt{12x}$ and $Y_2 = -\sqrt{12x}$. Do this and compare what you see with Figure 5. ∎

By reversing the steps we used to obtain equation (2), it follows that the graph of an equation of the form of equation (2), $y^2 = 4ax$, is a parabola; its vertex is at $(0, 0)$, its focus is at $(a, 0)$, its directrix is the line $x = -a$, and its axis of symmetry is the x-axis.

For the remainder of this section, the direction **"Analyze the equation"** will mean to find the vertex, focus, and directrix of the parabola and graph it.

EXAMPLE 2 | **Analyzing the Equation of a Parabola**

Figure 6

Analyze the equation: $y^2 = 8x$

Solution The equation $y^2 = 8x$ is of the form $y^2 = 4ax$, where $4a = 8$, so $a = 2$. Consequently, the graph of the equation is a parabola with vertex at $(0, 0)$ and focus on the positive x-axis at $(a, 0) = (2, 0)$. The directrix is the vertical line $x = -2$. The two points that determine the latus rectum are obtained by letting $x = 2$. Then $y^2 = 16$, so $y = \pm 4$. The points $(2, -4)$ and $(2, 4)$ determine the latus rectum. See Figure 6 for the graph.

Recall that we arrived at equation (2) after placing the focus on the positive *x*-axis. If the focus is placed on the negative *x*-axis, positive *y*-axis, or negative *y*-axis, a different form of the equation for the parabola results. The four forms of the equation of a parabola with vertex at $(0, 0)$ and focus on a coordinate axis a distance *a* from $(0, 0)$ are given in Table 1, and their graphs are given in Figure 7. Notice that each graph is symmetric with respect to its axis of symmetry.

Table 1

Equations of a Parabola: Vertex at (0, 0); Focus on an Axis; *a* > 0				
Vertex	**Focus**	**Directrix**	**Equation**	**Description**
$(0, 0)$	$(a, 0)$	$x = -a$	$y^2 = 4ax$	Axis of symmetry is the *x*-axis, opens right
$(0, 0)$	$(-a, 0)$	$x = a$	$y^2 = -4ax$	Axis of symmetry is the *x*-axis, opens left
$(0, 0)$	$(0, a)$	$y = -a$	$x^2 = 4ay$	Axis of symmetry is the *y*-axis, opens up
$(0, 0)$	$(0, -a)$	$y = a$	$x^2 = -4ay$	Axis of symmetry is the *y*-axis, opens down

Figure 7

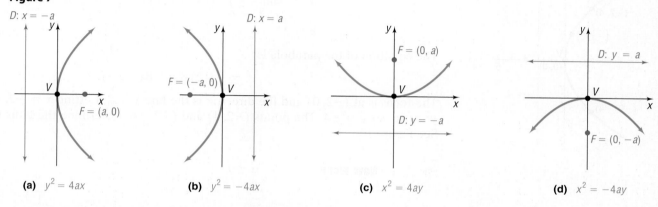

(a) $y^2 = 4ax$ (b) $y^2 = -4ax$ (c) $x^2 = 4ay$ (d) $x^2 = -4ay$

EXAMPLE 3 **Analyzing the Equation of a Parabola**

Analyze the equation: $x^2 = -12y$

Figure 8

Solution The equation $x^2 = -12y$ is of the form $x^2 = -4ay$, with $a = 3$. Consequently, the graph of the equation is a parabola with vertex at $(0, 0)$, focus at $(0, -3)$, and directrix the line $y = 3$. The parabola opens down, and its axis of symmetry is the *y*-axis. To obtain the points defining the latus rectum, let $y = -3$. Then $x^2 = 36$, so $x = \pm 6$. The points $(-6, -3)$ and $(6, -3)$ determine the latus rectum. See Figure 8 for the graph.

➡ **Now Work** PROBLEM 39

EXAMPLE 4 **Finding the Equation of a Parabola**

Find the equation of the parabola with focus at $(0, 4)$ and directrix the line $y = -4$. Graph the equation.

Figure 9

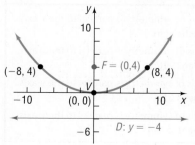

Solution A parabola whose focus is at $(0, 4)$ and whose directrix is the horizontal line $y = -4$ will have its vertex at $(0, 0)$. (Do you see why? The vertex is midway between the focus and the directrix.) Since the focus is on the positive *y*-axis at $(0, 4)$, the equation of this parabola is of the form $x^2 = 4ay$, with $a = 4$; that is,

$$x^2 = 4ay = 4(4)y = 16y$$
$$\uparrow$$
$$a = 4$$

The points $(8, 4)$ and $(-8, 4)$ determine the latus rectum. Figure 9 shows the graph of $x^2 = 16y$.

| EXAMPLE 5 | Finding the Equation of a Parabola |

Find the equation of a parabola with vertex at $(0,0)$ if its axis of symmetry is the x-axis and its graph contains the point $\left(-\frac{1}{2}, 2\right)$. Find its focus and directrix, and graph the equation.

Solution
The vertex is at the origin, the axis of symmetry is the x-axis, and the graph contains a point in the second quadrant, so the parabola opens to the left. We see from Table 1 that the form of the equation is

$$y^2 = -4ax$$

Because the point $\left(-\frac{1}{2}, 2\right)$ is on the parabola, the coordinates $x = -\frac{1}{2}$, $y = 2$ must satisfy $y^2 = -4ax$. Substituting $x = -\frac{1}{2}$ and $y = 2$ into this equation, we find that

$$4 = -4a\left(-\frac{1}{2}\right) \qquad y^2 = -4ax; \ x = -\frac{1}{2}, y = 2$$

$$a = 2$$

The equation of the parabola is

$$y^2 = -4(2)x = -8x$$

The focus is at $(-2, 0)$ and the directrix is the line $x = 2$. Letting $x = -2$, we find $y^2 = 16$, so $y = \pm 4$. The points $(-2, 4)$ and $(-2, -4)$ determine the latus rectum. See Figure 10.

Figure 10

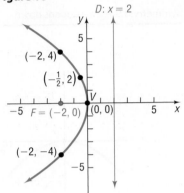

Now Work PROBLEM 27

2 Analyze Parabolas with Vertex at (h, k)

If a parabola with vertex at the origin and axis of symmetry along a coordinate axis is shifted horizontally h units and then vertically k units, the result is a parabola with vertex at (h, k) and axis of symmetry parallel to a coordinate axis. The equations of such parabolas have the same forms as those in Table 1, but with x replaced by $x - h$ (the horizontal shift) and y replaced by $y - k$ (the vertical shift). Table 2 gives the forms of the equations of such parabolas. Figures 11(a)–(d) illustrate the graphs for $h > 0, k > 0$.

NOTE It is not recommended that Table 2 be memorized. Rather use the ideas of transformations (shift horizontally h units, vertically k units) along with the fact that a represents the distance from the vertex to the focus to determine the various components of a parabola. It is also helpful to understand that parabolas of the form "$x^2 = $" will open up or down, while parabolas of the form "$y^2 = $" will open left or right. ∎

Table 2

Equations of a Parabola: Vertex at (h, k); Axis of Symmetry Parallel to a Coordinate Axis; $a > 0$				
Vertex	**Focus**	**Directrix**	**Equation**	**Description**
(h, k)	$(h + a, k)$	$x = h - a$	$(y - k)^2 = 4a(x - h)$	Axis of symmetry is parallel to the x-axis, opens right
(h, k)	$(h - a, k)$	$x = h + a$	$(y - k)^2 = -4a(x - h)$	Axis of symmetry is parallel to the x-axis, opens left
(h, k)	$(h, k + a)$	$y = k - a$	$(x - h)^2 = 4a(y - k)$	Axis of symmetry is parallel to the y-axis, opens up
(h, k)	$(h, k - a)$	$y = k + a$	$(x - h)^2 = -4a(y - k)$	Axis of symmetry is parallel to the y-axis, opens down

Figure 11

(a) $(y - k)^2 = 4a(x - h)$ **(b)** $(y - k)^2 = -4a(x - h)$ **(c)** $(x - h)^2 = 4a(y - k)$ **(d)** $(x - h)^2 = -4a(y - k)$

EXAMPLE 6   **Finding the Equation of a Parabola, Vertex Not at the Origin**

Find an equation of the parabola with vertex at $(-2, 3)$ and focus at $(0, 3)$. Graph the equation.

Solution

Figure 12

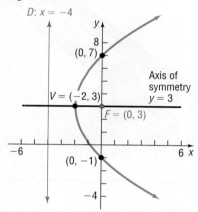

The vertex $(-2, 3)$ and focus $(0, 3)$ both lie on the horizontal line $y = 3$ (the axis of symmetry). The distance a from the vertex $(-2, 3)$ to the focus $(0, 3)$ is $a = 2$. Also, because the focus lies to the right of the vertex, we know that the parabola opens to the right. Consequently, the form of the equation is

$$(y - k)^2 = 4a(x - h)$$

where $(h, k) = (-2, 3)$ and $a = 2$. Therefore, the equation is

$$(y - 3)^2 = 4 \cdot 2[x - (-2)]$$
$$(y - 3)^2 = 8(x + 2)$$

To find the points that define the latus rectum, we let $x = 0$, so that $(y - 3)^2 = 16$. Then $y - 3 = \pm 4$, so $y = -1$ or $y = 7$. The points $(0, -1)$ and $(0, 7)$ determine the latus rectum; the line $x = -4$ is the directrix. See Figure 12.

Now Work PROBLEM 29

Polynomial equations define parabolas whenever they involve two variables that are quadratic in one variable and linear in the other. To analyze this type of equation, we first complete the square of the variable that is squared.

EXAMPLE 7   **Analyzing the Equation of a Parabola**

Analyze the equation: $x^2 + 4x - 4y = 0$

Solution

Figure 13

To analyze the equation $x^2 + 4x - 4y = 0$, we complete the square involving the variable x.

$$x^2 + 4x - 4y = 0$$
$$x^2 + 4x = 4y \qquad \text{Isolate the terms involving x on the left side.}$$
$$x^2 + 4x + 4 = 4y + 4 \qquad \text{Complete the square on the left side.}$$
$$(x + 2)^2 = 4(y + 1) \qquad \text{Factor.}$$

This equation is of the form $(x - h)^2 = 4a(y - k)$, with $h = -2$, $k = -1$, and $a = 1$. The graph is a parabola with vertex at $(h, k) = (-2, -1)$ that opens up. The focus is at $(-2, 0)$, and the directrix is the line $y = -2$. See Figure 13.

Now Work PROBLEM 47

3 Solve Applied Problems Involving Parabolas

Parabolas find their way into many applications. For example, as discussed in Section 2.4, suspension bridges have cables in the shape of a parabola. Another property of parabolas that is used in applications is their reflecting property.

Suppose that a mirror is shaped like a **paraboloid of revolution,** a surface formed by rotating a parabola about its axis of symmetry. If a light (or any other emitting source) is placed at the focus of the parabola, all the rays emanating from the light will reflect off the mirror in lines parallel to the axis of symmetry. This principle is used in the design of searchlights, flashlights, certain automobile headlights, and other such devices. See Figure 14.

Conversely, suppose that rays of light (or other signals) emanate from a distant source so that they are essentially parallel. When these rays strike the surface of a parabolic mirror whose axis of symmetry is parallel to these rays, they are reflected to a single point at the focus. This principle is used in the design of some solar energy devices, satellite dishes, and the mirrors used in some types of telescopes. See Figure 15.

Figure 14
Search Light

Figure 15
Telescope

EXAMPLE 8   **Satellite Dish**

A satellite dish is shaped like a paraboloid of revolution. The signals that emanate from a satellite strike the surface of the dish and are reflected to a single point, where the receiver is located. If the dish is 8 feet across at its opening and 3 feet deep at its center, at what position should the receiver be placed? That is, where is the focus?

Solution   Figure 16(a) shows the satellite dish. We draw the parabola used to form the dish on a rectangular coordinate system so that the vertex of the parabola is at the origin and its focus is on the positive *y*-axis. See Figure 16(b).

Figure 16

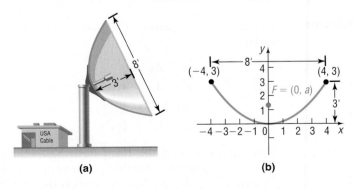

(a)   (b)

The form of the equation of the parabola is

$$x^2 = 4ay$$

and its focus is at $(0, a)$. Since $(4, 3)$ is a point on the graph, we have

$$4^2 = 4a(3) \quad \text{\small $x^2 = 4ay; x = 4, y = 3$}$$

$$a = \frac{4}{3} \quad \text{\small Solve for a.}$$

The receiver should be located $1\frac{1}{3}$ feet (1 foot, 4 inches) from the base of the dish, along its axis of symmetry.

Now Work PROBLEM 63

9.2 Assess Your Understanding

'Are You Prepared?' *Answers are given at the end of these exercises. If you get a wrong answer, read the pages listed in* red.

1. The formula for the distance d from $P_1 = (x_1, y_1)$ to $P_2 = (x_2, y_2)$ is $d =$ _____. (p. 3)

2. To complete the square of $x^2 - 4x$, add _____. (p. A40)

3. Use the Square Root Method to find the real solutions of $(x + 4)^2 = 9$. (pp. 142–143)

4. The point that is symmetric with respect to the x-axis to the point $(-2, 5)$ is _____. (pp. 12–14)

5. To graph $y = (x - 3)^2 + 1$, shift the graph of $y = x^2$ to the right _____ units and then _____ 1 unit. (pp. 90–93)

Concepts and Vocabulary

6. A(n) _____ is the collection of all points in the plane such that the distance from each point to a fixed point equals its distance to a fixed line.

Answer Problems 7–10 using the figure.

7. If $a \geq 0$, the equation of the parabola is of the form
   (a) $(y - k)^2 = 4a(x - h)$
   (b) $(y - k)^2 = -4a(x - h)$
   (c) $(x - h)^2 = 4a(y - k)$
   (d) $(x - h)^2 = -4a(y - k)$

8. The coordinates of the vertex are _____.

9. If $a = 4$, then the coordinates of the focus are _____.

10. If $a = 4$, then the equation of the directrix is _____.

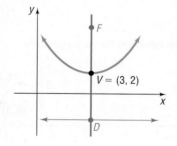

Skill Building

In Problems 11–18, the graph of a parabola is given. Match each graph to its equation.

(a) $y^2 = 4x$
(b) $x^2 = 4y$

(c) $y^2 = -4x$
(d) $x^2 = -4y$

(e) $(y - 1)^2 = 4(x - 1)$
(f) $(x + 1)^2 = 4(y + 1)$

(g) $(y - 1)^2 = -4(x - 1)$
(h) $(x + 1)^2 = -4(y + 1)$

11.

12.

13.

14.

15.

16.

17.

18.

In Problems 19–36, find the equation of the parabola described. Find the two points that define the latus rectum, and graph the equation.

19. Focus at $(4, 0)$; vertex at $(0, 0)$

20. Focus at $(0, 2)$; vertex at $(0, 0)$

21. Focus at $(0, -3)$; vertex at $(0, 0)$

22. Focus at $(-4, 0)$; vertex at $(0, 0)$

23. Focus at $(-2, 0)$; directrix the line $x = 2$

24. Focus at $(0, -1)$; directrix the line $y = 1$

25. Directrix the line $y = -\dfrac{1}{2}$; vertex at $(0, 0)$

26. Directrix the line $x = -\dfrac{1}{2}$; vertex at $(0, 0)$

27. Vertex at $(0, 0)$; axis of symmetry the y-axis; containing the point $(2, 3)$

28. Vertex at $(0, 0)$; axis of symmetry the x-axis; containing the point $(2, 3)$

29. Vertex at $(2, -3)$; focus at $(2, -5)$

30. Vertex at $(4, -2)$; focus at $(6, -2)$

31. Vertex at $(-1, -2)$; focus at $(0, -2)$

32. Vertex at $(3, 0)$; focus at $(3, -2)$

33. Focus at $(-3, 4)$; directrix the line $y = 2$

34. Focus at $(2, 4)$; directrix the line $x = -4$

35. Focus at $(-3, -2)$; directrix the line $x = 1$

36. Focus at $(-4, 4)$; directrix the line $y = -2$

In Problems 37–54, find the vertex, focus, and directrix of each parabola. Graph the equation.

37. $x^2 = 4y$

38. $y^2 = 8x$

39. $y^2 = -16x$

40. $x^2 = -4y$

41. $(y - 2)^2 = 8(x + 1)$

42. $(x + 4)^2 = 16(y + 2)$

43. $(x - 3)^2 = -(y + 1)$

44. $(y + 1)^2 = -4(x - 2)$

45. $(y + 3)^2 = 8(x - 2)$

46. $(x - 2)^2 = 4(y - 3)$

47. $y^2 - 4y + 4x + 4 = 0$

48. $x^2 + 6x - 4y + 1 = 0$

49. $x^2 + 8x = 4y - 8$

50. $y^2 - 2y = 8x - 1$

51. $y^2 + 2y - x = 0$

52. $x^2 - 4x = 2y$

53. $x^2 - 4x = y + 4$

54. $y^2 + 12y = -x + 1$

In Problems 55–62, write an equation for each parabola.

55.

56.

57.

58.

59.

60.

61.

62.

Applications and Extensions

63. Satellite Dish A satellite dish is shaped like a paraboloid of revolution. The signals that emanate from a satellite strike the surface of the dish and are reflected to a single point, where the receiver is located. If the dish is 10 feet across at its opening and 4 feet deep at its center, at what position should the receiver be placed?

64. Constructing a TV Dish A cable TV receiving dish is in the shape of a paraboloid of revolution. Find the location of the receiver, which is placed at the focus, if the dish is 6 feet across at its opening and 2 feet deep.

65. Constructing a Flashlight The reflector of a flashlight is in the shape of a paraboloid of revolution. Its diameter is 4 inches and its depth is 1 inch. How far from the vertex should the light bulb be placed so that the rays will be reflected parallel to the axis?

66. **Constructing a Headlight**   A sealed-beam headlight is in the shape of a paraboloid of revolution. The bulb, which is placed at the focus, is 1 inch from the vertex. If the depth is to be 2 inches, what is the diameter of the headlight at its opening?

67. **Suspension Bridge**   The cables of a suspension bridge are in the shape of a parabola, as shown in the figure. The towers supporting the cable are 600 feet apart and 80 feet high. If the cables touch the road surface midway between the towers, what is the height of the cable from the road at a point 150 feet from the center of the bridge?

68. **Suspension Bridge**   The cables of a suspension bridge are in the shape of a parabola. The towers supporting the cable are 400 feet apart and 100 feet high. If the cables are at a height of 10 feet midway between the towers, what is the height of the cable at a point 50 feet from the center of the bridge?

69. **Searchlight**   A searchlight is shaped like a paraboloid of revolution. If the light source is located 2 feet from the base along the axis of symmetry and the opening is 5 feet across, how deep should the searchlight be?

70. **Searchlight**   A searchlight is shaped like a paraboloid of revolution. If the light source is located 2 feet from the base along the axis of symmetry and the depth of the searchlight is 4 feet, what should the width of the opening be?

71. **Solar Heat**   A mirror is shaped like a paraboloid of revolution and will be used to concentrate the rays of the sun at its focus, creating a heat source. See the figure. If the mirror is 20 feet across at its opening and is 6 feet deep, where will the heat source be concentrated?

Sun's rays

72. **Reflecting Telescope**   A reflecting telescope contains a mirror shaped like a paraboloid of revolution. If the mirror is 4 inches across at its opening and is 3 inches deep, where will the collected light be concentrated?

73. **Parabolic Arch Bridge**   A bridge is built in the shape of a parabolic arch. The bridge has a span of 120 feet and a maximum height of 25 feet. See the illustration. Choose a suitable

rectangular coordinate system and find the height of the arch at distances of 10, 30, and 50 feet from the center.

25 ft
120 ft

74. **Parabolic Arch Bridge**   A bridge is to be built in the shape of a parabolic arch and is to have a span of 100 feet. The height of the arch a distance of 40 feet from the center is to be 10 feet. Find the height of the arch at its center.

75. **Gateway Arch**   The Gateway Arch in St. Louis is often mistaken to be parabolic in shape. In fact, it is a *catenary*, which has a more complicated formula than a parabola. The Arch is 625 feet high and 598 feet wide at its base.
    (a) Find the equation of a parabola with the same dimensions. Let x equal the horizontal distance from the center of the arc.
    (b) The table below gives the height of the Arch at various widths; find the corresponding heights for the parabola found in (a).

Width (ft)	Height (ft)
567	100
478	312.5
308	525

    (c) Do the data support the notion that the Arch is in the shape of a parabola?
    Source: Wikipedia, the free encyclopedia

76. Show that an equation of the form
$$Ax^2 + Ey = 0, \qquad A \neq 0, E \neq 0$$
is the equation of a parabola with vertex at $(0, 0)$ and axis of symmetry the y-axis. Find its focus and directrix.

77. Show that an equation of the form
$$Cy^2 + Dx = 0, \qquad C \neq 0, D \neq 0$$
is the equation of a parabola with vertex at $(0, 0)$ and axis of symmetry the x-axis. Find its focus and directrix.

78. Show that the graph of an equation of the form
$$Ax^2 + Dx + Ey + F = 0, \qquad A \neq 0$$
    (a) Is a parabola if $E \neq 0$.
    (b) Is a vertical line if $E = 0$ and $D^2 - 4AF = 0$.
    (c) Is two vertical lines if $E = 0$ and $D^2 - 4AF > 0$.
    (d) Contains no points if $E = 0$ and $D^2 - 4AF < 0$.

79. Show that the graph of an equation of the form
$$Cy^2 + Dx + Ey + F = 0, \qquad C \neq 0$$
    (a) Is a parabola if $D \neq 0$.
    (b) Is a horizontal line if $D = 0$ and $E^2 - 4CF = 0$.
    (c) Is two horizontal lines if $D = 0$ and $E^2 - 4CF > 0$.
    (d) Contains no points if $D = 0$ and $E^2 - 4CF < 0$.

'Are You Prepared?' Answers

1. $\sqrt{(x_2 - x_1)^2 + (y_2 - y_1)^2}$ 2. 4   3. $x + 4 = \pm 3$; $\{-7, -1\}$ 4. $(-2, -5)$ 5. 3; up

9.3 The Ellipse

PREPARING FOR THIS SECTION *Before getting started, review the following:*

- Distance Formula (Foundations, Section 1, p. 3)
- Completing the Square (Appendix A, Section A.4, p. A40)
- Intercepts (Foundations, Section 2, pp. 11–12)
- Symmetry (Foundations, Section 2, pp. 12–14)
- Circles (Foundations, Section 4, pp. 35–39)
- Graphing Techniques: Transformations (Section 1.5, pp. 90–99)

Now Work the '**Are You Prepared?**' problems on page 683.

> OBJECTIVES **1** Analyze Ellipses with Center at the Origin (p. 676)
>
> **2** Analyze Ellipses with Center at (*h, k*) (p. 680)
>
> **3** Solve Applied Problems Involving Ellipses (p. 682)

DEFINITION

An **ellipse** is the collection of all points in the plane, the sum of whose distances from two fixed points, called the **foci,** is a constant.

Figure 17

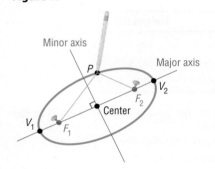

The definition contains within it a physical means for drawing an ellipse. Find a piece of string (the length of this string is the constant referred to in the definition). Then take two thumbtacks (the foci) and stick them into a piece of cardboard so that the distance between them is less than the length of the string. Now attach the ends of the string to the thumbtacks and, using the point of a pencil, pull the string taut. See Figure 17. Keeping the string taut, rotate the pencil around the two thumbtacks. The pencil traces out an ellipse, as shown in Figure 17.

In Figure 17, the foci are labeled F_1 and F_2. The line containing the foci is called the **major axis.** The midpoint of the line segment joining the foci is the **center** of the ellipse. The line through the center and perpendicular to the major axis is the **minor axis.**

The two points of intersection of the ellipse and the major axis are the **vertices,** V_1 and V_2, of the ellipse. The distance from one vertex to the other is the **length of the major axis.** The ellipse is symmetric with respect to its major axis, with respect to its minor axis, and with respect to its center.

1 Analyze Ellipses with Center at the Origin

With these ideas in mind, we are now ready to find the equation of an ellipse in a rectangular coordinate system. First, we place the center of the ellipse at the origin. Second, we position the ellipse so that its major axis coincides with a coordinate axis. Suppose that the major axis coincides with the *x*-axis, as shown in Figure 18. If *c* is the distance from the center to a focus, one focus will be at $F_1 = (-c, 0)$ and the other at $F_2 = (c, 0)$. If *a* is the distance from the center to a vertex, one vertex will be at $V_1 = (-a, 0)$ and the other at $V_2 = (a, 0)$. Recall, the sum of the distances from the two foci to *any* point on the ellipse is a constant. Suppose that we let the point on the ellipse be the vertex located at $V_2 = (a, 0)$. Then the sum of the distances from the foci to the vertex is

Figure 18

$$d(F_1, V_2) + d(F_2, V_2) = |-c - a| + |c - a|$$ On a real number line, $d(a, b) = |a - b|$.

$$= |-c - a| + |a - c|$$ $d(F_2, V_2) = d(V_2, F_2)$

$$= c + a + a - c$$ $|-c - a| = |-1(c + a)| = |-1||c + a| = c + a$ Since $a > c, |a - c| = a - c$.

$$= 2a$$

To see why $a > c$, look again at Figure 18. Then

$$d(F_1, P) + d(F_2, P) > d(F_1, F_2)$$ The sum of the lengths of two sides of a triangle is greater than the length of the third side.

$$2a > 2c$$ $d(F_1, P) + d(F_2, P) = 2a, \ d(F_1, F_2) = 2c$

$$a > c$$

So the constant sum referred to in the definition is $2a$.

Now, if $P = (x, y)$ is any point on the ellipse, we have

$$d(F_1, P) + d(F_2, P) = 2a$$ Sum of the distances from P to the foci equals a constant, $2a$.

$$\sqrt{(x + c)^2 + y^2} + \sqrt{(x - c)^2 + y^2} = 2a$$ Use the Distance Formula.

$$\sqrt{(x + c)^2 + y^2} = 2a - \sqrt{(x - c)^2 + y^2}$$ Isolate one radical.

$$(x + c)^2 + y^2 = 4a^2 - 4a\sqrt{(x - c)^2 + y^2}$$ Square both sides.
$$+ (x - c)^2 + y^2$$

$$x^2 + 2cx + c^2 + y^2 = 4a^2 - 4a\sqrt{(x - c)^2 + y^2}$$ Remove parentheses.
$$+ x^2 - 2cx + c^2 + y^2$$

$$4cx - 4a^2 = -4a\sqrt{(x - c)^2 + y^2}$$ Simplify; isolate the radical.

$$cx - a^2 = -a\sqrt{(x - c)^2 + y^2}$$ Divide each side by 4.

$$(cx - a^2)^2 = a^2[(x - c)^2 + y^2]$$ Square both sides again.

$$c^2x^2 - 2a^2cx + a^4 = a^2(x^2 - 2cx + c^2 + y^2)$$ Remove parentheses.

$$(c^2 - a^2)x^2 - a^2y^2 = a^2c^2 - a^4$$ Rearrange the terms.

$$(a^2 - c^2)x^2 + a^2y^2 = a^2(a^2 - c^2)$$ Multiply each side by -1; factor a^2 on the right side.   **(1)**

Since $a > c$, we also have $a^2 > c^2$, so $a^2 - c^2 > 0$. Let $b^2 = a^2 - c^2, b > 0$. Then $a > b$ and equation (1) can be written as

$$b^2x^2 + a^2y^2 = a^2b^2$$

$$\frac{x^2}{a^2} + \frac{y^2}{b^2} = 1$$ Divide each side by a^2b^2.

As you can verify, this equation is symmetric with respect to the x-axis, y-axis, and origin.

Because the major axis is the x-axis, we verify the vertices of this ellipse are at $(-a, 0)$ and $(a, 0)$ by letting $y = 0$. The vertices satisfy the equation $\frac{x^2}{a^2} = 1$, the solutions of which are $x = \pm a$. Consequently, the vertices of this ellipse are $V_1 = (-a, 0)$ and $V_2 = (a, 0)$. The y-intercepts of the ellipse, found by letting $x = 0$, have coordinates $(0, -b)$ and $(0, b)$. These four intercepts, $(a, 0)$, $(-a, 0)$, $(0, b)$, and $(0, -b)$, are used to graph the ellipse.

THEOREM

Figure 19

$V_1 = (-a, 0)$

$V_2 = (a, 0)$

$F_1 = (-c, 0)$ $F_2 = (c, 0)$

$(0, b)$

$(0, -b)$

Equation of an Ellipse: Center at (0, 0); Major Axis along the x-Axis

An equation of the ellipse with center at $(0, 0)$, foci at $(-c, 0)$ and $(c, 0)$, and vertices at $(-a, 0)$ and $(a, 0)$ is

$$\frac{x^2}{a^2} + \frac{y^2}{b^2} = 1$$ where $a > b > 0$ and $b^2 = a^2 - c^2$ **(2)**

The major axis is the x-axis. See Figure 19.

Notice in Figure 19 the right triangle formed with the points $(0, 0)$, $(c, 0)$, and $(0, b)$. Because $b^2 = a^2 - c^2$ (or $b^2 + c^2 = a^2$), the distance from the focus at $(c, 0)$ to the point $(0, b)$ is a.

This can be seen another way. Look at the two right triangles in Figure 19. They are congruent. Do you see why? Because the sum of the distances from the foci to a point on the ellipse is $2a$, it follows that the distance from $(c, 0)$ to $(0, b)$ is a.

EXAMPLE 1

Finding an Equation of an Ellipse

Find an equation of the ellipse with center at the origin, one focus at $(3, 0)$, and a vertex at $(-4, 0)$. Graph the equation.

Solution

If we plot the given points, we notice that the ellipse has its center at the origin and the given focus and vertex lie on the x-axis, so the major axis is the x-axis. The distance from the center, $(0, 0)$, to one of the foci, $(3, 0)$, is $c = 3$. The distance from the center, $(0, 0)$, to one of the vertices, $(-4, 0)$, is $a = 4$. From equation (2), it follows that

$$b^2 = a^2 - c^2 = 16 - 9 = 7$$

so an equation of the ellipse is

$$\frac{x^2}{16} + \frac{y^2}{7} = 1$$

Figure 20 shows the graph.

Figure 20

$$\frac{x^2}{16} + \frac{y^2}{7} = 1$$

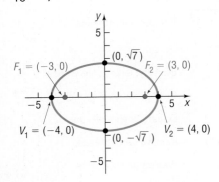

Notice in Figure 20 how we used the intercepts of the equation to graph the ellipse. Following this practice will make it easier for you to obtain an accurate graph of an ellipse when graphing.

COMMENT The intercepts of the ellipse also provide information about how to set the viewing rectangle for graphing an ellipse. To graph the ellipse

$$\frac{x^2}{16} + \frac{y^2}{7} = 1$$

discussed in Example 1, we set the viewing rectangle using a square screen that includes the intercepts, perhaps $-4.5 \le x \le 4.5, -3 \le y \le 3$. Then we proceed to solve the equation for y:

$$\frac{x^2}{16} + \frac{y^2}{7} = 1$$

$$\frac{y^2}{7} = 1 - \frac{x^2}{16} \qquad \text{Subtract } \frac{x^2}{16} \text{ from each side.}$$

$$y^2 = 7\left(1 - \frac{x^2}{16}\right) \qquad \text{Multiply both sides by 7.}$$

$$y = \pm\sqrt{7\left(1 - \frac{x^2}{16}\right)} \qquad \text{Take the square root of each side.}$$

Now graph the two functions

$$Y_1 = \sqrt{7\left(1 - \frac{x^2}{16}\right)} \quad \text{and} \quad Y_2 = -\sqrt{7\left(1 - \frac{x^2}{16}\right)}$$

Figure 21

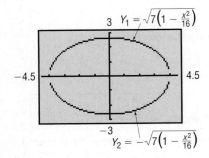

Figure 21 shows the result.

----**Now Work** PROBLEM **27**

An equation of the form of equation (2), with $a > b$, is the equation of an ellipse with center at the origin, foci on the x-axis at $(-c, 0)$ and $(c, 0)$, where $c^2 = a^2 - b^2$, and major axis along the x-axis.

For the remainder of this section, the direction **"Analyze the equation"** will mean to find the center, major axis, foci, and vertices of the ellipse and graph it.

EXAMPLE 2   **Analyzing the Equation of an Ellipse**

Analyze the equation: $\dfrac{x^2}{25} + \dfrac{y^2}{9} = 1$

Solution   The given equation is of the form of equation (2), with $a^2 = 25$ and $b^2 = 9$. The equation is that of an ellipse with center $(0, 0)$ and major axis along the x-axis. The vertices are at $(\pm a, 0) = (\pm 5, 0)$. Because $b^2 = a^2 - c^2$, we find that

$$c^2 = a^2 - b^2 = 25 - 9 = 16$$

The foci are at $(\pm c, 0) = (\pm 4, 0)$. Figure 22 shows the graph.

Figure 22

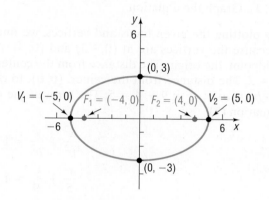

Now Work PROBLEM 17

If the major axis of an ellipse with center at $(0, 0)$ lies on the y-axis, the foci are at $(0, -c)$ and $(0, c)$. Using the same steps as before, the definition of an ellipse leads to the following result:

THEOREM

Equation of an Ellipse: Center at (0, 0); Major Axis along the y-Axis

An equation of the ellipse with center at $(0, 0)$, foci at $(0, -c)$ and $(0, c)$, and vertices at $(0, -a)$ and $(0, a)$ is

$$\frac{x^2}{b^2} + \frac{y^2}{a^2} = 1 \qquad \text{where } a > b > 0 \text{ and } b^2 = a^2 - c^2 \qquad (3)$$

The major axis is the y-axis.

Figure 23

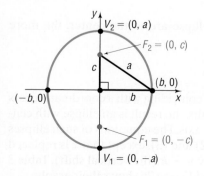

Figure 23 illustrates the graph of such an ellipse. Again, notice the right triangle formed by the points at $(0, 0)$, $(b, 0)$, and $(0, c)$, so that $a^2 = b^2 + c^2$ (or $b^2 = a^2 - c^2$).

Look closely at equations (2) and (3). Although they may look alike, there is a difference! In equation (2), the larger number, a^2, is in the denominator of the x^2-term, so the major axis of the ellipse is along the x-axis. In equation (3), the larger number, a^2, is in the denominator of the y^2-term, so the major axis is along the y-axis.

EXAMPLE 3

Analyzing the Equation of an Ellipse

Analyze the equation: $9x^2 + y^2 = 9$

Figure 24

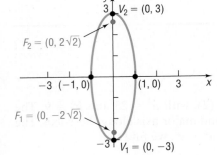

Solution

To put the equation in proper form, we divide each side by 9.

$$x^2 + \frac{y^2}{9} = 1$$

The larger denominator, 9, is in the y^2-term so, based on equation (3), this is the equation of an ellipse with center at the origin and major axis along the y-axis. Also, we conclude that $a^2 = 9$, $b^2 = 1$, and $c^2 = a^2 - b^2 = 9 - 1 = 8$. The vertices are at $(0, \pm a) = (0, \pm 3)$, and the foci are at $(0, \pm c) = (0, \pm 2\sqrt{2})$. Figure 24 shows the graph.

━━━━ **Now Work** PROBLEM 21

EXAMPLE 4

Finding an Equation of an Ellipse

Find an equation of the ellipse having one focus at $(0, 2)$ and vertices at $(0, -3)$ and $(0, 3)$. Graph the equation.

Solution

By plotting the given focus and vertices, we find that the major axis is the y-axis. Because the vertices are at $(0, -3)$ and $(0, 3)$, the center of this ellipse is at their midpoint, the origin. The distance from the center, $(0, 0)$, to one of the foci, $(0, 2)$, is $c = 2$. The distance from the center, $(0, 0)$, to one of the vertices, $(0, 3)$, is $a = 3$. So $b^2 = a^2 - c^2 = 9 - 4 = 5$. The form of the equation of this ellipse is given by equation (3).

Figure 25

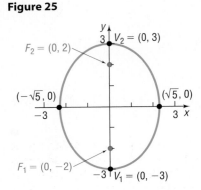

$$\frac{x^2}{b^2} + \frac{y^2}{a^2} = 1$$

$$\frac{x^2}{5} + \frac{y^2}{9} = 1$$

Figure 25 shows the graph.

━━━━ **Now Work** PROBLEM 29

The circle may be considered a special kind of ellipse. To see why, let $a = b$ in equation (2) or (3). Then

$$\frac{x^2}{a^2} + \frac{y^2}{a^2} = 1$$

$$x^2 + y^2 = a^2$$

This is the equation of a circle with center at the origin and radius a. The value of c is

$$c^2 = a^2 - b^2 = 0$$
$$\underset{a = b}{\uparrow}$$

We conclude that the closer the two foci of an ellipse are to the center, the more the ellipse will look like a circle.

2 Analyze Ellipses with Center at (h, k)

If an ellipse with center at the origin and major axis coinciding with a coordinate axis is shifted horizontally h units and then vertically k units, the result is an ellipse with center at (h, k) and major axis parallel to a coordinate axis. The equations of such ellipses have the same forms as those given in equations (2) and (3), except that x is replaced by $x - h$ (the horizontal shift) and y is replaced by $y - k$ (the vertical shift). Table 3 gives the forms of the equations of such ellipses, and Figure 26 shows their graphs.

Table 3

NOTE It is not recommended that Table 3 be memorized. Rather, use the ideas of transformations (shift horizontally h units, vertically k units) along with the fact that a represents the distance from the center to the vertices, c represents the distance from the center to the foci, and $b^2 = a^2 - c^2$ (or $c^2 = a^2 - b^2$). ■

Equations of an Ellipse: Center at (h, k); Major Axis Parallel to a Coordinate Axis

Center	Major Axis	Foci	Vertices	Equation
(h, k)	Parallel to the x-axis	$(h + c, k)$	$(h + a, k)$	$\dfrac{(x - h)^2}{a^2} + \dfrac{(y - k)^2}{b^2} = 1,$
		$(h - c, k)$	$(h - a, k)$	$a > b > 0$ and $b^2 = a^2 - c^2$
(h, k)	Parallel to the y-axis	$(h, k + c)$	$(h, k + a)$	$\dfrac{(x - h)^2}{b^2} + \dfrac{(y - k)^2}{a^2} = 1,$
		$(h, k - c)$	$(h, k - a)$	$a > b > 0$ and $b^2 = a^2 - c^2$

Figure 26

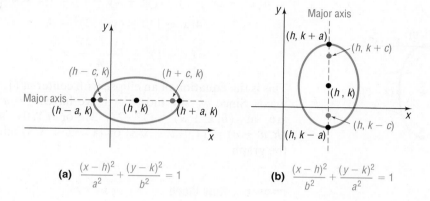

(a) $\dfrac{(x - h)^2}{a^2} + \dfrac{(y - k)^2}{b^2} = 1$

(b) $\dfrac{(x - h)^2}{b^2} + \dfrac{(y - k)^2}{a^2} = 1$

EXAMPLE 5   **Finding an Equation of an Ellipse, Center Not at the Origin**

Find an equation for the ellipse with center at $(2, -3)$, one focus at $(3, -3)$, and one vertex at $(5, -3)$. Graph the equation.

Solution   The center is at $(h, k) = (2, -3)$, so $h = 2$ and $k = -3$. If we plot the center, focus, and vertex, we notice that the points all lie on the line $y = -3$, so the major axis is parallel to the x-axis. The distance from the center $(2, -3)$ to a focus $(3, -3)$ is $c = 1$; the distance from the center $(2, -3)$ to a vertex $(5, -3)$ is $a = 3$. Then $b^2 = a^2 - c^2 = 9 - 1 = 8$. The form of the equation is

$$\frac{(x - h)^2}{a^2} + \frac{(y - k)^2}{b^2} = 1 \quad \text{where } h = 2, k = -3, a = 3, b = 2\sqrt{2}$$

$$\frac{(x - 2)^2}{9} + \frac{(y + 3)^2}{8} = 1$$

To graph the equation, we use the center $(h, k) = (2, -3)$ to locate the vertices. The major axis is parallel to the x-axis, so the vertices are $a = 3$ units left and right of the center $(2, -3)$. Therefore, the vertices are

$$V_1 = (2 - 3, -3) = (-1, -3) \quad \text{and} \quad V_2 = (2 + 3, -3) = (5, -3)$$

Since $c = 1$ and the major axis is parallel to the x-axis, the foci are 1 unit left and right of the center. Therefore, the foci are

$$F_1 = (2 - 1, -3) = (1, -3) \quad \text{and} \quad F_2 = (2 + 1, -3) = (3, -3)$$

Finally, we use the value of $b = 2\sqrt{2}$ to find the two points above and below the center.

$$\left(2, -3 - 2\sqrt{2}\right) \quad \text{and} \quad \left(2, -3 + 2\sqrt{2}\right)$$

Figure 27 shows the graph.

Figure 27

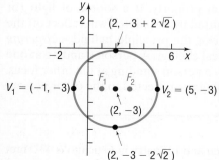

—**Now Work** PROBLEM 55

EXAMPLE 6 **Analyzing the Equation of an Ellipse**

Analyze the equation: $4x^2 + y^2 - 8x + 4y + 4 = 0$

Solution We proceed to complete the squares in x and in y.

$$4x^2 + y^2 - 8x + 4y + 4 = 0$$

$$4x^2 - 8x + y^2 + 4y = -4 \qquad \text{Group like variables; place the constant on the right side.}$$

$$4(x^2 - 2x) + (y^2 + 4y) = -4 \qquad \text{Factor out 4 from the first two terms.}$$

$$4(x^2 - 2x + 1) + (y^2 + 4y + 4) = -4 + 4 + 4 \qquad \text{Complete each square.}$$

$$4(x - 1)^2 + (y + 2)^2 = 4 \qquad \text{Factor.}$$

$$(x - 1)^2 + \frac{(y + 2)^2}{4} = 1 \qquad \text{Divide each side by 4.}$$

Figure 28

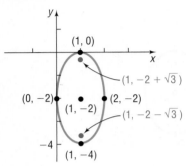

This is the equation of an ellipse with center at $(1, -2)$ and major axis parallel to the y-axis. Since $a^2 = 4$ and $b^2 = 1$, we have $c^2 = a^2 - b^2 = 4 - 1 = 3$. The vertices are at $(h, k \pm a) = (1, -2 \pm 2)$ or $(1, 0)$ and $(1, -4)$. The foci are at $(h, k \pm c) = (1, -2 \pm \sqrt{3})$ or $(1, -2 - \sqrt{3})$ and $(1, -2 + \sqrt{3})$. Figure 28 shows the graph.

●

━━━**Now Work** PROBLEM **47**

3 Solve Applied Problems Involving Ellipses

Ellipses are found in many applications in science and engineering. For example, the orbits of the planets around the Sun are elliptical, with the Sun's position at a focus. See Figure 29.

Figure 29

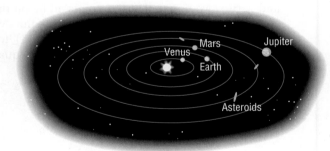

Stone and concrete bridges are often shaped as semielliptical arches. Elliptical gears are used in machinery when a variable rate of motion is required.

Ellipses also have an interesting reflection property. If a source of light (or sound) is placed at one focus, the waves transmitted by the source will reflect off the ellipse and concentrate at the other focus. This is the principle behind *whispering galleries*, which are rooms designed with elliptical ceilings. A person standing at one focus of the ellipse can whisper and be heard by a person standing at the other focus, because all the sound waves that reach the ceiling are reflected to the other person.

EXAMPLE 7 **A Whispering Gallery**

The whispering gallery in the Museum of Science and Industry in Chicago is 47.3 feet long. The distance from the center of the room to the foci is 20.3 feet. Find an equation that describes the shape of the room. How high is the room at its center?

Source: Chicago Museum of Science and Industry Web site; www.msichicago.org

Solution

We set up a rectangular coordinate system so that the center of the ellipse is at the origin and the major axis is along the x-axis. The equation of the ellipse is

$$\frac{x^2}{a^2} + \frac{y^2}{b^2} = 1$$

Since the length of the room is 47.3 feet, the distance from the center of the room to each vertex (the end of the room) will be $\frac{47.3}{2} = 23.65$ feet; so $a = 23.65$ feet. The distance from the center of the room to each focus is $c = 20.3$ feet. See Figure 30. Since $b^2 = a^2 - c^2$, we find $b^2 = 23.65^2 - 20.3^2 = 147.2325$. An equation that describes the shape of the room is given by

$$\frac{x^2}{23.65^2} + \frac{y^2}{147.2325} = 1$$

The height of the room at its center is $b = \sqrt{147.2325} \approx 12.1$ feet.

Figure 30

➤ **Now Work** PROBLEM 71

9.3 Assess Your Understanding

'Are You Prepared?' *Answers are given at the end of these exercises. If you get a wrong answer, read the pages listed in red.*

1. The distance d from $P_1 = (2, -5)$ to $P_2 = (4, -2)$ is $d =$ _____. (p. 3)

2. To complete the square of $x^2 - 3x$, add _____. (p. A40)

3. Find the intercepts of the equation $y^2 = 16 - 4x^2$. (p. 12)

4. The point that is symmetric with respect to the y-axis to the point $(-2, 5)$ is _____. (pp. 12–14)

5. To graph $y = (x + 1)^2 - 4$, shift the graph of $y = x^2$ to the (left/right) _____ unit(s) and then (up/down) _____ unit(s). (pp. 90–93)

6. The standard equation of a circle with center at $(2, -3)$ and radius 1 is _____. (p. 36)

Concepts and Vocabulary

7. A(n) _____ is the collection of all points in the plane the sum of whose distances from two fixed points is a constant.

8. For an ellipse, the foci lie on a line called the _____ axis.

9. In an ellipse whose major axis is parallel to the x-axis, b represents the distance to points on the graph of an ellipse directly above or below its center. If we know the values of a and c, we can find b using the fact that _____.

10. For the ellipse $\frac{x^2}{25} + \frac{y^2}{9} = 1$, the value of a is _____, the value of b is _____, and the major axis is the _____-axis.

11. If the center of an ellipse is $(2, -3)$, the major axis is parallel to the x-axis, and the distance from the center of the ellipse to its vertices is $a = 4$ units, then the coordinates of the vertices are _____ and _____.

12. If the foci of an ellipse are $(-4, 4)$ and $(6, 4)$, then the coordinates of the center of the ellipse are _____.

Skill Building

In Problems 13–16, the graph of an ellipse is given. Match each graph to its equation.

(A) $\dfrac{x^2}{4} + y^2 = 1$ (B) $x^2 + \dfrac{y^2}{4} = 1$ (C) $\dfrac{x^2}{16} + \dfrac{y^2}{4} = 1$ (D) $\dfrac{x^2}{4} + \dfrac{y^2}{16} = 1$

13.

14.

15.

16.

In Problems 17–26, find the vertices and foci of each ellipse. Graph each equation.

17. $\dfrac{x^2}{25} + \dfrac{y^2}{4} = 1$

18. $\dfrac{x^2}{9} + \dfrac{y^2}{4} = 1$

19. $\dfrac{x^2}{9} + \dfrac{y^2}{25} = 1$

20. $x^2 + \dfrac{y^2}{16} = 1$

21. $4x^2 + y^2 = 16$

22. $x^2 + 9y^2 = 18$

23. $4y^2 + x^2 = 8$

24. $4y^2 + 9x^2 = 36$

25. $x^2 + y^2 = 16$

26. $x^2 + y^2 = 4$

In Problems 27–38, find an equation for each ellipse. Graph the equation.

27. Center at $(0, 0)$;   focus at $(3, 0)$;   vertex at $(5, 0)$

28. Center at $(0, 0)$;   focus at $(-1, 0)$;   vertex at $(3, 0)$

29. Center at $(0, 0)$;   focus at $(0, -4)$;   vertex at $(0, 5)$

30. Center at $(0, 0)$;   focus at $(0, 1)$;   vertex at $(0, -2)$

31. Foci at $(\pm 2, 0)$;   length of the major axis is 6

32. Foci at $(0, \pm 2)$;   length of the major axis is 8

33. Focus at $(-4, 0)$;   vertices at $(\pm 5, 0)$

34. Focus at $(0, -4)$;   vertices at $(0, \pm 8)$

35. Foci at $(0, \pm 3)$; x-intercepts are ± 2

36. Vertices at $(\pm 4, 0)$; y-intercepts are ± 1

37. Center at $(0, 0)$;   vertex at $(0, 4)$; $b = 1$

38. Vertices at $(\pm 5, 0)$; $c = 2$

In Problems 39–42, write an equation for each ellipse.

39.

40.

41.

42.

In Problems 43–54, analyze each equation; that is, find the center, foci, and vertices of each ellipse. Graph each equation.

43. $\dfrac{(x-3)^2}{4} + \dfrac{(y+1)^2}{9} = 1$

44. $\dfrac{(x+4)^2}{9} + \dfrac{(y+2)^2}{4} = 1$

45. $(x+5)^2 + 4(y-4)^2 = 16$

46. $9(x-3)^2 + (y+2)^2 = 18$

47. $x^2 + 4x + 4y^2 - 8y + 4 = 0$

48. $x^2 + 3y^2 - 12y + 9 = 0$

49. $2x^2 + 3y^2 - 8x + 6y + 5 = 0$

50. $4x^2 + 3y^2 + 8x - 6y = 5$

51. $9x^2 + 4y^2 - 18x + 16y - 11 = 0$

52. $x^2 + 9y^2 + 6x - 18y + 9 = 0$

53. $4x^2 + y^2 + 4y = 0$

54. $9x^2 + y^2 - 18x = 0$

In Problems 55–64, find an equation for each ellipse. Graph the equation.

55. Center at $(2, -2)$;   vertex at $(7, -2)$;   focus at $(4, -2)$

56. Center at $(-3, 1)$;   vertex at $(-3, 3)$;   focus at $(-3, 0)$

57. Vertices at $(4, 3)$ and $(4, 9)$;   focus at $(4, 8)$

58. Foci at $(1, 2)$ and $(-3, 2)$;   vertex at $(-4, 2)$

59. Foci at $(5, 1)$ and $(-1, 1)$;   length of the major axis is 8

60. Vertices at $(2, 5)$ and $(2, -1)$; $c = 2$

61. Center at $(1, 2)$;   focus at $(4, 2)$;   contains the point $(1, 3)$

62. Center at $(1, 2)$;   focus at $(1, 4)$;   contains the point $(2, 2)$

63. Center at $(1, 2)$;   vertex at $(4, 2)$;   contains the point $(1, 5)$

64. Center at $(1, 2)$;   vertex at $(1, 4)$;   contains the point $(1 + \sqrt{3}, 3)$

In Problems 65–68, graph each function. Be sure to label all the intercepts.
[**Hint:** *Notice that each function is half an ellipse.*]

65. $f(x) = \sqrt{16 - 4x^2}$

66. $f(x) = \sqrt{9 - 9x^2}$

67. $f(x) = -\sqrt{64 - 16x^2}$

68. $f(x) = -\sqrt{4 - 4x^2}$

Applications and Extensions

69. Semielliptical Arch Bridge An arch in the shape of the upper half of an ellipse is used to support a bridge that is to span a river 20 meters wide. The center of the arch is 6 meters above the center of the river. See the figure. Write an equation for the ellipse in which the x-axis coincides with the water level and the y-axis passes through the center of the arch.

70. Semielliptical Arch Bridge The arch of a bridge is a semiellipse with a horizontal major axis. The span is 30 feet, and the top of the arch is 10 feet above the major axis. The roadway is horizontal and is 2 feet above the top of the arch. Find the vertical distance from the roadway to the arch at 5-foot intervals along the roadway.

71. Whispering Gallery A hall 100 feet in length is to be designed as a whispering gallery. If the foci are located 25 feet from the center, how high will the ceiling be at the center?

72. Whispering Gallery Jim, standing at one focus of a whispering gallery, is 6 feet from the nearest wall. His friend is standing at the other focus, 100 feet away. What is the length of this whispering gallery? How high is its elliptical ceiling at the center?

73. Semielliptical Arch Bridge A bridge is built in the shape of a semielliptical arch. The bridge has a span of 120 feet and a maximum height of 25 feet. Choose a suitable rectangular coordinate system and find the height of the arch at distances of 10, 30, and 50 feet from the center.

74. Semielliptical Arch Bridge A bridge is to be built in the shape of a semielliptical arch and is to have a span of 100 feet. The height of the arch, at a distance of 40 feet from the center, is to be 10 feet. Find the height of the arch at its center.

75. Racetrack Design Consult the figure. A racetrack is in the shape of an ellipse, 100 feet long and 50 feet wide. What is the width 10 feet from a vertex?

76. Semielliptical Arch Bridge An arch for a bridge over a highway is in the form of half an ellipse. The top of the arch is 20 feet above the ground level (the major axis). The highway has four lanes, each 12 feet wide; a center safety strip 8 feet wide; and two side strips, each 4 feet wide. What should the span of the bridge be (the length of its major axis) if the height 28 feet from the center is to be 13 feet?

77. Installing a Vent Pipe A homeowner is putting in a fireplace that has a 4-inch-radius vent pipe. He needs to cut an elliptical hole in his roof to accommodate the pipe. If the pitch of his roof is $\frac{5}{4}$, (a rise of 5, run of 4) what are the dimensions of the hole?
Source: www.pen.k12.va.us

78. Volume of a Football A football is in the shape of a **prolate spheroid,** which is simply a solid obtained by rotating an ellipse $\left(\frac{x^2}{a^2} + \frac{y^2}{b^2} = 1\right)$ about its major axis. An inflated NFL football averages 11.125 inches in length and 28.25 inches in center circumference. If the volume of a prolate spheroid is $\frac{4}{3}\pi ab^2$, how much air does the football contain? (Neglect material thickness).
Source: www.answerbag.com

In Problems 79–82, use the fact that the orbit of a planet about the Sun is an ellipse, with the Sun at one focus. The **aphelion** of a planet is its greatest distance from the Sun, and the **perihelion** is its shortest distance. The **mean distance** of a planet from the Sun is the length of the semimajor axis of the elliptical orbit. See the illustration.

79. Earth The mean distance of Earth from the Sun is 93 million miles. If the aphelion of Earth is 94.5 million miles, what is the perihelion? Write an equation for the orbit of Earth around the Sun.

80. Mars The mean distance of Mars from the Sun is 142 million miles. If the perihelion of Mars is 128.5 million miles, what is the aphelion? Write an equation for the orbit of Mars about the Sun.

81. Jupiter The aphelion of Jupiter is 507 million miles. If the distance from the center of its elliptical orbit to the Sun is 23.2 million miles, what is the perihelion? What is the mean distance? Write an equation for the orbit of Jupiter around the Sun.

82. Pluto The perihelion of Pluto is 4551 million miles, and the distance from the center of its elliptical orbit to the Sun is 897.5 million miles. Find the aphelion of Pluto. What is the mean distance of Pluto from the Sun? Write an equation for the orbit of Pluto about the Sun.

83. Show that an equation of the form
$$Ax^2 + Cy^2 + F = 0, \qquad A \neq 0, C \neq 0, F \neq 0$$
where A and C are of the same sign and F is of opposite sign,
(a) Is the equation of an ellipse with center at $(0, 0)$ if $A \neq C$.
(b) Is the equation of a circle with center $(0, 0)$ if $A = C$.

84. Show that the graph of an equation of the form

$$Ax^2 + Cy^2 + Dx + Ey + F = 0, \qquad A \neq 0, C \neq 0$$

where A and C are of the same sign,

(a) Is an ellipse if $\dfrac{D^2}{4A} + \dfrac{E^2}{4C} - F$ is the same sign as A.

(b) Is a point if $\dfrac{D^2}{4A} + \dfrac{E^2}{4C} - F = 0$.

(c) Contains no points if $\dfrac{D^2}{4A} + \dfrac{E^2}{4C} - F$ is of opposite sign to A.

Explaining Concepts: Discussion and Writing

85. The **eccentricity** e of an ellipse is defined as the number $\dfrac{c}{a}$, where a and c are the numbers given in equation (2) on page 677. Because $a > c$, it follows that $e < 1$. Write a brief paragraph about the general shape of each of the following ellipses. Be sure to justify your conclusions.

(a) Eccentricity close to 0   (b) Eccentricity = 0.5   (c) Eccentricity close to 1

'Are You Prepared?' Answers

1. $\sqrt{13}$ **2.** $\dfrac{9}{4}$ **3.** $(-2, 0), (2, 0), (0, -4), (0, 4)$ **4.** $(2, 5)$ **5.** left; 1; down: 4   **6.** $(x - 2)^2 + (y + 3)^2 = 1$

9.4 The Hyperbola

PREPARING FOR THIS SECTION   *Before getting started, review the following:*

- Distance Formula (Foundations, Section 1, p. 3)
- Completing the Square (Appendix A, Section A.4, p. A40)
- Intercepts (Foundations, Section 2, pp. 11–12)
- Symmetry (Foundations, Section 2, pp. 12–14)
- Asymptotes (Section 3.2, pp. 217–223)
- Graphing Techniques: Transformations (Section 1.5, pp. 90–99)
- Square Root Method (Section 2.3, pp. 142–143)

Now Work the **'Are You Prepared?'** problems on page 696.

OBJECTIVES   **1** Analyze Hyperbolas with Center at the Origin (p. 686)
  2 Find the Asymptotes of a Hyperbola (p. 691)
  3 Analyze Hyperbolas with Center at (h, k) (p. 693)
  4 Solve Applied Problems Involving Hyperbolas (p. 694)

DEFINITION

Figure 31

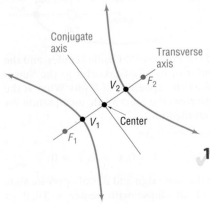

A **hyperbola** is the collection of all points in the plane, the difference of whose distances from two fixed points, called the **foci**, is a constant.

Figure 31 illustrates a hyperbola with foci F_1 and F_2. The line containing the foci is called the **transverse axis.** The midpoint of the line segment joining the foci is the **center** of the hyperbola. The line through the center and perpendicular to the transverse axis is the **conjugate axis.** The hyperbola consists of two separate curves, called **branches,** that are symmetric with respect to the transverse axis, conjugate axis, and center. The two points of intersection of the hyperbola and the transverse axis are the **vertices,** V_1 and V_2, of the hyperbola.

1 Analyze Hyperbolas with Center at the Origin

With these ideas in mind, we are now ready to find the equation of a hyperbola in the rectangular coordinate system. First, we place the center at the origin. Next, we position the hyperbola so that its transverse axis coincides with a coordinate axis. Suppose that the transverse axis coincides with the x-axis, as shown in Figure 32.

Figure 32

$d(F_1, P) - d(F_2, P) = \pm 2a$

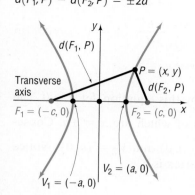

If a is the distance from the center to a vertex, one vertex will be at $V_1 = (-a, 0)$ and the other at $V_2 = (a, 0)$. If c is the distance from the center to a focus, one focus will be at $F_1 = (-c, 0)$ and the other at $F_2 = (c, 0)$. From Figure 32, we have that $c > a > 0 > -a > -c$. By definition, a hyperbola is the set of all points such that the difference in the distance from any point $P = (x, y)$ on the hyperbola to the foci is constant. Let the point P on the hyperbola have the coordinates $V_2 = (a, 0)$. Then the difference of the distances from the foci to the point is

$$
\begin{aligned}
d(F_1, V_2) - d(F_2, V_2) &= |-c - a| - |c - a| && \text{On a real number line } d(a, b) = |a - b| \\
&= c + a - (c - a) && |-c - a| = |-1(c + a)| = \\
& && |-1||c + a| = c + a; |c - a| = c - a \\
&= 2a && \text{since } c > a
\end{aligned}
$$

If we choose the point P on the hyperbola to have coordinates $V_1 = (-a, 0)$, then the difference of the distances from the foci to the point is

$$
\begin{aligned}
d(F_1, V_1) - d(F_2, V_1) &= |-c - (-a)| - |c - (-a)| && \text{On a real number line} \\
&= |a - c| - |c + a| && d(a, b) = |a - b| \\
&= -(a - c) - (c + a) && |a - c| = -(a - c) \\
&= -a + c - c - a && \text{since } c > a; |c + a| = c + a \\
&= -2a && \text{since } c > a > 0
\end{aligned}
$$

To see why $c > a$, look again at Figure 32.

$$
\begin{aligned}
d(F_2, P) + d(F_1, F_2) &> d(F_1, P) && \text{Use triangle } F_1PF_2. \\
d(F_1, F_2) &> d(F_1, P) - d(F_2, P) \\
2c &> 2a && P \text{ is on the right branch, so} \\
& && d(F_1, P) - d(F_2, P) = 2a; \\
& && d(F_1, F_2) = 2c. \\
c &> a
\end{aligned}
$$

In general, if the point P is on the left branch of the hyperbola, then $d(F_1, P) - d(F_2, P) = -2a$; if the point P is on the right branch of the hyperbola, then $d(F_1, P) - d(F_2, P) = 2a$. Therefore, the coordinates of P must satisfy the equation

$$
\begin{aligned}
d(F_1, P) - d(F_2, P) &= \pm 2a && \text{Difference of the distances from } P \\
& && \text{to the foci equals } \pm 2a. \\
\sqrt{(x + c)^2 + y^2} - \sqrt{(x - c)^2 + y^2} &= \pm 2a && \text{Use the Distance Formula.} \\
\sqrt{(x + c)^2 + y^2} &= \pm 2a + \sqrt{(x - c)^2 + y^2} && \text{Isolate one radical.} \\
(x + c)^2 + y^2 &= 4a^2 \pm 4a\sqrt{(x - c)^2 + y^2} && \text{Square both sides.} \\
& \quad + (x - c)^2 + y^2
\end{aligned}
$$

Next we remove the parentheses.

$$
\begin{aligned}
x^2 + 2cx + c^2 + y^2 &= 4a^2 \pm 4a\sqrt{(x - c)^2 + y^2} + x^2 - 2cx + c^2 + y^2 \\
4cx - 4a^2 &= \pm 4a\sqrt{(x - c)^2 + y^2} && \text{Simplify; isolate the radical.} \\
cx - a^2 &= \pm a\sqrt{(x - c)^2 + y^2} && \text{Divide each side by 4.} \\
(cx - a^2)^2 &= a^2[(x - c)^2 + y^2] && \text{Square both sides.} \\
c^2x^2 - 2ca^2x + a^4 &= a^2(x^2 - 2cx + c^2 + y^2) && \text{Simplify.} \\
c^2x^2 + a^4 &= a^2x^2 + a^2c^2 + a^2y^2 && \text{Remove parentheses and simplify.} \\
(c^2 - a^2)x^2 - a^2y^2 &= a^2c^2 - a^4 && \text{Rearrange terms.} \\
(c^2 - a^2)x^2 - a^2y^2 &= a^2(c^2 - a^2) && \text{Factor } a^2 \text{ on the right side.} \quad \textbf{(1)}
\end{aligned}
$$

Since $c > a$, we also have $c^2 > a^2$, so $c^2 - a^2 > 0$. Let $b^2 = c^2 - a^2, b > 0$. Then equation (1) can be written as

$$b^2x^2 - a^2y^2 = a^2b^2$$

$$\frac{x^2}{a^2} - \frac{y^2}{b^2} = 1 \qquad \text{Divide each side by } a^2b^2.$$

To find the vertices of the hyperbola defined by this equation, let $y = 0$. The vertices satisfy the equation $\frac{x^2}{a^2} = 1$, the solutions of which are $x = \pm a$. Consequently, the vertices of the hyperbola are $V_1 = (-a, 0)$ and $V_2 = (a, 0)$. Notice that the distance from the center $(0, 0)$ to either vertex is a.

THEOREM

Equation of a Hyperbola: Center at (0, 0); Transverse Axis along the x-Axis

An equation of the hyperbola with center at $(0, 0)$, foci at $(-c, 0)$ and $(c, 0)$, and vertices at $(-a, 0)$ and $(a, 0)$ is

$$\frac{x^2}{a^2} - \frac{y^2}{b^2} = 1, \qquad \text{where } b^2 = c^2 - a^2 \qquad \qquad (2)$$

The transverse axis is the x-axis.

Figure 33

$\frac{x^2}{a^2} - \frac{y^2}{b^2} = 1, \quad b^2 = c^2 - a^2$

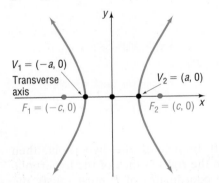

$V_1 = (-a, 0)$
Transverse axis
$F_1 = (-c, 0)$
$V_2 = (a, 0)$
$F_2 = (c, 0)$

See Figure 33. As you can verify, the hyperbola defined by equation (2) is symmetric with respect to the x-axis, y-axis, and origin. To find the y-intercepts, if any, let $x = 0$ in equation (2). This results in the equation $\frac{y^2}{b^2} = -1$, which has no real solution, so the hyperbola defined by equation (2) has no y-intercepts. In fact, since $\frac{x^2}{a^2} - 1 = \frac{y^2}{b^2} \geq 0$, it follows that $\frac{x^2}{a^2} \geq 1$. There are no points on the graph for $-a < x < a$.

EXAMPLE 1

Finding and Graphing an Equation of a Hyperbola

Find an equation of the hyperbola with center at the origin, one focus at $(3, 0)$, and one vertex at $(-2, 0)$. Graph the equation.

Solution

The hyperbola has its center at the origin. If we plot the center, focus, and vertex, we find that they all lie on the x-axis, so the transverse axis coincides with the x-axis. One focus is at $(c, 0) = (3, 0)$, so $c = 3$. One vertex is at $(-a, 0) = (-2, 0)$, so $a = 2$. From equation (2), it follows that $b^2 = c^2 - a^2 = 9 - 4 = 5$, so an equation of the hyperbola is

$$\frac{x^2}{4} - \frac{y^2}{5} = 1$$

To graph a hyperbola, it is helpful to locate and plot other points on the graph. For example, to find the points above and below the foci, we let $x = \pm 3$. Then

$$\frac{x^2}{4} - \frac{y^2}{5} = 1$$

$$\frac{(\pm 3)^2}{4} - \frac{y^2}{5} = 1 \qquad x = \pm 3$$

$$\frac{9}{4} - \frac{y^2}{5} = 1$$

Figure 34

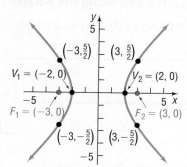

$$\frac{y^2}{5} = \frac{5}{4}$$

$$y^2 = \frac{25}{4}$$

$$y = \pm\frac{5}{2}$$

The points above and below the foci are $\left(\pm3, \frac{5}{2}\right)$ and $\left(\pm3, -\frac{5}{2}\right)$. These points determine the "opening" of the hyperbola. See Figure 34.

COMMENT To graph the hyperbola $\frac{x^2}{4} - \frac{y^2}{5} = 1$ discussed in Example 1, we need to graph the two functions $Y_1 = \sqrt{5}\sqrt{\frac{x^2}{4} - 1}$ and $Y_2 = -\sqrt{5}\sqrt{\frac{x^2}{4} - 1}$. Do this and compare what you see with Figure 34. ∎

Now Work PROBLEM 19

An equation of the form of equation (2) is the equation of a hyperbola with center at the origin, foci on the x-axis at $(-c, 0)$ and $(c, 0)$, where $c^2 = a^2 + b^2$, and transverse axis along the x-axis.

For the next two examples, the direction **"Analyze the equation"** will mean to find the center, transverse axis, vertices, and foci of the hyperbola and graph it.

EXAMPLE 2   **Analyzing the Equation of a Hyperbola**

Analyze the equation: $\dfrac{x^2}{16} - \dfrac{y^2}{4} = 1$

Solution   The given equation is of the form of equation (2), with $a^2 = 16$ and $b^2 = 4$. The graph of the equation is a hyperbola with center at $(0, 0)$ and transverse axis along the x-axis. Also, we know that $c^2 = a^2 + b^2 = 16 + 4 = 20$. The vertices are at $(\pm a, 0) = (\pm 4, 0)$, and the foci are at $(\pm c, 0) = (\pm 2\sqrt{5}, 0)$.

To locate the points on the graph above and below the foci, we let $x = \pm 2\sqrt{5}$. Then

Figure 35

$$\frac{x^2}{16} - \frac{y^2}{4} = 1$$

$$\frac{\left(\pm 2\sqrt{5}\right)^2}{16} - \frac{y^2}{4} = 1 \qquad x = \pm 2\sqrt{5}$$

$$\frac{20}{16} - \frac{y^2}{4} = 1$$

$$\frac{5}{4} - \frac{y^2}{4} = 1$$

$$\frac{y^2}{4} = \frac{1}{4}$$

$$y = \pm 1$$

The points above and below the foci are $(\pm 2\sqrt{5}, 1)$ and $(\pm 2\sqrt{5}, -1)$. See Figure 35.

The next result gives the form of the equation of a hyperbola with center at the origin and transverse axis along the y-axis.

THEOREM

Equation of a Hyperbola: Center at (0, 0); Transverse Axis along the y-Axis

An equation of the hyperbola with center at $(0, 0)$, foci at $(0, -c)$ and $(0, c)$, and vertices at $(0, -a)$ and $(0, a)$ is

$$\frac{y^2}{a^2} - \frac{x^2}{b^2} = 1, \qquad \text{where } b^2 = c^2 - a^2 \qquad \textbf{(3)}$$

The transverse axis is the y-axis.

Figure 36

$\frac{y^2}{a^2} - \frac{x^2}{b^2} = 1, b^2 = c^2 - a^2$

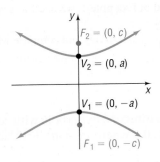

Figure 36 shows the graph of a typical hyperbola defined by equation (3).

An equation of the form of equation (2), $\frac{x^2}{a^2} - \frac{y^2}{b^2} = 1$, is the equation of a hyperbola with center at the origin, foci on the x-axis at $(-c, 0)$ and $(c, 0)$, where $c^2 = a^2 + b^2$, and transverse axis along the x-axis.

An equation of the form of equation (3), $\frac{y^2}{a^2} - \frac{x^2}{b^2} = 1$, is the equation of a hyperbola with center at the origin, foci on the y-axis at $(0, -c)$ and $(0, c)$, where $c^2 = a^2 + b^2$, and transverse axis along the y-axis.

Notice the difference in the forms of equations (2) and (3). When the y^2-term is subtracted from the x^2-term, the transverse axis is along the x-axis. When the x^2-term is subtracted from the y^2-term, the transverse axis is along the y-axis.

EXAMPLE 3

Analyzing the Equation of a Hyperbola

Analyze the equation: $y^2 - 4x^2 = 4$

Solution

To put the equation in proper form, we divide each side by 4:

$$\frac{y^2}{4} - x^2 = 1$$

Since the x^2-term is subtracted from the y^2-term, the equation is that of a hyperbola with center at the origin and transverse axis along the y-axis. Also, comparing the above equation to equation (3), we find $a^2 = 4$, $b^2 = 1$, and $c^2 = a^2 + b^2 = 5$. The vertices are at $(0, \pm a) = (0, \pm 2)$, and the foci are at $(0, \pm c) = (0, \pm\sqrt{5})$.

To locate other points on the graph, we let $x = \pm 2$. Then

Figure 37

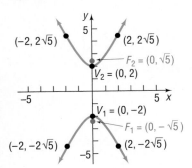

$$y^2 - 4x^2 = 4$$
$$y^2 - 4(\pm 2)^2 = 4 \qquad x = \pm 2$$
$$y^2 - 16 = 4$$
$$y^2 = 20$$
$$y = \pm 2\sqrt{5}$$

Four other points on the graph are $(\pm 2, 2\sqrt{5})$ and $(\pm 2, -2\sqrt{5})$. See Figure 37.

EXAMPLE 4

Finding an Equation of a Hyperbola

Find an equation of the hyperbola having one vertex at $(0, 2)$ and foci at $(0, -3)$ and $(0, 3)$. Graph the equation.

Solution

Since the foci are at $(0, -3)$ and $(0, 3)$, the center of the hyperbola, which is at their midpoint, is the origin. Also, the transverse axis is along the y-axis. The given

Figure 38

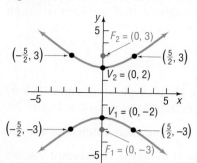

information also reveals that $c = 3$, $a = 2$, and $b^2 = c^2 - a^2 = 9 - 4 = 5$. The form of the equation of the hyperbola is given by equation (3):

$$\frac{y^2}{a^2} - \frac{x^2}{b^2} = 1$$

$$\frac{y^2}{4} - \frac{x^2}{5} = 1$$

Let $y = \pm 3$ to obtain points on the graph on either side of each foci. See Figure 38.

— Now Work PROBLEM 21

Look at the equations of the hyperbolas in Examples 2 and 4. For the hyperbola in Example 2, $a^2 = 16$ and $b^2 = 4$, so $a > b$; for the hyperbola in Example 4, $a^2 = 4$ and $b^2 = 5$, so $a < b$. We conclude that, for hyperbolas, there are no requirements involving the relative sizes of a and b. Contrast this situation to the case of an ellipse, in which the relative sizes of a and b dictate which axis is the major axis. Hyperbolas have another feature to distinguish them from ellipses and parabolas: Hyperbolas have asymptotes.

2   Find the Asymptotes of a Hyperbola

Recall from Section 3.2 that a horizontal or oblique asymptote of a graph is a line with the property that the distance from the line to points on the graph approaches 0 as $x \to -\infty$ or as $x \to \infty$. The asymptotes provide information about the end behavior of the graph of a hyperbola.

THEOREM

> **Asymptotes of a Hyperbola**
>
> The hyperbola $\dfrac{x^2}{a^2} - \dfrac{y^2}{b^2} = 1$ has the two oblique asymptotes
>
> $$y = \frac{b}{a}x \quad \text{and} \quad y = -\frac{b}{a}x \tag{4}$$

Proof   We begin by solving for y in the equation of the hyperbola.

$$\frac{x^2}{a^2} - \frac{y^2}{b^2} = 1$$

$$\frac{y^2}{b^2} = \frac{x^2}{a^2} - 1$$

$$y^2 = b^2\left(\frac{x^2}{a^2} - 1\right)$$

Since $x \neq 0$, we can rearrange the right side in the form

$$y^2 = \frac{b^2 x^2}{a^2}\left(1 - \frac{a^2}{x^2}\right)$$

$$y = \pm\frac{bx}{a}\sqrt{1 - \frac{a^2}{x^2}}$$

Now, as $x \to -\infty$ or as $x \to \infty$, the term $\dfrac{a^2}{x^2}$ approaches 0, so the expression under the radical approaches 1. So, as $x \to -\infty$ or as $x \to \infty$, the value of y approaches $\pm\dfrac{bx}{a}$; that is, the graph of the hyperbola approaches the lines

$$y = -\frac{b}{a}x \quad \text{and} \quad y = \frac{b}{a}x$$

These lines are oblique asymptotes of the hyperbola.   ■

Figure 39

$$\frac{x^2}{a^2} - \frac{y^2}{b^2} = 1$$

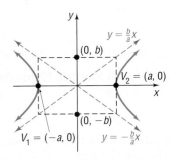

The asymptotes of a hyperbola are not part of the hyperbola, but they do serve as a guide for graphing a hyperbola. For example, suppose that we want to graph the equation

$$\frac{x^2}{a^2} - \frac{y^2}{b^2} = 1$$

We begin by plotting the vertices $(-a, 0)$ and $(a, 0)$. Then we plot the points $(0, -b)$ and $(0, b)$ and use these four points to construct a rectangle, as shown in Figure 39. The diagonals of this rectangle have slopes $\frac{b}{a}$ and $-\frac{b}{a}$, and their extensions are the asymptotes $y = \frac{b}{a}x$ and $y = -\frac{b}{a}x$ of the hyperbola. If we graph the asymptotes, we can use them to establish the "opening" of the hyperbola and avoid plotting other points.

THEOREM

Asymptotes of a Hyperbola

The hyperbola $\frac{y^2}{a^2} - \frac{x^2}{b^2} = 1$ has the two oblique asymptotes

$$y = \frac{a}{b}x \quad \text{and} \quad y = -\frac{a}{b}x \qquad \qquad \textbf{(5)}$$

You are asked to prove this result in Problem 84.

For the remainder of this section, the direction **"Analyze the equation"** will mean to find the center, transverse axis, vertices, foci, and asymptotes of the hyperbola and graph it.

EXAMPLE 5   **Analyzing the Equation of a Hyperbola**

Figure 40

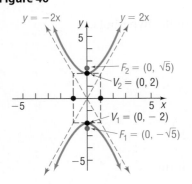

Analyze the equation: $\dfrac{y^2}{4} - x^2 = 1$

Solution   Since the x^2-term is subtracted from the y^2-term, the equation is of the form of equation (3) and is a hyperbola with center at the origin and transverse axis along the y-axis. Also, comparing this equation to equation (3), we find that $a^2 = 4$, $b^2 = 1$, and $c^2 = a^2 + b^2 = 5$. The vertices are at $(0, \pm a) = (0, \pm 2)$, and the foci are at $(0, \pm c) = (0, \pm\sqrt{5})$. Using equation (5) with $a = 2$ and $b = 1$, the asymptotes are the lines $y = \dfrac{a}{b}x = 2x$ and $y = -\dfrac{a}{b}x = -2x$. Form the rectangle containing the points $(0, \pm a) = (0, \pm 2)$ and $(\pm b, 0) = (\pm 1, 0)$. The extensions of the diagonals of this rectangle are the asymptotes. Now graph the rectangle, the asymptotes, and the hyperbola. See Figure 40.

●

EXAMPLE 6   **Analyzing the Equation of a Hyperbola**

Analyze the equation: $9x^2 - 4y^2 = 36$

Solution   Divide each side of the equation by 36 to put the equation in proper form.

$$\frac{x^2}{4} - \frac{y^2}{9} = 1$$

We now proceed to analyze the equation. The center of the hyperbola is the origin. Since the x^2-term is first in the equation, we know that the transverse axis is along the x-axis and the vertices and foci will lie on the x-axis. Using equation (2), we find $a^2 = 4$, $b^2 = 9$, and $c^2 = a^2 + b^2 = 13$. The vertices are $a = 2$ units left and right of

Figure 41

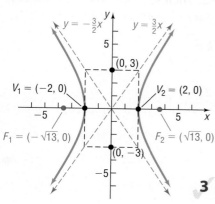

the center at $(\pm a, 0) = (\pm 2, 0)$, the foci are $c = \sqrt{13}$ units left and right of the center at $(\pm c, 0) = (\pm\sqrt{13}, 0)$, and the asymptotes have the equations

$$y = \frac{b}{a}x = \frac{3}{2}x \quad \text{and} \quad y = -\frac{b}{a}x = -\frac{3}{2}x$$

To graph the hyperbola, form the rectangle containing the points $(\pm a, 0)$ and $(0, \pm b)$, that is, $(-2, 0)$, $(2, 0)$, $(0, -3)$, and $(0, 3)$. The extensions of the diagonals of this rectangle are the asymptotes. See Figure 41 for the graph.

─ Now Work PROBLEM 31

3 Analyze Hyperbolas with Center at (h, k)

If a hyperbola with center at the origin and transverse axis coinciding with a coordinate axis is shifted horizontally h units and then vertically k units, the result is a hyperbola with center at (h, k) and transverse axis parallel to a coordinate axis. The equations of such hyperbolas have the same forms as those given in equations (2) and (3), except that x is replaced by $x - h$ (the horizontal shift) and y is replaced by $y - k$ (the vertical shift). Table 4 gives the forms of the equations of such hyperbolas. See Figure 42 for typical graphs.

Table 4

				Equation of a Hyperbola: Center at (h, k); Transverse Axis Parallel to a Coordinate Axis	
Center	**Transverse Axis**	**Foci**	**Vertices**	**Equation**	**Asymptotes**
(h, k)	Parallel to the x-axis	$(h \pm c, k)$	$(h \pm a, k)$	$\dfrac{(x - h)^2}{a^2} - \dfrac{(y - k)^2}{b^2} = 1$, $\ b^2 = c^2 - a^2$	$y - k = \pm\dfrac{b}{a}(x - h)$
(h, k)	Parallel to the y-axis	$(h, k \pm c)$	$(h, k \pm a)$	$\dfrac{(y - k)^2}{a^2} - \dfrac{(x - h)^2}{b^2} = 1$, $\ b^2 = c^2 - a^2$	$y - k = \pm\dfrac{a}{b}(x - h)$

Figure 42

NOTE It is not recommended that Table 4 be memorized. Rather use the ideas of transformations (shift horizontally h units, vertically k units) along with the fact that a represents the distance from the center to the vertices, c represents the distance from the center to the foci, and $b^2 = c^2 - a^2$ (or $c^2 = a^2 + b^2$). ■

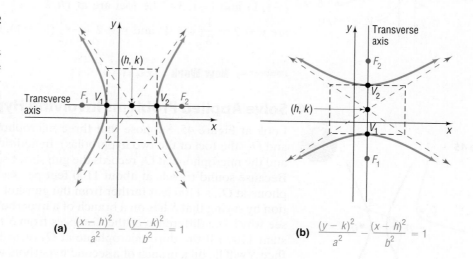

(a) $\dfrac{(x - h)^2}{a^2} - \dfrac{(y - k)^2}{b^2} = 1$

(b) $\dfrac{(y - k)^2}{a^2} - \dfrac{(x - h)^2}{b^2} = 1$

EXAMPLE 7 **Finding an Equation of a Hyperbola, Center Not at the Origin**

Find an equation for the hyperbola with center at $(1, -2)$, one focus at $(4, -2)$, and one vertex at $(3, -2)$. Graph the equation.

Solution The center is at $(h, k) = (1, -2)$, so $h = 1$ and $k = -2$. Since the center, focus, and vertex all lie on the line $y = -2$, the transverse axis is parallel to the x-axis. The distance from the center $(1, -2)$ to the focus $(4, -2)$ is $c = 3$; the distance from

Figure 43

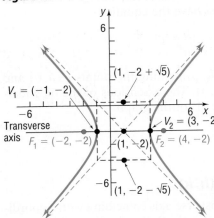

the center $(1, -2)$ to the vertex $(3, -2)$ is $a = 2$. Then $b^2 = c^2 - a^2 = 9 - 4 = 5$. The equation is

$$\frac{(x - h)^2}{a^2} - \frac{(y - k)^2}{b^2} = 1$$

$$\frac{(x - 1)^2}{4} - \frac{(y + 2)^2}{5} = 1$$

See Figure 43.

——— **Now Work** PROBLEM 41

EXAMPLE 8 **Analyzing the Equation of a Hyperbola**

Analyze the equation: $-x^2 + 4y^2 - 2x - 16y + 11 = 0$

Solution We complete the squares in x and in y.

$$-x^2 + 4y^2 - 2x - 16y + 11 = 0$$

$$-(x^2 + 2x) + 4(y^2 - 4y) = -11 \qquad \text{Group terms.}$$

$$-(x^2 + 2x + 1) + 4(y^2 - 4y + 4) = -11 - 1 + 16 \qquad \text{Complete each square.}$$

$$-(x + 1)^2 + 4(y - 2)^2 = 4 \qquad \text{Factor.}$$

$$(y - 2)^2 - \frac{(x + 1)^2}{4} = 1 \qquad \text{Divide each side by 4.}$$

Figure 44

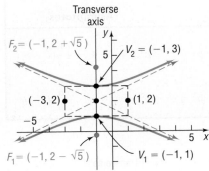

This is the equation of a hyperbola with center at $(-1, 2)$ and transverse axis parallel to the y-axis. Also, $a^2 = 1$ and $b^2 = 4$, so $c^2 = a^2 + b^2 = 5$. Since the transverse axis is parallel to the y-axis, the vertices and foci are located a and c units above and below the center, respectively. The vertices are at $(h, k \pm a) = (-1, 2 \pm 1)$, or $(-1, 1)$ and $(-1, 3)$. The foci are at $(h, k \pm c) = (-1, 2 \pm \sqrt{5})$. The asymptotes are $y - 2 = \frac{1}{2}(x + 1)$ and $y - 2 = -\frac{1}{2}(x + 1)$. Figure 44 shows the graph.

——— **Now Work** PROBLEM 55

4 Solve Applied Problems Involving Hyperbolas

Figure 45

Look at Figure 45. Suppose that three microphones are located at points O_1, O_2, and O_3 (the foci of the two hyperbolas). In addition, suppose that a gun is fired at S and the microphone at O_1 records the gun shot 1 second after the microphone at O_2. Because sound travels at about 1100 feet per second, we conclude that the microphone at O_1 is 1100 feet farther from the gunshot than O_2. We can model this situation by saying that S lies on a branch of a hyperbola with foci at O_1 and O_2. (Do you see why? The difference of the distances from S to O_1 and from S to O_2 is the constant 1100.) If the third microphone at O_3 records the gunshot 2 seconds after O_1, then S will lie on a branch of a second hyperbola with foci at O_1 and O_3. In this case, the constant difference will be 2200. The intersection of the two hyperbolas will identify the location of S.

EXAMPLE 9 **Lightning Strikes**

Suppose that two people standing 1 mile apart both see a flash of lightning. After a period of time, the person standing at point A hears the thunder. One second later, the person standing at point B hears the thunder. If the person at B is due west of the person at A and the lightning strike is known to occur due north of the person standing at point A, where did the lightning strike?

Solution   See Figure 46 in which the ordered pair (x, y) represents the location of the lightning strike. We know that sound travels at 1100 feet per second, so the person at point A is 1100 feet closer to the lightning strike than the person at point B. Since the difference of the distance from (x, y) to A and the distance from (x, y) to B is the constant 1100, the point (x, y) lies on a hyperbola whose foci are at A and B.

Figure 46

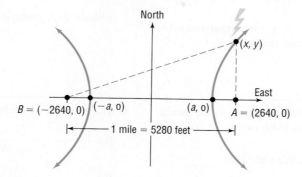

An equation of the hyperbola is

$$\frac{x^2}{a^2} - \frac{y^2}{b^2} = 1$$

where $2a = 1100$, so $a = 550$.

Because the distance between the two people is 1 mile (5280 feet) and each person is at a focus of the hyperbola, we have

$$2c = 5280$$

$$c = \frac{5280}{2} = 2640$$

Since $b^2 = c^2 - a^2 = 2640^2 - 550^2 = 6{,}667{,}100$, the equation of the hyperbola that describes the location of the lightning strike is

$$\frac{x^2}{550^2} - \frac{y^2}{6{,}667{,}100} = 1$$

Refer again to Figure 46. Since the lightning strikes due north of the individual at the point $A = (2640, 0)$, we let $x = 2640$ and solve the resulting equation.

$$\frac{2640^2}{550^2} - \frac{y^2}{6{,}667{,}100} = 1$$

$$-\frac{y^2}{6{,}667{,}100} = -22.04 \qquad \text{Subtract } \frac{2640^2}{550^2} \text{ from both sides.}$$

$$y^2 = 146{,}942{,}884 \qquad \text{Multiply both sides by } -6{,}667{,}100.$$

$$y = 12{,}122 \qquad \begin{array}{l}y > 0 \text{ since the lightning strike}\\ \text{occurred in quadrant I.}\end{array}$$

✓ **Check:** The difference between the distance from $(2640, 12122)$ to the person at the point $B = (-2640, 0)$ and the distance from $(2640, 12122)$ to the person at the point $A = (2640, 0)$ should be 1100. Using the distance formula, we find the difference in the distances is

$$\sqrt{[2640 - (-2640)]^2 + (12{,}122 - 0)^2} - \sqrt{(2640 - 2640)^2 + (12{,}122 - 0)^2} = 1100$$

as required.

The lightning strike is 12,122 feet north of the person standing at point A.

━━ **Now Work** PROBLEM 75

9.4 Assess Your Understanding

1. The distance d from $P_1 = (3, -4)$ to $P_2 = (-2, 1)$ is $d =$ _____. (p. 3)

2. To complete the square of $x^2 + 5x$, add _____. (p. A40)

3. Find the intercepts of the equation $y^2 = 9 + 4x^2$. (p. 12)

4. *True or False* The equation $y^2 = 9 + x^2$ is symmetric with respect to the x-axis, the y-axis, and the origin. (pp. 12–14)

5. To graph $y = (x - 5)^3 - 4$, shift the graph of $y = x^3$ to the (left/right) _____ unit(s) and then (up/down) _____ unit(s). (pp. 90–93)

6. Find the vertical asymptotes, if any, and the horizontal or oblique asymptote, if any, of $y = \dfrac{x^2 - 9}{x^2 - 4}$. (pp. 217–223)

Concepts and Vocabulary

7. A(n) _____ is the collection of points in the plane the difference of whose distances from two fixed points is a constant.

8. For a hyperbola, the foci lie on a line called the _____ _____.

Answer Problems 9–11 using the figure.

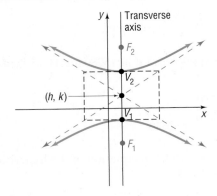

9. The equation of the hyperbola is of the form

(a) $\dfrac{(x - h)^2}{a^2} - \dfrac{(y - k)^2}{b^2} = 1$

(b) $\dfrac{(y - k)^2}{a^2} - \dfrac{(x - h)^2}{b^2} = 1$

10. If the center of the hyperbola is $(2, 1)$ and $a = 3$, then the coordinates of the vertices are _____ and _____.

11. If the center of the hyperbola is $(2, 1)$ and $c = 5$, then the coordinates of the foci are _____ and _____.

12. In a hyperbola, if $a = 3$ and $c = 5$, then $b =$ _____.

13. For the hyperbola $\dfrac{x^2}{4} - \dfrac{y^2}{9} = 1$, the value of a is _____, the value of b is _____, and the transverse axis is the _____-axis.

14. For the hyperbola $\dfrac{y^2}{16} - \dfrac{x^2}{81} = 1$, the asymptotes are _____ and _____.

Skill Building

In Problems 15–18, the graph of a hyperbola is given. Match each graph to its equation.

(A) $\dfrac{x^2}{4} - y^2 = 1$

(B) $x^2 - \dfrac{y^2}{4} = 1$

(C) $\dfrac{y^2}{4} - x^2 = 1$

(D) $y^2 - \dfrac{x^2}{4} = 1$

15.

16.

17.

18.

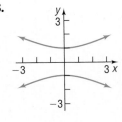

In Problems 19–28, find an equation for the hyperbola described. Graph the equation.

19. Center at $(0, 0)$; focus at $(3, 0)$; vertex at $(1, 0)$

20. Center at $(0, 0)$; focus at $(0, 5)$; vertex at $(0, 3)$

21. Center at $(0, 0)$; focus at $(0, -6)$; vertex at $(0, 4)$

22. Center at $(0, 0)$; focus at $(-3, 0)$; vertex at $(2, 0)$

23. Foci at $(-5, 0)$ and $(5, 0)$; vertex at $(3, 0)$

24. Focus at $(0, 6)$; vertices at $(0, -2)$ and $(0, 2)$

25. Vertices at $(0, -6)$ and $(0, 6)$; asymptote the line $y = 2x$

26. Vertices at $(-4, 0)$ and $(4, 0)$; asymptote the line $y = 2x$

27. Foci at $(-4, 0)$ and $(4, 0)$; asymptote the line $y = -x$

28. Foci at $(0, -2)$ and $(0, 2)$; asymptote the line $y = -x$

In Problems 29–36, find the center, transverse axis, vertices, foci, and asymptotes. Graph each equation.

29. $\dfrac{x^2}{25} - \dfrac{y^2}{9} = 1$ **30.** $\dfrac{y^2}{16} - \dfrac{x^2}{4} = 1$ **31.** $4x^2 - y^2 = 16$ **32.** $4y^2 - x^2 = 16$

33. $y^2 - 9x^2 = 9$ **34.** $x^2 - y^2 = 4$ **35.** $y^2 - x^2 = 25$ **36.** $2x^2 - y^2 = 4$

In Problems 37–40, write an equation for each hyperbola.

37.   **38.**   **39.**   **40.**

In Problems 41–48, find an equation for the hyperbola described. Graph the equation.

41. Center at $(4, -1)$; focus at $(7, -1)$; vertex at $(6, -1)$

42. Center at $(-3, 1)$; focus at $(-3, 6)$; vertex at $(-3, 4)$

43. Center at $(-3, -4)$; focus at $(-3, -8)$; vertex at $(-3, -2)$

44. Center at $(1, 4)$; focus at $(-2, 4)$; vertex at $(0, 4)$

45. Foci at $(3, 7)$ and $(7, 7)$; vertex at $(6, 7)$

46. Focus at $(-4, 0)$ vertices at $(-4, 4)$ and $(-4, 2)$

47. Vertices at $(-1, -1)$ and $(3, -1)$; asymptote the line $y + 1 = \dfrac{3}{2}(x - 1)$

48. Vertices at $(1, -3)$ and $(1, 1)$; asymptote the line $y + 1 = \dfrac{3}{2}(x - 1)$

In Problems 49–62, find the center, transverse axis, vertices, foci, and asymptotes. Graph each equation.

49. $\dfrac{(x - 2)^2}{4} - \dfrac{(y + 3)^2}{9} = 1$ **50.** $\dfrac{(y + 3)^2}{4} - \dfrac{(x - 2)^2}{9} = 1$ **51.** $(y - 2)^2 - 4(x + 2)^2 = 4$

52. $(x + 4)^2 - 9(y - 3)^2 = 9$ **53.** $(x + 1)^2 - (y + 2)^2 = 4$ **54.** $(y - 3)^2 - (x + 2)^2 = 4$

55. $x^2 - y^2 - 2x - 2y - 1 = 0$ **56.** $y^2 - x^2 - 4y + 4x - 1 = 0$ **57.** $y^2 - 4x^2 - 4y - 8x - 4 = 0$

58. $2x^2 - y^2 + 4x + 4y - 4 = 0$ **59.** $4x^2 - y^2 - 24x - 4y + 16 = 0$ **60.** $2y^2 - x^2 + 2x + 8y + 3 = 0$

61. $y^2 - 4x^2 - 16x - 2y - 19 = 0$ **62.** $x^2 - 3y^2 + 8x - 6y + 4 = 0$

In Problems 63–66, graph each function. Be sure to label any intercepts.
[Hint: *Notice that each function is half a hyperbola.]*

63. $f(x) = \sqrt{16 + 4x^2}$ **64.** $f(x) = -\sqrt{9 + 9x^2}$ **65.** $f(x) = -\sqrt{-25 + x^2}$ **66.** $f(x) = \sqrt{-1 + x^2}$

Mixed Practice

In Problems 67–74, analyze each conic.

67. $\dfrac{(x - 3)^2}{4} - \dfrac{y^2}{25} = 1$ **68.** $\dfrac{(y + 2)^2}{16} - \dfrac{(x - 2)^2}{4} = 1$ **69.** $x^2 = 16(y - 3)$

70. $y^2 = -12(x + 1)$ **71.** $25x^2 + 9y^2 - 250x + 400 = 0$ **72.** $x^2 + 36y^2 - 2x + 288y + 541 = 0$

73. $x^2 - 6x - 8y - 31 = 0$ **74.** $9x^2 - y^2 - 18x - 8y - 88 = 0$

Applications and Extensions

 75. Fireworks Display Suppose that two people standing 2 miles apart both see the burst from a fireworks display. After a period of time, the first person standing at point A hears the burst. One second later, the second person standing at point B hears the burst. If the person at point B is due west of the person at point A and if the display is known to occur due north of the person at point A, where did the fireworks display occur?

76. **Lightning Strikes** Suppose that two people standing 1 mile apart both see a flash of lightning. After a period of time, the first person standing at point A hears the thunder. Two seconds later, the second person standing at point B hears the thunder. If the person at point B is due west of the person at point A and if the lightning strike is known to occur due north of the person standing at point A, where did the lightning strike?

77. **Nuclear Power Plant** Some nuclear power plants utilize "natural draft" cooling towers in the shape of a **hyperboloid,** a solid obtained by rotating a hyperbola about its conjugate axis. Suppose that such a cooling tower has a base diameter of 400 feet and the diameter at its narrowest point, 360 feet above the ground, is 200 feet. If the diameter at the top of the tower is 300 feet, how tall is the tower?

Source: Bay Area Air Quality Management District

78. **An Explosion** Two recording devices are set 2400 feet apart, with the device at point A to the west of the device at point B. At a point between the devices, 300 feet from point B, a small amount of explosive is detonated. The recording devices record the time until the sound reaches each. How far directly north of point B should a second explosion be done so that the measured time difference recorded by the devices is the same as that for the first detonation?

79. **Rutherford's Experiment** In May 1911, Ernest Rutherford published a paper in *Philosophical Magazine*. In this article, he described the motion of alpha particles as they are shot at a piece of gold foil 0.00004 cm thick. Before conducting this experiment, Rutherford expected that the alpha particles would shoot through the foil just as a bullet would shoot through snow. Instead, a small fraction of the alpha particles bounced off the foil. This led to the conclusion that the nucleus of an atom is dense, while the remainder of the atom is sparse. Only the density of the nucleus could cause the alpha particles to deviate from their path. The figure shows a diagram from Rutherford's paper that indicates that the deflected alpha particles follow the path of one branch of a hyperbola.

(a) Find an equation of the asymptotes under this scenario.
(b) If the vertex of the path of the alpha particles is 10 cm from the center of the hyperbola, find a model that describes the path of the particle.

80. **Hyperbolic Mirrors** Hyperbolas have interesting reflective properties that make them useful for lenses and mirrors. For example, if a ray of light strikes a convex hyperbolic mirror on a line that would (theoretically) pass through its rear focus, it is reflected through the front focus. This property, and that of the parabola, were used to develop the *Cassegrain* telescope in 1672. The focus of the parabolic mirror and the rear focus of the hyperbolic mirror are the same point. The rays are collected by the parabolic mirror, reflected toward the (common) focus, and thus are reflected by the hyperbolic mirror through the opening to its front focus, where the eyepiece is located. If the equation of the hyperbola is $\dfrac{y^2}{9} - \dfrac{x^2}{16} = 1$ and the focal length (distance from the vertex to the focus) of the parabola is 6, find the equation of the parabola.

Source: www.enchantedlearning.com

81. The **eccentricity** e of a hyperbola is defined as the number $\dfrac{c}{a}$, where a and c are the numbers given in equation (2). Because $c > a$, it follows that $e > 1$. Describe the general shape of a hyperbola whose eccentricity is close to 1. What is the shape if e is very large?

82. A hyperbola for which $a = b$ is called an **equilateral hyperbola.** Find the eccentricity e of an equilateral hyperbola.

[**Note:** The eccentricity of a hyperbola is defined in Problem 81.]

83. Two hyperbolas that have the same set of asymptotes are called **conjugate.** Show that the hyperbolas

$$\frac{x^2}{4} - y^2 = 1 \quad \text{and} \quad y^2 - \frac{x^2}{4} = 1$$

are conjugate. Graph each hyperbola on the same set of coordinate axes.

84. Prove that the hyperbola

$$\frac{y^2}{a^2} - \frac{x^2}{b^2} = 1$$

has the two oblique asymptotes

$$y = \frac{a}{b}x \quad \text{and} \quad y = -\frac{a}{b}x$$

85. Show that the graph of an equation of the form

$$Ax^2 + Cy^2 + F = 0 \qquad A \neq 0, C \neq 0, F \neq 0$$

where A and C are of opposite sign, is a hyperbola with center at $(0, 0)$.

86. Show that the graph of an equation of the form

$$Ax^2 + Cy^2 + Dx + Ey + F = 0 \qquad A \neq 0, C \neq 0$$

where A and C are of opposite sign,

(a) is a hyperbola if $\dfrac{D^2}{4A} + \dfrac{E^2}{4C} - F \neq 0$.

(b) is two intersecting lines if $\dfrac{D^2}{4A} + \dfrac{E^2}{4C} - F = 0$.

'Are You Prepared?' Answers

1. $5\sqrt{2}$ 2. $\dfrac{25}{4}$ 3. $(0, -3), (0, 3)$ 4. True   5. right; 5; down: 4   6. Vertical: $x = -2, x = 2$; horizontal: $y = 1$

9.5 Rotation of Axes; General Form of a Conic

PREPARING FOR THIS SECTION *Before getting started, review the following:*

- Sum Formulas for Sine and Cosine (Section 6.5, pp. 514 and 517)
- Half-angle Formulas for Sine and Cosine (Section 6.6, p. 530)
- Double-angle Formulas for Sine and Cosine (Section 6.6, p. 526)

Now Work the *'Are You Prepared?'* problems on page 705.

OBJECTIVES 1 Identify a Conic (p. 699)
          2 Use a Rotation of Axes to Transform Equations (p. 700)
          3 Analyze an Equation Using a Rotation of Axes (p. 702)
          4 Identify Conics without a Rotation of Axes (p. 704)

In this section, we show that the graph of a general second-degree polynomial containing two variables x and y, that is, an equation of the form

$$Ax^2 + Bxy + Cy^2 + Dx + Ey + F = 0 \qquad \textbf{(1)}$$

where A, B, and C are not simultaneously 0, is a conic. We shall not concern ourselves here with the degenerate cases of equation (1), such as $x^2 + y^2 = 0$, whose graph is a single point $(0, 0)$; or $x^2 + 3y^2 + 3 = 0$, whose graph contains no points; or $x^2 - 4y^2 = 0$, whose graph is two lines, $x - 2y = 0$ and $x + 2y = 0$.

We begin with the case where $B = 0$. In this case, the term containing xy is not present, so equation (1) has the form

$$Ax^2 + Cy^2 + Dx + Ey + F = 0$$

where either $A \neq 0$ or $C \neq 0$.

1 Identify a Conic

We have already discussed the procedure for identifying the graph of this kind of equation; we complete the squares of the quadratic expressions in x or y, or both. Once this has been done, the conic can be identified by comparing it to one of the forms studied in Sections 9.2 through 9.4.

In fact, though, we can identify the conic directly from the equation without completing the squares.

THEOREM

Identifying Conics without Completing the Squares

Excluding degenerate cases, the equation

$$Ax^2 + Cy^2 + Dx + Ey + F = 0 \qquad \textbf{(2)}$$

where A and C cannot both equal zero:
(a) Defines a parabola if $AC = 0$.
(b) Defines an ellipse (or a circle) if $AC > 0$.
(c) Defines a hyperbola if $AC < 0$.

Proof
(a) If $AC = 0$, then either $A = 0$ or $C = 0$, but not both, so the form of equation (2) is either

$$Ax^2 + Dx + Ey + F = 0, \qquad A \neq 0$$

or

$$Cy^2 + Dx + Ey + F = 0, \qquad C \neq 0$$

Using the results of Problems 78 and 79 in Exercise 9.2, it follows that, except for the degenerate cases, the equation is a parabola.

(b) If $AC > 0$, then A and C are of the same sign. Using the results of Problem 84 in Exercise 9.3, except for the degenerate cases, the equation is an ellipse if $A \neq C$ or a circle if $A = C$.

(c) If $AC < 0$, then A and C are of opposite sign. Using the results of Problem 86 in Exercise 9.4, except for the degenerate cases, the equation is a hyperbola. ∎

We will not be concerned with the degenerate cases of equation (2). However, in practice, you should be alert to the possibility of degeneracy.

EXAMPLE 1   **Identifying a Conic without Completing the Squares**

Identify each equation without completing the squares.

(a) $3x^2 + 6y^2 + 6x - 12y = 0$ (b) $2x^2 - 3y^2 + 6y + 4 = 0$

(c) $y^2 - 2x + 4 = 0$

Solution   (a) We compare the given equation to equation (2) and conclude that $A = 3$ and $C = 6$. Since $AC = 18 > 0$, the equation is an ellipse.

(b) Here $A = 2$ and $C = -3$, so $AC = -6 < 0$. The equation is a hyperbola.

(c) Here $A = 0$ and $C = 1$, so $AC = 0$. The equation is a parabola.

●

──── **Now Work** PROBLEM 11

Although we can now identify the type of conic represented by any equation of the form of equation (2) without completing the squares, we will still need to complete the squares if we desire additional information about the conic, such as its graph.

2 Use a Rotation of Axes to Transform Equations

Now we turn our attention to equations of the form of equation (1), where $B \neq 0$. To discuss this case, we first need to investigate a new procedure: *rotation of axes.*

In a **rotation of axes,** the origin remains fixed while the x-axis and y-axis are rotated through an angle θ to a new position; the new positions of the x-axis and the y-axis are denoted by x' and y', respectively, as shown in Figure 47(a).

Now look at Figure 47(b). There the point P has the coordinates (x, y) relative to the xy-plane, while the same point P has coordinates (x', y') relative to the $x'y'$-plane. We seek relationships that will enable us to express x and y in terms of x', y', and θ.

As Figure 47(b) shows, r denotes the distance from the origin O to the point P, and α denotes the angle between the positive x'-axis and the ray from O through P. Then, using the definitions of sine and cosine, we have

$$x' = r \cos \alpha \qquad y' = r \sin \alpha \qquad \text{(3)}$$

$$x = r \cos(\theta + \alpha) \qquad y = r \sin(\theta + \alpha) \qquad \text{(4)}$$

Figure 47

(a)

(b)

Now

$$x = r \cos(\theta + \alpha)$$
$$= r(\cos \theta \cos \alpha - \sin \theta \sin \alpha) \qquad \text{Apply the Sum Formula for cosine.}$$
$$= (r \cos \alpha)(\cos \theta) - (r \sin \alpha)(\sin \theta)$$
$$= x' \cos \theta - y' \sin \theta \qquad \text{By equation (3)}$$

Similarly,

$$y = r \sin(\theta + \alpha)$$
$$= r(\sin \theta \cos \alpha + \cos \theta \sin \alpha) \qquad \text{Apply the Sum Formula for sine.}$$
$$= x' \sin \theta + y' \cos \theta \qquad \text{By equation (3)}$$

THEOREM

Rotation Formulas

If the x- and y-axes are rotated through an angle θ, the coordinates (x, y) of a point P relative to the xy-plane and the coordinates (x', y') of the same point relative to the new x'- and y'-axes are related by the formulas

$$x = x' \cos \theta - y' \sin \theta \qquad y = x' \sin \theta + y' \cos \theta \qquad (5)$$

EXAMPLE 2 **Rotating Axes**

Express the equation $xy = 1$ in terms of new $x'y'$-coordinates by rotating the axes through a $45°$ angle. Discuss the new equation.

Solution Let $\theta = 45°$ in equation (5). Then

$$x = x' \cos 45° - y' \sin 45° = x' \frac{\sqrt{2}}{2} - y' \frac{\sqrt{2}}{2} = \frac{\sqrt{2}}{2}(x' - y')$$

$$y = x' \sin 45° + y' \cos 45° = x' \frac{\sqrt{2}}{2} + y' \frac{\sqrt{2}}{2} = \frac{\sqrt{2}}{2}(x' + y')$$

Figure 48

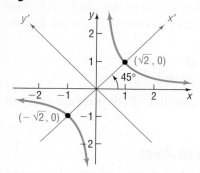

Substituting these expressions for x and y in $xy = 1$ gives

$$\left[\frac{\sqrt{2}}{2}(x' - y')\right]\left[\frac{\sqrt{2}}{2}(x' + y')\right] = 1$$

$$\frac{1}{2}(x'^2 - y'^2) = 1$$

$$\frac{x'^2}{2} - \frac{y'^2}{2} = 1$$

This is the equation of a hyperbola with center at $(0, 0)$ and transverse axis along the x'-axis. The vertices are at $(\pm\sqrt{2}, 0)$ on the x'-axis; the asymptotes are $y' = x'$ and $y' = -x'$ (which correspond to the original x- and y-axes). See Figure 48 for the graph.

●

As Example 2 illustrates, a rotation of axes through an appropriate angle can transform a second-degree equation in x and y containing an xy-term into one in x' and y' in which no $x'y'$-term appears. In fact, we will show that a rotation of axes through an appropriate angle will transform any equation of the form of equation (1) into an equation in x' and y' without an $x'y'$-term.

To find the formula for choosing an appropriate angle θ through which to rotate the axes, we begin with equation (1),

$$Ax^2 + Bxy + Cy^2 + Dx + Ey + F = 0 \qquad B \neq 0$$

Next we rotate through an angle θ using the rotation formulas (5).

$$A(x' \cos \theta - y' \sin \theta)^2 + B(x' \cos \theta - y' \sin \theta)(x' \sin \theta + y' \cos \theta)$$
$$+ C(x' \sin \theta + y' \cos \theta)^2 + D(x' \cos \theta - y' \sin \theta)$$
$$+ E(x' \sin \theta + y' \cos \theta) + F = 0$$

By expanding and collecting like terms, we obtain

$$(A \cos^2 \theta + B \sin \theta \cos \theta + C \sin^2 \theta)x'^2 + [B(\cos^2 \theta - \sin^2 \theta) + 2(C - A)(\sin \theta \cos \theta)]x'y'$$
$$+ (A \sin^2 \theta - B \sin \theta \cos \theta + C \cos^2 \theta)y'^2$$
$$+ (D \cos \theta + E \sin \theta)x'$$
$$+ (-D \sin \theta + E \cos \theta)y' + F = 0 \qquad (6)$$

In equation (6), the coefficient of $x'y'$ is

$$B(\cos^2 \theta - \sin^2 \theta) + 2(C - A)(\sin \theta \cos \theta)$$

Since we want to eliminate the $x'y'$-term, we select an angle θ so that this coefficient is 0.

$$B(\cos^2 \theta - \sin^2 \theta) + 2(C - A)(\sin \theta \cos \theta) = 0$$

$$B \cos(2\theta) + (C - A) \sin(2\theta) = 0 \qquad \text{Double-angle Formulas}$$

$$B \cos(2\theta) = (A - C) \sin(2\theta)$$

$$\cot(2\theta) = \frac{A - C}{B} \qquad B \neq 0$$

THEOREM

To transform the equation

$$Ax^2 + Bxy + Cy^2 + Dx + Ey + F = 0 \qquad B \neq 0$$

into an equation in x' and y' without an $x'y'$-term, rotate the axes through an angle θ that satisfies the equation

$$\cot(2\theta) = \frac{A - C}{B} \qquad \qquad \textbf{(7)}$$

WARNING Be careful if you use a calculator to solve equation (7).

1. If $\cot(2\theta) = 0$, then $2\theta = 90°$ and $\theta = 45°$.

2. If $\cot(2\theta) \neq 0$, first find $\cos(2\theta)$. Then use the inverse cosine function key(s) to obtain 2θ, $0° < 2\theta < 180°$. Finally, divide by 2 to obtain the correct acute angle θ. ∎

Equation (7) has an infinite number of solutions for θ. We shall adopt the convention of choosing the acute angle θ that satisfies (7). Then we have the following two possibilities:

If $\cot(2\theta) \geq 0$, then $0° < 2\theta \leq 90°$, so $0° < \theta \leq 45°$.

If $\cot(2\theta) < 0$, then $90° < 2\theta < 180°$, so $45° < \theta < 90°$.

Each of these results in a counterclockwise rotation of the axes through an acute angle θ.*

3 Analyze an Equation Using a Rotation of Axes

For the remainder of this section, the direction **"Analyze the equation"** will mean to transform the given equation so that it contains no xy-term and to graph the equation.

EXAMPLE 3

Analyzing an Equation Using a Rotation of Axes

Analyze the equation: $x^2 + \sqrt{3}xy + 2y^2 - 10 = 0$

Solution

Since an xy-term is present, we must rotate the axes. Using $A = 1$, $B = \sqrt{3}$, and $C = 2$ in equation (7), the appropriate acute angle θ through which to rotate the axes satisfies the equation

$$\cot(2\theta) = \frac{A - C}{B} = \frac{-1}{\sqrt{3}} = -\frac{\sqrt{3}}{3} \qquad 0° < 2\theta < 180°$$

Since $\cot(2\theta) = -\dfrac{\sqrt{3}}{3}$, we find $2\theta = 120°$, so $\theta = 60°$. Using $\theta = 60°$ in the rotation formulas (5), we find

$$x = x'\cos 60° - y'\sin 60° = \frac{1}{2}x' - \frac{\sqrt{3}}{2}y' = \frac{1}{2}\left(x' - \sqrt{3}y'\right)$$

$$y = x'\sin 60° + y'\cos 60° = \frac{\sqrt{3}}{2}x' + \frac{1}{2}y' = \frac{1}{2}\left(\sqrt{3}x' + y'\right)$$

---

*Any rotation through an angle θ that satisfies $\cot(2\theta) = \dfrac{A - C}{B}$ will eliminate the $x'y'$-term. However, the final form of the transformed equation may be different (but equivalent), depending on the angle chosen.

Substituting these values into the original equation and simplifying, we have

$$x^2 + \sqrt{3}xy + 2y^2 - 10 = 0$$

$$\frac{1}{4}(x' - \sqrt{3}y')^2 + \sqrt{3}\left[\frac{1}{2}(x' - \sqrt{3}y')\right]\left[\frac{1}{2}(\sqrt{3}x' + y')\right] + 2\left[\frac{1}{4}(\sqrt{3}x' + y')^2\right] = 10$$

Multiply both sides by 4 and expand to obtain

$$x'^2 - 2\sqrt{3}x'y' + 3y'^2 + \sqrt{3}(\sqrt{3}x'^2 - 2x'y' - \sqrt{3}y'^2) + 2(3x'^2 + 2\sqrt{3}x'y' + y'^2) = 40$$

$$10x'^2 + 2y'^2 = 40$$

$$\frac{x'^2}{4} + \frac{y'^2}{20} = 1$$

Figure 49

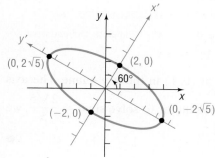

This is the equation of an ellipse with center at $(0, 0)$ and major axis along the y'-axis. The vertices are at $(0, \pm 2\sqrt{5})$ on the y'-axis. See Figure 49 for the graph.

●

━━━ **Now Work** PROBLEM 31

In Example 3, the acute angle θ through which to rotate the axes was easy to find because of the numbers that we used in the given equation. In general, the equation $\cot(2\theta) = \dfrac{A - C}{B}$ will not have such a "nice" solution. As the next example shows, we can still find the appropriate rotation formulas without using a calculator approximation by applying Half-angle Formulas.

EXAMPLE 4   **Analyzing an Equation Using a Rotation of Axes**

Analyze the equation: $4x^2 - 4xy + y^2 + 5\sqrt{5}x + 5 = 0$

Solution   Letting $A = 4$, $B = -4$, and $C = 1$ in equation (7), the appropriate angle θ through which to rotate the axes satisfies

$$\cot(2\theta) = \frac{A - C}{B} = \frac{3}{-4} = -\frac{3}{4}$$

To use the rotation formulas (5), we need to know the values of $\sin\theta$ and $\cos\theta$. Since we seek an acute angle θ, we know that $\sin\theta > 0$ and $\cos\theta > 0$. We use the Half-angle Formulas in the form

$$\sin\theta = \sqrt{\frac{1 - \cos(2\theta)}{2}} \qquad \cos\theta = \sqrt{\frac{1 + \cos(2\theta)}{2}}$$

Now we need to find the value of $\cos(2\theta)$. Since $\cot(2\theta) = -\dfrac{3}{4}$, then $90° < 2\theta < 180°$ (Do you know why?), so $\cos(2\theta) = -\dfrac{3}{5}$. Then

$$\sin\theta = \sqrt{\frac{1 - \cos(2\theta)}{2}} = \sqrt{\frac{1 - \left(-\frac{3}{5}\right)}{2}} = \sqrt{\frac{4}{5}} = \frac{2}{\sqrt{5}} = \frac{2\sqrt{5}}{5}$$

$$\cos\theta = \sqrt{\frac{1 + \cos(2\theta)}{2}} = \sqrt{\frac{1 + \left(-\frac{3}{5}\right)}{2}} = \sqrt{\frac{1}{5}} = \frac{1}{\sqrt{5}} = \frac{\sqrt{5}}{5}$$

With these values, the rotation formulas (5) are

$$x = \frac{\sqrt{5}}{5}x' - \frac{2\sqrt{5}}{5}y' = \frac{\sqrt{5}}{5}(x' - 2y')$$

$$y = \frac{2\sqrt{5}}{5}x' + \frac{\sqrt{5}}{5}y' = \frac{\sqrt{5}}{5}(2x' + y')$$

Substituting these values in the original equation and simplifying, we obtain

$$4x^2 - 4xy + y^2 + 5\sqrt{5}x + 5 = 0$$

$$4\left[\frac{\sqrt{5}}{5}(x' - 2y')\right]^2 - 4\left[\frac{\sqrt{5}}{5}(x' - 2y')\right]\left[\frac{\sqrt{5}}{5}(2x' + y')\right]$$

$$+ \left[\frac{\sqrt{5}}{5}(2x' + y')\right]^2 + 5\sqrt{5}\left[\frac{\sqrt{5}}{5}(x' - 2y')\right] = -5$$

Multiply both sides by 5 and expand to obtain

$$4(x'^2 - 4x'y' + 4y'^2) - 4(2x'^2 - 3x'y' - 2y'^2)$$

$$+ 4x'^2 + 4x'y' + y'^2 + 25(x' - 2y') = -25$$

$$25y'^2 - 50y' + 25x' = -25 \qquad \textit{Combine like terms.}$$

$$y'^2 - 2y' + x' = -1 \qquad \textit{Divide by 25.}$$

$$y'^2 - 2y' + 1 = -x' \qquad \textit{Complete the square in } y'.$$

$$(y' - 1)^2 = -x'$$

Figure 50

This is the equation of a parabola with vertex at $(0, 1)$ in the $x'y'$-plane. The axis of symmetry is parallel to the x'-axis. Using a calculator to solve $\sin\theta = \dfrac{2\sqrt{5}}{5}$, we find that $\theta \approx 63.4°$. See Figure 50 for the graph.

━━━━━**Now Work** PROBLEM 37

4 Identify Conics without a Rotation of Axes

Suppose that we are required only to identify (rather than analyze) an equation of the form

$$Ax^2 + Bxy + Cy^2 + Dx + Ey + F = 0 \qquad B \neq 0 \qquad \textbf{(8)}$$

If we apply the rotation formulas (5) to this equation, we obtain an equation of the form

$$A'x'^2 + B'x'y' + C'y'^2 + D'x' + E'y' + F' = 0 \qquad \textbf{(9)}$$

where A', B', C', D', E', and F' can be expressed in terms of A, B, C, D, E, F and the angle θ of rotation (see Problem 53). It can be shown that the value of $B^2 - 4AC$ in equation (8) and the value of $B'^2 - 4A'C'$ in equation (9) are equal no matter what angle θ of rotation is chosen (see Problem 55). In particular, if the angle θ of rotation satisfies equation (7), then $B' = 0$ in equation (9), and $B^2 - 4AC = -4A'C'$. Since equation (9) then has the form of equation (2),

$$A'x'^2 + C'y'^2 + D'x' + E'y' + F' = 0$$

we can identify it without completing the squares, as we did in the beginning of this section. In fact, now we can identify the conic described by any equation of the form of equation (8) without a rotation of axes.

THEOREM

Identifying Conics without a Rotation of Axes

Except for degenerate cases, the equation

$$\boxed{Ax^2 + Bxy + Cy^2 + Dx + Ey + F = 0}$$

(a) Defines a parabola if $B^2 - 4AC = 0$.
(b) Defines an ellipse (or a circle) if $B^2 - 4AC < 0$.
(c) Defines a hyperbola if $B^2 - 4AC > 0$.

You are asked to prove this theorem in Problem 56.

EXAMPLE 5	**Identifying a Conic without a Rotation of Axes**

Identify the equation: $8x^2 - 12xy + 17y^2 - 4\sqrt{5}x - 2\sqrt{5}y - 15 = 0$

Solution Here $A = 8$, $B = -12$, and $C = 17$, so $B^2 - 4AC = -400$. Since $B^2 - 4AC < 0$, the equation defines an ellipse.

●

—**Now Work** PROBLEM 43

9.5 Assess Your Understanding

'Are You Prepared?' *Answers are given at the end of these exercises. If you get a wrong answer, read the pages listed in* red.

1. The sum formula for the sine function is $\sin(A + B) =$ _____. (p. 517)

2. The Double-angle Formula for the sine function is $\sin(2\theta) =$ _____. (p. 526)

3. If θ is acute, the Half-angle Formula for the sine function is $\sin\dfrac{\theta}{2} =$ _____. (p. 530)

4. If θ is acute, the Half-angle Formula for the cosine function is $\cos\dfrac{\theta}{2} =$ _____. (p. 530)

Concepts and Vocabulary

5. To transform the equation
$$Ax^2 + Bxy + Cy^2 + Dx + Ey + F = 0, \qquad B \ne 0$$
into one in x' and y' without an $x'y'$-term, rotate the axes through an acute angle θ that satisfies the equation _____.

6. Except for degenerate cases, the equation $Ax^2 + Bxy + Cy^2 + Dx + Ey + F = 0$ defines a(n) _____ if $B^2 - 4AC = 0$.

7. Except for degenerate cases, the equation $Ax^2 + Bxy + Cy^2 + Dx + Ey + F = 0$ defines an ellipse if _____.

8. **True or False** The equation $ax^2 + 6y^2 - 12y = 0$ defines an ellipse if $a > 0$.

9. **True or False** The equation $3x^2 + Bxy + 12y^2 = 10$ defines a parabola if $B = -12$.

10. **True or False** To eliminate the xy-term from the equation $x^2 - 2xy + y^2 - 2x + 3y + 5 = 0$, rotate the axes through an angle θ, where $\cot\theta = B^2 - 4AC$.

Skill Building

In Problems 11–20, identify each equation without completing the squares.

11. $x^2 + 4x + y + 3 = 0$

12. $2y^2 - 3y + 3x = 0$

13. $6x^2 + 3y^2 - 12x + 6y = 0$

14. $2x^2 + y^2 - 8x + 4y + 2 = 0$

15. $3x^2 - 2y^2 + 6x + 4 = 0$

16. $4x^2 - 3y^2 - 8x + 6y + 1 = 0$

17. $2y^2 - x^2 - y + x = 0$

18. $y^2 - 8x^2 - 2x - y = 0$

19. $x^2 + y^2 - 8x + 4y = 0$

20. $2x^2 + 2y^2 - 8x + 8y = 0$

In Problems 21–30, determine the appropriate rotation formulas to use so that the new equation contains no xy-term.

21. $x^2 + 4xy + y^2 - 3 = 0$

22. $x^2 - 4xy + y^2 - 3 = 0$

23. $5x^2 + 6xy + 5y^2 - 8 = 0$

24. $3x^2 - 10xy + 3y^2 - 32 = 0$

25. $13x^2 - 6\sqrt{3}xy + 7y^2 - 16 = 0$

26. $11x^2 + 10\sqrt{3}xy + y^2 - 4 = 0$

27. $4x^2 - 4xy + y^2 - 8\sqrt{5}x - 16\sqrt{5}y = 0$

28. $x^2 + 4xy + 4y^2 + 5\sqrt{5}y + 5 = 0$

29. $25x^2 - 36xy + 40y^2 - 12\sqrt{13}x - 8\sqrt{13}y = 0$

30. $34x^2 - 24xy + 41y^2 - 25 = 0$

In Problems 31–42, rotate the axes so that the new equation contains no xy-term. Analyze and graph the new equation. Refer to Problems 21–30 for Problems 31–40.

31. $x^2 + 4xy + y^2 - 3 = 0$

32. $x^2 - 4xy + y^2 - 3 = 0$

33. $5x^2 + 6xy + 5y^2 - 8 = 0$

34. $3x^2 - 10xy + 3y^2 - 32 = 0$

35. $13x^2 - 6\sqrt{3}xy + 7y^2 - 16 = 0$

36. $11x^2 + 10\sqrt{3}xy + y^2 - 4 = 0$

37. $4x^2 - 4xy + y^2 - 8\sqrt{5}x - 16\sqrt{5}y = 0$

38. $x^2 + 4xy + 4y^2 + 5\sqrt{5}y + 5 = 0$

39. $25x^2 - 36xy + 40y^2 - 12\sqrt{13}x - 8\sqrt{13}y = 0$

40. $34x^2 - 24xy + 41y^2 - 25 = 0$

41. $16x^2 + 24xy + 9y^2 - 130x + 90y = 0$

42. $16x^2 + 24xy + 9y^2 - 60x + 80y = 0$

In Problems 43–52, identify each equation without applying a rotation of axes.

43. $x^2 + 3xy - 2y^2 + 3x + 2y + 5 = 0$

44. $2x^2 - 3xy + 4y^2 + 2x + 3y - 5 = 0$

45. $x^2 - 7xy + 3y^2 - y - 10 = 0$

46. $2x^2 - 3xy + 2y^2 - 4x - 2 = 0$

47. $9x^2 + 12xy + 4y^2 - x - y = 0$

48. $10x^2 + 12xy + 4y^2 - x - y + 10 = 0$

49. $10x^2 - 12xy + 4y^2 - x - y - 10 = 0$

50. $4x^2 + 12xy + 9y^2 - x - y = 0$

51. $3x^2 - 2xy + y^2 + 4x + 2y - 1 = 0$

52. $3x^2 + 2xy + y^2 + 4x - 2y + 10 = 0$

Applications and Extensions

In Problems 53–56, apply the rotation formulas (5) to
$$Ax^2 + Bxy + Cy^2 + Dx + Ey + F = 0$$
to obtain the equation
$$A'x'^2 + B'x'y' + C'y'^2 + D'x' + E'y' + F' = 0$$

53. Express A', B', C', D', E', and F' in terms of $A, B, C, D, E,$ F, and the angle θ of rotation.
  [Hint: Refer to equation (6).]

54. Show that $A + C = A' + C'$, and thus show that $A + C$ is **invariant;** that is, its value does not change under a rotation of axes.

55. Refer to Problem 54. Show that $B^2 - 4AC$ is invariant.

56. Prove that, except for degenerate cases, the equation
$$Ax^2 + Bxy + Cy^2 + Dx + Ey + F = 0$$
  (a) Defines a parabola if $B^2 - 4AC = 0$.
  (b) Defines an ellipse (or a circle) if $B^2 - 4AC < 0$.
  (c) Defines a hyperbola if $B^2 - 4AC > 0$.

57. Use the rotation formulas (5) to show that distance is invariant under a rotation of axes. That is, show that the distance from $P_1 = (x_1, y_1)$ to $P_2 = (x_2, y_2)$ in the xy-plane equals the distance from $P_1 = (x_1', y_1')$ to $P_2 = (x_2', y_2')$ in the $x'y'$-plane.

58. Show that the graph of the equation $x^{1/2} + y^{1/2} = a^{1/2}$ is part of the graph of a parabola.

Explaining Concepts: Discussion and Writing

59. Formulate a strategy for discussing and graphing an equation of the form
$$Ax^2 + Cy^2 + Dx + Ey + F = 0$$

60. How does your strategy change if the equation is of the following form?
$$Ax^2 + Bxy + Cy^2 + Dx + Ey + F = 0$$

'Are You Prepared?' Answers

1. $\sin A \cos B + \cos A \sin B$ **2.** $2 \sin \theta \cos \theta$ **3.** $\sqrt{\dfrac{1 - \cos \theta}{2}}$ **4.** $\sqrt{\dfrac{1 + \cos \theta}{2}}$

9.6 Polar Equations of Conics

PREPARING FOR THIS SECTION *Before getting started, review the following:*
- Polar Coordinates (Section 8.1, pp. 592–599)

Now Work the *'Are You Prepared?'* problems on page 711.

> OBJECTIVES **1** Analyze and Graph Polar Equations of Conics (p. 707)
>
> **2** Convert the Polar Equation of a Conic to a Rectangular Equation (p. 711)

1 Analyze and Graph Polar Equations of Conics

In Sections 9.2 through 9.4, we gave separate definitions for the parabola, ellipse, and hyperbola based on geometric properties and the distance formula. In this section, we present an alternative definition that simultaneously defines all these conics. As we shall see, this approach is well suited to polar coordinate representation. (Refer to Section 8.1.)

DEFINITION

Let D denote a fixed line called the **directrix;** let F denote a fixed point called the **focus,** which is not on D; and let e be a fixed positive number called the **eccentricity.** A **conic** is the set of points P in the plane such that the ratio of the distance from F to P to the distance from D to P equals e. That is, a conic is the collection of points P for which

$$\frac{d(F, P)}{d(D, P)} = e \qquad (1)$$

If $e = 1$, the conic is a **parabola.**
If $e < 1$, the conic is an **ellipse.**
If $e > 1$, the conic is a **hyperbola.**

Observe that if $e = 1$ the definition of a parabola in equation (1) is exactly the same as the definition used earlier in Section 9.2.

In the case of an ellipse, the **major axis** is a line through the focus perpendicular to the directrix. In the case of a hyperbola, the **transverse axis** is a line through the focus perpendicular to the directrix. For both an ellipse and a hyperbola, the eccentricity e satisfies

$$e = \frac{c}{a} \qquad (2)$$

where c is the distance from the center to the focus and a is the distance from the center to a vertex.

Just as we did earlier using rectangular coordinates, we derive equations for the conics in polar coordinates by choosing a convenient position for the focus F and the directrix D. The focus F is positioned at the pole, and the directrix D is either parallel or perpendicular to the polar axis.

Suppose that we start with the directrix D perpendicular to the polar axis at a distance p units to the left of the pole (the focus F). See Figure 51.

If $P = (r, \theta)$ is any point on the conic, then, by equation (1),

$$\frac{d(F, P)}{d(D, P)} = e \quad \text{or} \quad d(F, P) = e \cdot d(D, P) \qquad (3)$$

Figure 51

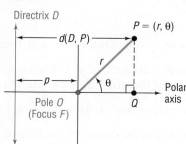

Now we use the point Q obtained by dropping the perpendicular from P to the polar axis to calculate $d(D, P)$.

$$d(D, P) = p + d(O, Q) = p + r \cos \theta$$

Using this expression and the fact that $d(F, P) = d(O, P) = r$ in equation (3), we get

$$d(F, P) = e \cdot d(D, P)$$
$$r = e(p + r \cos \theta)$$
$$r = ep + er \cos \theta$$
$$r - er \cos \theta = ep$$
$$r(1 - e \cos \theta) = ep$$
$$r = \frac{ep}{1 - e \cos \theta}$$

THEOREM

Polar Equation of a Conic; Focus at the Pole; Directrix Perpendicular to the Polar Axis a Distance p to the Left of the Pole

The polar equation of a conic with focus at the pole and directrix perpendicular to the polar axis at a distance p to the left of the pole is

$$r = \frac{ep}{1 - e \cos \theta} \qquad \text{(4)}$$

where e is the eccentricity of the conic.

EXAMPLE 1

Analyzing and Graphing the Polar Equation of a Conic

Analyze and graph the equation: $r = \dfrac{4}{2 - \cos \theta}$

Solution

The given equation is not quite in the form of equation (4), since the first term in the denominator is 2 instead of 1. We divide the numerator and denominator by 2 to obtain

$$r = \frac{2}{1 - \dfrac{1}{2} \cos \theta} \qquad r = \frac{ep}{1 - e \cos \theta}$$

This equation is in the form of equation (4), with

$$e = \frac{1}{2} \quad \text{and} \quad ep = 2$$

Then

$$\frac{1}{2}p = 2, \quad \text{so} \quad p = 4$$

We conclude that the conic is an ellipse, since $e = \dfrac{1}{2} < 1$. One focus is at the pole, and the directrix is perpendicular to the polar axis, a distance of $p = 4$ units to the left of the pole. It follows that the major axis is along the polar axis. To find the vertices, we let $\theta = 0$ and $\theta = \pi$. The vertices of the ellipse are $(4, 0)$ and $\left(\dfrac{4}{3}, \pi\right)$. The midpoint of the vertices, $\left(\dfrac{4}{3}, 0\right)$ in polar coordinates, is the center of the ellipse.

[Do you see why? The vertices $(4, 0)$ and $\left(\dfrac{4}{3}, \pi\right)$ in polar coordinates are $(4, 0)$ and $\left(-\dfrac{4}{3}, 0\right)$ in rectangular coordinates. The midpoint in rectangular coordinates is $\left(\dfrac{4}{3}, 0\right)$, which is also $\left(\dfrac{4}{3}, 0\right)$ in polar coordinates.] Then a = distance from the center to a vertex = $\dfrac{8}{3}$. Using $a = \dfrac{8}{3}$ and $e = \dfrac{1}{2}$ in equation (2), $e = \dfrac{c}{a}$, we find $c = ae = \dfrac{4}{3}$. Finally, using $a = \dfrac{8}{3}$ and $c = \dfrac{4}{3}$ in $b^2 = a^2 - c^2$, we have

$$b^2 = a^2 - c^2 = \frac{64}{9} - \frac{16}{9} = \frac{48}{9}$$

$$b = \frac{4\sqrt{3}}{3}$$

Figure 52 shows the graph.

Figure 52

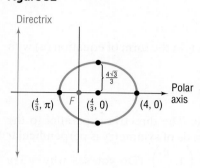

Directrix

$\left(\frac{4}{3}, \pi\right)$ F $\left(\frac{4}{3}, 0\right)$ $(4, 0)$

$\frac{4\sqrt{3}}{3}$

Polar axis

Exploration

Graph $r_1 = \dfrac{4}{2 + \cos\theta}$ and compare the result with Figure 52. What do you conclude? Clear the screen and graph $r_1 = \dfrac{4}{2 - \sin\theta}$ and then $r_1 = \dfrac{4}{2 + \sin\theta}$. Compare each of these graphs with Figure 52. What do you conclude?

Now Work PROBLEM 11

Equation (4) was obtained under the assumption that the directrix was perpendicular to the polar axis at a distance p units to the left of the pole. A similar derivation (see Problem 43), in which the directrix is perpendicular to the polar axis at a distance p units to the right of the pole, results in the equation

$$r = \frac{ep}{1 + e\cos\theta}$$

In Problems 44 and 45, you are asked to derive the polar equations of conics with focus at the pole and directrix parallel to the polar axis. Table 5 summarizes the polar equations of conics.

Table 5

Polar Equations of Conics (Focus at the Pole, Eccentricity e)	
Equation	**Description**
(a) $r = \dfrac{ep}{1 - e\cos\theta}$	Directrix is perpendicular to the polar axis at a distance p units to the left of the pole.
(b) $r = \dfrac{ep}{1 + e\cos\theta}$	Directrix is perpendicular to the polar axis at a distance p units to the right of the pole.
(c) $r = \dfrac{ep}{1 + e\sin\theta}$	Directrix is parallel to the polar axis at a distance p units above the pole.
(d) $r = \dfrac{ep}{1 - e\sin\theta}$	Directrix is parallel to the polar axis at a distance p units below the pole.
Eccentricity	
If $e = 1$, the conic is a parabola; the axis of symmetry is perpendicular to the directrix.	
If $e < 1$, the conic is an ellipse; the major axis is perpendicular to the directrix.	
If $e > 1$, the conic is a hyperbola; the transverse axis is perpendicular to the directrix.	

EXAMPLE 2   **Analyzing and Graphing the Polar Equation of a Conic**

Analyze and graph the equation: $r = \dfrac{6}{3 + 3\sin\theta}$

Solution   To place the equation in proper form, we divide the numerator and denominator by 3 to get

$$r = \dfrac{2}{1 + \sin\theta}$$

Referring to Table 5, we conclude that this equation is in the form of equation (c) with

$$e = 1 \quad \text{and} \quad ep = 2$$
$$p = 2 \quad e = 1$$

Figure 53

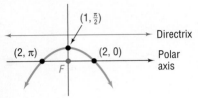

The conic is a parabola with focus at the pole. The directrix is parallel to the polar axis at a distance 2 units above the pole; the axis of symmetry is perpendicular to the polar axis. The vertex of the parabola is at $\left(1, \dfrac{\pi}{2}\right)$. (Do you see why?) See Figure 53 for the graph. Notice that we plotted two additional points, $(2, 0)$ and $(2, \pi)$, to assist in graphing.

━━━━━ **Now Work** PROBLEM 13

EXAMPLE 3   **Analyzing and Graphing the Polar Equation of a Conic**

Analyze and graph the equation: $r = \dfrac{3}{1 + 3\cos\theta}$

Solution   This equation is in the form of equation (b) in Table 5. We conclude that

$$e = 3 \quad \text{and} \quad ep = 3$$
$$p = 1 \quad e = 3$$

This is the equation of a hyperbola with a focus at the pole. The directrix is perpendicular to the polar axis, 1 unit to the right of the pole. The transverse axis is along the polar axis. To find the vertices, we let $\theta = 0$ and $\theta = \pi$. The vertices are $\left(\dfrac{3}{4}, 0\right)$ and $\left(-\dfrac{3}{2}, \pi\right)$. The center, which is at the midpoint of $\left(\dfrac{3}{4}, 0\right)$ and $\left(-\dfrac{3}{2}, \pi\right)$, is $\left(\dfrac{9}{8}, 0\right)$. Then $c = $ distance from the center to a focus $= \dfrac{9}{8}$. Since $e = 3$, it follows from equation (2), $e = \dfrac{c}{a}$, that $a = \dfrac{3}{8}$. Finally, using $a = \dfrac{3}{8}$ and $c = \dfrac{9}{8}$ in $b^2 = c^2 - a^2$, we find

Figure 54

$$b^2 = c^2 - a^2 = \dfrac{81}{64} - \dfrac{9}{64} = \dfrac{72}{64} = \dfrac{9}{8}$$

$$b = \dfrac{3}{2\sqrt{2}} = \dfrac{3\sqrt{2}}{4}$$

Figure 54 shows the graph. Notice that we plotted two additional points, $\left(3, \dfrac{\pi}{2}\right)$ and $\left(3, \dfrac{3\pi}{2}\right)$, on the left branch and used symmetry to obtain the right branch. The asymptotes of this hyperbola were found in the usual way by constructing the rectangle shown.

━━━━━ **Now Work** PROBLEM 17

2 Convert the Polar Equation of a Conic to a Rectangular Equation

> [!NOTE] EXAMPLE 4 — Converting a Polar Equation to a Rectangular Equation

Convert the polar equation

$$r = \frac{1}{3 - 3\cos\theta}$$

to a rectangular equation.

Solution　The strategy here is first to rearrange the equation and square each side before using the transformation equations.

$$r = \frac{1}{3 - 3\cos\theta}$$

$$3r - 3r\cos\theta = 1$$

$$3r = 1 + 3r\cos\theta \qquad \text{\textit{Rearrange the equation.}}$$

$$9r^2 = (1 + 3r\cos\theta)^2 \qquad \text{\textit{Square each side.}}$$

$$9(x^2 + y^2) = (1 + 3x)^2 \qquad \text{\textit{$x^2 + y^2 = r^2$; $x = r\cos\theta$}}$$

$$9x^2 + 9y^2 = 9x^2 + 6x + 1$$

$$9y^2 = 6x + 1$$

This is the equation of a parabola in rectangular coordinates.

── **Now Work** PROBLEM 25

9.6 Assess Your Understanding

'Are You Prepared?' *Answers are given at the end of these exercises. If you get a wrong answer, read the pages listed in red.*

1. If (x, y) are the rectangular coordinates of a point P and (r, θ) are its polar coordinates, then $x =$ _____ and $y =$ _____. (p. 594)

2. Transform the equation $r = 6\cos\theta$ from polar coordinates to rectangular coordinates. (p. 598)

Concepts and Vocabulary

3. A _____ is the set of points P in the plane such that the ratio of the distance from a fixed point called the _____ to P to the distance from a fixed line called the _____ to P equals e.

4. The eccentricity e of a parabola is _____, of an ellipse it is _____, and of a hyperbola it is _____.

5. *True or False* If (r, θ) are polar coordinates, the equation $r = \dfrac{2}{2 + 3\sin\theta}$ defines a hyperbola.

6. *True or False* The eccentricity e of any conic is $\dfrac{c}{a}$.

Skill Building

In Problems 7–12, identify the conic that each polar equation represents. Also, give the position of the directrix.

7. $r = \dfrac{1}{1 + \cos\theta}$

8. $r = \dfrac{3}{1 - \sin\theta}$

9. $r = \dfrac{4}{2 - 3\sin\theta}$

10. $r = \dfrac{2}{1 + 2\cos\theta}$

11. $r = \dfrac{3}{4 - 2\cos\theta}$

12. $r = \dfrac{6}{8 + 2\sin\theta}$

In Problems 13–24, analyze each equation and graph it.

13. $r = \dfrac{1}{1 + \cos\theta}$

14. $r = \dfrac{3}{1 - \sin\theta}$

15. $r = \dfrac{8}{4 + 3\sin\theta}$

16. $r = \dfrac{10}{5 + 4\cos\theta}$

17. $r = \dfrac{9}{3 - 6\cos\theta}$

18. $r = \dfrac{12}{4 + 8\sin\theta}$

19. $r = \dfrac{8}{2 - \sin\theta}$

20. $r = \dfrac{8}{2 + 4\cos\theta}$

21. $r(3 - 2\sin\theta) = 6$

22. $r(2 - \cos\theta) = 2$

23. $r = \dfrac{6\sec\theta}{2\sec\theta - 1}$

24. $r = \dfrac{3\csc\theta}{\csc\theta - 1}$

In Problems 25–36, convert each polar equation to a rectangular equation.

25. $r = \dfrac{1}{1 + \cos\theta}$

26. $r = \dfrac{3}{1 - \sin\theta}$

27. $r = \dfrac{8}{4 + 3\sin\theta}$

28. $r = \dfrac{10}{5 + 4\cos\theta}$

29. $r = \dfrac{9}{3 - 6\cos\theta}$

30. $r = \dfrac{12}{4 + 8\sin\theta}$

31. $r = \dfrac{8}{2 - \sin\theta}$

32. $r = \dfrac{8}{2 + 4\cos\theta}$

33. $r(3 - 2\sin\theta) = 6$

34. $r(2 - \cos\theta) = 2$

35. $r = \dfrac{6\sec\theta}{2\sec\theta - 1}$

36. $r = \dfrac{3\csc\theta}{\csc\theta - 1}$

In Problems 37–42, find a polar equation for each conic. For each, a focus is at the pole.

37. $e = 1$; directrix is parallel to the polar axis 1 unit above the pole.

38. $e = 1$; directrix is parallel to the polar axis 2 units below the pole.

39. $e = \dfrac{4}{5}$; directrix is perpendicular to the polar axis 3 units to the left of the pole.

40. $e = \dfrac{2}{3}$; directrix is parallel to the polar axis 3 units above the pole.

41. $e = 6$; directrix is parallel to the polar axis 2 units below the pole.

42. $e = 5$; directrix is perpendicular to the polar axis 5 units to the right of the pole.

Applications and Extensions

43. Derive equation (b) in Table 5:
$$r = \frac{ep}{1 + e\cos\theta}$$

44. Derive equation (c) in Table 5:
$$r = \frac{ep}{1 + e\sin\theta}$$

45. Derive equation (d) in Table 5:
$$r = \frac{ep}{1 - e\sin\theta}$$

46. Orbit of Mercury The planet Mercury travels around the Sun in an elliptical orbit given approximately by
$$r = \frac{(3.442)10^7}{1 - 0.206\cos\theta}$$

where r is measured in miles and the Sun is at the pole. Find the distance from Mercury to the Sun at *aphelion* (greatest distance from the Sun) and at *perihelion* (shortest distance from the Sun). See the figure. Use the aphelion and perihelion to graph the orbit of Mercury using a graphing utility.

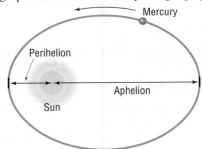

'Are You Prepared?' Answers

1. $r\cos\theta$; $r\sin\theta$

2. $x^2 + y^2 = 6x$ or $(x - 3)^2 + y^2 = 9$

9.7 Plane Curves and Parametric Equations

PREPARING FOR THIS SECTION *Before getting started, review the following:*

- Amplitude and Period of Sinusoidal Graphs (Section 5.6, pp. 439–440)

Now Work the 'Are You Prepared?' problem on page 722.

OBJECTIVES **1** Graph Parametric Equations (p. 713)

2 Find a Rectangular Equation for a Curve Defined Parametrically (p. 714)

3 Use Time as a Parameter in Parametric Equations (p. 716)

4 Find Parametric Equations for Curves Defined by Rectangular Equations (p. 719)

Equations of the form $y = f(x)$, where f is a function, have graphs that are intersected no more than once by any vertical line. The graphs of many of the conics and certain other, more complicated, graphs do not have this characteristic. Yet each graph, like the graph of a function, is a collection of points (x, y) in the xy-plane; that is, each is a *plane curve*. In this section, we discuss another way of representing such graphs.

> Let $x = f(t)$ and $y = g(t)$, where f and g are two functions whose common domain is some interval I. The collection of points defined by
>
> $$(x, y) = (f(t), g(t))$$
>
> is called a **plane curve.** The equations
>
> $$x = f(t) \qquad y = g(t)$$
>
> where t is in I, are called **parametric equations** of the curve. The variable t is called a **parameter.**

1 Graph Parametric Equations

Figure 55

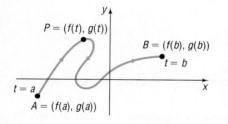

Parametric equations are particularly useful in describing movement along a curve. Suppose that a curve is defined by the parametric equations

$$x = f(t), \qquad y = g(t), \qquad a \le t \le b$$

where f and g are each defined over the interval $a \le t \le b$. For a given value of t, we can find the value of $x = f(t)$ and $y = g(t)$, obtaining a point (x, y) on the curve. In fact, as t varies over the interval from $t = a$ to $t = b$, successive values of t give rise to a directed movement along the curve; that is, the curve is traced out in a certain direction by the corresponding succession of points (x, y). See Figure 55. The arrows show the direction, or **orientation,** along the curve as t varies from a to b.

EXAMPLE 1 | **Graphing a Curve Defined by Parametric Equations**

Graph the curve defined by the parametric equations

$$x = 3t^2, \qquad y = 2t, \qquad -2 \le t \le 2 \qquad \textbf{(1)}$$

Solution

COMMENT Most graphing utilities have the capability of graphing parametric equations. See Section 9 in Appendix B. ∎

For each number t, $-2 \le t \le 2$, there corresponds a number x and a number y. For example, when $t = -2$, then $x = 3(-2)^2 = 12$ and $y = 2(-2) = -4$. When $t = 0$, then $x = 0$ and $y = 0$. We can set up a table listing various choices of the parameter t and the corresponding values for x and y, as shown in Table 6. Plotting these points and connecting them with a smooth curve leads to Figure 56. The arrows in Figure 56 are used to indicate the orientation.

Table 6

t	x	y	(x, y)
-2	12	-4	$(12, -4)$
-1	3	-2	$(3, -2)$
0	0	0	$(0, 0)$
1	3	2	$(3, 2)$
2	12	4	$(12, 4)$

Figure 56

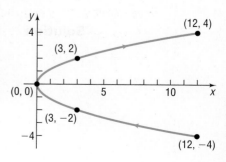

Exploration

Graph the following parametric equations using a graphing utility with $X\text{min} = 0, X\text{max} = 15$, $Y\text{min} = -5, Y\text{max} = 5$, and $T\text{step} = 0.1$:

1. $x = \dfrac{3t^2}{4}, y = t, -4 \leq t \leq 4$

2. $x = 3t^2 + 12t + 12, y = 2t + 4, -4 \leq t \leq 0$

3. $x = 3t^{\frac{2}{3}}, y = 2\sqrt[3]{t}, -8 \leq t \leq 8$

Compare these graphs to the graph in Figure 56. Conclude that parametric equations defining a curve are not unique; that is, different parametric equations can represent the same graph.

2 Find a Rectangular Equation for a Curve Defined Parametrically

The curve given in Example 1 should be familiar. To identify it accurately, we find the corresponding rectangular equation by eliminating the parameter t from the parametric equations given in Example 1:

$$x = 3t^2, \qquad y = 2t, \qquad -2 \leq t \leq 2$$

Noting that we can easily solve for t in $y = 2t$, obtaining $t = \dfrac{y}{2}$, we substitute this expression in the other equation.

$$x = 3t^2 = 3\left(\frac{y}{2}\right)^2 = \frac{3y^2}{4}$$

$$\uparrow$$
$$t = \frac{y}{2}$$

Exploration

In FUNCtion mode, graph $x = \dfrac{3y^2}{4}\left(Y_1 = \sqrt{\dfrac{4x}{3}} \text{ and } Y_2 = -\sqrt{\dfrac{4x}{3}}\right)$ with $X\text{min} = 0, X\text{max} = 15, Y\text{min} = -5, Y\text{max} = 5$. Compare this graph with Figure 56. Why do the graphs differ?

This equation, $x = \dfrac{3y^2}{4}$, is the equation of a parabola with vertex at $(0, 0)$ and axis of symmetry along the x-axis.

Note that the parameterized curve defined by equation (1) and shown in Figure 56 is only a part of the parabola $x = \dfrac{3y^2}{4}$. The graph of the rectangular equation obtained by eliminating the parameter will, in general, contain more points than the original parameterized curve. Care must therefore be taken when a parameterized curve is graphed after eliminating the parameter. Even so, the process of eliminating the parameter t of a parameterized curve to identify it accurately is sometimes a better approach than plotting points. However, the elimination process sometimes requires a little ingenuity.

EXAMPLE 2

Finding the Rectangular Equation of a Curve Defined Parametrically

Find the rectangular equation of the curve whose parametric equations are

$$x = a \cos t, \qquad y = a \sin t, \qquad -\infty < t < \infty$$

where $a > 0$ is a constant. Graph this curve, indicating its orientation.

Solution

The presence of sines and cosines in the parametric equations suggests that we use a Pythagorean Identity. In fact, since

$$\cos t = \frac{x}{a} \qquad \sin t = \frac{y}{a}$$

we find that

$$\cos^2 t + \sin^2 t = 1$$
$$\left(\frac{x}{a}\right)^2 + \left(\frac{y}{a}\right)^2 = 1$$
$$x^2 + y^2 = a^2$$

Figure 57

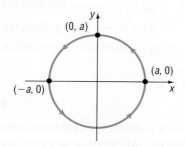

The curve is a circle with center at $(0, 0)$ and radius a. As the parameter t increases, say from $t = 0$ [the point $(a, 0)$] to $t = \dfrac{\pi}{2}$ [the point $(0, a)$] to $t = \pi$ [the point $(-a, 0)$], we see that the corresponding points are traced in a counterclockwise direction around the circle. The orientation is as indicated in Figure 57.

────── **Now Work** PROBLEMS 7 AND 19

Let's analyze the curve in Example 2 further. The domain of each parametric equation is $-\infty < t < \infty$. Thus, the graph in Figure 57 is actually being repeated each time that t increases by 2π.

If we wanted the curve to consist of exactly 1 revolution in the counterclockwise direction, we could write

$$x = a \cos t, \qquad y = a \sin t, \qquad 0 \le t \le 2\pi$$

This curve starts at $t = 0$ [the point $(a, 0)$] and, proceeding counterclockwise around the circle, ends at $t = 2\pi$ [also the point $(a, 0)$].

If we wanted the curve to consist of exactly three revolutions in the counterclockwise direction, we could write

$$x = a \cos t, \qquad y = a \sin t, \qquad -2\pi \le t \le 4\pi$$

or

$$x = a \cos t, \qquad y = a \sin t, \qquad 0 \le t \le 6\pi$$

or

$$x = a \cos t, \qquad y = a \sin t, \qquad 2\pi \le t \le 8\pi$$

EXAMPLE 3	**Describing Parametric Equations**

Find rectangular equations for the following curves defined by parametric equations. Graph each curve.

(a) $x = a \cos t, \quad y = a \sin t, \quad 0 \le t \le \pi, \quad a > 0$

(b) $x = -a \sin t, \quad y = -a \cos t, \quad 0 \le t \le \pi, \quad a > 0$

Solution   (a) We eliminate the parameter t using a Pythagorean Identity.

$$\cos^2 t + \sin^2 t = 1$$

$$\left(\frac{x}{a}\right)^2 + \left(\frac{y}{a}\right)^2 = 1$$

$$x^2 + y^2 = a^2$$

Figure 58

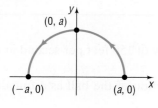

The curve defined by these parametric equations is a circle, with radius a and center at $(0, 0)$. The circle begins at the point $(a, 0)$, $t = 0$; passes through the point $(0, a)$, $t = \dfrac{\pi}{2}$; and ends at the point $(-a, 0)$, $t = \pi$.

The parametric equations define an upper semicircle of radius a with a counterclockwise orientation. See Figure 58. The rectangular equation is

$$y = \sqrt{a^2 - x^2} \qquad -a \le x \le a$$

(b) We eliminate the parameter t using a Pythagorean Identity.

$$\sin^2 t + \cos^2 t = 1$$

$$\left(\frac{x}{-a}\right)^2 + \left(\frac{y}{-a}\right)^2 = 1$$

$$x^2 + y^2 = a^2$$

Figure 59

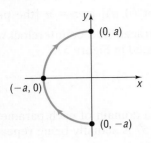

The curve defined by these parametric equations is a circle, with radius a and center at $(0, 0)$. The circle begins at the point $(0, -a)$, $t = 0$; passes through the point $(-a, 0)$, $t = \dfrac{\pi}{2}$; and ends at the point $(0, a)$, $t = \pi$. The parametric equations define a left semicircle of radius a with a clockwise orientation. See Figure 59. The rectangular equation is

$$x = -\sqrt{a^2 - y^2} \qquad -a \le y \le a$$

Example 3 illustrates the versatility of parametric equations for replacing complicated rectangular equations, while providing additional information about orientation. These characteristics make parametric equations very useful in applications, such as projectile motion.

3 Use Time as a Parameter in Parametric Equations

If we think of the parameter t as time, then the parametric equations $x = f(t)$ and $y = g(t)$ of a curve C specify how the x- and y-coordinates of a moving point vary with time.

For example, we can use parametric equations to model the motion of an object, sometimes referred to as **curvilinear motion.** Using parametric equations, we can specify not only where the object travels, that is, its location (x, y), but also when it gets there, that is, the time t.

When an object is propelled upward at an inclination θ to the horizontal with initial speed v_0, the resulting motion is called **projectile motion.** See Figure 60(a).

In calculus it is shown that the parametric equations of the path of a projectile fired at an inclination θ to the horizontal, with an initial speed v_0, from a height h above the horizontal are

$$x = (v_0 \cos \theta)t \qquad y = -\frac{1}{2}gt^2 + (v_0 \sin \theta)t + h \qquad \textbf{(2)}$$

where t is the time and g is the constant acceleration due to gravity (approximately 32 ft/sec/sec or 9.8 m/sec/sec). See Figure 60(b).

Seeing the Concept

Graph $\quad x = \cos t, y = \sin t \quad$ for $0 \le t \le 2\pi$. Compare to Figure 57. Graph $\quad x = \cos t, y = \sin t \quad$ for $0 \le t \le \pi$. Compare to Figure 58. Graph $x = -\sin t, y = -\cos t$ for $0 \le t \le \pi$. Compare to Figure 59.

Figure 60

(a)   (b)

EXAMPLE 4

Figure 61

Projectile Motion

Suppose that Jim hit a golf ball with an initial velocity of 150 feet per second at an angle of 30° to the horizontal. See Figure 61.

(a) Find parametric equations that describe the position of the ball as a function of time.

(b) How long was the golf ball in the air?

(c) When was the ball at its maximum height? Determine the maximum height of the ball.

(d) Determine the distance that the ball traveled.

(e) Using a graphing utility, simulate the motion of the golf ball by simultaneously graphing the equations found in part (a).

Solution   (a) We have $v_0 = 150$ ft/sec, $\theta = 30°$, $h = 0$ (the ball is on the ground), and $g = 32$ (since the units are in feet and seconds). Substituting these values into equations (2), we find that

$$x = (v_0 \cos \theta)t = (150 \cos 30°)t = 75\sqrt{3}\,t$$

$$y = -\frac{1}{2}gt^2 + (v_0 \sin \theta)t + h = -\frac{1}{2}(32)t^2 + (150 \sin 30°)t + 0$$

$$= -16t^2 + 75t$$

(b) To determine the length of time that the ball was in the air, we solve the equation $y = 0$.

$$-16t^2 + 75t = 0$$
$$t(-16t + 75) = 0$$
$$t = 0 \text{ sec} \quad \text{or} \quad t = \frac{75}{16} = 4.6875 \text{ sec}$$

The ball struck the ground after 4.6875 seconds.

(c) Notice that the height y of the ball is a quadratic function of t, so the maximum height of the ball can be found by determining the vertex of $y = -16t^2 + 75t$. The value of t at the vertex is

$$t = \frac{-b}{2a} = \frac{-75}{-32} = 2.34375 \text{ sec}$$

The ball was at its maximum height after 2.34375 seconds. The maximum height of the ball is found by evaluating the function y at $t = 2.34375$ seconds.

$$\text{Maximum height} = -16(2.34375)^2 + (75)2.34375 \approx 87.89 \text{ feet}$$

(d) Since the ball was in the air for 4.6875 seconds, the horizontal distance that the ball traveled is

$$x = \left(75\sqrt{3}\right)4.6875 \approx 608.92 \text{ feet}$$

 (e) We enter the equations from part (a) into a graphing utility with $T\text{min} = 0$, $T\text{max} = 4.7$, and $T\text{step} = 0.1$. We use ZOOM-SQUARE to avoid any distortion to the angle of elevation. See Figure 62.

Figure 62

 ## Exploration

Simulate the motion of a ball thrown straight up with an initial speed of 100 feet per second from a height of 5 feet above the ground. Use PARametric mode with $T\text{min} = 0$, $T\text{max} = 6.5$, $T\text{step} = 0.1$, $X\text{min} = 0$, $X\text{max} = 5$, $Y\text{min} = 0$, and $Y\text{max} = 180$. What happens to the speed with which the graph is drawn as the ball goes up and then comes back down? How do you interpret this physically? Repeat the experiment using other values for $T\text{step}$. How does this affect the experiment?

[**Hint:** In the projectile motion equations, let $\theta = 90°$, $v_0 = 100$, $h = 5$, and $g = 32$. We use $x = 3$ instead of $x = 0$ to see the vertical motion better.]

Result   See Figure 63. In Figure 63(a) the ball is going up. In Figure 63(b) the ball is near its highest point. Finally, in Figure 63(c) the ball is coming back down.

Figure 63

Notice that, as the ball goes up, its speed decreases, until at the highest point it is zero. Then the speed increases as the ball comes back down.

─══════── **Now Work** PROBLEM 49

A graphing utility can be used to simulate other kinds of motion as well. Let's work again Example 5 from Appendix A, Section A.9.

EXAMPLE 5 | **Simulating Motion**

Tanya, who is a long distance runner, runs at an average speed of 8 miles per hour. Two hours after Tanya leaves your house, you leave in your Honda and follow the same route. If your average speed is 40 miles per hour, how long will it be before you catch up to Tanya? See Figure 64. Use a simulation of the two motions to verify the answer.

Figure 64

Solution

We begin with two sets of parametric equations: one to describe Tanya's motion, the other to describe the motion of the Honda. We choose time $t = 0$ to be when Tanya leaves the house. If we choose $y_1 = 2$ as Tanya's path, then we can use $y_2 = 4$ as the parallel path of the Honda. The horizontal distances traversed in time t (Distance = Rate × Time) are

$$\text{Tanya: } x_1 = 8t \qquad \text{Honda: } x_2 = 40(t - 2)$$

The Honda catches up to Tanya when $x_1 = x_2$.

$$8t = 40(t - 2)$$
$$8t = 40t - 80$$
$$-32t = -80$$
$$t = \frac{-80}{-32} = 2.5$$

The Honda catches up to Tanya 2.5 hours after Tanya leaves the house.

In PARametric mode with Tstep $= 0.01$, we simultaneously graph

$$\text{Tanya: } x_1 = 8t \qquad \text{Honda: } x_2 = 40(t - 2)$$
$$y_1 = 2 \qquad\qquad y_2 = 4$$

for $0 \leq t \leq 3$.

Figure 65 shows the relative position of Tanya and the Honda for $t = 0$, $t = 2$, $t = 2.25$, $t = 2.5$, and $t = 2.75$.

Figure 65

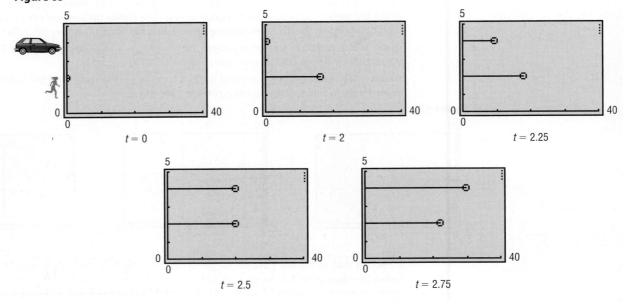

4 Find Parametric Equations for Curves Defined by Rectangular Equations

We now take up the question of how to find parametric equations of a given curve. If a curve is defined by the equation $y = f(x)$, where f is a function, one way of finding parametric equations is to let $x = t$. Then $y = f(t)$ and

$$x = t, \quad y = f(t), \qquad t \text{ in the domain of } f$$

are parametric equations of the curve.

EXAMPLE 6 | **Finding Parametric Equations for a Curve Defined by a Rectangular Equation**

Find two different parametric equations for the equation $y = x^2 - 4$.

Solution For the first parametric equation, let $x = t$. Then the parametric equations are

$$x = t, \quad y = t^2 - 4, \qquad -\infty < t < \infty$$

A second parametric equation is found by letting $x = t^3$. Then the parametric equations become

$$x = t^3, \quad y = t^6 - 4, \qquad -\infty < t < \infty$$

Care must be taken when using the second approach in Example 6, since the substitution for x must be a function that allows x to take on all the values stipulated by the domain of f. For example, letting $x = t^2$ so that $y = t^4 - 4$ does not result in equivalent parametric equations for $y = x^2 - 4$, since only points for which $x \geq 0$ are obtained; yet the domain of $y = x^2 - 4$ is $\{x \mid x \text{ is any real number}\}$.

➤ **Now Work** PROBLEM 33

EXAMPLE 7 | **Finding Parametric Equations for an Object in Motion**

Find parametric equations for the ellipse

$$x^2 + \frac{y^2}{9} = 1$$

where the parameter t is time (in seconds) and

(a) The motion around the ellipse is clockwise, begins at the point $(0, 3)$, and requires 1 second for a complete revolution.

(b) The motion around the ellipse is counterclockwise, begins at the point $(1, 0)$, and requires 2 seconds for a complete revolution.

Figure 66

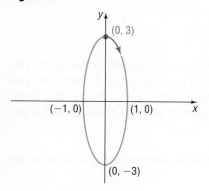

Solution

(a) See Figure 66. Since the motion begins at the point $(0, 3)$, we want $x = 0$ and $y = 3$ when $t = 0$. Furthermore, since the given equation is an ellipse, we begin by letting

$$x = \sin(\omega t) \quad y = 3 \cos(\omega t)$$

for some constant ω. These parametric equations satisfy the equation of the ellipse. Furthermore, with this choice, when $t = 0$, we have $x = 0$ and $y = 3$.
   For the motion to be clockwise, the motion will have to begin with the value of x increasing and y decreasing as t increases. This requires that $\omega > 0$. [Do you

know why? If $\omega > 0$, then $x = \sin(\omega t)$ is increasing when $t > 0$ is near zero and $y = 3\cos(\omega t)$ is decreasing when $t > 0$ is near zero.] See the red part of the graph in Figure 66.

Finally, since 1 revolution requires 1 second, the period $\dfrac{2\pi}{\omega} = 1$, so $\omega = 2\pi$. Parametric equations that satisfy the conditions stipulated are

$$x = \sin(2\pi t), \qquad y = 3\cos(2\pi t), \qquad 0 \le t \le 1 \qquad \textbf{(3)}$$

Figure 67

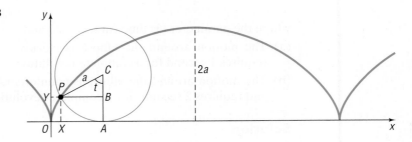

(0, 3)

(1, 0)

(−1, 0)

x

(0, −3)

(b) See Figure 67. Since the motion begins at the point $(1, 0)$, we want $x = 1$ and $y = 0$ when $t = 0$. The given equation is an ellipse, so we begin by letting

$$x = \cos(\omega t) \qquad y = 3\sin(\omega t)$$

for some constant ω. These parametric equations satisfy the equation of the ellipse. Furthermore, with this choice, when $t = 0$, we have $x = 1$ and $y = 0$.

For the motion to be counterclockwise, the motion will have to begin with the value of x decreasing and y increasing as t increases. This requires that $\omega > 0$. [Do you know why?] Finally, since 1 revolution requires 2 seconds, the period is $\dfrac{2\pi}{\omega} = 2$, so $\omega = \pi$. The parametric equations that satisfy the conditions stipulated are

$$x = \cos(\pi t), \qquad y = 3\sin(\pi t), \qquad 0 \le t \le 2 \qquad \textbf{(4)}$$

●

Either equation (3) or (4) can serve as parametric equations for the ellipse $x^2 + \dfrac{y^2}{9} = 1$ given in Example 7. The direction of the motion, the beginning point, and the time for 1 revolution give a particular parametric representation.

━━━▶ **Now Work** PROBLEM **39**

The Cycloid

Suppose that a circle of radius a rolls along a horizontal line without slipping. As the circle rolls along the line, a point P on the circle will trace out a curve called a **cycloid** (see Figure 68). We now seek parametric equations* for a cycloid.

Figure 68

y

C

a

P

t

B

Y

$2a$

O X A x

We begin with a circle of radius a and take the fixed line on which the circle rolls as the x-axis. Let the origin be one of the points at which the point P comes in contact with the x-axis. Figure 68 illustrates the position of this point P after the circle has rolled somewhat. The angle t (in radians) measures the angle through which the circle has rolled.

_____

*Any attempt to derive the rectangular equation of a cycloid would soon demonstrate how complicated the task is.

Since we require no slippage, it follows that

$$\text{Arc } AP = d(O, A)$$

The length of the arc AP is given by $s = r\theta$, where $r = a$ and $\theta = t$ radians. Then

$$at = d(O, A) \quad \text{\footnotesize $s = r\theta$, where $r = a$ and $\theta = t$}$$

The x-coordinate of the point P is

$$d(O, X) = d(O, A) - d(X, A) = at - a \sin t = a(t - \sin t)$$

The y-coordinate of the point P is equal to

$$d(O, Y) = d(A, C) - d(B, C) = a - a \cos t = a(1 - \cos t)$$

Exploration

Graph $\quad x = t - \sin t, y = 1 - \cos t,$
$0 \le t \le 3\pi$, using your graphing utility
with Tstep $= \dfrac{\pi}{36}$ and a square screen.
Compare your results with Figure 68.

The parametric equations of the cycloid are

$$x = a(t - \sin t) \qquad y = a(1 - \cos t) \tag{5}$$

Applications to Mechanics

If a is negative in equation (5), we obtain an inverted cycloid, as shown in Figure 69(a). The inverted cycloid occurs as a result of some remarkable applications in the field of mechanics. We shall mention two of them: the *brachistochrone* and the *tautochrone*.*

Figure 69

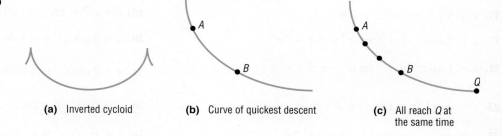

(a) Inverted cycloid   (b) Curve of quickest descent   (c) All reach Q at the same time

Figure 70

The **brachistochrone** is the curve of quickest descent. If a particle is constrained to follow some path from one point A to a lower point B (not on the same vertical line) and is acted on only by gravity, the time needed to make the descent is least if the path is an inverted cycloid. See Figure 69(b). This remarkable discovery, which is attributed to many famous mathematicians (including Johann Bernoulli and Blaise Pascal), was a significant step in creating the branch of mathematics known as the *calculus of variations*.

To define the **tautochrone,** let Q be the lowest point on an inverted cycloid. If several particles placed at various positions on an inverted cycloid simultaneously begin to slide down the cycloid, they will reach the point Q at the same time, as indicated in Figure 69(c). The tautochrone property of the cycloid was used by Christiaan Huygens (1629–1695), the Dutch mathematician, physicist, and astronomer, to construct a pendulum clock with a bob that swings along a cycloid (see Figure 70). In Huygen's clock, the bob was made to swing along a cycloid by suspending the bob on a thin wire constrained by two plates shaped like cycloids. In a clock of this design, the period of the pendulum is independent of its amplitude.

*In Greek, *brachistochrone* means "the shortest time" and *tautochrone* means "equal time."

9.7 Assess Your Understanding

'Are You Prepared?'   *Answers are given at the end of these exercises. If you get a wrong answer, read the page listed in red.*

1. The function $f(x) = 3\sin(4x)$ has amplitude _____ and period _____. (p. 440)

Concepts and Vocabulary

2. Let $x = f(t)$ and $y = g(t)$, where f and g are two functions whose common domain is some interval I. The collection of points defined by $(x, y) = (f(t), g(t))$ is called a(n) _____ _____. The variable t is called a(n) _____.

3. The parametric equations $x = 2\sin t, y = 3\cos t$ define a(n) _____.

4. If a circle rolls along a horizontal line without slippage, a point P on the circle will trace out a curve called a(n) _____.

5. *True or False*   Parametric equations defining a curve are unique.

6. *True or False*   Curves defined using parametric equations have an orientation.

Skill Building

In Problems 7–26, graph the curve whose parametric equations are given and show its orientation. Find the rectangular equation of each curve.

7. $x = 3t + 2,\quad y = t + 1;\quad 0 \le t \le 4$

8. $x = t - 3,\quad y = 2t + 4;\quad 0 \le t \le 2$

9. $x = t + 2,\quad y = \sqrt{t};\quad t \ge 0$

10. $x = \sqrt{2t},\quad y = 4t;\quad t \ge 0$

11. $x = t^2 + 4,\quad y = t^2 - 4;\quad -\infty < t < \infty$

12. $x = \sqrt{t} + 4,\quad y = \sqrt{t} - 4;\quad t \ge 0$

13. $x = 3t^2,\quad y = t + 1;\quad -\infty < t < \infty$

14. $x = 2t - 4,\quad y = 4t^2;\quad -\infty < t < \infty$

15. $x = 2e^t,\quad y = 1 + e^t;\quad t \ge 0$

16. $x = e^t,\quad y = e^{-t};\quad t \ge 0$

17. $x = \sqrt{t},\quad y = t^{3/2};\quad t \ge 0$

18. $x = t^{3/2} + 1,\quad y = \sqrt{t};\quad t \ge 0$

19. $x = 2\cos t,\quad y = 3\sin t;\quad 0 \le t \le 2\pi$

20. $x = 2\cos t,\quad y = 3\sin t;\quad 0 \le t \le \pi$

21. $x = 2\cos t,\quad y = 3\sin t;\quad -\pi \le t \le 0$

22. $x = 2\cos t,\quad y = \sin t;\quad 0 \le t \le \dfrac{\pi}{2}$

23. $x = \sec t,\quad y = \tan t;\quad 0 \le t \le \dfrac{\pi}{4}$

24. $x = \csc t,\quad y = \cot t;\quad \dfrac{\pi}{4} \le t \le \dfrac{\pi}{2}$

25. $x = \sin^2 t,\quad y = \cos^2 t;\quad 0 \le t \le 2\pi$

26. $x = t^2,\quad y = \ln t;\quad t > 0$

In Problems 27–34, find two different parametric equations for each rectangular equation.

27. $y = 4x - 1$

28. $y = -8x + 3$

29. $y = x^2 + 1$

30. $y = -2x^2 + 1$

31. $y = x^3$

32. $y = x^4 + 1$

33. $x = y^{3/2}$

34. $x = \sqrt{y}$

In Problems 35–38, find parametric equations that define the curve shown.

35.

36.

37.

38.

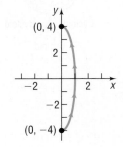

In Problems 39–42, find parametric equations for an object that moves along the ellipse $\dfrac{x^2}{4} + \dfrac{y^2}{9} = 1$ with the motion described.

39. The motion begins at $(2, 0)$, is clockwise, and requires 2 seconds for a complete revolution.

40. The motion begins at $(0, 3)$, is counterclockwise, and requires 1 second for a complete revolution.

41. The motion begins at $(0, 3)$, is clockwise, and requires 1 second for a complete revolution.

42. The motion begins at $(2, 0)$, is counterclockwise, and requires 3 seconds for a complete revolution.

In Problems 43 and 44, the parametric equations of four curves are given. Graph each of them, indicating the orientation.

43. C_1: $x = t$, $y = t^2$; $-4 \le t \le 4$

C_2: $x = \cos t$, $y = 1 - \sin^2 t$; $0 \le t \le \pi$

C_3: $x = e^t$, $y = e^{2t}$; $0 \le t \le \ln 4$

C_4: $x = \sqrt{t}$, $y = t$; $0 \le t \le 16$

44. C_1: $x = t$, $y = \sqrt{1 - t^2}$; $-1 \le t \le 1$

C_2: $x = \sin t$, $y = \cos t$; $0 \le t \le 2\pi$

C_3: $x = \cos t$, $y = \sin t$; $0 \le t \le 2\pi$

C_4: $x = \sqrt{1 - t^2}$, $y = t$; $-1 \le t \le 1$

In Problems 45–48, use a graphing utility to graph the curve defined by the given parametric equations.

45. $x = t \sin t$, $y = t \cos t$, $t > 0$

46. $x = \sin t + \cos t$, $y = \sin t - \cos t$

47. $x = 4 \sin t - 2 \sin(2t)$

$y = 4 \cos t - 2 \cos(2t)$

48. $x = 4 \sin t + 2 \sin(2t)$

$y = 4 \cos t + 2 \cos(2t)$

Applications and Extensions

49. Projectile Motion Bob throws a ball straight up with an initial speed of 50 feet per second from a height of 6 feet.
  (a) Find parametric equations that model the motion of the ball as a function of time.
  (b) How long is the ball in the air?
  (c) When is the ball at its maximum height? Determine the maximum height of the ball.
  (d) Simulate the motion of the ball by graphing the equations found in part (a).

50. Projectile Motion Alice throws a ball straight up with an initial speed of 40 feet per second from a height of 5 feet.
  (a) Find parametric equations that model the motion of the ball as a function of time.
  (b) How long is the ball in the air?
  (c) When is the ball at its maximum height? Determine the maximum height of the ball.
  (d) Simulate the motion of the ball by graphing the equations found in part (a).

51. Catching a Train Bill's train leaves at 8:06 AM and accelerates at the rate of 2 meters per second per second. Bill, who can run 5 meters per second, arrives at the train station 5 seconds after the train has left and runs for the train.
  (a) Find parametric equations that model the motions of the train and Bill as a function of time.
  [**Hint:** The position s at time t of an object having acceleration a is $s = \dfrac{1}{2} at^2$.]
  (b) Determine algebraically whether Bill will catch the train. If so, when?
  (c) Simulate the motion of the train and Bill by simultaneously graphing the equations found in part (a).

52. Catching a Bus Jodi's bus leaves at 5:30 PM and accelerates at the rate of 3 meters per second per second. Jodi, who can

run 5 meters per second, arrives at the bus station 2 seconds after the bus has left and runs for the bus.
  (a) Find parametric equations that model the motions of the bus and Jodi as a function of time.
  [**Hint:** The position s at time t of an object having acceleration a is $s = \dfrac{1}{2} at^2$.]
  (b) Determine algebraically whether Jodi will catch the bus. If so, when?
  (c) Simulate the motion of the bus and Jodi by simultaneously graphing the equations found in part (a).

53. Projectile Motion Ichiro throws a baseball with an initial speed of 145 feet per second at an angle of 20° to the horizontal. The ball leaves Ichiro's hand at a height of 5 feet.
  (a) Find parametric equations that model the position of the ball as a function of time.
  (b) How long is the ball in the air?
  (c) Determine the horizontal distance that the ball travels.
  (d) When is the ball at its maximum height? Determine the maximum height of the ball.
  (e) Using a graphing utility, simultaneously graph the equations found in part (a).

54. Projectile Motion Mark Texeira hit a baseball with an initial speed of 125 feet per second at an angle of 40° to the horizontal. The ball was hit at a height of 3 feet off the ground.
  (a) Find parametric equations that model the position of the ball as a function of time.
  (b) How long is the ball in the air?
  (c) Determine the horizontal distance that the ball travels.
  (d) When is the ball at its maximum height? Determine the maximum height of the ball.
  (e) Using a graphing utility, simultaneously graph the equations found in part (a).

55. Projectile Motion Suppose that Adam hits a golf ball off a cliff 300 meters high with an initial speed of 40 meters per second at an angle of 45° to the horizontal.
- (a) Find parametric equations that model the position of the ball as a function of time.
- (b) How long is the ball in the air?
- (c) Determine the horizontal distance that the ball travels.
- (d) When is the ball at its maximum height? Determine the maximum height of the ball.
- (e) Using a graphing utility, simultaneously graph the equations found in part (a).

56. Projectile Motion Suppose that Karla hits a golf ball off a cliff 300 meters high with an initial speed of 40 meters per second at an angle of 45° to the horizontal on the Moon (gravity on the Moon is one-sixth of that on Earth).
- (a) Find parametric equations that model the position of the ball as a function of time.
- (b) How long is the ball in the air?
- (c) Determine the horizontal distance that the ball travels.
- (d) When is the ball at its maximum height? Determine the maximum height of the ball.
- (e) Using a graphing utility, simultaneously graph the equations found in part (a).

57. Uniform Motion A Toyota Camry (traveling east at 40 mph) and a Chevy Impala (traveling north at 30 mph) are heading toward the same intersection. The Camry is 5 miles from the intersection when the Impala is 4 miles from the intersection. See the figure.

- (a) Find parametric equations that model the motion of the Camry and Impala.
- (b) Find a formula for the distance between the cars as a function of time.
- (c) Graph the function in part (b) using a graphing utility.
- (d) What is the minimum distance between the cars? When are the cars closest?
- (e) Simulate the motion of the cars by simultaneously graphing the equations found in part (a).

58. Uniform Motion A Cessna (heading south at 120 mph) and a Boeing 747 (heading west at 600 mph) are flying toward the same point at the same altitude. The Cessna is 100 miles from the point where the flight patterns intersect, and the 747 is 550 miles from this intersection point. See the figure.
- (a) Find parametric equations that model the motion of the Cessna and the 747.

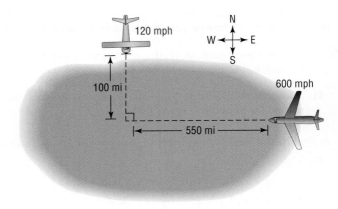

- (b) Find a formula for the distance between the planes as a function of time.
- (c) Graph the function in part (b) using a graphing utility.
- (d) What is the minimum distance between the planes? When are the planes closest?
- (e) Simulate the motion of the planes by simultaneously graphing the equations found in part (a).

59. The Green Monster The left field wall at Fenway Park is 310 feet from home plate; the wall itself (affectionately named the Green Monster) is 37 feet high. A batted ball must clear the wall to be a home run. Suppose a ball leaves the bat 3 feet off the ground, at an angle of 45°. Use $g = 32$ feet per second[2] as the acceleration due to gravity and ignore any air resistance.
- (a) Find parametric equations that model the position of the ball as a function of time.
- (b) What is the maximum height of the ball if it leaves the bat with a speed of 90 miles per hour? Give your answer in feet.
- (c) How far is the ball from home plate at its maximum height? Give your answer in feet.
- (d) If the ball is hit straight down the left field wall, will it clear the Green Monster? If it does, by how much does it clear the wall?

Source: The Boston Red Sox

60. Projectile Motion The position of a projectile fired with an initial velocity v_0 feet per second and at an angle θ to the horizontal at the end of t seconds is given by the parametric equations

$$x = (v_0 \cos \theta)t \qquad y = (v_0 \sin \theta)t - 16t^2$$

See the illustration.

- (a) Obtain the rectangular equation of the trajectory and identify the curve.
- (b) Show that the projectile hits the ground ($y = 0$) when $t = \dfrac{1}{16}v_0 \sin \theta$.
- (c) How far has the projectile traveled (horizontally) when it strikes the ground? In other words, find the range R.

(d) Find the time t when $x = y$. Then find the horizontal distance x and the vertical distance y traveled by the projectile in this time. Then compute $\sqrt{x^2 + y^2}$. This is the distance R, the range, that the projectile travels up a plane inclined at $45°$ to the horizontal $(x = y)$. See the following illustration. (See also Problem 97 in Section 6.6.)

61. Show that the parametric equations for a line passing through the points (x_1, y_1) and (x_2, y_2) are

$$x = (x_2 - x_1)t + x_1$$
$$y = (y_2 - y_1)t + y_1, \quad -\infty < t < \infty$$

What is the orientation of this line?

62. Hypocycloid The hypocycloid is a curve defined by the parametric equations

$$x(t) = \cos^3 t, \quad y(t) = \sin^3 t, \quad 0 \le t \le 2\pi$$

(a) Graph the hypocycloid using a graphing utility.
(b) Find a rectangular equation of the hypocycloid.

Explaining Concepts: Discussion and Writing

63. In Problem 62, we graphed the hypocycloid. Now graph the rectangular equations of the hypocycloid. Did you obtain a complete graph? If not, experiment until you do.

64. Look up the curves called *hypocycloid* and *epicycloid*. Write a report on what you find. Be sure to draw comparisons with the cycloid.

'Are You Prepared?' Answers

1. $3; \dfrac{\pi}{2}$

CHAPTER REVIEW

Things to Know

Equations

Parabola (pp. 667–673)	See Tables 1 and 2 (pp. 669 and 670).
Ellipse (pp. 676–683)	See Table 3 (p. 681).
Hyperbola (pp. 686–695)	See Table 4 (p. 693).
General equation of a conic (p. 704)	$Ax^2 + Bxy + Cy^2 + Dx + Ey + F = 0$ Parabola if $B^2 - 4AC = 0$
	Ellipse (or circle) if $B^2 - 4AC < 0$
	Hyperbola if $B^2 - 4AC > 0$
Polar equations of a conic with focus at the pole (pp. 707–711)	See Table 5 (p. 709).
Parametric equations of a curve (p. 713)	$x = f(t), y = g(t), t$ is the parameter

Definitions

Parabola (p. 667)	Set of points P in the plane for which $d(F, P) = d(P, D)$, where F is the focus and D is the directrix
Ellipse (p. 676)	Set of points P in the plane, the sum of whose distances from two fixed points (the foci) is a constant
Hyperbola (p. 686)	Set of points P in the plane, the difference of whose distances from two fixed points (the foci) is a constant
Conic in polar coordinates (p. 707)	$\dfrac{d(F, P)}{d(P, D)} = e$ Parabola if $e = 1$
	Ellipse if $e < 1$
	Hyperbola if $e > 1$

Formulas

Rotation formulas (p. 701)	$x = x' \cos \theta - y' \sin \theta$
	$y = x' \sin \theta + y' \cos \theta$
Angle θ of rotation that eliminates the $x'y'$-term (p. 702)	$\cot(2\theta) = \dfrac{A - C}{B}, \quad 0° < \theta < 90°$

Objectives

Section		You should be able to . . .	Example(s)	Review Exercises
9.1	1	Know the names of the conics (p. 666)		1–32
9.2	1	Analyze parabolas with vertex at the origin (p. 667)	1–5	1, 2, 21, 24
	2	Analyze parabolas with vertex at (h, k) (p. 670)	6, 7	7, 11, 12, 17, 18, 27, 30
	3	Solve applied problems involving parabolas (p. 672)	8	77, 78
9.3	1	Analyze ellipses with center at the origin (p. 676)	1–4	5, 6, 10, 22, 25
	2	Analyze ellipses with center at (h, k) (p. 680)	5, 6	14–16, 19, 28, 31
	3	Solve applied problems involving ellipses (p. 682)	7	79, 80
9.4	1	Analyze hyperbolas with center at the origin (p. 686)	1–4	3, 4, 8, 9, 23, 26
	2	Find the asymptotes of a hyperbola (p. 691)	5, 6	3, 4, 8, 9
	3	Analyze hyperbolas with center at (h, k) (p. 693)	7, 8	13, 20, 29, 32–36
	4	Solve applied problems involving hyperbolas (p. 694)	9	81
9.5	1	Identify a conic (p. 699)	1	37–40
	2	Use a rotation of axes to transform equations (p. 700)	2	47–52
	3	Analyze an equation using a rotation of axes (p. 702)	3, 4	47–52
	4	Identify conics without a rotation of axes (p. 704)	5	41–46
9.6	1	Analyze and graph polar equations of conics (p. 707)	1–3	53–58
	2	Convert the polar equation of a conic to a rectangular equation (p. 711)	4	59–62
9.7	1	Graph parametric equations (p. 713)	1	63–68
	2	Find a rectangular equation for a curve defined parametrically (p. 714)	2, 3	63–68
	3	Use time as a parameter in parametric equations (p. 716)	4, 5	82, 83
	4	Find parametric equations for curves defined by rectangular equations (p. 719)	6, 7	69–72

Review Exercises

In Problems 1–20, identify each equation. If it is a parabola, give its vertex, focus, and directrix; if it is an ellipse, give its center, vertices, and foci; if it is a hyperbola, give its center, vertices, foci, and asymptotes.

1. $y^2 = -16x$
2. $16x^2 = y$
3. $\dfrac{x^2}{25} - y^2 = 1$
4. $\dfrac{y^2}{25} - x^2 = 1$
5. $\dfrac{y^2}{25} + \dfrac{x^2}{16} = 1$
6. $\dfrac{x^2}{9} + \dfrac{y^2}{16} = 1$
7. $x^2 + 4y = 4$
8. $3y^2 - x^2 = 9$
9. $4x^2 - y^2 = 8$
10. $9x^2 + 4y^2 = 36$
11. $x^2 - 4x = 2y$
12. $2y^2 - 4y = x - 2$
13. $y^2 - 4y - 4x^2 + 8x = 4$
14. $4x^2 + y^2 + 8x - 4y + 4 = 0$
15. $4x^2 + 9y^2 - 16x - 18y = 11$
16. $4x^2 + 9y^2 - 16x + 18y = 11$
17. $4x^2 - 16x + 16y + 32 = 0$
18. $4y^2 + 3x - 16y + 19 = 0$
19. $9x^2 + 4y^2 - 18x + 8y = 23$
20. $x^2 - y^2 - 2x - 2y = 1$

In Problems 21–36, find an equation of the conic described. Graph the equation.

21. Parabola; focus at $(-2, 0)$; directrix the line $x = 2$
22. Ellipse; center at $(0, 0)$; focus at $(0, 3)$; vertex at $(0, 5)$
23. Hyperbola; center at $(0,0)$; focus at $(0,4)$; vertex at $(0, -2)$
24. Parabola; vertex at $(0, 0)$; directrix the line $y = -3$
25. Ellipse; foci at $(-3, 0)$ and $(3, 0)$; vertex at $(4, 0)$
26. Hyperbola; vertices at $(-2, 0)$ and $(2, 0)$; focus at $(4, 0)$
27. Parabola; vertex at $(2, -3)$; focus at $(2, -4)$
28. Ellipse; center at $(-1, 2)$; focus at $(0, 2)$; vertex at $(2, 2)$
29. Hyperbola; center at $(-2, -3)$; focus at $(-4, -3)$; vertex at $(-3, -3)$
30. Parabola; focus at $(3, 6)$; directrix the line $y = 8$
31. Ellipse; foci at $(-4, 2)$ and $(-4, 8)$; vertex at $(-4, 10)$
32. Hyperbola; vertices at $(-3, 3)$ and $(5, 3)$; focus at $(7, 3)$
33. Center at $(-1, 2)$; $a = 3$; $c = 4$; transverse axis parallel to the x-axis
34. Center at $(4, -2)$; $a = 1$; $c = 4$; transverse axis parallel to the y-axis
35. Vertices at $(0, 1)$ and $(6, 1)$; asymptote the line $3y + 2x = 9$
36. Vertices at $(4, 0)$ and $(4, 4)$; asymptote the line $y + 2x = 10$

In Problems 37–46, identify each conic without completing the squares and without applying a rotation of axes.

37. $y^2 + 4x + 3y - 8 = 0$

38. $2x^2 - y + 8x = 0$

39. $x^2 + 2y^2 + 4x - 8y + 2 = 0$

40. $x^2 - 8y^2 - x - 2y = 0$

41. $9x^2 - 12xy + 4y^2 + 8x + 12y = 0$

42. $4x^2 + 4xy + y^2 - 8\sqrt{5}x + 16\sqrt{5}y = 0$

43. $4x^2 + 10xy + 4y^2 - 9 = 0$

44. $4x^2 - 10xy + 4y^2 - 9 = 0$

45. $x^2 - 2xy + 3y^2 + 2x + 4y - 1 = 0$

46. $4x^2 + 12xy - 10y^2 + x + y - 10 = 0$

In Problems 47–52, rotate the axes so that the new equation contains no xy-term. Analyze and graph the new equation.

47. $2x^2 + 5xy + 2y^2 - \dfrac{9}{2} = 0$

48. $2x^2 - 5xy + 2y^2 - \dfrac{9}{2} = 0$

49. $6x^2 + 4xy + 9y^2 - 20 = 0$

50. $x^2 + 4xy + 4y^2 + 16\sqrt{5}x - 8\sqrt{5}y = 0$

51. $4x^2 - 12xy + 9y^2 + 12x + 8y = 0$

52. $9x^2 - 24xy + 16y^2 + 80x + 60y = 0$

In Problems 53–58, identify the conic that each polar equation represents and graph it.

53. $r = \dfrac{4}{1 - \cos\theta}$

54. $r = \dfrac{6}{1 + \sin\theta}$

55. $r = \dfrac{6}{2 - \sin\theta}$

56. $r = \dfrac{2}{3 + 2\cos\theta}$

57. $r = \dfrac{8}{4 + 8\cos\theta}$

58. $r = \dfrac{10}{5 + 20\sin\theta}$

In Problems 59–62, convert each polar equation to a rectangular equation.

59. $r = \dfrac{4}{1 - \cos\theta}$

60. $r = \dfrac{6}{2 - \sin\theta}$

61. $r = \dfrac{8}{4 + 8\cos\theta}$

62. $r = \dfrac{2}{3 + 2\cos\theta}$

In Problems 63–68, graph the curve whose parametric equations are given and show its orientation. Find the rectangular equation of each curve.

63. $x = 4t - 2$, $y = 1 - t$; $-\infty < t < \infty$

64. $x = 2t^2 + 6$, $y = 5 - t$; $-\infty < t < \infty$

65. $x = 3\sin t$, $y = 4\cos t + 2$; $0 \le t \le 2\pi$

66. $x = \ln t$, $y = t^3$; $t > 0$

67. $x = \sec^2 t$, $y = \tan^2 t$; $0 \le t \le \dfrac{\pi}{4}$

68. $x = t^{\frac{3}{2}}$, $y = 2t + 4$; $t \ge 0$

In Problems 69 and 70, find two different parametric equations for each rectangular equation.

69. $y = -2x + 4$

70. $y = 2x^2 - 8$

In Problems 71 and 72, find parametric equations for an object that moves along the ellipse $\dfrac{x^2}{16} + \dfrac{y^2}{9} = 1$ with the motion described.

71. The motion begins at $(4, 0)$, is counterclockwise, and requires 4 seconds for a complete revolution.

72. The motion begins at $(0, 3)$, is clockwise, and requires 5 seconds for a complete revolution.

---

73. Find an equation of the hyperbola whose foci are the vertices of the ellipse $4x^2 + 9y^2 = 36$ and whose vertices are the foci of this ellipse.

74. Find an equation of the ellipse whose foci are the vertices of the hyperbola $x^2 - 4y^2 = 16$ and whose vertices are the foci of this hyperbola.

75. Describe the collection of points in a plane so that the distance from each point to the point $(3, 0)$ is three-fourths of its distance from the line $x = \dfrac{16}{3}$.

76. Describe the collection of points in a plane so that the distance from each point to the point $(5, 0)$ is five-fourths of its distance from the line $x = \dfrac{16}{5}$.

77. Searchlight A searchlight is shaped like a paraboloid of revolution. If a light source is located 1 foot from the vertex

along the axis of symmetry and the opening is 2 feet across, how deep should the mirror be in order to reflect the light rays parallel to the axis of symmetry?

78. Parabolic Arch Bridge A bridge is built in the shape of a parabolic arch. The bridge has a span of 60 feet and a maximum height of 20 feet. Find the height of the arch at distances of 5, 10, and 20 feet from the center.

79. Semielliptical Arch Bridge A bridge is built in the shape of a semielliptical arch. The bridge has a span of 60 feet and a maximum height of 20 feet. Find the height of the arch at distances of 5, 10, and 20 feet from the center.

80. Whispering Gallery The figure on the following page shows the specifications for an elliptical ceiling in a hall designed to be a whispering gallery. Where are the foci located in the hall?

81. Calibrating Instruments In a test of their recording devices, a team of seismologists positioned two of the devices 2000 feet apart, with the device at point A to the west of the device at point B. At a point between the devices and 200 feet from point B, a small amount of explosive was detonated and a note made of the time at which the sound reached each device. A second explosion is to be carried out at a point directly north of point B. How far north should the site of the second explosion be chosen so that the measured time difference recorded by the devices for the second detonation is the same as that recorded for the first detonation?

82. Uniform Motion Mary's train leaves at 7:15 AM and accelerates at the rate of 3 meters per second per second. Mary, who can run 6 meters per second, arrives at the train station 2 seconds after the train has left.

(a) Find parametric equations that model the motion of the train and Mary as a function of time.
   [**Hint:** The position s at time t of an object having acceleration a is $s = \frac{1}{2}at^2$.]
(b) Determine algebraically whether Mary will catch the train. If so, when?
(c) Simulate the motions of the train and Mary by simultaneously graphing the equations found in part (a).

83. Projectile Motion Drew Brees throws a football with an initial speed of 80 feet per second at an angle of 35° to the horizontal. The ball leaves Brees's hand at a height of 6 feet.
(a) Find parametric equations that model the position of the ball as a function of time.
(b) How long is the ball in the air?
(c) When is the ball at its maximum height? Determine the maximum height of the ball.
(d) Determine the horizontal distance that the ball travels.
(e) Using a graphing utility, simultaneously graph the equations found in part (a).

84. Formulate a strategy for discussing and graphing an equation of the form
$$Ax^2 + Bxy + Cy^2 + Dx + Ey + F = 0$$

CHAPTER TEST

CHAPTER **Test Prep** VIDEOS

Step-by-step test solutions are found on the Chapter Test Prep Videos available via the Video Resources on DVD, in **MyMathLab** or on **You**Tube (search "Sullivan Precalculus: A Right Triangle Approach, Concepts Through Functions 2e).

In Problems 1–3, identify each equation. If it is a parabola, give its vertex, focus, and directrix; if an ellipse, give its center, vertices, and foci; if a hyperbola, give its center, vertices, foci, and asymptotes.

1. $\dfrac{(x+1)^2}{4} - \dfrac{y^2}{9} = 1$

2. $8y = (x-1)^2 - 4$

3. $2x^2 + 3y^2 + 4x - 6y = 13$

In Problems 4–6, find an equation of the conic described; graph the equation.

4. Parabola: focus $(-1, 4.5)$, vertex $(-1, 3)$

5. Ellipse: center $(0, 0)$, vertex $(0, -4)$, focus $(0, 3)$

6. Hyperbola: center $(2, 2)$, vertex $(2, 4)$, contains the point $\left(2 + \sqrt{10}, 5\right)$

In Problems 7–9, identify each conic without completing the square or rotating axes.

7. $2x^2 + 5xy + 3y^2 + 3x - 7 = 0$

8. $3x^2 - xy + 2y^2 + 3y + 1 = 0$

9. $x^2 - 6xy + 9y^2 + 2x - 3y - 2 = 0$

10. Given the equation $41x^2 - 24xy + 34y^2 - 25 = 0$, rotate the axes so that there is no xy-term. Analyze and graph the new equation.

11. Identify the conic represented by the polar equation $r = \dfrac{3}{1 - 2\cos\theta}$. Find the rectangular equation.

12. Graph the curve whose parametric equations are given and show its orientation. Find the rectangular equation for the curve.
$$x = 3t - 2, \quad y = 1 - \sqrt{t}, \quad 0 \le t \le 9$$

13. A parabolic reflector (paraboloid of revolution) is used by TV crews at football games to pick up the referee's announcements, quarterback signals, and so on. A microphone is placed at the focus of the parabola. If a certain reflector is 4 feet wide and 1.5 feet deep, where should the microphone be placed?

CUMULATIVE REVIEW

1. For $f(x) = -3x^2 + 5x - 2$, find
$$\frac{f(x+h) - f(x)}{h} \qquad h \ne 0$$

2. In the complex number system, solve the equation
$$9x^4 + 33x^3 - 71x^2 - 57x - 10 = 0$$

3. For what numbers x is $6 - x \ge x^2$?

4. (a) Find the domain and range of $y = 3^x + 2$.
   (b) Find the inverse of $y = 3^x + 2$ and state its domain and range.

5. $f(x) = \log_4(x - 2)$
   (a) Solve $f(x) = 2$.
   (b) Solve $f(x) \le 2$.

6. Find an equation for each of the following graphs:

(a) Line:

(b) Circle:

(c) Ellipse:

(d) Parabola:

(e) Hyperbola:

(f) Exponential:

7. Find all the solutions of the equation $\sin(2\theta) = 0.5$.

8. Find a polar equation for the line containing the origin that makes an angle of $30°$ with the positive x-axis.

9. Find a polar equation for the circle with center at the point $(0, 4)$ and radius 4. Graph this circle.

10. What is the domain of the function $f(x) = \dfrac{3}{\sin x + \cos x}$?

11. Solve the equation $\cot(2\theta) = 1$, where $0° < \theta < 90°$.

12. Find the rectangular equation of the curve

$$x = 5 \tan t, \quad y = 5 \sec^2 t, \quad -\frac{\pi}{2} < t < \frac{\pi}{2}$$

CHAPTER PROJECTS

I. The Orbits of Neptune and Pluto The orbit of a planet about the Sun is an ellipse, with the Sun at one focus. The **aphelion** of a planet is its greatest distance from the Sun and the **perihelion** is its shortest distance. The **mean distance** of a planet from the Sun is the length of the semimajor axis of the elliptical orbit. See the illustration.

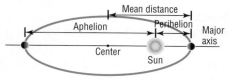

1. The aphelion of Neptune is 4532.2×10^6 kilometers (km) and its perihelion is 4458.0×10^6 km. Find a model for the orbit of Neptune around the Sun.

2. The aphelion of Pluto* is 7381.2×10^6 km and its perihelion is 4445.8×10^6 km. Find a model for the orbit of Pluto around the Sun.

3. Graph the orbits of Pluto and Neptune on a graphing utility. Knowing that the orbits of the planets intersect, what is wrong with the graphs you obtained?

4. The graphs of the orbits drawn in part 3 have the same center, but their foci lie in different locations. To see an accurate representation, the location of the Sun (a focus) needs to be the same for both graphs. This can be accomplished by shifting Pluto's orbit to the left. The shift amount is equal to Pluto's distance from the center (in the graph in part 3) to the Sun minus Neptune's distance from the center to the Sun. Find the new model representing the orbit of Pluto.

5. Graph the equation for the orbit of Pluto found in part 4 along with the equation of the orbit of Neptune. Do you see that Pluto's orbit is sometimes inside Neptune's?

6. Find the point(s) of intersection of the two orbits.

7. Do you think two planets will ever collide?

The following projects can be found on the Instructor's Resource Center (IRC):

II. Project at Motorola *Distorted Deployable Space Reflector Antennas* An engineer designs an antenna that will deploy in space to collect sunlight.

III. Constructing a Bridge over the East River The size of ships using a river and fluctuations in water height due to tides or flooding must be considered when designing a bridge that will cross a major waterway.

IV. Systems of Parametric Equations Choosing an approach to use when solving a system of equations depends on the form of the system and on the domains of the equations.

*Pluto's status was reduced to a dwarf planet in September 2006.

10 Systems of Equations and Inequalities

Economic Outcomes

Annual Earnings of Young Adults

For both males and females, earnings increase with education: full-time workers with at least a bachelor's degree have higher median earnings than those with less education. For example,

in 2003, male college graduates earned 93 percent more than male high school completers. Females with a bachelor's or higher degree earned 91 percent more than female high school completers. Males and females who dropped out of high school earned 37 and 39 percent less, respectively, than male and female high school completers.

The median earnings of young adults who have at least a bachelor's degree declined in the 1970s relative to their counterparts who were high school completers, before increasing between 1980 and 2003. Males with a bachelor's degree or higher had earnings 19 percent higher than male high school completers in 1980 and had earnings 93 percent higher in 2003. Among females, those with at least a bachelor's degree had earnings 34 percent higher than female high school completers in 1980, compared with earnings 91 percent higher in 2003.

—See Chapter Project I—

◁ **A Look Back** In Appendix A, Section A.8, and in Chapters 2, 3, and 4, we solved various kinds of equations and inequalities involving a single variable.

A Look Ahead ▷ In this chapter we take up the problem of solving equations and inequalities containing two or more variables. There are various ways to solve such problems.

The *method of substitution* for solving equations in several unknowns goes back to ancient times.

The *method of elimination*, although it had existed for centuries, was put into systematic order by Karl Friedrich Gauss (1777–1855) and by Camille Jordan (1838–1922).

The theory of *matrices* was developed in 1857 by Arthur Cayley (1821–1895), although only later were matrices used as we use them in this chapter. Matrices have become a very flexible instrument, useful in almost all areas of mathematics.

The method of *determinants* was invented by Takakazu Seki Kôwa (1642–1708) in 1683 in Japan and by Gottfried Wilhelm von Leibniz (1646–1716) in 1693 in Germany. *Cramer's Rule* is named after Gabriel Cramer (1704–1752) of Switzerland, who popularized the use of determinants for solving linear systems.

Section 10.5, on *partial fraction decomposition*, provides an application of systems of equations. This particular application is one that is used in integral calculus.

Section 10.8 introduces *linear programming*, a modern application of linear inequalities. This topic is particularly useful for students interested in operations research.

10.1 Systems of Linear Equations: Substitution and Elimination

PREPARING FOR THIS SECTION *Before getting started, review the following:*

- Linear Equations (Appendix A, Section A.8, pp. A65–A68)
- Lines (Foundations, Section 3, pp. 19–29)

Now Work the 'Are You Prepared?' problems on page 741.

OBJECTIVES 1 Solve Systems of Equations by Substitution (p. 733)
   2 Solve Systems of Equations by Elimination (p. 735)
   3 Identify Inconsistent Systems of Equations Containing Two Variables (p. 736)
   4 Express the Solution of a System of Dependent Equations Containing Two Variables (p. 737)
   5 Solve Systems of Three Equations Containing Three Variables (p. 738)
   6 Identify Inconsistent Systems of Equations Containing Three Variables (p. 739)
   7 Express the Solution of a System of Dependent Equations Containing Three Variables (p. 740)

We begin with an example.

EXAMPLE 1 **Movie Theater Ticket Sales**

A movie theater sells tickets for $8.00 each, with seniors receiving a discount of $2.00. One evening the theater took in $3580 in revenue. If x represents the number of tickets sold at $8.00 and y the number of tickets sold at the discounted price of $6.00, write an equation that relates these variables.

Solution Each nondiscounted ticket brings in $8.00, so x tickets will bring in $8x$ dollars. Similarly, y discounted tickets bring in $6y$ dollars. Since the total brought in is $3580, we must have

$$8x + 6y = 3580$$

In Example 1, suppose that we also know that 525 tickets were sold that evening. Then we have another equation relating the variables x and y:

$$x + y = 525$$

The two equations

$$8x + 6y = 3580$$
$$x + y = 525$$

form a *system* of equations.

In general, a **system of equations** is a collection of two or more equations, each containing one or more variables. Example 2 gives some illustrations of systems of equations.

EXAMPLE 2 **Examples of Systems of Equations**

(a) $\begin{cases} 2x + y = 5 & (1) \\ -4x + 6y = -2 & (2) \end{cases}$ Two equations containing two variables, x and y

(b) $\begin{cases} x + y^2 = 5 & (1) \\ 2x + y = 4 & (2) \end{cases}$ Two equations containing two variables, x and y

(c) $\begin{cases} x + y + z = 6 & (1) \\ 3x - 2y + 4z = 9 & (2) \\ x - y - z = 0 & (3) \end{cases}$ Three equations containing three variables, x, y, and z

(d) $\begin{cases} x + y + z = 5 & (1) \\ x - y \qquad = 2 & (2) \end{cases}$ Two equations containing three variables, x, y, and z

(e) $\begin{cases} x + y + z = 6 & (1) \\ 2x \qquad + 2z = 4 & (2) \\ \qquad y + z = 2 & (3) \\ x \qquad = 4 & (4) \end{cases}$ Four equations containing three variables, x, y, and z

We use a brace, as shown, to remind us that we are dealing with a system of equations. We also will find it convenient to number each equation in the system.

A **solution** of a system of equations consists of values for the variables that are solutions of each equation of the system. To **solve** a system of equations means to find all solutions of the system.

For example, $x = 2$, $y = 1$ is a solution of the system in Example 2(a), because

$\begin{cases} 2x + y = 5 & (1) \\ -4x + 6y = -2 & (2) \end{cases}$ $\begin{cases} 2(2) + 1 = 4 + 1 = 5 \\ -4(2) + 6(1) = -8 + 6 = -2 \end{cases}$

We may also write this solution as the ordered pair $(2, 1)$.

A solution of the system in Example 2(b) is $x = 1$, $y = 2$, because

$\begin{cases} x + y^2 = 5 & (1) \\ 2x + y = 4 & (2) \end{cases}$ $\begin{cases} 1 + 2^2 = 1 + 4 = 5 \\ 2(1) + 2 = 2 + 2 = 4 \end{cases}$

Another solution of the system in Example 2(b) is $x = \dfrac{11}{4}$, $y = -\dfrac{3}{2}$, which you can check for yourself.

A solution of the system in Example 2(c) is $x = 3$, $y = 2$, $z = 1$, because

$\begin{cases} x + y + z = 6 & (1) \\ 3x - 2y + 4z = 9 & (2) \\ x - y - z = 0 & (3) \end{cases}$ $\begin{cases} 3 + 2 + 1 = 6 & (1) \\ 3(3) - 2(2) + 4(1) = 9 - 4 + 4 = 9 & (2) \\ 3 - 2 - 1 = 0 & (3) \end{cases}$

We may also write this solution as the ordered triplet $(3, 2, 1)$.

Note that $x = 3$, $y = 3$, $z = 0$ is not a solution of the system in Example 2(c).

$\begin{cases} x + y + z = 6 & (1) \\ 3x - 2y + 4z = 9 & (2) \\ x - y - z = 0 & (3) \end{cases}$ $\begin{cases} 3 + 3 + 0 = 6 & (1) \\ 3(3) - 2(3) + 4(0) = 3 \neq 9 & (2) \\ 3 - 3 - 0 = 0 & (3) \end{cases}$

Although $x = 3$, $y = 3$, and $z = 0$ satisfy equations (1) and (3), they do not satisfy equation (2). Any solution of the system must satisfy *each* equation of the system.

Now Work PROBLEM 9

When a system of equations has at least one solution, it is said to be **consistent.** If a system of equations has no solution, it is called **inconsistent.**

An equation in n variables is said to be **linear** if it is equivalent to an equation of the form

$$a_1 x_1 + a_2 x_2 + \cdots + a_n x_n = b$$

where $x_1, x_2, \ldots, x_n$ are n distinct variables, $a_1, a_2, \ldots, a_n, b$ are constants, and at least one of the a's is not 0.

Some examples of linear equations are

$$2x + 3y = 2 \qquad 5x - 2y + 3z = 10 \qquad 8x + 8y - 2z + 5w = 0$$

If each equation in a system of equations is linear, we have a **system of linear equations.** The systems in Examples 2(a), (c), (d), and (e) are linear, whereas the system in Example 2(b) is nonlinear. In this chapter we shall solve linear systems in Sections 10.1 to 10.3. We discuss nonlinear systems in Section 10.6.

We begin by discussing a system of two linear equations containing two variables. We can view the problem of solving such a system as a geometry problem. The graph of each equation in such a system is a line. So a system of two linear equations containing two variables represents a pair of lines. The lines either (1) intersect or (2) are parallel or (3) are **coincident** (that is, identical).

1. If the lines intersect, the system of equations has one solution, given by the point of intersection. The system is **consistent** and the equations are **independent.** See Figure 1(a).

2. If the lines are parallel, the system of equations has no solution, because the lines never intersect. The system is **inconsistent.** See Figure 1(b).

3. If the lines are coincident (the lines lie on top of each other), the system of equations has infinitely many solutions, represented by the totality of points on the line. The system is **consistent** and the equations are **dependent.** See Figure 1(c).

Figure 1

(a) Intersecting lines; system has one solution

(b) Parallel lines; system has no solution

(c) Coincident lines; system has infinitely many solutions

EXAMPLE 3   **Graphing a System of Linear Equations**

Graph the system: $\begin{cases} 2x + y = 5 & (1) \\ -4x + 6y = 12 & (2) \end{cases}$

Solution   Equation (1) in slope–intercept form is $y = -2x + 5$, which has slope -2 and y-intercept 5. Equation (2) in slope–intercept form is $y = \frac{2}{3}x + 2$, which has slope $\frac{2}{3}$ and y-intercept 2. Figure 2 shows their graphs.

Figure 2

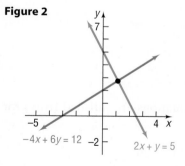

From the graph in Figure 2, we see that the lines intersect, so the system given in Example 3 is consistent. We can also use the graph as a means of approximating the solution. For this system, the solution appears to be close to the point $(1, 3)$. The actual solution, which you should verify, is $\left(\frac{9}{8}, \frac{11}{4}\right)$.

Seeing the Concept

Graph the lines $2x + y = 5$ $(Y_1 = -2x + 5)$ and $-4x + 6y = 12$ $\left(Y_2 = \frac{2}{3}x + 2\right)$ and compare what you see with Figure 2. Use INTERSECT to verify that the point of intersection is (1.125, 2.75).

1 Solve Systems of Equations by Substitution

Most of the time we must use algebraic methods to obtain exact solutions. A number of methods are available to us for solving systems of linear equations algebraically. In this section, we introduce two methods: *substitution* and *elimination.* We illustrate the **method of substitution** by solving the system given in Example 3.

| EXAMPLE 4 | **How to Solve a System of Linear Equations by Substitution** |

Solve: $\begin{cases} 2x + y = 5 & (1) \\ -4x + 6y = 12 & (2) \end{cases}$

Step-by-Step Solution

Step 1: Pick one of the equations and solve for one of the variables in terms of the remaining variable(s).

We solve equation (1) for y.

$$2x + y = 5 \qquad \text{Equation (1)}$$
$$y = -2x + 5 \quad \text{Subtract 2x from each side of (1).}$$

Step 2: Substitute the result into the remaining equation(s).

We substitute $-2x + 5$ for y in equation (2). The result is an equation containing just the variable x, which we can solve.

$$-4x + 6y = 12 \quad \text{Equation (2)}$$
$$-4x + 6(-2x + 5) = 12 \quad \text{Substitute } -2x + 5 \text{ for y in (2).}$$

Step 3: If one equation in one variable results, solve this equation. Otherwise, repeat Steps 1 and 2 until a single equation with one variable remains.

$$-4x - 12x + 30 = 12 \qquad \text{Distribute.}$$
$$-16x + 30 = 12 \qquad \text{Combine like terms.}$$
$$-16x = -18 \qquad \text{Subtract 30 from both sides.}$$
$$x = \frac{-18}{-16} \qquad \text{Divide both sides by } -16.$$
$$= \frac{9}{8} \qquad \text{Simplify.}$$

Step 4: Find the values of the remaining variables by back-substitution.

Because we know that $x = \frac{9}{8}$, we can find the value of y by **back-substitution,** that is, by substituting $\frac{9}{8}$ for x in one of the original equations. Equation (1) seems easier to work with, so we will back-substitute into equation (1).

$$2x + y = 5 \qquad \text{Equation (1)}$$
$$2\left(\frac{9}{8}\right) + y = 5 \qquad \text{Substitute } x = \frac{9}{8} \text{ into equation (1).}$$
$$\frac{9}{4} + y = 5 \qquad \text{Simplify.}$$
$$y = 5 - \frac{9}{4} \quad \text{Subtract } \frac{9}{4} \text{ from both sides.}$$
$$= \frac{11}{4} \qquad 5 - \frac{9}{4} = \frac{20}{4} - \frac{9}{4} = \frac{11}{4}$$

Step 5: Check the solution found.

We have $x = \frac{9}{8}$ and $y = \frac{11}{4}$. We verify that both equations are satisfied (true) for these values.

$$\begin{cases} 2x + y = 5: & 2\left(\frac{9}{8}\right) + \frac{11}{4} = \frac{9}{4} + \frac{11}{4} = \frac{20}{4} = 5 \\ -4x + 6y = 12: & -4\left(\frac{9}{8}\right) + 6\left(\frac{11}{4}\right) = -\frac{9}{2} + \frac{33}{2} = \frac{24}{2} = 12 \end{cases}$$

The solution of the system is $x = \frac{9}{8} = 1.125$ and $y = \frac{11}{4} = 2.75$. We can also write the solution as the ordered pair $\left(\frac{9}{8}, \frac{11}{4}\right)$.

●

Now Use Substitution to Work   PROBLEM 19

2 Solve Systems of Equations by Elimination

A second method for solving a system of linear equations is the *method of elimination*. This method is usually preferred over substitution if substitution leads to fractions or if the system contains more than two variables. Elimination also provides the necessary motivation for solving systems using matrices (the subject of Section 10.2).

The idea behind the **method of elimination** is to replace the original system of equations by an equivalent system so that adding two of the equations eliminates a variable. The rules for obtaining equivalent equations are the same as those studied earlier. However, we may also interchange any two equations of the system and/or replace any equation in the system by the sum (or difference) of that equation and a nonzero multiple of any other equation in the system.

> **In Words**
>
> When using elimination, we want to get the coefficients of one of the variables to be negatives of one another.

Rules for Obtaining an Equivalent System of Equations

1. Interchange any two equations of the system.
2. Multiply (or divide) each side of an equation by the same nonzero constant.
3. Replace any equation in the system by the sum (or difference) of that equation and a nonzero multiple of any other equation in the system.

An example will give you the idea. As you work through the example, pay particular attention to the pattern being followed.

EXAMPLE 5

How to Solve a System of Linear Equations by Elimination

Solve: $\begin{cases} 2x + 3y = 1 & (1) \\ -x + y = -3 & (2) \end{cases}$

Step-by-Step Solution

Step 1: Multiply both sides of one or both equations by a nonzero constant so that the coefficients of one of the variables are additive inverses.

We multiply both sides of equation (2) by 2 so that the coefficient of x in the two equations are additive inverses.

$$\begin{cases} 2x + 3y = 1 & (1) \\ -x + y = -3 & (2) \end{cases}$$

$$\begin{cases} 2x + 3y = 1 & (1) \\ 2(-x + y) = 2(-3) & (2) \quad \text{Multiply by 2.} \end{cases}$$

$$\begin{cases} 2x + 3y = 1 & (1) \\ -2x + 2y = -6 & (2) \end{cases}$$

Step 2: Add the equations to eliminate the variable. Solve the resulting equation for the remaining unknown.

$$\begin{cases} 2x + 3y = 1 & (1) \\ -2x + 2y = -6 & (2) \end{cases}$$
$$5y = -5 \quad \text{Add equations (1) and (2).}$$
$$y = -1 \quad \text{Divide both sides by 5.}$$

Step 3: Back-substitute the value of the variable found in Step 2 into one of the *original* equations to find the value of the remaining variable.

We back-substitute $y = -1$ into equation (1) and solve for x.

$$2x + 3y = 1 \quad \text{Equation (1)}$$
$$2x + 3(-1) = 1 \quad \text{Substitute } y = -1 \text{ into equation (1).}$$
$$2x - 3 = 1 \quad \text{Simplify.}$$
$$2x = 4 \quad \text{Add 3 to both sides.}$$
$$x = 2 \quad \text{Divide both sides by 2.}$$

Step 4: Check the solution found.

We leave the check to you.

The solution of the system is $x = 2$ and $y = -1$. We can also write the solution as the ordered pair $(2, -1)$.

Now Use Elimination to Work PROBLEM 19

EXAMPLE 6 | **Movie Theater Ticket Sales**

A movie theater sells tickets for $8.00 each, with seniors receiving a discount of $2.00. One evening the theater sold 525 tickets and took in $3580 in revenue. How many of each type of ticket were sold?

Solution | If x represents the number of tickets sold at $8.00 and y the number of tickets sold at the discounted price of $6.00, then the given information results in the system of equations

$$\begin{cases} 8x + 6y = 3580 & (1) \\ x + y = 525 & (2) \end{cases}$$

We use the method of elimination. First, multiply the second equation by -6, and then add the equations.

$$\begin{cases} 8x + 6y = 3580 \\ -6x - 6y = -3150 \end{cases}$$
$$ 2x = 430 \qquad \text{Add the equations.}$$
$$ x = 215$$

Since $x + y = 525$, then $y = 525 - x = 525 - 215 = 310$. So 215 nondiscounted tickets and 310 senior discount tickets were sold.

---

3 Identify Inconsistent Systems of Equations Containing Two Variables

The previous examples dealt with consistent systems of equations that had a single solution. The next two examples deal with two other possibilities that may occur, the first being a system that has no solution.

EXAMPLE 7 | **An Inconsistent System of Linear Equations**

Solve: $\begin{cases} 2x + y = 5 & (1) \\ 4x + 2y = 8 & (2) \end{cases}$

Solution | We choose to use the method of substitution and solve equation (1) for y.

$$2x + y = 5 \qquad (1)$$
$$y = -2x + 5 \qquad \text{Subtract 2x from each side.}$$

Now substitute $-2x + 5$ for y in equation (2) and solve for x.

$$4x + 2y = 8 \qquad (2)$$
$$4x + 2(-2x + 5) = 8 \qquad \text{Substitute } y = -2x + 5 \text{ in (2).}$$
$$4x - 4x + 10 = 8 \qquad \text{Remove parentheses.}$$
$$0 = -2 \qquad \text{Subtract 10 from both sides.}$$

This statement is false. We conclude that the system has no solution and is therefore inconsistent.

Figure 3

Figure 3 illustrates the pair of lines whose equations form the system in Example 7. Notice that the graphs of the two equations are lines, each with slope -2;

one has a y-intercept of 5, the other a y-intercept of 4. The lines are parallel and have no point of intersection. This geometric statement is equivalent to the algebraic statement that the system has no solution.

 Seeing the Concept
Graph the lines $2x + y = 5$ ($Y_1 = -2x + 5$) and $4x + 2y = 8$ ($Y_2 = -2x + 4$) and compare what you see with Figure 3. How can you be sure that the lines are parallel?

4 Express the Solution of a System of Dependent Equations Containing Two Variables

EXAMPLE 8 **Solving a System of Dependent Equations**

Solve: $\begin{cases} 2x + y = 4 & (1) \\ -6x - 3y = -12 & (2) \end{cases}$

Solution We choose to use the method of elimination.

$$\begin{cases} 2x + y = 4 & (1) \\ -6x - 3y = -12 & (2) \end{cases}$$

$$\begin{cases} 6x + 3y = 12 & (1) \quad \text{Multiply each side of equation (1) by 3.} \\ -6x - 3y = -12 & (2) \end{cases}$$

$$0 = 0 \qquad \text{Add equations (1) and (2).}$$

The statement $0 = 0$ is true. This means the equation $6x + 3y = 12$ is equivalent to $-6x - 3y = -12$. Therefore, the original system is equivalent to a system containing one equation, so the equations are dependent. This means that any values of x and y that satisfy $6x + 3y = 12$ or, equivalently, $2x + y = 4$ are solutions. For example, $x = 2$, $y = 0$; $x = 0$, $y = 4$; $x = -2$, $y = 8$; $x = 4$, $y = -4$; and so on, are solutions. There are, in fact, infinitely many values of x and y for which $2x + y = 4$, so the original system has infinitely many solutions. We will write the solution of the original system either as

$$y = -2x + 4 \quad \text{where } x \text{ can be any real number}$$

or as

$$x = -\frac{1}{2}y + 2 \quad \text{where } y \text{ can be any real number.}$$

We can also express the solution as $\{(x, y) \mid y = -2x + 4, x \text{ is any real number}\}$ or as $\{(x, y) \mid x = -\frac{1}{2}y + 2, y \text{ is any real number}\}$.

●

Figure 4

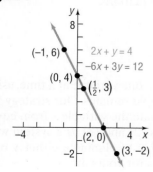

Figure 4 illustrates the situation presented in Example 8. Notice that the graphs of the two equations are lines, each with slope -2 and each with y-intercept 4. The lines are coincident. Notice also that equation (2) in the original system is -3 times equation (1), indicating that the two equations are dependent.

For the system in Example 8, we can write down some of the infinite number of solutions by assigning values to x and then finding $y = -2x + 4$.

If $x = -2$, then $y = -2(-2) + 4 = 8$.

If $x = 0$, then $y = 4$.

If $x = 2$, then $y = 0$.

The ordered pairs $(-2, 8)$, $(0, 4)$, and $(2, 0)$ are three of the points on the line in Figure 4.

> **Seeing the Concept**
>
> Graph the lines $2x + y = 4$ ($Y_1 = -2x + 4$) and $-6x - 3y = -12$ ($Y_2 = -2x + 4$) and compare what you see with Figure 4. How can you be sure that the lines are coincident?

→ **Now Work** PROBLEMS **25** AND **29**

5 Solve Systems of Three Equations Containing Three Variables

Just as with a system of two linear equations containing two variables, a system of three linear equations containing three variables has either (1) exactly one solution (a consistent system with independent equations), or (2) no solution (an inconsistent system), or (3) infinitely many solutions (a consistent system with dependent equations).

We can view the problem of solving a system of three linear equations containing three variables as a geometry problem. The graph of each equation in such a system is a plane in space. A system of three linear equations containing three variables represents three planes in space. Figure 5 illustrates some of the possibilities.

Figure 5

(a) Consistent system; one solution   Solutions **(b)** Consistent system; infinite number of solutions   Solutions **(c)** Inconsistent system; no solution

Recall that a **solution** to a system of equations consists of values for the variables that are solutions of each equation of the system. For example, $x = 3$, $y = -1$, $z = -5$ or, using an ordered triplet, $(3, -1, -5)$ is a solution to the system of equations

$$\begin{cases} x + y + z = -3 & (1) \\ 2x - 3y + 6z = -21 & (2) \\ -3x + 5y = -14 & (3) \end{cases} \qquad \begin{array}{l} 3 + (-1) + (-5) = -3 \\ 2(3) - 3(-1) + 6(-5) = 6 + 3 - 30 = -21 \\ -3(3) + 5(-1) = -9 - 5 = -14 \end{array}$$

because these values of the variables are solutions of each equation.

Typically, when solving a system of three linear equations containing three variables, we use the method of elimination. Recall that the idea behind the method of elimination is to form equivalent equations so that adding two of the equations eliminates a variable.

Let's see how elimination works on a system of three equations containing three variables.

EXAMPLE 9   **Solving a System of Three Linear Equations with Three Variables**

Use the method of elimination to solve the system of equations.

$$\begin{cases} x + y - z = -1 & (1) \\ 4x - 3y + 2z = 16 & (2) \\ 2x - 2y - 3z = 5 & (3) \end{cases}$$

Solution   For a system of three equations, we attempt to eliminate one variable at a time, using pairs of equations until an equation with a single variable remains. Our strategy for solving this system will be to use equation (1) to eliminate the variable x from equations (2) and (3). We can then treat the new equations (2) and (3) as a system with two unknowns. Alternatively, we could use equation (1) to eliminate either y or z from equations (2) and (3). Try one of these approaches for yourself.

We begin by multiplying each side of equation (1) by -4 and adding the result to equation (2). (Do you see why? The coefficients of x are now negatives of one another.) We also multiply equation (1) by -2 and add the result to equation (3). Notice that these two procedures result in the elimination of the variable x from equations (2) and (3).

$$\begin{aligned} x + y - z &= -1 \quad (1) \\ 4x - 3y + 2z &= 16 \quad (2) \end{aligned}$$ Multiply by -4

$$\begin{aligned} -4x - 4y + 4z &= 4 \quad (1) \\ \underline{4x - 3y + 2z} &= \underline{16} \quad (2) \\ -7y + 6z &= 20 \end{aligned}$$ Add.

$$\begin{aligned} x + y - z &= -1 \quad (1) \\ 2x - 2y - 3z &= 5 \quad (3) \end{aligned}$$ Multiply by -2

$$\begin{aligned} -2x - 2y + 2z &= 2 \quad (1) \\ \underline{2x - 2y - 3z} &= \underline{5} \quad (3) \\ -4y - z &= 7 \end{aligned}$$ Add.

$$\begin{cases} x + y - z = -1 & (1) \\ -7y + 6z = 20 & (2) \\ -4y - z = 7 & (3) \end{cases}$$

We now concentrate on the new equations (2) and (3), treating them as a system of two equations containing two variables. It is easiest to eliminate z. We multiply each side of equation (3) by 6 and add equations (2) and (3). The result is the new equation (3).

$$\begin{aligned} -7y + 6z &= 20 \quad (2) \\ -4y - z &= 7 \quad (3) \end{aligned}$$ Multiply by 6

$$\begin{aligned} -7y + 6z &= 20 \quad (2) \\ \underline{-24y - 6z} &= \underline{42} \quad (3) \\ -31y &= 62 \end{aligned}$$ Add.

$$\begin{cases} x + y - z = -1 & (1) \\ -7y + 6z = 20 & (2) \\ -31y = 62 & (3) \end{cases}$$

We now solve equation (3) for y by dividing both sides of the equation by -31.

$$\begin{cases} x + y - z = -1 & (1) \\ -7y + 6z = 20 & (2) \\ y = -2 & (3) \end{cases}$$

Back-substitute $y = -2$ in equation (2) and solve for z.

$$\begin{aligned} -7y + 6z &= 20 \quad (2) \\ -7(-2) + 6z &= 20 \quad \text{Substitute } y = -2 \text{ in (2).} \\ 6z &= 6 \quad \text{Subtract 14 from both sides of the equation.} \\ z &= 1 \quad \text{Divide both sides of the equation by 6.} \end{aligned}$$

Finally, we back-substitute $y = -2$ and $z = 1$ in equation (1) and solve for x.

$$\begin{aligned} x + y - z &= -1 \quad (1) \\ x + (-2) - 1 &= -1 \quad \text{Substitute } y = -2 \text{ and } z = 1 \text{ in (1).} \\ x - 3 &= -1 \quad \text{Simplify.} \\ x &= 2 \quad \text{Add 3 to both sides.} \end{aligned}$$

The solution of the original system is $x = 2$, $y = -2$, $z = 1$ or, using an ordered triplet, $(2, -2, 1)$. You should check this solution.

Look back over the solution given in Example 9. Note the pattern of removing one of the variables from two of the equations, followed by solving this system of two equations and two unknowns. Although which variables to remove is your choice, the methodology remains the same for all systems.

━━━━━▶ **Now Work** PROBLEM 43

6 Identify Inconsistent Systems of Equations Containing Three Variables

---

EXAMPLE 10   **Identify an Inconsistent System of Linear Equations**

Solve: $\begin{cases} 2x + y - z = -2 & (1) \\ x + 2y - z = -9 & (2) \\ x - 4y + z = 1 & (3) \end{cases}$

Solution   Our strategy is the same as in Example 9. However, in this system, it seems easiest to eliminate the variable z first. Do you see why?

Multiply each side of equation (1) by -1 and add the result to equation (2). Also, add equations (2) and (3).

$$
\begin{array}{ll}
-2x - y + z = 2 & \text{(1) Multiply by } -1. \\
\underline{x + 2y - z = -9} & \text{(2)} \\
-x + y = -7 & \text{Add.}
\end{array}
$$

$$
\begin{array}{ll}
x + 2y - z = -9 & \text{(2)} \\
\underline{x - 4y + z = 1} & \text{(3)} \\
2x - 2y = -8 & \text{Add.}
\end{array}
$$

$$
\left\{
\begin{array}{ll}
2x + y - z = -2 & \text{(1)} \\
-x + y = -7 & \text{(2)} \\
2x - 2y = -8 & \text{(3)}
\end{array}
\right.
$$

We now concentrate on the new equations (2) and (3), treating them as a system of two equations containing two variables. Multiply each side of equation (2) by 2 and add the result to equation (3).

$$
\begin{array}{ll}
-x + y = -7 & \text{(2) Multiply by 2.} \\
2x - 2y = -8 & \text{(3)}
\end{array}
$$

$$
\begin{array}{ll}
-2x + 2y = -14 & \text{(2)} \\
\underline{2x - 2y = -8} & \text{(3)} \\
0 = -22 & \text{Add.}
\end{array}
$$

$$
\left\{
\begin{array}{ll}
2x + y - z = -2 & \text{(1)} \\
-x + y = -7 & \text{(2)} \\
0 = -22 & \text{(3)}
\end{array}
\right.
$$

Equation (3) has no solution so the system is inconsistent.

    ●

7 Express the Solution of a System of Dependent Equations Containing Three Variables

---

EXAMPLE 11   **Solving a System of Dependent Equations**

Solve:
$$
\left\{
\begin{array}{ll}
x - 2y - z = 8 & \text{(1)} \\
2x - 3y + z = 23 & \text{(2)} \\
4x - 5y + 5z = 53 & \text{(3)}
\end{array}
\right.
$$

Solution   Our plan is to eliminate x from equations (2) and (3). Multiply each side of equation (1) by -2 and add the result to equation (2). Also, multiply each side of equation (1) by -4 and add the result to equation (3).

$$
\begin{array}{ll}
x - 2y - z = 8 & \text{(1) Multiply by } -2. \\
2x - 3y + z = 23 & \text{(2)}
\end{array}
$$

$$
\begin{array}{ll}
-2x + 4y + 2z = -16 & \text{(1)} \\
\underline{2x - 3y + z = 23} & \text{(2)} \\
y + 3z = 7 & \text{Add.}
\end{array}
$$

$$
\begin{array}{ll}
x - 2y - z = 8 & \text{(1) Multiply by } -4. \\
4x - 5y + 5z = 53 & \text{(3)}
\end{array}
$$

$$
\begin{array}{ll}
-4x + 8y + 4z = -32 & \text{(1)} \\
\underline{4x - 5y + 5z = 53} & \text{(2)} \\
3y + 9z = 21 & \text{Add.}
\end{array}
$$

$$
\left\{
\begin{array}{ll}
x - 2y - z = 8 & \text{(1)} \\
y + 3z = 7 & \text{(2)} \\
3y + 9z = 21 & \text{(3)}
\end{array}
\right.
$$

Treat equations (2) and (3) as a system of two equations containing two variables, and eliminate the variable y by multiplying both sides of equation (2) by -3 and adding the result to equation (3).

$$
\begin{array}{ll}
y + 3z = 7 & \text{Multiply by } -3. \\
3y + 9z = 21 &
\end{array}
$$

$$
\begin{array}{ll}
-3y - 9z = -21 & \\
\underline{3y + 9z = 21} & \text{Add.} \\
0 = 0 &
\end{array}
$$

$$
\left\{
\begin{array}{ll}
x - 2y - z = 8 & \text{(1)} \\
y + 3z = 7 & \text{(2)} \\
0 = 0 & \text{(3)}
\end{array}
\right.
$$

The original system is equivalent to a system containing two equations, so the equations are dependent and the system has infinitely many solutions. If we solve equation (2) for y, we can express y in terms of z as $y = -3z + 7$. Substitute this expression into equation (1) to determine x in terms of z.

$$x - 2y - z = 8 \quad \text{(1)}$$
$$x - 2(-3z + 7) - z = 8 \qquad \text{Substitute } y = -3z + 7 \text{ in (1).}$$
$$x + 6z - 14 - z = 8 \qquad \text{Remove parentheses.}$$
$$x + 5z = 22 \qquad \text{Combine like terms.}$$
$$x = -5z + 22 \qquad \text{Solve for } x.$$

We will write the solution to the system as

$$\begin{cases} x = -5z + 22 \\ y = -3z + 7 \end{cases} \quad \text{where } z \text{ can be any real number.}$$

This way of writing the solution makes it easier to find specific solutions of the system. To find specific solutions, choose any value of z and use the equations $x = -5z + 22$ and $y = -3z + 7$ to determine x and y. For example, if $z = 0$, then $x = 22$ and $y = 7$, and if $z = 1$, then $x = 17$ and $y = 4$.

Using ordered triplets, the solution is

$$\{(x, y, z) \mid x = -5z + 22, y = -3z + 7, z \text{ is any real number}\}$$

⬤

━━ **Now Work**   PROBLEM 45

Two distinct points in the Cartesian plane determine a unique line. Given three noncollinear points, we can find the (unique) quadratic function whose graph contains these three points.

EXAMPLE 12   **Curve Fitting**

Find real numbers a, b, and c so that the graph of the quadratic function $y = ax^2 + bx + c$ contains the points $(-1, -4)$, $(1, 6)$, and $(3, 0)$.

Solution   We require that the three points satisfy the equation $y = ax^2 + bx + c$.

For the point $(-1, -4)$ we have: $-4 = a(-1)^2 + b(-1) + c$ $-4 = a - b + c$
For the point $(1, 6)$ we have: $6 = a(1)^2 + b(1) + c$ $6 = a + b + c$
For the point $(3, 0)$ we have: $0 = a(3)^2 + b(3) + c$ $0 = 9a + 3b + c$

Figure 6

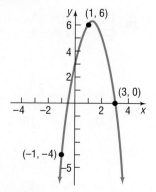

We wish to determine $a, b,$ and c so that each equation is satisfied. That is, we want to solve the following system of three equations containing three variables:

$$\begin{cases} a - b + c = -4 & \text{(1)} \\ a + b + c = 6 & \text{(2)} \\ 9a + 3b + c = 0 & \text{(3)} \end{cases}$$

Solving this system of equations, we obtain $a = -2, b = 5,$ and $c = 3$. So the quadratic function whose graph contains the points $(-1, -4)$, $(1, 6)$, and $(3, 0)$ is

$$y = -2x^2 + 5x + 3 \quad y = ax^2 + bx + c, \quad a = -2, b = 5, c = 3$$

Figure 6 shows the graph of the function along with the three points.

⬤

━━ **Now Work**   PROBLEM 69

10.1 Assess Your Understanding

'Are You Prepared?'   *Answers are given at the end of these exercises. If you get a wrong answer, read the pages listed in red.*

1. Solve the equation: $3x + 4 = 8 - x$. (p. A67)

2. (a) Graph the line: $3x + 4y = 12$.

  (b) What is the slope of a line parallel to this line? (pp. 19–28)

Concepts and Vocabulary

3. If a system of equations has no solution, it is said to be _____.

4. If a system of equations has one solution, the system is _____ and the equations are _____.

5. If the solution to a system of two linear equations containing two unknowns is $x = 3$, $y = -2$, then the lines intersect at _____.

6. If the lines that make up a system of two linear equations are coincident, then the system is _____ and the equations are _____.

Skill Building

In Problems 7–16, verify that the values of the variables listed are solutions of the system of equations.

7. $\begin{cases} 2x - y = 5 \\ 5x + 2y = 8 \end{cases}$
$x = 2, y = -1; (2, -1)$

8. $\begin{cases} 3x + 2y = 2 \\ x - 7y = -30 \end{cases}$
$x = -2, y = 4; (-2, 4)$

9. $\begin{cases} 3x - 4y = 4 \\ \frac{1}{2}x - 3y = -\frac{1}{2} \end{cases}$
$x = 2, y = \frac{1}{2}; \left(2, \frac{1}{2}\right)$

10. $\begin{cases} 2x + \frac{1}{2}y = 0 \\ 3x - 4y = -\frac{19}{2} \end{cases}$
$x = -\frac{1}{2}, y = 2; \left(-\frac{1}{2}, 2\right)$

11. $\begin{cases} x - y = 3 \\ \frac{1}{2}x + y = 3 \end{cases}$
$x = 4, y = 1; (4, 1)$

12. $\begin{cases} x - y = 3 \\ -3x + y = 1 \end{cases}$
$x = -2, y = -5; (-2, -5)$

13. $\begin{cases} 3x + 3y + 2z = 4 \\ x - y - z = 0 \\ 2y - 3z = -8 \end{cases}$
$x = 1, y = -1, z = 2;$
$(1, -1, 2)$

14. $\begin{cases} 4x - z = 7 \\ 8x + 5y - z = 0 \\ -x - y + 5z = 6 \end{cases}$
$x = 2, y = -3, z = 1;$
$(2, -3, 1)$

15. $\begin{cases} 3x + 3y + 2z = 4 \\ x - 3y + z = 10 \\ 5x - 2y - 3z = 8 \end{cases}$
$x = 2, y = -2, z = 2; (2, -2, 2)$

16. $\begin{cases} 4x - 5z = 6 \\ 5y - z = -17 \\ -x - 6y + 5z = 24 \end{cases}$
$x = 4, y = -3, z = 2; (4, -3, 2)$

In Problems 17–54, solve each system of equations. If the system has no solution, say that it is inconsistent.

17. $\begin{cases} x + y = 8 \\ x - y = 4 \end{cases}$

18. $\begin{cases} x + 2y = -7 \\ x + y = -3 \end{cases}$

19. $\begin{cases} 5x - y = 21 \\ 2x + 3y = -12 \end{cases}$

20. $\begin{cases} x + 3y = 5 \\ 2x - 3y = -8 \end{cases}$

21. $\begin{cases} 3x = 24 \\ x + 2y = 0 \end{cases}$

22. $\begin{cases} 4x + 5y = -3 \\ -2y = -8 \end{cases}$

23. $\begin{cases} 3x - 6y = 2 \\ 5x + 4y = 1 \end{cases}$

24. $\begin{cases} 2x + 4y = \frac{2}{3} \\ 3x - 5y = -10 \end{cases}$

25. $\begin{cases} 2x + y = 1 \\ 4x + 2y = 3 \end{cases}$

26. $\begin{cases} x - y = 5 \\ -3x + 3y = 2 \end{cases}$

27. $\begin{cases} 2x - y = 0 \\ 4x + 2y = 12 \end{cases}$

28. $\begin{cases} 3x + 3y = -1 \\ 4x + y = \frac{8}{3} \end{cases}$

29. $\begin{cases} x + 2y = 4 \\ 2x + 4y = 8 \end{cases}$

30. $\begin{cases} 3x - y = 7 \\ 9x - 3y = 21 \end{cases}$

31. $\begin{cases} 2x - 3y = -1 \\ 10x + y = 11 \end{cases}$

32. $\begin{cases} 3x - 2y = 0 \\ 5x + 10y = 4 \end{cases}$

33. $\begin{cases} 2x + 3y = 6 \\ x - y = \dfrac{1}{2} \end{cases}$

34. $\begin{cases} \dfrac{1}{2}x + y = -2 \\ x - 2y = 8 \end{cases}$

35. $\begin{cases} \dfrac{1}{2}x + \dfrac{1}{3}y = 3 \\ \dfrac{1}{4}x - \dfrac{2}{3}y = -1 \end{cases}$

36. $\begin{cases} \dfrac{1}{3}x - \dfrac{3}{2}y = -5 \\ \dfrac{3}{4}x + \dfrac{1}{3}y = 11 \end{cases}$

37. $\begin{cases} 3x - 5y = 3 \\ 15x + 5y = 21 \end{cases}$

38. $\begin{cases} 2x - y = -1 \\ x + \dfrac{1}{2}y = \dfrac{3}{2} \end{cases}$

39. $\begin{cases} \dfrac{1}{x} + \dfrac{1}{y} = 8 \\ \dfrac{3}{x} - \dfrac{5}{y} = 0 \end{cases}$

40. $\begin{cases} \dfrac{4}{x} - \dfrac{3}{y} = 0 \\ \dfrac{6}{x} + \dfrac{3}{2y} = 2 \end{cases}$

[**Hint:** Let $u = \dfrac{1}{x}$ and $v = \dfrac{1}{y}$, and solve for u and v. Then $x = \dfrac{1}{u}$ and $y = \dfrac{1}{v}$.]

41. $\begin{cases} x - y = 6 \\ 2x - 3z = 16 \\ 2y + z = 4 \end{cases}$

42. $\begin{cases} 2x + y = -4 \\ -2y + 4z = 0 \\ 3x - 2z = -11 \end{cases}$

43. $\begin{cases} x - 2y + 3z = 7 \\ 2x + y + z = 4 \\ -3x + 2y - 2z = -10 \end{cases}$

44. $\begin{cases} 2x + y - 3z = 0 \\ -2x + 2y + z = -7 \\ 3x - 4y - 3z = 7 \end{cases}$

45. $\begin{cases} x - y - z = 1 \\ 2x + 3y + z = 2 \\ 3x + 2y = 0 \end{cases}$

46. $\begin{cases} 2x - 3y - z = 0 \\ -x + 2y + z = 5 \\ 3x - 4y - z = 1 \end{cases}$

47. $\begin{cases} x - y - z = 1 \\ -x + 2y - 3z = -4 \\ 3x - 2y - 7z = 0 \end{cases}$

48. $\begin{cases} 2x - 3y - z = 0 \\ 3x + 2y + 2z = 2 \\ x + 5y + 3z = 2 \end{cases}$

49. $\begin{cases} 2x - 2y + 3z = 6 \\ 4x - 3y + 2z = 0 \\ -2x + 3y - 7z = 1 \end{cases}$

50. $\begin{cases} 3x - 2y + 2z = 6 \\ 7x - 3y + 2z = -1 \\ 2x - 3y + 4z = 0 \end{cases}$

51. $\begin{cases} x + y - z = 6 \\ 3x - 2y + z = -5 \\ x + 3y - 2z = 14 \end{cases}$

52. $\begin{cases} x - y + z = -4 \\ 2x - 3y + 4z = -15 \\ 5x + y - 2z = 12 \end{cases}$

53. $\begin{cases} x + 2y - z = -3 \\ 2x - 4y + z = -7 \\ -2x + 2y - 3z = 4 \end{cases}$

54. $\begin{cases} x + 4y - 3z = -8 \\ 3x - y + 3z = 12 \\ x + y + 6z = 1 \end{cases}$

Applications and Extensions

55. The perimeter of a rectangular floor is 90 feet. Find the dimensions of the floor if the length is twice the width.

56. The length of fence required to enclose a rectangular field is 3000 meters. What are the dimensions of the field if it is known that the difference between its length and width is 50 meters?

57. Orbital Launches In 2005 there was a total of 55 commercial and noncommercial orbital launches worldwide. In addition, the number of noncommercial orbital launches was one more than twice the number of commercial orbital launches. Determine the number of commercial and noncommercial orbital launches in 2005.

Source: Federal Aviation Administration

58. Movie Theater Tickets A movie theater charges $9.00 for adults and $7.00 for senior citizens. On a day when 325 people paid an admission, the total receipts were $2495. How many who paid were adults? How many were seniors?

59. Mixing Nuts A store sells cashews for $5.00 per pound and peanuts for $1.50 per pound. The manager decides to mix 30 pounds of peanuts with some cashews and sell the mixture for $3.00 per pound. How many pounds of cashews should be mixed with the peanuts so that the mixture will produce the same revenue as would selling the nuts separately?

60. Financial Planning A recently retired couple needs $12,000 per year to supplement their Social Security. They

have $150,000 to invest to obtain this income. They have decided on two investment options: AA bonds yielding 10% per annum and a Bank Certificate yielding 5%.
(a) How much should be invested in each to realize exactly $12,000?
(b) If, after 2 years, the couple requires $14,000 per year in income, how should they reallocate their investment to achieve the new amount?

61. **Computing Wind Speed** With a tail wind, a small Piper aircraft can fly 600 miles in 3 hours. Against this same wind, the Piper can fly the same distance in 4 hours. Find the average wind speed and the average airspeed of the Piper.

62. **Computing Wind Speed** The average airspeed of a single-engine aircraft is 150 miles per hour. If the aircraft flew the same distance in 2 hours with the wind as it flew in 3 hours against the wind, what was the wind speed?

63. **Restaurant Management** A restaurant manager wants to purchase 200 sets of dishes. One design costs $25 per set, while another costs $45 per set. If she only has $7400 to spend, how many of each design should be ordered?

64. **Cost of Fast Food** One group of people purchased 10 hot dogs and 5 soft drinks at a cost of $35.00. A second bought 7 hot dogs and 4 soft drinks at a cost of $25.25. What is the cost of a single hot dog? A single soft drink?

We paid $35.00.
How much is one hot dog?
How much is one soda?

We paid $25.25.
How much is one hot dog?
How much is one soda?

65. **Computing a Refund** The grocery store we use does not mark prices on its goods. My wife went to this store, bought three 1-pound packages of bacon and two cartons of eggs, and paid a total of $13.45. Not knowing that she went to the store, I also went to the same store, purchased two 1-pound packages of bacon and three cartons of eggs, and paid a total of $11.45. Now we want to return two 1-pound packages of bacon and two cartons of eggs. How much will be refunded?

66. **Finding the Current of a Stream** Pamela requires 3 hours to swim 15 miles downstream on the Illinois River. The

return trip upstream takes 5 hours. Find Pamela's average speed in still water. How fast is the current? (Assume that Pamela's speed is the same in each direction.)

67. **Pharmacy** A doctor's prescription calls for a daily intake containing 40 milligrams (mg) of vitamin C and 30 mg of vitamin D. Your pharmacy stocks two liquids that can be used: one contains 20% vitamin C and 30% vitamin D, the other 40% vitamin C and 20% vitamin D. How many milligrams of each compound should be mixed to fill the prescription?

68. **Pharmacy** A doctor's prescription calls for the creation of pills that contain 12 units of vitamin B_{12} and 12 units of vitamin E. Your pharmacy stocks two powders that can be used to make these pills: one contains 20% vitamin B_{12} and 30% vitamin E, the other 40% vitamin B_{12} and 20% vitamin E. How many units of each powder should be mixed in each pill?

69. **Curve Fitting** Find real numbers a, b, and c so that the graph of the function $y = ax^2 + bx + c$ contains the points $(-1, 4)$, $(2, 3)$, and $(0, 1)$.

70. **Curve Fitting** Find real numbers a, b, and c so that the graph of the function $y = ax^2 + bx + c$ contains the points $(-1, -2)$, $(1, -4)$, and $(2, 4)$.

71. **IS–LM Model in Economics** In economics, the IS curve is a linear equation that represents all combinations of income Y and interest rates r that maintain an equilibrium in the market for goods in the economy. The LM curve is a linear equation that represents all combinations of income Y and interest rates r that maintain an equilibrium in the market for money in the economy. In an economy, suppose that the equilibrium level of income (in millions of dollars) and interest rates satisfy the system of equations

$$\begin{cases} 0.06Y - 5000r = 240 \\ 0.06Y + 6000r = 900 \end{cases}$$

Find the equilibrium level of income and interest rates.

72. **IS–LM Model in Economics** In economics, the IS curve is a linear equation that represents all combinations of income Y and interest rates r that maintain an equilibrium in the market for goods in the economy. The LM curve is a linear equation that represents all combinations of income Y and interest rates r that maintain an equilibrium in the market for money in the economy. In an economy, suppose that the equilibrium level of income (in millions of dollars) and interest rates satisfy the system of equations

$$\begin{cases} 0.05Y - 1000r = 10 \\ 0.05Y + 800r = 100 \end{cases}$$

Find the equilibrium level of income and interest rates.

73. **Electricity: Kirchhoff's Rules** An application of Kirchhoff's Rules to the circuit shown results in the following system of equations:

$$\begin{cases} I_2 = I_1 + I_3 \\ 5 - 3I_1 - 5I_2 = 0 \\ 10 - 5I_2 - 7I_3 = 0 \end{cases}$$

Find the currents I_1, I_2, and I_3.

Source: Physics for Scientists & Engineers, 3rd ed., by Serway. © 1990 Brooks/Cole, a division of Thomson Learning.

74. **Electricity: Kirchhoff's Rules** An application of Kirchhoff's Rules to the circuit shown results in the following system of equations:

$$\begin{cases} I_3 = I_1 + I_2 \\ 8 = 4I_3 + 6I_2 \\ 8I_1 = 4 + 6I_2 \end{cases}$$

Find the currents I_1, I_2, and I_3.

Source: Physics for Scientists & Engineers, 3rd ed., by Serway. © 1990 Brooks/Cole, a division of Thomson Learning.

75. **Theater Revenues** A Broadway theater has 500 seats, divided into orchestra, main, and balcony seating. Orchestra seats sell for $50, main seats for $35, and balcony seats for $25. If all the seats are sold, the gross revenue to the theater is $17,100. If all the main and balcony seats are sold, but only half the orchestra seats are sold, the gross revenue is $14,600. How many are there of each kind of seat?

76. **Theater Revenues** A movie theater charges $8.00 for adults, $4.50 for children, and $6.00 for senior citizens. One day the theater sold 405 tickets and collected $2320 in receipts. Twice as many children's tickets were sold as adult tickets. How many adults, children, and senior citizens went to the theater that day?

77. **Nutrition** A dietitian wishes a patient to have a meal that has 66 grams (g) of protein, 94.5 g of carbohydrates, and

910 milligrams (mg) of calcium. The hospital food service tells the dietitian that the dinner for today is chicken, corn, and 2% milk. Each serving of chicken has 30 g of protein, 35 g of carbohydrates, and 200 mg of calcium. Each serving of corn has 3 g of protein, 16 g of carbohydrates, and 10 mg of calcium. Each glass of 2% milk has 9 g of protein, 13 g of carbohydrates, and 300 mg of calcium. How many servings of each food should the dietitian provide for the patient?

78. **Investments** Kelly has $20,000 to invest. As her financial planner, you recommend that she diversify into three investments: Treasury bills that yield 5% simple interest, Treasury bonds that yield 7% simple interest, and corporate bonds that yield 10% simple interest. Kelly wishes to earn $1390 per year in income. Also, Kelly wants her investment in Treasury bills to be $3000 more than her investment in corporate bonds. How much money should Kelly place in each investment?

79. **Prices of Fast Food** One group of customers bought 8 deluxe hamburgers, 6 orders of large fries, and 6 large colas for $26.10. A second group ordered 10 deluxe hamburgers, 6 large fries, and 8 large colas and paid $31.60. Is there sufficient information to determine the price of each food item? If not, construct a table showing the various possibilities. Assume that the hamburgers cost between $1.75 and $2.25, the fries between $0.75 and $1.00, and the colas between $0.60 and $0.90.

80. **Prices of Fast Food** Use the information given in Problem 79. Suppose that a third group purchased 3 deluxe hamburgers, 2 large fries, and 4 large colas for $10.95. Now is there sufficient information to determine the price of each food item? If so, determine each price.

81. **Painting a House** Three painters, Beth, Bill, and Edie, working together, can paint the exterior of a home in 10 hours (hr). Bill and Edie together have painted a similar house in 15 hr. One day, all three worked on this same kind of house for 4 hr, after which Edie left. Beth and Bill required 8 more hr to finish. Assuming no gain or loss in efficiency, how long should it take each person to complete such a job alone?

Explaining Concepts: Discussion and Writing

82. Make up a system of three linear equations containing three variables that has:
    (a) No solution
    (b) Exactly one solution
    (c) Infinitely many solutions
    Give the three systems to a friend to solve and critique.

83. Write a brief paragraph outlining your strategy for solving a system of two linear equations containing two variables.

84. Do you prefer the method of substitution or the method of elimination for solving a system of two linear equations containing two variables? Give reasons.

'Are You Prepared?' Answers

1. $\{1\}$ **2. (a)**   **(b)** $-\dfrac{3}{4}$

10.2 Systems of Linear Equations: Matrices

OBJECTIVES   **1** Write the Augmented Matrix of a System of Linear Equations (p. 747)
2 Write the System of Equations from the Augmented Matrix (p. 747)
3 Perform Row Operations on a Matrix (p. 748)
4 Solve a System of Linear Equations Using Matrices (p. 749)

The systematic approach of the method of elimination for solving a system of linear equations provides another method of solution that involves a simplified notation.
Consider the following system of linear equations:

$$\begin{cases} x + 4y = 14 \\ 3x - 2y = 0 \end{cases}$$

If we choose not to write the symbols used for the variables, we can represent this system as

$$\left[\begin{array}{cc|c} 1 & 4 & 14 \\ 3 & -2 & 0 \end{array}\right]$$

where it is understood that the first column represents the coefficients of the variable x, the second column the coefficients of y, and the third column the constants on the right side of the equal signs. The vertical line serves as a reminder of the equal signs. The large square brackets are used to denote a *matrix* in algebra.

DEFINITION

A **matrix** is defined as a rectangular array of numbers,

$$
\begin{array}{c}
\\
\text{Row 1} \\
\text{Row 2} \\
\vdots \\
\text{Row } i \\
\vdots \\
\text{Row } m
\end{array}
\begin{array}{cccccc}
\text{Column 1} & \text{Column 2} & & \text{Column } j & & \text{Column } n \\
\left[\begin{array}{cccccc}
a_{11} & a_{12} & \cdots & a_{1j} & \cdots & a_{1n} \\
a_{21} & a_{22} & \cdots & a_{2j} & \cdots & a_{2n} \\
\vdots & \vdots & & \vdots & & \vdots \\
a_{i1} & a_{i2} & \cdots & a_{ij} & \cdots & a_{in} \\
\vdots & \vdots & & \vdots & & \vdots \\
a_{m1} & a_{m2} & \cdots & a_{mj} & \cdots & a_{mn}
\end{array}\right]
\end{array}
\qquad (1)
$$

Each number a_{ij} of the matrix has two indexes: the **row index** i and the **column index** j. The matrix shown in display (1) has m rows and n columns. The numbers a_{ij} are usually referred to as the **entries** of the matrix. For example, a_{23} refers to the entry in the second row, third column.

1 Write the Augmented Matrix of a System of Linear Equations

Now we will use matrix notation to represent a system of linear equations. The matrix used to represent a system of linear equations is called an **augmented matrix.** In writing the augmented matrix of a system, the variables of each equation must be on the left side of the equal sign and the constants on the right side. A variable that does not appear in an equation has a coefficient of 0.

| EXAMPLE 1 | **Writing the Augmented Matrix of a System of Linear Equations** |

Write the augmented matrix of each system of equations.

(a) $\begin{cases} 3x - 4y = -6 & (1) \\ 2x - 3y = -5 & (2) \end{cases}$

(b) $\begin{cases} 2x - y + z = 0 & (1) \\ x + z - 1 = 0 & (2) \\ x + 2y - 8 = 0 & (3) \end{cases}$

Solution

(a) The augmented matrix is

$$\begin{bmatrix} 3 & -4 & | & -6 \\ 2 & -3 & | & -5 \end{bmatrix}$$

(b) Care must be taken that the system be written so that the coefficients of all variables are present (if any variable is missing, its coefficient is 0). Also, all constants must be to the right of the equal sign. We need to rearrange the given system as follows:

$$\begin{cases} 2x - y + z = 0 & (1) \\ x + z - 1 = 0 & (2) \\ x + 2y - 8 = 0 & (3) \end{cases}$$

$$\begin{cases} 2x - y + z = 0 & (1) \\ x + 0 \cdot y + z = 1 & (2) \\ x + 2y + 0 \cdot z = 8 & (3) \end{cases}$$

The augmented matrix is

$$\begin{bmatrix} 2 & -1 & 1 & | & 0 \\ 1 & 0 & 1 & | & 1 \\ 1 & 2 & 0 & | & 8 \end{bmatrix}$$

If we do not include the constants to the right of the equal sign, that is, to the right of the vertical bar in the augmented matrix of a system of equations, the resulting matrix is called the **coefficient matrix** of the system. For the systems discussed in Example 1, the coefficient matrices are

$$\begin{bmatrix} 3 & -4 \\ 2 & -3 \end{bmatrix} \quad \text{and} \quad \begin{bmatrix} 2 & -1 & 1 \\ 1 & 0 & 1 \\ 1 & 2 & 0 \end{bmatrix}$$

━━━━**Now Work** PROBLEM 7

2 Write the System of Equations from the Augmented Matrix

| EXAMPLE 2 | **Writing the System of Linear Equations from the Augmented Matrix** |

Write the system of linear equations corresponding to each augmented matrix.

(a) $\begin{bmatrix} 5 & 2 & | & 13 \\ -3 & 1 & | & -10 \end{bmatrix}$

(b) $\begin{bmatrix} 3 & -1 & -1 & | & 7 \\ 2 & 0 & 2 & | & 8 \\ 0 & 1 & 1 & | & 0 \end{bmatrix}$

Solution
(a) The matrix has two rows and so represents a system of two equations. The two columns to the left of the vertical bar indicate that the system has two variables. If x and y are used to denote these variables, the system of equations is

$$\begin{cases} 5x + 2y = 13 & (1) \\ -3x + y = -10 & (2) \end{cases}$$

(b) Since the augmented matrix has three rows, it represents a system of three equations. Since there are three columns to the left of the vertical bar, the system contains three variables. If x, y, and z are the three variables, the system of equations is

$$\begin{cases} 3x - y - z = 7 & (1) \\ 2x + 2z = 8 & (2) \\ y + z = 0 & (3) \end{cases}$$

3 Perform Row Operations on a Matrix

Row operations on a matrix are used to solve systems of equations when the system is written as an augmented matrix. There are three basic row operations.

> **Row Operations**
>
> 1. Interchange any two rows.
> 2. Replace a row by a nonzero multiple of that row.
> 3. Replace a row by the sum of that row and a constant nonzero multiple of some other row.

These three row operations correspond to the three rules given earlier for obtaining an equivalent system of equations. When a row operation is performed on a matrix, the resulting matrix represents a system of equations equivalent to the system represented by the original matrix.

For example, consider the augmented matrix

$$\begin{bmatrix} 1 & 2 & | & 3 \\ 4 & -1 & | & 2 \end{bmatrix}$$

Suppose that we want to apply a row operation to this matrix that results in a matrix whose entry in row 2, column 1 is a 0. The row operation to use is

> Multiply each entry in row 1 by -4 and add the result to the corresponding entries in row 2. **(2)**

If we use R_2 to represent the new entries in row 2 and r_1 and r_2 to represent the original entries in rows 1 and 2, respectively, we can represent the row operation in statement (2) by

$$R_2 = -4r_1 + r_2$$

Then

$$\begin{bmatrix} 1 & 2 & | & 3 \\ 4 & -1 & | & 2 \end{bmatrix} \underset{R_2 = -4r_1 + r_2}{\rightarrow} \begin{bmatrix} 1 & 2 & | & 3 \\ -4(1) + 4 & -4(2) + (-1) & | & -4(3) + 2 \end{bmatrix} = \begin{bmatrix} 1 & 2 & | & 3 \\ 0 & -9 & | & -10 \end{bmatrix}$$

As desired, we now have the entry 0 in row 2, column 1.

EXAMPLE 3 | **Applying a Row Operation to an Augmented Matrix**

Apply the row operation $R_2 = -3r_1 + r_2$ to the augmented matrix

$$\begin{bmatrix} 1 & -2 & | & 2 \\ 3 & -5 & | & 9 \end{bmatrix}$$

Solution The row operation $R_2 = -3r_1 + r_2$ tells us that the entries in row 2 are to be replaced by the entries obtained after multiplying each entry in row 1 by -3 and adding the result to the corresponding entries in row 2.

$$\begin{bmatrix} 1 & -2 & | & 2 \\ 3 & -5 & | & 9 \end{bmatrix} \rightarrow \begin{bmatrix} 1 & -2 & | & 2 \\ -3(1) + 3 & (-3)(-2) + (-5) & | & -3(2) + 9 \end{bmatrix} = \begin{bmatrix} 1 & -2 & | & 2 \\ 0 & 1 & | & 3 \end{bmatrix}$$

$$\uparrow$$
$$R_2 = -3r_1 + r_2$$

Now Work PROBLEM 17

EXAMPLE 4 | **Finding a Particular Row Operation**

Find a row operation that will result in the augmented matrix

$$\begin{bmatrix} 1 & -2 & | & 2 \\ 0 & 1 & | & 3 \end{bmatrix}$$

having a 0 in row 1, column 2.

Solution We want a 0 in row 1, column 2. Because there is a 1 in row 2, column 2, this result can be accomplished by multiplying row 2 by 2 and adding the result to row 1. That is, we apply the row operation $R_1 = 2r_2 + r_1$.

$$\begin{bmatrix} 1 & -2 & | & 2 \\ 0 & 1 & | & 3 \end{bmatrix} \rightarrow \begin{bmatrix} 2(0) + 1 & 2(1) + (-2) & | & 2(3) + 2 \\ 0 & 1 & | & 3 \end{bmatrix} = \begin{bmatrix} 1 & 0 & | & 8 \\ 0 & 1 & | & 3 \end{bmatrix}$$

$$\uparrow$$
$$R_1 = 2r_2 + r_1$$

A word about the notation that we have introduced. A row operation such as $R_1 = 2r_2 + r_1$ changes the entries in row 1. Note also that for this type of row operation we change the entries in a given row by multiplying the entries in some other row by an appropriate nonzero number and adding the results to the original entries of the row to be changed.

4 Solve a System of Linear Equations Using Matrices

To solve a system of linear equations using matrices, we use row operations on the augmented matrix of the system to obtain a matrix that is in *row echelon form*.

DEFINITION | A matrix is in **row echelon form** when the following conditions are met:

1. The entry in row 1, column 1 is a 1, and only 0's appear below it.

2. The first nonzero entry in each row after the first row is a 1, only 0's appear below it, and the 1 appears to the right of the first nonzero entry in any row above.

3. Any rows that contain all 0's to the left of the vertical bar appear at the bottom.

For example, for a system of three equations containing three variables, x, y, and z, with a unique solution, the augmented matrix is in row echelon form if it is of the form

$$\left[\begin{array}{ccc|c} 1 & a & b & d \\ 0 & 1 & c & e \\ 0 & 0 & 1 & f \end{array}\right]$$

where a, b, c, d, e, and f are real numbers. The last row of this augmented matrix states that $z = f$. We can then determine the value of y using back-substitution with $z = f$, since row 2 represents the equation $y + cz = e$. Finally, x is determined using back-substitution again.

Two advantages of solving a system of equations by writing the augmented matrix in row echelon form are the following:

1. The process is algorithmic; that is, it consists of repetitive steps that can be programmed on a computer.

2. The process works on any system of linear equations, no matter how many equations or variables are present.

The next example shows how to write a matrix in row echelon form.

EXAMPLE 5 | **How to Solve a System of Linear Equations Using Matrices**

Solve: $\begin{cases} 2x + 2y & = 6 \quad (1) \\ x + y + z = 1 \quad (2) \\ 3x + 4y - z = 13 \quad (3) \end{cases}$

Step-by-Step Solution

Step 1: Write the augmented matrix that represents the system.

We write the augmented matrix of the system.

$$\left[\begin{array}{ccc|c} 2 & 2 & 0 & 6 \\ 1 & 1 & 1 & 1 \\ 3 & 4 & -1 & 13 \end{array}\right]$$

Step 2: Perform row operations that result in the entry in row 1, column 1 becoming 1.

To get a 1 in row 1, column 1, we interchange rows 1 and 2. [Note that this is equivalent to interchanging equations (1) and (2) of the system.]

$$\left[\begin{array}{ccc|c} 1 & 1 & 1 & 1 \\ 2 & 2 & 0 & 6 \\ 3 & 4 & -1 & 13 \end{array}\right]$$

Step 3: Perform row operations that leave the entry in row 1, column 1 a 1, while causing the entries in column 1 below row 1 to become 0's.

Next, we want a 0 in row 2, column 1 and a 0 in row 3, column 1. We use the row operations $R_2 = -2r_1 + r_2$ and $R_3 = -3r_1 + r_3$ to accomplish this. Notice that row 1 is unchanged using these row operations. Also, do you see that performing these row operations simultaneously is the same as doing one followed by the other?

$$\left[\begin{array}{ccc|c} 1 & 1 & 1 & 1 \\ 2 & 2 & 0 & 6 \\ 3 & 4 & -1 & 13 \end{array}\right] \rightarrow \left[\begin{array}{ccc|c} 1 & 1 & 1 & 1 \\ 0 & 0 & -2 & 4 \\ 0 & 1 & -4 & 10 \end{array}\right]$$

$$\uparrow \begin{array}{l} R_2 = -2r_1 + r_2 \\ R_3 = -3r_1 + r_3 \end{array}$$

Step 4: Perform row operations that result in the entry in row 2, column 2 becoming 1 with 0's below it.

We want the entry in row 2, column 2 to be 1. We also want to have a 0 below the 1 in row 2, column 2. Interchanging rows 2 and 3 will accomplish both goals.

$$\left[\begin{array}{ccc|c} 1 & 1 & 1 & 1 \\ 0 & 0 & -2 & 4 \\ 0 & 1 & -4 & 10 \end{array}\right] \rightarrow \left[\begin{array}{ccc|c} 1 & 1 & 1 & 1 \\ 0 & 1 & -4 & 10 \\ 0 & 0 & -2 & 4 \end{array}\right]$$

Step 5: Repeat Step 4, placing a 1 in row 3, column 3.

To obtain a 1 in row 3, column 3, we use the row operation $R_3 = -\frac{1}{2}r_3$. The result is

$$\begin{bmatrix} 1 & 1 & 1 & | & 1 \\ 0 & 1 & -4 & | & 10 \\ 0 & 0 & -2 & | & 4 \end{bmatrix} \rightarrow \begin{bmatrix} 1 & 1 & 1 & | & 1 \\ 0 & 1 & -4 & | & 10 \\ 0 & 0 & 1 & | & -2 \end{bmatrix}$$

$$R_3 = -\frac{1}{2}r_3$$

Step 6: The matrix on the right in Step 5 is the row echelon form of the augmented matrix. Use back-substitution to solve the original system.

The third row of the augmented matrix represents the equation $z = -2$. Using $z = -2$, we back-substitute into the equation $y - 4z = 10$ (from the second row) and obtain

$$y - 4z = 10$$
$$y - 4(-2) = 10 \quad z = -2$$
$$y = 2 \quad \text{Solve for } y.$$

Finally, we back-substitute $y = 2$ and $z = -2$ into the equation $x + y + z = 1$ (from the first row) and obtain

$$x + y + z = 1$$
$$x + 2 + (-2) = 1 \quad y = 2, z = -2$$
$$x = 1 \quad \text{Solve for } x.$$

The solution of the system is $x = 1, y = 2, z = -2$ or, using an ordered triplet, $(1, 2, -2)$.

The steps that we used to solve the system of linear equations in Example 5 can be summarized as follows:

In Words
To obtain an augmented matrix in row echelon form:
- Add rows, exchange rows, or multiply a row by a nonzero constant.
- Work from top to bottom and left to right.
- Get 1's in the main diagonal with 0's below the 1's.
- Once the entry in row 1, column 1 is 1 with 0's below it, we do not use row 1 in our row operations. Once the entries in row 1, column 1 and row 2, column 2 are 1 with 0's below, we do not use rows 1 or 2 in our row operations (and so on).

Matrix Method for Solving a System of Linear Equations (Row Echelon Form)

STEP 1: Write the augmented matrix that represents the system.
STEP 2: Perform row operations that place the entry 1 in row 1, column 1.
STEP 3: Perform row operations that leave the entry 1 in row 1, column 1 unchanged, while causing 0's to appear below it in column 1.
STEP 4: Perform row operations that place the entry 1 in row 2, column 2, but leave the entries in columns to the left unchanged. If it is impossible to place a 1 in row 2, column 2, proceed to place a 1 in row 2, column 3. Once a 1 is in place, perform row operations to place 0's below it. (Place any rows that contain only 0's on the left side of the vertical bar, at the bottom of the matrix.)
STEP 5: Now repeat Step 4, placing a 1 in the next row, but one column to the right. Continue until the bottom row or the vertical bar is reached.
STEP 6: The matrix that results is the row echelon form of the augmented matrix. Analyze the system of equations corresponding to it to solve the original system.

In the next example, we solve a system of linear equations using these steps.

EXAMPLE 6 **Solving a System of Linear Equations Using Matrices (Row Echelon Form)**

Solve: $\begin{cases} x - y + z = 8 & (1) \\ 2x + 3y - z = -2 & (2) \\ 3x - 2y - 9z = 9 & (3) \end{cases}$

Solution   **STEP 1:** The augmented matrix of the system is

$$\begin{bmatrix} 1 & -1 & 1 & | & 8 \\ 2 & 3 & -1 & | & -2 \\ 3 & -2 & -9 & | & 9 \end{bmatrix}$$

STEP 2: Because the entry 1 is already present in row 1, column 1, we can go to step 3.

STEP 3: Perform the row operations $R_2 = -2r_1 + r_2$ and $R_3 = -3r_1 + r_3$. Each of these leaves the entry 1 in row 1, column 1 unchanged, while causing 0's to appear under it.

$$\begin{bmatrix} 1 & -1 & 1 & | & 8 \\ 2 & 3 & -1 & | & -2 \\ 3 & -2 & -9 & | & 9 \end{bmatrix} \longrightarrow \begin{bmatrix} 1 & -1 & 1 & | & 8 \\ 0 & 5 & -3 & | & -18 \\ 0 & 1 & -12 & | & -15 \end{bmatrix}$$

$$\uparrow$$
$$R_2 = -2r_1 + r_2$$
$$R_3 = -3r_1 + r_3$$

STEP 4: The easiest way to obtain the entry 1 in row 2, column 2 without altering column 1 is to interchange rows 2 and 3 (another way would be to multiply row 2 by $\frac{1}{5}$, but this introduces fractions).

$$\begin{bmatrix} 1 & -1 & 1 & | & 8 \\ 0 & 1 & -12 & | & -15 \\ 0 & 5 & -3 & | & -18 \end{bmatrix}$$

To get a 0 under the 1 in row 2, column 2, perform the row operation $R_3 = -5r_2 + r_3$.

$$\begin{bmatrix} 1 & -1 & 1 & | & 8 \\ 0 & 1 & -12 & | & -15 \\ 0 & 5 & -3 & | & -18 \end{bmatrix} \longrightarrow \begin{bmatrix} 1 & -1 & 1 & | & 8 \\ 0 & 1 & -12 & | & -15 \\ 0 & 0 & 57 & | & 57 \end{bmatrix}$$

$$\uparrow$$
$$R_3 = -5r_2 + r_3$$

STEP 5: Continuing, we obtain a 1 in row 3, column 3 by using $R_3 = \frac{1}{57}r_3$.

$$\begin{bmatrix} 1 & -1 & 1 & | & 8 \\ 0 & 1 & -12 & | & -15 \\ 0 & 0 & 57 & | & 57 \end{bmatrix} \longrightarrow \begin{bmatrix} 1 & -1 & 1 & | & 8 \\ 0 & 1 & -12 & | & -15 \\ 0 & 0 & 1 & | & 1 \end{bmatrix}$$

$$\uparrow$$
$$R_3 = \frac{1}{57}r_3$$

STEP 6: The matrix on the right is the row echelon form of the augmented matrix. The system of equations represented by the matrix in row echelon form is

$$\begin{cases} x - y + z = 8 & (1) \\ y - 12z = -15 & (2) \\ z = 1 & (3) \end{cases}$$

Using $z = 1$, we back-substitute to get

$$\begin{cases} x - y + 1 = 8 & (1) \\ y - 12(1) = -15 & (2) \end{cases} \xrightarrow[\text{Simplify.}]{} \begin{cases} x - y = 7 & (1) \\ y = -3 & (2) \end{cases}$$

We get $y = -3$ and, back-substituting into $x - y = 7$, we find that $x = 4$. The solution of the system is $x = 4$, $y = -3$, $z = 1$ or, using an ordered triplet, $(4, -3, 1)$.

Sometimes it is advantageous to write a matrix in **reduced row echelon form.** In this form, row operations are used to obtain entries that are 0 above (as well as below) the leading 1 in a row. For example, the row echelon form obtained in the solution to Example 6 is

$$\begin{bmatrix} 1 & -1 & 1 & | & 8 \\ 0 & 1 & -12 & | & -15 \\ 0 & 0 & 1 & | & 1 \end{bmatrix}$$

To write this matrix in reduced row echelon form, we proceed as follows:

$$\begin{bmatrix} 1 & -1 & 1 & | & 8 \\ 0 & 1 & -12 & | & -15 \\ 0 & 0 & 1 & | & 1 \end{bmatrix} \rightarrow \begin{bmatrix} 1 & 0 & -11 & | & -7 \\ 0 & 1 & -12 & | & -15 \\ 0 & 0 & 1 & | & 1 \end{bmatrix} \rightarrow \begin{bmatrix} 1 & 0 & 0 & | & 4 \\ 0 & 1 & 0 & | & -3 \\ 0 & 0 & 1 & | & 1 \end{bmatrix}$$
$$\uparrow \qquad\qquad\qquad\qquad\qquad \uparrow$$
$$R_1 = r_2 + r_1 \qquad\qquad\qquad R_1 = 11r_3 + r_1$$
$$R_2 = 12r_3 + r_2$$

The matrix is now written in reduced row echelon form. The advantage of writing the matrix in this form is that the solution to the system, $x = 4$, $y = -3$, $z = 1$, is readily found, without the need to back-substitute. Another advantage will be seen in Section 10.4, where the inverse of a matrix is discussed. The methodology used to write a matrix in reduced row echelon form is called **Gauss-Jordan elimination.**

━━━━━**Now Work** PROBLEMS **37** AND **47**

The matrix method for solving a system of linear equations also identifies systems that have infinitely many solutions and systems that are inconsistent. Let's see how.

EXAMPLE 7   **Solving a Dependent System of Linear Equations Using Matrices**

Solve: $\begin{cases} 6x - y - z = 4 & (1) \\ -12x + 2y + 2z = -8 & (2) \\ 5x + y - z = 3 & (3) \end{cases}$

Solution   We start with the augmented matrix of the system and proceed to obtain a 1 in row 1, column 1 with 0's below.

$$\begin{bmatrix} 6 & -1 & -1 & | & 4 \\ -12 & 2 & 2 & | & -8 \\ 5 & 1 & -1 & | & 3 \end{bmatrix} \rightarrow \begin{bmatrix} 1 & -2 & 0 & | & 1 \\ -12 & 2 & 2 & | & -8 \\ 5 & 1 & -1 & | & 3 \end{bmatrix} \rightarrow \begin{bmatrix} 1 & -2 & 0 & | & 1 \\ 0 & -22 & 2 & | & 4 \\ 0 & 11 & -1 & | & -2 \end{bmatrix}$$
$$\uparrow \qquad\qquad\qquad\qquad\qquad \uparrow$$
$$R_1 = -1r_3 + r_1 \qquad\qquad\qquad R_2 = 12r_1 + r_2$$
$$R_3 = -5r_1 + r_3$$

Obtaining a 1 in row 2, column 2 without altering column 1 can be accomplished by $R_2 = -\dfrac{1}{22}r_2$ or by $R_3 = \dfrac{1}{11}r_3$ and interchanging rows 2 and 3 or by $R_2 = \dfrac{23}{11}r_3 + r_2$. We shall use the first of these.

$$\begin{bmatrix} 1 & -2 & 0 & | & 1 \\ 0 & -22 & 2 & | & 4 \\ 0 & 11 & -1 & | & -2 \end{bmatrix} \rightarrow \begin{bmatrix} 1 & -2 & 0 & | & 1 \\ 0 & 1 & -\dfrac{1}{11} & | & -\dfrac{2}{11} \\ 0 & 11 & -1 & | & -2 \end{bmatrix} \rightarrow \begin{bmatrix} 1 & -2 & 0 & | & 1 \\ 0 & 1 & -\dfrac{1}{11} & | & -\dfrac{2}{11} \\ 0 & 0 & 0 & | & 0 \end{bmatrix}$$
$$\uparrow \qquad\qquad\qquad\qquad\qquad \uparrow$$
$$R_2 = -\dfrac{1}{22}r_2 \qquad\qquad\qquad R_3 = -11r_2 + r_3$$

This matrix is in row echelon form. Because the bottom row consists entirely of 0's, the system actually consists of only two equations.

$$\begin{cases} x - 2y = 1 & (1) \\ y - \dfrac{1}{11}z = -\dfrac{2}{11} & (2) \end{cases}$$

To make it easier to write down some of the solutions, we express both x and y in terms of z.

From the second equation, $y = \dfrac{1}{11}z - \dfrac{2}{11}$. Now back-substitute this solution for y into the first equation to get

$$x = 2y + 1 = 2\left(\dfrac{1}{11}z - \dfrac{2}{11}\right) + 1 = \dfrac{2}{11}z + \dfrac{7}{11}$$

The original system is equivalent to the system

$$\begin{cases} x = \dfrac{2}{11}z + \dfrac{7}{11} & (1) \\ y = \dfrac{1}{11}z - \dfrac{2}{11} & (2) \end{cases} \qquad \text{where } z \text{ can be any real number.}$$

Let's look at the situation. The original system of three equations is equivalent to a system containing two equations. This means that any values of x, y, z that satisfy both

$$x = \dfrac{2}{11}z + \dfrac{7}{11} \quad \text{and} \quad y = \dfrac{1}{11}z - \dfrac{2}{11}$$

will be solutions. For example, $z = 0, x = \dfrac{7}{11}, y = -\dfrac{2}{11}; z = 1, x = \dfrac{9}{11}, y = -\dfrac{1}{11};$ and $z = -1, x = \dfrac{5}{11}, y = -\dfrac{3}{11}$ are some of the solutions of the original system. There are, in fact, infinitely many values of $x, y,$ and z for which the two equations are satisfied. That is, the original system has infinitely many solutions. We will write the solution of the original system as

$$\begin{cases} x = \dfrac{2}{11}z + \dfrac{7}{11} \\ y = \dfrac{1}{11}z - \dfrac{2}{11} \end{cases} \qquad \text{where } z \text{ can be any real number}$$

or, using ordered triplets, as

$$\left\{ (x, y, z) \,\middle|\, x = \dfrac{2}{11}z + \dfrac{7}{11}, y = \dfrac{1}{11}z - \dfrac{2}{11}, z \text{ any real number} \right\}.$$

We can also find the solution by writing the augmented matrix in reduced row echelon form. Starting with the row echelon form, we have

$$\begin{bmatrix} 1 & -2 & 0 & \bigm| & 1 \\ 0 & 1 & -\dfrac{1}{11} & \bigm| & -\dfrac{2}{11} \\ 0 & 0 & 0 & \bigm| & 0 \end{bmatrix} \rightarrow \begin{bmatrix} 1 & 0 & -\dfrac{2}{11} & \bigm| & \dfrac{7}{11} \\ 0 & 1 & -\dfrac{1}{11} & \bigm| & -\dfrac{2}{11} \\ 0 & 0 & 0 & \bigm| & 0 \end{bmatrix}$$

$$\uparrow$$
$$R_1 = 2r_2 + r_1$$

The matrix on the right is in reduced row echelon form. The corresponding system of equations is

$$\begin{cases} x - \dfrac{2}{11}z = \dfrac{7}{11} & (1) \\[2mm] y - \dfrac{1}{11}z = -\dfrac{2}{11} & (2) \end{cases} \qquad \text{where } z \text{ can be any real number}$$

or, equivalently,

$$\begin{cases} x = \dfrac{2}{11}z + \dfrac{7}{11} \\[2mm] y = \dfrac{1}{11}z - \dfrac{2}{11} \end{cases} \qquad \text{where } z \text{ can be any real number}$$

──── **Now Work** PROBLEM 53

EXAMPLE 8   **Solving an Inconsistent System of Linear Equations Using Matrices**

Solve: $\begin{cases} x + y + z = 6 \\ 2x - y - z = 3 \\ x + 2y + 2z = 0 \end{cases}$

Solution   We proceed as follows, beginning with the augmented matrix.

$$\begin{bmatrix} 1 & 1 & 1 & | & 6 \\ 2 & -1 & -1 & | & 3 \\ 1 & 2 & 2 & | & 0 \end{bmatrix} \rightarrow \begin{bmatrix} 1 & 1 & 1 & | & 6 \\ 0 & -3 & -3 & | & -9 \\ 0 & 1 & 1 & | & -6 \end{bmatrix} \rightarrow \begin{bmatrix} 1 & 1 & 1 & | & 6 \\ 0 & 1 & 1 & | & -6 \\ 0 & -3 & -3 & | & -9 \end{bmatrix} \rightarrow \begin{bmatrix} 1 & 1 & 1 & | & 6 \\ 0 & 1 & 1 & | & -6 \\ 0 & 0 & 0 & | & -27 \end{bmatrix}$$

$$\begin{array}{c} R_2 = -2r_1 + r_2 \\ R_3 = -1r_1 + r_3 \end{array} \qquad \text{Interchange rows 2 and 3.} \qquad R_3 = 3r_2 + r_3$$

This matrix is in row echelon form. The bottom row is equivalent to the equation

$$0x + 0y + 0z = -27$$

which has no solution. The original system is inconsistent.

●

──── **Now Work** PROBLEM 27

The matrix method is especially effective for systems of equations for which the number of equations and the number of variables are unequal. Here, too, such a system is either inconsistent or consistent. If it is consistent, it will have either exactly one solution or infinitely many solutions.

Let's look at a system of four equations containing three variables.

EXAMPLE 9   **Solving a System of Linear Equations Using Matrices**

Solve: $\begin{cases} x - 2y + z = 0 & (1) \\ 2x + 2y - 3z = -3 & (2) \\ y - z = -1 & (3) \\ -x + 4y + 2z = 13 & (4) \end{cases}$

Solution   We proceed as follows, beginning with the augmented matrix.

$$\begin{bmatrix} 1 & -2 & 1 & | & 0 \\ 2 & 2 & -3 & | & -3 \\ 0 & 1 & -1 & | & -1 \\ -1 & 4 & 2 & | & 13 \end{bmatrix} \xrightarrow[\substack{R_2 = -2r_1 + r_2 \\ R_4 = r_1 + r_4}]{} \begin{bmatrix} 1 & -2 & 1 & | & 0 \\ 0 & 6 & -5 & | & -3 \\ 0 & 1 & -1 & | & -1 \\ 0 & 2 & 3 & | & 13 \end{bmatrix} \xrightarrow[\text{Interchange rows 2 and 3.}]{} \begin{bmatrix} 1 & -2 & 1 & | & 0 \\ 0 & 1 & -1 & | & -1 \\ 0 & 6 & -5 & | & -3 \\ 0 & 2 & 3 & | & 13 \end{bmatrix}$$

$$\xrightarrow[\substack{R_3 = -6r_2 + r_3 \\ R_4 = -2r_2 + r_4}]{} \begin{bmatrix} 1 & -2 & 1 & | & 0 \\ 0 & 1 & -1 & | & -1 \\ 0 & 0 & 1 & | & 3 \\ 0 & 0 & 5 & | & 15 \end{bmatrix} \xrightarrow[R_4 = -5r_3 + r_4]{} \begin{bmatrix} 1 & -2 & 1 & | & 0 \\ 0 & 1 & -1 & | & -1 \\ 0 & 0 & 1 & | & 3 \\ 0 & 0 & 0 & | & 0 \end{bmatrix}$$

We could stop here, since the matrix is in row echelon form, and back-substitute $z = 3$ to find x and y. Or we can continue and obtain the reduced row echelon form.

$$\begin{bmatrix} 1 & -2 & 1 & | & 0 \\ 0 & 1 & -1 & | & -1 \\ 0 & 0 & 1 & | & 3 \\ 0 & 0 & 0 & | & 0 \end{bmatrix} \xrightarrow[R_1 = 2r_2 + r_1]{} \begin{bmatrix} 1 & 0 & -1 & | & -2 \\ 0 & 1 & -1 & | & -1 \\ 0 & 0 & 1 & | & 3 \\ 0 & 0 & 0 & | & 0 \end{bmatrix} \xrightarrow[\substack{R_1 = r_3 + r_1 \\ R_2 = r_3 + r_2}]{} \begin{bmatrix} 1 & 0 & 0 & | & 1 \\ 0 & 1 & 0 & | & 2 \\ 0 & 0 & 1 & | & 3 \\ 0 & 0 & 0 & | & 0 \end{bmatrix}$$

The matrix is now in reduced row echelon form, and we can see that the solution is $x = 1$, $y = 2$, $z = 3$ or, using an ordered triplet, $(1, 2, 3)$.

●

──**Now Work** PROBLEM 69

───

EXAMPLE 10   **Financial Planning**

Adam and Michelle require an additional $25,000 in annual income (beyond their pension benefits). They are rather risk averse and have narrowed their investment choices down to Treasury notes that yield 3%, Treasury bonds that yield 5%, or corporate bonds that yield 6%. If they have $600,000 to invest and want the amount invested in Treasury notes to equal the total amount invested in Treasury bonds and corporate bonds, how much should be placed in each investment?

Solution   Let $n, b,$ and c represent the amount invested in Treasury notes, Treasury bonds, and corporate bonds, respectively. There is a total of $600,000 to invest, which means that the sum of the amount invested in Treasury notes, Treasury bonds, and corporate bonds should equal $600,000. The first equation is

$$n + b + c = 600,000 \quad (1)$$

If $100,000 was invested in Treasury notes, the income would be 0.03($100,000) = $3000. In general, if n dollars was invested in Treasury notes, the income would be $0.03n$. Since the total income is $25,000, we have the second equation as

$$0.03n + 0.05b + 0.06c = 25,000 \quad (2)$$

The amount invested in Treasury notes should equal the amount invested in Treasury bonds and corporate bonds, so the third equation is

$$n = b + c \quad \text{or} \quad n - b - c = 0 \quad (3)$$

We have the following system of equations:

$$\begin{cases} n + b + c = 600{,}000 & \text{(1)} \\ 0.03n + 0.05b + 0.06c = 25{,}000 & \text{(2)} \\ n - b - c = 0 & \text{(3)} \end{cases}$$

We begin with the augmented matrix and proceed as follows:

$$\begin{bmatrix} 1 & 1 & 1 & | & 600{,}000 \\ 0.03 & 0.05 & 0.06 & | & 25{,}000 \\ 1 & -1 & -1 & | & 0 \end{bmatrix} \rightarrow \begin{bmatrix} 1 & 1 & 1 & | & 600{,}000 \\ 0 & 0.02 & 0.03 & | & 7000 \\ 0 & -2 & -2 & | & -600{,}000 \end{bmatrix} \rightarrow$$

$$R_2 = -0.03r_1 + r_2$$
$$R_3 = -r_1 + r_2$$
$$R_2 = \frac{1}{0.02}r_2$$

$$\begin{bmatrix} 1 & 1 & 1 & | & 600{,}000 \\ 0 & 1 & 1.5 & | & 350{,}000 \\ 0 & -2 & -2 & | & -600{,}000 \end{bmatrix} \rightarrow \begin{bmatrix} 1 & 1 & 1 & | & 600{,}000 \\ 0 & 1 & 1.5 & | & 350{,}000 \\ 0 & 0 & 1 & | & 100{,}000 \end{bmatrix}$$

$$R_3 = 2r_2 + r_3$$

The matrix is now in echelon form. The final matrix represents the system

$$\begin{cases} n + b + c = 600{,}000 & \text{(1)} \\ b + 1.5c = 350{,}000 & \text{(2)} \\ c = 100{,}000 & \text{(3)} \end{cases}$$

 COMMENT Most graphing utilities have the capability to put an augmented matrix into row echelon form (ref) and also reduced row echelon form (rref). See Appendix B, Section B.7, for a discussion. ∎

From equation (3), we determine that Adam and Michelle should invest $100,000 in corporate bonds. Back-substitute $100,000 into equation (2) to find that $b = 200{,}000$, so Adam and Michelle should invest $200,000 in Treasury bonds. Back-substitute these values into equation (1) and find that $n = \$300{,}000$, so $300,000 should be invested in Treasury notes.

●

10.2 Assess Your Understanding

Concepts and Vocabulary

1. An m by n rectangular array of numbers is called a(n) _____.

2. The matrix used to represent a system of linear equations is called a(n) _____ matrix.

3. The notation a_{35} refers to the entry in the _____ row and _____ column of a matrix.

4. *True or False* The matrix $\begin{bmatrix} 1 & 3 & | & -2 \\ 0 & 1 & | & 5 \\ 0 & 0 & | & 0 \end{bmatrix}$ is in row echelon form.

Skill Building

In Problems 5–16, write the augmented matrix of the given system of equations.

5. $\begin{cases} x - 5y = 5 \\ 4x + 3y = 6 \end{cases}$

6. $\begin{cases} 3x + 4y = 7 \\ 4x - 2y = 5 \end{cases}$

7. $\begin{cases} 2x + 3y - 6 = 0 \\ 4x - 6y + 2 = 0 \end{cases}$

8. $\begin{cases} 9x - y = 0 \\ 3x - y - 4 = 0 \end{cases}$

9. $\begin{cases} 0.01x - 0.03y = 0.06 \\ 0.13x + 0.10y = 0.20 \end{cases}$

10. $\begin{cases} \dfrac{4}{3}x - \dfrac{3}{2}y = \dfrac{3}{4} \\ -\dfrac{1}{4}x + \dfrac{1}{3}y = \dfrac{2}{3} \end{cases}$

11. $\begin{cases} x - y + z = 10 \\ 3x + 3y = 5 \\ x + y + 2z = 2 \end{cases}$

12. $\begin{cases} 5x - y - z = 0 \\ x + y = 5 \\ 2x - 3z = 2 \end{cases}$

13. $\begin{cases} x + y - z = 2 \\ 3x - 2y = 2 \\ 5x + 3y - z = 1 \end{cases}$

14. $\begin{cases} 2x + 3y - 4z = 0 \\ x - 5z + 2 = 0 \\ x + 2y - 3z = -2 \end{cases}$

15. $\begin{cases} x - y - z = 10 \\ 2x + y + 2z = -1 \\ -3x + 4y = 5 \\ 4x - 5y + z = 0 \end{cases}$

16. $\begin{cases} x - y + 2z - w = 5 \\ x + 3y - 4z + 2w = 2 \\ 3x - y - 5z - w = -1 \end{cases}$

In Problems 17–24, write the system of equations corresponding to each augmented matrix. Then perform each row operation on the given augmented matrix.

17. $\begin{bmatrix} 1 & -3 & | & -2 \\ 2 & -5 & | & 5 \end{bmatrix}$ $R_2 = -2r_1 + r_2$

18. $\begin{bmatrix} 1 & -3 & | & -3 \\ 2 & -5 & | & -4 \end{bmatrix}$ $R_2 = -2r_1 + r_2$

19. $\begin{bmatrix} 1 & -3 & 4 & | & 3 \\ 3 & -5 & 6 & | & 6 \\ -5 & 3 & 4 & | & 6 \end{bmatrix}$
(a) $R_2 = -3r_1 + r_2$
(b) $R_3 = 5r_1 + r_3$

20. $\begin{bmatrix} 1 & -3 & 3 & | & -5 \\ -4 & -5 & -3 & | & -5 \\ -3 & -2 & 4 & | & 6 \end{bmatrix}$
(a) $R_2 = 4r_1 + r_2$
(b) $R_3 = 3r_1 + r_3$

21. $\begin{bmatrix} 1 & -3 & 2 & | & -6 \\ 2 & -5 & 3 & | & -4 \\ -3 & -6 & 4 & | & 6 \end{bmatrix}$
(a) $R_2 = -2r_1 + r_2$
(b) $R_3 = 3r_1 + r_3$

22. $\begin{bmatrix} 1 & -3 & -4 & | & -6 \\ 6 & -5 & 6 & | & -6 \\ -1 & 1 & 4 & | & 6 \end{bmatrix}$
(a) $R_2 = -6r_1 + r_2$
(b) $R_3 = r_1 + r_3$

23. $\begin{bmatrix} 5 & -3 & 1 & | & -2 \\ 2 & -5 & 6 & | & -2 \\ -4 & 1 & 4 & | & 6 \end{bmatrix}$
(a) $R_1 = -2r_2 + r_1$
(b) $R_1 = r_3 + r_1$

24. $\begin{bmatrix} 4 & -3 & -1 & | & 2 \\ 3 & -5 & 2 & | & 6 \\ -3 & -6 & 4 & | & 6 \end{bmatrix}$
(a) $R_1 = -r_2 + r_1$
(b) $R_1 = r_3 + r_1$

In Problems 25–36, the reduced row echelon form of a system of linear equations is given. Write the system of equations corresponding to the given matrix. Use x, y; or x, y, z; or x_1, x_2, x_3, x_4 as variables. Determine whether the system is consistent or inconsistent. If it is consistent, give the solution.

25. $\begin{bmatrix} 1 & 0 & | & 5 \\ 0 & 1 & | & -1 \end{bmatrix}$

26. $\begin{bmatrix} 1 & 0 & | & -4 \\ 0 & 1 & | & 0 \end{bmatrix}$

27. $\begin{bmatrix} 1 & 0 & 0 & | & 1 \\ 0 & 1 & 0 & | & 2 \\ 0 & 0 & 0 & | & 3 \end{bmatrix}$

28. $\begin{bmatrix} 1 & 0 & 0 & | & 0 \\ 0 & 1 & 0 & | & 0 \\ 0 & 0 & 0 & | & 2 \end{bmatrix}$

29. $\begin{bmatrix} 1 & 0 & 2 & | & -1 \\ 0 & 1 & -4 & | & -2 \\ 0 & 0 & 0 & | & 0 \end{bmatrix}$

30. $\begin{bmatrix} 1 & 0 & 4 & | & 4 \\ 0 & 1 & 3 & | & 2 \\ 0 & 0 & 0 & | & 0 \end{bmatrix}$

31. $\begin{bmatrix} 1 & 0 & 0 & 0 & | & 1 \\ 0 & 1 & 0 & 1 & | & 2 \\ 0 & 0 & 1 & 2 & | & 3 \end{bmatrix}$

32. $\begin{bmatrix} 1 & 0 & 0 & 0 & | & 1 \\ 0 & 1 & 0 & 2 & | & 2 \\ 0 & 0 & 1 & 3 & | & 0 \end{bmatrix}$

33. $\begin{bmatrix} 1 & 0 & 0 & 4 & | & 2 \\ 0 & 1 & 1 & 3 & | & 3 \\ 0 & 0 & 0 & 0 & | & 0 \end{bmatrix}$

34. $\begin{bmatrix} 1 & 0 & 0 & 0 & | & 1 \\ 0 & 1 & 0 & 0 & | & 2 \\ 0 & 0 & 1 & 2 & | & 3 \end{bmatrix}$

35. $\begin{bmatrix} 1 & 0 & 0 & 1 & | & -2 \\ 0 & 1 & 0 & 2 & | & 2 \\ 0 & 0 & 1 & -1 & | & 0 \\ 0 & 0 & 0 & 0 & | & 0 \end{bmatrix}$

36. $\begin{bmatrix} 1 & 0 & 0 & 0 & | & 1 \\ 0 & 1 & 0 & 0 & | & 2 \\ 0 & 0 & 1 & 0 & | & 3 \\ 0 & 0 & 0 & 1 & | & 0 \end{bmatrix}$

In Problems 37–72, solve each system of equations using matrices (row operations). If the system has no solution, say that it is inconsistent.

37. $\begin{cases} x + y = 8 \\ x - y = 4 \end{cases}$

38. $\begin{cases} x + 2y = 5 \\ x + y = 3 \end{cases}$

39. $\begin{cases} 2x - 4y = -2 \\ 3x + 2y = 3 \end{cases}$

40. $\begin{cases} 3x + 3y = 3 \\ 4x + 2y = \dfrac{8}{3} \end{cases}$

41. $\begin{cases} x + 2y = 4 \\ 2x + 4y = 8 \end{cases}$

42. $\begin{cases} 3x - y = 7 \\ 9x - 3y = 21 \end{cases}$

43. $\begin{cases} 2x + 3y = 6 \\ x - y = \dfrac{1}{2} \end{cases}$

44. $\begin{cases} \dfrac{1}{2}x + y = -2 \\ x - 2y = 8 \end{cases}$

45. $\begin{cases} 3x - 5y = 3 \\ 15x + 5y = 21 \end{cases}$

46. $\begin{cases} 2x - y = -1 \\ x + \dfrac{1}{2}y = \dfrac{3}{2} \end{cases}$

47. $\begin{cases} x - y = 6 \\ 2x - 3z = 16 \\ 2y + z = 4 \end{cases}$

48. $\begin{cases} 2x + y = -4 \\ -2y + 4z = 0 \\ 3x - 2z = -11 \end{cases}$

49. $\begin{cases} x - 2y + 3z = 7 \\ 2x + y + z = 4 \\ -3x + 2y - 2z = -10 \end{cases}$

50. $\begin{cases} 2x + y - 3z = 0 \\ -2x + 2y + z = -7 \\ 3x - 4y - 3z = 7 \end{cases}$

51. $\begin{cases} 2x - 2y - 2z = 2 \\ 2x + 3y + z = 2 \\ 3x + 2y = 0 \end{cases}$

52. $\begin{cases} 2x - 3y - z = 0 \\ -x + 2y + z = 5 \\ 3x - 4y - z = 1 \end{cases}$

53. $\begin{cases} -x + y + z = -1 \\ -x + 2y - 3z = -4 \\ 3x - 2y - 7z = 0 \end{cases}$

54. $\begin{cases} 2x - 3y - z = 0 \\ 3x + 2y + 2z = 2 \\ x + 5y + 3z = 2 \end{cases}$

55. $\begin{cases} 2x - 2y + 3z = 6 \\ 4x - 3y + 2z = 0 \\ -2x + 3y - 7z = 1 \end{cases}$

56. $\begin{cases} 3x - 2y + 2z = 6 \\ 7x - 3y + 2z = -1 \\ 2x - 3y + 4z = 0 \end{cases}$

57. $\begin{cases} x + y - z = 6 \\ 3x - 2y + z = -5 \\ x + 3y - 2z = 14 \end{cases}$

58. $\begin{cases} x - y + z = -4 \\ 2x - 3y + 4z = -15 \\ 5x + y - 2z = 12 \end{cases}$

59. $\begin{cases} x + 2y - z = -3 \\ 2x - 4y + z = -7 \\ -2x + 2y - 3z = 4 \end{cases}$

60. $\begin{cases} x + 4y - 3z = -8 \\ 3x - y + 3z = 12 \\ x + y + 6z = 1 \end{cases}$

61. $\begin{cases} 3x + y - z = \dfrac{2}{3} \\ 2x - y + z = 1 \\ 4x + 2y = \dfrac{8}{3} \end{cases}$

62. $\begin{cases} x + y = 1 \\ 2x - y + z = 1 \\ x + 2y + z = \dfrac{8}{3} \end{cases}$

63. $\begin{cases} x + y + z + w = 4 \\ 2x - y + z = 0 \\ 3x + 2y + z - w = 6 \\ x - 2y - 2z + 2w = -1 \end{cases}$

64. $\begin{cases} x + y + z + w = 4 \\ -x + 2y + z = 0 \\ 2x + 3y + z - w = 6 \\ -2x + y - 2z + 2w = -1 \end{cases}$

65. $\begin{cases} x + 2y + z = 1 \\ 2x - y + 2z = 2 \\ 3x + y + 3z = 3 \end{cases}$

66. $\begin{cases} x + 2y - z = 3 \\ 2x - y + 2z = 6 \\ x - 3y + 3z = 4 \end{cases}$

67. $\begin{cases} x - y + z = 5 \\ 3x + 2y - 2z = 0 \end{cases}$

68. $\begin{cases} 2x + y - z = 4 \\ -x + y + 3z = 1 \end{cases}$

69. $\begin{cases} 2x + 3y - z = 3 \\ x - y - z = 0 \\ -x + y + z = 0 \\ x + y + 3z = 5 \end{cases}$

70. $\begin{cases} x - 3y + z = 1 \\ 2x - y - 4z = 0 \\ x - 3y + 2z = 1 \\ x - 2y = 5 \end{cases}$

71. $\begin{cases} 4x + y + z - w = 4 \\ x - y + 2z + 3w = 3 \end{cases}$

72. $\begin{cases} -4x + y = 5 \\ 2x - y + z - w = 5 \\ z + w = 4 \end{cases}$

Applications and Extensions

73. Curve Fitting   Find the function $y = ax^2 + bx + c$ whose graph contains the points $(1, 2)$, $(-2, -7)$, and $(2, -3)$.

74. Curve Fitting   Find the function $y = ax^2 + bx + c$ whose graph contains the points $(1, -1)$, $(3, -1)$, and $(-2, 14)$.

75. Curve Fitting   Find the function $f(x) = ax^3 + bx^2 + cx + d$ for which $f(-3) = -112$, $f(-1) = -2$, $f(1) = 4$, and $f(2) = 13$.

76. Curve Fitting   Find the function $f(x) = ax^3 + bx^2 + cx + d$ for which $f(-2) = -10$, $f(-1) = 3$, $f(1) = 5$, and $f(3) = 15$.

77. Nutrition   A dietitian at Palos Community Hospital wants a patient to have a meal that has 78 grams (g) of protein, 59 g of carbohydrates, and 75 milligrams (mg) of vitamin A. The hospital food service tells the dietitian that the dinner for today is salmon steak, baked eggs, and acorn squash. Each serving of salmon steak has 30 g of protein, 20 g of carbohydrates, and 2 mg of vitamin A. Each serving of baked eggs contains 15 g of protein, 2 g of carbohydrates, and 20 mg of vitamin A. Each serving of acorn squash contains 3 g of protein, 25 g of carbohydrates, and 32 mg of vitamin A. How many servings of each food should the dietitian provide for the patient?

78. Nutrition   A dietitian at General Hospital wants a patient to have a meal that has 47 grams (g) of protein, 58 g of carbohydrates, and 630 milligrams (mg) of calcium. The hospital food service tells the dietitian that the dinner for today is pork chops, corn on the cob, and 2% milk. Each serving of pork chops has 23 g of protein, 0 g of carbohydrates, and 10 mg of calcium. Each serving of corn on the cob contains 3 g of protein, 16 g of carbohydrates, and 10 mg of calcium. Each glass of 2% milk contains 9 g of protein, 13 g of carbohydrates, and 300 mg of calcium. How many servings of each food should the dietitian provide for the patient?

79. Financial Planning   Carletta has $10,000 to invest. As her financial consultant, you recommend that she invest in Treasury bills that yield 6%, Treasury bonds that yield 7%, and corporate bonds that yield 8%. Carletta wants to have an annual income of $680, and the amount invested in corporate bonds must be half that invested in Treasury bills. Find the amount in each investment.

80. Landscaping   A landscape company is hired to plant trees in three new subdivisions. The company charges the developer for each tree planted, an hourly rate to plant the trees, and a fixed delivery charge. In one subdivision it took 166 labor hours to plant 250 trees for a cost of $7520. In a second subdivision it took 124 labor hours to plant 200 trees for a cost of $5945. In the final subdivision it took 200 labor hours to plant 300 trees for a cost of $8985. Determine the cost for each tree, the hourly labor charge, and the fixed delivery charge.

Sources: gurney.com; www.bx.org

81. Production   To manufacture an automobile requires painting, drying, and polishing. Epsilon Motor Company produces three types of cars: the Delta, the Beta, and the Sigma. Each Delta requires 10 hours (hr) for painting, 3 hr for drying, and 2 hr for polishing. A Beta requires 16 hr for painting, 5 hr for drying, and 3 hr for polishing, and a Sigma requires 8 hr for painting, 2 hr for drying, and 1 hr for polishing. If the company has 240 hr for painting, 69 hr for drying, and 41 hr for polishing per month, how many of each type of car are produced?

82. Production   A Florida juice company completes the preparation of its products by sterilizing, filling, and labeling bottles. Each case of orange juice requires 9 minutes (min) for sterilizing, 6 min for filling, and 1 min for labeling. Each case of grapefruit juice requires 10 min for sterilizing, 4 min for filling, and 2 min for labeling. Each case of tomato juice requires 12 min for sterilizing, 4 min for filling, and 1 min for labeling. If the company runs the sterilizing machine for 398 min, the filling machine for 164 min, and the labeling machine for 58 min, how many cases of each type of juice are prepared?

83. Electricity: Kirchhoff's Rules   An application of Kirchhoff's Rules to the circuit shown results in the following system of equations:

$$\begin{cases} -4 + 8 - 2I_2 = 0 \\ 8 = 5I_4 + I_1 \\ 4 = 3I_3 + I_1 \\ I_3 + I_4 = I_1 \end{cases}$$

Find the currents I_1, I_2, I_3, and I_4.

Source: Based on Raymond Serway. Physics, *3rd ed. (Philadelphia: Saunders, 1990), Prob. 34. p. 790.*

84. Electricity: Kirchhoff's Rules   An application of Kirchhoff's Rules to the circuit shown results in the following system of equations:

$$\begin{cases} I_1 = I_3 + I_2 \\ 24 - 6I_1 - 3I_3 = 0 \\ 12 + 24 - 6I_1 - 6I_2 = 0 \end{cases}$$

Find the currents I_1, I_2, and I_3.

Source: Ibid., Prob. 38, p. 791.

85. Financial Planning   Three retired couples each require an additional annual income of $2000 per year. As their financial consultant, you recommend that they invest some money in Treasury bills that yield 7%, some money in corporate bonds that yield 9%, and some money in junk bonds that yield 11%. Prepare a table for each couple showing the various ways that their goals can be achieved:
(a) If the first couple has $20,000 to invest.
(b) If the second couple has $25,000 to invest.
(c) If the third couple has $30,000 to invest.
(d) What advice would you give each couple regarding the amount to invest and the choices available?
[Hint: Higher yields generally carry more risk.**]**

86. Financial Planning   A young couple has $25,000 to invest. As their financial consultant, you recommend that they invest some money in Treasury bills that yield 7%, some money in corporate bonds that yield 9%, and some money in junk bonds that yield 11%. Prepare a table showing the various ways that this couple can achieve the following goals:
(a) $1500 per year in income
(b) $2000 per year in income
(c) $2500 per year in income
(d) What advice would you give this couple regarding the income that they require and the choices available?
[Hint: Higher yields generally carry more risk.**]**

87. Pharmacy   A doctor's prescription calls for a daily intake of a supplement containing 40 milligrams (mg) of vitamin C and 30 mg of vitamin D. Your pharmacy stocks three supplements that can be used: one contains 20% vitamin C and 30% vitamin D; a second, 40% vitamin C and 20% vitamin D; and a third, 30% vitamin C and 50% vitamin D. Create a table showing the possible combinations that could be used to fill the prescription.

88. Pharmacy   A doctor's prescription calls for the creation of pills that contain 12 units of vitamin B_{12} and 12 units of vitamin E. Your pharmacy stocks three powders that can be used to make these pills: one contains 20% vitamin B_{12} and 30% vitamin E; a second, 40% vitamin B_{12} and 20% vitamin E; and a third, 30% vitamin B_{12} and 40% vitamin E. Create a table showing the possible combinations of each powder that could be mixed in each pill.

Explaining Concepts: Discussion and Writing

89. Write a brief paragraph or two that outline your strategy for solving a system of linear equations using matrices.

90. When solving a system of linear equations using matrices, do you prefer to place the augmented matrix in row echelon form or in reduced row echelon form? Give reasons for your choice.

91. Make up a system of three linear equations containing three variables that has:
(a) No solution
(b) Exactly one solution
(c) Infinitely many solutions
Give the three systems to a friend to solve and critique.

10.3 Systems of Linear Equations: Determinants

OBJECTIVES **1** Evaluate 2 by 2 Determinants (p. 761)

2 Use Cramer's Rule to Solve a System of Two Equations Containing Two Variables (p. 762)

3 Evaluate 3 by 3 Determinants (p. 764)

4 Use Cramer's Rule to Solve a System of Three Equations Containing Three Variables (p. 766)

5 Know Properties of Determinants (p. 767)

In the preceding section, we described a method of using matrices to solve a system of linear equations. This section deals with yet another method for solving systems of linear equations; however, it can be used only when the number of equations equals the number of variables. Although the method will work for any system (provided that the number of equations equals the number of variables), it is most often used for systems of two equations containing two variables or three equations containing three variables. This method, called *Cramer's Rule*, is based on the concept of a *determinant*.

1 Evaluate 2 by 2 Determinants

DEFINITION If $a, b, c,$ and d are four real numbers, the symbol

$$D = \begin{vmatrix} a & b \\ c & d \end{vmatrix}$$

is called a **2 by 2 determinant.** Its value is the number $ad - bc$; that is,

$$D = \begin{vmatrix} a & b \\ c & d \end{vmatrix} = ad - bc \tag{1}$$

The following device may be helpful for remembering the value of a 2 by 2 determinant:

EXAMPLE 1 **Evaluating a 2 × 2 Determinant**

Evaluate: $\begin{vmatrix} 3 & -2 \\ 6 & 1 \end{vmatrix}$

Solution

$$\begin{vmatrix} 3 & -2 \\ 6 & 1 \end{vmatrix} = (3)(1) - (6)(-2)$$
$$= 3 - (-12)$$
$$= 15$$

═══ **Now Work** PROBLEM 7

2 Use Cramer's Rule to Solve a System of Two Equations Containing Two Variables

Let's now see the role that a 2 by 2 determinant plays in the solution of a system of two equations containing two variables. Consider the system

$$\begin{cases} ax + by = s & \text{(1)} \\ cx + dy = t & \text{(2)} \end{cases} \qquad \textbf{(2)}$$

We shall use the method of elimination to solve this system.

Provided $d \neq 0$ and $b \neq 0$, this system is equivalent to the system

$$\begin{cases} adx + bdy = sd & \text{(1)} \quad \text{Multiply by } d. \\ bcx + bdy = tb & \text{(2)} \quad \text{Multiply by } b. \end{cases}$$

Subtract the second equation from the first equation and obtain

$$\begin{cases} (ad - bc)x + 0 \cdot y = sd - tb & \text{(1)} \\ bcx + bdy = tb & \text{(2)} \end{cases}$$

Now the first equation can be rewritten using determinant notation.

$$\begin{vmatrix} a & b \\ c & d \end{vmatrix} x = \begin{vmatrix} s & b \\ t & d \end{vmatrix}$$

If $D = \begin{vmatrix} a & b \\ c & d \end{vmatrix} = ad - bc \neq 0$, we can solve for x to get

$$x = \frac{\begin{vmatrix} s & b \\ t & d \end{vmatrix}}{\begin{vmatrix} a & b \\ c & d \end{vmatrix}} = \frac{\begin{vmatrix} s & b \\ t & d \end{vmatrix}}{D} \qquad \textbf{(3)}$$

Return now to the original system (2). Provided that $a \neq 0$ and $c \neq 0$, the system is equivalent to

$$\begin{cases} acx + bcy = cs & \text{(1)} \quad \text{Multiply by } c. \\ acx + ady = at & \text{(2)} \quad \text{Multiply by } a. \end{cases}$$

Subtract the first equation from the second equation and obtain

$$\begin{cases} acx + bcy = cs & \text{(1)} \\ 0 \cdot x + (ad - bc)y = at - cs & \text{(2)} \end{cases}$$

The second equation can now be rewritten using determinant notation.

$$\begin{vmatrix} a & b \\ c & d \end{vmatrix} y = \begin{vmatrix} a & s \\ c & t \end{vmatrix}$$

If $D = \begin{vmatrix} a & b \\ c & d \end{vmatrix} = ad - bc \neq 0$, we can solve for y to get

$$y = \frac{\begin{vmatrix} a & s \\ c & t \end{vmatrix}}{\begin{vmatrix} a & b \\ c & d \end{vmatrix}} = \frac{\begin{vmatrix} a & s \\ c & t \end{vmatrix}}{D} \qquad \textbf{(4)}$$

Equations (3) and (4) lead us to the following result, called **Cramer's Rule.**

THEOREM

Cramer's Rule for Two Equations Containing Two Variables

The solution to the system of equations

$$\begin{cases} ax + by = s \quad (1) \\ cx + dy = t \quad (2) \end{cases} \qquad (5)$$

is given by

$$x = \frac{\begin{vmatrix} s & b \\ t & d \end{vmatrix}}{\begin{vmatrix} a & b \\ c & d \end{vmatrix}} \qquad y = \frac{\begin{vmatrix} a & s \\ c & t \end{vmatrix}}{\begin{vmatrix} a & b \\ c & d \end{vmatrix}} \qquad (6)$$

provided that

$$D = \begin{vmatrix} a & b \\ c & d \end{vmatrix} = ad - bc \neq 0$$

In the derivation given for Cramer's Rule, we assumed that none of the numbers a, b, c, and d was 0. In Problem 61 you will be asked to complete the proof under the less stringent condition that $D = ad - bc \neq 0$.

Now look carefully at the pattern in Cramer's Rule. The denominator in the solution (6) is the determinant of the coefficients of the variables.

$$\begin{cases} ax + by = s \\ cx + dy = t \end{cases} \qquad D = \begin{vmatrix} a & b \\ c & d \end{vmatrix}$$

In the solution for x, the numerator is the determinant, denoted by D_x, formed by replacing the entries in the first column (the coefficients of x) of D by the constants on the right side of the equal sign.

$$D_x = \begin{vmatrix} s & b \\ t & d \end{vmatrix}$$

In the solution for y, the numerator is the determinant, denoted by D_y, formed by replacing the entries in the second column (the coefficients in y) of D by the constants on the right side of the equal sign.

$$D_y = \begin{vmatrix} a & s \\ c & t \end{vmatrix}$$

Cramer's Rule then states that, if $D \neq 0$,

$$x = \frac{D_x}{D} \qquad y = \frac{D_y}{D} \qquad (7)$$

EXAMPLE 2

Solving a System of Linear Equations Using Determinants

Use Cramer's Rule, if applicable, to solve the system

$$\begin{cases} 3x - 2y = 4 \quad (1) \\ 6x + y = 13 \quad (2) \end{cases}$$

Solution The determinant D of the coefficients of the variables is

$$D = \begin{vmatrix} 3 & -2 \\ 6 & 1 \end{vmatrix} = (3)(1) - (6)(-2) = 15$$

Because $D \neq 0$, Cramer's Rule (7) can be used.

$$x = \frac{D_x}{D} = \frac{\begin{vmatrix} 4 & -2 \\ 13 & 1 \end{vmatrix}}{15} \qquad y = \frac{D_y}{D} = \frac{\begin{vmatrix} 3 & 4 \\ 6 & 13 \end{vmatrix}}{15}$$

$$= \frac{(4)(1) - (13)(-2)}{15} \qquad = \frac{(3)(13) - (6)(4)}{15}$$

$$= \frac{30}{15} \qquad\qquad = \frac{15}{15}$$

$$= 2 \qquad\qquad\qquad = 1$$

The solution is $x = 2$, $y = 1$ or, using an ordered pair, $(2, 1)$.

In attempting to use Cramer's Rule, if the determinant D of the coefficients of the variables is found to equal 0 (so that Cramer's Rule is not applicable), then the system is either inconsistent or has infinitely many solutions.

━━━━━**Now Work** PROBLEM 15

3 Evaluate 3 by 3 Determinants

To use Cramer's Rule to solve a system of three equations containing three variables, we need to define a 3 by 3 determinant.

A **3 by 3 determinant** is symbolized by

$$\begin{vmatrix} a_{11} & a_{12} & a_{13} \\ a_{21} & a_{22} & a_{23} \\ a_{31} & a_{32} & a_{33} \end{vmatrix} \tag{8}$$

in which $a_{11}, a_{12}, \ldots,$ are real numbers.

As with matrices, we use a double subscript to identify an entry by indicating its row and column numbers. For example, the entry a_{23} is in row 2, column 3.

The value of a 3 by 3 determinant may be defined in terms of 2 by 2 determinants by the following formula:

$$\begin{vmatrix} a_{11} & a_{12} & a_{13} \\ a_{21} & a_{22} & a_{23} \\ a_{31} & a_{32} & a_{33} \end{vmatrix} = a_{11} \overset{\text{Minus}}{\begin{vmatrix} a_{22} & a_{23} \\ a_{32} & a_{33} \end{vmatrix}} - a_{12} \begin{vmatrix} a_{21} & a_{23} \\ a_{31} & a_{33} \end{vmatrix} + a_{13} \overset{\text{Plus}}{\begin{vmatrix} a_{21} & a_{22} \\ a_{31} & a_{32} \end{vmatrix}} \tag{9}$$

| 2 by 2 determinant left after removing the row and column containing a_{11} | 2 by 2 determinant left after removing the row and column containing a_{12} | 2 by 2 determinant left after removing the row and column containing a_{13} |

The 2 by 2 determinants shown in formula (9) are called **minors** of the 3 by 3 determinant. For an n by n determinant, the **minor** M_{ij} of entry a_{ij} is the determinant resulting from removing the ith row and jth column.

─────────────────────────────────

EXAMPLE 3 **Finding Minors of a 3 by 3 Determinant**

For the determinant $A = \begin{vmatrix} 2 & -1 & 3 \\ -2 & 5 & 1 \\ 0 & 6 & -9 \end{vmatrix}$, find: (a) M_{12} (b) M_{23}

Solution   (a) M_{12} is the determinant that results from removing the first row and second column from A.

$$A = \begin{vmatrix} 2 & -1 & 3 \\ -2 & 5 & 1 \\ 0 & 6 & -9 \end{vmatrix} \qquad M_{12} = \begin{vmatrix} -2 & 1 \\ 0 & -9 \end{vmatrix} = (-2)(-9) - (0)(1) = 18$$

(b) M_{23} is the determinant that results from removing the second row and third column from A.

$$A = \begin{vmatrix} 2 & -1 & 3 \\ -2 & 5 & 1 \\ 0 & 6 & -9 \end{vmatrix} \qquad M_{23} = \begin{vmatrix} 2 & -1 \\ 0 & 6 \end{vmatrix} = (2)(6) - (0)(-1) = 12$$

Referring back to formula (9), we see that each element a_{ij} is multiplied by its minor, but sometimes this term is added and other times, subtracted. To determine whether to add or subtract a term, we must consider the *cofactor*.

DEFINITION

For an n by n determinant A, the **cofactor** of entry a_{ij}, denoted by A_{ij}, is given by

$$A_{ij} = (-1)^{i+j} M_{ij}$$

where M_{ij} is the minor of entry a_{ij}.

The exponent of $(-1)^{i+j}$ is the sum of the row and column of the entry a_{ij}, so if $i + j$ is even, $(-1)^{i+j}$ will equal 1, and if $i + j$ is odd, $(-1)^{i+j}$ will equal -1.

To find the value of a determinant, multiply each entry in any row or column by its cofactor and sum the results. This process is referred to as **expanding across a row or column.** For example, the value of the 3 by 3 determinant in formula (9) was found by expanding across row 1.

If we choose to expand down column 2, we obtain

$$\begin{vmatrix} a_{11} & a_{12} & a_{13} \\ a_{21} & a_{22} & a_{23} \\ a_{31} & a_{32} & a_{33} \end{vmatrix} = (-1)^{1+2} a_{12} \begin{vmatrix} a_{21} & a_{23} \\ a_{31} & a_{33} \end{vmatrix} + (-1)^{2+2} a_{22} \begin{vmatrix} a_{11} & a_{13} \\ a_{31} & a_{33} \end{vmatrix} + (-1)^{3+2} a_{32} \begin{vmatrix} a_{11} & a_{13} \\ a_{21} & a_{23} \end{vmatrix}$$

Expand down column 2.

If we choose to expand across row 3, we obtain

$$\begin{vmatrix} a_{11} & a_{12} & a_{13} \\ a_{21} & a_{22} & a_{23} \\ a_{31} & a_{32} & a_{33} \end{vmatrix} = (-1)^{3+1} a_{31} \begin{vmatrix} a_{12} & a_{13} \\ a_{22} & a_{23} \end{vmatrix} + (-1)^{3+2} a_{32} \begin{vmatrix} a_{11} & a_{13} \\ a_{21} & a_{23} \end{vmatrix} + (-1)^{3+3} a_{33} \begin{vmatrix} a_{11} & a_{12} \\ a_{21} & a_{22} \end{vmatrix}$$

Expand across row 3.

It can be shown that the value of a determinant does not depend on the choice of the row or column used in the expansion. However, expanding across a row or column that has an entry equal to 0 reduces the amount of work needed to compute the value of the determinant.

EXAMPLE 4   **Evaluating a 3 × 3 Determinant**

Find the value of the 3 by 3 determinant: $\begin{vmatrix} 3 & 0 & -1 \\ 4 & 6 & 2 \\ 8 & -2 & 3 \end{vmatrix}$

Solution   Because of the 0 in row 1, column 2, it is easiest to expand across row 1 or down row 2. We choose to expand across row 1.

$$\begin{vmatrix} 3 & 0 & -1 \\ 4 & 6 & 2 \\ 8 & -2 & 3 \end{vmatrix} = (-1)^{1+1} \cdot 3 \cdot \begin{vmatrix} 6 & 2 \\ -2 & 3 \end{vmatrix} + (-1)^{1+2} \cdot 0 \cdot \begin{vmatrix} 4 & 2 \\ 8 & 3 \end{vmatrix} + (-1)^{1+3} \cdot (-1) \cdot \begin{vmatrix} 4 & 6 \\ 8 & -2 \end{vmatrix}$$

$$= 3(18 - (-4)) - 0 + (-1)(-8 - 48)$$
$$= 3(22) + (-1)(-56)$$
$$= 66 + 56 = 122$$

●

—**Now Work** P R O B L E M **11**

4 Use Cramer's Rule to Solve a System of Three Equations Containing Three Variables

Consider the following system of three equations containing three variables.

$$\begin{cases} a_{11}x + a_{12}y + a_{13}z = c_1 \\ a_{21}x + a_{22}y + a_{23}z = c_2 \\ a_{31}x + a_{32}y + a_{33}z = c_3 \end{cases} \qquad \textbf{(10)}$$

It can be shown that if the determinant D of the coefficients of the variables is not 0, that is, if

$$D = \begin{vmatrix} a_{11} & a_{12} & a_{13} \\ a_{21} & a_{22} & a_{23} \\ a_{31} & a_{32} & a_{33} \end{vmatrix} \neq 0$$

the unique solution of system (10) is given by

THEOREM

> **Cramer's Rule for Three Equations Containing Three Variables**
>
> $$x = \frac{D_x}{D} \qquad y = \frac{D_y}{D} \qquad z = \frac{D_z}{D}$$
>
> where
>
> $$D_x = \begin{vmatrix} c_1 & a_{12} & a_{13} \\ c_2 & a_{22} & a_{23} \\ c_3 & a_{32} & a_{33} \end{vmatrix} \qquad D_y = \begin{vmatrix} a_{11} & c_1 & a_{13} \\ a_{21} & c_2 & a_{23} \\ a_{31} & c_3 & a_{33} \end{vmatrix} \qquad D_z = \begin{vmatrix} a_{11} & a_{12} & c_1 \\ a_{21} & a_{22} & c_2 \\ a_{31} & a_{32} & c_3 \end{vmatrix}$$

Do you see the similarity of this pattern and the pattern observed earlier for a system of two equations containing two variables?

EXAMPLE 5   **Using Cramer's Rule**

Use Cramer's Rule, if applicable, to solve the following system:

$$\begin{cases} 2x + y - z = 3 & (1) \\ -x + 2y + 4z = -3 & (2) \\ x - 2y - 3z = 4 & (3) \end{cases}$$

Solution   The value of the determinant D of the coefficients of the variables is

$$D = \begin{vmatrix} 2 & 1 & -1 \\ -1 & 2 & 4 \\ 1 & -2 & -3 \end{vmatrix} = (-1)^{1+1} \cdot 2 \cdot \begin{vmatrix} 2 & 4 \\ -2 & -3 \end{vmatrix} + (-1)^{1+2} \cdot 1 \cdot \begin{vmatrix} -1 & 4 \\ 1 & -3 \end{vmatrix} + (-1)^{1+3}(-1) \begin{vmatrix} -1 & 2 \\ 1 & -2 \end{vmatrix}$$

$$= 2(2) - 1(-1) + (-1)(0)$$
$$= 4 + 1 = 5$$

Because $D \neq 0$, we proceed to find the values of D_x, D_y, and D_z. To find D_x, we replace the coefficients of x with the constants and then evaluate the determinant.

$$D_x = \begin{vmatrix} 3 & 1 & -1 \\ -3 & 2 & 4 \\ 4 & -2 & -3 \end{vmatrix} = (-1)^{1+1} \cdot 3 \cdot \begin{vmatrix} 2 & 4 \\ -2 & -3 \end{vmatrix} + (-1)^{1+2} \cdot 1 \cdot \begin{vmatrix} -3 & 4 \\ 4 & -3 \end{vmatrix} + (-1)^{1+3}(-1) \begin{vmatrix} -3 & 2 \\ 4 & -2 \end{vmatrix}$$

$$= 3(2) - 1(-7) + (-1)(-2) = 15$$

$$D_y = \begin{vmatrix} 2 & 3 & -1 \\ -1 & -3 & 4 \\ 1 & 4 & -3 \end{vmatrix} = (-1)^{1+1} \cdot 2 \cdot \begin{vmatrix} -3 & 4 \\ 4 & -3 \end{vmatrix} + (-1)^{1+2} \cdot 3 \cdot \begin{vmatrix} -1 & 4 \\ 1 & -3 \end{vmatrix} + (-1)^{1+3}(-1) \begin{vmatrix} -1 & -3 \\ 1 & 4 \end{vmatrix}$$

$$= 2(-7) - 3(-1) + (-1)(-1) = -10$$

$$D_z = \begin{vmatrix} 2 & 1 & 3 \\ -1 & 2 & -3 \\ 1 & -2 & 4 \end{vmatrix} = (-1)^{1+1} \cdot 2 \cdot \begin{vmatrix} 2 & -3 \\ -2 & 4 \end{vmatrix} + (-1)^{1+2} \cdot 1 \cdot \begin{vmatrix} -1 & -3 \\ 1 & 4 \end{vmatrix} + (-1)^{1+3} \cdot 3 \cdot \begin{vmatrix} -1 & 2 \\ 1 & -2 \end{vmatrix}$$

$$= 2(2) - 1(-1) + 3(0) = 5$$

As a result,

$$x = \frac{D_x}{D} = \frac{15}{5} = 3 \qquad y = \frac{D_y}{D} = \frac{-10}{5} = -2 \qquad z = \frac{D_z}{D} = \frac{5}{5} = 1$$

The solution is $x = 3$, $y = -2$, $z = 1$ or, using an ordered triplet, $(3, -2, 1)$.

We already know that Cramer's Rule does not apply when the determinant of the coefficients on the variables, D, is 0. But can we learn anything about the system other than it is not a consistent and independent system if $D = 0$? The answer is yes!

Cramer's Rule with Inconsistent or Dependent Systems

- If $D = 0$ and at least one of the determinants D_x, D_y, or D_z is different from 0, then the system is inconsistent and the solution set is $\varnothing$ or $\{\ \}$.
- If $D = 0$ and all the determinants D_x, D_y, and D_z equal 0, then the system is consistent and dependent so that there are infinitely many solutions. The system must be solved using row reduction techniques.

 Now Work PROBLEM 33

5 Know Properties of Determinants

Determinants have several properties that are sometimes helpful for obtaining their value. We list some of them here.

THEOREM

The value of a determinant changes sign if any two rows (or any two columns) are interchanged.

(11)

Proof for 2 by 2 Determinants

$$\begin{vmatrix} a & b \\ c & d \end{vmatrix} = ad - bc \quad \text{and} \quad \begin{vmatrix} c & d \\ a & b \end{vmatrix} = bc - ad = -(ad - bc) \qquad \blacksquare$$

EXAMPLE 6 **Demonstrating Theorem (11)**

$$\begin{vmatrix} 3 & 4 \\ 1 & 2 \end{vmatrix} = 6 - 4 = 2 \qquad \begin{vmatrix} 1 & 2 \\ 3 & 4 \end{vmatrix} = 4 - 6 = -2$$

THEOREM

If all the entries in any row (or any column) equal 0, the value of the determinant is 0.

(12)

Proof Expand across the row (or down the column) containing the 0's. ∎

THEOREM If any two rows (or any two columns) of a determinant have corresponding entries that are equal, the value of the determinant is 0. **(13)**

You are asked to prove this result for a 3 by 3 determinant in which the entries in column 1 equal the entries in column 3 in Problem 64.

EXAMPLE 7 **Demonstrating Theorem (13)**

$$\begin{vmatrix} 1 & 2 & 3 \\ 1 & 2 & 3 \\ 4 & 5 & 6 \end{vmatrix} = (-1)^{1+1} \cdot 1 \cdot \begin{vmatrix} 2 & 3 \\ 5 & 6 \end{vmatrix} + (-1)^{1+2} \cdot 2 \cdot \begin{vmatrix} 1 & 3 \\ 4 & 6 \end{vmatrix} + (-1)^{1+3} \cdot 3 \cdot \begin{vmatrix} 1 & 2 \\ 4 & 5 \end{vmatrix}$$

$$= 1(-3) - 2(-6) + 3(-3) = -3 + 12 - 9 = 0$$

THEOREM If any row (or any column) of a determinant is multiplied by a nonzero number k, the value of the determinant is also changed by a factor of k. **(14)**

You are asked to prove this result for a 3 by 3 determinant using row 2 in Problem 63.

EXAMPLE 8 **Demonstrating Theorem (14)**

$$\begin{vmatrix} 1 & 2 \\ 4 & 6 \end{vmatrix} = 6 - 8 = -2$$

$$\begin{vmatrix} k & 2k \\ 4 & 6 \end{vmatrix} = 6k - 8k = -2k = k(-2) = k\begin{vmatrix} 1 & 2 \\ 4 & 6 \end{vmatrix}$$

THEOREM If the entries of any row (or any column) of a determinant are multiplied by a nonzero number k and the result is added to the corresponding entries of another row (or column), the value of the determinant remains unchanged. **(15)**

In Problem 65, you are asked to prove this result for a 3 by 3 determinant using rows 1 and 2.

EXAMPLE 9 **Demonstrating Theorem (15)**

$$\begin{vmatrix} 3 & 4 \\ 5 & 2 \end{vmatrix} = -14 \qquad \begin{vmatrix} 3 & 4 \\ 5 & 2 \end{vmatrix} \xrightarrow{\uparrow} \begin{vmatrix} -7 & 0 \\ 5 & 2 \end{vmatrix} = -14$$

Multiply row 2 by -2 and add to row 1.

10.3 Assess Your Understanding

Concepts and Vocabulary

1. $D = \begin{vmatrix} a & b \\ c & d \end{vmatrix} = \underline{\hspace{1.5cm}}$.

2. Using Cramer's Rule, the value of x that satisfies the system of equations $\begin{cases} 2x + 3y = 5 \\ x - 4y = -3 \end{cases}$ is $x = \dfrac{}{\begin{vmatrix} 2 & 3 \\ 1 & -4 \end{vmatrix}}$

3. **True or False** A determinant can never equal 0.

4. **True or False** When using Cramer's Rule, if $D = 0$, then the system of linear equations is inconsistent.

5. **True or False** The value of a determinant remains unchanged if any two rows or any two columns are interchanged.

6. **True or False** If any row (or any column) of a determinant is multiplied by a nonzero number k, the value of the determinant remains unchanged.

Skill Building

In Problems 7–14, find the value of each determinant.

7. $\begin{vmatrix} 6 & 4 \\ -1 & 3 \end{vmatrix}$

8. $\begin{vmatrix} 8 & -3 \\ 4 & 2 \end{vmatrix}$

9. $\begin{vmatrix} -3 & -1 \\ 4 & 2 \end{vmatrix}$

10. $\begin{vmatrix} -4 & 2 \\ -5 & 3 \end{vmatrix}$

11. $\begin{vmatrix} 3 & 4 & 2 \\ 1 & -1 & 5 \\ 1 & 2 & -2 \end{vmatrix}$

12. $\begin{vmatrix} 1 & 3 & -2 \\ 6 & 1 & -5 \\ 8 & 2 & 3 \end{vmatrix}$

13. $\begin{vmatrix} 4 & -1 & 2 \\ 6 & -1 & 0 \\ 1 & -3 & 4 \end{vmatrix}$

14. $\begin{vmatrix} 3 & -9 & 4 \\ 1 & 4 & 0 \\ 8 & -3 & 1 \end{vmatrix}$

In Problems 15–42, solve each system of equations using Cramer's Rule if it is applicable. If Cramer's Rule is not applicable, say so.

15. $\begin{cases} x + y = 8 \\ x - y = 4 \end{cases}$

16. $\begin{cases} x + 2y = 5 \\ x - y = 3 \end{cases}$

17. $\begin{cases} 5x - y = 13 \\ 2x + 3y = 12 \end{cases}$

18. $\begin{cases} x + 3y = 5 \\ 2x - 3y = -8 \end{cases}$

19. $\begin{cases} 3x = 24 \\ x + 2y = 0 \end{cases}$

20. $\begin{cases} 4x + 5y = -3 \\ -2y = -4 \end{cases}$

21. $\begin{cases} 3x - 6y = 24 \\ 5x + 4y = 12 \end{cases}$

22. $\begin{cases} 2x + 4y = 16 \\ 3x - 5y = -9 \end{cases}$

23. $\begin{cases} 3x - 2y = 4 \\ 6x - 4y = 0 \end{cases}$

24. $\begin{cases} -x + 2y = 5 \\ 4x - 8y = 6 \end{cases}$

25. $\begin{cases} 2x - 4y = -2 \\ 3x + 2y = 3 \end{cases}$

26. $\begin{cases} 3x + 3y = 3 \\ 4x + 2y = \dfrac{8}{3} \end{cases}$

27. $\begin{cases} 2x - 3y = -1 \\ 10x + 10y = 5 \end{cases}$

28. $\begin{cases} 3x - 2y = 0 \\ 5x + 10y = 4 \end{cases}$

29. $\begin{cases} 2x + 3y = 6 \\ x - y = \dfrac{1}{2} \end{cases}$

30. $\begin{cases} \dfrac{1}{2}x + y = -2 \\ x - 2y = 8 \end{cases}$

31. $\begin{cases} 3x - 5y = 3 \\ 15x + 5y = 21 \end{cases}$

32. $\begin{cases} 2x - y = -1 \\ x + \dfrac{1}{2}y = \dfrac{3}{2} \end{cases}$

33. $\begin{cases} x + y - z = 6 \\ 3x - 2y + z = -5 \\ x + 3y - 2z = 14 \end{cases}$

34. $\begin{cases} x - y + z = -4 \\ 2x - 3y + 4z = -15 \\ 5x + y - 2z = 12 \end{cases}$

35. $\begin{cases} x + 2y - z = -3 \\ 2x - 4y + z = -7 \\ -2x + 2y - 3z = 4 \end{cases}$

36. $\begin{cases} x + 4y - 3z = -8 \\ 3x - y + 3z = 12 \\ x + y + 6z = 1 \end{cases}$

37. $\begin{cases} x - 2y + 3z = 1 \\ 3x + y - 2z = 0 \\ 2x - 4y + 6z = 2 \end{cases}$

38. $\begin{cases} x - y + 2z = 5 \\ 3x + 2y = 4 \\ -2x + 2y - 4z = -10 \end{cases}$

39. $\begin{cases} x + 2y - z = 0 \\ 2x - 4y + z = 0 \\ -2x + 2y - 3z = 0 \end{cases}$

40. $\begin{cases} x + 4y - 3z = 0 \\ 3x - y + 3z = 0 \\ x + y + 6z = 0 \end{cases}$

41. $\begin{cases} x - 2y + 3z = 0 \\ 3x + y - 2z = 0 \\ 2x - 4y + 6z = 0 \end{cases}$

42. $\begin{cases} x - y + 2z = 0 \\ 3x + 2y = 0 \\ -2x + 2y - 4z = 0 \end{cases}$

In Problems 43–50, use properties of determinants to find the value of each determinant if it is known that

$$\begin{vmatrix} x & y & z \\ u & v & w \\ 1 & 2 & 3 \end{vmatrix} = 4$$

43. $\begin{vmatrix} 1 & 2 & 3 \\ u & v & w \\ x & y & z \end{vmatrix}$

44. $\begin{vmatrix} x & y & z \\ u & v & w \\ 2 & 4 & 6 \end{vmatrix}$

45. $\begin{vmatrix} x & y & z \\ -3 & -6 & -9 \\ u & v & w \end{vmatrix}$

46. $\begin{vmatrix} 1 & 2 & 3 \\ x - u & y - v & z - w \\ u & v & w \end{vmatrix}$

47. $\begin{vmatrix} 1 & 2 & 3 \\ x - 3 & y - 6 & z - 9 \\ 2u & 2v & 2w \end{vmatrix}$

48. $\begin{vmatrix} x & y & z - x \\ u & v & w - u \\ 1 & 2 & 2 \end{vmatrix}$

49. $\begin{vmatrix} 1 & 2 & 3 \\ 2x & 2y & 2z \\ u - 1 & v - 2 & w - 3 \end{vmatrix}$

50. $\begin{vmatrix} x + 3 & y + 6 & z + 9 \\ 3u - 1 & 3v - 2 & 3w - 3 \\ 1 & 2 & 3 \end{vmatrix}$

Mixed Practice

In Problems 51–56, solve for x.

51. $\begin{vmatrix} x & x \\ 4 & 3 \end{vmatrix} = 5$

52. $\begin{vmatrix} x & 1 \\ 3 & x \end{vmatrix} = -2$

53. $\begin{vmatrix} x & 1 & 1 \\ 4 & 3 & 2 \\ -1 & 2 & 5 \end{vmatrix} = 2$

54. $\begin{vmatrix} 3 & 2 & 4 \\ 1 & x & 5 \\ 0 & 1 & -2 \end{vmatrix} = 0$

55. $\begin{vmatrix} x & 2 & 3 \\ 1 & x & 0 \\ 6 & 1 & -2 \end{vmatrix} = 7$

56. $\begin{vmatrix} x & 1 & 2 \\ 1 & x & 3 \\ 0 & 1 & 2 \end{vmatrix} = -4x$

Applications and Extensions

57. Geometry: Equation of a Line An equation of the line containing the two points (x_1, y_1) and (x_2, y_2) may be expressed as the determinant

$$\begin{vmatrix} x & y & 1 \\ x_1 & y_1 & 1 \\ x_2 & y_2 & 1 \end{vmatrix} = 0$$

Prove this result by expanding the determinant and comparing the result to the two-point form of the equation of a line.

58. Geometry: Collinear Points Using the result obtained in Problem 57, show that three distinct points (x_1, y_1), (x_2, y_2), and (x_3, y_3) are collinear (lie on the same line) if and only if

$$\begin{vmatrix} x_1 & y_1 & 1 \\ x_2 & y_2 & 1 \\ x_3 & y_3 & 1 \end{vmatrix} = 0$$

59. Geometry: Area of a Triangle A triangle has vertices $(x_1, y_1), (x_2, y_2)$, and (x_3, y_3). The area of the triangle is given by the absolute value of D, where $D = \dfrac{1}{2}\begin{vmatrix} x_1 & x_2 & x_3 \\ y_1 & y_2 & y_3 \\ 1 & 1 & 1 \end{vmatrix}$.

Use this formula to find the area of a triangle with vertices $(2, 3), (5, 2)$, and $(6, 5)$.

60. Show that $\begin{vmatrix} x^2 & x & 1 \\ y^2 & y & 1 \\ z^2 & z & 1 \end{vmatrix} = (y - z)(x - y)(x - z)$.

61. Complete the proof of Cramer's Rule for two equations containing two variables.

[**Hint:** In system (5), page 763, if $a = 0$, then $b \neq 0$ and $c \neq 0$, since $D = -bc \neq 0$. Now show that equation (6) provides a solution of the system when $a = 0$. Then three cases remain: $b = 0$, $c = 0$, and $d = 0$.]

62. Interchange columns 1 and 3 of a 3 by 3 determinant. Show that the value of the new determinant is -1 times the value of the original determinant.

63. Multiply each entry in row 2 of a 3 by 3 determinant by the number $k, k \neq 0$. Show that the value of the new determinant is k times the value of the original determinant.

64. Prove that a 3 by 3 determinant in which the entries in column 1 equal those in column 3 has the value 0.

65. Prove that, if row 2 of a 3 by 3 determinant is multiplied by $k, k \neq 0$, and the result is added to the entries in row 1, there is no change in the value of the determinant.

10.4 Matrix Algebra

OBJECTIVES **1** Find the Sum and Difference of Two Matrices (p. 772)
   2 Find Scalar Multiples of a Matrix (p. 773)
   3 Find the Product of Two Matrices (p. 774)
   4 Find the Inverse of a Matrix (p. 779)
   5 Solve a System of Linear Equations Using an Inverse Matrix (p. 783)

In Section 10.2, we defined a matrix as a rectangular array of real numbers and used an augmented matrix to represent a system of linear equations. There is, however, a branch of mathematics, called **linear algebra,** that deals with matrices in such a way that an algebra of matrices is permitted. In this section, we provide a survey of how this **matrix algebra** is developed.

Before getting started, we restate the definition of a matrix.

DEFINITION

A **matrix** is defined as a rectangular array of numbers:

$$
\begin{array}{c}
\\
\text{Row 1} \\
\text{Row 2} \\
\vdots \\
\text{Row } i \\
\vdots \\
\text{Row } m
\end{array}
\begin{bmatrix}
a_{11} & a_{12} & \cdots & a_{1j} & \cdots & a_{1n} \\
a_{21} & a_{22} & \cdots & a_{2j} & \cdots & a_{2n} \\
\vdots & \vdots & & \vdots & & \vdots \\
a_{i1} & a_{i2} & \cdots & a_{ij} & \cdots & a_{in} \\
\vdots & \vdots & & \vdots & & \vdots \\
a_{m1} & a_{m2} & \cdots & a_{mj} & \cdots & a_{mn}
\end{bmatrix}
$$

with column labels Column 1, Column 2, Column j, Column n.

Each number a_{ij} of the matrix has two indexes: the **row index** i and the **column index** j. The matrix shown here has m rows and n columns. The numbers a_{ij} are usually referred to as the **entries** of the matrix. For example, a_{23} refers to the entry in the second row, third column.

Let's begin with an example that illustrates how matrices can be used to conveniently represent an array of information.

EXAMPLE 1 **Arranging Data in a Matrix**

In a survey of 900 people, the following information was obtained:

200 males	Thought federal defense spending was too high
150 males	Thought federal defense spending was too low
45 males	Had no opinion
315 females	Thought federal defense spending was too high
125 females	Thought federal defense spending was too low
65 females	Had no opinion

We can arrange these data in a rectangular array as follows:

	Too High	Too Low	No Opinion
Male	200	150	45
Female	315	125	65

or as the matrix

$$
\begin{bmatrix}
200 & 150 & 45 \\
315 & 125 & 65
\end{bmatrix}
$$

This matrix has two rows (representing male and female) and three columns (representing "too high," "too low," and "no opinion").

The matrix we developed in Example 1 has 2 rows and 3 columns. In general, a matrix with m rows and n columns is called an **m by n matrix.** The matrix we developed in Example 1 is a 2 by 3 matrix and contains $2 \cdot 3 = 6$ entries. An m by n matrix will contain $m \cdot n$ entries.

If an m by n matrix has the same number of rows as columns, that is, if $m = n$, then the matrix is referred to as a **square matrix.**

EXAMPLE 2 **Examples of Matrices**

(a) $\begin{bmatrix} 5 & 0 \\ -6 & 1 \end{bmatrix}$ A 2 by 2 square matrix

(b) $\begin{bmatrix} 1 & 0 & 3 \end{bmatrix}$ A 1 by 3 matrix

(c) $\begin{bmatrix} 6 & -2 & 4 \\ 4 & 3 & 5 \\ 8 & 0 & 1 \end{bmatrix}$ A 3 by 3 square matrix

1 Find the Sum and Difference of Two Matrices

We begin our discussion of matrix algebra by first defining what is meant by two matrices being equal and then defining the operations of addition and subtraction. It is important to note that these definitions require each matrix to have the same number of rows *and* the same number of columns as a condition for equality and for addition and subtraction.

We usually represent matrices by capital letters, such as A, B, and C.

DEFINITION

Two matrices A and B are said to be **equal,** written as

$$A = B$$

provided that A and B have the same number of rows and the same number of columns and each entry a_{ij} in A is equal to the corresponding entry b_{ij} in B.

For example,

$$\begin{bmatrix} 2 & 1 \\ 0.5 & -1 \end{bmatrix} = \begin{bmatrix} \sqrt{4} & 1 \\ \frac{1}{2} & -1 \end{bmatrix} \quad \text{and} \quad \begin{bmatrix} 3 & 2 & 1 \\ 0 & 1 & -2 \end{bmatrix} = \begin{bmatrix} \sqrt{9} & \sqrt{4} & 1 \\ 0 & 1 & \sqrt[3]{-8} \end{bmatrix}$$

$$\begin{bmatrix} 4 & 1 \\ 6 & 1 \end{bmatrix} \neq \begin{bmatrix} 4 & 0 \\ 6 & 1 \end{bmatrix} \qquad \text{Because the entries in row 1, column 2 are not equal}$$

$$\begin{bmatrix} 4 & 1 & 2 \\ 6 & 1 & 2 \end{bmatrix} \neq \begin{bmatrix} 4 & 1 & 2 & 3 \\ 6 & 1 & 2 & 4 \end{bmatrix} \qquad \text{Because the matrix on the left has 3 columns and the matrix on the right has 4 columns.}$$

Suppose that A and B represent two m by n matrices. We define their **sum** $A + B$ to be the m by n matrix formed by adding the corresponding entries a_{ij} of A and b_{ij} of B. The **difference $A - B$** is defined as the m by n matrix formed by subtracting the entries b_{ij} in B from the corresponding entries a_{ij} in A. Addition and subtraction of matrices are allowed only for matrices having the same number m of rows and the same number n of columns. For example, a 2 by 3 matrix and a 2 by 4 matrix cannot be added or subtracted.

EXAMPLE 3

Adding and Subtracting Matrices

Suppose that

$$A = \begin{bmatrix} 2 & 4 & 8 & -3 \\ 0 & 1 & 2 & 3 \end{bmatrix} \quad \text{and} \quad B = \begin{bmatrix} -3 & 4 & 0 & 1 \\ 6 & 8 & 2 & 0 \end{bmatrix}$$

Find: (a) $A + B$ (b) $A - B$

Solution (a) $A + B = \begin{bmatrix} 2 & 4 & 8 & -3 \\ 0 & 1 & 2 & 3 \end{bmatrix} + \begin{bmatrix} -3 & 4 & 0 & 1 \\ 6 & 8 & 2 & 0 \end{bmatrix}$

$$= \begin{bmatrix} 2 + (-3) & 4 + 4 & 8 + 0 & -3 + 1 \\ 0 + 6 & 1 + 8 & 2 + 2 & 3 + 0 \end{bmatrix} \qquad \text{Add corresponding entries.}$$

$$= \begin{bmatrix} -1 & 8 & 8 & -2 \\ 6 & 9 & 4 & 3 \end{bmatrix}$$

(b) $A - B = \begin{bmatrix} 2 & 4 & 8 & -3 \\ 0 & 1 & 2 & 3 \end{bmatrix} - \begin{bmatrix} -3 & 4 & 0 & 1 \\ 6 & 8 & 2 & 0 \end{bmatrix}$

$$= \begin{bmatrix} 2 - (-3) & 4 - 4 & 8 - 0 & -3 - 1 \\ 0 - 6 & 1 - 8 & 2 - 2 & 3 - 0 \end{bmatrix} \qquad \text{Subtract corresponding entries.}$$

$$= \begin{bmatrix} 5 & 0 & 8 & -4 \\ -6 & -7 & 0 & 3 \end{bmatrix}$$

Figure 7

 Seeing the Concept

Graphing utilities can make the sometimes tedious process of matrix algebra easy. In fact, most graphing calculators can handle matrices as large as 9 by 9, some even larger ones. Enter the matrices into a graphing utility. Name them [A] and [B]. Figure 7 shows the results of adding and subtracting [A] and [B].

━━━ **Now Work** PROBLEM 9

Many of the algebraic properties of sums of real numbers are also true for sums of matrices. Suppose that A, B, and C are m by n matrices. Then matrix addition is **commutative.** That is,

Commutative Property of Matrix Addition

$$A + B = B + A$$

Matrix addition is also **associative.** That is,

Associative Property of Matrix Addition

$$(A + B) + C = A + (B + C)$$

Although we shall not prove these results, the proofs, as the following example illustrates, are based on the commutative and associative properties for real numbers.

EXAMPLE 4   **Demonstrating the Commutative Property**

$$\begin{bmatrix} 2 & 3 & -1 \\ 4 & 0 & 7 \end{bmatrix} + \begin{bmatrix} -1 & 2 & 1 \\ 5 & -3 & 4 \end{bmatrix} = \begin{bmatrix} 2 + (-1) & 3 + 2 & -1 + 1 \\ 4 + 5 & 0 + (-3) & 7 + 4 \end{bmatrix}$$

$$= \begin{bmatrix} -1 + 2 & 2 + 3 & 1 + (-1) \\ 5 + 4 & -3 + 0 & 4 + 7 \end{bmatrix}$$

$$= \begin{bmatrix} -1 & 2 & 1 \\ 5 & -3 & 4 \end{bmatrix} + \begin{bmatrix} 2 & 3 & -1 \\ 4 & 0 & 7 \end{bmatrix}$$

A matrix whose entries are all equal to 0 is called a **zero matrix.** Each of the following matrices is a zero matrix.

$$\begin{bmatrix} 0 & 0 \\ 0 & 0 \end{bmatrix}$$ 2 by 2 square zero matrix $$\begin{bmatrix} 0 & 0 & 0 \\ 0 & 0 & 0 \end{bmatrix}$$ 2 by 3 zero matrix $$\begin{bmatrix} 0 & 0 & 0 \end{bmatrix}$$ 1 by 3 zero matrix

Zero matrices have properties similar to the real number 0. If A is an m by n matrix and 0 is an m by n zero matrix, then

$$A + 0 = 0 + A = A$$

In other words, the zero matrix is the additive identity in matrix algebra.

2 Find Scalar Multiples of a Matrix

We can also multiply a matrix by a real number. If k is a real number and A is an m by n matrix, the matrix kA is the m by n matrix formed by multiplying each entry a_{ij} in A by k. The number k is sometimes referred to as a **scalar,** and the matrix kA is called a **scalar multiple** of A.

EXAMPLE 5   **Operations Using Matrices**

Suppose that

$$A = \begin{bmatrix} 3 & 1 & 5 \\ -2 & 0 & 6 \end{bmatrix} \qquad B = \begin{bmatrix} 4 & 1 & 0 \\ 8 & 1 & -3 \end{bmatrix} \qquad C = \begin{bmatrix} 9 & 0 \\ -3 & 6 \end{bmatrix}$$

Find: (a) $4A$ (b) $\dfrac{1}{3}C$ (c) $3A - 2B$

Solution   (a) $4A = 4\begin{bmatrix} 3 & 1 & 5 \\ -2 & 0 & 6 \end{bmatrix} = \begin{bmatrix} 4 \cdot 3 & 4 \cdot 1 & 4 \cdot 5 \\ 4(-2) & 4 \cdot 0 & 4 \cdot 6 \end{bmatrix} = \begin{bmatrix} 12 & 4 & 20 \\ -8 & 0 & 24 \end{bmatrix}$

(b) $\dfrac{1}{3}C = \dfrac{1}{3}\begin{bmatrix} 9 & 0 \\ -3 & 6 \end{bmatrix} = \begin{bmatrix} \dfrac{1}{3} \cdot 9 & \dfrac{1}{3} \cdot 0 \\ \dfrac{1}{3}(-3) & \dfrac{1}{3} \cdot 6 \end{bmatrix} = \begin{bmatrix} 3 & 0 \\ -1 & 2 \end{bmatrix}$

(c) $3A - 2B = 3\begin{bmatrix} 3 & 1 & 5 \\ -2 & 0 & 6 \end{bmatrix} - 2\begin{bmatrix} 4 & 1 & 0 \\ 8 & 1 & -3 \end{bmatrix}$

$= \begin{bmatrix} 3 \cdot 3 & 3 \cdot 1 & 3 \cdot 5 \\ 3(-2) & 3 \cdot 0 & 3 \cdot 6 \end{bmatrix} - \begin{bmatrix} 2 \cdot 4 & 2 \cdot 1 & 2 \cdot 0 \\ 2 \cdot 8 & 2 \cdot 1 & 2(-3) \end{bmatrix}$

$= \begin{bmatrix} 9 & 3 & 15 \\ -6 & 0 & 18 \end{bmatrix} - \begin{bmatrix} 8 & 2 & 0 \\ 16 & 2 & -6 \end{bmatrix}$

$= \begin{bmatrix} 9 - 8 & 3 - 2 & 15 - 0 \\ -6 - 16 & 0 - 2 & 18 - (-6) \end{bmatrix}$

$= \begin{bmatrix} 1 & 1 & 15 \\ -22 & -2 & 24 \end{bmatrix}$

●

 ✓Check Enter the matrices $[A], [B]$, and $[C]$ into a graphing utility. Then find $4A, \dfrac{1}{3}C,$ and $3A - 2B$.

━ **Now Work**   PROBLEM 13

We list next some of the algebraic properties of scalar multiplication. Let h and k be real numbers, and let A and B be m by n matrices. Then

---

Properties of Scalar Multiplication

$$k(hA) = (kh)A$$
$$(k + h)A = kA + hA$$
$$k(A + B) = kA + kB$$

---

3 Find the Product of Two Matrices

Unlike the straightforward definition for adding two matrices, the definition for multiplying two matrices is not what we might expect. In preparation for this definition, we need the following definitions:

DEFINITION

A **row vector** R is a 1 by n matrix

$$R = [r_1 \quad r_2 \quad \cdots \quad r_n]$$

A **column vector** C is an n by 1 matrix

$$C = \begin{bmatrix} c_1 \\ c_2 \\ \vdots \\ c_n \end{bmatrix}$$

The **product** RC of R times C is defined as the number

$$RC = [r_1 \quad r_2 \cdots r_n] \begin{bmatrix} c_1 \\ c_2 \\ \vdots \\ c_n \end{bmatrix} = r_1 c_1 + r_2 c_2 + \cdots + r_n c_n$$

Notice that a row vector and a column vector can be multiplied only if they contain the same number of entries.

EXAMPLE 6   **The Product of a Row Vector and a Column Vector**

If $R = [3 \quad -5 \quad 2]$ and $C = \begin{bmatrix} 3 \\ 4 \\ -5 \end{bmatrix}$, then

$$RC = [3 \quad -5 \quad 2] \begin{bmatrix} 3 \\ 4 \\ -5 \end{bmatrix} = 3 \cdot 3 + (-5)4 + 2(-5) = 9 - 20 - 10 = -21$$

Let's look at an application of the product of a row vector by a column vector.

EXAMPLE 7   **Using Matrices to Compute Revenue**

A clothing store sells men's shirts for \$40, silk ties for \$20, and wool suits for \$400. Last month, the store had sales consisting of 100 shirts, 200 ties, and 50 suits. What was the total revenue due to these sales?

Solution   We set up a row vector R to represent the prices of each item and a column vector C to represent the corresponding number of items sold. Then

$$\begin{array}{cc} \text{Prices} & \text{Number} \\ \text{Shirts Ties Suits} & \text{sold} \end{array}$$

$$R = [40 \quad 20 \quad 400] \qquad C = \begin{bmatrix} 100 \\ 200 \\ 50 \end{bmatrix} \begin{array}{l} \text{Shirts} \\ \text{Ties} \\ \text{Suits} \end{array}$$

The total revenue obtained is the product RC. That is,

$$RC = [40 \ 20 \ 400] \begin{bmatrix} 100 \\ 200 \\ 50 \end{bmatrix}$$

$$= \underbrace{40 \cdot 100}_{\text{Shirt revenue}} + \underbrace{20 \cdot 200}_{\text{Tie revenue}} + \underbrace{400 \cdot 50}_{\text{Suit revenue}} = \underbrace{\$28,000}_{\text{Total revenue}}$$

The definition for multiplying two matrices is based on the definition of a row vector times a column vector.

DEFINITION

> Let A denote an m by r matrix and let B denote an r by n matrix. The **product** AB is defined as the m by n matrix whose entry in row i, column j is the product of the ith row of A and the jth column of B.

The definition of the product AB of two matrices A and B, in this order, requires that the number of columns of A equal the number of rows of B; otherwise, no product is defined.

$$\begin{array}{cc} A & B \\ m \text{ by } r & r \text{ by } n \end{array}$$

Must be same for AB to be defined
AB is m by n.

In Words

To find the product AB, the number of columns in A must equal the number of rows in B.

An example will help to clarify the definition.

EXAMPLE 8   Multiplying Two Matrices

Find the product AB if

$$A = \begin{bmatrix} 2 & 4 & -1 \\ 5 & 8 & 0 \end{bmatrix} \quad \text{and} \quad B = \begin{bmatrix} 2 & 5 & 1 & 4 \\ 4 & 8 & 0 & 6 \\ -3 & 1 & -2 & -1 \end{bmatrix}$$

Solution   First, we note that A is 2 by 3 and B is 3 by 4, so the product AB is defined and will be a 2 by 4 matrix.

Suppose that we want the entry in row 2, column 3 of AB. To find it, we find the product of the row vector from row 2 of A and the column vector from column 3 of B.

Column 3 of B

Row 2 of A
$$[5 \quad 8 \quad 0] \begin{bmatrix} 1 \\ 0 \\ -2 \end{bmatrix} = 5 \cdot 1 + 8 \cdot 0 + 0(-2) = 5$$

So far, we have

Column 3
$$AB = \begin{bmatrix} \underline{} & \underline{} & 5 & \underline{} \\ \underline{} & \underline{} & \underline{} & \underline{} \end{bmatrix} \leftarrow \text{Row 2}$$

Now, to find the entry in row 1, column 4 of AB, we find the product of row 1 of A and column 4 of B.

Column 4 of B

Row 1 of A
$$[2 \quad 4 \quad -1] \begin{bmatrix} 4 \\ 6 \\ -1 \end{bmatrix} = 2 \cdot 4 + 4 \cdot 6 + (-1)(-1) = 33$$

Continuing in this fashion, we find AB.

$$AB = \begin{bmatrix} 2 & 4 & -1 \\ 5 & 8 & 0 \end{bmatrix} \begin{bmatrix} 2 & 5 & 1 & 4 \\ 4 & 8 & 0 & 6 \\ -3 & 1 & -2 & -1 \end{bmatrix}$$

$$= \begin{bmatrix} \text{Row 1 of } A & \text{Row 1 of } A & \text{Row 1 of } A & \text{Row 1 of } A \\ \text{times} & \text{times} & \text{times} & \text{times} \\ \text{column 1 of } B & \text{column 2 of } B & \text{column 3 of } B & \text{column 4 of } B \\ \\ \text{Row 2 of } A & \text{Row 2 of } A & \text{Row 2 of } A & \text{Row 2 of } A \\ \text{times} & \text{times} & \text{times} & \text{times} \\ \text{column 1 of } B & \text{column 2 of } B & \text{column 3 of } B & \text{column 4 of } B \end{bmatrix}$$

$$= \begin{bmatrix} 2 \cdot 2 + 4 \cdot 4 + (-1)(-3) & 2 \cdot 5 + 4 \cdot 8 + (-1)1 & 2 \cdot 1 + 4 \cdot 0 + (-1)(-2) & 33(\text{from earlier}) \\ 5 \cdot 2 + 8 \cdot 4 + 0(-3) & 5 \cdot 5 + 8 \cdot 8 + 0 \cdot 1 & 5(\text{from earlier}) & 5 \cdot 4 + 8 \cdot 6 + 0(-1) \end{bmatrix}$$

$$= \begin{bmatrix} 23 & 41 & 4 & 33 \\ 42 & 89 & 5 & 68 \end{bmatrix}$$

 ✓**Check:** Enter the matrices $[A]$ and $[B]$. Then find AB. [See what happens if you try to find BA.)

➤ **Now Work**   PROBLEM 25

Notice that for the matrices given in Example 8 the product BA is not defined, because B is 3 by 4 and A is 2 by 3.

Another result that can occur when multiplying two matrices is illustrated in the next example.

EXAMPLE 9   **Multiplying Two Matrices**

If

$$A = \begin{bmatrix} 2 & 1 & 3 \\ 1 & -1 & 0 \end{bmatrix} \quad \text{and} \quad B = \begin{bmatrix} 1 & 0 \\ 2 & 1 \\ 3 & 2 \end{bmatrix}$$

find: (a) AB (b) BA

Solution   (a) $AB = \underset{2 \text{ by } 3}{\begin{bmatrix} 2 & 1 & 3 \\ 1 & -1 & 0 \end{bmatrix}} \underset{3 \text{ by } 2}{\begin{bmatrix} 1 & 0 \\ 2 & 1 \\ 3 & 2 \end{bmatrix}} = \underset{2 \text{ by } 2}{\begin{bmatrix} 13 & 7 \\ -1 & -1 \end{bmatrix}}$

(b) $BA = \underset{3 \text{ by } 2}{\begin{bmatrix} 1 & 0 \\ 2 & 1 \\ 3 & 2 \end{bmatrix}} \underset{2 \text{ by } 3}{\begin{bmatrix} 2 & 1 & 3 \\ 1 & -1 & 0 \end{bmatrix}} = \underset{3 \text{ by } 3}{\begin{bmatrix} 2 & 1 & 3 \\ 5 & 1 & 6 \\ 8 & 1 & 9 \end{bmatrix}}$

Notice in Example 9 that AB is 2 by 2 and BA is 3 by 3. It is possible for both AB and BA to be defined, yet be unequal. In fact, even if A and B are both n by n matrices so that AB and BA are each defined and n by n, AB and BA will usually be unequal.

EXAMPLE 10

Multiplying Two Square Matrices

If

$$A = \begin{bmatrix} 2 & 1 \\ 0 & 4 \end{bmatrix} \quad \text{and} \quad B = \begin{bmatrix} -3 & 1 \\ 1 & 2 \end{bmatrix}$$

find: (a) AB (b) BA

Solution   (a) $AB = \begin{bmatrix} 2 & 1 \\ 0 & 4 \end{bmatrix}\begin{bmatrix} -3 & 1 \\ 1 & 2 \end{bmatrix} = \begin{bmatrix} -5 & 4 \\ 4 & 8 \end{bmatrix}$

(b) $BA = \begin{bmatrix} -3 & 1 \\ 1 & 2 \end{bmatrix}\begin{bmatrix} 2 & 1 \\ 0 & 4 \end{bmatrix} = \begin{bmatrix} -6 & 1 \\ 2 & 9 \end{bmatrix}$

    ●

The preceding examples demonstrate that an important property of real numbers, the commutative property of multiplication, is not shared by matrices. In general:

THEOREM   Matrix multiplication is not commutative.

━━━━Now Work   PROBLEMS 15 AND 17

Next we give two of the properties of real numbers that are shared by matrices. Assuming that each product and sum is defined, we have the following:

Associative Property of Matrix Multiplication

$$A(BC) = (AB)C$$

Distributive Property

$$A(B + C) = AB + AC$$

For an n by n square matrix, the entries located in row i, column i, $1 \leq i \leq n$, are called the **diagonal entries** or **the main diagonal.** The n by n square matrix whose diagonal entries are 1's, while all other entries are 0's, is called the **identity matrix I_n.** For example,

$$I_2 = \begin{bmatrix} 1 & 0 \\ 0 & 1 \end{bmatrix} \qquad I_3 = \begin{bmatrix} 1 & 0 & 0 \\ 0 & 1 & 0 \\ 0 & 0 & 1 \end{bmatrix}$$

and so on.

EXAMPLE 11

Multiplication with an Identity Matrix

Let

$$A = \begin{bmatrix} -1 & 2 & 0 \\ 0 & 1 & 3 \end{bmatrix} \quad \text{and} \quad B = \begin{bmatrix} 3 & 2 \\ 4 & 6 \\ 5 & 2 \end{bmatrix}$$

Find: (a) AI_3 (b) I_2A (c) BI_2

Solution (a) $AI_3 = \begin{bmatrix} -1 & 2 & 0 \\ 0 & 1 & 3 \end{bmatrix}\begin{bmatrix} 1 & 0 & 0 \\ 0 & 1 & 0 \\ 0 & 0 & 1 \end{bmatrix} = \begin{bmatrix} -1 & 2 & 0 \\ 0 & 1 & 3 \end{bmatrix} = A$

(b) $I_2A = \begin{bmatrix} 1 & 0 \\ 0 & 1 \end{bmatrix}\begin{bmatrix} -1 & 2 & 0 \\ 0 & 1 & 3 \end{bmatrix} = \begin{bmatrix} -1 & 2 & 0 \\ 0 & 1 & 3 \end{bmatrix} = A$

(c) $BI_2 = \begin{bmatrix} 3 & 2 \\ 4 & 6 \\ 5 & 2 \end{bmatrix}\begin{bmatrix} 1 & 0 \\ 0 & 1 \end{bmatrix} = \begin{bmatrix} 3 & 2 \\ 4 & 6 \\ 5 & 2 \end{bmatrix} = B$

Example 11 demonstrates the following property:

Identity Property

If A is an m by n matrix, then

$$I_mA = A \quad \text{and} \quad AI_n = A$$

If A is an n by n square matrix,

$$AI_n = I_nA = A$$

An identity matrix has properties similar to those of the real number 1. In other words, the identity matrix is a multiplicative identity in matrix algebra.

4 Find the Inverse of a Matrix

DEFINITION Let A be a square n by n matrix. If there exists an n by n matrix A^{-1}, read "A inverse," for which

$$AA^{-1} = A^{-1}A = I_n$$

then A^{-1} is called the **inverse** of the matrix A.

NOTE If the determinant of A is zero, A is singular. (Refer to Section 10.3.) ■

As we shall soon see, not every square matrix has an inverse. When a matrix A does have an inverse A^{-1}, then A is said to be **nonsingular.** If a matrix A has no inverse, it is called **singular.**

EXAMPLE 12 **Multiplying a Matrix by Its Inverse**

Show that the inverse of

$$A = \begin{bmatrix} 3 & 1 \\ 2 & 1 \end{bmatrix} \quad \text{is} \quad A^{-1} = \begin{bmatrix} 1 & -1 \\ -2 & 3 \end{bmatrix}$$

Solution We need to show that $AA^{-1} = A^{-1}A = I_2$.

$$AA^{-1} = \begin{bmatrix} 3 & 1 \\ 2 & 1 \end{bmatrix}\begin{bmatrix} 1 & -1 \\ -2 & 3 \end{bmatrix} = \begin{bmatrix} 1 & 0 \\ 0 & 1 \end{bmatrix} = I_2$$

$$A^{-1}A = \begin{bmatrix} 1 & -1 \\ -2 & 3 \end{bmatrix}\begin{bmatrix} 3 & 1 \\ 2 & 1 \end{bmatrix} = \begin{bmatrix} 1 & 0 \\ 0 & 1 \end{bmatrix} = I_2$$

We now show one way to find the inverse of

$$A = \begin{bmatrix} 3 & 1 \\ 2 & 1 \end{bmatrix}$$

Suppose that A^{-1} is given by

$$A^{-1} = \begin{bmatrix} x & y \\ z & w \end{bmatrix} \qquad \textbf{(1)}$$

where x, y, z, and w are four variables. Based on the definition of an inverse, if A has an inverse,

$$AA^{-1} = I_2$$

$$\begin{bmatrix} 3 & 1 \\ 2 & 1 \end{bmatrix}\begin{bmatrix} x & y \\ z & w \end{bmatrix} = \begin{bmatrix} 1 & 0 \\ 0 & 1 \end{bmatrix}$$

$$\begin{bmatrix} 3x + z & 3y + w \\ 2x + z & 2y + w \end{bmatrix} = \begin{bmatrix} 1 & 0 \\ 0 & 1 \end{bmatrix}$$

Because corresponding entries must be equal, it follows that this matrix equation is equivalent to two systems of linear equations.

$$\begin{cases} 3x + z = 1 \\ 2x + z = 0 \end{cases} \qquad \begin{cases} 3y + w = 0 \\ 2y + w = 1 \end{cases}$$

The augmented matrix of each system is

$$\begin{bmatrix} 3 & 1 & | & 1 \\ 2 & 1 & | & 0 \end{bmatrix} \qquad \begin{bmatrix} 3 & 1 & | & 0 \\ 2 & 1 & | & 1 \end{bmatrix} \qquad \textbf{(2)}$$

The usual procedure would be to transform each augmented matrix into reduced row echelon form. Notice, though, that the left sides of the augmented matrices are equal, so the same row operations (see Section 10.2) can be used to reduce each one. We find it more efficient to combine the two augmented matrices (2) into a single matrix, as shown next, and then transform it into reduced row echelon form.

$$\begin{bmatrix} 3 & 1 & | & 1 & 0 \\ 2 & 1 & | & 0 & 1 \end{bmatrix}$$

We attempt to transform the left side into an identity matrix.

$$\begin{bmatrix} 3 & 1 & | & 1 & 0 \\ 2 & 1 & | & 0 & 1 \end{bmatrix} \xrightarrow[\uparrow]{} \begin{bmatrix} 1 & 0 & | & 1 & -1 \\ 2 & 1 & | & 0 & 1 \end{bmatrix}$$
$$R_1 = -1r_2 + r_1$$

$$\xrightarrow[\uparrow]{} \begin{bmatrix} 1 & 0 & | & 1 & -1 \\ 0 & 1 & | & -2 & 3 \end{bmatrix} \qquad \textbf{(3)}$$
$$R_2 = -2r_1 + r_2$$

Matrix (3) is in reduced row echelon form.

Now we reverse the earlier step of combining the two augmented matrices in (2) and write the single matrix (3) as two augmented matrices.

$$\begin{bmatrix} 1 & 0 & | & 1 \\ 0 & 1 & | & -2 \end{bmatrix} \quad \text{and} \quad \begin{bmatrix} 1 & 0 & | & -1 \\ 0 & 1 & | & 3 \end{bmatrix}$$

We conclude from these matrices that $x = 1$, $z = -2$, and $y = -1$, $w = 3$. Substituting these values into matrix (1), we find that

$$A^{-1} = \begin{bmatrix} 1 & -1 \\ -2 & 3 \end{bmatrix}$$

Notice in display (3) that the 2 by 2 matrix to the right of the vertical bar is, in fact, the inverse of A. Also notice that the identity matrix I_2 is the matrix that appears to the left of the vertical bar. These observations and the procedures followed to get display (3) will work in general.

Procedure for Finding the Inverse of a Nonsingular Matrix

To find the inverse of an n by n nonsingular matrix A, proceed as follows:

STEP 1: Form the matrix $[A|I_n]$.
STEP 2: Transform the matrix $[A|I_n]$ into reduced row echelon form.
STEP 3: The reduced row echelon form of $[A|I_n]$ will contain the identity matrix I_n on the left of the vertical bar; the n by n matrix on the right of the vertical bar is the inverse of A.

In other words, if A is nonsingular, we begin with the matrix $[A|I_n]$ and, after transforming it into reduced row echelon form, we end up with the matrix $[I_n|A^{-1}]$. Let's look at another example.

EXAMPLE 13

Finding the Inverse of a Matrix

The matrix

$$A = \begin{bmatrix} 1 & 1 & 0 \\ -1 & 3 & 4 \\ 0 & 4 & 3 \end{bmatrix}$$

is nonsingular. Find its inverse.

Solution   First, we form the matrix

$$[A|I_3] = \left[\begin{array}{ccc|ccc} 1 & 1 & 0 & 1 & 0 & 0 \\ -1 & 3 & 4 & 0 & 1 & 0 \\ 0 & 4 & 3 & 0 & 0 & 1 \end{array}\right]$$

Next, we use row operations to transform $[A|I_3]$ into reduced row echelon form.

$$\left[\begin{array}{ccc|ccc} 1 & 1 & 0 & 1 & 0 & 0 \\ -1 & 3 & 4 & 0 & 1 & 0 \\ 0 & 4 & 3 & 0 & 0 & 1 \end{array}\right] \rightarrow \left[\begin{array}{ccc|ccc} 1 & 1 & 0 & 1 & 0 & 0 \\ 0 & 4 & 4 & 1 & 1 & 0 \\ 0 & 4 & 3 & 0 & 0 & 1 \end{array}\right] \rightarrow \left[\begin{array}{ccc|ccc} 1 & 1 & 0 & 1 & 0 & 0 \\ 0 & 1 & 1 & \frac{1}{4} & \frac{1}{4} & 0 \\ 0 & 4 & 3 & 0 & 0 & 1 \end{array}\right]$$

$$\uparrow \qquad\qquad\qquad\qquad\qquad\qquad\qquad \uparrow$$
$$R_2 = r_1 + r_2 \qquad\qquad\qquad\qquad\qquad R_2 = \frac{1}{4} r_2$$

$$\rightarrow \left[\begin{array}{ccc|ccc} 1 & 0 & -1 & \frac{3}{4} & -\frac{1}{4} & 0 \\ 0 & 1 & 1 & \frac{1}{4} & \frac{1}{4} & 0 \\ 0 & 0 & -1 & -1 & -1 & 1 \end{array}\right] \rightarrow \left[\begin{array}{ccc|ccc} 1 & 0 & -1 & \frac{3}{4} & -\frac{1}{4} & 0 \\ 0 & 1 & 1 & \frac{1}{4} & \frac{1}{4} & 0 \\ 0 & 0 & 1 & 1 & 1 & -1 \end{array}\right]$$

$$\uparrow \qquad\qquad\qquad\qquad\qquad\qquad\qquad \uparrow$$
$$R_1 = -1r_2 + r_1 \qquad\qquad\qquad\qquad R_3 = -1r_3$$
$$R_3 = -4r_2 + r_3$$

$$\rightarrow \left[\begin{array}{ccc|ccc} 1 & 0 & 0 & \frac{7}{4} & \frac{3}{4} & -1 \\ 0 & 1 & 0 & -\frac{3}{4} & -\frac{3}{4} & 1 \\ 0 & 0 & 1 & 1 & 1 & -1 \end{array}\right]$$

$$\uparrow$$
$$R_1 = r_3 + r_1$$
$$R_2 = -1r_3 + r_2$$

The matrix $[A \mid I_3]$ is now in reduced row echelon form, and the identity matrix I_3 is on the left of the vertical bar. The inverse of A is

$$A^{-1} = \begin{bmatrix} \dfrac{7}{4} & \dfrac{3}{4} & -1 \\[2mm] -\dfrac{3}{4} & -\dfrac{3}{4} & 1 \\[2mm] 1 & 1 & -1 \end{bmatrix}$$

Figure 8

You can (and should) verify that this is the correct inverse by showing that

$$AA^{-1} = A^{-1}A = I_3$$

✓**Check:** Enter the matrix A into a graphing utility. Figure 8 shows A^{-1}.

Now Work PROBLEM 33

If transforming the matrix $[A \mid I_n]$ into reduced row echelon form does not result in the identity matrix I_n to the left of the vertical bar, A is singular and has no inverse. The next example demonstrates such a matrix.

EXAMPLE 14

Showing That a Matrix Has No Inverse

Show that the matrix $A = \begin{bmatrix} 4 & 6 \\ 2 & 3 \end{bmatrix}$ has no inverse.

Solution

Proceeding as in Example 13, we form the matrix

$$[A \mid I_2] = \begin{bmatrix} 4 & 6 & \vline & 1 & 0 \\ 2 & 3 & \vline & 0 & 1 \end{bmatrix}$$

Then we use row operations to transform $[A \mid I_2]$ into reduced row echelon form.

$$[A \mid I_2] = \begin{bmatrix} 4 & 6 & \vline & 1 & 0 \\ 2 & 3 & \vline & 0 & 1 \end{bmatrix} \underset{\substack{\uparrow \\ R_1 = \frac{1}{4}r_1}}{\rightarrow} \begin{bmatrix} 1 & \dfrac{3}{2} & \vline & \dfrac{1}{4} & 0 \\[2mm] 2 & 3 & \vline & 0 & 1 \end{bmatrix} \underset{\substack{\uparrow \\ R_2 = -2r_1 + r_2}}{\rightarrow} \begin{bmatrix} 1 & \dfrac{3}{2} & \vline & \dfrac{1}{4} & 0 \\[2mm] 0 & 0 & \vline & -\dfrac{1}{2} & 1 \end{bmatrix}$$

The matrix $[A \mid I_2]$ is sufficiently reduced for us to see that the identity matrix cannot appear to the left of the vertical bar. We conclude that A is singular and so has no inverse.

It can be shown that if the determinant of a matrix is 0 then the matrix is singular. The determinant of matrix A from Example 14 is

$$\begin{vmatrix} 4 & 6 \\ 2 & 3 \end{vmatrix} = 4 \cdot 3 - 6 \cdot 2 = 0$$

✓**Check:** Enter the matrix A. Try to find its inverse. What happens?

Now Work PROBLEM 61

5 Solve a System of Linear Equations Using an Inverse Matrix

Inverse matrices can be used to solve systems of equations in which the number of equations is the same as the number of variables.

EXAMPLE 15 **Using the Inverse Matrix to Solve a System of Linear Equations**

Solve the system of equations: $\begin{cases} x + y = 3 \\ -x + 3y + 4z = -3 \\ 4y + 3z = 2 \end{cases}$

Solution If we let

$$A = \begin{bmatrix} 1 & 1 & 0 \\ -1 & 3 & 4 \\ 0 & 4 & 3 \end{bmatrix} \quad X = \begin{bmatrix} x \\ y \\ z \end{bmatrix} \quad B = \begin{bmatrix} 3 \\ -3 \\ 2 \end{bmatrix}$$

the original system of equations can be written compactly as the matrix equation

$$AX = B \tag{4}$$

We know from Example 13 that the matrix A has the inverse A^{-1}, so we multiply each side of equation (4) by A^{-1}.

$$AX = B$$
$$A^{-1}(AX) = A^{-1}B \qquad \text{Multiply both sides by } A^{-1}.$$
$$(A^{-1}A)X = A^{-1}B \qquad \text{Associative Property of multiplication}$$
$$I_3 X = A^{-1}B \qquad \text{Definition of an inverse matrix}$$
$$X = A^{-1}B \qquad \text{Property of the identity matrix} \tag{5}$$

Now we use (5) to find $X = \begin{bmatrix} x \\ y \\ z \end{bmatrix}$.

$$X = \begin{bmatrix} x \\ y \\ z \end{bmatrix} = A^{-1}B = \underset{\substack{\uparrow \\ \text{Example 13}}}{\begin{bmatrix} \frac{7}{4} & \frac{3}{4} & -1 \\ -\frac{3}{4} & -\frac{3}{4} & 1 \\ 1 & 1 & -1 \end{bmatrix}} \begin{bmatrix} 3 \\ -3 \\ 2 \end{bmatrix} = \begin{bmatrix} 1 \\ 2 \\ -2 \end{bmatrix}$$

The solution is $x = 1$, $y = 2$, $z = -2$ or, using an ordered triplet, $(1, 2, -2)$.

The method used in Example 15 to solve a system of equations is particularly useful when it is necessary to solve several systems of equations in which the constants appearing to the right of the equal signs change, while the coefficients of the variables on the left side remain the same. See Problems 41–60 for some illustrations. Be careful; this method can only be used if the inverse exists. If it does not exist, row reduction must be used since the system is either inconsistent or dependent.

Now Work PROBLEM 45

Historical Feature

Arthur Cayley
(1821–1895)

Matrices were invented in 1857 by Arthur Cayley (1821–1895) as a way of efficiently computing the result of substituting one linear system into another (see Historical Problem 3). The resulting system had incredible richness, in the sense that a wide variety of mathematical systems could be mimicked by the matrices. Cayley and his friend James J. Sylvester (1814–1897) spent much of the rest of their lives elaborating the theory. The torch was then passed to Georg Frobenius (1849–1917), whose deep investigations established a central place for matrices in modern mathematics. In 1924, rather to the surprise of physicists, it was found that matrices (with complex numbers in them) were exactly the right tool for describing the behavior of atomic systems. Today, matrices are used in a wide variety of applications.

Historical Problems

1. **Matrices and Complex Numbers** Frobenius emphasized in his research how matrices could be used to mimic other mathematical systems. Here, we mimic the behavior of complex numbers using matrices. Mathematicians call such a relationship an *isomorphism*.

$$\text{Complex number} \longleftrightarrow \text{Matrix}$$

$$a + bi \longleftrightarrow \begin{bmatrix} a & b \\ -b & a \end{bmatrix}$$

   Note that the complex number can be read off the top line of the matrix. Thus,

$$2 + 3i \longleftrightarrow \begin{bmatrix} 2 & 3 \\ -3 & 2 \end{bmatrix} \text{ and } \begin{bmatrix} 4 & -2 \\ 2 & 4 \end{bmatrix} \longleftrightarrow 4 - 2i$$

   (a) Find the matrices corresponding to $2 - 5i$ and $1 + 3i$.

   (b) Multiply the two matrices.

   (c) Find the corresponding complex number for the matrix found in part (b).

   (d) Multiply $2 - 5i$ and $1 + 3i$. The result should be the same as that found in part (c).

   The process also works for addition and subtraction. Try it for yourself.

2. Compute $(a + bi)(a - bi)$ using matrices. Interpret the result.

3. **Cayley's Definition of Matrix Multiplication** Cayley invented matrix multiplication to simplify the following problem:

$$\begin{cases} u = ar + bs \\ v = cr + ds \end{cases} \quad \begin{cases} x = ku + lv \\ y = mu + nv \end{cases}$$

   (a) Find x and y in terms of r and s by substituting u and v from the first system of equations into the second system of equations.

   (b) Use the result of part (a) to find the 2 by 2 matrix A in

$$\begin{bmatrix} x \\ y \end{bmatrix} = A \begin{bmatrix} r \\ s \end{bmatrix}$$

   (c) Now look at the following way to do it. Write the equations in matrix form.

$$\begin{bmatrix} u \\ v \end{bmatrix} = \begin{bmatrix} a & b \\ c & d \end{bmatrix} \begin{bmatrix} r \\ s \end{bmatrix} \qquad \begin{bmatrix} x \\ y \end{bmatrix} = \begin{bmatrix} k & l \\ m & n \end{bmatrix} \begin{bmatrix} u \\ v \end{bmatrix}$$

   So

$$\begin{bmatrix} x \\ y \end{bmatrix} = \begin{bmatrix} k & l \\ m & n \end{bmatrix} \begin{bmatrix} a & b \\ c & d \end{bmatrix} \begin{bmatrix} r \\ s \end{bmatrix}$$

   Do you see how Cayley defined matrix multiplication?

10.4 Assess Your Understanding

Concepts and Vocabulary

1. A matrix that has the same number of rows as columns is called a(n) _____ matrix.

2. *True or False* Matrix addition is commutative.

3. To find the product AB of two matrices A and B, the number of _____ in matrix A must equal the number of _____ in matrix B.

4. *True or False* Matrix multiplication is commutative.

5. Suppose that A is a square n by n matrix that is nonsingular. The matrix B such that $AB = BA = I_n$ is called the _____ of the matrix A.

6. If a matrix A has no inverse, it is called _____.

7. *True or False* The identity matrix has properties similar to those of the real number 1.

8. If $AX = B$ represents a matrix equation where A is a nonsingular matrix, then we can solve the equation using $X =$ _____.

Skill Building

In Problems 9–24, use the following matrices to evaluate the given expression.

$$A = \begin{bmatrix} 0 & 3 & -5 \\ 1 & 2 & 6 \end{bmatrix} \qquad B = \begin{bmatrix} 4 & 1 & 0 \\ -2 & 3 & -2 \end{bmatrix} \qquad C = \begin{bmatrix} 4 & 1 \\ 6 & 2 \\ -2 & 3 \end{bmatrix}$$

9. $A + B$

10. $A - B$

11. $4A$

12. $-3B$

13. $3A - 2B$

14. $2A + 4B$

15. AC

16. BC

17. CA **18.** CB **19.** $C(A + B)$ **20.** $(A + B)C$

21. $AC - 3I_2$ **22.** $CA + 5I_3$ **23.** $CA - CB$ **24.** $AC + BC$

In Problems 25–30, find the product.

25. $\begin{bmatrix} 2 & -2 \\ 1 & 0 \end{bmatrix}\begin{bmatrix} 2 & 1 & 4 & 6 \\ 3 & -1 & 3 & 2 \end{bmatrix}$

26. $\begin{bmatrix} 4 & 1 \\ 2 & 1 \end{bmatrix}\begin{bmatrix} -6 & 6 & 1 & 0 \\ 2 & 5 & 4 & -1 \end{bmatrix}$

27. $\begin{bmatrix} 1 & 2 & 3 \\ 0 & -1 & 4 \end{bmatrix}\begin{bmatrix} 1 & 2 \\ -1 & 0 \\ 2 & 4 \end{bmatrix}$

28. $\begin{bmatrix} 1 & -1 \\ -3 & 2 \\ 0 & 5 \end{bmatrix}\begin{bmatrix} 2 & 8 & -1 \\ 3 & 6 & 0 \end{bmatrix}$

29. $\begin{bmatrix} 1 & 0 & 1 \\ 2 & 4 & 1 \\ 3 & 6 & 1 \end{bmatrix}\begin{bmatrix} 1 & 3 \\ 6 & 2 \\ 8 & -1 \end{bmatrix}$

30. $\begin{bmatrix} 4 & -2 & 3 \\ 0 & 1 & 2 \\ -1 & 0 & 1 \end{bmatrix}\begin{bmatrix} 2 & 6 \\ 1 & -1 \\ 0 & 2 \end{bmatrix}$

In Problems 31–40, each matrix is nonsingular. Find the inverse of each matrix.

31. $\begin{bmatrix} 2 & 1 \\ 1 & 1 \end{bmatrix}$ **32.** $\begin{bmatrix} 3 & -1 \\ -2 & 1 \end{bmatrix}$ **33.** $\begin{bmatrix} 6 & 5 \\ 2 & 2 \end{bmatrix}$ **34.** $\begin{bmatrix} -4 & 1 \\ 6 & -2 \end{bmatrix}$ **35.** $\begin{bmatrix} 2 & 1 \\ a & a \end{bmatrix}$ $a \neq 0$

36. $\begin{bmatrix} b & 3 \\ b & 2 \end{bmatrix}$ $b \neq 0$ **37.** $\begin{bmatrix} 1 & -1 & 1 \\ 0 & -2 & 1 \\ -2 & -3 & 0 \end{bmatrix}$ **38.** $\begin{bmatrix} 1 & 0 & 2 \\ -1 & 2 & 3 \\ 1 & -1 & 0 \end{bmatrix}$ **39.** $\begin{bmatrix} 1 & 1 & 1 \\ 3 & 2 & -1 \\ 3 & 1 & 2 \end{bmatrix}$ **40.** $\begin{bmatrix} 3 & 3 & 1 \\ 1 & 2 & 1 \\ 2 & -1 & 1 \end{bmatrix}$

In Problems 41–60, use the inverses found in Problems 31–40, to solve each system of equations.

41. $\begin{cases} 2x + y = 8 \\ x + y = 5 \end{cases}$ **42.** $\begin{cases} 3x - y = 8 \\ -2x + y = 4 \end{cases}$ **43.** $\begin{cases} 2x + y = 0 \\ x + y = 5 \end{cases}$ **44.** $\begin{cases} 3x - y = 4 \\ -2x + y = 5 \end{cases}$

45. $\begin{cases} 6x + 5y = 7 \\ 2x + 2y = 2 \end{cases}$ **46.** $\begin{cases} -4x + y = 0 \\ 6x - 2y = 14 \end{cases}$ **47.** $\begin{cases} 6x + 5y = 13 \\ 2x + 2y = 5 \end{cases}$ **48.** $\begin{cases} -4x + y = 5 \\ 6x - 2y = -9 \end{cases}$

49. $\begin{cases} 2x + y = -3 \\ ax + ay = -a \end{cases}$ $a \neq 0$ **50.** $\begin{cases} bx + 3y = 2b + 3 \\ bx + 2y = 2b + 2 \end{cases}$ $b \neq 0$ **51.** $\begin{cases} 2x + y = \dfrac{7}{a} \\ ax + ay = 5 \end{cases}$ $a \neq 0$ **52.** $\begin{cases} bx + 3y = 14 \\ bx + 2y = 10 \end{cases}$ $b \neq 0$

53. $\begin{cases} x - y + z = 0 \\ -2y + z = -1 \\ -2x - 3y = -5 \end{cases}$ **54.** $\begin{cases} x + 2z = 6 \\ -x + 2y + 3z = -5 \\ x - y = 6 \end{cases}$ **55.** $\begin{cases} x - y + z = 2 \\ -2y + z = 2 \\ -2x - 3y = \dfrac{1}{2} \end{cases}$ **56.** $\begin{cases} x + 2z = 2 \\ -x + 2y + 3z = -\dfrac{3}{2} \\ x - y = 2 \end{cases}$

57. $\begin{cases} x + y + z = 9 \\ 3x + 2y - z = 8 \\ 3x + y + 2z = 1 \end{cases}$ **58.** $\begin{cases} 3x + 3y + z = 8 \\ x + 2y + z = 5 \\ 2x - y + z = 4 \end{cases}$ **59.** $\begin{cases} x + y + z = 2 \\ 3x + 2y - z = \dfrac{7}{3} \\ 3x + y + 2z = \dfrac{10}{3} \end{cases}$ **60.** $\begin{cases} 3x + 3y + z = 1 \\ x + 2y + z = 0 \\ 2x - y + z = 4 \end{cases}$

In Problems 61–66, show that each matrix has no inverse.

61. $\begin{bmatrix} 4 & 2 \\ 2 & 1 \end{bmatrix}$ **62.** $\begin{bmatrix} -3 & \dfrac{1}{2} \\ 6 & -1 \end{bmatrix}$ **63.** $\begin{bmatrix} 15 & 3 \\ 10 & 2 \end{bmatrix}$

64. $\begin{bmatrix} -3 & 0 \\ 4 & 0 \end{bmatrix}$ **65.** $\begin{bmatrix} -3 & 1 & -1 \\ 1 & -4 & -7 \\ 1 & 2 & 5 \end{bmatrix}$ **66.** $\begin{bmatrix} 1 & 1 & -3 \\ 2 & -4 & 1 \\ -5 & 7 & 1 \end{bmatrix}$

In Problems 67–70, use a graphing utility to find the inverse, if it exists, of each matrix. Round answers to two decimal places.

67. $\begin{bmatrix} 25 & 61 & -12 \\ 18 & -2 & 4 \\ 8 & 35 & 21 \end{bmatrix}$

68. $\begin{bmatrix} 18 & -3 & 4 \\ 6 & -20 & 14 \\ 10 & 25 & -15 \end{bmatrix}$

69. $\begin{bmatrix} 44 & 21 & 18 & 6 \\ -2 & 10 & 15 & 5 \\ 21 & 12 & -12 & 4 \\ -8 & -16 & 4 & 9 \end{bmatrix}$

70. $\begin{bmatrix} 16 & 22 & -3 & 5 \\ 21 & -17 & 4 & 8 \\ 2 & 8 & 27 & 20 \\ 5 & 15 & -3 & -10 \end{bmatrix}$

In Problems 71–74, use the idea behind Example 15 with a graphing utility to solve the following systems of equations. Round answers to two decimal places.

71. $\begin{cases} 25x + 61y - 12z = 10 \\ 18x - 12y + 7y = -9 \\ 3x + 4y - z = 12 \end{cases}$

72. $\begin{cases} 25x + 61y - 12z = 15 \\ 18x - 12y + 7z = -3 \\ 3x + 4y - z = 12 \end{cases}$

73. $\begin{cases} 25x + 61y - 12z = 21 \\ 18x - 12y + 7z = 7 \\ 3x + 4y - z = -2 \end{cases}$

74. $\begin{cases} 25x + 61y - 12z = 25 \\ 18x - 12y + 7z = 10 \\ 3x + 4y - z = -4 \end{cases}$

Mixed Practice

In Problems 75–82, algebraically solve each system of equations using any method you wish.

75. $\begin{cases} 2x + 3y = 11 \\ 5x + 7y = 24 \end{cases}$

76. $\begin{cases} 2x + 8y = -8 \\ x + 7y = -13 \end{cases}$

77. $\begin{cases} x - 2y + 4z = 2 \\ -3x + 5y - 2z = 17 \\ 4x - 3y = -22 \end{cases}$

78. $\begin{cases} 2x + 3y - z = -2 \\ 4x + 3z = 6 \\ 6y - 2z = 2 \end{cases}$

79. $\begin{cases} 5x - y + 4z = 2 \\ -x + 5y - 4z = 3 \\ 7x + 13y - 4z = 17 \end{cases}$

80. $\begin{cases} 3x + 2y - z = 2 \\ 2x + y + 6z = -7 \\ 2x + 2y - 14z = 17 \end{cases}$

81. $\begin{cases} 2x - 3y + z = 4 \\ -3x + 2y - z = -3 \\ -5y + z = 6 \end{cases}$

82. $\begin{cases} -4x + 3y + 2z = 6 \\ 3x + y - z = -2 \\ x + 9y + z = 6 \end{cases}$

Applications and Extensions

83. College Tuition Nikki and Joe take classes at a community college, LCCC, and a local university, SIUE. The number of credit hours taken and the cost per credit hour (2009–2010 academic year, tuition only) are as follows:

	LCCC	SIUE
Nikki	6	9
Joe	3	12

Cost per Credit Hour	
LCCC	$80.00
SIUE	$277.80

(a) Write a matrix A for the credit hours taken by each student and a matrix B for the cost per credit hour.
(b) Compute AB and interpret the results.

Sources: www.lc.edu, www.siue.edu

84. School Loan Interest Jamal and Stephanie each have school loans issued from the same two banks. The amounts borrowed and the monthly interest rates are given next (interest is compounded monthly):

	Lender 1	Lender 2
Jamal	$4000	$3000
Stephanie	$2500	$3800

Monthly Interest Rate	
Lender 1	0.011 (1.1%)
Lender 2	0.006 (0.6%)

(a) Write a matrix A for the amounts borrowed by each student and a matrix B for the monthly interest rates.
(b) Compute AB and interpret the results.
(c) Let $C = \begin{bmatrix} 1 \\ 1 \end{bmatrix}$. Compute $A(C + B)$ and interpret the results.

85. Computing the Cost of Production The Acme Steel Company is a producer of stainless steel and aluminum containers. On a certain day, the following stainless steel containers were manufactured: 500 with 10-gallon (gal) capacity, 350 with 5-gal capacity, and 400 with 1-gal capacity. On the same day, the following aluminum containers were manufactured: 700 with 10-gal capacity, 500 with 5-gal capacity, and 850 with 1-gal capacity.
(a) Find a 2 by 3 matrix representing these data. Find a 3 by 2 matrix to represent the same data.
(b) If the amount of material used in the 10-gal containers is 15 pounds (lb), the amount used in the 5-gal containers is 8 lb, and the amount used in the 1-gal containers is 3 lb, find a 3 by 1 matrix representing the amount of material used.
(c) Multiply the 2 by 3 matrix found in part (a) and the 3 by 1 matrix found in part (b) to get a 2 by 1 matrix showing the day's usage of material.
(d) If stainless steel costs Acme $0.10 lb and aluminum costs $0.05 lb, find a 1 by 2 matrix representing cost.
(e) Multiply the matrices found in parts (c) and (d) to determine the total cost of the day's production.

86. Computing Profit Rizza Ford has two locations, one in the city and the other in the suburbs. In January, the city location sold 400 subcompacts, 250 intermediate-size cars, and 50 SUVs; in February, it sold 350 subcompacts, 100 intermediates, and 30 SUVs. At the suburban location in January, 450 subcompacts, 200 intermediates, and 140 SUVs were sold. In February, the suburban location sold 350 subcompacts, 300 intermediates, and 100 SUVs.
(a) Find 2 by 3 matrices that summarize the sales data for each location for January and February (one matrix for each month).

(b) Use matrix addition to obtain total sales for the 2-month period.

(c) The profit on each kind of car is $100 per subcompact, $150 per intermediate, and $200 per SUV. Find a 3 by 1 matrix representing this profit.

(d) Multiply the matrices found in parts (b) and (c) to get a 2 by 1 matrix showing the profit at each location.

87. Cryptography One method of encryption is to use a matrix to encrypt the message and then use the corresponding inverse matrix to decode the message. The encrypted matrix, E, is obtained by multiplying the message matrix, M, by a key matrix, K. The original message can be retrieved by multiplying the encrypted matrix by the inverse of the key matrix. That is, $E = M \cdot K$ and $M = E \cdot K^{-1}$.

(a) Given the key matrix $K = \begin{bmatrix} 2 & 1 & 1 \\ 1 & 1 & 0 \\ 1 & 1 & 1 \end{bmatrix}$, find its inverse, K^{-1}. [**Note:** This key matrix is known as the Q_2^3 Fibonacci encryption matrix.]

(b) Use your result from part (a) to decode the encrypted matrix $E = \begin{bmatrix} 47 & 34 & 33 \\ 44 & 36 & 27 \\ 47 & 41 & 20 \end{bmatrix}$.

(c) Each entry in your result for part (b) represents the position of a letter in the English alphabet ($A = 1$, $B = 2$, $C = 3$, and so on). What is the original message?

Source: goldenmuseum.com

88. Economic Mobility The relative income of a child (low, medium, or high) generally depends on the relative income of the child's parents. The matrix P, given by

Parent's Income

$$P = \begin{array}{c} \\ \end{array} \begin{array}{ccc} \text{L} & \text{M} & \text{H} \\ \begin{bmatrix} 0.4 & 0.2 & 0.1 \\ 0.5 & 0.6 & 0.5 \\ 0.1 & 0.2 & 0.4 \end{bmatrix} & \begin{array}{l} \text{L} \\ \text{M} \\ \text{H} \end{array} \end{array} \text{Child's income}$$

is called a *left stochastic transition matrix*. For example, the entry $P_{21} = 0.5$ means that 50% of the children of low relative income parents will transition to the medium level of income. The diagonal entry P_{ii} represents the percent of children who remain in the same income level as their parents. Assuming that the transition matrix is valid from one generation to the next, compute and interpret P^2.

Source: Understanding Mobility in America, April 2006

89. Consider the 2 by 2 square matrix

$$A = \begin{bmatrix} a & b \\ c & d \end{bmatrix}$$

If $D = ad - bc \neq 0$, show that A is nonsingular and that

$$A^{-1} = \frac{1}{D} \begin{bmatrix} d & -b \\ -c & a \end{bmatrix}$$

Explaining Concepts: Discussion and Writing

90. Create a situation different from any found in the text that can be represented by a matrix.

91. Explain why the number of columns in matrix A must equal the number of rows in matrix B when finding the product AB.

92. If $a, b,$ and $c \neq 0$ are real numbers with $ac = bc$, then $a = b$. Does this same property hold for matrices? In other words, if $A, B,$ and C, are matrices and $AC = BC$, must $A = B$?

10.5 Partial Fraction Decomposition

PREPARING FOR THIS SECTION *Before getting started, review the following:*

- Identity (Appendix A, Section A.8, p. A65)
- Proper and Improper Rational Functions (Section 3.2, p. 220)
- Factoring Polynomials (Appendix A, Section A.4, pp. A33–A39)
- Complex Zeros; Fundamental Theorem of Algebra (Section 3.6, p. 263)

Now Work the 'Are You Prepared?' problems on page 794.

OBJECTIVES 1 Decompose $\dfrac{P}{Q}$, Where Q Has Only Nonrepeated Linear Factors (p. 788)

2 Decompose $\dfrac{P}{Q}$, Where Q Has Repeated Linear Factors (p. 790)

3 Decompose $\dfrac{P}{Q}$, Where Q Has a Nonrepeated Irreducible Quadratic Factor (p. 792)

4 Decompose $\dfrac{P}{Q}$, Where Q Has a Repeated Irreducible Quadratic Factor (p. 793)

Consider the problem of adding two rational expressions:

$$\frac{3}{x + 4} \quad \text{and} \quad \frac{2}{x - 3}$$

The result is

$$\frac{3}{x + 4} + \frac{2}{x - 3} = \frac{3(x - 3) + 2(x + 4)}{(x + 4)(x - 3)} = \frac{5x - 1}{x^2 + x - 12}$$

The reverse procedure, starting with the rational expression $\dfrac{5x - 1}{x^2 + x - 12}$ and writing it as the sum (or difference) of the two simpler fractions $\dfrac{3}{x + 4}$ and $\dfrac{2}{x - 3}$, is referred to as **partial fraction decomposition,** and the two simpler fractions are called **partial fractions.** Decomposing a rational expression into a sum of partial fractions is important in solving certain types of calculus problems. This section presents a systematic way to decompose rational expressions.

 We begin by recalling that a rational expression is the ratio of two polynomials, say, P and $Q \neq 0$. We assume that P and Q have no common factors. Recall also that a rational expression $\dfrac{P}{Q}$ is called **proper** if the degree of the polynomial in the numerator is less than the degree of the polynomial in the denominator. Otherwise, the rational expression is termed **improper.**

Because any improper rational expression can be reduced by long division to a mixed form consisting of the sum of a polynomial and a proper rational expression, we shall restrict the discussion that follows to proper rational expressions.

The partial fraction decomposition of the rational expression $\dfrac{P}{Q}$ depends on the factors of the denominator Q. Recall from Section 3.6 that any polynomial whose coefficients are real numbers can be factored (over the real numbers) into products of linear and/or irreducible quadratic factors. This means that the denominator Q of the rational expression $\dfrac{P}{Q}$ will contain only factors of one or both of the following types:

1. *Linear factors* of the form $x - a$, where a is a real number.
2. *Irreducible quadratic factors* of the form $ax^2 + bx + c$, where a, b, and c are real numbers, $a \neq 0$, and $b^2 - 4ac < 0$ (which guarantees that $ax^2 + bx + c$ cannot be written as the product of two linear factors with real coefficients).

As it turns out, there are four cases to be examined. We begin with the case for which Q has only nonrepeated linear factors.

1 Decompose $\dfrac{P}{Q}$, Where Q Has Only Nonrepeated Linear Factors

Case 1:   Q has only nonrepeated linear factors.

Under the assumption that Q has only nonrepeated linear factors, the polynomial Q has the form

$$Q(x) = (x - a_1)(x - a_2) \cdots \cdot (x - a_n)$$

where none of the numbers $a_1, a_2, \ldots, a_n$ is equal. In this case, the partial fraction decomposition of $\dfrac{P}{Q}$ is of the form

$$\frac{P(x)}{Q(x)} = \frac{A_1}{x - a_1} + \frac{A_2}{x - a_2} + \cdots + \frac{A_n}{x - a_n} \qquad \textbf{(1)}$$

where the numbers $A_1, A_2, \ldots, A_n$ are to be determined.

We show how to find these numbers in the example that follows.

EXAMPLE 1 **Nonrepeated Linear Factors**

Write the partial fraction decomposition of $\dfrac{x}{x^2 - 5x + 6}$.

Solution First, we factor the denominator,

$$x^2 - 5x + 6 = (x - 2)(x - 3)$$

and conclude that the denominator contains only nonrepeated linear factors. Then we decompose the rational expression according to equation (1):

$$\frac{x}{x^2 - 5x + 6} = \frac{A}{x - 2} + \frac{B}{x - 3} \qquad \textbf{(2)}$$

where A and B are to be determined. To find A and B, we clear the fractions by multiplying each side by $(x - 2)(x - 3) = x^2 - 5x + 6$. The result is

$$x = A(x - 3) + B(x - 2) \qquad \textbf{(3)}$$

or

$$x = (A + B)x + (-3A - 2B)$$

This equation is an identity in x. We equate the coefficients of like powers of x to get

$$\begin{cases} 1 = A + B & \text{Equate coefficients of } x\text{: } 1x = (A + B)x. \\ 0 = -3A - 2B & \text{Equate the constants: } 0 = -3A - 2B. \end{cases}$$

This system of two equations containing two variables, A and B, can be solved using whatever method you wish. Solving it, we get

$$A = -2 \qquad B = 3$$

From equation (2), the partial fraction decomposition is

$$\frac{x}{x^2 - 5x + 6} = \frac{-2}{x - 2} + \frac{3}{x - 3}$$

✓**Check:** The decomposition can be checked by adding the rational expressions.

$$\frac{-2}{x - 2} + \frac{3}{x - 3} = \frac{-2(x - 3) + 3(x - 2)}{(x - 2)(x - 3)} = \frac{x}{(x - 2)(x - 3)}$$

$$= \frac{x}{x^2 - 5x + 6}$$

The numbers to be found in the partial fraction decomposition can sometimes be found more readily by using suitable choices for x (which may include complex numbers) in the identity obtained after fractions have been cleared. In Example 1, the identity after clearing fractions is equation (3):

$$x = A(x - 3) + B(x - 2)$$

If we let $x = 2$ in this expression, the term containing B drops out, leaving $2 = A(-1)$, or $A = -2$. Similarly, if we let $x = 3$, the term containing A drops out, leaving $3 = B$. As before, $A = -2$ and $B = 3$.

━━━━━**Now Work** PROBLEM **13**

2 Decompose $\dfrac{P}{Q}$, Where Q Has Repeated Linear Factors

Case 2: Q has repeated linear factors.

If the polynomial Q has a repeated linear factor, say $(x - a)^n$, $n \geq 2$ an integer, then, in the partial fraction decomposition of $\dfrac{P}{Q}$, we allow for the terms

$$\frac{A_1}{x - a} + \frac{A_2}{(x - a)^2} + \cdots + \frac{A_n}{(x - a)^n}$$

where the numbers $A_1, A_2, \ldots, A_n$ are to be determined.

| EXAMPLE 2 | **Repeated Linear Factors** |

Write the partial fraction decomposition of $\dfrac{x + 2}{x^3 - 2x^2 + x}$.

Solution First, we factor the denominator,

$$x^3 - 2x^2 + x = x(x^2 - 2x + 1) = x(x - 1)^2$$

and find that the denominator has the nonrepeated linear factor x and the repeated linear factor $(x - 1)^2$. By Case 1, we must allow for the term $\dfrac{A}{x}$ in the decomposition; and by Case 2 we must allow for the terms $\dfrac{B}{x - 1} + \dfrac{C}{(x - 1)^2}$ in the decomposition.

We write

$$\frac{x + 2}{x^3 - 2x^2 + x} = \frac{A}{x} + \frac{B}{x - 1} + \frac{C}{(x - 1)^2} \qquad \textbf{(4)}$$

Again, we clear fractions by multiplying each side by $x^3 - 2x^2 + x = x(x - 1)^2$. The result is the identity

$$x + 2 = A(x - 1)^2 + Bx(x - 1) + Cx \qquad \textbf{(5)}$$

If we let $x = 0$ in this expression, the terms containing B and C drop out, leaving $2 = A(-1)^2$, or $A = 2$. Similarly, if we let $x = 1$, the terms containing A and B drop out, leaving $3 = C$. Then equation (5) becomes

$$x + 2 = 2(x - 1)^2 + Bx(x - 1) + 3x$$

Now let $x = 2$ (any choice other than 0 or 1 will work as well). The result is

$$4 = 2(1)^2 + B(2)(1) + 3(2)$$
$$4 = 2 + 2B + 6$$
$$2B = -4$$
$$B = -2$$

We have $A = 2$, $B = -2$, and $C = 3$.
From equation (4), the partial fraction decomposition is

$$\frac{x + 2}{x^3 - 2x^2 + x} = \frac{2}{x} + \frac{-2}{x - 1} + \frac{3}{(x - 1)^2}$$

EXAMPLE 3   **Repeated Linear Factors**

Write the partial fraction decomposition of $\dfrac{x^3 - 8}{x^2(x - 1)^3}$.

Solution   The denominator contains the repeated linear factors x^2 and $(x - 1)^3$. The partial fraction decomposition takes the form

$$\frac{x^3 - 8}{x^2(x - 1)^3} = \frac{A}{x} + \frac{B}{x^2} + \frac{C}{x - 1} + \frac{D}{(x - 1)^2} + \frac{E}{(x - 1)^3} \quad (6)$$

As before, we clear fractions and obtain the identity

$$x^3 - 8 = Ax(x - 1)^3 + B(x - 1)^3 + Cx^2(x - 1)^2 + Dx^2(x - 1) + Ex^2 \quad (7)$$

Let $x = 0$. (Do you see why this choice was made?) Then

$$-8 = B(-1)$$
$$B = 8$$

Now let $x = 1$ in equation (7). Then

$$-7 = E$$

Use $B = 8$ and $E = -7$ in equation (7) and collect like terms.

$$x^3 - 8 = Ax(x - 1)^3 + 8(x - 1)^3$$
$$+ Cx^2(x - 1)^2 + Dx^2(x - 1) - 7x^2$$
$$x^3 - 8 - 8(x^3 - 3x^2 + 3x - 1) + 7x^2 = Ax(x - 1)^3$$
$$+ Cx^2(x - 1)^2 + Dx^2(x - 1)$$
$$-7x^3 + 31x^2 - 24x = x(x - 1)[A(x - 1)^2 + Cx(x - 1) + Dx]$$
$$x(x - 1)(-7x + 24) = x(x - 1)[A(x - 1)^2 + Cx(x - 1) + Dx]$$
$$-7x + 24 = A(x - 1)^2 + Cx(x - 1) + Dx \quad (8)$$

We now work with equation (8). Let $x = 0$. Then

$$24 = A$$

Now let $x = 1$ in equation (8). Then

$$17 = D$$

Use $A = 24$ and $D = 17$ in equation (8).

$$-7x + 24 = 24(x - 1)^2 + Cx(x - 1) + 17x$$

Now let $x = 2$ and simplify.

$$-14 + 24 = 24 + C(2) + 34$$
$$-48 = 2C$$
$$-24 = C$$

We now know all the numbers $A, B, C, D,$ and E, so, from equation (6), we have the decomposition

$$\frac{x^3 - 8}{x^2(x - 1)^3} = \frac{24}{x} + \frac{8}{x^2} + \frac{-24}{x - 1} + \frac{17}{(x - 1)^2} + \frac{-7}{(x - 1)^3}$$

●

━━━Now Work Example 3 by solving the system of five equations containing five variables that the expansion of equation (7) leads to.

━━━Now Work PROBLEM 19

The final two cases involve irreducible quadratic factors. A quadratic factor is irreducible if it cannot be factored into linear factors with real coefficients. A quadratic expression $ax^2 + bx + c$ is irreducible whenever $b^2 - 4ac < 0$. For example, $x^2 + x + 1$ and $x^2 + 4$ are irreducible.

3 Decompose $\dfrac{P}{Q}$, Where Q Has a Nonrepeated Irreducible Quadratic Factor

Case 3: Q contains a nonrepeated irreducible quadratic factor.

If Q contains a nonrepeated irreducible quadratic factor of the form $ax^2 + bx + c$, then, in the partial fraction decomposition of $\dfrac{P}{Q}$, allow for the term

$$\frac{Ax + B}{ax^2 + bx + c}$$

where the numbers A and B are to be determined.

EXAMPLE 4   **Nonrepeated Irreducible Quadratic Factor**

Write the partial fraction decomposition of $\dfrac{3x - 5}{x^3 - 1}$.

Solution   We factor the denominator,

$$x^3 - 1 = (x - 1)(x^2 + x + 1)$$

and find that it has a nonrepeated linear factor $x - 1$ and a nonrepeated irreducible quadratic factor $x^2 + x + 1$. We allow for the term $\dfrac{A}{x - 1}$ by Case 1, and we allow for the term $\dfrac{Bx + C}{x^2 + x + 1}$ by Case 3. We write

$$\frac{3x - 5}{x^3 - 1} = \frac{A}{x - 1} + \frac{Bx + C}{x^2 + x + 1} \qquad \textbf{(9)}$$

We clear fractions by multiplying each side of equation (9) by $x^3 - 1 = (x - 1)(x^2 + x + 1)$ to obtain the identity

$$3x - 5 = A(x^2 + x + 1) + (Bx + C)(x - 1) \qquad \textbf{(10)}$$

Expand the identity in (10) to obtain

$$3x - 5 = (A + B)x^2 + (A - B + C)x + (A - C)$$

This identity leads to the system of equations

$$\begin{cases} A + B & = & 0 & \text{(1)} \\ A - B + C & = & 3 & \text{(2)} \\ A \quad - C & = & -5 & \text{(3)} \end{cases}$$

The solution of this system is $A = -\dfrac{2}{3}, B = \dfrac{2}{3}, C = \dfrac{13}{3}$. Then, from equation (9), we see that

$$\frac{3x - 5}{x^3 - 1} = \frac{-\dfrac{2}{3}}{x - 1} + \frac{\dfrac{2}{3}x + \dfrac{13}{3}}{x^2 + x + 1}$$

━━━ **Now Work Example 4 using equation (10) and assigning values to x.**

━━━ **Now Work** PROBLEM 21

4 Decompose $\dfrac{P}{Q}$, Where Q Has a Repeated Irreducible Quadratic Factor

Case 4: Q contains a repeated irreducible quadratic factor.

If the polynomial Q contains a repeated irreducible quadratic factor $(ax^2 + bx + c)^n, n \geq 2, n$ an integer, then, in the partial fraction decomposition of $\dfrac{P}{Q}$, allow for the terms

$$\frac{A_1 x + B_1}{ax^2 + bx + c} + \frac{A_2 x + B_2}{(ax^2 + bx + c)^2} + \cdots + \frac{A_n x + B_n}{(ax^2 + bx + c)^n}$$

where the numbers $A_1, B_1, A_2, B_2, \ldots, A_n, B_n$ are to be determined.

EXAMPLE 5   **Repeated Irreducible Quadratic Factor**

Write the partial fraction decomposition of $\dfrac{x^3 + x^2}{(x^2 + 4)^2}$.

Solution   The denominator contains the repeated irreducible quadratic factor $(x^2 + 4)^2$, so we write

$$\frac{x^3 + x^2}{(x^2 + 4)^2} = \frac{Ax + B}{x^2 + 4} + \frac{Cx + D}{(x^2 + 4)^2} \qquad \textbf{(11)}$$

We clear fractions to obtain

$$x^3 + x^2 = (Ax + B)(x^2 + 4) + Cx + D$$

Collecting like terms yields the identity

$$x^3 + x^2 = Ax^3 + Bx^2 + (4A + C)x + 4B + D$$

Equating coefficients, we arrive at the system

$$\begin{cases} A = 1 \\ B = 1 \\ 4A + C = 0 \\ 4B + D = 0 \end{cases}$$

The solution is $A = 1, B = 1, C = -4, D = -4$. From equation (11),

$$\frac{x^3 + x^2}{(x^2 + 4)^2} = \frac{x + 1}{x^2 + 4} + \frac{-4x - 4}{(x^2 + 4)^2}$$

●

━━━━━**Now Work** PROBLEM 35

10.5 Assess Your Understanding

'Are You Prepared?' *Answers are given at the end of these exercises. If you get a wrong answer, read the pages listed in* red.

1. **True or False** The equation $(x - 1)^2 - 1 = x(x - 2)$ is an example of an identity. (p. A65)

2. **True or False** The rational expression $\dfrac{5x^2 - 1}{x^3 + 1}$ is proper. (p. 220)

3. Factor completely: $3x^4 + 6x^3 + 3x^2$ (pp. A33–A37)

4. **True or False** Every polynomial with real numbers as coefficients can be factored into products of linear and/or irreducible quadratic factors. (p. 263)

Skill Building

In Problems 5–12, tell whether the given rational expression is proper or improper. If improper, rewrite it as the sum of a polynomial and a proper rational expression.

5. $\dfrac{x}{x^2 - 1}$

6. $\dfrac{5x + 2}{x^3 - 1}$

7. $\dfrac{x^2 + 5}{x^2 - 4}$

8. $\dfrac{3x^2 - 2}{x^2 - 1}$

9. $\dfrac{5x^3 + 2x - 1}{x^2 - 4}$

10. $\dfrac{3x^4 + x^2 - 2}{x^3 + 8}$

11. $\dfrac{x(x - 1)}{(x + 4)(x - 3)}$

12. $\dfrac{2x(x^2 + 4)}{x^2 + 1}$

In Problems 13–46, write the partial fraction decomposition of each rational expression.

13. $\dfrac{4}{x(x - 1)}$

14. $\dfrac{3x}{(x + 2)(x - 1)}$

15. $\dfrac{1}{x(x^2 + 1)}$

16. $\dfrac{1}{(x + 1)(x^2 + 4)}$

17. $\dfrac{x}{(x - 1)(x - 2)}$

18. $\dfrac{3x}{(x + 2)(x - 4)}$

19. $\dfrac{x^2}{(x - 1)^2(x + 1)}$

20. $\dfrac{x + 1}{x^2(x - 2)}$

21. $\dfrac{1}{x^3 - 8}$

22. $\dfrac{2x + 4}{x^3 - 1}$

23. $\dfrac{x^2}{(x - 1)^2(x + 1)^2}$

24. $\dfrac{x + 1}{x^2(x - 2)^2}$

25. $\dfrac{x - 3}{(x + 2)(x + 1)^2}$

26. $\dfrac{x^2 + x}{(x + 2)(x - 1)^2}$

27. $\dfrac{x + 4}{x^2(x^2 + 4)}$

28. $\dfrac{10x^2 + 2x}{(x - 1)^2(x^2 + 2)}$

29. $\dfrac{x^2 + 2x + 3}{(x + 1)(x^2 + 2x + 4)}$

30. $\dfrac{x^2 - 11x - 18}{x(x^2 + 3x + 3)}$

31. $\dfrac{x}{(3x - 2)(2x + 1)}$

32. $\dfrac{1}{(2x + 3)(4x - 1)}$

33. $\dfrac{x}{x^2 + 2x - 3}$

34. $\dfrac{x^2 - x - 8}{(x + 1)(x^2 + 5x + 6)}$

35. $\dfrac{x^2 + 2x + 3}{(x^2 + 4)^2}$

36. $\dfrac{x^3 + 1}{(x^2 + 16)^2}$

37. $\dfrac{7x + 3}{x^3 - 2x^2 - 3x}$

38. $\dfrac{x^5 + 1}{x^6 - x^4}$

39. $\dfrac{x^2}{x^3 - 4x^2 + 5x - 2}$

40. $\dfrac{x^2 + 1}{x^3 + x^2 - 5x + 3}$

41. $\dfrac{x^3}{(x^2 + 16)^3}$

42. $\dfrac{x^2}{(x^2 + 4)^3}$

43. $\dfrac{4}{2x^2 - 5x - 3}$

44. $\dfrac{4x}{2x^2 + 3x - 2}$

45. $\dfrac{2x + 3}{x^4 - 9x^2}$

46. $\dfrac{x^2 + 9}{x^4 - 2x^2 - 8}$

'Are You Prepared?' Answers

1. True   2. True   3. $3x^2(x + 1)^2$ 4. True

10.6 Systems of Nonlinear Equations

PREPARING FOR THIS SECTION *Before getting started, review the following:*

- Lines (Foundations, Section 3, pp. 19–29)
- Circles (Foundations, Section 4, pp. 35–39)
- Parabolas (Section 9.2, pp. 667–671)
- Ellipses (Section 9.3, pp. 676–682)
- Hyperbolas (Section 9.4, pp. 686–694)

Now Work the 'Are You Prepared?' problems on page 800.

OBJECTIVES 1 Solve a System of Nonlinear Equations Using Substitution (p. 795)
2 Solve a System of Nonlinear Equations Using Elimination (p. 796)

1 Solve a System of Nonlinear Equations Using Substitution

In Section 10.1, we observed that the solution to a system of linear equations could be found geometrically by determining the point(s) of intersection (if any) of the equations in the system. Similarly, when solving systems of nonlinear equations, the solution(s) also represents the point(s) of intersection (if any) of the graphs of the equations.

There is no general methodology for solving a system of nonlinear equations. At times substitution is best; other times, elimination is best; and sometimes neither of these methods works. Experience and a certain degree of imagination are your allies here.

Before we begin, two comments are in order.

1. If the system contains two variables and if the equations in the system are easy to graph, then graph them. By graphing each equation in the system, you can get an idea of how many solutions a system has and approximately where they are located.

2. Extraneous solutions can creep in when solving nonlinear systems, so it is imperative that all apparent solutions be checked.

EXAMPLE 1 **Solving a System of Nonlinear Equations**

Solve the following system of equations:

$$\begin{cases} 3x - y = -2 & (1) \\ 2x^2 - y = 0 & (2) \end{cases}$$

Solution Using Substitution

Figure 9

First, we notice that the system contains two variables and that we know how to graph each equation. Equation (1) is the line $y = 3x + 2$ and equation (2) is the parabola $y = 2x^2$. See Figure 9. The system apparently has two solutions.

We will use substitution to solve the system. We choose to solve equation (1) for y.

$$3x - y = -2 \qquad \text{Equation (1)}$$
$$y = 3x + 2$$

We substitute this expression for y in equation (2). The result is an equation containing just the variable x, which we can then solve.

$$2x^2 - y = 0 \qquad \text{Equation (2)}$$
$$2x^2 - (3x + 2) = 0 \qquad \text{Substitute } 3x + 2 \text{ for } y.$$
$$2x^2 - 3x - 2 = 0 \qquad \text{Remove parentheses.}$$
$$(2x + 1)(x - 2) = 0 \qquad \text{Factor.}$$
$$2x + 1 = 0 \quad \text{or} \quad x - 2 = 0 \qquad \text{Apply the Zero-Product Property.}$$
$$x = -\frac{1}{2} \quad \text{or} \quad x = 2$$

Using these values for x in $y = 3x + 2$, we find

$$y = 3\left(-\frac{1}{2}\right) + 2 = \frac{1}{2} \quad \text{or} \quad y = 3(2) + 2 = 8$$

The apparent solutions are $x = -\frac{1}{2}$, $y = \frac{1}{2}$ and $x = 2$, $y = 8$.

✓ **Check:** For $x = -\frac{1}{2}$, $y = \frac{1}{2}$,

$$\begin{cases} 3\left(-\frac{1}{2}\right) - \frac{1}{2} = -\frac{3}{2} - \frac{1}{2} = -2 & (1) \\ 2\left(-\frac{1}{2}\right)^2 - \frac{1}{2} = 2\left(\frac{1}{4}\right) - \frac{1}{2} = 0 & (2) \end{cases}$$

For $x = 2$, $y = 8$,

$$\begin{cases} 3(2) - 8 = 6 - 8 = -2 & (1) \\ 2(2)^2 - 8 = 2(4) - 8 = 0 & (2) \end{cases}$$

Each solution checks. Now we know that the graphs in Figure 9 intersect at the points $\left(-\frac{1}{2}, \frac{1}{2}\right)$ and $(2, 8)$.

 ✓ **Check:** Graph $3x - y = -2$ ($Y_1 = 3x + 2$) and $2x^2 - y = 0$ ($Y_2 = 2x^2$) and compare what you see with Figure 9. Use INTERSECT (twice) to find the two points of intersection.

Now Work PROBLEM 15 USING SUBSTITUTION

2 Solve a System of Nonlinear Equations Using Elimination

Our next example illustrates how the method of elimination works for nonlinear systems.

EXAMPLE 2   **Solving a System of Nonlinear Equations**

Solve: $\begin{cases} x^2 + y^2 = 13 & (1) \\ x^2 - y = 7 & (2) \end{cases}$

Solution Using Elimination

Equation (1) is a circle and equation (2) is the parabola $y = x^2 - 7$. We graph each equation, as shown in Figure 10. Based on the graph, we expect four solutions. By subtracting equation (2) from equation (1), the variable x can be eliminated.

Figure 10

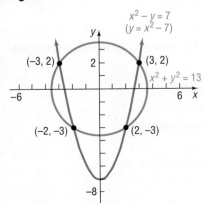

$$\begin{cases} x^2 + y^2 = 13 \\ x^2 - y = 7 \end{cases} \quad \text{Subtract.}$$
$$\overline{\quad\quad y^2 + y = 6}$$

This quadratic equation in y can be solved by factoring.

$$y^2 + y - 6 = 0$$
$$(y + 3)(y - 2) = 0$$
$$y = -3 \quad \text{or} \quad y = 2$$

We use these values for y in equation (2) to find x.

If $y = 2$, then $x^2 = y + 7 = 9$, so $x = 3$ or -3.
If $y = -3$, then $x^2 = y + 7 = 4$, so $x = 2$ or -2.

We have four solutions: $x = 3, y = 2$; $x = -3, y = 2$; $x = 2, y = -3$; and $x = -2, y = -3$.

You should verify that, in fact, these four solutions also satisfy equation (1), so all four are solutions of the system. The four points, $(3, 2)$, $(-3, 2)$, $(2, -3)$, and $(-2, -3)$, are the points of intersection of the graphs. Look again at Figure 10.

 ✓ **Check:** Graph $x^2 + y^2 = 13$ and $x^2 - y = 7$. (Remember that to graph $x^2 + y^2 = 13$ requires two functions: $Y_1 = \sqrt{13 - x^2}$ and $Y_2 = -\sqrt{13 - x^2}$.) Use INTERSECT to find the points of intersection.

━━━**Now Work** PROBLEM **13** USING ELIMINATION

EXAMPLE 3 **Solving a System of Nonlinear Equations**

Solve: $\begin{cases} x^2 - y^2 = 4 & (1) \\ y = x^2 & (2) \end{cases}$

Figure 11

Solution

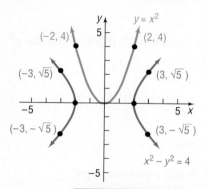

Equation (1) is a hyperbola and equation (2) is a parabola. See Figure 11. It appears the system has no solution. We verify this using substitution. Replace x^2 by y in equation (1). The result is

$$x^2 - y^2 = 4 \quad \text{Equation (1)}$$
$$y - y^2 = 4 \quad y = x^2$$
$$y^2 - y + 4 = 0 \quad \text{Place in standard form.}$$

This is a quadratic equation. Its discriminant is $(-1)^2 - 4 \cdot 1 \cdot 4 = 1 - 16 = -15 < 0$. The equation has no real solutions, so the system is inconsistent. The graphs of these two equations do not intersect.

EXAMPLE 4 **Solving a System of Nonlinear Equations**

Solve: $\begin{cases} x^2 + x + y^2 - 3y + 2 = 0 & (1) \\ x + 1 + \dfrac{y^2 - y}{x} = 0 & (2) \end{cases}$

Solution Using Elimination

First, we multiply equation (2) by x to eliminate the fraction. The result is an equivalent system because x cannot be 0. [Look at equation (2) to see why.]

$$\begin{cases} x^2 + x + y^2 - 3y + 2 = 0 & (1) \\ x^2 + x + y^2 - y = 0 & (2) \quad x \neq 0 \end{cases}$$

Now subtract equation (2) from equation (1) to eliminate x. The result is

$$-2y + 2 = 0$$
$$y = 1 \quad \text{Solve for y.}$$

To find x, we back-substitute $y = 1$ in equation (1):

$$x^2 + x + y^2 - 3y + 2 = 0 \quad \text{Equation (1)}$$
$$x^2 + x + 1 - 3 + 2 = 0 \quad \text{Substitute 1 for y in (1).}$$
$$x^2 + x = 0 \quad \text{Simplify.}$$
$$x(x + 1) = 0 \quad \text{Factor.}$$
$$x = 0 \quad \text{or} \quad x = -1 \quad \text{Apply the Zero-Product Property.}$$

Because x cannot be 0, the value $x = 0$ is extraneous, and we discard it.

✓**Check:** We now check $x = -1$, $y = 1$:

$$\begin{cases} (-1)^2 + (-1) + 1^2 - 3(1) + 2 = 1 - 1 + 1 - 3 + 2 = 0 & (1) \\ -1 + 1 + \dfrac{1^2 - 1}{-1} = 0 + \dfrac{0}{-1} = 0 & (2) \end{cases}$$

The solution is $x = -1$, $y = 1$. The point of intersection of the graphs of the equations is $(-1, 1)$.

●

In Problem 55 you are asked to graph the equations given in Example 4. Be sure to show holes in the graph of equation (2) for $x = 0$.

Now Work PROBLEMS 29 AND 49

EXAMPLE 5 **Solving a System of Nonlinear Equations**

Solve: $\begin{cases} 3xy - 2y^2 = -2 & (1) \\ 9x^2 + 4y^2 = 10 & (2) \end{cases}$

Solution We multiply equation (1) by 2 and add the result to equation (2) to eliminate the y^2 terms.

$$\begin{array}{rll} 6xy - 4y^2 &= -4 & (1) \\ 9x^2 + 4y^2 &= 10 & (2) \\ \hline 9x^2 + 6xy &= 6 & \text{Add.} \\ 3x^2 + 2xy &= 2 & \text{Divide each side by 3.} \end{array}$$

Since $x \neq 0$ (do you see why?), we can solve for y in this equation to get

$$y = \frac{2 - 3x^2}{2x} \qquad x \neq 0 \quad (3)$$

Now substitute for y in equation (2) of the system.

$$9x^2 + 4y^2 = 10 \qquad \text{Equation (2)}$$

$$9x^2 + 4\left(\frac{2 - 3x^2}{2x}\right)^2 = 10 \qquad \text{Substitute } y = \frac{2 - 3x^2}{2x} \text{ in (2).}$$

$$9x^2 + \frac{4 - 12x^2 + 9x^4}{x^2} = 10 \qquad \text{Expand and simplify.}$$

$$9x^4 + 4 - 12x^2 + 9x^4 = 10x^2 \qquad \text{Multiply both sides by } x^2.$$

$$18x^4 - 22x^2 + 4 = 0 \qquad \text{Subtract } 10x^2 \text{ from both sides.}$$

$$9x^4 - 11x^2 + 2 = 0 \qquad \text{Divide both sides by 2.}$$

This quadratic equation (in x^2) can be factored:

$$(9x^2 - 2)(x^2 - 1) = 0$$

$$9x^2 - 2 = 0 \qquad \text{or} \quad x^2 - 1 = 0$$

$$x^2 = \frac{2}{9} \qquad \text{or} \qquad x^2 = 1$$

$$x = \pm\sqrt{\frac{2}{9}} = \pm\frac{\sqrt{2}}{3} \quad \text{or} \qquad x = \pm 1$$

To find y, we use equation (3):

$$\text{If } x = \frac{\sqrt{2}}{3}: \quad y = \frac{2 - 3x^2}{2x} = \frac{2 - \dfrac{2}{3}}{2\left(\dfrac{\sqrt{2}}{3}\right)} = \frac{4}{2\sqrt{2}} = \sqrt{2}$$

$$\text{If } x = -\frac{\sqrt{2}}{3}: \quad y = \frac{2 - 3x^2}{2x} = \frac{2 - \dfrac{2}{3}}{2\left(-\dfrac{\sqrt{2}}{3}\right)} = \frac{4}{-2\sqrt{2}} = -\sqrt{2}$$

$$\text{If } x = 1: \quad y = \frac{2 - 3x^2}{2x} = \frac{2 - 3(1)^2}{2} = -\frac{1}{2}$$

$$\text{If } x = -1: \quad y = \frac{2 - 3x^2}{2x} = \frac{2 - 3(-1)^2}{-2} = \frac{1}{2}$$

The system has four solutions: $\left(\dfrac{\sqrt{2}}{3}, \sqrt{2}\right), \left(-\dfrac{\sqrt{2}}{3}, -\sqrt{2}\right), \left(1, -\dfrac{1}{2}\right), \left(-1, \dfrac{1}{2}\right)$. Check them for yourself.

Now Work PROBLEM 47

The next example illustrates an imaginative solution to a system of nonlinear equations.

EXAMPLE 6   **Running a Long Distance Race**

3 miles ───── 1 mile

In a 50-mile race, the winner crosses the finish line 1 mile ahead of the second-place runner and 4 miles ahead of the third-place runner. Assuming that each runner maintains a constant speed throughout the race, by how many miles does the second-place runner beat the third-place runner?

Solution   Let v_1, v_2, and v_3 denote the speeds of the first-, second-, and third-place runners, respectively. Let t_1 and t_2 denote the times (in hours) required for the first-place runner and second-place runner to finish the race. Then we have the system of equations

$$\begin{cases} 50 = v_1 t_1 & (1) \quad \textit{First-place runner goes 50 miles in } t_1 \textit{ hours.} \\ 49 = v_2 t_1 & (2) \quad \textit{Second-place runner goes 49 miles in } t_1 \textit{ hours.} \\ 46 = v_3 t_1 & (3) \quad \textit{Third-place runner goes 46 miles in } t_1 \textit{ hours.} \\ 50 = v_2 t_2 & (4) \quad \textit{Second-place runner goes 50 miles in } t_2 \textit{ hours.} \end{cases}$$

We seek the distance d of the third-place runner from the finish at time t_2. At time t_2, the third-place runner has gone a distance of $v_3 t_2$ miles, so the distance d remaining is $50 - v_3 t_2$. Now

$$d = 50 - v_3 t_2$$

$$= 50 - v_3 \left(t_1 \cdot \frac{t_2}{t_1}\right)$$

$$= 50 - (v_3 t_1) \cdot \frac{t_2}{t_1}$$

$$= 50 - 46 \cdot \dfrac{\dfrac{50}{v_2}}{\dfrac{50}{v_1}}$$

$$\begin{cases} \text{From (3), } v_3 t_1 = 46 \\[6pt] \text{From (4), } t_2 = \dfrac{50}{v_2} \\[6pt] \text{From (1), } t_1 = \dfrac{50}{v_1} \end{cases}$$

$$= 50 - 46 \cdot \dfrac{v_1}{v_2}$$

$$= 50 - 46 \cdot \dfrac{50}{49} \qquad \text{From the quotient of (1) and (2).}$$

$$\approx 3.06 \text{ miles}$$

Historical Feature

In the beginning of this section, we said that imagination and experience are important in solving systems of nonlinear equations. Indeed, these kinds of problems lead into some of the deepest and most difficult parts of modern mathematics. Look again at the graphs in Examples 1 and 2 of this section (Figures 9 and 10). We see that Example 1 has two solutions, and Example 2 has four solutions. We might conjecture that the number of solutions is equal to the product of the degrees of the equations involved. This conjecture was indeed made by Étienne Bézout (1730–1783), but working out the details took about 150 years. It turns out that, to arrive at the correct number of intersections, we must count not only the complex number intersections, but also those intersections that, in a certain sense, lie at infinity. For example, a parabola and a line lying on the axis of the parabola intersect at the vertex and at infinity. This topic is part of the study of algebraic geometry.

Historical Problem

A papyrus dating back to 1950 BC contains the following problem: "A given surface area of 100 units of area shall be represented as the sum of two squares whose sides are to each other as $1 : \dfrac{3}{4}$."

Solve for the sides by solving the system of equations

$$\begin{cases} x^2 + y^2 = 100 \\[4pt] x = \dfrac{3}{4} y \end{cases}$$

10.6 Assess Your Understanding

'Are You Prepared?' *Answers are given at the end of these exercises. If you get a wrong answer, read the pages listed in* red.

1. Graph the equation: $y = 3x + 2$ (pp. 21–22)
2. Graph the equation: $y + 4 = x^2$ (pp. 667–670)
3. Graph the equation: $y^2 = x^2 - 1$ (pp. 686–693)
4. Graph the equation: $x^2 + 4y^2 = 4$ (pp. 676–680)

Skill Building

In Problems 5–24, graph each equation of the system. Then solve the system to find the points of intersection.

5. $\begin{cases} y = x^2 + 1 \\ y = x + 1 \end{cases}$

6. $\begin{cases} y = x^2 + 1 \\ y = 4x + 1 \end{cases}$

7. $\begin{cases} y = \sqrt{36 - x^2} \\ y = 8 - x \end{cases}$

8. $\begin{cases} y = \sqrt{4 - x^2} \\ y = 2x + 4 \end{cases}$

9. $\begin{cases} y = \sqrt{x} \\ y = 2 - x \end{cases}$

10. $\begin{cases} y = \sqrt{x} \\ y = 6 - x \end{cases}$

11. $\begin{cases} x = 2y \\ x = y^2 - 2y \end{cases}$

12. $\begin{cases} y = x - 1 \\ y = x^2 - 6x + 9 \end{cases}$

13. $\begin{cases} x^2 + y^2 = 4 \\ x^2 + 2x + y^2 = 0 \end{cases}$

14. $\begin{cases} x^2 + y^2 = 8 \\ x^2 + y^2 + 4y = 0 \end{cases}$

15. $\begin{cases} y = 3x - 5 \\ x^2 + y^2 = 5 \end{cases}$

16. $\begin{cases} x^2 + y^2 = 10 \\ y = x + 2 \end{cases}$

17. $\begin{cases} x^2 + y^2 = 4 \\ y^2 - x = 4 \end{cases}$

18. $\begin{cases} x^2 + y^2 = 16 \\ x^2 - 2y = 8 \end{cases}$

19. $\begin{cases} xy = 4 \\ x^2 + y^2 = 8 \end{cases}$

20. $\begin{cases} x^2 = y \\ xy = 1 \end{cases}$

21. $\begin{cases} x^2 + y^2 = 4 \\ y = x^2 - 9 \end{cases}$

22. $\begin{cases} xy = 1 \\ y = 2x + 1 \end{cases}$

23. $\begin{cases} y = x^2 - 4 \\ y = 6x - 13 \end{cases}$

24. $\begin{cases} x^2 + y^2 = 10 \\ xy = 3 \end{cases}$

In Problems 25–54, solve each system. Use any method you wish.

25. $\begin{cases} 2x^2 + y^2 = 18 \\ \quad\ xy = 4 \end{cases}$

26. $\begin{cases} x^2 - y^2 = 21 \\ x + y = 7 \end{cases}$

27. $\begin{cases} \quad\quad\ y = 2x + 1 \\ 2x^2 + y^2 = 1 \end{cases}$

28. $\begin{cases} x^2 - 4y^2 = 16 \\ 2y - x = 2 \end{cases}$

29. $\begin{cases} x + y + 1 = 0 \\ x^2 + y^2 + 6y - x = -5 \end{cases}$

30. $\begin{cases} 2x^2 - xy + y^2 = 8 \\ \quad\quad\quad\ xy = 4 \end{cases}$

31. $\begin{cases} 4x^2 - 3xy + 9y^2 = 15 \\ \quad\quad\ 2x + 3y = 5 \end{cases}$

32. $\begin{cases} 2y^2 - 3xy + 6y + 2x + 4 = 0 \\ \quad\quad\quad\quad\ 2x - 3y + 4 = 0 \end{cases}$

33. $\begin{cases} x^2 - 4y^2 + 7 = 0 \\ 3x^2 + y^2 = 31 \end{cases}$

34. $\begin{cases} 3x^2 - 2y^2 + 5 = 0 \\ 2x^2 - y^2 + 2 = 0 \end{cases}$

35. $\begin{cases} 7x^2 - 3y^2 + 5 = 0 \\ 3x^2 + 5y^2 = 12 \end{cases}$

36. $\begin{cases} x^2 - 3y^2 + 1 = 0 \\ 2x^2 - 7y^2 + 5 = 0 \end{cases}$

37. $\begin{cases} x^2 + 2xy = 10 \\ 3x^2 - xy = 2 \end{cases}$

38. $\begin{cases} 5xy + 13y^2 + 36 = 0 \\ \quad\ xy + 7y^2 = 6 \end{cases}$

39. $\begin{cases} 2x^2 + y^2 = 2 \\ x^2 - 2y^2 + 8 = 0 \end{cases}$

40. $\begin{cases} y^2 - x^2 + 4 = 0 \\ 2x^2 + 3y^2 = 6 \end{cases}$

41. $\begin{cases} x^2 + 2y^2 = 16 \\ 4x^2 - y^2 = 24 \end{cases}$

42. $\begin{cases} 4x^2 + 3y^2 = 4 \\ 2x^2 - 6y^2 = -3 \end{cases}$

43. $\begin{cases} \dfrac{5}{x^2} - \dfrac{2}{y^2} + 3 = 0 \\ \dfrac{3}{x^2} + \dfrac{1}{y^2} = 7 \end{cases}$

44. $\begin{cases} \dfrac{2}{x^2} - \dfrac{3}{y^2} + 1 = 0 \\ \dfrac{6}{x^2} - \dfrac{7}{y^2} + 2 = 0 \end{cases}$

45. $\begin{cases} \dfrac{1}{x^4} + \dfrac{6}{y^4} = 6 \\ \dfrac{2}{x^4} - \dfrac{2}{y^4} = 19 \end{cases}$

46. $\begin{cases} \dfrac{1}{x^4} - \dfrac{1}{y^4} = 1 \\ \dfrac{1}{x^4} + \dfrac{1}{y^4} = 4 \end{cases}$

47. $\begin{cases} x^2 - 3xy + 2y^2 = 0 \\ x^2 + xy = 6 \end{cases}$

48. $\begin{cases} x^2 - xy - 2y^2 = 0 \\ xy + x + 6 = 0 \end{cases}$

49. $\begin{cases} y^2 + y + x^2 - x - 2 = 0 \\ y + 1 + \dfrac{x - 2}{y} = 0 \end{cases}$

50. $\begin{cases} x^3 - 2x^2 + y^2 + 3y - 4 = 0 \\ x - 2 + \dfrac{y^2 - y}{x^2} = 0 \end{cases}$

51. $\begin{cases} \log_x y = 3 \\ \log_x(4y) = 5 \end{cases}$

52. $\begin{cases} \log_x(2y) = 3 \\ \log_x(4y) = 2 \end{cases}$

53. $\begin{cases} \ln x = 4 \ln y \\ \log_3 x = 2 + 2 \log_3 y \end{cases}$

54. $\begin{cases} \ln x = 5 \ln y \\ \log_2 x = 3 + 2 \log_2 y \end{cases}$

55. Graph the equations given in Example 4.

56. Graph the equations given in Problem 49.

In Problems 57–64, use a graphing utility to solve each system of equations. Express the solution(s) rounded to two decimal places.

57. $\begin{cases} y = x^{2/3} \\ y = e^{-x} \end{cases}$

58. $\begin{cases} y = x^{3/2} \\ y = e^{-x} \end{cases}$

59. $\begin{cases} x^2 + y^3 = 2 \\ x^3 y = 4 \end{cases}$

60. $\begin{cases} x^3 + y^2 = 2 \\ x^2 y = 4 \end{cases}$

61. $\begin{cases} x^4 + y^4 = 12 \\ xy^2 = 2 \end{cases}$

62. $\begin{cases} x^4 + y^4 = 6 \\ xy = 1 \end{cases}$

63. $\begin{cases} xy = 2 \\ y = \ln x \end{cases}$

64. $\begin{cases} x^2 + y^2 = 4 \\ y = \ln x \end{cases}$

Mixed Practice

In Problems 65–70, graph each equation and find the point(s) of intersection, if any.

65. The line $x + 2y = 0$ and the circle $(x - 1)^2 + (y - 1)^2 = 5$

66. The line $x + 2y + 6 = 0$ and the circle $(x + 1)^2 + (y + 1)^2 = 5$

67. The circle $(x - 1)^2 + (y + 2)^2 = 4$ and the parabola $y^2 + 4y - x + 1 = 0$

68. The circle $(x + 2)^2 + (y - 1)^2 = 4$ and the parabola $y^2 - 2y - x - 5 = 0$

69. $y = \dfrac{4}{x - 3}$ and the circle $x^2 - 6x + y^2 + 1 = 0$

70. $y = \dfrac{4}{x + 2}$ and the circle $x^2 + 4x + y^2 - 4 = 0$

Applications and Extensions

71. The difference of two numbers is 2 and the sum of their squares is 10. Find the numbers.

72. The sum of two numbers is 7 and the difference of their squares is 21. Find the numbers.

73. The product of two numbers is 4 and the sum of their squares is 8. Find the numbers.

74. The product of two numbers is 10 and the difference of their squares is 21. Find the numbers.

75. The difference of two numbers is the same as their product, and the sum of their reciprocals is 5. Find the numbers.

76. The sum of two numbers is the same as their product, and the difference of their reciprocals is 3. Find the numbers.

77. The ratio of a to b is $\frac{2}{3}$. The sum of a and b is 10. What is the ratio of $a + b$ to $b - a$?

78. The ratio of a to b is $4:3$. The sum of a and b is 14. What is the ratio of $a - b$ to $a + b$?

79. Geometry The perimeter of a rectangle is 16 inches and its area is 15 square inches. What are its dimensions?

80. Geometry An area of 52 square feet is to be enclosed by two squares whose sides are in the ratio of $2:3$. Find the sides of the squares.

81. Geometry Two circles have circumferences that add up to 12π centimeters and areas that add up to 20π square centimeters. Find the radius of each circle.

82. Geometry The altitude of an isosceles triangle drawn to its base is 3 centimeters, and its perimeter is 18 centimeters. Find the length of its base.

83. The Tortoise and the Hare In a 21-meter race between a tortoise and a hare, the tortoise leaves 9 minutes before the hare. The hare, by running at an average speed of 0.5 meter per hour faster than the tortoise, crosses the finish line 3 minutes before the tortoise. What are the average speeds of the tortoise and the hare?

84. Running a Race In a 1-mile race, the winner crosses the finish line 10 feet ahead of the second-place runner and 20 feet ahead of the third-place runner. Assuming that each runner maintains a constant speed throughout the race, by how many feet does the second-place runner beat the third-place runner?

85. Constructing a Box A rectangular piece of cardboard, whose area is 216 square centimeters, is made into an open box by cutting a 2-centimeter square from each corner and turning up the sides. See the figure. If the box is to have a volume of 224 cubic centimeters, what size cardboard should you start with?

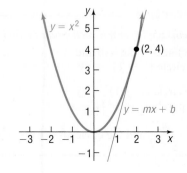

86. Constructing a Cylindrical Tube A rectangular piece of cardboard, whose area is 216 square centimeters, is made into a cylindrical tube by joining together two sides of the rectangle. See the figure. If the tube is to have a volume of 224 cubic centimeters, what size cardboard should you start with?

87. Fencing A farmer has 300 feet of fence available to enclose a 4500-square-foot region in the shape of adjoining squares, with sides of length x and y. See the figure. Find x and y.

88. Bending Wire A wire 60 feet long is cut into two pieces. Is it possible to bend one piece into the shape of a square and the other into the shape of a circle so that the total area enclosed by the two pieces is 100 square feet? If this is possible, find the length of the side of the square and the radius of the circle.

89. Geometry Find formulas for the length l and width w of a rectangle in terms of its area A and perimeter P.

90. Geometry Find formulas for the base b and one of the equal sides l of an isosceles triangle in terms of its altitude h and perimeter P.

91. Descartes's Method of Equal Roots Descartes's method for finding tangents depends on the idea that, for many graphs, the tangent line at a given point is the *unique* line that intersects the graph at that point only. We will apply his method to find an equation of the tangent line to the parabola $y = x^2$ at the point $(2, 4)$. See the figure.

First, we know that the equation of the tangent line must be in the form $y = mx + b$. Using the fact that the point $(2, 4)$ is on the line, we can solve for b in terms of m and get the equation $y = mx + (4 - 2m)$. Now we want $(2, 4)$ to be the *unique* solution to the system

$$\begin{cases} y = x^2 \\ y = mx + 4 - 2m \end{cases}$$

From this system, we get $x^2 - mx + (2m - 4) = 0$. By using the quadratic formula, we get

$$x = \frac{m \pm \sqrt{m^2 - 4(2m - 4)}}{2}$$

To obtain a unique solution for x, the two roots must be equal; in other words, the discriminant $m^2 - 4(2m - 4)$ must be 0. Complete the work to get m, and write an equation of the tangent line.

In Problems 92–98, use Descartes's method from Problem 91 to find the equation of the line tangent to each graph at the given point.

92. $x^2 + y^2 = 10$; at $(1, 3)$

93. $y = x^2 + 2$; at $(1, 3)$

94. $x^2 + y = 5$; at $(-2, 1)$

95. $2x^2 + 3y^2 = 14$; at $(1, 2)$

96. $3x^2 + y^2 = 7$; at $(-1, 2)$

97. $x^2 - y^2 = 3$; at $(2, 1)$

98. $2y^2 - x^2 = 14$; at $(2, 3)$

99. If r_1 and r_2 are two solutions of a quadratic equation $ax^2 + bx + c = 0$, it can be shown that

$$r_1 + r_2 = -\frac{b}{a} \quad \text{and} \quad r_1 r_2 = \frac{c}{a}$$

Solve this system of equations for r_1 and r_2.

Explaining Concepts: Discussion and Writing

100. A circle and a line intersect at most twice. A circle and a parabola intersect at most four times. Deduce that a circle and the graph of a polynomial of degree 3 intersect at most six times. What do you conjecture about a polynomial of degree 4? What about a polynomial of degree n? Can you explain your conclusions using an algebraic argument?

101. Suppose that you are the manager of a sheet metal shop. A customer asks you to manufacture 10,000 boxes, each box being open on top. The boxes are required to have a square

base and a 9-cubic-foot capacity. You construct the boxes by cutting out a square from each corner of a square piece of sheet metal and folding along the edges.

(a) What are the dimensions of the square to be cut if the area of the square piece of sheet metal is 100 square feet?

(b) Could you make the box using a smaller piece of sheet metal? Make a list of the dimensions of the box for various pieces of sheet metal.

'Are You Prepared?' Answers

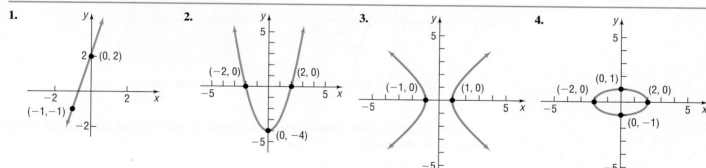

10.7 Systems of Inequalities

PREPARING FOR THIS SECTION *Before getting started, review the following:*

- Solving Linear Inequalities (Appendix A, Section A.10, pp. A83–A88)
- Lines (Foundations, Section 3, pp. 19–30)
- Circles (Foundations, Section 4, pp. 35–39)
- Graphing Techniques: Transformations (Section 1.5, pp. 90–99)

Now Work the 'Are You Prepared?' problems on page 809.

OBJECTIVES **1** Graph an Inequality (p. 804)
2 Graph a System of Inequalities (p. 806)

In the Appendix, Section A.10, we discussed inequalities in one variable. In this section, we discuss inequalities in two variables.

> EXAMPLE 1

Examples of Inequalities in Two Variables

(a) $3x + y \leq 6$ (b) $x^2 + y^2 < 4$ (c) $y^2 > x$

1 Graph an Inequality

An inequality in two variables x and y is **satisfied** by an ordered pair (a, b) if, when x is replaced by a and y by b, a true statement results. The **graph of an inequality in two variables** x and y consists of all points (x, y) whose coordinates satisfy the inequality. Let's look at an example.

> EXAMPLE 2

Graphing an Inequality

Graph the linear inequality: $3x + y \leq 6$

Solution

We begin by graphing the equation

$$3x + y = 6$$

formed by replacing (for now) the $\leq$ symbol with an $=$ sign. The graph of the equation is a line. See Figure 12(a). This line is part of the graph of the inequality that we seek because the inequality is nonstrict, so we draw the line as a solid line. (Do you see why? We are seeking points for which $3x + y$ is less than *or equal to* 6.)

Figure 12

(a) $3x + y = 6$ **(b)** Graph of $3x + y \leq 6$

Now let's test a few randomly selected points to see whether they belong to the graph of the inequality.

	$3x + y \leq 6$	**Conclusion**
$(4, -1)$	$3(4) + (-1) = 11 > 6$	Does not belong to the graph
$(5, 5)$	$3(5) + 5 = 20 > 6$	Does not belong to the graph
$(-1, 2)$	$3(-1) + 2 = -1 \leq 6$	Belongs to the graph
$(-2, -2)$	$3(-2) + (-2) = -8 \leq 6$	Belongs to the graph

Look again at Figure 12(a). Notice that the two points that belong to the graph both lie on the same side of the line, and the two points that do not belong to the graph lie on the opposite side. As it turns out, all the points that satisfy the inequality will lie on one side of the line, all the points that do not satisfy the inequality will lie on the other side. The graph we seek consists of all points that lie on the same side of the line as $(-1, 2)$ and $(-2, -2)$ and is shown as the shaded region in Figure 12(b).

━━━━**Now Work** PROBLEM 15

The graph of any inequality in two variables may be obtained in a like way. The steps to follow are given next.

Steps for Graphing an Inequality

NOTE The strict inequalities are $<$ and $>$. The nonstrict inequalities are $\leq$ and $\geq$. ∎

STEP 1: Replace the inequality symbol by an equal sign and graph the resulting equation. If the inequality is strict, use dashes; if it is nonstrict, use a solid mark. This graph separates the xy-plane into two or more regions.

STEP 2: In each region, select a test point P.

    (a) If the coordinates of P satisfy the inequality, so do all the points in that region. Indicate this by shading the region.

    (b) If the coordinates of P do not satisfy the inequality, none of the points in that region do.

EXAMPLE 3   Graphing an Inequality

Graph: $x^2 + y^2 \leq 4$

Solution

STEP 1: We graph the equation $x^2 + y^2 = 4$, a circle of radius 2, center at the origin. A solid circle will be used because the inequality is not strict.

STEP 2: We use two test points, one inside the circle, the other outside.

Inside $(0, 0)$: $x^2 + y^2 = 0^2 + 0^2 = 0 \leq 4$ *Belongs to the graph*

Outside $(4, 0)$: $x^2 + y^2 = 4^2 + 0^2 = 16 > 4$ *Does not belong to the graph*

All the points inside and on the circle satisfy the inequality. See Figure 13.

Figure 13

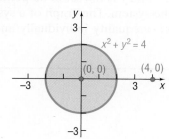

Now Work PROBLEM 17

Linear Inequalities

Linear inequalities are inequalities in one of the forms

$$Ax + By < C \qquad Ax + By > C \qquad Ax + By \leq C \qquad Ax + By \geq C$$

where A and B are not both zero.

    The graph of the corresponding equation of a linear inequality is a line that separates the xy-plane into two regions, called **half-planes.** See Figure 14.

    As shown, $Ax + By = C$ is the equation of the boundary line, and it divides the plane into two half-planes: one for which $Ax + By < C$ and the other for which $Ax + By > C$. Because of this, for linear inequalities, only one test point is required.

Figure 14

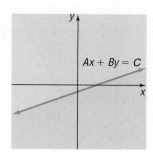

EXAMPLE 4

Graphing Linear Inequalities

Graph: (a) $y < 2$ (b) $y \geq 2x$

Solution

(a) The graph of the equation $y = 2$ is a dashed horizontal line and so is not part of the graph of the inequality. Since $(0, 0)$ satisfies the inequality, the graph consists of the half-plane below the line $y = 2$. See Figure 15.

(b) The graph of the equation $y = 2x$ is a solid line and so is part of the graph of the inequality. Using $(3, 0)$ as a test point, we find it does not satisfy the inequality $[0 < 2 \cdot 3]$. Points in the half-plane on the opposite side of $(3, 0)$ satisfy the inequality. See Figure 16.

Figure 15

Figure 16

COMMENT A graphing utility can be used to graph inequalities. To see how, read Appendix B, Section B.6 in the Appendix. ∎

Now Work PROBLEM 13

2 Graph a System of Inequalities

The **graph of a system of inequalities** in two variables x and y is the set of all points (x, y) that simultaneously satisfy *each* inequality in the system. The graph of a system of inequalities can be obtained by graphing each inequality individually and then determining where, if at all, they intersect.

EXAMPLE 5

Graphing a System of Linear Inequalities

Graph the system: $\begin{cases} x + y \geq 2 \\ 2x - y \leq 4 \end{cases}$

Solution

We begin by graphing the lines $x + y = 2$ and $2x - y = 4$ using a solid line since both inequalities are nonstrict. We use the test point $(0, 0)$ on each inequality. For example, $(0, 0)$ does not satisfy $x + y \geq 2$, so we shade above the line $x + y = 2$. See Figure 17(a). Also, $(0, 0)$ does satisfy $2x - y \leq 4$, so we shade above the line $2x - y = 4$. See Figure 17(b). The intersection of the shaded regions (in purple) gives us the result presented in Figure 17(c).

Figure 17

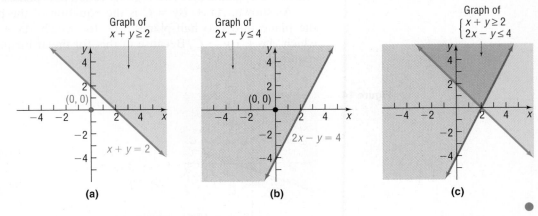

Now Work PROBLEM 23

EXAMPLE 6 | **Graphing a System of Linear Inequalities**

Graph the system: $\begin{cases} x + y \le 2 \\ x + y \ge 0 \end{cases}$

Solution See Figure 18. The overlapping purple-shaded region between the two boundary lines is the graph of the system.

Figure 18

$x + y = 0 \quad x + y = 2$

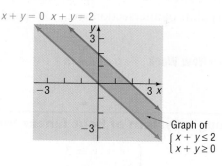

Graph of
$\begin{cases} x + y \le 2 \\ x + y \ge 0 \end{cases}$

Now Work PROBLEM 29

EXAMPLE 7 | **Graphing a System of Linear Inequalities**

Graph the systems:

(a) $\begin{cases} 2x - y \ge 0 \\ 2x - y \ge 2 \end{cases}$

(b) $\begin{cases} x + 2y \le 2 \\ x + 2y \ge 6 \end{cases}$

Solution (a) See Figure 19. The overlapping purple-shaded region is the graph of the system. Note that the graph of the system is identical to the graph of the single inequality $2x - y \ge 2$.

(b) See Figure 20. Because no overlapping region results, there are no points in the xy-plane that simultaneously satisfy each inequality. The system has no solution.

Figure 19

$2x - y = 0$

Graph of
$\begin{cases} 2x - y \ge 0 \\ 2x - y \ge 2 \end{cases}$

$2x - y = 2$

Figure 20

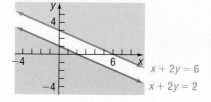

$x + 2y = 6$

$x + 2y = 2$

EXAMPLE 8 | **Graphing a System of Nonlinear Inequalities**

Graph the region below the graph of $x + y = 2$ and above the graph of $y = x^2 - 4$ by graphing the system:

$$\begin{cases} y \ge x^2 - 4 \\ x + y \le 2 \end{cases}$$

Label all points of intersection.

Solution Figure 21 on the following page shows the graph of the region above the graph of the parabola $y = x^2 - 4$ and below the graph of the line $x + y = 2$. The points of intersection are found by solving the system of equations

$$\begin{cases} y = x^2 - 4 \\ x + y = 2 \end{cases}$$

Figure 21

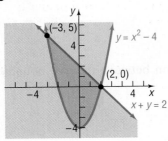

Using substitution, we find

$$x + (x^2 - 4) = 2$$
$$x^2 + x - 6 = 0$$
$$(x + 3)(x - 2) = 0$$
$$x = -3 \quad \text{or} \quad x = 2$$

The two points of intersection are $(-3, 5)$ and $(2, 0)$.

━━━━ **Now Work** PROBLEM 37

EXAMPLE 9 **Graphing a System of Four Linear Inequalities**

Graph the system:
$$\begin{cases} x + y \geq 3 \\ 2x + y \geq 4 \\ x \geq 0 \\ y \geq 0 \end{cases}$$

Figure 22

Solution See Figure 22. The two inequalities $x \geq 0$ and $y \geq 0$ require the graph of the system to be in quadrant I, which is shaded light gray. We concentrate on the remaining two inequalities. The intersection of the graphs of these two inequalities and quadrant I is shown in dark purple.

EXAMPLE 10 **Financial Planning**

A retired couple can invest up to $25,000. As their financial adviser, you recommend that they place at least $15,000 in Treasury bills yielding 6% and at most $5000 in corporate bonds yielding 9%.

(a) Using x to denote the amount of money invested in Treasury bills and y the amount invested in corporate bonds, write a system of linear inequalities that describes the possible amounts of each investment. We shall assume that x and y are in thousands of dollars.

(b) Graph the system.

Solution

(a) The system of linear inequalities is

$$\begin{cases} x \geq 0 & \text{\small x and y are nonnegative variables since they represent} \\ y \geq 0 & \text{\small money invested in thousands of dollars.} \\ x + y \leq 25 & \text{\small The total of the two investments, x + y, cannot} \\ & \text{\small exceed \$25,000.} \\ x \geq 15 & \text{\small At least \$15,000 in Treasury bills} \\ y \leq 5 & \text{\small At most \$5000 in corporate bonds} \end{cases}$$

Figure 23

(b) See the shaded region in Figure 23. Note that the inequalities $x \geq 0$ and $y \geq 0$ require that the graph of the system be in quadrant I.

The graph of the system of linear inequalities in Figure 23 is said to be **bounded,** because it can be contained within some circle of sufficiently large radius. A graph that cannot be contained in any circle is said to be **unbounded.** For example, the graph of the system of linear inequalities in Figure 22 is unbounded, since it extends indefinitely in the positive x and positive y directions.

Notice in Figures 22 and 23 that those points belonging to the graph that are also points of intersection of boundary lines have been plotted. Such points are referred to as **vertices** or **corner points** of the graph. The system graphed in Figure 22 has three corner points: $(0, 4)$, $(1, 2)$, and $(3, 0)$. The system graphed in Figure 23 has four corner points: $(15, 0)$, $(25, 0)$, $(20, 5)$, and $(15, 5)$.

These ideas will be used in the next section in developing a method for solving linear programming problems, an important application of linear inequalities.

—**Now Work** PROBLEM 45

10.7 Assess Your Understanding

'Are You Prepared?' *Answers are given at the end of these exercises. If you get a wrong answer, read the pages listed in* red.

1. Solve the inequality: $3x + 4 < 8 - x$ (pp. A86–A88)
2. Graph the equation: $3x - 2y = 6$ (p. 26)
3. Graph the equation: $x^2 + y^2 = 9$ (p. 36)
4. Graph the equation: $y = x^2 + 4$ (pp. 90–93)

5. *True or False* The lines $2x + y = 4$ and $4x + 2y = 0$ are parallel. (pp. 27–28)
6. The graph of $y = (x - 2)^2$ may be obtained by shifting the graph of _____ to the (left/right) a distance of _____ units. (pp. 90–93)

Concepts and Vocabulary

7. When graphing an inequality in two variables, if the inequality is strict use _____; if the inequality is nonstrict use a _____ mark.
8. The graph of the corresponding equation of a linear inequality is a line that separates the xy-plane into two regions. The two regions are called _____.

9. *True or False* The graph of a system of inequalities must have an overlapping region.
10. If a graph of a system of linear inequalities cannot be contained in any circle, then it is said to be _____.

Skill Building

In Problems 11–22, graph each inequality.

11. $x \geq 0$
12. $y \geq 0$
13. $x \geq 4$
14. $y \leq 2$

15. $2x + y \geq 6$
16. $3x + 2y \leq 6$
17. $x^2 + y^2 > 1$
18. $x^2 + y^2 \leq 9$

19. $y \leq x^2 - 1$
20. $y > x^2 + 2$
21. $xy \geq 4$
22. $xy \leq 1$

In Problems 23–34, graph each system of linear inequalities.

23. $\begin{cases} x + y \leq 2 \\ 2x + y \geq 4 \end{cases}$

24. $\begin{cases} 3x - y \geq 6 \\ x + 2y \leq 2 \end{cases}$

25. $\begin{cases} 2x - y \leq 4 \\ 3x + 2y \geq -6 \end{cases}$

26. $\begin{cases} 4x - 5y \leq 0 \\ 2x - y \geq 2 \end{cases}$

27. $\begin{cases} 2x - 3y \leq 0 \\ 3x + 2y \leq 6 \end{cases}$

28. $\begin{cases} 4x - y \geq 2 \\ x + 2y \geq 2 \end{cases}$

29. $\begin{cases} x - 2y \leq 6 \\ 2x - 4y \geq 0 \end{cases}$

30. $\begin{cases} x + 4y \leq 8 \\ x + 4y \geq 4 \end{cases}$

31. $\begin{cases} 2x + y \geq -2 \\ 2x + y \geq 2 \end{cases}$

32. $\begin{cases} x - 4y \leq 4 \\ x - 4y \geq 0 \end{cases}$

33. $\begin{cases} 2x + 3y \geq 6 \\ 2x + 3y \leq 0 \end{cases}$

34. $\begin{cases} 2x + y \geq 0 \\ 2x + y \geq 2 \end{cases}$

In Problems 35–42, graph each system of inequalities.

35. $\begin{cases} x^2 + y^2 \leq 9 \\ x + y \geq 3 \end{cases}$

36. $\begin{cases} x^2 + y^2 \geq 9 \\ x + y \leq 3 \end{cases}$

37. $\begin{cases} y \geq x^2 - 4 \\ y \leq x - 2 \end{cases}$

38. $\begin{cases} y^2 \leq x \\ y \geq x \end{cases}$

39. $\begin{cases} x^2 + y^2 \leq 16 \\ y \geq x^2 - 4 \end{cases}$

40. $\begin{cases} x^2 + y^2 \leq 25 \\ y \leq x^2 - 5 \end{cases}$

41. $\begin{cases} xy \geq 4 \\ y \geq x^2 + 1 \end{cases}$

42. $\begin{cases} y + x^2 \leq 1 \\ y \geq x^2 - 1 \end{cases}$

In Problems 43–52, graph each system of linear inequalities. Tell whether the graph is bounded or unbounded, and label the corner points.

43. $\begin{cases} x \geq 0 \\ y \geq 0 \\ 2x + y \leq 6 \\ x + 2y \leq 6 \end{cases}$

44. $\begin{cases} x \geq 0 \\ y \geq 0 \\ x + y \geq 4 \\ 2x + 3y \geq 6 \end{cases}$

45. $\begin{cases} x \geq 0 \\ y \geq 0 \\ x + y \geq 2 \\ 2x + y \geq 4 \end{cases}$

46. $\begin{cases} x \geq 0 \\ y \geq 0 \\ 3x + y \leq 6 \\ 2x + y \leq 2 \end{cases}$

47. $\begin{cases} x \geq 0 \\ y \geq 0 \\ x + y \geq 2 \\ 2x + 3y \leq 12 \\ 3x + y \leq 12 \end{cases}$

48. $\begin{cases} x \geq 0 \\ y \geq 0 \\ x + y \geq 2 \\ x + y \leq 8 \\ 2x + y \leq 10 \end{cases}$

49. $\begin{cases} x \geq 0 \\ y \geq 0 \\ x + y \geq 2 \\ x + y \leq 8 \\ 2x + y \leq 10 \end{cases}$

50. $\begin{cases} x \geq 0 \\ y \geq 0 \\ x + y \geq 2 \\ x + y \leq 8 \\ x + 2y \geq 1 \end{cases}$

51. $\begin{cases} x \geq 0 \\ y \geq 0 \\ x + 2y \geq 1 \\ x + 2y \leq 10 \end{cases}$

52. $\begin{cases} x \geq 0 \\ y \geq 0 \\ x + 2y \geq 1 \\ x + 2y \leq 10 \\ x + y \geq 2 \\ x + y \leq 8 \end{cases}$

In Problems 53–56, write a system of linear inequalities for the given graph.

53.

54.

55.

56.
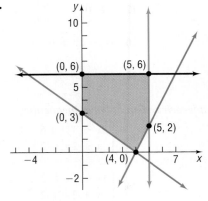

Applications and Extensions

57. **Financial Planning** A retired couple has up to $50,000 to invest. As their financial adviser, you recommend that they place at least $35,000 in Treasury bills yielding 7% and at most $10,000 in corporate bonds yielding 10%.
    (a) Using x to denote the amount of money invested in Treasury bills and y the amount invested in corporate bonds, write a system of linear inequalities that describes the possible amounts of each investment.
    (b) Graph the system and label the corner points.

58. **Manufacturing Trucks** Mike's Toy Truck Company manufactures two models of toy trucks, a standard model and a deluxe model. Each standard model requires 2 hours (hr) for painting and 3 hr for detail work; each deluxe model requires 3 hr for painting and 4 hr for detail work. Two painters and three detail workers are employed by the company, and each works 40 hr per week.
    (a) Using x to denote the number of standard-model trucks and y to denote the number of deluxe-model trucks,

write a system of linear inequalities that describes the possible number of each model of truck that can be manufactured in a week.

(b) Graph the system and label the corner points.

59. **Blending Coffee**   Bill's Coffee House, a store that specializes in coffee, has available 75 pounds (lb) of *A* grade coffee and 120 lb of *B* grade coffee. These will be blended into 1-lb packages as follows: An economy blend that contains 4 ounces (oz) of *A* grade coffee and 12 oz of *B* grade coffee and a superior blend that contains 8 oz of *A* grade coffee and 8 oz of *B* grade coffee.

(a) Using *x* to denote the number of packages of the economy blend and *y* to denote the number of packages of

the superior blend, write a system of linear inequalities that describes the possible number of packages of each kind of blend.

(b) Graph the system and label the corner points.

60. **Mixed Nuts**   Nola's Nuts, a store that specializes in selling nuts, has available 90 pounds (lb) of cashews and 120 lb of peanuts. These are to be mixed in 12-ounce (oz) packages as follows: a lower-priced package containing 8 oz of peanuts and 4 oz of cashews and a quality package containing 6 oz of peanuts and 6 oz of cashews.

(a) Use *x* to denote the number of lower-priced packages and use *y* to denote the number of quality packages. Write a system of linear inequalities that describes the possible number of each kind of package.

(b) Graph the system and label the corner points.

61. **Transporting Goods**   A small truck can carry no more than 1600 pounds (lb) of cargo nor more than 150 cubic (ft³) of cargo. A printer weighs 20 lb and occupies 3 ft³ of space. A microwave oven weighs 30 lb and occupies 2 ft³ of space.

(a) Using *x* to represent the number of microwave ovens and *y* to represent the number of printers, write a system of linear inequalities that describes the number of ovens and printers that can be hauled by the truck.

(b) Graph the system and label the corner points.

'Are You Prepared?' Answers

1. $\{x \mid x < 1\}$ or $(-\infty, 1)$

2.

3.

4.

5. True

6. $y = x^2$; right; 2

10.8 Linear Programming

OBJECTIVES   **1** Set up a Linear Programming Problem (p. 812)

2 Solve a Linear Programming Problem (p. 812)

Historically, linear programming evolved as a technique for solving problems involving resource allocation of goods and materials for the U.S. Air Force during World War II. Today, linear programming techniques are used to solve a wide variety of problems, such as optimizing airline scheduling and establishing telephone lines. Although most practical linear programming problems involve systems of several hundred linear inequalities containing several hundred variables, we will limit our discussion to problems containing only two variables, because we can solve such problems using graphing techniques.*

*The **simplex method** is a way to solve linear programming problems involving many inequalities and variables. This method was developed by George Dantzig in 1946 and is particularly well suited for computerization. In 1984, Narendra Karmarkar of Bell Laboratories discovered a way of solving large linear programming problems that improves on the simplex method.

1 Set up a Linear Programming Problem

We begin by returning to Example 10 of the previous section.

EXAMPLE 1 Financial Planning

A retired couple has up to $25,000 to invest. As their financial adviser, you recommend that they place at least $15,000 in Treasury bills yielding 6% and at most $5000 in corporate bonds yielding 9%. Develop a model that can be used to determine how much money should be placed in each investment so that income is maximized.

Solution The problem is typical of a *linear programming problem.* The problem requires that a certain linear expression, the income, be maximized. If I represents income, x the amount invested in Treasury bills at 6%, and y the amount invested in corporate bonds at 9%, then

$$I = 0.06x + 0.09y$$

We shall assume, as before, that I, x, and y are in thousands of dollars.

The linear expression $I = 0.06x + 0.09y$ is called the **objective function.** Further, the problem requires that the maximum income be achieved under certain conditions or **constraints,** each of which is a linear inequality involving the variables. (See Example 10 in Section 10.7.) The linear programming problem may be modeled as

$$\text{Maximize} \qquad I = 0.06x + 0.09y$$

subject to the conditions that

$$\begin{cases} x \geq 0, \quad y \geq 0 \\ x + y \leq 25 \\ x \geq 15 \\ y \leq 5 \end{cases}$$

In general, every linear programming problem has two components:

1. A linear objective function that is to be maximized or minimized
2. A collection of linear inequalities that must be satisfied simultaneously

---

DEFINITION

A **linear programming problem** in two variables x and y consists of maximizing (or minimizing) a linear objective function

$$z = Ax + By \qquad A \text{ and } B \text{ are real numbers, not both } 0$$

subject to certain conditions, or constraints, expressible as linear inequalities in x and y.

---

2 Solve a Linear Programing Problem

To maximize (or minimize) the quantity $z = Ax + By$, we need to identify points (x, y) that make the expression for z the largest (or smallest) possible. But not all points (x, y) are eligible; only those that also satisfy each linear inequality (constraint) can be used. We refer to each point (x, y) that satisfies the system of linear inequalities (the constraints) as a **feasible point.** In a linear programming problem, we seek the feasible point(s) that maximizes (or minimizes) the objective function.

Let's look again at the linear programming problem in Example 1.

EXAMPLE 2   **Analyzing a Linear Programming Problem**

Consider the linear programming problem

$$\text{Maximize} \qquad I = 0.06x + 0.09y$$

subject to the conditions that

$$\begin{cases} x \geq 0 \\ y \geq 0 \\ x + y \leq 25 \\ x \geq 15 \\ y \leq 5 \end{cases}$$

Graph the constraints. Then graph the objective function for $I = 0, 0.9, 1.35, 1.65,$ and 1.8.

Solution   Figure 24 shows the graph of the constraints. We superimpose on this graph the graph of the objective function for the given values of I.

For $I = 0$, the objective function is the line $0 = 0.06x + 0.09y$.
For $I = 0.9$, the objective function is the line $0.9 = 0.06x + 0.09y$.
For $I = 1.35$, the objective function is the line $1.35 = 0.06x + 0.09y$.
For $I = 1.65$, the objective function is the line $1.65 = 0.06x + 0.09y$.
For $I = 1.8$, the objective function is the line $1.8 = 0.06x + 0.09y$.

Figure 24

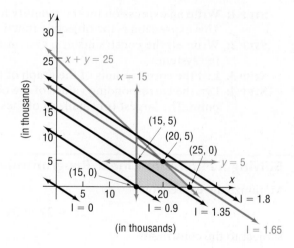

DEFINITION

> A **solution** to a linear programming problem consists of a feasible point that maximizes (or minimizes) the objective function, together with the corresponding value of the objective function.

One condition for a linear programming problem in two variables to have a solution is that the graph of the feasible points be bounded. (Refer to page 808.)

If none of the feasible points maximizes (or minimizes) the objective function or if there are no feasible points, the linear programming problem has no solution.

Consider the linear programming problem stated in Example 2, and look again at Figure 24. The feasible points are the points that lie in the shaded region. For example, $(20, 3)$ is a feasible point, as are $(15, 5)$, $(20, 5)$, $(18, 4)$, and so on. To find the solution of the problem requires that we find a feasible point (x, y) that makes $I = 0.06x + 0.09y$ as large as possible. Notice that, as I increases in value from

$I = 0$ to $I = 0.9$ to $I = 1.35$ to $I = 1.65$ to $I = 1.8$, we obtain a collection of parallel lines. Further, notice that the largest value of I that can be obtained using feasible points is $I = 1.65$, which corresponds to the line $1.65 = 0.06x + 0.09y$. Any larger value of I results in a line that does not pass through any feasible points. Finally, notice that the feasible point that yields $I = 1.65$ is the point $(20, 5)$, a corner point. These observations form the basis of the following result, which we state without proof.

THEOREM

Location of the Solution of a Linear Programming Problem

If a linear programming problem has a solution, it is located at a corner point of the graph of the feasible points.

If a linear programming problem has multiple solutions, at least one of them is located at a corner point of the graph of the feasible points.

In either case, the corresponding value of the objective function is unique.

We shall not consider here linear programming problems that have no solution. As a result, we can outline the procedure for solving a linear programming problem as follows:

Procedure for Solving a Linear Programming Problem

STEP 1: Write an expression for the quantity to be maximized (or minimized). This expression is the objective function.

STEP 2: Write all the constraints as a system of linear inequalities and graph the system.

STEP 3: List the corner points of the graph of the feasible points.

STEP 4: List the corresponding values of the objective function at each corner point. The largest (or smallest) of these is the solution.

EXAMPLE 3   **Solving a Minimum Linear Programming Problem**

Minimize the expression

$$z = 2x + 3y$$

subject to the constraints

$$y \leq 5 \qquad x \leq 6 \qquad x + y \geq 2 \qquad x \geq 0 \qquad y \geq 0$$

Solution   **STEP 1:** The objective function is $z = 2x + 3y$.

STEP 2: We seek the smallest value of z that can occur if x and y are solutions of the system of linear inequalities

$$\begin{cases} y \leq 5 \\ x \leq 6 \\ x + y \geq 2 \\ x \geq 0 \\ y \geq 0 \end{cases}$$

Figure 25

STEP 3: The graph of this system (the set of feasible points) is shown as the shaded region in Figure 25. We have also plotted the corner points.

STEP 4: Table 1 lists the corner points and the corresponding values of the objective function. From the table, we can see that the minimum value of z is 4, and it occurs at the point $(2, 0)$.

Objectives

Section		You should be able to . . .	Example(s)	Review Exercises
10.1	1	Solve systems of equations by substitution (p. 733)	4	1–14, 101, 102, 105–107
	2	Solve systems of equations by elimination (p. 735)	5, 6	1–14, 101, 102, 105–107
	3	Identify inconsistent systems of equations containing two variables (p. 736)	7	9, 10, 13, 98
	4	Express the solution of a system of dependent equations containing two variables (p. 737)	8	14, 97
	5	Solve systems of three equations containing three variables (p. 738)	9	15–18, 99, 100, 103
	6	Identify inconsistent systems of equations containing three variables (p. 739)	10	18
	7	Express the solution of a system of dependent equations containing three variables (p. 740)	11	17
10.2	1	Write the augmented matrix of a system of linear equations (p. 747)	1	35–44
	2	Write the system of equations from the augmented matrix (p. 747)	2	19, 20
	3	Perform row operations on a matrix (p. 748)	3, 4	35–44
	4	Solve a system of linear equations using matrices (p. 749)	5–10	35–44
10.3	1	Evaluate 2 by 2 determinants (p. 761)	1	45, 46
	2	Use Cramer's Rule to solve a system of two equations containing two variables (p. 762)	2	51–54
	3	Evaluate 3 by 3 determinants (p. 764)	4	47–50
	4	Use Cramer's Rule to solve a system of three equations containing three variables (p. 766)	5	55, 56
	5	Know properties of determinants (p. 767)	6–9	57, 58
10.4	1	Find the sum and difference of two matrices (p. 772)	3, 4	21, 22
	2	Find scalar multiples of a matrix (p. 773)	5	23, 24
	3	Find the product of two matrices (p. 774)	6–11	25–28
	4	Find the inverse of a matrix (p. 779)	12–14	29–34
	5	Solve a system of linear equations using an inverse matrix (p. 783)	15	35–44
10.5	1	Decompose $\dfrac{P}{Q}$, where Q has only nonrepeated linear factors (p. 788)	1	59, 60
	2	Decompose $\dfrac{P}{Q}$, where Q has repeated linear factors (p. 790)	2, 3	61, 62
	3	Decompose $\dfrac{P}{Q}$, where Q has a nonrepeated irreducible quadratic factor (p. 792)	4	63, 64, 67, 68
	4	Decompose $\dfrac{P}{Q}$, where Q has a repeated irreducible quadratic factor (p. 793)	5	65, 66
10.6	1	Solve a system of nonlinear equations using substitution (p. 795)	1, 3	69–78
	2	Solve a system of nonlinear equations using elimination (p. 796)	2, 4	69–78
10.7	1	Graph an inequality (p. 804)	2–4	79–82
	2	Graph a system of inequalities (p. 806)	5–10	83–92, 104
10.8	1	Set up a linear programming problem (p. 812)	1	108, 109
	2	Solve a linear programming problem (p. 812)	2–4	93–96, 108, 109

Review Exercises

In Problems 1–18, solve each system of equations using the method of substitution or the method of elimination. If the system has no solution, say that it is inconsistent.

1. $\begin{cases} 2x - y = 5 \\ 5x + 2y = 8 \end{cases}$

2. $\begin{cases} 2x + 3y = 2 \\ 7x - y = 3 \end{cases}$

3. $\begin{cases} 3x - 4y = 4 \\ x - 3y = \dfrac{1}{2} \end{cases}$

4. $\begin{cases} 2x + y = 0 \\ 5x - 4y = -\dfrac{13}{2} \end{cases}$

5. $\begin{cases} x - 2y - 4 = 0 \\ 3x + 2y - 4 = 0 \end{cases}$

6. $\begin{cases} x - 3y + 5 = 0 \\ 2x + 3y - 5 = 0 \end{cases}$

7. $\begin{cases} y = 2x - 5 \\ x = 3y + 4 \end{cases}$

8. $\begin{cases} x = 5y + 2 \\ y = 5x + 2 \end{cases}$

9. $\begin{cases} x - 3y + 4 = 0 \\ \dfrac{1}{2}x - \dfrac{3}{2}y + \dfrac{4}{3} = 0 \end{cases}$

10. $\begin{cases} x + \dfrac{1}{4}y = 2 \\ y + 4x + 2 = 0 \end{cases}$

11. $\begin{cases} 2x + 3y - 13 = 0 \\ 3x - 2y = 0 \end{cases}$

12. $\begin{cases} 4x + 5y = 21 \\ 5x + 6y = 42 \end{cases}$

13. $\begin{cases} 3x - 2y = 8 \\ x - \dfrac{2}{3}y = 12 \end{cases}$

14. $\begin{cases} 2x + 5y = 10 \\ 4x + 10y = 20 \end{cases}$

15. $\begin{cases} x + 2y - z = 6 \\ 2x - y + 3z = -13 \\ 3x - 2y + 3z = -16 \end{cases}$

16. $\begin{cases} x + 5y - z = 2 \\ 2x + y + z = 7 \\ x - y + 2z = 11 \end{cases}$

17. $\begin{cases} 2x - 4y + z = -15 \\ x + 2y - 4z = 27 \\ 5x - 6y - 2z = -3 \end{cases}$

18. $\begin{cases} x - 4y + 3z = 15 \\ -3x + y - 5z = -5 \\ -7x - 5y - 9z = 10 \end{cases}$

In Problems 19 and 20, write the system of equations corresponding to the given augmented matrix.

19. $\begin{bmatrix} 3 & 2 & | & 8 \\ 1 & 4 & | & -1 \end{bmatrix}$

20. $\begin{bmatrix} 1 & 2 & 5 & | & -2 \\ 5 & 0 & -3 & | & 8 \\ 2 & -1 & 0 & | & 0 \end{bmatrix}$

In Problems 21–28, use the following matrices to compute each expression.

$$A = \begin{bmatrix} 1 & 0 \\ 2 & 4 \\ -1 & 2 \end{bmatrix} \qquad B = \begin{bmatrix} 4 & -3 & 0 \\ 1 & 1 & -2 \end{bmatrix} \qquad C = \begin{bmatrix} 3 & -4 \\ 1 & 5 \\ 5 & 2 \end{bmatrix}$$

21. $A + C$

22. $A - C$

23. $6A$

24. $-4B$

25. AB

26. BA

27. CB

28. BC

In Problems 29–34, find the inverse, if there is one, of each matrix. If there is not an inverse, say that the matrix is singular.

29. $\begin{bmatrix} 4 & 6 \\ 1 & 3 \end{bmatrix}$

30. $\begin{bmatrix} -3 & 2 \\ 1 & -2 \end{bmatrix}$

31. $\begin{bmatrix} 1 & 3 & 3 \\ 1 & 2 & 1 \\ 1 & -1 & 2 \end{bmatrix}$

32. $\begin{bmatrix} 3 & 1 & 2 \\ 3 & 2 & -1 \\ 1 & 1 & 1 \end{bmatrix}$

33. $\begin{bmatrix} 4 & -8 \\ -1 & 2 \end{bmatrix}$

34. $\begin{bmatrix} -3 & 1 \\ -6 & 2 \end{bmatrix}$

In Problems 35–44, solve each system of equations using matrices. If the system has no solution, say that it is inconsistent.

35. $\begin{cases} 3x - 2y = 1 \\ 10x + 10y = 5 \end{cases}$

36. $\begin{cases} 3x + 2y = 6 \\ x - y = -\dfrac{1}{2} \end{cases}$

37. $\begin{cases} 5x - 6y - 3z = 6 \\ 4x - 7y - 2z = -3 \\ 3x + y - 7z = 1 \end{cases}$

38. $\begin{cases} 2x + y + z = 5 \\ 4x - y - 3z = 1 \\ 8x + y - z = 5 \end{cases}$

39. $\begin{cases} x - 2z = 1 \\ 2x + 3y = -3 \\ 4x - 3y - 4z = 3 \end{cases}$

40. $\begin{cases} x + 2y - z = 2 \\ 2x - 2y + z = -1 \\ 6x + 4y + 3z = 5 \end{cases}$

41. $\begin{cases} x - y + z = 0 \\ x - y - 5z - 6 = 0 \\ 2x - 2y + z - 1 = 0 \end{cases}$

42. $\begin{cases} 4x - 3y + 5z = 0 \\ 2x + 4y - 3z = 0 \\ 6x + 2y + z = 0 \end{cases}$

43. $\begin{cases} x - y - z - t = 1 \\ 2x + y - z + 2t = 3 \\ x - 2y - 2z - 3t = 0 \\ 3x - 4y + z + 5t = -3 \end{cases}$

44. $\begin{cases} x - 3y + 3z - t = 4 \\ x + 2y - z = -3 \\ x + 3z + 2t = 3 \\ x + y + 5z = 6 \end{cases}$

In Problems 45–50, find the value of each determinant.

45. $\begin{vmatrix} 3 & 4 \\ 1 & 3 \end{vmatrix}$

46. $\begin{vmatrix} -4 & 0 \\ 1 & 3 \end{vmatrix}$

47. $\begin{vmatrix} 1 & 4 & 0 \\ -1 & 2 & 6 \\ 4 & 1 & 3 \end{vmatrix}$

48. $\begin{vmatrix} 2 & 3 & 10 \\ 0 & 1 & 5 \\ -1 & 2 & 3 \end{vmatrix}$

49. $\begin{vmatrix} 2 & 1 & -3 \\ 5 & 0 & 1 \\ 2 & 6 & 0 \end{vmatrix}$

50. $\begin{vmatrix} -2 & 1 & 0 \\ 1 & 2 & 3 \\ -1 & 4 & 2 \end{vmatrix}$

In Problems 51–56, use Cramer's Rule, if applicable, to solve each system.

51. $\begin{cases} x - 2y = 4 \\ 3x + 2y = 4 \end{cases}$

52. $\begin{cases} x - 3y = -5 \\ 2x + 3y = 5 \end{cases}$

53. $\begin{cases} 2x + 3y - 13 = 0 \\ 3x - 2y = 0 \end{cases}$

54. $\begin{cases} 3x - 4y - 12 = 0 \\ 5x + 2y + 6 = 0 \end{cases}$

55. $\begin{cases} x + 2y - z = 6 \\ 2x - y + 3z = -13 \\ 3x - 2y + 3z = -16 \end{cases}$

56. $\begin{cases} x - y + z = 8 \\ 2x + 3y - z = -2 \\ 3x - y - 9z = 9 \end{cases}$

In Problems 57 and 58, use properties of determinants to find the value of each determinant if it is known that $\begin{vmatrix} x & y \\ a & b \end{vmatrix} = 8.$

57. $\begin{vmatrix} 2x & y \\ 2a & b \end{vmatrix}$

58. $\begin{vmatrix} y & x \\ b & a \end{vmatrix}$

In Problems 59–68, write the partial fraction decomposition of each rational expression.

59. $\dfrac{6}{x(x-4)}$

60. $\dfrac{x}{(x+2)(x-3)}$

61. $\dfrac{x-4}{x^2(x-1)}$

62. $\dfrac{2x-6}{(x-2)^2(x-1)}$

63. $\dfrac{x}{(x^2+9)(x+1)}$

64. $\dfrac{3x}{(x-2)(x^2+1)}$

65. $\dfrac{x^3}{(x^2+4)^2}$

66. $\dfrac{x^3+1}{(x^2+16)^2}$

67. $\dfrac{x^2}{(x^2+1)(x^2-1)}$

68. $\dfrac{4}{(x^2+4)(x^2-1)}$

In Problems 69–78, solve each system of equations.

69. $\begin{cases} 2x + y + 3 = 0 \\ x^2 + y^2 = 5 \end{cases}$

70. $\begin{cases} x^2 + y^2 = 16 \\ 2x - y^2 = -8 \end{cases}$

71. $\begin{cases} 2xy + y^2 = 10 \\ 3y^2 - xy = 2 \end{cases}$

72. $\begin{cases} 3x^2 - y^2 = 1 \\ 7x^2 - 2y^2 - 5 = 0 \end{cases}$

73. $\begin{cases} x^2 + y^2 = 6y \\ x^2 = 3y \end{cases}$

74. $\begin{cases} 2x^2 + y^2 = 9 \\ x^2 + y^2 = 9 \end{cases}$

75. $\begin{cases} 3x^2 + 4xy + 5y^2 = 8 \\ x^2 + 3xy + 2y^2 = 0 \end{cases}$

76. $\begin{cases} 3x^2 + 2xy - 2y^2 = 6 \\ xy - 2y^2 + 4 = 0 \end{cases}$

77. $\begin{cases} x^2 - 3x + y^2 + y = -2 \\ \dfrac{x^2 - x}{y} + y + 1 = 0 \end{cases}$

78. $\begin{cases} x^2 + x + y^2 = y + 2 \\ x + 1 = \dfrac{2 - y}{x} \end{cases}$

In Problems 79–82 graph each inequality.

79. $3x + 4y \leq 12$

80. $2x - 3y \geq 6$

81. $y \leq x^2$

82. $x \geq y^2$

In Problems 83–88, graph each system of inequalities. Tell whether the graph is bounded or unbounded, and label the corner points.

83. $\begin{cases} -2x + y \leq 2 \\ x + y \geq 2 \end{cases}$

84. $\begin{cases} x - 2y \leq 6 \\ 2x + y \geq 2 \end{cases}$

85. $\begin{cases} x \geq 0 \\ y \geq 0 \\ x + y \leq 4 \\ 2x + 3y \leq 6 \end{cases}$

86. $\begin{cases} x \geq 0 \\ y \geq 0 \\ 3x + y \geq 6 \\ 2x + y \geq 2 \end{cases}$

87. $\begin{cases} x \geq 0 \\ y \geq 0 \\ 2x + y \leq 8 \\ x + 2y \geq 2 \end{cases}$

88. $\begin{cases} x \geq 0 \\ y \geq 0 \\ 3x + y \leq 9 \\ 2x + 3y \geq 6 \end{cases}$

In Problems 89–92, graph each system of inequalities.

89. $\begin{cases} x^2 + y^2 \leq 16 \\ x + y \geq 2 \end{cases}$

90. $\begin{cases} y^2 \leq x - 1 \\ x - y \leq 3 \end{cases}$

91. $\begin{cases} y \leq x^2 \\ xy \leq 4 \end{cases}$

92. $\begin{cases} x^2 + y^2 \geq 1 \\ x^2 + y^2 \leq 4 \end{cases}$

In Problems 93–96, solve each linear programming problem.

93. Maximize $z = 3x + 4y$ subject to $x \geq 0, y \geq 0, 3x + 2y \geq 6, x + y \leq 8$

94. Maximize $z = 2x + 4y$ subject to $x \geq 0, y \geq 0, x + y \leq 6, x \geq 2$

95. Minimize $z = 3x + 5y$ subject to $x \geq 0, y \geq 0, x + y \geq 1, 3x + 2y \leq 12, x + 3y \leq 12$

96. Minimize $z = 3x + y$ subject to $x \geq 0, y \geq 0, x \leq 8, y \leq 6, 2x + y \geq 4$

---

97. Find A so that the system of equations has infinitely many solutions.

$$\begin{cases} 2x + 5y = 5 \\ 4x + 10y = A \end{cases}$$

98. Find A so that the system in Problem 97 is inconsistent.

99. Curve Fitting Find the quadratic function $y = ax^2 + bx + c$ that passes through the three points $(0, 1), (1, 0)$, and $(-2, 1)$.

100. Curve Fitting Find the general equation of the circle that passes through the three points $(0, 1), (1, 0)$, and $(-2, 1)$.

[**Hint:** The general equation of a circle is $x^2 + y^2 + Dx + Ey + F = 0$.]

101. Blending Coffee A coffee distributor is blending a new coffee that will cost \$6.90 per pound. It will consist of a blend of \$6.00 per pound coffee and \$9.00 per pound coffee. What amounts of each type of coffee should be mixed to achieve the desired blend?

[**Hint:** Assume that the weight of the blended coffee is 100 pounds.]

$6.00/lb

$6.90/lb

$9.00/lb

102. Farming A 1000-acre farm in Illinois is used to grow corn and soybeans. The cost per acre for raising corn is \$65, and the cost per acre for soybeans is \$45. If \$54,325 has been budgeted for costs and all the acreage is to be used, how many acres should be allocated for each crop?

103. Cookie Orders A cookie company makes three kinds of cookies, oatmeal raisin, chocolate chip, and shortbread, packaged in small, medium, and large boxes. The small box contains 1 dozen oatmeal raisin and 1 dozen chocolate chip; the medium box has 2 dozen oatmeal raisin, 1 dozen chocolate chip, and 1 dozen shortbread; the large box contains 2 dozen oatmeal raisin, 2 dozen chocolate chip, and 3 dozen shortbread. If you require exactly 15 dozen oatmeal raisin, 10 dozen chocolate chip, and 11 dozen shortbread, how many of each size box should you buy?

104. Mixed Nuts A store that specializes in selling nuts has available 72 pounds (lb) of cashews and 120 lb of peanuts. These are to be mixed in 12-ounce (oz) packages as follows: a lower-priced package containing 8 oz of peanuts and 4 oz of cashews and a quality package containing 6 oz of peanuts and 6 oz of cashews.

(a) Use x to denote the number of lower-priced packages and use y to denote the number of quality packages. Write a system of linear inequalities that describes the possible number of each kind of package.

(b) Graph the system and label the corner points.

105. Determining the Speed of the Current of the Aguarico River On a recent trip to the Cuyabeno Wildlife Reserve in the Amazon region of Ecuador, Mike took a 100-kilometer trip by speedboat down the Aguarico River from Chiritza to the Flotel Orellana. As Mike watched the Amazon unfold, he wondered how fast the speedboat was going and how fast the current of the white-water Aguarico River was. Mike timed the trip downstream at 2.5 hours and the return trip at 3 hours. What were the two speeds?

106. Finding the Speed of the Jet Stream On a flight between Midway Airport in Chicago and Ft. Lauderdale, Florida, a Boeing 737 jet maintains an airspeed of 475 miles per hour. If the trip from Chicago to Ft. Lauderdale takes 2 hours, 30 minutes and the return flight takes 2 hours, 50 minutes, what is the speed of the jet stream? (Assume that the speed of the jet stream remains constant at the various altitudes of the plane and that the plane flies with the jet stream one way and against it the other way.)

107. Constant Rate Jobs If Bruce and Bryce work together for 1 hour and 20 minutes, they will finish a certain job. If Bryce and Marty work together for 1 hour and 36 minutes, the same job can be finished. If Marty and Bruce work together, they can complete this job in 2 hours and 40 minutes. How long will it take each of them working alone to finish the job?

108. Maximizing Profit on Figurines A factory manufactures two kinds of ceramic figurines: a dancing girl and a mermaid. Each requires three processes: molding, painting, and glazing. The daily labor available for molding is no more than 90 work-hours, labor available for painting does not exceed 120 work-hours, and labor available for glazing is no more than 60 work-hours. The dancing girl requires 3 work-hours for molding, 6 work-hours for painting, and 2 work-hours for

glazing. The mermaid requires 3 work-hours for molding, 4 work-hours for painting, and 3 work-hours for glazing. If the profit on each figurine is $25 for dancing girls and $30 for mermaids, how many of each should be produced each day to maximize profit? If management decides to produce the number of each figurine that maximizes profit, determine which of these processes has work-hours assigned to it that are not used.

109. Minimizing Production Cost A factory produces gasoline engines and diesel engines. Each week the factory is obligated to deliver at least 20 gasoline engines and at least 15 diesel engines. Due to physical limitations, however, the factory cannot make more than 60 gasoline engines nor more than 40 diesel engines in any given week. Finally, to prevent layoffs, a total of at least 50 engines must be produced. If gasoline engines cost $450 each to produce and diesel engines cost $550 each to produce, how many of each should be produced per week to minimize the cost? What is the excess capacity of the factory; that is, how many of each kind of engine is being produced in excess of the number that the factory is obligated to deliver?

110. Describe four ways of solving a system of three linear equations containing three variables. Which method do you prefer? Why?

CHAPTER TEST

In Problems 1–4, solve each system of equations using the method of substitution or the method of elimination. If the system has no solution, say that it is inconsistent.

1. $\begin{cases} -2x + y = -7 \\ 4x + 3y = 9 \end{cases}$

2. $\begin{cases} \dfrac{1}{3}x - 2y = 1 \\ 5x - 30y = 18 \end{cases}$

3. $\begin{cases} x - y + 2z = 5 \\ 3x + 4y - z = -2 \\ 5x + 2y + 3z = 8 \end{cases}$

4. $\begin{cases} 3x + 2y - 8z = -3 \\ -x - \dfrac{2}{3}y + z = 1 \\ 6x - 3y + 15z = 8 \end{cases}$

5. Write the augmented matrix corresponding to the system of equations: $\begin{cases} 4x - 5y + z = 0 \\ -2x - y + 6 = -19 \\ x + 5y - 5z = 10 \end{cases}$

6. Write the system of equations corresponding to the augmented matrix: $\begin{bmatrix} 3 & 2 & 4 & | & -6 \\ 1 & 0 & 8 & | & 2 \\ -2 & 1 & 3 & | & -11 \end{bmatrix}$

In Problems 7–10, use the given matrices to compute each expression.

$$A = \begin{bmatrix} 1 & -1 \\ 0 & -4 \\ 3 & 2 \end{bmatrix} \quad B = \begin{bmatrix} 1 & -2 & 5 \\ 0 & 3 & 1 \end{bmatrix} \quad C = \begin{bmatrix} 4 & 6 \\ 1 & -3 \\ -1 & 8 \end{bmatrix}$$

7. $2A + C$

8. $A - 3C$

9. AC

10. BA

In Problems 11 and 12, find the inverse of each nonsingular matrix.

11. $A = \begin{bmatrix} 3 & 2 \\ 5 & 4 \end{bmatrix}$

12. $B = \begin{bmatrix} 1 & -1 & 1 \\ 2 & 5 & -1 \\ 2 & 3 & 0 \end{bmatrix}$

In Problems 13–16, solve each system of equations using matrices. If the system has no solution, say that it is inconsistent.

13. $\begin{cases} 6x + 3y = 12 \\ 2x - y = -2 \end{cases}$

14. $\begin{cases} x + \dfrac{1}{4}y = 7 \\ 8x + 2y = 56 \end{cases}$

15. $\begin{cases} x + 2y + 4z = -3 \\ 2x + 7y + 15z = -12 \\ 4x + 7y + 13z = -10 \end{cases}$

16. $\begin{cases} 2x + 2y - 3z = 5 \\ x - y + 2z = 8 \\ 3x + 5y - 8z = -2 \end{cases}$

In Problems 17 and 18, find the value of each determinant.

17. $\begin{vmatrix} -2 & 5 \\ 3 & 7 \end{vmatrix}$

18. $\begin{vmatrix} 2 & -4 & 6 \\ 1 & 4 & 0 \\ -1 & 2 & -4 \end{vmatrix}$

In Problems 19 and 20, use Cramer's Rule, if possible, to solve each system.

19. $\begin{cases} 4x + 3y = -23 \\ 3x - 5y = 19 \end{cases}$

20. $\begin{cases} 4x - 3y + 2z = 15 \\ -2x + y - 3z = -15 \\ 5x - 5y + 2z = 18 \end{cases}$

In Problems 21 and 22, solve each system of equations.

21. $\begin{cases} 3x^2 + y^2 = 12 \\ y^2 = 9x \end{cases}$

22. $\begin{cases} 2y^2 - 3x^2 = 5 \\ y - x = 1 \end{cases}$

23. Graph the system of inequalities: $\begin{cases} x^2 + y^2 \le 100 \\ 4x - 3y \ge 0 \end{cases}$

In Problems 24 and 25, write the partial fraction decomposition of each rational expression.

24. $\dfrac{3x + 7}{(x + 3)^2}$

25. $\dfrac{4x^2 - 3}{x(x^2 + 3)^2}$

26. Graph the system of inequalities. Tell whether the graph is bounded or unbounded, and label all corner points.

$$\begin{cases} x \ge 0 \\ y \ge 0 \\ x + 2y \ge 8 \\ 2x - 3y \ge 2 \end{cases}$$

27. Maximize $z = 5x + 8y$ subject to $x \ge 0, 2x + y \le 8$, and $x - 3y \le -3$.

28. Megan went clothes shopping and bought 2 pairs of flare jeans, 2 camisoles, and 4 T-shirts for $90.00. At the same store, Paige bought one pair of flare jeans and 3 T-shirts for $42.50, while Kara bought 1 pair of flare jeans, 3 camisoles, and 2 T-shirts for $62.00. Determine the price of each clothing item.

CUMULATIVE REVIEW

In Problems 1–6, solve each equation.

1. $2x^2 - x = 0$

2. $\sqrt{3x + 1} = 4$

3. $2x^3 - 3x^2 - 8x - 3 = 0$

4. $3^x = 9^{x+1}$

5. $\log_3(x - 1) + \log_3(2x + 1) = 2$

6. $3^x = e$

7. Determine whether the function $g(x) = \dfrac{2x^3}{x^4 + 1}$ is even, odd, or neither. Is the graph of g symmetric with respect to the x-axis, y-axis, or origin?

8. Find the center and radius of the circle $x^2 + y^2 - 2x + 4y - 11 = 0$. Graph the circle.

9. Graph $f(x) = 3^{x-2} + 1$ using transformations. What is the domain, range, and horizontal asymptote of f?

10. The function $f(x) = \dfrac{5}{x + 2}$ is one-to-one. Find f^{-1}. Find the domain and the range of f and the domain and the range of f^{-1}.

11. Graph each equation.
(a) $y = 3x + 6$
(b) $x^2 + y^2 = 4$
(c) $y = x^3$
(d) $y = \dfrac{1}{x}$
(e) $y = \sqrt{x}$
(f) $y = e^x$
(g) $y = \ln x$
(h) $2x^2 + 5y^2 = 1$
(i) $x^2 - 3y^2 = 1$
(j) $x^2 - 2x - 4y + 1 = 0$

12. $f(x) = x^3 - 3x + 5$
(a) Using a graphing utility, graph f and approximate the zero(s) of f.
(b) Using a graphing utility, approximate the local maxima and local minima.
(c) Determine the intervals on which f is increasing.

CHAPTER PROJECTS

I. **Markov Chains** A **Markov chain** (or process) is one in which future outcomes are determined by a current state. Future outcomes are based on probabilities. The probability of moving to a certain state depends only on the state previously occupied and does not vary with time. An example of a Markov chain is the maximum education achieved by children based on the highest education attained by their parents, where the states are (1) earned college degree, (2) high school diploma only, (3) elementary school only. If p_{ij} is the probability of moving from state i to state j, the **transition matrix** is the

$m \times m$ matrix

$$P = \begin{bmatrix} p_{11} & p_{12} & \cdots & p_{1m} \\ \vdots & \vdots & & \vdots \\ p_{m1} & p_{m2} & \cdots & p_{mm} \end{bmatrix}$$

The table represents the probabilities of the highest educational level of children based on the highest educational level of their parents. For example, the table shows that the probability p_{21} is 40% that parents with a high-school education (row 2) will have children with a college education (column 1).

Highest Educational Level of Parents	Maximum Education That Children Achieve		
	College	**High School**	**Elementary**
College	80%	18%	2%
High school	40%	50%	10%
Elementary	20%	60%	20%

1. Convert the percentages to decimals.
2. What is the transition matrix?

3. Sum across the rows. What do you notice? Why do you think that you obtained this result?
4. If P is the transition matrix of a Markov chain, the (i, j)th entry of P^n (nth power of P) gives the probability of passing from state i to state j in n stages. What is the probability that a grandchild of a college graduate is a college graduate?
5. What is the probability that the grandchild of a high school graduate finishes college?
6. The row vector $v^{(0)} = [0.288 \quad 0.569 \quad 0.143]$ represents the proportion of the U.S. population 25 years or older that has college, high school, and elementary school, respectively, as the highest educational level in 2007.* In a Markov chain the probability distribution $v^{(k)}$ after k stages is $v^{(k)} = v^{(0)}P^k$, where P^k is the kth power of the transition matrix. What will be the distribution of highest educational attainment of the grandchildren of the current population?
7. Calculate $P^3, P^4, P^5, \ldots$. Continue until the matrix does not change. This is called the long-run distribution. What is the long-run distribution of highest educational attainment of the population?

Source: U.S. Census Bureau.

The following projects are available at the Instructor's Resource Center (IRC):

II. Project at Motorola: *Error Control Coding* The high-powered engineering needed to assure that wireless communications are transmitted correctly is analyzed using matrices to control coding errors.

III. Using Matrices to Find the Line of Best Fit Have you wondered how our calculators get a line of best fit? See how to find the line by solving a matrix equation.

IV. CBL Experiment Simulate two people walking toward each other at a constant rate. Then solve the resulting system of equations to determine when and where they will meet.

11 Sequences; Induction; the Binomial Theorem

World Population Prospects

In July 2009, the world population will reach 6.8 billion, 313 million more than in 2005 or a gain of 78 million persons annually. Assuming that fertility levels continue to decline, the world population is expected to reach 9.1 billion in 2050 and to be increasing by about 33 million persons annually at that time.

Future population growth is highly dependent on the path that future fertility takes. Fertility is projected to decline from 2.56 children per woman in 2005–2010 to 2.02 children per woman in 2045–2050. If fertility were to remain about half a child above the levels projected, world population would reach 10.5 billion by 2050. A fertility path half a child below the levels projected would lead to a population of 8 billion by mid-century. Consequently, population growth until 2050 is inevitable even if the decline of fertility accelerates.

In the more developed regions, fertility has increased slightly in recent years so that its estimated level in 2005–2010, 1.64 children per woman according to the *2008 Revision*, is higher than the one reported in the *2006 Revision* (1.60 children per woman). As a result of the slightly higher projected fertility and a sustained net immigration averaging 2.4 million annually, the population of the more developed regions is expected to increase slightly from 1.23 billion in 2009 to 1.28 billion in 2050.

The population of the 49 least developed countries is still the fastest growing in the world, at 2.3 percent per year. Although its rate of increase is expected to moderate significantly over the next decades, the population of the least developed countries is projected to double, passing from 0.84 billion in 2009 to 1.7 billion in 2050. Growth in the rest of the developing world is also projected to be robust, though less rapid, with its population rising from 4.8 billion to 6.2 billion between 2009 and 2050 according to the medium variant.

Although the population of all countries is expected to age over the foreseeable future, the population will remain relatively young in countries where fertility is still high, many of which are experiencing very rapid population growth. High population growth rates prevail in many developing countries, most of which are least developed. Between 2010 and 2050, the populations of 31 countries, the majority of which are least developed, will double or more. Among them, the populations of Afghanistan, Burkina Faso, Niger, Somalia, Timor-Leste, and Uganda are projected to increase by 150 per cent or more.

Source: Adapted from Millennium Development Goals Report 2008, pp. IX–X. © United Nations, 2009. Reproduced with permission.

— See Chapter Project I —

◁ A Look Back, A Look Ahead ▷

This chapter may be divided into three independent parts: Sections 11.1–11.3, Section 11.4, and Section 11.5.

In Chapter 1, we defined a function and its domain, which was usually some set of real numbers. In Sections 11.1–11.3, we discuss a sequence, which is a function whose domain is the set of positive integers.

Throughout this text, where it seemed appropriate, we have given proofs of many of the results. In Section 11.4, a technique for proving theorems involving natural numbers is discussed.

In Appendix A, Section A.3, there are formulas for expanding $(x + a)^2$ and $(x + a)^3$. In Section 11.5, we discuss the Binomial Theorem, a formula for the expansion of $(x + a)^n$, where n is any positive integer.

The topics introduced in this chapter are covered in more detail in courses titled *Discrete Mathematics*. Applications of these topics can be found in the fields of computer science, engineering, business and economics, the social sciences, and the physical and biological sciences.

11.1 Sequences

PREPARING FOR THIS SECTION   *Before getting started, review the following concept:*

• Functions (Section 1.1, pp. 45–53)

Now Work the 'Are You Prepared?' problems on page 833.

OBJECTIVES   **1** Write the First Several Terms of a Sequence (p. 827)
  2 Write the Terms of a Sequence Defined by a Recursive Formula (p. 830)
  3 Use Summation Notation (p. 831)
  4 Find the Sum of a Sequence (p. 832)

When you hear the word *sequence* as in "a sequence of events," you likely think of something that happens first, then second, and so on. In mathematics, the word *sequence* also deals with outcomes that are first, second, and so on.

DEFINITION   A **sequence** is a function whose domain is the set of positive integers.

So in a sequence the inputs are $1, 2, 3, \ldots$. Because a sequence is a function, it will have a graph. In Figure 1(a), we show the graph of the function $f(x) = \dfrac{1}{x}, x > 0$. If all the points on this graph were removed except those whose x-coordinates are positive integers, that is, if all points were removed except $(1, 1), \left(2, \dfrac{1}{2}\right), \left(3, \dfrac{1}{3}\right)$, and so on, the remaining points would be the graph of the sequence $f(n) = \dfrac{1}{n}$, as shown in Figure 1(b). Notice that we use n to represent the independent variable in a sequence. This serves to remind us that n is a positive integer.

Figure 1

(a) $f(x) = \frac{1}{x}, x > 0$ (b) $f(n) = \frac{1}{n}$, n a positive integer

1 Write the First Several Terms of a Sequence

A sequence is usually represented by listing its values in order. For example, the sequence whose graph is given in Figure 1(b) might be represented as

$$f(1), f(2), f(3), f(4), \ldots \quad \text{or} \quad 1, \frac{1}{2}, \frac{1}{3}, \frac{1}{4}, \ldots$$

The list never ends, as the ellipsis indicates. The numbers in this ordered list are called the **terms** of the sequence.

In dealing with sequences, we usually use subscripted letters, such as a_1, to represent the first term, a_2 for the second term, a_3 for the third term, and so on. For the sequence $f(n) = \dfrac{1}{n}$, we write

$$\underbrace{a_1 = f(1) = 1,}_{\text{first term}} \quad \underbrace{a_2 = f(2) = \frac{1}{2},}_{\text{second term}} \quad \underbrace{a_3 = f(3) = \frac{1}{3},}_{\text{third term}} \quad \underbrace{a_4 = f(4) = \frac{1}{4}, \ldots}_{\text{fourth term}} \quad \underbrace{a_n = f(n) = \frac{1}{n}, \ldots}_{n^{\text{th}} \text{ term}}$$

In other words, we usually do not use the traditional function notation $f(n)$ for sequences. For this particular sequence, we have a rule for the nth term, which is $a_n = \dfrac{1}{n}$, so it is easy to find any term of the sequence.

When a formula for the nth term (sometimes called the **general term**) of a sequence is known, rather than write out the terms of the sequence, we usually represent the entire sequence by placing braces around the formula for the nth term.

For example, the sequence whose nth term is $b_n = \left(\dfrac{1}{2}\right)^n$ may be represented as

$$\{b_n\} = \left\{\left(\frac{1}{2}\right)^n\right\}$$

or by

$$b_1 = \frac{1}{2}, \quad b_2 = \frac{1}{4}, \quad b_3 = \frac{1}{8}, \dots, \quad b_n = \left(\frac{1}{2}\right)^n, \dots$$

EXAMPLE 1

Writing the First Several Terms of a Sequence

Write down the first six terms of the following sequence and graph it.

$$\{a_n\} = \left\{\frac{n-1}{n}\right\}$$

Figure 2

Solution The first six terms of the sequence are

$$a_1 = 0, \quad a_2 = \frac{1}{2}, \quad a_3 = \frac{2}{3}, \quad a_4 = \frac{3}{4}, \quad a_5 = \frac{4}{5}, \quad a_6 = \frac{5}{6}$$

See Figure 2 for the graph.

COMMENT Graphing utilities can be used to write the terms of a sequence and graph them. Figure 3 shows the sequence given in Example 1 generated on a TI-84 Plus graphing calculator. We can see the first few terms of the sequence on the viewing window. You need to press the right arrow key to scroll right to see the remaining terms of the sequence. Figure 4 shows a graph of the sequence. Notice that the first term of the sequence is not visible since it lies on the x-axis. TRACEing the graph will allow you to see the terms of the sequence. The TABLE feature can also be used to generate the terms of the sequence. See Table 1.

Figure 4

Figure 3

Table 1

⟋⟍⟋⟍⟋⟍**Now Work** PROBLEM **17**

EXAMPLE 2

Writing the First Several Terms of a Sequence

Write down the first six terms of the following sequence and graph it.

$$\{b_n\} = \left\{(-1)^{n+1}\left(\frac{2}{n}\right)\right\}$$

Figure 5

Solution   The first six terms of the sequence are

$$b_1 = 2, \quad b_2 = -1, \quad b_3 = \frac{2}{3}, \quad b_4 = -\frac{1}{2}, \quad b_5 = \frac{2}{5}, \quad b_6 = -\frac{1}{3}$$

See Figure 5 for the graph.

Notice in the sequence $\{b_n\}$ in Example 2 that the signs of the terms **alternate.** This occurs when we use factors such as $(-1)^{n+1}$, which equals 1 if n is odd and -1 if n is even, or $(-1)^n$, which equals -1 if n is odd and 1 if n is even.

EXAMPLE 3   **Writing the First Several Terms of a Sequence**

Write down the first six terms of the following sequence and graph it.

$$\{c_n\} = \begin{cases} n & \text{if } n \text{ is even} \\ \dfrac{1}{n} & \text{if } n \text{ is odd} \end{cases}$$

Figure 6

Solution   The first six terms of the sequence are

$$c_1 = 1, \quad c_2 = 2, \quad c_3 = \frac{1}{3}, \quad c_4 = 4, \quad c_5 = \frac{1}{5}, \quad c_6 = 6$$

See Figure 6 for the graph.

Now Work PROBLEM 19

Sometimes a sequence is indicated by an observed pattern in the first few terms that makes it possible to infer the makeup of the nth term. In the example that follows, a sufficient number of terms of the sequence is given so that a natural choice for the nth term is suggested.

EXAMPLE 4   **Determining a Sequence from a Pattern**

(a) $e, \dfrac{e^2}{2}, \dfrac{e^3}{3}, \dfrac{e^4}{4}, \ldots$ $\qquad a_n = \dfrac{e^n}{n}$

(b) $1, \dfrac{1}{3}, \dfrac{1}{9}, \dfrac{1}{27}, \ldots$ $\qquad b_n = \dfrac{1}{3^{n-1}}$

(c) $1, 3, 5, 7, \ldots$ $\qquad c_n = 2n - 1$

(d) $1, 4, 9, 16, 25, \ldots$ $\qquad d_n = n^2$

(e) $1, -\dfrac{1}{2}, \dfrac{1}{3}, -\dfrac{1}{4}, \dfrac{1}{5}, \ldots$ $\qquad e_n = (-1)^{n-1}\left(\dfrac{1}{n}\right)$

Now Work PROBLEM 27

The Factorial Symbol

Some sequences in mathematics involve a special product called a *factorial*.

DEFINITION   If $n \geq 0$ is an integer, the **factorial symbol** $n!$ is defined as follows:

$$0! = 1 \qquad 1! = 1$$
$$n! = n(n-1) \cdot \cdots \cdot 3 \cdot 2 \cdot 1 \qquad \text{if } n \geq 2$$

Table 2

n	$n!$
0	1
1	1
2	2
3	6
4	24
5	120
6	720

For example, $2! = 2 \cdot 1 = 2$, $3! = 3 \cdot 2 \cdot 1 = 6$, $4! = 4 \cdot 3 \cdot 2 \cdot 1 = 24$, and so on. Table 2 lists the values of $n!$ for $0 \leq n \leq 6$.

Because

$$n! = n\underbrace{(n-1)(n-2) \cdot \ldots \cdot 3 \cdot 2 \cdot 1}_{(n-1)!}$$

we can use the formula

$$\boxed{n! = n(n-1)!}$$

to find successive factorials. For example, because $6! = 720$, we have

$$7! = 7 \cdot 6! = 7(720) = 5040$$

and

$$8! = 8 \cdot 7! = 8(5040) = 40{,}320$$

Exploration

Your calculator has a factorial key. Use it to see how fast factorials increase in value. Find the value of 69!. What happens when you try to find 70!? In fact, 70! is larger than 10^{100} (a googol).

Now Work   PROBLEM 11

2 Write the Terms of a Sequence Defined by a Recursive Formula

A second way of defining a sequence is to assign a value to the first (or the first few) term(s) and specify the nth term by a formula or equation that involves one or more of the terms preceding it. Sequences defined this way are said to be defined **recursively,** and the rule or formula is called a **recursive formula.**

EXAMPLE 5

Writing the Terms of a Recursively Defined Sequence

Write down the first five terms of the following recursively defined sequence.

$$s_1 = 1, \qquad s_n = ns_{n-1}$$

Solution

The first term is given as $s_1 = 1$. To get the second term, we use $n = 2$ in the formula $s_n = ns_{n-1}$ to get $s_2 = 2s_1 = 2 \cdot 1 = 2$. To get the third term, we use $n = 3$ in the formula to get $s_3 = 3s_2 = 3 \cdot 2 = 6$. To get a new term requires that we know the value of the preceding term. The first five terms are

$$s_1 = 1$$
$$s_2 = 2 \cdot 1 = 2$$
$$s_3 = 3 \cdot 2 = 6$$
$$s_4 = 4 \cdot 6 = 24$$
$$s_5 = 5 \cdot 24 = 120$$

Do you recognize this sequence? $s_n = n!$

EXAMPLE 6

Writing the Terms of a Recursively Defined Sequence

Write down the first five terms of the following recursively defined sequence.

$$u_1 = 1, \qquad u_2 = 1, \qquad u_n = u_{n-2} + u_{n-1}$$

Solution

We are given the first two terms. To get the third term requires that we know both of the previous two terms. That is,

$$u_1 = 1$$
$$u_2 = 1$$
$$u_3 = u_1 + u_2 = 1 + 1 = 2$$
$$u_4 = u_2 + u_3 = 1 + 2 = 3$$
$$u_5 = u_3 + u_4 = 2 + 3 = 5$$

The sequence defined in Example 6 is called the **Fibonacci sequence,** and the terms of this sequence are called **Fibonacci numbers.** These numbers appear in a wide variety of applications (see Problems 85–88).

Now Work PROBLEMS 35 AND 43

3 Use Summation Notation

It is often important to find the sum of the first n terms of a sequence $\{a_n\}$, that is,

$$a_1 + a_2 + a_3 + \cdots + a_n$$

Rather than write down all these terms, we introduce a more concise way to express the sum, called **summation notation.** Using summation notation, we write the sum as

$$a_1 + a_2 + a_3 + \cdots + a_n = \sum_{k=1}^{n} a_k$$

The symbol Σ (the Greek letter sigma, which is an S in our alphabet) is simply an instruction to sum, or add up, the terms. The integer k is called the **index** of the sum; it tells you where to start the sum and where to end it. The expression

$$\sum_{k=1}^{n} a_k$$

is an instruction to add the terms a_k of the sequence $\{a_n\}$ starting with $k = 1$ and ending with $k = n$. We read the expression as "the sum of a_k from $k = 1$ to $k = n$."

EXAMPLE 7 **Expanding Summation Notation**

Write out each sum.

(a) $\displaystyle\sum_{k=1}^{n} \frac{1}{k}$ (b) $\displaystyle\sum_{k=1}^{n} k!$

Solution (a) $\displaystyle\sum_{k=1}^{n} \frac{1}{k} = 1 + \frac{1}{2} + \frac{1}{3} + \cdots + \frac{1}{n}$ (b) $\displaystyle\sum_{k=1}^{n} k! = 1! + 2! + \cdots + n!$

Now Work PROBLEM 51

EXAMPLE 8 **Writing a Sum in Summation Notation**

Express each sum using summation notation.

(a) $1^2 + 2^2 + 3^2 + \cdots + 9^2$ (b) $1 + \dfrac{1}{2} + \dfrac{1}{4} + \dfrac{1}{8} + \cdots + \dfrac{1}{2^{n-1}}$

Solution (a) The sum $1^2 + 2^2 + 3^2 + \cdots + 9^2$ has 9 terms, each of the form k^2, and starts at $k = 1$ and ends at $k = 9$:

$$1^2 + 2^2 + 3^2 + \cdots + 9^2 = \sum_{k=1}^{9} k^2$$

(b) The sum

$$1 + \frac{1}{2} + \frac{1}{4} + \frac{1}{8} + \cdots + \frac{1}{2^{n-1}}$$

has n terms, each of the form $\dfrac{1}{2^{k-1}}$, and starts at $k = 1$ and ends at $k = n$:

$$1 + \frac{1}{2} + \frac{1}{4} + \frac{1}{8} + \cdots + \frac{1}{2^{n-1}} = \sum_{k=1}^{n} \frac{1}{2^{k-1}}$$

The index of summation need not always begin at 1 or end at n; for example, we could have expressed the sum in Example 8(b) as

$$\sum_{k=0}^{n-1} \frac{1}{2^k} = 1 + \frac{1}{2} + \frac{1}{4} + \cdots + \frac{1}{2^{n-1}}$$

Letters other than k may be used as the index. For example,

$$\sum_{j=1}^{n} j! \quad \text{and} \quad \sum_{i=1}^{n} i!$$

each represent the same sum as the one given in Example 7(b).

Now Work   PROBLEM 61

4 Find the Sum of a Sequence

Next we list some properties of sequences using summation notation. These properties are useful for adding the terms of a sequence.

THEOREM

Properties of Sequences

If $\{a_n\}$ and $\{b_n\}$ are two sequences and c is a real number, then:

$$\sum_{k=1}^{n} (ca_k) = ca_1 + ca_2 + \cdots + ca_n = c(a_1 + a_2 + \cdots + a_n) = c\sum_{k=1}^{n} a_k \quad \textbf{(1)}$$

$$\sum_{k=1}^{n} (a_k + b_k) = \sum_{k=1}^{n} a_k + \sum_{k=1}^{n} b_k \quad \textbf{(2)}$$

$$\sum_{k=1}^{n} (a_k - b_k) = \sum_{k=1}^{n} a_k - \sum_{k=1}^{n} b_k \quad \textbf{(3)}$$

$$\sum_{k=j+1}^{n} a_k = \sum_{k=1}^{n} a_k - \sum_{k=1}^{j} a_k, \quad \text{where } 0 < j < n \quad \textbf{(4)}$$

The proof of property (1) follows from the distributive property of real numbers. The proofs of properties 2 and 3 are based on the commutative and associative properties of real numbers. Property (4) states that the sum from $j + 1$ to n equals the sum from 1 to n minus the sum from 1 to j. It can be helpful to employ this property when the index of summation begins at a number larger than 1.

Next we give some formulas for finding the sums of certain sequences.

THEOREM

Formulas for Sums of Sequences

$$\sum_{k=1}^{n} c = \underbrace{c + c + \cdots + c}_{n \text{ terms}} = cn \quad c \text{ is a real number} \quad \textbf{(5)}$$

$$\sum_{k=1}^{n} k = 1 + 2 + 3 + \cdots + n = \frac{n(n + 1)}{2} \quad \textbf{(6)}$$

$$\sum_{k=1}^{n} k^2 = 1^2 + 2^2 + 3^2 + \cdots + n^2 = \frac{n(n + 1)(2n + 1)}{6} \quad \textbf{(7)}$$

$$\sum_{k=1}^{n} k^3 = 1^3 + 2^3 + 3^3 + \cdots + n^3 = \left[\frac{n(n + 1)}{2}\right]^2 \quad \textbf{(8)}$$

The proof of formula (5) follows from the definition of summation notation. You are asked to prove formula (6) in Problem 92. The proofs of formulas (7) and (8) require mathematical induction, which is discussed in Section 11.4.

Notice the difference between formulas (5) and (6). In (5), the constant c is being summed from 1 to n, while in (6) the index of summation k is being summed from 1 to n.

EXAMPLE 9

Finding the Sum of a Sequence

Find the sum of each sequence.

(a) $\displaystyle\sum_{k=1}^{5}(3k)$

(b) $\displaystyle\sum_{k=1}^{10}(k^3+1)$

(c) $\displaystyle\sum_{k=1}^{24}(k^2-7k+2)$

(d) $\displaystyle\sum_{k=6}^{20}(4k^2)$

Solution

(a) $\displaystyle\sum_{k=1}^{5}(3k)=3\sum_{k=1}^{5}k$ Property (1)

$\qquad\qquad = 3\left(\dfrac{5(5+1)}{2}\right)$ Formula (6)

$\qquad\qquad = 3(15)$

$\qquad\qquad = 45$

(b) $\displaystyle\sum_{k=1}^{10}(k^3+1)=\sum_{k=1}^{10}k^3+\sum_{k=1}^{10}1$ Property (2)

$\qquad\qquad = \left(\dfrac{10(10+1)}{2}\right)^2+1(10)$ Formulas (8) and (5)

$\qquad\qquad = 3025+10$

$\qquad\qquad = 3035$

(c) $\displaystyle\sum_{k=1}^{24}(k^2-7k+2)=\sum_{k=1}^{24}k^2-\sum_{k=1}^{24}(7k)+\sum_{k=1}^{24}2$ Properties (2) and (3)

$\qquad\qquad = \displaystyle\sum_{k=1}^{24}k^2-7\sum_{k=1}^{24}k+\sum_{k=1}^{24}2$ Property (1)

$\qquad\qquad = \dfrac{24(24+1)(2\cdot24+1)}{6}-7\left(\dfrac{24(24+1)}{2}\right)+2(24)$ Formulas (7), (6), (5)

$\qquad\qquad = 4900-2100+48$

$\qquad\qquad = 2848$

(d) Notice that the index of summation starts at 6. We use property (4) as follows:

$$\sum_{k=6}^{20}(4k^2)=4\sum_{k=6}^{20}k^2=4\left[\sum_{k=1}^{20}k^2-\sum_{k=1}^{5}k^2\right]=4\left[\dfrac{20(21)(41)}{6}-\dfrac{5(6)(11)}{6}\right]$$

$\qquad\qquad\qquad$ Property (1) Property (4) $\qquad\qquad$ Formula (7)

$$= 4[2870-55]=11{,}260$$

Now Work PROBLEM 73

11.1 Assess Your Understanding

'Are You Prepared?' *Answers are given at the end of these exercises. If you get a wrong answer, read the pages listed in red.*

1. For the function $f(x)=\dfrac{x-1}{x}$, find $f(2)$ and $f(3)$. (pp. 46–49)

2. *True or False* A function is a relation between two sets D and R so that each element x in the first set D is related to exactly one element y in the second set R. (pp. 43–46)

Concepts and Vocabulary

3. A(n) _____ is a function whose domain is the set of positive integers.

4. *True or False* The notation a_5 represents the fifth term of a sequence.

5. If $n \geq 0$ is an integer, then $n! =$ _____ when $n \geq 2$.

6. The sequence $a_1 = 5$, $a_n = 3a_{n-1}$ is an example of a _____ sequence.

7. The notation $a_1 + a_2 + a_3 + \cdots + a_n = \sum_{k=1}^{n} a_k$ is an example of _____ notation.

8. *True or False* $\sum_{k=1}^{n} k = 1 + 2 + 3 + \cdots + n = \dfrac{n(n+1)}{2}$

Skill Building

In Problems 9–14, evaluate each factorial expression.

9. $10!$

10. $9!$

11. $\dfrac{9!}{6!}$

12. $\dfrac{12!}{10!}$

13. $\dfrac{3! \, 7!}{4!}$

14. $\dfrac{5! \, 8!}{3!}$

In Problems 15–26, write down the first five terms of each sequence.

15. $\{s_n\} = \{n\}$

16. $\{s_n\} = \{n^2 + 1\}$

17. $\{a_n\} = \left\{\dfrac{n}{n+2}\right\}$

18. $\{b_n\} = \left\{\dfrac{2n+1}{2n}\right\}$

19. $\{c_n\} = \{(-1)^{n+1}n^2\}$

20. $\{d_n\} = \left\{(-1)^{n-1}\left(\dfrac{n}{2n-1}\right)\right\}$

21. $\{s_n\} = \left\{\dfrac{2^n}{3^n + 1}\right\}$

22. $\{s_n\} = \left\{\left(\dfrac{4}{3}\right)^n\right\}$

23. $\{t_n\} = \left\{\dfrac{(-1)^n}{(n+1)(n+2)}\right\}$

24. $\{a_n\} = \left\{\dfrac{3^n}{n}\right\}$

25. $\{b_n\} = \left\{\dfrac{n}{e^n}\right\}$

26. $\{c_n\} = \left\{\dfrac{n^2}{2^n}\right\}$

In Problems 27–34, the given pattern continues. Write down the nth term of a sequence $\{a_n\}$ suggested by the pattern.

27. $\dfrac{1}{2}, \dfrac{2}{3}, \dfrac{3}{4}, \dfrac{4}{5}, \cdots$

28. $\dfrac{1}{1 \cdot 2}, \dfrac{1}{2 \cdot 3}, \dfrac{1}{3 \cdot 4}, \dfrac{1}{4 \cdot 5}, \cdots$

29. $1, \dfrac{1}{2}, \dfrac{1}{4}, \dfrac{1}{8}, \cdots$

30. $\dfrac{2}{3}, \dfrac{4}{9}, \dfrac{8}{27}, \dfrac{16}{81}, \cdots$

31. $1, -1, 1, -1, 1, -1, \ldots$

32. $1, \dfrac{1}{2}, 3, \dfrac{1}{4}, 5, \dfrac{1}{6}, 7, \dfrac{1}{8}, \cdots$

33. $1, -2, 3, -4, 5, -6, \ldots$

34. $2, -4, 6, -8, 10, \ldots$

In Problems 35–48, a sequence is defined recursively. Write down the first five terms.

35. $a_1 = 2$; $a_n = 3 + a_{n-1}$

36. $a_1 = 3$; $a_n = 4 - a_{n-1}$

37. $a_1 = -2$; $a_n = n + a_{n-1}$

38. $a_1 = 1$; $a_n = n - a_{n-1}$

39. $a_1 = 5$; $a_n = 2a_{n-1}$

40. $a_1 = 2$; $a_n = -a_{n-1}$

41. $a_1 = 3$; $a_n = \dfrac{a_{n-1}}{n}$

42. $a_1 = -2$; $a_n = n + 3a_{n-1}$

43. $a_1 = 1$; $a_2 = 2$; $a_n = a_{n-1} \cdot a_{n-2}$

44. $a_1 = -1$; $a_2 = 1$; $a_n = a_{n-2} + na_{n-1}$

45. $a_1 = A$; $a_n = a_{n-1} + d$

46. $a_1 = A$; $a_n = ra_{n-1}$, $r \neq 0$

47. $a_1 = \sqrt{2}$; $a_n = \sqrt{2 + a_{n-1}}$

48. $a_1 = \sqrt{2}$; $a_n = \sqrt{\dfrac{a_{n-1}}{2}}$

In Problems 49–58, write out each sum.

49. $\sum_{k=1}^{n} (k+2)$

50. $\sum_{k=1}^{n} (2k+1)$

51. $\sum_{k=1}^{n} \dfrac{k^2}{2}$

52. $\sum_{k=1}^{n} (k+1)^2$

53. $\sum_{k=0}^{n} \dfrac{1}{3^k}$

54. $\sum_{k=0}^{n} \left(\dfrac{3}{2}\right)^k$

55. $\sum_{k=0}^{n-1} \dfrac{1}{3^{k+1}}$

56. $\sum_{k=0}^{n-1} (2k+1)$

57. $\sum_{k=2}^{n} (-1)^k \ln k$

58. $\sum_{k=3}^{n} (-1)^{k+1} 2^k$

In Problems 59–68, express each sum using summation notation.

59. $1 + 2 + 3 + \cdots + 20$

60. $1^3 + 2^3 + 3^3 + \cdots + 8^3$

61. $\dfrac{1}{2} + \dfrac{2}{3} + \dfrac{3}{4} + \cdots + \dfrac{13}{13+1}$

62. $1 + 3 + 5 + 7 + \cdots + [2(12) - 1]$

63. $1 - \dfrac{1}{3} + \dfrac{1}{9} - \dfrac{1}{27} + \cdots + (-1)^6\left(\dfrac{1}{3^6}\right)$

64. $\dfrac{2}{3} - \dfrac{4}{9} + \dfrac{8}{27} - \cdots + (-1)^{12}\left(\dfrac{2}{3}\right)^{11}$

65. $3 + \dfrac{3^2}{2} + \dfrac{3^3}{3} + \cdots + \dfrac{3^n}{n}$

66. $\dfrac{1}{e} + \dfrac{2}{e^2} + \dfrac{3}{e^3} + \cdots + \dfrac{n}{e^n}$

67. $a + (a + d) + (a + 2d) + \cdots + (a + nd)$

68. $a + ar + ar^2 + \cdots + ar^{n-1}$

In Problems 69–80, find the sum of each sequence.

69. $\displaystyle\sum_{k=1}^{40} 5$

70. $\displaystyle\sum_{k=1}^{50} 8$

71. $\displaystyle\sum_{k=1}^{40} k$

72. $\displaystyle\sum_{k=1}^{24} (-k)$

73. $\displaystyle\sum_{k=1}^{20} (5k + 3)$

74. $\displaystyle\sum_{k=1}^{26} (3k - 7)$

75. $\displaystyle\sum_{k=1}^{16} (k^2 + 4)$

76. $\displaystyle\sum_{k=0}^{14} (k^2 - 4)$

77. $\displaystyle\sum_{k=10}^{60} (2k)$

78. $\displaystyle\sum_{k=8}^{40} (-3k)$

79. $\displaystyle\sum_{k=5}^{20} k^3$

80. $\displaystyle\sum_{k=4}^{24} k^3$

Applications and Extensions

81. Credit Card Debt John has a balance of $3000 on his Discover card that charges 1% interest per month on any unpaid balance. John can afford to pay $100 toward the balance each month. His balance each month after making a $100 payment is given by the recursively defined sequence

$$B_0 = \$3000 \qquad B_n = 1.01B_{n-1} - 100$$

Determine John's balance after making the first payment. That is, determine B_1.

82. Trout Population A pond currently has 2000 trout in it. A fish hatchery decides to add an additional 20 trout each month. In addition, it is known that the trout population is growing 3% per month. The size of the population after n months is given by the recursively defined sequence

$$p_0 = 2000 \qquad p_n = 1.03p_{n-1} + 20$$

How many trout are in the pond after two months? That is, what is p_2?

83. Car Loans Phil bought a car by taking out a loan for $18,500 at 0.5% interest per month. Phil's normal monthly payment is $434.47 per month, but he decides that he can afford to pay $100 extra toward the balance each month. His balance each month is given by the recursively defined sequence

$$B_0 = \$18,500 \qquad B_n = 1.005B_{n-1} - 534.47$$

Determine Phil's balance after making the first payment. That is, determine B_1.

84. Environmental Control The Environmental Protection Agency (EPA) determines that Maple Lake has 250 tons of pollutant as a result of industrial waste and that 10% of the pollutant present is neutralized by solar oxidation every year. The EPA imposes new pollution control laws that result in 15 tons of new pollutant entering the lake each year. The amount of pollutant in the lake after n years is given by the recursively defined sequence

$$p_0 = 250 \qquad p_n = 0.9p_{n-1} + 15$$

Determine the amount of pollutant in the lake after 2 years. That is, determine p_2.

85. Growth of a Rabbit Colony A colony of rabbits begins with one pair of mature rabbits, which will produce a pair of offspring (one male, one female) each month. Assume that all rabbits mature in 1 month and produce a pair of offspring (one male, one female) after 2 months. If no rabbits ever die, how many pairs of mature rabbits are there after 7 months?

[**Hint:** A Fibonacci sequence models this colony. Do you see why?]

1 mature pair
1 mature pair
2 mature pairs
3 mature pairs

86. Fibonacci Sequence Let

$$u_n = \dfrac{\left(1 + \sqrt{5}\right)^n - \left(1 - \sqrt{5}\right)^n}{2^n\sqrt{5}}$$

define the nth term of a sequence.
(a) Show that $u_1 = 1$ and $u_2 = 1$.
(b) Show that $u_{n+2} = u_{n+1} + u_n$.
(c) Draw the conclusion that $\{u_n\}$ is a Fibonacci sequence.

87. Pascal's Triangle Divide the triangular array shown (called Pascal's triangle) using diagonal lines as indicated. Find the sum of the numbers in each diagonal row. Do you recognize this sequence?

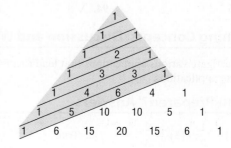

88. Fibonacci Sequence Use the result of Problem 86 to do the following problems:
(a) Write the first 11 terms of the Fibonacci sequence.
(b) Write the first 10 terms of the ratio $\dfrac{u_{n+1}}{u_n}$.
(c) As n gets large, what number does the ratio approach? This number is referred to as the **golden ratio**. Rectangles whose sides are in this ratio were considered pleasing to the eye by the Greeks. For example, the façade of the Parthenon was constructed using the golden ratio.
(d) Write down the first 10 terms of the ratio $\dfrac{u_n}{u_{n+1}}$.
(e) As n gets large, what number does the ratio approach? This number is referred to as the **conjugate golden ratio**. This ratio is believed to have been used in the construction of the Great Pyramid in Egypt. The ratio equals the sum of the areas of the four face triangles divided by the total surface area of the Great Pyramid.

89. Approximating $f(x) = e^x$ In calculus, it can be shown that

$$f(x) = e^x = \sum_{k=0}^{\infty} \frac{x^k}{k!}$$

We can approximate the value of $f(x) = e^x$ for any x using the following sum

$$f(x) = e^x \approx \sum_{k=0}^{n} \frac{x^k}{k!}$$

for some n.
(a) Approximate $f(1.3)$ with $n = 4$
(b) Approximate $f(1.3)$ with $n = 7$.
(c) Use a calculator to approximate $f(1.3)$.
(d) Using trial and error along with a graphing utility's SEQuence mode, determine the value of n required to approximate $f(1.3)$ correct to eight decimal places.

90. Approximating $f(x) = e^x$ Refer to Problem 89.
(a) Approximate $f(-2.4)$ with $n = 3$.
(b) Approximate $f(-2.4)$ with $n = 6$.
(c) Use a calculator to approximate $f(-2.4)$.
(d) Using trial and error along with a graphing utility's SEQuence mode, determine the value of n required to approximate $f(-2.4)$ correct to eight decimal places.

91. Bode's Law In 1772, Johann Bode published the following formula for predicting the mean distances, in astronomical units (AU), of the planets from the sun:

$$a_1 = 0.4 \qquad a_n = 0.4 + 0.3 \cdot 2^{n-2}, \, n \geq 2$$

where n is the number of the planet from the sun.
(a) Determine the first eight terms of this sequence.
(b) At the time of Bode's publication, the known planets were Mercury (0.39 AU), Venus (0.72 AU), Earth (1 AU), Mars (1.52 AU), Jupiter (5.20 AU), and Saturn (9.54 AU). How do the actual distances compare to the terms of the sequence?
(e) The planet Uranus was discovered in 1781 and the asteroid Ceres* was discovered in 1801. The mean orbital distances from the sun to Uranus and Ceres* are 19.2 AU and 2.77 AU, respectively. How well do these values fit within the sequence?
(d) Determine the ninth and tenth terms of Bode's sequence.
(e) The planets Neptune and Pluto* were discovered in 1846 and 1930, respectively. Their mean orbital distances from the sun are 30.07 AU and 39.44 AU, respectively. How do these actual distances compare to the terms of the sequence?
(f) On July 29, 2005, NASA announced the discovery of a dwarf planet* ($n = 11$), which has been named Eris. Use Bode's Law to predict the mean orbital distance of Eris from the sun. Its actual mean distance is not yet known, but Eris is currently about 97 astronomical units from the sun.

Sources: NASA.

92. Show that

$$1 + 2 + \cdots + (n - 1) + n = \frac{n(n+1)}{2}$$

[Hint: Let

$$S = 1 + 2 + \cdots + (n - 1) + n$$
$$S = n + (n - 1) + (n - 2) + \cdots + 1$$

Add these equations. Then

$$2S = \underbrace{[1 + n] + [2 + (n - 1)] + \cdots + [n + 1]}_{n \text{ terms in brackets}}$$

Now complete the derivation.]

*Ceres, Haumea, Makemake, Pluto, and Eris are referred to as dwarf planets.

Computing Square Roots A method for approximating $\sqrt{p}$ can be traced back to the Babylonians. The formula is given by the recursively defined sequence

$$a_0 = k \qquad a_n = \frac{1}{2}\left(a_{n-1} + \frac{p}{a_{n-1}}\right)$$

where k is an initial guess as to the value of the square root. Use this recursive formula to approximate the following square roots by finding a_5. Compare this result to the value provided by your calculator.

93. $\sqrt{5}$ **94.** $\sqrt{8}$ **95.** $\sqrt{21}$ **96.** $\sqrt{89}$

Explaining Concepts: Discussion and Writing

97. Investigate various applications that lead to a Fibonacci sequence, such as art, architecture, or financial markets. Write an essay on these applications.

'Are You Prepared?' Answers

1. $f(2) = \dfrac{1}{2}; f(3) = \dfrac{2}{3}$ **2.** True

11.2 Arithmetic Sequences

OBJECTIVES 1 Determine If a Sequence Is Arithmetic (p. 837)
2 Find a Formula for an Arithmetic Sequence (p. 838)
3 Find the Sum of an Arithmetic Sequence (p. 839)

1 Determine If a Sequence Is Arithmetic

When the difference between successive terms of a sequence is always the same number, the sequence is called **arithmetic**.

DEFINITION

An **arithmetic sequence*** may be defined recursively as $a_1 = a$, $a_n - a_{n-1} = d$, or as

$$a_1 = a, \qquad a_n = a_{n-1} + d \tag{1}$$

where $a_1 = a$ and d are real numbers. The number a is the first term, and the number d is called the **common difference**.

The terms of an arithmetic sequence with first term a_1 and common difference d follow the pattern

$$a_1, \quad a_1 + d, \quad a_1 + 2d, \quad a_1 + 3d, \ldots$$

EXAMPLE 1 **Determining If a Sequence Is Arithmetic**

The sequence

$$4, \quad 6, \quad 8, \quad 10, \ldots$$

is arithmetic since the difference of successive terms is 2. The first term is $a_1 = 4$, and the common difference is $d = 2$.

●

EXAMPLE 2 **Determining If a Sequence Is Arithmetic**

Show that the following sequence is arithmetic. Find the first term and the common difference.

$$\{s_n\} = \{3n + 5\}$$

Solution The first term is $s_1 = 3 \cdot 1 + 5 = 8$. The nth term and the $(n - 1)$st term of the sequence $\{s_n\}$ are

$$s_n = 3n + 5 \quad \text{and} \quad s_{n-1} = 3(n - 1) + 5 = 3n + 2$$

Their difference d is

$$d = s_n - s_{n-1} = (3n + 5) - (3n + 2) = 5 - 2 = 3$$

Since the difference of any two successive terms is the constant 3, the sequence $\{s_n\}$ is arithmetic and the common difference is 3.

●

EXAMPLE 3 **Determining If a Sequence Is Arithmetic**

Show that the sequence $\{t_n\} = \{4 - n\}$ is arithmetic. Find the first term and the common difference.

*Sometimes called an **arithmetic progression**.

Solution The first term is $t_1 = 4 - 1 = 3$. The nth term and the $(n - 1)$st term are

$$t_n = 4 - n \quad \text{and} \quad t_{n-1} = 4 - (n - 1) = 5 - n$$

Their difference d is

$$d = t_n - t_{n-1} = (4 - n) - (5 - n) = 4 - 5 = -1$$

Since the difference of any two successive terms is the constant -1, $\{t_n\}$ is an arithmetic sequence whose common difference is -1.

●

━━━**Now Work** PROBLEM 7

2 Find a Formula for an Arithmetic Sequence

Suppose that a is the first term of an arithmetic sequence whose common difference is d. We seek a formula for the nth term, a_n. To see the pattern, we write down the first few terms.

$$a_1 = a$$
$$a_2 = a_1 + d = a_1 + 1 \cdot d$$
$$a_3 = a_2 + d = (a_1 + d) + d = a_1 + 2 \cdot d$$
$$a_4 = a_3 + d = (a_1 + 2 \cdot d) + d = a_1 + 3 \cdot d$$
$$a_5 = a_4 + d = (a_1 + 3 \cdot d) + d = a_1 + 4 \cdot d$$
$$\vdots$$
$$a_n = a_{n-1} + d = [a_1 + (n - 2)d] + d = a_1 + (n - 1)d$$

We are led to the following result:

THEOREM

nth Term of an Arithmetic Sequence

For an arithmetic sequence $\{a_n\}$ whose first term is a_1 and whose common difference is d, the nth term is determined by the formula

$$a_n = a_1 + (n - 1)d \qquad \text{(2)}$$

EXAMPLE 4 **Finding a Particular Term of an Arithmetic Sequence**

Find the forty-first term of the arithmetic sequence: $2, 6, 10, 14, 18, \ldots$

Solution The first term of this arithmetic sequence is $a_1 = 2$, and the common difference is $d = 4$. By formula (2), the nth term is

$$a_n = 2 + (n - 1)4 \quad a_n = a_1 + (n - 1)d; \, a_1 = 2, d = 4$$

The forty-first term is

$$a_{41} = 2 + (41 - 1) \cdot 4 = 162$$

●

EXAMPLE 5 **Finding a Recursive Formula for an Arithmetic Sequence**

The eighth term of an arithmetic sequence is 75, and the twentieth term is 39.

(a) Find the first term and the common difference.
(b) Give a recursive formula for the sequence.
(c) What is the nth term of the sequence?

Solution   (a) By formula (2), we know that $a_n = a_1 + (n-1)d$. As a result,

$$\begin{cases} a_8 = a_1 + 7d = 75 \\ a_{20} = a_1 + 19d = 39 \end{cases}$$

This is a system of two linear equations containing two variables, a_1 and d, which we can solve by elimination. Subtracting the second equation from the first, we get

$$-12d = 36$$
$$d = -3$$

With $d = -3$, we use $a_1 + 7d = 75$ and find that $a_1 = 75 - 7d = 75 - 7(-3) = 96$. The first term is $a_1 = 96$, and the common difference is $d = -3$.

(b) Using formula (1), a recursive formula for this sequence is

$$a_1 = 96, \qquad a_n = a_{n-1} - 3$$

(c) Using formula (2), a formula for the nth term of the sequence $\{a_n\}$ is

$$a_n = a_1 + (n-1)d = 96 + (n-1)(-3) = 99 - 3n$$

●

> **Exploration**
>
> Graph the recursive formula from Example 5, $a_1 = 96$, $a_n = a_{n-1} - 3$, using a graphing utility. Conclude that the graph of the recursive formula behaves like the graph of a linear function. How is d, the common difference, related to m, the slope of a line?

Now Work PROBLEMS 23 AND 29

3 Find the Sum of an Arithmetic Sequence

The next result gives two formulas for finding the sum of the first n terms of an arithmetic sequence.

THEOREM

> **Sum of the First n Terms of an Arithmetic Sequence**
>
> Let $\{a_n\}$ be an arithmetic sequence with first term a_1 and common difference d. The sum S_n of the first n terms of $\{a_n\}$ may be found in two ways:
>
> $$S_n = a_1 + a_2 + a_3 + \cdots + a_n$$
>
> $$= \sum_{k=1}^{n} [a_1 + (k-1)d] = \frac{n}{2}[2a_1 + (n-1)d] \qquad (3)$$
>
> $$= \frac{n}{2}(a_1 + a_n) \qquad (4)$$

Proof

$$S_n = a_1 + a_2 + a_3 + \cdots + a_n \qquad \text{Sum of first } n \text{ terms}$$

$$= a_1 + (a_1 + d) + (a_1 + 2d) + \cdots + [a_1 + (n-1)d] \qquad \text{Formula (2)}$$

$$= \underbrace{(a_1 + a_1 + \cdots + a_1)}_{n \text{ terms}} + [d + 2d + \cdots + (n-1)d] \qquad \text{Rearrange terms}$$

$$= na_1 + d[1 + 2 + \cdots + (n-1)]$$

$$= na_1 + d\left[\frac{(n-1)n}{2}\right] \qquad \text{Formula 6, Section 11.1}$$

$$= na_1 + \frac{n}{2}(n-1)d$$

$$= \frac{n}{2}[2a_1 + (n-1)d] \qquad \text{Factor out } \frac{n}{2}; \text{ this is Formula (3).}$$

$$= \frac{n}{2}[a_1 + a_1 + (n-1)d]$$

$$= \frac{n}{2}(a_1 + a_n) \qquad \text{Use Formula (2); this is Formula (4).}$$

■

There are two ways to find the sum of the first n terms of an arithmetic sequence. Notice that formula (3) involves the first term and common difference, whereas formula (4) involves the first term and the nth term. Use whichever form is easier.

EXAMPLE 6

Finding the Sum of an Arithmetic Sequence

Find the sum S_n of the first n terms of the sequence $\{a_n\} = \{3n + 5\}$; that is, find

$$8 + 11 + 14 + \cdots + (3n + 5) = \sum_{k=1}^{n} (3k + 5)$$

Solution

The sequence $\{a_n\} = \{3n + 5\}$ is an arithmetic sequence with first term $a_1 = 8$ and nth term $a_n = 3n + 5$. To find the sum S_n, we use formula (4).

$$S_n = \sum_{k=1}^{n} (3k + 5) = \frac{n}{2}[8 + (3n + 5)] = \frac{n}{2}(3n + 13)$$

$$\uparrow$$
$$S_n = \frac{n}{2}(a_1 + a_n)$$

Now Work PROBLEM 37

EXAMPLE 7

Finding the Sum of an Arithmetic Sequence

Find the sum: $60 + 64 + 68 + 72 + \cdots + 120$

Solution

This is the sum S_n of an arithmetic sequence $\{a_n\}$ whose first term is $a_1 = 60$ and whose common difference is $d = 4$. The nth term is $a_n = 120$. We use formula (2) to find n.

$$
\begin{aligned}
a_n &= a_1 + (n - 1)d & \text{Formula (2)} \\
120 &= 60 + (n - 1) \cdot 4 & a_n = 120, a_1 = 60, d = 4 \\
60 &= 4(n - 1) & \text{Simplify.} \\
15 &= n - 1 & \text{Simplify.} \\
n &= 16 & \text{Solve for } n.
\end{aligned}
$$

Now we use formula (4) to find the sum S_{16}.

$$60 + 64 + 68 + \cdots + 120 = S_{16} = \frac{16}{2}(60 + 120) = 1440$$

$$\uparrow$$
$$S_n = \frac{n}{2}(a_1 + a_n)$$

Now Work PROBLEM 41

EXAMPLE 8

Creating a Floor Design

A ceramic tile floor is designed in the shape of a trapezoid 20 feet wide at the base and 10 feet wide at the top. See Figure 7. The tiles, 12 inches by 12 inches, are to be placed so that each successive row contains one less tile than the preceding row. How many tiles will be required?

Figure 7

Solution The bottom row requires 20 tiles and the top row, 10 tiles. Since each successive row requires one less tile, the total number of tiles required is

$$S = 20 + 19 + 18 + \cdots + 11 + 10$$

This is the sum of an arithmetic sequence; the common difference is -1. The number of terms to be added is $n = 11$, with the first term $a_1 = 20$ and the last term $a_{11} = 10$. The sum S is

$$S = \frac{n}{2}(a_1 + a_{11}) = \frac{11}{2}(20 + 10) = 165$$

In all, 165 tiles will be required.

11.2 Assess Your Understanding

Concepts and Vocabulary

1. In a(n) _____ sequence, the difference between successive terms is a constant.

2. *True or False* For an arithmetic sequence $\{a_n\}$ whose first term is a_1 and whose common difference is d, the nth term is determined by the formula $a_n = a_1 + nd$.

3. If the fifth term of an arithmetic sequence is 12 and the common difference is 5, then the sixth term of the sequence is _____.

4. *True or False* The sum S_n of the first n terms of an arithmetic sequence $\{a_n\}$ whose first term is a_1 can be found using the formula $S_n = \frac{n}{2}(a_1 + a_n)$.

Skill Building

In Problems 5–14, show that each sequence is arithmetic. Find the common difference and write out the first four terms.

5. $\{s_n\} = \{n + 4\}$ 6. $\{s_n\} = \{n - 5\}$ 7. $\{a_n\} = \{2n - 5\}$ 8. $\{b_n\} = \{3n + 1\}$ 9. $\{c_n\} = \{6 - 2n\}$

10. $\{a_n\} = \{4 - 2n\}$ 11. $\{t_n\} = \left\{\frac{1}{2} - \frac{1}{3}n\right\}$ 12. $\{t_n\} = \left\{\frac{2}{3} + \frac{n}{4}\right\}$ 13. $\{s_n\} = \{\ln 3^n\}$ 14. $\{s_n\} = \{e^{\ln n}\}$

In Problems 15–22, find the nth term of the arithmetic sequence $\{a_n\}$ whose initial term a and common difference d are given. What is the fifty-first term?

15. $a_1 = 2;\quad d = 3$ 16. $a_1 = -2;\quad d = 4$ 17. $a_1 = 5;\quad d = -3$ 18. $a_1 = 6;\quad d = -2$

19. $a_1 = 0;\quad d = \frac{1}{2}$ 20. $a_1 = 1;\quad d = -\frac{1}{3}$ 21. $a_1 = \sqrt{2};\quad d = \sqrt{2}$ 22. $a_1 = 0;\quad d = \pi$

In Problems 23–28, find the indicated term in each arithmetic sequence.

23. 100th term of $2, 4, 6, \ldots$

24. 80th term of $-1, 1, 3, \ldots$

25. 90th term of $1, -2, -5, \ldots$

26. 80th term of $5, 0, -5, \ldots$

27. 80th term of $2, \frac{5}{2}, 3, \frac{7}{2}, \ldots$

28. 70th term of $2\sqrt{5}, 4\sqrt{5}, 6\sqrt{5}, \ldots$

In Problems 29–36, find the first term and the common difference of the arithmetic sequence described. Give a recursive formula for the sequence. Find a formula for the nth term.

29. 8th term is 8;   20th term is 44

30. 4th term is 3;   20th term is 35

31. 9th term is -5;   15th term is 31

32. 8th term is 4;   18th term is -96

33. 15th term is 0;   40th term is -50

34. 5th term is -2;   13th term is 30

35. 14th term is -1;   18th term is -9

36. 12th term is 4;   18th term is 28

In Problems 37–54, find each sum.

37. $1 + 3 + 5 + \cdots + (2n - 1)$

38. $2 + 4 + 6 + \cdots + 2n$

39. $7 + 12 + 17 + \cdots + (2 + 5n)$

40. $-1 + 3 + 7 + \cdots + (4n - 5)$

41. $2 + 4 + 6 + \cdots + 70$

42. $1 + 3 + 5 + \cdots + 59$

43. $5 + 9 + 13 + \cdots + 49$

44. $2 + 5 + 8 + \cdots + 41$

45. $73 + 78 + 83 + 88 + \cdots + 558$

46. $7 + 1 - 5 - 11 - \cdots - 299$

47. $4 + 4.5 + 5 + 5.5 + \cdots + 100$

48. $8 + 8\frac{1}{4} + 8\frac{1}{2} + 8\frac{3}{4} + 9 + \cdots + 50$

49. $\sum_{n=1}^{80}(2n-5)$ **50.** $\sum_{n=1}^{90}(3-2n)$ **51.** $\sum_{n=1}^{100}\left(6-\frac{1}{2}n\right)$ **52.** $\sum_{n=1}^{80}\left(\frac{1}{3}n+\frac{1}{2}\right)$

53. The sum of the first 120 terms of the sequence 14, 16, 18, 20, ...

54. The sum of the first 46 terms of the sequence 2, −1, −4, −7, ...

Applications and Extensions

55. Find x so that $x+3$, $2x+1$, and $5x+2$ are consecutive terms of an arithmetic sequence.

56. Find x so that $2x$, $3x+2$, and $5x+3$ are consecutive terms of an arithmetic sequence.

57. How many terms must be added in an arithmetic sequence whose first term is 11 and whose common difference is 3 to obtain a sum of 1092?

58. How many terms must be added in an arithmetic sequence whose first term is 78 and whose common difference is −4 to obtain a sum of 702?

59. Drury Lane Theater The Drury Lane Theater has 25 seats in the first row and 30 rows in all. Each successive row contains one additional seat. How many seats are in the theater?

60. Football Stadium The corner section of a football stadium has 15 seats in the first row and 40 rows in all. Each successive row contains two additional seats. How many seats are in this section?

61. Creating a Mosaic A mosaic is designed in the shape of an equilateral triangle, 20 feet on each side. Each tile in the mosaic is in the shape of an equilateral triangle, 12 inches to a side. The tiles are to alternate in color as shown in the illustration. How many tiles of each color will be required?

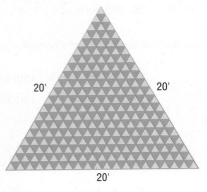

20' 20'

20'

62. Constructing a Brick Staircase A brick staircase has a total of 30 steps. The bottom step requires 100 bricks. Each successive step requires two less bricks than the prior step.
(a) How many bricks are required for the top step?
(b) How many bricks are required to build the staircase?

63. Cooling Air As a parcel of air rises (for example, as it is pushed over a mountain), it cools at the *dry adiabatic lapse rate* of 5.5°F per 1000 feet until it reaches its dew point. If the ground temperature is 67°F, write a formula for the sequence of temperatures, $\{T_n\}$, of a parcel of air that has risen n thousand feet. What is the temperature of a parcel of air if it has risen 5000 feet?

Source: National Aeronautics and Space Administration

64. Citrus Ladders Ladders used by fruit pickers are typically tapered with a wide bottom for stability and a narrow top for ease of picking. If the bottom rung of such a ladder is 49 inches wide and the top rung is 24 inches wide, how many rungs does the ladder have if each rung is 2.5 inches shorter than the one below it? How much material would be needed to make the rungs for the ladder described?

Source: www.stokesladders.com

65. Seats in an Amphitheater An outdoor amphitheater has 35 seats in the first row, 37 in the second row, 39 in the third row, and so on. There are 27 rows altogether. How many can the amphitheater seat?

66. Stadium Construction How many rows are in the corner section of a stadium containing 2040 seats if the first row has 10 seats and each successive row has 4 additional seats?

67. Salary If you received a job offer with a starting salary of $35,000 per year and a guaranteed raise of $1400 per year, how many years will it take before your aggregate salary is $280,000?

[Hint: Your aggregate salary after 2 years is $35,000 + ($35,000 + $1400).]

Explaining Concepts: Discussion and Writing

68. Make up an arithmetic sequence. Give it to a friend and ask for its twentieth term.

69. Describe the similarities and differences between arithmetic sequences and linear functions.

11.3 Geometric Sequences; Geometric Series

PREPARING FOR THIS SECTION   *Before getting started, review the following:*

• Compound Interest (Section 4.7, pp. 341–346)

Now Work the 'Are You Prepared?' problems on page 851.

OBJECTIVES   **1** Determine If a Sequence Is Geometric (p. 843)
   2 Find a Formula for a Geometric Sequence (p. 844)
   3 Find the Sum of a Geometric Sequence (p. 845)
   4 Determine Whether a Geometric Series Converges or Diverges (p. 846)
   5 Solve Annuity Problems (p. 849)

1 Determine If a Sequence Is Geometric

When the ratio of successive terms of a sequence is always the same nonzero number, the sequence is called **geometric.**

DEFINITION

A **geometric sequence*** may be defined recursively as $a_1 = a$, $\dfrac{a_n}{a_{n-1}} = r$, or as

$$a_1 = a, \qquad a_n = r a_{n-1} \tag{1}$$

where $a_1 = a$ and $r \neq 0$ are real numbers. The number a_1 is the first term, and the nonzero number r is called the **common ratio.**

The terms of a geometric sequence with first term a_1 and common ratio r follow the pattern

$$a_1, \quad a_1 r, \quad a_1 r^2, \quad a_1 r^3, \ldots$$

EXAMPLE 1

Determining If a Sequence Is Geometric

The sequence

$$2, \ 6, \ 18, \ 54, \ 162, \ldots$$

is geometric since the ratio of successive terms is 3; $\left(\dfrac{6}{2} = \dfrac{18}{6} = \dfrac{54}{18} = \cdots = 3 \right)$. The first term is $a_1 = 2$, and the common ratio is 3.

●

EXAMPLE 2

Determining If a Sequence Is Geometric

Show that the following sequence is geometric.

$$\{s_n\} = 2^{-n}$$

Find the first term and the common ratio.

Solution   The first term is $s_1 = 2^{-1} = \dfrac{1}{2}$. The nth term and the $(n-1)$st term of the sequence $\{s_n\}$ are

$$s_n = 2^{-n} \quad \text{and} \quad s_{n-1} = 2^{-(n-1)}$$

Their ratio is

$$\frac{s_n}{s_{n-1}} = \frac{2^{-n}}{2^{-(n-1)}} = 2^{-n+(n-1)} = 2^{-1} = \frac{1}{2}$$

*Sometimes called a **geometric progression.**

Because the ratio of successive terms is the nonzero constant $\frac{1}{2}$, the sequence $\{s_n\}$ is geometric with common ratio $\frac{1}{2}$.

EXAMPLE 3   **Determining If a Sequence Is Geometric**

Show that the following sequence is geometric.

$$\{t_n\} = \{3 \cdot 4^n\}$$

Find the first term and the common ratio.

Solution   The first term is $t_1 = 3 \cdot 4^1 = 12$. The nth term and the $(n-1)$st term are

$$t_n = 3 \cdot 4^n \quad \text{and} \quad t_{n-1} = 3 \cdot 4^{n-1}$$

Their ratio is

$$\frac{t_n}{t_{n-1}} = \frac{3 \cdot 4^n}{3 \cdot 4^{n-1}} = 4^{n-(n-1)} = 4$$

The sequence, $\{t_n\}$, is a geometric sequence with common ratio 4.

━━━━━**Now Work** PROBLEM 11

2 Find a Formula for a Geometric Sequence

Suppose that a_1 is the first term of a geometric sequence with common ratio $r \neq 0$. We seek a formula for the nth term a_n. To see the pattern, we write down the first few terms:

$$a_1 = a_1 \cdot 1 = a_1 r^0$$
$$a_2 = ra_1 = a_1 r^1$$
$$a_3 = ra_2 = r(a_1 r) = a_1 r^2$$
$$a_4 = ra_3 = r(a_1 r^2) = a_1 r^3$$
$$a_5 = ra_4 = r(a_1 r^3) = a_1 r^4$$
$$\vdots$$
$$a_n = ra_{n-1} = r(a_1 r^{n-2}) = a_1 r^{n-1}$$

We are led to the following result:

THEOREM   **nth Term of a Geometric Sequence**

For a geometric sequence $\{a_n\}$ whose first term is a_1 and whose common ratio is r, the nth term is determined by the formula

$$a_n = a_1 r^{n-1} \qquad r \neq 0 \tag{2}$$

EXAMPLE 4   **Finding a Particular Term of a Geometric Sequence**

(a) Find the nth term of the geometric sequence: $10, 9, \dfrac{81}{10}, \dfrac{729}{100} \cdots$

(b) Find the ninth term of this sequence.

(c) Find a recursive formula for this sequence.

Solution

(a) The first term of this geometric sequence is $a_1 = 10$ and the common ratio is $\dfrac{9}{10}$. (Use $\dfrac{9}{10}$ or $\dfrac{\frac{81}{10}}{9} = \dfrac{9}{10}$ or any two successive terms.) Then, by formula (2), the nth term is

$$a_n = 10\left(\frac{9}{10}\right)^{n-1} \qquad a_n = a_1 r^{n-1};\; a_1 = 10,\; r = \frac{9}{10}$$

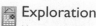

Exploration

Use a graphing utility to find the ninth term of the sequence given in Example 4. Use it to find the twentieth and fiftieth terms. Now use a graphing utility to graph the recursive formula found in Example 4(c). Conclude that the graph of the recursive formula behaves like the graph of an exponential function. How is r, the common ratio, related to a, the base of the exponential function $y = a^x$?

(b) The ninth term is

$$a_9 = 10\left(\frac{9}{10}\right)^{9-1} = 10\left(\frac{9}{10}\right)^{8} \approx 4.3046721$$

(c) The first term in the sequence is 10 and the common ratio is $r = \dfrac{9}{10}$. Using formula (1), the recursive formula is $a_1 = 10,\; a_n = \dfrac{9}{10}a_{n-1}$.

Now Work PROBLEMS 19, 27, AND 35

3 Find the Sum of a Geometric Sequence

The next result gives us a formula for finding the sum of the first n terms of a geometric sequence.

THEOREM

Sum of the First n Terms of a Geometric Sequence

Let $\{a_n\}$ be a geometric sequence with first term a_1 and common ratio r, where $r \neq 0, r \neq 1$. The sum S_n of the first n terms of $\{a_n\}$ is

$$S_n = a_1 + a_1 r + a_1 r^2 + \cdots + a_1 r^{n-1} = \sum_{k=1}^{n} a_1 r^{k-1}$$

$$= a_1 \cdot \frac{1 - r^n}{1 - r} \qquad r \neq 0, 1 \qquad (3)$$

Proof The sum S_n of the first n terms of $\{a_n\} = \{a_1 r^{n-1}\}$ is

$$S_n = a_1 + a_1 r + \cdots + a_1 r^{n-1} \qquad (4)$$

Multiply each side by r to obtain

$$rS_n = a_1 r + a_1 r^2 + \cdots + a_1 r^n \qquad (5)$$

Now, subtract (5) from (4). The result is

$$S_n - rS_n = a_1 - a_1 r^n$$
$$(1 - r)S_n = a_1(1 - r^n)$$

Since $r \neq 1$, we can solve for S_n.

$$S_n = a_1 \cdot \frac{1 - r^n}{1 - r}$$

∎

EXAMPLE 5 **Finding the Sum of the First n Terms of a Geometric Sequence**

Find the sum S_n of the first n terms of the sequence $\left\{\left(\dfrac{1}{2}\right)^n\right\}$; that is, find

$$\frac{1}{2} + \frac{1}{4} + \frac{1}{8} + \cdots + \left(\frac{1}{2}\right)^n = \sum_{k=1}^{n} \frac{1}{2}\left(\frac{1}{2}\right)^{k-1}$$

Solution The sequence $\left\{\left(\dfrac{1}{2}\right)^n\right\}$ is a geometric sequence with $a_1 = \dfrac{1}{2}$ and $r = \dfrac{1}{2}$. So we use formula (3) to get

$$S_n = \sum_{k=1}^{n} \frac{1}{2}\left(\frac{1}{2}\right)^{k-1} = \frac{1}{2} + \frac{1}{4} + \frac{1}{8} + \cdots + \left(\frac{1}{2}\right)^n$$

$$= \frac{1}{2}\left[\frac{1 - \left(\dfrac{1}{2}\right)^n}{1 - \dfrac{1}{2}}\right] \qquad \text{Formula (3); } a_1 = \frac{1}{2}, r = \frac{1}{2}$$

$$= \frac{1}{2}\left[\frac{1 - \left(\dfrac{1}{2}\right)^n}{\dfrac{1}{2}}\right]$$

$$= 1 - \left(\frac{1}{2}\right)^n$$

═══ **Now Work** PROBLEM 41

🖩 **EXAMPLE 6**

Using a Graphing Utility to Find the Sum of a Geometric Sequence

Use a graphing utility to find the sum of the first 15 terms of the sequence $\left\{\left(\dfrac{1}{3}\right)^n\right\}$; that is, find

$$\frac{1}{3} + \frac{1}{9} + \frac{1}{27} + \cdots + \left(\frac{1}{3}\right)^{15} = \sum_{k=1}^{15} \frac{1}{3}\left(\frac{1}{3}\right)^{k-1}$$

Figure 8

Solution Figure 8 shows the result obtained using a TI-84 Plus graphing calculator. The sum of the first 15 terms of the sequence $\left\{\left(\dfrac{1}{3}\right)^n\right\}$ is 0.4999999652.

```
sum(seq((1/3)^n,
n,1,15,1))
         .4999999652
```

═══ **Now Work** PROBLEM 47

4 Determine Whether a Geometric Series Converges or Diverges

DEFINITION An infinite sum of the form

$$a_1 + a_1 r + a_1 r^2 + \cdots + a_1 r^{n-1} + \cdots$$

with first term a_1 and common ratio r, is called an **infinite geometric series** and is denoted by

$$\sum_{k=1}^{\infty} a_1 r^{k-1}$$

Based on formula (3), the sum S_n of the first n terms of a geometric series is

$$S_n = a_1 \cdot \frac{1 - r^n}{1 - r} = \frac{a_1}{1 - r} - \frac{a_1 r^n}{1 - r} \qquad \textbf{(6)}$$

NOTE In calculus, we use limit notation and write

$$L = \lim_{n \to \infty} S_n = \lim_{n \to \infty} \sum_{k=1}^{n} a_1 r^{k-1} = \sum_{k=1}^{\infty} a_1 r^{k-1}$$ ∎

If this finite sum S_n approaches a number L as $n \to \infty$, we say the infinite geometric series $\sum_{k=1}^{\infty} a_1 r^{k-1}$ **converges.** We call L the **sum of the infinite geometric series** and we write

$$L = \sum_{k=1}^{\infty} a_1 r^{k-1}$$

If a series does not converge, it is called a **divergent series.**

THEOREM

Convergence of an Infinite Geometric Series

If $|r| < 1$, the infinite geometric series $\sum_{k=1}^{\infty} a_1 r^{k-1}$ converges. Its sum is

$$\sum_{k=1}^{\infty} a_1 r^{k-1} = \frac{a_1}{1 - r} \qquad (7)$$

Intuitive Proof   Since $|r| < 1$, it follows that $|r^n|$ approaches 0 as $n \to \infty$. Then, based on formula (6), the term $\dfrac{a_1 r^n}{1 - r}$ approaches 0, so the sum S_n approaches $\dfrac{a_1}{1 - r}$ as $n \to \infty$. ∎

EXAMPLE 7

Determining Whether a Geometric Series Converges or Diverges

Determine if the geometric series

$$\sum_{k=1}^{\infty} 2\left(\frac{2}{3}\right)^{k-1} = 2 + \frac{4}{3} + \frac{8}{9} + \cdots$$

converges or diverges. If it converges, find its sum.

Solution   Comparing $\sum_{k=1}^{\infty} 2\left(\dfrac{2}{3}\right)^{k-1}$ to $\sum_{k=1}^{\infty} a_1 r^{k-1}$, we see that the first term is $a_1 = 2$, and the common ratio is $r = \dfrac{2}{3}$. Since $|r| < 1$, the series converges. We use formula (7) to find its sum:

$$\sum_{k=1}^{\infty} 2\left(\frac{2}{3}\right)^{k-1} = 2 + \frac{4}{3} + \frac{8}{9} + \cdots = \frac{2}{1 - \dfrac{2}{3}} = 6$$

●

━━━ **Now Work** PROBLEM 53

EXAMPLE 8

Repeating Decimals

Show that the repeating decimal $0.999\ldots$ equals 1.

Solution   The decimal $0.999\ldots = 0.9 + 0.09 + 0.009 + \cdots = \dfrac{9}{10} + \dfrac{9}{100} + \dfrac{9}{1000} + \cdots$ is an infinite geometric series. We will write it in the form $\sum_{k=1}^{\infty} a_1 r^{k-1}$ so that we can use formula (7).

$$0.999\ldots = \frac{9}{10} + \frac{9}{100} + \frac{9}{1000} + \cdots = \sum_{k=1}^{\infty} \frac{9}{10^k} = \sum_{k=1}^{\infty} \frac{9}{10 \cdot 10^{k-1}} = \sum_{k=1}^{\infty} \frac{9}{10}\left(\frac{1}{10}\right)^{k-1}$$

Now we can compare this series to $\sum_{k=1}^{\infty} a_1 r^{k-1}$ and conclude that $a_1 = \dfrac{9}{10}$ and $r = \dfrac{1}{10}$. Since $|r| < 1$, the series converges and its sum is

$$0.999\ldots = \dfrac{\dfrac{9}{10}}{1 - \dfrac{1}{10}} = \dfrac{\dfrac{9}{10}}{\dfrac{9}{10}} = 1$$

The repeating decimal $0.999\ldots$ equals 1.

EXAMPLE 9

Pendulum Swings

Initially, a pendulum swings through an arc of 18 inches. See Figure 9. On each successive swing, the length of the arc is 0.98 of the previous length.

Figure 9

(a) What is the length of the arc of the 10th swing?
(b) On which swing is the length of the arc first less than 12 inches?
(c) After 15 swings, what total distance will the pendulum have swung?
(d) When it stops, what total distance will the pendulum have swung?

Solution

(a) The length of the first swing is 18 inches.
The length of the second swing is 0.98(18) inches.
The length of the third swing is $0.98(0.98)(18) = 0.98^2(18)$ inches.
The length of the arc of the 10th swing is

$$(0.98)^9(18) \approx 15.007 \text{ inches}$$

(b) The length of the arc of the nth swing is $(0.98)^{n-1}(18)$. For this to be exactly 12 inches requires that

$$(0.98)^{n-1}(18) = 12$$

$$(0.98)^{n-1} = \dfrac{12}{18} = \dfrac{2}{3} \qquad \textit{Divide both sides by 18.}$$

$$n - 1 = \log_{0.98}\left(\dfrac{2}{3}\right) \qquad \textit{Express as a logarithm.}$$

$$n = 1 + \dfrac{\ln\left(\dfrac{2}{3}\right)}{\ln 0.98} \approx 1 + 20.07 = 21.07 \qquad \textit{Solve for n; use the Change of Base Formula.}$$

The length of the arc of the pendulum exceeds 12 inches on the 21st swing and is first less than 12 inches on the 22nd swing.

(c) After 15 swings, the pendulum will have swung the following total distance L:

$$L = \underset{\text{1st}}{18} + \underset{\text{2nd}}{0.98(18)} + \underset{\text{3rd}}{(0.98)^2(18)} + \underset{\text{4th}}{(0.98)^3(18)} + \cdots + \underset{\text{15th}}{(0.98)^{14}(18)}$$

This is the sum of a geometric sequence. The common ratio is 0.98; the first term is 18. The sum has 15 terms, so

$$L = 18 \cdot \dfrac{1 - 0.98^{15}}{1 - 0.98} \approx 18(13.07) \approx 235.3 \text{ inches}$$

The pendulum will have swung through 235.3 inches after 15 swings.

(d) When the pendulum stops, it will have swung the following total distance T:

$$T = 18 + 0.98(18) + (0.98)^2(18) + (0.98)^3(18) + \cdots$$

This is the sum of an infinite geometric series. The common ratio is $r = 0.98$; the first term is $a_1 = 18$. Since $|r| < 1$, the series converges. Its sum is

$$T = \dfrac{a_1}{1 - r} = \dfrac{18}{1 - 0.98} = 900$$

The pendulum will have swung a total of 900 inches when it finally stops.

Now Work PROBLEM 87

5 Solve Annuity Problems

In Section 4.7 we developed the compound interest formula that gives the future value when a fixed amount of money is deposited in an account that pays interest compounded periodically. Often, though, money is invested in small amounts at periodic intervals. An **annuity** is a sequence of equal periodic deposits. The periodic deposits may be made annually, quarterly, monthly, or daily.

When deposits are made at the same time that the interest is credited, the annuity is called **ordinary.** We will only deal with ordinary annuities here. The **amount of an annuity** is the sum of all deposits made plus all interest paid.

Suppose that the interest rate that an account earns is i percent per payment period (expressed as a decimal). For example, if an account pays 12% compounded monthly (12 times a year), then $i = \dfrac{0.12}{12} = 0.01$. If an account pays 8% compounded quarterly (4 times a year), then $i = \dfrac{0.08}{4} = 0.02$.

To develop a formula for the amount of an annuity, suppose that P is deposited each payment period for n payment periods in an account that earns i percent per payment period. When the last deposit is made at the nth payment period, the first deposit of P has earned interest compounded for $n - 1$ payment periods, the second deposit of P has earned interest compounded for $n - 2$ payment periods, and so on. Table 3 shows the value of each deposit after n deposits have been made.

Table 3

Deposit	1	2	3	...	n − 1	n
Amount	$P(1 + i)^{n-1}$	$P(1 + i)^{n-2}$	$P(1 + i)^{n-3}$	...	$P(1 + i)$	P

The amount A of the annuity is the sum of the amounts shown in Table 3; that is,

$$A = P \cdot (1 + i)^{n-1} + P \cdot (1 + i)^{n-2} + \cdots + P \cdot (1 + i) + P$$
$$= P[1 + (1 + i) + \cdots + (1 + i)^{n-1}]$$

The expression in brackets is the sum of a geometric sequence with n terms and a common ratio of $(1 + i)$. As a result,

$$A = P[1 + (1 + i) + \cdots + (1 + i)^{n-2} + (1 + i)^{n-1}]$$
$$= P \frac{1 - (1 + i)^n}{1 - (1 + i)} = P \frac{1 - (1 + i)^n}{-i} = P \frac{(1 + i)^n - 1}{i}$$

We have established the following result:

THEOREM

Amount of an Annuity

Suppose that P is the deposit in dollars made at the end of each payment period for an annuity paying i percent interest per payment period. The amount A of the annuity after n deposits is

$$A = P \frac{(1 + i)^n - 1}{i} \tag{8}$$

NOTE In using formula (8), remember that when the nth deposit is made the first deposit has earned interest for $n - 1$ compounding periods. ■

EXAMPLE 10

Determining the Amount of an Annuity

To save for retirement, Brett decides to place $4000 into an Individual Retirement Account (IRA) each year for the next 30 years. What will the value of the IRA be when Brett makes his 30th deposit? Assume that the rate of return of the IRA is 10% per annum compounded annually. (This is the historical rate of return in the stock market.)

Solution   This is an ordinary annuity with $n = 30$ annual deposits of $P = \$4000$. The rate of interest per payment period is $i = \dfrac{0.10}{1} = 0.10$. The amount A of the annuity after 30 deposits is

$$A = \$4000\,\frac{(1 + 0.10)^{30} - 1}{0.10} = \$4000(164.494023) = \$657,976.09$$

●

EXAMPLE 11   **Determining the Amount of an Annuity**

To save for her daughter's college education, Ms. Miranda decides to put $50 aside every month in a credit union account paying 10% interest compounded monthly. She begins this savings program when her daughter is 3 years old. How much will she have saved by the time she makes the 180th deposit? How old is her daughter at this time?

Solution   This is an annuity with $P = \$50$, $n = 180$, and $i = \dfrac{0.10}{12}$. The amount A saved is

$$A = \$50\,\frac{\left(1 + \dfrac{0.10}{12}\right)^{180} - 1}{\dfrac{0.10}{12}} = \$50(414.47035) = \$20,723.52$$

Since there are 12 deposits per year, when the 180th deposit is made $\dfrac{180}{12} = 15$ years have passed and Ms. Miranda's daughter is 18 years old.

●

Now Work PROBLEM 91

Historical Feature

Fibonacci

Sequences are among the oldest objects of mathematical investigation, having been studied for over 3500 years. After the initial steps, however, little progress was made until about 1600.

Arithmetic and geometric sequences appear in the Rhind papyrus, a mathematical text containing 85 problems copied around 1650 BC by the Egyptian scribe Ahmes from an earlier work (see Historical Problem 1). Fibonacci (AD 1220) wrote about problems similar to those found in the Rhind papyrus, leading one to suspect that Fibonacci may have had material available that is now lost. This material would have been in the non-Euclidean Greek tradition of Heron (about AD 75) and Diophantus (about AD 250). One problem, again modified slightly, is still with us in the familiar puzzle rhyme "As I was going to St. Ives ..." (see Historical Problem 2).

The Rhind papyrus indicates that the Egyptians knew how to add up the terms of an arithmetic or geometric sequence, as did the Babylonians. The rule for summing up a geometric sequence is found in Euclid's *Elements* (Book IX, 35, 36), where, like all Euclid's algebra, it is presented in a geometric form.

Investigations of other kinds of sequences began in the 1500s, when algebra became sufficiently developed to handle the more complicated problems. The development of calculus in the 1600s added a powerful new tool, especially for finding the sum of an infinite series, and the subject continues to flourish today.

Historical Problems

1. *Arithmetic sequence problem from the Rhind papyrus (statement modified slightly for clarity)* One hundred loaves of bread are to be divided among five people so that the amounts that they receive form an arithmetic sequence. The first two together receive one-seventh of what the last three receive. How many loaves does each receive?

   [*Partial answer:* First person receives $1\frac{2}{3}$ loaves.]

2. The following old English children's rhyme resembles one of the Rhind papyrus problems.

   As I was going to St. Ives
   I met a man with seven wives

   Each wife had seven sacks
   Each sack had seven cats
   Each cat had seven kits [kittens]
   Kits, cats, sacks, wives
   How many were going to St. Ives?

   (a) Assuming that the speaker and the cat fanciers met by traveling in opposite directions, what is the answer?

   (b) How many kittens are being transported?

   (c) Kits, cats, sacks, wives; how many?

11.3 Assess Your Understanding

'Are You Prepared?' *Answers are given at the end of these exercises. If you get a wrong answer, read the pages listed in* red.

1. If $1000 is invested at 4% per annum compounded semiannually, how much is in the account after 2 years? (pp. 341–343)

2. How much do you need to invest now at 5% per annum compounded monthly so that in 1 year you will have $10,000? (pp. 345–346)

Concepts and Vocabulary

3. In a(n) _____ sequence the ratio of successive terms is a constant.

4. If $|r| < 1$, the sum of the geometric series $\sum_{k=1}^{\infty} ar^{k-1}$ is _____.

5. If a series does not converge, it is called a _____ _____.

6. *True or False* A geometric sequence may be defined recursively.

7. *True or False* In a geometric sequence the common ratio is always a positive number.

8. *True or False* For a geometric sequence with first term a_1 and common ratio r, where $r \neq 0, r \neq 1$, the sum of the first n terms is $S_n = a_1 \cdot \dfrac{1 - r^n}{1 - r}$.

Skill Building

In Problems 9–18, show that each sequence is geometric. Then find the common ratio and write out the first four terms.

9. $\{s_n\} = \{3^n\}$

10. $\{s_n\} = \{(-5)^n\}$

11. $\{a_n\} = \left\{-3\left(\dfrac{1}{2}\right)^n\right\}$

12. $\{b_n\} = \left\{\left(\dfrac{5}{2}\right)^n\right\}$

13. $\{c_n\} = \left\{\dfrac{2^{n-1}}{4}\right\}$

14. $\{d_n\} = \left\{\dfrac{3^n}{9}\right\}$

15. $\{e_n\} = \{2^{n/3}\}$

16. $\{f_n\} = \{3^{2n}\}$

17. $\{t_n\} = \left\{\dfrac{3^{n-1}}{2^n}\right\}$

18. $\{u_n\} = \left\{\dfrac{2^n}{3^{n-1}}\right\}$

In Problems 19–26, find the fifth term and the nth term of the geometric sequence whose initial term a_1 and common ratio r are given.

19. $a_1 = 2; \quad r = 3$

20. $a_1 = -2; \quad r = 4$

21. $a_1 = 5; \quad r = -1$

22. $a_1 = 6; \quad r = -2$

23. $a_1 = 0; \quad r = \dfrac{1}{2}$

24. $a_1 = 1; \quad r = -\dfrac{1}{3}$

25. $a_1 = \sqrt{2}; \quad r = \sqrt{2}$

26. $a_1 = 0; \quad r = \dfrac{1}{\pi}$

In Problems 27–32, find the indicated term of each geometric sequence.

27. 7th term of $1, \dfrac{1}{2}, \dfrac{1}{4}, \ldots$

28. 8th term of $1, 3, 9, \ldots$

29. 9th term of $1, -1, 1, \ldots$

30. 10th term of $-1, 2, -4, \ldots$

31. 8th term of $0.4, 0.04, 0.004, \ldots$

32. 7th term of $0.1, 1.0, 10.0, \ldots$

In Problems 33–40, find the nth term a_n of each geometric sequence. When given, r is the common ratio.

33. $7, 14, 28, 56, \ldots$

34. $5, 10, 20, 40, \ldots$

35. $-3, 1, -\dfrac{1}{3}, \dfrac{1}{9}, \ldots$

36. $4, 1, \dfrac{1}{4}, \dfrac{1}{16}, \ldots$

37. $a_6 = 243; \quad r = -3$

38. $a_2 = 7; \quad r = \dfrac{1}{3}$

39. $a_2 = 7; \quad a_4 = 1575$

40. $a_3 = \dfrac{1}{3}; \quad a_6 = \dfrac{1}{81}$

In Problems 41–46, find each sum.

41. $\dfrac{1}{4} + \dfrac{2}{4} + \dfrac{2^2}{4} + \dfrac{2^3}{4} + \cdots + \dfrac{2^{n-1}}{4}$

42. $\dfrac{3}{9} + \dfrac{3^2}{9} + \dfrac{3^3}{9} + \cdots + \dfrac{3^n}{9}$

43. $\sum_{k=1}^{n} \left(\dfrac{2}{3}\right)^k$

44. $\sum_{k=1}^{n} 4 \cdot 3^{k-1}$

45. $-1 - 2 - 4 - 8 - \cdots - (2^{n-1})$

46. $2 + \dfrac{6}{5} + \dfrac{18}{25} + \cdots + 2\left(\dfrac{3}{5}\right)^{n-1}$

For Problems 47–52, use a graphing utility to find the sum of each geometric sequence.

47. $\dfrac{1}{4} + \dfrac{2}{4} + \dfrac{2^2}{4} + \dfrac{2^3}{4} + \cdots + \dfrac{2^{14}}{4}$

48. $\dfrac{3}{9} + \dfrac{3^2}{9} + \dfrac{3^3}{9} + \cdots + \dfrac{3^{15}}{9}$

49. $\sum_{n=1}^{15} \left(\dfrac{2}{3}\right)^n$

50. $\sum_{n=1}^{15} 4 \cdot 3^{n-1}$

51. $-1 - 2 - 4 - 8 - \cdots - 2^{14}$

52. $2 + \dfrac{6}{5} + \dfrac{18}{25} + \cdots + 2\left(\dfrac{3}{5}\right)^{15}$

In Problems 53–68, determine whether each infinite geometric series converges or diverges. If it converges, find its sum.

53. $1 + \dfrac{1}{3} + \dfrac{1}{9} + \cdots$

54. $2 + \dfrac{4}{3} + \dfrac{8}{9} + \cdots$

55. $8 + 4 + 2 + \cdots$

56. $6 + 2 + \dfrac{2}{3} + \cdots$

57. $2 - \dfrac{1}{2} + \dfrac{1}{8} - \dfrac{1}{32} + \cdots$

58. $1 - \dfrac{3}{4} + \dfrac{9}{16} - \dfrac{27}{64} + \cdots$

59. $8 + 12 + 18 + 27 + \cdots$

60. $9 + 12 + 16 + \dfrac{64}{3} + \cdots$

61. $\displaystyle\sum_{k=1}^{\infty} 5\left(\dfrac{1}{4}\right)^{k-1}$

62. $\displaystyle\sum_{k=1}^{\infty} 8\left(\dfrac{1}{3}\right)^{k-1}$

63. $\displaystyle\sum_{k=1}^{\infty} \dfrac{1}{2}\,3^{k-1}$

64. $\displaystyle\sum_{k=1}^{\infty} 3\left(\dfrac{3}{2}\right)^{k-1}$

65. $\displaystyle\sum_{k=1}^{\infty} 6\left(-\dfrac{2}{3}\right)^{k-1}$

66. $\displaystyle\sum_{k=1}^{\infty} 4\left(-\dfrac{1}{2}\right)^{k-1}$

67. $\displaystyle\sum_{k=1}^{\infty} 3\left(\dfrac{2}{3}\right)^{k}$

68. $\displaystyle\sum_{k=1}^{\infty} 2\left(\dfrac{3}{4}\right)^{k}$

Mixed Practice

In Problems 69–82, determine whether the given sequence is arithmetic, geometric, or neither. If the sequence is arithmetic, find the common difference; if it is geometric, find the common ratio. If the sequence is arithmetic or geometric, find the sum of the first 50 terms.

69. $\{n + 2\}$

70. $\{2n - 5\}$

71. $\{4n^2\}$

72. $\{5n^2 + 1\}$

73. $\left\{3 - \dfrac{2}{3}n\right\}$

74. $\left\{8 - \dfrac{3}{4}n\right\}$

75. $1, 3, 6, 10, \ldots$

76. $2, 4, 6, 8, \ldots$

77. $\left\{\left(\dfrac{2}{3}\right)^{n}\right\}$

78. $\left\{\left(\dfrac{5}{4}\right)^{n}\right\}$

79. $-1, 2, -4, 8, \ldots$

80. $1, 1, 2, 3, 5, 8, \ldots$

81. $\{3^{n/2}\}$

82. $\{(-1)^{n}\}$

Applications and Extensions

83. Find x so that x, $x + 2$, and $x + 3$ are consecutive terms of a geometric sequence.

84. Find x so that $x - 1$, x, and $x + 2$ are consecutive terms of a geometric sequence.

85. Salary Increases If you have been hired at an annual salary of \$18,000 and expect to receive annual increases of 5%, what will your salary be when you begin your fifth year?

86. Equipment Depreciation A new piece of equipment cost a company \$15,000. Each year, for tax purposes, the company depreciates the value by 15%. What value should the company give the equipment after 5 years?

87. Pendulum Swings Initially, a pendulum swings through an arc of 2 feet. On each successive swing, the length of the arc is 0.9 of the previous length.
(a) What is the length of the arc of the 10th swing?
(b) On which swing is the length of the arc first less than 1 foot?
(c) After 15 swings, what total length will the pendulum have swung?
(d) When it stops, what total length will the pendulum have swung?

88. Bouncing Balls A ball is dropped from a height of 30 feet. Each time it strikes the ground, it bounces up to 0.8 of the previous height.

(a) What height will the ball bounce up to after it strikes the ground for the third time?
(b) How high will it bounce after it strikes the ground for the nth time?
(c) How many times does the ball need to strike the ground before its bounce is less than 6 inches?
(d) What total distance does the ball travel before it stops bouncing?

89. Retirement Christine contributes \$100 each month to her 401(k). What will be the value of Christine's 401(k) after the 360th deposit (30 years) if the per annum rate of return is assumed to be 12% compounded monthly?

90. Saving for a Home Jolene wants to purchase a new home. Suppose that she invests \$400 per month into a mutual fund. If the per annum rate of return of the mutual fund is assumed to be 10% compounded monthly, how much will Jolene have for a down payment after the 36th deposit (3 years)?

91. Tax Sheltered Annuity Don contributes \$500 at the end of each quarter to a tax-sheltered annuity (TSA). What will the value of the TSA be after the 80th deposit (20 years) if the per annum rate of return is assumed to be 8% compounded quarterly?

92. Retirement Ray contributes \$1000 to an Individual Retirement Account (IRA) semiannually. What will be the value of the IRA be when Ray makes his 30th deposit (after 15 years) if the per annum rate of return is assumed to be 10% compounded semiannually?

93. Sinking Fund Scott and Alice want to purchase a vacation home in 10 years and need \$50,000 for a down payment. How much should they place in a savings account each month if the per annum rate of return is assumed to be 6% compounded monthly?

94. Sinking Fund For a child born in 1996, the cost of a 4-year college education at a public university is projected to be \$150,000. Assuming an 8% per annum rate of return compounded monthly, how much must be contributed to a college fund every month to have $150,000 in 18 years when the child begins college?

95. **Grains of Wheat on a Chess Board** In an old fable, a commoner who had saved the king's life was told he could ask the king for any just reward. Being a shrewd man, the commoner said, "A simple wish, sire. Place one grain of wheat on the first square of a chessboard, two grains on the second square, four grains on the third square, continuing until you have filled the board. This is all I seek." Compute the total number of grains needed to do this to see why the request, seemingly simple, could not be granted. (A chessboard consists of $8 \times 8 = 64$ squares.)

96. Look at the figure. What fraction of the square is eventually shaded if the indicated shading process continues indefinitely?

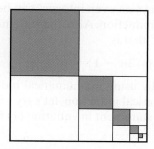

97. **Multiplier** Suppose that, throughout the U.S. economy, individuals spend 90% of every additional dollar that they earn. Economists would say that an individual's **marginal propensity to consume** is 0.90. For example, if Jane earns an additional dollar, she will spend $0.9(1) = \$0.90$ of it. The individual that earns $0.90 (from Jane) will spend 90% of it or $0.81. This process of spending continues and results in an infinite geometric series as follows:

$$1, 0.90, 0.90^2, 0.90^3, 0.90^4, \ldots$$

The sum of this infinite geometric series is called the **multiplier**. What is the multiplier if individuals spend 90% of every additional dollar that they earn?

98. **Multiplier** Refer to Problem 97. Suppose that the marginal propensity to consume throughout the U.S. economy is 0.95. What is the multiplier for the U.S. economy?

99. **Stock Price** One method of pricing a stock is to discount the stream of future dividends of the stock. Suppose that a stock pays $P per year in dividends and, historically, the dividend has been increased i% per year. If you desire an annual rate of return of r%, this method of pricing a stock states that the price that you should pay is the present value of an infinite stream of payments:

$$\text{Price} = P + P\frac{1 + i}{1 + r} + P\left(\frac{1 + i}{1 + r}\right)^2 + P\left(\frac{1 + i}{1 + r}\right)^3 + \cdots$$

The price of the stock is the sum of an infinite geometric series. Suppose that a stock pays an annual dividend of $4.00 and, historically, the dividend has been increased 3% per year. You desire an annual rate of return of 9%. What is the most you should pay for the stock?

100. **Stock Price** Refer to Problem 99. Suppose that a stock pays an annual dividend of $2.50 and, historically, the dividend has increased 4% per year. You desire an annual rate of return of 11%. What is the most that you should pay for the stock?

101. **A Rich Man's Promise** A rich man promises to give you $1000 on September 1, 2010. Each day thereafter he will give you $\frac{9}{10}$ of what he gave you the previous day. What is the first date on which the amount you receive is less than 1¢? How much have you received when this happens?

Explaining Concepts: Discussion and Writing

102. **Critical Thinking** You are interviewing for a job and receive two offers:
    A: $20,000 to start, with guaranteed annual increases of 6% for the first 5 years
    B: $22,000 to start, with guaranteed annual increases of 3% for the first 5 years
    Which offer is best if your goal is to be making as much as possible after 5 years? Which is best if your goal is to make as much money as possible over the contract (5 years)?

103. **Critical Thinking** Which of the following choices, A or B, results in more money?
    A: To receive $1000 on day 1, $999 on day 2, $998 on day 3, with the process to end after 1000 days
    B: To receive $1 on day 1, $2 on day 2, $4 on day 3, for 19 days

104. **Critical Thinking** You have just signed a 7-year professional football league contract with a beginning salary of $2,000,000 per year. Management gives you the following options with regard to your salary over the 7 years.
    1. A bonus of $100,000 each year
    2. An annual increase of 4.5% per year beginning after 1 year
    3. An annual increase of $95,000 per year beginning after 1 year

    Which option provides the most money over the 7-year period? Which the least? Which would you choose? Why?

105. **Critical Thinking** Suppose you were offered a job in which you would work 8 hours per day for 5 workdays per week for 1 month at hard manual labor. Your pay the first day would be 1 penny. On the second day your pay would be two pennies; the third day 4 pennies. Your pay would double on each successive workday. There are 22 workdays in the month. There will be no sick days. If you miss a day of work, there is no pay or pay increase. How much would you get

paid if you work all 22 days? How much do you get paid for the 22nd workday? What risks do you run if you take this job offer? Would you take the job?

106. Can a sequence be both arithmetic and geometric? Give reasons for your answer.

107. Make up a geometric sequence. Give it to a friend and ask for its 20th term.

108. Make up two infinite geometric series, one that has a sum and one that does not. Give them to a friend and ask for the sum of each series.

109. Describe the similarities and differences between geometric sequences and exponential functions.

'Are You Prepared?' Answers

1. $1082.43   **2.** $9513.28

11.4 Mathematical Induction

OBJECTIVE **1** Prove Statements Using Mathematical Induction (p. 854)

1 Prove Statements Using Mathematical Induction

Mathematical induction is a method for proving that statements involving natural numbers are true for all natural numbers.*

For example, the statement "$2n$ is always an even integer" can be proved for all natural numbers n by using mathematical induction. Also, the statement "the sum of the first n positive odd integers equals n^2," that is,

$$1 + 3 + 5 + \cdots + (2n - 1) = n^2 \qquad \textbf{(1)}$$

can be proved for all natural numbers n by using mathematical induction.

Before stating the method of mathematical induction, let's try to gain a sense of the power of the method. We shall use the statement in equation (1) for this purpose by restating it for various values of $n = 1, 2, 3, \ldots$.

$n = 1$ The sum of the first positive odd integer is 1^2; $1 = 1^2$.
$n = 2$ The sum of the first 2 positive odd integers is 2^2; $1 + 3 = 4 = 2^2$.
$n = 3$ The sum of the first 3 positive odd integers is 3^2; $1 + 3 + 5 = 9 = 3^2$.
$n = 4$ The sum of the first 4 positive odd integers is 4^2; $1 + 3 + 5 + 7 = 16 = 4^2$.

Although from this pattern we might conjecture that statement (1) is true for any choice of n, can we really be sure that it does not fail for some choice of n? The method of proof by mathematical induction will, in fact, prove that the statement is true for all n.

THEOREM   **The Principle of Mathematical Induction**

Suppose that the following two conditions are satisfied with regard to a statement about natural numbers:

CONDITION I: The statement is true for the natural number 1.
CONDITION II: If the statement is true for some natural number k, it is also true for the next natural number $k + 1$.

Then the statement is true for all natural numbers.

*Recall that the natural numbers are the numbers $1, 2, 3, 4, \ldots$. In other words, the terms *natural numbers* and *positive integers* are synonymous.

Figure 10

We shall not prove this principle. However, we can provide a physical interpretation that will help us to see why the principle works. Think of a collection of natural numbers obeying a statement as a collection of infinitely many dominoes. See Figure 10.

Now, suppose that we are told two facts:

1. The first domino is pushed over.

2. If one domino falls over, say the kth domino, so will the next one, the $(k + 1)$st domino.

Is it safe to conclude that *all* the dominoes fall over? The answer is yes, because if the first one falls (Condition I), the second one does also (by Condition II); and if the second one falls, so does the third (by Condition II); and so on.

Now let's prove some statements about natural numbers using mathematical induction.

EXAMPLE 1 | **Using Mathematical Induction**

Show that the following statement is true for all natural numbers n.

$$1 + 3 + 5 + \cdots + (2n - 1) = n^2 \tag{2}$$

Solution
We need to show first that statement (2) holds for $n = 1$. Because $1 = 1^2$, statement (2) is true for $n = 1$. Condition I holds.

Next, we need to show that Condition II holds. From statement (2), we assume that

$$1 + 3 + \cdots + (2k - 1) = k^2 \tag{3}$$

is true for some natural number k.

We wish to show that, based on equation (3), statement (2) holds for $k + 1$. We look at the sum of the first $k + 1$ positive odd integers to determine whether this sum equals $(k + 1)^2$.

$$1 + 3 + \cdots + (2k - 1) + [2(k + 1) - 1] = \underbrace{[1 + 3 + \cdots + (2k - 1)]}_{= \, k^2 \text{ by equation (3)}} + (2k + 1)$$

$$= k^2 + (2k + 1)$$

$$= k^2 + 2k + 1 = (k + 1)^2$$

Conditions I and II are satisfied; by the Principle of Mathematical Induction, statement (2) is true for all natural numbers n.

EXAMPLE 2 | **Using Mathematical Induction**

Show that the following statement is true for all natural numbers n.

$$2^n > n$$

Solution
First, we show that the statement $2^n > n$ holds when $n = 1$. Because $2^1 = 2 > 1$, the inequality is true for $n = 1$. Condition I holds.

Next, we assume, for some natural number k, that $2^k > k$. We wish to show that the formula holds for $k + 1$; that is, we wish to show that $2^{k+1} > k + 1$. Now

$$2^{k+1} = 2 \cdot 2^k \underset{\substack{\uparrow \\ \text{We know that} \\ 2^k > k.}}{>} 2 \cdot k = k + k \underset{\substack{\uparrow \\ k \ge 1}}{\ge} k + 1$$

If $2^k > k$, then $2^{k+1} > k + 1$, so Condition II of the Principle of Mathematical Induction is satisfied. The statement $2^n > n$ is true for all natural numbers n.

> [!NOTE] EXAMPLE 3

Using Mathematical Induction

Show that the following formula is true for all natural numbers n.

$$1 + 2 + 3 + \cdots + n = \frac{n(n+1)}{2} \tag{4}$$

Solution

First, we show that formula (4) is true when $n = 1$. Because

$$\frac{1(1+1)}{2} = \frac{1(2)}{2} = 1$$

Condition I of the Principle of Mathematical Induction holds.

Next, we assume that formula (4) holds for some k, and we determine whether the formula then holds for $k + 1$. We assume that

$$1 + 2 + 3 + \cdots + k = \frac{k(k+1)}{2} \quad \text{for some } k \tag{5}$$

Now we need to show that

$$1 + 2 + 3 + \cdots + k + (k+1) = \frac{(k+1)[(k+1)+1]}{2} = \frac{(k+1)(k+2)}{2}$$

We do this as follows:

$$1 + 2 + 3 + \cdots + k + (k+1) = \underbrace{[1 + 2 + 3 + \cdots + k]}_{= \frac{k(k+1)}{2} \quad \text{by equation (5)}} + (k+1)$$

$$= \frac{k(k+1)}{2} + (k+1)$$

$$= \frac{k^2 + k + 2k + 2}{2}$$

$$= \frac{k^2 + 3k + 2}{2} = \frac{(k+1)(k+2)}{2}$$

Condition II also holds. As a result, formula (4) is true for all natural numbers n.

═══ **Now Work** PROBLEM 1

> [!NOTE] EXAMPLE 4

Using Mathematical Induction

Show that $3^n - 1$ is divisible by 2 for all natural numbers n.

Solution

First, we show that the statement is true when $n = 1$. Because $3^1 - 1 = 3 - 1 = 2$ is divisible by 2, the statement is true when $n = 1$. Condition I is satisfied.

Next, we assume that the statement holds for some k, and we determine whether the statement then holds for $k + 1$. We assume that $3^k - 1$ is divisible by 2 for some k. We need to show that $3^{k+1} - 1$ is divisible by 2. Now

$$3^{k+1} - 1 = 3^{k+1} - 3^k + 3^k - 1 \qquad \text{Subtract and add } 3^k.$$
$$= 3^k(3 - 1) + (3^k - 1) = 3^k \cdot 2 + (3^k - 1)$$

Because $3^k \cdot 2$ is divisible by 2 and $3^k - 1$ is divisible by 2, it follows that $3^k \cdot 2 + (3^k - 1) = 3^{k+1} - 1$ is divisible by 2. Condition II is also satisfied. As a result, the statement "$3^n - 1$ is divisible by 2" is true for all natural numbers n.

WARNING The conclusion that a statement involving natural numbers is true for all natural numbers is made only after *both* Conditions I and II of the Principle of Mathematical Induction have been satisfied. Problem 28 demonstrates a statement for which only Condition I holds, but the statement is not true for all natural numbers. Problem 29 demonstrates a statement for which only Condition II holds, but the statement is not true for any natural number. ∎

11.4 Assess Your Understanding

Skill Building

In Problems 1–22, use the Principle of Mathematical Induction to show that the given statement is true for all natural numbers n.

1. $2 + 4 + 6 + \cdots + 2n = n(n + 1)$

2. $1 + 5 + 9 + \cdots + (4n - 3) = n(2n - 1)$

3. $3 + 4 + 5 + \cdots + (n + 2) = \frac{1}{2}n(n + 5)$

4. $3 + 5 + 7 + \cdots + (2n + 1) = n(n + 2)$

5. $2 + 5 + 8 + \cdots + (3n - 1) = \frac{1}{2}n(3n + 1)$

6. $1 + 4 + 7 + \cdots + (3n - 2) = \frac{1}{2}n(3n - 1)$

7. $1 + 2 + 2^2 + \cdots + 2^{n-1} = 2^n - 1$

8. $1 + 3 + 3^2 + \cdots + 3^{n-1} = \frac{1}{2}(3^n - 1)$

9. $1 + 4 + 4^2 + \cdots + 4^{n-1} = \frac{1}{3}(4^n - 1)$

10. $1 + 5 + 5^2 + \cdots + 5^{n-1} = \frac{1}{4}(5^n - 1)$

11. $\frac{1}{1 \cdot 2} + \frac{1}{2 \cdot 3} + \frac{1}{3 \cdot 4} + \cdots + \frac{1}{n(n + 1)} = \frac{n}{n + 1}$

12. $\frac{1}{1 \cdot 3} + \frac{1}{3 \cdot 5} + \frac{1}{5 \cdot 7} + \cdots + \frac{1}{(2n - 1)(2n + 1)} = \frac{n}{2n + 1}$

13. $1^2 + 2^2 + 3^2 + \cdots + n^2 = \frac{1}{6}n(n + 1)(2n + 1)$

14. $1^3 + 2^3 + 3^3 + \cdots + n^3 = \frac{1}{4}n^2(n + 1)^2$

15. $4 + 3 + 2 + \cdots + (5 - n) = \frac{1}{2}n(9 - n)$

16. $-2 - 3 - 4 - \cdots - (n + 1) = -\frac{1}{2}n(n + 3)$

17. $1 \cdot 2 + 2 \cdot 3 + 3 \cdot 4 + \cdots + n(n + 1) = \frac{1}{3}n(n + 1)(n + 2)$

18. $1 \cdot 2 + 3 \cdot 4 + 5 \cdot 6 + \cdots + (2n - 1)(2n) = \frac{1}{3}n(n + 1)(4n - 1)$

19. $n^2 + n$ is divisible by 2.

20. $n^3 + 2n$ is divisible by 3.

21. $n^2 - n + 2$ is divisible by 2.

22. $n(n + 1)(n + 2)$ is divisible by 6.

Applications and Extensions

In Problems 23–27, prove each statement.

23. If $x > 1$, then $x^n > 1$.

24. If $0 < x < 1$, then $0 < x^n < 1$.

25. $a - b$ is a factor of $a^n - b^n$.
    [**Hint:** $a^{k+1} - b^{k+1} = a(a^k - b^k) + b^k(a - b)$]

26. $a + b$ is a factor of $a^{2n+1} + b^{2n+1}$.

27. $(1 + a)^n \geq 1 + na$, for $a > 0$

28. Show that the statement "$n^2 - n + 41$ is a prime number" is true for $n = 1$, but is not true for $n = 41$.

29. Show that the formula

    $$2 + 4 + 6 + \cdots + 2n = n^2 + n + 2$$

    obeys Condition II of the Principle of Mathematical Induction. That is, show that if the formula is true for some k it is also true for $k + 1$. Then show that the formula is false for $n = 1$ (or for any other choice of n).

30. Use mathematical induction to prove that if $r \neq 1$ then

    $$a + ar + ar^2 + \cdots + ar^{n-1} = a\frac{1 - r^n}{1 - r}$$

31. Use mathematical induction to prove that

    $$a + (a + d) + (a + 2d)$$
    $$+ \cdots + [a + (n - 1)d] = na + d\frac{n(n - 1)}{2}$$

32. **Extended Principle of Mathematical Induction** The Extended Principle of Mathematical Induction states that if Conditions I and II hold, that is,
    (I) A statement is true for a natural number j.
    (II) If the statement is true for some natural number $k \geq j$, then it is also true for the next natural number $k + 1$.
    then the statement is true for all natural numbers $\geq j$. Use the Extended Principle of Mathematical Induction to show that the number of diagonals in a convex polygon of n sides is $\frac{1}{2}n(n - 3)$.

    [**Hint:** Begin by showing that the result is true when $n = 4$ (Condition I).]

33. **Geometry** Use the Extended Principle of Mathematical Induction to show that the sum of the interior angles of a convex polygon of n sides equals $(n - 2) \cdot 180°$.

Explaining Concepts: Discussion and Writing

34. How would you explain the Principle of Mathematical Induction to a friend?

11.5 The Binomial Theorem

OBJECTIVES **1** Evaluate $\binom{n}{j}$ (p. 858)

2 Use the Binomial Theorem (p. 860)

Formulas have been given for expanding $(x + a)^n$ for $n = 2$ and $n = 3$. The *Binomial Theorem** is a formula for the expansion of $(x + a)^n$ for any positive integer n. If $n = 1, 2, 3$, and 4, the expansion of $(x + a)^n$ is straightforward.

$$(x + a)^1 = x + a$$ Two terms, beginning with x^1 and ending with a^1

$$(x + a)^2 = x^2 + 2ax + a^2$$ Three terms, beginning with x^2 and ending with a^2

$$(x + a)^3 = x^3 + 3ax^2 + 3a^2x + a^3$$ Four terms, beginning with x^3 and ending with a^3

$$(x + a)^4 = x^4 + 4ax^3 + 6a^2x^2 + 4a^3x + a^4$$ Five terms, beginning with x^4 and ending with a^4

Notice that each expansion of $(x + a)^n$ begins with x^n and ends with a^n. As you read from left to right, the powers of x are decreasing by 1, while the powers of a are increasing by 1. Also, the number of terms equals $n + 1$. Notice, too, that the degree of each monomial in the expansion equals n. For example, in the expansion of $(x + a)^3$, each monomial $(x^3, 3ax^2, 3a^2x, a^3)$ is of degree 3. As a result, we might conjecture that the expansion of $(x + a)^n$ would look like this:

$$(x + a)^n = x^n + \underline{\quad} ax^{n-1} + \underline{\quad} a^2x^{n-2} + \cdots + \underline{\quad} a^{n-1}x + a^n$$

where the blanks are numbers to be found. This is, in fact, the case, as we shall see shortly.

Before we can fill in the blanks, we need to introduce the symbol $\binom{n}{j}$.

1 Evaluate $\binom{n}{j}$

COMMENT On a graphing calculator, the symbol $\binom{n}{j}$ may be denoted by the key $\boxed{nCr}$. ∎

We define the symbol $\binom{n}{j}$, read "n taken j at a time," as follows:

DEFINITION

If j and n are integers with $0 \leq j \leq n$, the symbol $\binom{n}{j}$ is defined as

$$\binom{n}{j} = \frac{n!}{j!(n-j)!} \qquad (1)$$

EXAMPLE 1

Evaluating $\binom{n}{j}$

Find:

(a) $\binom{3}{1}$ (b) $\binom{4}{2}$ (c) $\binom{8}{7}$ (d) $\binom{65}{15}$

*The name *binomial* is derived from the fact that $x + a$ is a binomial; that is, it contains two terms.

Solution (a) $\dbinom{3}{1} = \dfrac{3!}{1!(3-1)!} = \dfrac{3!}{1!\,2!} = \dfrac{3 \cdot 2 \cdot 1}{1(2 \cdot 1)} = \dfrac{6}{2} = 3$

(b) $\dbinom{4}{2} = \dfrac{4!}{2!(4-2)!} = \dfrac{4!}{2!\,2!} = \dfrac{4 \cdot 3 \cdot 2 \cdot 1}{(2 \cdot 1)(2 \cdot 1)} = \dfrac{24}{4} = 6$

(c) $\dbinom{8}{7} = \dfrac{8!}{7!(8-7)!} = \dfrac{8!}{7!\,1!} = \dfrac{8 \cdot 7!}{7! \cdot 1!} = \dfrac{8}{1} = 8$

$\underset{8! = 8 \cdot 7!}{\uparrow}$

Figure 11

65 nCr 15
  2.073746998ε14

(d) Figure 11 shows the solution using a TI-84 Plus graphing calculator. So

$$\dbinom{65}{15} \approx 2.073746998 \times 10^{14}$$

Now Work PROBLEM 5

Four useful formulas involving the symbol $\dbinom{n}{j}$ are

$$\dbinom{n}{0} = 1 \qquad \dbinom{n}{1} = n \qquad \dbinom{n}{n-1} = n \qquad \dbinom{n}{n} = 1$$

Proof $\dbinom{n}{0} = \dfrac{n!}{0!(n-0)!} = \dfrac{n!}{0!\,n!} = \dfrac{1}{1} = 1$

$\dbinom{n}{1} = \dfrac{n!}{1!(n-1)!} = \dfrac{n!}{(n-1)!} = \dfrac{n(n-1)!}{(n-1)!} = n$

You are asked to prove the remaining two formulas in Problem 45. ∎

Suppose that we arrange the values of the symbol $\dbinom{n}{j}$ in a triangular display, as shown next and in Figure 12.

$$\dbinom{0}{0}$$
$$\dbinom{1}{0} \quad \dbinom{1}{1}$$
$$\dbinom{2}{0} \quad \dbinom{2}{1} \quad \dbinom{2}{2}$$
$$\dbinom{3}{0} \quad \dbinom{3}{1} \quad \dbinom{3}{2} \quad \dbinom{3}{3}$$
$$\dbinom{4}{0} \quad \dbinom{4}{1} \quad \dbinom{4}{2} \quad \dbinom{4}{3} \quad \dbinom{4}{4}$$
$$\dbinom{5}{0} \quad \dbinom{5}{1} \quad \dbinom{5}{2} \quad \dbinom{5}{3} \quad \dbinom{5}{4} \quad \dbinom{5}{5}$$

Figure 12
Pascal's triangle

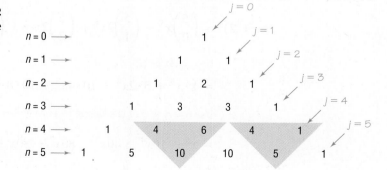

This display is called the **Pascal triangle,** named after Blaise Pascal (1623–1662), a French mathematician.

The Pascal triangle has 1's down the sides. To get any other entry, add the two nearest entries in the row above it. The shaded triangles in Figure 12 illustrate this feature of the Pascal triangle. Based on this feature, the row corresponding to $n = 6$ is found as follows:

$$n = 5 \rightarrow \quad 1 \quad 5 \quad 10 \quad 10 \quad 5 \quad 1$$
$$n = 6 \rightarrow \quad 1 \quad 6 \quad 15 \quad 20 \quad 15 \quad 6 \quad 1$$

Later we shall prove that this addition always works (see the theorem on page 862).

Although the Pascal triangle provides an interesting and organized display of the symbol $\binom{n}{j}$, in practice it is not all that helpful. For example, if you wanted to know the value of $\binom{12}{5}$, you would need to produce 13 rows of the triangle before seeing the answer. It is much faster to use definition (1).

2 Use the Binomial Theorem

Now we are ready to state the **Binomial Theorem.**

THEOREM

Binomial Theorem

Let x and a be real numbers. For any positive integer n, we have

$$(x + a)^n = \binom{n}{0}x^n + \binom{n}{1}ax^{n-1} + \cdots + \binom{n}{j}a^j x^{n-j} + \cdots + \binom{n}{n}a^n$$

$$= \sum_{j=0}^{n} \binom{n}{j}x^{n-j}a^j \qquad\qquad\qquad (2)$$

Now you know why we needed to introduce the symbol $\binom{n}{j}$; these symbols are the numerical coefficients that appear in the expansion of $(x + a)^n$. Because of this, the symbol $\binom{n}{j}$ is called a **binomial coefficient.**

EXAMPLE 2   **Expanding a Binomial**

Use the Binomial Theorem to expand $(x + 2)^5$.

Solution   In the Binomial Theorem, let $a = 2$ and $n = 5$. Then

$$(x + 2)^5 = \binom{5}{0}x^5 + \binom{5}{1}2x^4 + \binom{5}{2}2^2x^3 + \binom{5}{3}2^3x^2 + \binom{5}{4}2^4x + \binom{5}{5}2^5$$

↑
Use equation (2).

$$= 1 \cdot x^5 + 5 \cdot 2x^4 + 10 \cdot 4x^3 + 10 \cdot 8x^2 + 5 \cdot 16x + 1 \cdot 32$$

↑
Use row $n = 5$ of the Pascal triangle or formula (1) for $\binom{n}{j}$.

$$= x^5 + 10x^4 + 40x^3 + 80x^2 + 80x + 32$$

EXAMPLE 3

Expanding a Binomial

Expand $(2y - 3)^4$ using the Binomial Theorem.

Solution

First, we rewrite the expression $(2y - 3)^4$ as $[2y + (-3)]^4$. Now we use the Binomial Theorem with $n = 4$, $x = 2y$, and $a = -3$.

$$[2y + (-3)]^4 = \binom{4}{0}(2y)^4 + \binom{4}{1}(-3)(2y)^3 + \binom{4}{2}(-3)^2(2y)^2$$

$$+ \binom{4}{3}(-3)^3(2y) + \binom{4}{4}(-3)^4$$

$$= 1 \cdot 16y^4 + 4(-3)8y^3 + 6 \cdot 9 \cdot 4y^2 + 4(-27)2y + 1 \cdot 81$$

↑

Use row $n = 4$ of the Pascal triangle or formula (1) for $\binom{n}{j}$.

$$= 16y^4 - 96y^3 + 216y^2 - 216y + 81$$

In this expansion, note that the signs alternate due to the fact that $a = -3 < 0$.

●

━━ **Now Work** PROBLEM 21

EXAMPLE 4

Finding a Particular Coefficient in a Binomial Expansion

Find the coefficient of y^8 in the expansion of $(2y + 3)^{10}$.

Solution

We write out the expansion using the Binomial Theorem.

$$(2y + 3)^{10} = \binom{10}{0}(2y)^{10} + \binom{10}{1}(2y)^9(3)^1 + \binom{10}{2}(2y)^8(3)^2 + \binom{10}{3}(2y)^7(3)^3$$

$$+ \binom{10}{4}(2y)^6(3)^4 + \cdots + \binom{10}{9}(2y)(3)^9 + \binom{10}{10}(3)^{10}$$

From the third term in the expansion, the coefficient of y^8 is

$$\binom{10}{2}(2)^8(3)^2 = \frac{10!}{2!\,8!} \cdot 2^8 \cdot 9 = \frac{10 \cdot 9 \cdot 8!}{2 \cdot 8!} \cdot 2^8 \cdot 9 = 103{,}680$$

●

As this solution demonstrates, we can use the Binomial Theorem to find a particular term in an expansion without writing the entire expansion.

Based on the expansion of $(x + a)^n$, the term containing x^j is

$$\binom{n}{n - j}a^{n-j}x^j \qquad\qquad \textbf{(3)}$$

For example, we can solve Example 4 by using formula (3) with $n = 10$, $a = 3$, $x = 2y$, and $j = 8$. Then the term containing y^8 is

$$\binom{10}{10 - 8}3^{10-8}(2y)^8 = \binom{10}{2} \cdot 3^2 \cdot 2^8 \cdot y^8 = \frac{10!}{2!\,8!} \cdot 9 \cdot 2^8 y^8$$

$$= \frac{10 \cdot 9 \cdot 8!}{2 \cdot 8!} \cdot 9 \cdot 2^8 y^8 = 103{,}680 y^8$$

EXAMPLE 5 **Finding a Particular Term in a Binomial Expansion**

Find the sixth term in the expansion of $(x + 2)^9$.

Solution A We expand using the Binomial Theorem until the sixth term is reached.

$$(x + 2)^9 = \binom{9}{0}x^9 + \binom{9}{1}x^8 \cdot 2 + \binom{9}{2}x^7 \cdot 2^2 + \binom{9}{3}x^6 \cdot 2^3 + \binom{9}{4}x^5 \cdot 2^4$$
$$+ \binom{9}{5}x^4 \cdot 2^5 + \cdots$$

The sixth term is

$$\binom{9}{5}x^4 \cdot 2^5 = \frac{9!}{5!\,4!} \cdot x^4 \cdot 32 = 4032x^4$$

Solution B The sixth term in the expansion of $(x + 2)^9$, which has 10 terms total, contains x^4. (Do you see why?) By formula (3), the sixth term is

$$\binom{9}{9-4}2^{9-4}x^4 = \binom{9}{5}2^5 x^4 = \frac{9!}{5!\,4!} \cdot 32x^4 = 4032x^4$$

●

━━ **Now Work** PROBLEMS 29 AND 35

Next we show that the *triangular addition* feature of the Pascal triangle illustrated in Figure 12 always works.

THEOREM If n and j are integers with $1 \le j \le n$, then

$$\binom{n}{j-1} + \binom{n}{j} = \binom{n+1}{j} \tag{4}$$

Proof

$$\binom{n}{j-1} + \binom{n}{j} = \frac{n!}{(j-1)![n-(j-1)]!} + \frac{n!}{j!(n-j)!}$$

$$= \frac{n!}{(j-1)!(n-j+1)!} + \frac{n!}{j!(n-j)!}$$

$$= \frac{jn!}{j(j-1)!(n-j+1)!} + \frac{(n-j+1)n!}{j!(n-j+1)(n-j)!} \qquad \text{Multiply the first term by } \frac{j}{j} \text{ and}$$

$$\qquad\qquad\qquad\qquad\qquad\qquad\qquad\qquad\qquad \text{the second term by } \frac{n-j+1}{n-j+1}.$$

$$= \frac{jn!}{j!(n-j+1)!} + \frac{(n-j+1)n!}{j!(n-j+1)!} \qquad\qquad \text{Now the denominators are}$$
$$\qquad\qquad\qquad\qquad\qquad\qquad\qquad\qquad \text{equal.}$$

$$= \frac{jn! + (n-j+1)n!}{j!(n-j+1)!}$$

$$= \frac{n!(j + n - j + 1)}{j!(n-j+1)!}$$

$$= \frac{n!(n+1)}{j!(n-j+1)!} = \frac{(n+1)!}{j![(n+1)-j]!} = \binom{n+1}{j} \qquad ■$$

Historical Feature

Omar Khayyám (1050–1123)

The case $n = 2$ of the Binomial Theorem, $(a + b)^2$, was known to Euclid in 300 BC, but the general law seems to have been discovered by the Persian mathematician and astronomer Omar Khayyám (1050–1123), who is also well known as the author of the *Rubáiyát*, a collection of four-line poems making observations on the human condition. Omar Khayyám did not state the Binomial Theorem explicitly, but he claimed to have a method for extracting third, fourth, fifth roots, and so on. A little study shows that one must know the Binomial Theorem to create such a method.

The heart of the Binomial Theorem is the formula for the numerical coefficients, and, as we saw, they can be written in a symmetric triangular form. The Pascal triangle appears first in the books of Yang Hui (about 1270) and Chu Shih-chieh (1303). Pascal's name is attached to the triangle because of the many applications he made of it, especially to counting and probability. In establishing these results, he was one of the earliest users of mathematical induction.

Many people worked on the proof of the Binomial Theorem, which was finally completed for all n (including complex numbers) by Niels Abel (1802–1829).

11.5 Assess Your Understanding

Concepts and Vocabulary

1. _____ _____ is a triangular display of the binomial coefficients.

2. $\dbinom{n}{0} = $ _____ and $\dbinom{n}{1} = $ _____.

3. *True or False* $\dbinom{n}{j} = \dfrac{j!}{(n - j)! \, n!}$

4. The _____ _____ can be used to expand expressions like $(2x + 3)^6$.

Skill Building

In Problems 5–16, evaluate each expression.

5. $\dbinom{5}{3}$

6. $\dbinom{7}{3}$

7. $\dbinom{7}{5}$

8. $\dbinom{9}{7}$

9. $\dbinom{50}{49}$

10. $\dbinom{100}{98}$

11. $\dbinom{1000}{1000}$

12. $\dbinom{1000}{0}$

13. $\dbinom{55}{23}$

14. $\dbinom{60}{20}$

15. $\dbinom{47}{25}$

16. $\dbinom{37}{19}$

In Problems 17–28, expand each expression using the Binomial Theorem.

17. $(x + 1)^5$

18. $(x - 1)^5$

19. $(x - 2)^6$

20. $(x + 3)^5$

21. $(3x + 1)^4$

22. $(2x + 3)^5$

23. $(x^2 + y^2)^5$

24. $(x^2 - y^2)^6$

25. $(\sqrt{x} + \sqrt{2})^6$

26. $(\sqrt{x} - \sqrt{3})^4$

27. $(ax + by)^5$

28. $(ax - by)^4$

In Problems 29–42, use the Binomial Theorem to find the indicated coefficient or term.

29. The coefficient of x^6 in the expansion of $(x + 3)^{10}$

30. The coefficient of x^3 in the expansion of $(x - 3)^{10}$

31. The coefficient of x^7 in the expansion of $(2x - 1)^{12}$

32. The coefficient of x^3 in the expansion of $(2x + 1)^{12}$

33. The coefficient of x^7 in the expansion of $(2x + 3)^9$

34. The coefficient of x^2 in the expansion of $(2x - 3)^9$

35. The fifth term in the expansion of $(x + 3)^7$

36. The third term in the expansion of $(x - 3)^7$

37. The third term in the expansion of $(3x - 2)^9$

38. The sixth term in the expansion of $(3x + 2)^8$

39. The coefficient of x^0 in the expansion of $\left(x^2 + \dfrac{1}{x}\right)^{12}$

40. The coefficient of x^0 in the expansion of $\left(x - \dfrac{1}{x^2}\right)^9$

41. The coefficient of x^4 in the expansion of $\left(x - \dfrac{2}{\sqrt{x}}\right)^{10}$

42. The coefficient of x^2 in the expansion of $\left(\sqrt{x} + \dfrac{3}{\sqrt{x}}\right)^8$

Applications and Extensions

43. Use the Binomial Theorem to find the numerical value of $(1.001)^5$ correct to five decimal places.

[**Hint:** $(1.001)^5 = (1 + 10^{-3})^5$]

44. Use the Binomial Theorem to find the numerical value of $(0.998)^6$ correct to five decimal places.

45. Show that $\dbinom{n}{n-1} = n$ and $\dbinom{n}{n} = 1$.

46. Show that if n and j are integers with $0 \le j \le n$ then

$$\binom{n}{j} = \binom{n}{n-j}$$

Conclude that the Pascal triangle is symmetric with respect to a vertical line drawn from the topmost entry.

47. If n is a positive integer, show that

$$\binom{n}{0} + \binom{n}{1} + \cdots + \binom{n}{n} = 2^n$$

[**Hint:** $2^n = (1 + 1)^n$; now use the Binomial Theorem.]

48. If n is a positive integer, show that

$$\binom{n}{0} - \binom{n}{1} + \binom{n}{2} - \cdots + (-1)^n\binom{n}{n} = 0$$

49. $\dbinom{5}{0}\left(\dfrac{1}{4}\right)^5 + \dbinom{5}{1}\left(\dfrac{1}{4}\right)^4\left(\dfrac{3}{4}\right) + \dbinom{5}{2}\left(\dfrac{1}{4}\right)^3\left(\dfrac{3}{4}\right)^2$
$+ \dbinom{5}{3}\left(\dfrac{1}{4}\right)^2\left(\dfrac{3}{4}\right)^3 + \dbinom{5}{4}\left(\dfrac{1}{4}\right)\left(\dfrac{3}{4}\right)^4 + \dbinom{5}{5}\left(\dfrac{3}{4}\right)^5 = ?$

50. Stirling's Formula An approximation for $n!$, when n is large, is given by

$$n! \approx \sqrt{2n\pi}\left(\frac{n}{e}\right)^n\left(1 + \frac{1}{12n - 1}\right)$$

Calculate 12!, 20!, and 25! on your calculator. Then use Stirling's formula to approximate 12!, 20!, and 25!.

CHAPTER REVIEW

Things to Know

Sequence (p. 827)	A function whose domain is the set of positive integers
Factorials (p. 829)	$0! = 1, 1! = 1, n! = n(n - 1) \cdot \cdots \cdot 3 \cdot 2 \cdot 1$ if $n \ge 2$ is an integer
Arithmetic sequence (pp. 837 and 838)	$a_1 = a,\ a_n = a_{n-1} + d$, where $a_1 = a$ = first term, d = common difference $a_n = a_1 + (n - 1)d$
Sum of the first n terms of an arithmetic sequence (p. 839)	$S_n = \dfrac{n}{2}[2a_1 + (n - 1)d] = \dfrac{n}{2}(a_1 + a_n)$
Geometric sequence (pp. 843 and 844)	$a_1 = a,\quad a_n = ra_{n-1}$, where $a_1 = a$ = first term, r = common ratio $a_n = a_1 r^{n-1},\quad r \ne 0$
Sum of the first n terms of a geometric sequence (p. 845)	$S_n = a_1\dfrac{1 - r^n}{1 - r},\quad r \ne 0, 1$
Infinite geometric series (p. 846)	$a_1 + a_1 r + \cdots + a_1 r^{n-1} + \cdots = \displaystyle\sum_{k=1}^{\infty} a_1 r^{k-1}$
Sum of a convergent infinite geometric series (p. 847)	If $\lvert r \rvert < 1,\ \displaystyle\sum_{k=1}^{\infty} a_1 r^{k-1} = \dfrac{a_1}{1 - r}$
Principle of Mathematical Induction (p. 854)	If the following two conditions are satisfied, Condition I: The statement is true for the natural number 1. Condition II: If the statement is true for some natural number k, it is also true for $k + 1$. then the statement is true for all natural numbers.
Binomial coefficient (p. 858)	$\dbinom{n}{j} = \dfrac{n!}{j!(n - j)!}$
Pascal's triangle (p. 859)	See Figure 12.
Binomial Theorem (p. 860)	$(x + a)^n = \dbinom{n}{0}x^n + \dbinom{n}{1}ax^{n-1} + \cdots + \dbinom{n}{j}a^j x^{n-j} + \cdots + \dbinom{n}{n}a^n = \displaystyle\sum_{j=0}^{n}\dbinom{n}{j}x^{n-j}a^j$

Objectives

Section		You should be able to . . .	Example(s)	Review Exercises
11.1	1	Write the first several terms of a sequence (p. 827)	1–4	1–4
	2	Write the terms of a sequence defined by a recursive formula (p. 830)	5, 6	5–8
	3	Use summation notation (p. 831)	7, 8	9–12
	4	Find the sum of a sequence (p. 832)	9	25, 26
11.2	1	Determine if a sequence is arithmetic (p. 837)	1–3	13–24
	2	Find a formula for an arithmetic sequence (p. 838)	4, 5	31, 32, 35, 37–40, 65
	3	Find the sum of an arithmetic sequence (p. 839)	6–8	13, 14, 19, 20, 27, 28, 66
11.3	1	Determine if a sequence is geometric (p. 843)	1–3	13–24
	2	Find a formula for a geometric sequence (p. 844)	4	21, 22, 33, 34, 36, 70
	3	Find the sum of a geometric sequence (p. 845)	5, 6	17, 18, 21, 22, 29, 30, 67(a)–(c)
	4	Determine whether a geometric series converges or diverges (p. 846)	7–9	41–48, 67(d)
	5	Solve annuity problems (p. 849)	10, 11	68, 69
11.4	1	Prove statements using mathematical induction (p. 854)	1–4	49–54
11.5	1	Evaluate $\binom{n}{j}$ (p. 858)	1	55, 56
	2	Use the Binomial Theorem (p. 860)	2–5	57–64

Review Exercises

In Problems 1–8, write down the first five terms of each sequence.

1. $\{a_n\} = \left\{(-1)^n\left(\dfrac{n+3}{n+2}\right)\right\}$ **2.** $\{b_n\} = \{(-1)^{n+1}(2n+3)\}$ **3.** $\{c_n\} = \left\{\dfrac{2^n}{n^2}\right\}$ **4.** $\{d_n\} = \left\{\dfrac{e^n}{n}\right\}$

5. $a_1 = 3;\quad a_n = \dfrac{2}{3}a_{n-1}$ **6.** $a_1 = 4;\quad a_n = -\dfrac{1}{4}a_{n-1}$ **7.** $a_1 = 2;\quad a_n = 2 - a_{n-1}$ **8.** $a_1 = -3;\quad a_n = 4 + a_{n-1}$

In Problems 9 and 10, write out each sum.

9. $\displaystyle\sum_{k=1}^{4}(4k+2)$ **10.** $\displaystyle\sum_{k=1}^{3}(3-k^2)$

In Problems 11 and 12, express each sum using summation notation.

11. $1 - \dfrac{1}{2} + \dfrac{1}{3} - \dfrac{1}{4} + \cdots + \dfrac{1}{13}$ **12.** $2 + \dfrac{2^2}{3} + \dfrac{2^3}{3^2} + \cdots + \dfrac{2^{n+1}}{3^n}$

In Problems 13–24, determine whether the given sequence is arithmetic, geometric, or neither. If the sequence is arithmetic, find the common difference and the sum of the first n terms. If the sequence is geometric, find the common ratio and the sum of the first n terms.

13. $\{a_n\} = \{n+5\}$ **14.** $\{b_n\} = \{4n+3\}$ **15.** $\{c_n\} = \{2n^3\}$ **16.** $\{d_n\} = \{2n^2 - 1\}$

17. $\{s_n\} = \{2^{3n}\}$ **18.** $\{u_n\} = \{3^{2n}\}$ **19.** $0, 4, 8, 12, \ldots$ **20.** $1, -3, -7, -11, \ldots$

21. $3, \dfrac{3}{2}, \dfrac{3}{4}, \dfrac{3}{8}, \dfrac{3}{16}, \ldots$ **22.** $5, -\dfrac{5}{3}, \dfrac{5}{9}, -\dfrac{5}{27}, \dfrac{5}{81}, \ldots$ **23.** $\dfrac{2}{3}, \dfrac{3}{4}, \dfrac{4}{5}, \dfrac{5}{6}, \ldots$ **24.** $\dfrac{3}{2}, \dfrac{5}{4}, \dfrac{7}{6}, \dfrac{9}{8}, \dfrac{11}{10}, \ldots$

In Problems 25–30, find each sum.

25. $\displaystyle\sum_{k=1}^{50}(3k)$ **26.** $\displaystyle\sum_{k=1}^{30}k^2$ **27.** $\displaystyle\sum_{k=1}^{30}(3k-9)$

28. $\displaystyle\sum_{k=1}^{40}(-2k+8)$ **29.** $\displaystyle\sum_{k=1}^{7}\left(\dfrac{1}{3}\right)^k$ **30.** $\displaystyle\sum_{k=1}^{10}(-2)^k$

In Problems 31–36, find the indicated term in each sequence. [**Hint:** Find the general term first.]

31. 9th term of $3, 7, 11, 15, \ldots$

32. 8th term of $1, -1, -3, -5, \ldots$

33. 11th term of $1, \dfrac{1}{10}, \dfrac{1}{100}, \ldots$

34. 11th term of $1, 2, 4, 8, \ldots$

35. 9th term of $\sqrt{2}, 2\sqrt{2}, 3\sqrt{2}, \ldots$

36. 9th term of $\sqrt{2}, 2, 2^{3/2}, \ldots$

In Problems 37–40, find a general formula for each arithmetic sequence.

37. 7th term is 31; 20th term is 96

38. 8th term is -20; 17th term is -47

39. 10th term is 0; 18th term is 8

40. 12th term is 30; 22nd term is 50

In Problems 41–48, determine whether each infinite geometric series converges or diverges. If it converges, find its sum.

41. $3 + 1 + \dfrac{1}{3} + \dfrac{1}{9} + \cdots$

42. $2 + 1 + \dfrac{1}{2} + \dfrac{1}{4} + \cdots$

43. $2 - 1 + \dfrac{1}{2} - \dfrac{1}{4} + \cdots$

44. $6 - 4 + \dfrac{8}{3} - \dfrac{16}{9} + \cdots$

45. $\dfrac{1}{2} + \dfrac{3}{4} + \dfrac{9}{8} + \cdots$

46. $\displaystyle\sum_{k=1}^{\infty} 5 \cdot \left(-\dfrac{5}{4}\right)^{k-1}$

47. $\displaystyle\sum_{k=1}^{\infty} 4\left(\dfrac{1}{2}\right)^{k-1}$

48. $\displaystyle\sum_{k=1}^{\infty} 3\left(-\dfrac{3}{4}\right)^{k-1}$

In Problems 49–54, use the Principle of Mathematical Induction to show that the given statement is true for all natural numbers.

49. $3 + 6 + 9 + \cdots + 3n = \dfrac{3n}{2}(n + 1)$

50. $2 + 6 + 10 + \cdots + (4n - 2) = 2n^2$

51. $2 + 6 + 18 + \cdots + 2 \cdot 3^{n-1} = 3^n - 1$

52. $3 + 6 + 12 + \cdots + 3 \cdot 2^{n-1} = 3(2^n - 1)$

53. $1^2 + 4^2 + 7^2 + \cdots + (3n - 2)^2 = \dfrac{1}{2}n(6n^2 - 3n - 1)$

54. $1 \cdot 3 + 2 \cdot 4 + 3 \cdot 5 + \cdots + n(n + 2) = \dfrac{n}{6}(n + 1)(2n + 7)$

In Problems 55 and 56, evaluate each binomial coefficient.

55. $\dbinom{5}{2}$

56. $\dbinom{8}{6}$

In Problems 57–60, expand each expression using the Binomial Theorem.

57. $(x + 2)^5$

58. $(x - 3)^4$

59. $(2x + 3)^5$

60. $(3x - 4)^4$

---

61. Find the coefficient of x^7 in the expansion of $(x + 2)^9$.

62. Find the coefficient of x^3 in the expansion of $(x - 3)^8$.

63. Find the coefficient of x^2 in the expansion of $(2x + 1)^7$.

64. Find the coefficient of x^6 in the expansion of $(2x + 1)^8$.

65. Constructing a Brick Staircase A brick staircase has a total of 25 steps. The bottom step requires 80 bricks. Each successive step requires three less bricks than the prior step.
(a) How many bricks are required for the top step?
(b) How many bricks are required to build the staircase?

66. Creating a Floor Design A mosaic tile floor is designed in the shape of a trapezoid 30 feet wide at the base and 15 feet wide at the top. The tiles, 12 inches by 12 inches, are to be placed so that each successive row contains one less tile than the row below. How many tiles will be required?

67. Bouncing Balls A ball is dropped from a height of 20 feet. Each time it strikes the ground, it bounces up to three-quarters of the previous height.
(a) What height will the ball bounce up to after it strikes the ground for the third time?

(b) How high will it bounce after it strikes the ground for the nth time?
(c) How many times does the ball need to strike the ground before its bounce is less than 6 inches?
(d) What total distance does the ball travel before it stops bouncing?

68. Retirement Planning Chris gets paid once a month and contributes $200 each pay period into his 401(k). If Chris plans on retiring in 20 years, what will be the value of his 401(k) if the per annum rate of return of the 401(k) is 10% compounded monthly?

69. Retirement Planning Jacky contributes $500 every quarter to an IRA. If Jacky plans on retiring in 30 years, what will be the value of the IRA if the per annum rate of return of the IRA is 8% compounded quarterly?

70. Salary Increases Your friend has just been hired at an annual salary of $20,000. If she expects to receive annual increases of 4%, what will be her salary as she begins her fifth year?

CHAPTER TEST

CHAPTER
Test Prep
VIDEOS

Step-by-step test solutions are found on the Chapter Test Prep Videos available via the Video Resources on DVD, in **MyMathLab** or on **You Tube** (search "Sullivan Precalculus: A Right Triangle Approach, Concepts Through Functions 2e).

In Problems 1 and 2, write down the first five terms of each sequence.

1. $\{s_n\} = \left\{\dfrac{n^2 - 1}{n + 8}\right\}$

2. $a_1 = 4, a_n = 3a_{n-1} + 2$

5. Write the following sum using summation notation.
$$-\frac{2}{5} + \frac{3}{6} - \frac{4}{7} + \cdots + \frac{11}{14}$$

In Problems 3 and 4, write out each sum. Evaluate each sum.

3. $\displaystyle\sum_{k=1}^{3} (-1)^{k+1}\left(\frac{k+1}{k^2}\right)$

4. $\displaystyle\sum_{k=1}^{4}\left[\left(\frac{2}{3}\right)^k - k\right]$

In Problems 6–11, determine whether the given sequence is arithmetic, geometric, or neither. If the sequence is arithmetic, find the common difference and the sum of the first n terms. If the sequence is geometric, find the common ratio and the sum of the first n terms.

6. $6, 12, 36, 144, \ldots$

7. $\left\{-\dfrac{1}{2} \cdot 4^n\right\}$

8. $-2, -10, -18, -26, \ldots$

9. $\left\{-\dfrac{n}{2} + 7\right\}$

10. $25, 10, 4, \dfrac{8}{5}, \ldots$

11. $\left\{\dfrac{2n - 3}{2n + 1}\right\}$

12. Determine whether the infinite geometric series
$$256 - 64 + 16 - 4 + \cdots$$
converges or diverges. If it converges, find its sum.

13. Expand $(3m + 2)^5$ using the Binomial Theorem.

14. Use the Principle of Mathematical Induction to show that the given statement is true for all natural numbers.
$$\left(1 + \frac{1}{1}\right)\left(1 + \frac{1}{2}\right)\left(1 + \frac{1}{3}\right)\cdots\left(1 + \frac{1}{n}\right) = n + 1$$

15. A new car sold for \$31,000. If the vehicle loses 15% of its value each year, how much will it be worth after 10 years?

16. A weightlifter begins his routine by benching 100 pounds and increases the weight by 30 pounds for each set. If he does 10 repetitions in each set, what is the total weight lifted after 5 sets?

CUMULATIVE REVIEW

1. Find all the solutions, real and complex, of the equation $|x^2| = 9$

2. (a) Graph the circle $x^2 + y^2 = 100$ and the parabola $y = 3x^2$.

    (b) Solve the system of equations: $\begin{cases} x^2 + y^2 = 100 \\ y = 3x^2 \end{cases}$

    (c) Where do the circle and the parabola intersect?

3. Solve the equation $2e^x = 5$.

4. Find an equation of the line with slope 5 and x-intercept 2.

5. Find the standard equation of the circle whose center is the point $(-1, 2)$ if $(3, 5)$ is a point on the circle.

6. $f(x) = \dfrac{3x}{x - 2}, \quad g(x) = 2x + 1$

Find:

    (a) $(f \circ g)(2)$ (b) $(g \circ f)(4)$

    (c) $(f \circ g)(x)$ (d) The domain of $(f \circ g)(x)$

    (e) $(g \circ f)(x)$ (f) The domain of $(g \circ f)(x)$

    (g) The function g^{-1} and its domain

    (h) The function f^{-1} and its domain

7. Find the equation of an ellipse with center at the origin, a focus at $(0, 3)$, and a vertex at $(0, 4)$.

8. Find the equation of a parabola with vertex at $(-1, 2)$ and focus at $(-1, 3)$.

CHAPTER PROJECTS

I. Population Growth The size of the population of the United States essentially depends on its current population, the birth and death rates of the population, and immigration. Let b represent the birth rate of the U.S. population and d represent its death rate. Then $r = b - d$ represents the growth rate of the population, where r varies from year to year. The U.S. population after n years can be modeled using the recursive function

$$p_n = (1 + r)p_{n-1} + I$$

where I represents net immigration into the United States.

1. Using data from the CIA World Factbook at *www.cia.gov/ cia/publications/factbook/index.html*, determine the birth and death rates for all races for the most recent year that data are available. Birth rates and death rates are given as the number of live births per 1000 population. Each must be computed as the number of births (deaths) per individual. For example, in 2009, the birth rate was 13.82 per 1000 and the death rate was 8.38 per 1,000, so $b = \dfrac{13.82}{1000} = 0.01382$, while $d = \dfrac{8.38}{1,000} = 0.00838$.

Next, using data from the Immigration and Naturalization Service *www.fedstats.gov*, determine the net immigration to the United States for the same year used to obtain b and d in Problem 1.

2. Determine the value of r, the growth rate of the population.

3. Find a recursive formula for the population of the United States.

4. Use the recursive formula to predict the population of the United States in the following year. In other words, if data are available up to the year 2008, predict the U.S. population in 2009.

5. Compare your prediction to actual data.

6. Repeat Problems 1–5 for Uganda using the CIA World Factbook (in 2009, the birth rate was 47.84 per 1000 and the death rate was 12.09 per 1000).

7. Do your results for the United States (a developed country) and Uganda (a developing country) seem in line with the article in the chapter opener? Explain.

8. Do you think the recursive formula found in Problem 3 will be useful in predicting future populations? Why or why not?

The following projects are available at the Instructor's Resource Center (IRC):

II. Project at Motorola *Digital Wireless Communication* Cell phones take speech and change it into digital code using only zeros and ones. See how the code length can be modeled using a mathematical sequence.

III. Economics Economists use the current price of a good and a recursive model to predict future consumer demand and to determine future production.

IV. Standardized Tests Many tests of intelligence, aptitude, and achievement contain questions asking for the terms of a mathematical sequence.

Counting and Probability

12

Outline

Deal or No Deal

By LYNN ELBER, AP Television Writer—LOS ANGELES—The promise of an easy million bucks, a stage crowded with sexy models and the smoothly calibrated charm of host **Howie Mandel** made "Deal or No Deal" an unexpected hit in television's December dead zone. Based on a series that debuted in Holland in 2002 and became an international hit, "Deal or No Deal" is about luck and playing the odds. Contestants are faced with 26 briefcases held by 26 models, each case with a hidden value ranging from a penny to the top prize that will escalate by week's end to $3 million. As the game progresses and cases are eliminated, a contestant weighs the chance of snaring a big prize against lesser but still tempting offers made by the show's "bank," represented by an anonymous, silhouetted figure.

Source: Adapted from *Lynn Elber, "'Deal or No Deal' back with bigger prizes," Associated Press, February 24, 2006. © 2006 Associated Press.*

—*See Chapter Project I*—

◁ A Look Back   We introduced sets in Appendix A, Section A.1. We have been using sets to represent solutions of equations and inequalities and to represent the domain and range of functions.

A Look Ahead ▷   Here we discuss methods for counting the number of elements in a set and the role of sets in probability.

12.1 Counting

PREPARING FOR THIS SECTION *Before getting started, review the following:*

• Sets (Appendix A, Section A.1, pp. A1–A3)

Now Work the 'Are You Prepared?' problems on page 874.

OBJECTIVES 1 Find All the Subsets of a Set (p. 870)
2 Count the Number of Elements in a Set (p. 870)
3 Solve Counting Problems Using the Multiplication Principle (p. 873)

Counting plays a major role in many diverse areas, such as probability, statistics, and computer science; counting techniques are a part of a branch of mathematics called **combinatorics.**

1 Find All the Subsets of a Set

We begin by reviewing the ways that two sets can be compared.

If two sets A and B have precisely the same elements, we say that A and B are **equal** and write $A = B$.

If each element of a set A is also an element of a set B, we say that A is a **subset** of B and write $A \subseteq B$.

If $A \subseteq B$ and $A \neq B$, we say that A is a **proper subset** of B and write $A \subset B$.

If $A \subseteq B$, every element in set A is also in set B, but B may or may not have additional elements. If $A \subset B$, every element in A is also in B, and B has at least one element not found in A.

Finally, we agree that the empty set, $\emptyset$, is a subset of every set; that is,

$$\emptyset \subseteq A \qquad \text{for any set } A$$

EXAMPLE 1

Finding All the Subsets of a Set

Write down all the subsets of the set $\{a, b, c\}$.

Solution

To organize our work, we write down all the subsets with no elements, then those with one element, then those with two elements, and finally those with three elements. These will give us all the subsets. Do you see why?

0 Elements	**1 Element**	**2 Elements**	**3 Elements**
$\emptyset$	$\{a\}, \{b\}, \{c\}$	$\{a, b\}, \{b, c\}, \{a, c\}$	$\{a, b, c\}$

●

Now Work PROBLEM 9

2 Count the Number of Elements in a Set

As you count the number of students in a classroom or the number of pennies in your pocket, what you are really doing is matching, on a one-to-one basis, each object to be counted with the set of counting numbers, $1, 2, 3, \ldots, n$, for some number n. If a set A matched up in this fashion with the set $\{1, 2, \ldots, 25\}$, you would conclude that there are 25 elements in the set A. We use the notation $n(A) = 25$ to indicate that there are 25 elements in the set A.

Because the empty set has no elements, we write

$$n(\emptyset) = 0$$

If the number of elements in a set is a nonnegative integer, we say that the set is **finite.** Otherwise, it is **infinite.** We shall concern ourselves only with finite sets.

Look again at Example 1. A set with 3 elements has $2^3 = 8$ subsets. This result can be generalized.

In Words
We use the notation $n(A)$ to mean the number of elements in set A.

> If A is a set with n elements, A has 2^n subsets.

For example, the set $\{a, b, c, d, e\}$ has $2^5 = 32$ subsets.

| EXAMPLE 2 | **Analyzing Survey Data** |

In a survey of 100 college students, 35 were registered in College Algebra, 52 were registered in Computer Science I, and 18 were registered in both courses.

(a) How many students were registered in College Algebra or Computer Science I?

(b) How many were registered in neither course?

Solution (a) First, let A = set of students in College Algebra

B = set of students in Computer Science I

Then the given information tells us that

$$n(A) = 35 \qquad n(B) = 52 \qquad n(A \cap B) = 18$$

Figure 1

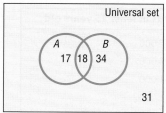

Refer to Figure 1. Since $n(A \cap B) = 18$, we know that the common part of the circles representing set A and set B has 18 elements. In addition, we know that the remaining portion of the circle representing set A will have $35 - 18 = 17$ elements. Similarly, we know that the remaining portion of the circle representing set B has $52 - 18 = 34$ elements. We conclude that $17 + 18 + 34 = 69$ students were registered in College Algebra or Computer Science I.

(b) Since 100 students were surveyed, it follows that $100 - 69 = 31$ were registered in neither course.

Now Work PROBLEMS 17 AND 27

The solution to Example 2 contains the basis for a general counting formula. If we count the elements in each of two sets A and B, we necessarily count twice any elements that are in both A and B, that is, those elements in $A \cap B$. To count correctly the elements that are in A or B, that is, to find $n(A \cup B)$, we need to subtract those in $A \cap B$ from $n(A) + n(B)$.

THEOREM

> **Counting Formula**
>
> If A and B are finite sets,
>
> $$n(A \cup B) = n(A) + n(B) - n(A \cap B) \qquad \text{(1)}$$

Refer back to Example 2. Using (1), we have

$$\begin{aligned} n(A \cup B) &= n(A) + n(B) - n(A \cap B) \\ &= 35 + 52 - 18 \\ &= 69 \end{aligned}$$

There are 69 students registered in College Algebra or Computer Science I.

A special case of the counting formula (1) occurs if A and B have no elements in common. In this case, $A \cap B = \varnothing$, so $n(A \cap B) = 0$.

THEOREM

> **Addition Principle of Counting**
>
> If two sets A and B have no elements in common, that is,
>
> $$\text{if } A \cap B = \varnothing, \text{ then } n(A \cup B) = n(A) + n(B) \qquad \text{(2)}$$

We can generalize formula (2).

THEOREM

General Addition Principle of Counting

If, for n sets $A_1, A_2, \ldots, A_n$, no two have elements in common,

$$n(A_1 \cup A_2 \cup \cdots \cup A_n) = n(A_1) + n(A_2) + \cdots + n(A_n) \qquad \textbf{(3)}$$

EXAMPLE 3 | Counting

Table 1 lists the level of education for all United States residents 25 years of age or older in 2007.

Table 1

Level of Education	Number of U.S. Residents at least 25 years old
Not a high school graduate	27,787,000
High school graduate	61,404,000
Some college, but no degree	32,451,000
Associate's degree	16,711,000
Bachelor's degree	36,726,000
Advanced degree	19,237,000

Source: Statistical Abstract of the United States, 2009

(a) How many U.S. residents 25 years of age or older had an associate's degree or bachelor's degree?

(b) How many U.S. residents 25 years of age or older had an associate's degree, a bachelor's degree, or an advanced degree?

Solution Let A represent the set of associate degree holders, B represent the set of bachelor degree holders, and C represent the set of advanced degree holders. No two of the sets A, B, or C have elements in common (while the holder of an advanced degree certainly also holds a bachelor's degree, the individual would be part of the set for which the highest degree has been conferred). Then

$$n(A) = 16,711,000 \quad n(B) = 36,726,000 \quad n(C) = 19,237,000$$

(a) Using formula (2), we have

$$n(A \cup B) = n(A) + n(B) = 16,711,000 + 36,726,000 = 53,437,000$$

There were 53,437,000 U.S. residents 25 years of age or older who had an associate's degree or bachelor's degree.

(b) Using formula (3), we have

$$\begin{aligned} n(A \cup B \cup C) &= n(A) + n(B) + n(C) \\ &= 16,711,000 + 36,726,000 + 19,237,000 \\ &= 72,674,000 \end{aligned}$$

There were 72,674,000 U.S. residents 25 years of age or older who had an associate's degree, bachelor's degree, or advanced degree.

Now Work PROBLEM 31

3 Solve Counting Problems Using the Multiplication Principle

We begin with an example that illustrates the multiplication principle.

EXAMPLE 4   **Counting the Number of Possible Meals**

The fixed-price dinner at Mabenka Restaurant provides the following choices:

Appetizer: soup or salad
Entree: baked chicken, broiled beef patty, baby beef liver, or roast
             beef au jus
Dessert: ice cream or cheese cake

How many different meals can be ordered?

Solution   Ordering such a meal requires three separate decisions:

Choose an Appetizer   **Choose an Entree**   **Choose a Dessert**

2 choices   4 choices   2 choices

Look at the **tree diagram** in Figure 2. We see that, for each choice of appetizer, there are 4 choices of entrees. And for each of these $2 \cdot 4 = 8$ choices, there are 2 choices for dessert. A total of

$$2 \cdot 4 \cdot 2 = 16$$

different meals can be ordered.

Figure 2

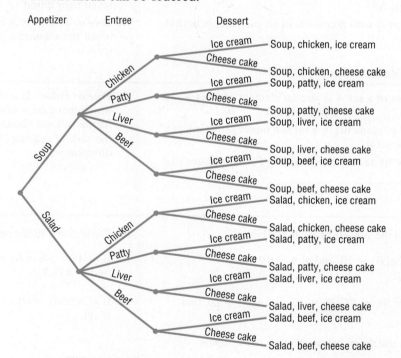

Example 4 demonstrates a general principle of counting.

THEOREM   **Multiplication Principle of Counting**

If a task consists of a sequence of choices in which there are p selections for the first choice, q selections for the second choice, r selections for the third choice, and so on, the task of making these selections can be done in

$$p \cdot q \cdot r \cdots$$

different ways.

EXAMPLE 5	**Forming Codes**

How many two-symbol code words can be formed if the first symbol is an uppercase letter and the second symbol is a digit?

Solution It sometimes helps to begin by listing some of the possibilities. The code consists of an uppercase letter followed by a digit, so some possibilities are A1, A2, B3, X0, and so on. The task consists of making two selections: the first selection requires choosing an uppercase letter (26 choices) and the second task requires choosing a digit (10 choices). By the Multiplication Principle, there are

$$26 \cdot 10 = 260$$

different codewords of the type described.

●

───── **Now Work** PROBLEM 23

12.1 Assess Your Understanding

'Are You Prepared?' *Answers are given at the end of these exercises. If you get a wrong answer, read the pages listed in* red.

1. The _____ of A and B consists of all elements in either A or B or both. (p. A2)

2. The _____ of A with B consists of all elements in both A and B. (p. A2)

3. **True or False** The intersection of two sets is always a subset of their union. (p. A2)

4. **True or False** If A is a set, the complement of A is the set of all the elements in the universal set that are not in A. (p. A2)

Concepts and Vocabulary

5. If each element of a set A is also an element of a set B, we say that A is a _____ of B and write A ____ B.

6. If the number of elements in a set is a nonnegative integer, we say that the set is _____.

7. If A and B are finite sets, the Counting Formula states that $n(A \cup B) = $ _____.

8. **True or False** If a task consists of a sequence of choices in which there are p selections for the first choice, q selections for the second choice, and r selections for the third choice, the task of making these selections can be done in $p \cdot q \cdot r$ different ways.

Skill Building

9. Write down all the subsets of $\{a, b, c, d\}$.

10. Write down all the subsets of $\{a, b, c, d, e\}$.

11. If $n(A) = 15$, $n(B) = 20$, and $n(A \cap B) = 10$, find $n(A \cup B)$.

12. If $n(A) = 30$, $n(B) = 40$, and $n(A \cup B) = 45$, find $n(A \cap B)$.

13. If $n(A \cup B) = 50$, $n(A \cap B) = 10$, and $n(B) = 20$, find $n(A)$.

14. If $n(A \cup B) = 60$, $n(A \cap B) = 40$, and $n(A) = n(B)$, find $n(A)$.

In Problems 15–22, use the information given in the figure.

15. How many are in set A?

16. How many are in set B?

17. How many are in A or B?

18. How many are in A and B?

19. How many are in A but not C?

20. How many are not in A?

21. How many are in A and B and C?

22. How many are in A or B or C?

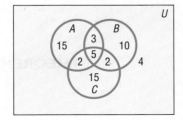

Applications and Extensions

23. **Shirts and Ties** A man has 5 shirts and 3 ties. How many different shirt and tie arrangements can he wear?

24. **Blouses and Skirts** A woman has 5 blouses and 8 skirts. How many different outfits can she wear?

25. Four-digit Numbers How many four-digit numbers can be formed using the digits 0, 1, 2, 3, 4, 5, 6, 7, 8, and 9 if the first digit cannot be 0? Repeated digits are allowed.

26. Five-digit Numbers How many five-digit numbers can be formed using the digits 0, 1, 2, 3, 4, 5, 6, 7, 8, and 9 if the first digit cannot be 0 or 1? Repeated digits are allowed.

27. Analyzing Survey Data In a consumer survey of 500 people, 200 indicated that they would be buying a major appliance within the next month, 150 indicated that they would buy a car, and 25 said that they would purchase both a major appliance and a car. How many will purchase neither? How many will purchase only a car?

28. Analyzing Survey Data In a student survey, 200 indicated that they would attend Summer Session I and 150 indicated Summer Session II. If 75 students plan to attend both summer sessions and 275 indicated that they would attend neither session, how many students participated in the survey?

29. Analyzing Survey Data In a survey of 100 investors in the stock market,

  50 owned shares in IBM
  40 owned shares in AT&T
  45 owned shares in GE
  20 owned shares in both IBM and GE
  15 owned shares in both AT&T and GE
  20 owned shares in both IBM and AT&T
  5 owned shares in all three

  (a) How many of the investors surveyed did not have shares in any of the three companies?
  (b) How many owned just IBM shares?
  (c) How many owned just GE shares?
  (d) How many owned neither IBM nor GE?
  (e) How many owned either IBM or AT&T but no GE?

30. Classifying Blood Types Human blood is classified as either Rh+ or Rh−. Blood is also classified by type: A, if it contains an A antigen but not a B antigen; B, if it contains a B antigen but not an A antigen; AB, if it contains both A and B antigens; and O, if it contains neither antigen. Draw a Venn diagram illustrating the various blood types. Based on this classification, how many different kinds of blood are there?

31. Demographics The following data represent the marital status of males 18 years old and older in 2007.
  (a) Determine the number of males 18 years old and older who are widowed or divorced.
  (b) Determine the number of males 18 years old and older who are married, widowed, or divorced.

Marital Status	Number (in thousands)
Married	63,318
Widowed	2,723
Divorced	9,200
Never married	29,608

Source: Current Population Survey

32. Demographics The following data represent the marital status of females 18 years old and older in 2007.
  (a) Determine the number of females 18 years old and older who are widowed or divorced.
  (b) Determine the number of females 18 years old and older who are married, widowed, or divorced.

Marital Status	Number (in thousands)
Married	63,971
Widowed	11,105
Divorced	12,932
Never married	24,318

Source: Current Population Survey

33. Stock Portfolios As a financial planner, you are asked to select one stock each from the following groups: 8 DOW stocks, 15 NASDAQ stocks, and 4 global stocks. How many different portfolios are possible?

Explaining Concepts: Discussion and Writing

34. Make up a problem different from any found in the text that requires the addition principle of counting to solve. Give it to a friend to solve and critique.

35. Investigate the notion of counting as it relates to infinite sets. Write an essay on your findings.

'Are You Prepared?' Answers

1. union   **2.** intersection   **3.** True   **4.** True

12.2 Permutations and Combinations

PREPARING FOR THIS SECTION *Before getting started, review the following:*

• Factorial (Section 11.1, pp. 829–830).

Now Work the 'Are You Prepared?' problems on page 881.

OBJECTIVES **1** Solve Counting Problems Using Permutations Involving *n* Distinct Objects (p. 876)

  2 Solve Counting Problems Using Combinations (p. 878)

  3 Solve Counting Problems Using Permutations Involving *n* Nondistinct Objects (p. 880)

1 Solve Counting Problems Using Permutations Involving n Distinct Objects

We begin with a definition.

DEFINITION

> A **permutation** is an ordered arrangement of r objects chosen from n objects.

We discuss three types of permutations:

1. The n objects are distinct (different), and repetition is allowed in the selection of r of them. [Distinct, with repetition]
2. The n objects are distinct (different), and repetition is not allowed in the selection of r of them, where $r \leq n$. [Distinct, without repetition]
3. The n objects are not distinct, and we use all of them in the arrangement. [Not distinct]

We take up the first two types here and deal with the third type at the end of this section.

The first type of permutation (n distinct objects, repetition allowed) is handled using the Multiplication Principle.

EXAMPLE 1   **Counting Airport Codes [Permutation: Distinct, with Repetition]**

The International Airline Transportation Association (IATA) assigns three-letter codes to represent airport locations. For example, the airport code for Ft. Lauderdale, Florida, is FLL. Notice that repetition is allowed in forming this code. How many airport codes are possible?

Solution   We are choosing 3 letters from 26 letters and arranging them in order. In the ordered arrangement a letter may be repeated. This is an example of a permutation with repetition in which 3 objects are chosen from 26 distinct objects.

The task of counting the number of such arrangements consists of making three selections. Each selection requires choosing a letter of the alphabet (26 choices). By the Multiplication Principle, there are

$$26 \cdot 26 \cdot 26 = 26^3 = 17{,}576$$

possible airport codes.

The solution given to Example 1 can be generalized.

THEOREM

> **Permutations: Distinct Objects with Repetition**
>
> The number of ordered arrangements of r objects chosen from n objects, in which the n objects are distinct and repetition is allowed, is n^r.

 Now Work   PROBLEM 33

Now let's consider permutations in which the objects are distinct and repetition is not allowed.

EXAMPLE 2   **Forming Codes [Permutation: Distinct, without Repetition]**

Suppose that we wish to establish a three-letter code using any of the 26 uppercase letters of the alphabet, but we require that no letter be used more than once. How many different three-letter codes are there?

Solution   Some of the possibilities are ABC, ABD, ABZ, ACB, CBA, and so on. The task consists of making three selections. The first selection requires choosing from 26 letters. Because no letter can be used more than once, the second selection requires choosing from 25 letters. The third selection requires choosing from 24 letters. (Do you see why?) By the Multiplication Principle, there are

$$26 \cdot 25 \cdot 24 = 15,600$$

different three-letter codes with no letter repeated.

For the second type of permutation, we introduce the following notation.

> The notation $P(n, r)$ represents the number of ordered arrangements of r objects chosen from n distinct objects, where $r \leq n$ and repetition is not allowed.

For example, the question posed in Example 2 asks for the number of ways that the 26 letters of the alphabet can be arranged in order using three nonrepeated letters. The answer is

$$P(26, 3) = 26 \cdot 25 \cdot 24 = 15,600$$

EXAMPLE 3   **Lining Up People**

In how many ways can 5 people be lined up?

Solution   The 5 people are distinct. Once a person is in line, that person will not be repeated elsewhere in the line; and, in lining up people, order is important. We have a permutation of 5 objects taken 5 at a time. We can line up 5 people in

$$P(5, 5) = \underbrace{5 \cdot 4 \cdot 3 \cdot 2 \cdot 1}_{5 \text{ factors}} = 120 \text{ ways}$$

Now Work PROBLEM 35

To arrive at a formula for $P(n, r)$, we note that the task of obtaining an ordered arrangement of n objects in which only $r \leq n$ of them are used, without repeating any of them, requires making r selections. For the first selection, there are n choices; for the second selection, there are $n - 1$ choices; for the third selection, there are $n - 2$ choices; ...; for the rth selection, there are $n - (r - 1)$ choices. By the Multiplication Principle, we have

$$
\begin{array}{ccccc}
& \text{1st} & \text{2nd} & \text{3rd} & \text{rth} \\
P(n, r) = & n \cdot & (n - 1) \cdot & (n - 2) \cdot & \cdots \cdot [n - (r - 1)]
\end{array}
$$

$$= n \cdot (n - 1) \cdot (n - 2) \cdot \cdots \cdot (n - r + 1)$$

This formula for $P(n, r)$ can be compactly written using factorial notation.*

$$P(n, r) = n \cdot (n - 1) \cdot (n - 2) \cdot \cdots \cdot (n - r + 1)$$

$$= n \cdot (n - 1) \cdot (n - 2) \cdot \cdots \cdot (n - r + 1) \cdot \frac{(n - r) \cdot \cdots \cdot 3 \cdot 2 \cdot 1}{(n - r) \cdot \cdots \cdot 3 \cdot 2 \cdot 1} = \frac{n!}{(n - r)!}$$

THEOREM

> **Permutations of r Objects Chosen from n Distinct Objects without Repetition**
>
> The number of arrangements of n objects using $r \leq n$ of them, in which
> 1. the n objects are distinct,
> 2. once an object is used it cannot be repeated, and
> 3. order is important,
>
> is given by the formula
>
> $$P(n, r) = \frac{n!}{(n - r)!} \qquad \textbf{(1)}$$

*Recall that $0! = 1$, $1! = 1$, $2! = 2 \cdot 1, \ldots, n! = n(n - 1) \cdot \cdots \cdot 3 \cdot 2 \cdot 1$.

EXAMPLE 4

Computing Permutations

Evaluate: (a) $P(7, 3)$ (b) $P(6, 1)$ (c) $P(52, 5)$

Solution

We shall work parts (a) and (b) in two ways.

(a) $P(7, 3) = \underbrace{7 \cdot 6 \cdot 5}_{3 \text{ factors}} = 210$

or

$$P(7, 3) = \frac{7!}{(7 - 3)!} = \frac{7!}{4!} = \frac{7 \cdot 6 \cdot 5 \cdot 4!}{4!} = 210$$

(b) $P(6, 1) = \underbrace{6}_{1 \text{ factor}} = 6$

or

$$P(6, 1) = \frac{6!}{(6 - 1)!} = \frac{6!}{5!} = \frac{6 \cdot 5!}{5!} = 6$$

Figure 3

(c) Figure 3 shows the solution using a TI-84 Plus graphing calculator. So

$$P(52, 5) = 311{,}875{,}200$$

━━━ **Now Work** PROBLEM 7

EXAMPLE 5

The Birthday Problem

All we know about Shannon, Patrick, and Ryan is that they have different birthdays. If we listed all the possible ways this could occur, how many would there be? Assume that there are 365 days in a year.

Solution

This is an example of a permutation in which 3 birthdays are selected from a possible 365 days, and no birthday may repeat itself. The number of ways that this can occur is

$$P(365, 3) = \frac{365!}{(365 - 3)!} = \frac{365 \cdot 364 \cdot 363 \cdot 362!}{362!} = 365 \cdot 364 \cdot 363 = 48{,}228{,}180$$

There are 48,228,180 ways in a group of three people for each to have a different birthday.

━━━ **Now Work** PROBLEM 47

2 Solve Counting Problems Using Combinations

In a permutation, order is important. For example, the arrangements ABC, CAB, BAC, ... are considered different arrangements of the letters A, B, and C. In many situations, though, order is unimportant. For example, in the card game of poker, the order in which the cards are received does not matter; it is the *combination* of the cards that matters.

DEFINITION

A **combination** is an arrangement, without regard to order, of r objects selected from n distinct objects without repetition, where $r \leq n$. The notation $C(n, r)$ represents the number of combinations of n distinct objects using r of them.

EXAMPLE 6

Listing Combinations

List all the combinations of the 4 objects a, b, c, d taken 2 at a time. What is $C(4, 2)$?

Solution

One combination of a, b, c, d taken 2 at a time is

$$ab$$

We exclude *ba* from the list because order is not important in a combination (this means that we do not distinguish *ab* from *ba*). The list of all combinations of a, b, c, d taken 2 at a time is

$$ab, \quad ac, \quad ad, \quad bc, \quad bd, \quad cd$$

so

$$C(4, 2) = 6$$

●

We can find a formula for $C(n, r)$ by noting that the only difference between a permutation of type 2 (distinct, without repetition) and a combination is that we disregard order in combinations. To determine $C(n, r)$, we need only eliminate from the formula for $P(n, r)$ the number of permutations that were simply rearrangements of a given set of r objects. This can be determined from the formula for $P(n, r)$ by calculating $P(r, r) = r!$. So, if we divide $P(n, r)$ by $r!$, we will have the desired formula for $C(n, r)$:

$$C(n, r) = \frac{P(n, r)}{r!} = \frac{\dfrac{n!}{(n - r)!}}{r!} = \frac{n!}{(n - r)!r!}$$

$\uparrow$
Use formula (1).

We have proved the following result:

THEOREM

> **Number of Combinations of *n* Distinct Objects Taken *r* at a Time**
>
> The number of arrangements of n objects using $r \le n$ of them, in which
>
> 1. the n objects are distinct,
> 2. once an object is used, it cannot be repeated, and
> 3. order is not important,
>
> is given by the formula
>
> $$C(n, r) = \frac{n!}{(n - r)!r!} \qquad (2)$$

Based on formula (2), we discover that the symbol $C(n, r)$ and the symbol $\binom{n}{r}$ for the binomial coefficients are, in fact, the same. Pascal's triangle (see Section 11.5) can be used to find the value of $C(n, r)$. However, because it is more practical and convenient, we will use formula (2) instead.

EXAMPLE 7 **Using Formula (2)**

Use formula (2) to find the value of each expression.
(a) $C(3, 1)$ (b) $C(6, 3)$ (c) $C(n, n)$ (d) $C(n, 0)$ (e) $C(52, 5)$

Solution (a) $C(3, 1) = \dfrac{3!}{(3 - 1)!1!} = \dfrac{3!}{2!1!} = \dfrac{3 \cdot 2 \cdot 1}{2 \cdot 1 \cdot 1} = 3$

(b) $C(6, 3) = \dfrac{6!}{(6 - 3)!3!} = \dfrac{6 \cdot 5 \cdot 4 \cdot 3!}{3! \cdot 3!} = \dfrac{6 \cdot 5 \cdot 4}{6} = 20$

(c) $C(n, n) = \dfrac{n!}{(n - n)!n!} = \dfrac{n!}{0!n!} = \dfrac{1}{1} = 1$

Figure 4

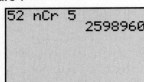

(d) $C(n, 0) = \dfrac{n!}{(n-0)!\,0!} = \dfrac{n!}{n!\,0!} = \dfrac{1}{1} = 1$

(e) Figure 4 shows the solution using a TI-84 Plus graphing calculator. So

$$C(52, 5) = 2{,}598{,}960$$

━━━━━ **Now Work** PROBLEM 15

EXAMPLE 8 | **Forming Committees**

How many different committees of 3 people can be formed from a pool of 7 people?

Solution The 7 people are distinct. More important, though, is the observation that the order of being selected for a committee is not significant. The problem asks for the number of combinations of 7 objects taken 3 at a time.

$$C(7, 3) = \frac{7!}{4!\,3!} = \frac{7 \cdot 6 \cdot 5 \cdot 4!}{4!\,3!} = \frac{7 \cdot 6 \cdot 5}{6} = 35$$

Thirty-five different committees can be formed.

EXAMPLE 9 | **Forming Committees**

In how many ways can a committee consisting of 2 faculty members and 3 students be formed if 6 faculty members and 10 students are eligible to serve on the committee?

Solution The problem can be separated into two parts: the number of ways that the faculty members can be chosen, $C(6, 2)$, and the number of ways that the student members can be chosen, $C(10, 3)$. By the Multiplication Principle, the committee can be formed in

$$C(6, 2) \cdot C(10, 3) = \frac{6!}{4!\,2!} \cdot \frac{10!}{7!\,3!} = \frac{6 \cdot 5 \cdot 4!}{4!\,2!} \cdot \frac{10 \cdot 9 \cdot 8 \cdot 7!}{7!\,3!}$$

$$= \frac{30}{2} \cdot \frac{720}{6} = 1800 \text{ ways}$$

━━━━━ **Now Work** PROBLEM 49

---

3 Solve Counting Problems Using Permutations Involving *n* Nondistinct Objects

We begin with an example.

EXAMPLE 10 | **Forming Different Words**

How many different words (real or imaginary) can be formed using all the letters in the word REARRANGE?

Solution Each word formed will have 9 letters: 3 R's, 2 A's, 2 E's, 1 N, and 1 G. To construct each word, we need to fill in 9 positions with the 9 letters:

$$\overline{1}\ \overline{2}\ \overline{3}\ \overline{4}\ \overline{5}\ \overline{6}\ \overline{7}\ \overline{8}\ \overline{9}$$

The process of forming a word consists of five tasks:

Task 1: Choose the positions for the 3 R's.

Task 2: Choose the positions for the 2 A's.

Task 3: Choose the positions for the 2 E's.

Task 4: Choose the position for the 1 N.
Task 5: Choose the position for the 1 G.

Task 1 can be done in $C(9, 3)$ ways. There then remain 6 positions to be filled, so Task 2 can be done in $C(6, 2)$ ways. There remain 4 positions to be filled, so Task 3 can be done in $C(4, 2)$ ways. There remain 2 positions to be filled, so Task 4 can be done in $C(2, 1)$ ways. The last position can be filled in $C(1, 1)$ way. Using the Multiplication Principle, the number of possible words that can be formed is

$$C(9, 3) \cdot C(6, 2) \cdot C(4, 2) \cdot C(2, 1) \cdot C(1, 1) = \frac{9!}{3! \cdot 6!} \cdot \frac{6!}{2! \cdot 4!} \cdot \frac{4!}{2! \cdot 2!} \cdot \frac{2!}{1! \cdot 1!} \cdot \frac{1!}{0! \cdot 1!}$$

$$= \frac{9!}{3! \cdot 2! \cdot 2! \cdot 1! \cdot 1!} = 15,120$$

15,120 possible words can be formed.

The form of the expression before the answer to Example 10 is suggestive of a general result. Had the letters in REARRANGE each been different, there would have been $P(9, 9) = 9!$ possible words formed. This is the numerator of the answer. The presence of 3 R's, 2 A's, and 2 E's reduces the number of different words, as the entries in the denominator illustrate. We are led to the following result:

THEOREM

Permutations Involving n Objects That Are Not Distinct

The number of permutations of n objects of which n_1 are of one kind, n_2 are of a second kind, ..., and n_k are of a kth kind is given by

$$\frac{n!}{n_1! \cdot n_2! \cdot \cdots \cdot n_k!} \tag{3}$$

where $n = n_1 + n_2 + \cdots + n_k$.

EXAMPLE 11

Arranging Flags

How many different vertical arrangements are there of 8 flags if 4 are white, 3 are blue, and 1 is red?

Solution

We seek the number of permutations of 8 objects, of which 4 are of one kind, 3 are of a second kind, and 1 is of a third kind. Using formula (3), we find that there are

$$\frac{8!}{4! \cdot 3! \cdot 1!} = \frac{8 \cdot 7 \cdot 6 \cdot 5 \cdot 4!}{4! \cdot 3! \cdot 1!} = 280 \text{ different arrangements}$$

━━━ **Now Work** PROBLEM 51

12.2 Assess Your Understanding

'Are You Prepared?' *Answers are given at the end of these exercises. If you get a wrong answer, read the pages listed in red.*

1. $0! = $ _____ ; $1! = $ _____ . (p. 829)

2. *True or False* $n! = \dfrac{(n + 1)!}{n}$. (p. 830)

Concepts and Vocabulary

3. A(n) _____ is an ordered arrangement of r objects chosen from n objects.

4. A(n) _____ is an arrangement of r objects chosen from n distinct objects, without repetition and without regard to order.

5. $P(n, r) = $ _____

6. $C(n, r) = $ _____

Skill Building

In Problems 7–14, find the value of each permutation.

7. $P(6, 2)$ **8.** $P(7, 2)$ **9.** $P(4, 4)$ **10.** $P(8, 8)$

11. $P(7, 0)$ **12.** $P(9, 0)$ **13.** $P(8, 4)$ **14.** $P(8, 3)$

In Problems 15–22, use formula (2) to find the value of each combination.

15. $C(8, 2)$ **16.** $C(8, 6)$ **17.** $C(7, 4)$ **18.** $C(6, 2)$

19. $C(15, 15)$ **20.** $C(18, 1)$ **21.** $C(26, 13)$ **22.** $C(18, 9)$

Applications and Extensions

23. List all the ordered arrangements of 5 objects a, b, c, d, and e choosing 3 at a time without repetition. What is $P(5, 3)$?

24. List all the ordered arrangements of 5 objects a, b, c, d, and e choosing 2 at a time without repetition. What is $P(5, 2)$?

25. List all the ordered arrangements of 4 objects 1, 2, 3, and 4 choosing 3 at a time without repetition. What is $P(4, 3)$?

26. List all the ordered arrangements of 6 objects 1, 2, 3, 4, 5, and 6 choosing 3 at a time without repetition. What is $P(6, 3)$?

27. List all the combinations of 5 objects a, b, c, d, and e taken 3 at a time. What is $C(5, 3)$?

28. List all the combinations of 5 objects a, b, c, d, and e taken 2 at a time. What is $C(5, 2)$?

29. List all the combinations of 4 objects 1, 2, 3, and 4 taken 3 at a time. What is $C(4, 3)$?

30. List all the combinations of 6 objects 1, 2, 3, 4, 5, and 6 taken 3 at a time. What is $C(6, 3)$?

31. Forming Codes How many two-letter codes can be formed using the letters A, B, C, and D? Repeated letters are allowed.

32. Forming Codes How many two-letter codes can be formed using the letters A, B, C, D, and E? Repeated letters are allowed.

33. Forming Numbers How many three-digit numbers can be formed using the digits 0 and 1? Repeated digits are allowed.

34. Forming Numbers How many three-digit numbers can be formed using the digits 0, 1, 2, 3, 4, 5, 6, 7, 8, and 9? Repeated digits are allowed.

35. Lining Up People In how many ways can 4 people be lined up?

36. Stacking Boxes In how many ways can 5 different boxes be stacked?

37. Forming Codes How many different three-letter codes are there if only the letters A, B, C, D, and E can be used and no letter can be used more than once?

38. Forming Codes How many different four-letter codes are there if only the letters A, B, C, D, E, and F can be used and no letter can be used more than once?

39. Stocks on the NYSE Companies whose stocks are listed on the New York Stock Exchange (NYSE) have their company name represented by either 1, 2, or 3 letters (repetition of letters is allowed). What is the maximum number of companies that can be listed on the NYSE?

40. Stocks on the NASDAQ Companies whose stocks are listed on the NASDAQ stock exchange have their company name represented by either 4 or 5 letters (repetition of letters is allowed). What is the maximum number of companies that can be listed on the NASDAQ?

41. Establishing Committees In how many ways can a committee of 4 students be formed from a pool of 7 students?

42. Establishing Committees In how many ways can a committee of 3 professors be formed from a department having 8 professors?

43. Possible Answers on a True/False Test How many arrangements of answers are possible for a true/false test with 10 questions?

44. Possible Answers on a Multiple-choice Test How many arrangements of answers are possible in a multiple-choice test with 5 questions, each of which has 4 possible answers?

45. Arranging Books Five different mathematics books are to be arranged on a student's desk. How many arrangements are possible?

46. Forming License Plate Numbers How many different license plate numbers can be made using 2 letters followed by 4 digits selected from the digits 0 through 9, if
(a) letters and digits may be repeated?
(b) letters may be repeated, but digits may not be repeated?
(c) neither letters nor digits may be repeated?

47. Birthday Problem In how many ways can 2 people each have different birthdays? Assume that there are 365 days in a year.

48. Birthday Problem In how many ways can 5 people each have different birthdays? Assume that there are 365 days in a year.

49. Forming a Committee A student dance committee is to be formed consisting of 2 boys and 3 girls. If the membership is to be chosen from 4 boys and 8 girls, how many different committees are possible?

50. Forming a Committee The student relations committee of a college consists of 2 administrators, 3 faculty members, and 5 students. Four administrators, 8 faculty members, and 20 students are eligible to serve. How many different committees are possible?

51. Forming Words How many different 9-letter words (real or imaginary) can be formed from the letters in the word ECONOMICS?

52. Forming Words How many different 11-letter words (real or imaginary) can be formed from the letters in the word MATHEMATICS?

53. Selecting Objects An urn contains 7 white balls and 3 red balls. Three balls are selected. In how many ways can the 3 balls be drawn from the total of 10 balls:
(a) If 2 balls are white and 1 is red?
(b) If all 3 balls are white?
(c) If all 3 balls are red?

54. Selecting Objects An urn contains 15 red balls and 10 white balls. Five balls are selected. In how many ways can the 5 balls be drawn from the total of 25 balls:
(a) If all 5 balls are red?
(b) If 3 balls are red and 2 are white?
(c) If at least 4 are red balls?

55. Senate Committees The U.S. Senate has 100 members. Suppose that it is desired to place each senator on exactly 1 of 7 possible committees. The first committee has 22 members, the second has 13, the third has 10, the fourth has 5, the fifth has 16, and the sixth and seventh have 17 apiece. In how many ways can these committees be formed?

56. Football Teams A defensive football squad consists of 25 players. Of these, 10 are linemen, 10 are linebackers, and 5 are safeties. How many different teams of 5 linemen, 3 linebackers, and 3 safeties can be formed?

57. Baseball In the American Baseball League, a designated hitter may be used. How many batting orders is it possible for a manager to use? (There are 9 regular players on a team.)

58. Baseball In the National Baseball League, the pitcher usually bats ninth. If this is the case, how many batting orders is it possible for a manager to use?

59. Baseball Teams A baseball team has 15 members. Four of the players are pitchers, and the remaining 11 members can play any position. How many different teams of 9 players can be formed?

60. World Series In the World Series the American League team (A) and the National League team (N) play until one team wins four games. If the sequence of winners is designated by letters (for example, $NAAAA$ means that the National League team won the first game and the American League won the next four), how many different sequences are possible?

61. Basketball Teams A basketball team has 6 players who play guard (2 of 5 starting positions). How many different teams are possible, assuming that the remaining 3 positions are filled and it is not possible to distinguish a left guard from a right guard?

62. Basketball Teams On a basketball team of 12 players, 2 only play center, 3 only play guard, and the rest play forward (5 players on a team: 2 forwards, 2 guards, and 1 center). How many different teams are possible, assuming that it is not possible to distinguish left and right guards and left and right forwards?

63. Combination Locks A combination lock displays 50 numbers. To open it, you turn to a number, then rotate clockwise to a second number, and then counterclockwise to the third number.
(a) How many different lock combinations are there?
(b) Comment on the description of such a lock as a *combination* lock.

Explaining Concepts: Discussion and Writing

64. Create a problem different from any found in the text that requires a permutation to solve. Give it to a friend to solve and critique.

65. Create a problem different from any found in the text that requires a combination to solve. Give it to a friend to solve and critique.

66. Explain the difference between a permutation and a combination. Give an example to illustrate your explanation.

'Are You Prepared?' Answers

1. $1; 1$ **2.** False

12.3 Probability

Probability is an area of mathematics that deals with experiments that yield random results, yet admit a certain regularity. Such experiments do not always produce the same result or outcome, so the result of any one observation is not predictable. However, the results of the experiment over a long period do produce regular patterns that enable us to predict with remarkable accuracy.

EXAMPLE 1 **Tossing a Fair Coin**

In tossing a fair coin, we know that the outcome is either a head or a tail. On any particular throw, we cannot predict what will happen, but, if we toss the coin many times, we observe that the number of times that a head comes up is approximately equal to the number of times that we get a tail. It seems reasonable, therefore, to assign a probability of $\frac{1}{2}$ that a head comes up and a probability of $\frac{1}{2}$ that a tail comes up.

●

1 Construct Probability Models

The discussion in Example 1 constitutes the construction of a **probability model** for the experiment of tossing a fair coin once. A probability model has two components: a sample space and an assignment of probabilities. A **sample space** S is a set whose elements represent all the possibilities that can occur as a result of the experiment. Each element of S is called an **outcome.** To each outcome, we assign a number, called the **probability** of that outcome, which has two properties:

1. The probability assigned to each outcome is nonnegative.

2. The sum of all the probabilities equals 1.

DEFINITION
A **probability model** with the sample space

$$S = \{e_1, e_2, \ldots, e_n\}$$

where $e_1, e_2, \ldots, e_n$ are the possible outcomes and $P(e_1), P(e_2), \ldots, P(e_n)$ are the respective probabilities of these outcomes, requires that

$$P(e_1) \geq 0, P(e_2) \geq 0, \ldots, P(e_n) \geq 0 \tag{1}$$

$$\sum_{i=1}^{n} P(e_i) = P(e_1) + P(e_2) + \cdots + P(e_n) = 1 \tag{2}$$

EXAMPLE 2 **Determining Probability Models**

In a bag of M&Ms,™ the candies are colored red, green, blue, brown, yellow, and orange. A candy is drawn from the bag and the color is recorded. The sample space of this experiment is {red, green, blue, brown, yellow, orange}. Determine which of the following are probability models.

(a)

Outcome	Probability
red	0.3
green	0.15
blue	0
brown	0.15
yellow	0.2
orange	0.2

(b)

Outcome	Probability
red	0.1
green	0.1
blue	0.1
brown	0.4
yellow	0.2
orange	0.3

(c)

Outcome	Probability
red	0.3
green	−0.3
blue	0.2
brown	0.4
yellow	0.2
orange	0.2

(d)

Outcome	Probability
red	0
green	0
blue	0
brown	0
yellow	1
orange	0

Solution

(a) This model is a probability model since all the outcomes have probabilities that are nonnegative and the sum of the probabilities is 1.

(b) This model is not a probability model because the sum of the probabilities is not 1.

(c) This model is not a probability model because $P(\text{green})$ is less than 0. Recall, all probabilities must be nonnegative.

(d) This model is a probability model because all the outcomes have probabilities that are nonnegative, and the sum of the probabilities is 1. Notice that $P(\text{yellow}) = 1$, meaning that this outcome will occur with 100% certainty each time that the experiment is repeated. This means that the bag of M&Ms™ has only yellow candies.

Now Work PROBLEM 7

EXAMPLE 3   **Constructing a Probability Model**

An experiment consists of rolling a fair die once. A die is a cube with each face having either 1, 2, 3, 4, 5, or 6 dots on it. See Figure 5. Construct a probability model for this experiment.

Figure 5

Solution   A sample space S consists of all the possibilities that can occur. Because rolling the die will result in one of six faces showing, the sample space S consists of

$$S = \{1, 2, 3, 4, 5, 6\}$$

Because the die is fair, one face is no more likely to occur than another. As a result, our assignment of probabilities is

$$P(1) = \frac{1}{6} \qquad P(2) = \frac{1}{6}$$

$$P(3) = \frac{1}{6} \qquad P(4) = \frac{1}{6}$$

$$P(5) = \frac{1}{6} \qquad P(6) = \frac{1}{6}$$

Now suppose that a die is loaded (weighted) so that the probability assignments are

$$P(1) = 0 \quad P(2) = 0 \quad P(3) = \frac{1}{3} \quad P(4) = \frac{2}{3} \quad P(5) = 0 \quad P(6) = 0$$

This assignment would be made if the die were loaded so that only a 3 or 4 could occur and the 4 is twice as likely as the 3 to occur. This assignment is consistent with the definition, since each assignment is nonnegative and the sum of all the probability assignments equals 1.

━━━━━**Now Work** PROBLEM 23

EXAMPLE 4   **Constructing a Probability Model**

An experiment consists of tossing a coin. The coin is weighted so that heads (H) is three times as likely to occur as tails (T). Construct a probability model for this experiment.

Solution   The sample space S is $S = \{H, T\}$. If x denotes the probability that a tail occurs,

$$P(T) = x \quad \text{and} \quad P(H) = 3x$$

Since the sum of the probabilities of the possible outcomes must equal 1, we have

$$P(T) + P(H) = x + 3x = 1$$
$$4x = 1$$
$$x = \frac{1}{4}$$

We assign the probabilities

$$P(T) = \frac{1}{4} \quad P(H) = \frac{3}{4}$$

●

━━━━━**Now Work** PROBLEM 27

In working with probability models, the term **event** is used to describe a set of possible outcomes of the experiment. An event E is some subset of the sample space S. The **probability of an event** E, $E \neq \varnothing$, denoted by $P(E)$, is defined as the sum of the probabilities of the outcomes in E. We can also think of the probability of an event E as the likelihood that the event E occurs. If $E = \varnothing$, then $P(E) = 0$; if $E = S$, then $P(E) = P(S) = 1$.

2 Compute Probabilities of Equally Likely Outcomes

When the same probability is assigned to each outcome of the sample space, the experiment is said to have **equally likely outcomes.**

THEOREM   **Probability for Equally Likely Outcomes**

If an experiment has n equally likely outcomes and if the number of ways that an event E can occur is m, then the probability of E is

$$P(E) = \frac{\text{Number of ways that } E \text{ can occur}}{\text{Number of possible outcomes}} = \frac{m}{n} \qquad (3)$$

If S is the sample space of this experiment,

$$P(E) = \frac{n(E)}{n(S)} \qquad (4)$$

EXAMPLE 5   **Calculating Probabilities of Events Involving Equally Likely Outcomes**

Calculate the probability that in a 3-child family there are 2 boys and 1 girl. Assume equally likely outcomes.

Solution   We begin by constructing a tree diagram to help in listing the possible outcomes of the experiment. See Figure 6, where B stands for boy and G for girl. The sample space S of this experiment is

$$S = \{BBB, BBG, BGB, BGG, GBB, GBG, GGB, GGG\}$$

Figure 6

1st child 2nd child 3rd child

so $n(S) = 8$.

We wish to know the probability of the event E: "having two boys and one girl." From Figure 6, we conclude that $E = \{BBG, BGB, GBB\}$, so $n(E) = 3$. Since the outcomes are equally likely, the probability of E is

$$P(E) = \frac{n(E)}{n(S)} = \frac{3}{8}$$

━━ **Now Work** PROBLEM 37

So far, we have calculated probabilities of single events. We will now compute probabilities of multiple events, called **compound probabilities.**

EXAMPLE 6   **Computing Compound Probabilities**

Consider the experiment of rolling a single fair die. Let E represent the event "roll an odd number," and let F represent the event "roll a 1 or 2."

(a) Write the event E and F. What is $n(E \cap F)$?
(b) Write the event E or F. What is $n(E \cup F)$?
(c) Compute $P(E)$. Compute $P(F)$.
(d) Compute $P(E \cap F)$.
(e) Compute $P(E \cup F)$.

Solution   The sample space S of the experiment is $\{1, 2, 3, 4, 5, 6\}$, so $n(S) = 6$. Since the die is fair, the outcomes are equally likely. The event E: "roll an odd number" is $\{1, 3, 5\}$, and the event F: "roll a 1 or 2" is $\{1, 2\}$, so $n(E) = 3$ and $n(F) = 2$.

(a) The word *and* in probability means the intersection of two events. The event E and F is

$$E \cap F = \{1, 3, 5\} \cap \{1, 2\} = \{1\} \qquad n(E \cap F) = 1$$

(b) The word *or* in probability means the union of the two events. The event E or F is

$$E \cup F = \{1, 3, 5\} \cup \{1, 2\} = \{1, 2, 3, 5\} \qquad n(E \cup F) = 4$$

(c) We use formula (4). Then

$$P(E) = \frac{n(E)}{n(S)} = \frac{3}{6} = \frac{1}{2} \qquad P(F) = \frac{n(F)}{n(S)} = \frac{2}{6} = \frac{1}{3}$$

(d) $P(E \cap F) = \dfrac{n(E \cap F)}{n(S)} = \dfrac{1}{6}$

(e) $P(E \cup F) = \dfrac{n(E \cup F)}{n(S)} = \dfrac{4}{6} = \dfrac{2}{3}$

3 Find Probabilities of the Union of Two Events

The next formula can be used to find the probability of the union of two events.

THEOREM

For any two events E and F,

$$P(E \cup F) = P(E) + P(F) - P(E \cap F) \qquad (5)$$

This result is a consequence of the Counting Formula discussed earlier in Section 12.1.

For example, we can use formula (5) to find $P(E \cup F)$ in Example 6(e). Then

$$P(E \cup F) = P(E) + P(F) - P(E \cap F) = \frac{1}{2} + \frac{1}{3} - \frac{1}{6} = \frac{3}{6} + \frac{2}{6} - \frac{1}{6} = \frac{4}{6} = \frac{2}{3}$$

as before.

EXAMPLE 7

Computing Probabilities of the Union of Two Events

If $P(E) = 0.2$, $P(F) = 0.3$, and $P(E \cap F) = 0.1$, find the probability of E or F. That is, find $P(E \cup F)$.

Solution

We use formula (5).

$$\text{Probability of } E \text{ or } F = P(E \cup F) = P(E) + P(F) - P(E \cap F)$$
$$= 0.2 + 0.3 - 0.1 = 0.4$$

     ●

A Venn diagram can sometimes be used to obtain probabilities. To construct a Venn diagram representing the information in Example 7, we draw two sets E and F. We begin with the fact that $P(E \cap F) = 0.1$. See Figure 7(a). Then, since $P(E) = 0.2$ and $P(F) = 0.3$, we fill in E with $0.2 - 0.1 = 0.1$ and F with $0.3 - 0.1 = 0.2$. See Figure 7(b). Since $P(S) = 1$, we complete the diagram by inserting $1 - (0.1 + 0.1 + 0.2) = 0.6$ outside the circles. See Figure 7(c). Now it is easy to see, for example, that the probability of F, but not E, is 0.2. Also, the probability of neither E nor F is 0.6.

Figure 7

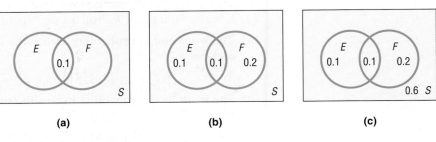

(a)   (b)   (c)

━━━**Now Work** PROBLEM 45

If events E and F are disjoint so that $E \cap F = \varnothing$, we say they are **mutually exclusive.** In this case, $P(E \cap F) = 0$, and formula (5) takes the following form:

THEOREM

Mutually Exclusive Events

If E and F are mutually exclusive events,

$$P(E \cup F) = P(E) + P(F) \qquad (6)$$

EXAMPLE 8   **Computing Probabilities of the Union of Two Mutually Exclusive Events**

If $P(E) = 0.4$ and $P(F) = 0.25$, and E and F are mutually exclusive, find $P(E \cup F)$.

Solution   Since E and F are mutually exclusive, we use formula (6).

$$P(E \cup F) = P(E) + P(F) = 0.4 + 0.25 = 0.65$$

━━━ Now Work   PROBLEM 47

4 Use the Complement Rule to Find Probabilities

Recall, if A is a set, the complement of A, denoted $\overline{A}$, is the set of all elements in the universal set U not in A. We similarly define the complement of an event.

DEFINITION   **Complement of an Event**

Let S denote the sample space of an experiment, and let E denote an event. The **complement of E,** denoted $\overline{E}$, is the set of all outcomes in the sample space S that are not outcomes in the event E.

The complement of an event E, that is, $\overline{E}$, in a sample space S has the following two properties:

$$E \cap \overline{E} = \varnothing \qquad E \cup \overline{E} = S$$

Since E and $\overline{E}$ are mutually exclusive, it follows from (6) that

$$P(E \cup \overline{E}) = P(S) = 1 \qquad P(E) + P(\overline{E}) = 1 \qquad P(\overline{E}) = 1 - P(E)$$

We have the following result:

THEOREM   **Computing Probabilities of Complementary Events**

If E represents any event and $\overline{E}$ represents the complement of E, then

$$P(\overline{E}) = 1 - P(E) \tag{7}$$

EXAMPLE 9   **Computing Probabilities Using Complements**

On the local news the weather reporter stated that the probability of rain tomorrow is 40%. What is the probability that it will not rain?

Solution   The complement of the event "rain" is "no rain."

$$P(\text{no rain}) = 1 - P(\text{rain}) = 1 - 0.4 = 0.6$$

There is a 60% chance of no rain tomorrow.

━━━ Now Work   PROBLEM 51

EXAMPLE 10   **Birthday Problem**

What is the probability that in a group of 10 people at least 2 people have the same birthday? Assume that there are 365 days in a year.

Solution   We assume that a person is as likely to be born on one day as another, so we have equally likely outcomes.

We first determine the number of outcomes in the sample space S. There are 365 possibilities for each person's birthday. Since there are 10 people in the group, there are 365^{10} possibilities for the birthdays. [For one person in the group, there are 365 days on which his or her birthday can fall; for two people, there are $(365)(365) = 365^2$ pairs of days; and, in general, using the Multiplication Principle, for n people there are 365^n possibilities.] So

$$n(S) = 365^{10}$$

We wish to find the probability of the event E: "at least two people have the same birthday." It is difficult to count the elements in this set; it is much easier to count the elements of the complementary event $\overline{E}$: "no two people have the same birthday."

We find $n(\overline{E})$ as follows: Choose one person at random. There are 365 possibilities for his or her birthday. Choose a second person. There are 364 possibilities for this birthday, if no two people are to have the same birthday. Choose a third person. There are 363 possibilities left for this birthday. Finally, we arrive at the tenth person. There are 356 possibilities left for this birthday. By the Multiplication Principle, the total number of possibilities is

$$n(\overline{E}) = 365 \cdot 364 \cdot 363 \cdot \cdots \cdot 356$$

The probability of the event $\overline{E}$ is

$$P(\overline{E}) = \frac{n(\overline{E})}{n(S)} = \frac{365 \cdot 364 \cdot 363 \cdot \cdots \cdot 356}{365^{10}} \approx 0.883$$

The probability of two or more people in a group of 10 people having the same birthday is then

$$P(E) = 1 - P(\overline{E}) \approx 1 - 0.883 = 0.117$$

The birthday problem can be solved for any group size. The following table gives the probabilities for two or more people having the same birthday for various group sizes. Notice that the probability is greater than $\frac{1}{2}$ for any group of 23 or more people.

	Number of People																
	5	10	15	20	21	22	23	24	25	30	40	50	60	70	80	90
Probability That Two or More Have the Same Birthday	0.027	0.117	0.253	0.411	0.444	0.476	0.507	0.538	0.569	0.706	0.891	0.970	0.994	0.99916	0.99991	0.99999

Now Work PROBLEM 71

Historical Feature

Blaise Pascal (1623–1662)

Set theory, counting, and probability first took form as a systematic theory in an exchange of letters (1654) between Pierre de Fermat (1601–1665) and Blaise Pascal (1623–1662). They discussed the problem of how to divide the stakes in a game that is interrupted before completion, knowing how many points each player needs to win. Fermat solved the problem by listing all possibilities and counting the favorable ones, whereas Pascal made use of the triangle that now bears his name. As mentioned in the text, the entries in Pascal's triangle are equivalent to $C(n, r)$. This recognition of the role of $C(n, r)$ in counting is the foundation of all further developments.

The first book on probability, the work of Christiaan Huygens (1629–1695), appeared in 1657. In it, the notion of mathematical expectation is explored. This allows the calculation of the profit or loss that a gambler might expect, knowing the probabilities involved in the game (see the Historical Problems that follow).

Although Girolamo Cardano (1501–1576) wrote a treatise on probability, it was not published until 1663 in Cardano's collected works, and this was too late to have any effect on the early development of the theory.

In 1713, the posthumously published *Ars Conjectandi* of Jakob Bernoulli (1654–1705) gave the theory the form it would have until 1900. Recently, both combinatorics (counting) and probability have undergone rapid development due to the use of computers.

A final comment about notation. The notations $C(n, r)$ and $P(n, r)$ are variants of a form of notation developed in England after 1830. The notation $\binom{n}{r}$ for $C(n, r)$ goes back to Leonhard Euler (1707–1783),

but is now losing ground because it has no clearly related symbolism of the same type for permutations. The set symbols $\cup$ and $\cap$ were introduced by Giuseppe Peano (1858-1932) in 1888 in a slightly different context. The inclusion symbol $\subset$ was introduced by E. Schroeder (1841–1902) about 1890. The treatment of set theory in the text is due to George Boole (1815-1864), who wrote $A + B$ for $A \cup B$ and AB for $A \cap B$ (statisticians still use AB for $A \cap B$).

Historical Problems

1. *The Problem Discussed by Fermat and Pascal*   A game between two equally skilled players, *A* and *B*, is interrupted when *A* needs 2 points to win and *B* needs 3 points. In what proportion would the stakes be divided?

   (a) *Fermat's solution*   List all possible outcomes that can occur as a result of four more plays. The probabilities for *A* to win and *B* to win then determine how the stakes should be divided.

   (b) *Pascal's solution*   Use combinations to determine the number of ways that the 2 points needed for *A* to win could occur in four plays. Then use combinations to determine the number of ways that the 3 points needed for *B* to win could occur. This is trickier than it looks, since *A* can win with 2 points in either two plays, three plays, or four plays. Compute the probabilities and compare with the results in part (a).

2. *Huygens's Mathematical Expectation*   In a game with *n* possible outcomes with probabilities $p_1, p_2, \ldots, p_n$, suppose that the *net* winnings are $w_1, w_2, \ldots, w_n$, respectively. Then the mathematical expectation is

$$E = p_1 w_1 + p_2 w_2 + \cdots + p_n w_n$$

The number *E* represents the profit or loss per game in the long run. The following problems are a modification of those of Huygens.

   (a) A fair die is tossed. A gambler wins $3 if he throws a 6 and $6 if he throws a 5. What is his expectation?
   [**Hint:** $w_1 = w_2 = w_3 = w_4 = 0$]

   (b) A gambler plays the same game as in part (a), but now the gambler must pay $1 to play. This means that $w_5 = \$5$, $w_6 = \$2$, and $w_1 = w_2 = w_3 = w_4 = -\$1$. What is the expectation?

12.3 Assess Your Understanding

Concepts and Vocabulary

1. When the same probability is assigned to each outcome of a sample space, the experiment is said to have _____ _____ outcomes.

2. The _____ of an event *E* is the set of all outcomes in the sample space *S* that are not outcomes in the event *E*.

3. *True or False*   The probability of an event can never equal 0.

4. *True or False*   In a probability model, the sum of all probabilities is 1.

Skill Building

5. In a probability model, which of the following numbers could be the probability of an outcome?

$$0 \quad 0.01 \quad 0.35 \quad -0.4 \quad 1 \quad 1.4$$

6. In a probability model, which of the following numbers could be the probability of an outcome?

$$1.5 \quad \frac{1}{2} \quad \frac{3}{4} \quad \frac{2}{3} \quad 0 \quad -\frac{1}{4}$$

7. Determine whether the following is a probability model.

Outcome	Probability
1	0.2
2	0.3
3	0.1
4	0.4

8. Determine whether the following is a probability model.

Outcome	Probability
Steve	0.4
Bob	0.3
Faye	0.1
Patricia	0.2

9. Determine whether the following is a probability model.

Outcome	Probability
Linda	0.3
Jean	0.2
Grant	0.1
Jim	0.3

10. Determine whether the following is a probability model.

Outcome	Probability
Erica	0.3
Joanne	0.2
Laura	0.1
Donna	0.5
Angela	-0.1

In Problems 11–16, construct a probability model for each experiment.

11. Tossing a fair coin twice

12. Tossing two fair coins once

13. Tossing two fair coins, then a fair die

14. Tossing a fair coin, a fair die, and then a fair coin

15. Tossing three fair coins once

16. Tossing one fair coin three times

In Problems 17–22, use the following spinners to construct a probability model for each experiment.

Spinner I   Spinner II   Spinner III
(4 equal areas)   (3 equal areas)   (2 equal areas)

17. Spin spinner I, then spinner II. What is the probability of getting a 2 or a 4, followed by Red?

18. Spin spinner III, then spinner II. What is the probability of getting Forward, followed by Yellow or Green?

19. Spin spinner I, then II, then III. What is the probability of getting a 1, followed by Red or Green, followed by Backward?

20. Spin spinner II, then I, then III. What is the probability of getting Yellow, followed by a 2 or a 4, followed by Forward?

21. Spin spinner I twice, then spinner II. What is the probability of getting a 2, followed by a 2 or a 4, followed by Red or Green?

22. Spin spinner III, then spinner I twice. What is the probability of getting Forward, followed by a 1 or a 3, followed by a 2 or a 4?

In Problems 23–26, consider the experiment of tossing a coin twice. The table lists six possible assignments of probabilities for this experiment. Using this table, answer the following questions.

Assignments	Sample Space			
	HH	**HT**	**TH**	**TT**
A	$\frac{1}{4}$	$\frac{1}{4}$	$\frac{1}{4}$	$\frac{1}{4}$
B	0	0	0	1
C	$\frac{3}{16}$	$\frac{5}{16}$	$\frac{5}{16}$	$\frac{3}{16}$
D	$\frac{1}{2}$	$\frac{1}{2}$	$-\frac{1}{2}$	$\frac{1}{2}$
E	$\frac{1}{4}$	$\frac{1}{4}$	$\frac{1}{4}$	$\frac{1}{8}$
F	$\frac{1}{9}$	$\frac{2}{9}$	$\frac{2}{9}$	$\frac{4}{9}$

23. Which of the assignments of probabilities is(are) consistent with the definition of a probability model?

24. Which of the assignments of probabilities should be used if the coin is known to be fair?

25. Which of the assignments of probabilities should be used if the coin is known to always come up tails?

26. Which of the assignments of probabilities should be used if tails is twice as likely as heads to occur?

27. Assigning Probabilities A coin is weighted so that heads is four times as likely as tails to occur. What probability should we assign to heads? to tails?

28. Assigning Probabilities A coin is weighted so that tails is twice as likely as heads to occur. What probability should we assign to heads? to tails?

29. Assigning Probabilities A die is weighted so that an odd-numbered face is twice as likely to occur as an even-numbered face. What probability should we assign to each face?

30. Assigning Probabilities A die is weighted so that a six cannot appear. The other faces occur with the same probability. What probability should we assign to each face?

For Problems 31–34, the sample space is $S = \{1, 2, 3, 4, 5, 6, 7, 8, 9, 10\}$. Suppose that the outcomes are equally likely.

31. Compute the probability of the event $E = \{1, 2, 3\}$.

32. Compute the probability of the event $F = \{3, 5, 9, 10\}$.

33. Compute the probability of the event E: "an even number."

34. Compute the probability of the event F: "an odd number."

For Problems 35 and 36, an urn contains 5 white marbles, 10 green marbles, 8 yellow marbles, and 7 black marbles.

35. If one marble is selected, determine the probability that it is white.

36. If one marble is selected, determine the probability that it is black.

In Problems 37–40, assume equally likely outcomes.

37. Determine the probability of having 3 boys in a 3-child family.

38. Determine the probability of having 3 girls in a 3-child family.

39. Determine the probability of having 1 girl and 3 boys in a 4-child family.

40. Determine the probability of having 2 girls and 2 boys in a 4-child family.

For Problems 41–44, two fair dice are rolled.

41. Determine the probability that the sum of the two dice is 7.

42. Determine the probability that the sum of the two dice is 11.

43. Determine the probability that the sum of the two dice is 3.

44. Determine the probability that the sum of the two dice is 12.

In Problems 45–48, find the probability of the indicated event if
$P(A) = 0.25$ and $P(B) = 0.45$.

45. $P(A \cup B)$ if $P(A \cap B) = 0.15$

46. $P(A \cap B)$ if $P(A \cup B) = 0.6$

47. $P(A \cup B)$ if A, B are mutually exclusive

48. $P(A \cap B)$ if A, B are mutually exclusive

49. If $P(A) = 0.60$, $P(A \cup B) = 0.85$, and $P(A \cap B) = 0.05$, find $P(B)$.

50. If $P(B) = 0.30$, $P(A \cup B) = 0.65$, and $P(A \cap B) = 0.15$, find $P(A)$.

51. **Automobile Theft**   According to the Insurance Information Institute, in 2004 there was a 13% probability that an automobile theft in the United States would be cleared by arrests. If an automobile theft case is randomly selected, what is the probability that it was not cleared by an arrest?

52. **Pet Ownership**   According to the American Pet Products Manufacturers Association's *2005-2006 National Pet Owners Survey*, there is a 63% probability that a U.S. household owns a pet. If a U.S. household is randomly selected, what is the probability that it does not own a pet?

53. **Cat Ownership**   According to the American Pet Products Manufacturers Association's *2005–2006 National Pet Owners Survey*, in 2004 there was a 34% probability that a U.S. pet owner owned a cat. If a U.S. pet owner is randomly selected, what is the probability that he or she does not own a cat?

54. **Doctorate Degrees**   According to the National Science Foundation, in 2004 there was a 13.7% probability that a doctoral degree awarded at a U.S. university was awarded in engineering. If a 2004 U.S. doctoral recipient is randomly selected, what is the probability that his or her degree was not in engineering?

55. **Online Gambling**   According to a Harris poll (January 12–17, 2006), 5% of U.S. adults admitted to having spent money gambling online. If a U.S. adult is selected at random, what is the probability that he or she has never spent any money gambling online?

56. **Girl Scout Cookies**   According to the Girl Scouts of America, in March 2006, 9% of all Girl Scout cookies sold are shortbread/trefoils. If a box of Girl Scout cookies is selected at random, what is the probability that it is not shortbread/trefoils?

For Problems 57–60, a golf ball is selected at random from a container. If the container has 9 white balls, 8 green balls, and 3 orange balls, find the probability of each event.

57. The golf ball is white or green.

58. The golf ball is white or orange.

59. The golf ball is not white.

60. The golf ball is not green.

61. On the "Price is Right" there is a game in which a bag is filled with 3 strike chips and 5 numbers. Let's say that the numbers in the bag are 0, 1, 3, 6, and 9. What is the probability of selecting a strike chip or the number 1?

62. Another game on the "Price is Right" requires the contestant to spin a wheel with numbers 5, 10, 15, 20, . . . , 100. What is the probability that the contestant spins 100 or 30?

Problems 63–66 are based on a consumer survey of annual incomes in 100 households. The following table gives the data.

Income	$0–9999	$10,000–19,999	$20,000–29,999	$30,000–39,999	$40,000 or more
Number of households	5	35	30	20	10

63. What is the probability that a household has an annual income of $30,000 or more?

64. What is the probability that a household has an annual income between $10,000 and $29,999, inclusive?

65. What is the probability that a household has an annual income of less than $20,000?

66. What is the probability that a household has an annual income of $20,000 or more?

67. **Surveys**   In a survey about the number of TV sets in a house, the following probability table was constructed:

Number of TV sets	0	1	2	3	4 or more
Probability	0.05	0.24	0.33	0.21	0.17

Find the probability of a house having:
(a) 1 or 2 TV sets
(b) 1 or more TV sets
(c) 3 or fewer TV sets
(d) 3 or more TV sets
(e) Fewer than 2 TV sets
(f) Fewer than 1 TV set
(g) 1, 2, or 3 TV sets
(h) 2 or more TV sets

68. **Checkout Lines**   Through observation, it has been determined that the probability for a given number of people waiting in line at the "5 items or less" checkout register of a supermarket is as follows:

Number waiting in line	0	1	2	3	4 or more
Probability	0.10	0.15	0.20	0.24	0.31

Find the probability of:
(a) At most 2 people in line
(b) At least 2 people in line
(c) At least 1 person in line

69. In a certain Algebra and Trigonometry class, there are 18 freshmen and 15 sophomores. Of the 18 freshmen, 10 are male, and of the 15 sophomores, 8 are male. Find the probability that a randomly selected student is:
(a) A freshman or female
(b) A sophomore or male

70. The faculty of the mathematics department at Joliet Junior College is composed of 4 females and 9 males. Of the 4 females, 2 are under age 40, and of the males 3 are under age 40. Find the probability that a randomly selected faculty member is:
(a) Female or under age 40
(b) Male or over age 40

71. Birthday Problem   What is the probability that at least 2 people have the same birthday in a group of 12 people? Assume that there are 365 days in a year.

72. Birthday Problem   What is the probability that at least 2 people have the same birthday in a group of 35 people? Assume that there are 365 days in a year.

73. Winning a Lottery   In a certain lottery, there are ten balls, numbered 1, 2, 3, 4, 5, 6, 7, 8, 9, 10. Of these, five are drawn in order. If you pick five numbers that match those drawn in the correct order, you win $1,000,000. What is the probability of winning such a lottery?

CHAPTER REVIEW

Things to Know

Counting formula (p. 871)	$n(A \cup B) = n(A) + n(B) - n(A \cap B)$
Addition Principle of Counting (p. 871)	If $A \cap B = \varnothing$, then $n(A \cup B) = n(A) + n(B)$.
Multiplication Principle of Counting (p. 873)	If a task consists of a sequence of choices in which there are p selections for the first choice, q selections for the second choice, and so on, the task of making these selections can be done in $p \cdot q \cdot \cdots$ different ways.
Permutation (p. 876)	An ordered arrangement of r objects chosen from n objects
Number of permutations: Distinct, with repetition (p. 876)	n^r The n objects are distinct (different), and repetition is allowed in the selection of r of them.
Number of permutations: Distinct, without repetition (p. 877)	$P(n, r) = n(n - 1) \cdot \cdots \cdot [n - (r - 1)] = \dfrac{n!}{(n - r)!}$ The n objects are distinct (different), and repetition is not allowed in the selection of r of them, where $r \le n$.
Combinations (p. 878)	An arrangement, without regard to order, of r objects selected from n distinct objects, where $r \le n$
Number of combinations (p. 879)	$C(n, r) = \dfrac{P(n, r)}{r!} = \dfrac{n!}{(n - r)! r!}$
Number of permutations: Not distinct (p. 881)	$\dfrac{n!}{n_1! n_2! \cdots n_k!}$ The number of permutations of n objects of which n_1 are of one kind, n_2 are of a second kind, ..., and n_k are of a kth kind, where $n = n_1 + n_2 + \cdots + n_k$
Sample space (p. 884)	Set whose elements represent the possible outcomes that can occur as a result of an experiment
Probability (p. 884)	A nonnegative number assigned to each outcome of a sample space; the sum of all the probabilities of the outcomes equals 1.
Probability for equally likely outcomes (p. 886)	$P(E) = \dfrac{n(E)}{n(S)}$ The same probability is assigned to each outcome.
Probability of the union of two events (p. 888)	$P(E \cup F) = P(E) + P(F) - P(E \cap F)$
Probability of the complement of an event (p. 889)	$P(\overline{E}) = 1 - P(E)$

Objectives

Section	You should be able to . . .	Example(s)	Review Exercises
12.1	**1** Find all the subsets of a set (p. 870)	1	1, 2
	2 Count the number of elements in a set (p. 870)	2, 3	3–10
	3 Solve counting problems using the Multiplication Principle (p. 873)	4, 5	15–18
12.2	**1** Solve counting problems using permutations involving n distinct objects (p. 876)	1–5	11, 12, 19, 20, 24–28, 33(a)
	2 Solve counting problems using combinations (p. 878)	6–9	13, 14, 21–23, 31, 32
	3 Solve counting problems using permutations involving n nondistinct objects (p. 880)	10, 11	29, 30
12.3	**1** Construct probability models (p. 884)	2–4	33(b)
	2 Compute probabilities of equally likely outcomes (p. 886)	5, 6	33(b), 34(a), 35(a), 36–39
	3 Find probabilities of the union of two events (p. 888)	7, 8	40
	4 Use the Complement Rule to find probabilities (p. 889)	9, 10	33(c), 34(b), 35(b), 36

Review Exercises

1. Write down all the subsets of the set {Dave, Joanne, Erica}.

2. Write down all the subsets of the set {Green, Blue, Red}.

3. If $n(A) = 8$, $n(B) = 12$, and $n(A \cap B) = 3$, find $n(A \cup B)$.

4. If $n(A) = 12$, $n(A \cup B) = 30$, and $n(A \cap B) = 6$, find $n(B)$.

In Problems 5–10, use the information supplied in the figure.

5. How many are in A?

6. How many are in A or B?

7. How many are in A and C?

8. How many are not in B?

9. How many are in neither A nor C?

10. How many are in B but not in C?

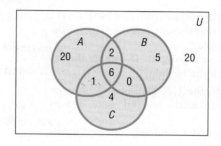

In Problems 11–14, compute the given expression.

11. $P(8, 3)$ **12.** $P(7, 3)$ **13.** $C(8, 3)$ **14.** $C(7, 3)$

15. Stocking a Store A clothing store sells pure wool and polyester-wool suits. Each suit comes in 3 colors and 10 sizes. How many suits are required for a complete assortment?

16. Wiring In connecting a certain electrical device, 5 wires are to be connected to 5 different terminals. How many different wirings are possible if 1 wire is connected to each terminal?

17. Baseball On a given day, the American Baseball League schedules 7 games. How many different outcomes are possible, assuming that each game is played to completion?

18. Baseball On a given day, the National Baseball League schedules 6 games. How many different outcomes are possible, assuming that each game is played to completion?

19. Choosing Seats If 4 people enter a bus having 9 vacant seats, in how many ways can they be seated?

20. Arranging Letters How many different arrangements are there of the letters in the word ROSE?

21. Choosing a Team In how many ways can a squad of 4 relay runners be chosen from a track team of 8 runners?

22. Writing a Test A professor has 10 similar problems to put on a test that has 3 problems. How many different tests can she design?

23. Baseball In how many ways can 2 teams from 14 teams in the American League be chosen without regard to which team is at home?

24. Arranging Books on a Shelf There are 5 different French books and 5 different Spanish books. How many ways are there to arrange them on a shelf if:
(a) Books of the same language must be grouped together, French on the left, Spanish on the right?
(b) French and Spanish books must alternate in the grouping, beginning with a French book?

25. Telephone Numbers Using the digits 0, 1, 2, ..., 9, how many 7-digit numbers can be formed if the first digit cannot be 0 or 9 and if the last digit is greater than or equal to 2 and less than or equal to 3? Repeated digits are allowed.

26. Home Choices A contractor constructs homes with 5 different choices of exterior finish, 3 different roof arrangements, and 4 different window designs. How many different types of homes can be built?

27. License Plate Possibilities A license plate consists of 1 letter, excluding O and I, followed by a 4-digit number that cannot have a 0 in the lead position. How many different plates are possible?

28. **Binary Codes** Using the digits 0 and 1, how many different numbers consisting of 8 digits can be formed?

29. **Forming Different Words** How many different words, real or imaginary, can be formed using all the letters in the word MISSING?

30. **Arranging Flags** How many different vertical arrangements are there of 10 flags if 4 are white, 3 are blue, 2 are green, and 1 is red?

31. **Forming Committees** A group of 9 people is going to be formed into committees of 4, 3, and 2 people. How many committees can be formed if:
    (a) A person can serve on any number of committees?
    (b) No person can serve on more than one committee?

32. **Forming Committees** A group consists of 5 men and 8 women. A committee of 4 is to be formed from this group, and policy dictates that at least 1 woman be on this committee.
    (a) How many committees can be formed that contain exactly 1 man?
    (b) How many committees can be formed that contain exactly 2 women?
    (c) How many committees can be formed that contain at least 1 man?

33. **Birthday Problem** For this problem, assume that a year has 365 days.
    (a) How many ways can 18 people have different birthdays?
    (b) What is the probability that nobody has the same birthday in a group of 18 people?
    (c) What is the probability in a group of 18 people that at least 2 people have the same birthday?

34. **Death Rates** According to the U.S. National Center for Health Statistics, 29% of all deaths in 2001 were due to heart disease.
    (a) What is the probability that a randomly selected death in 2001 was due to heart disease?
    (b) What is the probability that a randomly selected death in 2001 was not due to heart disease?

35. **Unemployment** According to the U.S. Bureau of Labor Statistics, 5.8% of the U.S. labor force was unemployed in 2002.
    (a) What is the probability that a randomly selected member of the U.S. labor force was unemployed in 2002?

    (b) What is the probability that a randomly selected member of the U.S. labor force was not unemployed in 2002?

36. From a box containing three 40-watt bulbs, six 60-watt bulbs, and eleven 75-watt bulbs, a bulb is drawn at random. What is the probability that the bulb is 40 watts? What is the probability that it is not a 75-watt bulb?

37. You have four $1 bills, three $5 bills, and two $10 bills in your wallet. If you pick a bill at random, what is the probability that it will be a $1 bill?

38. Each letter in the word ROSE is written on an index card and the cards are shuffled. When the cards are dealt, what is the probability that they spell the word ROSE?

39. Each of the numbers, 1, 2, . . . , 100 is written on an index card and the cards are shuffled. If a card is selected at random, what is the probability that the number on the card is divisible by 5? What is the probability that the card selected is either a 1 or names a prime number?

40. At the Milex tune-up and brake repair shop, the manager has found that a car will require a tune-up with a probability of 0.6, a brake job with a probability of 0.1, and both with a probability of 0.02.
    (a) What is the probability that a car requires either a tune-up or a brake job?
    (b) What is the probability that a car requires a tune-up but not a brake job?
    (c) What is the probability that a car requires neither a tune-up nor a brake job?

CHAPTER TEST

Step-by-step test solutions are found on the Chapter Test Prep Videos available via the Video Resources on DVD, in **MyMathLab** or on You Tube (search "Sullivan Precalculus: A Right Triangle Approach, Concepts Through Functions 2e).

In Problems 1–4, a survey of 70 college freshmen asked whether students planned to take biology, chemistry, or physics during their first year. Use the diagram to answer each question.

1. How many of the surveyed students plan to take physics during their first year?

2. How many of the surveyed students do not plan to take biology, chemistry, or physics during their first year?

3. How many of the surveyed students plan to take only biology and chemistry during their first year?

4. How many of the surveyed students plan to take physics or chemistry during their first year?

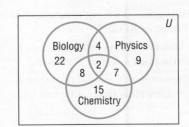

In Problems 5–7, compute the given expression.

5. $7!$ **6.** $P(10, 6)$ **7.** $C(11, 5)$

8. M&M's® offers customers the opportunity to create their own color mix of candy. There are 21 colors to choose from, and customers are allowed to select up to 6 different colors. How many different color mixes are possible, assuming that no color is selected more than once and 6 different colors are chosen?

9. How many distinct 8-letter words (real or imaginary) can be formed from the letters in the word REDEEMED?

10. In horse racing, an exacta bet requires the bettor to pick the first two horses in the exact order. If there are 8 horses in a race, in how many ways could you make an exacta bet?

11. On February 20, 2004, the Ohio Bureau of Motor Vehicles unveiled the state's new license plate format. The plate consists of three letters (A–Z) followed by 4 digits (0–9). Assume that all letters and digits may be used except that the third letter cannot be O, I, or Z. If repetitions are allowed, how many different plates are possible?

12. Kiersten applies for admission to the University of Southern California (USC) and Florida State University (FSU). She estimates that she has a 60% chance of being admitted to USC, a 70% chance of being admitted to FSU, and a 35% chance of being admitted to both universities.
(a) What is the probability that she will be admitted to either USC or FSU?

(b) What is the probability that she will not be admitted to FSU?

13. A cooler contains 8 bottles of Pepsi, 5 bottles of Coke, 4 bottles of Mountain Dew, and 3 bottles of IBC.
(a) What is the probability that a bottle chosen at random is Coke?
(b) What is the probability that a bottle chosen at random is either Pepsi or IBC?

14. A study on the age distribution of a community college gave the following table:

Age	17 and under	18–20	21–24	25–34	35–64	65 and over
Probability	0.03	???	0.23	0.29	0.25	0.01

What must be the probability that a randomly selected student at the college is between 18 and 20 years old?

15. Powerball is a multistate lottery in which 5 white balls from a drum with 53 balls and 1 red ball from a drum with 42 red balls are selected. For a $1 ticket, players get one chance at winning the jackpot by matching all 6 numbers. What is the probability of selecting the winning numbers on a $1 play?

16. If you roll a die five times, what is the probability that you obtain exactly 2 fours?

CUMULATIVE REVIEW

1. Solve: $3x^2 - 2x = -1$

2. Graph $f(x) = x^2 + 4x - 5$ by determining whether the graph opens up or down and by finding the vertex, axis of symmetry, and intercepts.

3. Graph $f(x) = 2(x + 1)^2 - 4$ using transformations.

4. Solve: $|x - 4| \le 0.01$

5. Find the complex zeros of
$$f(x) = 5x^4 - 9x^3 - 7x^2 - 31x - 6$$

6. Graph $g(x) = 3^{x-1} + 5$ using transformations. Determine the domain, the range, and horizontal asymptote of g.

7. What is the exact value of $\log_3 9$?

8. Solve: $\log_2(3x - 2) + \log_2 x = 4$

9. Solve the system: $\begin{cases} x - 2y + z = 15 \\ 3x + y - 3z = -8 \\ -2x + 4y - z = -27 \end{cases}$

10. What is the 33rd term in the sequence $-3, 1, 5, 9, \ldots$? What is the sum of the first 20 terms?

CHAPTER PROJECTS

I.   **The Monty Hall Game**   The *Monty Hall Game*, based on a segment from the game show *Let's Make a Deal*, is a classic probability problem that continues to stir debate. A more recent game show, *Deal or No Deal* (see the chapter opening vignette) has often been compared to the classic *Monty Hall Game*.

(a) Research the *Monty Hall Game* and *Deal or No Deal*.

(b) In the *Monty Hall Game*, what is the probability that a contestant wins if she does not switch? What is the probability of winning if she does switch? Perform a simulation to estimate the probabilities. Do the values agree with your research?

(c) Suppose the game *Deal or No Deal* is played with only three suitcases. Explain why this game is not the same as the *Monty Hall Game*.

(d) Suppose the game *Deal or No Deal* is played with 26 suitcases and the contestant is not allowed to switch at the end. What is the probability that the contestant will win the grand prize?

(e) Suppose the game *Deal or No Deal* is played with 26 suitcases and the contestant is allowed to switch at the end. Perform a simulation to estimate the probability that the contestant will win the grand prize if he does not switch at the end.

(f) Repeat part (e), but assume that the contestant will always switch at the end.

The following projects are available at the Instructor's Resource Center (IRC):

II. Project at Motorola ***Probability of Error Digital Wireless Communications*** Transmission errors in digital communications can often be detected by adding an extra digit of code to each transmitted signal. Investigate the probability of identifying an erroneous code using this simple coding method.

III. Surveys Polling (or taking a survey) is big business in the United States. Take and analyze a survey; then consider why different pollsters might get different results.

IV. Law of Large Numbers The probability that an event occurs, such as a head in a coin toss, is the proportion of heads you expect in the long run. A simulation is used to show that the more times a coin is flipped, the closer the probability that half of the outcomes is heads.

V. Simulation Electronic simulation of an experiment is often an economical way to investigate a theoretical probability. Develop a theory without leaving your desk.

Review

Outline

A.1 Algebra Essentials

PREPARING FOR THIS BOOK *Before getting started, read "To the Student" at the beginning of this book on page xxiv.*

OBJECTIVES   1   Work with Sets (p. A1)
              2   Graph Inequalities (p. A4)
              3   Find Distance on the Real Number Line (p. A5)
              4   Evaluate Algebraic Expressions (p. A6)
              5   Determine the Domain of a Variable (p. A7)
              6   Use the Laws of Exponents (p. A7)
              7   Evaluate Square Roots (p. A9)
              8   Use a Calculator to Evaluate Exponents (p. A10)

1 Work with Sets

A **set** is a well-defined collection of distinct objects. The objects of a set are called its **elements.** By **well-defined,** we mean that there is a rule that enables us to determine whether a given object is an element of the set. If a set has no elements, it is called the **empty set,** or **null set,** and is denoted by the symbol $\varnothing$.

For example, the set of *digits* consists of the collection of numbers 0, 1, 2, 3, 4, 5, 6, 7, 8, and 9. If we use the symbol D to denote the set of digits, then we can write

$$D = \{0, 1, 2, 3, 4, 5, 6, 7, 8, 9\}$$

In this notation, the braces $\{ \ \}$ are used to enclose the objects, or **elements,** in the set. This method of denoting a set is called the **roster method.** A second way to denote a set is to use **set-builder notation,** where the set D of digits is written as

$$D = \{ \quad x \quad | \quad x \text{ is a digit}\}$$

Read as "D is the set of all x such that x is a digit."

EXAMPLE 1

Using Set-builder Notation and the Roster Method

(a) $E = \{x \mid x \text{ is an even digit}\} = \{0, 2, 4, 6, 8\}$

(b) $O = \{x \mid x \text{ is an odd digit}\} = \{1, 3, 5, 7, 9\}$

●

Because the elements of a set are distinct, we never repeat elements. For example, we would never write $\{1, 2, 3, 2\}$; the correct listing is $\{1, 2, 3\}$. Because a set is a collection, the order in which the elements are listed does not matter. For example, $\{1, 2, 3\}, \{1, 3, 2\}, \{2, 1, 3\}$, and so on, all represent the same set.

If every element of a set A is also an element of a set B, then we say that A is a **subset** of B and write $A \subseteq B$. If two sets A and B have the same elements, then we say that A **equals** B and write $A = B$.

For example, $\{1, 2, 3\} \subseteq \{1, 2, 3, 4, 5\}$ and $\{1, 2, 3\} = \{2, 3, 1\}$.

DEFINITION

If A and B are sets, the **intersection** of A with B, denoted $A \cap B$, is the set consisting of elements that belong to both A and B. The **union** of A with B, denoted $A \cup B$, is the set consisting of elements that belong to either A or B, or both.

EXAMPLE 2

Finding the Intersection and Union of Sets

Let $A = \{1, 3, 5, 8\}$, $B = \{3, 5, 7\}$, and $C = \{2, 4, 6, 8\}$. Find:

(a) $A \cap B$ (b) $A \cup B$ (c) $B \cap (A \cup C)$

Solution

(a) $A \cap B = \{1, 3, 5, 8\} \cap \{3, 5, 7\} = \{3, 5\}$

(b) $A \cup B = \{1, 3, 5, 8\} \cup \{3, 5, 7\} = \{1, 3, 5, 7, 8\}$

(c) $B \cap (A \cup C) = \{3, 5, 7\} \cap (\{1, 3, 5, 8\} \cup \{2, 4, 6, 8\})$

$\qquad\qquad = \{3, 5, 7\} \cap \{1, 2, 3, 4, 5, 6, 8\} = \{3, 5\}$

●

━━━ **Now Work** PROBLEM 13

Usually, in working with sets, we designate a **universal set** U, the set consisting of all the elements that we wish to consider. Once a universal set has been designated, we can consider elements of the universal set not found in a given set.

DEFINITION

If A is a set, the **complement** of A, denoted $\overline{A}$, is the set consisting of all the elements in the universal set that are not in A.*

EXAMPLE 3

Finding the Complement of a Set

If the universal set is $U = \{1, 2, 3, 4, 5, 6, 7, 8, 9\}$ and if $A = \{1, 3, 5, 7, 9\}$, then $\overline{A} = \{2, 4, 6, 8\}$.

●

It follows from the definition of complement that $A \cup \overline{A} = U$ and $A \cap \overline{A} = \emptyset$. Do you see why?

Figure 1

━━━ **Now Work** PROBLEM 17

It is often helpful to draw pictures of sets. Such pictures, called **Venn diagrams,** represent sets as circles enclosed in a rectangle, which represents the universal set. Such diagrams often help us to visualize various relationships among sets. See Figure 1.

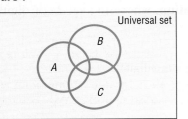

Universal set

*Some books use the notation A' or A^c for the complement of A.

If we know that $A \subseteq B$, we might use the Venn diagram in Figure 2(a). If we know that A and B have no elements in common, that is, if $A \cap B = \varnothing$, we might use the Venn diagram in Figure 2(b). The sets A and B in Figure 2(b) are said to be **disjoint.**

Figure 2

(a) $A \subseteq B$
subset

(b) $A \cap B = \varnothing$
disjoint sets

Figures 3(a), 3(b), and 3(c) use Venn diagrams to illustrate the definitions of intersection, union, and complement, respectively.

Figure 3

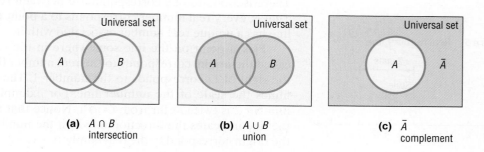

(a) $A \cap B$
intersection

(b) $A \cup B$
union

(c) $\bar{A}$
complement

Real Numbers

Real numbers are represented by symbols such as

$$25, \quad 0, \quad -3, \quad \frac{1}{2}, \quad -\frac{5}{4}, \quad 0.125, \quad \sqrt{2}, \quad \pi, \quad \sqrt[3]{-2}, \quad 0.666\ldots$$

The set of **counting numbers,** or **natural numbers,** contains the numbers in the set $\{1, 2, 3, 4, \ldots\}$. (The three dots, called an **ellipsis,** indicate that the pattern continues indefinitely.) The set of **integers** contains the numbers in the set $\{\ldots, -3, -2, -1, 0, 1, 2, 3, \ldots\}$. A **rational number** is a number that can be expressed as a *quotient* $\frac{a}{b}$ of two integers, where the integer b cannot be 0. Examples of rational numbers are $\frac{3}{4}, \frac{5}{2}, \frac{0}{4}$, and $-\frac{2}{3}$. Since $\frac{a}{1} = a$ for any integer a, every integer is also a rational number. Real numbers that are not rational are called **irrational.** Examples of irrational numbers are $\sqrt{2}$ and π (the Greek letter pi), which equals the constant ratio of the circumference to the diameter of a circle. See Figure 4.

Figure 4 $\pi = \dfrac{C}{d}$

Real numbers can be represented as **decimals.** Rational real numbers have decimal representations that either **terminate** or are nonterminating with **repeating** blocks of digits. For example, $\frac{3}{4} = 0.75$, which terminates; and $\frac{2}{3} = 0.666\ldots$, in which the digit 6 repeats indefinitely. Irrational real numbers have decimal representations that neither repeat nor terminate. For example, $\sqrt{2} = 1.414213\ldots$ and $\pi = 3.14159\ldots$. In practice, the decimal representation of an irrational number is given as an approximation. We use the symbol $\approx$ (read as "approximately equal to") to write $\sqrt{2} \approx 1.4142$ and $\pi \approx 3.1416$.

Two properties of real numbers that we shall use often are given next. Suppose that $a, b,$ and c are real numbers.

Distributive Property

$$a \cdot (b + c) = ab + ac$$

Zero-Product Property

If $ab = 0$, then either $a = 0$ or $b = 0$ or both equal 0.

The Distributive Property can be used to remove parentheses:

$$2(x + 3) = 2x + 2 \cdot 3 = 2x + 6$$

The Zero-Product Property will be used to solve equations (Section A.8). For example, if $2x = 0$, then $2 = 0$ or $x = 0$. Since $2 \neq 0$, it follows that $x = 0$.

The Real Number Line

The real numbers can be represented by points on a line called the **real number line.** There is a one-to-one correspondence between real numbers and points on a line. That is, every real number corresponds to a point on the line, and each point on the line has a unique real number associated with it.

Pick a point on the line somewhere in the center, and label it O. This point, called the **origin,** corresponds to the real number 0. See Figure 5. The point 1 unit to the right of O corresponds to the number 1. The distance between 0 and 1 determines the **scale** of the number line. For example, the point associated with the number 2 is twice as far from O as 1. Notice that an arrowhead on the right end of the line indicates the direction in which the numbers increase. Points to the left of the origin correspond to the real numbers $-1, -2,$ and so on. Figure 5 also shows the points associated with the rational numbers $-\dfrac{1}{2}$ and $\dfrac{1}{2}$ and with the irrational numbers $\sqrt{2}$ and π.

Figure 5 Real number line

DEFINITION

The real number associated with a point P is called the **coordinate** of P, and the line whose points have been assigned coordinates is called the **real number line.**

The real number line consists of three classes of real numbers, as shown in Figure 6.

Figure 6

1. The **negative real numbers** are the coordinates of points to the left of the origin O.

2. The real number **zero** is the coordinate of the origin O.

3. The **positive real numbers** are the coordinates of points to the right of the origin O.

Figure 7

(a) $a < b$

(b) $a = b$

(c) $a > b$

───── **Now Work** PROBLEM 21

2 Graph Inequalities

An important property of the real number line follows from the fact that, given two numbers (points) a and b, either a is to the left of b, or a is at the same location as b, or a is to the right of b. See Figure 7.

If a is to the left of b, we say that "a is less than b" and write $a < b$. If a is to the right of b, we say that "a is greater than b" and write $a > b$. If a is at the same location as b, then $a = b$. If a is either less than or equal to b, we write $a \leq b$. Similarly, $a \geq b$ means that a is either greater than or equal to b. Collectively, the symbols $<, >, \leq,$ and $\geq$ are called **inequality symbols.**

Note that $a < b$ and $b > a$ mean the same thing. It does not matter whether we write $2 < 3$ or $3 > 2$.

Furthermore, if $a < b$ or if $b > a$, then the difference $b - a$ is positive. Do you see why?

An **inequality** is a statement in which two expressions are related by an inequality symbol. The expressions are referred to as the **sides** of the inequality. Statements of the form $a < b$ or $b > a$ are called **strict inequalities,** whereas statements of the form $a \leq b$ or $b \geq a$ are called **nonstrict inequalities.**

Based on the discussion so far, we conclude that

$$a > 0 \quad \text{is equivalent to } a \text{ is positive}$$
$$a < 0 \quad \text{is equivalent to } a \text{ is negative}$$

We sometimes read $a > 0$ by saying that "a is positive." If $a \geq 0$, then either $a > 0$ or $a = 0$, and we may read this as "a is nonnegative."

Now Work PROBLEMS 25 AND 35

EXAMPLE 4

Graphing Inequalities

(a) On the real number line, graph all numbers x for which $x > 4$.
(b) On the real number line, graph all numbers x for which $x \leq 5$.

Solution

(a) See Figure 8. Notice that we use a left parenthesis to indicate that the number 4 is *not* part of the graph.
(b) See Figure 9. Notice that we use a right bracket to indicate that the number 5 *is* part of the graph.

Figure 8

Figure 9

Now Work PROBLEM 41

3 Find Distance on the Real Number Line

Figure 10

The *absolute value* of a number a is the distance from 0 to a on the number line. For example, -4 is 4 units from 0, and 3 is 3 units from 0. See Figure 10. Thus, the absolute value of -4 is 4, and the absolute value of 3 is 3.

A more formal definition of absolute value is given next.

DEFINITION

The **absolute value** of a real number a, denoted by the symbol $|a|$, is defined by the rules

$$|a| = a \quad \text{if } a \geq 0 \qquad \text{and} \qquad |a| = -a \quad \text{if } a < 0$$

For example, since $-4 < 0$, the second rule must be used to get $|-4| = -(-4) = 4$.

EXAMPLE 5

Computing Absolute Value

(a) $|8| = 8$ (b) $|0| = 0$ (c) $|-15| = -(-15) = 15$

Look again at Figure 10. The distance from -4 to 3 is 7 units. This distance is the difference $3 - (-4)$, obtained by subtracting the smaller coordinate from the larger. However, since $|3 - (-4)| = |7| = 7$ and $|-4 - 3| = |-7| = 7$, we can use absolute value to calculate the distance between two points without being concerned about which is smaller.

DEFINITION

If P and Q are two points on a real number line with coordinates a and b, respectively, the **distance between P and Q,** denoted by $d(P, Q)$, is

$$d(P, Q) = |b - a|$$

Since $|b - a| = |a - b|$, it follows that $d(P, Q) = d(Q, P)$.

EXAMPLE 6   **Finding Distance on a Number Line**

Let P, Q, and R be points on a real number line with coordinates -5, 7, and -3, respectively. Find the distance

(a) between P and Q

(b) between Q and R

Solution   See Figure 11.

Figure 11

(a) $d(P, Q) = |7 - (-5)| = |12| = 12$

(b) $d(Q, R) = |-3 - 7| = |-10| = 10$

Now Work PROBLEM 47

4 Evaluate Algebraic Expressions

In algebra we use letters such as x, y, a, b, and c to represent numbers. If a letter used is to represent *any* number from a given set of numbers, it is called a **variable.** A **constant** is either a fixed number, such as 5 or $\sqrt{3}$, or a letter that represents a fixed (possibly unspecified) number.

Constants and variables are combined using the operations of addition, subtraction, multiplication, and division to form *algebraic expressions*. Examples of algebraic expressions include

$$x + 3 \qquad \frac{3}{1 - t} \qquad 7x - 2y$$

To evaluate an algebraic expression, substitute a numerical value for each variable.

EXAMPLE 7   **Evaluating an Algebraic Expression**

Evaluate each expression if $x = 3$ and $y = -1$.

(a) $x + 3y$ (b) $5xy$ (c) $\dfrac{3y}{2 - 2x}$ (d) $|-4x + y|$

Solution   (a) Substitute 3 for x and -1 for y in the expression $x + 3y$.

$$x + 3y = 3 + 3(-1) = 3 + (-3) = 0$$

$$\uparrow$$
$$x = 3, y = -1$$

(b) If $x = 3$ and $y = -1$, then

$$5xy = 5(3)(-1) = -15$$

(c) If $x = 3$ and $y = -1$, then

$$\frac{3y}{2 - 2x} = \frac{3(-1)}{2 - 2(3)} = \frac{-3}{2 - 6} = \frac{-3}{-4} = \frac{3}{4}$$

(d) If $x = 3$ and $y = -1$, then

$$|-4x + y| = |-4(3) + (-1)| = |-12 + (-1)| = |-13| = 13$$

⬤

Now Work PROBLEMS 49 AND 57

5 Determine the Domain of a Variable

In working with expressions or formulas involving variables, the variables may be allowed to take on values from only a certain set of numbers. For example, in the formula for the area A of a circle of radius r, $A = \pi r^2$, the variable r is restricted to the positive real numbers since a radius cannot be negative or zero. In the expression $\frac{1}{x}$, the variable x cannot take on the value 0, since division by 0 is not defined.

DEFINITION

> The set of values that a variable may assume is called the **domain of the variable.**

EXAMPLE 8 **Finding the Domain of a Variable**

The domain of the variable x in the expression

$$\frac{5}{x - 2}$$

is $\{x | x \neq 2\}$, since, if $x = 2$, the denominator becomes 0, which is not defined.

⬤

EXAMPLE 9 **Circumference of a Circle**

In the formula for the circumference C of a circle of radius r,

$$C = 2\pi r$$

the domain of the variable r, representing the radius of the circle, is the set of positive real numbers, $\{r | r > 0\}$. The domain of the variable C, representing the circumference of the circle, is also the set of positive real numbers, $\{C | C > 0\}$.

⬤

In describing the domain of a variable, we may use either set notation or words, whichever is more convenient.

Now Work PROBLEM 67

6 Use the Laws of Exponents

Integer exponents provide a shorthand notation for representing repeated multiplications of a real number. For example,

$$2^3 = 2 \cdot 2 \cdot 2 = 8 \qquad 3^4 = 3 \cdot 3 \cdot 3 \cdot 3 = 81$$

DEFINITION

If a is a real number and n is a positive integer, then the symbol a^n represents the product of n factors of a. That is,

$$a^n = \underbrace{a \cdot a \cdot \ldots \cdot a}_{n\ factors} \qquad \text{(1)}$$

Here it is understood that $a^1 = a$.

Then $a^2 = a \cdot a$, $a^3 = a \cdot a \cdot a$, and so on. In the expression a^n, a is called the **base** and n is called the **exponent,** or **power.** We read a^n as "a raised to the power n" or as "a to the nth power." We usually read a^2 as "a squared" and a^3 as "a cubed."

In working with exponents, the operation of *raising to a power* is performed before any other operation. As examples,

WARNING Be careful with negatives and exponents.

$$-2^4 = -1 \cdot 2^4 = -16$$

whereas

$$(-2)^4 = (-2)(-2)(-2)(-2) = 16 \ \blacksquare$$

$$4 \cdot 3^2 = 4 \cdot 9 = 36 \qquad 2^2 + 3^2 = 4 + 9 = 13$$
$$-2^4 = -16 \qquad 5 \cdot 3^2 + 2 \cdot 4 = 5 \cdot 9 + 2 \cdot 4 = 45 + 8 = 53$$

Parentheses are used to indicate operations to be performed first. For example,

$$(-2)^4 = (-2)(-2)(-2)(-2) = 16 \qquad (2 + 3)^2 = 5^2 = 25$$

DEFINITION

If $a \neq 0$, we define

$$a^0 = 1 \quad \text{if } a \neq 0$$

DEFINITION

If $a \neq 0$ and if n is a positive integer, then we define

$$a^{-n} = \frac{1}{a^n} \quad \text{if } a \neq 0$$

Whenever you encounter a negative exponent, think "reciprocal."

EXAMPLE 10   **Evaluating Expressions Containing Negative Exponents**

(a) $2^{-3} = \dfrac{1}{2^3} = \dfrac{1}{8}$ (b) $-x^{-4} = \dfrac{-1}{x^4}$ (c) $\left(\dfrac{1}{5}\right)^{-2} = \dfrac{1}{\left(\dfrac{1}{5}\right)^2} = \dfrac{1}{\dfrac{1}{25}} = 25$

Now Work PROBLEMS 85 AND 105

The following properties, called the **Laws of Exponents,** can be proved using the preceding definitions. In the list, a and b are real numbers, and m and n are integers.

THEOREM

Laws of Exponents

$$a^m a^n = a^{m+n} \qquad (a^m)^n = a^{mn} \qquad (ab)^n = a^n b^n$$

$$\frac{a^m}{a^n} = a^{m-n} = \frac{1}{a^{n-m}} \quad \text{if } a \neq 0 \qquad \left(\frac{a}{b}\right)^n = \frac{a^n}{b^n} \quad \text{if } b \neq 0$$

EXAMPLE 11   **Using the Laws of Exponents**

Write each expression so that all exponents are positive.

(a) $\dfrac{x^5 y^{-2}}{x^3 y}$ $x \neq 0, \quad y \neq 0$ (b) $\left(\dfrac{x^{-3}}{3y^{-1}}\right)^{-2}$ $x \neq 0, \quad y \neq 0$

Solution   (a) $\dfrac{x^5 y^{-2}}{x^3 y} = \dfrac{x^5}{x^3} \cdot \dfrac{y^{-2}}{y} = x^{5-3} \cdot y^{-2-1} = x^2 y^{-3} = x^2 \cdot \dfrac{1}{y^3} = \dfrac{x^2}{y^3}$

(b) $\left(\dfrac{x^{-3}}{3y^{-1}}\right)^{-2} = \dfrac{(x^{-3})^{-2}}{(3y^{-1})^{-2}} = \dfrac{x^6}{3^{-2}(y^{-1})^{-2}} = \dfrac{x^6}{\dfrac{1}{9}y^2} = \dfrac{9x^6}{y^2}$

━ **Now Work** PROBLEMS 87 AND 97

7 Evaluate Square Roots

A real number is squared when it is raised to the power 2. The inverse of squaring is finding a **square root.** For example, since $6^2 = 36$ and $(-6)^2 = 36$, the numbers 6 and -6 are square roots of 36.

The symbol $\sqrt{}$, called a **radical sign,** is used to denote the **principal,** or nonnegative, square root. For example, $\sqrt{36} = 6$. We read $\sqrt{36}$ as "give me the nonnegative number whose square is 36."

DEFINITION

> If a is a nonnegative real number, the nonnegative number b, for which $b^2 = a$, is the principal square root of a and is denoted by $b = \sqrt{a}$.

The following comments are noteworthy:

In Words
The symbol $\sqrt{a}$ means "give me the nonnegative number whose square is a."

1. Negative numbers do not have square roots (in the real number system), because the square of any real number is *nonnegative*. For example, $\sqrt{-4}$ is not a real number, because there is no real number whose square is -4.
2. The principal square root of 0 is 0, since $0^2 = 0$. That is, $\sqrt{0} = 0$.
3. The principal square root of a positive number is positive.
4. If $c \geq 0$, then $(\sqrt{c})^2 = c$. For example, $(\sqrt{2})^2 = 2$ and $(\sqrt{3})^2 = 3$.

EXAMPLE 12   **Evaluating Square Roots**

(a) $\sqrt{64} = 8$ (b) $\sqrt{\dfrac{1}{16}} = \dfrac{1}{4}$ (c) $\left(\sqrt{1.4}\right)^2 = 1.4$

Examples 12(a) and (b) are examples of square roots of perfect squares, since $64 = 8^2$ and $\dfrac{1}{16} = \left(\dfrac{1}{4}\right)^2$.

Consider the expression $\sqrt{a^2}$. Since $a^2 \geq 0$, the principal square root of a^2 is defined whether $a > 0$ or $a < 0$. However, since the principal square root is nonnegative, we need an absolute value to ensure the nonnegative result. That is,

$$\sqrt{a^2} = |a| \qquad a \text{ any real number} \qquad (2)$$

EXAMPLE 13 **Simplifying Expressions Using Equation 2**

(a) $\sqrt{(2.3)^2} = |2.3| = 2.3$ (b) $\sqrt{(-2.3)^2} = |-2.3| = 2.3$

(c) $\sqrt{(x-5)^2} = |x-5|$

Now Work PROBLEM 93

Calculators

Calculators are finite machines. As a result, they are incapable of displaying decimals that contain a large number of digits. For example, some calculators are capable of displaying only eight digits. When a number requires more than eight digits, the calculator either truncates or rounds. To see how your calculator handles decimals, divide 2 by 3. How many digits do you see? Is the last digit a 6 or a 7? If it is a 6, your calculator truncates; if it is a 7, your calculator rounds.

There are different kinds of calculators. An **arithmetic** calculator can only add, subtract, multiply, and divide numbers; therefore, this type is not adequate for this course. **Scientific** calculators have all the capabilities of arithmetic calculators and also contain **function keys** labeled ln, log, sin, cos, tan, x^y, inv, and so on. **Graphing** calculators have all the capabilities of scientific calculators and contain a screen on which graphs can be displayed. We use the symbol whenever a graphing calculator is used. In this book, the graphing calculator is optional.

8 Use a Calculator to Evaluate Exponents

Your calculator has a caret key, $\boxed{\wedge}$, which is used for computations involving exponents.

EXAMPLE 14 **Exponents on a Graphing Calculator**

Evaluate: $(2.3)^5$

Solution Figure 12 shows the result using a TI-84 graphing calculator.

Now Work PROBLEM 123

Figure 12

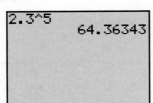

A.1 Assess Your Understanding

Concepts and Vocabulary

1. A(n) _____ is a letter used in algebra to represent any number from a given set of numbers.

2. On the real number line, the real number zero is the coordinate of the _____.

3. An inequality of the form $a > b$ is called a(n) _____ inequality.

4. In the expression 2^4, the number 2 is called the _____ and 4 is called the _____.

5. *True or False* The product of two negative real numbers is always greater than zero.

6. *True or False* The distance between two distinct points on the real number line is always greater than zero.

7. *True or False* The absolute value of a real number is always greater than zero.

8. *True or False* To multiply two expressions having the same base, retain the base and multiply the exponents.

Skill Building

In Problems 9–20, use U = universal set = {0, 1, 2, 3, 4, 5, 6, 7, 8, 9}, A = {1, 3, 4, 5, 9}, B = {2, 4, 6, 7, 8}, and C = {1, 3, 4, 6} to find each set.

9. $A \cup B$

10. $A \cup C$

11. $A \cap B$

12. $A \cap C$

13. $(A \cup B) \cap C$

14. $(A \cap B) \cup C$

15. $\overline{A}$

16. $\overline{C}$

17. $\overline{A \cap B}$

18. $\overline{B \cup C}$

19. $\overline{A} \cup \overline{B}$

20. $\overline{B} \cap \overline{C}$

21. On the real number line, label the points with coordinates 0, 1, -1, $\frac{5}{2}$, -2.5, $\frac{3}{4}$, and 0.25.

22. On the real number line, label the points with coordinates 0, -2, 2, -1.5, $\frac{3}{2}$, $\frac{1}{3}$, and $\frac{2}{3}$.

In Problems 23–32, replace the question mark by <, >, or =, whichever is correct.

23. $\frac{1}{2}$? 0

24. 5 ? 6

25. -1 ? -2

26 -3 ? $-\frac{5}{2}$

27. π ? 3.14

28. $\sqrt{2}$? 1.41

29. $\frac{1}{2}$? 0.5

30. $\frac{1}{3}$? 0.33

31. $\frac{2}{3}$? 0.67

32. $\frac{1}{4}$? 0.25

In Problems 33–38, write each statement as an inequality.

33. x is positive

34. z is negative

35. x is less than 2

36. y is greater than -5

37. x is less than or equal to 1

38. x is greater than or equal to 2

In Problems 39–42, graph the numbers x on the real number line.

39. $x \geq -2$

40. $x < 4$

41. $x > -1$

42. $x \leq 7$

In Problems 43–48, use the given real number line to compute each distance.

43. $d(C, D)$

44. $d(C, A)$

45. $d(D, E)$

46. $d(C, E)$

47. $d(A, E)$

48. $d(D, B)$

In Problems 49–56, evaluate each expression if x = −2 and y = 3.

49. $x + 2y$

50. $3x + y$

51. $5xy + 2$

52. $-2x + xy$

53. $\dfrac{2x}{x - y}$

54. $\dfrac{x + y}{x - y}$

55 $\dfrac{3x + 2y}{2 + y}$

56. $\dfrac{2x - 3}{y}$

In Problems 57–66, find the value of each expression if x = 3 and y = −2.

57. $|x + y|$

58. $|x - y|$

59. $|x| + |y|$

60. $|x| - |y|$

61. $\dfrac{|x|}{x}$

62. $\dfrac{|y|}{y}$

63. $|4x - 5y|$

64. $|3x + 2y|$

65. $||4x| - |5y||$

66. $3|x| + 2|y|$

In Problems 67–74, determine which of the value(s) (a) through (d), if any, must be excluded from the domain of the variable in each expression:

(a) $x = 3$ (b) $x = 1$ (c) $x = 0$ (d) $x = -1$

67. $\dfrac{x^2 - 1}{x}$

68. $\dfrac{x^2 + 1}{x}$

69. $\dfrac{x}{x^2 - 9}$

70. $\dfrac{x}{x^2 + 9}$

71. $\dfrac{x^2}{x^2 + 1}$

72. $\dfrac{x^3}{x^2 - 1}$

73. $\dfrac{x^2 + 5x - 10}{x^3 - x}$

74. $\dfrac{-9x^2 - x + 1}{x^3 + x}$

In Problems 75–78, determine the domain of the variable x in each expression.

75. $\dfrac{4}{x-5}$

76. $\dfrac{-6}{x+4}$

77. $\dfrac{x}{x+4}$

78. $\dfrac{x-2}{x-6}$

In Problems 79–82, use the formula $C = \dfrac{5}{9}(F - 32)$ for converting degrees Fahrenheit into degrees Celsius to find the Celsius measure of each Fahrenheit temperature.

79. $F = 32°$

80. $F = 212°$

81. $F = 77°$

82. $F = -4°$

In Problems 83–94, simplify each expression.

83. $(-4)^2$

84. -4^2

85. 4^{-2}

86. -4^{-2}

87. $3^{-6} \cdot 3^4$

88. $4^{-2} \cdot 4^3$

89. $(3^{-2})^{-1}$

90. $(2^{-1})^{-3}$

91. $\sqrt{25}$

92. $\sqrt{36}$

93. $\sqrt{(-4)^2}$

94. $\sqrt{(-3)^2}$

In Problems 95–104, simplify each expression. Express the answer so that all exponents are positive. Whenever an exponent is 0 or negative, we assume that the base is not 0.

95. $(8x^3)^2$

96. $(-4x^2)^{-1}$

97. $(x^2 y^{-1})^2$

98. $(x^{-1} y)^3$

99. $\dfrac{x^2 y^3}{x y^4}$

100. $\dfrac{x^{-2} y}{x y^2}$

101. $\dfrac{(-2)^3 x^4 (yz)^2}{3^2 x y^3 z}$

102. $\dfrac{4 x^{-2} (yz)^{-1}}{2^3 x^4 y}$

103. $\left(\dfrac{3x^{-1}}{4y^{-1}}\right)^{-2}$

104. $\left(\dfrac{5x^{-2}}{6y^{-2}}\right)^{-3}$

In Problems 105–116, find the value of each expression if $x = 2$ and $y = -1$.

105. $2xy^{-1}$

106. $-3x^{-1}y$

107. $x^2 + y^2$

108. $x^2 y^2$

109. $(xy)^2$

110. $(x + y)^2$

111. $\sqrt{x^2}$

112. $\left(\sqrt{x}\right)^2$

113. $\sqrt{x^2 + y^2}$

114. $\sqrt{x^2} + \sqrt{y^2}$

115. x^y

116. y^x

117. Find the value of the expression $2x^3 - 3x^2 + 5x - 4$ if $x = 2$. What is the value if $x = 1$?

118. Find the value of the expression $4x^3 + 3x^2 - x + 2$ if $x = 1$. What is the value if $x = 2$?

119. What is the value of $\dfrac{(666)^4}{(222)^4}$?

120. What is the value of $(0.1)^3 (20)^3$?

In Problems 121–128, use a calculator to evaluate each expression. Round your answer to three decimal places.

121. $(8.2)^6$

122. $(3.7)^5$

123. $(6.1)^{-3}$

124. $(2.2)^{-5}$

125. $(-2.8)^6$

126. -2.8^6

127. $(-8.11)^{-4}$

128. -8.11^{-4}

Applications and Extensions

In Problems 129–138, express each statement as an equation involving the indicated variables.

129. Area of a Rectangle   The area A of a rectangle is the product of its length l and its width w.

131. Circumference of a Circle   The circumference C of a circle is the product of π and its diameter d.

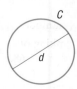

130. Perimeter of a Rectangle   The perimeter P of a rectangle is twice the sum of its length l and its width w.

132. Area of a Triangle The area A of a triangle is one-half the product of its base b and its height h.

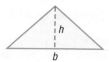

133. Area of an Equilateral Triangle The area A of an equilateral triangle is $\dfrac{\sqrt{3}}{4}$ times the square of the length x of one side.

134. Perimeter of an Equilateral Triangle The perimeter P of an equilateral triangle is 3 times the length x of one side.

135. Volume of a Sphere The volume V of a sphere is $\dfrac{4}{3}$ times π times the cube of the radius r.

136. Surface Area of a Sphere The surface area S of a sphere is 4 times π times the square of the radius r.

137. Volume of a Cube The volume V of a cube is the cube of the length x of a side.

138. Surface Area of a Cube The surface area S of a cube is 6 times the square of the length x of a side.

139. Manufacturing Cost The weekly production cost C of manufacturing x watches is given by the formula $C = 4000 + 2x$, where the variable C is in dollars.
(a) What is the cost of producing 1000 watches?
(b) What is the cost of producing 2000 watches?

140. Balancing a Checkbook At the beginning of the month, Mike had a balance of $210 in his checking account. During the next month, he deposited $80, wrote a check for $120, made another deposit of $25, and wrote two checks: one for $60 and the other for $32. He was also assessed a monthly service charge of $5. What was his balance at the end of the month?

In Problems 141 and 142, write an inequality using an absolute value to describe each statement.

141. x is at least 6 units from 4.

142. x is more than 5 units from 2.

143. U.S. Voltage In the United States, normal household voltage is 110 volts. It is acceptable for the actual voltage x to differ from normal by at most 5 volts. A formula that describes this is

$$|x - 110| \le 5$$

(a) Show that a voltage of 108 volts is acceptable.
(b) Show that a voltage of 104 volts is not acceptable.

144. Foreign Voltage In some countries, normal household voltage is 220 volts. It is acceptable for the actual voltage x to differ from normal by at most 8 volts. A formula that describes this is

$$|x - 220| \le 8$$

(a) Show that a voltage of 214 volts is acceptable.
(b) Show that a voltage of 209 volts is not acceptable.

145. Making Precision Ball Bearings The FireBall Company manufactures ball bearings for precision equipment. One of its products is a ball bearing with a stated radius of 3 centimeters (cm). Only ball bearings with a radius within 0.01 cm of this stated radius are acceptable. If x is the radius of a ball bearing, a formula describing this situation is

$$|x - 3| \le 0.01$$

(a) Is a ball bearing of radius $x = 2.999$ acceptable?
(b) Is a ball bearing of radius $x = 2.89$ acceptable?

146. Body Temperature Normal human body temperature is 98.6°F. A temperature x that differs from normal by at least 1.5°F is considered unhealthy. A formula that describes this is

$$|x - 98.6| \ge 1.5$$

(a) Show that a temperature of 97°F is unhealthy.
(b) Show that a temperature of 100°F is not unhealthy.

147. Does $\dfrac{1}{3}$ equal 0.333? If not, which is larger? By how much?

148. Does $\dfrac{2}{3}$ equal 0.666? If not, which is larger? By how much?

Explaining Concepts: Discussion and Writing

149. Is there a positive real number "closest" to 0?

150. Number game I'm thinking of a number! It lies between 1 and 10; its square is rational and lies between 1 and 10. The number is larger than π. Correct to two decimal places (that is, truncated to two decimal places) name the number. Now think of your own number, describe it, and challenge a fellow student to name it.

151. Write a brief paragraph that illustrates the similarities and differences between "less than" ($<$) and "less than or equal to" ($\le$).

152. Give a reason why the statement $5 < 8$ is true.

A.2 Geometry Essentials

OBJECTIVES   **1** Use the Pythagorean Theorem and Its Converse (p. A14)
2 Know Geometry Formulas (p. A15)
3 Understand Congruent Triangles and Similar Triangles (p. A16)

In this section we review some topics studied in geometry that we shall need throughout the course.

1 Use the Pythagorean Theorem and Its Converse

Figure 13

The *Pythagorean Theorem* is a statement about *right triangles*. A **right triangle** is one that contains a **right angle,** that is, an angle of 90°. The side of the triangle opposite the 90° angle is called the **hypotenuse;** the remaining two sides are called **legs.** In Figure 13 we have used c to represent the length of the hypotenuse and a and b to represent the lengths of the legs. Notice the use of the symbol $\ulcorner$ to show the 90° angle. We now state the Pythagorean Theorem.

PYTHAGOREAN THEOREM

In a right triangle, the square of the length of the hypotenuse is equal to the sum of the squares of the lengths of the legs. That is, in the right triangle shown in Figure 13,

$$c^2 = a^2 + b^2 \qquad\qquad (1)$$

A proof of the Pythagorean Theorem is given at the end of this section.

EXAMPLE 1   **Finding the Hypotenuse of a Right Triangle**

In a right triangle, one leg has length 4 and the other has length 3. What is the length of the hypotenuse?

Solution   Since the triangle is a right triangle, we use the Pythagorean Theorem with $a = 4$ and $b = 3$ to find the length c of the hypotenuse. From equation (1), we have

$$c^2 = a^2 + b^2$$
$$c^2 = 4^2 + 3^2 = 16 + 9 = 25$$
$$c = \sqrt{25} = 5$$

Now Work PROBLEM 13

The converse of the Pythagorean Theorem is also true.

CONVERSE OF THE PYTHAGOREAN THEOREM

In a triangle, if the square of the length of one side equals the sum of the squares of the lengths of the other two sides, the triangle is a right triangle. The 90° angle is opposite the longest side.

A proof is given at the end of this section.

EXAMPLE 2   **Verifying That a Triangle Is a Right Triangle**

Show that a triangle whose sides are of lengths 5, 12, and 13 is a right triangle. Identify the hypotenuse.

Solution   We square the lengths of the sides.

$$5^2 = 25, \qquad 12^2 = 144, \qquad 13^2 = 169$$

Figure 14

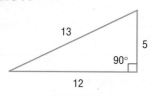

Notice that the sum of the first two squares (25 and 144) equals the third square (169). Hence, the triangle is a right triangle. The longest side, 13, is the hypotenuse. See Figure 14.

━━━━ **Now Work** PROBLEM 21

| EXAMPLE 3 | **Applying the Pythagorean Theorem** |

Figure 15

Excluding antenna, the tallest inhabited building in the world is Taipei 101 in Taipei, Taiwan. See Figure 15. If the indoor observation deck is 1437 feet above ground level, how far can a person standing on the observation deck see (with the aid of a telescope)? Use 3960 miles for the radius of Earth.

Source: Council on Tall Buildings and Urban Habitat, 2009.

Solution From the center of Earth, draw two radii: one through Taipei 101 and the other to the farthest point a person can see from the observation deck. See Figure 16. Apply the Pythagorean Theorem to the right triangle.

Since 1 mile = 5280 feet, then 1437 feet = $\dfrac{1437}{5280}$ mile. So we have

$$d^2 + 3960^2 = \left(3960 + \frac{1437}{5280}\right)^2$$

$$d^2 = \left(3960 + \frac{1437}{5280}\right)^2 - 3960^2 \approx 2155.57$$

$$d \approx 46.43$$

A person can see about 46 miles from the observation tower.

Figure 16

1437 ft   *d*

3960 mi

━━━━ **Now Work** PROBLEM 53

2 Know Geometry Formulas

Certain formulas from geometry are useful in solving problems. We list some of these next.

For a rectangle of length *l* and width *w*,

w

l

$$\text{Area} = lw \qquad \text{Perimeter} = 2l + 2w$$

For a triangle with base *b* and altitude *h*,

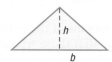

h

b

$$\text{Area} = \frac{1}{2}bh$$

For a circle of radius r (diameter $d = 2r$),

Area $= \pi r^2$ Circumference $= 2\pi r = \pi d$

For a closed rectangular box of length l, width w, and height h,

Volume $= lwh$ Surface area $= 2lh + 2wh + 2lw$

For a sphere of radius r,

Volume $= \dfrac{4}{3}\pi r^3$ Surface area $= 4\pi r^2$

For a right circular cylinder of height h and radius r,

Volume $= \pi r^2 h$ Surface area $= 2\pi r^2 + 2\pi rh$

Now Work PROBLEM 29

EXAMPLE 4 **Using Geometry Formulas**

A Christmas tree ornament is in the shape of a semicircle on top of a triangle. How many square centimeters (cm) of copper is required to make the ornament if the height of the triangle is 6 cm and the base is 4 cm?

Solution See Figure 17. The amount of copper required equals the shaded area. This area is the sum of the area of the triangle and the semicircle. The triangle has height $h = 6$ and base $b = 4$. The semicircle has diameter $d = 4$, so its radius is $r = 2$.

Figure 17

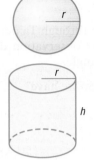

$$\text{Area} = \text{Area of triangle} + \text{Area of semicircle}$$

$$= \frac{1}{2}bh + \frac{1}{2}\pi r^2 = \frac{1}{2}(4)(6) + \frac{1}{2}\pi \cdot 2^2 \qquad {\scriptstyle b = 4;\ h = 6;\ r = 2}$$

$$= 12 + 2\pi \approx 18.28 \text{ cm}^2$$

About 18.28 cm² of copper is required.

Now Work PROBLEM 47

3 Understand Congruent Triangles and Similar Triangles

Throughout the text we will make reference to triangles. We begin with a discussion of *congruent* triangles. According to dictionary.com, the word **congruent** means coinciding exactly when superimposed. For example, two angles are congruent if they have the same measure and two line segments are congruent if they have the same length.

In Words
Two triangles are congruent if they have the same size and shape.

DEFINITION Two triangles are **congruent** if each of the corresponding angles is the same measure and each of the corresponding sides is the same length.

In Figure 18, corresponding angles are equal and the lengths of the corresponding sides are equal: $a = d$, $b = e$, and $c = f$. We conclude that these triangles are congruent.

Figure 18 Congruent Triangles

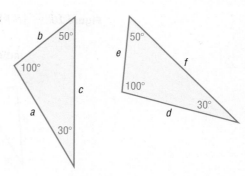

It is not necessary to verify that all three angles and all three sides are the same measure to determine whether two triangles are congruent.

Determining Congruent Triangles

1. **Angle–Side–Angle Case** Two triangles are congruent if two of the angles are equal and the lengths of the corresponding sides between the two angles are equal.

   For example, in Figure 19(a), the two triangles are congruent because two angles and the included side are equal.

2. **Side–Side–Side Case** Two triangles are congruent if the lengths of the corresponding sides of the triangles are equal.

   For example, in Figure 19(b), the two triangles are congruent because the three corresponding sides are all equal.

3. **Side–Angle–Side Case** Two triangles are congruent if the lengths of two corresponding sides are equal and the angles between the two sides are the same.

   For example, in Figure 19(c), the two triangles are congruent because two sides and the included angle are equal.

Figure 19

We contrast congruent triangles with *similar* triangles.

DEFINITION

Two triangles are **similar** if the corresponding angles are equal and the lengths of the corresponding sides are proportional.

For example, the triangles in Figure 20 on page A18 are similar because the corresponding angles are equal. In addition, the lengths of the corresponding sides are proportional because each side in the triangle on the right is twice as long as each corresponding side in the triangle on the left. That is, the ratio of the corresponding sides is a constant: $\dfrac{d}{a} = \dfrac{e}{b} = \dfrac{f}{c} = 2$.

Figure 20

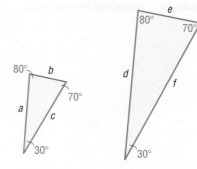

It is not necessary to verify that all three angles are equal and all three sides are proportional to determine whether two triangles are similar.

Determining Similar Triangles

1. **Angle–Angle Case**   Two triangles are similar if two of the corresponding angles are equal.

    For example, in Figure 21(a), the two triangles are similar because two angles are equal.

2. **Side–Side–Side Case**   Two triangles are similar if the lengths of all three sides of each triangle are proportional.

    For example, in Figure 21(b), the two triangles are similar because

$$\frac{10}{30} = \frac{5}{15} = \frac{6}{18} = \frac{1}{3}$$

3. **Side–Angle–Side Case**   Two triangles are similar if two corresponding sides are proportional and the angles between the two sides are equal.

    For example, in Figure 21(c), the two triangles are similar because $\frac{4}{6} = \frac{12}{18} = \frac{2}{3}$ and the angles between the sides are equal.

Figure 21

(a)   (b)   (c)

EXAMPLE 5

Using Similar Triangles

Given that the triangles in Figure 22 are similar, find the missing length x and the angles A, B, and C.

Figure 22

Solution   Because the triangles are similar, corresponding angles are equal. So $A = 90°$, $B = 60°$, and $C = 30°$. Also, the corresponding sides are proportional. That is, $\dfrac{3}{5} = \dfrac{6}{x}$. We solve this equation for x.

$$\frac{3}{5} = \frac{6}{x}$$

$$5x \cdot \frac{3}{5} = 5x \cdot \frac{6}{x} \qquad \text{Multiply both sides by 5x.}$$

$$3x = 30 \qquad \text{Simplify.}$$

$$x = 10 \qquad \text{Divide both sides by 3.}$$

The missing length is 10 units.

── **Now Work**   PROBLEM 41

Proof of the Pythagorean Theorem   We begin with a square, each side of length $a + b$. In this square, we can form four right triangles, each having legs equal in length to a and b. See Figure 23. All these triangles are congruent (two sides and their included angle are equal). As a result, the hypotenuse of each is the same, say c, and the pink shading in Figure 23 indicates a square with an area equal to c^2.

Figure 23

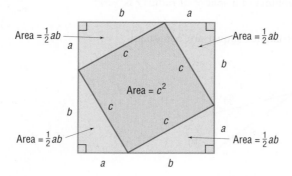

The area of the original square with sides $a + b$ equals the sum of the areas of the four triangles (each of area $\dfrac{1}{2}ab$) plus the area of the square with side c. That is,

$$(a + b)^2 = \frac{1}{2}ab + \frac{1}{2}ab + \frac{1}{2}ab + \frac{1}{2}ab + c^2$$

$$a^2 + 2ab + b^2 = 2ab + c^2$$

$$a^2 + b^2 = c^2$$

The proof is complete.   ∎

Figure 24

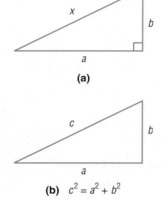

Proof of the Converse of the Pythagorean Theorem   We begin with two triangles: one a right triangle with legs a and b and the other a triangle with sides a, b, and c for which $c^2 = a^2 + b^2$. See Figure 24. By the Pythagorean Theorem, the length x of the third side of the first triangle is

$$x^2 = a^2 + b^2$$

But $c^2 = a^2 + b^2$. Hence,

$$x^2 = c^2$$

$$x = c$$

The two triangles have sides with the same length and are therefore congruent. This means corresponding angles are equal, so the angle opposite side c of the second triangle equals 90°.

The proof is complete.   ∎

A.2 Assess Your Understanding

Concepts and Vocabulary

1. A(n) _____ triangle is one that contains an angle of 90 degrees. The longest side is called the _____.

2. For a triangle with base b and altitude h, a formula for the area A is _____.

3. The formula for the circumference C of a circle of radius r is _____.

4. Two triangles are _____ if corresponding angles are equal and the lengths of the corresponding sides are proportional.

5. *True or False*   In a right triangle, the square of the length of the longest side equals the sum of the squares of the lengths of the other two sides.

6. *True or False*   The triangle with sides of length 6, 8, and 10 is a right triangle.

7. *True or False*   The volume of a sphere of radius r is $\frac{4}{3}\pi r^2$.

8. *True or False*   The triangles shown are congruent.

9. *True or False*   The triangles shown are similar.

10. *True or False*   The triangles shown are similar.

Skill Building

In Problems 11–16, the lengths of the legs of a right triangle are given. Find the length of the hypotenuse.

11. $a = 5, \quad b = 12$

12. $a = 6, \quad b = 8$

13. $a = 10, \quad b = 24$

14. $a = 4, \quad b = 3$

15. $a = 7, \quad b = 24$

16. $a = 14, \quad b = 48$

In Problems 17–24, the lengths of the sides of a triangle are given. Determine which are right triangles. For those that are, identify the hypotenuse.

17. $3, 4, 5$

18. $6, 8, 10$

19. $4, 5, 6$

20. $2, 2, 3$

21. $7, 24, 25$

22. $10, 24, 26$

23. $6, 4, 3$

24. $5, 4, 7$

25. Find the area A of a rectangle with length 4 inches and width 2 inches.

26. Find the area A of a rectangle with length 9 centimeters and width 4 centimeters.

27. Find the area A of a triangle with height 4 inches and base 2 inches.

28. Find the area A of a triangle with height 9 centimeters and base 4 centimeters.

29. Find the area A and circumference C of a circle of radius 5 meters.

30. Find the area A and circumference C of a circle of radius 2 feet.

31. Find the volume V and surface area S of a rectangular box with length 8 feet, width 4 feet, and height 7 feet.

32. Find the volume V and surface area S of a rectangular box with length 9 inches, width 4 inches, and height 8 inches.

33. Find the volume V and surface area S of a sphere of radius 4 centimeters.

34. Find the volume V and surface area S of a sphere of radius 3 feet.

35. Find the volume V and surface area S of a right circular cylinder with radius 9 inches and height 8 inches.

36. Find the volume V and surface area S of a right circular cylinder with radius 8 inches and height 9 inches.

In Problems 37–40, find the area of the shaded region.

37.

2
2

38.

2
2

39.

2
2

40.

2
2

In Problems 41–44, each pair of triangles is similar. Find the missing length x and the missing angles A, B, and C.

41.

60°
2
90°
4
30°

B x
A
8
C

42.

16
30°
75°
12
75°

x
B A
6 C

43.

20 60°
95°
45
25°

A
30
B
x
C

44.

10 125°
50°
50
5°

B
8
A
x
C

Applications and Extensions

45. How many feet does a wheel with a diameter of 16 inches travel after four revolutions?

46. How many revolutions will a circular disk with a diameter of 4 feet have completed after it has rolled 20 feet?

47. In the figure shown, $ABCD$ is a square, with each side of length 6 feet. The width of the border (shaded portion) between the outer square $EFGH$ and $ABCD$ is 2 feet. Find the area of the border.

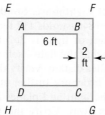
E   F
A   B
6 ft
2 ft
D   C
H   G

48. Refer to the figure. Square $ABCD$ has an area of 100 square feet; square $BEFG$ has an area of 16 square feet. What is the area of the triangle CGF?

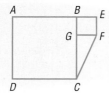
A   B E
G   F
D   C

49. Architecture A **Norman window** consists of a rectangle surmounted by a semicircle. Find the area of the Norman window shown in the illustration. How much wood frame is needed to enclose the window?

6'
4'

50. Construction A circular swimming pool, 20 feet in diameter, is enclosed by a wooden deck that is 3 feet wide. What is the area of the deck? How much fence is required to enclose the deck?

3"
20'

51. How Tall Is the Great Pyramid? The ancient Greek philosopher Thales of Miletus is reported on one occasion to have visited Egypt and calculated the height of the Great Pyramid of Cheops by means of shadow reckoning. Thales knew that each side of the base of the pyramid was 252 paces and that his own height was 2 paces. He measured the length of the pyramid's shadow to be 114 paces and determined the length of his shadow to be 3 paces. See the illustration. Using similar triangles, determine the height of the Great Pyramid in terms of the number of paces.

Source: www.anselm.edu/homepage/dbanach/thales.htm.
This site references another source: Selections, from Julia E. Diggins, *String, Straightedge, and Shadow,* Viking Press, New York, 1965, Illustrations by Corydon Bell.

52. The Bermuda Triangle Karen is doing research on the Bermuda Triangle which she defines roughly by Hamilton, Bermuda; San Juan, Puerto Rico; and Fort Lauderdale, Florida. On her atlas Karen measures the straight-line distances from Hamilton to Fort Lauderdale, Fort Lauderdale to San Juan, and San Juan to Hamilton to be approximately 57 millimeters (mm), 58 mm, and 53.5 mm, respectively. If the actual distance from Fort Lauderdale to San Juan is 1046 miles, approximate the actual distances from San Juan to Hamilton and from Hamilton to Fort Lauderdale.

Source: www.worldatlas.com

Source: www.en.wikipedia.org/wiki/Bermuda_Triangle.

In Problems 53–55, use the facts that the radius of Earth is 3960 miles and 1 mile = 5280 feet.

53. How Far Can You See? The conning tower of the U.S.S. *Silversides,* a World War II submarine now permanently stationed in Muskegon, Michigan, is approximately 20 feet above sea level. How far can you see from the conning tower?

54. How Far Can You See? A person who is 6 feet tall is standing on the beach in Fort Lauderdale, Florida, and looks out onto the Atlantic Ocean. Suddenly, a ship appears on the horizon. How far is the ship from shore?

55. How Far Can You See? The deck of a destroyer is 100 feet above sea level. How far can a person see from the deck? How far can a person see from the bridge, which is 150 feet above sea level?

56. Suppose that m and n are positive integers with $m > n$. If $a = m^2 - n^2$, $b = 2mn$, and $c = m^2 + n^2$, show that a, b, and c are the lengths of the sides of a right triangle. (This formula can be used to find the sides of a right triangle that are integers, such as 3, 4, 5; 5, 12, 13; and so on. Such triplets of integers are called Pythagorean triples.)

Explaining Concepts: Discussion and Writing

57. You have 1000 feet of flexible pool siding and wish to construct a swimming pool. Experiment with rectangular-shaped pools with perimeters of 1000 feet. How do their areas vary? What is the shape of the rectangle with the largest area? Now compute the area enclosed by a circular pool with a perimeter (circumference) of 1000 feet. What would be your choice of shape for the pool? If rectangular, what is your preference for dimensions? Justify your choice. If your only consideration is to have a pool that encloses the most area, what shape should you use?

58. The Gibb's Hill Lighthouse, Southampton, Bermuda, in operation since 1846, stands 117 feet high on a hill 245 feet high, so its beam of light is 362 feet above sea level. A brochure states that the light itself can be seen on the horizon about 26 miles distant. Verify the correctness of this information. The brochure further states that ships 40 miles away can see the light and planes flying at 10,000 feet can see it 120 miles away. Verify the accuracy of these statements. What assumption did the brochure make about the height of the ship?

A.3 Polynomials

OBJECTIVES
1 Recognize Monomials (p. A23)
2 Recognize Polynomials (p. A24)
3 Add and Subtract Polynomials (p. A25)
4 Multiply Polynomials (p. A26)
5 Know Formulas for Special Products (p. A27)
6 Divide Polynomials Using Long Division (p. A28)
7 Work with Polynomials in Two Variables (p. A31)

We have described algebra as a generalization of arithmetic in which letters are used to represent real numbers. From now on, we shall use the letters at the end of the alphabet, such as x, y, and z, to represent variables and the letters at the beginning of the alphabet, such as a, b, and c, to represent constants. In the expressions $3x + 5$ and $ax + b$, it is understood that x is a variable and that a and b are constants, even though the constants a and b are unspecified. As you will find out, the context usually makes the intended meaning clear.

1 Recognize Monomials

DEFINITION

NOTE The nonnegative integers are the integers 0, 1, 2, 3, ... ∎

A **monomial** in one variable is the product of a constant and a variable raised to a nonnegative integer power. A monomial is of the form

$$ax^k$$

where a is a constant, x is a variable, and $k \geq 0$ is an integer. The constant a is called the **coefficient** of the monomial. If $a \neq 0$, then k is called the **degree** of the monomial.

EXAMPLE 1 **Examples of Monomials**

Monomial	Coefficient	Degree	
(a) $6x^2$	6	2	
(b) $-\sqrt{2}x^3$	$-\sqrt{2}$	3	
(c) 3	3	0	Since $3 = 3 \cdot 1 = 3x^0, x \neq 0$
(d) $-5x$	-5	1	Since $-5x = -5x^1$
(e) x^4	1	4	Since $x^4 = 1 \cdot x^4$

Now let's look at some expressions that are not monomials.

EXAMPLE 2 **Examples of Nonmonomial Expressions**

(a) $3x^{1/2}$ is not a monomial, since the exponent of the variable x is $\frac{1}{2}$ and $\frac{1}{2}$ is not a nonnegative integer.

(b) $4x^{-3}$ is not a monomial, since the exponent of the variable x is -3 and -3 is not a nonnegative integer.

➡ Now Work PROBLEM 7

2 Recognize Polynomials

Two monomials with the same variable raised to the same power are called **like terms.** For example, $2x^4$ and $-5x^4$ are like terms. In contrast, the monomials $2x^3$ and $2x^5$ are not like terms.

We can add or subtract like terms using the Distributive Property. For example,

$$2x^2 + 5x^2 = (2 + 5)x^2 = 7x^2 \quad \text{and} \quad 8x^3 - 5x^3 = (8 - 5)x^3 = 3x^3$$

The sum or difference of two monomials having different degrees is called a **binomial.** The sum or difference of three monomials with three different degrees is called a **trinomial.** For example,

$$x^2 - 2 \text{ is a binomial}$$
$$x^3 - 3x + 5 \text{ is a trinomial}$$
$$2x^2 + 5x^2 + 2 = 7x^2 + 2 \text{ is a binomial}$$

DEFINITION

A **polynomial** in one variable is an algebraic expression of the form

$$a_n x^n + a_{n-1} x^{n-1} + \cdots + a_1 x + a_0 \tag{1}$$

where $a_n, a_{n-1}, \ldots, a_1, a_0$ are constants,* called the **coefficients** of the polynomial, $n \geq 0$ is an integer, and x is a variable. If $a_n \neq 0$, it is called the **leading coefficient,** and n is called the **degree** of the polynomial.

In Words
A polynomial is a sum of monomials.

The monomials that make up a polynomial are called its **terms.** If all the coefficients are 0, the polynomial is called the **zero polynomial,** which has no degree.

Polynomials are usually written in **standard form,** beginning with the nonzero term of highest degree and continuing with terms in descending order according to degree. If a power of x is missing, it is because its coefficient is zero.

EXAMPLE 3   **Examples of Polynomials**

Polynomial	Coefficients	Degree
$-8x^3 + 4x^2 - 6x + 2$	$-8, 4, -6, 2$	3
$3x^2 - 5 = 3x^2 + 0 \cdot x + (-5)$	$3, 0, -5$	2
$8 - 2x + x^2 = 1 \cdot x^2 + (-2)x + 8$	$1, -2, 8$	2
$5x + \sqrt{2} = 5x^1 + \sqrt{2}$	$5, \sqrt{2}$	1
$3 = 3 \cdot 1 = 3 \cdot x^0$	3	0
0	0	No degree

Although we have been using x to represent the variable, letters such as y or z are also commonly used.

$$3x^4 - x^2 + 2 \text{ is a polynomial (in } x) \text{ of degree 4}$$
$$9y^3 - 2y^2 + y - 3 \text{ is a polynomial (in } y) \text{ of degree 3}$$
$$z^5 + \pi \text{ is a polynomial (in } z) \text{ of degree 5}$$

Algebraic expressions such as

$$\frac{1}{x} \quad \text{and} \quad \frac{x^2 + 1}{x + 5}$$

*The notation a_n is read as "a sub n." The number n is called a **subscript** and should not be confused with an exponent. We use subscripts to distinguish one constant from another when a large or undetermined number of constants is required.

are not polynomials. The first is not a polynomial because $\dfrac{1}{x} = x^{-1}$ has an exponent that is not a nonnegative integer. Although the second expression is the quotient of two polynomials, the polynomial in the denominator has degree greater than 0, so the expression cannot be a polynomial.

Now Work PROBLEM 17

3 Add and Subtract Polynomials

Polynomials are added and subtracted by combining like terms.

EXAMPLE 4   **Adding Polynomials**

Find the sum of the polynomials:

$$8x^3 - 2x^2 + 6x - 2 \quad \text{and} \quad 3x^4 - 2x^3 + x^2 + x$$

Solution   We shall find the sum in two ways.

Horizontal Addition: The idea here is to group the like terms and then combine them.

$$(8x^3 - 2x^2 + 6x - 2) + (3x^4 - 2x^3 + x^2 + x)$$
$$= 3x^4 + (8x^3 - 2x^3) + (-2x^2 + x^2) + (6x + x) - 2$$
$$= 3x^4 + 6x^3 - x^2 + 7x - 2$$

Vertical Addition: The idea here is to vertically line up the like terms in each polynomial and then add the coefficients.

$$
\begin{array}{cccccc}
 & x^4 & x^3 & x^2 & x^1 & x^0 \\
 & & 8x^3 & - 2x^2 & + 6x & - 2 \\
+ & 3x^4 & - 2x^3 & + x^2 & + x & \\
\hline
 & 3x^4 & + 6x^3 & - x^2 & + 7x & - 2
\end{array}
$$

We can subtract two polynomials horizontally or vertically as well.

EXAMPLE 5   **Subtracting Polynomials**

Find the difference: $(3x^4 - 4x^3 + 6x^2 - 1) - (2x^4 - 8x^2 - 6x + 5)$

Solution   *Horizontal Subtraction:*

$$(3x^4 - 4x^3 + 6x^2 - 1) - (2x^4 - 8x^2 - 6x + 5)$$
$$= 3x^4 - 4x^3 + 6x^2 - 1 + \underbrace{(-2x^4 + 8x^2 + 6x - 5)}$$

Be sure to change the sign of each
term in the second polynomial.

$$= \underset{\uparrow}{(3x^4 - 2x^4)} + (-4x^3) + (6x^2 + 8x^2) + 6x + (-1 - 5)$$

Group like terms.

$$= x^4 - 4x^3 + 14x^2 + 6x - 6$$

COMMENT Vertical subtraction will be used when we divide polynomials. ∎

Vertical Subtraction: We line up like terms, change the sign of each coefficient of the second polynomial, and add vertically.

$$
\begin{array}{r}
x^4 \quad\ x^3 \quad\ x^2 \quad\ x^1 \quad\ x^0 \\
3x^4 - 4x^3 + 6x^2 \qquad\quad - 1 \ = \\
-\ [2x^4 \qquad\quad - 8x^2 - 6x + 5] = +
\end{array}
\qquad
\begin{array}{r}
x^4 \quad\ x^3 \quad\ x^2 \quad\ x^1 \quad\ x^0 \\
3x^4 - 4x^3 + 6x^2 \qquad\quad - 1 \\
-2x^4 \qquad\quad + 8x^2 + 6x - 5 \\
\hline
x^4 - 4x^3 + 14x^2 + 6x - 6
\end{array}
$$

The choice of which of these methods to use for adding and subtracting polynomials is left to you. To save space, we shall most often use the horizontal format.

Now Work PROBLEM 29

4 Multiply Polynomials

Two monomials may be multiplied using the Laws of Exponents and the Commutative and Associative Properties. For example,

$$(2x^3) \cdot (5x^4) = (2 \cdot 5) \cdot (x^3 \cdot x^4) = 10x^{3+4} = 10x^7$$

Products of polynomials are found by repeated use of the Distributive Property and the Laws of Exponents. Again, you have a choice of horizontal or vertical format.

EXAMPLE 6 **Multiplying Polynomials**

Find the product: $(2x + 5)(x^2 - x + 2)$

Solution **Horizontal Multiplication:**

$$(2x + 5)(x^2 - x + 2) = 2x(x^2 - x + 2) + 5(x^2 - x + 2)$$
↑
Distributive Property
$$= (2x \cdot x^2 - 2x \cdot x + 2x \cdot 2) + (5 \cdot x^2 - 5 \cdot x + 5 \cdot 2)$$
↑
Distributive Property
$$= (2x^3 - 2x^2 + 4x) + (5x^2 - 5x + 10)$$
↑
Law of Exponents
$$= 2x^3 + 3x^2 - x + 10$$
↑
Combine like terms.

Vertical Multiplication: The idea here is very much like multiplying a two-digit number by a three-digit number.

$$
\begin{array}{r}
x^2 - \quad x + 2 \\
2x + \quad 5 \\
\hline
5x^2 - 5x + 10 \\
(+)\ \underline{2x^3 - 2x^2 + 4x \qquad} \\
2x^3 + 3x^2 - \quad x + 10
\end{array}
$$

This line is $5(x^2 - x + 2)$.
This line is $2x(x^2 - x + 2)$.
Sum of the above two lines

Now Work PROBLEM 45

5 Know Formulas for Special Products

Certain products, which we call **special products,** occur frequently in algebra. We can calculate them using the **FOIL** (*First, Outer, Inner, Last*) method of multiplying two binomials.

$$(ax + b)(cx + d) = ax(cx + d) + b(cx + d)$$

$$= ax \cdot cx + ax \cdot d + b \cdot cx + b \cdot d$$
$$= acx^2 + adx + bcx + bd$$
$$= acx^2 + (ad + bc)x + bd$$

EXAMPLE 7 **Using FOIL**

(a) $(x - 3)(x + 3) = x^2 + 3x - 3x - 9 = x^2 - 9$
(b) $(x + 2)^2 = (x + 2)(x + 2) = x^2 + 2x + 2x + 4 = x^2 + 4x + 4$
(c) $(x - 3)^2 = (x - 3)(x - 3) = x^2 - 3x - 3x + 9 = x^2 - 6x + 9$
(d) $(x + 3)(x + 1) = x^2 + x + 3x + 3 = x^2 + 4x + 3$
(e) $(2x + 1)(3x + 4) = 6x^2 + 8x + 3x + 4 = 6x^2 + 11x + 4$

Now Work PROBLEMS 47 AND 55

Some products have been given special names because of their form. The following special products are based on Examples 7(a), (b), and (c).

Difference of Two Squares

$$(x - a)(x + a) = x^2 - a^2 \qquad (2)$$

Squares of Binomials, or Perfect Squares

$$(x + a)^2 = x^2 + 2ax + a^2 \qquad (3a)$$
$$(x - a)^2 = x^2 - 2ax + a^2 \qquad (3b)$$

EXAMPLE 8 **Using Special Product Formulas**

(a) $(x - 5)(x + 5) = x^2 - 5^2 = x^2 - 25$ — Difference of two squares
(b) $(x + 7)^2 = x^2 + 2 \cdot 7 \cdot x + 7^2 = x^2 + 14x + 49$ — Square of a binomial
(c) $(2x + 1)^2 = (2x)^2 + 2 \cdot 1 \cdot 2x + 1^2 = 4x^2 + 4x + 1$ — Notice that we used 2x in place of x in formula (3a).
(d) $(3x - 4)^2 = (3x)^2 - 2 \cdot 4 \cdot 3x + 4^2 = 9x^2 - 24x + 16$ — Replace x by 3x in formula (3b).

Now Work PROBLEMS 65, 67, AND 69

Let's look at some more examples that lead to general formulas.

EXAMPLE 9 **Cubing a Binomial**

(a) $(x + 2)^3 = (x + 2)(x + 2)^2 = (x + 2)(x^2 + 4x + 4)$ Formula (3a)

$$= x(x^2 + 4x + 4) + 2(x^2 + 4x + 4)$$
$$= (x^3 + 4x^2 + 4x) + (2x^2 + 8x + 8)$$
$$= x^3 + 6x^2 + 12x + 8$$

(b) $(x - 1)^3 = (x - 1)(x - 1)^2 = (x - 1)(x^2 - 2x + 1)$ Formula (3b)

$$= x(x^2 - 2x + 1) - 1(x^2 - 2x + 1)$$
$$= (x^3 - 2x^2 + x) + (-x^2 + 2x - 1)$$
$$= x^3 - 3x^2 + 3x - 1$$

Cubes of Binomials, or Perfect Cubes

$$(x + a)^3 = x^3 + 3ax^2 + 3a^2x + a^3 \qquad \textbf{(4a)}$$
$$(x - a)^3 = x^3 - 3ax^2 + 3a^2x - a^3 \qquad \textbf{(4b)}$$

Now Work PROBLEM 85

EXAMPLE 10 **Forming the Difference of Two Cubes**

$$(x - 1)(x^2 + x + 1) = x(x^2 + x + 1) - 1(x^2 + x + 1)$$
$$= x^3 + x^2 + x - x^2 - x - 1$$
$$= x^3 - 1$$

EXAMPLE 11 **Forming the Sum of Two Cubes**

$$(x + 2)(x^2 - 2x + 4) = x(x^2 - 2x + 4) + 2(x^2 - 2x + 4)$$
$$= x^3 - 2x^2 + 4x + 2x^2 - 4x + 8$$
$$= x^3 + 8$$

Examples 10 and 11 lead to two more special products.

Difference of Two Cubes

$$(x - a)(x^2 + ax + a^2) = x^3 - a^3 \qquad \textbf{(5)}$$

Sum of Two Cubes

$$(x + a)(x^2 - ax + a^2) = x^3 + a^3 \qquad \textbf{(6)}$$

---

6 Divide Polynomials Using Long Division

The procedure for dividing two polynomials is similar to the procedure for dividing two integers.

EXAMPLE 12   **Dividing Two Integers**

Divide 842 by 15.

Solution

$$
\begin{array}{r}
56 \quad \leftarrow \text{Quotient} \\
\text{Divisor} \rightarrow \quad 15\overline{)842} \quad \leftarrow \text{Dividend} \\
75 \quad \leftarrow 5 \cdot 15 \ (\text{subtract}) \\
\hline
92 \\
90 \quad \leftarrow 6 \cdot 15 \ (\text{subtract}) \\
\hline
2 \quad \leftarrow \text{Remainder}
\end{array}
$$

So, $\dfrac{842}{15} = 56 + \dfrac{2}{15}$.

In the long division process detailed in Example 12, the number 15 is called the **divisor,** the number 842 is called the **dividend,** the number 56 is called the **quotient,** and the number 2 is called the **remainder.**

To check the answer obtained in a division problem, multiply the quotient by the divisor and add the remainder. The answer should be the dividend.

$$
\boxed{(\text{Quotient})(\text{Divisor}) + \text{Remainder} = \text{Dividend}}
$$

For example, we can check the results obtained in Example 12 as follows:

$$(56)(15) + 2 = 840 + 2 = 842$$

To divide two polynomials, we first write each polynomial in standard form. The process then follows a pattern similar to that of Example 12. The next example illustrates the procedure.

EXAMPLE 13   **Dividing Two Polynomials**

Find the quotient and the remainder when

$$3x^3 + 4x^2 + x + 7 \quad \text{is divided by} \quad x^2 + 1$$

Solution

Each polynomial is in standard form. The dividend is $3x^3 + 4x^2 + x + 7$, and the divisor is $x^2 + 1$.

NOTE Remember, a polynomial is in standard form when the terms are written according to descending degree. ∎

STEP 1: Divide the leading term of the dividend, $3x^3$, by the leading term of the divisor, x^2. Enter the result, $3x$, over the term $3x^3$, as follows:

$$
\begin{array}{r}
3x \quad\quad\quad\quad\quad\quad \\
x^2 + 1\overline{)3x^3 + 4x^2 + x + 7}
\end{array}
$$

STEP 2: Multiply $3x$ by $x^2 + 1$ and enter the result below the dividend.

$$
\begin{array}{r}
3x \quad\quad\quad\quad\quad\quad\quad \\
x^2 + 1\overline{)3x^3 + 4x^2 + \ x + 7} \\
3x^3 \quad\quad + 3x \quad\quad \leftarrow 3x \cdot (x^2 + 1) = 3x^3 + 3x
\end{array}
$$

Notice that we align the 3x term under the x to make the next step easier.

STEP 3: Subtract and bring down the remaining terms.

$$
\begin{array}{r}
3x \quad\quad\quad\quad\quad\quad\quad \\
x^2 + 1\overline{)3x^3 + 4x^2 + \ x + 7} \\
3x^3 \quad\quad + 3x \quad\quad \leftarrow \text{Subtract (change the signs and add).} \\
\hline
4x^2 - 2x + 7 \quad \leftarrow \text{Bring down the } 4x^2 \text{ and the 7.}
\end{array}
$$

STEP 4: Repeat Steps 1–3 using $4x^2 - 2x + 7$ as the dividend.

$$
\begin{array}{r}
3x \;\; + 4 \\
x^2 + 1\overline{)3x^3 + 4x^2 + \;\; x + 7} \\
\underline{3x^3 \qquad\quad + 3x} \\
4x^2 - 2x + 7 \quad\leftarrow \text{Divide } 4x^2 \text{ by } x^2 \text{ to get 4.}\\
\underline{4x^2 \qquad\quad + 4} \quad\leftarrow \text{Multiply } x^2 + 1 \text{ by 4; subtract.}\\
-2x + 3
\end{array}
$$

Since x^2 does not divide into $-2x$ evenly (that is, the result is not a monomial), the process ends. The quotient is $3x + 4$, and the remainder is $-2x + 3$.

Check: (Quotient)(Divisor) + Remainder

$$
\begin{aligned}
&= (3x + 4)(x^2 + 1) + (-2x + 3) \\
&= 3x^3 + 3x + 4x^2 + 4 + (-2x + 3) \\
&= 3x^3 + 4x^2 + x + 7 = \text{Dividend}
\end{aligned}
$$

Then

$$
\frac{3x^3 + 4x^2 + x + 7}{x^2 + 1} = 3x + 4 + \frac{-2x + 3}{x^2 + 1}
$$

The next example combines the steps involved in long division.

EXAMPLE 14   **Dividing Two Polynomials**

Find the quotient and the remainder when

$$
x^4 - 3x^3 + 2x - 5 \quad \text{is divided by} \quad x^2 - x + 1
$$

Solution   In setting up this division problem, it is necessary to leave a space for the missing x^2 term in the dividend.

$$
\begin{array}{r}
x^2 - 2x \;\; - 3 \qquad\leftarrow \text{Quotient}\\
\text{Divisor} \rightarrow \quad x^2 - x + 1\overline{)x^4 - 3x^3 \qquad\quad + 2x - 5} \quad\leftarrow \text{Dividend}\\
\text{Subtract} \rightarrow \quad \underline{x^4 - x^3 + \;\; x^2} \qquad\qquad\\
-2x^3 - \;\; x^2 + 2x - 5\\
\text{Subtract} \rightarrow \quad \underline{-2x^3 + 2x^2 - 2x} \qquad\quad\\
-3x^2 + 4x - 5\\
\text{Subtract} \rightarrow \quad \underline{-3x^2 + 3x - 3}\\
x - 2 \quad\leftarrow \text{Remainder}
\end{array}
$$

Check: (Quotient)(Divisor) + Remainder

$$
\begin{aligned}
&= (x^2 - 2x - 3)(x^2 - x + 1) + (x - 2) \\
&= x^4 - x^3 + x^2 - 2x^3 + 2x^2 - 2x - 3x^2 + 3x - 3 + x - 2 \\
&= x^4 - 3x^3 + 2x - 5 = \text{Dividend}
\end{aligned}
$$

As a result,

$$
\frac{x^4 - 3x^3 + 2x - 5}{x^2 - x + 1} = x^2 - 2x - 3 + \frac{x - 2}{x^2 - x + 1}
$$

The process of dividing two polynomials leads to the following result:

THEOREM

Let Q be a polynomial of positive degree and let P be a polynomial whose degree is greater than or equal to the degree of Q. The remainder after dividing P by Q is either the zero polynomial or a polynomial whose degree is less than the degree of the divisor Q.

──Now Work PROBLEM 93

7 Work with Polynomials in Two Variables

A **monomial in two variables** x and y has the form $ax^n y^m$, where a is a constant, x and y are variables, and n and m are nonnegative integers. The **degree** of a monomial is the sum of the powers of the variables.

For example,

$$2xy^3, \quad x^2 y^2, \quad \text{and} \quad x^3 y$$

are monomials, each of which has degree 4.

A **polynomial in two variables** x and y is the sum of one or more monomials in two variables. The **degree of a polynomial** in two variables is the highest degree of all the monomials with nonzero coefficients.

EXAMPLE 15

Examples of Polynomials in Two Variables

$$3x^2 + 2x^3 y + 5 \qquad \pi x^3 - y^2 \qquad x^4 + 4x^3 y - xy^3 + y^4$$

Two variables, degree is 4.   Two variables, degree is 3.   Two variables, degree is 4.

Multiplying polynomials in two variables is handled in the same way as polynomials in one variable.

EXAMPLE 16

Using a Special Product Formula

To multiply $(2x - y)^2$, use the Square of a Binomial formula (3b) with $2x$ instead of x and y instead of a.

$$(2x - y)^2 = (2x)^2 - 2 \cdot y \cdot 2x + y^2$$
$$= 4x^2 - 4xy + y^2$$

──Now Work PROBLEM 79

A.3 Assess Your Understanding

Concepts and Vocabulary

1. The polynomial $3x^4 - 2x^3 + 13x^2 - 5$ is of degree _____. The leading coefficient is _____.

2. $(x - a)(x + a) = $ _____.

3. $(x - a)(x^2 + ax + a^2) = $ _____.

4. *True or False* $4x^{-2}$ is a monomial of degree -2.

5. *True or False* The degree of the product of two nonzero polynomials equals the sum of their degrees.

6. *True or False* $(x + a)(x^2 + ax + a) = x^3 + a^3$.

Skill Building

In Problems 7–16, tell whether the expression is a monomial. If it is, name the variable(s) and the coefficient and give the degree of the monomial. If it is not a monomial, state why not.

7. $2x^3$

8. $-4x^2$

9. $\dfrac{8}{x}$

10. $-2x^{-3}$

11. $-2xy^2$

12. $5x^2y^3$

13. $\dfrac{8x}{y}$

14. $-\dfrac{2x^2}{y^3}$

15. $x^2 + y^2$

16. $3x^2 + 4$

In Problems 17–26, tell whether the expression is a polynomial. If it is, give its degree. If it is not, state why not.

17. $3x^2 - 5$

18. $1 - 4x$

19. 5

20. $-\pi$

21. $3x^2 - \dfrac{5}{x}$

22. $\dfrac{3}{x} + 2$

23. $2y^3 - \sqrt{2}$

24. $10z^2 + z$

25. $\dfrac{x^2 + 5}{x^3 - 1}$

26. $\dfrac{3x^3 + 2x - 1}{x^2 + x + 1}$

In Problems 27–46, add, subtract, or multiply, as indicated. Express your answer as a single polynomial in standard form.

27. $(x^2 + 4x + 5) + (3x - 3)$

28. $(x^3 + 3x^2 + 2) + (x^2 - 4x + 4)$

29. $(x^3 - 2x^2 + 5x + 10) - (2x^2 - 4x + 3)$

30. $(x^2 - 3x - 4) - (x^3 - 3x^2 + x + 5)$

31. $(6x^5 + x^3 + x) + (5x^4 - x^3 + 3x^2)$

32. $(10x^5 - 8x^2) + (3x^3 - 2x^2 + 6)$

33. $(x^2 - 3x + 1) + 2(3x^2 + x - 4)$

34. $-2(x^2 + x + 1) + (-5x^2 - x + 2)$

35. $6(x^3 + x^2 - 3) - 4(2x^3 - 3x^2)$

36. $8(4x^3 - 3x^2 - 1) - 6(4x^3 + 8x - 2)$

37. $(x^2 - x + 2) + (2x^2 - 3x + 5) - (x^2 + 1)$

38. $(x^2 + 1) - (4x^2 + 5) + (x^2 + x - 2)$

39. $9(y^2 - 3y + 4) - 6(1 - y^2)$

40. $8(1 - y^3) + 4(1 + y + y^2 + y^3)$

41. $x(x^2 + x - 4)$

42. $4x^2(x^3 - x + 2)$

43. $-2x^2(4x^3 + 5)$

44. $5x^3(3x - 4)$

45. $(x + 1)(x^2 + 2x - 4)$

46. $(2x - 3)(x^2 + x + 1)$

In Problems 47–64, multiply the polynomials using the FOIL method. Express your answer as a single polynomial in standard form.

47. $(x + 2)(x + 4)$

48. $(x + 3)(x + 5)$

49. $(2x + 5)(x + 2)$

50. $(3x + 1)(2x + 1)$

51. $(x - 4)(x + 2)$

52. $(x + 4)(x - 2)$

53. $(x - 3)(x - 2)$

54. $(x - 5)(x - 1)$

55. $(2x + 3)(x - 2)$

56. $(2x - 4)(3x + 1)$

57. $(-2x + 3)(x - 4)$

58. $(-3x - 1)(x + 1)$

59. $(-x - 2)(-2x - 4)$

60. $(-2x - 3)(3 - x)$

61. $(x - 2y)(x + y)$

62. $(2x + 3y)(x - y)$

63. $(-2x - 3y)(3x + 2y)$

64. $(x - 3y)(-2x + y)$

In Problems 65–88, multiply the polynomials using the special product formulas. Express your answer as a single polynomial in standard form.

65. $(x - 7)(x + 7)$

66. $(x - 1)(x + 1)$

67. $(2x + 3)(2x - 3)$

68. $(3x + 2)(3x - 2)$

69. $(x + 4)^2$

70. $(x + 5)^2$

71. $(x - 4)^2$

72. $(x - 5)^2$

73. $(3x + 4)(3x - 4)$

74. $(5x - 3)(5x + 3)$

75. $(2x - 3)^2$

76. $(3x - 4)^2$

77. $(x + y)(x - y)$

78. $(x + 3y)(x - 3y)$

79. $(3x + y)(3x - y)$

80. $(3x + 4y)(3x - 4y)$

81. $(x + y)^2$

82. $(x - y)^2$

83. $(x - 2y)^2$

84. $(2x + 3y)^2$

85. $(x - 2)^3$

86. $(x + 1)^3$

87. $(2x + 1)^3$

88. $(3x - 2)^3$

In Problems 89–104, find the quotient and the remainder. Check your work by verifying that

$$(Quotient)(Divisor) + Remainder = Dividend$$

89. $4x^3 - 3x^2 + x + 1$ divided by $x + 2$

90. $3x^3 - x^2 + x - 2$ divided by $x + 2$

91. $4x^3 - 3x^2 + x + 1$ divided by x^2

92. $3x^3 - x^2 + x - 2$ divided by x^2

93. $5x^4 - 3x^2 + x + 1$ divided by $x^2 + 2$

94. $5x^4 - x^2 + x - 2$ divided by $x^2 + 2$

95. $4x^5 - 3x^2 + x + 1$ divided by $2x^3 - 1$

96. $3x^5 - x^2 + x - 2$ divided by $3x^3 - 1$

97. $2x^4 - 3x^3 + x + 1$ divided by $2x^2 + x + 1$

98. $3x^4 - x^3 + x - 2$ divided by $3x^2 + x + 1$

99. $-4x^3 + x^2 - 4$ divided by $x - 1$

100. $-3x^4 - 2x - 1$ divided by $x - 1$

101. $1 - x^2 + x^4$ divided by $x^2 + x + 1$

102. $1 - x^2 + x^4$ divided by $x^2 - x + 1$

103. $x^3 - a^3$ divided by $x - a$

104. $x^5 - a^5$ divided by $x - a$

Explaining Concepts: Discussion and Writing

105. Explain why the degree of the product of two nonzero polynomials equals the sum of their degrees.

106. Explain why the degree of the sum of two polynomials of different degrees equals the larger of their degrees.

107. Give a careful statement about the degree of the sum of two polynomials of the same degree.

108. Do you prefer adding two polynomials using the horizontal method or the vertical method? Write a brief position paper defending your choice.

109. Do you prefer to memorize the rule for the square of a binomial $(x + a)^2$ or to use FOIL to obtain the product? Write a brief position paper defending your choice.

A.4 Factoring Polynomials

OBJECTIVES
1 Factor the Difference of Two Squares and the Sum and Difference of Two Cubes (p. A34)
2 Factor Perfect Squares (p. A35)
3 Factor a Second-Degree Polynomial: $x^2 + Bx + C$ (p. A36)
4 Factor by Grouping (p. A37)
5 Factor a Second-Degree Polynomial: $Ax^2 + Bx + C, A \neq 1$ (p. A38)
6 Complete the Square (p. A40)

Consider the following product:

$$(2x + 3)(x - 4) = 2x^2 - 5x - 12$$

The two polynomials on the left side are called **factors** of the polynomial on the right side. Expressing a given polynomial as a product of other polynomials, that is, finding the factors of a polynomial, is called **factoring.**

We shall restrict our discussion here to factoring polynomials in one variable into products of polynomials in one variable, where all coefficients are integers. We call this **factoring over the integers.**

Any polynomial can be written as the product of 1 times itself or as -1 times its additive inverse. If a polynomial cannot be written as the product of two other polynomials (excluding 1 and -1), then the polynomial is said to be **prime.** When a polynomial has been written as a product consisting only of prime factors, it is said to be **factored completely.** Examples of prime polynomials (over the integers) are

COMMENT Over the real numbers, $3x + 4$ factors into $3\left(x + \dfrac{4}{3}\right)$. It is the noninteger $\dfrac{4}{3}$ that causes $3x + 4$ to be prime over the integers. ∎

$$2, \quad 3, \quad 5, \quad x, \quad x + 1, \quad x - 1, \quad 3x + 4, \quad x^2 + 4$$

The first factor to look for in a factoring problem is a common monomial factor present in each term of the polynomial. If one is present, use the Distributive Property to factor it out.

EXAMPLE 1   **Identifying Common Monomial Factors**

Polynomial	Common Monomial Factor	Remaining Factor	Factored Form
$2x + 4$	2	$x + 2$	$2x + 4 = 2(x + 2)$
$3x - 6$	3	$x - 2$	$3x - 6 = 3(x - 2)$
$2x^2 - 4x + 8$	2	$x^2 - 2x + 4$	$2x^2 - 4x + 8 = 2(x^2 - 2x + 4)$
$8x - 12$	4	$2x - 3$	$8x - 12 = 4(2x - 3)$
$x^2 + x$	x	$x + 1$	$x^2 + x = x(x + 1)$
$x^3 - 3x^2$	x^2	$x - 3$	$x^3 - 3x^2 = x^2(x - 3)$
$6x^2 + 9x$	$3x$	$2x + 3$	$6x^2 + 9x = 3x(2x + 3)$

●

Notice that, once all common monomial factors have been removed from a polynomial, the remaining factor is either a prime polynomial of degree 1 or a polynomial of degree 2 or higher. (Do you see why?)

✏️ **Now Work** PROBLEM 5

---

1 Factor the Difference of Two Squares and the Sum and Difference of Two Cubes

When you factor a polynomial, first check for common monomial factors. Then see whether you can use one of the special formulas discussed in the previous section.

Difference of Two Squares	$x^2 - a^2 = (x - a)(x + a)$
Perfect Squares	$x^2 + 2ax + a^2 = (x + a)^2$
	$x^2 - 2ax + a^2 = (x - a)^2$
Sum of Two Cubes	$x^3 + a^3 = (x + a)(x^2 - ax + a^2)$
Difference of Two Cubes	$x^3 - a^3 = (x - a)(x^2 + ax + a^2)$

EXAMPLE 2   **Factoring the Difference of Two Squares**

Factor completely: $x^2 - 4$

Solution   We notice that $x^2 - 4$ is the difference of two squares, x^2 and 2^2.

$$x^2 - 4 = (x - 2)(x + 2)$$

●

EXAMPLE 3	**Factoring the Difference of Two Cubes**

Factor completely: $x^3 - 1$

Solution Because $x^3 - 1$ is the difference of two cubes, x^3 and 1^3, we find that

$$x^3 - 1 = (x - 1)(x^2 + x + 1)$$

EXAMPLE 4	**Factoring the Sum of Two Cubes**

Factor completely: $x^3 + 8$

Solution Because $x^3 + 8$ is the sum of two cubes, x^3 and 2^3, we have

$$x^3 + 8 = (x + 2)(x^2 - 2x + 4)$$

EXAMPLE 5	**Factoring the Difference of Two Squares**

Factor completely: $x^4 - 16$

Solution Because $x^4 - 16$ is the difference of two squares, $x^4 = (x^2)^2$ and $16 = 4^2$, we have

$$x^4 - 16 = (x^2 - 4)(x^2 + 4)$$

But $x^2 - 4$ is also the difference of two squares. Thus,

$$x^4 - 16 = (x^2 - 4)(x^2 + 4) = (x - 2)(x + 2)(x^2 + 4)$$

Now Work PROBLEMS **15** AND **33**

2 Factor Perfect Squares

When the first term and third term of a trinomial are both positive and are perfect squares, such as x^2, $9x^2$, 1, and 4, check to see whether the trinomial is a perfect square.

EXAMPLE 6	**Factoring Perfect Squares**

Factor completely: $x^2 + 6x + 9$

Solution The first term, x^2, and the third term, $9 = 3^2$, are perfect squares. Because the middle term $6x$ is twice the product of x and 3, we have a perfect square.

$$x^2 + 6x + 9 = (x + 3)^2$$

EXAMPLE 7	**Factoring Perfect Squares**

Factor completely: $9x^2 - 6x + 1$

Solution The first term, $9x^2 = (3x)^2$, and the third term, $1 = 1^2$, are perfect squares. Because the middle term, $-6x$, is -2 times the product of $3x$ and 1, we have a perfect square.

$$9x^2 - 6x + 1 = (3x - 1)^2$$

EXAMPLE 8	**Factoring Perfect Squares**

Factor completely: $25x^2 + 30x + 9$

Solution The first term, $25x^2 = (5x)^2$, and the third term, $9 = 3^2$, are perfect squares. Because the middle term, $30x$, is twice the product of $5x$ and 3, we have a perfect square.

$$25x^2 + 30x + 9 = (5x + 3)^2$$

Now Work PROBLEM **25**

If a trinomial is not a perfect square, it may be possible to factor it using the technique discussed next.

3 Factor a Second-Degree Polynomial: $x^2 + Bx + C$

The idea behind factoring a second-degree polynomial like $x^2 + Bx + C$ is to see whether it can be made equal to the product of two, possibly equal, first-degree polynomials.

For example, we know that

$$(x + 3)(x + 4) = x^2 + 7x + 12$$

The factors of $x^2 + 7x + 12$ are $x + 3$ and $x + 4$. Notice the following:

$$x^2 + 7x + 12 = (x + 3)(x + 4)$$

12 is the product of 3 and 4
7 is the sum of 3 and 4

In general, if $x^2 + Bx + C = (x + a)(x + b) = x^2 + (a + b)x + ab$, then $ab = C$ and $a + b = B$.

To factor a second-degree polynomial $x^2 + Bx + C$, find integers whose product is C and whose sum is B. That is, if there are numbers a, b, where $ab = C$ and $a + b = B$, then

$$x^2 + Bx + C = (x + a)(x + b)$$

EXAMPLE 9

Factoring Trinomials

Factor completely: $x^2 + 7x + 10$

Solution First, determine all integers whose product is 10 and then compute their sums.

Integers whose product is 10	1, 10	−1, −10	2, 5	−2, −5
Sum	11	−11	7	−7

The integers 2 and 5 have a product of 10 and add up to 7, the coefficient of the middle term. As a result, we have

$$x^2 + 7x + 10 = (x + 2)(x + 5)$$

EXAMPLE 10

Factoring Trinomials

Factor completely: $x^2 - 6x + 8$

Solution First, determine all integers whose product is 8 and then compute each sum.

Integers whose product is 8	1, 8	−1, −8	2, 4	−2, −4
Sum	9	−9	6	−6

Since −6 is the coefficient of the middle term,

$$x^2 - 6x + 8 = (x - 2)(x - 4)$$

EXAMPLE 11   **Factoring Trinomials**

Factor completely: $x^2 - x - 12$

Solution   First, determine all integers whose product is -12 and then compute each sum.

Integers whose product is -12	1, -12	$-1, 12$	2, -6	$-2, 6$	3, -4	$-3, 4$
Sum	-11	11	-4	4	-1	1

Since -1 is the coefficient of the middle term,

$$x^2 - x - 12 = (x + 3)(x - 4)$$

●

EXAMPLE 12   **Factoring Trinomials**

Factor completely: $x^2 + 4x - 12$

Solution   The integers -2 and 6 have a product of -12 and have the sum 4. Thus,

$$x^2 + 4x - 12 = (x - 2)(x + 6)$$

●

To avoid errors in factoring, always check your answer by multiplying it out to see if the result equals the original expression.

When none of the possibilities works, the polynomial is prime.

EXAMPLE 13   **Identifying Prime Polynomials**

Show that $x^2 + 9$ is prime.

Solution   First, list the integers whose product is 9 and then compute their sums.

Integers whose product is 9	1, 9	$-1, -9$	3, 3	$-3, -3$
Sum	10	-10	6	-6

Since the coefficient of the middle term in $x^2 + 9 = x^2 + 0x + 9$ is 0 and none of the sums equals 0, we conclude that $x^2 + 9$ is prime.

●

Example 13 demonstrates a more general result:

THEOREM   Any polynomial of the form $x^2 + a^2$, a real, is prime.

━━━▶ **Now Work** PROBLEM 39

4 Factor by Grouping

Sometimes a common factor does not occur in every term of the polynomial, but in each of several groups of terms that together make up the polynomial. When this happens, the common factor can be factored out of each group by means of the Distributive Property. This technique is called **factoring by grouping.**

EXAMPLE 14   **Factoring by Grouping**

Factor completely by grouping: $(x^2 + 2)x + (x^2 + 2) \cdot 3$

Solution   Notice the common factor $x^2 + 2$. By applying the Distributive Property, we have

$$(x^2 + 2)x + (x^2 + 2) \cdot 3 = (x^2 + 2)(x + 3)$$

Since $x^2 + 2$ and $x + 3$ are prime, the factorization is complete.

●

⚠ The next example shows a factoring problem that occurs in calculus.

EXAMPLE 15 **Factoring by Grouping**

Factor completely by grouping: $3(x-1)^2(x+2)^4 + 4(x-1)^3(x+2)^3$

Solution Here, $(x-1)^2(x+2)^3$ is a common factor of $3(x-1)^2(x+2)^4$ and of $4(x-1)^3(x+2)^3$. As a result,

$$3(x-1)^2(x+2)^4 + 4(x-1)^3(x+2)^3 = (x-1)^2(x+2)^3[3(x+2) + 4(x-1)]$$
$$= (x-1)^2(x+2)^3[3x+6+4x-4]$$
$$= (x-1)^2(x+2)^3(7x+2)$$

●

EXAMPLE 16 **Factoring by Grouping**

Factor completely by grouping: $x^3 - 4x^2 + 2x - 8$

Solution To see if factoring by grouping will work, group the first two terms and the last two terms. Then look for a common factor in each group. In this example, we can factor x^2 from $x^3 - 4x^2$ and 2 from $2x - 8$. The remaining factor in each case is the same, $x - 4$. This means that factoring by grouping will work, as follows:

$$x^3 - 4x^2 + 2x - 8 = (x^3 - 4x^2) + (2x - 8)$$
$$= x^2(x-4) + 2(x-4)$$
$$= (x-4)(x^2 + 2)$$

Since $x^2 + 2$ and $x - 4$ are prime, the factorization is complete.

●

Now Work PROBLEMS 51 AND 127

5 Factor a Second-Degree Polynomial: $Ax^2 + Bx + C, A \neq 1$

To factor a second-degree polynomial $Ax^2 + Bx + C$, when $A \neq 1$ and A, B, and C have no common factors, follow these steps:

> **Steps for Factoring $Ax^2 + Bx + C$,**
> **when $A \neq 1$ and A, B, and C Have No Common Factors**
>
> **STEP 1:** Find the value of AC.
> **STEP 2:** Find integers whose product is AC that add up to B. That is, find a and b so that $ab = AC$ and $a + b = B$.
> **STEP 3:** Write $Ax^2 + Bx + C = Ax^2 + ax + bx + C$.
> **STEP 4:** Factor this last expression by grouping.

EXAMPLE 17 **Factoring Trinomials**

Factor completely: $2x^2 + 5x + 3$

Solution Comparing $2x^2 + 5x + 3$ to $Ax^2 + Bx + C$, we find that $A = 2$, $B = 5$, and $C = 3$.

STEP 1: The value of AC is $2 \cdot 3 = 6$.
STEP 2: Determine the integers whose product is $AC = 6$ and compute their sums.

Integers whose product is 6	1, 6	−1, −6	2, 3	−2, −3
Sum	7	−7	5	−5

STEP 3: The integers whose product is 6 that add up to $B = 5$ are 2 and 3.

$$2x^2 + 5x + 3 = 2x^2 + 2x + 3x + 3$$

STEP 4: Factor by grouping.

$$2x^2 + 2x + 3x + 3 = (2x^2 + 2x) + (3x + 3)$$
$$= 2x(x + 1) + 3(x + 1)$$
$$= (x + 1)(2x + 3)$$

As a result,

$$2x^2 + 5x + 3 = (x + 1)(2x + 3)$$

EXAMPLE 18   **Factoring Trinomials**

Factor completely: $2x^2 - x - 6$

Solution   Comparing $2x^2 - x - 6$ to $Ax^2 + Bx + C$, we find that $A = 2, B = -1$, and $C = -6$.

STEP 1: The value of AC is $2 \cdot (-6) = -12$.

STEP 2: Determine the integers whose product is $AC = -12$ and compute their sums.

Integers whose product is −12	1, −12	−1, 12	2, −6	−2, 6	3, −4	−3, 4
Sum	−11	11	−4	4	−1	1

STEP 3: The integers whose product is −12 that add up to $B = -1$ are −4 and 3.

$$2x^2 - x - 6 = 2x^2 - 4x + 3x - 6$$

STEP 4: Factor by grouping.

$$2x^2 - 4x + 3x - 6 = (2x^2 - 4x) + (3x - 6)$$
$$= 2x(x - 2) + 3(x - 2)$$
$$= (x - 2)(2x + 3)$$

As a result,

$$2x^2 - x - 6 = (x - 2)(2x + 3)$$

Now Work PROBLEM **57**

SUMMARY

Type of Polynomial	Method	Example
Any polynomial	Look for common monomial factors. (Always do this first!)	$6x^2 + 9x = 3x(2x + 3)$
Binomials of degree 2 or higher	Check for a special product: Difference of two squares, $x^2 - a^2$	$x^2 - 16 = (x - 4)(x + 4)$
	Difference of two cubes, $x^3 - a^3$	$x^3 - 64 = (x - 4)(x^2 + 4x + 16)$
	Sum of two cubes, $x^3 + a^3$	$x^3 + 27 = (x + 3)(x^2 - 3x + 9)$
Trinomials of degree 2	Check for a perfect square, $(x \pm a)^2$	$x^2 + 8x + 16 = (x + 4)^2$
		$x^2 - 10x + 25 = (x - 5)^2$
	Factoring $x^2 + Bx + C$ (p. A36)	$x^2 - x - 2 = (x - 2)(x + 1)$
	Factoring $Ax^2 + Bx + C$ (p. A38)	$6x^2 + x - 1 = (2x + 1)(3x - 1)$
Four or more terms	Grouping	$2x^3 - 3x^2 + 4x - 6 = (2x - 3)(x^2 + 2)$

6 Complete the Square

We now introduce the method of completing the square. The idea behind completing the square in one variable is to "adjust" an expression of the form $x^2 + bx$ to make it a perfect square. Perfect squares are trinomials of the form

$$x^2 + 2ax + a^2 = (x + a)^2 \text{ or } x^2 - 2ax + a^2 = (x - a)^2$$

For example, $x^2 + 6x + 9$ is a perfect square because $x^2 + 6x + 9 = (x + 3)^2$. And $p^2 - 12p + 36$ is a perfect square because $p^2 - 12p + 36 = (p - 6)^2$.

So how do we "adjust" $x^2 + bx$ to make it a perfect square? We do it by adding a number. For example, to make $x^2 + 6x$ a perfect square, we would add 9. But how do we know to add 9? If we divide the coefficient on the first-degree term, 6, by 2, and then square the result, we obtain 9. This approach works in general.

> **Completing the Square**
>
> Identify the coefficient of the first-degree term. Multiply this coefficient by $\dfrac{1}{2}$ and then square the result. That is, determine the value of b in $x^2 + bx$ and compute $\left(\dfrac{1}{2}b\right)^2$.

EXAMPLE 19   **Completing the Square**

Determine the number that must be added to each expression to complete the square. Then factor the expression.

Start	Add	Result	Factored Form
$y^2 + 8y$	$\left(\dfrac{1}{2} \cdot 8\right)^2 = 16$	$y^2 + 8y + 16$	$(y + 4)^2$
$x^2 + 12x$	$\left(\dfrac{1}{2} \cdot 12\right)^2 = 36$	$x^2 + 12x + 36$	$(x + 6)^2$
$a^2 - 20a$	$\left(\dfrac{1}{2} \cdot (-20)\right)^2 = 100$	$a^2 - 20a + 100$	$(a - 10)^2$
$p^2 - 5p$	$\left(\dfrac{1}{2} \cdot (-5)\right)^2 = \dfrac{25}{4}$	$p^2 - 5p + \dfrac{25}{4}$	$\left(p - \dfrac{5}{2}\right)^2$

●

Notice that the factored form of a perfect square is either

$$x^2 + bx + \left(\frac{b}{2}\right)^2 = \left(x + \frac{b}{2}\right)^2 \text{ or } x^2 - bx + \left(\frac{b}{2}\right)^2 = \left(x - \frac{b}{2}\right)^2$$

Figure 25

Now Work PROBLEM 69

Are you wondering why we call making an expression a perfect square "completing the square"? Look at the square in Figure 25. Its area is $(y + 4)^2$. The yellow area is y^2 and each orange area is $4y$ (for a total area of $8y$). The sum of these areas is $y^2 + 8y$. To complete the square, we need to add the area of the green region: $4 \cdot 4 = 16$. As a result, $y^2 + 8y + 16 = (y + 4)^2$.

A.4 Assess Your Understanding

Concepts and Vocabulary

1. **True or False** If factored completely, $3x^3 - 12x = 3x(x - 2)(x + 2)$.

2. If a polynomial cannot be written as the product of two other polynomials (excluding 1 and −1), then the polynomial is said to be _____.

3. **True or False** The polynomial $x^2 + 4$ is prime.

4. **True or False** $3x^3 - 2x^2 - 6x + 4 = (3x - 2)(x^2 + 2)$.

Skill Building

In Problems 5–14, factor each polynomial by removing the common monomial factor.

5. $3x + 6$

6. $7x - 14$

7. $ax^2 + a$

8. $ax - a$

9. $x^3 + x^2 + x$

10. $x^3 - x^2 + x$

11. $2x^2 - 2x$

12. $3x^2 - 3x$

13. $3x^2y - 6xy^2 + 12xy$

14. $60x^2y - 48xy^2 + 72x^3y$

In Problems 15–22, factor the difference of two squares.

15. $x^2 - 1$

16. $x^2 - 4$

17. $4x^2 - 1$

18. $9x^2 - 1$

19. $x^2 - 16$

20. $x^2 - 25$

21. $25x^2 - 4$

22. $36x^2 - 9$

In Problems 23–32, factor the perfect squares.

23. $x^2 + 2x + 1$

24. $x^2 - 4x + 4$

25. $x^2 + 4x + 4$

26. $x^2 - 2x + 1$

27. $x^2 - 10x + 25$

28. $x^2 + 10x + 25$

29. $4x^2 + 4x + 1$

30. $9x^2 + 6x + 1$

31. $16x^2 + 8x + 1$

32. $25x^2 + 10x + 1$

In Problems 33–38, factor the sum or difference of two cubes.

33. $x^3 - 27$

34. $x^3 + 125$

35. $x^3 + 27$

36. $27 - 8x^3$

37. $8x^3 + 27$

38. $64 - 27x^3$

In Problems 39–50, factor each polynomial.

39. $x^2 + 5x + 6$

40. $x^2 + 6x + 8$

41. $x^2 + 7x + 6$

42. $x^2 + 9x + 8$

43. $x^2 + 7x + 10$

44. $x^2 + 11x + 10$

45. $x^2 - 10x + 16$

46. $x^2 - 17x + 16$

47. $x^2 - 7x - 8$

48. $x^2 - 2x - 8$

49. $x^2 + 7x - 8$

50. $x^2 + 2x - 8$

In Problems 51–56, factor by grouping.

51. $2x^2 + 4x + 3x + 6$

52. $3x^2 - 3x + 2x - 2$

53. $2x^2 - 4x + x - 2$

54. $3x^2 + 6x - x - 2$

55. $6x^2 + 9x + 4x + 6$

56. $9x^2 - 6x + 3x - 2$

In Problems 57–68, factor each polynomial.

57. $3x^2 + 4x + 1$

58. $2x^2 + 3x + 1$

59. $2z^2 + 5z + 3$

60. $6z^2 + 5z + 1$

61. $3x^2 + 2x - 8$

62. $3x^2 + 10x + 8$

63. $3x^2 - 2x - 8$

64. $3x^2 - 10x + 8$

65. $3x^2 + 14x + 8$

66. $3x^2 - 14x + 8$

67. $3x^2 + 10x - 8$

68. $3x^2 - 10x - 8$

In Problems 69–74, determine the number that should be added to complete the square of each expression. Then factor each expression.

69. $x^2 + 10x$

70. $p^2 + 14p$

71. $y^2 - 6y$

72. $x^2 - 4x$

73. $x^2 - \dfrac{1}{2}x$

74. $x^2 + \dfrac{1}{3}x$

Mixed Practice

In Problems 75–122, factor completely each polynomial. If the polynomial cannot be factored, say it is prime.

75. $x^2 - 36$

76. $x^2 - 9$

77. $2 - 8x^2$

78. $3 - 27x^2$

79. $x^2 + 11x + 10$

80. $x^2 + 5x + 4$

81. $x^2 - 10x + 21$

82. $x^2 - 6x + 8$

83. $4x^2 - 8x + 32$

84. $3x^2 - 12x + 15$

85. $x^2 + 4x + 16$

86. $x^2 + 12x + 36$

87. $15 + 2x - x^2$

88. $14 + 6x - x^2$

89. $3x^2 - 12x - 36$

90. $x^3 + 8x^2 - 20x$

91. $y^4 + 11y^3 + 30y^2$

92. $3y^3 - 18y^2 - 48y$

93. $4x^2 + 12x + 9$

94. $9x^2 - 12x + 4$

95. $6x^2 + 8x + 2$ **96.** $8x^2 + 6x - 2$ **97.** $x^4 - 81$ **98.** $x^4 - 1$

99. $x^6 - 2x^3 + 1$ **100.** $x^6 + 2x^3 + 1$ **101.** $x^7 - x^5$ **102.** $x^8 - x^5$

103. $16x^2 + 24x + 9$ **104.** $9x^2 - 24x + 16$ **105.** $5 + 16x - 16x^2$ **106.** $5 + 11x - 16x^2$

107. $4y^2 - 16y + 15$ **108.** $9y^2 + 9y - 4$ **109.** $1 - 8x^2 - 9x^4$ **110.** $4 - 14x^2 - 8x^4$

111. $x(x + 3) - 6(x + 3)$ **112.** $5(3x - 7) + x(3x - 7)$ **113.** $(x + 2)^2 - 5(x + 2)$ **114.** $(x - 1)^2 - 2(x - 1)$

115. $(3x - 2)^3 - 27$ **116.** $(5x + 1)^3 - 1$ **117.** $3(x^2 + 10x + 25) - 4(x + 5)$

118. $7(x^2 - 6x + 9) + 5(x - 3)$ **119.** $x^3 + 2x^2 - x - 2$ **120.** $x^3 - 3x^2 - x + 3$

121. $x^4 - x^3 + x - 1$ **122.** $x^4 + x^3 + x + 1$

Applications and Extensions

✍ *In Problems 123–132, expressions that occur in calculus are given. Factor completely each expression.*

123. $2(3x + 4)^2 + (2x + 3) \cdot 2(3x + 4) \cdot 3$ **124.** $5(2x + 1)^2 + (5x - 6) \cdot 2(2x + 1) \cdot 2$

125. $2x(2x + 5) + x^2 \cdot 2$ **126.** $3x^2(8x - 3) + x^3 \cdot 8$

127. $2(x + 3)(x - 2)^3 + (x + 3)^2 \cdot 3(x - 2)^2$ **128.** $4(x + 5)^3(x - 1)^2 + (x + 5)^4 \cdot 2(x - 1)$

129. $(4x - 3)^2 + x \cdot 2(4x - 3) \cdot 4$ **130.** $3x^2(3x + 4)^2 + x^3 \cdot 2(3x + 4) \cdot 3$

131. $2(3x - 5) \cdot 3(2x + 1)^3 + (3x - 5)^2 \cdot 3(2x + 1)^2 \cdot 2$ **132.** $3(4x + 5)^2 \cdot 4(5x + 1)^2 + (4x + 5)^3 \cdot 2(5x + 1) \cdot 5$

133. Show that $x^2 + 4$ is prime. **134.** Show that $x^2 + x + 1$ is prime.

Explaining Conditions: Discussion and Writing

135. Make up a polynomial that factors into a perfect square. **136.** Explain to a fellow student what you look for first when presented with a factoring problem. What do you do next?

A.5 Synthetic Division

OBJECTIVE 1 Divide Polynomials Using Synthetic Division (p. A42)

1 Divide Polynomials Using Synthetic Division

To find the quotient as well as the remainder when a polynomial of degree 1 or higher is divided by $x - c$, a shortened version of long division, called **synthetic division,** makes the task simpler.

To see how synthetic division works, we will use long division to divide the polynomial $2x^3 - x^2 + 3$ by $x - 3$.

$$
\begin{array}{r}
2x^2 + 5x + 15 \qquad \leftarrow Quotient \\
x - 3 \overline{)2x^3 - x^2 \qquad\quad + 3} \\
\underline{2x^3 - 6x^2} \\
5x^2 \\
\underline{5x^2 - 15x} \\
15x + 3 \\
\underline{15x - 45} \\
48 \qquad \leftarrow Remainder
\end{array}
$$

✓**Check:** (Divisor) · (Quotient) + Remainder

$$
\begin{aligned}
&= (x - 3)(2x^2 + 5x + 15) + 48 \\
&= 2x^3 + 5x^2 + 15x - 6x^2 - 15x - 45 + 48 \\
&= 2x^3 - x^2 + 3
\end{aligned}
$$

The process of synthetic division arises from rewriting the long division in a more compact form, using simpler notation. For example, in the long division on page A42, the terms in blue are not really necessary because they are identical to the terms directly above them. With these terms removed, we have

$$
\begin{array}{r}
2x^2 + 5x + 15 \\
x - 3 \overline{)\,2x^3 - x^2 \qquad\quad + 3} \\
-6x^2 \\
\overline{5x^2} \\
-15x \\
\overline{15x} \\
-45 \\
\overline{48}
\end{array}
$$

Most of the x's that appear in this process can also be removed, provided that we are careful about positioning each coefficient. In this regard, we will need to use 0 as the coefficient of x in the dividend, because that power of x is missing. Now we have

$$
\begin{array}{r}
2x^2 + 5x + 15 \\
x - 3 \overline{)\,2 \quad -1 \qquad 0 \qquad 3} \\
-6 \\
\overline{5} \\
-15 \\
\overline{15} \\
-45 \\
\overline{48}
\end{array}
$$

We can make this display more compact by moving the lines up until the numbers in blue align horizontally.

$$
\begin{array}{r|rrrr}
& 2x^2 + 5x + 15 & & & \text{Row 1}\\
x - 3 \,)\,2 & -1 & 0 & 3 & \text{Row 2}\\
\hline
& -6 & -15 & -45 & \text{Row 3}\\
\bigcirc & 5 & 15 & 48 & \text{Row 4}
\end{array}
$$

Because the leading coefficient of the divisor is always 1, we know that the leading coefficient of the dividend will also be the leading coefficient of the quotient. So we place the leading coefficient of the quotient, 2, in the circled position. Now, the first three numbers in row 4 are precisely the coefficients of the quotient, and the last number in row 4 is the remainder. Thus, row 1 is not really needed, so we can compress the process to three rows, where the bottom row contains both the coefficients of the quotient and the remainder.

$$
\begin{array}{r|rrrl}
x - 3 \,)\,2 & -1 & 0 & 3 & \text{Row 1}\\
\hline
& -6 & -15 & -45 & \text{Row 2 (subtract)}\\
2 & 5 & 15 & 48 & \text{Row 3}
\end{array}
$$

Recall that the entries in row 3 are obtained by subtracting the entries in row 2 from those in row 1. Rather than subtracting the entries in row 2, we can change the sign of each entry and add. With this modification, our display will look like this:

$$
\begin{array}{r|rrrl}
x - 3 \,)\,2 & -1 & 0 & 3 & \text{Row 1}\\
\hline
& 6 & 15 & 45 & \text{Row 2 (add)}\\
2 & 5 & 15 & 48 & \text{Row 3}
\end{array}
$$

Notice that the entries in row 2 are three times the prior entries in row 3. Our last modification to the display replaces the $x - 3$ by 3. The entries in row 3 give the quotient and the remainder, as shown next.

$$
\begin{array}{r|rrrr}
3) & 2 & -1 & 0 & 3 \\
 & & 6 & 15 & 45 \\
\hline
 & 2 & 5 & 15 & 48
\end{array}
\quad
\begin{array}{l}
\text{Row 1} \\
\text{Row 2 (add)} \\
\text{Row 3}
\end{array}
$$

$$\underbrace{2x^2 + 5x + 15}_{\text{Quotient}} \quad \underbrace{48}_{\text{Remainder}}$$

Let's go through an example step by step.

EXAMPLE 1   **Using Synthetic Division to Find the Quotient and Remainder**

Use synthetic division to find the quotient and remainder when

$$x^3 - 4x^2 - 5 \text{ is divided by } x - 3$$

Solution   **STEP 1:** Write the dividend in descending powers of x. Then copy the coefficients, remembering to insert a 0 for any missing powers of x.

$$1 \quad -4 \quad 0 \quad -5 \quad \text{Row 1}$$

STEP 2: Insert the usual division symbol. In synthetic division, the divisor is of the form $x - c$, and c is the number placed to the left of the division symbol. Here, since the divisor is $x - 3$, we insert 3 to the left of the division symbol.

$$3)1 \quad -4 \quad 0 \quad -5 \quad \text{Row 1}$$

STEP 3: Bring the 1 down two rows, and enter it in row 3.

$$
\begin{array}{r|rrrr}
3) & 1 & -4 & 0 & -5 \\
 & \downarrow & & & \\
\hline
 & 1 & & &
\end{array}
\quad
\begin{array}{l}
\text{Row 1} \\
\text{Row 2} \\
\text{Row 3}
\end{array}
$$

STEP 4: Multiply the latest entry in row 3 by 3, and place the result in row 2, one column over to the right.

$$
\begin{array}{r|rrrr}
3) & 1 & -4 & 0 & -5 \\
 & & 3 & & \\
\hline
 & 1 & & &
\end{array}
\quad
\begin{array}{l}
\text{Row 1} \\
\text{Row 2} \\
\text{Row 3}
\end{array}
$$

STEP 5: Add the entry in row 2 to the entry above it in row 1, and enter the sum in row 3.

$$
\begin{array}{r|rrrr}
3) & 1 & -4 & 0 & -5 \\
 & & 3 & & \\
\hline
 & 1 & -1 & &
\end{array}
\quad
\begin{array}{l}
\text{Row 1} \\
\text{Row 2} \\
\text{Row 3}
\end{array}
$$

STEP 6: Repeat Steps 4 and 5 until no more entries are available in row 1.

$$
\begin{array}{r|rrrr}
3) & 1 & -4 & 0 & -5 \\
 & & 3 & -3 & -9 \\
\hline
 & 1 & -1 & -3 & -14
\end{array}
\quad
\begin{array}{l}
\text{Row 1} \\
\text{Row 2} \\
\text{Row 3}
\end{array}
$$

STEP 7: The final entry in row 3, the -14, is the remainder; the other entries in row 3, the $1, -1$, and -3, are the coefficients (in descending order) of a polynomial whose degree is 1 less than that of the dividend. This is the quotient. Thus,

$$\text{Quotient} = x^2 - x - 3 \qquad \text{Remainder} = -14$$

✓ **Check:** (Divisor)(Quotient) + Remainder

$$
\begin{aligned}
&= (x - 3)(x^2 - x - 3) + (-14) \\
&= (x^3 - x^2 - 3x - 3x^2 + 3x + 9) + (-14) \\
&= x^3 - 4x^2 - 5 = \text{Dividend}
\end{aligned}
$$

Let's do an example in which all seven steps are combined.

EXAMPLE 2	**Using Synthetic Division to Verify a Factor**

Use synthetic division to show that $x + 3$ is a factor of

$$2x^5 + 5x^4 - 2x^3 + 2x^2 - 2x + 3$$

Solution If the remainder is 0, then the divisor is a factor of the dividend, and the quotient is the other factor. The divisor is $x + 3 = x - (-3)$, so we place -3 to the left of the division symbol. Then the row 3 entries will be multiplied by -3, entered in row 2, and added to row 1.

$$
\begin{array}{r|rrrrrr}
-3 & 2 & 5 & -2 & 2 & -2 & 3 & \text{Row 1}\\
   &   & -6 & 3 & -3 & 3 & -3 & \text{Row 2}\\
\hline
   & 2 & -1 & 1 & -1 & 1 & 0 & \text{Row 3}
\end{array}
$$

Because the remainder is 0, we have

$$\text{Dividend} = (\text{Divisor})(\text{Quotient}) + \text{Remainder}$$

$$2x^5 + 5x^4 - 2x^3 + 2x^2 - 2x + 3 = (x + 3)(2x^4 - x^3 + x^2 - x + 1)$$

As we see, $x + 3$ is a factor of $2x^5 + 5x^4 - 2x^3 + 2x^2 - 2x + 3$.

As Example 2 illustrates, the remainder after division gives information about whether the divisor is, or is not, a factor. We shall have more to say about this in Chapter 3.

Now Work PROBLEMS 7 AND 17

A.5 Assess Your Understanding

Concepts and Vocabulary

1. To check division, the expression that is being divided, the dividend, should equal the product of the _____ and the _____ plus the _____.

2. To divide $2x^3 - 5x + 1$ by $x + 3$ using synthetic division, the first step is to write __ ⟌_____ .

3. *True or False* In using synthetic division, the divisor is always a polynomial of degree 1, whose leading coefficient is 1.

4. *True or False*
$$
\begin{array}{r|rrrr}
-2 & 5 & 3 & 2 & 1\\
   &   & -10 & 14 & -32\\
\hline
   & 5 & -7 & 16 & -31
\end{array}
$$
means $\dfrac{5x^3 + 3x^2 + 2x + 1}{x + 2} = 5x^2 - 7x + 16 + \dfrac{-31}{x + 2}$.

Skill Building

In Problems 5–16, use synthetic division to find the quotient and remainder when:

5. $x^3 - x^2 + 2x + 4$ is divided by $x - 2$

6. $x^3 + 2x^2 - 3x + 1$ is divided by $x + 1$

7. $3x^3 + 2x^2 - x + 3$ is divided by $x - 3$

8. $-4x^3 + 2x^2 - x + 1$ is divided by $x + 2$

9. $x^5 - 4x^3 + x$ is divided by $x + 3$

10. $x^4 + x^2 + 2$ is divided by $x - 2$

11. $4x^6 - 3x^4 + x^2 + 5$ is divided by $x - 1$

12. $x^5 + 5x^3 - 10$ is divided by $x + 1$

13. $0.1x^3 + 0.2x$ is divided by $x + 1.1$

14. $0.1x^2 - 0.2$ is divided by $x + 2.1$

15. $x^5 - 1$ is divided by $x - 1$

16. $x^5 + 1$ is divided by $x + 1$

In Problems 17–26, use synthetic division to determine whether $x - c$ is a factor of the given polynomial.

17. $4x^3 - 3x^2 - 8x + 4$; $x - 2$

18. $-4x^3 + 5x^2 + 8$; $x + 3$

19. $3x^4 - 6x^3 - 5x + 10$; $x - 2$

20. $4x^4 - 15x^2 - 4$; $x - 2$

21. $3x^6 + 82x^3 + 27$; $x + 3$

22. $2x^6 - 18x^4 + x^2 - 9$; $x + 3$

23. $4x^6 - 64x^4 + x^2 - 15$; $x + 4$

24. $x^6 - 16x^4 + x^2 - 16$; $x + 4$

25. $2x^4 - x^3 + 2x - 1$; $x - \dfrac{1}{2}$

26. $3x^4 + x^3 - 3x + 1$; $x + \dfrac{1}{3}$

Applications and Extensions

27. Find the sum of $a, b, c,$ and d if

$$\frac{x^3 - 2x^2 + 3x + 5}{x + 2} = ax^2 + bx + c + \frac{d}{x + 2}$$

Explaining Concepts: Discussion and Writing

28. When dividing a polynomial by $x - c$, do you prefer to use long division or synthetic division? Does the value of c make a difference to you in choosing? Give reasons.

A.6   Rational Expressions

OBJECTIVES   **1** Reduce a Rational Expression to Lowest Terms (p. A46)

2 Multiply and Divide Rational Expressions (p. A47)

3 Add and Subtract Rational Expressions (p. A48)

4 Use the Least Common Multiple Method (p. A50)

5 Simplify Complex Rational Expressions (p. A52)

1   Reduce a Rational Expression to Lowest Terms

If we form the quotient of two polynomials, the result is called a **rational expression.** Some examples of rational expressions are

(a) $\dfrac{x^3 + 1}{x}$ (b) $\dfrac{3x^2 + x - 2}{x^2 + 5}$ (c) $\dfrac{x}{x^2 - 1}$ (d) $\dfrac{xy^2}{(x - y)^2}$

Expressions (a), (b), and (c) are rational expressions in one variable, x, whereas (d) is a rational expression in two variables, x and y.

Rational expressions are described in the same manner as rational numbers. In expression (a), the polynomial $x^3 + 1$ is called the **numerator,** and x is called the **denominator.** When the numerator and denominator of a rational expression contain no common factors (except 1 and -1), we say that the rational expression is **reduced to lowest terms,** or **simplified.**

The polynomial in the denominator of a rational expression cannot be equal to 0 because division by 0 is not defined. For example, for the expression $\dfrac{x^3 + 1}{x}$, x cannot take on the value 0. The domain of the variable x is $\{x | x \neq 0\}$.

A rational expression is reduced to lowest terms by factoring completely the numerator and the denominator and canceling any common factors by using the Cancellation Property:

$$\frac{a\cancel{c}}{b\cancel{c}} = \frac{a}{b} \qquad \text{if } b \neq 0, c \neq 0 \tag{1}$$

EXAMPLE 1   **Reducing a Rational Expression to Lowest Terms**

Reduce to lowest terms: $\dfrac{x^2 + 4x + 4}{x^2 + 3x + 2}$

Solution   We begin by factoring the numerator and the denominator.

$$x^2 + 4x + 4 = (x + 2)(x + 2)$$
$$x^2 + 3x + 2 = (x + 2)(x + 1)$$

WARNING Apply the Cancellation Property only to rational expressions written in factored form. Be sure to cancel only common factors, not common terms! ∎

Since a common factor, $x + 2$, appears, the original expression is not in lowest terms. To reduce it to lowest terms, we use the Cancellation Property:

$$\frac{x^2 + 4x + 4}{x^2 + 3x + 2} = \frac{\cancel{(x + 2)}(x + 2)}{\cancel{(x + 2)}(x + 1)} = \frac{x + 2}{x + 1} \qquad x \neq -2, -1$$

EXAMPLE 2   **Reducing Rational Expressions to Lowest Terms**

Reduce each rational expression to lowest terms.

(a) $\dfrac{x^3 - 8}{x^3 - 2x^2}$

(b) $\dfrac{8 - 2x}{x^2 - x - 12}$

Solution   (a) $\dfrac{x^3 - 8}{x^3 - 2x^2} = \dfrac{\cancel{(x - 2)}(x^2 + 2x + 4)}{x^2 \cancel{(x - 2)}} = \dfrac{x^2 + 2x + 4}{x^2} \qquad x \neq 0, 2$

(b) $\dfrac{8 - 2x}{x^2 - x - 12} = \dfrac{2(4 - x)}{(x - 4)(x + 3)} = \dfrac{2(-1)\cancel{(x - 4)}}{\cancel{(x - 4)}(x + 3)} = \dfrac{-2}{x + 3} \qquad x \neq -3, 4$

━━━ **Now Work** PROBLEM 5

2 Multiply and Divide Rational Expressions

The rules for multiplying and dividing rational expressions are the same as the rules for multiplying and dividing rational numbers. If $\dfrac{a}{b}$ and $\dfrac{c}{d}$, $b \neq 0$, $d \neq 0$, are two rational expressions, then

$$\frac{a}{b} \cdot \frac{c}{d} = \frac{ac}{bd} \qquad \text{if } b \neq 0, d \neq 0 \qquad\qquad (2)$$

$$\frac{\dfrac{a}{b}}{\dfrac{c}{d}} = \frac{a}{b} \cdot \frac{d}{c} = \frac{ad}{bc} \qquad \text{if } b \neq 0, c \neq 0, d \neq 0 \qquad\qquad (3)$$

In using equations (2) and (3) with rational expressions, be sure first to factor each polynomial completely so that common factors can be canceled. Leave your answer in factored form.

EXAMPLE 3   **Multiplying and Dividing Rational Expressions**

Perform the indicated operation and simplify the result. Leave your answer in factored form.

(a) $\dfrac{x^2 - 2x + 1}{x^3 + x} \cdot \dfrac{4x^2 + 4}{x^2 + x - 2}$

(b) $\dfrac{\dfrac{x + 3}{x^2 - 4}}{\dfrac{x^2 - x - 12}{x^3 - 8}}$

Solution   (a) $\dfrac{x^2 - 2x + 1}{x^3 + x} \cdot \dfrac{4x^2 + 4}{x^2 + x - 2} = \dfrac{(x-1)^2}{x(x^2+1)} \cdot \dfrac{4(x^2+1)}{(x+2)(x-1)}$

$$= \dfrac{(x-1)^2(4)\cancel{(x^2+1)}}{x\cancel{(x^2+1)}(x+2)\cancel{(x-1)}}$$

$$= \dfrac{4(x-1)}{x(x+2)} \qquad x \neq -2, 0, 1$$

(b) $\dfrac{\dfrac{x+3}{x^2-4}}{\dfrac{x^2-x-12}{x^3-8}} = \dfrac{x+3}{x^2-4} \cdot \dfrac{x^3-8}{x^2-x-12}$

$$= \dfrac{x+3}{(x-2)(x+2)} \cdot \dfrac{(x-2)(x^2+2x+4)}{(x-4)(x+3)}$$

$$= \dfrac{\cancel{(x+3)}\cancel{(x-2)}(x^2+2x+4)}{\cancel{(x-2)}(x+2)(x-4)\cancel{(x+3)}}$$

$$= \dfrac{x^2+2x+4}{(x+2)(x-4)} \qquad x \neq -3, -2, 2, 4$$

Now Work   PROBLEMS **17** AND **25**

3 Add and Subtract Rational Expressions

The rules for adding and subtracting rational expressions are the same as the rules for adding and subtracting rational numbers. So, if the denominators of two rational expressions to be added (or subtracted) are equal, we add (or subtract) the numerators and keep the common denominator.

In Words

To add (or subtract) two rational expressions with the same denominator, keep the common denominator and add (or subtract) the numerators.

If $\dfrac{a}{b}$ and $\dfrac{c}{b}$ are two rational expressions, then

$$\dfrac{a}{b} + \dfrac{c}{b} = \dfrac{a+c}{b} \qquad \dfrac{a}{b} - \dfrac{c}{b} = \dfrac{a-c}{b} \qquad \text{if } b \neq 0 \qquad \textbf{(4)}$$

EXAMPLE 4   **Adding and Subtracting Rational Expressions with Equal Denominators**

Perform the indicated operation and simplify the result. Leave your answer in factored form.

(a) $\dfrac{2x^2 - 4}{2x + 5} + \dfrac{x + 3}{2x + 5} \qquad x \neq -\dfrac{5}{2}$ (b) $\dfrac{x}{x - 3} - \dfrac{3x + 2}{x - 3} \qquad x \neq 3$

Solution   (a) $\dfrac{2x^2 - 4}{2x + 5} + \dfrac{x + 3}{2x + 5} = \dfrac{(2x^2 - 4) + (x + 3)}{2x + 5}$

$$= \dfrac{2x^2 + x - 1}{2x + 5} = \dfrac{(2x - 1)(x + 1)}{2x + 5}$$

(b) $\dfrac{x}{x - 3} - \dfrac{3x + 2}{x - 3} = \dfrac{x - (3x + 2)}{x - 3} = \dfrac{x - 3x - 2}{x - 3}$

$$= \dfrac{-2x - 2}{x - 3} = \dfrac{-2(x + 1)}{x - 3}$$

| EXAMPLE 5 | **Adding Rational Expressions Whose Denominators Are Additive Inverses of Each Other** |

Perform the indicated operation and simplify the result. Leave your answer in factored form.

$$\frac{2x}{x-3} + \frac{5}{3-x} \qquad x \neq 3$$

Solution Notice that the denominators of the two rational expressions are different. However, the denominator of the second expression is the additive inverse of the denominator of the first. That is,

$$3 - x = -x + 3 = -1 \cdot (x - 3) = -(x - 3)$$

Then

$$\frac{2x}{x-3} + \frac{5}{3-x} \underset{\substack{\uparrow \\ 3-x = -(x-3)}}{=} \frac{2x}{x-3} + \frac{5}{-(x-3)} \underset{\substack{\uparrow \\ \frac{a}{-b} = \frac{-a}{b}}}{=} \frac{2x}{x-3} + \frac{-5}{x-3}$$

$$= \frac{2x + (-5)}{x-3} = \frac{2x-5}{x-3}$$

●

────**Now Work** PROBLEMS 37 AND 43

If the denominators of two rational expressions to be added or subtracted are not equal, we can use the general formulas for adding and subtracting rational expressions.

$$\frac{a}{b} + \frac{c}{d} = \frac{a \cdot d}{b \cdot d} + \frac{b \cdot c}{b \cdot d} = \frac{ad + bc}{bd} \qquad \text{if } b \neq 0, d \neq 0 \qquad \textbf{(5a)}$$

$$\frac{a}{b} - \frac{c}{d} = \frac{a \cdot d}{b \cdot d} - \frac{b \cdot c}{b \cdot d} = \frac{ad - bc}{bd} \qquad \text{if } b \neq 0, d \neq 0 \qquad \textbf{(5b)}$$

| EXAMPLE 6 | **Adding and Subtracting Rational Expressions with Unequal Denominators** |

Perform the indicated operation and simplify the result. Leave your answer in factored form.

(a) $\dfrac{x-3}{x+4} + \dfrac{x}{x-2} \qquad x \neq -4, 2$ (b) $\dfrac{x^2}{x^2-4} - \dfrac{1}{x} \qquad x \neq -2, 0, 2$

Solution (a) $\dfrac{x-3}{x+4} + \dfrac{x}{x-2} = \dfrac{x-3}{x+4} \cdot \dfrac{x-2}{x-2} + \dfrac{x+4}{x+4} \cdot \dfrac{x}{x-2}$

$$\underset{(5a)}{\uparrow}$$

$$= \frac{(x-3)(x-2) + (x+4)(x)}{(x+4)(x-2)}$$

$$= \frac{x^2 - 5x + 6 + x^2 + 4x}{(x+4)(x-2)} = \frac{2x^2 - x + 6}{(x+4)(x-2)}$$

(b) $\dfrac{x^2}{x^2 - 4} - \dfrac{1}{x} = \dfrac{x^2}{x^2 - 4} \cdot \dfrac{x}{x} - \dfrac{x^2 - 4}{x^2 - 4} \cdot \dfrac{1}{x} = \dfrac{x^2(x) - (x^2 - 4)(1)}{(x^2 - 4)(x)}$

$\uparrow$

(5b)

$= \dfrac{x^3 - x^2 + 4}{x(x - 2)(x + 2)}$

━━━━**Now Work** PROBLEM 47

4 Use the Least Common Multiple Method

If the denominators of two rational expressions to be added (or subtracted) have common factors, we usually do not use the general rules given by equations (5a) and (5b). Just as with fractions, we apply the **least common multiple (LCM) method.** The LCM method uses the polynomial of least degree that has each denominator polynomial as a factor.

The LCM Method for Adding or Subtracting Rational Expressions

The Least Common Multiple (LCM) Method requires four steps:

STEP 1: Factor completely the polynomial in the denominator of each rational expression.

STEP 2: The LCM of the denominator is the product of each of these factors raised to a power equal to the greatest number of times that the factor occurs in the polynomials.

STEP 3: Write each rational expression using the LCM as the common denominator.

STEP 4: Add or subtract the rational expressions using equation (4).

Let's work an example that only requires Steps 1 and 2.

EXAMPLE 7

Finding the Least Common Multiple

Find the least common multiple of the following pair of polynomials:

$$x(x - 1)^2(x + 1) \quad \text{and} \quad 4(x - 1)(x + 1)^3$$

Solution

STEP 1: The polynomials are already factored completely as

$$x(x - 1)^2(x + 1) \quad \text{and} \quad 4(x - 1)(x + 1)^3$$

STEP 2: Start by writing the factors of the left-hand polynomial. (Or you could start with the one on the right.)

$$x(x - 1)^2(x + 1)$$

Now look at the right-hand polynomial. Its first factor, 4, does not appear in the polynomial above, so we insert it.

$$4x(x - 1)^2(x + 1)$$

The next factor, $x - 1$, is already in our polynomial, so no change is necessary. The final factor is $(x + 1)^3$. Since our polynomial has $x + 1$ to the first power only, we replace $x + 1$ in the polynomial by $(x + 1)^3$. The LCM is

$$4x(x - 1)^2(x + 1)^3$$

Notice that the LCM is, in fact, the polynomial of least degree that contains $x(x-1)^2(x+1)$ and $4(x-1)(x+1)^3$ as factors.

Now Work PROBLEM 53

EXAMPLE 8

Using the Least Common Multiple to Add Rational Expressions

Perform the indicated operation and simplify the result. Leave your answer in factored form.

$$\frac{x}{x^2 + 3x + 2} + \frac{2x - 3}{x^2 - 1} \qquad x \neq -2, -1, 1$$

Solution

STEP 1: Factor completely the polynomials in the denominators.

$$x^2 + 3x + 2 = (x + 2)(x + 1)$$
$$x^2 - 1 = (x - 1)(x + 1)$$

STEP 2: The LCM is $(x + 2)(x + 1)(x - 1)$. Do you see why?

STEP 3: Write each rational expression using the LCM as the denominator.

$$\frac{x}{x^2 + 3x + 2} = \frac{x}{(x + 2)(x + 1)} = \frac{x}{(x + 2)(x + 1)} \cdot \frac{x - 1}{x - 1} = \frac{x(x - 1)}{(x + 2)(x + 1)(x - 1)}$$

Multiply numerator and denominator by $x - 1$ to get the LCM in the denominator.

$$\frac{2x - 3}{x^2 - 1} = \frac{2x - 3}{(x - 1)(x + 1)} = \frac{2x - 3}{(x - 1)(x + 1)} \cdot \frac{x + 2}{x + 2} = \frac{(2x - 3)(x + 2)}{(x - 1)(x + 1)(x + 2)}$$

Multiply numerator and denominator by $x + 2$ to get the LCM in the denominator.

STEP 4: Now we can add by using equation (4).

$$\frac{x}{x^2 + 3x + 2} + \frac{2x - 3}{x^2 - 1} = \frac{x(x - 1)}{(x + 2)(x + 1)(x - 1)} + \frac{(2x - 3)(x + 2)}{(x + 2)(x + 1)(x - 1)}$$

$$= \frac{(x^2 - x) + (2x^2 + x - 6)}{(x + 2)(x + 1)(x - 1)}$$

$$= \frac{3x^2 - 6}{(x + 2)(x + 1)(x - 1)} = \frac{3(x^2 - 2)}{(x + 2)(x + 1)(x - 1)}$$

EXAMPLE 9

Using the Least Common Multiple to Subtract Rational Expressions

Perform the indicated operations and simplify the result. Leave your answer in factored form.

$$\frac{3}{x^2 + x} - \frac{x + 4}{x^2 + 2x + 1} \qquad x \neq -1, 0$$

Solution

STEP 1: Factor completely the polynomials in the denominators.

$$x^2 + x = x(x + 1)$$
$$x^2 + 2x + 1 = (x + 1)^2$$

STEP 2: The LCM is $x(x + 1)^2$.

STEP 3: Write each rational expression using the LCM as the denominator.

$$\frac{3}{x^2 + x} = \frac{3}{x(x+1)} = \frac{3}{x(x+1)} \cdot \frac{x+1}{x+1} = \frac{3(x+1)}{x(x+1)^2}$$

$$\frac{x+4}{x^2 + 2x + 1} = \frac{x+4}{(x+1)^2} = \frac{x+4}{(x+1)^2} \cdot \frac{x}{x} = \frac{x(x+4)}{x(x+1)^2}$$

STEP 4: Subtract, using equation (4).

$$\frac{3}{x^2 + x} - \frac{x+4}{x^2 + 2x + 1} = \frac{3(x+1)}{x(x+1)^2} - \frac{x(x+4)}{x(x+1)^2}$$

$$= \frac{3(x+1) - x(x+4)}{x(x+1)^2}$$

$$= \frac{3x + 3 - x^2 - 4x}{x(x+1)^2}$$

$$= \frac{-x^2 - x + 3}{x(x+1)^2}$$

━━━ **Now Work** PROBLEM 63

5 Simplify Complex Rational Expressions

When sums and/or differences of rational expressions appear as the numerator and/or denominator of a quotient, the quotient is called a **complex rational expression.*** For example,

$$\frac{1 + \dfrac{1}{x}}{1 - \dfrac{1}{x}} \quad \text{and} \quad \frac{\dfrac{x^2}{x^2 - 4} - 3}{\dfrac{x-3}{x+2} - 1}$$

are complex rational expressions. To **simplify** a complex rational expression means to write it as a rational expression reduced to lowest terms. This can be accomplished in either of two ways.

> ### Simplifying a Complex Rational Expression
>
> **METHOD 1:** Treat the numerator and denominator of the complex rational expression separately, performing whatever operations are indicated and simplifying the results. Follow this by simplifying the resulting rational expression.
>
> **METHOD 2:** Find the LCM of the denominators of all rational expressions that appear in the complex rational expression. Multiply the numerator and denominator of the complex rational expression by the LCM and simplify the result.

We will use both methods in the next example. By carefully studying each method, you can discover situations in which one method may be easier to use than the other.

EXAMPLE 10 **Simplifying a Complex Rational Expression**

Simplify: $\dfrac{\dfrac{1}{2} + \dfrac{3}{x}}{\dfrac{x+3}{4}}$ $x \neq -3, 0$

*Some texts use the term **complex fraction.**

Solution   **Method 1:** First, we perform the indicated operation in the numerator, and then we divide.

$$\frac{\dfrac{1}{2}+\dfrac{3}{x}}{\dfrac{x+3}{4}}=\frac{\dfrac{1\cdot x+2\cdot 3}{2\cdot x}}{\dfrac{x+3}{4}}=\frac{\dfrac{x+6}{2x}}{\dfrac{x+3}{4}}=\frac{x+6}{2x}\cdot\frac{4}{x+3}$$

Rule for adding quotients   *Rule for dividing quotients*

$$=\frac{(x+6)\cdot 4}{2\cdot x\cdot(x+3)}=\frac{2\cdot 2\cdot(x+6)}{2\cdot x\cdot(x+3)}=\frac{2(x+6)}{x(x+3)}$$

Rule for multiplying quotients

Method 2: The rational expressions that appear in the complex rational expression are

$$\frac{1}{2}\qquad\frac{3}{x}\qquad\frac{x+3}{4}$$

The LCM of their denominators is $4x$. We multiply the numerator and denominator of the complex rational expression by $4x$ and then simplify.

$$\frac{\dfrac{1}{2}+\dfrac{3}{x}}{\dfrac{x+3}{4}}=\frac{4x\cdot\left(\dfrac{1}{2}+\dfrac{3}{x}\right)}{4x\cdot\left(\dfrac{x+3}{4}\right)}=\frac{4x\cdot\dfrac{1}{2}+4x\cdot\dfrac{3}{x}}{\dfrac{4x\cdot(x+3)}{4}}$$

Multiply the   *Use the Distributive Property*
numerator and   *in the numerator.*
denominator by 4x.

$$=\frac{2\cdot 2x\cdot\dfrac{1}{2}+4x\cdot\dfrac{3}{x}}{\dfrac{4x\cdot(x+3)}{4}}=\frac{2x+12}{x(x+3)}=\frac{2(x+6)}{x(x+3)}$$

Simplify.   *Factor.*

EXAMPLE 11   **Simplifying a Complex Rational Expression**

Simplify: $\dfrac{\dfrac{x^2}{x-4}+2}{\dfrac{2x-2}{x}-1}\qquad x\ne 0,2,4$

Solution   We will use Method 1.

$$\frac{\dfrac{x^2}{x-4}+2}{\dfrac{2x-2}{x}-1}=\frac{\dfrac{x^2}{x-4}+\dfrac{2(x-4)}{x-4}}{\dfrac{2x-2}{x}-\dfrac{x}{x}}=\frac{\dfrac{x^2+2x-8}{x-4}}{\dfrac{2x-2-x}{x}}$$

$$=\frac{\dfrac{(x+4)(x-2)}{x-4}}{\dfrac{x-2}{x}}=\frac{(x+4)\cancel{(x-2)}}{x-4}\cdot\frac{x}{\cancel{x-2}}=\frac{(x+4)\cdot x}{x-4}$$

Now Work PROBLEM 73

Application

EXAMPLE 12

Solving an Application in Electricity

Figure 26

An electrical circuit contains two resistors connected in parallel, as shown in Figure 26. If the resistance of each is R_1 and R_2 ohms, respectively, their combined resistance R is given by the formula

$$R = \frac{1}{\dfrac{1}{R_1} + \dfrac{1}{R_2}}$$

Express R as a rational expression; that is, simplify the right-hand side of this formula. Evaluate the rational expression if $R_1 = 6$ ohms and $R_2 = 10$ ohms.

Solution

We will use Method 2. If we consider 1 as the fraction $\dfrac{1}{1}$, then the rational expressions in the complex rational expression are

$$\frac{1}{1} \qquad \frac{1}{R_1} \qquad \frac{1}{R_2}$$

The LCM of the denominators is $R_1 R_2$. We multiply the numerator and denominator of the complex rational expression by $R_1 R_2$ and simplify.

$$R = \frac{1}{\dfrac{1}{R_1} + \dfrac{1}{R_2}} = \frac{1 \cdot R_1 R_2}{\left(\dfrac{1}{R_1} + \dfrac{1}{R_2}\right) \cdot R_1 R_2} = \frac{R_1 R_2}{\dfrac{1}{R_1} \cdot R_1 R_2 + \dfrac{1}{R_2} \cdot R_1 R_2} = \frac{R_1 R_2}{R_2 + R_1}$$

Thus,

$$R = \frac{R_1 R_2}{R_2 + R_1}$$

If $R_1 = 6$ and $R_2 = 10$, then

$$R = \frac{6 \cdot 10}{10 + 6} = \frac{60}{16} = \frac{15}{4} \text{ ohms}$$

●

A.6 Assess Your Understanding

Concepts and Vocabulary

1. When the numerator and denominator of a rational expression contain no common factors (except 1 and -1), the rational expression is in _____ _____.

2. LCM is an abbreviation for _____ _____ _____.

3. *True or False* The rational expression $\dfrac{2x^3 - 4x}{x - 2}$ is reduced to lowest terms.

4. *True or False* The LCM of $2x^3 + 6x^2$ and $6x^4 + 4x^3$ is $4x^3(x + 1)$.

Skill Building

In Problems 5–16, reduce each rational expression to lowest terms.

5. $\dfrac{3x + 9}{x^2 - 9}$

6. $\dfrac{4x^2 + 8x}{12x + 24}$

7. $\dfrac{x^2 - 2x}{3x - 6}$

8. $\dfrac{15x^2 + 24x}{3x^2}$

9. $\dfrac{24x^2}{12x^2 - 6x}$

10. $\dfrac{x^2 + 4x + 4}{x^2 - 4}$

11. $\dfrac{y^2 - 25}{2y^2 - 8y - 10}$

12. $\dfrac{3y^2 - y - 2}{3y^2 + 5y + 2}$

13. $\dfrac{x^2 + 4x - 5}{x^2 - 2x + 1}$

14. $\dfrac{x - x^2}{x^2 + x - 2}$

15. $\dfrac{x^2 + 5x - 14}{2 - x}$

16. $\dfrac{2x^2 + 5x - 3}{1 - 2x}$

In Problems 17–34, perform the indicated operation and simplify the result. Leave your answer in factored form.

17. $\dfrac{3x+6}{5x^2} \cdot \dfrac{x}{x^2-4}$

18. $\dfrac{3}{2x} \cdot \dfrac{x^2}{6x+10}$

19. $\dfrac{4x^2}{x^2-16} \cdot \dfrac{x^3-64}{2x}$

20. $\dfrac{12}{x^2+x} \cdot \dfrac{x^3+1}{4x-2}$

21. $\dfrac{4x-8}{-3x} \cdot \dfrac{12}{12-6x}$

22. $\dfrac{6x-27}{5x} \cdot \dfrac{2}{4x-18}$

23. $\dfrac{x^2-3x-10}{x^2+2x-35} \cdot \dfrac{x^2+4x-21}{x^2+9x+14}$

24. $\dfrac{x^2+x-6}{x^2+4x-5} \cdot \dfrac{x^2-25}{x^2+2x-15}$

25. $\dfrac{\dfrac{6x}{x^2-4}}{\dfrac{3x-9}{2x+4}}$

26. $\dfrac{\dfrac{12x}{5x+20}}{\dfrac{4x^2}{x^2-16}}$

27. $\dfrac{\dfrac{8x}{x^2-1}}{\dfrac{10x}{x+1}}$

28. $\dfrac{\dfrac{x-2}{4x}}{\dfrac{x^2-4x+4}{12x}}$

29. $\dfrac{\dfrac{4-x}{4+x}}{\dfrac{4x}{x^2-16}}$

30. $\dfrac{\dfrac{3+x}{3-x}}{\dfrac{x^2-9}{9x^3}}$

31. $\dfrac{\dfrac{x^2+7x+12}{x^2-7x+12}}{\dfrac{x^2+x-12}{x^2-x-12}}$

32. $\dfrac{\dfrac{x^2+7x+6}{x^2+x-6}}{\dfrac{x^2+5x-6}{x^2+5x+6}}$

33. $\dfrac{\dfrac{2x^2-x-28}{3x^2-x-2}}{\dfrac{4x^2+16x+7}{3x^2+11x+6}}$

34. $\dfrac{\dfrac{9x^2+3x-2}{12x^2+5x-2}}{\dfrac{9x^2-6x+1}{8x^2-10x-3}}$

In Problems 35–52, perform the indicated operation and simplify the result. Leave your answer in factored form.

35. $\dfrac{x}{2} + \dfrac{5}{2}$

36. $\dfrac{3}{x} - \dfrac{6}{x}$

37. $\dfrac{x^2}{2x-3} - \dfrac{4}{2x-3}$

38. $\dfrac{3x^2}{2x-1} - \dfrac{9}{2x-1}$

39. $\dfrac{x+1}{x-3} + \dfrac{2x-3}{x-3}$

40. $\dfrac{2x-5}{3x+2} + \dfrac{x+4}{3x+2}$

41. $\dfrac{3x+5}{2x-1} - \dfrac{2x-4}{2x-1}$

42. $\dfrac{5x-4}{3x+4} - \dfrac{x+1}{3x+4}$

43. $\dfrac{4}{x-2} + \dfrac{x}{2-x}$

44. $\dfrac{6}{x-1} - \dfrac{x}{1-x}$

45. $\dfrac{4}{x-1} - \dfrac{2}{x+2}$

46. $\dfrac{2}{x+5} - \dfrac{5}{x-5}$

47. $\dfrac{x}{x+1} + \dfrac{2x-3}{x-1}$

48. $\dfrac{3x}{x-4} + \dfrac{2x}{x+3}$

49. $\dfrac{x-3}{x+2} - \dfrac{x+4}{x-2}$

50. $\dfrac{2x-3}{x-1} - \dfrac{2x+1}{x+1}$

51. $\dfrac{x}{x^2-4} + \dfrac{1}{x}$

52. $\dfrac{x-1}{x^3} + \dfrac{x}{x^2+1}$

In Problems 53–60, find the LCM of the given polynomials.

53. $x^2-4, \quad x^2-x-2$

54. $x^2-x-12, \quad x^2-8x+16$

55. $x^3-x, \quad x^2-x$

56. $3x^2-27, \quad 2x^2-x-15$

57. $4x^3-4x^2+x, \quad 2x^3-x^2, \quad x^3$

58. $x-3, \quad x^2+3x, \quad x^3-9x$

59. $x^3-x, \quad x^3-2x^2+x, \quad x^3-1$

60. $x^2+4x+4, \quad x^3+2x^2, \quad (x+2)^3$

In Problems 61–72, perform the indicated operations and simplify the result. Leave your answer in factored form.

61. $\dfrac{x}{x^2-7x+6} - \dfrac{x}{x^2-2x-24}$

62. $\dfrac{x}{x-3} - \dfrac{x+1}{x^2+5x-24}$

63. $\dfrac{4x}{x^2-4} - \dfrac{2}{x^2+x-6}$

64. $\dfrac{3x}{x-1} - \dfrac{x-4}{x^2-2x+1}$

65. $\dfrac{3}{(x-1)^2(x+1)} + \dfrac{2}{(x-1)(x+1)^2}$

66. $\dfrac{2}{(x+2)^2(x-1)} - \dfrac{6}{(x+2)(x-1)^2}$

67. $\dfrac{x+4}{x^2-x-2} - \dfrac{2x+3}{x^2+2x-8}$

68. $\dfrac{2x-3}{x^2+8x+7} - \dfrac{x-2}{(x+1)^2}$

69. $\dfrac{1}{x} - \dfrac{2}{x^2+x} + \dfrac{3}{x^3-x^2}$

70. $\dfrac{x}{(x-1)^2} + \dfrac{2}{x} - \dfrac{x+1}{x^3-x^2}$

71. $\dfrac{1}{h}\left(\dfrac{1}{x+h} - \dfrac{1}{x}\right)$

72. $\dfrac{1}{h}\left[\dfrac{1}{(x+h)^2} - \dfrac{1}{x^2}\right]$

In Problems 73–84, perform the indicated operations and simplify the result. Leave your answer in factored form.

73. $\dfrac{1 + \dfrac{1}{x}}{1 - \dfrac{1}{x}}$

74. $\dfrac{4 + \dfrac{1}{x^2}}{3 - \dfrac{1}{x^2}}$

75. $\dfrac{2 - \dfrac{x+1}{x}}{3 + \dfrac{x-1}{x+1}}$

76. $\dfrac{1 - \dfrac{x}{x+1}}{2 - \dfrac{x-1}{x}}$

77. $\dfrac{\dfrac{x+4}{x-2} - \dfrac{x-3}{x+1}}{x+1}$

78. $\dfrac{\dfrac{x-2}{x+1} - \dfrac{x}{x-2}}{x+3}$

79. $\dfrac{\dfrac{x-2}{x+2} + \dfrac{x-1}{x+1}}{\dfrac{x}{x+1} - \dfrac{2x-3}{x}}$

80. $\dfrac{\dfrac{2x+5}{x} - \dfrac{x}{x-3}}{\dfrac{x^2}{x-3} - \dfrac{(x+1)^2}{x+3}}$

81. $1 - \dfrac{1}{1 - \dfrac{1}{x}}$

82. $1 - \dfrac{1}{1 - \dfrac{1}{1-x}}$

83. $\dfrac{2(x-1)^{-1} + 3}{3(x-1)^{-1} + 2}$

84. $\dfrac{4(x+2)^{-1} - 3}{3(x+2)^{-1} - 1}$

In Problems 85–92, expressions that occur in calculus are given. Reduce each expression to lowest terms.

85. $\dfrac{(2x+3)\cdot 3 - (3x-5)\cdot 2}{(3x-5)^2}$

86. $\dfrac{(4x+1)\cdot 5 - (5x-2)\cdot 4}{(5x-2)^2}$

87. $\dfrac{x\cdot 2x - (x^2+1)\cdot 1}{(x^2+1)^2}$

88. $\dfrac{x\cdot 2x - (x^2-4)\cdot 1}{(x^2-4)^2}$

89. $\dfrac{(3x+1)\cdot 2x - x^2\cdot 3}{(3x+1)^2}$

90. $\dfrac{(2x-5)\cdot 3x^2 - x^3\cdot 2}{(2x-5)^2}$

91. $\dfrac{(x^2+1)\cdot 3 - (3x+4)\cdot 2x}{(x^2+1)^2}$

92. $\dfrac{(x^2+9)\cdot 2 - (2x-5)\cdot 2x}{(x^2+9)^2}$

Applications and Extensions

93. **The Lensmaker's Equation** The focal length f of a lens with index of refraction n is

$$\frac{1}{f} = (n-1)\left[\frac{1}{R_1} + \frac{1}{R_2}\right]$$

where R_1 and R_2 are the radii of curvature of the front and back surfaces of the lens. Express f as a rational expression. Evaluate the rational expression for $n = 1.5$, $R_1 = 0.1$ meter, and $R_2 = 0.2$ meter.

94. **Electrical Circuits** An electrical circuit contains three resistors connected in parallel. If the resistance of each is R_1, R_2, and R_3 ohms, respectively, their combined resistance R is given by the formula

$$\frac{1}{R} = \frac{1}{R_1} + \frac{1}{R_2} + \frac{1}{R_3}$$

Express R as a rational expression. Evaluate R for $R_1 = 5$ ohms, $R_2 = 4$ ohms, and $R_3 = 10$ ohms.

Explaining Concepts: Discussion and Writing

95. The following expressions are called **continued fractions:**

$$1 + \frac{1}{x} \quad 1 + \frac{1}{1 + \dfrac{1}{x}} \quad 1 + \frac{1}{1 + \dfrac{1}{1 + \dfrac{1}{x}}} \quad 1 + \frac{1}{1 + \dfrac{1}{1 + \dfrac{1}{1 + \dfrac{1}{x}}}} \quad \cdots$$

Each simplifies to an expression of the form

$$\frac{ax + b}{bx + c}$$

Trace the successive values of a, b, and c as you "continue" the fraction. Can you discover the patterns that these values follow? Go to the library and research Fibonacci numbers. Write a report on your findings.

96. Explain to a fellow student when you would use the LCM method to add two rational expressions. Give two examples of adding two rational expressions, one in which you use the LCM and the other in which you do not.

97. Which of the two methods given in the text for simplifying complex rational expressions do you prefer? Write a brief paragraph stating the reasons for your choice.

A.7 *n*th Roots; Rational Exponents

PREPARING FOR THIS SECTION *Before getting started, review the following:*

- Exponents, Square Roots (Appendix A, Section A.1, pp. A7–A10)

Now Work the 'Are You Prepared?' problems on page A62.

OBJECTIVES 1 Work with *n*th Roots (p. A57)
2 Simplify Radicals (p. A58)
3 Rationalize Denominators (p. A59)
4 Simplify Expressions with Rational Exponents (p. A60)

1 Work with *n*th Roots

DEFINITION

The **principal *n*th root of a real number *a*,** $n \geq 2$ an integer, symbolized by $\sqrt[n]{a}$, is defined as follows:

$$\sqrt[n]{a} = b \quad \text{means} \quad a = b^n$$

where $a \geq 0$ and $b \geq 0$ if *n* is even and *a*, *b* are any real numbers if *n* is odd.

In Words
The symbol $\sqrt[n]{a}$ means "give me the number, when raised to the *n*th power, that results in *a*."

Notice that if *a* is negative and *n* is even then $\sqrt[n]{a}$ is not defined. When it is defined, the principal *n*th root of a number is unique.

The symbol $\sqrt[n]{a}$ for the principal *n*th root of *a* is called a **radical;** the integer *n* is called the **index,** and *a* is called the **radicand.** If the index of a radical is 2, we call $\sqrt[2]{a}$ the **square root** of *a* and omit the index 2 by writing $\sqrt{a}$. If the index is 3, we call $\sqrt[3]{a}$ the **cube root** of *a*.

EXAMPLE 1

Simplifying Principal *n*th Roots

(a) $\sqrt[3]{8} = \sqrt[3]{2^3} = 2$ The number whose cube is 8 is 2.   (b) $\sqrt[3]{-64} = \sqrt[3]{(-4)^3} = -4$

(c) $\sqrt[4]{\dfrac{1}{16}} = \sqrt[4]{\left(\dfrac{1}{2}\right)^4} = \dfrac{1}{2}$ (d) $\sqrt[6]{(-2)^6} = |-2| = 2$

These are examples of **perfect roots,** since each simplifies to a rational number. Notice the absolute value in Example 1(d). If *n* is even, the principal *n*th root must be nonnegative.

In general, if $n \geq 2$ is an integer and *a* is a real number, we have

$$\sqrt[n]{a^n} = a \qquad \text{if } n \geq 3 \text{ is odd} \tag{1a}$$
$$\sqrt[n]{a^n} = |a| \qquad \text{if } n \geq 2 \text{ is even} \tag{1b}$$

Now Work PROBLEM 7

Radicals provide a way of representing many irrational real numbers. For example, there is no rational number whose square is 2. Using radicals, we can say that $\sqrt{2}$ is the positive number whose square is 2.

EXAMPLE 2   **Using a Calculator to Approximate Roots**

Use a calculator to approximate $\sqrt[5]{16}$.

Solution   Figure 27 shows the result using a TI-84 Plus graphing calculator.

Figure 27

```
5×√16
        1.741101127
```

━━━━━▶ **Now Work** PROBLEM 101

2 **Simplify Radicals**

Let $n \geq 2$ and $m \geq 2$ denote positive integers, and let a and b represent real numbers. Assuming that all radicals are defined, we have the following properties:

> **Properties of Radicals**
>
> $$\sqrt[n]{ab} = \sqrt[n]{a}\ \sqrt[n]{b} \qquad\qquad \textbf{(2a)}$$
>
> $$\sqrt[n]{\dfrac{a}{b}} = \dfrac{\sqrt[n]{a}}{\sqrt[n]{b}} \qquad\qquad \textbf{(2b)}$$
>
> $$\sqrt[n]{a^m} = (\sqrt[n]{a})^m \qquad\qquad \textbf{(2c)}$$

When used in reference to radicals, the direction to "simplify" will mean to remove from the radicals any perfect roots that occur as factors.

EXAMPLE 3   **Simplifying Radicals**

(a) $\sqrt{32} = \sqrt{16 \cdot 2} = \sqrt{16} \cdot \sqrt{2} = 4\sqrt{2}$

    ↑   ↑
Factor out 16, (2a)
a perfect square.

(b) $\sqrt[3]{16} = \sqrt[3]{8 \cdot 2} = \sqrt[3]{8} \cdot \sqrt[3]{2} = \sqrt[3]{2^3} \cdot \sqrt[3]{2} = 2\sqrt[3]{2}$

    ↑   ↑
Factor out 8, (2a)
a perfect cube.

(c) $\sqrt[3]{-16x^4} = \sqrt[3]{-8 \cdot 2 \cdot x^3 \cdot x} = \sqrt[3]{(-8x^3)(2x)}$

    ↑   ↑
Factor perfect   Group perfect
cubes inside radical. cubes.

$$= \sqrt[3]{(-2x)^3 \cdot 2x} = \sqrt[3]{(-2x)^3} \cdot \sqrt[3]{2x} = -2x\sqrt[3]{2x}$$

                ↑
                (2a)

(d) $\sqrt[4]{\dfrac{16x^5}{81}} = \sqrt[4]{\dfrac{2^4 x^4 x}{3^4}} = \sqrt[4]{\left(\dfrac{2x}{3}\right)^4 \cdot x} = \sqrt[4]{\left(\dfrac{2x}{3}\right)^4} \cdot \sqrt[4]{x} = \left|\dfrac{2x}{3}\right| \sqrt[4]{x} = \dfrac{2}{3}|x|\sqrt[4]{x}$

━━━━━▶ **Now Work** PROBLEMS 11 AND 17

Two or more radicals can be added or subtracted, provided that they have the same index and the same radicand. Such radicals are called **like radicals.**

EXAMPLE 4   **Combining Like Radicals**

(a) $-8\sqrt{12} + \sqrt{3} = -8\sqrt{4 \cdot 3} + \sqrt{3}$

$\qquad\qquad\qquad = -8 \cdot \sqrt{4}\sqrt{3} + \sqrt{3}$

$\qquad\qquad\qquad = -16\sqrt{3} + \sqrt{3}$

$\qquad\qquad\qquad = -15\sqrt{3}$

(b) $\sqrt[3]{8x^4} + \sqrt[3]{-x} + 4\sqrt[3]{27x} = \sqrt[3]{2^3x^3x} + \sqrt[3]{-1 \cdot x} + 4\sqrt[3]{3^3x}$

$\qquad\qquad\qquad\qquad = \sqrt[3]{(2x)^3} \cdot \sqrt[3]{x} + \sqrt[3]{-1} \cdot \sqrt[3]{x} + 4\sqrt[3]{3^3} \cdot \sqrt[3]{x}$

$\qquad\qquad\qquad\qquad = 2x\sqrt[3]{x} - 1 \cdot \sqrt[3]{x} + 12\sqrt[3]{x}$

$\qquad\qquad\qquad\qquad = (2x + 11)\sqrt[3]{x}$

━━━**Now Work** PROBLEM 33

3 Rationalize Denominators

When radicals occur in quotients, it is customary to rewrite the quotient so that the new denominator contains no radicals. This process is referred to as **rationalizing the denominator.**

The idea is to multiply by an appropriate expression so that the new denominator contains no radicals. For example:

If a Denominator Contains the Factor	Multiply by	To Obtain a Denominator Free of Radicals
$\sqrt{3}$	$\sqrt{3}$	$(\sqrt{3})^2 = 3$
$\sqrt{3} + 1$	$\sqrt{3} - 1$	$(\sqrt{3})^2 - 1^2 = 3 - 1 = 2$
$\sqrt{2} - 3$	$\sqrt{2} + 3$	$(\sqrt{2})^2 - 3^2 = 2 - 9 = -7$
$\sqrt{5} - \sqrt{3}$	$\sqrt{5} + \sqrt{3}$	$(\sqrt{5})^2 - (\sqrt{3})^2 = 5 - 3 = 2$
$\sqrt[3]{4}$	$\sqrt[3]{2}$	$\sqrt[3]{4} \cdot \sqrt[3]{2} = \sqrt[3]{8} = 2$

In rationalizing the denominator of a quotient, be sure to multiply both the numerator and the denominator by the expression.

───────────────

EXAMPLE 5   **Rationalizing Denominators**

Rationalize the denominator of each expression:

(a) $\dfrac{1}{\sqrt{3}}$ (b) $\dfrac{5}{4\sqrt[3]{4}}$ (c) $\dfrac{\sqrt{2}}{\sqrt{3} - 3\sqrt{2}}$

Solution   (a) The denominator contains the factor $\sqrt{3}$, so we multiply the numerator and denominator by $\sqrt{3}$ to obtain

$$\frac{1}{\sqrt{3}} = \frac{1}{\sqrt{3}} \cdot \frac{\sqrt{3}}{\sqrt{3}} = \frac{\sqrt{3}}{(\sqrt{3})^2} = \frac{\sqrt{3}}{3}$$

(b) The denominator contains the factor $\sqrt[3]{4}$, so we multiply the numerator and denominator by $\sqrt[3]{2}$ to obtain

$$\frac{5}{4\sqrt[3]{4}} = \frac{5}{4\sqrt[3]{4}} \cdot \frac{\sqrt[3]{2}}{\sqrt[3]{2}} = \frac{5\sqrt[3]{2}}{4\sqrt[3]{8}} = \frac{5\sqrt[3]{2}}{8}$$

(c) The denominator contains the factor $\sqrt{3} - 3\sqrt{2}$, so we multiply the numerator and denominator by $\sqrt{3} + 3\sqrt{2}$ to obtain

$$\frac{\sqrt{2}}{\sqrt{3} - 3\sqrt{2}} = \frac{\sqrt{2}}{\sqrt{3} - 3\sqrt{2}} \cdot \frac{\sqrt{3} + 3\sqrt{2}}{\sqrt{3} + 3\sqrt{2}} = \frac{\sqrt{2}(\sqrt{3} + 3\sqrt{2})}{(\sqrt{3})^2 - (3\sqrt{2})^2}$$

$$= \frac{\sqrt{2}\sqrt{3} + 3(\sqrt{2})^2}{3 - 18} = \frac{\sqrt{6} + 6}{-15} = -\frac{6 + \sqrt{6}}{15}$$

━━━**Now Work** PROBLEM 47

4 Simplify Expressions with Rational Exponents

Radicals are used to define rational exponents.

DEFINITION

If a is a real number and $n \geq 2$ is an integer, then

$$a^{1/n} = \sqrt[n]{a} \qquad\qquad (3)$$

provided that $\sqrt[n]{a}$ exists.

Note that if n is even and $a < 0$ then $\sqrt[n]{a}$ and $a^{1/n}$ do not exist.

EXAMPLE 6 **Writing Expressions Containing Fractional Exponents as Radicals**

(a) $4^{1/2} = \sqrt{4} = 2$ (b) $8^{1/2} = \sqrt{8} = 2\sqrt{2}$

(c) $(-27)^{1/3} = \sqrt[3]{-27} = -3$ (d) $16^{1/3} = \sqrt[3]{16} = 2\sqrt[3]{2}$ •

DEFINITION

If a is a real number and m and n are integers containing no common factors, with $n \geq 2$, then

$$a^{m/n} = \sqrt[n]{a^m} = (\sqrt[n]{a})^m \qquad\qquad (4)$$

provided that $\sqrt[n]{a}$ exists.

We have two comments about equation (4):

1. The exponent $\dfrac{m}{n}$ must be in lowest terms and n must be positive.

2. In simplifying the rational expression $a^{m/n}$, either $\sqrt[n]{a^m}$ or $(\sqrt[n]{a})^m$ may be used, the choice depending on which is easier to simplify. Generally, taking the root first, as in $(\sqrt[n]{a})^m$, is easier.

EXAMPLE 7 **Using Equation (4)**

(a) $4^{3/2} = (\sqrt{4})^3 = 2^3 = 8$ (b) $(-8)^{4/3} = (\sqrt[3]{-8})^4 = (-2)^4 = 16$

(c) $(32)^{-2/5} = (\sqrt[5]{32})^{-2} = 2^{-2} = \dfrac{1}{4}$ (d) $25^{6/4} = 25^{3/2} = (\sqrt{25})^3 = 5^3 = 125$ •

Now Work PROBLEM 55

It can be shown that the Laws of Exponents (page A8) hold for rational exponents. The next example illustrates using the Law of Exponents to simplify.

EXAMPLE 8 **Simplifying Expressions Containing Rational Exponents**

Simplify each expression. Express your answer so that only positive exponents occur. Assume that the variables are positive.

(a) $(x^{2/3}y)(x^{-2}y)^{1/2}$ (b) $\left(\dfrac{2x^{1/3}}{y^{2/3}}\right)^{-3}$ (c) $\left(\dfrac{9x^2y^{1/3}}{x^{1/3}y}\right)^{1/2}$

Solution   (a) $(x^{2/3}y)(x^{-2}y)^{1/2} = (x^{2/3}y)[(x^{-2})^{1/2}y^{1/2}]$ $(ab)^n = a^nb^n$

$$= x^{2/3}yx^{-1}y^{1/2} \qquad (a^m)^n = a^{mn}$$

$$= (x^{2/3} \cdot x^{-1})(y \cdot y^{1/2})$$

$$= x^{-1/3}y^{3/2} \qquad a^m \cdot a^n = a^{m+n}$$

$$= \frac{y^{3/2}}{x^{1/3}} \qquad a^{-n} = \frac{1}{a^n}$$

(b) $\left(\dfrac{2x^{1/3}}{y^{2/3}}\right)^{-3} = \left(\dfrac{y^{2/3}}{2x^{1/3}}\right)^3 = \dfrac{(y^{2/3})^3}{(2x^{1/3})^3} = \dfrac{y^2}{2^3(x^{1/3})^3} = \dfrac{y^2}{8x}$

(c) $\left(\dfrac{9x^2y^{1/3}}{x^{1/3}y}\right)^{1/2} = \left(\dfrac{9x^{2-(1/3)}}{y^{1-(1/3)}}\right)^{1/2} = \left(\dfrac{9x^{5/3}}{y^{2/3}}\right)^{1/2} = \dfrac{9^{1/2}(x^{5/3})^{1/2}}{(y^{2/3})^{1/2}} = \dfrac{3x^{5/6}}{y^{1/3}}$

Now Work   PROBLEM 71

The next two examples illustrate some algebra that you will need to know for certain calculus problems.

EXAMPLE 9   **Writing an Expression as a Single Quotient**

Write the following expression as a single quotient in which only positive exponents appear.

$$(x^2 + 1)^{1/2} + x \cdot \frac{1}{2}(x^2 + 1)^{-1/2} \cdot 2x$$

Solution $(x^2 + 1)^{1/2} + x \cdot \dfrac{1}{2}(x^2 + 1)^{-1/2} \cdot 2x = (x^2 + 1)^{1/2} + \dfrac{x^2}{(x^2 + 1)^{1/2}}$

$$= \frac{(x^2 + 1)^{1/2}(x^2 + 1)^{1/2} + x^2}{(x^2 + 1)^{1/2}}$$

$$= \frac{(x^2 + 1) + x^2}{(x^2 + 1)^{1/2}}$$

$$= \frac{2x^2 + 1}{(x^2 + 1)^{1/2}}$$

Now Work   PROBLEM 77

EXAMPLE 10   **Factoring an Expression Containing Rational Exponents**

Factor: $\dfrac{4}{3}x^{1/3}(2x + 1) + 2x^{4/3}$

Solution   We begin by writing $2x^{4/3}$ as a fraction with 3 as denominator.

$$\frac{4}{3}x^{1/3}(2x + 1) + 2x^{4/3} = \frac{4x^{1/3}(2x + 1)}{3} + \frac{6x^{4/3}}{3} = \frac{4x^{1/3}(2x + 1) + 6x^{4/3}}{3}$$

↑ Add the two fractions.

$$= \frac{2x^{1/3}[2(2x + 1) + 3x]}{3} = \frac{2x^{1/3}(7x + 2)}{3}$$

↑ 2 and $x^{1/3}$ are common factors.   ↑ Simplify.

Now Work   PROBLEM 89

Historical Feature

The radical sign, $\sqrt{\ }$, was first used in print by Christoff Rudolff in 1525. It is thought to be the manuscript form of the letter r (for the Latin word *radix* = *root*), although this is not quite conclusively confirmed. It took a long time for $\sqrt{\ }$ to become the standard symbol for a square root and much longer to standardize $\sqrt[3]{\ }$, $\sqrt[4]{\ }$, $\sqrt[5]{\ }$ and so on. The indexes of the root were placed in every conceivable position, with

$$\sqrt[3]{8}, \quad \sqrt[3^7]{8}, \quad \text{and} \quad \underset{3}{\sqrt{}}8$$

all being variants for $\sqrt[3]{8}$. The notation $\sqrt{\sqrt{16}}$ was popular for $\sqrt[4]{16}$. By the 1700s, the index had settled where we now put it.

The bar on top of the present radical symbol, as follows,

$$\sqrt{a^2 + 2ab + b^2}$$

is the last survivor of the **vinculum,** a bar placed atop an expression to indicate what we would now indicate with parentheses. For example,

$$\overline{ab + c} = a(b + c)$$

A.7 Assess Your Understanding

'Are You Prepared?' *Answers are given at the end of these exercises. If you get a wrong answer, read the pages in* red.

1. $(-3)^2 =$ _____; $-3^2 =$ _____ (pp. A7–A8)

2. $\sqrt{16} =$ _____; $\sqrt{(-4)^2} =$ _____ (A9–A10)

Concepts and Vocabulary

3. In the symbol $\sqrt[n]{a}$, the integer n is called the _____.

4. **True or False** $\sqrt[5]{-32} = -2$

5. We call $\sqrt[3]{a}$ the _____ _____ of a.

6. **True or False** $\sqrt[4]{(-3)^4} = -3$

Skill Building

In Problems 7–42, simplify each expression. Assume that all variables are positive when they appear.

7. $\sqrt[3]{27}$

8. $\sqrt[4]{16}$

9. $\sqrt[3]{-8}$

10. $\sqrt[3]{-1}$

11. $\sqrt{8}$

12. $\sqrt[3]{54}$

13. $\sqrt[3]{-8x^4}$

14. $\sqrt[4]{48x^5}$

15. $\sqrt[4]{x^{12}y^8}$

16. $\sqrt[5]{x^{10}y^5}$

17. $\sqrt[4]{\dfrac{x^9 y^7}{xy^3}}$

18. $\sqrt[3]{\dfrac{3xy^2}{81x^4y^2}}$ —

19. $\sqrt{36x}$

20. $\sqrt{9x^5}$

21. $\sqrt{3x^2}\sqrt{12x}$

22. $\sqrt{5x}\sqrt{20x^3}$

23. $(\sqrt{5}\sqrt[3]{9})^2$

24. $(\sqrt[3]{3}\sqrt{10})^4$

25. $(3\sqrt{6})(2\sqrt{2})$

26. $(5\sqrt{8})(-3\sqrt{3})$

27. $3\sqrt{2} + 4\sqrt{2}$

28. $6\sqrt{5} - 4\sqrt{5}$

29. $-\sqrt{18} + 2\sqrt{8}$

30. $2\sqrt{12} - 3\sqrt{27}$

31. $(\sqrt{3} + 3)(\sqrt{3} - 1)$

32. $(\sqrt{5} - 2)(\sqrt{5} + 3)$

33. $5\sqrt[3]{2} - 2\sqrt[3]{54}$

34. $9\sqrt[3]{24} - \sqrt[3]{81}$

35. $(\sqrt{x} - 1)^2$

36. $(\sqrt{x} + \sqrt{5})^2$

37. $\sqrt[3]{16x^4} - \sqrt[3]{2x}$

38. $\sqrt[4]{32x} + \sqrt[4]{2x^5}$

39. $\sqrt{8x^3} - 3\sqrt{50x}$

40. $3x\sqrt{9y} + 4\sqrt{25y}$

41. $\sqrt[3]{16x^4y} - 3x\sqrt[3]{2xy} + 5\sqrt[3]{-2xy^4}$

42. $8xy - \sqrt{25x^2y^2} + \sqrt[3]{8x^3y^3}$

In Problems 43–54, rationalize the denominator of each expression. Assume that all variables are positive when they appear.

43. $\dfrac{1}{\sqrt{2}}$

44. $\dfrac{2}{\sqrt{3}}$

45. $\dfrac{-\sqrt{3}}{\sqrt{5}}$

46. $\dfrac{-\sqrt{3}}{\sqrt{8}}$

47. $\dfrac{\sqrt{3}}{5 - \sqrt{2}}$

48. $\dfrac{\sqrt{2}}{\sqrt{7} + 2}$

49. $\dfrac{2 - \sqrt{5}}{2 + 3\sqrt{5}}$

50. $\dfrac{\sqrt{3} - 1}{2\sqrt{3} + 3}$

51. $\dfrac{5}{\sqrt[3]{2}}$

52. $\dfrac{-2}{\sqrt[3]{9}}$

53. $\dfrac{\sqrt{x+h}-\sqrt{x}}{\sqrt{x+h}+\sqrt{x}}$

54. $\dfrac{\sqrt{x+h}+\sqrt{x-h}}{\sqrt{x+h}-\sqrt{x-h}}$

In Problems 55–66, simplify each expression.

55. $8^{2/3}$

56. $4^{3/2}$

57. $(-27)^{1/3}$

58. $16^{3/4}$

59. $16^{3/2}$

60. $25^{3/2}$

61. $9^{-3/2}$

62. $16^{-3/2}$

63. $\left(\dfrac{9}{8}\right)^{3/2}$

64. $\left(\dfrac{27}{8}\right)^{2/3}$

65. $\left(\dfrac{8}{9}\right)^{-3/2}$

66. $\left(\dfrac{8}{27}\right)^{-2/3}$

In Problems 67–74, simplify each expression. Express your answer so that only positive exponents occur. Assume that the variables are positive.

67. $x^{3/4}x^{1/3}x^{-1/2}$

68. $x^{2/3}x^{1/2}x^{-1/4}$

69. $(x^3y^6)^{1/3}$

70. $(x^4y^8)^{3/4}$

71. $\dfrac{(x^2y)^{1/3}(xy^2)^{2/3}}{x^{2/3}y^{2/3}}$

72. $\dfrac{(xy)^{1/4}(x^2y^2)^{1/2}}{(x^2y)^{3/4}}$

73. $\dfrac{(16x^2y^{-1/3})^{3/4}}{(xy^2)^{1/4}}$

74. $\dfrac{(4x^{-1}y^{1/3})^{3/2}}{(xy)^{3/2}}$

Applications and Extensions

In Problems 75–88, expressions that occur in calculus are given. Write each expression as a single quotient in which only positive exponents and/or radicals appear.

75. $\dfrac{x}{(1+x)^{1/2}}+2(1+x)^{1/2} \quad x>-1$

76. $\dfrac{1+x}{2x^{1/2}}+x^{1/2} \quad x>0$

77. $2x(x^2+1)^{1/2}+x^2\cdot\dfrac{1}{2}(x^2+1)^{-1/2}\cdot 2x$

78. $(x+1)^{1/3}+x\cdot\dfrac{1}{3}(x+1)^{-2/3} \quad x\neq -1$

79. $\sqrt{4x+3}\cdot\dfrac{1}{2\sqrt{x-5}}+\sqrt{x-5}\cdot\dfrac{1}{5\sqrt{4x+3}} \quad x>5$

80. $\dfrac{\sqrt[3]{8x+1}}{3\sqrt[3]{(x-2)^2}}+\dfrac{\sqrt[3]{x-2}}{24\sqrt[3]{(8x+1)^2}} \quad x\neq 2,\ x\neq-\dfrac{1}{8}$

81. $\dfrac{\sqrt{1+x}-x\cdot\dfrac{1}{2\sqrt{1+x}}}{1+x} \quad x>-1$

82. $\dfrac{\sqrt{x^2+1}-x\cdot\dfrac{2x}{2\sqrt{x^2+1}}}{x^2+1}$

83. $\dfrac{(x+4)^{1/2}-2x(x+4)^{-1/2}}{x+4} \quad x>-4$

84. $\dfrac{(9-x^2)^{1/2}+x^2(9-x^2)^{-1/2}}{9-x^2} \quad -3<x<3$

85. $\dfrac{\dfrac{x^2}{(x^2-1)^{1/2}}-(x^2-1)^{1/2}}{x^2} \quad x<-1 \ \text{ or } \ x>1$

86. $\dfrac{(x^2+4)^{1/2}-x^2(x^2+4)^{-1/2}}{x^2+4}$

87. $\dfrac{\dfrac{1+x^2}{2\sqrt{x}}-2x\sqrt{x}}{(1+x^2)^2} \quad x>0$

88. $\dfrac{2x(1-x^2)^{1/3}+\dfrac{2}{3}x^3(1-x^2)^{-2/3}}{(1-x^2)^{2/3}} \quad x\neq-1,\ x\neq 1$

In Problems 89–98, expressions that occur in calculus are given. Factor each expression. Express your answer so that only positive exponents occur.

89. $(x+1)^{3/2}+x\cdot\dfrac{3}{2}(x+1)^{1/2} \quad x\geq -1$

90. $(x^2+4)^{4/3}+x\cdot\dfrac{4}{3}(x^2+4)^{1/3}\cdot 2x$

91. $6x^{1/2}(x^2+x)-8x^{3/2}-8x^{1/2} \quad x\geq 0$

92. $6x^{1/2}(2x+3)+x^{3/2}\cdot 8 \quad x\geq 0$

93. $3(x^2 + 4)^{4/3} + x \cdot 4(x^2 + 4)^{1/3} \cdot 2x$

94. $2x(3x + 4)^{4/3} + x^2 \cdot 4(3x + 4)^{1/3}$

95. $4(3x + 5)^{1/3}(2x + 3)^{3/2} + 3(3x + 5)^{4/3}(2x + 3)^{1/2}$ $\quad x \geq -\dfrac{3}{2}$

96. $6(6x + 1)^{1/3}(4x - 3)^{3/2} + 6(6x + 1)^{4/3}(4x - 3)^{1/2}$ $\quad x \geq \dfrac{3}{4}$

97. $3x^{-1/2} + \dfrac{3}{2}x^{1/2}$ $\quad x > 0$

98. $8x^{1/3} - 4x^{-2/3}$ $\quad x \neq 0$

In Problems 99–106, use a calculator to approximate each expression. Round your answer to two decimal places.

99. $\sqrt{2}$

100. $\sqrt{7}$

101. $\sqrt[3]{4}$

102. $\sqrt[3]{-5}$

103. $\dfrac{2 + \sqrt{3}}{3 - \sqrt{5}}$

104. $\dfrac{\sqrt{5} - 2}{\sqrt{2} + 4}$

105. $\dfrac{3\sqrt[3]{5} - \sqrt{2}}{\sqrt{3}}$

106. $\dfrac{2\sqrt{3} - \sqrt[3]{4}}{\sqrt{2}}$

Applications and Extensions

107. Calculating the Amount of Gasoline in a Tank A Shell station stores its gasoline in underground tanks that are right circular cylinders lying on their sides. See the illustration. The volume V of gasoline in the tank (in gallons) is given by the formula

$$V = 40h^2\sqrt{\dfrac{96}{h} - 0.608}$$

where h is the height of the gasoline (in inches) as measured on a depth stick.

(a) If $h = 12$ inches, how many gallons of gasoline are in the tank?

(b) If $h = 1$ inch, how many gallons of gasoline are in the tank?

108. Inclined Planes The final velocity v of an object in feet per second (ft/sec) after it slides down a frictionless inclined plane of height h feet is

$$v = \sqrt{64h + v_0^2}$$

where v_0 is the initial velocity (in ft/sec) of the object.

(a) What is the final velocity v of an object that slides down a frictionless inclined plane of height 4 feet? Assume that the initial velocity is 0.

(b) What is the final velocity v of an object that slides down a frictionless inclined plane of height 16 feet? Assume that the initial velocity is 0.

(c) What is the final velocity v of an object that slides down a frictionless inclined plane of height 2 feet with an initial velocity of 4 ft/sec?

Problems 109–112 require the following information.

Period of a Pendulum *The period T, in seconds, of a pendulum of length l, in feet, may be approximated using the formula*

$$T = 2\pi\sqrt{\dfrac{l}{32}}$$

In Problems 109–112, express your answer both as a square root and as a decimal.

109. Find the period T of a pendulum whose length is 64 feet.

110. Find the period T of a pendulum whose length is 16 feet.

111. Find the period T of a pendulum whose length is 8 inches.

112. Find the period T of a pendulum whose length is 4 inches.

Explaining Concepts: Discussion and Writing

113. Give an example to show that $\sqrt{a^2}$ is not equal to a. Use it to explain why $\sqrt{a^2} = |a|$.

'Are You Prepared?' Answers

1. $9; -9$ **2.** $4; 4$

A.8 Solving Equations

PREPARING FOR THIS SECTION *Before getting started, review the following:*

- Factoring Polynomials (Appendix A, Section A.4, pp. A33–A39)
- Zero-Product Property (Appendix A, Section A.1, p. A4)
- Rational Expressions (Appendix A, Section A.6, pp. A46–A52)

- Square Roots (Appendix A, Section A.1, pp. A9–A10)
- *n*th Roots; Rational Exponents (Appendix A, Section A.7, pp. A57–A61)

Now Work the *'Are You Prepared?'* problems on page A72.

OBJECTIVES **1** Solve Linear Equations (p. A66)
 2 Solve Rational Equations (p. A68)
 3 Solve Equations by Factoring (p. A69)
 4 Solve Radical Equations (p. A70)

An **equation in one variable** is a statement in which two expressions, at least one containing the variable, are equal. The expressions are called the **sides** of the equation. Since an equation is a statement, it may be true or false, depending on the value of the variable. Unless otherwise restricted, the admissible values of the variable are those in the domain of the variable. Those admissible values of the variable, if any, that result in a true statement are called **solutions,** or **roots,** of the equation. To **solve an equation** means to find all the solutions of the equation.

For example, the following are all equations in one variable, x:

$$x + 5 = 9 \qquad x^2 + 5x = 2x - 2 \qquad \frac{x^2 - 4}{x + 1} = 0 \qquad \sqrt{x^2 + 9} = 5$$

The first of these statements, $x + 5 = 9$, is true when $x = 4$ and false for any other choice of x. Thus, 4 is a solution of the equation $x + 5 = 9$. We also say that 4 **satisfies** the equation $x + 5 = 9$, because, when we substitute 4 for x, a true statement results.

Sometimes an equation will have more than one solution. For example, the equation

$$\frac{x^2 - 4}{x + 1} = 0$$

has $x = -2$ and $x = 2$ as solutions.

Usually, we will write the solution of an equation in set notation. This set is called the **solution set** of the equation. For example, the solution set of the equation $x^2 - 9 = 0$ is $\{-3, 3\}$.

Some equations have no real solution. For example, $x^2 + 9 = 5$ has no real solution, because there is no real number whose square when added to 9 equals 5.

An equation that is satisfied for every choice of the variable for which both sides are defined is called an **identity.** For example, the equation

$$3x + 5 = x + 3 + 2x + 2$$

is an identity, because this statement is true for any real number x.

One method for solving equations requires that a series of *equivalent equations* be developed from the original equation until an obvious solution results.

For example, all the following equations are equivalent, because each has only the solution $x = 5$.

$$2x + 3 = 13$$
$$2x = 10$$
$$x = 5$$

The question, though, is "How do I obtain an equivalent equation?" In general, there are five ways to do so.

Procedures That Result in Equivalent Equations

1. Interchange the two sides of the equation:

   Replace $\quad 3 = x \quad$ by $\quad x = 3$

2. Simplify the sides of the equation by combining like terms, eliminating parentheses, and so on:

   Replace $\quad x + 2 + 6 = 2x + 3(x + 1)$

   by $\quad\quad\quad\quad x + 8 = 5x + 3$

3. Add or subtract the same expression on both sides of the equation:

   Replace $\quad\quad\quad 3x - 5 = 4$

   by $\quad (3x - 5) + 5 = 4 + 5$

4. Multiply or divide both sides of the equation by the same nonzero expression:

   Replace $\quad \dfrac{3x}{x - 1} = \dfrac{6}{x - 1} \quad\quad x \neq 1$

   by $\quad \dfrac{3x}{x - 1} \cdot (x - 1) = \dfrac{6}{x - 1} \cdot (x - 1)$

5. If one side of the equation is 0 and the other side can be factored, we may use the Zero-Product Property and set each factor equal to 0.

   Replace $\quad\quad\quad\quad x(x - 3) = 0$

   by $\quad x = 0 \quad$ or $\quad x - 3 = 0$

WARNING Squaring both sides of an equation does not necessarily lead to an equivalent equation. ∎

Whenever it is possible to solve an equation in your head, do so. For example:

The solution of $2x = 8$ is $x = 4$.

The solution of $3x - 15 = 0$ is $x = 5$.

Now Work PROBLEM 9

We now introduce specific types of equations that can be solved algebraically to obtain exact solutions. We start with **linear equations.**

1 Solve Linear Equations

Linear equations are equations such as

$$3x + 12 = 0 \quad\quad \frac{3}{4}x - \frac{1}{5} = 0 \quad\quad 0.62x - 0.3 = 0$$

A **linear equation in one variable** is equivalent to an equation of the form

$$ax + b = 0$$

where a and b are real numbers and $a \neq 0$.

Sometimes a linear equation is called a **first-degree equation,** because the left side is a polynomial in x of degree 1.

EXAMPLE 1

Solving a Linear Equation

Solve the equation: $3(x - 2) = 5(x - 1)$

Solution

$$3(x - 2) = 5(x - 1)$$

$3x - 6 = 5x - 5$ Use the Distributive Property.

$3x - 6 - 5x = 5x - 5 - 5x$ Subtract 5x from each side.

$-2x - 6 = -5$ Simplify.

$-2x - 6 + 6 = -5 + 6$ Add 6 to each side.

$-2x = 1$ Simplify.

$\dfrac{-2x}{-2} = \dfrac{1}{-2}$ Divide each side by −2.

$x = -\dfrac{1}{2}$ Simplify.

✓ **Check:** Let $x = -\dfrac{1}{2}$ in the expression in x on the left side of the equation and simplify. Let $x = -\dfrac{1}{2}$ in the expression in x on the right side of the equation and simplify. If the two expressions are equal, the solution checks.

$$3(x - 2) = 3\left(-\dfrac{1}{2} - 2\right) = 3\left(-\dfrac{5}{2}\right) = -\dfrac{15}{2}$$

$$5(x - 1) = 5\left(-\dfrac{1}{2} - 1\right) = 5\left(-\dfrac{3}{2}\right) = -\dfrac{15}{2}$$

Since the two expressions are equal, the solution $x = -\dfrac{1}{2}$ checks. The solution set is $\left\{-\dfrac{1}{2}\right\}$.

━━━━**Now Work**   PROBLEM 15

The next example illustrates the solution of an equation that does not appear to be linear, but leads to a linear equation upon simplification.

EXAMPLE 2

Solving an Equation That Leads to a Linear Equation

Solve the equation: $(2x - 1)(x - 1) = (x - 5)(2x - 5)$

Solution

$$(2x - 1)(x - 1) = (x - 5)(2x - 5)$$

$2x^2 - 3x + 1 = 2x^2 - 15x + 25$ Multiply and combine like terms.

$2x^2 - 3x + 1 - 2x^2 = 2x^2 - 15x + 25 - 2x^2$ Subtract 2x² from each side.

$-3x + 1 = -15x + 25$ Simplify.

$-3x + 1 - 1 = -15x + 25 - 1$ Subtract 1 from each side.

$-3x = -15x + 24$ Simplify.

$-3x + 15x = -15x + 24 + 15x$ Add 15x to each side.

$12x = 24$ Simplify.

$\dfrac{12x}{12} = \dfrac{24}{12}$ Divide each side by 12.

$x = 2$ Simplify.

✓Check: $(2x - 1)(x - 1) = (2 \cdot 2 - 1)(2 - 1) = (3)(1) = 3$
$(x - 5)(2x - 5) = (2 - 5)(2 \cdot 2 - 5) = (-3)(-1) = 3$

Since the two expressions are equal, the solution checks. The solution set is $\{2\}$. ●

Now Work PROBLEM 25

2 Solve Rational Equations

We now introduce another type of equation, the *rational equation*. A **rational equation** is an equation that contains a rational expression. Examples of rational equations are

$$\frac{3}{x + 1} = \frac{2}{x - 1} + 7 \quad \text{and} \quad \frac{x - 5}{x - 4} = \frac{3}{x + 2}$$

To solve a rational equation, multiply both sides of the equation by the least common multiple of the denominators of the rational expressions that make up the rational equation.

EXAMPLE 3 **Solving a Rational Equation**

Solve the equation: $\dfrac{3}{x - 2} = \dfrac{1}{x - 1} + \dfrac{7}{(x - 1)(x - 2)}$

Solution First, we note that the domain of the variable is $\{x | x \neq 1, x \neq 2\}$. We clear the equation of rational expressions by multiplying both sides by the least common multiple of the denominators of the three rational expressions, $(x - 1)(x - 2)$.

$$\frac{3}{x - 2} = \frac{1}{x - 1} + \frac{7}{(x - 1)(x - 2)}$$

$(x - 1)(x - 2)\dfrac{3}{x - 2} = (x - 1)(x - 2)\left[\dfrac{1}{x - 1} + \dfrac{7}{(x - 1)(x - 2)}\right]$ Multiply both sides by $(x - 1)(x - 2)$. Cancel on the left.

$3x - 3 = (x - 1)(x - 2)\dfrac{1}{x - 1} + (x - 1)(x - 2)\dfrac{7}{(x - 1)(x - 2)}$ Use the Distributive Property on each side; cancel on the right.

$3x - 3 = (x - 2) + 7$ Rewrite the equation.

$3x - 3 = x + 5$ Combine like terms.

$2x = 8$ Add 3 to each side; subtract x from each side.

$x = 4$ Divide by 2.

✓Check: $\dfrac{3}{x - 2} = \dfrac{3}{4 - 2} = \dfrac{3}{2}$

$\dfrac{1}{x - 1} + \dfrac{7}{(x - 1)(x - 2)} = \dfrac{1}{4 - 1} + \dfrac{7}{(4 - 1)(4 - 2)} = \dfrac{1}{3} + \dfrac{7}{3 \cdot 2} = \dfrac{2}{6} + \dfrac{7}{6} = \dfrac{9}{6} = \dfrac{3}{2}$

Since the two expressions are equal, the solution $x = 4$ checks. The solution set is $\{4\}$. ●

Now Work PROBLEM 35

Sometimes the process of creating equivalent equations leads to apparent solutions that are not solutions of the original equation. These are called **extraneous solutions.**

EXAMPLE 4

A Rational Equation with No Solution

Solve the equation: $\dfrac{3x}{x-1} + 2 = \dfrac{3}{x-1}$

Solution

First, we note that the domain of the variable is $\{x \mid x \neq 1\}$. Since the two rational expressions in the equation have the same denominator, $x - 1$, we can simplify by multiplying both sides by $x - 1$. The resulting equation is equivalent to the original equation, since we are multiplying by $x - 1$, which is not 0 (remember, $x \neq 1$).

$$\frac{3x}{x-1} + 2 = \frac{3}{x-1}$$

$$\left(\frac{3x}{x-1} + 2\right) \cdot (x-1) = \frac{3}{x-1} \cdot (x-1) \qquad \text{Multiply both sides by x − 1; cancel on the right.}$$

$$\frac{3x}{x-1} \cdot (x-1) + 2 \cdot (x-1) = 3 \qquad \text{Use the Distributive Property on the left side; cancel on the left.}$$

$$3x + 2x - 2 = 3 \qquad \text{Simplify.}$$

$$5x - 2 = 3 \qquad \text{Combine like terms.}$$

$$5x = 5 \qquad \text{Add 2 to each side.}$$

$$x = 1 \qquad \text{Divide both sides by 5.}$$

The solution appears to be 1. But recall that $x = 1$ is not in the domain of the variable, so $x = 1$ is an extraneous solution. The equation has no solution. The solution set is $\varnothing$.   ●

──**Now Work**   PROBLEM 37

3 Solve Equations By Factoring

Many equations can be solved by factoring and then using the Zero-Product Property.

EXAMPLE 5

Solving Equations by Factoring

Solve the equations:

(a) $x^2 = 4x$ (b) $x^3 - x^2 - 4x + 4 = 0$

Solution

(a) We begin by collecting all terms on one side. This results in 0 on one side and an expression to be factored on the other.

$$x^2 = 4x$$

$$x^2 - 4x = 0$$

$$x(x - 4) = 0 \qquad \text{Factor.}$$

$$x = 0 \quad \text{or} \quad x - 4 = 0 \qquad \text{Apply the Zero-Product Property.}$$

$$x = 4$$

The solution set is $\{0, 4\}$.

✓**Check:** $x = 0$: $0^2 = 4 \cdot 0$ *So 0 is a solution.*

$x = 4$: $4^2 = 4 \cdot 4$ *So 4 is a solution.*

(b) We group the terms of $x^3 - x^2 - 4x + 4 = 0$ as follows:

$$(x^3 - x^2) - (4x - 4) = 0$$

Factor out x^2 from the first grouping and 4 from the second.

$$x^2(x - 1) - 4(x - 1) = 0$$

This reveals the common factor $(x - 1)$, so we have

$$(x^2 - 4)(x - 1) = 0$$

$$(x - 2)(x + 2)(x - 1) = 0 \qquad \text{\textit{Factor again.}}$$

$x - 2 = 0$ or $x + 2 = 0$ or $x - 1 = 0$ *Set each factor equal to 0.*

$x = 2$ $x = -2$ $x = 1$ *Solve.*

The solution set is $\{-2, 1, 2\}$.

✓**Check:** $x = -2$: $(-2)^3 - (-2)^2 - 4(-2) + 4 = -8 - 4 + 8 + 4 = 0$ *−2 is a solution.*

$x = 1$: $1^3 - 1^2 - 4(1) + 4 = 1 - 1 - 4 + 4 = 0$ *1 is a solution.*

$x = 2$: $2^3 - 2^2 - 4(2) + 4 = 8 - 4 - 8 + 4 = 0$ *2 is a solution.*   ●

━━━━━**Now Work** PROBLEM 41

4 Solve Radical Equations

When the variable in an equation occurs in a square root, cube root, and so on, that is, when it occurs in a radical, the equation is called a **radical equation.** Sometimes a suitable operation will change a radical equation to one that is linear or quadratic. A commonly used procedure is to isolate the most complicated radical on one side of the equation and then eliminate it by raising each side to a power equal to the index of the radical. Care must be taken, however, because apparent solutions that are not, in fact, solutions of the original equation may result. Recall that these are called extraneous solutions. In radical equations, extraneous solutions may occur when the index of the radical is even. Therefore, we need to check all answers when working with radical equations.

EXAMPLE 6   **Solving a Radical Equation**

Find the real solutions of the equation: $\sqrt[3]{2x - 4} - 2 = 0$

Solution   The equation contains a radical whose index is 3. We isolate it on the left side:

$$\sqrt[3]{2x - 4} - 2 = 0$$

$$\sqrt[3]{2x - 4} = 2$$

Because the index of the radical is 3, raise each side to the third power and solve.

$$\left(\sqrt[3]{2x - 4}\right)^3 = 2^3 \qquad \text{\textit{Raise each side to the power 3.}}$$

$$2x - 4 = 8 \qquad \text{\textit{Simplify.}}$$

$$2x = 12 \qquad \text{\textit{Add 4 to both sides.}}$$

$$x = 6 \qquad \text{\textit{Divide both sides by 2.}}$$

✓**Check:** $\sqrt[3]{2(6) - 4} - 2 = \sqrt[3]{12 - 4} - 2 = \sqrt[3]{8} - 2 = 2 - 2 = 0$

The solution set is $\{6\}$.   ●

━━━━━**Now Work** PROBLEM 45

EXAMPLE 7   **Solving a Radical Equation**

Find the real solutions of the equation: $\sqrt{x-1} = x - 7$

Solution   We square both sides since the index of a square root is 2.

$$\sqrt{x-1} = x - 7$$

$$\left(\sqrt{x-1}\right)^2 = (x-7)^2 \qquad \text{Square both sides.}$$

$$x - 1 = x^2 - 14x + 49 \qquad \text{Remove parentheses.}$$

$$x^2 - 15x + 50 = 0 \qquad \text{Put in standard form.}$$

$$(x-10)(x-5) = 0 \qquad \text{Factor.}$$

$$x = 10 \quad \text{or} \quad x = 5 \qquad \text{Apply the Zero-Product Property and solve.}$$

✓ Check: $x = 10$: $\sqrt{x-1} = \sqrt{10-1} = \sqrt{9} = 3$ and $x - 7 = 10 - 7 = 3$
    $x = 5$: $\sqrt{x-1} = \sqrt{5-1} = \sqrt{4} = 2$ and $x - 7 = 5 - 7 = -2$

The apparent algebraic solution $x = 5$ is extraneous; the only solution of the equation is $x = 10$. The solution set is $\{10\}$.   ●

 Now Work PROBLEM 51

Sometimes, we need to raise each side to a power more than once in order to solve a radical equation.

EXAMPLE 8   **Solving a Radical Equation**

Find the real solutions of the equation: $\sqrt{2x+3} - \sqrt{x+2} = 2$

Solution   First, we choose to isolate the more complicated radical expression (in this case, $\sqrt{2x+3}$) on the left side:

$$\sqrt{2x+3} = \sqrt{x+2} + 2$$

Now square both sides (the index of the radical is 2).

$$\left(\sqrt{2x+3}\right)^2 = \left(\sqrt{x+2} + 2\right)^2$$

$$2x + 3 = \left(\sqrt{x+2}\right)^2 + 4\sqrt{x+2} + 4 \qquad \text{Remove parentheses.}$$

$$2x + 3 = x + 2 + 4\sqrt{x+2} + 4 \qquad \text{Simplify.}$$

$$2x + 3 = x + 6 + 4\sqrt{x+2} \qquad \text{Combine like terms.}$$

Because the equation still contains a radical, we isolate the remaining radical on the right side and again square both sides.

$$x - 3 = 4\sqrt{x+2} \qquad \text{Isolate the radical on the right side.}$$

$$(x-3)^2 = 16(x+2) \qquad \text{Square both sides.}$$

$$x^2 - 6x + 9 = 16x + 32 \qquad \text{Remove parentheses.}$$

$$x^2 - 22x - 23 = 0 \qquad \text{Put in standard form.}$$

$$(x-23)(x+1) = 0 \qquad \text{Factor.}$$

$$x = 23 \quad \text{or} \quad x = -1$$

✓ Check: $x = 23$: $\sqrt{2(23)+3} - \sqrt{23+2} = \sqrt{49} - \sqrt{25} = 7 - 5 = 2$
    $x = -1$: $\sqrt{2(-1)+3} - \sqrt{-1+2} = \sqrt{1} - \sqrt{1} = 1 - 1 = 0$

The apparent solution $x = -1$ is extraneous; the only solution is $x = 23$. The solution set of the equation is $\{23\}$.   ●

Now Work PROBLEM 57

A.8 Assess Your Understanding

'Are You Prepared?' *Answers are given at the end of these exercises. If you get a wrong answer, read the pages listed in red.*

1. Find the least common multiple of the denominators of
$\dfrac{3}{x^2 - 4}$ and $\dfrac{5}{x^2 - 3x + 2}$. (pp. A50–A52)

2. The solution set of the equation $(x - 3)(3x + 5) = 0$ is
_____. (p. A4)

3. Factor:
   (a) $2x^2 - 10x$ (pp. A33–A34)
   (b) $x^3 + 3x^2 - 4x - 12$ (pp. A37–A38)

4. (a) $\sqrt[3]{x^3} =$ _____. (p. A57)
   (b) $\left(\sqrt[4]{x - 3}\right)^4 =$ _____. (p. A57)

Concepts and Vocabulary

5. Two equations that have the same solution set are called _____ _____.

6. An equation that is satisfied for every choice of the variable for which both sides are defined is called a(n) _____.

7. When an apparent solution does not satisfy the original equation, it is called a(n) _____ solution.

8. *True or False:* Some equations have no solution.

Skill Building

In Problems 9–66, solve each equation.

9. $2x - 3 = 5$

10. $3x + 4 = -8$

11. $\dfrac{1}{3}x = \dfrac{5}{12}$

12. $\dfrac{2}{3}x = \dfrac{9}{2}$

13. $6 - x = 2x + 9$

14. $3 - 2x = 2 - x$

15. $2(3 + 2x) = 3(x - 4)$

16. $3(2 - x) = 2x - 1$

17. $8x - (2x + 1) = 3x - 10$

18. $5 - (2x - 1) = 10$

19. $\dfrac{1}{2}x - 4 = \dfrac{3}{4}x$

20. $1 - \dfrac{1}{2}x = 5$

21. $0.9t = 0.4 + 0.1t$

22. $0.9t = 1 + t$

23. $\dfrac{2}{y} + \dfrac{4}{y} = 3$

24. $\dfrac{4}{y} - 5 = \dfrac{5}{2y}$

25. $(x + 7)(x - 1) = (x + 1)^2$

26. $(x + 2)(x - 3) = (x - 3)^2$

27. $z(z^2 + 1) = 3 + z^3$

28. $w(4 - w^2) = 8 - w^3$

29. $x^2 = 9x$

30. $x^3 = 4x^2$

31. $t^3 - 9t = 0$

32. $4z^3 - 8z = 0$

33. $\dfrac{3}{2x - 3} = \dfrac{2}{x + 5}$

34. $\dfrac{-2}{x + 4} = \dfrac{-3}{x + 1}$

35. $\dfrac{2}{x - 2} = \dfrac{3}{x + 5} + \dfrac{10}{(x + 5)(x - 2)}$

36. $\dfrac{1}{2x + 3} + \dfrac{1}{x - 1} = \dfrac{1}{(2x + 3)(x - 1)}$

37. $\dfrac{x}{x - 2} + 3 = \dfrac{2}{x - 2}$

38. $\dfrac{2x}{x + 3} = \dfrac{-6}{x + 3} - 2$

39. $x^3 + x^2 - 20x = 0$

40. $x^3 + 6x^2 - 7x = 0$

41. $x^3 + x^2 - x - 1 = 0$

42. $x^3 + 4x^2 - x - 4 = 0$

43. $2x^3 + 4 = x^2 + 8x$

44. $3x^3 + 4x^2 = 27x + 36$

45. $\sqrt{2t - 1} = 1$

46. $\sqrt{3t + 4} = 2$

47. $\sqrt{3t + 4} = -6$

48. $\sqrt{5t + 3} = -2$

49. $\sqrt[3]{1 - 2x} - 3 = 0$

50. $\sqrt[3]{1 - 2x} - 1 = 0$

51. $\sqrt{15 - 2x} = x$

52. $\sqrt{12 - x} = x$

53. $3 + \sqrt{3x + 1} = x$

54. $2 + \sqrt{12 - 2x} = x$

55. $\sqrt{2x + 3} - \sqrt{x + 1} = 1$

56. $\sqrt{3x + 7} + \sqrt{x + 2} = 1$

57. $\sqrt{3x + 1} - \sqrt{x - 1} = 2$

58. $\sqrt{3x - 5} - \sqrt{x + 7} = 2$

59. $(3x + 1)^{1/2} = 4$

60. $(3x - 5)^{1/2} = 2$

61. $(5x - 2)^{1/3} = 2$

62. $(2x + 1)^{1/3} = -1$

63. $\dfrac{2}{y + 3} + \dfrac{3}{y - 4} = \dfrac{5}{y + 6}$

64. $\dfrac{5}{5z - 11} + \dfrac{4}{2z - 3} = \dfrac{-3}{5 - z}$

65. $\dfrac{x}{x^2 - 1} - \dfrac{x + 3}{x^2 - x} = \dfrac{-3}{x^2 + x}$

66. $\dfrac{x + 1}{x^2 + 2x} - \dfrac{x + 4}{x^2 + x} = \dfrac{-3}{x^2 + 3x + 2}$

Problem 67–72, list some formulas that occur in applications. Solve each formula for the indicated variable.

67. Electricity $\dfrac{1}{R} = \dfrac{1}{R_1} + \dfrac{1}{R_2}$ for R

68. Finance $A = P(1 + rt)$ for r

69. Mechanics $F = \dfrac{mv^2}{R}$ for R

70. Chemistry $PV = nRT$ for T

71. Mathematics $S = \dfrac{a}{1 - r}$ for r

72. Mechanics $v = -gt + v_0$ for t

Applications and Extensions

73. Physics: Using Sound to Measure Distance The distance to the surface of the water in a well can sometimes be found by dropping an object into the well and measuring the time elapsed until a sound is heard. If t_1 is the time (measured in seconds) that it takes for the object to strike the water, then t_1 will obey the equation $s = 16t_1^2$, where s is the distance (measured in feet). It follows that $t_1 = \dfrac{\sqrt{s}}{4}$. Suppose that t_2 is the time that it takes for the sound of the impact to reach your ears. Because sound travels at a speed of approximately 1100 feet per second, the time t_2 for the sound to

travel the distance s will be $t_2 = \dfrac{s}{1100}$. See the illustration. Now $t_1 + t_2$ is the total time that elapses from the moment that the object is dropped to the moment that a sound is heard. We have the equation

$$\text{Total time elapsed} = \frac{\sqrt{s}}{4} + \frac{s}{1100}$$

Find the distance to the water's surface if the total time elapsed from dropping a rock to hearing it hit water is 4 seconds.

Sound waves:
$t_2 = \dfrac{s}{1100}$

Falling object:
$t_1 = \dfrac{\sqrt{s}}{4}$

74. Crushing Load A civil engineer relates the thickness T, in inches, and height H, in feet, of a square wooden pillar to its crushing load L, in tons, using the model $T = \sqrt[4]{\dfrac{LH^2}{25}}$. If a square wooden pillar is 4 inches thick and 10 feet high, what is its crushing load?

75. Foucault's Pendulum The period of a pendulum is the time it takes the pendulum to make one full swing back and forth. The period T, in seconds, is given by the formula $T = 2\pi\sqrt{\dfrac{l}{32}}$, where l is the length, in feet, of the pendulum. In 1851, Jean-Bernard-Leon Foucault demonstrated the axial rotation of Earth using a large pendulum that he hung in the Pantheon in Paris. The period of Foucault's pendulum was approximately 16.5 seconds. What was its length?

Explaining Concepts: Discussion and Writing

76. Make up a radical equation that has no solution.

77. Make up a radical equation that has an extraneous solution.

78. Discuss the step in the solving process for radical equations that leads to the possibility of extraneous solutions. Why is there no such possibility for linear and quadratic equations?

79. What Went Wrong? On an exam, Jane solved the equation $\sqrt{2x + 3} - x = 0$ and wrote that the solution set was $\{-1, 3\}$. Jane received 3 out of 5 points for the problem. Jane asks you why she received 3 out of 5 points. Provide an explanation.

80. Which of the following pairs of equations are equivalent? Explain.

(a) $x^2 = 9$; $x = 3$

(b) $x = \sqrt{9}$; $x = 3$

(c) $(x - 1)(x - 2) = (x - 1)^2$; $x - 2 = x - 1$

81. The equation

$$\frac{5}{x + 3} + 3 = \frac{8 + x}{x + 3}$$

has no solution, yet when we go through the process of solving it we obtain $x = -3$. Write a brief paragraph to explain what causes this to happen.

82. Make up an equation that has no solution and give it to a fellow student to solve. Ask the student to write a critique of your equation.

'Are You Prepared?' Answers

1. $(x - 2)(x + 2)(x - 1)$

2. $\left\{-\dfrac{5}{3}, 3\right\}$

3. (a) $2x(x - 5)$ (b) $(x + 2)(x - 2)(x + 3)$

4. (a) x (b) $x - 3$

A.9 Problem Solving: Interest, Mixture, Uniform Motion, Constant Rate Jobs

OBJECTIVES 1 Translate Verbal Descriptions into Mathematical Expressions (p. A74)

2 Solve Interest Problems (p. A75)

3 Solve Mixture Problems (p. A76)

4 Solve Uniform Motion Problems (p. A77)

5 Solve Constant Rate Job Problems (p. A79)

NOTE The icon is a *Model It!* icon. It indicates that the discussion or problem involves modeling. ∎

Applied (word) problems do not come in the form "Solve the equation. . . . " Instead, they supply information using words, a verbal description of the real problem. So, to solve applied problems, we must be able to translate the verbal description into the language of mathematics. We do this by using variables to represent unknown quantities and then finding relationships (such as equations) that involve these variables. The process of doing all this is called **mathematical modeling.** The equation or inequality that represents the relationship among the variables is called a **model.**

Any solution to the mathematical problem must be checked against the mathematical problem, the verbal description, and the real problem. See Figure 28 for an illustration of the **modeling process.**

Figure 28

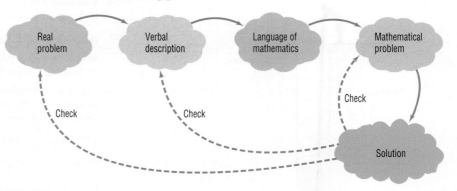

1 Translate Verbal Descriptions into Mathematical Expressions

Let's look at a few examples that will help you to translate certain words into mathematical symbols.

EXAMPLE 1

Translating Verbal Descriptions into Mathematical Expressions

(a) For uniform motion, the speed (rate) of an object equals the distance traveled divided by the time required.

Translation: If r is the speed (or rate), d the distance, and t the time, then $r = \dfrac{d}{t}$.

(b) Let x denote a number.

The number 5 times as large as x is $5x$.

The number 3 less than x is $x - 3$.

The number that exceeds x by 4 is $x + 4$.

The number that, when added to x, gives 5 is $5 - x$.

Now Work PROBLEM 7

Always check the units used to measure the variables of an applied problem. In Example 1(a), if r is measured in miles per hour, then the distance d must be expressed in miles and the time t must be expressed in hours. It is a good practice to check units to be sure that they are consistent and make sense.

2 Solve Interest Problems

Interest is money paid for the use of money. The total amount borrowed (whether by an individual from a bank in the form of a loan or by a bank from an individual in the form of a savings account) is called the **principal.** The **rate of interest,** expressed as a percent, is the amount charged for the use of the principal for a given period of time, usually on a yearly (that is, per annum) basis.

Simple Interest Formula

If a principal of P dollars is borrowed for a period of t years at a per annum interest rate r, expressed as a decimal, the interest I charged is

$$I = Prt \qquad\qquad\qquad (1)$$

Interest charged according to formula (1) is called **simple interest.** When using formula (1), be sure to express r as a decimal.

EXAMPLE 2   **Using the Simple Interest Formula:**
Computing Interest on a Loan

Suppose that Juanita borrows $500 for 6 months at the simple interest rate of 9% per annum. What is the interest that Juanita will be charged on this loan? How much does Juanita owe after 6 months?

Solution   The rate of interest is given per annum, so the actual time that the money is borrowed must be expressed in years. The interest charged would be the principal, $500, times the rate of interest (9% = 0.09) times the time in years, $\frac{1}{2}$:

$$\text{Interest charged} = I = Prt = (500)(0.09)\left(\frac{1}{2}\right) = \$22.50$$

After 6 months, Juanita will owe what she borrowed plus the interest: $500 + $22.50 = $522.50.

In the next example we illustrate the steps for solving applied problems.

EXAMPLE 3   **Financial Planning**

Candy has $70,000 to invest and requires an overall rate of return of 9%. She can invest in a safe, government-insured certificate of deposit, but it only pays 8%. To obtain 9%, she agrees to invest some of her money in noninsured corporate bonds paying 12%. How much should be placed in each investment to achieve her goal?

Solution   **STEP 1:** *Determine the question being asked.* The question is asking for two dollar amounts: the principal to invest in the corporate bonds and the principal to invest in the certificate of deposit.

STEP 2: *Assign a variable to the unknown.* We let b represent the amount (in dollars) to be invested in the bonds. Then $70,000 - b$ is the amount that will be invested in the certificate. (Do you see why?)

NOTE Don't forget to choose variables that remind you of what they represent, as in *b* for bonds. ∎

STEP 3: *Build the model.* We set up a table:

	Principal ($)	Rate	Time (yr)	Interest ($)
Bonds	b	12% = 0.12	1	$0.12b$
Certificate	$70{,}000 - b$	8% = 0.08	1	$0.08(70{,}000 - b)$
Total	$70{,}000$	9% = 0.09	1	$0.09(70{,}000) = 6300$

Since the total interest from the investments is equal to $0.09(70,000) = 6300$, we have the equation

$$0.12b + 0.08(70,000 - b) = 6300 \quad \text{The Model}$$

(Note that the units are consistent: the unit is dollars on each side.)

STEP 4: *Solve the equation.*

$$0.12b + 5600 - 0.08b = 6300$$
$$0.04b = 700$$
$$b = 17,500$$

Candy should place $17,500 in the bonds and $70,000 - $17,500 = $52,500 in the certificate.

STEP 5: *Check your answer.* The interest on the bonds after 1 year is $0.12(\$17,500) = \$2100$; the interest on the certificate after 1 year is $0.08(\$52,500) = \$4200$. The total annual interest is $6300, the required amount.

●

Now Work PROBLEM 17

3 Solve Mixture Problems

Oil refineries sometimes produce gasoline that is a blend of two or more types of fuel; bakeries occasionally blend two or more types of flour for their bread. These problems are referred to as **mixture problems** because they combine two or more quantities to form a mixture.

EXAMPLE 4   **Blending Coffees**

The manager of a Starbucks store decides to experiment with a new blend of coffee. She will mix some B grade Colombian coffee that sells for $5 per pound with some A grade Arabica coffee that sells for $10 per pound to get 100 pounds of the new blend. The selling price of the new blend is to be $7 per pound, and there is to be no difference in revenue from selling the new blend versus selling the other types. How many pounds of the B grade Colombian and A grade Arabica coffees are required?

Solution   Let c represent the number of pounds of the B grade Colombian coffee. Then $100 - c$ equals the number of pounds of the A grade Arabica coffee. See Figure 29.

Figure 29

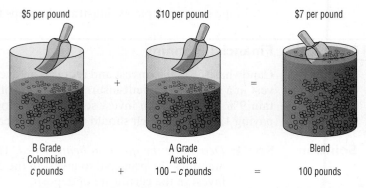

Since there is to be no difference in revenue between selling the A and B grades separately versus the blend, we have

$$\left\{\begin{array}{c}\text{Price per pound}\\\text{of B grade}\end{array}\right\}\left\{\begin{array}{c}\text{No. pounds}\\\text{B grade}\end{array}\right\} + \left\{\begin{array}{c}\text{Price per pound}\\\text{of A grade}\end{array}\right\}\left\{\begin{array}{c}\text{No. pounds}\\\text{A grade}\end{array}\right\} = \left\{\begin{array}{c}\text{Price per pound}\\\text{of blend}\end{array}\right\}\left\{\begin{array}{c}\text{No. pounds}\\\text{blend}\end{array}\right\}$$

$$\$5 \quad \cdot \quad c \quad + \quad \$10 \quad \cdot \quad (100 - c) \quad = \quad \$7 \quad \cdot \quad 100 \quad \text{The Model}$$

We have the equation

$$5c + 10(100 - c) = 7 \cdot (100)$$
$$5c + 1000 - 10c = 700$$
$$-5c = -300$$
$$c = 60$$

The manager should blend 60 pounds of B grade Colombian coffee with $100 - 60 = 40$ pounds of A grade Arabica coffee to get the desired blend.

✓**Check:** The 60 pounds of B grade coffee would sell for $(\$5)(60) = \$300$, and the 40 pounds of A grade coffee would sell for $(\$10)(40) = \$400$; the total revenue, $700, equals the revenue obtained from selling the blend, as desired.

●

➤**Now Work** PROBLEM 21

4 Solve Uniform Motion Problems

Objects that move at a constant speed are said to be in **uniform motion.** When the average speed of an object is known, it can be interpreted as its constant speed. For example, a bicyclist traveling at an average speed of 25 miles per hour is in uniform motion.

> **Uniform Motion Formula**
>
> If an object moves at an average speed (or rate) r, the distance d covered in time t is given by the formula
>
> $$d = r \cdot t \qquad\qquad (2)$$
>
> That is, Distance = Rate · Time.

| EXAMPLE 5 |

Physics: Uniform Motion

Tanya, who is a long-distance runner, runs at an average speed of 8 miles per hour (mi/hr). Two hours after Tanya leaves your house, you leave in your Honda and follow the same route. If your average speed is 40 mi/hr, how long will it be before you catch up to Tanya? How far will each of you be from your home?

Solution   Refer to Figure 30. We use t to represent the time (in hours) that it takes the Honda to catch up to Tanya. When this occurs, the total time elapsed for Tanya is $t + 2$ hours.

Figure 30

Set up the following table:

	Rate mi/hr	Time hr	Distance mi
Tanya	8	$t + 2$	$8(t + 2)$
Honda	40	t	$40t$

Since the distance traveled is the same, we are led to the following equation:

$$8(t + 2) = 40t \qquad \text{The Model}$$
$$8t + 16 = 40t$$
$$32t = 16$$
$$t = \frac{1}{2} \text{ hour}$$

It will take the Honda $\frac{1}{2}$ hour to catch up to Tanya. Each will have gone 20 miles.

✓ Check: In 2.5 hours, Tanya travels a distance of $(2.5)(8) = 20$ miles. In $\frac{1}{2}$ hour, the Honda travels a distance of $\left(\frac{1}{2}\right)(40) = 20$ miles.

EXAMPLE 6

Physics: Uniform Motion

A motorboat heads upstream a distance of 24 miles on a river whose current is running at 3 miles per hour (mi/hr). The trip up and back takes 6 hours. Assuming that the motorboat maintained a constant speed relative to the water, what was its speed?

Solution

See Figure 31. We use r to represent the constant speed of the motorboat relative to the water. Then the true speed going upstream is $r - 3$ mi/hr, and the true speed going downstream is $r + 3$ mi/hr. Since Distance = Rate × Time, then Time = $\dfrac{\text{Distance}}{\text{Rate}}$. We set up a table.

Figure 31

	Rate mi/hr	Distance mi	Time = $\dfrac{\text{Distance}}{\text{Velocity}}$ hr
Upstream	$r - 3$	24	$\dfrac{24}{r - 3}$
Downstream	$r + 3$	24	$\dfrac{24}{r + 3}$

Since the total time up and back is 6 hours, we have

$$\frac{24}{r - 3} + \frac{24}{r + 3} = 6 \qquad \text{The Model}$$

$$\frac{24(r + 3) + 24(r - 3)}{(r - 3)(r + 3)} = 6 \qquad \text{Add the quotients on the left.}$$

$$\frac{48r}{r^2 - 9} = 6 \qquad \text{Simplify.}$$

$$48r = 6(r^2 - 9)$$

$$6r^2 - 48r - 54 = 0 \qquad \text{Place in standard form.}$$

$$r^2 - 8r - 9 = 0 \qquad \text{Divide by 6.}$$

$$(r - 9)(r + 1) = 0 \qquad \text{Factor.}$$

$$r = 9 \quad \text{or} \quad r = -1 \qquad \text{Apply the Zero-Product Property and solve.}$$

We discard the solution $r = -1$ mi/hr, so the speed of the motorboat relative to the water is 9 mi/hr.

━━━━━━ **Now Work** PROBLEM 27

5 Solve Constant Rate Job Problems

This section involves jobs that are performed at a **constant rate.** Our assumption is that, if a job can be done in t units of time, $\dfrac{1}{t}$ of the job is done in 1 unit of time. So, if a job takes 4 hours, then $\dfrac{1}{4}$ of the job is done in 1 hour.

| EXAMPLE 7 | **Working Together to Do a Job** |

At 10 AM Danny is asked by his father to weed the garden. From past experience, Danny knows that this will take him 4 hours, working alone. His older brother, Mike, when it is his turn to do this job, requires 6 hours (Mike is more diligent than Dan). Since Mike wants to go golfing with Danny and has a reservation for 1 PM, he agrees to help Danny. Assuming no gain or loss of efficiency, when will they finish if they work together? Can they make the tee time?

Solution   We set up Table 1. If it takes Danny 4 hours to do the entire job, and he works at a constant rate, then in 1 hour, Danny does $\dfrac{1}{4}$ of the job. In 1 hour, Mike does $\dfrac{1}{6}$ of the job. Let t be the time (in hours) that it takes them to do the job together. In 1 hour, then, $\dfrac{1}{t}$ of the job is completed. We reason as follows:

Table 1

	Hours to Do Job	Part of Job Done in 1 Hour
Danny	4	$\dfrac{1}{4}$
Mike	6	$\dfrac{1}{6}$
Together	t	$\dfrac{1}{t}$

$$\left(\begin{array}{c}\text{Part done by Danny}\\ \text{in 1 hour}\end{array}\right) + \left(\begin{array}{c}\text{Part done by Mike}\\ \text{in 1 hour}\end{array}\right) = \left(\begin{array}{c}\text{Part done together}\\ \text{in 1 hour}\end{array}\right)$$

From Table 1,

$$\frac{1}{4} + \frac{1}{6} = \frac{1}{t} \qquad \text{The Model}$$

$$\frac{3}{12} + \frac{2}{12} = \frac{1}{t} \qquad \text{LCD = 12}$$

$$\frac{5}{12} = \frac{1}{t}$$

$$5t = 12 \qquad \text{Multiply both sides by 12}t.$$

$$t = \frac{12}{5} \qquad \text{Divide both sides by 5.}$$

Working together, the job can be done in $\dfrac{12}{5}$ hours, or 2 hours, 24 minutes. They should make the tee time, since they will finish at 12:24 PM.

●

Now Work PROBLEM 29

A.9 Assess Your Understanding

Concepts and Vocabulary

1. Using variables to represent unknown quantities and then finding relationships that involve these variables is referred to as _____ _____.

2. The money paid for the use of money is _____.

3. Objects that move at a constant speed are said to be in _____ _____.

4. *True or False*   The amount charged for the use of principal for a given period of time is called the rate of interest.

5. *True or False*   If an object moves at an average rate r, the distance d covered in time t is given by the formula $d = r \cdot t$.

6. Suppose that you want to mix two coffees to obtain 100 pounds of the blend. If x represents the number of pounds of coffee A, write an algebraic expression that represents the number of pounds of coffee B.

Applications and Extensions

In Problems 7–16, translate each sentence into a mathematical equation. Be sure to identify the meaning of all symbols.

7. **Geometry** The area of a circle is the product of the number π and the square of the radius.

8. **Geometry** The circumference of a circle is the product of the number π and twice the radius.

9. **Geometry** The area of a square is the square of the length of a side.

10. **Geometry** The perimeter of a square is four times the length of a side.

11. **Physics** Force equals the product of mass and acceleration.

12. **Physics** Pressure is force per unit area.

13. **Physics** Work equals force times distance.

14. **Physics** Kinetic energy is one-half the product of the mass and the square of the velocity.

15. **Business** The total variable cost of manufacturing x dishwashers is $150 per dishwasher times the number of dishwashers manufactured.

16. **Business** The total revenue derived from selling x dishwashers is $250 per dishwasher times the number of dishwashers sold.

17. **Financial Planning** Betsy, a recent retiree, requires $6000 per year in extra income. She has $84,000 to invest and can invest in B-rated bonds paying 8% per year or in a certificate of deposit (CD) paying 6% per year. How much money should be invested in each to realize exactly $6000 in interest per year?

18. **Financial Planning** After 2 years, Betsy (see Problem 17) finds that she will now require $6400 per year. Assuming that the remaining information is the same, how should the money be reinvested?

19. **Banking** A bank loaned out $120,000, part of it at the rate of 8% per year and the rest at the rate of 12% per year. If the interest received totaled $11,000, how much was loaned at 8%?

20. **Banking** Wendy, a loan officer at a bank, has $1,000,000 to lend and is required to obtain an average return of 8% per year. If she can lend at the rate of 7% or at the rate of 11%, how much can she lend at the 7% rate and still meet her requirement?

21. **Blending Teas** The manager of a store that specializes in selling tea decides to experiment with a new blend. She will mix some Earl Grey tea that sells for $5 per pound with some Orange Pekoe tea that sells for $3 per pound to get 100 pounds of the new blend. The selling price of the new blend is to be $4.50 per pound, and there is to be no difference in revenue from selling the new blend versus selling the other types. How many pounds of the Earl Grey tea and Orange Pekoe tea are required?

22. **Business: Blending Coffee** A coffee manufacturer wants to market a new blend of coffee that sells for $3.90 per pound by mixing two coffees that sell for $2.75 and $5 per pound, respectively. What amounts of each coffee should be blended to obtain the desired mixture?

[**Hint:** Assume that the total weight of the desired blend is 100 pounds.]

23. **Business: Mixing Nuts** A nut store normally sells cashews for $9.00 per pound and almonds for $3.50 per pound. But at the end of the month the almonds had not sold well, so, in order to sell 60 pounds of almonds, the manager decided to mix the 60 pounds of almonds with some cashews and sell the mixture for $7.50 per pound. How many pounds of cashews should be mixed with the almonds to ensure no change in the profit?

24. **Business: Mixing Candy** A candy store sells boxes of candy containing caramels and cremes. Each box sells for $12.50 and holds 30 pieces of candy (all pieces are the same size). If the caramels cost $0.25 to produce and the cremes cost $0.45 to produce, how many of each should be in a box to make a profit of $3?

25. **Physics: Uniform Motion** A motorboat can maintain a constant speed of 16 miles per hour relative to the water. The boat makes a trip upstream to a certain point in 20 minutes; the return trip takes 15 minutes. What is the speed of the current? (See the figure.)

26. **Physics: Uniform Motion** A motorboat heads upstream on a river that has a current of 3 miles per hour. The trip upstream takes 5 hours, and the return trip takes 2.5 hours. What is the speed of the motorboat? (Assume that the motorboat maintains a constant speed relative to the water.)

27. **Physics: Uniform Motion** A motorboat maintained a constant speed of 15 miles per hour relative to the water in going 10 miles upstream and then returning. The total time for the trip was 1.5 hours. Use this information to find the speed of the current.

28. **Physics: Uniform Motion** Two cars enter the Florida Turnpike at Commercial Boulevard at 8:00 AM, each heading for Wildwood. One car's average speed is 10 miles per hour more than the other's. The faster car arrives at Wildwood at 11:00 AM, $\frac{1}{2}$ hour before the other car. What is the average speed of each car? How far did each travel?

29. **Working Together on a Job** Trent can deliver his newspapers in 30 minutes. It takes Lois 20 minutes to do the same route. How long will it take them to deliver the newspapers if they work together?

30. **Working Together on a Job** Patrice, by himself, can paint four rooms in 10 hours. If he hires April to help, they can do the same job together in 6 hours. If he lets April work alone, how long will it take her to paint four rooms?

Mixed Practice

31. Enclosing a Garden   A gardener has 46 feet of fencing to be used to enclose a rectangular garden that has a border 2 feet wide surrounding it (see the figure).
  (a) If the length of the garden is to be twice its width, what are the dimensions of the garden?
  (b) What is the area of the garden?
  (c) If the length and width of the garden are to be the same, what are the dimensions of the garden?
  (d) What is the area of the square garden?

32. Construction   A pond is enclosed by a wooden deck that is 3 feet wide. The fence surrounding the deck is 100 feet long.
  (a) If the pond is square, what are its dimensions?
  (b) If the pond is rectangular and the length of the pond is to be three times its width, what are its dimensions?
  (c) If the pond is circular, what is its diameter?
  (d) Which pond has the most area?

33. Football   A tight end can run the 100-yard dash in 12 seconds. A defensive back can do it in 10 seconds. The tight end catches a pass at his own 20-yard line with the defensive back at the 15-yard line. (See the figure.) If no other players are nearby, at what yard line will the defensive back catch up to the tight end?

[**Hint:** At time $t = 0$, the defensive back is 5 yards behind the tight end.]

34. Computing Business Expense   Therese, an outside salesperson, uses her car for both business and pleasure. Last year she traveled 30,000 miles, using 900 gallons of gasoline. Her car gets 40 miles per gallon on the highway and 25 in the city. She can deduct all highway travel, but no city travel, on her taxes. How many miles should Therese be allowed as a business expense?

35. Mixing Water and Antifreeze   How much water should be added to 1 gallon of pure antifreeze to obtain a solution that is 60% antifreeze?

36. Mixing Water and Antifreeze   The cooling system of a certain foreign-made car has a capacity of 15 liters. If the system is filled with a mixture that is 40% antifreeze, how much of this mixture should be drained and replaced by pure antifreeze so that the system is filled with a solution that is 60% antifreeze?

37. Moving Walkways   The speed of a moving walkway is typically about 2.5 feet per second. Walking on such a moving walkway, it takes Karen a total of 40 seconds to travel 50 feet with the movement of the walkway and then back again against the movement of the walkway. What is Karen's normal walking speed?

Source: Answers.com

38. Moving Walkways   The Gare Montparnasse train station in Paris has a high-speed version of a moving walkway. If Jean Claude walks while riding this moving walkway, he can travel 200 meters in 30 seconds less time than if he stands still on the moving walkway. If he walks at a normal rate of 1.5 meters per second, what is the speed of the Gare Montparnasse walkway?

Source: Answers.com

39. Laser Printers   It takes an HP LaserJet 1300 laser printer 10 minutes longer to complete a 600-page print job by itself than it takes an HP LaserJet 2420 to complete the same job by itself. Together the two printers can complete the job in 12 minutes. How long does it take each printer to complete the print job alone? What is the speed of each printer?

Source: Hewlett-Packard

40. Tennis Anyone?   A regulation doubles tennis court has an area of 2808 square feet. If it is 6 feet longer than twice its width, determine the dimensions of the court.

Source: United States Tennis Association

41. Chemistry: Salt Solutions   How much water must be evaporated from 32 ounces of a 4% salt solution to make a 6% salt solution?

42. Chemistry: Salt Solutions   How much water must be evaporated from 240 gallons of a 3% salt solution to produce a 5% salt solution?

43. Purity of Gold   The purity of gold is measured in karats, with pure gold being 24 karats. Other purities of gold are expressed as proportional parts of pure gold. Thus, 18-karat gold is $\dfrac{18}{24}$, or 75% pure gold; 12-karat gold is $\dfrac{12}{24}$, or 50%

pure gold; and so on. How much 12-karat gold should be mixed with pure gold to obtain 60 grams of 16-karat gold?

44. **Chemistry: Sugar Molecules**   A sugar molecule has twice as many atoms of hydrogen as it does oxygen and one more atom of carbon than oxygen. If a sugar molecule has a total of 45 atoms, how many are oxygen? How many are hydrogen?

45. **Running a Race**   Mike can run the mile in 6 minutes, and Dan can run the mile in 9 minutes. If Mike gives Dan a head start of 1 minute, how far from the start will Mike pass Dan? How long does it take?

46. **Range of an Airplane**   An air rescue plane averages 300 miles per hour in still air. It carries enough fuel for 5 hours of flying time. If, upon takeoff, it encounters a head wind of 30 mi/hr, how far can it fly and return safely? (Assume that the wind remains constant.)

47. **Emptying Oil Tankers**   An oil tanker can be emptied by the main pump in 4 hours. An auxiliary pump can empty the tanker in 9 hours. If the main pump is started at 9 AM, when should the auxiliary pump be started so that the tanker is emptied by noon?

48. **Cement**   A 20-pound bag of Economy brand cement mix contains 25% cement and 75% sand. How much pure cement must be added to produce a cement mix that is 40% cement?

49. **Emptying a Tub**   A bathroom tub fills in 15 minutes with both faucets open and the stopper in place. With both faucets closed and the stopper removed, the tub empties in 20 minutes. How long will it take for the tub to fill if both faucets are open and the stopper is removed?

50. **Using Two Pumps**   A 5-horsepower (hp) pump can empty a pool in 5 hours. A smaller, 2-hp pump empties the same pool in 8 hours. The pumps are used together to begin emptying this pool. After 2 hours, the 2-hp pump breaks down. How long will it take the larger pump to empty the pool?

51. **A Biathlon**   Suppose that you have entered an 87-mile biathlon that consists of a run and a bicycle race. During your run, your average speed is 6 miles per hour, and during your bicycle race, your average speed is 25 miles per hour. You finish the race in 5 hours. What is the distance of the run? What is the distance of the bicycle race?

52. **Constructing a Coffee Can**   A 39-ounce can of Hills Bros.® coffee requires 188.5 square inches of aluminum. If its height is 7 inches, what is its radius? (The surface area S of a right cylinder is $S = 2\pi r^2 + 2\pi rh$, where r is the radius and h is the height.)

53. **Comparing Olympic Heroes**   In the 1984 Olympics, Carl Lewis of the United States won the gold medal in the 100-meter race with a time of 9.99 seconds. In the 1896 Olympics, Thomas Burke, also of the United States, won the gold medal in the 100-meter race in 12.0 seconds. If they ran in the same race repeating their respective times, by how many meters would Lewis beat Burke?

Explaining Concepts: Discussion and Writing

54. **Critical Thinking**   You are the manager of a clothing store and have just purchased 100 dress shirts for $20.00 each. After 1 month of selling the shirts at the regular price, you plan to have a sale giving 40% off the original selling price. However, you still want to make a profit of $4 on each shirt at the sale price. What should you price the shirts at initially to ensure this? If, instead of 40% off at the sale, you give 50% off, by how much is your profit reduced?

55. **Critical Thinking**   Make up a word problem that requires solving a linear equation as part of its solution. Exchange problems with a friend. Write a critique of your friend's problem.

56. **Critical Thinking**   Without solving, explain what is wrong with the following mixture problem: How many liters of 25% ethanol should be added to 20 liters of 48% ethanol to obtain a solution of 58% ethanol? Now go through an algebraic solution. What happens?

57. **Computing Average Speed**   In going from Chicago to Atlanta, a car averages 45 miles per hour, and in going from Atlanta to Miami, it averages 55 miles per hour. If Atlanta is halfway between Chicago and Miami, what is the average speed from Chicago to Miami? Discuss an intuitive solution. Write a paragraph defending your intuitive solution. Then solve the problem algebraically. Is your intuitive solution the same as the algebraic one? If not, find the flaw.

58. **Speed of a Plane**   On a recent flight from Phoenix to Kansas City, a distance of 919 nautical miles, the plane arrived 20 minutes early. On leaving the aircraft, I asked the captain, "What was our tail wind?" He replied, "I don't know, but our ground speed was 550 knots." How can you determine if enough information is provided to find the tail wind? If possible, find the tail wind. (1 knot = 1 nautical mile per hour)

A.10 Interval Notation; Solving Inequalities

PREPARING FOR THIS SECTION *Before getting started, review the following:*

- Algebra Essentials (Appendix A, Section A.1, pp. A4–A5)

Now Work the *'Are You Prepared?'* problems on page A88.

OBJECTIVES **1** Use Interval Notation (p. A83)
                2 Use Properties of Inequalities (p. A84)
                3 Solve Inequalities (p. A86)
                4 Solve Combined Inequalities (p. A87)

Suppose that a and b are two real numbers and $a < b$. We shall use the notation $a < x < b$ to mean that x is a number *between* a and b. The expression $a < x < b$ is equivalent to the two inequalities $a < x$ and $x < b$. Similarly, the expression $a \le x \le b$ is equivalent to the two inequalities $a \le x$ and $x \le b$. The remaining two possibilities, $a \le x < b$ and $a < x \le b$, are defined similarly.

Although it is acceptable to write $3 \ge x \ge 2$, it is preferable to reverse the inequality symbols and write instead $2 \le x \le 3$ so that, as you read from left to right, the values go from smaller to larger.

A statement such as $2 \le x \le 1$ is false because there is no number x for which $2 \le x$ and $x \le 1$. Finally, we never mix inequality symbols, as in $2 \le x \ge 3$.

1 Use Interval Notation

Let a and b represent two real numbers with $a < b$.

DEFINITION
> A **closed interval,** denoted by **[a, b],** consists of all real numbers x for which $a \le x \le b$.
> An **open interval,** denoted by **(a, b),** consists of all real numbers x for which $a < x < b$.
> The **half-open,** or **half-closed, intervals** are **(a, b],** consisting of all real numbers x for which $a < x \le b$, and **[a, b),** consisting of all real numbers x for which $a \le x < b$.

In each of these definitions, a is called the **left endpoint** and b the **right endpoint** of the interval.

The symbol ∞ (read as "infinity") is not a real number, but a notational device used to indicate unboundedness in the positive direction. The symbol $-\infty$ (read as "negative infinity") also is not a real number, but a notational device used to indicate unboundedness in the negative direction. Using the symbols ∞ and $-\infty$, we can define five other kinds of intervals:

$[a, \infty)$ Consists of all real numbers x for which $x \ge a$
(a, ∞) Consists of all real numbers x for which $x > a$
$(-\infty, a]$ Consists of all real numbers x for which $x \le a$
$(-\infty, a)$ Consists of all real numbers x for which $x < a$
$(-\infty, \infty)$ Consists of all real numbers x

Note that ∞ and $-\infty$ are never included as endpoints, since neither is a real number.

Table 2 summarizes interval notation, corresponding inequality notation, and their graphs.

Table 2

Interval	Inequality	Graph
The open interval (a, b)	$a < x < b$	
The closed interval $[a, b]$	$a \leq x \leq b$	
The half-open interval $[a, b)$	$a \leq x < b$	
The half-open interval $(a, b]$	$a < x \leq b$	
The interval $[a, \infty)$	$x \geq a$	
The interval (a, ∞)	$x > a$	
The interval $(-\infty, a]$	$x \leq a$	
The interval $(-\infty, a)$	$x < a$	
The interval $(-\infty, \infty)$	All real numbers	

EXAMPLE 1

Writing Inequalities Using Interval Notation

Write each inequality using interval notation.

(a) $1 \leq x \leq 3$ (b) $-4 < x < 0$ (c) $x > 5$ (d) $x \leq 1$

Solution

(a) $1 \leq x \leq 3$ describes all numbers x between 1 and 3, inclusive. In interval notation, we write $[1, 3]$.

(b) In interval notation, $-4 < x < 0$ is written $(-4, 0)$.

(c) $x > 5$ consists of all numbers x greater than 5. In interval notation, we write $(5, \infty)$.

(d) In interval notation, $x \leq 1$ is written $(-\infty, 1]$.

EXAMPLE 2

Writing Intervals Using Inequality Notation

Write each interval as an inequality involving x.

(a) $[1, 4)$ (b) $(2, \infty)$ (c) $[2, 3]$ (d) $(-\infty, -3]$

Solution

(a) $[1, 4)$ consists of all numbers x for which $1 \leq x < 4$.

(b) $(2, \infty)$ consists of all numbers x for which $x > 2$.

(c) $[2, 3]$ consists of all numbers x for which $2 \leq x \leq 3$.

(d) $(-\infty, -3]$ consists of all numbers x for which $x \leq -3$.

Now Work PROBLEMS 11, 23, AND 31

2 Use Properties of Inequalities

The product of two positive real numbers is positive, the product of two negative real numbers is positive, and the product of 0 and 0 is 0. For any real number a, the value of a^2 is 0 or positive; that is, a^2 is nonnegative. This is called the **nonnegative property.**

Nonnegative Property

In Words
The square of a real number is never negative.

For any real number a,

$$a^2 \geq 0 \qquad\qquad (1)$$

If we add the same number to both sides of an inequality, we obtain an equivalent inequality. For example, since $3 < 5$, then $3 + 4 < 5 + 4$ or $7 < 9$. This is called the **addition property** of inequalities.

Addition Property of Inequalities

For real numbers a, b, and c,

$$\text{If } a < b, \text{ then } a + c < b + c. \tag{2a}$$

$$\text{If } a > b, \text{ then } a + c > b + c. \tag{2b}$$

The addition property states that the sense, or direction, of an inequality remains unchanged if the same number is added to each side.

Now let's see what happens if we multiply each side of an inequality by a nonzero number. We begin with $3 < 7$ and multiply each side by 2. The numbers 6 and 14 that result obey the inequality $6 < 14$.

Now start with $9 > 2$ and multiply each side by -4. The numbers -36 and -8 that result obey the inequality $-36 < -8$.

Note that the effect of multiplying both sides of $9 > 2$ by the negative number -4 is that the direction of the inequality symbol is reversed. We are led to the following general **multiplication properties** for inequalities:

Multiplication Properties for Inequalities

For real numbers a, b, and c,

$$\text{If } a < b \text{ and if } c > 0, \text{ then } ac < bc.$$
$$\text{If } a < b \text{ and if } c < 0, \text{ then } ac > bc. \tag{3a}$$

$$\text{If } a > b \text{ and if } c > 0, \text{ then } ac > bc.$$
$$\text{If } a > b \text{ and if } c < 0, \text{ then } ac < bc. \tag{3b}$$

In Words

Multiplying by a negative number reverses the inequality.

The multiplication properties state that the sense, or direction, of an inequality *remains the same* if each side is multiplied by a *positive* real number, whereas the direction is *reversed* if each side is multiplied by a *negative* real number.

EXAMPLE 3   **Multiplication Property of Inequalities**

(a) If $2x < 6$, then $\dfrac{1}{2}(2x) < \dfrac{1}{2}(6)$ or $x < 3$.

(b) If $\dfrac{x}{-3} > 12$, then $-3\left(\dfrac{x}{-3}\right) < -3(12)$ or $x < -36$.

(c) If $-4x < -8$, then $\dfrac{-4x}{-4} > \dfrac{-8}{-4}$ or $x > 2$.

(d) If $-x > 8$, then $(-1)(-x) < (-1)(8)$ or $x < -8$.

═══Now Work PROBLEM 45

The **reciprocal property** states that the reciprocal of a positive real number is positive and that the reciprocal of a negative real number is negative.

Reciprocal Property for Inequalities

$$\text{If } a > 0, \text{ then } \frac{1}{a} > 0 \qquad \text{If } \frac{1}{a} > 0, \text{ then } a > 0 \qquad \textbf{(4a)}$$

$$\text{If } a < 0, \text{ then } \frac{1}{a} < 0 \qquad \text{If } \frac{1}{a} < 0, \text{ then } a < 0 \qquad \textbf{(4b)}$$

3 Solve Inequalities

An **inequality in one variable** is a statement involving two expressions, at least one containing the variable, separated by one of the inequality symbols $<$, $\leq$, $>$, or $\geq$. To **solve an inequality** means to find all values of the variable for which the statement is true. These values are called **solutions** of the inequality.

For example, the following are all inequalities involving one variable x:

$$x + 5 < 8 \qquad 2x - 3 \geq 4 \qquad x^2 - 1 \leq 3 \qquad \frac{x + 1}{x - 2} > 0$$

As with equations, one method for solving an inequality is to replace it by a series of equivalent inequalities until an inequality with an obvious solution, such as $x < 3$, is obtained. We obtain equivalent inequalities by applying some of the same properties as those used to find equivalent equations. The addition property and the multiplication properties form the basis for the following procedures.

Procedures That Leave the Inequality Symbol Unchanged

1. Simplify both sides of the inequality by combining like terms and eliminating parentheses:

   Replace $(x + 2) + 6 > 2x + 5(x + 1)$

   by $x + 8 > 7x + 5$

2. Add or subtract the same expression on both sides of the inequality:

   Replace $3x - 5 < 4$

   by $(3x - 5) + 5 < 4 + 5$

3. Multiply or divide both sides of the inequality by the same positive expression:

   Replace $4x > 16$ by $\dfrac{4x}{4} > \dfrac{16}{4}$

Procedures That Reverse the Sense or Direction of the Inequality Symbol

1. Interchange the two sides of the inequality:

   Replace $3 < x$ by $x > 3$

2. Multiply or divide both sides of the inequality by the same *negative* expression:

   Replace $-2x > 6$ by $\dfrac{-2x}{-2} < \dfrac{6}{-2}$

As the examples that follow illustrate, we solve inequalities using many of the same steps that we would use to solve equations. In writing the solution of an inequality, we may use either set notation or interval notation, whichever is more convenient.

EXAMPLE 4 **Solving an Inequality**

Solve the inequality: $4x + 7 \geq 2x - 3$
Graph the solution set.

Solution

$$4x + 7 \geq 2x - 3$$
$$4x + 7 - 7 \geq 2x - 3 - 7 \qquad \text{Subtract 7 from both sides.}$$
$$4x \geq 2x - 10 \qquad \text{Simplify.}$$
$$4x - 2x \geq 2x - 10 - 2x \qquad \text{Subtract 2x from both sides.}$$
$$2x \geq -10 \qquad \text{Simplify.}$$
$$\frac{2x}{2} \geq \frac{-10}{2} \qquad \text{Divide both sides by 2. (The direction of the inequality symbol is unchanged.)}$$
$$x \geq -5 \qquad \text{Simplify.}$$

Figure 32

The solution set is $\{x \mid x \geq -5\}$ or, using interval notation, all numbers in the interval $[-5, \infty)$. See Figure 32 for the graph.

─────── **Now Work** PROBLEM 5 7

4 Solve Combined Inequalities

EXAMPLE 5 **Solving Combined Inequalities**

Solve the inequality: $-5 < 3x - 2 < 1$
Graph the solution set.

Solution

Recall that the inequality

$$-5 < 3x - 2 < 1$$

is equivalent to the two inequalities

$$-5 < 3x - 2 \quad \text{and} \quad 3x - 2 < 1$$

We will solve each of these inequalities separately.

$-5 < 3x - 2$		$3x - 2 < 1$
$-5 + 2 < 3x - 2 + 2$	Add 2 to both sides.	$3x - 2 + 2 < 1 + 2$
$-3 < 3x$	Simplify.	$3x < 3$
$\dfrac{-3}{3} < \dfrac{3x}{3}$	Divide both sides by 3.	$\dfrac{3x}{3} < \dfrac{3}{3}$
$-1 < x$	Simplify.	$x < 1$

The solution set of the original pair of inequalities consists of all x for which

$$-1 < x \quad \text{and} \quad x < 1$$

Figure 33

This may be written more compactly as $\{x \mid -1 < x < 1\}$. In interval notation, the solution is $(-1, 1)$. See Figure 33 for the graph.

We observe in the preceding process that the two inequalities we solved required exactly the same steps. A shortcut to solving the original inequality algebraically is to deal with the two inequalities at the same time, as follows:

$$
\begin{aligned}
-5 &< 3x - 2 &&< 1 \\
-5 + 2 &< 3x - 2 + 2 &&< 1 + 2 \qquad \text{Add 2 to each part.}\\
-3 &< 3x &&< 3 \qquad \text{Simplify.}\\
\frac{-3}{3} &< \frac{3x}{3} &&< \frac{3}{3} \qquad \text{Divide each part by 3.}\\
-1 &< x &&< 1 \qquad \text{Simplify.}
\end{aligned}
$$

━━━ **Now Work**   PROBLEM 73

EXAMPLE 6   **Using the Reciprocal Property to Solve an Inequality**

Solve the inequality: $(4x - 1)^{-1} > 0$
Graph the solution set.

Solution   Since $(4x - 1)^{-1} = \dfrac{1}{4x - 1}$ and since the Reciprocal Property states that when

$\dfrac{1}{a} > 0$ then $a > 0$, we have

$$
(4x - 1)^{-1} > 0
$$

$$
\frac{1}{4x - 1} > 0
$$

$$
4x - 1 > 0 \qquad \text{Reciprocal Property}
$$

$$
4x > 1
$$

$$
x > \frac{1}{4}
$$

Figure 34

The solution set is $\left\{ x \middle| x > \dfrac{1}{4} \right\}$, that is, all x in the interval $\left(\dfrac{1}{4}, \infty \right)$. Figure 34 illustrates the graph.   ●

━━━ **Now Work**   PROBLEM 83

A.10 Assess Your Understanding

'Are You Prepared?'   *Answers are given at the end of these exercises. If you get a wrong answer, read the pages listed in red.*

1. Graph the inequality: $x \geq -2$. (pp. A4–A5)

2. Graph the inequality: $x < 1$. (p. A4–A5)

Concepts and Vocabulary

3. If each side of an inequality is multiplied by a(n) _____ number, then the direction of the inequality symbol is reversed.

4. A(n) _____ _____, denoted $[a, b]$, consists of all real numbers x for which $a \leq x \leq b$.

5. The _____ _____ state's that the sense, or direction, of an inequality remains the same if each side is multiplied by a positive number, whereas the direction is reversed if each side is multiplied by a negative number.

In Problems 6–9, determine if the statement is True or False if $a < b$ and $c < 0$.

6. $a + c < b + c$ **7.** $a - c < b - c$ **8.** $ac > bc$ **9.** $\dfrac{a}{c} < \dfrac{b}{c}$

10. *True or False:* The square of any real number is always nonnegative.

Skill Building

In Problems 11–16, express the graph shown in blue using interval notation. Also express each as an inequality involving x.

11.

12.

13.

14.

15.

16.

In Problems 17–22, an inequality is given. Write the inequality obtained by:
(a) *Adding 3 to each side of the given inequality.*
(b) *Subtracting 5 from each side of the given inequality.*
(c) *Multiplying each side of the given inequality by 3.*
(d) *Multiplying each side of the given inequality by −2.*

17. $3 < 5$ **18.** $2 > 1$ **19.** $4 > -3$ **20.** $-3 > -5$ **21.** $2x + 1 < 2$ **22.** $1 - 2x > 5$

In Problems 23–30, write each inequality using interval notation, and illustrate each inequality using the real number line.

23. $0 \le x \le 4$ **24.** $-1 < x < 5$ **25.** $4 \le x < 6$ **26.** $-2 < x < 0$

27. $x \ge 4$ **28.** $x \le 5$ **29.** $x < -4$ **30.** $x > 1$

In Problems 31–38, write each interval as an inequality involving x, and illustrate each inequality using the real number line.

31. $[2, 5]$ **32.** $(1, 2)$ **33.** $(-3, -2)$ **34.** $[0, 1)$

35. $[4, \infty)$ **36.** $(-\infty, 2]$ **37.** $(-\infty, -3)$ **38.** $(-8, \infty)$

In Problems 39–52, fill in the blank with the correct inequality symbol.

39. If $x < 5$, then $x - 5$ _____ 0. **40.** If $x < -4$, then $x + 4$ _____ 0.

41. If $x > -4$, then $x + 4$ _____ 0. **42.** If $x > 6$, then $x - 6$ _____ 0.

43. If $x \ge -4$, then $3x$ _____ -12. **44.** If $x \le 3$, then $2x$ _____ 6.

45. If $x > 6$, then $-2x$ _____ -12. **46.** If $x > -2$, then $-4x$ _____ 8.

47. If $x \ge 5$, then $-4x$ _____ -20. **48.** If $x \le -4$, then $-3x$ _____ 12.

49. If $2x > 6$, then x _____ 3. **50.** If $3x \le 12$, then x _____ 4.

51. If $-\dfrac{1}{2} x \le 3$, then x _____ -6. **52.** If $-\dfrac{1}{4} x > 1$, then x _____ -4.

In Problems 53–88, solve each inequality. Express your answer using set notation or interval notation. Graph the solution set.

53. $x + 1 < 5$ **54.** $x - 6 < 1$ **55.** $1 - 2x \le 3$

56. $2 - 3x \le 5$ **57.** $3x - 7 > 2$ **58.** $2x + 5 > 1$

59. $3x - 1 \geq 3 + x$

60. $2x - 2 \geq 3 + x$

61. $-2(x + 3) < 8$

62. $-3(1 - x) < 12$

63. $4 - 3(1 - x) \leq 3$

64. $8 - 4(2 - x) \leq -2x$

65. $\frac{1}{2}(x - 4) > x + 8$

66. $3x + 4 > \frac{1}{3}(x - 2)$

67. $\frac{x}{2} \geq 1 - \frac{x}{4}$

68. $\frac{x}{3} \geq 2 + \frac{x}{6}$

69. $0 \leq 2x - 6 \leq 4$

70. $4 \leq 2x + 2 \leq 10$

71. $-5 \leq 4 - 3x \leq 2$

72. $-3 \leq 3 - 2x \leq 9$

73. $-3 < \frac{2x - 1}{4} < 0$

74. $0 < \frac{3x + 2}{2} < 4$

75. $1 < 1 - \frac{1}{2}x < 4$

76. $0 < 1 - \frac{1}{3}x < 1$

77. $(x + 2)(x - 3) > (x - 1)(x + 1)$

78. $(x - 1)(x + 1) > (x - 3)(x + 4)$

79. $x(4x + 3) \leq (2x + 1)^2$

80. $x(9x - 5) \leq (3x - 1)^2$

81. $\frac{1}{2} \leq \frac{x + 1}{3} < \frac{3}{4}$

82. $\frac{1}{3} < \frac{x + 1}{2} \leq \frac{2}{3}$

83. $(4x + 2)^{-1} < 0$

84. $(2x - 1)^{-1} > 0$

85. $0 < \frac{2}{x} < \frac{3}{5}$

86. $0 < \frac{4}{x} < \frac{2}{3}$

87. $0 < (2x - 4)^{-1} < \frac{1}{2}$

88. $0 < (3x + 6)^{-1} < \frac{1}{3}$

Applications and Extensions

89. What is the domain of the variable in the expression $\sqrt{3x + 6}$?

90. What is the domain of the variable in the expression $\sqrt{8 + 2x}$?

91. A young adult may be defined as someone older than 21, but less than 30 years of age. Express this statement using inequalities.

92. Middle-aged may be defined as being 40 or more and less than 60. Express this statement using inequalities.

93. **Life Expectancy** According to the National Center for Health Statistics, an average 30-year-old male in 2004 could expect to live at least 46.9 more years, and an average 30-year-old female in 2004 could expect to live at least 51.5 more years.

(a) To what age can an average 30-year-old male expect to live? Express your answer as an inequality.

(b) To what age can an average 30-year-old female expect to live? Express your answer as an inequality.

(c) Who can expect to live longer, a male or a female? By how many years?

Source: National Vital Statistics Reports, Vol. 56, No. 9, December 2007.

94. **General Chemistry** For a certain ideal gas, the volume V (in cubic centimeters) equals 20 times the temperature T (in degrees Celsius). If the temperature varies from 80° to 120°C, inclusive, what is the corresponding range of the volume of the gas?

95. **Real Estate** A real estate agent agrees to sell an apartment complex according to the following commission schedule: $45,000 plus 25% of the selling price in excess of $900,000. Assuming that the complex will sell at some price between $900,000 and $1,100,000 inclusive, over what range does the agent's commission vary? How does the commission vary as a percent of selling price?

96. **Sales Commission** A used car salesperson is paid a commission of $25 plus 40% of the selling price in excess of owner's cost. The owner claims that used cars typically sell for at least owner's cost plus $200 and at most owner's cost plus $3000. For each sale made, over what range can the salesperson expect the commission to vary?

97. **Federal Tax Withholding** The percentage method of withholding for federal income tax (2009) states that a single person whose weekly wages, after subtracting withholding allowances, are over $696, but not over $1279, shall have $80.60 plus 25% of the excess over $696 withheld. Over what range does the amount withheld vary if the weekly wages vary from $800 to $1000 inclusive?

Source: Employer's Tax Guide. Department of the Treasury, Internal Revenue Service, Publication 2009.

98. **Exercising** Sue wants to lose weight. For healthy weight loss, the American College of Sports Medicine (ACSM) recommends 200 to 300 minutes of exercise per week. For the first six days of the week, Sue exercised 40, 45, 0, 50, 25, and 35 minutes. How long should Sue exercise on the seventh day in order to stay within the ACSM guidelines?

99. **Electricity Rates** Commonwealth Edison Company's charge for electricity in May 2009 is 7.598¢ per kilowatt-hour. In addition, each monthly bill contains a customer charge of \$8.23. If last year's bills ranged from a low of \$59.55 to a high of \$198.25, over what range did usage vary (in kilowatt-hours)?

Source: Commonwealth Edison Co., Chicago, Illinois, 2009.

100. Water Bills The Village of Oak Lawn charges homeowners \$28.84 per quarter-year plus \$2.28 per 1000 gallons for water usage in excess of 12,000 gallons. In 2009 one homeowner's quarterly bill ranged from a high of \$74.44 to a low of \$42.52. Over what range did water usage vary?

Source: Village of Oak Lawn, Illinois, April 2009.

101. Markup of a New Car The markup over dealer's cost of a new car ranges from 12% to 18%. If the sticker price is \$18,000, over what range will the dealer's cost vary?

102. IQ Tests A standard intelligence test has an average score of 100. According to statistical theory, of the people who take the test, the 2.5% with the highest scores will have scores of more than 1.96σ above the average, where σ (sigma, a number called the **standard deviation**) depends on the nature of the test. If $\sigma = 12$ for this test and there is (in principle) no upper limit to the score possible on the test, write the interval of possible test scores of the people in the top 2.5%.

103. Computing Grades In your Economics 101 class, you have scores of 68, 82, 87, and 89 on the first four of five tests. To get a grade of B, the average of the first five test scores must be greater than or equal to 80 and less than 90.
  (a) Solve an inequality to find the range of the score that you need on the last test to get a B.
  (b) What score do you need if the fifth test counts double?

What do I need to get a B?

104. "Light" Foods For food products to be labeled "light," the U.S. Food and Drug Administration requires that the altered product must either contain one-third or fewer calories than the regular product or it must contain one-half or less fat than the regular product. If a serving of Miracle Whip® Light contains 20 calories and 1.5 grams of fat, then what must be true about either the number of calories or the grams of fat in a serving of regular Miracle Whip®?

105. Arithmetic Mean If $a < b$, show that $a < \dfrac{a + b}{2} < b$. The number $\dfrac{a + b}{2}$ is called the **arithmetic mean** of a and b.

106. Refer to Problem 105. Show that the arithmetic mean of a and b is equidistant from a and b.

107. Geometric Mean If $0 < a < b$, show that $a < \sqrt{ab} < b$. The number $\sqrt{ab}$ is called the **geometric mean** of a and b.

108. Refer to Problems 105 and 107. Show that the geometric mean of a and b is less than the arithmetic mean of a and b.

109. Harmonic Mean For $0 < a < b$, let h be defined by

$$\frac{1}{h} = \frac{1}{2}\left(\frac{1}{a} + \frac{1}{b}\right)$$

Show that $a < h < b$. The number h is called the **harmonic mean** of a and b.

110. Refer to Problems 105, 107, and 109. Show that the harmonic mean of a and b equals the geometric mean squared divided by the arithmetic mean.

111. Another Reciprocal Property Prove that if $0 < a < b$ then $0 < \dfrac{1}{b} < \dfrac{1}{a}$.

Explaining Concepts: Discussion and Writing

112. Make up an inequality that has no solution. Make up one that has exactly one solution.

113. The inequality $x^2 + 1 < -5$ has no real solution. Explain why.

114. Do you prefer to use inequality notation or interval notation to express the solution to an inequality? Give your reasons. Are there particular circumstances when you prefer one to the other? Cite examples.

115. How would you explain to a fellow student the underlying reason for the multiplication properties for inequalities (page A85); that is, the sense or direction of an inequality remains the same if each side is multiplied by a positive real number, whereas the direction is reversed if each side is multiplied by a negative real number?

'Are You Prepared?' Answers

1.

A.11 Complex Numbers

PREPARING FOR THIS SECTION *Before getting started, review the following:*

- Classification of Numbers (Appendix A, Section A.1, p. A3)
- Rationalizing Denominators (Appendix A, Section A.7, p. A59)

Now Work the 'Are You Prepared?' problems on page A97.

OBJECTIVES **1** Add, Subtract, Multiply, and Divide Complex Numbers (p. A93)

Complex Numbers

One property of a real number is that its square is nonnegative (greater than or equal to 0). For example, there is no real number x for which

$$x^2 = -1$$

To remedy this situation, we introduce a number called the **imaginary unit,** which we denote by *i*, whose square is -1; that is,

$$i^2 = -1$$

This should not surprise you. If our universe were to consist only of integers, there would be no number x for which $2x = 1$. This unfortunate circumstance was remedied by introducing numbers such as $\frac{1}{2}$ and $\frac{2}{3}$, the *rational numbers*. If our universe were to consist only of rational numbers, there would be no x whose square equals 2. That is, there would be no number x for which $x^2 = 2$. To remedy this, we introduced numbers such as $\sqrt{2}$ and $\sqrt[3]{5}$, the *irrational numbers*. The *real numbers*, you will recall, consist of the rational numbers and the irrational numbers. Now, if our universe were to consist only of real numbers, then there would be no number x whose square is -1. To remedy this, we introduce a number i, whose square is -1.

In the progression outlined, each time that we encountered a situation that was unsuitable, we introduced a new number system to remedy this situation. And each new number system contained the earlier number system as a subset. The number system that results from introducing the number i is called the **complex number system.**

Complex numbers are numbers of the form $a + bi$, where a and b are real numbers. The real number a is called the **real part** of the number $a + bi$; the real number b is called the **imaginary part** of $a + bi$.

For example, the complex number $-5 + 6i$ has the real part -5 and the imaginary part 6.

When a complex number is written in the form $a + bi$, where a and b are real numbers, we say it is in **standard form.** However, if the imaginary part of a complex number is negative, such as in the complex number $3 + (-2)i$, we agree to write it instead in the form $3 - 2i$.

Also, the complex number $a + 0i$ is usually written simply as a. This serves to remind us that the real numbers are a subset of the complex numbers. The complex number $0 + bi$ is usually written as bi. Sometimes the complex number bi is called a **pure imaginary number.**

1 Add, Subtract, Multiply, and Divide Complex Numbers

Equality, addition, subtraction, and multiplication of complex numbers are defined so as to preserve the familiar rules of algebra for real numbers. Thus, two complex numbers are equal if and only if their real parts are equal and their imaginary parts are equal. That is,

Equality of Complex Numbers

$$a + bi = c + di \quad \text{if and only if} \quad a = c \text{ and } b = d \qquad (1)$$

Two complex numbers are added by forming the complex number whose real part is the sum of the real parts and whose imaginary part is the sum of the imaginary parts. That is,

Sum of Complex Numbers

$$(a + bi) + (c + di) = (a + c) + (b + d)i \qquad (2)$$

To subtract two complex numbers, we use this rule:

Difference of Complex Numbers

$$(a + bi) - (c + di) = (a - c) + (b - d)i \qquad (3)$$

EXAMPLE 1 | **Adding and Subtracting Complex Numbers**

(a) $(3 + 5i) + (-2 + 3i) = [3 + (-2)] + (5 + 3)i = 1 + 8i$

(b) $(6 + 4i) - (3 + 6i) = (6 - 3) + (4 - 6)i = 3 + (-2)i = 3 - 2i$ ●

▰▰▰ **Now Work** PROBLEM 13

Products of complex numbers are calculated as illustrated in Example 2.

EXAMPLE 2 | **Multiplying Complex Numbers**

$$
\begin{aligned}
(5 + 3i) \cdot (2 + 7i) &= 5(2 + 7i) + 3i(2 + 7i) && \text{Distributive Property} \\
&= 10 + 35i + 6i + 21i^2 && \text{Distributive Property} \\
&= 10 + 41i + 21(-1) && i^2 = -1 \\
&= -11 + 41i
\end{aligned}
$$

●

Based on the procedure of Example 2, we define the **product** of two complex numbers by the following formula:

Product of Complex Numbers

$$(a + bi) \cdot (c + di) = (ac - bd) + (ad + bc)i \qquad (4)$$

Do not bother to memorize formula (4). Instead, whenever it is necessary to multiply two complex numbers, follow the usual rules for multiplying two binomials, as in Example 2, remembering that $i^2 = -1$. For example,

$$(2i)(2i) = 4i^2 = -4$$
$$(2 + i)(1 - i) = 2 - 2i + i - i^2 = 3 - i$$

Now Work PROBLEM 19

Algebraic properties for addition and multiplication, such as the Commutative, Associative, and Distributive Properties, hold for complex numbers. However, the property that every nonzero complex number has a multiplicative inverse, or reciprocal, requires a closer look.

Conjugates

If $z = a + bi$ is a complex number, then its **conjugate**, denoted by $\bar{z}$, is defined as

$$\bar{z} = \overline{a + bi} = a - bi$$

For example, $\overline{2 + 3i} = 2 - 3i$ and $\overline{-6 - 2i} = -6 + 2i$.

EXAMPLE 3	**Multiplying a Complex Number by Its Conjugate**

Find the product of the complex number $z = 3 + 4i$ and its conjugate $\bar{z}$.

Solution Since $\bar{z} = 3 - 4i$, we have

$$z\bar{z} = (3 + 4i)(3 - 4i) = 9 - 12i + 12i - 16i^2 = 9 + 16 = 25$$

The result obtained in Example 3 has an important generalization.

THEOREM The product of a complex number and its conjugate is a nonnegative real number. That is, if $z = a + bi$, then

$$z\bar{z} = a^2 + b^2 \qquad\qquad (5)$$

Proof If $z = a + bi$, then

$$z\bar{z} = (a + bi)(a - bi) = a^2 - abi + abi - (bi)^2 = a^2 - b^2i^2 = a^2 + b^2 \qquad \blacksquare$$

To express the reciprocal of a nonzero complex number z in standard form, multiply the numerator and denominator of $\dfrac{1}{z}$ by its conjugate $\bar{z}$. That is, if $z = a + bi$ is a nonzero complex number, then

$$\frac{1}{a + bi} = \frac{1}{z} = \frac{1}{z} \cdot \frac{\bar{z}}{\bar{z}} = \frac{\bar{z}}{z\bar{z}} \underset{\substack{\uparrow \\ \text{Use (5).}}}{=} \frac{a - bi}{a^2 + b^2} = \frac{a}{a^2 + b^2} - \frac{b}{a^2 + b^2}i$$

EXAMPLE 4	**Writing the Reciprocal of a Complex Number in Standard Form**

Write $\dfrac{1}{3 + 4i}$ in standard form $a + bi$; that is, find the reciprocal of $3 + 4i$.

Solution

Multiply the numerator and denominator of $\dfrac{1}{3 + 4i}$ by the conjugate of $3 + 4i$, the complex number $3 - 4i$. The result is

$$\frac{1}{3 + 4i} = \frac{1}{3 + 4i} \cdot \frac{3 - 4i}{3 - 4i} = \frac{3 - 4i}{9 + 16} = \frac{3}{25} - \frac{4}{25}i$$

To express the quotient of two complex numbers in standard form, we multiply the numerator and denominator of the quotient by the conjugate of the denominator.

EXAMPLE 5

Writing the Quotient of Complex Numbers in Standard Form

Write each of the following in standard form.

(a) $\dfrac{1 + 4i}{5 - 12i}$ (b) $\dfrac{2 - 3i}{4 - 3i}$

Solution

(a) $\dfrac{1 + 4i}{5 - 12i} = \dfrac{1 + 4i}{5 - 12i} \cdot \dfrac{5 + 12i}{5 + 12i} = \dfrac{5 + 12i + 20i + 48i^2}{25 + 144}$

$= \dfrac{-43 + 32i}{169} = \dfrac{-43}{169} + \dfrac{32}{169}i$

(b) $\dfrac{2 - 3i}{4 - 3i} = \dfrac{2 - 3i}{4 - 3i} \cdot \dfrac{4 + 3i}{4 + 3i} = \dfrac{8 + 6i - 12i - 9i^2}{16 + 9}$

$= \dfrac{17 - 6i}{25} = \dfrac{17}{25} - \dfrac{6}{25}i$

Now Work PROBLEM 27

EXAMPLE 6

Writing Other Expressions in Standard Form

If $z = 2 - 3i$ and $w = 5 + 2i$, write each of the following expressions in standard form.

(a) $\dfrac{z}{w}$ (b) $\overline{z + w}$ (c) $z + \overline{z}$

Solution

(a) $\dfrac{z}{w} = \dfrac{z \cdot \overline{w}}{w \cdot \overline{w}} = \dfrac{(2 - 3i)(5 - 2i)}{(5 + 2i)(5 - 2i)} = \dfrac{10 - 4i - 15i + 6i^2}{25 + 4}$

$= \dfrac{4 - 19i}{29} = \dfrac{4}{29} - \dfrac{19}{29}i$

(b) $\overline{z + w} = \overline{(2 - 3i) + (5 + 2i)} = \overline{7 - i} = 7 + i$

(c) $z + \overline{z} = (2 - 3i) + (2 + 3i) = 4$

The conjugate of a complex number has certain general properties that we shall find useful later.

For a real number $a = a + 0i$, the conjugate is $\overline{a} = \overline{a + 0i} = a - 0i = a$. That is,

THEOREM

The conjugate of a real number is the real number itself.

Other properties that are direct consequences of the definition of the conjugate are given next. In each statement, z and w represent complex numbers.

THEOREM

The conjugate of the conjugate of a complex number is the complex number itself.

$$\overline{\overline{z}} = z \qquad\qquad (6)$$

The conjugate of the sum of two complex numbers equals the sum of their conjugates.

$$\overline{z + w} = \overline{z} + \overline{w} \qquad\qquad (7)$$

The conjugate of the product of two complex numbers equals the product of their conjugates.

$$\overline{z \cdot w} = \overline{z} \cdot \overline{w} \qquad\qquad (8)$$

We leave the proofs of equations (6), (7), and (8) as exercises. See Problems 60–62.

Powers of i

The powers of i follow a pattern that is useful to know.

$$i^1 = i \qquad\qquad i^5 = i^4 \cdot i = 1 \cdot i = i$$
$$i^2 = -1 \qquad\qquad i^6 = i^4 \cdot i^2 = -1$$
$$i^3 = i^2 \cdot i = -i \qquad\qquad i^7 = i^4 \cdot i^3 = -i$$
$$i^4 = i^2 \cdot i^2 = (-1)(-1) = 1 \qquad i^8 = i^4 \cdot i^4 = 1$$

And so on. The powers of i repeat with every fourth power.

EXAMPLE 7

Evaluating Powers of i

(a) $i^{27} = i^{24} \cdot i^3 = (i^4)^6 \cdot i^3 = 1^6 \cdot i^3 = -i$

(b) $i^{101} = i^{100} \cdot i^1 = (i^4)^{25} \cdot i = 1^{25} \cdot i = i$

EXAMPLE 8

Writing the Power of a Complex Number in Standard Form

Write $(2 + i)^3$ in standard form.

Solution

We use the special product formula for $(x + a)^3$.

$$(x + a)^3 = x^3 + 3ax^2 + 3a^2x + a^3$$

Using this special product formula,

$$(2 + i)^3 = 2^3 + 3 \cdot i \cdot 2^2 + 3 \cdot i^2 \cdot 2 + i^3$$
$$= 8 + 12i + 6(-1) + (-i)$$
$$= 2 + 11i$$

NOTE If you did not remember the special product formula for $(x + a)^3$, you could find $(2 + i)^3$ by simplifying $(2 + i)^2 (2 + i)$. ∎

━━**Now Work** PROBLEMS 33 AND 41

Square Roots of Negative Numbers

Now that we know $i^2 = -1$, we can define the square root of a negative number.

DEFINITION

If N is a positive real number, we define the **principal square root of** $-N$, denoted by $\sqrt{-N}$, as

$$\sqrt{-N} = \sqrt{N}i$$

where i is the imaginary unit and $i^2 = -1$.

EXAMPLE 9

Evaluating the Square Root of a Negative Number

(a) $\sqrt{-1} = \sqrt{1}i = i$ (b) $\sqrt{-16} = \sqrt{16}i = 4i$
(c) $\sqrt{-8} = \sqrt{8}i = 2\sqrt{2}i$

━━━━**Now Work** PROBLEM 47

A.11 Assess Your Understanding

'Are You Prepared?' *Answers are given at the end of these exercises. If you get a wrong answer, read the pages listed in red.*

1. Name the integers and the rational numbers in the set $\left\{-3, 0, \sqrt{2}, \dfrac{6}{5}, \pi\right\}$. (p. A3)

2. *True or False* Rational numbers and irrational numbers are in the set of real numbers. (p. A3)

3. Rationalize the denominator of $\dfrac{3}{2 + \sqrt{3}}$. (p. A59)

Concepts and Vocabulary

4. In the complex number $5 + 2i$, the number 5 is called the _____ part; the number 2 is called the _____ part; the number i is called the _____ _____.

5. $i^2 =$ _____ ; $i^3 =$ _____ ; $i^4 =$ _____ .

6. *True or False* The conjugate of $2 + 5i$ is $-2 - 5i$.

7. *True or False* All real numbers are complex numbers.

8. *True or False* The product of a complex number and its conjugate is a nonnegative real number.

Skill Building

In Problems 9–46, write each expression in the standard form $a + bi$.

9. $(2 - 3i) + (6 + 8i)$

10. $(4 + 5i) + (-8 + 2i)$

11. $(-3 + 2i) - (4 - 4i)$

12. $(3 - 4i) - (-3 - 4i)$

13. $(2 - 5i) - (8 + 6i)$

14. $(-8 + 4i) - (2 - 2i)$

15. $3(2 - 6i)$

16. $-4(2 + 8i)$

17. $2i(2 - 3i)$

18. $3i(-3 + 4i)$

19. $(3 - 4i)(2 + i)$

20. $(5 + 3i)(2 - i)$

21. $(-6 + i)(-6 - i)$

22. $(-3 + i)(3 + i)$

23. $\dfrac{10}{3 - 4i}$

24. $\dfrac{13}{5 - 12i}$

25. $\dfrac{2 + i}{i}$

26. $\dfrac{2 - i}{-2i}$

27. $\dfrac{6 - i}{1 + i}$

28. $\dfrac{2 + 3i}{1 - i}$

29. $\left(\dfrac{1}{2} + \dfrac{\sqrt{3}}{2} i\right)^2$ **30.** $\left(\dfrac{\sqrt{3}}{2} - \dfrac{1}{2} i\right)^2$ **31.** $(1 + i)^2$

32. $(1 - i)^2$ **33.** i^{23} **34.** i^{14}

35. i^{-15} **36.** i^{-23} **37.** $i^6 - 5$

38. $4 + i^3$ **39.** $6i^3 - 4i^5$ **40.** $4i^3 - 2i^2 + 1$

41. $(1 + i)^3$ **42.** $(3i)^4 + 1$ **43.** $i^7(1 + i^2)$

44. $2i^4(1 + i^2)$ **45.** $i^6 + i^4 + i^2 + 1$ **46.** $i^7 + i^5 + i^3 + i$

In Problems 47–52, perform the indicated operations and express your answer in the form $a + bi$.

47. $\sqrt{-4}$ **48.** $\sqrt{-9}$ **49.** $\sqrt{-25}$

50. $\sqrt{-64}$ **51.** $\sqrt{(3 + 4i)(4i - 3)}$ **52.** $\sqrt{(4 + 3i)(3i - 4)}$

Applications and Extensions

In Problems 53–56, $z = 3 - 4i$ and $w = 8 + 3i$. Write each expression in the standard form $a + bi$.

53. $z + \overline{z}$ **54.** $w - \overline{w}$ **55.** $z\overline{z}$ **56.** $\overline{z - w}$

57. Electrical Circuits   The impedance Z, in ohms, of a circuit element is defined as the ratio of the phasor voltage V, in volts, across the element to the phasor current I, in amperes, through the elements. That is, $Z = \dfrac{V}{I}$. If the voltage across a circuit element is $18 + i$ volts and the current through the element is $3 - 4i$ amperes, determine the impedance.

58. Parallel Circuits   In an ac circuit with two parallel pathways, the total impedance Z, in ohms, satisfies the formula $\dfrac{1}{Z} = \dfrac{1}{Z_1} + \dfrac{1}{Z_2}$, where Z_1 is the impedance of the first

pathway and Z_2 is the impedance of the second pathway. Determine the total impedance if the impedances of the two pathways are $Z_1 = 2 + i$ ohms and $Z_2 = 4 - 3i$ ohms.

59. Use $z = a + bi$ to show that $z + \overline{z} = 2a$ and that $z - \overline{z} = 2bi$.

60. Use $z = a + bi$ to show that $\overline{\overline{z}} = z$.

61. Use $z = a + bi$ and $w = c + di$ to show that $\overline{z + w} = \overline{z} + \overline{w}$.

62. Use $z = a + bi$ and $w = c + di$ to show that $\overline{z \cdot w} = \overline{z} \cdot \overline{w}$.

Explaining Concepts: Discussion and Writing

63. Explain to a friend how you would add two complex numbers and how you would multiply two complex numbers. Explain any differences in the two explanations.

64. Write a brief paragraph that compares the method used to rationalize denominators and the method used to write the quotient of two complex numbers in standard form.

65. Use an Internet search engine to investigate the origins of complex numbers. Write a paragraph describing what you find and present it to the class.

66. Explain how the method of multiplying two complex numbers is related to multiplying two binomials.

67. What Went Wrong?   A student multiplied $\sqrt{-9}$ and $\sqrt{-9}$ as follows:

$$\sqrt{-9} \cdot \sqrt{-9} = \sqrt{(-9)(-9)}$$
$$= \sqrt{81}$$
$$= 9$$

The instructor marked the problem incorrect. Why?

'Are You Prepared?' Answers

1. Integers: $\{-3, 0\}$; rational numbers: $\left\{-3, 0, \dfrac{6}{5}\right\}$ **2.** True   **3.** $3(2 - \sqrt{3})$

Graphing Utilities

Outline

B.1 The Viewing Rectangle

Figure 1
$y = 2x$

Figure 2

All graphing utilities, that is, all graphing calculators and all computer software graphing packages, graph equations by plotting points on a screen. The screen itself actually consists of small rectangles, called **pixels.** The more pixels the screen has, the better the resolution. Most graphing calculators have 2048 pixels per square inch; most computer screens have 4096 to 8192 pixels per square inch. When a point to be plotted lies inside a pixel, the pixel is turned on (lights up). The graph of an equation is a collection of pixels. Figure 1 shows how the graph of $y = 2x$ looks on a TI-84 Plus graphing calculator.

The screen of a graphing utility will display the coordinate axes of a rectangular coordinate system. However, you must set the scale on each axis. You must also include the smallest and largest values of x and y that you want included in the graph. This is called **setting the viewing rectangle** or **viewing window.** Figure 2 illustrates a typical viewing window.

To select the viewing window, we must give values to the following expressions:

Xmin: the smallest value of x
Xmax: the largest value of x
Xscl: the number of units per tick mark on the x-axis
Ymin: the smallest value of y
Ymax: the largest value of y
Yscl: the number of units per tick mark on the y-axis

Figure 3 illustrates these settings and their relation to the Cartesian coordinate system.

Figure 3

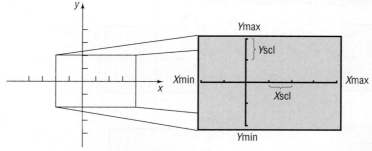

If the scale used on each axis is known, we can determine the minimum and maximum values of x and y shown on the screen by counting the tick marks. Look again at Figure 2. For a scale of 1 on each axis, the minimum and maximum values of x

are −10 and 10, respectively; the minimum and maximum values of y are also −10 and 10. If the scale is 2 on each axis, then the minimum and maximum values of x are −20 and 20, respectively; and the minimum and maximum values of y are −20 and 20, respectively.

Conversely, if we know the minimum and maximum values of x and y, we can determine the scales being used by counting the tick marks displayed. We shall follow the practice of showing the minimum and maximum values of x and y in our illustrations so that you will know how the viewing window was set. See Figure 4.

Figure 4

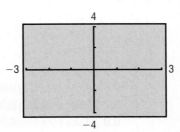

means

$X\min = -3 \quad Y\min = -4$
$X\max = 3 \quad Y\max = 4$
$X\text{scl} = 1 \quad Y\text{scl} = 2$

EXAMPLE 1 **Finding the Coordinates of a Point Shown on a Graphing Utility Screen**

Find the coordinates of the point shown in Figure 5. Assume that the coordinates are integers.

Figure 5

Solution First we note that the viewing window used in Figure 5 is

$X\min = -3 \quad Y\min = -4$
$X\max = 3 \quad Y\max = 4$
$X\text{scl} = 1 \quad Y\text{scl} = 2$

The point shown is 2 tick units to the left on the horizontal axis (scale = 1) and 1 tick up on the vertical axis (scale = 2). The coordinates of the point shown are $(-2, 2)$.

●

B.1 Exercises

In Problems 1–4, determine the coordinates of the points shown. Tell in which quadrant each point lies. Assume that the coordinates are integers.

1.

2.

3.

4.

In Problems 5–10, determine the viewing window used.

5.

6.

7.

8.

9.

10.

In Problems 11–16, select a setting so that each of the given points will lie within the viewing rectangle.

11. $(-10, 5), (3, -2), (4, -1)$

12. $(5, 0), (6, 8), (-2, -3)$

13. $(40, 20), (-20, -80), (10, 40)$

14. $(-80, 60), (20, -30), (-20, -40)$

15. $(0, 0), (100, 5), (5, 150)$

16. $(0, -1), (100, 50), (-10, 30)$

B.2 Using a Graphing Utility to Graph Equations

From Examples 2 and 3 in Foundations, Section 2, we see that a graph can be obtained by plotting points in a rectangular coordinate system and connecting them. Graphing utilities perform these same steps when graphing an equation. For example, the TI-84 Plus determines 95 evenly spaced input values,* starting at Xmin and ending at Xmax, uses the equation to determine the output values, plots these points on the screen, and finally (if in the connected mode) draws a line between consecutive points.

To graph an equation in two variables x and y using a graphing utility requires that the equation be written in the form $y = \{expression\ in\ x\}$. If the original equation is not in this form, replace it by equivalent equations until the form $y = \{expression\ in\ x\}$ is obtained.

Steps for Graphing an Equation Using a Graphing Utility

STEP 1: Solve the equation for y in terms of x.

STEP 2: Get into the graphing mode of your graphing utility. The screen will usually display $Y_1 =$, prompting you to enter the expression involving x that you found in Step 1. (Consult your manual for the correct way to enter the expression; for example, $y = x^2$ might be entered as $x^\wedge 2$ or as $x*x$ or as $x\ x^Y\ 2$).

STEP 3: Select the viewing window. Without prior knowledge about the behavior of the graph of the equation, it is common to select the **standard viewing window**** initially. The viewing window is then adjusted based on the graph that appears. In this text the standard viewing window is

$$X\text{min} = -10 \qquad Y\text{min} = -10$$
$$X\text{max} = 10 \qquad Y\text{max} = 10$$
$$X\text{scl} = 1 \qquad Y\text{scl} = 1$$

STEP 4: Graph.

STEP 5: Adjust the viewing window until a complete graph is obtained.

EXAMPLE 1

Graphing an Equation on a Graphing Utility

Graph the equation: $6x^2 + 3y = 36$

Solution   **STEP 1:** We solve for y in terms of x.

$$6x^2 + 3y = 36$$

$$3y = -6x^2 + 36 \quad \text{\small Subtract } 6x^2 \text{ from both sides of the equation.}$$

$$y = -2x^2 + 12 \quad \text{\small Divide both sides of the equation by 3 and simplify.}$$

---

*These input values depend on the values of Xmin and Xmax. For example, if Xmin $= -10$ and Xmax $= 10$, then the first input value will be -10 and the next input value will be $-10 + \dfrac{10 - (-10)}{94} = -9.7872$, and so on.

**Some graphing utilities have a ZOOM-STANDARD feature that automatically sets the viewing window to the standard viewing window and graphs the equation.

Step 2: From the $Y_1 =$ screen, enter the expression $-2x^2 + 12$ after the prompt.

Step 3: Set the viewing window to the standard viewing window.

Step 4: Graph. The screen should look like Figure 6.

Step 5: The graph of $y = -2x^2 + 12$ is not complete. The value of Ymax must be increased so that the top portion of the graph is visible. After increasing the value of Ymax to 12, we obtain the graph in Figure 7. The graph is now complete.

Figure 6

Figure 7

Figure 8

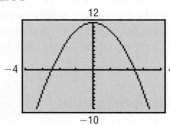

Look again at Figure 7. Although a complete graph is shown, the graph might be improved by adjusting the values of Xmin and Xmax. Figure 8 shows the graph of $y = -2x^2 + 12$ using Xmin $= -4$ and Xmax $= 4$. Do you think this is a better choice for the viewing window?

EXAMPLE 2   **Creating a Table and Graphing an Equation**

Create a table and graph the equation: $y = x^3$

Solution   Most graphing utilities have the capability of creating a table of values for an equation. (Check your manual to see if your graphing utility has this capability.) Table 1 illustrates a table of values for $y = x^3$ on a TI-84 Plus. See Figure 9 for the graph.

Table 1

Figure 9

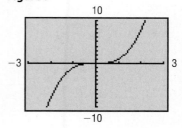

B.2 Exercises

In Problems 1–16, graph each equation using the following viewing windows:

(a) Xmin $= -5$
    Xmax $= 5$
     Xscl $= 1$
     Ymin $= -4$
     Ymax $= 4$
     Yscl $= 1$

(b) Xmin $= -10$
    Xmax $= 10$
     Xscl $= 2$
     Ymin $= -8$
     Ymax $= 8$
     Yscl $= 2$

1. $y = x + 2$ **2.** $y = x - 2$ **3.** $y = -x + 2$ **4.** $y = -x - 2$

5. $y = 2x + 2$ **6.** $y = 2x - 2$ **7.** $y = -2x + 2$ **8.** $y = -2x - 2$

9. $y = x^2 + 2$ **10.** $y = x^2 - 2$ **11.** $y = -x^2 + 2$ **12.** $y = -x^2 - 2$

13. $3x + 2y = 6$ **14.** $3x - 2y = 6$ **15.** $-3x + 2y = 6$ **16.** $-3x - 2y = 6$

17–32. *For each of the above equations, create a table, $-3 \le x \le 3$, and list points on the graph.*

B.3 Using a Graphing Utility to Locate Intercepts and Check for Symmetry

Value and Zero (or Root)

Most graphing utilities have an eVALUEate feature that, given a value of x, determines the value of y for an equation. We can use this feature to evaluate an equation at $x = 0$ to determine the y-intercept. Most graphing utilities also have a ZERO (or ROOT) feature that can be used to determine the x-intercept(s) of an equation.

EXAMPLE 1   **Finding Intercepts Using a Graphing Utility**

Use a graphing utility to find the intercepts of the equation $y = x^3 - 8$.

Solution   Figure 10(a) shows the graph of $y = x^3 - 8$.

Figure 10

(a)

(b)   (c)

The eVALUEate feature of a TI-84 Plus graphing calculator accepts as input a value of x and determines the value of y. If we let $x = 0$, we find that the y-intercept is -8. See Figure 10(b).

The ZERO feature of a TI-84 Plus is used to find the x-intercept(s). See Figure 10(c). The x-intercept is 2.

●

EXAMPLE 2   **Graphing the Equation** $y = \dfrac{1}{x}$

Graph the equation $y = \dfrac{1}{x}$. Based on the graph, infer information about intercepts and symmetry.

Solution   Figure 11 illustrates the graph. We infer from the graph that there are no intercepts; we may also infer that symmetry with respect to the origin is a possibility. The TABLE feature on a graphing utility can provide further evidence of symmetry with respect to the origin. Using a TABLE, we observe that for any ordered pair (x, y) the ordered pair $(-x, -y)$ is also a point on the graph.

Figure 11

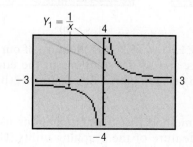

●

B.3 Exercises

In Problems 1–6, use ZERO (or ROOT) to approximate the smaller of the two x-intercepts of each equation. Express the answer rounded to two decimal places.

1. $y = x^2 + 4x + 2$

2. $y = x^2 + 4x - 3$

3. $y = 2x^2 + 4x + 1$

4. $y = 3x^2 + 5x + 1$

5. $y = 2x^2 - 3x - 1$

6. $y = 2x^2 - 4x - 1$

*In Problems 7–12, use ZERO (or ROOT) to approximate the **positive** x-intercepts of each equation. Express each answer rounded to two decimal places.*

7. $y = x^3 + 3.2x^2 - 16.83x - 5.31$

8. $y = x^3 + 3.2x^2 - 7.25x - 6.3$

9. $y = x^4 - 1.4x^3 - 33.71x^2 + 23.94x + 292.41$

10. $y = x^4 + 1.2x^3 - 7.46x^2 - 4.692x + 15.2881$

11. $y = x^3 + 19.5x^2 - 1021x + 1000.5$

12. $y = x^3 + 14.2x^2 - 4.8x - 12.4$

B.4 Using a Graphing Utility to Solve Equations

For many equations, there are no algebraic techniques that lead to a solution. For such equations, a graphing utility can often be used to investigate possible solutions. When a graphing utility is used to solve an equation, usually *approximate* solutions are obtained. Unless otherwise stated, we shall follow the practice of giving approximate solutions *rounded to two decimal places*.

The ZERO (or ROOT) feature of a graphing utility can be used to find the solutions of an equation when one side of the equation is 0. In using this feature to solve equations, we make use of the fact that the *x*-intercepts (or zeros) of the graph of an equation are found by letting $y = 0$ and solving the equation for *x*. Solving an equation for *x* when one side of the equation is 0 is equivalent to finding where the graph of the corresponding equation crosses or touches the *x*-axis.

EXAMPLE 1 **Using ZERO (or ROOT) to Approximate Solutions of an Equation**

Find the solution(s) of the equation $x^2 - 6x + 7 = 0$. Round answers to two decimal places.

Solution The solutions of the equation $x^2 - 6x + 7 = 0$ are the same as the *x*-intercepts of the graph of $Y_1 = x^2 - 6x + 7$. We begin by graphing the equation. See Figure 12(a).

From the graph there appear to be two *x*-intercepts (solutions to the equation): one between 1 and 2, the other between 4 and 5.

Figure 12

(a)   (b)   (c)

Using the ZERO (or ROOT) feature of our graphing utility, we determine that the *x*-intercepts, and so the solutions to the equation, are $x = 1.59$ and $x = 4.41$, rounded to two decimal places. See Figures 12(b) and (c).

A second method for solving equations using a graphing utility involves the INTERSECT feature of the graphing utility. This feature is used most effectively when one side of the equation is not 0.

EXAMPLE 2   **Using INTERSECT to Approximate Solutions of an Equation**

Find the solution(s) to the equation $3(x - 2) = 5(x - 1)$. Round answers to two decimal places.

Solution   We begin by graphing each side of the equation as follows: graph $Y_1 = 3(x - 2)$ and $Y_2 = 5(x - 1)$. See Figure 13(a).

Figure 13

(a)   (b)

At the point of intersection of the graphs, the value of the y-coordinate is the same. We conclude that the x-coordinate of the point of intersection represents the solution to the equation. Do you see why? The INTERSECT feature on a graphing utility determines the point of intersection of the graphs. Using this feature, we find that the graphs intersect at $(-0.5, -7.5)$. See Figure 13(b). The solution of the equation is therefore $x = -0.5$.

SUMMARY

The steps to follow for approximating solutions of equations are given next.

Steps for Approximating Solutions of Equations Using ZERO (or ROOT)

STEP 1: Write the equation in the form $\{expression\ in\ x\} = 0$.

STEP 2: Graph $Y_1 = \{expression\ in\ x\}$.
Be sure that the graph is complete. That is, be sure that all the intercepts are shown on the screen.

STEP 3: Use ZERO (or ROOT) to determine each x-intercept of the graph.

Steps for Approximating Solutions of Equations Using INTERSECT

STEP 1: Graph $Y_1 = \{expression\ in\ x\ on\ the\ left\ side\ of\ the\ equation\}$.
Graph $Y_2 = \{expression\ in\ x\ on\ the\ right\ side\ of\ the\ equation\}$.

STEP 2: Use INTERSECT to determine each x-coordinate of the point(s) of intersection, if any.
Be sure that the graphs are complete. That is, be sure that all the points of intersection are shown on the screen.

EXAMPLE 3   **Solving a Radical Equation**

Find the real solutions of the equation $\sqrt[3]{2x - 4} - 2 = 0$.

Solution   Figure 14 shows the graph of the equation $Y_1 = \sqrt[3]{2x - 4} - 2$. From the graph, we see one x-intercept near 6. Using ZERO (or ROOT), we find that the x-intercept is 6. The only solution is $x = 6$.

Figure 14

B.5 Square Screens

Figure 15

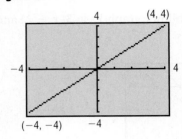

Most graphing utilities have a rectangular screen. Because of this, using the same settings for both x and y will result in a distorted view. For example, Figure 15 shows the graph of the line $y = x$ connecting the points $(-4, -4)$ and $(4, 4)$.

We expect the line to bisect the first and third quadrants, but it doesn't. We need to adjust the selections for Xmin, Xmax, Ymin, and Ymax so that a **square screen** results. On most graphing utilities, this is accomplished by setting the ratio of x to y at $3:2$. * For example, if

$$X\text{min} = -6 \qquad Y\text{min} = -4$$
$$X\text{max} = 6 \qquad Y\text{max} = 4$$

then the ratio of x to y is

$$\frac{X\text{max} - X\text{min}}{Y\text{max} - Y\text{min}} = \frac{6 - (-6)}{4 - (-4)} = \frac{12}{8} = \frac{3}{2}$$

for a ratio of $3:2$, resulting in a square screen.

EXAMPLE 1   **Examples of Viewing Rectangles That Result in Square Screens**

Figure 16

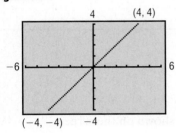

(a) $X\text{min} = -3$
$X\text{max} = 3$
$X\text{scl} = 1$
$Y\text{min} = -2$
$Y\text{max} = 2$
$Y\text{scl} = 1$

(b) $X\text{min} = -6$
$X\text{max} = 6$
$X\text{scl} = 1$
$Y\text{min} = -4$
$Y\text{max} = 4$
$Y\text{scl} = 1$

(c) $X\text{min} = -12$
$X\text{max} = 12$
$X\text{scl} = 1$
$Y\text{min} = -8$
$Y\text{max} = 8$
$Y\text{scl} = 2$

Figure 16 shows the graph of the line $y = x$ on a square screen using the viewing rectangle given in part (b). Notice that the line now bisects the first and third quadrants. Compare this illustration to Figure 15.

B.5 Exercises

In Problems 1–8, determine which of the given viewing rectangles result in a square screen.

1. $X\text{min} = -3$
$X\text{max} = 3$
$X\text{scl} = 2$
$Y\text{min} = -2$
$Y\text{max} = 2$
$Y\text{scl} = 2$

2. $X\text{min} = -5$
$X\text{max} = 5$
$X\text{scl} = 1$
$Y\text{min} = -4$
$Y\text{max} = 4$
$Y\text{scl} = 1$

3. $X\text{min} = 0$
$X\text{max} = 9$
$X\text{scl} = 3$
$Y\text{min} = -2$
$Y\text{max} = 4$
$Y\text{scl} = 2$

4. $X\text{min} = -6$
$X\text{max} = 6$
$X\text{scl} = 1$
$Y\text{min} = -4$
$Y\text{max} = 4$
$Y\text{scl} = 2$

5. $X\text{min} = -6$
$X\text{max} = 6$
$X\text{scl} = 1$
$Y\text{min} = -2$
$Y\text{max} = 2$
$Y\text{scl} = 0.5$

6. $X\text{min} = -6$
$X\text{max} = 6$
$X\text{scl} = 2$
$Y\text{min} = -4$
$Y\text{max} = 4$
$Y\text{scl} = 1$

7. $X\text{min} = 0$
$X\text{max} = 9$
$X\text{scl} = 1$
$Y\text{min} = -2$
$Y\text{max} = 4$
$Y\text{scl} = 1$

8. $X\text{min} = -6$
$X\text{max} = 6$
$X\text{scl} = 2$
$Y\text{min} = -4$
$Y\text{max} = 4$
$Y\text{scl} = 2$

9. If $X\text{min} = -4$, $X\text{max} = 8$, and $X\text{scl} = 1$, how should Ymin, Ymax, and Yscl be selected so that the viewing rectangle contains the point $(4, 8)$ and the screen is square?

10. If $X\text{min} = -6$, $X\text{max} = 12$, and $X\text{scl} = 2$, how should Ymin, Ymax, and Yscl be selected so that the viewing rectangle contains the point $(4, 8)$ and the screen is square?

*Some graphing utilities have a built-in function that automatically squares the screen. For example, the TI-84 has a ZSquare function that does this. Some graphing utilities require a ratio other than $3:2$ to square the screen. For example, the HP 48G requires the ratio of x to y to be $2:1$ for a square screen. Consult your manual.

B.6 Using a Graphing Utility to Graph Inequalities

It is easiest to begin with an example.

EXAMPLE 1   **Graphing an Inequality Using a Graphing Utility**

Use a graphing utility to graph: $3x + y - 6 \le 0$

Solution   We begin by graphing the equation $3x + y - 6 = 0$ ($Y_1 = -3x + 6$). See Figure 17.

As with graphing by hand, we need to test points selected from each region and determine whether they satisfy the inequality. To test the point $(-1, 2)$, for example, enter $3(-1) + 2 - 6 \le 0$. See Figure 18(a). The 1 that appears indicates that the statement entered (the inequality) is true. When the point $(5, 5)$ is tested, a 0 appears, indicating that the statement entered is false. Thus, $(-1, 2)$ is a part of the graph of the inequality and $(5, 5)$ is not. Figure 18(b) shows the graph of the inequality on a TI-84 Plus.*

Figure 17

Figure 18

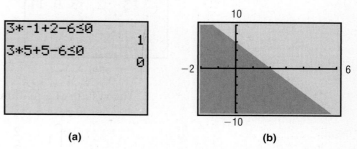

(a)   (b)

The steps to follow to graph an inequality using a graphing utility are given next.

Steps for Graphing an Inequality Using a Graphing Utility

STEP 1: Replace the inequality symbol by an equal sign, solve the equation for y, and graph the equation.

STEP 2: In each region, select a test point P and determine if the coordinates of P satisfy the inequality.

(a) If the test point satisfies the inequality, then so do all the points in the region. Indicate this by using the graphing utility to shade the region.

(b) If the coordinates of P do not satisfy the inequality, then none of the points in that region do.

B.7 Using a Graphing Utility to Solve Systems of Linear Equations

Most graphing utilities have the capability to put the augmented matrix of a system of linear equations in row echelon form. The next example, Example 6 from Section 10.2, demonstrates this feature using a TI-84 Plus graphing calculator.

*Consult your owner's manual for shading techniques.

EXAMPLE 1   **Solving a System of Linear Equations Using a Graphing Utility**

Solve: $\begin{cases} x - y + z = 8 & (1) \\ 2x + 3y - z = -2 & (2) \\ 3x - 2y - 9z = 9 & (3) \end{cases}$

Solution   The augmented matrix of the system is

$$\begin{bmatrix} 1 & -1 & 1 & | & 8 \\ 2 & 3 & -1 & | & -2 \\ 3 & -2 & -9 & | & 9 \end{bmatrix}$$

We enter this matrix into our graphing utility and name it A. See Figure 19(a). Using the REF (row echelon form) command on matrix A, we obtain the results shown in Figure 19(b). Since the entire matrix does not fit on the screen, we need to scroll right to see the rest of it. See Figure 19(c).

Figure 19

(a)   (b)   (c)

The system of equations represented by the matrix in row echelon form is

$$\begin{bmatrix} 1 & -\dfrac{2}{3} & -3 & | & 3 \\ 0 & 1 & \dfrac{15}{13} & | & -\dfrac{24}{13} \\ 0 & 0 & 1 & | & 1 \end{bmatrix} \qquad \begin{cases} x - \dfrac{2}{3}y - 3z = 3 & (1) \\ y + \dfrac{15}{13}z = -\dfrac{24}{13} & (2) \\ z = 1 & (3) \end{cases}$$

Using $z = 1$, we back-substitute to get

$$\begin{cases} x - \dfrac{2}{3}y - 3(1) = 3 & (1) \\ y + \dfrac{15}{13}(1) = -\dfrac{24}{13} & (2) \end{cases} \xrightarrow{\text{Simplify.}} \begin{cases} x - \dfrac{2}{3}y = 6 & (1) \\ y = \dfrac{-39}{13} = -3 & (2) \end{cases}$$

Solving the second equation for y, we find that $y = -3$. Back-substituting $y = -3$ into $x - \dfrac{2}{3}y = 6$, we find that $x = 4$. The solution of the system is $x = 4$, $y = -3$, $z = 1$.

●

Notice that the row echelon form of the augmented matrix using the graphing utility differs from the row echelon form in Chapter 10 (p. 752), yet both matrices provide the same solution! This is because the two solutions used different row operations to obtain the row echelon form. In all likelihood, the two solutions parted ways in Step 4 of the algebraic solution, where we avoided introducing fractions by interchanging rows 2 and 3.

Most graphing utilities also have the ability to put a matrix in reduced row echelon form. Figure 20 shows the reduced row echelon form of the augmented matrix from Example 1 using the RREF command on a TI-84 Plus graphing calculator. Using this command, we see that the solution of the system is $x = 4$, $y = -3$, $z = 1$.

Figure 20

B.8 Using a Graphing Utility to Graph a Polar Equation

Most graphing utilities require the following steps in order to obtain the graph of a polar equation. Be sure to be in POLar mode.

> **Graphing a Polar Equation Using a Graphing Utility**
>
> **STEP 1:** Set the mode to POLar. Solve the equation for r in terms of θ.
>
> **STEP 2:** Select the viewing rectangle in polar mode. Besides setting Xmin, Xmax, Xscl, and so forth, the viewing rectangle in polar mode requires setting the minimum and maximum values for θ and an increment setting for θ (θstep). In addition, a square screen and radian measure should be used.
>
> **STEP 3:** Enter the expression involving θ that you found in Step 1. (Consult your manual for the correct way to enter the expression.)
>
> **STEP 4:** Graph.

EXAMPLE 1 **Graphing a Polar Equation Using a Graphing Utility**

Use a graphing utility to graph the polar equation $r \sin \theta = 2$.

Solution **STEP 1:** We solve the equation for r in terms of θ.

$$r \sin \theta = 2$$

$$r = \frac{2}{\sin \theta}$$

STEP 2: From the POLar mode, select the viewing rectangle. We will use the one given next.

$$\theta\text{min} = 0 \qquad X\text{min} = -9 \qquad Y\text{min} = -6$$

$$\theta\text{max} = 2\pi \qquad X\text{max} = 9 \qquad Y\text{max} = 6$$

$$\theta\text{step} = \frac{\pi}{24} \qquad X\text{scl} = 1 \qquad Y\text{scl} = 1$$

θstep determines the number of points that the graphing utility will plot. For example, if θstep is $\frac{\pi}{24}$, the graphing utility will evaluate r at $\theta = 0(\theta\text{min})$, $\frac{\pi}{24}, \frac{2\pi}{24}, \frac{3\pi}{24}$, and so forth, up to $2\pi(\theta\text{max})$. The smaller θstep is, the more points that the graphing utility will plot. The student is encouraged to experiment with different values for θmin, θmax, and θstep to see how the graph is affected.

STEP 3: Enter the expression $\frac{2}{\sin \theta}$ after the prompt $r_1 =$.

STEP 4: Graph.

The graph is shown in Figure 21.

Figure 21

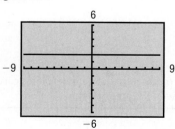

B.9 Using a Graphing Utility to Graph Parametric Equations

Most graphing utilities have the capability of graphing parametric equations. The following steps are usually required to obtain the graph of parametric equations. Check your owner's manual to see how yours works.

Graphing Parametric Equations Using a Graphing Utility

STEP 1: Set the mode to PARametric. Enter $x(t)$ and $y(t)$.

STEP 2: Select the viewing window. In addition to setting Xmin, Xmax, Xscl, and so on, the viewing window in parametric mode requires setting minimum and maximum values for the parameter t and an increment setting for t (Tstep).

STEP 3: Graph.

EXAMPLE 1

Graphing a Curve Defined by Parametric Equations Using a Graphing Utility

Graph the curve defined by the parametric equations

$$x = 3t^2, \qquad y = 2t, \qquad -2 \le t \le 2$$

Solution

STEP 1: Enter the equations $x(t) = 3t^2$, $y(t) = 2t$ with the graphing utility in PARametric mode.

STEP 2: Select the viewing window. The interval is $-2 \le t \le 2$, so we select the following square viewing window:

$$T\text{min} = -2 \qquad X\text{min} = 0 \qquad Y\text{min} = -5$$

$$T\text{max} = 2 \qquad X\text{max} = 15 \qquad Y\text{max} = 5$$

$$T\text{step} = 0.1 \qquad X\text{scl} = 1 \qquad Y\text{scl} = 1$$

We choose Tmin $= -2$ and Tmax $= 2$ because $-2 \le t \le 2$. Finally, the choice for Tstep will determine the number of points that the graphing utility will plot. For example, with Tstep at 0.1, the graphing utility will evaluate x and y at $t = -2, -1.9, -1.8$, and so on. The smaller the Tstep, the more points the graphing utility will plot. The reader is encouraged to experiment with different values of Tstep to see how the graph is affected.

STEP 3: Graph. Notice the direction in which the graph is drawn. This direction shows the orientation of the curve.

The graph shown in Figure 22 is complete. ●

Figure 22

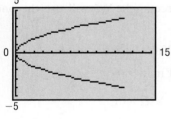

Exploration

Graph the following parametric equations using a graphing utility with Xmin $= 0$, Xmax $= 15$, Ymin $= -5$, Ymax $= 5$, and Tstep $= 0.1$.

1. $x = \dfrac{3t^2}{4}, \quad y = t, \quad -4 \le t \le 4$

2. $x = 3t^2 + 12t + 12, \quad y = 2t + 4, \quad -4 \le t \le 0$

3. $x = 3t^{2/3}, \quad y = 2\sqrt[3]{t}, \quad -8 \le t \le 8$

Compare these graphs to the graph in Figure 22. Conclude that parametric equations defining a curve are not unique; that is, different parametric equations can represent the same graph.

Exploration

In FUNCtion mode, graph $x = \dfrac{3y^2}{4}$ $\left(Y_1 = \sqrt{\dfrac{4x}{3}} \text{ and } Y_2 = -\sqrt{\dfrac{4x}{3}} \right)$ with Xmin $= 0$, Xmax $= 15$, Ymin $= -5$, Ymax $= 5$. Compare this graph with Figure 22. Why do the graphs differ?

Answers

Foundations: A Prelude to Functions

1 Assess Your Understanding (page 6)

7. abscissa; ordinate **8.** quadrants **9.** $\sqrt{(x_2 - x_1)^2 + (y_2 - y_1)^2}$ **10.** midpoint

11. (a) Quadrant II **(b)** Positive x-axis
(c) Quadrant III **(d)** Quadrant I
(e) Negative y-axis **(f)** Quadrant IV

13. The points will be on a vertical line that is 2 units to the right of the y-axis.

15. $\sqrt{5}$

17. $\sqrt{10}$

19. $2\sqrt{17}$

21. $\sqrt{85}$

23. $3\sqrt{5}$

25. $\sqrt{6.89} \approx 2.62$

27. $\sqrt{a^2 + b^2}$

29. $d(A, B) = \sqrt{13}$
$d(B, C) = \sqrt{13}$
$d(A, C) = \sqrt{26}$
$\left(\sqrt{13}\right)^2 + \left(\sqrt{13}\right)^2 = \left(\sqrt{26}\right)^2$
Area $= \dfrac{13}{2}$ square units

31. $d(A, B) = \sqrt{130}$
$d(B, C) = \sqrt{26}$
$d(A, C) = 2\sqrt{26}$
$\left(\sqrt{26}\right)^2 + \left(2\sqrt{26}\right)^2 = \left(\sqrt{130}\right)^2$
Area $= 26$ square units

33. $d(A, B) = 4$
$d(A, C) = 5$
$d(B, C) = \sqrt{41}$
$4^2 + 5^2 = 16 + 25 = \left(\sqrt{41}\right)^2$
Area $= 10$ square units

35. $(4, 0)$ **37.** $\left(\dfrac{3}{2}, 1\right)$ **39.** $\left(1, -\dfrac{7}{2}\right)$ **41.** $(1.05, 0.7)$ **43.** $\left(\dfrac{a}{2}, \dfrac{b}{2}\right)$

45. $(2, 2); (2, -4)$ **47.** $(0, 0); (8, 0)$ **49.** $\sqrt{17}; 2\sqrt{5}; \sqrt{29}$

51. $\left(\dfrac{s}{2}, \dfrac{s}{2}\right)$ **53.** $d(P_1, P_2) = 6; d(P_2, P_3) = 4; d(P_1, P_3) = 2\sqrt{13};$
right triangle **55.** $d(P_1, P_2) = 2\sqrt{17}; d(P_2, P_3) = \sqrt{34};$
$d(P_1, P_3) = \sqrt{34};$ isosceles right triangle **57.** $90\sqrt{2} \approx 127.28$ ft
59. (a) $(90, 0), (90, 90), (0, 90)$ **(b)** $5\sqrt{2161} \approx 232.43$ ft
(c) $30\sqrt{149} \approx 366.20$ ft **61.** $d = 50t$ mi **63. (a)** $(2.65, 1.6)$
(b) ≈ 1.285 units **65.** \$19,182; a slight overestimate **67.** $(5, -2)$

2 Assess Your Understanding (page 16)

3. intercepts **4.** zeros; roots **5.** y-axis **6.** True **7.** True **8.** False **9.** $(0, 0)$ is on the graph. **11.** $(0, 3)$ is on the graph. **13.** $(0, 2)$ and $\left(\sqrt{2}, \sqrt{2}\right)$ are on the graph.

15. $(-2, 0), (0, 2)$

17. $(-4, 0), (0, 8)$

19. $(-1, 0), (1, 0), (0, -1)$

21. $(-2, 0), (2, 0), (0, 4)$

23. $(3, 0), (0, 2)$

25. $(-2, 0), (2, 0), (0, 9)$

27.

29.

31.

33.

35.

37. (a) $(-1, 0), (1, 0)$
(b) Symmetric with respect to the x-axis, the y-axis and the origin

39. (a) $\left(-\dfrac{\pi}{2}, 0\right), (0, 1), \left(\dfrac{\pi}{2}, 0\right)$
(b) Symmetric with respect to the y-axis

41. (a) $(0, 0)$
(b) Symmetric with respect to the x-axis

43. (a) $(-2, 0), (0, 0), (2, 0)$
(b) Symmetric with respect to the origin

45. (a) x-intercepts: $[-2, 1]$; y-intercept: 0
(b) No symmetry

47. (a) No intercepts
(b) Symmetric with respect to the origin

49.

51.

53. $(-4, 0), (0, -2), (0, 2)$; symmetric with respect to the x-axis   **55.** $(0, 0)$; symmetric with respect to the origin   **57.** $(0, 9), (3, 0), (-3, 0)$; symmetric with respect to the y-axis
59. $(-2, 0), (2, 0), (0, -3), (0, 3)$; symmetric with respect to the x-axis, y-axis, and origin
61. $(0, -9), (3, 0), (-3, 0), (-1, 0)$; no symmetry   **63.** $(0, -4), (4, 0), (-4, 0)$; symmetric with respect to the y-axis   **65.** $(0, 0)$; symmetric with respect to the origin   **67.** $(0, 0)$; symmetric with respect to the origin

69.

71.

73. $b = 13$
75. -4 or 1

77. (a) $(0, -5), (-\sqrt{5}, 0), (\sqrt{5}, 0)$
(b) symmetric with respect to the y-axis
(c)

79. (a) $(-9, 0), (0, -3), (0, 3)$
(b) symmetric with respect to the x-axis
(c)

81. (a) $(0, 3), (0, -3), (-3, 0), (3, 0)$
(b) symmetric with respect to the x-axis, y-axis, and origin
(c)

83. (a) $(0, 0), (-2, 0), (2, 0)$
(b) symmetric with respect to the origin
(c)

85. $(-1, -2)$ **87.** 4
89. (a) $(0, 0), (2, 0), (0, 1), (0, -1)$
(b) x-axis symmetry
91. (a) $y = \sqrt{x^2}$ and $y = |x|$ have the same graph.
(b) $\sqrt{x^2} = |x|$
(c) $x \geq 0$ for $y = (\sqrt{x})^2$, while x can be any real numer for $y = x$
(d) $y \geq 0$ for $y = \sqrt{x^2}$

3 Assess Your Understanding (page 30)

1. undefined; 0   **2.** $\dfrac{y_2 - y_1}{x_2 - x_1}$ **3.** $y = b$; y-intercept   **4.** True   **5.** False   **6.** True   **7.** $m_1 = m_2$; y-intercepts; $m_1 m_2 = -1$; $m_1 = -\dfrac{1}{m_2}$

8. 2   **9.** $-\dfrac{1}{2}$ **10.** 3   **11. (a)** Slope $= \dfrac{1}{2}$ **(b)** If x increases by 2 units, y will increase by 1 unit.

13. (a) Slope $= -\dfrac{1}{3}$ **(b)** If x increases by 3 units, y will decrease by 1 unit.

15. Slope $= -\dfrac{3}{2}$ **17.** Slope $= -\dfrac{1}{2}$ **19.** Slope $= 0$ **21.** Slope undefined   **23.**   **25.**

27.

29.

31. $(2, 6); (3, 10); (4, 14)$ **33.** $(4, -7); (6, -10); (8, -13)$ **35.** $(-1, -5); (0, -7); (1, -9)$
37. $x - 2y = 0$ or $y = \dfrac{1}{2}x$ **39.** $x + y = 2$ or $y = -x + 2$ **41.** $2x - y = 3$ or $y = 2x - 3$
43. $x + 2y = 5$ or $y = -\dfrac{1}{2}x + \dfrac{5}{2}$ **45.** $3x - y = -9$ or $y = 3x + 9$

47. $2x + 3y = -1$ or $y = -\dfrac{2}{3}x - \dfrac{1}{3}$ **49.** $x - 2y = -5$ or $y = \dfrac{1}{2}x + \dfrac{5}{2}$ **51.** $3x + y = 3$ or $y = -3x + 3$ **53.** $x - 2y = 2$ or $y = \dfrac{1}{2}x - 1$
55. $x = 2$; no slope–intercept form   **57.** $y - 2 = 0$; $y = 2$ **59.** $2x - y = -4$ or $y = 2x + 4$ **61.** $2x - y = 0$ or $y = 2x$
63. $x = 4$; no slope–intercept form   **65.** $2x + y = 0$ or $y = -2x$ **67.** $x - 2y = -3$ or $y = \dfrac{1}{2}x + \dfrac{3}{2}$ **69.** $y = 4$

71. Slope $= 2$; y-intercept $= 3$ **73.** $y = 2x - 2$; slope $= 2$; y-intercept $= -2$ **75.** Slope $= \dfrac{1}{2}$; y-intercept $= 2$ **77.** Slope $= -\dfrac{1}{2}$; y-intercept $= 2$

79. Slope $= \dfrac{2}{3}$; y-intercept $= -2$ **81.** Slope $= -1$; y-intercept $= 1$ **83.** Slope undefined; no y-intercept   **85.** Slope $= 0$; y-intercept $= 5$

87. Slope $= 1$; y-intercept $= 0$ **89.** Slope $= \dfrac{3}{2}$; y-intercept $= 0$ **91. (a)** x-intercept: 3; y-intercept: 2   **93. (a)** x-intercept: -10; y-intercept: 8

(b)

(0, 0) (2, 2) (1, 1)

(0, 0) (2, 3)

(b)

(0, 2) (3, 0)

(b)

(−10, 0) (0, 8)

95. (a) x-intercept: 3; y-intercept: $\dfrac{21}{2}$ **97. (a)** x-intercept: 2; y-intercept: 3   **99. (a)** x-intercept: 5; y-intercept: -2 **101.** $y = 0$ **103.** Parallel

(b)

$\left(0, \dfrac{21}{2}\right)$ (3, 0)

(b)

(0, 3) (2, 0)

(b)

(5, 0) (0, −2)

105. Neither   **107. (b)**   **109. (d)**

111. $P_1 = (-2, 5)$, $P_2 = (1, 3)$, $m_1 = -\dfrac{2}{3}$; $P_2 = (1, 3)$, $P_3 = (-1, 0)$, $m_2 = \dfrac{3}{2}$; because $m_1 m_2 = -1$, the sides P_1, P_2 and P_2, P_3 are perpendicular
and the points $(-2, 5)$, $(1, 3)$, and $(-1, 0)$ are the vertices of a right triangle.

113. $P_1 = (-1, 0)$, $P_2 = (2, 3)$, $m = 1$; $P_3 = (1, -2)$, $P_4 = (4, 1)$, $m = 1$; $P_1 = (-1, 0)$, $P_3 = (1, -2)$, $m = -1$; $P_2 = (2, 3)$, $P_4 = (4, 1)$, $m = -1$;
opposite sides are parallel, and adjacent sides are perpendicular; the points are the vertices of a rectangle.   **115.** $C = 0.20x + 29$; \$51.00; \$75.00

117. $C = 0.15x + 1289$

119. (a) $C = 0.10701x + 10.47$, $0 \le x \le 400$ **(c)** \$21.17   **(d)** \$42.57   **(e)** Each additional kW-h used adds \$0.10701 to the bill.

(b)

(400, 53.27) (0, 10.47)

121. $^\circ C = \dfrac{5}{9}(^\circ F - 32)$; approximately 21.1°C   **123. (a)** $y = -\dfrac{2}{25}x + 30$ **(b)** x-intercept: 375;
the ramp meets the floor 375 in. (31.25 ft) from the base of the platform.   **(c)** No, because it is
31.25 feet long   **(d)** The only slope possible for the ramp to comply with the requirement is for it
to drop 1 in. for every 12-in. run.   **125. (a)** $A = \dfrac{1}{5}x + 20{,}000$ **(b)** \$80,000   **(c)** Each additional
box sold requires an additional \$0.20 in advertising.

127. All have the same slope, 2; the lines are parallel.

$2x - y = 0$
(0, 4)
(0, 0) $2x - y = 2$
$2x - y = -4$ (0, −2)

129. (b), (c), (e), (g)
131. (c)
137. No; consider $y = 0$; no
139. They are the same line.
141. Yes

4 Assess Your Understanding *(page 39)*

3. False   **4.** radius   **5.** $(x - h)^2 + (y - k)^2 = r^2$ **6.** unit   **7.** Center $(2, 1)$; radius $= 2$; $(x - 2)^2 + (y - 1)^2 = 4$

9. Center $\left(\dfrac{5}{2}, 2\right)$; radius $= \dfrac{3}{2}$; $\left(x - \dfrac{5}{2}\right)^2 + (y - 2)^2 = \dfrac{9}{4}$

11. $x^2 + y^2 = 4$;   **13.** $x^2 + (y - 2)^2 = 4$;   **15.** $(x - 4)^2 + (y + 3)^2 = 25$;   **17.** $(x + 2)^2 + (y - 1)^2 = 16$;   **19.** $\left(x - \dfrac{1}{2}\right)^2 + y^2 = \dfrac{1}{4}$;
   $x^2 + y^2 - 4 = 0$ $x^2 + y^2 - 4y = 0$ $x^2 + y^2 - 8x + 6y = 0$ $x^2 + y^2 + 4x - 2y - 11 = 0$ $x^2 + y^2 - x = 0$

(0, 0)

(0, 2)

(4, −3)

(−2, 1)

$\left(\dfrac{1}{2}, 0\right)$

21. (a) $(h, k) = (0, 0)$; $r = 2$ **23. (a)** $(h, k) = (3, 0)$; $r = 2$ **25. (a)** $(h, k) = (1, 2)$; $r = 3$ **27. (a)** $(h, k) = (-2, 2)$; $r = 3$

(b)

(0, 0)

(b)

(3, 0)

(b)

(1, 2)

(b)

(−2, 2)

(c) $(\pm 2, 0)$; $(0, \pm 2)$ **(c)** $(1, 0)$; $(5, 0)$ **(c)** $\left(1 \pm \sqrt{5}, 0\right)$; $\left(0, 2 \pm 2\sqrt{2}\right)$ **(c)** $\left(-2 \pm \sqrt{5}, 0\right)$; $\left(0, 2 \pm \sqrt{5}\right)$

29. (a) $(h, k) = \left(\frac{1}{2}, -1\right); r = \frac{1}{2}$ **31. (a)** $(h, k) = (3, -2); r = 5$ **33. (a)** $(h, k) = (-2, 0); r = 2$ **35.** $x^2 + y^2 = 13$

(b)

(b)

(b)

37. $(x - 2)^2 + (y - 3)^2 = 9$
39. $(x + 1)^2 + (y - 3)^2 = 5$
41. $(x + 1)^2 + (y - 3)^2 = 1$
43. (c) **45. (b)** **47.** 18 units^2
49. $x^2 + (y - 139)^2 = 15,625$

(c) $\left(3 \pm \sqrt{21}, 0\right); (0, -6), (0, 2)$ **(c)** $(0, 0), (-4, 0)$

(c) $(0, -1)$

51. $x^2 + y^2 + 2x + 4y - 4168.16 = 0$ **53.** $\sqrt{2}x + 4y = 9\sqrt{2}$ **55.** $(1, 0)$ **57.** $y = 2$ **59.** (b), (c), (e), (g)

CHAPTER 1 Functions and Their Graphs

1.1 Assess Your Understanding *(page 53)*

5. independent; dependent   **6.** range   **7.** $[0, 5]$ **8.** image   **9.** $(g - f)(x)$ **10.** F   **11.** T   **12.** T   **13.** F   **14.** F
15. Function; Domain: {Elvis, Colleen, Kaleigh, Marissa}, Range: {January 8, March 15, September 17}   **17.** Not a function   **19.** Not a function
21. Function; Domain: {1, 2, 3, 4}; Range: {3}   **23.** Not a function   **25.** Function; Domain: {−2, −1, 0, 1}, Range: {0, 1, 4}   **27.** Function
29. Function   **31.** Not a function   **33.** Not a function   **35.** Function   **37.** Not a function   **39. (a)** −4 **(b)** 1 **(c)** −3 **(d)** $3x^2 - 2x - 4$

(e) $-3x^2 - 2x + 4$ **(f)** $3x^2 + 8x + 1$ **(g)** $12x^2 + 4x - 4$ **(h)** $3x^2 + 6xh + 3h^2 + 2x + 2h - 4$ **41. (a)** 0 **(b)** $\frac{1}{2}$ **(c)** $-\frac{1}{2}$ **(d)** $\frac{-x}{x^2 + 1}$

(e) $\frac{-x}{x^2 + 1}$ **(f)** $\frac{x + 1}{x^2 + 2x + 2}$ **(g)** $\frac{2x}{4x^2 + 1}$ **(h)** $\frac{x + h}{x^2 + 2xh + h^2 + 1}$ **43. (a)** 4 **(b)** 5 **(c)** 5 **(d)** $|x| + 4$ **(e)** $-|x| - 4$ **(f)** $|x + 1| + 4$

(g) $2|x| + 4$ **(h)** $|x + h| + 4$ **45. (a)** $-\frac{1}{5}$ **(b)** $-\frac{3}{2}$ **(c)** $\frac{1}{8}$ **(d)** $\frac{2x - 1}{3x + 5}$ **(e)** $\frac{-2x - 1}{3x - 5}$ **(f)** $\frac{2x + 3}{3x - 2}$ **(g)** $\frac{4x + 1}{6x - 5}$ **(h)** $\frac{2x + 2h + 1}{3x + 3h - 5}$

47. All real numbers   **49.** All real numbers   **51.** $\{x | x \neq -4, x \neq 4\}$ **53.** $\{x | x \neq -3, x \neq 2, x \neq 3\}$ **55.** $\{x | x \geq 4\}$ **57.** $\{x | x > 9\}$

59. $\{x | x \leq 6\}$ **61.** $\{t | t \geq 4, t \neq 7\}$ **63. (a)** $(f + g)(x) = 5x + 1$; All real numbers   **(b)** $(f - g)(x) = x + 7$; All real numbers

(c) $(f \cdot g)(x) = 6x^2 - x - 12$; All real numbers   **(d)** $\left(\frac{f}{g}\right)(x) = \frac{3x + 4}{2x - 3}; \left\{x \,\middle|\, x \neq \frac{3}{2}\right\}$ **(e)** 16 **(f)** 11 **(g)** 10 **(h)** −7

65. (a) $(f + g)(x) = 2x^2 + x$; All real numbers **(b)** $(f - g)(x) = -2x^2 + x - 2$; All real numbers   **(c)** $(f \cdot g)(x) = 2x^3 - 2x^2 + x - 1$; All real

numbers **(d)** $\left(\frac{f}{g}\right)(x) = \frac{x - 1}{2x^2 + 1}$; All real numbers   **(e)** 21 **(f)** −30 **(g)** 9 **(h)** 0   **67. (a)** $(f + g)(x) = \sqrt{x + 1} + \sqrt{x - 1}; \{x | x \geq 1\}$

(b) $(f - g)(x) = \sqrt{x + 1} - \sqrt{x - 1}; \{x | x \geq 1\}$ **(c)** $(f \cdot g)(x) = \sqrt{x^2 - 1}; \{x | x \geq 1\}$ **(d)** $\left(\frac{f}{g}\right)(x) = \sqrt{\frac{x + 1}{x - 1}}; \{x | x > 1\}$ **(e)** $2 + \sqrt{2}$

(f) $\sqrt{5} - \sqrt{3}$ **(g)** $\sqrt{3}$ **(h)** Undefined   **69. (a)** $(f + g)(x) = \frac{6x + 3}{3x - 2}; \left\{x \,\middle|\, x \neq \frac{2}{3}\right\}$ **(b)** $(f - g)(x) = \frac{-2x + 3}{3x - 2}; \left\{x \,\middle|\, x \neq \frac{2}{3}\right\}$

(c) $(f \cdot g)(x) = \frac{8x^2 + 12x}{(3x - 2)^2}; \left\{x \,\middle|\, x \neq \frac{2}{3}\right\}$ **(d)** $\left(\frac{f}{g}\right)(x) = \frac{2x + 3}{4x}; \left\{x \,\middle|\, x \neq 0; x \neq \frac{2}{3}\right\}$ **(e)** 3 **(f)** $-\frac{1}{2}$ **(g)** $\frac{7}{2}$ **(h)** $\frac{5}{4}$

71. (a) $(f + g)(x) = \frac{2(2x^2 + 3x - 1)}{(x + 3)(x - 1)}; \{x | x \neq -3, x \neq 1\}$ **(b)** $(f - g)(x) = \frac{-2(x^2 + 5x - 2)}{(x + 3)(x - 1)}; \{x | x \neq -3, x \neq 1\}$

(c) $(f \cdot g)(x) = \frac{3x - 1}{x + 3}; \{x | x \neq -3, x \neq 1\}$ **(d)** $\left(\frac{f}{g}\right)(x) = \frac{(x - 1)^2}{(x + 3)(3x - 1)}; \left\{x \,\middle|\, x \neq -3, x \neq \frac{1}{3}, x \neq 1\right\}$ **(e)** $\frac{13}{3}$ **(f)** $-\frac{68}{21}$ **(g)** 1

(h) Undefined   **73.** $g(x) = 5 - \frac{7}{2}x$ **75.** 4   **77.** $2x + h - 1$ **79.** $6x + 3h - 2$ **81.** $3x^2 + 3xh + h^2$ **83.** $A = -\frac{7}{2}$ **85.** $A = -4$

87. $A = 8$; undefined at $x = 3$ **89.** $A(x) = \frac{1}{2}x^2$ **91.** $G(x) = 10x$ **93. (a)** P is the dependent variable; a is the independent variable.

(b) $P(20) = 197.34$; In 2005, there were 197.34 million people 20 years of age or older.   **(c)** $P(0) = 290.580$; In 2005, there were 290.580 million
people.   **95. (a)** 15.1 m, 14.071 m, 12.944 m, 11.719 m   **(b)** 1.01 s, 1.43 s, 1.75 s   **(c)** 2.02 s   **97. (a)** $222   **(b)** $225   **(c)** $220   **(d)** $230

99. $R(x) = \frac{L(x)}{P(x)}$ **101.** $H(x) = P(x) \cdot I(x)$ **103. (a)** $P(x) = -0.05x^3 + 0.8x^2 + 155x - 500$ **(b)** $P(15) = \$1836.25$ **(c)** When 15 hundred cell

phones are sold, the profit is $1836.25.   **105.** Only $h(x) = 2x$ **109.** Square the real number t and then subtract 2.

1.2 Assess Your Understanding *(page 62)*

3. vertical   **4.** 5; −3   **5.** $a = -2$ **6.** F   **7.** F   **8.** T   **9. (a)** $f(0) = 3; f(-6) = -3$ **(b)** $f(6) = 0; f(11) = 1$ **(c)** Positive   **(d)** Negative
(e) −3, 6, and 10   **(f)** $-3 < x < 6; 10 < x \leq 11$ **(g)** $\{x | -6 \leq x \leq 11\}$ **(h)** $\{y | -3 \leq y \leq 4\}$ **(i)** −3, 6, 10   **(j)** 3   **(k)** 3 times   **(l)** Once
(m) 0, 4   **(n)** −5, 8   **(o)** −3, 6, 10   **11.** Not a function   **13.** Function   **(a)** Domain: $\{x | -\pi \leq x \leq \pi\}$; Range: $\{y | -1 \leq y \leq 1\}$
(b) $\left(-\frac{\pi}{2}, 0\right), \left(\frac{\pi}{2}, 0\right), (0, 1)$ **(c)** y-axis   **15.** Not a function   **17.** Function   **(a)** Domain: $\{x | x > 0\}$; Range: all real numbers   **(b)** $(1, 0)$
(c) None   **19.** Function   **(a)** Domain: all real numbers; Range: $\{y | y \leq 2\}$ **(b)** $(-3, 0), (3, 0), (0, 2)$ **(c)** y-axis   **21.** Function   **(a)** Domain:
all real numbers; Range: $\{y | y \geq -3\}$ **(b)** $(1, 0), (3, 0), (0, 9)$ **(c)** None   **23. (a)** Yes   **(b)** $f(-2) = 9; (-2, 9)$ **(c)** $0, \frac{1}{2}; (0, -1), \left(\frac{1}{2}, -1\right)$
(d) All real numbers   **(e)** $-\frac{1}{2}, 1$ **(f)** −1   **(g)** $-\frac{1}{2}, 1$ **25. (a)** No   **(b)** $f(4) = -3; (4, -3)$ **(c)** $14; (14, 2)$ **(d)** $\{x | x \neq 6\}$ **(e)** −2   **(f)** $-\frac{1}{3}$

(g) -2 **27. (a)** Yes   **(b)** $f(2) = \dfrac{2}{5}; \left(2, \dfrac{2}{5}\right)$ **(c)** $1, 5; \left(1, \dfrac{5}{2}\right), \left(5, \dfrac{5}{2}\right)$ **(d)** All real numbers   **(e)** $0, -\dfrac{3}{2}$ **(f)** 0 **(g)** $0, -\dfrac{3}{2}$

29. (a) Approximately 10.4 ft high   **(b)** Approximately 9.9 ft high

(c)

31. (a) 230; 225 **(b)** $\{x \mid x > 0\}$

(c)

(d)

(d) The ball will not go through the hoop; $h(15) \approx 8.4$ ft.
If $v = 30$ ft/s, $h(15) = 10$ ft.

(e) 600 mi/hr

33. (a) 3 **(b)** -2 **(c)** -1 **(d)** 1 **(e)** 2 **(f)** $-\dfrac{1}{3}$ **35. (a)** 80; it costs $80 if you use 0 minutes.   **(b)** 80; it costs $80 if you use 1000 minutes.

(c) 210; it costs $210 if you use 2000 minutes.   **(d)** $\{m \mid 0 \le m \le 14{,}400\}$. There are at most 14,400 anytime minutes in a month.

37. The x-intercepts can number anywhere from 0 to infinitely many. There is at most one y-intercept.   **39. (a)** III   **(b)** IV   **(c)** I   **(d)** V   **(e)** II

41.

43. (a) 2 hr elapsed during which Kevin was between 0 and 3 mi from home.   **(b)** 0.5 hr elapsed during which Kevin was 3 mi from home.   **(c)** 0.3 hr elapsed during which Kevin was between 0 and 3 mi from home.
(d) 0.2 hr elapsed during which Kevin was 0 mi from home.   **(e)** 0.9 hr elapsed during which Kevin was between 0 and 2.8 mi from home.   **(f)** 0.3 hr elapsed during which Kevin was 2.8 mi from home.   **(g)** 1.1 hr elapsed during which Kevin was between 0 and 2.8 mi from home.   **(h)** 3 mi   **(i)** 2 times

45. No points whose x-coordinate is 5 or whose y-coordinate is 0 can be on the graph.

1.3 Assess Your Understanding *(page 75)*

6. increasing   **7.** even; odd   **8.** absolute maximum   **9.** local maximum   **10.** y-axis   **11.** Yes   **13.** No   **15.** $(-8, -2); (0, 2); (5, \infty)$ **17.** Yes; 10
19. $-2, 2; 6, 10$ **21. (a)** $(-2, 0), (0, 3), (2, 0)$ **(b)** Domain: $\{x \mid -4 \le x \le 4\}$ or $[-4, 4]$; Range: $\{y \mid 0 \le y \le 3\}$ or $[0, 3]$ **(c)** Increasing on $(-2, 0)$
and $(2, 4)$; Decreasing on $(-4, -2)$ and $(0, 2)$ **(d)** Even   **23. (a)** $(0, 1)$ **(b)** Domain: all real numbers; Range: $\{y \mid y > 0\}$ or $(0, \infty)$
(c) Increasing on $(-\infty, \infty)$ **(d)** Neither   **25. (a)** $(-\pi, 0), (0, 0), (\pi, 0)$ **(b)** Domain: $\{x \mid -\pi \le x \le \pi\}$ or $[-\pi, \pi]$; Range: $\{y \mid -1 \le y \le 1\}$
or $[-1, 1]$ **(c)** Increasing on $\left(-\dfrac{\pi}{2}, \dfrac{\pi}{2}\right)$; Decreasing on $\left(-\pi, -\dfrac{\pi}{2}\right)$ and $\left(\dfrac{\pi}{2}, \pi\right)$ **(d)** Odd   **27. (a)** $\left(0, \dfrac{1}{2}\right), \left(\dfrac{1}{3}, 0\right), \left(\dfrac{5}{2}, 0\right)$
(b) Domain: $\{x \mid -3 \le x \le 3\}$ or $[-3, 3]$; Range: $\{y \mid -1 \le y \le 2\}$ or $[-1, 2]$ **(c)** Increasing on $(2, 3)$; Decreasing on $(-1, 1)$; Constant on $(-3, -1)$
and $(1, 2)$ **(d)** Neither   **29. (a)** $0; 3$ **(b)** $-2, 2; 0, 0$ **31. (a)** $\dfrac{\pi}{2}; 1$ **(b)** $-\dfrac{\pi}{2}; -1$ **33.** Absolute maximum: $f(1) = 4$; Absolute minimum: $f(5) = 1$

35. Absolute maximum: none; Absolute minimum: $f(1) = 1$ **37.** Absolute maximum: none; Absolute minimum: $f(0) = 0$
39. Absolute maximum: none; Absolute minimum: none   **41.** Odd   **43.** Even   **45.** Odd   **47.** Neither   **49.** Even   **51.** Odd

53.

Increasing: $(-2, -1), (1, 2)$
Decreasing: $(-1, 1)$
Local maximum: 4 at $x = -1$
Local minimum: 0 at $x = 1$

55.

Increasing: $(-2, -0.77), (0.77, 2)$
Decreasing: $(-0.77, 0.77)$
Local maximum: 0.19 at $x = -0.77$
Local minimum: -0.19 at $x = 0.77$

57.

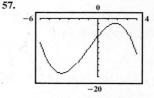

Increasing: $(-3.77, 1.77)$
Decreasing: $(-6, -3.77), (1.77, 4)$
Local maximum: -1.91 at $x = 1.77$
Local minimum: -18.89 at $x = -3.77$

59.

Increasing: $(-1.87, 0), (0.97, 2)$
Decreasing: $(-3, -1.87), (0, 0.97)$
Local maximum: 3 at $x = 0$
Local minimum: 0.95 at $x = -1.87$
Local minimum: 2.65 at $x = 0.97$

61. (a) -4 **(b)** -8 **(c)** -10 **63. (a)** 17 **(b)** -1 **(c)** 11 **65. (a)** 5 **(b)** $y = 5x - 2$ **67. (a)** -1
(b) $y = -x$ **69. (a)** 4 **(b)** $y = 4x - 8$ **71. (a)** Odd   **(b)** Local maximum: 54 at $x = -3$
73. (a) Even   **(b)** Local maximum: 24 at $x = -2$ **(c)** 47.4 sq. units

75. (a)

(b) 10 riding lawn mowers/hr
(c) $239/mower

77. (a) On average, the population is increasing at a rate of
0.036 g/hr from 0 to 2.5 hr.   **(b)** On average, from 4.5 to 6 hr,
the population is increasing at a rate of 0.1 g/hr.
(c) The average rate of change is increasing over time.
79. (a) 1 **(b)** 0.5 **(c)** 0.1 **(d)** 0.01 **(e)** 0.001
(f)

$y = x^2$ $y = x$
$y = 0.5x$
$y = 0.1x$
$y = 0.01x$
$y = 0.001x$

(g) They are getting closer to the tangent line at $(0, 0)$.
(h) They are getting closer to 0.

81. (a) 2
  (b) 2; 2; 2; 2
  (c) $y = 2x + 5$
  (d)

83. (a) $2x + h + 2$
  (b) 4.5; 4.1; 4.01; 4
  (c) $y = 4.01x - 1.01$
  (d)

85. (a) $4x + 2h - 3$
  (b) 2; 1.2; 1.02; 1
  (c) $y = 1.02x - 1.02$
  (d)

87. (a) $-\dfrac{1}{(x + h)x}$
  (b) $-\dfrac{2}{3}; -\dfrac{10}{11}; -\dfrac{100}{101}; -1$
  (c) $y = -\dfrac{100}{101}x + \dfrac{201}{101}$
  (d)

91. At most one   **93.** Yes; the function $f(x) = 0$ is both even and odd.   **95.** Not necessarily. It just means $f(5) > f(2)$.

1.4 Assess Your Understanding *(page 86)*

4. $(-\infty, 0)$ **5.** piecewise-defined   **6.** T   **7.** F   **8.** F   **9.** C   **11.** E   **13.** B   **15.** F

17.

19.

21.

23.

25. (a) 4   **(b)** 2   **(c)** 5   **27. (a)** -4 **(b)** -2 **(c)** 0   **(d)** 25

29. (a) All real numbers
  (b) $(0, 1)$
  (c)

  (d) $\{y | y \neq 0\}; (-\infty, 0) \cup (0, \infty)$
  (e) Discontinuous at $x = 0$

31. (a) All real numbers
  (b) $(0, 3)$
  (c)

  (d) $\{y | y \geq 1\}; [1, \infty)$
  (e) Continuous

33. (a) $\{x | x \geq -2\}; [-2, \infty)$
  (b) $(0, 3), (2, 0)$
  (c)

  (d) $\{y | y < 4, y = 5\};$
    $(-\infty, 4) \cup \{5\}$
  (e) Discontinuous at $x = 1$

35. (a) All real numbers
  (b) $(-1, 0), (0, 0)$
  (c)

  (d) All real numbers
  (e) Discontinuous at $x = 0$

37. (a) $\{x | x \geq -2, x \neq 0\};$
    $[-2, 0) \cup (0, \infty)$
  (b) No intercepts
  (c)

  (d) $\{y | y > 0\}; (0, \infty)$
  (e) Discontinuous at $x = 0$

39. (a) All real numbers
  (b) $(x, 0)$ for $0 \leq x < 1$
  (c)

  (d) Set of even integers
  (e) Discontinuous
    at $\{x | x$ is an integer$\}$

41. $f(x) = \begin{cases} -x & \text{if } -1 \leq x \leq 0 \\ \frac{1}{2}x & \text{if } 0 < x \leq 2 \end{cases}$ (Other answers are possible.)

43. $f(x) = \begin{cases} -x & \text{if } x \leq 0 \\ -x + 2 & \text{if } 0 < x \leq 2 \end{cases}$ (Other answers are possible.)

45. (a) 2   **(b)** 3   **(c)** -4 **47. (a)** \$39.99   **(b)** \$46.74   **(c)** \$40.44

49. (a) \$52.45   **(b)** \$277.36
  (c) $C(x) = \begin{cases} 0.73006x + 15.95 & \text{if } 0 \leq x \leq 50 \\ 0.4998x + 27.463 & \text{if } x > 50 \end{cases}$
  (d)

51. For schedule X:

$$f(x) = \begin{cases} 0.10x & \text{if} & 0 < x \leq 8350 \\ 835 + 0.15(x - 8350) & \text{if} & 8350 < x \leq 33{,}950 \\ 4675 + 0.25(x - 33{,}950) & \text{if} & 33{,}950 < x \leq 82{,}250 \\ 16{,}750 + 0.28(x - 82{,}250) & \text{if} & 82{,}250 < x \leq 171{,}550 \\ 41{,}754 + 0.33(x - 171{,}550) & \text{if} & 171{,}550 < x \leq 372{,}950 \\ 108{,}216 + 0.35(x - 372{,}950) & \text{if} & x > 372{,}950 \end{cases}$$

53. (a)

  (b) $C(x) = 10 + 0.4x$ **(c)** $C(x) = 70 + 0.25x$

55. $f(x) = \begin{cases} x & \text{if} & 0 \le x < 10 \\ 10 & \text{if} & 10 \le x < 500 \\ 30 & \text{if} & 500 \le x < 1000 \\ 50 & \text{if} & 1000 \le x < 1500 \\ 70 & \text{if} & x \ge 1500 \end{cases}$

57. (a) $10°C$
(b) $4°C$
(c) $-3°C$
(d) $-4°C$
(e) The wind chill is equal to the air temperature.
(f) At wind speed greater than 20 m/s, the wind chill factor depends only on the air temperature.

59.

$C(x) = \begin{cases} 1.17 & 0 < x \le 1 \\ 1.34 & 1 < x \le 2 \\ 1.51 & 2 < x \le 3 \\ 1.68 & 3 < x \le 4 \\ 1.85 & 4 < x \le 5 \\ 2.02 & 5 < x \le 6 \\ 2.19 & 6 < x \le 7 \\ 2.36 & 7 < x \le 8 \\ 2.53 & 8 < x \le 9 \\ 2.70 & 9 < x \le 10 \\ 2.87 & 10 < x \le 11 \\ 3.04 & 11 < x \le 12 \\ 3.21 & 12 < x \le 13 \end{cases}$

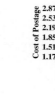

61. Each graph is that of $y = x^2$, but shifted horizontally. If $y = (x - k)^2$, $k > 0$, the shift is right k units; if $y = (x + k)^2$, $k > 0$, the shift is left k units.
63. The graph of $y = -f(x)$ is the reflection about the x-axis of the graph of $y = f(x)$.   **65.** Yes. The graph of $y = (x - 1)^3 + 2$ is the graph of $y = x^3$ shifted right 1 unit and up 2 units.   **67.** They all have the same general shape. All three go through the points $(-1, -1), (0, 0)$, and $(1, 1)$. As the exponent increases, the steepness of the curve increases (except near $x = 0$).

1.5 Assess Your Understanding (page 99)

1. horizontal; right **2.** y **3.** vertical; up **4.** T **5.** F **6.** T **7.** B **9.** H **11.** I **13.** L **15.** F **17.** G **19.** $y = (x - 4)^3$ **21.** $y = x^3 + 4$
23. $y = -x^3$ **25.** $y = 4x^3$ **27.** (1) $y = \sqrt{x} + 2$; (2) $y = -(\sqrt{x} + 2)$; (3) $y = -(\sqrt{-x} + 2)$ **29.** (1) $y = -\sqrt{x}$; (2) $y = -\sqrt{x} + 2$;
(3) $y = -\sqrt{x + 3} + 2$ **31.** (c) **33.** (c) **35.** (a) -7 and 1 (b) -3 and 5 (c) -5 and 3 (d) -3 and 5 **37.** (a) $(-3, 3)$ (b) $(4, 10)$
(c) Decreasing on $(-1, 5)$ (d) Decreasing on $(-5, 1)$

39.

Domain: $(-\infty, \infty)$;
Range: $[-1, \infty)$

41.
Domain: $(-\infty, \infty)$;
Range: $(-\infty, \infty)$

43.
Domain: $[2, \infty)$;
Range: $[0, \infty)$

45.
Domain: $(-\infty, \infty)$;
Range: $(-\infty, \infty)$

47.
Domain: $[0, \infty)$;
Range: $[0, \infty)$

49.

Domain: $(-\infty, 0) \cup (0, \infty)$
Range: $(-\infty, 0) \cup (0, \infty)$

51.

Domain: $(-\infty, \infty)$;
Range: $(-\infty, \infty)$

53.

Domain: $(-\infty, \infty)$;
Range: $(-\infty, \infty)$

55.

Domain: $(-\infty, \infty)$;
Range: $(-\infty, \infty)$

57.

Domain: $(-\infty, \infty)$;
Range: $[-3, \infty)$

59.

Domain: $[2, \infty)$;
Range: $[1, \infty)$

61.

Domain: $(-\infty, 0]$;
Range: $[-2, \infty)$

63.

Domain: $(-\infty, \infty)$;
Range: $(-\infty, \infty)$

65.

Domain: $(-\infty, \infty)$;
Range: $[0, \infty)$

67.

Domain: $(-\infty, \infty)$;
Range: $\{y \mid y \text{ is an even integer}\}$

69. (a) $F(x) = f(x) + 3$ **(b)** $G(x) = f(x + 2)$ **(c)** $P(x) = -f(x)$ **(d)** $H(x) = f(x + 1) - 2$ **(e)** $Q(x) = \dfrac{1}{2} f(x)$

(f) $g(x) = f(-x)$

(g) $h(x) = f(2x)$

71. (a) $F(x) = f(x) + 3$

(b) $G(x) = f(x + 2)$

(c) $P(x) = -f(x)$

(d) $H(x) = f(x + 1) - 2$

(e) $Q(x) = \dfrac{1}{2}f(x)$

(f) $g(x) = f(-x)$

(g) $h(x) = f(2x)$

73. $f(x) = (x + 1)^2 - 1$

75. $f(x) = (x - 4)^2 - 15$

77. $f(x) = \left(x + \dfrac{1}{2}\right)^2 + \dfrac{3}{4}$

79. $f(x) = 2(x - 3)^2 + 1$

81. $f(x) = -3(x + 2)^2 - 5$

83.

85. (a) 72°F; 65°F

(b) The temperature decreases by 2° to 70°F during the day and 63°F overnight.

(c) The time at which the temperature adjusts between the daytime and overnight settings is moved to 1 hr sooner. It begins warming up at 5:00 AM instead of 6:00 AM, and it begins cooling down at 8:00 PM instead of 9:00 PM.

87.

89. (a)

(b) 10% tax **(c)** Y_1 is the graph of $p(x)$ shifted down vertically 10,000 units. Y_2 is the graph of $p(x)$ vertically compressed by a factor of 0.9.

(d) 10% tax

91. (a)

(b)

93. (a) $(-4, 2)$
(b) $(1, -12)$
(c) $(-4, 5)$

95. The graph of $y = f(x) - 2$ is the graph of $y = f(x)$ shifted down 2 units. The graph of $y = f(x - 2)$ is the graph of $y = f(x)$ shifted right 2 units.

1.6 Assess Your Understanding *(page 106)*

1. (a) $d(x) = \sqrt{x^4 - 15x^2 + 64}$

(b) $d(0) = 8$ **(c)** $d(1) = \sqrt{50} \approx 7.07$

(d)

(e) d is smallest when $x \approx -2.74$ or $x \approx 2.74$

3. (a) $d(x) = \sqrt{x^2 - x + 1}$

(b)

(c) d is smallest when $x = 0.5$.

5. $A(x) = \dfrac{1}{2}x^4$

7. (a) $A(x) = x(16 - x^2)$

(b) Domain: $\{x \mid 0 < x < 4\}$

(c) The area is largest when $x \approx 2.31$.

9. (a) $A(x) = 4x\sqrt{4 - x^2}$ **(b)** $p(x) = 4x + 4\sqrt{4 - x^2}$ **11. (a)** $A(x) = x^2 + \dfrac{25 - 20x + 4x^2}{\pi}$

(c) A is largest when $x \approx 1.41$.   **(d)** p is largest when $x \approx 1.41$.

(b) Domain: $\{x | 0 < x < 2.5\}$

(c) A is smallest when $x \approx 1.40$ m.

13. (a) $C(x) = x$ **(b)** $A(x) = \dfrac{x^2}{4\pi}$ **15. (a)** $A(r) = 2r^2$ **(b)** $p(r) = 6r$ **17.** $A(x) = \left(\dfrac{\pi}{3} - \dfrac{\sqrt{3}}{4}\right)x^2$

19. (a) $d(t) = \sqrt{2500t^2 - 360t + 13}$ **21.** $V(r) = \dfrac{\pi H(R - r)r^2}{R}$ **23. (a)** $T(x) = \dfrac{12 - x}{5} + \dfrac{\sqrt{x^2 + 4}}{3}$ **25. (a)** $V(x) = x(24 - 2x)^2$

(b) d is smallest when $t \approx 0.07$ hr.

(b) $\{x | 0 \le x \le 12\}$ **(b)** 972 in.^{3} **(c)** 160 in.^{3}

(c) 3.09 hr   **(d)** V is largest when $x = 4$.

(d) 3.55 hr

1.7 Assess Your Understanding (page 112)

1. $y = kx$ **2.** F   **3.** $y = \dfrac{1}{5}x$ **5.** $A = \pi x^2$ **7.** $F = \dfrac{250}{d^2}$ **9.** $z = \dfrac{1}{5}(x^2 + y^2)$ **11.** $M = \dfrac{9d^2}{2\sqrt{x}}$ **13.** $T^2 = \dfrac{8a^3}{d^2}$ **15.** $V = \dfrac{4\pi}{3}r^3$ **17.** $A = \dfrac{1}{2}bh$

19. $F = 6.67 \times 10^{-11}\left(\dfrac{mM}{d^2}\right)$ **21.** $p = 0.00649B$; \$941.05   **23.** 144 ft; 2 s   **25.** 2.25   **27.** $R = 3.95g$; \$41.48   **29. (a)** $D = \dfrac{429}{p}$ **(b)** 143 bags

31. 450 cm^3 **33.** 124.76 lb   **35.** $V = \pi r^2 h$ **37.** 54.86 lb   **39.** $\sqrt[3]{6} \approx 1.82$ in.   **41.** 2812.5 joules   **43.** 384 psi

Review Exercises (page 116)

1. Function; domain $\{-1, 2, 4\}$, range $\{0, 3\}$ **3. (a)** 2   **(b)** -2 **(c)** $-\dfrac{3x}{x^2 - 1}$ **(d)** $-\dfrac{3x}{x^2 - 1}$ **(e)** $\dfrac{3(x - 2)}{x^2 - 4x + 3}$ **(f)** $\dfrac{6x}{4x^2 - 1}$

5. (a) 0   **(b)** 0   **(c)** $\sqrt{x^2 - 4}$ **(d)** $-\sqrt{x^2 - 4}$ **(e)** $\sqrt{x^2 - 4x}$ **(f)** $2\sqrt{x^2 - 1}$ **7. (a)** 0   **(b)** 0   **(c)** $\dfrac{x^2 - 4}{x^2}$ **(d)** $-\dfrac{x^2 - 4}{x^2}$ **(e)** $\dfrac{x(x - 4)}{(x - 2)^2}$ **(f)** $\dfrac{x^2 - 1}{x^2}$

9. $\{x | x \ne -3, x \ne 3\}$ **11.** $\{x | x \le 2\}$ **13.** $\{x | x > 0\}$ **15.** $\{x | x \ne -3, x \ne 1\}$

17. $(f + g)(x) = 2x + 3$; Domain: all real numbers

$(f - g)(x) = -4x + 1$; Domain: all real numbers

$(f \cdot g)(x) = -3x^2 + 5x + 2$; Domain: all real numbers

$\left(\dfrac{f}{g}\right)(x) = \dfrac{2 - x}{3x + 1}$; Domain: $\left\{x \,\middle|\, x \ne -\dfrac{1}{3}\right\}$

19. $(f + g)(x) = 3x^2 + 4x + 1$; Domain: all real numbers

$(f - g)(x) = 3x^2 - 2x + 1$; Domain: all real numbers

$(f \cdot g)(x) = 9x^3 + 3x^2 + 3x$; Domain: all real numbers

$\left(\dfrac{f}{g}\right)(x) = \dfrac{3x^2 + x + 1}{3x}$; Domain: $\{x | x \ne 0\}$

21. $(f + g)(x) = \dfrac{x^2 + 2x - 1}{x(x - 1)}$; Domain: $\{x | x \ne 0, x \ne 1\}$

$(f - g)(x) = \dfrac{x^2 + 1}{x(x - 1)}$; Domain: $\{x | x \ne 0, x \ne 1\}$

$(f \cdot g)(x) = \dfrac{x + 1}{x(x - 1)}$; Domain: $\{x | x \ne 0, x \ne 1\}$

$\left(\dfrac{f}{g}\right)(x) = \dfrac{x(x + 1)}{x - 1}$; Domain: $\{x | x \ne 0, x \ne 1\}$

23. $-4x + 1 - 2h$

25. (a) Domain: $\{x | -4 \le x \le 3\}$; Range: $\{y | -3 \le y \le 3\}$

(b) $(0, 0)$ **(c)** -1 **(d)** -4 **(e)** $\{x | 0 < x \le 3\}$

(f)   **(g)**   **(h)**

27. (a) Domain: $\{x | -4 \le x \le 4\}$ or $[-4, 4]$

Range: $\{y | -3 \le y \le 1\}$ or $[-3, 1]$

(b) Increasing on $(-4, -1)$ and $(3, 4)$; Decreasing on $(-1, 3)$

(c) Local maximum is 1 and occurs at $x = -1$.

Local minimum is -3 and occurs at $x = 3$.

(d) Absolute maximum: $f(-1) = 1$

Absolute minimum: $f(3) = -3$

(e) No symmetry   **(f)** Neither

(g) x-intercepts: $-2, 0, 4$; y-intercept: 0

29. Odd   **31.** Even   **33.** Neither   **35.** Odd

37.

Local maximum: 4.04 at $x = -0.91$
Local minimum: -2.04 at $x = 0.91$
Increasing: $(-3, -0.91)$; $(0.91, 3)$
Decreasing: $(-0.91, 0.91)$

39.

Local maximum: 1.53 at $x = 0.41$
Local minima: -0.54 at $x = -0.34$
and -3.56 at $x = 1.80$
Increasing: $(-0.34, 0.41)$; $(1.80, 3)$
Decreasing: $(-2, -0.34)$; $(0.41, 1.80)$

41. (a) 23 **(b)** 7 **(c)** 47 **43.** -5 **45.** -17 **47.** No **49.** Yes

51.

53.

55.

Intercepts: $(-4, 0), (4, 0), (0, -4)$
Domain: all real numbers
Range: $\{y | y \geq -4\}$ or $[-4, \infty)$

57.

Intercept: $(0, 0)$
Domain: all real numbers
Range: $\{y | y \leq 0\}$ or $(-\infty, 0]$

59.

Intercept: $(1, 0)$
Domain: $\{x | x \geq 1\}$ or $[1, \infty)$
Range: $\{y | y \geq 0\}$ or $[0, \infty)$

61.

Intercepts: $(0, 1), (1, 0)$
Domain: $\{x | x \leq 1\}$ or $(-\infty, 1]$
Range: $\{y | y \geq 0\}$ or $[0, \infty)$

63.

Intercept: $(0, 3)$
Domain: all real numbers
Range: $\{y | y \geq 2\}$ or $[2, \infty)$

65.

Intercepts: $(0, -2)$,
$\left(1 - \dfrac{\sqrt[3]{9}}{3}, 0\right)$ or about $(0.3, 0)$
Domain: all real numbers
Range: all real numbers

67. (a) $\{x | x > -2\}$ or $(-2, \infty)$
(b) $(0, 0)$
(c)

(d) $\{y | y > -6\}$ or $(-6, \infty)$
(e) Discontinuous at $x = 1$

69. (a) $\{x | x \geq -4\}$ or $[-4, \infty)$
(b) $(0, 1)$
(c)

(d) $\{y | -4 \leq y < 0 \text{ or } y > 0\}$
or $[-4, 0) \cup (0, \infty)$
(e) Discontinuous at $x = 0$

71. $A = 11$ **73. (a)** $A(x) = (8.5 - 2x)(11 - 2x)$
(b) $0 \leq x < 4.25, 0 < A \leq 93.5$
(c) $A(1) = 58.5$ in.^{2}, $A(1.2) = 52.46$ in.^{2},
$A(1.5) = 44$ in.^{2}
(d)

(e) $A(x) \approx 70$ when $x \approx 0.64$ in.;
$A(x) \approx 50$ when $x \approx 1.28$ in.

75. (a) $A(x) = 10x - x^3$
(b) The largest area that can be enclosed by the rectangle is
approximately 12.17 square units.
77. $R = 3.59g$; $\$40.21$
79. Approximately 36 million miles

Chapter Test *(page 119)*

1. (a) Function; domain: $\{2, 4, 6, 8\}$; range: $\{5, 6, 7, 8\}$ **(b)** Not a function **(c)** Not a function **(d)** Function; domain: all real numbers;
range: $\{y | y \geq 2\}$ **2.** Domain: $\left\{x | x \leq \dfrac{4}{5}\right\}$; $f(-1) = 3$ **3.** Domain: $\{x | x \neq -2\}$; $g(-1) = 1$ **4.** Domain: $\{x | x \neq -9, x \neq 4\}$; $h(-1) = \dfrac{1}{8}$

5. (a) Domain: $\{x | -5 \leq x \leq 5\}$; range: $\{y | -3 \leq y \leq 3\}$ **(b)** $(0, 2), (-2, 0)$, and $(2, 0)$ **(c)** $f(1) = 3$ **(d)** $x = -5$ and $x = 3$
(e) $\{x | -5 \leq x < -2 \text{ or } 2 < x \leq 5\}$ or $[-5, -2) \cup (2, 5]$ **6.** Local maxima: $f(-0.85) \approx -0.86$; $f(2.35) \approx 15.55$; local minima: $f(0) = -2$;
the function is increasing on the intervals $(-5, -0.85)$ and $(0, 2.35)$ and decreasing on the intervals $(-0.85, 0)$ and $(2.35, 5)$.

7. (a)

(b) $(0, -4), (4, 0)$
(c) $g(-5) = -9$
(d) $g(2) = -2$

8. 19
9. (a) $(f - g) = 2x^2 - 3x + 3$
(b) $(f \cdot g) = 6x^3 - 4x^2 + 3x - 2$
(c) $f(x + h) - f(x) = 4xh + 2h^2$

10. (a)

(b)

11. (a) 8.67% occurring in 1997 $(x \approx 5)$ **(b)** The model predicts that the interest rate will be -10.343%. This is not reasonable.

12. (a) $V(x) = \dfrac{x^2}{8} - \dfrac{5x}{4} + \dfrac{\pi x^2}{64}$ **(b)** 1297.61 ft^3 **13.** 14.69 ohms

CHAPTER 2 Linear and Quadratic Functions

2.1 Assess Your Understanding (page 130)

7. slope; y-intercept   **8.** $-4; 3$ **9.** positive   **10.** T   **11.** F   **12.** F

13. (a) $m = 2; b = 3$

(b)

(c) Domain and
    Range: $(-\infty, \infty)$
(d) 2   **(e)** Increasing

15. (a) $m = -3; b = 4$

(b)

(c) Domain and
    Range: $(-\infty, \infty)$
(d) -3 **(e)** Decreasing

17. (a) $m = \frac{1}{4}; b = -3$

(b)

(c) Domain and
    Range: $(-\infty, \infty)$
(d) $\frac{1}{4}$ **(e)** Increasing

19. (a) $m = 0; b = 4$

(b)

(c) Domain: $(-\infty, \infty)$
    Range: $\{y | y = 4\}$
(d) 0   **(e)** Constant

21. (a) 4

(b)

23. (a) 2

(b)

25. (a) 8

(b)

27. Linear; $y = -3x - 2$
29. Nonlinear
31. Nonlinear
33. Linear; $y = -4x + 9$

35. (a) $\frac{1}{4}$ **(b)** $\left\{x \middle| x > \frac{1}{4}\right\}$ or $\left(\frac{1}{4}, \infty\right)$ **(c)** 1
(d) $\{x | x \le 1\}$ or $(-\infty, 1]$ **(e)**

37. (a) 40   **(b)** 88   **(c)** -40
(d) $\{x | x > 40\}$ or $(40, \infty)$
(e) $\{x | x \le 88\}$ or $(-\infty, 88]$
(f) $\{x | -40 < x < 88\}$
    or $(-40, 88)$

39. (a) -4 **(b)** $\{x | x < -4\}$ or $(-\infty, -4)$ **41. (a)** -6 **(b)** $\{x | -6 \le x < 5\}$ or $[-6, 5)$
43. (a) \$45   **(b)** 180 mi   **(c)** 260 mi   **(d)** $\{x | x \ge 0\}$ or $[0, \infty)$ **45. (a)** \$16; 600 T-shirts   **(b)** $\$0 \le p < \$16$ **(c)** The price will increase.
47. (a) $\{x | 8350 \le x \le 33,950\}$ or $[8350, 33,950]$ **(b)** \$2582.50

49. (a) $x = 5000$ **(b)** $x > 5000$
(c) The independent variable is adjusted
    gross income, x.
(d) The dependent variable is the tax bill, T.

51. (a) $V(x) = -1000x + 3000$ **(b)** $\{x | 0 \le x \le 3\}$ or $[0, 3]$
(c)

(d) \$1000   **(e)** After 1 year

53. (a) $C(x) = 90x + 1800$
(b)

(c) \$3060   **(d)** 22 bicycles

(e) \$27,500

55. (a) $C(x) = 0.07x + 29$ **(b)** \$36.70; \$45.10
57. (a)

(e)

(b) Since each input (price) corresponds to a
    single output (quantity demanded), we
    know that quantity demanded is a function
    of price. Also, because the average rate of
    change is a constant -0.4 MP3 player per
    dollar, the function is linear.
(c) $q(p) = -0.4p + 160$
(d) $\{p | 0 \le p \le 400\}$ or $[0, 400]$

(f) If price increases by \$1, quantity demanded
    of MP3 players decreases by 0.4 player.
(g) q-intercept: When the price is \$0, 160 MP3
    players will be demanded. p-intercept: There
    will be 0 MP3 players demanded when the
    price is \$400.

59. (d), (e)   61. $b = 0$; yes

2.2 Assess Your Understanding *(page 137)*

3. scatter diagram   **4.** T   **5.** Linear relation, $m > 0$ **7.** Linear relation, $m < 0$ **9.** Nonlinear relation

11. (a)

(c)

(e)

    (b) Answers vary. Using $(4, 6)$ and $(8, 14)$, $y = 2x - 2$.   **(d)** $y = 2.0357x - 2.3571$

13. (a)

(c)

(e)

    (b) Answers will vary. Using $(-2, -4)$ and $(2, 5)$, $y = \dfrac{9}{4}x + \dfrac{1}{2}$.   **(d)** $y = 2.2x + 1.2$

15. (a)

(c)

(e)

    (b) Answers will vary. Using   **(d)** $y = 3.8613x + 180.2920$
        $(-20, 100)$ and $(-10, 140)$, $y = 4x + 180$

17. (a)

    (b) Linear with positive slope

(c) Answers will vary. Using the points $(39.52, 210)$ and $(66.45, 280)$, $y = 2.599x + 107.288$.

(d)

(e) 269 calories

(f) If the weight of a candy bar is increased by 1 gram, the number of calories will increase by 2.599, on average.

19. (a) The independent variable is the number of hours spent playing video games and cumulative grade-point average is the dependent variable because we are using number of hours playing video games to predict (or explain) cumulative grade-point average.

(b)

(c) $G(h) = -0.0942h + 3.2763$

(d) If the number of hours playing video games in a week increases by 1 hour, the cumulative grade-point average decreases 0.09, on average.

(e) 2.52

(f) Approximately 9.3 hours

21. (a) No

(b)

(c) $D = -1.3355p + 86.1974$

(d) If the price increases \$1, the quantity sold per day decreases by about 1.34 pairs of jeans, on average.

(e) $D(p) = -1.3355p + 86.1974$

(f) $\{p \mid 0 < p \leq 64\}$

(g) About 49 pairs

23.

No, the data do not follow a linear pattern.

25. No linear relation   **27.** 34.8 hours; A student whose GPA is 0 spends 34.8 hours each week playing video games.; $G(0) = 3.28$; The average GPA of a student who does not play video games is 3.28.

2.3 Assess Your Understanding (page 149)

7. repeated; multiplicity 2   **8.** discriminant; negative   **9.** 0, 1, 2   **10.** $x = \dfrac{-b \pm \sqrt{b^2 - 4ac}}{2a}$ **11.** Zeros: 0, 9; x-intercepts: 0, 9

13. Zeros: $-5, 5$; x-intercepts: $-5, 5$ **15.** Zeros: $-3, 2$; x-intercepts: $-3, 2$ **17.** Zeros: $-\dfrac{1}{2}, 3$; x-intercepts: $-\dfrac{1}{2}, 3$ **19.** Zeros: $-4, 4$; x-intercepts: $-4, 4$

21. Zeros: $-6, -2$; x-intercepts: $-6, -2$ **23.** Zero: $\dfrac{3}{2}$; x-intercept: $\dfrac{3}{2}$ **25.** Zeros: $-2\sqrt{2}, 2\sqrt{2}$; x-intercepts: $-2\sqrt{2}, 2\sqrt{2}$

27. Zeros: $-1, 3$; x-intercepts: $-1, 3$ **29.** Zeros: $\dfrac{-3 - 4\sqrt{2}}{2}, \dfrac{-3 + 4\sqrt{2}}{2}$; x-intercepts: $\dfrac{-3 - 4\sqrt{2}}{2}, \dfrac{-3 + 4\sqrt{2}}{2}$

31. Zeros: $-2 - 2\sqrt{3}, -2 + 2\sqrt{3}$; x-intercepts: $-2 - 2\sqrt{3}, -2 + 2\sqrt{3}$ **33.** Zeros: $-\dfrac{1}{4}, \dfrac{3}{4}$; x-intercepts: $-\dfrac{1}{4}, \dfrac{3}{4}$

35. Zeros: $\dfrac{-1 - \sqrt{7}}{6}, \dfrac{-1 + \sqrt{7}}{6}$; x-intercepts: $\dfrac{-1 - \sqrt{7}}{6}, \dfrac{-1 + \sqrt{7}}{6}$ **37.** Zeros: $2 - \sqrt{2}, 2 + \sqrt{2}$; x-intercepts: $2 - \sqrt{2}, 2 + \sqrt{2}$

39. Zeros: $2 - \sqrt{5}, 2 + \sqrt{5}$; x-intercepts: $2 - \sqrt{5}, 2 + \sqrt{5}$ **41.** Zeros: $1, \dfrac{3}{2}$; x-intercepts: $1, \dfrac{3}{2}$ **43.** No real zeros; no x-intercepts

45. Zeros: $\dfrac{-1 - \sqrt{5}}{4}, \dfrac{-1 + \sqrt{5}}{4}$; x-intercepts: $\dfrac{-1 - \sqrt{5}}{4}, \dfrac{-1 + \sqrt{5}}{4}$ **47.** Zeros: $\dfrac{-2 - \sqrt{10}}{2}, \dfrac{-2 + \sqrt{10}}{2}$; x-intercepts: $\dfrac{-2 - \sqrt{10}}{2}, \dfrac{-2 + \sqrt{10}}{2}$

49. Zero: $\dfrac{1}{3}$; x-intercept: $\dfrac{1}{3}$ **51.** $\{-6, 0\}$; $(-6, 3), (0, 3)$ **53.** $\left\{-1, -\dfrac{1}{2}\right\}$; $(-1, -1), \left(-\dfrac{1}{2}, \dfrac{1}{2}\right)$ **55.** $\{-3, 5\}$; $(-3, 13), (5, 21)$

57. Zeros: $-2\sqrt{2}, 2\sqrt{2}$; x-intercepts: $-2\sqrt{2}, 2\sqrt{2}$ **59.** Zeros: $-2, -1, 1, 2$; x-intercepts: $-2, -1, 1, 2$ **61.** Zeros: $-1, 1$; x-intercepts: $-1, 1$

63. Zeros: $-2, 1$; x-intercepts: $-2, 1$ **65.** Zeros: $-6, -5$; x-intercepts: $-6, -5$ **67.** Zero: $-\dfrac{1}{3}$; x-intercept: $-\dfrac{1}{3}$ **69.** Zeros: $-\dfrac{3}{2}, 2$; x-intercepts: $-\dfrac{3}{2}, 2$

71. Zeros: 0, 16; x-intercepts: 0, 16   **73.** Zero: 16; x-intercept: 16   **75.** Zeros: $-5\sqrt{2}, 5\sqrt{2}$; x-intercepts: $-5\sqrt{2}, 5\sqrt{2}$ **77.** Zero: $\dfrac{1}{4}$; x-intercept: $\dfrac{1}{4}$

79. Zeros: $-\dfrac{3}{5}, \dfrac{5}{2}$; x-intercepts: $-\dfrac{3}{5}, \dfrac{5}{2}$ **81.** Zeros: $-\dfrac{1}{2}, \dfrac{2}{3}$; x-intercepts: $-\dfrac{1}{2}, \dfrac{2}{3}$ **83.** Zeros: $\dfrac{-\sqrt{2} + 2}{2}, \dfrac{-\sqrt{2} - 2}{2}$; x-intercepts: $\dfrac{-\sqrt{2} + 2}{2}, \dfrac{-\sqrt{2} - 2}{2}$

85. Zeros: $\dfrac{-1 - \sqrt{17}}{2}, \dfrac{-1 + \sqrt{17}}{2}$; x-intercepts: $\dfrac{-1 - \sqrt{17}}{2}, \dfrac{-1 + \sqrt{17}}{2}$

87. (a)   **(b)** $-1, 3$ **89. (a)**   **(b)** $-2, -6$ **91. (a)**   **(b)** $3 - \sqrt{2}, 3 + \sqrt{2}$

93. $\left\{-\dfrac{1}{4}, \dfrac{2}{3}\right\}$; $\left(-\dfrac{1}{4}, \dfrac{25}{16}\right), \left(\dfrac{2}{3}, -\dfrac{10}{9}\right)$ **95.** $\{-8\}$; $(-8, 180)$ **97.** $\left\{\dfrac{5}{3}\right\}$; $\left(\dfrac{5}{3}, -\dfrac{45}{88}\right)$ **99. (a)** $-2 - \sqrt{13}, -2 + \sqrt{13}$ **(b)** 5   **(c)** $-7, -4, 1, 2$

101. 11 ft by 13 ft   **103.** 4 ft by 4 ft

105. (a) The ball strikes the ground after 6 sec.   **(b)** The ball passes the top of the building on its way down after 5 sec.

107. 20 consecutive integers

109. $\dfrac{-b + \sqrt{b^2 - 4ac}}{2a} + \dfrac{-b - \sqrt{b^2 - 4ac}}{2a} = \dfrac{-2b}{2a} = -\dfrac{b}{a}$ **111.** $k = \dfrac{1}{2}$ or $k = -\dfrac{1}{2}$

113. $ax^2 + bx + c = 0, x = \dfrac{-b \pm \sqrt{b^2 - 4ac}}{2a}$; $ax^2 - bx + c = 0, x = \dfrac{b \pm \sqrt{(-b)^2 - 4ac}}{2a} = \dfrac{b \pm \sqrt{b^2 - 4ac}}{2a} = -\dfrac{-b \pm \sqrt{b^2 - 4ac}}{2a}$

115. (b)

2.4 Assess Your Understanding (page 161)

5. parabola   **6.** axis or axis of symmetry   **7.** $-\dfrac{b}{2a}$ **8.** T   **9.** T   **10.** T   **11.** C   **13.** F   **15.** G   **17.** H

19.   **21.**   **23.**   **25.** $f(x) = -(x + 1)^2 + 1$

27. $f(x) = 2(x + 1)^2 - 4$

29. (a)

(b) Domain: $(-\infty, \infty)$
Range: $[-1, \infty)$
(c) Decreasing: $(-\infty, -1)$
Increasing: $(-1, \infty)$

31. (a)

(b) Domain: $(-\infty, \infty)$
Range: $(-\infty, 9]$
(c) Increasing: $(-\infty, -3)$
Decreasing: $(-3, \infty)$

33. (a)

(b) Domain: $(-\infty, \infty)$
Range: $[-9, \infty)$
(c) Decreasing: $(-\infty, -1)$
Increasing: $(-1, \infty)$

35. (a)

(b) Domain: $(-\infty, \infty)$
Range: $[0, \infty)$
(c) Decreasing: $(-\infty, -1)$
Increasing: $(-1, \infty)$

37. (a)

(b) Domain: $(-\infty, \infty)$

Range: $\left[\dfrac{15}{8}, \infty\right)$

(c) Decreasing: $\left(-\infty, \dfrac{1}{4}\right)$

Increasing: $\left(\dfrac{1}{4}, \infty\right)$

39. (a)

(b) Domain: $(-\infty, \infty)$

Range: $\left(-\infty, -\dfrac{5}{2}\right]$

(c) Increasing: $\left(-\infty, \dfrac{1}{2}\right)$

Decreasing: $\left(\dfrac{1}{2}, \infty\right)$

41. (a)

(b) Domain: $(-\infty, \infty)$

Range: $\left(-\infty, \dfrac{17}{4}\right]$

(c) Increasing: $\left(-\infty, -\dfrac{3}{4}\right)$

Decreasing: $\left(-\dfrac{3}{4}, \infty\right)$

43. $f(x) = (x + 1)^2 - 2 = x^2 + 2x - 1$ **45.** $f(x) = -(x + 3)^2 + 5 = -x^2 - 6x - 4$ **47.** $f(x) = 2(x - 1)^2 - 3 = 2x^2 - 4x - 1$
49. Minimum value; -18 **51.** Minimum value; -21 **53.** Maximum value; 21 **55.** Maximum value; 13

57. (a)

(b) Domain: $(-\infty, \infty)$
Range: $[-16, \infty)$
(c) Decreasing: $(-\infty, 1)$
Increasing: $(1, \infty)$

59. (a)

(b) Domain: $(-\infty, \infty)$
Range: $(-\infty, \infty)$
(c) Increasing: $(-\infty, \infty)$

61. (a)

(b) Domain: $(-\infty, \infty)$
Range: $(-\infty, 2]$
(c) Increasing: $(-\infty, 3)$
Decreasing: $(3, \infty)$

63. (a)

(b) Domain: $(-\infty, \infty)$

Range: $\left[\dfrac{7}{8}, \infty\right)$

(c) Decreasing: $\left(-\infty, -\dfrac{1}{4}\right)$

Increasing: $\left(-\dfrac{1}{4}, \infty\right)$

65. (a)

(b) Domain: $(-\infty, \infty)$
Range: $(-\infty, \infty)$
(c) Decreasing: $(-\infty, \infty)$

67. (a)

(b) Domain: $(-\infty, \infty)$
Range: $(-\infty, 0]$

(c) Increasing: $\left(-\infty, -\dfrac{1}{2}\right)$

Decreasing: $\left(-\dfrac{1}{2}, \infty\right)$

69. $a = 6, b = 0, c = 2$
71. (a), (c), (d)

(b) $\{-1, 3\}$

73. (a), (c), (d)

(b) $\{-1, 3\}$

75. (a), (c), (d)

(b) $\{-1, 2\}$

77. (a) $a = 1$: $f(x) = (x + 3)(x - 1) = x^2 + 2x - 3$
$a = 2$: $f(x) = 2(x + 3)(x - 1) = 2x^2 + 4x - 6$
$a = -2$: $f(x) = -2(x + 3)(x - 1) = -2x^2 - 4x + 6$
$a = 5$: $f(x) = 5(x + 3)(x - 1) = 5x^2 + 10x - 15$
(b) The value of a does not affect the x-intercepts, but it changes the y-intercept by a
factor of a. **(c)** The value of a does not affect the axis of symmetry. It is $x = -1$
for all values of a. **(d)** The value of a does not affect the x-coordinate of the
vertex. However, the y-coordinate of the vertex is multiplied by a.
(e) The mean of the x-intercepts is the x-coordinate of the vertex.

79. (a) $(-2, -25)$ **(b)** $-7, 3$
(c) $-4, 0; (-4, -21), (0, -21)$
(d)

81. $500; $1,000,000 **83.** **(a)** 70,000 mp3 players **(b)** $2500 **85.** **(a)** 171 ft **(b)** 49 mph **(c)** Reaction time **87.** **(a)** 187 or 188 watches; $7031.20
(b) $P(x) = -0.2x^2 + 43x - 1750$ **(c)** 107 or 108 watches; $561.20 **89.** $f(x) = 2(x + 4)(x - 2)$ **91.** If x is even, then ax^2 and bx are even and
$ax^2 + bx$ is even, which means that $ax^2 + bx + c$ is odd. If x is odd, then ax^2 and bx are odd and $ax^2 + bx$ is even, which means that $ax^2 + bx + c$ is
odd. In either case, $f(x)$ is odd. **93.** **95.** $b^2 - 4ac < 0$ **97.** No

2.5 Assess Your Understanding *(page 166)*

3. **(a)** $\{x | x < -2 \text{ or } x > 2\}$; $(-\infty, -2) \cup (2, \infty)$ **(b)** $\{x | -2 \le x \le 2\}$; $[-2, 2]$ **5.** **(a)** $\{x | -2 \le x \le 1\}$; $[-2, 1]$
 (b) $\{x | x < -2 \text{ or } x > 1\}$; $(-\infty, -2) \cup (1, \infty)$ **7.** $\{x | -2 \le x \le 5\}$; $[-2, 5]$ **9.** $\{x | x < 0 \text{ or } x > 4\}$; $(-\infty, 0) \cup (4, \infty)$

11. $\{x | x < -4 \text{ or } x > 3\}$; $(-\infty, -4) \cup (3, \infty)$ **13.** $\{-3\}$ **15.** $\left\{x \left| -\dfrac{1}{2} \le x \le 3\right.\right\}$; $\left[-\dfrac{1}{2}, 3\right]$ **17.** $\{x | x < -1 \text{ or } x > 8\}$; $(-\infty, -1) \cup (8, \infty)$

19. No real solution **21.** All real numbers; $(-\infty, \infty)$ **23.** $\{x | x \le -4 \text{ or } x \ge 4\}$; $(-\infty, -4] \cup [4, \infty)$ **25.** **(a)** $\{-1, 1\}$ **(b)** $\{-1\}$ **(c)** $\{-1, 4\}$
 (d) $\{x | x < -1 \text{ or } x > 1\}$; $(-\infty, -1) \cup (1, \infty)$ **(e)** $\{x | x \le -1\}$; $(-\infty, -1]$ **(f)** $\{x | x < -1 \text{ or } x > 4\}$; $(-\infty, -1) \cup (4, \infty)$
 (g) $\left\{x | x \le -\sqrt{2} \text{ or } x \ge \sqrt{2}\right\}$; $\left(-\infty, -\sqrt{2}\right] \cup \left[\sqrt{2}, \infty\right)$ **27.** **(a)** $\{-1, 1\}$ **(b)** $\left\{-\dfrac{1}{4}\right\}$ **(c)** $\{-4, 0\}$ **(d)** $\{x | -1 < x < 1\}$; $(-1, 1)$
 (e) $\left\{x \left| x \le -\dfrac{1}{4}\right.\right\}$ or $\left(-\infty, -\dfrac{1}{4}\right]$ **(f)** $\{x | -4 < x < 0\}$; $(-4, 0)$ **(g)** $\{0\}$ **29.** **(a)** $\{-2, 2\}$ **(b)** $\{-2, 2\}$ **(c)** $\{-2, 2\}$
 (d) $\{x | x < -2 \text{ or } x > 2\}$; $(-\infty, -2) \cup (2, \infty)$ **(e)** $\{x | x \le -2 \text{ or } x \ge 2\}$; $(-\infty, -2] \cup [2, \infty)$ **(f)** $\{x | x < -2 \text{ or } x > 2\}$; $(-\infty, -2) \cup (2, \infty)$
 (g) $\left\{x | x \le -\sqrt{5} \text{ or } x \ge \sqrt{5}\right\}$; $\left(-\infty, -\sqrt{5}\right] \cup \left[\sqrt{5}, \infty\right)$ **31.** **(a)** $\{-1, 2\}$ **(b)** $\{-2, 1\}$ **(c)** $\{0\}$ **(d)** $\{x | x < -1 \text{ or } x > 2\}$; $(-\infty, -1) \cup (2, \infty)$
 (e) $\{x | -2 \le x \le 1\}$; $[-2, 1]$ **(f)** $\{x | x < 0\}$; $(-\infty, 0)$ **(g)** $\left\{x \left| x \le \dfrac{1 - \sqrt{13}}{2} \text{ or } x \ge \dfrac{1 + \sqrt{13}}{2}\right.\right\}$; $\left(-\infty, \dfrac{1 - \sqrt{13}}{2}\right] \cup \left[\dfrac{1 + \sqrt{13}}{2}, \infty\right)$

33. **(a)** 5 sec **(b)** The ball is more than 96 ft above the ground for time t between 2 and 3 sec, $2 < t < 3$. **35.** **(a)** $0, $1000 **(b)** The revenue is
more than $800,000 for prices between $276.39 and $723.61, $276.39 < p < $723.61. **41.** If the inequality is non-strict.

2.6 Assess Your Understanding *(page 173)*

3. **(a)** $R(x) = -\dfrac{1}{6}x^2 + 100x$ **(b)** $\{x | 0 \le x < 600\}$ **(c)** $13,333.33 **(d)** 300; $15,000 **(e)** $50 **5.** **(a)** $R(x) = -\dfrac{1}{5}x^2 + 20x$ **(b)** $255
 (c) 50; $500 **(d)** $10 **(e)** Between $8 and $12 **7.** **(a)** $A(w) = -w^2 + 200w$ **(b)** A is largest when $w = 100$ yd. **(c)** 10,000 yd^2 **9.** 2,000,000 m^2
11. **(a)** $\dfrac{625}{16} \approx 39$ ft **(b)** $\dfrac{7025}{32} \approx 219.5$ ft **(c)** About 170 ft **13.** 18.75 m **15.** **(a)** 3 in. **(b)** Between 2 in. and 4 in.
 (d) **(f)** When the height is 100 ft,
the projectile is about
135.7 ft from the cliff. **17.** $\dfrac{750}{\pi} \approx 238.73$ m by 375 m
19. $x = \dfrac{a}{2}$ **21.** $\dfrac{38}{3}$ **23.** $\dfrac{248}{3}$

25. **(a)**
The data appear to follow a
quadratic relation with $a < 0$.

(b) $I(x) = -45.466x^2 + 4314.374x - 55,961.675$ **(e)**
(c) About 47.4 years of age
(d) Approximately $46,388

27. **(a)**

The data appear to be linearly related with positive slope.
(b) $R(x) = 0.836x + 1032.273$ **(c)** $1743

29. **(a)**

The data appear to follow a quadratic relation with $a > 0$.
(b) $P(a) = 0.015a^2 - 1.332a + 39.823$ **(c)** 11.6%

2.7 Assess Your Understanding (page 180)

7. $2 + 3i$ **8.** T

9. $-2i, 2i$

11. $-4, 4$

13. $3 - 2i, 3 + 2i$

15. $3 - i, 3 + i$

17. $2 - \sqrt{3}, 2 + \sqrt{3}$

19. $-\frac{1}{2} - \frac{1}{2}i, -\frac{1}{2} + \frac{1}{2}i$

21. $-\frac{1}{2} - \frac{\sqrt{3}}{2}i, -\frac{1}{2} + \frac{\sqrt{3}}{2}i$

23. $\frac{4 - 3\sqrt{2}}{2}, \frac{4 + 3\sqrt{2}}{2}$

25. Two complex solutions that are conjugates of each other **27.** Two unequal real solutions **29.** A repeated real solution **31.** $-2, 2, -2i, 2i$

33. $-\frac{1}{2} - \frac{\sqrt{3}}{2}i, -\frac{1}{2} + \frac{\sqrt{3}}{2}i, -1 - \sqrt{3}i, -1 + \sqrt{3}i, 1, 2$

2.8 Assess Your Understanding (page 184)

7. $-a; a$ **8.** $-a < u < a$ **9.** $\leq$ **10.** T **11.** F **12.** F **3. (a)** $\{-9, 3\}$ **(b)** $\{x | -9 \leq x \leq 3\}; [-9, 3]$ **(c)** $\{x | x < -9 \text{ or } x > 3\}$;
$(-\infty, -9) \text{ or } (3, \infty)$ **15. (a)** $\{-2, 3\}$ **(b)** $\{x | x \leq -2 \text{ or } x \geq 3\}; (-\infty, -2] \text{ or } [3, \infty)$ **(c)** $\{x | -2 < x < 3\}; (-2, 3)$ **17.** $\{-6, 6\}$ **19.** $\{-4, 1\}$

21. $\left\{-1, \frac{3}{2}\right\}$ **23.** $\{-4, 4\}$ **25.** $\left\{-\frac{1}{2}, \frac{1}{2}\right\}$ **27.** $\left\{-\frac{27}{2}, \frac{27}{2}\right\}$ **29.** $\left\{-\frac{36}{5}, \frac{24}{5}\right\}$ **31.** No real solution **33.** $\{-3, 3\}$ **35.** $\{-1, 3\}$ **37.** $\{-2, -1, 0, 1\}$

39. $\{x | -6 < x < 6\}; (-6, 6)$

41. $\{x | x < -4 \text{ or } x > 4\}$;
$(-\infty, -4) \text{ or } (4, \infty)$

43. $\{x | -4 < x < 4\}; (-4, 4)$

45. $\{x | x < -4 \text{ or } x > 4\}$;
$(-\infty, -4) \text{ or } (4, \infty)$

47. $\{x | 1 < x < 3\}; (1, 3)$

49. $\left\{t \left| -\frac{2}{3} \leq t \leq 2\right.\right\}; \left[-\frac{2}{3}, 2\right]$

51. $\{x | x \leq 1 \text{ or } x \geq 5\}$;
$(-\infty, 1] \text{ or } [5, \infty)$

53. $\left\{x \left| -1 < x < \frac{3}{2}\right.\right\}; \left(-1, \frac{3}{2}\right)$

55. $\{x | x < -1 \text{ or } x > 2\}$;
$(-\infty, -1) \text{ or } (2, \infty)$

57. No real solution; $\varnothing$

59. $\left\{x \left| \frac{17}{6} < x < \frac{19}{6}\right.\right\}; \left(\frac{17}{6}, \frac{19}{6}\right)$

61. $\{x | -2 < x < 4\}; (-2, 4)$

63. (a) $\left\{-\frac{1}{5}, 1\right\}$ **(b)** $\left\{x \left| -\frac{1}{5} < x < 1\right.\right\}; \left(-\frac{1}{5}, 1\right)$ **(c)** $\left\{x \left| x \leq -\frac{1}{5} \text{ or } x \geq 1\right.\right\}; \left(-\infty, -\frac{1}{5}\right] \cup [1, \infty)$ **65. (a)** $\{-2, 6\}$
(b) $\{x | x \leq -2 \text{ or } x \geq 6\}; (-\infty, -2] \cup [6, \infty)$ **(c)** $\{x | -2 < x < 6\}; (-2, 6)$ **67.** $|x - 10| < 2; \{x | 8 < x < 12\}$ **69.** $|2x + 1| > 5$;
$\{x | x < -3 \text{ or } x > 2\}$

71. Between 5.6995 in. and 5.7005 in. **73.** IQs less than 71 or greater than 129 **75.** $|5x + 1| + 7 = 5$ is equivalent to $|5x + 1| = -2$. Because $|u| \geq 0$ for any real number u, the equation has no real solution. **77.** $|2x - 1|$ is greater than or equal to 0 for all real numbers x. The only real number x such that $|2x - 1| \leq 0$ is the value of x for which $2x - 1 = 0$, which is $\frac{1}{2}$.

Review Exercises (page 187)

1. (a) $m = 2; b = -5$
(b)

(c) Domain and
Range: $(-\infty, \infty)$
(d) 2 **(e)** Increasing

3. (a) $m = \frac{4}{5}; b = -6$
(b)

(c) Domain and
Range: $(-\infty, \infty)$
(d) $\frac{4}{5}$ **(e)** Increasing

5. (a) $m = 0; b = 4$
(b)

(c) Domain: $(-\infty, \infty)$
Range: $\{y | y = 4\}$
(d) 0 **(e)** Constant

7. Zero: -7; y-intercept: 14

9. Linear; $y = 5x + 3$　**11.** Zeros: $-9, 8$; x-intercepts: $-9, 8$　**13.** Zeros: $-\dfrac{1}{3}, \dfrac{5}{2}$; x-intercepts: $-\dfrac{1}{3}, \dfrac{5}{2}$　**15.** Zeros: $1, 5$; x-intercepts: $1, 5$

17. Zero: $-\dfrac{1}{3}$; x-intercept: $-\dfrac{1}{3}$　**19.** Zeros: $\dfrac{2 - \sqrt{6}}{2}, \dfrac{2 + \sqrt{6}}{2}$; x-intercepts: $\dfrac{2 - \sqrt{6}}{2}, \dfrac{2 + \sqrt{6}}{2}$　**21.** Zeros: $-\dfrac{1}{2}, 1$; x-intercepts: $-\dfrac{1}{2}, 1$

23. $\{-1, 7\}$　**25.** $\{-2, 2\}$　**27.** Zeros: $-2, -1, 1, 2$; x-intercepts: $-2, -1, 1, 2$　**29.** Zeros: $-3, 11$; x-intercepts: $-3, 11$

31. Zero: 25; x-intercept: 25

33.

35.

37.

39. (a)

(b) Domain: $(-\infty, \infty)$
Range: $[2, \infty)$

(c) Decreasing: $(-\infty, 2)$
Increasing: $(2, \infty)$

41. (a)

(b) Domain: $(-\infty, \infty)$
Range: $[-16, \infty)$
(c) Decreasing: $(-\infty, 0)$
Increasing: $(0, \infty)$

43. (a)

(b) Domain: $(-\infty, \infty)$
Range: $(-\infty, 1]$
(c) Increasing: $\left(-\infty, \dfrac{1}{2}\right)$
Decreasing: $\left(\dfrac{1}{2}, \infty\right)$

45. (a)

(b) Domain: $(-\infty, \infty)$
Range: $\left[\dfrac{1}{2}, \infty\right)$
(c) Decreasing: $\left(-\infty, -\dfrac{1}{3}\right)$
Increasing: $\left(-\dfrac{1}{3}, \infty\right)$

47. (a)

(b) Domain: $(-\infty, \infty)$
Range: $\left[-\dfrac{7}{3}, \infty\right)$
(c) Decreasing: $\left(-\infty, -\dfrac{2}{3}\right)$
Increasing: $\left(-\dfrac{2}{3}, \infty\right)$

49. Minimum value; 1　**51.** Maximum value; 12　**53.** Maximum value; 16

55. $\{x \mid -8 < x < 2\}$; $(-8, -2)$　**57.** $\left\{x \mid x \le -\dfrac{1}{3} \text{ or } x \ge 5\right\}$; $\left(-\infty, -\dfrac{1}{3}\right]$ or $[5, \infty)$

59. $-2\sqrt{2}i, 2\sqrt{2}i$　**61.** $-1 - \sqrt{5}, -1 + \sqrt{5}$　**63.** $1 - \dfrac{\sqrt{2}}{2}i, 1 + \dfrac{\sqrt{2}}{2}i$　**65.** $-\dfrac{1}{2} - \dfrac{\sqrt{2}}{2}i, -\dfrac{1}{2} + \dfrac{\sqrt{2}}{2}i$

67. $\{-5, 2\}$　**69.** $\left\{-\dfrac{5}{3}, 3\right\}$

71. $\left\{x \mid -\dfrac{3}{2} < x < -\dfrac{7}{6}\right\}$; $\left(-\dfrac{3}{2}, -\dfrac{7}{6}\right)$

73. $\{x \mid x \le -2 \text{ or } x \ge 7\}$;
$(-\infty, -2]$ or $[7, \infty)$

75. $\left\{x \mid 0 \le x \le \dfrac{4}{3}\right\}$; $\left[0, \dfrac{4}{3}\right]$

77. $\left\{x \mid x < -1 \text{ or } x > \dfrac{7}{3}\right\}$;
$(-\infty, -1)$ or $\left(\dfrac{7}{3}, \infty\right)$

79. (a) Company A: $C(x) = 0.06x + 7$; Company B: $C(x) = 0.08x$　**(b)** 350 min　**(c)** $0 \le x < 350$　**81. (a)** $R(p) = -10p^2 + 1500p$
(b) $\{p \mid 0 < p \le 150\}$　**(c)** \$75　**(d)** \$56,250　**(e)** 750 units　**(f)** Between \$70 and \$80　**83.** The length of the plasterboard should be cut down
to about 6.47 ft.　**85.** The width is 6 in. and the length is 8 in.　**87.** 50 by 50 ft　**89.** 25 sq units

91. (a)

(b) Yes; the two variables appear to be linearly related.
(c) $y = 1.3902x + 1.114$　**(d)** 38.0 mm

Chapter Test *(page 190)*

1. (a) Slope: -4; y-intercept: 3
 (b) Decreasing
 (c)

2. Linear; $y = f(x) = -5x + 2$ **3.** $-\dfrac{4}{3}, 2$ **4.** $\dfrac{2 - \sqrt{6}}{2}, \dfrac{2 + \sqrt{6}}{2}$

5. $\{-1, 3\}$ **6.** $-3, 0$ **7.**

8. (a) Opens up
 (b) $(2, -8)$
 (c) $x = 2$
 (d) x-intercepts: $\dfrac{6 - 2\sqrt{6}}{3}, \dfrac{6 + 2\sqrt{6}}{3}$;
   y-intercept: 4

(e)

(f) Domain: All real numbers; $(-\infty, \infty)$
   Range: $\{y \mid y \geq -8\}$; $[-8, \infty)$
 (g) Decreasing: $(-\infty, 2)$; Increasing: $(2, \infty)$

9. (a) Opens down
 (b) $(1, -3)$
 (c) $x = 1$
 (d) No x-intercepts: y-intercept: -5

(e)

(f) Domain: All real numbers; $(-\infty, \infty)$
   Range: $\{y \mid y \leq -3\}$; $(-\infty, -3]$
 (g) Increasing: $(-\infty, 1)$; Decreasing: $(1, \infty)$

10. $f(x) = 2x^2 - 4x - 30$ **11.** Maximum value; 21   **12.** $\{x \mid x \leq 4 \text{ or } x \geq 6\}$; $(-\infty, 4]$ or $[6, \infty)$ **13.** $-1 - \dfrac{\sqrt{6}}{2}i, -1 + \dfrac{\sqrt{6}}{2}i$ **14.** $\left\{-3, \dfrac{7}{3}\right\}$

15. $\{x \mid -11 < x < 5\}$; $(-11, 5)$ **16.** $\{x \mid x \leq -5 \text{ or } x \geq 2\}$; $(-\infty, -5]$ or $[2, \infty)$ **17. (a)** $C(m) = 0.15m + 129.50$
 (b) $258.50   **(c)** 562 miles

18. (a) $R(x) = -\dfrac{1}{10}x^2 + 1000x$ **(b)** $384,000   **(c)** 5000 units; $2,500,000   **(d)** $500

19. (a)

Linear with negative slope   Quadratic that opens up

(b) $y = -4.234x - 2.362$ **(c)** $y = 1.993x^2 + 0.289x + 2.503$

Cumulative Review *(page 191)*

1. $5\sqrt{2}$; $(1.5, 0.5)$ **2.** $(-2, -1)$ and $(2, 3)$ are on the graph.

3. $\left\{x \mid x \geq -\dfrac{3}{5}\right\}$ or $\left[-\dfrac{3}{5}, \infty\right)$ **4.** $y = -2x + 2$ **5.** $y = -\dfrac{1}{2}x + \dfrac{13}{2}$ **6.** $(x - 2)^2 + (y + 4)^2 = 25$

7. Yes, a function   **8. (a)** -3 **(b)** $x^2 - 4x - 2$ **(c)** $x^2 + 4x + 1$ **(d)** $-x^2 + 4x - 1$ **(e)** $x^2 - 3$ **(f)** $2x + h - 4$ **9.** $\left\{z \mid z \neq \dfrac{7}{6}\right\}$
 10. Yes, a function   **11. (a)** No   **(b)** -1; $(-2, -1)$ is on the graph.   **(c)** -8; $(-8, 2)$ is on the graph.   **12.** Neither
 13. Local maximum is 5.30 and occurs at $x = -1.29$.   **14. (a)** -4 **(b)** $\{x \mid x > -4\}$ or $(-4, \infty)$
   Local minimum is -3.30 and occurs at $x = 1.29$.
   Increasing: $(-4, -1.29)$ or $(1.29, 4)$
   Decreasing: $(-1.29, 1.29)$

15. (a) Domain: $\{x \mid -4 \leq x \leq 4\}$; Range: $\{y \mid -1 \leq y \leq 3\}$ **(b)** $(-1, 0), (0, -1), (1, 0)$ **(c)** y-axis   **(d)** 1   **(e)** -4 and 4   **(f)** $\{x \mid -1 < x < 1\}$
 (g)   **(h)**   **(i)**

(j) Even   **(k)** $(0, 4)$

CHAPTER 3 Polynomial and Rational Functions

3.1 Assess Your Understanding *(page 210)*

7. smooth; continuous **8.** touches **9.** $(-1, 1)$; $(0, 0)$; $(1, 1)$ **10.** r is a real zero of f; r is an x-intercept of the graph of f; $x - r$ is a factor of f.
11. turning points **12.** $y = 3x^4$ **13.** ∞; $-\infty$ **14.** As x increases in the positive direction, $f(x)$ decreases without bound. **15.** Yes; degree 3

17. Yes; degree 2 **19.** No; x is raised to the -1 power. **21.** No; x is raised to the $\frac{3}{2}$ power. **23.** Yes; degree 4 **25.** Yes; degree 4

27.   **29.**   **31.**   **33.**

35.   **37.**   **39.**

41. $f(x) = x^3 - 3x^2 - x + 3$ for $a = 1$
43. $f(x) = x^3 - x^2 - 12x$ for $a = 1$
45. $f(x) = x^4 - 15x^2 + 10x + 24$ for $a = 1$
47. $f(x) = x^3 - 5x^2 + 3x + 9$ for $a = 1$

49. (a) 7, multiplicity 1; -3, multiplicity 2 **(b)** Graph touches the x-axis at -3 and crosses it at 7. **(c)** Near -3: $f(x) \approx -30(x + 3)^2$;
Near 7: $f(x) \approx 300(x - 7)$ **(d)** 2 **(e)** $y = 3x^3$ **51. (a)** 2, multiplicity 3 **(b)** Graph crosses the x-axis at 2 **(c)** Near 2: $f(x) \approx 20(x - 2)^3$

(d) 4 **(e)** $y = 4x^5$ **53. (a)** $-\dfrac{1}{2}$, multiplicity 2; -4, multiplicity 3 **(b)** Graph touches x-axis at $-\dfrac{1}{2}$ and crosses at -4

(c) Near $-\dfrac{1}{2}$: $f(x) \approx -85.75\left(x + \dfrac{1}{2}\right)^2$; Near -4: $f(x) \approx -24.5(x + 4)^3$ **(d)** 4 **(e)** $y = -2x^5$. **55. (a)** 5, multiplicity 3; -4, multiplicity 2

(b) Graph touches the x-axis at -4 and crosses it at 5. **(c)** Near -4: $f(x) \approx -729(x + 4)^2$; Near 5: $f(x) \approx 81(x - 5)^3$ **(d)** 4 **(e)** $y = x^5$
57. (a) No real zeros **(b)** Graph neither crosses nor touches the x-axis. **(c)** No real zeros **(d)** 5 **(e)** $y = 3x^6$ **59. (a)** 0, multiplicity 2;
$-\sqrt{2}, \sqrt{2}$, multiplicity 1 **(b)** Graph touches the x-axis at 0 and crosses at $-\sqrt{2}$ and $\sqrt{2}$. **(c)** Near $-\sqrt{2}$: $f(x) \approx 11.31(x + \sqrt{2})$;
Near 0: $f(x) \approx 4x^2$; Near $\sqrt{2}$: $f(x) \approx -11.31(x - \sqrt{2})$ **(d)** 3 **(e)** $y = -2x^4$ **61.** Could be; zeros: $-1, 1, 2$; Least degree is 3

63. Cannot be the graph of a polynomial; gap at $x = -1$ **65.** $f(x) = x(x - 1)(x - 2)$ **67.** $f(x) = -\dfrac{1}{2}(x + 1)(x - 1)^2(x - 2)$

69. Step 1: $y = x^3$
Step 2: x-intercepts: $0, 3$; y-intercept: 0
Step 3: 0: multiplicity 2, touches; 3: multiplicity 1, crosses
Step 4: At most 2 turning points
Step 5: Near 0: $f(x) \approx -3x^2$; Near 3: $f(x) \approx 9(x - 3)$
Step 6:

71. Step 1: $y = x^3$
Step 2: x-intercepts: $-4, 2$; y-intercept: 16
Step 3: -4: multiplicity 1, crosses; 2: multiplicity 2, touches
Step 4: At most 2 turning points
Step 5: Near -4: $f(x) \approx 36(x + 4)$; Near 2: $f(x) \approx 6(x - 2)^2$
Step 6:

73. Step 1: $y = -2x^4$
Step 2: x-intercepts: $-2, 2$; y-intercept: 32
Step 3: -2: multiplicity 1, crosses; 2: multiplicity 3, crosses
Step 4: At most 3 turning points
Step 5: Near -2: $f(x) \approx 128(x + 2)$; Near 2: $f(x) \approx -8(x - 2)^3$
Step 6:

75. Step 1: $y = x^3$
Step 2: x-intercepts: $-4, -1, 2$; y-intercept: -8
Step 3: $-4, -1, 2$: multiplicity 1, crosses
Step 4: At most 2 turning points
Step 5: Near -4: $f(x) \approx 18(x + 4)$; Near -1: $f(x) \approx -9(x + 1)$;
Near 2: $f(x) \approx 18(x - 2)$
Step 6:

77. Step 1: $y = x^4$
Step 2: x-intercepts: $-2, 0, 2$; y-intercept: 0
Step 3: $-4, 3$, multiplicity 1, crosses; 0, multiplicity 2, touches
Step 4: At most 3 turning points
Step 5: Near -2: $f(x) \approx -16(x + 2)$; Near 0: $f(x) \approx -4x^2$;
Near 2: $f(x) \approx 16(x - 2)$
Step 6:

79. Step 1: $y = x^4$
Step 2: x-intercepts: $-1, 2$; y-intercept: 4
Step 3: $-1, 2$: multiplicity 2, touches
Step 4: At most 3 turning points
Step 5: Near -1: $f(x) \approx 9(x + 1)^2$; Near 2: $f(x) \approx 9(x - 2)^2$
Step 6:

81. Step 1: $y = x^4$
   Step 2: x-intercepts: $-1, 0, 3$; y-intercept: 0
   Step 3: $-1, 3$: multiplicity 1, crosses; 0: multiplicity 2, touches
   Step 4: At most 3 turning points
   Step 5: Near -1: $f(x) \approx -4(x + 1)$; Near 0: $f(x) \approx -3x^2$;
   Near 3: $f(x) \approx 36(x - 3)$
   Step 6:

83. Step 1: $y = x^4$
   Step 2: x-intercepts: $-2, 4$; y-intercept: 64
   Step 3: $-2, 4$, multiplicity 2, touches
   Step 4: At most 3 turning points
   Step 5: Near -2: $f(x) \approx 36(x + 2)^2$; Near 4: $f(x) \approx 36(x - 4)^2$
   Step 6:

85. Step 1: $y = x^5$
   Step 2: x-intercepts: $0, 2$; y-intercept: 0
   Step 3: 2, multiplicity 1, crosses;
   0, multiplicity 2, touches

   Step 4: At most 4 turning points
   Step 5: Near 0: $f(x) \approx -6x^2$; Near 2: $f(x) \approx 28(x - 2)$
   Step 6:

87. Step 1: $y = x^3$
   Step 2:

   Step 3: x-intercepts: $-1.26, -0.20, 1.26$
   y-intercept: -0.31752
   Step 4:

   Step 5: $(-0.80, 0.57)$; $(0.66, -0.99)$
   Step 6:

   Step 7: Domain: $(-\infty, \infty)$;
   Range: $(-\infty, \infty)$
   Step 8: Increasing on $(-\infty, -0.80)$
   and $(0.66, \infty)$
   Decreasing on $(-0.80, 0.66)$

89. Step 1: $y = x^3$
   Step 2:

   Step 3: x-intercepts: $-3.56, 0.50$
   y-intercept: 0.89
   Step 4:

   Step 5: $(-2.21, 9.91)$; $(0.50, 0)$
   Step 6:

   Step 7: Domain: $(-\infty, \infty)$; Range: $(-\infty, \infty)$
   Step 8: Increasing on $(-\infty, -2.21)$ and $(0.50, \infty)$
   Decreasing on $(-2.21, 0.50)$

91. Step 1: $y = x^4$
   Step 2:

   Step 3: x-intercepts: $-1.5, -0.5, 0.5, 1.5$
   y-intercept: 0.5625
   Step 4:

   Step 5: $(-1.12, -1)$; $(1.12, -1)$; $(0, 0.56)$
   Step 6:

   Step 7: Domain: $(-\infty, \infty)$;
   Range: $[-1, \infty)$
   Step 8: Increasing on $(-1.12, 0)$
   and $(1.12, \infty)$
   Decreasing on $(-\infty, -1.12)$ and $(0, 1.12)$

93. Step 1: $y = 2x^4$
   Step 2:

   Step 3: x-intercepts: $-1.07, 1.62$;
   y-intercept: -4
   Step 4:

   Step 5: $(-0.42, -4.64)$
   Step 6:

   Step 7: Domain: $(-\infty, \infty)$;
   Range: $[-4.64, \infty)$
   Step 8: Increasing on $(-0.42, \infty)$
   Decreasing on $(-\infty, -0.42)$

95. $f(x) = -x(x + 2)(x - 2)$
   Step 1: $y = -x^3$
   Step 2: x-intercepts: $-2, 0, 2$;
   y-intercept: 0
   Step 3: $-2, 0, 2$: multiplicity 1, crosses
   Step 4: At most 2 turning points
   Step 5: Near -2: $f(x) \approx -8(x + 2)$;
   Near 0: $f(x) \approx 4x$;
   Near 2: $f(x) \approx -8(x - 2)$
   Step 6:

97. $f(x) = x(x + 4)(x - 3)$
  Step 1: $y = x^3$
  Step 2: x-intercepts: $-4, 0, 3$; y-intercept: 0
  Step 3: $-4, 0, 3$: multiplicity 1, crosses
  Step 4: At most 2 turning points
  Step 5: Near -4: $f(x) \approx 28(x + 4)$; Near 0: $f(x) \approx -12x$;
    Near 3: $f(x) \approx 21(x - 3)$
  Step 6:

99. $f(x) = 2x(x + 6)(x - 2)(x + 2)$
  Step 1: $y = 2x^4$
  Step 2: x-intercepts: $-6, -2, 0, 2$; y-intercept: 0
  Step 3: $-6, -2, 0, 2$: multiplicity 1, crosses
  Step 4: At most 3 turning points
  Step 5: Near -6: $f(x) \approx -384(x + 6)$; Near -2: $f(x) \approx 64(x + 2)$;
    Near 0: $f(x) \approx -48x$; Near 2: $f(x) \approx 128(x - 2)$
  Step 6:

101. $f(x) = -x^2(x + 1)^2(x - 1)$
  Step 1: $y = -x^5$
  Step 2: x-intercepts: $-1, 0, 1$; y-intercept: 0
  Step 3: 1: multiplicity 1, crosses; $-1, 0$: multiplicity 2, touches
  Step 4: At most 4 turning points

Step 5: Near -1: $f(x) \approx 2(x + 1)^2$; Near 0: $f(x) \approx x^2$;
  Near 1: $-4(x - 1)$
Step 6:

103. $f(x) = 3(x + 3)(x - 1)(x - 4)$ **105.** $f(x) = -2(x + 5)^2(x - 2)(x - 4)$ **107. (a)** $-3, 2$ **(b)** $-6, -1$
109. (a)
$$T(r) = \underbrace{500(1 + r)(1 + r)}_{\text{Account value of deposit 1}} + \underbrace{500(1 + r)}_{\text{Account value of deposit 2}} + \underbrace{500}_{\text{Deposit 3}} \quad \textbf{(b)}\ \$2155.06$$

$$= 500(1 + 2r + r^2) + 500 + 500r + 500 = 500 + 1000r + 500r^2 + 500 + 500r + 500 = 500r^2 + 1500r + 1500$$

111. (a)

The relation appears to be cubic

  (b) $H(x) = 0.1591x^3 - 2.3203x^2 + 9.3301x - 2.2143$
  (c) About 6 major hurricanes

  (d)

  (e) Approximately 10 major hurricanes

113. (a)

The relation appears to be cubic.

  (b) Average rate of change: $2.27°$/h
  (c) Average rate of change: $0°$/h
  (d) $T(x) = -0.0103x^3 + 0.3174x^2 - 1.3742x + 45.3929$; The predicted temperature at 5 PM is $63.2°$F.

  (e)

  (f) The y-intercept, $45.4°$, is the predicted temperature at midnight.

119. (a)–(d)

3.2 Assess Your Understanding *(page 223)*

5. F **6.** horizontal asymptote **7.** vertical asymptote **8.** proper **9.** T **10.** F **11.** $y = 0$ **12.** T **13.** All real numbers except 3; $\{x | x \neq 3\}$

15. All real numbers except 2 and -4; $\{x | x \neq 2, x \neq -4\}$ **17.** All real numbers except $-\dfrac{1}{2}$ and 3; $\left\{x \middle| x \neq -\dfrac{1}{2}, x \neq 3\right\}$

19. All real numbers except 2; $\{x | x \neq 2\}$ **21.** All real numbers **23.** All real numbers except -3 and 3; $\{x | x \neq -3, x \neq 3\}$

25. (a) Domain: $\{x | x \neq 2\}$; Range: $\{y | y \neq 1\}$ **(b)** $(0, 0)$ **(c)** $y = 1$ **(d)** $x = 2$ **(e)** None

27. (a) Domain: $\{x | x \neq 0\}$; Range: all real numbers **(b)** $(-1, 0), (1, 0)$ **(c)** None **(d)** $x = 0$ **(e)** $y = 2x$

29. (a) Domain: $\{x | x \neq -2, x \neq 2\}$; Range: $\{y | y \leq 0, y > 1\}$ **(b)** $(0, 0)$ **(c)** $y = 1$ **(d)** $x = -2, x = 2$ **(e)** None

31.

33.

35.

37.

39.

41.

43. Vertical asymptote: $x = -4$; horizontal asymptote: $y = 3$ **45.** Vertical asymptote: $x = 3$; oblique asymptote: $y = x + 5$

47. Vertical asymptotes: $x = 1$, $x = -1$; horizontal asymptote: $y = 0$ **49.** Vertical asymptote: $x = -\dfrac{1}{3}$; horizontal asymptote: $y = \dfrac{2}{3}$

51. Vertical asymptote: none; oblique asymptote: $y = 2x - 1$ **53.** Vertical asymptote: $x = 0$; no horizontal or oblique asymptote

55. (a) 9.8208 m/sec^2 **(b)** 9.8195 m/sec^2 **(c)** 9.7936 m/sec^2 **(d)** h-axis **(e)** $\varnothing$

57. (a) **(b)** Horizontal: $R_{tot} = 10$; as the resistance of R_2 increases without bound, the total resistance approaches 10 ohms, the resistance R_1. **(c)** $R_1 \approx 103.5$ ohms

3.3 Assess Your Understanding *(page 237)*

3. in lowest terms **4.** vertical **5.** True **6.** True **7. 1.** Domain: $\{x \mid x \neq 0, x \neq -4\}$ **2.** R is in lowest terms **3.** no y-intercept; x-intercept: -1
4. R is in lowest terms; vertical asymptotes: $x = 0$, $x = 4$ **5.** Horizontal asymptote: $y = 0$, intersected at $(-1, 0)$

6.

Interval	$(-\infty, -4)$	$(-4, -1)$	$(-1, 0)$	$(0, \infty)$
Number Chosen	-5	-2	$-\frac{1}{2}$	1
Value of R	$R(-5) = -\frac{4}{5}$	$R(-2) = \frac{1}{4}$	$R\left(-\frac{1}{2}\right) = -\frac{2}{7}$	$R(1) = \frac{2}{5}$
Location of Graph	Below x-axis	Above x-axis	Below x-axis	Above x-axis
Point on Graph	$\left(-5, -\frac{4}{5}\right)$	$\left(-2, \frac{1}{4}\right)$	$\left(-\frac{1}{2}, -\frac{2}{7}\right)$	$\left(1, \frac{2}{5}\right)$

7/8.

9. 1. $R(x) = \dfrac{3(x + 1)}{2(x + 2)}$; domain: $\{x \mid x \neq -2\}$ **2.** R is in lowest terms **3.** y-intercept: $\dfrac{3}{4}$; x-intercept: -1
4. R is in lowest terms; vertical asymptote: $x = -2$ **5.** Horizontal asymptote: $y = \dfrac{3}{2}$, not intersected

6.

Interval	$(-\infty, -2)$	$(-2, -1)$	$(-1, \infty)$
Number Chosen	-3	$-\frac{3}{2}$	0
Value of R	$R(-3) = 3$	$R\left(-\frac{3}{2}\right) = -\frac{3}{2}$	$R(0) = \frac{3}{4}$
Location of Graph	Above x-axis	Below x-axis	Above x-axis
Point on Graph	$(-3, 3)$	$\left(-\frac{3}{2}, -\frac{3}{2}\right)$	$\left(0, \frac{3}{4}\right)$

7/8.

11. 1. $R(x) = \dfrac{3}{(x + 2)(x - 2)}$; domain: $\{x \mid x \neq -2, x \neq 2\}$ **2.** R is in lowest terms **3.** y-intercept: $-\dfrac{3}{4}$; no x-intercept
4. R is in lowest terms; vertical asymptotes: $x = 2$, $x = -2$ **5.** Horizontal asymptote: $y = 0$, not intersected

6.

Interval	$(-\infty, -2)$	$(-2, 2)$	$(2, \infty)$
Number Chosen	-3	0	3
Value of R	$R(-3) = \frac{3}{5}$	$R(0) = -\frac{3}{4}$	$R(3) = \frac{3}{5}$
Location of Graph	Above x-axis	Below x-axis	Above x-axis
Point on Graph	$\left(-3, \frac{3}{5}\right)$	$\left(0, -\frac{3}{4}\right)$	$\left(3, \frac{3}{5}\right)$

7/8.

13. 1. $P(x) = \dfrac{(x^2 + x + 1)(x^2 - x + 1)}{(x + 1)(x - 1)}$; domain: $\{x \mid x \neq -1, x \neq 1\}$ **2.** P is in lowest terms **3.** y-intercept: -1; no x-intercept
4. P is in lowest terms; vertical asymptotes: $x = -1$, $x = 1$ **5.** No horizontal or oblique asymptote

6.

Interval	$(-\infty, -1)$	$(-1, 1)$	$(1, \infty)$
Number Chosen	-2	0	2
Value of P	$P(-2) = 7$	$P(0) = -1$	$P(2) = 7$
Location of Graph	Above x-axis	Below x-axis	Above x-axis
Point on Graph	$(-2, 7)$	$(0, -1)$	$(2, 7)$

7/8.

15. 1. $H(x) = \dfrac{(x - 1)(x^2 + x + 1)}{(x + 3)(x - 3)}$; domain: $\{x \mid x \neq -3, x \neq 3\}$ **2.** H is in lowest terms **3.** y-intercept: $\dfrac{1}{9}$; x-intercept: 1
4. H is in lowest terms; vertical asymptotes: $x = 3$, $x = -3$ **5.** Oblique asymptote: $y = x$, intersected at $\left(\dfrac{1}{9}, \dfrac{1}{9}\right)$

6.

Interval	$(-\infty, -3)$	$(-3, 1)$	$(1, 3)$	$(3, \infty)$
Number Chosen	-4	0	2	4
Value of H	$H(-4) \approx -9.3$	$H(0) = \frac{1}{9}$	$H(2) = -1.4$	$H(4) = 9$
Location of Graph	Below x-axis	Above x-axis	Below x-axis	Above x-axis
Point on Graph	$(-4, -9.3)$	$\left(0, \frac{1}{9}\right)$	$(2, -1.4)$	$(4, 9)$

7/8.

17. 1. $R(x) = \dfrac{x^2}{(x+3)(x-2)}$; domain: $\{x \neq -3, x \neq 2\}$ **2.** R is in lowest terms   **3.** y-intercept: 0; x-intercept: 0

4. R is in lowest terms; vertical asymptotes: $x = 2$, $x = -3$ **5.** Horizontal asymptote: $y = 1$, intersected at $(6, 1)$

6.

Interval	$(-\infty, -3)$	$(-3, 0)$	$(0, 2)$	$(2, \infty)$
Number Chosen	-6	-1	1	3
Value of R	$R(-6) = 1.5$	$R(-1) = -\frac{1}{6}$	$R(1) = -0.25$	$R(3) = 1.5$
Location of Graph	Above x-axis	Below x-axis	Below x-axis	Above x-axis
Point on Graph	$(-6, 1.5)$	$\left(-1, -\frac{1}{6}\right)$	$(1, -0.25)$	$(3, 1.5)$

7/8.

19. 1. $G(x) = \dfrac{x}{(x+2)(x-2)}$; domain: $\{x \mid x \neq -2, x \neq 2\}$ **2.** G is in lowest terms   **3.** y-intercept: 0; x-intercept: 0

4. G is in lowest terms; vertical asymptotes: $x = -2$, $x = 2$ **5.** Horizontal asymptote: $y = 0$, intersected at $(0, 0)$

6.

Interval	$(-\infty, -2)$	$(-2, 0)$	$(0, 2)$	$(2, \infty)$
Number Chosen	-3	-1	1	3
Value of G	$G(-3) = -\frac{3}{5}$	$G(-1) = \frac{1}{3}$	$G(1) = -\frac{1}{3}$	$G(3) = \frac{3}{5}$
Location of Graph	Below x-axis	Above x-axis	Below x-axis	Above x-axis
Point on Graph	$\left(-3, -\frac{3}{5}\right)$	$\left(-1, \frac{1}{3}\right)$	$\left(1, -\frac{1}{3}\right)$	$\left(3, \frac{3}{5}\right)$

7/8.

21. 1. $R(x) = \dfrac{3}{(x-1)(x+2)(x-2)}$; domain: $\{x \mid x \neq 1, x \neq -2, x \neq 2\}$ **2.** R is in lowest terms   **3.** y-intercept: $\frac{3}{4}$; no x-intercept

4. R is in lowest terms; vertical asymptotes: $x = -2$, $x = 1$, $x = 2$ **5.** Horizontal asymptote: $y = 0$, not intersected

6.

Interval	$(-\infty, -2)$	$(-2, 1)$	$(1, 2)$	$(2, \infty)$
Number Chosen	-3	0	1.5	3
Value of R	$R(-3) = -\frac{3}{20}$	$R(0) = \frac{3}{4}$	$R(1.5) = -\frac{24}{7}$	$R(3) = \frac{3}{10}$
Location of Graph	Below x-axis	Above x-axis	Below x-axis	Above x-axis
Point on Graph	$\left(-3, -\frac{3}{20}\right)$	$\left(0, \frac{3}{4}\right)$	$\left(1.5, -\frac{24}{7}\right)$	$\left(3, \frac{3}{10}\right)$

7/8.

23. 1. $H(x) = \dfrac{(x+1)(x-1)}{(x^2+4)(x+2)(x-2)}$; domain: $\{x \mid x \neq -2, x \neq 2\}$ **2.** H is in lowest terms   **3.** y-intercept: $\frac{1}{16}$; x-intercepts: $-1, 1$

4. H is in lowest terms; vertical asymptotes: $x = -2$, $x = 2$ **5.** Horizontal asymptote: $y = 0$, intersected at $(-1, 0)$ and $(1, 0)$

6.

Interval	$(-\infty, -2)$	$(-2, -1)$	$(-1, 1)$	$(1, 2)$	$(2, \infty)$
Number Chosen	-3	-1.5	0	1.5	3
Value of H	$H(-3) \approx 0.12$	$H(-1.5) \approx -0.11$	$H(0) = \frac{1}{16}$	$H(1.5) \approx -0.11$	$H(3) \approx 0.12$
Location of Graph	Above x-axis	Below x-axis	Above x-axis	Below x-axis	Above x-axis
Point on Graph	$(-3, 0.12)$	$(-1.5, -0.11)$	$\left(0, \frac{1}{16}\right)$	$(1.5, -0.11)$	$(3, 0.12)$

7/8.

25. 1. $F(x) = \dfrac{(x+1)(x-4)}{x+2}$; domain: $\{x \mid x \neq -2\}$ **2.** F is in lowest terms   **3.** y-intercept: -2; x-intercept: $-1, 4$

4. F is in lowest terms; vertical asymptote: $x = -2$ **5.** Oblique asymptote: $y = x - 5$, not intersected

6.

Interval	$(-\infty, -2)$	$(-2, -1)$	$(-1, 4)$	$(4, \infty)$
Number Chosen	-3	-1.5	0	5
Value of F	$F(-3) = -14$	$F(-1.5) = 5.5$	$F(0) = -2$	$F(5) \approx 0.86$
Location of Graph	Below x-axis	Above x-axis	Below x-axis	Above x-axis
Point on Graph	$(-3, -14)$	$(-1.5, 5.5)$	$(0, -2)$	$(5, 0.86)$

7/8.

27. 1. $R(x) = \dfrac{(x+4)(x-3)}{x-4}$; domain: $\{x \mid x \neq 4\}$ **2.** R is in lowest terms   **3.** y-intercept: 3; x-intercepts: $-4, 3$

4. R is in lowest terms; vertical asymptote: $x = 4$ **5.** Oblique asymptote: $y = x + 5$, not intersected

6.

Interval	$(-\infty, -4)$	$(-4, 3)$	$(3, 4)$	$(4, \infty)$
Number Chosen	-5	0	3.5	5
Value of R	$R(-5) = -\frac{8}{9}$	$R(0) = 3$	$R(3.5) = -7.5$	$R(5) = 18$
Location of Graph	Below x-axis	Above x-axis	Below x-axis	Above x-axis
Point on Graph	$\left(-5, -\frac{8}{9}\right)$	$(0, 3)$	$(3.5, -7.5)$	$(5, 18)$

7/8.

29. 1. $F(x) = \dfrac{(x+4)(x-3)}{x+2}$; domain: $\{x | x \neq -2\}$ **2.** F is in lowest terms **3.** y-intercept: -6; x-intercepts: $-4, 3$

4. F is in lowest terms; vertical asymptote: $x = -2$ **5.** Oblique asymptote: $y = x - 1$, not intersected

6.

Interval	$(-\infty, -4)$	$(-4, -2)$	$(-2, 3)$	$(3, \infty)$
Number Chosen	-5	-3	0	4
Value of F	$F(-5) = -\frac{8}{3}$	$F(-3) = 6$	$F(0) = -6$	$F(4) = \frac{4}{3}$
Location of Graph	Below x-axis	Above x-axis	Below x-axis	Above x-axis
Point on Graph	$\left(-5, -\frac{8}{3}\right)$	$(-3, 6)$	$(0, -6)$	$\left(4, \frac{4}{3}\right)$

7/8.

31. 1. Domain: $\{x | x \neq -3\}$ **3.** y-intercept: 0; x-intercepts: $0, 1$ **4.** Vertical asymptote: $x = -3$

5. Horizontal asymptote: $y = 1$, not intersected

6.

Interval	$(-\infty, -3)$	$(-3, 0)$	$(0, 1)$	$(1, \infty)$
Number Chosen	-4	-1	$\frac{1}{2}$	2
Value of R	$R(-4) = 100$	$R(-1) = -0.5$	$R\left(\frac{1}{2}\right) \approx 0.003$	$R(2) = 0.016$
Location of Graph	Above x-axis	Below x-axis	Above x-axis	Above x-axis
Point on Graph	$(-4, 100)$	$(-1, -0.5)$	$\left(\frac{1}{2}, 0.003\right)$	$(2, 0.016)$

7/8.

See enlarged view at right.

Enlarged view

33. 1. $R(x) = \dfrac{(x+4)(x-3)}{(x+2)(x-3)}$; domain: $\{x | x \neq -2, x \neq 3\}$ **2.** In lowest terms, $R(x) = \dfrac{x+4}{x+2}$ **3.** y-intercept: 2; x-intercept: -4

4. Vertical asymptote: $x = -2$; hole at $\left(3, \dfrac{7}{5}\right)$ **5.** Horizontal asymptote: $y = 1$, not intersected

6.

Interval	$(-\infty, -4)$	$(-4, -2)$	$(-2, 3)$	$(3, \infty)$
Number Chosen	-5	-3	0	4
Value of R	$R(-5) = \frac{1}{3}$	$R(-3) = -1$	$R(0) = 2$	$R(4) = \frac{4}{3}$
Location of Graph	Above x-axis	Below x-axis	Above x-axis	Above x-axis
Point on Graph	$\left(-5, \frac{1}{3}\right)$	$(-3, -1)$	$(0, 2)$	$\left(4, \frac{4}{3}\right)$

7/8.

35. 1. $R(x) = \dfrac{(3x+1)(2x-3)}{(x-2)(2x-3)}$; domain: $\left\{x \middle| x \neq \dfrac{3}{2}, x \neq 2\right\}$ **2.** In lowest terms, $R(x) = \dfrac{3x+1}{x-2}$ **3.** y-intercept: $-\dfrac{1}{2}$; x-intercept: $-\dfrac{1}{3}$

4. Vertical asymptote: $x = 2$; hole at $\left(\dfrac{3}{2}, -11\right)$ **5.** Horizontal asymptote: $y = 3$, not intersected

6.

Interval	$\left(-\infty, -\frac{1}{3}\right)$	$\left(-\frac{1}{3}, \frac{3}{2}\right)$	$\left(\frac{3}{2}, 2\right)$	$(2, \infty)$
Number Chosen	-1	0	1.7	6
Value of R	$R(-1) = \frac{2}{3}$	$R(0) = -\frac{1}{2}$	$R(1.7) \approx -20.3$	$R(6) = 4.75$
Location of Graph	Above x-axis	Below x-axis	Below x-axis	Above x-axis
Point on Graph	$\left(-1, \frac{2}{3}\right)$	$\left(0, -\frac{1}{2}\right)$	$(1.7, -20.3)$	$(6, 4.75)$

7/8.

37. 1. $R(x) = \dfrac{(x+3)(x+2)}{x+3}$; domain: $\{x | x \neq -3\}$ **2.** In lowest terms, $R(x) = x + 2$ **3.** y-intercept: 2; x-intercept: -2

4. Vertical asymptote: none; hole at $(-3, -1)$ **5.** Oblique asymptote: $y = x + 2$ intersected at all points except $x = -3$

6.

Interval	$(-\infty, -3)$	$(-3, -2)$	$(-2, \infty)$
Number Chosen	-4	$-\frac{5}{2}$	0
Value of R	$R(-4) = -2$	$R\left(-\frac{5}{2}\right) = -\frac{1}{2}$	$R(0) = 2$
Location of Graph	Below x-axis	Below x-axis	Above x-axis
Point on Graph	$(-4, -2)$	$\left(-\frac{5}{2}, -\frac{1}{2}\right)$	$(0, 2)$

7/8.

39. 1. $f(x) = \dfrac{x^2 + 1}{x}$; domain: $\{x | x \neq 0\}$ **2.** f is in lowest terms **3.** no y-intercept; no x-intercepts **4.** f is in lowest terms; vertical asymptote: $x = 0$

5. Oblique asymptote: $y = x$, not intersected

6.

Interval	$(-\infty, 0)$	$(0, \infty)$
Number Chosen	-1	1
Value of f	$f(-1) = -2$	$f(1) = 2$
Location of Graph	Below x-axis	Above x-axis
Point on Graph	$(-1, -2)$	$(1, 2)$

7/8.

41. 1. $f(x) = \dfrac{x^3 + 1}{x} = \dfrac{(x+1)(x^2 - x + 1)}{x}$; domain: $\{x | x \neq 0\}$ **2.** f is in lowest terms   **3.** no y-intercept; x-intercept: -1

4. f is in lowest terms; vertical asymptote: $x = 0$ **5.** No horizontal or oblique asymptote

6.

Interval	$(-\infty, -1)$	$(-1, 0)$	$(0, \infty)$
Number Chosen	-2	$-\frac{1}{2}$	1
Value of f	$f(-2) = \frac{7}{2}$	$f\left(-\frac{1}{2}\right) = -\frac{7}{4}$	$f(1) = 2$
Location of Graph	Above x-axis	Below x-axis	Above x-axis
Point on Graph	$\left(-2, \frac{7}{2}\right)$	$\left(-\frac{1}{2}, -\frac{7}{4}\right)$	$(1, 2)$

7/8.

43. 1. $f(x) = \dfrac{x^4 + 1}{x^3}$; domain: $\{x | x \neq 0\}$ **2.** f is in lowest terms   **3.** no y-intercept; no x-intercepts

4. f is in lowest terms; vertical asymptote: $x = 0$ **5.** Oblique asymptote: $y = x$, not intersected

6.

Interval	$(-\infty, 0)$	$(0, \infty)$
Number Chosen	-1	1
Value of f	$f(-1) = -2$	$f(1) = 2$
Location of Graph	Below x-axis	Above x-axis
Point on Graph	$(-1, -2)$	$(1, 2)$

7/8.

45. Minimum value: 2.00 at $x = 1.00$

47. Minimum value: 1.89 at $x = 0.79$

49. Minimum value: 1.75 at $x = 1.32$

51. One possibility: $R(x) = \dfrac{x^2}{x^2 - 4}$ **53.** One possibility: $R(x) = \dfrac{(x-1)(x-3)\left(x^2 + \dfrac{4}{3}\right)}{(x+1)^2(x-2)^2}$

55. (a) t-axis; $C(t) \to 0$
(b)

(c) 0.71 h after injection

57. (a) $C(x) = 16x + \dfrac{5000}{x} + 100$
(b) $x > 0$
(c)

(d) Approximately 17.7 ft by 56.6 ft (longer side parallel to river)

59. (a) $S(x) = 2x^2 + \dfrac{40,000}{x}$
(b)

(c) 2784.95 in.^{2}
(d) 21.54 in. $\times$ 21.54 in. $\times$ 21.54 in.
(e) To minimize the cost of materials needed for construction

61. (a) $C(r) = 12\pi r^2 + \dfrac{4000}{r}$
(b)

The cost is smallest when $r = 3.76$ cm.

63. No. Each function is a quotient of polynomials, but it is not written in lowest terms. Each function is undefined for $x = 1$; each graph has a hole at $x = 1$.
69. If there is a common factor between the numerator and the denominator, and the factor yields a real zero, then the graph will have a hole.

3.4 Assess Your Understanding *(page 245)*

3. T   **4.** F   **5. (a)** $\{x | 0 < x < 1 \text{ or } x > 2\}$; $(0, 1) \cup (2, \infty)$ **(b)** $\{x | x \leq 0 \text{ or } 1 \leq x \leq 2\}$; $(-\infty, 0] \cup [1, 2]$ **7. (a)** $\{x | -1 < x < 0 \text{ or } x > 1\}$; $(-1, 0) \cup (1, \infty)$
(b) $\{x | x < -1 \text{ or } 0 \leq x < 1\}$; $(-\infty, -1) \cup [0, 1)$ **9.** $\{x | x < 0 \text{ or } 0 < x < 3\}$; $(-\infty, 0) \cup (0, 3)$ **11.** $\{x | x \geq -4\}$; $[-4, \infty)$
13. $\{x | x \leq -2 \text{ or } x \geq 2\}$; $(-\infty, -2] \cup [2, \infty)$ **15.** $\{x | -4 < x < -1 \text{ or } x > 0\}$; $(-4, -1) \cup (0, \infty)$ **17.** $\{x | -2 < x \leq -1\}$; $(-2, -1]$
19. $\{x | x < -2\}$; $(-\infty, -2)$ **21.** $\{x | x > 4\}$; $(4, \infty)$ **23.** $\{x | -4 < x < 0 \text{ or } x > 0\}$; $(-4, 0) \cup (0, \infty)$ **25.** $\{x | x \leq 1 \text{ or } 2 \leq x \leq 3\}$; $(-\infty, 1] \cup [2, 3]$

27. $\{x | -1 < x < 0 \text{ or } x > 3\}; (-1, 0) \cup (3, \infty)$ **29.** $\{x | x < -1 \text{ or } x > 1\}; (-\infty, -1) \cup (1, \infty)$ **31.** $\{x | x < -1 \text{ or } x > 1\}; (-\infty, -1) \cup (1, \infty)$

33. $\{x | x < -1 \text{ or } x > 1\}; (-\infty, -1) \cup (1, \infty)$ **35.** $\{x | x \le -1 \text{ or } 0 < x \le 1\}; (-\infty, -1] \cup (0, 1]$ **37.** $\{x | x < -1 \text{ or } x > 1\}; (-\infty, -1) \cup (1, \infty)$

39. $\{x | x < 2\}; (-\infty, 2)$ **41.** $\{x | -2 < x \le 9\}; (-2, 9]$ **43.** $\{x | x < 2 \text{ or } 3 < x < 5\}; (-\infty, 2) \cup (3, 5)$

45. $\{x | x < -5 \text{ or } -4 \le x \le -3 \text{ or } x = 0 \text{ or } x > 1\}; (-\infty, -5) \cup [-4, -3] \cup \{0\} \cup (1, \infty)$ **47.** $\left\{x \left| -\frac{1}{2} < x < 1 \text{ or } x > 3 \right.\right\}; \left(-\frac{1}{2}, 1\right) \cup (3, \infty)$

49. $\{x | -1 < x < 3 \text{ or } x > 5\}; (-1, 3) \cup (5, \infty)$ **51.** $\left\{x \left| x \le -4 \text{ or } x \ge \frac{1}{2} \right.\right\}; (-\infty, -4] \cup \left[\frac{1}{2}, \infty\right)$ **53.** $\{x | x < 3 \text{ or } x \ge 7\}; (-\infty, 3) \cup [7, \infty)$

55. $\{x | x < 2\}; (-\infty, 2)$ **57.** $\left\{x \left| x < -\frac{2}{3} \text{ or } 0 < x < \frac{3}{2} \right.\right\}; \left(-\infty, -\frac{2}{3}\right) \cup \left(0, \frac{3}{2}\right)$ **59.** $\{x | x \le -3 \text{ or } 0 \le x \le 3\}; (-\infty, -3] \cup [0, 3]$

61. $\{x | x > 4\}; (4, \infty)$ **63.** $\{x | x \le -2 \text{ or } x \ge 2\}; (-\infty, -2] \cup [2, \infty)$ **65.** $\{x | x < -4 \text{ or } x \ge 2\}; (-\infty, -4) \cup [2, \infty)$

67.

$f(x) \le g(x)$ if $-1 \le x \le 1$

69.

$f(x) \le g(x)$ if $-2 \le x \le 2$

71. Produce at least 250 bicycles
73. (a) The stretch is less than 39 ft.
   (b) The ledge should be at least 84 ft above the ground for a 150-lb jumper.
75. At least 50 students must attend.

Historical Problems (page 259)

1.
$$\left(x - \frac{b}{3}\right)^3 + b\left(x - \frac{b}{3}\right)^2 + c\left(x - \frac{b}{3}\right) + d = 0$$

$$x^3 - bx^2 + \frac{b^2 x}{3} - \frac{b^3}{27} + bx^2 - \frac{2b^2 x}{3} + \frac{b^3}{9} + cx - \frac{bc}{3} + d = 0$$

$$x^3 + \left(c - \frac{b^2}{3}\right)x + \left(\frac{2b^3}{27} - \frac{bc}{3} + d\right) = 0$$

Let $p = c - \frac{b^2}{3}$ and $q = \frac{2b^3}{27} - \frac{bc}{3} + d$. Then $x^3 + px + q = 0$.

2.
$$(H + K)^3 + p(H + K) + q = 0$$
$$H^3 + 3H^2 K + 3HK^2 + K^3 + pH + pK + q = 0$$
Let $3HK = -p$.
$$H^3 - pH - pK + K^3 + pH + pK + q = 0, \qquad H^3 + K^3 = -q$$

3.
$$3HK = -p$$
$$K = -\frac{p}{3H}$$
$$H^3 + \left(-\frac{p}{3H}\right)^3 = -q$$
$$H^3 - \frac{p^3}{27H^3} = -q$$
$$27H^6 - p^3 = -27qH^3$$
$$27H^6 + 27qH^3 - p^3 = 0$$
$$H^3 = \frac{-27q \pm \sqrt{(27q)^2 - 4(27)(-p^3)}}{2 \cdot 27}$$
$$H^3 = \frac{-q}{2} \pm \sqrt{\frac{27^2 q^2}{2^2(27^2)} + \frac{4(27)p^3}{2^2(27^2)}}$$
$$H^3 = \frac{-q}{2} \pm \sqrt{\frac{q^2}{4} + \frac{p^3}{27}}$$
$$H = \sqrt[3]{\frac{-q}{2} + \sqrt{\frac{q^2}{4} + \frac{p^3}{27}}}$$

Choose the positive root for now.

4. $H^3 + K^3 = -q$
$$K^3 = -q - H^3$$
$$K^3 = -q - \left[\frac{-q}{2} + \sqrt{\frac{q^2}{4} + \frac{p^3}{27}}\right]$$
$$K^3 = \frac{-q}{2} - \sqrt{\frac{q^2}{4} + \frac{p^3}{27}}$$
$$K = \sqrt[3]{\frac{-q}{2} - \sqrt{\frac{q^2}{4} + \frac{p^3}{27}}}$$

5. $x = H + K$
$$x = \sqrt[3]{\frac{-q}{2} + \sqrt{\frac{q^2}{4} + \frac{p^3}{27}}} + \sqrt[3]{\frac{-q}{2} - \sqrt{\frac{q^2}{4} + \frac{p^3}{27}}}$$
(Note that had we used the negative root in 3 the result would be the same.)

6. $x = 3$ **7.** $x = 2$ **8.** $x = 2$

3.5 Assess Your Understanding (page 259)

5. Remainder; dividend   **6.** $f(c)$ **7.** -4 **8.** F   **9.** 0   **10.** T   **11.** $R = f(2) = 8$; no   **13.** $R = f(2) = 0$; yes   **15.** $R = f(-3) = 0$; yes

17. $R = f(-4) = 1$; no   **19.** $R = f\left(\frac{1}{2}\right) = 0$; yes   **21.** 7; 3 or 1 positive; 2 or 0 negative   **23.** 6; 2 or 0 positive; 2 or 0 negative

25. 3; 2 or 0 positive; 1 negative   **27.** 4; 2 or 0 positive; 2 or 0 negative   **29.** 5; 0 positive; 3 or 1 negative   **31.** 6; 1 positive; 1 negative

33. $\pm 1, \pm \frac{1}{3}$ **35.** $\pm 1, \pm 3$ **37.** $\pm 1, \pm 2, \pm \frac{1}{4}, \pm \frac{1}{2}$ **39.** $\pm 1, \pm 3, \pm 9, \pm \frac{1}{2}, \pm \frac{1}{3}, \pm \frac{1}{6}, \pm \frac{3}{2}, \pm \frac{9}{2}$ **41.** $\pm 1, \pm 2, \pm 3, \pm 4, \pm 6, \pm 12, \pm \frac{1}{2}, \pm \frac{3}{2}$

43. $\pm 1, \pm 2, \pm 4, \pm 5, \pm 10, \pm 20, \pm \frac{1}{2}, \pm \frac{5}{2}, \pm \frac{1}{3}, \pm \frac{2}{3}, \pm \frac{4}{3}, \pm \frac{5}{3}, \pm \frac{10}{3}, \pm \frac{20}{3}, \pm \frac{1}{6}, \pm \frac{5}{6}$ **45.** $-3, -1, 2; f(x) = (x + 3)(x + 1)(x - 2)$

47. $\frac{1}{2}; f(x) = 2\left(x - \frac{1}{2}\right)(x^2 + 1)$ **49.** $2, \sqrt{5}, -\sqrt{5}; f(x) = 2(x - 2)(x - \sqrt{5})(x + \sqrt{5})$

51. $-1, \frac{1}{2}, \sqrt{3}, -\sqrt{3}; f(x) = 2(x + 1)\left(x - \frac{1}{2}\right)(x - \sqrt{3})(x + \sqrt{3})$ **53.** 1, multiplicity 2; $-2, -1; f(x) = (x + 2)(x + 1)(x - 1)^2$

55. $-1, -\frac{1}{4}; f(x) = 4(x + 1)\left(x + \frac{1}{4}\right)(x^2 + 2)$ **57.** $\{-1, 2\}$ **59.** $\left\{\frac{2}{3}, -1 + \sqrt{2}, -1 - \sqrt{2}\right\}$ **61.** $\left\{\frac{1}{3}, \sqrt{5}, -\sqrt{5}\right\}$ **63.** $\{-3, -2\}$ **65.** $\left\{-\frac{1}{3}\right\}$

67. $\left\{\frac{1}{2}, 2, 5\right\}$ **69.** 5 **71.** 2 **73.** 5 **75.** $\frac{3}{2}$ **77.** $f(0) = -1; f(1) = 10$ **79.** $f(-5) = -58; f(-4) = 2$ **81.** $f(1.4) = -0.17536; f(1.5) = 1.40625$

83. 0.21 **85.** −4.04 **87.** 1.15 **89.** 2.53

91.

93.

95.

97.

99.

101.

103. $k = 5$ **105.** −7 **107.** If $f(x) = x^n - c^n$, then $f(c) = c^n - c^n = 0$, so $x - c$ is a factor of f. **109.** 5 **111.** 7 in. **113.** All the potential rational zeros are integers, so r is either an integer or is not a rational zero (and is therefore irrational). **117.** No; by the Rational Zeros Theorem, $\frac{1}{3}$ is not a potential rational zero. **119.** No; by the Rational Zeros Theorem, $\frac{2}{3}$ is not a potential rational zero.

3.6 Assess Your Understanding (page 266)

3. one **4.** $3 - 4i$ **5.** T **6.** F **7.** $4 + i$ **9.** $-i, 1 - i$ **11.** $-i, -2i$ **13.** $-i$ **15.** $2 - i, -3 + i$ **17.** $f(x) = x^4 - 14x^3 + 77x^2 - 200x + 208; a = 1$

19. $f(x) = x^5 - 4x^4 + 7x^3 - 8x^2 + 6x - 4; a = 1$ **21.** $f(x) = x^4 - 6x^3 + 10x^2 - 6x + 9; a = 1$ **23.** $-2i, 4$ **25.** $2i, -3, \frac{1}{2}$

27. $3 + 2i, -2, 5$ **29.** $4i, -\sqrt{11}, \sqrt{11}, -\frac{2}{3}$ **31.** $1, -\frac{1}{2} - \frac{\sqrt{3}}{2}i, -\frac{1}{2} + \frac{\sqrt{3}}{2}i; f(x) = (x - 1)\left(x + \frac{1}{2} + \frac{\sqrt{3}}{2}i\right)\left(x + \frac{1}{2} - \frac{\sqrt{3}}{2}i\right)$

33. $2, 3 - 2i, 3 + 2i; f(x) = (x - 2)(x - 3 + 2i)(x - 3 - 2i)$ **35.** $-i, i, -2i, 2i; f(x) = (x + i)(x - i)(x + 2i)(x - 2i)$

37. $-5i, 5i, -3, 1; f(x) = (x + 5i)(x - 5i)(x + 3)(x - 1)$ **39.** $-4, \frac{1}{3}, 2 - 3i, 2 + 3i; f(x) = 3(x + 4)\left(x - \frac{1}{3}\right)(x - 2 + 3i)(x - 2 - 3i)$

41. Zeros that are complex numbers must occur in conjugate pairs; or a polynomial with real coefficients of odd degree must have at least one real zero.
43. If the remaining zero were a complex number, its conjugate would also be a zero, creating a polynomial of degree 5.

Review Exercises (page 269)

1. Polynomial of degree 5 **3.** Neither

5.

7.

9.

11. **1.** $y = x^3$
2. x-intercepts: $-4, -2, 0$; y-intercept: 0
3. $-4, -2, 0$, multiplicity 1, crosses
4. 2
5. Near -4: $f(x) \approx 8(x + 4)$
  Near -2: $f(x) \approx -4(x + 2)$
  Near 0: $f(x) \approx 8x$
6.

13. **1.** $y = x^3$
2. x-intercepts: $-4, 2$; y-intercept: 16
3. -4, multiplicity 1, crosses; 2, multiplicity 2, touches
4. 2
5. Near -4: $f(x) \approx 36(x + 4)$
  Near 2: $f(x) \approx 6(x - 2)^2$
6.

15. **1.** $y = -2x^3$
2. $f(x) = -2x^2(x - 2)$
  x-intercepts: 0, 2; y-intercept: 0
3. 0, multiplicity 2, touches;
  2, multiplicity 1, crosses
4. 2
5. Near 0: $f(x) \approx 4x^2$
  Near 2: $f(x) \approx -8(x - 2)$
6.

17. 1. $y = x^4$
2. x-intercepts: $-3, -1, 1$; y-intercept: 3
3. $-3, -1$, multiplicity 1, crosses; 1, multiplicity 2, touches
4. 3
5. Near -3: $f(x) \approx -32(x + 3)$
Near -1: $f(x) \approx 8(x + 1)$; Near 1: $f(x) \approx 8(x - 1)^2$

6.

19. Domain: $\{x | x \neq -3, x \neq 3\}$; horizontal asymptote: $y = 0$; vertical asymptotes: $x = -3, x = 3$

21. Domain: $\{x | x \neq -2\}$; horizontal asymptote: $y = 1$; vertical asymptote: $x = -2$

23. 1. $R(x) = \dfrac{2(x - 3)}{x}$; domain: $\{x | x \neq 0\}$ **2.** R is in lowest terms   **3.** no y-intercept; x-intercept: 3
4. R is in lowest terms; vertical asymptote: $x = 0$ **5.** Horizontal asymptote: $y = 2$; not intersected

6.

Interval	$(-\infty, 0)$	$(0, 3)$	$(3, \infty)$
Number Chosen	-2	1	4
Value of R	$R(-2) = 5$	$R(1) = -4$	$R(4) = \frac{1}{2}$
Location of Graph	Above x-axis	Below x-axis	Above x-axis
Point on Graph	$(-2, 5)$	$(1, -4)$	$\left(4, \frac{1}{2}\right)$

7/8.

25. 1. Domain: $\{x | x \neq 0, x \neq 2\}$ **2.** H is in lowest terms   **3.** no y-intercept; x-intercept: -2 **4.** H is in lowest terms; vertical asymptote: $x = 0, x = 2$
5. Horizontal asymptote: $y = 0$; intersected at $(-2, 0)$

6.

Interval	$(-\infty, -2)$	$(-2, 0)$	$(0, 2)$	$(2, \infty)$
Number Chosen	-3	-1	1	3
Value of H	$H(-3) = -\frac{1}{15}$	$H(-1) = \frac{1}{3}$	$H(1) = -3$	$H(3) = \frac{5}{3}$
Location of Graph	Below x-axis	Above x-axis	Below x-axis	Above x-axis
Point on Graph	$\left(-3, -\frac{1}{15}\right)$	$\left(-1, \frac{1}{3}\right)$	$(1, -3)$	$\left(3, \frac{5}{3}\right)$

7/8.

27. 1. $R(x) = \dfrac{(x + 3)(x - 2)}{(x - 3)(x + 2)}$; domain: $\{x | x \neq -2, x \neq 3\}$ **2.** R is in lowest terms   **3.** y-intercept: 1; x-intercepts: $-3, 2$
4. R is in lowest terms; vertical asymptotes: $x = -2, x = 3$ **5.** Horizontal asymptote: $y = 1$; intersected at $(0, 1)$

6.

Interval	$(-\infty, -3)$	$(-3, -2)$	$(-2, 2)$	$(2, 3)$	$(3, \infty)$
Number Chosen	-4	$-\frac{5}{2}$	0	$\frac{5}{2}$	4
Value of R	$R(-4) = \frac{3}{7}$	$R\left(-\frac{5}{2}\right) = -\frac{9}{11}$	$R(0) = 1$	$R\left(\frac{5}{2}\right) = -\frac{11}{9}$	$R(4) = \frac{7}{3}$
Location of Graph	Above x-axis	Below x-axis	Above x-axis	Below x-axis	Above x-axis
Point on Graph	$\left(-4, \frac{3}{7}\right)$	$\left(-\frac{5}{2}, -\frac{9}{11}\right)$	$(0, 1)$	$\left(\frac{5}{2}, -\frac{11}{9}\right)$	$\left(4, \frac{7}{3}\right)$

7/8.

29. 1. $F(x) = \dfrac{x^3}{(x + 2)(x - 2)}$; domain: $\{x | x \neq -2, x \neq 2\}$ **2.** F is in lowest terms   **3.** y-intercept: 0; x-intercept: 0
4. F is in lowest terms; vertical asymptote: $x = -2, x = 2$ **5.** Oblique asymptote: $y = x$; intersected at $(0, 0)$

6.

Interval	$(-\infty, -2)$	$(-2, 0)$	$(0, 2)$	$(2, \infty)$
Number Chosen	-3	-1	1	3
Value of F	$F(-3) = -\frac{27}{5}$	$F(-1) = \frac{1}{3}$	$F(1) = -\frac{1}{3}$	$F(3) = \frac{27}{5}$
Location of Graph	Below x-axis	Above x-axis	Below x-axis	Above x-axis
Point on Graph	$\left(-3, -\frac{27}{5}\right)$	$\left(-1, \frac{1}{3}\right)$	$\left(1, -\frac{1}{3}\right)$	$\left(3, \frac{27}{5}\right)$

7/8.

31. 1. Domain: $\{x | x \neq 1\}$ **2.** R is in lowest terms   **3.** y-intercept: 0; x-intercept: 0   **4.** R is in lowest terms; vertical asymptote: $x = 1$
5. No oblique or horizontal asymptote

6.

Interval	$(-\infty, 0)$	$(0, 1)$	$(1, \infty)$
Number Chosen	-2	$\frac{1}{2}$	2
Value of R	$R(-2) = \frac{32}{9}$	$R\left(\frac{1}{2}\right) = \frac{1}{2}$	$R(2) = 32$
Location of Graph	Above x-axis	Above x-axis	Above x-axis
Point on Graph	$\left(-2, \frac{32}{9}\right)$	$\left(\frac{1}{2}, \frac{1}{2}\right)$	$(2, 32)$

7/8.

33. 1. $G(x) = \dfrac{(x+2)(x-2)}{(x+1)(x-2)}$; domain: $\{x\,|\,x \neq -1, x \neq 2\}$ **2.** In lowest terms, $G(x) = \dfrac{x+2}{x+1}$ **3.** y-intercept: 2; x-intercept: -2

4. Vertical asymptote: $x = -1$; hole at $\left(2, \dfrac{4}{3}\right)$ **5.** Horizontal asymptote: $y = 1$, not intersected

6.

Interval	$(-\infty, -2)$	$(-2, -1)$	$(-1, 2)$	$(2, \infty)$
Number Chosen	-3	$-\frac{3}{2}$	0	3
Value of G	$G(-3) = \frac{1}{2}$	$G\left(-\frac{3}{2}\right) = -1$	$G(0) = 2$	$G(3) = \frac{5}{4}$
Location of Graph	Above x-axis	Below x-axis	Above x-axis	Above x-axis
Point on Graph	$\left(-3, \frac{1}{2}\right)$	$\left(-\frac{3}{2}, -1\right)$	$(0, 2)$	$\left(3, \frac{5}{4}\right)$

7/8.

35. $\{x\,|\,x < -2 \text{ or } -1 < x < 2\}$; $(-\infty, -2) \cup (-1, 2)$

37. $\{x\,|\,-3 < x \le 3\}$; $(-3, 3]$

39. $\{x\,|\,x < 1 \text{ or } x > 2\}$; $(-\infty, 1) \cup (2, \infty)$

41. $\{x\,|\,1 \le x \le 2 \text{ or } x > 3\}$; $[1, 2] \cup (3, \infty)$

43. $\{x\,|\,x < -4 \text{ or } 2 < x < 4 \text{ or } x > 6\}$; $(-\infty, -4) \cup (2, 4) \cup (6, \infty)$

45. $R = 10$; g is not a factor of f.   **47.** $R = 0$; g is a factor of f.   **49.** $f(4) = 47{,}105$ **51.** 4, 2, or 0 positive; 2 or 0 negative

53. $\pm 1, \pm 3, \pm \dfrac{1}{2}, \pm \dfrac{3}{2}, \pm \dfrac{1}{3}, \pm \dfrac{1}{4}, \pm \dfrac{3}{4}, \pm \dfrac{1}{6}, \pm \dfrac{1}{12}$ **55.** $-2, 1, 4$; $f(x) = (x+2)(x-1)(x-4)$ **57.** $\dfrac{1}{2}$, multiplicity 2; -2; $f(x) = 4\left(x - \dfrac{1}{2}\right)^2 (x+2)$

59. 2, multiplicity 2; $f(x) = (x-2)^2(x^2 + 5)$ **61.** $\{-3, 2\}$ **63.** $\left\{-3, -1, -\dfrac{1}{2}, 1\right\}$ **65.** 5   **67.** $\dfrac{37}{2}$ **69.** $f(0) = -1$; $f(1) = 1$

71. $f(0) = -1$; $f(1) = 1$ **73.** 1.52   **75.** 0.93   **77.** $4 - i$; $f(x) = x^3 - 14x^2 + 65x - 102$ **79.** $-i, 1 - i$; $f(x) = x^4 - 2x^3 + 3x^2 - 2x + 2$

81. $-2, 1, 4$; $f(x) = (x+2)(x-1)(x-4)$ **83.** $-2, \dfrac{1}{2}$ (multiplicity 2); $f(x) = 4(x+2)\left(x - \dfrac{1}{2}\right)^2$

85. 2 (multiplicity 2), $-\sqrt{5}i, \sqrt{5}i$; $f(x) = (x + \sqrt{5}i)(x - \sqrt{5}i)(x-2)^2$ **87.** $-3, 2, -\dfrac{\sqrt{2}}{2}i, \dfrac{\sqrt{2}}{2}i$; $f(x) = 2(x+3)(x-2)\left(x + \dfrac{\sqrt{2}}{2}i\right)\left(x - \dfrac{\sqrt{2}}{2}i\right)$

89. (a) $A(r) = 2\pi r^2 + \dfrac{500}{r}$

(b) 223.22 cm^2

(c) 257.08 cm^2

(d)

A is smallest when $r \approx 3.41$ cm.

93. (a) Even

(b) Positive

(c) Even

(d) The graph touches the x-axis at $x = 0$, but does not cross it there.

(e) 8

Chapter Test *(page 272)*

1.

2. (a) 3

(b) Every zero of g lies between -15 and 15.

(c) $\dfrac{p}{q}$: $\pm\dfrac{1}{2}, \pm 1, \pm\dfrac{3}{2}, \pm\dfrac{5}{2}, \pm 3, \pm 5, \pm\dfrac{15}{2}, \pm 15$

(d) $-5, -\dfrac{1}{2}, 3$; $g(x) = (x+5)(2x+1)(x-3)$

(e) y-intercept: -15; x-intercepts: $-5, -\dfrac{1}{2}, 3$

(f) Crosses at $-5, -\dfrac{1}{2}, 3$

(g) $y = 2x^3$

(h) Near -5: $g(x) \approx 72(x+5)$

Near $-\dfrac{1}{2}$: $g(x) \approx -\dfrac{63}{4}(2x+1)$

Near 3: $g(x) \approx 56(x-3)$

(i)

3. $4, -5i, 5i$ **4.** $\left\{1, \dfrac{5 - \sqrt{61}}{6}, \dfrac{5 + \sqrt{61}}{6}\right\}$ **5.** Domain: $\{x\,|\,x \neq -10, x \neq 4\}$; asymptotes: $x = -10$, $y = 2$

6. Domain: $\{x\,|\,x \neq -1\}$; asymptotes: $x = -1$, $y = x + 1$

7.

8. Answers may vary. One possibility is $f(x) = x^4 - 4x^3 - 2x^2 + 20x$.

9. Answers may vary. One possibility is $r(x) = \dfrac{2(x - 9)(x - 1)}{(x - 4)(x - 9)}$.

10. $f(0) = 8$; $f(4) = -36$

Since $f(0) = 8 > 0$ and $f(4) = -36 < 0$, the Intermediate Value Theorem guarantees that there is at least one real zero between 0 and 4.

11. $\{x | x < 3 \text{ or } x > 8\}$; $(-\infty, 3) \cup (8, \infty)$

Cumulative Review *(page 272)*

1. $\sqrt{26}$

2. $\{x | x \leq 0 \text{ or } x \geq 1\}$; $(-\infty, 0]$ or $[1, \infty)$

3. $\{x | -1 < x < 4\}$; $(-1, 4)$

4. $f(x) = -3x + 1$

5. $y = 2x - 1$

6.

7. Not a function; 3 has two images.

8. $\{0, 2, 4\}$

9. $\left\{ x \middle| x \geq \dfrac{3}{2} \right\}$; $\left[\dfrac{3}{2}, \infty \right)$

10. Center: $(-2, 1)$; radius: 3

11. x-intercepts: $-3, 0, 3$; y-intercept: 0; symmetric with respect to the origin

12. $y = -\dfrac{2}{3}x + \dfrac{17}{3}$

13. Not a function; it fails the Vertical Line Test.

14. **(a)** 22 **(b)** $x^2 - 5x - 2$ **(c)** $-x^2 - 5x + 2$ **(d)** $9x^2 + 15x - 2$ **(e)** $2x + h + 5$

15. **(a)** $\{x | x \neq 1\}$ **(b)** No; $(2, 7)$ is on the graph. **(c)** 4; $(3, 4)$ is on the graph. **(d)** $\dfrac{7}{4}$; $\left(\dfrac{7}{4}, 9 \right)$ is on the graph.
(e) Rational

16.

17.

18. 6; $y = 6x - 1$

19. **(a)** x-intercepts: $-5, -1, 5$; y-intercept: -3

(b) No symmetry

(c) Neither

(d) Increasing: $(-\infty, -3)$ and $(2, \infty)$; decreasing: $(-3, 2)$

(e) Local maximum is 5 and occurs at $x = -3$.

(f) Local minimum is -6 and occurs at $x = 2$.

20. Odd

21. **(a)** Domain: $\{x | x > -3\}$ or $(-3, \infty)$

(b) x-intercept: $-\dfrac{1}{2}$; y-intercept: 1

(c)

(d) Range: $\{y | y < 5\}$ or $(-\infty, 5)$

22.

23. **(a)** $(f + g)(x) = x^2 - 9x - 6$; domain: all real numbers

(b) $\left(\dfrac{f}{g} \right)(x) = \dfrac{x^2 - 5x + 1}{-4x - 7}$; domain: $\left\{ x \middle| x \neq -\dfrac{7}{4} \right\}$

24. **(a)** $R(x) = -\dfrac{1}{10}x^2 + 150x$

(b) $\$14,000$ **(c)** 750; $\$56,250$ **(d)** $\$75$

CHAPTER 4 Exponential and Logarithmic Functions

4.1 Assess Your Understanding *(page 280)*

4. composite function; $f(g(x))$ **5.** F **6.** F **7.** **(a)** -1 **(b)** -1 **(c)** 8 **(d)** 0 **(e)** 8 **(f)** -7 **9.** **(a)** 4 **(b)** 5 **(c)** -1 **(d)** -2 **11.** **(a)** 98
(b) 49 **(c)** 4 **(d)** 4 **13.** **(a)** 97 **(b)** $-\dfrac{163}{2}$ **(c)** 1 **(d)** $-\dfrac{3}{2}$ **15.** **(a)** $2\sqrt{2}$ **(b)** $2\sqrt{2}$ **(c)** 1 **(d)** 0 **17.** **(a)** $\dfrac{1}{17}$ **(b)** $\dfrac{1}{5}$ **(c)** 1 **(d)** $\dfrac{1}{2}$

19. **(a)** $\dfrac{3}{\sqrt[3]{4} + 1}$ **(b)** 1 **(c)** $\dfrac{6}{5}$ **(d)** 0 **21.** $\{x | x \neq 0, x \neq 2\}$ **23.** $\{x | x \neq -4, x \neq 0\}$ **25.** $\left\{ x \middle| x \geq -\dfrac{3}{2} \right\}$ **27.** $\{x | x \geq 1\}$

29. **(a)** $(f \circ g)(x) = 6x + 3$; all real numbers **31.** **(a)** $(f \circ g)(x) = 3x^2 + 1$; all real numbers **33.** **(a)** $(f \circ g)(x) = x^4 + 8x^2 + 16$; all real numbers
(b) $(g \circ f)(x) = 6x + 9$; all real numbers **(b)** $(g \circ f)(x) = 9x^2 + 6x + 1$; all real numbers
(c) $(f \circ f)(x) = 4x + 9$; all real numbers numbers **(b)** $(g \circ f)(x) = x^4 + 4$; all real numbers
(d) $(g \circ g)(x) = 9x$; all real numbers **(c)** $(f \circ f)(x) = 9x + 4$; all real numbers **(c)** $(f \circ f)(x) = x^4$; all real numbers
(d) $(g \circ g)(x) = x^4$; all real numbers **(d)** $(g \circ g)(x) = x^4 + 8x^2 + 20$; all real numbers

35. **(a)** $(f \circ g)(x) = \dfrac{3x}{2 - x}$; $\{x | x \neq 0, x \neq 2\}$ **(b)** $(g \circ f)(x) = \dfrac{2(x - 1)}{3}$; $\{x | x \neq 1\}$ **(c)** $(f \circ f)(x) = \dfrac{3(x - 1)}{4 - x}$; $\{x | x \neq 1, x \neq 4\}$

(d) $(g \circ g)(x) = x$; $\{x | x \neq 0\}$ **37.** **(a)** $(f \circ g)(x) = \dfrac{4}{4 + x}$; $\{x | x \neq -4, x \neq 0\}$ **(b)** $(g \circ f)(x) = \dfrac{-4(x - 1)}{x}$; $\{x | x \neq 0, x \neq 1\}$

(c) $(f \circ f)(x) = x; \{x|x \neq 1\}$ **(d)** $(g \circ g)(x) = x; \{x|x \neq 0\}$ **39. (a)** $(f \circ g)(x) = \sqrt{2x+3}; \left\{x|x \geq -\dfrac{3}{2}\right\}$

(b) $(g \circ f)(x) = 2\sqrt{x}+3; \{x|x \geq 0\}$ **(c)** $(f \circ f)(x) = \sqrt[4]{x}; \{x|x \geq 0\}$ **(d)** $(g \circ g)(x) = 4x+9$; all real numbers
41. (a) $(f \circ g)(x) = x; \{x|x \geq 1\}$ **(b)** $(g \circ f)(x) = |x|$; all real numbers **(c)** $(f \circ f)(x) = x^4 + 2x^2 + 2$; all real numbers

(d) $(g \circ g)(x) = \sqrt{\sqrt{x-1}-1}; \{x|x \geq 2\}$ **43. (a)** $(f \circ g)(x) = -\dfrac{4x-17}{2x-1}; \left\{x|x \neq 3; x \neq \dfrac{1}{2}\right\}$

(b) $(g \circ f)(x) = -\dfrac{3x-3}{2x+8}; \{x|x \neq -4; x \neq -1\}$ **(c)** $(f \circ f)(x) = -\dfrac{2x+5}{x-2}; \{x|x \neq -1; x \neq 2\}$

(d) $(g \circ g)(x) = -\dfrac{3x-4}{2x-11}; \left\{x|x \neq \dfrac{11}{2}; x \neq 3\right\}$ **45.** $(f \circ g)(x) = f(g(x)) = f\left(\dfrac{1}{2}x\right) = 2\left(\dfrac{1}{2}x\right) = x; (g \circ f)(x) = g(f(x)) = g(2x) = \dfrac{1}{2}(2x) = x$

47. $(f \circ g)(x) = f(g(x)) = f(\sqrt[3]{x}) = (\sqrt[3]{x})^3 = x; (g \circ f)(x) = g(f(x)) = g(x^3) = \sqrt[3]{x^3} = x$

49. $(f \circ g)(x) = f(g(x)) = f\left(\dfrac{1}{2}(x+6)\right) = 2\left[\dfrac{1}{2}(x+6)\right] - 6 = x + 6 - 6 = x;$

$(g \circ f)(x) = g(f(x)) = g(2x-6) = \dfrac{1}{2}(2x-6+6) = \dfrac{1}{2}(2x) = x$

51. $(f \circ g)(x) = f(g(x)) = f\left(\dfrac{1}{a}(x-b)\right) = a\left[\dfrac{1}{a}(x-b)\right] + b = x; (g \circ f)(x) = g(f(x)) = g(ax+b) = \dfrac{1}{a}(ax+b-b) = x$

53. $f(x) = x^4; g(x) = 2x+3$ (Other answers are possible.) **55.** $f(x) = \sqrt{x}; g(x) = x^2 + 1$ (Other answers are possible.)
57. $f(x) = |x|; g(x) = 2x+1$ (Other answers are possible.) **59.** $(f \circ g)(x) = 11; (g \circ f)(x) = 2$ **61.** $-3, 3$
63. (a) $(f \circ g)(x) = acx + ad + b$ **(b)** $(g \circ f)(x) = acx + bc + d$ **(c)** The domains of both $f \circ g$ and $g \circ f$ are all real numbers.

(d) $f \circ g = g \circ f$ when $ad + b = bc + d$ **65.** $S(t) = \dfrac{16}{9}\pi t^6$ **67.** $C(t) = 15,000 + 800,000t - 40,000t^2$

69. $C(p) = \dfrac{2\sqrt{100-p}}{25} + 600, 0 \leq p \leq 100$ **71.** $V(r) = 2\pi r^3$ **73. (a)** $f(x) = 0.7143x$ **(b)** $g(x) = 137.402x$

(c) $g(f(x)) = g(0.7143x) = 98.1462486x$ **(d)** 98,146.2486 yen **75. (a)** $f(p) = p - 200$ **(b)** $g(p) = 0.8p$
(c) $(f \circ g)(p) = 0.8p - 200; (g \circ f)(p) = 0.8p - 160$; The 20% discount followed by the $200 rebate is the better deal.
77. f is an odd function, so $f(-x) = -f(x)$. g is an even function, so $g(-x) = g(x)$. Then $(f \circ g)(-x) = f(g(-x)) = f(g(x)) = (f \circ g)(x)$. So $f \circ g$
is even. Also, $(g \circ f)(-x) = g(f(-x)) = g(-f(x)) = g(f(x)) = (g \circ f)(x)$, so $g \circ f$ is even.

4.2 Assess Your Understanding (page 292)

5. $f(x_1) \neq f(x_2)$ **6.** one-to-one **7.** 3 **8.** $y = x$ **9.** $[4, \infty)$ **10.** T **11.** one-to-one **13.** not one-to-one **15.** not one-to-one **17.** one-to-one
19. one-to-one **21.** not one-to-one **23.** one-to-one

25.

Annual Rainfall (inches)	Location
460.00	→ Mt Waialeale, Hawaii
202.01	→ Monrovia, Liberia
196.46	→ Pago Pago, American Samoa
191.02	→ Moulmein, Burma
182.87	→ Lae, Papua New Guinea

Domain: {460.00, 202.01, 196.46, 191.02, 182.87}
Range: {Mt Waialeale, Monrovia, Pago Pago, Moulmein, Lae}
29. $\{(5, -3), (9, -2), (2, -1), (11, 0), (-5, 1)\}$
Domain: $\{5, 9, 2, 11, -5\}$
Range: $\{-3, -2, -1, 0, 1\}$
31. $\{(1, -2), (2, -3), (0, -10), (9, 1), (4, 2)\}$
Domain: $\{1, 2, 0, 9, 4\}$
Range: $\{-2, -3, -10, 1, 2\}$
33. $f(g(x)) = f\left(\dfrac{1}{3}(x-4)\right) = 3\left[\dfrac{1}{3}(x-4)\right] + 4 = (x-4) + 4 = x$

$g(f(x)) = g(3x+4) = \dfrac{1}{3}[(3x+4) - 4] = \dfrac{1}{3}(3x) = x$

35. $f(g(x)) = f\left(\dfrac{x}{4} + 2\right) = 4\left[\dfrac{x}{4} + 2\right] - 8 = (x + 8) - 8 = x$

$g(f(x)) = g(4x-8) = \dfrac{4x-8}{4} + 2 = (x - 2) + 2 = x$

37. $f(g(x)) = f(\sqrt[3]{x+8}) = (\sqrt[3]{x+8})^3 - 8 = (x+8) - 8 = x$
$g(f(x)) = g(x^3 - 8) = \sqrt[3]{(x^3-8)+8} = \sqrt[3]{x^3} = x$

27.

Monthly Cost of Life Insurance	Age
$7.09	→ 30
$8.40	→ 40
$11.29	→ 45

Domain: {$7.09, $8.40, $11.29}
Range: {30, 40, 45}

39. $f(g(x)) = f\left(\dfrac{1}{x}\right) = \dfrac{1}{\left(\dfrac{1}{x}\right)} = x; x \neq 0, g(f(x)) = g\left(\dfrac{1}{x}\right) = \dfrac{1}{\left(\dfrac{1}{x}\right)} = x, x \neq 0$

41. $f(g(x)) = f\left(\dfrac{4x-3}{2-x}\right) = \dfrac{2\left(\dfrac{4x-3}{2-x}\right) + 3}{\dfrac{4x-3}{2-x} + 4}$

$= \dfrac{2(4x-3) + 3(2-x)}{4x-3 + 4(2-x)} = \dfrac{5x}{5} = x, x \neq 2$

$g(f(x)) = g\left(\dfrac{2x+3}{x+4}\right) = \dfrac{4\left(\dfrac{2x+3}{x+4}\right) - 3}{2 - \dfrac{2x+3}{x+4}}$

$= \dfrac{4(2x+3) - 3(x+4)}{2(x+4) - (2x+3)} = \dfrac{5x}{5} = x, x \neq -4$

43. **45.** **47.**

49. $f^{-1}(x) = \dfrac{1}{3}x$

$f(f^{-1}(x)) = f\left(\dfrac{1}{3}x\right) = 3\left(\dfrac{1}{3}x\right) = x$

$f^{-1}(f(x)) = f^{-1}(3x) = \dfrac{1}{3}(3x) = x$

51. $f^{-1}(x) = \dfrac{x}{4} - \dfrac{1}{2}$

$f(f^{-1}(x)) = f\left(\dfrac{x}{4} - \dfrac{1}{2}\right) = 4\left(\dfrac{x}{4} - \dfrac{1}{2}\right) + 2$

$= (x - 2) + 2 = x$

$f^{-1}(f(x)) = f^{-1}(4x + 2) = \dfrac{4x + 2}{4} - \dfrac{1}{2}$

$= \left(x + \dfrac{1}{2}\right) - \dfrac{1}{2} = x$

53. $f^{-1}(x) = \sqrt[3]{x + 1}$

$f(f^{-1}(x)) = f\left(\sqrt[3]{x + 1}\right)$

$= \left(\sqrt[3]{x + 1}\right)^3 - 1 = x$

$f^{-1}(f(x)) = f^{-1}(x^3 - 1)$

$= \sqrt[3]{(x^3 - 1) + 1} = x$

55. $f^{-1}(x) = \sqrt{x - 4}, \ x \geq 4$

$f(f^{-1}(x)) = f(\sqrt{x - 4}) = (\sqrt{x - 4})^2 + 4 = x$

$f^{-1}(f(x)) = f^{-1}(x^2 + 4) = \sqrt{(x^2 + 4) - 4} = \sqrt{x^2} = x, \ x \geq 0$

57. $f^{-1}(x) = \dfrac{4}{x}$

$f(f^{-1}(x)) = f\left(\dfrac{4}{x}\right) = \dfrac{4}{\left(\dfrac{4}{x}\right)} = x$

$f^{-1}(f(x)) = f^{-1}\left(\dfrac{4}{x}\right) = \dfrac{4}{\left(\dfrac{4}{x}\right)} = x$

59. $f^{-1}(x) = \dfrac{2x + 1}{x}$

$f(f^{-1}(x)) = f\left(\dfrac{2x + 1}{x}\right) = \dfrac{1}{\dfrac{2x + 1}{x} - 2} = \dfrac{x}{(2x + 1) - 2x} = x$

$f^{-1}(f(x)) = f^{-1}\left(\dfrac{1}{x - 2}\right) = \dfrac{2\left(\dfrac{1}{x - 2}\right) + 1}{\dfrac{1}{x - 2}} = \dfrac{2 + (x - 2)}{1} = x$

61. $f^{-1}(x) = \dfrac{2 - 3x}{x}$

$f(f^{-1}(x)) = f\left(\dfrac{2 - 3x}{x}\right) = \dfrac{2}{3 + \dfrac{2 - 3x}{x}} = \dfrac{2x}{3x + 2 - 3x} = \dfrac{2x}{2} = x$

$f^{-1}(f(x)) = f^{-1}\left(\dfrac{2}{3 + x}\right) = \dfrac{2 - 3\left(\dfrac{2}{3 + x}\right)}{\dfrac{2}{3 + x}} = \dfrac{2(3 + x) - 3 \cdot 2}{2} = \dfrac{2x}{2} = x$

Domain of f = range of f^{-1} = all real numbers except -3

Range of f = domain of f^{-1} = all real numbers except 0

63. $f^{-1}(x) = \dfrac{-2x}{x - 3}$

$f(f^{-1}(x)) = f\left(\dfrac{-2x}{x - 3}\right) = \dfrac{3\left(\dfrac{-2x}{x - 3}\right)}{\dfrac{-2x}{x - 3} + 2}$

$= \dfrac{3(-2x)}{-2x + 2(x - 3)} = \dfrac{-6x}{-6} = x$

$f^{-1}(f(x)) = f^{-1}\left(\dfrac{3x}{x + 2}\right) = \dfrac{-2\left(\dfrac{3x}{x + 2}\right)}{\dfrac{3x}{x + 2} - 3}$

$= \dfrac{-2(3x)}{3x - 3(x + 2)} = \dfrac{-6x}{-6} = x$

Domain of f = range of f^{-1} = all real numbers except -2

Range of f = domain of f^{-1} = all real numbers except 3

65. $f^{-1}(x) = \dfrac{x}{3x - 2}$

$f(f^{-1}(x)) = f\left(\dfrac{x}{3x - 2}\right) = \dfrac{2\left(\dfrac{x}{3x - 2}\right)}{3\left(\dfrac{x}{3x - 2}\right) - 1}$

$= \dfrac{2x}{3x - (3x - 2)} = \dfrac{2x}{2} = x$

$f^{-1}(f(x)) = f^{-1}\left(\dfrac{2x}{3x - 1}\right) = \dfrac{\dfrac{2x}{3x - 1}}{3\left(\dfrac{2x}{3x - 1}\right) - 2}$

$= \dfrac{2x}{6x - 2(3x - 1)} = \dfrac{2x}{2} = x$

Domain of f = range of f^{-1} = all real numbers except $\dfrac{1}{3}$

Range of f = domain of f^{-1} = all real numbers except $\dfrac{2}{3}$

67. $f^{-1}(x) = \dfrac{3x + 4}{2x - 3}$

$$f(f^{-1}(x)) = f\left(\dfrac{3x + 4}{2x - 3}\right) = \dfrac{3\left(\dfrac{3x + 4}{2x - 3}\right) + 4}{2\left(\dfrac{3x + 4}{2x - 3}\right) - 3}$$

$$= \dfrac{3(3x + 4) + 4(2x - 3)}{2(3x + 4) - 3(2x - 3)}$$

$$= \dfrac{17x}{17} = x$$

$$f^{-1}(f(x)) = f^{-1}\left(\dfrac{3x + 4}{2x - 3}\right) = \dfrac{3\left(\dfrac{3x + 4}{2x - 3}\right) + 4}{2\left(\dfrac{3x + 4}{2x - 3}\right) - 3}$$

$$= \dfrac{3(3x + 4) + 4(2x - 3)}{2(3x + 4) - 3(2x - 3)}$$

$$= \dfrac{17x}{17} = x$$

Domain of f = range of f^{-1} = all real numbers except $\dfrac{3}{2}$

Range of f = domain of f^{-1} = all real numbers except $\dfrac{3}{2}$

69. $f^{-1}(x) = \dfrac{-2x + 3}{x - 2}$

$$f(f^{-1}(x)) = f\left(\dfrac{-2x + 3}{x - 2}\right) = \dfrac{2\left(\dfrac{-2x + 3}{x - 2}\right) + 3}{\dfrac{-2x + 3}{x - 2} + 2}$$

$$= \dfrac{2(-2x + 3) + 3(x - 2)}{-2x + 3 + 2(x - 2)} = \dfrac{-x}{-1} = x$$

$$f^{-1}(f(x)) = f^{-1}\left(\dfrac{2x + 3}{x + 2}\right) = \dfrac{-2\left(\dfrac{2x + 3}{x + 2}\right) + 3}{\dfrac{2x + 3}{x + 2} - 2}$$

$$= \dfrac{-2(2x + 3) + 3(x + 2)}{2x + 3 - 2(x + 2)} = \dfrac{-x}{-1} = x$$

Domain of f = range of f^{-1} = all real numbers except -2

Range of f = domain of f^{-1} = all real numbers except 2

71. $f^{-1}(x) = \dfrac{2}{\sqrt{1 - 2x}}$

$$f(f^{-1}(x)) = f\left(\dfrac{2}{\sqrt{1 - 2x}}\right) = \dfrac{\dfrac{4}{1 - 2x} - 4}{2 \cdot \dfrac{4}{1 - 2x}} = \dfrac{4 - 4(1 - 2x)}{2 \cdot 4}$$

$$= \dfrac{8x}{8} = x$$

$$f^{-1}(f(x)) = f^{-1}\left(\dfrac{x^2 - 4}{2x^2}\right) = \dfrac{2}{\sqrt{1 - 2\left(\dfrac{x^2 - 4}{2x^2}\right)}} = \dfrac{2}{\sqrt{\dfrac{4}{x^2}}} = \sqrt{x^2}$$

$$= x, \text{ since } x > 0$$

Domain of f = range of f^{-1} = $(0, \infty)$

Range of f = domain of f^{-1} = $\left(-\infty, \dfrac{1}{2}\right)$

73. (a) 0 **(b)** 2 **(c)** 0 **(d)** 1
75. 7 **77.** Domain of f^{-1}: $[-2, \infty)$; range of f^{-1}: $[5, \infty)$
79. Domain of g^{-1}: $[0, \infty)$; range of g^{-1}: $(-\infty, 0]$
81. Increasing on the interval $(f(0), f(5))$
83. $f^{-1}(x) = \dfrac{1}{m}(x - b), m \neq 0$ **85.** Quadrant I
87. Possible answer: $f(x) = |x|, x \geq 0$, is one-to-one; $f^{-1}(x) = x, x \geq 0$
89. (a) $r(d) = \dfrac{d + 90.39}{6.97}$

(b) $r(d(r)) = \dfrac{6.97r - 90.39 + 90.39}{6.97} = \dfrac{6.97r}{6.97} = r$

$$d(r(d)) = 6.97\left(\dfrac{d + 90.39}{6.97}\right) - 90.39 = d + 90.39 - 90.39 = d$$

(c) 56 miles per hour

91. (a) 77.6 kg **(b)** $h(W) = \dfrac{W - 50}{2.3} + 60 = \dfrac{W + 88}{2.3}$ **(c)** $h(W(h)) = \dfrac{50 + 2.3(h - 60) + 88}{2.3} = \dfrac{2.3h}{2.3} = h$ **(d)** 73 inches

$$W(h(W)) = 50 + 2.3\left(\dfrac{W + 88}{2.3} - 60\right)$$

$$= 50 + W + 88 - 138 = W$$

93. (a) $\{g \mid 33,950 \leq g \leq 82,250\}$

(b) $\{T \mid 4675 \leq T \leq 16,750\}$

(c) $g(T) = \dfrac{T - 4675}{0.25} + 33,950$

Domain: $\{T \mid 4675 \leq T \leq 16,750\}$
Range: $\{g \mid 33,950 \leq g \leq 82,250\}$

95. (a) t represents time, so $t \geq 0$.

(b) $t(H) = \sqrt{\dfrac{H - 100}{-4.9}} = \sqrt{\dfrac{100 - H}{4.9}}$

(c) 2.02 seconds

97. $f^{-1}(x) = \dfrac{-dx + b}{cx - a}; f = f^{-1}$ if $a = -d$ **101.** No

4.3 Assess Your Understanding *(page 307)*

6. Exponential function; growth factor; initial value **7.** $a; a$ **8.** T **9.** F **10.** T **11.** $\left(-1, \dfrac{1}{a}\right); (0, 1); (1, a)$ **12.** 1 **13.** 4 **14.** F

15. (a) 11.212 **(b)** 11.587 **(c)** 11.664 **(d)** 11.665 **17. (a)** 8.815 **(b)** 8.821 **(c)** 8.824 **(d)** 8.825 **19. (a)** 21.217 **(b)** 22.217 **(c)** 22.440
(d) 22.459 **21.** 3.320 **23.** 0.427 **25.** Neither **27.** Exponential; $H(x) = 4^x$ **29.** Exponential; $f(x) = 3(2^x)$ **31.** Linear; $H(x) = 2x + 4$
33. B **35.** D **37.** F **39.** E

41.

Domain: All real numbers
Range: $\{y \mid y > 1\}$ or $(1, \infty)$
Horizontal asymptote: $y = 1$

43.

Domain: All real numbers
Range: $\{y \mid y > 0\}$ or $(0, \infty)$
Horizontal asymptote: $y = 0$

45.

Domain: All real numbers
Range: $\{y \mid y > 0\}$ or $(0, \infty)$
Horizontal asymptote: $y = 0$

47.

Domain: All real numbers
Range: $\{y \mid y > -2\}$ or $(-2, \infty)$
Horizontal asymptote: $y = -2$

49.

Domain: All real numbers
Range: $\{y \mid y > 2\}$ or $(2, \infty)$
Horizontal asymptote: $y = 2$

51.

Domain: All real numbers
Range: $\{y \mid y > 2\}$ or $(2, \infty)$
Horizontal asymptote: $y = 2$

53.

Domain: All real numbers
Range: $\{y \mid y > 0\}$ or $(0, \infty)$
Horizontal asymptote: $y = 0$

55.

Domain: All real numbers
Range: $\{y \mid y > 0\}$ or $(0, \infty)$
Horizontal asymptote: $y = 0$

57.

Domain: All real numbers
Range: $\{y \mid y < 5\}$ or $(-\infty, 5)$
Horizontal asymptote: $y = 5$

59.

Domain: All real numbers
Range: $\{y \mid y < 2\}$ or $(-\infty, 2)$
Horizontal asymptote: $y = 2$

61. $\{3\}$ **63.** $\{-4\}$ **65.** $\{2\}$ **67.** $\left\{\dfrac{3}{2}\right\}$ **69.** $\{-\sqrt{2}, 0, \sqrt{2}\}$ **71.** $\{6\}$

73. $\{-1, 7\}$ **75.** $\{-4, 2\}$ **77.** $\{-4\}$ **79.** $\{1, 2\}$ **81.** $\dfrac{1}{49}$ **83.** $\dfrac{1}{4}$

85. $f(x) = 3^x$ **87.** $f(x) = -6^x$ **89.** $f(x) = 3^x + 2$

91. (a) $16; (4, 16)$ **(b)** $-4; \left(-4, \dfrac{1}{16}\right)$ **93. (a)** $\dfrac{9}{4}; \left(-1, \dfrac{9}{4}\right)$

(b) $3; (3, 66)$ **95. (a)** $60; (-6, 60)$ **(b)** $-4: (-4, 12)$ **(c)** -2

97.

Domain: $(-\infty, \infty)$
Range: $[1, \infty)$
Intercept: $(0, 1)$

99.

Domain: $(-\infty, \infty)$
Range: $[-1, 0)$
Intercept: $(0, -1)$

101. (a) 74% **(b)** 47%
103. (a) \$12,123 **(b)** \$6443
105. 3.35 mg; 0.45 mg

107. (a) 0.632 **(b)** 0.982 **(c)** 1
(d)
(e) About 7 min

109. (a) 0.0516 **(b)** 0.0888 **111. (a)** 70.95% **(b)** 72.62% **(c)** 100%
113. (a) 5.41 amp, 7.59 amp, 10.38 amp **(b)** 12 amp
(d) 3.34 amp, 5.31 amp, 9.44 amp **(e)** 24 amp
(c), (f)

115. 36

117.

Final Denominator	Value of Expression	Compare Value to $e \approx$ 2.718281828
1 + 1	2.5	2.5 < e
2 + 2	2.8	2.8 > e
3 + 3	2.7	2.7 < e
4 + 4	2.721649485	2.721649485 > e
5 + 5	2.717770035	2.717770035 < e
6 + 6	2.718348855	2.718348855 > e

119. $f(A + B) = a^{A+B} = a^A \cdot a^B = f(A) \cdot f(B)$ **121.** $f(\alpha x) = a^{\alpha x} = (a^x)^\alpha = [f(x)]^\alpha$

123. (a) $f(-x) = \dfrac{1}{2}(e^{-x} + e^{(-x)})$
$= \dfrac{1}{2}(e^{-x} + e^x)$
$= \dfrac{1}{2}(e^x + e^{-x})$
$= f(x)$

(b)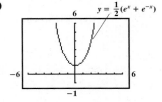

(c) $(\cosh x)^2 - (\sinh x)^2$
$= \left[\dfrac{1}{2}(e^x + e^{-x})\right]^2 - \left[\dfrac{1}{2}(e^x - e^{-x})\right]^2$
$= \dfrac{1}{4}[e^{2x} + 2 + e^{-2x} - e^{2x} + 2 - e^{-2x}]$
$= \dfrac{1}{4}(4) = 1$

125. 59 minutes

4.4 Assess Your Understanding (page 321)

4. $\{x \mid x > 0\}$ or $(0, \infty)$ **5.** $\left(\dfrac{1}{a}, -1\right), (1, 0), (a, 1)$ **6.** 1 **7.** F **8.** T **9.** $2 = \log_3 9$ **11.** $2 = \log_a 1.6$ **13.** $x = \log_2 7.2$ **15.** $x = \ln 8$

17. $2^3 = 8$ **19.** $a^6 = 3$ **21.** $3^x = 2$ **23.** $e^x = 4$ **25.** 0 **27.** 2 **29.** -4 **31.** $\dfrac{1}{2}$ **33.** 4 **35.** $\dfrac{1}{2}$ **37.** $\{x \mid x > 3\}; (3, \infty)$

39. All real numbers except 0; $\{x \mid x \neq 0\}$ **41.** $\{x \mid x > 10\}$; $(10, \infty)$ **43.** $\{x \mid x > -1\}$; $(-1, \infty)$ **45.** $\{x \mid x < -1 \text{ or } x > 0\}$; $(-\infty, -1) \cup (0, \infty)$
47. $\{x \mid x \geq 1\}$; $[1, \infty)$ **49.** 0.511 **51.** 30.099 **53.** 2.303 **55.** -53.991 **57.** $\sqrt{2}$

59.

61.

63. B **65.** D **67.** A **69.** E

71. (a) Domain: $(-4, \infty)$
(b)
(c) Range: $(-\infty, \infty)$
Vertical asymptote: $x = -4$
(d) $f^{-1}(x) = e^x - 4$
(e) Range of f: $(-\infty, \infty)$
Range of f^{-1}: $(-4, \infty)$
(f)

73. (a) Domain: $(0, \infty)$
(b)
(c) Range: $(-\infty, \infty)$
Vertical asymptote: $x = 0$
(d) $f^{-1}(x) = e^{x-2}$
(e) Range of f: $(-\infty, \infty)$
Range of f^{-1}: $(0, \infty)$
(f)

75. (a) Domain: $(0, \infty)$
(b)
(c) Range: $(-\infty, \infty)$
Vertical asymptote: $x = 0$
(d) $f^{-1}(x) = \dfrac{1}{2} e^{x+3}$
(e) Range of f: $(-\infty, \infty)$
Range of f^{-1}: $(0, \infty)$
(f)

77. (a) Domain: $(4, \infty)$
(b)
(c) Range: $(-\infty, \infty)$
Vertical asymptote: $x = 4$
(d) $f^{-1}(x) = 10^{x-2} + 4$
(e) Range of f: $(-\infty, \infty)$
Range of f^{-1}: $(4, \infty)$
(f)

79. (a) Domain: $(0, \infty)$
(b)
(c) Range: $(-\infty, \infty)$
Vertical asymptote: $x = 0$
(d) $f^{-1}(x) = \dfrac{1}{2} \cdot 10^{2x}$
(e) Range of f: $(-\infty, \infty)$
Range of f^{-1}: $(0, \infty)$
(f)

81. (a) Domain: $(-2, \infty)$
(b)
(c) Range: $(-\infty, \infty)$
Vertical asymptote: $x = -2$
(d) $f^{-1}(x) = 3^{x-3} - 2$
(e) Range of f: $(-\infty, \infty)$
Range of f^{-1}: $(-2, \infty)$
(f)

83. (a) Domain: $(-\infty, \infty)$
(b)
(c) Range: $(-3, \infty)$
Horizontal asymptote: $y = -3$
(d) $f^{-1}(x) = \ln(x + 3) - 2$
(e) Range of f: $(-3, \infty)$
Range of f^{-1}: $(-\infty, \infty)$
(f)

85. (a) Domain: $(-\infty, \infty)$
(b)
(c) Range: $(4, \infty)$
Horizontal asymptote: $y = 4$
(d) $f^{-1}(x) = 3 \log_2 (x - 4)$
(e) Range of f: $(4, \infty)$
Range of f^{-1}: $(-\infty, \infty)$
(f)

87. $\{9\}$ **89.** $\left\{\dfrac{7}{2}\right\}$ **91.** $\{2\}$ **93.** $\{5\}$
95. $\{3\}$ **97.** $\{2\}$ **99.** $\left\{\dfrac{\ln 10}{3}\right\}$
101. $\left\{\dfrac{\ln 8 - 5}{2}\right\}$ **103.** $\{-2\sqrt{2}, 2\sqrt{2}\}$
105. $\{-1\}$ **107.** $\left\{5 \ln \dfrac{7}{5}\right\}$
109. $\left\{2 - \log \dfrac{5}{2}\right\}$
111. (a) $\left\{x \mid x > -\dfrac{1}{2}\right\}$; $\left(-\dfrac{1}{2}, \infty\right)$
(b) 2; (40, 2) **(c)** 121; (121, 3) **(d)** 4

113.

Domain: $\{x \mid x \neq 0\}$
Range: $(-\infty, \infty)$
Intercepts: $(-1, 0), (1, 0)$

115.

Domain: $\{x \mid x > 0\}$
Range: $\{y \mid y \geq 0\}$
Intercept: $(1, 0)$

117. (a) 1 **(b)** 2 **(c)** 3
(d) It increases. **(e)** 0.000316
(f) 3.981×10^{-8}
119. (a) 5.97 km **(b)** 0.90 km
121. (a) 6.93 min **(b)** 16.09 min
123. $h \approx 2.29$, so the time between
injections is about 2 h, 17 min.

125. 0.2695 s
0.8959 s

127. 50 decibels (dB) **129.** 90 dB **131.** 8.1 **133. (a)** $k \approx 11.216$ **(b)** 6.73 **(c)** 0.41% **(d)** 0.14%
135. Because $y = \log_1 x$ means $1^y = 1 = x$, which cannot be true for $x \neq 1$

4.5 Assess Your Understanding *(page 332)*

1. 0 **2.** 1 **3.** M **4.** r **5.** $\log_a M; \log_a N$ **6.** $\log_a M; \log_a N$ **7.** $r \log_a M$ **8.** 6 **9.** 7 **10.** F **11.** F **12.** F **13.** 71 **15.** -4 **17.** 7 **19.** 1

21. 1 **23.** 3 **25.** $\dfrac{5}{4}$ **27.** 4 **29.** $a + b$ **31.** $b - a$ **33.** $3a$ **35.** $\dfrac{1}{5}(a + b)$ **37.** $2 + \log_5 x$ **39.** $3 \log_2 z$ **41.** $1 + \ln x$ **43.** $\ln x - x$

45. $2 \log_a u + 3 \log_a v$ **47.** $2 \ln x + \dfrac{1}{2}\ln(1 - x)$ **49.** $3 \log_2 x - \log_2(x - 3)$ **51.** $\log x + \log(x + 2) - 2 \log(x + 3)$

53. $\dfrac{1}{3}\ln(x - 2) + \dfrac{1}{3}\ln(x + 1) - \dfrac{2}{3}\ln(x + 4)$ **55.** $\ln 5 + \ln x + \dfrac{1}{2}\ln(1 + 3x) - 3 \ln(x - 4)$ **57.** $\log_5 u^3 v^4$ **59.** $\log_3\left(\dfrac{1}{x^{5/2}}\right)$ **61.** $\log_4\left[\dfrac{x - 1}{(x + 1)^4}\right]$

63. $-2 \ln(x - 1)$ **65.** $\log_2[x(3x - 2)^4]$ **67.** $\log_a\left(\dfrac{25x^6}{\sqrt{2x + 3}}\right)$ **69.** $\log_2\left[\dfrac{(x + 1)^2}{(x + 3)(x - 1)}\right]$ **71.** 2.771 **73.** -3.880 **75.** 5.615 **77.** 0.874

79. $y = \dfrac{\log x}{\log 4}$

81. $y = \dfrac{\log(x + 2)}{\log 2}$

83. $y = \dfrac{\log(x + 1)}{\log(x - 1)}$

85. (a) $(f \circ g)(x) = x$; $\{x \mid x$ is any real number$\}$ or $(-\infty, \infty)$
(b) $(g \circ f)(x) = x$; $\{x \mid x > 0\}$ or $(0, \infty)$ **(c)** 5
(d) $(f \circ h)(x) = \ln x^2$; $\{x \mid x \neq 0\}$ or $(-\infty, 0) \cup (0, \infty)$ **(e)** 2

87. $y = Cx$ **89.** $y = Cx(x + 1)$ **91.** $y = Ce^{3x}$ **93.** $y = Ce^{-4x} + 3$

95. $y = \dfrac{\sqrt[3]{C}(2x + 1)^{1/6}}{(x + 4)^{1/9}}$ **97.** 3 **99.** 1

101. $\log_a\left(x + \sqrt{x^2 - 1}\right) + \log_a\left(x - \sqrt{x^2 - 1}\right) = \log_a\left[\left(x + \sqrt{x^2 - 1}\right)\left(x - \sqrt{x^2 - 1}\right)\right] = \log_a[x^2 - (x^2 - 1)] = \log_a 1 = 0$
103. $\ln(1 + e^{2x}) = \ln[e^{2x}(e^{-2x} + 1)] = \ln e^{2x} + \ln(e^{-2x} + 1) = 2x + \ln(1 + e^{-2x})$
105. $y = f(x) = \log_a x; a^y = x$ implies $a^y = \left(\dfrac{1}{a}\right)^{-y} = x$, so $-y = \log_{1/a} x = -f(x)$.

107. $f(x) = \log_a x; f\left(\dfrac{1}{x}\right) = \log_a \dfrac{1}{x} = \log_a 1 - \log_a x = -f(x)$

109. $\log_a \dfrac{M}{N} = \log_a(M \cdot N^{-1}) = \log_a M + \log_a N^{-1} = \log_a M - \log_a N$, since $a^{\log_a N^{-1}} = N^{-1}$ implies $a^{-\log_a N^{-1}} = N$; i.e., $\log_a N = -\log_a N^{-1}$.

4.6 Assess Your Understanding *(page 338)*

5. $\{16\}$ **7.** $\left\{\dfrac{16}{5}\right\}$ **9.** $\{6\}$ **11.** $\{16\}$ **13.** $\left\{\dfrac{1}{3}\right\}$ **15.** $\{3\}$ **17.** $\{5\}$ **19.** $\left\{\dfrac{21}{8}\right\}$ **21.** $\{-6\}$ **23.** $\{-2\}$ **25.** $\{-1 + \sqrt{1 + e^4}\} \approx \{6.456\}$

27. $\left\{\dfrac{-5 + 3\sqrt{5}}{2}\right\} \approx \{0.854\}$ **29.** $\{2\}$ **31.** $\left\{\dfrac{9}{2}\right\}$ **33.** $\{8\}$ **35.** $\{\log_2 10\} = \left\{\dfrac{\ln 10}{\ln 2}\right\} \approx \{3.322\}$ **37.** $\{-\log_8 1.2\} = \left\{-\dfrac{\ln 1.2}{\ln 8}\right\} \approx \{-0.088\}$

39. $\left\{\dfrac{1}{3}\log_2 \dfrac{8}{5}\right\} = \left\{\dfrac{\ln \dfrac{8}{5}}{3 \ln 2}\right\} \approx \{0.226\}$ **41.** $\left\{\dfrac{\ln 3}{2 \ln 3 + \ln 4}\right\} \approx \{0.307\}$ **43.** $\left\{\dfrac{\ln 7}{\ln 0.6 + \ln 7}\right\} \approx \{1.356\}$ **45.** $\{0\}$ **47.** $\left\{\dfrac{\ln \pi}{1 + \ln \pi}\right\} \approx \{0.534\}$

49. $\left\{\dfrac{\ln 3}{\ln 2}\right\} \approx \{1.585\}$ **51.** $\{0\}$ **53.** $\left\{\log_4\left(-2 + \sqrt{7}\right)\right\} \approx \{-0.315\}$ **55.** $\{\log_5 4\} \approx \{0.861\}$ **57.** No real solution **59.** $\{\log_4 5\} \approx \{1.161\}$

61. $\{2.79\}$ **63.** $\{-0.57\}$ **65.** $\{-0.70\}$ **67.** $\{0.57\}$ **69.** $\{0.39, 1.00\}$ **71.** $\{1.32\}$ **73.** $\{1.31\}$ **75.** $\{1\}$ **77.** $\{16\}$ **79.** $\left\{-1, \dfrac{2}{3}\right\}$ **81.** $\{0\}$

83. $\left\{\ln\left(2 + \sqrt{5}\right)\right\} \approx \{1.444\}$ **85.** $\left\{e^{\frac{\ln 5 \cdot \ln 3}{\ln 15}}\right\} \approx \{1.921\}$ **87. (a)** $\{5\}; (5, 3)$ **(b)** $\{5\}; (5, 4)$ **(c)** $\{1\}$; yes, at $(1, 2)$ **(d)** $\{5\}$ **(e)** $\left\{-\dfrac{1}{11}\right\}$

89. (a), (b)

(c) $\{x \mid x > 0.710\}$ or $(0.710, \infty)$

91. (a), (b), (c)

93. (a), (b), (c)

95. (a)

97. (a) 2025 **(b)** 2043

99. (a) After 3.1 yr
(b) After 7.1 yr
(c) After 10.6 yr

(b) 2
(c) $\{x \mid x < 2\}$ or $(-\infty, 2)$

4.7 Assess Your Understanding *(page 347)*

3. principal **4.** *I; Prt;* simple interest **5.** 4 **6.** effective rate of interest **7.** $108.29 **9.** $609.50 **11.** $697.09 **13.** $1246.08 **15.** $88.72

17. $860.72 **19.** $554.09 **21.** $59.71 **23.** 5.095% **25.** 5.127% **27.** $6\frac{1}{4}$% compounded annually **29.** 9% compounded monthly **31.** 25.992%

33. 24.573% **35. (a)** About 8.69 yr **(b)** About 8.66 yr **37.** 6.823% **39.** 5.09 yr; 5.07 yr **41.** 15.27 yr or 15 yr, 3 mo **43.** $104,335

45. $12,910.62 **47.** About $30.17 per share or $3017 **49.** Not quite. Jim will have $1057.60. The second bank gives a better deal, since Jim will have $1060.62 after 1 yr. **51.** Will has $11,632.73; Henry has $10,947.89. **53. (a)** $79,129 **(b)** $38,516 **55.** About $1019 billion; about $232 billion

57. $940.90 **59.** 2.53% **61.** 34.31 yr **63. (a)** $1364.62 **(b)** $1353.35 **65.** $4631.93

67. (a) 6.12 yr
(b) 18.45 yr

(c) $mP = P\left(1 + \dfrac{r}{n}\right)^{nt}$

$$m = \left(1 + \dfrac{r}{n}\right)^{nt}$$

$$\ln m = \ln\left(1 + \dfrac{r}{n}\right)^{nt} = nt \ln\left(1 + \dfrac{r}{n}\right)$$

$$t = \dfrac{\ln m}{n \ln\left(1 + \dfrac{r}{n}\right)}$$

69. (a) 2.82% **(b)** In 2020 or after 12 yr **71.** 22.7 yr

4.8 Assess Your Understanding *(page 359)*

1. (a) 500 insects **(b)** $0.02 = 2\%$ per day **(c)** About 611 insects **(d)** After about 23.5 days **(e)** After about 34.7 days

3. (a) $-0.0244 = -2.44\%$ per year **(b)** About 391.7 g **(c)** After about 9.1 yr **(d)** 28.4 yr **5. (a)** $N(t) = N_0 e^{kt}$ **(b)** 5832 **(c)** 3.9 days

7. (a) $N(t) = N_0 e^{kt}$ **(b)** 25,198 **9.** 9.797 g **11.** 9727 yr ago **13. (a)** 5:18 PM **(b)** About 14.3 min **(c)** The temperature of the pan approaches 70°F. **15.** 18.63°C; 25.1°C **17.** 1.7 ppm; 7.17 days or 172 hr **19.** 0.26 M; 6.58 hr or 395 min **21.** 26.6 days **23. (a)** 1000 g **(b)** 43.9% **(c)** 30 g
(d) 616.6 g **(e)** After 9.85 h **(f)** About 7.9 h **25. (a)** 9.23×10^{-3}, or about 0 **(b)** 0.81, or about 1 **(c)** 5.01, or about 5
(d) 57.91°, 43.99°, 30.07°

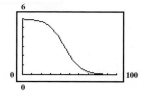

4.9 Assess Your Understanding *(page 365)*

1. (a)

(d)

3. (a)

(d)

(b) $y = 0.0903(1.3384)^x$
(c) $N(t) = 0.0903e^{0.2915t}$

(e) 0.69
(f) After about 7.26 hr

(b) $y = 100.326(0.8769)^x$
(c) $A(t) = 100.326e^{-0.1314t}$

(e) 5.3 weeks **(f)** 0.14 g
(g) After about 12.3 weeks

5. (a)   **(b)** $y = 32{,}741.02 - 6070.96 \ln x$ **(c)**   **(d)** Approximately 168 computers

7. (a)

(b) $y = \dfrac{799{,}475{,}916.5}{1 + 9.1968e^{-0.0160x}}$

(c)

(d) 799,475,917
(e) Approximately 291,599,733
(f) 2006

9. (a)   **(c)**

(b) $y = \dfrac{67{,}856.6}{1 + 19.844e^{-0.2029x}}$

(d) About 67,856,600 subscribers
(e) About 67,711,000 subscribers

11. (a)

(b) Quadratic with $a < 0$ because of the "upside down U-shape" of the data
(c) $y = -0.0311x^2 + 3.4444x + 118.2493$

(d)   **(e)** 201

13. (a)

(b) Exponential because depreciation of a car is described by exponential models in the theory of finance.
(c) $y = 31{,}808.51(0.8474)^x$

(d)   **(e)** $13,899

Review Exercises *(page 371)*

1. (a) -26 **(b)** -241 **(c)** 16 **(d)** -1 **3. (a)** $\sqrt{11}$ **(b)** 1 **(c)** $\sqrt{\sqrt{6} + 2}$ **(d)** 19   **5. (a)** e^4 **(b)** $3e^{-2} - 2$ **(c)** e^{e^4} **(d)** -17

7. $(f \circ g)(x) = 1 - 3x$, all real numbers; $(g \circ f)(x) = 7 - 3x$, all real numbers;
$(f \circ f)(x) = x$, all real numbers; $(g \circ g)(x) = 9x + 4$, all real numbers

9. $(f \circ g)(x) = 27x^2 + 3|x| + 1$, all real numbers; $(g \circ f)(x) = 3|3x^2 + x + 1|$,
all real numbers; $(f \circ f)(x) = 27x^4 + 18x^3 + 24x^2 + 7x + 5$, all real numbers; $(g \circ g)(x) = 9|x|$, all real numbers

11. $(f \circ g)(x) = \dfrac{1 + x}{1 - x}$, $\{x | x \neq 0, x \neq 1\}$; $(g \circ f)(x) = \dfrac{x - 1}{x + 1}$, $\{x | x \neq -1, x \neq 1\}$; $(f \circ f)(x) = x$, $\{x | x \neq 1\}$; $(g \circ g)(x) = x$, $\{x | x \neq 0\}$

13. (a) one-to-one   **(b)** $\{(2, 1), (5, 3), (8, 5), (10, 6)\}$

15.

17. $f^{-1}(x) = \dfrac{2x + 3}{5x - 2}$

$$f(f^{-1}(x)) = \dfrac{2\left(\dfrac{2x + 3}{5x - 2}\right) + 3}{5\left(\dfrac{2x + 3}{5x - 2}\right) - 2} = x$$

$$f^{-1}(f(x)) = \dfrac{2\left(\dfrac{2x + 3}{5x - 2}\right) + 3}{5\left(\dfrac{2x + 3}{5x - 2}\right) - 2} = x$$

Domain of f = range of f^{-1} = all real numbers except $\dfrac{2}{5}$

Range of f = domain of f^{-1} = all real numbers except $\dfrac{2}{5}$

19. $f^{-1}(x) = \dfrac{x + 1}{x}$

$$f(f^{-1}(x)) = \dfrac{1}{\dfrac{x + 1}{x} - 1} = x$$

$$f^{-1}(f(x)) = \dfrac{\dfrac{1}{x - 1} + 1}{\dfrac{1}{x - 1}} = x$$

Domain of f = range of f^{-1} = all real numbers except 1
Range of f = domain of f^{-1} = all real numbers except 0

21. $f^{-1}(x) = \dfrac{27}{x^3}$

$$f(f^{-1}(x)) = \dfrac{3}{\left(\dfrac{27}{x^3}\right)^{1/3}} = x$$

$$f^{-1}(f(x)) = \dfrac{27}{\left(\dfrac{3}{x^{1/3}}\right)^3} = x$$

Domain of f = range of f^{-1} = all real numbers except 0
Range of f = domain of f^{-1} = all real numbers except 0

23. (a) 81 (b) 2 (c) $\dfrac{1}{9}$ (d) -3 **25.** $\log_5 z = 2$ **27.** $5^{13} = u$

29. $\left\{x \middle| x > \dfrac{2}{3}\right\}; \left(\dfrac{2}{3}, \infty\right)$ **31.** $\{x | x < 1 \text{ or } x > 2\}; (-\infty, 1) \cup (2, \infty)$

33. -3 **35.** $\sqrt{2}$ **37.** 0.4 **39.** $\log_3 u + 2\log_3 v - \log_3 w$

41. $2\log x + \dfrac{1}{2}\log(x^3 + 1)$ **43.** $\ln x + \dfrac{1}{3}\ln(x^2 + 1) - \ln(x - 3)$ **45.** $\dfrac{25}{4}\log_4 x$

47. $-2\ln(x + 1)$ **49.** $\log\left(\dfrac{4x^3}{[(x + 3)(x - 2)]^{1/2}}\right)$ **51.** 2.124

53.

55. (a) Domain of f: $(-\infty, \infty)$
(b)

(c) Range of f: $(0, \infty)$
Horizontal asymptote: $y = 0$
(d) $f^{-1}(x) = 3 + \log_2 x$
(e) Range of f: $(0, \infty)$
(f)

57. (a) Domain of f: $(-\infty, \infty)$
(b)

(c) Range of f: $(0, \infty)$
Horizontal asymptote: $y = 0$
(d) $f^{-1}(x) = -\log_3(2x)$
(e) Range of f: $(0, \infty)$
(f)

59. (a) Domain of f: $(-\infty, \infty)$
(b)

(c) Range of f: $(-\infty, 1)$
Horizontal asymptote: $y = 1$
(d) $f^{-1}(x) = -\ln(1 - x)$
(e) Range of f: $(-\infty, \infty)$
(f)

61. (a) Domain of f: $(-3, \infty)$
(b)

(c) Range of f: $(-\infty, \infty)$
Vertical asymptote: $x = -3$
(d) $f^{-1}(x) = e^{2x} - 3$
(e) Range of f: $(-\infty, \infty)$
(f)

63. $\left\{\dfrac{1}{4}\right\}$ **65.** $\left\{\dfrac{-1 - \sqrt{3}}{2}, \dfrac{-1 + \sqrt{3}}{2}\right\} \approx \{-1.366, 0.366\}$ **67.** $\left\{\dfrac{1}{4}\right\}$ **69.** $\left\{\dfrac{2\ln 3}{\ln 5 - \ln 3}\right\} \approx \{4.301\}$ **71.** $\left\{\dfrac{12}{5}\right\}$ **73.** $\{83\}$ **75.** $\left\{\dfrac{1}{2}, -3\right\}$

77. $\{-1\}$ **79.** $\{1 - \ln 5\} \approx \{-0.609\}$ **81.** $\left\{\log_3\left(-2 + \sqrt{7}\right)\right\} = \left\{\dfrac{\ln\left(-2 + \sqrt{7}\right)}{\ln 3}\right\} \approx \{-0.398\}$

83. (a), (e)

(b) 3; (6, 3) (c) 10; (10, 4)
(d) $\left\{x \middle| x > \dfrac{5}{2}\right\}$ or $\left(\dfrac{5}{2}, \infty\right)$
(e) $f^{-1}(x) = 2^{x-1} + 2$

85. 3229.5 m **87.** (a) 37.3 W (b) 6.9 dB
89. (a) 9.85 yr (b) 4.27 yr **91.** \$41,668.97 **93.** 24,203 yr ago
95. 7,237,271,501 **97.** \$483.67 billion

99. (a) 165.1

(c) 165.1

101. (a) 50

(c) 50

(b) $y = 165.73(0.9951)^x$

(d) Approximately 83 s

(b) $C = \dfrac{46.93}{1 + 21.273e^{-0.7306t}}$

(d) About 47 people; 50 people
(e) 2.4 days; during the tenth hour of day 3
(f) 9.5 days

Chapter Test *(page 375)*

1. (a) $f \circ g = \dfrac{2x + 7}{2x + 3}$; domain: $\left\{x \mid x \neq -\dfrac{3}{2}\right\}$ **(b)** $(g \circ f)(-2) = 5$ **(c)** $(f \circ g)(-2) = -3$

2. (a) The function is not one-to-one. **(b)** The function is one-to-one.

3. $f^{-1}(x) = \dfrac{2 + 5x}{3x}$; domain of $f = \left\{x \mid x \neq \dfrac{5}{3}\right\}$, range of $f = \{y \mid y \neq 0\}$; domain of $f^{-1} = \{x \mid x \neq 0\}$; range of $f^{-1} = \left\{y \mid y \neq \dfrac{5}{3}\right\}$

4. The point $(-5, 3)$ must be on the graph of f^{-1}. **5.** $x = 5$ **6.** $b = 4$ **7.** $x = 625$ **8.** $e^3 + 2 \approx 22.086$ **9.** $\log 20 \approx 1.301$

10. $\log_3 21 = \dfrac{\ln 21}{\ln 3} \approx 2.771$ **11.** $\ln 133 \approx 4.890$

12. (a) Domain of f: $\{x \mid -\infty < x < \infty\}$ or $(-\infty, \infty)$ **(e)** Range of f: $\{y \mid y > -2\}$ or $(-2, \infty)$ **13. (a)** Domain of f: $\{x \mid x > 2\}$ or $(2, \infty)$ **(e)** Range of f: $\{y \mid -\infty < y < \infty\}$ or $(-\infty, \infty)$

(b)

(f)

(b)

(f)

(c) Range of f: $\{y \mid y > -2\}$ or $(-2, \infty)$;
Horizontal asymptote: $y = -2$
(d) $f^{-1}(x) = \log_4(x + 2) - 1$

(c) Range of f: $\{y \mid -\infty < y < \infty\}$ or $(-\infty, \infty)$; vertical asymptote: $x = 2$
(d) $f^{-1}(x) = 5^{1-x} + 2$

14. $\{1\}$ **15.** $\{91\}$ **16.** $\{-\ln 2\} \approx \{-0.693\}$ **17.** $\left\{\dfrac{1 - \sqrt{13}}{2}, \dfrac{1 + \sqrt{13}}{2}\right\} \approx \{-1.303, 2.303\}$ **18.** $\left\{\dfrac{3 \ln 7}{1 - \ln 7}\right\} \approx \{-6.172\}$

19. $\{2\sqrt{6}\} \approx \{4.899\}$ **20.** $2 + 3\log_2 x - \log_2(x - 6) - \log_2(x + 3)$ **21.** About 250.39 days **22. (a)** \$1033.82 **(b)** \$963.42 **(c)** 11.9 yr

23. (a) About 83 dB **(b)** The pain threshold will be exceeded if 31,623 people shout at the same time.

Cumulative Review *(page 375)*

1. Yes; no **2. (a)** 10 **(b)** $2x^2 + 3x + 1$ **(c)** $2x^2 + 4xh + 2h^2 - 3x - 3h + 1$ **3.** $\left(\dfrac{1}{2}, \dfrac{\sqrt{3}}{2}\right)$ is on the graph. **4.** $\{-26\}$

5.

6. (a)

7. $f(x) = 2(x - 4)^2 - 8 = 2x^2 - 16x + 24$

8.

(b) $\{x \mid -\infty < x < \infty\}$

9. $f(g(x)) = \dfrac{4}{(x - 3)^2} + 2$; domain: $\{x \mid x \neq 3\}$; 3

10. (a) Zeros: $-4, -\dfrac{1}{4}, 2$

(b) x-intercepts: $-4, -\dfrac{1}{4}, 2$; y-intercept: -8

(c) Local maximum of 60.75 occurs at $x = -2.5$.
Local minimum of -25 occurs at $x = 1$.

(d)

11. (a), (c)

Domain $g = $ range $g^{-1} = (-\infty, \infty)$
Range $g = $ domain $g^{-1} = (2, \infty)$
(b) $g^{-1}(x) = \log_3(x - 2)$

12. $\left\{-\dfrac{3}{2}\right\}$ **13.** $\{2\}$ **14. (a)** $\{-1\}$ **(b)** $\{x|x > -1\}$ or $(-1, \infty)$ **(c)** $\{25\}$

15. (a)   **(b)** Logarithmic; $y = 49.293 - 10.563 \ln x$ **(c)** Highest value of $|r|$

CHAPTER 5 Trigonometric Functions

5.1 Assess Your Understanding (page 387)

3. standard position   **4.** central angle   **5.** radian   **6.** $r\theta; \dfrac{1}{2}r^2\theta$ **7.** π **8.** $\dfrac{s}{t}; \dfrac{\theta}{t}$ **9.** T   **10.** F

11.   **13.**   **15.**   **17.**   **19.**   **21.**

23. $40.17°$ **25.** $1.03°$ **27.** $9.15°$ **29.** $40°19'12''$ **31.** $18°15'18''$ **33.** $19°59'24''$ **35.** $\dfrac{\pi}{6}$ **37.** $\dfrac{4\pi}{3}$ **39.** $-\dfrac{\pi}{3}$ **41.** π **43.** $-\dfrac{3\pi}{4}$ **45.** $-\dfrac{\pi}{2}$ **47.** $60°$

49. $-225°$ **51.** $90°$ **53.** $15°$ **55.** $-90°$ **57.** $-30°$ **59.** 0.30 **61.** -0.70 **63.** 2.18 **65.** $179.91°$ **67.** $114.59°$ **69.** $362.11°$ **71.** 5 m   **73.** 6 ft

75. 0.6 radian   **77.** $\dfrac{\pi}{3} \approx 1.047$ in.   **79.** 25 m^2 **81.** $2\sqrt{3} \approx 3.464$ ft   **83.** 0.24 radian   **85.** $\dfrac{\pi}{3} \approx 1.047 \text{ in.}^2$ **87.** $s = 2.094$ ft; $A = 2.094 \text{ ft}^2$

89. $s = 14.661$ yd; $A = 87.965 \text{ yd}^2$ **91.** $3\pi \approx 9.42$ in; $5\pi \approx 15.71$ in.   **93.** $2\pi \approx 6.28 \text{ m}^2$ **95.** $\dfrac{675\pi}{2} \approx 1060.29 \text{ ft}^2$

97. $\omega = \dfrac{1}{60}$ radian/s; $v = \dfrac{1}{12}$ cm/s   **99.** ≈ 452.5 rpm   **101.** ≈ 359 mi   **103.** ≈ 898 mi/h   **105.** ≈ 2292 mi/h   **107.** $\dfrac{3}{4}$ rpm   **109.** ≈ 2.86 mi/h

111. ≈ 31.47 rpm   **113.** ≈ 1037 mi/h   **115.** Radius ≈ 3979 mi; circumference $\approx 25,000$ mi   **117.** $v_1 = r_1\omega_1$, $v_2 = r_2\omega_2$, and $v_1 = v_2$, so
$r_1\omega_1 = r_2\omega_2$ Q $\dfrac{r_1}{r_2} = \dfrac{\omega_2}{\omega_1}$.

5.2 Assess Your Understanding (page 400)

3. complementary   **4.** cosine   **5.** $62°$ **6.** 1   **7.** T   **8.** F   **9.** T   **10.** F

11. $\sin\theta = \dfrac{5}{13}$; $\cos\theta = \dfrac{12}{13}$; $\tan\theta = \dfrac{5}{12}$; $\csc\theta = \dfrac{13}{5}$; $\sec\theta = \dfrac{13}{12}$; $\cot\theta = \dfrac{12}{5}$ **13.** $\sin\theta = \dfrac{2\sqrt{13}}{13}$; $\cos\theta = \dfrac{3\sqrt{13}}{13}$; $\tan\theta = \dfrac{2}{3}$; $\csc\theta = \dfrac{\sqrt{13}}{2}$; $\sec\theta = \dfrac{\sqrt{13}}{3}$;

$\cot\theta = \dfrac{3}{2}$ **15.** $\sin\theta = \dfrac{\sqrt{3}}{2}$; $\cos\theta = \dfrac{1}{2}$; $\tan\theta = \sqrt{3}$; $\csc\theta = \dfrac{2\sqrt{3}}{3}$; $\sec\theta = 2$; $\cot\theta = \dfrac{\sqrt{3}}{3}$ **17.** $\sin\theta = \dfrac{\sqrt{6}}{3}$; $\cos\theta = \dfrac{\sqrt{3}}{3}$; $\tan\theta = \sqrt{2}$;

$\csc\theta = \dfrac{\sqrt{6}}{2}$; $\sec\theta = \sqrt{3}$; $\cot\theta = \dfrac{\sqrt{2}}{2}$ **19.** $\sin\theta = \dfrac{\sqrt{5}}{5}$; $\cos\theta = \dfrac{2\sqrt{5}}{5}$; $\tan\theta = \dfrac{1}{2}$; $\csc\theta = \sqrt{5}$; $\sec\theta = \dfrac{\sqrt{5}}{2}$; $\cot\theta = 2$ **21.** $\tan\theta = \dfrac{\sqrt{3}}{3}$; $\csc\theta = 2$;

$\sec\theta = \dfrac{2\sqrt{3}}{3}$; $\cot\theta = \sqrt{3}$ **23.** $\tan\theta = \dfrac{2\sqrt{5}}{5}$; $\csc\theta = \dfrac{3}{2}$; $\sec\theta = \dfrac{3\sqrt{5}}{5}$; $\cot\theta = \dfrac{\sqrt{5}}{2}$ **25.** $\cos\theta = \dfrac{\sqrt{2}}{2}$; $\tan\theta = 1$; $\csc\theta = \sqrt{2}$; $\sec\theta = \sqrt{2}$; $\cot\theta = 1$

27. $\sin\theta = \dfrac{2\sqrt{2}}{3}$; $\tan\theta = 2\sqrt{2}$; $\csc\theta = \dfrac{3\sqrt{2}}{4}$; $\sec\theta = 3$; $\cot\theta = \dfrac{\sqrt{2}}{4}$ **29.** $\sin\theta = \dfrac{\sqrt{5}}{5}$; $\cos\theta = \dfrac{2\sqrt{5}}{5}$; $\csc\theta = \sqrt{5}$; $\sec\theta = \dfrac{\sqrt{5}}{2}$; $\cot\theta = 2$

31. $\sin\theta = \dfrac{2\sqrt{2}}{3}$; $\cos\theta = \dfrac{1}{3}$; $\tan\theta = 2\sqrt{2}$; $\csc\theta = \dfrac{3\sqrt{2}}{4}$; $\cot\theta = \dfrac{\sqrt{2}}{4}$ **33.** $\sin\theta = \dfrac{\sqrt{6}}{3}$; $\cos\theta = \dfrac{\sqrt{3}}{3}$; $\csc\theta = \dfrac{\sqrt{6}}{2}$; $\sec\theta = \sqrt{3}$; $\cot\theta = \dfrac{\sqrt{2}}{2}$

35. $\sin\theta = \dfrac{1}{2}$; $\cos\theta = \dfrac{\sqrt{3}}{2}$; $\tan\theta = \dfrac{\sqrt{3}}{3}$; $\sec\theta = \dfrac{2\sqrt{3}}{3}$; $\cot\theta = \sqrt{3}$ **37.** 1   **39.** 1   **41.** 0   **43.** 0   **45.** 1   **47.** 0   **49.** 0   **51.** 1   **53.** 1

55. (a) $\dfrac{1}{2}$ **(b)** $\dfrac{3}{4}$ **(c)** 2   **(d)** 2   **57. (a)** 17   **(b)** $\dfrac{1}{4}$ **(c)** 4   **(d)** $\dfrac{17}{16}$ **59. (a)** $\dfrac{1}{4}$ **(b)** 15   **(c)** 4   **(d)** $\dfrac{16}{15}$ **61. (a)** 0.78   **(b)** 0.79   **(c)** 1.27

(d) 1.28   **(e)** 1.61   **(f)** 0.78   **(g)** 0.62   **(h)** 1.27   **63.** 0.6   **65.** $20°$

67. (a) 10 min   **(d)** Approximately 15.8 min   **(f)** $70.5°$; 177 ft,
(b) 20 min   **(e)** Approximately 10.4 min   9.7 min
(c) $T(\theta) = 5\left(1 - \dfrac{1}{3\tan\theta} + \dfrac{1}{\sin\theta}\right)$

69. (a) $Z = 200\sqrt{13} \approx 721.1$ ohms **(b)** $\tan\phi = \dfrac{2}{3}$; $\sin\phi = \dfrac{2\sqrt{13}}{13}$; $\cos\phi = \dfrac{3\sqrt{13}}{13}$; $\cot\phi = \dfrac{3}{2}$; $\csc\phi = \dfrac{\sqrt{13}}{2}$; $\sec\phi = \dfrac{\sqrt{13}}{3}$

71. (a) $|OA| = |OC| = 1$; angle OAC = angle OCA; angle OAC + angle OAC + $180° - \theta = 180°$; angle $OAC = \dfrac{\theta}{2}$

  (b) $\sin\theta = \dfrac{|CD|}{|OC|} = |CD|$; $\cos\theta = \dfrac{|OD|}{|OC|} = |OD|$ **(c)** $\tan\dfrac{\theta}{2} = \dfrac{|CD|}{|AD|} = \dfrac{\sin\theta}{1 + |OD|} = \dfrac{\sin\theta}{1 + \cos\theta}$

73. $h = x\tan\theta$ and $h = (1 - x)\tan n\theta$; thus, $x\tan\theta = (1 - x)\tan n\theta$, so $x = \dfrac{\tan n\theta}{\tan\theta + \tan n\theta}$.

75. (a) Area $\triangle OAC = \dfrac{1}{2}|AC||OC| = \dfrac{1}{2}\cdot\dfrac{|AC|}{1}\cdot\dfrac{|OC|}{1} = \dfrac{1}{2}\sin\alpha\cos\alpha$

  (b) Area $\triangle OCB = \dfrac{1}{2}|BC||OC| = \dfrac{1}{2}|OB|^2\dfrac{|BC|}{|OB|}\cdot\dfrac{|OC|}{|OB|} = \dfrac{1}{2}|OB|^2\sin\beta\cos\beta$

  (c) Area $\triangle OAB = \dfrac{1}{2}|BD||OA| = \dfrac{1}{2}|OB|\dfrac{|BD|}{|OB|} = \dfrac{1}{2}|OB|\sin(\alpha + \beta)$

  (d) $\dfrac{\cos\alpha}{\cos\beta} = \dfrac{\dfrac{|OC|}{1}}{\dfrac{|OC|}{|OB|}} = |OB|$

  (e) Area$\triangle OAB$ = area $\triangle OAC$ + area $\triangle OCB$

  $\dfrac{1}{2}|OB|\sin(\alpha + \beta) = \dfrac{1}{2}\sin\alpha\cos\alpha + \dfrac{1}{2}|OB|^2\sin\beta\cos\beta$

  $\sin(\alpha + \beta) = \dfrac{\sin\alpha\cos\alpha + |OB|^2\sin\beta\cos\beta}{|OB|}$

  $\sin(\alpha + \beta) = \dfrac{\sin\alpha(|OB|\cos\beta) + |OB|^2\sin\beta\left(\dfrac{\cos\alpha}{|OB|}\right)}{|OB|}$

  $\sin(\alpha + \beta) = \sin\alpha\cos\beta + \cos\alpha\sin\beta$

77. $\sin\alpha = \tan\alpha\cos\alpha = \cos\beta\cos\alpha = \cos\beta\tan\beta = \sin\beta$;

  $\sin^2\alpha + \cos^2\alpha = 1$

  $\sin^2\alpha + \tan^2\beta = 1$

  $\sin^2\alpha + \dfrac{\sin^2\beta}{\cos^2\beta} = 1$

  $\sin^2\alpha + \dfrac{\sin^2\alpha}{1 - \sin^2\alpha} = 1$

  $\sin^2\alpha - \sin^4\alpha + \sin^2\alpha = 1 - \sin^2\alpha$

  $\sin^4\alpha - 3\sin^2\alpha + 1 = 0$

  $\sin^2\alpha = \dfrac{3 \pm \sqrt{5}}{2}$

  $\sin^2\alpha = \dfrac{3 - \sqrt{5}}{2}$

  $\sin\alpha = \sqrt{\dfrac{3 - \sqrt{5}}{2}}$

79. Since $a^2 + b^2 = c^2, a > 0, b > 0$, then $0 < a^2 < c^2$ or $0 < a < c$. Thus $0 < \dfrac{a}{c} < 1$ and $0 < \cos\theta < 1$ or $\dfrac{\cos\theta}{\cos\theta} < \dfrac{1}{\cos\theta}$; therefore, $\sec\theta > 1$.

5.3 Assess Your Understanding *(page 410)*

1. $\dfrac{3}{2}$ **2.** 0.91 **3.** T **4.** F **5.** $\sin 45° = \dfrac{\sqrt{2}}{2}$; $\cos 45° = \dfrac{\sqrt{2}}{2}$; $\tan 45° = 1$; $\csc 45° = \sqrt{2}$; $\sec 45° = \sqrt{2}$; $\cot 45° = 1$ **7.** $\dfrac{\sqrt{3}}{2}$ **9.** $\dfrac{1}{2}$ **11.** $\dfrac{3}{4}$

13. $\sqrt{3}$ **15.** $\dfrac{\sqrt{3}}{4}$ **17.** $\sqrt{2}$ **19.** 2 **21.** $\sqrt{2} + \dfrac{4\sqrt{3}}{3}$ **23.** $-\dfrac{8}{3}$ **25.** $\dfrac{1}{2}$ **27.** 0 **29.** 0.47 **31.** 0.38 **33.** 1.33 **35.** 0.31 **37.** 3.73 **39.** 1.04

41. 0.84 **43.** 0.02 **45.** 0.31 **47.** $\dfrac{1 + \sqrt{3}}{2}$ **49.** $\dfrac{1}{2}$ **51.** $\dfrac{\sqrt{3}}{2}$ **53.** $\dfrac{\sqrt{2}}{4}$ **55. (a)** $\dfrac{\sqrt{2}}{2}$; $\left(\dfrac{\pi}{4}, \dfrac{\sqrt{2}}{2}\right)$ **(b)** $\left(\dfrac{\sqrt{2}}{2}, \dfrac{\pi}{4}\right)$ **(c)** $\left(\dfrac{\pi}{4}, -2\right)$

57. $R \approx 310.56$ ft; $H \approx 77.64$ ft **59.** $R \approx 19{,}541.95$ m; $H \approx 2278.14$ m **61. (a)** 1.20 s **(b)** 1.12 s **(c)** 1.20 s

63. (a) $T(\theta) = 1 + \dfrac{2}{3\sin\theta} - \dfrac{1}{4\tan\theta}$ **(b)** 1.9 h; 0.57 h **(c)** 1.69 h; 0.75 h **(d)** 1.63 h; 0.86 h **(e)** 1.67 h **(f)** 2.75 h

  (g)

65. 4.59 in.; 6.55 in. **67. (a)** 5.52 in. or 11.83 in. **69.** 70.02 ft **71.** 985.91 ft **73.** $50\dfrac{\sqrt{3}}{3} \approx 28.87$ m

75. 1978.09 ft **77.** 60.27 ft **79.** $190\sqrt{2} \approx 268.70$ ft **81.** 554.52 ft **83.** 3.83 mi

85. The white ball should hit the top cushion 4.125 ft from the upper-left corner.

67.98°; 1.62 h; 0.9 h

87.

θ	0.5	0.4	0.2	0.1	0.01	0.001	0.0001	0.00001
$f(\theta) = \dfrac{\sin\theta}{\theta}$	0.9589	0.9735	0.9933	0.9983	1.0000	1.0000	1.0000	1.0000

$\dfrac{\sin\theta}{\theta}$ approaches 1 as θ : 0.

89. 1 **91.** $\dfrac{\sqrt{2}}{2}$

5.4 Assess Your Understanding (page 422)

1. tangent; cotangent **2.** coterminal **3.** 60° **4.** F **5.** T **6.** T **7.** 60° **8.** I, IV **9.** $\dfrac{\pi}{2}, \dfrac{3\pi}{2}$ **10.** $\dfrac{\pi}{3}$ **11.** $\sin\theta = \dfrac{4}{5}$; $\cos\theta = -\dfrac{3}{5}$;

$\tan\theta = -\dfrac{4}{3}$; $\csc\theta = \dfrac{5}{4}$; $\sec\theta = -\dfrac{5}{3}$; $\cot\theta = -\dfrac{3}{4}$ **13.** $\sin\theta = -\dfrac{3\sqrt{13}}{13}$; $\cos\theta = \dfrac{2\sqrt{13}}{13}$; $\tan\theta = -\dfrac{3}{2}$; $\csc\theta = -\dfrac{\sqrt{13}}{3}$; $\sec\theta = \dfrac{\sqrt{13}}{2}$; $\cot\theta = -\dfrac{2}{3}$

15. $\sin\theta = -\dfrac{\sqrt{2}}{2}$; $\cos\theta = -\dfrac{\sqrt{2}}{2}$; $\tan\theta = 1$; $\csc\theta = -\sqrt{2}$; $\sec\theta = -\sqrt{2}$; $\cot\theta = 1$

17. $\sin\theta = \dfrac{1}{2}$; $\cos\theta = \dfrac{\sqrt{3}}{2}$; $\tan\theta = \dfrac{\sqrt{3}}{3}$; $\csc\theta = 2$; $\sec\theta = \dfrac{2\sqrt{3}}{3}$; $\cot\theta = \sqrt{3}$ **19.** $\sin\theta = -\dfrac{\sqrt{2}}{2}$; $\cos\theta = \dfrac{\sqrt{2}}{2}$; $\tan\theta = -1$; $\csc\theta = -\sqrt{2}$;

$\sec\theta = \sqrt{2}$; $\cot\theta = -1$ **21.** $\dfrac{\sqrt{2}}{2}$ **23.** 1 **25.** 1 **27.** $\sqrt{3}$ **29.** $\dfrac{\sqrt{2}}{2}$ **31.** 0 **33.** II **35.** IV **37.** IV **39.** III **41.** 30° **43.** 60° **45.** 60° **47.** $\dfrac{\pi}{4}$

49. $\dfrac{\pi}{3}$ **51.** 45° **53.** $\dfrac{\pi}{3}$ **55.** 80° **57.** $\dfrac{\pi}{4}$ **59.** $\dfrac{1}{2}$ **61.** $\dfrac{1}{2}$ **63.** $\dfrac{\sqrt{2}}{2}$ **65.** -2 **67.** $-\sqrt{3}$ **69.** $\dfrac{\sqrt{2}}{2}$ **71.** $-\dfrac{\sqrt{2}}{2}$ **73.** $-\dfrac{\sqrt{3}}{2}$ **75.** $-\sqrt{3}$ **77.** 0 **79.** 0

81. -1 **83.** $\cos\theta = -\dfrac{5}{13}$; $\tan\theta = -\dfrac{12}{5}$; $\csc\theta = \dfrac{13}{12}$; $\sec\theta = \dfrac{13}{5}$; $\cot\theta = -\dfrac{5}{12}$ **85.** $\sin\theta = -\dfrac{3}{5}$; $\tan\theta = \dfrac{3}{4}$; $\csc\theta = -\dfrac{5}{3}$; $\sec\theta = -\dfrac{5}{4}$; $\cot\theta = \dfrac{4}{3}$

87. $\cos\theta = -\dfrac{12}{13}$; $\tan\theta = -\dfrac{5}{12}$; $\csc\theta = \dfrac{13}{5}$; $\sec\theta = -\dfrac{13}{12}$; $\cot\theta = -\dfrac{12}{5}$ **89.** $\sin\theta = -\dfrac{2\sqrt{2}}{3}$; $\tan\theta = 2\sqrt{2}$; $\csc\theta = -\dfrac{3\sqrt{2}}{4}$; $\sec\theta = -3$; $\cot\theta = \dfrac{\sqrt{2}}{4}$

91. $\cos\theta = -\dfrac{\sqrt{5}}{3}$; $\tan\theta = -\dfrac{2\sqrt{5}}{5}$; $\csc\theta = \dfrac{3}{2}$; $\sec\theta = -\dfrac{3\sqrt{5}}{5}$; $\cot\theta = -\dfrac{\sqrt{5}}{2}$ **93.** $\sin\theta = -\dfrac{\sqrt{3}}{2}$; $\cos\theta = \dfrac{1}{2}$; $\tan\theta = -\sqrt{3}$; $\csc\theta = -\dfrac{2\sqrt{3}}{3}$;

$\cot\theta = -\dfrac{\sqrt{3}}{3}$ **95.** $\sin\theta = -\dfrac{3}{5}$; $\cos\theta = -\dfrac{4}{5}$; $\csc\theta = -\dfrac{5}{3}$; $\sec\theta = -\dfrac{5}{4}$; $\cot\theta = \dfrac{4}{3}$ **97.** $\sin\theta = \dfrac{\sqrt{10}}{10}$; $\cos\theta = -\dfrac{3\sqrt{10}}{10}$;

$\csc\theta = \sqrt{10}$; $\sec\theta = -\dfrac{\sqrt{10}}{3}$; $\cot\theta = -3$ **99.** $\sin\theta = -\dfrac{1}{2}$; $\cos\theta = -\dfrac{\sqrt{3}}{2}$; $\tan\theta = \dfrac{\sqrt{3}}{3}$; $\sec\theta = -\dfrac{2\sqrt{3}}{3}$; $\cot\theta = \sqrt{3}$ **101.** 0

103. (a) $-\dfrac{\sqrt{2}}{2}$; $\left(315°, -\dfrac{\sqrt{2}}{2}\right)$ **(b)** $\sqrt{2}$; $(315°, \sqrt{2})$ **(c)** -1; $(315°, -1)$ **105. (a)** $-\dfrac{\sqrt{3}}{2}$; $\left(\dfrac{7\pi}{6}, -\dfrac{\sqrt{3}}{2}\right)$ **(b)** -2; $\left(\dfrac{7\pi}{6}, -2\right)$ **(c)** 1; $(-315°, 1)$

107. -0.2 **109.** 3 **111.** -5 **113.** 0 **115. (a)** Approximately 16.6 ft **(b)** **(c)** 67.5°

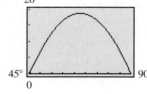

5.5 Assess Your Understanding (page 433)

4. 2π; π **5.** b; a **6.** $\dfrac{b}{r}$; $\dfrac{a}{r}$ **7.** -0.2; 0.2 **8.** T

9. $\sin t = -\dfrac{1}{2}$; $\cos t = \dfrac{\sqrt{3}}{2}$; $\tan t = -\dfrac{\sqrt{3}}{3}$; $\csc t = -2$; $\sec t = \dfrac{2\sqrt{3}}{3}$; $\cot t = -\sqrt{3}$ **11.** $\sin t = -\dfrac{\sqrt{2}}{2}$; $\cos t = -\dfrac{\sqrt{2}}{2}$; $\tan t = 1$;

$\csc t = -\sqrt{2}$; $\sec t = -\sqrt{2}$; $\cot t = 1$ **13.** $\sin t = \dfrac{2}{3}$; $\cos t = \dfrac{\sqrt{5}}{3}$; $\tan t = \dfrac{2\sqrt{5}}{5}$; $\csc t = \dfrac{3}{2}$; $\sec t = \dfrac{3\sqrt{5}}{5}$; $\cot t = \dfrac{\sqrt{5}}{2}$

15. $\sin\theta = -\dfrac{4}{5}$; $\cos\theta = \dfrac{3}{5}$; $\tan\theta = -\dfrac{4}{3}$; $\csc\theta = -\dfrac{5}{4}$; $\sec\theta = \dfrac{5}{3}$; $\cot\theta = -\dfrac{3}{4}$

17. $\sin\theta = \dfrac{3\sqrt{13}}{13}$; $\cos\theta = -\dfrac{2\sqrt{13}}{13}$; $\tan\theta = -\dfrac{3}{2}$; $\csc\theta = \dfrac{\sqrt{13}}{3}$; $\sec\theta = -\dfrac{\sqrt{13}}{2}$; $\cot\theta = -\dfrac{2}{3}$

19. $\sin\theta = -\dfrac{\sqrt{2}}{2}$; $\cos\theta = -\dfrac{\sqrt{2}}{2}$; $\tan\theta = 1$; $\csc\theta = -\sqrt{2}$; $\sec\theta = -\sqrt{2}$; $\cot\theta = 1$ **21.** $\dfrac{\sqrt{2}}{2}$ **23.** 1 **25.** 1 **27.** $\sqrt{3}$ **29.** $\dfrac{\sqrt{2}}{2}$ **31.** 0 **33.** $\sqrt{2}$

35. $\dfrac{\sqrt{3}}{3}$ **37.** $-\dfrac{\sqrt{3}}{2}$ **39.** $-\dfrac{\sqrt{3}}{3}$ **41.** 2 **43.** -1 **45.** -1 **47.** $\dfrac{\sqrt{2}}{2}$ **49.** 0 **51.** $-\sqrt{2}$ **53.** $\dfrac{2\sqrt{3}}{3}$ **55.** -1

57. -2 **59.** $\dfrac{2-\sqrt{2}}{2}$ **61.** $(-\infty, \infty)$ **63.** At odd multiples of $\dfrac{\pi}{2}$ **65.** At odd multiples of $\dfrac{\pi}{2}$ **67.** $[-1, 1]$ **69.** $(-\infty, \infty)$ **71.** $(-\infty, -1] \cup [1, \infty)$

73. Odd; yes; origin **75.** Odd; yes; origin **77.** Even; yes; y-axis **79.** 0.9 **81.** 9 **83. (a)** $-\dfrac{1}{3}$ **(b)** 1 **85. (a)** -2 **(b)** 6 **87. (a)** -4 **(b)** -12

89. (a) Using graph: $\sin 1 \approx 0.8$, $\cos 1 \approx 0.5$, $\tan 1 \approx 1.6$, $\csc 1 \approx 1.3$, $\sec 1 \approx 2$, $\cot 1 \approx 0.6$;

Using calculator: $\sin 1 \approx 0.8$, $\cos 1 \approx 0.5$, $\tan 1 \approx 1.6$, $\csc 1 \approx 1.2$, $\sec 1 \approx 1.9$, $\cot 1 \approx 0.6$

(b) Using graph: $\sin 5.1 \approx -0.9$, $\cos 5.1 \approx 0.4$, $\tan 5.1 \approx -2.3$, $\csc 5.1 \approx -1.1$, $\sec 5.1 \approx 2.5$, $\cot 5.1 \approx -0.4$;

Using calculator: $\sin 5.1 \approx -0.9$, $\cos 5.1 \approx 0.4$, $\tan 5.1 \approx -2.4$, $\csc 5.1 \approx -1.1$, $\sec 5.1 \approx 2.6$, $\cot 5.1 \approx -0.4$

91. Let $P = (x, y)$ be the point on the unit circle that corresponds to t. Consider the equation $\tan t = \dfrac{y}{x} = a$. Then $y = ax$.

But $x^2 + y^2 = 1$, so $x^2 + a^2x^2 = 1$. Thus $x = \pm\dfrac{1}{\sqrt{1 + a^2}}$ and $y = \pm\dfrac{a}{\sqrt{1 + a^2}}$; that is, for any real number a, there is a point $P = (x, y)$ on the unit circle for which $\tan t = a$. In other words, $-\infty < \tan t < \infty$, and the range of the tangent function is the set of all real numbers.

93. Suppose that there is a number p, $0 < p < 2\pi$, for which $\sin(\theta + p) = \sin \theta$ for all θ. If $\theta = 0$, then $\sin(0 + p) = \sin p = \sin 0 = 0$; so $p = \pi$. If $\theta = \dfrac{\pi}{2}$, then $\sin\left(\dfrac{\pi}{2} + p\right) = \sin\left(\dfrac{\pi}{2}\right)$. But $p = \pi$. Thus $\sin\left(\dfrac{3\pi}{2}\right) = -1 = \sin\left(\dfrac{\pi}{2}\right) = 1$. This is impossible. Therefore, the smallest positive number p for which $\sin(\theta + p) = \sin \theta$ for all θ is $p = 2\pi$.

95. $\sec \theta = \dfrac{1}{\cos \theta}$; since $\cos \theta$ has period 2π, so does $\sec \theta$.

97. If $P = (a, b)$ is the point on the unit circle corresponding to θ, then $Q = (-a, -b)$ is the point on the unit circle corresponding to $\theta + \pi$. Thus $\tan(\theta + p) = \dfrac{(-b)}{(-a)} = \dfrac{b}{a} = \tan \theta$. If there exists a number p, $0 < p < \pi$, for which $\tan(\theta + p) = \tan \theta$ for all θ, then, if $\theta = 0$, $\tan(p) = \tan 0 = 0$. But this means that p is a multiple of π. Since no multiple of π exists in the interval $(0, \pi)$, this is impossible. Therefore, the fundamental period of $f(\theta) = \tan \theta$ is π.

99. $m = \dfrac{\sin \theta - 0}{\cos \theta - 0} = \dfrac{\sin \theta}{\cos \theta} = \tan \theta$

5.6 Assess Your Understanding (page 445)

3. $1; \dfrac{\pi}{2}$ **4.** $3; \pi$ **5.** $3; \dfrac{\pi}{3}$ **6.** T   **7.** F   **8.** T   **9. (a)** 0 **(b)** $-\dfrac{\pi}{2} < x < \dfrac{\pi}{2}$ **(c)** 1 **(d)** $0, \pi, 2\pi$

(e) $f(x) = 1$ for $x = -\dfrac{3\pi}{2}, \dfrac{\pi}{2}$; $f(x) = -1$ for $x = -\dfrac{\pi}{2}, \dfrac{3\pi}{2}$ **(f)** $-\dfrac{5\pi}{6}, -\dfrac{\pi}{6}, \dfrac{7\pi}{6}, \dfrac{11\pi}{6}$ **(g)** $\{x \mid x = k\pi, k \text{ an integer}\}$ **11.** Amplitude $= 2$; period $= 2\pi$

13. Amplitude $= 4$; period $= \pi$ **15.** Amplitude $= 6$; period $= 2$ **17.** Amplitude $= \dfrac{1}{2}$; period $= \dfrac{4\pi}{3}$ **19.** Amplitude $= \dfrac{5}{3}$; period $= 3$

21. F   **23.** A   **25.** H   **27.** C   **29.** J   **31.** A   **33.** B

35.

Domain: $(-\infty, \infty)$
Range: $[-4, 4]$

37.
Domain: $(-\infty, \infty)$
Range: $[-4, 4]$

39.

Domain: $(-\infty, \infty)$
Range: $[-1, 1]$

41.

Domain: $(-\infty, \infty)$
Range: $[-1, 1]$

43.

Domain: $(-\infty, \infty)$
Range: $[-2, 2]$

45.
Domain: $(-\infty, \infty)$
Range: $\left[-\dfrac{1}{2}, \dfrac{1}{2}\right]$

47.
Domain: $(-\infty, \infty)$
Range: $[1, 5]$

49.

Domain: $(-\infty, \infty)$
Range: $[-8, 2]$

51.
Domain: $(-\infty, \infty)$
Range: $[-2, 10]$

53.

Domain: $(-\infty, \infty)$
Range: $[2, 8]$

55.
Domain: $(-\infty, \infty)$
Range: $\left[-\dfrac{5}{3}, \dfrac{5}{3}\right]$

57.
Domain: $(-\infty, \infty)$
Range: $[-1, 2]$

59. $y = \pm 3 \sin(2x)$ **61.** $y = \pm 3 \sin(\pi x)$ **63.** $y = 5 \cos\left(\dfrac{\pi}{4}x\right)$ **65.** $y = -3 \cos\left(\dfrac{1}{2}x\right)$ **67.** $y = \dfrac{3}{4}\sin(2\pi x)$ **69.** $y = -\sin\left(\dfrac{3}{2}x\right)$

71. $y = -\cos\left(\dfrac{4\pi}{3}x\right) + 1$ **73.** $y = 3 \sin\left(\dfrac{\pi}{2}x\right)$ **75.** $y = -4 \cos(3x)$ **77.** $\dfrac{2}{\pi}$ **79.** $\dfrac{\sqrt{2}}{\pi}$

81. $(f \circ g)(x) = \sin(4x)$

$(g \circ f)(x) = 4 \sin x$

83. $(f \circ g)(x) = -2 \cos x$

$(g \circ f)(x) = \cos(-2x)$

85. Period $= \dfrac{1}{30}$ s; amplitude $= 220$ amp

87. (a) Amplitude $= 220$ V; period $= \dfrac{1}{60}$ s

(b), (e)

(c) $I(t) = 22 \sin(120\pi t)$

(d) Amplitude $= 22$ amp; period $= \dfrac{1}{60}$ s

89. (a) $P(t) = \dfrac{[V_0 \sin(2\pi f t)]^2}{R} = \dfrac{V_0^2}{R}\sin^2(2\pi f t)$ **(b)** Since the graph of P has amplitude $\dfrac{V_0^2}{2R}$ and period $\dfrac{1}{2f}$ and is of the form $y = A \cos(\omega t) + B$,

then $A = -\dfrac{V_0^2}{2R}$ and $B = \dfrac{V_0^2}{2R}$. Since $\dfrac{1}{2f} = \dfrac{2\pi}{\omega}$, then $\omega = 4\pi f$. Therefore, $P(t) = -\dfrac{V_0^2}{2R}\cos(4\pi f t) + \dfrac{V_0^2}{2R} = \dfrac{V_0^2}{2R}[1 - \cos(4\pi f t)]$.

91. (a) Physical potential: $\omega = \dfrac{2\pi}{23}$; emotional potential: $\omega = \dfrac{\pi}{14}$; intellectual potential: $\omega = \dfrac{2\pi}{33}$

(b)

(c) No   **(d)** Physical potential peaks at 15 days after 20th birthday. Emotional potential is 50% at 17 days, with a maximum at 10 days and a minimum at 24 days. Intellectual potential starts fairly high, drops to a minimum at 13 days, and rises to a maximum at 29 days.

93.

95. Answers may vary. $\left(-\dfrac{5\pi}{3}, \dfrac{1}{2}\right), \left(-\dfrac{\pi}{3}, \dfrac{1}{2}\right), \left(\dfrac{\pi}{3}, \dfrac{1}{2}\right), \left(\dfrac{5\pi}{3}, \dfrac{1}{2}\right)$ **97.** Answers may vary. $\left(-\dfrac{3\pi}{4}, 1\right), \left(\dfrac{\pi}{4}, 1\right), \left(\dfrac{5\pi}{4}, 1\right), \left(\dfrac{9\pi}{4}, 1\right)$

5.7 Assess Your Understanding (page 456)

3. origin; odd multiples of $\dfrac{\pi}{2}$ **4.** y-axis; odd multiples of $\dfrac{\pi}{2}$ **5.** $y = \cos x$ **6.** T   **7.** 0   **9.** 1

11. $\sec x = 1$ for $x = -2\pi, 0, 2\pi$; $\sec x = -1$ for $x = -\pi, \pi$ **13.** $-\dfrac{3\pi}{2}, -\dfrac{\pi}{2}, \dfrac{\pi}{2}, \dfrac{3\pi}{2}$ **15.** $-\dfrac{3\pi}{2}, -\dfrac{\pi}{2}, \dfrac{\pi}{2}, \dfrac{3\pi}{2}$

17.

Domain: $\left\{ x \;\middle|\; x \neq \dfrac{k\pi}{2}, k \text{ is an odd integer} \right\}$
Range: $(-\infty, \infty)$

19.

Domain: $\{x | x \neq k\pi, k \text{ is an integer}\}$
Range: $(-\infty, \infty)$

21.

Domain: $\{x | x \text{ does not equal an odd integer}\}$
Range: $(-\infty, \infty)$

23.

Domain: $\{x | x \neq 4k\pi, k \text{ is an integer}\}$
Range: $(-\infty, \infty)$

25.

Domain: $\left\{ x \;\middle|\; x \neq \dfrac{k\pi}{2}, k \text{ is an odd integer} \right\}$
Range: $\{y | y \leq -2 \text{ or } y \geq 2\}$

27.

Domain: $\{x | x \neq k\pi, k \text{ is an integer}\}$
Range: $\{y | y \leq -3 \text{ or } y \geq 3\}$

29.

Domain: $\{x | x \neq k\pi, k \text{ is an odd integer}\}$
Range: $\{y | y \leq -4 \text{ or } y \geq 4\}$

31.

Domain: $\{x | x \text{ does not equal an integer}\}$
Range: $\{y | y \leq -2 \text{ or } y \geq 2\}$

33.

Domain: $\{x | x \neq 2\pi k, k \text{ is an odd integer}\}$
Range: $(-\infty, \infty)$

35.

Domain: $\left\{ x \middle| x \neq \dfrac{3}{4}k, k \text{ is an odd integer} \right\}$
Range: $\{y | y \leq 1 \text{ or } y \geq 3\}$

37.

Domain: $\{x | x \neq 2\pi k, k \text{ is an odd integer}\}$
Range: $(-\infty, \infty)$

39.

Domain: $\{x | x \neq 3\pi k, k \text{ is an integer}\}$
Range: $\{y | y \leq -3 \text{ or } y \geq 1\}$

41. $\dfrac{2\sqrt{3}}{\pi}$ **43.** $\dfrac{6\sqrt{3}}{\pi}$ **45.** $(f \circ g)(x) = \tan(4x)$ $(g \circ f)(x) = 4 \tan x$ **47.** $(f \circ g)(x) = -2 \cot x$ $(g \circ f)(x) = \cot(-2x)$

49. (a) $L(\theta) = \dfrac{3}{\cos \theta} + \dfrac{4}{\sin \theta}$
$= 3 \sec \theta + 4 \csc \theta$
(c) ≈ 0.83 **(d)** ≈ 9.86 ft

(b)

51.

$y = \tan x$ $y = -\cot\left(x + \dfrac{\pi}{2}\right)$

5.8 Assess Your Understanding (page 466)

1. phase shift   **2.** F
3. Amplitude $= 4$
Period $= \pi$
Phase shift $= \dfrac{\pi}{2}$

5. Amplitude $= 2$
Period $= \dfrac{2\pi}{3}$
Phase shift $= -\dfrac{\pi}{6}$

7. Amplitude $= 3$
Period $= \pi$
Phase shift $= -\dfrac{\pi}{4}$

9. Amplitude $= 4$
Period $= 2$
Phase shift $= -\dfrac{2}{\pi}$

11. Amplitude $= 3$
Period $= 2$
Phase shift $= \dfrac{2}{\pi}$

13. Amplitude $= 3$
Period $= \pi$
Phase shift $= \dfrac{\pi}{4}$

15. $y = 2\sin\left[2\left(x - \dfrac{1}{2}\right)\right]$ or $y = 2\sin(2x - 1)$ **17.** $y = 3\sin\left[\dfrac{2}{3}\left(x + \dfrac{1}{3}\right)\right]$ or $y = 3\sin\left(\dfrac{2}{3}x + \dfrac{2}{9}\right)$

19.   **21.**   **23.**   **25.**

27. Period $= \dfrac{1}{15}$ s;
amplitude $= 120$ amp;
phase shift $= \dfrac{1}{90}$ s

29. (a)   **(b)** $y = 15.9\sin\left[\dfrac{\pi}{6}(x - 4)\right] + 40.1$ or $y = 15.9\sin\left(\dfrac{\pi}{6}x - \dfrac{2\pi}{3}\right) + 40.1$

(c)   **(d)** $y = 15.62\sin(0.517x - 2.096) + 40.377$

(e)

31. (a)   **(b)** $y = 24.95\sin\left[\dfrac{\pi}{6}(x - 4)\right] + 50.45$ or $y = 24.95\sin\left(\dfrac{\pi}{6}x - \dfrac{2\pi}{3}\right) + 50.45$

(c)   **(d)** $y = 25.693\sin(0.476x - 1.814) + 49.854$

(e)

33. (a) 11:55 PM   **(b)** $y = 3.105\sin\left[\dfrac{24\pi}{149}(x - 8.3958)\right] + 2.735$ or $y = 3.105\sin\left[\dfrac{24\pi}{149}x - 4.2485\right] + 2.735$ **(c)** 2.12 ft

35. (a) $y = 1.6\sin\left(\dfrac{2\pi}{365}x - 1.39\right) + 12.15$
(b) 12.43 h
(c)

(d) The actual hours of sunlight on April 1, 2010, were 12.43 hours. This is the same as the predicted amount.

37. (a) $y = 6.97\sin\left(\dfrac{2\pi}{365}x - 1.39\right) + 12.45$
(b) 13.67 h
(c)

(d) The actual hours of sunlight on April 1 2010, were 13.38 hours. This is close to the predicted amount of 13.67 hours.

Review Exercises (page 471)

1. $\dfrac{3\pi}{4}$ **3.** $\dfrac{\pi}{10}$ **5.** $135°$ **7.** $-450°$ **9.** $\dfrac{1}{2}$ **11.** $\dfrac{3\sqrt{2}}{2} - \dfrac{4\sqrt{3}}{3}$ **13.** $-3\sqrt{2} - 2\sqrt{3}$ **15.** 3 **17.** 0 **19.** 0 **21.** 1 **23.** 1 **25.** 1 **27.** -1 **29.** 1

31. $\cos\theta = \dfrac{3}{5}$; $\tan\theta = \dfrac{4}{3}$; $\csc\theta = \dfrac{5}{4}$; $\sec\theta = \dfrac{5}{3}$; $\cot\theta = \dfrac{3}{4}$ **33.** $\sin\theta = -\dfrac{12}{13}$; $\cos\theta = -\dfrac{5}{13}$; $\csc\theta = -\dfrac{13}{12}$; $\sec\theta = -\dfrac{13}{5}$; $\cot\theta = \dfrac{5}{12}$

35. $\sin\theta = \dfrac{3}{5}$; $\cos\theta = -\dfrac{4}{5}$; $\tan\theta = -\dfrac{3}{4}$; $\csc\theta = \dfrac{5}{3}$; $\cot\theta = -\dfrac{4}{3}$ **37.** $\cos\theta = -\dfrac{5}{13}$; $\tan\theta = -\dfrac{12}{5}$; $\csc\theta = \dfrac{13}{12}$; $\sec\theta = -\dfrac{13}{5}$; $\cot\theta = -\dfrac{5}{12}$

39. $\cos\theta = \dfrac{12}{13}$; $\tan\theta = -\dfrac{5}{12}$; $\csc\theta = -\dfrac{13}{5}$; $\sec\theta = \dfrac{13}{12}$; $\cot\theta = -\dfrac{12}{5}$ **41.** $\sin\theta = -\dfrac{\sqrt{10}}{10}$; $\cos\theta = -\dfrac{3\sqrt{10}}{10}$; $\csc\theta = -\sqrt{10}$; $\sec\theta = -\dfrac{\sqrt{10}}{3}$; $\cot\theta = 3$

43. $\sin\theta = -\dfrac{2\sqrt{2}}{3}$; $\cos\theta = \dfrac{1}{3}$; $\tan\theta = -2\sqrt{2}$; $\csc\theta = -\dfrac{3\sqrt{2}}{4}$; $\cot\theta = -\dfrac{\sqrt{2}}{4}$ **45.** $\sin\theta = \dfrac{\sqrt{5}}{5}$; $\cos\theta = -\dfrac{2\sqrt{5}}{5}$; $\tan\theta = -\dfrac{1}{2}$; $\csc\theta = \sqrt{5}$; $\sec\theta = -\dfrac{\sqrt{5}}{2}$

47.
Domain: $(-\infty,\infty)$
Range: $[-2, 2]$

49.
Domain: $(-\infty,\infty)$
Range: $[-2, 2]$

51.
Domain: $\left\{x \,\middle|\, x \neq \dfrac{k\pi}{2}, k \text{ is an odd integer}\right\}$
Range: $(-\infty, \infty)$

53.
Domain: $\left\{x \,\middle|\, x \neq \dfrac{\pi}{6} + k \cdot \dfrac{\pi}{3}, k \text{ is an integer}\right\}$
Range: $(-\infty, \infty)$

55.
Domain: $\left\{x \,\middle|\, x \neq -\dfrac{\pi}{4} + k\pi, k \text{ is an integer}\right\}$
Range: $(-\infty, \infty)$

57.
Domain: $\left\{x \,\middle|\, x \neq \dfrac{k\pi}{4}, k \text{ is an odd integer}\right\}$
Range: $\{y|y \leq -4 \text{ or } y \geq 4\}$

59.
Domain: $(-\infty,\infty)$
Range: $[-6, 2]$

61.
Domain: $\left\{x \,\middle|\, x \neq \dfrac{\pi}{2} + k \cdot 2\pi, k \text{ is an integer}\right\}$
Range: $(-\infty, \infty)$

63. Amplitude $= 4$; period $= 2\pi$

65. Amplitude $= 8$; period $= 4$

67. Amplitude $= 4$
Period $= \dfrac{2\pi}{3}$
Phase shift $= 0$

69. Amplitude $= 2$
Period $= \pi$
Phase shift $= \dfrac{\pi}{2}$

71. Amplitude $= \dfrac{1}{2}$
Period $= \dfrac{4\pi}{3}$
Phase shift $= \dfrac{2\pi}{3}$

73. Amplitude $= \dfrac{2}{3}$
Period $= 2$
Phase shift $= \dfrac{6}{\pi}$

75. $y = 5\cos\dfrac{x}{4}$ **77.** $y = -6\cos\left(\dfrac{\pi}{4}x\right)$ **79.** $\sin\theta = \dfrac{3}{7}$; $\cos\theta = \dfrac{2\sqrt{10}}{7}$; $\tan\theta = \dfrac{3\sqrt{10}}{20}$; $\csc\theta = \dfrac{7}{3}$; $\sec\theta = \dfrac{7\sqrt{10}}{20}$; $\cot\theta = \dfrac{2\sqrt{10}}{3}$

81. Sine, cosine, cosecant and secant: negative; tangent and cotangent: positive

83. $\sin\theta = \dfrac{2\sqrt{2}}{3}$; $\cos\theta = -\dfrac{1}{3}$; $\tan\theta = -2\sqrt{2}$; $\csc\theta = \dfrac{3\sqrt{2}}{4}$; $\sec\theta = -3$; $\cot\theta = -\dfrac{\sqrt{2}}{4}$

85. Domain: $\left\{x \,\middle|\, x \neq \text{odd multiple of } \dfrac{\pi}{2}\right\}$; range: $\{y||y| \geq 1\}$; period $= 2\pi$ **87.** $\dfrac{\pi}{3} \approx 1.05$ ft; $\dfrac{\pi}{3} \approx 1.05$ ft^2 **89.** ≈ 114.59 revolutions/hr

91. 0.1 revolution/sec $= \dfrac{\pi}{5}$ radian/sec **93.** 839.10 ft **95.** 23.32 ft **97.** 2.15 mi

99. (a)

(c)

(e)

(b) $y = 19.5\sin\left[\dfrac{\pi}{6}(x - 4)\right] + 70.5$ or
$y = 19.5\sin\left(\dfrac{\pi}{6}x - \dfrac{2\pi}{3}\right) + 70.5$

(d) $y = 19.52\sin(0.54x - 2.28) + 71.01$

Chapter Test *(page 474)*

1. $\dfrac{13\pi}{9}$ **2.** $-\dfrac{20\pi}{9}$ **3.** $\dfrac{13\pi}{180}$ **4.** $-22.5°$ **5.** $810°$ **6.** $135°$ **7.** $\dfrac{1}{2}$ **8.** 0 **9.** $-\dfrac{1}{2}$ **10.** $-\dfrac{\sqrt{3}}{3}$ **11.** 2 **12.** $\dfrac{3(1-\sqrt{2})}{2}$ **13.** 0.292 **14.** 0.309

15. -1.524 **16.** 2.747 **17.**

18. $-\dfrac{3}{5}$

	$\sin\theta$	$\cos\theta$	$\tan\theta$	$\sec\theta$	$\csc\theta$	$\cot\theta$
θ in QI	+	+	+	+	+	+
θ in QII	+	−	−	−	+	−
θ in QIII	−	−	+	−	−	+
θ in QIV	−	+	−	+	−	−

19. $\cos\theta = -\dfrac{2\sqrt{6}}{7}$; $\tan\theta = -\dfrac{5\sqrt{6}}{12}$; $\csc\theta = \dfrac{7}{5}$; $\sec\theta = -\dfrac{7\sqrt{6}}{12}$; $\cot\theta = -\dfrac{2\sqrt{6}}{5}$ **20.** $\sin\theta = -\dfrac{\sqrt{5}}{3}$; $\tan\theta = -\dfrac{\sqrt{5}}{2}$; $\csc\theta = -\dfrac{3\sqrt{5}}{5}$;

$\sec\theta = \dfrac{3}{2}$; $\cot\theta = -\dfrac{2\sqrt{5}}{5}$ **21.** $\sin\theta = \dfrac{12}{13}$; $\cos\theta = -\dfrac{5}{13}$; $\csc\theta = \dfrac{13}{12}$; $\sec\theta = -\dfrac{13}{5}$; $\cot\theta = -\dfrac{5}{12}$ **22.** $\dfrac{7\sqrt{53}}{53}$ **23.** $-\dfrac{5\sqrt{146}}{146}$ **24.** $-\dfrac{1}{2}$

25.

26.

27. $y = -3\sin\left(3x + \dfrac{3\pi}{4}\right)$ **28.** 78.93 ft^2 **29.** 143.5 rpm

30. The ship is about 838 ft from the statue.

31. The building is 122.4 ft tall.

Cumulative Review *(page 475)*

1. $\left\{-1, \dfrac{1}{2}\right\}$ **2.** $y - 5 = -3(x + 2)$ or $y = -3x - 1$ **3.** $x^2 + (y + 2)^2 = 16$

4. A line; slope $\dfrac{2}{3}$; intercepts $(6, 0)$ and $(0, -4)$

5. A circle; center $(1, -2)$; radius 3

6.

7. (a)   **(b)**   **(c)**   **(d)**   **(e)**   **(f)**

8. $f^{-1}(x) = \dfrac{1}{3}(x + 2)$ **9.** -2 **10.**

11. $3 - \dfrac{3\sqrt{3}}{2}$ **12.** $y = 2(3^x)$ **13.** $y = 3\cos\left(\dfrac{\pi}{6}x\right)$

14. (a) $f(x) = -3x - 3$;
$m = -3$; $(-1, 0)$, $(0, -3)$

(b) $f(x) = (x - 1)^2 - 6$; $(0, -5)$,
$\left(-\sqrt{6} + 1, 0\right), \left(\sqrt{6} + 1, 0\right)$

(c) We have that $y = 3$ when $x = -2$ and $y = -6$ when $x = 1$. Both points satisfy $y = ae^x$. Therefore, for $(-2, 3)$ we have $3 = ae^{-2}$, which implies that $a = 3e^2$. But for $(1, -6)$ we have $-6 = ae^1$, which implies that $a = -6e^{-1}$. Therefore, there is no exponential function $y = ae^x$ that contains $(-2, 3)$ and $(1, -6)$.

15. (a) $f(x) = \dfrac{1}{6}(x + 2)(x - 3)(x - 5)$ **(b)** $R(x) = -\dfrac{(x + 2)(x - 3)(x - 5)}{3(x - 2)}$

CHAPTER 6 Analytic Trigonometry

6.1 Assess Your Understanding *(page 487)*

7. $x = \sin y$ **8.** $0 \le x \le \pi$ **9.** $-\infty < x < \infty$ **10.** F **11.** T **12.** T **13.** 0 **15.** $-\dfrac{\pi}{2}$ **17.** 0 **19.** $\dfrac{\pi}{4}$ **21.** $\dfrac{\pi}{3}$ **23.** $\dfrac{5\pi}{6}$ **25.** 0. 10 **27.** 1.37

29. 0.51 **31.** -0.38 **33.** -0.12 **35.** 1.08 **37.** $\dfrac{4\pi}{5}$ **39.** $-\dfrac{3\pi}{8}$ **41.** $-\dfrac{\pi}{8}$ **43.** $-\dfrac{\pi}{5}$ **45.** $\dfrac{1}{4}$ **47.** 4 **49.** Not defined **51.** π

53. $f^{-1}(x) = \sin^{-1}\dfrac{x - 2}{5}$

Range of f = Domain of f^{-1} = $[-3, 7]$

Range of f^{-1} = $\left[-\dfrac{\pi}{2}, \dfrac{\pi}{2}\right]$

55. $f^{-1}(x) = \dfrac{1}{3}\cos^{-1}\left(-\dfrac{x}{2}\right)$

Range of f = Domain of f^{-1} = $[-2, 2]$

Range of f^{-1} = $\left[0, \dfrac{\pi}{3}\right]$

57. $f^{-1}(x) = -\tan^{-1}(x + 3) - 1$

Range of f = Domain of f^{-1} = $(-\infty, \infty)$

Range of f^{-1} = $\left(-1 - \dfrac{\pi}{2}, \dfrac{\pi}{2} - 1\right)$

59. $f^{-1}(x) = \dfrac{1}{2}\left[\sin^{-1}\left(\dfrac{x}{3}\right) - 1\right]$

Range of f = Domain of f^{-1} = $[-3, 3]$

Range of f^{-1} = $\left[-\dfrac{1}{2} - \dfrac{\pi}{4}, -\dfrac{1}{2} + \dfrac{\pi}{4}\right]$

61. $\left\{\dfrac{\sqrt{2}}{2}\right\}$ **63.** $\left\{-\dfrac{1}{4}\right\}$ **65.** $\{\sqrt{3}\}$ **67.** $\{-1\}$

69. (a) 13.92 h or 13 h, 55 min **(b)** 12 h **(c)** 13.85 h or 13 h, 51 min

71. (a) 13.3 h or 13 h, 18 min **(b)** 12 h **(c)** 13.26 h or 13 h, 15 min

73. (a) 12 h **(b)** 12 h **(c)** 12 h **(d)** It is 12 h. **75.** 3.28 min **77. (a)** $\dfrac{\pi}{3}$ square units **(b)** $\dfrac{5\pi}{12}$ square units **79.** 4250 mi

6.2 Assess Your Understanding *(page 494)*

4. $x = \sec y; \ge 1; 0; \pi$ **5.** cosine **6.** F **7.** T **8.** T **9.** $\dfrac{\sqrt{2}}{2}$ **11.** $-\dfrac{\sqrt{3}}{3}$ **13.** 2 **15.** $\sqrt{2}$ **17.** $-\dfrac{\sqrt{2}}{2}$ **19.** $\dfrac{2\sqrt{3}}{3}$ **21.** $\dfrac{3\pi}{4}$ **23.** $\dfrac{\pi}{6}$ **25.** $\dfrac{\sqrt{2}}{4}$

27. $\dfrac{\sqrt{5}}{2}$ **29.** $-\dfrac{\sqrt{14}}{2}$ **31.** $-\dfrac{3\sqrt{10}}{10}$ **33.** $\sqrt{5}$ **35.** $-\dfrac{\pi}{4}$ **37.** $\dfrac{\pi}{6}$ **39.** $-\dfrac{\pi}{2}$ **41.** $\dfrac{\pi}{6}$ **43.** $\dfrac{2\pi}{3}$ **45.** 1.32 **47.** 0.46 **49.** -0.34 **51.** 2.72 **53.** -0.73

55. 2.55 **57.** $\dfrac{1}{\sqrt{1 + u^2}}$ **59.** $\dfrac{u}{\sqrt{1 - u^2}}$ **61.** $\dfrac{\sqrt{u^2 - 1}}{|u|}$ **63.** $\dfrac{\sqrt{u^2 - 1}}{|u|}$ **65.** $\dfrac{1}{u}$ **67.** $\dfrac{5}{13}$ **69.** $\dfrac{3\pi}{4}$ **71.** $-\dfrac{3}{4}$ **73.** $\dfrac{5}{13}$ **75.** $\dfrac{\pi}{6}$ **77.** $-\sqrt{15}$

79. (a) $\theta = 31.89°$ **(b)** 54.64 ft in diameter **(c)** 37.96 ft high **81. (a)** $\theta = 22.3°$ **(b)** $v_0 = 2940.23$ ft/s

83.

6.3 Assess Your Understanding *(page 502)*

7. $\dfrac{\pi}{6}; \dfrac{5\pi}{6}$ **8.** $\left\{\theta \middle| \theta = \dfrac{\pi}{6} + 2\pi k, \theta = \dfrac{5\pi}{6} + 2\pi k, k \text{ is any integer}\right\}$ **9.** F **10.** F **11.** $\left\{\dfrac{7\pi}{6}, \dfrac{11\pi}{6}\right\}$ **13.** $\left\{\dfrac{\pi}{3}, \dfrac{2\pi}{3}, \dfrac{4\pi}{3}, \dfrac{5\pi}{3}\right\}$ **15.** $\left\{\dfrac{\pi}{4}, \dfrac{3\pi}{4}, \dfrac{5\pi}{4}, \dfrac{7\pi}{4}\right\}$

17. $\left\{\dfrac{\pi}{2}, \dfrac{7\pi}{6}, \dfrac{11\pi}{6}\right\}$ **19.** $\left\{\dfrac{\pi}{3}, \dfrac{2\pi}{3}, \dfrac{4\pi}{3}, \dfrac{5\pi}{3}\right\}$ **21.** $\left\{\dfrac{4\pi}{9}, \dfrac{8\pi}{9}, \dfrac{16\pi}{9}\right\}$ **23.** $\left\{\dfrac{7\pi}{6}, \dfrac{11\pi}{6}\right\}$ **25.** $\left\{\dfrac{3\pi}{4}, \dfrac{7\pi}{4}\right\}$ **27.** $\left\{\dfrac{2\pi}{3}, \dfrac{4\pi}{3}\right\}$ **29.** $\left\{\dfrac{3\pi}{4}, \dfrac{5\pi}{4}\right\}$

31. $\left\{\dfrac{3\pi}{4}, \dfrac{7\pi}{4}\right\}$ **33.** $\left\{\dfrac{11\pi}{6}\right\}$ **35.** $\left\{\theta \middle| \theta = \dfrac{\pi}{6} + 2k\pi, \theta = \dfrac{5\pi}{6} + 2k\pi\right\}; \dfrac{\pi}{6}, \dfrac{5\pi}{6}, \dfrac{13\pi}{6}, \dfrac{17\pi}{6}, \dfrac{25\pi}{6}, \dfrac{29\pi}{6}$

37. $\left\{\theta \middle| \theta = \dfrac{5\pi}{6} + k\pi\right\}; \dfrac{5\pi}{6}, \dfrac{11\pi}{6}, \dfrac{17\pi}{6}, \dfrac{23\pi}{6}, \dfrac{29\pi}{6}, \dfrac{35\pi}{6}$ **39.** $\left\{\theta \middle| \theta = \dfrac{\pi}{2} + 2k\pi, \theta = \dfrac{3\pi}{2} + 2k\pi\right\}; \dfrac{\pi}{2}, \dfrac{3\pi}{2}, \dfrac{5\pi}{2}, \dfrac{7\pi}{2}, \dfrac{9\pi}{2}, \dfrac{11\pi}{2}$

41. $\left\{\theta \middle| \theta = \dfrac{\pi}{3} + k\pi, \theta = \dfrac{2\pi}{3} + k\pi\right\}; \dfrac{\pi}{3}, \dfrac{2\pi}{3}, \dfrac{4\pi}{3}, \dfrac{5\pi}{3}, \dfrac{7\pi}{3}, \dfrac{8\pi}{3}$ **43.** $\left\{\theta \middle| \theta = \dfrac{8\pi}{3} + 4k\pi, \theta = \dfrac{10\pi}{3} + 4k\pi\right\}; \dfrac{8\pi}{3}, \dfrac{10\pi}{3}, \dfrac{20\pi}{3}, \dfrac{22\pi}{3}, \dfrac{32\pi}{3}, \dfrac{34\pi}{3}$

45. $\{0.41, 2.73\}$ **47.** $\{1.37, 4.51\}$ **49.** $\{2.69, 3.59\}$ **51.** $\{1.82, 4.46\}$ **53.** $\{2.08, 5.22\}$ **55.** $\{0.73, 2.41\}$ **57.** $\left\{\dfrac{\pi}{2}, \dfrac{2\pi}{3}, \dfrac{4\pi}{3}, \dfrac{3\pi}{2}\right\}$

59. $\left\{\dfrac{\pi}{2}, \dfrac{7\pi}{6}, \dfrac{11\pi}{6}\right\}$ **61.** $\left\{0, \dfrac{\pi}{4}, \dfrac{5\pi}{4}\right\}$ **63.** $\left\{\dfrac{\pi}{2}, \dfrac{2\pi}{3}, \dfrac{4\pi}{3}, \dfrac{3\pi}{2}\right\}$ **65.** $\{\pi\}$ **67.** $\left\{\dfrac{\pi}{4}, \dfrac{5\pi}{4}\right\}$ **69.** $\left\{0, \dfrac{\pi}{3}, \pi, \dfrac{5\pi}{3}\right\}$ **71.** $\left\{\dfrac{\pi}{6}, \dfrac{5\pi}{6}, \dfrac{3\pi}{2}\right\}$ **73.** $\left\{\dfrac{\pi}{2}\right\}$ **75.** $\{0\}$

77. $\left\{\dfrac{\pi}{3}, \dfrac{5\pi}{3}\right\}$ **79.** No real solution **81.** $-1.31, 1.98, 3.84$ **83.** 0.52 **85.** 1.26 **87.** $-1.02, 1.02$ **89.** 0, 2.15 **91.** 0.76, 1.35 **93.** $\dfrac{\pi}{3}, \dfrac{2\pi}{3}, \dfrac{4\pi}{3}, \dfrac{5\pi}{3}$

95. (a) $-2\pi, -\pi, 0, \pi, 2\pi, 3\pi, 4\pi$ **(b)**

(c) $\left\{ -\dfrac{11\pi}{6}, -\dfrac{7\pi}{6}, \dfrac{\pi}{6}, \dfrac{5\pi}{6}, \dfrac{13\pi}{6}, \dfrac{17\pi}{6} \right\}$

(d) $\left\{ x \,\middle|\, -\dfrac{11\pi}{6} < x < -\dfrac{7\pi}{6} \text{ or } \dfrac{\pi}{6} < x < \dfrac{5\pi}{6} \text{ or } \dfrac{13\pi}{6} < x < \dfrac{17\pi}{6} \right\}$

97. (a) $\left\{ x \,\middle|\, x = -\dfrac{\pi}{4} + k\pi, k \text{ is any integer} \right\}$ **(b)** $-\dfrac{\pi}{2} < x < -\dfrac{\pi}{4} \text{ or } \left(-\dfrac{\pi}{2}, -\dfrac{\pi}{4} \right)$

99. (a), (d)

(b) $\left\{ \dfrac{\pi}{12}, \dfrac{5\pi}{12} \right\}$

(c) $\left\{ x \,\middle|\, \dfrac{\pi}{12} < x < \dfrac{5\pi}{12} \right\} \text{ or } \left(\dfrac{\pi}{12}, \dfrac{5\pi}{12} \right)$

101. (a), (d)

(b) $\left\{ \dfrac{2\pi}{3}, \dfrac{4\pi}{3} \right\}$

(c) $\left\{ x \,\middle|\, \dfrac{2\pi}{3} < x < \dfrac{4\pi}{3} \right\} \text{ or } \left(\dfrac{2\pi}{3}, \dfrac{4\pi}{3} \right)$

103. (a) 0 s, 0.43 s, 0.86 s **(b)** 0.21 s **(c)** $[0, 0.03] \cup [0.39, 0.43] \cup [0.86, 0.89]$
105. (a) 150 mi **(b)** 6.06, 8.44, 15.72, 18.11 min **(c)** Before 6.06 min, between 8.44 and 15.72 min, and after 18.11 min **(d)** No
107. $2.03, 4.91$

109. (a) $30°, 60°$ **(b)** 123.6 m **(c)**

111. 28.90° **113.** Yes; it varies from 1.25 to 1.34. **115.** 1.47
117. If θ is the original angle of incidence and ϕ is the angle of refraction, then $\dfrac{\sin\theta}{\sin\phi} = n_2$. The angle of incidence of the emerging beam is also ϕ, and the index of refraction is $\dfrac{1}{n_2}$. Thus, θ is the angle of refraction of the emerging beam.

6.4 Assess Your Understanding *(page 511)*

3. identity; conditional **4.** -1 **5.** 0 **6.** T **7.** F **8.** T **9.** $\dfrac{1}{\cos\theta}$ **11.** $\dfrac{1 + \sin\theta}{\cos\theta}$ **13.** $\dfrac{1}{\sin\theta \cos\theta}$ **15.** 2 **17.** $\dfrac{3\sin\theta + 1}{\sin\theta + 1}$

19. $\csc\theta \cdot \cos\theta = \dfrac{1}{\sin\theta} \cdot \cos\theta = \dfrac{\cos\theta}{\sin\theta} = \cot\theta$ **21.** $1 + \tan^2(-\theta) = 1 + (-\tan\theta)^2 = 1 + \tan^2\theta = \sec^2\theta$

23. $\cos\theta(\tan\theta + \cot\theta) = \cos\theta\left(\dfrac{\sin\theta}{\cos\theta} + \dfrac{\cos\theta}{\sin\theta} \right) = \cos\theta\left(\dfrac{\sin^2\theta + \cos^2\theta}{\cos\theta \sin\theta} \right) = \cos\theta\left(\dfrac{1}{\cos\theta \sin\theta} \right) = \dfrac{1}{\sin\theta} = \csc\theta$

25. $\tan u \, \cot u - \cos^2 u = \tan u \cdot \dfrac{1}{\tan u} - \cos^2 u = 1 - \cos^2 u = \sin^2 u$ **27.** $(\sec\theta - 1)(\sec\theta + 1) = \sec^2\theta - 1 = \tan^2\theta$

29. $(\sec\theta + \tan\theta)(\sec\theta - \tan\theta) = \sec^2\theta - \tan^2\theta = 1$ **31.** $\cos^2\theta(1 + \tan^2\theta) = \cos^2\theta \sec^2\theta = \cos^2\theta \cdot \dfrac{1}{\cos^2\theta} = 1$

33. $(\sin\theta + \cos\theta)^2 + (\sin\theta - \cos\theta)^2 = \sin^2\theta + 2\sin\theta\cos\theta + \cos^2\theta + \sin^2\theta - 2\sin\theta\cos\theta + \cos^2\theta$
$\qquad = \sin^2\theta + \cos^2\theta + \sin^2\theta + \cos^2\theta = 1 + 1 = 2$

35. $\sec^4\theta - \sec^2\theta = \sec^2\theta(\sec^2\theta - 1) = (1 + \tan^2\theta)\tan^2\theta = \tan^4\theta + \tan^2\theta$

37. $\sec u - \tan u = \dfrac{1}{\cos u} - \dfrac{\sin u}{\cos u} = \dfrac{1 - \sin u}{\cos u} \cdot \dfrac{1 + \sin u}{1 + \sin u} = \dfrac{1 - \sin^2 u}{\cos u(1 + \sin u)} = \dfrac{\cos^2 u}{\cos u(1 + \sin u)} = \dfrac{\cos u}{1 + \sin u}$

39. $3\sin^2\theta + 4\cos^2\theta = 3\sin^2\theta + 3\cos^2\theta + \cos^2\theta = 3(\sin^2\theta + \cos^2\theta) + \cos^2\theta = 3 + \cos^2\theta$

41. $1 - \dfrac{\cos^2\theta}{1 + \sin\theta} = 1 - \dfrac{1 - \sin^2\theta}{1 + \sin\theta} = 1 - \dfrac{(1 + \sin\theta)(1 - \sin\theta)}{1 + \sin\theta} = 1 - (1 - \sin\theta) = \sin\theta$

43. $\dfrac{1 + \tan v}{1 - \tan v} = \dfrac{1 + \dfrac{1}{\cot v}}{1 - \dfrac{1}{\cot v}} = \dfrac{\dfrac{\cot v + 1}{\cot v}}{\dfrac{\cot v - 1}{\cot v}} = \dfrac{\cot v + 1}{\cot v - 1}$ **45.** $\dfrac{\sec\theta}{\csc\theta} + \dfrac{\sin\theta}{\cos\theta} = \dfrac{\dfrac{1}{\cos\theta}}{\dfrac{1}{\sin\theta}} + \tan\theta = \dfrac{\sin\theta}{\cos\theta} + \tan\theta = \tan\theta + \tan\theta = 2\tan\theta$

47. $\dfrac{1 + \sin\theta}{1 - \sin\theta} = \dfrac{1 + \dfrac{1}{\csc\theta}}{1 - \dfrac{1}{\csc\theta}} = \dfrac{\dfrac{\csc\theta + 1}{\csc\theta}}{\dfrac{\csc\theta - 1}{\csc\theta}} = \dfrac{\csc\theta + 1}{\csc\theta - 1}$

49. $\dfrac{1 - \sin v}{\cos v} + \dfrac{\cos v}{1 - \sin v} = \dfrac{(1 - \sin v)^2 + \cos^2 v}{\cos v(1 - \sin v)} = \dfrac{1 - 2\sin v + \sin^2 v + \cos^2 v}{\cos v(1 - \sin v)} = \dfrac{2 - 2\sin v}{\cos v(1 - \sin v)} = \dfrac{2(1 - \sin v)}{\cos v(1 - \sin v)} = \dfrac{2}{\cos v} = 2\sec v$

51. $\dfrac{\sin\theta}{\sin\theta - \cos\theta} = \dfrac{1}{\dfrac{\sin\theta - \cos\theta}{\sin\theta}} = \dfrac{1}{1 - \dfrac{\cos\theta}{\sin\theta}} = \dfrac{1}{1 - \cot\theta}$

53. $(\sec\theta - \tan\theta)^2 = \sec^2\theta - 2\sec\theta\tan\theta + \tan^2\theta = \dfrac{1}{\cos^2\theta} - \dfrac{2\sin\theta}{\cos^2\theta} + \dfrac{\sin^2\theta}{\cos^2\theta} = \dfrac{1 - 2\sin\theta + \sin^2\theta}{\cos^2\theta} = \dfrac{(1-\sin\theta)^2}{1-\sin^2\theta} = \dfrac{(1-\sin\theta)^2}{(1-\sin\theta)(1+\sin\theta)}$

$= \dfrac{1-\sin\theta}{1+\sin\theta}$

55. $\dfrac{\cos\theta}{1-\tan\theta} + \dfrac{\sin\theta}{1-\cot\theta} = \dfrac{\cos\theta}{1 - \dfrac{\sin\theta}{\cos\theta}} + \dfrac{\sin\theta}{1 - \dfrac{\cos\theta}{\sin\theta}} = \dfrac{\cos\theta}{\dfrac{\cos\theta - \sin\theta}{\cos\theta}} + \dfrac{\sin\theta}{\dfrac{\sin\theta - \cos\theta}{\sin\theta}} = \dfrac{\cos^2\theta}{\cos\theta - \sin\theta} + \dfrac{\sin^2\theta}{\sin\theta - \cos\theta}$

$= \dfrac{\cos^2\theta - \sin^2\theta}{\cos\theta - \sin\theta} = \dfrac{(\cos\theta - \sin\theta)(\cos\theta + \sin\theta)}{\cos\theta - \sin\theta} = \sin\theta + \cos\theta$

57. $\tan\theta + \dfrac{\cos\theta}{1+\sin\theta} = \dfrac{\sin\theta}{\cos\theta} + \dfrac{\cos\theta}{1+\sin\theta} = \dfrac{\sin\theta(1+\sin\theta) + \cos^2\theta}{\cos\theta(1+\sin\theta)} = \dfrac{\sin\theta + \sin^2\theta + \cos^2\theta}{\cos\theta(1+\sin\theta)} = \dfrac{\sin\theta + 1}{\cos\theta(1+\sin\theta)} = \dfrac{1}{\cos\theta} = \sec\theta$

59. $\dfrac{\tan\theta + \sec\theta - 1}{\tan\theta - \sec\theta + 1} = \dfrac{\tan\theta + (\sec\theta - 1)}{\tan\theta - (\sec\theta - 1)} \cdot \dfrac{\tan\theta + (\sec\theta - 1)}{\tan\theta + (\sec\theta - 1)} = \dfrac{\tan^2\theta + 2\tan\theta(\sec\theta - 1) + \sec^2\theta - 2\sec\theta + 1}{\tan^2\theta - (\sec^2\theta - 2\sec\theta + 1)}$

$= \dfrac{\sec^2\theta - 1 + 2\tan\theta(\sec\theta - 1) + \sec^2\theta - 2\sec\theta + 1}{\sec^2\theta - 1 - \sec^2\theta + 2\sec\theta - 1} = \dfrac{2\sec^2\theta - 2\sec\theta + 2\tan\theta(\sec\theta - 1)}{-2 + 2\sec\theta}$

$= \dfrac{2\sec\theta(\sec\theta - 1) + 2\tan\theta(\sec\theta - 1)}{2(\sec\theta - 1)} = \dfrac{2(\sec\theta - 1)(\sec\theta + \tan\theta)}{2(\sec\theta - 1)} = \tan\theta + \sec\theta$

61. $\dfrac{\tan\theta - \cot\theta}{\tan\theta + \cot\theta} = \dfrac{\dfrac{\sin\theta}{\cos\theta} - \dfrac{\cos\theta}{\sin\theta}}{\dfrac{\sin\theta}{\cos\theta} + \dfrac{\cos\theta}{\sin\theta}} = \dfrac{\dfrac{\sin^2\theta - \cos^2\theta}{\cos\theta\sin\theta}}{\dfrac{\sin^2\theta + \cos^2\theta}{\cos\theta\sin\theta}} = \dfrac{\sin^2\theta - \cos^2\theta}{1} = \sin^2\theta - \cos^2\theta$

63. $\dfrac{\tan u - \cot u}{\tan u + \cot u} + 1 = \dfrac{\dfrac{\sin u}{\cos u} - \dfrac{\cos u}{\sin u}}{\dfrac{\sin u}{\cos u} + \dfrac{\cos u}{\sin u}} + 1 = \dfrac{\dfrac{\sin^2 u - \cos^2 u}{\cos u \sin u}}{\dfrac{\sin^2 u + \cos^2 u}{\cos u \sin u}} + 1 = \sin^2 u - \cos^2 u + 1 = \sin^2 u + (1 - \cos^2 u) = 2\sin^2 u$

65. $\dfrac{\sec\theta + \tan\theta}{\cot\theta + \cos\theta} = \dfrac{\dfrac{1}{\cos\theta} + \dfrac{\sin\theta}{\cos\theta}}{\dfrac{\cos\theta}{\sin\theta} + \cos\theta} = \dfrac{\dfrac{1+\sin\theta}{\cos\theta}}{\dfrac{\cos\theta + \cos\theta\sin\theta}{\sin\theta}} = \dfrac{1+\sin\theta}{\cos\theta} \cdot \dfrac{\sin\theta}{\cos\theta(1+\sin\theta)} = \dfrac{\sin\theta}{\cos\theta} \cdot \dfrac{1}{\cos\theta} = \tan\theta\sec\theta$

67. $\dfrac{1-\tan^2\theta}{1+\tan^2\theta} + 1 = \dfrac{1-\tan^2\theta + 1 + \tan^2\theta}{1+\tan^2\theta} = \dfrac{2}{1+\tan^2\theta} = \dfrac{2}{\sec^2\theta} = 2\cos^2\theta$

69. $\dfrac{\sec\theta - \csc\theta}{\sec\theta\csc\theta} = \dfrac{\sec\theta}{\sec\theta\csc\theta} - \dfrac{\csc\theta}{\sec\theta\csc\theta} = \dfrac{1}{\csc\theta} - \dfrac{1}{\sec\theta} = \sin\theta - \cos\theta$

71. $\sec\theta - \cos\theta = \dfrac{1}{\cos\theta} - \cos\theta = \dfrac{1-\cos^2\theta}{\cos\theta} = \dfrac{\sin^2\theta}{\cos\theta} = \sin\theta \cdot \dfrac{\sin\theta}{\cos\theta} = \sin\theta\tan\theta$

73. $\dfrac{1}{1-\sin\theta} + \dfrac{1}{1+\sin\theta} = \dfrac{1+\sin\theta + 1 - \sin\theta}{(1+\sin\theta)(1-\sin\theta)} = \dfrac{2}{1-\sin^2\theta} = \dfrac{2}{\cos^2\theta} = 2\sec^2\theta$

75. $\dfrac{\sec\theta}{1-\sin\theta} = \dfrac{\sec\theta}{1-\sin\theta} \cdot \dfrac{1+\sin\theta}{1+\sin\theta} = \dfrac{\sec\theta(1+\sin\theta)}{1-\sin^2\theta} = \dfrac{\sec\theta(1+\sin\theta)}{\cos^2\theta} = \dfrac{1+\sin\theta}{\cos^3\theta}$

77. $\dfrac{(\sec v - \tan v)^2 + 1}{\csc v(\sec v - \tan v)} = \dfrac{\sec^2 v - 2\sec v\tan v + \tan^2 v + 1}{\dfrac{1}{\sin v}\left(\dfrac{1}{\cos v} - \dfrac{\sin v}{\cos v}\right)} = \dfrac{2\sec^2 v - 2\sec v\tan v}{\dfrac{1}{\sin v}\left(\dfrac{1-\sin v}{\cos v}\right)} = \dfrac{\dfrac{2}{\cos^2 v} - \dfrac{2\sin v}{\cos^2 v}}{\dfrac{1-\sin v}{\sin v\cos v}} = \dfrac{2 - 2\sin v}{\cos^2 v} \cdot \dfrac{\sin v\cos v}{1-\sin v}$

$= \dfrac{2(1-\sin v)}{\cos v} \cdot \dfrac{\sin v}{1-\sin v} = \dfrac{2\sin v}{\cos v} = 2\tan v$

79. $\dfrac{\sin\theta + \cos\theta}{\cos\theta} - \dfrac{\sin\theta - \cos\theta}{\sin\theta} = \dfrac{\sin\theta}{\cos\theta} + 1 - 1 + \dfrac{\cos\theta}{\sin\theta} = \dfrac{\sin^2\theta + \cos^2\theta}{\cos\theta\sin\theta} = \dfrac{1}{\cos\theta\sin\theta} = \sec\theta\csc\theta$

81. $\dfrac{\sin^3\theta + \cos^3\theta}{\sin\theta + \cos\theta} = \dfrac{(\sin\theta + \cos\theta)(\sin^2\theta - \sin\theta\cos\theta + \cos^2\theta)}{\sin\theta + \cos\theta} = \sin^2\theta + \cos^2\theta - \sin\theta\cos\theta = 1 - \sin\theta\cos\theta$

83. $\dfrac{\cos^2\theta - \sin^2\theta}{1 - \tan^2\theta} = \dfrac{\cos^2\theta - \sin^2\theta}{1 - \dfrac{\sin^2\theta}{\cos^2\theta}} = \dfrac{\cos^2\theta - \sin^2\theta}{\dfrac{\cos^2\theta - \sin^2\theta}{\cos^2\theta}} = \cos^2\theta$

85. $\dfrac{(2\cos^2\theta - 1)^2}{\cos^4\theta - \sin^4\theta} = \dfrac{[2\cos^2\theta - (\sin^2\theta + \cos^2\theta)]^2}{(\cos^2\theta - \sin^2\theta)(\cos^2\theta + \sin^2\theta)} = \dfrac{(\cos^2\theta - \sin^2\theta)^2}{\cos^2\theta - \sin^2\theta} = \cos^2\theta - \sin^2\theta = (1 - \sin^2\theta) - \sin^2\theta = 1 - 2\sin^2\theta$

87. $\dfrac{1 + \sin\theta + \cos\theta}{1 + \sin\theta - \cos\theta} = \dfrac{(1+\sin\theta) + \cos\theta}{(1+\sin\theta) - \cos\theta} \cdot \dfrac{(1+\sin\theta) + \cos\theta}{(1+\sin\theta) + \cos\theta} = \dfrac{1 + 2\sin\theta + \sin^2\theta + 2(1+\sin\theta)\cos\theta + \cos^2\theta}{1 + 2\sin\theta + \sin^2\theta - \cos^2\theta}$

$= \dfrac{1 + 2\sin\theta + \sin^2\theta + 2(1+\sin\theta)(\cos\theta) + (1 - \sin^2\theta)}{1 + 2\sin\theta + \sin^2\theta - (1 - \sin^2\theta)} = \dfrac{2 + 2\sin\theta + 2(1+\sin\theta)(\cos\theta)}{2\sin\theta + 2\sin^2\theta}$

$= \dfrac{2(1+\sin\theta) + 2(1+\sin\theta)(\cos\theta)}{2\sin\theta(1+\sin\theta)} = \dfrac{2(1+\sin\theta)(1+\cos\theta)}{2\sin\theta(1+\sin\theta)} = \dfrac{1+\cos\theta}{\sin\theta}$

89. $(a \sin \theta + b \cos \theta)^2 + (a \cos \theta - b \sin \theta)^2 = a^2 \sin^2 \theta + 2ab \sin \theta \cos \theta + b^2 \cos^2 \theta + a^2 \cos^2 \theta - 2ab \sin \theta \cos \theta + b^2 \sin^2 \theta$

$= a^2(\sin^2 \theta + \cos^2 \theta) + b^2(\cos^2 \theta + \sin^2 \theta) = a^2 + b^2$

91. $\dfrac{\tan \alpha + \tan \beta}{\cot \alpha + \cot \beta} = \dfrac{\tan \alpha + \tan \beta}{\dfrac{1}{\tan \alpha} + \dfrac{1}{\tan \beta}} = \dfrac{\tan \alpha + \tan \beta}{\dfrac{\tan \beta + \tan \alpha}{\tan \alpha \tan \beta}} = (\tan \alpha + \tan \beta) \cdot \dfrac{\tan \alpha \tan \beta}{\tan \alpha + \tan \beta} = \tan \alpha \tan \beta$

93. $(\sin \alpha + \cos \beta)^2 + (\cos \beta + \sin \alpha)(\cos \beta - \sin \alpha) = (\sin^2 \alpha + 2 \sin \alpha \cos \beta + \cos^2 \beta) + (\cos^2 \beta - \sin^2 \alpha)$

$= 2 \cos^2 \beta + 2 \sin \alpha \cos \beta = 2 \cos \beta (\cos \beta + \sin \alpha) = 2 \cos \beta (\sin \alpha + \cos \beta)$

95. $\ln|\sec \theta| = \ln|\cos \theta|^{-1} = -\ln|\cos \theta|$

97. $\ln|1 + \cos \theta| + \ln|1 - \cos \theta| = \ln(|1 + \cos \theta||1 - \cos \theta|) = \ln|1 - \cos^2 \theta| = \ln|\sin^2 \theta| = 2 \ln|\sin \theta|$

99. $g(x) = \sec x - \cos x = \dfrac{1}{\cos x} - \cos x = \dfrac{1}{\cos x} - \dfrac{\cos^2 x}{\cos x} = \dfrac{1 - \cos^2 x}{\cos x} = \dfrac{\sin^2 x}{\cos x} = \sin x \cdot \dfrac{\sin x}{\cos x} = \sin x \cdot \tan x = f(x)$

101. $f(\theta) = \dfrac{1 - \sin \theta}{\cos \theta} - \dfrac{\cos \theta}{1 + \sin \theta} = \dfrac{1 - \sin \theta}{\cos \theta} \cdot \dfrac{1 + \sin \theta}{1 + \sin \theta} - \dfrac{\cos \theta}{1 + \sin \theta} \cdot \dfrac{\cos \theta}{\cos \theta} = \dfrac{1 - \sin^2 \theta}{\cos \theta(1 + \sin \theta)} - \dfrac{\cos^2 \theta}{\cos \theta(1 + \sin \theta)}$

$= \dfrac{\cos^2 \theta}{\cos \theta(1 + \sin \theta)} - \dfrac{\cos^2 \theta}{\cos \theta(1 + \sin \theta)} = 0 = g(\theta)$

103. $1200 \sec \theta (2 \sec^2 \theta - 1) = 1200 \dfrac{1}{\cos \theta}\left(\dfrac{2}{\cos^2 \theta} - 1\right) = 1200 \dfrac{1}{\cos \theta}\left(\dfrac{2}{\cos^2 \theta} - \dfrac{\cos^2 \theta}{\cos^2 \theta}\right) = 1200 \dfrac{1}{\cos \theta}\left(\dfrac{2 - \cos^2 \theta}{\cos^2 \theta}\right) = \dfrac{1200 \, (1 + 1 - \cos^2 \theta)}{\cos^3 \theta}$

$= \dfrac{1200 \, (1 + \sin^2 \theta)}{\cos^3 \theta}$

6.5 Assess Your Understanding (page 523)

5. – **6.** – **7.** F **8.** F **9.** F **10.** T **11.** $\dfrac{1}{4}(\sqrt{6} + \sqrt{2})$ **13.** $\dfrac{1}{4}(\sqrt{2} - \sqrt{6})$ **15.** $-\dfrac{1}{4}(\sqrt{2} + \sqrt{6})$ **17.** $2 - \sqrt{3}$ **19.** $-\dfrac{1}{4}(\sqrt{6} + \sqrt{2})$

21. $\sqrt{6} - \sqrt{2}$ **23.** $\dfrac{1}{2}$ **25.** 0 **27.** 1 **29.** -1 **31.** $\dfrac{1}{2}$ **33. (a)** $\dfrac{2\sqrt{5}}{25}$ **(b)** $\dfrac{11\sqrt{5}}{25}$ **(c)** $\dfrac{2\sqrt{5}}{5}$ **(d)** 2

35. (a) $\dfrac{4 - 3\sqrt{3}}{10}$ **(b)** $\dfrac{-3 - 4\sqrt{3}}{10}$ **(c)** $\dfrac{4 + 3\sqrt{3}}{10}$ **(d)** $\dfrac{25\sqrt{3} + 48}{39}$ **37. (a)** $-\dfrac{5 + 12\sqrt{3}}{26}$ **(b)** $\dfrac{12 - 5\sqrt{3}}{26}$ **(c)** $\dfrac{-5 + 12\sqrt{3}}{26}$ **(d)** $\dfrac{-240 + 169\sqrt{3}}{69}$

39. (a) $-\dfrac{2\sqrt{2}}{3}$ **(b)** $\dfrac{-2\sqrt{2} + \sqrt{3}}{6}$ **(c)** $\dfrac{-2\sqrt{2} + \sqrt{3}}{6}$ **(d)** $\dfrac{9 - 4\sqrt{2}}{7}$ **41.** $\dfrac{1 - 2\sqrt{6}}{6}$ **43.** $\dfrac{\sqrt{3} - 2\sqrt{2}}{6}$ **45.** $\dfrac{8\sqrt{2} - 9\sqrt{3}}{5}$

47. $\sin\left(\dfrac{\pi}{2} + \theta\right) = \sin\dfrac{\pi}{2}\cos \theta + \cos\dfrac{\pi}{2}\sin \theta = 1 \cdot \cos \theta + 0 \cdot \sin \theta = \cos \theta$

49. $\sin(\pi - \theta) = \sin \pi \cos \theta - \cos \pi \sin \theta = 0 \cdot \cos \theta - (-1) \sin \theta = \sin \theta$

51. $\sin(\pi + \theta) = \sin \pi \cos \theta + \cos \pi \sin \theta = 0 \cdot \cos \theta + (-1) \sin \theta = -\sin \theta$

53. $\tan(\pi - \theta) = \dfrac{\tan \pi - \tan \theta}{1 + \tan \pi \tan \theta} = \dfrac{0 - \tan \theta}{1 + 0 \cdot \tan \theta} = -\tan \theta$ **55.** $\sin\left(\dfrac{3\pi}{2} + \theta\right) = \sin\dfrac{3\pi}{2}\cos \theta + \cos\dfrac{3\pi}{2}\sin \theta = (-1)\cos \theta + 0 \cdot \sin \theta = -\cos \theta$

57. $\sin(\alpha + \beta) + \sin(\alpha - \beta) = \sin \alpha \cos \beta + \cos \alpha \sin \beta + \sin \alpha \cos \beta - \cos \alpha \sin \beta = 2 \sin \alpha \cos \beta$

59. $\dfrac{\sin(\alpha + \beta)}{\sin \alpha \cos \beta} = \dfrac{\sin \alpha \cos \beta + \cos \alpha \sin \beta}{\sin \alpha \cos \beta} = \dfrac{\sin \alpha \cos \beta}{\sin \alpha \cos \beta} + \dfrac{\cos \alpha \sin \beta}{\sin \alpha \cos \beta} = 1 + \cot \alpha \tan \beta$

61. $\dfrac{\cos(\alpha + \beta)}{\cos \alpha \cos \beta} = \dfrac{\cos \alpha \cos \beta - \sin \alpha \sin \beta}{\cos \alpha \cos \beta} = \dfrac{\cos \alpha \cos \beta}{\cos \alpha \cos \beta} - \dfrac{\sin \alpha \sin \beta}{\cos \alpha \cos \beta} = 1 - \tan \alpha \tan \beta$

63. $\dfrac{\sin(\alpha + \beta)}{\sin(\alpha - \beta)} = \dfrac{\sin \alpha \cos \beta + \cos \alpha \sin \beta}{\sin \alpha \cos \beta - \cos \alpha \sin \beta} = \dfrac{\dfrac{\sin \alpha \cos \beta + \cos \alpha \sin \beta}{\cos \alpha \cos \beta}}{\dfrac{\sin \alpha \cos \beta - \cos \alpha \sin \beta}{\cos \alpha \cos \beta}} = \dfrac{\dfrac{\sin \alpha \cos \beta}{\cos \alpha \cos \beta} + \dfrac{\cos \alpha \sin \beta}{\cos \alpha \cos \beta}}{\dfrac{\sin \alpha \cos \beta}{\cos \alpha \cos \beta} - \dfrac{\cos \alpha \sin \beta}{\cos \alpha \cos \beta}} = \dfrac{\tan \alpha + \tan \beta}{\tan \alpha - \tan \beta}$

65. $\cot(\alpha + \beta) = \dfrac{\cos(\alpha + \beta)}{\sin(\alpha + \beta)} = \dfrac{\cos \alpha \cos \beta - \sin \alpha \sin \beta}{\sin \alpha \cos \beta + \cos \alpha \sin \beta} = \dfrac{\dfrac{\cos \alpha \cos \beta - \sin \alpha \sin \beta}{\sin \alpha \sin \beta}}{\dfrac{\sin \alpha \cos \beta + \cos \alpha \sin \beta}{\sin \alpha \sin \beta}} = \dfrac{\dfrac{\cos \alpha \cos \beta}{\sin \alpha \sin \beta} - \dfrac{\sin \alpha \sin \beta}{\sin \alpha \sin \beta}}{\dfrac{\sin \alpha \cos \beta}{\sin \alpha \sin \beta} + \dfrac{\cos \alpha \sin \beta}{\sin \alpha \sin \beta}} = \dfrac{\cot \alpha \cot \beta - 1}{\cot \beta + \cot \alpha}$

67. $\sec(\alpha + \beta) = \dfrac{1}{\cos(\alpha + \beta)} = \dfrac{1}{\cos \alpha \cos \beta - \sin \alpha \sin \beta} = \dfrac{\dfrac{1}{\sin \alpha \sin \beta}}{\dfrac{\cos \alpha \cos \beta - \sin \alpha \sin \beta}{\sin \alpha \sin \beta}} = \dfrac{\dfrac{1}{\sin \alpha} \cdot \dfrac{1}{\sin \beta}}{\dfrac{\cos \alpha \cos \beta}{\sin \alpha \sin \beta} - \dfrac{\sin \alpha \sin \beta}{\sin \alpha \sin \beta}} = \dfrac{\csc \alpha \csc \beta}{\cot \alpha \cot \beta - 1}$

69. $\sin(\alpha - \beta) \sin(\alpha + \beta) = (\sin \alpha \cos \beta - \cos \alpha \sin \beta)(\sin \alpha \cos \beta + \cos \alpha \sin \beta) = \sin^2 \alpha \cos^2 \beta - \cos^2 \alpha \sin^2 \beta$

$= (\sin^2 \alpha)(1 - \sin^2 \beta) - (1 - \sin^2 \alpha)(\sin^2 \beta) = \sin^2 \alpha - \sin^2 \beta$

71. $\sin(\theta + k\pi) = \sin \theta \cos k\pi + \cos \theta \sin k\pi = (\sin \theta)(-1)^k + (\cos \theta)(0) = (-1)^k \sin \theta$, k any integer

73. $\dfrac{\sqrt{3}}{2}$ **75.** $-\dfrac{24}{25}$ **77.** $-\dfrac{33}{65}$ **79.** $\dfrac{63}{65}$ **81.** $\dfrac{48 + 25\sqrt{3}}{39}$ **83.** $\dfrac{4}{3}$ **85.** $u\sqrt{1 - v^2} - v\sqrt{1 - u^2}$; $-1 \le u \le 1$; $-1 \le v \le 1$

87. $\dfrac{u\sqrt{1 - v^2} - v}{\sqrt{1 + u^2}}$: $-\infty < u < \infty$; $-1 \le v \le 1$ **89.** $\dfrac{uv - \sqrt{1 - u^2}\sqrt{1 - v^2}}{v\sqrt{1 - u^2} + u\sqrt{1 - v^2}}$: $-1 \le u \le 1$; $-1 \le v \le 1$ **91.** $\left\{\dfrac{\pi}{2}, \dfrac{7\pi}{6}\right\}$ **93.** $\left\{\dfrac{\pi}{4}\right\}$ **95.** $\left\{\dfrac{11\pi}{6}\right\}$

97. Let $\alpha = \sin^{-1} v$ and $\beta = \cos^{-1} v$. Then $\sin \alpha = \cos \beta = v$, and since $\sin \alpha = \cos\left(\dfrac{\pi}{2} - \alpha\right)$, $\cos\left(\dfrac{\pi}{2} - \alpha\right) = \cos \beta$.

If $v \geq 0$, then $0 \leq \alpha \leq \dfrac{\pi}{2}$, so $\left(\dfrac{\pi}{2} - \alpha\right)$ and β both lie on $\left[0, \dfrac{\pi}{2}\right]$. If $v < 0$, then $-\dfrac{\pi}{2} \leq \alpha < 0$, so $\left(\dfrac{\pi}{2} - \alpha\right)$ and β both lie on $\left(\dfrac{\pi}{2}, \pi\right]$.

Either way, $\cos\left(\dfrac{\pi}{2} - \alpha\right) = \cos \beta$ implies $\dfrac{\pi}{2} - \alpha = \beta$, or $\alpha + \beta = \dfrac{\pi}{2}$.

99. Let $\alpha = \tan^{-1}\dfrac{1}{v}$ and $\beta = \tan^{-1} v$. Because $v \neq 0$, $\alpha, \beta \neq 0$. Then $\tan \alpha = \dfrac{1}{v} = \dfrac{1}{\tan \beta} = \cot \beta$, and since

$\tan \alpha = \cot\left(\dfrac{\pi}{2} - \alpha\right)$, $\cot\left(\dfrac{\pi}{2} - \alpha\right) = \cot \beta$. Because $v > 0$, $0 < \alpha < \dfrac{\pi}{2}$, and so $\left(\dfrac{\pi}{2} - \alpha\right)$ and β both lie on $\left(0, \dfrac{\pi}{2}\right)$.

Then $\cot\left(\dfrac{\pi}{2} - \alpha\right) = \cot \beta$ implies $\dfrac{\pi}{2} - \alpha = \beta$, or $\alpha = \dfrac{\pi}{2} - \beta$.

101. $\sin(\sin^{-1} v + \cos^{-1} v) = \sin(\sin^{-1} v)\cos(\cos^{-1} v) + \cos(\sin^{-1} v)\sin(\cos^{-1} v) = (v)(v) + \sqrt{1 - v^2}\sqrt{1 - v^2} = v^2 + 1 - v^2 = 1$

103. $\dfrac{\sin(x + h) - \sin x}{h} = \dfrac{\sin x \cos h + \cos x \sin h - \sin x}{h} = \dfrac{\cos x \sin h - \sin x(1 - \cos h)}{h} = \cos x \cdot \dfrac{\sin h}{h} - \sin x \cdot \dfrac{1 - \cos h}{h}$

105. (a) $\tan(\tan^{-1} 1 + \tan^{-1} 2 + \tan^{-1} 3) = \tan((\tan^{-1} 1 + \tan^{-1} 2) + \tan^{-1} 3) = \dfrac{\tan(\tan^{-1} 1 + \tan^{-1} 2) + \tan(\tan^{-1} 3)}{1 - \tan(\tan^{-1} 1 + \tan^{-1} 2)\tan(\tan^{-1} 3)}$

$= \dfrac{\dfrac{\tan(\tan^{-1} 1) + \tan(\tan^{-1} 2)}{1 - \tan(\tan^{-1} 1)\tan(\tan^{-1} 2)} + 3}{1 - \dfrac{\tan(\tan^{-1} 1) + \tan(\tan^{-1} 2)}{1 - \tan(\tan^{-1} 1)\tan(\tan^{-1} 2)} \cdot 3} = \dfrac{\dfrac{1 + 2}{1 - 1 \cdot 2} + 3}{1 - \dfrac{1 + 2}{1 - 1 \cdot 2} \cdot 3} = \dfrac{\dfrac{3}{-1} + 3}{1 - \dfrac{3}{-1} \cdot 3} = \dfrac{-3 + 3}{1 + 9} = \dfrac{0}{10} = 0$

(b) From the definition of the inverse tangent function $0 < \tan^{-1} 1 < \dfrac{\pi}{2}$, $0 < \tan^{-1} 2 < \dfrac{\pi}{2}$, and $0 < \tan^{-1} 3 < \dfrac{\pi}{2}$,

so $0 < \tan^{-1} 1 + \tan^{-1} 2 + \tan^{-1} 3 < \dfrac{3\pi}{2}$.

On the interval $\left(0, \dfrac{3\pi}{2}\right)$, $\tan \theta = 0$ if and only if $\theta = \pi$. Therefore, from part (a), $\tan^{-1} 1 + \tan^{-1} 2 + \tan^{-1} 3 = \pi$.

107. $\tan \theta = \tan(\theta_2 - \theta_1) = \dfrac{\tan \theta_2 - \tan \theta_1}{1 + \tan \theta_1 \tan \theta_2} = \dfrac{m_2 - m_1}{1 + m_1 m_2}$

109. $2\cot(\alpha - \beta) = \dfrac{2}{\tan(\alpha - \beta)} = 2\left(\dfrac{1 + \tan \alpha \tan \beta}{\tan \alpha - \tan \beta}\right) = 2\left(\dfrac{1 + (x + 1)(x - 1)}{(x + 1) - (x - 1)}\right) = 2\left(\dfrac{1 + x^2 - 1}{x + 1 - x + 1}\right) = \dfrac{2x^2}{2} = x^2$

111. $\tan \dfrac{\pi}{2}$ is not defined; $\tan\left(\dfrac{\pi}{2} - \theta\right) = \dfrac{\sin\left(\dfrac{\pi}{2} - \theta\right)}{\cos\left(\dfrac{\pi}{2} - \theta\right)} = \dfrac{\cos \theta}{\sin \theta} = \cot \theta$.

6.6 Assess Your Understanding *(page 533)*

1. $\sin^2 \theta$; $2\cos^2 \theta$; $2\sin^2 \theta$ **2.** $1 - \cos \theta$ **3.** $\sin \theta$ **4.** T **5.** F **6.** F **7. (a)** $\dfrac{24}{25}$ **(b)** $\dfrac{7}{25}$ **(c)** $\dfrac{\sqrt{10}}{10}$ **(d)** $\dfrac{3\sqrt{10}}{10}$

9. (a) $\dfrac{24}{25}$ **(b)** $-\dfrac{7}{25}$ **(c)** $\dfrac{2\sqrt{5}}{5}$ **(d)** $-\dfrac{\sqrt{5}}{5}$ **11. (a)** $-\dfrac{2\sqrt{2}}{3}$ **(b)** $\dfrac{1}{3}$ **(c)** $\sqrt{\dfrac{3 + \sqrt{6}}{6}}$ **(d)** $\sqrt{\dfrac{3 - \sqrt{6}}{6}}$

13. (a) $\dfrac{4\sqrt{2}}{9}$ **(b)** $-\dfrac{7}{9}$ **(c)** $\dfrac{\sqrt{3}}{3}$ **(d)** $\dfrac{\sqrt{6}}{3}$ **15. (a)** $-\dfrac{4}{5}$ **(b)** $\dfrac{3}{5}$ **(c)** $\sqrt{\dfrac{5 + 2\sqrt{5}}{10}}$ **(d)** $\sqrt{\dfrac{5 - 2\sqrt{5}}{10}}$

17. (a) $-\dfrac{3}{5}$ **(b)** $-\dfrac{4}{5}$ **(c)** $\dfrac{1}{2}\sqrt{\dfrac{10 - \sqrt{10}}{5}}$ **(d)** $-\dfrac{1}{2}\sqrt{\dfrac{10 + \sqrt{10}}{5}}$ **19.** $\dfrac{\sqrt{2 - \sqrt{2}}}{2}$ **21.** $1 - \sqrt{2}$ **23.** $-\dfrac{\sqrt{2 + \sqrt{3}}}{2}$

25. $\dfrac{2}{\sqrt{2 + \sqrt{2}}} = (2 - \sqrt{2})\sqrt{2 + \sqrt{2}}$ **27.** $-\dfrac{\sqrt{2 - \sqrt{2}}}{2}$ **29.** $-\dfrac{4}{5}$ **31.** $\dfrac{\sqrt{10(5 - \sqrt{5})}}{10}$ **33.** $\dfrac{4}{3}$ **35.** $-\dfrac{7}{8}$ **37.** $\dfrac{\sqrt{10}}{4}$ **39.** $-\dfrac{\sqrt{15}}{3}$

41. $\sin^4 \theta = (\sin^2 \theta)^2 = \left(\dfrac{1 - \cos(2\theta)}{2}\right)^2 = \dfrac{1}{4}[1 - 2\cos(2\theta) + \cos^2(2\theta)] = \dfrac{1}{4} - \dfrac{1}{2}\cos(2\theta) + \dfrac{1}{4}\cos^2(2\theta)$

$= \dfrac{1}{4} - \dfrac{1}{2}\cos(2\theta) + \dfrac{1}{4}\left(\dfrac{1 + \cos(4\theta)}{2}\right) = \dfrac{1}{4} - \dfrac{1}{2}\cos(2\theta) + \dfrac{1}{8} + \dfrac{1}{8}\cos(4\theta) = \dfrac{3}{8} - \dfrac{1}{2}\cos(2\theta) + \dfrac{1}{8}\cos(4\theta)$

43. $\cos(3\theta) = 4\cos^3 \theta - 3\cos \theta$ **45.** $\sin(5\theta) = 16\sin^5 \theta - 20\sin^3 \theta + 5\sin \theta$ **47.** $\cos^4 \theta - \sin^4 \theta = (\cos^2 \theta + \sin^2 \theta)(\cos^2 \theta - \sin^2 \theta) = \cos(2\theta)$

49. $\cot(2\theta) = \dfrac{1}{\tan(2\theta)} = \dfrac{1 - \tan^2 \theta}{2\tan \theta} = \dfrac{1 - \dfrac{1}{\cot^2 \theta}}{2\left(\dfrac{1}{\cot \theta}\right)} = \dfrac{\dfrac{\cot^2 \theta - 1}{\cot^2 \theta}}{\dfrac{2}{\cot \theta}} = \dfrac{\cot^2 \theta - 1}{\cot^2 \theta} \cdot \dfrac{\cot \theta}{2} = \dfrac{\cot^2 \theta - 1}{2\cot \theta}$

51. $\sec(2\theta) = \dfrac{1}{\cos(2\theta)} = \dfrac{1}{2\cos^2 \theta - 1} = \dfrac{1}{\dfrac{2}{\sec^2 \theta} - 1} = \dfrac{1}{\dfrac{2 - \sec^2 \theta}{\sec^2 \theta}} = \dfrac{\sec^2 \theta}{2 - \sec^2 \theta}$ **53.** $\cos^2(2u) - \sin^2(2u) = \cos[2(2u)] = \cos(4u)$

55. $\dfrac{\cos(2\theta)}{1+\sin(2\theta)} = \dfrac{\cos^2\theta - \sin^2\theta}{1 + 2\sin\theta\cos\theta} = \dfrac{(\cos\theta - \sin\theta)(\cos\theta + \sin\theta)}{\sin^2\theta + \cos^2\theta + 2\sin\theta\cos\theta} = \dfrac{(\cos\theta - \sin\theta)(\cos\theta + \sin\theta)}{(\sin\theta + \cos\theta)(\sin\theta + \cos\theta)} = \dfrac{\cos\theta - \sin\theta}{\cos\theta + \sin\theta}$

$= \dfrac{\dfrac{\cos\theta - \sin\theta}{\sin\theta}}{\dfrac{\cos\theta + \sin\theta}{\sin\theta}} = \dfrac{\dfrac{\cos\theta}{\sin\theta} - \dfrac{\sin\theta}{\sin\theta}}{\dfrac{\cos\theta}{\sin\theta} + \dfrac{\sin\theta}{\sin\theta}} = \dfrac{\cot\theta - 1}{\cot\theta + 1}$

57. $\sec^2\dfrac{\theta}{2} = \dfrac{1}{\cos^2\left(\dfrac{\theta}{2}\right)} = \dfrac{1}{\dfrac{1+\cos\theta}{2}} = \dfrac{2}{1+\cos\theta}$

59. $\cot^2\dfrac{v}{2} = \dfrac{1}{\tan^2\left(\dfrac{v}{2}\right)} = \dfrac{1}{\dfrac{1-\cos v}{1+\cos v}} = \dfrac{1+\cos v}{1-\cos v} = \dfrac{1 + \dfrac{1}{\sec v}}{1 - \dfrac{1}{\sec v}} = \dfrac{\dfrac{\sec v + 1}{\sec v}}{\dfrac{\sec v - 1}{\sec v}} = \dfrac{\sec v + 1}{\sec v}\cdot\dfrac{\sec v}{\sec v - 1} = \dfrac{\sec v + 1}{\sec v - 1}$

61. $\dfrac{1 - \tan^2\left(\dfrac{\theta}{2}\right)}{1 + \tan^2\left(\dfrac{\theta}{2}\right)} = \dfrac{1 - \dfrac{1-\cos\theta}{1+\cos\theta}}{1 + \dfrac{1-\cos\theta}{1+\cos\theta}} = \dfrac{\dfrac{1+\cos\theta - (1-\cos\theta)}{1+\cos\theta}}{\dfrac{1+\cos\theta + 1 - \cos\theta}{1+\cos\theta}} = \dfrac{2\cos\theta}{1+\cos\theta}\cdot\dfrac{1+\cos\theta}{2} = \cos\theta$

63. $\dfrac{\sin(3\theta)}{\sin\theta} - \dfrac{\cos(3\theta)}{\cos\theta} = \dfrac{\sin(3\theta)\cos\theta - \cos(3\theta)\sin\theta}{\sin\theta\cos\theta} = \dfrac{\sin(3\theta - \theta)}{\dfrac{1}{2}(2\sin\theta\cos\theta)} = \dfrac{2\sin(2\theta)}{\sin(2\theta)} = 2$

65. $\tan(3\theta) = \tan(\theta + 2\theta) = \dfrac{\tan\theta + \tan(2\theta)}{1 - \tan\theta\tan(2\theta)} = \dfrac{\tan\theta + \dfrac{2\tan\theta}{1-\tan^2\theta}}{1 - \dfrac{\tan\theta(2\tan\theta)}{1-\tan^2\theta}} = \dfrac{\tan\theta - \tan^3\theta + 2\tan\theta}{1 - \tan^2\theta - 2\tan^2\theta} = \dfrac{3\tan\theta - \tan^3\theta}{1 - 3\tan^2\theta}$

67. $\dfrac{1}{2}(\ln|1 - \cos(2\theta)| - \ln 2) = \ln\left(\dfrac{|1 - \cos(2\theta)|}{2}\right)^{1/2} = \ln|\sin^2\theta|^{1/2} = \ln|\sin\theta|$ **69.** $\left\{\dfrac{\pi}{3}, \dfrac{2\pi}{3}, \dfrac{4\pi}{3}, \dfrac{5\pi}{3}\right\}$ **71.** $\left\{0, \dfrac{2\pi}{3}, \dfrac{4\pi}{3}\right\}$

73. $\left\{0, \dfrac{\pi}{3}, \dfrac{\pi}{2}, \dfrac{2\pi}{3}, \pi, \dfrac{4\pi}{3}, \dfrac{3\pi}{2}, \dfrac{5\pi}{3}\right\}$ **75.** No real solution **77.** $\left\{0, \dfrac{\pi}{3}, \pi, \dfrac{5\pi}{3}\right\}$ **79.** $\dfrac{\sqrt{3}}{2}$ **81.** $\dfrac{7}{25}$ **83.** $\dfrac{24}{7}$ **85.** $\dfrac{24}{25}$ **87.** $\dfrac{1}{5}$ **89.** $\dfrac{25}{7}$ **91.** $0, \dfrac{\pi}{3}, \pi, \dfrac{5\pi}{3}$

93. $\dfrac{\pi}{2}, \dfrac{3\pi}{2}$ **95. (a)** $W = 2D(\csc\theta - \cot\theta) = 2D\left(\dfrac{1}{\sin\theta} - \dfrac{\cos\theta}{\sin\theta}\right) = 2D\dfrac{1-\cos\theta}{\sin\theta} = 2D\tan\dfrac{\theta}{2}$ **(b)** $\theta = 24.45°$

97. (a) $R = \dfrac{v_0^2\sqrt{2}}{16}\cos\theta(\sin\theta - \cos\theta)$ **(b)** $\dfrac{3\pi}{8}$ or $67.5°$ **(d)**

$= \dfrac{v_0^2\sqrt{2}}{32}(2\cos\theta\sin\theta - 2\cos^2\theta)$ **(c)** $32(2-\sqrt{2}) \approx 18.75$ ft

$= \dfrac{v_0^2\sqrt{2}}{32}[\sin(2\theta) - \cos(2\theta) - 1]$

$\theta = 67.5°$

$\left(\dfrac{3\pi}{8}\text{ radians}\right)$ makes R largest. $R = 18.75$ ft

99. $A = \dfrac{1}{2}h(\text{base}) = h\left(\dfrac{1}{2}\text{base}\right) = s\cos\dfrac{\theta}{2}\cdot s\sin\dfrac{\theta}{2} = \dfrac{1}{2}s^2\sin\theta$ **101.** $\sin(2\theta) = \dfrac{4x}{4+x^2}$ **103.** $-\dfrac{1}{4}$

105. $\dfrac{2z}{1+z^2} = \dfrac{2\tan\left(\dfrac{\alpha}{2}\right)}{1+\tan^2\left(\dfrac{\alpha}{2}\right)} = \dfrac{2\tan\left(\dfrac{\alpha}{2}\right)}{\sec^2\left(\dfrac{\alpha}{2}\right)} = \dfrac{\dfrac{2\sin\left(\dfrac{\alpha}{2}\right)}{\cos\left(\dfrac{\alpha}{2}\right)}}{\dfrac{1}{\cos^2\left(\dfrac{\alpha}{2}\right)}} = 2\sin\left(\dfrac{\alpha}{2}\right)\cos\left(\dfrac{\alpha}{2}\right) = \sin\left(2\cdot\dfrac{\alpha}{2}\right) = \sin\alpha$ **107.**

109. $\sin\dfrac{\pi}{24} = \dfrac{\sqrt{2}}{4}\sqrt{4 - \sqrt{6} - \sqrt{2}}$

$\cos\dfrac{\pi}{24} = \dfrac{\sqrt{2}}{4}\sqrt{4 + \sqrt{6} + \sqrt{2}}$

111. $\sin^3\theta + \sin^3(\theta + 120°) + \sin^3(\theta + 240°) = \sin^3\theta + (\sin\theta\cos 120° + \cos\theta\sin 120°)^3 + (\sin\theta\cos 240° + \cos\theta\sin 240°)^3$

$= \sin^3\theta + \left(-\dfrac{1}{2}\sin\theta + \dfrac{\sqrt{3}}{2}\cos\theta\right)^3 + \left(-\dfrac{1}{2}\sin\theta - \dfrac{\sqrt{3}}{2}\cos\theta\right)^3$

$= \sin^3\theta + \dfrac{1}{8}\left(3\sqrt{3}\cos^3\theta - 9\cos^2\theta\sin\theta + 3\sqrt{3}\cos\theta\sin^2\theta - \sin^3\theta\right) - \dfrac{1}{8}\left(\sin^3\theta + 3\sqrt{3}\sin^2\theta\cos\theta + 9\sin\theta\cos^2\theta + 3\sqrt{3}\cos^3\theta\right)$

$= \dfrac{3}{4}\sin^3\theta - \dfrac{9}{4}\cos^2\theta\sin\theta = \dfrac{3}{4}[\sin^3\theta - 3\sin\theta(1 - \sin^2\theta)] = \dfrac{3}{4}(4\sin^3\theta - 3\sin\theta) = -\dfrac{3}{4}\sin(3\theta)$ (from Example 2)

6.7 Assess Your Understanding *(page 538)*

1. $\frac{1}{2}[\cos(2\theta) - \cos(6\theta)]$ **3.** $\frac{1}{2}[\sin(6\theta) + \sin(2\theta)]$ **5.** $\frac{1}{2}[\cos(2\theta) + \cos(8\theta)]$ **7.** $\frac{1}{2}[\cos\theta - \cos(3\theta)]$ **9.** $\frac{1}{2}[\sin(2\theta) + \sin\theta]$ **11.** $2\sin\theta\cos(3\theta)$

13. $2\cos(3\theta)\cos\theta$ **15.** $2\sin(2\theta)\cos\theta$ **17.** $2\sin\theta\sin\frac{\theta}{2}$ **19.** $\dfrac{\sin\theta + \sin(3\theta)}{2\sin(2\theta)} = \dfrac{2\sin(2\theta)\cos\theta}{2\sin(2\theta)} = \cos\theta$

21. $\dfrac{\sin(4\theta) + \sin(2\theta)}{\cos(4\theta) + \cos(2\theta)} = \dfrac{2\sin(3\theta)\cos\theta}{2\cos(3\theta)\cos\theta} = \dfrac{\sin(3\theta)}{\cos(3\theta)} = \tan(3\theta)$ **23.** $\dfrac{\cos\theta - \cos(3\theta)}{\sin\theta + \sin(3\theta)} = \dfrac{2\sin(2\theta)\sin\theta}{2\sin(2\theta)\cos\theta} = \dfrac{\sin\theta}{\cos\theta} = \tan\theta$

25. $\sin\theta[\sin\theta + \sin(3\theta)] = \sin\theta[2\sin(2\theta)\cos\theta] = \cos\theta[2\sin(2\theta)\sin\theta] = \cos\theta\left[2\cdot\frac{1}{2}[\cos\theta - \cos(3\theta)]\right] = \cos\theta[\cos\theta - \cos(3\theta)]$

27. $\dfrac{\sin(4\theta) + \sin(8\theta)}{\cos(4\theta) + \cos(8\theta)} = \dfrac{2\sin(6\theta)\cos(2\theta)}{2\cos(6\theta)\cos(2\theta)} = \dfrac{\sin(6\theta)}{\cos(6\theta)} = \tan(6\theta)$

29. $\dfrac{\sin(4\theta) + \sin(8\theta)}{\sin(4\theta) - \sin(8\theta)} = \dfrac{2\sin(6\theta)\cos(-2\theta)}{2\sin(-2\theta)\cos(6\theta)} = \dfrac{\sin(6\theta)}{\cos(6\theta)}\cdot\dfrac{\cos(2\theta)}{-\sin(2\theta)} = \tan(6\theta)[-\cot(2\theta)] = -\dfrac{\tan(6\theta)}{\tan(2\theta)}$

31. $\dfrac{\sin\alpha + \sin\beta}{\sin\alpha - \sin\beta} = \dfrac{2\sin\dfrac{\alpha+\beta}{2}\cos\dfrac{\alpha-\beta}{2}}{2\sin\dfrac{\alpha-\beta}{2}\cos\dfrac{\alpha+\beta}{2}} = \dfrac{\sin\dfrac{\alpha+\beta}{2}}{\cos\dfrac{\alpha+\beta}{2}}\cdot\dfrac{\cos\dfrac{\alpha-\beta}{2}}{\sin\dfrac{\alpha-\beta}{2}} = \tan\dfrac{\alpha+\beta}{2}\cot\dfrac{\alpha-\beta}{2}$

33. $\dfrac{\sin\alpha + \sin\beta}{\cos\alpha + \cos\beta} = \dfrac{2\sin\dfrac{\alpha+\beta}{2}\cos\dfrac{\alpha-\beta}{2}}{2\cos\dfrac{\alpha+\beta}{2}\cos\dfrac{\alpha-\beta}{2}} = \dfrac{\sin\dfrac{\alpha+\beta}{2}}{\cos\dfrac{\alpha+\beta}{2}} = \tan\dfrac{\alpha+\beta}{2}$

35. $1 + \cos(2\theta) + \cos(4\theta) + \cos(6\theta) = [1 + \cos(6\theta)] + [\cos(2\theta) + \cos(4\theta)] = 2\cos^2(3\theta) + 2\cos(3\theta)\cos(-\theta)$
$= 2\cos(3\theta)[\cos(3\theta) + \cos\theta] = 2\cos(3\theta)[2\cos(2\theta)\cos\theta] = 4\cos\theta\cos(2\theta)\cos(3\theta)$

37. $\left\{0, \dfrac{\pi}{3}, \dfrac{\pi}{2}, \dfrac{2\pi}{3}, \pi, \dfrac{4\pi}{3}, \dfrac{3\pi}{2}, \dfrac{5\pi}{3}\right\}$ **39.** $\left\{0, \dfrac{\pi}{5}, \dfrac{2\pi}{5}, \dfrac{3\pi}{5}, \dfrac{4\pi}{5}, \pi, \dfrac{6\pi}{5}, \dfrac{7\pi}{5}, \dfrac{8\pi}{5}, \dfrac{9\pi}{5}\right\}$

41. (a) $y = 2\sin(2061\pi t)\cos(357\pi t)$
(b) $y_{\max} = 2$
(c)

43. $I_u = I_x\cos^2\theta + I_y\sin^2\theta - 2I_{xy}\sin\theta\cos\theta = I_x\cos^2\theta + I_y\sin^2\theta - I_{xy}\sin 2\theta$
$= I_x\left(\dfrac{\cos 2\theta + 1}{2}\right) + I_y\left(\dfrac{1 - \cos 2\theta}{2}\right) - I_{xy}\sin 2\theta$
$= \dfrac{I_x}{2}\cos 2\theta + \dfrac{I_x}{2} + \dfrac{I_y}{2} - \dfrac{I_y}{2}\cos 2\theta - I_{xy}\sin 2\theta$
$= \dfrac{I_x + I_y}{2} + \dfrac{I_x - I_y}{2}\cos 2\theta - I_{xy}\sin 2\theta$

$I_v = I_x\sin^2\theta + I_y\cos^2\theta + 2I_{xy}\sin\theta\cos\theta = I_x\left(\dfrac{1 - \cos 2\theta}{2}\right) + I_y\left(\dfrac{\cos 2\theta + 1}{2}\right) + I_{xy}\sin 2\theta$
$= \dfrac{I_x}{2} - \dfrac{I_x}{2}\cos 2\theta + \dfrac{I_y}{2}\cos 2\theta + \dfrac{I_y}{2} + I_{xy}\sin 2\theta$
$= \dfrac{I_x + I_y}{2} - \dfrac{I_x - I_y}{2}\cos 2\theta + I_{xy}\sin 2\theta$

45. $\sin(2\alpha) + \sin(2\beta) + \sin(2\gamma) = 2\sin(\alpha + \beta)\cos(\alpha - \beta) + \sin(2\gamma) = 2\sin(\alpha + \beta)\cos(\alpha - \beta) + 2\sin\gamma\cos\gamma$
$= 2\sin(\pi - \gamma)\cos(\alpha - \beta) + 2\sin\gamma\cos\gamma = 2\sin\gamma\cos(\alpha - \beta) + 2\sin\gamma\cos\gamma = 2\sin\gamma[\cos(\alpha - \beta) + \cos\gamma]$
$= 2\sin\gamma\left(2\cos\dfrac{\alpha - \beta + \gamma}{2}\cos\dfrac{\alpha - \beta - \gamma}{2}\right) = 4\sin\gamma\cos\dfrac{\pi - 2\beta}{2}\cos\dfrac{2\alpha - \pi}{2} = 4\sin\gamma\cos\left(\dfrac{\pi}{2} - \beta\right)\cos\left(\alpha - \dfrac{\pi}{2}\right)$
$= 4\sin\gamma\sin\beta\sin\alpha = 4\sin\alpha\sin\beta\sin\gamma$

47. $\sin(\alpha - \beta) = \sin\alpha\cos\beta - \cos\alpha\sin\beta$
       $\sin(\alpha + \beta) = \sin\alpha\cos\beta + \cos\alpha\sin\beta$
$\sin(\alpha - \beta) + \sin(\alpha + \beta) = 2\sin\alpha\cos\beta$
           $\sin\alpha\cos\beta = \dfrac{1}{2}[\sin(\alpha + \beta) + \sin(\alpha - \beta)]$

49. $2\cos\dfrac{\alpha + \beta}{2}\cos\dfrac{\alpha - \beta}{2} = 2\cdot\dfrac{1}{2}\left[\cos\left(\dfrac{\alpha + \beta}{2} + \dfrac{\alpha - \beta}{2}\right) + \cos\left(\dfrac{\alpha + \beta}{2} - \dfrac{\alpha - \beta}{2}\right)\right] = \cos\dfrac{2\alpha}{2} + \cos\dfrac{2\beta}{2} = \cos\alpha + \cos\beta$

Review Exercises *(page 541)*

1. $\dfrac{\pi}{2}$ **3.** $\dfrac{\pi}{4}$ **5.** $\dfrac{5\pi}{6}$ **7.** $\dfrac{\pi}{4}$ **9.** $\dfrac{3\pi}{8}$ **11.** $-\dfrac{\pi}{3}$ **13.** $\dfrac{\pi}{7}$ **15.** 0.9 **17.** -0.3 **19.** Not defined   **21.** $-\dfrac{\pi}{6}$ **23.** $-\dfrac{\pi}{4}$ **25.** $-\sqrt{3}$ **27.** $\dfrac{2\sqrt{3}}{3}$ **29.** $\dfrac{4}{5}$

31. $-\dfrac{4}{3}$ **33.** $f^{-1}(x) = \dfrac{1}{3}\sin^{-1}\left(\dfrac{x}{2}\right)$; Range of f = Domain of f^{-1} = $[-2, 2]$; Range of f^{-1} = $\left[-\dfrac{\pi}{6}, \dfrac{\pi}{6}\right]$ **35.** $f^{-1}(x) = \cos^{-1}(3 - x)$;

Range of f = Domain of f^{-1} = $[2, 4]$; Range of f^{-1} = $[0, \pi]$ **37.** $\sqrt{1 - u^2}$ **39.** $\dfrac{1}{u}$ **41.** $\tan\theta\cot\theta - \sin^2\theta = 1 - \sin^2\theta = \cos^2\theta$

43. $\sin^2\theta(1 + \cot^2\theta) = \sin^2\theta\csc^2\theta = 1$ **45.** $5\cos^2\theta + 3\sin^2\theta = 2\cos^2\theta + 3(\cos^2\theta + \sin^2\theta) = 3 + 2\cos^2\theta$

47. $\dfrac{1 - \cos\theta}{\sin\theta} + \dfrac{\sin\theta}{1 - \cos\theta} = \dfrac{(1 - \cos\theta)^2 + \sin^2\theta}{\sin\theta(1 - \cos\theta)} = \dfrac{1 - 2\cos\theta + \cos^2\theta + \sin^2\theta}{\sin\theta(1 - \cos\theta)} = \dfrac{2(1 - \cos\theta)}{\sin\theta(1 - \cos\theta)} = 2\csc\theta$

49. $\dfrac{\cos\theta}{\cos\theta - \sin\theta} = \dfrac{\frac{\cos\theta}{\cos\theta}}{\frac{\cos\theta - \sin\theta}{\cos\theta}} = \dfrac{1}{1 - \frac{\sin\theta}{\cos\theta}} = \dfrac{1}{1 - \tan\theta}$ **51.** $\dfrac{\csc\theta}{1 + \csc\theta} = \dfrac{\frac{1}{\sin\theta}}{1 + \frac{1}{\sin\theta}} = \dfrac{1}{1 + \sin\theta} = \dfrac{1}{1 + \sin\theta} \cdot \dfrac{1 - \sin\theta}{1 - \sin\theta} = \dfrac{1 - \sin\theta}{1 - \sin^2\theta} = \dfrac{1 - \sin\theta}{\cos^2\theta}$

53. $\csc\theta - \sin\theta = \dfrac{1}{\sin\theta} - \sin\theta = \dfrac{1 - \sin^2\theta}{\sin\theta} = \dfrac{\cos^2\theta}{\sin\theta} = \cos\theta \cdot \dfrac{\cos\theta}{\sin\theta} = \cos\theta\cot\theta$

55. $\dfrac{1 - \sin\theta}{\sec\theta} = \cos\theta(1 - \sin\theta) \cdot \dfrac{1 + \sin\theta}{1 + \sin\theta} = \dfrac{\cos\theta(1 - \sin^2\theta)}{1 + \sin\theta} = \dfrac{\cos^3\theta}{1 + \sin\theta}$ **57.** $\cot\theta - \tan\theta = \dfrac{\cos\theta}{\sin\theta} - \dfrac{\sin\theta}{\cos\theta} = \dfrac{\cos^2\theta - \sin^2\theta}{\sin\theta\cos\theta} = \dfrac{1 - 2\sin^2\theta}{\sin\theta\cos\theta}$

59. $\dfrac{\cos(\alpha + \beta)}{\cos\alpha\sin\beta} = \dfrac{\cos\alpha\cos\beta - \sin\alpha\sin\beta}{\cos\alpha\sin\beta} = \dfrac{\cos\alpha\cos\beta}{\cos\alpha\sin\beta} - \dfrac{\sin\alpha\sin\beta}{\cos\alpha\sin\beta} = \cot\beta - \tan\alpha$

61. $\dfrac{\cos(\alpha - \beta)}{\cos\alpha\cos\beta} = \dfrac{\cos\alpha\cos\beta + \sin\alpha\sin\beta}{\cos\alpha\cos\beta} = \dfrac{\cos\alpha\cos\beta}{\cos\alpha\cos\beta} + \dfrac{\sin\alpha\sin\beta}{\cos\alpha\cos\beta} = 1 + \tan\alpha\tan\beta$

63. $(1 + \cos\theta)\left(\tan\dfrac{\theta}{2}\right) = (1 + \cos\theta) \cdot \dfrac{\sin\theta}{1 + \cos\theta} = \sin\theta$

65. $2\cot\theta\cot 2\theta = 2\left(\dfrac{\cos\theta}{\sin\theta}\right)\left(\dfrac{\cos 2\theta}{\sin 2\theta}\right) = \dfrac{2\cos\theta(\cos^2\theta - \sin^2\theta)}{2\sin^2\theta\cos\theta} = \dfrac{\cos^2\theta - \sin^2\theta}{\sin^2\theta} = \cot^2\theta - 1$

67. $1 - 8\sin^2\theta\cos^2\theta = 1 - 2(2\sin\theta\cos\theta)^2 = 1 - 2\sin^2(2\theta) = \cos(4\theta)$ **69.** $\dfrac{\sin(2\theta) + \sin(4\theta)}{\cos(2\theta) + \cos(4\theta)} = \dfrac{2\sin(3\theta)\cos(-\theta)}{2\cos(3\theta)\cos(-\theta)} = \tan(3\theta)$

71. $\dfrac{\cos(2\theta) - \cos(4\theta)}{\cos(2\theta) + \cos(4\theta)} - \tan\theta\tan(3\theta) = \dfrac{-2\sin(3\theta)\sin(-\theta)}{2\cos(3\theta)\cos(-\theta)} - \tan\theta\tan(3\theta) = \tan(3\theta)\tan\theta - \tan\theta\tan(3\theta) = 0$ **73.** $\dfrac{1}{4}\left(\sqrt{6} - \sqrt{2}\right)$

75. $\dfrac{1}{4}\left(\sqrt{6} - \sqrt{2}\right)$ **77.** $\dfrac{1}{2}$ **79.** $\sqrt{2} - 1$ **81. (a)** $-\dfrac{33}{65}$ **(b)** $-\dfrac{56}{65}$ **(c)** $-\dfrac{63}{65}$ **(d)** $\dfrac{33}{56}$ **(e)** $\dfrac{24}{25}$ **(f)** $\dfrac{119}{169}$ **(g)** $\dfrac{5\sqrt{26}}{26}$ **(h)** $\dfrac{2\sqrt{5}}{5}$

83. (a) $-\dfrac{16}{65}$ **(b)** $-\dfrac{63}{65}$ **(c)** $-\dfrac{56}{65}$ **(d)** $\dfrac{16}{63}$ **(e)** $\dfrac{24}{25}$ **(f)** $\dfrac{119}{169}$ **(g)** $\dfrac{\sqrt{26}}{26}$ **(h)** $-\dfrac{\sqrt{10}}{10}$

85. (a) $-\dfrac{63}{65}$ **(b)** $\dfrac{16}{65}$ **(c)** $\dfrac{33}{65}$ **(d)** $-\dfrac{63}{16}$ **(e)** $\dfrac{24}{25}$ **(f)** $-\dfrac{119}{169}$ **(g)** $\dfrac{2\sqrt{13}}{13}$ **(h)** $-\dfrac{\sqrt{10}}{10}$

87. (a) $\dfrac{-\sqrt{3} - 2\sqrt{2}}{6}$ **(b)** $\dfrac{1 - 2\sqrt{6}}{6}$ **(c)** $\dfrac{-\sqrt{3} + 2\sqrt{2}}{6}$ **(d)** $\dfrac{8\sqrt{2} + 9\sqrt{3}}{23}$ **(e)** $-\dfrac{\sqrt{3}}{2}$ **(f)** $-\dfrac{7}{9}$ **(g)** $\dfrac{\sqrt{3}}{3}$ **(h)** $\dfrac{\sqrt{3}}{2}$

89. (a) 1 **(b)** 0 **(c)** $-\dfrac{1}{9}$ **(d)** Not defined **(e)** $\dfrac{4\sqrt{5}}{9}$ **(f)** $-\dfrac{1}{9}$ **(g)** $\dfrac{\sqrt{30}}{6}$ **(h)** $-\dfrac{\sqrt{6}\sqrt{3 - \sqrt{5}}}{6}$ **91.** $\dfrac{4 + 3\sqrt{3}}{10}$ **93.** $-\dfrac{48 + 25\sqrt{3}}{39}$

95. $-\dfrac{24}{25}$ **97.** $\left\{\dfrac{\pi}{3}, \dfrac{5\pi}{3}\right\}$ **99.** $\left\{\dfrac{3\pi}{4}, \dfrac{5\pi}{4}\right\}$ **101.** $\left\{\dfrac{3\pi}{4}, \dfrac{7\pi}{4}\right\}$ **103.** $\left\{0, \dfrac{\pi}{2}, \pi, \dfrac{3\pi}{2}\right\}$ **105.** $\left\{\dfrac{\pi}{3}, \dfrac{2\pi}{3}, \dfrac{4\pi}{3}, \dfrac{5\pi}{3}\right\}$ **107.** $\{0.25, 2.89\}$

109. $\left\{0, \dfrac{2\pi}{3}, \pi, \dfrac{4\pi}{3}\right\}$ **111.** $\left\{0, \dfrac{\pi}{6}, \dfrac{5\pi}{6}\right\}$ **113.** $\left\{\dfrac{\pi}{6}, \dfrac{\pi}{2}, \dfrac{5\pi}{6}\right\}$ **115.** $\left\{\dfrac{\pi}{3}, \dfrac{5\pi}{3}\right\}$ **117.** $\left\{\dfrac{\pi}{4}, \dfrac{\pi}{2}, \dfrac{3\pi}{4}, \dfrac{3\pi}{2}\right\}$ **119.** $\left\{\dfrac{\pi}{2}, \pi\right\}$ **121.** 0.78 **123.** -1.11

125. 1.23 **127.** $\{1.11\}$ **129.** $\{0.87\}$ **131.** $\{2.22\}$ **133.** $\left\{-\dfrac{\sqrt{3}}{2}\right\}$

135. $\sin 15° = \sqrt{\dfrac{1 - \cos 30°}{2}} = \sqrt{\dfrac{1 - \frac{\sqrt{3}}{2}}{2}} = \sqrt{\dfrac{2 - \sqrt{3}}{4}} = \sqrt{\dfrac{2 - \sqrt{3}}{2}};$

$\sin 15° = \sin(45° - 30°) = \sin 45°\cos 30° - \cos 45°\sin 30° = \dfrac{\sqrt{2}}{2} \cdot \dfrac{\sqrt{3}}{2} - \dfrac{\sqrt{2}}{2} \cdot \dfrac{1}{2} = \dfrac{\sqrt{6}}{4} - \dfrac{\sqrt{2}}{4} = \dfrac{\sqrt{6} - \sqrt{2}}{4};$

$\left[\dfrac{\sqrt{2 - \sqrt{3}}}{2}\right]^2 = \dfrac{2 - \sqrt{3}}{4} = \dfrac{4(2 - \sqrt{3})}{4 \cdot 4} = \dfrac{8 - 4\sqrt{3}}{16} = \dfrac{6 - 2\sqrt{12} + 2}{16} = \left(\dfrac{\sqrt{6} - \sqrt{2}}{4}\right)^2$

Chapter Test (page 544)

1. $\dfrac{\pi}{6}$ **2.** $-\dfrac{\pi}{4}$ **3.** $\dfrac{\pi}{5}$ **4.** $\dfrac{7}{3}$ **5.** 3 **6.** $-\dfrac{4}{3}$ **7.** 0.39 **8.** 0.78 **9.** 1.25 **10.** 0.20

11. $\dfrac{\csc\theta + \cot\theta}{\sec\theta + \tan\theta} = \dfrac{\csc\theta + \cot\theta}{\sec\theta + \tan\theta} \cdot \dfrac{\csc\theta - \cot\theta}{\csc\theta - \cot\theta} = \dfrac{\csc^2\theta - \cot^2\theta}{(\sec\theta + \tan\theta)(\csc\theta - \cot\theta)} = \dfrac{1}{(\sec\theta + \tan\theta)(\csc\theta - \cot\theta)}$

$= \dfrac{1}{(\sec\theta + \tan\theta)(\csc\theta - \cot\theta)} \cdot \dfrac{\sec\theta - \tan\theta}{\sec\theta - \tan\theta} = \dfrac{\sec\theta - \tan\theta}{(\sec^2\theta - \tan^2\theta)(\csc\theta - \cot\theta)} = \dfrac{\sec\theta - \tan\theta}{\csc\theta - \cot\theta}$

12. $\sin\theta\tan\theta + \cos\theta = \sin\theta \cdot \dfrac{\sin\theta}{\cos\theta} + \cos\theta = \dfrac{\sin^2\theta}{\cos\theta} + \dfrac{\cos^2\theta}{\cos\theta} = \dfrac{\sin^2\theta + \cos^2\theta}{\cos\theta} = \dfrac{1}{\cos\theta} = \sec\theta$

13. $\tan\theta + \cot\theta = \dfrac{\sin\theta}{\cos\theta} + \dfrac{\cos\theta}{\sin\theta} = \dfrac{\sin^2\theta}{\sin\theta\cos\theta} + \dfrac{\cos^2\theta}{\sin\theta\cos\theta} = \dfrac{\sin^2\theta + \cos^2\theta}{\sin\theta\cos\theta} = \dfrac{1}{\sin\theta\cos\theta} = \dfrac{2}{2\sin\theta\cos\theta} = \dfrac{2}{\sin(2\theta)} = 2\csc(2\theta)$

14. $\dfrac{\sin(\alpha + \beta)}{\tan\alpha + \tan\beta} = \dfrac{\sin\alpha\cos\beta + \cos\alpha\sin\beta}{\dfrac{\sin\alpha}{\cos\alpha} + \dfrac{\sin\beta}{\cos\beta}} = \dfrac{\sin\alpha\cos\beta + \cos\alpha\sin\beta}{\dfrac{\sin\alpha\cos\beta}{\cos\alpha\cos\beta} + \dfrac{\cos\alpha\sin\beta}{\cos\alpha\cos\beta}} = \dfrac{\sin\alpha\cos\beta + \cos\alpha\sin\beta}{\dfrac{\sin\alpha\cos\beta + \cos\alpha\sin\beta}{\cos\alpha\cos\beta}}$

$= \dfrac{\sin\alpha\cos\beta + \cos\alpha\sin\beta}{1} \cdot \dfrac{\cos\alpha\cos\beta}{\sin\alpha\cos\beta + \cos\alpha\sin\beta} = \cos\alpha\cos\beta$

15. $\sin(3\theta) = \sin(\theta + 2\theta) = \sin\theta\cos(2\theta) + \cos\theta\sin(2\theta) = \sin\theta \cdot (\cos^2\theta - \sin^2\theta) + \cos\theta \cdot 2\sin\theta\cos\theta = \sin\theta\cos^2\theta - \sin^3\theta + 2\sin\theta\cos^2\theta$
$= 3\sin\theta\cos^2\theta - \sin^3\theta = 3\sin\theta(1 - \sin^2\theta) - \sin^3\theta = 3\sin\theta - 3\sin^3\theta - \sin^3\theta = 3\sin\theta - 4\sin^3\theta$

16. $\dfrac{\tan\theta - \cot\theta}{\tan\theta + \cot\theta} = \dfrac{\dfrac{\sin\theta}{\cos\theta} - \dfrac{\cos\theta}{\sin\theta}}{\dfrac{\sin\theta}{\cos\theta} + \dfrac{\cos\theta}{\sin\theta}} = \dfrac{\dfrac{\sin^2\theta - \cos^2\theta}{\sin\theta\cos\theta}}{\dfrac{\sin^2\theta + \cos^2\theta}{\sin\theta\cos\theta}} = \dfrac{\sin^2\theta - \cos^2\theta}{\sin^2\theta + \cos^2\theta} = \dfrac{(1 - \cos^2\theta) - \cos^2\theta}{1} = 1 - 2\cos^2\theta$ **17.** $\dfrac{1}{4}\left(\sqrt{6} + \sqrt{2}\right)$

18. $2 + \sqrt{3}$ **19.** $\dfrac{\sqrt{5}}{5}$ **20.** $\dfrac{12\sqrt{85}}{49}$ **21.** $\dfrac{2\sqrt{13}\left(\sqrt{5} - 3\right)}{39}$ **22.** $\dfrac{2 + \sqrt{3}}{4}$ **23.** $\dfrac{\sqrt{6}}{2}$ **24.** $\dfrac{\sqrt{2}}{2}$ **25.** $\left\{\dfrac{\pi}{3}, \dfrac{2\pi}{3}, \dfrac{4\pi}{3}, \dfrac{5\pi}{3}\right\}$ **26.** $\{0, 1.911, \pi, 4.373\}$

27. $\left\{\dfrac{3\pi}{8}, \dfrac{7\pi}{8}, \dfrac{11\pi}{8}, \dfrac{15\pi}{8}\right\}$ **28.** $\{0.285, 3.427\}$ **29.** $\{0.253, 2.889\}$ **30.** The change in elevation during the time trial was about 1.18 km.

Cumulative Review *(page 544)*

1. $\left\{\dfrac{-1 - \sqrt{13}}{6}, \dfrac{-1 + \sqrt{13}}{6}\right\}$ **2.** $y + 1 = -1(x - 4)$ or $x + y = 3; 6\sqrt{2}; (1, 2)$ **3.** x-axis symmetry; $(0, -3), (0, 3), (3, 0)$

4.

5.

6.

7. (a)

(b)

(c)

(d)

8. (a) $-\dfrac{2\sqrt{2}}{3}$ **(b)** $\dfrac{\sqrt{2}}{4}$ **(c)** $\dfrac{4\sqrt{2}}{9}$ **(d)** $\dfrac{7}{9}$ **(e)** $\sqrt{\dfrac{3 + 2\sqrt{2}}{6}}$ **(f)** $-\sqrt{\dfrac{3 - 2\sqrt{2}}{6}}$

9. $\dfrac{\sqrt{5}}{5}$ **10. (a)** $-\dfrac{2\sqrt{2}}{3}$ **(b)** $-\dfrac{2\sqrt{2}}{3}$ **(c)** $\dfrac{7}{9}$ **(d)** $\dfrac{4\sqrt{2}}{9}$ **(e)** $\dfrac{\sqrt{6}}{3}$

11. (a) $f(x) = (2x - 1)(x - 1)^2(x + 1)^2;$ **(b)** $(0, -1); \left(\dfrac{1}{2}, 0\right); (-1, 0); (1, 0)$ **(c)** $y = 2x^5$
$\dfrac{1}{2}$ multiplicity 1; 1 and -1 multiplicity 2

(d)

(e) Minima -1.33 at $x = -0.29$, Minima 0 at $x = 1$
Maxima 0 at $x = -1$, Maxima 0.10 at $x = 0.69$

(g) Increasing: $(-\infty, -1), (-0.29, 0.69), (1, \infty)$
Decreasing: $(-1, -0.29), (0.69, 1)$

(f)

12. (a) $\left\{-1, -\dfrac{1}{2}\right\}$ **(b)** $\{-1, 1\}$

(c) $(-\infty, -1) \cup \left(-\dfrac{1}{2}, \infty\right)$

(d) $(-\infty, -1] \cup [1, \infty)$

CHAPTER 7 Applications of Trigonometric Functions

7.1 Assess Your Understanding *(page 550)*

5. 90 **6.** direction; bearing **7.** T **8.** F **9.** $a \approx 13.74, c \approx 14.62, A = 70°$ **11.** $b \approx 5.03, c \approx 7.83, A = 50°$ **13.** $a \approx 0.71, c \approx 4.06, B = 80°$
15. $b \approx 10.72, c \approx 11.83, B = 65°$ **17.** $b \approx 3.08, a \approx 8.46, A = 70°$ **19.** $c \approx 5.83, A \approx 59.0°, B \approx 31.0°$ **21.** $b \approx 4.58, A \approx 23.6°, B \approx 66.4°$
23. $23.6°$ and $66.4°$ **25.** $80.5°$ **27. (c)** Under $18.8°$ **29. (b)** $0.000214°$ **31.** S76.6°E **33.** The embankment is 30.5 m high.
35. The buildings are 7984 ft apart. **37.** $69.0°$ **39.** $38.9°$

7.2 Assess Your Understanding *(page 559)*

4. oblique **5.** $\dfrac{\sin A}{a} = \dfrac{\sin B}{b} = \dfrac{\sin C}{c}$ **6.** F **7.** F **8.** ambiguous case **9.** $a \approx 3.23, b \approx 3.55, A = 40°$ **11.** $a \approx 3.25, c \approx 4.23, B = 45°$
13. $C = 95°, c \approx 9.86, a \approx 6.36$ **15.** $A = 40°, a = 2, c \approx 3.06$ **17.** $C = 120°, b \approx 1.06, c \approx 2.69$ **19.** $A = 100°, a \approx 5.24, c \approx 0.92$
21. $B = 40°, a \approx 5.64, b \approx 3.86$ **23.** $C = 100°, a \approx 1.31, b \approx 1.31$ **25.** One triangle; $B \approx 30.7°, C \approx 99.3°, c \approx 3.86$ **27.** One triangle;
$C \approx 36.2°, A \approx 43.8°, a \approx 3.51$ **29.** No triangle **31.** Two triangles; $C_1 \approx 30.9°, A_1 \approx 129.1°, a_1 \approx 9.07$ or $C_2 \approx 149.1°, A_2 \approx 10.9°, a_2 \approx 2.20$
33. No triangle **35.** Two triangles; $A_1 \approx 57.7°, B_1 \approx 97.3°, b_1 \approx 2.35$ or $A_2 \approx 122.3°, B_2 \approx 32.7°, b_2 \approx 1.28$ **37. (a)** Station Able is about 143.33 mi
from the ship: Station Baker is about 135.58 mi from the ship. **(b)** Approximately 41 min **39.** 1490.48 ft **41.** 381.69 ft **43.** The tree is 39.4 ft high.
45. Adam receives 88.3 more frequent flyer miles. **47.** $84.7°; 183.72$ ft **49.** 2.64 mi **51.** 38.5 in. **53.** 449.36 ft **55.** 187,600,000 km or 101,440,000 km
57. The diameter is 252 ft.

59. $\dfrac{a-b}{c} = \dfrac{a}{c} - \dfrac{b}{c} = \dfrac{\sin A}{\sin C} - \dfrac{\sin B}{\sin C} = \dfrac{\sin A - \sin B}{\sin C} = \dfrac{2\sin\left(\dfrac{A-B}{2}\right)\cos\left(\dfrac{A+B}{2}\right)}{2\sin\dfrac{C}{2}\cos\dfrac{C}{2}} = \dfrac{\sin\left(\dfrac{A-B}{2}\right)\cos\left(\dfrac{\pi}{2} - \dfrac{C}{2}\right)}{\sin\dfrac{C}{2}\cos\dfrac{C}{2}} = \dfrac{\sin\left(\dfrac{A-B}{2}\right)}{\cos\dfrac{C}{2}}$

61. $\dfrac{a-b}{a+b} = \dfrac{\dfrac{a-b}{c}}{\dfrac{a+b}{c}} = \dfrac{\dfrac{\sin\left[\dfrac{1}{2}(A-B)\right]}{\cos\dfrac{C}{2}}}{\dfrac{\cos\left[\dfrac{1}{2}(A-B)\right]}{\sin\dfrac{C}{2}}} = \dfrac{\tan\left[\dfrac{1}{2}(A-B)\right]}{\cot\dfrac{C}{2}} = \dfrac{\tan\left[\dfrac{1}{2}(A-B)\right]}{\tan\left(\dfrac{\pi}{2} - \dfrac{C}{2}\right)} = \dfrac{\tan\left[\dfrac{1}{2}(A-B)\right]}{\tan\left[\dfrac{1}{2}(A+B)\right]}$

7.3 Assess Your Understanding (page 566)

3. Cosines   **4.** Sines   **5.** Cosines   **6.** F   **7.** F   **8.** T   **9.** $b \approx 2.95$, $A \approx 28.7°$, $C \approx 106.3°$ **11.** $c \approx 3.75$, $A \approx 32.1°$, $B \approx 52.9°$
13. $A \approx 48.5°$, $B \approx 38.6°$, $C \approx 92.9°$ **15.** $A \approx 127.2°$, $B \approx 32.1°$, $C \approx 20.7°$ **17.** $c \approx 2.57$, $A \approx 48.6°$, $B \approx 91.4°$
19. $a \approx 2.99$, $B \approx 19.2°$, $C \approx 80.8°$ **21.** $b \approx 4.14$, $A \approx 43.0°$, $C \approx 27.0°$ **23.** $c \approx 1.69$, $A \approx 65.0°$, $B \approx 65.0°$ **25.** $A \approx 67.4°$, $B \approx 90°$, $C \approx 22.6°$
27. $A = 60°$, $B = 60°$, $C = 60°$ **29.** $A \approx 33.6°$, $B \approx 62.2°$, $C \approx 84.3°$ **31.** $A \approx 97.9°$, $B \approx 52.4°$, $C \approx 29.7°$ **33.** $A = 85°$, $a = 14.56$, $c = 14.12$
35. $A = 40.8°$, $B = 60.6°$, $C = 78.6°$ **37.** $A = 80°$, $b = 8.74$, $c = 13.80$ **39.** Two triangles: $B_1 = 35.4°$, $C_1 = 134.6°$, $c_1 = 12.29$;
$B_2 = 144.6°$, $C_2 = 25.4°$, $c_2 = 7.40$ **41.** $B = 24.5°$, $C = 95.5°$, $a = 10.44$ **43.** 165 yd   **45. (a)** 26.4°   **(b)** 30.8 h   **47. (a)** 63.7 ft   **(b)** 66.8 ft
(c) 92.8°   **49. (a)** 492.6 ft   **(b)** 269.3 ft   **51.** 342.33 ft   **53.** The footings should be 7.65 ft apart.

55. Suppose $0 < \theta < \pi$. Then, by the Law of Cosines, $d^2 = r^2 + r^2 - 2r^2\cos\theta = 4r^2\left(\dfrac{1-\cos\theta}{2}\right) \Rightarrow d = 2r\sqrt{\dfrac{1-\cos\theta}{2}} = 2r\sin\dfrac{\theta}{2}$.

Since, for any angle in $(0, \pi)$, d is strictly less than the length of the arc subtended by θ, that is, $d < r\theta$, then $2r\sin\dfrac{\theta}{2} < r\theta$, or $2\sin\dfrac{\theta}{2} < \theta$.

Since $\cos\dfrac{\theta}{2} < 1$, then, for $0 < \theta < \pi$, $\sin\theta = 2\sin\dfrac{\theta}{2}\cos\dfrac{\theta}{2} < 2\sin\dfrac{\theta}{2} < \theta$. If $\theta \geq \pi$, then, since $\sin\theta \leq 1$, $\sin\theta < \theta$. Thus $\sin\theta < \theta$ for all $\theta > 0$.

57. $\sin\dfrac{C}{2} = \sqrt{\dfrac{1-\cos C}{2}} = \sqrt{\dfrac{1 - \dfrac{a^2 + b^2 - c^2}{2ab}}{2}} = \sqrt{\dfrac{2ab - a^2 - b^2 + c^2}{4ab}} = \sqrt{\dfrac{c^2 - (a-b)^2}{4ab}} = \sqrt{\dfrac{(c+a-b)(c+b-a)}{4ab}}$

$= \sqrt{\dfrac{(2s-2b)(2s-2a)}{4ab}} = \sqrt{\dfrac{(s-a)(s-b)}{ab}}$

7.4 Assess Your Understanding (page 572)

2. $\dfrac{1}{2}ab\sin C$ **3.** $\sqrt{s(s-a)(s-b)(s-c)}$; $\dfrac{1}{2}(a+b+c)$ **4.** T   **5.** 2.83   **7.** 2.99   **9.** 14.98   **11.** 9.56   **13.** 3.86   **15.** 1.48   **17.** 2.82   **19.** 30

21. 1.73   **23.** 19.90   **25.** $K = \dfrac{1}{2}ab\sin C = \dfrac{1}{2}a\sin C\left(\dfrac{a\sin B}{\sin A}\right) = \dfrac{a^2\sin B\sin C}{2\sin A}$ **27.** 0.92   **29.** 2.27   **30.** 5.44   **33.** 9.03 sq ft   **35.** \$5446.38

37. The area of home plate is about 216.5 in.^{2} **39.** $K = \dfrac{1}{2}r^2(\theta + \sin\theta)$ **41.** The ground area is 7517.4 ft^2.

43. (a) Area $\triangle OAC = \dfrac{1}{2}|OC||AC| = \dfrac{1}{2}\cdot\dfrac{|OC|}{1}\cdot\dfrac{|AC|}{1} = \dfrac{1}{2}\sin\alpha\cos\alpha$

(b) Area $\triangle OCB = \dfrac{1}{2}|BC||OC| = \dfrac{1}{2}|OB|^2\dfrac{|BC|}{|OB|}\cdot\dfrac{|OC|}{|OB|} = \dfrac{1}{2}|OB|^2\sin\beta\cos\beta$

(c) Area $\triangle OAB = \dfrac{1}{2}|BD||OA| = \dfrac{1}{2}|OB|\dfrac{|BD|}{|OB|} = \dfrac{1}{2}|OB|\sin(\alpha+\beta)$

(d) $\dfrac{\cos\alpha}{\cos\beta} = \dfrac{1}{\dfrac{|OC|}{|OB|}} = |OB|$

(e) Area $\triangle OAB = $ Area $\triangle OAC + $ Area $\triangle OCB$

$\dfrac{1}{2}|OB|\sin(\alpha+\beta) = \dfrac{1}{2}\sin\alpha\cos\alpha + \dfrac{1}{2}|OB|^2\sin\beta\cos\beta$

$\sin(\alpha+\beta) = \dfrac{1}{|OB|}\sin\alpha\cos\alpha + |OB|\sin\beta\cos\beta$

$\sin(\alpha+\beta) = \dfrac{\cos\beta}{\cos\alpha}\sin\alpha\cos\alpha + \dfrac{\cos\alpha}{\cos\beta}\sin\beta\cos\beta$

$\sin(\alpha+\beta) = \sin\alpha\cos\beta + \cos\alpha\sin\beta$

45. 31,145 ft^2 **47. (a)** The perimeter and area are both 36.   **(b)** The perimeter and area are both 60.

49. $K = \dfrac{1}{2}ah = \dfrac{1}{2}ab\sin C \Rightarrow h = b\sin C = \dfrac{a\sin B\sin C}{\sin A}$

51. $\cot\dfrac{C}{2} = \dfrac{\cos\dfrac{C}{2}}{\sin\dfrac{C}{2}} = \dfrac{c\sin\dfrac{A}{2}\sin\dfrac{B}{2}}{r} = \dfrac{c\sqrt{\dfrac{(s-b)(s-c)}{bc}}\sqrt{\dfrac{(s-a)(s-c)}{ac}}}{r\sqrt{\dfrac{(s-a)(s-b)}{ab}}} = \dfrac{c}{r}\sqrt{\dfrac{(s-c)^2}{c^2}} = \dfrac{s-c}{r}$

53. $K = $ area of triangle $QOR + $ area of $ROP + $ area of $POQ = \dfrac{1}{2}ar + \dfrac{1}{2}br + \dfrac{1}{2}cr = r\dfrac{1}{2}(a+b+c) = rs$, so

$r = \dfrac{K}{s} = \dfrac{\sqrt{s(s-a)(s-b)(s-c)}}{s} = \sqrt{\dfrac{(s-a)(s-b)(s-c)}{s}}$.

7.5 Assess Your Understanding *(page 582)*

2. Simple harmonic; amplitude   **3.** Simple harmonic motion; damped   **4.** T   **5.** $d = -5\cos(\pi t)$ **7.** $d = -6\cos(2t)$ **9.** $d = -5\sin(\pi t)$

11. $d = -6\sin(2t)$ **13. (a)** Simple harmonic   **(b)** 5 m   **(c)** $\dfrac{2\pi}{3}$ sec   **(d)** $\dfrac{3}{2\pi}$ oscillation/sec   **15. (a)** Simple harmonic   **(b)** 6 m   **(c)** 2 sec

(d) $\dfrac{1}{2}$ oscillation/sec   **17. (a)** Simple harmonic   **(b)** 3 m   **(c)** 4π sec   **(d)** $\dfrac{1}{4\pi}$ oscillation/sec   **19. (a)** Simple harmonic   **(b)** 2 m

(c) 1 sec   **(d)** 1 oscillation/sec

21.   **23.**   **25.**   **27.**   **29.**

31.

33. (a) $f(x) = \dfrac{1}{2}\left[\cos x - \cos(3x)\right]$

35. (a) $G(x) = \dfrac{1}{2}\left[\cos(6x) + \cos(2x)\right]$

(b)

(b)

(b)

(b)

37. (a) $H(x) = \sin(4x) + \sin(2x)$

(b)

39. (a) $d = -10e^{-0.7t/50}\cos\left(\sqrt{\dfrac{4\pi^2}{25} - \dfrac{0.49}{2500}}\,t\right)$

(b)

41. (a) $d = -18e^{-0.6t/60}\cos\left(\sqrt{\dfrac{\pi^2}{4} - \dfrac{0.36}{3600}}\,t\right)$

(b)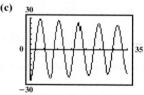

43. (a) $d = -5e^{-0.8t/20}\cos\left(\sqrt{\dfrac{4\pi^2}{9} - \dfrac{0.64}{400}}\,t\right)$

(b)

45. (a) The motion is damped. The bob has mass $m = 20$ kg with a damping factor of 0.7 kg/sec.

(b) 20 m leftward

(c)

(d) 18.33 m   **(e)** $d \to 0$

47. (a) The motion is damped. The bob has mass $m = 40$ kg with a damping factor of 0.6 kg/sec.

(b) 30 m leftward

(c)

(d) 28.47 m   **(e)** $d \to 0$

49. (a) The motion is damped. The bob has mass $m = 15$ kg with a damping factor of 0.9 kg/sec.

(b) 15 m leftward

(c)

(d) 12.53 m   **(e)** $d \to 0$

51. $\omega = 1040\pi$; $d = 0.80\cos(1040\pi t)$

53. $\omega = 880\pi$; $d = 0.01\sin(880\pi t)$

55. (a)

(b) At $t = 0, 2$; at $t = 1, t = 3$

(c) During the approximate intervals $0.35 < t < 0.67$, $1.29 < t < 1.75$, and $2.19 < t \le 3$

57.

61.

63. $y = \dfrac{1}{x}\sin x$

$y = \dfrac{1}{x^2}\sin x$

$y = \dfrac{1}{x^3}\sin x$

Review Exercises *(page 585)*

1. $A = 70°$, $b \approx 3.42$, $a \approx 9.40$ **3.** $a \approx 4.58$, $A \approx 66.4°$, $B \approx 23.6°$ **5.** $C = 100°$, $b \approx 0.65$, $c \approx 1.29$ **7.** $B \approx 56.8°$, $C \approx 23.2°$, $b \approx 4.25$
9. No triangle   **11.** $b \approx 3.32$, $A \approx 62.8°$, $C \approx 17.2°$ **13.** No triangle   **15.** $c \approx 2.32$, $A \approx 16.1°$, $B \approx 123.9°$ **17.** $B \approx 36.2°$, $C \approx 63.8°$, $c \approx 4.55$
19. $A \approx 39.6°$, $B \approx 18.6°$, $C \approx 121.9°$ **21.** Two triangles: $B_1 \approx 13.4°$, $C_1 \approx 156.6°$, $c_1 \approx 6.86$ or $B_2 \approx 166.6°$, $C_2 \approx 3.4°$, $c_2 \approx 1.02$
23. $a \approx 5.23$, $B \approx 46.0°$, $C \approx 64.0°$ **25.** 1.93   **27.** 18.79   **29.** 6   **31.** 3.80   **33.** 0.32   **35.** 12.7°   **37.** 29.97 ft   **39.** 6.22 mi   **41.** 71.12 ft

43. $222,983.51   **45.** S4.0°E   **47.** 79.69 in.   **49.** $d = -5\cos\left(\dfrac{\pi}{3}t\right)$ **51. (a)** simple harmonic   **(b)** 2 ft   **(c)** $\dfrac{\pi}{2}$ s   **(d)** $\dfrac{2}{\pi}$ oscillation/s

53. (a) simple harmonic   **(b)** 3 ft   **(c)** 4 s   **(d)** $\dfrac{1}{4}$ oscillation/s

55. (a) $d = -13e^{-0.65t/50}\cos\left(\sqrt{\dfrac{\pi^2}{4} - \dfrac{0.4225}{2500}}\,t\right)$

(b)

57. (a) The motion is damped. The bob has mass
$m = 30$ kg with a damping factor of 0.5 kg/s.
(b) 20 m leftward
(c)

(d) 19.51 m leftward   **(e)** $d \to 0$

59.

Chapter Test *(page 588)*

1. 61.0°   **2.** 1.3°   **3.** $a = 15.88$, $B \approx 57.5°$, $C \approx 70.5°$ **4.** $b \approx 6.85$, $C = 117°$, $c \approx 16.30$ **5.** $A \approx 52.4°$, $B \approx 29.7°$, $C \approx 97.9°$
6. $b \approx 4.72$, $c \approx 1.67$, $B = 105°$ **7.** No triangle   **8.** $c \approx 7.62$, $A \approx 80.5°$, $B \approx 29.5°$ **9.** 15.04 square units   **10.** 19.81 square units
11. The area of the shaded region is 9.26 cm².   **12.** 54.15 square units   **13.** Madison will have to swim about 2.23 miles.   **14.** 12.63 square units
15. The lengths of the sides are 15, 18, and 21.   **16.** $d = 5(\sin 42°)\sin\left(\dfrac{\pi t}{3}\right)$ or $d \approx 3.346\sin\left(\dfrac{\pi t}{3}\right)$

Cumulative Review *(page 589)*

1. $\left\{\dfrac{1}{3}, 1\right\}$ **2.** $(x + 5)^2 + (y - 1)^2 = 9$ **3.** $\{x\,|\,x \le -1 \text{ or } x \ge 4\}$ **4.**   **5.**

6. (a) $-\dfrac{2\sqrt{5}}{5}$ **(b)** $\dfrac{\sqrt{5}}{5}$ **(c)** $-\dfrac{4}{5}$ **(d)** $-\dfrac{3}{5}$ **(e)** $\sqrt{\dfrac{5 - \sqrt{5}}{10}}$ **(f)** $-\sqrt{\dfrac{5 + \sqrt{5}}{10}}$
7. (a)   **(b)**   **(c)**   **(d)**

8. (a)   **(b)**   **(c)**   **(d)**   **(e)**   **(f)**

(g)   **(h)**   **(i)**

9. Two triangles: $A_1 \approx 59.0°$, $B_1 \approx 81.0°$, $b_1 \approx 23.05$
or $A_2 \approx 121.0°$, $B_2 \approx 19.0°$, $b_2 \approx 7.59$

10. $\left\{-2i, 2i, \dfrac{1}{3}, 1, 2\right\}$

11. $R(x) = \dfrac{(2x + 1)(x - 4)}{(x + 5)(x - 3)}$; domain: $\{x\,|\,x \ne -5, x \ne 3\}$

Intercepts: $\left(-\dfrac{1}{2}, 0\right)$, $(4, 0)$, $\left(0, \dfrac{4}{15}\right)$

No symmetry
Vertical asymptotes: $x = -5$, $x = 3$
Horizontal asymptote: $y = 2$

Intersects: $\left(\dfrac{26}{11}, 2\right)$

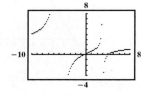

12. $\{2.26\}$ **13.** $\{1\}$ **14. (a)** $\left\{-\dfrac{5}{4}\right\}$ **(b)** $\{2\}$ **(c)** $\left\{\dfrac{-1-3\sqrt{13}}{2}, \dfrac{-1+3\sqrt{13}}{2}\right\}$ **(d)** $\left\{x\,\middle|\,x > -\dfrac{5}{4}\right\}$ or $\left(-\dfrac{5}{4}, \infty\right)$

(e) $\{x\,|\,-8 \le x \le 3\}$ or $[-8, 3]$ **(f)**

(g)

CHAPTER 8 Polar Coordinates; Vectors

8.1 Assess Your Understanding (page 599)

5. pole; polar axis **6.** True **7.** False **8.** $r\cos\theta$; $r\sin\theta$ **9.** A **11.** C **13.** B **15.** A

17. **19.** **21.** **23.** **25.** **27.**

29. **31.**

(a) $\left(5, -\dfrac{4\pi}{3}\right)$ **(b)** $\left(-5, \dfrac{5\pi}{3}\right)$ **(c)** $\left(5, \dfrac{8\pi}{3}\right)$ **33.**

(a) $(2, -2\pi)$ **(b)** $(-2, \pi)$

(c) $(2, 2\pi)$

35. **37.**

39. $(0, 3)$ **41.** $(-2, 0)$ **43.** $(-3\sqrt{3}, 3)$ **45.** $(\sqrt{2}, -\sqrt{2})$ **47.** $\left(-\dfrac{1}{2}, \dfrac{\sqrt{3}}{2}\right)$ **49.** $(2, 0)$

(a) $\left(1, -\dfrac{3\pi}{2}\right)$

(b) $\left(-1, \dfrac{3\pi}{2}\right)$

(c) $\left(1, \dfrac{5\pi}{2}\right)$

(a) $\left(3, -\dfrac{5\pi}{4}\right)$

(b) $\left(-3, \dfrac{7\pi}{4}\right)$

(c) $\left(3, \dfrac{11\pi}{4}\right)$

51. $(-2.57, 7.05)$ **53.** $(-4.98, -3.85)$ **55.** $(3, 0)$ **57.** $(1, \pi)$ **59.** $\left(\sqrt{2}, -\dfrac{\pi}{4}\right)$ **61.** $\left(2, \dfrac{\pi}{6}\right)$

63. $(2.47, -1.02)$ **65.** $(9.30, 0.47)$ **67.** $r^2 = \dfrac{3}{2}$ or $r = \dfrac{\sqrt{6}}{2}$ **69.** $r^2\cos^2\theta - 4r\sin\theta = 0$

71. $r^2\sin 2\theta = 1$ **73.** $r\cos\theta = 4$ **75.** $x^2 + y^2 - x = 0$ or $\left(x - \dfrac{1}{2}\right)^2 + y^2 = \dfrac{1}{4}$

77. $(x^2 + y^2)^{3/2} - x = 0$ **79.** $x^2 + y^2 = 4$ **81.** $y^2 = 8(x + 2)$

83. (a) $(-10, 36)$ **(b)** $\left(2\sqrt{349}, 180° + \tan^{-1}\left(-\dfrac{18}{5}\right)\right) \approx (37.36, 105.5°)$ **(c)** $(-3, -35)$ **(d)** $\left(\sqrt{1234}, 180° + \tan^{-1}\left(\dfrac{35}{3}\right)\right) \approx (35.13, 265.1°)$

8.2 Assess Your Understanding (page 613)

7. polar equation **8.** False **9.** $-\theta$ **10.** $\pi - \theta$ **11.** True **12.** $2n$; n

13. $x^2 + y^2 = 16$; circle, radius 4, center at pole

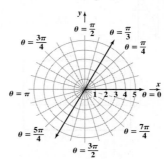

15. $y = \sqrt{3}x$; line through pole, making an angle of $\dfrac{\pi}{3}$ with polar axis

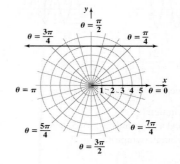

17. $y = 4$; horizontal line 4 units above the pole

19. $x = -2$; vertical line 2 units to the left of the pole

21. $(x - 1)^2 + y^2 = 1$; circle, radius 1, center $(1, 0)$ in rectangular coordinates

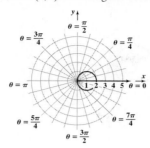

23. $x^2 + (y + 2)^2 = 4$; circle, radius 2, center at $(0, -2)$ in rectangular coordinates

25. $(x - 2)^2 + y^2 = 4$, $x \neq 0$; circle, radius 2, center at $(2, 0)$ in rectangular coordinates, hole at $(0, 0)$

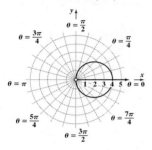

27. $x^2 + (y + 1)^2 = 1$, $x \neq 0$; circle, radius 1, center at $(0, -1)$ in rectangular coordinates, hole at $(0, 0)$

29. E **31.** F **33.** H **35.** D
37. Cardioid

39. Cardioid

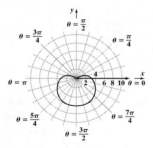

41. Limaçon without inner loop

43. Limaçon without inner loop

45. Limaçon with inner loop

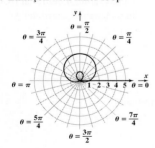

47. Limaçon with inner loop

49. Rose

51. Rose

53. Lemniscate

55. Spiral

57. Cardioid

59. Limaçon with inner loop

61.

63.

65.

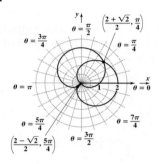

67. $r = 3 + 3\cos\theta$ **69.** $r = 4 + \sin\theta$

71.

73.

75.

77.

79.

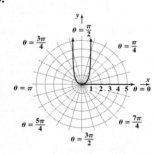

81. $r\sin\theta = a$
$y = a$

83.
$$r = 2a\sin\theta$$
$$r^2 = 2ar\sin\theta$$
$$x^2 + y^2 = 2ay$$
$$x^2 + y^2 - 2ay = 0$$
$$x^2 + (y - a)^2 = a^2$$
Circle, radius a, center at $(0, a)$
in rectangular coordinates

85.
$$r = 2a\cos\theta$$
$$r^2 = 2ar\cos\theta$$
$$x^2 + y^2 = 2ax$$
$$x^2 - 2ax + y^2 = 0$$
$$(x - a)^2 + y^2 = a^2$$
Circle, radius a, center at $(a, 0)$
in rectangular coordinates

87. **(a)** $r^2 = \cos\theta$; $r^2 = \cos(\pi - \theta)$
$$r^2 = -\cos\theta$$
Not equivalent; test fails.
$$(-r)^2 = \cos(-\theta)$$
$$r^2 = \cos\theta$$
New test works.

(b) $r^2 = \sin\theta$: $r^2 = \sin(\pi - \theta)$
$$r^2 = \sin\theta$$
Test works.
$$(-r)^2 = \sin(-\theta)$$
$$r^2 = -\sin\theta$$
Not equivalent; new test fails.

Historical Problems *(page 621)*
1. (a) $1 + 4i, 1 + i$ **(b)** $-1, 2 + i$

8.3 Assess Your Understanding *(page 622)*

5. real; imaginary   **6.** magnitude; modulus; argument   **7.** $r_1 r_2$; $\theta_1 + \theta_2$; $\theta_1 + \theta_2$ **8.** r^n; $n\theta$; $n\theta$ **9.** three   **10.** False

11.

$$\sqrt{2}(\cos 45° + i\sin 45°)$$

13.

$$2(\cos 330° + i\sin 330°)$$

15.

$$3(\cos 270° + i\sin 270°)$$

17.

$$4\sqrt{2}(\cos 315° + i \sin 315°)$$

19.

$$5(\cos 306.9° + i \sin 306.9°)$$

21.

$$\sqrt{13}(\cos 123.7° + i \sin 123.7°)$$

23. $-1 + \sqrt{3}i$ **25.** $2\sqrt{2} - 2\sqrt{2}i$ **27.** $-3i$ **29.** $-0.035 + 0.197i$ **31.** $1.970 + 0.347i$

33. $zw = 8(\cos 60° + i \sin 60°); \dfrac{z}{w} = \dfrac{1}{2}(\cos 20° + i \sin 20°)$ **35.** $zw = 12(\cos 40° + i \sin 40°); \dfrac{z}{w} = \dfrac{3}{4}(\cos 220° + i \sin 220°)$

37. $zw = 4\left(\cos \dfrac{9\pi}{40} + i \sin \dfrac{9\pi}{40}\right); \dfrac{z}{w} = \cos \dfrac{\pi}{40} + i \sin \dfrac{\pi}{40}$ **39.** $zw = 4\sqrt{2}(\cos 15° + i \sin 15°); \dfrac{z}{w} = \sqrt{2}(\cos 75° + i \sin 75°)$

41. $-32 + 32\sqrt{3}i$ **43.** $32i$ **45.** $\dfrac{27}{2} + \dfrac{27\sqrt{3}}{2}i$ **47.** $-\dfrac{25\sqrt{2}}{2} + \dfrac{25\sqrt{2}}{2}i$ **49.** $-4 + 4i$ **51.** $-23 + 14.142i$

53. $\sqrt[6]{2}(\cos 15° + i \sin 15°), \sqrt[6]{2}(\cos 135° + i \sin 135°), \sqrt[6]{2}(\cos 255° + i \sin 255°)$

55. $\sqrt[4]{8}(\cos 75° + i \sin 75°), \sqrt[4]{8}(\cos 165° + i \sin 165°), \sqrt[4]{8}(\cos 255° + i \sin 255°), \sqrt[4]{8}(\cos 345° + i \sin 345°)$

57. $2(\cos 67.5° + i \sin 67.5°), 2(\cos 157.5° + i \sin 157.5°), 2(\cos 247.5° + i \sin 247.5°), 2(\cos 337.5° + i \sin 337.5°)$

59. $\cos 18° + i \sin 18°, \cos 90° + i \sin 90°, \cos 162° + i \sin 162°, \cos 234° + i \sin 234°, \cos 306° + i \sin 306°$

61. $1, i, -1, -i$

63. Look at formula (8); $|z_k| = \sqrt[n]{r}$ for all k.

65. Look at formula (8). The z_k are spaced apart by an angle of $\dfrac{2\pi}{n}$.

67. (a)

$z = a_0$	a_1	a_2	a_3	a_4	a_5	a_6
$0.1 - 0.4i$	$-0.05 - 0.48i$	$-0.13 - 0.35i$	$-0.01 - 0.31i$	$0.004 - 0.395i$	$-0.06 - 0.0i$	$-0.06 - 0.35i$
$0.5 + 0.8i$	$0.11 + 1.6i$	$-2.05 + 1.15i$	$3.37 - 3.92i$	$-3.52 - 25.6i$	$-641.7 + 180.0i$	$379073 - 232071i$
$-0.9 + 0.7i$	$-0.58 - 0.56i$	$-0.88 + 1.35i$	$-1.95 - 1.67i$	$0.13 + 7.21i$	$-52.88 + 2.56i$	$2788.5 - 269.6i$
$-1.1 + 0.1i$	$0.1 - 0.12i$	$-1.10 + 0.76i$	$0.11 - 0.068i$	$-1.09 + 0.085i$	$0.085 - 0.85i$	$-1.10 + 0.086i$
$0 - 1.3i$	$-1.69 - 1.3i$	$1.17 + 3.09i$	$-8.21 + 5.92i$	$32.46 - 98.47i$	$-8643.6 - 6393.7i$	$33833744 + 110529134.4i$
$1 + 1i$	$1 + 3i$	$-7 + 7i$	$1 - 97i$	$-9407 - 193i$	$88454401 + 3631103i$	$7.8 \times 10^{15} + 6.4 \times 10^{14}i$

(b) z_1 and z_4 are in the Mandlebrot set. a_6 for the complex numbers not in the set have very large components.

(c)

z	$|z|$	$|a_6|$
$0.1 - 0.4i$	0.4	0.4
$0.5 + 0.8i$	0.9	444470
$-0.9 + 0.7i$	1.1	2802
$-1.1 + 0.1i$	1.1	1.1
$0 - 1.3i$	1.3	115591573
$1 + 1i$	1.4	7.8×10^{15}

The numbers that are in the Mandlebrot set satisfy the condition $|a_n| \le 2$.

8.4 Assess Your Understanding (page 634)

1. vector **2.** **0** **3.** unit **4.** position **5.** horizontal; vertical **6.** resultant **7.** True **8.** False

9.

11.

13.

15.

17. T **19.** F **21.** F **23.** T **25.** 12 **27.** $\mathbf{v} = 3\mathbf{i} + 4\mathbf{j}$ **29.** $\mathbf{v} = 2\mathbf{i} + 4\mathbf{j}$ **31.** $\mathbf{v} = 8\mathbf{i} - \mathbf{j}$ **33.** $\mathbf{v} = -\mathbf{i} + \mathbf{j}$ **35.** 5 **37.** $\sqrt{2}$ **39.** $\sqrt{13}$ **41.** $-\mathbf{j}$

43. $\sqrt{89}$ **45.** $\sqrt{34} - \sqrt{13}$ **47.** $\mathbf{i}$ **49.** $\dfrac{3}{5}\mathbf{i} - \dfrac{4}{5}\mathbf{j}$ **51.** $\dfrac{\sqrt{2}}{2}\mathbf{i} - \dfrac{\sqrt{2}}{2}\mathbf{j}$ **53.** $\mathbf{v} = \dfrac{8\sqrt{5}}{5}\mathbf{i} + \dfrac{4\sqrt{5}}{5}\mathbf{j}$ or $\mathbf{v} = -\dfrac{8\sqrt{5}}{5}\mathbf{i} - \dfrac{4\sqrt{5}}{5}\mathbf{j}$

55. $\left\{-2 + \sqrt{21}, -2 - \sqrt{21}\right\}$ **57.** $\mathbf{v} = \dfrac{5}{2}\mathbf{i} + \dfrac{5\sqrt{3}}{2}\mathbf{j}$ **59.** $\mathbf{v} = -7\mathbf{i} + 7\sqrt{3}\mathbf{j}$ **61.** $\mathbf{v} = \dfrac{25\sqrt{3}}{2}\mathbf{i} - \dfrac{25}{2}\mathbf{j}$ **63.** 45° **65.** 150° **67.** 333.4° **69.** 258.7°

71. (a) $(-1, 4)$
(b)

73. $\mathbf{F} = 20\sqrt{3}\mathbf{i} + 20\mathbf{j}$ **75.** $\mathbf{F} = \left(20\sqrt{3} + 30\sqrt{2}\right)\mathbf{i} + \left(20 - 30\sqrt{2}\right)\mathbf{j}$

77. (a) $\mathbf{v}_a = 550\mathbf{j}$; $\mathbf{v}_w = 50\sqrt{2}\mathbf{i} + 50\sqrt{2}\mathbf{j}$ **(b)** $\mathbf{v}_g = 50\sqrt{2}\mathbf{i} + (550 + 50\sqrt{2})\mathbf{j}$
(c) $\|\mathbf{v}_g\| = 624.7$ mph; S 80.6°W

91.

79. $\mathbf{v} = (250\sqrt{2} - 30)\mathbf{i} + (250\sqrt{2} + 30\sqrt{3})\mathbf{j}$; 518.8 km/h; N38.6°E
81. Approximately 4031 lb **83.** 8.6° left of direct heading across the river; 1.52 min
85. Tension in right cable: 1000 lb; tension in left cable: 845.2 lb **87.** Tension in right part:
1088.4 lb; tension in left part: 1089.1 lb **89.** The truck must pull with a force of 4635.2 lb.

Historical Problem (page 642)

$(a\mathbf{i} + b\mathbf{j}) \cdot (c\mathbf{i} + d\mathbf{j}) = ac + bd$

Real part $[(\overline{a + bi})(c + di)]$ = real part$[(a - bi)(c + di)]$ = real part$[ac + adi - bci - bdi^2] = ac + bd$

8.5 Assess Your Understanding (page 643)

2. dot product **3.** orthogonal **4.** parallel **5.** T **6.** F **7. (a)** 0 **(b)** 90° **(c)** orthogonal **9. (a)** 0 **(b)** 90° **(c)** orthogonal
11. (a) $\sqrt{3} - 1$ **(b)** 75° **(c)** neither **13. (a)** -50 **(b)** 180° **(c)** parallel **15. (a)** 0 **(b)** 90° **(c)** orthogonal
17. $\dfrac{2}{3}$ **19.** $\mathbf{v}_1 = \dfrac{5}{2}\mathbf{i} - \dfrac{5}{2}\mathbf{j}, \mathbf{v}_2 = -\dfrac{1}{2}\mathbf{i} - \dfrac{1}{2}\mathbf{j}$ **21.** $\mathbf{v}_1 = -\dfrac{1}{5}\mathbf{i} - \dfrac{2}{5}\mathbf{j}, \mathbf{v}_2 = \dfrac{6}{5}\mathbf{i} - \dfrac{3}{5}\mathbf{j}$ **23.** $\mathbf{v}_1 = \dfrac{14}{5}\mathbf{i} + \dfrac{7}{5}\mathbf{j}, \mathbf{v}_2 = \dfrac{1}{5}\mathbf{i} - \dfrac{2}{5}\mathbf{j}$ **25.** 9 ft-lb
27. (a) $\|\mathbf{I}\| \approx 0.022$; the intensity of the sun's rays is approximately 0.022 W/cm². $\|\mathbf{A}\| = 500$; the area of the solar panel is 500 cm².
(b) $W = 10$; ten watts of energy is collected. **(c)** Vectors $\mathbf{I}$ and $\mathbf{A}$ should be parallel with the solar panels facing the sun.
29. Force required to keep Sienna from rolling down the hill: 737.6 lb; force perpendicular to the hill: 5248.4 lb
31. Timmy must exert 85.5 lb. **33.** 60° **35.** Let $\mathbf{v} = a\mathbf{i} + b\mathbf{j}$. Then $\mathbf{0} \cdot \mathbf{v} = 0a + 0b = 0$.
37. $\mathbf{v} = \cos \alpha \mathbf{i} + \sin \alpha \mathbf{j}, 0 \le \alpha \le \pi; \mathbf{w} = \cos \beta \mathbf{i} + \sin \beta \mathbf{j}, 0 \le \beta \le \pi$. If θ is the angle between $\mathbf{v}$ and $\mathbf{w}$, then $\mathbf{v} \cdot \mathbf{w} = \cos\theta$, since $\|\mathbf{v}\| = 1$ and $\|\mathbf{w}\| = 1$.
Now $\theta = \alpha - \beta$ or $\theta = \beta - \alpha$. Since the cosine function is even, $\mathbf{v} \cdot \mathbf{w} = \cos(\alpha - \beta)$. Also, $\mathbf{v} \cdot \mathbf{w} = \cos \alpha \cos \beta + \sin \alpha \sin \beta$. So $\cos(\alpha - \beta)$
$= \cos \alpha \cos \beta + \sin \alpha \sin \beta$.
39. (a) If $\mathbf{u} = a_1\mathbf{i} + b_1\mathbf{j}$ and $\mathbf{v} = a_2\mathbf{i} + b_2\mathbf{j}$, then, since $\|\mathbf{u}\| = \|\mathbf{v}\|, a_1^2 + b_1^2 = \|\mathbf{u}\|^2 = \|\mathbf{v}\|^2 = a_2^2 + b_2^2$,
$(\mathbf{u} + \mathbf{v}) \cdot (\mathbf{u} - \mathbf{v}) = (a_1 + a_2)(a_1 - a_2) + (b_1 + b_2)(b_1 - b_2) = (a_1^2 + b_1^2) - (a_2^2 + b_2^2) = 0$.
(b) The legs of the angle can be made to correspond to vectors $\mathbf{u} + \mathbf{v}$ and $\mathbf{u} - \mathbf{v}$.
41. $(\|\mathbf{w}\|\mathbf{v} + \|\mathbf{v}\|\mathbf{w}) \cdot (\|\mathbf{w}\|\mathbf{v} - \|\mathbf{v}\|\mathbf{w}) = \|\mathbf{w}\|^2\mathbf{v}\cdot\mathbf{v} - \|\mathbf{w}\|\|\mathbf{v}\|\mathbf{v}\cdot\mathbf{w} + \|\mathbf{v}\|\|\mathbf{w}\|\mathbf{w}\cdot\mathbf{v} - \|\mathbf{v}\|^2\mathbf{w}\cdot\mathbf{w} = \|\mathbf{w}\|^2\mathbf{v}\cdot\mathbf{v} - \|\mathbf{v}\|^2\mathbf{w}\cdot\mathbf{w} = \|\mathbf{w}\|^2\|\mathbf{v}\|^2 - \|\mathbf{v}\|^2\|\mathbf{w}\|^2 = 0$
43. $\|\mathbf{u} + \mathbf{v}\|^2 - \|\mathbf{u} - \mathbf{v}\|^2 = (\mathbf{u} + \mathbf{v}) \cdot (\mathbf{u} + \mathbf{v}) - (\mathbf{u} - \mathbf{v}) \cdot (\mathbf{u} - \mathbf{v}) = (\mathbf{u}\cdot\mathbf{u} + \mathbf{u}\cdot\mathbf{v} + \mathbf{v}\cdot\mathbf{u} + \mathbf{v}\cdot\mathbf{v}) - (\mathbf{u}\cdot\mathbf{u} - \mathbf{u}\cdot\mathbf{v} - \mathbf{v}\cdot\mathbf{u} + \mathbf{v}\cdot\mathbf{v})$
$= 2(\mathbf{u}\cdot\mathbf{v}) + 2(\mathbf{v}\cdot\mathbf{u}) = 4(\mathbf{u}\cdot\mathbf{v})$

8.6 Assess Your Understanding (page 652)

2. xy-plane **3.** components **4.** 1 **5.** F **6.** T **7.** All points of the form $(x, 0, z)$ **9.** All points of the form $(x, y, 2)$
11. All points of the form $(-4, y, z)$ **13.** All points of the form $(1, 2, z)$ **15.** $\sqrt{21}$ **17.** $\sqrt{33}$ **19.** $\sqrt{26}$
21. $(2, 0, 0); (2, 1, 0); (0, 1, 0); (2, 0, 3); (0, 1, 3); (0, 0, 3)$ **23.** $(1, 4, 3); (3, 2, 3); (3, 4, 3); (3, 2, 5); (1, 4, 5); (1, 2, 5)$
25. $(-1, 2, 2); (4, 0, 2); (4, 2, 2); (-1, 2, 5); (4, 0, 5); (-1, 0, 5)$ **27.** $\mathbf{v} = 3\mathbf{i} + 4\mathbf{j} - \mathbf{k}$ **29.** $\mathbf{v} = 2\mathbf{i} + 4\mathbf{j} + \mathbf{k}$ **31.** $\mathbf{v} = 8\mathbf{i} - \mathbf{j}$ **33.** 7 **35.** $\sqrt{3}$
37. $\sqrt{22}$ **39.** $-\mathbf{j} - 2\mathbf{k}$ **41.** $\sqrt{105}$ **43.** $\sqrt{38} - \sqrt{17}$ **45.** $\mathbf{i}$ **47.** $\dfrac{3}{7}\mathbf{i} - \dfrac{6}{7}\mathbf{j} - \dfrac{2}{7}\mathbf{k}$ **49.** $\dfrac{\sqrt{3}}{3}\mathbf{i} + \dfrac{\sqrt{3}}{3}\mathbf{j} + \dfrac{\sqrt{3}}{3}\mathbf{k}$ **51.** $\mathbf{v} \cdot \mathbf{w} = 0; \theta = 90°$
53. $\mathbf{v} \cdot \mathbf{w} = -2, \theta \approx 100.3°$ **55.** $\mathbf{v} \cdot \mathbf{w} = 0; \theta = 90°$ **57.** $\mathbf{v} \cdot \mathbf{w} = 52; \theta = 0°$
59. $\alpha \approx 64.6°; \beta \approx 149.0°; \gamma \approx 106.6°; \mathbf{v} = 7(\cos 64.6°\mathbf{i} + \cos 149.0°\mathbf{j} + \cos 106.6°\mathbf{k})$
61. $\alpha = \beta = \gamma \approx 54.7°; \mathbf{v} = \sqrt{3}(\cos 54.7°\mathbf{i} + \cos 54.7°\mathbf{j} + \cos 54.7°\mathbf{k})$ **63.** $\alpha = \beta = 45°; \gamma = 90°; \mathbf{v} = \sqrt{2}(\cos 45°\mathbf{i} + \cos 45°\mathbf{j} + \cos 90°\mathbf{k})$
65. $\alpha \approx 60.9°; \beta \approx 144.2°; \gamma \approx 71.1°; \mathbf{v} = \sqrt{38}(\cos 60.9°\mathbf{i} + \cos 144.2°\mathbf{j} + \cos 71.1°\mathbf{k})$ **67. (a)** $\mathbf{d} = \mathbf{a} + \mathbf{b} + \mathbf{c} = <7, 1, 5>$ **(b)** 8.66 ft
69. $(x - 3)^2 + (y - 1)^2 + (z - 1)^2 = 1$ **71.** Radius = 2, center $(-1, 1, 0)$ **73.** Radius = 3, center $(2, -2, -1)$

75. Radius = $\dfrac{3\sqrt{2}}{2}$, center $(2, 0, -1)$ **77.** 2 newton-meters = 2 joules **79.** 9 newton-meters = 9 joules

8.7 Assess Your Understanding (page 658)

1. T **2.** T **3.** T **5.** F **4.** F **6.** T **7.** 2 **9.** 4 **11.** $-11A + 2B + 5C$ **13.** $-6A + 23B - 15C$ **15. (a)** $5\mathbf{i} + 5\mathbf{j} + 5\mathbf{k}$ **(b)** $-5\mathbf{i} - 5\mathbf{j} - 5\mathbf{k}$
(c) 0 **(d)** 0 **17. (a)** $\mathbf{i} - \mathbf{j} - \mathbf{k}$ **(b)** $-\mathbf{i} + \mathbf{j} + \mathbf{k}$ **(c)** 0 **(d)** 0 **19. (a)** $-\mathbf{i} + 2\mathbf{j} + 2\mathbf{k}$ **(b)** $\mathbf{i} - 2\mathbf{j} - 2\mathbf{k}$ **(c)** 0 **(d)** 0 **21. (a)** $3\mathbf{i} - \mathbf{j} + 4\mathbf{k}$
(b) $-3\mathbf{i} + \mathbf{j} - 4\mathbf{k}$ **(c)** 0 **(d)** 0 **23.** $-9\mathbf{i} - 7\mathbf{j} - 3\mathbf{k}$ **25.** $9\mathbf{i} + 7\mathbf{j} + 3\mathbf{k}$ **27.** 0 **29.** $-27\mathbf{i} - 21\mathbf{j} - 9\mathbf{k}$ **31.** $-18\mathbf{i} - 14\mathbf{j} - 6\mathbf{k}$ **33.** 0 **35.** -25
37. 25 **39.** 0 **41.** Any vector of the form $c(-9\mathbf{i} - 7\mathbf{j} - 3\mathbf{k})$, where c is a nonzero scalar **43.** Any vector of the form $c(-\mathbf{i} + \mathbf{j} + 5\mathbf{k})$, where c is a
nonzero scalar **45.** $\sqrt{166}$ **47.** $\sqrt{555}$ **49.** $\sqrt{34}$ **51.** $\sqrt{998}$ **53.** $\dfrac{11\sqrt{19}}{57}\mathbf{i} + \dfrac{\sqrt{19}}{57}\mathbf{j} + \dfrac{7\sqrt{19}}{57}\mathbf{k}$ or $-\dfrac{11\sqrt{19}}{57}\mathbf{i} - \dfrac{\sqrt{19}}{57}\mathbf{j} - \dfrac{7\sqrt{19}}{57}\mathbf{k}$

57. $\mathbf{u} \times \mathbf{v} = \begin{vmatrix} \mathbf{i} & \mathbf{j} & \mathbf{k} \\ a_1 & b_1 & c_1 \\ a_2 & b_2 & c_2 \end{vmatrix} = (b_1c_2 - b_2c_1)\mathbf{i} - (a_1c_2 - a_2c_1)\mathbf{j} + (a_1b_2 - a_2b_1)\mathbf{k}$

$\|\mathbf{u} \times \mathbf{v}\|^2 = \left(\sqrt{(b_1c_2 - b_2c_1)^2 + (a_1c_2 - a_2c_1)^2 + (a_1b_2 - a_2b_1)^2} \right)^2$

$= b_1^2c_2^2 - 2b_1b_2c_1c_2 + b_2^2c_1^2 + a_1^2c_2^2 - 2a_1a_2c_1c_2 + a_2^2c_1^2 + a_1^2b_2^2 - 2a_1a_2b_1b_2 + a_2^2b_1^2$

$\|\mathbf{u}\|^2 = a_1^2 + b_1^2 + c_1^2, \|\mathbf{v}\|^2 = a_2^2 + b_2^2 + c_2^2$

$\|\mathbf{u}\|^2\|\mathbf{v}\|^2 = (a_1^2 + b_1^2 + c_1^2)(a_2^2 + b_2^2 + c_2^2) = a_1^2a_2^2 + a_1^2b_2^2 + a_1^2c_2^2 + b_1^2a_2^2 + b_1^2b_2^2 + b_1^2c_2^2 + a_2^2c_1^2 + b_2^2c_1^2 + c_1^2c_2^2$

$(\mathbf{u} \cdot \mathbf{v})^2 = (a_1a_2 + b_1b_2 + c_1c_2)^2 = (a_1a_2 + b_1b_2 + c_1c_2)(a_1a_2 + b_1b_2 + c_1c_2)$

$= a_1^2a_2^2 + a_1a_2b_1b_2 + a_1a_2c_1c_2 + b_1b_2c_1c_2 + b_1b_2a_1a_2 + b_1^2b_2^2 + b_1b_2c_1c_2 + a_1a_2c_1c_2 + c_1^2c_2^2$

$= a_1^2a_2^2 + b_1^2b_2^2 + c_1^2c_2^2 + 2a_1a_2b_1b_2 + 2b_1b_2c_1c_2 + 2a_1a_2c_1c_2$

$\|\mathbf{u}\|^2\|\mathbf{v}\|^2 - (\mathbf{u} \cdot \mathbf{v})^2 = a_1^2b_2^2 + a_1^2c_2^2 + b_1^2a_2^2 + a_2^2c_1^2 + b_2^2c_1^2 + b_1^2c_2^2 - 2a_1a_2b_1b_2 - 2b_1b_2c_1c_2 - 2a_1a_2c_1c_2$, which equals $\|\mathbf{u} \times \mathbf{v}\|^2$.

59. By Problem 58, since $\mathbf{u}$ and $\mathbf{v}$ are orthogonal, $\|\mathbf{u} \times \mathbf{v}\| = \|\mathbf{u}\|\|\mathbf{v}\|$. If, in addition, $\mathbf{u}$ and $\mathbf{v}$ are unit vectors, $\|\mathbf{u} \times \mathbf{v}\| = 1 \cdot 1 = 1$.

61. Assume that $\mathbf{u} = a\mathbf{i} + b\mathbf{j} + c\mathbf{k}$, $\mathbf{v} = d\mathbf{i} + e\mathbf{j} + f\mathbf{k}$, and $\mathbf{w} = l\mathbf{i} + m\mathbf{j} + n\mathbf{k}$. Then $\mathbf{u} \times \mathbf{v} = (bf - ec)\mathbf{i} - (af - dc)\mathbf{j} + (ae - db)\mathbf{k}$,

$\mathbf{u} \times \mathbf{w} = (bn - mc)\mathbf{i} - (an - lc)\mathbf{j} + (am - lb)\mathbf{k}$, and $\mathbf{v} + \mathbf{w} = (d + l)\mathbf{i} + (e + m)\mathbf{j} + (f + n)\mathbf{k}$. Therefore,

$(\mathbf{u} \times \mathbf{v}) + (\mathbf{u} \times \mathbf{w}) = (bf - ec + bn - mc)\mathbf{i} - (af - dc + an - lc)\mathbf{j} + (ae - db + am - lb)\mathbf{k}$ and $\mathbf{u} \times (\mathbf{v} + \mathbf{w})$

$= [b(f + n) - (e + m)c]\mathbf{i} - [a(f + n) - (d + l)c]\mathbf{j} + [a(e + m) - (d + l)b]\mathbf{k}$

$= (bf - ec + bn - mc)\mathbf{i} - (af - dc + an - lc)\mathbf{j} + (ae - db + am - lb)\mathbf{k}$, which equals $(\mathbf{u} \times \mathbf{v}) + (\mathbf{u} \times \mathbf{w})$.

Review Exercises *(page 660)*

1. $\left(\dfrac{3\sqrt{3}}{2}, \dfrac{3}{2} \right)$

3. $(1, \sqrt{3})$

5. $(0, 3)$

7. $\left(3\sqrt{2}, \dfrac{3\pi}{4} \right), \left(-3\sqrt{2}, -\dfrac{\pi}{4} \right)$

9. $\left(2, -\dfrac{\pi}{2} \right), \left(-2, \dfrac{\pi}{2} \right)$

11. $(5, 0.93), (-5, 4.07)$

13. **(a)** $x^2 + (y - 1)^2 = 1$ **(b)** circle, radius 1, center $(0, 1)$ in rectangular coordinates

15. **(a)** $x^2 + y^2 = 25$ **(b)** circle, radius 5, center at pole

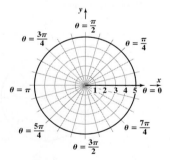

17. **(a)** $x + 3y = 6$ **(b)** line through $(6, 0)$ and $(0, 2)$ in rectangular coordinates

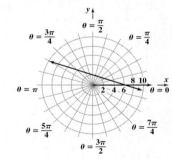

19. Circle; radius 2, center at $(2, 0)$ in rectangular coordinates; symmetric with respect to the polar axis

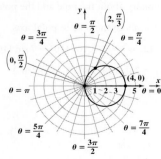

21. Cardioid; symmetric with respect to the line $\theta = \dfrac{\pi}{2}$

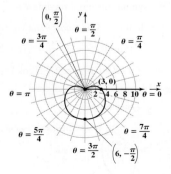

23. Limaçon without inner loop; symmetric with respect to the polar axis

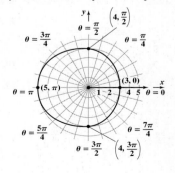

25. $\sqrt{2}(\cos 225° + i \sin 225°)$ **27.** $5(\cos 323.1° + i \sin 323.1°)$

29. $-\sqrt{3} + i$

31. $-\dfrac{3}{2} + \dfrac{3\sqrt{3}}{2}i$

33. $0.10 - 0.02i$

35. $zw = \cos 130° + i \sin 130°$; $\dfrac{z}{w} = \cos 30° + i \sin 30°$ **37.** $zw = 6(\cos 0 + i \sin 0) = 6$; $\dfrac{z}{w} = \dfrac{3}{2}\left(\cos \dfrac{8\pi}{5} + i \sin \dfrac{8\pi}{5}\right)$

39. $zw = 5(\cos 5° + i \sin 5°)$; $\dfrac{z}{w} = 5(\cos 15° + i \sin 15°)$ **41.** $\dfrac{27}{2} + \dfrac{27\sqrt{3}}{2}i$ **43.** $4i$ **45.** 64 **47.** $-527 - 336i$

49. $3, 3(\cos 120° + i \sin 120°), 3(\cos 240° + i \sin 240°)$ or $3, -\dfrac{3}{2} + \dfrac{3\sqrt{3}}{2}i, -\dfrac{3}{2} - \dfrac{3\sqrt{3}}{2}i$

51.

53.

55. $\mathbf{v} = 2\mathbf{i} - 4\mathbf{j}$; $\|\mathbf{v}\| = 2\sqrt{5}$ **57.** $\mathbf{v} = -\mathbf{i} + 3\mathbf{j}$; $\|\mathbf{v}\| = \sqrt{10}$ **59.** $2\mathbf{i} - 2\mathbf{j}$ **61.** $-20\mathbf{i} + 13\mathbf{j}$

63. $\sqrt{5}$ **65.** $\sqrt{5} + 5 \approx 7.24$ **67.** $-\dfrac{2\sqrt{5}}{5}\mathbf{i} + \dfrac{\sqrt{5}}{5}\mathbf{j}$ **69.** $\mathbf{v} = \dfrac{3}{2}\mathbf{i} + \dfrac{3\sqrt{3}}{2}\mathbf{j}$ **71.** $120°$

73. $\sqrt{43} \approx 6.56$ **75.** $\mathbf{v} = 3\mathbf{i} - 5\mathbf{j} + 3\mathbf{k}$ **77.** $21\mathbf{i} - 2\mathbf{j} - 5\mathbf{k}$ **79.** $\sqrt{38}$

81. 0 **83.** $3\mathbf{i} + 9\mathbf{j} + 9\mathbf{k}$ **85.** $\dfrac{3\sqrt{14}}{14}\mathbf{i} + \dfrac{\sqrt{14}}{14}\mathbf{j} - \dfrac{\sqrt{14}}{7}\mathbf{k}$; $-\dfrac{3\sqrt{14}}{14}\mathbf{i} - \dfrac{\sqrt{14}}{14}\mathbf{j} + \dfrac{\sqrt{14}}{7}\mathbf{k}$

87. $\mathbf{v} \cdot \mathbf{w} = -11$; $\theta \approx 169.7°$ **89.** $\mathbf{v} \cdot \mathbf{w} = -4$; $\theta \approx 153.4°$ **91.** $\mathbf{v} \cdot \mathbf{w} = 1$; $\theta \approx 70.5°$

93. $\mathbf{v} \cdot \mathbf{w} = 0$; $\theta = 90°$ **95.** Parallel   **97.** Parallel   **99.** Orthogonal   **101.** $\mathbf{v}_1 = \dfrac{4}{5}\mathbf{i} - \dfrac{3}{5}\mathbf{j}$; $\mathbf{v}_2 = \dfrac{6}{5}\mathbf{i} + \dfrac{8}{5}\mathbf{j}$ **103.** $\mathbf{v}_1 = \dfrac{9}{10}(3\mathbf{i} + \mathbf{j})$; $\mathbf{v}_2 = -\dfrac{7}{10}\mathbf{i} + \dfrac{21}{10}\mathbf{j}$

105. $a \approx 56.1°$; $\beta \approx 138°$; $\gamma \approx 68.2°$ **107.** $2\sqrt{83}$ **109.** $-2\mathbf{i} + 3\mathbf{j} - \mathbf{k}$ **111.** $\sqrt{29} \approx 5.39$ mi/hr; 0.4 mi
113. Left cable: 1843.21 lb; right cable: 1630.41 lb   **115.** A force of 697.2 lb is needed to keep the van from rolling down the hill. The magnitude of the force on the hill is 7969.6 lb.

Chapter Test (page 663)

1–3.

4. $\left(4, \dfrac{\pi}{3}\right)$ **5.** $x^2 + y^2 = 49$ **6.** $\dfrac{y}{x} = 3$ or $y = 3x$

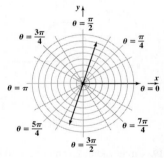

7. $8y = x^2$

8. $r^2 \cos \theta = 5$ is symmetric about the pole, the polar axis, and the line $\theta = \dfrac{\pi}{2}$.

9. $r = 5 \sin \theta \cos^2 \theta$ is symmetric about the line $\theta = \dfrac{\pi}{2}$. The tests for symmetry about the pole and the polar axis fail, so the graph of $r = 5 \sin \theta \cos^2 \theta$ may or may not be symmetric about the pole or polar axis.

10. $z \cdot w = 6(\cos 107° + i \sin 107°)$ **11.** $\dfrac{w}{z} = \dfrac{3}{2}(\cos 297° + i \sin 297°)$ **12.** $w^5 = 243(\cos 110° + i \sin 110°)$

13. $z_0 = 2\sqrt[3]{2}(\cos 40° + i \sin 40°)$, $z_1 = 2\sqrt[3]{2}(\cos 160° + i \sin 160°)$, $z_2 = 2\sqrt[3]{2}(\cos 280° + i \sin 280°)$

14. $\mathbf{v} = \langle 5\sqrt{2}, -5\sqrt{2} \rangle$ **15.** $\|\mathbf{v}\| = 10$ **16.** $\mathbf{u} = \dfrac{\mathbf{v}}{\|\mathbf{v}\|} = \left\langle \dfrac{\sqrt{2}}{2}, -\dfrac{\sqrt{2}}{2} \right\rangle$ **17.** $315°$ off the positive x-axis **18.** $\mathbf{v} = 5\sqrt{2}\mathbf{i} - 5\sqrt{2}\mathbf{j}$

19. $\mathbf{v}_1 + 2\mathbf{v}_2 - \mathbf{v}_3 = \langle 6, -10 \rangle$ **20.** Vectors $\mathbf{v}_1$ and $\mathbf{v}_4$ are parallel. **21.** Vectors $\mathbf{v}_2$ and $\mathbf{v}_3$ are orthogonal. **22.** $172.87°$ **23.** $-9\mathbf{i} - 5\mathbf{j} + 3\mathbf{k}$

24. $\alpha \approx 57.7°, \beta \approx 143.3°, \gamma \approx 74.5°$ **25.** $\sqrt{115}$ **26.** The cable must be able to endure a tension of approximately 670.82 lb.

Cumulative Review *(page 663)*

1. $\{-3, 3\}$ **2.** $y = \dfrac{\sqrt{3}}{3}x$ **3.** $x^2 + (y - 1)^2 = 9$ **4.** $\left\{ x \,\middle|\, x < \dfrac{1}{2} \right\}$ or $\left(-\infty, \dfrac{1}{2} \right)$ **5.** Symmetry with respect to the y-axis **6.**

7.   **8.**   **9.** $-\dfrac{\pi}{6}$ **10.**   **11.**   **12.** Amplitude: 4; period: 2

CHAPTER 9 Analytic Geometry

9.2 Assess Your Understanding *(page 673)*

6. parabola **7.** (c) **8.** $(3, 2)$ **9.** $(3, 6)$ **10.** $y = -2$ **11.** B **13.** E **15.** H **17.** C

19. $y^2 = 16x$ **21.** $x^2 = -12y$ **23.** $y^2 = -8x$ **25.** $x^2 = 2y$

27. $x^2 = \dfrac{4}{3}y$ **29.** $(x - 2)^2 = -8(y + 3)$ **31.** $(y + 2)^2 = 4(x + 1)$ **33.** $(x + 3)^2 = 4(y - 3)$

35. $(y + 2)^2 = -8(x + 1)$ **37.** Vertex: $(0, 0)$; focus: $(0, 1)$; directrix: $y = -1$ **39.** Vertex: $(0, 0)$; focus: $(-4, 0)$; directrix: $x = 4$ **41.** Vertex: $(-1, 2)$; focus: $(1, 2)$; directrix: $x = -3$

43. Vertex: $(3, -1)$; focus: $\left(3, -\dfrac{5}{4} \right)$; directrix: $y = -\dfrac{3}{4}$ **45.** Vertex: $(2, -3)$; focus: $(4, -3)$; directrix: $x = 0$ **47.** Vertex: $(0, 2)$; focus: $(-1, 2)$; directrix: $x = 1$ **49.** Vertex: $(-4, -2)$; focus: $(-4, -1)$; directrix: $y = -3$

51. Vertex: $(-1, -1)$; focus: $\left(-\dfrac{3}{4}, -1\right)$;   **53.** Vertex: $(2, -8)$; focus: $\left(2, -\dfrac{31}{4}\right)$;

directrix: $x = -\dfrac{5}{4}$ directrix: $y = -\dfrac{33}{4}$

55. $(y - 1)^2 = x$ **57.** $(y - 1)^2 = -(x - 2)$

59. $x^2 = 4(y - 1)$ **61.** $y^2 = \dfrac{1}{2}(x + 2)$

63. 1.5625 ft from the base of the dish, along the axis of symmetry
65. 1 in. from the vertex, along the axis of symmetry
67. 20 ft   **69.** 0.78125 ft
71. 4.17 ft from the base, along the axis of symmetry
73. 24.31 ft, 18.75 ft, 7.64 ft

75. (a) $y = -\dfrac{625}{(299)^2}x^2 + 625$

    (b) 567 ft: 63.12 ft; 478 ft: 225.67 ft; 308 ft: 459.2 ft   **(c)** No

77. $Cy^2 + Dx = 0, C \neq 0, D \neq 0$ This is the equation of a parabola with vertex at $(0, 0)$ and axis of symmetry the x-axis.

$Cy^2 = -Dx$ The focus is $\left(-\dfrac{D}{4C}, 0\right)$; the directrix is the line $x = \dfrac{D}{4C}$. The parabola opens to the right if

$y^2 = -\dfrac{D}{C}x$ $-\dfrac{D}{C} > 0$ and to the left if $-\dfrac{D}{C} < 0$.

79. $Cy^2 + Dx + Ey + F = 0, C \neq 0$ **(a)** If $D \neq 0$, then the equation may be written as

$Cy^2 + Ey = -Dx - F$ $\left(y + \dfrac{E}{2C}\right)^2 = -\dfrac{D}{C}\left(x - \dfrac{E^2 - 4CF}{4CD}\right).$

$y^2 + \dfrac{E}{C}y = -\dfrac{D}{C}x - \dfrac{F}{C}$ This is the equation of a parabola with vertex at $\left(\dfrac{E^2 - 4CF}{4CD}, -\dfrac{E}{2C}\right)$

$\left(y + \dfrac{E}{2C}\right)^2 = -\dfrac{D}{C}x - \dfrac{F}{C} + \dfrac{E^2}{4C^2}$ and axis of symmetry parallel to the x-axis.

$\left(y + \dfrac{E}{2C}\right)^2 = -\dfrac{D}{C}x + \dfrac{E^2 - 4CF}{4C^2}$ **(b)–(d)** If $D = 0$, the graph of the equation contains no points if $E^2 - 4CF < 0$, is a single horizontal line if $E^2 - 4CF = 0$, and is two horizontal lines if $E^2 - 4CF > 0$.

9.3 Assess Your Understanding *(page 683)*

7. ellipse   **8.** major   **9.** $b^2 = a^2 - c^2$ **10.** 5; 3; x **11.** $(-2, -3)$; $(6, -3)$ **12.** $(1, 4)$ **13.** C   **15.** B

17. Vertices: $(-5, 0)$, $(5, 0)$
    Foci: $(-\sqrt{21}, 0)$, $(\sqrt{21}, 0)$

19. Vertices: $(0, -5)$, $(0, 5)$
    Foci: $(0, -4)$, $(0, 4)$

21. $\dfrac{x^2}{4} + \dfrac{y^2}{16} = 1$

    Vertices: $(0, -4)$, $(0, 4)$
    Foci: $(0, -2\sqrt{3})$, $(0, 2\sqrt{3})$

23. $\dfrac{x^2}{8} + \dfrac{y^2}{2} = 1$

    Vertices: $\left(-2\sqrt{2}, 0\right)$, $\left(2\sqrt{2}, 0\right)$
    Foci: $\left(-\sqrt{6}, 0\right)$, $\left(\sqrt{6}, 0\right)$

25. $\dfrac{x^2}{16} + \dfrac{y^2}{16} = 1$

    Vertices: $(-4, 0)$, $(4, 0)$,
    $(0, -4)$, $(0, 4)$; Focus: $(0, 0)$

27. $\dfrac{x^2}{25} + \dfrac{y^2}{16} = 1$

29. $\dfrac{x^2}{9} + \dfrac{y^2}{25} = 1$

31. $\dfrac{x^2}{9} + \dfrac{y^2}{5} = 1$

33. $\dfrac{x^2}{25} + \dfrac{y^2}{9} = 1$

35. $\dfrac{x^2}{4} + \dfrac{y^2}{13} = 1$

37. $x^2 + \dfrac{y^2}{16} = 1$

39. $\dfrac{(x + 1)^2}{4} + (y - 1)^2 = 1$

41. $(x - 1)^2 + \dfrac{y^2}{4} = 1$

43. Center: $(3, -1)$; vertices: $(3, -4), (3, 2)$;
foci: $(3, -1 - \sqrt{5}), (3, -1 + \sqrt{5})$

45. $\dfrac{(x + 5)^2}{16} + \dfrac{(y - 4)^2}{4} = 1$
Center: $(-5, 4)$; vertices: $(-9, 4), (-1, 4)$;
foci: $(-5 - 2\sqrt{3}, 4), (-5 + 2\sqrt{3}, 4)$

47. $\dfrac{(x + 2)^2}{4} + (y - 1)^2 = 1$
Center: $(-2, 1)$; vertices: $(-4, 1), (0, 1)$;
foci: $(-2 - \sqrt{3}, 1), (-2 + \sqrt{3}, 1)$

49. $\dfrac{(x - 2)^2}{3} + \dfrac{(y + 1)^2}{2} = 1$
Center: $(2, -1)$; vertices: $(2 - \sqrt{3}, -1)$;
$(2 + \sqrt{3}, -1)$; foci: $(1, -1), (3, -1)$

51. $\dfrac{(x - 1)^2}{4} + \dfrac{(y + 2)^2}{9} = 1$
Center: $(1, -2)$; vertices: $(1, -5), (1, 1)$;
foci: $(1, -2 - \sqrt{5}), (1, -2 + \sqrt{5})$

53. $x^2 + \dfrac{(y + 2)^2}{4} = 1$
Center: $(0, -2)$; vertices: $(0, -4), (0, 0)$;
foci: $(0, -2 - \sqrt{3}), (0, -2 + \sqrt{3})$

55. $\dfrac{(x - 2)^2}{25} + \dfrac{(y + 2)^2}{21} = 1$

57. $\dfrac{(x - 4)^2}{5} + \dfrac{(y - 6)^2}{9} = 1$

59. $\dfrac{(x - 2)^2}{16} + \dfrac{(y - 1)^2}{7} = 1$

61. $\dfrac{(x - 1)^2}{10} + (y - 2)^2 = 1$

63. $\dfrac{(x - 1)^2}{9} + \dfrac{(y - 2)^2}{9} = 1$

65.

67.

69. $\dfrac{x^2}{100} + \dfrac{y^2}{36} = 1$ **71.** 43.3 ft   **73.** 24.65 ft, 21.65 ft, 13.82 ft   **75.** 30 ft   **77.** The elliptical hole will have a major axis of length $2\sqrt{41}$ in. and a minor axis of length 8 in.   **79.** 91.5 million mi; $\dfrac{x^2}{(93)^2} + \dfrac{y^2}{8646.75} = 1$

81. Perihelion: 460.6 million mi; mean distance: 483.8 million mi; $\dfrac{x^2}{(483.8)^2} + \dfrac{y^2}{233,524.2} = 1$

83. (a) $Ax^2 + Cy^2 + F = 0$ If A and C are of the same sign and F is of opposite sign, then the equation takes the form
$$Ax^2 + Cy^2 = -F \qquad \dfrac{x^2}{\left(-\dfrac{F}{A}\right)} + \dfrac{y^2}{\left(-\dfrac{F}{C}\right)} = 1, \text{ where } -\dfrac{F}{A} \text{ and } -\dfrac{F}{C} \text{ are positive. This is the equation of an ellipse with center at } (0, 0).$$

(b) If $A = C$, the equation may be written as $x^2 + y^2 = -\dfrac{F}{A}$.

This is the equation of a circle with center at $(0, 0)$ and radius equal to $\sqrt{-\dfrac{F}{A}}$.

9.4 Assess Your Understanding *(page 696)*

7. hyperbola   **8.** transverse axis   **9.** b
10. $(2, 4); (2, -2)$ **11.** $(2, 6); (2, -4)$
12. 4   **13.** $2; 3; x$
14. $y = -\dfrac{4}{9}x; y = \dfrac{4}{9}x$ **15.** B   **17.** A

19. $x^2 - \dfrac{y^2}{8} = 1$

21. $\dfrac{y^2}{16} - \dfrac{x^2}{20} = 1$

23. $\dfrac{x^2}{9} - \dfrac{y^2}{16} = 1$

25. $\dfrac{y^2}{36} - \dfrac{x^2}{9} = 1$

27. $\dfrac{x^2}{8} - \dfrac{y^2}{8} = 1$

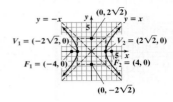

29. $\dfrac{x^2}{25} - \dfrac{y^2}{9} = 1$

Center: $(0, 0)$
Transverse axis: x-axis
Vertices: $(-5, 0), (5, 0)$
Foci: $\left(-\sqrt{34}, 0\right), \left(\sqrt{34}, 0\right)$
Asymptotes: $y = \pm\dfrac{3}{5}x$

31. $\dfrac{x^2}{4} - \dfrac{y^2}{16} = 1$

Center: $(0, 0)$
Transverse axis: x-axis
Vertices: $(-2, 0), (2, 0)$
Foci: $\left(-2\sqrt{5}, 0\right), \left(2\sqrt{5}, 0\right)$
Asymptotes: $y = \pm 2x$

33. $\dfrac{y^2}{9} - x^2 = 1$

Center: $(0, 0)$
Transverse axis: y-axis
Vertices: $(0, -3), (0, 3)$
Foci: $\left(0, -\sqrt{10}\right), \left(0, \sqrt{10}\right)$
Asymptotes: $y = \pm 3x$

35. $\dfrac{y^2}{25} - \dfrac{x^2}{25} = 1$

Center: $(0, 0)$
Transverse axis: y-axis
Vertices: $(0, -5), (0, 5)$
Foci: $\left(0, -5\sqrt{2}\right), \left(0, 5\sqrt{2}\right)$
Asymptotes: $y = \pm x$

37. $x^2 - y^2 = 1$

39. $\dfrac{y^2}{36} - \dfrac{x^2}{9} = 1$

41. $\dfrac{(x - 4)^2}{4} - \dfrac{(y + 1)^2}{5} = 1$

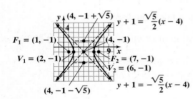

43. $\dfrac{(y + 4)^2}{4} - \dfrac{(x + 3)^2}{12} = 1$

45. $(x - 5)^2 - \dfrac{(y - 7)^2}{3} = 1$

47. $\dfrac{(x - 1)^2}{4} - \dfrac{(y + 1)^2}{9} = 1$

49. $\dfrac{(x - 2)^2}{4} - \dfrac{(y + 3)^2}{9} = 1$

Center: $(2, -3)$
Transverse axis: parallel to x-axis
Vertices: $(0, -3), (4, -3)$
Foci: $\left(2 - \sqrt{13}, -3\right), \left(2 + \sqrt{13}, -3\right)$
Asymptotes: $y + 3 = \pm\dfrac{3}{2}(x - 2)$

51. $\dfrac{(y-2)^2}{4} - (x+2)^2 = 1$

Center: $(-2, 2)$
Transverse axis: parallel to y-axis
Vertices: $(-2, 0), (-2, 4)$
Foci: $\left(-2, 2 - \sqrt{5}\right), \left(-2, 2 + \sqrt{5}\right)$
Asymptotes: $y - 2 = \pm 2(x + 2)$

53. $\dfrac{(x+1)^2}{4} - \dfrac{(y+2)^2}{4} = 1$

Center: $(-1, -2)$
Transverse axis: parallel to x-axis
Vertices: $(-3, -2), (1, -2)$
Foci: $\left(-1 - 2\sqrt{2}, -2\right), \left(-1 + 2\sqrt{2}, -2\right)$
Asymptotes: $y + 2 = \pm(x + 1)$

55. $(x-1)^2 - (y+1)^2 = 1$

Center: $(1, -1)$
Transverse axis: parallel to x-axis
Vertices: $(0, -1), (2, -1)$
Foci: $\left(1 - \sqrt{2}, -1\right), \left(1 + \sqrt{2}, -1\right)$
Asymptotes: $y + 1 = \pm(x - 1)$

57. $\dfrac{(y-2)^2}{4} - (x+1)^2 = 1$

Center: $(-1, 2)$
Transverse axis: parallel to y-axis
Vertices: $(-1, 0), (-1, 4)$
Foci: $\left(-1, 2 - \sqrt{5}\right), \left(-1, 2 + \sqrt{5}\right)$
Asymptotes: $y - 2 = \pm 2(x + 1)$

59. $\dfrac{(x-3)^2}{4} - \dfrac{(y+2)^2}{16} = 1$

Center: $(3, -2)$
Transverse axis: parallel to x-axis
Vertices: $(1, -2), (5, -2)$
Foci: $\left(3 - 2\sqrt{5}, -2\right), \left(3 + 2\sqrt{5}, -2\right)$
Asymptotes: $y + 2 = \pm 2(x - 3)$

61. $\dfrac{(y-1)^2}{4} - (x+2)^2 = 1$

Center: $(-2, 1)$
Transverse axis: parallel to y-axis
Vertices: $(-2, -1), (-2, 3)$
Foci: $\left(-2, 1 - \sqrt{5}\right), \left(-2, 1 + \sqrt{5}\right)$
Asymptotes: $y - 1 = \pm 2(x + 2)$

63.

65.

67. Center: $(3, 0)$

Transverse axis: parallel to x-axis
Vertices: $(1, 0), (5, 0)$
Foci: $(3 - \sqrt{29}, 0), (3 + \sqrt{29}, 0)$
Asymptotes: $y = \pm \dfrac{5}{2}(x - 3)$

69. Vertex: $(0, 3)$; focus: $(0, 7)$;
directrix: $y = -1$

71. $\dfrac{(x-5)^2}{9} + \dfrac{y^2}{25} = 1$

Center: $(5, 0)$; vertices: $(5, 5), (5, -5)$;
foci: $(5, -4), (5, 4)$

73. $(x-3)^2 = 8(y+5)$
Vertex: $(3, -5)$; foci: $(3, -3)$;
directrix: $y = -7$

75. The fireworks display is 50,138 ft north of the person at point A.   **77.** The tower is 592.4 ft tall.   **79. (a)** $y = \pm x$ **(b)** $\dfrac{x^2}{100} - \dfrac{y^2}{100} = 1, x \geq 0$

81. If the eccentricity is close to one, the "opening" of the hyperbola is very small. As e increases, the opening gets bigger.

83. $\dfrac{x^2}{4} - y^2 = 1$; asymptotes $y = \pm\dfrac{1}{2}x$

$y^2 - \dfrac{x^2}{4} = 1$; asymptotes $y = \pm\dfrac{1}{2}x$

85. $Ax^2 + Cy^2 + F = 0$ If A and C are of opposite sign and $F \neq 0$, this equation may be written as $\dfrac{x^2}{\left(-\dfrac{F}{A}\right)} + \dfrac{y^2}{\left(-\dfrac{F}{C}\right)} = 1,$

$Ax^2 + Cy^2 = -F$ where $-\dfrac{F}{A}$ and $-\dfrac{F}{C}$ are opposite in sign. This is the equation of a hyperbola with center $(0,0)$.
The transverse axis is the x-axis if $-\dfrac{F}{A} > 0$; the transverse axis is the y-axis if $-\dfrac{F}{A} < 0$.

9.5 Assess Your Understanding *(page 705)*

5. $\cot(2\theta) = \dfrac{A-C}{B}$ **6.** Parabola **7.** $B^2 - 4AC < 0$ **8.** T **9.** T **10.** F **11.** Parabola **13.** Ellipse **15.** Hyperbola

17. Hyperbola **19.** Circle **21.** $x = \dfrac{\sqrt{2}}{2}(x' - y'), y = \dfrac{\sqrt{2}}{2}(x' + y')$ **23.** $x = \dfrac{\sqrt{2}}{2}(x' - y'), y = \dfrac{\sqrt{2}}{2}(x' + y')$

25. $x = \dfrac{1}{2}(x' - \sqrt{3}y'), y = \dfrac{1}{2}(\sqrt{3}x' + y')$ **27.** $x = \dfrac{\sqrt{5}}{5}(x' - 2y'), y = \dfrac{\sqrt{5}}{5}(2x' + y')$ **29.** $x = \dfrac{\sqrt{13}}{13}(3x' - 2y'), y = \dfrac{\sqrt{13}}{13}(2x' + 3y')$

31. $\theta = 45°$ (see Problem 21)

$x'^2 - \dfrac{y'^2}{3} = 1$

Hyperbola
Center at origin
Transverse axis is the x'-axis.
Vertices at $(\pm 1, 0)$

33. $\theta = 45°$ (see Problem 23)

$x'^2 + \dfrac{y'^2}{4} = 1$

Ellipse
Center at $(0,0)$
Major axis is the y'-axis.
Vertices at $(0, \pm 2)$

35. $\theta = 60°$ (see Problem 25)

$\dfrac{x'^2}{4} + y'^2 = 1$

Ellipse
Center at $(0,0)$
Major axis is the x'-axis.
Vertices at $(\pm 2, 0)$

37. $\theta \approx 63°$ (see Problem 27)

$y'^2 = 8x'$

Parabola
Vertex at $(0,0)$
Focus at $(2,0)$

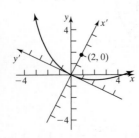

39. $\theta \approx 34°$ (see Problem 29)

$\dfrac{(x'-2)^2}{4} + y'^2 = 1$

Ellipse
Center at $(2,0)$
Major axis is the x'-axis.

Vertices at $(4,0)$ and $(0,0)$

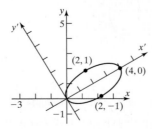

41. $\cot(2\theta) = \dfrac{7}{24}$;

$\theta = \sin^{-1}\left(\dfrac{3}{5}\right) \approx 37°$

$(x'-1)^2 = -6\left(y' - \dfrac{1}{6}\right)$

Parabola
Vertex at $\left(1, \dfrac{1}{6}\right)$

Focus at $\left(1, -\dfrac{4}{3}\right)$

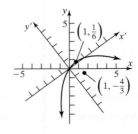

43. Hyperbola **45.** Hyperbola **47.** Parabola **49.** Ellipse **51.** Ellipse

53. Refer to equation (6): $A' = A\cos^2\theta + B\sin\theta\cos\theta + C\sin^2\theta$
$B' = B(\cos^2\theta - \sin^2\theta) + 2(C - A)(\sin\theta\cos\theta)$
$C' = A\sin^2\theta - B\sin\theta\cos\theta + C\cos^2\theta$
$D' = D\cos\theta + E\sin\theta$
$E' = -D\sin\theta + E\cos\theta$
$F' = F$

55. Use Problem 53 to find $B'^2 - 4A'C'$. After much cancellation, $B'^2 - 4A'C' = B^2 - 4AC$.

57. The distance between P_1 and P_2 in the $x'y'$-plane equals $\sqrt{(x_2' - x_1')^2 + (y_2' - y_1')^2}$.
Assuming that $x' = x\cos\theta - y\sin\theta$ and $y' = x\sin\theta + y\cos\theta$, then
$(x_2' - x_1')^2 = (x_2\cos\theta - y_2\sin\theta - x_1\cos\theta + y_1\sin\theta)^2$
$= \cos^2\theta(x_2 - x_1)^2 - 2\sin\theta\cos\theta(x_2 - x_1)(y_2 - y_1) + \sin^2\theta(y_2 - y_1)^2$, and
$(y_2' - y_1')^2 = (x_2\sin\theta + y_2\cos\theta - x_1\sin\theta - y_1\cos\theta)^2 = \sin^2\theta(x_2 - x_1)^2 + 2\sin\theta\cos\theta(x_2 - x_1)(y_2 - y_1) + \cos^2\theta(y_2 - y_1)^2$.
Therefore, $(x_2' - x_1')^2 + (y_2' - y_1')^2 = \cos^2\theta(x_2 - x_1)^2 + \sin^2\theta(x_2 - x_1)^2 + \sin^2\theta(y_2 - y_1)^2 + \cos^2\theta(y_2 - y_1)^2$
$= (x_2 - x_1)^2(\cos^2\theta + \sin^2\theta) + (y_2 - y_1)^2(\sin^2\theta + \cos^2\theta) = (x_2 - x_1)^2 + (y_2 - y_1)^2$.

9.6 Assess Your Understanding *(page 711)*

3. conic; focus; directrix **4.** $1; <1; >1$ **5.** T **6.** T **7.** Parabola; directrix is perpendicular to the polar axis 1 unit to the right of the pole.

9. Hyperbola; directrix is parallel to the polar axis $\dfrac{4}{3}$ units below the pole.

11. Ellipse; directrix is perpendicular to the polar axis $\dfrac{3}{2}$ units to the left of the pole.

13. Parabola; directrix is perpendicular to the polar axis 1 unit to the right of the pole; vertex is at $\left(\dfrac{1}{2}, 0\right)$.

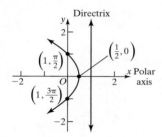

15. Ellipse; directrix is parallel to the polar axis $\dfrac{8}{3}$ units above the pole; vertices are at $\left(\dfrac{8}{7}, \dfrac{\pi}{2}\right)$ and $\left(8, \dfrac{3\pi}{2}\right)$.

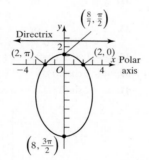

17. Hyperbola; directrix is perpendicular to the polar axis $\dfrac{3}{2}$ units to the left of the pole; vertices are at $(-3, 0)$ and $(1, \pi)$.

19. Ellipse; directrix is parallel to the polar axis 8 units below the pole; vertices are at $\left(8, \dfrac{\pi}{2}\right)$ and $\left(\dfrac{8}{3}, \dfrac{3\pi}{2}\right)$.

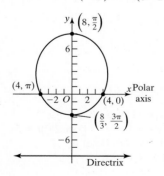

21. Ellipse; directrix is parallel to the polar axis 3 units below the pole; vertices are at $\left(6, \dfrac{\pi}{2}\right)$ and $\left(\dfrac{6}{5}, \dfrac{3\pi}{2}\right)$.

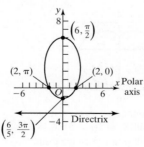

23. Ellipse; directrix is perpendicular to the polar axis 6 units to the left of the pole; vertices are at $(6, 0)$ and $(2, \pi)$.

25. $y^2 + 2x - 1 = 0$ **27.** $16x^2 + 7y^2 + 48y - 64 = 0$ **29.** $3x^2 - y^2 + 12x + 9 = 0$ **31.** $4x^2 + 3y^2 - 16y - 64 = 0$

33. $9x^2 + 5y^2 - 24y - 36 = 0$ **35.** $3x^2 + 4y^2 - 12x - 36 = 0$ **37.** $r = \dfrac{1}{1 + \sin\theta}$ **39.** $r = \dfrac{12}{5 - 4\cos\theta}$ **41.** $r = \dfrac{12}{1 - 6\sin\theta}$

43. Use $d(D, P) = p - r\cos\theta$ in the derivation of equation (a) in Table 5.

45. Use $d(D, P) = p + r\sin\theta$ in the derivation of equation (a) in Table 5.

9.7 Assess Your Understanding *(page 722)*

2. plane curve; parameter **3.** ellipse **4.** cycloid **5.** F **6.** T

7.

$x - 3y + 1 = 0$

9.

$y = \sqrt{x - 2}$

11.

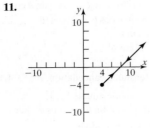

$y = x - 8$

13.

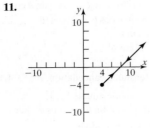

$x = 3(y - 1)^2$

15.

$2y = 2 + x$

17.

$y = x^3$

19.

$\dfrac{x^2}{4} + \dfrac{y^2}{9} = 1$

21.

$\dfrac{x^2}{4} + \dfrac{y^2}{9} = 1$

23.

$(\sqrt{2}, 1)$

$x^2 - y^2 = 1$

25.

back and forth twice

$x + y = 1$

27. $x = t$ or $x = \dfrac{t + 1}{4}$
$y = 4t - 1$ $y = t$

29. $x = t$ or $x = t^3$
$y = t^2 + 1$ $y = t^6 + 1$

31. $x = t$ or $x = \sqrt[3]{t}$
$y = t^3$ $y = t$

33. $x = t$ or $x = t^3$
$y = t^{2/3}, t \geq 0$ $y = t^2, t \geq 0$

35. $x = t + 2, y = t, 0 \leq t \leq 5$ **37.** $x = 3\cos t, y = 2\sin t, 0 \leq t \leq 2\pi$

39. $x = 2\cos(\pi t), y = -3\sin(\pi t), 0 \leq t \leq 2$

41. $x = 2\sin(2\pi t), y = 3\cos(2\pi t), 0 \leq t \leq 1$

43.

$(-4, 16)$ $(4, 16)$ 16
$(-1, 1)$ $(1, 1)$
-4 $(0, 0)$ C_1 4

$(-1, 1)$ $(1, 1)$ 16
-4 C_2 4

$(4, 16)$ 16
$(1, 1)$
-4 C_3 4

$(4, 16)$ 16
$(1, 1)$
-4 C_4 4

45.

47.

49. (a) $x = 3$
$y = -16t^2 + 50t + 6$
(b) 3.24 s
(c) 1.56 s; 45.06 ft
(d)

51. (a) Train: $x_1 = t^2, y_1 = 1$;
Bill: $x_2 = 5(t - 5), y_2 = 3$
(b) Bill won't catch the train.
(c)

Bill
Train
0 $t = 9.5$ 100

53. (a) $x = (145\cos 20°)t$
$y = -16t^2 + (145\sin 20°)t + 5$
(b) 3.20 s **(c)** 435.61 ft
(d) 1.55 s; 43.43 ft
(e)

170
0 440
-120

55. (a) $x = (40 \cos 45°)t$

$y = -4.9t^2 + (40 \sin 45°)t + 300$

(b) 11.23 s   **(c)** 317.52 m

(d) 2.89 s; 340.82 m

(e)

57. (a) Camry: $x = 40t - 5$, $y = 0$; Chevy Impala: $x = 0$, $y = 30t - 4$

(b) $d = \sqrt{(40t - 5)^2 + (30t - 4)^2}$

(c)

(d) 0.2 mi; 7.68 min

(e) Turn axes off to see the graph:

59. (a) $x = \dfrac{\sqrt{2}}{2} v_0 t$, $y = -16t^2 + \dfrac{\sqrt{2}}{2} v_0 t + 3$ **(b)** Maximum height is 139.1 ft.   **(c)** The ball is 272.25 ft from home plate.   **(d)** Yes, the ball will clear the wall by about 99.5 ft.   **61.** The orientation is from (x_1, y_1) to (x_2, y_2).

Review Exercises (page 726)

1. Parabola; vertex $(0, 0)$, focus $(-4, 0)$, directrix $x = 4$ **3.** Hyperbola; center $(0, 0)$, vertices $(5, 0)$ and $(-5, 0)$, foci $\left(\sqrt{26}, 0\right)$ and $\left(-\sqrt{26}, 0\right)$, asymptotes $y = \dfrac{1}{5}x$ and $y = -\dfrac{1}{5}x$ **5.** Ellipse; center $(0, 0)$, vertices $(0, 5)$ and $(0, -5)$, foci $(0, 3)$ and $(0, -3)$ **7.** $x^2 = -4(y - 1)$: Parabola; vertex $(0, 1)$, focus $(0, 0)$, directrix $y = 2$ **9.** $\dfrac{x^2}{2} - \dfrac{y^2}{8} = 1$: Hyperbola; center $(0, 0)$, vertices $\left(\sqrt{2}, 0\right)$ and $\left(-\sqrt{2}, 0\right)$, foci $\left(\sqrt{10}, 0\right)$ and $\left(-\sqrt{10}, 0\right)$, asymptotes $y = 2x$ and $y = -2x$ **11.** $(x - 2)^2 = 2(y + 2)$: Parabola; vertex $(2, -2)$, focus $\left(2, -\dfrac{3}{2}\right)$, directrix $y = -\dfrac{5}{2}$

13. $\dfrac{(y - 2)^2}{4} - (x - 1)^2 = 1$: Hyperbola; center $(1, 2)$, vertices $(1, 4)$ and $(1, 0)$, foci $\left(1, 2 + \sqrt{5}\right)$ and $\left(1, 2 - \sqrt{5}\right)$, asymptotes $y - 2 = \pm 2(x - 1)$

15. $\dfrac{(x - 2)^2}{9} + \dfrac{(y - 1)^2}{4} = 1$: Ellipse; center $(2, 1)$, vertices $(5, 1)$ and $(-1, 1)$, foci $\left(2 + \sqrt{5}, 1\right)$ and $\left(2 - \sqrt{5}, 1\right)$

17. $(x - 2)^2 = -4(y + 1)$: Parabola; vertex $(2, -1)$, focus $(2, -2)$, directrix $y = 0$

19. $\dfrac{(x - 1)^2}{4} + \dfrac{(y + 1)^2}{9} = 1$: Ellipse; center $(1, -1)$, vertices $(1, 2)$ and $(1, -4)$, foci $\left(1, -1 + \sqrt{5}\right)$ and $\left(1, -1 - \sqrt{5}\right)$

21. $y^2 = -8x$

23. $\dfrac{y^2}{4} - \dfrac{x^2}{12} = 1$

25. $\dfrac{x^2}{16} + \dfrac{y^2}{7} = 1$

27. $(x - 2)^2 = -4(y + 3)$

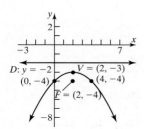

29. $(x + 2)^2 - \dfrac{(y + 3)^2}{3} = 1$

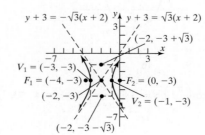

31. $\dfrac{(x + 4)^2}{16} + \dfrac{(y - 5)^2}{25} = 1$

AN78 **ANSWERS** Chapter 9 Review Exercises

33. $\dfrac{(x+1)^2}{9} - \dfrac{(y-2)^2}{7} = 1$

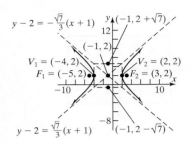

35. $\dfrac{(x-3)^2}{9} - \dfrac{(y-1)^2}{4} = 1$

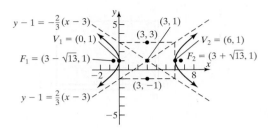

37. Parabola   **39.** Ellipse
41. Parabola   **43.** Hyperbola
45. Ellipse

47. $x'^2 - \dfrac{y'^2}{9} = 1$
Hyperbola
Center at the origin
Transverse axis the x'-axis
Vertices at $(\pm 1, 0)$

49. $\dfrac{x'^2}{2} + \dfrac{y'^2}{4} = 1$
Ellipse
Center at origin
Major axis the y'-axis
Vertices at $(0, \pm 2)$

51. $y'^2 = -\dfrac{4\sqrt{13}}{13}x'$
Parabola
Vertex at the origin
Focus on the x'-axis at $\left(-\dfrac{\sqrt{13}}{13}, 0\right)$

53. Parabola; directrix is perpendicular to the polar axis 4 units to the left of the pole; vertex is $(2, \pi)$.

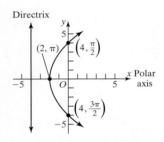

55. Ellipse; directrix is parallel to the polar axis 6 units below the pole; vertices are $\left(6, \dfrac{\pi}{2}\right)$ and $\left(2, \dfrac{3\pi}{2}\right)$.

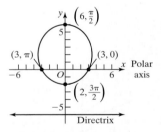

57. Hyperbola; directrix is perpendicular to the polar axis 1 unit to the right of the pole; vertices are $\left(\dfrac{2}{3}, 0\right)$ and $(-2, \pi)$.

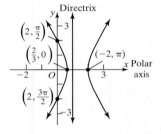

59. $y^2 - 8x - 16 = 0$ **61.** $3x^2 - y^2 - 8x + 4 = 0$

63.

$x + 4y = 2$

65.

$\dfrac{x^2}{9} + \dfrac{(y-2)^2}{16} = 1$

67.

$1 + y = x$

69. $x = t, y = -2t + 4, -\infty < t < \infty$ **71.** $x = 4\cos\left(\dfrac{\pi}{2}t\right), y = 3\sin\left(\dfrac{\pi}{2}t\right), 0 \le t \le 4$ **73.** $\dfrac{x^2}{5} - \dfrac{y^2}{4} = 1$ **75.** The ellipse $\dfrac{x^2}{16} + \dfrac{y^2}{7} = 1$

$x = \dfrac{t-4}{-2}, y = t, -\infty < t < \infty$

77. $\frac{1}{4}$ ft or 3 in.   **79.** 19.72 ft, 18.86 ft, 14.91 ft   **81.** 450 ft

83. (a) $x = (80\cos 35°)t$

$y = -16t^2 + (80\sin 35°)t + 6$

(b) 2.9932 s
(c) 1.4339 s; 38.9 ft
(d) 196.15 ft

(e)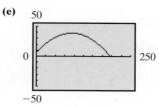

Chapter Test *(page 728)*

1. Hyperbola; center: $(-1, 0)$; vertices: $(-3, 0)$ and $(1, 0)$; foci: $\left(-1 - \sqrt{13}, 0\right)$ and $\left(-1 + \sqrt{13}, 0\right)$; asymptotes: $y = -\frac{3}{2}(x + 1)$ and $y = \frac{3}{2}(x + 1)$

2. Parabola; vertex: $\left(1, -\frac{1}{2}\right)$; focus: $\left(1, \frac{3}{2}\right)$; directrix: $y = -\frac{5}{2}$

3. Ellipse; center: $(-1, 1)$; foci: $\left(-1 - \sqrt{3}, 1\right)$ and $\left(-1 + \sqrt{3}, 1\right)$; vertices: $(-4, 1)$ and $(2, 1)$

4. $(x + 1)^2 = 6(y - 3)$

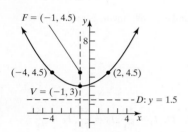

5. $\dfrac{x^2}{7} + \dfrac{y^2}{16} = 1$

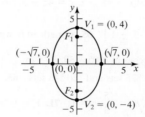

6. $\dfrac{(y - 2)^2}{4} - \dfrac{(x - 2)^2}{8} = 1$

7. Hyperbola   **8.** Ellipse   **9.** Parabola

10. $x'^2 + 2y'^2 = 1$. This is the equation of an ellipse with center at $(0, 0)$ in the $x'y'$-plane. The vertices are at $(-1, 0)$ and $(1, 0)$ in the $x'y'$-plane.

11. Hyperbola; $(x + 2)^2 - \dfrac{y^2}{3} = 1$ **12.** $y = 1 - \sqrt{\dfrac{x + 2}{3}}$

13. The microphone should be located $\frac{2}{3}$ ft from the base of the reflector, along its axis of symmetry.

Cumulative Review *(page 728)*

1. $-6x + 5 - 3h$ **2.** $\left\{-5, -\frac{1}{3}, 2\right\}$ **3.** $\{x \mid -3 \le x \le 2\}$ or $[-3, 2]$ **4. (a)** Domain: $(-\infty, \infty)$; range: $(2, \infty)$

(b) $y = \log_3(x - 2)$; domain: $(2, \infty)$; range: $(-\infty, \infty)$ **5. (a)** $\{18\}$ **(b)** $(2, 18]$ **6. (a)** $y = 2x - 2$ **(b)** $(x - 2)^2 + y^2 = 4$ **(c)** $\dfrac{x^2}{9} + \dfrac{y^2}{4} = 1$

(d) $y = 2(x - 1)^2$ **(e)** $y^2 - \dfrac{x^2}{3} = 1$ **(f)** $y = 4^x$ **7.** $\theta = \dfrac{\pi}{12} \pm \pi k$, k is any integer; $\theta = \dfrac{5\pi}{12} \pm \pi k$, k is any integer   **8.** $\theta = \dfrac{\pi}{6}$

9. $r = 8\sin\theta$ **10.** $\left\{x \mid x \neq \dfrac{3\pi}{4} \pm \pi k, k \text{ is an integer}\right\}$ **11.** $\{22.5°\}$ **12.** $y = \dfrac{x^2}{5} + 5$

CHAPTER 10   Systems of Equations and Inequalities

10.1 Assess Your Understanding *(page 741)*

3. inconsistent   **4.** consistent; independent   **5.** $(3, -2)$ **6.** consistent; dependent

7. $\begin{cases} 2(2) - (-1) = 5 \\ 5(2) + 2(-1) = 8 \end{cases}$ **9.** $\begin{cases} 3(2) - 4\left(\dfrac{1}{2}\right) = 4 \\ \dfrac{1}{2}(2) - 3\left(\dfrac{1}{2}\right) = -\dfrac{1}{2} \end{cases}$ **11.** $\begin{cases} 4 - 1 = 3 \\ \dfrac{1}{2}(4) + 1 = 3 \end{cases}$ **13.** $\begin{cases} 3(1) + 3(-1) + 2(2) = 4 \\ 1 - (-1) - 2 = 0 \\ 2(-1) - 3(2) = -8 \end{cases}$

15. $\begin{cases} 3(2) + 3(-2) + 2(2) = 4 \\ 2 - 3(-2) + 2 = 10 \\ 5(2) - 2(-2) - 3(2) = 8 \end{cases}$ **17.** $x = 6, y = 2;\ (6, 2)$ **19.** $x = 3, y = -6;\ (3, -6)$ **21.** $x = 8, y = -4;\ (8, -4)$

23. $x = \dfrac{1}{3}, y = -\dfrac{1}{6};\ \left(\dfrac{1}{3}, -\dfrac{1}{6}\right)$ **25.** Inconsistent   **27.** $x = \dfrac{3}{2}, y = 3;\ \left(\dfrac{3}{2}, 3\right)$

29. $\{(x, y)\,|\,x = 4 - 2y, y \text{ is any real number}\}$ or $\left\{(x, y)\,\middle|\,y = \dfrac{4 - x}{2}, x \text{ is any real number}\right\}$ **31.** $x = 1, y = 1;\ (1, 1)$ **33.** $x = \dfrac{3}{2}, y = 1;\ \left(\dfrac{3}{2}, 1\right)$

35. $x = 4, y = 3;\ (4, 3)$ **37.** $x = \dfrac{4}{3}, y = \dfrac{1}{5};\ \left(\dfrac{4}{3}, \dfrac{1}{5}\right)$ **39.** $x = \dfrac{1}{5}, y = \dfrac{1}{3};\ \left(\dfrac{1}{5}, \dfrac{1}{3}\right)$ **41.** $x = 8, y = 2, z = 0;\ (8, 2, 0)$

43. $x = 2, y = -1, z = 1;\ (2, -1, 1)$ **45.** Inconsistent   **47.** $\{(x, y, z)\,|\,x = 5z - 2, y = 4z - 3; z \text{ is any real number}\}$ **49.** Inconsistent

51. $x = 1, y = 3, z = -2;\ (1, 3, -2)$ **53.** $x = -3, y = \dfrac{1}{2}, z = 1;\ \left(-3, \dfrac{1}{2}, 1\right)$ **55.** Length 30 ft; width 15 ft

57. There were 18 commercial launches and 37 noncommercial launches in 2005.   **59.** 22.5 lb

61. Average wind speed 25 mph; average airspeed 175 mph   **63.** 80 \$25 sets and 120 \$45 sets   **65.** \$9.96

67. Mix 50 mg of first compound with 75 mg of second.   **69.** $a = \dfrac{4}{3}, b = -\dfrac{5}{3}, c = 1$ **71.** $Y = 9000, r = 0.06$ **73.** $I_1 = \dfrac{10}{71}, I_2 = \dfrac{65}{71}, I_3 = \dfrac{55}{71}$

75. 100 orchestra, 210 main, and 190 balcony seats   **77.** 1.5 chicken, 1 corn, 2 milk

79. If $x = $ price of hamburgers, $y = $ price of fries, and $z = $ price of colas, then

$x = 2.75 - z, y = \dfrac{41}{60} + \dfrac{1}{3}z, \$0.60 \le z \le \$0.90.$

x	\$2.13	\$2.01	\$1.86
y	\$0.89	\$0.93	\$0.98
z	\$0.62	\$0.74	\$0.89

There is not sufficient information:

81. It will take Beth 30 hr, Bill 24 hr, and Edie 40 hr.

10.2 Assess Your Understanding *(page 757)*

1. matrix   **2.** augmented   **3.** third; fifth   **4.** T

5. $\begin{bmatrix} 1 & -5 & | & 5 \\ 4 & 3 & | & 6 \end{bmatrix}$ **7.** $\begin{bmatrix} 2 & 3 & | & 6 \\ 4 & -6 & | & -2 \end{bmatrix}$ **9.** $\begin{bmatrix} 0.01 & -0.03 & | & 0.06 \\ 0.13 & 0.10 & | & 0.20 \end{bmatrix}$ **11.** $\begin{bmatrix} 1 & -1 & 1 & | & 10 \\ 3 & 3 & 0 & | & 5 \\ 1 & 1 & 2 & | & 2 \end{bmatrix}$ **13.** $\begin{bmatrix} 1 & 1 & -1 & | & 2 \\ 3 & -2 & 0 & | & 2 \\ 5 & 3 & -1 & | & 1 \end{bmatrix}$ **15.** $\begin{bmatrix} 1 & -1 & -1 & | & 10 \\ 2 & 1 & 2 & | & -1 \\ -3 & 4 & 0 & | & 5 \\ 4 & -5 & 1 & | & 0 \end{bmatrix}$

17. $\begin{cases} x - 3y = -2 & (1) \\ 2x - 5y = 5 & (2) \end{cases}$ $\begin{bmatrix} 1 & -3 & | & -2 \\ 0 & 1 & | & 9 \end{bmatrix}$ **19.** $\begin{cases} x - 3y + 4z = 3 & (1) \\ 3x - 5y + 6z = 6 & (2) \\ -5x + 3y + 4z = 6 & (3) \end{cases}$ **(a)** $\begin{bmatrix} 1 & -3 & 4 & | & 3 \\ 0 & 4 & -6 & | & -3 \\ -5 & 3 & 4 & | & 6 \end{bmatrix}$ **(b)** $\begin{bmatrix} 1 & -3 & 4 & | & 3 \\ 3 & -5 & 6 & | & 6 \\ 0 & -12 & 24 & | & 21 \end{bmatrix}$

21. $\begin{cases} x - 3y + 2z = -6 & (1) \\ 2x - 5y + 3z = -4 & (2) \\ -3x - 6y + 4z = 6 & (3) \end{cases}$ **(a)** $\begin{bmatrix} 1 & -3 & 2 & | & -6 \\ 0 & 1 & -1 & | & 8 \\ -3 & -6 & 4 & | & 6 \end{bmatrix}$ **(b)** $\begin{bmatrix} 1 & -3 & 2 & | & -6 \\ 2 & -5 & 3 & | & -4 \\ 0 & -15 & 10 & | & -12 \end{bmatrix}$

23. $\begin{cases} 5x - 3y + z = -2 & (1) \\ 2x - 5y + 6z = -2 & (2) \\ -4x + y + 4z = 6 & (3) \end{cases}$ **(a)** $\begin{bmatrix} 1 & 7 & -11 & | & 2 \\ 2 & -5 & 6 & | & -2 \\ -4 & 1 & 4 & | & 6 \end{bmatrix}$ **(b)** $\begin{bmatrix} 1 & -2 & 5 & | & 4 \\ 2 & -5 & 6 & | & -2 \\ -4 & 1 & 4 & | & 6 \end{bmatrix}$

25. $\begin{cases} x = 5 \\ y = -1 \end{cases}$
Consistent; $x = 5, y = -1$ or $(5, -1)$

27. $\begin{cases} x = 1 \\ y = 2 \\ 0 = 3 \end{cases}$
Inconsistent

29. $\begin{cases} x + 2z = -1 \\ y - 4z = -2 \\ 0 = 0 \end{cases}$
Consistent:
$\begin{cases} x = -1 - 2z \\ y = -2 + 4z \\ z \text{ is any real number or} \end{cases}$
$\{(x, y, z) \mid x = -1 - 2z,$
$y = -2 + 4z, z \text{ is any real}$
number$\}$

31. $\begin{cases} x_1 = 1 \\ x_2 + x_4 = 2 \\ x_3 + 2x_4 = 3 \end{cases}$
Consistent:
$\begin{cases} x_1 = 1, x_2 = 2 - x_4 \\ x_3 = 3 - 2x_4 \\ x_4 \text{ is any real number or} \end{cases}$
$\{(x_1, x_2, x_3, x_4) \mid x_1 = 1,$
$x_2 = 2 - x_4, x_3 = 3 - 2x_4,$
$x_4 \text{ is any real number}\}$

33. $\begin{cases} x_1 + 4x_4 = 2 \\ x_2 + x_3 + 3x_4 = 3 \\ 0 = 0 \end{cases}$
Consistent:
$\begin{cases} x_1 = 2 - 4x_4 \\ x_2 = 3 - x_3 - 3x_4 \\ x_3, x_4 \text{ are any real numbers or} \end{cases}$
$\{(x_1, x_2, x_3, x_4) \mid x_1 = 2 - 4x_4,$
$x_2 = 3 - x_3 - 3x_4, x_3, x_4 \text{ are}$
any real numbers$\}$

35. $\begin{cases} x_1 + x_4 = -2 \\ x_2 + 2x_4 = 2 \\ x_3 - x_4 = 0 \\ 0 = 0 \end{cases}$
Consistent:
$\begin{cases} x_1 = -2 - x_4 \\ x_2 = 2 - 2x_4 \\ x_3 = x_4 \\ x_4 \text{ is any real number or} \end{cases}$
$\{(x_1, x_2, x_3, x_4) \mid x_1 = -2 - x_4,$
$x_2 = 2 - 2x_4, x_3 = x_4, x_4 \text{ is any}$
real number$\}$

37. $x = 6, y = 2; (6, 2)$ **39.** $x = \dfrac{1}{2}, y = \dfrac{3}{4}; \left(\dfrac{1}{2}, \dfrac{3}{4}\right)$ **41.** $x = 4 - 2y, y$ is any real number; $\{(x, y) \mid x = 4 - 2y, y \text{ is any real number}\}$

43. $x = \dfrac{3}{2}, y = 1; \left(\dfrac{3}{2}, 1\right)$ **45.** $x = \dfrac{4}{3}, y = \dfrac{1}{5}; \left(\dfrac{4}{3}, \dfrac{1}{5}\right)$ **47.** $x = 8, y = 2, z = 0; (8, 2, 0)$ **49.** $x = 2, y = -1, z = 1; (2, -1, 1)$

51. Inconsistent **53.** $x = 5z - 2, y = 4z - 3$, where z is any real number; $\{(x, y, z) \mid x = 5z - 2, y = 4z - 3, z \text{ is any real number}\}$

55. Inconsistent **57.** $x = 1, y = 3, z = -2; (1, 3, -2)$ **59.** $x = -3, y = \dfrac{1}{2}, z = 1; \left(-3, \dfrac{1}{2}, 1\right)$ **61.** $x = \dfrac{1}{3}, y = \dfrac{2}{3}, z = 1; \left(\dfrac{1}{3}, \dfrac{2}{3}, 1\right)$

63. $x = 1, y = 2, z = 0, w = 1; (1, 2, 0, 1)$ **65.** $y = 0, z = 1 - x, x$ is any real number; $\{(x, y, z) \mid y = 0, z = 1 - x, x \text{ is any real number}\}$

67. $x = 2, y = z - 3, z$ is any real number; $\{(x, y, z) \mid x = 2, y = z - 3, z \text{ is any real number}\}$ **69.** $x = \dfrac{13}{9}, y = \dfrac{7}{18}, z = \dfrac{19}{18}; \left(\dfrac{13}{9}, \dfrac{7}{18}, \dfrac{19}{18}\right)$

71. $x = \dfrac{7}{5} - \dfrac{3}{5}z - \dfrac{2}{5}w, y = -\dfrac{8}{5} + \dfrac{7}{5}z + \dfrac{13}{5}w$, where z and w are any real numbers; $\left\{(x, y, z, w) \,\middle|\, x = \dfrac{7}{5} - \dfrac{3}{5}z - \dfrac{2}{5}w, y = -\dfrac{8}{5} + \dfrac{7}{5}z + \dfrac{13}{5}w,\right.$

z and w are any real numbers$\Big\}$ **73.** $y = -2x^2 + x + 3$ **75.** $f(x) = 3x^3 - 4x^2 + 5$ **77.** 1.5 salmon steak, 2 baked eggs, 1 acorn squash

79. $4000 in Treasury bills, $4000 in Treasury bonds, $2000 in corporate bonds **81.** 8 Deltas, 5 Betas, 10 Sigmas **83.** $I_1 = \dfrac{44}{23}, I_2 = 2, I_3 = \dfrac{16}{23}, I_4 = \dfrac{28}{23}$

85. (a)

Amount Invested At		
7%	9%	11%
0	10,000	10,000
1000	8000	11,000
2000	6000	12,000
3000	4000	13,000
4000	2000	14,000
5000	0	15,000

(b)

Amount Invested At		
7%	9%	11%
12,500	12,500	0
14,500	8500	2000
16,500	4500	4000
18,750	0	6250

(c) All the money invested at 7% provides $2100, more than what is required.

87.

First Supplement	Second Supplement	Third Supplement
50 mg	75 mg	0 mg
36 mg	76 mg	8 mg
22 mg	77 mg	16 mg
8 mg	78 mg	24 mg

10.3 Assess Your Understanding *(page 768)*

1. $ad - bc$ **2.** $\begin{vmatrix} 5 & 3 \\ -3 & -4 \end{vmatrix}$ **3.** F **4.** F **5.** F **6.** F **7.** 22 **9.** -2 **11.** 10 **13.** -26 **15.** $x = 6, y = 2; (6, 2)$ **17.** $x = 3, y = 2; (3, 2)$

19. $x = 8, y = -4; (8, -4)$ **21.** $x = 4, y = -2; (4, -2)$ **23.** Not applicable **25.** $x = \dfrac{1}{2}, y = \dfrac{3}{4}; \left(\dfrac{1}{2}, \dfrac{3}{4}\right)$ **27.** $x = \dfrac{1}{10}, y = \dfrac{2}{5}; \left(\dfrac{1}{10}, \dfrac{2}{5}\right)$

29. $x = \dfrac{3}{2}, y = 1; \left(\dfrac{3}{2}, 1\right)$ **31.** $x = \dfrac{4}{3}, y = \dfrac{1}{5}; \left(\dfrac{4}{3}, \dfrac{1}{5}\right)$ **33.** $x = 1, y = 3, z = -2; (1, 3, -2)$ **35.** $x = -3, y = \dfrac{1}{2}, z = 1; \left(-3, \dfrac{1}{2}, 1\right)$

37. Not applicable **39.** $x = 0, y = 0, z = 0; (0, 0, 0)$ **41.** Not applicable **43.** -4 **45.** 12 **47.** 8 **49.** 8 **51.** -5 **53.** $\dfrac{13}{11}$ **55.** 0 or -9

57. $(y_1 - y_2)x - (x_1 - x_2)y + (x_1y_2 - x_2y_1) = 0$ **59.** The triangle has an area of 5 square units.

$(y_1 - y_2)x + (x_2 - x_1)y = x_2y_1 - x_1y_2$

$(x_2 - x_1)y - (x_2 - x_1)y_1 = (y_2 - y_1)x + x_2y_1 - x_1y_2 - (x_2 - x_1)y_1$

$(x_2 - x_1)(y - y_1) = (y_2 - y_1)x - (y_2 - y_1)x_1$

$y - y_1 = \dfrac{y_2 - y_1}{x_2 - x_1}(x - x_1)$

61. If $a = 0$, we have

$$by = s$$
$$cx + dy = t$$

Thus, $y = \dfrac{s}{b}$ and

$$x = \dfrac{t - dy}{c} = \dfrac{tb - ds}{bc}$$

Using Cramer's Rule, we get

$$x = \dfrac{sd - tb}{-bc} = \dfrac{tb - sd}{bc}$$

$$y = \dfrac{-sc}{-bc} = \dfrac{s}{b}$$

If $b = 0$, we have

$$ax = s$$
$$cx + dy = t$$

Since $D = ad \neq 0$, then $a \neq 0$ and $d \neq 0$.

Thus, $x = \dfrac{s}{a}$ and

$$y = \dfrac{t - cx}{d} = \dfrac{ta - cs}{ad}$$

Using Cramer's Rule, we get

$$x = \dfrac{sd}{ad} = \dfrac{s}{a}$$

$$y = \dfrac{ta - cs}{ad}$$

If $c = 0$, we have

$$ax + by = s$$
$$dy = t$$

Since $D = ad \neq 0$, then $a \neq 0$ and $d \neq 0$.

Thus, $y = \dfrac{t}{d}$ and

$$x = \dfrac{s - by}{a} = \dfrac{sd - bt}{ad}$$

Using Cramer's Rule, we get

$$x = \dfrac{sd - bt}{ad}$$

$$y = \dfrac{at}{ad} = \dfrac{t}{d}$$

If $d = 0$, we have

$$ax + by = s$$
$$cx = t$$

Since $D = -bc \neq 0$, then $b \neq 0$ and $c \neq 0$.

Thus, $x = \dfrac{t}{c}$ and

$$y = \dfrac{s - ax}{b} = \dfrac{sc - at}{bc}$$

Using Cramer's Rule, we get

$$x = \dfrac{-tb}{-bc} = \dfrac{t}{c}$$

$$y = \dfrac{at - sc}{-bc} = \dfrac{sc - at}{bc}$$

63. $\begin{vmatrix} a_{11} & a_{12} & a_{13} \\ ka_{21} & ka_{22} & ka_{23} \\ a_{31} & a_{32} & a_{33} \end{vmatrix} = -ka_{21}(a_{12}a_{33} - a_{32}a_{13}) + ka_{22}(a_{11}a_{33} - a_{31}a_{13}) - ka_{23}(a_{11}a_{32} - a_{31}a_{12})$

$$= k[-a_{21}(a_{12}a_{33} - a_{32}a_{13}) + a_{22}(a_{11}a_{33} - a_{31}a_{13}) - a_{23}(a_{11}a_{32} - a_{31}a_{12})] = k\begin{vmatrix} a_{11} & a_{12} & a_{13} \\ a_{21} & a_{22} & a_{23} \\ a_{31} & a_{32} & a_{33} \end{vmatrix}$$

65. $\begin{vmatrix} a_{11} + ka_{21} & a_{12} + ka_{22} & a_{13} + ka_{23} \\ a_{21} & a_{22} & a_{23} \\ a_{31} & a_{32} & a_{33} \end{vmatrix} = (a_{11} + ka_{21})(a_{22}a_{33} - a_{32}a_{23}) - (a_{12} + ka_{22})(a_{21}a_{33} - a_{31}a_{23}) + (a_{13} + ka_{23})(a_{21}a_{32} - a_{31}a_{22})$

$$= a_{11}a_{22}a_{33} - a_{11}a_{32}a_{23} + \cancel{ka_{21}a_{22}a_{33}} - \cancel{ka_{21}a_{32}a_{23}} - a_{12}a_{21}a_{33} + a_{12}a_{31}a_{23}$$
$$- \cancel{ka_{22}a_{21}a_{33}} + \cancel{ka_{22}a_{31}a_{23}} + a_{13}a_{21}a_{32} - a_{13}a_{31}a_{22} + \cancel{ka_{23}a_{21}a_{32}} - \cancel{ka_{23}a_{31}a_{22}}$$
$$= a_{11}a_{22}a_{33} - a_{11}a_{32}a_{23} - a_{12}a_{21}a_{33} + a_{12}a_{31}a_{23} + a_{13}a_{21}a_{32} - a_{13}a_{31}a_{22}$$
$$= a_{11}(a_{22}a_{33} - a_{32}a_{23}) - a_{12}(a_{21}a_{33} - a_{31}a_{23}) + a_{13}(a_{21}a_{32} - a_{31}a_{22})$$
$$= \begin{vmatrix} a_{11} & a_{12} & a_{13} \\ a_{21} & a_{22} & a_{23} \\ a_{31} & a_{32} & a_{33} \end{vmatrix}$$

Historical Problems *(page 784)*

1. (a) $2 - 5i \longleftrightarrow \begin{bmatrix} 2 & -5 \\ 5 & 2 \end{bmatrix}$, $1 + 3i \longleftrightarrow \begin{bmatrix} 1 & 3 \\ -3 & 1 \end{bmatrix}$ **(b)** $\begin{bmatrix} 2 & -5 \\ 5 & 2 \end{bmatrix}\begin{bmatrix} 1 & 3 \\ -3 & 1 \end{bmatrix} = \begin{bmatrix} 17 & 1 \\ -1 & 17 \end{bmatrix}$ **(c)** $17 + i$ **(d)** $17 + i$

2. $\begin{bmatrix} a & b \\ -b & a \end{bmatrix}\begin{bmatrix} a & -b \\ b & a \end{bmatrix} = \begin{bmatrix} a^2 + b^2 & 0 \\ 0 & b^2 + a^2 \end{bmatrix}$; the product is a real number.

3. (a) $x = k(ar + bs) + l(cr + ds) = r(ka + lc) + s(kb + ld)$ **(b)** $A = \begin{bmatrix} ka + lc & kb + ld \\ ma + nc & mb + nd \end{bmatrix}$

$\quad\quad y = m(ar + bs) + n(cr + ds) = r(ma + nc) + s(mb + nd)$

10.4 Assess Your Understanding *(page 784)*

1. square **2.** T **3.** columns; rows **4.** F **5.** inverse **6.** singular **7.** T **8.** $A^{-1}B$

9. $\begin{bmatrix} 4 & 4 & -5 \\ -1 & 5 & 4 \end{bmatrix}$ **11.** $\begin{bmatrix} 0 & 12 & -20 \\ 4 & 8 & 24 \end{bmatrix}$ **13.** $\begin{bmatrix} -8 & 7 & -15 \\ 7 & 0 & 22 \end{bmatrix}$ **15.** $\begin{bmatrix} 28 & -9 \\ 4 & 23 \end{bmatrix}$ **17.** $\begin{bmatrix} 1 & 14 & -14 \\ 2 & 22 & -18 \\ 3 & 0 & 28 \end{bmatrix}$ **19.** $\begin{bmatrix} 15 & 21 & -16 \\ 22 & 34 & -22 \\ -11 & 7 & 22 \end{bmatrix}$ **21.** $\begin{bmatrix} 25 & -9 \\ 4 & 20 \end{bmatrix}$

23. $\begin{bmatrix} -13 & 7 & -12 \\ -18 & 10 & -14 \\ 17 & -7 & 34 \end{bmatrix}$ **25.** $\begin{bmatrix} -2 & 4 & 2 & 8 \\ 2 & 1 & 4 & 6 \end{bmatrix}$ **27.** $\begin{bmatrix} 5 & 14 \\ 9 & 16 \end{bmatrix}$ **29.** $\begin{bmatrix} 9 & 2 \\ 34 & 13 \\ 47 & 20 \end{bmatrix}$ **31.** $\begin{bmatrix} 1 & -1 \\ -1 & 2 \end{bmatrix}$ **33.** $\begin{bmatrix} 1 & -\dfrac{5}{2} \\ -1 & 3 \end{bmatrix}$ **35.** $\begin{bmatrix} 1 & -\dfrac{1}{a} \\ -1 & \dfrac{2}{a} \end{bmatrix}$

37. $\begin{bmatrix} 3 & -3 & 1 \\ -2 & 2 & -1 \\ -4 & 5 & -2 \end{bmatrix}$ **39.** $\begin{bmatrix} -\dfrac{5}{7} & \dfrac{1}{7} & \dfrac{3}{7} \\ \dfrac{9}{7} & \dfrac{1}{7} & -\dfrac{4}{7} \\ \dfrac{3}{7} & -\dfrac{2}{7} & \dfrac{1}{7} \end{bmatrix}$ **41.** $x = 3, y = 2; (3, 2)$ **43.** $x = -5, y = 10; (-5, 10)$ **45.** $x = 2, y = -1; (2, -1)$

47. $x = \dfrac{1}{2}, y = 2; \left(\dfrac{1}{2}, 2\right)$ **49.** $x = -2, y = 1; (-2, 1)$ **51.** $x = \dfrac{2}{a}, y = \dfrac{3}{a}; \left(\dfrac{2}{a}, \dfrac{3}{a}\right)$ **53.** $x = -2, y = 3, z = 5; (-2, 3, 5)$

55. $x = \frac{1}{2}, y = -\frac{1}{2}, z = 1; \left(\frac{1}{2}, -\frac{1}{2}, 1\right)$ **57.** $x = -\frac{34}{7}, y = \frac{85}{7}, z = \frac{12}{7}; \left(-\frac{34}{7}, \frac{85}{7}, \frac{12}{7}\right)$ **59.** $x = \frac{1}{3}, y = 1, z = \frac{2}{3}; \left(\frac{1}{3}, 1, \frac{2}{3}\right)$

61. $\begin{bmatrix} 4 & 2 & | & 1 & 0 \\ 2 & 1 & | & 0 & 1 \end{bmatrix} \rightarrow \begin{bmatrix} 1 & \frac{1}{2} & | & \frac{1}{4} & 0 \\ 2 & 1 & | & 0 & 1 \end{bmatrix} \rightarrow \begin{bmatrix} 1 & \frac{1}{2} & | & \frac{1}{4} & 0 \\ 0 & 0 & | & -\frac{1}{2} & 1 \end{bmatrix}$ **63.** $\begin{bmatrix} 15 & 3 & | & 1 & 0 \\ 10 & 2 & | & 0 & 1 \end{bmatrix} \rightarrow \begin{bmatrix} 1 & \frac{1}{5} & | & \frac{1}{15} & 0 \\ 10 & 2 & | & 0 & 1 \end{bmatrix} \rightarrow \begin{bmatrix} 1 & \frac{1}{5} & | & \frac{1}{15} & 0 \\ 0 & 0 & | & -\frac{2}{3} & 1 \end{bmatrix}$

65. $\begin{bmatrix} -3 & 1 & -1 & | & 1 & 0 & 0 \\ 1 & -4 & -7 & | & 0 & 1 & 0 \\ 1 & 2 & 5 & | & 0 & 0 & 1 \end{bmatrix} \rightarrow \begin{bmatrix} 1 & 2 & 5 & | & 0 & 0 & 1 \\ 1 & -4 & -7 & | & 0 & 1 & 0 \\ -3 & 1 & -1 & | & 1 & 0 & 0 \end{bmatrix} \rightarrow \begin{bmatrix} 1 & 2 & 5 & | & 0 & 0 & 1 \\ 0 & -6 & -12 & | & 0 & 1 & -1 \\ 0 & 7 & 14 & | & 1 & 0 & 3 \end{bmatrix}$

$\rightarrow \begin{bmatrix} 1 & 2 & 5 & | & 0 & 0 & 1 \\ 0 & 1 & 2 & | & 0 & -\frac{1}{6} & \frac{1}{6} \\ 0 & 1 & 2 & | & \frac{1}{7} & 0 & \frac{3}{7} \end{bmatrix} \rightarrow \begin{bmatrix} 1 & 2 & 5 & | & 0 & 0 & 1 \\ 0 & 1 & 2 & | & 0 & -\frac{1}{6} & \frac{1}{6} \\ 0 & 0 & 0 & | & \frac{1}{7} & \frac{1}{6} & \frac{11}{42} \end{bmatrix}$

67. $\begin{bmatrix} 0.01 & 0.05 & -0.01 \\ 0.01 & -0.02 & 0.01 \\ -0.02 & 0.01 & 0.03 \end{bmatrix}$ **69.** $\begin{bmatrix} 0.02 & -0.04 & -0.01 & 0.01 \\ -0.02 & 0.05 & 0.03 & -0.03 \\ 0.02 & 0.01 & -0.04 & 0.00 \\ -0.02 & 0.06 & 0.07 & 0.06 \end{bmatrix}$ **71.** $x = 4.57, y = -6.44, z = -24.07$ or $(4.57, -6.44, -24.07)$

73. $x = -1.19, y = 2.46, z = 8.27$ or $(-1.19, 2.46, 8.27)$ **75.** $x = -5, y = 7; (-5, 7)$ **77.** $x = -4, y = 2, z = \frac{5}{2}; \left(-4, 2, \frac{5}{2}\right)$

79. Inconsistent; $\varnothing$ **81.** $x = -\frac{1}{5}z + \frac{1}{5}, y = \frac{1}{5}z - \frac{6}{5}$, where z is any real number; $\left\{(x, y, z) \mid x = -\frac{1}{5}z + \frac{1}{5}, y = \frac{1}{5}z - \frac{6}{5}, z \text{ is any real number}\right\}$

83. (a) $A = \begin{bmatrix} 6 & 9 \\ 3 & 12 \end{bmatrix}; B = \begin{bmatrix} 80.00 \\ 277.80 \end{bmatrix}$ (b) $AB = \begin{bmatrix} 2980.20 \\ 3573.60 \end{bmatrix}$; Nikki's total tuition is \$2980.20, and Joe's total tuition is \$3573.60.

85. (a) $\begin{bmatrix} 500 & 350 & 400 \\ 700 & 500 & 850 \end{bmatrix}; \begin{bmatrix} 500 & 700 \\ 350 & 500 \\ 400 & 850 \end{bmatrix}$ (b) $\begin{bmatrix} 15 \\ 8 \\ 3 \end{bmatrix}$ (c) $\begin{bmatrix} 11,500 \\ 17,050 \end{bmatrix}$ (d) $[0.10 \quad 0.05]$ (e) \$2002.50

87. (a) $K^{-1} = \begin{bmatrix} 1 & 0 & -1 \\ -1 & 1 & 1 \\ 0 & -1 & 1 \end{bmatrix}$ (b) $M = \begin{bmatrix} 13 & 1 & 20 \\ 8 & 9 & 19 \\ 6 & 21 & 14 \end{bmatrix}$ (c) Math is fun.

89. If $D = ad - bc \neq 0$, then $a \neq 0$ and $d \neq 0$, or $b \neq 0$ and $c \neq 0$. Assuming the former,

$$\begin{bmatrix} a & b & | & 1 & 0 \\ c & d & | & 0 & 1 \end{bmatrix} \rightarrow \begin{bmatrix} 1 & \frac{b}{a} & | & \frac{1}{a} & 0 \\ c & d & | & 0 & 1 \end{bmatrix} \rightarrow \begin{bmatrix} 1 & \frac{b}{a} & | & \frac{1}{a} & 0 \\ 0 & \frac{D}{a} & | & -\frac{c}{a} & 1 \end{bmatrix} \rightarrow \begin{bmatrix} 1 & \frac{b}{a} & | & \frac{1}{a} & 0 \\ 0 & 1 & | & -\frac{c}{D} & \frac{a}{D} \end{bmatrix} \rightarrow \begin{bmatrix} 1 & 0 & | & \frac{d}{D} & -\frac{b}{D} \\ 0 & 1 & | & -\frac{c}{D} & \frac{a}{D} \end{bmatrix}$$

$$R_1 = \frac{1}{a}r_1 \qquad R_2 = -cr_1 + r_2 \qquad R_2 = \frac{a}{D}r_2 \qquad R_1 = -\frac{b}{a}r_2 + r_1$$

10.5 Assess Your Understanding (page 794)

5. Proper **7.** Improper; $1 + \dfrac{9}{x^2 - 4}$ **9.** Improper; $5x + \dfrac{22x - 1}{x^2 - 4}$ **11.** Improper; $1 + \dfrac{-2(x - 6)}{(x + 4)(x - 3)}$ **13.** $\dfrac{-4}{x} + \dfrac{4}{x - 1}$ **15.** $\dfrac{1}{x} + \dfrac{-x}{x^2 + 1}$

17. $\dfrac{-1}{x - 1} + \dfrac{2}{x - 2}$ **19.** $\dfrac{\frac{1}{4}}{x + 1} + \dfrac{\frac{3}{4}}{x - 1} + \dfrac{\frac{1}{2}}{(x - 1)^2}$ **21.** $\dfrac{\frac{1}{12}}{x - 2} + \dfrac{-\frac{1}{12}(x + 4)}{x^2 + 2x + 4}$ **23.** $\dfrac{\frac{1}{4}}{x - 1} + \dfrac{\frac{1}{4}}{(x - 1)^2} + \dfrac{-\frac{1}{4}}{x + 1} + \dfrac{\frac{1}{4}}{(x + 1)^2}$

25. $\dfrac{-5}{x + 2} + \dfrac{5}{x + 1} + \dfrac{-4}{(x + 1)^2}$ **27.** $\dfrac{\frac{1}{4}}{x} + \dfrac{1}{x^2} + \dfrac{-\frac{1}{4}(x + 4)}{x^2 + 4}$ **29.** $\dfrac{\frac{2}{3}}{x + 1} + \dfrac{\frac{1}{3}(x + 1)}{x^2 + 2x + 4}$ **31.** $\dfrac{\frac{2}{7}}{3x - 2} + \dfrac{\frac{1}{7}}{2x + 1}$ **33.** $\dfrac{\frac{3}{4}}{x + 3} + \dfrac{\frac{1}{4}}{x - 1}$

35. $\dfrac{1}{x^2 + 4} + \dfrac{2x - 1}{(x^2 + 4)^2}$ **37.** $\dfrac{-1}{x} + \dfrac{2}{x - 3} + \dfrac{-1}{x + 1}$ **39.** $\dfrac{4}{x - 2} + \dfrac{-3}{x - 1} + \dfrac{-1}{(x - 1)^2}$ **41.** $\dfrac{x}{(x^2 + 16)^2} + \dfrac{-16x}{(x^2 + 16)^3}$

43. $\dfrac{-\frac{8}{7}}{2x + 1} + \dfrac{\frac{4}{7}}{x - 3}$ **45.** $\dfrac{-\frac{2}{9}}{x} + \dfrac{-\frac{1}{3}}{x^2} + \dfrac{\frac{1}{6}}{x - 3} + \dfrac{\frac{1}{18}}{x + 3}$

Historical Problem *(page 800)*

$x = 6$ units, $y = 8$ units

10.6 Assess Your Understanding *(page 800)*

5.

7.

9.

11.

13.

15.

17.

19.

21. No points of intersection

23.

25. $x = 1, y = 4; x = -1, y = -4; x = 2\sqrt{2}, y = \sqrt{2}; x = -2\sqrt{2}, y = -\sqrt{2}$ or $(1, 4), (-1, -4), (2\sqrt{2}, \sqrt{2}), (-2\sqrt{2}, -\sqrt{2})$

27. $x = 0, y = 1; x = -\dfrac{2}{3}, y = -\dfrac{1}{3}$ or $(0, 1), \left(-\dfrac{2}{3}, -\dfrac{1}{3}\right)$ **29.** $x = 0, y = -1; x = \dfrac{5}{2}, y = -\dfrac{7}{2}$ or $(0, -1), \left(\dfrac{5}{2}, -\dfrac{7}{2}\right)$

31. $x = 2, y = \dfrac{1}{3}; x = \dfrac{1}{2}, y = \dfrac{4}{3}$ or $\left(2, \dfrac{1}{3}\right), \left(\dfrac{1}{2}, \dfrac{4}{3}\right)$ **33.** $x = 3, y = 2; x = 3, y = -2; x = -3, y = 2; x = -3, y = -2$ or

$(3, 2), (3, -2), (-3, 2), (-3, -2)$ **35.** $x = \dfrac{1}{2}, y = \dfrac{3}{2}; x = \dfrac{1}{2}, y = -\dfrac{3}{2}; x = -\dfrac{1}{2}, y = \dfrac{3}{2}; x = -\dfrac{1}{2}, y = -\dfrac{3}{2}$ or $\left(\dfrac{1}{2}, \dfrac{3}{2}\right), \left(\dfrac{1}{2}, -\dfrac{3}{2}\right), \left(-\dfrac{1}{2}, \dfrac{3}{2}\right), \left(-\dfrac{1}{2}, -\dfrac{3}{2}\right)$

37. $x = \sqrt{2}, y = 2\sqrt{2}; x = -\sqrt{2}, y = -2\sqrt{2}$ or $(\sqrt{2}, 2\sqrt{2}), (-\sqrt{2}, -2\sqrt{2})$ **39.** No real solution exists.

41. $x = \dfrac{8}{3}, y = \dfrac{2\sqrt{10}}{3}; x = -\dfrac{8}{3}, y = \dfrac{2\sqrt{10}}{3}; x = \dfrac{8}{3}, y = -\dfrac{2\sqrt{10}}{3}; x = -\dfrac{8}{3}, y = -\dfrac{2\sqrt{10}}{3}$ or $\left(\dfrac{8}{3}, \dfrac{2\sqrt{10}}{3}\right), \left(-\dfrac{8}{3}, \dfrac{2\sqrt{10}}{3}\right), \left(\dfrac{8}{3}, -\dfrac{2\sqrt{10}}{3}\right), \left(-\dfrac{8}{3}, -\dfrac{2\sqrt{10}}{3}\right)$

43. $x = 1, y = \dfrac{1}{2}; x = -1, y = \dfrac{1}{2}; x = 1, y = -\dfrac{1}{2}; x = -1, y = -\dfrac{1}{2}$ or $\left(1, \dfrac{1}{2}\right), \left(-1, \dfrac{1}{2}\right), \left(1, -\dfrac{1}{2}\right), \left(-1, -\dfrac{1}{2}\right)$ **45.** No real solution exists.

47. $x = \sqrt{3}, y = \sqrt{3}; x = -\sqrt{3}, y = -\sqrt{3}; x = 2, y = 1; x = -2, y = -1$ or $(\sqrt{3}, \sqrt{3}), (-\sqrt{3}, -\sqrt{3}), (2, 1), (-2, -1)$

49. $x = 0, y = -2; x = 0, y = 1; x = 2, y = -1$ or $(0, -2), (0, 1), (2, -1)$ **51.** $x = 2, y = 8$ or $(2, 8)$ **53.** $x = 81, y = 3$ or $(81, 3)$

55.

57. $x = 0.48, y = 0.62$ **59.** $x = -1.65, y = -0.89$

61. $x = 0.58, y = 1.86; x = 1.81, y = 1.05; x = 0.58, y = -1.86; x = 1.81, y = -1.05$

63. $x = 2.35, y = 0.85$

65.

67.

69.

71. 3 and 1; -1 and -3 **73.** 2 and 2; -2 and -2 **75.** $\dfrac{1}{2}$ and $\dfrac{1}{3}$ **77.** 5 **79.** 5 in. by 3 in. **81.** 2 cm and 4 cm **83.** tortoise: 7 m/hr, hare: $7\dfrac{1}{2}$ m/hr

85. 12 cm by 18 cm **87.** $x = 60$ ft; $y = 30$ ft **89.** $l = \dfrac{P + \sqrt{P^2 - 16A}}{4}; w = \dfrac{P - \sqrt{P^2 - 16A}}{4}$ **91.** $y = 4x - 4$ **93.** $y = 2x + 1$

95. $y = -\dfrac{1}{3}x + \dfrac{7}{3}$ **97.** $y = 2x - 3$ **99.** $r_1 = \dfrac{-b + \sqrt{b^2 - 4ac}}{2a}; r_2 = \dfrac{-b - \sqrt{b^2 - 4ac}}{2a}$ **101. (a)** 4.274 ft by 4.274 ft or 0.093 ft by 0.093 ft

10.7 Assess Your Understanding *(page 809)*

7. dashes; solid **8.** half-planes **9.** F **10.** unbounded

11.

13.

15.

17.

19.
$y \leq x^2 - 1$

21.
$xy \geq 4$

23.
$x + y = 2$ $2x + y = 4$

25.
$3x + 2y = -6$ $2x - y = 4$

27.
$2x - 3y = 0$ $3x + 2y = 6$

29.
$2x - 4y = 0$ $x - 2y = 6$

31.
$2x + y = -2$ $2x + y = 2$

33. No solution

$2x + 3y = 6$ $2x + 3y = 0$

35.
$x^2 + y^2 = 9$ $x + y = 3$

37.
$y = x^2 - 4$ $y = x - 2$

39.
$y = x^2 - 4$ $x^2 + y^2 = 16$

41.
$y = x^2 + 1$ $xy = 4$

43. Bounded

$2x + y = 6$ $(0, 3)$ $(2, 2)$ $(0, 0)$ $(3, 0)$ $x + 2y = 6$

45. Unbounded

$x + y = 2$ $(0, 4)$ $(2, 0)$ $2x + y = 4$

47. Bounded

$3x + y = 12$ $(0, 4)$ $\left(\frac{24}{7}, \frac{12}{7}\right)$ $x + y = 2$ $(0, 2)$ $(2, 0)$ $(4, 0)$ $2x + 3y = 12$

49. Bounded

$2x + y = 10$ $(0, 8)$ $(2, 6)$ $x + y = 2$ $x + y = 8$ $(0, 2)$ $(2, 0)$ $(5, 0)$

51. Bounded

$x + 2y = 10$ $(0, 5)$ $(10, 0)$ $\left(0, \frac{1}{2}\right)$ $(1, 0)$ $x + 2y = 1$

53. $\begin{cases} x \leq 4 \\ x + y \leq 6 \\ x \geq 0 \\ y \geq 0 \end{cases}$

55. $\begin{cases} x \leq 20 \\ y \geq 15 \\ x + y \leq 50 \\ x - y \leq 0 \\ x \geq 0 \end{cases}$

57. (a) $\begin{cases} x + y \leq 50{,}000 \\ x \geq 35{,}000 \\ y \leq 10{,}000 \\ y \geq 0 \end{cases}$

(b)
$x = 35{,}000$ $40{,}000$ $(40{,}000, 10{,}000)$ $(35{,}000, 10{,}000)$ $y = 10{,}000$ $-20{,}000$ $(50{,}000, 0)$ $(35{,}000, 0)$ $x + y = 50{,}000$

59. (a) $\begin{cases} x \geq 0 \\ y \geq 0 \\ x + 2y \leq 300 \\ 3x + 2y \leq 480 \end{cases}$

(b)
$3x + 2y = 480$ 400 $(0, 150)$ $(90, 105)$ 400 $(0, 0)$ $(160, 0)$ $x + 2y = 300$

61. (a) $\begin{cases} 3x + 2y \leq 160 \\ 2x + 3y \leq 150 \\ x \geq 0 \\ y \geq 0 \end{cases}$

(b)
80 $3x + 2y = 160$ $(0, 50)$ $(36, 26)$ $(0, 0)$ 80 $2x + 3y = 150$ $\left(\frac{160}{3}, 0\right)$

10.8 Assess Your Understanding (page 816)

1. objective function   **2.** T   **3.** Maximum value is 11; minimum value is 3.   **5.** Maximum value is 65; minimum value is 4.   **7.** Maximum value is 67; minimum value is 20.   **9.** The maximum value of z is 12, and it occurs at the point $(6, 0)$.   **11.** The minimum value of z is 4, and it occurs at the point $(2, 0)$.   **13.** The maximum value of z is 20, and it occurs at the point $(0, 4)$.   **15.** The minimum value of z is 8, and it occurs at the point $(0, 2)$.   **17.** The maximum value of z is 50, and it occurs at the point $(10, 0)$.   **19.** Produce 8 downhill and 24 cross-country; \$1760; \$1920 which is the profit when producing 16 downhill and 16 cross-country.   **21.** Rent 15 rectangular tables and 16 round tables for a minimum cost of \$1252.   **23. (a)** \$10,000 in a junk bond and \$10,000 in Treasury bills   **(b)** \$12,000 in a junk bond and \$8000 in Treasury bills   **25.** 100 lb of ground beef should be mixed with 50 lb of pork.   **27.** Manufacture 10 racing skates and 15 figure skates.   **29.** Order 2 metal samples and 4 plastic samples; \$34   **31. (a)** Configure with 10 first class seats and 120 coach seats.   **(b)** Configure with 15 first class seats and 120 coach seats.

Review Exercises (page 819)

1. $x = 2, y = -1$ or $(2, -1)$ **3.** $x = 2, y = \frac{1}{2}$ or $\left(2, \frac{1}{2}\right)$ **5.** $x = 2, y = -1$ or $(2, -1)$ **7.** $x = \frac{11}{5}, y = -\frac{3}{5}$ or $\left(\frac{11}{5}, -\frac{3}{5}\right)$ **9.** Inconsistent

11. $x = 2, y = 3$ or $(2, 3)$ **13.** Inconsistent   **15.** $x = -1, y = 2, z = -3$ or $(-1, 2, -3)$

17. $x = \frac{7}{4}z + \frac{39}{4}, y = \frac{9}{8}z + \frac{69}{8}$, where z is any real number or $\left\{ (x, y, z) \middle| x = \frac{7}{4}z + \frac{39}{4}, y = \frac{9}{8}z + \frac{69}{8}, z \text{ is any real number} \right\}$

19. $\begin{cases} 3x + 2y = 8 \\ x + 4y = -1 \end{cases}$ **21.** $\begin{bmatrix} 4 & -4 \\ 3 & 9 \\ 4 & 4 \end{bmatrix}$ **23.** $\begin{bmatrix} 6 & 0 \\ 12 & 24 \\ -6 & 12 \end{bmatrix}$ **25.** $\begin{bmatrix} 4 & -3 & 0 \\ 12 & -2 & -8 \\ -2 & 5 & -4 \end{bmatrix}$ **27.** $\begin{bmatrix} 8 & -13 & 8 \\ 9 & 2 & -10 \\ 22 & -13 & -4 \end{bmatrix}$ **29.** $\begin{bmatrix} \frac{1}{2} & -1 \\ -\frac{1}{6} & \frac{2}{3} \end{bmatrix}$

31. $\begin{bmatrix} -\frac{5}{7} & \frac{9}{7} & \frac{3}{7} \\ \frac{1}{7} & \frac{1}{7} & -\frac{2}{7} \\ \frac{3}{7} & -\frac{4}{7} & \frac{1}{7} \end{bmatrix}$ **33.** Singular   **35.** $x = \frac{2}{5}, y = \frac{1}{10}$ or $\left(\frac{2}{5}, \frac{1}{10}\right)$ **37.** $x = 9, y = \frac{13}{3}, z = \frac{13}{3}$ or $\left(9, \frac{13}{3}, \frac{13}{3}\right)$

39. $x = -\frac{1}{2}, y = -\frac{2}{3}, z = -\frac{3}{4}$, or $\left(-\frac{1}{2}, -\frac{2}{3}, -\frac{3}{4}\right)$

41. $z = -1, x = y + 1$, where y is any real number or $\left\{(x, y, z) \mid x = y + 1, z = -1, y \text{ is any real number}\right\}$

43. $x = 4, y = 2, z = 3, t = -2$ or $(4, 2, 3, -2)$ **45.** 5   **47.** 108   **49.** -100 **51.** $x = 2, y = -1$ or $(2, -1)$ **53.** $x = 2, y = 3$ or $(2, 3)$

55. $x = -1, y = 2, z = -3$ or $(-1, 2, -3)$ **57.** 16   **59.** $\dfrac{-\frac{3}{2}}{x} + \dfrac{\frac{3}{2}}{x - 4}$ **61.** $\dfrac{-3}{x - 1} + \dfrac{3}{x} + \dfrac{4}{x^2}$ **63.** $\dfrac{-\frac{1}{10}}{x + 1} + \dfrac{\frac{1}{10}x + \frac{9}{10}}{x^2 + 9}$

65. $\dfrac{x}{x^2 + 4} + \dfrac{-4x}{(x^2 + 4)^2}$ **67.** $\dfrac{\frac{1}{2}}{x^2 + 1} + \dfrac{\frac{1}{4}}{x - 1} + \dfrac{-\frac{1}{4}}{x + 1}$ **69.** $x = -\frac{2}{5}, y = -\frac{11}{5}; x = -2, y = 1$ or $\left(-\frac{2}{5}, -\frac{11}{5}\right), (-2, 1)$

71. $x = 2\sqrt{2}, y = \sqrt{2}; x = -2\sqrt{2}, y = -\sqrt{2}$ or $(2\sqrt{2}, \sqrt{2}), (-2\sqrt{2}, -\sqrt{2})$ **73.** $x = 0, y = 0; x = -3, y = 3; x = 3, y = 3$ or

$(0, 0), (-3, 3), (3, 3)$ **75.** $x = \sqrt{2}, y = -\sqrt{2}; x = -\sqrt{2}, y = \sqrt{2}; x = \frac{4}{3}\sqrt{2}, y = -\frac{2}{3}\sqrt{2}; x = -\frac{4}{3}\sqrt{2}, y = \frac{2}{3}\sqrt{2}$ or

$(\sqrt{2}, -\sqrt{2}), (-\sqrt{2}, \sqrt{2}), \left(\frac{4}{3}\sqrt{2}, -\frac{2}{3}\sqrt{2}\right), \left(-\frac{4}{3}\sqrt{2}, \frac{2}{3}\sqrt{2}\right)$ **77.** $x = 1, y = -1$ or $(1, -1)$

79.   **81.**   **83.** Unbounded   **85.** Bounded

87. Bounded   **89.**   **91.**

93. The maximum value is 32 when $x = 0$ and $y = 8$.   **95.** The minimum value is 3 when $x = 1$ and $y = 0$.   **97.** 10

99. $y = -\frac{1}{3}x^2 - \frac{2}{3}x + 1$ **101.** Mix 70 lb of \$6.00 coffee and 30 lb of \$9.00 coffee.   **103.** Buy 1 small, 5 medium, and 2 large.

105. Speedboat: 36.67 km/hr; Aguarico River: 3.33 km/hr   **107.** Bruce: 4hr; Bryce: 2hr; Marty: 8hr

109. Produce 35 gasoline engines and 15 diesel engines; the factory is producing an excess of 15 gasoline engines and 0 diesel engines.

Chapter Test *(page 823)*

1. $x = 3, y = -1$ or $(3, -1)$ **2.** Inconsistent   **3.** $x = -z + \frac{18}{7}, y = z - \frac{17}{7}$, where z is any real number or

$\left\{(x, y, z) \Big| x = -z + \frac{18}{7}, y = z - \frac{17}{7}, z \text{ is any real number}\right\}$ **4.** $x = \frac{1}{3}, y = -2, z = 0$ or $\left(\frac{1}{3}, -2, 0\right)$ **5.** $\begin{bmatrix} 4 & -5 & 1 & | & 0 \\ -2 & -1 & 0 & | & -25 \\ 1 & 5 & -5 & | & 10 \end{bmatrix}$

6. $\begin{cases} 3x + 2y + 4z = -6 \\ 1x + 0y + 8z = 2 \\ -2x + 1y + 3z = -11 \end{cases}$ or $\begin{cases} 3x + 2y + 4z = -6 \\ x + 8z = 2 \\ -2x + y + 3z = -11 \end{cases}$ **7.** $\begin{bmatrix} 6 & 4 \\ 1 & -11 \\ 5 & 12 \end{bmatrix}$ **8.** $\begin{bmatrix} -11 & -19 \\ -3 & 5 \\ 6 & -22 \end{bmatrix}$ **9.** The operation cannot be performed.

10. $\begin{bmatrix} 16 & 17 \\ 3 & -10 \end{bmatrix}$ **11.** $\begin{bmatrix} 2 & -1 \\ -\frac{5}{2} & \frac{3}{2} \end{bmatrix}$ **12.** $\begin{bmatrix} 3 & 3 & -4 \\ -2 & -2 & 3 \\ -4 & -5 & 7 \end{bmatrix}$ **13.** $x = \frac{1}{2}, y = 3$ or $\left(\frac{1}{2}, 3\right)$ **14.** $x = -\frac{1}{4}y + 7$, where y is any real number or

$\left\{(x, y) \Big| x = -\frac{1}{4}y + 7, y \text{ is any real number}\right\}$ **15.** $x = 1, y = -2, z = 0$ or $(1, -2, 0)$ **16.** Inconsistent   **17.** -29 **18.** -12

19. $x = -2, y = -5$ or $(-2, -5)$ **20.** $x = 1, y = -1, z = 4$ or $(1, -1, 4)$ **21.** $(1, -3)$ and $(1, 3)$ **22.** $(3, 4)$ and $(1, 2)$

23.

24. $\dfrac{3}{x+3} + \dfrac{-2}{(x+3)^2}$

25. $\dfrac{-\dfrac{1}{3}}{x} + \dfrac{\dfrac{1}{3}x}{(x^2+3)} + \dfrac{5x}{(x^2+3)^2}$

26. Unbounded

27. The maximum value of z is 64, and it occurs at the point $(0, 8)$.   **28.** Flare jeans cost \$24.50, camisoles cost \$8.50, and T-shirts cost \$6.00.

Cumulative Review *(page 824)*

1. $\left\{0, \dfrac{1}{2}\right\}$ **2.** $\{5\}$ **3.** $\left\{-1, -\dfrac{1}{2}, 3\right\}$ **4.** $\{-2\}$ **5.** $\left\{\dfrac{5}{2}\right\}$ **6.** $\left\{\dfrac{1}{\ln 3}\right\}$ **7.** Odd; symmetric with respect to the origin

8. Center: $(1, -2)$; radius $= 4$

9. Domain: all real numbers
Range: $\{y | y > 1\}$
Horizontal asymptote: $y = 1$

10. $f^{-1}(x) = \dfrac{5}{x} - 2$
Domain of f: $\{x | x \neq -2\}$
Range of f: $\{y | y \neq 0\}$
Domain of f^{-1}: $\{x | x \neq 0\}$
Range of f^{-1}: $\{y | y \neq -2\}$

11. (a)

(b)

(c)

(d)

(e)

(f)

(g)

(h)

(i)

(j)

12. (a)
; -2.28
(b) Local maximum of 7 at $x = -1$;
local minimum of 3 at $x = 1$
(c) $(-\infty, -1), (1, \infty)$

CHAPTER 11 Sequences; Induction; the Binomial Theorem

11.1 Assess Your Understanding *(page 833)*

3. sequence   **4.** True   **5.** $n(n-1) \cdots 3 \cdot 2 \cdot 1$ **6.** recursive   **7.** summation   **8.** True   **9.** 3,628,800   **11.** 504   **13.** 1260   **15.** $s_1 = 1, s_2 = 2, s_3 = 3,$
$s_4 = 4, s_5 = 5$ **17.** $a_1 = \dfrac{1}{3}, a_2 = \dfrac{1}{2}, a_3 = \dfrac{3}{5}, a_4 = \dfrac{2}{3}, a_5 = \dfrac{5}{7}$ **19.** $c_1 = 1, c_2 = -4, c_3 = 9, c_4 = -16, c_5 = 25$ **21.** $s_1 = \dfrac{1}{2}, s_2 = \dfrac{2}{5}, s_3 = \dfrac{2}{7}, s_4 = \dfrac{8}{41}, s_5 = \dfrac{8}{61}$

23. $t_1 = -\dfrac{1}{6}, t_2 = \dfrac{1}{12}, t_3 = -\dfrac{1}{20}, t_4 = \dfrac{1}{30}, t_5 = -\dfrac{1}{42}$ **25.** $b_1 = \dfrac{1}{e}, b_2 = \dfrac{2}{e^2}, b_3 = \dfrac{3}{e^3}, b_4 = \dfrac{4}{e^4}, b_5 = \dfrac{5}{e^5}$ **27.** $a_n = \dfrac{n}{n+1}$ **29.** $a_n = \dfrac{1}{2^{n-1}}$

31. $a_n = (-1)^{n+1}$ **33.** $a_n = (-1)^{n+1}n$ **35.** $a_1 = 2, a_2 = 5, a_3 = 8, a_4 = 11, a_5 = 14$ **37.** $a_1 = -2, a_2 = 0, a_3 = 3, a_4 = 7, a_5 = 12$

39. $a_1 = 5, a_2 = 10, a_3 = 20, a_4 = 40, a_5 = 80$ **41.** $a_1 = 3, a_2 = \dfrac{3}{2}, a_3 = \dfrac{1}{2}, a_4 = \dfrac{1}{8}, a_5 = \dfrac{1}{40}$ **43.** $a_1 = 1, a_2 = 2, a_3 = 2, a_4 = 4, a_5 = 8$

45. $a_1 = A, a_2 = A + d, a_3 = A + 2d, a_4 = A + 3d, a_5 = A + 4d$

47. $a_1 = \sqrt{2}, a_2 = \sqrt{2 + \sqrt{2}}, a_3 = \sqrt{2 + \sqrt{2 + \sqrt{2}}},$ **49.** $3 + 4 + \cdots + (n+2)$ **51.** $\dfrac{1}{2} + 2 + \dfrac{9}{2} + \cdots + \dfrac{n^2}{2}$ **53.** $1 + \dfrac{1}{3} + \dfrac{1}{9} + \cdots + \dfrac{1}{3^n}$
$a_4 = \sqrt{2 + \sqrt{2 + \sqrt{2 + \sqrt{2}}}},$
$a_5 = \sqrt{2 + \sqrt{2 + \sqrt{2 + \sqrt{2 + \sqrt{2}}}}}$ **55.** $\dfrac{1}{3} + \dfrac{1}{9} + \cdots + \dfrac{1}{3^n}$ **57.** $\ln 2 - \ln 3 + \ln 4 - \cdots + (-1)^n \ln n$ **59.** $\displaystyle\sum_{k=1}^{20} k$

61. $\sum_{k=1}^{13} \dfrac{k}{k+1}$ **63.** $\sum_{k=0}^{6}(-1)^k\left(\dfrac{1}{3^k}\right)$ **65.** $\sum_{k=1}^{n}\dfrac{3^k}{k}$ **67.** $\sum_{k=0}^{n}(a+kd)$ or $\sum_{k=1}^{n+1}[a+(k-1)d]$ **69.** 200 **71.** 820 **73.** 1110 **75.** 1560 **77.** 3570

79. 44,000 **81.** $2930 **83.** $18,058.03 **85.** 21 pairs **87.** Fibonacci sequence **89. (a)** 3.630170833 **(b)** 3.669060828 **(c)** 3.669296668 **(d)** 12

91. (a) $a_1 = 0.4; a_2 = 0.7; a_3 = 1; a_4 = 1.6; a_5 = 2.8; a_6 = 5.2; a_7 = 10; a_8 = 19.6$

(b) Except for term 5, which has no match, Bode's formula provides excellent approximations for the mean distances of the planets from the Sun.

(c) The mean distance of Ceres from the Sun is approximated by $a_5 = 2.8$ and that of Uranus is $a_8 = 19.6$.

(d) $a_9 = 38.8; a_{10} = 77.2$ **(e)** Pluto's distance is approximated by a_9, but no term approximates Neptune's mean distance from the Sun.

(f) According to Bode's Law, the mean orbital distance of Eris will be 154 AU from the Sun.

93. $a_0 = 2; a_5 = 2.236067977; 2.236067977$ **95.** $a_0 = 4; a_5 = 4.582575695; 4.582575695$

11.2 Assess Your Understanding (page 841)

1. arithmetic **2.** F **3.** 17 **4.** T **5.** $s_n - s_{n-1} = (n+4) - [(n-1)+4] = n+4-(n+3) = n+4-n-3 = 1$, a constant;
$d = 1; s_1 = 5, s_2 = 6, s_3 = 7, s_4 = 8$

7. $a_n - a_{n-1} = (2n-5) - [2(n-1)-5] = 2n-5-(2n-2-5) = 2n-5-(2n-7) = 2n-5-2n+7 = 2$, a constant;
$d = 2; a_1 = -3, a_2 = -1, a_3 = 1, a_4 = 3$

9. $c_n - c_{n-1} = (6-2n) - [6-2(n-1)] = 6-2n-(6-2n+2) = 6-2n-(8-2n) = 6-2n-8+2n = -2$, a constant;
$d = -2; c_1 = 4, c_2 = 2, c_3 = 0, c_4 = -2$

11. $t_n - t_{n-1} = \left(\dfrac{1}{2} - \dfrac{1}{3}n\right) - \left[\dfrac{1}{2} - \dfrac{1}{3}(n-1)\right] = \dfrac{1}{2} - \dfrac{1}{3}n - \left(\dfrac{1}{2} - \dfrac{1}{3}n + \dfrac{1}{3}\right) = \dfrac{1}{2} - \dfrac{1}{3}n - \left(\dfrac{5}{6} - \dfrac{1}{3}n\right) = \dfrac{1}{2} - \dfrac{1}{3}n - \dfrac{5}{6} + \dfrac{1}{3}n = -\dfrac{1}{3}$, a constant;
$d = -\dfrac{1}{3}; t_1 = \dfrac{1}{6}, t_2 = -\dfrac{1}{6}, t_3 = -\dfrac{1}{2}, t_4 = -\dfrac{5}{6}$

13. $s_n - s_{n-1} = \ln 3^n - \ln 3^{n-1} = n\ln 3 - (n-1)\ln 3 = n\ln 3 - (n\ln 3 - \ln 3) = n\ln 3 - n\ln 3 + \ln 3 = \ln 3$, a constant;
$d = \ln 3; s_1 = \ln 3, s_2 = 2\ln 3, s_3 = 3\ln 3, s_4 = 4\ln 3$

15. $a_n = 3n - 1; a_{51} = 152$ **17.** $a_n = 8 - 3n; a_{51} = -145$ **19.** $a_n = \dfrac{1}{2}(n-1); a_{51} = 25$ **21.** $a_n = \sqrt{2}n; a_{51} = 51\sqrt{2}$ **23.** 200 **25.** -266 **27.** $\dfrac{83}{2}$

29. $a_1 = -13; d = 3; a_n = a_{n-1} + 3; a_n = -16 + 3n$ **31.** $a_1 = -53; d = 6; a_n = a_{n-1} + 6; a_n = -59 + 6n$

33. $a_1 = 28; d = -2; a_n = a_{n-1} - 2; a_n = 30 - 2n$ **35.** $a_1 = 25; d = -2; a_n = a_{n-1} - 2; a_n = 27 - 2n$ **37.** n^2 **39.** $\dfrac{n}{2}(9 + 5n)$ **41.** 1260

43. 324 **45.** 30,919 **47.** 10,036 **49.** 6080 **51.** -1925 **53.** 15,960 **55.** $-\dfrac{3}{2}$ **57.** 24 terms **59.** 1185 seats

61. 210 beige and 190 blue **63.** $\{T_n\} = \{-5.5n + 67\}; T_5 = 39.5°F$ **65.** The amphitheater has 1647 seats. **67.** 8 yr

Historical Problems (page 850)

1. $1\dfrac{2}{3}$ loaves, $10\dfrac{5}{6}$ loaves, 20 loaves, $29\dfrac{1}{6}$ loaves, $38\dfrac{1}{3}$ loaves **2. (a)** 1 person **(b)** 2401 kittens **(c)** 2800

11.3 Assess Your Understanding (page 851)

3. geometric **4.** $\dfrac{a}{1-r}$ **5.** divergent series **6.** T **7.** F **8.** T **9.** $r = 3; s_1 = 3, s_2 = 9, s_3 = 27, s_4 = 81$

11. $r = \dfrac{1}{2}; a_1 = -\dfrac{3}{2}, a_2 = -\dfrac{3}{4}, a_3 = -\dfrac{3}{8}, a_4 = -\dfrac{3}{16}$ **13.** $r = 2; c_1 = \dfrac{1}{4}, c_2 = \dfrac{1}{2}, c_3 = 1, c_4 = 2$ **15.** $r = 2^{1/3}; e_1 = 2^{1/3}, e_2 = 2^{2/3}, e_3 = 2, e_4 = 2^{4/3}$

17. $r = \dfrac{3}{2}; t_1 = \dfrac{1}{2}, t_2 = \dfrac{3}{4}, t_3 = \dfrac{9}{8}, t_4 = \dfrac{27}{16}$ **19.** $a_5 = 162; a_n = 2 \cdot 3^{n-1}$ **21.** $a_5 = 5; a_n = 5 \cdot (-1)^{n-1}$ **23.** $a_5 = 0; a_n = 0$

25. $a_5 = 4\sqrt{2}; a_n = (\sqrt{2})^n$ **27.** $a_7 = \dfrac{1}{64}$ **29.** $a_9 = 1$ **31.** $a_8 = 0.00000004$ **33.** $a_n = 7 \cdot 2^{n-1}$ **35.** $a_n = -3 \cdot \left(-\dfrac{1}{3}\right)^{n-1} = \left(-\dfrac{1}{3}\right)^{n-2}$

37. $a_n = -(-3)^{n-1}$ **39.** $a_n = \dfrac{7}{15}(15)^{n-1} = 7 \cdot 15^{n-2}$ **41.** $-\dfrac{1}{4}(1 - 2^n)$ **43.** $2\left[1 - \left(\dfrac{2}{3}\right)^n\right]$ **45.** $1 - 2^n$

47.
```
(1/4)sum(seq(2^n
,n,0,14,1))
         8191.75
```

49.
```
sum(seq((2/3)^n,
n,1,15,1))
         1.995432683
```

51.
```
-1sum(seq(2^n,n,
0,14,1))
             -32767
```

53. Converges; $\dfrac{3}{2}$ **55.** Converges; 16 **57.** Converges; $\dfrac{8}{5}$ **59.** Diverges **61.** Converges; $\dfrac{20}{3}$ **63.** Diverges **65.** Converges; $\dfrac{18}{5}$ **67.** Converges; 6

69. Arithmetic; $d = 1; 1375$ **71.** Neither **73.** Arithmetic; $d = -\dfrac{2}{3}; -700$ **75.** Neither **77.** Geometric; $r = \dfrac{2}{3}; 2\left[1 - \left(\dfrac{2}{3}\right)^{50}\right]$

79. Geometric; $r = -2; -\dfrac{1}{3}[1 - (-2)^{50}]$ **81.** Geometric; $r = 3^{1/2}; -\dfrac{\sqrt{3}}{2}(1 + \sqrt{3})(1 - 3^{25})$ **83.** -4 **85.** $21,879.11 **87. (a)** 0.775 ft **(b)** 8th

(c) 15.88 ft **(d)** 20 ft **89.** $349,496.41 **91.** $96,885.98 **93.** $305.10 **95.** 1.845×10^{19} **97.** 10 **99.** $72.67 per share **101.** December 20, 2007; $9999.92 **103.** Option B results in more money ($524,287 versus $500,500). **105.** Total pay: $41,943.03; pay on day 22: $20,971.52

11.4 Assess Your Understanding *(page 857)*

1. (I) $n = 1: 2(1) = 2$ and $1(1 + 1) = 2$

(II) If $2 + 4 + 6 + \cdots + 2k = k(k + 1)$, then $2 + 4 + 6 + \cdots + 2k + 2(k + 1) = (2 + 4 + 6 + \cdots + 2k) + 2(k + 1)$
$= k(k + 1) + 2(k + 1) = k^2 + 3k + 2 = (k + 1)(k + 2) = (k + 1)[(k + 1) + 1]$.

3. (I) $n = 1: 1 + 2 = 3$ and $\dfrac{1}{2}(1)(1 + 5) = \dfrac{1}{2}(6) = 3$

(II) If $3 + 4 + 5 + \cdots + (k + 2) = \dfrac{1}{2}k(k + 5)$, then $3 + 4 + 5 + \cdots + (k + 2) + [(k + 1) + 2]$

$= [3 + 4 + 5 + \cdots + (k + 2)] + (k + 3) = \dfrac{1}{2}k(k + 5) + k + 3 = \dfrac{1}{2}(k^2 + 7k + 6) = \dfrac{1}{2}(k + 1)(k + 6) = \dfrac{1}{2}(k + 1)[(k + 1) + 5]$.

5. (I) $n = 1: 3(1) - 1 = 2$ and $\dfrac{1}{2}(1)[3(1) + 1] = \dfrac{1}{2}(4) = 2$

(II) If $2 + 5 + 8 + \cdots + (3k - 1) = \dfrac{1}{2}k(3k + 1)$, then $2 + 5 + 8 + \cdots + (3k - 1) + [3(k + 1) - 1]$

$= [2 + 5 + 8 + \cdots + (3k - 1)] + (3k + 2) = \dfrac{1}{2}k(3k + 1) + (3k + 2) = \dfrac{1}{2}(3k^2 + 7k + 4) = \dfrac{1}{2}(k + 1)(3k + 4)$

$= \dfrac{1}{2}(k + 1)[3(k + 1) + 1]$.

7. (I) $n = 1: 2^{1-1} = 1$ and $2^1 - 1 = 1$

(II) If $1 + 2 + 2^2 + \cdots + 2^{k-1} = 2^k - 1$, then $1 + 2 + 2^2 + \cdots + 2^{k-1} + 2^{(k+1)-1} = (1 + 2 + 2^2 + \cdots + 2^{k-1}) + 2^k$
$= 2^k - 1 + 2^k = 2(2^k) - 1 = 2^{k+1} - 1$.

9. (I) $n = 1: 4^{1-1} = 1$ and $\dfrac{1}{3}(4^1 - 1) = \dfrac{1}{3}(3) = 1$

(II) If $1 + 4 + 4^2 + \cdots + 4^{k-1} = \dfrac{1}{3}(4^k - 1)$, then $1 + 4 + 4^2 + \cdots + 4^{k-1} + 4^{(k+1)-1} = (1 + 4 + 4^2 + \cdots + 4^{k-1}) + 4^k$

$= \dfrac{1}{3}(4^k - 1) + 4^k = \dfrac{1}{3}[4^k - 1 + 3(4^k)] = \dfrac{1}{3}[4(4^k) - 1] = \dfrac{1}{3}(4^{k+1} - 1)$.

11. (I) $n = 1: \dfrac{1}{1 \cdot 2} = \dfrac{1}{2}$ and $\dfrac{1}{1 + 1} = \dfrac{1}{2}$

(II) If $\dfrac{1}{1 \cdot 2} + \dfrac{1}{2 \cdot 3} + \dfrac{1}{3 \cdot 4} + \cdots + \dfrac{1}{k(k + 1)} = \dfrac{k}{k + 1}$, then $\dfrac{1}{1 \cdot 2} + \dfrac{1}{2 \cdot 3} + \dfrac{1}{3 \cdot 4} + \cdots + \dfrac{1}{k(k + 1)} + \dfrac{1}{(k + 1)[(k + 1) + 1]}$

$= \left[\dfrac{1}{1 \cdot 2} + \dfrac{1}{2 \cdot 3} + \dfrac{1}{3 \cdot 4} + \cdots + \dfrac{1}{k(k + 1)}\right] + \dfrac{1}{(k + 1)(k + 2)} = \dfrac{k}{k + 1} + \dfrac{1}{(k + 1)(k + 2)} = \dfrac{k(k + 2) + 1}{(k + 1)(k + 2)}$

$= \dfrac{k^2 + 2k + 1}{(k + 1)(k + 2)} = \dfrac{(k + 1)^2}{(k + 1)(k + 2)} = \dfrac{k + 1}{k + 2} = \dfrac{k + 1}{(k + 1) + 1}$.

13. (I) $n = 1: 1^2 = 1$ and $\dfrac{1}{6} \cdot 1 \cdot 2 \cdot 3 = 1$

(II) If $1^2 + 2^2 + 3^2 + \cdots + k^2 = \dfrac{1}{6}k(k + 1)(2k + 1)$, then $1^2 + 2^2 + 3^2 + \cdots + k^2 + (k + 1)^2$

$= (1^2 + 2^2 + 3^2 + \cdots + k^2) + (k + 1)^2 = \dfrac{1}{6}k(k + 1)(2k + 1) + (k + 1)^2 = \dfrac{1}{6}(2k^3 + 9k^2 + 13k + 6)$

$= \dfrac{1}{6}(k + 1)(k + 2)(2k + 3) = \dfrac{1}{6}(k + 1)[(k + 1) + 1][2(k + 1) + 1]$.

15. (I) $n = 1: 5 - 1 = 4$ and $\dfrac{1}{2}(1)(9 - 1) = \dfrac{1}{2} \cdot 8 = 4$

(II) If $4 + 3 + 2 + \cdots + (5 - k) = \dfrac{1}{2}k(9 - k)$, then $4 + 3 + 2 + \cdots + (5 - k) + [5 - (k + 1)]$

$= [4 + 3 + 2 + \cdots + (5 - k)] + 4 - k = \dfrac{1}{2}k(9 - k) + 4 - k = \dfrac{1}{2}(9k - k^2 + 8 - 2k) = \dfrac{1}{2}(-k^2 + 7k + 8)$

$= \dfrac{1}{2}(k + 1)(8 - k) = \dfrac{1}{2}(k + 1)[9 - (k + 1)]$.

17. (I) $n = 1: 1 \cdot (1 + 1) = 2$ and $\dfrac{1}{3} \cdot 1 \cdot 2 \cdot 3 = 2$

(II) If $1 \cdot 2 + 2 \cdot 3 + 3 \cdot 4 + \cdots + k(k + 1) = \dfrac{1}{3}k(k + 1)(k + 2)$, then $1 \cdot 2 + 2 \cdot 3 + 3 \cdot 4 + \cdots + k(k + 1)$

$+ (k + 1)[(k + 1) + 1] = [1 \cdot 2 + 2 \cdot 3 + 3 \cdot 4 + \cdots + k(k + 1)] + (k + 1)(k + 2)$

$= \dfrac{1}{3}k(k + 1)(k + 2) + \dfrac{1}{3} \cdot 3(k + 1)(k + 2) = \dfrac{1}{3}(k + 1)(k + 2)(k + 3) = \dfrac{1}{3}(k + 1)[(k + 1) + 1][(k + 1) + 2]$.

19. (I) $n = 1: 1^2 + 1 = 2$, which is divisible by 2.

(II) If $k^2 + k$ is divisible by 2, then $(k + 1)^2 + (k + 1) = k^2 + 2k + 1 + k + 1 = (k^2 + k) + 2k + 2$. Since $k^2 + k$ is divisible by 2 and $2k + 2$ is divisible by 2, $(k + 1)^2 + (k + 1)$ is divisible by 2.

21. (I) $n = 1: 1^2 - 1 + 2 = 2$, which is divisible by 2.
  (II) If $k^2 - k + 2$ is divisible by 2, then $(k + 1)^2 - (k + 1) + 2 = k^2 + 2k + 1 - k - 1 + 2 = (k^2 - k + 2) + 2k$. Since $k^2 - k + 2$ is divisible by 2 and $2k$ is divisible by 2, $(k + 1)^2 - (k + 1) + 2$ is divisible by 2.

23. (I) $n = 1$: If $x > 1$, then $x^1 = x > 1$.
  (II) Assume, for an arbitrary natural number k, that if $x > 1$ then $x^k > 1$. Multiply both sides of the inequality $x^k > 1$ by x. If $x > 1$, then $x^{k+1} > x > 1$.

25. (I) $n = 1: a - b$ is a factor of $a^1 - b^1 = a - b$.
  (II) If $a - b$ is a factor of $a^k - b^k$, then $a^{k+1} - b^{k+1} = a(a^k - b^k) + b^k(a - b)$.
  Since $a - b$ is a factor of $a^k - b^k$ and $a - b$ is a factor of $a - b$, then $a - b$ is a factor of $a^{k+1} - b^{k+1}$.

27. (I) $n = 1: (1 + a)^1 = 1 + a \geq 1 + 1 \cdot a$
  (II) Assume that there is an integer k for which the inequality holds. So $(1 + a)^k \geq 1 + ka$. We need to show that $(1 + a)^{k+1} \geq 1 + (k + 1)a$. $(1 + a)^{k+1} = (1 + a)^k (1 + a) \geq (1 + ka)(1 + a) = 1 + ka^2 + a + ka = 1 + (k + 1)a + ka^2 \geq 1 + (k + 1)a$.

29. If $2 + 4 + 6 + \cdots + 2k = k^2 + k + 2$, then $2 + 4 + 6 + \cdots + 2k + 2(k + 1)$
  $= (2 + 4 + 6 + \cdots + 2k) + 2k + 2 = k^2 + k + 2 + 2k + 2 = k^2 + 3k + 4 = (k^2 + 2k + 1) + (k + 1) + 2$
  $= (k + 1)^2 + (k + 1) + 2.$
  But $2 \cdot 1 = 2$ and $1^2 + 1 + 2 = 4$. The fact is that $2 + 4 + 6 + \cdots + 2n = n^2 + n$, not $n^2 + n + 2$ (Problem 1).

31. (I) $n = 1: [a + (1 - 1)d] = a$ and $1 \cdot a + d\dfrac{1 \cdot (1 - 1)}{2} = a.$

  (II) If $a + (a + d) + (a + 2d) + \cdots + [a + (k - 1)d] = ka + d\dfrac{k(k - 1)}{2}$, then

  $a + (a + d) + (a + 2d) + \cdots + [a + (k - 1)d] + [a + ((k + 1) - 1)d] = ka + d\dfrac{k(k - 1)}{2} + a + kd$

  $= (k + 1)a + d\dfrac{k(k - 1) + 2k}{2} = (k + 1)a + d\dfrac{(k + 1)(k)}{2} = (k + 1)a + d\dfrac{(k + 1)[(k + 1) - 1]}{2}.$

33. (I) $n = 3$: The sum of the angles of a triangle is $(3 - 2) \cdot 180° = 180°$.
  (II) Assume for some $k \geq 3$ that the sum of the angles of a convex polygon of k sides is $(k - 2) \cdot 180°$. A convex polygon of $k + 1$ sides consists of a convex polygon of k sides plus a triangle (see the illustration). The sum of the angles is
  $(k - 2) \cdot 180° + 180° = (k - 1) \cdot 180° = [(k + 1) - 2] \cdot 180°.$

k sides
$k + 1$ sides

11.5 Assess Your Understanding (page 863)

1. Pascal's triangle **2.** $1; n$ **3.** F **4.** Binomial Theorem **5.** 10 **7.** 21 **9.** 50 **11.** 1 **13.** $\approx 1.8664 \times 10^{15}$ **15.** $\approx 1.4834 \times 10^{13}$
17. $x^5 + 5x^4 + 10x^3 + 10x^2 + 5x + 1$ **19.** $x^6 - 12x^5 + 60x^4 - 160x^3 + 240x^2 - 192x + 64$ **21.** $81x^4 + 108x^3 + 54x^2 + 12x + 1$
23. $x^{10} + 5x^8y^2 + 10x^6y^4 + 10x^4y^6 + 5x^2y^8 + y^{10}$ **25.** $x^3 + 6\sqrt{2}x^{5/2} + 30x^2 + 40\sqrt{2}x^{3/2} + 60x + 24\sqrt{2}x^{1/2} + 8$
27. $a^5x^5 + 5a^4bx^4y + 10a^3b^2x^3y^2 + 10a^2b^3x^2y^3 + 5ab^4xy^4 + b^5y^5$ **29.** 17,010 **31.** $-101,376$ **33.** 41,472 **35.** $2835x^3$
37. $314,928x^7$ **39.** 495 **41.** 3360 **43.** 1.00501

45. $\dbinom{n}{n - 1} = \dfrac{n!}{(n - 1)! [n - (n - 1)]!} = \dfrac{n!}{(n - 1)! \, 1!} = \dfrac{n \cdot (n - 1)!}{(n - 1)!} = n; \dbinom{n}{n} = \dfrac{n!}{n! \, (n - n)!} = \dfrac{n!}{n! 0!} = \dfrac{n!}{n!} = 1$

47. $2^n = (1 + 1)^n = \dbinom{n}{0}1^n + \dbinom{n}{1}(1)^{n-1}(1) + \cdots + \dbinom{n}{n}1^n = \dbinom{n}{0} + \dbinom{n}{1} + \cdots + \dbinom{n}{n}$ **49.** 1

Review Exercises (page 865)

1. $a_1 = -\dfrac{4}{3}, a_2 = \dfrac{5}{4}, a_3 = -\dfrac{6}{5}, a_4 = \dfrac{7}{6}, a_5 = -\dfrac{8}{7}$ **3.** $c_1 = 2, c_2 = 1, c_3 = \dfrac{8}{9}, c_4 = 1, c_5 = \dfrac{32}{25}$ **5.** $a_1 = 3, a_2 = 2, a_3 = \dfrac{4}{3}, a_4 = \dfrac{8}{9}, a_5 = \dfrac{16}{27}$

7. $a_1 = 2, a_2 = 0, a_3 = 2, a_4 = 0, a_5 = 2$ **9.** $6 + 10 + 14 + 18 = 48$ **11.** $\displaystyle\sum_{k=1}^{13}(-1)^{k+1}\dfrac{1}{k}$ **13.** Arithmetic; $d = 1$; $S_n = \dfrac{n}{2}(n + 11)$ **15.** Neither

17. Geometric; $r = 8$; $S_n = \dfrac{8}{7}(8^n - 1)$ **19.** Arithmetic; $d = 4$; $S_n = 2n(n - 1)$ **21.** Geometric; $r = \dfrac{1}{2}$; $S_n = 6\left[1 - \left(\dfrac{1}{2}\right)^n\right]$ **23.** Neither

25. 3825 **27.** 1125 **29.** $\dfrac{1093}{2187} \approx 0.49977$ **31.** 35 **33.** $\dfrac{1}{10^{10}}$ **35.** $9\sqrt{2}$ **37.** $\{a_n\} = \{5n - 4\}$ **39.** $\{a_n\} = \{n - 10\}$ **41.** Converges; $\dfrac{9}{2}$

43. Converges; $\dfrac{4}{3}$ **45.** Diverges **47.** Converges; 8

49. (I) $n = 1: 3 \cdot 1 = 3$ and $\dfrac{3 \cdot 1}{2}(1 + 1) = 3$

  (II) If $3 + 6 + 9 + \cdots + 3k = \dfrac{3k}{2}(k + 1)$, then $3 + 6 + 9 + \cdots + 3k + 3(k + 1) = (3 + 6 + 9 + \cdots + 3k) + (3k + 3)$

  $= \dfrac{3k}{2}(k + 1) + (3k + 3) = \dfrac{3k^2}{2} + \dfrac{3k}{2} + \dfrac{6k}{2} + \dfrac{6}{2} = \dfrac{3}{2}(k^2 + 3k + 2) = \dfrac{3}{2}(k + 1)(k + 2) = \dfrac{3(k + 1)}{2}[(k + 1) + 1].$

51. (I) $n = 1: 2 \cdot 3^{1-1} = 2$ and $3^1 - 1 = 2$
(II) If $2 + 6 + 18 + \cdots + 2 \cdot 3^{k-1} = 3^k - 1$, then $2 + 6 + 18 + \cdots + 2 \cdot 3^{k-1} + 2 \cdot 3^{(k+1)-1} = (2 + 6 + 18 + \cdots + 2 \cdot 3^{k-1}) + 2 \cdot 3^k$
$= 3^k - 1 + 2 \cdot 3^k = 3 \cdot 3^k - 1 = 3^{k+1} - 1.$

53. (I) $n = 1: (3 \cdot 1 - 2)^2 = 1$ and $\dfrac{1}{2} \cdot 1 \cdot [6(1)^2 - 3(1) - 1] = 1$

(II) If $1^2 + 4^2 + 7^2 + \cdots + (3k - 2)^2 = \dfrac{1}{2}k(6k^2 - 3k - 1)$, then $1^2 + 4^2 + 7^2 + \cdots + (3k - 2)^2 + [3(k + 1) - 2]^2$

$= [1^2 + 4^2 + 7^2 + \cdots + (3k - 2)^2] + (3k + 1)^2 = \dfrac{1}{2}k(6k^2 - 3k - 1) + (3k + 1)^2 = \dfrac{1}{2}(6k^3 - 3k^2 - k) + (9k^2 + 6k + 1)$

$= \dfrac{1}{2}(6k^3 + 15k^2 + 11k + 2) = \dfrac{1}{2}(k + 1)(6k^2 + 9k + 2) = \dfrac{1}{2}(k + 1)[6(k + 1)^2 - 3(k + 1) - 1].$

55. 10 **57.** $x^5 + 10x^4 + 40x^3 + 80x^2 + 80x + 32$ **59.** $32x^5 + 240x^4 + 720x^3 + 1080x^2 + 810x + 243$ **61.** 144 **63.** 84

65. (a) 8 bricks **(b)** 1100 bricks **67. (a)** $20\left(\dfrac{3}{4}\right)^3 = \dfrac{135}{16}$ ft **(b)** $20\left(\dfrac{3}{4}\right)^n$ ft **(c)** 13 times **(d)** 140 ft **69.** \$244,129.08

Chapter Test *(page 867)*

1. $0, \dfrac{3}{10}, \dfrac{8}{11}, \dfrac{5}{4}, \dfrac{24}{13}$ **2.** $4, 14, 44, 134, 404$ **3.** $2 - \dfrac{3}{4} + \dfrac{4}{9} = \dfrac{61}{36}$ **4.** $-\dfrac{1}{3} - \dfrac{14}{9} - \dfrac{73}{27} - \dfrac{308}{81} = -\dfrac{680}{81}$ **5.** $\displaystyle\sum_{k=1}^{10}(-1)^k\left(\dfrac{k + 1}{k + 4}\right)$ **6.** Neither

7. Geometric; $r = 4$; $S_n = \dfrac{2}{3}(1 - 4^n)$ **8.** Arithmetic: $d = -8$; $S_n = n(2 - 4n)$ **9.** Arithmetic; $d = -\dfrac{1}{2}$; $S_n = \dfrac{n}{4}(27 - n)$

10. Geometric; $r = \dfrac{2}{5}$; $S_n = \dfrac{125}{3}\left[1 - \left(\dfrac{2}{5}\right)^n\right]$ **11.** Neither **12.** Converges; $\dfrac{1024}{5}$ **13.** $243m^5 + 810m^4 + 1080m^3 + 720m^2 + 240m + 32$

14. First we show that the statement holds for $n = 1$. $\left(1 + \dfrac{1}{1}\right) = 1 + 1 = 2$. The equality is true for $n = 1$, so Condition I holds. Next we assume that

$\left(1 + \dfrac{1}{1}\right)\left(1 + \dfrac{1}{2}\right)\left(1 + \dfrac{1}{3}\right)\cdots\left(1 + \dfrac{1}{n}\right) = n + 1$ is true for some k, and we determine whether the formula then holds for $k + 1$. We assume that

$\left(1 + \dfrac{1}{1}\right)\left(1 + \dfrac{1}{2}\right)\left(1 + \dfrac{1}{3}\right)\cdots\left(1 + \dfrac{1}{k}\right) = k + 1$. Now we need to show that $\left(1 + \dfrac{1}{1}\right)\left(1 + \dfrac{1}{2}\right)\left(1 + \dfrac{1}{3}\right)\cdots\left(1 + \dfrac{1}{k}\right)\left(1 + \dfrac{1}{k + 1}\right)$

$= (k + 1) + 1 = k + 2$. We do this as follows:

$\left(1 + \dfrac{1}{1}\right)\left(1 + \dfrac{1}{2}\right)\left(1 + \dfrac{1}{3}\right)\cdots\left(1 + \dfrac{1}{k}\right)\left(1 + \dfrac{1}{k + 1}\right) = \left[\left(1 + \dfrac{1}{1}\right)\left(1 + \dfrac{1}{2}\right)\left(1 + \dfrac{1}{3}\right)\cdots\left(1 + \dfrac{1}{k}\right)\right]\left(1 + \dfrac{1}{k + 1}\right)$

$= (k + 1)\left(1 + \dfrac{1}{k + 1}\right)$(induction assumption) $= (k + 1) \cdot 1 + (k + 1) \cdot \dfrac{1}{k + 1} = k + 1 + 1 = k + 2$

Condition II also holds. Thus, the formula holds true for all natural numbers.

15. After 10 years, the Durango will be worth \$6103.11. **16.** The weightlifter will have lifted a total of 8000 pounds after 5 sets.

Cumulative Review *(page 867)*

1. $\{-3, 3, -3i, 3i\}$ **2. (a)**

(b) $\left\{\left(\sqrt{\dfrac{-1 + \sqrt{3601}}{18}}, \dfrac{-1 + \sqrt{3601}}{6}\right), \left(-\sqrt{\dfrac{-1 + \sqrt{3601}}{18}}, \dfrac{-1 + \sqrt{3601}}{6}\right)\right\}$

(c) The circle and the parabola intersect at

$\left(\sqrt{\dfrac{-1 + \sqrt{3601}}{18}}, \dfrac{-1 + \sqrt{3601}}{6}\right), \left(-\sqrt{\dfrac{-1 + \sqrt{3601}}{18}}, \dfrac{-1 + \sqrt{3601}}{6}\right).$

3. $\left\{\ln\left(\dfrac{5}{2}\right)\right\}$ **4.** $y = 5x - 10$ **5.** $(x + 1)^2 + (y - 2)^2 = 25$ **6. (a)** 5 **(b)** 13 **(c)** $\dfrac{6x + 3}{2x - 1}$ **(d)** $\left\{x\middle| x \neq \dfrac{1}{2}\right\}$ **(e)** $\dfrac{7x - 2}{x - 2}$ **(f)** $\{x|x \neq 2\}$

(g) $g^{-1}(x) = \dfrac{1}{2}(x - 1)$; all reals **(h)** $f^{-1}(x) = \dfrac{2x}{x - 3}$; $\{x|x \neq 3\}$ **7.** $\dfrac{x^2}{7} + \dfrac{y^2}{16} = 1$ **8.** $(x + 1)^2 = 4(y - 2)$

CHAPTER 12 Counting and Probability

12.1 Assess Your Understanding *(page 874)*

5. subset; $\subseteq$ **6.** finite **7.** $n(A) + n(B) - n(A \cap B)$ **8.** T **9.** $\varnothing$, $\{a\}$, $\{b\}$, $\{c\}$, $\{d\}$, $\{a, b\}$, $\{a, c\}$, $\{a, d\}$, $\{b, c\}$, $\{b, d\}$, $\{c, d\}$, $\{a, b, c\}$, $\{b, c, d\}$, $\{a, c, d\}$,
$\{a, b, d\}$, $\{a, b, c, d\}$ **11.** 25 **13.** 40 **15.** 25 **17.** 37 **19.** 18 **21.** 5 **23.** 15 different arrangements **25.** 9000 numbers **27.** 175; 125
29. (a) 15 **(b)** 15 **(c)** 15 **(d)** 25 **(e)** 40 **31. (a)** 11,923 thousand **(b)** 75,241 thousand **33.** 480 portfolios

12.2 Assess Your Understanding *(page 881)*

3. permutation **4.** combination **5.** $\dfrac{n!}{(n - r)!}$ **6.** $\dfrac{n!}{(n - r)!r!}$ **7.** 30 **9.** 24 **11.** 1 **13.** 1680 **15.** 28 **17.** 35 **19.** 1 **21.** 10,400,600

23. {abc, abd, abe, acb, acd, ace, adb, adc, ade, aeb, aec, aed, bac, bad, bae, bca, bcd, bce, bda, bdc, bde, bea, bec, bed, cab, cad, cae, cba, cbd, cbe, cda, cdb, cde, cea, ceb, ced, dab, dac, dae, dba, dbc, dbe, dca, dcb, dce, dea, deb, dec, eab, eac, ead, eba, ebc, ebd, eca, ecb, ecd, eda, edb, edc}; 60

25. {123, 124, 132, 134, 142, 143, 213, 214, 231, 234, 241, 243, 312, 314, 321, 324, 341, 342, 412, 413, 421, 423, 431, 432}; 24   **27.** {abc, abd, abe, acd, ace, ade, bcd, bce, bde, cde}; 10   **29.** {123, 124, 134, 234}; 4   **31.** 16   **33.** 8   **35.** 24   **37.** 60   **39.** 18,278   **41.** 35   **43.** 1024   **45.** 120   **47.** 132,860

49. 336   **51.** 90,720   **53. (a)** 63   **(b)** 35   **(c)** 1   **55.** 1.157×10^{76} **57.** 362,880   **59.** 660   **61.** 15   **63. (a)** 125,000; 117,600   **(b)** A better name for a *combination* lock would be a *permutation* lock because the order of the numbers matters.

Historical Problems *(page 891)*

1. (a) {AAAA, AAAB, AABA, AABB, ABAA, ABAB, ABBA, ABBB, BAAA, BAAB, BABA, BABB, BBAA, BBAB, BBBA, BBBB}

(b) $P(A \text{ wins}) = \dfrac{C(4,2) + C(4,3) + C(4,4)}{2^4} = \dfrac{6 + 4 + 1}{16} = \dfrac{11}{16}$; $P(B \text{ wins}) = \dfrac{C(4,3) + C(4,4)}{2^4} = \dfrac{4 + 1}{16} = \dfrac{5}{16}$

2. (a) $\$\dfrac{3}{2} = \$1.50$ **(b)** $\$\dfrac{1}{2} = \$0.50$

12.3 Assess Your Understanding *(page 891)*

1. equally likely   **2.** complement   **3.** F   **4.** T   **5.** 0, 0.01, 0.35, 1   **7.** Probability model   **9.** Not a probability model

11. $S = \{HH, HT, TH, TT\}$; $P(HH) = \dfrac{1}{4}$, $P(HT) = \dfrac{1}{4}$, $P(TH) = \dfrac{1}{4}$, $P(TT) = \dfrac{1}{4}$ **13.** $S = \{HH1, HH2, HH3, HH4, HH5, HH6, HT1, HT2, HT3,$

HT4, HT5, HT6, TH1, TH2, TH3, TH4, TH5, TH6, TT1, TT2, TT3, TT4, TT5, TT6}; each outcome has the probability of $\dfrac{1}{24}$.

15. $S = \{HHH, HHT, HTH, HTT, THH, THT, TTH, TTT\}$; each outcome has the probability of $\dfrac{1}{8}$.

17. $S = \{1 \text{ Yellow}, 1 \text{ Red}, 1 \text{ Green}, 2 \text{ Yellow}, 2 \text{ Red}, 2 \text{ Green}, 3 \text{ Yellow}, 3 \text{ Red}, 3 \text{ Green}, 4 \text{ Yellow}, 4 \text{ Red}, 4 \text{ Green}\}$; each outcome has the probability

of $\dfrac{1}{12}$; thus, $P(2 \text{ Red}) + P(4 \text{ Red}) = \dfrac{1}{12} + \dfrac{1}{12} = \dfrac{1}{6}$.

19. $S = \{1 \text{ Yellow Forward}, 1 \text{ Yellow Backward}, 1 \text{ Red Forward}, 1 \text{ Red Backward}, 1 \text{ Green Forward}, 1 \text{ Green Backward}, 2 \text{ Yellow Forward}, 2 \text{ Yellow}$

Backward, 2 Red Forward, 2 Red Backward, 2 Green Forward, 2 Green Backward, 3 Yellow Forward, 3 Yellow Backward, 3 Red Forward, 3 Red

Backward, 3 Green Forward, 3 Green Backward, 4 Yellow Forward, 4 Yellow Backward, 4 Red Forward, 4 Red Backward, 4 Green Forward, 4 Green

Backward}; each outcome has the probability of $\dfrac{1}{24}$; thus, $P(1 \text{ Red Backward}) + P(1 \text{ Green Backward}) = \dfrac{1}{24} + \dfrac{1}{24} = \dfrac{1}{12}$.

21. $S = \{11 \text{ Red}, 11 \text{ Yellow}, 11 \text{ Green}, 12 \text{ Red}, 12 \text{ Yellow}, 12 \text{ Green}, 13 \text{ Red}, 13 \text{ Yellow}, 13 \text{ Green}, 14 \text{ Red}, 14 \text{ Yellow}, 14 \text{ Green}, 21 \text{ Red}, 21 \text{ Yellow}, 21 \text{ Green},$

22 Red, 22 Yellow, 22 Green, 23 Red, 23 Yellow, 23 Green, 24 Red, 24 Yellow, 24 Green, 31 Red, 31 Yellow, 31 Green, 32 Red, 32 Yellow, 32 Green, 33 Red,

33 Yellow, 33 Green, 34 Red, 34 Yellow, 34 Green, 41 Red, 41 Yellow, 41 Green, 42 Red, 42 Yellow, 42 Green, 43 Red, 43 Yellow, 43 Green, 44 Red,

44 Yellow, 44 Green}; each outcome has the probability of $\dfrac{1}{48}$; thus, $E = \{22 \text{ Red}, 22 \text{ Green}, 24 \text{ Red}, 24 \text{ Green}\}$; $P(E) = \dfrac{n(E)}{n(S)} = \dfrac{4}{48} = \dfrac{1}{12}$.

23. A, B, C, F   **25.** B   **27.** $P(H) = \dfrac{4}{5}$; $P(T) = \dfrac{1}{5}$ **29.** $P(1) = P(3) = P(5) = \dfrac{2}{9}$; $P(2) = P(4) = P(6) = \dfrac{1}{9}$ **31.** $\dfrac{3}{10}$ **33.** $\dfrac{1}{2}$ **35.** $\dfrac{1}{6}$ **37.** $\dfrac{1}{8}$

39. $\dfrac{1}{4}$ **41.** $\dfrac{1}{6}$ **43.** $\dfrac{1}{18}$ **45.** 0.55   **47.** 0.70   **49.** 0.30   **51.** 0.87   **53.** 0.66   **55.** 0.95   **57.** $\dfrac{17}{20}$ **59.** $\dfrac{11}{20}$ **61.** $\dfrac{1}{2}$ **63.** $\dfrac{3}{10}$ **65.** $\dfrac{2}{5}$

67. (a) 0.57   **(b)** 0.95   **(c)** 0.83   **(d)** 0.38   **(e)** 0.29   **(f)** 0.05   **(g)** 0.78   **(h)** 0.71   **69. (a)** $\dfrac{25}{33}$ **(b)** $\dfrac{25}{33}$ **71.** 0.167   **73.** 0.000033069

Review Exercises *(page 895)*

1. $\varnothing$, {Dave}, {Joanne}, {Erica}, {Dave, Joanne}, {Dave, Erica}, {Joanne, Erica}, {Dave, Joanne, Erica}

3. 17   **5.** 29   **7.** 7   **9.** 25   **11.** 336   **13.** 56   **15.** 60   **17.** 128   **19.** 3024   **21.** 1680   **23.** 91   **25.** 1,600,000   **27.** 216,000   **29.** 1260

31. (a) 381,024   **(b)** 1260   **33. (a)** $8.634628387 \times 10^{45}$ **(b)** 0.6531   **(c)** 0.3469   **35. (a)** 0.058   **(b)** 0.942   **37.** $\dfrac{4}{9}$ **39.** 0.2; 0.26

Chapter Test *(page 896)*

1. 22   **2.** 3   **3.** 8   **4.** 45   **5.** 5040   **6.** 151,200   **7.** 462   **8.** There are 54,264 ways to choose 6 different colors from the 21 available colors.

9. There are 840 distinct arrangements of the letters in the word REDEEMED.   **10.** There are 56 different exacta bets for an 8-horse race.

11. There are 155,480,000 possible license plates using the new format.   **12. (a)** 0.95   **(b)** 0.30   **13. (a)** 0.25   **(b)** 0.55   **14.** 0.19

15. $P(\text{win on } \$1 \text{ play}) = \dfrac{1}{120,526,770} \approx 0.0000000083$ **16.** $P(\text{exactly 2 fours}) = \dfrac{1250}{7776} \approx 0.1608$

Cumulative Review *(page 897)*

1. $\left\{ \dfrac{1}{3} - \dfrac{\sqrt{2}}{3}i, \dfrac{1}{3} + \dfrac{\sqrt{2}}{3}i \right\}$ **2.** [graph]   **3.** [graph]   **4.** $\{x \mid 3.99 \le x \le 4.01\}$ or $[3.99, 4.01]$

5. $\left\{-\dfrac{1}{2}+\dfrac{\sqrt{7}}{2}i, -\dfrac{1}{2}-\dfrac{\sqrt{7}}{2}i, -\dfrac{1}{5}, 3\right\}$ **6.**   **7.** 2   **8.** $\left\{\dfrac{8}{3}\right\}$ **9.** $x = 2, y = -5, z = 3$ **10.** 125; 700

Domain: all real numbers
Range: $\{y | y > 5\}$
Horizontal asymptote: $y = 5$

APPENDIX A Review

A.1 Assess Your Understanding *(page A10)*

1. variable   **2.** origin   **3.** strict   **4.** base; exponent or power   **5.** T   **6.** T   **7.** F   **8.** F   **9.** $\{1, 2, 3, 4, 5, 6, 7, 8, 9\}$ **11.** $\{4\}$ **13.** $\{1, 3, 4, 6\}$
15. $\{0, 2, 6, 7, 8\}$ **17.** $\{0, 1, 2, 3, 5, 6, 7, 8, 9\}$ **19.** $\{0, 1, 2, 3, 5, 6, 7, 8, 9\}$ **21.** ⟵•──•─○•●──•⟶   **23.** >   **25.** >   **27.** >   **29.** =   **31.** <

33. $x > 0$ **35.** $x < 2$ **37.** $x \le 1$ **39.** ⟶   **41.** ⟶   **43.** 1   **45.** 2   **47.** 6   **49.** 4   **51.** -28
53. $\dfrac{4}{5}$ **55.** 0   **57.** 1   **59.** 5   **61.** 1   **63.** 22   **65.** 2   **67.** $x = 0$ **69.** $x = 3$ **71.** None   **73.** $x = 0, x = 1, x = -1$ **75.** $\{x | x \ne 5\}$ **77.** $\{x | x \ne -4\}$
79. 0°C   **81.** 25°C   **83.** 16   **85.** $\dfrac{1}{16}$ **87.** $\dfrac{1}{9}$ **89.** 9   **91.** 5   **93.** 4   **95.** $64x^6$ **97.** $\dfrac{x^4}{y^2}$ **99.** $\dfrac{x}{y}$ **101.** $-\dfrac{8x^3z}{9y}$ **103.** $\dfrac{16x^2}{9y^2}$ **105.** -4 **107.** 5   **109.** 4
111. 2   **113.** $\sqrt{5}$ **115.** $\dfrac{1}{2}$ **117.** 10; 0   **119.** 81   **121.** 304,006.671   **123.** 0.004   **125.** 481.890   **127.** 0.000   **129.** $A = lw$ **131.** $C = \pi d$
133. $A = \dfrac{\sqrt{3}}{4}x^2$ **135.** $V = \dfrac{4}{3}\pi r^3$ **137.** $V = x^3$ **139. (a)** \$6000   **(b)** \$8000   **141.** $|x - 4| \ge 6$ **143. (a)** $2 \le 5$ **(b)** $6 > 5$ **145. (a)** Yes   **(b)** No
147. No; $\dfrac{1}{3}$ is larger; 0.000333 …   **149.** No

A.2 Assess Your Understanding *(page A20)*

1. right; hypotenuse   **2.** $A = \dfrac{1}{2}$ bh   **3.** $C = 2\pi r$ **4.** similar   **5.** T   **6.** T   **7.** F   **8.** T   **9.** T   **10.** F   **11.** 13   **13.** 26   **15.** 25   **17.** Right triangle; 5
19. Not a right triangle   **21.** Right triangle; 25   **23.** Not a right triangle   **25.** 8 in.^{2} **27.** 4 in.^{2} **29.** $A = 25\pi$ m^2; $C = 10\pi$ m
31. $V = 224$ ft^3; $S = 232$ ft^2 **33.** $V = \dfrac{256}{3}\pi$ cm^3; $S = 64\pi$ cm^2 **35.** $V = 648\pi$ in.^{3}; $S = 306\pi$ in.^{2} **37.** π square units   **39.** 2π square units
41. $x = 4$ units; $A = 90°$; $B = 60°$; $C = 30°$ **43.** $x = 67.5$ units; $A = 60°$; $B = 95°$; $C = 25°$ **45.** About 16.8 ft   **47.** 64 ft^2
49. $24 + 2\pi \approx 30.28$ ft^2; $16 + 2\pi \approx 22.28$ ft   **51.** 160 paces   **53.** About 5.477 mi   **55.** From 100 ft: 12.2 mi; From 150 ft: 15.0 mi

A.3 Assess Your Understanding *(page A31)*

1. 4; 3   **2.** $x^2 - a^2$ **3.** $x^3 - a^3$ **4.** F   **5.** T   **6.** F   **7.** Monomial; variable: x; coefficient: 2; degree: 3   **9.** Not a monomial; the exponent of the variable is not a nonnegative integer   **11.** Monomial; variables: x, y; coefficient: -2; degree: 3   **13.** Not a monomial; the exponent of one of the variables is not a nonnegative integer   **15.** Not a monomial; it has more than one term   **17.** Yes; 2   **19.** Yes; 0   **21.** No; the variable of one of the terms is not a nonnegative integer   **23.** Yes; 3   **25.** No; the polynomial of the denominator has a degree greater than 0   **27.** $x^2 + 7x + 2$ **29.** $x^3 - 4x^2 + 9x + 7$
31. $6x^5 + 5x^4 + 3x^2 + x$ **33.** $7x^2 - x - 7$ **35.** $-2x^3 + 18x^2 - 18$ **37.** $2x^2 - 4x + 6$ **39.** $15y^2 - 27y + 30$ **41.** $x^3 + x^2 - 4x$
43. $-8x^5 - 10x^2$ **45.** $x^3 + 3x^2 - 2x - 4$ **47.** $x^2 + 6x + 8$ **49.** $2x^2 + 9x + 10$ **51.** $x^2 - 2x - 8$ **53.** $x^2 - 5x + 6$ **55.** $2x^2 - x - 6$
57. $-2x^2 + 11x - 12$ **59.** $2x^2 + 8x + 8$ **61.** $x^2 - xy - 2y^2$ **63.** $-6x^2 - 13xy - 6y^2$ **65.** $x^2 - 49$ **67.** $4x^2 - 9$ **69.** $x^2 + 8x + 16$
71. $x^2 - 8x + 16$ **73.** $9x^2 - 16$ **75.** $4x^2 - 12x + 9$ **77.** $x^2 - y^2$ **79.** $9x^2 - y^2$ **81.** $x^2 + 2xy + y^2$ **83.** $x^2 - 4xy + 4y^2$
85. $x^3 - 6x^2 + 12x - 8$ **87.** $8x^3 + 12x^2 + 6x + 1$ **89.** $4x^2 - 11x + 23$; remainder -45 **91.** $4x - 3$; remainder $x + 1$
93. $5x^2 - 13$; remainder $x + 27$ **95.** $2x^2$; remainder $-x^2 + x + 1$ **97.** $x^2 - 2x + \dfrac{1}{2}$; remainder $\dfrac{5}{2}x + \dfrac{1}{2}$ **99.** $-4x^2 - 3x - 3$; remainder -7
101. $x^2 - x - 1$; remainder $2x + 2$ **103.** $x^2 + ax + a^2$; remainder 0

A.4 Assess Your Understanding *(page A41)*

1. $3x(x - 2)(x + 2)$ **2.** prime   **3.** T   **4.** F   **5.** $3(x + 2)$ **7.** $a(x^2 + 1)$ **9.** $x(x^2 + x + 1)$ **11.** $2x(x - 1)$ **13.** $3xy(x - 2y + 4)$
15. $(x + 1)(x - 1)$ **17.** $(2x + 1)(2x - 1)$ **19.** $(x + 4)(x - 4)$ **21.** $(5x + 2)(5x - 2)$ **23.** $(x + 1)^2$ **25.** $(x + 2)^2$ **27.** $(x - 5)^2$
29. $(2x + 1)^2$ **31.** $(4x + 1)^2$ **33.** $(x - 3)(x^2 + 3x + 9)$ **35.** $(x + 3)(x^2 - 3x + 9)$ **37.** $(2x + 3)(4x^2 - 6x + 9)$ **39.** $(x + 2)(x + 3)$
41. $(x + 6)(x + 1)$ **43.** $(x + 5)(x + 2)$ **45.** $(x - 8)(x - 2)$ **47.** $(x - 8)(x + 1)$ **49.** $(x + 8)(x - 1)$ **51.** $(x + 2)(2x + 3)$
53. $(x - 2)(2x + 1)$ **55.** $(2x + 3)(3x + 2)$ **57.** $(3x + 1)(x + 1)$ **59.** $(z + 1)(2z + 3)$ **61.** $(x + 2)(3x - 4)$ **63.** $(x - 2)(3x + 4)$
65. $(x + 4)(3x + 2)$ **67.** $(x + 4)(3x - 2)$ **69.** 25; $(x + 5)^2$ **71.** 9; $(y - 3)^2$ **73.** $\dfrac{1}{16}$; $\left(x - \dfrac{1}{4}\right)^2$ **75.** $(x + 6)(x - 6)$ **77.** $2(1 + 2x)(1 - 2x)$
79. $(x + 1)(x + 10)$ **81.** $(x - 7)(x - 3)$ **83.** $4(x^2 - 2x + 8)$ **85.** Prime   **87.** $-(x - 5)(x + 3)$ **89.** $3(x + 2)(x - 6)$ **91.** $y^2(y + 5)(y + 6)$
93. $(2x + 3)^2$ **95.** $2(3x + 1)(x + 1)$ **97.** $(x - 3)(x + 3)(x^2 + 9)$ **99.** $(x - 1)^2(x^2 + x + 1)^2$ **101.** $x^5(x - 1)(x + 1)$ **103.** $(4x + 3)^2$
105. $-(4x - 5)(4x + 1)$ **107.** $(2y - 5)(2y - 3)$ **109.** $-(3x - 1)(3x + 1)(x^2 + 1)$ **111.** $(x + 3)(x - 6)$ **113.** $(x + 2)(x - 3)$

115. $(3x - 5)(9x^2 - 3x + 7)$ **117.** $(x + 5)(3x + 11)$ **119.** $(x - 1)(x + 1)(x + 2)$ **121.** $(x - 1)(x + 1)(x^2 - x + 1)$
123. $2(3x + 4)(9x + 13)$ **125.** $2x(3x + 5)$ **127.** $5(x + 3)(x - 2)^2(x + 1)$ **129.** $3(4x - 3)(4x - 1)$ **131.** $6(3x - 5)(2x + 1)^2(5x - 4)$
133. The possibilities are $(x \pm 1)(x \pm 4) = x^2 \pm 5x + 4$ or $(x \pm 2)(x \pm 2) = x^2 \pm 4x + 4$, none of which equals $x^2 + 4$.

A.5 Assess Your Understanding *(page A45)*

1. quotient; divisor; remainder **2.** $-3\overline{)2\ \ 0\ \ -5\ \ 1}$ **3.** T **4.** T **5.** $x^2 + x + 4$; remainder 12 **7.** $3x^2 + 11x + 32$; remainder 99
9. $x^4 - 3x^3 + 5x^2 - 15x + 46$; remainder -138 **11.** $4x^5 + 4x^4 + x^3 + x^2 + 2x + 2$; remainder 7 **13.** $0.1x^2 - 0.11x + 0.321$; remainder -0.3531
15. $x^4 + x^3 + x^2 + x + 1$; remainder 0 **17.** No **19.** Yes **21.** Yes **23.** No **25.** Yes **27.** -9

A.6 Assess Your Understanding *(page A54)*

1. lowest terms **2.** least common multiple **3.** T **4.** F **5.** $\dfrac{3}{x - 3}$ **7.** $\dfrac{x}{3}$ **9.** $\dfrac{4x}{2x - 1}$ **11.** $\dfrac{y + 5}{2(y + 1)}$ **13.** $\dfrac{x + 5}{x - 1}$ **15.** $-(x + 7)$ **17.** $\dfrac{3}{5x(x - 2)}$

19. $\dfrac{2x(x^2 + 4x + 16)}{x + 4}$ **21.** $\dfrac{8}{3x}$ **23.** $\dfrac{x - 3}{x + 7}$ **25.** $\dfrac{4x}{(x - 2)(x - 3)}$ **27.** $\dfrac{4}{5(x - 1)}$ **29.** $-\dfrac{(x - 4)^2}{4x}$ **31.** $\dfrac{(x + 3)^2}{(x - 3)^2}$ **33.** $\dfrac{(x - 4)(x + 3)}{(x - 1)(2x + 1)}$ **35.** $\dfrac{x + 5}{2}$

37. $\dfrac{(x - 2)(x + 2)}{2x - 3}$ **39.** $\dfrac{3x - 2}{x - 3}$ **41.** $\dfrac{x + 9}{2x - 1}$ **43.** $\dfrac{4 - x}{x - 2}$ **45.** $\dfrac{2(x + 5)}{(x - 1)(x + 2)}$ **47.** $\dfrac{3x^2 - 2x - 3}{(x + 1)(x - 1)}$ **49.** $\dfrac{-(11x + 2)}{(x + 2)(x - 2)}$ **51.** $\dfrac{2(x^2 - 2)}{x(x - 2)(x + 2)}$

53. $(x - 2)(x + 2)(x + 1)$ **55.** $x(x - 1)(x + 1)$ **57.** $x^3(2x - 1)^2$ **59.** $x(x - 1)^2(x + 1)(x^2 + x + 1)$ **61.** $\dfrac{5x}{(x - 6)(x - 1)(x + 4)}$

63. $\dfrac{2(2x^2 + 5x - 2)}{(x - 2)(x + 2)(x + 3)}$ **65.** $\dfrac{5x + 1}{(x - 1)^2(x + 1)^2}$ **67.** $\dfrac{-x^2 + 3x + 13}{(x - 2)(x + 1)(x + 4)}$ **69.** $\dfrac{x^3 - 2x^2 + 4x + 3}{x^2(x + 1)(x - 1)}$ **71.** $\dfrac{-1}{x(x + h)}$ **73.** $\dfrac{x + 1}{x - 1}$

75. $\dfrac{(x - 1)(x + 1)}{2x(2x + 1)}$ **77.** $\dfrac{2(5x - 1)}{(x - 2)(x + 1)^2}$ **79.** $\dfrac{-2x(x^2 - 2)}{(x + 2)(x^2 - x - 3)}$ **81.** $\dfrac{-1}{x - 1}$ **83.** $\dfrac{3x - 1}{2x + 1}$ **85.** $\dfrac{19}{(3x - 5)^2}$ **87.** $\dfrac{(x + 1)(x - 1)}{(x^2 + 1)^2}$

89. $\dfrac{x(3x + 2)}{(3x + 1)^2}$ **91.** $-\dfrac{(x + 3)(3x - 1)}{(x^2 + 1)^2}$ **93.** $f = \dfrac{R_1 \cdot R_2}{(n - 1)(R_1 + R_2)}; \dfrac{2}{15}$ m

A.7 Assess Your Understanding *(page A62)*

3. index **4.** T **5.** cube root **6.** F **7.** 3 **9.** -2 **11.** $2\sqrt{2}$ **13.** $-2x\sqrt[3]{x}$ **15.** x^3y^2 **17.** x^2y **19.** $6\sqrt{x}$ **21.** $6x\sqrt{x}$ **23.** $15\sqrt[3]{3}$ **25.** $12\sqrt{3}$

27. $7\sqrt{2}$ **29.** $\sqrt{2}$ **31.** $2\sqrt{3}$ **33.** $-\sqrt[3]{2}$ **35.** $x - 2\sqrt{x} + 1$ **37.** $(2x - 1)\sqrt[3]{2x}$ **39.** $(2x - 15)\sqrt{2x}$ **41.** $-(x + 5y)\sqrt[3]{2xy}$ **43.** $\dfrac{\sqrt{2}}{2}$

45. $-\dfrac{\sqrt{15}}{5}$ **47.** $\dfrac{(5 + \sqrt{2})\sqrt{3}}{23}$ **49.** $\dfrac{8\sqrt{5} - 19}{41}$ **51.** $\dfrac{5\sqrt[3]{4}}{2}$ **53.** $\dfrac{2x + h - 2\sqrt{x^2 + xh}}{h}$ **55.** 4 **57.** -3 **59.** 64 **61.** $\dfrac{1}{27}$ **63.** $\dfrac{27\sqrt{2}}{32}$ **65.** $\dfrac{27\sqrt{2}}{32}$

67. $x^{7/12}$ **69.** xy^2 **71.** $x^{2/3}y$ **73.** $\dfrac{8x^{5/4}}{y^{3/4}}$ **75.** $\dfrac{3x + 2}{(1 + x)^{1/2}}$ **77.** $\dfrac{x(3x^2 + 2)}{(x^2 + 1)^{1/2}}$ **79.** $\dfrac{22x + 5}{10\sqrt{(x - 5)(4x + 3)}}$ **81.** $\dfrac{2 + x}{2(1 + x)^{3/2}}$ **83.** $\dfrac{4 - x}{(x + 4)^{3/2}}$

85. $\dfrac{1}{x^2(x^2 - 1)^{1/2}}$ **87.** $\dfrac{1 - 3x^2}{2\sqrt{x}(1 + x^2)^2}$ **89.** $\dfrac{1}{2}(5x + 2)(x + 1)^{1/2}$ **91.** $2x^{1/2}(3x - 4)(x + 1)$ **93.** $(x^2 + 4)^{1/3}(11x^2 + 12)$

95. $(3x + 5)^{1/3}(2x + 3)^{1/2}(17x + 27)$ **97.** $\dfrac{3(x + 2)}{2x^{1/2}}$ **99.** 1.41 **101.** 1.59 **103.** 4.89 **105.** 2.15 **107. (a)** 15,660.4 gal **(b)** 390.7 gal

109. $2\pi\sqrt{2} \approx 8.89$ s **111.** $\dfrac{\pi\sqrt{3}}{6} \approx 0.91$ s

A.8 Assess Your Understanding *(page A72)*

5. equivalent equations **6.** identity **7.** extraneous **8.** T **9.** {4} **11.** $\left\{\dfrac{5}{4}\right\}$ **13.** $\{-1\}$ **15.** $\{-18\}$ **17.** $\{-3\}$ **19.** $\{-16\}$ **21.** $\{0.5\}$ **23.** $\{2\}$

25. {2} **27.** {3} **29.** $\{0, 9\}$ **31.** $\{-3, 0, 3\}$ **33.** {21} **35.** {6} **37.** $\varnothing$ or $\{\ \}$ **39.** $\{-5, 0, 4\}$ **41.** $\{-1, 1\}$ **43.** $\left\{-2, \dfrac{1}{2}, 2\right\}$ **45.** {1} **47.** No real solution

49. $\{-13\}$ **51.** {3} **53.** {8} **55.** $\{-1, 3\}$ **57.** $\{1, 5\}$ **59.** {5} **61.** {2} **63.** $\left\{-\dfrac{11}{6}\right\}$ **65.** $\{-6\}$ **67.** $R = \dfrac{R_1 R_2}{R_1 + R_2}$ **69.** $R = \dfrac{mv^2}{F}$ **71.** $r = \dfrac{S - a}{S}$

73. The distance is approximately 229.94 ft. **75.** approx. 221 ft **79.** The apparent solution -1 is extraneous. **81.** When multiplying both sides by $x + 3$, we are actually multiplying both sides by 0 when $x = -3$. This violates the Multiplication Property of Equality.

A.9 Assess Your Understanding *(page A80)*

1. mathematical modeling **2.** interest **3.** uniform motion **4.** F **5.** T **6.** $100 - x$ **7.** $A = \pi r^2$; r = radius, A = area
9. $A = s^2$; A = area, s = length of a side **11.** $F = ma$; F = force, m = mass, a = acceleration **13.** $W = Fd$; W = work, F = force, d = distance
15. $C = 150x$; C = total variable cost, x = number of dishwashers **17.** Invest \$48,000 in bonds and \$36,000 in CDs. **19.** \$85,000 was loaned out at
8%. **21.** Mix 75 lb of Earl Grey tea with 25 lb of Orange Pekoe tea. **23.** Mix 160 lb of cashews with the almonds. **25.** The speed of the current is
2.286 mi/hr. **27.** The speed of the current is 5 mi/hr. **29.** Working together, it takes 12 min. **31. (a)** The dimensions are 10 ft by 5 ft. **(b)** The area
is 50 sq ft. **(c)** The dimensions would be 7.5 ft by 7.5 ft. **(d)** The area would be 56.25 sq ft. **33.** The defensive back catches up to the tight end at
the tight end's 45-yd line. **35.** Add $\dfrac{2}{3}$ gal of water. **37.** Approximately 4.05 ft/s. **39.** hp 1300; 30 min; 20 pages/min; hp 2420; 20 min; 30 pages/min

41. Evaporate 10.67 oz of water. **43.** 40 g of 12-karat gold should be mixed with 20 g of pure gold. **45.** Mike passes Dan $\dfrac{1}{3}$ mi from the start,

2 min from the time Mike started to race.   **47.** Start the auxiliary pump at 9:45 A.M.   **49.** The tub will fill in 1 hr.   **51.** The distance of the run is 12 mi.
The distance of the bicycle race is 75 mi.   **53.** Lewis would beat Burke by 16.75 m.   **57.** The average speed is 49.5 mi/hr.

A.10 Assess Your Understanding (page A88)

3. negative   **4.** closed interval   **5.** Multiplication Property   **6.** T   **7.** T   **8.** T   **9.** F   **10.** T   **11.** $[0,2]; 0 \le x \le 2$ **13.** $[2,\infty); x \ge 2$
15. $[0,3); 0 \le x < 3$ **17. (a)** $6 < 8$ **(b)** $-2 < 0$ **(c)** $9 < 15$ **(d)** $-6 > -10$ **19. (a)** $7 > 0$ **(b)** $-1 > -8$ **(c)** $12 > -9$ **(d)** $-8 < 6$
21. (a) $2x + 4 < 5$ **(b)** $2x - 4 < -3$ **(c)** $6x + 3 < 6$ **(d)** $-4x - 2 > -4$

23. $[0,4]$

25. $[4,6)$

27. $[4,\infty)$

29. $(-\infty, -4)$

31. $2 \le x \le 5$

33. $-3 < x < -2$

35. $x \ge 4$

37. $x < -3$

39. $<$ **41.** $>$ **43.** $\ge$ **45.** $<$ **47.** $\le$ **49.** $>$ **51.** $\ge$

53. $\{x|x < 4\}$ or $(-\infty, 4)$

55. $\{x|x \ge -1\}$ or $[-1,\infty)$

57. $\{x|x > 3\}$ or $(3,\infty)$

59. $\{x|x \ge 2\}$ or $[2,\infty)$

61. $\{x|x > -7\}$ or $(-7,\infty)$

63. $\left\{x \middle| x \le \dfrac{2}{3}\right\}$ or $\left(-\infty, \dfrac{2}{3}\right)$

65. $\{x|x < -20\}$ or $(-\infty, -20)$

67. $\left\{x \middle| x \ge \dfrac{4}{3}\right\}$ or $\left[\dfrac{4}{3}, \infty\right)$

69. $\{x|3 \le x \le 5\}$ or $[3,5]$

71. $\left\{x \middle| \dfrac{2}{3} \le x \le 3\right\}$ or $\left[\dfrac{2}{3}, 3\right]$

73. $\left\{x \middle| -\dfrac{11}{2} < x < \dfrac{1}{2}\right\}$ or $\left(-\dfrac{11}{2}, \dfrac{1}{2}\right)$ **75.** $\{x|-6 < x < 0\}$ or $(-6,0)$

77. $\{x|x < -5\}$ or $(-\infty, -5)$

79. $\{x|x \ge -1\}$ or $[-1,\infty)$

81. $\left\{x \middle| \dfrac{1}{2} \le x < \dfrac{5}{4}\right\}$ or $\left[\dfrac{1}{2}, \dfrac{5}{4}\right)$

83. $\left\{x \middle| x < -\dfrac{1}{2}\right\}$ or $\left(-\infty, -\dfrac{1}{2}\right)$

85. $\left\{x \middle| x > \dfrac{10}{3}\right\}$ or $\left(\dfrac{10}{3}, \infty\right)$

87. $\{x|x > 3\}$ or $(3,\infty)$

89. $\{x|x \ge -2\}$ **91.** $21 < \text{Age} < 30$
93. (a) Male ≥ 76.9 years   **(b)** Female ≥ 81.5 years
(c) A female can expect to live 4.6 years longer.

95. The agent's commission ranges from $45,000 to $95,000, inclusive. As a percent of selling price, the commission ranges from 5% to 8.6%,
inclusive.   **97.** The amount withheld varies from $106.60 to $156.60, inclusive.   **99.** The usage varies from 675.44 kW · hr to 2500.92 kW · hr, inclusive.
101. The dealer's cost varies from $15,254.24 to $16,071.43, inclusive.   **103. (a)** You need at least a 74 on the fifth test.   **(b)** You need at least a 77 on
the fifth test.
105. $\dfrac{a+b}{2} - a = \dfrac{a+b-2a}{2} = \dfrac{b-a}{2} > 0$; therefore, $a < \dfrac{a+b}{2}$. $b - \dfrac{a+b}{2} = \dfrac{2b-a-b}{2} = \dfrac{b-a}{2} > 0$; therefore, $b > \dfrac{a+b}{2}$.
107. $(\sqrt{ab})^2 - a^2 = ab - a^2 = a(b - a) > 0$; thus $(\sqrt{ab})^2 > a^2$ and $\sqrt{ab} > a$.
$\quad b^2 - (\sqrt{ab})^2 = b^2 - ab = b(b - a) > 0$; thus $b^2 > (\sqrt{ab})^2$ and $b > \sqrt{ab}$.
109. $h - a = \dfrac{2ab}{a+b} - a = \dfrac{ab - a^2}{a+b} = \dfrac{a(b-a)}{a+b} > 0$; thus $h > a$. $b - h = b - \dfrac{2ab}{a+b} = \dfrac{b^2 - ab}{a+b} = \dfrac{b(b-a)}{a+b} > 0$; thus $h < b$.
111. Since $0 < a < b$, then $a - b < 0$ and $\dfrac{a-b}{ab} < 0$. So $\dfrac{a}{ab} - \dfrac{b}{ab} < 0$, or $\dfrac{1}{b} - \dfrac{1}{a} < 0$. Therefore, $\dfrac{1}{b} < \dfrac{1}{a}$. And $0 < \dfrac{1}{b}$ because $b > 0$.
113. $x^2 + 1 \ge 1$ for all real numbers x.

A.11 Assess Your Understanding (page A97)

4. real; imaginary; imaginary unit   **5.** $\{-2i, 2i\}$ **6.** F   **7.** T   **8.** T   **9.** $8 + 5i$ **11.** $-7 + 6i$ **13.** $-6 - 11i$ **15.** $6 - 18i$ **17.** $6 + 4i$ **19.** $10 - 5i$
21. 37   **23.** $\dfrac{6}{5} + \dfrac{8}{5}i$ **25.** $1 - 2i$ **27.** $\dfrac{5}{2} - \dfrac{7}{2}i$ **29.** $-\dfrac{1}{2} + \dfrac{\sqrt{3}}{2}i$ **31.** $2i$ **33.** $-i$ **35.** i **37.** -6 **39.** $-10i$ **41.** $-2 + 2i$ **43.** 0   **45.** 0   **47.** $2i$
49. $5i$ **51.** $5i$ **53.** 6   **55.** 25   **57.** $2 + 3i$ **59.** $z + \bar{z} = (a + bi) + (a - bi) = 2a$; $z - \bar{z} = (a + bi) - (a - bi) = 2bi$
61. $\overline{z + w} = \overline{(a + bi) + (c + di)} = \overline{(a + c) + (b + d)i} = (a + c) - (b + d)i = (a - bi) + (c - di) = \bar{z} + \bar{w}$
67. $\sqrt{a} \cdot \sqrt{b} = \sqrt{ab}$ only when $\sqrt{a}, \sqrt{b}$ are real numbers.

APPENDIX B Graphing Utilities

B.1 Exercises *(page B2)*

1. $(-1, 4)$; II   **3.** $(3, 1)$; I   **5.** $X\min = -6, X\max = 6, X\mathrm{scl} = 2, Y\min = -4, Y\max = 4, Y\mathrm{scl} = 2$
7. $X\min = -6, X\max = 6, X\mathrm{scl} = 2, Y\min = -1, Y\max = 3, Y\mathrm{scl} = 1$ **9.** $X\min = 3, X\max = 9, X\mathrm{scl} = 1, Y\min = 2, Y\max = 10, Y\mathrm{scl} = 2$
11. $X\min = -11, X\max = 5, X\mathrm{scl} = 1, Y\min = -3, Y\max = 6, Y\mathrm{scl} = 1$
13. $X\min = -30, X\max = 50, X\mathrm{scl} = 10, Y\min = -90, Y\max = 50, Y\mathrm{scl} = 10$
15. $X\min = -10, X\max = 110, X\mathrm{scl} = 10, Y\min = -10, Y\max = 160, Y\mathrm{scl} = 10$

B.2 Exercises *(page B4)*

1. (a) **(b)** **3. (a)** **(b)**

5. (a) **(b)** **7. (a)** **(b)**

9. (a) **(b)** **11. (a)** **(b)**

13. (a) **(b)** **15. (a)** **(b)**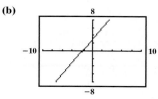

17. $Y_1 = X + 2$ **19.** $Y_1 = -X + 2$ **21.** $Y_1 = 2X + 2$ **23.** $Y_1 = -2X + 2$

25. $Y_1 = X^2 + 2$ **27.** $Y_1 = -X^2 + 2$ **29.** $Y_1 = -(3/2)X + 3$ **31.** $Y_1 = (3/2)X + 3$

B.3 Exercises *(page B6)*

1. -3.41 **3.** -1.71 **5.** -0.28 **7.** 3.00 **9.** 4.50 **11.** $1.00, 23.00$

B.5 Exercises *(page B8)*

1. Yes   **3.** Yes   **5.** No   **7.** Yes   **9.** Answers may vary. A possible answer is $Y\min = 4, Y\max = 12,$ and $Y\mathrm{scl} = 1$.

Index

CONICS

Parabola

$$y^2 = 4ax \qquad\qquad y^2 = -4ax \qquad\qquad x^2 = 4ay \qquad\qquad x^2 = -4ay$$

Ellipse

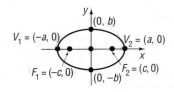

$$\frac{x^2}{a^2} + \frac{y^2}{b^2} = 1, \quad a > b, \quad c^2 = a^2 - b^2 \qquad\qquad \frac{x^2}{b^2} + \frac{y^2}{a^2} = 1, \quad a > b, \quad c^2 = a^2 - b^2$$

Hyperbola

 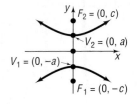

$$\frac{x^2}{a^2} - \frac{y^2}{b^2} = 1, \quad c^2 = a^2 + b^2 \qquad\qquad \frac{y^2}{a^2} - \frac{x^2}{b^2} = 1, \quad c^2 = a^2 + b^2$$

$$\text{Asymptotes:} \quad y = \frac{b}{a}x, \quad y = -\frac{b}{a}x \qquad\qquad \text{Asymptotes:} \quad y = \frac{a}{b}x, \quad y = -\frac{a}{b}x$$

PROPERTIES OF LOGARITHMS

$$\log_a(MN) = \log_a M + \log_a N$$

$$\log_a\left(\frac{M}{N}\right) = \log_a M - \log_a N$$

$$\log_a M^r = r \log_a M$$

$$\log_a M = \frac{\log M}{\log a} = \frac{\ln M}{\ln a}$$

PERMUTATIONS/COMBINATIONS

$$0! = 1 \qquad 1! = 1$$
$$n! = n(n-1)\cdot\ldots\cdot(3)(2)(1)$$

$$P(n,r) = \frac{n!}{(n-r)!}$$

$$C(n,r) = \binom{n}{r} = \frac{n!}{(n-r)!\,r!}$$

BINOMIAL THEOREM

$$(a+b)^n = a^n + \binom{n}{1}ba^{n-1} + \binom{n}{2}b^2 a^{n-2}$$
$$+ \cdots + \binom{n}{n-1}b^{n-1}a + b^n$$

ARITHMETIC SEQUENCE

$$a_1 + (a_1 + d) + (a_1 + 2d) + \cdots + [a_1 + (n-1)d]$$
$$= \frac{n}{2}[2a_1 + (n-1)d] = \frac{n}{2}[a_1 + a_n]$$

GEOMETRIC SEQUENCE

$$a_1 + a_1 r + a_1 r^2 + \cdots + a_1 r^{n-1} = a_1 \cdot \frac{1 - r^n}{1 - r}$$

GEOMETRIC SERIES

$$\text{If } |r| < 1,\ a_1 + a_1 r + a_1 r^2 + \cdots = \sum_{k=1}^{\infty} a_1 r^{k-1}$$
$$= \frac{a_1}{1 - r}$$